The British Library General Subject Catalogue 1975 to 1985

The British Library General Subject Catalogue 1975 to 1985

1

A

Airports

K · G · Saur
London · München · New York · Paris 1986

Project Consultant:

Judi Vernau B.A., A.L.A.

British Library Cataloguing in Publication Data

British Library
The British Library general subject catalogue
1975 to 1985.
1: A – Airports
1. British Library *Department of Printed Books*
– Catalogs
I. Title
017'.5 Z921.L553

ISBN 0-86291-651-8
ISBN 0-86291-650-x Set of 75 vols

Digiset Data preparation and automatic data processing
by Computaprint Limited, London
Printed in the Federal Republic of Germany
ISBN 3-598-32191-0

'1' GAUGE MODEL RAILWAYS
1895-1926 - Catalogues
BRISTOL INDUSTRIAL MUSEUM
Gauge 1 railway models in Bristol Industrial Museum. - [Bristol] : City of Bristol Museum and Art Gallery, 1983
64p : ill
0-900199-22-9 L.49/2385
3086461

1:2500 MAPS. Great Britain
Ordnance Survey 1:2500 maps. Mapping of land use. 1855-1918
HARLEY, John Brian
The Ordnance Survey and land-use mapping. - Norwich : Geo Abstracts [for] Historical Geography Research Group, 1979
[6],59p. - (Historical geography research series ; no.2)
0-86094-037-3 Map Library
2311127

1ST AIRBORNE RECONNAISSANCE SQUADRON
See Great Britain. Army. 1st Airborne Reconnaissance Squadron

1ST BRAMHALL SCOUT GROUP
to 1977
ROSE, J. R.
The official history of the 1st Bramhall Scout Group, 1909-1977. - Bramhall : 1st Bramhall Scout Group, [1980]
[136]p : ill
B80-40608 X.629/14260 Woolwich
2289113

1ST CHISWICK SCOUT TROOP
1908-1939
DAWSON, Alwyn
The story of the 1st Chiswick early years, 1908-1939. - [Ipswich] : The author, 1978
[1],87p
B78-25953 X.702/5725
1522310

1ST OF MAY GROUP
See Grupo 1o de Mayo

1ST ST AUSTELL BOY SCOUT & NIGERIAN BOY SCOUT CANOE EXPEDITION
TO THE NIGER DELTA 1981
SHEEN, Brian
1st. St. Austell Scout & Nigerian Boy Scout Canoe expedition to the Niger Delta 1981. - Par : B. Sheen, c1981
iv,48p : ill
B82-24858 X.809/52006
273656x

1ST TONBRIDGE B.P. SCOUT GROUP
to 1983
1ST TONBRIDGE B.P. SCOUT GROUP
1st Tonbridge Scouts 1908-1983. - [Rev. and updated]. - [Tunbridge Wells] : [1st Tonbridge B.P. Scout Group], [1983]
35p : ill
B84-04421 X.808/38870
2995042

2 OU 3 CHOSE QUE JE SAIS D'ELLE
GUZZETTI, Alfred
Two or three things I know about her. - Cambridge, Mass. ; London : Harvard University Press, 1981
x,366p : ill. - (Harvard film studies)
0-674-91500-3 X.955/547
2409305

2,4,5-T
See Trichlorophenoxyacetic acid
2,4,5-Trichlorophenoxyacetic acid

2E DIVISION BLINDÉE
See France. Armée. 2e Division Blindée

2M WAVEBAND RADIO EQUIPMENT
2m waveband frequency modulation radio equipment: Receivers. Antennas
JUDD, F. C. (Frederick Charles)
Two-metre antenna handbook. - London : Newnes Technical, 1980
157p : ill. - Bibliography: p154. - Includes index
0-408-00402-9 SRL
1823450

2ND QUEEN'S OWN CAMERON HIGHLANDERS
See Great Britain. Army. 2nd Queen's Own Cameron Highlanders

3-D
See Three-dimensional

3RD EYE CENTRE
See Third Eye Centre

3RD EYE CENTRE
Exhibits: English sculptures. Neagu, Paul - Catalogues
NEAGU, Paul
Paul Neagu. - Glasgow : Third Eye Centre, 1979
12,[4]p : chiefly ill
0-906474-05-1 X.425/1512 Woolwich
2060248

3RD READING SCOUT GROUP
to 1983
3RD READING SCOUT GROUP
3rd Reading Scout Group, St Peters, Caversham. - Reading : The Group, 1984
32p : ill
B85-02948 X.809/61847
3212351

4F IONS. Crystalline solids
Electrons - Conference proceedings
INTERNATIONAL CONFERENCE ON CRYSTALLINE ELECTRIC FIELD EFFECTS IN F-ELECTRON SYSTEMS (4th : 1981 : Wroclaw)
Crystalline electric field effects in f-electron magnetism. - New York ; London : Plenum, c1982
xvii,586p : ill. - Includes index
0-306-41004-4 SRL
2209993

5-0-2 MOTOR AMBULANCE CONVOY
See Great Britain. Army. Auxiliary Territorial Service. 5-0-2 Ambulance Convoy

5TH DRAGOON GUARDS
See Great Britain. Army. 5th Dragoon Guards

6 P.M. RADIO NEWS. News programmes. Radio programmes. British Broadcasting Corporation
- Indexes - Serials
BRITISH BROADCASTING CORPORATION
BBC 6 p.m. radio news index. - Vol.1, no.1- ; 1 Jan.-31 Mar. 1978-. - Cambridge : Chadwyck-Healey Ltd [etc.], 1978-
B79-00009 P.901/3118
1732978
- [Another issue, bound in annual volumes] P.903/596

7TH DUKE OF EDINBURGH'S OWN GURKHA RIFLES
See Great Britain. Army. 7th Duke of Edinburgh's Own Gurkha Rifles

8 1/2. Italian cinema films
PERRY, Ted
Filmguide to '8 1/2'. - Bloomington ; London : Indiana University Press, 1975
v,89p. - (Indiana University Press filmguide series). - Bibliography: p.73-86. - List of films: p.71-72. - List of gramophone records: p.72
0-253-39321-3 X.989/50782
0683655

8 BORE SHOTGUNS
Cartridges. Loading
MCDOUGALL, Douglas
8 bore guns and their cartridges. - Gt Yarmouth : [The author], 1979
68p : ill
0-9506710-0-2 X.629/13330 Woolwich
2073463

8MM CINEMATOGRAPHY
Super 8mm cinematography
ARROWSMITH, Frank
Beginner's guide to Super 8 film making. - London : Newnes Technical Books, 1981
162p,[8]p of plates : ill(some col.). - Includes index
0-408-00509-2 X.629/16857 Woolwich
1175831

8MM CINEMATOGRAPHY
Super 8mm cinematography
FILM maker's guide to super-8 / by the editors of Super-8 filmmaker magazine. - San Francisco : Sheptow, c1980
357p : ill
0-442-23344-2 X.629/22256
063641x

MCCLAIN, Bebe Ferrell
Super 8 filmmaking from scratch. - Englewood Cliffs ; London : Prentice-Hall, 1978
xiv,226p : ill. - Includes index
0-13-876128-0 X.900/25651
063641x

MATZKIN, Myron Archie
The super 8 film maker's handbook. - London : Focal Press, 1976
x,298p : ill. - Includes index
0-240-50755-x X.629/10487 Woolwich
063641x

SCHIFF, Lillian
Getting started in film-making. - New York : Sterling [etc.] ; London : Distributed by Ward Lock, 1978
96p : ill. - Includes index
0-7061-2564-9 X.989/52599
1175831

WATKINS, Derek
Cine photography made easy. - Newton Abbot : David and Charles, 1977
126p : ill. - (David & Charles leisure & travel series). - Bibliography: p.124. - Includes index
0-7153-7437-0 X.629/11382 Woolwich
1175831

8MM CINEMATOGRAPHY. Curriculum subjects. Schools
Super 8mm cinematography. Techniques - For teaching
HUXLEY, D. R.
Making films in Super Eight. - Cambridge : Cambridge University Press, 1978
vii,88p : ill. - Bibliography: p.85
0-521-29300-6 X.619/19246 Woolwich
1757830

8MM FILMS. United States
8mm cinema films: Feature films - Directories
LIMBACHER, James L. (James Louis)
Feature films on 8mm, 16mm, and videotape. - 7th ed. - New York ; London : Bowker, 1982
ix,481p. - Bibliography: p471. - Includes index
0-8352-1486-9 X.622/13783
0346667

LIMBACHER, James Louis
Feature films on 8mm, 16mm and videotape. - 6th ed. - New York ; London : Bowker, 1979
ix,447p. - Bibliography: p.437. - Includes index
0-8352-1112-6 X.985/1213
0346667

Feature films on 8mm and 16mm. - 5th ed. - New York ; London : Bowker, 1977
ix,422p. - Bibliography: p.411. - Includes index
0-8352-0919-9 X.985/1093
0346667

11 PLUS EXAMINATIONS. Entrance requirements. Secondary schools. England
SOCIALIST EDUCATIONAL ASSOCIATION
Return of 11 plus. - [London] : [The Association], [1970?]
Folder([4]p)
r6031724 X.510/15441
1090453

12 BORE SHOTGUNS
Cartridges. End caps. Collecting & display
SMITH, Colin, 1930-
Collecting & displaying 12 bore cartridge end caps. - [Hornchurch] : The Hobby Horse, 1979
[9]p(2 fold) : ill
B80-41925 X.421/21270 Woolwich
2320339

13TH MASSACHUSETTS INFANTRY
See United States. Army. 13th Massachusetts Infantry

16-BIT MICROPROCESSORS
WHITWORTH, Ian R.
16-bit microprocessors. - London : Granada, 1984
vii,381p : ill. - Includes index
0-246-11572-6 SRL
3133915

16-BIT MICROPROCESSORS
Use in design of microprocessor systems
BOWEN, B. A.
The logical design of multiple-microprocessor systems. - Englewood Cliffs ; London : Prentice-Hall, c1980
xiv,310p : ill. - Includes bibliographies and index
0-13-539908-4 SRL
1925733

GARLAND, Harry
Introduction to microprocessor system design. - New York ; London : McGraw-Hill, 1979
xi,192p : ill. - Includes index
0-07-022871-x X.619/19994 Woolwich
1925733

MCGLYNN, Daniel R. (Daniel Robert)
Modern microprocessor system design. - New York ; Chichester : Wiley, c1980
viii,295p,[2] folded leaves : ill. - Bibliography: p283-284. - Includes index
0-471-06492-0 SRL
1925733

16MM FILMS
Cutting

BURDER, John
16mm film cutting. – London : Focal Press, 1975
[1],158p : ill(chiefly col). – (Media manuals). –
Bibliography: p.151
0–240–50857–2 X.619/15389 Woolwich
0566381

16MM FILMS
Editing

BURDER, John
The technique of editing 16mm. films. – 4th ed. – London
: Focal Press [etc.], 1979
152p : ill. – (The library of communication techniques).
– Includes index
0–240–51019–4 SRL
1490354

16MM FILMS. Adaptations. Books. United States
16mm cinema films: Short films – Bibliographies

PARLATO, Salvatore J. (Salvatore Joseph)
Films ex libris. – Jefferson, N.C. : McFarland, 1980
xii,271p : ill. – Includes index
0–89950–006–4 X.950/7795
2561328

16MM FILMS. Audio–visual aids. Schools. Great Britain
Use in teaching of science

GEE, Brian
The 16mm film in school science and technology. –
Plymouth : College of St Mark and St John, Science
Education Unit, 1979
36,[48]p : ill. – Bibliography: p.14
0–9506525–0–4 X.622/8738 Woolwich
2303493

16MM FILMS. Audio–visual aids. Schools. Great Britain
Use in teaching of technology

GEE, Brian
The 16mm film in school science and technology. –
Plymouth : College of St Mark and St John, Science
Education Unit, 1979
36,[48]p : ill. – Bibliography: p.14
0–9506525–0–4 X.622/8738 Woolwich
2303507

16MM FILMS. Stock. National Audio–Visual Aids Library
– Catalogues

NATIONAL AUDIO–VISUAL AIDS LIBRARY
16mm films in the National Audio–Visual Aids Library. –
[London] : Educational Foundation for Visual Aids, 1976
x,288p. – Includes index
B77–29127 2725.aa.31
124504x

16MM FILMS. Teaching aids. Schools. United States

GAFFNEY, Maureen
What to do when the lights go on. – Phoenix : Oryx, 1981
xxiii,268p : ill. – Bibliography: p253–264. – Includes
index
0–912700–65–3 X.622/15172
2878038

16MM FILMS. United States
16mm cinema films: Feature films – Directories

LIMBACHER, James L. (James Louis)
Feature films on 8mm, 16mm, and videotape. – 7th ed. –
New York ; London : Bowker, 1982
ix,481p. – Bibliography: p471. – Includes index
0–8352–1486–9 X.622/13783
0346667

LIMBACHER, James Louis
Feature films on 8mm, 16mm and videotape. – 6th ed. – New
York ; London : Bowker, 1979
ix,447p. – Bibliography: p.437. – Includes index
0–8352–1112–6 X.985/1213
0346667

Feature films on 8mm and 16mm. – 5th ed. – New York ;
London : Bowker, 1977
ix,422p. – Bibliography: p.411. – Includes index
0–8352–0919–9 X.985/1093
0346667

19TH MICHIGAN VOLUNTEER INFANTRY
See United States. Army. 19th Michigan Volunteer Infantry

20 CROWE STREET. Museum of East Anglian Life. Cottages .
Architecture, to 1975

MUSEUM OF EAST ANGLIAN LIFE
Abbot's Hall cottages. – [Stowmarket] : [The Museum],
[1976]
Folder ([6]p) : 1 ill
0–905596–06–4 X.419/3204 Woolwich
071545x

21 CLUB

KAYTOR, Marilyn
"21". – New York : The Viking Press, [1975]
191p : ill
0–670–73460–8 X.622/10980 Woolwich
8743010

21ST INDEPENDENT PARACHUTE COMPANY
See Great Britain. Army. 21st Independent Parachute Company

21ST MISSOURI VOLUNTEER INFANTRY
See United States. Army. Twnety-first Missouri Volunteer
Infantry

22 SPECIAL AIR SERVICE REGIMENT
See Great Britain. Army. 22 Special Air Service Regiment

24 PULK ULANOW
See Poland. Wojsko. 24 Pulk Ulanow

24TH POLISH LANCERS
See Poland. Wojsko. 24 Pulk Ulanow

27 MHZ WAVEBAND
See also
Citizens' band

28TH ARMY. JAPAN
See Japan. Army. 28th Army

30 HOUR CLOCKS
30 hour long case clocks. Automatic winding mechanisms.
Making

WILDING, John, b.1924
Automatic winding for a 30 hour longcase clock. –
Ashford, Kent : Brant Wright Associates Ltd, 1975
[2],12p : ill
0–903512–03–3 SRL
0795941

30 HOUR CLOCKS
30 hour weight driven alarm clocks. Making

WILDING, John, b.1924
A 30-hour weight driven alarm clock. – Ashford, Kent :
Brant Wright Associates Ltd, [1976]
[4],71p : ill
0–903512–04–1 SRL
079595x

35MM CAMERAS
See also
Canon cameras
Contax cameras
Fujica cameras
Konica cameras
Leica cameras
Minolta cameras
Nikon cameras
Pentax cameras
Praktica cameras

35MM CAMERAS

CALDER, Julian
The 35mm photographer's handbook. – London : Pan Books,
1979
240p : ill(some col). – Bibliography: p.232–233. –
Includes index
0–330–26106–1 X.629/12863 Woolwich
0578061

FOOTHORAP, Robert
Independent photography. – European ed.. – New York ;
London : Omnibus Press, 1976
[8],227p : ill. – Bibliography: p.221–222. – Includes
index
0–86001–294–8 X.619/17599 Woolwich
0578061

FREEMAN, Michael, 1945–
The 35mm handbook. – London : Windward, c1980
320p : ill(some col.). – Bibliography: p320. – Includes
index
0–7112–0046–7 X.622/9173 Woolwich
0578061

GAUNT, Leonard
The photoguide to 35mm. – London : Focal Press, 1975
227p : ill(some col). – (The Focal photoguides). –
Includes index
0–240–50925–0 WP.13581/107
0578061

LANGFORD, Michael, 1933–
Michael Langford's 35mm handbook. – London : Ebury, 1983
224p : ill(some col.). – Includes index
0–85223–218–7 X.429/14811
0578061

YULSMAN, Jerry
The complete book of 35mm photography. – London : Hale,
1979
224p : ill
0–7091–7347–4 X.629/12347 Woolwich
0578061

35MM CAMERAS
35mm single lens reflex cameras

BROWN, Dolores
PSL guide to 35mm SLR photography. – Cambridge :
Stephens, c1981
157p,[8]p of plates : ill(some col.). – Includes index
0–85059–549–5 X.629/16161
077443x

The CANON camera handbook / by the editors of Curtin &
London. – Somerville, Mass. : Curtin & London, 1981
ix,163p : ill(some col.)
0–930764–32–3 X.629/26680
1906763

GAUNT, Leonard
The photoguide to the 35mm single lens reflex. – London :
Focal Press, 1977
237p : ill(some col). – ([Focal] photoguides). – Includes
index
0–240–50768–1 X.629/13135 Woolwich
0080454

GRILL, Tom
[Taking better pictures with 35mm SLR]. Taking better
pictures with your 35mm SLR. – London : Hale, 1982, c1981
ix,142p : ill(some col.). – Includes index
0–7091–9626–1 X.425/3245 Woolwich
077443x

KILPATRICK, David
Improve your SLR photography. – London : Collins, 1983
96p : ill(some col.). – (You and your camera photography
series). – Includes index
0–00–411770–0 X.622/16120
077443x

LONDON, Barbara
A short course in Canon photography. – Marblehead, Mass.
: Curtin and London : London : Newnes Technical Books,
1979
ix,2–136p : ill(some col). – Bibliography: p.118. –
Includes index
0–408–00422–3 X.410/10376 Woolwich
1906763

A short course in Minolta photography. – Marblehead,
Mass. : Curtin and London : London : Newnes Technical,
1979
ix,2–136p : ill(some col). – Bibliography: p.118. –
Includes index
0–408–00423–1 X.410/10377 Woolwich
1906771

A short course in Nikon photography. – Marblehead, Mass.
: Curtin and London : London : Newnes Technical Books,
1979
ix,2–136p : ill(some col). – Bibliography: p.118. –
Includes index
0–408–00421–5 X.421/21069 Woolwich
190678x

A short course in Nikon photography. – Rev. ed. –
Somerville, Mass : Curtin & London, 1983
vii,138p : ill(some col.)
0–930764–54–4 X.622/17626
190678x

A short course in photography. – Marblehead, Mass. :
Curtin and London ; London : Newnes Technical, 1979
ix,2–136p : ill(some col). – Includes index
0–408–00419–3 X.410/72(6
077443x

A short course in photography. – Rev. ed. – Somerville,
Mass. : Curtin & London, 1983
vii,135p : ill(some col.)
0–930764–51–x X.622/17627
077443x

MASON, R. H. (Reg Herbert)
Photography using a 35mm camera. – [London] : Chancerel :
Barrie and Jenkins, 1977
2–91p : ill. – (Action books)
0–905703–01–4 X.620/16944 Woolwich
077443x

SLR handbook / [edited by Michael Langford]. – London :
Ebury Press, 1980
2–92p : chiefly ill. – (Basic photography series). –
Includes index
0–85223–171–7 X.629/13380 Woolwich
077443x

WATKINS, Derek
SLR photography. – Newton Abbot : David and Charles, 197(
128p : ill. – Bibliography: p.124. – Includes index
0–7153–7301–3 X.620/16812 Woolwich
077443x

35MM CAMERAS
35mm single lens reflex cameras, to 1967

HICKS, Roger, 19—
A history of the 35mm still camera. – London : Focal,
1984
198p : ill. – Bibliography: p188–189. – Includes index
0–240–51233–2 X.622/224.6
3053695

35MM CAMERAS
35mm single lens reflex cameras. Accessories

EXANDING your SLR system / consultant editor David Kilpatrick. - London : Collins, 1981
96p : ill(some col.). - (You and your camera photography series). - Includes index
0-00-411685-2 X.421/22255 Woolwich
563967

35MM CAMERAS
Automatic 35mm single lens reflex cameras

CURTIN, Dennis P.
What are you doing wrong? with your automatic camera. - Somerville : Curtin & London ; London : Newnes Technical, 1980
v.138p : ill(some col.). - Includes index
0-408-00542-4 X.629/15660 Woolwich
287829

Your automatic camera. - Somerville : Curtin & London ; London : Newnes Technical, 1980
v.122p : ill(some col.). - Includes index
0-408-00541-6 X.629/15662 Woolwich
287829

PARTRIDGE, Joe
One touch photography. - London : Pan, 1984
13p : ill(some col.). - Includes index
0-330-28629-3 X.421/26548
287829

35MM CAMERAS
Automatic 35mm single lens reflex cameras. Indoor photography

LONDON, Barbara
Photographing indoors with your automatic camera. - Somerville : Curtin & London ; London : Newnes Technical, 1981
vi137p : ill(some col.). - Includes index
0408-00544-0 X.629/15663 Woolwich
287837

35MM CAMERAS
Automatic 35mm single lens reflex cameras. Outdoor photography

LONDON, Barbara
Photographing outdoors with your automatic camera. - Somerville : Curtin & London ; London : Newnes Technical, 1981
vi137p : ill(some col.). - Includes index
0408-00543-2 X.629/15661 Woolwich
287845

35MM CAMERAS. Great Britain
Consumer behaviour - For marketing

35MM camera survey. - [London] : NOP Market Research Limited, 1980
1 v.,3leaves
B1-38715 SRL
257261

35MM FILMSTRIPS ON GEOGRAPHY. Stock. National Audio-Visual Aids Library
- Catalogues

NATIONAL AUDIO-VISUAL AIDS LIBRARY
Filmstrips for regional geography. - [London] : Educational Foundation for Visual Aids, [1978]
[2]43p
B9-08459 2725.aa.79
185822

41ST (WELSH) REGIMENT OF FOOT
See Great Britain. Army. 41st (Welsh) Regiment of Foot

43RD (MONMOUTHSHIRE) REGIMENT OF FOOT
See Great Britain. Army. 43rd (Monmouthshire) Regiment of Foot

50 NAUTICAL MILE LIMIT. Restrictions. Fishing. Coastal waters. Iceland

ANDERSEN, Hans G.
The Icelandic fishery limits and the concept of the exclusive economic zone. - Reykjavík : Orator, Law Students' Association, University of Iceland, 1974
18p.. - (Úlfljótur ; vol. 27, no. 3, suppl)
r601233 X.311/4511 Woolwich
09035X

JÓNSON, Hannes
Iceland's 50 miles and the reasons why. - 2nd print, slightly rev. - Reykjavík : Government of Iceland, 1973
15p : ill
r603722 S.B.462/1
09035X

51ST HIGHLAND DIVISION
See Great Britain. Army. 51st Highland Division

56 GROUP WALES. British visual arts

56 GROUP WALES
The artist and how to employ him. - [Cardiff] : [The Group], 1976
64p : ill(some col)
0-9505005-0-x X.415/5118 Woolwich
0681059

78 R.P.M. SOUND DISCS. Speech in English 1888-1953 - Bibliographies

RUST, Brian
Discography of historical records on cylinders and 78s. - Westport, Conn. ; London : Greenwood Press, 1979
ix,327,[1]p. - Includes index
0-313-20561-2 2725.ff.99
194455x

83-ĪĀ GVARDEĪSKAĪĀ STRELKOVAĪĀ GORODOKSKAĪĀ KRASNOZNAMENNAĪĀ DIVIZIĪĀ
See Union of Soviet Socialist Republics. Armiīā. 83-īā gvardeiskaīā strelkovaīā Gorodokskaīā Krasnoznamennaīā diviziīā †

94TH REGIMENT OF FOOT
See Great Britain. Army. 94th Regiment of Foot

95TH (RIFLE) REGIMENT
See Great Britain. Army. 95th (Rifle) Regiment

112 WORKSHOP

112 Workshop, 112 Greene Street / edited by Robyn Brentano with Mark Savitt ; designed by Barbara Du Pree Knowles. - New York : New York University Press, 1981
xii,385p : ill
0-8147-1037-9 L.42/1644
2792087

200 NAUTICAL MILE LIMIT. Restrictions. Fishing. Coastal waters. Iceland

ICELAND. Utanríksraduneytid
The fishery limits off Iceland. - Reykjavík : Ministry for Foreign Affairs, 1975
37 p.
r6001229 S.B.460/4
0909289

"291" (GALLERY)
Role in American dadaism, to 1923

LEAVENS, Ileana B.
From '291' to Zurich. - Epping : Bowker, c1983
x,196p. - (Studies in fine arts. The avant-garde ; no.39). - Bibliography: p185-189. - Includes index
0-8357-1447-0 X.421/28434
3046478

457 HEAVY AIR DEFENCE REGIMENT. ISLE OF WIGHT (PRINCESS BEATRICE'S) BATTERY
See Great Britain. Army. 457 Heavy Air Defence Regiment. Isle of Wight (Princess Beatrice's) Battery

727 AEROPLANES
Boeing 727 aeroplanes

The BOEING 727 / edited by John A. Whittle, in collaboration with ... [others]. - Saffron Walden : Air-Britain, 1976
82p,[18]p of plates : ill. - (Monographs / Air-Britain). - Includes index
0-85130-047-2 X.611/8738 Woolwich
183651x

830 SQUADRON RN
See Great Britain. Royal Navy. 830 Squadron

1766 ARS ALLMÄNNA FISKESTADGA

PLAENGE JACOBSON, Sigbrit
[Fiskestadga 1766]. 1766 års allmänna fiskestadga. - Uppsala ; Stockholm : Distributor : Almqvist & Wiksell International, 1978
196p. - (Studia historica upsaliensia ; 98). - Includes bibliographies
91-554-0722-6 Ac.1077.ng.[vol.98]
1977792

1976 BRITISH HOGGAR MOUNTAINS AND BILMA SANDS EXPEDITION. Bilma Sands. Niger

1976 BRITISH HOGGAR MOUNTAINS AND BILMA SANDS EXPEDITION
1976 British Hoggar Mountains and Bilma Sands Expedition. - London : [The Expedition], 1977
[7],69[i.e.80]p,[5] leaves of plates : ill. - Bibliography: p.68-69
B80-09015 X.702/6221
2044579

1976 BRITISH HOGGAR MOUNTAINS AND BILMA SANDS EXPEDITION. Hoggar Mountains. Algeria

1976 BRITISH HOGGAR MOUNTAINS AND BILMA SANDS EXPEDITION
1976 British Hoggar Mountains and Bilma Sands Expedition. - London : [The Expedition], 1977
[7],69[i.e.80]p,[5] leaves of plates : ill. - Bibliography: p.68-69
B80-09015 X.702/6221
2044560

A-AB-BA HU-LUH-HA. Sumerian laments

KUTSCHER, Raphael
Oh angry sea (a-ab-ba hu-luh-ha). - New Haven ; London : Yale University Press, 1975 [i.e. 1976]
xvi,178p : ill. - (Yale Near Eastern researches ; 6). - Bibliography: p.xi-xvi. - Includes index
0-300-01579-8 X.200/20404
0656488

À BECKET, THOMAS, SAINT
See Becket, Thomas, Saint

A. DE GRUCHY & CO. LTD.
See De Gruchys (Firm)

A. DONATANTONIO & SONS LIMITED
- Serials

[AD ... (London)]. AD .... - 1980, no.1-. - London : Donatantonio, 1980-
v. : ill
B81-38524 P.513/2618
2565633

A. E.. English literature

DAVIS, Robert Bernard
George William Russell ("AE"). - London : Prior [etc.], 1977
163p. - (Twayne's English authors series ; 208). - Bibliography: p.155-158. - Includes index
0-86043-089-8 X.989/52049
1291114

KAIN, Richard Morgan
George Russell (A.E.). - Lewisburg : Bucknell University Press ; London : Associated University Presses, 1976
93p. - (The Irish writers series). - Bibliography: p.91-93
0-8387-1101-4 X.989/52980
1291114

A. E.. English literature
- Biographies

DAVIS, Robert Bernard
George William Russell ("AE"). - Boston, Mass. : Twayne, 1977
163p. - (Twayne's English authors series ; 208)
0-8057-6677-4 X.989/82850
0610208

SUMMERFIELD, Henry, 1935-
That myriad-minded man. - Gerrards Cross : Smythe, 1975
xiii,354p,[16]p of plates : ill. - Includes index
0-900675-69-1 X.809/40726
0610208

À FORFAIT TRANSACTIONS
See Forfaiting

A KEMPIS, THOMAS
See Thomas, à Kempis

A LA RECHERCHE DU TEMPS PERDU. Proust, Marcel. Fiction in French

ALDEN, Douglas
Marcel Proust's Grasset proofs. - Chapel Hill : UNC Department of Romance Languages, 1978
561p. - (North Carolina studies in the romance languages and literatures ; no 193)
0-8078-9193-2 AC.2685.K/7(193)
0122831

BENOIST-MÉCHIN, Jacques
Avec Marcel Proust. - Paris : Albin Michel, 1977
183p,8p of plates : ill
2-226-00501-3 X.989/72293
0122831

BOWIE, Malcolm
Proust, jealousy, knowledge. - London : University of London, Queen Mary College, 1978
[4],24p. - (Inaugural lectures / Queen Mary College)
0-902238-04-3 X.529/35827
0122831

CROSMAN, Inge Karalus
Metaphoric narration. - Chapel Hill : University of North Caroline, Department of Romance Languages, 1978
230p. - (North Carolina studies in the Romance languages and literatures ; no. 204)
0-8078-9204-1 Ac.2685.K/7(204)
0122831

A LA RECHERCHE DU TEMPS PERDU. Proust, Marcel. Fiction in French–continued

GUARALDO, Enrico
Lo specchio della differenza. – Roma : Bulzoni, 1977
239p. – (Biblioteca di cultura ; 110)
r6100524 X.909/83282
0122831

MAGNANI, Luigi
La musica in Proust. – Torino : Einaudi, 1978
134p. – (Saggi ; 594)
r6195347 X.909/88366
0122831

MILLY, Jean
La phrase de Proust. – Paris : Librairie Larousse. – (Collection L) – Bibliography : p. 211–217
des phrases de Bergotte aux phrases de Vinteuil. – [1975]
223p
2-03-036014-7 X.909/36494
0122831

SPALDING, Philip Anthony
A reader's handbook to Proust. – Revised ed.. – London : George Prior Publishers, 1975
xii,303p
0-904000-05-2 X.989/40387
0561800

A LA RECHERCHE DU TEMPS PERDU. Proust, Marcel. Fiction in French
– Indexes

KILMARTIN, Terence
A guide to Proust. – London : Chatto & Windus, 1983
viii,193p
0-7011-2636-1 X.809/58851
0561800

A LA RECHERCHE DU TEMPS PERDU. Proust, Marcel. Fiction in French
Composition

WINTON, Alison
Proust's additions. – Cambridge : Cambridge University Press, 1977
2v.(x,393p; 209p). – Bibliography: p.364–381. – Includes index
0-521-21612-5 X.981/20885
1285203

A LA RECHERCHE DU TEMPS PERDU. Proust, Marcel. Fiction in French
Special subjects: Communication & interpersonal relationships

GHINSTE, J. van de
Rapports humains et communication dans "A la recherche du temps perdu". – Paris : Éditions A.-G. Nizet, 1975
285 p.. – Bibliography : p. 277–280
r6001118 X.510/9358
0924377

A LA RECHERCHE DU TEMPS PERDU. Proust, Marcel. Fiction in French
Special themes: Man. Sexuality

RIVERS, J. E. (Julius Edwin)
Proust & the art of love. – New York ; Guildford : Columbia University Press, c1980
xiii,327p. – Bibliography: p297–309. – Includes index
B85-02902 X.529/65716
2366460

Proust & the art of love. – New York ; Guildford : Columbia University Press, 1980
xii,327p. – Bibliography: p297–309. – Includes index
0-231-05036-4 X.520/23150
2366460

A LA RECHERCHE DU TEMPS PERDU. Proust, Marcel. Fiction in French
Special themes: Space

POULET, Georges
Proustian space. – Baltimore ; London : Johns Hopkins University Press, 1977
[5],114p
0-8018-1921-0 X.989/52019
1285068

A. LANGE & SÖHNE (GLASHÜTTE)

HUBER, Martin
Die Uhren von A. Lange & Söhne Glashütte / Sachsen. – 3. erw. Aufl. – München : Callwey, 1979
111,15p,[4]p of plates : ill(some col.)
3-7667-0499-0 X.622/11105 Woolwich
8729816

A LEVEL
See G.C.E. (A level)

A PRIORI KNOWLEDGE
Theories of Kant, Immanuel

BECK, Lewis White
Kant's theory of knowledge. – Dordrecht : D. Reidel Publishing Company, [1974]
vi,217p.
90-277-0529-1 X.519/38684
112773X

HENRICH, DIETER
Identität und Objektivität. – Heidelberg : Winter, 1976
112p. – (Sitzungsberichte der Heidelberger Akademie der Wissenschaften. Phil.-hist. Klasse ; Jahrg. 1976, Abh. 1)
3-533-02484-9 Ac.892/2 (1976.Abh.1)
112773X

A&AEE
See Aeroplane and Armament Experimental Establishment

A9 ROAD REGION. Scotland

The A9. – Gartocharn, Alexandria : Famedram, [1983?]
1v. : ill
B83-42302 X.809/56549
2969114

A74 ROAD. South-west Scotland
Improvement

TRANSPORT ACTION SCOTLAND
Scotland's missing link. – [Glasgow] : [Royal Automobile Club], [1978]
5,[8]leaves : 1 ill
B79-32436 X.512/6960
1955187

A350 ROAD
See also
Countryside adjacent to A350 road

AA
See Automobile Association

AABEL, PER. Acting. Theatre. Norway
– Biographies

AABEL, Per
Den stundesløse Per Aabel. – 4. oppl. – Oslo : Gyldendal, 1981
210p,plates
82-05-12409-4 X.955/1737
8991456

AABENRAA
See Åbenraa

AABENRAA FJORD. Denmark
Pollution

BISGAARD, Agnete
Fjordliv. – Aabenraa : Miljøforeningen Aabenraa, [1975]
13 leaves : ill
r6078850 X.313/782 Woolwich
1389734

AACHEN
See also
Battle of Aachen

AACHEN. West Germany
to date

GIELEN, Viktor
Aachen im Vormärz. – Eupen : Grenz Echo, 1982
216p,[16]p of plates : ill. – Includes bibliography
3-923099-13-4 X.809/59042
8769702

KAEMMERER, Walter
Aachener Quellentexte. – Aachen : Mayer, 1980
viii,347p,17p of plates : ill. – Bibliography
3-87519-008-4 X.0800/1623(1)
8769702

AACHEN. West Germany
to date – Serials

VERÖFFENTLICHUNGEN des Stadtarchivs Aachen. – Bd.1–. – Aachen : Mayer, 1980–
r6301785 X.0800/1623
8769702

AACHEN. West Germany
1500–1900

ALTE Aachener Stadtansichten / herausgegeben von Helmut A. Crous. – Aachen : Mayer, 1977– : ill
r6062430 Tab.583.f.32
1427164

AACHEN. West Germany
Antiquities. Archaeology, to date

AQUAE Granni / von Heinz Cüppers ... [et al.]. – Köln : Rheinland-Verlag, 1982
x,213p.,98p.of plates(9leaves folded) : ill. – (Rheinische Ausgrabungen ; Bd.22)
3-7927-0313-0 X.0421/158(22) Woolwich
8820848

AACHEN. West Germany
Cathedrals: Aachener Dom

KREMER, Jacob
Der Aachener Dom. – Monchengladbach : Kuhlen, [1979]
72p : ill
3-87448-099-2 X.421/24654
8824738

AACHEN. West Germany
Convents: Kloster Burtscheid

WURZEL, Thomas
Die Reichsabtei Burtscheid von der Grundung bis zur frühen Neuzeit. – Aachen : Mayer, 1984
245p,plates. – (Veröffentlichungen des Stadtarchivs Aachen ; Bd.4). – Includes bibliography
3-87519-102-1 X.0800/1623(4)
916619x

AACHEN. West Germany
Museums: Suermondt-Ludwig-Museum

SUERMONDT-LUDWIG-MUSEUM Aachen / [Autoren Ernst Gunther Grimme ... et al.]. – Braunschweig : Westermann, 1982
128p : ill(some col). – (Museum ; Sept. 1982)
r6305875 P.421/392[Sept.1982]
8913307

AACHEN. West Germany
Religious institutions: Aachener Marienstift

FALKENSTEIN, Ludwig
Karl der Grosse und die Entstehung des Aachener Marienstiftes. – Paderborn : Schöningh, 1981
149p. – (Quellen und Forschungen aus dem Gebiet der Geschichte ; n.F., Hft.3). – Bibliography
3-506-73253-6 X.0520/634(3)
8793271

AACHEN. West Germany
Sculptures, to date

CORR, Elfriede
Aachener Brunnen und Denkmäler. – 3., erw. Aufl. – Aachen : Mayer, [1981]
153p : chiefly ill
3-87519-018-1 X.421/22015 Woolwich
8751218

AACHEN (GEMEINDE). West Germany
Christian church, 1933–1945

GEMEINDE Aachen im Dritten Reich / herausgegeben von Hermann Korth. – Köln : Rheinland, 1980
xxv,252p. – (Schriftenreihe des Vereins für Rheinische Geschichte ; Bd 58)
3-7927-0499-4 X.200/36344
8773572

AACHEN REGION. West Germany
Catholic Church, to date

LEBENSRAUM Bistum Aachen / Philipp Boonen (Herausgeber). – Aachen : Einhard, 1982
373p. – (Aachener Beiträge zu Pastoral- und Bildungsfragen ; 10)
3-920284-09-7 X.200/42849
9081062

AACHEN REGION. West Germany
Imperial estates. Financial management, 800–1350

FLACH, Dietmar
Untersuchungen zur Verfassung und Verwaltung des Aachener Reichsgutes von der Karlingerzeit bis zur Mitte des 14. Jahrhunderts. – Göttingen : Vandenhoeck & Ruprecht, 1976
431p. – (Veröffentlichungen des Max-Planck-Instituts für Geschichte ; 46). – Includes bibliographies
3-525-35355-3 9044.aa1/46
1330438

AACHENER DOM

KREMER, Jacob
Der Aachener Dom. – Mönchengladbach : Kühlen, [1979]
72p : ill
3-87448-099-2 X.421/24654
8824738

AACHENER MARIENSTIFT

FALKENSTEIN, Ludwig
Karl der Grosse und die Entstehung des Aachener Marienstiftes. – Paderborn : Schöningh, 1981
149p. – (Quellen und Forschungen aus dem Gebiet der Geschichte ; n.F., Hft.3). – Bibliography
3-506-73253-6 X.0520/634(3)
8793271

AACR 2
See Anglo-American cataloguing rules
2nd ed

AAFJES, BERTUS. Dutch literature
- Biographies

MOLIN, Rob
Bertus Aafjes. - Beveren : Orion, [1981]
103p,[12]p of plates - (Grote ontmoetingen ; 58)
90-264-3351-4 X.950/20334
8723699

AAKJÆR, JEPPE. Fiction in Danish

BJERRE, Solvejg
Den hvide rug, den blide rug-. - Herning, Kristensen, 1980
63p : ill
87-7468-113-3 L.49/1810
815189x

HJORDT-VETLESEN, Inger-Lise
Flugtens fængsel. - [Odense] : Odense Universitetsforlag, 1981
136p. - (Dansk kulturhistorie og bevidsthedsdannelse 1880-1920 ; 11)
87-7492-354-4 X.0709/929(11)
815189x

JEPPE Aakjær / Aakjær-selskabet. - 1-. - [København] : Selskabet, 1982-
r6610409 X.0950/499
815189x

AALBORG. Denmark
to date

WITT, Torben
Aalborg og fabrikkerne - byvækst i 1890erne. - [Aalborg] : Selskabet for Aalborgs Historie, [1980]
95p : ill. - (Aalborg-bogen ; 1979-80)
r6610894 X.809/55800
8978344

AALBORG. Denmark
to date - Bibliographies

JENSEN, Ole Steen
Lokalbibliografi for Aalborg by 1482-1978. - Aalborg : Aalborg Universitetsforlag, 1980
225p
87-7307-072-6 X.809/56026
8979677

AALBORG. Denmark
ca 1150-1550

AALBORG HISTORISKE MUSEUM
Aalborg og middelalderen. - [Aalborg] : Aalborg Historiske Museum, [1973]
24 p. : ill. - Includes bibliographies
87-980257-0-8 X.709/20457
0855111

AALBORG. Denmark
Chemists' shops, to 1976

HASSING POVLSEN, Holger
Aalborgs apotekere. - [Aalborg] : Selskabet for Aalborgs Historie, [1976]
114p : ill. - (Aalborg-bogen ; 1976)
r6034283 X.519/23040
1228250

AALBORG. Denmark
Grammar schools: Aalborg Katedralskole, 1940-1945

STEGMANN, Hother
Erindringer fra Aalborg Katedralskole under besættelsen. - [Aalborg] : [H. Stegmann], [ca.1975]
31 leaves
r6060864 X.512/3097
1860313

AALBORG. Denmark
Residences. Removal, to date

SPØHR, Hanne
Flyttekæder pa boligmarkedet. - Hørsholm : Statens Byggeforskningsinstut, 1981
168p. - (SBI-rapport ; 129)
87-563-0396-3 S.B.218/2[no.129]
8873461

AALBORG. Denmark
Social conditions, to date

DAHL, Robert
Med brede strøg. - Aalborg : Lokalhistorisk Arkiv for Aalborg kommune, 1982
216p : ill
87-981065-1-1 X.525/10383
8079404

AALBORG. Denmark
Social conditions, 1900-1945

UNG i Aalborg / redaktion : Aage Svendstorp, Kaj Hansen, Johs. Juncker... - [Aalborg] : Lions Club, 1976
168p : ill
r6028997 X.700/18459
1230433

AALBORG. Denmark
Unemployment, 1926-1936

CHRISTENSEN, Allan
Krise og arbejdsløshed. - Aalborg : Aalborg Universitetsforlag, 1979
160p. - (Serie om arbejdsforhold og arbejdsmarked ; nr 6)
87-7307-014-9 X.709/50358
2292661

AALBORG HARBOUR. Denmark

AALBORG havn kort fortalt. - [Aalborg] : [s.n.], [1976?]
11p : ill
r6033775 X.708/42752
142064X

[AALBORG havn kort fortalt. English]. Port of Aalborg in short. - [Aalborg] : [s.n.], [1976?]
11p : ill
r6033776 X.708/42753
142064X

SIMONSEN, Jørgen D.
[Aalborg viser vej. English]. Aalborg shows the way - its harbour the sea way. - [Aalborg] : Port of Aalborg Committee, [1976]
28p : ill
r6033774 X.708/42754
142064X

Aalborg viser vej- havnen søvejen. - [Aalborg] : Aalborg Havnevæsen, [1976]
40p : ill
r6033773 X.708/42751
142064X

AALBORG HARBOUR. Denmark
1476-1976

AALBORG. Havneudvalget
Aalborg havn. - [Ålborg] : Aalborg havneudvalg, [1976]
238p.,[3]fold. leaves of plates : ill
r6034144 X.800/12833
1327828

AALBORG KATEDRALSKOLE
1940-1945

STEGMANN, Hother
Erindringer fra Aalborg Katedralskole under besættelsen. - [Aalborg] : [H. Stegmann], [ca.1975]
31 leaves
r6060864 X.512/3097
1860313

AALBORG REGION. Denmark
to 1977

STOKHOLM, Torben
Landsbymiljøer ved Aalborg. - [Aalborg] : Selskabet for Aalborgs Historie, [1977]
92p : ill. - (Aalborg-bogen ; 1977)
r6061162 X.709/29525
2134624

AALBORG TEATER
1878-1978

AALBORG Teater 100 år / redigeret af Svend M. Clausen, Poul Krabbe og Pia Leonhardt. - [Ålborg] : E. Ruhländer, 1978
96p : chiefly ill
87-980723-0-7 X.902/3451
2027879

AALBORG Teater 1878-1978 / udgivet af Selskabet for Aalborgs Historie ; [redaktion : Ivan Rønn]. - [Ålborg] : Selskabet for Aalborgs Historie, [1978]
123p : ill. - (Aalborg-bogen ; 1978)
r6113305 X.989/87222
2027879

AALBORG UNIVERSITETSCENTER
See Alborg Universitetscenter

AALST
See Alost

AALTO, ALVAR. Architectural design. Finland

CRESTI, Carlo
Alvar Aalto. - Firenze : Sansoni, 1975
97p : ill. - (I maestri del Novecento ; 25). - Bibliography : p.94-96
r6024431 X.415/2165 Woolwich
029232X

PEARSON, Paul David
Alvar Aalto and the international style. - New York : Whikney Library of Design, 1978
240p : ill. - Bibliography : p. 236-237 -Includes index
0-8230-7023-9 X.423/10778 Woolwich
029232X

PORPHYRIOS, Demetri
Sources of modern eclecticism. - London : Academy Editions, 1982
xvii,138p : ill. - Includes index
0-312-74673-3 X.622/13314
029232x

QUANTRILL, Malcolm
Alvar Aalto. - London : Secker & Warburg, 1983
xii,307,[8]p of plates : ill(some col.). - Bibliography: p282-297. - Includes index
0-436-39400-6 X.622/16728
029232x

SCHILDT, Göran
Det vita bordet. - [Stockholm] : Arkitektur förlag, 1982
292p : ill
91-86050-03-6 X.421/25896
8890382

AALTO, ALVAR. Architectural design. Finland
- Bibliographies

MILLER, William C. (William Charles), 1945-
Alvar Aalto. - New York ; London : Garland, 1984
xxix,244p,[19]p of plates : ill. - (Garland bibliographies in architecture and planning ; v.4) (Garland reference library of the humanities ; v.390). - Includes index
0-8240-9188-4 X.950/35069
3170632

AALTO, ALVAR. Architectural design. Finland
- Serials

ALVAR Aallon arkkitehtuuria = Architecture by Alvar Aalto. - no. 1-. - Jyväskylä : Alvar Aalto museo, 1976-
r6017714 X.0419/98 Woolwich
029232X

AALTO, ALVAR. Finnish design

BLASER, Werner
[Il design di Alvar Aalto. German]. Alvar Aalto als Designer. - Stuttgart : Deutsche Verlagsanstalt, 1982
143p : ill
3-421-02589-4 L.49/2271
9060758

AAM
See Association of Assistant Mistresses

AAR
See Ammonia absorption refrigeration

AARAU. Switzerland
Buildings. Architecture, to date

AARAUER Stadtbilder aus fünf Jahrhunderten / herausgegeben von der Ortsbürgergemeinde zur Feier "700 Jahre Stadtrecht Aarau" ; gesammelt und kommentiert von Theo Elsasser. - Aarau : Sauerländer, 1983
79p : ill
3-7941-2400-6 X.525/9216
9083014

AARAU. Switzerland
Libraries. Stock: Manuscripts - Catalogues

SCARPATETTI, Beat Matthias von
Die Handschriften der Bibliotheken von Aarau, Appenzell und Basel. - Dietikon-Zürich : Graf, 1977
2v : ill. - (Katalog der datierten Handschriften in der Schweiz in lateinischer Schrift vom Anfang des Mittelalters bis 1550 ; Bd 1)
3-85951-125-4 2702.f.16
1512005

AARDVARK AEROPLANES
See F-111 aeroplanes

AARESTRUP, EMIL. Poetry in Danish

JACOBSEN, Hans Henrik
Emil Aarestrup. - [Odense] : Odense Universitetsforlag, 1975
314 p. : ill. - Includes bibliographies
87-7492-122-3 X.709/20416
0852910

AARESTRUP, EMIL. Poetry in Danish
- Biographies

ZERUNEITH, Keld
Den frigjorte. - [København] : Gyldendal, 1981
452p. - Bibliography
87-00-77031-0 X.322/11336
8349541

AARGAN. Switzerland
Rural craft industries, to date

HAENGGLI, Adrian
Die Freiaemter Strohindustrie im 19. Jahrhundert. - Fribourg, Suisse : Éditions universitaires, 1973
163leaves. - (Études et recherches d'histoire contemporaine. Série Mémoires de licence ; 16). - Bibliography
r6305538 X.0702/161c(16)
8818231

AARGAU. Switzerland
Agricultural colleges, to date - Serials

BERICHT ... über die landwirtschaftliche Berufsbildung / Kanton Aargau, Finanzdepartement, Abteilung Landwirtschaft. - 1979/81-. - [Aarau] . [die Abteilung], [1981]-
r6303923 S.P.2/6
8805989

LANDWIRTSCHAFTLICHE Berufsbildung / Kanton Aargau. Finanzdepartement, Abteilung Landwirtschaft. - 1977/79. - Aarau : die Abteilung, [1979]
1v.
r6303922 S.P.2/6
8805989

AARGAU. Switzerland
Economic conditions, to date

BIRCHER, Silvio
Politik und Volkswirtschaft des Aargaus. - Aarau : AT Verlag, [1979]
194p. : ill
3-85502-056-6 X.529/48920
8854025

AARGAU. Switzerland
Politics, to date

BIRCHER, Silvio
Politik und Volkswirtschaft des Aargaus. - Aarau : AT Verlag, [1979]
194p. : ill
3-85502-056-6 X.529/48920
8854033

AARGAU. Switzerland
Public administration, to date

BIRCHER, Silvio
Politik und Volkswirtschaft des Aargaus. - Aarau : AT Verlag, [1979]
194p. : ill
3-85502-056-6 X.529/48920
8854041

AARGAU. Switzerland
Public roads. Law

MÜLLER, Walter, of Boswil
Die öffentliche Strasse und ihre Benutzung nach aargauischem Verwaltungsrecht. - Zürich, 1973
xi, 180 p.
r6003663 X.100/14897
0882593

AARHUS
See Arhus

AARHUS UNIVERSITET

AARHUS Universitet 1928-1978 / udgivet af Det lærde Selskab ; under redaktion of Gustav Albeck. - [Århus] : Universitetsforlaget i Aarhus, 1978
648p. - (Acta jutlandica ; 51)
87-87671-07-7 Ac.1022[vol.51]
8921970

AARON. Bible. O.T. Exodus
Ceremonial clothing

TILLER, Lawrence V.
Garments for glory and for beauty. - Cheltenham : [Grenehurst Press], c1981
80p : ill
B82-08684 X.208/7077
261510x

AARON WARD (SHIP). Minelayers. United States. Navy. Fifth Fleet
Naval operations during Battle of Okinawa in World War 2

LOTT, Arnold S.
Brave ship, brave men. - New York : Dale Books ; [London] : Distributed by Magnum Books, 1978
[13],268p : ill
0-458-93540-9 X.708/21901
1892401

AARSCHOT. Belgium
to date

MILLET, Charles
De kroniek van Aarschot van Charles Millet (1597). - [Aarschot] : Aarschotse Kring voor Heemkunde, 1983
168p : ill. - (Bijdragen tot de geschiedenis van het land van Aarschot)
r6306960 X.520/35512
9127232

AARSDALE. Denmark
Houses, to date

CHRISTENSEN, Hanne
Historiske huse i Aarsdale fiskerleje. - [København] : Nationalmuseet, [1980]
192p : ill
87-480-0313-1 X.425/3791
893276x

AARTRIJKE. Belgium
to date

POLLET, Jules
Bijdragen tot de geschiedenis van Aartrijke. - Zedelgem : Heemkundige Kring 'Pastoor Ronse', 1982
216p : ill
r6330739 X.805/5019
9040447

AAS, ALF-JØRGEN. Norwegian paintings

ALF-JØRGEN Aas 1915-1981 / [katalog redigert av Anne Grete Ljøsne]. - Oslo : Oslo kunstforening, [1983]
45p : ill
Catalogue of the exhibition inserted in this copy
r6612193 X.421/26685
8329710

AASEN, IVAR. Linguistics. Norwegian language

HOVDEN, Anders
Ivar Aasen i kvardagslaget. - 3. uppl. - [Oslo] : Fonna, 1978
60p
82-513-0037-1 X.989/84102
2219980

MYHREN, Magne
Ei bok om Ivar Aasen, sprakgranskaren og malreisaren. - Oslo : Det Norske Samlaget, 1975
220 p.. - (Orion-bøkene ; no. 170)
82-521-0425-8 X.908/33685
0935751

AASKOV. Denmark
ca 1650-ca 1950

VEJEN LARSEN, Jens
Brogede blade af Aaskov kommunes historie, med enkelte udsnit fra nabosognet Arnborg. - [Kibæk] : Aaskov kommune, 1974
55 p.
r6001505 X.709/20796
0877050

ABACA
See Manila hemp

ABACKA. Sweden
Sjögren, Gunda - Biographies

SJÖGREN, Gunda
My roots survive the flood. - London : Grosvenor Books, 1978
107p,[8]p of plates : ill. - (A Grosvenor biography)
0-901269-31-x X.708/21139
1413570

ABACUS. Software packages. ACT Sirius microcomputer systems

MORRIS, Stephen, 1955 Nov. 14-
Using Abacus on the ACT Sirius. - London : Duckworth, 1985
197p : ill. - (Duckworth business computing). - Includes index
0-7156-1941-1 X.629/26686
320166x

ABAD GÓMEZ, HÉCTOR. Intellectual life. Colombia - Biographies

ABAD GÓMEZ, Héctor
Cartas desde Asia. - [Medellín] : Universidad de Antioquia, [1977?]
127p. - (Ediciones Universidad de Antioquia ; vol.1)
r6701785 X.529/43848
8741492

ABADIA DE TIBÃES
Ecclesiastical estates. Social conditions, 1630-1680

OLIVEIRA, Aurélio de
A Abadia de Tibães e o seu domínio, 1630-1680. - Porto : Faculdade de Letras do Porto, 1974
406p.,5 leaves of plates. - Bibliography : p.367-389
r6037928 X.510/10104
1162675

ABAELANDUS, PETRUS. Scholastic philosophy - Conference proceedings

ABÉLARD en son temps. - Paris : Les Belles Lettres, 1981
237p : ill. - ([Classiques de l'histoire de France au Moyen-Age])
2-251-34302-4 WP.1576/32
8387958

ABAELARDUS, PETRUS
Theories of Christian doctrine compared with theories of Bernard of Clairvaux, Saint

PADELLARO DE ANGELIS, Rosa
Filosofia e Mistica nel XII secolo. - Roma : ELIA, 1976
233p
r6100274 X.100/28267
2429578

ABAELARDUS, PETRUS
Theories of universals

TWEEDALE, Martin Middleton
Abailard on universals. - Amsterdam ; Oxford : North-Holland Publishing Co., 1976
x,337p. - Bibliography: p.329-330. - Includes index
0-7204-8040-x X.519/26297
0763284

ABAELARDUS, PETRUS. French philosophy - Conference proceedings

ABÉLARD. - Genève : Revue de théologie et de philosophie, 1981
131p. - (Cahiers de la Revue de théologie et de philosophie ; 6)
r6209025 X.529/4973?
817282x

ABAELARDUS, PETRUS. Moral philosophy

WILLIAMS, Paul L.
The moral philosophy of Peter Abelard. - Lanham : University Press of America, [1980]
iii,187p. - Bibliography
0-8191-1137-6 X.529/5685?
873819x

ABAELARDUS, PETRUS. Scholastic philosophy

ABAELARDUS, Petrus
[Correspondence. Selections]. Peter Abelard. Letters IX-XIV. - Groningen : [E.R. Smits], 1983
xii,315p. - Includes bibliography
90-608-8085-4 X.529/6243?
8387958

ABAELARDUS, Petrus
[Historia calamitatum. English]. The story of his misfortunes ; and, The personal letters. - London : Folio Society, 1977
130p : ill
B77-28376 X.981/2076?
0075671

CROCCO, Antonio
Abelardo. - Napoli : Liguori, 1979
178p. - (Antichi e moderni. Sezione medievale). - Bibliography
88-207-0872-8 X.529/4917?
8387958

INTERNATIONAL CONFERENCE ON PETER ABELARD (1971 : Louvain)
Peter Abelard. - Leuven : Leuven University Press ; Hague : Nijhoff, 1974
xiv,181p. - (Mediaevalia Lovaniensia. Series 1 Studia 2)
90-618-6005-9 X.0702/33?a.(2
1475126

PERNOUD, Régine
Héloïse et Abélard. - Nouvelle éd. - Paris : Albin Michel, [1981]
298p. - Includes bibliography
2-226-01170-6 X.529/?866?
8387958

URBANI ULIVI, Lucia
La psicologia di Abelardo e il Tractatus de intellectibus. - Roma : Edizioni di storia e letteratura, 1976
137p. - (Edizioni di storia e letteratura. Opere varie) - Includes bibliographies
r6069180 X.510/?178?
2094207

ABAELARDUS, PETRUS. Scholastic philosophy - Biographies

LUSCOMBE, David Edward
Peter Abelard. - London : Historical Association, 1979
40p. - (General series / Historical Association ; 95) - Bibliography: p.38-40
0-85278-202-0 WP.31?5/9?
1912828

ABAELARDUS, PETRUS. Scholastic philosophy
– Conference proceedings

COLLOQUE ABÉLARD (1979 : Neuchâtel)
Abélard. – Lausanne : Revue de théologie et de Philosophie, 1981
131p. – (Cahiers de la Revue de théologie et de philosophie ; 6)
r6207244 X.529/47739
8387958

ABAELARDUS, PETRUS. Scholastic philosophy
– Conference proceedings – Biographies

PETRUS Abaelardus (1079–1142) / herausgegeben von Rudolf Thomas ... [et al.]. – Trier : Paulinus-Verlag, 1980
339p. – (Trierer theologische Studien ; Bd.38)
3–7902–0041–7 X.520/24926
8387958

ABAELARDUS, PETRUS. Scholastic philosophy
– Correspondence, diaries

ABAELARDUS, Petrus
[Historia calamitatum. Old French]. Le lettere di Abelardo ed Eloisa nella traduzione di Jean de Meun. – Modena : STEM Mucchi, 1977
2v.. – (Studi, testi e manuali / Istituto di filologia romanza dell'Università di Roma ; 5)
r6507830 X.200/40766
8387958

ABAELARDUS, PETRUS. Scholastic philosophy
Correspondence with Héloïse, Abbess, the Paraclete. Authenticity

DRONKE, Peter
Abelard and Heloise in medieval testimonies. – [Glasgow] : University of Glasgow Press, 1976
61p. – (W.P. Ker memorial lecture ; 26th)
0–85261–136–6 X.909/41278
1064045

ABAELARDUS, PETRUS. Scholastic philosophy
Interpersonal relationships with Héloïse, Abbess of the Paraclete

ROBERTSON, Durant Waite
Abelard and Heloise. – London : Millington, 1974
iii–xvi,238p. – Bibliography: p.227–228. – Includes index
0–86000–005–2 X.529/30175
0352969

ABAFA
See Association of British Adoption and Fostering Agencies

ABAILARD, PIERRE
See Abaelardus, Petrus †

ABAKUÁ
Rituals & symbols

CABRERA, Lydia
Anaforuana. – Madrid : Ediciones R, 1975
498p : ill. – (Los de hoy)
84–399–4611–2 X.700/22849
2440644

ABALLI, ANGEL ARTURO. Paediatrics. Cuba
– Biographies

ALONSO MENÉNDEZ, Daniel
Profesor Angel Arturo Aballí, maestro de la pediatría cubana. – La Habana : Academia de Ciencias de Cuba, Centro de Estudios de Historia y Organización de la Ciencia Carlos J. Finlay, 1979
18leaves. – (Conferencias y estudios de historia y organización de la ciencia ; 12)
r6799020 LAS.D.300/15 Woolwich
2479931

ABANDONED CANALS. West Midlands (Metropolitan County)

CHESTER-BROWNE, Richard
The other sixty miles. – Stafford : Birmingham Canal Navigations Society, 1981
[42]p : ill. – Bibliography: p42
B82–41034 X.622/13887
2801205

ABANDONED VEHICLES. England
Abandoned motor vehicles. Disposal. Law

DISPOSAL of abandoned vehicles and other refuse. – London : Shaw, 1981
29p. – (The Shaway guides ; no.4)
0–7219–0875–6 X.0529/584.(4)
257232x

ABANDONMENT. Aircraft
Parachuting, to 1945 – Collections

BOWMAN, Gerald
Jump for it. – London : White Lion Publishers, 1975
198p,[12]p of plates : ill
0–85617–739–3 X.629/10484 Woolwich
0636142

ABANDONMENT. Aircraft
Parachuting, to 1976 – Collections

MACKERSEY, Ian
Into the silk. – Revised ed.. – London : Mayflower, 1978
232p,[8]p of plates : ill. – Includes index
0–583–12793–2 X.708/21102
1400045

ABANDONMENT. Projects, 1945–1972. Aeronautical engineering. Great Britain

WOOD, Derek
Project cancelled. – London : Macdonald and Jane's, 1975
253p : ill
0–356–08109–5 X.620/15158
0501441

ABANYAMBO. Karagwe. Tanzania
to date

KATOKE, Israel K.
The Karagwe Kingdom. – Nairobi : East African Publishing House, 1975
xx,182p : ill. – (Peoples of East Africa ; 5)
r6052924 X.809/50933
8806179

ABARCA DE BOLEA Y XIMÉNEZ DE URREA, PEDRO PABLO, CONDE DE ARANDA
See Aranda, Pedro Pablo Abarca de Bolea y Ximénez de Urrea, Conde de

ABARTH CARS
Abarth racing cars

BRADEN, Pat
Abarth. – Santa Ana : Newport Press ; London : Osprey, 1983
160p : ill
0–85045–517–0 X.622/20817
3139069

ABASHEVO ANTIQUITIES

PRÏÂKHIN, Anatoliĭ Dmitrievich
Pogrebal'nye abashevskie pamîâtniki. – Voronezh : Izdatel'stvo Voronezhskogo universiteta, 1977
166p : ill
r6094858 X.419/8248 Woolwich
1229265

Poseleniîâ abashevskoĭ obshchnosti. – Voronezh : Izdatel'stvo Voronezhskogo universiteta, 1976
167p,[2]fold p of plates : ill. – Includes bibliographies
r6046271 X.709/23410
1229265

ABATTOIR INDUSTRIES
See Slaughterhouse industries

ABATTOIRS
See Slaughterhouses

ABAZA FOLKLORE
[NARTSKIĬ ėpos. Abaza]. Narty / [sbor, sostavlenie, perevod i kommentarii Vladimira Meremkulova]. – Cherkessk : Stavropol'skoe knizhnoe izdatel'stvo, Karachaevo–Cherkesskoe otdelenie, 1975
335p
r6096832 X.989/79415
1974653

ABAZA LANGUAGE
See Abkhaz language

ABBA. Pop music

SCHULMAN, Leif
Succé på världs-scenen. – [Helsingborg] : Allerbok, 1979
95p : ill
91–7506–041–8 X.435/1085
8365539

ABBA. Pop music
– Serials

ABBA magazine. – No.1– ; [Dec. 1977]–. – Knutsford : Poster Plus Ltd : London : Distributed by Seymour Press, 1977– : ill
B78–30832 P.431/436
1569023

ABBA (GROUP). Singing. Pop music
to 1976

KOFOED, Rud
Bogen om ABBA. – København : Erichsen, [1977]
143p : ill
87–555–0475–2 X.431/2454
1391062

ABBA (GROUP). Singing. Pop music
to 1977 – Biographies

EDGINGTON, Harry
Abba. – Revised ed. – London : Magnum Books, 1978
176p,[16]p of plates
0–417–03370–2 X.439/8525
1449397

LINDVALL, Marianne
Abba. – London : Pop Universal Ltd : Souvenir Press, 1977
96p,fold plate : ill
0–285–62312–5 X.431/10470
1449397

ABBADIA SAN SALVATORE, MONTE AMIATA
736–1198

ABBADIA SAN SALVATORE (Monte Amiata)
Codex diplomaticus Amiatinus. – Tübingen : M. Niemeyer, 1974–
Bd. 1–
r6004004 X.0800/694
0906778

'ABBAS 'ABD AL-BAHÂ'
See 'Abd al-Bahâ' ibn Bahâ' Allâh

'ABBAS EFENDI 'ABD AL-BAHÂ'
See 'Abd al-Bahâ' ibn Bahâ' Allâh

ABBAS, ZAHEER. Cricket
– Biographies

ABBAS, Zaheer
Zed. – Kingswood : World's Work, 1983
159p : ill. – Bibliography: p154. – Includes index
0–437–05420–9 X.629/21296
2961083

ABBASID KHALIFATE
750–ca 1000

LASSNER, Jacob
The shaping of 'Abbâsid rule. – Princeton ; Guildford : Princeton University Press, 1980
xvii,329p : ill. – (Princeton studies on the Near East). – Bibliography: p.305–316. – Includes index
0–691–05281–6 X.800/29188
2176041

ABBASID REVOLUTION

SHABAN, M. A.
The 'Abbâsid Revolution. – Cambridge : Cambridge University Press, 1979
xxii,182p. – Bibliography: p.169–176. – Includes index
B79–14312 X.809/44211
1847899

SHARON, Moshe
Black banners from the east. – Jerusalem : Magnes Press, 1983
265p,[2]leaves of plates. – (Max Schloessinger memorial series. monographs ; 2). – Includes bibliography
9–652–23501–6 X.800/39731
1847899

ABBASOV, SHUKHRAT. Directing. Uzbek cinema films
– Biographies

ABUL-KASYMOVA, Khandzhara
Shukhrat Abbasov. – Tashkent : Izdatel'stvo literatury i iskusstva im. Gafura Guliâma, 1976
112p : ill
r6037020 X.907/15857
1215175

ABBATIA LUBINENSIS
See Opactwo benedyktynów w Lubiniu

ABBAYE DE BÉGARD

LE GOFF, Hervé
Bégard ou, Le petit Cîteaux de l'Armorique. – Guipavas : Éditions Kelenn, 1980
394p,1leaf of plates,6p of plates : ill
r6206844 X.200/41597
8995567

ABBAYE DE CAMBRON
See also
Manuscripts written at Abbaye de Cambron

ABBAYE DE CLUNY

ERLANDE-BRANDENBURG, Alain
L'abbaye de Cluny. – Paris : Caisse nationale des monuments historiques et des sites, 1981
63p : ill. – (Petites notes sur les grands édifices)
2–85822–041–7 X.808/38939
9018530

ROSENWEIN, Barbara H.
Rhinoceros bound. – Philadelphia : University of Pennsylvania Press, 1982
xxii,173p. – (The middle ages). – Includes bibliography
0–8122–7830–5 X.800/39928
9018530

ABBAYE DE FLEURY-SAINT-BENOÎT
Social conditions, ca 1220
DAVRIL, Anselme, dom
La vie à l'abbaye de Fleury-saint-Benoît au XIIIe siècle. - Orléans : Société Archéologique et Historique de l'Orléanais, 1976
30 p. : ill. - (Bulletin de la Société archéologique et historique de l'Orléanais ; n.s. tom. 6, no. 45 bis 1974)
r6012290 Ac.5324[nouv.sér.tom.6.no.45 bis]
0869317

ABBAYE DE FONTAINE-LE-COMTE
RECUEIL des documents de l'Abbaye de Fontaine-le-Comte (XIIe-XIIIe siècles) / publié par Georges Pon ; Archives de la Vienne, Poitiers. - Poitiers : Société des archives historiques du Poitou, 1982
xxix,499p. - (Archives historiques du Poitou ; 61)
r6209664 Ac.6887[61]
8934487

ABBAYE DE LA TRINITÉ (VENDÔME)
1032-1187
JOHNSON, Penelope D. (Penelope Delafield)
Prayer, patronage, and power. - New York : New York University Press, 1981
xii,213p : ill. - Bibliography: p194-207. - Includes index
0-8147-4162-2 X.800/34030
278873x

ABBAYE DE MONT-SAINT-MICHEL
Architecture
ADAMS, Henry
Mont-Saint-Michel and Chartres. - London : Hamlyn, 1980
192p : ill(some col.). - (A Bison book). - Includes index
0-600-34182-8 L.45/833
2350289
Mont-Saint-Michel and Chartres. - Princeton ; Guildford : Princeton University Press, 1981, c1933
xiv,401p,[12]leaves of plates : ill. - Includes index
0-691-03971-2 X.800/32475
2350289

ABBAYE DE SAINT-ÉTIENNE DE DIJON
Cartularies
RIANDEY, Paul
L'organisation financière de la Bourgogne sous Philippe le Hardi. Chartes de l'abbaye de Saint-Étienne de Dijon de 1280 à 1285. - Genève : Megariotis Reprints, 1981
xi,162p. - Includes bibliography
r6210583 X.529/58161
9059415

ABBAYE DE SAINT-HUBERT
CHRONIQUE de l'Abbaye de St-Hubert dite Cantatorium / traduite par A.L.P. de Robaulx de Soumoy ... ; suivie du texte. - Bruxelles : Éditions culture et civilisation, 1982
352p
r6209886 X.200/41248
8977097

ABBAYE DE SAINT-MAURICE D'AGAUNE
Treasure
BOUFFARD, Pierre
Saint-Maurice d'Agaune. - Genève : Les Éditions de Bonvent, 1974
210p. : ill. - (Orfèvrerie médiévale ; 1). - Bibliography : p.199-200
r6000730 LR.420.d.7
0861529

ABBAYE DE ST. GERMAIN DES PRÉS
See St. Germain des Prés (Abbey)

ABBAYE D'OIGNY
Community life
Le COUTUMIER de l'abbaye d'Oigny en Bourgogne au XIIe siècle / introduction, texte critique et tables par Pl.F.Lefèvre...et A.H. Thomas... - Leuven : Spicilegium Sacrum Lovaniense, 1976
lxxxi,124p.,2 leaves of plates : ill. - (Spicilegium sacrum lovaniense. études et documents ; fasc39). - Bibliography : p.lxxx-lxxxi
r6026841 W.P.7396/39
0395323

ABBAYE D'ORVAL
Influence of Jansenism to 1741
DEMOULIN, Louis
Le jansénisme et l'abbaye d'Orval. - Bruxelles, Rome : Institut Historique Belge de Rome, 1976
332p, 1 leaf of plates : ill. - (Bibliothèque de l'Institut historique belge de Rome ; fasc. 23)
r6058905 Ac.6544/5[fasc.23]
1539205

ABBAYE DU BEC
Role in society in Normandy, 1034-1136
VAUGHN, Sally N.
The Abbey of Bec and the Anglo-Norman state, 1034-1136. - Woodbridge : Boydell Press, c1981
168p. - Includes index
0-85115-140-x X.800/30562
2349574

ABBAYE ROYALE DE SAINT-DENIS
LE MOËL, Michel
L'abbaye royale de Saint-Denis. - Paris : Éditions de la Tourelle-Maloine, [1980]
56p : ill. - (Histoire et arts)
r6210142 X.429/14620
8971986

ABBAYE ROYALE DE SAINT-DENIS
- Catalogues
The ROYAL abbey of Saint-Denis in the time of abbot Suger (1122-1151) / Sumner McKnight Crosby ... [et al.] ; [Catalogue of an exhibition held at the Metropolitan Museum of Art, 1981]. - New York : The Metropolitan Museum of Art, [1981]
127p : ill. - Includes bibliography
0-87099-261-9 X.205/1371
8971986

ABBAYE ROYALE DE ST. GERMAIN DES PRÉS
See St. Germain des Prés (Abbey)

ABBAYE SAINT-YVED DE BRAINE
KLEIN, Bruno
Saint-Yved in Braine und die Anfänge der hochgotischen Architektur in Frankreich. - Köln : Kunsthistorisches Institut der Universität, 1984
ii,282p,plates. - (Veröffentlichung der Abteilung Architektur des Kunsthistorischen Instituts der Universität zu Köln ; 28). - Includes bibliography
r6310134 X.429/16570
9159150

ABBAYE SAINTE TRINITÉ DE CAEN
Ecclesiastical estates in England, ca 1100-ca 1300
ABBAYE SAINTE TRINITÉ DE CAEN
Charters and custumals of the Abbey of Holy Trinity, Caen. - London : Published for the British Academy by the Oxford University Press, c1982
liv,163p. - (Records of social and economic history. new series ; 5). - Bibliography: pxv-xvi. - Includes index
0-19-726009-8 Ac.1186/5(5)
2782278

ABBAYE ST-BÉNIGNE, DIJON
Architecture, 1001-1018
SCHLINK, Wilhelm
Saint-Bénigne in Dijon. - Berlin : Mann, 1978
205p,[8,]48p of plates : ill,plans. - (Frankfurter Forschungen zur Architekurgeschichte ; Bd 5)
3-7861-1139-1 7823:hh.4./5 Woolwich
2429330

ABBAZIA DI FERENTILLO
See Abbazia di San Pietro in Valle

ABBAZIA DI MONTECASSINO
DORMEIER, Heinrich
Montecassino und die Laien im 11. und 12. Jahrhundert. - Stuttgart : Hiersemann, 1979
xxxvii,296p. - (Schriften der Monumenta Germaniae historica ; Bd 27). - Bibl : p.viii-xxxvii
3-7772-7901-3 Ac.7003/5[Bd.27]
2423510

ABBAZIA DI OGNISSANTI DI CUTI
to 1811
SADA, Luigi
L'abbazia benedettina d'Ognissanti di Cuti in Terra di Bari. - Bari : Società di Storia Patria per la Puglia, 1974
126 p. : ill. - (Quaderni dell'Archivio storico pugliese ; n. 13)
r6024149 X.700/17539
1019538

ABBAZIA DI SAN PIETRO IN VALLE
ORAZI, Anna Maria
L'Abbazia di Ferentillo centro politico, religioso, culturale dell' Alto Medio Evo. - 2a ed. - Roma : Bulzoni, 1977
83p : ill
r6504070 X.421/22960 Woolwich
8819734

ABBAZIA DI SANTA MARIA DI CHIARAVALLE DI FIASTRA
GENTILI, Otello
L'Abbazia di S.M. di Chiaravalle di Fiastra. - Roma : Herder, 1978
xii,307p : ill(some col.)
r6505880 X.429/13967
8896488

ABBAZIA DI SESTO IN SYLRIS
L'ABBAZIA di Sesto in Sylvis dalle origini alla fine del '200. - [Udine] : La Nuova Base, 1979
xiv,397p. - Includes bibliography
r6512644 X.800/40777
9161384

ABBERLEY HALL
to 1961
ASHTON, Gilbert
Abberley Hall 1921-1961. - Kidderminster : Tomkinson, 1980
244p,[9]p of plates : ill
0-907083-02-1 X.809/46541
2356465

ABBEVILLE. France
Capture by German military forces, 1940
WAILLY, Henri de
Le coup de faux. - Paris : Copernic, 1980
255p,16p of plates : ill,map,plan
2-85984-051-6 X.709/516310
2439964

ABBEY DORE. Hereford and Worcester
Churches: Church of Abbey Dore
MORGAN, Frederick Charles
A short account of the Church of Abbey Dore. - 8th ed. revised. - [Abbey Dore] : The Church, 1977
8p,[4]p of plates : ill. - Bibliography: p.8
B78-10545 X.100/17397
1372866

ABBEY, EDWARD. American writers. Fiction in English
RONALD, Ann
The New West of Edward Abbey. - Albuquerque : University of New Mexico Press, [1982]
xvi,255p. - Includes bibliography
0-8263-0635-7 X.800/40696
8153272

ABBEY HOUSE MUSEUM
ABBEY House Museum Kirkstall. - Leeds : Abbey House Museum, [1984]
15p : col.ill
B85-10689 X.809/62747
1877887

ABBEY HOUSE MUSEUM, KIRKSTALL
KIRKSTALL Abbey House Museum. - St Ives, Cambs. : Photo Precision, [1979?]
16,[4]p : col ill
B79-16313 X.708/21876
1877887

ABBEY OF BEC
See Abbaye du Bec

ABBEY OF COUPAR ANGUS
to 1562
SIEVWRIGHT, Martha Jane
The Abbey of Coupar Angus. - Coupar Angus : Culross, 1983
25p
0-900323-58-2 X.629/21713
2987287

ABBEY OF LA TRINITÉ (VENDÔME)
See Abbaye de la Trinité (Vendôme)

ABBEY OF OUR LADY OF GETHSEMANI
Community life, 1946-1952 - Biographies
MERTON, Thomas
The sign of Jonas. - London : Sheldon Press, 1976
[7],362p
0-85969-081-4 X.100/16581
0741981

ABBEY OF WALTHAM HOLY CROSS AND ST LAWRENCE
BASCOMBE, Kenneth Norman
Notes on the Abbey of Waltham Holy Cross and St Lawrence, Essex. - [Waltham Abbey] : [Waltham Abbey Historical Society], [1967]
[1],6 leaves
r6032141 X.102/2090
1193694

ABBEY ROAD (STUDIOS)
to 1982
SOUTHALL, Brian
Abbey Road. - Cambridge : Stephens, 1982
217p : ill. - Includes index
0-85059-601-7 X.431/11819
2830302

ABBEY THEATRE
to 1909 - Correspondence, diaries
YEATS, W. B. (William Butler)
Theatre business. - Gerrards Cross : Smythe, 1982
330p,[8]p of plates : ill. - Bibliography: p301-307. - Includes index
0-86140-042-9 X.950/15375
2176297

ABBEY THEATRE
to 1978
HUNT, Hugh
The Abbey. - Dublin : Gill and Macmillan, 1979
xiii,306p,[16]p of plates : ill. - Includes index
0-7171-0919-4 X.989/54145
2021439

ABBEY THEATRE
Horniman, Annie Elizabeth Fredericka
FLANNERY, James W.
Miss Annie F. Horniman and the Abbey Theatre. - Dublin : Dolmen Press, 1970 [i.e. 1976]
40p,[2] leaves of plates. - (The Irish theatre series ; 3)
B76-12807 X.0909/458(3)
0626252

ABBEYDALE. Sheffield. South Yorkshire (Metropolitan County)
Watermills, to ca 1850
MEREDITH, Rosamond
The watermills of Abbeydale. - 2nd enlarged ed. - [Sheffield] : [Sheffield City Libraries], 1974
Folder ([6]p) : ill. - (Local studies leaflets / Sheffield City Libraries)
0-900660-22-8 X.0702/348(9)
0309729

ABBEYFEALE. Limerick (County)
Social conditions, to 1979
GLÓR na Féile / in eagar ag Pádraig Ó Fiannachta. - Mainistir na Féile : Coiste na Feile, 1979
80p : ill
B85-27211 X.950/43488
3299619

ABBEYS
Benedictine abbeys. Architecture
PRICE, Lorna
The Plan of St Gall. - Berkeley ; London : University of California Press, 1982
xii,99p : ill(some col.)
0-520-04736-2 L.45/2288
2976218

ABBEYS
Cistercian abbeys, 1050-1200
DUBY, Georges
Saint Bernard. - Paris : Arts et Métiers Graphiques, 1976
220p : ill(some col.). - (Les grands bâtisseurs)
2-7004-0020-8 LR.416.q.4
1158147
- [Another copy.] L.45/509

ABBEYS. Admont. Austria
Stift Admont, to 1974
LIST, Rudolf
Stift Admont 1074-1974. - Ried im Innkreis, 1974
xxiii, 559 p. : ill
r6002373 X.421/8779 Woolwich
0881007

ABBEYS. Amesbury. Wiltshire
Amesbury Abbey. Clock
BACKINSELL, William G. C.
The medieval clock in Amesbury Abbey. - Salisbury : South Wiltshire Industrial Archaeology Society, 1979
Folder ([4]p) : ill. - (Historical monographs / South Wiltshire Industrial Archaeology Society ; 3)
0-906195-02-0 X.702/6420
2175150

ABBEYS. Arnsburg. West Germany
Kloster Arnsburg. Archives. Historical sources for Hesse personal names, 1200-1600
MULCH, Roland
Arnsburger Personennamen. - Darmstadt, 1974
420 p.. - (Quellen und Forschungen zur hessischen Geschichte ; 29)
r6003661 Ac7354[no.29]
0863599

ABBEYS. Assisi. Italy
Badia San Nicolò di Campolango, to 1972
FALCINELLI, Vittorio
Badia S. Nicolò di Campolongo. - Assisi : Tipografia Porziuncola Santa Maria degli Angeli, 1973
135 p., 1 fold. leaf of plates, 10 p. of plates : ill
r6013981 X.700/18904
0863564

ABBEYS. Bad Gandersheim. West Germany
Libraries: Stiftskirche Gandersheim. Bibliothek. Stock: Manuscripts - Catalogues
HARTEL, Helmar
Die Handschriften der Stiftsbibliothek zu Gandersheim. - Wiesbaden : Harrassowitz, 1978
83p. - (Mittelalterliche Handschriften in Niedersachsen ; Hft 2). - Includes bibliographies
3-447-01910-7 2719.b.60/2
1537253

ABBEYS. Bardney. Lincolnshire
Bardney Abbey, to 1976
A GUIDE to the Abbey and Parish of Bardney / [by Peter Clarke et al.]. - [Gainsborough] : Association of the Friends of Bardney Abbey, [1976?]
20p : ill
B79-03298 X.108/17275
1756494

ABBEYS. Bardney. Lincolnshire
Bardney Abbey. Community life, to 1538
OWEN, Dorothy Mary
Life at Bardney Abbey. - Bardney : Bardney Branch of the Workers' Educational Association, 1974
[1],5p. - (Kitching memorial lecture ; 1972)
0-9505165-0-3 X.102/2043
0709964

ABBEYS. Bardney. Lincolnshire
Bardney Abbey. Refounding, 1087
ROGERS, Alan, 1933-
The re-founding of Bardney Abbey by Gilbert of Ghent. - [Bardney] : Association of Friends of Bardney Abbey, 1979
9p : ill
B83-31079 X.205/1284
2928647

ABBEYS. Bari. Italy
Abbazia di Ognissanti di Cuti, to 1811
SADA, Luigi
L'abbazia benedettina d'Ognissanti di Cuti in Terra di Bari. - Bari : Società di Storia Patria per la Puglia, 1974
126 p. : ill. - (Quaderni dell'Archivio storico pugliese ; n. 13)
r6024149 X.700/17539
1019538

ABBEYS. Battle. East Sussex
Battle Abbey
EAST SUSSEX. Planning Department
Battle Abbey. - [Lewes] : The Department, [1979]
12,[4]p(4 fold) : ill(chiefly col). - Bibliography: p.[4]
B79-18798 X.709/33174
1888900

SOMERSET FRY, Plantagenet
Battle Abbey and the Battle of Hastings. - London : Pitkin Pictorials, c1984
23p : ill(some col.). - Bibliography: p23
1-85074-004-6 X.805/6617
3280381

ABBEYS. Battle. East Sussex
Battle Abbey, to 1176 - Translations
BATTLE ABBEY
[Chronicon Monasterii de Bello. English & Latin]. The chronicle of Battle Abbey. - Oxford : Clarendon Press, 1980
xiii,357p. - (Oxford medieval texts). - Bibliography: p.xi-xiii. - Includes index
0-19-822238-6 WP.3189/43
1402838

ABBEYS. Beauchief. South Yorkshire (Metropolitan County)
Beauchief Abbey, to 1974
WALTON, Mary
Beauchief Abbey past and present. - 2nd ed. - Sheffield : Sheffield City Libraries, 1975
Folder([6]p) : ill. - (Local studies leaflets / Sheffield City Libraries). - Bibliography: p.[6]
0-900660-24-4 X.0702/348(5)
0573760

ABBEYS. Beaulieu. Hampshire
Beaulieu Abbey, to 1538
HOCKEY, Frederick
Beaulieu, King John's abbey. - [London] : Pioneer Publications Ltd, 1976
xiv,251p : ill. - Bibliography: p.234-240. - Includes index
0-9502786-1-0 X.200/30816
1063146

ABBEYS. Beaulieu. Hampshire
Beaulieu Abbey. Ecclesiastical estates
BEAULIEU ABBEY
The Beaulieu Cartulary. - Southampton : Southampton University Press, 1974
lxx,281p,IIp of plates. - (Southampton records series ; 17). - Includes index
0-85432-118-7 Ac.2659.da(17)
0358223

ABBEYS. Beaulieu. Hampshire
Beaulieu Abbey. Ecclesiastical estates, 1269-1270
BEAULIEU ABBEY
The account-book of Beaulieu Abbey. - London : The Society, 1975
vi,348p. - (Camden fourth series ; vol.16). - Includes index
0-901050-27-x AC.8118/8[ser.4.vol.16.]
0652490

ABBEYS. Bec-Hellouin. France
to date
BOÜARD, Michel de
L'abbaye du Bec-Hellouin. - Paris : Caisse nationale des monuments historiques et des sites, 1981
63p : ill. - (Monographies)
2-85822-039-5 X.200/36942
8773920

ABBEYS. Belgium
to date
REMOORTERE, Julien van
Abbayes et béguinages de Belgique. - Bruxelles : Meddens, 1981
136p : ill
2-87013-045-7 L.42/2557
9052712

ABBEYS. Belgium
Norbertijnenabdij Tongerlo
TONGERLO. - Westerlo : De Oude Liuden, 1980
80p,[4]leaves of plates : ill
r6303176 X.200/36587
8744289

ABBEYS. Billy-sur-Seine. France
Abbaye d'Oigny. Community life
Le COUTUMIER de l'abbaye d'Oigny en Bourgogne au XIIe siècle / introduction, texte critique et tables par Pl.F.Lefèvre...et A.H. Thomas... - Leuven : Spicilegium Sacrum Lovaniense, 1976
lxxxi,124p.,2 leaves of plates : ill. - (Spicilegium sacrum lovaniense. études et documents ; fasc39). - Bibliography : p.lxxx-lxxxi
r6026841 W.P.7396/39
0395323

ABBEYS. Braine. France
Abbaye Saint-Yved de Braine
KLEIN, Bruno
Saint-Yved in Braine und die Anfänge der hochgotischen Architektur in Frankreich. - Köln : Kunsthistorisches Institut der Universität, 1984
ii,282p,plates. - (Veröffentlichung der Abteilung Architektur des Kunsthistorischen Instituts der Universität zu Köln ; 28). - Includes bibliography
r6310134 X.429/16570
9159150

ABBEYS. Buildwas. Shropshire
Buildwas Abbey
GREAT BRITAIN. Department of the Environment
Buildswas Abbey. - 2nd ed. - [London] : [The Department], 1978
8p
0-86056-047-3 BS.414/2011
111526x

ABBEYS. Burton upon Trent. Staffordshire
Burton Abbey. Ecclesiastical estates
BURTON ABBEY
Charters of Burton Abbey. - Oxford : Oxford University Press for the British Academy, 1979
lix,93p,[4]p of plates : 1 ill. - (Anglo-Saxon charters ; 2). - Includes index
0-19-725940-5 X.0800/527(2)
1876562

ABBEYS. Bury St Edmunds. Suffolk
Bury St Edmunds Abbey. Community life, 1173–1202

SCARFE, Norman
A monk named Jocelin. – Edinburgh : Chambers, c1976
48p : ill. – (The Way it was. Norman invasions). – Bibliography: p48. – Includes index
B82–25840 X.0700/1008(31)
2728192

ABBEYS. Bury St Edmunds. Suffolk
Bury St Edmunds Abbey. Ecclesiastical estates. Archives

The ARCHIVES of the Abbey of Bury St Edmunds / edited by Rodney M. Thomson. – Woodbridge : Published for the Suffolk Records Society by Boydell, c1980
ix, 173p. – (Suffolk charters) (Suffolk Records Society ; vol.21). – Includes index
0–85115–087–x Ac.8139 [vol.21]
2153254

ABBEYS. Caen. France
Abbaye Sainte Trinité de Caen. Ecclesiastical estates in England, ca 1100–ca 1300

ABBAYE SAINTE TRINITÉ DE CAEN
Charters and custumals of the Abbey of Holy Trinity, Caen. – London : Published for the British Academy by the Oxford University Press, c1982
liv,163p. – (Records of social and economic history. new series ; 5). – Bibliography: pxv–xvi. – Includes index
0–19–726009–8 Ac.1186/5(5)
2782278

ABBEYS. Caldey Island. Dyfed
Caldey Abbey. Community life, to 1975

HOWELLS, Roscoe
Total community. – Tenby : H.G. Walters (Publishers) Ltd, 1975
224p : ill
0–901906–11–5 X.200/20408
0657786

ABBEYS. Canterbury. Kent
St Augustine's Abbey (Canterbury)

CLAPHAM, Sir Alfred William
St Augustine's Abbey, Canterbury, Kent. – London : H.M.S.O., 1975
31p,[2]fold leaves : ill. – (Ancient monuments and historic buildings) (Department of the Environment official handbook)
0–11–670159–5 BS.46/2(371)
036293x

ABBEYS. Cava de'Tirreni. Italy
Badia della Santissima Trinità, Cava de'Tirreni. Stock: Illuminated manuscripts – Catalogues

BADIA DELLA SANTISSIMA TRINITÀ (Cava de' Tirreni)
La miniatura nella Badia di Cava. – Napoli : Di Mauro, 1976–1978
2v : facsims
r6068753 2702.p.9
2040069

ABBEYS. Cluny. France
Abbaye de Cluny

ERLANDE-BRANDENBURG, Alain
L'abbaye de Cluny. – Paris : Caisse nationale des monuments historiques et des sites, 1981
63p : ill. – (Petites notes sur les grands édifices)
2–85822–041–7 X.808/38939
9018530

ROSENWEIN, Barbara H.
Rhinoceros bound. – Philadelphia : University of Pennsylvania Press, 1982
xxii,173p. – (The middle ages). – Includes bibliography
0–8122–7830–5 X.800/39928
9018530

ABBEYS. Cluny. France
Benedictine abbeys: Cluny III (Abbey). Design & construction

ARMI, C. Edson
Masons and sculptors in Romanesque Burgundy. – University Park ; London : Pennsylvania State University Press. Illustrations. – c1983
1v. : all ill
0–271–00338–3 X.425/4877
3189996

ABBEYS. Coupar Angus. Tayside Region. Scotland
Abbey of Coupar Angus, to 1562

SIEVWRIGHT, Martha Jane
The Abbey of Coupar Angus. – Coupar Angus : Culross, 1983
25p
0–900323–58–2 X.629/21713
2987287

ABBEYS. Crowland. Lincolnshire
Crowland Abbey. Ecclesiastical estates, ca 970–ca 1535

RABAN, Sandra
The estates of Thorney and Crowland. – Cambridge : University of Cambridge Department of Land Economy, 1977
[9],106p : ill. – (Occasional papers / University of Cambridge. Department of Land Economy ; no.7). – Bibliography: p.102–106
B78–02064 X.0329/97(7) Woolwich
1285327

ABBEYS. Cuneo. Italy
to date

PIOVANO, Anita
Abbazie e certose. – Cavallermaggiore : Gribaudo, 1979
191p,[48]p of col plates : ill
r6504867 L.45/1522
8820007

ABBEYS. Dale. Derbyshire
Dale Abbey, to ca 1270

DE MUSCA, Thomas
The foundation of Dale Abbey in legend and chronicle. – Ilkeston : Ilkeston and District Local History Society, 1977
iv,44p,leaf of plate,[2]p of plates : ill. – (Occasional papers / Ilkeston and District Local History Society ; no.6). – Includes index
0–902165–09–7 X.0802/151(6)
1117386

ABBEYS. Dijon. France
Abbaye de Saint-Étienne de Dijon. Cartularies

RIANDEY, Paul
L'organisation financière de la Bourgogne sous Philippe le Hardi. Chartes de l'abbaye de Saint-Étienne de Dijon de 1280 à 1285. – Genève : Megariotis Reprints, 1981
xi,162p. – Includes bibliography
r6210583 X.529/58161
9059415

ABBEYS. Dijon. France
Abbaye St-Bénigne, Dijon. Architecture, 1001–1018

SCHLINK, Wilhelm
Saint-Bénigne in Dijon. – Berlin : Mann, 1978
205p,[8,]48p of plates : ill,plans. – (Frankfurter Forschungen zur Architekurgeschichte ; Bd 5)
3–7861–1139–1 7823:hh.4./5 Woolwich
2429330

ABBEYS. Dunkeswell. Devon
Dunkeswell Abbey. Community life, to 1539

SPARKS, J. A.
In the shadow of the Blackdowns. – Bradford-on-Avon : Moonraker Press, 1978
viii,133p,plate : ill(some col). – Includes index
0–239–00178–8 X.800/27218
1594923

ABBEYS. Elgin. Grampian Region. Scotland
Libraries: Pluscarden Library. Stock: Books printed before 1801 – Catalogues

PLUSCARDEN ABBEY
Benedictine Abbey of Our Lady and St John the Baptist and St Andrew, Pluscarden. – St Andrews : University Library, 1980
xviii,48p : ill. – Bibliography: pxiv–xviii. – Includes index
0–900897–03–1 2725.c.100
2623617

ABBEYS. Elgin. Grampian Region. Scotland
Pluscarden Abbey

SKINNER, Basil C. (Basil Chisholm)
Pluscarden Abbey, near Elgin, Moray, Scotland. – Derby : Pilgrim, c1981
[24]p : ill(some col.)
0–900594–54–3 X.208/6266
0647624

ABBEYS. England

JONES, David, 1936– (Nov.)
Your book of abbeys. – London : Faber, 1979
52p,[12]p of plates : ill. – (The your book series). – Bibliography: p.50. – Includes index
0–571–10924–1 X.200/31815
1569139

NEW, Anthony S. B. (Anthony Sherwood Brooks)
A guide to the abbeys of England and Wales. – London : Constable, 1985
[466]p : ill
0–09–463520–x X.329/19638
3235157

ABBEYS. England
Abbeys & cathedrals

THURLOW, Gilbert
Cathedrals and abbeys of England. – Norwich : Jarrold & Sons, c1981
[32]p : ill(some col.). – (Jarrold Cotman house series)
0–85306–934–4 X.200/35261
2376342

ABBEYS. England
Abbeys & cathedrals, 1970–1979

THURLOW, Gilbert
Cathedrals & abbeys of England. – Norwich : Jarrold and Sons, 1975
[34]p : chiefly ill(chiefly col)
0–85306–583–7 X.100/16166
0575038

ABBEYS. England
Abbeys & cathedrals. Architecture, 600–1540

MORRIS, Richard, 1947–
Cathedrals and abbeys of England and Wales. – London Dent, 1979
[10],294p,[8]p of plates : ill(some col). – Bibliography: p.283–289. – Includes index
0–460–04334–x X.421/10952
156840x
– [Another copy: paperback.] ISBN 0–460–02239–3 X.200/41030

ABBEYS. England
Abbeys & cathedrals. Flower arrangement, 1966–1976

CLEMENTS, Julia
Flowers in praise. – London : Batsford, 1977
96p,[8]p of plates : chiefly ill(some col)
0–7134–3260–8 X.320/10887 Woolwich
1193422

Flowers in praise. – London : Batsford, 1979
96p,[8]p of plates : ill
0–7134–3328–0 X.423/10928 Woolwich
1193422

ABBEYS. England
Abbeys & priories, to 1978

LITTLE, Bryan
Abbeys and priories in England and Wales. – London : Batsford, 1979
216p : ill. – Bibliography: p.212. – Includes index
0–7134–1712–9 X.423/10821
1936638

ABBEYS. Europe
Religious life, ca 1400–ca 1500

RUTLAND, Jonathan
See inside an abbey. – London : Hutchinson, 1978
4–29p : ill(chiefly col). – (See inside). – Includes index
0–09–128680–8 X.203/2129
1457829

ABBEYS. Evesham. Hereford and Worcester
Evesham Abbey

COX, D. C. (David C.)
Evesham Abbey and the parish churches. – Evesham : Vale of Evesham Historical Society, c1980
25p : ill
0–907353–01–0 X.421/21590 Woolwich
2373130

ABBEYS. Ferentillo. Italy
Abbazia di San Pietro in Valle

ORAZI, Anna Maria
L'Abbazia di Ferentillo centro politico, religioso, culturale dell' Alto Medio Evo. – 2a ed. – Roma : Bulzoni, 1977
83p : ill
r6504070 X.421/22960 Woolwich
8819734

ABBEYS. Fontevrault. France
to date

MELOT, Michel
Fontevrault. – Paris : Éditions Jacques Lanore, 1971
115p : ill. – (Petites monographies des grands édifices de la France ; 90)
r6206981 X.429/13526
8852014

ABBEYS. France
Abbaye de Bégard

LE GOFF, Hervé
Bégard ou, Le petit Cîteaux de l'Armorique. – Guipavas : Éditions Kelenn, 1980
394p,1leaf of plates,6p of plates : ill
r6206844 X.200/41597
8995567

ABBEYS. Garendon. Leicestershire
Cistercian Abbey of St Mary (Garendon). Ecclesiastical estates
HUMPHREY, Wallace
The Cistercian Abbey of St Mary, Garendon. - [Loughborough] : East Midlands Studies Unit, Loughborough University with the assistance of Charnwood Borough Council, [1982?]
78p : ill
0-946207-01-1 X.809/55135
2889811

ABBEYS. Germany
Cistercian abbeys. Churches, to date
COESTER, Ernst
Die einschiffigen Cistercienserinnenkirchen West- und Süddeutschlands von 1200 bis 1350. - Mainz : Gesellschaft für Mittelrheinische Kirchengeschichte, 1984
xx,473p : ill. - (Quellen und Abhandlungen zur mittelrheinischen Kirchengeschichte ; Bd.46)
r6309175 X.0100/505(46)
9076891

ABBEYS. Glastonbury. Somerset
Glastonbury Abbey, to ca 1350
JOHN OF GLASTONBURY
Cronica sive antiquitates Glastoniensis ecclesie. - Oxford : British Archaeological Reports, 1978
2v.([15],lii,417p). - (British archaeological reports ; 47). - Bibliography: [9-15]. - Includes index
0-86054-009-x X.702/6177
1964933

WILLIAM, of Malmesbury
[De antiquitate Glastonie Ecclesie. English]. The early history of Glastonbury. - Woodbridge : Boydell, c1981
viii,224p. - Bibliography: p211-215. - Includes index
0-85115-154-x X.800/33212
2735296

ABBEYS. Glastonbury. Somerset
Glastonbury Abbey, to 1965
WATKIN, Aelred
The story of Glastonbury. - London : Catholic Truth Society, 1982
16p : ill
0-85183-506-6 X.808/38090
2906457

ABBEYS. Glastonbury. Somerset
Glastonbury Abbey. Architecture
BOND, Frederick Bligh
An architectural handbook of Glastonbury Abbey. - London : Research into Lost Knowledge Organization, 1981
88p,[17]p of plates : ill. - Bibliography: p88. - Includes index
0-902103-06-7 X.200/35500
1291564

GAFFNEY, Nicholas J. W.
Glastonbury Abbey. - Henley on Thames : Cressrelles Publishing Co., 1977
19p : ill
0-85956-031-7 Cup.1285/257
1291564

ABBEYS. Glastonbury. Somerset
Glastonbury Abbey. Geomantic aspects
PENNICK, Nigel
The geomancy of Glastonbury Abbey. - [Cambridge] : Fenris-Wolf Publications, 1976
[1],8p : ill. - (Megalithic visions antiquarian papers ; no.11)
0-9505403-7-4 X.0708/1524(11)
1091964
The geomancy of Glastonbury Abbey. - [Cambridge] : Fenris-Wolf, [1983]
8p : ill
B83-22637 X.200/40336
1091964

ABBEYS. Gloucester
Monastery of St Peter's, Gloucester. Ecclesiastical estates, to ca 1400
WALKER, David, 1923-
Register of the churches of the monastery of St Peter's, Gloucester.. - (In. An ecclesiastical miscellany. - Bristol, 1976. - pp.1-58,133-144) (Bristol and Gloucestershire Archaeological Society. Records Section. Publications ; vol.11)
r6051003 Ac.5650.b[vol.11]
106049X

ABBEYS. Graignamanagh. Kilkenny (County)
Duiske Abbey, to 1975
CARVILLE, Geraldine
Norman splendour. - Belfast : Blackstaff Press, 1979
119p : ill. - Bibliography: p.111-112. - Includes index
0-85640-171-4 X.809/49672
1939017

ABBEYS. Great Britain
Abbeys & priories
BOTTOMLEY, Frank, 1920-
The abbey explorer's guide. - London : Kaye & Ward, c1981
vii,284p : ill. - Bibliography: p194
0-7182-1280-0 X.808/35777
2692767

ABBEYS. Great Britain
Abbeys & priories, to 1979
WRIGHT, Geoffrey N. (Geoffrey Norman)
Discovering abbeys and priories. - 2nd ed. - Aylesbury : Shire Publications, 1979
64p,[16]p of plates : ill. - (Discovering series ; no.57). - Includes index
0-85263-454-4 X.429/11704 Woolwich
234890x

ABBEYS. Great Britain
Abbeys, cathedrals & priories, to 1977
NEWCOMBE, Lisa
Cathedrals and abbeys of Britain. - Norwich : Jarrold and Sons, [1978?]
[68]p : ill(chiefly col). - (Jarrold Brancaster book series). - Includes index
0-85306-674-4 X.702/5932
1809164

ABBEYS. Haddington. Lothian Region. Scotland
Nunraw Abbey, to 1981
NUNRAW, past and present. - [Haddington] : [Nunraw Abbey], [1982]
[8]p : ill
B82-26636 X.808/36713
2746956

ABBEYS. Horning. Norfolk
St. Benet's Abbey (Horning)
SNELLING, Joan M.
St. Benet's Abbey, Norfolk. - [New and rev. ed.]. - Norwich : St. Benet's Abbey Committee of Management, [1983]
20p : ill. - Bibliography: p20
0-9509043-0-9 X.809/60139
3138356

ABBEYS. Italy
Cistercian abbeys. Architecture, to date
FARINA, Federico
L'architettura cistercense e l'abbazia di Casamari. - Casamari : Edizioni Casamari, 1978
xi,152p : ill,maps,plans. - Bibliography
r6500918 X.205/815
8728194

NEGRI, Daniele
Abbazie cistercensi in Italia. - Pistoia : Tellini, c1981
246p : ill
r6512306 X.425/5332
8728194

ABBEYS. Keynsham. Avon
Keynsham Abbey. Pottery paving tiles, ca 1350-ca 1500
LOWE, Barbara J.
Medieval floor tiles of Keynsham Abbey. - [Keynsham] : [The author], [1978]
[1],iii,151p : ill. - Bibliography: p.149
B78-33915 X.419/34726 Woolwich
1595334

ABBEYS. Kirkstall. West Yorkshire (Metropolitan County)
Kirkstall Abbey
BREARS, Peter C. D. (Peter Charles David)
Kirkstall Abbey. - [Leeds] : Leeds City Museum, [1982?]
16p : ill(some col.)
B83-27304 X.421/24122
187005x

KIRKSTALL Abbey, Leeds. - St Ives [Cambs.] : Photo Precision, 1973
16,[4]p : col ill. - Bibliography: p.16
B79-15125 X.708/21846
187005x

ABBEYS. Kirkstall. West Yorkshire (Metropolitan County)
Kirkstall Abbey, to 1925
THORNTON, David, 1935-
The picture story of Kirkstall Abbey. - Leeds : D. and J. Thornton, [1978]
16p : of ill
B78-13997 X.709/32119
1404938

ABBEYS. Kornelimünster. West Germany
Kloster Kornelimünster
KÜHN, Norbert
Die Reichsabtei Kornelimünster im Mittelalter. - Aachen : Mayer, 1982
178p.,16p.of plates,[1]folded leaf of plates : ill. - (Veröffentlichungen des Stadtarchivs Aachen ; Bd.3). - Bibliography
3-87519-000-9 X.0800/1623(3)
8977089

ABBEYS. Lacock. Wiltshire
Lacock Abbey. Ecclesiastical estates
LACOCK Abbey charters / edited by Kenneth H. Rogers. - Devizes : Wiltshire Record Society, 1979
[3],vi,139p. - (Publications / Wiltshire Record Society ; vol.34 ; 1978). - Includes index
0-901333-11-5 Ac.5740/14(34)
1994840

ABBEYS. Le Bec-Hellouin. France
Abbaye du Bec. Role in society in Normandy, 1034-1136
VAUGHN, Sally N.
The Abbey of Bec and the Anglo-Norman state, 1034-1136. - Woodbridge : Boydell Press, c1981
168p. - Includes index
0-85115-140-x X.800/30562
2349574

ABBEYS. Leiston. Suffolk
Leiston Abbey. Ecclesiastical estates
LEISTON Abbey cartulary and Butley Priory charters / edited by Richard Mortimer. - Ipswich : Boydell Press for the Suffolk Record Society, 1979
xiv,174p. - (Suffolk charters ; 1). - Bibliography: p.xi-xii. - Includes index
0-85115-106-x X.800/35254(1)
1588419

ABBEYS. Liège (Province). Belgium
Abbaye de Saint-Hubert
CHRONIQUE de l'Abbaye de St-Hubert dite Cantatorium / traduite par A.L.P. de Robaulx de Soumoy ... ; suivie du texte. - Bruxelles : Éditions culture et civilisation, 1982
352p
r6209886 X.200/41248
8977097

ABBEYS. Lilienfeld. Austria
Stift Lilienfeld
MÜLLER, Eugen Hermann
Geschichtlicher Abriss des Stiftes Lilienfeld seit 1700. - Lilienfeld : Verlag des Stiftes Lilienfeld, 1979
440p
r6301717 X.200/36655
9080562

ABBEYS. Llantarnam. Gwent
Llantarnam Abbey, to 1979
MAHONEY, Teresa A.
Llantarnam Abbey. - [Cwmbran] : [Sisters of St Joseph], [1981]
vii,238p. - Bibliography: p195-205
B81-40665 X.200/37051
2575566

ABBEYS. Llanthony. Gwent
Llanthony Abbey, to ca 1850
HOPKINS, Gordon
Llanthony Abbey and Walter Savage Landor. - [s.l.] : C.H.G. Hopkins, 1979
48p : 1ill
0-905928-04-0 X.809/46811
2366762

ABBEYS. Logum. Denmark
Logum Kloster, to 1973
BARTHOLDY, Olga
Munkeliv i Løgum kloster. - 2. opl. - Løgumkloster : Refugiets Kulturforening, 1974
89p : ill. - Includes bibliographies
87-980229-1-1 X.415/2411 Woolwich
1977253

ABBEYS. Loughrea. Galway (County)
Old Abbey (Loughrea), to 1650
MONAHAN, Phelim
The Old Abbey Loughrea, 1300-1650. - [Loughrea] : [P. Monahan], [1983?]
16p : ill
B84-21548 X.809/59179
3083667

ABBEYS. Lubiń. Poland
Opactwo benedyktynów w Lubiniu, 1200–1500
OPACTWO BENEDYKTYNÓW W LUBINIU
Dokumenty opactwa benedyktynów w Lubiniu z XIII–XV wieku = [Diplomata Abbatiae Lubinensis saec. XIII–XV]. – Warszawa : Państwowe Wydawnictwo Naukowe, 1975
430 p.
r6008269 Ac.915/13[tom.17]
0935298

ABBEYS. Macerata (Province). Italy
Abbazia di Santa Maria di Chiaravalle di Fiastra
GENTILI, Otello
L'Abbazia di S.M. di Chiaravalle di Fiastra. – Roma : Herder, 1978
xii,307p : ill(some col.)
r6505880 X.429/13967
8896488

ABBEYS. Malmesbury. Wiltshire
Malmesbury Abbey, to 1832
LUCE, Sir Richard H.
The history of the abbey and town of Malmesbury. – [Chippenham] : Friends of Malmesbury Abbey, 1979
iv–x,206[i.e.212]p,[2]leaves of plates(1 fold) : ill. – Includes index
B80–20065 X.800/28859
2081245

ABBEYS. Milton Abbas. Dorset
Milton Abbey, to 1540
TRASKEY, J. P.
Milton Abbey. – Tisbury : Compton Press, 1978 [i.e. 1979]
xi,268p,[18]p of plates : ill. – Bibliography: p.251–254. – Includes index
0–900193–73–5 X.800/27628
1820907

ABBEYS. Milton Keynes. Buckinghamshire
Bradwell Abbey
BRADWELL Abbey. – Milton Keynes : Milton Keynes Development Corporation, [1980]
16,[3]p : ill(some col). – Bibliography: p.16
0–903379–04–x X.700/26546
2282550

ABBEYS. Mogilno. Poland
CHUDZIAKOWA, Jadwiga
Opactwo benedyktynów w Mogilnie. – Warszawa : Państwowe Wydawnictwo Naukowe, 1977
93p : ill. – (Towarzystwo Naukowe w Toruniu. Prace popularnonaukowe ; nr.31. Zabytki Polski północnej ; nr.3)
r6016429 Ac.766/11[nr.31]
1148893

ABBEYS. Mont-Saint-Michel. France
Abbaye de Mont-Saint-Michel. Architecture
ADAMS, Henry
Mont-Saint-Michel and Chartres. – London : Hamlyn, 1980
192p : ill(some col.). – (A Bison book). – Includes index
0–600–34182–8 L.45/833
2350289

Mont-Saint-Michel and Chartres. – Princeton ; Guildford : Princeton University Press, 1981, c1933
xiv,401p,[12]leaves of plates : ill. – Includes index
0–691–03971–2 X.800/32475
2350289

ABBEYS. Monte Amiata. Italy
Abbadià San Salvatore, Monte Amiata, 736–1198
ABBADIA SAN SALVATORE (Monte Amiata)
Codex diplomaticus Amiatinus. – Tübingen : M. Niemeyer, 1974–
Bd. 1–
r6004004 X.0800/694
0906778

ABBEYS. Montserrat. Spain
Libraries: Montserrat (Benedictine Abbey). Biblioteca. Stock: Manuscripts – Catalogues
MONTSERRAT (Benedictine Abbey). Biblioteca
Catàleg dels manuscrits de la Biblioteca del Monestir de Montserrat. – Montserrat : Monestir de Montserrat, 1977
xxiv,562p.. – (Scripta et documenta ; 25)
84–7202–144–0 2719.g.55
2133873

ABBEYS. Montserrat. Spain
Montserrat (Benedictine Abbey), 1800–1979
MASSOT I MUNTANER, Josep
Els creadors del Montserrat modern. – [Montserrat] : Abadia de Montserrat, 1979
285p. – (Biblioteca Serra d'or ; 23)
84–7202–337–0 X.808/33560
2481596

ABBEYS. Nelson County. Kentucky
Abbey of Our Lady of Gethsemani. Community life, 1946–1952 – Biographies
MERTON, Thomas
The sign of Jonas. – London : Sheldon Press, 1976
[7],362p
0–85969–081–4 X.100/16581
0741981

ABBEYS. Om. Denmark
Om Kloster, to ca 1550
MCGUIRE, Brian Patrick
Conflict and continuity at øm Abbey. – Copenhagen : Museum Tusculanum, 1976
151p.,7p.of plates : ill. – (Opuscula graecolatina ; v.8). – Includes bibliographies
r6028928 X.0900/517(8)
1977369

ABBEYS. Orval. Belgium
Abbaye d'Orval. Influence of Jansenism to 1741
DEMOULIN, Louis
Le jansénisme et l'abbaye d'Orval. – Bruxelles, Rome : Institut Historique Belge de Rome, 1976
332p, 1 leaf of plates : ill. – (Bibliothèque de l'Institut historique belge de Rome ; fasc. 23)
r6058905 Ac.6544/5[fasc.23]
1539205

ABBEYS. Paris. France
St. Germain des Prés (Abbey), 1600–1700
ULTEE, Maarten
The Abbey of St. Germain des Prés in the seventeenth century. – New Haven ; London : Yale University Press, c1981
ix,210p,[7]p of plates : ill. – Bibliography: p193–204. – Includes index
0–300–02562–9 X.200/37397
2596822

ABBEYS. Peterborough. Cambridgeshire
Peterborough Abbey. Cartularies & registers
MARTIN, Janet D. (Janet Douglas)
The court and account rolls of Peterborough Abbey. – [Leicester] : University of Leicester, History Department, 1980
ix,51p. – (Occasional publication / University of Leicester, History Department ; no.2)
0–906696–01–1 X.0700/1083(2)
1882163

PETERBOROUGH CATHEDRAL. Chapter
The cartularies and registers of Peterborough Abbey. – [Peterborough] : Northamptonshire Record Society [for] the Dean and Chapter of Peterborough, 1978
xx,52p,4p of plates : 1 ill. – (Publications / Anthony Mellows Memorial Trust ; vol.1) (Publications / Northamptonshire Record Society ; vol.28). – Includes index
0–901275–39–5 Ac.8128[vol.28]
1882163

ABBEYS. Poitou. France
Abbaye de Fontaine-le-Comte
RECUEIL des documents de l'Abbaye de Fontaine-le-Comte (XIIe–XIIIe siècles) / publié par Georges Pon ; Archives de la Vienne, Poitiers. – Poitiers : Société des archives historiques du Poitou, 1982
xxix,499p. – (Archives historiques du Poitou ; 61)
r6209664 Ac.6887[61]
8934487

ABBEYS. Pontrhydfendigaid. Dyfed
Strata Florida Abbey
SMITH, Jenky Beverley
Abaty Ystrad Fflur, Dyfed. – Llundain [i.e. London] : H.M.S.O., 1977
32p,fold plate : ill. – (Official handbooks / Great Britain. Department of the Environment). – Bibliography: p.32
0–11–670341–5 BS.46/2(386)
1265814

ABBEYS. Provence. France
to date
GERBAULT, Pierre
Les abbayes provençales. – Rennes : Ouest-France, 1980
57p : ill
2–85882–288–3 X.205/887
8754616

ABBEYS. Prüfening. Regensburg. West Germany
Kloster Prüfening, 1100–1200
SCHMITZ, Hans-Georg
Kloster Prüfening im 12. Jahrhundert. – München, 1975
xii, 462 p.. – (Miscellanea Bavarica Monacensia ; Hft/49)
3–87913–049–3 WP.8420/67
0891916

ABBEYS. Ramsey. Cambridgeshire
Ramsey Abbey, to 1540
WISE, John, fl.1881
Ramsey Abbey. – Fenstanton : Grasshopper Press, 1981
226,viip,[5]leaves of plates : ill. – Includes index
0–904701–10–7 X.208/6538
2526948

ABBEYS. Redditch. Hereford and Worcester
Bordesley Abbey. Archaeology, to 1976
RAHTZ, Philip Arthur
Bordesley Abbey, Redditch. – Birmingham : School of History, University of Birmingham, [1976]
[2],25p,[3] leaves of plates : ill(some col)
B76–30593 Cup.1281/327
0728276

Bordesley Abbey, Redditch, Hereford-Worcestershire – Oxford : British Archaeological Reports, 1976
[4],277p,[3]leaves of plates,[17]p of plates : ill(incl 3 col). – (British archaeological reports ; 23). – Bibliography: p.272–276
0–904531–26–0 Cup.911/34
0728276

ABBEYS. Reichenau. West Germany
Church of St George, Reichenau. Frescoes
HUKLOW, Karl
The working of Christ in man. – Edinburgh : Floris Books, 1979
62p : ill
0–903540–28–2 X.200/19082
1911910

ABBEYS. Ripon. North Yorkshire
Fountains Abbey (Ripon)
ROBSON, Peter
Fountains Abbey. – York : P. Robson, c1983
44p : ill
0–9508518–0–9 X.200/41886
1269089

ABBEYS. Ripon. North Yorkshire
Fountains Abbey (Ripon). Ecclesiastical estates. Leases, 1495–1539
The FOUNTAINS Abbey lease book / edited by D.J.H. Michelmore. – [Leeds] : [Yorkshire Archaeological Society], 1981
lxxxv,368p,[1]folded leaf of plates. – (Record series / Yorkshire Archaeological Society ; v.140 for the years 1979 and 1980). – Includes index
B83–07251 Ac.5652/8
2830612

ABBEYS. Ripon. North Yorkshire
Fountains Abbey. Ecclesiastical estates in Craven (District), 1150–1780
RAISTRICK, Arthur
Monks & shepherds in the Yorkshire Dales. – [Leyburn] : Yorkshire Dales National Park Committee, 1976
[3],22,[1]p : ill
0–905455–04–5 BS.116/153
103846x

ABBEYS. Rogate. West Sussex
Durford Abbey. Ecclesiastical estates, to 1536
YATES, E. M. (Edward Marshall)
Durford Abbey and its lands. – [Petersfield] : Published for the Harting Society, 1980
33p : ill. – (The Harting papers)
B83–20317 X.809/54897
2888149

ABBEYS. Saint-Benoît-sur-Loire. France
Abbaye de Fleury-Saint-Benoît. Social conditions, ca 1220
DAVRIL, Anselme, dom
La vie à l'abbaye de Fleury-saint-Benoît au XIIIe siècle. – Orléans : Société Archéologique et Historique de l'Orléanais, 1976
30 p. : ill. – (Bulletin de la Société archéologique et historique de l'Orléanais ; n.s. tom. 6, no. 45 bis 1974)
r6012290 Ac.5324[nouv.sér.tom.6.no.45 bis]
0869317

ABBEYS. Saint-Denis. France
Abbaye royale de Saint-Denis
LE MOËL, Michel
L'abbaye royale de Saint-Denis. – Paris : Éditions de la Tourelle-Maloine, [1980]
56p : ill. – (Histoire et arts)
r6210142 X.429/14620
8971986

ABBEYS. Saint-Denis. France
Abbaye royale de Saint-Denis - Catalogues

The ROYAL abbey of Saint-Denis in the time of abbot Suger (1122-1151) / Sumner McKnight Crosby ... [et al.] : [Catalogue of an exhibition held at the Metropolitan Museum of Art, 1981]. - New York : The Metropolitan Museum of Art, [1981]
127p : ill. - Includes bibliography
0-87099-261-9 X.205/1371
8971986

ABBEYS. Saint-Maurice. Switzerland
Abbaye de Saint-Maurice d'Agaune. Treasure

BOUFFARD, Pierre
Saint-Maurice d'Agaune. - Genève : Les Éditions de Bonvent, 1974
210p. : ill. - (Orfèvrerie médiévale ; 1). - Bibliography : p.199-200
r6000730 LR.420.d.7
0861529

ABBEYS. Scotland
Abbeys & cathedrals. Architecture, ca 400-1977

FENWICK, Hubert
Scotland's abbeys and cathedrals. - London : Hale, 1978
288p : ill. - Includes index
0-7091-6374-6 X.429/10446 Woolwich
1689800

ABBEYS. Sesto al Rhéghena. Italy
Abbazia di Sesto in Sylris

L'ABBAZIA di Sesto in Sylvis dalle origini alla fine del '200. - [Udine] : La Nuova Base, 1979
xiv,397p. - Includes bibliography
r6512644 X.800/40777
9161384

ABBEYS. Settimo. Florence (Province). Italy
Badia a Settimo

CALZOLAI, Carlo Celso
La storia della Badia a Settimo. - 2.ed completamente rifatta. - Firenze : Libreria Editrice Fiorentina, 1976
174p : ill
r6067424 X.709/25431
152951X

ABBEYS. Sherborne. Dorset
Sherborne Abbey. Architecture

ROYAL COMMISSION ON HISTORICAL MONUMENTS (ENGLAND)
Sherborne Abbey, the early Church. - London : H.M.S.O., 1975
[2]p,p.li-lvii : ill
0-11-700750-1 Ac.8072[Dorset - vol.1. Add.]
0589977

ABBEYS. Shrewsbury. Shropshire
Shrewsbury Abbey

SHREWSBURY ABBEY
The cartulary of Shrewsbury Abbey. - Aberystwyth : National Library of Wales, 1975
2v.([4],xxxi,514,[5]p). - Includes index
0-901833-70-3 X.200/20350
0637661

ABBEYS. Stein am Rhein. Switzerland
Kloster Sankt Georgen, Stein am Rhein. Buildings, to 1926

WALDVOGEL, Heinrich
Die Baugeschichte des Klosters St. Georgen in Stein am Rhein. - [Zurich], 1973
120 p. : ill
r6002364 X.709/20823
087888X

ABBEYS. Stoke-on-Trent. Staffordshire
Hulton Abbey, to 1980

LANCASTER, Thelma W.
Hulton Abbey. - Stoke-on-Trent : City Museum and Art Gallery, 1981
iv,25p : ill. - Bibliography: p25
B82-11576 X.200/37684
2626683

Hulton Abbey. - Rev. 2nd ed. - Stoke-on-Trent : City Museum and Art Gallery, 1984
iv,25p : ill. - Bibliography: p25
B84-40062 X.809/61422
2626683

ABBEYS. Stoke-on-Trent. Staffordshire
Hulton Abbey. Archaeology, 1960-1980

HULTON Abbey, Stoke-on-Trent. - [S.l] : [S.n], [1981?]
3p
B81-37074 X.808/35004
2558394

ABBEYS. Stratton-on-the-Fosse. Somerset
Downside Abbey

FITZGERALD-LOMBARD, Charles
Downside Abbey. - Bath : Downside Abbey Trustees, c1981
24p : ill(some col.)
B82-25960 X.800/33261
274502x

ABBEYS. Sumpetar. Dalmatia. Yugoslavia
Sveti Petar u Selu u Poljicima. Ecclesiastical estates, 1080-1187

The CARTULARY of the Benedictine Abbey of St Peter of Gumay (Croatia) 1080-1187 / edited by Edo Pivčević. - Bristol : Arthur, 1984
112p : ill. - Bibliography: p112. - Includes index
B85-15316 X.800/40550
3256480

ABBEYS. Sunderland. Tyne and Wear (Metropolitan County)
St Peter's, Wearmouth, to ca 730

ROBSON, H. L.
The great days of Wearmouth. - [Sunderland] : J.R. Salkeld, [1978]
[1],40p : ill
B79-04011 X.709/32771
1765655

ABBEYS. Sweden
Vreta kloster

ANDRÆ, Gustaf-Adolf
Det hände i klostret. - Borensberg : Noteria, 1981
96p : ill
91-85694-12-6 X.429/15581
9090541

ABBEYS. Thorney. Cambridgeshire
Thorney Abbey. Ecclesiastical estates, ca 970-ca 1535

RABAN, Sandra
The estates of Thorney and Crowland. - Cambridge : University of Cambridge Department of Land Economy, 1977
[9],106p : ill. - (Occasional papers / University of Cambridge. Department of Land Economy ; no.7). - Bibliography: p.102-106
B78-02064 X.0329/97(7) Woolwich
1285319

ABBEYS. Thoronet. France
to date

BERENGUIER, Raoul
L'abbaye du Thoronet. - Paris : Caisse nationale des monuments historiques et des sites, 1981
61p : ill. - (Monographies)
2-85822-038-7 X.200/36925
8773912

ABBEYS. Thurles. Tipperary (County)
Holy Cross Abbey, Thurles. Restoration, 1971-1977

HOLY Cross Abbey restoration = Athchóiriu na Mainistreach. - [Thurles] : Restoration Committee, Holy Cross Abbey, 1978
16,[4]p : of ill
B78-33912 X.702/5762
1594184

ABBEYS. Tibães. Portugal
Abadia de Tibães. Ecclesiastical estates. Social conditions, 1630-1680

OLIVEIRA, Aurélio de
A Abadia de Tibães e o seu dominio, 1630-1680. - Porto : Faculdade de Letras do Porto, 1974
406p.,5 leaves of plates. - Bibliography : p.367-389
r6037928 X.510/10104
1162675

ABBEYS. Tintern. Gwent
Tintern Abbey, to 1981

KELLEY, Frank
850th anniversary of the foundation of Tintern Abbey. - [Abergavenny : [Tintern Festival Organising Committee], [1981?]
8,[4]p : ill
B81-31982 X.200/36386
253830x

ABBEYS. Tintern. Gwent. Local associations. Romanticism
Tintern Abbey

KNIGHT, Jeremy
Tintern and the Romantic movement. - [London] : Department of the Environment, 1977
[1],13p : ill(some col). - Bibliography: p.12
0-86056-002-3 BS.46/2(398)
138287x

ABBEYS. Torquay. Devon
Torre Abbey, to 1539

SEYMOUR, Deryck John
Torre Abbey. - [Torquay] : [The author], 1977
xvi,303p,[40]p of plates : ill. - Bibliography: p.281. - Includes index
0-9505949-0-3 X.800/26546
1282875

ABBEYS. Torquay. Devon
Torre Abbey. Ecclesiastical estates in Devon, to 1539

SEYMOUR, Deryck John
Torre Abbey. - [Torquay] : [The author], 1977
xvi,303p,[40]p of plates : ill. - Bibliography: p.281. - Includes index
0-9505949-0-3 X.800/26546
1282867

ABBEYS. Trzemeszno. Poland

JÓZEFOWICZÓWNA, Krystyna
Trzemeszno. - Warszawa : Państwowe Wydawnictwo Naukowe, 1978
153p,[3] fold. leaves of plates : ill. - (Poznańskie Towarzystwo Przyjaciół Nauk. Prace Komisji Historii Sztuki ; t. 10)
r6095324 AC915/30[t.10]
1978829

ABBEYS. Vale Royal (District). Cheshire
Vale Royal Abbey

VALE Royal Abbey. - [Derby] : English Life, c1984
[12]p : ill
0-85101-219-1 X.808/41172
3251314

ABBEYS. Vallbona. Spain
Monestir de Vallbona, to 1977

PIQUER I JOVER, Josep Joan
Abaciologi de Vallbona, 1153-1977. - Santes Creus : Fundació Roger de Belfort, 1978
440p.,42p.of plates : ill. - (Publicació ; n.5). - Includes bibliographies
84-400-5204-9 X.100/28913
216910X

ABBEYS. Vendôme. France
Abbaye de la Trinité (Vendôme), 1032-1187

JOHNSON, Penelope D. (Penelope Delafield)
Prayer, patronage, and power. - New York : New York University Press, 1981
xii,213p : ill. - Bibliography: p194-207. - Includes index
0-8147-4162-2 X.800/34030
278873x

ABBEYS. Venosa. Italy
Chiesa della Ss. Trinità di Venosa

BOZZONI, Corrado
Saggi di architettura medievale. - Roma : Università degli studi di Roma, Istituto di fondamenti dell'architettura, 1979
141p : ill. - (Saggi di storia dell'architettura ; 3). - Includes bibliography
r6510609 X.200/43633
9093001

ABBEYS. Vreta. Sweden
Vreta Kloster. Community life, 1200-1300

ANDRÆ, Gustaf-Adolf
Klosterbrev från Vreta. - [Stockholm] : Verbum, 1975
69 p. : ill
r6001529 X.419/1810 Woolwich
0889482

ABBEYS. Wales

KING, D. J. Cathcart
Castles and abbeys of Wales. - London : H.M.S.O., 1975
40p : ill(some col). - Bibliography: p.38. - Includes index
0-11-670440-3 BS.46/2(372)
0641421

ABBEYS. Waltham Abbey. Essex
Abbey of Waltham Holy Cross and St Lawrence

BASCOMBE, Kenneth Norman
Notes on the Abbey of Waltham Holy Cross and St Lawrence, Essex. - [Waltham Abbey] : [Waltham Abbey Historical Society], [1967]
[1],6 leaves
r6032141 X.102/2090
1193694

ABBEYS. Waverley (District). Surrey
Waverley Abbey. Community life, to 1536
WARE, Gwen
The White Monks of Waverley. – Farnham : Farnham and District Museum Society, 1976
[4],44,[1]p,plate : 2 ill. – Bibliography: p.42
0–901638–06–4 X.108/16537
1042882

ABBEYS. Wealden (District). East Sussex
Bayham Abbey. Antiquities. Archaeology
STREETEN, Anthony
Bayham Abbey. – Lewes : Sussex Archaeological Society, 1983
viii,149p,[10]p of plates : ill. – (Sussex Archaeological Society monograph ; 2). – Bibliography: p138–142. – Includes index
B83–40637 X.800/36278
2962144

ABBEYS. Wealden (District). East Sussex
Bayham Abbey. Turnham, Robert de
ELVINS, Mark Turnham
Bayham Abbey 1182-1982. – Hove : Chichester Diocesan Fund and Board of Finance, [1981]
5p
B82–05934 X.200/37381
2603160

ABBEYS. Weltenburg. West Germany
Kloster Weltenburg, 1626–1803
RIESS, Otmar
Die Abtei Weltenburg zwischen Dreissigjährigem Krieg und Säkularisation (1626–1803). – Regensburg, 1975
20,514 p. : ill. – (Beiträge zur Geschichte des Bistums Regensburg ; Bd 9)
r6003822 P.101/423[Bd.9]
0876917

ABBEYS. Witham Valley. Lincolnshire
WHITE, Andrew, 1948–
Abbeys of the Witham Valley between Lincoln and Tattershall. – Lincoln : Lincolnshire Museums, City and County Museum, 1978
Folder([1],11p) : ill. – Bibliography: p.1
B79–28244 X.410/10388 Woolwich
193337x

ABBEYS. Zwettl. Austria
Stift Zwettl
LIBER fundatorum Zwetlensis Monasterii. – Graz : Akademische Druck- u. Verlagsanstalt, [1981]
[390]p : ill. – (Codices selecti ; 73)
3–201–01165–7 Ms.Facs.581 (73, 73a)
9080554

ABBOT HALL ART GALLERY
Exhibits: Northern English decorative arts – Catalogues
The MAKERS. – Sunderland : Ceolfrith Press, [1975]
84p : ill. – Includes bibliographies
B79–17741 X.419/4119
1882910

ABBOT, RICK. Musical plays in English
– Librettos
ABBOT, Rick
Beauty and the beast really. – New York ; London : French, c1983
70p
0–573–61906–9 X.958/21890
3076407

ABBOTS. Benedictines. Congregación de San Benito el Real de Valladolid
1499–1568 – Biographies
ZARAGOZA PASCUAL, Ernesto
Los generales de la Congregación de San Benito de Valladolid. – Silos : [Monasterio de Santo Domingo de Silos]. – (Studia silensia ; n.2) – Bibliography : p.455–478
vol.II: Los abades trienales, 1499–1568. – 1976
534p.14p.of plates,1 leaf of plates,1 fold.leaf
84–7009–177–8 X.100/26054
1315013

ABBOTS. Monestir de Santes Creus
Bonanat de Vila-seca, abat de Santes Creus
FORT I COGUL, Eufemià
Un gran vila-secà del segle XIII : Fra Bonanat, abat de Santes Creus. – Santes Creus : Arxiu Bibliogràfic, 1975
95p.,8p.of plates. – (Publicacions de l'Arxiu Bibliogràfic de Santes Creus ; n.33)
84–400–8958–9 X.100/27112
1534459

ABBOTS BROMLEY. Staffordshire
Country houses: Blithfield Hall
BAGOT, Nancy Constance, Lady
Blithfield Hall. – [New ed.]. – Derby : English Life Publications, 1979
16p : ill(some col)
B79–10700 X.708/21756
182287x

ABBOTS BROMLEY. Staffordshire
Wildlife. Conservation
DRABBLE, Phil
My beloved wilderness. – [London] : Fontana, 1977
127p,[8]p of plates : ill. – Includes index
0–00–634892–0 X.319/17803 Woolwich
1298623

ABBOT'S HALL BARN. Museum of East Anglian Life
Architecture
HEWITT, C. A. (Cecil Alec)
Abbot's Hall Barn. – Stowmarket : Museum of East Anglian Life, 1975
8p : ill
0–905596–02–1 X.419/3203 Woolwich
0715433

ABBOT'S HALL MUSEUM
See also
Museum of East Anglian Life

ABBOT'S HALL MUSEUM
Alton Watermill
MUSEUM OF EAST ANGLIAN LIFE
Alton Mill. – Stowmarket : Museum of East Anglian Life, [1977]
10p : chiefly ill
B78–09083 Cup.1281/769
1351605

ABBOT'S HALL MUSEUM
Alton Watermill. Flour. Milling, to 1973
ABBOT'S HALL MUSEUM
The Alton watermill. – [Abbot's Hall] : Friends of the Abbot's Hall Museum, [1974]
23p : ill
0–9501979–1–2 X.419/3229 Woolwich
0737674

ABBOTS LANGLEY. Hertfordshire
Salvatorians, to 1978
SALVATORIAN golden jubilee, Abbots Langley, 1928–1978. – [Watford] : [Salvatorian Fathers], [1978]
40,[3]p : ill(incl 1 col)
B80–04667 X.100/18789
2020807

ABBOTS RIPTON. Cambridgeshire
Nature reserves: Monks Wood National Nature Reserve
NATURE CONSERVANCY COUNCIL
Monks Wood National Nature Reserve. – Deutsche Ausg. – London : The Council, [1977?]
[16]p : ill(chiefly col)
0–901204–55–2 BS.32/150
1754130

ABBOTSLEY. Cambridgeshire
to 1978
WISSON, F. M.
Abbotsley. – [Abbotsley] : [The author], [1978?]
[7],207p : ill. – Bibliography: p.206–207
B79–03301 X.709/32715
1755625

ABBOTSWOOD
Architecture
CROOT, Arnold
Grove Farm, Christchurch, Dorset. – Bournemouth : Bournemouth Local Studies, 1984
22p,[4]p of plates : ill
0–906287–53–7 X.809/62590
3240703

ABBOTT, BERENICE. American photography
O'NEAL, Hank
Berenice Abbott. – London : Thames & Hudson, 1982
255p : ill. – Bibliography: p255
0–500–54086–1 L.45/1866
2849259

ABBOTT, BUD. Acting. American cinema films
Abbott, Bud & Costello, Lou – Biographies
THOMAS, Bob
Bud & Lou. – London : W.H. Allen, 1977
224p : ill. – List of films: p.203–209. – Includes index
0–491–02212–3 X.981/20841
1273159

ABBOTT, EDITH. Social reform. United States
Abbott, Edith & Abbott, Grace – Biographies
COSTIN, Lela B.
Two sisters for social justice. – Urbana : University of Illinois Press, c1983
xv,315p,[8]p of plates : ill
0–252–01013–2 X.950/33688
3071987

ABBOTT, GRACE. Social reform. United States
Abbott, Edith & Abbott, Grace – Biographies
COSTIN, Lela B.
Two sisters for social justice. – Urbana : University of Illinois Press, c1983
xv,315p,[8]p of plates : ill
0–252–01013–2 X.950/33688
3071987

ABBOTT, JACK HENRY. Crimes. United States
– Biographies
ABBOTT, Jack Henry
In the belly of the beast. – London : Arrow, 1982, c1981
xvi,166p
0–09–928140–6 X.529/49770
2783835

ABBOTTS ANN. Hampshire
Church of England. Best, Samuel
GEDDES, Alastair
Samuel Best & the Hampshire labourer. – Andover : Andover Local History Society, [1981?]
64p : 1ill
0–903755–11–4 X.809/50448
2620316

ABBREVIATED HANDSTAMPS. Postmarks
1459–1951
HEINS, H. H. (Henry Hardy)
Numeral cancellations of the British Empire. – 4th ed. – London : Robson Lowe, 1979
64,20p : ill
0–85397–101–3 X.520/22979
2355973

ABBREVIATIONS
See also
Acronyms

ABBREVIATIONS
KOS, Zdeněk
Zkratky, značky, akronymy. – Praha : Horizont, 1983
289p
r6441332 X.950/32790
8958823

ABBREVIATIONS
– Dictionaries
ALEKSEEV, Dmitriĭ Ivanovich
Slovar' sokrashchenii russkogo i͡azyka. – Izd. 2., ispr. i dop. – Moskva : Russkiĭ i͡azyk, 1977
414p
r6055888 X.989/77550
1862804

BANCA COMMERCIALE ITALIANA. Nucleo editoriale
Il siglario mondiale di Enti e Imprese economiche. – Milano : la Banca, 1977
xlii,1253p
r6068990 X.981/12595
006419X

BRID, Alfred R.
Aide-mémoire des sigles et abréviations. – Paris : La Pensée Universelle, [1975]
143 p.
r6001035 X.908/33111
006419X

ABBREVIATIONS
– Dictionaries
COLLINDER, Björn
Förkortningsordbok. – Stockholm : Liber förlag, 1981
101p
91–380–5992–4 X.950/20818
8958823

DE SOLA, Ralph
Abbreviations dictionary .... – New international 5th ed. – New York ; Oxford : Elsevier ; Limerick : [Distributed by] Thomond Books, 1978
xiv,654p
0–444–00213–8 X.950/13544
006419x
– [Another copy] X.981/21556
Abbreviations dictionary. – Expanded international 6th ed. – New York ; Oxford : Elsevier, c1981
xvi,966p
0–444–00380–0 X.950/7530
006419x

ABREVIATIONS
-Dictionaries-continued

GLOSSAIRE d'abréviations multilingue / Conseil des communautés européennes. - Luxembourg : Office des publications officielles des Communautés européennes. 1983
xxxii,589p
92-8240-115-4 066.0148 OPL
8958823

JONES, David J.
The Australian dictionary of acronyms & abbreviations. - Manly : Second Back Row Press, 1977
156p
0-909325-08-1 2050.a Reading Room
006419X

KOBLISCHKE, Heinz
Grosses Abkürzungsbuch. - 2., durchgesehene Aufl. - Leipzig : Bibliographisches Institut, 1980
508p
r6308769 X.950/33589
8958823

LARSSON, William
[Förkortningsordboken]. Nya förkortn. ordboken. - Ny, helt omarb. och utvidgad uppl. - Halmstad : Spektra, 1975
167 p.
91-7136-166-9 X.981/10334
006419X

LEWIS, Harriette
The dictionary of initials. - Kingswood : Elliot Right Way, c1983
128p. - (Paperfronts)
0-7160-0702-9 X.958/20820
006419x
Initial-mania concise initial dictionary. - London : Initials Pocket-Dictionary Ltd ; Sunninghill : Distributed by Ken Dickson (Marketing) Ltd, 1978
[2],144p
0-9506098-0-3 X.907/25434
006419x

LØLAND, Stale
Norsk forkortingsbok. - 2. oppl. - [Oslo] : Cappelen, 1977
68p. - (Norsk sprakrad. Skrifter ; 14)
82-02-01325-9 X.0909/442(14)
1465139

MARSHALL, Nancy
Dictionary of abbreviations. - London : Collins, 1980
[6],314p. - (Gem dictionaries)
0-00-458324-8 X.907/25878
006419X

MILÉA, Antonino Paolo
Dicionário de siglas e abreviaturas. - 3.ed.. - São Paulo : Nova Época Editorial, 1975
280p
r6042981 X.909/33733
006419X

MOLINA CORREA, Gilberto
1001 siglas .... - Quito : Editorial Ecuatoriana, 1975
153p : ill
r6157568 X.909/85629
2292491

PAXTON, John
Everyman's dictionary of abbreviations. - Reprinted with revisions and supplement. - London : Dent, 1981
xiii,384,[8]p. - (Everyman's reference library)
0-460-04538-5 WP.474/36
006419x

PUGH, Eric
Third dictionary of acronyms & abbreviations. - London : Bingley [etc.], 1977
208p. - Includes index
0-85157-224-3 X.989/51226
006419x

REVERSE acronyms, initialisms, & abbreviations dictionary : a companion volume to Acronyms, initialisms, & abbreviations dictionary, with terms arranged alphabetically by meaning of acronym, initialism, or abbreviation ... / edited by Ellen T. Crowley [et al.]. - 5th ed. - Detroit : Gale Research, [1976]
x,754p
0-8103-0515-1 OPL
006419X

UNITED NATIONS. Economic Commission for Latin America. Library
Lista de siglas latinoamericanas = Latin American initialisms and acronyms with English translations. - Detroit : Blaine Ethridge-Books, 1974
157p. - (E/CN.12/LIB.3)
0-87917-035-2 2050.a. Reading Room
006419X

ABBREVIATIONS
Nigerian abbreviations

OSOMUGHA, Chukwunedu N.C
Nigerian book of abbreviations and acronyms. - Onitsha : ABIC Publishers, 1978
183p.
978-226-900-x X.902/3797
2418630

ABBREVIATIONS. Bibliographies in Eastern European languages

GÖRNER, Franz
Osteuropäische bibliographische Abkürzungen = East European bibliographical abbreviations. - 3. neu bearb. und erg. Aufl. - München, 1975-
T.1-
3-7940-3196-2 SF.514/4(2)
116239X

ABBREVIATIONS. Brazilian manuscripts
to date

FLEXOR, Maria Helena Ochi
Abreviaturas. - São Paulo : Governo do Estado de São Paulo, Secretaria da Cultura, Divisão de Arquivo do Estado, 1979
xii,391p
r6706869 LAS.B.275/230
8968403

ABBREVIATIONS. Coins
to date - Dictionaries

SCHLICKEYSEN, F. W. A. (Friedrich Wilhelm Adolf)
Erklärung der Abkürzungen auf Münzen der neueren Zeit, des Mittelalters und des Altertums, sowie auf Denkmünzen und münzartigen Zeichen. - 3. unveränd. Aufl. - Berlin : Transpress, 1981
xxxii,511,9p.,iip.of plates : ill
r6306158 X.429/13429 Woolwich
884271x

ABBREVIATIONS. Czechoslovakia

HRBÁČEK, Josef
Jazykové zkratky v češtině. - Praha : Univerzita Karlova, 1979
126p. - (Acta Universitatis Carolinae. Philologica monographia ; 78). - Bibliography
r6441114 Ac.798/8g(78)
9014705

ABBREVIATIONS. Education. Great Britain

ACRONYMS and initialisms in education. - 2nd ed.. - Leicester : LISE, 1984
iii,[24]p
0-901922-13-7 X.525/10248
321799x
- [Another copy.] 370.942 OPL

ABBREVIATIONS. Electrical engineering

INSTITUTION OF ELECTRICAL ENGINEERS
Symbols and abbreviations for electrical and electronic engineering. - 3rd ed. - London : I.E.E., [1979]
16p : ill
0-85296-197-9 X.619/20017 Woolwich
1934732

ABBREVIATIONS. Electrical engineering
- Dictionaries

BROWN, Phil, 19—
Dictionary of electrical, electronics and computer abbreviations. - London : Butterworth, 1985
[6],237p : ill. - Bibliography: 6th prelim. page
0-408-01210-2 SRL
1934732

ABBREVIATIONS. English language
- Dictionaries

POPA, Constantin
Culegere de abrevieri în limba engleză. - Bucureşti : Editura tehnică, 1978
576p
r6183588 X.989/85197
2265389

ABBREVIATIONS. French language

MENZEL, Hans-Bernd
Abkürzungen im heutigen Französisch. - Rheinfelden : Schäuble, c1983
2v. : ill. - (Reihe Romanistik ; Nr.35). - Includes bibliography
3-87718-735-8 X.0900/328(35)
8976287

ABBREVIATIONS. Insurance. Air transport. Great Britain

SPURWAY, M. J.
Aviation insurance abbreviations, organisations and institutions. - London : Witherby, 1983
63p. - (Monument series)
0-900886-77-3 X.629/23920
3169278

ABBREVIATIONS. Italian language
- Dictionaries

CAPPELLI, Adriano
Lexicon abbreviaturarum. - 6. ed. - Milano : Hoepli, 1979
lxxiii,531p,ixleaves of plates(some fold.) : facsims. - (Manuali Hoepli)
88-203-1100-3 X.989/88321
2475464

ABBREVIATIONS. Labels. Matchboxes
to 1979 - Dictionaries

GWYN-SMITH, Sid
AZ. - 3rd expanded ed. - [Oxford?] : British Matchbox Label Society Bookshop, 1980
36p
0-906937-01-9 X.529/48279
2754436

ABBREVIATIONS. Latin language
- Dictionaries

CAPPELLI, Adriano
Lexicon abbreviaturarum. - 6. ed. - Milano : Hoepli, 1979
lxxiii,531p,ixleaves of plates(some fold.) : facsims. - (Manuali Hoepli)
88-203-1100-3 X.989/88321
2475472

ABBREVIATIONS. Librarianship & information science
- Dictionaries

SLOVAR' sokrashchenii po informatike... = Dictionary of abbreviations in informatics... / [nauchnyĭ redaktor...A.I. Mikhaĭlov]. - Moskva : Mezhdunarodnyĭ t͡sentr nauchnoĭ i tekhnicheskoĭ informat͡sii, 1976
405p
r6021469 X.989/39143
1131265

ABBREVIATIONS. Names. International organisations
to date

GERMANY (Federal Republic). Statistisches Bundesamt
Verzeichnis wichtiger internationaler Abkürzungen. - Ausg. 1981. - Stuttgart : Kohlhammer, 1981
67p.
3-17-003241-0 X.950/37913
9141006

ABBREVIATIONS. National health services. Great Britain
- Dictionaries

HAWLEY, Fred
Initially speaking - all (or nearly all) you ever wanted to know about Health Service abbreviations but were too shy to ask. - [Birmingham] : National Association of Health Authorities in England and Wales, [1978]
[32]p. - Includes index
B79-18207 X.519/29682
1892169

ABBREVIATIONS. Norwegian language
- Dictionaries

LØLAND, Ståle
Norsk forkortingsbok. - [Oslo] : Cappelen, 1976
68p. - (Norsk språkråd. Skrifter ; 14)
82-02-00892-1 X.909/37019
1220136

ABBREVIATIONS. Russian language
- Dictionaries

ALEKSEEV, Dmitriĭ Ivanovich
Slovar' sokrashcheniĭ russkogo i͡azyka. - Izd. 3-e, s prilozheniem novykh sokrashchenii. - Moskva : Russkiĭ i͡azyk, 1983
r6448167 X.950/30451
9076034

ABBREVIATIONS. Système International d'Unités. Measurement. Biology

UNITS, symbols, and abbreviations / editor D.N. Baron. - 3rd ed. - London : Royal Society of Medicine, 1977
48p
0-905958-00-4 SRL
1147102

ABBREVIATIONS. Système International d'Unités. Measurement. Medicine

UNITS, symbols, and abbreviations / editor D.N. Baron. - 3rd ed. - London : Royal Society of Medicine, 1977
48p
0-905958-00-4 SRL
1147110

ABBREVIATIONS. Système International d'Unités. Measurement. Technology

COUNCIL OF TECHNICAL EXAMINING BODIES
SI symbols, abbreviations and conventions. - London : City and Guilds of London Institute, 1976
24p
B78-10958 SRL
1382128

ABBREVIATIONS. Terminology. Jurisprudence. Benelux countries
to date - Dictionaries

SPRUDZS, Adolf
Benelux abbreviations and symbols. - Dobbs Ferry : Oceana, 1971
iii,129p
0-379-00120-9 OPL
892614x

ABBREVIATIONS. Terminology. Seafaring
- Dictionaries
FAVOROV, P. A. (Petr Alekseevich)
Anglo-russkiĭ morskoi slovar'sokrashchenii. - Moskva : Voennoe izdatel'stvo, 1983
632p
r6448165 X.950/30448
9063919

ABBREVIATIONS. Terminology. Textile industries
- Dictionaries
FISCHER-BOBSIEN, C -H.
Internationales Abkürzungslexikon Textil- und Bekleidungsindustrie aus 120 Ländern. - Dülmen / Westfalen : Laumann, c1981
xiv,557p
3-87466-020-6 X.629/24909
9166831

ABBREVIATIONS. Titles. Serials
- Standards
BRITISH STANDARDS INSTITUTION
Specification, the abbreviation of titles of periodicals. - London : B.S.I.
Part 2: Word-abbreviation list. - 1st revision. - 1975
[2],71p. - (BS4148. Part 2. 1975)
0-580-08596-1 Enquiry Desk Reading Room
0601055
- [Another copy] 20091.b. Woolwich
Specification...the abbreviation of titles of periodicals. - 1st revision, incorporating amendment issued October 1975. - London : B.S.I.. - (BS4148. part 1. 1970)
Part 1: Principles. - [1975]
11p
0-580-05957-x Enquiry Desk Reading Room
0601055
- [Another copy] 20091.b Woolwich

ABBREVIATIONS. Titles. Serials
to date
ALKIRE, Leland G.
Periodical title abbreviations .... - 3rd ed. - Detroit : Gale, 1981.
By abbreviation
xix,706p
0-8103-0337-x X.950/31197
8825238

ABBREVIATIONS. Titles. Serials
to date - Serials
INTERNATIONAL CENTER FOR THE REGISTRATION OF SERIALS
Liste d'abréviations de mots des titres de périodiques. supplément. - (jan-juin 1981)-. - Paris : CIEPS, 1981-
r6299156 L.49/1259
8825238

ABBREVIATIONS. United States. Army
- Serials
DEFENSE LOGISTICS SERVICES CENTER
Master requirements directory. - Mar. 1977-. - Battle Creek : The Center, 1977-. - (DLAH 4140-3) (DA Pam 708-6)
r6196040 SPR Mic.E.268
2244683

ABBREVIATIONS. United States. Navy
WEDERTZ, Bill
Dictionary of naval abbreviations. - 3rd ed. - Annapolis : Naval Institute Press, c1984
viii,330p
0-87021-155-2 X.800/41280
3328392

ABC
See American Broadcasting Company †
Australian Broadcasting Commission

'ABC'
Reporting on political events in Soviet Union, 1917
LAZO DÍAZ, Alfonso
La revolución rusa en el diario ABC de la época. - Sevilla : Universidad de Sevilla, 1975
174p. - (Colección de bolsillo ; n.34)
84-600-1752-4 X.708/41475
120937X

ABC MICROCOMPUTER SYSTEMS
ABC 80 microcomputer systems. Applications
WESTH, Ake
Measurement and control with a small computer system. - Lund : Studentlitteratur ; Bromley : Chartwell-Bratt, 1981
125p : ill. - Bibliography: p123. - Includes index
0-86238-026-x SRL
2781093

'ABD AL-AZIZ AL-RAHMAN AL SA'UD
See al-Rahman Al Sa'ud, 'Abd al-Aziz

'ABD AL BAHĀ' IBN BAHĀ' ALLĀH. Bahaism
HOFMAN, David
A commentary on the Will and testament of 'Abdu'l-Bahá. - Rev.ed. - Oxford : Ronald, 1982
54p. - Bibliography: p54
0-85398-158-2 X.808/38046
2899167

ABD AL-BAHĀ IBN BAHĀ ALLĀH. Bahaism
- Texts
MEHRABI, Jacqueline
Stories of 'Abdu'l-Behá. - London : Bahá'í Publishing Trust, c1984
48p
0-900125-54-3 X.208/10600
3013057

'ABD AL-MU'MIN IBN 'ALI. FIRST ALMOHAD CALIPH. Almohad Khalifate
BOUROUIBA, Rachid
Abdelmoumène. - Alger : Ministère de l'information et de la culture, 1976
134p : ill. - (Art et culture)
84-399-6079-4 S.V.58/50
8725055

'ABD AL-NASIR, JAMAL
See Nasser, Gamal Abul

'ABD AL-QĀDIR, EMĪR OF MASCARA. Algeria
- Biographies
CHURCHILL, Charles Henry
[The LIFE of Abdel Kader. French]. La vie d'Abdel Kader. - 3e éd. - Alger : S.N.E.D., 1981
353p
r6207523 X.809/53096
8738351

ABD AL-QĀDIR AL JĪLANĪ. Islam
- Biographies
URFI, Aziz
Ghous ul Azam Dastgir. - Karachi : Gaylani Publishers, 1973
xii,242p. 2p. of plates.. - Bibliography : p. 230-234
r6026034 X.200/10808
0392758

ABD AL-QĀHIR AL-JURJĀNĪ
See Al-jurjāni 'Abd al-Qāhir

'ABD AL-RAHMĀN IBRAHIMA
See Ibrahima, 'Abd al-Raḥmān

'ABD AL-RAHMĀN KHĀN. Afghanistan
- Biographies
'ABD AL-RAHMĀN KHĀN
The life of Abdur Rahman. - Karachi ; Oxford : Oxford University Press, 1980
2v. : ill. - (Oxford in Asia historical reprints). - Includes index
0-19-577258-x X.809/49745
2579448

'ABD AL-WAHHĀB IBN AHMAD AL-SHA'RĀNĪ
See Al-Sha'rānī, 'Abd al-Wahhāb ibn Ahmad

'ABD ALLĀH, SĪ (FAMILY)
See Si 'Abd Allāh (Family)

'ABD AR-RAHMĀN IBN MUHAMMAD IBN KHALDŪN
See Ibn Khaldūn, 'Abd ar-Raḥmān ibn Muḥammad

ABD-RU-SHIN. Messianic movements. Germany
AHLBÄCK, Tore
Människosonen - en självgjord messias i Tyskland efter första världskriget. - Abo : Abo akademi, 1983
297p. - (Meddelanden fran Stiftelsens för Abo akademi forskningsinstitut ; nr 81). - Includes bibliography
951-648-891-9 X.0519/601(81)
9091467

'ABD UL-BAHĀ IBN BAHĀ ULLĀH. Bahaism
FOREL, Auguste
Auguste Forel and the Bahá'í Faith. - Oxford : Ronald, 1978
[3],60p
0-85398-076-4 X.108/17858
2051680
The MYSTERY of God / compiled by Mrs Īrán Fúrútan Muhájir. - Rev. ed. - London : Bahá'í Publishing Trust, 1979, c1971
327p : ill
0-900125-44-6 X.200/40088
2886634
WARD, Allan L.
239 days. - Wilmette : Bahá'í Publishing Trust, 1979
xii,218p : ill
0-87743-129-9 X.809/55206
8023956

'ABD UL-BAHĀ IBN BAHĀ ULLĀH. Bahaism
- Biographies
HONNOLD, Annamarie
Vignettes from the life of 'Abdu'l-Bahá. - Oxford George Ronald, c1982
xi,199p,[12]p of plates : ill. - Bibliography: p175-179. - Includes index to anecdotes
0-85398-128-0 X.200/40437
2904454

ABDALLAH, KING OF JORDAN
- Biographies
ABDALLAH, King of Jordan
[al-Takmilah. English]. My memoirs completed, 'al-Takmilah'. - London : Longman, 1978
xxvii,102p,plate
0-582-78082-9 14561.ddd.452 OMPB
1303597

ABDEL KADER, ÉMIR DE MASCARA
See 'Abd al-Qādir Emir of Mascara †

ABDELMOUMÈNE
See 'Abd al-Mu'min ibd 'Ali First Almohad Caliph †

ABDICATED EMPERORS
See also
Insei

ABDOMEN
See also
Omentum

ABDOMEN. Cats. Pets
Diagnosis. Radiography
O'BRIEN, Timothy R.
Radiographic diagnosis of abdominal disorders in the dog and cat. - Philadelphia ; London : Saunders, 1978
xi,682p : ill(some col). - Includes bibliographies and index
0-7216-6898-4 SRL
1566679

ABDOMEN. Children
Diagnosis. Gray scale ultrasonography
GATES, Gary F.
Atlas of abdominal ultrasonography in children. - New York ; Edinburgh : Churchill Livingstone, 1978
[1],xvii,283p : ill
0-443-08015-1 SRL
1774913

ABDOMEN. Children
Pain. Diagnosis
APLEY, John
The child with abdominal pains. - 2nd ed. - Oxford : Blackwell Scientific, 1975
x,117p : ill. - Bibliography: p.108-113. - Includes index
0-632-00641-2 X.319/10180
0515493

ABDOMEN. Dogs. Livestock
Diagnosis. Radiography
O'BRIEN, Timothy R.
Radiographic diagnosis of abdominal disorders in the dog and cat. - Philadelphia ; London : Saunders, 1978
xi,682p : ill(some col). - Includes bibliographies and index
0-7216-6898-4 SRL
1566660

ABDOMEN. Man
Acute abdomen
ANGELL, James Cyril
The acute abdomen for the man on the spot. - 3rd ed. - Tunbridge Wells : Pitman Medical, 1978
116p
0-272-79519-4 X.319/18307 Woolwich
1568361
SHEPHERD, John A. (John Alfred)
Management of the acute abdomen. - Oxford : Oxford University Press, 1982
vii,366p : ill. - (Oxford medical publications). - Bibliography: p359-360. - Includes index
0-19-261322-7 SRL
1568361

ABDOMEN. Man
Acute abdomen. Diagnosis
BEAL, John Mann
Diagnosis of acute abdominal disease. - Philadelphia : Lee and Febiger ; London : Kimpton, 1979
ix,169p : ill. - Bibliography - Includes index
0-8121-0678-4 X.329/12362 Woolwich
1190687

ABDOMEN. Man
Acute abdomen. Diagnosis–continued

BOTSFORD, Thomas Winston
The acute abdomen. – 2nd ed. – Philadelphia ; London : Saunders, 1977
xvi,325p : ill. – Bibliography: p.313–315. – Includes index
0–7216–1886–3 SRL
1190687

COPE, Sir Zachary
Cope's early diagnosis of the acute abdomen. – 16th ed.. – New York : Oxford University Press, 1983
xiv,284p,14p of plates : ill. – Includes index
0–19–503294–2 SRL
1190687

DE DOMBAL, F. T.
Diagnosis of acute abdominal pain. – Edinburgh : Churchill Livingstone, 1980
170p : ill. – Includes bibliographies and index
0–443–01901–0 SRL
1190687

ABDOMEN. Man
Cancer. Diagnosis. Ultrasonography

BRASCHO, Donn J.
Abdominal ultrasound in the cancer patient. – New York ; Chichester : Wiley, c1980
xv,414p : ill. – (Wiley series in diagnostic and therapeutic radiology). – Includes index
0–471–01742–6 SRL
2330814

ABDOMEN. Man
Communicable diseases

WILSON, Samuel E. (Samuel Eric)
Intra-abdominal infection. – New York ; London : McGraw-Hill, c1982
xvi,495p : ill(some col.). – Includes index
0–07–070815–0 SRL
269736x

ABDOMEN. Man
Diagnosis

ABDOMINAL diagnosis / edited by Stewart Wolf. – Philadelphia : Lea and Febiger ; London : Kimpton, 1979
x,209p : ill. – Bibliography: p.187–189. – Includes index
0–8121–0659–8 X.319/19055 Woolwich
1946358

ABDOMEN. Man
Diagnosis. Gray scale ultrasonography

ABDOMINAL gray scale ultrasonography / edited by Barry B. Goldberg. – New York ; London : Wiley, 1977
[9],372p,[8]p of plates : ill(some col). – (Wiley series in diagnostic and therapeutic radiology). – Includes index
0–471–01510–5 SRL
1300032

ABDOMEN. Man
Diagnosis. Laparoscopy

BECK, Kurt
Color atlas of laporoscopy. – Philadelphia ; London : Saunders, 1984
x,508p : ill(some col.). – Includes bibliographies and index
0–7216–1612–7 SRL
0494283

ABDOMEN. Man
Diagnosis. Radiography

ABDOMEN / edited by Abass Alavi, Peter H. Arger. – New York ; London : Grune & Stratton, c1980
xi,448p : ill. – (Multiple imaging procedures ; v.3). – Includes bibliographies and index
0–8089–1306–9 SRL
0593842

ABDOMINAL radiology / James J. McCort, editor ... [et al.]. – Baltimore ; London : Williams & Wilkins, c1981
x,344p : ill. – Includes bibliographies and index
0–683–05751–0 SRL
0593842

CHESNEY, Dorothy Noreen
Radiographic anatomy of the chest and abdomen. – Oxford : Blackwell Scientific, 1976
x,278p : ill. – Includes index
0–632–09440–0 SRL
0593842

INTERVENTIONAL radiology of the abdomen / [edited by] Joseph T. Ferrucci, Jack Wittenberg. – Baltimore ; London : Williams & Wilkins, c1981
xviii,243p : ill. – Includes index
0–683–03174–0 SRL
0593842

ABDOMEN. Man
Diagnosis. Radiography. Anatomical aspects

WHALEN, Joseph P.
Radiology of the abdomen. – Philadelphia : Lea and Febiger ; London : Kimpton, 1976
xv,310p : ill. – Bibliography: p.285–286. – Includes index
0–8121–0569–9 SRL
1089552

ABDOMEN. Man
Diagnosis. Tomography. Applications of digital computer systems

HAAGA, John
Computed tomography of abdominal abnormalities. – Saint Louis : Mosby ; London : Distributed by Kimpton, 1978
xiii,430p : ill. – Includes index
0–8016–2006–6 SRL
1566652

ABDOMEN. Man
Diagnosis. Ultrasonography

DIAGNOSTIC ultrasound of the lower abdomen / A. Trenta ... [et al.]. – Padua : Piccin ; London : Butterworths, 1985
[199]p : ill. – Bibliography: p183–196. – Includes index
0–407–01055–6 SRL
0178659

LEOPOLD, George R. (George Robert)
Fundamentals of abdominal and pelvic ultrasonography. – Philadelphia ; London : Saunders, 1975 [i.e. 1976]
ii–x,267p,[24]p of plates,leaf of plate : ill(some col). – (Saunders monographs in clinical radiology ; vol.6). – Includes bibliographies and index
0–7216–5731–1 SRL
0178659

METREWELI, Constantine
Practical abdominal ultrasound. – London : Heinemann Medical, 1978
xi,123p,[19]p of plates : ill
0–433–21311–6 SRL
0178659

ABDOMEN. Man
Diagnosis. Ultrasonography. Ultrasonograms. Interpretation

BARTRUM, Royal J.
Case studies in ultrasound. – Philadelphia ; London : Saunders, 1979
xiii,435p : ill. – Includes index
0–7216–1553–8 X.622/6899
1904531

ABDOMEN. Man
Hernias. Surgery

PONKA, Joseph L.
Hernias of the abdominal wall. – Philadelphia ; London : Saunders, 1980
xix,644p : ill. – Includes bibliographies and index
0–7216–7274–4 SRL
2253127

ABDOMEN. Man
Muscles. Exercises

DUGAN, Ann
Flatten your stomach for women & men. – London : Century, c1983
64p : col.ill
0–7126–0278–x X.622/17669
1911376

EVERROAD, Jim
How to flatten your stomach. – London : Pan Books, 1979
47p : ill
0–330–25877–x X.319/18933 Woolwich
1911376

ABDOMEN. Man
Pain. Diagnosis

CURRIE, Donald J.
Abdominal pain. – Washington ; London : Hemisphere Publishing Corporation ; New York ; London : McGraw-Hill, 1979
xiii,233p : ill. – Includes bibliographies and index
0–07–014942–9 SRL
1932985

ABDOMEN. Man
Surgery

EDWARDS, Edward Allen
Operative anatomy of abdomen and pelvis. – Philadelphia : Lea and Febiger ; London : Distributed by Kimpton, 1975
xxii,418p : ill(some col). – Includes bibliographies and index
0–8121–0491–9 SRL
0007552

MAINGOT, Rodney
Abdominal operations. – 7th ed. – New York : Appleton-Century-Crofts, c1980
2v(xxxii,2577,75p,[6]p of plates) : ill(some col.). – Includes bibliographies
r6864775 X.322/10839
0007552

NARDI, George Lionel
Abdominal surgery. – Philadelphia ; London : Saunders, 1976
iii–xiii,290p : ill(chiefly col). – Includes index
0–7216–8950–7 SRL
0007552

SCHEIN, Clarence J. (Clarence Jacob)
Introduction to abdominal surgery. – Hagerstown ; London : Harper & Row, c1981
xiv,521p : ill. – Includes bibliographies and index
0–06–142381–5 SRL
2491842

ABDOMEN. Man
Veins. Diseases – Conference proceedings

PELVIC VEIN SYMPOSIUM (1980 : Igls–Innsbruck)
Pelvic and abdominal veins. – Amsterdam ; Oxford : Excerpta Medica, 1981
xvi,374p : ill(some col.). – (International congress series ; no.550). – Includes bibliographies and index
90–219–0494–2 SRL
2552477

ABDOMEN. Newborn babies
Acute abdomen. Diagnosis. Radiography

CIPEL, Louis
Radiology of the acute abdomen in the newborn. – New York ; London : Grune and Stratton ; London : Distributed by Academic Press, 1978
xvii,294p : ill. – Includes bibliographies and index
0–8089–1071–x X.320/11644 Woolwich
1840975

ABDOMENS. Insects
Evolution

MATSUDA, Ryuichi
Morphology and evolution of the insect abdomen. – Oxford : Pergamon, 1976
viii,534p : ill. – (International series in pure and applied biology. zoology division ; vol.56). – Bibliography: p.431–501. – Includes index
0–08–018753–6 SRL
0760625

ABDUCTION
See Kidnapping

'ABDU'L-BAHÁ
See 'Abd al-Bahā' ibn Bahā' Allāh

ABDUL BOKAR KAN. Futa Toro. Senegal
Role in political events, 1853–1891

ROBINSON, David, 1938–
Chiefs and clerics. – Oxford : Clarendon Press, 1975
xiv,239p,[1],vi p of plates : ill. – (Oxford studies in African affairs). – Bibliography: p.206–221. – Includes index
0–19–822701–9 WP.D.16/44
0568295

ABDUL RAHMAN PUTRA, TUNKU. Malaysia
– Biographies

HEALY, Allen
Tunku Abdul RahmanrdA.M. Healy. – St Lucia, Qld. : University of Queensland Press, [1982]
33p. – (Leaders of Asia ; 9)
0–7022–1790–5 X.809/57399
8295085

ABDULLAEV, MIKAIL GUSEIN-OGLY. Azerbaijani paintings

ABDULLAEV, Mikail Guseĭn-ogly
Zhenshchiny Indii. – Moskva : Sovetskiĭ khudozhnik, 1976
[32]p : of col ill
r6173519 X.415/2902 Woolwich
2117231
– [Another copy] X.415/3087 Woolwich

ABEILARD, PIERRE
See Abaelardus, Petrus †

ABEL, BOBBY. Cricket
– Biographies

KYNASTON, David
Bobby Abel. – London : Secker & Warburg, 1982
x,164p,[16]p of plates : ill. – Bibliography: p157–159. – Includes index
0–436–23951–5 X.622/14854
2864568

ÁBEL, JENŐ. Classical studies
- Biographies
BORZSÁK, István
Ábel Jenő. - Budapest : Akadémiai Kiadó, 1981
198p. - (A múlt magyar tudósai). - Bibliography
963-05-2485-6 X.958/13694
8856753

ABEL', RUDOL'F. Soviet espionage
KHENKIN, Kirill
Okhotnik vverkh nogami. - Frankfurt / Main : Possev, [1980]
311p[5]leaves of plates : ill,ports
r6182587 X.708/49184
2508087

ABEL', RUDOL'F. Soviet espionage
- Biographies
KHENKIN, Kirill
[Okhotnik vverkh nogami. French]. L'espionnage soviétique. - Paris : Fayard, 1981
371p,4p of plates
2-213-00983-x X.800/32482
8819807

ABÉLARD, PIERRE
See Abaelardus, Petrus †

ABELARDO, PETRUS
See Abaelardus, Petrus

ABELARDO, PIETRO
See Abaelardus, Petrus †

ABELE (FAMILY). Bavaria. West Germany
HÄUPLER, Hans-Joachim
Das Drama von Ludwigsthal. - Sauerlach : H.-J. Häupler, 1982
121p.,50p.of plates : ill. - (Glashistorische Forschungshefte ; Hft.1)
r6306391 X.0802/1616(1)
8907285

ABELIAN GROUPS
See also
Modules. Algebra
Rings. Algebra

ABELIAN GROUPS
Finite abelian groups. Smooth actions on differentiable manifolds
KOSNIOWSKI, Czes
Actions of finite abelian groups. - London : Pitman, 1978
[9],230p : ill. - (Research notes in mathematics ; 18). - Bibliography: p.222-226. - Includes index
0-273-08405-4 SRL
1338803

ABELIAN GROUPS
Locally compact abelian groups. Pontryagin duality
MORRIS, Sidney A.
Pontryagin duality and the structure of locally compact abelian groups. - Cambridge : Cambridge University Press, 1977
viii,128p : ill. - (Lecture note series / London Mathematical Society ; 29). - Bibliography: p.119-120. - Includes index
0-521-21543-9 SRL
1147080

ABELIAN VARIETIES
MUMFORD, David
Abelian varieties. - 2nd ed.. - London : Oxford University Press for the Tata Institute of Fundamental Research, 1974 [i.e. 1975]
x,279p : ill. - (Studies in mathematics / Tata Institute of Fundamental Research ; no.5). - Bibliography: p.276-277. - Includes index
0-19-560528-4 SRL
0553220

ABELL, KJELD. Danish paintings
BREDSDORFF, Elias
Kjeld Abells billedkunst. - [København] : Gyldendal, 1979
97p : chiefly ill(some col)
87-01-69011-6 X.429/10757 Woolwich
2212587

ABELL, KJELD. Drama in Danish
BJØRNSEN, Johan Faltin
Kjeld Abell. - [Lysaker] : Solum ; [København] : Gyldendal, 1978
161p
82-560-0113-5 X.909/85437
2136465

ABELOUR TRUST
to 1975
ABERLOUR TRUST
The Aberlour Trust centenery [i.e. centenary] year, 1875-1975. - [Stirling] : The Trust, [1975]
[1],43p : ill
0-9504502-0-0 X.519/25939
0686751

ABENCERRAJE, EL. Fiction in Spanish
CARRASCO, María Soledad
The Moorish novel. - Boston [mass.] : Twayne, 1976
197 p.. - (Twayne's world authors series. Spain). - Bibliography: p. 177-187. - Includes index
0-8057-6178-0 X.989/51104
0768448

ABENDS. Programs written in Cobol language. Digital computer systems
Debugging
BOAR, B. H.
Abend debugging for COBOL programmers. - New York : London : Wiley, 1976
xiii,321p : ill. - Bibliography: p.314-315. - Includes index
0-471-08413-1 SRL
1074350

ÅBENRAA. Denmark
to date
BUCHREITZ, Georg
Hjelmhaverne i Aabenraa. - [Aabenraa] : Historisk Samfund for Sønderjylland, 1981
115p : ill
87-7406-036-8 X.809/53185
8860394

ÅBENRAA. Denmark
1850-1920
DRAGSBO, Peter
Mennesker og huse i Aabenraa. - [Esbjerg] : Sydjysk Universitetsforlag, 1978
131p : ill
87-85198-42-0 X.421/20473 Woolwich
2264587

ÅBENRAA. Denmark
Buildings of historical importance
PETERSEN, Viggo
Det gamle Åbenrå. - [København] : Nationalmuseet, c1982
74p : ill
87-480-0298-4 X.429/14619
8953333

ÅBENRAA. Denmark
Buildings of historical importance. Conservation
ENGQVIST, Hans Henrik
Historiske huse i Aabenraa. - [København] : Nationalmuseet, 1975
121p,[1]fold leaf of plates : ill
87-480-0075-2 X.415/3050 Woolwich
2139251

ÅBENRAA. Denmark
Trade licences, 1686-1867
KAMPHÖVENER, Morten
Borgerskaber i Åbenrå 1686-1867. - Åbenrå, 1974
2 bd
87-7016-085-6 X.702/2967
0875589

ABER. Gwynedd
Nature reserves: Coedydd Aber National Nature Reserve. Nature trails
NATURE CONSERVANCY COUNCIL
Coedydd Aber National Nature Reserve nature trail. - Revised [ed.]. - Bangor : The Council, 1977
[12]p : ill
0-901204-36-6 BS.32/146
0322040

ABERAVON. West Glamorgan
Presbyterian Church of Wales. Bethlehem Forward Movement Church (Aberavon), 1927-1938 - Biographies
LLOYD-JONES, Bethan
Memories of Sandfields 1927-1938. - Edinburgh : Banner of Truth Trust, 1983
96p : ill
0-85151-366-2 X.208/9060
2977095

ABERCONWY (DISTRICT). Gwynedd
Gardens open to the public: Bodnant (Garden). Pochin, Henry Davis
MILLIKEN, Harold Turner
The road to Bodnant. - Manchester : Morten, 1975
viii,111p : ill. - Bibliography: p.110-111
0-85972-021-7 X.319/10100
0372188

ABERCROMBIE, LASCELLES. Poetry in English
RAJU, Anand Kumar
Georgian romantic. - Bombay : Blackie, 1983
x,326p. - Includes bibliography
r6808078 X.950/41653
8349738

SASTRI, C. L.
Abercrombie's poems. - Waltair : Andhra University Press, 1971
vii,181p. - (Andhra University series ; no95). - Bibliography : p.175-176. - List of works p. : 177-181
r6071651 X.989/52300
1379895

ABERCYNON. Mid Glamorgan
to 1975
EVANS, Thomas
The story of Abercynon. - 3rd ed. revised and enlarged. - Risca : Starling Press, 1976
82p,leaf of plate,[54]p of plates : ill
0-903434-16-4 X.809/41035
0673765

ABERCYNON. Mid Glamorgan
Bowden, Wilfred George - Biographies
BOWDEN, Wilfred George
Abercynon to Flanders and back. - Newport : Starling, 1984
71p,[9]p of plates : ill
0-903434-71-7 X.809/62906
3246795

ABERDARE. Mid Glamorgan
1870-1920
OLD Aberdare & Merthyr Tydfil in photographs / with commentaries by R. Ivor Parry and Tom Whitney. - Barry : Stewart Williams, Publishers, 1976
[120]p : of ill
0-900807-18-0 X.800/11594
0632422

ABERDEEN
Adolescents. Health
SURVIVAL guide for school-leavers. - [1978]. - [Aberdeen] : [Workers' Educational Association], [1978]
[5],39p : ill
Subsequent volumes, when received, will be placed at this shelfmark
B79-01749 P.523/878
1753053

ABERDEEN
Scotland. First employment
SURVIVAL guide for school-leavers. - [1978]. - [Aberdeen] : [Workers' Educational Association], [1978]
[5],39p : ill
Subsequent volumes, when received, will be placed at this shelfmark
B79-01749 P.523/878
1753045

ABERDEEN. Grampian Region. Scotland
ABERDEEN. - [Aberdeen] : [City of Aberdeen, Department of Development & Tourism], [1981]
32p : col.ill. - (City of Aberdeen tourist guide ; v.8, 1981)
B81-16998 X.805/1389
2392720

ABERDEEN. Grampian Region. Scotland
- Walkers' guides
GRAHAM, Cuthbert
Aberdeen. - [Aberdeen] : [City of Aberdeen Department of Development and Tourism], [1983?]
120p : ill
B83-26243 X.809/55272
2908336

ABERDEEN. Grampian Region. Scotland
ca1850-1979
WALKIN' the mat / compiled by Andrew Cluer. - Aberdeen : Lantern Books, 1980
viii,184p : chiefly ill. - Includes index
B80-40038 L.49/493
2272407

ABERDEEN. Grampian Region. Scotland
Airports: Aberdeen Airport. Development
The DEVELOPMENT of Aberdeen airport 1981-1995. - Glasgow : Scottish Airports, 1981
28p,[7]p,[5]leaves of plates(3 folded) : ill
B81-37626 BS.401/29
2560542

ABERDEEN. Grampian Region. Scotland
Airports: Dyce Airport. Industrial relations. Disputes between Bristow Helicopters Limited & British Air Line Pilots Association

MCDONALD, Robert Howat, Lord McDonald
Report of a court of inquiry into a trade dispute at Dyce Airport, Aberdeen, between Bristow Helicopters Limited and members of the British Air Line Pilots Association, conducted by the Hon. Lord McDonald MC ... - London : H.M.S.O., [1977]
[1],54p. - (Cmnd.6951)
0-10-169510-1 BS.23/866
1257854

ABERDEEN. Grampian Region. Scotland
Architectural design. Simpson, Archibald

ARCHIBALD Simpson. - [Aberdeen] : Aberdeen Civic Society with the financial assistance of the Clydesdale Bank Ltd. and the Scottish Arts Council, 1978
47p : ill
B83-15252 X.429/14164
2869489

ABERDEEN. Grampian Region. Scotland
Arts - Serials

GALLERY. - Sept. 1973. - Aberdeen : The Gallery, 1973
r6051250 P.423/233
0165840

ABERDEEN. Grampian Region. Scotland
Bacteriology. Research organisations: Torry Research Station. Stock: Industrial bacteria. Cultures. Collections: National Collection of Industrial Bacteria

NATIONAL COLLECTION OF INDUSTRIAL BACTERIA
The National Collection of Industrial Bacteria. - 3rd ed.. - London : H.M.S.O., 1975
v,271p. - Includes index
0-11-240540-1 SRL
0604585

ABERDEEN. Grampian Region. Scotland
Bishop, Edith - Biographies

BISHOP, Edith
For you I remember. - Aberdeen : Gourdas House, 1980
103p : ill
0-907301-01-0 X.529/56082
294197x

ABERDEEN. Grampian Region. Scotland
Cathedrals: St. Andrew's Cathedral (Aberdeen). Musical instruments: Pipe organs, to 1981

MORRISSON, A. R. (Andrew Russell)
The organ and organists of St. Andrew's Cathedral, Aberdeen. - Aberdeen : Armusic, 1981
16p. - Includes list of sound recordings
0-9507337-0-9 X.439/11768
2623927

ABERDEEN. Grampian Region. Scotland
Cathedrals: St Machar's Cathedral (Aberdeen), ca 1150-ca 1500

COWAN, Ian B. (Ian Borthwick)
St Machar's Cathedral in the early Middle Ages. - Aberdeen : Friends of St Machar's Cathedral, 1980
12p. - (Occasional papers ; no.6)
B85-05951 X.200/45417
3219690

ABERDEEN. Grampian Region. Scotland
Cathedrals: St Machar's Cathedral (Aberdeen), 1560-1690

STEVENSON, David, 1942-
St Machar's Cathedral and the Reformation. - Aberdeen : Friends of St Machar's Cathedral, 1981
17p. - (Occasional papers ; no.7)
B85-05952 X.200/45415
3219704

ABERDEEN. Grampian Region. Scotland
Cathedrals: St Machar's Cathedral (Aberdeen). Churchyards

SHORT, Agnes
The Kirkyard of St Machar's Cathedral Aberdeen. - Aberdeen : Friends of St Machar's Cathedral, 1982
14p. - (Occasional papers ; no.9)
B85-05950 X.200/45416
3219682

ABERDEEN. Grampian Region. Scotland
Cathedrals: St Machar's Cathedral (Aberdeen). Stained glass windows

MCCARTNEY, Crear
The stained glass windows of St Machar's Cathedral Aberdeen. - Aberdeen : Friends of St Machar's Cathedral, 1979
6p. - (Occasional papers ; no.5)
B85-05949 X.200/45414
3219674

ABERDEEN. Grampian Region. Scotland
Central areas. Environment planning

CENTRAL District local plan. - Aberdeen : Department of Planning and Building Control, City of Aberdeen District Council, 1980
92p : ill
B81-19796 X.525/4409
2406683

ABERDEEN. Grampian Region. Scotland
Central institutions. Curriculum subjects: Home economics: Robert Gordon's Institute of Technology. School of Home Economics, to 1978

ABERDEEN School of Domestic Science / edited by Robert A. Bayliss. - Aberdeen : Robert Gordon's Institute of Technology, 1979
75p : ill. - Bibliography: p.51-52
0-9506289-1-3 X.708/22050
1939823

ABERDEEN. Grampian Region. Scotland
Central institutions: Robert Gordon's Institute of Technology, to 1977

ROBERT GORDON'S INSTITUTE OF TECHNOLOGY
RGIT past, present & future. - Aberdeen : Robert Gordon's Institute of Technology, [1978]
16p : ill(chiefly col)
B80-09654 X.619/22341 Woolwich
2045028

ABERDEEN. Grampian Region. Scotland
Church of Scotland. Mannofield Church (Aberdeen), to 1982

MANNOFIELD CHURCH (Aberdeen)
A century at Mannofield, 1882-1982. - [Aberdeen] : [Mannofield Church?], [1982]
28p : ill
B84-43593 X.805/5986
319387x

ABERDEEN. Grampian Region. Scotland
Community information centres

KOHN, Paul
Community information and advice services in Aberdeen. - Aberdeen : St Katherine's Welfare Rights Project, 1979
iii,49p : ill
B79-16499 X.519/29573
1880853

ABERDEEN. Grampian Region. Scotland
Economic conditions. Effects of exploitation of petroleum deposits in North Sea

MACKAY, G. A. (George Anthony)
North Sea oil and the Aberdeen economy. - London : Social Science Research Council, c1980
105p : ill. - (Occasional paper / North Sea Oil Panel ; no.3). - Bibliography: p104-105
0-86226-036-1 BS.379/65(3)
2390779

ABERDEEN. Grampian Region. Scotland
Environment planning

ABERDEEN (Grampian). Department of Planning and Building Control
City North local plan. - Aberdeen : The Department, 1984
77p,[6]leaves : ill(some col.)
B85-14776 X.805/6419
306297x

ABERDEEN (Grampian). District Council
City north local plan. - Aberdeen : City of Aberdeen District Council, 1983
68p
B84-15363 X.805/5029
306297x

ABERDEEN. Grampian Region. Scotland
Fascist movements. Conflict with anti-fascist movements, 1936-1939

KIBBLEWHITE, Liz
Fascism in Aberdeen. - Aberdeen : Aberdeen People's Press ; London : Distributed by Publications Distribution Coop, 1978
48p : ill. - (APP labour history series ; no.2). - Bibliography: p.48
B79-33959 X.709/33437
1954172

ABERDEEN. Grampian Region. Scotland
Fishing industries, 1870-1880

WATERMAN, J. J.
Aberdeen and the fishing industry in the eighteen seventies. - Aberdeen : [University of Aberdeen. Centre for Scottish Studies], [1983?]
50p. - (Local history pamphlet series)
B84-09842 X.809/57584
3041557

ABERDEEN. Grampian Region. Scotland
Freedom lands. Boundary stones

ABERDEEN (Grampian). Department of Planning and Building Control
The freedom lands and marches of Aberdeen. - Aberdeen : [The Department, [1977]
[9]p : ill
B80-41422 X.805/1285
2321610

ABERDEEN. Grampian Region. Scotland
Goldsmiths, 1450-1850

JAMES, I. E. (Isabel Elizabeth)
The goldsmiths of Aberdeen. - Aberdeen : Bieldside Books, c1981
156p : ill. - Bibliography: p151. - Inlcudes index
B82-22470 X.805/2617
2725746

ABERDEEN. Grampian Region. Scotland
Green, The , to 1900

FRASER, George Milne
Historical Aberdeen. - Finzean : Robin Callander, 1980
[5],46[i.e.52],iv p : ill. - Includes index
B80-40037 X.708/22633
228961x

ABERDEEN. Grampian Region. Scotland
Housing - Serials

ABERDEEN (Grampian). District Council
Housing plan. - 1978-1983-. - [Aberdeen] : [The Council], [1978?]
v.
Holdings: 1980-1985-
B82-19884 P.623/1187
2716968

ABERDEEN. Grampian Region. Scotland
Information services

KOHN, Paul
Community information and advice services in Aberdeen. - Aberdeen : St Katherine's Welfare Rights Project, 1979
iii,49p : ill
B79-16499 X.519/29573
1880861

ABERDEEN. Grampian Region. Scotland
Lower Deeside. Environment planning

LOWER Deeside draft local plan / Department of Planning and Building Control, City of Aberdeen District Council. - Aberdeen : [The Department], 1982
51p,[6]leaves of plates (some folded) : ill
B82-30648 X.805/2930
2759144

ABERDEEN. Grampian Region. Scotland
Lower Deeside. Social conditions - For environment planning

ABERDEEN (Grampian). Department of Planning and Building Control
Lower Deeside local plan, report of survey. - Aberdeen : The Department, 1980
[1],34p,[7]fold leaves of plates : ill
B80-38860 X.525/3788
2302489

ABERDEEN. Grampian Region. Scotland
Maritime aspects, to 1980

DUTHIE, John Lowe
Aberdeen Maritime Museum. - Aberdeen : Aberdeen Art Gallery and Museums, City of Aberdeen District Council, c1984
[30]p : ill(some col.)
B85-00744 X.429/16529
3195872

ABERDEEN. Grampian Region. Scotland
Medicine. Fraser, James Fowler - Biographies

FRASER, James Fowler
Doctor Jimmy. - Aberdeen : Aberdeen University Press, 1980
x,150p
0-08-025737-2 X.950/39490
3249115

ABERDEEN. Grampian Region. Scotland
Middle classes. Social conditions, 1700-1800

EVANS, Godfrey, 1953-
Polite society in Aberdeen in the eighteenth century. - [Aberdeen] : Aberdeen Art Gallery and Museums, [1980]
[1],19p : ill
B80-35565 X.708/22577
2274191

ABERDEEN. Grampian Region. Scotland
Museums: James Dun's House. Exhibits: Scottish antiquities from National Museum of Antiquities of Scotland - Catalogues
NATIONAL MUSEUM OF ANTIQUITIES OF SCOTLAND
Ancient treasures of Scotland. - [Aberdeen] : [The Department], [1978]
[1],47p : ill. - Bibliography: p.46
B78-15925 X.419/34598 Woolwich
1435108

ABERDEEN. Grampian Region. Scotland
Newspapers: 'Evening express' & 'Press and journal', to 1978
FRASER, George
The northern lights. - London : Hamilton, 1978
[6],137p,[16]p of plates : ill. - Includes index
0-241-10073-9 X.989/52954
1450832

ABERDEEN. Grampian Region. Scotland
Organisms
MARREN, Peter
A natural history of Aberdeen. - Finzean : Callander, 1982
184p : ill. - Bibliography: p171-174. - Includes index
0-907184-03-0 X.809/51692
2722003

ABERDEEN. Grampian Region. Scotland
Parks, to date
ABERDEEN. Department of Leisure & Recreation
[Miscellaneous publications not catalogued separately]. - Aberdeen : the Department, [1981]-
r6800311 X.0802/1519
8781907

ABERDEEN. Grampian Region. Scotland
Political parties: South Aberdeen Conservative Association - Serials
ABERDEEN record. - Issue no.1 (Mar./May 1981)-. - Aberdeen : [South Aberdeen Conservative Association], 1981-
v.
B81-38525 P.523/1562
2565641

ABERDEEN. Grampian Region. Scotland
Political strikes: General Strike, 1926
KIBBLEWHITE, Liz
Aberdeen in the General Strike. - Aberdeen : Aberdeen People's Press ; London : [Distributed by] Publications Distribution Co-op, 1977
33p : ill. - Bibliography: p.32
B77-29231 X.709/31584
1249916

ABERDEEN. Grampian Region. Scotland
Public libraries: Aberdeen City Libraries. Commercial and Technical Department. Stock: Documents on natural gas & petroleum - Catalogues
BIBLIOGRAPHY of material on oil and gas. - 3rd ed.. - [Aberdeen] : [City of Aberdeen, Libraries Department], 1981
101p
B82-01497 X.322/10381 Woolwich
2113872

ABERDEEN. Grampian Region. Scotland
Shopping facilities
DRIVERS JONAS (Firm)
City and Royal Burgh of Aberdeen, Shopping Study, 1976. - [London] : [Drivers Jonas for Aberdeen District Council].
Part 1: Report of studies. - [1977]
81,33,[8]p,10 leaves of plates : ill
B79-27262 X.512/6774
1930575
Part 2: Summary and recommendations. - [1977]
[7],42,v p,plate : 1 ill
B79-27263 X.512/6775
1930575

ABERDEEN. Grampian Region. Scotland
Social conditions, 1875-1910
WALKIN' the mat / [compiled by Andrew Cluer]. - Aberdeen : Lantern Books, 1976
[164]p : chiefly ill
0-9502853-1-5 Cup.1281/520
1109030

ABERDEEN. Grampian Region. Scotland
Special schools: Camphill Rudolf Steiner Schools. Faculty of Curative Education. Courses in professional education of teachers in special schools
CURATIVE education. - [Aberdeen] : Published for the Faculty of Curative Education on behalf of the Association of Camphill Communities by Aberdeen University Press, 1981
x,30p
0-08-028440-x X.529/43679
2559080

ABERDEEN. Grampian Region. Scotland
Town houses: Provost Skene's House - Catalogues
ABERDEEN ART GALLERY AND MUSEUMS
Provost Skene's House. - [Aberdeen] : The Gallery, 1980
[24]p : ill(some in col)
B80-41192 X.421/21176 Woolwich
2312182

ABERDEEN. Grampian Region. Scotland
Universities. Libraries: University of Aberdeen. Library - Directories
UNIVERSITY OF ABERDEEN
Guide to the Library. - [No. 1]- ; [1900?]. - [Aberdeen] : [The University], [1900?]-
Subsequent volumes, when received, will be placed at this shelfmark
r6010738 2719.aa.56
1051083

ABERDEEN. Grampian Region. Scotland
Universities. Libraries: University of Aberdeen. Library. Catalogues
UNIVERSITY OF ABERDEEN. Library
Guide to the catalogues. - [Aberdeen] : [The Library], [1976]
[1], 12 p.
r6010594 2719ff.60
1040286

ABERDEEN. Grampian Region. Scotland
Universities. Libraries: University of Aberdeen. Library. Stock: Books with European imprints: Books printed between 1500 & 1600
UNIVERSITY OF ABERDEEN. Library
A short-title catalogue of books printed on the continent of Europe, 1501-1600, in Aberdeen University Library. - Oxford : Oxford University Press for the University of Aberdeen, 1979
xii,313p. - (Studies / University of Aberdeen ; no.156). - Bibliography: p.xi-xii. - Includes index
0-19-714106-4 Ac.1482(156)
1772686

ABERDEEN. Grampian Region. Scotland
Universities: University of Aberdeen
UNIVERSITY OF ABERDEEN
An introduction to the University of Aberdeen. - [Aberdeen] : [The University], [1975]
[2],21p
0-902604-24-4 X.510/15068
0530875

ABERDEEN. Grampian Region. Scotland
Universities: University of Aberdeen - Serials
UNIVERSITY OF ABERDEEN
Staff lists. - 1982-1983-. - Aberdeen : The University, 1983-
v.
B84-09645 P.861/363
0228141

ABERDEEN. Grampian Region. Scotland
Universities: University of Aberdeen. Institute of South-East Asian Biology - Serials
[BULLETIN (University of Aberdeen. Institute of South-East Asian Biology)]. Bulletin / University of Aberdeen Institute of South-East Asian Biology. - No.1 (Summer term 1969)-. - Aberdeen : The Institute, 1969-
v.
r6865251 P.523/940
2794047

ABERDEEN. Grampian Region. Scotland
Universities: University of Aberdeen. Institute of South-East Asian Biology, to 1979
UNIVERSITY OF ABERDEEN. Institute of South-East Asian Biology
The Institute of South-East Asian Biology. - Aberdeen : University of Aberdeen, 1980
ii,12p. - Includes bibliographies
B82-18390 X.322/10749 Woolwich
2708507

ABERDEEN. Grampian Region. Scotland
Universities: University of Aberdeen. Graduates, to 1970
UNIVERSITY OF ABERDEEN
Roll of the graduates of the University of Aberdeen 1956-1970. - [Aberdeen] : Published for the University of Aberdeen by Aberdeen University Press, 1982
xxi,749p
0-08-028469-8 X.520/29796
2893398

ABERDEEN. Grampian Region. Scotland
Welfare services - For retired persons
ABERDEEN OLD PEOPLE'S WELFARE COUNCIL
Retirement. - [Aberdeen] : The Council, 1980
32,[2]p
B80-39158 X.529/37998
2288540

ABERDEEN. Grampian Region. Scotland
Woodside. Public libraries: Sir John Anderson Branch Library, to 1983 - Biographies
HENDERSON, Moira
The admirable mechanic. - [Aberdeen] : Aberdeen City Libraries, [c1983]
25p : ill. - Bibliography: p25
0-946920-00-1 2719.e.938
3251756

ABERDEEN. Hong Kong
Social conditions
HAHN, Werner
Aberdeen. - Hong Kong : Perennial Press, 1974
201 p. : ill
r6000001 X.802/3508
0900435

ABERDEEN (DISTRICT). Grampian Region. Scotland
Bus services. Routes
GRAMPIAN (Region). Department of Public Transportation
Bus service network in Aberdeen City District. - Aberdeen : The Department, 1978
iii[i.e. ii],41,[1] leaves(2 fold),7 leaves of plates(4 fold)
B80-11830 X.512/7200
2051281

ABERDEEN (DISTRICT). Grampian Region. Scotland
Dormer windows. Design
ABERDEEN. Department of Planning and Building Control
Dormer windows. - Aberdeen : [The Department], 1978
15p : ill. - Bibliography: p.15
B80-13474 X.415/6016 Woolwich
2053411

ABERDEEN (DISTRICT). Grampian Region. Scotland
Northern Aberdeen (District). Environment planning
ABERDEEN (Grampian). Department of Planning and Building Control
City North draft local plan. - Aberdeen : The Department, 1982
69p,[1]fold leaf of plates
B82-38916 X.805/3185
2788578

ABERDEEN (DISTRICT). Grampian Region. Scotland
Single persons. Housing
BUCHANAN, Glen
Aberdeen district report. - Edinburgh : Scottish Council for Single Homeless, 1981
ix,58p : ill
0-907050-04-2 X.525/5801
2731967

ABERDEEN (DISTRICT). ART GALLERY AND MUSEUMS DEPARTMENT. JAMES DUN'S HOUSE
See James Dun's House

ABERDEEN (GRAMPIAN). DISTRICT COUNCIL
- Serials
ABERDEEN (Grampian). District Council
Financial statement for the year ended 31st March. - [19-]-. - Aberdeen : The Council, [19-]-
v
Holdings: 1983-
r6863983 P.515/624
3183831

ABERDEEN (GRAMPIAN). DISTRICT COUNCIL
Trade unions: National and Local Government Officers Association. Aberdeen City Branch - Serials
[NALGONEWS (Aberdeen : 1982)]. Nalgonews. - No.37 (Feb. 1982)-. - [Aberdeen] : Aberdeen City Branch of NALGO, 1982-
v.
r6864912 P.523/1683
2742004

ABERDEEN AIRPORT
Development
The DEVELOPMENT of Aberdeen airport 1981-1995. - Glasgow : Scottish Airports, 1981
28p,[7]p,[5]leaves of plates(3 folded) : ill
B81-37626 BS.401/29
2560542

ABERDEEN AND NORTH EAST SCOTLAND FAMILY HISTORY SOCIETY
Members. Interests - Directories

ABERDEEN AND NORTH EAST SCOTLAND FAMILY HISTORY SOCIETY
Aberdeen & North East Scotland Family History Society. - [Aberdeen] : [The Society], 1981
33p
0-9507828-0-7 X.529/45175
2606690

ABERDEEN AND ORKNEY (DIOCESE). Episcopal Church in Scotland
Churches

STRANRAER-MULL, Gerald
A view of the diocese of Aberdeen and Orkney. - [Ellon] [The author], 1977
23[i.e. 24]p : ill
B80-40226 X.200/34230
2308916

ABERDEEN CITY LIBRARIES. COMMERCIAL AND TECHNICAL DEPARTMENT
Stock: Documents on natural gas & petroleum - Catalogues

BIBLIOGRAPHY of material on oil and gas. - 3rd ed.. - [Aberdeen] : [City of Aberdeen, Libraries Department], 1981
101p
B82-01497 X.322/10381 Woolwich
2113872

ABERDEEN CITY LIBRARIES. SIR JOHN ANDERSON BRANCH LIBRARY
See Sir John Anderson Branch Library

ABERDEEN FOOTBALL CLUB
to 1978

WEBSTER, Jack
The Dons. - London : Paul, 1978
224p,[16]p of plates : ill. - Includes index
0-09-133730-5 X.620/17889 Woolwich
1415816

ABERDEEN, GEORGE HAMILTON GORDON, EARL OF. Great Britain
- Biographies

CHAMBERLAIN, M. E. (Muriel Evelyn)
Lord Aberdeen. - London : Longman, 1983
xv,583p : ill. - Bibliography: p535-542
0-582-50462-7 X.520/31206
158247x

IREMONGER, Lucille
Lord Aberdeen. - London : Collins, 1978
384p,[8]p of plates : ill. - Bibliography: p.365-371. - Includes index
0-00-216082-x X.809/43526
158247x

ABERDEEN-GLASGOW ROUTE. Scotland
Road freight transport. Costs. Effects of road design standards

FREIGHT TRANSPORT ASSOCIATION. Scottish Region
The Aberdeen journey. - Glasgow : The Association, [1978]
[2],8p,plate : 1 ill
B79-13657 X.615/2308 Woolwich
1838113

ABERDEEN HARBOUR. Grampian Region. Scotland

ABERDEEN harbour / published by the authority of Aberdeen Harbour Board. - London : Burrow, [1983]
92p : ill(some col.)
B84-19468 X.0800/1902
0591432

ABERDEEN REGION. Grampian Region. Scotland
Environment planning

GRAMPIAN (Region). Council
Aberdeen area structure plan. - [Aberdeen] : [The Council], [1979?]
[3],55p : ill
B80-39780 X.525/3723
2300389

ABERDEEN REGION. Grampian Region. Scotland
Shopping centres. Planning

DRIVERS JONAS (Firm)
City of Aberdeen, district and local centres shopping study, 1978. - [London] : [Drivers Jonas for Aberdeen District Council], [1978]
[10],95[i.e.97]p,[22] leaves,14[i.e.13] leaves of plates
B79-28851 X.415/5823 Woolwich
1939467

ABERDEEN SKI CLUB
- Serials

ABERDEEN Ski Club journal. - [19-]-. - [Aberdeen] : The Club, [19-]-
v. : ill
r6810581 P.623/1271
2919699

ABERDEEN UNIVERSITY
See University of Aberdeen

ABERDEENSHIRE. Grampian Region. Scotland
- For motoring

CAR tours. - Aberdeen : Department of Development & Tourism, [1982?]
64p : ill
B82-27496 X.809/52327
2750287

ABERDEENSHIRE. Grampian Region. Scotland
Cricket. Clubs: Aberdeenshire Cricket Club, 1857-1982

ABERDEENSHIRE Cricket Club / H.O. Smith ... [et al.]. - [Aberdeen] : [Aberdeenshire Cricket Club], [1983]
103p : ill
B83-38300 X.622/16720
295317x

ABERDEENSHIRE. Grampian Region. Scotland
Murray (Family)

MURRAY, Caroline
The Murray family register 1794-1977. - [Pretoria] : [C. Murray], [1978]
122p
r6807293 X.529/59507
9066438

ABERDEENSHIRE. Grampian Region. Scotland
Turnpike roads

PATRICK, John, 19—
The coming of turnpikes to Aberdeenshire. - [Aberdeen] : Centre for Scottish Studies], [1982?]
55p : ill. - (Local history pamphlet series)
B82-31665 X.809/52709
2763141

ABERDEENSHIRE CRICKET CLUB
1857-1982

ABERDEENSHIRE Cricket Club / H.O. Smith ... [et al.]. - [Aberdeen] : [Aberdeenshire Cricket Club], [1983]
103p : ill
B83-38300 X.622/16720
295317x

ABERDOUR. Fife Region. Scotland
1880-1980

MACDONALD, Niall
Aberdour. - [Aberdour] : [N. MacDonald], 1981
30p : ill
B83-43320 X.809/56689
2976013

ABERDOVEY. Gwynedd
1884-1945

ABERDOVEY / assembled by Hugh M. Lewis. - [Aberdovey] : [Hugh M. Lewis], [1975]
[64]p : of ill
0-9504457-0-3 X.708/20164
0542326

ABERDOVEY. Gwynedd
Literary institutions: Aberdovey Literary Institute, to 1982

BIRCH, Henry Kenneth
The pages of time. - [Aberdovey] : [Aberdovey Literary Institute], [1982?]
47p : ill
B84-30365 X.950/33382
3138690

ABERDOVEY LITERARY INSTITUTE
to 1982

BIRCH, Henry Kenneth
The pages of time. - [Aberdovey] : [Aberdovey Literary Institute], [1982?]
47p : ill
B84-30365 X.950/33382
3138690

ABERFELDY REGION. Tayside Region. Scotland
Self-sufficiency

The ABERFELDY project / editor Howard Liddell ; authors Judy Appleby ... [et al.] ; illustrations Judy Appleby, Howard Liddell, Mark Standley. - Hull : Polygon Press, [1979]
[5],68p : ill
0-903057-03-4 X.512/6757
1927116

ABERFOYLE REGION. Central Region. Scotland
to 1980

DICKIE, D. M.
An introduction to Aberfoyle. - 4th ed. - Stirling : Educational Resources Unit, 1981
27p : ill
0-906098-06-8 X.809/62233
3221113

ABERG, ERIC. Seafaring. Sweden
- Biographies

ABERG, Eric
Aren da jag seglade. - Stockholm : Liber förlag, 1977
221p : ill
91-380-3552-9 X.629/7947 Woolwich
1538799

ÅBERG, Eric
Aren pa bryggan. - Stockholm : Liber förlag, 1978
160p : ill
91-380-4367-x X.629/8113 Woolwich
1538799

ABERGAVENNY. Gwent

ABERGAVENNY, Gwent, illustrated official guide. - [New ed.]. - Gloucester : British Publishing Co., [1978]
44p : ill
0-7140-1483-4 X.708/21663
1089498

ABERGAVENNY. Gwent
Social conditions, 1860-1960

TUCKER, Anna
Abergavenny in old photographs. - [Abergavenny] : Monmouth District Museum Service, [1983?]
96p : ill. - Bibliography: p96
0-9503295-5-x X.809/56211
2955040

ABERGAVENNY. Gwent
Social conditions, 1900-1960

OLD Abergavenney in photographs / [compiled] by Albert Lyons. - Barry : Williams, 1983
[118]p : chiefly ill
0-900807-56-3 X.800/35966
2944448

ABERGAVENNY REGION. Gwent
- Walkers' guides

BRECON BEACONS NATIONAL PARK COMMITTEE
30 walks in the Abergavenny area. - [Brecon] : The Committee, 1976
[16]p
0-905293-01-0 BS.116/109
0660981

ABERGAVENNY REGION. Gwent
Description & travel

STINCHCOMBE, M. C.
Portraits of the gateway to Wales. - Bath : Kingsmead, 1978
48p : ill
0-906230-35-7 X.800/26938
1485997

ABERGAVENNY REGION. Gwent
Welsh language. Organisations: Cymdeithas Cymreigyddion Y Fenni, to 1854

THOMAS, Mair Elvet
Afiaith yng Ngwent. - Caerdydd [i.e. Cardiff] : Gwasg Prifysgol Cymru, 1978
[9],150p : ill. - Bibliography: p.147-149
0-7083-0685-3 X.709/32939
1575325

ABERGAVENNY WELSH SOCIETY
See Cymdeithas Cymreigyddion y Fenni

ABERHART, WILLIAM. Alberta

WILLIAM Aberhart and social credit in Alberta / edited by Lewis H. Thomas. - Vancouver : Copp Clark Publishing, c1977
175p. - (Issues in Canadian history)
0-7730-3128-6 X.529/58210
9024220

ABERLADY BAY LOCAL NATURE RESERVE

ABERLADY Bay L.N.R. descriptive management plan. - [Haddington] : [East Lothian Department of Physical Planning], 1977
94p,[10]folded leaves of plates. - Bibliography: p92-93
B83-32824 L.49/1846
2933349

ABERLE, ERNST. Jews
- Biographies

ABERLE, Ernst
Vi må ikke glemme. - [Oslo] : Cappelen, 1980
74p,[8]p of plates
82-02-04521-5 X.809/54762
8273103

ABERMULE. Powys
Castles: Dolforwyn Castle. Archaeology

INTERIM reports on excavations at three castles in Wales 1981-1982 / by C. Arnold ... [et al.]. - Cardiff : Department of Extra-Mural Studies, University College, 1983
29p. - Includes bibliographies
B83-41050 X.425/4086
296385x

ABERNATHY, JEWELL. Christian missions. China
Abernathy, John & Abernathy, Jewell - Biographies

FLETCHER, Jesse C.
Living sacrifices. - Nashville, Tenn : Broadman Press, 1974
x,13-157p
0-8054-7210-x X.200/31107
1256793

ABERNATHY, JOHN. Christian missions. China
Abernathy, John & Abernathy, Jewell - Biographies

FLETCHER, Jesse C.
Living sacrifices. - Nashville, Tenn : Broadman Press, 1974
x,13-157p
0-8054-7210-x X.200/31107
1256793

ABERNETHY. Tayside Region. Scotland
to 1971

CAMERON, Christina
A short history of Abernethy. - Abernethy : Abernethy Women's Rural Institute, [1972?]
31p : ill
B78-07980 X.709/31972
1342991

ABERRATIONS. Butterflies. Great Britain

RUSSWURM, A. D. A. (Alexander Donald Abbot)
Aberrations of British butterflies. - Faringdon : Classey, 1978
151p : col ill. - Includes index
0-86096-002-1 SRL
1702742

ABERSOCH. Gwynedd
Primary education, ca 1920-ca 1960

ROBERTS, Megan
Puryd a mân us. - [Caernarfon] : Gwasanaeth Llyfrgell Gwynedd, 1983
40p. - (Darlith flynyddol Llŷn)
0-904852-37-7 X.950/35687
3188310

ABERTILLERY. Gwent
1890-1972

OLD Abertillery in photographs / [compiled] by Keith Thomas. - Barry : Stewart Williams, 1981
[121]p : chiefly.ill
0-900807-45-8 X.805/1556
2498189

ABERYSTWYTH. Dyfed
to 1977

ABERYSTWYTH, 1277-1977 / edited by Ieuan Gwynedd Jones. - Llandysul : Gomer Press, 1977
ix,142p,[54]p of plates : ill. - Includes index
0-85088-497-7 X.800/26697
1351095

ABERYSTWYTH. Dyfed
1864-1964

ABERYSTWYTH yesterday / [compiled] by Howard C. Jones. - Barry : S. Williams, 1980
[120]p : of ill
0-900807-38-5 X.800/28826
2160447

ABERYSTWYTH. Dyfed
Congregational churches. Seion Congregational Church, Aberystwyth, to 1978

JONES, E. D. (Evan David)
Trem ar ganrif yn hanes Eglwys Gynulleidfaol Baker Street, Aberystwyth. - Aberystwyth : [Yr awdur], 1978
[1],38p
B79-15811 X.100/18350
1875280

ABERYSTWYTH. Dyfed
Mints: Aberystwyth Mint, to 1667

BOON, George C. (George Counsell)
Cardiganshire silver and the Aberystwyth Mint in peace & war. - Cardiff : National Museum of Wales, 1981
xiii,287p,[3]leaves of plates(some folded) : ill(some col.). - Bibliography: p145-150. - Includes index
0-7200-0240-0 X.421/22616 Woolwich
2625504

ABERYSTWYTH. Dyfed
Schools of librarianhsip: College of Librarianship, Wales

COLLEGE of Librarianship Wales-higher degree studies. - Aberystwyth : University of Wales School of Librarianship and Information Studies, 1982
40p : col.ill
r6815700 2719.e.828
3071804

ABERYSTWYTH. Dyfed
Schools of librarianship: College of Librarianship, Wales. Curriculum subjects: Audiovisual materials

COLLEGE OF LIBRARIANSHIP, WALES
Non-book media at CLW. - Aberystwyth : College of Librarianship Wales, [1982?]
[12]p : ill
B82-27865 2719.e.369
2751429

ABERYSTWYTH. Dyfed
Universities. Colleges. Libraries: Hugh Owen Library. Architecture

PERCY THOMAS PARTNERSHIP
The Hugh Owen Library. - Aberystwyth : University College of Wales, 1977
[2],12[i.e.13]p : ill
0-904222-04-7 2719.aa.99
1248502

ABERYSTWYTH. Dyfed
Universities. Colleges. Libraries: Library University College of Wales, to 1976

LLOYD, T. G.
Llyfrgell y Coleg, 1872-1976. - Aberystwyth : University College of Wales, 1977
[2],18p : ill
0-904222-09-8 2719.aa.96
1196332

ABERYSTWYTH. Dyfed
Universities. Colleges. Libraries: University College of Wales. Library

UNIVERSITY COLLEGE OF WALES. Library
Cyfarwyddyd i'r Llyfrgell. - Aberystwyth : Coleg Prifysgol Cymru, 1976
[1],27p,[2]leaves
r6032157 X.519/27164
1194690
Library guide. - Aberystwyth : The College, 1977
[1],28p,[2]leaves. - Includes index
r6032158 2719.aa.93
1194739

ABERYSTWYTH. Dyfed
Universities. Colleges. Libraries: University College of Wales. Library, to 1976

LLOYD, T. G.
The College Library, 1872-1976. - Aberystwyth : University College of Wales, 1977
[2],18p : ill
0-904222-08-x 2719.aa.95
1196324

ABERYSTWYTH. Dyfed
Universities. Colleges. Libraries: University College of Wales. Library. Exhibits: Items associated with Dürrenmatt, Friedrich - Catalogues

UNIVERSITY COLLEGE OF WALES. Library
Friedrich Dürrenmatt. - [Aberystwyth] : [The Library], 1976
[1],14p. - Bibliography: p.12-14
0-904222-03-9 X.909/41139
0798126

ABERYSTWYTH. UNIVERSITY COLLEGE OF WALES
See University College of Wales

ABERYSTWYTH MINT
to 1667

BOON, George C. (George Counsell)
Cardiganshire silver and the Aberystwyth Mint in peace & war. - Cardiff : National Museum of Wales, 1981
xiii,287p,[3]leaves of plates(some folded) : ill(some col.). - Bibliography: p145-150. - Includes index
0-7200-0240-0 X.421/22616 Woolwich
2625504

ABHIDHAMMAPITAKA
See Tipiṭaka. Abhidhammapiṭaka

ABHIDHAMMATTHA SANGAHA. Buddhism

[ABHIDHAMMATTHA sangaha. English & Pali]. A manual of abhidhamma / edited in the original Pali text with English translation and explanatory notes Narada Maha Thera. - 3rd revised ed. - Kandy : Buddhist Publication Society, 1975
[4], vii, 451 p., fold. plate. - Includes index
r6010333 14099.z.5
079886X

ABHIDHARMAPIṬAKA
See Tripiṭaka. Abhidharmapiṭaka

ABHISHIKTANANDA, SWAMI. Benedictines

DAVY, Marie-Madeleine
Henri le Saux. - Paris : Les Éditions du Cerf, 198?
206p : ill. - (Témoins spirituels d'aujourd'hui). - Bibliography
2-204-01667-5 X.200/37672
8027714

ABHISHIKTANANDA, SWAMI. Hinduism
- Biographies

ABHISHIKTANANDA, Swami
Souvenirs d'Arunâchala. - Paris : Épi, c1978
177p
2-7045-0082-7 X.200/43742
8242178

ABIDJAN. Ivory Coast
Muslims, to date

DELVAL, Raymond
Les musulmans d'Abidjan. - Paris : Centre de hautes études sur l'Afrique et l'Asie moderne, 1980
106leaves,3leaves of plates : ill. - (Cahiers du C.H.E.A.M. ; no.10)
2-903182-04-3 X.205/897
8760098

ABIDJAN. Ivory Coast
Social conditions

GIBBAL, Jean Marie
Citadins et paysans dans la ville africaine. - Grenoble : Presses Universitaires de Grenoble, 1974
398 p.; 1 leaf of plates, 2 p. of plates : ill. - (Bibliothèque d'anthropologie)
2-7061-0036-2 X.611/4506 Woolwich
0905283

ABILDGAARD, N A. Danish drawings

SKOVGAARD, Bente
N.A. Abildgaard. - København : Den kgl. Kobberstiksamling, Statens Museum for Kunst, 1978
136p : ill. - (Kobberstiksamlingens billedhæfter ; 4)
87-87273-18-7 X.410/6952 Woolwich
179342X

ABILDGAARD, N. A.. Danish drawings
- Catalogues

AARHUS KUNSTMUSEUM
N.A. Abildgaard. - Arhus : Aarhus Kunstmuseum, 1977.
[36]p : ill. - (Billedhæfte ; 1)
r6112739 X.421/21219
2462761

ABILDGAARD, N A. Danish drawings
- Catalogues

CHRISTENSEN, Charlotte
N.A. Abildgaard. - København : Thorvaldsens Museum, [1978]
69p : ill
87-7521-046-0 X.415/3023 Woolwich
179342X

NANNESTAD, Kirsten
Nicolai Abildgaard 1743-1809. - [Hillerød] : Det Nationalhistoriske Museum pa Frederiksborg, 1979
56p : chiefly ill
87-87237-25-3 X.419/4096 Woolwich
179342X

ABILITY
See also
Aptitude
Competence
Creativity
Genius
Learning ability
Mixed ability
Musical ability
Spatial ability
Verbal ability

ABILITY. Girls. Students. Fifth forms. Secondary schools. Great Britain
compared with aspirations

RAUTA, Irene
Fifth form girls. - London : H.M.S.O., 1975
iii,104p : ill
0-11-700693-9 BS.81/15(173)
052073x

ABILIT'. Mn
Theorie
WERDEL. Hans
Zur Prolem der Begabung und Intelligenz. - Basel ; Lodox. Karger, c1980
193 - Psychologische Praxis ; Bd 55). - Bibliography p17-19
3-0551123-x X.529/39691
23025

ABILIT' Stdents. Universities
Learnig abity. Improvement. Role of counsellors
MAZZETI, Roberto
Quae scola secondaria? - Roma : Armando Armando. [19-]
15 - Controcampo ; 6)
r67856 X.519/39589
04733

ABILIT' TETS
ABILTY sting / Alexandra K. Wigfor and Wendell R. Garer, ditors ; Committee on Ability Testing, Assembly of Behavoral and Social Sciences, National Research Council. Washington, D.C. : National Academy Press. Pt1: Roort of the Committee. - 1982
x,22p. Includes bibliographies
0-309-3228-8 X.529/48170
27916
ABILTY sting / Alexandra K. Wigdor and Wendell R. Garer, ditors ; Committee on Ability Testing, Assembly of Behavoral and Social Sciences, National Research Council. Washington, D.C. : National Academy Press. Part 2: Documentation section. - 1982
x,44p ill. - Includes bibliographies
0-309-3229-6 X.529/48168
27916

ABINGDON Oxfordshire
ABINGDO on Thames today / [text by Harry Knights]. - Abingdo : Harry Knights, 1982
48p: allll. - Bibliography on inside back cover
0-9461-00-6 L.49/1511
11372
LIVERSIGE, W. H. A.
A guide Abingdon on Thames, Berkshire. - [Revised ed.] - Aingdon : Abbey Press (Oxon.), [1979]
48pfoldplate : ill. - Bibliography: p.2
B79-3391 X.708/22127
11372

ABINGDON Oxfordshire
to 1977
HAMMOD, Nigel
The book f Abingdon. - Buckingham : Barracuda Books, 1979
2-140p :l. - Bibliography: p.136-137. - Includes index
0-8602:067-8 X.802/10887
189944

ABINGDON Oxfordshire
to date
COX, Mieke
Abingdc. - [Abingdon?] : [M. Cox?], [1974?]
108p : i;
r681101 X.809/56338
899901:
- Signe by the author X.809/56338

ABINGDON Oxfordshire
1850-1950
ABINGDO in camera / [compiled by] M.J. Thomas. - [Abingdo:: Abingdon Area Archaeological and Historical Society, 979
[50]p : ciefly ill
0-9000-41-2 LR.421/233
195650

ABINGDON Oxfordshire
Ashville Trang Estate. Antiquities. Archaeology, 1974-1976
PARRINGTON, Michael
The excation of an Iron Age settlement, Bronze Age ring-dites and Roman features at Ashville Trading Estate, Aingdon (Oxfordshire) 1974-76. - London : Council r British Archaeology [for] the Oxfordshire Archaeogical Unit, 1978
viii,139p,0]p of plates : ill. - (Reports / Oxfordshire Archaeoical Unit ; 1) (Research reports / Council for British Achaeology ; 28). - Includes bibliographies
0-9003-50-5 X.0410/35(28) Woolwich
158863

ABINGDON Oxfordshire
Boys' public schools: Abingdon School, 1939-1976 - Biograpes
WILLIS, Donald
A song a bugle blown. - Cumnor : Kenton, 1983
xii,232p ill. - Includes index
0-95085-0-6 X.808/37968
289269

ABINGDON. Oxfordshire
Boys' public schools: Radley College. Buildings, 1720-1900
CHERNIAVSKY, M. T. (Michael Theodore)
Looking at Radley. - [Abingdon] : Radley College, [1981]
61p : ill(some col.)
B82-02888 X.421/22412 Woolwich
2590166

ABINGDON. Oxfordshire
Charities: Christ's Hospital (Abingdon), to 1900
CARTER, John, 19—
Give & take. - [Abingdon] : [J. Carter and J. Smith], [c1981]
112p : ill. - Bibliography p105-107. - Includes index
0-9507680-0-6 X.329/14498 Woolwich
2562022

ABINGDON. Oxfordshire
Churches: Church of St Helen, Abingdon. Painted ceilings
LIVERSIDGE, M. J. H.
The fourteenth century decorated ceiling in St Helen's, Abingdon. - Great Britain : [s.n.], 1973
[2],17p : ill. - Bibliography : p.17
B75-28624 X.419/1731 Woolwich
0517828

ABINGDON. Oxfordshire
Museums: Abingdon Borough Museum
COX, Mieneke
Abingdon Museum. - [Abingdon] : [Abingdon Borough Museum], [1979]
18p : ill
B80-32830 X.429/11385 Woolwich
2256169

ABINGDON. Oxfordshire
Primary schools: St. Edmund's School, to 1982
LET'S celebrate 1982. - [Abingdon] : St. Edmund's School, [1982?]
65p : ill
B82-33643 X.520/27859
2769158

ABINGDON. Oxfordshire
Primary schools: St Edmund's School, Abingdon - Serials
NEWSDESK / St Edmunds School, Abingdon. - No. 1- ; Nov. 16th, 1976-. - [Abingdon] : [The School], 1976-
r6031307 P.523/744
1067990

ABINGDON (ROYAL AIR FORCE STATION)
See Royal Air Force, Abingdon

ABINGDON BOROUGH MUSEUM
COX, Mieneke
Abingdon Museum. - [Abingdon] : [Abingdon Borough Museum], [1979]
18p : ill
B80-32830 X.429/11385 Woolwich
2256169

ABINGDON REGION. Oxfordshire
Sand deposits & gravel deposits
CORSER, C. E.
The sand and gravel resources of the country around Abingdon, Oxfordshire. - London : H.M.S.O., 1978
iii,105p : ill. - (Mineral assessment report ; 38). - Bibliography: p.104-105
0-11-884084-3 SRL
1902083

ABINGDON SCHOOL. Oxfordshire
1939-1976 - Biographies
WILLIS, Donald
A song on a bugle blown. - Cumnor : Kenton, 1983
xii,232p : ill. - Includes index
0-9508503-0-6 X.808/37968
2892693

ABJECTION. Special themes. Literature, to 1979
KRISTEVA, Julia
Powers of horror. - New York ; Guildford : Columbia University Press, 1982
x,219p. - (European perspectives)
0-231-05346-0 X.520/36941
2914239

ABKHAZ ASSR. Georgia (Soviet Socialist Republic)
PACHULĪĀ, V. P. (Vianor Pandzhovich)
Abkhaziīā. - Sukhumi : Izdatel'stvo "Alashara", 1976
159p,[8]p of plates : ill
r6044556 X.708/41766
1314831

ABKHAZ ASSR. Georgia (Soviet Socialist Republic)
1800-1900
DZIDZARIĪĀ, G. A. (Georgiĭ Alekseevich)
Makhadzhirstvo i problemy istorii Abkhazii XIX stoletiīā. - Sukhumi : Alashara, 1975
525p : ill
r6094021 X.809/26595
1972820

ABKHAZ ASSR. Georgia (Soviet Socialist Republic)
Antiquities
BGAZHBA, Oleg Khukhutovich
Ocherki po remeslu srednevekovoi Abkhazii (VIII-XIV vv.). - Sukhumi : Alashara, 1977
79p : ill
r6176147 X.419/4250 Woolwich
1860828
GUNBA, Mikhail Mikhaĭlovich
Novye pamīātniki T͡sebel'dinskoĭ Kul'tury. - Tbilisi : Met͡sniereba, 1978
123p,LIIp of plates : ill. - Bibliography : p.115-118
r6094112 X.429/10723 Woolwich
1860828
MATERIALY po arkheologii Abkhazii / [nauchnaīā redakt͡siīā Otara Lordkipanidze]. - Tbilisi : Met͡sniereba, 1979
87p,63p of plates : ill
r6183233 X.421/20550
1860828
SHAMBA, G. K. (Georgiĭ Kuchievich)
Ėsherskoe gorodishche. - Tbilisi : "Met͡sniereba", 1980
68, lxxvi p. : ill.. - Bibliography: p. 61-[66]
r6465752 X.421/26264
1860828
VORONOV, ĪUriĭ Nikolaevich
Dioskuriada—Sebastopolis—T͡Skh um. - Moskva : Nauka, 1980
127 p., [8] p. of plates : ill.. - (Stranit͡sy istorii nasheĭ Rodiny). - Includes bibliographical references
r6490015 X.529/67423
1860828
Drevnosti Voenno-Sukhumskoĭ dorogi. - Sukhumi : Alashara, 1977
74p. : ill
r6094887 X.419/3758
1860828

ABKHAZ ASSR. Georgia (Soviet Socialist Republic)
Architecture
VORONOV, ĪUriĭ Nikolaevich
V mire arkhitekturnykh pamīātnikov Abkhazii. - Moskva : Iskusstvo, 1978
172p : ill. - (Dorogi k prekrasnomu)
r6096780 X.419/3696 Woolwich
2094932

ABKHAZ ASSR. Georgia (Soviet Socialist republic)
Bronze Age antiquities
T͡SVINARIĪĀ, Igor' Ivanovich
Poselenie Guandra. - Tbilisi : Met͡sniereba, 1978
56p,xlviip of plates : ill. - Bibliography : p.51-53
r6177645 X.700/22370
2265036

ABKHAZ ASSR. Georgia (Soviet Socialist Republic)
Coins, to 1500
SHAMBA, S. M. (Sergeĭ Mironovich)
O chem govorīāt monety. - Sukhumi : Alashara, 1982
53p : ill
r6448217 X.429/15739
9082565

ABKHAZ ASSR. Georgia (Soviet Socialist Republic)
Constitution
ABKHAZIA (Abkhaz ASSR)
[Constitution, 1978]. Konstitut͡siīā (Osnovnoĭ Zakon) Abkhazskoĭ Avtonomnoĭ Sovetskoĭ Sot͡sialisticheskoĭ Respubliki. - Sukhumi : Alashara, 1979
42p
r6176936 S.NLA.1
2297728

ABKHAZ ASSR. Georgia (Soviet Socialist Republic)
Cultural processes
KHASHBA, Arseniĭ Konstantinovich
[Selections. Abkhaz and Russian]. Izbrannye raboty. - Sukhumi : Alashara, 1972
230p,[1]leaf of plates,[4]p of plates : ill,ports
r6178255 X.989/85978
2446952

ABKHAZ ASSR. Georgia (Soviet Socialist Republic)
Cultural processes, to date
ZVANBA, S. T. (Solomon Teĭmurkovich)
Abkhazkie ėtnograficheskie ėtīūdy. - Sukhumi : Alashara, 1982
89p : ill
r6448250 X.808/39851
9087540

ABKHAZ ASSR. Georgia (Soviet Socialist Republic)
Cultural relations with Russia (RSFSR)

TARBA, Boris G.
Russkiĭ îazyk v razvitii kul'tury i prosveshcheniîa abkhazov. - Sukhumi : Izdatel'stvo "Alashara", 1976
115p
r6021660 X.708/19781
1124420

ABKHAZ ASSR. Georgia (Soviet Socialist Republic)
Government

SOVETY Abkhazii (1922-1937 gg.) / [sostaviteli : A. O. Tulumdzhîan, R. M. Shvet͡sova, E. V. Osipova]. - Sukhumi : Izdatel'stvo "Alashara", 1976
397 p.
r6020720 X.809/23618
0899283

ABKHAZ ASSR. Georgia (Soviet Socialist Republic)
Gulripshi region. Antiquities. Archaeology, to date

ARKHEOLOGICHESKIE issledovaniîa v T͡Sebel'de / [redaktor : O. Lordkipanidze]. - Tbilisi : Met͡sniereba, 1983
131p : ill
r6489962 X.421/26235
9125493

ABKHAZ ASSR. Georgia (Soviet Socialist Republic)
Intellectuals, 1800-1917

DZIDZARIÎA, G. A. (Georgiĭ Alekseevich)
Formirovanie dorevoliut͡sionnoĭ abkhazskoĭ intelligent͡sii. - Sukhumi : Alashara, 1979
353p
r6177862 X.809/28789
2441071

ABKHAZ ASSR. Georgia (Soviet Socialist Republic)
Metallurgy, to date

BGAZHBA (Oleg Khukhutovich)
Chernaîa metallurgiîa i metalloobrabotka v drevneĭ i srednevekovoĭ Abkhazii (VIII v. do n.ė. - XV v.n.e.). - Tbilisi : Met͡sniereba, 1983
102p,plates
r6489965 X.629/25029
9155813

BGAZHBA, Oleg Khukhutovich
Po sledam kuznet͡sa Aĭnara. - Sukhumi : Alashara, 1982
58,20p : ill
r6492159 X.629/26495
9155813

ABKHAZ ASSR. Georgia (Soviet Socialist Republic)
Migration, to 1918

DZIDZARIÎA, G. A. (Georgiĭ Alekseevich)
Makhadzhirstvo i problemy istorii Abkhazii XIX stoletiîa. - Izd. 2-e, dop. - Sukhumi : Alashara, 1982
529p,plates
r6448314 X.529/60406
9074449

ABKHAZ ASSR. Georgia (Soviet Socialist Republic)
Mosaics, to date

SHERVASHIDZE, L. A. (Leo Alekseevich)
Srednevekovaîa monumental'naîa zhivopis' v Abkhazii. - Tbilisi : Met͡sniereba, 1980
252p,plates. - Includes bibliography
r6464661 X.421/25229
9074341

ABKHAZ ASSR. Georgia (Soviet Socialist Republic)
Religious buildings. Wall paintings, to date

SHERVASHIDZE, L. A. (Leo Alekseevich)
Srednevekovaîa monumental'naîa zhivopis' v Abkhazii. - Tbilisi : Met͡sniereba, 1980
252p,plates. - Includes bibliography
r6464661 X.421/25229
9074333

ABKHAZ ASSR. Georgia (Soviet Socialist Republic)
Theatre

ARGUN, Alyk'sa Kh
St͡sena i zhizn'. - Tbilisi : Khelovneba, 1977
227p
r6087989 X.989/77638
1002759

ABKHAZ ASSR. Georgia (Soviet Socialist Republic)
Theatre - Biographies

ARGUN, Alyk'sa Kh.
Talant i vdokhnovenie. - Sukhumi : Izdatel'stvo "Alashara", 1976
197p : ill
r6021175 X.529/20017
1002759

ABKHAZ ASSR. Georgia (Soviet Socialist Republic)
Theatre, 1917-

ARGUN, Alyk'sa Kh.
Istoriîa abkhazskogo teatra. - Sukhumi : Alashara, 1978
440p,[16]p of plates : ill
r6498076 X.958/3223
8713804

ABKHAZ CIVILIZATION

INAL-IPA, Shalva Denisovich
Voprosy ėtno-kul'turnoĭ istorii abkhazov. - Sukhumi : Izdatel'stvo "Alashara", 1976
463p
r6020894 X.529/20257
0899399

ABKHAZ CULTURE
Influence on Georgian literature, to date

BGAZHBA, Khukut Solomonovich
Stranit͡sy iz letopisi druzhby. - Tbilisi : Met͡sniereba, 1983
115p
r6448358 X.950/30716
9076050

ABKHAZ CULTURE
Influence on Russian literature, to date

BGAZHBA, Khukut Solomonovich
Stranit͡sy iz letopisi druzhby. - Tbilisi : Met͡sniereba, 1983
115p
r6448358 X.950/30716
9076069

ABKHAZ FOLKLORE

ANSHBA, A. A. (Artur Artemovich)
Abkhazskiĭ fol'klor i deĭstvitel'nost'. - Tbilisi : Met͡sniereba, 1982
279p
r6464765 X.950/31248
9073388

INAL-IPA, Shalva Denisovich
Pamîatniki abkhazskogo fol'klora. - Sukhumi : Alashara, 1977
197p
r6095063 X.809/26542
1976796

ABKHAZ LANGUAGE

ARISTAVA, Sh. K. (Shota Konstantinovich)
Problema prostogo predlozheniîa v abkhazskom îazyke. - Tbilisi : Izdatel'stvo Tbilisskogo universiteta, 1982
253p. - Includes bibliography
r6464753 X.950/31448
2430525

GET͡SADZE, Irina Okropirovna
Ocherki po sintaksisu abkhazskogo îazyka. - Leningrad : Nauka, Leningradskoe otdelenie, 1979
133p
r6173630 X.909/84350
2430525

HEWITT, B. G.
Abkhaz. - Amsterdam : North-Holland, 1979
v,283p. - (Lingua descriptive studies ; v2). - Includes bibliography
r6306982 X.950/3835(2.)
2430525

KVARCHIÎA, V. E. (Valeriĭ Ermeevich)
Zhivotnovodcheskaîa (pastusheskaîa) leksika v abkhazskom îazyke. - Sukhumi : Alashara, 1981
132p
r6489039 X.958/19134
2430525

SHAGIROV, Amin Kabt͡suevich
Material'nye i strukturnye obshchnosti leksiki abkhazo-adygskikh îazykov. - Moskva : Nauka, 1982
163p
r6496923 X.950/30288
2430525

ABKHAZ LEGENDS
Abkhaz tales & legends

KHASHBA, Arseniĭ Konstantinovich
[Selections. Abkhaz and Russian]. Izbrannye raboty. - Sukhumi : Alashara, 1972
230p,[1]leaf of plates,[4]p of plates : ill,ports
r6178255 X.989/85978
2446960

ABKHAZ LEGENDS
Abkhaz tales & legends - Anthologies

ABKHAZSKIE narodnye skazki / [sostavitel' i avtor primechaniĭ : K.S. Shakryl]. - Moskva : Nauka, Glavnaîa redakt͡siîa vostochnoĭ literatury, 1975
469p. - (Skazki i mify narodov Vostoka)
r6176041 X.989/81276
1125257

ABKHAZSKIE skazki / [sostavil, obrabotal i perevel s abkhazskogo Kh.S. Bgazhba]. - Izd.4. - Sukhumi : Izdatel'stvo "Alashara", 1976
313p : ill
r6021732 X.989/38782
1125257

ABKHAZ LEGENDS
Characters: Narts

SALAKAÎA, Shota Khichovich
Abkhazskiĭ nartskiĭ ėpos. - Tbilisi : Izdatel'stvo "Met͡sniereba", 1976
233p
r6045908 X.809/25016
1384139

ABKHAZ LITERATURE
See also
Fiction in Abkhaz
Poetry in Abkhaz

ABKHAZ LITERATURE
to date

INAL-IPA, Shalva Denisovich
Stranit͡sy abkhazskoĭ literatury. - Sukhumi : Alashara, 1980
249 p.. - Includes bibliographical references
r6482838 X.958/25111
9110852

ABKHAZ MYTHOLOGY

AKABA, Lili Kh
Iz mifologii abkhazov. - Sukhumi : Izdatel'stvo "Alashara", 1976
103 p.
r6020826 X.908/36214
0898376

ABKHAZ TALES
Abkhaz tales & legends

KHASHBA, Arseniĭ Konstantinovich
[Selections. Abkhaz and Russian]. Izbrannye raboty. - Sukhumi : Alashara, 1972
230p,[1]leaf of plates,[4]p of plates : ill,ports
r6178255 X.989/85978
2446960

ABKHAZ TALES
Abkhaz tales & legends - Anthologies

ABKHAZSKIE narodnye skazki / [sostavitel' i avtor primechaniĭ : K.S. Shakryl]. - Moskva : Nauka, Glavnaîa redakt͡siîa vostochnoĭ literatury, 1975
469p. - (Skazki i mify narodov Vostoka)
r6176041 X.989/81276
1125257

ABKHAZSKIE skazki / [sostavil, obrabotal i perevel s abkhazskogo Kh.S. Bgazhba]. - Izd.4. - Sukhumi : Izdatel'stvo "Alashara", 1976
313p : ill
r6021732 X.989/38782
1125257

ABKHAZIAN PEOPLE

ANCHABADZE, Zurab V.
Ocherk ėtnicheskoĭ istorii abkhazskogo naroda. - Sukhumi : Izdatel'stvo "Alashara", 1976
165p
r6044035 X.709/23570
1328530

ABKHAZIANS
Religious beliefs

AKABA, Lili Kh.
U istokov religii abkhazov. - Sukhumi : Alashara, 1979
121p
r6465128 X.208/10288
9162259

ABKHAZIANS. Georgia (Soviet Socialist Republic)
Cultural processes

INAL-IPA, Shalva Denisovich
Tradit͡sii i sovremennost'. - Izd. 2-oe, dop. i perer. - Sukhumi : Alashara, 1978
110p
r6098537 X.519/41535
011832X

ABKHAZIANS. World War 2
Role in Russian campaign

MINASÎAN, Ė. G. (Ėduard Gareginovich)
Istrebitel'nye batal'ony Abkhazii v Velikoĭ Otechestvennoĭ voĭne (1941-1945 gg.). - Sukhumi : Alashara, 1980
91p
r6462187 X.808/37895
8944849

ABKHAZO-ADYGHIAN LANGUAGES
See North-west Caucasian languages

ABLATIVE CASE. Latin language

PRAT, Louis C.
Morphosyntaxe de l'ablatif en latin archaique. - Paris Société d'Édition Les Belles Lettres, [1975]
xxxvii, 444 p. : ill. - (Collection d'études latines. série scientifique ; fasc. 32). - Bibliography : p. xii-xxxvii
r6012629 Ac.5336.c/2a[fasc.32]
0890928

ABLINGA. Lithuania

KUNČIUS, Algimantas
Ablinga. - Vilnius : Vaga. 1975
[112]p : chiefly ill
r6014914 X.421/9249 Woolwich
1319094

ABMS
See Anti-ballistic missile systems

ABNAKI
to date

SÉVIGNY, P.-André
Les Abénaquis : habitat et migrations (17e et 18e siècles). - Montréal : Les Éditions Bellarmin, 1976
247p,4fold,leaves of plates. - (Cahiers d'histoire des Jésuites ; 3). - Bibliography
0-88502-212-2 X.808/35077
8778108

ABNEY. Derbyshire
to 1980

TOMLINSON, Tom D.
Ye ancient parish of Habenai (Abney). - [Hathersage] : [Hathersage Parochial Church Council], [1980]
12p : ill
B81-18660 X.809/47675
2401053

ABNORMAL DEVELOPMENT
See also
Teratology

ABNORMAL DEVELOPMENT. Children
- Conference proceedings

GATLINBURG (TENNESSEE) CONFERENCE ON RESEARCH AND THEORY IN MENTAL RETARDATION (7th : 1974)
Aberrant development in infancy. - Hillsdale : Erlbaum ; New York ; London : Distributed by Wiley, 1975 [i.e. 1977]
viii,287p : ill. - Includes bibliographies and index
0-470-23859-3 SRL
1079840

ABNORMAL PSYCHOLOGY
See also
Behavioural disorders
Mental disorders
Mentally disordered
Psychiatry

ABNORMAL PSYCHOLOGY

DIMENSIONS in abnormal psychology / edited by Glenn Shean. - 2nd ed. - Chicago : Rand McNally, 1976
xi,533p : ill. - (Rand McNally psychology series). - Includes bibliographies and index
0-528-62049-5 X.510/16090
0301507

FARINA, Amerigo
Abnormal psychology. - Englewood Cliffs ; London : Prentice-Hall, 1976
xv,207p. - (The Prentice-Hall series in personality). - Includes bibliographies and index
0-13-001164-9 X.319/16392 Woolwich
0115797

HARMATZ, Morton G.
Abnormal psychology. - Englewood Cliffs ; London : Prentice-Hall, 1978
xxiv,584p : ill(some col). - Bibliography: p.541-565. - Includes index
0-13-000885-0 X.520/12518
0115797

KLEINMUNTZ, Benjamin
Essentials of abnormal psychology. - 2nd ed. - San Francisco ; London : Harper and Rows, 1980
xii,660p : ill(some col). - Bibliography : p.588-641. - Includes index
0-06-043711-1 X.520/20833
0115797

LAING, R. D. (Ronald David)
The politics of experience ; and, The bird of paradise. - Harmondsworth : Penguin, 1967
155p. - (A Pelican book)
0-14-022410-6 X.529/65060
0115797

MAHONEY, Michael J. (Michael John)
Abnormal psychology - San Francisco ; London : Harper and Row, 1980
xvi,590p : ill(some col) - Bibliography: p. 525-575. - Includes index
0-397-47410-5 X.520/21481
0115797

MARTIN, Barclay
Abnormal psychology. - New York ; London : Holt, Rinehart and Winston, 1977
xiv,2-642p,[4]p of plates : ill(some col). - Bibliography: p.601-630. - Includes index
0-03-014896-0 X.322/8370 Woolwich
0115797
Abnormal psychology. - 2nd ed. - New York ; London : Holt, Rinehart and Winston, c1981
xiv,578p : ill(some col.). - Bibliography: p533-565. - Includes index
0-03-050721-9 X.520/23092
0115797

RIMM, David C.
Abnormal psychology. - New York ; London : Academic Press, 1977
xvi,696p : ill. - Bibliography: p.615-662. - Includes index
0-12-588840-6 X.520/12004
0115797

ROSENHAN, David L.
Abnormal psychology. - New York ; London : Norton, c1984
xxiii,728p : ill(some col.). - Bibliography: p678-704. - Includes index
0-393-95277-0 X.525/10131
0115797

SARASON, Irwin G. (Irwin Gerald)
Abnormal psychology. - 4th ed. - Englewood Cliffs ; London : Prentice-Hall, c1984
xviii,557p : ill. - Bibliography: p515-538. - Includes index
0-13-000638-6 X.525/9807
0115797

SARASON, Irwin Gerald
Abnormal psychology. - 2nd ed. - Englewood Cliffs ; London : Prentice-Hall, 1976
xiv,687p,2p of plates : ill(some col). - (The century psychology series). - Bibliography: p.633-660. - Includes index
0-13-001123-1 WP.9/65
0115797
Abnormal psychology. - 3rd ed. - Englewood Cliffs ; London : Prentice-Hall, 1980
xvi,573p : ill(some col). - Bibliography: p.527-552. - Includes index
0-13-001107-x X.525/3553
0115797

THEORY and research in abnormal psychology / edited by David L. Rosenhan, Perry London. - 2nd ed. - New York ; London : Holt, Rinehart and Winston, 1975
xix,454p : ill. - Includes bibliographies and index
0-03-084294-8 X.311/8035
0301507

WHITE, Robert W. (Robert Winthrop)
The abnormal personality. - 5th ed. - New York ; Chichester : Wiley, c1981
xiii,793p : ill. - Bibliography: p677-732. - Includes index
0-471-04599-3 X.520/23400
0115797

ZAX, Melvin
Abnormal psychology. - 2nd ed. - New York ; London : Holt, Rinehart and Winston, 1976
xii,643p : ill. - Bibliography: p.577-611. - Includes index
0-03-089517-0 X.320/10599 Woolwich
0115797

ABNORMAL PSYCHOLOGY
to 1983 - Dictionaries

HOWELLS, John G. (John Gwilym)
A reference companion to the history of abnormal psychology. - Westport, Conn. ; London : Greenwood Press, 1984
2v.(xviii,1141p) : ill. - Includes index
0-313-22183-9 X.520/36086
3228134

ABNORMAL PSYCHOLOGY. Special subjects. European literatures, 1900-

AMO, Javier del
Literatura y neurosis. - Madrid : Editora Nacional, [1974]
144 p.. - (Escalada)
84-276-1157-9 X.319/8330 Woolwich
0912166

ABNORMALITIES
See also
Aberrations
Congenital abnormalities
Freaks

ABNORMALITIES. Chromosomes. Man

GROUCHY, J. de
Clinical atlas of human chromosomes. - New York ; Chichester : Wiley, 1977
xxv,319p : ill. - Includes bibliographies
0-471-01704-3 SRL
0202533

NEW chromosomal syndromes / edited by Jorge J. Yunis. - New York ; London : Academic Press, 1977
xi,404p : ill. - (Chromosomes in biology and medicine). - Includes bibliographies and index
0-12-775165-3 SRL
0202533

SYMPOSIUM ON HUMAN POPULATION CYTOGENETICS 1975 Albany)
Population cytogenetics. - New York ; London : Academic Press, 1977
xv,374p : ill. - (Symposia / Birth Defects Institute ; 6th). - Includes index
0-12-355460-8 SRL
0202533

ZELLWEGER, Hans
Chromosomes of man. - London : Heinemann Medical [etc. for] Spastics International Medical Publications, 1977
xii,228p : ill. - (Clinics in developmental medicine ; no.65-66). - Bibliography: p.191-221. - Includes index
0-433-39886-8 SRL
0202533

ABNORMALITIES. Chromosomes. Man
Psychological aspects

NYBORG, Helmuth
Sex chromosome abnormalities and cognitive performance. - Risskov : University of Aarhus, Institute of Psychology, 1976-
1-. - (Psychological reports Aarhus ; vol.1, no2 etc). - Includes bibliographies
r6017668 X.0512/454(1,ii-)
0549185

ABNORMALITIES. Haemoglobins. Blood. Man

BUNN, H. Franklin
Hemoglobinopathies. - Philadelphia ; London Saunders, 1977
ix,308p : ill. - (Major problems in internal medicine ; vol.12). - Bibliography: p.251-291. - Includes index
0-7216-2179-1 SRL
0562971

ABNORMALITIES. Haemoglobins. Blood. Man
- Conference proceedings

ABNORMAL haemoglobins and thalassaemia / edited by R.M. Schmidt. - New York ; London : Academic Press, 1975
lvi,371p : ill. - Includes bibliographies and index
0-12-627150-x SRL
0562971

BIOCHEMICAL and chemical aspects of henoglobin abnormalities / edited by Winslow S. Caughey .. - New York ; London : Academic Press, 1978
xx,725p : ill. - Includes index
0-12-164350-6 SRL
0562971

COMPREHENSIVE SICKLE CELL CENTER SYMPOSIUM ON THE MOLECULAR BASIS OF MUTANT HEMOGLOBIN DYSFUNCTION (1979 : University of Chicago)
The molecular basis of mutant hemoglobin dysfunction. - New York ; Amsterdam : Elsevier/North Holland, c1981
xii,338p : ill. - (The University of Chicago Sickle Cell Center Hemoglobin Symposia ; v.1). - Includes index
0-444-00631-1 SRL
0562971

ABNORMALITIES. Mammals

PARÉ, Ambroise
On monsters and marvels. - Chicago ; London : University of Chicago Press, 1982
xxxii,224p : ill. - Bibliography: p208-212. - Includes index
0-226-64562-2 X.550/18583
2846853

ABNORMALITIES. Menisci. Joints. Knees. Man

PAVLOV, Helene
Atlas of knee menisci. - Norwalk, Conn. ; London : Appleton-Century-Crofts, c1983
ix,166p : ill(some col.). - Bibliography: p16[illegible]
0-8385-0444-2 SRL
3076679

ABNORMALITIES. Reproductive system. Women
Surgery

JONES, Howard W. (Howard Wilbur)
Reparative and constructive surgery of the female generative tract. - Baltimore ; London : Williams & Wilkins, c1983
ix,350p : ill(some col.). - Bibliography: p325-340. - Includes index
0-683-04470-2 SRL
2933470

ABNORMALITIES DIAGNOSTIC SERVICES. Great Britain
Congenital foetal abnormalities diagnostic services

CLINICAL GENETICS SOCIETY. Working Party on Prenatal Diagnosis in Relation to Genetic Counselling
The provision of services for the prenatal diagnosis of fetal abnormality in the United Kingdom. - London : Eugenics Society, [1978]
iii,32p. - (Supplements to the 'Bulletin of the Eugenics Society' / Eugenics Society ; no.3). - Bibliography: p.28
B79-19990 X.319/18870 Woolwich
1898280

'IT all depends where you live'. - London : The Maternity Alliance, 1982
26p
B83-14748 X.322/11636
1898280

ABO
See Turku

ABO AKADEMI

NALLINMAA-LUOTO, Terhi
Ylioppilaitten ja kaupunkilaisten suhteet Turun vanhan akatemian alkuaikoina. - [Turku] : Turun yliopisto, Suomen historian laitos, 1973
93 leaves. - (Monistesarja / Turun yliopiston Suomen historian laitos. B, Tutkimuksia ; 5)
r6600037 X.0525/514(5)
9030522

ABO AKADEMI
Curriculum subjects: Sociology. Teaching methods, 1973-1974

GRÖNDAHL, Aulis
Sociologistudier i smagrupper. - Abo : Stiftelsens för Abo akademi forskningsinstitut, 1976
97p. - (Meddelanden fran Stiftelsens för Abo akademi forskningsinstitut ; nr 5)
951-648-208-2 X.0519/601(5)
1866273

ABO AKADEMI
Curriculum subjects: Theology

KURTÉN, Tage
Teori och praxis inom teologin. - Abo : Institutionen för systematisk teologi, 1976
iv,148p. - (Studier utgivna av Institutionen för systematisk teologi vid Abo Akademi ; 12). - Includes bibliographies
951-648-284-8 X.0102/136(12)
1980459

ABO AKADEMI. BIBLIOTEK
- Bibliographies

MUSTELIN, Olof
Publikationer rörande Abo akademis bibliotek. - Abo : AAB, 1976
108 leaves. - (Abo akademis bibliotek, Stencilserie ; 14)
r6028859 P.1393/1 [vol.14]
151024X

ABO AKADEMI. BIBLIOTEK
1919-1979

BÖCKER, samlingar, bibliotekarier. - Abo, 1979
120p : ill. - (Skrifter utgivna av Abo akademis bibliotek ; 14). - Includes bibliographies
951-648-461-1 2762.t.5/14
2218968

ABO AKADEMI. BIBLIOTEK
Stock. Inter-library loans, 1973

RAJALIN, Marita
Fjärrlaningen till Abo akademis bibliotek. - Abo : Abo akademis bibliotek, 1977
50 leaves. - (Stencilserie ; 15)
951-648-274-0 P.1393/1[no.15]
1859811

ABOLITION. Allowances. Personal income tax. Married men. Great Britain
Mathematical models

ZABALZA, Antoni
Efficiency and equity effects of reforming the British system of direct taxation. - London : London School of Economics, Centre for Labour Economics, 1983
50p : ill. - (Discussion paper / Centre for Labour Economics. London School of Economics ; no.175). - Bibliography: p50
B84-30828 X.529/63030
3139557

ABOLITION. Capital punishment
Conferences - Conference proceedings

CONFERENCE ON THE ABOLITION OF THE DEATH PENALTY (1977 Stockholm)
Report of the Amnesty International conference on the death penalty, Stockholm, Sweden, 10-11 December 1977. - London : Amnesty International, 1978
[5],32p
B79-07182 X.702/5914
180118x

ABOLITION. Capital transfer tax. Great Britain

BRACEWELL-MILNES, Barry
A fairer tax on transfers? - London [Selsdon Group], [1981?]
23leaves. - (Selson brief : no.2)
B82-08019 X.525/5284
2611074

A tax for the axe. - London : CUT, the taxpayer Union, 1979
iii,10leaves
B82-08018 X.525/5296
2611074

ABOLITION. Clerical clothing. Christian church

BROWN, Roger Lee
How to rid your church of vestments. - [Cardiff] : R.L. Brown, 1982
7p
B82-35599 X.200/38932
2775891

ABOLITION. Corporal punishment. Schools. Europe

The EUROPEAN example. - Croydon : Society of Teachers Opposed to Physical Punishment, [1981?]
28p. - Bibliography: p28
B82-14281 X.529/46333
2639459

ABOLITION. Corporal punishment. Schools. Great Britain

ABOLITION handbook / edited for STOPP, Society of Teachers Opposed to Physical Punishment, by Peter Newell. - [Croydon] : [STOPP], [1981?]
40p : ill. - Bibliography: p38
B82-14280 X.529/46332
2639440

AFTER Strasbourg. - London : AMMA, [1982?]
24p
0-902983-18-0 X.529/50605
2639440

NATIONAL UNION OF TEACHERS
Corporal punishment. - 2nd ed. - [London] : [Education Department, NUT], [1984]
24p
B84-39779 X.200/44608
2639440

ABOLITION. Corporal punishment. Schools. Great Britain
- Serials

STOPP news / Society of Teachers Opposed to Physical Punishment. - Oct. 1979-. - Croydon : The Society, 1979-
v.
B82-23591 P.523/1650
2736004

ABOLITION. Corporal punishment. Schools. Scotland

MAKING the change / C.E. Cumming ... [et al.]. - Dunton Green : Hodder and Stoughton for the Scottish Council for Research in Education, c1981
viii,48p : 1ill. - (SCRE publication ; 76)
0-340-28180-4 X.529/45975
262771x

ABOLITION. Dumping waste materials from coal industries on beaches. Durham (County)

GREAT BRITAIN. Working Party on Colliery Waste Tipping on Durham Beaches
Interim report. - [London] : Department of the Environment, [Planning Regional and Minerals Directorate, Division 3], 1975
6 leaves
0-7184-0046-1 BS.414/257
0702048

ABOLITION. Flogging. Corporal punishment. Military offenders. Great Britain. Army
Effects of inquest into death of White, Frederick John, 1846

HOPKINS, Harry
The strange death of Private White. - London : Weidenfeld and Nicolson, 1977
273p,[8]p of plates : ill. - Bibliography: p.257-266. - Includes index
0-297-77293-7 X.809/42046
1135643

ABOLITION. Greater London Council

RATE capping & abolition / Labour Research Department. - London : LRD Publications, 1984
24p : ill
0-900508-70-1 X.529/68233
3202100

ABOLITION. Greater London Council
Proposals

GREAT BRITAIN
Abolition of The Greater London Council and The Metropolitan County Councils. - London : Department of the Environment, 1984
[48]p
B85-11821 BS.414/2189
3192253

GREATER LONDON COUNCIL
The future of the Greater London Council. - London : Greater London Council, c1984
48p : ill(some col.)
0-7168-1408-0 X.525/10002
3192253

Responses by GLC Committees to the Government's White Paper 'Streamlining the Cities'. - London : Greater London Council, [1984]
[340]p in various pagings : ill
B84-43632 X.525/10024
3192253

ABOLITION. Greater London Council
Proposals. Attitudes of voluntary organisations

KNIGHT, Barry
Government and the GLC. - London : London Voluntary Service Council, 1984
5p
0-901171-48-4 X.525/10172
3215016

ABOLITION. Industrial training boards. Great Britain

ELLIOT, Clive
Industrial training boards. - London : Centre for Policy Studies, 1981
27leaves
0-905880-35-8 X.525/4317
2397056

ABOLITION. Infanticide of daughters & suttee to ca 1860

SAXENA, R. K.
Social reforms. - New Delhi : Trimurti Publications, 1975
xi,155p. - Bibliography: p.149-152.- Includes index
r6091839 X.529/33143
1890182

ABOLITION. Intensive production. Livestock
- Serials

NATIONAL SOCIETY AGAINST FACTORY FARMING
NSAFF news. - Vol. 1, no. 1-; 197-?-. - London : Smith East Associates Limited, 197-?- : ill
Holdings: vol.15, no.1; April 1979, etc
B80-20806 P.313 / 496
2062798

ABOLITION. Interest. Finance. Islamic countries

MAHMUD AHMAD, Shaikh
Social justice in Islam. - 2nd revised and enlarged ed. - Lahore : Institute of Islamic Culture, 1975
xll, 120 p.. - Includes index
r6031317 X.519/26638
106827X

SIDDIQI, Muhammad Nejatullah
Banking without interest. - Leicester : The Islamic Foundation, c1983
192p. - (Islamic economics series ; 5). - Includes index
0-86037-119-0 X.529/54304
106827x

ABOLITION. Jurisdiction. Obsolete local courts. England

GREAT BRITAIN. Law Commission
Jurisdiction of certain ancient courts. - London : H.M.S.O., [1976]
ii,16p. - (Law Com. ; no.72) (Cmnd.6385)
0-10-163850-7 BS.24f/108
0605875

ABOLITION. Metropolitan counties. England

RATE capping & abolition / Labour Research Department. - London : LRD Publications, 1984
24p : ill
0-900508-70-1 X.529/68233
3172562

STREAMLINING the cities. - [London] : Coopers & Lybrand Associates, 1983
28p : ill
B84-39215 X.529/64936
3172562

ABOLITION. Metropolitan counties. England –continued

STREAMLINING the cities. - [London] : Coopers & Lybrand Associates, 1984
52p : ill
B84–39214 X.529/64935
3172562

ABOLITION. Monarchy. Italy
1943–1946

DALLA monarchia alla repubblica. 1943–1946 / a cura di Enzo Santarelli. - Roma : Editori Riuniti, 1974
276 p. - (Strumenti per la ricerca interdisciplinare ; n. 6). - Bibliography : p. 272–276
r6014170 X.708/18087
0859672

ABOLITION. Negro slavery. Southern states. United States
Role of Lincoln, Abraham

OATES, Stephen B.
Builders of the dream. - Fort Wayne, Ind. : Louis A. Warren Lincoln Library and Museum, 1982
24p,4p of plates
r6818509 X.800/39193
9119353

ABOLITION. Road Haulage Wages Council
Proposals

ADVISORY, CONCILIATION AND ARBITRATION SERVICE
Road Haulage Wages Council. - [London] : [The Service], [1977]
64p. - (ACAS reports / Advisory, Conciliation and Arbitration Service ; no.6)
0–906073–04–9 BS.427/3
1111590

ABOLITION. Serfdom. Poland
1750–1900

GRONIOWSKI, Krzysztof
Uwłaszczenie chłopów w Polsce. - Warszawa : P.W. "Wiedza Powszechna", 1976
363 p., 24 p. of plates. - (Biblioteka Wiedzy historycznej. Historia Polski)
r6014902 X.708/18447
0888257

ABOLITION. Serfdom. Russia
1855–1861

FIELD, Daniel
The end of serfdom. - Cambridge, Mass. ; London : Harvard University Press, 1976
xv,472p. - (Studies / Harvard University. Russian Research Center ; 75). - Bibliography: p.449–464. - Includes index
0–674–25240–3 Ac.2692.bx(75)
0764000

ABOLITION. Serfdom. Transylvania. Romania
to 1918

EGYED, Ákos
Falu, város, civilizáció. - Bukarest : Kriterion, 1981
359p.. - Includes bibliographical references and indexes
r6455075 X.529/33993
9101004

ABOLITION. Sex discrimination. Age of eligibility. State superannuation schemes. Great Britain

GREAT BRITAIN. Equal Opportunities Commission
Equalising the pension age. - [London] : [The Commission], 1978
[3],43p : 1 ill
B78–19049 BS.433/49
1103172

Sex equality and the pension age. - Manchester : The Commission, [1977]
23p : 1 ill
0–905829–00–x BS.433
1103172

NATIONAL ASSOCIATION OF PENSION FUNDS
Towards equality in retirement ages. - Croydon : The Association, [1977]
[1],12p. - (Notes on pensions ; no.5)
B78–10813 X.0100/1015(5)
1103172

ABOLITION. Sex discrimination. Law. Industrial health & industrial safety. Great Britain

GREAT BRITAIN. Equal Opportunities Commission
Health & safety legislation. - Manchester : The Commission, 1979
[5],261[i.e.262]p
0–905829–20–4 BS.433/7
1959379
- [Another copy] 614.852 OPL

ABOLITION. Slave trade
- Conference proceedings

The ABOLITION of the Atlantic slave trade / edited by David Eltis and James Walvin with the collaboration of Svend E. Green-Pedersen. - Madison ; London : University of Wisconsin Press, 1981
xiii,314p : ill. - Bibliography: p303–306. - Includes index
0–299–08490–6 X.520/26807
2711796

ABOLITION. Slave trade. Cuba
Foreign relations between Great Britain & Spain, 1770–1870

MURRAY, David R.
Odious commerce. - Cambridge : Cambridge University Press, 1980
xi,423p. - (Cambridge Latin American studies ; 37). - Bibliography: p400–413. - Includes index
0–521–22867–0 X.809/47723
1963384

ABOLITION. Slave trade. Great Britain
1800–1837

GREAT BRITAIN
[Treaties, etc. Spain, 1817 Sept.23]. Tratado entre Su Majestad el Rey de España y de las Indias y Su Majestad el Rey del Reino Unido de la Gran Bretaña e Irlanda, para la abolición del tráfico de negros (1817). - Ed. facsimilar. - Windsor, Ontario : Rolston-Bain, 1983
ii,81p. - (Colección Documenta Novae Hispaniae ; v.B–3)
0–88653–001–6 X.525/9000
9079890

ABOLITION. Slave trade. Great Britain
Causes. Theories

DRESCHER, Seymour
Econocide. - Pittsburgh : University of Pittsburgh Press ; London : Feffer and Simons, 1977
xiv,279p : ill. - Bibliography: p.261–271. - Includes index
0–8229–3344–6 X.800/26913
1477625

ABOLITION. Slave trade. Portugal
1810–1842

CAPELA, José
As burguesias portuguesas e a abolição do tráfico da escravatura, 1810–1842. - Porto : Afrontamento, 1979
254p. - (Colecção As armas e os varões ; 8). - Bibliography
r6160828 X.708/49334
2466953

ABOLITION. Slave trade. Spain
1814–1868

GREAT BRITAIN
[Treaties, etc. Spain, 1817 Sept.23]. Tratado entre Su Majestad el Rey de España y de las Indias y Su Majestad el Rey del Reino Unido de la Gran Bretaña e Irlanda, para la abolición del tráfico de negros (1817). - Ed. facsimilar. - Windsor, Ontario : Rolston-Bain, 1983
ii,81p. - (Colección Documenta Novae Hispaniae ; v.B–3)
0–88653–001–6 X.525/9000
9079904

ABOLITION. Slavery. Brazil

NABUCO, Joaquim
O abolicionismo. - 4.ed. - Petrópolis : Editora Vozes, 1977
204p. - (Coleção Dimensões do Brasil ; 4)
r6099694 X.519/41580
2042738

[O abolicionismo. English]. Abolitionism. - Urbana ; London : University of Illinois Press, 1977
xxvi,186p,[12]p of plates. - (Blacks in the New World). - Bibliography: p.174–178. - Includes index
0–252–00601–1 X.520/12422
1433652

ABOLITION. Slavery. Brazil
1822–1889

BEIGUELMAN, Paula
A crise do escravismo e a grande imigração. - 2a ed. - São Paulo : Brasiliense, 1981
62p : ill. - (Tudo é história ; 2)
r6706327 X.808/39803
8624313

NABUCO, Joaquim
[Selections]. Joaquim Nabuco. - São Paulo : Atica, 1982
192p. - (Coleção Grandes cientistas sociais ; 23)
r6710124 X.809/64679
8624313

SANTOS, Ronaldo Marcos dos
Resistência e superação do escravismo na província de São Paulo, 1885–1888. - São Paulo : Instituto de Pesquisas Econômicas, 1980
142p. - (Série Ensaios econômicos ; v.5)
r6705973 X.529/51775
8624313

ABOLITION. Slavery. Brazil
1871

CENTENÁRIO da Lei do Ventre Livre. - [Brasília] : Conselho Federal de Cultura, 1976
240p.,5 leaves of plates
r6025111 X.900/18513
1152408

ABOLITION. Slavery. Brazil
Patrocínio, José do - Biographies

ORICO, Osvaldo
O Tigre da Abolição. - Rio de Janeiro : Civilização Brasileira, 1977
308p. - (Coleção Vera cruz ; literatura brasileira ; v.238)
r6101838 X.519/44011
2297973

ABOLITION. Slavery. Mauritius
1810–1968

RIVALTZ QUENETTE, L
La fin d'une légende. - Port-Louis [Mauritius] : Monoprint, [1979?]
iii,116p,1leaf of plates : ill
r6206384 X.520/24692
8717826

ABOLITION. Slavery. Mossoró. Brazil
- Serials

MOMENTO. - Ano 1, no.0 (30.09.76). - Mossoró : ASTECAM, 1976
30p : chiefly ill
r6159778 X.709/52124
2474506

ABOLITION. Slavery. Puerto Rico
to 1898

CIBES VIADÉ, Alberto
El Gobernador Pezuela y el abolicionismo puertorriqueño, 1848–1873. - Río Piedras : Edil, [1978?]
156p. - (Colección Historia). - Bibliography
84–399–7275–x X.529/47715
8834881

El PROCESO abolicionista en Puerto Rico / [recopilados por el] Centro de Investigaciones Históricas ... Universidad de Puerto Rico [y el] Instituto de Cultura Puertorriqueña. - San Juan, Puerto Rico : El Instituto.
v.2: Proceso y efectos de la abolición : 1866–1896. - 1978
xxvi,492p
r6704254 X.0700/988(2)
8834881

ABOLITION. Slavery. Puerto Rico
Role of Pezuela, Juan de la

CIBES VIADÉ, Alberto
Don Juan de la Pezuela inicia el abolicionismo puertorriqueño. - Río Piedras : Editorial Madre Isla, 1975
33p
r6705863 X.800/34296
8971854

ABOLITION. Taxation. Great Britain

MACMURCHIE, David A.
For God's sake set the people free and end legalised theft! a blue-print for full employment and national prosperity. - Dundee : [D.A. MacMurchie], [1980]
[7]p
B81–06453 X.529/39686
2358816

ABOLITION. Tied cottages. Agriculture. Great Britain

GREAT BRITAIN. Department of the Environment
Abolition of the tied cottage system in agriculture. - [London] : [The Department], [1976]
33p
0–7184–0060–7 BS.414/273
0701262

ABOLITION. Tuition fees for first two years. Colleges. United States

CARNEGIE COUNCIL ON POLICY STUDIES IN HIGHER EDUCATION
Low or no tuition. - San Francisco ; London : Jossey-Bass ; [London] : [Distributed by Dent], 1975
viii,88p : 1 ill. - ([Carnegie Council series]). - Bibliography: p.83–85. - Includes index
0–87589–257–4 X.0520/501(18)
0362506

ABOLITION. Use by private patients of national health services facilities. Hospitals. Great Britain

HEALTH SERVICES BOARD
Withdrawal of authorisations for the use of NHS hospital accommodation and services by private patients. - London : H.M.S.O., [1977]
23p. - (Cmnd.6963)
0–10–169630–2 BS.385/306
126852x

ABOLITION. Use by private patients of national health services facilities. Hospitals. Great Britain–continued

Withdrawal of authorisations for the use of NHS hospital accommodation and services by private patients - Health Services Board's third set of proposals ... - London : H.M.S.O., 1978
16p. - ([1978–79 H.C.] 4)
0–10–200479–x BS.385/487
126852x

Withdrawal of authorisations for the use of NHS hospital accommodation and services by private patients - Health Services Board's sixth set of proposals ... - London : H.M.S.O., [1979]
6p. - ([1979–80] H.C. 148)
0–10–214880–5 BS.385/497
126852x

Withdrawal of authorisations for the use of NHS hospital accommodation and services by private patients - Health Services Board's fourth set of proposals ... - London : H.M.S.O., [1979]
19p. - ([1978–79 H.C.] 188)
0–10–218879–3 BS.385/473
126852x

Withdrawal of authorisations for the use of NHS hospital accommodation and services by private patients - Health Services Board's seventh set of proposals. - London : H.M.S.O., [1979]
14p. - ([1979–80] H.C. 212)
0–10–221280–5 BS.385/573
126852x

Withdrawal of authorisations for the use of NHS hospital accommodation and services by private patients - Health Services Board's second proposals ... - London : H.M.S.O., [1978]
10p. - ([1977–78 H.C.] 358)
0–10–235878–8 BS.91/161
126852x

ABOLITION. Use of torture by governments

AGAINST torture. - London : Amnesty International, 1984
12p : ill. - (An Amnesty International briefing)
0–86210–069–0 X.525/9998
0334030

AMNESTY INTERNATIONAL
Cinquieme Congres des Nations Unies pour la Prevention du Crime et le Traitement des Delinquants, Toronto, 1-15 Septembre 1975, propositions presentées par Amnesty International. - London : Amnesty International, [1975]
[1],18p
0–900058–18–8 X.102/2010
0570494

Quinto Congreso de las Naciones Unidas sobre Prevencion del Delito y Tratamiento del Delincuente, Toronto, 1-15 Septiembre de 1975, propuestas presentadas por Amnesty International. - London : Amnistia International, [1975]
[1],18p
0–900058–19–6 X.102/2011
0570508

TORTURE in the eighties. - Oxford : Robertson and Amnesty International, 1984
263p : ill
0–85520–781–7 X.529/67095
0334030

TORTURE in the eighties. - London : Amnesty International, 1984
263p : ill. - (An Amnesty International report)
0–86210–066–6 X.529/65367
0334030

ABOLITION. Value-added tax. Financial transactions between value-added tax registered traders. Great Britain Proposals

ELIMINATION of VAT on transactions between registered traders / report by a Working Party. - London : Consultative Committee of Accountancy Bodies, 1979
[1],7p
B80–41447 X.200/34274
2306786

ABOLITION. Vehicle excise duty. Great Britain

GREAT BRITAIN. Department of Transport
The future of vehicle excise duty. - [London] : [The Department], 1978
14,[6]p
B80–40390 BS.43/547
2321386

ABOLITION. Vehicle excise duty. Great Britain Proposals

COUSINS, Steve
Road tax changes. - Milton Keynes : New Towns Study Unit, the Open University, 1979
[3],18p. - Bibliography: p.14
B80–40391 X.625/620 Woolwich
2321378

ABOLITIONIST MOVEMENT. Slavery
American abolitionist movement. Garrison, William Lloyd - Correspondence, diaries

GARRISON, William Lloyd
The letters of William Lloyd Garrison. - Cambridge, Mass. ; London : Belknap Press of Harvard University Press.
Vol.6: To rouse the slumbering land 1868-1879. - 1981
xx,637p. - Includes index
0–674–52666–x X.520/25116
0555991

ABOLITIONIST MOVEMENTS. Negro slavery. United States
American abolitionist movements. Douglas, Frederick - Biographies

PRESTON, Dickson J.
Young Frederick Douglas. - Baltimore ; London : Johns Hopkins University Press, c1980
xvii,242p : ill. - Includes index
0–8018–2439–7 X.520/23389
2379554

ABOLITIONIST MOVEMENTS. Negro slavery. United States
British abolitionist movements. Role of American negroes, 1830–1860

BLACKETT, R. J. M.
Building on antislavery wall. - Baton Rouge ; London : Louisiana State University Press, c1983
xii,237p. - Bibliography p217-231. - Includes index
0–8071–1082–5 X.520/34061
3136523

ABOLITIONIST MOVEMENTS. Slavery
1700–1900

COCHIN, Augustin
L'abolition de l'esclavage. - Fort-de-France : Desormeaux, 1979
407p,plates. - (Histoire de l'esclavage aux Antilles)
2–85275–106–2 X.809/61080
8405816

LUDWIG, Charles
He freed Britain's slaves. - Scottdale, Pa. : Herald Press, 1977
206p
0–8361–1822–7 X.808/34550
8405816

ABOLITIONIST MOVEMENTS. Slavery
ca 1770–1861 - Conference proceedings

ANTI-SLAVERY, religion, and reform / edited by Christine Bolt, Seymour Drescher. - Folkestone : Dawson [etc.], 1980
xii,377p,plate : 1 ill
0–7129–0945–1 X.809/45638
1996770

ABOLITIONIST MOVEMENTS. Slavery
American abolitionist movements, to 1865 - Indexes

ANTI-SLAVERY newspapers and periodicals / edited by John W. Blassingame and Mae G. Henderson. - Boston, Mass. : G.K. Hall, c1980–81
3v.. - (Reference publications in Afro-American studies)
r6808145 X.955/1432
8573263

ABOLITIONIST MOVEMENTS. Slavery
American abolitionist movements, 1830–1865

FRIEDMAN, Lawrence J. (Lawrence Jacob)
Gregarious saints. - Cambridge : Cambridge University Press, 1982
xi,344p. - Bibliography: p333-336. - Includes index
0–521–24429–3 X.520/27939
1147188

LACY, Dan
The abolitionists. - New York : McGraw-Hill, c1978
viii,148p. - Bibliography : p141-142. - Includes index
0–07–035753–6 X.809/28737
1147188

MCKIVIGAN, John R.
The war against proslavery religion. - Ithaca ; London : Cornell University Press, 1984
327p. - Bibliography: p288-318. - Includes index
0–8014–1589–6 X.200/45617
1147188

STANGE, Douglas C.
Patterns of antislavery among American Unitarians, 1831-1860. - Rutherford, N.J : Fairleigh Dickenson University Press, [1977]
308p. - Includes bibliographies
0–8386–1797–2 X.529/34946
1147188

WALTERS, Ronald G.
The antislavery appeal. - Baltimore ; London : Johns Hopkins University Press, 1976
xvii,196,[2]p : ill. - Includes index
0–8018–1861–3 X.520/11757
1147188

ABOLITIONIST MOVEMENTS. Slavery
American abolitionist movements, 1832–1970

TAYLOR, Clare
British and American abolitionists. - Edinburgh : Edinburgh University Press [etc.], 1974
[5],586p. - Bibliography: p.549-558. - Includes index
0–85224–161–5 X.520/10120
0505587

ABOLITIONIST MOVEMENTS. Slavery
American abolitionist movements, 1837–1860

SEWELL, Richard Herbert
Ballots for freedom. - New York : Oxford University Press, 1976
xvi, 379 p.. - Includes index
0–19–501997–0 X.520/10907
069729X

ABOLITIONIST MOVEMENTS. Slavery
American abolitionist movements, 1841–1895

DOUGLASS, Frederick
The Frederick Douglass papers. - New Haven, [Conn.] ; London : Yale University Press.
Ser.1: Speeches, debates and interviews
Vol.2: 1847-54. - c1982
xxxvii,613p. - Includes index
0–300–02661–7 X.520/30456
2060450
- New Haven ; London : Yale University Press. - Includes index
Series 1 : Speeches, debates, and interviews. Vol.1: 1841–46. - 1979
cii,530p, plate
0–300–02246–8 X.981/22340
2060450

ABOLITIONIST MOVEMENTS. Slavery
American abolitionist movements, 1870–1910

MCPHERSON, James M. (James Munro)
The abolitionist legacy. - Princeton ; Guildford : Princeton University Press, [1976]
xiii,438p. - Bibliography: p.417-422. - Includes index
0–691–04637–9 X.520/10526
0612774

ABOLITIONIST MOVEMENTS. Slavery
American abolitionist movements. Brown, John, b.1800

FRIED, Albert
John Brown's journey. - Garden City : Anchor Press/Doubleday, 1978
293p.;[12]p. of plates
0–385–05511–0 X.529/21943
2191032

ABOLITIONIST MOVEMENTS. Slavery
American abolitionist movements. Channing, William Ellery - Biographies

DELBANCO, Andrew
William Ellery Channing. - Cambridge, Mass. ; London : Harvard University Press, 1981
xviii,203p. - Includes index
0–674–95335–5 X.529/42601
2528347

ABOLITIONIST MOVEMENTS. Slavery
American abolitionist movements. Garnet, Henry Highland, to 1865

SCHOR, Joel
Henry Highland Garnet. - Westport, Conn. ; London : Greenwood Press, 1977
xii,250,[1]p. - (Contributions in American history ; no.54). - Bibliography: p.221-229. - Includes index
0–8371–8937–3 X.809/42592
1275038

ABOLITIONIST MOVEMENTS. Slavery
American abolitionist movements. Grimké, Angelina - Biographies

LUMPKIN, Katharine Du Pre
The emancipation of Angelina Grimké. - Chapel Hill : University of North Carolina Press, [1974]
xv, 265 p.. - Bibliography : p. 247-255
0–8078–1232–3 X.520/9262
1009699

ABOLITIONIST MOVEMENTS. Slavery
American abolitionist movements. Members: Women. Role, 1830–1850

TAYLOR, Clare
Romantic reform and Anglo-American women, London, the 1840 Convention and its aftermath. - Aberystwyth : [Department of History, The University College of Wales], 1981
24leaves
B81–34254 X.805/1808
2547120

ABOLITIONIST MOVEMENTS. Slavery
American abolitionist movements. Phillips, Ann Terry Greene

BARTLETT, Irving H. (Irving Henry)
Wendell and Ann Phillips. - New York : Norton, c1979
249p
0-393-01426-6 X.529/64177
8405824
- [Another copy] X.809/61496

ABOLITIONIST MOVEMENTS. Slavery
American abolitionist movements. Phillips, Wendell

BARTLETT, Irving H. (Irving Henry)
Wendell and Ann Phillips. - New York : Norton, c1979
249p
0-393-01426-6 X.529/64177
8405824
- [Another copy.] X.809/61496

ABOLITIONIST MOVEMENTS. Slavery
American abolitionist movements. Role of Lane Seminary, 1828-1860

LESICK, Lawrence Thomas
The Lane rebels. - Metuchen ; London : Scarecrow, 1980
ix,278p. - (Studies in evangelicalism ; no.2). - Bibliography: p239-262. - Includes index
0-8108-1372-6 X.529/41243
2395797

ABOLITIONIST MOVEMENTS. Slavery
American abolitionist movements. Weld, Theodore Dwight - Biographies

ABZUG, Robert H.
Passionate liberator. - New York : Oxford University Press, 1980
xi,370p. - Includes bibliography
0-19-502771-x X.529/57391
8405824

ABOLITIONIST MOVEMENTS. Slavery
Bristol abolitionist movements, 1830-1833

MARSHALL, Peter, 1926-
Bristol and the abolition of slavery. - Bristol : Bristol Branch of the Historical Association, The University, 1975
28,iv p,[4]p of plates : ill. - (Local history pamphlets / Historical Association. Bristol Branch ; 37)
0-901388-15-7 10099.y.3/37
0682624

ABOLITIONIST MOVEMENTS. Slavery
British abolitionist movements, to 1833

BRADY, Terence
The fight against slavery. - London : British Broadcasting Corporation, 1975
144p : ill
0-563-12868-2 X.520/10315
0530492

SLAVERY and British society 1776-1846 / edited by James Walvin. - London : Macmillan, 1982
272p : ill. - (Problems in focus series). - Includes index
0-333-28073-3 X.529/49352
0530492

ABOLITIONIST MOVEMENTS. Slavery
British abolitionist movements, 1832-1970

TAYLOR, Clare
British and American abolitionists. - Edinburgh : Edinburgh University Press [etc.], 1974
[5],586p. - Bibliography: p.549-558. - Includes index
0-85224-161-5 X.520/10120
0369209

ABOLITIONIST MOVEMENTS. Slavery
British abolitionist movements, 1833-1870

TEMPERLEY, Howard
British antislavery, 1833-1870. - London : Longman, 1976
xvii,298p : 1 ill. - Bibliography: p.277-284. - Includes index
0-582-50126-1 X.709/30560
068001x

ABOLITIONIST MOVEMENTS. Slavery
British abolitionist movements. Members: Women. Role, 1830-1850

TAYLOR, Clare
Romantic reform and Anglo-American women, London, the 1840 Convention and its aftermath. - Aberystwyth : [Department of History, The University College of Wales], 1981
24leaves
B81-34254 X.805/1808
2547112

ABOLITIONIST MOVEMENTS. Slavery
British abolitionist movements. Ramsay, James - Biographies

SHYLLON, Folarin Olawale
James Ramsay. - Edinburgh : Canongate Publishing, 1977
viii,144p,plate. - Bibliography: p.134-139. - Includes index
0-903937-11-5 X.809/42140
1164996

ABOLITIONIST MOVEMENTS. Slavery
British abolitionist movements. Wilberforce, William - Biographies

CORMACK, Patrick (1939-)
Wilberforce. - Basingstoke : Pickering, 1983
128p. - Bibliography: p122. - Includes index
0-7208-0544-9 X.200/46149
010129x

LEAN, Garth
God's politician. - London : Darton, Longman & Todd, 1980
xii,180p. - Bibliography: p175-176. - Includes index
0-232-51505-0 X.809/46586
2362740

POLLOCK, John, 1923-
Wilberforce. - Berkhamsted : Lion Publishing, 1978
xv,368p. - (An Aslan lion book). - Bibliography: p.351-359. - Includes index
0-85648-089-4 X.708/21236
010129x

Wilberforce. - Tring : Lion, 1978, c1977
368p. - Bibliography: p355-359. - Includes index
0-86760-377-1 X.808/37347
010129x

Wilferforce. - London : Constable, 1977
xvi,368p,[2]leaves of plates(1 fold),[8]p of plates : ill. - Bibliography: p.355-359. - Includes index
0-09-460780-x X.800/26138
010129x

ABOLITIONIST MOVEMENTS. Slavery
Scottish abolitionist movements, 1833-1861

RICE, C. Duncan (Charles Duncan)
The Scots abolitionists 1833-1861. - Baton Rouge ; London : Louisiana State University Press, c1981
xii,221p. - Includes index
0-8071-0861-8 X.800/33895
2772361
- [ANOTHER COPY : PAPERBACK.] ISBN 0-8071-0898-7 X.800/33896

ABOLITIONIST MOVEMENTS. Slavery. Brazil
1850-1888

CONRAD, Robert
[The destruction of Brazilian slavery, 1850-1888. Portuguese]. Os últimos anos da escravatura no Brasil, 1850-1888. - Rio de Janeiro : Civilização Brasileira, 1975
xxiii,394p.,8p. of plates : ill. - (Coleção Retratos do Brasil ; v.90). - Bibliography : p. 379-394
r6033084 X.709/22699
1124951

ABOLITIONIST MOVEMENTS. Slavery. Demerara. Guyana
British abolitionist movements. Smith, John, b.1790 - Biographies

NORTHCOTT, Cecil
Slavery's martyr. - London : Epworth Press, 1976
136p. - Bibliography: p.131-133. - Includes index
0-7162-0269-7 X.519/26429
0793361

ABOLITIONIST MOVEMENTS. Slavery. East Africa
British abolitionist movements. Sulivan, George Lydiard - Biographies

COLLISTER, Peter
The Sulivans and the slave trade. - London : Collins, 1980, c1979
197p,[5]p of plates : ill. - Bibliography: p184-186. - Includes index
0-86036-121-7 X.529/39605
2354446

ABOLITIONIST MOVEMENTS. Slavery. North America
American abolitionist movements, 1793-1867

ANTISLAVERY reconsidered / edited by Lewis Perry and Michael Fellman. - Baton Rouge ; London : Louisiana State University Press, 1979
xvi,348p. - Includes index
0-8071-0479-5 X.800/29120
2160161

ANTISLAVERY reconsidered / edited by Lewis Perry and Michael Fellman. - Baton Rouge ; London : Louisiana State University Press, 1981, c1979
xvi,348p. - Includes index
0-8071-0889-8 X.529/45658
2160161

ABOLITIONIST MOVEMENTS. Slavery. North America
British abolitionist movements, ca 1790-1863

ANTISLAVERY reconsidered / edited by Lewis Perry and Michael Fellman. - Baton Rouge ; London : Louisiana State University Press, 1979
xvi,348p. - Includes index
0-8071-0479-5 X.800/29120
216017x

ABOLITIONIST MOVEMENTS. Slavery. North America
British abolitionist movements, ca 1790-1863

ANTISLAVERY reconsidered / edited by Lewis Perry and Michael Fellman. - Baton Rouge ; London : Louisiana State University Press, 1981, c1979
xvi,348p. - Includes index
0-8071-0889-8 X.529/45658
216017x

ABOLITIONIST MOVEMENTS. Slavery. North America
British abolitionist movements. Clarkson, John - Biographies

WILSON, Ellen Gibson
John Clarkson and the African adventure. - London : Macmillan, 1980
x,236p,8p of plates : ill. - Bibliography : p.206-225. - Includes index
0-333-26594-7 X.809/46138
1888315

ABOLITIONIST MOVEMENTS. Slavery. United States
American abolitionist movements, ca 1840-1900 - Biographies

WALKER, Peter F.
Moral choices. - Baton Rouge ; London : Louisiana State University Press, 1978
xxii,387p,[6]p of plates. - Bibliography: p.373-382. - Includes index
0-8071-0262-8 X.800/27782
1885901

ABOLITIONIST MOVEMENTS. Slavery. United States
American abolitionist movements. Attitudes to law of slavery, 1760-1848

WIECEK, William Michael
The sources of antislavery constitutionalism in America, 1760-1848. - Ithaca ; London : Cornell University Press, 1977
307p. - Includes index
0-8014-1089-4 X.529/32229
1433024

ABOLITIONIST MOVEMENTS. Slavery. United States
Role of women, 1830-ca 1860

HERSH, Blanche Glassman
The slavery of sex. - Urbana ; London : University of Illinois Press, 1978
xv,280p. - Bibliography: p.257-269. - Includes index
0-252-00695-x X.520/13949
1895583

ABOLITIONIST VIEWPOINTS
See also
National Campaign for the Abolition of Capital Punishment viewpoints

ABOMINABLE SNOWMEN
See Yetis

ABONY. Hungary
Women's movements, to date

FEHÉR, Jánosné
A nőmozgalom története Abonyban. - Abony : [s.n.], 1976
67leaves
r6449260 X.525/6032
8851735

ABORGINES
Australian aborgines. Social conditions

HARNEY, William Edward
Brimming billabongs. - Adelaide : Seal Books, 1976
174p,8p of plates : ill
0-85179-335-5. X.709/31719
1269577

ABORIGINAL ANTIQUITIES
Australian aboriginal antiquities

FLOOD, Josephine
Archaeology of the dreamtime. - Sydney ; London : Collins, 1983
288p,[8]p of plates : ill(some col.). - Bibliography: p275-277. - Includes index
0-00-216444-2 X.800/35943
1297929
- [Another copy] X.421/25266

MCCOURT, Tom
Aboriginal artefacts. - Adelaide : Rigby, 1975
[10],154p : ill(chiefly col). - Includes index
0-85179-867-5 X.421/10547 Woolwich
1297929

ABORIGINAL ANTIQUITIES. Sydney region. New South Wales
Australian aboriginal antiquities. Archaeology, 1964–1967

The RECENT archaeology of the Sydney district / edited by J. V. S. Megaw. – Canberra : Australian Institute of Aboriginal Studies, 1974
viii, 40 p., [31] p. of plates : ill. – ([Australian aboriginal studies. research and regional studies ; No. 1]). – Bibliography : p. 39–40
0–85575–034–0 X.415/5252 Woolwich
1041991

ABORIGINAL ARTEFACTS
Australian aboriginal artefacts – Catalogues

ROYAL ALBERT MEMORIAL MUSEUM
Australia. – [Exeter] : The City Museum, 1982
41p
0–86114–353–1 X.425/3423
2774763

ABORIGINAL ARTS
Australian aboriginal arts, to date

BERNDT, Ronald M. (Ronald Murray)
Aboriginal Australian art. – Sydney : Methuen Australia, 1982
176p : ill
0–454–00276–9 X.421/25003
9043772

ABORIGINAL ARTS
Australian aboriginal visual arts – Conference proceedings

SCHEMATISATION IN ART (Conference : 1974 : Canberra)
Form in indigenous art. – Canberra : Australian Institute of Aboriginal Studies ; London : Duckworth [etc.], 1977
[6],486p : ill(some col). – (Prehistory and material culture series ; no.13). – Includes bibliographies and index
0–7156–1305–7 Cup.1281/1022
176747x

ABORIGINAL ARTS. Stock. Western Australian Art Gallery
Australian aboriginal visual arts – Catalogues

WESTERN AUSTRALIAN ART GALLERY
Aboriginal art. – [Perth] : The Gallery, 1975
[2],66p : ill
r6031761 X.421/10392 Woolwich
1096435

ABORIGINAL CHILDREN
Australian aboriginal children, to date

THOMSON, Donald (Donald Ferguson)
Children of the wilderness. – South Yarra, Vic. : Currey O'Neil, 1983
x,85p : ill. – Includes bibliography
0–85902–329–x L.42/3228
914367x

ABORIGINAL CHILDREN. Central Australia
Australian aboriginal children. Cognitive development. Cultural factors

SEAGRIM, Gavin
Furnishing the mind. – Sydney ; London : Academic Press, 1980
xii,242p : ill. – (Behavioural development ; 1). – Bibliography: p223–232. – Includes index
0–12–634340–3 X.520/23766
2396246

ABORIGINAL CIVILIZATION
Australian aboriginal civilization, to ca 1800

BLAINEY, Geoffrey
Triumph of the nomads. – London : Macmillan, 1976
x,285p. – Bibliography: p.255–257. – Includes index
0–333–17583–2 X.809/40668
0598097

ABORIGINAL CIVILIZATION
Australian aboriginal civilization, to 1851

BLAINEY, Geoffrey
Triumph of the nomads. – Rev. ed. – Melbourne : Sun Books, 1983
x,285p
0–7251–0412–0 X.529/61326
8500959

ABORIGINAL HATCHETS
Australian aboriginal stone hatchets

DICKSON, F. P.
Australian stone hatchets. – Sydney ; London : Academic Press, 1981
viii,240p : ill. – (Studies in archaeological sciences). – Bibliography: p221–227. – Includes index
0–12–215220–4 X.421/23090
2766744

ABORIGINAL LANGUAGES
See also
Alyawarra language
Diyari language
Djaru language
Djingili language
Jindjibarndi language
Mara language
Ngiyambaa language
Nunggubuju language
Wangkatja language
Warndarang language
Wik-Munkan languages
Yidin language
Yuwaalaraay language

ABORIGINAL LANGUAGES
Australian aboriginal languages

BLAKE, Barry J.
Australian aboriginal languages. – London : Angus & Robertson, 1981
137p. – Bibliography: p135–137
0–207–14044–8 X.950/22656
2283379

Australian Aboriginal languages. – Sydney ; London : Angus & Robertson, 1981
137p
0–207–14150–9 X.950/22656
2283379

DIXON, Robert M. W. (Robert Malcolm Ward)
The languages of Australia. – Cambridge : Cambridge University Press, 1980
xxii,547p : ill. – (Cambridge language surveys). – Bibliography: p.515–533. – Includes index
0–521–22329–6 X.981/22502
2283379

ABORIGINAL LANGUAGES
Australian aboriginal languages. Grammar

GRAMMATICAL categories in Australian languages / edited by R.M.W. Dixon. – Canberra : Australian Institute of Aboriginal Studies [etc.], 1976
[7],776p : ill. – (Linguistic series ; no.22). – Includes bibliographies
0–85575–055–3 X.0800/39(71)
137267X

ABORIGINAL LEGENDS
Australian aboriginal tales & legends

ROBERTS, Ainslie
Dreamtime. – Adelaide : Rigby, 1981
144p
0–7270–1496–x L.45/1969
8021015

ABORIGINAL LEGENDS
Australian aboriginal tales & legends – Anthologies

HARNEY, William Edward
Tales from the Aborigines. – Adelaide : Seal Books, 1976
189p,[12]p of plates : ill
0–85179–217–0 H.78/6 Woolwich
118668X

REED, A. W. (Alexander Wyclif)
Myths and legends of Australia. – Sydney ; London : A.H. and A.W. Reed, 1976
256p : ill. – Bibliography: p.255–256
0–589–07041–x X.990/9036 Woolwich
118668x

ABORIGINAL LITERATURES
See also
Children's stories in Ngaanyatjarra

ABORIGINAL MYTHS
Australian aboriginal myths

ALLEN, Louis A. (Louis Alexander)
Time before morning. – Adelaide : Rigby, 1976
xii,304p,plates
0–7270–0076–4 X.421/24677
0049816

ABORIGINAL PAINTINGS
Australian aboriginal paintings, to date

BRANDL, E. J. (Eric Joseph)
Australian Aboriginal paintings in Western and Central Arnhem Land. – Repr. with amendments. – Canberra : Australian Institute of Aboriginal Studies, 1982
214p : ill. – (AIAS new series ; no.35)
0–85575–130–4 L.45/2970
9062998

ABORIGINAL RELIGIONS
Australian aboriginal religions

BERNDT, Ronald M. (Ronald Murray)
Australian Aboriginal religion. – Leiden : E. J. Brill, 1974
1 vol. (various pagings), [175] p. of plates, [5] fold leaves. : ill. – (Iconography of religions ; section 5). – Includes bibliographies
90–04–03861–2 X.203/673
0900869

ABORIGINAL TALES
Australian aboriginal tales – Anthologies

AUSTRALIAN legendary tales / collected from various tribes by Mrs K. Langloh Parker ; with introductions by Andrew Lang and Wandjuk Marika ; illustrated by Rex Backhaus-Smith. – London : Bodley Head, 1978
191p : ill. – (Bodley Head source books of fairy tales and folklore) ([The Bodley Head source books of fairy tales and folklore]). – Includes index
0–370–10841–8 X.981/20980
1137875

LAND of the Rainbow Snake / arranged and translated [from the Gunwinggu] by Catherine H. Berndt ; illustrated by Djoki Yunupingu. – Sydney ; London : Collins, 1979
96p : ill
0–00–184384–2 X.990/16092 Woolwich
2088401

WALKER, Kath
Stradbroke dreamtime. – Re-illustrated ed.. – London : Angus & Robertson, 1982
64p : ill(some col.)
0–207–14770–1 X.992/5257
2088401

ABORIGINAL TALES
Australian aboriginal tales – Texts

CLARKE, Mollie, 1916–
Walloobahl. – London : Collins Educational, c1984
24p : ill(some col.). – (A World of folk tales)
0–247–13402–3 X.990/23257
2566656

The LEGEND of the frogs. – [London] : Save the Children Fund, 1981
32p : ill(some col.). – (Round the world folk tales)
0–333–30760–7 X.990/17820 Woolwich
2566656

ABORIGINAL TALES
Australian aboriginal tales & legends

ROBERTS, Ainslie
Dreamtime. – Adelaide : Rigby, 1981
144p
0–7270–1496–x L.45/1969
8021015

ABORIGINAL TALES
Australian aboriginal tales & legends – Anthologies

HARNEY, William Edward
Tales from the Aborigines. – Adelaide : Seal Books, 1976
189p,[12]p of plates : ill
0–85179–217–0 H.78/6 Woolwich
118668X

REED, A. W. (Alexander Wyclif)
Myths and legends of Australia. – Sydney ; London : A.H. and A.W. Reed, 1976
256p : ill. – Bibliography: p.255–256
0–589–07041–x X.990/9036 Woolwich
118668x

ABORIGINES
See also
Batek
Jiman
Mah Meri
Mardudjara
Orang Asli
Pitjandjar
Tiwi

ABORIGINES
Attitudes to colonisation by Europe, to 1978

The WHITE men / [compiled by] Julia Blackburn. – London : Orbis Books, 1979
192p : ill(some col). – Bibliography: p.190. – Includes index
0–85613–399–x X.800/28175
1992821

ABORIGINES
Australian aborigines

ABORIGINES / consultant editor Henry Pluckrose ; illustrated by Maurice Wilson. – London : Hamilton, 1981
28p : col.ill. – (Small world). – Includes index
0–241–10630–3 X.808/35021
2081156

ABORIGINES
Australian aborigines–continued

AUSTRALIAN aboriginal concepts / edited by L.R. Hiatt. – Canberra : Australian Institute of Aboriginal Studies, 1978
198p.
0–391–00887–0 X.0800/39
133266X

EDWARDS, Robert, b.1930
Australian Aborigines. – Adelaide : Rigby, 1976
2–30p : col ill. – (Opal young)
0–85179–744–x X.800/26622
133266X

HUGHES, Jill
A closer look at aborigines. – London : H. Hamilton, 1979
32p : chiefly col ill. – (A closer look book). – Includes index
0–241–10333–9 X.802/11159
2113252

MATHEWS, Janet
Wurley & wommera. – Sydney : London : Collins, 1979
128p : ill. – Includes index
0–00–195032–0 X.802/11100
2081156
– [Another copy] X.525/9058

MOUNTFORD, Charles Pearcy
The aborigines and their country. – Adelaide : Rigby, 1975
143p : chiefly ill(some col)
0–7270–0027–6 X.802/10420
1246828

ABORIGINES
Australian aborigines – Serials

ABORIGINAL history. – vol.1– ; 1977–. – Canberra, 1977–
r6089456 P.701/1154
133266X

ABORIGINES
Australian aborigines, to date

BERNDT, Catherine H. (Catherine Helen)
The aboriginal Australians. – 2nd ed. – Carlton, Vict. : Pitman, 1983
xii,151p : ill. – Includes bibliography
0–85896–916–5 X.809/60497
8532915

DAWSON, James, 1806–1900
Australian Aborigines. – Canberra : Australian Institute of Aboriginal Studies, 1981
viii,[113]p,2leaves of plates. – (AIAS new series ; no.26)
0–85575–118–5 X.805/5836
8532915

NGABIDJ, Grant
My country of the pelican dreaming. – Canberra : Australian Institute of Aboriginal Studies, 1981
202p,1 leaf of plates : ill,maps. – Bibliography
0–85575–088–x X.800/39140
8532915

ROWLEY, C. D.
The destruction of aboriginal society. – [New ed.]. – Ringwood, Vict. : Penguin Books Australia, 1972
xv,430p. – Includes bibliography
0–14–021452–6 X.529/62926
8532915

ABORIGINES
Australian aborigines, to date – Serials

AUSTRALIAN aboriginal studies. – No.1 (1983)–. – Melbourne : Australian Institute of Aboriginal Studies, 1983–
v. : ill
r6814276 X.0525/685
8532915

ABORIGINES
Australian aborigines, ca 1805

TETLEY, Captain
Australian aborigines. – St Peter Port : Toucan Press, 1977
[14]leaves : chiefly ill
B78–02910 X.419/3451 Woolwich
1288423

ABORIGINES
Australian aborigines. Alcoholism, to date

GOODLUCK, Jack
The grog-breakers. – Casuarina, N.T. : Nungalinya Publications, [1980]
40p
r6803436 L.49/1013
8768838

ABORIGINES
Australian aborigines. Attitudes to colonisation of Australia by Europeans, 1788–1900

REYNOLDS, Henry, 1938–
The other side of the frontier. – Ringwood ; Harmondsworth : Penguin, 1982
xii,255p. – (Pelican books). – Bibliography: p222-247 – Includes index
0–14–022475–0 X.808/38137
2912236

ABORIGINES
Australian aborigines: Bennelong – Biographies

PHIPSON, Joan
Bennelong. – Sydney ; London : Collins, 1975
64p : ill. – (Australians in history)
0–00–195024–x X.992/3093 Woolwich
0587621

ABORIGINES
Australian aborigines. Cognition – Conference proceedings

COGNITION SYMPOSIUM (1974 : Canberra)
Aboriginal cognition. – Canberra : Australian Institute of Aboriginal Studies [etc], 1976
[6],418p : ill. – (Australian Institute of Aboriginal Studies. Psychology series ; no.1). – Includes bibliographies and index
0–85575–057–x X.520/12973
1687816

ABORIGINES
Australian aborigines. Cultural processes, to 1973

GIBBS, R. M.
The aborigines. – Melbourne : Longman Cheshire, 1977
[8],112,R7p : ill. – Bibliography : p.106–107. – Includes index
0–582–68120–0 X.700/25706
1494317

TINDALE, Norman Barnett
The Australian Aborigines. – [Windsor, Victoria] : Lloyd O'Neil, 1979
62p. : ill. – (Budget Books)
0–908505–16–7 X.802/4752
1494317

ABORIGINES
Australian aborigines. Cultural processes, to date

KIRK, R. L.
Aboriginal man adapting. – Melbourne : Oxford University Press, [1983]
vii,229p. – (Research monographs on human population biology). – Includes bibliography
0–19–554408–0 X.520/34001
8534071

TURNER, David H.
Australian aboriginal social organization. – Atlantic Highlands, N.J. : Humanities Press, 1980
xii,163p. – ([Publications] / [Australian Institute of Aboriginal Studies]). – Bibliography
0–391–01008–5 X:0800/39
8534071

ABORIGINES
Australian aborigines. Ecological aspects

KIRK, R. L.
Aboriginal man adapting. – Oxford : Clarendon, 1981
vii,229p : ill. – ([Research monographs on human population biology]). – Includes index
0–19–857532–7 X.325/1280 Woolwich
2205025

ABORIGINES
Australian aborigines. Economic status

ALTMAN, Jon C.
The economic status of Australian aborigines. – Cambridge : Cambridge University Press, 1979
xxi,230p : ill. – Bibliography: p.211–220. – Includes index
0–521–22421–7 X.520/13948
1896067

ABORIGINES
Australian aborigines. Effects of colonisation of Australia by Great Britain, to 1977

ROBERTS, Janine
From massacres to mining. – London : CIMRA : London : War on Want, 1978
xi,212p : ill. – Bibliography: p.187–188
0–905990–05–6 X.519/30439
2001543

ABORIGINES
Australian aborigines: Jackey Jackey – Biographies

PAICE, Margaret
Jackey Jackey. – Sydney : Collins, 1976
63p : ill. – (Australians in history series)
0–00–195028–2 X.800/27212
1592025

ABORIGINES
Australian aborigines. Kinship. Classification

SCHEFFLER, Harold Walter
Australian kin classification. – Cambridge : Cambridge University Press, 1978
xxi,567p : ill. – (Cambridge studies in social anthropology ; 23). – Bibliography: p.545–555. – Includes index
0–521–21906–x X.520/13045
1702653

ABORIGINES
Australian aborigines. Land tenure. Rights, to date

ABORIGINES, land and land rights / edited by Nicolas Peterson and Marcia Langton. – Canberra : Australian Institute of Aboriginal Studies, 1983
viii,468p. – (AIAS new series ; no.42)
0–85575–145–2 X.800/39134
9090533

ABORIGINES
Australian aborigines. Nicholls, Sir Douglas – Biographies

CLARK, Mavis Thorpe
The boy from Cumeroogunga. – Sydney : London : Hodder and Stoughton, 1979
179p,[8]p of plates : 1 ill
0–340–23933–6 X.809/45762
1845853

Pastor Doug. – Adelaide : Rigby, 1975
[7],259,[12]p of plates : ill. – (Seal books)
0–85179–890–x X.708/21836
1845853

ABORIGINES
Australian aborigines. Policies of Australian government, 1967–1977

COOMBS, H. C.
Australia's policy towards aborigines, 1967–1977. – London : Minority Rights Group, 1978
20,[2]p. – (Reports / Minority Rights Group ; no.35). – Bibliography: p.18. – List of films: p.19–20
B78–18013 P.803/299
1456687

ABORIGINES
Australian aborigines. Religious beliefs

MOL, Hans
The firm and the formless. – Waterloo, Ont. : Wilfrid Laurier University Press, c1982
viii,103p. – (Religion and identity ; 2). – Bibliography: p94–99. – Includes index
0–88920–117–x X.200/40320
2898136

ABORIGINES
Australian aborigines. Religious beliefs. Effects of social change, to date

KOLIG, Erich
The silent revolution. – Philadelphia : Institute for the Study of Human Issues, [1981]
xii,192p. – Includes bibliography
0–89727–020–7 X.200/45829
9166394

ABORIGINES
Australian aborigines: Roberts, Phillip – Biographies

LOCKWOOD, Douglas
I, the aboriginal. – Adelaide : Rigby, 1980
200p
0–7270–1365–3 X.800/33200
8894906

ABORIGINES
Australian aborigines. Sign languages

ABORIGINAL sign languages of the Americas and Australia / edited by D. Jean Umiker-Sebeok and Thomas A. Sebeok. – New York ; London : Plenum Press.
Vol.1: North America. – 1978
xxxii,442p : ill. – Includes bibliographies and index
0–306–31073–2 X.0520/545(1)
1565796

ABORIGINAL sign languages of the Americas and Australia / edited by D. Jean Umiker-Sebeok and Thomas A. Sebeok. – New York ; London : Plenum Press.
Vol.2: The Americas and Australia. – 1978
xxxiv,445p : ill. – Includes bibliographies and index
0–306–31081–3 X.0520/545(2)
1565796

ABORIGINES
Australian aborigines. Social conditions

BROWNE, Rollo
Aboriginal family. – London : A. and C. Black, c1983
25p : col.ill. – (Beans. Geography)
0–7136–2293–8 X.529/55620
2933829

ABORIGINES
Australian aborigines. Social conditions–continued

BURT, Ben
Aborigines. – London : British Museum Publications Ltd [for] Museum of Mankind, 1977
[2],16p : ill. – (Discovering other cultures). – Bibliography: p.16
0–7141–0060–9 BS.4/269(1)
1293389

MATHEWS, Janet
Totem & taboo. – Sydney ; London : Collins, 1979
112p : ill. – Includes index
0–00–195033–9 X.805/4521
2933829

SUTER, Keith
Aboriginal Australians. – London : Minority Rights Group, 1982
20p. – (Report / Minority Rights Group ; no.35). – Bibliography: p20
0–903114–77–1 P.803/299(35)
3150658

WOMAN'S role in Aboriginal society / Fay Gale ; editor. – 2nd ed. – Canberra : Australian Institute of Aboriginal Studies, 1974
viii,84p. – (Australian Aboriginal studies ; no 36) (Social anthropology series ; no. 6)
0–85575–033–2 X.0800/39(36a)
1293389

ABORIGINES
Australian aborigines. Social conditions – Librettos

WALKER, Kath
Stradbroke dreamtime. – Re-illustrated ed.. – London : Angus & Robertson, 1982
64p : ill(some col.)
0–207–14770–1 X.992/5257
305814x

ABORIGINES
Australian aborigines. Social conditions, to 1978

LULING, Virginia
Aborigines. – London : Macdonald Educational, 1979
48p : ill(chiefly col). – (Surviving peoples). – Bibliography: p.47. – Includes index
0–356–05952–9 X.802/10934
1935607

ABORIGINES
Australian aborigines. Social conditions, to date

BROOME, Richard
Aboriginal Australians. – 2nd impression. – Sydney : Allen & Unwin, 1982
227p : ill. – (The Australian experience ; no.4)
0–86861–051–8 X.809/63871
8021066

ABORIGINES
Australian aborigines. Social conditions, 1803–1835

MORGAN, John, fl.1852
The life and adventures of William Buckley. – Firle : Caliban, 1979
238p : ill
0–904573–12–5 X.808/34711
2540762

ABORIGINES
Australian aborigines. Social conditions, 1932–1964

DRYSDALE, Ingrid
The end of dreaming. – London : Hale [etc.], 1975
[25],194p,[16]p of plates : ill
0–7091–5040–7 X.809/40117
0362387

ABORIGINES
Australian aborigines. Social conditions, 1967–1976

AUSTRALIA. Parliament. Senate. Standing Committee on Social Environment
Aborigines and Torres Strait Islanders. – Canberra : Government Printer of Australia, [1977?]
6v.(3548p.)
r6051819 CS.G.588/182
1205757

COOMBS, H. C.
Aboriginal Australians, 1967–1976. – Perth, Western Australia : Murdoch University, 1976
[19]p
r6032237 X.700/25520
1205757

ABORIGINES
Australian aborigines. Social problems – Conference proceedings

ABORIGINES and change / edited by R.M. Berndt. – Canberra : Australian Institute of Aboriginal Studies [etc.], 1977
424p. – (Social anthropology series ; no. 11)
0–85575–059–6 X.0800/39.(11)
0190276

ABORIGINES
Australian aborigines. Totemism

BRANDENSTEIN, C. G. von (Carl Georg)
Names and substance of the Australian subsection system. – Chicago ; London : University of Chicago Press, 1982
199p,[1]leaf of plates : ill(some col.). – Bibliography: p176–193. – Includes index
0–226–86481–2 X.800/35848
2926687

ABORIGINES
Australian aborigines. Traditional activities, to date

MOUNTFORD, Charles Pearcy
Brown men and red sand. – Sydney ; London : Angus & Robertson, 1981
192p,16p of plates
0–207–14373–0 X.950/14252
8532869

ABORIGINES
Australian aborigines. Treatment by criminal law

EGGLESTON, Elizabeth
Fear, favour or affection. – Canberra : Australian National University Press, 1976
xx, 399 p., 4 p. of plates : ill. – (Aborigines in Australian society ; 13). – Bibliography : p. 378–385. – Includes index
0–7081–0191–7 X.200/30538
0711381

ABORIGINES
Australian aborigines. Tribes

TINDALE, Norman Barnett
Aboriginal tribes of Australia. – Berkeley ; London : University of California Press, 1974
xii,404p,leaf of plate,[52]p of plates : ill(some col). – Bibliography: p.357–386. – Includes index
0–520–02005–7 X.802/10030
0500380

ABORIGINES
Australian aborigines. Tribes. Evolution. Genetic factors

BALAKRISHNAN, V.
Genetic diversity among Australian aborigines. – Canberra : Australian Institute of Aboriginal Studies, 1975
ix, 115 p. : ill. – (Australian aboriginal studies. research and regional studies ; no. 3). – Bibliography : p. 106–115
0–85575–043–x X.311/8377 Woolwich
1042009

ABORIGINES
Indian aborigines. Tribes. Cultural processes

FUCHS, Stephen
The aboriginal tribes of India. – London : Macmillan, 1977
vii,308p. – Includes bibliographies and index
0–333–19782–8 X.809/42791
1333127

ABORIGINES
Rights. International law

BENNETT, Gordon
Aboriginal rights in international law. – London : Royal Anthropological Institute [for] Survival International, 1978
[5],88p. – (Occasional papers / Royal Anthropological Institute ; no.37). – Bibliography: p.69
B78–39531 Ac.6236/3[no.37]
1734571

ABORIGINES. Arnhem Land. Northern Territory. Australia
Australian aborigines. Social conditions, ca 1947

MARSHALL, Alan, b.1902
These were my tribesmen. – 3rd ed. – Adelaide : Rigby, 1976
[8],214,[22]p of plates : ill. – (Seal books)
0–85179–408–4 X.908/42374
1881965

ABORIGINES. Brisbane. Queensland
Australian aborigines. Social conditions, 1965–1966

SMITH, Hazel M.
Look forward, not back. – Canberra : Australian National University Press, 1975
xix, 163 p.. – (Aborigines in Australian society ; 12)
0–7081–0697–8 X.809/22314
0905542

ABORIGINES. Central Australia
Australian aborigines. Cultural processes

AUSTRALIA. Parliament. Senate. Select Committee on Aborigines and Torres Strait Islanders
The environmental conditions of Aborigines and Torres Strait Islanders and the preservation of their sacred sites. – Canberra : Australian Government Publishing Service, 1976
xxix,314p
0–642–02236–4 CS.G.588/230
1058851

RÓHEIM, Géza
Children of the desert. – New York ; London : Harper and Row.
Vol.1. – 1976
xix,262p : ill. – Includes index
0–06–131906–6 X.0700/1028(1)
1058851

ABORIGINES. Deserts. Western Australia
Australian aborigines. Adult education – Serials

WESTERN AUSTRALIA. Adult Aboriginal Education
Annual report. – 1976–. – West Perth : Education Department of Western Australia. Technical Education Division, Adult Aboriginal Education, 1977–
r6051911 CS.G.485/109
1319779

ABORIGINES. Deserts. Western Australia
Australian aborigines. Making of stone tools

HAYDEN, Brian
Palaeotothic reflections. – Canberra : Australian Institute of Aboriginal Studies., 1979
viii,181p,31leaves of plates : ill. – (A.I.A.S. new series no.5.)
r6194015 X.805/5838
0436550

ABORIGINES. Groote Eylandt. Northern Territory. Australia
Use of plants, to date

LEVITT, Dulcie
Plants and people. – Canberra : Australian Institute of Aboriginal Studies, 1981
viii,166p,100p of plates : ill. – (AIAS new series ; no.22)
0–85575–115–0 X.805/5837
8865434

ABORIGINES. New South Wales
Australian aborigines, to date

BLOMFIELD, Geoffrey
Baal Belbora. – Chippendale, N.S.W. : APCOL, 1981
xviii,148p : ill
Signed by the author
0–909188–57–2 X:809/61911
9007229

ROWLEY, C. D.
Equality by instalments. – Canberra : Australian Institute of Aboriginal Studies, 1982
33p. – (AIAS new series ; no.40)
0–85575–137–1 X.800/39136
9007229

ABORIGINES. New South Wales
Australian aborigines. Barker, Jimmie – Biographies

BARKER, Jimmie
The two worlds of Jimmie Baker. – Canberra : Australian Institute of Aborginal Studies, 1977
xiv,218p : 2 ill. – (Australian Aboriginal studies. ethnohistory series ; no.4)
0–85575–056–1 X.0800/39(65)
1882996

ABORIGINES. Northern Territory. Australia
Australian aborigines. Social conditions, 1919–1956

HARNEY, William Edward
Life among the Aborigines. – Adelaide : Rigby, 1976
219p,[1] leaf of plate,[20]p of plates : ill. – (Seal books)
0–85179–215–4 X.708/21828
1845918

ABORIGINES. Port Keats. Northern Territory. Australia
Australian aborigines. Kinship

FALKENBERG, Aslaug
The affinal relationship system. – Oslo : Universitetsforlaget, c1981
206p : ill. – (Oslo studies in social anthropology ; no.1). – Bibliography: p201–202. – Includes index
82–00–05563–9 X.529/44458
2583895

ABORIGINES. Port Keats. Northern Territory. Australia
Australian aborigines. Marriage

FALKENBERG, Aslaug
The affinal relationship system. - Oslo : Universitetsforlaget, c1981
206p : ill. - (Oslo studies in social anthropology ; no.1). - Bibliography: p201-202. - Includes index
82-00-05563-9 X.529/44458
2583887

ABORIGINES. Queensland
Australian aborigines. Cultural processes, 1882-1883

LUMHOLTZ, Carl
Among cannibals. - Firle : Caliban Books, 1979
[24],383p,3 leaves of plates(1 fold) : ill. - Includes index
0-904573-13-3 X.520/9901
1873393

ABORIGINES. Special themes. Australian writers, to date. English literature
Australian aborigines

HEALY, J. J. (John Joseph)
Literature and the aborigine in Australia, 1770-1975. - St. Lucia : University of Queensland Press, c1978
x,305p
0-7022-1197-4 X:950/35111
9113622

ABORIGINES. Sydney region. New South Wales
Australian aborigines, 1788-1860

WILLEY, Keith
When the sky fell down. - Sydney ; London : Collins, 1979
231p,[12]p of plates : ill(some col.). - Bibliography: p221-223. - Includes index
0-00-216434-5 X.809/46369
2350327

ABORIGINES. Tasmania
Australian aborigines, to 1876

PLOMLEY, Norman James Brian
The Tasmanian aborigines. - Launceston : The author with the Adult Education Division, Tasmania, 1977
viii,72p : ill. - Bibliography : p.70-72
r6032698 X.709/32023
1353543

ABORIGINES. Tasmania
Australian aborigines, to 1980

RYAN, Lyndall
The Aboriginal Tasmanians. - St Lucia ; London : University of Queensland Press, 1982, c1981
xiv,315p : ill. - Bibliography: p278-302. - Includes index
0-7022-1544-9 X.809/63951
3281612

ABORIGINES. Warrabri. Northern Territory. Australia
Australian women aborigines. Cultural processes

BELL, Diane
Daughters of the dreaming. - Melbourne : McPhee Gribble, 1983
297p : ill. - Bibliography: p273-282
0-86861-472-6 X.520/34237
301780x

ABORIGINES. Western Australia
Behaviour. Adaptation to environment expounding adaptation to environment of behaviour of prehistoric man

GOULD, Richard Allan
Living archaeology. - Cambridge : Cambridge University Press, 1980
xv,270p : ill. - (New studies in archaeology). - Bibliography: p.255-264. - Includes index
0-521-23093-4 X.421/21048 Woolwich
2237849

ABORTION
See also
Anti-abortion movements
Women undergoing abortion

ABORTION

ABORTION / report of the Public Morals Committee of the Evangelical Presbyterian Church. - Belfast : Public Morals Committee of the Evangelical Presbyterian Church, 1981
[16]p
B81-28456 X.329/14274 Woolwich
2527162

ABORTION & law. - Dublin : Dominican, 1983
130p. - (A Doctrine and life special)
0-907271-19-7 X.200/41180
0287245

ABORTION and the right to live. - Abbots Langley : Catholic Information Services, [1980?]
[16]p
0-85183-313-6 X.329/15821
0287245

ABORTION research / Henry P. David, editor. - Lexington, Mass. ; London : D.C. Heath, 1974
xix,249p : ill. - Includes bibliographies
0-669-94896-9 X.320/10099
050002x

BARNES, Peter, 1953 Nov. 13-
Open your mouth for the dumb. - Edinburgh : Banner of Truth Trust, 1984
31p
0-85151-390-5 X.208/10207
0056960

BRENT AGAINST CORRIE PAMPHLET GROUP
Mixed feelings. - London : B.A.C.P.G. [etc.], 1980
30p
B80-36681 X.629/14104 Woolwich
227549x

BRODY, Baruch A. (Baruch Alter)
Abortion and the sanctity of human life. - Cambridge, Mass. ; London : MIT Press, 1975
viii,162p. - Includes index
0-262-02116-1 X.329/10214
0542016

DE MARCO, Donald
Abortion in perspective. - Cincinnati, Ohio : Hiltz and Hayes, 1974
xiv,192p : ill
r6051557 X.519/30290
0542016

DOHERTY, Peter, 1924-
Abortion - is this your choice?. - Wallington, Surrey : Faith Pamphlets, [1983?]
20p
B84-00019 X.329/17041
0287245

FINNIS, J. M. (John Mitchell)
What do you know about abortion?. - [Oxford] : Oxford School Publications, 1976
iii,37p : ill. - Bibliography: p.37
0-905697-00-6 Cup.351/65
1056735

What do you know about abortion?. - Rev. ed. - Witney : Oxford School Publications, 1980
iii,37p : ill
0-905697-02-2 Cup.355/23
1056735

GODFROY, Raymonde
Non aux camps de la mort, secours aux innocentes victimes, non à l'avortement. - Paris : La Pensée Universelle, 1974
63 p.
r6000941 X.108/14383
0287245

HARRIS, Brian, 1932-
Abortion. - [Bridgend] : Evangelical Movement of Wales [Literature Committee], [1977]
15p. - (God has spoken)
B78-36596 X.100/17963
0056960

MIURA, Domyo
The forgotten child. - Henley-on-Thames : Ellis, 1983
124p
0-85628-130-1 X.529/62253
3132102
- [Another copy: paperback.] ISBN 0-85628-133-6 X.529/62254

NILSEN, Per Agnar
Seksualetikk og abort. - [Oslo] : Det Norske misjonsforbunds ungdom, [1974]
14p
r6034098 X.319/8673 Woolwich
0056960

O'DONOVAN, Oliver
The Christian and the unborn child. - 2nd ed. - Bramcote : Grove Books, 1975
24p. - (Grove booklets on ethics ; no.1)
0-901710-67-9 X.100/16169
0056960

POWELL, John, 1925-
Abortion. - Allen, Texas : Argus Communications, c1981
183p : col ill. - Bibliography: p181-182
0-89505-063-3 X.329/16340
0056960

SHAPIRO, Rose
Abortion advice. - London : Thames Television, [1983?]
9p
B84-03675 X.322/12640
2992949

Abortion advice. - London : Thames Television, [1983?]
9p. - (A Help! booklet)
B84-37602 X.525/9794
2992949

STIRRAT, Gordon A.
Legalised abortion. - London : Christian Medical Fellowship, 1979
32p. - (Topic booklet)
0-85111-966-2 X.108/17535
0056960

ABORTION

VIOLA, Carmelo Rosario
Aborto : perchè deve decidere la donna. - Cosenza : Pellegrini, 1977
215p. - (Collana di. "Cultura politica e sociale")
r6085554 X.519/41117
2029235

ABORTION
- For teaching

LIFE must go on / compiled by Noirin Hynes ... [et al.]. - Dublin : Helicon, 1982
104p
B83-18437 X.200/40124
2881454

ABORTION
to 1974

SARDI, Paolo
L'aborto ieri e oggi. - Brescia : Paideia, 1975
389 p.. - (Biblioteca di cultura religiosa ; 26)
r6024359 x.100/15385
1024957

ABORTION
to 1976

POTTS, Malcolm
Abortion. - Cambridge : Cambridge University Press, 1977
x,575p : ill. - Includes index
0-521-21442-4 X.320/10930 Woolwich
1251899

ABORTION
to date

DAVIDSON, Barbro
Abort förr och nu. - Stockholm : Röda rummet, 1982
118p : ill
91-7362-084-x X.329/17352
8619190

GARCÍA-VELUTINI, Oscar
Sobre derechos personales y la dignidad humana. - Caracas : Editorial Sucre, 1980
200p
r6706729 X.329/16280
8619190

RETT til liv / Norsk Pro Vita ; redaktør : Bjørn Roksand. - [Oslo] : Lunde, 1978
119p
82-520-3640-6 X.329/16389
8619190

THÉVENET, Emilio F
Aborto en el Uruguay. - Montevideo : [s.n.], 1978
85p. - Bibliography
r6700939 X.329/14284 Woolwich
8619190

ABORTION
to date. Psychological aspects

TROST, Anne-Christine
Abort och psykiska besvär. - Västerås : International Library, 1982
173p. - Bibliography
91-85898-03-1 X.529/49577
8852006

ABORTION
1800-1982

FRANCOME, Colin
Abortion freedom. - London : Allen & Unwin, 1984
241p. - Bibliography: p225-234. - Includes index
0-04-179001-4 X.529/62678
3138542

ABORTION
Ethical aspects

ABORTION / edited by Sidney Callahan and Daniel Callahan. - New York ; London : Plenum, c1984
xxii,338p. - (The Hastings Center series in ethics). - Includes index
0-306-41640-9 X.520/36854
2501678

ABORTION and the status of the fetus / edited by William B. Bondeson ... [et al.]. - Dordrecht ; Lancaster : Reidel, c1983
xxxii,349p : ill. - (Philosophy and medicine ; v.13). - Includes bibliographies and index
90-277-1493-2 X.329/17036
2501678

SUMNER, L. W.
Abortion and moral theory. - Princeton ; Guildford : Princeton University Press, c1981
x,246p. - Bibliography: p229-239. - Includes index
0-691-07262-0 X.329/13969 Woolwich
2501678

ABORTION
Ethical aspects–continued

TOOLEY, Michael
Abortion and infanticide. - Oxford : Clarendon, 1983
ix,441p. - Bibliography: p426–433. - Includes index
0–19–824674–9 X.529/61136
2501678

ABORTION
Ethics

CHURCH OF ENGLAND. Board for Social Responsibility
Abortion. - London : Church Information Office for the Board, 1973
70p
0–7151–6502–x X.529/65396
0047236

SALVONI, Fausto
Abominevole e contrario alla morale. - Genova : Lanterna, [1976]
100p. - (Attualità biblica ; 8)
r6085276 X.519/40780
1985191

WOODY, Jeanine
Abortion?. - Houston : Hunter Ministries, [1977]
95p
0–917726–10–3 X.529/34948
2250365

ABORTION
Law related to fertility of women

VAN DER TAK, Jean
Abortion, fertility, and changing legislation. - Lexington ; London : Lexington Books, [1974]
xxi, 141 p. : ill
0–669–94029–1 X.320/5159 Woolwich
1020021

ABORTION
Legal aspects

The HUMAN problem of abortion / edited by Rebecca J. Cook, Pramilla Senanayake. - London : International Planned Parenthood Federation, c1979
56p : col.ill. - Bibliography: p49–56
0–86089–022–8 X.525/10533
1837753

ABORTION
Medical aspects

HERN, Warren M. (Warren Martin)
Abortion practice. - Philadelphia ; London : Lippincott, c1984
xviii,340p : ill. - Includes index
0–397–50607–4 SRL
1837761

The HUMAN problem of abortion / edited by Rebecca J. Cook, Pramilla Senanayake. - London : International Planned Parenthood Federation, c1979
56p : col.ill. - Bibliography: p49–56
0–86089–022–8 X.525/10533
1837761

ABORTION
Medical aspects - Conference proceedings

WORKSHOP AND POSTGRADUATE COURSE ON PREGNANCY TERMINATION : PROCEDURES, SAFETY, AND NEW DEVELOPMENTS (1978 : Nassau)
Pregnancy termination. - Hagerstown ; Cambridge : Harper and Row, 1979
xxvii,447p : ill. - (PARFR series on fertility regulation / Program for Applied Research on Fertility Regulation). - Includes bibliographies and index
0–06–142900–7 SRL
1837761

ABORTION
Menstrual regulation

INTERNATIONAL PLANNED PARENTHOOD FEDERATION. Central Medical Committee
Menstrual regulation. - London : International Planned Parenthood Federation, 1976
34p : ill. - Bibliography: p.33–34
0–900924–81–0 SRL
0786411

ABORTION
Second-trimester abortion. Medical aspects

SECOND-TRIMESTER abortion / edited by Gary S. Berger, William E. Brenner, Louis G. Keith. - The Hague ; London : Nijhoff, 1981
xxii,340p : ill. - Includes index
B82–06415 SRL
2593742

SECOND trimester pregnancy termination / edited by Marc J.N.C. Keirse ... [et al.]. - The Hague ; London : Leiden University Press, 1982
x,210p : ill. - (Boerhaave series for postgraduate medical education ; v.22). - Includes index
90–602–1490–0 X.0320/59(22)
2593742

ABORTION
Sex relations, divorce & abortion

INDEBITE euforie / a cura di Annarosa Martelli. - Milano : Sugar Co, 1974
90 p.. - (Fatti e misfatti ; n. 10)
r6024088 X.519/21369
1005863

ABORTION
Social aspects

ABORTION and sterilization / edited by Jane E. Hodgson. - London : Academic Press, 1981
x,594p : ill. - Includes bibliographies and index
0–12–792030–7 SRL
028338x

WATTERS, Wendell W.
Compulsory parenthood. - Toronto : McClelland and Stewart, 1976
304p
0–7710–8880–9 X.320/11876 Woolwich
028338X

ABORTION. Adolescent girls
Psychosocial aspects

ALLEN, James E. (James Edward)
Managing teenage pregnancy. - New York : Praeger, 1980
xxvii,300p : ill. - (Praeger special studies). - Bibliography: p290–295. - Includes index
0–03–053991–9 X.322/9480 Woolwich
055376X

ABORTION. Australia
to date

WAINER, Bertram
It isn't nice. - [Sydney?] : Alpha Books, 1972
224p.,4p of plates : ill
r6813403 X.200/41435
9023674

ABORTION. Birmingham. West Midlands (Metropolitan County) 1967–1977

ABORTION in Birmingham - ten years on. - [Birmingham] : [Birmingham National Abortion Campaign], [1977]
16p : ill
B81–10682 X.525/4086
2372517

ABORTION. Commonwealth countries
Law

COOK, Rebecca J.
Emerging issues in commonwealth Abortion Laws, 1982. - London : Commonwealth Secretariat, c1983
135p. - Bibliography: p92–110
0–85092–230–5 X.525/9631
3151158

ABORTION. Denmark

USSING, Jytte
Om abort = Facts about legally induced abortion. - København : Socialforskningsinstituttet : I kommission hos Teknisk Forlag, 1979
87p. - (Socialforskningsinstituttet. Publikation ; 87)
87–7487–153–6 S.B.210/9(87)
2165716

ABORTION. Denmark
Opposition

MICHELET, Else
Vet de hva de snakker om? Abortsmotstanderne og virkeligheten. - Oslo : Pax, [1975]
[32]p : ill. - (Pax rapport)
82–530–0719–1 X.319/8697 Woolwich
0394955

ABORTION. England
- Serials

ABORTION review. - No.1 (Oct./Nov. 1981)–. - London : Birth Control Trust, 1981–
v.
B82–11836 P.523/1620
2627655

ABORTION. England
- Statistics - Serials

GREAT BRITAIN
[Office of Population Censuses and Surveys]. OPCS Monitor. Legal abortions. - AB 75/1–. - London : OPCS, [1975]–
r6000595 BS.34/70(2)
1316613

ABORTION. England
Legal abortion - Statistics

ABORTION statistics / Office of Population Censuses and Surveys. - 1974. - London : H.M.S.O, 1977
vi,56p
Subsequent volumes, when received, will be placed at this shelfmark
0–11–690635–9 BS.34/90
126267x

ABORTION. Europe
Ethics

INTERNATIONAL PLANNED PARENTHOOD FEDERATION. Europe Region
Ethical aspects of abortion. - London : International Planned Parenthood Federation, Europe Region, 1978
[2],82p. - Bibliography: p.79–80
0–904983–03–x X.319/18229 Woolwich
1516892

ABORTION. Finland
to date. Social aspects

KOTIRANTA, Maija-Liisa
Aborttipyynnön sosiaaliset perusteet. - Helsinki : Väestöntutkimuslaitos, 1975
iii,110 leaves. - (Väestöntutkimuslaitoksen julkaisuja. Sarja B ; n:o 39)
951–90481–8–9 X.0512/156b(39)
8945071

ABORTION. Foetuses. Man

VOLUNTARY termination of pregnancy / editor E.S.E. Hafez. - Lancaster : MTP, c1984
ix,178p : ill. - (Advances in reproductive health care). - Includes bibliographies and index
0–85200–483–4 SRL
1762079

WORLD HEALTH ORGANIZATION
Induced abortion. - Geneva : World Health Organization ; [London] : [H.M.S.O.], 1978
65p : ill. - (Technical report series / World Health Organization ; 623)
B79–02793 SRL
1762079

ABORTION. Foetuses with congenital abnormalities. Man

HARRIS, Harry, 1919–
Prenatal diagnosis and selective abortion. - London : Nuffield Provincial Hospitals Trust, 1974
ix,101p : ill. - (The Rock Carling Fellowship ; 1974)
0–900574–23–2 SRL
0748617

ABORTION. France

FIGUERAS, André
Avorteurs et avortons. - Paris : A. Figueras, 1980
159p
r6127916 X.329/13127 Woolwich
2463466

LAISSEZ-LES vivre / Jérôme Lejeune [et al.] .... - Paris : Éditions P. Lethielleux, 1975
351 p.
2–249–60100–3 X.319/8411 Woolwich
0938025

ABORTION. France
Law. Code de la santé publique. Chapitre III bis. Interruption volontaire de la grossesse. Enactment, 1973–1974

L'AVORTEMENT / rédigé sous la direction de Bernard Pingaud. - [Paris] : Flammarion, 1975
292 p.
2–08–060816–9 X.708/17578
0917060

ABORTION. Great Britain

GREAT BRITAIN. Parliament. House of Commons. Select Committee on Abortion
Report from the Select Committee on Abortion, session 1975–76. - London : H.M.S.O.
1st. Vol.1: Report. - [1976]
v,43p. - ([1975–76] H.C.573–I)
0–10–297476–4 BS.92/220
0729876
1st. Vol.2: Minutes of evidence and appendices. - [1976]
iv,322p : ill. - ([1975–76] H.C.573–II)
0–10–267176–1 BS.92/220
0729876
2nd. - [1976]
xix,27p. - ([1975–76 H.C.]737)
0–10–009676–x BS.92/220
0729876

ABORTION. Great Britain
- Serials

ABORTION today. - No.1–; Autumn 1976–. - Solihull : British Pregnancy Advisory Service, [1978]–
B79–08538 P.521/3211
1804049

ABORTION. Great Britain
Law. Great Britain. Parliament. House of Commons. Abortion (Amendment) Bill. Session 1974-75. Protest movements by women. Organisation

ALRA
A woman's right to choose action guide. - London : [Alra], [1975]
[1],51p : ill
0-9500063-3-5 X.319/10229
0523844

ABORTION. Great Britain
Law. Reform - Serials

BREAKING chains - No 1- ; May-June 1977- - [London] : [A.L.R.A.], 1977- : ill
B78-04903 P.323/321
006839x

ABORTION. Great Britain
Law. Social aspects, 1967-1975

GREENWOOD, Victoria
Abortion in demand. - London : Pluto Press, 1976
153p. - Bibliography: p.144-150. - Includes index
0-904383-11-3 X.319/16933 Woolwich
0761524

ABORTION. Great Britain
Legislation. Voting by members of Great Britain. Parliament. House of Commons, 1967-1981

HOW M.P.'s vote on abortion / [compiled] by Joanna Chambers and Dilys Cossey. - London : Birth Control Campaign, 1982
38p. - Bibliography: p38
B84-03563 X.200/41969
298542x

ABORTION. Great Britain
Rights of women

ABORTION rights. - London : Relgocrest Ltd, [1979]
32,[2]p : ill. - ('Socialist woman' pamphlet)
0-85612-200-9 X.519/29907
1920847

ABORTION. Hungary
1945- - Statistics

HUNGARY. Központi Statisztikai Hivatal. Népesedésstatisztikai főosztály. Népesedésstatisztikai osztály
Egységes szülészeti adatszolgáltatási rendszer. - Budapest, 1983
114p
r6449367 S.461/223
891947x

ABORTION. India (Republic)
- Conference proceedings

SEMINAR ON MEDICAL AND SOCIO-ECONOMIC ASPECTS OF ABORTION (1972 : Calcutta)
Medical and socio-economic aspects of abortion. - Calcutta : The Association, [1974?]
[2], xxii, 172 p., [2] p. of plates : 1 ill
r6010773 X.319/17125 Woolwich
105337X

ABORTION. Ireland (Republic)

NO more chains. - Dublin : [People's Democracy], [1982?]
28p : ill. - (A People's Democracy pamphlet)
B83-30212 X.529/55205
2926040

ABORTION. Italy

SESSO amaro / a cura di Fausta Cecchini...[et al.]. - Roma : Editori Riuniti, 1977
vi,195p. - (La questione femminile ; 5)
r6078352 X.519/38978
1711164

ABORTION. Italy
- Conference proceedings

L'ABORTO in Italia / autori vari. - Bologna : Edizioni Dehoniane, 1975
164p. - (Quaderni di teologia morale. Dibattito "aperto" ; 2)
r6067402 X.319/9356 Woolwich
1474472

ABORTION. Italy
to 1975

FACCIO, Adele
Il reato di massa. - Milano : Sugar Co Edizioni, 1975
257p : ill. - (Fatti e misfatti ; no.21). - Bibliography : p.257
r6011989 X.319/8668 Woolwich
0388602

ABORTION. Italy
to date

L'INTERRUZIONE volontaria di gravidanza / UICEMP - Firenze : Nuova Guaraldi, 1981
xiv,164p. - (Contributi salute ; 2)
r6509714 X.329/16948
857605x

ABORTION. Italy
Attitudes of women undergoing abortion

FACCIO, Adele
Le mie ragioni. - Milano : Feltrinelli, 1975
45p. - (Libelli)
r6068063 X.319/9705 Woolwich
178742X

ABORTION. Italy
Law

LIBRO bianco sull'aborto / Gruppo democratico cristiano della Camera dei Deputati ; a cura di Paolo Piccoli. - Milano : Rusconi, 1977
383p. - (Problemi attuali)
r6068714 X.319/20021 Woolwich
2036592

ABORTION. Italy
Law - Conference proceedings

DIFESA del diritto alla nascita / [relazioni di] Adriano Bompiani, Ignazio Marcello Gallo [e] Vittorio Mathieu. - Roma : Ed. Giuffrè, [1975]
xv, 257 p., 5 leaves of plates, 4 p. of plates : ill. - (Quaderni di "Iustitia" ; n. 25)
r6024112 X.100/15326
1020013

ABORTION. Italy
Law, to date

L'ABORTO / Paolo Martini ... [et al.]. - Milano : Giuffrè, 1979
iv,93p
r6503253 X.200/37530
8265399

CONTI, Laura
Il tormento e lo scudo. - Milano : Mazzotta, 1981
143p. - (Biblioteca di nuova cultura ; 81)
88-202-0474-6 X.208/10659
8265399

MARINUCCI, Elena
Guida all'aborto legale. - Venezia : Marsilio, 1978
vi,157p. - (Materiali Marsilio ; 14)
r6500218 X.200/35871
8265399

ABORTION. Italy
Law. Reform

FUSARO, Carlo
Aborto libero : ecco perché. - Livorno : la Nuova Frontiera, 1975
111 p.
r6006558 X.108/14439
0933007

ABORTION. Maryland
Attitudes of women undergoing abortion

MCCORMICK, Evelyn Patricia
Attitudes toward abortion. - Lexington, Mass. ; London : D.C. Heath, 1975
xiv,161p : ill. - Bibliography: p.143-151. - Includes index
0-669-96594-4 X.320/10121
0506206

ABORTION. Methods. Birth control
Social aspects

LUKER, Kristin
Taking chances. - Berkeley ; London : University of California Press, [1976]
xii,207p : ill. - Bibliography: p.195-204. - Includes index
0-520-02872-4 X.329/10700 Woolwich
0759716

ABORTION. Mexico

El ABORTO en México / Mariclaire Acosta ... [et al.]. - México : Fondo de Cultura Económica, 1976
83p. - (Archivo del Fondo ; n.57)
r6067128 X.519/39585
1785362

ABORTION. Norway

MENINGER om abort / redigert av Aase Kagge. - [Oslo] : Grøndahl, [1972]
100p
r6034203 X.319/8775 Woolwich
1221264

ABORTION. Norway
Law - Conference proceedings

ABORTHØRINGA (1976 : Oslo)
Hvem er de umoralske?. - Oslo : Brød og roser, [1976]
47p : ill
r6033821 X.313/806 Woolwich
1421670

ABORTION. Peru

CASTILLO MURRUGARRA, Victoria del
El aborto, su legalización en el Perú. - Lima : Ediciones Raíz, 1976
180p
r6053654 X.319/9078 Woolwich
1390570

ABORTION. Portugal

METRASS, Célia
Aborto. - Lisboa : Editorial Futura, 1975
360p. - (Colecção Mulheres em luta)
r6032951 X.319/8716 Woolwich
0398829

ABORTION. Slovenia. Yugoslavia
to date

SVOBODNO odločanje o rojstvu otrok. - [Ljubljana] : Uradni list SRS, 1972
80p
r6435512 X.529/52863
8890838

ABORTION. South Africa
Social aspects - Conference proceedings

SYMPOSIUM ON ABORTION (1973 : Durban)
The great debate. - Cape Town : Timmins, 1974
251 p. - (Human Sciences Research Council. Publication series ; no. 47). - Includes bibliographies
0-86978-098-0 X.529/31313
1073451

ABORTION. Spain
to date

IGLESIAS DE USSEL, Julio
El aborto. - Madrid : Centro de Investigaciones Sociológicas, 1979
127p. - (Colección Monografías ; n.26). - Bibliography
84-7476-020-8 X.529/43110
8724806

ABORTION. Sweden
to date

TROST, Jan
Aborter i länen. - Uppsala : Uppsala universitet, 1980
116p
91-506-0254-3 X.529/49465
8851999

ABORTION. Sweden
Attitudes of women undergoing abortion

VALFRIDSSON, Inga-Lill
Aborten. - Stockholm : Liber förlag, 1979
124p
91-380-4562-1 X.519/44010
2216515

ABORTION. Sweden
Law

LINDQVIST, Inge
Abortlagen med kommentarer. - Stockholm : Natur och Kultur, 1975
85 p.
91-270-0152-0 X.319/8338 Woolwich
0914746

ABORTION. United States

ABORTION / edited by Edward Manier, William Liu, David Solomon. - Notre Dame ; London : University of Notre Dame Press, 1977
xi,186p. - Includes index
0-268-00582-6 X.320/11367 Woolwich
0280062

FRANCKE, Linda Bird
The ambivalence of abortion. - London : Allen Lane, 1979
[7],261p. - Includes index
0-7139-1213-8 X.529/33258
1872311

The ambivalence of abortion. - Harmondsworth : Penguin, 1980
[9],261p. - Includes index
0-14-005440-5 Cup.358/70
1872311

JAFFE, Frederick S.
Abortion politics. - New York ; London : McGraw-Hill, c1981
viii,216p. - Includes index
0-07-032189-2 X.520/23157
0280062

ABORTION. United States –continued

MERTON, Andrew H.
Enemies of choice. – Boston, [Mass.] : Beacon Press, c1981
ix,241p
0–8070–0485–5 X.529/54451
0280062

ABORTION. United States
to date

NATHANSON, Bernard N.
Aborting America. – Garden City : Doubleday, c1979
x,320p
0–385–14461–x X.329/14950 Woolwich
8724814

NOONAN, John T.
A private choice. – New York : The Free Press : London : Collier MacMillan, [1979]
vii,244p
0–02–923160–4 X.520/24614
8724814

RICE, Charles E.
Beyond abortion. – Chicago : Franciscan Herald Press, c1979
vii,159p
0–8199–0696–4 X.529/54797
8724814
– [Another copy] X.529/65417

ABORTION. United States
Ethical aspects

RAMSEY, Paul, 1913–
Ethics at the edges of life. – New Haven ; London : Yale University Press, 1978
xvii,353p. – (The Bampton lectures in America). – Includes index
0–300–02137–2 WP.1578/21
1480405
Ethics at the edges of life. – New Haven ; London : Yale University Press, 1980
xvii,353p. – (Bampton lectures in America). – Includes index
B80–40170 X.325/978 Woolwich
1480405

ABORTION. United States
Law, 1800–1900. Social aspects

MOHR, James Crail
Abortion in America. – New York : Oxford University Press, 1978
xii,331p : ill. – Includes bibliographies and index
0–19–502249–1 X.329/11464 Woolwich
1484494

ABORTION. United States
Legal aspects

The LAW and politics of abortion / edited by Carl E. Schneider, Maris A. Vinovskis. – Lexington : Lexington, c1980
xlvii,268p : ill. – Includes index
0–669–03386–3 X.200/35465
1480421

RAMSEY, Paul, 1913–
Ethics at the edges of life. – New Haven ; London : Yale University Press, 1978
xvii,353p. – (The Bampton lectures in America). – Includes index
0–300–02137–2 WP.1578/21
1480421

RUBIN, Eva R.
Abortion, politics and the courts. – Westport, Conn. ; London : Greenwood, 1982
viii,211p. – (Contributions in American studies ; no.57). – Bibliography: p193–201. – Includes index
0–313–23018–8 X.529/51624
1480421

ABORTION. United States
Political aspects

FALIK, Marilyn
Ideology and abortion policy politics. – New York : Praeger, 1983
228p. – (Landmark dissertations in women's studies series). – Bibliography: p220–228
0–03–062813–x X.520/31672
2975211

TATALOVICH, Raymond
The politics of abortion. – New York : Praeger, 1981
vi,248p. – Bibliography: p234–240
0–03–058036–6 X.520/32478
2975211

ABORTION. United States
Social aspects, 1914

'WOMAN rebel' / Alex Baskin, advisory editor & historical consultant. – New York : Archives of Social History, 1976
[2],xxii,56p
0–914924–02–8 X.525/2126
1073818

ABORTION. United States
Sociopolitical aspects

LUKER, Kristin
Abortion and the politics of motherhood. – Berkeley : London : University of California Press, c1984
xvi,324. – Bibliography: p291–309. – Includes index
0–520–04314–6 X.529/69159
3211282

PETEHESKY, Rosalind Pollack
Abortion and woman's choice. – New York ; London : Longman, c1984
xii,444p. – (Longman series in feminist theory). – Includes index
0–582–28215–2 X.322/14286
3211282

ABORTION. Uruguay
to date

THÉVENET, Emilio F
Aborto en el Uruguay. – Montevideo : [s.n.], 1978
85p. – Bibliography
r6700939 X.329/14284 Woolwich
8724822

ABORTION. Venezuela

GARCÍA-VELUTINI, Oscar
Un foro sobre el aborto. – Caracas . [s.n.], 1977
40p
r6084785 X.319/9774 Woolwich
1791362

ABORTION. West Germany
to date

MATERIALIEN zum Bericht der Kommission zur Auswertung der Erfahrungen mit dem reformierten [Art.] 218 StGB. – Stuttgart : Kohlhammer, 1981–. – (Schriftenreihe des Bundesministers für Jugend, Familie und Gesundheit ; Bd.92). – Includes bibliographies
r6305251 S.F.467/13[Bd.92]
8840431

ABORTION (AMENDMENT) BILL. Great Britain. Parliament. House of Commons
Session 1974–75. Protest movements by women. Organisation

ALRA
A woman's right to choose action guide. – London : [Alra], [1975]
[1],51p : ill
0–9500063–3–5 X.319/10229
0523844

ABORTION SERVICES. Camden (London Borough)
Attitudes of women undergoing abortion

CLARKE, Lynda
Camden abortion study. – Solihull : British Pregnancy Advisory Service, c1983
xv,98p : ill
0–9508643–0–7 X.322/12170
2937948

ABORTION SERVICES. National health services. Great Britain
Day care abortion services

DAY-CARE abortion and the NHS. – London : The Birth Control Trust, 1980
68,[4]p : ill. – Includes bibliographies
B80–40573 X.329/13133 Woolwich
2322021

ABOUKIR (SHIP). Cruisers. Great Britain. Royal Navy
Naval operations during North Sea campaign of World War 1. Sinking by U-9 (Submarine)

COLES, Alan
Three before breakfast. – Havant : K. Mason, 1979
192p,[8]p of plates : ill. – Bibliography: p.185–186. – Includes index
0–85937–168–9 X.809/44780
1901567

ABOVILLE, GÉRARD D'. Single-handed voyages. Atlantic Ocean – Biographies

ABOVILLE, Gérard d'
L'Atlantique à bout de bras. – Paris : Arthaud, 1980
180p,16p of plates
2–7003–0347–4 X.800/31059
8761205

ABOYNE REGION. Grampian region. Scotland
Organisations – Directories

DIRECTORY of societies and organisations, Aboyne area. – [Aberdeen] : North East Scotland Library Service, [1982?]
25p
B83–19601 2725.g.144
2886057

ABRAHAM. Bible. O.T.
Abraham & David, King of Israel. Local associations: Mediterranean region

BRUCE, F. F. (Frederick Fyvie)
Abraham and David. – London : Scripture Union, 1984
128p : ill(some col.). – (Places they knew)
0–86201–111–6 X.205/1720
3284344

ABRAHAM. Bible. O.T. Genesis

[BIBLE O.T. Genesis. Welsh. Selections 1983]. Abraham / addasiad o storiau o'r Beibl gan Emir Jones ; darluniau gan Harry Bishop. – Abertawe [Swansea] : Gwasg Mynydd Mawr, c1983
28p : ill(some col.)
1–85012–005–6 X.208/9907
3144569

FRANK, Penny
Abraham friend of God. – Tring : Lion, 1984
23p : col.ill. – (The Lion story Bible ; 4)
0–85648–729–5 X.208/10083
0350699

MCMEEKIN, Barbara
Abraham's big family. – Leigh-on-Sea : Palm Tree, c1983
[29]p : col.ill. – (Palm Tree Bible stories)
0–86208–038–x X.208/8799
0350699

MEYER, F. B. (Frederick Brotherton)
Abraham, or, The obedience of faith. – London : Lakeland, 1978
159p
0–551–00032–5 W.P.D.404/354
0752614

RÉCRÉO
Abraham, founder of a nation. – Tring : Lion Publishing, 1979
46p : chiefly col ill
0–85648–151–3 X.102/2235
1994492

ROBERTSON, Jenny
Abraham's family. – London : Scripture Union/Ladybird, 1980
[28]p : col.ill. – (A Ladybird Bible book)
0–85421–894–7 X.205/968
0350699

VAN SETERS, John
Abraham in history and tradition. – New Haven ; London : Yale University Press, 1975 [i.e. 1976]
xiii,335p. – Bibliography: p.xi–xiii. – Includes index
0–300–01792–8 X.200/30621
0752614

WILLIAMS-ELLIS, Virginia
Day by day stories about Abraham. – London : Dean, 1982
28p : ill(some col.)
0–603–00319–2 X.990/19090
0350699

ABRAHAM. Bible. O.T. Genesis
Abraham & Joseph (Biblical patriarch)

COX, Molly
The family of Abraham. – London : Collins, 1977
[32]p : chiefly col ill
0–00–183704–4 Cup.1281/766
1350064
– [Another copy: stapled]
ISBN 0–00–661316–0 Cup.1281/767

ABRAHAM. Characters. Bible. O.T.

FORSBERG, Juhani
Das Abrahambild in der Theologie Luthers. – Stuttgart : Steiner, 1984
xii,193p. – (Veröffentlichungen des Instituts für europäische Geschichte Mainz, Abteilung für abendländische Religionsgeschichte ; Bd.117). – Includes bibliography
3–515–04316–0 Ac.6431[Bd.117]
9099050

ABRAHAM, ASHLEY. Rock climbing. Special subjects. English photography
Abraham, Ashley & Abraham, George

ABRAHAM, George
Camera on the crags. – London : Heinemann Educational, 1975
vi,42p,[94]p of plates : ill. – Includes index
0–435–86000–3 X.802/10057
0526193

ABRAHAM, GEORGE. Rock climbing. Special subjects. English photography
Abraham, Ashley & Abraham, George

ABRAHAM, George
Camera on the crags. – London : Heinemann Educational, 1975
vi,42p,[94]p of plates : ill. – Includes index
0–435–86000–3 X.802/10057
0526193

ABRAHAMS, HAROLD. Sprinting. Running. Athletics
Abrahams, Harold & Liddell, Eric
WEATHERBY, W. J. (William John)
Chariots of fire. - London : Granada, 1982, c1981
126p : ill
0-586-05776-5 H.84/999
3087700

ABRAHAMS, IVOR. English sculptures. Exhibits. Art galleries. Great Britain
- Catalogues
ABRAHAMS, Ivor
Ivor Abrahams. - London : Mayor Gallery, [1979 or 1980]
[16]p : ill
B81-37518 X.421/22217 Woolwich
2561123

ABRAHAMS, PETER, 1919- South African writers. Fiction in English
OGUNGBESAN, Kolawole
The writing of Peter Abrahams. - London : Hodder and Stoughton, 1979
[4],156p. - Bibliography: p.151-156
0-340-23601-9 X.809/44301
1585827

ABRAHAMS, PETER, 1919-. South African writers. Fiction in English
Biographies: Tell freedom
DANIELS, Russell
Tell freedom. - Harlow : Longman, 1981
vii,86p. - (Longman guides to literature)
0-582-65534-x X.958/2267
2382776

ABRAHAMS, PETER, 1919-. South African writers. Fiction in English
Mine boy
DANIELS, Russell
'Mine boy'. - [London] : Longman, 1978
vi,57p. - (Longman guides to literature)
0-582-60184-3 X.908/41691
1478613

ABRAHAMS, PETER W.. Church of England
ABRAHAMS, Peter W.
New creation. - London : Mowbray, 1982
xi,130p. - (Mowbray's popular Christian paperbacks)
0-264-66867-7 X.208/8112
2838672

ABRAMS TANKS. Armoured combat vehicles
M1 Abrams tanks, to 1980
ZALOGA, Steven J.
The M1 Abrams battle tank. - London : Osprey, 1985
40p,A-Hp of plates : ill(some col.). - (Vanguard series ; 41)
0-85045-584-7 X.622/22621
3116255

ABRANTÈS, LAURE PERMON, DUCHESSE D'. Historiography. France
- Biographies
GUNN, Peter
Napoleon's 'Little pest'. - London : H. Hamilton, 1979
ix,214p,[4]p of plates : ill. - Bibliography: p.205. - Includes index
0-241-10183-2 X.809/44385
1593226

ABRASION
See also
Scuffing

ABRASIVE MACHINERY. Materials
MCKEE, Richard L.
Machining with abrasives. - New York ; London : Van Nostrand Reinhold, c1982
ix,304p : ill. - Includes index
0-442-25281-1 SRL
2723239

ABRASIVES
Coated abrasives. Safety measures
SAFETY - a code of practice / Coated Abrasives Manufacturers Association. - [Bracknell] : CAMA, 1983
8p
B83-40643 X.529/56909
2962918

ABREU, CASIMIRO DE. Brazilian writers. Poetry in Portuguese
MAGALHÃES JÚNIOR, R. (Raimundo)
Poesia e vida de Casimiro de Abreu. - 3a. ed. revista e aumentada. - Rio de Janeiro : Civilização Brasileira, 1980
293p. - (Coleção Vera Cruz ; 298)
r6705602 X.950/13709
8358745

ABREU E LIMA, JOSÉ INÁCIO DE
See Lima, José Inácio de Abreu e

ABREU E SILVA, FLORÊNCIO CARLOS DE
See Abreu, Florêncio de +

ABREU, FLORÊNCIO DE. Politics. Brazil
SILVA, Florêncio de Abreu e
O Senador do Império Florêncio de Abreu e a política do seu tempo. - [Porto Alegre] : Escola Superior de Teologia São Lourenço de Brindes, 1979
109p. - ([Coleção "Temas gaúchos"])
r6701692 X.809/48782
8034680

ABRIACHAN. Highland Region. Scotland
to 1977
INVERNESS HIGH SCHOOL
Abriachan. - [Inverness] : The School, [1978]
18,[2]p : ill. - Bibliography: p.17
B79-14286 X.708/21838
1847953

ABRIACHAN. Highland Region. Scotland
Social conditions, ca 1950-ca 1960
STEWART, Katharine
A croft in the hills. - [1st ed. reprinted]. - Perth : Melven Press, 1979
[1],xiii,185p : ill
0-9505884-6-6 X.909/43872
1953001

ABRIDGEMENT OR SURVEY OF POPERIE. Sutcliffe, Matthew
KELLISON, Matthew
A reply to Sotcliffes answer. - Ilkley : Scolar Press, 1977
[57],439,[31]p. - (English recusant literature, 1558-1640 ; vol.330). - Includes index
0-85967-346-4 1654/330
1117963

ABRUZZI. Italy
to date
COLAPIETRA, Raffaele
Itinerari storici abruzzesi. - Lanciano : RAI, sede regionale d'Abruzzo, 1979
106p,[16]p of plates : ill. - (Testi e monografie ; 4) (Quaderno ; n.5 / RAI, sede regionale per l'Abruzzo)
r6504698 X.809/51420
8265402

ABRUZZI. Italy
Antiquities
COARELLI, Filippo
Abruzzo Molise. - Roma : Laterza, 1984
346p : ill. - (Guide archeologiche Laterza. L'Italia ; 9). - Includes bibliography
88-420-2519-4 X.809/63216
887655x

SGATTONI, Giammario
L'Abruzzo antico. - Lanciano : RAI, Sede regionale d'Abruzzo, 1979
xv,285p,[64]p of plates : ill. - (Testi e monografie ; 6)
r6505509 X.429/13822
887655x

ABRUZZI. Italy
Antiquities, to ca B.C.700
RADMILLI, Antonio Mario
Storia dell'Abruzzo dalle origini all'età del bronzo. - Pisa : Giardini, 1977
457p : ill. - (Studi paletnologici ; 1). - Bibliography : p.435-450
r6065405 X.410/6587 Woolwich
1505378

ABRUZZI. Italy
Archaeology - Bibliographies
FUCINESE, Damiano Venanzio
Arte e archeologia in Abruzzo. - Roma : Officina, 1978
xxii,381p
r6100557 X.311/5037 Woolwich
211996X

ABRUZZI. Italy
Castles, to 1974
PEROGALLI, Carlo
Castelli dell' Abruzzo e del Molise. - Milano : Görlich, 1975
151p : ill. - (Castelli delle regioni italiane)
r6034805 LR.416.pp.9
1161709

ABRUZZI. Italy
Croats, to date
SCOTTI, Giacomo
Z one bane mora. - Rijeka : Otokar Keršovani, 1980
81p
r6435267 X.958/10547
8845891

ABRUZZI. Italy
Description & travel
LEVI, Primo
Abruzzo forte e gentile. - Sulmona : Libreria editrice A. Di Cioccio, 1976
141p.,[16]p. of plates. - (Scrittori abruzzesi e scritti sull'Abruzzo : 3)
r6106925 X.700/21650
0149292

ABRUZZI. Italy
Politics. Social aspects
WHITE, Caroline
Patrons and partisans. - Cambridge : Cambridge University Press, 1980
xii,195p : ill. - (Cambridge studies in social anthropology ; 31). - Bibliography: p.180-184. - Includes index
0-521-22872-7 X.800/15446
2309769

ABRUZZI. Italy
Public administration, 1946-
QUARANTA, Alfonso
La regione Abruzzo. - Milano : Giuffrè, 1980
78p. - (L'ordinamento amministrativo delle regioni : 13)
r6507559 X.529/54638
8952302

ABRUZZI. Italy
Social conditions
La MODERNIZZAZIONE sperata / S. Acquaviva ... [et al.]. - Napoli : Guida, 1978
288p. - (La terra deportata ; 8)
r6500083 X.529/40045
2477297

ABRUZZI. Italy
Social conditions, 1790-1806
D'ANDREA, Uberto
Gli avvenimenti dal 1791 al 1806 nelle valli dell'Alto Sangro e del Sagittario ed in alcune zone della Marsica e della Conca Peligna. - [s.l.] : [s.n.], 1974
67p
r6067582 X.709/26358
1711881

ABRUZZI. Italy
Wines industries
GIULIANI, Guido
Il vino in Abruzzo. - L'Aquila : L.U. Japadre, 1975
246p : ill
r6052848 X.319/9020 Woolwich
1510622

ABRUZZI ARTS. Italian arts
- Bibliographies
FUCINESE, Damiano Venanzio
Arte e archeologia in Abruzzo. - Roma : Officina, 1978
xxii,381p
r6100557 X.311/5037 Woolwich
2119978

ABRUZZI CUSTOMS. Italian customs
FINAMORE, Gennaro
Tradizioni popolari abruzzesi. - Palermo : Edikronos, 1981
xiii,239p. - (Biblioteca delle tradizioni popolari italiane)
r6510068 X.808/39063
9038566

ABRUZZI DIALECTS. Italian language
PROFILO dei dialetti italiani / a cura di Manlio Cortelazzo. - Pisa : Pacini. - (Consiglio nazionale delle ricerche s Centro di studio per la dialettologia italiana ; 5)
13: Abruzzo. - 1979
282p,[1]folded leaf of plates
r6502443 X.0900/596(5/13)
879555x
- [Accompanying gramophone record] Cup.575.ee.26

ABRUZZI DISHES. Food
FABRIS, Ida
Le ricette della mia cucina marchigiana umbra e abruzzese. - Firenze : Edizioni del Riccio, 1979
159p. - (Vita e tradizioni)
r6500185 X.629/15209 Woolwich
213957x

ABRUZZI DISHES. Food –continued

NATALI, Carlo
Abruzzi e Molise in bocca. – Palermo : Il Vespro, 1978
201p : ill
r6188251 X.620/8886 Woolwich
213957X

ABRUZZI FOLKLORE. Italian folklore

GIANCRISTOFARO, Emiliano
Totemàjje. – Lanciano : RAI. Sede regionale d'Abruzzo, 1978
255p,[32]p of plates : ill. – (Testi e monografie ; 3) (Quaderno / RAI, Sede regionale per l'Abruzzo ; n.4). – Bibliography
r6504995 X.809/52436
8845336

ABRUZZI WRITERS. Italian literature
to date

CIRCEO, Ermanno
Da Croce a Silone. – Roma : Ateneo, 1981
141p. – (Collana di cultura ; 34)
r6510524 X.950/29746
9058273

ABRUZZI WRITERS. Italian literature
to date – Anthologies

COLONIA aternina de' velati (1816–1841) / a cura di Francesco Di Gregorio. – Roma : Palombi, 1979–1980
4v.. – (Biblioteca di cultura abruzzese. Testi ; 1)
r6510526 X.950/29575
9058273

ABRUZZO
See Abruzzi

ABRUZZO. Italy
Architecture

ROSSI, Sara
Abruzzo e Molise. – [Roma] : Officina, 1976
106p : ill(some col.)
r6100447 X.421/9920 Woolwich
2043831

ABS
See Association of Broadcasting Staff

ABSALOM AND ACHITOPHEL. Dryden, John, 1631–1700. Poetry in English

THOMAS, W. K.
The crafting of 'Absalom and Architophel'. – Waterloo, Ont. : Wilfred Laurier University Press, 1978
239p
0–88920–059–9 X.981/22147
0035831

ABSE, DANNIE. Poetry in English

CURTIS, Tony, 146–
Dannie Abse. – [Cardiff] : University of Wales Press on behalf of Welsh Arts Council, 1985
123p. – (Writers of Wales) (Writers of Wales). – Bibliography: p115–119
0–7083–0896–1 X.950/43044
3060918

ABSE, DANNIE. Poetry in English
– Biographies

ABSE, Dannie
Ash on a young man's sleeve. – Harmondsworth : Penguin, 1982, c1954
167p. – (A King Penguin)
0–14–005153–8 X.958/11710
0171816

A poet in the family. – London : Robson, 1984, c1974
198p
0–86051–280–0 X.950/37354
0171816

ABSE, Dannie
A strong dose of myself. – London : Hutchinson, 1983
220p
0–09–151260–3 X.529/55351
0171816

ABSE, DANNIE. Poetry in English
– Texts

The POETRY of Dannie Abse / edited and with an introduction by Joseph Cohen. – London : Robson, 1983
187p. – Bibliography: p182–183
0–86051–243–6 X.950/28882
3060918

ABSENCE. Nurses. Psychiatric hospitals. Great Britain
Unauthorised absence from work

CORMACK, Desmond
Sickness and absence amongst nursing staffs in a psychiatric, mental deficiency hospital group. – Dundee : Royal Dundee Liff Hospital ; [Dundee] : [Distributed by the author], 1973
[3],27p. – Bibliography: p.25–26
0–9505369–0–3 X.311/8334 Woolwich
0787140

ABSENCE. Students. Schools. England
– Conference proceedings

NATIONAL ASSOCIATION OF CHIEF EDUCATION WELFARE OFFICERS
These we serve. – [Halifax] : [The Association], [1975]
43p
0–9504540–0–1 X.529/63662
0510475

ABSENTEE OWNERS. British sugar plantations. Jamaica
Phillips, Nathaniel

TAYLOR, Clare
The journal of an absentee proprietor. – Aberystwyth : C. Taylor, 1979
21leaves
B83–00371 X.805/3278
2803933

ABSENTEEISM
See also
Dismissal for absenteeism

ABSENTEEISM. Personnel

Il CAPO e l'assenteismo / Guglielmo Elia [et al.]. – Milano : Franco Angeli, [1975]
116p. – Includes bibliographies
r6059392 X.519/24761
1465775

CHADWICK-JONES, J. K. (John K)
Social psychology of absenteeism. – New York : Praeger, 1982
xiv,161p : ill. – Bibliography: p149–159
0–03–056652–5 X.520/27718
2742268

ABSENTEEISM. Personnel
Management aspects

KELLY, Brian, 19—
Preventing absenteeism. – Dublin : Irish Productivity Centre, Labour/Management Service, [1981]
22p : ill
0–902939–13–0 X.529/51434
2532948

ABSENTEEISM. Personnel
Monitoring. Techniques

BEHREND, Hilde
How to monitor absence from work. – London : Institute of Personnel Management, 1978
[6],iv,62p : ill. – Bibliography: p.61
0–85292–209–4 X.615/2164 Woolwich
1740458

ABSENTEEISM. Personnel. Great Britain
Management aspects

BARLOW, David H. (David Henry)
An employer's guide to absenteeism and sick-pay. – London : Kogan Page, 1982
174p : ill. – Includes index
0–85038–582–2 X.529/50229
2789450

GARDNER, A. Ward (Archibald Ward)
How to prevent absenteeism. – Fawley : Applecross, 1983
xi,72p : 2ill
0–9508632–0–3 X.529/55346
2789450

ROBERTS, Cory
Absent from work. – London : Duncan Publishing, 1982
viii,104p : ill. – Includes index
0–946093–00–8 X.529/55770
2789450

ABSENTEEISM. Personnel. Italy
1970–1975

ALBANI, Paolo
L'assenteismo operaio. – Roma : Coines, 1976
159p. – ("Universale Coines". Economia e lavoro ; 53)
r6052507 X.519/24834
1469150

ABSENTEEISM. Personnel. Poland

KALINA-PRASZNIC, Urszula
Uwarunkowania rozmiarów absencji w pracy. – Wrocław : Wydawnictwa Uniwersytetu Wrocławskiego, 1978
168p. – (Acta Universitatis Wratislaviensis : no 442. Prawo ; 77)
r6185981 Ac868c/3[no.442]
2439387

ABSENTEEISM. Students
See also
Truancy

ABSOLUTE CONFIGURATIONS. Organic compounds

KLYNE, William
Atlas of stereochemistry. – 2nd ed. – London : Chapman and Hall.
Vol.1. – 1978
xix,236p : ill
0–412–15450–1 SRL
0232890
Vol.2. – 1978
xi,241p : ill. – Includes index
0–412–15460–9 SRL
0232890

ABSOLUTISM
See also
Enlightened despotism
Tyranny

ABSOLUTISM. China
Role in politics, 1723–1735

HUANG, Pei
Autocracy at work. – Bloomington ; London : Indiana University Press for the International Affairs Center, 1974
xii,500p : ill. – (East Asian studies). – Bibliography: p.410–492. – Includes index
0–253–39103–2 X.0800/1502(1)
0353442

ABSOLUTISM. Europe
to ca 1920

ANDERSON, Perry
Lineages of the absolutist state. – London : Verso Editions, 1979
573p. – Includes index
0–86091–710–x X.709/33108
031563x

ABSOLUTISM. France
to 1789

EMMANUELLI, François-Xavier
Un mythe de l'absolutisme bourbonien. – Aix en Provence : Université de Provence, 1981
xiii,199p. – (Études historiques ; no.6)
r6208727 X.529/53776
8985332

ABSOLUTISM. France
1600–1700

PARKER, David, 1941-
The making of French absolutism. – London : Arnold, 1983
xvi,160p. – Bibliography: p152–155. – Includes index
0–7131–6382–8 X.800/38704
3141659

ABSOLUTISM. France
Theories, 1489–1519

SCIACCA, Enzo
Le radici teoriche dell' assolutismo nel pensiero politico francese ... 1498–1519. – Milano : A. Giuffrè, 1975
vi, 180 p.. – (Università di Catania, Pubblicazioni della Facoltà di Giurisprudenza ; n. 79)
r6013769 X.702/2940
0854417

ABSOLUTISM. Germany
Influence of Pietism, 1680–1780

FULBROOK, Mary
Piety and politics. – Cambridge : Cambridge University Press, 1983
viii,215p. – Bibliography: p190–210. – Includes index
0–521–25612–7 X.200/43183
3087603

ABSOLUTISM. Great Britain
Influence of Puritanism, 1550–1660

FULBROOK, Mary
Piety and politics. – Cambridge : Cambridge University Press, 1983
viii,215p. – Bibliography: p190–210. – Includes index
0–521–25612–7 X.200/43183
308759x

ABSOLUTISM. Russia
1613–1801

DUKES, Paul, 1934-
The making of Russian absolutism 1613–1801. - London : Longman, 1982
197p. - (Longman history of Russia). - Bibliography: p187–188. - Includes index
0–582–48684–x X.800/32476
2641445

ABSOLUTISM. Russia
1796–1917

EROSHKIN, N. P. (Nikolai Petrovich)
Krepostnicheskoe samoderzhavie i ego politicheskie instituty. - Moskva : Mysl', 1981
251p
r6400894 X.808/35215
8524165

ABSOLUTISM. Russia
1800–1917

DAVIDOVICH, Aleksandr Mikhaĭlovich
Samoderzhavie v epokhu imperializma. - Moskva : Nauka, 1975
348p
r6077156 X.808/11735
1974475

ABSOLUTISM. Russia
1900–1917

DĪAKIN, V. S. (Valentin Semenovich)
Samoderzhavie, burzhuaziīa i dvorīanstvo, v 1907–1911gg.. - Leningrad : Nauka, Leningradskoe otdelenie, 1978
244p
r6093842 X.709/52289
2467186

ABSOLUTISM. Spain
1598–1868

EMMANUELLI, François-Xavier
Un mythe de l'absolutisme bourbonien. - Aix en Provence : Université de Provence, 1981
xiii,199p. - (Études historiques ; no.6)
r6208727 X.529/53776
8985340

ABSOLUTISM. Special subjects. Drama in German. 1625–1750

REICHELT, Klaus
Barockdrama und Absolutismus. - Frankfurt am Main : Lang, 1981
400p.. - (Arbeiten zur mittleren deutschen Literatur und Sprache ; Bd.8). - Bibliography
3–8204–6201–5 X.0909/918(8)
8714908

ABSOLUTISM. Special themes. Tragedies, 1558–1625. Drama in English

LEVER, J. W. (Julius Walter)
The tragedy of state. - London : Methuen, 1980, c1971
ix,100p. - Includes index
0–416–30550–4 X.950/1382
2336421

ABSOLUTISM. Sweden
1680–1693

BARUDIS, Günter
Absolutismus - Zerstörung der "libertären Verfassung". - Wiesbaden : Steiner, 1976
xi,248p. - (Frankfurter historische Abhandlungen ; Bd 13). - Bibliography : p.239–248
3–515–02362–3 P.701/604[Bd.13]
1155768

ABSORPTION. Chemical engineering
- Conference proceedings

WHAT'S new in absorption with chemical reaction? / edited by J.E. Gillett. - [Manchester] : Institution of Chemical Engineers, North Western Branch, c1982
1v.(various pagings) : ill. - (Symposium papers 1981 ; no.6)
0–906636–13–2 SRL
2770687

ABSORPTION. Drugs. Gastrointestinal tract. Man
- Conference proceedings

INTERNATIONAL CONFERENCE ON DRUG ABSORPTION (1979 : Edinburgh)
Drug absorption. - Lancaster : MTP Press, c1981
xi,353p : ill. - Includes bibliographies and index
0–85200–528–8 SRL
2361922

ABSORPTION. Gases. Atmosphere

MCCARTNEY, Earl J.
Absorption and emission by atmospheric gases. - New York : Chichester : Wiley, c1983
xii,320p : ill. - (Wiley series in pure and applied optics). - Bibliography: p293–308. - Includes index
0–471–04817–8 SRL
2970163

ABSORPTION. Intestines
See also
Malabsorption. Intestines

ABSORPTION. Intestines. Mammals
Of antigens

ANTIGEN absorption by the gut / edited by W.A. Hemmings. - Lancaster : M.T.P. Press, 1978
ix,226p : ill. - Includes bibliographies and index
0–85200–101–0 SRL
1684159

ABSORPTION. Intestines. Man

INTESTINAL absorption in man / edited by Ian McColl and G.E. Sladen. - London : Academic Press, 1975
ix,363p : ill. - Includes bibliographies and index
0–12–482150–2 SRL
0369462

ABSORPTION. Intestines. Man
- Conference proceedings

INTERNATIONAL CONFERENCE ON INTESTINAL ABSORPTION AND SECRETION (1983 : Titisee)
Intestinal absorption and secretion. - Lancaster : MTP, 1984
xxiv,567p : ill. - (Falk symposium ; 36). - Includes index
0–85200–765–5 SRL
3160513

ABSORPTION. Lipids. Mammals
- Conference proceedings

LIPID absorption / edited by K. Rommel and H. Goebell, co-editor R. Böhmer. - Lancaster : M.T.P. Press, 1976
xi,352p : ill. - Includes bibliographies
0–85200–145–2 SRL
0751243

ABSORPTION. Low-frequency mechanical ground-borne vibration
Use of natural rubber. Engineering aspects

ROSE, I. G. (I Graham)
Vibration isolation with natural rubber. - [Brickendonbury] : Malaysian Rubber Producers' Research Association, [1982?]
ii,10p : ill. - (NR background ; 5)
B82–38738 X.622/13671
2787989

ABSORPTION. Manufacture. Sculphuric acid
Waste heat. Recovery. Techniques

BERK Spencer Acids Ltd. / [Energy Conservation Demonstration Projects Scheme]. - [London] : [Information Division, Distribution Unit], [1981]
7leaves : col.ill
B83–14718 BS.149/208
286696x

ABSORPTION. Mechanical vibration
- Conference proceedings

DYNAMIC vibration isolation and absorption. - Worthing : Mechanical Engineering Publications for Publication Services, 1982
147p : ill
0–85298–506–1 SRL
282471x

ABSORPTION. Mechanical vibration
Dynamic absorption

HUNT, J. B.
Dynamic vibration absorbers. - London : Mechanical Engineering Publications, 1979
[7],117p : ill. - Includes index
0–85298–417–0 SRL
192561x

ABSORPTION. Mineral ions
By plant roots

BOWLING, Dudley James Francis
Uptake of ions by plant roots. - London : Chapman and Hall, 1976
xii,212p,[4]p of plates : ill. - Bibliography: p.183–207. - Includes index
0–412–12000–3 SRL
064899x

ABSORPTION. Solutes. Soils
By roots of crops

NYE, Peter Hague
Solute movement in the soil-root system. - Oxford : Blackwell Scientific, 1977
xiv,342p : ill. - (Studies in ecology ; vol.4). - Bibliography: p.290–323. - Includes index
0–632–09730–2 SRL
1047698

ABSORPTION. Steroid hormones. Skin. Man
- Conference proceedings

PERCUTANEOUS absorption of steroids / edited by P. Mauvais-Jarvis, C.F.H. Vickers, J. Wepierre. - London : Academic Press, 1980
xviii,294p : ill. - Includes index
0–12–480680–5 SRL
2085038

ABSORPTION EDGES. Spectra. X-ray spectroscopy

CAUCHOIS, Y.
Wavelengths of x-ray emission lines and absorption edges. - Oxford : Pergamon, 1978
[19],320p. - (International tables of selected constants ; 18). - Bibliography: p.299–314. - Includes index
0–08–022448–2 SRL
1433253

ABSORPTION SPECTROSCOPY
Atomic absorption spectroscopy

EBDON, L. (Les)
An introduction to atomic absorption spectroscopy. - London : Heyden, c1982
xiii,138p : ill. - Bibliography: p130–133. - Includes index
0–85501–714–7 X.622/23794
0349380

THOMPSON, K. C. (Kenneth Clive)
Atomic absorption, fluorescence and flame emission spectroscopy. - 2nd ed.. - London : Griffin, 1978
x,319p : ill. - Includes index
0–85264–246–6 SRL
0349380

ABSORPTION SPECTROSCOPY
Extended x-ray absorption fine structure spectroscopy - Conference proceedings

EXAFS spectroscopy / [based on the proceedings of a symposium on the Applications of EXAFS to Materials Science, held at the 1979 meeting of the Materials Research Society, November 26–30, 1979, in Boston, Massachusetts] ; edited by B.K. Teo and D.C. Joy. - New York ; London : Plenum, c1981
viii,275p : ill. - Includes index
0–306–40654–3 SRL
2495341

ABSORPTION SPECTROSCOPY. Anti-knock additives. Petrol. Motor vehicles
Atomic absorption spectroscopy. Use in determination of lead

The DETERMINATION of lead in gasoline by atomic absorption spectrophotometry. - London : Institute of Petroleum, [1982?]
14p. - (IP ; 82–008)
B84–31401 X.622/19935
3140563

ABSORPTION SPECTROSCOPY. Chemical analysis
Atomic absorption spectroscopy

HASSAN, S. S. M.
Organic analysis using AAS. - Chichester : Horwood, Dec.1983
[250]p. - (Ellis Horwood series in analytical chemistry)
0–85312–559–7 SRL
1871749

SLAVIN, Morris, 1901-
Atomic absorption spectroscopy. - 2nd ed.. - New York ; Chichester : Wiley, 1978
xv,193p : ill. - (Chemical analysis ; vol.25). - Includes index
0–471–79652–2 SRL
1871749

ABSORPTION SPECTROSCOPY. Chemical analysis
Atomic absorption spectroscopy. Techniques

ATOMIC absorption spectrometry / edited by John Edward Cantle. - Amsterdam ; Oxford : Elsevier Scientific, 1982
xvi,448p : ill. - (Techniques and instrumentation in analytical chemistry ; v.5). - Includes index
0–444–42015–0 SRL
2485664

PRACTICAL absorption spectrometry / Ultraviolet Spectrometry Group ; edited by A. Knowles and C. Burgess. - London : Chapman and Hall, 1984
xxii,234p : ill. - (Techniques in visible and ultraviolet spectrometry ; v.3). - Includes index
0–412–24390–3 SRL
2485664

ABSORPTION SPECTROSCOPY. Chemical analysis
Atomic absorption spectroscopy. Techniques–continued

VAN LOON, J. C. (Jon Clement)
Analytical atomic absorption spectroscopy. – New York : London : Academic Press, 1980
xi,337p : ill. – Includes index
0–12–714050–6 SRL
2485664

ABSORPTION SPECTROSCOPY. Chemical analysis
Standards

STANDARDS in absorption spectrometry / Ultraviolet Spectrometry Group ; edited by C. Burgess and A. Knowles. – London : Chapman and Hall, 1981
x,142p : ill. – (Techniques in visible and ultraviolet spectrometry ; v.1). – Includes index
0–412–22470–4 SRL
2185202

ABSORPTION SPECTROSCOPY. Chemical analysis. Cadmium. Pollutants of drinking water
Atomic absorption spectroscopy

STANDING COMMITTEE OF ANALYSTS
Cadmium in potable waters by atomic absorption spectrophotometry, 1976 tentative method. – London : H.M.S.O., 1977
15,[1]p. – (Methods for the examination of waters and associated materials)
0–11–751136–6 SRL
1376950

ABSORPTION SPECTROSCOPY. Chemical analysis. Silver. Contaminants. Water
Atomic Absorption spectroscopy

SILVER in water, sewages and effluents by atomic absorption spectrophotometry. – London : H.M.S.O., 1983
18p. – (Methods for the examination of waters and associated materials)
0–11–751678–3 SRL
298203x

ABSORPTION SPECTROSCOPY. Chemical analysis. Water
Atomic absorption spectroscopy

THOMPSON, K. C. (Kenneth Clive)
Atomic absorption spectrophotometry, 1979 version. – London : H.M.S.O., [1980]
50p : ill. – (Methods for the examination of waters and associated materials). – Bibliography: p.40
0–11–751461–6 SRL
2281252

ABSORPTIVE CAPACITY. Economies
Measurement

KASIM, Sola
The measurement of absorptive capacity. – [Aberdeen] : University of Aberdeen, Department of Political Economy, [1982?]
26leaves : 3ill. – (Discussion paper ; no.82–08). – Bibliography: leaf 26
B83–26065 X.525/7427
2907453

ABSTRACT ALGEBRA

ALLENBY, R. B. J. T.
Rings, fields and groups. – London : Edward Arnold, 1983
xxvi,294p : ill. – Bibliography: p277–282. – Includes index
0–7131–3476–3 X.622/17340
0061204

BIRKHOFF, Garrett
A survey of modern algebra. – 4th ed. – New York : Macmillan ; London : Collier Macmillan, 1977
xi,500p : ill. – Bibliography: p.438–485. – Includes index
0–02–310070–2 X.620/16936 Woolwich
0061204

BLYTH, T. S. (Thomas Scott)
Set theory and abstract algebra. – London : Longman, 1975
ix,222p : ill. – (Longman mathematical texts). – Includes index
0–582–44284–2 X.619/15190
0061204

BRISLEY, Warren
A basis for linear algebra. – Sydney ; London : Wiley, 1973
viii,189p : ill. – Includes index
0–471–10460–4 X.620/17060 Woolwich
0061204
– [Another copy : paperback]
ISBN 0–471–10461–2 X.611/7745 Woolwich

BUDDEN, Francis James
An introduction to algebraic structures. – London : Longman, 1975
viii,136p : ill. – (Mathematical topics). – Includes index
0–582–35218–5 X.0619/88(10)
0061204

CHOW, Yutze
Modern abstract algebra. – New York : London : Gordon and Breach, 1976
2v.([40],742p). – Includes index
0–677–03880–1 X.620/16497 Woolwich
0061204

COHN, P. M. Paul Moritz
Algebra. – Chichester : Wiley
Vol.1. – 2nd ed. – c1982
xv,410p : ill. – Bibliography: p400–401. – Includes index
0–471–10169–9 X.629/18401
0061204

DAVIDSON, Neil
Abstract algebra. – Boston [Mass.] . London : Houghton Mifflin, 1976
xvi,336p : ill. – Bibliography: p.329–330. – Includes index
0–395–20663–4 X.620/17486 Woolwich
0061204

DORNHOFF, Larry L.
Applied modern algebra. – New York : Macmillan ; London : Collier-Macmillan, 1978
xi,500p : ill. – Bibliography: p.486–490. – Includes index
0–02–329980–0 X.620/17780 Woolwich
0061204

DURBIN, John R.
Modern algebra. – New York ; Chichester : Wiley, 1979
xviii,329p : ill. – Includes bibliographies and index
0–471–02158–x X.620/18531 Woolwich
0061204

FISHER, James L.
Application-oriented algebra. – New York : IEP, 1977
xix,362p : ill. – (IEP series in mathematics). – Includes index
0–7002–2504–8 X.620/17973 Woolwich
0061204

FRALEIGH, John Blackman
A first course in abstract algebra. – 2nd ed. – Reading, Mass. ; London : Addison-Wesley, 1976
xviii,455p : ill. – (Addison-Wesley series in mathematics). – Includes bibliographies and index
0–201–01984–1 WP.16967/128
0061204

GARDINER, C. F. (Cyril Frederick)
Modern algebra. – Chichester : Horwood, 1981
288p : ill. – (Ellis Horwood series in mathematics and its applications). – Bibliography: p283. – Includes index
0–85312–285–7 SRL
0061204

GILBERT, William J.
Modern algebra with applications. – New York ; London : Wiley, [1977]
xiii,348p : ill. – Bibliography: p.319–323. – Includes index
0–471–29891–3 X.620/16657 Woolwich
0061204

HERSTEIN, Israel Nathan
Topics in algebra. – 2nd ed. – New York ; London : Wiley, 1975
xi,388p : ill. – Includes bibliographies and index
0–471–01090–1 SRL
0061204

JONES, Burton Wadsworth
An introduction to modern algebra. – New York : Macmillan ; London : Collier Macmillan, 1975
xiii,366p : ill. – Bibliography: p.325–326. – Includes index
0–02–361250–9 X.622/5411 Woolwich
0061204

KIM, Ki Hang
Applied abstract algebra. – Chichester : Horwood, 1983
265p : ill. – (Ellis Horwood series in mathematics and its applications). – Bibliography: p259–260. – Includes index
0–85312–563–5 SRL
0061204

MCCOY, Neal Henry
Algebra. – Boston, Mass. ; London : Allyn and Bacon, 1977
xiii,658p : ill. – Bibliography: p.633–645. – Includes index
0–205–05699–7 X.620/17653 Woolwich
0061204

MACLANE, Saunders
Algebra. – 2nd ed. – New York : Macmillan ; London : Collier Macmillan, 1979
xv,586p : ill. – Bibliography: p.561–563. – Includes index
0–02–374310–7 X.620/18602 Woolwich
0061204
– [Another copy: paperback]
ISBN 0–02–978830–7 X.611/9159 Woolwich

MOORE, John Thomas
Introduction to abstract algebra. – New York ; London : Academic Press, 1975
xi,291p : ill. – Includes index
0–12–505750–4 X.620/8312 Woolwich
0061204

SAWYER, Walter Warwick
A concrete approach to abstract algebra. – New York : Dover Publications [etc.] ; London : Constable, 1978
[3],234p : ill. – Includes index
0–486–63647–x X.611/8311 Woolwich
0061204

SCHILLING, Otto Franz Georg
Basic abstract algebra. – Boston, Mass. : Allyn and Bacon, 1975
xv, 394 p. : ill. – (Allyn and Bacon series in advanced mathematics). – Bibliography : p. 363–365. – Includes index
r6009907 X.620/16313 Woolwich
0061204

SHAPIRO, Louis
Introduction to abstract algebra. – New York ; London : McGraw-Hill, 1975
x,340p : ill. – (International series in pure and applied mathematics). – Bibliography: p.282–286. – Includes index
0–07–056415–9 WP.3170/54
0061204

WHITELAW, Thomas A.
An introduction to abstract algebra. – Glasgow : Blackie, 1978
ix,166p : ill. – Includes index
0–216–90488–9 X.619/18522 Woolwich
0061204

ABSTRACT ALGEBRA
– For computer sciences

GILL, Arthur
Applied algebra for the computer sciences. – Englewood Cliffs ; London : Prentice-Hall, 1976
xv,432p : ill. – (Prentice-Hall series in automatic computation). – Includes bibliographies and index
0–13–039222–7 X.622/5228 Woolwich
0018643

SIMS, Charles C.
Abstract algebra. – New York ; Chichester : Wiley, c1984
xv,491p : ill. – Bibliography: p480–481. – Includes index
0–471–09846–9 X.622/20400
0018643

ABSTRACT ARTS
See also
Op art

ABSTRACT ARTS
to date

BALLO, Guido
Origini dell'astrattismo, 1885–1919. – Milano : Electa, c1980
263p : chiefly ill
r6507704 X.421/24131
8953147

ABSTRACT ARTS
1900–1977

OSBORNE, Harold
Abstraction and artifice in twentieth-century art. – Oxford : Clarendon Press, 1979
vi,193p,[16]p of plates : ill. – Includes index
0–19–817359–8 X.429/10503 Woolwich
1402579

ABSTRACT ARTS
Abstract graphic arts, 1900–

KRAHL, Ilse
Technische Medien als Herausforderung. – Sankt Augustin : Richarz, 1983
63p : ill. – (Duisburger Studien. Beiheft ; 1)
3–88345–608–x X.421/24733
9033017

ABSTRACT ARTS
Abstract graphic arts. Techniques

BROOKS, Leonard, 1911–
Painting and understanding abstract art. – New York ; London : Van Nostrand Reinhold, 1980, c1964
143p : ill(some col.)
0–442–24334–0 X.421/21619 Woolwich
2375923

ABSTRACT ARTS
Brazilian abstract arts, to date

BENTO, Antonio
Abstração na arte dos indios brasileiros. – Rio de Janeiro : Spala Editora, c1979
176p : ill(some col.)
r6704127 L.42/1479
8820600

ABSTRACT ARTS
European abstract arts, 1875–1935

BARR, Alfred Hamilton
Cubism and abstract art .... – London : Secker and Warburg, 1975
249p : ill. – Bibliography: p.234–249
0–436–03356–9 X.410/10018
0513008

ABSTRACT ARTS
European abstract arts, 1900– – Catalogues

ABSTRAKT kunst i en dansk privatsamling / [redaktion. Lise Funder og Nanna Merete Gøtzsche]. – København : [Glyptoteket], [1982]
56p : ill
87–7452–056–3 X.429/16198
9103805

ABSTRACT ARTS
European abstract arts, 1910–1920

TOWARDS a new art. – London : Tate Gallery, 1980
239p : ill(some col)
0–905005–22–8 X.421/20738 Woolwich
2155176

ABSTRACT ARTS
European abstract arts, 1950–1975. Special themes: Eroticism

CRISPOLTI, Enrico
Erotismo nell'arte astratta. – [Trapani] : Celebes, 1977
361p.,[56]p.of plates : ill
r6085301 Cup.821.d.13
1988948

ABSTRACT ARTS
German abstract arts. Blaue Reiter, Der (Movement)

VOGT, Paul
The Blue Rider. – Woodbury, N.Y. ; London : Barron's, c1980
vii,138p : ill. – Bibliography: p127
0–8120–2100–2 X.429/14037
2844117

ABSTRACT ARTS
Italian abstract arts, to 1938

BELLI, Carlo
Lettera sulla nascita dell'astrattismo in Italia. – Milano : All'insegna del pesce d'oro, 1978
48p
r6107742 X.419/4503 Woolwich
2215950

ABSTRACT ARTS
Italian abstract arts, to date

RADICE, Mario
Memoire del primo astrattismo italiano degli anni '30 e '40. – Lugano : Edizioni Pantarei, 1979
51p,[35]p of plates : ill. – (Collezione dell' "Acero")
r6505135 X.421/23384
8546177

SARTORIS, Alberto
La lunga marcia dell'arte astratta in Italia. – Milano : All'insegna del pesce d'oro, 1980
62p : ill. – (Documenti d'arte astratta ; n.4)
r6510200 X.429/15346
8546177

ABSTRACT ARTS
Italian abstract arts. Bonfanti, Arturo

BONFANTI, Arturo
Bonfanti. – [Venezia] : Alfieri, [1979]
237p : chiefly ill(some col.)
r6505903 L.42/1811
8889716

ABSTRACT CODING THEORY

FOUNDATIONS of coding theory / edited by William E. Hartnett. – Dordrecht : Reidel, 1974
xiii,216p : ill. – (Episteme ; vol.1). – Bibliography : p.207–209.– Includes index
90–277–0536–4 X.629/12061 Woolwich
1587064

ABSTRACT CONCEPTS
Role in Western civilization

DUPLESSY, Lucien
Bilan du monde moderne. – Paris : Le Courrier du Livre, 1974
317 p.
r6012107 X.519/20583
0945986

ABSTRACT CONCEPTS
Theories of Broad, Charles Dunbar

KIVINEN, Seppo
C.D. Broad's views on abstracta. – [Helsinki] : Helsingin yliopisto, Filosofian laitos, 1977
6p. – (Helsingin yliopiston filosofian laitoksen julkaisuja = Reports from the Department of Philosophy, University of Helsinki ; 1977, no.1)
951–451–008–9 X.0512/385[1977, no.1]
8961778

ABSTRACT DRAWINGS. Exhibits. LYC Museum and Gallery
Chinese abstract drawings. Li Yuan Chia

LI YUAN–CHIA
Water + colour = 56/7. – Brampton : LYC Museum, 1977
[36]p : ill
B78–37841 X.908/42024
1706888

ABSTRACT EXPRESSIONISM. American paintings to 1950

GUILBAUT, Serge
How New York stole the idea of modern art. – Chicago ; London : University of Chicago Press, 1983
x,277p,[16]p of plates : ill. – Bibliography: p251–266. – Includes index
0–226–31038–8 X.421/25421
3130320

HOBBS, Robert Carleton
Abstract expressionism. – Ithaca, N.Y. ; London : Cornell University Press, 1981, c1978
140p : ill(some col.)
0–8014–1365–6 X.425/3101 Woolwich
2628031

SEITZ, William C. (William Chapin)
Abstract expressionist painting in America. – Cambridge, Mass. ; London : Published for the National Gallery of Art by Harvard University Press, 1983
xxii,490p : ill(some col.). – (The Ailsa Mellon Bruce studies in American art). – Includes index
0–674–00215–6 X.421/25981
2628031

ABSTRACT EXPRESSIONISM. American paintings to 1970

COX, Annette
Art–as–politics. – Ann Arbor : UMI Research ; Epping : Bowker, c1982
x,206p : ill. – (Studies in the fine arts. The avant–garde ; no.26). – Bibliography: p185–199. – Includes index
0–8357–1318–0 X.421/24136
0529230

EVERITT, Anthony
Abstract expressionism. – London : Thames and Hudson, 1975
64p,[64]p of plates : ill(chiefly col). – (A dolphin art book)
0–500–41059–3 X.419/3051
0529230

SANDLER, Irving
The triumph of American painting. – New York[; London : Harper and Row, [1977]
xv,301p : ill. – Bibliography: p.281–298. – Includes index
0–06–430075–7 X.410/10141 Woolwich
0529230

ABSTRACT EXPRESSIONISM. American paintings to date

BRADBURY, Ray
The ghosts of forever. – New York, : Rizzoli International Publications,, 1981
130p,plates : ill
0–8478–0358–9 L.45/3146
8008361

ABSTRACT EXPRESSIONISM. Paintings to date

MEEWIS, Wim
Iconologie van de action painting. – Brussel : AWLSK, 1983
235p,[12]p of plates. – (Verhandelingen van de Koninklijke Academie voor Wetenschappen, Letteren en Schone Kunsten van België. Klasse der Schone Kunsten ; jg.45 ; nr.37). – Includes bibliography
90–656–9335–1 Ac.988ea/2.[vol.37.]
9097864

ABSTRACT FILMS
to 1975

LE GRICE, Malcolm
Abstract film and beyond. – London : Studio Vista, 1977
160p : ill. – Bibliography: p.156–158. – Includes index
0–289–70591–6 X.989/52020
1285394

ABSTRACT GROUP THERAPY
1800–1900

WUSSING, Hans
The genesis of the abstract group concept. – Cambridge, Mass. ; London : MIT, c1984
331p. – Bibliography: p295–323. – Includes index
0–262–23109–3 X.520/36207
3246558

ABSTRACT PAINTINGS
1910–1960

THOMAS, Denis
Abstract painting. – Oxford : Phaidon, 1976
16p.48p of plates. – ([Phaidon colour plate series])
0–7148–1746–5 X.425/1242 Woolwich
069858x

ABSTRACT PAINTINGS
1910–1978

POHRIBNY, Arsén
Abstract painting. – Oxford : Phaidon, 1979
[120]p : col ill. – Bibliography: p.40
0–7148–1956–5 LR.421/173
1942565

ABSTRACT PAINTINGS
American abstract paintings. Rothko, Mark

CLEARWATER, Bonnie
Mark Rothko. – New York : Hudson Hills, c1984
144p : ill(some col.). – Bibliography: p140–141
0–933920–53–9 L.42/2859
1833421

ABSTRACT PAINTINGS
American abstract paintings. Rothko, Mark – Biographies

ROTHKO, Mark
Mark Rothko. – London : Thames and Hudson, 1978
296p : ill(some col). – Bibliography: p.292–295
0–500–09131–5 X.423/10750 Woolwich
1833421

ABSTRACT PAINTINGS
Canadian abstract paintings. Gaucher, Yves – Catalogues

GAUCHER, Yves
Yves Gaucher. – [Paris] : [Le Centre], [1983]
16p : ill(some col.)
2–900434–65–3 L.49/2075
3030725

ABSTRACT PAINTINGS
English abstract paintings. Canney, Michael – Catalogues

CANNEY, Michael
Michael Canney. – Penzance : Newly Orion, c1983
16p : ill(some col.)
0–9506579–3–x X.429/16305
316098x

ABSTRACT PAINTINGS
French abstract paintings. Staël, Nicolas de – Correspondence, diaries

STAËL, Nicolas de
Lettres à Jacques Duborg. – London : Taranman, 1981
[107]p
0–906499–11–9 X.955/638
2547546

ABSTRACT PAINTINGS
Italian abstract paintings, to date – Catalogues

ASTRATTISMO CLASSICO (Exhibition) (1980–1981 : Florence)
Astrattismo classico. – Firenze : Vallecchi, 1980
179p : ill
r6510101 X.421/24759
9058095

ABSTRACT PAINTINGS. Exhibits. Annely Juda Fine Art 1914–1939 – Catalogues

ANNELY JUDA FINE ART
The non–objective world. – London : Annely Juda Fine Art, 1978
96p : chiefly ill
B78–33148 X.419/34719 Woolwich
1589121

ABSTRACT PAINTINGS. Exhibits. Arts Council of Great Britain
English abstract paintings, 1973–1977 – Catalogues

ARTS COUNCIL OF GREAT BRITAIN
A free hand. – [London] : [The Council], 1978
32p : ill
0–7287–0155–3 X.419/34597 Woolwich
1413996

ABSTRACT PAINTINGS. Exhibits. Talbot Rice Art Centre
British abstract paintings, 1910–1960 – Catalogues
ANNELY JUDA FINE ART
Aspects of abstract painting in Britain, 1910–1960. – London : [Annely Juda Fine Art] ; London : [Alexander Postan Fine Art], [1974]
[72]p : ill
0–9504121–3–9 X.419/3053
0530948

ABSTRACT PAINTINGS. Exhibits. Tate Gallery
to 1920
ABSTRACTION. – London : Tate Gallery Publications, 1980
128p,[8]p of plates : ill(some col)
0–905005–12–0 X.423/10965 Woolwich
2115735

ABSTRACT PHOTOGRAPHY
English abstract photography, 1950– – Collections
HILLIARD, John
Black depths, white expanse, grey extent, and four other works, 1974 and 1975. – [London] : Robert Self (Publications), 1976
[18]p : of ill
B78–30823 X.410/10249 Woolwich
157504x

ABSTRACT PHOTOGRAPHY
New Zealand abstract photography, 1950– – Collections
WEBB, Boyd
Tableaux. – [London] : Robert Self, 1978
[24]p : of col ill
B78–30824 X.410/10248 Woolwich
1575058

ABSTRACT PHOTOGRAPHY. Exhibits. Arts Council of Great Britain
English abstract photography, 1975–1978 – Catalogues
ARTS COUNCIL OF GREAT BRITAIN
Nature as material. – London : Arts Council of Great Britain, 1980
[16]p : ill
0–7287–0244–4 X.325/720 Woolwich
2318539

ABSTRACT SCULPTURES
KNIGHT, Vivien
Have you seen sculpture from the body?. – London : Tate Gallery, 1984
56p : ill. – Bibliography: p56
0–946590–13–3 X.421/26433
325450x

ABSTRACT SCULPTURES. Exhibits. Arts Council of Great Britain
English abstract sculptures, 1973–1977 – Catalogues
ARTS COUNCIL OF GREAT BRITAIN
A free hand. – [London] : [The Council], 1978
32p : ill
0–7287–0155–3 X.419/34597 Woolwich
1414003

ABSTRACT SCULPTURES. Exhibits. Arts Council of Great Britain
English abstract sculptures, 1975–1978 – Catalogues
ARTS COUNCIL OF GREAT BRITAIN
Nature as material. – London : Arts Council of Great Britain, 1980
[16]p : ill
0–7287–0244–4 X.325/720 Woolwich
2318520

ABSTRACT SCULPTURES. Exhibits. David Mirvish Gallery
Canadian abstract metal sculptures. Murray, Robert, b.1936
DAVID MIRVISH GALLERY
Robert Murray. – [Toronto] : [David Mirvish Gallery], [1975]
[18] p. : ill. – Bibliography : p. [14]–[15]
r6009703 Cup.1281/301
0701114

ABSTRACT SPACES
See also
Bornological spaces
Ordered spaces
Topological spaces

ABSTRACT SPACES
Statistical inference
GRENANDER, Ulf
Abstract inference. – New York ; Chichester : Wiley, c1981
ix,526p : ill. – (Wiley series in probability and mathematical statistics). – Bibliography: p511–521. – Includes index
0–471–08267–8 SRL
2395800

ABSTRACT STRUCTURES. Mathematics
See also
Classes. Mathematics
Sets. Mathematics
Steiner systems

ABSTRACTING
– Bibliographies
WELLISCH, Hans H.
Indexing and abstracting. – Santa Barbara ; Oxford : Published in cooperation with the American Society of Indexers and the Society of Indexers (U.K.) [by] ABC–Clio, c1980
xxi,308p. – Includes index
0–87436–300–4 X.525/4135
237949x

ABSTRACTING. Documents
BORKO, Harold
Abstracting concepts and methods. – New York ; London : Academic Press, 1975
x,250p : 1 ill. – (Library and information science). – Bibliography: p.219–223. – Includes index
0–12–118650–4 2719.aa.25
0680877

ROWLEY, J. E. (Jennifer E)
Abstracting and indexing. – London : Bingley, 1982
155p : ill. – (Outlines of modern librarianship). – Bibliography: p147–151. – Includes index
0–85157–336–3 2719.e.323
0680877

ABSTRACTING. Documents. Soviet Union
– Standards
UNION OF SOVIET SOCIALIST REPUBLICS. Gosudarstvennyi standart
Sistema informat͡sionno – bibliograficheskoĭ dokumentat͡sii. – Izd. ofit͡sial'noe. – Moskva : Izdatel'stvo standartov, 1978
5p
r6172590 S.N.176/10
2098547

ABSTRACTING SERVICES. Great Britain
Abstracting services & indexing services
BURGESS, G.
Inventory of abstracting and indexing services produced in the UK. – London : British Library, 1978
iv,103p. – (Research and development reports / British Library ; no.5420). – Includes index
0–905984–11–0 Bar.T.18.c.11. Reading Room
129217x
– [Another copy.] BS.10/747
– [Another copy] OPL

ABSTRACTING SERVICES. Science. Great Britain
– Directories
BRITISH COUNCIL. Education Information Unit
British scientific documentation services. – [Harlow] : Longman, 1976
104p. – Includes index
0–582–01262–7 X.619/16819 Woolwich
0601292

ABSTRACTING SERVICES. Social sciences
Coverage & overlap
DESIGN OF INFORMATION SYSTEMS IN THE SOCIAL SCIENCES
The planning of indexing and abstracting services in the social sciences. – [Bath] : Bath University Library, 1976
vi,107,[60]p. – (Research reports. series A ; no.5). – Bibliography: p.97–106
0–900843–55–1 X.512/5320
0711551

ABSTRACTING SERVICES. Technology. Great Britain
– Directories
BRITISH COUNCIL. Education Information Unit
British scientific documentation services. – [Harlow] : Longman, 1976
104p. – Includes index
0–582–01262–7 X.619/16819 Woolwich
0601306

ABSTRACTNESS. Generative phonology
Theories
GUSSMANN, Edmund
Studies in abstract phonology. – Cambridge, Mass. ; London : MIT Press, c1980
xii,160p. – (Linguistic inquiry monographs ; 4). – Bibliography: p147–149. – Includes index
0–262–07081–2 X.622/10797 Woolwich
2528843

ABSTRACTS
See also
Abstracts under subject, or consult list of books under Abstracts in author index

ABSTRACTS. Serial articles. Biology
Proposed standards
LEWINGTON, R. J.
Testing and revision of guidelines for writing abstracts of papers in the biological sciences. – [Slough] : Commonwealth Agricultural Bureaux, 1980
42leaves. – (BL R & D report ; no.5568)
B83–40522 X.622/16868
2962187

ABSTRACTS. Stock. Members. Humberside Libraries Technical Interloan Scheme
– Bibliographies – Catalogues
HUMBERSIDE LIBRARIES TECHNICAL INTERLOAN SCHEME
A guide to abstracts and indexes in HULTIS members libraries. – Hull : Library of Science, Technology and Commerce, 1977
[7],40p. – Includes index
0–904451–06–2 2725.ff.15
1043986

ABSTRACTS JOURNALS
to 1920
MANZER, Bruce M.
The abstract journal, 1790–1920. – Metuchen ; London : Scarecrow Press ; [Folkestone] : [Distributed by Bailey and Swinfen], 1977
ix,312p : ill. – Bibliography: p.277–286. – Includes index
0–8108–1047–6 X.989/52351
1409352

ABSTRACTS JOURNALS
Publishing. Applications of on-line computer systems
SZCZEPANSKI, Adam J.
Synoptic publishing and on-line full text. – Leicester : Primary Communications Research Centre, University of Leicester, 1983
68p : ill. – (British Library R. & D. report ; 5809). – Includes bibliography
B85–10366 2708h.86
3241904

ABSTRACTS JOURNALS. Chemistry
Chemical abstracts
STANDARD abbreviations, acronyms, special characters and symbols in CAS Computer-readable files and publications. – [United States] : Chemical Abstracts Service, 1982
23p
r6863058 X.622/15102
2856484

ABSTRACTS JOURNALS. Documents on biology
'Biological abstracts', to 1976
STEERE, William Campbell
Biological abstracts–BIOSIS. – New York ; London : Plenum Press, 1976
xvii,233p : ill. – Bibliography: p.219–224. – Includes index
0–306–30915–7 SRL
1282468

ABSTRACTS JOURNALS. Documents on mechanical engineering. Mechanical engineering
Feasibility
MILLSON, R. J.
An experiment in synopsis publishing in the field of mechanical engineering. – London : British Library Research and Development Department, 1979
viii,112p : ill. – (Research and development reports / British Library ; 5498)
0–905984–41–2 X.622/9067 Woolwich
1922610

ABSTRACTS JOURNALS. Documents on medicine
– Bibliographies
DALBY, Andrew
Medical abstracts and indexes, 1975. – Cambridge : Cambridge University Library, 1975
86p. – (Librarianship series / University of Cambridge. Library ; 2). – Includes index
0–902205–17–x SRL
0593923

ABSTRACTS JOURNALS. Documents on science
Referativnyĭ Zhurnal. Use
COPLEY, Eric James
A guide to 'Referativnyi Zhurnal'. – 3rd ed. – [London] : British Library, Science Reference Library, 1975
[2],26p. – (Occasional publications / Science Reference Library)
0–902914–13–8 BS.10/688
0587311

ABSTRACTS JOURNALS. Science
- Bibliographies

SCIENCE REFERENCE LIBRARY
Abstracting and indexing periodicals in the Science Reference Library. - 2nd ed.. - London : Science Reference Library, c1982
54p. - Includes index
0-902914-69-3 L.49/1473
060741x

ABSTRACTS JOURNALS. Science
- Directories

SCIENCE REFERENCE LIBRARY
Abstracting & indexing periodicals in the Science Reference Library. - London : Science Reference Library, 1975
iv,71p. - Includes index
0-902914-17-0 BS.10/437
060741x
- [Another copy] O.P.L.

ABSTRACTS JOURNALS. Serial articles. Humanities
- Bibliographies

UNIVERSITY OF BIRMINGHAM. Library
Bibliographical information in the social sciences and humanities. - Birmingham : [The Library], 1978
32[i.e. 33]p,[7] leaves. - (Special guides / University of Birmingham. Library ; no. 18)
0-7044-0323-4 X.0512/235
2163470

ABSTRACTS JOURNALS. Serial articles. Humanities
- Catalogues

HARZFELD, Lois A.
Periodical indexes in the social sciences and humanities. - Metuchen ; London : Scarecrow ; [Folkestone] : [Distributed by Bailey and Swinfen], 1978
xiv,174p. - Includes index
0-8108-1133-2 Bar.T.16.bb.17 Reading Room
1586513
- [Another copy] OPL
- [Another copy.] 2725.aa.50

ABSTRACTS JOURNALS. Serial articles on drug therapy
National drug information abstracts. Indexing. Applications of digital computer systems

BISHOP, J. M. (Jacqueline Margaret)
The application of computer indexing to the operation of an N.H.S. pharmacy-based regional drug information centre. - [London] : S.E. Thames Regional Drug Information Centre, Pharmacy Dept., Guy's Hospital, 1981
[60]leaves
B82-09555 SRL
2618842

ABSTRACTS JOURNALS. Serial articles. Science
Abstracts journals in stock of University College of North Wales. Science Library. Use

MADLEY, M.
How to attack periodicals. - Bangor : Science Library, University College of North Wales, 1979
[4],13p : ill. - (Publications / University College of North Wales. Science Library ; no.4)
B80-37595 X.909/45061
225817x

ABSTRACTS JOURNALS. Serial articles. Social sciences
- Bibliographies

UNIVERSITY OF BIRMINGHAM. Library
Bibliographical information in the social sciences and humanities. - Birmingham : [The Library], 1978
32[i.e. 33]p,[7] leaves. - (Special guides / University of Birmingham. Library ; no. 18)
0-7044-0323-4 X.0512/235
2163489

ABSTRACTS JOURNALS. Serial articles. Social sciences
- Catalogues

HARZFELD, Lois A.
Periodical indexes in the social sciences and humanities. - Metuchen ; London : Scarecrow ; [Folkestone] : [Distributed by Bailey and Swinfen], 1978
xiv,174p. - Includes index
0-8108-1133-2 Bar.T.16.bb.17 Reading Room
1586505
- [Another copy] OPL
- [Another copy.] 2725.aa.50

ABSTRACTS JOURNALS IN ENGLISH
Coverage of British patents, 1973

MANN, Henry
The coverage of UK patent specifications by abstracting journals. - London : British Library, Science Reference Library, 1976
[2],26p : ill. - (Reports / Science Reference Library)
0-902914-29-4 X.615/1713 Woolwich
0669792

ABSURD
See also
Theatre of the absurd

ABSURD. Literature, to date

TOMOVIĆ, Slobodan
Junak apsurda. - Titograd : Pobjeda, 1980
290p. - (Studije - kritike - eseji)
r6435577 X.958/17430
8996970

ABSURD HEROES. Special themes. American writers, 1945-. Fiction in English

GALLOWAY, David, 1937-
The absurd hero in American fiction. - 2nd rev. ed. - Austin : University of Texas Press, c1981
xv,265p. - Bibliography: p239-258
0-292-70356-2 X.950/23860
2919907

ABTEI AMORBACH

GORENFLO, Roger M.
Die mittelalterliche Baugeschichte der ehemaligen Abteikirche Amorbach. - Frankfurt am Main : Kunstgeschichtliches Institut der Johann Wolfgang Goethe-Universität, 1983
457p : ill. - (Frankfurter Fundamente der Kunstgeschichte ; Bd.3)
3-923813-02-3 X.200/42657
9039325

ABTEI FULDA
See Kloster Fulda †

ABTEI IBURG
See Kloster Iburg †

ABTEI PFÄFERS
See Kloster Pfäfers †

ABTEI ROMMERSDORF

SCHULZE, Heiko K. L
Die ehemalige Prämonstratenser-Abtei Rommersdorf. - Mainz : Selbstverlag der Gesellschaft für Mittelrheinische Kirchengeschichte, 1983
l,365p : ill. - (Quellen und Abhandlungen zur mittelrheinischen Kirchengeschichte ; Bd.44). - Includes bibliography
r6308038 X.0100/505(44)
897408x

ABTEI ST. GEORGENBERG-FIECHT

Das ÄLTESTE Urbar der Benediktinerabtei St. Georgenberg zu Fiecht von 1361/70 und das Weinzinsregister von 1420 und 1422 / bearbeitet von Hanns Bachmann. - Innsbruck : Wagner, 1981
239p,4p of plates,[1] folded leaf of plates. - (Österreichische Urbare. Abt.3., Urbare geistlicher Grundherrschaften. Bd.5, Die mittelalterlichen Stiftsurbare des Bistums Brixen ; Tl.4)
3-7030-0091-0 Ac.811/7 [Abt.3, Bd.5, Tl.4]
8809747

ABTEI WADGASSEN
See Kloster Wadgassen †

ABU (MOUNT)
See Mount Abu

ABU AL-RAIḤAN MUḤAMMAD IBN AḤMAD, AL-BĪRŪNĪ
See al-Birūni, Abu al-Raiḥān Muḥammad ibn Aḥmad †

ABU ALI IBN SINA
See Avicenna †

ABŪ BAKR IBN AL-ṬUFAIL, ABŪ JA'FAR, AL-ISHBĪLĪ. Islam
- Biographies

FEROZE, Muḥammad Rašid
Abu Bakr. - Leicester : Islamic Foundation, c1976
44p : ill. - (Glimpses of Islamic history)
B83-32524 X.200/40945
0726710
- [Another copy.] X.709/30710

HASAN, Masudul
Sidiq-i-Akbar Hazrat Abu Bakr. - Lahore : Ferozsons, [1976?]
vi,241p
r6040743 X.100/27526
0726710

ABU DHABI. United Arab Emirates
Social conditions

O'BRIEN, Edna
Arabian days. - London : Quartet Books, 1977
160p : chiefly ill(same col.)
0-7043-2150-5 X.805/1042
1273442
- [Another copy: paperback]
ISBN 0-7043-3369-4 X.805/1334

ABŪ ḤAYYĀN AL-TAWḤĪDĪ
See At-Tawhidi, Abū Ḥayyān †

ABŪ ḤAYYĀN AL-TAWḤĪDĪ
See At-Tawhidi, Abū Ḥayyān †

ABU KLEA MEDAL. Medals
Awards to British military forces, 1885

WEBB, J. V. (John Vernon)
The Abu Klea medal rolls. - London : J.V. Webb, 1981
xviii,158p
B81-37384 X.622/11239 Woolwich
2559412

ABŪ L-FATḤ MAḤMŪD IBN ḤUSAYN KUŠĀĞIM
See Kušāğim, Abū l-Fatḥ maḥmūd ibn Ḥusayn †

ABŪ L-HASAN 'ABD AL-JABBĀR BN AHMAD AL-HAMADĀNĪ QĀDĪ L-QUDĀT
Theories of Islamic doctrine

PETERS, Johannes Reinier Theodorus Maria
God's created speech. - Leiden : Brill, 1976
xii,447,[1]p. - Bibliography : p.422-429. - Includes index
r6031753 X.100/16895
109064X

ABŪ L-HASAN 'ALI AL-MAS'ŪDĪ
See Al-Mas'ūdi

ABU SALABIKH. Iraq
Antiquities. Archaeology

POSTGATE, J. N. (John Nicholas)
The West Mound surface clearance. - [London?] : British School of Archaeology in Iraq, 1983
vii,111p,T3p of plates(some folded) : ill. - (Abu Salabikh excavations ; v.1). - Bibliography: p105-106
0-903472-06-6 X.425/4942
3198782

ABU SIMBEL. Egypt
Antiquities. Archaeology, to date

CENTRE DE DOCUMENTATION ET D'ÉTUDES SUR L'ANCIENNE ÉGYPTE
Grand temple d'Abou Simbel. - Le Caire : Centre de Documentation et d'Études sur l'Ancienne Égypte, 1971
vi,63leaves,plates. - (Collection scientifique)
r6210821 L.42/3253
9146571

GRAND temple d'Abou Simbel / S. Donadoni .. [et al.]. - Le Caire : Centre d'Études et de Documentation sur l'Ancienne Égypte, 1975
2v.in 1(plates) : ill. - (Collection Scientifique)
r6211818 X.421/25662
9127933

ABUA-OGBIA LANGUAGES
See Central Delta languages

ABUAN LANGUAGE
- Dictionaries

GARDNER, Ian D.
Abuan-English, English-Abuan dictionary. - [Port Harcourt] : University of Port Harcourt, 1980
xv,112p. - (Deltaseries ; 1 (1980))
978-232-108-7 X.950/19766
8916845

ABULAFIA, MEIR BEN TODROS. Rabbis. Judaism. Spain

SEPTIMUS, Bernard
Hispano-Jewish culture in transition. - Cambridge, Mass. ; London : Harvard University Press, 1982
ix,180p. - (Harvard Judaic monographs ; 4). - Includes index
0-674-39230-2 X520/29091
2846047

ABUNDANCE MODELS. Diversity. Population. Organisms

ENGEN, S.
Stochastic abundance models. - London : Chapman and Hall, 1978
x,126p : ill. - (Monographs on applied probability and statistics). - Bibliography: p.111-115. - Includes index
0-412-15240-1 X.629/12298 Woolwich
1808591

ABUSCH, ALEXANDER. Communist movements. Germany
- Biographies

ABUSCH, Alexander
Memoiren. - Berlin : Dietz, 1981- : ill
r6305148 X.0808/2030
8856125

ABUSE
See also
Physically abused children
Sexual abuse

ABUSE. Children
By adults
ASSAULT against children / edited by John H. Meier. - London : Taylor & Francis, c1985
xxxii,304p : ill. - Bibliography: p271-292. - Includes index
0-85066-507-8 X.529/69972
1736116

ABUSE. Children
By adults - Bibliographies
WELLS, Dorothy P.
Child abuse. - Metuchen ; London : Scarecrow Press ; [Folkestone] : [Distributed by Bailey and Swinfen], 1980
viii,450p. - Includes index
0-8108-1264-9 X.529/36170
2181495

ABUSE. Children
By adults - Conference proceedings
CHILD abuse. - [Dundee] : [Tayside Regional Council Department of Social Work?], [1979?]
36p
B83-32828 X.525/7708
1736116

CHILD abuse. - [Dundee] : [Tayside Regional Council Department of Social Work?], [1982]
29p
B83-32829 X.525/7706
1736116

DECISION making in child abuse. - [Dundee] : [Tayside Regional Council Department of Social Work?], [1981]
31p
B83-32827 X.525/7709
1736116

INTERNATIONAL CONGRESS ON CHILD ABUSE AND NEGLECT (2nd : 1978 : London)
The abused child in the family and in the community. - Oxford : Pergamon, 1980
2v.(1092,xli p) : ill. - Includes bibliographies and index
Subsequent volumes, when received, will be placed at this shelfmark
0-08-023430-5 X.0525/355
1736116

INTERNATIONAL CONGRESS ON CHILD ABUSE AND NEGLECT (2nd : 1978 : London)
Second International Congress on Child Abuse & Neglect, London, 12-15 September 1978. - [Oxford, etc.] : Pergamon for the Organizing Committee of the Congress. Abstracts. - [1979]
iv[i.e. vi],278,[1],11p. - Includes bibliographies and index
0-08-023438-0 X.519/29453
1736116

NATO ADVANCED STUDY INSTITUTE ON RESEARCH AND INNOVATION IN CHILD ABUSE AND NEGLECT (1982 : Les Arcs)
Child abuse and neglect. - The Hague ; Lancaster : Nijhoff in cooperation with NATO Scientific Affairs Division, 1983
xi,340p : ill. - (NATO advanced study institutes series. Series D, Behavioural and social sciences ; no.16). - Includes index
90-247-2862-2 X.520/32624
1736116

ABUSE. Children
By adults. Cultural aspects - Conference proceedings
CHILD abuse and neglect / edited by Jill E. Korbin. - Berkeley ; London : University of California Press, c1981
xiv,217p. - Includes bibliographies and index
0-520-04432-0 X.520/26209
2635178

ABUSE. Children
By parents
CHILD abuse. - Boston, [Mass.] : Little, Brown, 1982
282p : ill. - (Little, Brown series in clinical pediatrics). - Includes bibliographies
0-316-59794-5 X.520/29443
1589059

INGLIS, Ruth
Sins of the fathers. - London : Owen, 1978
220p. - Bibliography: p.214-220
0-7206-0507-5 X.529/32565
1589059

SHEPPARD, Michael
Perceptions of child abuse. - Norwich : University of East Anglia in association with Social Work Today, 1982
56leaves. - (Social work monographs ; 2). - Bibliography: leaves 45-56
B84-00257 X.525/8314
1589059

ABUSE. Children
By parents - Conference proceedings
ASSESSMENT and social work practice in child abuse. - [Cardiff] : Welsh Office, Social Work Service, 1984
iii,27p
B85-10995 BS.128A/387
1589059

ABUSE. Children
By parents. Prevention. Psychological aspects
PSYCHOLOGICAL approaches to child abuse / edited by Neil Frude. - London : Batsford Academic and Educational, 1980
240p : ill. - Includes bibliographies and index
0-7134-3714-6 X.529/40506
2149028

ABUSE. Children. Canada
to date
SCHLESINGER, Benjamin
Child abuse in Canada. - Toronto : Guidance Centre, Faculty of Education, University of Toronto, c1977
36p. - (Social problems in Canada)
0-7713-0040-9 X.529/56657
9075119

ABUSE. Children. Canada
By adults, to date
VAN STOLK, Mary
The battered child in Canada. - Rev. ed.. - Toronto : McClelland and Stewart, 1979
xii,178p
0-7710-8712-8 X.529/62247
9088768

ABUSE. Children. England
REPORT on the death of Shirley Woodcock. - [London] : London Borough of Hammersmith & Fulham, 1984
iii,66p : ill. - Includes bibliography
B85-26692 X.529/70474
3299791

ABUSE. Children. England
1837-1901
BEHLMER, George K.
Child abuse and moral reform in England, 1870-1908. - Stanford, Calif. : Stanford University Press, 1982
viii,320p : ill. - Includes bibliography
0-8047-1127-5 X.529/66822
915390x

ABUSE. Children. Great Britain
By parents. Prevention
DINGWALL, Robert
The protection of children. - Oxford : Basil Blackwell, 1983
x,286p. - Bibliography: p267-282. - Includes index
0-631-13105-1 X.520/32105
2988208

ABUSE. Children. Great Britain
Prevention. Cooperation between welfare services
CHILD abuse / Department of Health and Social Security. - London : H.M.S.O., 1982
vi,93p
0-11-320788-3 BS.385/748
1930818

ABUSE. Children. United States
The BATTERED child / edited by C. Henry Kempe and Ray E. Helfer. - 3rd ed., rev. and expanded. - Chicago ; London : University of Chicago Press, 1980
xvii,440p : ill. - Includes index
0-226-43038-3 X.520/23462
2381745

DAMAGED parents / Norman A. Blansky ... [et al.]. - Chicago ; London : University of Chicago Press, 1981
xii,271p. - Bibliography: p255-264. - Includes index
0-226-67221-2 X.520/24544
2381745

The SOCIAL context of child abuse and neglect / edited by Leroy H. Pelton. - New York ; London : Human Sciences, c1981
331p. - Includes bibliographies and index
0-87705-504-1 X.529/53403
2381745

ABUSE. Children. United States
- For welfare work
MAYHALL, Pamela D.
Child abuse and neglect. - New York ; Chichester : Wiley, c1983
xiv,400p : ill. - Includes bibliographies and index
0-471-09929-5 X.520/32200
2939312

ABUSE. Children. United States
By adults, to date
HERBRUCK, Christine Comstock
Breaking the cycle of child abuse. - Minneapolis : Winston Press, c1979
205p
0-03-045691-6 X.529/67723
9164057

ABUSE. Children. United States
By parents
GABARINO, James
Understanding abusive families. - [Lexington, Mass.] : Lexington Books, [1981], c1980
xvi,263p : ill. - Includes index
0-669-03621-8 X.520/23939
2280272

GIOVANNONI, Jeanne M.
Defining child abuse. - New York : Free Press ; London : Collier Macmillan, 1979
xviii,302p : 2 ill. - Bibliography: p 287-295. - Includes index
0-02-911750-x X.520/21968
2131757

The MALTREATMENT of the school-aged child / edited by Richard Volpe and Margot Breton, Judith Mitton. - Lexington : Lexington, c1980
xiii,210p. - Includes bibliographies and index
0-669-03463-0 X.520/23716
2280272

TRAUMATIC abuse and neglect of children at home / edited by Gertrude J. Williams and John Money. - Baltimore ; London : Johns Hopkins University Press, 1980
xv,616p : ill. - Includes bibliographies and index
0-8018-2231-9 X.529/37866
2280272

ABUSE. Civilians. Saverne. France
By officers of Germany. Heer, 1913
SCHOENBAUM, David
Zabern 1913. - London : Allen & Unwin, 1982
197p. - Bibliography: p187-192. - Includes index
0-04-943025-4 X.529/49432
2771519

ABUSE. Old persons. United States
ABUSE and maltreatment of the elderly / edited by Jordan I. Kosberg. - Boston ; Bristol : John Wright, 1983
xiv,448p : ill. - Includes bibliographies and index
0-7236-7025-0 X.520/31459
2963051

ABUSE. Wives. Great Britain
By husbands
MARITAL violence / edited by Norman Johnson. - London : Routledge & Kegan Paul, 1985
ix,195p. - (Sociological review monograph ; 31)
0-7102-0448-5 Ac.2263/3(1)/31
315663x

ABUSED CHILDREN
See also
Welfare work with abused children

ABUSIVE LETTERS IN ENGLISH
to 1978 - Anthologies
DEAR sir, drop dead! / edited by Donald Carroll. - London : Eyre Methuen, 1979
152p
0-413-39560-x X.981/22614
1883887

ABWEHR
See Germany. Abwehr

ABXAZO LANGUAGE
See Abkhaz language

ABXAZO LITERATURE
See Abkhaz literature

ABYDOS. Egypt
Ancient Egyptian temples: Temple of Seti I (Abydos). Relief sculptures. Special subjects: Ancient Egyptian religion. Rituals
DAVID, Rosalie
A guide to religious ritual at Abydos. - Warminster : Aris & Philips, c1981
xiv,182p,[6]p of plates : ill. - (Modern Egyptology). - Bibliography: pvii-xiii
0-85668-060-5 L.49/1053
2584425

ABYSSAL ZONE. Atlantic Ocean
Circulation
WÜST, Georg
The stratosphere of the Atlantic Ocean. - New Delhi : Amervid, c1978
xv,112p,[25]leaves of plates (some folded) : ill
0-86249-375-7 X.622/21744
3221628

ABYSSAL ZONE. Oceans
Animals
KNIGHT, Dennis
Monsters of the deep. - London : Royal Sovereign Group Ltd, 1976
[2],18p,[2]leaves of plates : ill(chiefly col). - (Activity books / Dennis Knight)
0-900862-44-0 X.319/17098 Woolwich
1042106

ABYSSAL ZONE. Oceans
Pressure. Adaptation by marine animals. Biochemistry
BIOCHEMISTRY at depth / edited and organized by P.W. Hochachka. - Oxford : Pergamon, 1976
[5],202p,plate : ill. - Includes bibliographies and index
0-08-019960-7 X.622/5532 Woolwich
0555703

ABYSSINIA
See Ethiopia

ABYSSINIAN
See Ethiopian

ABYSSINIAN EXPEDITION
Army operations by Great Britain. Army, 1867-1868
BATES, Sir Darrell
The Abyssinian difficulty. - Oxford : Oxford University Press, 1979
xvi,240p,[16]p of plates : ill. - Bibliography: p.229-231. - Includes index
0-19-211747-5 X.809/45020
1841580

ABYSSINIANS
See Ethiopians

A.C.
See Alternating current

AC
See Alternating current

A.C.
See Alternating current

AC
See Alternating current

A.C.
See Alternating current

AC
See Alternating current

A.C.
See Alternating current

AC
See Alternating current

A.C.
See Alternating current

AC
See Alternating current

A.C.
See Alternating current

AC
See Alternating current

A.C. BUTCHER COLLECTION. Collections. Social conditions. 1890-1910. Gundagai. New South Wales. Special subjects. Australian photography. Stock. National Library of Australia
GABRIEL, Charles Louis
Gundagai album. - Canberra : National Library of Australia, 1976
xx,[8]p,118[i.e.135]p of plates : ill. - Bibliography (2p.)
0-642-99058-1 Cup.1281/796
1382705

AC CARS
to 1975
WATKINS, Martyn
AC. - [Yeovil] : [Haynes], [1976]
160p : ill. - (Foulis mini marque history series)
0-85429-204-7 X.620/16393 Woolwich
0757098

AC CARS
AC Cobra racing cars
GRAINGER, Rod
AC/Ford/Shelby Cobra. - [Yeovil] : Foulis, 1984
56p : ill(some col.). - (Super profile) (A Foulis motoring book)
0-85429-381-7 X.622/21740
3112276

AC CARS
AC Cobra racing sports cars
LEGATE, Trevor
Cobra. - Yeovil : Haynes, 1984
256p : ill(some col.). - (A Foulis motoring book). - Includes index
0-85429-424-4 X.622/23451
3112276
MCCOMB, F. Wilson
AC (Shelby) Cobra. - London : Osprey, 1984
135p : ill(some col.). - (Autohistory) (Osprey autohistory). - Includes index
0-85045-554-5 X.629/26016
3112276

AC CARS
AC Cobra racing sports cars. Racing, 1962-1965
FRIEDMAN, Dave
Carroll Shelby's Racing Cobra. - London : Osprey, c1982
207p : ill
0-85045-457-3 X.622/15215
2880180

ACA FORM OF BUILDING AGREEMENT. Association of Consultant Architects
1984
POWELL-SMITH, Vincent
AJ contracts guide to: ASCA Form of Building Agreement 1982, second edition 1984, BPF/ACA Form of Building Agreement 1984. - London : Architectural Press, 1984
14p : ill
0-85139-756-5 X.525/10970
320457x

ACACIAS
Phyllodia
WILKINSON, H. P.
The anatomy of the pulvinus in various Mimosoideae in relation to function ; Acacia phyllodes. A bibliography on the anatomy of organs connected with nyctinastic and seismonastic movements in higher plants. - Kew : Jodrell Laboratory, Royal Botanic Gardens, 1983
30p : ill. - (Notes from the Jodrell Laboratory ; 10)
B84-01523 BS.36/23
2983176

ACADEMIA ALTORFIA
Prizes. Oration medals, 1577-1626
STOPP, Frederick John
The emblems of the Altdorf Academy. - London : Modern Humanities Research Association, 1974
xxii,226p,22p of plates,leaf of plate : ill. - (Publications / Modern Humanities Research Association ; vol.6). - Bibliography: p.xiii-xxii. - Includes index
0-900547-32-4 X.0525/110(6)
0738859

ACADEMIA COLOMBIANA DE LA HISTORIA
to 1972
ACADEMIA COLOMBIANA DE LA HISTORIA
Academia Colombiana de Historia. - Bogotá : Editorial Kelly, 1972
96p : ill
r6025395 X.700/18864
1329898

ACADEMIA DE CIENCIAS DE CUBA
Relations with Soviet research organisations, 1962-1979
SÁENZ, Tirso W
El papel de la colaboración con las instituciones científicas soviéticas en la creación y desarrollo de la Academia de Ciencias de Cuba. - La Habana : Editorial Academia, 1979
35p
r6160784 X.622/9275 Woolwich
2470101
El papel de la colaboración con las instituciones científicas soviéticas en la creación y desarrollo de la Academia de Ciencias de Cuba. - La Habana : Editorial Academia, 1979
35p
r6160784 X.622/9275 Woolwich

ACADEMIA DE CIENCIAS MÉDICAS, FÍSICAS Y NATURALES DE LA HABANA
BUSTAMANTE O'LEARY, Jose A
La Academia de Ciencias Médicas, Físicas y Naturales de La Habana. - La Habana : Academia de Ciencias de Cuba, Centro de Estudios de Historia y Organización de la Ciencia Carlos J. Finlay, 1979
19leaves. - (Conferencias y estudios de historia y organización de la ciencia ; 11)
r6701134 LAS.D.300/15 Woolwich
8744297

ACADÈMIA DE CULTURA VALENCIANA
PÉREZ MORAGON, Francesc
L'Acadèmia de Cultura Valenciana. - València : Eliseu Climent, editor, 1982
78p. - (Quaderns 3 i 4 ; 23)
84-7502-053-4 X.529/51287
8883351

ACADEMIA DE SAN CARLOS
1785-1810
BROWN, Thomas Anthony
La Academia de San Carlos de la Nueva España. - México : Secretaría de Educación Pública, 1976
2v : ill. - (SepSetentas ; 299-300). - Includes bibliographies
r6084771 X.708/45143
1860348

ACADEMIA DE SÃO PAULO
See Universidade de São Paulo. Faculdade de Direito †

ACADEMIA ECUATORIANA
Academicians, 1875-1975 - Biographies
TOBAR DONOSO, Julio
Los Miembros de Número de la Academia Ecuatoriana muertos en el primer siglo de su existencia, 1875-1975. - Quito : Editorial Ecuatoriana, 1976
495p.,17 leaves of plates
r6101580 X.709/29237
2166186

ACADEMIA LUGDUNO BATAVA. FACULTEIT DER RECHTSGELEERDHEID
Professors: Bronchorst, Everardus; Vinnius, Arnoldus & Voet, Johannes. Influence on European civil law
FEENSTRA, R. (Robert)
Seventeenth-century Leyden law professors and their influence on the development of the civil law. - Amsterdam ; Oxford : North-Holland Publishing Co., 1975
124p,xii p of plates. - (Verhandelingen : nieuwe reeks / Nederlandse Akademie van Wetenschappen. Afdeling Letterkunde ; deel 90). - Bibliography: p.111-115. - Includes index
0-7204-8296-8 Ac.944/3[nieuwe reeks.dl.90]
0573949
- [Another copy] X.205/776

ACADEMIA NACIONAL DE LA HISTORIA (ARGENTINA)
- Conference proceedings
CONGRESO DE HISTORIA ARGENTINA Y REGIONAL (2nd : 1973 : Comodoro Rivadavia)
Crónica del segundo Congreso de Historia Argentina y Regional, Comodoro Rivadavia, 12-15 de enero de 1973. - Buenos Aires : Academia Nacional de la Historia, 1973
79 p., 9 leaves of plates
r6003595 X.702/2927
0850519

ACADEMIA NACIONAL DE LA HISTORIA (VENEZUELA)
Stock: Serials
BRICEÑO PEROZO, Mario
La hemeroteca de la Academia Nacional de la Historia. - Caracas : [s.n.], 1972
79p : ill. - (Biblioteca venezolana de historia ; 17)
r6099971 X.709/29343
2193361

ACADEMIA OLIMPICA, VICENZA
Exhibits: Photographs. Special subjects. Brazil. Immigration from Veneto, Italy - Catalogues
I VENETI in Brasile nel centenario dell'emigrazione, 1876-1976 / a cura di Mario Sabbatini e Emilio Franzina. - Vicenza : Edizioni dell' Accademia olimpica, 1977
xv,157p,[3]fold leaves,[120]p of plates
r6500580 X.529/39946
2478765

ACADEMIA OLIMPICA, VICENZA
Exhibits: Photographs. Special subjects: Italy. Veneto. Emigration to Brazil - Catalogues
I VENETI in Brasile nel centenario dell'emigrazione, 1876-1976 / a cura di Mario Sabbatini e Emilio Franzina. - Vicenza : Edizioni dell' Accademia olimpica, 1977
xv,157p,[3]fold leaves,[120]p of plates
r6500580 X.529/39946
2478765

ACADEMIA REPUBLICII SOCIALISTE ROMÂNIA

MIHOC, Gheorghe
The Academy of the Socialist Republic of Romania. - București : Editura Academiei Republicii Socialiste România, 1981
52p
r6457182 X.529/49152
8858381

ACADEMIA VIADRINA
See Universität Frankfurt an der Oder †

ACADEMIC ACHIEVEMENT
See also
Great Britain. Assessment of Performance Unit

ACADEMIC ACHIEVEMENT. Children
related to character of parents

FINNÄS, Leif
Nagra föräldrakarakteristikas samband med relativ skolprestation. - Abo : Abo akademi, 1978
300p. - (Meddelanden fran Stiftelsens för Abo akademi forskningsinstitut ; nr 26). - Includes bibliographies
951-648-364-x X.0519/601(26)
2026384

ACADEMIC ACHIEVEMENT. Children
related to psychology

GIL MUÑOZ, Carlos
Causas del fracaso escolar. - Madrid : CVS Ediciones, 1975
162 p.. - (Colección Ateneo ; no. 6)
84-354-0015-8 X.519/20357
0931470

ACADEMIC ACHIEVEMENT. Children. Ethnic minorities. Australia

MARJORIBANKS, Kevin
Ethnic families and children's achievements. - Sydney ; London : Allen and Unwin, 1980
172p. - Bibliography: p.158-167
0-86861-305-3 X.529/36279
2175118

ACADEMIC ACHIEVEMENT. Ethnic minority students. Schools. United States
Effects of family life

CLARK, Reginald
Family life and school achievement. - Chicago ; London : University of Chicago Press, 1983
xiii,249p. - Bibliography: p229-249
0-226-10769-8 X.520/33976
313413x

ACADEMIC ACHIEVEMENT. Girls. Students. Secondary schools
In mathematics

MATHEMATICS education and girls / Sheffield City Polytechnic Department of Mathematics, Statistics and Operational Research ; sponorship by the British Petroleum Company. - Sheffield : The Department, 1982
iiileaves,198p : ill. - Bibliography: p193-198
0-903761-45-9 X.525/5494
2692856

ACADEMIC ACHIEVEMENT. Negroes. Langa. Cape Town. South Africa
Effects on employment

HUBBARD, Michael
Education for underemployment. - [Cape Town] : South African Institute of Race Relations, Cape Western Region, [1976]
[7], 26 p.
r6031203 X.512/5535
1053345

ACADEMIC ACHIEVEMENT. Personnel. Organisation for Economic Co-operation and Development countries
related to remuneration

PSACHAROPOULOS, George
Earnings and education in OECD countries. - Paris : Organisation for Economic Co-operation and Development ; [London] : [H.M.S.O.], 1975
[2],194p. - Includes bibliographies
B76-05495 UN.P.106/538
0582476

ACADEMIC ACHIEVEMENT. Personnel. United States
Effects on upward job mobility

LEIGH, Duane E.
An analysis of the determinants of occupational upgrading. - New York ; London : Academic Press, 1978
xv,185p : ill. - (Monograph series / University of Wisconsin. Institute for Research on Poverty). - Bibliography: p.181-185
0-12-442850-9 X.520/9905
1822020

ACADEMIC ACHIEVEMENT. School leavers. England
Great Britain. Parliament. House of Commons. Expenditure Committee. 'Attainments of the school leaver'

GREAT BRITAIN. Department of Education and Science Expenditure Committee. Government observations on tenth report, session 1976-77, 'The attainments of the school leaver'. - London : H.M.S.O., [1978]
[1],13p. - (Cmnd.7124)
0-10-171240-5 BS.10/713
1433857

ACADEMIC ACHIEVEMENT. School leavers. Great Britain
Assessment

BURGESS, Tyrrell
Outcomes of education. - Basingstoke : Macmillan, 1980
xvii,181p : ill. - Bibliography: p.177-181
0-333-27800-3 X.520/22435
2154587

GREAT BRITAIN. Department of Education and Science
Records of achievement. - [London] : [Department of Education and Science], 1984
11p
B85-13783 BS.10/1052
2154587

ACADEMIC ACHIEVEMENT. School leavers. Multiracial secondary schools. Great Britain

DRIVER, Geoffrey
Beyond under-achievement. - London : Commission for Racial Equality, 1980
83p
0-902355-90-2 BS.400/51
2311232

ACADEMIC ACHIEVEMENT. Secondary schools. Scotland
Diagnostic tests

BLACK, H. D. (Harold David)
Diagnostic assessment in secondary schools. - [Edinburgh] : The Scottish Council for Research in Education, c1980
ix,86p : ill. - Bibliography: p86
0-901116-26-2 X.329/13478 Woolwich
2549719
- [Another copy] X.529/47100
Diagnostic assessment in secondary schools. - Sevenoaks : Published by Hodder & Stoughton [for] Scottish Council for Research in Education, 1981, c1980
x,86p. - (SCRE publication ; 77). - Bibliography: p86
0-340-28240-1 X.529/55419
2549719

ACADEMIC ACHIEVEMENT. Slow learning children. West Germany
related to social class

MEIDEL, Klaus
Analyse der Wirkungen von sozialen Umweltfaktoren bei lernbehinderten Sonderschülern unter Berücksichtigung ihrer Schichtzugehörigkeit. - Erlangen : Offsetdruck-Fotodruck, [1975]
xxiii [i.e. xxxiii], 500, 27 p.. - Bibliography : p. xv-xxiii [i.e. xxxiii]
r6002085 X.519/20276
0927562

ACADEMIC ACHIEVEMENT. Students
Assessment

AHMANN, J. Stanley (John Stanley)
Test items based on Evaluating student progress. - Boston, Mass. ; London : Allyn and Bacon, c1981
81p : 1ill
0-205-06562-7 X.529/40305
2373017

BEJAR, Isaac I.
Achievement testing. - Beverly Hills ; London : Sage, c1983
88p : ill. - (Quantitative applications in the social sciences ; no.07-036). - Bibliography: p77-87
0-8039-2047-4 X.529/62954
0180009

CHASE, Clinton I. (Clinton Irvin)
Measurement for educational evaluation. - 2nd ed. - Reading, Mass. ; London : Addison-Wesley, 1978
xviii,361p : ill. - (Addison-Wesley series in education). - Includes bibliographies and index
0-201-01019-4 X.520/12703
0180009

EBEL, Robert Louis
Essentials of educational measurement. - 3rd ed. - Englewood Cliffs ; London : Prentice-Hall, 1979
xi,388p : ill. - Includes index
0-13-286013-9 X.520/14236
0180009

GRONLUND, Norman E. (Norman Edward)
Measurement and evaluation in teaching. - 4th ed. - New York : Macmillan ; London : Collier Macmillan, c1981
vii,597p : ill. - Includes bibliographies and index
0-02-348020-3 X.520/23885
0180009
- [Another copy: paperback]
ISBN 0-02-978700-5 X.520/24448

GRONLUND, Norman Edward
Measurement and evaluation in teaching. - 3rd ed. - New York : Macmillan ; London : Collier Macmillan, 1976
x,590p : ill. - Includes index
0-02-348050-5 X.520/10759
0180009

GRONLUND, Norman Edward
Measurement and evaluation in teaching. - 3rd ed. - New York : Macmillan . London : Collier Macmillan, 1976
x,590p : ill. - Includes bibliographies and index
0-02-979400-5 X.510/15521
0180009

MEHRENS, William A. (William Arthur)
Measurement and evolution in education and psychology. - 3rd ed. - New York ; London : Holt, Rinehart and Winston, c1984
xiv,658p : ill. - Bibliography: p599-639. - Includes index
0-03-062491-6 X.520/34060
0180009

MEHRENS, William Arthur
Measurement and evaluation in education and psychology. - 2nd ed. - New York . London : Holt, Rinehart and Winston, 1978
viii,759p : ill. - Bibliography: p.705-743. - Includes index
0-03-089966-4 X.520/12738
0180009

POPHAM, W. James (William James)
Modern educational measurement. - Englewood Cliffs : Prentice-Hall, c1981
xiii,441p : ill. - Includes bibliographies and index
0-13-591982-7 X.520/24016
0180009

READINGS in measurement and evaluation in education and psychology / [edited by] William A. Mehrens. - New York ; London : Holt, Rinehart and Winston, 1976
x,374p : ill. - Includes bibliographies and index
0-03-089680-0 X.519/26392
0180009

The RISE and fall of national test scores / edited by Gilbert R. Austin, Herbert Garber. - New York ; London : Academic Press, c1982
xvii,270p : ill. - (Educational psychology). - Includes bibliographies and index
0-12-068580-9 X.520/27495
0180009

WORNER, Roger B.
Student diagnosis, placement, and prescription. - Bloomington ; London : Indiana University Press, 1977
xiii,237p : ill. - Bibliography: p.233. - Includes index
0-253-35526-5 X.520/12273
0180009

ACADEMIC ACHIEVEMENT. Students
Assessment. Analysis

ROWNTREE, Derek
Assessing students. - London : Harper and Row, 1977
ix,269p : ill. - Bibliography: p.244-258. - Includes index
0-06-318062-6 X.529/31755
1063987

ACADEMIC ACHIEVEMENT. Students
Assessment. Criterion-referenced tests

BROWN, Sally
What do they know?. - Edinburgh : H.M.S.O., 1980
x,127p. - (Occasional papers / Great Britain. Scottish Education Department). - Bibliography: p.113-121. - Includes index
0-11-491659-4 BS.152/222(3)
1449893

POPHAM, William James
Criterion-referenced measurement. - Englewood Cliffs ; London : Prentice-Hall, 1978
xii,260p : ill. - Includes bibliographies and index
0-13-193607-7 X.611/8031 Woolwich
1449893

ROID, Gale H.
A technology for test-item writing. - New York ; London : Academic Press, 1982
xii,247p : ill. - (The Educational technology series). - Includes bibliographies and index
0-12-593250-2 X.622/13069 Woolwich
1449893

ACADEMIC ACHIEVEMENT. Students
Assessment. Criterion-referenced tests - Bibliographies

TURNER, Eileen, 1939-
Criterion referenced assessment. - [Stirling] : University of Stirling, Department of Education, 1980
[4]leaves, viii,230p. - (Stirling technical reports in education ; no.1). - Includes index
0-901636-28-2 X.525/3629
2252333

ACADEMIC ACHIEVEMENT. Students
Assessment. Criterion-referenced tests - Conference proceedings
JOHNS HOPKINS UNIVERSITY NATIONAL SYMPOSIUM ON EDUCATIONAL RESEARCH (1st : 1978 : Washington, D.C.)
Criterion-referenced measurement. - Baltimore ; London : Johns Hopkins University Press, 1980
x,234,[1]p : ill. - Includes bibliographies and index
0-8018-2264-5 X.520/21334
217460x

ACADEMIC ACHIEVEMENT. Students
Assessment. Norm-referenced tests. Standard scores
CHILDS, Roy
Norm referenced testing and the standard scores. - [Windsor] : NFER, [1978]
[1],11p : ill. - Bibliography: p.10
B78-25959 X.319/18237 Woolwich
1522132

ACADEMIC ACHIEVEMENT. Students
Assessment. Techniques
MARTUZA, Victor R.
Applying norm-referenced and criterion-referenced measurement in education. - Boston, Mass. ; London : Allyn and Bacon, 1977
viii,352p : ill. - Includes bibliographies and index
0-205-05545-1 X.520/11698
1122606

SUMNER, Raymond
Criterion-referenced measurement and criterion-referenced tests. - Windsor : NFER, 1977
iii,134p. - Bibliography: p.125-127. - Includes index
0-85633-116-3 X.512/5780
131131x

ACADEMIC ACHIEVEMENT. Students
Assessment. Techniques - Conference proceedings
INTERNATIONAL SYMPOSIUM ON EDUCATIONAL TESTING (2nd : 1975 : Montreux)
Advances in psychological and educational measurement. - London : Wiley, 1976
x,320p : ill. - Includes bibliographies and index
0-471-01817-1 X.520/11291
1122606

ACADEMIC ACHIEVEMENT. Students
Assessment. Tests. Scores. Evaluation
LYMAN, Howard Burbeck
Test scores and what they mean. - 3rd ed. - Englewood Cliffs ; London : Prentice-Hall, 1978
xv,190p : ill. - Bibliography: p.166-169. - Includes index
0-13-903823-x X.519/28467
1585312

ACADEMIC ACHIEVEMENT. Students
Assessment. Tests. Standardisation. Effects
KELLAGHAN, Thomas
The effects of standardized testing. - Boston, Mass. ; London : Kluwer-Nijhoff, c1982
xiii,284p : ill. - (Evaluation in education and human services). - Bibliography: p263-273. - Includes index
0-89838-076-6 X.520/26753
2710277

ACADEMIC ACHIEVEMENT. Students
Effects of expectations of teachers
NASH, Roy
Teacher expectations and pupil learning. - London : Routledge and Kegan Paul, 1976
ix,89p. - (Students library of education). - Bibliography: p.85-89
0-7100-8338-6 X.529/30695
0651354

ACADEMIC ACHIEVEMENT. Students
Objective tests
DAVIS, Terence N.
Objective tests in theory and practice. - Réduit : Mauritius Institute of Education, 1975
121p : ill. - Bibliography:p.110-113. - Includes index
r6051167 X.619/19723 Woolwich
1887696

ACADEMIC ACHIEVEMENT. Students
Psychosocial factors
CHILD, Dennis
Affective influences on academic performance. - [Newcastle upon Tyne] : School of Education, University of Newcastle upon Tyne, 1977
22p : 1 ill. - (Inaugural lectures / University of Newcastle upon Tyne)
0-7017-0006-8 X.519/28174
149385x

ACADEMIC ACHIEVEMENT. Students
related to sex differences
KELLY, Alison, of the University of Manchester
Girls and science. - Stockholm : Almqvist & Wiksell International, 1978
ix,152p. - (IEA monograph studies ; no.9). - Includes bibliographies
r6193763 X.520/20227
0845418

ACADEMIC ACHIEVEMENT. Students
Testing - Conference proceedings
INTERNATIONAL SYMPOSIUM ON EDUCATIONAL TESTING (1973 : The Hague)
Contemporary issues in educational testing. - The Hague : Mouton, 1974
[4], 344 p. : ill. - (Psychological studies ; 9). - Includes bibliographies and index
90-279-7521-3 X.520/11493
1080741

ACADEMIC ACHIEVEMENT. Students. 10-13 years. Schools. Sweden
DRAKENBERG, Margareth
Förändringar i prestationsprocessen. - [Lund] : Gleerup, 1978
95p
91-400-4649-4 X.519/38574
1535986

ACADEMIC ACHIEVEMENT. Students. 11-12 years. Primary schools. Great Britain
Assessment. Tests: Transitional Assessment Modules
TRANSITIONAL assessment. - Windsor : NFER Publishing Co. Ltd., [1980]
20p : ill
B83-09219 X.525/6704
2837935

ACADEMIC ACHIEVEMENT. Students, 14 years. Secondary schools
Expectations of students
LITTLE, Angela
The occupational and educational expectations of students in developed and developing countries. - Brighton : Institute of Development Studies, 1978
[1], ix, 69p. - (IDS research reports / Institute of Development Studies) (Education report ; no.3). - Bibliography: p.66-69
0-903354-64-0 X.622/9110 Woolwich
1549839
- [Another copy] X.525/9055

ACADEMIC ACHIEVEMENT. Students, 16-19 years. Educational institutions. England
In French language. Assessment - For teaching
FRENCH 16-19 STUDY GROUP
French 16-19. - London : Hodder and Stoughton, c1981
v,101p : ill
0-340-25528-5 X.950/4379
2259508

ACADEMIC ACHIEVEMENT. Students, 18 years-. Secondary schools. Great Britain
In G.C.E. (A level) examinations
MILES, H. B.
Some factors affecting attainment at 18+. - Oxford : Pergamon, 1979
xix,137p,leaf : ill. - Bibliography: p.134-137
0-08-024678-8 X.529/33735
1804081

ACADEMIC ACHIEVEMENT. Students awaiting transfer. Schools. England
Assessment
SUMNER, Raymond
Assessment for transition. - Windsor : NFER, 1977
[3],iii,92p : ill. - Bibliography: p.91-92
0-85633-132-5 X.512/5779
1311328

ACADEMIC ACHIEVEMENT. Students. Co-educational secondary schools. England
compared with academic achievement of students in single-sex secondary schools
STEEDMAN, Jane
Examination results in mixed and single sex schools. - Manchester : Equal Opportunities Commission, 1983
37p
B85-08041 BS.433/63
3227081

ACADEMIC ACHIEVEMENT. Students. Courses. Adult literacy education. Great Britain
A SURVEY of attainment and progress of students in adult literacy schemes / ALBSU. - London : Adult Literacy and Basic Skills Unit, [1983?]
11p
B84-06636 X.529/58451
3036782

ACADEMIC ACHIEVEMENT. Students. Diploma in Management Studies courses. Higher education institutions. England
Assessment
POLYTECHNIC OF CENTRAL LONDON. School of Management Studies. Research Unit
An analysis of the assessment methods used by colleges offering the Diploma in Management Studies. - [London] : [The Research Unit], [1976]
[1],30 leaves
0-905410-03-3 X.512/5166
0637343

ACADEMIC ACHIEVEMENT. Students. Educational institutions
In chemistry. Assessment. Techniques - Conference proceedings
RESEARCH in assessment / edited by R. Kempa, D.E. Hoare, R.J.D. Rutherford. - London : Education Division, Chemical Society, [1976]
[7],94p : ill. - Includes bibliographies
0-85186-739-1 X.619/17339 Woolwich
1040693

ACADEMIC ACHIEVEMENT. Students. Educational institutions
In medicine. Assessment. Multiple-choice tests
ANDERSON, John, 1921-
The multiple choice question in medicine. - Tunbridge Wells : Pitman Medical, 1976
viii,152p. - Bibliography: p.57-58. - Includes index
0-272-79376-0 X.319/17305 Woolwich
1112376

The multiple choice question in medicine. - 2nd ed. - London : Pitman, 1982
xii,216p. - Bibliography: p85-88. - Includes index
0-272-79642-5 X.329/15281 Woolwich
1112376

ACADEMIC ACHIEVEMENT. Students. Educational institutions
In modern languages. Testing
CURRENT developments in language testing / edited by Arthur Hughes and Don Porter. - London : Academic, 1983
x,167p. - (Applied language studies). - Bibliography: p154-162. - Includes index
0-12-360880-5 X.950/34529
1256041

FOREIGN language testing. - 2nd ed.. - London : Centre for Information on Language Teaching and Research, 1981
108p. - (Specialised bibliography ; 1)
0-903466-32-5 X.950/17897
169958x

VALETTE, Rebecca M.
Modern language testing. - 2nd ed. - New York : Harcourt Brace Jovanovich, 1977
xiv,349p : ill. - Bibliography : p.335-336.-Includes index
0-15-561926-8 X.900/25454
1256041

ACADEMIC ACHIEVEMENT. Students. Educational institutions
In modern languages. Testing - Bibliographies
CENTRE FOR INFORMATION ON LANGUAGE TEACHING AND RESEARCH
Foreign language testing. - [New ed.]. - [London] : [C.I.L.T.], 1977
71p. - (Specialised bibliographies / Centre for Information on Language Teaching and Research). - Includes index
0-903466-16-3 X.909/42964
169958x

ACADEMIC ACHIEVEMENT. Students. Europe
Assessment
INGENKAMP, Karlheinz
Educational assessment. - Windsor : NFER [for the Council of Europe, Committee for Educational Research], 1977
141p. - (European trend reports on educational research). - Bibliography: p.114-135. - Includes index
0-85633-130-9 X.510/15667
137141x

ACADEMIC ACHIEVEMENT. Students from social minorities. Schools. United States
Assessment
ASSESSING minority group children / edited by Beeman N. Phillips. - New York : Behavioral Publications ; London : Distributed by Eurospan Ltd, [1978]
[4]p,p291-413. - Includes bibliographies and index
0-87705-154-2 X.520/13547
1821342

ACADEMIC ACHIEVEMENT. Students. Further education institutions
In business studies. Assessment
MACDOUGALL, Margaret J.
Handbook on testing and evaluation in business education. - Toronto : Copp Clark Pitman, c1983
203p : ill. - Bibliography: p201-203
0-7730-4320-9 X.520/34265
3176010

ACADEMIC ACHIEVEMENT. Students. Further education institutions. Great Britain
Assessment – For teaching

WARD, Christine
Designing a scheme of assessment. – Cheltenham : Thornes, 1980
vii,159p : ill. – (ST(P) handbooks for further education). – Bibliography: p153. – Includes index
0–85950–428–x X.529/39002
2104652

ACADEMIC ACHIEVEMENT. Students. Higher education institutions
Assessment

CLIFT, John C.
Assessing students, appraising teaching. – London : Croom Helm, c1981
176p : ill. – (New patterns of learning series). – Bibliography: p162–170. – Includes index
0–7099–0230–1 X.529/39802
1100165

HEYWOOD, John, 1930–
Assessment in higher education. – London : Wiley, 1977
xiii,289p : ill. – Bibliography: p.258–284. – Includes index
0–471–99404–9 X.520/11580
1100165

ACADEMIC ACHIEVEMENT. Students. Higher education institutions
Assessment. Tests

SHERMAN, Thomas M.
Proven strategies for successful test taking. – Columbus, Ohio ; London : Merrill, c1982
xiii,138p. – Includes index
0–675–09843–2 X.529/63017
3143309

ACADEMIC ACHIEVEMENT. Students. Higher education institutions. Great Britain

ENTWISTLE, Noel James
Degrees of excellence. – London : Hodder and Stoughton, 1977
ix,226p : ill. – Bibliography: p.169–177. – Includes index
0–340–20901–1 X.520/11829
1171232

ACADEMIC ACHIEVEMENT. Students. Higher education institutions. Great Britain
Assessment

KLUG, Brian
The grading game. – [2nd ed.]. – London : NUS Publications, 1977
94p : ill. – Bibliography: p.83–86. – Includes index
0–900544–63–5 X.529/65324
0155624

ACADEMIC ACHIEVEMENT. Students. Higher education institutions. Ireland (Republic)
Assessment – Conference proceedings

STUDENT ASSESSMENT IN THIRD LEVEL EDUCATION (Conference : 1975 : Dublin)
Student assessment in third level education. – [Dublin] : N.C.E.A., [1977]
83p : ill. – (Publications / National Council for Educational Awards ; 4). – Includes bibliographies
B77–21061 X.519/28612
1144219

ACADEMIC ACHIEVEMENT. Students. Polytechnic of North London
In sociology. Assessment

DEGREES of choice?. – London : Polytechnic of North London Students' Union, 1978
45p : ill
0–906513–00–6 X.909/44511
2054469

ACADEMIC ACHIEVEMENT. Students. Primary schools
Assessment related to language skills

LANGUAGE use and school performance. – New York ; London : Academic Press, 1974
ix,368p : ill. – Bibliography: p.352–365. – Includes index
0–12–174950–9 X.981/20044
0379123

ACADEMIC ACHIEVEMENT. Students. Primary schools
Assessment. Techniques

CARTWRIGHT, Carol A.
Developing observation skills. – 2nd ed. – New York ; London : McGraw-Hill, c1984
xii,163p : ill. – (McGraw-Hill series in special education). – Bibliography: p153–159. – Includes index
0–07–010185–x X.520/36427
1138022

SPOONCER, Frank
Testing for teaching. – London : Hodder and Stoughton, c1983
148p : ill. – Includes bibliographies
0–340–27689–4 X.520/35348
1138022

ACADEMIC ACHIEVEMENT. Students. Primary schools. Dumfries and Galloway Region. Scotland
In mathematics

TURNBULL, G. H.
Survey of attainment in reading and mathematics in primary schools in the Dumfries and Galloway region. – Ayr : Craigie College of Education, 1980
42leaves
B83–14686 X.525/6841
2867052

ACADEMIC ACHIEVEMENT. Students. Primary schools. Dumfries and Galloway Region. Scotland
In reading

TURNBULL, G. H.
Survey of attainment in reading and mathematics in primary schools in the Dumfries and Galloway region. – Ayr : Craigie College of Education, 1980
42leaves
B83–14686 X.525/6841
2867044

ACADEMIC ACHIEVEMENT. Students. Primary schools. England
Effects of teaching methods

BENNETT, Neville
Teaching styles and pupil progress. – London : Open Books Publishing Ltd, 1976
xiii,201p : ill. – Bibliography: p.189–197. – Includes index
0–7291–0098–7 X.529/30677
0647640

KITWOOD, Tom
Mind that child!. – London : Writers and Readers Publishing Cooperative, 1977
54p
0–904613–44–5 X.519/30574
2022052

PROGRESS and performance in the primary classroom / edited by Maurice Galton and Brian Simon. – London : Routledge & Kegan Paul, 1980
xiii,257p : ill. – Bibliography: p246–250. – Includes index
0–7100–0669–1 X.529/38912
2022052

ACADEMIC ACHIEVEMENT. Students. Primary schools. England
In English language. Assessment by Great Britain. Assessment of Performance Unit

ROSEN, Harold
The language monitors. – London : Institute of Education, University of London, 1982
51p. – (Bedford Way papers ; 11). – Bibliography: p50–51
0–85473–134–2 X.529/53564
2886499

ACADEMIC ACHIEVEMENT. Students. Primary schools. England
In mathematics. Assessment

TARGET attainments. – [Hertford] : [Hertfordshire County Council], 1977
5leaves. – (Hertfordshire guidelines in mathematics, supplementary booklet)
B84–03689 X.625/852
2819449

ACADEMIC ACHIEVEMENT. Students. Primary schools. Ireland (Republic)
In spoken Irish language. Assessment

HARRIS, John, 19—
Spoken Irish in primary schools. – Dublin : Institiúid Teangeolaíochta Éireann, c1984
xiv,153p. – Bibliography: p145–148
0–946452–02–4 X.950/38409
3223779

ACADEMIC ACHIEVEMENT. Students. Primary schools. United States

ENTWISLE, Doris R.
Early schooling. – Baltimore ; London : The Johns Hopkins University Press, c1982
xv,215p : ill. – Bibliography: p204–211. – Includes index
0–8018–2761–2 X.525/7043
2882019

ACADEMIC ACHIEVEMENT. Students. Primary schools. United States
Effects of socioeconomic expectations

ENTWISLE, Doris R.
Too great expectations. – Baltimore ; London : Johns Hopkins University Press, 1978
xiv,193,[1]p : ill. – Bibliography: p.187–190. – Includes index
0–8018–1986–5 X.520/12822
1572210

ACADEMIC ACHIEVEMENT. Students. Schools

MATTHEWS, Kenneth M.
Managing classroom learning. – Lanham ; London : University Press of America, c1984
viii,69p : ill
0–8191–3620–4 X.529/65032
3184447

ACADEMIC ACHIEVEMENT. Students. Schools
Assessment

BLOOM, Benjamin S. (Benjamin Samuel)
Evaluation to improve learning. – New York ; London McGraw-Hill, c1981
ix,356p : ill. – Includes bibliographies and index
0–07–006109–2 X.520/24960
2554755
– Reprinted – 1983 X.520/30967

ISSUES in evaluation and accountability / edited by Colin Lacey and Denis Lawton. – London : Methuen, 1981
x,248p. – Includes bibliographies and index
0–416–74740–x X.529/46866
2554755

SHIPMAN, M. D. (Marten Dorrington)
Assessment in primary and middle schools. – London : Croom Helm, c1983
xiii,168p : ill. – (Croom Helm teaching 5–13 series). – Bibliography: p160–165. – Includes index
0–7099–2344–9 X.529/59697
2554755

TESTS in education / compiled by Philip Levy, Harvey Goldstein. – London : Academic Press, 1984
xxix,718p. – Includes bibliographies and index
0–12–445880–7 X.520/36346
2554755

ACADEMIC ACHIEVEMENT. Students. Schools
Assessment – For teaching

BLACK, Harry
Keeping track of teaching. – London : Routledge & Kegan Paul, 1982
x,154p. – (Routledge education books). – Includes index
0–7100–9017–x X.529/50199
2791595

TESTING children / Caroline Gipps ... [et al.]. – London : Heinmann Educational, 1983
188p. – Bibliography: p166–169. – Includes index
0–435–82323–x X.529/61984
2791595

ACADEMIC ACHIEVEMENT. Students. Schools
Assessment. Comparative studies: International Association for the Evaluation of Academic Achievement. Six subject survey. Mathematics

PEAKER, Gilbert Fawcett
An empirical study of education in twenty-one countries. – Stockholm : Almqvist och Wiksell ; New York : London : Wiley, 1975
230p : ill. – (International studies in evaluation ; no.8). – Bibliography: p.219–220
0–470–67456–3 X.519/25630
0616850

ACADEMIC ACHIEVEMENT. Students. Schools
Assessment. Techniques

PEDDIE, Bill
Testing in practice. – Auckland : Heinemann Educational, 1978
iv,67p : ill. – Bibliography : p.66–67
0–86863–258–9 X.519/29309
0679283

SCANNELL, Dale Paul
Testing and measurement in the classroom. – Boston [Mass.] ; London : Houghton Mifflin, 1975
xi,288p : ill. – Includes bibliographies and index
0–395–18608–0 X.519/25918
0679283

ACADEMIC ACHIEVEMENT. Students. Schools
Effects of self-image

COVINGTON, Martin
Self-worth and school learning. – New York ; London : Holt, Rinehart and Winston, [1977]
ix,166p : ill. – (The principles of educational psychology series). – Bibliography: p.149–162. – Includes index
0–03–015286–0 X.519/26705
1079492

ACADEMIC ACHIEVEMENT. Students. Schools
Grading. Applications of Apple II microcomputer systems

BLANKENSHIP, John
The gradebook system. – Englewood Cliffs ; London : Prentice-Hall, c1984
v,68p : ill
0–13–362526–5 X.622/23163
3277623

ACADEMIC ACHIEVEMENT. Students. Schools
In mathematics. Assessment

BENTLEY, Charles
Mathematics. - London : Macmillan Education, 1983
vii,94p : ill. - (Guides to assessment in education). - Bibliography: p89-93. - Includes index
0-333-31912-5 X.629/21443
2967219

ACADEMIC ACHIEVEMENT. Students. Schools
In mathematics. Effects of social class of students & educational resources of schools

NOONAN, Richard D.
School resources, social class and student achievement. - Stockholm : Almqvist and Wiksell International ; New York ; London : Wiley, 1976
xiv,127p : ill. - (Monograph studies / International Association for the Evaluation of Educational Achievement ; no.5). - Bibliography: p.122-127
0-470-15091-2 X.519/26231
0750441

ACADEMIC ACHIEVEMENT. Students. Schools
In reading. Assessment

The MACMILLAN group reading test / Macmillan Test Unit.
23p : ill
0-333-39458-5 X.955/3046
2967189

RABAN, Bridie
Reading. - London : Macmillan Education, 1983
vii,63p : ill. - (Guides to assessment in education). - Bibliography: p61-62. - Includes index
0-333-31913-3 X.529/57224
2967189

ACADEMIC ACHIEVEMENT. Students. Schools
In science. Assessment

HARLEN, Wynne
Science. - London : Macmillan Education, 1983
viii,82p : ill. - (Guides to assessment in education). - Bibliography: p80-81. - Includes index
0-333-31916-8 X.629/21444
2967200

ACADEMIC ACHIEVEMENT. Students. Schools
Influence of home environment. Research

MARJORIBANKS, Kevin
Families and their learning environments. - London : Routledge and Kegan Paul, 1979
xv,255p : ill. - Bibliography: p.231-247. - Includes index
0-7100-0167-3 X.529/34078
1826549

ACADEMIC ACHIEVEMENT. Students. Schools
Monitoring - Conference proceedings

EUROPEAN CONTACT WORKSHOP (1st : 1976 : Windsor)
Monitoring national standards of attainment in schools. - Windsor : NFER, 1977
274p : ill. - Includes bibliographies
0-85633-144-9 X.512/5807
134465x

ACADEMIC ACHIEVEMENT. Students. Schools
related to bilingualism

CUMMINS, Jim, 19—
Bilingualism and special education. - Clevedon : Multilingual Matters Ltd., c1984
ix,306p. - (Multilingual matters ; 6). - Bibliography: p276-297. - Includes index
0-905028-14-7 X.950/36511
0851507

LANGUAGE proficiency and academic achievement / edited by Charlene Rivera. - Clevedon : Multilingual Matters, c1984
xiv,80p. - (Multilingual matters ; 10). - Includes bibliographies and index
0-905028-24-4 X.529/64719
0851507
- [Another copy: paperback] ISBN 0-905028-23-6 X.529/64720

SKUTNABB-KANGAS, Tove
Om tvasprakighet och skolframgang. - Abo : Svenska Litteratursällskapets i Finland Nämnd för Samhällsforskning, 1975
120 leaves. - (Forskningsrapporter ; no. 20). - Bibliography p.87-120
r6000655 P.703/129(20)
0851507

ACADEMIC ACHIEVEMENT. Students. Schools
Testing

HOPKINS, Kenneth D.
Educational and psychological measurement and evaluation. - 6th ed. - Englewood Cliffs ; London : Prentice-Hall, c1981
xxi,517p : ill. - (Prentice-Hall series in educational measurement, research and statistics). - Bibliography: p472-505. - Includes index
0-13-236273-2 X.520/24269
2505274

ACADEMIC ACHIEVEMENT. Students. Schools
Testing - For teaching

BROWN, Frederick (Frederick Gramm)
Measuring classroom achievement. - New York ; London : Holt, Rinehart and Winston, c1981
x,224p : ill. - Bibliography: p214-216. - Includes index
0-03-052421-0 X.520/23383
1054821

TUCKMAN, Bruce W.
Measuring educational outcomes. - New York : Harcourt Brace Jovanovich, 1975
xix, 527 p. : ill. - Bibliography : p. 508-519. - Includes index
0-15-557692-5 X.520/11409
1054821
- [Another copy : paperback] X.510/15394

ACADEMIC ACHIEVEMENT. Students. Schools. Denmark
Assessment

PRØVER! Karakterer! Pædagogik? / redigeret af Jørgen Gregersen. - Copenhagen : Gyldendal, 1977
120p. - (Gyldendals pædagogiske bibliotek)
87-01-54081-5 X.510/10192
1312634

ACADEMIC ACHIEVEMENT. Students. Schools. England
Assessment

EXPOSURE and performance. - [London] : Assessment of Performance Unit, c1983
28p
B84-23021 X.525/9021
3087174

ACADEMIC ACHIEVEMENT. Students. Schools. England
In mathematics. Testing

KYLES, I.
Tests of attainment in mathematics in schools, continuation of monitoring feasibility study. - Windsor : NFER, 1977
[2],75p : ill
0-85633-135-x X.615/1914 Woolwich
0605204

SUMNER, R. (Raymond)
Tests of attainment in mathematics in schools. - London : National Foundation for Educational Research in England and Wales, 1975
vii,113,[106]p : ill. - (Occasional reports : second series / National Foundation for Educational Research in England and Wales)
0-85633-084-1 X.615/1584 Woolwich
0605204

ACADEMIC ACHIEVEMENT. Students. Schools. England
In science. Assessment

SCIENCE in schools. - London : Department of Education and Science.
Age 11, report no.3. - c1984
v.347p : ill. - Bibliography: p301-302
B85-03649 X.622/21462
3214540

ACADEMIC ACHIEVEMENT. Students. Schools. England
Standards

BOGDANOR, Vernon
Standards in schools. - Esher : National Council for Educational Standards, [1979?]
v,22p. - (Kay-Shuttleworth papers on education ; no. 1)
0-9506883-0-4 X.529/37873
2280086

ACADEMIC ACHIEVEMENT. Students. Schools. Great Britain
In mathematics. Tests

TESTS. - Leicester : Mathematical Association, c1979
42p
B85-23870 X.622/23893
3285200

ACADEMIC ACHIEVEMENT. Students. Schools. Great Britain
In mathematics. Tests: Profile of Mathematical Skills - For teaching

FRANCE, Norman
Profile of mathematical skills. - Sunbury-on-Thames : Nelson, 1979
31p
0-17-420193-1 X.622/11228 Woolwich
2563215

ACADEMIC ACHIEVEMENT. Students. Schools. Great Britain
In science. Assessment

MURPHY, Patricia
Science. - [London?] : Assessment of Performance Unit, c1984
48p : ill. - (Science report for teachers ; 2)
0-86357-008-9 X.629/25123
319051x

Science at age 13. - [Stanmore] : DES, c1984
44p : ill. - (Science report for teachers ; 3)
0-86357-009-7 X.629/24330
319051x

ACADEMIC ACHIEVEMENT. Students. Schools. Hungary
Assessment, 1945-

VESZPRÉMI, László
Az értékelés és osztályozás korszerüsítésének kérdései. - Budapest : Akadémiai Kiadó, 1981
155p : ill. - Bibliography
963-05-2353-1 X.520/25469
8766649

ACADEMIC ACHIEVEMENT. Students. Schools. Indonesia
to date

MANI, A.
Determinants of educational aspirations among Indonesian youth. - Hong Kong : Maruzen Asia, 1983
vii,51p. - (Occasional paper / Institute of Southeast Asian Studies ; no.71)
962-220-125-3 X.0702/265(71)
9142584

ACADEMIC ACHIEVEMENT. Students. Schools. Scotland
Assessment. Methods: Scottish Council for Research in Education Profile Assessment System

SCOTTISH COUNCIL FOR RESEARCH IN EDUCATION
SCRE profile assessment system. - [Edinburgh] : The Council, 1977
[2],34p
0-901116-10-6 X.519/27042
1167650

ACADEMIC ACHIEVEMENT. Students. Schools. Slovenia. Yugoslavia
related to social background

TOLIČIČ, Ivan
Okolje in uspešnost učencev. - Ljubljana : Državna založba Slovenije, 1977
359p. - Bibliography : p.338-340
r6178007 X.529/35494
2219425

ACADEMIC ACHIEVEMENT. Students. Schools. United States
to date

COPPERMAN, Paul
The literary hoax. - New York, 1978
323p.. - Bibliography
0-688-03353-9 X.520/24463
8785236

GARBARINO, James
Successful schools and competent students. - Lexington, Mass. : Lexington Books, c1981
iv,170p. - Includes index
0-669-04526-8 X.520/26279
8785236

ACADEMIC ACHIEVEMENT. Students. Schools. United States
Assessment. Minimum competency tests

LAZARUS, Mitchell
Goodbye to excellence. - Epping : Bowker, c1981
xv,206p : Iill. - (NAESP studies in education and public policy). - Bibliography: p197-200. - Includes index
0-89158-771-3 X.520/31275
2956683

ACADEMIC ACHIEVEMENT. Students. Schools. United States
Assessment. Minimum competency tests. Validity. Law

The COURTS, validity, and minimum competency testing / George F. Madaus. - Boston ; London : Kluwer-Nijhoff, c1983
xi,289p : ill. - Includes bibliographies
0-89838-113-4 X.520/29290
2848104

ACADEMIC ACHIEVEMENT. Students. Secondary schools
Assessment. Techniques

MCGUIRE, James
An outline of assessment methods in secondary education in selected countries. - Edinburgh : Scottish Council for Research in Education, [1976]
[1],11p. - (Educational issues review ; no.2). - Bibliography: p.11
0-901116-07-6 X.512/5192
0655449

ACADEMIC ACHIEVEMENT. Students. Secondary schools
In crafts. Assessment – For teaching

BLACK, H. D. (Harold David)
Diagnostic assessment in technical education. – [Edinburgh] : Scottish Council for Research in Education, c1980
x,75p : ill
0–901116–25–4 X.529/42972
2539721

ACADEMIC ACHIEVEMENT. Students. Secondary schools
In economics. Assessment. Multiple–choice tests

TEST DEVELOPMENT AND RESEARCH UNIT
The 1974 inter board multiple choice economics test, report and teachers' guide. – Cambridge : The Unit, 1974
[2],41[i.e. 43]p : ill. – (Occasional publications / Test Development and Research Unit ; no.1)
0–905631–00–5 X.0515/305(1)
0739685

Inter-Board multiple choice economics 1976. – Cambridge : The Unit, 1977
66p : ill. – (Occasional publications / Test Development and Research Unit ; no.6)
B78–02040 X.0515/305(6)
0739685

ACADEMIC ACHIEVEMENT. Students. Secondary schools
In English language. Assessment – For teaching

CHATER, Pauline
Marking and assessment in English. – London : Methuen, 1984
xiii,213p : ill. – (Teaching secondary English). – Bibliography: p198–207
0–416–33180–7 X.950/39437
3112640
– [Another copy : paperback] ISBN 0–416–33190–4 X.950/39438

ACADEMIC ACHIEVEMENT. Students. Secondary schools
In geography. Assessment – For teaching

BLACK, H. D. (Harold David)
Diagnostic assessment in geography. – [Edinburgh] : Scottish Council for Research in Education, c1980
x,80p : ill
0–901116–24–6 X.529/42970
253973x

ACADEMIC ACHIEVEMENT. Students. Secondary schools
In home economics. Assessment – For teaching

BLACK, H. D. (Harold David)
Diagnostic assessment in home economics. – [Edinburgh] : Scotland Council for Research in Education, c1981
x,76p : ill
0–901116–27–0 X.529/42971
2539713

ACADEMIC ACHIEVEMENT. Students. Secondary schools. England

STEEDMAN, Jane
Progress in secondary schools. – London : National Children's Bureau, 1980
[6],vi,265p : ill. – Bibliography: p.265
0–902817–17–5 X.525/3780
2286440

ACADEMIC ACHIEVEMENT. Students. Secondary schools. England
In biology. Assessment – For teaching

SANDS, M. K. (Margaret Kathleen)
Practical biology. – London : Bell & Hyman, 1984
iv,124p : ill. – (Modern teaching series)
0–7135–1405–1 X.329/19686
3300226

ACADEMIC ACHIEVEMENT. Students. Secondary schools. England
In examinations compared with academic achievement in examinations of secondary school students in Wales

GREAT BRITAIN. Welsh Education Office
Examination achievement in Wales. – [Cardiff] : [The Office], 1978
[1],8p
B78–37616 BS.442/7
1706683

ACADEMIC ACHIEVEMENT. Students. Secondary schools. England
In mathematics. Assessment

MATHEMATICAL development / by D.D. Foxman ... [et al.]. – London : H.M.S.O., 1980
xiii,131p : ill. – Includes index
Subsequent surveys are also placed at this shelfmark
0–11–270516–2 BS.10/927
2700360

ACADEMIC ACHIEVEMENT. Students. Secondary schools. England
related to choice of questions in G.C.E. (A level) examinations in English literature

FAIN, Malcolm Thomas
An investigation into the effect of question–choice in Advanced Level English. – Aldershot : Associated Examining Board for General Certificate of Education, 1975
iv,39p : ill
0–901893–04–8 X.909/40152
052428x

ACADEMIC ACHIEVEMENT. Students. Secondary schools. France
related to social class, 1600–1800

FRIJHOFF, Willem
École et société dans la France d'Ancien Régime. – Paris : Librairie Armand Colin, 1975
118 p.. – (Cahiers des Annales ; 35)
r6012074 X.510/9423
0940453

ACADEMIC ACHIEVEMENT. Students. Secondary schools. Great Britain
Assessment

ENGEL CLOUGH, Elizabeth
Assessing pupils. – Windsor : NFER-Nelson, 1984
xi,148p : ill. – Bibliography: p148
0–7005–0664–0 X.529/70139
3138925

ACADEMIC ACHIEVEMENT. Students. Secondary schools. Great Britain
Assessment – Conference proceedings

ASSESSMENT and the curriculum / edited by R.W. Fairbrother. – London : Chelsea College, University of London, c1980
vi,202p
B84–30672 X.529/62730
3138925

ACADEMIC ACHIEVEMENT. Students. Secondary schools. Great Britain
Assessment – For teaching

MACINTOSH, Henry Gordon
Assessment and the secondary school teacher. – London : Routledge and Kegan Paul, 1976
xiii,131p : ill. – (Students library of education). – Bibliography: p.129–131
0–7100–8472–2 X.529/31166
079371x

ACADEMIC ACHIEVEMENT. Students. Secondary schools. Great Britain
In G.C.E. (A level) examinations related to academic achievement at university

WANKOWSKI, J.A. (Janislaw Adam)
G.C.E.'s and degrees. – 3rd ed. [i.e. 1st ed. reprinted]. – Birmingham : University of Birmingham, Educational Survey, 1978
[2],iv,28,ii,37 leaves. – Bibliography: leaf 28
0–7044–0294–7 X.525/3716
2234831

ACADEMIC ACHIEVEMENT. Students. Secondary schools. Great Britain
related to results of applications for admission to higher education institutions – Serials

DEGREE course offers. – 1973– / edited by Brian W. Heap. – London : Career Consultants, 1973–
r6000258 P.521/3006
0055565

ACADEMIC ACHIEVEMENT. Students. Secondary schools. Inner London
In C.S.E. examinations & G.C.E. (O level) examinations, 1980. Sex differences

MARTINI, Richard
Sex differences and achievement. – [London] : Inner London Education Authority, 1982
18p. – (Report / ILEA Research and Statistics ; RS823/82)
B84–05342 X.525/8483
299609x

ACADEMIC ACHIEVEMENT. Students. Secondary schools. Ireland (Republic)
In mathematics. Assessment

HEYWOOD, John, 1930–
Assessment in mathematics, twelve to fifteen. – Dublin : School of Education, Trinity College, Dublin, 1976
[3],vii,88p : ill. – (Reports / Public Examinations Evaluation Project ; no.2). – Bibliography: p.80–88
B78–12740 X.615/2205
1398997

ACADEMIC ACHIEVEMENT. Students. Secondary schools. Northern Ireland
In mathematics. Assessment

WELLS, Ian F.
School mathematics in perspective. – [Belfast] : Northern Ireland Council for Educational Research, c1982
xii,73p : ill. – (Publications of the Northern Ireland Council for Educational Research ; 22)
0–903478–21–8 X.622/21295
3197093

ACADEMIC ACHIEVEMENT. Students. Secondary schools. Scotland
Assessment by teachers

SCOTTISH COUNCIL FOR RESEARCH IN EDUCATION
Pupils in profile. – London : Hodder and Stoughton, 1977
xii,172p : ill. – (Publications / Scottish Council for Research in Education). – Bibliography: p.169–170. – Includes index
0–340–22059–7 WP.338/67
1122533

ACADEMIC ACHIEVEMENT. Students. Secondary schools. Scotland
In S.C.E. (O grade) examinations. Effects of non–intellectual factors

MACIVER, Luis
A study of non-intellectual factors affecting pupils O–Grade performance. – [Dundee] : Dundee College of Education, 1974
90p,fold leaf : ill
0–903765–03–9 X.512/5015
0350362

ACADEMIC ACHIEVEMENT. Students. Secondary schools. Sweden
Assessment

SWEDEN. Utbildningsdepartementet
Betyg 76. – Stockholm : Liber, 1976
272p. – (Ds. U. 1976. 4)
91–380–2949–9 X.510/9940
0393894

ACADEMIC ACHIEVEMENT. Students. Secondary schools. United States
Effects of social class of students

REHBERG, Richard A.
Class and merit in the American high school. – New York ; London : Longman, 1978
xiii,337p : ill. – (Educational policy, planning and theory). – Includes index
0–582–28013–3 X.510/15857
1575465

ACADEMIC ACHIEVEMENT. Students. Secondary schools. Wales
Assessment, to date

LOOSMORE, F. A.
Curriculum and assessment in Wales. – Cardiff : Schools Council Committee for Wales, 1981
vii,103p
r6806897 X.525/7056
8944024

ACADEMIC ACHIEVEMENT. Students. Secondary schools. Wales
In examinations compared with academic achievement in examinations of secondary school students in England

GREAT BRITAIN. Welsh Education Office
Examination achievement in Wales. – [Cardiff] : [The Office], 1978
[1],8p
B78–37616 BS.442/7
1706675

ACADEMIC ACHIEVEMENT. Students. Secondary schools. Wales
In mathematics. Assessment

MATHEMATICAL development / D.D. Foxman ... [et al.]. – Cardiff : H.M.S.O., 1980
iii,63p : ill
Subsequent surveys are also placed at this shelfmark
0–11–790150–4 BS.10/928
2700352

ACADEMIC ACHIEVEMENT. Students. Secondary schools. West Germany
Assessment, 1945–

NAUCK, Joachim
Erfolgskontrollen in der Orientierungsstufe. – [Hildesheim] : [J. Nauck], [1981?]
209p.. – Bibliography
r6303723 X.529/46029
8926530

ACADEMIC ACHIEVEMENT. Students. United States
Assessment – Serials

NEW directions for testing and measurement. – No.1– ; 1979–. – San Francisco ; London : Jossey-Bass, 1979–
B80–28171 P.521/3463
2103605

ACADEMIC ACHIEVEMENT Students Universities Austria
Effects of social class. 1945-

SCHMIERER, Christian
Soziale Bestimmungsgrunde des Studienerfolges. - Wien : VWGÖ, 1976
184p. - (Dissertationen der Johannes Kepler-Universitat Linz ; 5). - Bibliography
r6301493 X.529/46030
8816670

ACADEMIC ACHIEVEMENT Students Universities Great Britain related to academic achievement in G C E (A level) examinations

WANKOWSKI, J A (Janislaw Adam)
G C E s and degrees. - 3rd ed [i e 1st ed reprinted] - Birmingham : University of Birmingham. Educational Survey, 1978
[2],iv,28,ii,37 leaves. - Bibliography: leaf 28
0-7044-0294-7 X.525/3716
223484x

ACADEMIC ACHIEVEMENT Students. University of Birmingham related to personality

HOLDER, Roger
Personality & academic performance of students at university. - Guildford : Society for Research into Higher Education, 1980
[9],103p : ill. - (Research into higher education monographs). - Includes bibliographies
0-900868-72-4 X.529/37437
0027553

ACADEMIC ACHIEVEMENT. Undergraduates. Great Britain related to motivation

WANKOWSKI, J. A. (Janislaw Adam)
Random sample analysis. - 2nd ed. - Birmingham : University of Birmingham Educational Survey & Counselling Unit, 1978
24,40leaves. - Bibliography: leaf 24
0-7044-0327-7 X.525/4054
2367947

ACADEMIC ACHIEVEMENT. West Indian students. Schools. Great Britain

GREAT BRITAIN. Committee of Inquiry into the Education of Children from Ethnic Minority Groups
West Indian children in our schools. - London : H.M.S.O., [1981]
v,119p : ill. - (Cmnd. ; 8273)
0-10-182730-x BS.10/916
2534614

ACADEMIC ACHIEVEMENT. West Indian students. Schools. Great Britain
Effects of English-Creole languages

EDWARDS, Viv
West Indian language. - Revised ed. - Derby : National Association for Multi-racial Education, 1977
[2],14p. - (NAME handbooks / National Association for Multi-racial Education)
B78-18181 X.909/42596
1044494

The West Indian language in British schools. - London : Routledge and Kegan Paul, 1979
vii,168p. - (Routledge education books). - Bibliography: p.155-163. - Includes index
0-7100-0172-x X.989/53275
1044494

ACADEMIC ACHIEVEMENT. West Indian students. Schools. Great Britain
Theories of Great Britain. Parliament. House of Commons. Select Committee on Race Relations and Immigration. Criticism. Great Britain. Department of Education and Science. Consultative document

GREAT BRITAIN. Commission for Racial Equality
Education of ethnic minorities. - London : The Commission, 1977
[2],6p. - (Occasional paper ; no.1)
B78-38993 BS.400/30(1)
1728032

ACADEMIC ACHIEVEMENT. West Indian students. Schools. Redbridge (London Borough)

BLACK PEOPLES PROGRESSIVE ASSOCIATION
Cause for concern. - Ilford : The Association ; [Ilford] : The Council, 1978
[3],21p
B78-40121 X.512/6186
1732676

ACADEMIC ACHIEVEMENT. Women
In mathematics

WOMEN AND MATHEMATICS (Conference : 1976 : Boston)
Women and the mathematical mystique. - Baltimore ; London : Johns Hopkins University Press, c1980
viii,211p : ill. - Includes bibliographies and index
0-8018-2341-2 X.520/22892
233495x

ACADEMIC BOOKS. Non-fiction
Books with British imprints. Authors. Relations with publishers

MANN, Peter H. (Peter Henry)
Author-publisher relationships in scholarly publishing. - London : British Library, 1978
[4],79p. - (Research and development reports / British Library ; no.5416). - Bibliography: p.74-79
0-905984-12-9 BS.10/743
1305980

ACADEMIC BOOKS. Non-fiction
Books with British imprints. Prices

WOOD, Lawraine
Average prices of British academic books. - [Loughborough] : Centre for Library and Information Management, c1984
13p : 1ill. - (Report ; no.37)
0-904924-61-0 2708.h.101
2767058

ACADEMIC BOOKS. Non-fiction. Great Britain
Publishing. Economic aspects

The ECONOMICS of university and college textbook publishing in the United Kingdom. - [London] : University College & Professional Publishing Council, [1984?]
[8]p : ill
B85-06153 X.529/66629
3220966

ACADEMIC BOOKS. Non-fiction. United States
Marketing

BODIAN, Nat G.
Book marketing handbook. - New York ; London : Bowker, 1980
xxvii,482p : ill. - Includes index
0-8352-1286-6 2708.g.72
2336081

Book marketing handbook. - New York ; London : Bowker. Vol.2. - 1983
xxxv,607p : ill. - Includes index
0-8352-1685-3 X.955/1987
2336081

ACADEMIC BOOKS. Non-fiction. United States
Printing, publishing & marketing by university presses

ONE book/five ways. - Los Altos : William Kaufmann ; Farnham : [Distributed by] Omnibus Book Service, 1978
[10],337p : ill. - Bibliography: p.331-337
0-913232-53-x X.902/5408
1818252

ACADEMIC COSTUME. University of Oxford

VENABLES, D. R. (Dennis Roy)
Academic dress of the University of Oxford. - 4th ed. - Oxford : [The University], 1975
[2],39p : col ill
0-903690-03-9 X.629/19857
056883x

Academic dress of the University of Oxford. - 5th ed. - [Oxford] : [Oxford University Registry], 1979
38p : col.ill
B82-11920 X.529/54027
056883x
- [Another copy] X.529/54027

ACADEMIC DOCUMENTS
Publishing & dissemination

MACHLUP, Fritz
Information through the printed word. - New York ; London : Praeger. - (Praeger special studies)
Vol.1: Book publishing. - 1978
xviii,301,[1]p
0-03-047401-9 X.981/21770
1843214

Vol.2: Journals. - 1978
xxiii,338,[1]p : ill
0-03-047406-x X.520/13681
1843214

Vol.3: Libraries. - 1978
xv,201,[1]p
0-03-047411-6 2719.cc.80
1843214

ACADEMIC DOCUMENTS
Refereeing - For publishing

GORDON, Michael, 1952-
Running a refereeing system. - Leicester : Primary Communications Research Centre, University of Leicester, c1983
58p : ill. - (Aids to scholarly communication). - Bibliography: p57-58
0-906083-25-7 X.529/67921
3250512

ACADEMIC DOCUMENTS. Europe
Publishing. Innovation

UNIVERSITY OF LEICESTER. Primary Communications Research Centre
Scholarly publishers guide. new methods and techniques. - Leicester : The Centre, 1977
1v.
0-906083-00-1 X.985/1205
1743155

ACADEMIC DOCUMENTS. South-east Asia
Publishing, to date - Conference proceedings

SEMINAR ON SCHOLARLY PUBLISHING IN SOUTH-EAST ASIA (1975 Kuala Lumpur)
Scholarly publishing in South-East Asia. - Kuala Lumpur : Association of Southeast Asian Institutions of Higher Learning, 1975
240p : ill
r6814240 2708.e.503
9153640

ACADEMIC FREEDOM. Higher education. United States

REGULATING the intellectuals / edited by Craig Kaplan and Ellen Schrecker. - New York ; Eastbourne : Praeger, 1983
ix,260p. - (Praeger special studies). - Includes index
0-03-063943-3 X.520/35437
0354597

ACADEMIC FREEDOM. Higher education. United States - Conference proceedings

CONFERENCE ON THE CONCEPT OF ACADEMIC FREEDOM (1972 : University of Texas at Austin)
The concept of academic freedom. - Austin ; London : University of Texas Press, 1975
xiv,272p. - Bibliography: p.265-267. - Includes index
0-292-71016-x X.520/10068
0354597

ACADEMIC FREEDOM. Universities

HALSEY, Albert Henry
Academic freedom and the idea of a university.... - [Cape Town] : University of Cape Town, 1976
[3], 16 p.. - (T. B. Davie memorial lecture ; 17th)
0-7992-0180-4 X.519/26535
1043625

ACADEMIC FREEDOM. Universities
1945-

MACEDO, Jorge Borges de
A independência da universidade. - Lisboa : Instituto Democracia e Liberdade, 1977
p.11-36
r6704699 X.520/26572
8622663

ACADEMIC FREEDOM. Universities admitting students of all races and colours. South Africa
1957-1974

UNIVERSITY OF CAPE TOWN. Academic Freedom Committee
The open universities in South Africa and academic freedom, 1957-1974. - Cape Town : Juta, 1974
x,49p
0-7021-0588-0 X.709/32913
1803026

ACADEMIC FREEDOM. Universities. Brazil
to date

LIBERDADE acadêmica e opção totalitária / [introdução e compilação de] Antonio Ferreira Paim. - [Rio de Janeiro] : Editora Artenova, c1979
172p
r6706264 X.529/61613
9085688

ACADEMIC FREEDOM. Universities. Germany
to date

SCHREINER, Klaus
Disziplinierte Wissenschaftsfreiheit. - Tübingen : Mohr, 1981
151p. - (Contubernium ; Bd.22)
3-16-444453-2 P.521/1282[Bd.22]
8740437

ACADEMIC FREEDOM. Universities. Poland

STRUŚ, Adam
Gaudeamus. - Londyn [i.e. London] : The author, 1978
71p
B79-14618 X.519/29501
1869620

ACADEMIC FREEDOM. Universities. Sweden 1809–1832

SEGERSTEDT, Torgny T.
Den akademiska friheten 1809–1832. – Uppsala : Distributör Almqvist & Wiksell, 1976
320p. – (Skrifter rörande Uppsala universitet. C, organisation och historia ; 35) (Skrifter utgivna till Uppsala universitets 500–arsjubileum. 2, studier ; 9)
91–554–0474–x Ac.1075/6(11)[35]
0851426

ACADEMIC FREEDOM. Universities. West Germany 1945–

ZUM Beispiel Peter Brückner / herausgegeben von Alfred Krovoza, Axel R. Oestmann und Klaus Ottomeyer. – Frankfurt am Main : Europäische Verlagsanstalt, 1981
352p. – Bibliography
3–434–00430–0 X.809/49330
8737991

ACADEMIC FREEDOM. Washington State University to date

SANDERS, Jane
Cold war on the campus. – Seattle : University of Washington Press, c1979
xii,243p.;[8]p.of plates
0–295–95652–6 X.520/29899
8941270

ACADEMIC JOURNALS
Serial articles. Readability

KATZEN, May
The visual impact of scholarly journal articles. – Leicester : Primary Communications Research Centre, University of Leicester, 1977
[iii],i,99[i.e.103]p. – Bibliography: p.81–99
0–906083–02–8 X.515/593
1753509
– [Another copy] BS.10/843

ACADEMIC JOURNALS. Great Britain
Publishing. Cooperation between learned societies & publishing industries

SINGLETON, A. K. J. (Alan K. J.)
Learned societies, journals and collaboration with publishers. – Leicester : Primary Communications Research Centre, University of Leicester, c1980
vi,164p : ill. – (A Primary Communications Research publication) (BL (R&D) report ; no.5565)
0–906083–14–1 X.525/4960
2223309

ACADEMIC JOURNALS IN GERMAN
Book reviews. Authors, to 1800 – Indexes

SCHIMPF, Wolfgang
Die Rezensenten der Göttingischen Gelehrten Anzeigen 1760–1768. – Göttingen : Vandenhoeck & Ruprecht, 1982
100p. – (Arbeiten aus der Niedersächsischen Staats- und Universitätsbibliothek Göttingen ; Bd.18)
3–525–85270–3 Ac.671.fa.[Bd.18]
8975965

ACADEMIC LIBRARIES
See Libraries. Higher education institutions

ACADEMIC LIBRARIES. Denmark
Stock. Cooperative cataloguing. Machine–readable files

FORSKNINGSBIBLIOTEKERNES EDB–UDVALG
SAMKAT Juni 1978. – København : EDB–Udvalget, 1978
204p
87–87521–01–6 X.902/3596
2168901

ACADEMIC LIBRARIES. Great Britain
– Conference proceedings

GEMEINSAME Probleme von Staats- und Hochschulbibliotheken in Großbritannien und der Bundesrepublik Deutschland = Common themes in academic librarianship in the Federal Republic of Germany and in the United Kingdom. – Berlin : Deutsches Bibliotheksinstitut ; London : Standing Conference of National and University Libraries, 1981
285p. – (Dbi–materialien ; 10)
0–900210–06–0 X.809/50667
2631652

ACADEMIC LIBRARIES. Netherlands
to date

BAAS, L. J. C
Depot van publikaties, nationale bibliografie en catalogusautomatisering bij wetenschappelijke bibliotheken in Nederland. – 's–Gravenhage : Staatsuitgeverij, 1978
122p. – (Publikaties / Bibliotheekraad ; nr.2)
90–12–02382–3 2719.e.122
8756864

ACADEMIC LIBRARIES. Poland
to date

ERGETOWSKI, Ryszard
Universitätsbibliotheken und Bibliotheken der Akademie der Wissenschaften in Polen. – Frankfurt am Main : Lang, 1982
81p.. – (Symbolae Slavicae ; Bd.14). – Includes bibliographies
3–8204–5953–7 X.950/13450
8820783

ACADEMIC LIBRARIES. Soviet Union
1945–

ÉFFEKTIVNOST' informatsionno–bibliograficheskoi raboty akademicheskikh bibliotek / [glavnyi redaktor M.V. Varfolomeeva]. – Moskva : Bibliotechnyi sovet po estestvennym naukam pri Prezidiume Akademii nauk SSSR, 1981
101p. – (Biblioteki Akademii nauk SSSR i akademii nauk soiuznykh respublik ; vyp.3)
r6463187 2719.e.24/3
8890854

ACADEMIC LIBRARIES. Soviet Union
Stock, 1917–

SOVREMENNYE problemy fondov nauchnykh bibliotek SSSR / [pod redaktsiei V.V. Serova]. – Moskva : Kniga, 1975
118p : ill. – Bibliography
r6402684 2719.e.283
8781990

ACADEMIC LIBRARIES. United States

ESSAYS from the New England Academic Librarians' Writing Seminar / edited by Norman D. Stevens. – Metuchen ; London : Scarecrow, 1980
vi,224p. – Bibliography: p194–196
0–8108–1365–3 X.950/3883
2395339

ACADEMIC LIBRARIES. United States
– Conference proceedings

ASSOCIATION OF COLLEGE AND RESEARCH LIBRARIES. National Conference (1st : 1978 : Boston, Mass.)
New horizons for academic libraries. – München : London : Saur, 1979
viii,583p : ill
B83–15955 2719.e.470
2395339

ACADEMIC LIBRARIES. West Germany
1945– – Conference proceedings

DEUTSCHER BIBLIOTHEKARTAG (71st : 1981 : Regensburg)
Bestände in wissenschaftlichen Bibliotheken. – Frankfurt am Main : Klostermann, 1982
269p : ill. – (Zeitschrift für Bibliothekswesen und Bibliographie. Sonderheft ; 34)
3–465–01511–8 PP.6492.gab(3)[no.34]
8806160

ACADEMIC PERSONNEL
See also
Proctors. Universities
Senior fellows
Teachers

ACADEMIC PERSONNEL. Colleges of education. England
Trade unions: Association of Teachers in Colleges and Departments of Education, 1978

BROWNE, Joan D. (Joan Dillon)
Teachers of teachers. – London : Hodder and Stoughton [for] the National Association of Teachers in Further and Higher Education, 1979
x,277p. – Bibliography: p.268–269. – Includes index
0–340–24460–7 X.529/34355
1883895

ACADEMIC PERSONNEL. Education departments. Universities. England
Trade unions: Association of Teachers in Colleges and Departments of Education, to 1978

BROWNE, Joan D. (Joan Dillon)
Teachers of teachers. – London : Hodder and Stoughton [for] the National Association of Teachers in Further and Higher Education, 1979
x,277p. – Bibliography: p.268–269. – Includes index
0–340–24460–7 X.529/34355
1883909

ACADEMIC PERSONNEL. Higher education institutions
Professional development – Serials

ASTD newsletter / Loughborough University of Technology. – Issue no.1 (autumn 1977)–. – [Loughborough] : The University, 1977–
v. : ill
Holdings: Issue no.20 (Spring 1984)–
r6863836 P.525/634
3182282

ACADEMIC PERSONNEL. Humanities. United States
– Biographies – Dictionaries

DIRECTORY of American scholars. – 8th ed.. – New York ; London : Bowker, 1982
4v.
0–8352–1476–1 X.0525/512(1)–512(4)
2892472

ACADEMIC PERSONNEL. Learned institutions. Belorussia
Publications – Bibliographies – Serials

AKADEMIÎA NAVUK BELARUSKAÎ SSR
Bibliografiîa trudov sotrudnikov uchrezhdenii Akademii nauk BSSR. – 1970/1971–. – Minsk : Akademiîa nauk Belorusskoi SSSR, Fundamental'naîa biblioteka im ÎA. Kolasa, 1973–
r6186383 X.529/21823
2191164

ACADEMIC PERSONNEL. Learned institutions. Belorussia
Publications, 1917– – Bibliographies

BIBLIOGRAFIÎA pechatnykh rabot professorsko–prepodavatel'skogo sostava Minskogo gosudarstvennogo pedagogicheskogo instituta inostrannykh îazykov. – Minsk : Minskii gosudarstvennyi pedagogicheskii institut inostrannykh îazykov, 1973–
v
r6403932 X.0529/989
8027544

ACADEMIC PERSONNEL. Learned institutions. Bonn. West Germany
Publications, 1945– – Directories

BONNER Literaturkalender 1980 / [Redaktion : Arno Hillebrand]. – Bonn : Bouvier, 1979
211p
3–416–01535–5 X.529/41895
8710120

ACADEMIC PERSONNEL. Learned institutions. Düsseldorf. West Germany
Publications, 1945– – Bibliographies

RHEINISCH-WESTFÄLISCHE AKADEMIE DER WISSENSCHAFTEN
Gesamtverzeichnis der Veröffentlichungen. – Opladen : Westdeutscher Verlag, [1979]
53p
r6301532 X.520/24135
8710112

ACADEMIC PERSONNEL. Learned institutions. Ljubljana. Yugoslavia
– Biographies

SLOVENSKA akademija znanosti in umetnosti ob štiridesetletnici, 1978 / [urednik : Tine Logar]. – Ljubljana : Slovenska akademija znanosti in umetnosti, [1978]
266p,[6]leaves of plates : ill
r6172899 X.520/9985
2031248

ACADEMIC PERSONNEL. Lunds universitet
1834–1847

GERLE, Eva
Lunds universitetets matrikel 1834–47. – Lund : Gleerup, 1978
134p
91–400–4670–2 X.529/34970
229575X

ACADEMIC PERSONNEL. Universität Leipzig
to date – Biographies – Serials

NAMHAFTE Hochschullehrer der Karl–Marx–Universität Leipzig. – 1–. – Leipzig : Die Universität, 1982–
r6308799 X.0529/937
9085327

ACADEMIC PERSONNEL. Universities. Italy
Organisation structure, to date

GIGLIOLI, Pier Paolo
Baroni e burocrati. – Bologna : Il Mulino, 1979
232p. – (Studi e ricerche ; 104)
r6511155 X.529/67491
9130284

ACADEMIC PERSONNEL. Universities. Netherlands
Training, to date

AUWERA, M. van der
Docententrainingen aan Nederlandse universiteiten en hogescholen. – Gent : [Rijksuniversiteit, Centrum voor de vergelijkende studie van het hoger onderwijs], 1980
86p. – (Bijdragen van het Centrum voor de vergelijkende studie van het hoger onderwijs ; nr.12)
r6330266 X.520/29218
8922020

ACADEMIC PERSONNEL. Universities. United States
Careers. Development. Econometric models
CORNFORD, Tony
A life-cycle model of career experience. - London : Centre for Labour Economics, 1980
23p : ill. - (Discussion paper / Centre for Labour Economics, London School of Economics ; no.68). - Bibliography: p12
B83-38509 X.525/7963
2953773

ACADEMIC PERSONNEL. University College, Cardiff
Trade unions. Association of University Teachers University College Cardiff Branch - Serials
ASSOCIATION OF UNIVERSTITY TEACHERS
Autotype. - No. 1 - : Feb. 1977- - [Cardiff] : [The Association], [1977]-
r6072594 P.523/863
1720120

ACADEMIC PERSONNEL. University of Leicester. Department of English Local History
Publications - Bibliographies
EVERITT, Alan
English local history at Leicester 1948-1978. - Leicester : Department of English Local History, University of Leicester, c1981
87p. - Bibliography: p87. - Includes index
0-907647-00-6 X.809/49954
2592703

ACADEMIC PERSONNEL. Uppsala universitet
1951-1960 - Biographies
DINTLER, ake
Uppsala universitets matrikel 1951-1960. - Uppsala : Almqvist & Wiksell, 1975
680p
r6043240 X.320/3162 Woolwich
0941948

ACADEMIC PERSONNEL. Users. Information sources on patents. Great Britain
JALLOQ, Monica C.
Use of patent literature by academics in chemistry and physics, electrical & mechanical engineering. - Aberystwyth : College of Librarianship Wales. - (British Library R & D report ; 5771)
Vol.2: Report to the British Library Research & Development Department on Project SI/G/321. - 1982
iii,180leaves
B84-36862 2719.k.259
3162346

ACADEMIC PUBLISHING. Great Britain
- Conference proceedings
WORKSHOP ON TRENDS IN SCHOLARLY PUBLISHING (1976 : Leicester)
Trends in scholarly publishing. - London : The Library, 1976
[2],iv,86p. - (Reports / British Library. Research and Development Department ; no.5299 HC). - Bibliography: p.80
0-85350-141-6 BS.10/676
0741639

ACADEMIC PUBLISHING. Great Britain
Direct-mail marketing
COLLINS, Joyce
Direct mail marketing for scholarly publishers. - Leicester : Primary Communications Research Centre, University of Leicester, 1983
ii,91p : ill. - (Aids to scholarly communication)
0-906083-26-5 X.629/26659
3286495

ACADEMIC REQUIREMENTS. Occupations. England
SMITH, Michael, 1931-
Your choice at 13+. - [New ed.]. - Cambridge : Hobsons Press [for] Careers Research and Advisory Centre, 1976
[4],216p : ill. - Bibliography: p.210-212. - Includes index
0-86021-130-4 X.519/28404
157552x

Your choice at 13+. - 4th ed. - [Cambridge] : CRAC, 1983
315p : ill
0-86021-542-3 X.629/23479
157552x

Your choice at 15+. - New ed. - Cambridge : Hobsons, 1982
159p : ill. - (A CRAC publication)
0-86021-472-9 X.529/54017
157552x

ACADEMIC UNDERACHIEVEMENT. Students. Secondary schools. England
- Conference proceedings
BOURNE, Eric
Under-achievement. - Sheffield : Association of Colleges for Further and Higher Education, [1978]
16,[2]p
B79-36265 X.519/30373
1993496

ACADEMICIANS. Academia Ecuatoriana
1875-1975 - Biographies
TOBAR DONOSO, Julio
Los Miembros de Numero de la Academia Ecuatoriana muertos en el primer siglo de su existencia, 1875-1975. - Quito : Editorial Ecuatoriana, 1976
495p.,17 leaves of plates
r6101580 X.709/29237
2166186

ACADEMICIANS. Magyar Tudományos Akadémia
1825-1975
A MAGYAR Tudományos Akadémia tagjai 1825-1973 / összeállította : Fekete Gézáné. - Budapest : Magyar Tudományos Akadémia Könyvtára, 1975
609p
r6044732 X.510/10262
0940984

SŐTÉR, István
A sas és a serleg. - Budapest : Akadémiai Kiadó, 1975
203 p.
963-05-0614-9 X.519/20531
0940984

ACADEMICS
Women academics. Equality of opportunity
WOMEN in academe / edited by Resa L. Dudovitz. - Oxford : Pergamon, 1984
viii,p.131-244. - Includes bibliographies
0-08-030819-8 X.520/37438
3329828

ACADEMICS. Soviet Union
PUCHKOV, Igor' Semenovich
Sot͡sial'no-demograficheskai͡a kharakteristika nauchnykh kadrov. - Moskva : Statistika, 1976
78p
r6183197 X.529/35638
2426277

ACADEMICS. United States
Influence on government, 1960-1969
BOM, Philip C
Academocracy. - [Amsterdam] : [P.C. Bom], [1976]
176p. - Includes bibliography and index
r6303264 X.805/1963
8780935

ACADÉMIE DE GENÈVE
Curriculum subjects: Philosophy. Teaching, 1660-1705
HEYD, Michael
Between orthodoxy and the Enlightenment. - The Hague ; London : Nijhoff, 1982
xii,308p : ill. - (Archives internationale d'histoire des idées = International archives of the history of ideas ; 96). - Bibliography: p289-302. - Includes index
90-247-2508-9 X.520/30142
2907615

ACADÉMIE DE LANGUE ARABE DU CAIRE
1932-1975
HAMZAOUI, Rached
L'Académie de langue arabe du Caire. - Tunis : Publications de l'Université de Tunis, 1975
661p. - (Faculté des lettres et sciences humaines de Tunis ; sér.6, vol.8). - Bibliography : p.607-627
r6028045 Ac.1997.f/3f(vol.8)
1126423

ACADEMIE DE STRASBOURG. UNIVERSITE DES SCIENCES JURIDIQUES, POLITIQUES ET SOCIALES
See Université de Strasbourg III

ACADÉMIE DES SCIENCES
FONTENELLE, Bernard le Bovier de
[Éloges]. Choix d'éloges de savants. - Paris : Librairie philosophique J. Vrin, 1981
2v.
r6208863 X.629/20890
8985456

ACADÉMIE DES SCIENCES
Members. Eulogies, 1699-1791
PAUL, Charles B.
Science and immortality. - Berkeley ; London : University of California Press, c1980
x,202p. - Bibliography: p181-197. - Includes index
0-520-03986-6 X.622/9609 Woolwich
2355035

ACADÉMIE DES SCIENCES
Research on music, to ca 1790
COHEN, Albert
Cohen. - Princeton ; Guildford : Princeton University Press, c1981
xvii,150p,[8]p of plates : ill. - Bibliography: p123-139. - Includes index
0-691-09127-7 X.431/11581
2636492

ACADÉMIE DU PALAIS DU LOUVRE
SEALY, Robert J.
The Palace Academy of Henry III [of France]. - Genève : Librairie Droz, 1981
x,209p,1leaf of plates. - (Travaux d'humanisme et renaissance ; no 184)
r6206299 WP.A.31/184
8729387

ACADÉMIE FRANÇAISE
DUTOURD, Jean
Discours de réception de M. Jean Dutourd à l'Académie française et réponse de M. Maurice Schumann. - Paris : Flammarion, 1980
103p,1leaf of plates
2-08-064311-8 X.958/4844
872685x

Les FEMMES et l'Academie française. - Paris : Les Éditions de l'Opale, 1980
118p
2-86283-009-7 X.529/44738
872685x

JAUBERT, Jacques
L'académie épinglée. - Paris : Éditions Sylvie Messinger, 1981
236p
2-86583-008-x X.950/16827
8172498

SCHUMANN, Maurice
Discours de réception de M. Maurice Schumann à l'Académie Française et réponse de M. le duc de Castries. - Paris : Librairie Arthème Fayard, 1975
146 p.
2-213-00233-9 X.708/17576
0916552

ACADÉMIE FRANÇAISE
Members. Elections. Canvassing, 1970 - Biographies
ISORNI, Jacques
La fièvre verte. - Paris : Flammarion, 1975
244 p.
2-08-060831-2 X.519/20577
0945021

ACADÉMIE FRANÇAISE
Relations with Voltaire, 1712-1778
RACEVSKIS, Karlis
Voltaire and the French Academy. - Chapel Hill : U.N.C. Department of Romance Languages, 1975
142p. - (North Carolina studies in the Romance languages and literatures. essays ; no.4). - Bibliography : p.137-142
0-88438-004-1 Ac.2685.k/7a(4)
1499726

ACADÉMIE GONCOURT
to 1974
ROBICHON, Jacques
Le défi des Goncourt. - Paris : Denoël, 1975
377 p., 16 leaves of plates : ill
r6012038 X.909/33363
0938017

ACADÉMIE ROYALE DES SCIENCES
See Académie des sciences †

ACADÉMIE ROYALE DES SCIENCES, DES LETTRES ET DES BEAUX-ARTS DE BELGIQUE. BIBLIOTHÈQUE
- Catalogues
ACADÉMIE ROYALE DES SCIENCES, DES LETTRES ET DES BEAUX-ARTS DE BELGIQUE. Bibliothèque
Catalogue des périodiques de la bibliothèque, Académie royale des sciences, des lettres et des beaux-arts de Belgique, Bibliothèque. - Bruxelles : [l'Académie], 1982
xiii,1311p
2-8031-0032-0 2719.e.437
8929890

ACADEMIES
See also
Flemish speaking academies
Military academies
Naval academies

ACADEMIES. Architecture. Soviet Union
MASTERSKAĪA monumental'noĭ zhivopisi pri Akademii arkhitektury SSSR / [sostavitel'sbornika i avtor vstupitel'noĭ stat'i E.V. Shunkova]. - Moskva : Sovetskiĭ khudozhnik, 1978
215p,[210]p of plates : chiefly ill(some col). - Bibliography : p.201-202
r6173658 X.429/7824 Woolwich
2165589

ACADEMIES. Arezzo. Italy
1600-1800
GAZZOLA STACCHINI, Vanna
Le accademie dell'Aretino nel XVII e XVIII secolo. - Firenze : Leo S. Olschki, 1978
xv,595p. - (Biblioteca dell'Archivum romanicum. Ser. 1 ; v. 141). - Includes bibliographies
r6068282 PP.4884.dba[Ser.1.vol.141]
1863223

ACADEMIES. Arts. Rome. Italy
Accademia di Francia
CORRESPONDANCE des directeurs de l'Académie de France à Rome. - Roma : Edizioni dell'elefante, 1979-
r6506670 X.0529/881
9113142

ACADEMIES. Brussels. Belgium
Libraries: Académie royale des sciences, des lettres et des beaux-arts de Belgique. Bibliothèque - Catalogues
ACADÉMIE ROYALE DES SCIENCES, DES LETTRES ET DES BEAUX-ARTS DE BELGIQUE. Bibliothèque
Catalogue des périodiques de la bibliothèque, Académie royale des sciences, des lettres et des beaux-arts de Belgique, Bibliothèque. - Bruxelles : [l'Académie], 1982
xiii,1311p
2-8031-0032-0 2719.e.437
8929890

ACADEMIES. Europe
1500-1700
CONRADS, Norbert
Ritterakademien der frühen Neuzeit. - Göttingen : Vandenhoeck & Ruprecht, 1982
414p,4p of plates : ill. - (Schriftenreihe der Historischen Kommission bei der Bayerischen Akademie der Wissenschaften ; Schrift 21). - Bibliography: p.333-400
3-525-35918-7 Ac.713.b/3[no.21]
8994323

ACADEMIES. Europe
1600-1800 - Conference proceedings
Der AKADEMIEGEDANKE im 17. und 18. Jahrhundert / herausgegeben von Fritz Hartmann und Rudolf Vierhaus. - Bremen : Jacobi, 1977
192p. - (Wolfenbütteler Forschungen ; Bd 3)
3-87447-102-0 2719.c.6(3)
1529935

ACADEMIES. Italy
1600-1800
ACCADEMIE e cultura / [saggi di Bruno Neveu ... et al.]. - Firenze : Olschki, 1979
289p. - (Centro di studi muratoriani, Modena. Biblioteca dell'edizione nazionale del carteggio di L.A. Muratori ; 5)
88-222-2874-x X.950/22945
2441837

ACADEMIES. Law. Düsseldorf. West Germany
Rechtsakademie zu Düsseldorf
FISCHER, Guntram
Düsseldorf und seine Rechtsakademie. - Düsseldorf : Triltsch, c1983
307p.. - Includes bibliography
3-7998-0024-7 X.200/45018
9170170

ACADEMIES. Rome. Italy
Accademia di Francia, Rome. Exhibits: French paintings. Poussin, Nicolas - Catalogues
ACCADEMIA DI FRANCIA
Nicolas Poussin, 1594-1665. - Roma : Edizioni dell'elefante, 1977
251p.,[8]p.of col.plates : ill
r6195393 X.421/21188
2462389

ACADEMIES. Theology. Rome. Italy
Accademia di religione cattolica
PIOLANTI, Antonio
L'Accademia di religione cattolica. - Città del Vaticano : Libreria editrice vaticana, 1977
xi,545p,[2]leaves of plates : ill. - (Biblioteca per la storia del Tomismo ; 9)
r6500110 X.200/35002
8794073

ACADEMIES. Vicenza. Italy
Academia Olimpica, Vicenza. Exhibits: Photographs. Special subjects. Brazil. Immigration from Veneto. Italy - Catalogues
I VENETI in Brasile nel centenario dell'emigrazione. 1876-1976 / a cura di Mario Sabbatini e Emilio Franzina. - Vicenza : Edizioni dell' Accademia olimpica. 1977
xv,157p,[3]fold leaves,[120]p of plates
r6500580 X.529/39946
2478765

ACADEMIES. Vicenza. Italy
Academia Olimpica. Vicenza. Exhibits: Photographs. Special subjects: Italy. Veneto. Emigration to Brazil - Catalogues
I VENETI in Brasile nel centenario dell'emigrazione. 1876-1976 / a cura di Mario Sabbatini e Emilio Franzina. - Vicenza : Edizioni dell' Accademia olimpica, 1977
xv,157p,[3]fold leaves,[120]p of plates
r6500580 X.529/39946
2478765

ACADEMIES. Visual arts. Düsseldorf. West Germany
Staatliche Kunstakademie Düsseldorf - Catalogues
DÜSSELDORFI KOOLKOND EESTI KUNSTIS (Exhibition)
Düsseldorfi koolkond eesti kunstis. - Tallinn : ENSV Riiklik Kunstimuuseum, 1980
29p : chiefly ill
r6461595 X.429/13446 Woolwich
8867577

ACADEMIES. Visual arts. Düsseldorf. West Germany
Staatliche Kunstakademie Düsseldorf, to 1973
ZWEIHUNDERT Jahre Kunstakademie Dusseldorf / herausgegeben für die Ernst-Forberg-Stiftung von Eduard Trier, unter Mitwirkung von Paul Böhringer...[et al.]. - Düsseldorf, 1973
xv,229p : ill
r6016886 X.423/2584 Woolwich
116225X

ACADEMIES. Visual arts. Italy
to date
CARAMEL, Luciano
L'arte bella. - Milano : Feltrinelli, 1979
541p. - (Strumenti per una nuova scuola ; 35). - Bibliography
r6500454 X.429/12202 Woolwich
8720630

ACADEMIES. Visual arts. Leningrad. Russia (RSFSR)
Akademiīa khudozhestv SSSR
KEMENOV, Vladimir Semenovich
[Akademiīa Khudozhestv SSSR. English]. The USSR Academy of Arts. - Leningrad : Aurora Art Publishers, 1982
413p,plates : ill(some col.)
r6463741 L.42/2070
8940053

LISOVSKIĬ, Vladimir Grigor'evich
Akademiīa khudozhestv. - 2-e izd. perer. i dop. - Leningrad : Lenizdat, 1982
222p : ill(some col.)
r6464039 X.429/14221
8940053

ACADEMIES. Visual arts. Leningrad. Russia (RSFSR)
Akademiīa khudozhestv SSSR - Catalogues
225 let Akademii khudozhestv SSSR / [sostaviteli I toma I.A. Bartenev (otvetstvennyĭ redaktor) ... i dr.] ; [sostaviteli II toma B.V. Veĭmarn (otvetstvennyĭ redaktor) ... i dr.]. - Moskva : Izobrazitel'noe iskusstvo, 1983
2v : ill
r6492525 X.421/26370
8940053

ACADEMIES. Visual arts. Leningrad. Russia (RSFSR)
Akademiīa khudozhestv SSSR - Serials
TRUDY Akademii khudozhestv SSSR. - Vyp.1-. - Moskva : Izobrazitel'noe iskusstvo, 1983-
v.
r6496850 P.421/844
8940053

ACADEMIES. Visual arts. Mantua. Italy
Accademia di Mantova - Catalogues
ARCHITETTURA E PITTURA ALL'ACCADEMIA DI MANTOVA (1752-1802)
(Exhibition) (1980 : Mantua)
Architettura e pittura all'Accademia di Mantova (1752-1802). - Firenze : Centro Di, 1980
111p : ill
88-7038-002-5 X.421/24160
8974985

ACADEMIES. Visual arts. New Zealand
New Zealand Academy of Fine Arts
KAY, Robin
Portrait of a century. - Wellington, N.Z. : Millwood Press, 1983
219p : ill
0-908582-60-9 X.425/4617
9118225

ACADEMIES. Visual arts. Rome. Italy
Archives: Accademia nazionale di San Luca. Archivio storico
Stock: Architectural drawings
ACCADEMIA NAZIONALE DI SAN LUCA. Archivio storico
I disegni di architettura dell'Archivio storico dell'Accademia di San Luca. - Roma : De Luca, 1974
2v : chiefly ill
r6042441 LR.413.x.38
1394371

ACADEMIES OF SCIENCE. Cuba
Academia de Ciencias Médicas, Físicas y Naturales de La Habana
BUSTAMANTE O'LEARY, Jose A
La Academia de Ciencias Médicas, Físicas y Naturales de La Habana. - La Habana : Academia de Ciencias de Cuba. Centro de Estudios de Historia y Organización de la Ciencia Carlos J. Finlay, 1979
19leaves. - (Conferencias y estudios de historia y organización de la ciencia ; 11)
r6701134 LAS.D.300/15 Woolwich
8744297

ACADEMIES OF SCIENCE. France
Académie des sciences
FONTENELLE, Bernard le Bovier de
[Éloges]. Choix d'éloges de savants. - Paris : Librairie philosophique J. Vrin, 1981
2v.
r6208863 X.629/20890
8985456

ACADEMIES OF SCIENCES. Cuba
Academia de Ciencias de Cuba. Relations with Soviet research organisations, 1962-1979
SÁENZ, Tirso W
El papel de la colaboración con las instituciones científicas soviéticas en la creación y desarrollo de la Academia de Ciencias de Cuba. - La Habana : Editorial Academia, 1979
35p
r6160784 X.622/9275 Woolwich
2470101
El papel de la colaboración con las instituciones científicas soviéticas en la creación y desarrollo de la Academia de Ciencias de Cuba. - La Habana : Editorial Academia, 1979
35p
r6160784 X.622/9275 Woolwich

ACADEMIES OF SCIENCES. Soviet Union
DOROFEEVA, Vera
Vremīa, uchenye, svershenīa. - Moskva : Izdatel'stvo politicheskoĭ literatury, 1975
286p.,[16]p.of plates : ill
r6196108 X.510/11358
1983814

ACADEMY FOR RURAL DEVELOPMENT
See also
Bangladesh Academy for Rural Development

ACADEMY OF FINE ARTS
See Baghdad University. Academy of Fine Arts

ACADEMY OF GENEVA
See Académie de Genève

ACADEMY OF KOREAN STUDIES
- Serials
ACADEMY OF KOREAN STUDIES
Academy news. - Vol.1, no.1 (Autumn 1983)-. - Kyonggi-do : The Academy, 1983-
v.
r6813371 X.0805/153
9092021

ACADEMY OF MUSIC (NEW YORK)
1883-1886
CONE, John Frederick
First rival of the Metropolitan Opera. - New York ; Guildford, Surrey : Columbia University Press, 1983
xvi,357p : ill. - Bibliography: p239-245. - Includes index
0-231-05748-2 X.431/12414
3172007

ACADEMY OF SCIENCES OF THE USSR
See Akademiīa nauk SSSR

ACADEMY OF ST MARTIN IN THE FIELDS
to 1980

HARRIES, Meirion
The Academy of St. Martin in the Fields. - London : Joseph, 1981
ix,278p : ill. - List of sound recordings: p241-264. - Includes index
0-7181-2049-3 X.520/25567
2591294

ACADIA
to date

BRUN, Régis
De Grand-Pre a Kouchibougouac. - [Moncton] : Editions d'Acadie, 1982
175p.plates
2-7600-0078-8 X.809/63850
8977070

DU PONT DUVIVIER, François
Course à l'Accadie. - Moncton, N.B. : Éditions d'Acadie, 1982
195p : ill
2-7600-0074-5 X.809/55905
8977070

ROY, Michel
L'Acadie, des origines à nos jours. - Montréal : Québec / Amérique, 1981
340p : ill. - Includes bibliography
2-89037-085-2 X.809/58336
8977070

RUMILLY, Robert
L'Acadie française (1497-1713). - Montréal : Fides, 1981
253p
2-7621-1111-0 X.809/57663
8977070

ACADIA
Description & travel, to date

CORMIER, Roméo
Image of Acadia. - Toronto : Oxford University Press, 1980
85p : col.ill
0-19-540330-4 L.49/936
8724792
- [Another copy] L:49/2236

Images de l'Acadie. - Montréal : Hurtubise HMH, 1980
86p : col.ill
2-89045-315-4 L.49/1665
8724792

GALLANT, Melvin
Le pays d'Acadie. - Moncton : Éditions d'Acadie, 1980
206p : ill
2-7600-0051-6 L.49/1940
8724792

ACADIA
Social conditions, to date

ARSENAULT, Georges
Courir la Chandeleur. - [Moncton] : Éditions d'Acadie, 1982
116p : ill. - Includes bibliography
2-7600-0079-6 X.529/68981
8876738

DUPONT, Jean-Claude
Histoire populaire de l'Acadie. - [Montréal] : Leméac, 1979
440p : ill. - Bibliography
2-7609-5278-9 X.529/51361
8876738

THÉRIAULT, Léon
La question du pouvoir en Acadie. - 2nd ed.. - Moncton, N.B. : Éditions d'Acadie, 1982
256p. - Includes bibliography
2-7600-0067-2 X.520/34097
8876738

ACADIAN LEGENDS
Acadian tales & legends - Texts

JOLICOEUR, Catherine
Les plus belles légendes acadiennes. - Montréal : Stanké, [1981]
280p : ill
2-7604-0111-1 X.950/24901
9008241

ACADIAN SONGS
Acadian folk songs: Songs from Shippegan Peninsula - Collections

CHANSONS de Shippagan / recueillies par Dominique Gauthier. - Québec : Les Presses de l'Université Laval, 1975
xxviii, 178 p.. - (Les archives de folklore ; 16)
0-7746-6730-3 Ac.2702.g/3[16]
0848980

ACADIAN TALES
Acadian tales & legends - Texts

JOLICOEUR, Catherine
Les plus belles légendes acadiennes. - Montréal : Stanké, [1981]
280p : ill
2-7604-0111-1 X.950/24901
9008241

ACADIANS
to date

ARSENAULT, Bona
History of the Acadians. - [Montreal] : Lemeac, [1978]
265p. - Includes bibliography
0-7761-5034-0 X.809/55641
8963576

LAPIERRE, Jean-William
Les Acadiens. - [Paris] : Presses Universitaires de France, 1983
127p. - (Que sais-je? ; no.2078)
2-13-037842-0 X.808/40832
8963576

ACAHUALINCA. Nicaragua
Prehistoric antiquities. Archaeology

MATILLÓ VILA, Joaquín
Las huellas de Acahualinca en el panorama arqueológico de Nicaragua. - [Managua] : Ministerio de Educación Pública, Departamento de Extensión Cultural, 1975
45p : ill. - (Publicaciones del Museo Nacional. serie arqueológica ; n.1)
r6084028 X.419/8273 Woolwich
1983768

ACALI (RAFT). Rafts
Voyages in Atlantic Ocean, 1973. Psychological aspects

GENOVÉS, Santiago
Acali. - 7.ed. - Barcelona : Editorial Planeta, 1975
455p.,64p. of plates (mainly col.). - (Espejo del mundo ; 6). - Bibliography : p. 443-449
84-320-6606-0 X.800/12336
1124153

AÇÃO INTEGRALISTA BRASILEIRA

TRINDADE, Hélgio
Integralismo. - 2.ed. revista e ampliada. - São Paulo : Difel, 1979
xxvii,379p,20p of plates. - (Corpo e alma do Brasil ; 40)
r6703338 X.809/50036
8793859

ACAPULCO. Mexico

BROSNAHAN, Tom
Frommer's 1983-84 guide to Mexico City & Acapulco. - New York : Frommer / Pasmantier, c1983
218p : ill
0-671-45298-3 X.808/38273
2918587

ACAPULCO. Mexico
to 1970

OTEIZA IRIARTE, Tomás
Acapulco, la ciudad de las naos de oriente y de las sirenas modernas. - México : Editorial Diana, 1973
400p : ill
r6052234 X.708/42132
1323601

ACAPULCO. Mexico. Special subjects. Photography

MOLDVAY, Albert
Photographing Mexico City & Acapulco. - Cambridge : Stephens, 1981, c1980
128p : col.ill. - (PSL travel guide). - Includes index
0-85059-520-7 X.800/30182
2384000

ACARBOSE. Drug therapy. Metabolic disorders. Man
- Conference proceedings

INTERNATIONAL SYMPOSIUM ON ACARBOSE (1st : 1981 : Montreux)
First International Symposium on Acarbose. - Amsterdam ; Oxford : Excerpta Medica, 1982
xiii,573p : ill. - (International congress series ; no.594). - Includes index
90-219-9540-9 SRL
2788772

ACARDA
See Asociaţiei Culturale a Românilor din Anglia

ACARINA
See also
Mites
Ticks

ACAROLOGY
See also
Acarina

ACAS
See Advisory, Conciliation and Arbitration Service

ACAST
See United Nations. Advisory Committee on the Application of Science and Technology for Development

ACATZINGO. Mexico
Population, to date

CALVO, Thomas
Acatzingo. - México : Instituto Nacional de Antropología e Historia, 1973
124p : ill. - (Colección científica. Historia ; 6). - Bibliography
r6702340 X.525/5085
8759871

ACAYUCÁN. Mexico
Rebellions, 1906

PASQUEL, Leonardo
La rebelión agraria de Acayucán en 1906. - México : Citlaltepetl, 1976
66p.,4 leaves of plates,22p. of plates : ill. - (Colección Suma veracruzana. Serie historiografía)
r6084624 X.709/27189
1794280

ACCADEMIA DEI CONCORDI
- Catalogues

ACCADEMIA DEI CONCORDI
La Pinacoteca dell'Accademia dei Concordi. - Rovigo : Accademia dei Concordi, 1981
517p,plates. - (Studi e ricerche ; 2). - Includes bibliography
r6509346 X.421/25450
9083243

ACCADEMIA DELL'ARCADIA
- Indexes

GIORGETTI VICHI, Anna Maria
Gli Arcadi dal 1690 al 1800. - Roma : Arcadia, 1977
xviii,408p. - Bibl
r6195611 X.902/3896
2421836

ACCADEMIA DELLE ARTI DEL DISEGNO
Exhibits: Italian drawings. Guttuso, Renato - Catalogues

GUTTUSO, Renato
Da Michelangelo. - Firenze : Stiav, 1975
[14]p.,[29]leaves of plates : chiefly ill(some col)
r6050547 X.410/6396 Woolwich
1391321

ACCADEMIA DELLE ARTI DEL DISEGNO
Exhibits: Italian visual arts - Catalogues

ACCADEMIA DELLE ARTI DEL DISEGNO
Mostra degli Accademici delle arti del disegno, 20 dicembre 1975-20 gennaio 1976. - Firenze : l'Accademia, 1975
x,179p : ill
r6048761 X.421/9273 Woolwich
132568X

ACCADEMIA DI FRANCIA

CORRESPONDANCE des directeurs de l'Académie de France à Rome. - Roma : Edizioni dell'elefante, 1979-
r6506670 X.0529/881
9113142

LE NORMAND, Antoinette
La tradition classique & l'esprit romantique. - Roma : Edizioni dell'elefante, 1981
378p : ill. - (Collection Académie de France à Rome ; 3)
r6510341 X.525/8729
9039619

ACCADEMIA DI FRANCIA
Exhibits: French visual arts. Derain, André - Catalogues

ACCADEMIA DI FRANCIA
Derain. - Roma : Edizioni dell' elefante, 1976
162p : ill(some col.). - Includes bibliographies
r6068353 X.410/7010 Woolwich
1866370

ACCADEMIA DI FRANCIA, ROME
Exhibits: French paintings. Poussin, Nicolas - Catalogues

ACCADEMIA DI FRANCIA
Nicolas Poussin, 1594-1665. - Roma : Edizioni dell'elefante, 1977
251p.,[8]p.of col.plates : ill
r6195393 X.421/21188
2462389

ACCADEMIA DI MANTOVA
- Catalogues

ARCHITETTURA E PITTURA ALL'ACCADEMIA DI MANTOVA (1752-1802)
(Exhibition) (1980 : Mantua)
Architettura e pittura all'Accademia di Mantova (1752-1802). - Firenze : Centro Di, 1980
111p : ill
88-7038-002-5 X.421/24160
8974985

ACCADEMIA DI RELIGIONE CATTOLICA

PIOLANTI, Antonio
L'Accademia di religione cattolica. - Città del Vaticano : Libreria editrice vaticana, 1977
xi,545p,[2]leaves of plates : ill. - (Biblioteca per la storia del Tomismo ; 9)
r6500110 X.200/35002
8794073

ACCADEMIA ETRUSCA DI CORTONA. MUSEO
Exhibits: Ancient Roman & Etruscan coins - Catalogues

PANCRAZZI, Orlanda
Le monete dell'Accademia di Cortona fino a Teodosio. - Pisa : Pacini, 1975
457 p., 39 p. of plates : ill
r6014359 X.410/5810 Woolwich
0880582

ACCADEMIA NAVALE

SANTINI, Aldo
Livorno ammiraglia. - Livorno : Belforte, 1981
360p : ill
r6509685 X.629/21396
9010890

ACCADEMIA NAZIONALE DEI LINCEI
- Bibliographies

SCHETTINI PIAZZA, Enrica
Bibliografia storica dell'Accademia nazionale dei Lincei. - Firenze : Olschki, 1980
180p. - (Biblioteconomia e bibliografia ; 12)
88-222-2885-5 2715.bbb.1/12
2463113

ACCADEMIA NAZIONALE DEI LINCEI
Members - Biographies - Bibliographies

ACCADEMIA NAZIONALE DEI LINCEI
Biografie e bibliografie degli Accademici Lincei. - Roma : l'Accademia, 1976
lxi,1319p.,ca.250 leaves of plates
r6100646 X.512/3240
2038951

ACCADEMIA NAZIONALE DEI LINCEI. BIBLIOTECA
Stock: Periodicals - Catalogues

ACCADEMIA NAZIONALE DEI LINCEI. Biblioteca
Catalogo dei periodici e atti accademici posseduti dalla biblioteca [dell'Accademia nazionale dei Lincei] con elenco degli Enti e bibliografia. - Roma : l'Accademia, 1975-. - (Indici e sussidi bibliografici della Biblioteca ; 7)
r6068593 Ac.102/28[vol.7 etc.]
2027534

ACCADEMIA NAZIONALE DEI LINCEI. BIBLIOTECA
Stock: Persian manuscripts - Catalogues

ACCADEMIA NAZIONALE DEI LINCEI. Biblioteca
I manoscritti persiani dell'Accademia nazionale dei Lincei. - Roma : l'Accademia, 1974
xxvi,82p.,[4]p.of plates : ill. - (Indici e sussidi bibliografici della Biblioteca ; 8)
r6100638 Ac.102/28[vol.8]
2040131

ACCADEMIA NAZIONALE DEI QUARANTA
to 1977

PENSO, Giuseppe
Scienziati italiani e unità d'Italia. - Roma : Bardi, 1978
xiv,505p.,[1]leaf of col.plates,[4]p.of col.plates : ill,facsims,ports. - Bibl
r6197617 X.520/22083
2460068

ACCADEMIA NAZIONALE DI SAN LUCA. ARCHIVIO STORICO
Stock: Architectural drawings

ACCADEMIA NAZIONALE DI SAN LUCA. Archivio storico
I disegni di architettura dell'Archivo storico dell'Accademia di San Luca. - Roma : De Luca, 1974
2v : chiefly ill
r6042441 LR.413.s.38
1394371

ACCADEMICI INTRONATI DI SIENA. Drama in Italian
- Texts

ACCADEMICI INTRONATI DI SIENA
[Gl'ingannati]. La commedia degli Ingannati. - Firenze : Olschki, 1980
246p. - (Biblioteca dell' "Archivum romanicum". Serie 1, storia, letteratura, paleografia ; v 156). - Bibliography
88-222-2958-4 P.P.4884dba[vol.156]
8094489

ACCADIAN LANGUAGE
See Akkadian language

ACCELERATED CURING. Elastomers
Use of organic compounds

BLOKH, G. A.
Organic accelerators and curing systems for elastomers. - Shrewsbury : Rubber and Plastics Research Association of Great Britain, 1981
xiii,324p : ill. - (International polymer science and technology monograph ; 3). - Bibliography: p323-324
0-902348-22-1 SRL
2550229

ACCELERATED DRAGON VARIATIONS. Sicilian Defence. Openings. Chess

LEVY, D. N. L. (David Neil Laurence)
The Sicilian accelerated dragons. - London : Batsford, 1975
x,164p : ill. - (Batsford chess books) ([Contemporary chess openings]). - Includes index
0-7134-3031-1 X.629/10241
053174x

ACCELERATED PROCEDURE. Criminal courts. Poland

GABERLE, Andrzej
Postępowanie przyśpieszone w polskim procesie karnym. - [Cracow] : Nakładem Uniwersytetu Jagiellońskiego, 1975
210 p.. - (Zeszyty naukowe Uniwersytetu Jagiellońskiego ; tom 410. Prace prawnicze ; zesz. 72). - Bibliography: p. 195-201
r6008739 Ac.749/27c [zesz. 72]
0868647

ACCELERATION
See also
Deceleration

ACCELERATION. Convergence. Infinite series
Techniques

HALLIDAY, A. S. M.
A review of methods for the acceleration of convergence of infinite series. - Cardiff : Department of Mathematics, University of Wales Institute of Science and Technology, c1980
38p. - (Math report ; 80-3). - Bibliography: p37-38
B81-15098 X.622/9992 Woolwich
2387557

ACCELERATORS. Nuclear physics
See Particle accelerators

ACCENTS
See also
Diacritics

ACCENTS. Pronunciation. Ukrainian language

SKLÎARENKO, V. H. (Vitaliĭ Hryhor'evych)
Narysy z istorychnoï aktsentologiï ukraïns'koï movy. - Kyïv : Naukova dumka, 1983
239p
r6489168 X.958/22542
9085661

ACCEPTING HOUSES
See Merchant banks

ACCESS. Arts. Great Britain
By physically-handicapped persons

GREATER LONDON ARTS ASSOCIATION
Needing a push. - London : Greater London Arts Association ; [London] : Greater London Association for the Disabled, [1979]
27p : ill
0-9502447-4-0 X.520/21782
2184109

ACCESS. Birth certificates. England
Rights of adopted persons

DAY, Cyril
Access to birth records. - London : Association of British Adoption & Fostering Agencies, 1980
36p. - (Research series ; 1)
0-903534-29-0 X.529/53326
2881683
- [Another copy] 347.426 OPL

ACCESS. Broadcasting services. European Community countries
By political parties

ROBERTS, G. K. (Geoffrey Keith), 1936-
Access aux émissions politiques dans la CEE. - [Manchester] : [For] Institut Européen de la Communication [by] Manchester University Press, [c1984]
30p
0-7190-1482-4 X.520/35301
3192555

Access to political broadcasting in the EEC. - Manchester : [For] European Institute for the Media [by] Manchester University Press, c1984
29p
0-7190-1481-6 X.520/35300
3192563

ACCESS. Buildings
By physically handicapped persons - Conference proceedings

INTO the eighties - access to the built environment / Greater London Association for the Disabled ; photographs Maria Bartha ; editor John Keet ; assistant editor Virginia Armstrong. - London : The Association, [1982?]
24p : ill
B82-23266 X.525/5755
2730014

ACCESS. Buildings. Hove. East Sussex
By physically handicapped persons - For architectural design

HOVE. Planning Department
Access for the disabled. - Hove : Borough of Hove Planning Department, 1977
[3],14p : ill
B77-30460 SRL
1252550

ACCESS. Buildings. Wales
By physically handicapped persons - For architectural design

GREAT BRITAIN. Central Office of Information
Access for the disabled. - [London] : H.M.S.O., 1978
[2],17p : ill(chiefly col). - Bibliography: p.17
B80-38411 BS.128A/263
2282526

ACCESS. Canadian government publications in libraries. Canada
By public

JARVI, Edith
Access to Canadian government publications in Canadian academic and public libraries. - Ottawa : Canadian Library Association, [1976]
116p : ill
B83-17515 2719.e.511
2878356

ACCESS. Canadian government publications in libraries of higher education institutions. Canada
By public

JARVI, Edith
Access to Canadian government publications in Canadian academic and public libraries. - Ottawa : Canadian Library Association, 1976
[5],iii,116p : ill. - Bibliography : p.77-93. - Includes index
0-88802-113-5 2719.bb.110
1721577
- [Another copy] OPL

ACCESS. Canadian government publications in public libraries. Canada
By public

JARVI, Edith
Access to Canadian government publications in Canadian academic and public libraries. - Ottawa : Canadian Library Association, 1976
[5],iii,116p : ill. - Bibliography : p.77-93. - Includes index
0-88802-113-5 2719.bb.110
1721569
- [Another copy] OPL

ACCESS. Children. Divorced persons. England
Law

LEVY, Allan
Custody and access. - London : Oyez Longman, 1983
xxiv,147p : ill. - (Oyez Longman practice notes). - Includes index
0-85120-748-0 WP.1357
303948x

ACCESS. Children. England
By divorced parents

DIVIDED children. - [London] : Gingerbread, 1982
53p : ill
B83-44039 X.529/57423
2978733

ACCESS. Children. England
Rights of divorced fathers – Serials
FAMILIES NEED FATHERS (Organisation)
Newsletter. – [1]- ; Autumn 1977-. – London . Families Need Fathers, 1977 – : ill
Holdings: Spring 1980, etc
B80–29376 P.521 / 3409
2204355

ACCESS. Children in care. England
By parents
FOSTERING parental contact. – London Family Rights Group. 1982
75p. – Includes bibliographies
B83–36886 X.529/56307
2184214

ACCESS. Children in care. England
By parents – Conference proceedings
TERMINATING parental contact / editors Margaret Adcock, Richard White. – London : Association of British Adoption and Fostering Agencies, 1980
56p. – (Discussion series / Association of British Adoption and Fostering Agencies ; 2)
0–903534–26–6 X.529/36265
2184214

ACCESS. Confidential records. Students. Schools
NATIONAL UNION OF TEACHERS
Confidentiality & school records. – London : National Union of Teachers, 1984
40p
B85–07513 X.529/67327
3224317

ACCESS. Confidential records. Students. Schools
Rights of parents
HODGES, Lucy
Out in the open?. – London : Writers and Readers in association with Chameleon, [1981]
88p. – Bibliography: p83. – Includes index
0–906495–58–x X.529/51247
2817691

ACCESS. Countryside. Great Britain
By public, to 1979
HILL, Howard
Freedom to roam. – Ashbourne : Moorland, 1980
139p : ill. – Includes index
0–903485–77–x X.809/46112
196626x

ACCESS. Countryside. Sweden
By public
ROSÉN, Bo
The right of common access in Sweden. – Stockholm : Svenska institutet, 1979
11p. – (Current Sweden ; no.235)
r6062116 X.0512/423(235)
2213338

ACCESS. Health services. Great Britain
Inequalities. Effects of national health services, 1948–1978
WALTERS, Vivienne
Class inequality and health care. – London : Croom Helm, 1980
175p. – Bibliography: p.163–171. – Includes index
0–85664–685–7 X.329/12461 Woolwich
1929143

ACCESS. Health services. United States
ADAY, Lu Ann
Health care in the U.S.. – Beverly Hills ; London : Sage Publications, 1980
415p : 1 ill. – Bibliography: p.249–253
0–8039–1373–7 X.329/12637 Woolwich
2208040

IN search of equity / edited by Ronald Bayer, Arthur L. Caplan and Norman Daniels. – New York ; London : Plenum, c1983
xxvi,240p. – (The Hastings Center series in ethics). – Includes index
0–306–41212–8 X.329/17138
2208040

ACCESS. Health services. United States
By patients, 1969–1975. Economic aspects
SLOAN, Frank A.
Access to ambulatory care and the U.S. economy. – Lexington, Mass. : Heath ; [Farnborough, Hants.] : [Distributed by] Teakfield Limited, 1979
xxi,175p : 1 ill. – Includes index
0–669–02510–0 X.520/21246
2063972

ACCESS. Higher education. Denmark
DENMARK. Undervisningsministeriet
Nye veje til en højere uddannelse. – København : [Undervisningsministeriet], 1976
142p. – (Betænkning ; nr 778)
87–503–2055–6 S.B.9/4(778)
1329278

ACCESS. Higher education. Great Britain
ACCESS to higher education / Oliver Fulton editor. – Guildford : Society for Research into Higher Education, 1981
202p : ill. – (Research into higher education monographs ; 44). – Includes bibliographies
0–900868–89–9 X.429/15059
2734370

FULTON, Oliver
Demand, supply and access. – Doncaster : Association of Colleges for Further and Higher Education, [1984]
16p. – Bibliography: p16
B85–09089 X.529/67242
2734370

ACCESS. Higher education. Great Britain
– Serials
ACCESS to higher education forum. – No.1 (autumn 1983)-. – Ealing : Access Forum, 1983–
v.
B85–13195 P.823/596
2734370

ACCESS. Higher education. South Yorkshire (Metropolitan County)
By mature students – Conference proceedings
RECENT developments in mature student access to education in South Yorkshire / edited by R.H. Mercer. – [Sheffield] : Sheffield City Polytechnic Department of Education Management, 1980
117p. – (Sheffield papers in education management ; 8)
B84–23007 X.529/61258
3087344

ACCESS. Kinder Scout. Derbyshire
Restrictions. Protests by walkers, 1932
ROTHMAN, Benny
The 1932 Kinder trespass. – Altrincham : Willow Publishing, 1982
48p : ill
0–9506043–7–2 X.800/33360
2747839

ACCESS. Neighbouring land. England
Rights of landowners. Law
GREAT BRITAIN. Law Commission
Rights of access to neighbouring land. – London : H.M.S.O., 1980
v,70p. – (Working paper ; no.78)
0–11–730158–2 BS.24f/47
2641968

ACCESS. Official documents
Rights of public – Serials
INTERNATIONAL FREEDOM OF INFORMATION COMMISSION
Newsletter. – [Vol.1, no.1]- ; [1979]-. – London : The Commission, 1979– : ill
B80–24126 P.703/660
2067412

ACCESS. Official documents. Developed countries
Rights of public
ADMINISTRATIVE secrecy in developed countries / edited [and translated from the French] by Donald C. Rowat. – London : Macmillan, 1979
xxiv,364p. – (Studies in administrative procedures). – Includes index
0–333–24636–5 X.529/33189
1439464

GREAT BRITAIN. Civil Service Department
Disclosure of official information. – London : H.M.S.O., 1979
v,55,150p. – Bibliography: p.143–150
0–11–630415–4 BS.6/131
1901621

ACCESS. Official documents. Great Britain
Rights of public
BARNES, Trevor
Open up!. – London : Fabian Society, 1980
[2],24p. – (Fabian tract ; 467)
0–7163–0467–8 8275.dd.7/467
1186485

GREAT BRITAIN. Lord Privy Seal
Open government. – London : H.M.S.O., [1979]
18p. – (Cmnd.7520)
0–10–175200–8 BS.91/170
1186485

JUSTICE
Freedom of information. – London : JUSTICE, 1978
[6],18p
0–9505247–5–1 X.100/17810
1551329

MICHAEL, James
The politics of secrecy. – London : National Council for Civil Liberties, 1979
50p. – Bibliography: p.50
0–901108–80–4 X.200/33837
1551329

WRAITH, Ronald Edward
Open government. – London : Royal Institute of Public Administration, 1977
74p. – (Research booklets / Royal Institute of Public Administration)
0–900628–15–4 X.709/31431
1186485

ACCESS. Official documents. Great Britain
Rights of public, 1978–1979
OUTER CIRCLE POLICY UNIT
A consumers' guide to open government. – London : The Unit, 1980
[3],92p
B80–24125 X.702/6353
2185466
– [Another copy] OPL

ACCESS. Official documents. Great Britain
Rights of public. Law
PUBLIC SECRETS (Group)
An Official Information Act. – London : Outer Circle Policy Unit, [1979?]
[2],78p
B79–24322 X.100/18471
191832x
Official information bill. – London : Outer Circle Policy Unit, [1978]
[85]p in various pagings
B79–25009 X.102/2209
191832x

ACCESS. Official documents. Great Britain, South Africa & United States
Rights of public. Law
MATHEWS, Anthony Stuart
The darker reaches of government. – Berkeley ; London : University of California Press, 1978
xxi,245p. – (Perspectives on southern Africa ; 27). – Bibliography: p.ix–xii. – Includes index
0–520–03803–7 X.800/27938
1918125

ACCESS. Official documents. Norway
Rights of public. Law
LYNGDAL, Lars Erik
Undersøkelse om offentlighetsloven. – Oslo : Universitetet i Oslo, Institutt for presseforskning, [1972–] : forms. – (Universitetet i Oslo. Institutt for presseforskning. Stensil ; nr 23, etc)
r6112621 P.903/113[no.23etc.]
2473933

ACCESS. Ombudsmen. Great Britain
WHICH Ombudsman? / [prepared by the Office of the Parliamentary Ombudsman ... et al.]. – [London] : H.M.S.O., 1982
[16]p
B83–09241 BS.585/1
2837528

ACCESS. Personal records
to date – Conference proceedings
PROCEEDINGS of a symposium on personal integrity and the need for data in the social sciences / and sponsored by the Swedish Council for Social Science Research ; editors of proceedings : Tore Dalenius, Anders Klevmarken. – [Stockholm] : The Council, [1977]
198p
91–970044–6–4 X.520/29821
8937346

ACCESS. Personal records. Great Britain
Rights of subjects
COHEN, Ruth N. (Ruth Naomi)
Whose file is it anyway?. – London : National Council for Civil Liberties, 1982
77p. – Bibliography: p77
0–901108–97–9 X.529/53146
2877686

ACCESS. Primary health services. England
SIMPSON, Robin
Access to primary care. – London : H.M.S.O., 1979
x,62p. – (Research papers / Great Britain. Royal Commission on the National Health Service ; no.6)
0–11–730116–7 BS.77/108(6)
1878077

ACCESS. Primary health services. England –continued
Access to primary care. – London : H.M.S.O., 1979
x,62p. – (Research papers / Great Britain. Royal Commission on the National Health Service ; no.6)
0–11–730116–7 BS.77/108(6)
1878085

ACCESS. Private land. Peak District. England
By public, to 1981

RICKWOOD, Pat
The story of access in the Peak District. – Bakewell : Peak Park Joint Planning Board, c1982
19p : ill. – (A Peak National Park publication). – Bibliography: p19
0–901428–98–1 BS.554/8
2759659

ACCESS. Private land. Rural regions. Scotland
By public. Agreements

ACCESS agreements. – Redgorton : Countryside Commission for Scotland, 1982
11p
0–902226–61–4 BS.182/45
0710091

COUNTRYSIDE COMMISSION FOR SCOTLAND
Access agreements. – Redgorton : The Commission, [1971?]
i,10p. – (Information sheet ; no.2)
0–902226–06–1 BS.116/115
0710091

ACCESS. Private land. Rural regions. Sweden
By public

BOLLVIK, Rune
Gäst i naturen. – Stockholm : Svenska naturskyddsföreningen, 1977
28p : ill
91–558–7221–2 X.709/25356
1474588

ROSÉN, Bo
Allemansrätt – allemansskyldighet. – [Stockholm] : Rabén & Sjögren, 1976
176p : ill
91–294–6629–6 X.808/11566
1474588

ACCESS. Public places
By physically handicapped persons – For architectural design

KAMM, Jonathan
Designing for disabled people. – [Winchester] : The Department in association with the District Councils of Hampshire, 1981
13p : ill
B82–17403 X.525/5574
2702525

ACCESS. Public records. Great Britain
Rights of public

MODERN public records / Chairman Sir Duncan Wilson ; presented to Parliament by the Lord High Chancellor. – London : H.M.S.O., 1981
xiii,256p,[1]leaf of plates : 1ill. – (Cmnd. ; 8204)
0–10–182040–2 BS24F/168
2411172

ACCESS. Public records. Great Britain
Rights of public. Modern public records

MODERN public records. – London : H.M.S.O., [1982]
ii,29p. – (Cmnd. ; 8531)
0–10–185310–6 BS.24F/183
2763354

ACCESS. Publications
By public – Conference proceedings

LIBRARY ASSOCIATION. Conference (1983 : Torquay)
Papers: access to published information. – London : The Association, c1984
vi,122p. – Includes index
0–85365–876–5 2719.e.814
3022242

ACCESS. Publications. Local authorities. Great Britain
By public

ACCESS to local-government documentation / Capital Planning Information. – [London] : British Library, 1981
vii,73p. – (British Library research & development reports ; no.5619)
0–905984–68–4 X.525/4892
2528525
– [Another copy] OPL

ACCESS. Recreation facilities. Rural regions. England
Role of public transport services

DARTINGTON AMENITY RESEARCH TRUST
Public transport for countryside recreation. – [Cheltenham] : [Countryside Commission], 1976
iv,52p : ill. – (DART publication ; no.21) (CCP94)
0–902590–44–8 BS.116/51
0735043

ACCESS. Recreation facilities. Rural regions. England
Role of public transport services – Conference proceedings

OUT and about / editors Chris Banister & David Groome. – Manchester : Department of Town and Country Planning, University of Manchester, 1984
73p : ill. – (Occasional paper ; no.12). – Bibliography: p69–71
B85–27461 X.525/11028
0735043

ACCESS. Residences
Social aspects

UNIVERSITY OF MANCHESTER. School of Architecture
Access in dwellings. – Manchester : The School, [1975]
[7],160p : ill. – Bibliography: p.156
0–905230–00–0 X.415/5123 Woolwich
0686921

ACCESS. Residents. Rural regions. England
To transport

BANISTER, David
Transport mobility and deprivation in inter-urban areas. – Reading : University of Reading, Department of Geography, 1980
[2],iv,62p : ill. – (Geographical papers ; no. 71). – Bibliography: p.60–62
0–7049–0659–7 X.0519/1008(71)
2307421

ACCESS. Sequential machine-readable files. Magnetic discs. Digital computer systems
Direct access. Logarithmic techniques

HANSON, Owen
Statistical searching of sequential disc files. – London : City University Business School, [1980]
[21] leaves. – (Working paper series / City University (London). Business School ; no.15)
B80–41257 SRL
2321246

ACCESS. Shopping facilities. Great Britain
By physically handicapped persons

CENTRE ON ENVIRONMENT FOR THE HANDICAPPED
Access in the High Street. – London : The Centre, 1981
16p : ill. – Bibliography: on inside cover
0–903976–04–8 X.525/4807
2548933

ACCESS. Stonehenge
Restrictions. Attitudes of visitors

BAINBRIDGE, Sheila
Restrictions at Stonehenge. – London : H.M.S.O., 1979
[4],36p : 1 ill
0–11–700834–6 BS.34/114
2142813

ACCESS COURSES. Further education institutions. England

A SURVEY of "access" courses in the North and Midlands / edited by Susan Lucas and Derek Crossland. – Lancaster : School of Education, University of Lancaster, 1983
iv,48p. – Includes index
0–901272–01–9 X.529/62707
3138240

ACCESS EQUIPMENT
High-rise access equipment – Serials

ACCESS today. – Issue no.1– ; Apr. 1977–. – Edgware : MW Publishers, 1977– : ill
B77–20348 SRL
1132547

ACCESS EQUIPMENT. Cargo ships

BUXTON, Ian Lyon
Cargo access equipment for merchant ships. – London : Spon, 1978
xviii,366p : ill. – Includes index
0–419–11490–4 SRL
1527347

ACCESS EQUIPMENT INDUSTRIES. Hemel Hempstead. Hertfordshire
Access Equipment Limited – Serials

ORBIT. – Vol.1, no.1– ; Feb.1976–. – Hemel Hempstead : Access Equipment Ltd, [1976]– : ill
r6032742 P.625/220
1368125

ACCESS EQUIPMENT LIMITED
– Serials

ORBIT. – Vol.1, no.1– ; Feb.1976–. – Hemel Hempstead : Access Equipment Ltd, [1976]– : ill
r6032742 P.625/220
1368125

ACCESS METHODS. IBM 360 & 370 digital computer systems
Access methods & job control languages

HANNULA, Reino
System 360–370 job control language and the access methods. – Reading, Mass. ; London : Addison-Wesley, 1977
xii,340p : ill. – Includes index
0–201–02755–0 SRL
1304372

ACCESS SURGERY. Veins. Man
– Conference proceedings

INTERNATIONAL CONGRESS ON ACCESS SURGERY (1982 : University of Limburg)
Access surgery. – Lancaster : MTP Press, c1983
xxviii,337p : ill. – Includes index
0–85200–453–2 SRL
292823x

ACCESSIBILITY. Amenities. Rural regions
Measurement

NUTLEY, Stephen D.
The evaluation of accessibility levels in rural areas. – [Cardiff] : Welsh Office, c1981
ii,43p : ill. – Bibliography: p28–30
0–904251–47–0 BS.128A/288
257490x

ACCESSIBILITY. Amenities. Urban regions
Accessibility as criteria for assessment of influence of planners' theories of social justice on urban planning

LOODMER, Tisha
Social justice and planning. – [London] : London School of Economics and Political Science, Graduated Geography Department, 1976
[1],16[i.e. 17]p : ill. – (Discussion papers / London School of Economics and Political Science. Graduate School of Geography ; no.57). – Bibliography: p.15–16
0–85328–054–1 X.0702/279(57)
0774952

ACCESSIBILITY. Arts. Great Britain
Role of universities

FORSTER, W. (William)
The universities, the arts and the public. – London : Arts Council of Great Britain, 1979
[12]p
B80–34538 X.0708/1537(3)
224022x

ACCESSIBILITY. Community hospitals. Rural regions. England

HAYNES, R. M.
Community hospitals and rural accessibility. – Farnborough, Hants. : Saxon House, 1979
vii,200p : ill. – Bibliography: p.196–197. – Includes index
0–566–00271–x X.329/11860 Woolwich
1727133

ACCESSIBILITY. Employment. Cumbria
Implications for unemployed unskilled persons in Cleator Moor & Frizington

SUGDEN, Robert
Unskilled and unemployed in West Cumbria. – Heslington : University of York, Department of Social Administration and Social Work, CDP Unit, 1978
[6],51p : ill. – (Papers in community studies ; no.3)
B79–06307 X.512/6315
1767518

ACCESSIBILITY. Hospitals. Great Britain

WALTERS, Jean
Access to hospitals for visitors. – Cardiff : Transport & Traffic Studies Unit, Department of Town Planning, UWIST, 1982
62,[24]leaves : ill. – Bibliography: p60–62
B84–33632 X.322/13754
3147312

ACCESSIBILITY. Local shopping facilities. Reading. Berkshire

GUY, Clifford M.
The assessment of access to local shopping opportunities. – [Cardiff] : Department of Town Planning, University of Wales Institute of Science and Technology, 1981
i,44p : 1ill. – (Papers in planning research ; 41). – Bibliography: p41–42
B84–08654 X.525/8554
3034720

ACCESSION. Cape Verde
To ACP-EEC Convention of Lomé

CAPE VERDE
[Treaties, etc. European Economic Community, 1977 Mar.28]. Agreement on the accession of the Republic of Cape Verde to the ACP-EEC Convention of Lomé (with final act and exchange of letters), Brussels, 28 March 1977 ... – London : H.M.S.O., [1977]
8p. – (European Communities ; no.30(1977)) (Cmnd.6955)
0–10–169550–0 BS.14/3066
1289810

ACCESSION. Greece
To ACP-EEC Convention of Lomé

[The SECOND ACP-EEC Convention of Lome and Agreement on Products within the Province of the European Coal and Steel Community and Final Act (with Related Documents) (1979). Protocols, etc. 1981 Oct.8]. Protocol to the Second ACP-EEC Convention consequent on the accession of the Hellenic Republic to the Community. - London : H.M.S.O., [1983]
31p. - (European Communities : no.30 (1983)) (Cmnd. : 8889)
0-10-188890-2 BS.14/3678
2987341

ACCESSION. Papua New Guinea
To ACP-EEC Convention of Lomé

EUROPEAN ECONOMIC COMMUNITY
[Treaties, etc. Papua New Guinea. 1977 Mar.28]. Agreement on the accession of Papua New Guinea to the ACP-EEC Convention of Lomé (with final act and exchange of letters). Brussels, 28 March 1977 ... - London : H.M.S.O., [1977]
8p. - (European Communities : no.31 (1977)) (Cmnd.6957)
0-10-169570-5 BS.14/3064
1289837

ACCESSION. São Tomé e Principe
To ACP-EEC Convention of Lomé

EUROPEAN ECONOMIC COMMUNITY
[Treaties, etc. São Tomé e Principe. 1977 Mar.28]. Agreement on the accession of the Democratic Republic of São Tomé and Principe to the ACP-EEC Convention of Lomé (with final act and exchange of letters). Brussels, 28 March 1977 ... - London : H.M.S.O., [1977]
8p. - (European Communities : no.32(1977)) (Cmnd.6958)
0-10-169580-2 BS.14/3062
1289829

ACCESSION. Zimbabwe
To ACP-EEC Convention of Lomé

AGREEMENT on the accession of the Republic of Zimbabwe to the Second ACP-EEC Convention signed at Lomé on 31 October 1979 with final act, Luxembourg, 4 November 1980. - London : H.M.S.O., 1981
10p : ill. - (Cmnd. : 8150) (European Communities ; no.9 (1981))
0-10-181500-x BS.14/3390
2384477

ACCESSION NUMBERS. Patents
- Indexes

WPI accession number index (V) P + Q. - 1976 (X01-X52). - London : Derwent Publications, [1977]
2v.([2]leaves,812p)
Subsequent volumes, when received, will be placed in SRL
B78-02929 SRL
1290029

ACCESSION NUMBERS. Patents on chemical products
- Indexes

CPI/WPI accession number index (V) A-M/Ch. - 1976 (X01-X52). - London : Derwent Publications, [1977]
2v.([2]leaves,1054p)
Subsequent volumes, when received, will be found at SRL
B78-25782 SRL
1520970

ACCESSION NUMBERS. Patents on electrical products
- Indexes

WPI accession number index (V) R. - 1976 (X01-X52). - London : Derwent Publications, [1977]
[2]leaves,483p
Subsequent volumes, when received, will be placed in SRL
B78-02930 SRL
1516914

ACCESSION NUMBERS. Patents on mechanical products
- Indexes

WPI accession number index (V) P + Q. - 1976 (X01-X52). - London : Derwent Publications, [1977]
2v.([2]leaves,812p)
Subsequent volumes, when received, will be placed in SRL
B78-02929 SRL
1516906

ACCESSIONS. Norwegian paintings. Exhibits. Nasjonalgalleriet 1968-1978 - Catalogues

NASJONALGALLERIET
Katalog over norsk malerkunst. - Oslo : Nasjonalgalleriet, 1979
112p
82-990576-0-4 SB.741/4
2452820

ACCESSORIES
See also
Travelling accessories

ACCESSORIES. 35mm single lens reflex cameras

EXPANDING your SLR system / consultant editor David Kilpatrick. - London : Collins, 1981
96p : ill(some col.). - (You and your camera photography series). - Includes index
0-00-411685-2 X.421/22255 Woolwich
2563967

ACCESSORIES. British firearms
1760-1860

NEAL, W. Keith (William Keith)
British gunmakers. - Warminster : Compton Press, c1980
300 166p : ill, 26cm. - Include index
0-900193-58-1 X.622/9579 Woolwich
2360721

ACCESSORIES. Cameras

GAUNT, Leonard
The Focalguide to camera accessories. - London : Focal Press [etc.], 1979
176p : ill(some col). - (Focalguides). - Includes index
0-240-51043-7 WP.13581/123
1879162

ACCESSORIES. Cars
Installation

The TOTAL book of car improvements and accessories. - London : Marshall Cavendish, 1980
245p : col.ill. - Includes index
0-85685-764-5 X.622/11791 Woolwich
2603985

ACCESSORIES. Cars. Exports from Great Britain to Japan
Marketing - For British businessmen

INTERNATIONAL BUSINESS INFORMATION
The auto-accessory market in Japan. - [London] : British Overseas Trade Board : [Distributed by Department of Industry Central Library], 1975
[1],18 leaves : 1 ill
0-85605-171-3 SRL
0602256

ACCESSORIES. Cars. Imports to Japan from Great Britain
Marketing

INTERNATIONAL BUSINESS INFORMATION
The auto-accessory market in Japan. - [London] : British Overseas Trade Board : [Distributed by Department of Industry Central Library], 1975
[1],18 leaves : 1 ill
0-85605-171-3 SRL
0602256

ACCESSORIES. Citizens' Band radio equipment
Construction

BENREY, Ronald M.
CB accessories you can build. - Rochelle Park, N.J. : Hayden Book Co., c1979
110 : ill
0-8104-0854-6 X.629/19915
2605066

PENFOLD, R. A.
CB projects. - London : Babani, 1981
83p : ill. - (BP ; 96)
0-85934-071-6 SRL
2605066

ACCESSORIES. Fashion
- Serials

FASHION extras. - Vol.1, no.1 (June '83)-. - Tonbridge : Benn, 1983-
v. : ill(some col.)
B84-08767 P.443/488
304064x

ACCESSORIES. Fashion
1800-1914

JOHNSON, Eleanor, 1921-
Fashion accessories. - Aylesbury : Shire, 1980
32p : ill. - (Shire album ; 58). - Bibliography: p32
0-85263-530-3 X.0972/143(58)
2359405

ACCESSORIES. Leica cameras

ERNST LEITZ (Firm)
Leica, instructions for use of reproduction devices including visoflexes and bellows etc.. - [Hove] : [Hove Camera Company], [1980]
[78]p : ill
0-906447-04-6 X.629/14277 Woolwich
2312514

ACCESSORIES. Motor vehicles
See also
Motor vehicle accessories industries

ACCESSORIES. Motor vehicles
See also
Motor vehicle accessories shops
Motor vehicle accessories trades
Trim. Motor vehicles

ACCESSORIES. Sporting firearms
1883 - Catalogues

The DIXON catalogue of 1883. - [London] : Ken Trotman, [1984]
58p : ill. - (Arms & armour monographs ; 1). - Includes index
0-946879-05-2 X.809/61514
3175758

ACCESSORY ORGANS. Reproductive system. Male mammals
Physiology

MALE accessory sex organs / edited by David Brandes. - New York ; London : Academic Press, 1974
xi,527p : ill. - Includes bibliographies and index
0-12-125650-2 SRL
0361534

ACCIDENT VICTIMS
See also
Avalanche victims
Fire victims
Hit & run victims

ACCIDENT VICTIMS
Emergency treatment

RESCUE emergency care / edited by Ken Easton. - London : Heinemann Medical, 1977
xix,490p : ill. - Includes bibliographies and index
0-433-08000-0 X.329/11005 Woolwich
1140744

ACCIDENT VICTIMS
Road accident victims: Adolescents. Severe injuries. Recovery

MILLER, Barbara
[Kathy]. All I need is time. - Eastbourne : Kingsway, 1981, c1980
126p
0-86065-106-1 X.958/2848
240589x

ACCIDENT VICTIMS. Australia
Personal injuries. Compensation. Law. Reform, 1973-1979

PALMER, Geoffrey, 19—
Compensation for incapacity. - Wellington [N.Z.] ; Oxford : Oxford University Press, c1979
460p. - Includes index
0-19-558045-1 X.200/34682
2340283

ACCIDENT VICTIMS. England
Personal injuries. Damages. Claims. Law

PERSONAL injury claims. - Guildford : College of Law, c1981
vii,88p. - (College of Law lectures)
B82-11339 X.200/37667
2624966

PERSONAL injury claims. - Guildford : College of Law, c1981
vii,88p. - (College of Law lectures)
B82-22003 X.200/38323
2624966

ACCIDENT VICTIMS. England
Personal injuries. Damages. Claims. Law. Procedure

GOLDREIN, Iain S. (Iain Saville)
Key to personal injury practice and compensation. - London : Oyez, c1981
25p
0-85120-615-8 X.200/37243
0795836

PRITCHARD, John M.
Personal injury litigation. - London : Oyez Publishing, 1976
xvi,264p : ill. - (Oyez practice notes ; 63). - Includes index
0-85120-285-3 WP.1357/63
0795836

PRITCHARD, John M. (John Michael)
Personal injury litigation. - 2nd ed. - London : Oyez Publishing, 1978
iii-xvii,319p : ill. - (Oyez practitioner series). - Includes index
0-85120-371-x X.100/18160
0795836

Personal injury litigation. - 3rd ed. - London : Oyez, 1981
xxi,274p : ill. - (Oyez practitioner series). - Includes index
0-85120-523-2 X.200/35589
0795836
- [Another copy] OPL

ACCIDENT VICTIMS. England
Personal injuries. Damages. Claims. Law. Procedure-continued
Personal injury litigation. - 4th ed. - London : Oyez Longman, 1983
xvi,276p : ill. - Includes index
0-85120-788-x X.200/42993
0795836

ACCIDENT VICTIMS. England
Road accident victims. Personal injuries. Damages. Claims. Law
BINGHAM, Leonard
Bingham's motor claims cases. - 7th ed. - London : Butterworth.
2nd (cumulative) supplement. - 1978
xiv,104p
0-406-11808-6 X.200/7736b
0619353

ACCIDENT VICTIMS. Great Britain
Personal injuries. Compensation
GREAT BRITAIN. Royal Commission on Civil Liability and Compensation for Personal Injury
Report .... - London : H.M.S.O.
Vol.1. - [1978]
xvi,545p. - (Cmnd.7054-I). - Includes index
0-10-170540-9 BS.77/105
144249x
Vol.2: Statistics and costings. - [1978]
ix,259p. - (Cmnd.7054-II)
0-10-170542-5 BS.77/105
144249x
Vol.3: Overseas systems of compensation. - [1978]
vii,280p. - (Cmnd.7054-III). - Includes index
0-10-170543-3 BS.77/105
144249x

ACCIDENT VICTIMS. Great Britain
Personal injuries. Compensation. Great Britain. Royal Commission on Civil Liability and Compensation for Personal Injuries. Reports
ACCIDENT compensation after Pearson / edited by David K. Allen, C.J. Bourn, Jon Holyoak. - London : Sweet and Maxwell, 1979
x,262p. - Includes index
0-421-25810-1 X.100/18901
2050757

ACCIDENT VICTIMS. Great Britain
Personal injuries. Compensation. Great Britain. Royal Commission on Civil Liability and Compensation for Personal Injury. Report. Implications for personal injury liability insurance - Conference proceedings
PEARSON REPORT: THE IMPLICATIONS FOR INSURERS (Conference : 1978 : London)
The Pearson Report. - [London] : [Lloyd's of London Press Ltd], [1978]
53 leaves in various pagings
0-904093-55-7 X.525/3397
2065592

ACCIDENT VICTIMS. Great Britain
Personal injuries. Economic aspects
GHOSH, Debapriya
The economics of personal injury. - Farnborough, Hants. : Saxon House [etc.], 1976
viii,137p : ill. - ([Saxon House studies]). - Bibliography: p.129-133. - Includes index
0-347-01111-x X.520/10517
0610658

ACCIDENT VICTIMS. Great Britain
Road accident victims: Children. Severe injuries. Recovery
COLLINS, Joan, 1933-
Katy. - Large print ed. - Bath : Chivers, 1983
viii,206p. - (A Lythway memoir)
0-85046-923-6 Nov.51790
2739682

ACCIDENT VICTIMS. Great Britain
Road accident victims: Children. Severe injuries. Recovery - Correspondence, diaries
COLLINS, Joan, 1933-
Katy. - London : Gollancz, 1982
156p,[24]p of plates
0-575-03111-5 X.329/15304 Woolwich
2739682
Katy. - London : Corgi, 1983, c1982
153p,[24]p of plates : ill
0-552-12210-6 X.958/20171
2739682

ACCIDENT VICTIMS. Great Britain
Road accident victims: Cyclists & pedestrians
RESEARCH INSTITUTE FOR CONSUMER AFFAIRS
Knocked down. - London : Consumers' Association, 1980
41leaves : ill
B83-08196 X.525/6661
2834693

ACCIDENT VICTIMS. New Zealand
Personal injuries. Compensation. Law
ISON, Terence G. (Terence George)
Accident compensation. - London : Croom Helm, c1980
201p. - Includes index
0-7099-0249-2 X.529/39898
2141035

ACCIDENT VICTIMS. New Zealand
Personal injuries. Compensation. Law. Reform, 1967-1979
PALMER, Geoffrey, 19—
Compensation for incapacity. - Wellington [N.Z.] ; Oxford : Oxford University Press, c1979
460p. - Includes index
0-19-558045-1 X.200/34682
2340275

ACCIDENTAL DEATHS. England
Accidental deaths & violent deaths - Statistics
MORTALITY statistics, accidents and violence / Office of Population Censuses and Surveys. - 1974. - London : H.M.S.O, 1976
viii,35p
Subsequent volumes, when received, will be placed at this shelfmark
0-11-690555-7 BS.34/74
0778567

ACCIDENTAL FIRES. Residences. Special subjects. Comprehension. English language
Safety aspects
ENGLISH. - London : Macmillan Education for the Home Office, 1981
1portfolio : ill(some col.). - (11 to 16+ project fire)
0-333-31771-8 Cup.900.Zb.38
2566893

ACCIDENTS
See also
Accident victims
Collisions
Disasters

ACCIDENTS
See also
Emergencies
Information services on domestic accidents
Injuries
Injuries caused by accidents
Life saving
Petroleum spilled during accident to Christos Bitas (Ship)
Petroleum spilled during accidents to ships
Rescue
Rescue operations
Rescue services
Rescues
Safety measures
Salvage operations
Sinking
Spillage
Spillages
Spillages on land
Wrecks

ACCIDENTS
ROBERTS, John L. (John Laing)
If only-. - Sunbury-on-Thames : Nelson, 1977
64p : ill. - (Nelson health education). - Bibliography: p.60-63
0-17-438224-3 X.311/8702 Woolwich
1399934

ACCIDENTS
1837-1900
ACCIDENTAL Times / compiled by Jane Lambert ; illustrated by Bill Tidy. - London : Allen & Unwin, 1984
136p : ill. - Includes index
0-04-808041-1 X.529/66168
3199096

ACCIDENTS
Astrological aspects
CARTER, Charles Ernest Owen
The astrology of accidents. - [2nd ed.]. - London : Theosophical Publishing House, 1977
123p : ill
0-7229-5045-4 X.529/32030
1335685
- [Another copy: paperback]
ISBN 0-7229-5059-4 X.519/27611

ACCIDENTS
Causes & safety measures
PERROW, Charles
Normal accidents. - New York : Basic Books, c1984
x,386p : ill. - Bibliography: p367-375
0-465-05143-x X.622/23379
1375652

THYGERSON, Alton Luie
Accidents & disasters. - Englewood Cliffs ; London : Prentice-Hall, 1977
xix,411p : ill. - Includes index
0-13-000968-7 X.620/16691 Woolwich
1086820

ACCIDENTS
Emergency action & prevention
MCGEE, Eddie
The emergency handbook. - London : Carousel, 1984
171p : ill. - Includes index
0-552-54252-0 X.329/19441
3274101

ACCIDENTS
Prevention - For children
CHAPMAN, Jean
Beware, take care!. - Sydney ; London : Hodder and Stoughton, 1978
77p : ill(chiefly col)
0-340-23247-1 X.322/8734 Woolwich
1778803

ACCIDENTS
Prevention - For Girl Guides
CAWKELL, Rosemary
Accident prevention. - London : Girl Guides Association, 1984
16p : ill. - (Help with guide badges)
B85-01093 X.629/24488
2240440

ACCIDENTS. Acrylonitrile. Industrial chemicals
Prevention - For management
CHEMICAL INDUSTRY SAFETY AND HEALTH COUNCIL. Safety Committee
The safe design, construction and use of plants producing or consuming acrylonitrile. - London : The Council, 1978
[4],23p : 1 ill. - (Codes of practice for chemicals with major hazards)
B79-30680 X.611/9035 Woolwich
194701x

ACCIDENTS. Aeroplanes
The BLACK box / edited by Malcolm MacPherson. - London : Panther, c1984
156p
0-586-06362-5 X.629/24460
1371371
PILOT error / by the editors of 'Flying magazine'. - New York ; London : Van Nostrand Reinhold, 1977
160p. - Includes index
0-442-22451-6 X.620/17408
1371371
- [Another copy: paperback] ISBN 0-442-22481-8 X.629/18874

ACCIDENTS. Aeroplanes
Survival
BARKER, Ralph
Survival in the sky. - London : Kimber, 1976
186p,[12]p of plates : ill
0-7183-0354-7 X.620/16078 Woolwich
0659134
Survival in the sky. - London : Pan Books, 1977
206p,[8]p of plates : ill
0-330-25109-0 X.708/20947
0659134

ACCIDENTS. Aeroplanes
Survival. Techniques
NESBITT, Paul Homer
A pilot's survival manual. - New York ; London : Van Nostrand Reinhold, 1978
v,338p : ill. - Bibliography: p.322-323. - Includes index
0-442-25992-1 X.620/18415 Woolwich
1816268
- [Another copy: paperback]
ISBN 0-442-25994-8 X.619/19462 Woolwich

ACCIDENTS. Aeroplanes. Mount Erebus. Antarctic
1979
HICKSON, Ken
Flight 901 to Erebus. - Christchurch [N.Z.] ; London : Whitcoulls, 1980
278p,[16]p of plates : ill(some col.)
0-7233-0641-9 X.809/47335
2385775

ACCIDENTS. Aeroplanes. Munich. West Germany
Accidents involving Manchester United Football Club players, 1958. Survivors' experiences
TAYLOR, Frank, 1920-
The day a team died. - London : Souvenir, 1983
192p,[16]p of plates : ill
0-285-62563-2 X.629/20038
2889757

ACCIDENTS. Aeroplanes. Salmon River Mountains. Idaho 1979. Survivors experiences

GZOWSKI, Peter
The Sacrament. - London : Hale, 1981, c1980
203p : ill
0-7091-8916-8 X.622/9829 Woolwich
2377837

ACCIDENTS. Aeroplanes. Tuskar Rock. Wexford (County) 1968

WALSH, Dermot
Tragedy at Tuskar Rock. - Dublin : Mercier, c1983
89p
0-85342-696-1 X.529/59561
3060411

ACCIDENTS. Agricultural tractors. Great Britain
Overturning

HUNTER, A. G. M.
A discussion of tractor overturning accidents on slopes due to stability or control loss. - Penicuik : [Scottish Institute of Agricultural Engineering], 1981
25p : ill. - (Technical report / Scottish Institute of Agricultural Engineering ; no.3)
0-7084-0187-2 SRL
2905329

Tractor accidents on slopes. - Penicuik : [Scottish Institute of Agricultural Engineering], 1981
30p in various pagings : ill. - (Technical report / Scottish Institute of Agricultural Engineering ; no.4)
0-7084-0188-0 SRL
2905329

ACCIDENTS. Aircraft
See also
Players killed in Munich air disaster

ACCIDENTS. Aircraft

BEATY, David
Strange encounters. - London : Methuen, 1982
160p,[12]p of plates : ill. - Bibliography: p152-153. - Includes index
0-413-39760-2 X.520/29288
2848082

GREAT BRITAIN. Accidents Investigation Branch
BAC One-Eleven Type 518 G-AXMG, report on the accident at Stansted Airport, Essex on 27 February 1978. - London : H.M.S.O., 1979
[3],12p. - (Aircraft accident report ; 7/78)
0-11-512447-0 SRL
0000469

Boeing 707 321C G-BEBP, report on the accident near Lusaka International Airport, Zambia, on 14 May 1977. - London : H.M.S.O., 1979
38,[4]p,[29] leaves of plates(4 fold) : ill(some col). - (Aircraft accident report ; 9/78)
0-11-512449-7 SRL
0000469

Boeing 707 Series 436 G-APFK, report on the accident at Prestwick Airport, Scotland on 17 March 1977. - London : H.M.S.O., 1978
[3],39p(4 fold) : ill. - (Aircraft accident report ; 6/78)
0-11-512446-2 SRL
0000469

!Pipe!Piper!. - London : H.M.S.O., 1979
[3],9p. - (Aircraft accident report ; 1/79)
0-11-512450-0 SRL
0000469

Piper PA-34 (Seneca II) G-BFKO. - London : H.M.S.O., c1979
29p,3leaves of plates(1folded) : ill. - (Aircraft accident report ; 7/80). - Bibliography: p29
0-11-513175-2 X.625/693 Woolwich
0000469

Sikorsky S.6 IN helicopter, G-BBHN, report on the accident in the North Sea, north east of Aberdeen on 1 October 1977. - London : H.M.S.O., 1979
[3],28p : ill. - (Aircraft accident report ; 8/78)
0-11-512448-9 SRL
0000469

SPAIN. Civil Aviation Accident Commission
British Airways Trident 1E G-AVYD, report on the accident at Bilboa Airport, Spain on 15 September 1975. - London : H.M.S.O., 1978
[2],22p,2p of plates,[2]leaves of plates(1 fold) : ill. - (Aircraft accident report ; 8/77)
0-11-511812-8 SRL
0000469

ACCIDENTS. Aircraft
Investigation

BARLAY, Stephen
Aircrash detective. - [Revised ed.]. - London : Coronet, 1975
395p. - Bibliography: p.381-382. - Includes index
0-340-19890-7 X.619/15297 Woolwich
0539066

ACCIDENTS. Aircraft
Investigation - Conference proceedings

SYMPOSIUM ON INTERNATIONAL AIRCRAFT ACCIDENTS INVESTIGATION
(1973 : London)
Symposium on International Aircraft Accidents Investigation, 15th January 1973. - London : Royal Aeronautical Society, [1975?]
58 leaves in various pagings
0-903409-01-1 X.105/500
0539066

ACCIDENTS. Aircraft. Andes. Mendoza Province Argentina 1972. Survivors' experiences

READ, Piers Paul
Alive. - London : Pan Books, 1975
xii,318p,[16]p of plates : ill
0-330-24392-6 X.708/20174
0097020

ACCIDENTS. Aircraft. Antarctic
to date

VETTE, Gordon
Impact Erebus. - Auckland : Hodder and Stoughton, 1983
347p : ill
0-340-32010-9 X.622/22256
9130438

ACCIDENTS. Aircraft. Barbados
1976 accidents

LEÓN COTAYO, Nicanor
Crimen en Barbados. - 2a ed. - La Habana : Editorial de Ciencias Sociales, 1978
295p : ill
r6700525 X.629/14876 Woolwich
2481936

ACCIDENTS. Aircraft. Nairobi. Kenya
1974. Survivors' experiences

MOORHOUSE, Earl
Wake Up, it's a crash!. - London : Corgi, 1982, c1980
172p,[16]p of plates : ill. - Includes index
0-552-11932-6 X.808/36395
2149141

ACCIDENTS. Aircraft. Tulare County. California
1976. Survivors' experiences

ELDER, Lauren
And I alone survived. - London : Collins, 1978
xiv,188p
0-00-216781-6 X.809/43372
1544322

And I alone survived. - London : Pan Books, 1978
159p
0-330-25911-3 X.708/22262
1544322

ACCIDENTS. Albion Colliery
Explosions, 1894

EVANS, R. Meurig (Richard Meurig)
One Saturday afternoon. - [Cardiff] : National Museum of Wales, 1984
87p : ill. - Bibliography: p87
0-7200-0293-1 X.809/64572
3284964

ACCIDENTS. American civil aircraft
Investigations by United States. Bureau of Aviation Safety

DORMAN, Michael
Detectives of the sky. - New York ; London : F. Watts, 1976
[9],106,[1]p : ill. - Bibliography: p.97. - Includes index
0-531-00342-6 X.620/16818 Woolwich
113342x

ACCIDENTS. Bicycles & motorcycles. Road traffic

ORGANISATION FOR ECONOMIC CO-OPERATION AND DEVELOPMENT
Safety of two-wheelers. - Paris : Organisation for Economic Co-operation and Development ; [London] : [H.M.S.O.], 1978
123p : ill. - (Road research). - Bibliography: p.111-120
B78-37635 SRL
1701215

ACCIDENTS. Boilers. Steam locomotives. Great Britain
Explosions, to 1962

HEWISON, Christian H.
Locomotive boiler explosions. - Newton Abbot : David & Charles, c1983
144p : ill. - Includes index
0-7153-8305-1 X.629/21193
2920956

ACCIDENTS. Buses. Great Britain
Accidents involving full-time drivers compared with accidents involving part-time drivers

FULL time/part time PSV driver accident study. - [London] : Department of Transport, 1982
8p
B84-05293 BS.43/664
2995832

ACCIDENTS. Cargo handling. Ports. Great Britain
Costs

BEARHAM, John
The cost of accidents within the port transport industry. - London : National Ports Council, 1976
vi,34p : ill
0-901058-89-0 BS.43/407
0770221

ACCIDENTS. Cars

KOSSMANN, Alfred
Laatst ging ik spelevaren. - Amsterdam : Querido, 1973
82 p.
90-214-1079-6 X.709/20390
0849790

ACCIDENTS. Cars
Effects on child passengers. Limitation by seat belts & booster cushions

BACON, D. G. C.
The restraint of children in car impacts using booster cushions and adult seat belts. - [Nuneaton] : [Motor Industry Research Association], 1982
72p,[39]leaves of plates : ill. - (Report ; no.K42703)
r6810632 P.523/1928
2972514

ACCIDENTS. Chemical engineering plants. Great Britain
Explosions

The EXPLOSION and fire at Chemstar Limited 6 September 1981 / a report by HM Factory Inspectorate. - London : H.M.S.O., 1982
17p,[3]leaves of plates,(some folded) : ill
0-11-883666-8 BS.438/156.
0787043

GREAT BRITAIN. Factory Inspectorate
The explosion at Laporte Industries Ltd, Ilford, 5 April 1975. - [London] : H.M.S.O., 1976
24p : ill
0-11-880333-6 SRL
0787043

The explosion at the Dow chemical factory, King's Lynn, 27 June 1976. - [London] : H.M.S.O., 1977
19p : ill
0-11-883003-1 BS.438/93
0787043
- [Another copy] SRL

ACCIDENTS. Chemical engineering plants. Seveso. Italy
Explosions, 1976. Effects

CONTI, Laura
Visto da Seveso. - Milano : Feltrinelli, 1977
218p. - (Attualità)
r6106838 X.519/43230
1852450

ICMESA / Giovanni Cerruti ... [et al.]. - Milano : G. Mazzotta, 1976
119p. - (Nuova informazione ; 55)
r6068271 X.619/10563 Woolwich
1852450

MARGERISON, Tom
The superpoison. - London : Macmillan, 1979
236p. - Bibliography; p.227-229. - Includes index
0-333-22797-2 X.809/44958
1852450

The superpoison 1976-1978. - London : Macmillan, 1981
236p. - Bibliography: p227-229. - Includes index
B81-23262 X.629/15607
1852450

ACCIDENTS. Children
Prevention

PLAY it safe!. - Dublin : Health Education Bureau, [1984?]
36p : ill(some col.)
B85-01853 X.322/14243
3197999

ACCIDENTS. Children
Prevention - Serials

The TUFTY times. - Spring 1979-. - Birmingham : The Royal Society for the Prevention of Accidents, 1979-
v. : ill
B81-39739 P.993/311
2574020

ACCIDENTS. Children. Great Britain
Causes & safety measures

CHILDREN, the environment and accidents / edited by R.H. Jackson. - Tunbridge : Pitman Medical, 1977
[11],164p : ill. - Includes bibliographies and index
0-272-79406-6 X.319/17519 Woolwich
1198785

ACCIDENTS. Children. Pedestrians. Manchester & Salford. Greater Manchester (Metropolitan County)
1969

PRESTON, Barbara, 1915-
Child pedestrian accidents in Manchester and Salford. - Manchester : Institute of Advanced Studies, 1976
[1],59p : ill. - (Manchester studies). - Bibliography: p.58-59
B78-05150 X.512/5778
1297740

ACCIDENTS. Children. Residences. Great Britain
1977

GREAT BRITAIN. Consumer Safety Unit
The home accident surveillance system. - London : The Unit, 1979
[7],66p : 1 ill
B79-20971 BS.431/24
1901699

ACCIDENTS. Chlorine. Industrial chemicals. Great Britain
Emergency action

CHEMICAL INDUSTRIES ASSOCIATION
Inter-company collaboration for chlorine emergencies. - London : C.I.A., 1978
14, [5]p : 1 col ill
B80-40793 X.529/38282
2314290

ACCIDENTS. Civil aircraft. Great Britain
Investigation. Methods - For police

GREAT BRITAIN. Accidents Investigation Branch
Memorandum on the investigation of civil air accidents. - [Revised ed.]. - London : [The Department], 1976
[1],iv,25p : ill
0-905660-00-5 BS.41/1137
0740004

ACCIDENTS. Coal mines. Great Britain

The EXPLOSION at Cardowan colliery, Stepps, Strathclyde Region 27 January 1982 / a report by HM Inspectorate of Mines and Quarries. - [London] : H.M.S.O., 1982
vi,9p,[2]folded leaves of plates
0-11-883644-7 SRL
002080x

GREAT BRITAIN. Inspectorate of Mines and Quarries
The accident at Bentley Colliery, South Yorkshire, 21 November 1978. - [London] : H.M.S.O., 1979
18p : ill
0-11-883287-5 BS.438/107
002080X

The explosion at Golborne Colliery, Greater Manchester County, 18 March 1979. - [London] : H.M.S.O., 1979
v,15p,3 leaves of plates : ill(some col)
0-11-883288-3 BS.438/101
002080x

ACCIDENTS. Coal mines. Great Britain
to date

ACCIDENTS and disasters in coal mines / [compiled by the] National Coal Board, Headquarters Library. - London : National Coal Board, Headquarters Library, 1978
iii,70p. - Includes bibliography
r6814455 BS82/181
9140271

ACCIDENTS. Coal mines. Leigh region. Greater Manchester (Metropolitan County)
1856-1910

MINING disasters in the Leigh area / edited by Tim Macquiban and Norma Ackers. - Leigh : Leigh Local History Society, 1975
[4],24p : 1 ill. - (Publications / Leigh Local History Society ; no.3). - Bibliography: p.24
0-905235-02-9 X.0700/1005(3)
0606146

ACCIDENTS. Coal mines. Most. Czechoslovakia
to 1918

KOŠEK, Zdeněk
Zborcení města. - V Ústí nad Labem : Severočeské nakladatelství, c1980
74p,[8]p of plates : ill
r6439941 X.629/16500 Woolwich
8778493

ACCIDENTS. Coal mines. Sacriston. Durham (County)
1903

The SACRISTON mine disaster / [compiled] by Gavin Purdon. - Beamish : [The compiler], [1979]
56,[4]p : ill
B80-34356 X.708/22553
2220938

ACCIDENTS. Commercial aircraft
Role of pilot error

PILOT error / edited by Ronald Hurst. - London : Crosby Lockwood Staples, 1976
282p : ill. - Includes bibliographies and index
0-258-97072-3 X.620/16426 Woolwich
0764604

PILOT error. - 2nd ed.. - London : Granada, 1982
x,185p : ill. - Includes bibliographies and index
0-246-11499-1 X.622/15139
0764604

ACCIDENTS. Compressed gases. Great Britain
Emergency action

BRITISH COMPRESSED GASES ASSOCIATION
Guidance notes on the preparation of major procedures. - London : British Compressed Gases Association, 1977
16p
B81-25950 X.329/13981 Woolwich
2501686

ACCIDENTS. Construction industries. Great Britain
Fatal accidents

GREAT BRITAIN. Factory Inspectorate
Fatal accidents in construction, 1977. - London : H.M.S.O., 1979
16p
0-11-883238-7 SRL
1813390

ACCIDENTS. Construction industries. Scotland
Fatal accidents, 1969-1975

GREAT BRITAIN. Factory Inspectorate
One hundred fatal accidents in construction. - [London] : H.M.S.O., 1978
19p
0-11-883071-6 BS.438/61
1480596

ACCIDENTS. Construction sites. Power stations. Great Britain

The HOIST accident at Littlebrook 'D' power station, 9 January 1978 / a report by HM Factory Inspectorate. - London : H.M.S.O., 1981
16p : ill
0-11-883273-5 SRL
2172321

ACCIDENTS. Consumer goods. Organisation for Economic Co-operation and Development countries
Data. Recording

ORGANISATION FOR ECONOMIC CO-OPERATION AND DEVELOPMENT.
Committee on Consumer Policy
Data collection systems related to injuries involving consumer products. - Paris : O.E.C.D. ; [London] : [H.M.S.O.], 1978
72p : ill
B79-08761 UNP.108/185
1814060

ACCIDENTS. Container depots. Great Britain
Explosions

GREAT BRITAIN. Factory Inspectorate
The fire and explosion at Braehead container depot, Renfrew, 4 January 1977. - London : H.M.S.O., 1979
27p : ill
0-11-883220-4 SRL
1737996

ACCIDENTS. Cosmos 954. McKenzie District. Northwest Territories
1978

HEAPS, Leo
Operation Morning Light. - New York ; London : Paddington Press, 1978
208p : ill. - Bibliography: p.201-203. - Includes index
0-7092-0323-3 X.620/18391 Woolwich
1871617

ACCIDENTS. DeHavilland Mosquito aeroplanes
1942-1964

SMITH, David J.
De Havilland Mosquito. - 2nd ed. (fully rev.). - Earl Shilton : Midland Counties Publications, c1980
48p : ill
0-904597-33-4 X.629/18371
2773384

ACCIDENTS. Educational institutions
First aid

IRELAND, Kenneth
Illness and injury at school. - Kettering : Schoolmaster Publishing Company Ltd, 1978
89p : ill. - Includes index
0-900642-29-7 X.319/17977 Woolwich
1405969

ACCIDENTS. England
Causes & safety measures

CLIFF, Kenneth Stephenson
Accidents. - London : Croom Helm, c1984
301p : ill. - Bibliography: p274-296. - Includes index
0-7099-0792-3 X.529/65044
3050831

ACCIDENTS. England
Compensation. Law

ATIYAH, Patrick Selim
Accidents, compensation and the law. - 2nd ed. - London : Weidenfeld and Nicolson, 1975
xxii,646p. - (Law in context). - Includes index
0-297-76964-2 X.100/16245
061520x

ACCIDENTS. England
Fatal accidents - Statistics - Serials

GREAT BRITAIN. Office of Population Censuses and Surveys
OPCS monitor Deaths from accidents. - DH 4 75/1- . - London : OPCS, 1975-
r6000613 BS.34/70(10)
1316575

ACCIDENTS. England
Law

VANDYK, Neville David
Accidents and the law. - 2nd ed. - London : Oyez Publishing, 1979
xxi,162p. - (It's your law). - Includes index
0-85120-463-5 X.108/17863
029084x

ACCIDENTS. Ethylene oxide. Industrial chemicals
Prevention - For management

GUIDELINES for bulk handling of ethylene oxide. - London : Chemical Industries Association, c1983
28p : 2ill
B84-31752 X.622/19987
1037714

ACCIDENTS. Extraction. Offshore natural gas & petroleum deposits
Blowouts. Prevention & control

WELL control and blowout prevention / Drilling & Production Technology Training Centre. - Montrose : Offshore Petroleum Industry Training Board, 1984
1v(looseleaf) : ill
B85-12529 L.45/3149
3244008

ACCIDENTS. Extraction. Petroleum deposits
Blowouts. Prevention & control

BLOWOUT prevention and well control / Chambre syndicale de la recherche et de la production du pétrole et du gaz naturel, Comité des techniciens, Commission exploitation sous-commission forage ; translation from the French by Paul W. Ellis. - London : Graham & Trotman, 1981
xvi,164p,[2]folded leaves of plates : ill
0-86010-337-4 SRL
2548704

ACCIDENTS. Extraction. Petroleum deposits
Blowouts. Prevention & control - Conference proceedings

MANAGING technological accidents / David W. Fischer, editor. - Oxford : Pergamon, 1982
xii,234p : ill. - (IIASA proceedings series ; v.16). - Bibliography: p90-91. - Includes index
0-08-029346-8 X.622/13311
2548704

ACCIDENTS. Factories
Investigation. Methods

FERRY, Ted S.
Modern accident investigation and analysis. - New York ; Chichester : Wiley, c1981
xxiv,273p : ill. - Includes index
0-471-07776-3 X.520/23960
2490242

ACCIDENTS. Factories
Role of alcohol

ARGYROPOULOS-GRISANOS, M. A.
Alcohol and industrial accidents. - London : Christian Economic Research Foundation, [1980?]
52p : ill. - Bibliography: p51-52
B81-31116 X.529/42928
2535459

ACCIDENTS. Factories. Great Britain
Causes & prevention

FACTORY accidents. - 13th ed. - Birmingham : Royal Society for the Prevention of Accidents, c1982
28p : col.ill
0-900635-59-2 X.629/19891
2882221

ACCIDENTS. Factories. Great Britain
Causes & prevention - Conference proceedings

SAFETY at work / edited by Jennifer Phillips. - Oxford : Centre for Socio-Legal Studies, Wolfson College ; [London] : Social Science Research Council, 1977
[2],iv,86p : ill. - (Conference papers / Wolfson College Centre for Socio-Legal Studies ; no.1)
0-900296-65-8 BS.379/46
2882221

ACCIDENTS. Factories. Great Britain
Economic aspects

LEES, Dennis
Economic aspects of industrial accidents in Great Britain. - [Stockport] : Manchester Statistical Society, [1977?]
32p : 1 ill
B78-13364 X.519/27820
1405985

ACCIDENTS. Factories. Great Britain
Prevention

GREAT BRITAIN. Accident Prevention Advisory Unit
Success and failure in accident prevention. - London : H.M.S.O., 1976
[3],44p : ill. - Bibliography: p.43-44
0-11-880330-1 BS.385/261
1037765

ACCIDENTS. Factories. United States
Causes & prevention

READINGS in industrial accident prevention / edited by Dan Petersen and Jerry Goodale. - New York ; London : McGraw-Hall, 1980
xiv,393p : ill. - Includes bibliographies
0-07-049591-2 X.620/8273 Woolwich
2235013

ACCIDENTS. Gas industries
Explosions. Prevention & control - Conference proceedings

The CONTROL and prevention of gas explosions. - London : Oyez Scientific and Technical Services, c1984
iv,251p : ill
0-907822-38-x SRL
3049736

ACCIDENTS. Halifax. Nova Scotia
Explosions, 1917

The HALIFAX explosion December 6, 1917 / compiled and edited by Graham Metson ; including the complete text of 'The Halifax disaster' by Archibald MacMechan. - Toronto ; London : McGraw-Hill, 1978
173p : ill
0-07-082797-4 X.802/11214
2109409

ACCIDENTS. Handley Page Hampden aeroplanes
1939-1944

ROBERTS, Nicholas
Handley Page Hampden & Hereford. - Earl Shilton : Midland Counties Publications, c1980
56p : ill. - Bibliography: p56
0-904597-34-2 X.629/18370
2773376

ACCIDENTS. Immigrant personnel. Factories. Great Britain

LEE, Gloria L.
Accidents are colour-blind. - [London] : Community Relations Commission for the Birmingham Community Relations Council, 1977
24p : ill. - Bibliography: p.20
B77-18802 BS.412/56
1132067

ACCIDENTS. India (Republic)
Law

SHRIVASTVA, H. P.
Law relating to accidents. - Indore : National Law House, 1974-75
2 v.
r6028601 X.200/10650
1015249

ACCIDENTS. Industrial chemicals
Prevention - Conference proceedings

MAJOR CHEMICAL HAZARDS (Conference : 1978 : Lorch Foundation)
Proceedings of the seminar on major chemical hazards held at the Lorch Foundation 26 and 27 April 1978. - [Abingdon] : Harwell, 1978
[1],x,243p : ill. - (Harwell environmental seminar)
B79-20968 SRL
190180x

ACCIDENTS. Industrial motor vehicles. Great Britain

TRANSPORT kills : a report by HM Factory Inspectorate. - London : H.M.S.O., 1982
79p : ill
0-11-883659-5 SRL
2901994
- [Another copy] X.622/20679

ACCIDENTS. Lancaster Pit. Universal Colliery
1913

BROWN, John H.
The valley of the shadow. - Port Talbot : Alun Books, c1981
171p : ill. - Bibliography: p163-164. - Includes index
0-907117-07-4 X.809/49998
2592312

ACCIDENTS. McDonnell Douglas DC10 aeroplanes
to 1974

GODSON, John
The rise and fall of the DC-10. - London : New English Library, 1975
351p,[8]p of plates : ill
0-450-02355-9 X.629/10173
0514136

ACCIDENTS. McDonnell Douglas DC10 aeroplanes. Ermonville. France
1974. Implications for safety measures in design of McDonnell Douglas DC10 aeroplanes

EDDY, Paul
Destination disaster. - London : Hart-Davis, 1976
x,435p,[32]p of plates : ill. - Includes index
0-246-10883-5 X.620/16407 Woolwich
0762105

ACCIDENTS. Mines. Germany
to date

PABST, Wolfgang
350 Männer starben. - Herne : Wolf, 1982
154p. : ill
3-923400-01-2 X.629/23800
8832048

VÖLKENING, Ulrich
Unfallentwicklung und -verhütung im Bergbau des Deutschen Kaiserreiches. - Bremerhaven : Wirtschaftsverlag NW, 1980
127p,[14]p of plates. - (Schriftenreihe Arbeitsschutz ; Nr.23). - Bibliography
3-88314-111-9 S.F.583/67[no.23]
8832048

ACCIDENTS. Mont Blanc (Ship). Merchant ships. Halifax. Nova Scotia
Explosions, 1917

BIRD, Michael, b.1928
The town that died. - London : New English Library, 1977
144p
0-450-03470-4 X.708/21009
1343610

ACCIDENTS. Norway
Role of alcohol & drugs affecting nervous system

REIGSTAD, Astor
Ulykker, alkohol og nervemedisin. - Oslo : Universitetsforlaget, 1977
104p. - (Statens institutt for alkoholforskning. Skrifter ; nr 16). - Includes bibliographies
82-00-09445-6 S.B.703
1427806

ACCIDENTS. Nuclear engineering industries
1945-1977

TIBALDI, Ettore
Il disastro nucleare. - Milano : Collettivo editoriale Librirossi, 1978
84p
r6107414 X.619/11326 Woolwich
245016X

ACCIDENTS. Nuclear power industries
Medical aspects - Conference proceedings

MEDICAL BASIS FOR RADIATION ACCIDENT PREPAREDNESS (Conference : 1979 : Oak Ridge)
The medical basis for radiation accident preparedness. - New York ; Oxford : Elsevier/North-Holland, c1980
xxxviii,545p : ill. - Includes index
0-444-00431-9 SRL
2360314

ACCIDENTS. Nuclear power industries. United States

BURLESON, Clyde W.
The day the bomb fell. - London : Sphere, 1980
316p : 1 ill
0-7221-2060-5 H.80/302 Woolwich
2111063

ACCIDENTS. Nuclear power stations. Three Mile Island. Pennsylvania

ANSWERS to frequently ashed questions about cleanup activities at Three Mile Island, Unit 2 / [prepared by] US Nuclear Regulatory Commission, Office of Nuclear Reactor Regulation, Three Mile Island Program Office. - Washington, D.C. : Nuclear Regulatory Commission, Office of Nuclear Reactor Regulation, Three Mile Island Program Office, 1980
34p. - (NUREG ; 0732)
r6801008 AS:783/374
2020440

ENVIRONMENTAL assessment of radiological effluents from data gathering and maintenance operation on Three Mile Island Unit 2 / [prepared by] US Nuclear Regulatory Commission, Office of Nuclear Reactor Regulation Three Mile Island Program Office. - Washington, D.C. : Nuclear Regulatory Commission, Office of Nuclear Reactor Regulation, Three Mile Island Program Office, 1980
6p. - (NUREG ; 0681)
r6801007 AS:783/359
2020440

NRC Action Plan developed as a result of the TMI-2 accident / [prepared by] US Nuclear Regulatory Commission. - Washington, D.C. : The Commission, 1980
2v.. - (NUREG-0660)
r6202859 AS:783/358
2020440

PRESIDENT'S COMMISSION ON THE ACCIDENT AT THREE MILE ISLAND
Report of the President's Commission on the Accident at Three Mile Island. - New York ; Oxford : Pergamon, [1979]
201p : ill
0-08-025949-9 X.622/14025
2020440

TMI-1 restart : evaluation of licensee's compliance with the short-and long-term items of Section II of the NRC Order dated August 9, 1979, Metropolitan Edison Company, et al., Three Mile Island Nuclear Station, Unit 1, Docket 50-289 / [prepared by] US Nuclear Regulatory Commission, Office of Nuclear Reactor Regulation, Division of Project Management. - Washington, D.C. : The Commission, 1980
1v.(various pagings). - (NUREG-0680)
r6202851 AS:783/362
2020440
- Supplement no.2-

TMI-RELATED requirements for new operating licenses / [prepared by] US Nuclear Regulatory Commission. - Washington, D.C. : The Commission, 1980
38p. - (NUREG-0694)
r6202902 AS:783/377
2020440

UNITED STATES. Congress. Senate. Committee on Environment and Public Works. Subcommittee on Nuclear Regulation
Nuclear accident and recovery at Three Mile Island. - Washington, DC : Govt. Print. Off., 1980
vii,423p : ill. - (96th Congress, 2nd Session, Committee Print) (Serial no.96-14)
r6201529 As.50/67(30)
2020440

UNITED STATES. Nuclear Regulatory Commission. Three Mile Island Program Office
Answers to questions about removing krypton from the Three Mile Island, Unit 2 Reactor Building. - Washington, DC : The Commission, 1980
ii,17p.. - (NUREG-0673)
r6201955 AS:783/352
2020440

UNITED STATES. Office of Nuclear Reactor Regulation. TMI-2 Lessons Learned Task Force
Final report. - Washington, D.C. : US Nuclear Regulatory Commission, Office of Nuclear Reactor Regulation, 1979
1v([various pagings]). - (NUREG ; 0585)
r6801009 AS:783/316
2020440

ACCIDENTS. Nuclear power stations. Three Mile Island. Pennsylvania
- Serials

UNITED STATES. President's Commission on the Accident at Three Mile Island
[Staff reports of various task forces] to the President's Commission on the Accident at Three Mile Island. - 1979-
- Washington, DC : Govt Print. Off., 1980-
r6200878 AS.403/559
2020440

ACCIDENTS. Nuclear power stations. Three Mile Island. Pennsylvania
to date

ACCIDENT at Three Mile Island / edited by David L. Sills, C.P. Wolf and Vivien B. Shelanski. - Boulder : Westview, 1982
xviii,258p. - (A Westview special study). - Includes bibliographies
0-86531-187-0 X.529/47182
8872392

CANTELON, Philip L. (Philip Louis)
Crisis contained. - Carbondale : Southern Illinois University Press, [1982]
xxi,213p,19p of plates : ill. - (Science and international affairs series)
0-8093-1079-1 X.622/17497
8872392

DEL TREDICI, Robert
The people of Three Mile Island. - San Francisco : Sierra Club Books, c1980
127p : ill
0-87156-237-5 X.805/4066
8872392

FORD, Daniel F.
Three mile island. - Harmondsworth : Penguin Books, 1982
271p
0-14-006048-0 X.529/69478
8872392

MARTIN, Daniel, 1948-
Three Mile Island. - Cambridge, Mass. : Ballinger, c1980
ix,253p
r6807876 X.622/13885
8872392

ACCIDENTS. Nuclear power stations. Three Mile Island. Pennsylvania
1978

GRAY, Mike
The warning. - New York ; London : Norton, c1982
287p : ill. - Includes index
0-393-01522-x X.809/56469
2897148

ACCIDENTS. Nuclear power stations. Three Mile Island. Pennsylvania
1978

STEPHENS, Mark
Three Mile Island. - London : Junction, c1980
x,245p. - Includes index
0-86245-034-9 X.800/35272
2897148
- [Another copy] X.800/35271

ACCIDENTS. Nypro (UK) Limited
Explosions, 1974. Implications for safety measures in British industries

TAYLOR, Henry Dennis
Flixborough. - London : Keith Shipton Developments Limited ; Cambridge : Woodhead-Faulkner Limited, 1975
ii,30p : 1 ill. - (Special studies / Keith Shipton Developments ; no.3)
0-85941-028-5 X.519/25134
0365351

ACCIDENTS. Oil tankers. Ships

The TARPENBEK incident / Department of Trade. - London : The Department, 1979
31p
B83-08290 BS.41/1342
2835096

ACCIDENTS. Pedestrians. Great Britain
Psychological factors

PEDESTRIAN accidents / edited by Antony J. Chapman, Frances M. Wade and Hugh C. Foot. - Chichester : Wiley, c1982
xii,354p : ill. - Bibliography: p293-339. - Includes index
0-471-10057-9 X.520/27785
2764784

ACCIDENTS. Pedestrians. Ireland (Republic)

GOLDEN, J. M.
Pedestrian accidents. - Dublin : Foras Forbartha, 1980
[1],10 leaves. - (RS225A)
B80-35039 X.525/3607
2274566

Pedestrian accidents = Timpistí coisithe. - Dublin : Foras Forbartha, 1980
v,145p : ill. - (RS 225)
0-906120-32-2 X.525/3606
2274566

ACCIDENTS. Personnel
Role of stress

RADL, Gerald W.
Psychische Beanspruchung und Arbeitsunfall. - Bremerhaven : Wirtschaftsverlag NW, 1975
317p.. - (Forschungsbericht / Bundesanstalt für Arbeitsschutz und Unfallforschung ; Nr.145). - Bibliography: p.241-311
r6303744 S.F.583/283[Nr.145]
8823553

ACCIDENTS. Personnel. Health services. England
Law

FARNDALE, W. A. J. (William Arthur James)
Law on accidents to health service staff and volunteers. - Beckenham : Ravenswood Publications Ltd, 1977
72p. - (Case studies on law and practice for health service management ; no.7)
0-901812-23-4 X.0319/108(7) Woolwich
1301306

ACCIDENTS. Platforms. North Sea
to date

["ALEXANDER L. Kielland"-ulykken. English]. The "Alexander L. Kielland"-accident / from a commission appointed by Royal Decree of 28th March 1980 [chairman : Thor Næsheim]. - Oslo : [s.n.], 1981
vi,472p : ill
r6611683 X.525/10285
8938113

ENGHAUG, Aage
Katastrofe i Nordsjøen. - Oslo : Dreyers, 1980
172p : ill
82-09-01825-6 X.529/53035
8938113

ACCIDENTS. Railways. Grantham. Lincolnshire
1906

BONNETT, Harold
The Grantham rail crash of 1906. - [Grantham] : Bygone Grantham, 1978
40p : ill. - (The bygone series)
0-906338-05-0 X.709/33280
1917749

ACCIDENTS. Railways. Great Britain

GREAT BRITAIN. Railway Inspectorate
Railway accident, report on the collision that occurred on 3rd August 1977 near New Cross in the Southern Region, British Railways. - London : H.M.S.O., 1979
[3],6p : 1 ill
0-11-550495-8 SRL
0000450

Railway accident, report on the collision that occurred on 4th July 1977 near Hall Road Station in the London Midland Region, British Railways. - London : H.M.S.O., 1979
[2],14p
0-11-550500-8 SRL
0000450

Railway accident, report on the collision that occurred on 5th September 1977 at Farnley Junction, Leeds in the Eastern Region, British Railways. - London : H.M.S.O., 1979
[2],21p
0-11-550498-2 SRL
0000450

Railway accident, report on the collision that occurred on 6th August 1975 at Weaver Junction in the London Midland Region, British Railways. - London : H.M.S.O., 1978
[2],16p,fold plate : ill
0-11-550480-x SRL
0000450

Railway accident, report on the collision that occurred on 16th February 1977 at Darlington in the Eastern Region, British Railways. - London : H.M.S.O., 1978
11p,[2]fold leaves of plates : ill
0-11-550485-0 SRL
0000450

Railway accident, report on the collision that occurred on 26th October 1975 at Lunan Bay, between Montrose and Arbroath in the Scottish Region, British Railways. - London : H.M.S.O., 1976
11p,fold plate : ill
0-11-550410-9 SRL
0000450

Railway accident, report on the derailment of a parcels train that occurred on 2nd July 1977 between Wolverton and Bletchley in the London Midland Region, British Railways. - London : H.M.S.O., 1978
[2],10p,fold plate : ill
0-11-550484-2 SRL
0000450

Railway accident, report on the derailment that occurred on 2nd June 1976 near Reading station in the Western Region, British Railways. - London : H.M.S.O., 1978
[2],13p,plate : 2 ill
0-11-550493-1 SRL
0000450

Railway accident, report on the derailment that occurred on 6th May, 1977 at Borough Market Junction in the Southern Region, British Railways. - London : H.M.S.O., 1978
[1],10p,plate : ill
0-11-550488-5 SRL
0000450

TRAINS in trouble. - St Day : Atlantic.
Vol.1. - 1980
[48]p : ill
0-906899-01-x X.629/18471
2783681

Vol.2. - 1981
[48]p : ill
0-906899-03-6 X.629/18489
2783681

ACCIDENTS. Railways. Great Britain
to 1968

NOCK, O. S. (Oswald Stevens)
Historic railway disasters. - [2nd ed.]. - London : Arrow Books, 1978
255p,[32]p of plates : ill. - Includes index
0-09-907720-5 X.708/20992
1335456

ACCIDENTS. Railways. Great Britain
to 1975

ROLT, L. T. C. (Lionel Thomas Caswall)
Red for danger. - 3rd ed.. - Newton Abbot : David and Charles, 1976
5-297p,12p of plates : ill. - Bibliography: p.293. - Includes index
0-7153-7292-0 X.629/11070 Woolwich
1055003

Red for danger. - 3rd (revised and extended) ed.. - London : Pan Books, 1978
301p,[8]p of plates : ill. - Bibliography: p.294. - Includes index
0-330-25555-x X.619/18430 Woolwih
1055003

ACCIDENTS. Railways. Great Britain
to 1980

NOCK, O. S. (Oswald Stevens)
Historic railway disasters. - 3rd ed. - London : Ian Allan, 1983
294p : ill. - Includes index
0-7110-1205-9 X.622/16736
2784157

ROLT, L. T. C. (Lionel Thomas Caswall)
Red for danger. - 4th ed.. - Newton Abbot : David & Charles, 1982
308p,[8]p of plates : ill. - Bibliography: p308. - Includes index
0-7153-8362-0 X.629/18566
2784157

ACCIDENTS. Railways. Great Britain
1906-1960

GERARD, Malcolm
Rails to disaster. - London : Allen & Unwin, 1984
124p : ill. - (Steam past). - Includes index
0-04-385103-7 X.809/58160
2537443

HAMILTON, J. A. B. (James Alan Bousfield)
Trains to nowhere. - 2nd ed.. - London : Allen & Unwin, 1981
96p : ill. - (Steam past). - Includes index
0-04-385084-7 X.629/16088 Woolwich
2537443

ACCIDENTS. Railways. Harrow (London Borough)
1952

COOMBS, L. F. E.
The Harrow railway disaster 1952. - Newton Abbot : David and Charles, 1977
164p : ill. - Bibliography: p.156-158. - Includes index
0-7153-7409-5 X.629/11474 Woolwich
1244213

ACCIDENTS. Residences

GARDNER, Mary
People get hurt. - Hamilton : Lanarkshire Health Board, [1978]
[16]p : ill(chiefly col). - (Health education booklets)
B78-16799 X.319/18046 Woolwich
1445944

ACCIDENTS. Residences
Prevention

EWART, Neil
Unsafe as houses. - Poole : Blandford, 1981
157p : ill
0-7137-1090-x X.629/15109 Woolwich
0613975

FOSTER, Susan
Home safety. - Havant : K. Mason, 1979
[18]p : col ill. - (Keep well series)
0-85937-172-7 LR.421/147
1897764

ACCIDENTS. Residences
Research - Conference proceedings

SOCIAL, ECONOMIC, AND LEGAL ASPECTS OF ACCIDENTS IN THE HOME (Conference : 1976 : Wolfson College)
Accidents in the home. - London : Croom Helm for [the] Centre for Socio-Legal Studies, 1977
140p : ill. - Includes bibliographies and index
0-85664-452-8 X.329/11010 Woolwich
1141465

ACCIDENTS. Residences. England
Fatal electrical accidents

A COMMENTARY on electrical fatalities in the home / Department of Prices and Consumer Protection. - 1974-75. - London : [The Department], 1976
[4],51p : ill
Subsequent volumes, when received, will be placed at this shelfmark
0-86058-003-2 BS.431/25
0796700

A COMMENTARY on electrical fatalities in the home / Department of Prices and Consumer Protection. - 1976. - London : [The Department], 1977
6,[32]p
B78-27384 BS.431/28
0796700

ACCIDENTS. Residences. England
Human factors

POYNER, Barry
Personal factors in domestic accidents. - London : Department of Trade, Consumer Safety Unit, 1980
viii,28p : ill
B80-41747 SRL
2314223

ACCIDENTS. Residences. Great Britain

The HOME accident surveillance system. - London : Consumer Safety Unit, Department of Trade, 1979
v,43p
B83-04453 BS.431/39(5)
2819503

The HOME accident surveillance system. - London : Department of Prices and Consumer Protection, Consumer Safety Unit.
[No.1]-. - 1977-
[1],v[i.e.vii],17p : 1 ill
B78-02375 BS.431/39
1289896

ACCIDENTS. Road traffic
Causes

IRK, Ferenc
Közúti balesetek. - Budapest : Közgazdasági és Jogi Könyvkiadó, 1979
471p. - Bibliography : p.429-460
963-220-772-6 X.529/35812
1828320

ACCIDENTS. Road traffic
Role of alcohol

WAGENAAR, Alexander C.
Alcohol, young drivers, and traffic accidents. - [Lexington, Mass.] : Lexington Books, 1983
xii,151p : ill. - Bibliography: p135-145. - Includes index
0-669-06696-6 X.529/63148
3143120

ACCIDENTS. Road traffic
Role of alcohol, to date - Conference proceedings

INTERNATIONAL CONFERENCE ON ALCOHOL, DRUGS AND TRAFFIC SAFETY (8th : 1980 : Stockholm)
Abstracts. - [Stockholm?] : [s.n.], [1980?]
134p
r6611224 X.329/17317
9009809

Alcohol, drugs and traffic safety. - Stockholm : Almqvist & Wiksell International, 1981
3v.
r6611226 X.322/12851
9009809

ACCIDENTS. Road traffic
Role of alcohol & drugs - Conference proceedings

SATELLITE SYMPOSIUM OF THE 6TH INTERNATIONAL CONGRESS OF PHARMACOLOGY ON ALCOHOL, DRUGS AND DRIVING (1975 : Helsinki)
Alcohol, drugs and driving. - Basel ; London : Karger ; [Chichester] : [Distributed by Wiley], 1976
vii,102p : ill. - (Modern problems of pharmacopsychiatry ; vol. 11). - Includes bibliographies and index
B76-33996 SRL
0752606

ACCIDENTS. Road traffic
Role of behaviour of motorists

NÄÄTÄNEN, Risto
Road-user behavior and traffic accidents. - Amsterdam ; Oxford : North-Holland Publishing Co. [etc.], 1976
ix,270p : ill. - Bibliography: p.244-258. - Includes index
0-7204-0338-3 X.529/30415
0583855

ACCIDENTS. Road traffic. Austria
1945- - Statistics - Serials

STRASSENVERKEHRSSICHERHEIT / bearbeitet im Österreichischen Statistischen Zentralamt. - 1980-. - Wien : das Zentralamt, 1981-. - (Beiträge zur österreichischen Statistik)
r6303597 S.36/30[Hft.612,690,726, etc.]
8718083

ACCIDENTS. Road traffic. Bristol. Avon

URBAN safety area project / Avon County Council, Transport and Road Research Laboratory. - Bristol : County of Avon Public Relations and Publicity Department, 1983
[28]p,1folded leaf of plates
0-86063-209-1 X.525/9648
3163016

ACCIDENTS. Road traffic. Buckinghamshire

BUCKINGHAMSHIRE. Engineer's Department
Accident statistics & road traffic flows. - 1975. - [Aylesbury] : [The Council], 1976
33[i.e. 35]p : ill
Subsequent volumes, when received, will be placed at this shelfmark
0-902805-33-9 X.512/5259
0685836

ACCIDENTS. Road traffic. Dundalk. Louth (County)

CROWLEY, F. (Finbarr)
Dundalk traffic accident study. - Dublin : An Foras Forbartha, 1983
vii,92p,11folded leaves of plates : ill
0-906120-96-9 Map Library
3175979

ACCIDENTS. Road traffic. France
to date

EWALD, François
L'accident nous attend au coin de la rue. - Paris : La Documentation française, 1982
162p
2-11-000929-2 S.E.400/1160
8929823

ACCIDENTS. Road traffic. Great Britain
- Statistics - Serials

ROAD accident statistics English regions / Department of Transport. - 1981-. - London : The Department, 1983-
v. : ill
B85-05556 P.803/1508
0029262

ACCIDENTS. Road traffic. Great Britain
Effects of motor vehicle handling characteristics

JONES, Ian Shore
The effect of vehicle characteristics on road accidents. - Oxford : Pergamon, 1976
xiii,219p : ill. - Bibliography: p.210-212. - Includes index
0-08-018963-6 X.629/10579 Woolwich
0551821

ACCIDENTS. Road traffic. Great Britain
Investigation

BYATT, R.
Manual of road accident investigation. - London : Pitman.
Vol.1. - 1980
[7],151p : ill. - Includes index
Subsequent volumes, when received, will be placed at this shelfmark
0-273-01325-4 X.0629/539 Woolwich
207494x
Vol.2. - 1981
vi,167p : ill. - Includes index
0-273-01600-8 SRL
207494x

ACCIDENTS. Road traffic. Great Britain
Notification of Great Britain. Department of Transport by local authorities & police. Data. Presentation

GREAT BRITAIN. Department of Transport
Procedures for submitting road accident data to the Department of Transport. - London : [Department of Transport] Statistics Transport G Division, [1977]
[3],68p. - (Stats ; 21(1977)). - Includes index
B79-29354 BS.43/455
1942093

ACCIDENTS. Road traffic. Great Britain
Post-accident procedure - For motorists

DAVIES, Michael D.
A motorist's guide through the experience of having an accident. - Nuneaton : The author, 1979
[1],19 leaves
B80-09660 X.512/7154
2046202

ACCIDENTS. Road traffic. Great Britain
Report forms: Great Britain. Department of Transport. Stats 19. Filling

GREAT BRITAIN. Department of Transport
Instructions for the completion of road accident reports. - [2nd ed.]. - London : Statistics Transport G Division, [1978]
[133]p : ill. - (STATS ; 20)
B79-19263 BS.43/453
1893335

ACCIDENTS. Road traffic. Great Britain
Statistics. Official recording systems

GREAT BRITAIN. Department of Transport
Road accident statistics. - London : [The Department], 1977
[56]p
B78-25062 BS.43/433
1517317

ACCIDENTS. Road traffic. Hertfordshire
Effects of improvement of blacksites, 1971-1978

HERTFORDSHIRE. Highways Department
Before & after accident studies at blacksites. - [Hertford] : [The Department], 1978
[3],14 leaves,plate : 2 ill. - (Report ; no.H.C.C./T270/1). - Bibliography: leaf 13
B79-20123 X.512/6639
189580x

ACCIDENTS. Road traffic. Hertfordshire
Effects of improvement of blacksites, 1971-1980

BEFORE and after study. - [Hertford] : Hertfordshire County Council, Highways Dept., [1981]
[22]leaves. - (H.C.C. T/287/1)
B81-38043 X.525/4913
2563444

ACCIDENTS. Road traffic. Ireland (Republic)

HIGH accident locations. - Dublin : An Foras Forbartha [National Institute for Physical Planning and Construction Research]. - (RS 234) - Bibliography: p.6
[5]: National routes, 1974-1976. - [1978]
[5],127p : chiefly ill
B80-33403 X.525/3517
223792x

REDUCING the frequency and severity of traffic accidents. - Dublin : An Foras Forbartha, 1983
90p in various pagings : ill
B84-40678 X.525/9990
3176002

ACCIDENTS. Road traffic. Ireland (Republic)
- Statistics

ROAD accident facts, Ireland = Fíorais thimpistí bóthair, Eire / An Foras Forbartha. - 1977. - Dublin : National Institute for Physical Planning and Construction Research, 1978
vi,33p : ill
Subsequent volumes, when received, will be placed at this shelfmark
B79-18254 P.523/1201
0251224

ACCIDENTS. Road traffic. Ireland (Republic)
Reduction. Effects of compulsory wearing of motor vehicle seat belts

HEARNE, R. (Rory)
The initial impact of the safety-belt legislation in Ireland. - Dublin : An Foras Forbartha, 1981
14p : ill. - (RS ; 255)
0-906120-46-2 X.205/976
2693828

ACCIDENTS. Road traffic. National Primary Routes. Ireland (Republic)
HEARNE, R. (Rory)
The national routes (1977-1979). - Dublin : An Foras Forbartha, 1982
120p,leaf of plate : ill. - (High accident locations ; 6)
0-906120-69-1 Map library
2899299

ACCIDENTS. Road traffic. Organisation for Economic Co-operation and Development countries
Role of alcohol & drugs
ORGANISATION FOR ECONOMIC CO-OPERATION AND DEVELOPMENT
New research on the role of alcohol and drugs in road accidents. - Paris : Organisation for Economic Co-operation and Development : [London] : [H.M.S.O.], 1978
151p : ill. - (Road research)
B79-09546 SRL
1816071

ACCIDENTS. Road traffic. Sweden
to date
SWEDEN. Statistiska centralbyran
Trafikant på väg. - Stockholm : Statistiska centralbyrån, 1983
59p : ill. - (Sveriges officiella statistik)
91-7142-001-0 X.622/18360
9059555

ACCIDENTS. Road traffic. United States
Investigation
CLARK, Warren E.
Traffic management and collision investigation. - Englewood Cliffs ; London : Prentice-Hall, c1982
xiv,332p : ill. - (Prentice-Hall series in criminal justice). - Bibliography: p325-326. - Includes index
0-13-926162-1 X.520/26898
2715201

ACCIDENTS. Road traffic. Wales
- Statistics - Serials
ROAD accidents, Wales / Welsh Office = Damweiniau ffyrdd, Cymru / y Swyddfa Gymreig. - No.1 (1979/80)-. - Cardiff : Welsh Office, 1983-
v.
Holdings: no 2, 1981-
B84-28571 P.525/485
314156x

ACCIDENTS. Road traffic. West Germany
to date. Law
PFANNMÜLLER, Norbert
Das Verhalten nach einem Verkehrsunfall. - Giessen : N. Pfannmüller, 1981
xxx,342p. - Includes bibliography
r6399227 X.529/55360
8981159

ACCIDENTS. Shipping. Freight transport
Liability. Implications of due diligence - Conference proceedings
DUE diligence. - [London] : [Lloyd's of London Press], [1980]
[52]leaves
0-904093-89-1 X.525/7058
288335x

ACCIDENTS. Shipping of dangerous goods
Emergency action
CARRIAGE of dangerous goods / Department of Trade, Marine Division. - London : [Department of Trade], Marine Library, 1980
178p
B83-00496 BS.41/1317
2805421

INTERNATIONAL MARITIME ORGANIZATION. Maritime Safety Committee
Emergency procedures for ships carrying dangerous goods .... - [London?] : [International Maritime Organization?], [1982?]
6p
B83-42839 X.629/21482
2805421

ACCIDENTS. Ships
Liability of shipping services. Limitation
[INTERNATIONAL Convention relating to the Limitation of the Liability of Owners of Sea-going Ships. Protocols, amendments, etc., December 21 1979]. Protocol amending the International Convention relating to the Limitation of the Liability of Owners of Sea-going ships dated 10 October 1957, Brussels, 21 Decvember 1979 ... - London : H.M.S.O., [1980]
6p. - (Miscellaneous ; no.15 (1980)) (Cmnd.7926)
0-10-179260-3 BS.14 / 3338
1337572

ACCIDENTS. Ships
Liability of shipping services. Limitation - Conference proceedings
INTERNATIONAL CONFERENCE ON LIMITATION OF LIABILITY FOR MARITIME CLAIMS (1976 : London)
Final act of the International Conference on Limitation of Liability for Maritime Claims, 1976, with text of convention and understanding of the Conference, London, 1 to 19 November 1976 ... - London : H.M.S.O., 1977
18p. - (Miscellaneous ; no.31(1977)) (Cmnd.7035)
0-10-170350-3 BS.14/3108
1337572

International Conference on Limitation of Liability for Maritime Claims, 1976. - London : Inter-Governmental Maritime Consultative Organization, 1977
47p
B78-10768 X.100/17418
1337572

ACCIDENTS. Ships registered in Iceland
- Serials
RANNSÓKNARNEFND SJÓLYSA
Skýrsla Rannsóknarnefndar sjólysa. - 1971-. - Reykjavík : Rannsóknarnefnd sjólysa og Siglingamálastofnun ríkisins, 1973-
r6033720 S.B.456/2
1792571

ACCIDENTS. Skateboarding. England
1977-1978
GREAT BRITAIN. Department of Prices and Consumer Protection
The Home Accident Surveillance System, analysis of skateboard accidents. - London : Department of Trade, Consumer Safety Unit, 1979
v,31p
B80-15835 BS.41/1320
1573551

ROYAL SOCIETY FOR THE PREVENTION OF ACCIDENTS
Skateboarding accident statistics. - [Birmingham] : [The Society], [1978]
[3],25p. - (Document ; SE95)
B78-32188 X.512/6153
1573551

ACCIDENTS. Special schools. West Germany
1945-
VERNOOIJ, Monika A.
Statistische Analyse des Unfallgeschehens an Sonderschulen. - Bremerhaven : Wirtschaftsverlag NW, 1981
377p. : ill. - (Forschungsbericht / Bundesanstalt für Arbeitsschutz und Unfallforschung ; Nr.258). - Bibliography
3-88314-147-x S.F.583/283[Nr.258]
8765200

ACCIDENTS. Sports
WELCH, Ann
Accidents happen. - London : J. Murray, 1978
xii,240p : ill. - Includes index
0-7195-3545-x X.619/18851 Woolwich
1578553

ACCIDENTS. Steelworks. Great Britain
Explosions
GREAT BRITAIN. Factory Inspectorate
The explosion at the Appleby-Frodingham steelworks, Scunthorpe, 4 November 1975. - [London] : H.M.S.O., 1976
59p,fold plate : ill(some col)
0-11-880331-x SRL
0658162

ACCIDENTS. Tankers. Ships
1968-1977 - Statistics - Statistics
UNITED KINGDOM TANKER SAFETY GROUP. Sub-Group on Tanker Casualties
Tanker casualties report. - London : Inter-Governmental Maritime Consultative Organization, 1978
30p
B79-15459 SRL
1874411

ACCIDENTS. Transport. Hull region. Humberside
1837-1927
ULYATT, Michael E.
Five Hull tragedies. - Hull : Bradley, 1977
28p : ill
B82-09373 X.958/7368
2618699

ACCIDENTS. Transport. Nuclear fuels containers. London
Emergency action
MATTHEWS, R. R.
Notes on the contingency arrangements for irradiated fuel flask incidents. - London : Central Electricity Generating Board, 1980
[12]p
B80-40794 BS.85/196
2313863

ACCIDENTS. Tynewydd Colliery
1877
LLEWELLYN, Ken
Disaster at Tynewydd. - Cardiff : ap Dafydd Publications Ltd, 1975
92p(2fold) : ill
0-905041-00-3 X.319/10307 Woolwich
0570826

ACCIDENTS. Warehouses. Great Britain
Explosions
The FIRE and explosions at Permaflex Ltd, Trubshaw Cross, Longport, Stoke on Trent, 11 February 1980 / Health and Safety Executive. - Newcastle-under-Lyme : The Executive 1981
[8]p
0-7176-0073-4 BS.438/117
2395975

ACCIDENTS. Water traffic. Ports. Great Britain
NATIONAL MARITIME INSTITUTE
Analysis of marine incidents in ports and harbours. - London : The Council, 1976
[68]leaves : ill
0-901058-93-9 BS.414/397
0754439

ACCIDENTS. West Germany
1945- - Serials
[FORSCHUNGSBERICHT (Bundesanstalt für Arbeitsschutz und Unfallforschung)]. Forschungsbericht / Bundesanstalt für Arbeitsschutz und Unfallforschung. - Nr.142-. - Bremerhaven : Wirtschaftsverlag NW, 1975-
r6303742 S.F.583/283
8799873

ACCIDENTS. Workplaces
- Conference proceedings
INTERNATIONAL SEMINAR ON OCCUPATIONAL ACCIDENT RESEARCH
(1983 : Saltsjöbaden)
Occupational accident research. - Oxford : Elsevier Science, Oct.1984
1v.
0-444-42415-6 X.622/22748
3119319

ACCIDENTS. Workplaces. Great Britain
Notification to Great Britain. Health and Safety Executive
GREAT BRITAIN. Health and Safety Commission
Proposals for notification of accidents, dangerous occurrences and ill health regulations, and guidance notes. - London : H.M.S.O., 1979
[2],24p
0-11-883239-5 BS.438/84
1728091

GREAT BRITAIN. Health and Safety Executive
Proposals for the notification of accidents and dangerous occurrences. - London : H.M.S.O., 1977
[2],6p
0-11-883009-0 BS.438/53
1728091

ACCIDENTS. Workplaces. Great Britain
Notification to Great Britain. Health and Safety Executive. Law: Notification of Accidents and Dangerous Occurrences Regulations 1980
The NOTIFICATION of accidents and dangerous occurrences regulations, 1980. - London : Industrial Relations Briefing, c1981
56p. - (IRB question & answer series ; no.10)
0-86231-006-7 X.529/42746
2534045

ACCIDENTS. Workplaces. United States
Prevention
HEINRICH, H. W. (Herbert William)
Industrial accident prevention. - 5th ed. - New York ; London : McGraw-Hill, c1980
xii,468p : ill. - Bibliography: p455-461. - Includes index
0-07-028061-4 SRL
2737655

ACCIDENTS RESULTING IN BURNS. Residences
Prevention
THOMSON, William, 1925-
People get burned. - Hamilton [Lanarkshire] : Lanarkshire Health Board, [1976]
[12]p : col ill
0-905453-00-x X.519/25765
0651435

ACCIÓ CATALANA

BARAS I GOMEZ, Montserrat
Accio Catalana, 1922–1936. – Barcelona : Curial, 1984
470p. – (Biblioteca de cultura catalana ; 53). – Includes bibliography
84–7256–221–2 X.958/24166
8419698

ACCION CATOLICA (GUATEMALA) Quiche (Department) Guatemala 1948–1970

FALLA, Ricardo
Quiche rebelde. – Guatemala : Editorial Universitaria, 1978
574p. – (Coleccion "Realidad nuestra" ; v.7)
r6162369 X.208/5906
2474204

ACCION CATÓLICA (SPAIN) 1934–1939

TABOADA LAGO, Jose Maria
Por una España mejor. – Madrid : G. del Toro Editor, 1977
305p : ill
84–312–0215–7 X.200/11785
1860127

ACCIÓN DEMOCRÁTICA (VENEZUELA)

GONZÁLEZ HERRERA, Luis
Romulo en Berna. – Caracas : Centauro, 1978
2v.(64p of plates)
r6701230 X.809/48113
8462399

ACCIÓN REPUBLICANA

ESPÍN, Eduardo
Azaña en el poder. – Madrid : Centro de Investigaciones Sociológicas, 1980
401p. – (Colección "Monografías" ; no.35). – Includes bibliography
84–7476–033–x X.529/56782
8421676

ACCIPITRIDAE
See also
Buzzards
Eagles
Hawks
Honey buzzards
Red kites. Birds
Vultures

ACCIUS, LUCIUS. Drama in Latin – Indexes

DE ROSALIA, Antonino
Lexicon Accianum. – Hildesheim : Olms, 1982
xxiii,202p. – (Alpha–Omega. Reihe A Lexika, Indizes, Konkordanzen zur klassischen Philologie ; 53). – Includes bibliographies
3–487–07261–0 X.950/32122
8094705

ACCLAIM CARS
Triumph Acclaim cars. Maintenance & repair

LEGG, A. K.
Triumph Acclaim owners workshop manual. – Yeovil : Haynes, c1983
206p : ill. – Includes index
0–85696–792–0 X.622/16917
2963175

TRIUMPH Acclaim / by the Autobooks team of writers and illustrators. – Wakefield : Autobooks, 1983
160p : ill. – (Owners workshop manual / Autobooks ; 996). – Includes index
0–85146–232–4 X.622/18481
2963175

ACCOMMODATION
See also
Bed & breakfast accommodation
Boarding houses
Cabins
Guest houses
Halls of residence
Hostels
Hotels
Housing
Inns
Lodging houses
Lodging schemes
Low price accommodation
Rented accommodation
Residences
Self-catering accommodation
Supported accommodation

ACCOMMODATION
See also
Temporary accommodation
Time-sharing accommodation
Transport cafés providing overnight accommodation
Vacation accommodation

ACCOMMODATION. Adult education. England

PRIME use accommodation for adult education / by Keith Percy ... [et al.] ; [for] ACACE. – Leicester : ACACE, [1982]
71p : 2ill. – (Report / Advisory Council for Adult and Continuing Education)
0–906436–15–x X.529/50235
2791137

ACCOMMODATION. Colleges of education. Great Britain
Financing. Use of student cash maintenance allowances

GREAT BRITAIN. Department of Education and Science
Catering and residence for students. – London : The Department, 1976
[1],59p : ill
0–85522–076–7 BS.10/915
0715034

ACCOMMODATION. Crews. British fishing boats
Surveying

GREAT BRITAIN. Department of Trade
Survey of crew accommodation in fishing vessels. – London : H.M.S.O., 1976
1v.. – Includes index
0–11–511716–4 BS.41/1141
0734543

ACCOMMODATION. Crews. British merchant ships
Surveying

GREAT BRITAIN. Department of Trade
Survey of crew accommodation in merchant ships. – London : H.M.S.O., 1979
1v.. – Includes index
0–11–512207–9 BS.41/1194
1966774

ACCOMMODATION. Dermatology services. Hospitals. Great Britain
Design

GREAT BRITAIN. Department of Health and Social Security
Accommodation for hospital dermatology services. – [London] : D.H.S.S., 1976
[2],12p
B80–30074 BS.385/516
2088169

ACCOMMODATION. England
Self-catering vacation accommodation. Conversion from farm buildings – For farmers

ENGLISH TOURIST BOARD
Self-catering on the farm. – London : The Board, 1979
48p : ill. – Bibliography: p.40
B79–21115 BS.411/268
1901443

ACCOMMODATION. Ex-patients. Mental illness hospitals. Scotland

SCOTTISH ASSOCIATION FOR MENTAL HEALTH
Accommodation, an aspect of community care for the mentally ill. – Edinburgh : S.A.M.H., 1979
[1],18p
B80–25411 X.525/3304
218110x

ACCOMMODATION. Ex-patients. Psychiatric hospitals. England

MIND
The next step. – London : MIND, [1977?]
[3],24p. – (Reports / MIND ; 16)
B78–08185 X.0313/510(16) Woolwich
1348671

ACCOMMODATION. Homeless persons. England
Provision by local authorities – Statistics – Serials

HOMELESSNESS statistics. Actuals / CIPFA Statistical Information Service. – 1978–79–. – London : Chartered Institute of Public Finance and Accountancy, 1980–
v.
B82–32153 P.523/1689
2770423

ACCOMMODATION. Homeless single persons. Great Britain
Provision by housing associations

CURRIE, Hector
Single initiatives. – Edinburgh : Scottish Council for Single Homeless, 1980
44p : ill
0–907050–01–8 X.525/4151
2376873

ACCOMMODATION. Homeless single persons. Great Britain
Provision by local authorities

DRAKE, Madeline
Policy and provision for the single homeless. – London : Personal Social Services Council, 1977
vi,49,[1]p : 1 ill. – Bibliography: p.47–49
B78–00831 X.512/5728
1277200

ACCOMMODATION. Homeless single persons. Scotland
Emergency accommodation

ALTERNATIVES to night shelters. – Edinburgh : Scottish Council for Single Homeless, 1984
ii,46p : ill. – Bibliography: p46
0–907050–09–3 X.525/9220
313282x

ACCOMMODATION. Homeless single young persons. Cities. Great Britain

AKEHURST, Michael D. (Michael David)
Running to the city. – Nuneaton : National Association of Youth Clubs, [1977]
[1],14,[1]p : 1 ill. – Bibliography: p.11–13
0–901528–77–3 X.519/27351
1264303

ACCOMMODATION. Homeless single young persons. Great Britain

BEYOND the hostel / text by Community Action Projects ; graphics by Dave Rogers. – Manchester : CAP, c1982
128p : ill
0–907095–33–x X.529/51699
1199323

ACCOMMODATION. Homeless women. Hungary
1945–

MÁTYUS, Aliz
Holnapon innen, tegnapon túl. – Budapest : Szépirodalmi Könyvkiadó, 1980
318p,[32]p of plates : ill. – (Magyarország felfedezése)
963–15–1664–4 X.520/27007
8822107

ACCOMMODATION. Motorists. Tourists. United States
1910–1945

BELASCO, Warren James
Americans on the road. – Cambridge, Mass. ; London : M.I.T. Press, 1979
xi,212p : ill. – Bibliography: p.195–205. – Includes index
0–262–02123–4 X.800/29010
2105306

ACCOMMODATION. Old persons. Great Britain

WILSON, Monica
The College of Health guide to homes for elderly people. – London : College of Health, 1984
52p,12p
B85–16787 X.529/68683
3261417

ACCOMMODATION. Patients. Hospitals. Great Britain
Administration

BURROUGH, Evan Jerome Ridgway
A scandalous impromptu. – Oxford : The author, 1976
23p
0–9505008–0–1 X.319/16555 Woolwich
0658480

ACCOMMODATION. Personnel. Bethlem Royal Hospital
Land. Leasing by Great Britain. Department of Health and Social Security: Leasing to Bethlem Royal Hospital and the Maudsley Hospital Board of Governors

GREAT BRITAIN. Treasury
Treasury minute dated 13 July 1977 relative to the leasing at a nominal rent of land at the Bethlem Royal Hospital, Beckenham, to the Board of Governors of the Bethlem Royal and Maudsley Hospitals. – London : H.M.S.O., [1977]
Sheet([2]p). – (Cmnd.6868)
0–10–168680–3 BS.44/470
1191594

ACCOMMODATION. Pet owners. West Germany
1945–. Legal aspects

TIERHALTUNG in Miet- und Eigentumswohnungen in der Rechtsprechung / Herausgeber, Bund gegen den Missbrauch der Tiere ; [Vorwort und Zusammenstellung, Eduard Heppl]. – 3. Aufl., erg. und verb. – München : Der Bund, 1978
115p
r6301354 X.529/44124
8785295

ACCOMMODATION. Pilgrims
to date

GNESDA, Luciano
Gli "ospizi" nelle Dolomiti. - Firenze : Olschki, 1979
81p,[1]folded leaf,[9]p of plates. - (Comitato dei geografi italiani ; 4)
88-222-2852-9 X.800/32170
8832846

ACCOMMODATION. Pilgrims. Mecca. Saudi Arabia

MAKKY, Ghazy Abdul Wahed
Mecca, the pilgrimage city. - London : Croom Helm for the Hajj Research Centre, King Abdul Aziz University, Jeddah, 1978
95[i.e.97]p,[8]p of plates : ill. - (Studies / King Abdul Aziz University. Hajj Research Centre ; 1). - Bibliography: p.86-88. - Includes index
0-85664-591-5 X.809/42953
120386x

ACCOMMODATION. Playgroups. Great Britain
Architectural design. Plans

PRE-SCHOOL playgroup designs. - [London] : International Timber Corporation, 1976
24p : ill
r6813794 X.425/4962
3126641

ACCOMMODATION. Single persons. Scotland
Conversion from buildings

NATIONAL BUILDING AGENCY
Conversion for single people. - Edinburgh : The Agency, 1979
[2],51 leaves : 1 ill
B80-39761 X.525/3793
2303752

ACCOMMODATION. Single persons. Scotland
Planning

CURRIE, Hector
Single initiatives II. - Edinburgh : Scottish Council for Single Homeless, 1981
iv,43p : ill
0-907050-03-4 X.525/5797
2731916

ACCOMMODATION. Sociology
See also
Intergroup accommodation

ACCOMMODATION. Stock. Science Reference Library

SCIENCE REFERENCE LIBRARY
The arrangement of SRL stock in the 1980s. - [London] : [The Library], [1979]
[1],11p. - (Discussion papers / Science Reference Library ; no.1)
B80-09559 2719.hh.12
2017474

ACCOMMODATION. Students. Slovenia. Yugoslavia
Architectural design - Catalogues

SLOVENSKI ŠOLSKI MUZEJ
Dijaški in študentski domovi. - Ljubljana, 1976
45p : ill
r6074184 X.709/25697
1542257

ACCOMMODATION. Students. University of Birmingham

UNIVERSITY OF BIRMINGHAM. Lodgings Warden and Student Welfare Adviser
Living in Birmingham. - 5th ed. - Birmingham : University of Birmingham, 1979
[2],21p : ill
0-7044-0348-x X.519/30090
1938614

ACCOMMODATION. Tourists. London
Demand - Forecasts

HORWATH AND HORWATH (UK) LTD
London accommodation, demand forecasts. - London : B.T.A., 1978
118,[2]p
0-7095-0140-4 BS.460/37
2061392

ACCOMMODATION. Tourists. West Germany
1945- - Statistics - Serials

HANDEL, Gastgewerbe, Reiseverkehr. Reihe 7.1, Übernachtungen in Beherbergungsstätten / Herausgegeber : Statistisches Bundesamt Wiesbaden. - Jan. 1977-. - Stuttgart : Kohlhammer, 1977-. - (Fachserie 6 / Statistisches Bundesamt)
Issues under the earlier title are entered in the General Catalogue
r6308783 S.F.555/102(10)
906950x

ACCOMMODATION. Young persons. France
1945-

BONNAUD, Jacques
Foyers de jeunes travailleurs en Midi-Pyrénées. - Toulouse : Direction régionale du travail et de l'emploi de la Région Midi-Pyrénées, 1982
2v.
r6210450 SE.400/1200
8966826

Foyers de jeunes travailleurs en Midi-Pyrénées. - Toulouse : Direction régionale du travail et de l'emploi de la Region Midi-Pyrénées, 1982
2v.
r6299167 SE.400/1200
8966826

ACCOMMODATION AGENCIES
Great Britain. Race relations. Law. Race Relations Act 1976

GREAT BRITAIN. Commission for Racial Equality
A guide for accommodation agencies and landlords. - London : The Commission, 1980
20p : 1 col ill
B80-36811 BS.400/48
2276437

ACCOMMODATION AGENCIES. England
Illegal payments by clients. Recovery - For unrepresented litigants

MANCHESTER LAW CENTRE
Deposits, premiums and accommodation agency fees. - [Manchester] : [The Centre], [1979]
[3],22p : 1 ill
B80-40524 X.529/38276
0265128

PADDINGTON NEIGHBOURHOOD LAW CENTRE
Got your money back?. - London : The Centre, [1974]
[1],10 leaves
0-9504000-0-9 X.102/1107
0265128

ACCOMMODATION FOR RETIRED PERSONS. Great Britain
Choice

CONSUMERS' ASSOCIATION
Where to live after retirement. - Revised [ed.]. - London : The Association, 1978
[5],165p : ill. - (A consumer publication). - Includes index
0-85202-157-7 X.519/29736
1279734

ACCOMMODATION SERVICES
See also
Housing services

ACCOMMODATION SERVICES. Great Britain
Management

ALLEN, David M.
Accommodation and cleaning services. - London : Hutchinson. - ([Catering and hotel management books])
Vol.1: Operations. - 1983
307p : ill. - Includes index
0-09-150990-4 X.622/18708
0092177

Vol.2: Management. - 1983
102p : ill. - Includes index
0-09-151000-7 X.622/18707
0092177

ACCOMPANIMENT. Christmas carols in English
Piano playing. Techniques

VARNUM, Brooke Minarik
Play & sing - It's Christmas!. - New York : Macmillan ; London : Collier Macmillan, 1980
48p : col.ill
0-02-045420-1 X.431/11508
2582066

ACCOMPANIMENT. Folk songs in Czech
- Conference proceedings

DOPROVODY lidových písní a jejich improvizace / redakce Karel Padrta. - České Budějovice, 1971
50 leaves,[8]leaves of plates
r6015641 X435/420
1718258

ACCOMPANIMENT. Music
See also
Songs with orchestral accompaniment

ACCOMPANIMENT. Music
Piano playing

MOORE, Gerald, 1899-
Furthermoore. - London : Hamilton, 1983
149p. - Includes index
0-241-10909-4 X.439/12370
2902907

ACCOMPANIMENT. Music
Piano playing. Moore, Gerald, 1899-

MOORE, Gerald, 1899-
The unashamed accompanist. - Rev. ed. - London : MacRae, 1984
125p : ill
0-86203-181-8 X.431/12360
0158860

ACCOMPANIMENT. Music
Piano playing. Moore, Gerald, 1899- - Biographies

MOORE, Gerald
[Am I too loud?. German]. Bin ich zu laut?. - Kassel : Bärenreiter, 1976
267p
3-7618-0560-8 X.439/4539
0158860

ACCOMPANIMENT. Music
Piano playing. Moore, Gerald, b.1899, to 1961 - Biographies

MOORE, Gerald, b.1899
Am I too loud?. - Harmondsworth : Penguin, 1979
286p,[8]p of plates. - Includes index
B80-12463 X.439/8798
2078252

Farewell recital. - Harmondsworth : Penguin, 1979
208p,8p of plates : ill. - Includes index
0-14-004941-x X.439/8797
2078252

ACCOMPANIMENT. Music
Piano playing. Moore, Gerald, b.1899, 1967-1977 - Biographies

MOORE, Gerald, b.1899
Farewell recital. - London : Hamilton, 1978
[11],178p,[8]p of plates : ill. - Includes index
0-241-89817-x X.439/8436
1343890

ACCOMPANIMENT. Music
Piano playing. Performance

ADLER, Kurt
The art of accompanying and coaching. - Corrected ed. - New York : Da Capo Press, 1976
[11],260p : ill. - Bibliography: p.243-245.- Includes index
0-306-80027-6 X.431/10649
1893610

ACCOMPANIMENT. Practising. Ballet
Piano playing

LISHKA, Gerald R.
A handbook for the ballet accompanist. - Bloomington ; London : Indiana University Press, 1979
xi,136p. - Includes index
0-253-32704-0 X.622/8118 Woolwich
2206900

ACCOMPANIMENT. Songs
Piano playing

DUCKENFIELD, Bridget
The care and nurture of your pianist partner or lesser spotted accompanist. - Ross-on-Wye : Hydatum, 1978
[2],18p : ill(chiefly col)
0-905682-10-6 X.431/10536
1586742

Hazards of the pianist partner. - Epsom : Rosslyn, 1984
45p : ill
0-9508969-1-8 X.439/13175
1586742

The ideal accompanist. - Epsom : Rosslyn, 1983
41p : ill
0-9508969-0-x X.950/28104
1586742

ACCORD CARS
Honda Accord cars. Maintenance & repair

BALL, Kenneth
Accord 1976-77 autobook. - Brighton : Autobooks, [1978]
152p : ill. - (The autobook series of workshop manuals). - Includes index
0-85147-740-2 X.620/17687 Woolwich
1382187

ACCORDION MUSIC
to date - Catalogues

GOSUDARSTVENNAÎA BIBLIOTEKA SSSR IM. V.I. LENINA. Otdel notnykh izdaniĭ i zvukozapiseĭ
Notnye izdaniîa v fondakh Gosudarstvennoĭ biblioteki SSSR im. V.I. Lenina. Garmonika, baîan. - Moskva : Gosudarstvennaîa biblioteka SSSR im. V.I. Lenina, Otdel notnykh izdaniĭ i zvukozapiseĭ, 1978
4v
r6464204 X.0439/351
8977062

ACCORDION PLAYING
Shand, Jimmy - Biographies
PHILLIPS, D. (David), 1914-
Jimmy Shand. - Dundee : D. Winter and Son Ltd, 1976
166p.[32]p of plates : ill. - List of sound discs: p.132-143
0-902804-05-7 X.439/8263
1047949

ACCORDION PLAYING
Two row chromatic button accordion playing
MCDOWELL, John, 1943-
Button keyed accordeon. - [Gosport] : Metrorange Publications, [1977]
[3],26p
B77-27283 X.439/8345
0737887

ACCORDION PLAYING. Italy
to date
GIANNATTASIO, Francesco
L'organetto. - Roma : Bulzoni, [1979]
135p,[12]p of plates,1folded leaf of plates : ill. - (Etnomusicologia ; 3. Strumenti). - Includes bibliography and discography
r6501164 X.439/11920
9004386

ACCORDION PLAYING. Norwegian folk music
FAUKSTAD, Jon
Ein-raderen i norsk folkemusikk. - Oslo : Universitesforlaget, [1978]
183p,[4]p of plates : ill
82-00-01832-6 X.439/10720
2217155

ACCORDION PLAYING. Scottish folk music
Piano accordion playing. Organisations: National Association of Accordion and Fiddle Clubs - Serials
BOX and fiddle. - No.1- ; Oct. 1977-. - [Stranraer] : National Association of Accordion and Fiddle Clubs, [1977]- : ill
B78-03581 P.2000/643
1295403

ACCOUNT OF THE RISE AND PROGRESS OF MAHOMETANISM. Stubbe, Henry. Comparative religion
HOLT, Peter Malcolm
A seventeenth-century defender of Islam. - London : Dr. Williams's Trust, 1972
29p. - (Friends of Dr. Williams's Library. lecture ; 26)
r6196018 PP.6481.bcf[no.26]
2483033

ACCOUNTABILITY. Boards of directors. Companies - Conference proceedings
MANAGEMENT accountability and corporate governance / edited by Kenneth Midgley. - London : Macmillan, 1982
xvii,229p. - Includes index
0-333-31200-7 X.529/48348
2753634

ACCOUNTABILITY. Companies. India (Republic)
related to ownership
GUPTA, L.C.
Corporate management and accountability. - [Madras] : Macmillan [for] Institute for Financial Management and Research, 1974
xi,128p
r6025829 X.520/9328
1027336

ACCOUNTABILITY. Educational institutions. England
ACCOUNTABILITY in the English educational system / edited by Hugh Sockett. - London : Hodder and Stoughton, c1980
117p. - (Unibooks). - Bibliography: p112-117. - Includes index
0-340-24459-3 X.529/47055
1929151

ACCOUNTABILITY. Educational institutions. Great Britain
ACCOUNTABILITY in education / edited by Tony Becher and Stuart Maclure. - Windsor : NFER [for the Educational Research Board of the Social Science Research Council], 1978
256p : 1 ill. - Includes bibliographies and index
0-85633-167-8 X.519/29610
1883453

CALLING education to account / edited by Robert McCormick ... [et al.]. - London : Heinemann Educational in association with the Open University Press, 1982
376p : ill. - (Open University set book). - Includes bibliographies and index
0-435-80629-7 X.529/47354
1883453

ACCOUNTABILITY. Government. Athenian Empire
ROBERTS, Jennifer Tolbert
Accountability in Athenian government. - Madison, Wis. ; London : University of Wisconsin Press, 1982
xi,270p. - (Wisconsin studies in classics). - Bibliography: p235-251. - Includes index
0-299-08680-1 X.529/54860
2920220

ACCOUNTABILITY. Leaders. Labour Party (Great Britain)
DEMOCRACY and accountability in the Labour Party / edited by John B. Burnell. - [Nottingham] : Spokesman for the Institute for Workers' Control, 1980
87p
0-85124-299-5 X.529/46222
264035x

ACCOUNTABILITY. Local government. Great Britain
REGAN, David Edward
A headless state. - [Nottingham] : [University of Nottingham], [1980]
26p. - (Inaugural lectures / University of Nottingham). - Bibliography: p.25-26
0-900572-60-4 X.529/38154
2308258

ACCOUNTABILITY. Members. Great Britain. Parliament. House of Commons
AVEBURY, Eric Lubbock, Baron
Authority and accountability. - [London] : Birkbeck College, 1980
19p
B83-16517 X.520/29386
2874474

ACCOUNTABILITY. Multinational companies. Pharmaceutical industries. Developing countries
MEDAWAR, Charles
Drug diplomacy. - London : Social Audit, [c 1982]
vii,119p : ill. - Bibliography: p112-115
0-9503392-9-6 X.329/19338
3258246

ACCOUNTABILITY. National health services. England
ELCOCK, Howard
The buck stops where?. - [Hull] : University of Hull, Institute for Health Studies, 1980
iv leaves, 134 [i.e. 139]p : ill
B80-31239 X.525/4189
2242117

ACCOUNTABILITY. Personnel. Education. Great Britain
ACCOUNTABILITY in education / edited by John Lello. - London : Ward Lock, 1979
[6],192p : 2 ill. - Bibliography: p.184-186. - Includes index
0-7062-3813-3 X.519/29802
1912992

ACCOUNTABILITY. Police. Great Britain
REGAN, David
Are the police under control?. - London : Social Affairs Unit, c1983
16p. - (Research reports ; [1])
0-907631-06-1 X.529/61404
3121038

ACCOUNTABILITY. Police. Great Britain
to 1981
BROGDEN, Michael
The police. - London : Academic Press, 1982
xv,265p : ill. - (Law, state and society series ; 7). - Bibliography: p251-259. - Includes index
0-12-135180-7 X.200/40267
2892138

ACCOUNTABILITY. Schools. Great Britain
BECHER, Tony
Policies for educational accountability. - London : Heinemann Educational, 1981
168p. - Bibliography: p158-163. - Includes index
0-435-80060-4 X.529/63803
2003767

SCHOOL accountability / John Elliott ... [et al.]. - London : McIntyre, 1981
xvii,247p : ill. - Bibliography: p239-244. - Includes index
0-86216-074-x X.529/46086
2003767

ACCOUNTABILITY. Schools. Great Britain
Evaluation by schools
GUIDELINES for review and internal development in schools / Agnes McMahon ... [et al.]. - York : Longman. - (Schools Council programme. 1, Purpose and planning in schools) Secondary school handbook. - 1984
78p
0-582-17219-5 X.525/11014
2993244

ACCOUNTABILITY. Schools. Great Britain
Evaluation by schools - Conference proceedings
NUTTALL, Desmond L. (Desmond Lewes)
School self-evaluation. - London : Schools Council, c1981
53p : 1ill. - (Schools Council programme. 1, Purpose and planning in schools). - Bibliography: p48-49
0-582-39674-3 X.525/8438
2993244

ACCOUNTABILITY. Schools. Great Britain
related to planning of curriculum
OLLERENSHAW, Dame Kathleen
Accountability and the curriculum. - London : Southern Executive of the College of Preceptors, [1978]
11p. - (Joseph Payne memorial lecture)
B79-04236 X.519/29049
1195824

WHITFIELD, Richard C. (Richard Charles)
Curriculum planning, teaching and education accountability. - [Birmingham] : University of Aston in Birmingham, 1976
28p. - (Aston educational enquiry monographs ; no.6)
B84-33109 X.0519/1018
1195824

ACCOUNTABILITY. Teachers. Higher education institutions. Great Britain
ACCOUNTABILITY or freedom for teachers? / Donald Bligh (editor) ; [contributors] Norman Lindop ... [et al.]. - Guildford : Society for Research into Higher Education, 1982
137p. - (Programme of study into the future of higher education ; 1 7) (SRHE monograph ; 50). - Includes bibliographies
0-900868-88-0 X.529/58070
299366x

ACCOUNTABILITY. Welfare workers. Great Britain
FLETCHER, Harry
No accountability, no redress. - London : One Parent Families, 1982
10p
B82-32071 X.525/6066
276489x

ACCOUNTANCY
MARCH, Robert T.
[Running Press glossary of accounting language]. The language of accountancy. - London : W.H. Allen, 1981, c1978
77p. - (A Star book)
0-352-30957-1 X.529/44732
2591081

NOBES, Christopher
The economist pocket accountant. - Oxford : Blackwell, 1985
171p : ill
0-631-13965-6 X.629/26118
2582511

STUDIES in accountancy. - 1976. - London : Institute of Chartered Accountants in England and Wales, [1977?]
[4],160p : ill
Subsequent volumes, when received, will be placed at this shelfmark
B80-08846 P.513/2397
0253456

ACCOUNTANCY
- Dictionaries
ABDEEN, Adnan
English-Arabic dictionary of accounting and finance. - Chichester : Wiley, c1981
xi,226p,[42]p
0-471-27673-1 X.950/12297
1092251
- [Another copy] OPL

ABU-GHAZALEH, Talal
The Abu-Ghazaleh English-Arabic dictionary of accountancy. - London : Macmillan, 1978
[9],210p
0-333-21468-4 X.525/2382
1092251

ROBB, Alan J.
A dictionary of accounting terms. - Christchurch, [N.Z.] ; London : Whitcoulls, 1981
72p
0-7233-0665-6 X.629/16672 Woolwich
2582511
- [Another copy] OPL

ACCOUNTANCY
- For Irish students

DOWLING, Noel
Leaving Certificate accountancy. - Dublin : Helicon, 1977
[4],337p. - (Study assessment series)
B77-29549 X.519/27248
1251937

ACCOUNTANCY
- For law

BERGER, Robert O.
Practical accounting for lawyers. - New York : Chichester : Wiley, c1981
xiv,357p. - (Modern accounting perspectives and practices series). - Bibliography: p345-346. - Includes index
0-471-08486-7 X.520/25698
0370363

WILLIAMS, Donald B. (Donald Brian)
A solicitor's introduction to accounts. - London : Sweet and Maxwell, 1975
ix,150p. - (Concise college texts)
0-421-16340-2 X.100/16055
0370363

ACCOUNTANCY
- For undergraduates

HOPKINS, Leon
Accountancy and law. - 2nd ed. - London : Haymarket Publishing, 1975
127p : ill. - (DOG careers guides). - Includes bibliographies
0-86024-025-8 X.100/16595
0747696

Accountancy and law. - 4th ed. - London : Haymarket Publishing, 1977
118p : ill. - (DOG careers guides. 1978). - Includes bibliographies
0-86024-069-x X.100/18280
0747696

ACCOUNTANCY
- For West African students

AJAYI, C. A. (Catherine A)
Principles of accounts for WASC/GCE and RSA Accounting Stage 2. - London : Edward Arnold, 1983
xi,307p : ill. - Includes index
0-7131-8141-9 X.529/63226
3128040

ACCOUNTANCY
- Serials

CHART key. - 1977-. - London : Chart Tutors Ltd., 1977-
v. : ill
r6865284 P.513/1271
2794403

ACCOUNTANCY
to date - Encyclopaedias

HANDWÖRTERBUCH des Rechnungswesens. - 2., völlig neu gestaltete Aufl.. - Stuttgart : Poeschel, 1981
xlviii,2064columns. - (Enzyklopädie der Betriebswirtschaftslehre ; Bd.3)
3-7910-8018-0 X.0525/475(3a)
9011013

ACCOUNTANCY
Companies. Management - For accountancy

OWENS, Tom
A student guide to business management. - [Dublin] : [Leinster Students' Society of the Association of Certified Accountants], [1977]
[3],46p : ill
B78-14439 X.519/27838
1408186

ACCOUNTANCY
Computer systems - For accountancy

VENABLES, Delia
Computers & word processors for accountants. - Uppingham : European Study Conferences, [1981]
75p
B82-23280 X.525/5744
2728559

Computers & word processors for accountants. - Uppingham : European Study Conferences, 1982
6leaves : ill
B82-23281 X.525/5745
2728559

ACCOUNTANCY
Concepts: Revenue. Definition

ACCOUNTANTS INTERNATIONAL STUDY GROUP
Revenue recognition. - [London] : The Group, 1978
[22]p
B78-33097 X.519/28493
157681x

ACCOUNTANCY
Digital computer systems - For accountancy

BHASKAR, Krish
Management information systems and data processing. - London : Heinemann in association with the Institute of Cost and Management Accountants, 1985, c1984
[x,350]p. - Bibliography: pix. - Includes index
0-434-90794-4 X.520/34951
2774631

PERRY, William E.
The accountants' guide to computer systems. - New York ; Chichester : Wiley, c1982
xi,286p : ill. - (Modern accounting perspectives and practice). - Includes index
0-471-08992-3 X.622/13444
2774631

ACCOUNTANCY
Digital computer systems. Programming languages: Basic language - For accountancy

PARKER, Alan J. (Alan Jay)
Accountant's BASIC programming for the APPLE II. - Reston, Va. : Reston, c1981
x,301p : ill
0-8359-0047-9 X.520/30460
286584x

SCORGIE, Michael E. (Michael Edwin)
BASIC for accountants. - Sydney ; London : Prentice-Hall, 1982
191p : ill. - Includes index
0-7248-0098-0 X.622/14915
286584x

ACCOUNTANCY
England. Companies. Creditors' voluntary liquidation. Procedure - For accountancy

DIQUE, Trevor
Creditors' voluntary liquidation. - London : Institute of Chartered Accountants in England and Wales, 1984
92leaves. - (Work pack)
0-85291-364-8 X.525/9207
0602590

INSTITUTE OF CHARTERED ACCOUNTANTS IN ENGLAND AND WALES
Creditors voluntary liquidation. - London : ICAEW, 1975
44 leaves in various pagings. - (Work pack)
0-85291-130-0 X.512/5114
0602590

ACCOUNTANCY
England. Companies. Liquidation. Receivership. Procedure - For accountancy

INSTITUTE OF CHARTERED ACCOUNTANTS IN ENGLAND AND WALES
Receivership. - London : ICAEW, 1975
39 leaves in various pagings. - (Work pack)
0-85291-129-7 X.515/509
0601918

ACCOUNTANCY
England. Companies. Members' voluntary liquidation. Procedure - For accountancy

DIQUE, Trevor
Members' voluntary liquidation. - London : Institute of Chartered Accountants in England and Wales, 1984
86leaves. - (Work pack)
0-85291-363-x X.525/9206
059783x

INSTITUTE OF CHARTERED ACCOUNTANTS IN ENGLAND AND WALES
Members voluntary liquidation. - London : ICAEW, 1975
37 leaves in various pagings. - (Work pack)
0-85291-131-9 X.512/5104
059783x

ACCOUNTANCY
England. Executorship. Law - For accountancy

RANKING, D. F. de l'Hoste (Devey Fearon de l'Hoste)
Ranking, Spicer and Pegler's executorship law and accounts. - 22nd ed.. - London : H.F.L., 1982
1v(loose-leaf). - Includes index
0-372-30005-7 X.200/46171
062828x

THOMPSON, Jack
Executorship law and accounts. - London : Data Study Books, [1975]
50p
0-905419-00-6 X.100/16276
062828x

ACCOUNTANCY
England. Inherited property: Estates. Administration. Law - For accountancy

SHERRING, Tony
Law and accounts for executors and trustees. - London : Gee, 1981
xvi,257p. - Bibliography: p248. - Includes index
0-85258-195-5 X.200/35773
0017205

VICKERY, Bertram George
Law and accounts of executors, administrators and trustees. - 19th ed.. - Reading : Donnington Press ; London : Distributed by Cassell and Collier Macmillan, 1975
xii,496p. - Includes index
0-304-29629-5 X.200/20231
0017205

Law and accounts of executors, administrators, and trustees. - 20th ed. - London : Cassell, 1980
xii,371p. - (A Cassell professional handbook)
0-304-30452-2 X.200/33713
0017205

ACCOUNTANCY
England. Insolvency. Procedure - For accountancy

CHAMBERS, Bernard
Insolvency. - [London] : [Certified Accountants Educational Trust], [1978]
118p. - (Certified Accountants technical publication)
B78-31262 X.510/15871
1574094

ACCOUNTANCY
England. Law - For accountancy

CARD, Richard
Law for accountancy students. - London : Butterworth, 1978
xlii,660p : 1 ill. - Includes index
0-406-70881-9 X.100/17934
1689150

Law for accountancy students. - 2nd ed. - London : Butterworths, 1982
xl,547p. - Includes index
0-406-70882-7 X.200/38961
1689150

LEAL, A. R. (Andrew Richard)
Essential law for accountancy students. - London : Edward Arnold, 1982
viii,278p : ill. - Includes index
0-7131-0605-0 X.200/38040
1689150

ACCOUNTANCY
Ethical aspects

WINDAL, Floyd W.
The accounting professional. - Englewood Cliffs ; London : Prentice-Hall, 1980
xv,456p. - Includes bibliographies and index
0-13-003020-1 X.520/21791
2155621

ACCOUNTANCY
Ethics

ETHICS in the accounting profession / edited by Stephen E. Loeb. - Santa Barbara ; Chichester : Wiley, 1978
xii,479p : ill. - (Wiley/Hamilton series in accounting and information systems). - Includes bibliographies and index
0-471-54331-4 X.519/28773
1733044

ACCOUNTANCY
Great Britain. Companies. Taxation. Forms. Filling - For accountancy

ROBSON RHODES (Firm)
Chartac tax and company law forms. - London : Gee in associaition with the Institute of Chartered Accountants in England and Wales, 1984
1v.(looseleaf). - Includes index
0-85258-813-5 X.205/1714
3023419

ACCOUNTANCY
Great Britain. Income tax. Back duty - For accountancy

CERTIFIED ACCOUNTANTS EDUCATIONAL TRUST
Back duty practice notes. - 3rd ed.. - London : The Trust, 1976
[3],16p. - (Certified Accountants technical publication)
B79-02453 X.510/15958
0157856

ACCOUNTANCY
Great Britain. Prices. Control. Policies of government: Price Code, 1977-1978 - For accountancy

CONRAN, C. C.
Guide to the 1977 counter-inflation legislation. - London : Institute of Chartered Accountants in England and Wales, 1978
24p. - (Professional briefs for practitioners / Institute of Chartered Accountants in England and Wales. Smaller Practitioner Advisory Committee). - Bibliography: p.24
0-85291-191-2 X.100/17582
1457322

ACCOUNTANCY
Great Britain. Sex discrimination. Law. Sex Discrimination Act 1975 - For accountancy

TRESMAN, John Henry
Sex Discrimination Act 1975. - Birmingham : Society of Company and Commercial Accountants, [1976]
[2],13p
0-9503913-2-8 X.100/16268
0628344

ACCOUNTANCY
Great Britain. Social security benefits - For accountancy

BARTLETT, K. D
Employment, unemployment and disability. - London : Institute of Chartered Accountants in England and Wales, 1978
33p. - (Professional briefs for practitioners / Institute of Chartered Accountants in England and Wales. Smaller Practitioner Advisory Committee)
0-85291-210-2 X.519/25909
1848003

Family related benefits. - London : Institute of Chartered Accountants in England and Wales, 1978
34p. - (Professional briefs for practitioners / Institute of Chartered Accountants in England and Wales. Smaller Practitioner Advisory Committee)
0-85291-209-9 X.519/29510
1848003

ACCOUNTANCY
Marketing - For accountancy

COWLEY, Simon
Practice development. - London : Butterworths, 1984
viii,119p : ill. - Includes index
0-406-26040-0 X.622/23010
2281872

ACCOUNTANCY
Materials management - For accountancy

HARRIS, N. D. (Neville D)
Materials management. - [London] : Institute of Chartered Accountants in England and Wales, 1980
vii,77p : ill. - (The Accountant in industry and commerce). - Bibliography: p76-77
0-85291-279-x X.629/19023
2814102

ACCOUNTANCY
Organisations: Association of Certified Accountants

ASSOCIATION OF CERTIFIED ACCOUNTANTS
An association of ideas. - [London] : The Association, [1978]
[3],17p
B78-35400 X.519/28587
1689312

ACCOUNTANCY
Public speaking - For accountancy

HOLGATE, John, 1911-
Effective speaking for accountants. - London : Institute of Chartered Accountants in England and Wales, 1981
viii,88p : 2ill
0-85291-308-7 X.950/10538
2626845

ACCOUNTANCY
Standards - Conference proceedings

FINANCIAL reporting and accounting standards. - [Glasgow] : University of Glasgow Press, 1978
[7],80p : ill. - Bibliography: p.37-39
0-85261-150-1 X.519/29138
1780336

ACCOUNTANCY
Terminology - Dictionaries

LEXIQUE UEC / Union européenne des experts comptables économiques et financiers (UEC). - 2. éd. - Düsseldorf : IdW-Verlag, 1974
1v.(various pagings)
3-8021-0073-5 X.529/47023
8724784

SVENSK-ENGELSK ordlista för redovisning och revision / [utgiven av Föreningen Auktoriserade revisorer]. - Stockholm : FAR, 1981
120p
91-85992-07-0 X.950/32797
8724784

UNION EUROPÉENNE DES EXPERTS COMPTABLES, ÉCONOMIQUES ET FINANCIERS
International accounting lexicon. - Special ed. - Düsseldorf : IdW-Verlag, 1980
[184]p
3-8021-0192-8 X.529/43234
8724784

WORTERBUCH der Deutsch-englischen Fachausdrücke des Rechnungswesens und des Geschäftsverkehrs = Dictionary of German-English accounting and business terms / herausgegeben von Arthur Andersen & Co. - 2. Aufl. - Stuttgart : Schäffer, 1981
255p
3-8202-0193-9 X.520/30819
8724784

ACCOUNTANCY
United States. Law courts. Expert evidence - For accountancy

DYKEMAN, Francis C.
Forensic accounting. - New York ; Chichester : Wiley, c1982
ix,236p. - Includes index
0-471-08395-x X.520/26654
2707284

ACCOUNTANCY. Curriculum subjects. Higher education institutions. Great Britain
Degree courses - Serials

APPROVED courses for accountancy education. - 1981 ed.-. - [London] : [Accounting Education Consultative Board], [1981?]-
v.
B82-18543 P.513/2586
2379716

DEGREE studies and the accountancy profession / Accounting Education Consultative Board. - 1980 ed. - London : The Board, [1980?]
27p
Subsequent volumes, when received, will be placed at this shelf mark
B81-13087 P.513/2586
2379716

ACCOUNTANCY. Developing countries

ENTHOVEN, Adolf Jan Henri
Accounting systems in Third World economies .... - Amsterdam ; Oxford : North-Holland Publishing Co., 1977
viii,359p,[5]leaves. - Includes index
0-7204-0721-4 X.520/11886
1188119

ACCOUNTANCY. England

A GUIDE to professional practice regulations. - London : The Institute of Chartered Accountants in England and Wales, 1983
111p. - Bibliography: p111
0-85291-338-9 X.520/31407
0621625

ACCOUNTANCY. England
Applications of digital computer systems

SHEFFIELD AND DISTRICT SOCIETY OF CHARTERED ACCOUNTANTS.
Technical Advisory Committee. Computer Sub-Committee
The use of computers and computer bureaux in practising accountants' offices. - London : Institute of Chartered Accountants in England and Wales, 1978
32p. - (Professional briefs for practitioners / Institute of Chartered Accountants in England and Wales. Smaller Practitioner Advisory Committee)
0-85291-193-9 X.519/27984
1451081

ACCOUNTANCY. England
Fees

ALFORD, Mervyn H.
How to reduce your accountancy and audit fees. - Barry : Vale Publications, 1983
71p. - Includes index
0-946389-00-4 X.529/56931
0013269

ACCOUNTANCY. England
Organisations: Institute of Chartered Accountants in England and Wales, to 1965

HOWITT, Sir Harold
The history of the Institute of Chartered Accountants in England and Wales 1870-1965. - New York ; London : Garland, 1984, c1966
xiv,269p,[16]p of plates : ill. - (Accounting history and the development of a profession). - Bibliography: p260-262. - Includes index
0-8240-6329-5 X.809/63802
3263541

ACCOUNTANCY. England
Organisations: Institute of Chartered Accountants in England and Wales, to 1980

HOPKINS, Leon
The hundredth year. - Plymouth : Macdonald and Evans, 1980
vii,168p : ill
0-7121-0816-5 X.529/36801
2207044

ACCOUNTANCY. England
Organisations: Institute of Chartered Accountants in England and Wales. Organisation

INSTITUTE OF CHARTERED ACCOUNTANTS IN ENGLAND AND WALES
Governing the Institute. - London : The Institute, 1983
193p : ill. - Bibliography: p191-193
0-85291-342-7 X.520/31215
295432x

ACCOUNTANCY. Finland
- Bibliographies

VAISTO, Erkki
Laskentatoimen kirjallisuus Suomessa vuosilta 1945-1971 = Bibliography of accountancy in Finland 1945-1971. - Helsinki : Kauppakorkeakoulu : Jakelu : Kyriiri, 1973
244p. - (Kauppakorkeakoulun julkaisuja : sarja C : 2 ; 15)
951-699-008-8 P.511/1681[pt.2,no.15]
243931X

ACCOUNTANCY. Great Britain

ACCOUNTANCY / [edited by SGS Associates (Education) Limited]. - London : Careers and Occupational Information Centre, 1978
16p : ill. - (Career profiles ; 8)
0-86110-075-1 BS.421/27(8)
2238799

MAIN services. - 3rd ed. - London : Institute of Chartered Accountants in England and Wales, 1984
7p
0-85291-375-3 X.525/10480
0777196

NOBES, Christopher
Becoming an accountant. - London : Longman, 1983
x,123p : ill. - (Longman accounting and finance). - Includes index
0-582-29640-4 X.529/60660
2238799

PARKER, R. H. (Robert Henry)
Men of account. - [Exeter] : University of Exeter, 1977
22p. - (Inaugural lectures / University of Exeter)
0-85989-018-x X.519/28080
0777196

TAYLOR, Felicity
Careers in accountancy. - London : Kogan Page, 1982
110p
0-85038-520-2 X.529/63436
2238799

ACCOUNTANCY. Great Britain
- Directories

1984 training opportunities / The Institute of Chartered Accountants in England and Wales. - London : The Institute, [1983]
43p
B84-13285 X.529/59144
2577216

ACCOUNTANCY. Great Britain
- For graduates

The DOG guide to accountancy & finance / [editor Iris Rosier]. - [London] : [VNU Business Publications], [1981?]
52p : ill
0-86271-017-0 X.520/25682
2589052

ACCOUNTANCY. Great Britain
- Serials

The ACCOUNTING bulletin. - Issue no.1 (Oct. 1982)-. - London : Michael Lafferty Publications, 1982-
v. : ill
B83-09844 P.513/3029
0042781

INSTITUTE OF CHARTERED ACCOUNTANTS IN ENGLAND AND WALES
Technical bulletin. - Vol. 1, no. 1-; Dec. 1979-. - London : The Institute, 1979-
B80-23446 P.511/5215
0042781

NEWSLETTER for Chartac technical manual users. - 1-. - London : Gee, [1983]-
v. : ill
B84-22217 P.625/288
0042781

[UPDATE (London : 1984)]. Update. - Vol.1, no.1 (Jan. 1984)-. - London : Institute of Chartered Accountants in England and Wales, 1984-
v.
B85-03236 P.823/388
0042781

ACCOUNTANCY. Great Britain
Historical sources: Books: Subscription lists
WALLIS, Peter John
Book subscription lists and the accounting historian. - [Newcastle upon Tyne] : University of Newcastle upon Tyne, 1980
10,15p. - (PHIBB ; 234)
0-7017-0024-6 X.805/3729
2866420

ACCOUNTANCY. Great Britain
Organisations. Libraries. Stock - Catalogues
CURRENT accounting literature. - London : Mansell. 1971-. - [1971]-
r6086473 OPL
001561X
- [Another copy.] 2740.c.9

ACCOUNTANCY. Great Britain
Organisations: Society of Incorporated Accountants, to 1957
GARRETT, A. A. (Alexander Adnett)
History of the Society of Incorporated Accountants 1885-1957. - New York ; London : Garland, 1984, c1961
xviii,360p,[14]leaves of plates : ill. - (Accounting history and the development of a profession). - Includes index
0-8240-6327-9 X.629/26048
326355x

ACCOUNTANCY. Great Britain
Partnerships: Touche Ross & Co., to 1981
TOUCHE ROSS & CO.
Touche Ross & Co. 1899-1981. - London : Touche Ross & Co., c1981
xiii,145p : ill
B82-12189 X.800/32322
2629828

ACCOUNTANCY. Great Britain
Smallpeice, Sir Basil - Biographies
SMALLPEICE, Sir Basil
Of comets and queens. - Shrewsbury : Airlife, 1981, c1980
274p,18p of plates : ill. - Includes index
0-906393-10-8 X.629/15589 Woolwich
2485087

ACCOUNTANCY. Hungary
to date
BALÁS, Alajos
Könyvszakértők kézikönyve. - Budapest : Közgazdasági és Jogi Könyvkiadó, 1981
444p
963-221-004-2 X.520/29279
8553181

ACCOUNTANCY. Ireland (Republic)
CHARTERED accountant 80. - [Dublin] : The Institute of Chartered Accountants in Ireland, [1980?]
68p : ill(some col.)
B82-30397 X.529/48620
2397137

ACCOUNTANCY. Netherlands
Organisations: Nederlands Instituut van Registeraccountants
BRUYNE, A. L. de
De geschiedenis van de beroepsorganisatie der accountants in Nederland 1895-1975. - Alphen aan den Rijn : Samsom, 1980
151p : ill
90-14-03021-5 X.520/24112
870662x

ACCOUNTANCY. Norway
- For agricultural industries - Serials
MODELLBRUKSBEREGNINGER / Budsjettnemnda for jordbruket. - 1978-. - Oslo : Budsjettnemnda, 1980-
r6112771 SB.795/32
2465388

ACCOUNTANCY. Norway
Ethics
PROFESSIONAL ethics for statsautoriserte revisorer in Norway / a study undertaken by the Institute of Chartered Accountants in England and Wales, Norges Statsautoriserte Revisorers Forening. - London : Published on behalf of the Anglo-Nordic Liaison Committee by the Institute of Chartered Accountants in England and Wales, 1980
24p
0-85291-288-9 X.525/7137
2890372

ACCOUNTANCY. Scotland
INSTITUTE OF CHARTERED ACCOUNTANTS OF SCOTLAND
The C.A. student guide. - [Edinburgh] : [The Institute], 1982
64p : 1ill
B83-04733 X.529/51491
256176x

ACCOUNTANCY. Scotland
1854-1879 - Biographies
STEWART, James C.
Pioneers of a profession. - Edinburgh : Scottish Committee on Accounting History, the Institute of Chartered Accountants of Scotland, 1977
xii,181p : ill. - Includes index
0-85099-026-2 X.529/32055
1342673

ACCOUNTANCY. Scotland
Organisations: Institute of Chartered Accountants of Scotland, 1954-1979
MCDOUGALL, E. H. V.
Fifth quarter-century. - Edinburgh : Accountants' Publishing Co. for the Institute of Chartered Accountants of Scotland, 1980
xv,101p,xvip of plates : ill
0-85099-027-0 X.529/37748
2275414

ACCOUNTANCY. United States
ABRAHAM, Stanley Charles
The public accounting profession. - Lexington, Mass. : Heath ; Farnborough [Hants.] : [Distributed by] Teakfield Ltd, 1978
xx,295p : ill. - Includes index
0-669-01606-3 X.520/13355
1777904

RANDALL, Michael
How to prepare for the C.P.A. examination. - New York ; London : McGraw-Hill, 1977
[10],436p : ill
0-07-051182-9 X.525/3289
2220687

ACCOUNTANCY. United States
- Dictionaries
KOHLER, Eric L. (Eric Louis)
Kohler's dictionary for accountants. - 6th ed. - Englewood Cliffs, N.J. ; London : Prentice-Hall, c1983
xi,574p : ill. - Includes index
0-13-516658-6 X.520/30242
0682985

ACCOUNTANCY. United States
- For businessmen
HAYES, Rick Stephan
Simplified accounting for non-accountants. - New York ; Chichester : Willey, 1980
ix,291p : ill. - (Small business management series). - Includes index
0-471-04977-8 X.529/36708
221069x

ACCOUNTANCY. United States
- Indexes - Dictionaries
KOHLER, Eric Louis
A dictionary for accountants. - 5th ed. - Englewood Cliffs ; London : Prentice-Hall, 1975
xi,498p : ill
0-13-209783-4 OPL
0682985
- [Another copy] X.520/10826

ACCOUNTANCY FIRMS
Efficiency. Measurement. Use of regression analysis
DEWHURST, Roland Frederick James
Costing the routine outputs of an accountant's department. - [London] : Institute of Chartered Accountants in England and Wales, Research Committee, 1978
[9],96p : ill. - (Occasional papers / Institute of Chartered Accountants in England and Wales. Research Committee ; no.15). - Bibliography: p.96
0-85291-204-8 X.0512/168(15)
1572857

ACCOUNTANCY FIRMS. Great Britain
Coopers & Lybrand (Firm) - Serials
C & L journal. - No.1 ([19-])-. - London : Coopers & Lybrand, [19-]-
v : ill(some col.)
Holdings: No.41 (Dec. 1983)-
r6816766 P.903/968
3328430

ACCOUNTANCY FIRMS. Great Britain
Ernest & Whinney, to 1980
JONES, Edgar
Accountancy and the British economy 1840-1980. - London : Batsford, 1981
288p,[16]p of plates : ill. - Bibliography: p271-281. - Includes index
0-7134-3776-6 X.529/46276
2638304

ACCOUNTANCY FIRMS. Great Britain
Organisation
WALDRON, Robert Sydney
Building a profitable practice. - London : Certified Accountants Educational Trust, 1974
73p : ill. - Bibliography: p.73
0-900094-25-7 X.510/11267
0165905

ACCOUNTANCY FIRMS. United States
Haskins & Sells, to 1970
HASKINS & SELLS
Our first seventy-five years. - New York ; London : Garland, 1984, c1970
192p. - (Accounting history and the development of a profession). - Includes index
0-8240-6326-0 X.800/40878
3263592

ACCOUNTANCY FIRMS. United States
Management
FOX, Jack
Starting and building your own accounting business. - New York ; Chichester : Wiley, c1984
xvi,257p : ill. - (A Ronald Press publication) (The Wiley/Ronald - National association of Accountants professional book series). - Includes index
0-471-80053-8 X.622/22713
325738x

ACCOUNTANCY SERVICES. Great Britain
Supply. Effects of restrictions on advertising by accountants
GREAT BRITAIN. Monopolies and Mergers Commission
Accountancy services. - London : H.M.S.O., [1976]
[4],104p. - (Cmnd. 6573). - Includes index
0-10-165730-7 BS.83/93
0736295

ACCOUNTANCY SERVICES. United States
Marketing
DENNEY, Robert W.
Marketing accounting services. - New York ; London : Van Nostrand Reinhold, c1983
xi,269p. - Includes index
0-442-22003-0 X.520/32500
1778455

MAHON, James J.
The marketing of professional accounting services. - New York ; Chichester : Wiley, 1978
xv,189p : 1 ill. - Includes index
0-471-04480-6 X.520/13351
1778455

MAHON, James J. (James Joseph)
The marketing of professional accounting services. - 2nd ed. - New York ; Chichester : Wiley, c1982
xviii,199p : 1ill. - (Modern accounting perspectives and practice). - Includes index
0-471-86731-4 X.520/29332
1778455

ACCOUNTANCY SERVICES. United States
Quality control
PATTILLO, James W.
Quality control and peer revew. - New York ; Chichester : Wiley, c1984
xviii,301p. - (Modern accounting perspectives and practices series). - Includes index
0-471-86986-4 X.520/35737
3217086

ACCOUNTANTS
See also
Chartered accountants
Trainee accountants

ACCOUNTANTS
Professional education - Serials
CERTIFIED Accountants Students' Newsletter. - Vol.1, no.1- ; Jan. 1976-. - London : Certified Accountants Educational Trust, 1976- : ill
B76-11451 P.515/400
0616095

ACCOUNTANTS. Australia
Communication skills. Development
GALVIN, Michael
Communication skills. - [West mead, N.S.W.] : [M. Galvin?], c1983
60pin various pagings. - Bibliography: p1-3
0-908078-03-x X.520/32422
3031586

ACCOUNTANTS. Australia
Professional education. Curriculum

FLANAGAN, Jack
The Core of the curriculum for accounting undergraduates in Australia. - [Westmead, N.S.W.] : [J. Flanagan?], c1982
ca.194p
0-908078-02-1 X.520/32423
3031578

ACCOUNTANTS. Belgium
Professional education

ANGLO-BELGIAN LIAISON COMMITTEE OF THE INSTITUTE OF CHARTERED ACCOUNTANTS IN ENGLAND AND WALES. COLLEGE NATIONAL DES EXPERTS COMPTABLES DE BELGIQUE. INSTITUT DES REVISEURS D'ENTREPRISES
Education, training and examination requirements for admission to the profession of expert comptable (CNECB) in Belgium. - London : The Institute, 1975
[4],27p
0-85291-138-6 X.515/507
059623x

ACCOUNTANTS. Developing countries
Professional education. Role in economic development of developing countries

ENTHOVEN, Adolf J. H. (Adolf Jan Henri)
Accounting education in economic development management. - Amsterdam ; Oxford : North-Holland, c1981
ix,448p : ill. - Bibliography: p433. - Includes index
0-444-86195-5 X.520/24210
2500213

ACCOUNTANTS. England
Professional education. Practical experience. Self-assessment - For trainee accountants

STUDENTS'S training record file / Institute of Chartered Accountants in England and Wales. - [London] : [The Institute], [1981?]
1v.(loose-leaf)
B82-07138 L.45/1386
2608057

ACCOUNTANTS. Great Britain
Advertising. Restrictions. Effects on supply of accountancy services

GREAT BRITAIN. Monopolies and Mergers Commission
Accountancy services. - London : H.M.S.O., [1976]
[4],104p. - (Cmnd. 6573). - Includes index
0-10-165730-7 BS.83/93
0736295

ACCOUNTANTS. Great Britain
Negligence. Liability. Law

POCKSON, Jonathan R. H. H.
Accountant's professional negligence. - London : Macmillan, 1982
xxi,212p : ill. - Bibliography: p192-208. - Includes index
0-333-27845-3 X.200/38235
271714x

WOOLF, Emile
Legal liabilities of practising accountants. - London : Butterworths, 1984
xx,457p. - Bibliography: p445. - Includes index
0-406-25960-7 X.205/1677
271714x

ACCOUNTANTS. Great Britain
Remuneration

SMALL, John, 1933-
Accountants' remuneration survey 1983-4. - Croydon : Tolley, 1984
70p : ill
0-85459-145-1 X.525/10076
0323977

ACCOUNTANTS. Industries
Chief accountants. Leadership

MCKENNA, Eugene F.
The management style of the chief accountant. - Farnborough, Hants. : Saxon House, 1978
xiii,307p : ill. - Bibliography: p.280-300. - Includes index
0-566-00216-7 X.529/33717
1403516

ACCOUNTANTS. Industries. Great Britain
Training

SOCIETY OF COMPANY AND COMMERCIAL ACCOUNTANTS
Training the accountant in commerce and industry. - Bristol : The Society, [1975]
27p
0-9503913-1-x X.519/25926
0680834

ACCOUNTANTS. United States
Negligence. Liability. Law

DAVIS, Jonathan J.
CPA liability. - New York ; Chichester : Wiley, c1983
ix,212p. - (Modern accounting perspectives and practice). - Includes index
0-471-06290-1 X.520/31355
2717158

POCKSON, Jonathan R. H. H.
Accountant's professional negligence. - London : Macmillan, 1982
xxi,212p : ill. - Bibliography: p192-208. - Includes index
0-333-27845-3 X.200/38235
2717158

ACCOUNTANTS. United States
Professional education. C.P.A. examinations

DELANEY, Patrick R.
CPA examination review. - 1984 ed. - New York ; London : Plenum, c1984
3v. - Includes index
B85-06992 X.205/1619-1621
2582457

STENZEL, Alvin M.
Approaching the CPA examination. - New York ; Chichester : Wiley, 1981
102p. - Bibliography: p97-102
0-471-08699-1 X.529/44412
2582457

ACCOUNTING
See also
Auditing
Book-keeping
Budgeting
Cost value reconciliation
Financial accounting
Human asset accounting
Inflation accounting
International accounting
Management accounting
National accounting
Profit-centre accounting
Replacement-value accounting
Social accounting
Variance accounting

ACCOUNTING

ACCOUNTING / edited by Gerhard G. Mueller, Charles H. Smith. - 2nd ed. - Hinsdale, Ill. : Dryden Press, 1977
x, 310 p. : ill. - Bibliography : p. 57
0-03-089908-7 X.512/5560
0164259

ACCOUNTING theory / edited by Francis A. Bird. - Richmond, Va. : Dame, c1981
xiii,543p
0-936328-09-6 X.529/68267
0045543

ADVANCED accounting / prepared by Chart Foulks Lynch. - London : Holt, Rinehart and Winston, c1984
x,481p. - (100 questions and answers). - Includes index
0-03-910503-2 X.529/59986
0630950

BARTON, A. D.
The anatomy of accounting. - 2nd ed. - St Lucia : University of Queensland Press ; Hemel Hempstead : Distributed by Prentice-Hall, 1977
xv,593p : ill. - Includes bibliographies and index
0-7022-1482-5 X.520/12029
0045543

BASIC accounting / prepared by Chart Foulks Lynch. - London : Holt, Rinehart and Winston, c1983
x,359p. - (100 questions and answers)
0-03-910437-0 X.529/56357
0630950

BASTON, Andrew
Elements of accounts. - 10th ed., completely revised. - London : Cassell, 1980
vi,417p : ill
0-304-30557-x X.529/35477
0045543

BEAMS, Floyd A.
Advanced accounting. - Englewood Cliffs ; London : Prentice-Hall, 1979
xiv,816p : ill. - Includes bibliographies and index
0-13-010124-9 X.520/14635
0294624

BEAMS, Floyd A.
Advanced accounting. - 2nd ed. - Englewood Cliffs ; London : Prentice-Hall, c1982
xvii,904p : ill. - Includes bibliographies and index
0-13-010157-5 X.520/26838
0045543

BEDFORD, Norton Moore
Advanced accounting. - 4th ed. - New York ; Chichester : Wiley, 1979
xix,892p : ill. - (Wiley series in accounting and information systems). - Includes index
0-471-02927-0 X.520/13856
0294624

BISHOP, K. R. (Kenneth Richard)
Numeracy and accounting. - London : Cassell, 1980
x,230p : ill. - (Cassell's BEC series. National level)
0-304-30333-x X.629/14559 Woolwich
0045543

BLAKE, G.
Numeracy and accounting. - London : Pitman, 1979
[6],296p : ill. - (Pitman business education). - Includes index
0-273-01323-8 X.519/30245
0304018

BRAY, F. Sewell (Frank Sewell)
Frank Sewell Bray. - Glasgow : Strathclyde Convergencies, 1982
141p : ill. - List of writings: p30-32
0-906161-08-8 X.622/14879
0045543

BRIGHT, George
Practical accounts 2. - London : Pan, 1983
237p : ill. - (Pan breakthrough books). - Includes index
0-330-28058-9 X.529/58695
0045543

BRISTON, Richard J. (Richard Jeremy)
Introduction to accountancy and finance. - London : Macmillan in association with the Institute of Cost and Management Accountants, 1981
xiv,663p : ill. - Includes bibliographies and index
0-333-24669-1 X.622/10358 Woolwich
1198300

BUTT, Safdar Ali
A practice manual in accounting. - London : Cassell.
Vol.1. - 1978
[5],204p
0-304-29896-4 X.0519/1034
0630950
Vol.2. - 1978
[5],314p
0-304-29947-2 X.0519/1034
0630950

CALLAM, A. A.
A foundation in business accounting. - London : Macmillan, 1977
viii,403p : ill. - Includes index
0-333-18452-1 X.510/15494
0045543

CAMERON, James B.
Advanced accounting. - Boston [Mass.] ; London : Houghton Mifflin, 1979
xv,816p : ill. - Includes index
0-395-27446-x X.520/13798
0294624

CAPITAL, income & decision making / [edited by] Ian Tilley / Peter Jubb. - Sydney ; London : Holt, Rinehart and Winston, 1978
[9],313p : ill. - Includes index
0-03-900126-1 X.510/16438
0045543

CARRINGTON, Athol S.
Data and information. - [Glasgow] : Department of Accountancy, University of Glasgow, 1983
20leaves : ill. - (Arthur Young lecture ; no.5). - Bibliography: p18-20
0-85261-176-5 X.525/8035
0045543

CARTER, Roger Neale
Carter's advanced accounts. - 7th ed. revised. - London : Pitman. - (A Pitman international text)
Key to Carter's advanced accounts. - 1978
v p,p101-2939
0-273-01093-x X.519/30262
0294624

CASHIN, James A.
Schaum's outline of theory and problems of accounting 1. - 2nd ed. - New York ; London : McGraw-Hill, c1980
278p : ill. - (Schaum's outline series). - Includes index
0-07-010251-1 X.525/4407
0045543
Schaum's outline of theory and problems of intermediate accounting. - New York ; London : McGraw-Hill. - (Schaum's outline series)
1. - [1976]
[9],306p : ill. - Includes index
0-07-010202-3 X.512/5500
0085650

CASTLE, E. F. (Eric Frank)
Principles of accounts. - 6th ed. - Plymouth : Macdonald and Evans, c1981
viii,440p. - (The M & E handbook series). - Includes index
0-7121-1692-3 X.529/41426
0045543

ACCOUNTING -continued

CHAHIN, M. (Mekerdich)
Key to Favell's Practical bookkeeping and accounts. - 2nd ed. - Slough : University Tutorial Press, 1980
218p
0-7231-0804-8 X.529/49011
0630950

CHAHIN, Mekerdich
Accounting for first examinations. - London : University Tutorial Press, 1977
[6],337p
0-7231-0747-5 X.510/15491
0304018

CLAY, Raymond J.
Student study guide to accompany Intermediate accounting, fourth edition, by Donald E. Kieso and Jerry J. Weygandt. - New York ; Chichester : Wiley, c1983
222p
0-471-08883-8 X.525/9637
0630950

COLINSON, Robert
Idea rationaria, or The perfect accomptant, 1683. - London : Scolar Press ; Tokyo : Yushodo Press, 1978
[260]p. - (Historic accounting literature ; vol.7)
r6112456 X.0625/102(7)
1198300

CONTEMPORARY accounting thought / edited by M.J.R. Gaffikin. - Sydney : Prentice-Hall of Australia ; London : Prentice-Hall International, c1984
346p. - Includes bibliographies
0-7248-0201-0 X.529/70006
0045543

COOPER, J. M. John Macneill
Objective tests in accounts 1. - London : Longman, 1981
81p. - (Longman objective tests series)
0-582-41577-2 X.0619/402 Woolwich
0033863

CORREA, Hector
Integrated economic accounting. - Lexington, Mass. : D.C. Heath, 1977
xxi,223p. - Includes bibliographies and index
0-669-00779-x X.520/12302
0045543

COULDERY, Frederick Alan James
Understanding accounts. - London : Gee and Co., 1979
88p : ill(chiefly col). - (A Gee's study book). - Includes index
0-85258-183-1 X.520/21299
0045543

COY, John
Accounting case problems. - 2nd ed.. - Stockport : Polytech, 1983
60p
0-85505-072-1 X.525/10569
1966944

CROWLEY, T. Pearse
Modern accounting. - Dublin : Gill and Macmillan, c1984
540p
0-7171-1023-0 X.520/35874
0045543

CURRENT issues in accounting / editors Bryan Carsberg and Tony Hope. - Oxford : P. Allan, 1977
[9],293p. - (Philip Allan textbooks in business studies)
0-86003-503-4 X.529/31878
0045543

CURRENT issues in accounting / editors Bryan Carsberg and Tony Hope. - 2nd ed. - Oxford : Philip Allen, 1984
xi,317p
0-86003-522-0 X.529/67478
0045543

DAFF, Trevor
Accounting workbook 2. - London : Pitman, 1982
187p
0-273-01730-6 X.529/48140
0630950
Quantitative and accounting methods. - London : Pitman, 1983
vii,303p : ill. - Includes index
0-273-01825-6 X.529/58683
1052098

DAFFORNE, Richard
The merchants mirrour, 1635/[by] Richard Dafforne. An introduction to merchants accounts, 1653/[by] John Collins. - London : Scolar Press ; Tokyo : Yushodo Press, 1978
[26],55,[332]p. - (Historic accounting literature ; vol.4)
r6112455 X.0625/102(4)
1198300

DAVIDSON, Sidney
Intermediate accounting. - 3rd ed. - Chicago ; London : Dryden Press, c1982
[1100]p. - (The Dryden Press series in accounting). - Includes index
0-03-058916-9 X.520/28121
0045543

DAWBER, Ron
Clerical accounting. - London : Pitman, 1976
vi,122p : ill
0-273-00443-3 X.512/5371
0033863
Clerical accounting. - London : Pitman.
Teacher's manual. - 1976
viii,67p
0-273-00961-3 X.512/5371a
0033863

DIXON, Robert L. (Robert Livingston)
The executive's accounting primer. - 2nd ed. - New York ; London : McGraw-Hill, c1982
xv,395p. - Includes index
0-07-017079-7 X.520/27363
0045543
The McGraw-Hill 36-hour accounting course. - New York ; London : McGraw-Hill, 1976
xxiii,344p
0-07-017090-8 X.520/11257
0045543
The McGraw-Hill 36-hour accounting course. - 2nd ed. - New York ; London : McGraw-Hill, c1982
xxiii,412p. - Includes index
0-07-017091-6 X.520/27358
0045543

DOWNIE, John J.
Numeracy and accounting. - London : McGraw-Hill, 1979
xii,168p : ill. - (McGraw-Hill business education courses)
0-07-084600-6 X.510/16119
0045543

DUFFY, Leo
Introductory accounting. - Melbourne : Longman Cheshire, 1977
x,379p. - Bibliography : p.369. - Includes index
0-582-68525-7 X.510/15697
0045543

EDEY, Harold Cecil
Introduction to accounting. - 4th ed. - London : Hutchinson, 1978
192p : ill. - Bibliography: p.185-187. - Includes index
0-09-132580-3 WP.1413/451
0045543

ETHERIDGE, Richard Alfred
Numeracy and accounting. - Cheltenham : Thornes, 1980
viii, 232p : ill. - Includes index
0-85950-082-9 X.520/22457
0045543

EVERINGHAM, G. K.
Generally accepted accounting practice. - [Cape Town] : Juta, 1979
140p
0-7021-0998-3 X.520/14274
0045543

FINNEY, Harry Anson
Finney and Miller's principles of accounting. - 8th ed.. - Englewood Cliffs ; London : Prentice-Hall, 1980
xiv,754p : ill. - Includes index
0-13-317370-4 X.520/20697
1198300

FORBES numbers game / edited and with an introduction by Lawrence Minard and David A. Wilson. - Englewood Cliffs ; London : Prentice-Hall, c1980
xvi,175p. - Includes index
0-13-325100-4 X.629/15340 Woolwich
0045543

FOUNDATION in accounting. - London : Prentice-Hall.
1. - 1976
x,166p : ill. - Includes index
B76-20642 X.519/25853
0304018
2. - 1977
xvi,511p : ill. - Bibliography: p.505-506. - Includes index
0-13-329672-5 X.519/27107
0045543

FRANCIS, D. Pitt (David Pitt)
Numeracy and accounting. - London : Holt, Rinehart and Winston, c1981
viii,342p : ill. - (Holt business text). - Includes index
0-03-910301-3 X.629/15639 Woolwich
0045543

FRED learns accounts ... - [London] : Continua Productions, 1975
96p : ill. - (A Continua Fredbook). - Includes index
0-7227-0102-0 X.510/15083
0045543

GAMBLING, Trevor
Positive accounting. - London : Macmillan, 1984
xviii,184p : ill. - Bibliography: p175-179. - Includes index
0-333-26263-8 X.529/66626
0045543

GARBUTT, Douglas
Accounting foundations. - London : Pitman, 1980
xiii,439p : ill. - Includes index
0-273-01369-6 X.520/27541
1198300
Guide to accounting foundations. - London : Pitman, 1980
xi,348p : ill
0-273-01370-x X.520/27540
0630950

GLAUTIER, M. W. E. (Michael William Edgard)
Accounting theory and practice. - London : Pitman, 1976
xvi,741p : ill. - ([Accounting theory and practice series]). - Includes bibliographies and index
0-273-00705-x X.510/15311
0045543

GLAUTIER, M. W. E. (Michael William Edgard)
Accounting theory and practice. - 2nd ed. - London : Pitman, 1982
xxii,641p : ill. - (Accounting theory and practice series)
0-273-01541-9 X.520/29161
0045543
Basic accounting practice. - London : Pitman, 1978
xvii,426p : ill. - (Accounting theory and practice series) (A Pitman international text). - Includes index
0-273-01058-1 X.510/15705
0045543
Basic accounting practice. - 2nd ed. - London : Pitman, 1980
xv,426p : ill. - (A Pitman international text). - Includes index
0-273-01597-4 X.520/22560
0045543
Basic accounting practice. - 2nd ed. - London : Pitman. - (A Pitman international text)
Teacher's guide. - 1980
243p
0-273-01521-4 X.520/25789
0045543

GRANOF, Michael H.
Accounting for managers and investors. - Englewood Cliffs ; London : Prentice-Hall, c1983
xii,706p : ill. - Includes index
0-13-002725-1 X.520/31346
0045543

HALL, Arthur Gerald
An introduction to modern accounting. - 2nd ed. - London : Heinemann Educational, 1974
vi,177p : ill
0-435-45675-x X.519/26018
0045543

HANDBOOK of modern accounting / Sidney Davidson, editor ; Roman L. Weil, associate editor. - 2nd ed. - New York ; London : McGraw-Hill, 1977
1408p in various pagings : ill. - Includes bibliographies and index
0-07-015451-1 X.520/11949
1198300

HANDBOOK of modern accounting / edited by Sidney Davison and Roman L. Weil. - 3rd ed. - New York ; London : McGraw-Hill, c1983
1(various pagings) : ill. - Includes bibliographies and index
0-07-015492-9 X.520/34349
1198300

HARIED, Andrew A.
Advanced accounting. - New York ; Chichester : Wiley, 1979
xxiii,717p. - (Wiley series in accounting and information systems). - Includes index
0-471-02374-4 X.520/13793
0294624
Advanced accounting. - 2nd ed. - New York ; Chichester : Wiley, c1982
xxiv,912p. - Includes index
0-471-08717-3 X.520/26582
0045543
Instructor's manual to accompany Advanced accounting, second edition. - New York ; Chichester : Wiley, c1982
552p
0-471-86340-8 X.525/10296
0045543

HARRISON, John, b.1931
Accounting. - 3rd ed.. - Melbourne : Longman Cheshire, 1977
x,644p : ill. - Includes index
0-7015-2072-8 X.520/12417
0045543
Accounting. - 3rd ed. - Melbourne : Longman Cheshire. solutions manual. - 1978
[5],258p
0-582-71027-8 X.520/12418
0045543

HARVEY, Mike
The accounting framework in practice. - London : Certified Accountants Educational Trust, c1984
xiv,291p. - Bibliography: p281-282. - Includes index
0-900094-55-9 X.629/25693
0045543

HELMKAMP, John G.
Instructor's manual to accompany Principles of accounting. - New York ; Chichester : Wiley, c1983
182p
0-471-86287-8 X.525/7908
0045543

Principles of accounting. – New York : Chichester : Wiley, c1983
xx,1113,[46]p : ill(some col.). – Includes index
0–471–08510–3 X.520/30561
0045543

HERBERT, Michael, 19—
Practical accounts 1. – London : Pan, 1983
224p : ill. – (Pan breakthrough books). – Includes index
0–330–28057–0 X.529/58696
0045543

HINDMARCH, Arthur
Accounting. – London : Macmillan, 1977
xvii,314p : ill. – ([Macmillan business management and administration series]). – Includes bibliographies and index
0–333–15023–6 X.520/11718
0045543

HISTORIC accounting literature / edited by M.F. Bywater. – London : Scolar Press [etc.], 1978
51v.
B79–01286 X.0625/102 Woolwich
1751654

HITCHING, Claude
Understand accounting!. – London : Pitman, 1984
xiv,402p : ill. – Includes index
0–273–01883–3 X.529/66920
0045543

HOFFMAN, Frank
Accounting fundamentals. – 3rd ed. – New York ; London : McGraw–Hill, 1980
3v : ill. – (Continuing education series). – Includes index
0–07–029208–6 X.525/3122
0304018

HORNGREN, Charles T. (Charles Thomas)
Introduction to the financial accounting. – London : Prentice–Hall International, c1981
xviii,745p. – (Prentice–Hall series in accounting). – Includes index
0–13–483685–5 X.520/35483
1198300

HOUGHTON, Diane
Accounting terms and book–keeping procedures explained. – Farnborough, Hants. : Gower, c1980
x,268p
0–566–00393–7 X.529/40593
2090619

Students' accounting vocabulary. – Farnborough, Hants. : Gower, c1980
x,268p
0–566–00330–9 X.529/40594
2090619

HOWARD, Leslie R. (Leslie Reginald)
Advanced accounting practice. – London : Gee, 1985
[272]p. – Includes index
0–85258–241–2 X.525/10974
0045543

HOYLE, Ken
Business statistics and accounting made simple. – London : W.H. Allen, 1979
xii,350p : ill. – (Made simple books). – Includes index
0–491–02299–9 X.629/14267 Woolwich
0045543

INTERNATIONAL accounting standard / issued for comment by the International Accounting Standards Committee. – [London?] : [The Committee], 1983
[16]p. – (Exposure draft ; 25)
B84–27377 P.881/315
3131378

ISTVAN, Donald F.
Accounting principles. – New York : Harcourt Brace Jovanovich, c1979
xxi,1095p : ill(some col.)
0–15–500420–4 X.520/26969
1198300

Accounting principles. – New York : Harcourt Brace Jovanovich.
Study guide .... – 1979
xii,467p : ill
0–15–500423–9 X.512/6833
0045543

JOHNSON, G. B. (Gary B)
Accounting. – Wetherby : Michael Benn and Associates Limited. – (Revision cards)
1. – 3rd ed. – 1977
[2],iii,98p
0–904096–05–x X.519/27914
0045543
2. – 3rd ed. – 1977
[3],69,[7]p : ill
0–904096–06–8 X.519/27915
0045543
– 4th ed. – London : Pitman. – (Pitman revision cards)
2. – 1979
99p : ill
B83–04505 X.629/19050
0045543

JOHNSON, Harry
A practical foundation in accounting. – London : Allen & Unwin, 1982
xi,447p. – Includes index
0–04–332082–1 X.520/27885
1198300

A practical foundation in accounting. – London : Allen & Unwin, 1982
202p
0–04–332085–6 X.520/27544
0630950

JOHNSON, Harry, 1931–
A practical foundation in accounting. – 2nd ed. – London : Allen & Unwin, 1984
x,468p : ill. – Includes index
0–04–332095–3 X.622/20958
1198300

A practical foundation in accounting. – 2nd ed. – London : Allen & Unwin, 1984
210p : 1ill
0–04–332096–1 X.622/20950
0630950

JOPSON, Harold
Jopson's accounts. – London : Pitman Education, 1981
iv,284p : ill. – Includes index
0–273–01491–9 X.529/40400
0045543

KALUZA, Henry J.
Accounting projects and simulations. – Toronto ; London : McGraw–Hill, 1977
iii,161p
0–07–082470–3 X.512/5665
0630950

KEISTER, Orville R.
Study guide to accompany Haried, Imdieke, Smith Advanced accounting, second edition. – New York ; Chichester : Wiley, c1982
284p
0–471–09235–5 X.525/10295
0045543

KELLOCK, John
Elements of accounting. – London : Heinemann, 1978
xi,229p. – (The Heinemann accountancy and administration series). – Includes index
0–434–91046–5 X.519/28922
0045543

KIESO, Donald E. (Donald Ervin)
Intermediate accounting. – 4th ed. – New York ; Chichester : Wiley, c1983
xxii,1297p : ill. – Includes index
0–471–08871–4 X.520/30559
0045543

KIESO, Donald E. (Donald Ervin)
Overhead transparencies to accompany Intermediate accounting, fourth edition. – [New York] : Wiley, [1983]
414transparencies : b.&w.
0–471–08877–3 Cup.918/255
0045543

Solutions manual to accompany Intermediate accounting, fourth edition. – New York ; Chichester : Wiley, c1983
1148p
0–471–08881–1 X.525/9636
0630950

KIESO, Donald Ervin
Intermediate accounting. – 2nd ed. – Santa Barbara ; London : Wiley, 1977
xvii,1165p : ill. – Includes index
0–471–47466–5 X.520/11690
0085650

Intermediate accounting. – 3rd ed. – New York ; London : Wiley, 1980
xix,1278p : ill. – Includes index
0–471–04819–4 X.520/20823
0085650

LANGDON, Telles
Learning the basic principles and structure of accounts. – [2nd ed]. – [Croydon] : T. Langdon, [1976]
21p. – (The Complete answer series)
r6816701 X.529/69750
0045543

Variations in accounts. – [Croydon] : T. Langdon.
No.1. – [3rd ed.]. – [1975]
47p. – (The Complete answer series)
r6816708 X.529/69751
0045543
– [Croydon] : T. Langdon. – [1976].
No.2. – [4th ed.]
43p. – (The Complete answer series)
r6816704 X.529/69752
0045543

LANGLEY, F. P. (Frederick Percy)
Workbook in accounting. – 3rd ed. – London : Butterworths, 1981
273p : ill
0–408–10680–8 X.520/26262
0630950
– Teacher's guide X.520/26261

Workbook in accounting. – 3rd ed. – London : Butterworths, Oct.1981
[280]p
0–408–10823–1 X.520/26261
0085650

LANGLEY, Frederick Percy
Introduction to accounting for business studies. – 3rd ed. – London : Butterworth, 1978
xi,411p : ill. – Bibliography : p.405–406. – Includes index
0–406–75043–2 X.519/28279
0045543

LEE, G. A. (Geoffrey Alan)
Modern financial accounting. – 3rd ed. – Walton–on–Thames : Nelson, 1981
ix,543p. – (Nelson series in accounting and finance). – Includes index
0–17–761057–3 X.622/10185
0045543

LEE, Nanci
Elementary accounting. – Chicago ; London : Dryden, c1984
xxi,504p : ill. – Includes index
0–03–060446–x X.520/34530
0045543

MA, Ronald
The accounting framework. – Sydney : Longman Cheshire, 1979
xiv,732p : ill. – Includes bibliographies
0–582–71094–4 X.520/26722
0045543

MA, Ronald A.
The accounting framework. – Melbourne : Longman Cheshire.
Students' companion. – 1979
[5],306p : ill
0–582–71080–4 X.510/16393
0630950

MCCULLERS, Levis D.
Accounting theory. – Santa Barbara ; Chichester : Wiley, 1978
xiii,577p : ill. – (Wiley/Hamilton series in accounting and information systems)
0–471–58364–2 X.520/12702
0045543

Accounting theory. – 2nd ed. – New York ; Chichester : Wiley, c1982
xii,686p : ill. – Includes bibliographies and index
0–471–06029–1 X.520/26826
0045543

MCNAMARA, Terry M.
Numeracy and accounting. – Plymouth : Macdonald and Evans, 1979
viii,350p : ill. – (M & E BEC books). – Includes index
0–7121–1411–4 X.519/30396
0045543

MCQUAIG, Douglas J.
Career accounting fundamentals. – Boston [Mass.] ; London : Houghton Mifflin, 1977
xvi,442p : ill. – Includes index
0–395–18979–9 X.520/22133
1198300

College accounting fundamentals. – Boston [Mass.] ; London : Houghton Mifflin, 1977
xxiv,936p : ill. – Includes index
0–395–18986–1 X.520/22134
1198300

MAGEE, Charles Ciaran
Accounting. – [Cardiff] : University College, Cardiff, [1972]
[1],23p. – (Inaugural lectures / University College (Cardiff))
B76–03286 X.519/19876
0045543

Framework of accountancy. – 2nd ed. – Plymouth : Macdonald and Evans, 1979
x,323p. – Bibliography: p.297–298. – Includes index
0–7121–0631–6 X.519/29330
0045543

MAGEE, J. O. (James Owen)
Basic accounting. – 2nd ed. – Plymouth : Macdonald and Evans, 1979
x,340p. – (M & E handbooks). – Bibliography: p.vii. – Includes index
0–7121–0284–1 X.519/30613
0045543

MALCOLM, Alexander
A treatise of book–keeping, 1731. – London : Scolar Press ; Tokyo : Yushodo Press, 1979
vi,145,[36],20,[20],34,[2]p. – (Historic accounting literature ; vol.11)
r6112470 X.0625/102(11) Woolwich
0045543

MARTIN, E. R. (Ernest Richard)
Multiple choice questions on the principles of accounts. – London : Edward Arnold, 1980
[3],87p
0–7131–0441–4 X.629/13973 Woolwich
0630950

MAY, Robert George
A new introduction to financial accounting. - Englewood Cliffs ; London : Prentice-Hall, 1975
xix,645p. - Includes index
0-13-615021-7 X.520/10287
0304018

MEIGS, Walter B. (Walter Berkeley)
Accounting. - 5th ed.. - New York ; London : McGraw-Hill, c1981
xxii,1090p : ill. - Includes index
0-07-041551-x X.520/23349
0045543

Accounting. - 6th ed.. - New York ; London : McGraw-Hill, c1984
xxiii,1104p : ill. - Includes index
0-07-041587-0 X.520/34344
0045543

Modern advanced accounting. - 2nd ed.. - Tokyo ; London : McGraw-Hill Kogakusha, c1979
xix,744p : ill. - Includes index
0-07-066403-x X.529/40301
1198300

MEIGS, Walter Berkeley
Accounting. - 4th ed.. - New York ; London : McGraw-Hill, 1977
xix,1034p : ill(chiefly col). - Includes index
0-07-041241-3 X.520/12026
0045543

Intermediate accounting. - 4th ed.. - New York ; London : McGraw-Hill, 1978
xx,1132p : ill. - Includes index
0-07-041255-3 X.520/12468
0085650

METCALF, Richard W.
Principles of accounting. - Philadelphia ; London : Saunders, 1976
xxvi,1023p : ill(some col). - Includes index
0-7216-6313-3 X.520/11196
0045543

MILLER, Derek P.
Objective tests in the principles of accounts. - London : Hodder and Stoughton, 1979
iv,76p
0-340-23442-3 X.519/29447
0033863

MILLMORE, Lionel
Accounting through numeracy. - Walton-on-Thames : Nelson, 1982
188p : ill. - (Nelson BEC books). - Includes index
0-17-741131-7 X.520/27640
1198300

MONTEAGE, Stephen
Debtor and creditor made easie ... to which is added a maiden scholars advice, [second edition,] 1682 ; Instructions for rent-gatherers accompts, 1683. - London : Scolar Press ; Tokyo : Yushodo Press, 1979
[182],36,[2],36p. - (Historic accounting literature ; vol.6)
r6112471 X.0625/102(6) Woolwich
0045543

MOSICH, A. N.
The CPA examination. - New York ; London : McGraw-Hill, 1978
xvii,860p : ill
0-07-043435-2 X.520/12944
0630950

Intermediate accounting. - 5th ed.. - New York ; London : McGraw-Hill, c1982
xxii,1114p. - Includes index
0-07-041580-3 X.520/26491
0045543

NATHAN, Peter, 1938-
Profitable accounting. - Bath : Sewell's Profit and Information Unit, [1976]
[2],iii,90[i.e. 91] leaves,[14] leaves of plates. - (Sewell management accounting series ; book 3) (Sewell's management guide books)
0-86016-010-6 Cup.1285/93
0045543

NORKETT, P. T. C. (Paul T. C.)
Accountancy. - London : Longman, [1977]
[7],167p. - (Questions and answers)
0-582-41555-1 X.519/27559
0630950

NORKETT, P. T. C. (Paul T C)
Financial accounting. - Harlow : Longman, 1981
xvi,335p : ill. - (Accountancy for non-accountants ; v.1). - Includes index
0-582-41207-2 X.529/45519
0045543

NORTH, Roger, 1651-1734
The gentleman accomptant, 1714. - London : Scolar Press ; Tokyo : Yushodo Press, 1979
vi,263,[35]p,1 fold.leaf of plates. - (Historic accounting literature ; vol.9)
r6112469 X.0625/102(9) Woolwich
0045543

OFFICE accounting / Rocco M. Santovo ... [et al.]. - New York ; Chichester : Wiley, c1984
xiv,590p : ill. - Includes index
0-471-86413-7 X.520/35104
0045543

ORACLE. - [London] : Association of Independent Tutors in Accountancy, [1982]
127p : ill
B83-11359 X.525/6680
0630950

ORACLE. - [London] : Association of Independent Tutors in Accountancy, [1982]
95p : ill
B83-11360 X.622/14635
0630950

ORACLE. - [London] : Association of Independent Tutors in Accountancy, [1983]
102,18p
0-86277-078-5 X.525/9729
0630950

ORACLE. - [London] : Association of Independent Tutors in Accountancy, [1983]
121,21p : ill
0-86277-079-3 X.525/9730
0630950

OWLER, L. W. J. (Leonard William John)
Intermediate accounts. - 3rd ed. - London : Macdonald and Evans, 1975
vi,390p. - (The M. & E. handbook series). - Includes index
0-7121-0936-6 X.519/25062
0085650

PALMER, Alfred
Palmer's examination notebook for accountancy and secretarial students. - 13th ed.. - London : Gee and Co., 1977
[7],208p. - (A Gee's study book). - Includes index
0-85258-152-1 X.519/26783
0045543

PERKINS, James H.
Computer-assisted practice set and problems. - New York ; Chichester : Wiley, c1983
118p
0-471-89372-2 X.622/23342
0630950

PERSON, Samuel
Intermediate accounting. - 1981/82 ed. - Chicago ; London : Dryden, c1982
xv,151p. - (CPA exam supplement)
0-03-059793-5 X.525/5764
0630950

PETERSON, Robert C.
Understand accounting - fast. - New York ; London : McGraw-Hill, 1976
xiii,107p : ill. - Includes index
0-07-049615-3 X.520/12526
0045543

PICKLES, William, b.1893
Accountancy. - 4th ed.. - London : Pitman, 1974
ix,1490p in various pagings : ill. - Includes index
0-273-31538-2 X.519/25109
0045543

Accountancy, questions and key. - 4th ed.. - London : Pitman, 1976
viii,729p. - Includes index
0-273-00837-4 X.519/25681
0630950

PICKLES, William, 1893-
Accountancy. - 5th ed.. - London : Pitman, 1982
1v.(various pagings) : ill. - Includes index
0-273-01256-8 X.529/56384
0045543

PIZZEY, Alan
Accounting and finance. - London : Holt, Rinehart and Winston, 1980
xii,391p. - (Holt business texts). - Includes index
0-03-910275-0 X.520/21152
0304018

PIZZEY,, Alan
Accounting and finance. - 2nd ed. - London : Holt, Rinehart and Winston, c1985
ix,460p : ill. - Includes index
0-03-910586-5 X.622/23492
0045543

PRICE WATERHOUSE INTERNATIONAL
Accounting principles and reporting practices. - [New ed.]. - [London] : Price Waterhouse [International], 1975
[91],264,[13]p. - Includes index
0-9502558-1-5 X.529/53815
0096113

A survey in 46 countries. - [New ed.]. - London : Institute of Chartered Accountants in England and Wales, 1976
[91],264,[13]p. - Includes index
0-85291-159-9 X.519/26555
0096113

REXROAD, W. Max
Professional examination manual to accompany Intermediate accounting, fourth edition, by Donald E. Kieso and Jerry J. Weygandt. - New York ; Chichester : Wiley, c1983
ix,252p
0-471-87165-6 X.622/23343
0630950

REYNOLDS, Isaac N.
Principles of accounting. - 3rd ed. - Chicago ; London : Dryden, c1984
xxv,1036,I-5,CL-4p : ill. - Includes index
0-03-063313-3 X.520/33906
0045543

RIGGS, Henry E.
Accounting. - New York ; London : McGraw-Hill, 1981
xvii,552p : ill. - Includes index
0-07-052851-9 X.520/23830
0045543

ROBINSON, David F. (David Foster)
Key definitions in finance and accounting. - London : Muller, 1980
125p : ill. - (A language of its own)
0-584-10546-0 X.529/39377
2090619
- [Another copy: Paperback] ISBN 0-584-10556-8 X.529/39378

ROWE, Arthur, 1921-
Accounting reports today. - London : Gee and Co., 1980
[15],206p. - Includes index
0-85258-178-5 X.529/36420
0045543

An outline of accounting method. - London : Gee, 1979
189p : ill. - (GeeBEC series of study books). - Includes index
0-85258-171-8 X.529/54154
0045543

ROWLAND, Stanley William
Rowland and Magee's accounting. - 9th ed.. - London : Gee and Co., 1977
xiii,390p : 2 ill. - (A Gee's study book). - Includes index
0-85258-158-0 X.519/27340
0045543

RUNDLE, Cyril
Accountancy for everyone. - Newton Abbot : David and Charles, 1977
173p : ill. - Includes index
0-7153-6864-8 X.529/31651
0045543

SIEGEL, Joel G.
Schaum's outline of theory and problems of financial accounting. - New York ; London : McGraw-Hill, c1983
310p. - (Schaum's outline series). - Includes index
0-07-057304-2 X.525/7341
0630950

SIMINI, Joseph Peter
Accounting made simple. - 2nd ed.. - London : Heinemann, 1984
xiii,369p. - (Made simple books). - Includes index
0-434-98469-8 X.529/62445
0045543

SLAVIN, Albert
Financial accounting. - Hinsdale, Ill. : Dryden Press, 1980
689p
0-03-048906-7 X.520/21814
0045543

SMITH, Jack L.
Accounting principles. - New York ; London : McGraw-Hill, c1983
xxv,1143,17p : ill(some col.). - Includes index
0-07-059060-5 X.520/32499
0045543

SMYTH, W. M.
Junior accounting. - Christchurch, N.Z. ; London : Whitcoulls, 1977
133p : ill. - Includes index
0-7233-0530-7 X.529/32020
1052098

SMYTH, W. M. (William Maclurg)
Junior accounting. - Rev. ed. - Christchurch ; London : Whitcoulls, 1979, c1977
137p : ill. - Includes index
0-7233-0593-5 X.529/44922
1052098

School certificate accounting. - Christchurch ; London : Whitcoulls, 1982
271p : ill. - Includes index
0-7233-0681-8 X.529/49896
1052098

SOLOMON, Lanny M.
Accounting principles. - New York ; London : Harper & Row, c1983
xix,1127,[49]p : ill(some col.). - Includes index
0-06-046348-1 X.520/30641
0045543

SOLUTION to practice set 1B. - New York ; Chichester : Wiley, c1983
[56]p
0-471-80215-8 X.525/10976
0045543

SPICER, E. E. (Ernest Evan)
Spicer and Pegler's Book-keeping and accounts. - 18th ed.. - London : HFL, 1980
viii,576p. - Includes index
0-372-30034-0 X.622/23195
0045543

STALLINGS, Warren D.
Computerized practice set. - New York ; Chichester : Wiley, c1983
34p,[60]p of plates : ill
0-471-87938-x X.525/10978
0045543

STEPHEN, Alan
Accounts. - Sunbury on Thames : Celtic Revision Aids, 1980
[7],81p. - (Model answers. O Level)
0-17-751120-6 X.529/35481
0033863
Accounts. - Sunbury on Thames : Celtic Revision Aids, 1980
[2],93p. - (Worked examples. A Level)
0-17-751163-x X.529/35475
0033863

STEPHEN, Alan
Accounts. - Sunbury on Thames : Celtic Revision Aids, 1980
[3],91p. - (Multiple choice. O Level)
0-17-751188-5 X.519/30963
2108380
Accounts. - Sunbury on Thames : Celtic Revision Aids, 1980
[1],85p. - (Test yourself)
0-17-751241-5 X.529/36348
2175487

STOTT, J. Randall (John Randall)
Mastering principles of accounts. - London : Macmillan, 1982
237p. - (Macmillan master series). - Includes index
0-333-31289-9 X.529/46759
0045543

STOTT, John Randall
Basic accounts. - London : Edward Arnold, 1977
167p
0-7131-0166-0 X.510/15553
0045543
Basic accounts. - London : Heinemann Educational.
Work book 1. - 1976
[2], 90, [7] p.
0-435-45800-0 X.510/9210
0045543
Work book 2. - 1976
[2], 90, [7] p.
0-435-45801-9 X.510/9210
0045543
Work book 3. - 1976
[2], 83, [14] p.
0-435-45802-7 X.510/9210
0045543

STUDIES in accounting. - 3rd ed.. - London : Institute of Chartered Accountants in England and Wales, 1977
xii,426p : ill. - Includes index
0-85291-179-3 X.520/12169
0045543

THACKER, Ronald J.
Accounting principles. - 2nd ed. - Englewood Cliffs ; London : Prentice-Hall, 1979
xxvii,996p : ill. - Includes index
0-13-002766-9 X.520/14526
0045543

THOMAS, D. J. (Derek John)
That's the way the money goes!. - London : Pitman, 1976
iii,115p : ill
0-273-00329-1 X.510/15388
1052098

TURNER, D. E.
Accounting for 'O' level. - London : Edward Arnold, 1982
ix,303p : ill. - Includes index
0-7131-0719-7 X.529/49759
1052098

VICKERY, Bertram George
Principles and practice of book-keeping and accounts. - 21st ed.. - London : Cassell, 1979
[7],914p : ill. - Includes index
B79-32642 X.519/30215
0045543

WALGENBACH, Paul H.
Principles of accounting. - New York : Harcourt Brace Jovanovich, 1976
xxi, 934 p. : ill. - Includes index
0-15-571320-5 X.520/11371
0045543

Principles of accounting. - 2nd ed. - New York ; London : Harcourt Brace Jovanovich, 1980
xxi,1065p : ill(some col). - Includes index
0-15-571336-1 X.520/20736
0045543

WALGENBACH, Paul H.
Principles of accounting. - New York : HarcourtBrace Jovanovich.
Solutions manual. - 1976
xi,635p : ill
0-15-571322-1 X.615/2270 Woolwich
0045543

WILKINSON-RIDDLE, G. J.
Accounting Level III. - Plymouth : Macdonald and Evans, 1982
viii,261p : ill. - (The M & E BEC book series). - Includes index
0-7121-0175-6 X.529/49959
0045543

WILSON, David A.
Contemporary financial reporting. - Englewood Cliffs ; London : Prentice-Hall, 1979
iii-x,385p : ill
0-13-170332-3 X.510/16396
1966944

WISEMAN, Joseph A.
Schaum's outline of theory and problems of advanced accounting. - New York ; London : McGraw-Hill, c1982
402p : 1ill. - (Schaum's outline series). - Includes index
0-07-071138-0 X.622/14426
0630950

WOELFEL, Charles J.
An introduction to financial accounting. - Santa Monica : Goodyear Publishing, c1980
x,662p : ill. - Includes index
0-8302-4059-4 X.520/22449
1198300

WOLTZ, Phebe M.
College accounting. - New York ; London : McGraw-Hill, 1979
xvii,413p. - Includes index
0-07-071551-3 X.520/13724
1198300

WOOD, Frank
Accounting 2. - Stockport : Polytech, 1980
310p : ill. - (Business Education Council national award courses). - Includes index
0-85505-041-1 X.529/41658
0045543
Business accounting. - 3rd ed. - London : Longman.
1. - 1979
[8],484p. - Includes index
Subsequent volumes, when received, will be placed at this shelfmark
0-582-41560-8 X.0519/1081
0045543
- 2nd ed. - [London] : English Language Book Society [etc.].
2. - 1978
ix,p501-1095 : ill. - Includes index
0-582-41564-0 X.519/29780
0045543

ACCOUNTING
- Bibliographies - Catalogues

INSTITUTE OF CHARTERED ACCOUNTANTS IN ENGLAND AND WALES.
Library
Historical accounting literature. - London : Mansell Information Publishing, 1975
xxvi,360p : ill. - Includes index
0-7201-0519-6 BB.C.b.6 Reading Room
1410385
- [Another copy] OPL
- [Another copy] X.525/2008
- Supplement updated to 31st August 1975. - 15p. ; 29cm.
- ISBN 0-85291-134-3

ACCOUNTING
- Catalogues

INSTITUTE OF CHARTERED ACCOUNTANTS IN ENGLAND AND WALES.
Library
Reading lists on accounting finance & taxation. - London : The Library, [1977]
v[i.e.ix],271p
B78-14026 X.515/578
1410385
- [Another copy] OPL

INSTITUTE OF CHARTERED ACCOUNTANTS OF SCOTLAND
An accountants' book collection 1494-1930. - 3rd ed.. - Edinburgh : Institute of Chartered Accountants of Scotland, Scottish Committee on Accounting History, 1976
xviii,120,6p : ill
0-85099-025-4 X.955/1484
1410385

ACCOUNTING
- Dictionaries

ESTES, Ralph
Dictionary of accounting. - Cambridge, Mass. ; London : MIT Press, c1981
161p : ill
0-262-05024-2 X.629/15951 Woolwich
2090619

PARKER, R. H. (Robert Henry)
Macmillan dictionary of accounting. - London : Macmillan, 1984
xi,185p. - (Macmillan reference books)
0-333-36327-2 X.520/36420
025472x

ACCOUNTING
- For banking

EDWARDS, J. R. (John Richard)
Accountancy for banking students. - London : Institute of Bankers, 1983
xxii,495p : ill. - Includes index
0-85297-079-x X.529/70038
1147137

EGGINTON, Don A. (Don Albert)
Accounting for the banker. - 2nd ed. - London : Longman, 1982
vii,280p : ill. - Includes index
0-582-29633-1 X.520/29033
1147137

EGGINTON, Don Albert
Accounting for the banker. - London ; New York : Longman, 1977
x,272p : ill. - Includes index
0-582-45055-1 X.510/15498
1147137

NIBBS, Colin
Accountancy in banking. - Cambridge : Woodhead-Faulkner, 1983
ix,244p : ill. - Includes index
0-85941-192-3 X.520/31532
1147137
- [Another Copy: paperback.]
ISBN 0-85941-193-1 X.520/31533

ACCOUNTING
- For business studies

CLARET, Jake
Accounting 2. - London : McGraw-Hill, c1981
xiv,290p : ill. - (McGraw-Hill business education courses). - Includes index
0-07-084621-9 X.520/25389
2152460

COUNSELL, Robert E.
Accounting II. - London : Cassell, 1982
281p. - (Cassell BEC series. National level)
0-304-30337-2 X.529/46378
2152460

EDWARDS, Bernard, 1921-
Quantitative and accounting methods. - Plymouth : Macdonald and Evans, 1980
ix,242p : ill. - (M & E BEC books). - Includes index
0-7121-1704-0 X.629/13881 Woolwich
2152460

EVE, John
Accounting. - London : Pitman, 1979
[7],184p : ill. - (Pitman business education). - Includes index
0-273-01260-6 X.529/36193
2152460

FRANCIS, D. Pitt (David Pitt)
Accounting concepts and methods. - London : Holt, Rinehart and Winston, c1981
100p. - (Holt business texts)
0-03-910310-2 X.525/4506
2152460
Accounting concepts and methods. - London : Holt, Rinehart and Winston, c1982
xi,354p : ill. - (Holt business texts). - Includes index
0-03-910342-0 X.529/47139
2152460

JONES, Graham, 1947-
Financial practice and control. - Plymouth : Macdonald & Evans, 1982
viii,148p : ill. - (The M. & E. Beckook series). - Includes index
0-7121-0640-5 X.529/51394
2152460

NEWCOMB, V. N.
Practical accounting for business studies. - Chichester : Wiley, c1983
viii,372p : ill. - Includes index
0-471-90007-9 X.520/31104
2152460

STOTT, J. Randall (John Randall)
Basic accounting. - London : Hodder and Stoughton, 1985
282p. - (Teach yourself books). - Includes index
0-340-36513-7 WP.706/859
2152460

ACCOUNTING
- For business studies–continued

TURNER, D. E.
Accounting and numeric methods for business studies students. – London : Edward Arnold, 1981
ix,278p : ill. – Includes index
0–7131–0590–9 X.529/44274
2152460

WILKINSON, Gill
Numeracy and accounting. – Amersham : Hulton Educational, 1981
252p : ill. – (Hulton BEC books). – Includes index
0–7175–0841–2 X.622/9899 Woolwich
2152460

ACCOUNTING
- For Caribbean students

WOOD, Frank
Principles of accounts for the Caribbean. – Harlow : Longman, 1982
312p. – Includes index
0–582–76596–x X.529/49768
2782367

ACCOUNTING
- For Irish students

ACCOUNTING 1981 (Ordinary and Higher Level). – [Dublin] : Folens.
Solutions. – [1981?]
32p
B82–18013 X.520/26661
2582805

O'CONNOR, John, 1949–
Accounting assignments. – [Dublin] : Folens, [1979?]. – (Leaving certificate)
1: Ordinary level
48p
0–86121–054–9 X.0520/662
2582805
– Subsequent volumes when recieved will be placed at this shelfmark X.0520/662

ACCOUNTING
- For management

ACCOUNTING & finance for managers / the Open University, Open Business School. – [Milton Keynes] : [Open University Press], [1984]
1case : ill. – (P671)
0–335–10540–8 Cup 930/34
1244086

FINANCIAL accounting for management / Almand R. Coleman, E. Richard Brownlee II, C. Ray Smith. – Richmond, Va. : Dame, c1981
xiii,441p
0–936328–03–7 X.520/37412
1244086

FLESHER, Dale L. (Dale Lee)
Accounting for the middle manager. – New York ; London : Van Nostrand Reinhold, 1980
ix,454p : ill. – Includes index
0–442–23875–4 X.525/3490
1244086

HARTLEY, Walter Colin Farrar
An introduction to business accounting for managers. – 3rd ed. – Oxford : Pergamon, 1980
xiv,221p : ill. – (Pergamon international library). – Includes index
0–08–024061–5 X.529/38704
1244086

JONES, D. M. C. (David Maurice Clifford)
Management accounting for non-financial managers. – London : Research and Publications Division, BIM Foundation, c1978
47p : ill. – (Management guide ; no.7). – Bibliography: p48
0–85946–094–0 X.0615/96(7)
1244086

ROBSON, Alan Peel
Essential accounting for managers. – 4th ed. – London : Cassell, 1979
ix,130p : ill. – Includes index
0–304–30385–2 X.519/30227
1244086

SLATER, H. W.
Accounting for managers. – Henley : Administrative Staff College, 1977
[5],105p(4 fold) : ill
B77–29550 X.512/5692
1244086

WALKER, James William, b.1912
Making sense of finance and accounts in business. – London : Bodley Head : H.F.L., 1977
xv,187p : ill. – Includes index
0–372–30006–5 X.519/27548
1244086

ACCOUNTING
- For pharmacy

MARTIN, Francis A.
Principles of pharmaceutical accounting. – Philadelphia : Lea and Febiger ; London : Kimpton, 1980
xi,241p : ill. – Includes index
0–8121–0634–2 X.525/3691
2304643

ACCOUNTING
- For South African students

SELECTED questions in accounting / edited by B. J. S. Wimble and T. Cairns. – 4th ed.. – Cape Town : Juta.
2: Intermediate. – 1976
[7], 283 p.
0–7021–0678–x X.519/25971
0695289

ACCOUNTING
- For West African students

OKOLO, M. N. (Michael N.)
Key to book-keeping and accounts. – London : Longman, 1980
204p. – (Longman commercial studies)
0–582–65029–1 X.520/23418
237983x

OKOLO, Michael N.
Questions and answers, O-Level principles of accounts. – London : Longman, 1979
vii,216p. – (Study for success)
0–582–60692–6 X.529/36196
2154226

WOOD, Frank
Business accounting 1. – West African ed.. – Harlow : Longman, 1982
424p. – Includes index
0–582–65602–8 X.0529/729(1)
237983x

Business accounting 2. – West African ed.. – Harlow : Longman, 1983
498p. – Includes index
0–582–65603–6 X.0529/729(2)
237983x

ACCOUNTING
- Programmed texts

ANTHONY, Robert Newton
Essentials of accounting. – 2nd ed. – Reading, Mass. ; London : Addison-Wesley, 1976
v,175[i.e.176],[54]p
0–201–00252–3 X.512/5600
1105027

ACCOUNTING
- Serials

ACCOUNTING history. – Vol.1, no.1 (Feb. 1976)–. – Canterbury : Accounting History Society, 1976–
v
r6816767 P.705/373
3328422

ACCOUNTING, Organizations and Society. – Vol.1, no.1– ; [June 4] 1976–. – Oxford : Pergamon, 1976– : ill
B76–32532 P.521/3065
0672920

ANNUAL accounting review. – Vol.1 (1979)–. – Chur ; London : Harwood Academic, 1979–
v.
Holdings: Vol 4 (1982)–
B83–12098 P.511/5378
0672920

'FINANCIAL Times' World Accounting Report. – [No.1]– ; May 1976–. – London : 'Financial Times', 1976–. – (FT business newsletters)
B76–21968 P.513/2022
0672920

INTERNATIONAL accounting & financial report. – Vol. C77, no.1– ; May 24, 1976–. – London : Institute for International Research Ltd, [1976]–
B78–03493 P.513/2139
0672920

INTERNATIONAL accounting bulletin. – Issue no.1 (July 1983)–. – London : Lafferty Publications, 1983–
v.
B84–05581 P.823/167
0672920

INTERNATIONAL finance & accounting report. – No.1 (Jan. 1984)–. – London : Stonehart Research International, 1984–
v.
Holdings: No.2 February 1984 etc
B85–13179 P.823/469
0672920

ACCOUNTING
to 1983

PARKER, R. H. (Robert Henry)
Papers on accounting history. – New York ; London : Garland, 1984
xi,184p. – (Accounting history and the development of a profession). – Includes bibliographies
0–8240–6310–4 X.520/37534
329711x

ACCOUNTING
to date

DELLA GATTA, Giacomo
Nuova prattica d'aritmetica mercantile, 1774. – London : Scolar Press : Tokyo : Yushodo Press, 1978
475p. – (Historic accounting literature ; v.27)
r6198019 X.0625/102(27) Woolwich
8800804

MANZONI, Domenico
Quaderno doppio col suo giornale, 1540. – 2nd ed reproduced. – London : Scolar Press ; Tokyo : Yushodo Press, 1978
1v.(various pagings). – (Historic accounting literature ; v.23)
r6198011 X.0625/102(23) Woolwich
8800804

PIETRA, Angelo
Indrizzo de gli economi, 1586. – London : Scolar Press ; Tokyo : Yushodo Press, 1978
1v.(various pagings). – (Historic accounting literature ; v.24)
r6198016 X.0625/102(24) Woolwich
8800804

ACCOUNTING
to date – Conference proceedings

COMMONWEALTH CONFERENCE OF ACCOUNTANTS (1975 : New Delhi)
[Papers]. – [s.l.] : [Institute of Chartered Accountants of India, [1975]
216p
r6804724 X.525/4997
8752079

ACCOUNTING
1500–1850. Early works

BYWATER, M. F.
Historic accounting literature. – London : Scolar, 1982
255p. – (Historic accounting literature). – Bibliography: p242–249. – Includes index
0–85967–647–1 X.0625/102a
2756102
– [Another copy.] OPL

ACCOUNTING
Applications of behavioural sciences

The EVOLUTION of behavioral accounting research / edited by Robert H. Ashton. – New York ; London : Garland, 1984
x,121p : ill. – (Accounting history and the development of a profession). – Bibliography: p120–121
0–8240–6302–3 X.520/36567
3255948

ACCOUNTING
Applications of digital computer systems

COWE, Roger
Computers for accountants. – London : Sinclair Browne, 1983
186p : ill. – (A Sinclair computerguide). – Includes index
0–946195–08–0 X.629/23696
0379204

MCRAE, Thomas Watson
Computers and accounting. – London : Wiley, 1976
xi,167p : ill. – Includes bibliographies and index
0–471–58985–3 X.620/16831 Woolwich
0379204

The impact of computers on accounting. – London : Wiley, 1964 [i.e. 1973]
xiv,304p : ill. – Bibliography: p.287–290. – Includes index
0–471–58730–3 X.520/10405
0379204

VAN LINT, Wennie
Introduction to computer-based accounting, with applications for Apple II, TR8 80, Commodore Pet. – Sydney ; London : Prentice-Hall of Australia, c1982
x,202p : ill. – Includes index
0–7248–0651–2 X.520/30828
0379204

WEIS, Steven S.
Computer applications guide for accountants. – Reston : Reston Publishing, c1984
xv,335p : ill
0–8359–0850–x X.622/19874
0379204

ACCOUNTING
Applications of digital computer systems. Software packages
ACCOUNTING software controls. - Manchester : NCC Publications, 1983
21p. - Bibliography: p19-21
0-85012-430-1 X.529/58791
3043371

ACCOUNTING
Applications of measure theory
MOCK, Theodore J
Measurement, accounting, and organizational information. - New York ; Chichester : Wiley, 1979
xviii,238p : ill. - (Wiley series in accounting and information systems). - Bibliography: p.229-234. - Includes index
0-471-61202-2 X.520/14422
1946285

ACCOUNTING
Applications of microcomputer systems
NEEDLEMAN, Theodore
Microcomputers for accountants. - Englewood Cliffs ; London : Prentice-Hall, c1983
v,183p : ill. - (A Spectrum book). - Includes index
0-13-580696-8 X.622/16681
2936569

ACCOUNTING
Applications of small digital computer systems
HAYES, Rick Stephan
Simplified accounting for the computer industry. - New York ; Chichester : Wiley, c1981
vii,191p : ill. - (Small business management series) (A Ronald Press publication). - Includes index
0-471-05703-7 X.525/4536
2493934

ACCOUNTING
Channel Islands. Income tax. Law - For accounting
PIPER, G.S.F
The Channel Islands and the Isle of Man. - 2nd ed. - London : Institute of Chartered Accountants in England and Wales, 1978
1 sound cassette(ca 35min.)
39p.. - (Current taxation problems)
r6196163 Cup.580.cc.17
2008246

ACCOUNTING
Companies. Corporate planning - For accounting
HUSSEY, D. E. (David Edward)
Corporate planning. - London : Institute of Chartered Accountants in England and Wales, c1981
viii,102p : ill. - (The Accountant in industry and commerce). - Bibliography: p100-102
0-85291-299-4 X.629/17038 Woolwich
2615878

ACCOUNTING
Companies. Corporate planning. Effects of inflation - For accounting
HUSSEY, David Edward
Corporate planning in inflationary conditions. - London : Institute of Cost and Management Accountants, 1978
v,99p : ill. - (Management accounting in inflation series). - Bibliography: p.98-99
0-901308-51-x X.529/37792
2229285

ACCOUNTING
Companies. Financial statements, to 1978 - For accounting
The EVOLUTION of corporate financial reporting / [edited by] T.A. Lee and R.H. Parker. - Sunbury-on-Thames : Nelson, 1979
xi,307p. - (Nelson series in accounting and finance). - Bibliography: p.299-300. - Includes index
0-17-761047-6 X.529/34158
1903411

ACCOUNTING
Companies. Prospectuses. Preparation - For accounting
PROSPECTUSES and the reporting accountant. - London : Auditing Practices Committee, 1984
22p. - (Auditing guideline ; 5330)
B84-40279 X.529/65203
0320439

ACCOUNTING
Computer services. Charges - For accounting
CHARGING for computer services. - London : Institute of Cost and Management Accountants, 1982
iii,31p. - (Management accounting guidelines ; 4)
0-901308-66-8 X.629/26310
2058561

PORTER, James G
Integrating the computer with your business. - London : H.F.L., 1980
308p : ill. - Includes index
0-372-30032-4 X.629/25990
2058561

ACCOUNTING
Decision making
LIBBY, Robert
Accounting and human information processing - Englewood Cliffs ; London : Prentice-Hall, c1981
xiii,203p : ill. - (Prentice-Hall contemporary topics in accounting series) - Bibliography p185-198. - Includes index
0-13-001818-x X.529/44211
2573199

ACCOUNTING
Effects of inflation
BRÖMS, Jan
Inflationsredovisning. - Stockholm : Sveriges industriförbund, [1974]
122p. - (Ekonomiska utredningsrapporter. serie A ; 13)
r6029106 X.519/40061
0099023

CROSS, Stephen M.
Economic decisions under inflation. - Greenwich, Conn. ; London : Jai, c1982
xii,158p. - (Contemporary studies in economic and financial analysis ; v.40). - Bibliography: p147-152. - Includes index
0-89232-240-3 X.520/31474
0099023

ACCOUNTING
England. Consumer goods. Physical distribution - For accounting
NEWTON, Keith
Distribution. - [London] : Institute of Chartered Accountants in England and Wales, 1981
vii,84p : ill. - (The Accountant in industry and commerce). - Bibliography: p80-81
0-85291-291-9 X.529/41236
240088x

ACCOUNTING
European Community - For accounting
EVANS, Dennis, 1921-
Accountants guide to the European Communities. - Plymouth : Macdonald and Evans, 1981
xix,361p,[1]leaf of plates : ill. - Includes index
0-7121-0156-x X.520/23720
2395460
- [Another copy] Enquiry desk OPL

ACCOUNTING
Finance. Inflation. Effects on corporate planning in business firms - For accounting
HUSSEY, David Edward
Corporate planning in inflationary conditions. - London : Institute of Cost and Management Accountants, 1978
v,99p : ill. - (Management accounting in inflation series). - Bibliography: p.98-99
0-901308-51-x X.529/37792
2229285

ACCOUNTING
Financial management - For accounting
BROCKINGTON, Raymond
Financial management. - 2nd ed. - Winchester : D.P. Publications, 1981
v,294p. - Includes index
0-905435-16-8 X.529/41029
2392313

ACCOUNTING
Great Britain. Businessmen. Foreign income. Taxation. Law - For accounting
ALLAN, R.J
U.K. executives' foreign earnings. - London : Institute of Chartered Accountants in England and Wales, 1977
1 sound cassette(ca 20min.)
43p. - (Current taxation problems)
r6196175 Cup.580.ee.25
2008289

ACCOUNTING
Great Britain. Capital transfer tax. Liability of trusts. Law - For accounting
SHERRING, F.A
Trusts after capital transfer tax. - 2nd ed. - London : Institute of Chartered Accountants in England and Wales, 1979
1 sound cassette(ca 27min.)
31p.. - (Current taxation problems)
r6196177 Cup.580.ee.27
2008130

ACCOUNTING
Great Britain. Close companies. Income. Law - For accounting
PACKER, W.R
Close companies, apportionments. - London : Institute of Chartered Accountants in England and Wales, 1975
1 sound cassette(ca 20min.)
52p.. - (Current taxation problems)
r6196178 Cup.580.ee.28
2008181

ACCOUNTING
Great Britain. Companies. Occupational superannuation schemes. Financial information. Disclosure - For accounting
DISCLOSURE of pension information in company accounts. - London : Accounting Standards Committee, 1983
[20]p. - (Exposure draft ; 32)
B84-22552 X.529/61283
3086062

ACCOUNTING
Great Britain. Employment. Termination. Payments. Taxation. Law - For accounting
LANCH, D.
Termination payments. - London : Institute of Chartered Accountants in England and Wales, 1977
1 sound cassette(ca 25min.)
43p.. - (Current taxation problems)
r6196168 Cup.580.cc.22
2008262

ACCOUNTING
Great Britain. Foreign personnel. Taxation. Law - For accounting
ALLAN, R.J
Earnings of foreign executives in the U.K.. - London : Institute of Chartered Accountants in England and Wales, 1977
1 sound cassette(ca 25min.)
49p.. - (Current taxation problems)
r6196166 Cup.580.cc.20
2008173

ACCOUNTING
Great Britain. Manufacturing industries. Productivity. Improvement. Role of added value - For accounting
COX, Bernard
Value added. - London : Heinemann [for] the Institute of Cost and Management Accountants, 1979
xii,220p : ill. - Bibliography: p.212-213. - Includes index
0-434-90276-4 X.529/34085
1961233

ACCOUNTING
Great Britain. Occupational superannuation schemes. Funds. Investments: Insurance policies. Valuation - For accounting
VALUING insurance policies for accounts purposes / Pensions Research Accountants Group. - [s.l.] : The Group, c1981
15p. - (Notes on pensions ; no.10)
0-907110-01-0 X.529/44283
2562561

ACCOUNTING
Great Britain. Personal income: Interest. Taxation. Allowances. Law - For accounting
STOCKWELL, R.A.O
Interest relief for individuals. - London : Institute of Chartered Accountants in England and Wales, 1977
1 sound cassette(ca 25min.)
55p.. - (Current taxation problems)
r6196165 Cup.580.cc.19
2008157

ACCOUNTING
Great Britain. Personal income tax. Allowances. Law - For accounting
WALTERS, R.M
Income tax losses and the individual. - London : Institute of Chartered Accountants in England and Wales, 1977
1 sound cassette(ca 20min)
50p.. - (Current taxation problems)
r6196174 Cup.580.ee.24
2008300

ACCOUNTING
Great Britain. Personal taxation. Implications of capital transfer tax - For accounting
SHERRING, F.A
Living with capital transfer tax. - 2nd ed. - London : Institute of Chartered Accountants in England and Wales, 1979
1 sound cassette(ca 40min.)
40p.. - (Current taxation problems)
r6196164 Cup.580.cc.18
2008203

ACCOUNTING
Great Britain. Physical assets. Taxation. Law - For accounting
MACLEOD, K.A.D
Recent tax cases on 'plant'. - London : Institute of Chartered Accountants in England and Wales, 1977
1 sound cassette(ca 25min.)
29p.. - (Current taxation problems)
r6196162 Cup.580.cc.16
200819X

ACCOUNTING
Great Britain. Private companies. Implications of capital transfer tax - For accounting
SHERRING, F.A
Capital transfer tax and the family company. - 2nd ed. - London : Institute of Chartered Accountants in England and Wales, 1979
1 sound cassette(ca 27min.)
27p.. - (Current taxation problems)
r6196169 Cup.580.cc.23
2008165

ACCOUNTING
Great Britain. Securities. Prices. Determination by securities markets. Efficiency - For accounting
KEANE, Simon M.
The efficient market hypothesis. - London : Published for the Institute of Chartered Accountants of Scotland by Gee, 1980
37p. - Bibliography: p35. - Includes index
0-85258-198-x X.529/39557
2351455

ACCOUNTING
Great Britain. Superannuation schemes - For accounting
NAPIER, Christopher J.
Accounting for the cost of pensions. - London : Institute of Chartered Accountants in England and Wales, 1983
v,202p : ill. - Bibliography: p199-202
0-85291-343-5 X.520/32092
2518694

ACCOUNTING
Great Britain. Superannuation schemes. Financial reports - For accounting
PENSIONS RESEARCH ACCOUNTANTS GROUP
Financial reports for pension funds. - Sheffield : The Group, 1978
32p
B80-39668 X.529/38004
2302918

ACCOUNTING
Great Britain. Taxation. Law. Finance (No.5) Act 1976 - For accounting
VIELER, G.H
Stock increase relief. - New ed. - London : Institute of Chartered Accountants in England and Wales, 1978
1 sound cassette(ca 25min.)
54p.. - (Current taxation problems)
r6196161 Cup.580.cc.15
2008211

ACCOUNTING
Great Britain. Taxation. Liability of British emigrants - For accounting
PIPER, G.S.F
Emigration. - London : Institute of Chartered Accountants in England and Wales, 1978
1 sound cassette(ca 30min)
43p.. - (Current taxation problems)
r6196167 Cup.580.cc.21
2008556

ACCOUNTING
Historiography
PARKER, R. H. (Robert Henry)
The study of accounting history. - [Exeter] : [Exeter Research Group], 1980
22p. - (Discussion paper / Exeter Research Group ; 8). - Bibliography: p17-22
0-907021-03-4 X.809/53944
2821230

ACCOUNTING
Implications of variation in foreign exchange rates
WAINMAN, David
Currency fluctuation. - Cambridge : Woodhead-Faulkner, 1976
72p. - Includes index
0-85941-039-0 X.520/11163
0764132

ACCOUNTING
Information systems
CUSHING, Barry E.
Accounting information systems and business organizations. - 2nd ed. - Reading, Mass. ; London : Addison-Wesley, 1978
[1],xi,557p : ill. - Includes bibliographies and index
0-201-01016-x X.520/12941
1123440

LEITCH, Robert A.
Accounting information systems. - Englewood Cliffs ; London : Prentice-Hall, c1983
xiv,652p : ill. - Includes bibliographies and index
0-13-002949-1 X.520/33067
1123440

PAGE, John, 1949-
Accounting and information systems. - 2nd ed. - Englewood Cliffs ; London : Prentice-Hall, c1982
xiii,607p : ill. - Includes index
0-8359-0095-9 X.520/27843
1123440

WU, Frederick H.
Accounting information systems. - New York ; London : McGraw-Hill, c1983
xviii,605p : ill. - Includes bibliographies and index
0-07-072121-1 X.520/31550
1123440

ACCOUNTING
Information systems. Applications of automatic data processing
CUSHING, Barry E.
Accounting information systems and business organizations. - Reading, Mass. ; London : Addison-Wesley, 1974
x,469p : ill. - Includes index
0-201-01263-4 X.520/12232
1341375

ACCOUNTING
Information systems. Applications of digital computer systems
ACCOUNTING information systems / Robert G. Murdick ... [et al.]. - Englewood Cliffs ; London : Prentice-Hall, 1978
ix,499p : ill. - Includes bibliographies and index
0-13-002014-1 X.520/13180
1756850

BODNAR, George H.
Accounting information systems. - Boston, Mass. ; London : Allyn and Bacon, 1980
xvii,476p : ill. - Bibliography - Includes index
0-205-06884-7 X.520/20870
1756850
- Solutions manual
ISBN 0-205-06885-5 X.520/20870a

Accounting information systems. - 2nd ed. - Boston [Mass.] ; London : Allyn and Bacon, c1983
xi,676p : ill. - Includes bibliographies and index
0-205-07929-6 X.520/32267
1756850

MOSCOVE, Stephen A.
Accounting information systems. - New York ; Chichester : Wiley, c1981
xx,616p : ill. - (Wiley series in accounting and information systems). - Includes index
0-471-03369-3 X.520/22916
1756850

Accounting information systems. - 2nd ed. - New York ; Chichester : Wiley, c1984
xxi,721p : ill. - Includes index
0-471-88354-9 X.520/33739
1756850

Instructor's manual to accompany Accounting information systems - concepts and practice for effective decision making. - New York ; Chichester : Wiley, c1981
467,[100]p : ill
0-471-03371-5 X.622/13767
2789779

ROBINSON, Leonard A.
Accounting information systems. - New York ; London : Harper & Row, c1982
xv,584p : ill. - Includes bibliographies and index
0-06-045509-8 X.622/13203
1756850

WILKINSON, Joseph W.
Accounting and information systems. - New York Chichester : Wiley, c1982
xvi,845p : ill. - Includes bibliographies and index
0-471-04986-7 X.520/26653
1756850

ACCOUNTING
Information systems. Applications of digital computer systems, to date
PUHL, Werner
Entwurf und Realisierung eines computergestützten Kosten- und Erlösinformationssystems auf der Basis einer Datenbank und einer Methodensammlung. - [Nürnberg] [W Puhl], [1983]
iii,169,v,viip.. - Includes bibliography
r6307952 X.529/60462
9076883

ACCOUNTING
International standards
INTERNATIONAL ACCOUNTING STANDARDS COMMITTEE
Preface to statements of international accounting standards. - London : International Accounting Standard Committee, 1983
[6]p
B84-03824 X.529/58036
0272779

ACCOUNTING
International standards. Organisations: International Accounting Standards Committee
INTERNATIONAL ACCOUNTING STANDARDS COMMITTEE
Objectives and procedures. - London : The Committee, 1983
[29]p
B84-03826 X.520/32026
2992302

ACCOUNTING
Isle of Man. Income tax. Law - For accounting
PIPER, G.S.F
The Channel Islands and the Isle of Man. - 2nd ed. - London : Institute of Chartered Accountants in England and Wales, 1978
1 sound cassette(ca 35min.)
39p.. - (Current taxation problems)
r6196163 Cup.580.cc.17
2008254

ACCOUNTING
Mathematical models
MEPHAM, Michael J.
Accounting models. - Stockport : Polytech, 1980
624p : ill. - Includes bibliographies and index
0-85505-040-3 X.529/41585
2487918

ACCOUNTING
Mathematics
NUMERACY and accounting / contributing authors Ron Jones ... [et al.]. - Stockport : Polytech, 1979
349p : ill. - (Business Education Council national award courses). - Includes index
0-85505-033-0 X.529/47472
248790x

ACCOUNTING
Quantitative methods
LEININGER, Wayne E.
Quantitative methods in accounting. - New York ; London : D. Van Nostrand, 1980
xii,353p : ill. - Includes bibliographies and index
0-442-80503-9 X.520/21206
2072874

LUCEY, T. (Terence)
Quantitative techniques. - Winchester : D.P. Publications, 1979, c1980
iv,339p : ill. - Includes index
0-905435-09-5 X.529/44985
2072874

QUANTITATIVE and accounting methods / contributing authors Frank Wood ... [et al.]. - Stockport : Polytech, 1980
394p : ill. - (Business Education Council national award courses). - Includes index
0-85505-042-x X.529/41584
2072874

ACCOUNTING
Research
TRICKER, Robert Ian
Research in accounting. - [Glasgow] : University of Glasgow Press, 1978
35 leaves : ill. - (Arthur Young lecture ; no.1)
0-85261-155-2 X.512/6765
1933752

ACCOUNTING
Standards
ACCOUNTING standards setting / edited by Michael Bromwich and Anthony G. Hopwood. - London : Pitman, 1983
xxiv,184p. - Includes bibliographies
0-273-01850-7 X.200/41315
2616157

ACCOUNTING
Standards–comtinued

BLAKE, John, 1950–
Accountimg standards. – London : Longman, 1981
276p : ill. – (Longman professional education series). – Includes imdex
0–582–41211–0 X.529/45467
2616157

PREFACE to Statements of international accounting standards. – London International Accounting Standards Committee, 1978
[11]p
B84–09425 X.529/58735
2616157

ACCOUNTING
Standards. Political aspects

SOLOMONS, David
The political implications of accounting and accounting standard setting. – [Glasgow] : University of Glasgow Press, 1980
26leaves. – (Arthur Young lecture ; no.3). – Bibliography: p24–26
0–85261–165–x X.622/11058 Woolwich
2550008

ACCOUNTING
Statistical mathematics – For accounting

DICKINSON, John Philip
Statistics for business, finance, and accounting. – Plymouth : Macdonald and Evans, 1976
xiv,312p : ill. – Bibliography: p.297–299. – Includes index
0–7121–1939–6 X.529/30687
0187682

FRANCIS, David Pitt
Statistical method for accounting students. – London : Heinemann, 1978
xii,293p : ill. – (The Heinemann accountancy and administration series). – Bibliography: p.284–286. – Includes index
0–434–90580–1 X.519/28291
0187682

STATISTICS and quantitative methods / prepared by Chart Foulks Lynch. – London : Holt, Rinehart and Winston, c1984
x,289p : ill. – (100 questions and answers). – Includes index
0–03–910499–0 X.629/23542
0187682

ACCOUNTING
Statistical mathematics. Sampling

VANASSE, Robert W.
Statistical sampling for auditing and accounting decisions. – 2nd ed. – New York ; London : McGraw-Hill, 1976
viii,108p : ill. – Bibliography: p.59–60
0–07–066851–5 X.512/5457
0792438

ACCOUNTING
Teaching aids

CITY OF LONDON POLYTECHNIC. Library and Learning Resources Service
Media resources for lectures in accountancy, taxation and financial management. – [London] : [The Polytechnic Library and Learning Resources Service], 1976
[1],5p
0–904264–16–5 X.512/5297
0701653

ACCOUNTING
Theories, to 1965

DEINZER, Harvey T. (Harvey Theodore)
Development of accounting thought. – New York ; London : Garland, 1984, c1965
viii,184p : 1ill. – (Accounting history and the development of a profession). – Bibliography: p166–174. – Includes index
0–8240–6321–x X.529/69098
3263533

ACCOUNTING
United States. Capital markets. Financial management. Efficient market hypothesis – For accounting

DYCKMAN, Thomas Richard
Efficient capital markets and accounting. – Englewood Cliffs ; London : Prentice-Hall, 1975
xiii,130p : ill. – (Prentice-Hall contemporary topics in accounting series). – Bibliography: p.115–124. – Includes index
0–13–246975–8 X.519/25980
0697133

ACCOUNTING. ASEAN countries

COMPARATIVE accounting practices in ASEAN / [prepared by the SGV Group]. – London : Oyez Longman, 1984
58p
0–85120–814–2 L.49/2990
3212955

ACCOUNTING. Assets. Companies

ELLERMAN, David P.
Economics, accounting and property theory. – Lexington, Mass. : Lexington Books, 1983, c1982
xv,203p : ill. – Bibliography. p191–195. – Includes index
0–669–05552–2 X.520/30891
2922592

ACCOUNTING. Austria
to date

KOCH, Friedrich
Das Buchsachverständigenwesen und die Wirtschaftsprüfung in Österreich von ihren Anfängen im 18. Jahrhundert bis 1938. – Linz : Trauner, c1983
vii,238p. – (Linzer Schriften zur Sozial- und Wirtschaftsgeschichte ; Bd.8). – Includes bibliography
Wanting pp.v–vi
3–85320–288–8 X.0525/376(8)
9000917

ACCOUNTING. Banks. Germany
to 1945

ETTLINGER, Stefan
Die Gewinn- und Verlustrechnung der deutschen Kreditbanken. – Frankfurt am Main : Keip ; Tokyo : Yushodo, 1980
188p. – (Betriebs- und finanzwirtschaftliche Forschungen. Serie 2 ; Hft.61). – Bibliography
r6305146 X.529/47152
8820880

ACCOUNTING. Book trades. Soviet Union
1917–

BUKHGALTERSKIĬ uchet i kontrol' v knizhnoĭ torgovle / [pod obshcheĭ redaktsieĭ G.A. Kovaleva]. – Moskva : Kniga, 1974
425p
r6402111 2708.e.148
8739609

IVANOVA, E. N.
Izderzhki, obrashcheniia, torgovye skidki i rentabel'nost'v knizhnoĭ torgovle. – Moskva : Moskovskiĭ poligraficheskiĭ institut, 1982
63p
r6448388 2708.e.429
8739609

ACCOUNTING. Building industries. Great Britain

HUGHES, D. A. (David Anthony)
Basic accounting for builders. – London : Longman, 1983
xi,161p. – (Longman technician series. Construction and civil engineering). – Includes index
0–582–41351–6 X.629/24270
314450x

ACCOUNTING. Canada, Great Britain & United States

ACCOUNTANTS INTERNATIONAL STUDY GROUP
Comparative glossary of accounting terms in Canada, the United Kingdom and the United States. – [London] : [Institute of Chartered Accountants in England and Wales for] the Group, 1975
[11],72p
0–85291–132–7 X.519/25288
0523917

ACCOUNTING. Clubs. Great Britain
– For treasurers

PETERS, M. A. (Myles Anthony)
The club treasurer's handbook. – Bristol : Rose/Jordan, c1980
vii,127p : ill. – Includes index
0–907313–00–0 X.529/40154
2367173

ACCOUNTING. Companies

ACCOUNTING for business combinations / issued for comment by the International Accounting Standards Committee. – London : The Committee, 1981
[19]p. – (International accounting standard) (Exposure draft ; 22)
B84–34774 X.0529/853
1393065

ACCOUNTING for government grants and disclosure of government assistance / issued for comment by the International Accounting Standards Committee. – London : The Committee, 1981
[11]p. – (International accounting standard) (Exposure draft ; 21)
B84–34778 X.0529/853
1393065

ACCOUNTING for leases / issued for comment by the International Accounting Standards Committee. – London : The Committee, 1980
[23]p : ill. – (International accounting standard) (Exposure draft ; 19)
B84–34775 X.0529/853
1393065

GUIDANCE notes on SSAP 21 / The Accounting Standards Committee. – [London] : [Institute of Charterd Accountants in England and Wales], 1984
58p. – (Guidance notes)
0–85291–573–x X.529/67457
1393065

KELLOCK, John
A modern approach to company accounts. – London : Heinemann, 1978
xi,308p. – (The Heinemann accountancy and administration series). – Includes index
0–434–91047–3 X.519/28748
1393065

MARTYN, Simon
Company accounts and finance. – Wantage : Shaws Linton Publications, [1979]
vii,144p : ill. – Includes index
0–906653–00–2 X.529/33346
1393065

OLOFSSON, Torbjörn
Konstruktion av ekonomisystem. – [Lund : Studentlitteratur], 1976
304p. – Includes bibliographies
91–444–0911–7 X.519/23962
1393065

REVENUE recognition / issued for comment by the International Accounting Standards Committee. – London : The Committee, 1981
[18]p. – (International accounting standard) (Exposure draft ; 20)
B84–34776 X.0529/853
1393065

ACCOUNTING. Companies
to date

HENNIG, Bettina
Bilanzierung latenter Steurn. – Bochum : Brockmeyer, 1982
xiii,329p. – (Bochumer wirtschaftswissenschaftliche Studien ; Nr.83). – Includes bibliography
3–88339–219–7 X.529/58149
9026487

ACCOUNTING. Companies
Mathematical models

HARLAN, Kenneth M. (Kenneth Mark)
Accounting distortions and the consistency of firms' performance measures. – Epping : Bowker, c1983
xv,180p. – (Research for business decisions ; no.65). – Bibliography: p175–178. – Includes index
0–8357–1486–1 X.520/34688
3170195

ACCOUNTING. Companies. Construction industries. Developing countries

MILES, Derek
Accounting and book-keeping for the small building contractor. – London : Intermediate Technology Publications, 1978
190p : ill
0–903031–54–x X.519/29491
1846760

ACCOUNTING. Companies. European Community countries
to date

BRUINESSEN, W. van
Richtlijnen voor jaarrekeningen .... – Deventer : Kluwer, [1983]
19p
90–267–0851–3 X.520/30678
8967121

ACCOUNTING. Companies. Great Britain

ASHTON, R. K. (Ray K)
Financial accounting for managers. – London : Armstrong, 1982
151p
0–946291–00–4 X.529/53620
1745158

BRADBURY, Farel
Accountability & company administration. – Ross-on-Wye : Hydatum ; London : Distributed by Bradbury Controls Ltd, 1976
[2],9p
0–905682–02–5 X.510/15354
1031287

COMPANIES Act 1981. – London : Binder Hamlyn, [1982]
121p
B82–36400 X.525/616
1745158

ACCOUNTING. Companies. Great Britain –continued

HELLINGS, Jack
Accounting for small businesses. – [London] : Sphere, 1983
ix,129p. – (Business guidebooks) (Sphere study aids)
0–7221–4843–7 X.529/59042
2503549

MAGEE, J. O. (James Owen)
Company accounts. – 2nd ed. – Plymouth : Macdonald and Evans, 1978
viii,423p. – (The M. & E. handbook series). – Bibliography: p.vi. – Includes index
0–7121–0384–8 X.519/28852
1745158

Company accounts. – 3rd ed. – Plymouth : Macdonald and Evans, 1982
x,445p. – (The M & E handbook series). – Includes index
0–7121–0470–4 X.529/51122
1745158

MILLINGTON, I. E. (Ian Edward)
Direct accounting for the self-employed and small businesses. – Stockport : Direct Accounting, c1981
[53]leaves : ill
B81–26441 X.525/4614
2503549

MOTT, Graham
Accounting for non-accountants. – London : Pan, 1984
222p : ill. – Includes index
0–330–28235–2 X.529/64419
1745158

ACCOUNTING. Companies. Great Britain
– For banking

LESLIE, James
Business accounting 'A'. – Edinburgh : Institute of Bankers in Scotland, 1978
161p
B81–07686 X.525/3977
235960x

SCOTT, I. W. S.
Business accounting 'B'. – [Edinburgh] : Institute of Bankers in Scotland, [1979?]
95p
B81–07687 X.525/3974
235960x

ACCOUNTING. Companies. Great Britain
Law: Companies Act 1981

ACCOUNTING and reporting requirements of the Companies Acts / Arthur Andersen & Co. – London : Arthur Andersen & Co., [1983?]
146p. – Includes index
B83–27702 X.205/1261
269767x

The COMPANIES Act 1981 handbook / edited by Alan Hardcastle and Michael Renshall. – London : Institute of Chartered Accountants in England and Wales, 1981
xxv,230p. – Includes index
0–85291–303–6 X.529/47753
269767x

GODFREY, M. (Michael)
The Companies Act 1981 and the small family company. – [Guildford] : [The College of Law], [1982]
21p. – (Crash course lecture / The College of Law ; 1982)
B82–40233 X.200/39119
269767x

MCMONNIES, Peter N.
The Companies Act 1981. – London : Oyez, 1982
viii,184p. – Includes index
0–85120–609–3 X.529/46563
269767x

SUTCLIFFE, Charles E. T.
Tolley's accounting problems of the Companies Acts. – Croydon : Tolley, 1984
xii,123p. – Includes index
0–85459–147–8 X.529/65261
269767x

ACCOUNTING. Companies. Great Britain
Policies. Effects on liquidity

LAWSON, Gerald H. (Gerald Hartley)
Company profitability and the U.K. stock market. – [Manchester] : Centre for Business Research in association with the Manchester Business School, [197–?]
65p. – (Research report / Centre for Business Research). – Bibliography: p52
B83–00735 X.525/6341
2806592

ACCOUNTING. Companies. Ireland (Republic)

CALLENDER, Rosheen
Account for yourself. – [Dublin] : Kincora Books [for] the Irish Transport and General Workers Union, 1977
240p : ill. – Bibliography: p.235–236. – Includes index
B78–38538 X.519/28725
1723537

ACCOUNTING. Companies. Italy
Law, 1946–

JAEGER, Pier Giusto
Il bilancio d'esercizio delle società per azioni. – Milano : Giuffrè, 1980
vii,135p. – (Quaderni di giurisprudenza commerciale ; 27)
r6508815 X.200/41198
8982953

ACCOUNTING. Companies. Manufacturing industries. United States

BAKER, C. Richard
Accounting for small manufacturers. – New York ; Chichester : Wiley, 1980
ix,197p : ill. – (Small business management series). – Bibliography: p.173. – Includes index
0–471–05704–5 X.525/3735
2272571

ACCOUNTING. Companies. North-eastern United States
Questionable practices

KENNEDY, Tom, 1950–
An examination of questionable payments and practices. – New York ; London : Praeger, 1978
xiii,416p. – (Praeger special studies)
0–03–046321–1 X.520/13708
187070x

ACCOUNTING. Companies. Northern Ireland
Law: Companies (Northern Ireland) Order 1982

SWINSON, Chris
Companies (Northern Ireland) order 1982. – Belfast : Institute of Chartered Accountants in Ireland, 1983
v,121p
0–903854–05–8 X.525/8202
297133x

ACCOUNTING. Companies. Sweden
to date

THORELL, Per
Skattelag och affärssed. – Stockholm : Norstedts, 1984
413p. – Includes bibliography
91–18–47061–0 X.529/66130
9129103

ACCOUNTING. Companies. United States

DOYLE, Dennis M.
Efficient accounting and record-keeping. – New York : McKay, 1977
vi,114p. – (Small business profits program). – Includes index
0–679–50738–8 X.529/32357
149337X

Efficient accounting and record-keeping. – New York ; Chichester : Wiley, c1978
vii,116p : ill. – (The Small business series). – Bibliography: p109–110. – Includes index
0–471–04026–6 X.529/52379
149337x

Efficient accounting and record-keeping. – New York ; Chichester : Wiley, 1978
vii,116p. – (Small business series). – Bibliography: p.109–110. – Includes index
0–471–05044–x X.519/30491
149337x
– [Another copy] X.529/47977

KIRSNER, Laura
Practical accounting for small businesses. – 2nd ed.. – New York ; London : Van Nostrand Reinhold, 1983
xv,254p : ill. – Includes index
0–442–28420–9 X.520/33923
149337x

LIPAY, Raymond J.
Accounting services for your small business. – New York ; Chichester : Wiley, c1983
xii,258p. – Includes index
0–471–09160–x X.520/30583
149337x

ACCOUNTING. Companies. West Germany
1945–

OECHSLE, Eberhard
Die aktienrechtliche Forderung nach Vermittlung eines den tatsächlichen Verhältnissen entsprechenden Bildes der Vermögens-, Finanz- und Ertragslage, untersucht anhand ausgewählter Rechnungslegungsfragen. – [Nürnberg] : [E. Oechsle], [1980]
xlii,203p. – Bibliography: p.xiv–xlii
r6300771 X.529/42202
8714258

ACCOUNTING. Companies. West Germany
1945– – Statistics – Serials

UNTERNEHMEN und Arbeitsstätten. Reihe 2.1, Abschlüsse der Aktiengesellschaften / Herausgeber : Statistisches Bundesamt Wiesbaden. – Geschäftsjahr 1975–. – Stuttgart : Kohlhammer, 1978–. – (Fachserie 2 / Statistisches Bundesamt)
r6310698 S.F.555/98(2)
8714258

ACCOUNTING. Construction industries

COOMBS, William E.
Construction accounting and financial management. – 2nd ed.. – New York ; London : McGraw-Hill, 1977
ix,558p : ill. – (McGraw-Hill series in modern structures). – Includes index
0–07–012610–0 X.520/11903
0200700

ACCOUNTING. Construction industries. United States

COOMBS, William E.
The handbook of construction accounting and financial management. – 3rd ed. – New York ; London : McGraw-Hill, c1984
ix,572p. – Includes index
0–07–012611–9 X.520/34366
3144356

LUCAS, Paul D.
Modern construction accounting methods and controls. – Englewood Cliffs ; London : Prentice-Hall, c1984
xv,252p : 3ill. – Includes index
0–13–590241–x X.520/36073
3144356

ACCOUNTING. Costs. Superannuation schemes. Companies. Great Britain

NAPIER, Christopher J.
Accounting for pension costs. – London : The Committee, c1982
15p
B83–39531 X.529/56827
2958007

PENSION scheme accounts. – London : Accounting Standards Committee, 1984
27p. – (Exposure draft ; 34)
B84–40280 X.529/65170
2958007

ACCOUNTING. Credit institutions. Italy

FORESTIERI, Giancarlo
Analisi dei bilanci delle aziende di credito. – Milano : Giuffrè, 1977
xvi,114p. – (Istituto di economia aziendale dell'Università commerciale "L. Bocconi" – Milano. Ser.9 ; n.8)
r6100143 X.510/11530
204238X

ACCOUNTING. Curriculum subjects. London School of Economics and Political Science
to 1980

DEV, Susan
Accounting and the L.S.E. tradition. – [London] : London School of Economics and Political Science, 1980
23p
0–85328–068–1 X.529/50916
2814315

ACCOUNTING. Curriculum subjects. Universities

JOHNSON, H. Thomas
The role of accounting history in the education of prospective accountants. – [Glasgow] : Department of Accountancy, University of Glasgow, 1984
28leaves. – (Arthur Young lecture ; no.6). – Bibliography: leaves 26–28
0–85261–184–6 X.525/10087
2761831

PERKS, R. W.
Accounting as a university subject. – [Belfast] : Queen's University of Belfast, c1982
18p. – (New lecture series ; no.130)
0–85389–209–1 Ac.1522/2(130)
2761831

ACCOUNTING. Deferred taxation. Companies. Great Britain

ACCOUNTING for deferred tax. – London : Accounting Standards Committee, 1983
26p. – (Exposure draft ; 33)
B84–22554 X.529/61292
3086046

STITT, Iain P. A.
Practical aspects of deferred tax accounting. – London : Institute of Chartered Accountants in England and Wales, 1976
ix,193p. – Includes index
0–85291–141–6 X.200/30450
0675822

ACCOUNTING. Depreciation. Assets

INTERNATIONAL ACCOUNTING STANDARDS COMMITTEE
International accounting standard, depreciation accounting. – London : The Committee, 1976
[11]p. – (International accounting standard ; no.4)
0–905625–01–3 X.519/26829
1103202

ACCOUNTING. Depreciation. Assets. Companies

DEPRECIATION and capital maintenance / edited by Richard P Brief. - New York ; London : Garland, 1984
137p : ill. - (Accounting history and the development of a profession)
0-8240-6304-x X.520/36568
3255956

ACCOUNTING. Depreciation. Assets. Companies. Great Britain

BAXTER, W. T. (William Threipland)
Depreciating assets. - London : Published for The Institute of Chartered Accountants of Scotland by Gee, 1981
iii,61p : ill. - Includes index
0-85258-204-8 X.529/54912
2914972

ACCOUNTING. England
Extraordinary items & prior period adjustment. Standards. Institute of Chartered Accountants in England and Wales. Extraordinary items and prior year adjustments

A REVIEW of SSAP6 - extraordinary items and prior year adjustments. - London : Accounting Standards Committee, c1983
36p
B83-38828 X.529/56740
2954885

ACCOUNTING. England
Research

TRICKER, Robert Ian
Research in accountancy, a strategy for further work. - London : Social Science Research Council, 1975
[3],55,xvii[i.e. xix]p. - Bibliography: p.xiii-xvii
0-900296-45-3 BS.379/35
0643866

ACCOUNTING. England
Standards. Institute of Chartered Accountants in England and Wales. Statements of standard accounting practice

ASHTON, R. K. (Raymond K)
UK financial accounting standard. - Cambridge : Woodhead-Faulkner, 1983
vi,234p. - Includes index
0-85941-201-6 X.529/54549
1486594
- [Another copy: paperback.] ISBN 0-85941-202-4 X.529/54550

COULDERY, Frederick A. J. (Frederick Alan James)
Accounting standards study book. - 3rd ed. - London : Gee, 1981
100p. - (A Gee's study book)
0-85258-207-2 X.520/23784
1486594

Accounting standards study book. - 4th ed. - Wokingham : Van Nostrand Reinhold, 1983
vi,89p. - (A Gee study book)
0-85258-232-3 X.520/31516
1486594

COULDERY, Frederick Alan James
Accounting standards study book. - London : Gee and Co., 1978
62p. - (A Gee's study book)
0-85258-164-5 X.510/15769
1486594

Accounting standards study book. - 2nd ed. - London : Gee and Co., 1979
100p. - (A Gee's study book)
0-85258-173-4 X.622/9582 Woolwich
1486594

ACCOUNTING. Europe

PERCIVAL, C. T.
Glossary of European accounting charts. - Durham : Flambard (European).
Vol.1: France, Plan Comptable Général [PCG], Germany BDI Industrie-Kontenrahmen (IKR), Spain, Plan General de Contabilidad (PGC). - c1982
ii,170p. - Includes index
0-9507949-0-2 X.0525/448(1)
2717131

PRICE WATERHOUSE AND COMPANY
Accounting principles and practices in European countries. - Revised [ed.]. - [London] : Price Waterhouse and Co., 1976
[2],104p
B77-10440 X.519/26635
1067907

ACCOUNTING. European Community countries

LAFFERTY, Michael
Accounting in Europe. - Cambridge : Woodhead-Faulkner Ltd [for] National Westminster Bank, 1975
xiii,425p : ill. - Bibliography: p.394-406. - Includes index
0-85941-013-7 X.519/25268
0519952

ACCOUNTING. European Community countries
- Statistics - Serials

REGIONAL accounts : economic aggregates / [issued by] Statistical Office of the European Communities. - 1970-. - Luxembourg : Office for Official Publications of the European Communities, 1976-
r6077712 UN.P.131/249
0519952

ACCOUNTING. European Community countries
European Community law

The IMPACT of the Seventh Directive / Ernst & Whinney. - London : Financial Times Business Information, 1984
111p : ill
0-903199-74-2 X.525/10026
3191117

ACCOUNTING. Farms. Agriculture. Great Britain

CRITCHLEY, R.
Understanding your farm accounts. - [Aberdeen] : [North of Scotland College of Agriculture Agriculture Economics Division School of Agriculture], 1982
i,45p : ill
B83-17097 X.525/7007
2877112

ACCOUNTING. Farms. Agriculture. Great Britain
Applications of digital computer systems - Conference proceedings

EFFECTIVE use of computing for farm accounting / edited by Ann Foster. - Aldershot : Gower, c1981
72p : ill. - (Computing in business report)
0-566-03028-4 SRL
2727420

ACCOUNTING. Financial transactions. Related parties. Canada, Great Britain & United States

ACCOUNTANTS INTERNATIONAL STUDY GROUP
Related party transactions. - [London] : The Group, 1978
[17]p
B78-38539 X.439/8582
1723324

ACCOUNTING. France
1815-1958

PAYEN, Manufacturier
Essai sur la tenue des livres d'un manufacturier. De la comptabilité dans une entreprise industrielle-. - London : Scolar Press, 1980
111,40p. - (Historic accounting literature ; vol.40)
r6208878 X.0625/102(40)
8977054

ACCOUNTING. France
1958-

PESQUEUX, Yvon
Le nouveau plan comptable. - Paris : La Documentation française, [1981]
303p. - (Pratique de-)
2-11-000552-1 X.622/14015
887672x

ACCOUNTING. Great Britain

ACCOUNTING STANDARDS COMMITTEE
Accounting standards 1978. - London : The Institute, 1978
v,294p. - Includes bibliographies and index
0-85291-196-3 X.519/30121
1946293

ACCOUNTING. Great Britain
- Serials

[IN touch (London : 1979)]. In touch. - Issue no.1979/1 (12th June 1979)-. - [London] : Touche Ross & Co. (United Kingdom), 1979-
v.
B83-41497 P.513/3177
2970996

ACCOUNTING. Great Britain
to 1976

KITCHEN, Jack
Accounting. - [Hull] : University of Hull, 1978
[1],32p. - (Inaugural lectures / University of Hull)
0-85958-418-6 X.519/29966
1928627

ACCOUNTING. Great Britain
1800-1901

CORY, Isaac Preston
A practical treatise on accounts, mercantile, partnership, solicitor's, private, steward's, receiver's, executor's, trustee's &c. - 1839. - London : Scolar, 1980
v,340p. - (Historic accounting literature ; v.20)
r6812065 X.0625/102(20)
9052739

ACCOUNTING. Great Britain
Contingency planning. Standards

ACCOUNTING STANDARDS COMMITTEE
Accounting for contingencies. - London : The Committee, 1978
[9]p. - (ASC exposure draft ; 23)
B80-08844 X.519/30591
2044021

ACCOUNTING. Great Britain
Funds statements

KNOX, R. W.
Statements of source and application of funds. - London : Institute of Chartered Accountants in England and Wales, 1977
[8],117,7p
0-85291-178-5 X.519/27785
1383213

STATEMENTS of source and application of funds. - London : Institute of Chartered Accountants in England and Wales, 1976
1 sound cassette(ca.36min.)
71p.. - (Finance managers' guidelines ; 5)
r6196141 Cup.580.ff.19
2008653

ACCOUNTING. Great Britain
Law

CARR, J. Graham
Accountancy law and practice manual. - Farnborough, Hants. : Gower Press, 1978
1v.(loose-leaf)
Supplements, when received, will be placed at this shelfmark
0-566-02090-4 X.0200/1190
1567314

CURRENT accounting law & practice. - London : Sweet and Maxwell.
[1976]. - 1976
xxi,394p. - Includes index
Subsequent volumes, when received, will be placed at this shelfmark
0-421-21490-2 X.0100/1005
0603473

ACCOUNTING. Great Britain
Standards. Formulation

ACCOUNTING STANDARDS COMMITTEE
Setting accounting standards. - London : Institute of Chartered Accountants in England and Wales, 1978
[4],47p
0-85291-212-9 X.519/29068
1773658

Setting accounting standards. - [London] : [Institute of Chartered Accountants in England & Wales], c1981
56p
0-85291-300-1 X.529/41657
1773658

ACCOUNTING. Great Britain
Standards. Formulation. Proposals: Accounting Standards Committee, 1970-1980

BRITISH accounting standards / edited by Sir Ronald Leach and Edward Stamp. - Cambridge : Woodhead-Faulkner, 1981
viii,247p. - Bibliography: p81-84
0-85941-149-4 X.520/26913
2719487

ACCOUNTING. Great Britain
Standards. Formulation. Proposals: Accounting Standards Committee. Setting accounting standards : a consultative document

SUBMISSIONS on the Accounting Standards Committee's consultative document 'Setting accounting standards'. - [London] : Accounting Standards Committee, 1979
2v.([6],i,678p[5]leaves)
0-85291-236-6 X.512/6959
1956248

ACCOUNTING. Great Britain
Standards. Setting

ACCOUNTING STANDARDS COMMITTEE
Review of the standard setting process. - [London] : [The Committee], c1983
23p
0-85291-348-6 X.529/57796
2990695

ACCOUNTING. Groups. Companies

PENDLEBURY, M. W.
An international comparison of group accounting in the Federal Republic of Germany, France, the United Kingdom and the United States of America. - [London] : [Institute of Chartered Accountants in England and Wales], [1978]
[5],97p. - (Occasional papers / Institute of Chartered Accountants in England and Wales. Research Committee ; no.16). - Bibliography: p.92-95
0-85291-211-0 X.0512/168(16)
1773054

ACCOUNTING. Groups. Companies. Great Britain

FRANKEL, M. R.
The preparation of group accounts. - London : Financial Training with HFL (Publishers), 1979
259p. - Includes index
0-372-30031-6 X.525/5581
2703769

ACCOUNTING. Higher education institutions. United States

SCHEPS, Clarence
Accounting for colleges and universities. - 3rd ed. - Baton Rouge ; London : Louisiana State University Press, 1978
xv,384p. - Includes index
0-8071-0388-8 X.629/12164 Woolwich
1737465

ACCOUNTING. Hospitals

BROYLES, Robert W.
Hospital accounting practice. - Rockville, Ma. ; London : Aspen, 1982
2v. : ill. - Includes indexes
B83-33734 X.520/30863-30864
2937344

ACCOUNTING. Hotel industries & catering industries. Great Britain

HARRIS, Peter J. (Peter James)
Accounting and financial management in the hotel and catering industry. - 3rd ed. - London : Hutchinson, 1983
2v. : ill. - Includes index
B83-23686 X.520/32864-32865
2506939

KOTAS, Richard
Accounting in the hotel and catering industry. - 4th ed. - London : International Textbook Co., 1981
344p : ill. - Includes index
0-7002-0279-x X.520/24312
2506939

WALTON, Peter, 19—
Modern financial accounting in the hospitality industry. - London : Hutchinson, 1983
138p : ill. - (Hutchinson catering and hotel management series). - Includes index
0-09-153100-4
2506939
- [Another Copy : paperback.] ISBN 0-09-153101-2 X.520/32503

ACCOUNTING. Hotel industries. Great Britain

FENTON, Lawrence S.
Hotel accounts and their audit. - London : Institute of Chartered Accountants in England and Wales, 1978
xii,287p : ill. - (Industry accounting and auditing guides). - Bibliography: p.265. - Includes index
0-85291-202-1 X.510/16009
1777378

ACCOUNTING. Hotel industries. United States

HORWATH, Ernest B.
Hotel accounting. - 4th ed.. - New York ; Chichester : Wiley, 1977
vi,424p : ill. - Includes index
0-471-07247-8 X.520/14587
1964275

ACCOUNTING. Income tax. Companies
International standards

INTERNATIONAL ACCOUNTING STANDARDS COMMITTEE
International accounting standard, proposed statement, accounting for taxes on income. - London : International Accounting Standards Committee, 1978
[15]p. - (Exposure draft ; 13)
B78-18474 X.519/28007
1452088

ACCOUNTING. Insolvency. Companies. United States

NEWTON, Grant W.
Bankruptcy and insolvency accounting. - 2nd ed. - New York ; Chichester : Wiley, c1981
xii,676p. - (The Wiley/Ronald series in professional accounting and business). - Bibliography: p641-650. - Includes index
0-471-07992-8 X.200/36777
2562731

ACCOUNTING. Interest. Finance

ANTHONY, Robert Newton
Accounting for the cost of interest. - Lexington, Mass. ; London : D.C. Heath, 1975
xiii,135p. - Includes index
0-669-00027-2 X.520/10522
0612197

ACCOUNTING. Interest. Finance. Great Britain

RICKWOOD, C. P. (Colin P)
Accounting treatment of capitalised interest. - London : Published for the Association of Certified Accountants by Certified Accountant Publications, 1983
32p
1-85011-000-x X.529/68748
3258971

ACCOUNTING. Local authorities. England
Standards

DRAFT guidance notes on the application of statements of standard accounting practice to local authorities in England and Wales / ... prepared by the Chartered Institute of Public Finance and Accountancy on behalf of the Accounting Standards Committee ... ; issued by the Accounting Standards Committee for comment by persons interested in financial reporting. - London : The Committee, 1982
10p
B83-39530 X.529/56825
295799x

ACCOUNTING. Local authorities. England
Standards. Institute of Chartered Accountants in England and Wales. Statements of standard accounting practice

GUIDANCE notes on the application of statements of standard accounting practice to local authorities in England and Wales / issued to preparers, auditors an users of local authority accounts in England and Wales by the Accounting Standards Committee of the Institute of Chartered Accountants in England and Wales ... [et al.]. - London : Accounting Standards Committee, 1983
18p
B84-21604 X.529/61005
2534185

ACCOUNTING. Local authorities. Great Britain

ROCKLEY, L. E. (Lawrence Edwin)
Public and local authority accounts. - 2nd ed. - London : Heinemann, 1978
xi,280p. - Includes index
0-434-91736-2 X.629/12088 Woolwich
0275808

ACCOUNTING. Local authorities. Ireland (Republic)

GOLDEN, T. P.
Local authority accounting in Ireland. - Dublin : Institute of Public Administration, [1977]
xii,276p : ill. - Includes index
0-902173-77-4 X.519/27141
1188194

ACCOUNTING. Local authorities. United States

TIERNEY, Cornelius E.
Governmental accounting. - New York ; Oxford : Elsevier, c1983
ix,223p. - Includes index
0-444-00792-x X.525/9699
316330x

ACCOUNTING. Local authority housing. England

GREAT BRITAIN. Department of the Environment
Manual on local authority housing subsidies and accounting. - London : H.M.S.O., 1975
viii,108p. - Includes index
0-11-751001-7 BS.414/704
056799x

HOUSING subsidies and accounting manual / Department of the Environment, the Welsh Office. - 1981 ed. - London : H.M.S.O., 1982
i,78p. - Includes index
0-11-751612-0 BS.414/2101
056799x
- [Another copy.] X.525/10830

ACCOUNTING. Local councils. England

DAVIES, Richard Ungoed
Parish council and town council accounts. - London : C. Knight, 1977
160p. - Includes index
0-85314-288-2 X.519/27140
1188186

ACCOUNTING. Manufacturing industries

WHITMORE, John
Factory accounts. - New York ; London : Garland, 1984
123p : Iill. - (Accounting history and the development of a profession)
0-8240-6314-7 X.622/23344
327974x

ACCOUNTING. Manufacturing industries. Great Britain

MONAGHAN, M. W. (Michael William)
Basic factory accounting and administration. - Aldershot : Gower, c1983
x,191p : ill. - Includes index
0-566-02281-8 X.520/30799
2938766

ACCOUNTING. Multinational companies

ARPAN, Jeffrey S.
International accounting and multinational enterprises. - Boston [Mass.] : Warren, Gorham & Lamont, c1981
xv,400p : ill. - Bibliography: p375-392
0-88262-539-x X.520/29422
1968343

MILLER, Elwood L.
Accounting problems of multinational enterprises. - [Lexington, Mass. etc.] : Heath] ; [Farnborough, Hants.] : [Distributed by] Teakfield Limited, 1979
xiii,289p : 2 ill. - Bibliography: p.275-277. - Includes index
0-669-02712-x X.520/14614
1968343

ACCOUNTING. Multinational companies
- For management

CHOI, Frederick D. S.
International accounting. - [Rev. ed.]. - Englewood Cliffs ; London : Prentice-Hall, c1984
xvi,525p : ill. - Includes bibliographies and index
0-13-470931-4 X.520/36067
1512080

CHOI, Frederick D. S.
An introduction to multinational accounting. - Englewood Cliffs ; London : Prentice-Hall, 1978
xiii,365p : ill. - Includes bibliographies and index
0-13-489302-6 X.520/12673
1512080

ACCOUNTING. Multinational companies
to date. International legal aspects

BÖNKOST, Klaus Jürgen
Weltabschlüsse multinationaler Konzerne. - München : Florentz, 1980
vi,463p. - (Wirtschaftswissenschaftliche Forschung und Entwicklung ; Bd 59). - Bibliography: p435-463
3-88259-064-5 X.529/51325
8891753

ACCOUNTING. New Zealand

WHEELER, R. B. (Robert Brian)
Advanced accounting in New Zealand. - 2nd ed. - Christchurch ; London : Whitcoulls, 1978
436p. - Includes index
0-7233-0560-9 X.520/27484
2756765

ACCOUNTING. Non-profit making organisations. Great Britain

PRESENTING your group's annual accounts. - [London] : [Gingerbread], [1983]
[6]p. - (Gingerbread group leaflet)
B84-30198 X.525/9251
3137945

ACCOUNTING. Non-profit making organisations. United States

GROSS, Malvern J.
Principles of accounting and financial reporting for nonprofit organizations. - New York ; Chichester : Wiley, 1979
xv,415p. - Bibliography: p.395-400. - Includes index
0-471-05719-3 X.322/9240 Woolwich
2141205

HENKE, Emerson O.
Accounting for nonprofit organizations. - 3rd ed. - Boston, Mass. : Kent, c1983
viii,274p : ill. - Bibliography: p263-267
0-534-01429-1 X.520/37395
2052393

POWELL, Ray M.
Accounting procedures for institutions. - Notre Dame ; London : University of Notre Dame Press, 1978
xx,441p. - (Notre Dame studies in the management of not-for-profit institutions). - Includes index
0-268-00590-7 X.525/2973
2052393

SMITH, G. Stevenson
Accounting for librarians. - Chicago : American Library Association, 1983
ix,470p : ill
0-8389-0385-1 2719e.798
2052393

ACCOUNTING. Offshore natural gas & petroleum industries. Great Britain

NORTON, John C.
Accounting and auditing guide for United Kingdom oil and gas exploration and production. - London : Institute of Chartered Accountants in England and Wales, 1978
x,326p. - (Industry accounting and auditing guides). - Bibliography: p.323. - Includes index
0-85291-206-4 X.611/8327 Woolwich
1729268

ACCOUNTING. Organisations
Information systems

MACINTOSH, Norman B.
The social software of accounting and information systems. – Chichester : Wiley, c1985
xvi,294p : ill. – Includes bibliographies and index
0-471-90543-7 X.520/37351
3055272

ACCOUNTING. Partnerships. General practice. Great Britain

PARKER, E. J. (Ernest James)
The doctor's guide to partnership accounts. – London : Regency, c1984
76p. – Includes index
B85-02238 X.529/65579
3142035

ACCOUNTING. Partnerships. Great Britain

MAGEE, J. O. (James Owen)
Partnership accounts. – 2nd ed. – Plymouth : Macdonald and Evans, 1977
x,370p. – (The M. & E. handbook series). – Includes index
0-7121-1666-4 X.519/27479
1288105

ACCOUNTING. Public administration
to date

PARENZAN, Tullio
Profili del sistema dei controlli esterni sulla pubblica amministrazione. – Milano : Giuffrè, 1983
297p. – (Università di Trieste, Facoltà di scienze politiche ; 23)
r6509974 X.520/31962
9020152

ACCOUNTING. Public bodies. Great Britain

ISSUES in public sector accounting / edited by Anthony Hopwood and Cyril Tomkins. – Deddington : Philip Allan, 1984
ix,247p : ill. – Includes bibliographies
0-86003-523-9 X.529/64642
2984121

JONES, Rowan
Public sector accounting. – London : Pitman, 1984
viii,309p : ill. – Bibliography: p300-306. – Includes index
0-273-01857-4 X.529/64228
2984121

PUBLIC sector accounting and financial control / Sir Douglas Henley ... [et al.] ; sponsored by the Chartered Institute of Public Finance and Accountancy. – Wokingham : Van Nostrand Reinhold, c1983
xiv,251p : ill. – (The VNR series in accounting and finance). – Includes bibliographies and index
0-442-30543-5 X.529/57999
2984121

ACCOUNTING. Public bodies. United States

LYNN, Edward S.
Fund accounting. – 2nd ed. – Englewood Cliffs ; London : Prentice-Hall, c1983
xiv,930p : ill. – Includes index
0-13-332411-7 X.520/30494
2926806

ACCOUNTING. Public finance. Italy
Legal aspects

PEREZ, Rita
Le procedure finanziarie dello Stato. – Milano : Franco Angeli, [1978]
200p. – (Studi e documenti sul settore pubblico dell'economia) (Collana Ciriec ; 20)
r6160083 X.529/35307
2263157

ACCOUNTING. Public utilities. Great Britain

ROCKLEY, L. E. (Lawrence Edwin)
Public and local authority accounts. – 2nd ed. – London : Heinemann, 1978
xi,280p. – Includes index
0-434-91736-2 X.629/12088 Woolwich
0275816

ACCOUNTING. Real property

KLINK, James J.
Real estate accounting and reporting. – New York ; Chichester : Wiley, 1980
xi,227p. – Bibliography: p.217-219. – Includes index
0-471-06041-0 X.520/21801
222755x

ACCOUNTING. Real property investment companies. Great Britain

MILNES, G. A.
Property company accounts. – London : Institute of Chartered Accountants in England and Wales, 1978
xi,185p : 1 ill. – (Industry accounting and auditing guides). – Includes index
0-85291-207-2 X.510/16019
179793x

ACCOUNTING. Real property. United States

WILEY, Robert J.
Real estate accounting and mathematics handbook. – New York ; Chichester : Wiley, c1980
xxiv,310p : ill. – (Real estate for professional practitioners). – Includes index
0-471-04812-7 X.520/23065
2354810

ACCOUNTING. Remuneration. Executives. Companies. United States

FINANCIAL considerations of executive compensation and retirement plans / Peter T. Chingos ... [et al.]. – New York ; Chichester : Wiley, c1984
vii,236p. – (The Wiley/Ronald series in professional accounting and business). – Includes index
0-471-87489-2 X.520/35736
3216993

ACCOUNTING. Remuneration. Personnel. United States

KEELING, B. Lewis (Billy Lewis)
Payroll accounting. – 1981 ed. – Cincinnati : South-Western, 1981
ix,346p. – Includes index
0-538-01461-x X.525/7047
2882205

ACCOUNTING. Research & development by companies

INTERNATIONAL ACCOUNTING STANDARDS COMMITTEE
International accounting standard, proposed statement, accounting for research and development costs. – London : Institute of Chartered Accountants in England and Wales, 1977
[11]p. – (Exposure draft ; 9)
0-905625-02-1 X.519/26828
1103180

ACCOUNTING. Research & development by companies. Great Britain

BATTY, J. (Joseph)
Accounting for research and development. – London : Business Books, 1976
x,237p : ill. – Includes index
B76-25880 X.520/10939
0702390

LOTHIAN, Niall
How companies manage R & D. – London : Institute of Cost and Management Accountants, 1984
ii,118p. – Bibliography: p117-118
0-901308-74-9 X.629/24545
3210103

ACCOUNTING. Risks. Foreign exchange

WALKER, David P.
An economic analysis of foreign exchange risk. – [London] : Institute of Chartered Accountants in England and Wales, 1978
[7],77p. – (Occasional papers / Institute of Chartered Accountants in England and Wales. Research Committee ; no.14)
0-85291-195-5 X.0512/168(14)
1476939

ACCOUNTING. Romania
– Serials

REVISTA de contabilitate. – [1927 nr. 1-]. – București : Ministerul Finanțelor, [1927-]
Holdings: an. 43, nr.1. ian. 1979-
r6174473 S.M.20/18
229334X

ACCOUNTING. Sale of local authority houses. Great Britain

WILSON, John Pearson
Accounting for the sale of council houses. – London : Institute of Cost and Management Accountants, 1977
44p. – (Occasional papers / Institute of Cost and Management Accountants)
0-901308-41-2 X.519/28084
1476750

ACCOUNTING. Scotland
to 1920 – Bibliographies

PRYCE-JONES, Janet
Accounting in Scotland. – 2nd ed. – Edinburgh : Institute of Chartered Accountants of Scotland : Scottish Committee on Accounting History, 1976
[1],107,xviii p,plate. – Includes index
B76-26792 OPL
0209139
– [Another copy] X.512/5317

Accounting in Scotland. – [2nd ed.]. – New York ; London : Garland, 1984
x,107p. – (Accounting history and the development of a profession). – Includes index
0-8240-6333-3 X.950/41469
0209139

ACCOUNTING. Slovenia. Yugoslavia

ZDRUŽENJE RAČUNOVODSKIH IN FINANČNIH DELAVCEV SR SLOVENIJE
Almanah ob dvajsetletnici organizacije računovodskih in finančnih delavcev v SR Sloveniji. – Ljubljana : [Združenje računovodskih in finančnih delavcev SR Slovenije], 1977
196p : ill
r6172897 X.520/9986
2097427

ACCOUNTING. Soviet Union

BAILEY, David, 1929-
Some reflections on Soviet accounting. – [Birmingham] : University of Birmingham, Faculty of Commerce and Social Science, 1973
[3]leaves,35,9p : 2 ill. – (Series B ; no.29) (Series RC/B ; no.5) (Discussion paper)
r6072857 X.510/15984
1770853

ACCOUNTING. Stock & work in progress. Companies. Great Britain

ACCOUNTING for stocks and work in progress. – London : Institute of Chartered Accountants in England and Wales, 1976
1 sound cassette(ca.60min.)
42p.. – (Finance managers' guidelines ; 4). – Includes bibliographies
r6196140 Cup.580.ff.20
0542644

ACCOUNTING. Sweden
Law – Texts

BÖKMARK, Jan
Bokföringslagen. – Stockholm : Liber förlag : Jurist- och Samhälls-vetareförbundet, 1977
162p. – (Publica/Rätt och samhälle)
91-380-2795-x X.200/11517
1539906

ACCOUNTING. Taxation. Companies. Great Britain

SCRIMGEOUR, J. Larry
Accounting for UK company taxation. – London : Longman, 1984
260p. – Bibliography: p255. – Includes index
0-582-29621-8 X.622/19491
3132269

ACCOUNTING. Taxation. Companies. United States

MILES, Catherine E.
Business and personal taxes. – 4th revision. – Boston [Mass.] ; London : Allyn and Bacon, 1977
372p in various pagings. – Includes index
0-205-05808-6 X.520/14447
194889x

Business and personal taxes 1981. – [New ed.]. – Boston [Mass.] ; London : Allyn and Bacon, c1981
x,421p. – Includes index
0-205-07163-5 X.200/34501
194889x

MILES, Catherine E.
Instructor's manual to accompany Business and personal taxes 1981. – Boston, [Mass.] ; London : Allyn and Bacon, c1980
166p
0-205-07164-3 X.529/52953
2873206

ACCOUNTING. Taxation. United States

MOSS, Morris H.
Schaum's outline of theory and problems of tax accounting. – New York ; London : McGraw-Hill, 1980
[7],216p. – (Schaum's outline series in accounting ; 5). – Includes index
0-07-043470-0 X.525/3692
2287188

ACCOUNTING. Trusts. Scotland
Legal aspects

MORLEY, Michael F. (Michael Francis)
Accounting For Scottish executries and trusts. – Edinburgh : Law Society of Scotland, 1984
vii,95p
0-902023-13-6 X.529/69106
3278972

ACCOUNTING. United Nations

ANSON-MEYER, Monique
La nouvelle comptabilité des Nations-Unies en Afrique. - Paris : L'Harmattan, 1982
421p. - (Bibliothèque du développement). - Includes bibliography
2-85802-253-4 X.520/35699
9139419

ACCOUNTING. United States

ACCOUNTANT'S handbook. - 6th ed.. - New York ; Chichester : Wiley, c1981
2v.(2004p in various pagings) : ill. - (The Wiley/Ronald series in professional accounting and business). - Includes bibliographies and index
0-471-05505-0 X.520/24140
2497077

BIRD, Francis A.
Accounting theory. - Richmond, Va. : Dame, c1981
viii,391p : ill
0-936328-02-9 X.622/19886
2497077

DANOS, Paul
Intermediate accounting. - Englewood Cliffs, N.J. ; London : Prentice-Hall, 1983
xxviii,1122p : ill. - (Prentice-Hall series in accounting). - Includes index
0-13-469338-8 X.520/30241
2497077

FLEGM, Eugene H.
Accounting. - New York ; Chichester : Wiley, c1984
xviii,261p. - (Modern accounting perspectives and practice). - Includes index
0-471-09326-2 X.520/34954
2497077

ACCOUNTING. United States
- For engineering

HAYES, Rick Stephan
Simplified accounting for engineering and technical consultants. - New York ; Chichester : Wiley, 1980
ix,171p : ill. - (Small business management series). - Includes index
0-471-05708-8 X.525/3374
2225425

ACCOUNTING. United States
to 1978

PREVITS, Gary John
A history of accounting in America. - New York ; Chichester : Wiley, 1979
xiii,378p : ill. - Bibliography: p.336-368. - Includes index
0-471-05172-1 X.800/28955
206913x

ACCOUNTING. United States
C.P.A. practice

HUBBARD, Thomas Dean
CPA review. - 2nd ed. - Santa Barbara ; Chichester : Wiley, 1978
2v.(ix,1059;iii,340p) : ill. - Includes bibliographies and index
0-471-02299-3 X.520/12600
0066060

LANGENDERFER, Harold Q.
C.P.A. examination. - Columbus, Ohio ; London : Merrill.
Vol.1: Problems. - 3rd ed. - c1979
xii,1089p : ill
0-675-08298-6 X.0520/672(1)
0066060
- 3rd ed. - Columbus, Ohio ; London : Merrill.
Vol.2: Solutions. - c1979
vi,566p
0-675-08297-8 X.0520/672(2)
0066060

ACCOUNTING. United States
Theories, 1957-1964

The ACCOUNTING postulates and principles controversy of the 1960s / compiled and edited by Stephen A. Zeff. - New York ; London : Garland, 1982
1v.(various pagings). - (Accountancy in transition)
0-8240-5341-9 X.520/32304
3034321

ACCOUNTING. Valuation. Foreign stock. Companies. United States

BINDON, Kathleen Ranney
Inventories and foreign currency translation requirements. - Epping : Bowker, c1983
ix,145p : ill. - (Research for business decisions ; no.56). - Bibliography: p135-142. - Includes index
0-8357-1391-1 X.520/31310
2954427

ACCOUNTING. Western Europe

OLDHAM, K. Michael (Kenneth Michael)
Accounting systems and practice in Europe. - 2nd ed. - Aldershot : Gower, 1981, c1975
xiv,271p. - Bibliography: p263-264. - Includes index
0-566-02147-1 X.520/26596
0543233

OLDHAM, Kenneth Michael
Accounting systems and practice in Europe. - Epping : Gower Press, 1975
xi,183p : ill. - Includes index
0-7161-0267-6 X.520/10341
0543233

ACCOUNTING. Yugoslavia
to date

KRONIKA ob desetem simpoziju o sodobnih metodah v računovodstvu in poslovnih financah / Koordinacijski odbor Zveze ekonomistov Slovenije ; [Koordinacijski odbor] Zveze računovodskih in finančnih delavcev SR Slovenije. - Ljubljana : [Zveza ekonomistov Slovenije], 1978
97p : ill
r6435525 X.529/52862
8927707

ACCOUNTING FOR THE EFFECTS OF CHANGING PRICES. Accounting Standards Committee

JOHNSON, Barry M.
Accounting for the effects of changing prices. - London : Deloitte, Haskins & Sells, 1984
90p
0-86349-040-9 X.529/68065
3254453

ACCOUNTING PRICES. Rice. Thailand
Conversion of market prices. Conversion factors. Calculation

VEITCH, Mike
The accounting price of rice in Thailand. - Bradford : Project Planning Centre, University of Bradford, [1979]
ii,33,[3]p. - (Occasional papers / University of Bradford. Project Planning Centre for Developing Countries ; no.4). - Bibliography: p.32-33
0-901945-36-6 X.622/9051 Woolwich
2278316

ACCOUNTING STANDARDS COMMITTEE
1970-1980

BRITISH accounting standards / edited by Sir Ronald Leach and Edward Stamp. - Cambridge : Woodhead-Faulkner, 1981
viii,247p. - Bibliography: p81-84
0-85941-149-4 X.520/26913
2719487

ACCOUNTING STANDARDS COMMITTEE
Accounting for the effects of changing prices

JOHNSON, Barry M.
Accounting for the effects of changing prices. - London : Deloitte, Haskins & Sells, 1984
90p
0-86349-040-9 X.529/68065
3254453

ACCOUNTING STANDARDS COMMITTEE
Setting accounting standards : a consultative document

SUBMISSIONS on the Accounting Standards Committee's consultative document 'Setting accounting standards'. - [London] : Accounting Standards Committee, 1979
2v.([6],i,678p[5]leaves)
0-85291-236-6 X.512/6959
1956248

ACCOUNTING SYSTEMS
See also
BEA accounting system

ACCOUNTING SYSTEMS
Computerised accounting systems. Control

EDWARDS, John D.
Accounting and management controls for computer systems. - North Ryde : CCH Australia ; Bicester : CCH Editions, 1980
xi,236p : ill. - Bibliography: p225. - Includes index
0-86903-285-2 X.629/19729
2871831

ACCOUNTING SYSTEMS. Banks

PATTEN, James A.
Fundamentals of bank accounting. - Reston : Reston, c1983
xv,285p. - Bibliography: p275-278
0-8359-2119-0 X.520/32292
303044x

ACCOUNTING SYSTEMS. Farms. Agriculture. European Community countries

KERR, Hugh William Thomas
Farm costing in the European Economic Community. - [Loughborough] : University of Nottingham, Department of Agriculture and Horticulture, 1974
[3],33,[7]leaves(2fold) : ill
0-900573-39-2 X.313/1273 Woolwich
0621137

ACCOUNTING SYSTEMS. Income. Companies. Great Britain

ARNOLD, John, 1944-
A study of the relative usefulness of six accounting measures of income. - [London] : Institute of Chartered Accountants in England and Wales, 1978
[7],96p : ill. - (Occasional papers / Institute of Chartered Accountants in England and Wales. Research Committee ; no.13)
0-85291-192-0 X.512/5913
1413589

ACCOUNTING SYSTEMS. Organisations

GEE, Kenneth P. (Kenneth Philip)
Financial control. - [Salford] : University of Salford, [1980?]
[14]p : ill. - (Inaugural lecture)
B82-30529 X.529/48621
2758903

ACCOUNTING SYSTEMS. Organisations
On-line accounting systems: ADNDA

ADNDA user guide. - 3rd ed. - Maidenhead : Lipman Management Resources, 1982
104p in various pagings. - Includes index
0-906168-04-x SRL
2895781

ACCOUNTING UNITS
See also
Green money

ACCOUNTING UNITS. Civil liability for pollution by petroleum
- Conference proceedings

FINAL act of the conference to revise the unit of account provisions of the International Convention on Civil Liability for Oil Pollution Damage, 1969 (with protocol and resolution), London 19 November 1976 ... - London : H.M.S.O., [1977]
10p. - (Miscellaneous ; no.26 (1977)) (Cmnd.7028)
0-10-170280-9 BS.14/3101
1432338

ACCOUNTING UNITS. European Investment Bank

TREATY amending certain provisions of the Protocol on the statute of the European Investment Bank, Brussels, 10 July 1975 ... - London : H.M.S.O., [1975]
Folder (4p). - (European Communities ; no.139 (1975)) (Cmnd.6253)
0-10-162530-8 BS.14/137
0668559

ACCOUNTS
See also
Audits
Balance sheets
Budgets
Consolidated accounts
Deposit accounts
Financial statements
International accounts
Municipal accounts
National accounts

ACCOUNTS

MUNRO, Andrew
Munro's book-keeping and accountancy. - 23rd ed.. - London : Pitman, 1975
viii,563p. - Includes index
0-273-00924-9 X.519/25296
0000566

ACCOUNTS
- For Asian students

LI, Betsy
Principles of accounts. - Kuala Lumpur ; Oxford : Oxford University Press, 1978
iv,168p. - (Modern certificate guides)
0-19-581147-x X.520/27951
2774488

ACCOUNTS
- For Nigerian students

BASTON, Andrew
Elements of accounts for elementary and intermediate stage students. - Nigerian ed.. - London : Cassell, 1979
[6],378p : ill
0-304-30253-8 X.519/30274
195833x

ACCOUNTS
- For Nigerian students–continued

Key to elements of accounts. – Nigerian ed. – London : Cassell, 1980
229p
0–304–30254–6 X.529/42162
2525127

ACCOUNTS
Computerised accounts. Auditing

AUDITING in a computer environment. – London : Auditing Practices Committee, 1982
31p. – (Auditing guidelines) (APC publications ; no 5080)
B83–37526 X.529/56514
1781626

BEST, Peter J. (Peter John)
Auditing computer–based accounting systems. – Sydney : Prentice–Hall of Australia ; London : Prentice–Hall International, c1983
247p : ill. – Bibliography: p241–243. – Includes index
0–7248–0045–x X.622/20007
1781626

CHAMBERS, Andrew D.
Computer auditing. – London : Pitman, 1981
xv,237p : ill. – Bibliography: p229–231. – Includes index
0–273–01633–4 X.622/23916
1781626

DAVIS, Keagle W.
Auditing computer applications. – New York ; Chichester : Wiley, c1982
xvi,601p : ill
0–471–05482–8 X.622/13762
1781626

DAWLEY, Donald L. (Donald Lee)
What auditors should know about data processing. – Epping : Bowker, c1983
ix,236p : ill. – (Research for business decisions ; no.63). – Bibliography: p225–234. – Includes index
0–8357–1483–7 X.520/34589
1781626

DOUGLAS, I. J. (Ian J)
Audit and control of mini– and microcomputers. – Manchester : NCC Publications, 1982
112p. – Bibliography: p95–97. – Includes index
0–85012–368–2 X.629/20218
1781626

JANCURA, Elise G. (Elise Geraldine)
Establishing controls and auditing the computerized accounting system. – New York ; London : Van Nostrand Reinhold, c1981
vii,288p : ill. – Includes index
0–442–80507–1 X.525/3911
1781626

JENKINS, Brian G.
An audit approach to computers. – [2nd ed.]. – London : Institute of Chartered Accountants in England and Wales, 1978
xxii,435p : ill. – Includes index
0–85291–187–4 X.525/2315
1781626

ROBERTS, Martin B.
EDP controls. – New York ; Chichester : Wiley, c1983
1v.(loose–leaf) : ill. – (Modern accounting perspectives and practices). – Includes index
0–471–86599–0 L.42/3146
1781626

WATNE, Donald A.
Auditing EDP systems. – Englewood Cliffs ; London : Prentice–Hall, c1984
xxiii,584p : ill. – Bibliography: p564–567. – Includes index
0–13–051631–7 X.622/19828
1781626

WEBER, Ron
EDP auditing. – New York ; London : McGraw–Hill, c1982
xiv,642p : ill. – (McGraw–Hill series in management information systems). – Includes bibliographies and index
0–07–068830–3 SRL
1781626

ACCOUNTS. Banks
Confidentiality

CHAMBOST, Edouard
Bank accounts. – Chichester : Wiley, 1983
xii,322p. – Includes index
0–471–90076–1 X.520/31838
298332x

ACCOUNTS. Brazil. Presidência da Republica
1974–

BRAZIL. Congresso. Câmara dos Deputados. Comissão de Fiscalização Financeira e Tomada de Contas
Relatório sobre as contas do Presidente da República. – Brasília : Câmara dos Deputados, Diretoria Legislativa, Centro de Documentação e Informação, Coordenação de Publicações.
exercício de 1974–. – 1975–
r6066156 LAS.B.275/95 Woolwich
178840X

ACCOUNTS. Companies

PRICE, A. St. J.
Understand your accounts. – Revised ed. – London : Kogan Page, 1979
96p : ill
0–85038–205–x X.519/29747
0189286

REID, Walter
The meaning of company accounts. – 3rd ed. – Aldershot : Gower, c1982
ix,353p. – (A Gower workbook). – Includes index
0–566–02284–2 L.49/1297
0189286
– [Another copy: paperback] ISBN 0–566–02285–0 L.49/1340

ACCOUNTS. Companies. Great Britain

BIRD, Peter, 1934–
Understanding company accounts. – London : Pitman, 1979
vii,111p : ill. – Bibliography: p.93–94. – Includes index
0–273–01289–4 X.519/29434
1106805

FANNING, David
Company accounts. – London : Allen & Unwin, 1984
223p. – Bibliography: p219–220. – Includes index
0–04–332092–9 X.520/33749
1106805

FLINT, David
A true and fair view in company accounts. – London : Published for the Institute of Chartered Accountants of Scotland by Gee, 1982
vi,47p. – Bibliography: p46–47
0–85258–223–4 X.529/49556
1106805

JONES, Frank H. (Frank Horace)
Guide to company balance sheets and profit & loss accounts. – 8th ed., revised. – Bournemouth : Barkeley Book Co., 1977
xxii,640p. – Includes index
0–900132–08–6 X.529/31434
1106805

WARREN, Roy
How to understand and use company accounts. – London : Business Books, 1983
xxii,221p : ill. – Includes index
0–09–145890–0 X.520/30424
1106805
– [Another copy: paperback.] ISBN 0–09–145891–9 X.520/30425

WOOD, E. G. (Edward Geoffrey)
Practical exercise in interpretation of company accounts. – [Sheffield] : [Centre for Innovation and Productivity, Sheffield City Polytechnic], [1984?]
[14]leaves
0–86339–049–8 X.525/9550
1106805

ACCOUNTS. Companies. Great Britain
Added value statements

GRAY, Sidney
Value added reporting. – London : Association of Certified Accountants, c1980
71p. – Bibliography: p64–71
0–900094–39–7 X.525/5050
1447742

MORLEY, Michael Francis
The Value Added Statement. – London : Gee and Co. for the Institute of Chartered Accounts of Scotland, 1978
x,150p. – Bibliography: p.149–150
0–85258–162–9 X.519/27950
1447742

ACCOUNTS. Companies. Great Britain
Law

CHECK schedules for annual accounts and directors' reports. – St Albans : Fiscal, [1982?]
24p
B83–00628 X.200/39170
0271853

COMPANIES accounts check list / Peat, Marwick, Mitchell & Co.. – Croydon : Tolley, c1981
1v.(loose–leaf)
0–85459–050–1 X.525/5464
0271853

COMPANIES accounts check list no.1 / Peat, Marwick, Mitchell & Co.. – Croydon : Tolley, c1982
24p
0–85459–066–8 X.622/14181
0271853

COMPANIES accounts check list no.2 / Peat, Marwick, Mitchell & Co.. – Croydon : Tolley, c1982
29p
0–85459–073–0 X.622/14180
0271853

FORM and content of company accounts / Coopers & Lybrand. – London : Financial Training, 1982
vii,227p. – Includes index
0–906322–14–6 X.520/31458
0271853

GREAT BRITAIN
[Companies Acts]. The Companies Acts. – London : Institute of Chartered Accountants in England and Wales, 1984
148p
0–85291–377–x X.525/10079
0271853

JONES, Frank H. (Frank Horace)
Accounting requirements of companies. – 11th ed. – Bournemouth : Barkeley Book Co., 1977
47p. – Includes index
0–900132–07–8 X.519/26706
0271853

STILLING, P. J.
Company accounting requirements. – London : Oyez, 1981
78p
0–85120–619–0 X.520/26435
0271853

Company accounting requirements. – Rev. reprint. – London : Oyez Longman, 1983
78p
0–85120–755–3 X.520/31025
0271853

TOUCHE, Anthony George
Accounting requirements of the Companies Acts. – 2nd ed. (including the accounting requirements of the 4th Directive, Statements of Standard Accounting Practice, and the 1979 Stock Exchange Listing Agreement). – London : Butterworth, 1979
iv,172p
0–406–40203–5 X.200/32854
0271853

ACCOUNTS. Companies. Great Britain
With banks. Information. Requesting by auditors

BANK reports for audit purposes. – [England?] : [s.n.], [1982]
14p. – (Auditing guideline ; 401)
B83–04280 X.529/51273
2818809

ACCOUNTS. Companies. Great Britain
With banks. Law

MATHER, L. C.
The accounts of limited company customers. – 5th rev. ed. – London : Waterlow, 1983
184p. – Includes index
0–900791–83–7 X.529/57651
2316528

MATHER, Leonard Charles
The accounts of limited company customers. – 4th (revised) ed. – London : Waterlow, 1978
160p. – Includes index
0–900791–41–1 X.529/38337
2316528

ACCOUNTS. Companies. Italy
Law, to date

Il BILANCIO per gli anni '80. – Roma : Angeli, c1981
147p. – (Collana di economia e direzione d'impresa ; 7)
88–204–2120–8 X.200/45536
9155120

ACCOUNTS. England. Exchequer
1163–1214. Pipe rolls published by Pipe Roll Society. Introductions – Indexes

LIBER memorialis. – London : Pipe Roll Society, 1976
x,220p,plate. – (Publications : new series / Pipe Roll Society ; vol.41). – Bibliography: p.33–34
B77–24649 2082.b Reading Room
1185489
– [Another copy] X.800/26840

ACCOUNTS. Evans, David, b. ca. 1825. Shipping. Freight transport. Mawddach River. Gwynedd
1846–1869 – Biographies

LLOYD, Lewis
Sails on the Mawddach. – Harlech : L. Lloyd, c1981
48p : ill. – (Maritime Merioneth ; 7)
0–9504022–1–4 X.808/35517
2618265

ACCOUNTS. Farms. Agriculture. East Midlands. England
1708–1950 – Catalogues

UNIVERSITY OF NOTTINGHAM. Manuscripts Department
Farm records c.1708–1950. – [Nottingham] : University of Nottingham, Manuscripts Department, [c1984]
7p
B85–02916 X.805/6068
321172x

ACCOUNTS. Government departments. Russia
to 1917
PRIKHODO-RASKHODNYE knigi moskovskikh prikazov 1619-1621gg. / [sostavil: S.B. Veselovskii]. - Moskva : Nauka, 1983
477p
r6489583 X.525/8607
900968x

ACCOUNTS. Groups. Companies. Great Britain
ROBERTSON, A. T. (Alan T)
Group accounts. - London : Gee, 1980
182p : ill. - Includes index
0-85258-190-4 X.529/39532
2350904

ACCOUNTS. Historical sources. America, to 1645
Polish accounts, 1505-1645
WRÓBLEWSKI, Romuald
Znajomość Ameryki w Polsce okresu Odrodzenia. - Warszawa : Państwowe Wydawnictwo Naukowe, 1977
152p : ill. - Bibliography : p.145-150
r6016150 X.700/19494
1501100

ACCOUNTS. Historical sources. Royal residences. Cracow. Poland
BONER, Jan
Rachunki wielkorządowe Jana Bonera 1558. - Kraków : Nakładem Ministerstwa Kultury i Sztuki
xiii,208p. - (Państwowe zbiory sztuki na Wawelu. Źródła do dziejów Wawelu t. VII)
r6044383 AC.7190a/4[tom 7]
1431439

ACCOUNTS. Historical sources. South German merchants, 1288-1370. Tyrol
Tyrolean accounts
BASTIAN, Franz
Oberdeutsche Kaufleute in den älteren Tiroler Raitbüchern (1288-1370). - Aalen : Scientia, 1973
vii, 193 p.. - (Schriftenreihe zur bayerischen Landesgeschichte ; Bd 10)
3-511-06510-0 X.800/11717
0873586

ACCOUNTS. Housing associations. England
Auditing
GRAY, Desmond J.
Housing association accounts and their audit. - London : Institute of Chartered Accountants in England and Wales, 1980
xvi,221p. - (Industry accounting and auditing guides). - Includes index
0-85291-260-9 X.520/23031
2359936

HOUSING associations. - London : Auditing Practices Committee, 1983
36p. - (Auditing guideline) (APC publications ; 5098)
B83-38900 X.529/56745
2359936

ACCOUNTS. Local authorities. England
Auditing. Law
JONES, Reginald
Local government audit law. - London : H.M.S.O., 1981
xxxvii,380p. - Includes index
0-11-751486-1 X.200/37527
2607913

ACCOUNTS. Merchants. Ancient Mesopotamia
SNELL, Daniel C.
Ledgers and prices. - New Haven ; London : Yale University Press, c1982
xx,282p,xliiplates : ill. - (Yale Near Eastern researches ; 8). - Bibliography: p279-282
0-300-02517-3 X.520/27793
2752891

ACCOUNTS. Mildmay, Benjamin, Earl Fitzwalter. England
1724-1754
EDWARDS, Arthur Charles
The account books of Benjamin Mildmay, Earl Fitzwalter. - London : Regency Press, 1977
[4],213,[11]p,plate : ill. - Includes index
0-7212-0480-5 X.809/41781
1075535

ACCOUNTS. Nationalised industries. Great Britain
LIKIERMAN, Andrew
The reports and accounts of nationalised industries. - London : H.M.S.O., 1979
iii,31p. - (Handbooks / Civil Service College ; no.20)
0-11-630426-x BS.44/330
2317850

ACCOUNTS. Petre, John, Baron Petre. England
Collections: Essex Record Office
EDWARDS, Arthur Charles
John Petre. - London : Regency Press, 1975
156p,[4]p of plates : ill. - Includes index
0-7212-0431-7 X.809/40715
0606448

ACCOUNTS. Taxpayers. Great Britain
Investigation by Great Britain. Board of Inland Revenue
READER, M. (Michael)
Tax investigation handbook. - London : Butterworth, 1981
xvi,158p
0-406-25410-9 X.200/35836
250071x

ACCOUNTS. Unit trusts. Great Britain
UNIT TRUST ASSOCIATION
Unit trust accounts. - London : Department of Trade, 1981
28p
B81-28723 BS.41/1265
252824x
Unit trust accounts. - Repr. with amendments. - London : Department of Trade, 1981
28p
B82-40100 BS.41/1323
252824x

ACCOUNTS. University of Essex
- Serials
UNIVERSITY OF ESSEX
Accounts. - [19-]-. - [Colchester] : The University, [19-]-
v.
Holdings: 1979-80-
r6865252 P.513/1257
2793903

ACCRA. Ghana
Central Accra. Ga. Cultural processes
KILSON, Marion
African urban kinsmen. - New York : St Martin's Press, 1974
122p
r6189240 X.809/45376
0291498

ACCRA. Ghana
Ga. Social change. Role of family life
AZU, Diana Gladys
The Ga family and social change. - Leiden : Afrika-Studiecentrum ; Cambridge : African Studies Centre, 1974
vii,137,fold plate : ill. - (African social research documents ; vol.5). - Bibliography: p.117-119
B77-25401 X.0311/143(5) Woolwich
1190202

ACCRA. Ghana
Trade exhibitions: Ghana International Trade Fair, 2nd, 1971
OFFICIAL catalogue and directory of the Second Ghana International Trade Fair, [held] 1st-14th February 1971. - [Ghana] : [s.n.], [1971]
207 p. : ill
r6031420 X.519/26668
1072862

ACCREDITATION. Higher education institutions. United States
ORLANS, Harold
Private accreditation and public eligibility. - Lexington, Mass. ; London : D.C. Heath, 1975
xxi,263p. - Includes index
0-669-99267-4 X.520/10250
0509299

UNDERSTANDING accreditation / Kenneth E. Young ... [et al.] with the assistance of Ruth Cargo. - San Francisco ; London : Jossey-Bass, 1983
xxiv,502p. - (The Jossey-Bass higher education series). - Bibliography: p459-483. - Includes index
0-87589-570-0 X.520/34289
0509299

ACCREDITATION. Testing laboratories
- Serials
NATLAS newsletter / National Testing Laboratory Accreditation Scheme. - Issue no.1 (Nov. 1983)-. - Teddington : National Physical Laboratory, 1983-
v.
B85-03971 P.625/494
2558599

ACCREDITATION. Testing laboratories. Great Britain
NATIONAL PHYSICAL LABORATORY
The accreditation of test houses. - Teddington : The Laboratory, 1979
[2],14p
B79-30043 BS.424/182
1946455

ACCREDITATION. Welfare workers. Great Britain
MALHERBE, Madelaine
Accreditation in social work. - London : Central Council for Education and Training in Social Work, [1979]
89,4p
B83-00863 X.525/6344
2807041

ACCRETION. Binary stars
FRANK, J.
Accretion power in astrophysics. - Cambridge : Cambridge University Press, 1985
[422]p : ill. - (Cambridge astrophysics series). - Includes bibliography and index
0-521-24530-3 SRL
305456x

ACCRINGTON. Lancashire
ca 1880-1936
ACCRINGTON as it was / [compiled] by Brian Ashton. - Nelson : Hendon Publishing Co., 1975
[46]p : of ill
0-902907-58-1 Cup.1281/180
0612731

ACCRINGTON REGION. Lancashire
Tram services, 1886-1932
TRAMWAY treasury / [compiled] by Trevor Dawson. - [Accrington] : [The compiler], 1975
65p : ill
0-9505310-0-6 X.619/17165 Woolwich
0757403

ACCRINGTON WRITERS. Lancashire writers. English literature
1900-1970 - Anthologies
An ACCRINGTON miscellany / compiled by Ronald Y. Digby and Alice Miller ; edited by J.C. Goddard. - [Accrington] : Lancashire County Council Library and Leisure Committee, 1975
243p,[12] leaves of plates : ill. - Includes index
0-902228-21-8 X.708/20402
0668699

ACCRUAL. Capital gains. Great Britain
Accrual & realisation, 1965-1975
KING, J. R.
Realisations and accruals of capital gains (with particular reference to company securities). - London : Central Division, Inland Revenue, 1978
[1],29p : ill. - (Government Economic Service working paper ; no.15)
B79-27050 BS.44/505(15)
1934228

ACCULTURATION
See also
Assimilation. Racial groups

ACCULTURATION. Bronze Age settlements. Western Europe
Archaeology - Conference proceedings
ATLANTIC COLLOQUIUM (4th : 1975 : Ghent)
Acculturation and continuity in Atlantic Europe, mainly during the Neolithic period and the Bronze Age. - Brugge : De Tempel, 1976
3-310 p., 2 fold leaves of plates : ill. - (Dissertationes archaeologicae Gandenses ; vol 16). - Includes bibliographies
r6010129 Ac.2647/5[16]
0768553

ACCULTURATION. German Jews. Germany
1800-1923. Effects of immigration of Eastern European Jews
ASCHHEIM, Steven E.
Brothers and strangers. - Madison ; London : University of Wisconsin Press, 1982
xiv,331p : ill. - Bibliography: p307-323. - Includes index
0-299-09110-4 X.800/40730
3263444

ACCULTURATION. Migrants
to date
ABOU, Selim
L'identité culturelle. - Paris : Éditions Anthropos, 1981
235p
2-7157-1025-9 X.529/52304
8913633

ACCULTURATION. Neolithic settlements. Western Europe
Archaeology - Conference proceedings
ATLANTIC COLLOQUIUM (4th : 1975 : Ghent)
Acculturation and continuity in Atlantic Europe, mainly during the Neolithic period and the Bronze Age. - Brugge : De Tempel, 1976
3-310 p., 2 fold leaves of plates : ill. - (Dissertationes archaeologicae Gandenses ; vol 16). - Includes bibliographies
r6010129 Ac.2647/5[16]
0768561

ACCUMULATION. Capital
1492-1789
FRANK, Andre Gunder
World accumulation, 1492-1789. - London : Macmillan, 1978
303p : 1 ill. - Bibliography: p.273-289. - Includes index
0-333-23883-4 X.529/33237
145885x

ACCUMULATION. Capital
By capitalist countries, to 1930 related to dependency of developing countries
FRANK, Andre Gunder
Dependent accumulation and underdevelopment. - London : Macmillan, 1978
xx,226p : 1 ill. - Bibliography: p.209-219. - Includes index
0-333-23950-4 X.529/33199
1351419

ACCUMULATION. Capital. Brazil
related to dependency
EVANS, Peter, 1944-
Dependent development. - Princeton ; Guildford : Princeton University Press, 1979
xviii,362,[1]p. - Bibliography: p.331-351. - Includes index
0-691-07606-5 X.529/33639
1918982

ACCUMULATION. Capital. Capitalist countries
HARRIS, Donald J.
Capital accumulation and income distribution. - London : Routledge and Kegan Paul, 1978
[1],ix,313p : ill. - Bibliography: p.297-307. - Includes index
0-7100-8923-6 X.520/13128
152108x

ACCUMULATION. Capital. Developing countries. Africa
POL'SHIKOV, P. I. (Petr Ivanovich)
Capital accumulation and economic growth in developing Africa. - Moscow : Progress, c1981
206p. - (Problems of the developing countries)
0-7147-1738-x X.529/48520
2755645

ACCUMULATION. Capital. Mexico
Effects of private investment
FITZGERALD, Edmund Valpy Knox
The fiscal deficit and development finance. - Cambridge : Centre of Latin American Studies, University of Cambridge, 1979
[3],28p. - (Working papers / University of Cambridge. Centre of Latin American Studies ; no.35). - Bibliography: p.25-28
B80-30930 X.525/3492
2243210

ACCUMULATION. Capital. United States
related to imperfect competition, 1869-1939
STEINDL, Josef
Maturity and stagnation in American capitalism. - [1st ed. reprinted]. - New York ; London : Monthly Review Press, 1976
xix,248p : ill. - Includes index
0-85345-318-7 X.529/34179
1992511

ACCUMULATION. Paperwork. Office practices
Control measures
GROSSMAN, Lee
Fat paper. - New York ; London : McGraw-Hill, 1976
ix,160p : ill. - Includes index
0-07-024981-4 X.529/32224
1409972

Fat paper. - New York ; London : McGraw-Hill, 1978
viii,160p : ill. - Includes index
0-07-024983-0 X.519/29549
1409972

ACCUMULATION. Stocks. Consumer goods. Eastern Europe
related to discretionary expenditure by consumers
ANDERSON, Evan Eugene
The growth of income, consumer discretion, and the accumulation of stocks in socialist economics. - Oxford : St Antony's College, Centre for Soviet and East European Studies, [1975]
[1],30[i.e. 33]p : ill. - (Papers in East European economics / St Antony's College. Centre for Soviet and East European Studies ; 45). - Bibliography: p.25
0-904908-05-4 X.0515/69(45)
0687863

ACCUMULATORS
See Storage batteries

ACCURACY
See also
Errors
Validation

ACCURACY. Construction. Buildings
RYAN, Nicholas M. (Nicholas Murth)
A survey of building accuracy. - Dublin : An Foras Forbartha.
3: Surface regularity of concrete floors. - 1984 : ill
1-85053-005-x X.805/6354
2234475

ACCURACY. Dimensions. Building components
Control
RYAN, Nicholas Murth
Accuracy and dimensional control in building. - Dublin : National Institute for Physical Planning and Construction Research, 1980
[4],32p : ill. - Bibliography: p.32
0-906120-33-0 SRL
2316897

ACCURACY. Electric measuring instruments
Effects of electrical noise
VAN DER ZIEL, Aldert
Noise in measurements. - New York ; London : Wiley, 1976
xi,228p : ill. - Includes index
0-471-89895-3 SRL
1145134

ACCURACY. Government propaganda on nuclear weapons. Educational institutions. Great Britain
SMITH, Andrew, 1943-
Not the whole truth. - [Coalville] : [The Commission], [1983]
37p
B83-40836 X.529/57035
2963825

ACCURACY. Illustrations, ca 1130-ca 1940. Antiquities
PIGGOTT, Stuart
Antiquity depicted. - London : Thames and Hudson, 1978
64p : ill. - (Walter Neurath memorial lectures ; 10)
0-500-55010-7 X.429/10484 Woolwich
1810928

ACCURACY. Measuring instruments. Engineering
Testing
HAYWARD, A. T. J.
Repeatability and accuracy. - London : Mechanical Engineering Publications, 1977
63p : ill. - Bibliography: p.56. - Includes index
0-85298-351-4 SRL
120193x

ACCURACY. Setting out. Construction sites
Testing
RYAN, Nicholas M. (Nicholas Murth)
Accuracy of setting out in construction work. - Dublin : An Foras Forbartha, 1981
iv,18p : ill
0-906120-45-4 X.525/4929
256341x

ACCURACY. Statistical mathematics
GETTING it right / [Schools Council Project on Statistical Education]. - Slough : Published for the Schools Council by Foulsham Educational, c1981
19p : ill. - (Statistics in your world. [Level 2])
0-572-01078-8 X.529/41078
2393638
- [Teachers' notes] X.529/41077

ACCUSED PERSONS. Soviet Union
Innocence, to date. Legal aspects
LARIN, A. M. (Aleksandr Mikhaĭlovich)
Prezumpt͡sii͡a nevinovnosti. - Moskva : Nauka, 1982
150p. - (Konstitut͡sii͡a SSSR / Akademii͡a nauk SSSR)
r6464336 X.208/9008
9039600

ACDA
See United States. Arms Control and Disarmament Agency

ACEH. Indonesia
Rural regions. Economic conditions. Effects of development of agricultural industries
GIBBONS, David S.
Agricultural modernization, poverty and inequality. - Farnborough, Hants. : Saxon House, 1980
xi,225p : ill. - Bibliography: p.213-219. -Includes index
0-566-00331-7 X.329/13187 Woolwich
1933558

ACER PSEUDO-PLATANUS
See Sycamore trees

ACES. Fighter aeroplanes
to 1975. Air operations
SHORES, Christopher
Fighter aces. - London : Hamlyn, 1975
160p : ill(some col). - Bibliography: p.158. - Includes index
0-600-30230-x X.802/10100
058861x

ACES. French fighter aeroplanes
Air operations during World War I
PORRET, D.
Les "AS" français de la Grande Guerre. - Paris : CEDOCAR, [1983]
2v. : ill
2-7170-0741-5 X.800/36686
9032851

ACETABULARIA
- Conference proceedings
INTERNATIONAL MEETING ON ACETABULARIA (4th : 1976 : University of Massachusetts)
Progress in acetabularia research. - New York ; London : Academic Press, 1977
xv,341p : ill. - Includes index
0-12-763750-8 SRL
1520040

ACETABULARIA
Cells. Development - Conference proceedings
ROUND TABLE ON ACETABULARIA (1978 : Moscow)
Developmental biology of acetabularia. - Amsterdam ; Oxford : Elsevier/North Holland Biomedical Press, 1979
xxiv,312p : ill. - (Developments in cell biology ; vol.3). - Includes index
0-444-80098-0 SRL
1878239

ACETALDEHYDE MANUFACTURING INDUSTRIES. Minamata. Japan
Chisso Corporation. Role in mercury poisoning of inhabitants, 1956-1973
SMITH, W. Eugene
Minamata. - London : Chatto and Windus, 1975
192p : ill
0-7011-2131-9 Cup.1285/11
0521450

ACETAMIDE
See also
Fluoroacetamide

ACETATES
See also
Sodium fluoroacetate

ACETYLCHOLINE. Man
Role in nervous diseases
BRAIN acetylcholine and neuropsychiatric disease / edited by Kenneth L. Davis and Philip A. Berger. - New York ; London : Plenum Press, 1979
xvi,597p : ill. - Includes bibliographies and index
0-306-40157-6 SRL
1949543

ACETYLCHOLINE. Neurons. Mammals
Biosynthesis
TUČEK, S.
Acetylcholine synthesis in neurons. - London : Chapman and Hall [etc.], 1978
xiii,259p : ill. - Bibliography: p.199-254. - Includes index
0-412-15030-1 SRL
1585290

ACETYLENE MANUFACTURED BY BRITISH OXYGEN COMPANY. Great Britain
Dissolved acetylene manufactured by British Oxygen Company. Prices. Increases

GREAT BRITAIN. Price Commission
BOC Limited - compressed permanent gases and dissolved acetylene sold in cylinders, cylinder rentals and fixed charges .... - London : H.M.S.O., [1979]
vii,42p : 2 ill. - ([1978-79 H.C.] 223)
0-10-222379-3 BS.416/16(32)
1887629

ACETYLENES
See Alkynes

ACEVEDO DÍAZ, EDUARDO. Uruguayan writers. Fiction in Spanish - Biographies

DEUS, Sergio
Eduardo Acevedo Díaz. - Montevideo : Acali, 1978
261p
r6161062 X.989/86843
2451395

ACEVEDO DÍAZ, EDUARDO. Uruguayan writers. Fiction in Spanish - Catalogues

EDUARDO ACEVEDO DÍAZ 1851-1921 (Exhibition) (1981 : Montevideo)
Eduardo Acevedo Díaz, 1851-1921. - Montevideo : La Biblioteca, 1981
86p,plates
r6708493 X.950/41448
8158908

ACEVEDO DÍAZ, EDUARDO. Uruguayan writers. Fiction in Spanish - Correspondence, diaries

ACEVEDO DÍAZ, Eduardo
[Correspondence. Selections]. Correspondencia familiar e intima de Eduardo Acevedo Díaz, 1880-1898. - Montevideo : Biblioteca Nacional, 1979
66p : ill
r6705727 X.808/36774
8158908

ACEVEDO, EDUARDO. Politics. Uruguay

GIL, Felipe
En torno a Eduardo Acevedo y otros rubros orientales. - Montevideo : Barreiro y Ramos, 1980
22p
r6705573 X.809/52023
8461244

ACEVEDO, EVARISTO. Journalism. Spain - Biographies

ACEVEDO, Evaristo
Un humorista en la España de Franco. - Barcelona : Editorial Planeta, 1976
241p : ill. - (Espejo de España. Serie La España de la posguerra ; no.24)
84-320-5624-3 X.981/11784
139102X

ACHACACHI. Bolivia
Peasants. Political movements, to date

ALBÓ, Javier
Achacachi. - La Paz : CIPCA, 1979
173p. - (Cuaderno de investigación ; 19)
r6703638 X.800/31624
8770514

ACHAEMENIAN ANTIQUITIES. Pasargadae. Ancient Iran
Archaeology, 1961-1963

STRONACH, David
Pasargadae. - Oxford : Clarendon Press, 1978
xiv,326p,[8] fold leaves of plates,192p of plates : ill. - Bibliography: p.xi-xii. - Includes index
0-19-813190-9 Cup.1285/296
0733431

ACHAEMENIAN SEALS
Inscriptions in Old Persian

ALTPERSISCHE Siegel-Inschriften / [bearbeitet von] Rüdiger Schmitt. - Wien : Österreichische Akademie der Wissenschaften, 1981
49,[4]p of plates : ill. - (Sitzungsberichte / Österreichische Akademie der Wissenschaften. Philosophisch-historische Klasse ; Bd.381) (Veröffentlichungen der Iranischen Kommission ; Bd 10)
3-7001-0390-5 Ac.810/6a[Bd.381]
8750467

ACHALASIA. Cardia. Oesophagus. Man

HARLEY, Hugh Rosborough Swanzy
Achalasia of the cardia. - Bristol : J. Wright, 1978
[5],174p : ill. - Bibliography: p.155-165. - Includes index
0-7236-0488-6 SRL
1284312

ACHEBE, CHINUA. Nigerian writers. Fiction in English

CARROLL, David, 1932-
Chinua Achebe. - London : Macmillan, 1980
ix,192p. - (Macmillan Commonwealth writers series). - Bibliography: p187-190. - Includes index
0-333-25574-7 X.958/7857
1196618
- [Another copy: paperback.] ISBN 0-333-25575-5 X.958/9236

CRITICAL perspectives on Chinua Achebe / edited by C.L. Innes, Bernth Lindfors. - London : Heinemann Educational, 1979
[4],316p. - Bibliography: p.294-310. - Includes index
0-435-91612-2 X.909/43741
1925261

DUNN, T. A. (Thomas Alexander)
Things fall apart. - Harlow : Longman, 1981
64p. - (York notes ; 96)
0-582-78204-x X.950/4091
2406187

KILLAM, G. D. (Gordon Douglas)
The writings of Chinua Achebe. - Revised ed. - London : Heinemann Educational, 1977
viii,132p. - (Studies in African literature)
0-435-91665-3 X.909/41675
1196618

PROBYN, Clive T.
A man of the people. - Harlow : Longman, 1981
64p. - (York notes ; 116)
0-582-78179-5 X.950/6888
2546523

WREN, Robert M. (Robert Meriwether)
Achebe's world. - Harlow : Longman, 1981, c1980
xxv,166p. - (Longman studies in African literature). - Bibliography: p140-159. - Includes index
0-582-64251-5 X.950/9737
1196618

ACHEBE, CHINUA. Nigerian writers. Fiction in English
Arrow of God

CARROLL, David, 1932-
Arrow of God. - Harlow : Longman, 1980
80p. - (York notes ; 92). - Bibliography: p74
0-582-78145-0 X.950/4191
2405768

ACHEBE, CHINUA. Nigerian writers. Fiction in English
Thing fall apart - For West African students

MBOCK, Charly Gabriel
'Le monde s'effrondre' de Chinua Achebe. - Yaoundé : Éditions Buma Kor, [1978]
108p. - (Chemin facile)
r6104224 X.908/90021
2283514
- [Another copy.] X.907/20759

WREN, Robert M. (Robert Meriwether)
Chinua Achebe, 'Things fall apart'. - [London] : Longman, 1980
vi,56p. - (Longman guides to literature)
0-582-60109-6 X.908/43632
2283514

ACHEBE, CHINUA. Nigerian writers. Fiction in English
Things fall apart

TURKINGTON KATE
Chinua Achebe, 'Things fall apart'. - London : Edward Arnold, 1977
63,[1]p. - (Studies in English literature ; no.66). - Bibliography: p.63. - Includes index
0-7131-5932-4 011879.de.2/66
1102028

ACHEEN
See Aceh

ACHEH
See Aceh

ACHEL. Belgium
Monasteries: Achelse Kluis

JONG, Domien de
Momenten uit drie eeuwen kluishistorie. - [Achel] : Achelse Kluis, 1973
xxiv,216p : ill. - Includes bibliography
r6330624 X.421/24450
9024573

ACHELSE KLUIS

JONG, Domien de
Momenten uit drie eeuwen kluishistorie. - [Achel] : Achelse Kluis, 1973
xxiv,216p : ill. - Includes bibliography
r6330624 X.421/24450
9024573

ACHERON (SHIP). Steamships
Voyages round New Zealand, 1848-1852

NATUSCH, Sheila
The cruise of the 'Acheron'. - Christchurch, N.Z. ; London : Whitcoulls, 1978
190p,[8]p of plates : ill(some col). - Bibliography: p.175-179. - Includes index
0-7233-0550-1 X.809/44221
1868357
- [Another copy] X.809/44495

ACHEULEAN ANTIQUITIES. Caddington. Bedfordshire

PALEOECOLOGY and archeology of an Acheulian site at Caddington, England / edited by C. Garth Sampson. - Dallas : Department of Anthropology, Institute for the Study of Earth and Man, Southern Methodist University, c1978
x,158p : ill. - Bibliography: p.153-158
r6804836 X.425/2286 Woolwich
8778809

ACHIEVEMENT
See also
Academic achievement
Occupational attainment

ACHIEVEMENT. Man

COOPER, Wendy
Human potential. - Newton Abbot : David and Charles, c1981
223p
0-7153-7925-9 X.325/990 Woolwich
2381559

ACHIEVEMENT. Military self-reliance in developing countries
Role of multinational companies in military equipment industries

TUOMI, Helena
Transnational corporations, armaments and development. - Aldershot : Gower, 1982
xv,312p
0-566-00506-9 X.529/47622
2726033

ACHIEVEMENT. Personal power. Executives. Companies. United States
Use of manipulative behaviour

JONES, William M.
Survival. - Englewood Cliffs ; London : Prentice-Hall, 1981, c1979
261p. - (A Reward book). - Includes index
0-13-879098-1 X.529/42387
2527014

ACHIEVEMENT. Personnel. Industries. Soviet Union
1945-

KOSTROMSKOE sozvezdie / [sostavitel' : A.L. Lipatov]. - ÎAroslavl' : Verkhne-Volzhskoe knizhnoe izdatel'stvo, 1975
219p : ill
r6093573 X.809/53507
8882797

RÎABOV, ÎA. P. (ÎAkov Petrovich)
Iniťsiativa i tvorchestvo mass. - Moskva : Izdatel'stvo politicheskoĭ literatury, 1973
189p,plates
r6465034 X.529/64071
8882797

ACHIEVEMENT. Persons. United States
to date

EPSTEIN, Joseph
Ambition. - New York : Dutton, [1980]
312p. - Bibliography
0-525-05280-1 X.529/55449
9075216

ACHIEVEMENT. Success. Man
Effects of population size

EASTERLIN, Richard A.
Birth & fortune. - London : Great McIntyre, 1980
xi,205p : ill. - Includes index
0-86216-036-7 X.529/38712
2149206

ACHIEVEMENT. Success. Personnel. Companies
Applications of transactional analysis of interpersonal relationships

JONGEWARD, Dorothy
Choosing success. - New York ; Chichester : Wiley, 1978
[5],362p : ill. - Bibliography: p.360. - Includes index
0-471-02884-3 X.510/15981
1768468

ACHIEVEMENT MOTIVATION
related to economic development

MCCLELLAND, David Clarence
The achieving society. – [1st ed. reprinted]. – New York : Irvington ; New York ; London : Distributed by Wiley, 1976
xv,512,A–G p : ill. – Bibliography: p.439–449. – Includes index
0–470–01397–4 X.520/11866
1185721

ACHIEVEMENT MOTIVATION. Entrants. Higher education institutions. Great Britain

BEARD, Ruth Mary
Motivating students. – London : Routledge and Kegan Paul, 1980
viii,107p. – (Routledge education books). – Bibliography : p.92–102. – Includes index
0–7100–0594–6 X.529/36282
2109646

ACHIEVEMENT MOTIVATION. Man

ATKINSON, John W. (John William)
Personality, motivation, and action. – New York : Praeger, 1983
xvi,432p : ill. – (Centennial psychology series). – Bibliography: p397–417
0–03–060541–5 X.520/31667
1491008

ATKINSON, John William
Personality, motivation and achievement. – [Abridged ed.]. – Washington, D.C. ; London : Hemisphere ; New York ; London : Wiley, 1978
x,272p : ill. – Bibliography: p.243–254. – Includes index
0–470–99336–7 X.519/28166
1491008

RAYNOR, Joel O.
Motivation, career striving and aging. – Washington [D.C.] ; London : Hemispher, c1982
x,406p : ill. – Bibliography: p379–392. – Includes index
0–07–051274–4 X.520/28342
1491008

SPENCE, Janet T. (Janet Taylor)
Achievement and achievement motives. – San Francisco : Freeman, c1983
x,381p. – Includes bibliographies
0–7167–1396–9 X.520/31240
1491008

ACHIEVEMENT MOTIVATION. Man
– Conference proceedings

ACHIEVEMENT motivation. – New York ; London : Plenum Press, 1980
ix,471p : ill. – Includes bibliographies and index
0–306–40549–0 X.520/22102
1491008

ACHIEVEMENT MOTIVATION. Man
– Serials

ADVANCES in motivation and achievement. – Vol.1 (1984)–. – Greenwich, Conn. ; London : JAI, 1984–
v. : ill
B85–24377 P.861/823
1491008

ACHIEVEMENT MOTIVATION. Man
Measurement

The ACHIEVEMENT motive / by David C. McClelland ... [et al.]. – [1st ed. reprinted]. – New York : Irvington ; New York ; London : Distributed by Wiley, 1976
[12],xxiii,386p,[2]p plates : ill. – (The century psychology series). – Bibliography: p.377–384. – Includes index
0–470–01390–7 WP.9/71
1057995

SMITS, Bernardus Wilhelmus Gerardus Maria
Motivatie en meetmethode. – Lisse : Swets & Zeitlinger, 1982
xi,316p. – Includes bibliography
90–265–0402–0 X.520/29449
8933731

ACHIEVEMENT MOTIVATION. Man
Persistence

NYGARD, Roald
Personality, situation, and persistence. – Oslo : Universitetsforlaget ; Henley on Thames : [Distributed by] Global Book Resources [etc.], 1977
282p : ill. – (Norwegian studies in education ; no.6). – Bibliography: p.255–266. – Includes index
B77–28352 X.519/27244
1207741

ACHIEVEMENT MOTIVATION. Students. Primary schools. Wales
Effects of family life compared with effects of family life on achievement motivation in primary school students in Malaysia

WAN-RAFAEI, A. R.
Achievement motivation in Malaysia and Great Britain. – Cardiff : [UWIST Department of Applied Psychology], 1979
17leaves : ill. – (Occasional paper / UWIST Department of Applied Psychology ; no.10). – Bibliography: leaf 9
B81–24801 X.525/4560
2496836

ACHIEVEMENT MOTIVATION. Women
related to role of women in society

WOMEN and achievement / Martha Tamara Shuch Mednick, Sandra Schwartz Tangri, Lois Wladis Hoffman, editors. – Washington ; London : Hemisphere ; New York ; London : Distributed by Wiley, 1975
xv,447p : ill. – Includes bibliographies and index
0–470–59025–4 X.520/10571
0624179

WOMEN and achievement / Martha Tamara Shuch Mednick, Sandra Schwartz Tangri, Lois Wladis Hoffman, editors. – Washington, D.C. ; London : Hemisphere ; New York ; London : Distributed by Wiley, [1977]
xv,447p. – Includes bibliographies and index
0–470–99150–x X.510/15619
0624179

ACHILL ISLAND DIALECT. Mayo (County) dialect. Irish language
– Collections

The IRISH of Achill, Co. Mayo / [compiled by] Gerard Stockman. – Belfast : Institute of Irish Studies, The Queen's University of Belfast, 1974
[7],ix,466p. – (Studies in Irish language and literature ; vol.2). – Includes index
B79–20862 X.981/12408
1904574

ACHILLES. Characters. Homer. Epic poetry in Greek

MACCARY, W. Thomas (William Thomas)
Childlike Achilles. – New York ; Guildford : Columbia University Press, 1982
xiii,276p : ill. – Includes index
0–231–05504–8 X.950/19942
2882469

ACHILLES. Special subjects. Ancient Greek visual arts & ancient Roman visual arts

KEMP-LINDEMANN, Dagmar
Darstellungen des Achilleus in griechischer und römischer Kunst. – Bern : Lang, 1975
V, 287 p., 1 leaf of plates : ill. – (Archäologische Studien ; No. 3)
3–261–01770–8 X.0519/647(3)
085204X

ACHILLES (SHIP). Cruisers. New Zealand. Royal New Zealand Navy
Naval operations during World War 2

HARKER, Jack S. (Jack Stanley)
HMNZS Achilles. – Auckland ; London : Collins, 1980
264p,[24]p of plates : ill. – Bibliography: p259. – Includes index
0–00–216961–4 X.809/46360
2335107

ACHIMOTA. Ghana
Brewing industries: Tata Brewery, to 1973

BENIBENGOR-BLAY, John
The story of Tata. – Aboso : Benibengor Book Agency, [1976]
3–44p : 1 ill
r6091778 X.708/21899
1893157

ACHIN
See Aceh

ACHMORE. Isle of Lewis. Western Isles. Scotland
Megalithic circles. Archaeology

PONTING, Margaret
Achmore stone circle. – Callanish : G. & M. Ponting, 1981
57p,[1]folded leaf of plates : ill. – Bibliography: p57
0–9505998–3–2 X.425/3099 Woolwich
2628910

ACHNASHELLACH. Ross and Cromarty (District). Highland Region. Scotland
to 1981

BANKS, George, 1912–
Achnashellach saga. – [Scotland] : G. Banks, c1982
36p : col.ill
B85–22612 X.809/64497
3284425

ACHOLI. Uganda
to date

OCITTI, J. P.
African indigenous education. – Nairobi : East African Literature Bureau, 1973
x,116p
r6808341 X.529/53445
8916853

ACHOLI LITERATURE
See also
Poetry in Acholi

ACHTERBERG, GERRIT. Poetry in Dutch
GERRIT Achterberg. – 's-Gravenhage : Nederlands Letterkundig Museum en Documentatiecentrum, 1981
54p : ill. – (Schrijvers prentenboek ; 21)
r6306605 11876.cc.71/21
8700060

HOUWINK, Roel
Het raadsel "Achterberg". – Brugge : Orion, c1973
106 p.. – (Orions literair atelier)
90–264–3520–7 X.909/33394
0941743

JAGENEAU, Lambert
O gij die ik had omgebracht. – 2de, herz. druk. – [Antwerp] : Soethoudt, 1983
87p : ill
90–637–2067–x X.958/25813
8700060

ROEST, Hans
Schaatsenrijder. – Heemstede : Lojen Deur, 1982
29p : ill
r6330961 Cup.408y.32
8700060

ACHTERBERG, GERRIT. Poetry in Dutch
– Biographies

HAZEU, Wim
Gerrit Achterberg. – Apeldoorn : Walraven, [1982]
102p. – (Zilverschoonreeks ; nr.12)
90–604–9340–0 X.958/16308
8700060

ACHTERNBUSCH, HERBERT. Directing. German cinema films

HERBERT Achternbusch / mit Beiträgen von Wolfgang Jacobsen ... [et al.]. – München : Hanser, c1984
190p. : ill. – (Reihe Film ; 32)
3–446–14133–2 X.629/25891
8705798

ACHTERNBUSCH, HERBERT. Fiction in German

HERBERT Achternbusch / herausgegeben von Jörg Drews. – Frankfurt am Main : Suhrkamp, 1982
318p,plates. – (Suhrkamp Taschenbuch ; 2015. Materialien). – Includes bibliography
3–518–38515–1 X.958/25457
8156220

ACHTRUP. West Germany
Population. Language

RYGE PETERSEN, Søren
Dansk eller tysk?. – Flensborg : Studieafdelingen ved Dansk Centralbibliotek for Sydslesvig, 1975
131 p.
87–980193–2–5 X.709/20795
0877042

ACHUALE LEGENDS
Achuale tales & legends – Anthologies

CUENTOS folklóricos de los achual / recopilados por Gerhard Fast Mowitz. – 2a ed. – [Yarinacocha] : Instituto Lingüístico de Verano, 1978
242p. – (Comunidades y culturas peruanas ; no.3)
r6708747 X.958/25676
9124187

ACHUALE TALES
Achuale tales & legends – Anthologies

CUENTOS folklóricos de los achual / recopilados por Gerhard Fast Mowitz. – 2a ed. – [Yarinacocha] : Instituto Lingüístico de Verano, 1978
242p. – (Comunidades y culturas peruanas ; no.3)
r6708747 X.958/25676
9124187

ACHUAR. Ecuador
Cultural processes, to date – Correspondence, diaries

ARNALOT, José
Lo que los Achuar me han enseñado. – Sucua (Morona Santiago), Ecuador : Ediciones Mundo Shuar, 1978
266p,plates : ill
r6708222 X.200/44752
9165002

ACI
See Azione cattolica †

ACID–BASE BALANCE
See also
Acidosis

ACID–BASE BALANCE
Theories

FINSTON, Harman L.
A new view of current acid–base theories. – New York ; Chichester : Wiley, c1982
viii,216p : ill. – (A Wiley-Interscience publication). – Includes index
0–471–08472–7 SRL
2754304

ACID–BASE BALANCE. Man

EASTHAM, R. D. (Robert Duncan)
A guide to water, electrolyte and acid–base metabolism. – Bristol : Wright, 1983
ix,230p : ill. – Includes index
0–7236–0662–5 SRL
051179x

GAMBLE, James L.
Acid–base physiology. – Baltimore ; London : Johns Hopkins University Press, c1982
xv,125p : ill. – Includes index
0–8018–2629–2 SRL
1266535

GARDNER, Michael L. G.
Medical acid–base balance. – London : Baillière Tindall, 1978
viii,125p : ill. – Bibliography: p.117–118. – Includes index
0–7020–0687–4 X.311/8867 Woolwich
1266535

LI, Arthur K. C.
Fluid, electrolytes, acid–base and nutrition. – London : Academic Press, 1980
ix,80p : ill. – Includes index
0–12–448150–7 SRL
051179x

MASORO, Edward Joseph
Acid–base regulation. – 2nd ed. – Philadelphia ; London : Saunders, 1977
viii,169p : ill. – Includes bibliographies and index
0–7216–6144–0 SRL
1266535

ROOTH, Gösta
Acid–base and electrolyte balance. – [Lund] : Studentlitteratur ; London : Distributed by Wolfe Medical, 1975
118p : ill. – Bibliography: p.112–118
0–7234–0678–2 X.619/15188
051179x

STEWART, Peter A.
How to understand acid–base. – London : Edward Arnold, 1981
xii,186p : ill. – Includes index
0–7131–4390–8 SRL
051179x

WILLATTS, Sheila M.
Lecture notes on fluid and electrolyte balance. – Oxford : Blackwell Scientific, 1982
ix,308p : ill. – (Lecture notes). – Bibliography: p296–297. – Includes index
0–632–00862–8 S.R.L.
051179x

ACID–BASE BALANCE. Man
Disorders

ACID–BASE and potassium homeostasis / edited by Barry M. Brenner and Jay H. Stein. – New York ; Edinburgh : Churchill Livingstone, 1978
ix,300p : ill. – (Contemporary issues in nephrology ; vol.2). – Includes bibliographies and index
0–443–08017–8 SRL
1904043

FLUID, electrolyte and acid–base disturbances / edited by R.H. Loblay, D.J. Tiller, W.J. Hensley. – Aylesbury : HM+M, 1976
[6],136p : ill. – Includes index
0–85602–065–6 X.619/17439 Woolwich
1067583

MOYER, Carl A. (Carl Alfred)
Moyer's fluid balance. – 3rd ed. – Chicago ; London : Year Book Medical, c1982
xiv,149p : ill. – Includes index
0–8151–8963–x X.329/16299
1067583

ROSE, Burton David
Clinical physiology of acid–base and electrolyte disorders. – New York ; London : McGraw-Hill, 1977
viii,548p : ill. – Includes index
0–07–053621–x SRL
1067583
– [Another copy] X:329/17525

ACID–BASE BALANCE. Man
Disorders – Conference proceedings

ACID–BASE and electrolyte balance / edited by Allan B. Schwartz, Harvey Lyons. – New York ; London : Grune and Stratton, 1977
xv,320p : ill. – Includes bibliographies and index
0–8089–0991–6 SRL
1269682

ACID–BASE BALANCE. Man
Disorders. Syndromes

GOLDBERGER, Emanuel
A primer of water, electrolyte and acid–base syndromes. – 5th ed. – Philadelphia : Lea and Febiger ; London : [Distributed by] Kimpton, 1975
xi,604p : ill. – Includes bibliographies and index
0–8121–0525–7 SRL
0518581

ACID PRECIPITATION
See Acid rain

ACID PRECIPITATION
International law, to date

VAN LIER, Irene H.
Acid rain and international law. – Toronto : Bunsel, [1981?]
xxii,278p
0–7703–0185–1 OPL
8967822

ACID PROTEINASES. Organisms
– Conference proceedings

CONFERENCE ON ACID PROTEASES – STRUCTURE, FUNCTION, AND BIOLOGY (1976 : University of Oklahoma)
Acid proteases. – New York ; London : Plenum Press, 1977
ix,355p : ill. – (Advances in experimental medicine and biology ; vol.95). – Includes index
0–306–32695–7 SRL
1432222

ACID RAIN

OWENS, Russell V.
Acidification of the environment including acid rain. – Stoke on Trent : Pyramid, 1983
35p. – Bibliography: p35
0–946924–00–7 X.329/17825
2966034

ACID RAIN
– Conference proceedings

ACID deposition / edited by S. Beilke and A.J. Elshout. – Dordrecht ; Lancaster : Reidel, c1983
x,235p : ill. – Includes bibliographies and index
90–277–1588–2 SRL
2966034

ACID RAIN. Europe
Effects on surface water

MASON, B. J. (Basil John)
The current status of research of acidification of surface waters. – London : Royal Society, 1984
23p : ill. – Bibliography: p23
B85–19906 X.622/23202
3272397

ACID RAIN. Europe
Pollution of environment

ELSWORTH, Steve
Acid rain. – London : Pluto, c1984
154p. – Includes index
0–86104–791–5 X.529/64925
3024857

ACID RAIN. Great Britain

UNITED KINGDOM REVIEW GROUP ON ACID RAIN
Acid deposition in the United Kingdom. – Stevenage : Warren Spring Laboratory, 1983
72p : ill
0–85624–323–x SRL
3173887

ACIDAEMIAS. Man
Organic acidaemias

CHALMERS, R. A. (Ronald Aubrey)
Organic acids in man. – London : Chapman and Hall, 1982
xii,523p : ill. – Includes bibliographies and index
0–412–14890–0 SRL
2880261

ACIDIC ROCKS
See also
Granite

ACIDITY. Soils. Flood plains. Rivers. Norfolk Broads

ACID sulphate soils in Broadland / Consultant Environmental Resource Management Ltd. – Norwich : Broads Authority, 1981
36p,2leaves of plates : ill. – Bibliography: p35
B83–36392 X.622/16494
2946114

ACIDOSIS
See also
Metabolic acidosis

ACIDS
See also
Aluminium compounds soluble in acids
Dichlorophenoxyacetic acid
Fluosilicic acid
Hard acids
Inorganic acids
Methychlorophenoxyacetic acid
Nitric acid
Organic acids
Oxalic acid
Pantothenic acid
Phosphoric acids
Soft acids
Sorbic acid
Sulphuric acid
Trichlorophenoxyacetic acid

ACIDS
Experiments

ACIDS and alkalis. – St. Albans : Hart-Davis Educational, 1981, c1980
31p : ill. – (Centre science)
0–247–13123–7 X.629/15712 Woolwich
2495724

ACIDS. Catalysts

CATALYSIS by acids and bases / editors, B. Imelik ... [et al.]. – Amsterdam ; Oxford : Elsevier, 1985
xiv,445p : ill. – (Studies in surface science and catalysis ; 20)
0–444–42449–0 SRL
3233537

ACIDULANTS. Additives of food

ARNOLD, Michael Henry Miller
Acidulants for foods and beverages. – London : Food Trade Press, 1975
vi,137p : ill. – Includes index
0–900379–19–7 SRL
0355119

ACIDURIAS. Man
Organic acidurias

CHALMERS, R. A. (Ronald Aubrey)
Organic acids in man. – London : Chapman and Hall, 1982
xii,523p : ill. – Includes bibliographies and index
0–412–14890–0 SRL
2880253

ACIDURIAS. Man
Organic acidurias – Conference proceedings

SOCIETY FOR THE STUDY OF INBORN ERRORS OF METABOLISM. Symposium (21st : 1983 : Lyon)
Organic acidurias. – Lancaster : MTP, c1984
x,160p : ill. – Includes bibliographies and index
0–85200–875–9 X.322/14569
2880253

ACINONYX
See Cheetahs

ACIREALE. Italy
Town halls: Palazzo Comunale, Acireale. Exhibits: Visual arts. Irony – Catalogues

RASSEGNA INTERNAZIONALE D.ARTE ACIREALE TURISTICO TERMALE (8th : 1974 : Acireale)
Ironia come alternativa. – [s.l.] : [s.n.], [1974]
[69]p.36p.of plates : ill
r6024307 X.410/6005 Woolwich
1026925

ACKERLEY, J. R.. English literature
– Biographies

ACKERLEY, J. R. (Joe Randolph)
My father and myself. – Harmondsworth : Penguin, 1971, c1968
191p
B84–33192 X.958/24574
3146707

ACKERLEY, J. R.. English literature
- Correspondence, diaries

ACKERLEY, J. R. (Joe Randolph)
My sister and myself. - London : Hutchinson, 1982
216p,[8]p of plates. - Includes index
0-09-147020-x X.950/12646
0268763

ACKERMANN (FIRM)
1783-1983

ACKERMANN (Firm)
Ackermann 1783-1983. - London : Ackermann, 1983
256p : ill(some col.). -- Bibliography: p233-236. -
Includes index
0-946186-03-0 X.425/4813
3170098

ACKERMANN-GEMEINDE

IM Geiste des "Ackermann aus Böhmen" / herausgegeben von
Franz Kubin. - Bamberg : Ackermann-Gemeinde, 1981
111p : ill
r6310085 X.529/64260
8861862

ACKLAND, VALENTINE. English literature
- Biographies

ACKLAND, Valentine
For Sylvia. - London : Chatto & Windus, 1985
134p
0-7011-3941-2 X.950/40162
3112845

ACKWORTH. West Yorkshire (Metropolitan County)
- Walkers' guides

CAMPLIN, C. E.
Ackworth Trail. - [Pontefract] : [The author], [1976]
[2],17 leaves,fold plate
0-9505679-0-6 X.709/31172
1096575

ACKWORTH. West Yorkshire (Metropolitan County)
Society of Friends schools: Ackworth School, to 1979

ACKWORTH SCHOOL
So numerous a family. - Ackworth : The School, 1979
60,[3]p : ill(incl 1 col)
0-85245-141-5 X.512/6666
1907719

ACKWORTH OLD SCHOLARS' ASSOCIATION
Members - Directories

ACKWORTH OLD SCHOLARS' ASSOCIATION
List of members. - [Pontefract] : [The Association].
[No. 1]-. - [19-?]-
r6031258 P.521/3097
1055496

ACKWORTH SCHOOL. West Yorkshire (Metropolitan County)
to 1979

ACKWORTH SCHOOL
So numerous a family. - Ackworth : The School, 1979
60,[3]p : ill(incl 1 col)
0-85245-141-5 X.512/6666
1907719

ACLAND (FAMILY). Devon
- Biographies

ACLAND, Lady
A Devon family. - London : Phillimore, 1981
xviii,170p,xvip of plates : ill. - Bibliography:
p157-158. - Includes index
0-85033-356-3 X.809/55715
2936240

ACLI
See Associazione Cristiane Lavoratori Italiani

ACLU
See American Civil Liberties Union †

ACNE. Man
- For adolescents

HOEHN, Gustave H.
Acne can be cured. - New York : Arco, 1978
144p
0-668-04153-6 X.319/19285 Woolwich
1352482

MURRAY, David, 1943-
The anti-acne book. - London : Arlington, 1981
64p : ill. - (Arlington pocket books)
0-85140-282-8 X.329/13892 Woolwich
1352482

NOURSE, Alan Edward
Clear skin, healthy skin. - [New] ed. - New York ; London
: F. Watts, 1978
[7],59p : ill. - (A concise guide). - Includes index
0-85166-661-2 X.320/11119 Woolwich
1352482

ACNE. Man
Therapy. Diet

MOYLE, Alan
Diets to help acne. - Wellingborough : Thorsons, 1979
47p
B80-24511 X.329/12600
217930x

ACOA
See Administrative and Clerical Officers Association

ACOL BIDDING. Contract bridge

BEARD, John
Develop your bridge the Acol way. - Revised ed. -
Plymouth : Develop Your Bridge, 1979
63p : ill
0-906827-00-0 X.619/22215 Woolwich
0202045

COHEN, Ben
All about Acol. - 3rd ed.. - London : Allen and Unwin,
1976
3-264p. - Includes index
0-04-793027-6 X.629/10887 Woolwich
0202045

COHEN, Ben
All about Acol. - 4th ed.. - London : Unwin Paperbacks,
1978
266p. - Includes index
0-04-793032-2 X.619/19014 Woolwich
0202045

Basic Acol. - 3rd revised ed.. - London : Unwin
Paperbacks, 1979
112p
0-04-793038-1 X.629/14343 Woolwich
0202045

Basic Acol. - 4th rev. ed.. - London : Unwin Paperbacks,
1981
112p : ill
0-04-793049-7 X.629/24131
0202045

KLINGER, Ron
Basic bridge. - London : Gollancz in association with
Peter Crawley, c1978
127p. - (Master bridge series)
0-575-02637-5 X.622/9672 Woolwich
0202045

Basic bridge. - London : Ward Lock, 1978
127p. - (Master bridge series)
0-7063-5424-9 X.622/6325 Woolwich
0202045

LEDERER, Rhoda
Acol-ite's quiz. - 3rd ed. - London : Unwin Paperbacks,
1982
197p
0-04-793058-6 X629/19880
0202045

Acol-ite's quiz, including the basic rules of the Acol
system, quiz exercises and competition quizzes. - London
: Unwin Paperbacks, 1978
197p : ill
0-04-793033-0 X.619/19043 Woolwich
1402609

REESE, Terence
Bridge, the Acol system of bidding. - Fully revised ed. -
London : Pan Books, 1978
156p
0-330-25540-1 X.619/19092 Woolwich
0202045

Bridge, the Acol system of bidding. - Fully revised ed. -
London : Hale, 1979
156p
0-7091-7382-2 X.629/12309 Woolwich
0202045

ACOL BIDDING. Contract bridge
Acol competitive bidding

CROWHURST, Eric
Acol in competition. - London : Pelham, 1980
383p. - Includes index
0-7207-1273-4 X.629/14612 Woolwich
2326299

ACOL in competition. - London : Pelham, 1980
383p : ill. - Includes index
0-7207-1525-3 X629/22495
2326299

ACOL BIDDING. Contract bridge
Acol precision bidding

CROWHURST, Eric
Precision bidding in Acol. - London : Pelham, 1974
256p. - Includes index
0-7207-1488-5 X.629/22285
3062252

ACOLI
See Acholi

ACOMB. York. North Yorkshire
Manors: Acomb (Manor). Law courts

ACOMB (Manor)
Court rolls of the manor of Acomb. - [Leeds] : Yorkshire
Archaeological Society.
Vol.2. - 1978
ix,xv,253-468p. - (Record series / Yorkshire
Archaeological Society ; vol.137). - Includes index
B79-22292 Ac.5652/8[vol.137]
190793x

ACOMB (MANOR)
Law courts

ACOMB (Manor)
Court rolls of the manor of Acomb. - [Leeds] : Yorkshire
Archaeological Society.
Vol.2. - 1978
ix,xv,253-468p. - (Record series / Yorkshire
Archaeological Society ; vol.137). - Includes index
B79-22292 Ac.5652/8[vol.137]
190793x

ACONMORE. Hammersmith and Fulham (London Borough)
Young persons, to 19 years. Leisure activities

NOBLE, Paul
Recreational needs. - [London] : Hammersmith & Fulham,
1981
[72]p. - (Research report ; 50)
B82-06132 X.525/5243
260423x

ACORN MICROCOMPUTER SYSTEMS
Acorn Atom microcomputer systems

39 tested programs for the Acorn Atom / edited by Tim
Hartnell. - London : INTERFACE, 1982
48p : ill
0-907563-12-0 X.622/16684
2733331

HARTNELL, Tim
The book of listings. - London : British Broadcasting
Corporation, 1982
156p. - (BBC soft)
0-563-16534-0 X.629/20128
2733331

INGLIS, Jonathan
Simple maths, tables and numbers. - London : Granada,
1984
64p : col.ill. - (Micro mates) (Micro mates ; 1). -
Includes index
0-246-12268-4 X.629/25497
2733331

JOHNSON-DAVIES, David
Practical programs. - Wilmslow : Sigma Technical, c1982
119p : ill. - Bibliography: p119
0-905104-14-5 X.629/17722 Woolwich
2733331

ACORN MICROCOMPUTER SYSTEMS
Acorn Atom microcomputer systems - Serials

ACORN user. - No.1 (July/Aug. 1982)-. - London :
Addison-Wesley, 1982-
v. : ill(some col.)
B83-09838 SRL
2839679

ACORN MICROCOMPUTER SYSTEMS
Acorn Atom microcomputer systems. Programming - Serials

[INTERFACE (London : 1981)]. Interface. - Vol.1, issue 7
(Mar. 1981)-. - London : National ZX80 & ZX81 Users'
Club, 1981-
v. : ill
B82-15773 P.623/1174
2698161

ZX computing. - Vol.1, no.1 (summer 1982)-. - London :
Argus Specialist Publications, 1982-
v. : ill
B83-04809 SRL
2698161

ACORN MICROCOMPUTER SYSTEMS
Acorn Electron microcomputer systems

AUGHTON, Jeff
Invaluable utilities for the Electron. - London : Pan,
1984
222p : ill. - (Pan/Personal computer news computer
library)
0-330-28675-7 X.622/23074
3021726

BISHOP, O. N. (Owen Neville)
Practical programme for the electron. - London : Granada,
1984
165p
0-246-12362-1 SRL
3021726

ACORN MICROCOMPUTER SYSTEMS
Acorn Electron microcomputer systems–continued

BLACKWELL, Grahame
The Electron micro guide. - London : Longman, [1984]
103p : ill. - (Longman micro guides). - Includes index
0-582-36207-5 X.629/26282
3048217

CRYER, Andrew
Learning to use the Electron computer. - Aldershot : Gower, c1984
x,84p : ill. - Includes index
0-566-03509-x X.629/24215
3022064

DEWHIRST, John
A child's guide to the Electron Micro. - Cambridge : Cambridge University Press, 1984
95p : ill
0-521-31844-0 X.622/21977
3021513

DUNN, Seamus
The Acorn Electron for beginners. - Englewood Cliffs : London : Prentice-Hall, 1984
xi,307p : ill. - (Prentice-Hall International personal computer book). - Bibliography: p300-301. - Includes index
0-13-003211-5 X.629/24853
3022064

GORDON, John, 1952-
100 programs for the Acorn Electron. - Englewood Cliffs : London : Prentice/Hall International, c1984
ix,208p : ill. - (Prentice Hall International personal computer books)
0-13-634858-0 X.629/23886
3021726

INGLIS, Jonathan
The Electron. - London : Granada, 1984
61p : col.ill. - (Beginners' Micro guides). - Includes index
0-246-12381-8 X.622/22831
3022064

MCGREGOR, James J.
The Electron book. - London : Addison-Wesley, c1983
vii,322p : ill. - Includes index
0-201-14514-6 X.622/23245
3022064

MORSE, Peter
Microguide for the Electron. - London : Century Communications, 1984
40p : ill
0-7126-0505-3 X.629/25637
3022064

MURRAY, Ian, 1951-
Educational programs for the Electron. - London : Century Communications, 1984
196p : ill
0-7126-0405-7 X.629/25664
3021726

SCRIVEN, John
The working Electron. - London : Sunshine, 1984
vii,177p : ill. - Includes index
0-946408-52-1 X.622/22620
3021726

SIMON
Mastering the Electron. - Chichester : Horwood, 1984
262p : ill. - (Ellis Horwood books in computing). - Includes index
0-85312-679-8 X.622/21963
3022064

Quality programs for the Electron. - Tunbridge Wells : Micro, 1984
197p : ill. - Includes index
0-7447-0004-3 X.622/22664
3021726

SMALLEY, Brian
Electron magic. - London : Foulsham, c1984
137p,[4]p of plates : ill(some col.). - Includes index
0-572-01270-5 X.629/25842
301567x

SOLOMON, Meyer
My Electron and me. - London : Duckworth, 1984
93p : ill. - (Duckworth home computing). - Includes index
0-7156-1847-4 X.629/24671
301567x

ACORN MICROCOMPUTER SYSTEMS
Acorn Electron microcomputer systems - Serials

ACORN programs. - Dec. 1983/Jan. 1984-. - London : ECC Publications, 1984-
v. : ill
B85-14363 P.843/289
3252469

ACORN MICROCOMPUTER SYSTEMS
Acorn Electron microcomputer systems. Applications in artificial intelligence

BRAIN, K. R. (Keith Roger)
Artificial intelligence on the BBC and Electron. - London : Sunshine, 1984
viii,136p : ill
0-946408-36-x X.622/22071
3017974

VINE, Jeremy
On the road to artificial intelligence. - Nantwich : Shiva, c1984
102p : ill. - (Shiva's friendly micro series)
1-85014-045-6 X.622/24005
3017974

ACORN MICROCOMPUTER SYSTEMS
Acorn Electron microcomputer systems. Applications in electronic games

APPS, Vince
40 educational games for the Electron. - London : Granada, 1984
196p : ill
0-246-12404-0 X.622/21436
3126277

BRADBURY, A. J.
Adventure games for the Electron. - London : Granada, 1984
234p : ill. - Includes index
0-246-12417-2 X.622/20706
3022056

BRYANT, Martin
White Knight mk 11. - London : British Broadcasting Corporation, 1984
1program
0-563-21096-6 Cup.930/69
3126277

EWBANK, K. (Kay)
The Electron gamesmaster. - London : Granada, 1984
161p : ill
0-246-12514-4 X.622/21520
3126277

FIRST moves. - Harlow : Longman, c1984
1program. - (Longman software)
0-582-83590-9 Cup.930/71
3232220

FROST, Jean
Instant arcade games for the Electron. - London : Pan, 1983
199p. - (Pan/Personal Computer News computer library)
0-330-28406-1 X.629/23437
3022056

GAMES and other programs for the Acorn Electron. - Harmondsworth : Penguin, 1983
123p : ill. - (Penguin Acorn computer library)
0-14-007807-x X.629/23438
3022056

JAMES, M. (Mike)
21 games for the Electron. - London : Granada, 1984
145p : ill
0-246-12344-3 X.622/19288
3126277

PCW games collection for the Electron. - London : Century Communications, 1984
194p : ill
0-7126-0623-8 X.629/26244
3115968

ACORN MICROCOMPUTER SYSTEMS
Acorn Electron microcomputer systems. Applications in music

RITCHIE, Ian
BBC Micro music masterclass. - London : Pan, 1984
240p : ill. - (Pan/Personal computer news computer library). - Bibliography: p235. - Includes index
0-330-28673-0 X.622/23076
3112519

ACORN MICROCOMPUTER SYSTEMS
Acorn Electron microcomputer systems. Assembly languages

HOLMES, Peter, 1939-
Beginners assembly language programming for the Electron. - [Barnet] : [Glentop], [c1984]
215p in various pagings : ill. - (Dr. Watson series). - Includes index
0-907792-23-5 X.629/24793
3230740

ACORN MICROCOMPUTER SYSTEMS
Acorn Electron microcomputer systems. Assembly programming

GRIFFITHS, Jonathan
Creative assembler. - Harmondsworth : Penguin, 1984
191p,[4]p of plates : ill(some col.)
0-14-007809-6 X.629/25395
3243699

ACORN MICROCOMPUTER SYSTEMS
Acorn Electron microcomputer systems. Graphic displays. Programs

ALLAN, Boris
Graphic art for the Electron computer. - London : Sunshine, 1984
viii,107p : ill. - Includes index
0-946408-20-3 X.622/22678
3020150

MCGREGOR, James J.
Graphics programming on your Electron. - London : Corgi/Addison-Wesley, 1984
168p : ill
0-552-99102-3 X.629/25360
3020150

ACORN MICROCOMPUTER SYSTEMS
Acorn Electron microcomputer systems. Machine codes

SINCLAIR, Ian R. (Ian Robertson)
Electron machine code for beginners. - London : Granada, 1984
viii,156p : ill. - Includes index
0-246-12152-1 SRL
3020045

STEPHENSON, A. P.
Advanced electron machine code techniques. - London : Granada, 1984
208p. - Includes index
0-246-12403-2 SRL
3020045

ACORN MICROCOMPUTER SYSTEMS
Acorn Electron microcomputer systems. Programming

DEESON, Eric
Easy programming for the Electron. - Nantwich : Shiva, c1984
164p : ill. - (Shiva's friendly micro series). - Includes index
0-906812-66-6 X.622/20394
3049582

GEE, S. M.
The Electron programmer. - London : Granada, 1984
viii,183p : ill. - Includes index
0-246-12340-0 SRL
3049582

JAMES, M. (Mike)
Advanced programming for the Electron. - London : Granada, 1984
viii,192p : ill. - Bibliography: p.190. - Includes index
0-246-12402-4 SRL
3049582

MCGREGOR, James J.
Advanced programming techniques for the Electron. - London : Addison-Wesley, 1984
335p : ill. - Includes index
0-201-14517-0 X.622/23640
3049582

PENFOLD, R. A.
An introduction to programming the Acorn Electron. - London : Babani, 1984
134p : ill
0-85934-117-8 X.629/24023
3049582

PROGRAMMING the Electron. - London : Newnes Technical, Oct.1984
[176]p
0-408-01540-3 X.629/25126
3049582

SEAL, Peter
Advancing with the Electron. - Tunbridge Wells : Micro, 1984
109p : ill. - Includes index
0-7447-0012-4 X.622/22333
3049582

ACORN MICROCOMPUTER SYSTEMS
Acorn Electron microcomputer systems. Programming languages: Basic language

BAYLEY, Bruce
Building blocks for BBC games. - [Tring] : Melbourne House, [c1984]
140p : ill
0-86161-141-1 X.629/26768
3026183

BOYDE-SHAW, Brian
The Acorn Electron puzzle book. - Wilmslow : Sigma, c1984
126p
0-905104-79-x X.629/26059
3026183

CRYER, Neil
Basic programming on the Acorn Electron. - Englewood Cliffs ; London : Prentice-Hall, c1984
xiv,312p : ill. - Includes index
0-13-066259-3 SRL
3026183

ACORN MICROCOMPUTER SYSTEMS
Acorn Electron microcomputer systems. Programming languages: Basic language–continued

MCGREGOR, James J
Better BASIC for your Electron. - London : Corgi Addison-Wesley, 1984
184p : ill
0-552-99103-1 X.629/25361
3026183

ACORN MICROCOMPUTER SYSTEMS
Acorn Electron microcomputer systems. Programs. Applications in mathematics

KOSNIOWSKI, Czes
Essential maths on the BBC and Electron computers. - London : Sunshine, 1984
viii,212p : ill
0-946408-34-3 X.622/22062
3021114

ACORN MICROCOMPUTER SYSTEMS
Acorn Electron microcomputer systems. Read only memory

PLUMBLEY, Mark D.
The BASIC ROM user guide. - Cambridge : Adder, 1984
359p : ill. - Bibliography: p345. - Includes index
0-947929-04-5 X.629/25976
3287173

ACORN MICROCOMPUTER SYSTEMS
Acorn Electron microcomputer systems. Software packages

BILBOUL, Roger R.
The good software guide. - [London] : Fontana, 1984
222p. - Includes index
0-00-636847-6 X.629/25131
3240126

ACOSTA, AGUSTIN. Cuban writers. Poetry in Spanish

FORÉS, Aldo R.
La poesia de Agustin Acosta, poeta nacional de Cuba. - Miami : Universal, 1976
xv,254p. - (Colección Polymita). - Includes bibliographies
r6099214 X.909/83072
2028794
- [Another copy] X.909/88239

ACOSTA, CECILIO. Intellectual life. Venezuela
- Correspondence, diaries

ACOSTA, Cecilio
[Correspondence]. Epistolario. - Bogotá : Instituto Caro y Cuervo, 1981
lix,289p,13p of plates,10leaves of plates. - (Publicaciones / del Instituto Caro y Cuervo ; 15)
r6704941 X.520/27202
8840385

ACOSTA, CECILIO. Venezuela
- Biographies

DÍAZ SEIJAS, Pedro
Cecilio Acosta. - Los Teques : Gobernación del Estado Miranda, 1977
115p. - (Biblioteca popular mirandina)
r6706615 X.809/54511
8461937

ACOSTA GARCIA, JULIO. Costa Rica

ACADEMIA DE GEOGRAFÍA E HISTORIA DE COSTA RICA
Homenaje al Lic. Don Julio Acosta García en el centenario de su nacimiento el día 23 de mayo de 1872. - [San José] : la Academia, 1972
69p.,[4]leaves of plates : ill
r6085904 X.709/28078
2033356

ACOSTA, MIGUEL MARIA DE LOS SANTOS
See Santos Acosta, Miguel Maria de los

ACOSTA, URIEL
See Costa, Uriel da †

ACOT
See Council for Development in Agriculture

ACOUSTIC EMISSION

WILLIAMS, R. V. (Royson Vaughan)
Acoustic emission. - Bristol : Hilger, c1980
xiii,118p : ill. - Includes bibliographies and index
0-85274-359-9 SRL
2153017

ACOUSTIC EMISSION
- Bibliographies

DROUILLARD, Thomas F.
Acoustic emission. - New York ; London : IFI/Plenum, 1979
xix,787p. - (IFI data base library). - Includes index
0-306-65179-3 SRL
1871765

ACOUSTIC EMISSION. Testing. Pressure vessels
- Conference proceedings

COLLOQUIUM ON ACOUSTIC EMISSION (1975 : Tel Aviv)
Acoustic emission. - London : Applied Science Publishers, 1976
ix,121p : ill. - Includes bibliographies and index
0-85334-681-x SRL
0702463

ACOUSTIC ENCLOSURES. Chemical engineering equipment
- For petroleum industries

GENERAL specification for silencers and acoustic enclosures. - London : Heyden on behalf of the Institute of Petroleum in association with the Oil Companies Materials Association, 1979
iii,16p. - (Specification / Oil Companies Materials Association ; no.NWG-4)
0-85501-397-4 SRL
2580470
- [Another copy.] X.525/10618

ACOUSTIC ENGINEERING
See also
Sound recording
Sound reproduction

ACOUSTIC ENGINEERING

DOAK, Philip Ellis
Acoustical engineering. - [Southampton] : University of Southampton, 1975
32p : ill. - (Inaugural lectures / University of Southampton)
0-85432-144-6 X.619/15330 Woolwich
0566918

EVEREST, F. Alton (Frederick Alton)
The master handbook of acoustics. - Blue Ridge Summit, Pa. : Tab, c1981
352p : ill. - Includes index
0-8306-0008-6 X.629/22735
0566918

HEMOND, Conrad J.
Engineering acoustics and noise control. - Englewood Cliffs ; London : Prentice-Hall, c1983
xiv,192p : ill. - Includes index
0-13-278911-6 X.622/15451
0566918

REYNOLDS, Douglas D.
Engineering principles of acoustics. - Boston, Mass. ; London : Allyn and Bacon, c1981
641p : ill. - (Allyn and Bacon series in mechanical engineering and applied mechanics). - Bibliography: p632-633. - Includes index
0-205-07271-2 X.622/9750
0566918
- [Another copy: paperback]
ISBN 0-205-07283-6 X.622/9751

ACOUSTIC ENGINEERING
- Serials

MOA. - Vol.1, no.1- ; Sept. 1976-. - Stevenage : Institution of Electrical Engineers, 1976- : ill
B77-12499 SRL
1085360

ACOUSTIC EQUIPMENT
See also
Electro-acoustic equipment

ACOUSTIC EQUIPMENT
Underwater acoustic navigation equipment

MILNE, P. H. (Peter Henry)
Underwater acoustic positioning systems. - London : Spon, 1983
x,284p : ill. - Bibliography: p274-280. - Includes index
0-419-12100-5 SRL
2990822

ACOUSTIC EQUIPMENT
Underwater acoustic navigation equipment - Serials

SONARDYNE sensor. - Autumn 1983-. - Fleet, Hants. : Sonardyne, 1983-
v. : ill
r6813671 P.625/275
3068676

ACOUSTIC IMAGING

ACOUSTIC imaging / edited by Glen Wade. - New York ; London : Plenum Press, 1976
xiv,325p : ill. - Includes index
0-306-30914-9 X.620/16855 Woolwich
1139827

GREGUSS, Pál
Ultrasonic imaging. - London : Focal, 1980
224p : ill. - Includes index
0-240-51039-9 SRL
1139827

ACOUSTIC IMAGING
- Conference proceedings

ACOUSTICAL holography. - New York ; London : Plenum Press.
Vol.3. - 1971
xv,399p : ill. - Includes index
0-306-37723-3 X.620/18903 Woolwich
1958119

Vol.4. - 1972
xii,740p : ill. - Includes index
0-306-37724-1 X.620/18902 Woolwich
1958119

Vol.5. - 1974
xiv,738p : ill. - Includes index
0-306-37725-x X.620/18901 Woolwich
1958119

Vol.6. - 1975
xi,760p : ill. - Includes index
0-306-37726-8 X.620/18900 Woolwich
1958119

Vol.7. - 1977
xiii,654p : ill. - Includes bibliographies and index
0-306-37727-6 SRL
1407058

ACOUSTIC PHONETICS. Pulkau Valley usage. Austrian dialects. German language

HRACHOVEC, Herbert
Sonographische Untersuchungen zur Mundart des mittleren Paulkautales, N.Ö.. - Wien : VWGÖ, 1975
250p,[96]p of plates. - (Dissertationen der Universität Wien ; 124). - Includes bibliographies
r6009299 P.521/573[no.124]
1149555

ACOUSTIC-REFLEX STIMULATION. Measurement. Hearing. Man

HEARING assessment with the acoustic reflex / edited by Gerald R. Popelka. - New York ; London : Grune & Stratton, c1981
vii,168p : ill. - Bibliography: p161-168
0-8089-1361-1 SRL
2570513

ACOUSTIC SPECTROSCOPY
See also
Photoacoustic spectroscopy

ACOUSTIC SPECTROSCOPY. Polymers

PEREPECHKO, Igor' Ivanovich
Acoustic methods of investigating polymers. - Moscow : Mir, 1975
312 p. : ill. - Includes index
r6031588 X.629/11118 Woolwich
1082507

ACOUSTIC WAVE DEVICES
Interdigital surface acoustic wave devices

KEY papers on surface-acoustic-wave passive interdigital devices / edited by David P. Morgan. - Stevenage : Peregrinus [for] the Institution of Electrical Engineers, 1976
xi,372p : ill. - (IEE reprint series / Institution of Electrical Engineers ; 2). - Bibliography: p.347-370. - Includes index
0-901223-82-4 X.0615/502(2)
079838x

ACOUSTIC WAVE DEVICES
Surface acoustic wave devices. Design. Applications of computer systems - Conference proceedings

EUROPEAN WORKSHOP ON COMPUTER AIDED DESIGN OF SAW DEVICES
(1976 : Bologna)
Computer-aided design of surface acoustic wave devices. - Amsterdam ; Oxford : Elsevier, 1976
vii,308p : ill
0-444-41476-2 X.620/16326 Woolwich
073859x

ACOUSTIC WAVE FILTERS
Surface acoustic wave filters

SURFACE wave filters / Herbert Matthews, editor. - New York ; London : Wiley, 1977
xiii,521p : ill. - Includes index
0-471-58030-9 SRL
136880x

ACOUSTIC WAVEGUIDES. Microelectronic devices
- Conference proceedings

SYMPOSIUM ON OPTICAL AND ACOUSTICAL MICRO-ELECTRONICS (1974 : New York)
Proceedings of the Symposium on Optical and Acoustical Micro-electronics, New York, N.Y., April 16-18, 1974. - Brooklyn, N.Y. : Polytechnic Press of the Polytechnic Institute of New York ; New York ; London : Distributed by Wiley, 1975
[1],xxxvii,646p : ill. - (Symposia series / Microwave Research Institute ; vol.23). - Bibliography: p.xxxi-xxxii. - Includes index
0-470-27440-9 X.629/10570 Woolwich
0579351

ACOUSTIC WAVES
See Sound waves

ACOUSTICS
See also
Aeroacoustics
Electro-acoustics
Sound

ACOUSTICS

PHYSICAL acoustics / edited by Warren P. Mason and R.N. Thurston. - New York ; London : Academic Press. Vol.11. - 1975
xiii,364p : ill. - Includes bibliographies and index
0-12-477911-5 SRL
0156191

PIERCE, Allan D.
Acoustics. - New York ; London : McGraw-Hill, c1981
xxii,642p : ill. - (McGraw-Hill series in mechanical engineering). - Includes index
0-07-049961-6 SRL
0156191

WICKS, Keith
Sound and recording. - Harlow : Longman, 1982
61p : ill(some col.). - (Understanding science). - Includes index
0-582-39169-5 X.622/12784 Woolwich
0156191

ACOUSTICS
- For architecture

SILVA, Pérides
Acústica arquitetônica. - [3a ed.]. - Belo Horizonte : Edições Engenharia e Arquitetura, 1971
220 p., 2 fold. leaves of plates, 4 p. of plates : ill
Author's presentation copy to the British Library
r6032820 X.611/7541 Woolwich
1024256

ACOUSTICS
- Serials

ACOUSTICS Bulletin. - Vol.1, no.1- ; [Apr. 1976]-. - London : Institute of Acoustics, 1976-
B76-17491 SRL
065020x

ACOUSTICS letters. - Vol.1- ; 1977-. - [London] : Multi-Science Publishing Co. Ltd, [1977]- : ill
B77-24910 SRL
065020x
- Supplement, 1 ; July 1980-, etc SRL

ACOUSTICS
to ca 1800

HUNT, Frederick Vinton
Origins in acoustics. - New Haven ; London : Yale University Press, 1978
xv,196p,[6]p of plates : ill. - Includes index
0-300-02220-4 X.629/12194 Woolwich
1746693

ACOUSTICS
Applications in oceanography

CLAY, Clarence Samuel
Acoustical oceanography. - New York ; London : Wiley, 1977
xix,544p : ill. - (Ocean engineering). - Bibliography: p.526-536. - Includes index
0-471-16041-5 X.620/17268 Woolwich
1284991

ACOUSTICS
Applied acoustics

PORGES, G.
Applied acoustics. - London : Edward Arnold, 1977
ix,180p : ill. - Bibliography: p.175-176. - Includes index
0-7131-2658-2 X.619/18236 Woolwich
105175x

ACOUSTICS
Sokolov, Sergeĭ Iakovlevich - Biographies

IOFE, Viktor Kivovich
Sergeĭ Īākovlevich Sokolov (1897-1957). - Leningrad : Izdatel'stvo "Nauka", Leningradskoe otdelenie, 1976
149p. : ill. - (Akademiīā nauk SSSR. Nauchno-biograficheskaīā seriīā). - Bibliography : p.145-147
r6021009 Ac.1125/331(181)
1148451

ACOUSTICS
Underwater acoustics

BURDIC, William S.
Underwater acoustic system analysis. - Englewood Cliffs ; London : Prentice-Hall, c1984
xiv,445p : ill. - (Prentice-Hall signal processing series). - Includes index
0-13-936716-0 SRL
0073725

URICK, Robert Joseph
Principles of underwater sound. - [Revised ed.] - New York ; London : McGraw-Hill, 1975
xiii,384p : ill. - Includes index
0-07-066086-7 SRL
0073725

ACOUSTICS
Underwater acoustics - Conference proceedings

UNDERWATER acoustics and signal processing / edited by Leif Bjorno. - Dordrecht ; London : Reidel published in cooperation with NATO Scientific Affairs Division, c1981
xv,736p : ill. - (NATO advanced study institutes series. Series C, Mathematical and physical sciences ; v.66). - Includes index
90-277-1255-7 X.622/10599 Woolwich
0073725

ACOUSTICS. Bassoons, ca 1600-1975

VOIGT, Wolfgang
Untersuchungen zur Formantbildung in Klängen von Fagott und Dulzianen. - Regensburg, 1975
249, 7, 26 p. : ill. - (Kölner Beiträge zur Musikforschung ; Bd 80. Akustische Reihe ; Bd 5)
3-7649-2588-4 WP.13067/80
1005383

ACOUSTICS. Buildings
- For architectural design

MOORE, John Edwin
Design for good acoustics and noise control. - London : Macmillan, 1978
[8],213p. - Bibliography: p.205-207. - Includes index
0-333-24292-0 SRL
1349740

ACOUSTICS. Classical techniques. Guitar playing

TAYLOR, John, 1951-
Tone production on the classical guitar. - London : Musical News Services Ltd, 1978
80p : ill. - Includes index
B80-19863 X.431/11066
211173x

ACOUSTICS. Fluids

NONLINEAR acoustics in fluids / edited by Robert T. Beyer. - New York ; Wokingham : Van Nostrand Reinhold, c1984
xv,384p : ill. - (Benchmark papers in acoustics ; v.18). - Includes bibliographies and index
0-442-21182-1 SRL
3305120

ACOUSTICS. Guitars

STRASSENBURG, Arnold Adolph
The guitar. - New York ; London : McGraw-Hill, [1976]
[4],81p : ill. - (Physics of technology)
0-07-001716-6 X.431/10324
1066420

ACOUSTICS. Interiors. Auditoriums
- Conference proceedings

INTERNATIONAL SYMPOSIUM ON ARCHITECTURAL ACOUSTICS (1974 : Heriot-Watt University)
Auditorium acoustics. - London : Applied Science Publishers, 1975
xii,231p : ill. - Includes index
0-85334-646-1 SRL
0518204

ACOUSTICS. Interiors. Buildings

EVEREST, F. Alton (Frederick Alton)
Acoustic techniques for home & studio. - 2nd ed. - Blue Ridge Summit : Tab, c1984
viii,344p : ill. - Includes index
0-8306-0696-3 X.629/24597
0182745

PARKIN, Peter Hubert
Acoustics, noise and buildings. - 4th ed.. - London : Faber, 1979
297p : ill. - Includes index
0-571-04952-4 SRL
0182745

SOUND RESEARCH LABORATORIES
Practical building acoustics. - Sudbury : S.R.L. . London : Distributed by Spon, 1976
[5],226,[32]p : ill. - Bibliography (4p.)
B77-13763 X.615/2025 Woolwich
0182745

ACOUSTICS. Interiors. Buildings
Architectural design. Detail drawings

TEMPLETON, Duncan
Detailing for acoustics. - London : Architectural, 1983
191p : ill. - Bibliography: p188-189. - Includes index
0-85139-794-8 L.49/2045
299822x

ACOUSTICS. Interiors. Buildings
Mathematical models

KUTTRUFF, Heinrich
Room acoustics. - 2nd ed. - London : Applied Science Publishers, 1979
xxii,309p : ill. - Includes index
0-85334-813-8 SRL
0022179

ACOUSTICS. Music

ASKILL, John
Physics of musical sounds. - New York ; London : D. Van Nostrand, 1979
xiii,370p : ill. - Bibliography: p.359-362. - Includes index
0-442-20381-0 X.431/10654
0685593

BENADE, Arthur Henry
Fundamentals of musical acoustics. - New York ; London : Oxford University Press, 1975
xii,596p. - Includes index
0-19-502030-8 X.431/10243
0685593

LEVARIE, Siegmund
Tone. - 2nd ed. - Kent, Ohio : Kent State University Press, c1980
xvii,256p
0-87338-250-1 X.439/12828
879491x

LIPSCHUTZ, Frank
Zu Musikelementen. - Madrid : [El Autor], 1982
30p
84-300-7049-4 X.439/13475
879491x

LLOYD, Llewelyn Southworth
Intervals, scales and temperaments .... - [New] ed. - London : Macdonald and Jane's, 1978
xix,322p : ill. - Bibliography: p.112-114. - Includes index
0-354-04229-7 X.439/8656
1845187

MUSIC room and acoustics / papers given at a seminar, organized at the Royal Institute of Technology in Stockholm by the Royal Swedish Academy of Music, the Center for Human Technology and the Center for Speech Communication Research and Musical Acoustics in April 1975. - [Stockholm] : [Musikaliska akademien], 1977
187p : ill. - (Publications issued by the Royal Swedish Academy of Music ; 17). - Includes bibliographies
91-85428-03-5 X.0431/56(17)
0685593

N.A. Garbuzov-muzykant, issledovatel', pedagog / [sostaviteli O.E. Sakhaltueva, O.I. Sokolova]. - Moskva : Muzyka, 1980
302p,[8]p of plates : ill. - Bibliography
r6400918 X.439/11687
879491x

PIERCE, John R. (John Robinson)
The science of musical sound. - New York : Scientific American, c1983
xii,242p : ill(some col.). - (Scientific American library). - Bibliography: p221-222
0-7167-1508-2 X.431/12153
0685593

RIGDEN, John S.
Physics and the sound of music. - New York ; London : Wiley, 1977
xiii,286p : ill. - Includes index
0-471-02433-3 X.431/10418
0685593

ACOUSTICS. Music –continued

TIEFENSTRUKTUR der Musik / [herausgegeben vom Fachbereich 1. Fachgebiet Kommunikationswissenschaft der Technischen Universität Berlin in Verbindung mit der Akademie der Künste Berlin ; Redaktion : Carl Dahlhaus et al.]. – Berlin : TU, 1982
xv,232p : ill. – Includes bibliographies
3-7983-0821-7 X.439/13099
879491x

WOOD, Alexander
Alexander Wood's the physics of music. – 7th ed.. – London : Chapman and Hall [etc.], 1975
xiv,258p,xvi p of plates : ill. – Bibliography: p.252. – Includes index
0-412-13250-8 SRL
0685593
– [Another copy.] X.439/8276

ACOUSTICS. Musical instruments

SOUND generation in winds, strings, computers / papers by [Arthur Henry] Benade ... [Carleen Maley] Hutchins ... [et al.] ; given at seminars of The Committee for the Acoustics of Music. – Stockholm : Royal Swedish Academy of Music, 1980
248p : ill. – (Publications issued by the Royal Swedish Academy of Music ; no.29)
r6611193 X.0431/56(29)
8970165

ACOUSTICS. Pipe organs. Musical instruments to date

LOTTERMOSER, Werner
Orgeln, Kirchen und Akustik. – 2. Aufl. – Frankfurt am Main : Bochinsky, 1983
2v. : ill. – (Fachbuchreihe Das Musikinstrument ; Bd.16)
r6309038 X.0439/37(16)
9054073

ACOUSTICS. Religious buildings. Great Britain – For architectural design

ALLEN, William, 1914–
Acoustic treatment for places of worship. – [Newcastle upon Tyne] : Ecclesiastical Architects' and Surveyors' Associaton, c1981
22p : ill. – (EASA papers ; 2)
0-907866-01-8 X.425/3086 Woolwich
2623633

ACOUSTICS. Speech. Man

ACOUSTIC phonetics / [compiled and] edited by D.B. Fry. – Cambridge : Cambridge University Press, 1976
469p,[32]p of plates : ill. – Includes bibliographies
0-521-21393-2 X.989/51068
0761451

CONTEMPORARY issues in experimental phonetics / edited by Norman J. Lass. – New York ; London : Academic Press, 1976
xiii,498p : ill. – (Perspectives in neurolinguistics and psycholinguistics). – Includes bibliographies and index
0-12-437150-7 SRL
077331x

FRY, Dennis Butler
The physics of speech. – Cambridge : Cambridge University Press, 1979
vi,148p : ill. – (Cambridge textbooks in linguistics). – Includes index
0-521-22173-0 X.620/18401 Woolwich
077331x

PICKETT, J.M.
The sounds of speech communication. – Baltimore : University Park Press, c1980
xi,249p : ill. – (Perspectives in audiology). – Includes bibliographies and index
0-8391-1533-4 X.981/13415
077331X

ACOUSTICS. Violin family instruments

MUSICAL acoustics / edited by Carleen M. Hutchins. – Stroudsburg : Dowden, Hutchinson and Ross. – (Benchmark papers in acoustics ; 6) – Includes bibliographies and index
Part 2: Violin family functions. – 1976
xvii, 379 p. : ill
0-87933-153-4 X.431/10241
0698326

ACOUSTICS INSULATING MATERIALS. Great Britain
Asbestos acoustics insulating materials. Installation & removal. Industrial health measures

GREAT BRITAIN. Advisory Committee on Asbestos
Asbestos, work on thermal and acoustic insulation and sprayed coatings. – London : H.M.S.O., 1978
16p. – Bibliography: p.16
0-11-883083-x SRL
1842994

ACOUSTO-OPTICS

SAPRIEL, J.
Acousto-optics. – Chichester : Wiley, 1979
xv,126p : ill. – Includes index
0-471-99700-5 SRL
1818635

ACP COUNTRIES
Economic relations with European Community countries

The POLITICAL economy of EEC relations with African, Caribbean and Pacific States / edited by Frank Long. – Oxford : Pergamon, 1980
xii,193p. – (Pergamon international library). – Includes index
0-08-024077-1 X.809/46775
2362775

ACP COUNTRIES
Economic relations with European Community countries, 1970–1979

TWITCHETT, Carol Cosgrove
A framework for development. – London : Allen & Unwin, 1981
160p. – Bibliography: p149–153. – Includes index
0-04-338094-8 X.529/39885
2228653

ACP COUNTRIES
Exports to European Community countries: Cane sugar

AGREEMENT in the form of an exchange of letters between the Euopean Economic Community and Bardados, the People's Republic of the Congo, Fiji, the Co-operative Republic of Guyana, Jamaica, the Republic of Kenya, the Democratic Republic of Madagascar, the Republic of Malawi, Mauritius, the Republic of Suriname, the Kingdom of Swaziland, the United Republic of Tanzania, Trinidad and Tobago, the Republic of Uganda and Belize on the accession of the latter country to Protocol no.7 on ACP sugar annexed to the Second ACP-EEC Convention. – London : H.M.S.O., [1982]
3p. – (European Communities ; no.46 (1982)) (Cmnd. ; 8723)
0-10-187230-5 BS.14/3667
1308858

AGREEMENT in the form of an exchange of letters between the European Economic Community and Barbados, the People's Republic of the Congo, Fiji, the Co-operative Republic of Guyana, Jamaica, the Republic of Kenya, the Democratic Republic of Madagascar, the Republic of Malawi, Mauritius, the Republic of Suriname, the Kingdom of Swaziland, the United Republic of Tanzania, Trinidad and Tobago, the Republic of Uganda and the Republic of Zimbabwe on the accession of the latter country to the Protocol no.7 on ACP sugar annexed to the Second ACP-EEC Convention. – London : H.M.S.O., [1982]
3p. – (European Communities ; no.45 (1982)) (Cmnd. ; 8722)
0-10-187220-8 BS.14/3641
1308858

ACP COUNTRIES
Exports to European Community countries: Cane sugar – Texts

AGREEMENT in the form of an exchange of letters between the European Economic Community and Barbados, the People's Republic of the Congo, Fiji, the Co-operative Republic of Guyana, Jamaica, the Republic of Kenya, the Democratic Republic of Madagascar, the Republic of Malawi, Mauritius, the Republic of Surinam, the Kingdom of Swaziland, the United Republic of Tanzania, Trinidad and Tobago and the Republic of Uganda concerning the agreed quantities of cane sugar for certain ACP states, Brussels, 29 June 1977 ... – London : H.M.S.O., [1977]
Folder (4p). – (European Communities ; no.46(1977)) (Cmnd.7021)
0-10-170210-8 BS.14/3103
1308858

ACP COUNTRIES
Exports to European Community countries: Cane sugar. Guaranteed prices

AGREEMENT in the form of an exchange of letters between the European Economic Community and Barbados, the People's Republic of the Congo, Fiji, the Cooperative Republic of Guyana, Jamaica, the Republic of Kenya, the Democratic Republic of Madagascar, the Republic of Malawi, Mauritius, the Republic of Suriname, the Kingdom of Swaziland, the United Republic of Tanzania, Trinidad and Tobago and the Republic of Uganda on the guaranteed prices for cane sugar for 1981/1982. – London : H.M.S.O., [1982]
3p. – (European Communities ; no.44 (1982)) (Cmnd. ; 8721)
0-10-187210-0 BS.14/3642
1931458

ACP COUNTRIES
Exports to European Community countries: Cane sugar. Guaranteed prices – Texts

AGREEMENT in the form of an exchange of letters between the European Economic Community and Barbados, the People's Republic of the Congo, Fiji, the Co-operative Republic of Guyana, Jamaica, the Republic of Kenya, the Democratic Republic of Madagascar, the Republic of Malawi, Mauritius, the Republic of Suriname, the Kingdom of Swaziland, the United Republic of Tanzania, Trinidad and Tobago and the Republic of Uganda on the guaranteed prices for cane sugar for 1978/79, Brussels, 24 July 1978 ... – London : H.M.S.O., [1979]
Folder(4p). – (European Communities ; no.1(1979)) (Cmnd.7483)
0-10-174830-2 BS.14/3223
1931458

AGREEMENT in the form of an exchange of letters between the European Economic Community and Barbados, the People's Republic of the Congo, Fiji, the Co-operative Republic of Guyana, Jamaica, the Republic of Kenya, the Democratic Republic of Madagascar, the Republic of Malawi, Mauritius, the Republic of Suriname, the Kingdom of Swaziland, the United Republic of Tanzania, Trinidad and Tobago and the Republic of Uganda on the guaranteed prices for cane sugar for 1980/81, Brussels, 17 December 1980. – London : H.M.S.O., 1981
3p. – (European communities ; no.41(1981)) (Cmnd. ; 8306)
0-10-183060-2 BS.14/3423
1931458

ACP COUNTRIES
Exports to European Community countries: Coal. Import duties

DECISION of the representatives of the Governments of the member states of the European Coal and Steel Community, meeting within the Council, extending the decision of 24 June 1975 opening tariff preferences for products within the province of that Community originating in the African, Caribbean and Pacific States and in the Overseas countries and territories associated with the Community, Brussels, 23 February 1976 ... – London : H.M.S.O., [1976]
folder (4p). – (European Communities ; no.19 (1976)) (Cmnd. 6595)
0-10-165950-4 BS.14/1772
0760870

ACP COUNTRIES
Foreign trade with European Community countries

ACP-EEC Convention of Lomé and Agreement on Products within the Province of the European Coal and Steel Community, Lomé, 28 February 1975 (with related documents) ... – London : H.M.S.O., [1975]
187p. – (European Communities ; no.128(1975)) (Cmnd.6220)
0-10-162200-7 BS.14/1879
0543004

COUNCIL OF MINISTERS OF THE AFRICAN, CARIBBEAN AND PACIFIC STATES
[Treaties, etc. Council of the European Communities, 1975 Feb.28. Protocols, etc., 1976 Feb.26]. Exchange of letters between the Council of the European Communities and the Council of Ministers of the African, Caribbean and Pacific States extending application of the provisions set out in paragraph 2 of the exchange of letters of 28 February 1975, Brussels, 26 February 1976 ... – London : H.M.S.O., [1976]
Folder(4p). – (European Communities ; no.16(1976)) (Cmnd.6551)
0-10-165510-x BS.14/1754
0543004

The SECOND ACP-EEC Convention of Lomé and Agreement on products within the province of the European Coal and Steel Community and final act, Lomé, 31 October 1979 (with related documents) ... – London : H.M.S.O., [1980]
iv,232p. – (European Communities ; no.16(1980)) (Cmnd.7895)
0-10-178950-5 BS.14/3318
0543004

ACP COUNTRIES
Foreign trade with European Community countries. Treaties: ACP-EEC Convention of Lomé

The LOMÉ Convention and a new international economic order / edited by Frans A.M. Alting von Geusau ; with contributions from J.C. Anyiwo [et al.]. – Leyden : Sijthoff, 1977
xvi,249p. – (Publications of the John F. Kennedy Institute, Center for International Studies ; nr.11)
90-286-0217-8 X.520/9640
0621099

WALL, David
The European Community's Lomé Convention. – London : Trade Policy Research Centre, [1976]
[1],i,22 leaves. – (Guest papers / Trade Policy Research Centre ; no.4)
0-902926-04-7 X.0512/512(4)
0621099

ACP COUNTRIES
Foreign trade with European Community countries. Treaties: ACP-EEC Convention of Lomé. Accession of Cape Verde

CAPE VERDE
[Treaties, etc. European Economic Community, 1977 Mar.28]. Agreement on the accession of the Republic of Cape Verde to the ACP-EEC Convention of Lomé (with final act and exchange of letters), Brussels, 28 March 1977 ... - London : H.M.S.O., [1977]
8p. - (European Communities ; no.30(1977)) (Cmnd.6955)
0-10-169550-0 BS.14/3066
1289810

ACP COUNTRIES
Foreign trade with European Community countries. Treaties: ACP-EEC Convention of Lomé. Accession of Greece

[The SECOND ACP-EEC Convention of Lomé and Agreement on Products within the Province of the European Coal and Steel Community and Final Act (with Related Documents) (1979). Protocols, etc. 1981 Oct.8]. Protocol to the Second ACP-EEC Convention consequent on the accession of the Hellenic Republic to the Community. - London : H.M.S.O., [1983]
31p. - (European Communities : no.30 (1983)) (Cmnd. : 8889)
0-10-188890-2 BS.14/3678
2987341

ACP COUNTRIES
Foreign trade with European Community countries. Treaties: ACP-EEC Convention of Lomé. Accession of Papua New Guinea

EUROPEAN ECONOMIC COMMUNITY
[Treaties, etc. Papua New Guinea. 1977 Mar.28]. Agreement on the accession of Papua New Guinea to the ACP-EEC Convention of Lomé (with final act and exchange of letters), Brussels, 28 March 1977 ... - London : H.M.S.O., [1977]
8p. - (European Communities ; no.31 (1977)) (Cmnd.6957)
0-10-169570-5 BS.14/3064
1289837

ACP COUNTRIES
Foreign trade with European Community countries. Treaties: ACP-EEC Convention of Lomé. Accession of São Tomé e Príncipe

EUROPEAN ECONOMIC COMMUNITY
[Treaties, etc. São Tomé e Príncipe, 1977 Mar.28]. Agreement on the accession of the Democratic Republic of São Tomé and Príncipe to the ACP-EEC Convention of Lomé (with final act and exchange of letters), Brussels, 28 March 1977 ... - London : H.M.S.O., [1977]
8p. - (European Communities ; no.32(1977)) (Cmnd.6958)
0-10-169580-2 BS.14/3062
1289829

ACP COUNTRIES
Foreign trade with European Community countries. Treaties: ACP-EEC Convention of Lomé. Accession of Zimbabwe

AGREEMENT on the accession of the Republic of Zimbabwe to the Second ACP-EEC Convention signed at Lomé on 31 October 1979 with final act, Luxembourg, 4 November 1980. - London : H.M.S.O., 1981
10p : ill. - (Cmnd. ; 8150) (European Communities ; no.9 (1981))
0-10-181500-x BS.14/3390
2384477

ACP COUNTRIES
Foreign trade with European Community countries. Treaties: ACP-EEC Convention of Lomé. Renegotiation

RUBIN, Abby
Lomé II. - London : Catholic Institute for International Relations ; [Dublin] : Trócaire, Catholic Agency for World Development, 1978
25,[3] leaves. - (A CIIR / TRÓCAIRE position paper)
B80-37849 X.525/3682
1581309

SUCCESSOR to Lomé / memorandum by the Foreign and Commonwealth Office. - [London] : [Foreign and Commonwealth Office], [1983]
[28]p. - (Foreign policy document / Great Britain. Foreign and Commonwealth Office ; no.97)
B85-27297 X.805/6839
1581309

ACP COUNTRIES
Foreign trade with European Community countries. Treaties: ACP-EEC Convention of Lomé. Renegotiation - Conference proceedings

RENEGOTIATION OF THE LOME CONVENTION (Conference : 1978 : London)
The renegotiation of the Lomé Convention. - London : Catholic Institute for International Relations ; Dublin : Trócaire, Catholic Agency for World Development, 1978
[3],87p
B80-37850 X.525/3683
2076853

ACP COUNTRIES
Foreign trade with European Community countries. Treaties: ACP-EEC Convention of Lomé. Renegotiation, to 1984

The LOMÉ negotiations. - London : Catholic Institute for International Relations, c1984
25p. - (Comment). - Bibliography: p25
0-946848-35-1 X.529/65442
319597x

ACP COUNTRIES
Foreign trade with European Community countries. Treaties. ACP-EEC Convention of Lomé. Renegotiation. Great Britain. Parliament. House of Commons. Select Committee on Overseas Development. Renegotiation of the Lome Convention

The RENEGOTIATION of the Lome Convention. - London : H.M.S.O., 1978
16p. - (Miscellaneous ; no.34(1978)) (Cmnd.7397)
0-10-173970-2 BS.14/3201
1845497

ACP COUNTRIES
Foreign trade with United States. Implications of ACP-EEC Convention of Lomé

MOSS, Joanna
The Lome Conventions and their implications for the United States. - Epping, [1982]
xvii,218p : ill. - (Westview replica editions). - Bibliography: p207-214. - Includes index
0-86531-935-9 X.529/55581
2936631

ACP COUNTRIES
Social conditions - Directories - Serials

ANNUAIRE des Etats ACP = ACP States yearbook. - 80/81- . - Bruxelles : Editions Delta, 1980-
v.
B81-04496 P.523/1445
2339420
- [Another copy.] P.703/837

ACP COUNTRIES
Treaties with members of European Coal and Steel Community: Treaties on entry of Greece into European Coal and Steel Community

[The SECOND ACP-EEC Convention of Lomé and Agreement on Products within the Province of the European Coal and Steel Community and Final Act (with Related Documents) (1979). Protocols, etc. 1981 Oct.8]. Protocol to the agreement between the member states of the European Coal and Steel Community and the ACP States consequent on the accession of the Hellenic Republic to the Community. - London : H.M.S.O., [1983]
11p. - (European Communities) (Cmnd. ; 8882)
0-10-188820-1 BS.14/3682
2987333

ACP-EEC CONVENTION OF LOMÉ

The LOMÉ Convention and a new international economic order / edited by Frans A.M. Alting von Geusau ; with contributions from J.C. Anyiwo [et al.]. - Leyden : Sijthoff, 1977
xvi,249p. - (Publications of the John F. Kennedy Institute, Center for International Studies ; nr.11)
90-286-0217-8 X.520/9640
0621099

WALL, David
The European Community's Lomé Convention. - London : Trade Policy Research Centre, [1976]
[1],i,22 leaves. - (Guest papers / Trade Policy Research Centre ; no.4)
0-902926-04-7 X.0512/512(4)
0621099

ACP-EEC CONVENTION OF LOMÉ
Accession of Cape Verde

CAPE VERDE
[Treaties, etc. European Economic Community, 1977 Mar.28]. Agreement on the accession of the Republic of Cape Verde to the ACP-EEC Convention of Lomé (with final act and exchange of letters), Brussels, 28 March 1977 ... - London : H.M.S.O., [1977]
8p. - (European Communities ; no.30(1977)) (Cmnd.6955)
0-10-169550-0 BS.14/3066
1289810

ACP-EEC CONVENTION OF LOMÉ
Accession of Greece

[The SECOND ACP-EEC Convention of Lome and Agreement on Products within the Province of the European Coal and Steel Community and Final Act (with Related Documents) (1979). Protocols, etc. 1981 Oct.8]. Protocol to the Second ACP-EEC Convention consequent on the accession of the Hellenic Republic to the Community. - London : H.M.S.O., [1983]
31p. - (European Communities ; no.30 (1983)) (Cmnd. ; 8889)
0-10-188890-2 BS.14/3678
2987341

ACP-EEC CONVENTION OF LOMÉ
Accession of Papua New Guinea

EUROPEAN ECONOMIC COMMUNITY
[Treaties, etc. Papua New Guinea. 1977 Mar.28]. Agreement on the accession of Papua New Guinea to the ACP-EEC Convention of Lomé (with final act and exchange of letters), Brussels, 28 March 1977 ... - London : H.M.S.O., [1977]
8p. - (European Communities ; no.31 (1977)) (Cmnd.6957)
0-10-169570-5 BS.14/3064
1289837

ACP-EEC CONVENTION OF LOMÉ
Accession of São Tomé e Príncipe

EUROPEAN ECONOMIC COMMUNITY
[Treaties, etc. São Tomé e Príncipe, 1977 Mar.28] Agreement on the accession of the Democratic Republic of São Tomé and Príncipe to the ACP-EEC Convention of Lomé (with final act and exchange of letters), Brussels, 28 March 1977 ... - London : H.M.S.O., [1977]
8p. - (European Communities ; no.32(1977)) (Cmnd.6958)
0-10-169580-2 BS.14/3062
1289829

ACP-EEC CONVENTION OF LOMÉ
Accession of Zimbabwe

AGREEMENT on the accession of the Republic of Zimbabwe to the Second ACP-EEC Convention signed at Lomé on 31 October 1979 with final act, Luxembourg, 4 November 1980 - London : H.M.S.O., 1981
10p : ill. - (Cmnd. ; 8150) (European Communities ; no.9 (1981))
0-10-181500-x BS.14/3390
2384477

ACP-EEC CONVENTION OF LOMÉ
Implications for foreign trade between United States & ACP countries

MOSS, Joanna
The Lome Conventions and their implications for the United States. - Epping, [1982]
xvii,218p : ill. - (Westview replica editions). - Bibliography: p207-214. - Includes index
0-86531-935-9 X.529/55581
2936623

ACP-EEC CONVENTION OF LOMÉ
Renegotiation

RUBIN, Abby
Lomé II. - London : Catholic Institute for International Relations ; [Dublin] : Trócaire, Catholic Agency for World Development, 1978
25,[3] leaves. - (A CIIR / TRÓCAIRE position paper)
B80-37849 X.525/3682
1581309

SUCCESSOR to Lomé / memorandum by the Foreign and Commonwealth Office. - [London] : [Foreign and Commonwealth Office], [1983]
[28]p. - (Foreign policy document / Great Britain. Foreign and Commonwealth Office ; no.97)
B85-27297 X.805/6839
1581309

ACP-EEC CONVENTION OF LOMÉ
Renegotiation - Conference proceedings

RENEGOTIATION OF THE LOMÉ CONVENTION (Conference : 1978 London)
The renegotiation of the Lomé Convention. - London : Catholic Institute for International Relations ; Dublin : Trócaire, Catholic Agency for World Development, 1978
[3],87p
B80-37850 X.525/3683
2076853

ACP-EEC CONVENTION OF LOMÉ
Renegotiation, to 1984

The LOMÉ negotiations. - London : Catholic Institute for International Relations, c1984
25p. - (Comment). - Bibliography: p25
0-946848-35-1 X.529/65442
319597x

ACP-EEC CONVENTION OF LOMÉ
Renegotiation. Great Britain. Parliament. House of Commons. Select Committee on Overseas Development. Renegotiation of the Lomé Convention

The RENEGOTIATION of the Lomé Convention. - London : H.M.S.O., 1978
16p. - (Miscellaneous ; no.34(1978)) (Cmnd.7397)
0-10-173970-2 BS.14/3201
1845497

ACQUAINTANCE. Interpersonal relationships

DUCK, Steve
The study of acquaintance. - Farnborough, Hants. : Saxon House, 1977
ix,236p. - Bibliography: p.200-229. - Includes index
0-566-00160-8 X.529/31954
1111663

ACQUAROSSA. Italy
Antiquities. Archaeology, to date

SVENSKA INSTITUTET I ROM
Acquarossa. - Stockholm : The Institute, 1981-
v. - (Skrifter utgivna av Svenska institutet i Rom. 4° ; 38)
r6610565 Ac.5234.b[no.38]
8890226

ACQUI. Italy
Cathedrals. Cattedrale dell'Assunta. Acqui. Stock. Documents in Latin

Le CARTE medievali della Chiesa d'Acqui / [a cura di] Romeo Pavoni. - Bordighera : Istituto internazionale di studi liguri, 1977
641p. - (Collana storica di fonti e studi ; 22) (Collana storica dell'Oltregiogo ligure ; 2)
r6069580 X.421/21405 Woolwich
2467380

ACQUI TERME. Italy
Buildings. Palazzo Liceo Saracco. Exhibits: Italian paintings. Guttuso, Renato - Catalogues

GUTTUSO, Renato
Renato Guttuso. - Acqui Terme : Azienda Autonoma di Cura, 1974
[28] p., 21 leaves of plates, 22 p. of plates : chiefly ill
r6013897 X.419/1660 Woolwich
0860808

ACQUIRED IMMUNE DEFICIENCY SYNDROME. Man

The AIDS epidemic / Kevin M. Cahill. - London : Hutchinson, 1984, c1983
xvi,173p : ill
0-09-148651-3 X.329/17546
3014282

MAYER, Kenneth H.
The AIDS fact book. - Toronto ; London : Bantam, 1983
[xii],135p : ill. - Bibliography: p123-125. - Includes index
0-553-23870-1 X.329/17905
3014282

ACQUIRED IMMUNE DEFICIENCY SYNDROME. Man
Aetiology. Nutritional factors

FOOD imprint as cause of AIDS. - Dover : Collation Research, 1984
[14]p
0-906470-05-6 X.329/18352
3167860

ACQUISITION. Agricultural land. Great Britain

GREAT BRITAIN. Committee of Inquiry into the Acquisition and Occupancy of Agricultural Land
Report of the Committee of Inquiry into the Acquisition and Occupancy of Agricultural Land. - London : H.M.S.O., [1979]
xiii,378p : ill. - (Cmnd.7599). - Bibliography: p.370-373
0-10-175990-8 BS.3/665
196514x

ROYAL INSTITUTION OF CHARTERED SURVEYORS
The acquisition and occupancy of agricultural land. - London : The Institution, 1977
[3],37 leaves
B78-14143 X.313/1527 Woolwich
1408275

ACQUISITION. American citizenship
By negro slaves owned by Cherokees in Southern states, 1863-1902

LITTLEFIELD, Daniel F.
The Cherokee freedmen. - Westport, Conn. ; London : Greenwood Press, 1978
xii,283p : ill. - (Contributions in Afro-American and African studies ; no.4). - Bibliography: p.259-265. - Includes index
0-313-20413-6 X.809/44466
1906143

ACQUISITION. American citizenship, 1900-1917
By Puerto Ricans, 1900-1917

CABRANES, José A.
Citizenship and the American empire. - New Haven, [Conn.] ; London : Yale University Press, 1979
viii,101p
0-300-02325-1 X.800/31866
2588099

ACQUISITION. American government publications. Stock. Libraries. United States

HERNON, Peter
Development collections of U.S. government publications. - Greenwich, Conn. ; London : Jai, c1982
xxiii,289p : ill. - (Foundations in library and information science ; v.12). - Bibliography: p275-279. - Includes index
0-89232-135-0 2725.c.184
2963604

MCCLURE, Charles R.
Improving the quality of reference service for government publications. - Chicago : American Library Association, 1983
xiii,270p. - (ALA studies in librarianship ; no.10). - Bibliography: p257-265
0-8389-0388-6 X.520/32763
2963604
- [Another copy] OPL

ACQUISITION. Art objects
Ethics

BURNHAM, Bonnie
The art crisis. - London : Collins, 1975
256p. - Includes index
0-00-211884-x X.429/10069
0583669

ACQUISITION. Books & manuscripts. Stock. Kungliga biblioteket ca 1630-ca 1650

CALLMER, Christian
Königin Christina, ihre Bibliothekare und ihre Handschriften. - Stockholm : [Kungl. biblioteket], 1977
270p. - (Acta Bibliothecæ Regiæ Stockholmiensis ; 30)
91-7000-066-2 Ac.9233.zk[vol.30]
2441985

ACQUISITION. Books in Punjabi. Stock. Public libraries. London

RAIT, S. K.
Acquisition, cataloguing and transliteration of Punjabi literature in some London libraries. - Leeds : School of Librarianship, Leeds Polytechnic, 1981
43p
0-900738-30-8 2725.g.135
2878399

ACQUISITION. Books. Stock. Doncaster Borough Library
Selection, ca 1880-1885

BLOOMFIELD, M. A. (Mark A)
About 'About books'. - Alsager : M.A. Bloomfield, c1982
15p : 1ill
0-9502955-1-5 2719.k.53.
2816938

ACQUISITION. Books. Stock. Libraries
From American publishers

KIM, Ung Chon
Policies of publishers. - 1982 ed. - Metuchen ; London : Scarecrow, 1982
xi,161p
0-8108-1527-3 2702.f.72
2763885

ACQUISITION. Books. Stock. Libraries
Selection

LUNATI, Rinaldo
Book selection. - Metuchen : Scarecrow Press, 1975
vii, 167 p.. - Includes bibliographies
0-8108-0846-3 2719ff.39
0779415

ACQUISITION. Books. Stock. Libraries. Universities. United States
Selection. Approval plans

MCCULLOUGH, Kathleen
Approval plans and academic libraries. - Phoenix : Oryx Press ; London : Distributed by Mansell Information Publishing, 1977
xi,154p. - Bibliography: p.149-154
0-912700-05-x 2719.bb.24
1300970

ACQUISITION. Books. Stock. Libraries. Universities. United States
Selection. Approval plans. Administration

CARGILL, Jennifer S.
Practical approval plan management. - Phoenix : Oryx Press ; London : Distributed by Mansell Information Publishing, 1979
viii,95p. - Bibliography: p.89-91. - Includes index
0-912700-52-1 X.520/21081
2130319

ACQUISITION. Books. Stock. Public libraries. Finland
Selection

TIIHONEN, Pertti
Yleisten kirjastojen kirjanvalinnasta. - Tampere : Tampereen yliopisto, Tiedotusopin laitos, 1976
66,[14]p : charts,forms. - (Tampereen yliopisto. Tiedotusopin laitos. Julkaisuja = Reports ; n:o 30). - Bibl
951-440-441-6 P.905/73(30)
2428318

ACQUISITION. Books. Stock. Public libraries. Great Britain
Selection

SPILLER, David
Book selection. - 3rd ed. - New York : Saur ; London : Bingley, 1980
206p. - Includes index
0-85157-305-3 2719.e.100
0085189

ACQUISITION. Books. Stock. Schools. Great Britain

NATIONAL BOOK LEAGUE
Books for schools. - London : The League, 1979
70p : ill. - Bibliography: p.63
0-85353-328-8 X.519/29887
1919016

ACQUISITION. Brazilian citizenship
By Portuguese persons. Law

XAVIER, Alberto
Estatuto jurídico dos portugueses no Brasil. - Rio de Janeiro : Liber Juris, 1979
117p
r6159728 X.529/38068
2432846

ACQUISITION. Brazilian citizenship
Law

LEITE, Yára Muller
Como ser brasileiro, não nascendo no Brasil. - São Paulo : Edição Saraiva, 1974
142 p.
r6009028 X.100/14579
0919888

ACQUISITION. British citizenship
- Statistics

STATISTICS of persons acquiring citizenship of the United Kingdom and colonies ... / Home Office. - 1977. - London : H.M.S.O, [1978]
11p
0-10-173100-0 BS.18/230
0173959

ACQUISITION. British citizenship
Fees

GREAT BRITAIN. Parliament. House of Commons. Home Affairs Committee
Third report from the Home Affairs Committee, Session 1982-83. - London : H.M.S.O., [1983]
xxi,130p : ill. - (H.C. ; 248)
0-10-224883-4 BS.17/784
2992345

ACQUISITION. Bury St Edmunds cross. Exhibits. Metropolitan Museum of Art

HOVING, Thomas
King of the Confessors. - London : Hamilton, 1981
365p : ill(some col.). - Bibliography: p353-354. - Includes index
0-241-10674-5 X.800/31994
2604434

ACQUISITION. Canadian provincial government publications. Stock. Libraries

PROSS, Catherine A.
A guide to the identification and acquisition of Canadian provincial government publications. - London : The Vine Press, 1977
[2]p,i,[67]leaves. - (Dalhousie University Libraries and Dalhousie University School of Library Service occasional papers / Dalhousie University ; 16). - Includes bibliographies
0-7703-0153-3 X.0512/507(16)
1457187

ACQUISITION. Children's books. Stock. Libraries
Selection

BUYING books / edited by Vivien Griffiths. - Towcester : Library Association Youth Libraries Group, c1983
64p. - (Practical guide ; no.2)
0-946581-01-0 2708.e.428
3079775

ACQUISITION. Children's stories. Stock. Children's libraries. Canada
Effects on publishing of children's stories in Canada

WILKINSON, J. P.
Canadian juvenile fiction and the library market. - Ottawa : Canadian Library Association, 1976
iii-viii,87p. - Includes bibliographies
0-88802-112-7 X.900/25364
1087797

ACQUISITION. Cinema film rights
Fiction in English, 1945–: Books with British imprints – Serials

FILM rights. – [Issue 127(1)?]–. – London : Jaguar, c1978–
v.
Holdings: issue 128(5), etc
B81–06778 P.2000/871
2353741

ACQUISITION. Cognitive skills
Simulations. Applications of digital computer systems. Programs: HACKER program

SUSSMAN, Gerald Jay
A computer model of skill acquisition. – New York : American Elsevier ; London : Elsevier, 1975
xv,133p : ill. – (Elsevier computer science library. artificial intelligence series ; 1). – Bibliography: p.128–133
0–444–00163–8 X.620/8319 Woolwich
0511994

ACQUISITION. Communication skills. Children
Role in development of children

COMMUNICATION in development / edited by W.P. Robinson. – London : Published in cooperation with European Association of Experimental Social Psychology by Academic Press, 1981
xiv,292p : ill. – (European monographs in social psychology ; 24). – Includes bibliographies and index
0–12–590140–2 X.520/25605
2595516

ACQUISITION. Companies. European Community countries

CHOWN, John F. (John Francis)
The acquisition of assets, companies and real estate in Belgium, France, Germany, The Netherlands, United Kingdom. – London : 'Financial Times', 1974
[8],157p
0–903199–03–3 X.512/5060
052249x

ACQUISITION. Companies. Great Britain
Financial management

BEAN, David Geoffrey
Financial strategy in the acquisition decision. – Epping : Gower Press, 1975
xii,175p : ill. – (A Gower Press special study). – Bibliography: p.164–167. – Includes index
0–7161–0268–4 X.520/10390
0578088

ACQUISITION. Companies. Great Britain
Financial management aspects

ACCOUNTANTS' reports on take-overs, mergers, acquisitions and realisations / prepared by Deloitte Haskins & Sells. – London : Deloitte Haskins & Sells, 1984
392p
0–86349–035–2 X.529/67626
3250075

ACQUISITION. Companies. Great Britain
Management aspects

JONES, C. Stuart
Successful management of acquisitions. – London : Derek Beattie, 1982
xviii,173p. – Includes index
0–907591–02–7 X.529/50378
0264601

ACQUISITION. Companies. United States
Management aspects

BING, Gordon
Corporate acquisitions. – Houston ; London : Gulf Publishing, c1980
xi,248p. – Includes index
0–87201–009–0 X.520/23193
237272x

BRADLEY, James W.
Acquisition and corporate development. – Lexington, [Mass.] : Lexington Books, 1981
xiv,252p : ill. – (Arthur D. Little books). – Includes bibliographies and index
0–669–03170–4 X.520/26119
237272x

ACQUISITION. Computer systems. Government departments. Great Britain
Procedure

CENTRAL COMPUTER AND TELECOMMUNICATIONS AGENCY
CCTA procedures review for the development and procurement of computer based systems. – London : H.M. Treasury, Central Computer and Telecommunications Agency, 1982
[42]p
0–7115–0043–6 BS.6/213
2900270

ACQUISITION. Concept of conservation
By children

PINARD, Adrien
The conservation of conservation. – Chicago ; London : University of Chicago Press, 1981
xii,199p. – Bibliography: p181–194. – Includes index
0–226–66834–7 X.529/44030
2571439

ACQUISITION. Concept of number
By children. Projects: Project on Children's Learning and Development

GONCHAR, Arthur J.
A study in the nature and development of the natural number concept. – Madison, Wis. : Wisconsin Research and Development Center for Cognitive Learning, The University of Wisconsin, 1975
x, 135 p. : ill. – (Technical report ; no. 340). – Bibliography : p. 95–97
r6010365 X.615/1760 Woolwich
102843X

ACQUISITION. Concept of probabilities
By children

FISCHBEIN, E.
The intuitive sources of probabilistic thinking in children. – Dordrecht : Reidel, 1975
xiii, 204 p. : ill. – (Synthese library ; vol. 85). – Includes bibliographies and index
90–277–0626–3 X.529/48663
0734209

ACQUISITION. Concept of society
By children in England

FURTH, Hans Gerald
The world of grown-ups. – New York ; Oxford : Elsevier New York ; Limerick : Distributed by Thomond Books, 1980
xiii,221p. – Bibliography: p.215–216. – Includes index
0–444–99065–8 X.520/21746
2160919

ACQUISITION. Concepts

STENILD, Mariane
Concept learning. – Risskov : University of Aarhus, Institute of Psychology, 1977
31,5p. – (Psychological reports Aarhus ; vol.2, no.2). – Includes bibliographies
r6017670 X.0512/454
1986848

ACQUISITION. Concepts
– Conference proceedings

MEANING and understanding / edited by Herman Parret and Jacques Bouveresse. – Berlin : de Gruyter, 1981
x,442p.. – (Foundations of communication). – Bibliography
3–11–008116–4 X.950/11373
8776458

ACQUISITION. Contemporary objects. Exhibits. Museums. Sweden
to date

TODAY for tomorrow / edited by Göran Rosander. – Stockholm : SAMDOK, 1980
63p : ill
91–7108–190–9 X.429/14865
8999554

ACQUISITION. Development land. Great Britain
Acquisition & disposal. Role of local authorities

BARRETT, Susan, 1942–
Local authorities and land supply. – Bristol : School for Advanced Urban Studies, University of Bristol, c1983
88p. – (Occasional paper ; no.10)
0–86292–103–1 X.529/62384
313346x

ACQUISITION. Development land. Wales
Acquisition & disposal. Policies of Land Authority for Wales

LAND policy statement and rolling programme / Land Authority for Wales. – 1978. – Cardiff : The Authority, [1978]
[7],89p : ill(incl 1 col)
Subsequent volumes, when received, will be placed at this shelfmark
B79–01755 BS446/3
1743112

ACQUISITION. Documents in foreign languages. Documents for ethnic minorities. Stock. Public libraries. Great Britain

COOKE, Madeleine
Public library provision for ethnic minorities in the United Kingdom. – Leicester : Distributed by Leicestershire Libraries and Information Service, 1979
[5],46p
0–85022–057–2 X.525/3432
2243016

ACQUISITION. Documents on Caribbean countries. Stock. Libraries
Cooperation – Conference proceedings

SHARING Caribbean resources / compiled by Robert V. Vaughn. – Kingshill, St Croix, V.I. : Aye-Aye Press, 1975
[2], 6 leaves
r6010784 2719.ff.70
1052004

ACQUISITION. Documents on South East Asia. Stock. Australian National University. Library
– Bibliographies

AUSTRALIAN NATIONAL UNIVERSITY. Library
Southeast Asian materials in the Australian National University Library. – Canberra : The Library, 1975
[2], iv, 180 p.. – Bibliography : p. 177–180
0–7081–1333–8 X.702/5331
1050400

ACQUISITION. Documents with European imprints. Stock. Libraries. Great Britain
– Conference proceedings

EUROPEAN acquisitions and their bibliographical control / edited by Colin Steele and Gregory Walker. – [Lancaster] : University of Lancaster Library, 1975
[4].129p. – (Occasional papers / University of Lancaster Library ; 9)
0–901699–32–2 2719.ff.19
0334634

ACQUISITION. Documents with foreign imprints. Stock. Libraries. United States

ACQUISITION of foreign materials for U.S. libraries / compiled and edited by Theodore Samore. – 2nd ed. – Metuchen ; London : Scarecrow, 1982
xi,218p. – Includes index
0–8108–1560–5 2719.e.465
2842327

ACQUISITION. Documents with Latin American imprints. Stock. Libraries
to date

SEMINAR on the Acquistion of Latin American Library Materials. – Reprinted in microfiche. – Washington, D.C. : Brookhaven Press, 1976
75microfiche
r6805477 Mic.F.202/1–75
8884234

ACQUISITION. Ecuadorian citizenship
Law, to date

ECUADOR
Ley de naturalización. – 2.a ed., actualizada a Julio de 1977. – Quito : Corporación de Estudios y Publicaciones, 1977
12p
r6701238 X.200/36592
8828342

ACQUISITION. Firearms. Western Europe
By individuals

EUROPEAN Convention on the Control of the Acquisition and Possession of Firearms by Individuals (with explanatory report), Strasbourg, June 28, 1978. – London : H.M.S.O., [1979]
29p. – (Miscellaneous ; no. 19 (1979)) (Cmnd. 7683)
0–10–176830–3 BS.14/3276
2255952

ACQUISITION. Foreign language skills

KRASHEN, Stephen
The input hypothesis. – London : Longman, 1985
[224]p. – Includes bibliography and index
0–582–55381–4 X.629/26517
3202305

ACQUISITION. Foreign language skills
– Conference proceedings

INTERLANGUAGE / Alan Davies, C. Criper, A.P.R. Howatt, editors. – Edinburgh : Edinburgh University Press, c1984
xvi,384p : ill. – Bibliography: p346–380. – Includes index
0–85224–499–1 X.950/41146
3202305

ACQUISITION. Foreign language skills
Psychological aspects

MCDONOUGH, Steven H.
Psychology in foreign language teaching. – London : Allen & Unwin, 1981
168p : ill. – Bibliography: p157–166. – Includes index
0–04–418002–0 X.529/40929
2390396

ACQUISITION. Foreign language skills
Sociolinguistic aspects

LOVEDAY, Leo
The sociolinguistics of learning and using a non-native language. - Oxford : Pergamon, 1982
196p : ill. - (Language teaching methodology series). - Bibliography: p179-196
0-08-028668-2 X.950/16098
279196x

ACQUISITION. Foreign language skills. Adults
Theories

KRASHEN, Stephen
Second language acquisition and second language learning. - Oxford : Pergamon, 1981
vii,151p : ill. - (Language teaching methodology series) (Pergamon Institute of English (Oxford)). - Bibliography: p139-151
0-08-025338-5 X.950/8784
2152770

ACQUISITION. French citizenship
By Alsace & Lorraine persons, 1871-1872

WAHL, Alfred
L'option et l'émigration des Alsaciens-Lorrains (1871-1872). - Paris : Éditions Ophrys, 1974
276 p.. - (Association des publications près les Universités de Strasbourg ; 151). - Bibliography: p. 265-271
2-7080-0407-7 Ac.2633 e (151)
0854360

ACQUISITION. Knowledge

STEMMER, Nathan
The roots of knowledge. - Oxford : Basil Blackwell, 1983
viii,200p : ill. - Bibliography: p193-196. - Includes index
0-631-13382-8 X.520/32620
3056341

ACQUISITION. Knowledge
Attitudes of Catholic Church

CROWLEY, Aleister
The banned lecture. - [Carnforth] : [Society for the Propagation of Religious Truth], [c1982]
[24]p
B82-20932 X.958/9244
2716887

ACQUISITION. Knowledge
Effects of sex discrimination - Conference proceedings

PRISM OF SEX : TOWARD ON EQUITABLE PURSUIT OF KNOWLEDGE (Conference : 1977 : Madison)
The prism of sex. - Madison ; London : University of Wisconsin Press, 1979
vii,286,[1]p. - Includes index
0-299-08010-2 X.520/21200
2144360

ACQUISITION. Knowledge
Methodology. Theories of Descartes, René & Locke, John, 1632-1704

SCHOULS, Peter A.
The imposition of method. - Oxford : Clarendon Press, 1980
x,271p. - Bibliography: p.261-265. - Includes index
0-19-824613-7 X.529/38694
1902849

ACQUISITION. Knowledge
Role of education

BROWNHILL, R. J. (Robert J)
Education and the nature of knowledge. - London : Croom Helm, c1983
135p. - (New patterns of learning series). - Includes index
0-7099-0654-4 X.529/53654
2890143

ACQUISITION. Knowledge
Role of education - Conference proceedings

CONFERENCE ON SCHOOLING AND THE ACQUISITION OF KNOWLEDGE (1975 : Shelter Island)
Schooling and the acquisition of knowledge. - Hillsdale : Erlbaum ; New York ; London : Distributed by Wiley, 1977
xi,448p : ill. - Includes bibliographies and index
0-470-99293-x X.520/12388
2890143

ACQUISITION. Knowledge
Role of human cognition & perception

DRETSKE, Fred I. (Fred Irwin)
Knowledge and the flow of information. - Oxford : Blackwell, 1981
xiv,273p : ill. - Includes index
0-631-12765-8 X.520/25282
2572273

RIEDL, Rupert
Biology of knowledge. - Chichester : Wiley, c1984
lx,252p : ill. - Bibliography: p229-240. - Includes index
0-471-10309-8 X.622/19484
2572273

ACQUISITION. Land. England
Acquisition & disposal by local authorities. Law

RATCLIFFE, John
Land acquisition and disposal. - Edinburgh : Capital Planning Information, 1982
65p : ill. - (Planning reviews). - Bibliography: p64-65
0-906011-13-2 X.322/12947
3046788

ACQUISITION. Land. Palestine
By Jews, 1948-1980

HADAWI, Sami
Zionism and the lands of Palestine. - London : EAFORD, [1982?]
19p. - (Paper ; no.17)
B84-06260 X.529/58317
2999722

ACQUISITION. Land. Proposed trunk roads. Scotland
Application. Plans. Preparation

LAND acquisition for motorways and trunk roads. - Rev. - [Edinburgh] : Scottish Development Department (Roads Division), 1983
15p
B84-09797 X.629/22013
3041387

ACQUISITION. Land. United States
By railways. Subsidies. Policies of government, to date

RAE, John Bell
The development of railway land subsidy policy in the United States. - New York : Arno, 1979
337p. - (The development of public land law in the United States). - Includes bibliography
0-405-11366-8 X.622/22546
9143696

ACQUISITION. Language skills. Adolescents deprived of personal contact

CURTISS, Susan
Genie. - New York ; London : Academic Press, 1977
xvi,288p : ill. - (Perspectives in neurolinguistics and psycholinguistics). - Bibliography: p.277-283. - Includes index
0-12-196350-0 X.981/20986
1377353

ACQUISITION. Language skills. Aphasic children
LANGUAGE development and aphasia in children / edited by R.W. Rieber. - New York ; London : Academic Press, 1980
xv,236p. - (Perspectives in neurolinguistics and psycholinguistics). - Includes index
0-12-588280-7 X.322/9527 Woolwich
2227002

ACQUISITION. Language skills. Babies
Use of holophrases

RODGON, Maris Monitz
Single-word usage, cognitive development, and the beginnings of combinatorial speech. - Cambridge : Cambridge University Press, 1976
ix,163p : ill. - Bibliography: p.156-158. - Includes index
0-521-20884-x X.981/20292
0670324

ACQUISITION. Language skills. Babies, to 18 months

HALLIDAY, Michael Alexander Kirkwood
Learning how to mean. - London : Edward Arnold, 1975
[12],164p. - (Explorations in language study). - Bibliography: p.159-164
0-7131-1872-5 X.909/40082
0371270

ACQUISITION. Language skills. Blind children - Conference proceedings

LANGUAGE acquisition in the blind child / edited by Anne E. Mills. - London : Croom Helm, c1983
235p : ill. - Bibliography: p197-214. - Includes index
0-7099-1768-6 X.529/56975
296368x

ACQUISITION. Language skills. Children

BLOOM, Lois
Language development and language disorders. - New York ; London : Wiley, 1978
xii,689p : ill. - (Wiley series on communication disorders). - Bibliography: p.616-654. - Includes index
0-471-08220-1 X.981/21189
0530514

BRUNER, Jerome S. (Jerome Seymour)
Child's talk. - Oxford : Oxford University Press, 1983
144p : ill. - Includes index
0-19-857613-7 X.950/33824
0530514
- [Another Copy: paperback.]
ISBN 0-19-857511-4 X.950/33825

Entry into early language. - [Swansea] : University College of Swansea, 1975
30p. - (Charles Gittins memorial lecture ; 1973). - Bibliography: p.28-30
0-901626-92-9 X.909/40172
0530514

BRYEN, Diane Nelson
Inquiries into child language. - Boston, Mass. ; London : Allyn and Bacon, [1982?]
x,436p : ill. - Bibliography: p411-426. - Includes index
0-205-07642-4 X.950/13865
0530514

The CHILD'S construction of language / edited by Werner Deutsch. - London : Academic Press, 1981
x,393p : ill. - (Behavioural development). - Includes bibliographies and index
0-12-213580-6 X.520/26418
0530514

CRUTTENDEN, Alan
Language in infancy and childhood. - Manchester : Manchester University Press, 1979
xiv,193p : ill. - Bibliography: p.171-188. - Includes index
0-7190-0706-2 X.989/78928
0530514

DE VILLIERS, Jill G.
Language acquisition. - Cambridge, Mass. ; London : Harvard University Press, 1978
xi,312p : ill. - Bibliography: p.281-303. - Includes index
0-674-50931-5 X.981/21181
0530514

FRANCESCATO, Giuseppe
Il linguaggio infantile, strutturazione e apprendimento. - 3. ed. riveduta. - Torino : Giulio Einaudi, 1975
285 p.. - (Piccola biblioteca Einaudi ; n. 131)
r6006531 X.908/33530
0530514

GEEST, Ton van der
Some aspects of communicative competence and their implications for language acquisition. - Assen : Van Gorcum, 1975
[6], ii, 263 p. : ill. - Bibliography : p. 256-261. - Includes index
90-232-1261-4 X.900/25252
0530514

LANGUAGE acquisition / edited by Eric Wanner and Lila R. Gleitman. - Cambridge : Cambridge University Press, 1982
x,532p : ill. - Bibliography: p487-520. - Includes index
0-521-23817-x X.520/33780
0530514

LIGHTFOOT, David
The language lottery. - Cambridge, Mass. ; London : MIT Press, c1982.
xii,224p : ill. - Bibliography: p213-219. - Includes index
0-262-12096-8 X.950/23235
0530514

The LOGICAL problem of language acquisition / edited by C.L. Baker and John J. McCarthy. - Cambridge, Mass. ; London : MIT, c1981
xii,358p : ill. - (MIT Press series on cognitive theory and mental representation). - Bibliography: p330-344. - Includes index
0-262-02159-5 X.950/12679
0530514

MARINI MARIUCCI, Fatima
Creatività e linguaggi nella prescuola e a scuola. - Roma : Armando Armando, [1976]
183p. - (Serie di linguistica teorica e applicata ; 8)
r6078392 X.519/39099
0530514

PETERS, Ann M.
The units of language acquisition. - Cambridge : Cambridge University Press, 1983
xii,131p. - (Cambridge monographs and texts in applied psycholinguistics). - Bibliography: p115-121. - Includes index
0-521-27071-5 X.950/32812
0530514

RAMGE, Hans
Spracherwerb. - Tübingen : Niemeyer, 1973
x, 110 p.. - (Germanistische Arbeitshefte ; 14). - Includes bibliographies
3-484-25016-x X.900/17078
0530514
- [Another copy] X:900/18281

ACQUISITION. Language skills. Children –continued

SOLAN, Lawrence
Pronominal reference. – Dordrecht ; Lancaster : Reidel, c1983
xii,237p : ill. – (Studies in theoretical psycholinguistics). – Bibliography: p224–232. – Includes index
90–277–1495–9 X.950/23953
0530514

TOUGH, Joan
The development of meaning. – London : Allen and Unwin, 1977
[9],198p. – (Unwin education books ; 33). – Bibliography: p.193–195. – Includes index
0–04–372020–x X.989/51346
0530514

WEXLER, Kenneth
Formal principles of language acquisition. – Cambridge, Mass. ; London : M.I.T. Press, 1980
xix,647p : ill. – Bibliography: p.615–629. – Includes index
0–262–23099–2 X.981/22485
0530514

ACQUISITION. Language skills. Children
– Bibliographies

ABRAHAMSEN, Adele A.
Child language. – Baltimore : London : University Park Press, 1977
xxv,381p. – Includes index
0–8391–1128–2 X.909/41797
1244809

ACQUISITION. Language skills. Children
– Conference proceedings

EARLY language / edited by Richard L. Schiefelbusch and Diane D. Bricker. – Baltimore : University Park Press, c1981
x,606p. – (Language intervention series ; 6)
0–8391–1618–7 X.950/8239
0253227

INTERNATIONAL CHILD LANGUAGE SYMPOSIUM (3rd : 1975 : School of Oriental and African Studies)
The development of communication. – Chichester : Wiley, 1978
xxv,498p : ill. – Includes bibliographies and index
0–471–99628–9 X.981/21060
0530514

INTERNATIONAL CONGRESS FOR THE STUDY OF CHILD LANGUAGE (2nd : 1981 : Vancouver)
Proceedings of the Second International Congress for the Study of Child Language. – Lanham ; London : University Press of America.
Vol.2. – c1984
viii,520p : ill. – Includes bibliographies
0–8191–3694–8 X.950/37173
0530514

LANGUAGE and language acquisition / edited by F. Lowenthal, F. Vandamme and J. Cordier. – New York ; London : Plenum, c1982
xvi,379p : ill. – Includes bibliographies and index
0–306–41128–8 X.955/1618
0253227

LANGUAGE behavior in infancy and early childhood / editor Rachel E. Stark ; sponsored by Johnson & Johnson Baby Products Company. – New York ; Oxford : Elsevier/North-Holland, c1981
xix,479p : ill. – Includes bibliographies and index
0–444–00627–3 X.520/25717
0253227

LANGUAGE INPUT AND ACQUISITION (Conference : 1974 : American Academy of Arts and Sciences)
Talking to children. – Cambridge : Cambridge University Press, 1977
x,369p : ill. – Bibliography: p.341–369
0–521–21318–5 X.981/20852
0253227

ACQUISITION. Language skills. Children
– For teaching

DERRICK, June
The child's acquisition of language. – Windsor : NFER, 1977
[5],53p. – Bibliography: p.50–53
0–85633–110–4 X.909/41526
056740x

DISTEFANO, Philip
Elementary language arts. – New York ; Chichester : Wiley, c1984
xiii,423p : ill. – Includes bibliographies and index
0–471–86658–x X.950/35934
056740x

FISHER, Carol J.
Children's language and the language arts. – 2nd ed. – New York ; London : McGraw-Hill, c1982
xv,361p : ill. – Includes index
0–07–021108–6 X.950/13868
056740x

FRANCIS, Hazel
Language in teaching and learning. – London : Allen and Unwin, 1977
3–139p. – (Unwin education books ; 36). – Bibliography: p.131–136. – Includes index
0–04–407003–9 X.909/42566
056740x

LINDFORS, Judith Wells
Children's language and learning. – Englewood Cliffs : London : Prentice-Hall, 1980
xv,447p : ill. – Bibliography: p.428–439. – Includes index
0–13–131953–1 X.520/20968
056740x

MOYLE, Donald
Children's words. – London : Grant McIntyre, 1982
viii,212p : ill. – Bibliography: p207–210. – Includes index
0–86216–032–4 X.958/20197
056740x

ACQUISITION. Language skills. Children
– Serials

FIRST language. – Vol.1, pt.1, no.1 (Feb.1980)–. – Chalfont St Giles : Alpha Academic, 1980–
v.
B81–02341 P.901/3228
0326992

ACQUISITION. Language skills. Children
Neurophysiological aspects

LANGUAGE development and neurological theory / edited by Sidney J. Segalowitz, Frederic A. Gruber. – New York ; London : Academic Press, 1977
xv,376p : ill. – (Perspectives in neurolinguistics and psycholinguistics). – Includes bibliographies and index
0–12–635650–5 X.981/20957
1348442

ACQUISITION. Language skills. Children
Neuropsychological aspects

NEUROPSYCHOLOGY of language, reading, and spelling / edited by Ursula Kirk. – New York ; London : Academic Press, 1983
xviii,283p : ill. – (Educational psychology). – Includes bibliographies and index
0–12–409680–8 X.322/12097
2929775

ACQUISITION. Language skills. Children
Phonological aspects – Conference proceedings

CHILD phonology / edited by Grace H. Yeni-Komshian, James F. Kavanagh, Charles A. Ferguson. – New York ; London : Academic Press. – (Perspectives in neurolinguistics, neuropsychology and psycholinguistics)
Vol.1: Production. – 1980
xiv,304p : ill. – Includes bibliographies and index
0–12–770601–1 X.950/3926
2393824

ACQUISITION. Language skills. Children
Psychological aspects

KARMILOFF-SMITH, Annette
A functional approach to child language. – Cambridge : Cambridge University Press, 1979
ix,258p : ill. – (Cambridge studies in linguistics ; 24). – Bibliography: p.242–251. – Includes index
0–521–22416–0 X.981/21645
191975x

A functional approach to child language. – Cambridge : Cambridge University Press, 1979
vi,258p : ill. – Bibliography: p242–251. – Includes index
0–521–28549–6 X.950/8094
191975x

ACQUISITION. Language skills. Children
related to cognitive development

BEILIN, Harry
Studies in the cognitive basis of language development. – New York ; London : Academic Press, 1975
xi,420p : ill. – (The child psychology series). – Bibliography: p.399–413. – Includes index
0–12–085650–6 X.981/20155
0586951

CHILDREN thinking through language / edited by Michael Beveridge. – London : Edward Arnold, 1982
x,272p : ill. – Includes bibliographies and index
0–7131–6352–6 X.529/47240
0586951

The FUNCTIONS of language and cognition / edited by Grover J. Whitehurst, Barry J. Zimmerman. – New York ; London : Academic Press, 1979
xiii,313p : ill. – Includes bibliographies and index
0–12–747050–6 X.981/21912
0586951

HAMLYN, D. W. (David Walter)
Experience and the growth of understanding. – London : Routledge and Kegan Paul, 1978
xii,159p. – (International library of the philosophy of education). – Bibliography: p.154–156. – Includes index
0–7100–8735–7 X.529/32027
0586951

Experience and the growth of understanding. – London : Routledge and Kegan Paul, 1979
xii,159p. – (International library of the philosophy of education). – Bibliography: p.154–156. – Includes index
0–7100–0336–6 X.519/30072
0586951

LANGUAGE and operational thought / edited by Barbara Z Presseisen, David Goldstein, Marilyn H. Appel. – New York ; London : Plenum Press, 1978
xv,210p : ill. – (Topics in cognitive development ; vol.2). – Includes bibliographies and index
0–306–33002–4 X.981/21069
0586951

LANGUAGE learning and thought / edited by John Macnamara. – New York ; London : Academic Press, 1977
xii,296p : ill. – (Perspectives in neurolinguistics and psycholinguistics). – Includes bibliographies and index
0–12–464750–2 X.520/12114
0586951

OLÉRON, Pierre
Language and mental development. – Hillsdale : Erlbaum ; New York ; London : Distributed by Wiley, 1977
ix,182p : ill. – Bibliography: p.167–177. – Includes index
0–470–99027–9 X.981/20884
0586951

ACQUISITION. Language skills. Children
related to linguistic change

BARON, Naomi S.
Language acquisition and historical change. – Amsterdam ; Oxford : North-Holland Publishing Co., 1977
xiv,320p : ill. – (North-Holland linguistic series ; 36). – Bibliography: p.291–308. – Includes index
0–444–85077–5 X.989/52232
1347918

ACQUISITION. Language skills. Children
related to social class

BERNSTEIN, Basil
Class, codes and control. – 2nd ed. – London : Routledge and Kegan Paul. – (Primary socialization, language and education ; 4)
Vol.1: Theoretical studies towards a sociology of language. – 1977
xiv,266p : ill. – Includes bibliographies and index
0–7100–8654–7 X.0529/254(12)
0296813

Vol.3: Towards a theory of educational transmissions. – 1975 [i.e. 1976]
vii,167p : ill. – (Set books / Open University). – Includes bibliographies and index
0–7100–8488–9 X.0529/254(9)
0296813

Vol.3: Towards a theory of educational transmissions. – Revised ed. – 1977
viii,207p : ill. – Includes bibliographies and index
0–7100–8666–0 X.0529/254(3)
0296813

ACQUISITION. Language skills. Children
related to socialisation

CHILDREN and language / edited by Sinclair Rogers. – London : Oxford University Press, 1975
ix,346p : ill. – Includes bibliographies
0–19–437007–0 X.989/40339
0534943

ACQUISITION. Language skills. Children
Research. Applications of linguistics

CRYSTAL, David
Child language, learning and linguistics. – London : Edward Arnold, 1976
106p : ill. – Includes index
0–7131–5890–5 X.909/41327
1074296

ACQUISITION. Language skills. Children
Role of families

TAYLOR, Denny
Family literacy. – Exeter, N.H. ; London : Heinemann Educational, 1983
120p,[1]p of plates : ill. – Bibliography: p115–120
0–435–08204–3 X.950/29783
3073793

ACQUISITION. Language skills. Children
Role of listening

SLADE, Peter
Listening and language development. - [Bromley] : Educational Drama Association, [1977]
[1],16p
B77-29356 X.909/41789
1245244

ACQUISITION. Language skills. Children
Role of mothers. Implications for language disordered children

CLEZY, Gillian
Modification of the mother-child interchange in language, speech and hearing. - London : Edward Arnold, 1979
xiii,162p : ill. - Includes bibliographies and index
0-7131-6226-0 X.520/14705
1998854

ACQUISITION. Language skills. Children
Semantic aspects

LEONARD, Laurence B.
Meaning in child language. - New York ; London : Grune and Stratton, 1976
xii,259p : ill. - Bibliography: p.245-252. - Includes index
0-8089-0977-0 X.981/22152
2051613

ACQUISITION. Language skills. Children
Theories

ATKINSON, Martin
Explanations in the study of child language development. - Cambridge : Cambridge University Press, 1982
viii,289p. - (Cambridge studies in linguistics ; 35). - Bibliography: p265-280. - Includes index
0-521-24302-5 X.520/26615
2705621

ACQUISITION. Language skills. Children, 4-6 years

FERREIRO, Emilia
Literacy before schooling. - Exeter, N.H. ; London : Heinemann Educational, 1983, c1982
xii,289p : ill. - Bibliography: p287-289
0-435-80474-x X.950/21204
2903350

ACQUISITION. Language skills. Children, 7-13 years. Great Britain
- For teaching

SCHOOLS COUNCIL COMMUNICATION SKILLS PROJECT : 7-13
Talk for teaching and learning. - London : Ward Lock ; Cardiff : Drake Educational Associates, 1979
vii,247p,6 leaves of plates : ill. - Bibliography: p.238-239. - Includes index
0-7062-3905-9 X.520/21244
2068648

ACQUISITION. Language skills. Children, to 6 years

DE VILLIERS, Peter A.
Early language. - London : Open Books, 1979
157p. - (The developing child). - Bibliography: p.151-152. - Includes index
0-7291-0009-x X.529/33280
1767356

ACQUISITION. Language skills. Children, to 7 years
- For teaching

SCHOOLS COUNCIL COMMUNICATION SKILLS IN EARLY CHILDHOOD PROJECT
Talking and learning. - London : Ward Lock ; Cardiff : Drake Educational Associates, 1977
viii,327p : ill. - Bibliography: p.308-309. - Includes index
0-7062-3610-6 X.520/12489
1452835

ACQUISITION. Language skills. Children, to 10 years
Teaching

MEERS, Hilda J.
Helping our children talk. - London : Longman, 1976
v,162p : ill. - (Longman early childhood education). - Bibliography: p.152-157. - Includes index
0-582-25014-5 X.989/51432
1085840
- [Another copy: paperback]
ISBN 0-582-25010-2 X.909/41374

ACQUISITION. Language skills. Children, to 12 years
- For teaching

ANDERSON, N. V.
Words, meanings and classrooms. - Ely : Ely Area Resource Organisation, 1978
60p in various pagings. - Bibliography(5p.)
0-904463-51-6 X.900/25690
1799231

ACQUISITION. Language skills. Dyslexic children
Teaching

HICKEY, Kathleen
Dyslexia. - London : The author, 1977
x,238p(8 fold) : ill. - Includes bibliographies
0-9505386-0-4 Cup.915/107
1883585

ACQUISITION. Language skills. German speaking children

BIELEFELD, Renate
Untersuchungen zum Spracherwerb 2-4 jähriger Kinder unter besonderer Berucksichtigung semantischer Aspekte. - Hamburg : R. Bielefeld, 1972
iv,431p. - Bibliography
r6305439 X.958/12390
8884021

ACQUISITION. Language skills. Hearing disordered children

QUIGLEY, Stephen P.
Language and deafness. - San Diego : College-Hill ; London : Croom Helm, c1984
xvi,277p. - Bibliography: p237-263. - Includes index
0-7099-2103-9 X.322/13616
0062200

STRENG, Alice H.
Language, learning and deafness. - New York ; London : Grune and Stratton, 1978
viii,232p : ill. - Includes bibliographies and index
0-8089-1060-4 X.520/9647
0062200

ACQUISITION. Language skills. Man

PAPERS on language acquisition, language learning and language teaching / edited by Henning Wode. - Heidelberg : Groos, 1983
223p. - (Studies in descriptive linguistics ; v.11). - Includes bibliography
3-87276-260-5 X.0900/613(11)
9065636

ACQUISITION. Language skills. Man
Individual variation

INDIVIDUAL differences in language ability and language behaviour / edited by Charles J. Fillmore, Daniel Kempler, William S-Y. Wang. - New York ; London : Academic Press, 1979
xv,346p : ill. - (Perspectives in neurolinguistics and psycholinguistics). - Includes bibliographies and index
0-12-255950-9 X.981/22398
2105764

ACQUISITION. Language skills. Pre-school children

BROWN, Roger, 1925-
A first language. - Harmondsworth : Penguin, 1976
493p : ill. - (Penguin education). - Bibliography: p.467-475. - Includes index
0-14-081054-4 X.908/40470
006002x

FELDBUSCH, Elisabeth
Sprachförderung im Vorschulalter. - Marburg, 1976
xvi,359p. - (Deutsche Dialektgeographie ; Bd 81. Kindersprache und Schule in Hessen ; 1)
3-7708-0554-2 12964.k.1/81 Woolwich
1212400

IWAMURA, Susan Grohs
The verbal games of pre-school children. - London : Croom Helm, 1980
237p. - Bibliography: p.226-234. - Includes index
0-7099-0211-5 X.989/54267
1931547

LANGUAGE development in pre-school children / research director Gordon Wells. - Bristol : University of Bristol School of Education.
Transcripts of boys aged 15 months [A]. - 1975
149p in various pagings
0-903024-11-x X.0902/1006(4)
0599166

Transcripts of boys aged 15 months [B]. - 1975
156p in various pagings
0-903024-12-8 X.0902/1006(3)
0599166

LANGUAGE development in pre-school children / research director Gordon Wells. - Bristol : University of Bristol School of Education.
Transcripts of boys aged 15 months [C]. - 1975
147p in various pagings
0-903024-13-6 X.0902/1006(5)
0599166

Transcripts of boys aged 15 months [D]. - 1975
114p in various pagings
0-903024-14-4 X.0902/1006(10)
0599166

Transcripts of boys aged 39 months [A]. - 1975
191p in various pagings
0-903024-19-5 X.0902/1006(9)
0599166

Transcripts of boys aged 39 months [B]. - 1975
205p in various pagings
0-903024-20-9 X.0902/1006(8)
0599166

Transcripts of boys aged 39 months [C]. - 1975
181p in various pagings
0-903024-21-7 X.0902/1006(7)
0599166

Transcripts of boys aged 39 months [D]. - 1975
124p in various pagings
0-903024-22-5 X.0902/1006(6)
0599166

Transcripts of girls aged 15 months [A]. - 1975
171p in various pagings
0-903024-07-1 X.0902/1006(16)
0599166

Transcripts of girls aged 15 months [B]. - 1975
154p in various pagings
0-903024-08-x X.0902/1006(15)
0599166

Transcripts of girls aged 15 months [C]. - 1975
119p in various pagings
0-903024-09-8 X.0902/1006(1)
0599166

Transcripts of girls aged 15 months [D]. - 1975
161p in various pagings
0-903024-10-1 X.0902/1006(2)
0599166

Transcripts of girls aged 39 months [A]. - 1975
196p in various pagings
0-903024-15-2 X.0902/1006(11)
0599166

Transcripts of girls aged 39 months [B]. - 1975
195p in various pagings
0-903024-16-0 X.0902/1006(12)
0599166

Transcripts of girls aged 39 months [C]. - 1975
176p in various pagings
0-903024-17-9 X.0902/1006(13)
0599166

Transcripts of girls aged 39 months [D]. - 1975
206p in various pagings
0-903024-18-7 X.0902/1006(14)
0599166

OKSAAR, Els
Language acquisition in the early years. - London : Batsford Academic and Educational, 1983
vi,232p : ill. - Bibliography: p205-220. - Includes index
0-7134-3694-8 X.950/21874
006002x

ACQUISITION. Language skills. Pre-school children. Sweden
Semantic aspects

LINDHAGEN, Karin
Semantic relations in Swedish children's early sentences. - Uppsala : Department of Psychology, 1976
172p. - (Studia psychologica upsaliensia ; 3). - Includes bibliographies
91-554-0382-4 X.0510/257(3)
1428535

ACQUISITION. Language skills. Socially disadvantaged children. United States
- For teaching

ALDER, Sol
Poverty children and their language. - New York ; London : Grune and Stratton, 1979
xv,320p : ill. - Includes bibliographies and index
0-8089-1194-5 X.529/37190
210394x

ANASTASIOW, Nicholas J.
Language & reading strategies for poverty children. - Baltimore, Md. : University Park Press, c1982
xv,222p : ill,music. - Includes bibliographies
0-8391-1709-4 X.529/47522
210394x

ACQUISITION. Language skills. Students. Schools. United States

DESTEFANO, Johanna S.
Language, the learner & the school. - New York ; Chichester : Wiley, 1978
xi,221p. - Includes bibliographies and index
0-471-02378-7 X.909/42522
1482394

ACQUISITION. Language skills. Twins, 1-3 years

SAVIĆ, Svenka
How twins learn to talk. - London : Academic Press, 1980
xiv,195p : ill. - Bibliography: p182-190. - Includes index
0-12-619580-3 X.520/22590
2126133

ACQUISITION. Mathematical concepts
By school students

ACQUISITION of mathematics concepts and processes / edited by Richard Lesh, Marsha Landau. - New York ; London : Academic, 1983
xii,409p : ill. - (Developmental psychology series). - Includes bibliographies and index
0-12-444220-x X.622/20892
0559520

NUMBERS and the world of things / [edited by] E.A. Lunzer, A.W. Bell and C.M. Shiu. - Nottingham : University of Nottingham, School of Education, Shell Centre for Mathematical Education, 1976
viii,228p : ill
0-85359-040-0 X.615/1702 Woolwich
0559520

ACQUISITION. Monographs in print. Stock. Libraries. United States

AMERICAN LIBRARY ASSOCIATION. Bookdealer-Library Relations Committee
Guidelines for handling library orders for in-print monographic publications. - [Chicago] : A.L.A., 1973
[2],16p
0-8389-3148-0 2719.bb.25
130271X

ACQUISITION. Motor skills. Children, 4-12 years related to cognitive development

PIAGET, Jean
The grasp of consciousness. - London : Routledge and Kegan Paul, 1977
viii,360p : ill. - Includes index
0-7100-8400-5 X.529/31514
0531294
Success and understanding. - London : Routledge and Kegan Paul, 1978
viii,236p : ill. - Includes index
0-7100-8946-5 X.529/32700
0531294

ACQUISITION. Number skills. Children

CHILDREN'S counting types / Leslie P. Steffe ... [et al.]. - New York : Praeger, 1983
xviii,152p : ill. - Bibliography: p137-144
0-03-063744-9 X.520/32581
3066568

ACQUISITION. Photosynthesis. Concepts. Biology. Curriculum subjects. Secondary schools. Scotland
By students

ARNOLD, Brian
An investigation of the development of the concept photosynthesis to Scottish Certificate of Education 'O' grade. - [Aberdeen] : Aberdeen College of Education, 1980
221p in various pagings : ill. - Bibliography: p101-104
B81-21924 X.525/4488
248434x

ACQUISITION. Pragmatics. Language skills. Children

DEVELOPMENT pragmatics / edited by Elinor Ochs, Bambi B. Schieffelin. - New York ; London : Academic Press, 1979
xv,437p : ill. - Bibliography: p.415-429. - Includes index
0-12-524550-5 X.981/22372
2081393

ACQUISITION. Pragmatics. Language skills. Children, to 6 years

BATES, Elizabeth
Language and context. - New York ; London : Academic Press, 1976
xv,375p : ill. - (Language, thought and culture). - Bibliography: p.355-365. - Includes index
0-12-081550-8 X.981/20472
0788856

ACQUISITION. Printed ephemera. Local history. England

STURGES, R. P. (Rodney Paul)
An investigation of local publications. - [Loughborough] : Loughborough University, Department of Library and Information Studies, 1983
66p : ill. - (BLRDD report ; no.5645)
0-904641-20-1 X.955/2168
3072533

ACQUISITION. Private companies. England
Law

KNIGHT, W. J. L. (William John Langford)
The acquisition of private companies. - London : Oyez Publishing, 1975
xvi,153p. - Includes index
0-85120-260-8 X.100/16340
0648817
The acquisition of private companies. - 2nd ed. - [London] : Oyez Publishing, 1979
xxii,265p : 2 ill. - Includes index
0-85120-411-2 X.519/29896
0648817
The acquisition of private companies. - 3rd ed. - London : Oyez Longman, 1982
xviii,228p : ill. - Includes index
B82-30501 X.200/38639
0648817

ACQUISITION. Provincial Building Society mortgages
By women. Equality of opportunity

MORTGAGE lending and practice. - Manchester : Equal Opportunities Commission, [1982]
[16]p
0-905829-61-1 BS.433/33
2891123

ACQUISITION. Reading habits. Children

CHAMBERS, Aidan
Introducing books to children. - 2nd ed., completely rev. and expanded. - London : Heinemann Educational, 1984
vii,223p. - Bibliography: p215-216. - Includes index
0-435-80180-5 X.950/35790
3172155

ACQUISITION. Reading skills. Children, to 10 years
- Conference proceedings

HYMAN BLUMBERG SYMPOSIUM ON RESEARCH IN EARLY CHILDHOOD EDUCATION (5th : 1974 : Johns Hopkins University)
Aspects of reading acquisition. - Baltimore ; London : Johns Hopkins University Press, 1976
xvii,222p : ill. - Includes bibliographies and index
0-8018-1800-1 X.981/20438
0766321

ACQUISITION. Reading skills. Children, to 10 years
Theories

SMITH, Donald Eugene Phillip
A technology of reading and writing. - New York ; London : Academic Press. - (Educational psychology)
Vol.1: Learning to read and write. - 1976
xi,163p : ill. - Bibliography: p.155-158. - Includes index
0-12-651701-0 X.0981/1511
0748552
Vol.2: Criterion-referenced tests for reading and writing. - 1977
xii,279p. - (Educational psychology). - Includes bibliographies and index
0-12-651702-9 X.0981/1511
0748552
Vol.3: The adaptive classroom. - 1976
x,320p : ill. - Bibliography: p.313-316. - Includes index
0-12-651703-7 X.981/20735
0748552
Vol.4: Designing instructional tasks. - 1978
x,212p : ill. - Bibliography: p.197. - Includes index
0-12-651704-5 X.0981/1511
0748552

ACQUISITION. Reading skills. Pre-school chilren

PAYTON, Shirley
Developing awareness of print. - Birmingham : Educational review, 1984
vi,117p : ill. - (Off-set publication ; no.2). - Bibliography: p111-117
0-7044-0700-0 X.525/10235
3216136

ACQUISITION. Real property
By foreigners. Law

LEGAL aspects of alien acquisition of real property / edited by Dennis Campbell. - Deventer ; London : Kluwer, 1980
206p. - Includes index
90-268-1169-1 OPL
282793x
- [Another copy] X.520/28837

ACQUISITION. Real property. England
By Christian church. Law. Statutes, 1279-1500

RABAN, Sandra
Mortmain legislation and the English church 1279-1500. - Cambridge : Cambridge University Press, 1982
xiii,216p. - (Cambridge studies in medieval life and thought. 3rd series ; v.17). - Bibliography: p195-204. - Includes index
0-521-24233-9 X.0909/360(17)
2787024

ACQUISITION. Real property. European Community countries

CHOWN, John F. (John Francis)
The acquisition of assets, companies and real estate in Belgium, France, Germany, The Netherlands, United Kingdom. - London : 'Financial Times', 1974
[8],157p
0-903199-03-3 X.512/5060
0522503

ACQUISITION. Scientific knowledge
Role of cognition

MEY, Marc de
The cognitive paradigm. - Dordrecht ; London : Reidel, 1982
xx,314p. - (Sociology of the sciences). - Bibliography: p283-303. - Includes index
90-277-1382-0 X.622/13871
280039x

ACQUISITION. Scientific knowledge
Theories of Kuhn, Thomas, S

YOUNG, R. H. (Roy Harold)
The methodological limitations of Kuhn's model of science. - Salford : University of Salford, Department of Geography, 1978
15p. - (Discussion paper in geography / Department of Geography, University of Salford ; no.8). - Bibliography: p15
B83-02751 X.805/3357
2813718

ACQUISITION. Sculptures from Parthenon. Exhibits. British Museum. Department of Greek and Roman Antiquities

ST. CLAIR, William
Lord Elgin and the Marbles. - [New ed.]. - Oxford : Oxford University Press, 1983
x,311p,[9]p of plates : ill. - (Oxford paperbacks). - Bibliography: p279-281. - Includes index
0-19-285140-3 X.808/39590
308678x

ACQUISITION. Sculptures from Parthenon. Exhibits. British Museum. Department of Greek and Roman Antiquities
Acquisition & appreciation, 1799-1816

ROTHENBERG, Jacob
'Descensus ad terram'. - New York ; London : Garland, 1977
[4],xii[i.e.xv],469p,76p of plates : ill. - (Outstanding dissertations in the fine arts). - Bibliography: p.456-468
0-8240-2726-4 X.429/10616 Woolwich
2053470

ACQUISITION. Serials. Stock. Libraries
Selection - Bibliographies

MARSHALL, Joan K.
Serial for libraries. - New York : Neal/Schuman ; Santa Barbara ; Oxford : ABC-Chio, [1979]
xv,494p.
0-87436-280-6 2719.i.66
2425513

ACQUISITION. Serials. Stock. Libraries. Austria
1945-

UNIVERSITÄT GRAZ. Bibliothek
Regeln für die Erfassung der Zeitschriften und Serien ('Spiegel'-Kartei). - Graz : Universitätsbibliothek, 1978
96leaves. - (Schriftenreihe EDV-Projekt / Universitätsbibliothek Graz ; 1)
r6399222 X.955/917
8939462

ACQUISITION. Serials. Stock. Libraries. United States
From serials agents - Directories

KATZ, William Armstrong
Guide to magazine and serial agents. - New York ; London : Bowker, 1975
xii,239p. - Includes index
0-8352-0789-7 X.981/20173
0598046

ACQUISITION. Skills in sports. Man

KNAPP, Barbara
Skill in sport. - London : Routledge and Kegan Paul, 1977
xii,204p,8p of plates : ill. - Includes index
0-7100-1691-3 X.619/18195 Woolwich
1304631

ACQUISITION. Skills. Skilled personnel. Industries. Great Britain
1870-1914. Social aspects

MORE, Charles
Skill and the English working class, 1870-1914. - London : Croom Helm, c1980
252p. - (Croom Helm social history series). - Bibliography: p240-247. - Includes index
0-7099-0327-8 X.529/39449
2090066

ACQUISITION. Software. Computer systems
Management aspects

GLASEMAN, Steven
Comparative studies in software acquisition. - [Lexington] : Lexington Books, 1982
xii,131p. - (Lexington Books series in computer science). - Bibliography: p123-126. - Includes index
0-669-05422-4 SRL
2879964

ACQUISITION. Speech skills. Children, to 3 years
Role of mothers

SCHACHTER, Frances Fuchs
Everyday mother talk to toddlers. - New York ; London : Academic Press, 1979
xix,211p : ill. - Bibliography: p.199-205. - Includes index
0-12-621360-7 X.520/14667
1995057

ACQUISITION. Speech skills. Man
Psycholinguistic aspects

OCHS, Elinor
Acquiring conversational competence. - London : Routledge & Kegan Paul, 1983
vi,208p. - (Language, education and society). - Bibliography: p192-202. - Includes index
0-7100-9459-0 X.950/27566
3034380

ACQUISITION. Statistical publications. Stock. Libraries. Universities. Camden (London Borough)
Committees: University of London. Library Resources Co-ordinating Committee. Subject Sub-Committee in Statistics

UNIVERSITY OF LONDON. Library Resources Co-ordinating Committee. Subject Sub-Committee in Statistics
Minutes of first meeting, held at Senate House on 5 July 1977. - London : The Sub-Committee, [1977]
[5] leaves
r6072632 2719.gg.54
1720090

ACQUISITION. Stock. Libraries

BACKGROUND readings in building library collections. - 2nd ed.. - Metuchen ; London : Scarecrow Press ; [Folkestone] : [Distributed by Bailey and Swinfen], 1979
ix,417p : 1 ill. - Includes index
0-8108-1200-2 2719.cc.90
1936484

ACQUISITION. Stock. Libraries
- Serials

LIBRARY acquisitions. - Vol.1, no.1- ; Jan. 1977-. - New York ; Oxford : Pergamon, 1977- : ill
B77-23059 2719.aa.89
1170317

ACQUISITION. Stock. Libraries
Selection

BONK, Wallace John
Building library collections. - 5th ed.. - Metuchen ; London : Scarecrow Press ; [Folkestone] : [Distributed by Bailey and Swinfen], 1979
[1],ix,380p. - Includes bibliographies and index
0-8108-1214-2 2719.dd.7
1321854

CABECEIRAS, James
The multimedia library. - New York ; London : Academic Press, 1978
xv,275p : ill. - (Library and information science series). - Includes bibliographies and index
0-12-153950-4 2719.bb.96
1321854

The multimedia library. - 2nd ed. - New York ; London : Academic Press, 1982
xvi,290p : ill. - (Library and information science). - Includes bibliographies and index
0-12-153952-0 2719.e.552
1321854

CARTER, Mary Duncan
Building library collections. - 4th ed.. - Metuchen, N.J. : Scarecrow Press, 1974
xvi,415p.
0-8108-0730-0 2719 f.51
1321854

KATZ, William Armstrong
Collection development. - New York : Holt, Rinehart and Winston, 1980
352p
0-03-050266-7 2719.e59
1321854

ACQUISITION. Stock. Libraries
Selection - Conference proceedings

SELECTION at work / compiled and edited by Frank Baguley. - [Winchester] : Library Association. South Western Branch, 1978
71,[4]p
0-85365-511-1 2719.gg.44
158264x

ACQUISITION. Stock. Libraries
Selection. National cooperation

COLLINS, Judith
National acquisition policies and systems. - Wetherby : IFLA International Office for UAP, 1982
vi,219p. - Bibliography: p160-221
0-85350-185-8 X.529/47951
2736195

ACQUISITION. Stock. Libraries. England
Selection. Policies of management

APTED, Sheila
Information materials selection. - [Birmingham] : City of Birmingham Polytechnic. Department of Librarianship, 1978
[1],12 leaves. - Bibliography(1p.)
B79-01674 2719.gg.59
1744550

ACQUISITION. Stock. Libraries holding documents on Latin America
to date - Conference proceedings

SEMINAR ON THE ACQUISITION OF LATIN AMERICAN LIBRARY MATERIALS (25th : 1980 : Albuquerque)
Library resources on Latin America. - Madison : SALALM Secretariat, 1981
vii,318p
r6805068 2717.c.2
8934940

ACQUISITION. Stock. Libraries. Schools

MASON, Robert S.
Free and cheap resources for schools. - London : The Library Association, 1984
167p. - Bibliography: p131-134. - Includes index
0-85365-886-2 X.520/36210
3231291

ACQUISITION. Stock. Libraries. Schools. United States

SCHOOL library and media center acquisitions policies and procedures / edited by Mary M. Taylor. - Phoenix, Ariz. : Oryx, 1981
xxxiii,272p. - Bibliography: pxvi-xvii. - Includes index
0-912700-70-x 2719.e.519
287654x

ACQUISITION. Stock. Libraries. Soviet Union

INSTRUKT͡SII͡A po otboru sovetskikh knig, periodicheskikh i prodolzhai͡ushchikhsi͡a izdaniĭ, podlezhashchikh gruppovoĭ obrabotke / [sostaviteli : L.B. Borilin, Ė.A. Bervechenko]. - Izd. 3-e, ispr. i dop. - Moskva : Gosudarstvennai͡a biblioteka SSSR im. V.I. Lenina, Otdel katalogizat͡sii i altavitnykh katalogov, 1979
27p
r6178695 2719.h.99
2042924

KOMPLEKTOVANIE fondov t͡sentralizovannoĭ bibliotechnoĭ sistemy / pod redakt͡sieĭ A.P. Seligerskogo i V.V. Serova. - Moskva : Kniga, 1979
121p. - Bibliography : p.119-121
r6097836 2719.d.99
2042924

RAZRABOTKA profili͡a i tematicheskogo plana komplektovanii͡a / [sostavitel' Gorkina N.P. ; redaktor Kanevskiĭ B.P.]. - Moskva : Gosudarstvennai͡a biblioteka SSSR im. V.I. Lenina, Otdel inostrannogo komplektovanii͡a i mezhdunarodnogo knigoobmena, 1978
19p
r6056774 2719.c.14
2098644

VAĬNTRAUB, I. A.
Mezhdunarodnyĭ abonement v bibliotekakh SSSR. - Moskva : Gosudarstvennai͡a biblioteka SSSR im. V.I. Lenina, 1978
28p,[1]fold.leaf
r6187318 2719.e.77
2042924

ACQUISITION. Stock. Libraries. Soviet Union
1945-

AREST, I͡A. I. (I͡Akov Iserovich)
Bibliotechnyĭ kollektor. - 2-e izd., pererab. i dop. - Moskva : Kniga, 1983
155p
r6463718 2719.e.593
8538433

DINAMIKA formirovanii͡a i ispol'zovanii͡a fondov nauchnykh bibliotek / [redakt͡sionnai͡a kollegii͡a : I͡U.A. Grikhanov (otvetstvennyĭ redaktor) ... i dr.]. - Moskva : Gosudarstvennai͡a biblioteka SSSR imeni V.I. Lenina, Nauchno-issledovatel'skiĭ otdel bibliotekovedenii͡a, 1978
70p
r6498113 2719.e.216
877854x

INSTRUKT͡SII͡A po otboru sovetskikh knig i periodicheskikh izdaniĭ, podlezhashchikh gruppovoĭ obrabotke / [sostaviteli : L.B. Borilin, Ė.A. Bervechenko]. - Izd. 2-e, ispr. i dop.. - Moskva : Gosudarstvennai͡a biblioteka SSSR imeni V.I. Lenina, Otdel katalogizat͡sii i alfavitnykh katalogov, 1973
36p
r6498124 2719.e.196
8720894

ACQUISITION. Stock. Libraries. Soviet Union
Selection, to date

MODELIROVANIE bibliotechnykh fondov / [redakt͡sionnai͡a kollegii͡a: N.S. Kartashov (predsedatel')]. - Moskva : Gosudarstvennai͡a biblioteka SSSR im. V.I. Lenina, Nauchno-metodicheskiĭ otdel, 1983
159p
r6448076 2719.e.801
8914389

TEORII͡A i praktika formirovanii͡a knizhnogo i͡adra bibliotechnogo fonda / [sostavitel' D.V. Evseev]. - Leningrad : Gosudarstvennai͡a publichnai͡a biblioteka im. M.E. Saltykova-Shchedrina, 1980
210p. - (Problema "Sistema komplektovanii͡a i fondy massovykh bibliotek")
r6463129 2719.e.436
8914389

ACQUISITION. Stock. Libraries. United States
- Bibliographies

GODDEN, Irene P.
Collection development and acquisitions, 1970-80. - Metuchen ; London : Scarecrow, 1982
vii,138p. - Includes index
0-8108-1499-4 X.529/47914
2735547

ACQUISITION. Stock. Libraries. United States
Selection

COLLECTION development in libraries / edited by Robert D. Stueart, George B. Miller, Jr.. - Greenwich, Conn. : Jai, c1980
2v.(602p) : ill. - (Foundations in library and information science ; v.10). - Includes bibliographies and index
B82-05919 2719.e.313-314
1297511

LIBRARY acquisition policies and procedures / edited by Elizabeth Futas. - Phoenix : Oryx Press ; London : Distributed by Mansell Information Publishing, 1977
xli,406p : ill. - Bibliography: p.xli. - Includes index
0-912700-02-5 2719.e.307
1297511

ACQUISITION. Stock. Libraries. United States
Selection - Conference proceedings

SHAPING library collections for the 1980s / edited by Peter Spyers-Duran and Thomas Mann, Jr. - Phoenix, Ariz. : Oryx, 1980
xii,235p : ill. - Includes index
0-912700-58-0 2719.e.518
1297511

ACQUISITION. Stock. Libraries. Universities. United States

GRIEDER, Ted
Acquisitions. - Westport, Conn. ; London : Greenwood Press, 1978
x,277,[1]p : ill. - (Contributions in librarianship and information science ; no.22). - Includes index
0-8371-9890-9 2719.cc.17
1799304

ACQUISITION. Stock. Metropolitan Museum of Art

HESS, John L.
The grand aquisitors. - Boston : Houghton Mifflin Co., 1974
178p.,[32] pages of plates
0-395-18013-9 X.429/7227 Woolwich
1422588

ACQUISITION. Stock. National libraries. Great Britain

WILSON, Alexander, 1921 Feb. 12-
The national published record. - Manchester : John Rylands University Library of Manchester, 1982
p246-264
r6813043 X.950/29189
3068293

ACQUISITION. Stock. Public libraries. Italy
- Bibliographies

La CONSULTAZIONE nelle biblioteche pubbliche / [diretta da Alessandro Barbetta e Lilli Dalle Nogare]. - [Milan] : Arnoldo Mondadori, 1975
xxii,572p
r6029661 2719 f.50
1318837

ACQUISITION. Stock. Public libraries. Sweden
Selection

SVENSSON, Gunnar, 1932-
Att välja är politik. - Lund : Bibliotekstjänst, 1978
102p. - (Biblioteksdebatt ; 9). - Includes bibliographies
91-7018-169-1 2715.ss.9/9
2119463

ACQUISITION. Stock. Saltley Action Centre. Library
Procedure

KELLY, Lesley U.
Saltley action centre. - [Birmingham] : Birmingham Library School Co-operative, 1980
[3],20 leaves. - (Occasional publications / Birmingham Library School Co-operative). - Bibliography: p.20
0-906945-01-1 2719i18
231357x

ACQUISITION. Stock. University of Lancaster. Library

HINDLE, Anthony
Developing an acquistions system for a university library. - London : British Library Research and Development Department, 1977
iii,45p : ill. - (Research and development reports / British Library ; no.5351 HC). - Bibliography: p.41
0-905984-03-x BS.10/630
113714x

ACQUISITION. Sub-standard rented residences. England
Policies of local authorities

STUDY GROUP ON PROGRAMMES OF SOCIAL OWNERSHIP AND RENOVATION OF COUNCIL DWELLINGS
Study Group on Programmes of Social Ownership and Renovation of Council Dwellings, first report. - [London] : Department of the Environment : Association of District Councils : Association of Metropolitan Authorities : Greater London Council : London Boroughs Association, [1976]
[50]p in various pagings
0-7184-0076-3 BS.414/384
1061267

ACQUISITION. Symbolic skills. Children
Role in development of children - Conference proceedings

NATO CONFERENCE ON THE ACQUISITION OF SYMBOLIC SKILLS (1982 : University of Keele)
The acquisition of symbolic skills. - New York ; London : Published in cooperation with NATO Scientific Affairs Division [by] Plenum, c1983
xii,623p : ill. - (NATO conference series. III, Human factors ; v.22). - Includes bibliographies and index
0-306-41368-x U.N.P.142/10
3058247

ACQUISITION. Syntactic skills. German speaking children

CLAHSEN, Harald
Spracherwerb in der Kindheit. - Tübingen : Narr, 1982
183p. - (Tübinger Beiträge zur Linguistik. Ser. A, Language development ; 4). - Includes bibliography
3-87808-548-6 X.0950/202(4)
8992312

ACQUISITION. Syntactic skills. Man

MOULTON, Janice
The organization of language. - Cambridge : Cambridge University Press, 1981
xvi,389p : ill. - Includes index
0-521-23129-9 X.950/3998
2402025

ACQUISITION. Trident I nuclear weapons systems
By Great Britain - Texts

The BRITISH strategic nuclear force, July 1980. - London : H.M.S.O., [1981]
3p. - (Cmnd. ; 7979)
0-10-179790-7 BS.102/100
2340054

ACQUISITION. Unquoted companies. Great Britain
Management aspects

PEARSON, Barrie
Successful acquisition of unquoted companies. - Aldershot : Gower, in association with the Institute of Cost and Management Accountants, c1983
viii,99p. - Includes index
0-566-02418-7 X.520/31856
2988526

ACQUISITION. Use of deictic words. Language skills. Children

TANZ, Christine
Studies in the acquisition of deictic terms. - Cambridge : Cambridge University Press, 1980
ix,184p : ill. - (Cambridge studies in linguistics ; 26). - Bibliography: p.174-179. - Includes index
0-521-22740-2 X.520/22027
2141078

ACQUISITION. Use of direct & indirect reference. Language skills. Children

MARATSOS, Michael Philip
The use of definite and indefinite reference in young children. - Cambridge : Cambridge University Press, 1976
xiv,144p : 1 ill. - Bibliography: p.139-141. - Includes index
0-521-20924-2 X.809/40924
0647160

ACQUISITION. Virtue
Philosophical perspectives. Plato. Meno

STERNFELD, Robert
Plato's 'Meno'. - Carbondale : Southern Illinois University Press ; London : Feffer and Simons, 1978
xvi,176p : ill. - (Philosophical explorations). - Includes index
0-8093-0838-x X.520/12929
1590456

ACQUISITION. Visual arts, to 1977. Stock. Great Britain. Department of the Environment
- Catalogues

PROPERTY SERVICES AGENCY. Library Service
The collection of the Department of the Environment, pictures and other works of art, acquisitions for 1977-78. - Croydon : The Service, 1978
[5],38p. - (Current papers / Property Services Agency. Library Service)
0-86177-019-6 BS.414/1821(4)
2301342

ACQUISITION. Vocabulary. Children

GARCÍA HOZ, Víctor
Estudios experimentales sobre el vocabulario. - Madrid : Instituto de Pedagogía "San José de Calasanz", 1977
126p
84-00-03712-x X.900/21579
1367382

ACQUISITION. Vocabulary. Foreign languages

TRAMPE, Peter af
An experiment in foreign language vocabulary learning. - Stockholm : University of Stockholm, Institute of Linguistics, 1982
121p : ill. - (Papers from the Institute of Linguistics, University of Stockholm ; 45). - Includes bibliography
r6610879 X.0905/48(45)
8970815

ACQUISITION. Weapons. Military equipment. African countries
International political aspects

ARLINGHAUS, Bruce E.
Military development in Africa. - Epping : Bowker, c1984
xiv,152p. - (Westview special studies on Africa). - Includes index
0-86531-434-9 X.520/37477
3286959

ACQUISITION. Weapons. Military equipment. Developing countries
International political aspects

BRZOSKA, Michael
Rüstung und Dritte Welt. - München ; London : Weltforum Verlag, c1981
x,209p : ill. - (Weltwirtschaft und internationale Beziehungen ; 25)
3-8039-0197-9 X.809/50611
2629607

JONES, Rodney W. (Rodney William)
Modern weapons and Third World powers. - Boulder ; London : Westview, c1984
xii,125p. - (CSIS significant issues series ; v.6, no.4)
0-86531-871-9 X.629/25425
3248097

ACQUISITION. Writing skills. Children
Drawings

ALIPRANDI, Cecilia
Images. - London : Good Reading, 1978, c1971
128p : all ill(some col.)
0-904223-35-3 X.421/21788 Woolwich
2399784

ACQUISITION. Writing skills. Children, to 10 years
Theories

SMITH, Donald Eugene Phillip
A technology of reading and writing. - New York ; London : Academic Press. - (Educational psychology)
Vol.1: Learning to read and write. - 1976
xi,163p : ill. - Bibliography: p.155-158. - Includes index
0-12-651701-0 X.0981. 1511
0748560
Vol.2: Criterion-referenced tests for reading and writing. - 1977
xii,279p. - (Educational psychology). - Includes bibliographies and index
0-12-651702-9 X.0981 '1511
0748560
Vol.3: The adaptive classroom. - 1976
x,320p : ill. - Bibliography: p.313-316. - Includes index
0-12-651703-7 X.981/20735
0748560
Vol.4: Designing instructional tasks. - 1978
x,212p : ill. - Bibliography: p.197. - Includes index
0-12-651704-5 X.0981/1511
0748560

ACQUISITION SYSTEMS. University of Southampton. Library
Automated acquisition systems, 1971-1978

FRANCIS, B. (Bernice)
The acquisition system, 1971-1978. - [Southampton] : [University of Southampton], 1979
[122]p. - (Automation project reports / University of Southampton. Library ; no.6)
0-85432-191-8 P.623/336 [no.6]
2131595

ACQUISITIONS
See also
Accession numbers

ACQUISITIONS. Books. Stock. Svenska barnbokinstitutet. Biblioteket
1967-1973 - Bibliographies

SVENSKA BARNBOKSINSTITUTET
Svenska Barnboksinstitutets förvärvskatalog över referenslitteratur 1967-1973 = The Swedish Institute for Children's Books catalogue of the reference collection acquisitions 1967-1973. - Stockholm : Svenska Barnboksinstitutet, 1975
177 p. : ill
91-7260-059-4 X.900/18085
1026585

ACQUISITIONS. Exhibits. Musée national d'art moderne. Cabinet d'art graphique
1971-1976 - Catalogues

MUSÉE NATIONAL D'ART MODERNE. Cabinet d'art graphique
Musée national d'art moderne. - [Paris] : Centre national d'art et de Culture Georges Pompidou : Musée national d'art moderne, 1977
ca.200p : ill
2-85850-027-4 X.419/7640 Woolwich
1468138

ACQUISITIONS. Limited companies
Taxation. Law

TAX aspects of acquisitions and mergers / edited by Philip Cooke, Jan M. van der Beek. - 2nd ed. - Deventer ; London : Kluwer Law and Taxation, c1983
150p : 2ill
90-654-4151-4 X.520/33701
3127184

ACQUISITIONS. Rare books & books with fine bindings. Stock. Badische Landesbibliothek
1955-1974 - Catalogues

SCHMITT, Franz Anselm
Kostbare Einbände-seltene Drucke. - Karlsruhe, [1974]
96p : ill
3-7617-0056-3 667.r.87
0392448

ACQUISITIONS. Stock. Archives municipales de Bordeaux
1965-1975

BORDEAUX. Archives municipales
Dix ans d'acquisitions aux Archives municipales de Bordeaux. - Bordeaux : Archives Municipales, 1975
47p.,16p. of plates : ill
r6012782 S.E.123/600
0396303

ACQUISITIONS. Stock. Biblioteca nazionale universitaria
- Catalogues

BIBLIOTECA NAZIONALE UNIVERSITARIA
Primo [ecc.] catalogo di recenti acquisti ordinati secondo la Classificazione decimale Dewey. - Torino : la Biblioteca, 1976-
r6068548 2725p.21
2039974

ACQUISITIONS. Stock. British Library. Department of Manuscripts
1756–1782 – Catalogues

BRITISH LIBRARY
Catalogue of additions to the manuscripts. 1756–1782. additional manuscripts 4101–5017. – London : British Museum Publications Ltd for the British Library, 1977
ix,706p
0–7141–0490–6 2179.b.6
1195905
– [Another copy.] BS.10/745
– [Another copy.] BS.4/282

ACQUISITIONS. Stock. British Museum. Department of Medieval and Later Antiquities

BRITISH MUSEUM. Department of Medieval and Later Antiquities
Department of Medieval and Later Antiquities. – London : British Museum. – (Occasional paper / British Museum ; no.10)
Pt.1: Medieval. – 1980
68p : ill
0–86159–009–0 BS.4/246(10)
2864533

ACQUISITIONS. Stock. British Museum. Department of Prints and Drawings

BRITISH MUSEUM. Department of Prints and Drawings
New acquisitions. – London : British Museum. – (Occasional papers / British Museum ; no.8)
No.1 (1976–78). – 1979
39p
0–86159–007–4 BS.4/246(8)
2821095

ACQUISITIONS. Stock. Chartered Insurance Institute. Library
– Serials

CHARTERED INSURANCE INSTITUTE. Library
A selection of books, periodicals and pamphlets added to the library. – [List no. 1]–. – London : [The Library], [19–]–
Holdings : no 31, etc. 1 May 1974–30 April 1975, etc
r6031564 P.513/2092
1082272

ACQUISITIONS. Stock. Libraries. Universities. Sweden
to date

HAGERLID, Jan
Förvärvsstatistik ur Libris. – Uppsala : Universitetsbiblioteket, 1983
31 leaves. – (Carolina-rapport ; 1)
r6611132 2725.g.194(1)
9044051

ACQUISITIONS. Stock. Museums. Naples. Italy
Exhibitions in Villa Pignatelli – Catalogues

ACQUISIZIONI 1960–1975 / [redazione del catalogo...curata da Nicola Spinosa]. – Napoli : Società editrice napoletana, 1975
xxx,272p : ill. – (Mostre e musei ; 1)
r6067303 X.419/7495 Woolwich
1464191

ACQUISITIONS. Stock. National Gallery of Ireland
– Catalogues

NATIONAL GALLERY OF IRELAND
Acquisitions 1982–83. – Dublin : The National Gallery, 1984
94p : ill(some col.). – Includes index
0–903162–12–1 X.429/15774
3022005

ACQUISITIONS. Stock. Nehru Memorial Museum and Library
– Serials

NEHRU MEMORIAL MUSEUM AND LIBRARY
List of additions. – [No.1]–. – New Delhi : [The Library], [19–]–
r6072552 2725.ff.54
170172X

ACQUISITIONS. Stock. Stanford University. Museum of Art

STANFORD UNIVERSITY. Museum of Art
The Stanford Museum. – Vols 4–5 : 1975. – Stanford, Calif. : Committee for Art, Stanford University Museum of Art, 1976
56,[3]p : ill
Subsequent volumes, when received, will be placed at this shelfmarkHoldings : vols. 4–5, etc : 1975, etc
r6073355 P.441/864
1800442

ACQUISTION. Stock. Special libraries. Bulgaria

PROBLEMI na spetsialnite biblioteki / [redaktsionna Kolegiia : E. Savova (otg. redaktor), M. Kadūnkova-Radoslavova, M. Argirova]. – Sofiia : Būlgarska akademiia na naukite, Tsentūr za nauchna informatsiia s tsentralna biblioteka i nauchen arkhiv, 1978
410p,[5]fold.leaves of plates. – Includes bibliographies
r6097975 2719.d.77
2042169

ACQUSITIONS. Stock. British Museum. Department of Greek and Roman Antiquities

BRITISH MUSEUM. Department of Greek and Roman Antiquities
Department of Greek and Roman Antiquities. – London : British Museum, 1981
iv,51p : ill. – (Occasional paper / British Museum ; no.22). – Includes bibliographies
B82–11551 BS.4/246(22)
2626411

ACRASIALES
See also
Myxomycetes

ACRE. Brazil
to date

COSTA, Craveiro
A conquista do deserto occidental. – 2a ed. – São Paulo : Companhia Editora Nacional, 1974
213p,plates. – (Brasiliana ; v.191)
r6707403 X.809/57980
9039678

ACRE. Brazil
Castro, Plácido de, 1902–1903

CASTILHOS GOYCOCHÊA, Luiz Felippe de
O espírito militar na questão acreana. – 2.ed. – Rio de Janeiro : Comissão Nacional do Centenário de Plácido de Castro, 1973
159p.
r6024996 X.700/18307
1151886

GOYCOCHÊA, Luiz Felippe de Castilhos
O espírito militar na questão acreana. – 2. ed. – [Rio de Janeiro] : Comissão Nacional do Centenário de Plácido de Castro, 1973
159p.,[2]fold leaves of plates
r6065627 X.700/19631
1151886

ACRE. Israel
Manuscripts. Illumination by Crusaders, 1275–1291

FOLDA, Jaroslav
Crusader manuscript illumination at Saint-Jean d'Acre, 1275–1291. – Princeton ; Guildford : Princeton University Press, 1976
xxix,234p,leaf of plate,[96]p of plates : ill(some col). – Includes index
0–691–03907–0 2702.ff.1
0640263

ACREMANT, GERMAINE. Fiction in French
– Biographies

ACREMANT, Germaine
Hier que j'aimais. – Paris : Plon, 1980
216p
2–259–00667–1 X.958/13154
8155453

ACRIDOIDEA

UVAROV, Sir Boris
Grasshoppers and locusts. – London : Centre for Overseas Pest Research.
Vol.2: Behaviour, ecology, biogeography, population dynamics. – 1977
ix,613p : ill. – Bibliography: p.533–588. – Includes index
0–85135–072–0 SRL
1276778

ACRIDOMORPHOIDEA
Taxonomy

DIRSH, Vitaly Michailovitsh
Classification of the acridomorphoid insects. – Faringdon : Classey, 1975
vii,171p : ill. – Bibliography: p.167–170. – Includes index
0–900848–82–0 SRL
0357995

ACRITARCHS. Strata. Great Britain

DOWNIE, C. (Charles)
Acritarchs in British stratigraphy. – Oxford : Blackwell Scientific, 1984
26p : ill. – (Special report ; no.17). – Bibliography: p23–26
0–632–01225–0 X.322/15160
3022641

ACROBATICS

GEDNEY, Judith M.
Tumbling and balancing. – Englewood Cliffs ; London : Prentice-Hall, 1977
xvii,172p : ill. – Includes index
0–13–932798–3 X.620/17040 Woolwich
1187074

ACROBATICS
– Serials

ACROBATICS Magazine. – No.1– ; Aug. 1976–. – Feltham : Technimedia for G. Garrard, [1976]– : ill
B76–35376 P.441/817
076132x

ACROBATICS
1945–

GUREVICH, Z. B. (Zinoviĭ Bonich)
Ėkvilibristika. – Moskva : Iskusstvo, 1982
168p,plates : ill
r6490213 X.629/25028
8001634

ACROBATICS. Sports

COULTON, Jill
Sports acrobatics. – Wakefield : EP Sport, 1981
120p : ill. – Bibliography: p120
0–7158–0770–6 X.629/16089 Woolwich
2537176

SMITH, Morgan
An introduction to sports acrobatics. – London : Stanley Paul, 1982
89p : ill
0–09–146230–4 X.622/13928
2537176
– [Another copy: paperback] ISBN 0–09–146231–2 X.622/13529

ACRONYMS
– Dictionaries

ARDÓN, Juan Ramón
Enciclopedia de las siglas. – Tegucigalpa : [s.n.], [1978]
396p
r6162125 X.909/88899
2458039

ACROPOLIS. Athens. Greece

MELETZÊS, Spyros
[Akropolis kai mouseion]. Acropolis and Museum. – 6.ed. – Munich & Zurich : Schnell & Steiner, 1973
44p.,ca 110p. of plates
3–7954–0554–8 X.410/6258 Woolwich
1314912

MULRYNE, T. W.
The Acropolis. – London : Bell and Hyman, 1978
32p : ill. – (The ancient world in action). – Bibliography: p.32
0–7135–1978–9 X.419/34656 Woolwich
1514717

ACROPOLIS. Athens. Greece
to date

DONTAS, George
The Acropolis and its museum. – Athens : Clio editions, 1979
135p,[1]leaf of plates : ill
r6598131 X.421/25026
8546320

ACROPOLIS. Athens. Greece
Ancient Greek temples: Parthenon

THEOCHARÊS, Periklês S.
Peiramatikê meletê tês eustatheias tou Parthenônos. – Athênai : Grapheion Dêmosieumatôn tês Akadêmias Athênôn, 1979
74p,[33]folded leaves,[50]p of plates. – (Pragmateiai tês Akadêmias Athênôn ; 44). – Bibliography
r6501543 Ac.23/5 [vol.44]
8747288

WOODFORD, Susan
The Parthenon. – Cambridge : Cambridge University Press, 1981
48p : ill. – (Cambridge introduction to the history of mankind. topic books)
0–521–22629–5 X.809/48365
8747288

ACROPOLIS. Athens. Greece
Ancient Greek temples: Parthenon. Architecture

ORLANDOS, Anast. K.
Hē architektonikē tou Parthenōnos. - Athēnai : Archaiologikē Hetaireia, 1976-78
3v.(incl.portfolio),ca 160 leaves of plates(part fold.) : ill. - (Vivliothēkē tēs en Athēnais Archaiologikēs Hetaireias ; 86)
r6069348 N.Tab.2025/10
241774X
- Vol 1. N.Tab.2025/10
- Vol. 2 and 3 Ac.5202/9(86)

ACROPOLIS. Athens. Greece
Ancient Greek temples: Parthenon. Friezes

MELETZĒS, Spyros
[Akropolis kai mouseion. Italian]. L'Acropoli. - 10a ed., completamente di nuovo rappresentata. - Monaco : Schnell & Steiner, 1979
104p : ill(some col)
3-7954-0809-1 X.805/3553
0287636

ACROPOLIS. Athens. Greece
Ancient Greek temples: Parthenon. Sculptures

BROMMER, Frank
The sculptures of the Parthenon. - London : Thames and Hudson, 1979
219p in various pagings,[1]fold leaf : chiefly ill. - Bibliography: p.[71]
0-500-23296-2 LR.425/466
1994581

DESPINĒS, Giōrgos
Parthenōneia. - Athēnai : Hē en Athēnais Archaiologikē Hetaireia, 1982
119p,71p of plates. - (Vivliothēkē tēs en Athēnais Archaiologikēs Hetaireias ; ar.97)
r6509079 Ac.5202/9[vol.97]
823289x

ACROPOLIS. Athens. Greece
Ancient Greek temples: Parthenon. Sculptures: Exhibits at British Museum. Department of Greek and Roman Antiquities

BRITISH MUSEUM
An historical guide to the sculptures of the Parthenon. - [London] : British Museum Publications Ltd for the Trustees, 1975[i.e.1976]
[2],48p : ill. - Bibliography: p.47-48
B76-24517 X.410/10074 Woolwich
0689602
- [Another copy: hardback] BS.10/644

COOK, B. F. (Brian F)
The Elgin marbles. - London : published for the Trustees of the British Museum by British Museum, c1984
72p : ill(some col.). - Bibliography: p71. - Includes index
0-7141-2026-x X.429/16806
0689602

ACROPOLIS. Athens. Greece
Ancient Greek temples: Parthenon. Sculptures: Exhibits at British Museum. Department of Greek and Roman Antiquities. Acquisition

ST. CLAIR, William
Lord Elgin and the Marbles. - [New ed.]. - Oxford : Oxford University Press, 1983
x,311p,[9]p of plates : ill. - (Oxford paperbacks). - Bibliography: p279-281. - Includes index
0-19-285140-3 X.808/39590
308678x

ACROPOLIS. Athens. Greece
Ancient Greek temples: Parthenon. Sculptures: Exhibits at British Museum. Department of Greek and Roman Antiquities. Acquisition & appreciation, 1799-1816

ROTHENBERG, Jacob
'Descensus ad terram'. - New York ; London : Garland, 1977
[4],xii[i.e.xv],469p,76p of plates : ill. - (Outstanding dissertations in the fine arts). - Bibliography: p.456-468
0-8240-2726-4 X.429/10616 Woolwich
2053470

ACROTERIA. Temples. Attica. Ancient Greece B.C.500-B.C.400

DELIVORRIAS, Angelos
Attische Giebelskulpturen und Akrotere des fünften Jahrhunderts. - Tübingen : Wasmuth, c1974
xvi, 208 p., 64 p. of plates, 5 fold. leaves of plates. - (Tübinger Studien zur Archäologie und Kunstgeschichte ; Bd. 1)
3-8030-1900-1 X.0415/219(1) Woolwich
0915963

ACROW GROUP
to 1976

ACROW GROUP
Acrow, the success story of achievement through team spirit. - [London] : [Acrow], [1976]
108p : ill(chiefly col)
0-901983-01-2 X.625/516 Woolwich
0551554

ACRYLIC PAINTINGS
Acrylic landscape paintings. Techniques

BLAKE, Wendon
Landscapes in acrylic. - London : Pitman House, 1980
80p : ill(some col.). - (The Artist's painting library)
0-273-01359-9 X.425/1875 Woolwich
0756296

PELLEW, John Clifford
Acrylic landscape painting. - New York : Watson-Guptill ; London : Pitman, 1976
158,[1]p : ill(some col). - Includes index
0-273-00239-2 X.415/5191 Woolwich
0756296

ACRYLIC PAINTINGS
Acrylic portrait paintings. Techniques

VAN WYK, Helen
Acrylic portrait painting. - New York : Watson-Guptill ; London : Pitman, 1976 [i.e. 1977]
174p : chiefly ill(some col). - (Art instruction paperbacks). - Includes index
0-273-00975-3 X.415/5321
1038052

ACRYLIC PAINTINGS
Acrylic seascape paintings. Techniques

BLAKE, Wendon
Seascapes in acrylic. - London : Pitman, 1979[i.e.1980]
80p : ill(chiefly col). - (Artist's painting library / Wendon Blake)
0-273-01364-5 X.415/6052 Woolwich
2110814

ACRYLIC PAINTINGS
English acrylic paintings. Torok, Karl - Catalogues

TOROK, Karl
In a summer garden. - Bradford : Bradford Art Galleries and Museums, [1978]
[11]p : col ill
B79-08118 X.419/34842 Woolwich
1808206

ACRYLIC PAINTINGS
Techniques

BLAKE, Wendon
Acrylic painting. - London : Pitman, 1979 [i.e. 1980]
80p : ill(chiefly col). - (Artist's painting library / Wendon Blake)
0-273-01417-x X.421/20500 Woolwich
0074853

The acrylic painting book. - New York : Watson-Guptill ; London : Pitman, 1978
256p : ill(chiefly col)
0-273-01257-6 X.423/10736 Woolwich
0074853

CATCHPOLE, Joan
Painting in oils and acrylics. - Guildford : Lutterworth Press, 1978
152p,iv p of plates : ill(some col). - Bibliography: p.147-148. - Includes index
0-7188-2342-7 X.629/12059 Woolwich
0074853

CRAWSHAW, Alwyn
Learn to paint with acrylic colours. - Glasgow : Collins, 1979
64p : ill(chiefly col)
0-00-411867-7 X.415/5879 Woolwich
0074853

DANIELS, Alfred
Enjoying acrylics. - London : Luscombe, 1976
128p,8p of plates : ill(some col). - Bibliography: p.126. - Includes index
0-86002-142-4 X.429/10076
0074853

MILLS, John FitzMaurice
Materials & techniques of acrylic painting. - London : Warne, 1981
48p : ill(some col.). - (An Observer's guide. Art and craft)
0-7232-2476-5 X.429/12669 Woolwich
0074853

TAUBES, Frederic
Acrylic painting for the beginner. - New York : Watson-Guptill, 1971
143p : ill(some col.)
0-8230-0061-3 X.421/22842
0074853

ACRYLIC PAINTINGS. Exhibits. Bluecoat Gallery
English acrylic paintings. Farthing, Stephen - Catalogues

FARTHING, Stephen
The construction of a monument. - Liverpool : Bluecoat Gallery, 1981
1sheet : ill
B82-09033 X.429/12894 Woolwich
2617145

ACRYLIC PLASTICS
See also
Handicrafts using acrylic plastics

ACRYLONITRILE. Industrial chemicals
Accidents. Prevention - For management

CHEMICAL INDUSTRY SAFETY AND HEALTH COUNCIL Safety Committee
The safe design, construction and use of plants producing or consuming acrylonitrile. - London : The Council, 1978
[4],23p : 1 ill. - (Codes of practice for chemicals with major hazards)
B79-30680 X.611/9035 Woolwich
194701x

ACRYLONITRILE INDUSTRIES

BRITISH SULPHUR CORPORATION
World survey of acrylonitrile. - London : The Corporation, 1975
43p : col ill
0-902777-33-5 SRL
0528005

ACSÁDY, IGNÁC. Historiography. Hungary
- Biographies

GUNST, Péter
Acsády Ignác. - Budapest : Akadémiai Kiadó, 1973
236p,[1]leaf of plates. - (A múlt magyar tudósai)
r6015833 X.808/10792
1243004

ACSET
Theories of Great Britain. Adolescents, 16-19 years. Vocational preparation. Great Britain. Manpower Services Commission. Courses. Effects on career development of teachers in further education institutions. ACSET. Teacher training and staff development implications of the expansion of MSC-related programmes

FEU response to ACSET advice 'Teacher training and staff development implications of the expansion of MSC-related programmes'. - London : Further Education Curriculum Review and Development Unit, 1982
[7]p
B83-13761 X.525/6791
2863421

ACSET. Development. Careers. Teachers. Further education institutions. Great Britain
Theories of effects of courses of Great Britain. Manpower Services Commission for vocational preparation of adolescents, 16-19 years

FEU response to ACSET advice 'Teacher training and staff development implications of the expansion of MSC-related programmes'. - London : Further Education Curriculum Review and Development Unit, 1982
[7]p
B83-13761 X.525/6791
2863421

ACT
See American College Testing Program

ACT MICROCOMPUTER SYSTEMS
ACT Apricot microcomputer systems

GOSLING, P. E. (Peter Edward)
The Apricot. - London : Pitman, 1985
117p : ill. - (Microcomputers handbooks). - Includes index
0-273-02317-9 X.629/26711
3234444

MORRIS, Stephen, 19—
Getting to know your ACT Apricot. - London : Duckworth, 1984
159p : ill. - Includes index
0-7156-1839-3 X.629/25399
3240266

ACT MICROCOMPUTER SYSTEMS
ACT Sirius microcomputer systems

JARRETT, Dennis
The Sirius and Victor 9000 computer book. - London : Hutchinson, 1985
[320]p : ill. - Includes index
0-09-155001-7 X.622/22749
3159531

ACT MICROCOMPUTER SYSTEMS
ACT Sirius microcomputer systems–continued
LANE, John B.
The ACT Sirius 1 computer. - Englewood Cliffs ; London : Prentice-Hall, c1985
x,180p : ill. – (Prentice-Hall International personal computer book). – Includes index
0–13–003567–x X.629/26588
3159531

MORRIS, Stephen
Getting to know your ACT Sirius. - London : Duckworth, 1984
143p : ill. – Includes index
0–7156–1838–5 X.629/25188
3159531

ACT MICROCOMPUTER SYSTEMS
ACT Sirius microcomputer systems. Graphic displays. Applications in companies. Software packages: Easel
MORRIS, Stephen, 1955 Nov. 14–
Using Easel on the ACT Sirius. – London : Duckworth, 1985
185p : ill. – (Duckworth business computing)
0–7156–1940–3 X.629/26703
3201643

ACT MICROCOMPUTER SYSTEMS
ACT Sirius microcomputer systems. Machine-readable files. Management. Software packages: Archive
MORRIS, Stephen, 1955 Nov. 14–
Using Archive on the ACT Sirius. – London : Duckworth, 1985
192p : ill. - (Duckworth business computing). – Includes index
0–7156–1939–x X.629/26685
3201627

ACT MICROCOMPUTER SYSTEMS
ACT Sirius microcomputer systems. Software packages: Abacus
MORRIS, Stephen, 1955 Nov. 14–
Using Abacus on the ACT Sirius. – London : Duckworth, 1985
197p : ill. - (Duckworth business computing). – Includes index
0–7156–1941–1 X.629/26686
320166x

ACT MICROCOMPUTER SYSTEMS
ACT Sirius microcomputer systems. Software packages: Quill. Applications in word processing
MORRIS, Stephen, 1955 Nov. 14–
Using Quill on the ACT Sirius. – London : Duckworth, 1985
187p : ill. - (Duckworth business computing). – Includes index
0–7156–1938–1 X.629/26704
3201635

ACT OF CONSECRATION OF MAN
See Eucharist. Christian Community (1922)

ACTIN. Cells. Organisms
– Conference proceedings
ACTIN / edited by Cristobal G. dos Remedios, Julian A. Barden. – Sydney ; London : Academic, 1983
xvi,336p : ill. – Includes bibliographies and index
0–12–221180–4 SRL
3171752

ACTING
See also
Actors
Actresses
Impersonation
Representation. Drama
Roles. Theatre
Stage fights

ACTING
BARKWORTH, Peter
More about acting. – London : Secker & Warburg, 1984
210p. – Includes index
0–436–03293–7 X.950/35778
0748366
– [Another copy: paperback] ISBN 0–436–03294–5 X.950/35779

BOWSKILL, Derek
Acting and stagecraft. – London : Heinemann, 1973
xviii,362p,[16]p of plates : ill. – (Made simple books). – Includes index
0–434–98574–0 X.950/22667
0748366

COHEN, Robert, 1938–
Acting one. – [Palo Alto, Calif.] : Mayfield, c1984
xviii,204p : ill
0–87484–669–2 X.950/33639
0748366

GLENN, Stanley L.
The complete actor. – Boston, Mass. ; London : Allyn and Bacon, 1977
xii,339p : ill. – Includes bibliographies and index
0–205–05580–x X.981/20645
0748366

GREEN, Michael, 1927–
The art of coarse acting. – London : Arrow Books, 1975
128p,[8]p of plates : ill. – Includes index
0–09–907530–x X.958/581
0263044

MCGAW, Charles
Acting is believing. – 4th ed. – New York ; London : Holt, Rinehart and Winston, 1980
viii,264p : ill. – Bibliography: p.253–257. – Includes index
0–03–021671–0 X.981/22269
0748366
Working a scene. – New York ; London : Holt, Rinehart and Winston, 1977
ix,209p : ill. – Includes index
0–03–080086–2 X.900/25477
0748366

RIZZO, Raymond
The total actor. – Indianapolis : Odyssey Press, 1975
xv, 247 p. : ill. – Bibliography: p. 243–244. – Includes index
0–672–63276–4 X.909/41065
0748366

SWIFT, Clive
The job of acting. – London : Harrap, 1976
x,133p : ill. – Bibliography: p.132–133
0–245–52782–6 X.908/40627
0748366
The job of acting. – 2nd ed. rev. – London : Harrap, 1984
xi,131p : ill
0–245–54056–3 X.529/64140
0748366
The performing world of the actor. – London : Hamilton, 1981
114p : ill
0–241–10585–4 X.622/23454
2339188

TING : THEATRE OF MISTAKES
Elements of performance art. – Revised ed. – London : The Ting, Theatre of Mistakes, 1977
83p : ill
B77–25808 X.619/17791 Woolwich
0748366

WOOD, Joan
The casting couch. – London : Elek, 1975
[8],216p
0–236–31037–2 X.989/40102
036794x

ACTING
– Texts
SHARKEY, Jack
Jack Sharkey's audition pieces and classroom exercises. – New York ; London : French, c1984
xvi,138p
0–573–69036–7 X.950/38412
0748366

ACTING
Bankhead, Tallulah – Biographies
BRIAN, Dennis
Tallulah, darling. – London : Sidgwick & Jackson, 1980
ix,292p,[12]p of plates : ill. – Includes index
0–283–98709–x X.989/90006
0031682

ACTING
Berlioz, Harriet Smithson – Biographies
RABY, Peter
'Fair Ophelia'. – Cambridge : Cambridge University Press, 1982
xiii,216p : ill. – Bibliography: p204–209. – Includes index
0–521–24421–8 X.800/34510
2821893

ACTING
Bloom, Claire – Biographies
BLOOM, Claire
Limelight and after. – Harmondsworth : Penguin, 1983, c1982
187p,[8]p of plates
0–14–006529–6 X.958/19532
8001693

ACTING
Brown, Joan Winmill – Biographies
BROWN, Joan Winmill
No longer alone. – London : Hodder and Stoughton, 1976
187p,[8]p of plates : ill. – (Hodder Christian paperbacks)
0–340–20123–1 X.100/16416
0678716

ACTING
Carmichael, Ian – Biographies
CARMICHAEL, Ian
Will the real Ian Carmichael–. – London : Macmillan, 1979
400p : ill. – Includes index
0–333–25476–7 X.981/21693
1935313
Will the real Ian Carmichael –. – London : Futura Publications, 1980
444p,[8]p of plates : ill
0–7088–1860–9 X.907/26004
1935313

ACTING
Crosby, Kathryn. Interpersonal relationships with Crosby, Bing – Biographies
CROSBY, Kathryn
My life with Bing. – London : Quartet, 1983
xi,351p : ill
0–7043–2379–6 X.955/2095
3063852

ACTING
Culver, Roland – Biographies
CULVER, Roland
Not quite a gentleman. – London : Kimber, 1979
192p,[16]p of plates : ill. – Includes index
0–7183–0107–2 X.981/22219
2057409

ACTING
Donlan, Yolande, ca 1920–1946 – Biographies
DONLAN, Yolande
Shake the stars down. – London : Hodder and Stoughton, 1976
223p,[8]p of plates : ill
0–340–20726–4 X.989/50876
0715522

ACTING
Fyodorova, Victoria – Biographies
FYODOROVA, Victoria
The admiral's daughter. – London : Joseph, 1979
[7],372p,[16]p of plates. – Includes index
0–7181–1832–4 X.629/12713 Woolwich
1966103
The admiral's daughter. – Feltham : Hamlyn, 1980, c1979
372p,[8]p of plates. – Includes index
0–600–20334–4 X.529/39894
1966103

ACTING
Harrison, Elizabeth – Biographies
HARRISON, Elizabeth
Love, honour and dismay. – London : Weidenfeld and Nicolson, 1976
viii,168p,[12]p of plates : ill. – Includes index
0–297–77245–7 X.809/41334
0755796
Love, honour and dismay. – London : Star Books, 1978
[8],168p,[8]p of plates : ill. – Includes index
0–352–30112–0 X.908/41495
0755796

ACTING
Harrison, Rex – Biographies
HARRISON, Rex
Rex. – Large print [ed.]. – Boston, Mass. : G.K. Hall ; London : Prior, 1976
[7],371p. – List of plays and films: p.367–371
0–904000–99–0 X.981/20422
0224278

ACTING
Howard, Leslie – Biographies
HOWARD, Leslie
Trivial fond records. – London : Kimber, 1982
187p,[16]p of plates : ill. – Includes index
0–7183–0418–7 X.800/33860
2775077

HOWARD, Ronald
In search of my father. – London : Kimber, 1981
255p,[16]p of plates : ill. – List of films: p245–246. – Includes index
0–7183–0168–4 X.950/6089
2531003

ACTING
Improvisation
JOHNSTONE, Keith
Impro. – London : Faber, 1979
3–208p
0–571–10989–6 X.989/53518
0107174
Impro. – Corr. ed. – London : Eyre Methuen, 1981
208p
0–413–46430–x X.950/1882
0107174

ACTING
Improvisation–continued

POLSKY, Milton E.
Let's improvise. – Englewood Cliffs ; London : Prentice-Hall, c1980
xvi,316p : ill. – (A Spectrum book). – Bibliography: p309–310. – Includes index
0–13–532069–0 X.900/27911
0107174

ACTING
Improvisation – For teaching

SPOLIN, Viola
Improvisation for the theatre. – London : Pitman, 1977
xxv,395p : ill
B79–02996 WP.1442/51
1759965

ACTING
Jackson, Glenda – Biographies

NATHAN, David
Glenda Jackson. – Tunbridge Wells : Spellmount, 1984
95p : ill. – (Film and theatre stars)
0–946771–35–9 X.958/24816
3012409

ACTING
Lawrence, Gertrude – Biographies

MORLEY, Sheridan
Gertude Lawrence. – London : Weidenfeld and Nicolson, c1981
xii,228p[16]p of plates : ill. – Bibliography: p203–205. – List of films: p209. – List of sound discs: p210–216. – Includes index
0–297–77882–x X.950/1385
2350238

ACTING
Lindfors, Viveca – Biographies

LINDFORS, Viveca
Viveka – Viveca –. – [Stockholm] : Bonniers, 1978
271p,[8]p.of plates : ill
91–00–43690–9 X.989/78811
2026058

ACTING
Marquet, Mary

MARQUET, Mary
Mes noces d'or avec la poésie. – Paris : Jacques Granger, 1976
tom.1–
r6043582 X.0908/847
1235044

ACTING
Martin, Mary, b.1913 – Biographies

MARTIN, Mary, b.1913
My heart belongs. – London : W.H. Allen, 1977
320p. – Includes index
0–491–01668–9 X.981/20588
1083597

ACTING
Massey, Raymond, to 1921 – Biographies

MASSEY, Raymond
When I was young. – London : Robson, 1977
271p : ill
0–903895–85–4 X.981/20672
1139398

ACTING
Massey, Raymond, 1922–1976 – Biographies

MASSEY, Raymond
A hundred different lives. – London : Robson, 1979
447p : ill. – Includes index
0–86051–066–2 X.981/21702
1811894

ACTING
Maynard, Bill – Biographies

MAYNARD, Bill
The yo-yo man. – London : Golden Eagle Press, 1975
189p : ill
0–901482–21–8 X.989/50607
0622214

ACTING
McCowen, Alec – Biographies

MCCOWEN, Alec
Double bill. – London : Elm Tree Books, 1980
[10],198p : ill
0–241–10395–9 X.989/89221
158359X
Young gemini. – London : Elm Tree Books, 1979
87p,plate
0–241–10188–3 X.809/44200
158359x

ACTING
Mills, John, 1908– – Biographies

MILLS, John, 1908–
Up in the clouds, gentlemen please. – London : Weidenfeld and Nicolson, 1980
vii,290p,[24]p of plates : ill. – List of films and theatrical productions: p276–280. – Includes index
0–297–77841–2 X.981/22610
2326167
Up in the clouds, gentlemen please. – Harmondsworth : Penguin, 1981, c1980
427p,[16]p of plates : ill. – Includes index
0–14–005827–3 X.958/5893
2326167

ACTING
More, Kenneth – Biographies

MORE, Kenneth
More or less. – London : Hodder and Stoughton, 1978
249p,[16]p of plates : ill. – Includes index
0–340–22603–x X.620/18328 Woolwich
1485318
More or less. – Bolton-By-Bowland : Magna Print, 1980, c1978
547p
0–86009–218–6 Nov.42216 Woolwich
1485318

ACTING
Neagle, Anna – Biographies

NEAGLE, Anna
Anna Neagle says 'There's always tomorrow'. – Large print [ed.]. – Litton : Magna Print Books, 1976
403p. – List of films: p.398–399
0–86009–047–7 X.989/52166
0227226
'There's always tomorrow'. – [1st ed. reprinted with new epilogue]. – London : Futura Publications, 1979
224p,[16]p of plates : 1 ill. – Includes index
0–7088–1616–9 X.907/25781
0227226

ACTING
Palmer, Lilli – Biographies

PALMER, Lilli
Change lobsters, and dance. – London : W.H. Allen, 1976
[8],310p,[24]p of plates : ill
0–491–01714–6 X.981/20201
0627925

ACTING
Preparation

STANISLAVSKII, K. S. (Konstantin Sergeevich)
An actor prepares. – London : Eyre Methuen, 1980, c1936
313p. – (An Eyre Methuen dramabook)
0–413–46190–4 X.958/3169
2494159

ACTING
Roberts, Rachel – Biographies

ROBERTS, Rachel
No bells on Sunday. – London : Pavilion, 1984
x,246p,[16]p of plates : ill
0–907516–26–2 X.950/41963
323150x

ACTING
Stars, ca 1900–1972

The RISE and fall of the matinée idol / edited by Anthony Curtis. – London : New English Library, 1976
[8],215p,[24]p of plates : ill. – Includes index
0–450–02662–0 X.900/25232
0073687

ACTING
Techniques

BARKWORTH, Peter
About acting. – London : Secker and Warburg, 1980
xvi,160p
0–436–03290–2 X.989/85501
2179873
– [Another copy : paperback]
ISBN 0–436 03290 2 X.989/85502

BENEDETTI, Robert L.
The actor at work. – 3rd ed. – Englewood Cliffs ; London : Prentice-Hall, c1981
xvii,286p : ill. – (Prentice-Hall series in theatre and drama). – Bibliography: p279–281. – Includes index
0–13–003673–0 X.950/3793
2179873

COHEN, Robert, 1938–
Acting power. – Palo Alto, Calif. : Mayfield, 1978
xiv,266p : ill. – Bibliography : p.249–259. – Includes index
0–87484–408–8 X.981/21412
1180487

KNEBEL', M. (Mariīa)
O deistvennom analize p'esy i roli. – 3–e izd. – Moskva : Iskusstvo, 1982
117p
r6464307 X.958/18397
9000909

STANISLAVSKII, K. S. (Konstantin Sergeevich)
Stanislavski's legacy. – Rev. and expanded ed. – London : Eyre Methuen, 1981, c1968
x,209p. – Includes index
0–413–47770–3 X.958/14959
2179873

ACTING
Twiggy – Biographies

TWIGGY
Twiggy. – London : Hart-Davis MacGibbon, 1975
[4],156p,leaf of plate,[32]p of plates : ill
0–246–10895–9 X.809/40352
0521418
Twiggy. – St Albans : Mayflower, 1976
160p,[16]p of plates : ill
0–583–12564–6 X.709/30869
0521418

ACTING
Ullmann, Liv – Biographies

ULLMANN, Liv
Changing. – London : Weidenfeld and Nicolson, 1977
[7],244p
0–297–77285–6 X.989/51465
1096397
Changing. – London : Star Books, 1978
[7],244p
0–352–30163–5 H.79/1819 Woolwich
1096397
Forandringen. – [Oslo] : H. Erichsen, [1976]
235p
82–545–0013–4 X.429/7162 Woolwich
1096397

ACTING
Warner, Jack – Biographies

WARNER, Jack
'Evening all. – London : Star Books, 1979
xi,226,8p of plates : ill
0–352–30342–5 X.619/19667 Woolwich
0301981

ACTING
West, Mae

EELLS, George
Mae West. – London : Robson, 1984
351p : ill. – Includes index
0–86051–260–6 X.950/32771
0302589

WEST, Mae
Peel me a grape. – London : Futura Publications, 1975
[1],96p,[32]p of plates : ill
0–86007–240–1 X.908/40161
0538302

ACTING
West, Mae – Biographies

CASHIN, Fergus
Mae West. – London : W.H. Allen, c1981
197p,[8]p of plates : ill. – List of plays and films: p190. – Includes index
0–491–02984–5 X.950/5296
0302589
Mae West. – London : W.H. Allen, 1982, c1981
197p : ill. – (A Star book)
0–352–31094–4 X.808/36943
0302589

ACTING
White, Carol, 1943– – Biographies

WHITE, Carol, 1943–
Carol comes home. – Sevenoaks : New English Library, 1982
284p,[16]p of plates : ill
0–450–05528–0 X.958/26358
3196801

ACTING. Activities. Children
Teaching

MALKIN, Michael R.
Training the young actor. – South Brunswick ; New York : Barnes ; London : Yoseloff, 1979
128p : ill. – Includes index
0–498–01957–8 X.981/22284
2141086

ACTING. Advertisements. Television
Cats: Arthur

MONTGOMERY, John
Arthur. – London : W.H. Allen, 1975
128p : ill
0–491–01705–7 X.329/1016
0524255

ACTING. All my children. Television drama series in English
Warwick, Ruth - Biographies

WARRICK, Ruth
The confessions of Phoebe Tyler. - Englewood Cliffs ; London : Prentice-Hall, 1980
x,227p.[32]p of plates : ill
0-13-167403-x X.950/1243
2347091

ACTING. American cinema films

BARR, Tony
Acting for the camera. - Boston [Mass.] ; London : Allyn and Bacon, c1982
ix,310p : ill. - Includes index
0-205-07368-9 X.950/9474
2239019

NOOSE, Theodore
Hollywood film acting. - South Brunswick ; New York : A.S. Barnes ; London : Yoseloff, 1979
176p. - Includes index
0-498-02207-2 X.989/88354
2239019

ACTING. American cinema films
1937-1972

HOPE, Bob, b.1904
The road to Hollywood. - London : W.H. Allen, 1977
2-271p : ill
0-491-02150-x X.981/20837
1271873

ACTING. American cinema films
Abbott, Bud & Costello, Lou - Biographies

THOMAS, Bob
Bud & Lou. - London : W.H. Allen, 1977
224p : ill. - List of films: p.203-209. - Includes index
0-491-02212-3 X.981/20841
1273159

ACTING. American cinema films
Astaire, Fred - Biographies

CARRICK, Peter
A tribute to Fred Astaire. - London : Hale, 1984
188p,[24]p of plates : ill. - List of films: p166-177. - Includes index
0-7090-2023-6 X.950/42768
0309222

FREEDLAND, Michael
Fred Astaire. - London : W.H. Allen, 1976
ix,277p,[16]p of plates. - Includes index
0-491-01786-3 X.989/50871
0309222

GREEN, Benny
Fred Astaire. - London : Hamlyn, 1979
176p : ill(some col). - Bibliography: p.174. - List of films: p.156-168. - List of sound discs: p.169-173. - Includes index
0-600-33703-0 X.435/659
0309222
Fred Astaire. - London : Hamlyn, 1979
176p : ill(some col.). - (Gondola). - Includes index
0-600-34707-9 X.435/1178
0309222

ACTING. American cinema films
Autry, Gene - Biographies

AUTRY, Gene
Back in the saddle again. - Garden City, N.Y. : Doubleday, 1978
252p,plates
0-385-03234-x X.950/41012
8738505

ACTING. American cinema films
Bacall, Lauren - Biographies

BACALL, Lauren
By myself. - London : Cape, 1979
[6],377,32p of plates
0-224-01692-x X.981/21388
1573721

ACTING. American cinema films
Bacall, Lauren & Bogart, Humphrey - Biographies

GREENBERGER, Howard
Bogey's baby. - London : W.H. Allen, 1976
viii,216p,[24]p of plates : ill
0-491-01953-x X.809/40733
0578282

HYAMS, Joe
Bogart & Bacall. - London : Joseph, 1976
[5],245p,[32]p of plates. - Includes index
0-7181-1484-1 X.989/40440
0578282
Bogart and Bacall. - London : Sphere, 1976
205p,[16]p of plates : ill. - Includes index
0-7221-0416-2 X.907/25151
0578282

ACTING. American cinema films
Ball, Lucille - Biographies

ANDREWS, Bart
Loving Lucy. - London : Robson, 1980
226p : ill. - Includes index
0-86051-127-8 X.955/806
0052450

ACTING. American cinema films
Barrymore (Family) - Biographies

KOTSILIBAS-DAVIS, James
The Barrymores. - London : Arthur Barker, 1981
376p : ill. - List of films: p338-366. - Bibliography: p367-368. - Includes index
0-213-16805-7 X.950/4686
2498642

ACTING. American cinema films
Baxter, Anne - Biographies

BAXTER, Anne
Intermission. - London : W.H. Allen, 1977
384p
0-491-02480-0 Nov.34795 Woolwich
1172786

ACTING. American cinema films
Beatty, Warren

QUIRK, Lawrence J.
The films of Warren Beatty. - Secaucas : The Citadel Press, c1979
222p : ill
0-8065-0670-9 X.955/954
8738505

ACTING. American cinema films
Beatty, Warren - Biographies

MUNSHOWER, Suzanne L.
Warren Beatty. - London : Everest, 1976
138p,[8]p of plates
0-905018-27-3 X.908/41351
1276425

ACTING. American cinema films
Belushi, John - Biographies

WOODWARD, Bob
Wired. - London : Faber, 1985
461p,[14]p of plates : ill. - Includes index
0-571-13577-3 X.950/40912
3152669

ACTING. American cinema films
Bogart, Humphrey

THOMPSON, Verita
Bogie and me. - London : W.H. Allen, 1983, c1982
xiv,240p,[16]p of plates : ill
0-491-03230-7 X.950/39122
0533432

ACTING. American cinema films
Bogart, Humphrey - Biographies

BENCHLEY, Nathaniel
Humphrey Bogart. - London : Hutchinson, 1975
[8],242p : ill. - List of films: p.237-239
0-09-124990-2 X.981/20095
0533432

ACTING. American cinema films
Boyer, Charles - Biographies

SWINDELL, Larry
Charles Boyer. - London : Weidenfeld and Nicolson, 1983
vii,280p,[24]p of plates. - List of films: p260-269. - Includes index
0-297-78255-x X.950/23100
293857x

ACTING. American cinema films
Brando, Marlon - Biographies

BRANDO, Anna Kashfi
Brando for breakfast. - New York : Crown : [1979]
viii,273p
0-517-53686-2 X.950/8089
8738505

ACTING. American cinema films
Bronson, Charles - Biographies

DOWNING, David
Charles Bronson. - London : W.H. Allen, 1982
156p,[8]p of plates : ill. - (A Star book). - Includes index
0-352-31448-6 X.958/24575
0628573
Charles Bronson. - London : W.H. Allen, 1982
156p,[8]p of plates. - List of films: p139-152. - Includes index
0-491-02937-3 X.950/30492
0628573

HARBINSON, W. A. (William Allen)
Bronson!. - London : W.H. Allen, 1976
157p,[12]p of plates : ill. - List of films: p.151-157
0-491-01536-4 X.989/50626
0628573

WHITNEY, Steven
Charles Bronson. - London : Hale, 1978
272p,[8]p of plates : 1 ill
0-7091-7134-x X.989/52876
0628573

ACTING. American cinema films
Brooks, Louise - Biographies

BROOKS, Louise
Lulu in Hollywood. - London : Hamilton, 1982
viii,109p,[40]p of plates : ill
0-241-10761-x X.950/15516
2779080
Lulu in Hollywood. - London : Hamish Hamilton, 1982
viii,113p,[40]p of plates : ill. - List of films: p112-113
0-241-11022-x X.950/23261
2779080

ACTING. American cinema films
Cagney, James - Biographies

CAGNEY, James
Cagney by Cagney. - London : New English Library, 1976
[11],202p,[24]p of plates
0-450-02839-9 X.989/50629
0228591

MCGILLIGAN, Patrick
Cagney. - South Brunswick ; New York : Barnes ; London : Tantivy Press, 1975
240p : ill. - Bibliography: p.236-238. - List of films: p.214-234. - Includes index
0-904208-45-1 X.981/20794
1253190

WARREN, Doug
James Cagney. - London : Robson, 1983
xv,239p,[32]p of plates : ill. - Filmography: p211-232. - Includes index
0-86051-252-5 X.950/28923
0228591

ACTING. American cinema films
Chaplin, Charles, 1889-1977

CHAPLIN. - Manchester : World Distributors, 1978
32p : ill(some col)
0-7235-0864-x X.902/5728
030249x

The LEGEND of Charlie Chaplin / collected and introduced by Peter Haining. - London : W.H. Allen, 1982
[224]p : ill
0-491-02608-0 X.805/5751
030249x

ACTING. American cinema films
Chaplin, Charles, 1889-1977 - Bibliographies

ERIKSSON, Lennart
Books on / by Chaplin. - 2nd ed. - Västerås : L. Eriksson, c1981
41,ivp
r6611403 L.45/2691
8738505

LYONS, Timothy J.
Charles Chaplin. - London : Prior [etc.], 1979
iii-xvii,232p,plate : 1 ill. - (A reference publication in film). - Includes index
0-86043-177-0 X.985/706
1894951

ACTING. American cinema films
Chaplin, Charles, 1889-1977 - Biographies

CHAPLIN, Charles, 1889-1977
My life in pictures. - New York : Grossek and Dunlap, 1975
320p
0-448-12037-2 X.985/1249
030249X

HOYT, Edwin P. (Edwin Palmer)
Sir Charlie. - London : Hale, 1977
206p,[16]p of plates : ill. - Bibliography: p.197-198. - Includes index
0-7091-6192-1 X.989/51875
030249x

ISAAC, Frederick
When the moon shone bright on Charlie Chaplin. - Melksham : Venton, 1978
108p,12p of plates : ill. - List of films: p.102-103. - Bibliography: p.105-106. - Includes index
0-85475-098-3 X.989/52375
030249x

KANE, Tom
Charlie, Sir. - Bognor Regis : New Horizon, c1982
92p
0-86116-606-x Nov.47663
030249x

ACTING. American cinema films
Chaplin, Charles, 1889–1977 – Biographies–continued

MCCABE, John, b.1920
Charlie Chaplin. – London : Robson, 1978
xiii,297p,[6]p of plates : ill. – Bibliography: p.245–257. – List of films: p.261–279. – Includes index
0–86051–039–5 X.989/52570
030249x

Charlie Chaplin. – London : Magnum Books, 1979
xiii,297p,[8]p of plates : ill. – Bibliography: p.245–257. – List of films: p.261–279. – Includes index
0–417–03340–0 X.708/21703
030249x

ROBINSON, David, 1930–
Chaplin. – London : Secker & Warburg, 1983, c1984
viii,205p,[1]p of plates : ill. – Bibliography: p197–198. – Includes index
0–436–42053–8 X.950/32335
030249x
– [Another Copy : paperback.] ISBN 0–253–21160–3 X.950/32336

SACRANIE, Raj
Charlie Chaplin. – London : H. Hamilton, 1980
64p : ill. – (Profiles)
0–241–10479–3 X.989/88386
2148811

ACTING. American cinema films
Chaplin, Charles, 1889–1977, to 1916 – Biographies

CHAPLIN, Charles, 1889–1977
My early years. – London : Bodley Head, 1979
[11],192p,[16]p of plates : ill
0–370–30230–3 X.989/54041
1837680

ACTING. American cinema films
Chaplin, Charles, 1889–1977, to 1917 – Biographies

CHAPLIN, Charles, 1889–1977
My early years. – Large print ed. – Bath : Chivers, 1982, c1964
273p. – (A Lythway autobiography)
0–85119–843–0 X.950/16666
1296663

SOBEL, Raoul
Chaplin. – London : Quartet Books, 1977
253p : ill. – Bibliography: p.223–227. – List of films: p.229–242. – Includes index
0–7043–2105–x X.989/52065
1296663
– [Another copy: paperback]
ISBN 0–7043–3134–9 X.909/42040

ACTING. American cinema films
Clift, Montgomery – Biographies

BOSWORTH, Patricia
Montgomery Clift. – New York ; London : Harcourt Brace Jovanovich ; [London] : [Sidgwick and Jackson], 1978 [i.e.1979]
ix,438p,[24]p of plates : ill. – List of films: p.424–431. – Includes index
0–283–98575–5 X.800/27897
118847x

Montgomery Clift. – Toronto ; London : Bantam, 1979
xi,433,[1],leaf of plate,[32]p of plates : ill. – List of films: p.419–426. – Includes index
0–553–17002–3 X.907/25668
118847x

KASS, Judith M.
The films of Montgomery Clift. – Secaucus : Citadel Press, [1979]
223p : ill
0–8065–0717–9 X.955/792
8738505

LAGUARDIA, Robert
Montgomery Clift. – London : W.H. Allen, 1977
xi,304p : ill. – List of films: p.285–287. – Includes index
0–491–02320–0 X.981/20722
118847x

ACTING. American cinema films
Cody, Iron Eyes – Biographies

CODY, Iron Eyes
Iron Eyes. – London : Muller, 1982
290p,[32]p of plates : ill. – Includes index
0–584–11050–2 X.950/18392
2844400

ACTING. American cinema films
Cooper, Gary – Biographies

ARCE, Hector
Gary Cooper. – New York : Morrow, 1979
288p
0–688–03604–x X.950/13351
8738505

SWINDELL, Larry
The last hero. – London : Robson, 1981
xiv,343,[32]p of plates : ill. – Includes index
0–86051–132–4 X.950/2486
1134574

ACTING. American cinema films
Crawford, Christina. Interpersonal relationships with Crawford, Joan

CRAWFORD, Christina
Mommie dearest. – London : Hart-Davis, MacGibbon, 1979
282p,[16]p of plates
0–246–11110–0 X.981/12572
2066394

CRAWFORD, Christina
Mommie dearest. – London : Mayflower, 1980
282p,[8]p of plates
0–583–13066–6 X.907/25910
2066394

ACTING. American cinema films
Crawford, Joan – Biographies

CASTLE, Charles, 1939–
Joan Crawford. – London : New English Library, 1977
207p : ill. – Bibliography: p.188. – List of films: p.189–201. – Includes index
0–450–03685–5 X.981/21080
1476629

HOUSTON, David, 1938–
Jazz baby. – London : Robson, 1984, c1983
xvi,203p,[8]p of plates : ill. – Bibliography: p195–203
0–86051–261–4 X.950/34264
1476629

NEWQUIST, Roy
Conversations with Joan Crawford. – Secaucus, N.J : Citadel Press, [1980]
175p : ill
0–8065–0720–9 X.950/11469
8738505

THOMAS, Bob
Joan Crawford. – London : Weidenfeld and Nicolson, 1979
315p,[32]p of plates : ill. – Bibliography: p.299–300. – List of films: p.277–295. – Includes index
0–297–77617–7 X.981/13312
1476629

WALKER, Alexander
Joan Crawford. – London : Weidenfeld and Nicolson, 1983
192p : ill. – Bibliography: p188. – Includes index. – Filmography: p182–187
0–297–78216–9 X.955/1820
1476629

ACTING. American cinema films
Crawford, Joan; Dietrich, Marlene; Garbo, Greta & Swanson, Gloria

FOUR fabulous faces / [compiled by] Larry Carr. – Harmondsworth : Penguin, 1978
[9],492p : ill
0–14–004988–6 X.615/2169 Woolwich
1746359

ACTING. American cinema films
Crawford, Joan. Interpersonal relationships with Crawford, Christina

CRAWFORD, Christina
Mommie dearest. – London : Hart-Davis, MacGibbon, 1979
282p,[16]p of plates
0–246–11110–0 X.981/12572
2066394
Mommie dearest. – London : Mayflower, 1980
282p,[8]p of plates
0–583–13066–6 X.907/25910
2066394

ACTING. American cinema films
Davies, Marion – Biographies

DAVIES, Marion
The times we had. – London : Angus and Robertson, 1976
[10],276p : ill. – Includes index
0–207–95671–5 X.981/20198
0620815

ACTING. American cinema films
Davis, Bette, 1908– – Biographies

HIGHAM, Charles, 1931–
Bette. – New York : Macmillan, c1981
xviii,316p
0–02–551500–4 X.950/34994
8738505
Bette. – London : New English Library, 1981
vi,250p : ill. – Includes index
0–450–04875–6 X.950/10429
2610957

HIGHAM, Charles, 1931–
Bette. – Sevenoaks : New English Library, 1981
392p,[16]p of plates : ill. – Includes index
0–450–05509–4 X.958/23310
2610957

STINE, Whitney
Mother Goddam. – London : W.H. Allen, 1975
[7],374p : ill. – List of films: p.343–360. – Includes index
0–491–01504–6 X.981/20021
0366129

Mother Goddam. – London : W.H. Allen, 1982, c1974
308p,[4]p of plates : ill. – (A Star book)
0–352–31142–8 X.808/36942
2610957

ACTING. American cinema films
De Havilland, Olivia & Fontaine, Joan – Biographies

HIGHAM, Charles, 1931–
[Sisters]. Olivia and Joan. – Sevenoaks : New English Library, 1984
217p,[16]p of plates : ill. – Includes index
0–450–06052–7 X.950 38083
3115453

ACTING. American cinema films
Dean, James, 1931–1955 – Biographies

BOURGET, Jean-Loup
James Dean. – Paris : Henri Veyrier, 1983
152p : ill
2–85199–279–1 X.955/2997
8738505

DALTON, David
James Dean. – London : Plexus, 1983, c1974
356p : ill. – Bibliography: p344–350. – Includes index
0–85965–067–7 X.950/38814
0224480

James Dean. – London : Sidgwick & Jackson, 1985, c1984
285p : ill(some col.)
0–283–99159–3 L.45/3283
0224480

HERNDON, Venable
James Dean. – London : Futura Publications, 1974
288p,[16]p of plates : ill. – Includes index
0–86007–600–8 X.989/51756
0224480

HOWLETT, John, 1940–
James Dean. – London : Plexus Publishing, 1975
191p : ill. – Bibliography: p.191
0–85965–011–1 X.985/1021
0224480

SCHATT, Roy
James Dean. – London : Sidgwick & Jackson, 1982
146p : ill
0–283–98877–0 X.955/1161
2764504

ACTING. American cinema films
Del Rio, Dolores

WOLL, Allen L.
The films of Dolores Del Rio. – New York : Gordon Press, 1978
129p. – (Gordon Press film series)
0–8490–1364–x X.950/18729
8738505

ACTING. American cinema films
Dietrich, Marlene

NAVACELLE, Thierry de
Sublime Marlene. – Paris : Éditions Ramsay, 1982
159p : ill
2–85956–307–5 L.42/2874
8738505

ACTING. American cinema films
Douglas, Kirk – Biographies

MUNN, Michael
Kirk Douglas. – London : Robson, 1985
224p : ill
0–86051–300–9 X.950/42536
3202607

ACTING. American cinema films
Eastwood, Clint

COLE, Gerald
Clint Eastwood. – London : W.H. Allen, c1983
224p,[4]p of plates : ill(some col.)
0–491–03241–2 X.955/2795
1123513

DOWNING, David
Clint Eastwood, all-American anti-hero. – London : Omnibus Press, 1977
144p. – List of films: p.139–143
0–86001–412–6 X.902/5298
1123513

ACTING. American cinema films
Eastwood, Clint - Biographies

AGAN, Patrick
Clint Eastwood. the man behind the myth. - London : Hale, 1977
189p.[12]p of plates : ill. - List of films: p.171-189
0-7091-6159-x X.989/51953
1123513

JOHNSTONE, Iain
The man with no name. - London : Plexus, c1981
144p : ill. - List of films: p141-144
0-85965-025-1 X.955/2968
1123513

ACTING. American cinema films
Eastwood, Clint - Texts - Biographies

AGAN, Patrick
Clint Eastwood. - Sevenoaks : Coronet, 1978
172p,[12]p of plates : ill(some col). - List of films: p.153-172
0-340-23090-8 H.78/1530 Woolwich
1123513

ACTING. American cinema films
Evans, Dale - Biographies

ROGERS, Roy
Happy trails. - Waco, Tex. : Word Books, [1979]
213p,plates
0-8499-0086-7 X.950/43244
8738505

ACTING. American cinema films
Fairbanks, Douglas, b.1883 - Biographies

SCHICKEL, Richard
Douglas Fairbanks. - London : Elm Tree Books, 1976
160p : ill. - List of films : p.149-152. - Bibliography: p.153-155. - Includes index
0-241-89443-3 X.981/20222
0554979

ACTING. American cinema films
Farmer, Frances - Biographies

ARNOLD, William, 1945-
Shadowland. - New York ; London : McGraw-Hill, 1978
260p : ill
0-07-002311-5 X.809/44080
1810677

FARMER, Frances
Will there really be a morning?. - London : Allison & Busby, 1974
318p,[4]p of plates
B82-26839 X.809/52191
1810677

FARMER, Frances
Will there really be a morning?. - [London] : Fontana Paperbacks, 1983, c1972
318p,[4]p of plates
0-00-636526-4 X.958/16468
1810677

ACTING. American cinema films
Fields, William Claude

'GODFREY Daniels!' / edited by Richard J. Anobile. - London : Joseph, 1975
3-224p : of ill
0-7181-1419-1 X.985/1011
0021008

ACTING. American cinema films
Fields, William Claude, 1932-1946 - Biographies

MONTI, Carlotta
W.C. Fields & me. - London : New English Library, 1975
207p,[8]p of plates : ill
0-450-02423-7 X.908/40224
0046124

ACTING. American cinema films
Flynn, Errol - Biographies

FLYNN, Errol
My wicked, wicked ways. - London : Pan Books in association with Heinemann, 1961, c1959
379p
0-330-23343-2 X.958/2894
1259296

FREEDLAND, Michael
Errol Flynn. - London : A. Barker, 1978
x,258p,16p of plates : ill
0-213-16652-6 X.989/53070
1259296

GODFREY, Lionel
The life and crimes of Errol Flynn. - London : Hale [etc.], 1977
176p,[24]p of plates : ill
0-7091-6091-7 X.981/20807
1259296

HIGHAM, Charles
Errol Flynn. - London : Granada, 1980
416p,[16]p of plates : ill. - List of films: p.402-404
0-246-11153-4 X.989/88295
1259296

HIGHAM, Charles, 1931-
Errol Flynn. the untold story. - London : Granada, 1980
416p,[8]p of plates. - (A Mayflower book)
0-583-13440-8 X.958/3260
1259296

ACTING. American cinema films
Flynn, Errol, 1958-1959 - Biographies

CONRAD, Earl
Errol Flynn. - London : Hale, 1979
222p,[16]p of plates : ill. - List of films.: p.221-222
0-7091-7405-5 X.989/53334
1875000

ACTING. American cinema films
Fontaine, Joan - Biographies

FONTAINE, Joan
No bed of roses. - London : W.H. Allen, 1978
319p : ill. - Includes index
0-491-02016-3 X.800/27578
1810685

ACTING. American cinema films
Gable, Clark & Lombard, Carole - Biographies

HARRIS, Warren Gene
Gable and Lombard. - London : Cassell, 1976
189p,[16]p of plates : ill. - Includes index
0-304-29397-0 X.981/20306
0682314

HARRIS, Warren Gene
Gable and Lombard. - London : Corgi, 1977
207p : ill. - Includes index
0-552-10436-1 X.908/41032
0682314

ACTING. American cinema films
Gable, Clark; Harlow, Jean; Lombard, Carole & Powell, William - Biographies

MORELLA, Joe
Gable & Lombard & Powell & Harlow. - 1st British ed. - London : W.H. Allen, 1976
204p,[8]p of plates : ill
0-491-01975-0 X.989/50601
0620270

ACTING. American cinema films
Garbo, Greta - Biographies

PAYNE, Robert, 1911-1983
The great Garbo. - London : W.H. Allen, 1976
[7],198p,[24]p of plates : ill. - Bibliography: p.191-192. - Includes index
0-491-01538-0 X.989/51115
0772739

SANDS, Frederick
The divine Garbo. - London : Sidgwick and Jackson, 1979
243p : ill. - Includes index
0-283-98545-3 LR.421/295
0772739
- [Another copy: paperback]
ISBN 0-283-98752-9 X.958/2520
The divine Garbo. - London : Sidgwick & Jackson in association with New English Library, 1979
160p,16p of plates : ill. - Includes index
0-283-98752-9 X.958/2649
0772739

WALKER, Alexander
Garbo. - London : Weidenfeld and Nicolson, 1980
191p : ill. - Bibliography: p.188. - List of films: p.183-187. - Includes index
0-297-77799-8 X.985/1286
0772739
Garbo. - London : Sphere, 1982, c1980
191p : ill. - Bibliography: p188. - Includes index. - List of film: p183-187
0-7221-8867-6 X.622/14452
0772739

ACTING. American cinema films
Gardel, Carlos - Biographies

GALLARDO, Juan Carlos
Carlos Gardel. - [Bogotá?] : Antonio Malaver Ruiz, [1977]
117p : ill. - (Colección Idolos)
r6099201 X.908/86982
2028735

ACTING. American cinema films
Garland, Judy - Biographies

DIORIO, Al
Little girl lost. - London : Coronet, 1976
190p,[32]p of plates : ill
0-340-20747-7 H.76/1229 Woolwich
0290807

EDWARDS, Anne
Judy Garland. - London : Corgi, 1976
350p,[8]p of plates. - List of films: p.315-317. - List of records: p.318-324. - Includes index
0-552-10063-3 H.76/662 Woolwich
0290807

EDWARDS, Anne, 1909-, 1927-
Judy Garland. - London : Constable, 1975
350p,[32]p of plates : ill. - List of films: p.315-317. - List of records: p.318-324. - Includes index
0-09-461000-2 X.989/40356
0290807

FINCH, Christopher
Rainbow. - London : Joseph, 1975
255p : ill. - Includes index
0-7181-1439-6 X.902/5016
0290807

FRANK, Gerold
Judy. - London : W.H. Allen, 1975
xviii,654p,[32]p of plates. - Includes index
0-491-01735-9 X.981/20041
0290807

SMITH, Lorna
Judy, with love. - London : Hale, 1975
208p,[16]p of plates. - List of films: p.199-201. - Includes index
0-7091-5257-4 X.989/40146
0290807

ACTING. American cinema films
Garland, Judy, to 1935 - Biographies

DAHL, David
Young Judy. - London : Hart-Davis MacGibbon, 1976
xxvii,250p : ill. - Bibliography: p.247-250
0-246-10917-3 X.981/20370
0726036
Young Judy. - St Albans : Mayflower, 1977
238p,[8]p of plates : ill. - Bibliography: p.235-238
0-583-12592-1 X.908/41077
0726036

ACTING. American cinema films
Garland, Judy & Minnelli, Liza - Biographies

SPADA, James
Judy and Liza. - London : Sidgwick & Jackson, 1983
216p : ill. - Bibliography: p213
0-283-98993-9 X.435/1128
3080218

ACTING. American cinema films
Granger, Stewart - Biographies

GRANGER, Stewart
Sparks fly upward. - London : Granada, 1981
416p,[24]p of plates : ill. - Includes index
0-246-11403-7 X.950/2857
2377616
Sparks fly upwards. - London : Granada, 1981
416p,[32]p of plates : ill. - (A Panther book). - Includes index
0-586-05599-1 X.958/12193
2377616

ACTING. American cinema films
Grant, Cary - Biographies

BRITTON, Andrew
Cary Grant. - Newcastle-upon-Tyne : TC, 1983
[24]p. - Bibliography: p22
0-946493-00-6 X.950/21955
2920093

GODFREY, Lionel
Cary Grant. - London : Hale, 1981
224p,[24]p of plates : ill. - Bibliography: p203-204. - Includes index
0-7091-9038-7 X.950/5291
2486423

MCINTOSH, William Currie
The private Cary Grant. - London : Sidgwick & Jackson, c1983
158p,[12]p of plates. - Includes index
0-283-98989-0 X.950/28122
2486423

SCHICKEL, Richard
Cary Grant. - London : Pavilion, 1983
190p : ill. - List of films: p175-190
0-907516-25-4 X.955/2213
2920093

WANSELL, Geoffrey
Cary Grant. - London : Collins, 1983
350p,[12]p of plates : ill. - Includes index
0-00-216371-3 X.950/31482
2486423
Cary Grant. - [London] : Fontana, 1984, c1983
350p,[8]p of plates : ill. - Includes index
0-00-636799-2 X.958/28673
2486423

ACTING. American cinema films
Hardy, Oliver

GIUSTI, Marco
Laurel & Hardy. - Firenze : La nuova Italia, 1978 [i.e. 1979]
140p. - (Il castoro cinema ; 57). - Includes filmography and bibliography
r6509487 X.629/24745
8738505

ACTING. American cinema films
Hardy, Oliver & Laurel, Stan - Biographies

MCCABE, John, b.1920
Mr Laurel and Mr Hardy. - [Revised ed.]. - London : Robson Books, 1976
vii,259p,[16]p of plates : ill. - List of films: p.249-252. - Includes index
0-903895-63-3 X.989/50578
0611956

SCAGNETTI, Jack
The Laurel and Hardy scrapbook. - Middle Village, N.Y. : Jonathan David, c1976
131p,[29]p of plates : ill. - List of films : p122-131
0-8246-0207-2 X.985/793
0611956

ACTING. American cinema films
Havoc, June - Biographies

HAVOC, June
More Havoc. - New York : Harper & Row, 1980
277p : ill
0-06-011811-3 X.950/26719
8738505

ACTING. American cinema films
Hayward, Susan - Biographies

LINET, Beverly
Susan Hayward. - 3rd print. - New York : Atheneum, 1980
xiii,336p
0-689-11079-0 X.950/28155
8738505

MORENO, Eduardo, 1936-
The films of Susan Hayward. - Secaucus, N.J. : Citadel Press, [1979]
285p : chiefly ill
0-8065-0682-2 X.955/848
8738505

ACTING. American cinema films
Hayworth, Rita - Biographies

HILL, James
Rita Hayworth. - London : Robson, 1983
238p,16p of plates
0-86051-239-8 X.950/28113
3045978

KOBAL, John
Rita Hayworth. - London : W.H. Allen, 1977
416p : ill. - List of films: p.340-407. - Includes index
0-491-02330-8 X.981/20726
1188488

ACTING. American cinema films
Hepburn, Katharine - Biographies

CAREY, Gary
Katharine Hepburn. - Sevenoaks : New English Library, 1985, c1983
287p,[8]p of plates : ill. - List of films: p277-281. - Includes index
0-450-05808-5 X.958/29267
0366153

Katherine Hepburn. - London : Robson, 1983
284p,[12]p of plates : ill. - Includes index
0-86051-210-x X.950/28115
0366153

HIGHAM, Charles, b.1931
Kate. - London : W.H. Allen, 1975
xix,244p,[12]p of plates : ill. - Includes index
0-491-01983-1 X.989/40099
0366153

MORLEY, Sheridan
Katharine Hepburn. - London : Pavilion, 1984
190p : ill. - List of films: p177-190
0-907516-51-3 X.955/2772
0366153

ACTING. American cinema films
Heston, Charlton, 1956-1976 - Biographies

HESTON, Charlton
The actor's life. - London : Allen Lane, 1979
xvii,482p,[12]p of plates : ill
0-7139-1252-9 X.950/2902
1845616

The actor's life. - Harmondsworth : Penguin, 1980
xvii,482p,[12]p of plates : ill
0-14-005173-2 X.908/43255
1845616

ACTING. American cinema films
Holden, William, 1918-1981 - Biographies

THOMAS, Bob
Golden boy. - London : Weidenfeld & Nicolson, 1983
xi,276p,[32]p of plates : ill. - Includes index
0-297-78344-0 X.950/26957
2999099

ACTING. American cinema films
Holliday, Judy - Biographies

CAREY, Gary
Judy Holliday. - London : Robson, 1983
xii,271p,[8]p of plates. - Includes index
0-86051-169-3 X.950/22637
2915057

ACTING. American cinema films
Influence of Stanislavskii, K. S.

BLUM, Richard A.
American film acting. - Epping : Bowker, c1984
xvi,115p. - (Studies in cinema ; no.28). - Bibliography: p103-111. - Includes index
0-8357-1609-0 X.950/43186
3286967

ACTING. American cinema films
Jones, Dean, 1933- - Biographies

JONES, Dean, 1933-
Under running laughter. - Eastbourne : Kingsway, 1983, c1982
185p
0-86065-227-0 X.958/17868
2942402

ACTING. American cinema films
Keaton, Buster

BENAYOUN, Robert
The look of Buster Keaton. - London : Pavilion, c1984
206p : ill. - Bibliography: p206
0-907516-40-8 L.42/3002
1408232

CREMONINI, Giorgio
Buster Keaton. - Firenze : La nuova Italia, 1976
127p. - (Il castoro cinema ; 28). - Includes filmography and bibliography
r6509472 X.629/24741
8738505

DARDIS, Tom
Keaton. - Harmondsworth : Penguin, 1980, c1979
xi,340p : ill. - Bibliography: p329-331. - List of films: p285-317. - Includes index
0-14-005701-3 X.958/2984
1993909

WEAD, George
Buster Keaton and the dynamics of visual wit. - New York : Arno Press, 1976
370p. - Includes bibliographies
0-405-07538-3 X.981/13235
1408232

The film career of Buster Keaton. - London : Prior [etc.], 1977
iii-x,174p,plate. - (A reference publication in film). - Includes bibliographies and index
0-86043-143-6 X.981/21023
1408232

ACTING. American cinema films
Keaton, Buster - Biographies

DARDIS, Tom
Keaton. - London : Deutsch, 1979
xi,340p : ill. - Bibliography: p.329-331. - List of films: p.285-317. - Includes index
0-233-97137-8 X.981/21841
1993909

TICHY, Wolfram
Buster Keaton. - Reinbek bei Hamburg : Rowohlt Taschenbuch, c1983
155p : ill. - (Rowohlts Monographien ; 318). - Includes bibliography and filmography
3-499-50318-2 X.629/24914
8738505

ACTING. American cinema films
Keyes, Evelyn - Biographies

KEYES, Evelyn
Scarlett O'Hara's younger sister. - London : W.H. Allen, 1978
318p,[16]p of plates : ill. - Includes index
0-491-02054-6 X.981/20959
1350102

ACTING. American cinema films
Ladd, Alan - Biographies

LINET, Beverly
Ladd. - London : Robson, 1980
[2], xxiii, 294p, [32] p of plates : ill. - List of films: p.271-280. - Includes index
0-86051-097-2 X.981/22296
2005182

ACTING. American cinema films
Lamour, Dorothy - Biographies

LAMOUR, Dorothy
My side of the road. - London : Robson : 1981, c1980
244p,[24]p of plates : ill. - List of films: p227-2.5. - Includes index
0-86051-126-x X.800/.4007
2220261

ACTING. American cinema films
Lancaster, Burt - Biographies

CLINCH, Minty
Burt Lancaster. - London : Barker, 1984
vii,184p,[16]p of plates : ill - Includes index
0-213-16891-x X.950/.3807
3141861

ACTING. American cinema films
Langdon, Harry - Biographies

SCHELLY, William
Harry Langdon. - Methuen : London : Scarecrow, 982
vii,239p : ill. - (Filmmakers ; no.3). - Bibliography: p212-214. - Includes index
0-8108-1567-2 X.950.20416
2890100

ACTING. American cinema films
Laurel, Stan

GIUSTI, Marco
Laurel & Hardy. - Firenze : La nuova Italia, 1978 i.e. 1979]
140p. - (Il castoro cinema ; 57). - Includes filmography and bibliography
r6509487 X.629 24745
8738505

ACTING. American cinema films
Laurel, Stan - Biographies

GUILES, Fred Lawrence
Stan. - London : Joseph, 1980
240p,[32]p of plates : ill. - Bibliography: p.235-23[illegible]. - List of films: p.228-234. - Includes index
0-7181-1908-8 X.98[illegible]/88995
0506893

MCCABE, John, b.1920
The comedy world of Stan Laurel. - London : Robson Books, 1975
xvii,221p,[24]p of plates : ill. - Includes index
0-903895-48-x X.98[illegible]/40198
0506893

OWEN-PAWSON, Jenny
Laurel before Hardy. - Kendal : Westmorland Gazette 1984
142p : ill
0-902272-51-9 X.80[illegible]/62445
0506893

ACTING. American cinema films
Lemmon, Jack - Biographies

WIDENER, Don
Lemmon. - London : W.H. Allen, 1977
vii,247p,[32]p of plates : ill. - Includes index
0-491-01578-x X.981/20533
1054163

ACTING. American cinema films
Lewis, Jerry

CREMONINI, Giorgio
Jerry Lewis. - Firenze : La nuova Italia, 1979 [i.e. 1980]
117p. - (Il castoro cinema ; 72). - Includes filmography and bibliography
r6509481 X.629/24782
8738505

ACTING. American cinema films
Love, Bessie - Biographies

LOVE, Bessie
From Hollywood with love. - London : Elm Tree Books, 1977
160p : ill. - List of films: p.148-156. - Includes index
0-241-89342-9 X.981/20782
1246054

ACTING. American cinema films
Lovelace, Linda

LOVELACE, Linda
Inside Linda Lovelace. - New York : Pinnacle Books, 1973
184p,[16]p of plates, fold leaf of plate
r6080995 Cup.358/38
0246832

ACTING. American cinema films
Lovelace, Linda - Biographies

LOVELACE, Linda
Ordeal. - [London] : Star, 1981, c1980
206p
0-352-30857-5 Cup.358/103
0246832

ACTING. American cinema films
Lovelace, Linda - Biographies-continued
Ordeal. - London : W.H. Allen, 1981, c1980
206p
0-491-02903-9 Cup.351/284
0246832

ACTING. American cinema films
Lugosi, Bela
BOJARSKI, Richard
The films of Bela Lugosi. - Secaucus : Citadel Press, [1980]
256p : ill
0-8065-0716-0 X.955/2906
8738505

ACTING. American cinema films
MacLaine, Shirley, 1970-1974 - Biographies
MACLAINE, Shirley
You can get there from here. - London : Bodley Head, 1975
3-249p
0-370-10384-x X.989/40230
0513962
You can get there from here. - Large print [ed.]. - London : George Prior Publishers [etc.], 1975
[5],377p
0-904000-36-2 X.981/20135
0513962
You can get there from here. - London : Corgi, 1977
[5],218p
0-552-10411-6 H.77/553 Woolwich
0513962

ACTING. American cinema films
Management. Pero, Taylor. Interpersonal relationships with Turner, Lana
PERO, Taylor
Always, Lana. - Toronto ; London : Bantam Books, 1982
274p. - (A Bantam book)
0-553-20805-5 X.958/15856
2903083

ACTING. American cinema films
Mansfield, Jayne - Biographies
MANN, May
Jayne Mansfield. - St Albans : Mayflower, 1975
271p,[8]p of plates : ill
0-583-12408-9 X.708/41354
0235008
SAXTON, Martha
Jayne Mansfield and the American fifties. - Toronto ; London : Bantam, 1976
xvii,202p,[16]p of plates : ill
0-552-62309-1 X.909/41015
0235008
STRAIT, Raymond
The tragic secret life of Jayne Mansfield. - London : Hale, 1976
xvi,207p,[24]p of plates : ill
0-7091-5543-3 X.989/50715
0235008

ACTING. American cinema films
Marvin, Lee - Biographies
ZEC, Donald
Marvin. - London : New English Library, 1979
252p,[16]p of plates : ill. - List of films: p.240-243. - Includes index
0-450-04115-8 X.989/54099
2003678

ACTING. American cinema films
Marx (Brothers)
MARTINI, Andrea
I Marx. - Firenze : La nuova Italia, 1980 [i.e. 1981]
121p. - (Il castoro cinema ; 79-80). - Includes bibliography
r6509488 X.629/24879
8738505

ACTING. American cinema films
Marx Brothers
BENAYOUN, Robert
Les frères Marx. - [Paris] : Seghers, 1979
165p,4p of plates : ill. - (Seghers / humour ; 9)
r6129007 X:907/20843
0007439

ACTING. American cinema films
Marx, Groucho - Biographies
ARCE, Hector
Groucho. - New York : Putnam, c1979
541p,[16]p of plates
0-399-12046-7 X.950/23495
8738505

CHANDLER, Charlotte
Hello, I must be going. - London : Robson, 1979
vii,568p,[24]p of plates : ill
0-86051-059-x X.989/53180
1591347

MARX, Groucho
The Groucho letters. - London : Sphere, 1976
252p
0-7221-5858-0 X.908/41272
009353x
The Groucho phile. - Indianapolis : Bobbs-Merrill, 1976
xvi,384p,[16]p of plates : ill(some col). - Includes index
0-672-52239-x X.981/21856
1398385
The Groucho phile. - London : W.H. Allen, 1978
xvi,384p,[16]p of plates : ill(some col). - Includes index
0-491-02134-8 X.985/1175
1398385
The secret word is Groucho. - New York : Putnam, 1976
217p
0-399-11690-7 X.989/82688
1398385

ACTING. American cinema films
Marx, Harpo - Biographies
MARX, Harpo
Harpo speaks!. - London : Robson Books, 1976
451p : ill
0-903895-62-5 X.989/50618
0628409
Harpo speaks. - Sevenoaks : Coronet, 1978
478p,[8]p of plates : ill
0-340-22308-1 X.907/25368
0628409

ACTING. American cinema films
McQueen, Steve - Biographies
MCCOY, Malachy
Steve McQueen. - [New ed.]. - Sevenoaks : Coronet, 1981, c1980
224p,[12]p of plates : ill
0-340-27059-4 X.958/3097
0198161
NOLAN, William F. (William Francis)
McQueen. - London : Barker, c1984
230p,[32]p of plates : ill. - Includes index
0-213-16898-7 X.950/34478
0198161
SATCHELL, Tim
McQueen. - London : Sidgwick & Jackson, 1981
137p : ill. - List of films: p129-135. - Includes index
0-283-98778-2 X.955/732
0198161

ACTING. American cinema films
Minnelli, Liza - Biographies
PARISH, James Robert
Liza. - London : W.H. Allen, 1975
vi,186p,[16]p of plates : ill. - List of films: p.178-182. - List of gramophone records: p.183-186
0-491-01574-7 X.989/40129
0373575

ACTING. American cinema films
Mix, Tom - Biographies
NICHOLAS, John H.
Tom Mix. - Oklahoma City : National Cowboy Hall of Fame and Western Heritage Center, [1980]
xi,95p : ill. - (A Persimmon Hill book). - Includes bibliographies
0-932154-05-0 L.45/2451
8738505

ACTING. American cinema films
Monroe, Marilyn
CONWAY, Michael
The films of Marilyn Monroe. - 9th paperbound printing. - Secaucus : Citadel Press, 197?
160p
0-8065-0145-6 X.955/2548
8738505

ACTING. American cinema films
Monroe, Marilyn - Biographies
GUILES, Fred Lawrence
Norma Jeane. - London : Granada, 1985, c1984
xii,377p,[16]p of plates : ill. - Bibliography: p359-361. - Includes index
0-246-12307-9 X.950/43066
0155705
MARILYN Monroe / introduction by David Robinson ; compiled and edited by John Kobal. - London : Hamlyn, 1974, c1983
176p : ill(some col.). - (Gondola). - Bibliography: p174. - Includes index
0-600-37310-x X.425/4366
0155705

MONROE, Marilyn
My story. - London : W.H. Allen, 1975
143p
0-491-01754-5 X.989/40267
0155705

SHAW, Sam
Marilyn Monroe in the camera eye. - London : Hamlyn, 1979
160p : ill
0-600-34156-9 X.421/20764 Woolwich
2205173

SPADA, James
Monroe. - London : Sidgwick & Jackson, 1982
194p
0-283-98871-1 X.955/1831
2205173

STERN, Bert
The last sitting. - London : Orbis, [1982]
188p
0-85613-468-6 L.45/1820
2205173

WEATHERBY, William John
Conversations with Marilyn. - London : Sphere, 1977
[6],229p
0-7221-8982-6 X.709/31769
0155705

ACTING. American cinema films
Monroe, Marilyn, 1958-1962 - Biographies
PEPITONE, Lena
Marilyn Monroe confidential. - London : Sidgwick and Jackson, 1979
3-251p,[16]p of plates : 1 ill
0-283-98537-2 X.989/53508
1868721

ACTING. American cinema films
Monroe, Marilyn, 1960 - Biographies
LEMBOURN, Hans Jorgen
Forty days with Marilyn. - London : Hutchinson, 1979
[7],214p
0-09-139010-9 X.629/12616 Woolwich
1835165
Forty days with Marilyn. - London : Arrow Books, 1980
[7],214p
0-09-922690-1 X.907/25921
1835165

ACTING. American cinema films
Monroe, Marilyn, 1960-1962 - Biographies
WEATHERBY, William John
Conversations with Marilyn. - London : Robson Books, 1976
[6],229p,plate
0-903895-68-4 X.989/50718
0658081

ACTING. American cinema films
Novak, Kim
KLENO, Larry
Kim Novak on camera. - 2. [print.]. - San Diego : A.S. Barnes, c1980
320p
r6807786 X.425/3688
8738505

ACTING. American cinema films
Peck, Gregory - Biographies
FREEDLAND, Michael
Gregory Peck. - London : W.H. Allen, 1980
252p,[24]p of plates : ill. - List of films: p.251-252
0-491-02178-x X.981/22425
2210371

ACTING. American cinema films
Pickford, Mary, b.1893 & Fairbanks, Douglas, b.1883 - Biographies
HERNDON, Booton
Mary Pickford and Douglas Fairbanks. - London : W.H. Allen, 1978
xii,324p : ill. - Bibliography: p.310-317. - Includes index
0-491-02354-5 X.981/21160
1546090

ACTING. American cinema films
Poitier, Sidney
MARILL, Alvin H.
The films of Sidney Poitier. - Secaucus : The Citadel Press, c1978
222p : ill
0-8065-0612-1 X.955/896
8738505

ACTING. American cinema films
Poitier, Sidney - Biographies

POITIER, Sidney
This life. - London : Hodder and Stoughton, 1980
374p,[24]p of plates
0-340-25654-0 X.950/1379
214882x

ACTING. American cinema films
Power, Tyrone

BELAFONTE, Dennis
The films of Tyrone Power. - Secancus, N.J. : Citadel Press, c1979
224p
0-8065-0477-3 X.955/1195
8738505

ACTING. American cinema films
Power, Tyrone - Biographies

GUILES, Fred Lawrence
Tyrone Power. - London : Granada, 1980
406p,32p of plates : ill. - Bibliography: p.391-393. - List of films: p.334-390. - Includes index
0-246-11311-1 X.989/88214
2221268

ACTING. American cinema films
Presley, Elvis, to 1972 - Biographies

LICHTER, Paul
Elvis in Hollywood. - London : Hale, 1977
188p : ill. - List of sound discs: p.183-185
0-7091-5733-9 X.431/10335
1091212

ACTING. American cinema films
Pryor, Richard - Biographies

ROVIN, Jeff
Richard Pryor. - London : Orbis, 1984, c1983
248p,[16]p of plates : ill
0-85613-697-2 X.958/28003
3204707

ACTING. American cinema films
Reagan, Ronald

THOMAS, Tony, 1927-
The films of Ronald Reagan. - Secaucus, N.J. : Citadel Press, c1980
224p
0-8065-0751-9 X.955/1586
8738505

ACTING. American cinema films
Redford, Robert

SPADA, James
The films of Robert Redford. - Secaucus : The Citadel Press, 1978
255p : mainly ill
0-8065-0654-7 X.955/959
8738505

ACTING. American cinema films
Redford, Robert - Biographies

DOWNING, David
Robert Redford. - London : W.H. Allen, 1982
224p,[8]p of plates : ill(some col.). - List of films: p217-219. - Includes index
0-491-02977-2 X.805/4293
2948869

ACTING. American cinema films
Reynolds, Burt - Biographies

ELIOT, Marc
Burt!. - [Sevenoaks] : Coronet, 1983, c1982
224p,[8]p of plates : ill
0-340-32804-5 X.958/18876
2991187

HURWOOD, Bernhardt J.
Burt Reynolds. - New York ; London : Quick Fox, 1979
112p : ill
0-8256-3942-5 X.950/30911
8738505

RESNICK, Sylvia Safran
Burt Reynolds. - London : W.H. Allen, c1983
222p,[4]p of plates : ill(some.col.)
0-491-03340-0 X.955/2167
2991187

WHITLEY, Dianna
Burt Reynolds. - New York : Grosset & Dunlap, 1979
112p : ill
0-448-15479-x X.955/2240
8738505

ACTING. American cinema films
Ritter, Tex - Biographies

BOND, Johnny
The Tex Ritter story. - New York : Chappell Music Company, 1976
[14], 397 p. : ill. - List of sound discs : p. 275-305. - List of films : p. 307-376
r6031527 X.431/10331
1080431

ACTING. American cinema films
Robinson, Edward G - Biographies

GANSBERG, Alan L.
Little Caesar. - Sevenoaks : New English Library, 1983
334p,[8]p of plates : ill. - List of films: p295-311. - Includes index
0-450-05779-8 X.958/28488
0132225

Little Caesar. - London : New English Library, 1983
248p,[32]p of plates : ill. - List of films: p222-235. - Includes index
0-450-06004-7 X.950/31530
0132225

ACTING. American cinema films
Rogers, Roy - Biographies

ROGERS, Roy
Happy trails. - Waco, Tex. : Word Books, [1979]
213p,plates
0-8499-0086-7 X.950/43244
8738505

ACTING. American cinema films
Roth, Lillian - Biographies

ROTH, Lillian
I'll cry tomorrow. - Large print ed. - Bath : Chivers, 1978
xi,451p
0-85997-279-8 Nov.36122 Woolwich
1378015

ACTING. American cinema films
Russell, Rosalind - Biographies

RUSSELL, Rosalind
Life is a banquet. - London : W.H. Allen, 1978
xxii,260p,[32]p of plates : ill. - List of films: p.245-246. - Includes index
0-491-02024-4 X.981/21073
1456326

ACTING. American cinema films
Sedgwick, Edith - Biographies

EDIE / [compiled by] Jean Stein ; edited with George Plimpton. - London : Panther, 1984, c1982
455p : ill
0-586-05785-4 X.958/25241
3163520

ACTING. American cinema films
Smith, C. Aubrey - Biographies

ALLEN, David Rayvern
Sir Aubrey. - London : Elm Tree, 1982
xiii,172p,[8]p of plates : ill. - Bibliography: p167-168. - Includes index
0-241-10590-0 X.809/52634
2763451

ACTING. American cinema films
Stars, to 1930 - Biographies

BRUNDIDGE, Harry T.
Twinkle, twinkle, movie star!. - New York ; London : Garland, 1977
[4],xiii,284p,[32]leaves of plates. - (The Garland classics of film literature ; 6)
0-8240-2868-6 X.989/88439
2073293

ACTING. American cinema films
Stars, to 1945

DODY, Sandford
Giving up the ghost. - New York : M. Evans and Company, c1980
334p
0-87131-142-9 X.950/14114
8007004

EELLS, George
Ginger, Loretta and Irene who?. - New York : G.P. Putnam's Sons, c1976
393p
0-399-11822-5 X.950/20505
8007004

ACTING. American cinema films
Stars, to 1980 - Biographies

QUINLAN, David, 1941-
The illustrated directory of film stars. - London : Batsford, 1981
499p. - Bibliography: p498-499
0-7134-3891-6 X.955/752
2584255

ACTING. American cinema films
Stars, to date

GERMANOS, Phrenty
Kalėnychta theie Oskar-. - [Athena] : Kaktos, c1982
191p,[1]leaf of plates : ill
r6597071 X.950/38049
8007012

ACTING. American cinema films
Stars, 1930-1940

MOVIE-STAR portraits of the forties / edited by John Kobal. - New York : Dover Publications, [etc.], 1977
ii-xii,162p of ports : ill
0-486-23546-7 X.415/5463
1434616

ACTING. American cinema films
Streisand, Barbra - Biographies

BRADY, Frank, 1934-
Barbra. - New York : Grosset & Dunlap, 1980
151p
0-448-16534-1 X.955/2239
8738505

ACTING. American cinema films
Supporting actors & supporting actresses - Biographies

JONES, Ken D.
Character people. - South Brunswick ; New York : Barnes ; London : Yoseloff, 1976
3-209p : ill
0-498-01697-8 X.985/1192
1549391

PARISH, James Robert
Hollywood character actors. - New York : Arlington House, [1978]
542p
0-87000-384-4 X.955/557
8007039

ACTING. American cinema films
Swanson, Gloria - Biographies

SWANSON, Gloria
Swanson on Swanson. - Feltham : Hamlyn, 1981, c1980
vi,519p,[32]p of plates : ill. - Includes index
0-600-20496-0 X.958/8903
2387751

Swanson on Swanson. - London : Joseph, 1981, c1980
vi,535p,[48]p of plates : ill. - Includes index
0-7181-1990-8 X.950/3501
2387751

ACTING. American cinema films
Temple, Shirley

WINDELER, Robert
The films of Shirley Temple. - Secaucus : Citadel Press, c1978
256p : ill
0-8065-0615-6 X.425/3530
8738505

ACTING. American cinema films
Temple, Shirley - Biographies

DAVID, Lester
The Shirley Temple story. - London : Robson, 1984
224p,[16]p of plates. - Bibliography: p215. - Includes index
0-86051-279-7 X.950/37184
0584541

WINDELER, Robert
Shirley Temple. - London : W.H. Allen, 1976
160p,[24]p of plates : ill. - List of films: p.146-153. - Includes index
0-491-01524-0 X.989/40468
0584541

ACTING. American cinema films
Travolta, John - Biographies

MUNSHOWER, Suzanne L.
The John Travolta scrapbook. - London : Pop Universal Ltd : [Distributed by] Souvenir Press, 1978
[5],121p : ill
0-285-62382-6 X.902/5314
1571052

TRAVOLTA superstar special!. - [London] : [IPC Magazines], [1978]
[48]p : ill(some col)
B79-01359 X.902/5363
1571052

ACTING. American cinema films
Turner, Lana - Biographies

TURNER, Lana
Lana. - Sevenoaks : New English Library, 1983, c1982
255p,[40]p of plates
0-450-06019-5 X.950/22398
2926458

ACTING. American cinema films
Turner, Lana. Interpersonal relationships with Pero, Taylor

PERO, Taylor
Always, Lana. - Toronto ; London : Bantam Books, 1982
274p. - (A Bantam book)
0-553-20805-5 X.958/15856
2903083

ACTING. American cinema films
Valentino, Rudolph

TRINCHERO, Sergio
Rodolfo Valentino. - [Ivrea] : Priuli e Verlucca, 1975
175 p. : ill. - Filmografia e bibliografia : p. 164-172
r6024206 X.900/17820
022927X

ACTING. American cinema films
Valentino, Rudolph - Biographies

BOTHAM, Noel
Valentino. - London : Everest, 1976
248p,[32]p of plates : ill. - List of films : p.245-248
0-903925-49-4 X.908/40794
022927x

SCAGNETTI, Jack
The intimate life of Rudolph Valentino. - Middle Village : Jonathan David, 1975
160p
0-8246-0197-1 X.802/4730
022927X

STEIGER, Brad
Valentino. - London : Corgi, 1976
203p,8p of plates : ill
0-552-10235-0 X.908/40518
022927x

WALKER, Alexander
Rudolph Valentino. - London : Elm Tree Books, 1976
127p : ill. - Bibliography: p.124. - List of films : p.121-123. - Includes index
0-241-89349-6 X.981/20286
022927x

Rudolph Valentino. - New York : Penguin, 1977
127p : ill. - Bibliography : p.124. - List of films : p.121-123.- Includes index
0-14-003615-6 X.909/43552
022927X

Rudolph Valentino. - London : Sphere, 1977
92p,[32]p of plates : ill. - Bibliography: p.88. - List of films: p.83-87. - Includes index
0-7221-8868-4 X.907/25275
022927x

ACTING. American cinema films
Varconi, Victor - Biographies

VARCONI, Victor
It's not enough to be Hungarian. - Denver : Graphic Impression, c1976
192p : ill
0-914628-09-7 X.950/11927
8738505

ACTING. American cinema films
Wayne, John

STACY, Pat
Duke. - London : Souvenir, 1983
xvii,233p,[20]p plates
0-285-62606-x X.950/28470
0162736

ACTING. American cinema films
Wayne, John, 1907-1979 - Biographies

GOLDSTEIN, Norm
John Wayne. - [New York] : Associated Press ; [London] : Windward, c1979
159p : ill. - List of films: p135-159
0-7112-0018-1 X.955/751
2053039

The LIFE and times of John Wayne / [by David Hanna]. - London : Hamlyn, 1979
[88]p : ill. - List of films (8p.)
0-600-39499-9 X.985/1272
2053039

ACTING. American cinema films
Wayne, John, b.1907

SHAW, Sam
John Wayne in the camera eye. - London : Hamlyn, 1980
160p : ill
0-600-34160-7 X.421/20763 Woolwich
2205203

ACTING. American cinema films
Wayne, John, b.1907 - Biographies

CARPOZI, George
The John Wayne story. - Revised ed. - Sevenoaks : Coronet, 1977
190p,[8]p of plates : ill. - List of films: p.164-180. - Includes index
0-340-22012-0 X.907/25321
0126551
- [Another copy.] X.907/25321

ACTING. American cinema films
Welles, Orson - Biographies

ORSON Uells / [sostavitel' : Kirill Ėmil'evich Razlogov]. - Moskva : "Iskusstvo", 1975
270p,[30]p of plates : ill. - (Mastera zarubezhnogo kinoiskusstva)
r6056293 X.907/17272
1501259

ACTING. American cinema films
Winters, Shelley - Biographies

WINTERS, Shelley
Shelley. - London : Granada, 1981
511p : ill
0-246-11496-7 X.950/3650
2389185

Shelley. - London : Granada, 1982, c1980
500p,[32]p of plates : 2ill. - (A Mayflower book)
0-583-13506-4 X.958/12318
2389185

ACTING. American cinema films. Special subjects. American photography
Stars, 1950-1960 - Collections

FILM-STAR portraits of the fifties / edited with an introduction by John Kobal. - Leicester : Windward, c1980
xii,163p : chiefly ill
0-7112-0200-1 X.955/695
2563622

ACTING. American cinema films. Special subjects. American portrait photography, 1900-1950
Stars, 1916-1942 - Collections

KOBAL, John
The art of the great Hollywood portrait photographers 1925-1940. - London : Allen Lane, 1980
ix,291p. - Bibliography: p287. - Includes index
0-7139-1393-2 L.45/827
2347407

ACTING. American cinema films. Special subjects. American portrait photography, 1900-1950
Stars, 1930-1970

ENGSTEAD, John
Star shots. - London : Columbus Books, c1978
250p. - Includes index
0-86287-036-4 X.425/4127
2979276

ACTING. American cinema films. Special subjects. American portrait photography, 1900-1950
Stars, 1934-1960 - Collections

HOLLYWOOD colour portraits / [selected by] John Kobal. - London : Aurum, c1981
156p
0-906053-27-7 X.425/3594
2816776
- [Another issue]. - New York: William Morrow, 1981. - ISBN: 0688007538 X.425/5331

ACTING. American silent cinema films
Lloyd, Harold

REILLY, Adam
Harold Lloyd. - London : Deutsch, 1978
xv,240p : ill. - Bibliography: p.229. - Includes index
0-233-96995-0 X.981/21086
1479334

ACTING. American silent cinema films
Stars - Biographies

SLIDE, Anthony
The idols of silence. - South Brunswick ; New York : Barnes ; London : Yoseloff, 1976
208p : ill. - Bibliography: p.119-207
0-498-01611-0 X.985/1163
1254677

ACTING. Argentinian cinema films
Sarli, Isabel

BO, Armandó
Los films de Armando Bó con Isabel Sarli. - Buenos Aires : Corregidor, 1981
174p : ill. - (Serie mayor)
r6709023 X.955/2543
9153837

ACTING. Armenian cinema films
to date

KALANTAR, K. (Karen)
Akter v armi͡anskom kino. - Erevan : Sovetakan grokh, 1982
321p
r6403164 X.950/27826
9032169

ACTING. Bohemia. Czechoslovakia
Filipovský, František

TVRZNÍK, J. (Jiří)
Šest dýmek Františka Filipovského. - Praha : Novinář, 1981
348p : ill
r6440717 X.622/15423
8851980

ACTING. Bohemia. Czechoslovakia
Mošna, Jindřich - Biographies

PÁLENÍČEK, Ludvík
Jindřich Mošna. - 1. vyd. - Praha : Melantrich, 1983
271 p., [60] p. of plates : ill.. - (Odkazy pokrokových osobností naší minulosti ; sv. 69). - Includes indexes. - Bibliography: p. 263-264
r6441541 X.950/37089
8851980

ACTING. Bohemia. Czechoslovakia
Werich, Jan

ŠKUTINA, Vladimír
Tak už jsem tady s tím vápnem, pane Werichu!. - Zürich : Magazín, 1981
160p : ill
r6440589 X.329/15530
8851980

ACTING. British cinema films
Caine, Michael - Biographies

HALL, William, 1935-
Raising Caine. - London : Sidgwick & Jackson, 1981
260p,[16]p of plates. - List of films: p231-256. - Includes index
0-283-98777-4 X.950/8565
2581981

Raising Caine. - London : Arrow Books, 1982
vi,344p,[16]p of plates : ill. - List of films: p298-333. - Includes index
0-09-929200-9 X.808/38092
2581981

ACTING. British cinema films
Fox, James, 1939- - Biographies

FOX, James, 1939-
Comeback. - London : Hodder and Stoughton, 1983
151p,[16]p of plates : ill
0-340-27733-5 X.809/55530
2928299

FOX, James, 1939-
Comeback. - London : Hodder and Stoughton, 1983
151p,[16]p of plates
0-340-35357-0 X.958/24746
2928299

ACTING. Bulgaria
Kabakchiev, Liubomir - Biographies

G'OROVA, Sevelina Tsokova
Li͡ubomir Kabakchiev. - Sofii͡a : Nauka i izkustvo, 1976
86p, 16p of plates : ill. - (Tvortsi na sti͡senata i ekvana)
r6043472 X.909/37530
1208780

ACTING. Characters. Drama

APPEL, Libby
Mask characterization. - Carbondale : Southern Illinois University Press, c1982
xv,120p : ill
0-8093-1039-2 X.950/37548
8090645

STANISLAVSKIĬ, K. S. (Konstantin Sergeevich)
Creating a role. - London : Eyre Methuen, 1981, c1961
xiv,271p
0-413-47760-6 X.808/33399
2334682

Creating a role. - London : Eyre Methuen, 1981, c1961
xiv,271p
0-413-47760-6 X.808/33399
2334690

ACTING. Cinema films
Astaire, Fred - Biographies

GREEN, Stanley
Starring Fred Astaire. - Garden City : Doubleday, 1977
501p : ill. - Bibliography of radio and television appearances and discography: p.462-482
0-385-11604-7 X.955/492
8063850

ACTING. Cinema films
Baker, Sir Stanley, b.1928 – Biographies

STOREY, Anthony
Stanley Baker. – London : W.H. Allen, 1977
160p,[16]p of plates : ill. – List of films: p.139–160
0–491–02230–1 X.989/51728
118654x

ACTING. Cinema films
Bergen, Candice – Biographies

BERGEN, Candice
Knock wood. – London : Hamilton, 1984
xii,225p,[16]p of plates : ill
0–241–11358–x X.950/38424
3024156

ACTING. Cinema films
Bergman, Ingrid, 1915– – Biographies

BERGMAN, Ingrid
Ingrid Bergman. – London : Joseph, 1980
480p,[64]p of plates : ill. – Includes index
0–7181–1946–0 X.981/22644
2326574

BERGMAN, Ingrid, 1915–
Ingrid Bergman. – London : Sphere, 1981, c1980
xvii,556p,[16]p of plates : ill. – List of films: p534–545. – Includes index
0–7221–1631–4 X.958/6789
2326574

ACTING. Cinema films
Bergman, Ingrid, 1915–1982 – Biographies

BROWN, Curtis F.
Ingrid Bergman. – London : Star Books, 1975
157p : ill. – (The Star illustrated history of the movies). – Bibliography: p.145. – List of films: p.146–150. – Includes index
0–352–39804–3 X.908/40246
0577987

TAYLOR, John Russell
Ingrid Bergman. – London : Elm Tree, 1983
127p,[8]p of plates : ill(some col.)
0–241–11106–4 X.955/2214
0577987

ACTING. Cinema films
Bogarde, Dirk, to ca 1960 – Biographies

BOGARDE, Dirk
A postillion struck by lightning. – London : Chatto and Windus, 1977
[12],268p,leaf of plate,10p of plates : ill. – Includes index
0–7011–2207–2 X.989/51336
0668974
– [2nd impression.] X.950/42185

A postillion struck by lightning. – St Albans : Triad, 1978
302p,[8]p of plates : ill
0–586–04569–4 X.907/25449
0668974

A postillion struck by lightning. – Large print [ed.]. – Bolton-by-Bowland : Magna Print Books, 1979
568p : ill
0–86009–125–2 X.629/12645 Woolwich
0668974

ACTING. Cinema films
Bogarde, Dirk, 1939–1972 – Biographies

BOGARDE, Dirk
Snakes and ladders. – London : Chatto and Windus, 1978
xi,339p,leaf of plate,[22]p of plates : ill. – List of films: p.331–332. – Includes index
0–7011–2377–x X.809/43693
1441019

Snakes and ladders. – St Albans : Triad, 1979
395p,[16]p of plates : ill. – List of films: p.385–386. – Includes index
0–586–04946–0 X.907/26019
1441019

ACTING. Cinema films
Bogarde, Dirk, 1972–1982 – Biographies

BOGARDE, Dirk
An orderly man. – London : Chatto & Windus, 1983
291p,[17]p of plates : ill(some col.). – Includes index
0–7011–2659–0 Nov.48544
2914794

An orderly man. – London : Triad, 1984, c1983
352p,[16]p of plates : ill. – Includes index
0–586–05893–1 X.958/25786
2914794

An orderly man. – Large print ed. – Long Preston : Magna, Feb.1985
[550]p
0–86009–697–1 Nov.53482
2914794

ACTING. Cinema films
Brando, Marlon – Biographies

FIORE, Carlo
The Brando I knew. – London : Hart-Davis MacGibbon, 1975
207p,[32]p of plates : ill
0–246–10838–x X.989/40193
0253200

The Brando I knew. – St Albans : Mayflower, 1976
207p,[8]p of plates : ill
0–583–12527–1 X.907/25166
0253200

SHAW, Sam
Brando in the camera eye. – London : Hamlyn, 1979
160p : ill
0–600–34157–7 X.421/20762 Woolwich
2205181

ACTING. Cinema films
Burton, Richard, 1925– – Biographies

CASHIN, Fergus
Richard Burton. – London : W.H. Allen, 1982
191p,[16]p of plates. – Includes index
0–491–02868–7 X.950/22638
0187259

FERRIS, Paul
Richard Burton. – London : Weidenfeld and Nicolson, 1981
xi,212p : ill. – Bibliography: p203–204. – Includes index
0–297–77966–4 X.800/34949
0187259

ACTING. Cinema films
Burton, Richard, 1925–. Interpersonal relationships with Taylor, Elizabeth, 1932–

DAVID, Lester
Richard and Elizabeth. – London : Star Books, 1979
[11],226p,[8]p of plates : ill. – Includes index
0–352–30330–1 X.908/42328
1253247

ACTING. Cinema films
Burton, Richard, 1925–1984 – Biographies

CASHIN, Fergus
Richard Burton. – London : W.H. Allen, 1982
191p,[8]p of plates : ill. – (A Star book). – Includes index
0–352–31311–0 X.958/25609
0187259

FERRIS, Paul
Richard Burton. – Long Preston : Magna, 1985, c1981
480p,[16]p of plates : ill
0–86009–713–7 X.950/42092
0187259

ACTING. Cinema films
Burton, Richard, 1925–1984. Interpersonal relationships with Taylor, Elizabeth, 1932– – Biographies

DAVID, Lester
Richard & Elizabeth. – London : A. Barker, 1977
[11],242p : ill. – Includes index
0–213–16660–7 X.989/51895
1253247

ACTING. Cinema films
Chevalier, Maurice – Biographies

FREEDLAND, Michael
Maurice Chevalier. – London : Arthur Barker, c1981
287p : ill
0–213–16789–1 X.800/32453
2641054

HARDING, James, 1929–
Maurice Chevalier. – London : Secker & Warburg, 1982
xv,220p,[32]p of plates. – Bibliography: p212–214. – Includes index
0–436–19107–5 X.800/34468
2641054

SABATÈS, Fabien
Maurice Chevalier. – Paris : Olivier Orban, [1981]
203p : ill
2–85565–170–0 X.955/1627
2641054

ACTING. Cinema films
Collins, Joan, 1933– – Biographies

COLLINS, Joan, 1933–
Past imperfect. – London : W.H. Allen, 1978
ix,252p,[32]p of plates : ill. – List of films: p.250–252
0–491–02014–7 X.981/21109
1485709

Past imperfect. – Sevenoaks : Coronet, 1979
301p,[8]p of plates. – List of films: p.297–301
0–340–23828–3 X.908/43791
1485709

ACTING. Cinema films
Colman, Ronald – Biographies

COLMAN, Juliet Benita
Ronald Colman. – London : W.H. Allen, 1975
ix,294p,[16]p of plates : ill. – Bibliography: p.287–288. – List of films: p.281–282. – Includes index
0–491–01785–5 X.989/40388
0562106

ACTING. Cinema films
Connery, Sean – Biographies

PASSINGHAM, Kenneth
Sean Connery. – London : Sidgwick & Jackson, 1983
160p,[16]p of plates : ill. – Includes index
0–283–98921–1 X.800/35351
290439x

Sean Connery. – London : Sidgwick & Jackson, 1983
160p,[8]p of plates : ill. – Includes index
0–450–05712–7 X.958/26196
290439x

RISSIK, Andrew
The James Bond man. – London : Elm Tree, 1983
217p,[8]p of plates : ill. – List of films: p197–208. – Includes index
0–241–11067–x X.950/27963
304422x

ACTING. Cinema films
Day, Doris – Biographies

DAY, Doris
Doris Day. – London : W.H. Allen, 1976
365p[36]p of plates : ill
0–491–01767–7 X.981/20382
0730181

Doris Day. – New York : Morrow, 1976
305p
0–688–02968–x X.981/21938
1517945

Doris Day. – London : Star Books, 1977
365p,[8]p of plates : ill
0–352–39568–0 X.908/41776
1517945

ACTING. Cinema films
Delon, Alain – Biographies

RODE, Henri
Alain Delon. – Nouvelle édition revue et augmentée. – Paris : Éditions Pac, 1982
344p,plates
r6209821 X.629/20280
8063850

ACTING. Cinema films
Delon, Alain, 1957–1974 – Biographies

RODE, Henri
Le fascinant Monsieur Delon. – Paris : PAC, [1974]
149 p. : ill. – (Têtes d'affiche)
2–85336–006–7 X.909/32318
0924725

ACTING. Cinema films
Dietrich, Marlene – Biographies

HIGHAM, Charles
Marlene. – London : Mayflower, 1978
320p,[16]p of plates : ill. – List of films: p.299–300. – Includes index
0–583–12916–1 X.708/21739
0152935

HIGHAM, Charles, b.1931
Marlene. – London : Hart-Davis MacGibbon, 1978
319p : ill. – Includes index
0–246–11051–1 X.989/52306
0152935

MORLEY, Sheridan
Marlene Dietrich. – London : Elm Tree Books, 1976
128p : ill. – List of films: p.119–123. – Bibliography: p.124–125. – Includes index
0–241–89348–8 X.981/20330
0152935

Marlene Dietrich. – London : Sphere, 1978
93p,[32]p of plates : ill. – Bibliography: p.84–87. – List of films: p.76–83. – Includes index
0–7221–6163–8 X.907/25372
0152935

NAVACELLE, Thierry de
Sublime Marlene. – London : Sidgwick & Jackson, 1984
158p : ill. – Includes index
0–283–99042–2 X.425/4871
0152935

WALKER, Alexander
Dietrich. – London : Thames and Hudson, 1984
207p : ill. – Bibliography: p203. – Includes index
0–500–01340–3 X.955/2988
0152935

ACTING. Cinema films
Dors, Diana - Biographies

DORS, Diana
Behind closed Dors. - London : W.H. Allen, 1979
208p : ill. - (A Star book)
0-352-30335-2 X.958/3035
1355368

Behind closed Dors. - London : W.H. Allen, 1979
208p : ill
0-491-02387-1 X.629/12320 Woolwich
1355368

Dors by Diana. - London : Macdonald Futura, 1981
317p,[16]p of plates : ill
0-362-00549-4 X.950/8195
1355368
- [Another copy: paperback]
ISBN 0708820255 X.950/8196

For adults only. - London : Star Books, 1978
256p : ill
0-352-30158-9 X.908/41526
1355368

For adults only. - London : W.H. Allen, 1978
256p : ill
0-491-02372-3 X.989/52253
1355368

ACTING. Cinema films
Douglas, Angela - Biographies

DOUGLAS, Angela
Swings and roundabouts. - London : Elm Tree, c1983
255p,[16]p of plates
0-241-10953-1 X.950/29138
3063283

Swings and roundabouts. - London : Corgi, 1985, c1983
330p,[16]p of plates : ill
0-552-12553-9 X.958/30090
3063283

ACTING. Cinema films
Ekland, Britt - Biographies

EKLAND, Britt
True Britt. - London : Sphere, 1980
245p,[16]p of plates. - List of films: p244-245
0-7221-3270-0 X.958/2717
2397374

ACTING. Cinema films
Fairbanks, Douglas, b.1883 & Fairbanks, Douglas, b.1909 - Biographies

The FAIRBANKS album / drawn from the family archives by Douglas Fairbanks, Jr ; introduction and narrative by Richard Schickel. - London : Secker and Warburg, 1976
287p : chiefly ill
0-436-44415-1 X.985/1040
0643955

ACTING. Cinema films
Female stars, 1912-1974 - Biographies

MANVELL, Roger
Love goddesses of the movies. - London : Hamlyn, 1975
176p : ill(some col). - Bibliography: p.176. - List of films: p.172-176
0-600-35485-7 Cup.815/2
0350982

ACTING. Cinema films
Finch, Peter - Biographies

DUNDY, Elaine
Finch, bloody Finch. - London : Joseph, 1980
352p,[32]p of plates : ill. - Includes index
0-7181-1904-5 X.981/13208
2148145

Finch, bloody Finch. - [London] : Magnum, 1981, c1980
354p,[16]p of plates. - Includes index
0-417-06600-7 X.958/8628
2148145

ACTING. Cinema films
Finch, Peter, 1916 Sept. 28-1977

FINCH, Yolande
Finchy. - London : Arrow, 1980
215p,[16]p of plates : ill
0-09-924190-0 X.958/12764
2086522

ACTING. Cinema films
Finch, Peter, 1916 Sept. 28-1977 - Biographies

FAULKNER, Trader
Peter Finch. - London : Angus and Robertson, 1979
312p,[32]p of plates : ill. - List of films: p.293-303. - Includes index
0-207-95831-9 Cup.408.l.16
2086522

Peter Finch. - London : Pan Books, 1980
364p,[8]p of plates. - List of films and plays: p.337-352. - Includes index
0-330-26120-7 X.908/43376
2086522

ACTING. Cinema films
Fonda, Henry - Biographies

FONDA, Henry
Fonda. - London : W.H. Allen, 1982, c1981
xii,372p,[32]p of plates : ill. - Includes index
0-491-02866-0 X.950/22639
2866595

GOLDSTEIN, Norm
Henry Fonda. - London : Joseph, 1982
123p : ill
0-7181-2245-3 X.955/1471
2866595

ACTING. Cinema films
Fonda, Jane - Biographies

GUILES, Fred Lawrence
Jane Fonda. - London : Michael Joseph, 1981
278p,[32]p of plates : ill. - Bibliography: p277-278. - Includes index
0-7181-2071-x X.950/8558
0185051

KIERNAN, Thomas
Jane Fonda. - St Albans : Mayflower, 1976
xviii,430p
0-583-12717-7 X.908/40774
0185051

Jane Fonda. - London : Granada, 1982
368p,[16]p of plates : ill. - List of films : p368
0-246-11921-7 X.520/30298
0185051

Jane Fonda. - London : Granada, 1982
367p,[16]p of plates : ill
0-583-13681-8 X.958/22897
0185051

ACTING. Cinema films
Fröbe, Gert - Biographies

BALL, Gregor
Gert Fröbe. - München : Heyne, 1982
238p. - (Heyne Filmbibliothek ; 37). - Includes filmography
3-453-86041-1 X.429/16194
8063850

ACTING. Cinema films
Gable, Clark - Biographies

TORNABENE, Lyn
Long live the king. - London : W.H. Allen, 1977
396p,[32]p of plates : ill. - Includes index
0-491-02470-3 X.981/20752
1198319

ACTING. Cinema films
Gardner, Ava - Biographies

FLAMINI, Roland
Ava. - London : Hale, 1984, c1983
269p : ill
0-7090-1555-0 X.950/34082
0289027

ACTING. Cinema films
Grable, Betty - Biographies

WARREN, Doug
Betty Grable. - London : Robson, 1982, c1981
237p,[24]p of plates : ill. - List of films: p219-232. - Includes index
0-86051-165-0 X.950/13027
2727129

ACTING. Cinema films
Guinness, Alec

HUNTER, Allan
Alec Guinness on screen. - Edinburgh : Polygon, c1982
103p : ill. - List of films: p95-100. - Includes index
0-904919-58-7 X.529/53392
2885867

ACTING. Cinema films
Harvey, Laurence - Biographies

HICKEY, Des
The prince. - London : Frewin, 1975
272p,[24]p of plates : ill. - Includes index
0-85632-121-4 X.989/50577
0529346

ACTING. Cinema films
Hawkins, Jack - Biographies

HAWKINS, Jack
Anything for a quiet life. - London : Coronet, 1975
223p,[8]p of plates : ill. - Includes index
0-340-19866-4 X.908/40028
0356514

ACTING. Cinema films
Hayward, Susan

MORENO, Eduardo
The films of Susan Hayward. - Secausus : Citadel, c1979
285p : ill
0-8065-0757-8 X.955/1417
2853566

ACTING. Cinema films
Hoffman, Dustin - Biographies

JOHNSTONE, Iain
Dustin Hoffman. - Tunbridge Wells : Spellmount, 1984
94p : ill. - (Film and theatre stars)
0-946771-25-1 X.958/26146
3012395

ACTING. Cinema films
Jürgens, Curd - Biographies

BALL, Gregor
Curd Jürgens. - München : Heyne, 1982
286p. - (Heyne Filmbibliothek ; 45). - Includes filmography
3-453-86045-4 X.429/16196
8063850

ACTING. Cinema films
Kerr, Deborah - Biographies

BRAUN, Eric
Deborah Kerr. - London : W.H. Allen, 1977
264p,[32]p of plates : ill. - Bibliography: p.236. - List of films: p.237-249. - Includes index
0-491-02341-3 X.981/20897
1290126

ACTING. Cinema films
Knef, Hildegard - Biographies

KNEF, Hildegard
Der geschenkte Gaul. - Hamburg : Knaus, c1982
382p.
3-8135-0069-1 X.950/32826
8063850

[Der geschenkte Gaul. English]. The gift horse. - London : Panther, 1978
410p
B79-08135 X.907/25585
062263x

[Das Urteil. English]. The verdict. - London : Weidenfeld and Nicolson, 1976
[5],377p
0-297-77118-3 X.809/40800
062263x

ACTING. Cinema films
Laughton, Charles - Biographies

HIGHAM, Charles, b.1931
Charles Laughton. - London : W.H. Allen, 1976
xxi,239p,[16]p of plates : ill. - Includes index
0-491-01696-4 X.989/50872
0714518

Charles Laughton. - Sevenoaks : Coronet, 1979
271p,[4]p of plates : ill. - Includes index
0-340-24493-3 X.908/43392
0714518

ACTING. Cinema films
Lee, Bruce

BRUCE Lee. - Hollywood : Unique, c1980
76p : ill
r6864763 X.622/12638
0232076

LESTER, Paul
The meaning of Bruce Lee. - [Birmingham] : Running Horse Press Publications, c1975
8p
r6815595 X.529/61380
0232076

ACTING. Cinema films
Lee, Bruce - Biographies

LEE, Linda
The life and tragic death of Bruce Lee. - London : Star Books, 1975
188p,[8]p of plates : ill
0-352-30043-4 X.708/20672
0232076

ACTING. Cinema films
Lee, Christopher, 1922-

POHLE, Robert W.
The films of Christopher Lee. - Metuchen, N.J. ; London : Scarecrow, 1983
xxi,227p,[56]p of plates : ill. - Bibliography: p211-218. - Includes index
0-8108-1573-7 X.955/1724
2938596

ACTING. Cinema films
Lee, Christopher, 1922- - Biographies
LEE, Christopher, 1922-
Tall, dark and gruesome. - London : W.H. Allen, 1977
284p,leaf of plate,[32]p of plates : ill. - Includes index
0-491-01968-8 X.989/51790
1198297
Tall, dark and gruesome. - London : Mayflower, 1978
286p,[8]p of plates : ill. - Includes index
0-583-12853-x X.907/25494
1198297

ACTING. Cinema films
Lester, Mark - Biographies
KIDD, Paul
Mark Lester. - Ilfracombe : Stockwell, 1975
81p,[8]p of plates : ill
0-7223-0697-0 X.989/40073
0355712

ACTING. Cinema films
Loder, John
LODER, John
Hollywood hussar. - London : H. Baker, 1977
178p,[16]p of plates : ill. - Includes index
0-7030-0121-3 X.989/52675
1571257

ACTING. Cinema films
Loren, Sophia - Biographies
LEVY, Alan
Forever, Sophia. - London : Magnum Books, 1979
xii,227p,[16]p of plates : ill
0-417-05500-5 X.908/42836
0303488
Forever, Sophia. - London : Hale, 1980, c1979
xi,227p,[16]p of plates : ill
0-7091-8594-4 X.981/22572
0303488
LOREN, Sophia
Sophia. - London : Joseph, 1979
216p,[64]p of plates
0-7181-1741-7 X.989/53262
1838555
Sophia. - London : Corgi, 1980
x,245p,[32]p of plates : ill
0-552-11301-8 X.708/22382
1838555
SHAW, Sam
Sophia Loren in the camera eye. - London : Hamlyn, 1980
160p : ill
0-600-34155-0 X.421/20761 Woolwich
220519x
ZEC, Donald
Sophia. - Large print ed. - London : Prior [etc.], 1976
xxix,463p. - List of films: p.454-459
0-86043-006-5 X.981/20426
0303488

ACTING. Cinema films
MacLaine, Shirley - Biographies
MACLAINE, Shirley
'Don't fall off the mountain'. - Toronto ; London : Bantam, 1971, c1970
292p
0-553-23662-8 X.958/21935
0513962
Out on a limb. - London : Elm Tree, 1983
372p
0-241-11105-6 X.950/28632
0513962
PICKARD, Roy
Shirley MacLaine. - Tunbridge Wells : Spellmount, 1985
95p : ill. - List of films: p89-93
0-946771-86-3 X.950/43151
0513962

ACTING. Cinema films
Manfredi, Nino
BERNARDINI, Aldo
Nino Manfredi. - Roma : Gremese, 1979
224p : ill. - (Le stelle filanti ; 5)
r6502702 X.955/725
8063850

ACTING. Cinema films
Mason, James, 1909- - Biographies
MASON, James, 1909-1984
Before I forget. - London : Hamilton, 1981
x,345p,16p of plates : ill. - Includes index
0-241-10677-x X.950/8377
256209x
MASON, James, 1909-1984
Before I forget. - London : Sphere, 1982, c1981
466p,[8]p of plates : ill
0-7221-5763-0 X.958/13011
256209x
Before I forget. - Long Preston : Magna, 1983, c1981
692p
0-86009-513-4 Nov.49645
256209x

ACTING. Cinema films
Milland, Ray - Biographies
MILLAND, Ray
Wide-eyed in Babylon. - London : Coronet, 1977
251p,[12]p of plates : ill
0-340-21315-9 X.907/25320
027108x

ACTING. Cinema films
Miranda, Isa
CALDIRON, Orio
Isa Miranda. - Roma : Gremese, 1978
236p : ill. - (Le stelle filanti ; 1)
r6194903 X.902/3802
2298988

ACTING. Cinema films
Mitchum, Robert - Biographies
EELLS, George
Robert Mitchum. - London : Robson, 1984
328p : ill. - List of films: p295-319. - Includes index
0-86051-309-2 X.950/37362
016254x
MALCOLM, Derek
Robert Mitchum. - Tunbridge Wells : Spellmount, [1984]
95p : ill. - (Film and theatre stars)
0-946771-30-8 X.958/24814
016254x

ACTING. Cinema films
Moore, Roger, 1927- - Biographies
DONOVAN, Paul
Roger Moore. - London : W.H. Allen, 1983
224p,[4]p of plates : ill. - Includes index
0-491-03051-7 X.955/2443
3141683
MOSELEY, Roy
Roger Moore. - Sevenoaks : New English Library, 1985
249p,[16]p of plates : ill. - Includes index
0-450-06114-0 X.950/42540
3141683

ACTING. Cinema films
Morgan, Michèle - Biographies
MORGAN, Michèle
With those eyes. - London : W.H. Allen, 1978
[8],305p,[12]p of plates : ill
0-491-02185-2 X.809/43670
1697226

ACTING. Cinema films
Newman, Paul, 1925- - Biographies
GODFREY, Lionel
Paul Newman, superstar. - New York : St Martin's Press ; London : Hale, 1978
208p,[16]p of plates : ill. - Bibliography: p.189-190. - List of films: p.191-197. - Includes index
0-7091-6914-0 X.808/33230
0249823
KERBEL, Michael
Paul Newman. - London : Star Books, 1975
158p : ill. - (The Star illustrated history of the movies). - Bibliography: p.147-148. - List of films: p.149-153. - Includes index
0-352-39803-5 X.908/40241
0577979
LANDRY, J. C.
Paul Newman. - London : Sidgwick & Jackson, 1983
144p : ill. - List of films: p142. - Includes index
0-283-98898-3 X.955/1730
0249823

ACTING. Cinema films
Nicholson, Jack
DOWNING, David
Jack Nicholson. - London : W.H. Allen, 1983
192p,[8]p of plates : ill. - Includes index
0-491-03010-x X.950/35808
3112942

ACTING. Cinema films
Nielsen, Asta - Biographies
ASTA Nielsen / herausgegeben von Renate Seydel und Allan Hagedorff. - Berlin : Henschelverlag, 1981
263p : ill
r6305014 L.42/1369
8063850

ACTING. Cinema films
Niven, David
HUTCHINSON, Tom
Niven's Hollywood. - London : Macmillan, 1984
192p : ill. - Includes index
0-333-37025-2 X.955/2626
1335499
NIVEN, David
Bring on the empty horses. - London : Hamilton, 1975
xv,362p,[32]p of plates : 1 ill. - Includes index
0-241-89273-2 X.989/40169
1335499

ACTING. Cinema films
Niven, David - Biographies
NIVEN, David
Bring on the empty horses. - London : Prior [etc.], 1976
2v.([22],730p)
0-86043-007-3 X.981/20425
1335499
The moon's a balloon. - Large print ed. - Litton : Magna Print Books, 1976
2v.([9],672p)
B78-06269 X.809/42794
1335499
Niven. - London : Hamilton, 1984
541p : ill
0-241-11360-1 X.950/40877
1335499

ACTING. Cinema films
Oberon, Merle - Biographies
HIGHAM, Charles, 1931-
Merle. - Sevenoaks : New English Library, 1983
xii,227p,[30]p of plates : ill. - Filmography: p218-220. - Includes index
0-450-06009-8 X.800/36541
2983753

ACTING. Cinema films
O'Toole, Peter - Biographies
FREEDLAND, Michael
Peter O'Toole. - London : W.H. Allen, 1983
237p,[16]p of plates : ill. - Includes index
0-491-02648-x X.950/30425
3075435
WAPSHOTT, Nicholas
Peter O'Toole. - London : New English Library, 1983
299p,[32]p of plates : ill. - Includes index
0-450-06003-9 X.950/32333
3127168
Peter O'Toole. - London : New English Library, 1984, c1983
239p,[8]p of plates : ill. - Includes index
0-450-05754-2 X.958/27073
3075435

ACTING. Cinema films
Rathbone, Basil
DRUXMAN, Michael B.
Basil Rathbone. - South Brunswick : Barnes ; London : Yoseloff, 1976
359p : ill
0-498-01471-1 X.981/20771
1249150

ACTING. Cinema films
Redgrave, Michael - Biographies
REDGRAVE, Michael
In my mind's eye. - London : Weidenfeld and Nicolson, 1983
viii,256p,[24]p of plates : ill. - Includes index
0-297-78278-9 X.950/28479
3056503
In my mind's eye. - Sevenoaks : Coronet, 1984, c1983
316p,[8]p of plates : ill. - Includes index
0-340-36382-7 X.958/28734
3056503

ACTING. Cinema films
Reed, Oliver - Biographies
REED, Oliver
Reed all about me. - London : W.H. Allen, 1979
x,261p,[16]p of plates : ill. - Includes index
0-491-02039-2 X.989/54120
2017199

ACTING. Cinema films
Schneider, Romy
KNEF, Hildegard
Romy. - Hamburg : Knaus, c1983
204p
3-8135-0443-3 X.958/29128
8063850

ACTING. Cinema films
Sedgwick, Edie – Biographies

STEIN, Jean
Edie. – New York : Knopf, 1982
455p : ill
0–394–48819–9 X.950/37754
8063850

ACTING. Cinema films
Sellers, Peter – Biographies

EVANS, Peter, 1933–
Peter Sellers. – Rev. and updated ed. – London : Severn House, 1981, c1980
256,[16]p of plates : ill. – List of films: p248–250. – Includes index
0–7278–0688–2 X.950/3329
2382628

SELLERS, Michael, 1954–
P.S. I love you. – London : Collins, 1981
238p,[24]p of plates : ill
0–00–216649–6 X.950/9086
2382628

P.S. I love you. – [London] : Fontana/Collins, 1982, c1981
238p,[8]p of plates : ill
0–00–636515–9 X.958/11832
2382628

SYLVESTER, Derek
Peter Sellers. – London : Proteus, 1981
128p ill(some col.). – List of films: p119–128
0–906071–51–8 X.955/688
2382628
– (Another copy: paperback)
ISBN. 0906 071 518 X.955/689

WALKER, Alexander
Peter Sellers. – London : Weidenfeld and Nicolson, 1981
xvi,240p,[24]p of plates : ill. – Includes index
0–297–77965–6 X.950/19495
2382628

ACTING. Cinema films
Sharif, Omar – Biographies

SHARIF, Omar
The eternal male. – London : W.H. Allen, 1977
155p,[16]p of plates : ill
0–491–02431–2 X.989/52036
1290800

ACTING. Cinema films
Signoret, Simone, to 1975 – Biographies

SIGNORET, Simone
Nostalgia isn't what it used to be. – London : Weidenfeld and Nicolson, 1978
ix,403p. – Includes index
0–297–77417–4 X.981/21156
1519964

Nostalgia isn't what it used to be. – London : Panther, 1979
446p. – Includes index
0–586–04836–7 X.619/22260 Woolwich
1519964

ACTING. Cinema films
Sordi, Alberto – Biographies

GOVERNI, Giancarlo
Alberto Sordi. – [Milano] : Milano libri, 1979
219p. – (Prima pagina. Spettacolo)
r6500146 X.950/2990
8063850

ACTING. Cinema films
Stars – Biographies

NORMAN, Barry
The movie greats. – London : Hodder and Stoughton, 1981
319p : ill. – Includes index
0–340–25972–8 X.950/5949
2531070

ACTING. Cinema films
Stars, to 1978 – Encyclopaedias

WLASCHIN, Ken
The illustrated encyclopedia of the world's great movie stars and their films. – London : Salamander Books : [Distributed by New English Library], 1979
[3],235p : col ill. – Includes index
0–86101–030–2 X.985/1231
1916750

ACTING. Cinema films
Stars, to 1982

DAVIS, Daphne
Stars!. – Bromley : Columbus, c1983
277p : chiefly ill(some col.). – Bibliography: p272. – List of films: p270–271. – Includes index
0–86287–048–8 X.955/2291
3084876

ACTING. Cinema films
Stars, ca 1910–1977

SUPERSTARS / selected and introduced by Alexander Walker. – Oxford : Phaidon [etc.], 1978
[8],64p
0–7148–1851–8 Cup.1285/335
1588273

ACTING. Cinema films
Stars, 1967–1980 – Biographies

CASTELL, David
Screen stars of the 70's. – Godalming : LSP, c1981
141p : ill(some col.). – List of films: p128–136
0–85321–082–9 X.955/601
2504081

ACTING. Cinema films
Stars, 1970–1974

CASTELL, David
Superstars of the '70s. – London : Barnden Castell Williams Ltd, 1974
96p : ill(some col). – Bibliography: p.94–95
0–904159–16–7 X.900/25195
067771x

ACTING. Cinema films
Streep, Meryl – Biographies

MAYCHICK, Diana
Meryl Streep. – London : Robson, 1984
x,166p,[30]p of plates : ill. – Includes index
0–86051–308–4 X.950/39386
3114872

ACTING. Cinema films
Taylor, Elizabeth, 1932– – Biographies

KELLEY, Kitty
Elizabeth Taylor. – London : Joseph, c1981
ix,346p : ill. – Bibliography: p309–310. – List of films: p311–336. – Includes index
0–7181–2075–2 X.950/9060
0271098

Elizabeth Taylor. – [Sevenoaks] : Coronet, 1982, c1981
448p,[16]p of plates. – Includes index
0–340–28345–9 X.958/23297
0271098

MADDOX, Brenda
Who's afraid of Elizabeth Taylor?. – London : Hart-Davis MacGibbon, 1977
256p,[8]p of plates : 2 ill. – Includes index
0–246–10916–5 X.989/52403
0271098

Who's afraid of Elizabeth Taylor?. – London : Mayflower, 1979
256p,[8]p of plates : 2 ill. – List of films: p.248–249. – Includes index
0–583–12585–9 X.619/19446 Woolwich
0271098

NICKENS, Christopher
Elizabeth Taylor. – London : Hutchinson, 1984
186p,[8]p of plates : ill. – (A James Spada associates book)
0–09–159101–5 X.955/2856
0271098

WATERBURY, Ruth
Elizabeth Taylor. – Toronto ; London : Bantam, 1982
293p,[14]p of plates : ill
0–553–22613–4 X.808/37143
0271098

ZEC, Donald
Liz. – London : Mirror Books, 1982
48p : ill(some col.)
0–85939–316–x X.525/6112
0271098

ACTING. Cinema films
Todd, Ann – Biographies

TODD, Ann
The eighth veil. – London : Kimber, 1980
173p,[24]p of plates : ill. – Includes index
0–7183–0387–3 X.981/22647
2322668

ACTING. Cinema films
Tognazzi, Ugo

BERNARDINI, Aldo
Ugo Tognazzi. – Roma : Gremese, 1978
251p : ill. – (Le stelle filanti ; 2)
r6194904 X.902/3803
229897X

ACTING. Cinema films
Topol, Chaim – Biographies

TOPOL, Chaim
Topol. – London : Weidenfeld and Nicolson, c1981
222p,[12]p of plates : ill. – Includes index
0–297–77919–2 X.950/8376
2562103

ACTING. Cinema films
Trintignant, Jean Louis – Biographies

TRINTIGNANT, Jean Louis
Un homme à sa fenêtre. – Paris : Jean–Claude Simoën, 1977
199p,32p of plates : ill. – (L'illusion d'optique)
r6059917 X.900/20331
1536079

ACTING. Cinema films
Wilding, Michael, 1912–1979 – Biographies

WILDING, Michael, 1912–1979
Apple sauce. – London : Allen & Unwin, 1982
xiii,190p,[32]p of plates : ill. – List of films: p182–187. – Includes index
0–04–920064–x X.809/52879
2770733

ACTING. Cinema films. Hollywood. Los Angeles. California
British stars, to 1983

MORLEY, Sheridan
Tales from the Hollywood raj. – London : Weidenfeld and Nicolson, 1983
237p,[16]p of plates : ill. – Includes index
0–297–78289–4 X.950/35533
3167887

ACTING. Cinema films. Hollywood. Los Angeles. California
Stars, 1915–1979 – Biographies

CLOSE-UPS / by their co-stars, directors, screenwriters and friends ; edited by Danny Peary. – Leicester : Windward, c1978
xvi,606p : ill. – Includes index
0–7112–0051–3 X.955/215
2342812

ACTING. Cinema films. Hollywood. Los Angeles. California
Stars, 1930–1950

BROWN, Peter Harry
The MGM girls. – London : Harrap, 1984, c1983
xvi,288p : ill. – Bibliography: p279–282. – Includes index
0–245–54220–5 X.950/36186
1877100

ACTING. Cinema films. Hollywood. Los Angeles. California
Stars, 1930–1950 – Biographies

NORMAN, Barry
The Hollywood greats. – London : Hodder and Stoughton : British Broadcasting Corporation, 1979
272p. – Includes index
0–340–23332–x X.981/21823
1877100

The Hollywood greats. – London : Arrow, 1980, c1979
271p,[47]p of plates. – Includes index
0–09–924030–0 X.907/26036
1877100

ACTING. Cinema films. Hollywood. Los Angeles. California
Stars, 1930–1964

MCCLURE, Arthur Frederick
Star quality. – South Brunswick ; New York : Barnes ; London : Yoseloff, 1974
285p : chiefly ill
0–498–01374–x X.985/1009
0377600

ACTING. Cinema films. Hollywood. Los Angeles. California
Stars, ca 1930–1970

HARRIS, Radie
Radie's world. – London : W.H. Allen, 1975
x,225p,[32]p of plates. – Includes index
0–491–01514–3 X.809/40119
0361089

ACTING. Cinema films. Hollywood. Los Angeles. California
Stars. Social conditions, ca 1930–ca 1940

FINCH, Christopher
Gone Hollywood. – London : Weidenfeld and Nicolson, 1980, c1979
xii,396p : ill. – Bibliography: p369–376. – Includes index
0–297–77860–9 X.950/1855
2362686

ACTING. Cinema films. Hollywood. Los Angeles. California
Supporting actors, 1937–ca 1970

BARRIS, Alex
Hollywood's other men. – South Brunswick ; New York : Barnes ; London : Yoseloff, 1975
223p : ill. – Includes index
0–498–01428–2 X.985/1022
0535788

ACTING. Cinema films. Hollywood. Los Angeles. California
Supporting actresses, 1930–1973 – Biographies

BARRIS, Alex
Hollywood's other women. – South Brunswick : New York : Barnes ; London : Yoseloff, 1975
212p : ill. – Includes index
0–498–01488–6 X.985/1166
1264257

ACTING. Cinema films. Special subjects. American portrait photography, 1900–1950
Stars, 1926–1949 – Collections

HOLLYWOOD glamor portraits / edited by John Kobal. – New York : Dover Publications [etc.] ; London : Constable, 1976
xiv,144p : of ill. – Includes index
0–486–23352–9 X.415/5260 Woolwich
1061712

ACTING. Comedies. American silent cinema films
Lloyd, Harold

MCCAFFREY, Donald W.
Three classic silent screen comedies starring Harold Lloyd. – Rutherford : Fairleigh Dickinson University Press ; London : Associated University Presses, 1976
264p : ill. – Bibliography: p.258–260. – List of films: p.240–247. – Includes index
0–8386–1455–8 X.981/21346
1764292

ACTING. Coronation Street. Television drama series in English
Phoenix, Pat – Biographies

PHOENIX, Pat
All my burning bridges. – London : Star Books, 1976
188p,[8]p of plates : ill
0–352–39839–6 X.908/40884
0260150

All my burning bridges. – London : Panther, 1984, c1974
245p,[8]p of plates : ill
0–586–06243–2 X.958/25788
0260150

Love, curiosity, freckles and doubt. – London : Arlington, 1983
190p,[16]p of plates
0–85140–623–8 X.950/28114
0260150

Love, curiosity, freckles and doubt. – London : Panther, 1984, c1983
190p,[16]p of plates : ill
0–586–06201–7 X.958/25787
0260150

ACTING. Czech cinema films
Baarová, Lída – Biographies

BAAROVÁ, Lída
Útěky. – Toronto : Sixty-Eight Publishers, 1983
319p
0–88781–133–7 X.958/16709
888014x

ACTING. Czech cinema films
Mandlová, Adina – Biographies

MANDLOVÁ, Adina
Dneska už se tomu směju. – Toronto : Sixty-Eight Publishers, 1977
333p : ill
r6440671 X.629/19138
888014x

ACTING. Czechoslovakia
Pešek, Ladislav – Biographies

PEŠEK, Ladislav
Tvář bez masky. – vyd. druhé. – Praha : Odeon, 1979
309p,[72]p of plates : ill,ports
r6177524 X.989/84410
2414333

ACTING. Denmark
Legal aspects

ROTHE, Bendt
Skuespillerkunst og skuespillerret. – [København] : Juristforbundets Forlag, 1980
528p. – Bibl
87–574–3120–5 X.622/8398 Woolwich
2460513

ACTING. Estonia
Ots, Georg – Biographies

TÕNSON, Helga
Georg Ots. – Teine, täiendatud trükk. – Tallinn : Eesti Raamat, 1976
319p, 72p of plates
r6046783 X.989/70680
1317652

ACTING. Estonian cinema films
Järvet, Jüri – Biographies

ORAV, Õie
Jüri Järvet. – Tallinn : Eesti Raamat, 1977
126p : ill
r6021983 X.909/39161
1421573

ACTING. France
to date

MAZEAU, Jacques
Les grands acteurs français contemporains. – Paris : Presses Universitaires de France, 1982
127p. – (Que sais-je? : 887)
2–13–037493–x X.429/15959
9090878

ACTING. France
Fresnay, Pierre – Biographies

FRESNAY, Pierre
Pierre Fresnay. – Paris : La Table Ronde, 1975
188 p., 4 p. of plates : ill
r6012158 X.900/17229
0859869

ACTING. France
Grey, Denise

GREY, Denise
D'une loge à l'autre. – Paris : Flammarion, 1980
219p.,16p.of plates : ill,ports
2–08–064264–2 X.989/88872
2437260

ACTING. France
Marais, Jean – Biographies

MARAIS, Jean
Histoires de ma vie. – Paris : Albin Michel, 1975
315 p., 16 pages of plates : ill
2–226–00153–0 X.900/17145
0943878

ACTING. France
Printemps, Yvonne – Biographies

YVONNE Printemps, 1894–1977. – [Richmond, Surrey] : [Patrick O'Connor], [1978]
[22]p,[8]p of plates : ill. – List of sound discs: p.[18]–[21]
B80–03271 X.902/5554
2019752

ACTING. French cinema films
Bardot, Brigitte

ROBERTS, Glenys
Bardot. – London : Sidgwick & Jackson, 1984
290p,[24]p of plates : ill. – Bibliography: p284–285. – Includes index
0–283–99103–8 X.950/40137
1590731

ACTING. French cinema films
Bardot, Brigitte – Biographies

CRAWLEY, Tony
Bébé. – London : LSP Books, 1975
256p
0–85321–065–9 X.985/1032
0611689

ECKARDT, Bernard d'
Brigitte Bardot. – München : Heyne, 1982
263p. – (Heyne Filmbibliothek)
3–453–86050–0 X.958/23906
8793638

FRISCHAUER, Willi
Bardot. – London : Joseph, 1978
222p,[32]p of plates : ill. – List of films: p.207–217. – Includes index
0–7181–1744–1 X.809/43562
1590731

Bardot. – London : Sphere, 1979
[5],214p,[8]p of plates. – Includes index
0–7221–3686–2 Cup 719/49
1590731

ACTING. French cinema films
Bardot, Brigitte – Biographies

HAINING, Peter
The legend of Brigitte Bardot. – London : W.H. Allen, c1983
224p,[8]p of plates : ill
0–491–03360–5 X.955/2477
1590731

LEVIN, Sam
Brigitte Bardot. – London : Sidgwick and Jackson, 1983
[125]p : chiefly ill(some col.). – List of films: p73. – List of sound recordings: p120–122
0–283–99080–5 L.45/2762
3136280

ACTING. French cinema films
Bourvil, André – Biographies

BERRUER, Pierre
Bourvil. – Paris : Presses de la Cité, 1975
248 p., 16 p. of plates
r6001114 X.908/33447
0925721

ACTING. French cinema films
Depardieu, Gérard

CHAZAL, Robert
Gérard Depardieu. – Renens : 5 Continents, [1983]
160p. – (Bibliothèque du cinéma)
2–88003–033–1 X.429/15451
8793638

ACTING. French cinema films
Fernandel

JELOT-BLANC, Jean-Jacques
Fernandel. – Nice : Éditions Alain Lefeuvre, 1981
255p : ill
2–902639–62–7 X.955/891
8793638

ACTING. French cinema films
Feuillère, Edwige – Biographies

FEYDEAU, Alain
Edwige Feuillère. – Paris : PAC, 1983
146p,plates. – (Grand écran). – Includes bibliography and discography
r6298332 X.955/3006
8793638

ACTING. French cinema films
Gabin, Jean – Biographies

COLIN, Gerty
Jean Gabin. – Paris : Presses de la Cité, [1983]
207p,plates. – Includes bibliography
2–258–01205–8 X.950/40463
8793638

ACTING. French cinema films
Hanin, Roger – Biographies

HANIN, Roger
L'ours en lambeaux. – [Paris] : Encre, 1983
219p
2–86418–132–0 X.950/40896
8793638

ACTING. French cinema films
Le Vigan, Robert

VANDROMME, Pol
Robert Le Vigan. – Kessel-Lo : La Revue Célinienne, 1980
204p
2–87067–000–1 X.958/19035
8793638

ACTING. French cinema films
Montand, Yves – Biographies

SEMPRUN, Jorge
Montand, la vie continue. – Paris : Denoël, [1983]
310p
2–207–22876–2 X.950/40894
8793638

ACTING. French cinema films
Morgan, Michèle – Biographies

BOUNIQ-MERCIER, Claude
Michèle Morgan. – Paris : Colona, 1983
207p : ill
2–904232–03–6 X.955/2995
8793638

ACTING. German cinema films
Werner, Ilse – Biographies

WERNER, Ilse
So wird's nie wieder sein –. – 5. Aufl. – Bayreuth : Hestia, 1982
322p,[36]p of plates : ill. – Filmography
3–7770–0217–8 X.950/16370
8858799

ACTING. Germany
Gründgens, Gustaf

SPANGENBERG, Eberhard
Karriere eines Romans. – München : Ellermann, c1982
238p : ill. – (Edition Spangenberg)
3–7707–0186–0 X.950/24401
8798591

ACTING. Germany
Hellberg, Martin – Biographies

HELLBERG, Martin
Die bunte Lüge. – Berlin : Henschelverlag, 1974
305 p., [16] p. of plates : ill
r6019490 X.989/37198
8798591

ACTING. Germany
Hellberg, Martin - Biographies–continued
Die bunte Lüge. - 2., durchges. Aufl. - Berlin : Henschelverlag, 1977
289p.[16]p of plates : ill
r6063707 X.989/76246
8798591

Mit scharfer Optik. - Berlin : Henschelverlag, 1982
411p.,plates
r6309681 X.808/39624
8798591

ACTING. Germany
Ruhmann, Heinz - Biographies

BALL, Gregor
Heinz Rühmann. - München : Heyne, 1981
239p. : ill. - (Heyne Filmbibliothek ; 24)
3-453-86024-1 X.429/12936 Woolwich
8798591

RUHMANN, Heinz
Das war's. - 3. Aufl. - Frankfurt/M. : Ullstein, 1982
311p,plates
3-550-06472-1 X.950/22780
8798591

ACTING. Great Britain
1558-1979

FORBES, Bryan
That despicable race. - London : Elm Tree, 1980
326p,[8]p of plates : ill(some col.). - Bibliography: p311-316. - Includes index
0-241-10164-6 X.981/22580
2005735

ACTING. Great Britain
Adams, Shirley. Interpersonal relationships with Adams, Peter

ADAMS, Peter, 1940-
Knockback. - London : Duckworth, 1982
176p
0-7156-1677-3 X.529/52330
2849763

ACTING. Great Britain
Bennett, Jill, 1931-. Interpersonal relationships with Tearle, Sir Godfrey

BENNETT, Jill, 1931-
Godfrey. - London : Hodder and Stoughton, 1983
186p : ill. - Includes index
0-340-33160-7 X.809/56237
2956632

ACTING. Great Britain
Bennett, Jill, 1931-. Interpersonal relationships with Tearle, Sir Godfrey - Biographies

BENNETT, Jill, 1931-
Godfrey. - Large print ed. - Bath : Chivers, 1984, c1983
180p. - (A New Portway large print book)
0-85119-284-x Nov.52444
2956632

Godfrey. - Sevenoaks : Coronet, 1985, c1983
172p,[8]p of plates : ill. - Includes index
0-340-36652-4 X.958/29776
2956632

ACTING. Great Britain
Brambell, Wilfrid - Biographies

BRAMBELL, Wilfrid
All above board. - London : W.H. Allen, 1976
189p.[24]p of plates : ill
0-491-01695-6 X.989/50561
0607495

ACTING. Great Britain
Cooklin, Shirley. Interpersonal relationships with Adams, Peter, 1940-

ADAMS, Peter, 1940-
Knockback. - London : Ariel, 1984, c1982
183p
0-563-20260-2 X.958/28897
2849763

ACTING. Great Britain
Drake, Fabia - Biographies

DRAKE, Fabia
Blind fortune. - London : Kimber, 1978
207p.[24]p of plates : ill. - Includes index
0-718?-0455-1 X.981/21097
1483722

ACTING. Great Britain
Ellis, Mary - Biographies

ELLIS, Mary
Those dancing years. - London : Murray, 1982
ix,182p[28]p of plates : ill. - Includes index
0-7195-3984-6 X.950/19022
286605x

ACTING. Great Britain
England, John - Biographies

ENGLAND, John, 1902-
Across my path. - Weybridge : Pennant Publishing Co., 1975
[7],239p,[20]p of plates : ill. - List of films. - Includes index
0-904141-05-5 X.809/40838
0632600

ACTING. Great Britain
Evans, Dame Edith

FORBES, Bryan
Ned's girl. - London : Elm Tree Books, 1977
xvii,297p,[24]p of plates. - Includes index
0-241-89600-2 X.981/20754
1201980

ACTING. Great Britain
Evans, Dame Edith, 1946-1971 - Biographies

BATTERS, Jean
Edith Evans. - London : Hart-Davis MacGibbon, 1977
159p,[8]p of plates : ill. - Includes index
0-246-10994-7 X.989/51805
1201999

ACTING. Great Britain
Guinness, Alec - Biographies

TAYLOR, John Russell
Alec Guinness. - London : Pavilion, 1984
184p : ill. - List of films: p175-178. - Bibliography: p179. - Includes index
0-907516-41-6 X.955/2544
3173208

ACTING. Great Britain
Hampshire, Susan - Biographies

HAMPSHIRE, Susan
Susan's story. - London : Sidgwick & Jackson, 1981
167p,[24]p of plates : ill. - Bibliography: p166-167
0-283-98735-9 X.809/49912
2588129

Susan's story. - Large print ed. - Bath : Chivers, 1982
192p. - (A Lythway book)
0-85119-876-7 X.950/19431
2588129

Susan's story. - London : Sphere, 1983, c1981
181p,[8]p of plates : ill
0-7221-4219-6 X.958/19161
2588129

ACTING. Great Britain
Hird, Thora - Biographies

HIRD, Thora
Scene and Hird. - London : W.H. Allen, 1976
ix,238p,leaf of plate,[24]p of plates : ill
0-491-01965-3 X.989/50564
0609757

ACTING. Great Britain
Konstam, Phyllis. Interpersonal relationships with Austin, Henry Wilfred - Biographies

AUSTIN, Henry Wilfred
To Phyll with love. - London : Grosvenor Books, 1979
[2],105p
0-901269-42-5 X.208/5503
2071142

ACTING. Great Britain
Le Mesurier, John - Biographies

LE MESURIER, John
A jobbing actor. - London : Elm Tree, 1984
159p,[8]p of plates
0-241-11063-7 X.950/32801
305201x

A jobbing actor. - Large print ed. - Bath : Chivers, 1984
246p. - (A Lythway book)
0-7451-0051-1 Nov.52850
305201x

ACTING. Great Britain
Leigh, Vivien - Biographies

EDWARDS, Anne, 1909-, 1927-
Vivien Leigh. - London : Coronet, 1978
334p,[16]p of plates : ill. - List of films: p.319-328. - Includes index
0-340-23024-x X.708/21507
1186965

EDWARDS, Anne, 1927-
Vivien Leigh. - London : W.H. Allen, 1977
318p,[32]p of plates : ill. - Includes index
0-491-02140-2 X.981/20718
1186965

Vivien Leigh. - Large print [ed.]. - Bolton-by-Bowland : Magna Print Books, 1979
579p
0-86009-127-9 X.629/12646 Woolwich
1186965

TAYLOR, John Russell
Vivien Leigh. - London : Elm Tree, 1984
128p,[8]p of plates : ill(some col.)
0-241-11333-4 X.955/2753
1186965

UTILOV, V. (Vladimir)
Vivien Li. - Moskva : Iskusstvo, 1980
208 p, [25] p. of plates : ill.. - (Serii͡a "Zhizn' v iskusstve"). - Includes bibliography: p. 208
r6465657 X.950/36924
8183252

ACTING. Great Britain
Leigh, Vivien. Interpersonal relationships with Olivier, Laurence Olivier, Baron

LASKY, Jesse Louis, b.1910
Love scene. - London : Sphere, 1980
iii-xiii,270p,[16]p of plates. - Bibliography: p.263-266. - Includes index
0-7221-5448-8 X.907/25966
1944339

O'CONNOR, Garry
Darlings of the gods. - London : Hodder and Stoughton, 1984
192p,[8]p of plates : ill. - Bibliography: p181-183. - Includes index
0-340-34346-x X.950/32904
1944339

ACTING. Great Britain
Leigh, Vivien. Interpersonal relationships with Olivier, Laurence Olivier, Baron - Biographies

LASKY, Jesse Louis
Love scene. - Brighton : Angus and Robertson, 1978
256p : ill. - Bibliography: p.252-253. - Includes index
0-207-95824-6 X.981/21731
1944339

O'CONNOR, Garry
Darlings of the gods. - Large print ed. - Bath : Chivers, 1985, c1984
vii,253p. - (A Lythway book). - Bibliography: p245-248
0-7451-0149-6 X.950/42931
1944339

ACTING. Great Britain
McKern, Leo - Biographies

MCKERN, Leo
Just resting. - London : Methuen, 1983
222p,[16]p of plates : 1ill
0-413-52120-6 X.950/26939
3036332

ACTING. Great Britain
Olivier, Laurence Olivier, Baron - Biographies

HIRSCH, Foster
Laurence Olivier. - Boston [Mass.] : Twayne Publishers, 1979
190p : ill. - (Twayne's Theatrical Arts series). - Bibliography. - Filmography
0-8057-9260-0 X.950/14636
8183252

ACTING. Great Britain
Pertwee, Jon - Biographies

PERTWEE, John
Moon boots and dinner suits. - London : Elm Tree, 1984
x,225p[16]p of plates : ill
0-241-11337-7 X.950/38821
311595x

ACTING. Great Britain
Reid, Beryl - Biographies

REID, Beryl
So much love. - London : Hutchinson, 1984
xii. 294p,[16]p of plates : ill
0-09-155730-5 X.950/38427
3117014

ACTING. Great Britain
Robson, Frederick - Biographies

SANDS, Mollie
Robson of the Olympic. - London : Society for Theatre Research, 1979
[2],155p 12p of plates : ill. - Bibliography: p.147-149. - Includes index
0-85430-029-5 X.989/88231
2114186

ACTING. Great Britain
Rutherford, Dame Margaret, 1892-1972 - Biographies

SIMMONS, Dawn Langley
Margaret Rutherford. - London : Barker, c1983
xxii,196p,[8]p of plates : ill
0-213-16890-1 X.950/34492
3148785

ACTING. Great Britain
Sher, Antony

SHER, Antony
Year of the King. - London : Chatto & Windus, 1985
249p : ill
0-7011-2926-3 X.950/43056
3237486

ACTING. Great Britain
Sinden, Donald - Biographies

SINDEN, Donald
A touch of the memoirs. - London : Hodder and Stoughton, 1982
256p,[16]p of plates : ill. - Includes index
0-340-26235-4 X.800/33081
2725029
A touch of the memoirs. - London : Futura, 1983, c1982
312p,[8]p of plates : ill. - (A Futura book). - Includes index
0-7088-2285-1 X.958/16012
2725029

ACTING. Great Britain
Williams, Charles, 1915- - Biographies

WILLIAMS, Charles, 1915-
Wel dyma fo - Charles Williams. - Pengroes, Caernarfon : Cyhoeddiadau Mei, 1983
143p : ill
0-905775-05-8 X.958/25685
3190269

ACTING. Great Britain
Witcombe, Peggy - Biographies

WITCOMBE, Peggy
Both sides of the curtain. - Pontypool : Hughes and Son, 1975
142p : ill
0-904069-04-4 X.989/40461
058228x

ACTING. Great Britain
Withers, Googie - Biographies

MCCALLUM, John, b.1914
Life with Googie. - London : Heinemann, 1979
[7],275p,[20]p of plates : ill. - Includes index
0-434-40020-3 X.985/668
1889486

ACTING. Hungarian cinema films
Törőcsik, Mari

MOLNÁR GÁL, Péter
Törőcsik Mari. - Budapest : Magyar Filmtudományi Intézet és Filmarchívum, [1980]
102p : ill. - (Filmbarátok kiskönyvtára)
963-562-731-9 X.958/18244
9005951

ACTING. Hungary
Latinovits, Zoltán

HEGYI, Béla
Latinovits. - Budapest : Gondolat, 1983
373p,plates : ill
963-281-271-9 X.950/28017
9007644

ACTING. Indian cinema films
Stars - Serials

FILMWORLD. - vol.1- ; [Jan. 1964?]-. - Bombay : T.M. Ramachandran, [1964?]-
Holdings: vol.1, vol.15, no.1-
r6098907 P.901/1753
0269697

ACTING. Italian cinema films
Sordi, Alberto

FAVA, Claudio G.
Alberto Sordi. - Roma : Gremese, 1979
260p : ill. - (Le stelle filanti ; 6)
r6502247 X.421/22384 Woolwich
8772576

ACTING. Italian cinema films
Totò

GOVERNI, Giancarlo
Vita di Totò. - Milano : Rusconi, 1980
257p,plates
r6507751 X.950/22073
8772576

ACTING. Italian cinema films
Totò - Biographies

FALDINI, Franca
Totò : l'uomo e la maschera. - 2a ed. - Milano : Feltrinelli, 1977
269p.,[32]p.of plates : ill. - (Universale economica ; 788)
r6100276 X.909/83625
203994X

ACTING. Latvia
Lācis, Teodors - Biographies

GUDRIĶE, Biruta
Teodors Lācis. - Riga : Liesma, 1977
196p,[24]p of plates : ill
r6177613 X.989/85136
226210X

ACTING. Latvia
Ṫsilinskiĭ, Gunar Al'fredovich - Biographies

KOKOREVICH, Boris Augustovich
Gunar Ṫsilinskiĭ. - Moskva : Iskusstvo, 1978
110p,[30]p of plates : ill. - (Mastera sovetskogo teatra i kino)
r6089162 X.708/45522
197971X

ACTING. Lithuania
Babkauskas, Bronius

AB, Evgeniĭ
Broniūs Babkauskas. - Leningrad : Iskusstvo, Leningradskoe otdelenie, 1979
196p,[32]p of plates. - (Mastera sovetskogo teatra i kino)
r6485556 X.958/14536
8890862

ACTING. Lithuania
Vaičiūniene, Teofilija - Biographies

VAIČIŪNIENĖ, Teofilija
Scena ir gyvenimas. - Vilnius : Vaga, 1975
265p,[1] leaf of plates, [32] p of plates : ill
r6016651 X.989/70271
1243551

ACTING. Macedonia. Yugoslavia
Džuvalekovski, Ilija

DŽUVALEKOVSKI / [tekst : Gane Todorovski, Branko Varošlija]. - Skopje : Dramski teatar, [1975]
[141]p : chiefly ill
r6057078 X.900/19534
1466879

ACTING. Majorca. Spain
Valls, Cristina - Biographies

JANÉ MANILA, Gabriel
Implicació social i humana del teatre. - Barcelona : Dopesa, 1975
244 p.. - (Pinya de rosa ; no. 23)
84-7235-245-5 X.908/36512
1019333

ACTING. Netherlands
Mauhs, Else - Biographies

HARTEN, Jaap
Else Mauhs. - Amsterdam : Querido, 1984
152p
90-214-6541-8 X.950/36585
9159215

ACTING. Poland
Ćwiklińska, Mieczyslawa - Biographies

OKOŃSKA, Alicja
Rozmowyz panią Miecią. - [Warsaw] : Ludowa Spółdzielnia Wydawnicza, 1976
443p[48]p of plates : ill
r6044226 X.989/70186
1312677

ACTING. Poland
Gogolewski, Ignacy - Biographies

FILLER, Witold
Ignacy Gogolewski. - Warszawa : Wydawnictwa Artystyczne i Filmowe, 1979
73p,[24]p of plates. - List of roles : p.62-74
83-221-0035-3 X.958/2317
8783217

ACTING. Poland
Jaracz, Stefan

KRASIŃSKI, Edward
Stefan Jaracz. - Warszawa : Państwowe Wydawnictwo Naukowe, 1976
205p,1 leaf of plates : ill. - (Sylwetki warszawskie)
r6036579 X.907/15820
1215817

ACTING. Poland
Komorowska, Maja - Biographies

ŻÓRAWSKI, Kazimierz
Maja Komorowska. - Warszawa : Wydawnictwo Artystyczne i Filmowe, 1977
70p,32p of plates : ill
r6037189 X.908/38921
122672X

ACTING. Poland
Krafftówna, Barbara

KUCHTÓWNA, Lidia
Barbara Krafftówna. - Warszawa : Wydawnictwa Artystyczne i Filmowe, 1975
46 p., 24 p. of plates
r6015220 X.908/36350
1015192

ACTING. Poland
Kucówna, Zofia - Biographies

HAUSBRANDT, Andrzej
Zofia Kucówna. - Warszawa : Wydawnictwa Artystyczne i Filmowe, 1977
114p,24p of plates
r6057208 X.908/81552
1388096

ACTING. Poland
Lomnicki, Tadeusz - Biographies

FILLER, Witold
Tadeusz Lomnicki. - Warszawa : Wydawnictwa Artystyczne i Filmowe, 1976
58p,24p of plates : ill
r6036583 X.908/38190
1215833

ACTING. Poland
Pszoniak, Wojciech - Biographies

KARPIŃSKI, Maciej
Wojciech Pszoniak. - Warszawa : Wydawnictwo Artystyczne i Filmowe, 1976
66p
r6044218 X.908/39561
1312707

ACTING. Poland
Zapasiewicz, Zbigniew - Biographies

SZCZYGIELSKA, Zofia
Zbigniew Zapasiewicz. - Warszawa : Wydawnictwa Artystyczne i Filmowe, 1978
73p,[24]p of plates : ill
r6076855 X.908/84759
1715860

ACTING. Polish cinema films
Cybulski, Zbigniew - Biographies

EBERHARDT, Konrad
Zbigniew Cybulski. - Warszawa : Wydawnictwa Artystyczne i Filmowe, 1976
89p,[32]p of plates. - (Aktorzy filmowi świata)
r6015908 X.908/38550
1150367

ACTING. Radio & television programmes. Great Britain
Daniels, Bebe & Lyon, Ben - Biographies

ALLGOOD, Jill
Bebe and Ben. - London : Hale, 1975
192p,[12]p of plates : ill
0-7091-4942-5 X.989/40098
0363723

ACTING. Russia (RSFSR)
Batalov, A. - Biographies

BATALOV, A. (Alekseĭ)
Sud'ba i remeslo. - Moskva : Iskusstvo, 1984
253p : ill. - (Tvorcheskaīa laboratoriīa kinematografista)
r6465803 X.950/37628
8851972

ACTING. Russia (RSFSR)
Bosilashvili, Oleg

ĪĀSNEṪS, Ė (Ėmil')
Oleg Basilashvili. - Leningrad : Iskusstvo, Leningradskoe otdelenie, 1981
213p,[31]p of plates : ill. - (Mastera sovetskogo teatra i kino)
r6486722 X.429/13498
8851972

ACTING. Russia (RSFSR)
Livanov, Boris - Biographies

BORIS Livanov / [sostavitel' E.K. Livanova]. - Moskva : Vserossiĭskoe teatral'noe obshchestvo, 1983
214p : ill
r6403978 X.950/34961
8851972

ACTING. Russia (RSFSR)
Strzhel'chik, Vladislav - Biographies

ZABOZLAEVA, T. B. (Tat'īana Borisovna)
Vladislav Strzhel'chik. - Leningrad : Iskusstvo, Leningradskoe otdelenie, 1979
188,plates. - (Mastera sovetskogo teatra i kino)
r6465112 X.808/40817
8851972

ACTING Russia (RSFSR)
T͡Sarev, Mikhail

MIKHAIL T͡Sarev / [otvetstvennyi redaktor P.A. Markov] - Moskva : Vserossiiskoe teatral'noe obshchestvo, 1983
638p : illl
r64039779 X.950/34962
8851972

ACTING Russian cinema films
Bernes, Mark - Biographies

MARK Bernes / [sostavitel' L.M. Bernes-Bodrova] ; [vstupitel'nai͡a stat'i͡a i obshchai͡a redakt͡sii͡a N.I. Smirnovoi] - Moskva : Iskusstvo, 1980
215p,plates
r6462288 X.958/15454
8725748

ACTING. Russian cinema films
Chirkov, Boris - Biographies

CHIRKOV, Boris
Azorskie ostrova. - Moskva : Sovetskai͡a Rossii͡a, 1978
280p,[32]p of plates : ill
r6498446 X.950/6640
8725748

ACTING. Russian cinema films
Grovskaia, Ol'ga Vladimirovna - Biographies

OL'GA Vladimirovna Gzovskai͡a / [sostavitel' V.G. Gaidarov]. - Moskva : Vserossiiskoe teatral'noe obshchestvo, 1976
429p,[44]p of plates : ill
r6088840 X.989/76791
1849832

ACTING. Russian cinema films
Lapikov, Ivan

SMELKOV, I͡U. (I͡Uliĭ)
Ivan Lapikov. - Moskva : Iskusstvo, 1976
151p,plates. - (Mastera sovetskogo teatra i kino)
r6462797 X.958/21243
8725748

ACTING. Russian cinema films
Munzuk, Maksim

VASIL'EV, Vladimir
Maksim Munzuk - Dersu Uzala. - Kyzyl : Tuvinskoe knizhnoe izdatel'stvo, 1979
169p : ill
r6487243 X.958/23348
8725748

ACTING. Russian cinema films
Pereverzev, Ivan - Biographies

ANDREEV, F. (Feliks)
Ivan Pereverzev. - Moskva : Iskusstvo, 1982
176p,[31]p plates
r6448709 X.629/18713
8725748

ACTING. Silent cinema films
Stars, to 1926

PULLA, Armas J.
Mykän filmin sankarit. - Helsingissä : Otava, 1974
120 p. : ill
951-10-1488-9 X.989/35281
0876178

ACTING. Silent cinema films. Hollywood. Los Angeles. California
Stars - Biographies

CLOSE-UPS / John Richard Finch, planning, design, and direction ; Paul Elby, historical research and liaison ; Roland Liot, photographic consultant. - South Brunswick ; New York : Barnes ; London : Tantivy Press, 1978
[1],566p,[8]p of plates
0-904208-41-9 X.981/12428
1897772

ACTING. Soviet cinema films
Cherkasov, Nikolaĭ Konstantinovich - Biographies

CHERKASOVA, Nina Nikolaevna
Ri͡adom c Cherkasovym. - Leningrad : Lenizdat, 1978
253p,[1]leaf of plates,[48]p of plates : ill
r6184138 X.989/84058
1430602

GERASIMOV, I͡Uriĭ Konstantinovich
Cherkasov. - Moskva : Molodai͡a gvardii͡a, 1976
346p,[1]leaf of plates,[48]p of plates : ill. - (Zhizn' zamechatel'nykh li͡udeĭ ; vyp.11(563))
r6444859 W.P.16340/160
1430602

NIKOLAĬ Cherkasov / [sostavitel' N.N. Cherkasova]. - Moskva : Vserossiiskoe teatral'noe obshchestvo, 1976
429p,[71]p of plates : ill
r6088062 X.981/12430
1430602

ACTING. Soviet cinema films
Khokhlova, Aleksandra Sergeevna - Biographies

KULESHOV, L. V. (Lev Vladimirovich)
50 let v kino. - Moskva : Iskusstvo, 1975
301p,[64]p of plates : ill. - (Memuary kinematografistov)
r6095211 X.989/80663
2121247

ACTING. Soviet cinema films
Kuleshov, Lev Vladimirovich - Biographies

KULESHOV, L. V. (Lev Vladimirovich)
50 let v kino. - Moskva : Iskusstvo, 1975
301p,[64]p of plates : ill. - (Memuary kinematografistov)
r6095211 X.989/80663
2121239

ACTING. Soviet Union

O vremeni i o sebe / [redaktor-sostavitel' M.L. Rogachevskiĭ]. - Moskva : Vserossiiskoe teatral'noe obshchestvo, 1977
284p
r6077146 X.809/26306
0899003

UL'I͡ANOV, Mikhail Aleksandrovich
Moi͡a professii͡a. - 2-e izd.. - Moskva : Molodai͡a gvardii͡a, 1976
252 p., 32 p. of plates : ill. - (Mastera iskusstv molodezhi)
r6020887 X.907/15201
0899003

ACTING. Soviet Union
to date

KARA, S. (Sokrat)
Vkus pravdy. - Leningrad : Lenizdat, 1979
334p,plates
r6464611 X.958/23285
8536104

ACTING. Soviet Union
Knipper, Ol'ga Leonardovna - Biographies

PITCHER, Harvey John
Chekhov's leading lady. - London : J. Murray, 1979
viii,288p,[12]p of plates : ill. - Bibliography: p.280-282. - Includes index
0-7195-3681-2 X.989/54093
1840452

ACTING. Soviet Union
Leonov, Evgeniĭ

ISMAILOVA, Ninel' Khazbulatovna
Evgeniĭ Leonov. - Moskva : Iskusstvo, 1979
158p,[60]p of plates : ill. - (Mastera sovetskogo teatra i kino)
r6177920 X.989/86964
2417979

ACTING. Soviet Union
Mikhalovskaia, Nina Valerianovna - Biographies

MIKHALOVSKAI͡A, Nina Valerianovna
Glazami aktrisy. - Moskva : Iskusstvo, 1978
116p,[8]p of plates : ill
r6096778 X.907/19136
2094800

ACTING. Soviet Union
Mordvinov, Nikolaĭ Dmitrievich

NAPALKOV, Fedor Mikhaĭlovich
Ot Ti͡umeni do Kirkenesa. - Sverdlovsk : Sredne-Ural'skoe knizhnoe izdatel'stvo, 1976
175 p., 16 p. of plates : ill
r6020868 X.808/10252
0897671

PI͡ATKOV, Vasiliĭ Karpovich
Tret'i͡a udarnai͡a. - Moskva : Voennoe izdatel'stvo Ministerstva oborony SSSR, 1976
254p : ill
r6021445 X.809/23981
0897671

V dni grozovye / [sostavitel'...Vasiliĭ Georgievich Korolev]. - Moskva : Voennoe izdatel'stvo Ministerstva oborony SSSR, 1975
108 p. : ill
r6021376 X.908/36490
0897671

ACTING. Soviet Union
Mordvinov, Nikolaĭ Dmitrievich - Biographies

MORDVINOV, Nikolaĭ Dmitrievich
Dnevniki, 1938-1966. - Moskva : Vserossiĭskoe teatral'noe obshchestvo, 1976
679 p., 1 leaf of plates, 72 p. of plates : ill
r6020526 X.989/35955
0897671

VERI͡U v nashu pobedu- / [sostaviteli i avtory rasskazov o geroi͡akh E.D. Drobiz, A.A. Zolotov]. - Cheli͡abinsk : I͡Uzhne-ural'skoe knizhnoe izdatel'stvo, 1976
221p,20p. of plates : ill
r6021602 X.808/10470
0897671

ACTING. Soviet Union
Papanov, Anatolii Dmitrievich - Biographies

LINET͡SKAI͡A, Marta I͡Akovlevna
Anatoliĭ Papanov. - Moskva : Iskusstvo, 1972
198p,[48]p of plates : ill. - (Mastera sovetskogo teatra)
r6175569 X.907/19592
213473X

ACTING. Soviet Union
Shtraukh, Maksim Maksimovich - Biographies

STRAUKH, Maksim Maksimovich
Glavnai͡a rol'. - Moskva : Vserossiĭskoe teatral'noe obshchestvo, 1977
273p : ill
r6087023 X.989/76887
1794973

ACTING. Soviet Union
Solov'ev, Ivan Ivanovich - Biographies

SOLOV'EV, Ivan Ivanovich
Monolog pod zanaves. - Moskva : Iskusstvo, 1979
222p,[1]leaf of plates[46]p of plates : ill,port
r6175819 X.429/7949 Woolwich
2449501

ACTING. Soviet Union
Sverdlin, Lev Naumovich

LEV Sverdlin / [sostaviteli N.A. Velekhova i A.G. Obrazt͡sova]. - Moskva : Iskusstvo, 1979
316p,[1]leaf of plates,[46]p of plates : ill
r6175924 X.989/83627
2260832

ACTING. Soviet Union
T͡Sarev, Mikhail - Biographies

ZUBKOV, I͡Uriĭ Aleksandrovich
Mikhail T͡Sarev v Malom teatre. - Moskva : Iskusstvo, 1978
181p,[42]p of plates : ill
r6098138 X.989/79296
1975854

ACTING. Starsky & Hutch. Television drama series in English
Glaser, Paul Michael & Soul, David

TRIBUTE to Starsky & Hutch. - [London] : [IPC Magazines], [1976]
Sheet ([8]) p.
r6010672 X.902/5158
1048120

ACTING. Swedish cinema films
Carlsson, Sickan - Biographies

CARLSSON, Sickan
Sickan. - [Stockholm] : Bonniers, 1977
235p,[32]p of plates : ill
91-00-41937-0 X.989/74516
1538780

ACTING. Swedish cinema films
Rodin, Gösta - Biographies

RODIN, Gösta
Den svenska filmens glada 30-tal. - [Stockholm] : Nordisk bokindustri, 1976
339p : ill
91-970184-0-6 X.909/36470
0931357

ACTING. Television drama
Timothy, Christopher - Biographies

TIMOTHY, Christopher
Vet behind the ears. - London : Pan Books, 1979
122p,[8]p of plates : ill
0-330-26075-8 X.907/25888
2065533

Vet behind the ears. - London : Severn House, 1981, c1979
121p,[4]p of plates : ill
0-7278-0698-x X.989/90113
2065533

ACTING. Theatre

GOLDING, Alfred Siemon
Classicistic acting. to which is appended an annotated translation of the Lessons on the principles of gesticulation and mimic expression. - Lanham, Md. ; London : University Press of America, c1984
xv,627p : ill
0-8191-3680-8 X.950/38267
1411195

ACTING. Theatre -continued
GOLDMAN, Michael
The actor's freedom. - New York : Viking Press, 1975
180p
0-670-10347-0 X.989/54156
1411195

MAROWITZ, Charles
The act of being. - London : Secker and Warburg, 1978
x,196,[16]p of plates : ill
0-436-27326-8 X.989/52360
1411195

ACTING. Theatre
REDGRAVE, Sir Michael
The actor's ways and means. - London : Heinemann Educational [etc.], 1979
96p : ill
0-435-18762-7 X.909/43713
1411195

ACTING. Theatre
to date
BOAL, Augusto
200 exercícios e jogos para o ator e o não-ator com vontade de dizer algo através do teatro. - 2. ed. - Rio de Janeiro : Civilização Brasileira, 1979
123p : ill. - (Coleção Teatro hoje ; vol.30)
r6701962 X.958/7972
8660255

CAUNE, Jean
La dramatisation. - Louvain-la-neuve : Édition des Cahiers Théâtre Louvain, [1981]
368p. - (Arts du spectacle)
r6207312 P.901/1319
8660255

HELBO, André
Les mots et les gestes. - Lille : Presses Universitaires de Lille, 1983
117p. - (Littérature Générale). - Includes bibliography
2-85939-226-2 X.950/32128
8660255

KÖRMENDI, János
Levelek az urológiáról. - Budapest : Magvető Könyvkiadó, [1982]
199p
963-271-687-6 X.958/18212
8660255

Il SIGNORE della scena / introduzione e cura di Maria Grazia Gregori. - Milano : Feltrinelli, 1979
237p. - (SC10 ; 91)
r6503238 X.958/7081
8660255

STANISLAVSKIĬ, K. S. (Konstantin Sergeevich)
[The art of the actor and the art of the director. Swedish]. Skadespelarkonsten och regissörskonsten. - Norrköping : IOR, 1982
29p. - (Dramatens skriftserie ; nr 5)
91-970467-0-1 X.0909/748(5)
8660255

ACTING. Theatre
Adler, Jacob - Biographies
ROSENFELD, Lulla
Bright star of exile. - London : Barrie and Jenkins, 1978
xvi,368p[16]p of plates : ill. - Bibliography: p.359-360. - Includes index
0-214-20590-8 X.981/22391
2127857

ACTING. Theatre
Bloom, Claire - Biographies
BLOOM, Claire
Limelight and after. - London : Weidenfeld & Nicolson, 1982
187p,[40]p of plates : ill
0-297-78051-4 X.809/51455
2711729

Limelight and after. - Large print ed. - Bath : Chivers, 1983, c1982
221p. - (Lythway book)
0-85119-954-2 Nov.49536
2711729

ACTING. Theatre
Draper, Ruth - Biographies
DRAPER, Ruth
The letters of Ruth Draper, 1920-1956. - London : H. Hamilton, 1979
xxii,362p : ill. - Bibliography: p.348-349. - List of sound discs: p.349. - Includes index
0-241-10229-4 X.981/21677
1733419

ACTING. Theatre
Korian, Tola
TOLA Korian / [medalcgà Katarzyna Fejcher]. - Londyn : Oficyna Poetów i Malarzy, 1984
167p,[18]p of plates : ill
B85-10311 X.429/16860
3241688

ACTING. Theatre
Moissi, Alexander - Biographies
KOSOVA, Bardhyl
Aleksandër Moisiu. - Prishtine : Rilindja, 1971
117p,[2]leaves of plates,[28]p of plates : ill. - (Portrete-monografi)
r6176761 X.989/83322
221685X

ACTING. Theatre
Nonverbal communication, to date
CREATIVITÀ, gesto, comportamento / con i contributi di Ennio Castaldini ... [et al.]. - Bologna : Pàtron, 1980
337p : ill. - Includes bibliography
r6508531 X.950/22673
8977135

ACTING. Theatre
Techniques
FO, Dario
Dario Fo and Franca Rame. - London : Red Notes, 1983
xviii,75p : ill
0-906305-08-x X.955/2219
3079848

KNIGHT, George Wilson
Symbol of man. - London : Regency Press, 1979
192p,[i],xv leaves of plates : ill. - Bibliography: p.7-8. - Includes index
B80-06515 X.989/54529
2088304

ACTING. Theatre. Arab countries
to date
HADDAD, Youssef Rachid
Art du conteur, art de l'acteur. - Louvain-la-Neuve : Édition des Cahiers théâtre louvain, [1982]
147p,[8]p of plates : ill. - (Cahiers théâtre louvain ; 47) (Arts du spectacle). - Bibliography
r6209691 P.901/1319[47]
8940479

ACTING. Theatre. Austria
Krones, Therese - Catalogues
DEUTSCHMANN, Wilhelm
Therese Krones. - Wien : Museen der Stadt, [1980]
104p. : ill
r6303644 X.950/9627
8772398

ACTING. Theatre. Austria
Schratt, Katharina - Biographies
MARKUS, Georg
Katharina Schratt. - 2. Aufl. - Wien : Amalthea, 1982
298p.,plates
3-85002-162-9 X.950/39693
8772398

ACTING. Theatre. Austria
Schratt, Katharina. Interpersonal relationships with Franz Joseph, Emperor of Austria
HASLIP, Joan
The Emperor & the actress. - London : Weidenfeld and Nicolson, 1982
x,284p,[8]p of plates : ill. - Bibliography: p273-274. - Includes index
0-297-78102-2 X.809/53286
2788381

ACTING. Theatre. Azerbaijan
Rukhulla, Sidgi - Biographies
SHEĬKHZADE, Ėmin
Sidgi Rukhulla. - Baku : Ishyg, 1978
90p : ill
r6097170 X.909/84075
2098431

ACTING. Theatre. Bavaria. West Germany
Esslair, Ferdinand - Biographies
OHLEN, Joachim von, Freiherr
Ferdinand Esslair, Kgl bayerischer Hofschauspieler (1772-1840). - München : Kommissionsbuchhandlung R. Wölfle, 1972
158 p., 12 leaves of plates : ill. - (Genealogia boica ; Bd 1)
r6000350 X.0700/785(1)
0916102

ACTING. Theatre. Bohemia. Czechoslovakia
Marvan, Jaroslav - Biographies
MARVAN, Jaroslav
Jaroslav Marvan vypravuje. - Praha : Novinář, 1975
224p,plates
r6439202 X.950/28829
8875774

ACTING. Theatre. Bohemia. Czechoslovakia
Vydra, Václav - Biographies
VYDRA, Václav
Má pouť životem a uměním. - Praha : Melantrich, 1976
433p,[48]p of plates : ill
r6440584 X.329/15525
8875774

ACTING. Theatre. Brazil
Costa, Beatriz - Biographies
COSTA, Beatriz
Sem papas na língua. - 4. ed. - [Lisboa?] : Publicações Europa-América, 1976
291 p.
r6027410 X.709/21829
1019627

ACTING. Theatre. Bulgaria
Atanasov, Petko
PŬRVANOVA, Donka Marinova
Petko Atanasov. - Sofiia : Nauka i izkustvo, 1979
114p[16]p of plates : ill. - (Maistori na bŭlgarskata stsena)
r6173815 X.908/89444
2136252

ACTING. Theatre. Chuvash ASSR. Russia (RSFSR)
Alekseev, B. A.
ROMANOVA, Faina Aleksandrovna
Boris Alekseev. - Cheboksary : Chuvashskoe knizhnoe izdatel'stvo, 1980
150p
r6485646 X.958/8613
8809143

ACTING. Theatre. Czechoslovakia
Hlavatý, Vladimír
HLAVATÝ, Vladimír
Monolog herce z Vinohrad. - Praha : Melantrich, 1984
189p,plates
r6442627 X.950/42817
8711666

ACTING. Theatre. Czechoslovakia
Myron, Jiří - Biographies
ZBAVITEL, Miloš
Jiří Myron. - Ostrava : Profil, 1980
235p,[32]p of plates : ill. - Bibliography: p.221-223
r6499092 X.950/4898
8711666

ACTING. Theatre. Debrecen. Hungary
to date
A DEBRECENI színészet története / [szerkesztette és lektorálta Katona Ferenc] ; [munkatársak Béber László ... et al.]. - Debrecen : Kiadja Debrecen megyei Városi Tanács V.B. Művelődésügyi Osztálya, 1976
513p. - Bibliography
963-03-0400-7 X.520/26463
881564x

ACTING. Theatre. Denmark
Andreasen, Dagmar - Biographies
ANDREASEN, Dagmar
Teaterarbejder. - København : Reitzels, 1980
84p : ill
87-412-3658-0 X.958/15674
8925542

ACTING. Theatre. Denmark
Crone, Maria - Biographies
CRONE, Maria
Han hed Othello. - [Lynge] : Bogan, [1980]
113p,[16]leaves of plates
87-87533-83-9 X.950/18269
8925542

ACTING. Theatre. Denmark
Heiberg, Johanne Luise
FORSSBERGER, Annalisa
Johanne Luise Heiberg. - København : Nordisk Bogforlag, [1973]
149p : ill. - (Skrifter udgivet af Heiberg-Selskabet ; 1)
87-85083-32-1 X.0909/1062
1859390

ACTING. Theatre. Denmark
Nielsen. Carl. 1933–

BAKGAARD. Hans
[Correspondence]. Breve. – Hurup Thy : Hans Bakgaards Venners Forlag. [1980]
141p : ill
r6600225 X.950/22034
8925542

ACTING. Theatre. Denmark
Pio. Elith – Biographies

PIO. Elith
At være – og – ikke være. – København : Berlingske. 1980
87p : ill
87–19–01359–0 X.950/21047
8925542

ACTING. Theatre. Denmark
Rode. Ebbe

HEIN. Piet
Privat tryk. – [Virum] : Greens Boghandel. 1980
17p
No.302 of a numbered ed. of 500 copies
87–87831–08–2 X.950/19892
8925542

ACTING. Theatre. Denmark
Storm Petersen. Robert

BOYE. Ib
Skuespilleren Storm P.. – [København] : Carit Andersen. 1982
100p
87–424–4353–9 X.622/20085
8925542

ACTING. Theatre. Denmark
Walter. Caroline – Biographies

NEIIENDAM. Klaus
Caroline Walter. – [København] : Nyt Nordisk Forlag. 1983
171p : ill. – Includes bibliography
87–17–05171–1 X.950/38573
8925542

ACTING. Theatre. East Germany
Heinz. Wolfgang – Biographies

WAACK. Renate
Wolfgang Heinz. – Berlin : Henschelverlag. 1980
207p.[48]p of plates : ill
r6305121 X.950/12012
8779228

ACTING. Theatre. East Germany
Weigel. Helene – Biographies
Die WEIGEL / fotografiert von Vera Tenschert. – Berlin : Henschelverlag. 1981
[216]p : chiefly ill(one col.)
r6399188 L.42/1367
8779228

ACTING. Theatre. England
to 1980

MILES. Bernard
Curtain calls. – Guildford : Lutterworth. 1981
191p : ill. – Includes index
0–7188–2476–8 X.950/8956
2591537

ACTING. Theatre. England
to 1980 – Texts

A NIGHT at the theatre / edited by Ronald Harwood. – London : Methuen. 1982
200p
0–413–49950–2 X.950/18057
2591537

ACTING. Theatre. England
1890–1983

SANDERSON. Michael. 1939 Jan. 23–
From Irving to Olivier. – London : Athlone. 1984
xii,375p.[8]p of plates : ill. – Bibliography: p336–351. – Includes index
0–485–11252–3 X.950/39651
3055930

ACTING. Theatre. Estonia
Ird. Kaarel

VANEMUISLANE Kaarel Ird = Kaarel iz teatra "Vanemuine" = Kaarel Ird aus dem Theater "Vanemuine" / [koostanud K. Kask ja M. Säre]. – Tallinn : Eesti Raamat. 1981
187p : chiefly ill
r6461625 L.49/1289
8867941

ACTING. Theatre. Estonia
Nuude. Rudolf – Biographies

LOTT. Jüri
Rudolf Nuude. – Tallinn.
55p.[24]p of plates. – (Teatriteed)
r6461001 X.950/1844
2481677

ACTING. Theatre. Estonia
Ratassepp. Valdeko – Biographies

SAAREPERA. Maimu
Valdeko Ratassepp. – Tallinn : Eesti Raamat. 1981
63p.plates. – (Teatriteed)
r6461663 X.950/22438
8867941

ACTING. Theatre. Estonia
Särev. Andres – Biographies

KUNINGAS. O.
Andres Särev. – Tallinn : Eesti Raamat. 1977
91p.[32]p of plates. – (Teatriteed)
r6461003 X.950/1837
2481669

ACTING. Theatre. Estonia
Üksip. Albert

ÜKSIP. Albert
Mälestused. – Tallinn : Eesti Raamat. 1975
341p.[32]p of plates : ill
r6045552 X.989/70841
1323393

ACTING. Theatre. Estonia
Ulla. Ülle – Biographies

TAMM. Helle
Ülle Ulla. – Tallinn : Eesti Raamat. 1976
72p.[24]p of plates. – (Teatriteed)
r6461002 X.950/1836
2481650

ACTING. Theatre. Europe
1700–1900

CHRISTIANSEN. Svend
Klassisk skuespilkunst. – [Copenhagen] : Akademisk Forlag. 1975
376 p. : ill. – (Teatervidenskabelige studier ; bd. 4). – Bibliography : p. 358–365
87–500–1572–9 X.0900/183(4)
0903671

ACTING. Theatre. Finland
Salonen. Sylvi – Biographies

SEPPÄLÄ. Raimo
Sylvi. – [Espoo] : Weilin + Göös. 1981
159p.plates
951–352–612–7 X.950/24866
9008101

ACTING. Theatre. Finland
Siivola. Vilho – Biographies

SIIVOLA. Vilho
Myllykylästä maailmalle. – 2.painos. – Helsinki : Tammi. 1975
268p.,[24]p.of plates : ill
951–303–070–9 X.809/23687
1027387

ACTING. Theatre. France
Bernhardt. Sarah

BERNHARDT and the theatre of her time / edited by Eric Salmon. – Westport. Conn. ; London : Greenwood Press. 1984
x,287p : ill. – (Contributions in drama and theatre studies ; no.6). – Bibliography: p269–272. – Includes index
0–313–23755–7 X.950/38103
0529311

SARAH Bernhardt. 1844–1923 / [images et photographies. réalisation et collection : André Fildier]. – Paris : Éditions Fildier Cartophilie. 1980
112p : ill. – (Collection Im. phot)
r6205726 X.955/414
8725969

SPIVAKOFF. Pierre
Sarah Bernhardt vue par les Nadar. – Paris : Herscher. 1982
[112]p
2–7335–0039–2 L.45/2630
816942x

ACTING. Theatre. France
Bernhardt. Sarah – Biographies

BERNHARDT. Sarah
My double life. – London : Owen. 1977
vi,453p.[8]p of plates. – Includes index
0–7206–0502–4 X.989/51472
0529311

My double life. – London : Zenith. 1984
vi,453p.[16]p of plates : ill. – Includes index
0–09–933560–3 X.958/25155
0529311

Die GOTTLICHE Sarah Bernhardt / herausgegeben und mit einem Vorwort von Maria Moog-Grunewald. – Dortmund : Harenberg. c1982
130p : ill. – (Die bibliophilen Taschenbücher ; Nr. 338)
3–88379–338–8 X.958/20098
816942x

RICHARDSON. Joanna
Sarah Bernhardt and her world. – London : Weidenfeld and Nicolson. 1977
232p : ill(some col). – Bibliography: p.225. – Includes index
0–297–77253–8 X.981/20575
0529311

ACTING. Theatre. France
Charon. Jacques – Biographies

CHARON. Jacques
Moi, un comédien. – Paris : Albin Michel. 1975
283 p., 16 p. of plates
2–226–00134–4 X.900/17057
0926671

ACTING. Theatre. France
Goulebenéze – Biographies

LABODINIÈRE. Georges
Goulebenéze. – 3e éd. – La Rochelle : Éditions Rupella. 1976
265p : ill
r6055137 X.909/39687
1502670

ACTING. Theatre. France
Guitry. Sacha – Biographies

JADOUX. Henri
Sacha Guitry. – Paris : Librairie Académique Perrin. 1982
318p.plates
2–262–00260–6 X.950/26087
816942x

ACTING. Theatre. France
Huster. Francis – Biographies

ROGALE. Jean-Yves
Francis Huster. – Paris : Éditions Mengès. 1982
316p
2–85620–145–8 X.421/25296
816942x

ACTING. Theatre. France
Le Poulain. Jean – Biographies

LE POULAIN. Jean
Je rirai le dernier. – Paris : Éditions Robert Laffont. 1977
245p.[8p of plates] : ill. – (A jeu découvert)
r6055157 X.909/39682
1507397

ACTING. Theatre. France
Maréchal. Marcel – Biographies

PIERRON. Agnès
Maréchal. – [Lausanne] : L'Âge d'Homme. 1977
236p.16p of plates : ill
r6055110 X:709/25483
1507419

ACTING. Theatre. France
Pierre-Louis – Biographies

PIERRE-LOUIS
Mes bonnes fréquentations. – Paris : France-Empire. [1983]
256p.plates
2–7048–0282–3 X.950/41339
816942x

ACTING. Theatre. France
Rachel

TOUSSAINT DU WAST. Nicole
Rachel, amours et tragédie. – Paris : Stock. 1980
316p.,16p.of plates : ill. – (Femmes dans leur temps)
2–234–01234–1 X.909/88788
0204684

ACTING. Theatre. Georgia (Soviet Socialist Republic)
Andzhaparidze. Veriko – Biographies

URUSHADZE. N. (Natela)
Veriko Andzhaparidze. – Moskva : Iskusstvo. 1972
192p.[48]p of plates : ill. – (Mastera sovetskogo teatra)
r6178880 X.907/21035
2467623

ACTING. Theatre. German speaking countries
to date

Die STERNE dürfet ihr verschwenden / herausgegeben und kommentiert von Barbara Albrecht und Günter Albrecht. - 2. Aufl. - Berlin : Buchverlag der Morgen, 1982
544p.,plates. - Includes bibliography
r6307780 X.958/19470
9028722

ACTING. Theatre. Germany
Giehse, Therese

GIEHSE, Therese
"Ich hab nichts zum Sagen". - München, c 1973
223 p. : ill
3-570-08405-1 X.981/11178
1014129

ACTING. Theatre. Germany
Hahn, Elise

BÜRGERS Liebe / herausgegeben und mit einem Nachwort von Hermann Kinder. - Frankfurt am Main : Insel, 1981
188p. : ill. - (Insel Taschenbuch ; 564)
3-458-32264-7 X.429/13812
8876398

ACTING. Theatre. Great Britain
1837-1901 - Dictionaries

VICTORIAN actors and actresses in review / compiled and edited with an introduction by Donald Mullin. - Westport, Conn. ; London : Greenwood Press, 1983
xxxvi,571p : ill. - Bibliography: p526-529. - Includes index
0-313-23316-0 X.950/36370
3209083

ACTING. Theatre. Great Britain
Baddeley, Hermione - Biographies

BADDELEY, Hermione
The unsinkable Hermione Baddeley. - London : Collins, 1984
240p,[32]p of plates : ill. - Includes index
0-00-216493-0 X.950/39186
3024334

ACTING. Theatre. Great Britain
Barkworth, Peter, to 1964 - Biographies

BARKWORTH, Peter
First houses. - London : Secker & Warburg, 1983
132p,[16]p of plates. - Includes index
0-436-03292-9 X.950/28679
306008x

ACTING. Theatre. Great Britain
Callow, Simon

CALLOW, Simon
Being an actor. - London : Methuen, 1984
ix,190p,[8]p of plates : ill. - Includes index
0-413-52440-x X.950/36039
3186296

ACTING. Theatre. Great Britain
Campbell, Mrs Patrick, 1865-1940 - Biographies

PETERS, Margot
Mrs Pat. - London : Bodley Head, 1984
531p,[44]p of plates : ill(some col.). - Includes index
0-370-30567-1 X.950/36349
182838x

ACTING. Theatre. Great Britain
Casson, Sir Lewis - Biographies

DEVLIN, Diana
A speaking part. - London : Hodder and Stoughton, 1982
vi,277p,[8]p of plates : ill. - Bibliography: p267-268. - Includes index
0-340-28090-5 X.800/34267
2804336

ACTING. Theatre. Great Britain
Cibber, Susannah Maria - Biographies

NASH, Mary
The provoked wife. - London : Hutchinson, 1977
xii,369p : ill. - Bibliography: p.353-358. - Includes index
0-09-129380-4 X.800/26376
1207369

ACTING. Theatre. Great Britain
Cooke, George Frederick - Biographies

HARE, Arnold
George Frederick Cooke. - London : Society for Theatre Research, c1980
viii,255p,[8]p of plates : ill. - Bibliography: p214-218. - Includes index
0-85430-031-7 X.950/1231
2341239
- [Another copy.] X.950/1857

WILMETH, Don B. (Don Burton)
George Frederick Cooke. - Westport, Conn. ; London : Greenwood Press, 1980
xv,364p : ill. - (Contributions in drama and theatre studies ; no.2). - Bibliography: p345-350. - Includes index
0-313-21487-5 X.950/1308
2337800

ACTING. Theatre. Great Britain
Cooper, Gladys - Biographies

MORLEY, Sheridan
Gladys Cooper. - London : Heinemann, 1979
xxii,314p : ill. - Bibliography: p.301-303. - Includes index
0-434-47896-2 X.800/27790
1889567

ACTING. Theatre. Great Britain
Du Maurier, Sir Gerald - Biographies

DU MAURIER, Daphne
Gerald. - Harmondsworth : Penguin, 1977
252p
B77-09186 X.908/40863
1061100

ACTING. Theatre. Great Britain
Farr, Florence - Biographies

JOHNSON, Josephine
Florence Farr. - Gerrards Cross : Smythe, 1975
xiv,222p,leaf of plate,[16]p of plates : ill. - Includes index
0-901072-15-x X.529/30243
0519960

ACTING. Theatre. Great Britain
Furse, Jill - Biographies

WHISTLER, Laurence
The initials in the heart. - 3rd ed.. - London : J. Murray, 1975
264p,[7] leaves of plates : ill
0-7195-3265-5 X.809/40853
0636061

ACTING. Theatre. Great Britain
Garrick, David, 1717-1779 - Biographies

SMITH, Helen R.
David Garrick, 1717-1779. - London : British Library, c1979
80p : ill. - (British Library monograph ; no.1). - Bibliography: p76-77
0-904654-40-0 X.0950/388(1)
1937529

STONE, George Winchester
David Garrick. - Carbondale : Southern Illinois University Press, [1979]
xx,771p : ill
0-8093-0931-9 X.981/13222
1937529

ACTING. Theatre. Great Britain
Gielgud, John - Biographies

The AGES of Gielgud / edited by Ronald Harwood. - London : Hodder and Stoughton, c1984
182p,[24]p of plates : ill. - Includes index
0-340-34828-3 X.950/29610
2022311

BRANDRETH, Gyles
John Gielgud. - London : I Pavilion, 1984
186p : ill. - Bibliography: p181. - Includes index
0-907516-38-6 X.525/8874
2022311

GIELGUD, Sir John
An actor and his time. - London : Sidgwick and Jackson, 1979
253p,[8]p of plates : ill(some col). - Includes index
0-283-98573-9 X.981/21949
2022311
An actor and his time. - Harmondsworth : Penguin, 1981, c1979
228p,[16]pof plates. - Includes index
0-14-005636-x X.958/2975
2022311
- [A re-issue]. - 1984 X.958/26372

ACTING. Theatre. Great Britain
Irving, Henry - Biographies

BINGHAM, Madeleine, Baroness Clanmorris
Henry Irving. - New York : Stein and Day, 1978
312p,[8]p of plates : ill. - Bibliography : p.302-305. - Includes index
0-8128-2160-2 X.981/21858
1451324
Henry Irving and the Victorian theatre. - London : Allen and Unwin, 1978
312p,8p of plates : ill. - Bibliography: p.302-305. - Includes index
0-04-928038-4 X.981/21068
1451324

ACTING. Theatre. Great Britain
Kean, Edmund - Biographies

FITZSIMONS, Raymund
Edmund Kean. - London : Hamilton, 1976
xiii,255p,8p of plates : ill. - Bibliography: p.240-244. - Includes index
0-241-89337-2 X.989/50562
0608726

PLAYFAIR, Giles
The flash of lightning. - London : Kimber, 1983
188p,[6]p of plates : ill. - Includes index
0-7183-0399-7 X.950/32274
0608726

ACTING. Theatre. Great Britain
Kemble, Fanny - Biographies

FURNAS, Joseph Chamberlain
Fanny Kemble. - New York : Dial Press, 1982
xiv,494p
Signed by the author
0-385-27208-1 X.950/26807
822661x

KEMBLE, Fanny
The terrific Kemble. - London : Hamilton, 1978
xv,272p,[8]p of plates. - Bibliography: p.262-263. - Includes index
0-241-89884-6 X.989/52442
1200089

MARSHALL, Dorothy
Fanny Kemble. - London : Weidenfeld and Nicolson, 1977
vii,280p,[8]p of plates : ill. - Bibliography: p.4-5. - Includes index
0-297-77282-1 X.989/51616
0128082

ACTING. Theatre. Great Britain
Kemble, John Philip & Siddons, Sarah - Biographies

KELLY, Linda, 1936-
The Kemble era. - London : Bodley Head, 1980
221p,[32]p of plates : ill. - Bibliography: p.213-216. - Includes index
0-370-10466-8 X.989/88279
1818015

ACTING. Theatre. Great Britain
Kortner, Fritz

BRAND, Matthias
Fritz Kortner in der Weimarer Republik. - Rheinfelden : Schäuble, [1981]
332p.. - (Theater unserer Zeit ; Nr.17). - Bibliography and list of plays
3-87718-008-6 X.958/6077
822661x

ACTING. Theatre. Great Britain
Langtry, Lillie - Biographies

BIRKETT, Jeremy
Lillie Langtry. - Poole : Blandford Press [for] Rupert Shuff and the Société Jersiaise, 1979
64p : ill
0-7137-1073-x X.908/42737
1992740
Lillie Langtry. - St. Clements Bay, Jersey : Shuff in association with the Société Jersiase, 1979
62p : ill
0-9506604-1-8 X.529/55502
1992740

ACTING. Theatre. Great Britain
Langtry, Lillie, to 1924 - Biographies

LANGTRY, Lillie
The days that I knew. - [Abridged ed.]. - London : Futura Publications, 1978
256p,8p of plates : ill
0-7088-1444-1 X.907/25658
1900498

ACTING. Theatre. Great Britain
Langtry, Lillie. Interpersonal relationships with Edward VII, King of Great Britain

BROUGH, James
The prince & the lily. - London : Hodder and Stoughton, 1975
352p,[32]p of plates : ill. - Includes index
0-340-16835-8 X.800/25160
050405x

ACTING. Theatre. Great Britain
Langtry, Lillie. Interpersonal relationships with Edward VII, King of Great Britain - Biographies

BROUGH, James
The prince and the lily. - Sevenoaks : Coronet, 1978, c1975
332p. - Includes index
0-340-23384-2 X.808/33548
050405x

ACTING. Theatre. Great Britain
Morley, Robert - Biographies

MORLEY, Margaret
Larger than life. - London : Robson Books, 1979
206p,[12]p of plates : ill - Bibliography p.195-196. - Includes index
0-86051-064-6 X.989/53543
1801414

MORLEY, Robert
The best of Robert Morley. - London : Robson, 1981
378p
0-86051-146-4 X.950/11909
1801414

ACTING. Theatre. Great Britain
Nesbitt, Cathleen - Biographies

NESBITT, Cathleen
A little love & good company. - [New ed.] - Owings Mills, Md : Stemmer House Publishers, 1977
262p
0-916144-10-0 X.981/13099
0294438

ACTING. Theatre. Great Britain
Olivier, Laurence Olivier, Baron - Biographies

LEFÈVRE, Raymond
Sir Laurence Olivier. - Paris : Éditions PAC, [1980]
290p : ill
r6298192 X.950/39653
822661x

ACTING. Theatre. Great Britain
Pritchard, Hannah - Biographies

VAUGHAN, Anthony
Born to please. - London : Society for Theatre Research, 1979
xi,195p,12p of plates : ill(incl 1 col). - Bibliography: p.179-184. - Includes index
0-85430-030-9 X.981/22312
2144565

ACTING. Theatre. Great Britain
Richardson, Sir Ralph

RALPH Richardson / devised and compiled by Robert Tanitch. - London : Evans, 1982
128p : chiefly ill. - List of films and plays: p122-127. - Includes index
0-237-45680-x X.525/6692
2838427

ACTING. Theatre. Great Britain
Richardson, Sir Ralph - Biographies

O'CONNOR, Garry
Ralph Richardson. - London : Hodder and Stoughton, 1982
260p,[16]p of plates : ill. - Bibliography: p245-248. - Includes index
0-340-27041-1 X.950/17597
2822377

ACTING. Theatre. Great Britain
Rix, Brian - Biographies

RIX, Brian
My farce from my elbow. - London : Secker and Warburg, 1975
[7],248p,[16]p of plates : ill. - Includes index
0-436-41500-3 X.989/40131
0372803

ACTING. Theatre. Great Britain
Terry, Ellen - Biographies

PRIDEAUX, Tom
Love or nothing. - London : Millington, 1976
xi,288p,[16]p of plates : ill. - Bibliography: p.273-277. - Includes index
0-86000-074-5 X.989/51785
1196820

ACTING. Theatre. Great Britain
Terry, Ellen - Biographies

TERRY, Ellen
The story of my life. - Woodbridge : Boydell, 1982
viii,240p. - (Bookmarks)
0-85115-204-x X.950/14230
1196820

ACTING. Theatre. Great Britain
Terry, Ellen - Correspondence, diaries

TERRY, Ellen
[Correspondence. Selections. Polish]. Listy Ellen Terry i Bernarda Shaw. - Warszawa : Państwowy Instytut Wydawniczy, 1981
491p,[12]p of plates
83-06-00415-9 X.958/4735
822661x

ACTING. Theatre. Great Britain
Thorndike, Dame Sybil - Biographies

MORLEY, Sheridan
Sybil Thorndike. - London : Weidenfeld and Nicolson, 1977
183p. - Includes index
0-297-77388-7 X.981/20788
1252801

ACTING. Theatre. Great Britain
Vestris, Lucia Elizabeth - Biographies

APPLETON, William W.
Madame Vestris and the London stage. - New York ; London : Columbia University Press, 1974
ix,230p. : ill. - Bibliography: p.217-221
0-231-03794-5 X.950/5678
822661x

ACTING. Theatre. Great Britain
Wolfit, Sir Donald - Biographies

HARWOOD, Ronald
Sir Donald Wolfit C.B.E.. - Oxford : Amber Lane, 1983
xviii,302p. - Bibliography: p291-292. - Includes index
0-906399-43-2 X.958/24329
3139670

ACTING. Theatre. Great Britain
Wyndham, Sir Charles & Moore, Mary, 1862-1931 - Biographies

TREWIN, Wendy
All on stage. - London : Harrap, 1980
x,244,[16]p of plates : ill
0-245-53444-x X.981/22699
2337126

ACTING. Theatre. Greece
Chalkousē, Helenē - Biographies

CHALKOUSĒ, Helenē
Theatriko hēmerologio. - [Athēnai] : Kaktos, c1981
309p : ill
r6598234 X.429/15805
9090355

ACTING. Theatre. Hungary
Jászai, Mari

FÖLDES, Anna
Jászai Mari és a magyar színház. - Budapest : Magvető Kiadó, 1983
475p
963-271-850-x X.958/19419
8925577

ACTING. Theatre. Hungary
Jávor, Pál - Biographies

BÁNOS, Tibor
Jávor Pál. - Budapest : Gondolat Könyvkiadó, 1978
287p : ill. - (Szemtől szemben)
963-280-481-3 X.958/14287
8925577

ACTING. Theatre. Hungary
Ódry, Árpád - Biographies

CSILLAG, Ilona
Ódry Árpád. - Budapest : Gondolat, 1982
261p : ill. - (Szemtől szemben)
963-281-014-7 X.958/18248
8925577

ACTING. Theatre. Italy
Capodaglio, Wanda - Biographies

BRAGAGLIA, Leonardo
Wanda Capodaglio attrice. - Roma : Bulzoni, c1981
143p : ill
r6510392 X.950/29276
8908753

ACTING. Theatre. Italy
Degli Esposti, Piera

DEGLI ESPOSTI, Piera
Storia di Piera. - 2a ed. - Milano : Bompiani, 1980
135p
r6599120 X.950/18782
8908753

ACTING. Theatre. Italy
Duse, Eleonora - Biographies

DUSE, Eleonora
Lettere d'amore. - Milano : Il saggiatore, 1979
xxix,982p,[16]p of plates : ill. - (Il filo di Arianna ; 9)
r6505237 X.439/12172
8908753

SETTI, Dora
La Duse com'era. - Milano : Pan, 1978
233p : ill. - (Collana storica ; 6)
r6197821 X.909/89346
2468549

WEAVER, William, 1923-
Duse. - London : Thames and Hudson, c1984
382p : ill. - Bibliography: p372-375. - Includes index
0-500-01341-1 X.950/40088
2468549

ACTING. Theatre. Italy
Duse, Eleonora. Items associated with Duse, Eleonora exhibited at Museo di Asolo - Catalogues

ELEONORA Duse e il suo tempo, 1858-1924 / [acura] di Gerardo Guerrieri. - Treviso : Editrice Canova, [1974?]
[208] p., 106 p. of plates : ill
r6013937 X.909/34248
089205X

ACTING. Theatre. Italy
Duse, Eleonora. Tours of South America, 1906-1907 - Correspondence, diaries

NOCCIOLI, Guido
Duse on tour. - Manchester : Manchester University Press, c1982
x,178p : ill. - Bibliography: p174-178
0-7190-0847-6 X.950/13770
2748266

ACTING. Theatre. Italy
Gassman, Vittorio - Biographies

GASSMAN, Vittorio
Un grande avvenire dietro le spalle. - 6 ed. - Milano : Longanesi, 1982
244p. - (La gaja scienza ; v.40)
r6512124 X.950/38116
8908753

ACTING. Theatre. Italy
Novelli, Ermete

GALLI, Quirino
Un capocomico nell'Italia borghese. - Napoli : Conte, 1979
226p. - (Riscontri ; 6)
r6506273 X.950/19564
8942455

ACTING. Theatre. Italy
Ruggeri, Ruggero - Catalogues

"CESTA" DI RUGGERO RUGGERI (Exhibition) (1980 : Milan)
La "cesta" di Ruggero Ruggeri. - Milano : De Carlo, 1980
84p : ill
r6508523 X.958/18565
8908753

ACTING. Theatre. Italy
Tòfano, Sergio - Catalogues

STORIA LUNGA UN MILIONE (Exhibition) (1979-1981 : Genoa, etc)
Una storia lunga un milione. - Roma : Bulzoni, 1980
53p : ill
r6506550 X.950/24806
8908753

ACTING. Theatre. Italy
Valenti, Osvaldo & Ferida, Luisa. Role in political events, 1945

LUALDI, Aldo
Morire a Salò. - Milano : Sugarco, 1975
280p.,[4]p. of plates : ill. - (Fatti e misfatti ; 23)
r6042795 X.909/37588
1235540

ACTING. Theatre. Italy
Viviani, Raffaele

RAO, Angela Maria
Raffaele Viviani, o, Della miseriacoatta. - Poggibonsi : A. Lalli, 1981
116p. - (Materiali di letteratura)
r6511182 X.950/38901
8908753

ACTING. Theatre. Komi ASSR. Russia (RSFSR)
Avramov, Ivan - Biographies

KLEĬN, Aleksandr
Ivan Avramov. - Syktyvkar : Komi knizhnoe izdatel'stvo, 1980
126p : ill
r6462193 X.958/14976
8942447

ACTING. Theatre. Netherlands
Dalsum, Albert van - Biographies

PLEKKER, Jan
Albert van Dalsum, man van het toneel. - Zutphen : Walburg, [1983]
287p
90-601-1240-7 X.950/34237
8908737

ACTING. Theatre. Netherlands
Royaards-Sandberg, Jacqueline - Correspondence, diaries

ROYAARDS-SANDBERG, Jacqueline
[Correspondence. Selections]. Ik heb je zoveel te vertellen. - Baarn : Rap, 1981
xix,501p : ill
90-600-5164-5 X.950/16898
8908737

ACTING. Theatre. Netherlands
Ulsen, Henk van

DOUDART DE LA GRÉE, Marie-Louise
De speler. - Alphen aan den Rijn : GEKA, [1974]
155p,plates
90-617-6990-6 X.958/16899
8908737

ACTING. Theatre. Norway
Aabel, Per - Biographies

AABEL, Per
Den stundesløse Per Aabel. - 4. oppl. - Oslo : Gyldendal, 1981
210p,plates
82-05-12409-4 X.955/1737
8991456

ACTING. Theatre. Norway
Bjørnson, Bjørn

ERBE, Berit
Bjørn Bjørnsons vej mod realismens teater. - Bergen : Universitetsforlaget, 1976
339p : ill. - (Bergens Teatermuseums skrifter; 3)
82-00-01464-9 X.909/38659
9005269

ACTING. Theatre. Norway
Heide-Steen, Harald - Biographies

HEIDE-STEEN, Harald
Sett og hørt. - 4. oppl. - Oslo : Gyldendal, 1981
215p : ill
82-05-13008-6 X.950/24998
900808x

ACTING. Theatre. Norway
Maurstad, Tordis

BRODAL, Svein Erik
Tordis Maurstad. - Oslo : Det Norske samlaget, 1981
172p,plates
82-521-2056-3 X.950/28734
8991456

ACTING. Theatre. Oradea. Romania

NAGYVÁRADI színikritikák a Holnap évtizedében / Ady Endre, Biró Lajos, Dutka Ákos-és Juhász Gyula írásai a színházról ; válogatta, a bevezető tanulmányt írta és a jegyzeteket összeállította Indig Otto. - Bukarest : Kriterion Könyvkiadó, 1975
169p.,[4]p.of plates
963-07-0657-1 X.989/71542
1327941

ACTING. Theatre. Plzeň. Czechoslovakia

HOŘEC, Petr
Herecká ohlédnutí. - Plzeň : Západočeské nakladatelství, 1977
205p,[64]p of plates : ill
r6096422 X.989/78753
2024233

ACTING. Theatre. Poland
Jaracz, Stefan - Biographies

KRASIŃSKI, Edward
Stefan Jaracz. - Warszawa : Państwowy Instytut Wydawniczy, 1983
575p,plates
83-06-00542-2 X.950/32241
8363595

SZLETYŃSKI, Henryk
Stefana Jaracza wzloty i zabłądzenia. - Kraków : Wydawnictwo Literackie, 1984
197p. - Includes bibliography
83-08-01273-6 X.958/28894
8363595

ACTING. Theatre. Poland
Kaminska, Ruth Turkow - Biographies

KAMINSKA, Ruth Turkow
Mink coats and barbed wire. - London : Collins : Harvill Press, 1979
xv,255p,[12]p of plates : ill. - Includes index
0-00-262310-2 X.989/53689
1935720

ACTING. Theatre. Poland
Kotarbiński, Józef

KOTARBIŃSKA, Lucyna
[Correspondence. Selected correspondence]. Listy Lucyny i Józefa Kotarbińskich. - Warszawa : Państwowy Instytut Wydawniczy, 1978
301p,[32]p of plates : ill
r6186881 X.989/83721
221573X

ACTING. Theatre. Poland
Kotarbiński, Lucyna

KOTARBIŃSKA, Lucyna
[Correspondence. Selected correspondence]. Listy Lucyny i Józefa Kotarbińskich. - Warszawa : Państwowy Instytut Wydawniczy, 1978
301p,[32]p of plates : ill
r6186881 X.989/83721
2215748

ACTING. Theatre. Poland
Malkowska, Hanna - Biographies

MAŁKOWSKA, Hanna
Teatr mojego życia. - Łódź : Wydawnictwo Łódzkie, 1976
462 p., 24 p. of plates : ill
r6008391 X.989/34780
0856835

ACTING. Theatre. Poland
Modrzejewska, Helena - Biographies

SZCZUBLEWSKI, Józef
Żywot Modrzejewskiej. - Warszawa : Państwowy Instytut Wydawniczy, 1975
721 p., 1 leaf of plates, 24 p. of plates : ill. - (Żywoty)
r6014873 X.989/35199
0873373

ACTING. Theatre. Poland
Rudzki, Kazimierz - Biographies

WSPOMNIENIA o Kazimierzu Rudzkim / [redaktor Helena Głowacka]. - Warszawa : Czytelnik, 1981
417p : ill
83-07-00273-7 X.958/23086
8363595

ACTING. Theatre. Poland
Solska, Irena

SOLSKA, Irena
Pamiętnik. - Warszawa : Wydawnictwo Artystyczne i Filmowe, 1978
201p,[32]p of plates,[1] leaf of plates : ill
r6095362 X.989/78197
1981692

ACTING. Theatre. Poland
Solska, Irena - Biographies

KUCHTÓWNA, Lidia
Irena Solska. - Warszawa : Pavistwowy Instytut Wydawniczy, 1980
280p,[65]leaves of plates : ill
83-06-00121-4 X.950/15240
8363595

ACTING. Theatre. Poland
Wysocka, Stanisława - Biographies

WILSKI, Zbigniew
Wielka tragiczka. - Kraków : Wydawnictwo Literackie, 1982
352p[30]p of plates : ill
83-08-00576-4 X.958/13237
8363595

ACTING. Theatre. Romania
1945-1973

SĂCEANU, Amza
Fața văzută și nevăzută a teatrului. - București : Editura Eminescu, 1974
279 p. : ill. - (Colecția ""Masca")
r6007997 X.908/35058
0873829

ACTING. Theatre. Romania
Solomon Callimachi, Dida - Biographies

SOLOMON CALLIMACHI, Dida
Amintirile domnișoare Iulia. - București : Editura Cartea Românească, 1974
167p,16p of plates : ill
r6000699 X.908/35454
0890111

ACTING. Theatre. Russia
to 1917

MOROZOV, Mikhail
[Selections]. Izbrannoe. - Moskva : Iskusstvo, 1979
668p,[1]leaf of plates : ill. - Bibliography: p.656-666
r6497537 X.950/7827
873190x

ACTING. Theatre. Russia
Andreev-Burlak, Vasiliĭ Nikolaevich

NEL'S, S. (Sof'ia)
Andreev-Burlak. - Moskva : Iskusstvo, 1971
285p,[24]p of plates : ill. - (Korifei russkoi i zarubezhnoi sts̄eny)
r6448697 X.958/12095
8876401

ACTING. Theatre. Russia
IAkovlev, Alekseĭ Semenovich - Biographies

KULIKOVA, Kira Fedorovna
Alekseĭ ÎAkovlev. - Leningrad : Iskusstvo, Leningradskoe otdelenie, 1977
220p.[1]leaf of plates,[24]p of plates : ill. - (Seriia "Zhizn' v iskusstve")
r6088495 X.989/77312
185755X

ACTING. Theatre. Russia
Klimova, Aleksandra

BUR'ĨAN, Boris
Sud'ba chuzhaia-kak svoia. - Minsk : Mastatskaia litaratura, 1982
189p,[9]p.of plates : ill
r6448731 X.958/20308
8876401

ACTING. Theatre. Russia
Mochalov, Pavel Stepanovich

ALPERS, Boris Vladimirovich
Teatr Mochalova i Shchepkina. - Moskva : Iskusstvo, 1979
630p : ill
r6184170 X.989/84148
2245558

ACTING. Theatre. Russia
Orlenev, Pavel Nikolaevich - Biographies

MA+T# SKIN, Aleksandr Petrovich
Orlenev. - Moskva : Iskusstvo, 1977
380p,[47]p of plates : ill. - (Zhizn. i iskusstvo)
r6187284 X.989/79811
1990861

ACTING. Theatre. Russia
Savina, M. G - Biographies

SAVINA, M. G. (Mariia Gavrilovna)
Goresti i skitaniia (zapiski 1854-1877). - Izd. 2-e, dop.. - Leningrad : Iskusstvo, Leningradskoe otdelenie, 1983
221p,plates
r6448111 X.950/30963
8876401

ACTING. Theatre. Russia
Shchepkin, Mikhail Semenovich

ALPERS, Boris Vladimirovich
Teatr Mochalova i Shchepkina. - Moskva : Iskusstvo, 1979
630p : ill
r6184170 X.989/84148
2245566

ACTING. Theatre. Russia
Shchepkin, Mikhail Semenovich - Biographies

SENELICK, Laurence
Serf actor. - Westport, Conn. ; London : Greenwood, 1984
xvi,305p : ill. - (Contributions in drama and theatre studies ; no.9). - Bibliography: p283-290. - Includes index
0-313-22494-3 X.950/39123
2245566

ACTING. Theatre. Russia
Sumbatov-ÎUzhin, A. I.

CHKHIKVISHVILI, D. I.
Aleksandr Ivanovich Sumbatov-ÎUzhin. - Tbilisi : Izdatel'stvo Tbilisskogo universiteta, 1982
456p,plates
r6448286 X.950/30213
8876401

ACTING. Theatre. Russia
Vasil'ev, Pavel Vasil'evich - Biographies

MORDISON, Grigoriĭ Zalmanovich
Zhizn' aktera Pavla Vasil'eva .... - Leningrad : Iskusstvo, Leningradskoe otdelenie, 1972
199p,[40]p of plates. - (Korifei russkoĭ i zarubezhnoĭ sts̄eny)
r6174326 X.989/81862
2190575

ACTING. Theatre. Russia (RSFSR)
Gribov, Aleksei Nikolaevich - Biographies

ROGACHEVSKIĬ, Mikhail Lazarevich
Alekseĭ Gribov. - Moskva : Iskusstvo, 1979
262p,[1]leaf of plates,[30]p of plates : ill. - (Mastera sovetskogo teatra i kino)
r6184146 X.989/84077
2244276

ACTING. Theatre. Soviet Union
GARIN, Erast Pavlovich
S Meierkhol'dom. - Moskva : Iskusstvo, 1974
289p.[1]leaf of plates.[48]p of plates : ill
r6186516 X.989/81209
2191490

ACTING. Theatre. Soviet Union
1917-
ABALKIN, N. (Nikolai)
Dialog s akterom. - Moskva : Vserossiiskoe teatral'noe obshchestvo, 1982
317p
r6489170 X.950/31333
8580464

SOLOV'EV, Ivan Ivanovich
Po sobstvennomu opytu. - Moskva : Vserossiiskoe teatral'noe obshchestvo, 1982
205p.plates
r6489171 X.950/31329
8580464

SOVETSKOE akterskoe iskusstvo 50-70 gody / [otvetstvennyi redaktor D.N. Kaidalova]. - Moskva : Iskusstvo, 1982
301p.plates
r6463614 X.950/20996
8580464

ACTING. Theatre. Soviet Union
Dobronravov, B. G. - Biographies
BORIS Georgievich Dobronravov / [sostavitel', avtor letopisi i primechaniĭ V.S. Davydov]. - Moskva : Iskusstvo, 1983
207 p, [76] p. of plates : ports.
r6465767 X.950/38480
8711658

ACTING. Theatre. Soviet Union
Gaideburov, Pavel Pavlovich
GAĬDEBUROV, Pavel Pavlovich
Literaturnoe nasledie. - Moskva : Vserossiĭskoe teatral'noe obshchestvo, 1977
461p,[1]leaf of plates,[32]p of plates : ill
r6172808 X.989/80135
2119315

ACTING. Theatre. Soviet Union
Glumov, A. - Biographies
GLUMOV, A. (Aleksandr)
Nestertye stroki. - Moskva : Vserossiĭskoe teatral'noe obshchestvo, 1977
421p,[24]leaves of plates : ill
r6497504 X.950/4879
8711658

ACTING. Theatre. Soviet Union
IAkhontov, Vladimir Nikolaevich - Biographies
KRYMOVA, Natal'i͡a Anatol'evna
Vladimir I͡Akhontov. - Moskva : Iskusstvo, 1978
317p,[2]leaves of plates,[46]p of plates : ill. - (Zhizn' v iskusstve)
r6173256 X.989/80631
2120208

ACTING. Theatre. Soviet Union
Il'inskii, Igor' Vladimirovich - Biographies
IL'INSKIĬ, Igor' Vladimirovich
Sam o sebe. - Izd. 2., ispr. i dop. - Moskva : Iskusstvo, 1973
527p.,[64]p.of plates,[1]leaf of plates
r6097292 X.989/78615
2033054

ACTING. Theatre. Soviet Union
Konstantinov, P. A. - Biographies
RYZHOVA, V.
Petr Konstantinov. - Moskva : Vserossiiskoe teatral'noe obshchestvo, 1981
86p,[48]p of plates : ill. - (V tvorcheskoĭ laboratorii masterov sovetskoĭ s͡tseny)
r6462213 X.950/20050
8418225

ACTING. Theatre. Soviet Union
Ktorov, Anatolii Petrovich - Biographies
POLEZHAEVA, Elena Aleksandrovna
Anatoliĭ Ktorov. - Moskva : Iskusstvo, 1978
222p,[31]p of plates : ill. - (Mastera sovetskogo teatra i kino)
r6184013 X.907/20166
2195208

ACTING. Theatre. Soviet Union
Leshkovskai͡a, E. K. - Biographies
SLONOVA, N. (Nadezhda)
Leshkovskai͡a. - Moskva : Iskusstvo, 1979
270p,[32]p of plates : ill. - (Teatral'nye imena)
r6498288 X.958/3458
8418225

ACTING. Theatre. Soviet Union
Mordvinov, Nikolaĭ Dmitrievich - Biographies
PARFENOV, Lev Aleksandrovich
Nikolaĭ Mordvinov. - Moskva : Iskusstvo, 1981
199p,[30]p of plates. - (Mastera sovetskogo teatra i kino)
r6448936 X.958/15908
8418225

ACTING. Theatre. Soviet Union
Mysov, P.A. - Biographies
POPOV, Ėrast
Narodnyi artist P.A. Mysov. - Syktyvkar : Komi knizhnoe izdatel'stvo, 1981
118p : ill
r6448536 X.958/8624
8418225

ACTING. Theatre. Soviet Union
Polonskai͡a, Veronika
CHERTOK, Semen Markovich
Posledni͡ai͡a li͡ubov' Mai͡akovskogo. - Ann Arbor : Ėrmitazh, 1983
123p : ill
0-938920-31-6 X.958/20294
8418225

ACTING. Theatre. Soviet Union
Ryzhov, N. I.
RYZHOV, N. I. (Nikolaĭ Ivanovich)
Ryzhov o Ryzhovoĭ .... - Moskva : Vserossiĭskoe teatral'noe obshchestvo, 1984
405p : ill
r6490542 X.950/42329
8418225

ACTING. Theatre. Soviet Union
Ryzhova, V. N.
RYZHOV, N. I. (Nikolaĭ Ivanovich)
Ryzhov o Ryzhovoĭ .... - Moskva : Vserossiĭskoe teatral'noe obshchestvo, 1984
405p : ill
r6490542 X.950/42329
8418225

ACTING. Theatre. Soviet Union
Sadovskii, Mikhail Mikhailovich - Biographies
SADOVSKIĬ, Mikhail Mikhaĭlovich
Zapiski aktera. - Moskva : Iskusstvo, 1975
301p,[48]p of plates : ill
r6095217 X.989/80682
2120496

ACTING. Theatre. Soviet Union
Sashin-Nikol'skii, Aleksandr Ivanovich
A.I. Sashin-Nikol'skiĭ / [sostavitel' V.V. Temkina]. - Moskva : Vserossiĭskoe teatral'noe obshchestvo, 1979
271p,[46]p of plates : ill
r6184082 X.989/83808
2261200

ACTING. Theatre. Soviet Union
Sukharevskaia, Lidiia Pavlovna - Biographies
VLADIMIROVA, Zoi͡a Vladimirovna
Lidii͡a Sukharevskai͡a. - Moskva : Iskusstvo, 1977
263p,[30]p. of plates : ill. - (Mastera sovetskogo teatra i kino)
r6087949 X.907/18560
1856367

ACTING. Theatre. Soviet Union
Tarasova, Alla Konstantinovna - Biographies
ALLA Konstantinovna Tarasova / [redaktory-sostaviteli O.A. Radishcheva I E.A. Shingareva]. - Moskva : Iskusstvo, 1978
444p,[49]p of plates : ill
r6095101 X.989/77774
1973193

ACTING. Theatre. Soviet Union
T͡Sarev, Mikhail
KORZHEVICH, L. (Li͡udmila)
Put' aktera. - Moskva : Vserossiĭskoe teatral'noe obshchestvo, 1981
238p,[80]p of plates : ill
r6483496 X.950/14610
8418225

ACTING. Theatre. Soviet Union
Zakhava, Boris - Biographies
ZAKHAVA, Boris
[Selections]. Vospominanii͡a ; Spektakli i roli ; Stat'i. - Moskva : Vserossiĭskoe teatral'noe obshchestvo, 1982
397p,[64]p of plates : ill
r6484199 X.950/19623
8418225

ACTING. Theatre. Special subjects. American portrait photography, 1950-
Stars, 1940-1967
FEHL, Fred
Stars of the Broadway stage 1940-1967. - New York : Dover ; London : Constable, c1983
122p. - Includes index
0-486-24398-2 L.45/2873
3170977

ACTING. Theatre. Special subjects. English graphic arts
Doggett, Thomas. Authenticity
LEON, Walter
Thomas Doggett pictur'd. - [London] : The Company of Watermen & Lightermen of the River Thames, 1980
52p : ill
B81-39376 X.955/707
2569264

ACTING. Theatre. Special subjects. Gainsborough, Thomas. English portrait paintings
Garrick, David, 1717-1779. Authentication - Texts
CORRI, Adrienne
The search for Gainsborough. - London : Cape, 1984
278p. - Bibliography: p266-268. - Includes index
0-224-02162-1 X.429/16863
3113817

ACTING. Theatre. Sweden
De Wahl, Anders - Biographies
WENNERHOLM, Eric
Anders de Wahl. - [Stockholm] : Bonniers, 1974
230 p., 16 p. of plates : ill
91-00-39394-0 X.809/22490
0923230

ACTING. Theatre. Sweden
Kulle, Jarl - Biographies
WENNERHOLM
Jag Kulle. - Stockholm : Bonniers, 1979
309p : ill,ports
91-00-44120-1 X.989/85660
241693X

ACTING. Theatre. Tatar ASSR. Russia (RSFSR)
ILʹI͡ALOVA, Il'tani Iskhakovna
Akterskoe iskusstvo sovremennogo tatarskogo teatra. - Kazan' : Tatarskoe knizhnoe izdatel'stvo, 1978
141p : ill
r6097782 X.907/19071
2035243

ACTING. Theatre. Ukraine
Belousov, Mykhaĭlo - Biographies
SENNIKOVA, R. D. (Raisa Denisovna)
Mikhail Belousov. - Kiev : Mistet͡stvo, 1980
131p,plates. - (Mastera s͡tseny i ėkrana)
r6462298 X.629/20159
8942439

ACTING. Theatre. Ukraine
Zan'kovets'ka, Marii͡a
DURYLIN, Sergeĭ Nikolaevich
Marii͡a Zan'kovet͡skai͡a, 1854-1934. - Kiev : Mistet͡stvo, 1982
445p,32p.of plates : ill. - (Bibliotechnai͡a serii͡a)
r6448272 X.950/30209
8942439

ACTING. Theatre. United States
1837-1901 - Dictionaries
VICTORIAN actors and actresses in review / compiled and edited with an introduction by Donald Mullin. - Westport, Conn. ; London : Greenwood Press, 1983
xxxvi,571p : ill. - Bibliography: p526-529. - Includes index
0-313-23316-0 X.950/36370
3209091

ACTING. Theatre. United States
Booth, John Wilkes - Biographies
SAMPLES, Gordon
Lust for fame. - Jefferson ; London : McFarland, 1982
xii,238p : 1ill. - Bibliography: p191-195. - Includes index
0-89950-042-0 X.950/21397
2905647

ACTING. Theatre. United States
Menken, Adah Isaacs - Biographies
MANKOWITZ, Wolf
Mazeppa. - London : Blond & Briggs, 1982
270p : ill. - Bibliography: p259-261. - Includes index
0-85634-119-3 X.800/34697
2824558

ACTING. Theatre. United States
Quinn, Elizabeth – Biographies

QUINN, Elizabeth
Listen to me. – London : Joseph, 1984
xii,208p,[8]p of plates : ill
0-7181-2288-7 X.950/34761
3173151

ACTING. Theatre. Uzbekistan
Burkhanov, Shukur – Biographies

RAKHMANOV, Mamadzhan
Shukur Burkhanov. – Tashkent : Izdatel'stvo literatury i iskusstva im. Gafura Guli͡ama, 1974
171p,plates
r6465433 X.958/29125
9065911

ACTING. Theatre. Uzbekistan
Khodzhaev, Alim

FEL'DMAN, ÎA. S. (ÎAkov Solomonovich)
Besedy S Alimom Khodzhaevym, vydai͡ushchimsi͡a uzbekskim akterom. – Tashkent : Izdatel'stvo literatury i isskustva imeni Gafura Guli͡ama, 1981
158p
r6489155 X.958/21040
9065911

ACTING. Ukraine
Matveev, Evgeniĭ – Biographies

ISAEVA, K. (Klara)
Evgeniĭ Matveev. – Moskva : Iskusstvo, 1980
189p,[31]p of plates : ill. – (Mastera sovetskogo teatra i kino)
r6484295 X.958/14938
8916861

ACTING. United States

COHEN, Robert, 1938–
Acting professionally. – 3rd ed. – New York ; London : Barnes & Noble, 1983, c1981
xiii,122p
0-06-463570-8 X.950/22641
2932652

KALTER, Joanmarie
Actors on acting. – New York : Sterling ; London : Oak Tress Press : Distributed by Ward Lock, 1979
256p. – Includes index
0-7061-2660-2 X.981/22313
2151871

ACTING. United States
Strasberg, Lee

ADAMS, Cindy
Lee Strasberg. – New York : Doubleday, 1980
398p,[20]p of plates
0-385-12496-1 X.629/19966
891687x

ACTING. Vienna. Austria
Bardach, Emilie. Interpersonal relationships with Ibsen, Henrik

BARDACH, Emilie
[Tagebuch. T.1, 1889]. Nova über Henrik Ibsen und sein Alterswerk. – Oslo : Edizione "a", 1977
94p : ill
r6060852 X.950/7043
1790544

ACTING. West German cinema films
Schygulla, Hanna – Biographies

SCHYGULLA, Hanna
Hanna Schygulla. – München : Schirmer / Mosel, 1981
189p : ill(some col.)
3-921375-83-5 X.950/19680
892483x

ACTING. West Germany
Frank, Horst – Biographies

FRANK, Horst
Leben heisst Leben. – München : Herbig, c1981
343p,plates
3-7766-1169-3 X.950/23246
8725640

ACTING. West Germany
Juhnke, Harald – Biographies

JUHNKE, Harald
Die Kunst ein Mensch zu sein. – 2. Aufl. – München : Herbig, 1980
352p.,[32]p.of plates
3-7766-1043-3 X.950/6399
8725640

ACTING. West Germany
Lüders, Günther – Biographies

GÜNTHER Lüders / zusammengestellt von Hans-Gerd Kästner und Wolf-Rüdiger Ohlhoff. – Lübeck : Amt für Kultur des Senats, 1985
94p : ill. – (Veröffentlichungen des Senats der Hansestadt Lübeck. Amt für Kultur. Reihe B ; Hft.5)
3-924214-26-3 X.0805/188(5)
8725640

ACTINIDE COMPOUNDS
Organic actinide compounds – Conference proceedings

ORGANOMETALLICS of the f-elements / edited by Tobin J Marks and R. Dieter Fischer. – Dordrecht ; London : Reidel, c1979
xv,517p : ill. – (Nato advanced study institutes series. Series C, Mathematical and Physical Sciences ; v.44). – Includes index
90-277-0990-4 X.622/9976 Woolwich
2388200

ACTINIDE COMPOUNDS
Physical properties

ERDÖS, Paul
The physics of actinide compounds. – New York ; London : Plenum, 1983
xi,213p : ill. – (Physics of solids and liquids). – Includes index
0-306-41150-4 SRL
2992329

ACTINIDE ELEMENTS
See also
Americium
Curium
Plutonium
Transuranium elements
Uranium

ACTINIDE ELEMENTS
– Conference proceedings

ACTINIDES – 1981 CONFERENCE (Pacific Grove)
Actinides in perspective. – Oxford : Pergamon, 1982
ix,610p : ill. – Includes bibliographies and index
0-08-029193-7 SRL
1839829

HEAVY element properties / editors W. Müller, H. Blank. – Amsterdam ; Oxford : North-Holland Publishing Co. [etc.], 1976
vii,125p : ill. – Includes index
0-7204-0402-9 SRL
1839829

INTERNATIONAL CONFERENCE ON PLUTONIUM AND OTHER ACTINIDES
(5th : 1975 : Baden Baden)
Plutonium 1975 and other actinides. – Amsterdam ; Oxford : North-Holland Publishing Co. [etc.], 1976
xviii,978p : ill. – Includes index
0-7204-0404-5 SRL
1839829

ACTINIDE ELEMENTS
Chemical properties

The CHEMISTRY of the actinides / S. Ahrland ... [et al.]. – Oxford : Pergamon, 1975
635,xix p : ill. – (Pergamon international library) ([Pergamon texts in inorganic chemistry ; vol.10])
0-08-018794-3 X.620/15145
0374067

ACTINIDE ELEMENTS
Electrons. Properties

The ACTINIDES / edited by A.J. Freeman, J.B. Darby, Jr. – New York ; London : Academic Press. – (Materials science and technology)
Vol.1. – 1974
xxi,360p : ill. – Includes bibliographies and index
0-12-266701-8 SRL
0505676
Vol.2. – 1974
xiv,386p : ill. – Includes bibliographies and index
0-12-266702-6 SRL
0505676

ACTINOBACILLUS. Pathogens. Mammals
– Conference proceedings

HAEMOPHILUS influenzae / edited by Sarah H. Sell and Peter F. Wright. – New York ; Oxford : Elsevier Biomedical, c1982
xix,321p : ill. – Includes index
0-444-00683-4 SRL
2609819

HAEMOPHILUS, pasteurella and actinobacillus / edited by M. Kilian, W. Frederiksen, E.L. Biberstein. – London : Academic Press, 1981
xiv,294p : ill. – Includes index
0-12-406780-8 SRL
2609819

ACTINOMYCETES
See also
Nocardiae

ACTINOZOA
See Anthozoa

ACTION AT A DISTANCE. Particles
Relativistic theories

BROWN, G. Burniston
Retarded action-at-a-distance. – Luton : Cortney, 1982
vi,145p : ill. – Includes index
0-904378-14-4 X.629/18145
0350443

The THEORY of action-at-a-distance in relativistic particle dynamics / edited by Edward H. Kerner. – New York ; London : Gordon and Breach, 1972
xxi,209p : ill. – (International science review series ; vol.11)
0-677-13990-x X.620/8267 Woolwich
0350443

ACTION CATHOLIQUE OUVRIÈRE

DEBÈS, Joseph
Naissance de l'Action catholique ouvrière. – Paris : Les Éditions ouvrières, [1982]
285p. – (Mémoires du futur)
2-7082-2262-7 X.800/36976
9056610

ACTION-CENTRED LEADERSHIP. Organisations

ADAIR, John, 1934–
Action-Centred Leadership. – Farnborough, Hants. : Gower Press, 1979
xi,186p : ill. – Bibliography: p.179–180. – Includes index
0-566-02143-9 X.520/14084
1907999

ACTION FRANÇAISE
to 1928

TROFIMENKOFF, Susan Mann
Action française. – Toronto : University of Toronto Press, 1975
xi,157p. – Bibliography : p.121–122. – Includes index
r6043190 X.800/26102
1116088

ACTION FRANÇAISE
Relations with syndicalism, 1870–1940

MAZGAJ, Paul
The Action française and revolutionary syndicalism. – Chapel Hill : University of North Carolina Press, [1979]
ix,281p
0-8078-1316-8 X.529/54855
8964041

ACTION GROUPS
See also
Community action groups
Ecology action groups
Housing action groups
Squatters' action groups

ACTION GROUPS. Norway

KOLBENSTVEDT, Marika
Lokale aksjonsgrupper. – Oslo : Norsk institutt for by- og regionforskning, 1978
100p : ill. – (NIBR rapport ; 44)
r6060234 X.0802/100.(44)
1542133

ACTION GROUPS. Norway
Organisation

HVINDEN, Bjørn
Grasrothåndbok. – Oslo : Pax, [1974]
141 p. : ill. – (Pax-bok ; 406)
82-530-0631-4 X.708/19816
086997X

ACTION GROUPS. Reorganisation. Secondary education. Ferguslie Park. Paisley. Strathclyde Region. Scotland

ROBERTSON, H. B.
The Schools Action Group. – [Paisley] : Community Development Project, Ferguslie Park ; [Glasgow] : [Distributed by Administration Department, Strathclyde Regional Council], 1977
[2],10 leaves
B80-25453 X.525/3301
2144689

ACTION LEARNING. Teaching methods. Training. Managers

ACTION learning in practice / edited by Mike Pedler. – Aldershot : Gower, c1983
xvi,297p : ill. – Bibliography: p277–282. – Includes index
0-566-02383-0 X.529/57669
2900742

ACTION LEARNING. Teaching methods. Training. Managers -continued

ASHMAWY, Saad
The Nile project. - [Southport] : Action Learning Projects International, [1971]
ii,42p : 2ill
r6810586 X.520/31187
2900742

BAQUER, Ali
"But surely, that is their job?". - Southport : Action Learning Projects International, 1973
71p : 1ill. - Bibliography: p70-71
r6810588 X.520/31185
2900742

CORTAZZI, Diana
Illuminative incident analysis. - [Southport] : Action Learning Projects International, [1973]
16p : ill
r6810585 X.520/31188
2900742

MUSSCHOOT, F.
Action learning in small enterprises. - [Southport] : Action Learning Projects International, [1973]
19p
r6810587 X.520/31186
2900742

PART time action learning in the IWM. - Luton : Institution of Works Managers, 1978
6p
B83-23839 X.529/54188
2900742

REVANS, R. W. (Reginald William)
Studies in action learning. - [Altrincham] : [R.W. Revans], [1983]
31p
B84-23730 X.622/19316
2900742

ACTION MINIMA. Numerical solution. Equations. Two-dimensional Euclidean scalar fields

SEN, Siddharta
Action minima among solutions to the two dimensional Euclidean [phi]4 scalar field equations. - Dublin : Trinity College, Dublin, School of Mathematics, [1978]
15,[3]leaves : 1 ill
B80-41708 X.622/9073 Woolwich
2315521

ACTION PHOTOGRAPHY

HODGSON, David, 1942-
All about action photography. - London : Pelham, 1976
190p : ill. - Includes index
0-7207-0888-5 X.629/10450 Woolwich
0623636

MORLEY, Don
The focalguide to action photography. - London : Focal Press [etc.], 1978
189p : ill(some col). - Includes index
0-240-50984-6 WP.13581/117
0623636

ACTION PHOTOGRAPHY
Techniques

GREAT action photography / edited by Bryn Campbell. - London : Ebury, 1983
160p : ill(some col.). - Includes index
0-85223-325-6 X.421/25137
3078167

ACTION RESEARCH
Applications in improvement of efficiency in manufacturing industries - Conference proceedings

NATO ADVANCED RESEARCH INSTITUTE ON THE EFFICIENCY OF MANUFACTURING SYSTEMS (1982 : Amsterdam)
Efficiency of manufacturing systems. - New York ; London : Published in cooperation with NATO Scientific Affairs Division [by] Plenum, c1983
ix,336p : ill. - (NATO conference series. II, Systems science ; v.14). - Includes bibliographies and index
0-306-41256-x X.520/31529
2968568

ACTION RESEARCH
Applications in management of irrigation of agricultural land

BOTTRALL, Anthony
The action research approach to problem solving, with illustrations from irrigation management. - London : Overseas Development Institute, 1982
20p. - (ODI working paper ; no.9). - Bibliography: p19-20
0-85003-083-8 X.525/6155
2776774

Action research towards improved water distribution. - London : Overseas Development Institute, Agricultural Administration Unit, [1981]
20p. - (Network paper ; 1/81/2)
B84-30331 X.329/17995
2776774

ACTION RESEARCH
Role in formulations of government policies on social planning in urban regions - Conference proceedings

The USE of action planning in developing urban planning policy. - London : Department of the Environment, Directorate of Planning Intelligence, [1976]
[3],127p : ill
0-7184-0057-7 BS.414/285
0701238

ACTION RESEARCH. Schools
- For teachers

A TEACHER'S guide to action research / edited by Jon Nixon - London : Grant McIntyre, 1981
209p : ill
0-86216-040-5 X.529/41748
2497026

ACTION RESEARCH. Social services. South-east London

SOUTH EAST LONDON CONSORTIUM. Action Research Team
The Action Research experiment. - London : The Consortium, 1978
[5],129p
B78-17284 X.519/27969
1449095

ACTION RESOURCE CENTRE

ACTION RESOURCE CENTRE
ARC assessment study. - [London?] : The Centre, 1976
8p
B83-12757 X.529/52478
2851342

ACTION RESOURCE CENTRE
- Serials

[ARC news (London : 1979)]. ARC news / Action Resource Centre. - No.14 (Oct.1979)-. - London : The Centre, 1979-
v. : ill
B81-04865 P.523/1415
2342715

ARCLIGHT. - Apr.1976-July 1979. - London : The Centre, 1976-1979
[13?]v. : ill
Holdings : Jan. 1977, etc
B81-04864 P.523/1415
2342715

ACTION SONGS
Children's action songs - Collections

NELSON, Esther L.
Musical games for children of all ages. - New York : Sterling [etc.] ; London : Distributed by Ward Lock, 1977
72p : ill. - Includes index
0-7061-2520-7 Cup.1281/577
1186051

ACTION SONGS IN ENGLISH
Children's action songs in English - Collections

ACTION rhymes / [compiled] by Dorothy Taylor ; with illustrations by Kathy Layfield & Brian Price Thomas and photographs by John Moyes. - Loughborough : Ladybird Books, 1974
52p : col ill. - (Learning with traditional rhymes ; book 5)
0-7214-0439-1 X.439/8179
072534x

SINGING rhymes / [compiled] by Dorothy Taylor ; with illustrations by Brian Price Thomas and photographs by John Moyes. - Loughborough : Ladybird Books, 1979
52p : chiefly col ill. - (Learning with traditional rhymes ; book 7)
0-7214-0537-1 X.439/8622
072534x

ACTION SONGS IN ENGLISH
Children's action songs in English - For parents - Collections

LANGUAGE through song. - [Twickenham] : [Cameron Forbes Nedale], [1983]
17p : ill(some col.)
B84-00552 Cup.580.h.35
2980681

ACTION SONGS IN ENGLISH
Children's action songs in English: Caribbean songs. Words - Anthologies

BENJAMIN, Joe
A musical journey for littleys. - [London] : Thomas Joseph, c1982
31p : ill
B82-27243 X.992/4845 Woolwich
2749343

ACTION SONGS IN ENGLISH
Children's action songs in English. Words - Anthologies

EVANS, Delphine
Fingers, feet and fun!. - London : Hutchinson, 1984
106p : ill. - Includes index
0-09-156040-3 X.990/23399
0745774

OKKI - tokki - unga / chosen by Beatrice Harrop ; drawings by David McKee. - Words ed. - London : A. and C. Black, 1976
[64]p : ill. - Includes index
0-7136-1683-0 X.439/8193
0745774

OKKI-TOKKI-UNGA / chosen by Beatrice Harrop ; with piano accompaniments, chords for guitar, actions by Linda Friend and drawings by David McKee. - London : A. and C. Black, 1976
[96] p. : ill. - Includes index
0-7136-1685-7 X.435/538
0745774
- Schools ed.
ISBN 0-7136-1684-9 X.435/539

ACTION SPACE. England
to 1982

TURNER, Mary, 1933-
Play, education and the arts. - London : NPFA, c1983
[16]p : ill
0-946085-02-1 X.529/58421
3038858

ACTION THEORY. Sociology

BURT, Ronald S.
Toward a structural theory of action. - New York ; London : Academic Press, 1982
xv,381p : ill. - (Quantitative studies in social relations). - Bibliography: p357-353. - Includes index
0-12-147150-0 X.520/28969
130934x

PARSONS, Talcott
Action theory and the human condition. - New York : Free Press ; London : Collier Macmillan, c1978
xi,464p : ill. - Bibliography: p434-450. - Includes index
0-02-923990-7 X.520/23214
130934x
Social systems and the evolution of action theory. - New York : Free Press ; London : Collier Macmillan, 1977
xi,420p : ill. - Includes bibliographies and index
0-02-924800-0 X.520/12170
130934x

SCHUTZ, Alfred
The theory of social action. - Bloomington ; London : Indiana University Press, 1978
xxvi,145p. - (Studies in phenomenology and existential philosophy). - Includes index
0-253-35957-0 X.520/13360
177932x

ACTION THEORY. Sociology
to date. Philosophical aspects

COLLECTIVE behavior and social movements / Louis E. Genevie, editor. - Itasca, Ill. : F.E. Peacock Publishers, c1978
xxii,531p. - Includes bibliography
0-87581-228-7 X.520/37247
9100199

PHILOSOPHISCHE Probleme der Handlungstheorie / Beiträge von Holger van den Boom ... [et al.] ; herausgegeben von Hans Poser. - Freiburg [im Breisgau] : Alber, 1982
361p. - (Praktische Philosophie ; Bd.17)
3-495-47488-9 X.529/62963
9100199

ACTION THEORY. Sociology
Theories of Parsons, Talcott

ADRIAANSENS, Hans P. M.
Talcott Parsons and the conceptual dilemma. - London : Routledge and Kegan Paul, 1980
vii,200p : ill. - (International library of sociology). - Bibliography: p.189-197. - Includes index
0-7100-0519-9 WP.8084/447
2070804

ACTIONS
See also
Symbolic actions

ACTIONS
Role in human cognition

Le RÔLE de l'action dans la formation des connaissances / editors: A. Morf, L. Apostel, M. de Mey. - Ghent : Communication & Cognition, [1982]
156p. - (Studies in action theory)
r6330830 X.529/62157
9091351

ACTIONS. Law
See Legal proceedings

ACTIONS. Man

The ANALYSIS of action / edited by Mario von Cranach and Rom Harré. - Cambridge : Cambridge University Press, 1982
404p : ill. - (European studies in social psychology). - Includes bibliographies and index
0-521-24229-0 X.520/28615
2382342

ARNÁIZ, José María
Antropología del obrar humano. - Fribourg : Facultad de Letras de la Universidad de Fribourg, 1971
295,72,24p. - Includes bibliographies
r6037973 X.512/2546
0712582

BRAND, Myles
Intending and acting. - Cambridge, Mass. ; London : MIT, c1984
xiii,296p : ill. - (A Bradford book). - Bibliography: p279-288. - Includes index
0-262-02202-8 X.520/36208
0712582

DAVIS, Lawrence H.
Theory of action. - Englewood Cliffs ; London : Prentice-Hall, 1979
viii,152p. - (Prentice-Hall foundations of philosophy series). - Bibliography: p.142-147. - Includes index
0-13-913152-3 X.519/29460
0712582

GOLDMAN, Alvin Ira
A theory of human action. - Princeton ; Guildford : Princeton University Press, 1976
[1],xi,230p : ill. - Includes index
0-691-07216-7 X.519/26768
0712582

HAMPSHIRE, Stuart
Thought and action. - New ed. - London : Chatto & Windus, 1982
298p. - Includes index
0-7011-2604-3 X.529/50369
0712582

HOLLIS, Martin
Models of man. - Cambridge : Cambridge University Press, 1977
vii,198p : ill. - Bibliography: p.191-195. - Includes index
0-521-21546-3 X.529/31873
0712582

HORNSBY, Jennifer
Actions. - London : Routledge and Kegan Paul, 1980
vi,146p. - (International library of philosophy). - Bibliography: p.138-143. - Includes index
0-7100-0451-6 X.989/88254
0712582

JOHN PAUL II, Pope
The acting person. - Dordrecht ; London : Reidel, c1979
xxiii,367p. - (Analecta Husserliana ; v.10). - Includes index
90-277-0985-8 X.200/35343
0712582

PRAXIOLOGICAL studies / edited by Wojciech Gasparski and Tadeusz Pszczołowski. - Dordrecht ; London : Reidel, c1983
xiv,418p : ill. - (Theory and decision library ; v.34). - Includes bibliographies and index
90-277-1258-1 X.629/22536
0712582

REID, Thomas
Essays on the active powers of man, 1788. - New York ; London : Garland, 1977
[1],vii,493p. - (British philosophers and theologians of the 17th & 18th centuries)
0-8240-1802-8 X.529/33311
0712582

ROSENBERG, Harold
Act and the actor. - Chicago ; London : University of Chicago Press, 1983, c1970
ix,214p. - Includes index
0-226-72675-4 X.958/19507
0712582

THALBERG, Irving
Perception, emotion and action. - Oxford : Blackwell, 1977
[8],142p. - (Library of philosophy and logic). - Bibliography: p.131-137. - Includes index
0-631-17600-4 X.529/31707
0712582

THOMSON, Judith Jarvis
Acts and other events. - Ithaca ; London : Cornell University Press, 1977
274,[1]p : ill. - (Contemporary philosophy). - Includes index
0-8014-1050-9 WP.14759/12
0712582

TUOMELA, Raimo
A theory of social action. - Dordrecht ; Lancaster : Reidel, 1984
x,534p : ill. - (Synthese library ; 171). - Bibliography: p500-510. - Includes index
90-277-1703-6 X.0529/4(152)
0712582

ACTIONS. Man
Causes

DAVIDSON, Donald, 1917-
Essays on actions and events. - Oxford : Clarendon, 1980
xvi,304p. - Bibliography: p291-299. - Includes index
0-19-824529-7 X.529/67347
2071975

ACTIONS. Man
Causes. Theories of Davidson, Donald, 1917-

ESSAYS on Davidson / edited by Bruce Vermazen and Merrill B. Hintikka. - Oxford : Clarendon, 1985
viii, 57p. - Includes index
0-19-824749-4 X.950/42038
3153363

ACTIONS. Man
Decision making

CASTAÑEDA, Hector Neri
Thinking and doing. - Dordrecht : Reidel, 1975
xviii,366p : ill. - (Philosophical studies series in philosophy ; vol.7)
90-277-0610-7 X.0529/321(7)
1194720

ACTIONS. Man
Explanation - Conference proceedings

ACCOUNTS and action / [edited by] G. Nigel Gilbert, Peter Abell. - Aldershot : Gower, c1983
vii,190p : ill. - (Surrey conferences on sociological theory and method ; 1). - Includes bibliographies and index
0-566-00620-0 X.529/64623
1259652

WINNIPEG CONFERENCE ON HUMAN ACTION (1975)
Action theory. - Dordrecht : Reidel, 1976
vi,345p. - (Synthese library ; vol 97). - Bibliography-Includes index
90-277-0671-9 X.0529/4(133)
1259652

ACTIONS. Man
Norms

RAZ, Joseph
Practical reason and norms. - London : Hutchinson, 1975
192p. - (Hutchinson university library). - Includes index
0-09-123981-8 WP.1413/434
0565164

ACTIONS. Man
Philosophical perspectives, 1845-1971

BERNSTEIN, Richard J. (Richard Jacob)
Praxis and action. - London : Duckworth, 1972 [i.e. 1976]
xv,344p. - Bibliography: p.321-333. - Includes index
0-7156-0779-0 X.519/25756
0645710

ACTIONS. Man
Psychological aspects

HARRÉ, Rom
Motives and mechanisms. - London : Methuen, 1985
x,161p : ill. - Bibliography: p149-154. - Includes index
0-416-36230-3 X.529/69816
1920219
- [Another copy: paperback] ISBN 0-416-36240-0 X.529/69817

PHILOSOPHICAL problems in psychology / edited by Neil Bolton. - London : Methuen, 1979
xiii,207p : ill. - (Psychology in progress). - Includes bibliographies and index
0-416-70980-x X.529/33636
1920219

ACTIONS. Man
Reasons

BOND, E. J.
Reason and value. - Cambridge : Cambridge University Press, 1983
ix,166p. - (Cambridge studies in philosophy). - Bibliography: p162-163. - Includes index
0-521-24571-0 X.529/54519
1393219

PARFIT, Derek
Reasons and persons. - Oxford : Clarendon, 1984
xv,543p : ill. - Bibliography: p533-540. - Includes index
0-19-824615-3 X.520/33223
1393219

TUOMELA, Raimo
A causal theory of complex actions. - [Helsinki] : Helsingin yliopiston filosofian laitos, 1975
57p. - (Helsingin yliopiston filosofian laitoksen julkaisuja ; 1975, no.1)
951-450-706-1 X.0512/385(11)
1393219
Human action and its explanation. - Helsinki : Helsingin yliopiston filosofian laitos, 1974
115,4 leaves. - (Helsingin yliopiston filosofian laitoksen julkaisuja ; 1974, no.2)
951-450-427-5 X.0512/385(8)
1393219

ACTIONS. Man
Reasons. Moral aspects

WESTON, Michael, 19—
Morality and the self. - Oxford : Blackwell, 1975
vii,95p. - (Library of philosophy and logic). - Includes index
0-631-15700-x X.529/30213
0511471

ACTIONS. Man
related to cognition

The EXPRESSION of knowledge / edited by Robert L. Isaacson and Norman E. Spear. - New York ; London : Plenum, c1982
xviii,423p : ill. - Includes index
0-306-40927-5 X.520/29302
1211986

LINHART, Josef
Činnost a poznávání. - Praha : Academia, 1976
574p. - Bibliography : p. 520-552
r6015408 X.325/144 Woolwich
1211986

ACTIONS. Man
related to cognition - Conference proceedings

INTERNATIONAL CONGRESS OF PSYCHOLOGY (22nd : 1980 : Leipzig)
Cognitive and motivational aspects of action. - Amsterdam ; Oxford : North-Holland, 1982
173p : ill. - Includes bibliographies
0-444-86348-6 X.520/30208
1211986

ACTIONS. Man
related to contemplation in religion

CONTEMPLATION and action in world religions / edited by Yusuf Ibish and Ileana Marculescu. - [Houston] : Rothko Chapel ; Seattle, c1978
274p
0-295-95634-8 X.200/35111
1585649

ACTIONS. Man
related to contemplation in religion - Conference proceedings

TRADITIONAL modes of contemplation and action / edited by Yusuf Ibish and Peter Lamborn Wilson. - Tehran : Imperial Iranian Academy of Philosophy, 1977
[5],viii,[i.e.ix],477p. - (Publication ; no.24)
r6072224 X.200/31527
1585649

ACTIONS. Man
Responsibility

HARRIS, John, 1945 Aug.21-
Violence and responsibility. - London : Routledge and Kegan Paul, 1980
viii,177p. - Bibliography: p.166-174. - Includes index
0-7100-0448-6 X.529/38722
2162202

ACTIONS. Man
Role of reasoning

AUNE, Bruce
Reason and action. - Dordrecht : Reidel, 1977
xi,206p. - (Philosophical studies series in philosophy ; vol.9). - Bibliography: p.198-201. - Includes index
90-277-0805-3 X.0529/321(9)
0682950

CASTAÑEDA, Hector-Neri
Thinking and doing. - Dordrecht ; London : Reidel, 1975
xviii,366p. - (Pallas paperbacks). - Includes index
90-277-1375-8 X.529/45884
0682950

DREHER, Eva
Handlungsplanung als Komponente kognitiver Sozialisation. - [Augsburg] : [E. Dreher], 1980
v,287p. - Bibliography
r6305166 X.529/47250
8739897

EBELING, Hans
Die ideale Sinndimension. - Freiburg : Alber, 1982
128p.. - (Alber-Broschur Philosophie)
3-495-47502-8 X.529/63515
8739897

ACTIONS. Man
Role of reasoning–continued

KÖRNER, Stephan
Experience and conduct. – Cambridge : Cambridge University Press, 1976
ix,268p. – Bibliography: p.262–263. – Includes index
0–521–21075–5 X.520/10825
0682950

KÖRNER, Stephen
Experience and conduct. – Cambridge : Cambridge University Press, 1976
ix,268p. – Bibliography: p262–263. – Includes index
0–521–29943–8 X.529/38978
0682950

MILLIGAN, David, 1930–
Reasoning and the explanation of actions. – Brighton : Harvester, 1980
xi,194p. – ([Studies in philosophy ; no.23]). – Includes index
0–85527–433–6 X.529/39689
1402765

PERSSON, Ingmar
Reasons and reason-governed actions. – Lund : [The University], 1981
194p. – (Lund monographs in practical philosophy ; 4). – Bibliography
91–7222–360–x X.529/42466
8739897

PRACTICAL reasoning / edited by Joseph Raz. – Oxford : Oxford University Press, 1978
[5],202p. – (Oxford readings in philosophy). – Bibliography: p.199–201. – Includes index
0–19–875041–2 X.519/29500
1402765

RATIONAL action / edited by Ross Harrison. – Cambridge : Cambridge University Press, 1979 [i.e. 1980]
xii,176p : ill. – Bibliography – Includes index
0–521–22714–3 X.520/20915
0682950

ACTIONS. Man
Role of rules

HANSON, Bo
Application of rules in new situations. – [Lund] : Gleerup, 1977
159p. – (Studia philosophiae religionis ; 3)
91–400–4537–4 X.0100/617(3)
8933707

ACTIONS. Man
Sociobiological aspects

REYNOLDS, Vernon
The biology of human action. – 2nd ed.. – Oxford : W.H. Freeman, 1980
xxiv,303p : ill. – Bibliography: p278–291. – Includes index
0–7167–1239–3 X.325/862 Woolwich
2359863

ACTIONS. Man
Teleological explanation

TAYLOR, Richard, 1919–
Action and purpose. – Atlantic Highlands : Humanities Press ; Brighton : Harvester Press, 1980
xiv,269p. – Includes index
0–85527–187–6 X.529/36046
2149842

ACTIONS. Man
Theories of Aristotle

CHARLES, David, 19—
Aristotle's philosophy of action. – London : Duckworth, 1984
xi,282p. – Bibliography: p266–271. – Includes index
0–7156–1005–8 X.520/36470
3250849

ACTIVATED SLUDGE
Activated sewage sludge. Bulking – Conference proceedings

BULKING of activated sludge / editors B. Chambers and E.J. Tomlinson. – Chichester : Published by Ellis Horwood for the Water Research Centre, 1982
279p : ill. – Includes index
0–85312–350–0 SRL
2697807

ACTIVATION ANALYSIS
Neutron activation analysis. Applications in agriculture

CAMPBELL, J. A. (Jacqueline Ann)
Neutron activation analysis. – Slough : Commonwealth Agricultural Bureaux, 1978
36p. – (Special publication / Commonwealth Bureau of Soils ; no.7) – Bibliography: p20–36
0–85198–438–x X.322/11903
2923211

ACTIVATION ANALYSIS
Non-destructive neutron activation analysis

NONDESTRUCTIVE activation analysis / edited by Saadia Amiel. – Amsterdam ; Oxford : Elsevier Scientific, 1981
xvi,369p : ill. – (Studies in analytical chemistry ; 3). – Includes index
0–444–41942–x SRL
2388278

ACTIVATION ANALYSIS. Chemical analysis. Trace elements

VALKOVIĆ, Vlado
Nuclear microanalysis. – New York ; London : Garland Publishing, 1977
xii,415p : ill. – (Garland reference library of science and technology ; vol.10). – Bibliography: p.368–373. – Includes index
0–8240–9898–6 SRL
1398121

ACTIVE FILTERS

BOWRON, P.
Active filters for communications and instrumentation. – London : McGraw-Hill, 1979
viii,285p,2p of plates : ill(incl 2 col). – Includes index
0–07–084086–5 SRL
0209759
– [Another copy : paperback]
ISBN 0–07–084085–7 X.629/13542 Woolwich

ACTIVE FILTERS
Design

DARYANANI, Gobind
Principles of active network synthesis and design. – New York ; London : Wiley, 1976
xv,495p : ill. – Includes bibliographies and index
0–471–19545–6 X.620/10586 Woolwich
057600x

JOHNSON, David Edsel
Rapid practical designs of active filters. – New York ; London : Wiley-Interscience, 1975
xi,264p : ill. – Includes index
0–471–44304–2 SRL
057600x

VAN VALKENBURG, M. E. (Mac Elwyn)
Analog filter design. – New York ; London : Holt, Rinehart and Winston, c1982
xi,608p : ill. – (HRW series in electrical and computer engineering). – Bibliography: p601–604. – Includes index
0–03–059246–1 SRL
057600x
– [Another copy: paperback.]
ISBN 4–8338–0091–3 X.629/18567

WILLIAMS, Arthur B. (Arthur Bernard)
Active filter design. – London : Adtech, 1978, c1975
189p : ill. – (Modern frontiers in applied science). – Bibliography: p186. – Includes index
0–89006–044–4 X.622/16621
057600x

ACTIVE FILTERS
Design. Applications of minicomputer systems & pocket programmable electronic calculators

MOSCHYTZ, G. S. (George Samson)
Active filter design handbook. – Chichester : Wiley, c1981
viii,316p : ill. – Bibliography: p177–179. – Includes index
0–471–27850–5 SRL
2158884

ACTIVE GAMES

HINDMAN, Darwin Alexander
Kick the can. – Englewood Cliffs ; London : Prentice-Hall, [1979]
[4],430p : ill. – Includes index
0–13–515163–5 X.629/13490 Woolwich
2068532

ACTIVE HEPATITIS. Liver. Man
Chronic active hepatitis

CHRONIC active liver disease / edited by Sidney Cohen and Roger D. Soloway. – New York ; Edinburgh : Churchill Livingstone, 1983
xvi,266p : ill. – (Contemporary issues in gastroenterology ; v.2). – Includes index
0–443–08223–5
2913488

ACTIVE NUCLEI. Galaxies
– Conference proceedings

ACTIVE galactic nuclei / edited by C. Hazard and S. Mitton. – Cambridge : Cambridge University Press, 1979
vii,317p : ill. – (Cambridge astrophysics series) (NATO Advanced Study Institutes). – Includes bibliographies and index
0–521–22494–2 SRL
200111x

ACTIVE OXYGEN. Organisms
– Conference proceedings

BIOCHEMICAL and medical aspects of active oxygen / edited by Osamu Hayaishi, Kozi Asada. – Baltimore ; London : University Park Press, 1977
xiii,313p : ill. – Includes index
0–8391–1204–1 SRL
1435558

ACTIVE SITES. Biopolymers
Research. Use of affinity labelling

THEORY and practice in affinity techniques / edited by P.V. Sundaram and F. Eckstein. – London : Academic Press, 1978
x,246p : ill. – Includes index
0–12–677150–2 SRL
1873717

ACTIVE TUTORIAL WORK PROJECT

BOLAM, Ray
An evaluation of the training and dissemination methods used by the Active Tutorial Work Project team. – Bristol : University of Bristol School of Education, 1983
1v.(various pagings). – Includes bibliography
B85–27794 X.525/11034
3302946

ACTIVITIES
See also
Christmas activities
Cultural activities
Dangerous activities
Demonstrations
Easter activities
Educational activities
Extra-curricular activities
Government activities
Group activities
Indoor activities
Musical activities
Outdoor activities
Overseas activities
Physical activities
Resource-based activities
Seaside activities
Seasonal activities
Social activities

ACTIVITIES
Drama – For non-English speaking students

DIXEY, James
Get up & do it!. – London : Longman.
Students' book. – 1978
[3],36p : ill
0–582–55088–2 X.900/25533
1409468

Get up and do it!. – London : Longman.
Teacher's book. – 1978
[5],43p : ill
0–582–55089–0 X.0900/1046
1409468

ACTIVITIES
Drama – For non-English speaking students – Texts

SWAN, H. A. (Hazel Ann)
Act one in English. – Amersham : Hulton Educational, 1983
3v. : ill
B84–08399 X.950/27562–27564
1409468

ACTIVITIES
Drama – Texts

NEWTON, Bill
The miracle man. – London : Stainer and Bell, 1976
63p : ill
0–85249–312–6 X.900/25183
0676195

ACTIVITIES
Projects

NICHOLSON, Simon
Interactive art and play. – Milton Keynes : Open University Press, 1976
64p,[12]p of plates : ill(some col). – ([Technology/arts/social sciences], a second level interdisciplinary course. art and environment ; unit 10) (TAD292 ; unit 10)
0–335–06209–1 X.615/1757 Woolwich
1028642

ACTIVITIES. American usage. English language. Curriculum subjects. Secondary schools
– For teaching

BERGMAN, Floyd L.
The English teacher's activities handbook. – 2nd ed. – Boston [Mass.] ; London : Allyn and Bacon, c1982
xviii,342p : ill. – Bibliography: p311–332. – Includes index
0–205–07383–2 X.520/25733
2606119

ACTIVITIES. Botany

BISHOP, O.N. (Owen Neville)
Adventures with small plants. – London : Murray, 1983
56p : ill. – Bibliography: p56
0–7195–3953–6 X.322/12315
2953161

ACTIVITIES. Boys' Brigade
Award schemes

BOYS' BRIGADE
Award regulations (Company Section). – London : The Brigade, 1976
61p : ill
B79–02353 X.519/28944
1761552

ACTIVITIES. Boys' Brigade
Award schemes: Target Award

BOYS' BRIGADE
Target award. – London : The Brigade, [1978?]
13,[1]p. – (Awards pamphlets ; no.2)
B79–11435 X.519/29400
1835017

ACTIVITIES. Boys' public schools. England
Sports & games, 1837–1910

MANGAN, J. A. (James Anthony)
Athleticism in the Victorian and Edwardian public school. – Cambridge : Cambridge University Press, 1981
xv,345p : ill. – Bibliography: p309–336. – Includes index
0–521–23388–7 X.520/24606
2257491

ACTIVITIES. Brownie Guides

The BOOK of Brownie skills. – Maidenhead : Purnell, 1976
192p : col ill
0–361–03577–2 X.619/17382 Woolwich
0156442

BRAMBLEBY, Ailsa
The Brownie Guide handbook. – 3rd ed., revised. – London : Girl Guides Association, 1980
[2],189p : ill(chiefly col). – Includes index
B80–31295 X.529/37285
0156442

BROWN, Rosalie
Ideas for packs. – Wendover : Goodchild in conjunction with the Girl Guides Association, 1977
92p : ill
0–903445–27–1 X.529/51034
0156442

The BROWNIE gift book. – Maidenhead : Purnell, 1981
61p : ill(some col.)
0–361–05074–7 X.622/10757 Woolwich
0156442

The BROWNIE Guide handbook / illustrated by Jennetta Vise. – 3rd ed., rev.. – London : Girl Guides Association, 1980
189p : col.ill. – Includes index
B82–28247 X.529/48404
0156442

BROWNIE guide record book. – [London] : [Girl Guides Association], [1980]
[2],23p : ill
B80–33324 X.529/37375
0156442

BROWNIE guide record book. – [London] : [Girl Guides Association].
2. – [1980]
[2],24p : ill
B80–33325 X.529/37376
0156442
3. – [1980]
[2],23p : ill(some col)
B80–33326 X.529/37377
0156442

A HANDBOOK for guiders. – London : Girl Guides Association.
Brownie Guide section. – 1975
2–141p : ill(some col). – Includes index
0–85260–004–6 X.519/25360
0156442
Brownie guide section. – 1980
179,x,189p : ill(some col.). – Includes indexes
B83–02915 X.629/18969
0156442

HASLAM, S. M. (Sylvia Mary)
Games for Brownies. – [Cambridge] : Cambridgeshire East Girl Guides, [1984]
29p
B84–43623 X.629/24407
0156442

ACTIVITIES. Brownie Guides. Great Britain

BROWN, Rosalie
Brownie journey colouring book. – Wendover : John Goodchild, Publishers [for] the Girl Guides Association.
No.1. – 1976
[24]p : of ill
0–903445–23–9 X.992/3325 Woolwich
1098306
No.2. – 1976
[24]p : of ill
0–903445–24–7 X.992/3326 Woolwich
1098306
No.3. – 1976
[24]p : of ill
0–903445–25–5 X.992/3327 Woolwich
1098306
No.4. – 1976
[24]p : of ill
0–903445–26–3 X.992/3328 Woolwich
1098306

ACTIVITIES. Children
Acting. Teaching

MALKIN, Michael R.
Training the young actor. – South Brunswick ; New York : Barnes ; London : Yoseloff, 1979
128p : ill. – Includes index
0–498–01957–8 X.981/22284
2141086

ACTIVITIES. Children. Special subjects. Bewick, Thomas. English wood engravings

BEWICK, Thomas
Children's pastimes. – London : Merrion Press, 1975
Fold leaf : of ill. – (Wood engravings by the English engraver Thomas Bewick, 1753–1828 / Thomas Bewick)
0–903560–08–9 Cup.510.a.9
0649333

ACTIVITIES. Children, to 5 years
– For parents

CRAFT, Ruth
Play school play ideas. – London : BBC : Knight, 1983, c1971
91p : ill
0–340–33299–9 X.529/54384
024208x

ACTIVITIES. Children. United Reformed Church

JUNIOR work book. – Rev. ed. – [London] : United Reformed Church, Church Life Department, 1981
39p : ill
0–902256–50–5 X.529/47261
2714914

ACTIVITIES. Christian religious education. Children, to 10 years
– Serials

IDEAS for under-tens leaders. – 1981–. – London : Covenanters, 1981–
v.
B84–14817 P.101/2141
3057127

ACTIVITIES. Christian religious education. Primary schools
– For teaching

FREEMAN, Marjorie
Steps and stones. – London : CIO.
Bk.1. – c1983
80p : ill
0–7151–0413–6 X.200/41952
2985896
– London : For the General Synod Board of Education by CIO.
Bk.2. – 1984
82p : ill
0–7151–0421–7 X.200/44961
2985896

ACTIVITIES. Christian religious education. Schools
– For Christian students – Serials

IS annual / [Inter-School Christian Fellowship]. – [No. 1]–. – London : ISCF, 1980–
v. : ill
B81–15289 P.103/596
2388227

ACTIVITIES. Christian religious groups
– For parish priests

MALLISON, John
Creative ideas. – London : SU Specialist Publications, 1981, c1978
249p : ill. – (Abridged small group series ; no.2). – Includes bibliographies
0–85421–906–4 X.0200/1130(2)
2533456

ACTIVITIES. Coaching. Hockey

CADMAN, John, 1934–
Games for hockey training. – London : Pelham, 1981
158p : ill
0–7207–1246–7 X.629/15537 Woolwich
2410036

ACTIVITIES. Colfe's School
Rugby football, 1925–1980

COLFE'S SCHOOL. Rubgy Club
The jubilee history. – [Great Britain] : [Colfe's School], [1980?]
19p : ill
B85–06065 X.629/24942
3220761

ACTIVITIES. Community & youth work. Liverpool. Merseyside (Metropolitan County)
English literature. Projects: Windows Project, to 1978

WINDOWS PROJECT
City of poems. – [Liverpool] : [The project], 1979
[58]p : ill
B82–14766 X.958/7999
2641631

ACTIVITIES. Community & youth work. Nottingham
Drama. Projects, 1973–1980

HESCOTT, Bob
The feast of fools. – London : UK Branch, Calouste Gulbenkian Foundation, 1983
75p : ill
0–903319–30–6 X.629/21658
2984733

ACTIVITIES. Cub Scouts

The BEST of the cub scout supplements from 'Scouting' magazine / [general editor Ron Jeffries ; editor Paul Moynihan]. – London : Scout Association, 1978
30,[3]p : ill
0–85165–152–6 X.992/4958
2022419

MORE cub scout ideas sheets. – London : Scouting, 1983
39p : ill
0–85165–176–3 X.622/16815
2022419

ACTIVITIES. Cub scouts
Planning

ADDIS, Hazel Iris
Programme planning in the Cub Scout pack. – London : Scout Association, 1975
160p : ill
0–85165–107–0 X.619/15236
0525235

ACTIVITIES. Cub Scouts
Planning

BARROW, Jean
A pack scouter's activity book. – Rev. ed.. – Glasgow : Brown, Son & Ferguson, c1981
vii,149p : ill
0–85174–415–x X.529/44715
0525235

ACTIVITIES. Cub Scouts. Ireland (Republic)

ELLIS, Charles E.
My cub scout handbook. – Dublin : Scout Association of Ireland, 1982
63p : ill(some col.)
B83–34074 X.629/20836
2938065

ACTIVITIES. Deaf children
Music. Teaching

KEINER, Rita
Music for deaf children. – London : R. Keiner, c1984
30p : ill. – Bibliography: p30
B85–25255 X.435/1228
3287130

ACTIVITIES. Denmark. Ministeriet for Kulturelle Anliggender
KULTUR nyt. – Nr 1– , 1977–. – København : Ministeriet for Kulturelle Anliggender, 1977–
r6060211 S.B.286/45
242911X

ACTIVITIES. Druids. Stonehenge
to 1977

GRINSELL, L. V. (Leslie Valentine)
The Druids and Stonehenge. – St Peter Port : Toucan Press, 1978
2–29p : ill. – (West Country folklore ; no.11). – Bibliography: p.26–29
0–85694–147–6 X.708/21515
1702564

ACTIVITIES. Ecumenical movement. Christian church

UNITED in Christ / general editor Josephine Clemson ; contributors and advisers P. Ryan ... [et al.]. – Slough : St. Paul, 1981
122p. – Includes index
0–85439–196–7 X.200/37298
260177x

ACTIVITIES. Education. Children
Drama - Bibliographies

KENNEDY, Carol Jean
Child drama. - [Washington] : Children's Theatre Association of America, 1981
iv,51p
r6812566 X.950/28095
3064514

KLOCK, Mary Eileen
Creative drama:. - [Washington] : Children's Association of America, 1975
vi,26p
r6812567 X.950/28094
3064514

ACTIVITIES. Educational institutions
Drama

COLSON, Greta
Drama skills. - London : Barrie and Jenkins, 1980
160p : ill
0-214-20571-1 L.49/645
1950398

ACTIVITIES. Educational institutions
Drama. Improvisation. Teaching methods. Theories of Heathcote, Dorothy

WAGNER, Betty Jane
Dorothy Heathcote. - London : Hutchinson, 1979
238p. - Bibliography: p.235. - List of films: p.236-237. - List of videotapes: p.238
0-09-138851-1 X.909/43606
1760327

ACTIVITIES. Educational institutions. United States
Athletics. Fund raising

BRONZAN, Robert T.
Public relations, promotions, and fund-raising for athletic and physical education programs. - New York ; London : Wiley, 1977
xiv,268p : ill. - Bibliography: p.259-262. - Includes index
0-471-01540-7 X.620/16678 Woolwich
1080172

ACTIVITIES. Educational institutions. United States
Athletics. Public relations

BRONZAN, Robert T.
Public relations, promotions, and fund-raising for athletic and physical education programs. - New York ; London : Wiley, 1977
xiv,268p : ill. - Bibliography: p.259-262. - Includes index
0-471-01540-7 X.620/16678 Woolwich
1080164

ACTIVITIES. Educational institutions. United States
Sports & games. Administration

VANDERZWAAG, Harold J.
Sport management in schools and colleges. - New York ; Chichester : Wiley, c1984
xviii,265p : ill. - Includes bibliographies and index
0-471-87135-4 X.622/20961
3186849

ACTIVITIES. Environmental studies. Curriculum subjects. Schools. England
- For teaching

ADAMS, Eileen
Art and the built environment : study activities. - York : Longman for Schools Council, 1982
64p : ill. - (Schools Council programme. 3, Developing the curriculum for a changing world)
0-582-26168-6 L.49/1973
2971496

ACTIVITIES. Globe Theatre
ca 1600

THOMSON, Peter, 1938-
Shakespeare's theatre. - London : Routledge & Kegan Paul, 1983
xiii,190p,[8]p of plates : ill. - (Theatre production studies). - Bibliography: p180-184. - Includes index
0-7100-9480-9 X.950/24583
1796402

ACTIVITIES. Group therapy. Speech disorders. Stroke victims

BOLITHO, Jane
Speech and language activities for use with stroke groups. - London : Chest, Heart and Stroke Association, 1979
108p : ill
0-901548-45-6 X.909/44146
2021951

ACTIVITIES. Group worship. Christian church

BEALL, Patricia
God's family at worship. - Poole : Celebration, 1983
144p : ill. - (Expressing his life). - Bibliography: p136-142
0-906309-26-3 X.0108/518(3)
2340291

HODGETTS, Colin
Exploring worship. - London : Mowbray, 1980
vi,153p : ill. - (Mowbray's popular Christian paperbacks)
0-264-66460-4 X.208/5995
2340291

ACTIVITIES. Handicapped children
Movement. Teaching

HOLLIS, Katy
Progress to standing. - New ed. - Kidderminster : British Institute of Mental Handicap, 1980
v,16p : ill. - Bibliography: pv
0-906054-16-8 X.329/15486
1192213

MORTIMORE, Frederick
Making movement fun. - London : National Society for Mentally Handicapped Children, 1977
[24]p : ill. - (Advisory leaflets / National Society for Mentally Handicapped Children)
B77-25505 X.519/27155
1192213

ACTIVITIES. Handicapped persons. Great Britain
Arts

The ARTS and disabilities / edited by Geoffrey Lord. - Edinburgh : Macdonald, 1981
135p : ill. - Includes bibliographies
0-904265-62-5 X.421/22833
2709155
- [Another copy: paperback]
I.S.B.N. 0-904265-61-7 X.421/22834

ACTIVITIES. Higher education institutions. United States
Sports & games - Directories

CALLAHAN, Timothy R.
Callahan's college guide to athletics and academics in America. - New York ; London : Harper & Row, c1984
257p
0-06-015249-4 X.525/10271
3220087

ACTIVITIES. Libraries. Primary schools

MARSHALL, Karen K.
Back to books. - Jefferson, N.C. ; [London] : McFarland, 1983
v,138p : ill. - Includes index
0-89950-092-7 X.950/34034
314528x

ACTIVITIES. Libraries. United States

68 great ideas / a project of Elizabeth W. Stone ; editor, Peggy Barber ; review committee Melvin George, Loretta O'Brien, Esther Perica ; with suggestions from the John Cotton Dana Public Relations Award Committee. - Chicago : American Library Association, 1982
66p : ill. - Includes index
0-8389-0376-2 2719.e.555
2600285

ROBOTHAM, John S.
Library programs. - 2nd ed. - Metuchen, N.J. ; London : Scarecrow, 1981
xii,352p : ill. - Bibliography: p332-341. - Includes index
0-8108-1422-6 2719.e.303
2600285

ACTIVITIES. Mathematics

WILSON, Des, 1934-
Maths activity in the early years. - [Sheffield] : Eigen.
4: Patterns of three. - c1984
12p : ill
1-85039-123-8 X.629/25720
2907291
5: Number lines. - c1984
12p : ill
1-85039-124-6 X.629/25717
2907291
6: Number games. - c1984
13p : ill
1-85039-125-4 X.629/25718
2907291
7: Shapes. - c1984
12p : ill
1-85039-126-2 X.629/25719
2907291
8: Mirrors and patterns. - c1984
12p : ill
1-85039-127-0 X.629/25721
2907291
9: Logic. - c1984
12p : ill
1-85039-128-9 X.629/25722
2907291

ACTIVITIES. Members. Trade unions. Great Britain
Time off. Permission by employers

ADVISORY, CONCILIATION AND ARBITRATION SERVICE
Time off for trade union duties and activities. - [London] : H.M.S.O., 1977
[2],8,[1]p. - (Code of practice ; 3)
0-11-885002-4 BS.427/12(3)
1289934

GUIDE to time-off arrangements / [IDS]. - London : Incomes Data Services, 1981
92p. - Includes index
B82-09131 X.622/11990 Woolwich
1289934

ACTIVITIES. Middle schools

BAKER, Peter, 1943-
Interplay. - London : Macmillan.
1: Animalia. - 1976
1v. : ill(some col)
0-333-18023-2 Cup.911/31
0702706
2: Fantasia. - 1976
1v. : ill(some col)
0-333-18024-0 Cup.911/31
0702706

ACTIVITIES. Middle schools
Drama

NORRIS, Jay
Drama resource cards. - London : Longman, 1977
Portfolio : ill
0-582-36204-0 Cup.911/53
025715x

PORTER, Sue
Action pack. - [London] : Edward Arnold, 1976
1v. : ill
0-7131-0076-1 Cup.911/64
025715x
- [Another copy] Cup900.ss.10

PRIOR, John, 1933-
Act it out. - London : Macmillan, 1978
1v. : ill
0-333-18793-8 Cup.900.za.93
025715x

SLADE, Peter
Drama and the middle school. - 2nd ed. - [Birmingham] : Educational Drama Association, [1976?]
17p
B77-06223 X.909/41199
1040898

ACTIVITIES. Middle schools. Great Britain
Creative music. Projects

OSBORNE, Nigel
Creative projects. - London : Chappell Music, c1981
32p : ill. - (Ways with music)
B81-21967 L.49/837
2484897

STURMAN, Paul
Creating music. - Harlow : Longman, 1982
64p : ill
0-582-20001-6 L.49/1206
2484897

ACTIVITIES. Middle schools. United States

HECK, Shirley F.
All the classroom is a stage. - New York ; Oxford : Pergamon, 1978
xvii,157p : ill. - Includes bibliographies and index
0-08-022248-x X.519/29356
1824988

ACTIVITIES. Morning assembly. Primary schools. Great Britain
- For teaching

BARRATT, Sylvia
The tinder-box assembly book. - London : Black, 1982
112p : ill. - Includes bibliographies and index
0-7136-2169-9 L.49/1583
0205079

FARNCOMBE, Anne
It's our assembly. - Redhill : National Christian Education Council, 1979
128p
0-7197-0220-8 X.100/18218
0205079

GRIFFIN, Warwick
Exploring primary assemblies. - London : Macmillan Education, 1984
x,165p : ill. - Includes bibliographies
0-333-36588-7 X.520/35463
0205079

JUNIOR assemblies / edited by Geraldine Witcher ; illustrations by Elizabeth Norman-Clarke. - London : Scripture Union, c1982
160p : ill. - Includes bibliography
0-85421-949-8 X.200/38544
0205079

ACTIVITIES. Morning assembly. Primary schools. Great Britain
- For teaching - Serials

HANDS together. - Issue no.1 (Oct. 1983)-. - Leamington Spa : Scholastic Publications (Magazines) Ltd., 1983-
v. : ill
B84-28554 P.525/486
0205079

ACTIVITIES. Morning assembly. Schools
- For teaching

SMITH, Harry, 19—
Assemblies. - London : Heinemann Educational, 1981
x,245p. - Includes bibliographies
0-435-01830-2 X.529/44700
0364770

TAYLOR, Dorothy J. (Dorothy Jean)
Explorations in assembly with children. - Guildford : Lutterworth Press, 1973
94p. - Bibliography: p.85-91
0-7188-2042-8 X.108/16531
0364770

WATERS, Derek
A book of assemblies. - London : Mills and Boon, 1975
128p : 1 ill. - Bibliography: p.110-124. - List of gramophone records and music: p.125-128
0-263-05782-8 X.200/20052
0364770

ACTIVITIES. Morning assembly. Schools
Special themes: Special occasions

MORGAN, Diana, 19—
Let's celebrate. - London : Mowbray, 1984
ix,85p
0-264-67024-8 X.200/45640
2384183

TAYLOR, Dorothy J. (Dorothy Jean)
Exploring red letter days. - Guildford : Lutterworth Educational, 1981
x,150p. - Bibliography: p139-147
0-7188-2479-2 X.520/23511
2384183

ACTIVITIES. Morning assembly. Secondary schools
Special themes: Harvest

CHRISTIAN AID
The harvesters. - London : Christian Aid ; London : Christian Education Movement, 1976
14p. - Bibliography: p.5-5
B77-03486 X.100/16675
0790206

ACTIVITIES. Morning assembly. Secondary schools. Great Britain
- For teaching

CASTLE, Tony
Tuesday again!. - Great Wakering : Mayhew-McCrimmon, 1978
103p. - Includes bibliographies. - List of films
0-85597-241-6 X.100/17904
1594524

GENT, William A.
Living centre. - London : Christian Education Movement, c1984
36p : ill
0-905022-97-1 X.200/45974
1594524

ACTIVITIES. Natural history

ARNOLD, Neil, 1939-
The young naturalist. - London : Ward Lock, 1983
48p : ill(some col.). - Bibliography: p48
0-7063-6219-5 X.322/15156
0361267

CHINERY, Michael
The family naturalist. - London : Macdonald and Jane's, 1977
192p : ill(some col). - Bibliography: p.188-189. - Includes index
0-354-04155-x X.322/8476 Woolwich
1275305
- [Second impression] X.325/558 Woolwich

HAVE fun with nature / [co-ordinating editor Lesley Firth]. - London : Macdonald Educational.
1. - 1976
32p : ill
0-356-05512-4 X.311/8216(1) Woolwich
0361267
2. - 1976
32p : ill
0-356-05513-2 X.311/8216(2) Woolwich
0361267
3. - 1976
32p : ill
0-356-05514-0 X.311/8216(3) Woolwich
0361267
4. - 1976
32p : ill
0-356-05515-9 X.311/8216(4) Woolwich
0361267

MITCHELL, Andrew
The young naturalist. - London : Usborne, 1982
32p : ill(some col.). - (Usborne hobby guides). - Includes index
0-86020-653-x X.322/11474
0361267

VAN GELDEREN, Gerrit
Doings. - Dublin : Educational Company of Ireland, 1975
32p : chiefly ill. - (Environmental project book ; 1)
0-904916-10-3 X.319/10063
0361267

WILLSON, Robina Beckles
Eyes wide open. - London : Heimemann, 1981
95p : ill. - Bibliography: p91. - Includes index
0-434-97259-2 X.329/14447 Woolwich
0361267

ACTIVITIES. Natural history. Australia

BUTLER, Harry, 1930-
Looking at the wild. - Sydney ; London : Hodder and Stoughton, 1981
142p,[32]p of plates : ill(some col.). - Bibliography: p139. - Includes index
0-340-26619-8 X.322/14794
3260194

ACTIVITIES. Netherhall School
Cricket. Staff teams, 1933-1982

GREY, D. S. (Duncan S)
Netherhall staff cricket, 1933-83. - Cambridge : Golden Bustard, c1983
[28]p
B83-38298 Cup.408.r.38
2953145

ACTIVITIES. Nursery schools. United States
Arts

BRITTAIN, W. Lambert
Creativity, art, and the young child. - New York : Macmillan ; London : Collier Macmillan, 1979
xiii,242p,[4]p of plates : ill(some col). - Bibliography: p.233-237. - Includes index
0-02-314990-6 X.421/10986 Woolwich
1937863

ACTIVITIES. Organisations. Conservation. Environment. Great Britain

A CATALOGUE of conservation activities / [editor J.T. Williams]. - [London] : Professional Institutions Council for Conservation, 1975
[7],30p. - Bibliography: p.17
0-905020-03-0 X.313/1285 Woolwich
0683779

ACTIVITIES. Personnel. Welfare rights services. Local authorities. Great Britain

SIMPSON, Tony, 1941-
Advocacy and social change. - London : National Institute for Social Work, 1978
[7],113p. - (Papers / National Institute for Social Work ; no.8). - Bibliography: p.106-113
0-902789-13-9 X.512/6772
193449x

ACTIVITIES. Playgroups

BROAD, Laura Peabody
The playgroup handbook. - London : Souvenir Press, 1976
[13],306p : ill. - Includes index
0-285-64802-0 X.520/10559
0125563

The playgroup handbook. - Updated ed. - London : Sphere, 1977
[13],306p : ill. - Includes index
0-7221-1874-0 X.619/18520 Woolwich
0125563

COLLIN, Margaret
Interest tables. - Glasgow : Scottish Pre-School Playgroups Association, [1982?]
16p : ill. - (Activity booklet)
0-907662-01-3 X.629/18855
0125563

GAMWELL, Sue
Serendipity. - Glasgow : Scottish Pre-school Playgroups Association, 1982
16p : ill. - (Activity booklet)
0-907662-07-2 X.529/51835
0125563

SERENDIPITY 2. - Glasgow : Scottish Pre-school Playgroups Association, 1983
16p : ill
0-907662-10-2 X.529/62025
0125563

ACTIVITIES. Playgroups
Music

SHEPHARD, Margaret
Playgroup music. - London : Pre-School Playgroups Association, c1982
[12]p : ill. - (PPA publication). - Bibliography: p11-12
0-901755-34-6 X.529/51338
0125571

ACTIVITIES. Pre-school children

TAETZSCH, Sandra Zeitlin
Preschool games and activities. - Belmont, Calif. : Fearon Publishers, 1974
xvii, 88 p. : ill. - (A Fearon teacher-aid book). - Bibliography : p. 87-88
0-8224-5605-2 X.619/17456 Woolwich
1082159

ACTIVITIES. Pre-school children
- For parents

MENLOVE, Coleen Kent
Ready, set, go!. - Englewood Cliffs ; London : Prentice-Hall, 1978
xii,244p : ill. - Bibliography: p.223-240. - Includes index
0-13-762286-4 X.519/29348
1824392

MOLLOY, Cora
Playtime handbook. - Tallaght : Folens, c1979
iv,20p
B83-32418 X.520/30687
1824392

ACTIVITIES. Pre-school children, 3-5 years
Music

ANDRESS, Barbara
Music experiences in early childhood. - New York ; London : Holt, Rinehart and Winston, 1980
ii-x,198p : ill. - Bibliography: p.193-195. - Includes index
0-03-021771-7 X.431/11149
223548x

ACTIVITIES. Pre-school Playgroups Association
- Conference proceedings

PRE-SCHOOL PLAYGROUPS ASSOCIATION. Annual Conference (15th : 1976 : University of Warwick)
Our voluntary contribution. - London : P.P.A., 1977
24p
0-901755-16-8 X.519/27498
129203x

ACTIVITIES. Primary schools
- For teaching

BERNSTEIN TARROW, Norma
Activities and resources for guiding young children's learning. - New York ; London : McGraw-Hill, c1981
xiii,290p : ill. - Bibliography: p280-285
0-07-039108-4 X.525/7774
264357x

GILBERT, Jean
Topic anthologies for young children. - London : Oxford University Press, 1979.
1: Water, machines, toys
viii,85p : ill
0-19-321280-3 X.431/11845
264357x

LONG, Jan
[Doing social studies]. Looking, making, displaying. - [Driffield] : [Studies in Education], 1980
69p : ill. - (New primary ideas)
0-905484-28-2 X.629/17358 Woolwich
264357x

YARDLEY, Alice
Early concepts. - London : Macmillan.
1: Being. - 1978
1v. : ill(some col)
0-333-24357-9 Cup.915/103
1824783

2: Discovering. - 1978
1v. : ill(some col)
0-333-24358-7 Cup.915/102
1824783

ACTIVITIES. Primary schools
- Texts

CARRUTH, Jane
What will I do at school?. - [Maidenhead] : [Purnell], [1978]
[28]p : chiefly col ill. - (My first colour library)
0-361-03849-6 X.990/10036 Woolwich
1478230

ACTIVITIES. Primary schools
Books. Making – For teaching

BENNETT, Leslie
Children making books. – London : A. and C. Black, 1978
x,118p : ill. – Bibliography: p.112. – Includes index
0–7136–1893–0 2708.bb.5
1404806

ACTIVITIES. Primary schools
Drama – For teaching

FULLEYLOVE, Eve
Starting points for drama. – Banbury : Kemble, 1984
viii,120p : ill
0–906835–22–4 X.950/42937
0098043

JAMES, Ronald
Developing drama. – Rev. ed. – Sunbury-on-Thames : Nelson, 1983
vii,51p. – Includes index
0–906835–10–0 X.950/25924
0098043

SILVERMAN, Eleanor
Dramatics for children. – Metuchen, N.J. ; London : Scarecrow, 1983
x,149p : ill. – Includes index
0–8108–1612–1 X.955/1970
0098043

ACTIVITIES. Primary schools
Drama. Improvisation – For teaching

ERRINGTON, Eddie
A time and a place. – Driffield : Studies in Education, 1980
91p : ill. – Bibliography: p91
0–905484–27–4 X.529/39486
234808x

ACTIVITIES. Primary schools
Drama. Teaching

DAVIES, G. C. (Geoffrey Charles)
Practical primary drama. – London : Heinemann Educational, 1983
63p
0–435–18236–6 X.529/53986
2177331

HEINIG, Ruth Beall
Creative drama for the classroom teacher. – 2nd ed. – Englewood Cliffs ; London : Prentice-Hall, c1981
xii,256p : ill. – Includes bibliographies and index
0–13–189415–3 X.950/3966
2177331

ACTIVITIES. Primary schools
Hockey

CAVILL, Gwen
Make hockey fun. – [London] : Edward Arnold, 1985
75p : ill
0–7131–8246–6 X.629/25705
3118789

ACTIVITIES. Primary schools
Music. Projects – For teaching

BIRKENSHAW, Lois
Music for fun, music for learning. – 2nd ed. – Toronto : Holt, Rinehart and Winston ; Eastbourne : Holt-Saunders, c1977
xii, : ill. – Includes index
0–03–920001–9 X.431/11843
2643596

LEASK, Jeff
Listening, creating, moving. – [Driffield] : Studies in Education, 1980
59p : ill. – Includes index
0–905484–23–1 X.529/52900
2643596

ACTIVITIES. Primary schools
Performing arts. Special subjects: Europe, 1250–1400

BAGENAL, Alison
This merry company. – London : Oxford University Press.
Pupils' book. 1. – 1979
[2],33p : ill(chiefly col)
0–19–321041–x L.49/557
2257238
Teacher's book. – 1979
viii,40p : ill. – Bibliography: p.38–39. – List of sound discs : p.39
0–19–321040–1 X.431/11194
2257238

ACTIVITIES. Primary schools
Themes – For group teaching

TOMOS, Eirlys Wynn
Camau cywaith. – [Caerdydd] : Gwasg Prifysgol Cymru, 1981
94p : ill
0–7083–0804–x X.520/24946
255397x

ACTIVITIES. Primary schools. Great Britain
Sports & games – For teaching

CASTLE, R. N.
Games teaching to children in the 7–11 age group. – Ipswich : [Suffolk County Council], [1982]
135p : ill
B82–36652 X.525/6204
2289989

CASTLE, R. N. (Robert Norman)
Games and small apparatus skills for infants. – Ipswich : [Suffolk county Council], [1979]
[3],85[i.e.88]p : ill. – Bibliography: p.84
0–86055–054–0 X.622/8769 Woolwich
2289989

LENEL, R. M. (Renate Milly)
Games in the primary school. – 2nd ed. – London : Hodder and Stoughton, 1984
160p : ill. – Bibliography: p158. – Includes index
0–340–35964–1 X.629/24896
2289989

MAULDON, E. (Elizabeth)
Games teaching. – 2nd ed. – Plymouth : Macdonald and Evans, 1981
ix,131p. – Includes index
0–7121–0739–8 X.529/41619
2289989

WISE, W. M. (William Mark)
Games and sports. – 2nd ed. – London : Heinemann Educational, 1983
viii,129p : ill. – (Activity in the primary school). – Includes index
0–435–80930–x X.629/21946
2289989

ACTIVITIES. Primary schools. Multicultural societies
Drama – For teaching

DEMMERY, Sylvia
Drama in a multi-cultural society. – [Birmingham] : Educational Drama Association, [1981?]
16p. – Bibliography: p16
0–904027–10–4 X.950/4193
2406470

ACTIVITIES. Primary schools. Scotland
Performing arts

TOWARDS a policy for expressive arts in the primary school. – Edinburgh : Scottish Committee on Expressive Arts in the Primary School, 1983
8p. – Bibliography: p8
B84–11962 X.525/8646
3046001

ACTIVITIES. Primary schools. Scotland
Visual arts

TOWARDS a policy for expressive arts in the primary school. – Edinburgh : Scottish Committee on Expressive Arts in the Primary School, 1983
8p. – Bibliography: p8
B84–11962 X.525/8646
304601x

ACTIVITIES. Primary schools. United States

HECK, Shirley F.
All the classroom is a stage. – New York ; Oxford : Pergamon, 1978
xvii,157p : ill. – Includes bibliographies and index
0–08–022248–x X.519/29356
182497x

ACTIVITIES. Primary schools. United States
– For teaching

STEPHENS, Lillian S. (Lillian Sklaire)
Developing thinking skills through real-life activities. – Boston [Mass.] ; London : Allyn and Bacon, c1983
xii,276p : ill
0–205–07977–6 X.525/8528
3038173

ACTIVITIES. Primary schools. Wales
Themes – For teaching

CYNLLUN ADDYSG DDWYIEITHOG
Gweithgareddau i'r plant bach. – London : Cyn-gor Ysgolion.
Gweithgareddau atodol. Llawlyfr i'r athrawes. – 1975
5v : ill
r6071769 X.519/27816
1404628

ACTIVITIES. Public worship. Christian church
Drama

BURBRIDGE, Paul
Time to act. – London : Hodder and Stoughton, 1979
127p. – (Hodder Christian paperbacks)
0–340–24699–5 X.208/7408
0320722

[Time to act. Welsh]. 'Gymrwch chi ran?. – Abertawe [Swansea] : Tŷ John Penry, 1982, c1979
138p
0–903701–52–9 X.958/13477
2831260

CLAPSON, Ged
Come, celebrate!. – London : Triangle, 1984
x,131p
0–281–04104–0 X.208/10278
0320722

GRINHAM, Gillian
Know how to use drama in church. – [London] : Scripture Union, 1979
[3],71p : ill. – Bibliography: p.70–71
0–85421–819–x 4194.m.1/13
0320722

ACTIVITIES. Public worship. Christian church
Drama & dancing

BLOGG, Martin H.
Time to dance. – London : Collins, 1984
72p : ill. – Bibliography: p72
0–00–599777–1 X.622/21299
0791601

GRAINGER, Roger
Presenting drama in church. – London : Epworth, 1985
[vii,128]p
0–7162–0408–8 X.950/42196
0791601

LAMONT, Gordon
Move yourselves. – Swindon : Bible Society, 1983
112p : ill. – (Creative resources). – Bibliography: p107. – Includes index
0–564–07292–3 P.101/2146
0791601

LONG, Anne
Praise Him in the dance. – London : Hodder and Stoughton, 1976
192p,[8]p of plates : ill. – (Ecclesia books). – Bibliography: p.192
0–340–20976–3 X.439/8248
0791601

ACTIVITIES. Public worship. Christian church
Folk arts

BARKER, Martha Keys
Building worship together. – [Poole] : Celebration, 1981
126p. – List of sound recordings: p125–126
0–906309–21–2 X.208/7011
1930427

BEALL, Patricia
The folk arts in God's family. – Rev. ed.. – London : Hodder and Stoughton, 1984
318p : ill. – (Hodder Christian paperbacks). – Bibliography: p311–316
0–340–35511–5 X.208/10401
1930427

ACTIVITIES. Public worship. Christian church
Performing arts – Serials

GALLERY. – No.1– ; Feb. 1974–. – London : Christian Arts Project (London), 1974– : ill
B76–12346 P.425/100
062098x

ACTIVITIES. Public worship. Christian church
Story-telling

WARNE, Clifford
Know how to tell a story. – London : Scripture Union, 1979
[5],53p : ill
0–85421–818–1 4194.m.1/15
2020734

ACTIVITIES. Public worship for families. Christian church
Performing arts

FISHERFOLK, THE
All aboard the ark. – Lytchett Minster : Celebration Publishing, 1978
1v. : ill. – (Fisherfolk kit ; 1). – Bibliography (1p.). – List of sound discs (1p.)
0–906309–00–x Cup.915/98
1779451

ACTIVITIES. Public worship for families. Christian church
Performing arts. Special subjects: Jesus Christ. Nativity

FISHERFOLK, THE
Watch, look, listen. – Lytchett Minster : Celebration Publishing, 1978
1v. : ill. – (Fisherfolk kit ; 2). – Bibliography (1p.). – List of sound discs/sound tapes (1p.)
0–906309–06–9 Cup.915/99
177946x
– Accompanying sound tape Cup.575.l.54

ACTIVITIES. Schools
Association football – For teaching

JARMAN, John
Junior soccer. – London : Faber, 1976
77p : ill. – Includes index
0–571–10846–6 X.629/10935 Woolwich
029439x

WADE, Allen
The F.A. guide to teaching football. – London : Heinemann [for] the Football Association, 1978
xvi,171p : ill
0–434–92210–2 X.620/17933 Woolwich
029439x

WORTHINGTON, Eric
Teaching soccer skill. – 2nd ed. – London : Lepus Books, 1980
[7],218p : ill(some col). – Bibliography: p.218
0–86019–039–0 X.629/14390 Woolwich
029439x

ACTIVITIES. Schools
Drama

ADLAND, David
Group approach to drama 1. – 2nd ed. – Harlow : Longman, 1981
92p
0–582–21960–4 X.950/9731
2608820

ADLAND, David
Group approach to drama 2. – 2nd ed. – Harlow : Longman, 1981
92p
0–582–21961–2 X.950/9732
2608820

Group approach to drama 3. – 2nd ed. – Harlow : Longman, 1981
108p : ill
0–582–21962–0 X.950/9733
2608820

FRANCIS, Hilton
The vocabulary of educational drama. – 2nd ed. – Banbury : Kemble Press, [1979]
[2], ix, 51p. – Bibliography: p.49–50. – Includes index
0–906835–00–3 X.909/45173
1916521

LIMB, Sue
[Action pack]. Action packed. – London : Arnold, 1985
[64]p : ill
0–7131–7437–4 X.950/41489
2608820

ACTIVITIES. Schools
Drama – Bibliographies

PICKERING, Kenneth
Select bibliography of drama and education. – London : Commission for a British Theatre Institute, 1975
ii,14p. – (BTI bibliographic series ; no.3)
0–904512–03–7 P.903/252
0589829

REDINGTON, Christine
Select bibliography of drama and education. – [Rev.] ed.. – London : British Theatre Institute, 1980
19p. – (BTI bibliographic series ; no.3)
B81–16269 X.950/4319
0589829

ACTIVITIES. Schools
Drama. Improvisation – For teaching

STANLEY, Susan M.
Drama without script. – London : Hodder and Stoughton, 1980
[4],139p : ill. – (Unibooks)
0–340–23768–6 X.909/45212
202103X

ACTIVITIES. Schools
Drama. Teaching

MCCASLIN, Nellie
Creative drama in the classroom. – 4th ed. – New York ; London : Longman, c1984
xxiv,423p : ill. – Bibliography: p397–417. – Includes index
0–582–28508–9 X.950/36755
0245917

Creative dramatics in the classroom. – 2nd ed. – New York ; London : Longman, 1977
xviii,236p : ill. – Bibliography: p.221–231. – List of films: p.231–234. – Includes index
0–582–28007–9 X.909/42620
0245917

PEACHMENT, Brian
Educational drama. – Plymouth : Macdonald and Evans, 1976
xvi,215p : ill. – Includes bibliographies and index
0–7121–0552–2 X.909/40890
0245917

ACTIVITIES. Schools
Improvised drama, to date

FRIBERT, Søren
Ideer til dramatik. – [Grasten] : Drama, 1979
xvi,88p
87–7419–102–0 X.529/55799
8987793

ACTIVITIES. Schools
Music. Projects

DENMAN, Christopher
Ten scenarios. – Cambridge : Cambridge University Press, 1984
48p : ill. – (Music in the making)
0–521–27491–5 L.49/3037
3219739

ACTIVITIES. Schools. England
Commonwealth Day celebrations – For teaching

COMMONWEALTH INSTITUTE. Education Department
Commonwealth Day in schools. – [London] : The Institute, [1978]
14,[4]p : ill
B79–03952 X.702/5842
1761714

ACTIVITIES. Schools. England
Drama

ALLEN, John, 1912–
Drama in schools. – London : Heinemann Educational, 1979
xii,196p. – Bibliography: p184–191. – Includes index
0–435–18033–9 X.950/4097
0794538

SLADE, Peter
New trends. – [Birmingham] : Educational Drama Association, [1976?]
14p
B77–04354 X.909/41120
0794538

ACTIVITIES. Schools. England
Drama – Serials

LEEDS drama. – Issue 1 (winter 1982/83)–. – Leeds : The Association, 1982–
v. : ill
B84–24955 P.903/935
0794538

ACTIVITIES. Schools. England
Drama. Teaching

ALLEN, John, 1912–
Drama in schools. – London : Heinemann Educational, 1979
xii,196p. – Includes index
0–435–18031–2 X.989/53130
1564870

ACTIVITIES. Schools. England
Volleyball. Teaching

JAMES, Dave
Volleyball for schools. – London : Pelham, 1976
141p : ill
0–7207–0886–9 X.629/10891 Woolwich
0754226

ACTIVITIES. Schools. Great Britain
Drama – Conference proceedings

DRAMA AND THEATRE IN EDUCATION (Conference : 1969 : Bristol)
Drama and theatre in education. – London : Heinemann Educational, 1973
[7],175p
0–435–18271–4 X.909/43589
1895443

ACTIVITIES. Schools. Great Britain
Drama. Influence of professional theatre – Conference proceedings

THEATRE-EDUCATION, AN EXPLORATION (Conference : 1978 : Riverside Studios, London)
Exploring theatre and education. – London : Heinemann Educational, 1980
viii,191p : ill. – Bibliography : p.189–191
0–435–18780–5 X.529/36786
2050242

ACTIVITIES. Schools. Great Britain
Drama. Role in development of students – For teaching – Conference proceedings

SLADE, Peter
Child drama and its value in education. – [Bromley] : Educational Drama Association, 1974
18p
0–9501295–6–9 X.900/25316
104088x

ACTIVITIES. Schools. Great Britain
Drama. Teaching. Organisations: Educational Drama Association – Serials

[NEWSLETTER (Educational Drama Association)]. Newsletter / Educational Drama Association. – [No.1]–. – [Birmingham] : The Association, [197–?]–
v.
Holdings: Feb. 1980, etc
B81–06388 P.523/1452
2358913

ACTIVITIES. Schools. Great Britain
Sports & games – For teaching – Serials

SCHOOL Sport / [National Council for Schools' Sports]. – Vol.1, no.1– ; Jan.–Feb. 1976–. – London : Simavi Publishing Co. Ltd, 1976– : ill
B76–25987 P.443/310
0697788

ACTIVITIES. Science. Curriculum subjects. Infant schools – For teaching

HARLAN, Jean
Science experience for the early childhood years. – 2nd ed. – Columbus ; London : Merrill, c1980
xi,237p : ill. – Includes bibliographies and index
0–675–08155–6 X.622/9931 Woolwich
2383691

ACTIVITIES. Scouts

BARROW, Jean
A Pack Scouter's activity book. – Glasgow : Brown and Ferguson, 1975
[4],vii,147p : ill
0–85174–230–0 X.619/15265
0365335

CALAMO
Scout spare time activities. – London : Scout Association, 1974
32p : ill. – (The patrol books ; no.25)
0–85165–052–x WP.3046/25
0365335

EUROPEAN SCOUT FEDERATION. British Association
Handbook. – Manchester : The Federation.
Vol.1: Wolf cubs. – 1978
[2],51p : ill
B80–00649 X.0619/15009(1) Woolwich
0365335

ACTIVITIES. Secondary schools
Drama

HANRATTY, Jerome
Drama cards. – Cambridge : Cambridge University Press, 1977
1v. : ill(some col)
0–521–20926–9 Cup.911/55
1334514

KING, Neil
Classical beginnings. – Amersham : Hulton Educational, 1984
96p : ill. – (Hulton drama series)
0–7175–1230–4 X.950/36332
1453793

LEWIS, Rosemary
Active drama. – Richmond, Vic. : Heinemann Educational Australia, 1981
viii,109p : ill(some col.)
0–85859–244–4 X.950/21345
1453793

ACTIVITIES. Secondary schools
Drama – Texts

BOWSKILL, Derek
Drama in action. – London : Hutchinson, 1979
174p
0–09–140191–7 X.909/44168
0122386

KING, Neil
Mystery and morality. – Amersham : Hulton Educational, 1984
96p : ill. – (Hulton drama series)
0–7175–1231–2 X.950/36333
1453793

ACTIVITIES. Secondary schools
Drama. Improvisation

NASH, S. G.
Outlines. – Huddersfield : Schofield & Sims, 1981
48p
0–7217–0428–x X.950/10280
1413899

ACTIVITIES. Secondary schools
Drama. Improvisation – For teaching

JAMES, Ronald
A guide to improvisation. – Banbury : Kemble, c1980
x,108p
0–906835–05–4 X.529/39408
2125617

ACTIVITIES. Secondary schools
Drama. Improvisation - Texts
PORTER, Sue
Play it again. - London : Arnold, 1984
60p
0-7131-0698-0 X.950/32885
1413899
SEELY, John
In role. - London : Edward Arnold, 1978
vi,56p
0-7131-0295-0 X.909/43317
1413899

ACTIVITIES. Secondary schools
Drama. Improvisation. Role in development of students' language skills
SEELY, John
In context. - London : Oxford University Press, 1976
[9],170p : ill. - (Oxford studies in education ; 12). - Bibliography: p.168-170
0-19-913222-4 X.909/40579
0644455

ACTIVITIES. Secondary schools
Drama. Improvisation. Teaching
CHILVER, Peter
Teaching improvised drama. - London : Batsford, 1978
176p. - Bibliography: p.175-176
0-7134-1036-1 X.989/52688
1551922

ACTIVITIES. Secondary schools
Drama. Role-playing
RAMSEY, Gaynor
Play your part. - London : Longman, 1978
vi,33p : ill
0-582-55250-8 X.611/8390 Woolwich
1755560

ACTIVITIES. Secondary schools. United States
Sports & games. Management
FUOSS, Donald E.
Creative management techniques in interscholastic athletics. - New York ; London : Wiley, 1977
xviii,494p : ill. - Includes bibliographies and index
0-471-28815-2 X.620/16942 Woolwich
1166409

ACTIVITIES. Socially disadvantaged pre-school children
- For teaching
CURTIS, Audrey
My world. - Windsor : NFER, 1978
167p : ill. - Bibliography: p.149-151. - Includes index
0-85633-156-2 X.510/16143
1879316

ACTIVITIES. Sunday schools
Handicrafts using paper - For teaching
WATNEY, Myrtle
Activities for Bible lessons. - London : Oliphants.
2: Things to make with cardboard. - 1974
[1],33p : ill. - Bibliography: inside cover
0-551-00514-9 X.0108/506(2)
0724629
3: Things to make with cardboard. - 1974
[1],33p : ill. - Bibliography: inside cover
0-551-00515-7 X.0108/506(3)
0724629

ACTIVITIES. Teaching aids. Development. Babies
- For parents
[BABY progress guides]. - [Hoylandswaine] : Psycho-Educational Products, c1982
6pieces
B83-12726 X.529/52545
2851172

ACTIVITIES. Teaching aids. Development. Children, to 5 years
BAILEY, Rebecca Anne
The dynamic self. - St. Louis ; London : Mosby, 1982
viii,195p : ill. - Bibliography: p188-189. - Includes index
0-8016-0438-9 X.520/25839
2613271

ACTIVITIES. Teaching aids. Development. Reading skills. Children
FOWLER, Zinita
Monster magic. - Phoenix : Oryx, 1983
65p : ill. - (A Fun with reading book)
0-89774-044-0 X.525/8074
2958368
KIMBALL, Judith A.
Children's caravan. - Phoenix : Oryx, 1983
x,75p : ill. - (A Fun with reading book). - Includes bibliographies
0-89774-043-2 X.525/8073
2958368
THOMAS, Carol H.
Sports splash. - Phoenix : Oryx, c1983
xi,109p : ill. - (A Fun with reading book)
0-89774-000-9 X.525/8072
2958368

ACTIVITIES. Teaching aids. English language. Curriculum subjects. Non-English speaking students
GREENALL, Simon
Language games and activities. - Amersham : Hulton, 1984
95p : 1ill. - Includes index
0-7175-1272-x X.950/36680
3209660

ACTIVITIES. Teaching aids. Mathematics. Curriculum subjects. Primary schools
SMITH, Peter, 1924-
Activities for primary mathematics. - London : Macmillan Education, 1983
v,49p : ill. - (Classroom guides)
0-333-30694-5 X.629/21634
2983117

ACTIVITIES. Teaching aids. Mathematics. Curriculum subjects. Secondary schools
Sports
KIRKBY, Dave
Mathematics in sport. - Sheffield : Eigen.
6: The marathon. - c1984
12p : 2ill
1-85039-165-3 X.629/25620
3254763

ACTIVITIES. Teaching aids. Reading skills. Students. Primary schools
SMITH, Peter, 1924-
Reading skills and reference work. - London : Macmillan, 1983
v,32p : ill. - (Classroom guides). - Bibliography: p32
0-333-30695-3 X.529/57791
2983125

ACTIVITIES. Teaching. Handwriting. Curriculum subjects. Children
GOURDIE, Tom
Fun and games. - Edinburgh : Macdonald Publishers, c1981
32p : ill. - (Teaching children to write ; bk.1)
0-904265-63-3 L.49/1087(1)
2600161

ACTIVITIES. United Nations
Coordination
HILL, Martin
The United Nations system. - Cambridge : Cambridge University Press, 1978
xv,252p : 1 ill. - Includes index
0-521-21674-5 X.520/21073
2130734

ACTIVITIES. United Nations
Political aspects
GRIFFIN, G. Edward
The fearful master. - Boston, Mass. : Western Islands, c1964
ix,244p,[64]p of plates : ill
r6865447 X.809/54245
2854171

ACTIVITIES. West Germany
Traditional seasonal activities, to date
OESCHGER, Bernhard
Zwischen Santiklaus und Martinsritt. - Frankfurt am Main : Lang, 1981
347p : ill. - (Artes populares ; Bd.5). - Bibliography
3-8204-6096-9 X.0709/839(5)
8750394

ACTIVITIES. Young persons
Award schemes: Duke of Edinburgh's Award
COPISAROW, Alcon
A study of the Duke of Edinburgh's Award Scheme. - [London] : [The Duke of Edinburgh's Award], [1976]
[1],16p : ill
0-905425-00-6 X.512/5165
0637254

ACTIVITIES. Young persons
Award schemes: Duke of Edinburgh's Award - Serials
AWARD news / the Duke of Edinburgh's Award Scheme. - [No.1]-. - London : Duke of Edinburgh's Award, [19-]- : ill
Holdings: Jan.1971-
r6072487 P.443/351
1701592

ACTIVITIES. Young persons organisations. Romania
ACTIVITĂTI tehnice - stiintifice pionierești / [volumul a fost pregătit pentru tipar de Gheorghita Fleancu]. - București : Editura politică, 1971
486p
r6074827 X.519/35304
1510924

ACTIVITIES. Young persons. Union of Welsh Independents
- Texts
NICHOLAS, W. Rhys (William Rhys)
Oedfa'r ifanc. - 3 argraffiad. - Abertawe [Swansea] : Gwasg John Penry, 1982
96p
B83-29631 X.950/22162
2923890

ACTIVITIES. Youth clubs
CURRICULUM development in the youth club. - Leicester : National Youth Bureau, 1983
8p
0-86155-074-9 X.525/10020
0331120

ACTIVITIES. Youth groups
- Bibliographies
TREE, Keith S.
A book list for youth leaders. - [Orpington] : Baden-Powell Scouts Association, [c1984]
[17]p
0-9501609-1-1 2725.g.302
3273024

ACTIVITIES BASED ON STORY-TELLING BY STUDENTS. Primary schools
WALKER, Brenda
We made a story. - London : Miller [etc.], 1975
vii,136p : ill. - Bibliography: p.129-135. - List of gramophone records: p.125-129
0-85343-535-9 X.989/50749
0668540

ACTIVITIES CONNECTED WITH ANIMALS. Activities for pre-school children
- For playgroups
GOODAY, Margaret
Animals. - Glasgow : Scottish Pre-school Playgroups Association, 1982
12p : ill. - (Activity booklet)
0-907662-08-0 X.329/15929
2835568

ACTIVITIES CONNECTED WITH BEACHES. Great Britain
Safety measures
A SAFETY code for all who use our beaches / [by Tony Bailey et al.]. - St Austell : Haven Marketing, [1976]
[1],22p : ill
0-9505226-0-0 X.619/17091 Woolwich
0739774

ACTIVITIES CONNECTED WITH BED
WIK, Lars
Bedtime. - London : Methuen, 1979
[15]p : of col ill. - (Look and see books)
0-416-88000-2 X.990/16277 Woolwich
2130955

ACTIVITIES CONNECTED WITH BIRTHDAY PARTIES
GIBBONS, Gail
Things to make and do for your birthday. - New York ; London : F. Watts, 1978
48p : col ill. - (A things to make and do book)
0-531-02380-x X.992/3582 Woolwich
1589237

ACTIVITIES CONNECTED WITH CHRISTIAN OVERSEAS MISSIONS. Christian youth groups
LENNON, Dennis
Young world. - [Revised] ed. - Sevenoaks : OMF Books, 1979
44p : ill. - (A Belmont booklet)
0-85363-125-5 X.108/17602
1935003

ACTIVITIES CONNECTED WITH CHURCHES
Activities for children - Texts
HENCHER, John
A place to dream - a new way of looking at churches and cathedrals. - [London] : Church Information Office, 1976
viii,71p : ill
0-7151-0352-0 X.100/16596
0747947

ACTIVITIES CONNECTED WITH CREATION. Christian religious education. Children, 9-13 years
DURRAN, Maggie
Creation. - Poole : Post Green Press, c1983
118p : ill. - (A Take time to grow course). - Bibliography: p116-117
0-906309-23-9 L.49/1768
290909x

ACTIVITIES CONNECTED WITH DANCING. Activities for educationally subnormal children
LUDUS NORTH WEST DANCE IN EDUCATION (Group)
The adventures of Mr Lear and his cat Foss. - Lancaster : Ludus North West Dance in Education.
Follow up material. - [1978]
Portfolio : ill
B79-03508 Cup.900.w.29
1765485
Preparatory material. - [1978]
Portfolio : ill
B79-03509 Cup.900.w.28
1765485

ACTIVITIES CONNECTED WITH DIGITAL COMPUTER SYSTEMS. Activities for children
DIAMOND, Dorothy
Pre-computer activities. - Amersham : Hulton, 1983
71p : ill. - (Hulton practical teaching series). - Includes index
0-7175-1154-5 X.629/20970
2946378

ACTIVITIES CONNECTED WITH ENVIRONMENT
NORTON, Michael
Colour in your environment. - London : CSV Advisory Service, c1981
31p : ill
0-9501450-7-6 L.49/1784
2903105

ACTIVITIES CONNECTED WITH FESTIVALS
CORDELLO, Becky Stevens
Celebrations. - New York : Butterick, c1977
ix,311p32p of plates : ill(some col.). - Includes index
0-88421-034-0 X.622/6378 Woolwich
1521500
- [Another copy] X.622/9559 Woolwich

ACTIVITIES CONNECTED WITH FESTIVALS
Special subjects: Egypt
FARNAY, Josie
Egypt. - New York : Sterling [etc.] ; London : Distributed by Ward Lock, 1979
96p : ill. - Includes index
0-7061-2615-7 X.808/33359
2062410

ACTIVITIES CONNECTED WITH FESTIVALS
Special subjects: India
SOLEILLANT, Claude
Activities & projects, India in color. - New York : Sterling [etc.] ; London : Distributed by Ward Lock, 1977
96p : ill(chiefly col). - Includes index
0-7061-2559-2 X.629/11972 Woolwich
1544209

ACTIVITIES CONNECTED WITH FESTIVALS
Special subjects: Mexico
SOLEILLANT, Claude
Activities & projects, Mexico in color. - New York : Sterling [etc.] ; London : Distributed by Ward Lock, 1977
96p : ill(chiefly col). - Includes index
0-7061-2560-6 X.629/11965 Woolwich
1545566

ACTIVITIES CONNECTED WITH GARDENS
- Collections
GUNDREY, Elizabeth
Fun in the garden. - London : Pan Books, 1976
124p : ill. - (A piccolo original)
0-330-24654-2 X.619/16612 Woolwich
0619922
Fun in the garden. - London : Severn House : [Distributed by Hutchinson], 1978
127p : ill
0-7278-0334-4 X.990/10327 Woolwich
0619922

ACTIVITIES CONNECTED WITH GROUPING
Activities for pre-school children
GRENDER, Iris
Sorting & matching. - [Kenley] : Three Four Five, [1976]
[4].p,12 leaves : chiefly ill(chiefly col). - (Pre-school activity books)
0-903016-41-9 X.619/17499 Woolwich
1092111

ACTIVITIES CONNECTED WITH HANDS
HARRISON, Jeanne
First themes, hands. - London : Longman, 1977
5v : ill(some col). - (Aware). - Includes bibliographies
0-582-18863-6 Cup.1281/918
1585452

ACTIVITIES CONNECTED WITH INTERPERSONAL RELATIONSHIPS
Families
HENDRICKS, Gay
The family centering book. - Englewood Cliffs ; London : Prentice-Hall, 1979
xii,146p : 1 ill. - (The transformation series). - Includes index
0-13-301846-6 X.519/30790
2075822

ACTIVITIES CONNECTED WITH LEARNING. Activities for pre-school children
- For parents
HARTMAN, Harriet
Let's play and learn. - New York : Human Sciences Press ; London : Distributed by Eurospan Ltd, 1976
138p : ill. - (Early childhood series). - Bibliography: p.17
0-87705-241-7 X.519/29459
1844318

ACTIVITIES CONNECTED WITH MATHEMATICAL RELATIONS. Middle schools
- For teaching
SPARROW, Len
Relationships. - Nottingham : University of Nottingham, School of Education, 1979
[3],47[i.e.58]p : ill. - Bibliography: p.47
0-85359-060-5 X.512/6677
1915479

ACTIVITIES CONNECTED WITH MUSIC. Activities for primary school children
THOMPSON, Diana
Pompaleerie jig. - Leeds : E.J. Arnold, 1978
47p : ill(chiefly col)
0-560-00319-6 Cup.1281/1016
1771558

ACTIVITIES CONNECTED WITH NUMERATION
Activities for pre-school children
GRENDER, Iris
Counting games. - [Kenley] : Three Four Five, [1976]
[4]p,12 leaves : chiefly ill(chiefly col). - (Pre-school activity books)
0-903016-42-7 X.619/17479 Woolwich
1089196

ACTIVITIES CONNECTED WITH ORGANISMS. Activities for children
KATZ, Adrienne
Naturewatch. - London : Piatkus, 1985
128p : ill. - Includes index
0-86188-498-1 X.329/19690
3118665

MELLANBY, Jean
Wonder why book of nature fun. - London : Transworld, 1978
[2],33p : chiefly col ill
0-552-57028-1 X.992/3684 Woolwich
1772562

ACTIVITIES CONNECTED WITH ORGANISMS. Activities for children. Countryside
CORNELL, Joseph Bharat
Sharing nature with children. - [Watford] : Exley in association with Inter-Action Inprint, [1982?]
143p : ill
0-905521-36-6 X.329/16338
290084x

ACTIVITIES CONNECTED WITH PARTS
Activities for pre-school children
EHRENMARK, Michele
Bits & pieces. - [London] : Macdonald Educational, [1979]
[4]p,12 leaves : chiefly ill(chiefly col). - (Pre-school activity books) (Macdonald 345 play and learn book)
0-903016-82-6 X.619/19758 Woolwich
1891294

ACTIVITIES CONNECTED WITH PEACE. Society of Friends
- Serials
QUAPS. - No.1- ; Mar. 1979-. - London : Quaker Peace and Service, [1979]- : ill
B80-18594 P.103/565
2051206

ACTIVITIES CONNECTED WITH PICNICS
MANLEY, Deborah
The piccolo picnic book. - London : Pan Books, 1976
142p : ill. - (A piccolo original)
0-330-24717-4 X.619/16862 Woolwich
0675792

ACTIVITIES CONNECTED WITH RHYTHMS. Teaching aids. Primary schools
JOHNSON, Mary E
You and rhythm. - Lanham ; London : University Press of America, c1984
xii,87p : ill
0-8191-3670-0 X.49/294
3186776

ACTIVITIES CONNECTED WITH SEASONAL FESTIVALS
Activities for families - Collections
CAREY, Diana
Festivals, family and food. - Stroud : Hawthorn, c1982
viii,216p : ill. - (Who's bringing them up?) - Includes index
0-9507062-3-x X.622/1536
2890720

ACTIVITIES CONNECTED WITH SEASONAL FESTIVALS. Primary schools
- For teaching
BRANDLING, Redvers
Festive occasions in the primary school. - London : Ward Lock, 1978
iv,236p : ill. - Includes bibliographies
0-7062-3746-3 X.519/2881
1736647

ACTIVITIES CONNECTED WITH SHIPS OF ISLE OF MAN STEAM PACKET COMPANY LIMITED. Activities for children. Irish Sea
HUGHES, R. D.
Young spotters' guide to Isle of Man ships. - Chorley : The author, 1978
[1],26p : ill
B78-29841 X.709/3245
1570277

ACTIVITIES CONNECTED WITH SICKNESS. Primary schools
- For teaching
MICHAEL, Bill
Something to say about feeling poorly. - London : Evans Bros, 1977
48p : ill. - (Practical language development)
0-237-29174-6 X.900/2540
117018x

ACTIVITIES CONNECTED WITH STORY-TELLING. Activities for children, 2-8 years
PETERSON, Carolyn Sue
Story programs. - Metuchen ; London : Scarecrow, 1980
v,294p : ill. - Bibliography: p264-266. - Includes index
0-8108-1317-3 X.622/9449 Woolwic
2345633

ACTIVITIES CONNECTED WITH TIME
Activities for pre-school children
EHRENMARK, Michele
Talking about time. - [London] : Macdonald Educationa
[1979]
[4]p,12 leaves : chiefly ill(chiefly col). - (Pre-school activity books) (Macdonald 345 play and learn books)
0-903016-81-8 X.619/197
189241x

ACTIVITIES CONNECTED WITH TOWNS. Activities for childre
CRUSH, Margaret
Town trail. - London : Dragon, 1985
144p : ill. - Bibliography: p144. - Includes index
0-583-30691-8 X.808/417
3206548

ACTIVITIES CONNECTED WITH TRAVEL. Primary schools
- For teaching
MICHAEL, Bill
Something to say about going places. - London : Evans Bros, 1977
47p : ill. - (Practical language development)
0-237-29175-4 X.900/254
1170171

ACTIVITIES CONNECTED WITH TROOP NIGHTS. Scouts. Great Brit
TROOP night in pictures / [general editor Ron Jeffries ; editor Peter Brooks]. - London : Scout Association, 197
40,[2]p : chiefly ill(some col)
0-85165-147-x X.622/8908 Woolwi
2319721

CTIVITIES CONNECTED WITH WATER Primary schools - For teaching

MICHAEL, Bill
Something to say about water - London Evans Bros. 1977
48p : ill. - (Practical language development)
0-237-29176-2 X.900/25406
1170163

CTIVITIES CONNECTED WITH WINDS Primary schools - For teaching

MICHAEL, Bill
Something to say about the wind - London Evans Bros. 1977
48p : ill. - (Practical language development)
0-237-29173-8 X.900/25409
1170848

CTIVITIES CONNECTED WITH WOODLANDS Girl Guides & Brownie Guides. Great Britain

MORTON, Brenda
The woodland book for Guides and Brownies - Revised ed - Glasgow : Brown and Ferguson, 1976
vii,121p : ill. - Includes index
0-85174-256-4 X.319/16909
0756458
- [Another copy: reprint] X.529/49176

CTIVITIES FOR ADOLESCENT GIRLS

QUANT, Mary
Mary Quant's daisy chain of things to make and do. - Glasgow : Collins, 1975
3-46p : col ill
0-00-103351-4 X.622/5107
0529524

CTIVITIES FOR CATHOLIC FAMILIES
Christmas activities & Easter activities - Collections

PRATT, Oliver
The Christmas and Easter ideas book. - London : Sheed and Ward, 1977
ix,85p : ill. - Bibliography: p.70-72
0-7220-7714-9 X.619/17740 Woolwich
1168142

CTIVITIES FOR CHILDREN
See also
Christian activities for children
Church of England activities for children
Seasonal activities for children

CTIVITIES FOR CHILDREN

215 ideas to keep children happy / [edited by Magda Gray and Yvonne Deutch]. - [London] : [Marshall Cavendish], [1975]
[92]p : ill(some col)
0-85685-137-x X.622/5675 Woolwich
0342955

AXWORTHY, Ann
Cinderella's indoor activity book. - London : n Carousel Books, 1979
[49]p : chiefly ill(some col). - (Famous character activity books)
0-552-54150-8 X.992/3948 Woolwich
0342955
Monster mix. - [London] : Hippo, 1981
48p : col.ill
0-590-70028-6 X.990/19972
0342955

AXWORTHY, Ann
Robin Hood's outdoor activity book. - London : n Carousel Books, 1979
[49]p. - (Famous character activity books)
0-552-54149-4 X.992/3949 Woolwich
0342955
Robinson Crusoe's holiday activity book. - London : n Carousel Books, 1979
[49]p : chiefly ill(some col). - (Famous character activity books)
0-552-54151-6 X.992/3950 Woolwich
0342955
Sinbad's round the world activity book. - London : n Carousel Books, 1979
[49]p : Chiefly ill(some col). - (Famous character activity books)
0-552-54148-6 X.992/3947 Woolwich
0342955

BARBER, Janet
Magpie lots more make & do. - London : Independent Television Books Ltd : Arrow Books, 1979
125p : ill. - (Look-in books)
0-09-917390-5 X.629/13465 Woolwich
0342955

BARWELL, Eve
Free fun. - London : Dragon Books, 1978
92p : ill
B78-13719 X.619/18419 Woolwich
0342955

BENJAMIN, Floella
Floella's fun book. - London : Methuen, 1984
64p : ill. - (A Magnet book)
0-416-46110-7 X.629/24223
0342955

BOND, Michael
Fun and games with Paddington - Glasgow Collins, 1977
4-61p : ill(chiefly col)
0-00-138045-1 X.992/3392 Woolwich
0342955

BRANDRETH, Gyles
Challenge. - London : Transworld, 1981
159p : ill. - (A Carousel book)
0-552-54194-x X.629/17095 Woolwich
0342955
Fun & games for every day of the year. - Leicester : Knight Books, 1976
128p : ill
0-340-20225-4 X.619/17214 Woolwich
0342955
Knight book of fun and games for a rainy day. - Leicester : Knight Books, 1976
126p : ill
0-340-19781-1 X.619/17356 Woolwich
0342955

CHETWOOD, Doreen
What's in a game?. - Edinburgh : Oliver and Boyd, 1976
32p : ill(chiefly col). - (You at home). - Bibliography: p.32
0-05-003006-x X.990/8550 Woolwich
0342955

DUNN, Roger
Things to do with junk. - London : Macdonald Educational, 1980
[24]p : col.ill. - (Things to do books ; 6)
0-356-07057-3 X.990/19496
0342955

GILL, Rosemary
The multi-coloured swap shop book. - London : British Broadcasting Corporation, 1978
80p : ill(chiefly col)
0-563-17603-2 X.431/10554
0342955
The second multi-coloured swap shop book. - London : British Broadcasting Corporation, 1979
80p : ill(chiefly col)
0-563-17741-1 X.622/8010 Woolwich
0342955

GILL, Rosemary
Swap shop. - London : British Broadcasting Corporation, c1981
128p : ill(some col.)
0-563-17989-9 X.622/11881 Woolwich
0342955

GOLDSTEIN-JACKSON, Kevin
Activities with everyday objects. - London : Souvenir Press, 1980
143p : ill
0-285-62439-3 X.529/38453
0342955
Activities with everyday objects. - London : Granada, 1981, c1980
128p : ill
0-583-30464-8 X.990/17433 Woolwich
0342955

GOODLAND, Patrick
300 games for all occasions. - London : Scripture Union, 1979
143p : ill
0-85421-678-2 X.619/19882 Woolwich
057614x

GORDON, Jennie
Difyrrwch hamdden i blant. - Llandysul : Gwasg Gomer, 1978
45p : ill
0-85088-560-4 X.619/18767 Woolwich
1189018

HAYWARD, Stan
The Henry's cat funtime book. - [Bristol] : Purnell, c1983
77p : col.ill
0-361-05927-2 X.992/5261
0342955

HOME and away / compiled by Eva Wilson ; with illustrations by Brian Robins. - London : Collins, 1975
125p : ill. - (Lions)
0-00-670994-x X.619/17307 Woolwich
0342955

HUNTER, Norman
Professor Branestawm's do-if-yourself handbook. - Harmondsworth : Puffin Books, 1979
155,[1]p : ill. - (Puffin books)
0-14-031148-3 X.619/19910 Woolwich
0342955
Professor Branestawm's do-it-yourself handbook. - London : Bodley Head, 1976
133p : ill
0-370-10847-7 X.629/10932 Woolwich
0342955

ICHIKAWA, Satomi
Let's play. - London : Heinemann, 1981
[32]p : chiefly col.ill
0-434-94365-7 L.49/939
2540088

The KNOW How make & do encyclopedia / written and devised by Heather Amery ... [et al.] ; compiled by Christopher Rawson ; designed by John Jamieson ... [et al.] ; illustrated by Colin King ... [et al.]. - London : Usborne, 1976
4-157p : col ill. - Bibliography: p.157. - Includes index
0-86020-035-3 X.622/5831 Woolwich
0342955

The KNOWHOW omnibus. - London : Usborne, 1979
192p : col.ill
0-86020-357-3 X.622/12194 Woolwich
0342955

LEWIS, Shari
Spooky stuff. - Kingswood : Kaye & Ward, 1982, c1979
96p : ill. - Includes index
0-7182-3751-x X.990/19624
0342955

MACDONALD, Maryann
Moving, doing, building, being. - London : Andersen, 1980
[25]p : col.ill
0-905478-79-7 X.990/17098 Woolwich
2336804

MCPHEE GRIBBLE PUBLISHERS
Strange things to do and make. - Harmondsworth : Puffin Books, 1976
[34]p : chiefly ill(chiefly col). - (Practical puffins)
0-14-049137-6 X.619/16947 Woolwich
0342955
Strange things to do and make. - Harmondsworth : Kestrel Books, 1976
[32]p : ill(some col). - (Practical puffin)
0-7226-5340-9 X.629/10708 Woolwich
0342955
Strange things to do and make. - Revised ed. - Harmondsworth : Puffin Books, 1978
[34]p : ill(chiefly col). - (Practical puffins)
B79-07469 X.990/10989 Woolwich
0342955
Strange things to do and make. - Revised ed. - Harmondsworth : Kestrel Books, 1978
[32]p : ill(chiefly col). - (Practical puffin)
B79-11139 X.629/12304 Woolwich
0342955

MANLEY, Deborah
Fun for one. - London : Dragon Books, 1979
96p : ill
0-583-30293-9 X.619/19863 Woolwich
0342955

MARZOLLO, Jean
Superkids. - London : Unwin Paperbacks, 1984, c1981
xi,184p : ill. - Includes index
0-04-649028-0 X.629/23710
0342955

MILES, Vanessa
What can I do?. - London : Severn House, 1984
212p : ill
0-7278-1071-5 X.629/24845
0342955

MURPHY, Hilary, 19—
Why don't you? 2. - London : British Broadcasting Corporation, 1983
96p : ill. - Includes index
0-563-20187-8 X.629/21587
0342955

NEVER a dull moment / [editor Sue Tarsky]. - London : Walker Books, 1983
93p : col.ill
0-7445-0000-1 X.622/17483
0342955

OWEN, Elina
Wrth droed yr enfys. - [Llanrwst] : Gwasg Carreg Gwalch, 1984
[64]p : ill
B85-26412 X.622/24079
1189018

PACK of fun / Islington Bus Co., CSV Advisory Service. - London : Community Services Volunteers & Islington Bus Co., c1984
1portfolio : ill
0-907829-95-3 Cup.920/91
057614x

The PUFFIN do it yourself book / written and produced by McPhee Gribble Publishers ; with illustrations by David Lancashire and Jenny Elliott. - Ringwood, Vic. ; Harmondsworth : Puffin, 1984
120p : ill(some col.)
0-14-031723-6 X.629/25401
0342955

RHYS, Ieuan
Hwyl a joio. - Talybont, Ceredigion : Y Lolfa, 1981
128p : ill
0-86243-013-5 X.629/19502
1189018

ACTIVITIES FOR CHILDREN -continued

Joio mas draw. - Talybont, Ceredigion : Y Lolfa, 1983
156p : ill
0-86243-052-6 X.990/22040
1189018

SATURDAY SuperStore. - London : British Broadcasting Corporation, 1984
96p : ill(some col.)
0-563-20319-6 X.950/41165
0342955

SCARLETT, Clare
Make and share. - London : Piccolo, 1984
48p : ill(some col.)
0-330-28514-9 X.629/25982
0342955

SCARRY, Richard
Richard Scarry's best make-it book ever!. - Glasgow : Collins, 1978
143p : ill(chiefly col)
0-00-138224-1 X.611/8041 Woolwich
0342955

STAPLETON, Marjorie
Child's play. - [London] : Star, 1981, c1980
144p : ill. - Includes index
0-352-30680-7 X.629/15234 Woolwich
057614x

STEVENS, Catrin
Sgorio 1,000. - Talybont, Ceredigion : Y Lolfa.
6: Cartrefi Cymru. - 1982
63p : ill
0-86243-035-6 X.808/41063
2946084

9: Gwaith a hamdden. - 1983
96p : ill
0-86243-062-3 X.958/25274
2946084

10: Y byd pop. - 1983
96p : ill
0-86243-063-1 X.529/65627
2946084

Sgorio 1000. - Talybont, Ceredigion : Y Lolfa.
2: Ar y ffordd. - 1981
64p : ill
0-86243-017-8 X.629/20963
2946084

4: Ar y traeth. - 1981
96p : ill
0-86243-024-0 X.329/16733
2946084

STOLL, Martin
Fun for free. - London : Armada, 1981
128p : ill. - Includes index
0-00-691866-2 X.629/15050 Woolwich
0342955

SUPERBOOK / [edited and designed by the Reader's Digest Association Limited]. - London : Published by the Reader's Digest Association Limited in conjunction with Mothercare Limited, c1981
223p : col.ill
B81-34999 X.622/11059 Woolwich
0342955

TANAKA, Béatrice
Build-it workshop. - London : Pepper, 1983
47p : ill(some col.)
0-237-45721-0 X.622/23006
0342955

The TISWAS phantom flan flinger's fun book / compiled by Helen Piddock ; illustrated by Mike Miller. - London : Carousel, 1981
136p : ill
0-552-54167-2 X.990/18347 Woolwich
2336804

VANCE, Eleanor Graham
A treasury of things to make and do. - London : Hamlyn, 1975
4-140p : ill(chiefly col). - Includes index
0-600-33122-9 X.622/5189 Woolwich
057614x

WHY don't you - ? / compiled by Hilary Murphy. - London : British Broadcasting Corporation, 1982
95p : ill. - Includes index
0-563-20051-0 X.629/18537
0342955

WILLIAMS, Cen
Ffau'r Llewod!. - Talybont, Ceredigion : Y Lolfa, 1977
95p : ill. - (Pocedlyfrau'r Lolfa)
0-904864-43-x X.619/17803 Woolwich
1189018

WILLIAMS, Eilir M.
Anrhegion syml i blant. - Penygroes, Caernarfon : Cyhoeddiadau Mei, 1982
24p : ill. - (Llyfr sboncyn)
B83-25729 X.990/20320
1189018

ACTIVITIES FOR CHILDREN
- For parents

BECKETT, Geoffrey
Amusing the children. - Maidenhead : Sampson Low, 1975
96p : ill(chiefly col). - ([Leisuretime])
0-562-00018-6 X.622/5120
0534501

JOHNSON, June
555 ways to amuse a child. - U.K. ed. abridged and adapted. - Kingswood : Elliot Right Way, c1981
173p : ill. - (Paperfronts). - Includes index
0-7160-0668-5 X.629/15708 Woolwich
0534501

MAKING and using. - London : Mitchell Beazley, 1982
72p : col.ill. - (Let's learn). - Includes index
0-85533-425-8 L.45/2041
0534501

MILLER, Maureen
To share with your children. - Niles, Ill. : Argus Communciations, 1978
107p : col,ill
r6051256 X.519/29778
0534501

ACTIVITIES FOR CHILDREN
- For parents - Serials

LEARNING together. - 1-4. - London : Evans Brothers, 1980-1980
v. : ill
Holdings: no 2, 1980 etc
B81-36557 P.523/1552
2555409

ACTIVITIES FOR CHILDREN
- Texts

MOSCHOPS playtime book. - Bristol : Purnell, 1983
77p : col.ill
0-361-05953-1 X.992/5367
0342955

PORTLAND Bill's sunshine book. - Bristol : Purnell, 1983
78p : col.ill
0-361-05951-5 X.992/5368
0342955

ACTIVITIES FOR CHILDREN
Activities connected with digital computer systems

DIAMOND, Dorothy
Pre-computer activities. - Amersham : Hulton, 1983
71p : ill. - (Hulton practical teaching series). - Includes index
0-7175-1154-5 X.629/20970
2946378

ACTIVITIES FOR CHILDREN
Activities connected with organisms

KATZ, Adrienne
Naturewatch. - London : Piatkus, 1985
128p : ill. - Includes index
0-86188-498-1 X.329/19690
3118665

MELLANBY, Jean
Wonder why book of nature fun. - London : Transworld, 1978
[2],33p : chiefly col ill
0-552-57028-1 X.992/3684 Woolwich
1772562

ACTIVITIES FOR CHILDREN
Activities connected with towns

CRUSH, Margaret
Town trail. - London : Dragon, 1985
144p : ill. - Bibliography: p144. - Includes index
0-583-30691-8 X.808/41757
3206548

ACTIVITIES FOR CHILDREN
Organisations

CAMPBELL, Helen F. (Helen Farrar)
Children at play. - New ed. - Belfast : Northern Ireland Council of Social Service, 1981
32p : ill
0-903087-08-1 X.529/56553
2937719

ACTIVITIES FOR CHILDREN. Activities connected with churches
- Texts

HENCHER, John
A place to dream - a new way of looking at churches and cathedrals. - [London] : Church Information Office, 1976
viii,71p : ill
0-7151-0352-0 X.100/16596
0747947

ACTIVITIES FOR CHILDREN. Brighton. East Sussex

CHILDREN'S Brighton / compiled and written by members and friends of the Brighton Society ; edited by Robert Logan ; illustrated by Selma Nankivell. - Bristol : Redcliffe Press Ltd, 1977
112p : ill. - Includes index
0-905459-02-4 X.708/21533
1726420

ACTIVITIES FOR CHILDREN. Bristol region. Avon

CHILDREN'S Bristol / compiled and written by Janet Bishop ... [et al.] ; edited by John Sansom ; illustrated by Alan Hurst. - Bristol : Redcliffe Press Ltd, 1976
120p : ill. - Includes index
0-905459-00-8 X.708/20412
0672580

CHILDREN'S Bristol / compiled and written by Merlyn Chambers ... [et al.] ; illustrated by Alan Hurst. - Rev. and enl. ed. - Bristol : Redcliffe Press, 1979
152p : ill. - Includes index
0-905459-14-8 X.808/33641
0672580

ACTIVITIES FOR CHILDREN. Countryside
Activities connected with organisms

CORNELL, Joseph Bharat
Sharing nature with children. - [Watford] : Exley in association with Inter-Action Inprint, [1982?]
143p : ill
0-905521-36-6 X.329/16338
290084x

ACTIVITIES FOR CHILDREN. England
Information sources

CONGDON, Peter J. (Peter John)
Which and where?. - [Solihull] : Gifted Children's Information Centre, [1980]
[3],75p. - Includes index
B80-41115 X.529/38222
2311488

ACTIVITIES FOR CHILDREN. Eucharist. Church of England

FRANCIS, Leslie
The child & the Eucharist. - 2nd and rev. ed. - Great Wakering : Mayhew McCrimmon, 1982
32p : ill. - Bibliography: p16
0-85597-320-x X.200/40017
0696811

FRANCIS, Leslie Roose
His spirit is with us. - London : Collins, 1981
238p : ill
0-00-599684-8 X.200/37311
0696811

SHARE the Word / by the Wadderton Group. - [London] : CIO.
2. - Rev. - 1980
viii,159p : ill
0-7151-0392-x X.200/35719
0696811

WADDERTON GROUP
Share the word. - [London] : Church Information Office [for] the General Synod Board of Education, 1976
viii,144p : ill
0-7151-0344-x X.100/16470
0696811

WADDERTON GROUP
Share the word. - [London] : CIO Publishing for the General Synod Board of Education.
2. - 1977
viii,165p : ill
0-7151-0359-8 X.100/17363
0696811

ACTIVITIES FOR CHILDREN. Gardening

ROSS, Alison
Gardening with children. - London : Faber, 1980
176p : ill. - Bibliography: p161-163. - Includes index
0-571-11564-0 X.329/13309 Woolwich
2076934

ACTIVITIES FOR CHILDREN. Gardens

SMITH, Roger, 1930 Mar.15-
Garden fun. - London : Armada, 1980
128p : ill
0-00-691650-3 X.329/12977 Woolwich
2242613

ACTIVITIES FOR CHILDREN. Harvest festivals. Church of England

TOGETHER for harvest. - London : Church Information Office, 1976
78p
B83-19152 X.200/40147
0721123

TOGETHER for harvest. - London : Church Information Office for the General Synod, Board of Education, 1976
[3],78p
0-7151-0345-8 X.909/40829
0721123

ACTIVITIES FOR CHILDREN. Irish Sea
Activities connected with ships of Isle of Man Steam Packet Company Limited

HUGHES, R. D.
Young spotters' guide to Isle of Man ships. - Chorley : The author, 1978
[1],26p : ill
B78-29841 X.709/32454
1570277

ACTIVITIES FOR CHILDREN. Methodist Church

FACE to face with children & young people on Sundays / [editor : Beryl Underhill] ; [illustration : Jim Moss]. - London : Methodist Church, Division of Education and Youth, 1982
60p : ill
0-7192-0149-7 X.205/1295
2921979

FACE to face with young people / [editor : David Pendle] ; [illustration : Jim Moss]. - London : Methodist Church, Division of Education and Youth, 1982
32p : ill
0-7192-0148-9 X.525/7483
2921979

ACTIVITIES FOR CHILDREN. Religious festivals. Church of England

TOGETHER for festivals. - Westminster : Church Information Office, 1975
184p : ill. - Bibliography: p.169
0-7151-0333-4 X.909/40174
0531111

ACTIVITIES FOR CHILDREN, 2-5 YEARS
- For parents

WEBB, Alison
Fun for kids. - Wellington [N.Z.] ; London : A.H. and A.W. Reed, 1976
[2],72p : ill(some col)
0-589-00911-7 Cup.1281/742
1336177

ACTIVITIES FOR CHILDREN, 2-8 YEARS
Activities connected with story-telling

PETERSON, Carolyn Sue
Story programs. - Metuchen ; London : Scarecrow, 1980
v,294p : ill. - Bibliography: p264-266. - Includes index
0-8108-1317-3 X.622/9449 Woolwich
2345633

ACTIVITIES FOR CHILDREN, 4-7 YEARS

EVANS, Delphine
What shall we do today. - London : Hutchinson, 1985
111p : ill
0-09-160400-1 X.990/24476
1097296

PRESLAND, John, 1935-
How many?. - Purton : Child's Play (International) Ltd, 1975
[14]p,[2]leaves of plates : chiefly col ill. - (Child's Play motivation books ; 3)
0-85953-045-0 X.0990/407(3) Woolwich
1097296

Letters and words. - Purton : Child's Play (International) Ltd, 1975
[14]p,[2]leaves of plates : chiefly col ill. - (Child's Play motivation books ; 4)
0-85953-046-9 X.0990/407(4) Woolwich
1097296

Same and different. - Purton : Child's Play (International) Ltd, 1975
[14]p,[2]leaves of plates : chiefly col ill. - (Child's Play motivation books ; 1)
0-85953-043-4 X.0990/407(1) Woolwich
1097296

What is it?. - Purton : Child's Play (International) Ltd, 1975
[14]p,[2] leaves of plates : chiefly col ill. - (Child's Play motivation books ; 2)
0-85953-044-2 X.0990/407(2) Woolwich
1097296

ACTIVITIES FOR CHILDREN, 7-14 YEARS
Organisation

CATHOLIC YOUTH COUNCIL
In it together. - Dublin : C.Y.C., [1978?]
33,[3]p : ill
B79-14018 X.619/19613 Woolwich
1846450

ACTIVITIES FOR CHILDREN ON CAR JOURNEYS
- Collections

DANBY, Mary
Fun on wheels. - London : Armada, 1979
128p : ill
0-00-691651-1 X.619/20018 Woolwich
1905295

MILLER, A. M. L.
For children on wheels. - London : Royal Automobile Club, 1979
118p : ill
0-902628-49-6 X.519/42858
1905295

ACTIVITIES FOR CHILDREN, TO 5 YEARS
- For parents

CORP, Prue
Enquire within upon entertaining the under 5s. - London : Barrie and Jenkins, 1976
vii,91p
0-214-20286-0 X.629/10687 Woolwich
0077720

WARD, Carole
Play school play ideas 2. - London : British Broadcasting Corporation, 1977
127p : ill. - Includes index
0-563-17212-6 BS.129/42
0077720

WHAT to do when 'there's nothing to do' / members of the staff of the Boston Children's Medical Center and Elizabeth M. Gregg ; foreword by Philip Evans ; illustrations by Marc Simont. - London : Arrow Books, 1976
189p : ill. - Bibliography: p.157-170. - List of sound discs: p.171-179. - Includes index
B77-21978 X.619/17711 Woolwich
0077720

ACTIVITIES FOR CHILDREN, TO 5 YEARS
- For parents & for teaching

CROFT, Doreen J.
An activities handbook for teachers of young children. - 2nd ed. - Boston [Mass.] ; London : Houghton Mifflin, 1975
xiv,209p : ill. - Includes bibliographies
0-395-19821-6 X.512/5236
0052574

An activities handbook for teachers of young children. - 3rd ed. - Boston [Mass.] ; London : Houghton Mifflin, 1980
xi,256p : ill. - Includes bibliographies and index
0-395-28698-0 X.525/3100
0052574

ACTIVITIES FOR CHILDREN, TO 6 YEARS
- For parents

BRAGA, Joseph
Children and adults. - Englewood Cliffs ; London : Prentice-Hall, 1976
xi,310p : ill. - (Human development books). - Bibliography: p.283-295. - Includes index
0-13-130351-1 X.520/10576
062523x
- [Another copy] X.512/6234

ACTIVITIES FOR CHILDREN, TO 6 YEARS
- For parents & for teaching

BELTON, Sandra
Sparks. - New York : Human Sciences Press ; London : Distributed by Eurospan Ltd, 1976[i.e.1978]
120,[5]p : ill
0-87705-710-9 LR.421/48
1869329

JORDE, Paula
Living & learning with children. - Rev., expanded and updated ed. - Washington, D.C. : Acropolis Books, c1981
ix,70p : ill. - (Bright ideas)
0-912078-47-2 X.525/7835
1869329

ACTIVITIES FOR CHILDREN, TO 9 YEARS

CHETWOOD, Doreen
Where shall we go today?. - Edinburgh : Oliver and Boyd, 1976
32p : ill(some col). - (You at home). - Includes bibliographies
0-05-003007-8 X.700/25352
1058991

ACTIVITIES FOR EDUCATIONALLY SUBNORMAL CHILDREN
Activities connected with dancing

LUDUS NORTH WEST DANCE IN EDUCATION (Group)
The adventures of Mr Lear and his cat Foss. - Lancaster : Ludus North West Dance in Education.
Follow up material. - [1978]
Portfolio : ill
B79-03508 Cup.900.w.29
1765485

Preparatory material. - [1978]
Portfolio : ill
B79-03509 Cup.900.w.28
1765485

ACTIVITIES FOR FAMILIES. Activities connected with seasonal festivals
- Collections

CAREY, Diana
Festivals, family and food. - Stroud : Hawthorn, c1982
viii,216p : ill. - (Who's bringing them up?). - Includes index
0-9507062-3-x X.622/15363
2890720

ACTIVITIES FOR GIFTED CHILDREN
- For teaching

CONGDON, Peter J. (Peter John)
Teaching able children. - Solihull : Gifted Children's Information Centre, 1978
[2],42p : 1 ill
B79-08646 X.519/29246
1811134

ACTIVITIES FOR HANDICAPPED CHILDREN
- For parents

LEAR, Roma
Play helps. - London : Heinemann Medical, 1977
xi,158p : ill. - (Heinemann health books). - Includes bibliographies and index
0-433-19085-x X.611/7595 Woolwich
108030x

ACTIVITIES FOR MENTALLY HANDICAPPED CHILDREN
Teaching

PHYSICAL and creative activities for the mentally handicapped / edited by Graham Upton. - Cambridge : Cambridge University Press, 1979
xiii,178p : ill. - Bibliography: p.170-175. - Includes index
0-521-21778-4 X.520/14563
1961829

ACTIVITIES FOR ONE PERSON

ERSKINE, Jim
Fold a banana - today!. - London : Muller, 1979
[96]p : chiefly ill
0-584-10360-3 X.421/21289 Woolwich
1929054

ACTIVITIES FOR PRE-SCHOOL CHILDREN

MARTIN, Toy
The pre-school craft book. - Poole : Blandford Press, 1983, c1981
64p : col.ill
0-7137-1330-5 X.629/21142
0165743

ACTIVITIES FOR PRE-SCHOOL CHILDREN
- For teaching - Serials

[ACTIVITY booklets (Scottish Pre-School Playgroups Association)]. Activity booklets / Scottish Pre-school Playgroups Association. - No.1 (Oct. 1980)-. - Glasgow : The Association, 1980-
v. : ill
B82-29055 P.441/974
2759306

ACTIVITIES FOR PRE-SCHOOL CHILDREN
Activities connected with animals - For playgroups

GOODAY, Margaret
Animals. - Glasgow : Scottish Pre-school Playgroups Association, 1982
12p : ill. - (Activity booklet)
0-907662-08-0 X.329/15929
2835568

ACTIVITIES FOR PRE-SCHOOL CHILDREN
Activities connected with learning - For parents

HARTMAN, Harriet
Let's play and learn. - New York : Human Sciences Press ; London : Distributed by Eurospan Ltd, 1976
138p : ill. - (Early childhood series). - Bibliography: p.17
0-87705-241-7 X.519/29459
1844318

ACTIVITIES FOR PRE-SCHOOL CHILDREN
Activities using percussion instruments - For teaching

BLADES, James
Ready to play. - London : British Broadcasting Corporation, 1978
64p : ill
0-563-17610-5 X.431/10557
1720260

ACTIVITIES FOR PRE-SCHOOL CHILDREN. Activities connected with grouping

GRENDER, Iris
Sorting & matching. - [Kenley] : Three Four Five, [1976]
[4],p,12 leaves : chiefly ill(chiefly col). - (Pre-school activity books)
0-903016-41-9 X.619/17499 Woolwich
1092111

ACTIVITIES FOR PRE-SCHOOL CHILDREN. Activities connected with numeration

GRENDER, Iris
Counting games. - [Kenley] : Three Four Five, [1976]
[4]p,12 leaves : chiefly ill(chiefly col). - (Pre-school activity books)
0-903016-42-7 X.619/17479 Woolwich
1089196

ACTIVITIES FOR PRE-SCHOOL CHILDREN. Activities connected with parts

EHRENMARK, Michele
Bits & pieces. - [London] : Macdonald Educational, [1979]
[4]p,12 leaves : chiefly ill(chiefly col). - (Pre-school activity books) (Macdonald 345 play and learn book)
0-903016-82-6 X.619/19758 Woolwich
1891294

ACTIVITIES FOR PRE-SCHOOL CHILDREN. Activities connected with time

EHRENMARK, Michele
Talking about time. - [London] : Macdonald Educational, [1979]
[4]p,12 leaves : chiefly ill(chiefly col). - (Pre-school activity books) (Macdonald 345 play and learn books)
0-903016-81-8 X.619/19780
189241x

ACTIVITIES FOR PRIMARY SCHOOL CHILDREN
Activities connected with music

THOMPSON, Diana
Pompaleerie jig. - Leeds : E.J. Arnold, 1978
47p : ill(chiefly col)
0-560-00319-6 Cup.1281/1016
1771558

ACTIVITIES FOR SCHOOL LEAVERS

CAREERS RESEARCH AND ADVISORY CENTRE
Time between. - [New ed.]. - Cambridge : Hobsons Press, 1977
[1],52p. - Includes bibliographies and index
0-86021-135-5 X.519/28366
1570730

Time between. - 2nd ed. - Cambridge : Hobsons Press, 1980
63p. - Includes bibliographies and index
0-86021-334-x X.529/37145
1570730

ACTIVITIES FOR STUDENTS. Great Britain
- Directories

YOUNG visitors to Britain. - 1976. - London : Central Bureau for Educational Visits and Exchanges, [1976]
36p : ill
0-900087-32-3 X.702/5106
0061492

YOUNG visitors to Britain / the Central Bureau for Educational Visits and Exchanges in association with the British Tourist Authority. - 1978. - London : The Bureau, [1978]
35p : ill
0-900087-41-2 BS.411/246
0061492

ACTIVITIES FOR WOMEN

CONRAN, Shirley
Superwoman 2. - London : Sidgwick and Jackson, 1977
xv,440p : ill. - Includes index
0-283-98389-2 X.520/24573
1516582

Superwoman in action. - Harmondsworth : Penguin, 1979
xv,487p : ill. - Includes index
0-14-004972-x X.519/42862
1516582

ACTIVITIES FOR YOUNG PERSONS
- For welfare work

DEARLING, Alan
The youth arts and craft book. - [Bridge of Weir] : I.T. Resource Centre, 1982
247p : ill. - Bibliography: p242-247
0-9506951-4-9 X.429/14377
2894408

ACTIVITIES PRODUCING SOUND

KIRK, Helen
Making noises. - London : Studio Vista, 1976
44p : chiefly ill. - (Something to do). - Includes index
0-289-70685-8 X.620/16172 Woolwich
0686557

ACTIVITIES USING DIGITAL COMPUTER SYSTEMS. Chemistry
Curriculum subjects. Secondary schools
- For teaching

COMPUTERS IN THE CURRICULUM (Project)
Computers in the chemistry curriculum. - [London] : Schools Council Publications : Distributed by Edward Arnold, 1978
vii,138p : ill. - Includes bibliographies
0-7131-0226-8 X.615/2291 Woolwich
1836463

ACTIVITIES USING DIGITAL COMPUTER SYSTEMS. Economics
Curriculum subjects. Secondary schools
- For teaching

COMPUTERS IN THE CURRICULUM (Project)
Computers in the economics curriculum. - [London] : Schools Council : Distributed by Edward Arnold, 1978
vii,187p : ill. - Includes bibliographies
0-7131-0229-2 X.615/2293 Woolwich
1836455

ACTIVITIES USING ELASTIC BANDS

KENNEWAY, Eric
The rubber band book. - London : Beaver, 1977
93p : ill. - (A Beaver original)
0-09-933790-8 X.629/23642
1275232

The rubber band book. - London : Beaver Books, 1977
93p : chiefly ill
0-600-38753-4 X.619/18064 Woolwich
1275232

ACTIVITIES USING PAINTS

MCGLASHON, Angela
Fun with paints and paper. - London : Carousel, 1981
112p : ill
0-552-54190-7 X.990/17795 Woolwich
2558920

ACTIVITIES USING PAPER

CAKET, Colin
The paper book. - London : Macmillan Children's, 1981
96p : ill. - Includes index
0-333-34108-2 X.629/19133
060609x

ELDIN, Peter
Fun and games with paper. - London : Pan Books, 1976
110p : ill. - (A piccolo original)
0-330-24626-7 X.619/16573 Woolwich
060609x

KENNEWAY, Eric
Paper fun. - London : Beaver Books, 1978
109p : ill
0-600-33630-1 X.619/18363
060609x
- Reprinted. - 1983 X.990/20341

MCGLASHON, Angela
Fun with paints and paper. - London : Carousel, 1981
112p : ill
0-552-54190-7 X.990/17795 Woolwich
060609x

ACTIVITIES USING PERCUSSION INSTRUMENTS. Activities for pre-school children
- For teaching

BLADES, James
Ready to play. - London : British Broadcasting Corporation, 1978
64p : ill
0-563-17610-5 X.431/10557
1720260

ACTIVITIES USING POCKET ELECTRONIC CALCULATORS. Mathematics.
Curriculum subjects. Schools
- For teaching

BITTER, Gary G.
Activities handbook for teaching with the hand-held calculator. - Boston, Mass. ; London : Allyn and Bacon, 1980
viii,336p : ill. - Bibliography: p.322-333. - Includes index
0-205-06713-1 X.525/3340
2152061

MOURSURD, David G.
Calculators in the classroom. - New York ; Chichester : Wiley, c1981
vi,202p : ill. - Bibliography: p174-177
0-471-08113-2 X.629/16372 Woolwich
2152061

ACTIVITIES USING TORCHLIGHT

MACLEAN, Colin
Playing with torchlight. - London : Studio Vista, 1976
44p : chiefly ill. - (Something to do). - Includes index
0-289-70684-x X.620/16171 Woolwich
0686565

ACTIVITIES USING WATER

ARDLEY, Neil
Working with water. - London : Watts, 1983
32p : col.ill. - (Action science). - Includes index
0-86313-021-6 X.622/16852
2813750

DUNN, Roger
Things to do with water. - London : Macdonald Educational, 1980
[24]p : col.ill. - (Things to do books ; 8)
0-356-07058-1 X.990/19495
2813750

ACTIVITY. Concepts. Soviet theories. Psychology

The CONCEPT of activity in Soviet psychology / translated and edited by James V. Wertsch. - Armonk, N.Y. : Sharpe, c1981
xi,441p : ill. - Includes bibliographies
0-87332-158-8 X.529/46748
2701111

ACTIVITY ANALYSIS
Applications in computation of equilibria in general equilibrium theory

GINSBURGH, Victor A.
Activity analysis and general equilibrium modelling. - Amsterdam ; Oxford : North-Holland, c1981
xvi,373p : ill. - (Contributions to economic analysis ; 125). - Bibliography: p357-364. - Includes index
0-444-86011-8 WPC19/125
2557630

ACTIVITY ANALYSIS. Central areas. Copenhagen. Denmark

KØBENHAVNS bykerne, indhold og aktivitet. - København : Reitzel, 1975
ii, 216 p.. - (Kulturgeografiske skrifter ; bd 10). - Bibliography : p. 193-200
87-421-0507-2 Ac.6109b/2 [bd. 10]
0909416

ACTIVITY GROUPS. Intermediate treatment. Juvenile delinquency. England
Attitudes of participants

JONES, Ray
Fun and therapy. - Leicester : National Youth Bureau, 1979
44,[3]p : ill. - (IT papers). - Bibliography: p.43-44
0-86155-014-5 X.512/7191
2048795

ACTIVITY METHOD. Teaching methods. Schools

GREIG, Thomas Ogilvie
Activity methods in the middle years. - Edinburgh : Oliver and Boyd, 1975
v,88p : ill. - (Teaching in focus). - Bibliography: p.67-69
0-05-002857-x X.519/25258
0517143

ACTIVITY SPACES. Adolescents. Leeds. West Yorkshire (Metropolitan County)

LEACH, Bridget
Youth and spatial poverty. - Bristol : S.S.R.C. Research Unit on Ethnic Relations, 1978
[4],78p. - (Working papers on ethnic relations / Great Britain. Research Unit on Ethnic Relations ; no.9). - Bibliography: p.76-78
B80-21431 X.525/3215
2068451

ACTIVITY WEEKS. Speech disordered children
Planning

ASSOCIATION FOR ALL SPEECH IMPAIRED CHILDREN
Activity week guidelines. - London : A.F.A.S.I.C., 1978
[2],35p
B78-26683 X.512/6035
1525859

ACTON. Ealing (London Borough)
1872-1928

ACTON as it was / compiled and annotated by London Borough of Ealing Library Service and R.N.G. Rowland. - Nelson : Hendon Publishing, c1981
[44]p : chiefly ill
0-86067-067-8 L.49/1120
2613603

ACTON. Ealing (London Borough)
Street names

ROWLAND, R. N. G.
The street-names of Acton, Middlesex. - 2nd ed., revised and enlarged. - London : London Borough of Ealing Library Service, 1977[i.e.1978]
viii,46p : 1 ill
0-9502264-3-2 X.702/5638
1406299

ACTON AERODROME
to 1917
LONDON BOROUGH OF EALING LIBRARY SERVICE
The story of Acton Aerodrome and the Alliance Factory. with notes on other early aviation in the Borough. - 2nd ed. - London : The Service, 1978
[7],31p
0-9502264-4-0 X.702/5915
1799185

ACTON FAMILY
to 1977 - Biographies
GUNN, Peter
The Actons. - London : H. Hamilton, 1978
[9],212p,[4]p of plates : ill - Bibliography p.198-200 - Includes index
0-241-89977-x X.809/43678
1351478

ACTON HALL
Household financial management, 1412-1413
The HOUSEHOLD book of Dame Alice de Bryene of Acton Hall, Suffolk / translated by Miss M.K. Dale ; edited by Vincent B. Redstone. - Bungay : Paradigm, 1984
viii,148p. - Includes index
0-9506104-7-x X.629/23527
3025128

ACTON, HAROLD. English literature
- Bibliographies
RITCHIE, Neil
Harold Acton. - Florence : [s.n.], 1984
185p,plates
r6509405 X.950/35225
8714614

ACTON, HAROLD. Historiography
- Biographies
ACTON, Harold
Memoirs of an aesthete. - London : Hamilton, 1984, c1948
415p,[8]p of plates : ill. - Includes index
0-241-11373-3 X.958/27532
3244393

ACTON, JOHN EMERICH DAHLBERG-ACTON, BARON. Historiography
Interpersonal relationships with Gladstone, W. E. (William Ewart)
CHADWICK, Owen
Acton and Gladstone. - London : Athlone Press, 1976
56p. - (Creighton lectures in history ; 1975)
0-485-14122-1 WP.4049/18
0737704

ACTOR-MANAGERS. Theatre. England
Barrett, Wilson - Biographies
THOMAS, James, 1943-
The art of the actor-manager. - Epping : Bowker, c1984
x,203p : ill. - (Theater and dramatic studies ; no.15). - Bibliography: p193-199. - Includes index
0-8357-1492-6 X.950/36052
318546x

ACTORS
See also
Actor-managers
Cinema films featuring coloured actors & actresses
Itinerant actors

ACTORS
Physical fitness. Exercises
PISK, Litz
The actor and his body. - London : Harrap, 1975
95p,[1],7p of plates : ill
0-245-52707-9 X.620/15050
0351164

ACTORS
Superstitions
HUGGETT, Richard
The curse of Macbeth. - [Chippenham] : Picton, 1981
259p : ill. - Includes index
0-902633-72-4 X.950/4391
2484811

ACTORS
Training
BARKER, Clive
Theatre games. - London : Eyre Methuen, 1977
xi,226p : ill. - Bibliography: p.225-226
0-413-45370-7 X.989/51902
1254529

STANISLAVSKIĬ, K. S. (Konstantin Sergeevich)
Building a character. - London : Methuen, 1979
[4],300p
0-413-36720-7 X.908/42198
1799835

ACTORS
Training. Theories of Stanislavskii, K. S.
BENEDETTI, Jean
Stanislavski. - London : Methuen, 1982
xii,79p : ill. - Includes index
0-413-50030-6 X.958/12334
110179x

MOORE, Sonia
The Stanislavski system. - New revised ed. - Harmondsworth : Penguin, 1976
xxi,106p. - (Penguin handbooks). - Includes index
0-14-046259-7 WP.4003/220
110179x

ACTORS
Voice. Medical aspects
PUNT, Norman Arthur
The singer's and actor's throat. - 3rd ed. - London : Heinemann Medical, 1979
vii,100p : ill. - Bibliography: p.93-96. - Includes index
0-433-26451-9 X.319/18838 Woolwich
1889508

ACTORS. Acting. American cinema films
Supporting actors - Biographies
JONES, Ken D.
Character people. - South Brunswick ; New York : Barnes ; London : Yoseloff, 1976
3-209p : ill
0-498-01697-8 X.985/1192
1549391

PARISH, James Robert
Hollywood character actors. - New York : Arlington House, [1978]
542p
0-87000-384-4 X.955/557
8007039

ACTORS. Acting. Cinema films. Hollywood. Los Angeles. California
Supporting actors, 1937-ca 1970
BARRIS, Alex
Hollywood's other men. - South Brunswick ; New York : Barnes ; London : Yoseloff, 1975
223p : ill. - Includes index
0-498-01428-2 X.985/1022
0535788

ACTORS. American cinema films
- Biographies
NORMAN, Barry
The movie greats. - London : Arrow, 1982, c1981
288p,[16]p of plates : ill. - Includes index
0-09-929170-3 X.629/19241
2591995

SHIPMAN, David
The great movie stars. - New rev. ed. - London : Angus & Robertson, 1979
592p : ill. - Bibliography: p586-588. - Includes index
0-207-95829-7 X.955/771
2591995

ACTORS. American cinema films
to date - Biographies
PARISH, James Robert
Hollywood character actors. - New York : Arlington House, [1978]
542p
0-87000-384-4 X.955/557
8899029

ACTORS. American cinema films
1948-1977
PITTS, Michael R.
Hollywood on record. - Metuchen ; London : Scarecrow Press ; [Folkestone] : [Distributed by Bailey and Swinfen], 1978
xi,411p
0-8108-1093-x X.439/8524
1551620

ACTORS. American musical shows
1860-1950
STARS of the American musical theater in historic photographs / edited by Stanley Appelbaum and James Camner. - New York ; Dover ; London : Constable, 1981
168p. - Includes index
0-486-24209-9 X.805/3110
2783231

ACTORS. Characters. Drama in German, 1625-1830
PIETSCH-EBERT, Lilly
Die Gestalt des Schauspielers auf der deutschen Bühne des 17. und 18. Jahrh.. - Nendeln/Liechtenstein : Kraus Reprint, 1977
135p. - (Theatergeschichtliche Forschungen ; Hft 46)
r6090700 011795.i.55/46
2038099

ACTORS. Cinema films
- Bibliographies
SKINNER, J. W.
Skinner's bibliography of screen celebrities. - Worthing : I.F.C., 1978
[12]p
B83-07262 X.955/1386
2831112

ACTORS. Cinema films
- Biographies
COHEN, Daniel
Screen goddesses. - London : Hamlyn, c1984
128p : ill(some col.). - (A Bison book). - Includes index
0-600-34738-9 L.45/3036
0215066

The MOVIE stars story. - London : Octopus, 1984
287p : ill(some col.). - Includes index
0-7064-2092-6 L.45/3252
0215066

NORMAN, Barry
The film greats. - Sevenoaks : Hodder and Stoughton, 1985
262p : ill. - Includes index
0-340-35270-1 X.950/40123
0215066

PARISH, James Robert
The debonairs. - New Rochelle : Arlington House, c1975
511p : ill. - Includes index
0-87000-293-7 X.985/789
0215066

SHIPMAN, David
The great movie stars. - New revised ed. - London : Angus and Robertson, 1980
ix,646p : ill. - Bibliography: p.639-640
0-207-95858-0 X.981/22416
0215066

SKINNER, John Walter
Who's who on the screen. - Worthing : Madeleine, c1983
127p : ill
0-905380-02-9 X.525/10813
0215066

TRUITT, Evelyn Mack
Who was who on screen. - 2nd ed. - New York ; London : Bowker, 1977
vi,505p. - Bibliography: p.505
0-8352-0914-8 BB.E.dd.48 Reading Room
0215066
- [Another copy] X.622/6445 Woolwich

ACTORS. Cinema films
- Biographies - Dictionaries
TRUITT, Evelyn Mack
Who was who on screen. - 3rd ed. - New York ; London : Bowker, 1983
lx,788p. - Bibliography: p787-788
0-8352-1578-4 X.955/1785
0215066
- [Another copy.] BB.E.dd.48 Reading Room

Who was who on screen. - Illustrated ed. - New York ; London : Bowker, 1984
ix,438p. - Bibliography: p437-438
0-8352-1906-2 X.955/2658
0215066

ACTORS. Cinema films
- Biographies - Directories
TRUITT, Evelyn Mack
Who was who on screen. - New York ; London : Bowker, 1974
vii,363p. - Bibliography: p.363
0-8352-0719-6 X.985/815
0215066
- [Another copy.] X.902/2462

ACTORS. Cinema films
British actors
PALMER, Scott
A who's who of British film actors. - Metuchen ; London : Scarecrow, 1981
561p
0-8108-1388-2 X.950/7403
2551241

ACTORS. Cinema films
British character actors - Biographies
PETTIGREW, Terence
British film character actors. - Newton Abbot : David & Charles, 1982
208p. - Includes lists of films and index
0-7153-8270-5 X.950/16042
2790432

ACTORS. Cinema films
German actors, to 1945

HILCHENBACH, Maria
Kino im Exil. - München : Saur, 1982
286p.. - (Kommunikation und Politik ; Bd.14). - Includes bibliography
3-598-20544-9 X.529/61568
9081836

ACTORS. Cinema films
Interpersonal relationships, to date

HEINZLMEIER, Adolf
Happy-End. - Frankfurt am Main : Fischer, 1981
239p.. - (Fischer Taschenbücher ; 3668)
3-596-23668-1 X.429/13648
8870381

ACTORS. Cinema films, to 1975
Western European actors - Directories

PARISH, James Robert
Film actors guide. - Metuchen ; London : Scarecrow Press ; [Folkestone] : [Distributed by Bailey and Swinfen], 1977
xv,606p
0-8108-1044-1 X.989/52445
1451898

ACTORS. Cinema films, to date

DELVILLE, Olivier
Mes grandes heures de cinéma. - Bruxelles : Legrain, [1982]
168p : ill
r6298231 X.950/41532
9049258

ACTORS. Cinema films, to date
- Biographies

DUREAU, Christian
Dictionnaire mondial des comédiens. - [Paris?] : L.S.P. éditions, 1982
593p,plates
2-903927-00-6 X.950/29800
9049258

WERNER, Paul, 1951-
Göttinnen der Leinwand. - Dortmund : Harenberg, c1983
180p. : chiefly ill.. - (Die bibliophilen Taschenbücher ; Nr.371)
3-88379-371-x X.958/24488
9049258

ACTORS. Cinema films produced by Selznick, David O., 1936-1957

BOWERS, Ronald L.
The Selznick players. - South Brunswick ; New York : Barnes ; London : Yoseloff, 1976
255p : ill. - Bibliography: p.246. - Includes index
0-498-01375-8 X.985/1161
1253581

ACTORS. Comedies. American cinema films
1920-1930

MANCHEL, Frank
The talking clowns. - New York ; London : F. Watts, 1976
[10],129p : ill. - Bibliography: p.119-121. - Includes index
0-531-01153-4 X.981/20650
1119907

ACTORS. Denmark
Organisations: Fællesforbundet for Teater, Film & TV

FÆLLES forbundets overenskomst-avis / udgivet af Fællesforbundet for Teater, Film & TV. - Februar 1978. - København : Fællesforbundet, 1978
11p : ill
r6112761 P.2000/528
2463733

ACTORS. French cinema films
to date

CADARS, Pierre
Les séducteurs du cinéma français (1928-1958). - Paris : Henri Veyrier, 1982
221p. - (Collection Cinéma)
2-85199-220-1 X.955/2245
9080899

ACTORS. Great Britain
Trade unions, 1800-1951

MACLEOD, Joseph, 1903-
The actor's right to act. - London : Lawrence & Wishart for the friends of Equity, 1981
205p,[8]p of plates. - Includes index
0-85315-538-0 X.950/2850
2379449

ACTORS. Horror films. Cinema films

BECK, Calvin Thomas
Heroes of the horrors. - New York : Collier Books ; London : Collier Macmillan, 1975
xiv,353p : ill. - List of films: p.339-353
0-02-508190-x X.902/5153
0594172

PITTS, Michael R.
Horror film stars. - Jefferson, N.C. : McFarland, 1981
vii,324p : ill. - Bibliography: p303-306. - Includes index
0-89950-003-x X.950/7590
0594172

ACTORS. Horror films. Cinema films
- Biographies

WELSH, Paul, 1953-
The spine chillers. - [Ilfracombe] : [Stockwell], [1975]
[1],68p
0-7223-0849-3 X.908/40302
0594172

ACTORS. Musicals. 1934-1959. Westerns. American cinema films
- Biographies

ROTHEL, David
The singing cowboys. - South Brunswick ; New York : Barnes ; London : Yoseloff, 1978
272p : ill. - Bibliography: p.264-265. - Includes index
0-498-02163-7 X.435/717
2211742
- [Another copy] X.981/13667

ACTORS. National theatres. Czechoslovakia

NEJEDLÝ, Zdeněk
O divadle a jeho umělcích. - Praha : Panorama, 1978
221
r6187266 X.989/80270
2134136

ACTORS. Netherlands
to date - Directories

HONIG, Piet Hein
Acteurs- en kleinkunstenaarslexicon. - Diepenveen : P. Honig, [1984]
1079p
90-900051-1-0 X.520/36915
9162119

ACTORS. Polish cinema films
- Biographies

EBERHARDT, Konrad
Album aktorów polskiego filmu i telewizji. - Warszawa : Wydawnictwa Artystyczne i Filmowe, 1975
158 p. : ill
r6014945 X.900/17619
0889156

ACTORS. Russia (RSFSR)

ÈSKIN, A. M. (Aleksandr Moiseevich)
V nashem Dome. - Izd.2-e, dop. - Moskva : Vserossiĭskoe teatral'noe obshchestvo, 1980
318p : ill
r6400559 X.950/2603
2482444

ACTORS. Swedish cinema films
- Biographies

WINQUIST, Sven G.
Svenskt filmskådespelarlexikon. - Stockholm : Forum, i samarbete med Svenska Filminstitutet, 1973
946p : ill. - (Skrifter från Dokumentationsavdelningen ; nr 15). - Includes bibliographies
91-370-5247-0 X.0907/103(15)
0390518

ACTORS. Swedish cinema films
to 1977

COWIE, Peter
Stars and players. - London : Tantivy Press [etc.] [for] the Swedish Film Institute, Stockholm, 1977
128p : ill. - (Film in Sweden). - List of films: p.118-126. - Includes index
0-498-02013-4 X.989/52801
1688936

ACTORS. Television drama. United States
Performances, 1950-

PARISH, James Robert
Actors' television credits. - Metuchen ; London : Scarecrow Press.
Suppl.2: 1977-1981. - 1982
ix,327p
0-8108-1559-1 X.950/20448
1498339

Actors' television credits, 1950-1972. - Metuchen ; London : Scarecrow Press ; [Folkestone] : [Distributed by Bailey and Swinfen].
Supplement 1. - 1978
x,423p : ill
0-8108-1053-0 X.0989/2041(1)
1498339

ACTORS. Theatre. Ancient Greece
to ca B.C.300

GHIRON-BISTAGNE, Paulette
Recherches sur les acteurs dans la Grèce antique. - Paris : Société d'Édition 'Les Belles Lettres', 1976
xxviii,404p : ill. - Bibliography p.xv-xxviii
r6028336 X.981/11308
0927384
- [Another copy] X.981/10382

ACTORS. Theatre. Békéscsaba. Hungary
to 1945

HÉT évtized a békéscsabai színészet multjából 1875-1944 / [a kötet anyagát válogatta, sajtó alá rendezte és jegyzetekkel ellátta Papp János]. - Békéscsaba, 1980
66p,[8]p of plates : ill. - (Bibliotheca Bekesiensis ; 18)
r6449090 X.0808/883(18)
8752915

ACTORS. Theatre. Brazil
Training. Exercises

BOAL, Augusto
200 exercícios e jogos para o ator e o não-ator com vontade de dizer algo através do teatro. - Rio de Janeiro : Civilização Brasileira, 1977
123p : ill. - (Coleção Teatro hoje ; v.30)
r6101871 X.909/86899
2260956

ACTORS. Theatre. Denmark

HOLM, Kai
-som jeg husker dem. - [København] : Delta, 1980
157p : ill
87-7449-053-2 X.950/20889
8942463

ACTORS. Theatre. England
Professional actors, 1590-1642

BENTLEY, Gerald Eades
The profession of player in Shakespeare's time, 1590-1642. - Princeton ; Guildford : Princeton University Press, c1984
xiv,315p. - Includes index
0-691-06596-9 X.950/37808
3219879

ACTORS. Theatre. England
Professional actors. Social status, 1830-1890

BAKER, Michael, 1948-
The rise of the Victorian actor. - London : Croom Helm [etc.], 1978
249p. - (Croom Helm social history series). - Bibliography: p.226-240. - Includes index
0-85664-542-7 X.989/52392
1266446

ACTORS. Theatre. Europe
1900- - Biographies

GARDEROBENGESPRÄCHE / herausgegeben von Ingeborg Pietzsch. - Berlin : Henschelverlag, 1982
215p : ill. - (Dialog)
r6307294 X.958/17174
8988528

ACTORS. Theatre. Great Britain
Actors & actresses, 1925-1982 - Biographies

FINDLATER, Richard
These our actors. - London : Elm Tree in association with the Theatre Museum, the Victoria and Albert Museum, 1983
191p : ill. - Bibliography: p182-183. - Includes index
0-241-11060-2 X.950/32203
3122530

ACTORS. Theatre. Great Britain
Actors & actresses, 1950-1980 - Biographies

MARKS, Claude
Theatre sketchbook. - Ambergate : Amber Lane, 1982
127p
0-906399-33-5 X.950/17759
282650x

ACTORS. Theatre. Hungary

A BORZASZTÓ torony / válogatta, szerkesztette és a jegyzeteket írta Kerényi Ferenc. - Budapest : Magvető Könyvkiadó, 1979
412p,[25]p of plates : ill. - (Magyar tallózó)
963-270-925-x X.989/80471
2165988

ACTORS. Theatre. Hungary –continued

NÁDUDVARI, Anna
Honnét lesz a tűz? - Budapest : Kozmosz Konyvek, 1981
318p
963-211-428-0 X.529/69223
8592748

ACTORS. Theatre. Italy

GAUDIO, Silvana
Dietro le quinte. - Roma : La rassegna, [1979]
224p,xiip of plates : ill. - (Immagini allo specchio ; 3)
r6501236 X.950/13156
8838372

ACTORS. Theatre. Netherlands
Actors & actresses, to date

MOL, Albert
Het doek viel te vroeg. - Amsterdam : Tiebosch, [1977]
159p
90-627-8521-2 X.950/29083
9039236

ACTORS. Theatre. Norway
to date - Biographies

JENSSON, Liv
Biografisk skuespillerleksikon. - Oslo : Universitetsforlaget, c1981
193p
82-00-05622-8 X.950/24712
9008071

ACTORS. Theatre. Poland
1945- - Biographies

ŻUROWSKI, Andrzej
Ludzie w reflektorach. - Warszawa : Czytelnik, 1982
295p
83-07-00787-9 X.958/18141
9008055

ACTORS. Theatre. Prague. Czechoslovakia

ČERNÝ, František
Měnivá tvář divadla aneb Dvě století s pražskými herci. - Praha : Mladá fronta, 1978
319p[48]p of plates : ill
r6096493 X.981/12631
2134152

ACTORS. Theatre. Slovenia. Yugoslavia
- Biographies

KALAN, Filip
Veliki igralci. - Ljubljana : Zveza kulturno-provetnih organizacij Slovenije, 1975
34p : ports. - (Umetnost in kultura ; 99 : Gledališki večer)
r6176688 X.989/82640
2448602

ACTORS. Theatre. Soviet Union
1917-

MASTERA khudozhestvennogo slova / [redaktor T.P. Bazhenova]. - Moskva : Iskusstvo, 1983
269 p. : ill., ports.
r6482922 X.958/26234
8839417

ACTORS. Theatre. Soviet Union
1917- - Biographies

PIMENOV, Vladimir Fedorovich
Vstrechi posle spektakleĭ. - Moskva : Iskusstvo, 1981
140p
r6485693 X.958/8894
8839417

ACTORS. Theatre. Special subjects. Paintings. Exhibits. National Theatre
Actors & actresses: Maugham Collection - Catalogues

NATIONAL THEATRE
Guide to the Maugham collection of theatrical paintings. - [London] : Heinemann [for] the National Theatre, 1980
68p : ill(some col.). - Includes index
0-435-18591-8 X.425/2169 Woolwich
2542293

ACTORS. Theatre. United States
to 1974 - Biographies

FAMOUS actors and actresses on the American stage / [compiled by] William C. Young. - New York ; London : Bowker, 1975
2v.(xxi,1298p). - Bibliography: p.1215-1227. - Includes index
0-8352-0821-4 X.981/20172
0594601

ACTORS. Theatre. United States
Actors & actresses, 1850-1950 - Biographies

GREAT actors and actresses of the American stage in historic photographs / edited by Stanley Appelbaum. - New York : Dover ; London : Constable, c1983
v,136p. - Includes index
0-486-24555-1 X.955/2849
3254739

ACTORS. Travelling theatre. England
Children. Home-based education

KIDDLE, Catherine
What shall we do with the children?. - Barnstaple : Spindlewood, 1981
109p
0-907349-05-6 X.520/26159
2632101

ACTORS. Westerns, 1920-1976. American cinema films

HORWITZ, James
They went thataway. - London : W.H. Allen, 1977
v,281p : ill
0-491-01818-5 X.989/51526
1113682

ACTORS. Westerns, 1935-1955. American cinema films

ROTHEL, David
Those great cowboy sidekicks. - Metuchen ; London : Scarecrow, 1984
ix,325p : ill. - Bibliography: p306-308. - Includes index
0-8108-1707-1 X.955/3057
3302156

ACTRESSES
See also
Cinema films featuring coloured actors & actresses
Dolls representing actresses dressed in costumes from American cinema films

ACTRESSES
Voice. Medical aspects

PUNT, Norman Arthur
The singer's and actor's throat. - 3rd ed. - London : Heinemann Medical, 1979
vii,100p : ill. - Bibliography: p.93-96. - Includes index
0-433-26451-9 X.319/18838 Woolwich
1889508

ACTRESSES. Acting. American cinema films
Supporting actresses - Biographies

JONES, Ken D.
Character people. - South Brunswick ; New York : Barnes ; London : Yoseloff, 1976
3-209p : ill
0-498-01697-8 X.985/1192
1549391

PARISH, James Robert
Hollywood character actors. - New York : Arlington House, [1978]
542p
0-87000-384-4 X.955/557
8007039

ACTRESSES. Acting. Cinema films. Hollywood. Los Angeles. California
Supporting actresses, 1930-1973 - Biographies

BARRIS, Alex
Hollywood's other women. - South Brunswick : New York : Barnes ; London : Yoseloff, 1975
212p : ill. - Includes index
0-498-01488-6 X.985/1166
1264257

ACTRESSES. American cinema films
- Biographies

NORMAN, Barry
The movie greats. - London : Arrow, 1982, c1981
288p,[16]p of plates : ill. - Includes index
0-09-929170-3 X.629/19241
2591995

SHIPMAN, David
The great movie stars. - New rev. ed. - London : Angus & Robertson, 1979
592p : ill. - Bibliography: p586-588. - Includes index
0-207-95829-7 X.955/771
2591995

ACTRESSES. American cinema films
1948-1977

PITTS, Michael R.
Hollywood on record. - Metuchen ; London : Scarecrow Press ; [Folkestone] : [Distributed by Bailey and Swinfen], 1978
xi,411p
0-8108-1093-x X.439/8524
1551620

ACTRESSES. American musical shows
1860-1950

STARS of the American musical theater in historic photographs / edited by Stanley Appelbaum and James Camner. - New York ; Dover ; London : Constable, 1981
168p. - Includes index
0-486-24209-9 X.805/3110
2783231

ACTRESSES. Cinema films
- Bibliographies

SKINNER, J. W.
Skinner's bibliography of screen celebrities. - Worthing : I.F.C., 1978
[12]p
B83-07262 X.955/1386
2831112

ACTRESSES. Cinema films
- Biographies

COHEN, Daniel
Screen goddesses. - London : Hamlyn, c1984
128p : ill(some col.). - (A Bison book). - Includes index
0-600-34738-9 L.45/3036
0215066

The MOVIE stars story. - London : Octopus, 1984
287p : ill(some col.). - Includes index
0-7064-2092-6 L.45/3252
0215066

NORMAN, Barry
The film greats. - Sevenoaks : Hodder and Stoughton, 1985
262p : ill. - Includes index
0-340-35270-1 X.950/40123
0215066

PARISH, James Robert
The debonairs. - New Rochelle : Arlington House, c1975
511p : ill. - Includes index
0-87000-293-7 X.985/789
0215066

SHIPMAN, David
The great movie stars. - New revised ed. - London : Angus and Robertson, 1980
ix,646p : ill. - Bibliography: p.639-640
0-207-95858-0 X.981/22416
0215066

SKINNER, John Walter
Who's who on the screen. - Worthing : Madeleine, c1983
127p : ill
0-905380-02-9 X.525/10813
0215066

TRUITT, Evelyn Mack
Who was who on screen. - 2nd ed. - New York ; London : Bowker, 1977
vi,505p. - Bibliography: p.505
0-8352-0914-8 BB.E.dd.48 Reading Room
0215066
- [Another copy] X.622/6445 Woolwich

ACTRESSES. Cinema films
- Biographies - Dictionaries

TRUITT, Evelyn Mack
Who was who on screen. - 3rd ed. - New York ; London : Bowker, 1983
lx,788p. - Bibliography: p787-788
0-8352-1578-4 X.955/1785
0215066
- [Another copy.] BB.E.dd.48 Reading Room
Who was who on screen. - Illustrated ed. - New York ; London : Bowker, 1984
ix,438p. - Bibliography: p437-438
0-8352-1906-2 X.955/2658
0215066

ACTRESSES. Cinema films
- Biographies - Directories

TRUITT, Evelyn Mack
Who was who on screen. - New York ; London : Bowker, 1974
vii,363p. - Bibliography: p.363
0-8352-0719-6 X.985/815
0215066
- [Another copy.] X.902/2462

ACTRESSES. Cinema films
British actresses

PALMER, Scott
A who's who of British film actors. - Metuchen ; London : Scarecrow, 1981
561p
0-8108-1388-2 X.950/7403
2551241

ACTRESSES. Cinema films, to date

DELVILLE, Olivier
Mes grandes heures de cinéma. - Bruxelles : Legrain, [1982]
168p : ill
r6298231 X.950/41532
9049258

ACTRESSES. Cinema films, to date
- Biographies

DUREAU, Christian
Dictionnaire mondial des comédiens. - [Paris?] : L.S.P. éditions, 1982
593p,plates
2-903927-00-6 X.950/29800
9049258

WERNER, Paul, 1951-
Göttinnen der Leinwand. - Dortmund : Harenberg, c1983
180p. : chiefly ill.. - (Die bibliophilen Taschenbücher ; Nr.371)
3-88379-371-x X.958/24488
9049258

ACTRESSES. Cinema films produced by Republic Pictures Corporation
- Biographies

SWANN, Thomas Burnett
The heroine or the horse. - South Brunswick ; New York : Barnes ; London : Yoseloff, 1977
134p : ill. - Bibliography: p.130-131. - Includes index
0-498-01962-4 X.985/1235
1937227

ACTRESSES. Cinema films produced by Selznick, David O., 1936-1957

BOWERS, Ronald L.
The Selznick players. - South Brunswick ; New York : Barnes ; London : Yoseloff, 1976
255p : ill. - Bibliography: p.246. - Includes index
0-498-01375-8 X.985/1161
1253581

ACTRESSES. Comedies. American cinema films
1920-1930

MANCHEL, Frank
The talking clowns. - New York ; London : F. Watts, 1976
[10],129p : ill. - Bibliography: p.119-121. - Includes index
0-531-01153-4 X.981/20650
1119907

ACTRESSES. Great Britain
Trade unions, 1800-1951

MACLEOD, Joseph, 1903-
The actor's right to act. - London : Lawrence & Wishart for the friends of Equity, 1981
205p,[8]p of plates. - Includes index
0-85315-538-0 X.950/2850
2379449

ACTRESSES. Horror films. Cinema films

BECK, Calvin Thomas
Scream queens. - New York : Collier Books ; London : Collier Macmillan, [1980]
viii,344p : ill. - Bibliography: p.335-338. - index
0-02-508170-5 X.902/5715
2282275

ACTRESSES. Polish cinema films
- Biographies

EBERHARDT, Konrad
Album aktorów polskiego filmu i telewizji. - Warszawa : Wydawnictwa Artystyczne i Filmowe, 1975
158 p. : ill
r6014945 X.900/17619
0889156

ACTRESSES. Swedish cinema films
- Biographies

WINQUIST, Sven G.
Svenskt filmskadespelarlexikon. - Stockholm : Forum, i samarbete med Svenska Filminstitutet, 1973
946p : ill. - (Skrifter fran Dokumentationsavdelningen ; nr 15). - Includes bibliographies
91-370-5247-0 X.0907/103(15)
0390518

ACTRESSES. Swedish cinema films
to 1977

COWIE, Peter
Stars and players. - London : Tantivy Press [etc.] [for] the Swedish Film Institute, Stockholm, 1977
128p : ill. - (Film in Sweden). - List of films: p.118-126. - Includes index
0-498-02013-4 X.989/52801
1688936

ACTRESSES. Television drama. United States
Performances, 1950-

PARISH, James Robert
Actors' television credits. - Metuchen ; London : Scarecrow Press.
Suppl.2: 1977-1981. - 1982
ix,327p
0-8108-1559-1 X.950/20448
1498339

Actors' television credits, 1950-1972. - Metuchen ; London : Scarecrow Press : [Folkestone] : [Distributed by Bailey and Swinfen].
Supplement 1. - 1978
x,423p : ill
0-8108-1053-0 X.0989/2041(1)
1498339

ACTRESSES. Theatre. Denmark
to date

HOLM, Kai
-som jeg husker dem. - [København] : Delta, 1980
157p : ill
87-7449-053-2 X.950/20889
8942463

ACTRESSES. Theatre. England
Professional actresses. Social status, 1830-1890

BAKER, Michael, 1948-
The rise of the Victorian actor. - London : Croom Helm [etc.], 1978
249p. - (Croom Helm social history series). - Bibliography: p.226-240. - Includes index
0-85664-542-7 X.989/52392
1266446

ACTRESSES. Theatre. Great Britain
1660-1975

FINDLATER, Richard
The player queens. - London : Weidenfeld and Nicolson, 1976
vi,250p,[8]p of plates. - Bibliography: p.238-243. - Includes index
0-297-77158-2 X.989/51085
0762989

ACTRESSES. Theatre. Great Britain
Actors & actresses, 1925-1982 - Biographies

FINDLATER, Richard
These our actors. - London : Elm Tree in association with the Theatre Museum, the Victoria and Albert Museum, 1983
191p : ill. - Bibliography: p182-183. - Includes index
0-241-11060-2 X.950/32203
3122530

ACTRESSES. Theatre. Great Britain
Actors & actresses, 1950-1980 - Biographies

MARKS, Claude
Theatre sketchbook. - Ambergate : Amber Lane, 1982
127p
0-906399-33-5 X.950/17759
282650x

ACTRESSES. Theatre. Hungary

A BORZASZTÓ torony / válogatta, szerkesztette és a jegyzeteket írta Kerényi Ferenc. - Budapest : Magvető Könyvkiadó, 1979
412p,[25]p of plates : ill. - (Magyar tallózó)
963-270-925-x X.989/80471
2165988

ACTRESSES. Theatre. Hungary
to date

NÁDUDVARI, Anna
Honnét lesz a tűz?. - Budapest : Kozmosz Könyvek, 1981
318p
963-211-428-0 X.529/69223
8592748

ACTRESSES. Theatre. Italy
to date

GAUDIO, Silvana
Dietro le quinte. - Roma : La rassegna, [1979]
224p,xiip of plates : ill. - (Immagini allo specchio ; 3)
r6501236 X.950/13156
8838372

ACTRESSES. Theatre. Italy
1800-1945

CIOTTI CAVALLETTO, Giovanna
Attrici e società nell'Ottocento italiano. - Milano : Mursia, [1978]
181p. - (Problemi di storia dello spettacolo ; 6)
r6069574 X.950/4666
8706638

ACTRESSES. Theatre. Netherlands
Actors & actresses, to date

MOL, Albert
Het doek viel te vroeg. - Amsterdam : Tiebosch, [1977]
159p
90-627-8521-2 X.950/29083
9039236

ACTRESSES. Theatre. Special subjects. Paintings. Exhibits. National Theatre
Actors & actresses: Maugham Collection - Catalogues

NATIONAL THEATRE
Guide to the Maugham collection of theatrical paintings. - [London] : Heinemann [for] the National Theatre, 1980
68p : ill(some col.). - Includes index
0-435-18591-8 X.425/2169 Woolwich
2542293

ACTRESSES. Theatre. United States
to 1974 - Biographies

FAMOUS actors and actresses on the American stage / [compiled by] William C. Young. - New York ; London : Bowker, 1975
2v.(xxi,1298p). - Bibliography p.1215-1227. - Includes index
0-8352-0821-4 X.981/20172
0594601

ACTRESSES. Theatre. United States
1890-1920

AUSTER, Albert
Actresses and suffragists. - New York ; Eastbourne : Praeger, 1984
ix,177p,[6]p of plates : ill. - (Praeger special studies). - Bibliography: p149-164. - Includes index
0-03-069778-6 X.950/36366
319809x

ACTRESSES. Theatre. United States
Actors & actresses, 1850-1950 - Biographies

GREAT actors and actresses of the American stage in historic photographs / edited by Stanley Appelbaum. - New York : Dover ; London : Constable, c1983
v,136p. - Includes index
0-486-24555-1 X.955/2849
3254739

ACTRESSES. Travelling theatre. England
Children. Home-based education

KIDDLE, Catherine
What shall we do with the children?. - Barnstaple : Spindlewood, 1981
109p
0-907349-05-6 X.520/26159
2632101

ACTS. Bible
See Bible. N.T. Acts

ACTS OF JOHN

ACTA Iohannis / cura Eric Junod et Jean-Daniel Kaestli. - Turnhout : Brepols, 1983
426-949p. - (Corpus Christianorum. Series Apocryphorum ; 2)
r6211641 X.0200/1317
9073884

ACTS OF PARLIAMENT
See Statutes

'ACTUAL'
- Indexes

HERNÁNDEZ, Manuel
Actual. - Mérida : Universidad de los Andes, Biblioteca "Gonzalo Rincón Gutiérrez, 1977
v,32leaves
r6159578 P.801/1353
2419491

ACTUARIAL SCIENCE
See also
Rating. Risks. Life insurance

ACTUARIAL SCIENCE

POLLARD, J. H. (John Hurlstone)
The analysis of mortality and other actuarial statistics. - 2nd ed. - London : Heinemann for the Institute of Actuaries and the Faculty of Actuaries, 1980
xiv,466p : ill. - Includes bibliographies and index
0-434-90137-7 X.529/37150
2230607

ACTUARIAL SCIENCE
- Serials

ASTIN bulletin. - Vol.1, no.1 (1958)-. - Clevedon : Tieto Ltd., 1958-
v. : ill
Holdings: Vol.13, no.2 (Dec. 1982)-
r6813691 P.803/1409
2230607

ACTUARIAL SCIENCE. Great Britain

The ACTUARIAL profession. - [London] : [Institute of Actuaries], [1984]
8p : col.ill
B85-13610 X.525/10554
0760676

ACTUARIAL SCIENCE. Great Britain –continued
FACULTY OF ACTUARIES
The actuarial profession as a career. – Edinburgh : The Faculty, 1976
16p
0–901019–08–9 X.619/17177 Woolwich
0760676

ACTUARIES. Great Britain
Role in superannuation schemes
HARRIS GRAHAM AND PARTNERS (Firm)
The role of the actuary in a pension plan. – London : Harris Graham and Partners, [1978]
[2],31p : col ill
B79–02877 X.519/28928
1757520

ACTUATOR DISCS
Use in simulation of regulation of fluid flow
HORLOCK, John Harold
Actuator disk theory. – New York ; London : McGraw-Hill, 1978
xviii,242p : ill. – Includes index
0–07–030360–6 SRL
1271172

ACTUATORS
See also
Valve actuators

ACU-YOGA
GACH, Michael Reed
Acu-yoga. – Tokyo : Japan Publications, 1981
247p : ill. – Bibliography: p239–241. – Includes index
0–87040–489–x X.525/5475
2639947

ACUERDO DE CARTAGENA
See Andean Pact (1969)

ACUFF, ROY. Singing. Country music
SCHLAPPI, Elizabeth
Roy Acuff. – Gretna, La. : Pelican Publishing Co., c1978
viii,289p,[16]p.of plates : ill
0–88289–144–8 X.431/11778
8073465

ACULUM. Suffixes. Latin language
Development in Romance languages
TUTTLE, Edward Fowler
Studies in the derivational suffix –āculum. – Tübingen : Max Niemeyer, 1975
viii,128p. : 2 ill. – (Beihefte zur Zeitschrift für romanische Philologie ; Bd 146). – Bibliography : p. 103–114. – Includes index
3–484–52051–5 PP.5044.ac/3[Bd.146]
1049909

ACUÑA DE FIGUEROA, FRANCISCO. Composition. Uruguayan national anthems
– Biographies
URUGUAY. Consejo de Educación Primaria. Inspección Nacional de Educación Musical
Autores del himno nacional. – [Montevideo] : el Consejo, 1974
6 f.
r6003565 X.510/9420
0939358

ACUÑA GALÉ, JULIÁN. Research. Agriculture. Cuba
MONCADA FERRERA, Milagros
Recordando a Julián Acuña. – La Habana : Academia de Ciencias de Cuba, Centro de Estudios de Historia y Organización de la Ciencia Carlos J. Finlay, 1980
10leaves. – (Conferencias y estudios de historia y organización de la ciencia ; 22)
r6701145 LAS.D.300/15 Woolwich
8747482

ACUÑA GALÉ, JULIÁN. Research. Agriculture. Cuba
– Biographies
MARTÍNEZ VIERA, Rafael
Julián Acuña Galé en el 80 aniversario de su nacimiento. – La Habana : Academia de Ciencias de Cuba, Centro de Estudios de Historia y Organización de la Ciencia Carlos J. Finlay, 1980
13 leaves. – (Conferencias y estudios de historia y organización de la ciencia ; 21)
r6701147 LAS.D.300/15 Woolwich
8747482

ACUÑA GALÉ, JULIÁN BALDOMERO
See Acuña, Julián †

ACUÑA, GUIDO. Venezuelan writers. Short stories in Spanish
VELÁSQUEZ, Ramón J
Discurso del Doctor Ramón J. Velásquez con motivo de bautizar el libro de Guido Acuña "Guatire, crónicas del pueblo". – Guatire : Concejo Municipal, [1976?]
[22]p
r6701554 X.950/5180
9064958

ACUÑA, JULIAN
See Acuña Galé, Julian †

ACUÑA, JULIAN. Botany
MUÑIZ, Onaney
Cinco hipótesis acerca de la obra científica de Julián Acuña. – La Habana : Academia de Ciencias de Cuba, Centro de Estudios de Historia y Organización de la Ciencia Carlos J. Finlay, 1980
15p
r6703122 X.329/14467 Woolwich
8032858

ACUÑA, MAURO FERNÁNDEZ
See Fernández, Mauro

ACUPRESSURE
See Shiatzu

ACUPUNCTURE
See also
Electro-acupuncture therapy
Shiatzu

ACUPUNCTURE
ACADEMY OF TRADITIONAL CHINESE MEDICINE
An outline of Chinese acupuncture. – Pei-ching [i.e. Peking] : [Foreign Languages Press] ; Oxford : [Distributed by] Pergamon, [1978]
[4],xii,306p,[4]p of plates,[9]leaves of plates(2 fold) : ill(chiefly col). – Includes index
0–08–021545–9 X.322/8579 Woolwich
0077704

AUSTIN, Mary, 1914–
Acupuncture therapy. – [Abridged and rev. ed]. – Wellingborough : Turnstone, 1981
192p. – Includes index
0–85500–142–9 X.329/13802 Woolwich
0077704

ESSENTIALS of Chinese acupuncture / compiled by Beijing College of Traditional Chinese Medicine ... [et al.]. – Oxford : Pergamon, 1981
xiii,432p : ill. – Includes index
0–08–027995–3 X.322/11693
0077704

JAYASURIYA, Anton
Traditional Chinese acupuncture. – [Sri Lanka] : Acupuncture Foundation of Sri Lanka, [198?]
ii,572p
r6815275 X.322/15054
9090525

LEGER, J. P.
The little red book of acupuncture. – Wellingborough : Thorsons, 1978
64p : ill. – (Nature's way)
0–7225–0430–6 SRL
0077704

LEWITH, G. T. (George T)
The acupuncture treatment of internal disease. – Wellingborough : Thorsons, 1985
[160]p : ill. – Includes index
0–7225–1103–5 X.322/15105
0077704

LOW, Royston H.
The celestial stems. – Wellingborough : Thorsons, 1985
96p : ill. – Bibliography: p93–94. – Includes index
0–7225–0788–7 X.329/19552
0077704
The secondary vessels of acupuncture. – Wellingborough : Thorsons, 1983
192p : ill(some col.). – Bibliography: p187. – Includes index
0–7225–0736–4 X.322/13946
0077704

MACDONALD, Alexander, 19––
Acupuncture. – London : Allen & Unwin, 1982
xiv,184p,7p of plates : ill. – (The Medicine today series). – Includes index
0–04–616023–x X.329/15531
0077704

MACDONALD, Alexander, 1941–
Acupuncture. – London : Unwin Paperbacks, 1984, c1982
xiv,184p,[8]p of plates : ill. – Includes index
0–04–616026–4 X.329/17449
0077704

MANN, Felix
Acupuncture. – London : Pan Books, 1978
xiii,124p : ill
B78–15000 X.319/18019 Woolwich
0077704

Acupuncture. – 3rd ed. – London : Heinemann Medical, 1978
xi,200p : ill. – Bibliography: p.199–200
0–433–20310–2 SRL
0077704

Scientific aspects of acupuncture. – London : Heinemann Medical, 1977
viii,77p : ill
0–433–20309–9 SRL
0077704

MARCUS, Paul
Acupuncture. – Wellingborough : Thorsons, 1984
112p. – Includes index
0–7225–0921–9 X.329/18141
0077704

MEERAN, Munsif H.
Acupuncture. – London : Mehar, [1984]
64p : ill
0–9509035–0–7 X.329/18191
0077704

OMURA, Yoshiaki
Acupuncture medicine. – Tokyo : Japan Publications, 1982
287p : ill. – Includes index
0–87040–491–1 SRL
0077704

PLAYER, Graham
Disease and diagnosis for the acupuncturist. – Wellingborough : Thorsons, 1984
144p : ill. – Bibliography: p143. – Includes index
0–7225–0822–0 X.329/18254
0077704

ROBERTS, Michael J.
Acupuncture therapy. – Leicester : Eresus Publications, c1981
17p
0–9507758–0–0 X.329/14550 Woolwich
0077704

ROSE-NEIL, Sidney
Acupuncture and the life energies. – London : British Acupuncture Association, 1979
160p : 1 ill. – Includes index
0–9504160–5–3 X.320/11808 Woolwich
0077704

The TREATMENT of 100 common diseases by new acupuncture. – Hong Kong : Medicine and Health Publishing Co., [1977]
89p
r6092718 X.322/8884 Woolwich
0077704

WARREN, Frank Z.
Handbook of medical acupuncture. – New York ; London : Van Nostrand Reinhold, 1976
[11],273p : ill. – Includes index
0–442–29198–1 SRL
0077704

WOOLLERTON, Henry
Acupuncture energy in health and disease. – Wellingborough : Thorsons, 1979
126p : ill. – Bibliography: p.121. – Includes index
0–7225–0521–3 X.319/18879 Woolwich
0077704

WORSLEY, J. R.
Talking about acupuncture. – Royal Leamington Spa : College of Traditional Chinese Acupuncture with Element Books, c1982
112p : ill
0–906540–24–0 X.329/18208
0077704

ACUPUNCTURE
– Serials
BRITISH ACUPUNCTURE ASSOCIATION
British Acupuncture Association journal. – Vol.1, no.1– ; 1978–. – [London] : [The Association], [1978]–
B79–04374 P.323/332
1768530

The JOURNAL of traditional acupuncture. – Vol.1 no.1–. – Columbia, Md. : Traditional Acupuncture Foundation, 1977–
v.
Holdings: Vol.6, no.1 (Spring / Summer 1982)–
r6813170 X.322/292
9090525

ACUPUNCTURE
to 1979
LU, Gwei-djen
Celestial lancets. – Cambridge : Cambridge University Press, 1980
xxi,427p : ill. – Bibliography: p324–381. – Includes index
0–521–21513–7 X.622/9321 Woolwich
2050196

ACUPUNCTURE
Man. Organs - For acupuncture

ROSS, Jeremy
Zang fu. - Bristol : J. Ross, c1984
xvii,241p : ill. - Bibliography: p234-235. - Includes index
0-9509953-0-4 X.322/15129
3284557

ACUPUNCTURE
Points

CHINESE TRADITIONAL MEDICAL COLLEGE OF SHANGHAI
Anatomical charts of the acupuncture points and 14 meridians. - Shang-hai : [People's Publishing House] ; Oxford : [Distributed by] Pergamon, [1978]
[2],viii,65p : ill
0-08-021546-7 15593.c.3 OMPB
0133272

LOW, Royston H.
Acupuncture atlas and reference book. - Wellingborough : Thorsons, 1985
56p : ill(some col.)
0-7225-0898-0 SRL
0133272

ACUPUNCTURE
Use in relief of pain in man

CHAITOW, Leon
The acupuncture treatment of pain. - Wellingborough : Thorsons, 1976
160p : ill(chiefly col). - Includes index
0-7225-0356-3 SRL
1033921
The acupuncture treatment of pain. - 2nd ed. - Wellingborough : Thorsons, 1983
188p : ill. - Includes index
0-7225-0811-5 X.322/12243
1033921

ACUPUNCTURE
Use in therapy of horses

WESTERMAYER, Erwin
The treatment of horses by acupuncture. - Holsworthy : Health Science Press, 1979
40p : ill. - Includes index
0-85032-161-1 SRL
1951009

ACUPUNCTURE. Auricles. Ears. Man
Use in suppression of appetite for slimming

WARREN, Frank Z.
Acupuncture diet. - New York : St Martin's Press, 1976
[7], 136 p. : ill
0-900997-85-0 X.329/10772 Woolwich
1029436

ACUPUNCTURE. China

WEWITH, G. T. (George T)
Modern Chinese acupuncture. - 2nd ed., rev. and reset. - Wellingborough : Thorsons, 1983
144p : ill. - Bibliography: p142. - Includes index
0-7225-0924-3 X.329/18550
3162834

ACUPUNCTURE. China
to 1974

FU WEI-KANG
The story of Chinese acupuncture and moxibustion. - Peking : Foreign Languages Press, 1975
[3], 40 p., [4] p. of plates : ill
r6031363 X.619/17432 Woolwich
1070975

ACUPUNCTURE. China
to date

CHINESE acupuncture handbook / edited and translated by Samuel M. Wong, T.W. Law. - Hong Kong : Light Publishing Co., [1982]
150p
r6810785 X.329/17211
8584575

ACUPUNCTURE. China
1975

ROSE-NEIL, Sidney
An acupuncturist visits China. - London : British Acupuncture Association, 1979
192p
0-9504160-6-1 X.320/11841 Woolwich
1932977

ACUPUNCTURE. China
1979

LEWITH, G. T. (George T.)
Modern Chinese acupuncture. - Wellingborough : Thorsons, 1980
12mp : ill. - Includes index
0-7225-0662-7 SRL
2158558

ACUPUNCTURE. Great Britain
Organisations: British Acupuncture Association

BRITISH ACUPUNCTURE ASSOCIATION
The British Acupuncture Association and Register. - [Leicester] : British Acupuncture Association, [1983?]
[21]leaves
B84-30433 X.322/13611
3138461

ACUPUNCTURE. Great Britain
Organisations: British Acupuncture Association - Directories

BRITISH ACUPUNCTURE ASSOCIATION
Register and year book. - 1977. - [London] : [The Association], [1978]
[1],56p
0-9504160-3-7 SRL
1588982

Register and year book. - 1979-1980. - [Manchester] : The Association, [1979]
[1],64p
0-9504160-7-x SRL
1588982

ACUPUNCTURE. Great Britain
Organisations: British Acupuncture Association - Serials

[NEWSLETTER (British Acupuncture Association)]. Newsletter / the British Acupuncture Association. - Manchester : The Association, 1976-
v.
Holdings: Winter 1981, etc
B81-10252 P.321/524
2371081

ACUPUNCTURE ASSOCIATION
See also
British Acupuncture Association

ACUPUNCTURE POINT MASSAGE
See Shiatzu

ACUTE ABDOMEN. Man

ANGELL, James Cyril
The acute abdomen for the man on the spot. - 3rd ed. - Tunbridge Wells : Pitman Medical, 1978
116p
0-272-79519-4 X.319/18307 Woolwich
1568361

SHEPHERD, John A. (John Alfred)
Management of the acute abdomen. - Oxford : Oxford University Press, 1982
vii,366p : ill. - (Oxford medical publications). - Bibliography: p359-360. - Includes index
0-19-261322-7 SRL
1568361

ACUTE ABDOMEN. Man
Diagnosis

BEAL, John Mann
Diagnosis of acute abdominal disease. - Philadelphia : Lee and Febiger ; London : Kimpton, 1979
ix,169p : ill. - Bibliography - Includes index
0-8121-0678-4 X.329/12362 Woolwich
1190687

BOTSFORD, Thomas Winston
The acute abdomen. - 2nd ed. - Philadelphia ; London : Saunders, 1977
xvi,325p : ill. - Bibliography: p.313-315. - Includes index
0-7216-1886-3 SRL
1190687

COPE, Sir Zachary
Cope's early diagnosis of the acute abdomen. - 16th ed.. - New York : Oxford University Press, 1983
xiv,284p,14p of plates : ill. - Includes index
0-19-503294-2 SRL
1190687

DE DOMBAL, F. T.
Diagnosis of acute abdominal pain. - Edinburgh : Churchill Livingstone, 1980
170p : ill. - Includes bibliographies and index
0-443-01901-0 SRL
1190687

ACUTE ABDOMEN. Newborn babies
Diagnosis. Radiography

CIPEL, Louis
Radiology of the acute abdomen in the newborn. - New York ; London : Grune and Stratton ; London : Distributed by Academic Press, 1978
xvii,294p : ill. - Includes bibliographies and index
0-8089-1071-x X.320/11644 Woolwich
1840975

ACUTE ALLERGIES. Man
Immunological aspects - Conference proceedings

MOLECULAR and biological aspects of the acute allergic reaction / edited by S.G.O. Johansson and Kjell Strandberg and Börje Uvnäs. - New York : London Plenum Press, 1976
xi,463p. - Includes bibliographies and index
0-306-33703-7 SRL
1965603

ACUTE DISEASES
See also
Patients with acute diseases

ACUTE DISEASES. Digestive system. Man
Emergency treatment - Conference proceedings

GASTROINTESTINAL emergencies / editors Harris R. Clearfield and Vicente P. Dinoso, Jr ; Edith Schwager, executive editor. - New York ; London : Grune and Stratton, 1976
xiii,305p : ill. - (Hahnemann symposium ; 34). - Includes index
0-8089-0713-1 SRL
0673617

INTERNATIONAL SYMPOSIUM ON GASTROINTESTINAL EMERGENCIES
(1st : 1975 : Stockholm)
Gastrointestinal emergencies. - Oxford : Pergamon, 1977
xvii,428p : ill. - (International symposium series / Wenner-Gren Center ; vol.26). - Includes bibliographies and index
0-08-020494-5 SRL
0673617

ACUTE DISEASES. Heart. Man
Acute coronary diseases. Emergency treatment

The ACUTE coronary attack / J.F. Pantridge ... [et al.]. - Tunbridge Wells : Pitman Medical, 1975
x,141p : ill. - Bibliography: p.132-136. - Includes index
0-272-79370-1 X.329/10197
0535575

ACUTE phase of ischemic heart disease and myocardial infarction / edited by A.A. Jennifer Adgey. - The Hague ; London : Nijhoff, 1982
xii,201p : ill. - (Developments in cardiovascular medicine ; v.21). - Includes index
90-247-2675-1 SRL
0535575

ACUTE DISEASES. Heart. Man
Diagnosis. Electrocardiography

LYON, Leonard J.
Basic electrocardiography handbook. - New York ; London : Van Nostrand Reinhold, 1977
xi,175p : ill. - Includes index
0-442-24960-8 SRL
1375067

ACUTE DISEASES. Man

CONCEPTS common to acute illness / edited by Laura K. Hart, Jean L. Reese, Margery O. Fearing. - St. Louis, Mo. ; London : Mosby, 1981
ix,371p : ill. - Includes index
0-8016-2117-8 X.322/10859 Woolwich
2727218

ACUTE DISEASES. Man
Acute bacterial diseases

GARDNER, Pierce
Manual of acute bacterial infections. - 2nd ed. - Boston, [Mass.] : Little, Brown, c1984
xi,445p : ill. - (A Little, Brown spiral manual). - Includes bibliographies and index
0-316-30389-5 X.329/19557
3281892

ACUTE DISEASES. Man
Recovery. Psychological factors

COUSINS, Norman
Anatomy of an illness as perceived by the patient. - New York ; London : Norton, c1979
173p. - Bibliography: p163-173
0-393-01252-2 X.329/13535 Woolwich
235862x

Anatomy of an illness as perceived by the patient. - Toronto ; London : Bantam, 1981, c1979
173p. - Bibliography: p163-173
0-553-01293-2 X.329/14227 Woolwich
235862x

WILSON-BARNETT, Jenifer
Recovery from illness. - Chichester : Wiley, c1982
vii,143p : ill. - (Wiley series on developments in nursing research ; v.1). - Includes bibliographies and index
0-471-10408-6 X.329/16024
235862x

ACUTE DISEASES. Man
Treatment. Psychosocial aspects
GARFIELD, Charles A.
Stress and survival. - St. Louis ; London : Mosby, 1979
xviii,388p : ill. - Includes bibliographies and index
0-8016-1743-x X.325/942 Woolwich
2378086

ACUTE DISEASES. Old persons
Emergency treatment
ACUTE geriatric medicine / edited by Davis Coakley. - London : Croom Helm, c1981
290p : ill. - Includes bibliographies and index
0-7099-0150-x SRL
2153157

ACUTE DISEASES. Urogenital system. Man
Emergency treatment
HANDBOOK of urologic emergencies / ... chief editor Robert Kessler, associate editor Rodney U. Anderson. - London : Kimpton, 1976
154p : ill. - Includes bibliographies and index
0-87488-647-3 SRL
0763322

ACUTE DISEASES CAUSED BY DRUG ABUSE. Man
Emergency treatment
ACUTE drug abuse emergencies / edited by Peter G. Bourne. - New York ; London : Academic Press, 1976 [i.e. 1977]
xx,361p : ill. - Includes bibliographies and index
0-12-119560-0 SRL
1095838

ACUTE drug abuse emergencies / edited by Peter G. Bourne. - New York ; London : Academic Press, 1976
xx,361p : ill. - Includes bibliographies and index
0-12-119565-1 X.319/17343
1095838
- [Another copy.] SRL

ACUTE DISSECTION. Aorta. Man
Surgery
ANAGNOSTOPOULOS, Constantine Efthymios
Acute aortic dissections. - Baltimore ; London : University Park Press ; [Lancaster] : [Distributed by Medical and Technical Publishing Co.], 1975
xv,255p : ill. - Includes index
0-8391-0692-0 SRL
0361410

ACUTE INFARCTION. Muscles. Heart. Man
ACUTE myocardial infarction / Rue L. Cromwell ... [et al.]. - Saint Louis : Mosby ; London : Distributed by Kimpton, 1977
xvi,224p : ill. - Includes index
0-8016-1079-6 SRL
1245392

MCBRIDE, Helena
Acute myocardial infarction. - New York : Appleton-Century-Crofts [for the American Heart Association Council on Cardiovascular Nursing] ; London : Prentice-Hall, 1979
xi,79p : ill. - (Continuing education in cardiovascular nursing. series 1, myocardial infarction ; unit 3). - Includes bibliographies
0-8385-0049-8 X.329/13022 Woolwich
1245392

ACUTE INFARCTION. Muscles. Heart. Man
Diagnosis. Use of radioisotope scanning: Thallium-201 imaging & technetium-99m-pyrophosphate imaging
MYOCARDIAL imaging in the coronary care unit / edited by Frans J.Th. Wackers. - The Hague ; London : Nijhoff, 1980
xii,255p : ill. - (Developments in cardiovascular medicine ; v.9). - Includes index
90-247-2396-5 X.325/898 Woolwich
2356562

ACUTE INFLAMMATION. Tissues. Man
HURLEY, J. V. (John Victor)
Acute inflammation. - 2nd ed. - Edinburgh : Churchill Livingstone, 1983
157p : ill. - Includes bibliographies and index
0-443-02507-x SRL
0464430

ACUTE INJURIES. Hands. Man
Surgery
MUTILATING injuries of the hand / edited by D.A. Campbell Reid and the late J. Gosset in collaboration with ... [others] ; [translated from the French]. - Edinburgh : Churchill Livingstone, 1979
[9],134p : ill. - (Monographs / Groupe d'étude de la main ; no.3). - Includes bibliographies and index
0-443-01684-4 SRL
008364x

ACUTE INJURIES. Hands. Man
Treatment
MUTILATING injuries of the hand. - 2nd ed. - Edinburgh : Churchill Livingstone, 1984
viii,212p : ill. - (Gem monograph series). - Includes bibliographies and index
0-443-02369-7 SRL
240253x

WEEKS, Paul M.
Acute bone and joint injuries of the hand and wrist. - St. Louis ; London : Mosby, 1981
viii,299p : ill. - Bibliography: p283-290. - Includes index
0-8016-5373-8 SRL
240253x

ACUTE INJURIES. Head. Man
HAYWARD, Richard David
Management of acute head injuries. - Oxford : Blackwell Scientific, 1980
vi,106p : ill. - Bibliography: p.101-103. - Includes index
0-632-00515-7 SRL
2085135

ACUTE INJURIES. Wrists. Man
Treatment
WEEKS, Paul M.
Acute bone and joint injuries of the hand and wrist. - St. Louis ; London : Mosby, 1981
viii,299p : ill. - Bibliography: p283-290. - Includes index
0-8016-5373-8 SRL
2402548

ACUTE LEUKAEMIA. Children
Therapy
PINKEL, Donald
Treatment of acute lymphocytic leukemia. - London : Leukaemia Research Fund, 1973
51p : ill. - (Annual guest lectures / Leukaemia Research Fund ; 8th). - Bibliography: p.49-51
0-9502230-3-4 SRL
0762725

ACUTE POISONING. Man
Therapy
PROUDFOOT, A. T. (Alexander Thompson)
Diagnosis and management of acute poisoning. - Oxford : Blackwell Scientific, 1982
viii,237p. - Includes bibliographies and index
0-632-00584-x SRL
181480x

ACUTE POSTERIOR GANGLIONITIS
See Shingles

ACUTE RENAL FAILURE. Man
ACUTE renal failure / edited by Antoine Chapman. - Edinburgh : Churchill Livingstone, 1980
vi,178p : ill. - (Clinics in critical care medicine ; 1). - Includes bibliographies and index
0-443-01930-4 SRL
0216240

ACUTE renal failure / edited by Barry M. Brenner and Jay H. Stein. - New York ; Edinburgh : Churchill Livingstone, 1980
viii,296p : ill. - (Contemporary issues in nephrology ; v.6). - Includes bibliographies and index
0-443-08116-6 SRL
0216240

KUEHNEL, Edward
Acute renal failure. - London : Kimpton, 1976
121p : ill. - Includes index
0-87488-878-6 SRL
0216240

ACUTE RENAL FAILURE. Man
- Conference proceedings
ACUTE renal failure / edited by H.E. Eliahou. - London : Libbey, 1982
x,310p : ill. - Includes index
0-86196-013-0 SRL
0216240

ACUTE RENAL FAILURE. Man
Therapy
The TREATMENT of renal failure / edited by J.E. Castro. - Lancaster : MTP, c1982
vii,392p : ill. - Includes index
0-85200-336-6 SRL
0563668

YUILL, George Martin
The treatment of renal failure. - Manchester : Manchester University Press, 1975
vi,33p : 2 ill. - Includes index
0-7190-0628-7 X.319/10283 Woolwich
0563668

ACUTE RESPIRATORY FAILURE. Man
Emergency treatment
NELSON, Ellan J.
Critical care respiratory therapy. - Boston [Mass.] : Little, Brown, c1983
xvii,394p : ill
0-316-60314-7 X.322/13060
139651x

RESPIRATORY emergencies / edited by Elaine M. Shibel, Kenneth M. Moser. - Saint Louis : Mosby ; London : Distributed by Kimpton, 1977
xi,290p : ill. - Includes bibliographies and index
0-8016-4583-2 SRL
139651x

RESPIRATORY emergencies. - 2nd ed. - St. Louis ; London : Mosby, 1982
xi,316p : ill. - Includes bibliographies and index
0-8016-4584-0 SRL
139651x

ZAGELBAUM, Gary L.
Manual of acute respiratory care. - Boston : Little, Brown, c1982
xii,300p : ill
0-316-98467-1 X.329/15551
139651x

ACUTE RESPIRATORY FAILURE. Man
Therapy
TRACHEOSTOMY and artificial ventilation in the treatment of respiratory failure. - 3rd ed.. - London : Edward Arnold, 1977
x,212p : ill. - Includes bibliographies and index
0-7131-4288-x SRL
0191477

ACUTE RESPIRATORY FAILURE. Man
Therapy. Use of artificial lungs - Conference proceedings
INTERNATIONAL CONFERENCE ON MEMBRANE LUNG TECHNOLOGY AND PROLONGED EXTRACORPOREAL PERFUSION (1975 : Rungstedgaard)
Artificial lungs for acute respiratory failure. - New York ; London : Academic Press ; Washington ; London : Hemisphere Publishing Corporation, 1976
xii,557p : ill. - Includes bibliographies and index
0-12-775450-4 SRL
105810x

ACUTELY INJURED ADULTS
See also
Day hospitals for acutely injured adults

ACUTELY INJURED PATIENTS. Rehabilitation centres. England
Acutely head injured patients. Rehabilitation
REHABILITATION after severe head injury / edited by C.D. Evans. - Edinburgh : Churchill Livingstone, 1981
vi,218p : ill. - Includes bibliographies and index
0-443-01693-3 X.329/14622 Woolwich
2573423

ACUTT (FAMILY). South Africa
MILLER, Yvonne
Acutts in Africa. - [Durban?] : Yvonne Miller, c1978
xiv,206p : ill. - Bibliography: p.205
r6864674 X.800/33070
8410062
- No.162 of an edition of 500 copies X.800/33070

ACVS
See Hovercraft

AD-'ANDALUSÎ, IBN HÂNI. Poetry in Arabic
YALAOUI, Mohammed
Un poète chiite d'Occident au IVème/Xème siècle : Ibn Hânî al-Andalusî. - [Tunis] : Publications de l'Université de Tunis, 1976
474p. - (Faculté des lettres et Sciences humaines [de l'Université de Tunis] ; sér, 6, vol.9). - Bibliography : p.459-468
r6028058 Ac.1997.f/3f.[vol.9]
1158244

AḌ-ḌARÎḤ AL-MALAKÎ AL-MÛRÎTÂNÎ. Algeria
BOUCHENAKI, Mounir
Le Mausolée royal de Maurétanie. - Alger : Direction des Musées de l'archéologie et des monuments et sites historiques, 1979
32,22p : ill
r6209283 X.421/23859
8947376

AD HOC COMMITTEES OF INQUIRY. Administration. Local authorities. Great Britain
Procedure
AD hoc inquiries in local government. - Lewes : SOLACE, 1978
[2],viii,57p. - (Occasional papers / Society of Local Authority Chief Executives)
0-905566-02-5 X.709/32167
1434063

AD INTERNATIONAL LIMITED
Mergers with Dentsply International Incorporated
GREAT BRITAIN. Monopolies and Mergers Commission
Dentsply International Incorporated and AD International Limited. - London : H.M.S.O., [1975]
v,65p : ill. - ([1974-75 H.C.]394)
0-10-023945-5 BS.83/79
0364908

ADA. Nabokov, Vladimir. American writers. Fiction in English
Special themes: Nature
MASON, Bobbie Ann
Nabokov's garden. - Ann Arbor : Ardis, 1974
196p. - Bibliography : p.193-196
0-88233-052-7 X.981/12033
1494287

ADA LANGUAGE. Programming languages. Computer systems
LE VERRAND, D. (Dominique)
Evaluating ADA. - Rev. and updated. - Oxford : North Oxford Academic, c1985
vii,288p : ill. - (Studies in computer science). - Bibliography: p281-286. - Includes index
0-946536-15-5 SRL
2142198

ADA LANGUAGE. Programming languages. Digital computer systems
ADA / edited by M.W. Rogers. - Cambridge : Cambridge University Press on behalf of the Commission of the European Communities, 1984
1v.(various pagings). - (The Ada companion series). - Bibliography: p41p
0-521-26464-2 SRL
2142198

ADA-BASED system development methodology study report. - London : Department of Industry, 1981
3v. : ill
B82-10039 SRL
2142198

BARNES, J. G. P. (John Gilbert Presslie)
Programming in Ada. - London : Addison-Wesley, c1982
x,340p : ill. - (International computer science series). - Includes index
0-201-13793-3 X.622/12226 Woolwich
2142198
Programming in Ada. - 2nd ed. - London : Addison-Wesley, c1984
xv,367p : ill. - (International computer science series). - Includes index
0-201-13799-2 SRL
2142198

BUHR, R. J. A.
System design with Ada. - Englewood Cliffs : Prentice-Hall ; London : Prentice-Hall International, c1984
xv,256p : ill. - Includes index
0-13-881623-9 SRL
2142198

CLARK, Robert G.
Programming in Ada. - Cambridge : Cambridge University Press, 1985
[217]p : ill. - Includes index
0-521-25728-x X.622/23544
2142198

DOWNES, Valerie A.
Programming embedded systems with Ada. - Englewood Cliffs ; London : Prentice-Hall, c1982
xv,377p : ill. - Includes index
0-13-730010-7 SRL
2142198

HILFINGER, Paul N.
Abstraction mechanisms and language design. - Cambridge, Mass. ; London : MIT Press, c1983
xi,176p. - (ACM distinguished dissertations)
0-262-08134-2 X.950/26284
2142198

KATZAN, Harry
Invitation to Ada. - Condensed version. - New York : PBI, c1984
x,173p : ill. - (Petrocelli invitation to series). - Bibliography: p167-169
0-89433-239-2 X.622/23661
2142198

PYLE, I. C. (Ian C.)
The ADA programming language. - Englewood Cliffs ; London : Prentice-Hall, c1981
x,293p
0-13-003921-7 SRL
2142198

REFERENCE manual for the Ada programming language / United States Department of Defense. - Tunbridge Wells : Castle House, 1983
[350]p. - Includes index
0-7194-0097-x SRL
2142198

SAXON, James A. (James Anthony)
Beginning programming with Ada. - Englewood Cliffs ; London : Prentice-Hall, c1983
xv,218p. - Includes index
0-13-071688-x SRL
2142198

SHUMATE, Kenneth C.
Understanding Ada. - New York ; London : Harper & Row, c1984
xii,387p. - Includes index
0-06-046133-0 X.622/22279
2142198

YOUNG, Stephen J. (Stephen John)
An introduction to Ada. - Chichester : Horwood, 1983
400p : ill. - (Ellis Horwood series in computers and their applications). - Includes index
0-85312-535-x SRL
2142198

ADA LANGUAGE. Programming languages. Digital computer systems - Serials
ADA UK news. - Vol.1, no.1 (Mar. 1980)-. - London : Ada Language UK Ltd, 1980-
v. : ill
B83-24913 SRL
2905515

ADA LANGUAGE. Programming languages. Digital computer systems
Applications in problem solving
MAYOH, Brian
Problem solving with ADA. - Chichester : Wiley, 1982
viii,233p : ill. - Includes index
0-471-10025-0 SRL
2764717

ÅDALEN. Sweden
Bridges. Collapse, to date
NENZÉN, Börje
Ådalen -39. - [Bjästa] : CeWe, 1979
133p : ill
91-7542-015-5 X.809/54540
8892830

ÅDALEN. Sweden
Political events, 1905-1945
ÅDALEN / [redaktion: Hans Arvidsson, Jan-Erik Olson och Uno Palmström]. - [Stockholm] : Arbetarkultur, 1981
95p : ill
91-7014-132-0 X.800/36135
8874816

ELMBRANT, Björn
Hungermarschen 1917. - Bjästa : CEWE, 1980
133p : ill
91-7542-035-x X.808/38232
8874816

NORMAN, Birger
Ådalen 31. - Ny utvidgad uppl.. - [Stockholm] : Rabén & Sjögren, 1981
204p,[24]p of plates : ill
91-295-4856-x X.809/53164
8874816

ÅDALEN. Sweden
Social conditions
EDSTRÖM, Mauritz
Mitt Ådalen. - Stockholm : Liber Förlag, 1978
260p : ill
91-380-4279-7 X.809/26975
2026902

ÅDALEN. Sweden
Stevedores. Social conditions, ca 1870-1975
BJÖRKLUND, Anders, 1948-
Lundebusar och bondesjaare. - Stockholm : Liber förlag, 1978
252p : ill. - (Sjöhistoriska museet. Rapport ; nr 9). - Includes bibliographies
91-380-4403-x X.0709/717(9)
1984020

ADAM. Bible. O.T. Genesis
BUTCHER, Geoffrey
Adam. - London : Hodder and Stoughton, 1982
[28]p : col.ill. - (First Bible stories)
0-340-28185-5 X.208/7330
2714507

ADAM. Bible. O.T. Genesis
Adam & Eve
STORR, Catherine
Adam and Eve. - London : Methuen Children's Books, 1983
[32]p : col.ill. - (People of the Bible)
0-416-42960-2 X.990/21340
2765462

Adam and Eve. - London : Watts, in association with Belitha, 1983
[30]p : col.ill. - (People of the Bible)
0-86313-004-6 X.200/4 265
2765462

ADAM. Bible. O.T. Genesis
Adam & Eve - Texts
CASELEY, Judith
The garden of Eden. - [London] : Abelard, 1982
[26]p : col.ill
0-200-72746-x X.990/18934
2765462

ADAM. Bible. O.T. Genesis
Adam & Eve. Theories
CHURN, Charles Reginald
For discussion, the Adam and Eve story. - Truro : [The author], [1976]
19p
0-9505325-0-9 X.100/15638
0766925

ADAM, DE LA HALLE. Songs in Old French
BARTH-WEHRENALP, Renate
Studien zu Adan de le Hale. - Tutzing : Schneider, 1982
108p. - (Wiener Veröffentlichungen zur Musikwissenschaft ; Bd.18). - Includes bibliography
3-7952-0204-3 X.431/22220
9061258

ADAM, ONKEL
See Wetterbergh, C. A. †

ADAM, ADOLPHE. Ballet
Performance by Pavlova, Nadezhda & Solov'ev, Iūrii
GREGORY, John, 1914-
Giselle immortal. - London : Robson, 1982
58p : ill. - Bibliography: p58
0-86051-177-4 X.622/13845
2801655

ADAM, ADOLPHE. Ballet
Performances
HERING, Doris
Giselle & Albrecht. - New York : Dance Horizons, 1981
191p : ill. - Bibliography: p187-188
0-87127-124-9 X.622/15178
2856549

ADAM, AUGUSTE VILLIERS DE L'ISLE-
See Villiers de l'Isle-Adam, Auguste

ADAM BEDE. Eliot, George. Fiction in English
ADAM, Ian
George Eliot, 'Adam Bede'. - London : Longman [etc.], 1980
79p. - (York notes ; 14). - Bibliography: p.76-78
0-582-78143-4 X.909/44648
2149591

ADAM DE LA HALLE. Songs in Old French
MAILLARD, Jean
Adam de la Halle. - Paris : Éditions Honoré Champion, 1982
221p. - (Librairie Honoré Champion). - Includes bibliography
2-85203-101-9 X.439/12538
9012265

ADAM DE ORLETON
See also
Worcester (Diocese). Bishop, 1327-1333 (De Orleton)

ADAM HOUSES. Great Britain
Adam country houses
BEARD, Geoffrey
Robert Adam's country houses. - Edinburgh : Bartholomew, 1981
31p : ill(some col.). - Bibliography: on inside cover
0-7028-8061-2 X.421/22415 Woolwich
2592509

ADAM, JAMES
See also
Buildings designed by Adam, James

ADAM, JAMES. Architectural design
Adam, Robert & Adam, James
ADAM, Robert
The works in architecture of Robert & James Adam. - New York : Dover Publications ; London : Constable, 1980
xii,23p,105p of plates : ill. - Bibliography: pviii
0-486-23810-5 N.Tab 2025/31
2378698

ADAM, JAMES Scottish architectural drawings. Exhibits. Sir John Soane's Museum
Adam, Robert & Adam, James - Catalogues
SIR JOHN SOANE'S MUSEUM
Catalogue of the drawings and designs of Robert and James Adam in Sir John Soane's Museum. - Cambridge : Chadwyck-Healey [etc.], 1979
82,[4],92p
0-85964-055-8 X.421/11101 Woolwich
1701827

ADAM, JOHN Iron industries. United States
HOWELL, Kenneth T
Men of iron. - Lakeville : Pocketknife Press, c1980
157p : ill
Signed by Kenneth T. Howell
r6810126 X.629/20505
8966745

ADAM, PAUL. French literature
- Biographies
ADAM, Paul
[Correspondence. Selections]. L'époque symboliste et le monde proustien a travers la correspondance de Paul Adam 1884-1920. - Paris : Éditions A.-G. Nizet, 1982
140p. - "Bibliographie - œuvres de Paul Adam" : p.137
r6208609 X.950/17558
8171866

ADAM, ROBERT
See also
Adam houses
Buildings designed by Adam, Robert

ADAM, ROBERT. Architectural design
BEARD, Geoffrey
The work of Robert Adam. - Edinburgh : J. Bartholomew, 1978
xi,244p : chiefly ill(some col). - Bibliography: p.28-31. - Includes index
0-7028-1087-8 X.421/10674 Woolwich
1301748

ADAM, ROBERT. Architectural design
- Biographies - Biographies
MACLEAN, Colin C.
Robert Adam F.R.S.E. (1728-1792). - Edinburgh : History of Medicine and Science Unit, University of Edinburgh, [1982]
[20]p : ill. - (Scottish men of letters)
0-907692-40-0 X.329/16533
0381276

ADAM, ROBERT. Architectural design
Adam, Robert & Adam, James
ADAM, Robert
The works in architecture of Robert & James Adam. - New York : Dover Publications ; London : Constable, 1980
xii,23p,105p of plates : ill. - Bibliography: pviii
0-486-23810-5 N.Tab 2025/31
2378698

ADAM, ROBERT. Scottish architectural drawings. Exhibits. Sir John Soane's Museum
Adam, Robert & Adam, James - Catalogues
SIR JOHN SOANE'S MUSEUM
Catalogue of the drawings and designs of Robert and James Adam in Sir John Soane's Museum. - Cambridge : Chadwyck-Healey [etc.], 1979
82,[4],92p
0-85964-055-8 X.421/11101 Woolwich
1701827

ADAM SMITH INSTITUTE
- Serials
INSIDER bulletin / Adam Smith Institute. - Feb. 1981-. - London : The Institute, 1981-
v.
B83-41484 P.523/1959
296807x

ADAM SMITH, PATSY. Sailing. Australia
- Texts
ADAM SMITH, Patsy
Trader to the islands. - Adelaide : Seal Books, 1977
[8],220p,[8]p of plates
0-7270-0240-6 X.708/21402
1586793

ADAM, THEO. Singing. Opera
- Biographies
ADAM, Theo
Seht, hier ist Tinte, Feder, Papier -. - München : Max Hieber, 1981
225p,plates. - (MH ; 9001)
r6307267 X.439/12453
8332991

ADAM VON BREMEN. Christian church. Scandinavia
Manuscripts
KRISTENSEN, Arne K G
Studien zur Adam von Bremen Überlieferung. - København : Københavns Universitet, Historiske Institut, 1975
218p. - (Skrifter udgivet af det Historiske Institut ved Københavns Universitet ; bd 5). - Bibliography
87-503-1759-8 Ac.1025.i[bd.5]
247090X

ADAM, WILLIAM, B.1689
See also
Buildings designed by Adam, William, b.1689

ADAM WILSON & SONS
to 1979
ADAM WILSON & SONS
Adam Wilson & Sons. - Ayr : Adam Wilson & Sons Ltd, 1980
40p : ill
B82-29241 X.809/52456
2755270

ADAMANSHUIS (ZUTPHEN)
DOORNINK-HOGENRAAD, M. M.
Adamanshuis. - Zutphen : Walburg, [1983]
112p : ill
90-601-1227-x X.200/44849
9166165

ADAMAWA. Nigeria
to date
ABUBAKAR, Sa'ad
The Lāmībe of Fombina. - [Zaria] : Ahmadu Bello University Press, [1977]
x,190p. - (Ahmadu Bello University history series). - Includes bibliography
978-12-5011-9 X.800/35903
8985952

ADAMAWA ARTS
Adamawa decorative arts
CHAPPEL, T. J. H.
Decorated gourds in north-eastern Nigeria. - London : Ethnographica [etc.], 1977
vii,222p,[14]p of plates : ill(some col). - Bibliography: p.216-218. - Includes index
0-905788-02-8 X.423/5243 Woolwich
1147420

ADAMAWA LANGUAGES
See also
Mumuye language

ADAMCLISI. Romania
Ancient Roman monuments
RĂDULESCU, A.V.
The triumphal monument Tropaeum at Adamclisi. - [Constanța] : Museum of National History and Archaeology, [1978]
35p : ill
r6183512 X.708/48138
2412748

ADAMGARH. India (Republic)
Antiquities
JOSHI, Ramchandra Vinayak
Stone Age cultures of Central India. - Poona : Deccan College, Post-Graduate and Research Institute, 1978
vii,96p,18p of plates,1 leaves of plates : ill. - (Publication ; DC:M-181)
r6803895 X.800/32587
8809682

ADAMIČ, LOUIS. American writers. English literature
PETRIČ, Jerneja
Svetovi Louisa Adamiča. - Ljubljana : Cankarjeva založba, 1981
235p,plates
r6494959 X.958/26524
8131120

ADAMIČ, LOUIS. American writers. Fiction in English
ADAMIČ, France
Spomini in pričevonjo o življenju in delu Louiso Adomičo. - Ljubljana : Prešernova družba, 1983
206p
r6494949 X.958/26543
8153272

ADAMIČ, LOUIS. American writers. Fiction in English
- Correspondence, diaries
ADAMIČ, Louis
[Correspondence. Slovene. Selections]. Izbrana pisma Louisa Adamiča. - v Ljubljani : Cankarjeva založba, 1981
473p
r6435363 X.958/11982
8153272

ADAMO, PAOLA. Italy
- Biographies
L'ARCO, Adolfo
Una meteora viva. - Napoli : Edizioni dehoniane, 1981
116p : ill
r6510943 X.808/40195
8265283

ADAMOV. Moravia. Czechoslovakia
Iron industries, 1600-1900 & engineering industries, 1600-1900
KREPS, Miloš
Dějiny adamovských železáren a strojíren do roku 1905. - v Brně : Blok, 1976
254p,lvip of plates : ill
r6183371 X.622/7840 Woolwich
2431874

ADAMOV, ARTHUR. Drama in French
CHAHINE, Samia Assad
Regards sur le théâtre d'Arthur Adamov. - Paris : A.G. Nizet, 1981
246p
r6210102 X.950/21201
870564x

MCCANN, John J.
The theater of Arthur Adamov. - Chapel Hill : U.N.C. Department of Romance Languages, 1975
168p : ill. - (North Carolina studies in the Romance languages and literatures. essays ; 13). - Bibliography : p.160-165.-Includes index
0-88438-013-0 Ac.2685.k/7a(13)
1513621

MODERNES französisches Theater / herausgegeben von Karl Alfred Blüher. - Darmstadt : Wissenschaftliche Buchgesellschaft, 1982
469p. - (Wege der Forschung ; Bd.532). - Includes bibliography
3-534-07599-4 X.958/25453
870564x

REILLY, John H.
Arthur Adamov. - New York : Twayne Publishers, 1974
177 p.. - (Twayne's World authors series ; 318)
0-8057-2005-7 X.989/32191
0900095

ADAMOV, ARTHUR. Drama in French
- Conference proceedings
LECTURES d'Adamov / édités par Robert Abirached, Ernstpeter Ruhe, Richard Schwaderer. - Tübingen : Narr, 1983
181p. - (Études littéraires françaises ; 27)
3-87808-592-3 X.0909/1049(27)
870564x

ADAMOVICH, GEORGIĬ. Russian literature
- Bibliographies
HAGGLUND, Roger
Georgy Adamovich. - Ann Arbor : Ardis, 1985
102p
0-88233-996-6 X.950/42997
8383936

ADAMOVICH, GEORGIĬ. Russian literature
- Biographies
HAGGLUND, Roger
A vision of unity. - Ann Arbor : Ardis, 1985
119p. - Includes bibliography
0-88233-410-7 X.950/42998
8383936

ADAMOVICH, GEORGY
See Adamovich, Georgiĭ †

ADAMS (FAMILY). United States
1735-1927 - Biographies
NAGEL, Paul C.
Descent from glory. - New York ; Oxford : Oxford University press, 1983
xiv,400p,[16]p of plates : ill. - Includes index
0-19-503172-5 X.800/35636
2925958

ADAMS, ABIGAIL. United States
- Biographies
WITHEY, Lynne
Dearest friend. - New York : Free Press ; London : Collier Macmillan, c1981
xiv,369p,[8]p of plates : ill. - Bibliography: p347-356. - Includes index
0-02-934760-2 X.800/32893
2709651

ADAMS, ABIGAIL. United States
Adams, John, 1735–1826 & Adams, Abigail – Texts
ADAMS (Family)
The book of Abigail and John. – Cambridge, Mass. ; London : Harvard University Press, 1975
ix,411p,plate : ill. – Includes index
0–674–07855–1 X.800/25353
0581585

The book of Abigail and John. – Cambridge, Mass. ; London : Harvard University Press, 1976
ix,411p : ill. – Includes index
0–674–07854–3 X.709/30924
0581585

ADAMS, ABIGAIL SMITH. United States – Biographies
AKERS, Charles W.
Abigail Adams. – Boston, Mass. : Little, Brown & Co., [1980]
207p. – (Library of American biography)
0–316–02040–0 X.808/35841
8446814

ADAMS, ALFRED. Zululand. South Africa – Biographies
MIKULA, M. (Maggie)
The Adams story. – [Durban?] : [M. Mikula?], [1981?]
153p
Copy no.250 of an edition of 400
0–620–05404–2 X.809/55381
8975647

ADAMS, ANSEL. American photography – Biographies
NEWHALL, Nancy
Ansel Adams. – Rev. ed. – New York : Aperture, c1980
175p : ill
0–89381–066–5 L.45/1016
2405113

ADAMS, ANSEL. American photography. Exhibits. Moderna museet. Fotografiska museet – Catalogues
MODERNA MUSEET. Fotografiska museet
Ansel Adams. – [Stockholm] : Moderna museet, 1977
27p : ill. – (Moderna museets utställningskatalog ; nr 153)
91–7100–131–x WP.16291/153
1533363

ADAMS, CHARLIE. Zululand. South Africa – Biographies
MIKULA, M. (Maggie)
The Adams story. – [Durban?] : [M. Mikula?], [1981?]
153p
Copy no.250 of an edition of 400
0–620–05404–2 X.809/55381
8975647

ADAMS, CHRIS. Children's musical plays in English – Librettos
ADAMS, Chris
The evacuees. – Kingston upon Thames : Youngsong Music, c1981
41p. – Bibliography: p41
B82–32924 X.950/14975
2307839

ADAMS, CHRIS. Children's musical plays in English – Texts
ADAMS, Chris
The lollipop lady. – Kingston upon Thames : Youngsong Music, [1979]
[2],29p
B80–41103 X.902/5726
2307839

ADAMS COUNTY. Mississippi
Negro sharecroppers. Social conditions, 1860–1890
DAVIS, Ronald L. F. (Ronald Lee Franklin)
Good and faithful labor. – Westport, Conn. ; London : Greenwood, 1982
xv,225p : ill. – (Contributions in American history ; no.100). – Bibliography: p207–220. – Includes index
0–313–23134–6 X.529/54591
2908891

ADAMS, GEORGE, 1894–1963. Anthroposophy – Correspondence, diaries
ADAMS, George, 1894–1963
A letter from George Adams. – London : Steiner, 1978
47p
0–85440–339–6 X.525/6489
2820609

ADAMS, HENRY
Theories of American thought
ROWE, John Carlos
Henry Adams and Henry James. – Ithaca ; London : Cornell University Press, 1976
255p. – Bibliography: p.243–250. – Includes index
0–8014–0954–3 X.989/50728
0661880

ADAMS, HENRY. Historiography
BYRNES, Joseph F.
The Virgin of Chartres. – East Brunswick, N.J. : Associated University Presses, c1981
194p
0–8386–2369–7 X.950/16418
8242496
– [Another copy.] X.809/59694

ADAMS, HENRY. Historiography – Biographies
DUSINBERRE, William
Henry Adams. – Charlottesville : University Press of Virginia, 1980
250p
0–8139–0833–7 X.800/33286
8242496

ADAMS, HENRY. Historiography. United States
BLACKMUR, R. P. (Richard Palmer)
Henry Adams. – London : Secker and Warburg, 1980
xxvii,354p,[1] leaves of plates. – Bibliography: p.346–348. – Includes index
0–436–05060–9 X.950/1722
2359472

ADAMS, HENRY. Historiography. United States – Biographies
ADAMS, Henry
The letters of Henry Adams. – Cambridge, Mass. : Belknap Press, c1982
3v. : ill(some col.). – Includes index
0–674–52685–6 X.950/42906
3284395

CONTOSTA, David R.
Henry Adams and the American experiment. – Boston : Little, Brown & Co., [1980]
viii,159p. – (Library of American biography). – Includes bibliography
0–316–15400–8 X.809/61836
9093303

ADAMS, JOHN, 1735–1826
Theories on American Independence, 1775–1789 – Biographies
MORGAN, Edmund Sears
The meaning of independence. – New York : W.W. Norton, 1978
85p. – (Norton Library)
0–393–00896–7 X.808/38177
9018492

ADAMS, JOHN, 1735–1826. United States
EAST, Robert A. (Robert Abraham)
John Adams. – Boston : Twayne, c1979
126p. – (Twayne's world leaders series ; 78)
0–8057–7723–7 X.529/54051
8446814

ADAMS, JOHN, 1735–1826. United States – Biographies
ADAMS, John, 1735–1826
Papers of John Adams. – Cambridge, Mass. ; London : Belknap Press. – (The Adams papers. series 3)
Vol.3 : May 1775–Jan. 1776 ; [and], Vol.4 : Feb.–Aug. 1776. – 1979
2v.(xxxv,417p;xi,550p) : ill
0–674–65442–0 10712.p.1 / 3a
2061813

ADAMS, JOHN, 1735–1826. United States – Correspondence, diaries
ADAMS, John, 1735–1826
Papers of John Adams. – Cambridge, Mass. ; London : Belknap Press of Harvard University Press. – (The Adams papers. Series 3, General correspondence and other papers of the Adams statesmen)
Vol.5: August 1776–March 1778. – 1983
xliii,410p : ill
B84–37263 10712.p.1/3a
2061813

ADAMS, JOHN, 1735–1826. United States
Adams, John, 1735–1826 & Adams, Abigail – Texts
ADAMS (Family)
The book of Abigail and John. – Cambridge, Mass. ; London : Harvard University Press, 1975
ix,411p,plate : ill. – Includes index
0–674–07855–1 X.800/25353
0581585

The book of Abigail and John. – Cambridge, Mass. ; London : Harvard University Press, 1976
ix,411p : ill. – Includes index
0–674–07854–3 X.709/30924
0581585

ADAMS, JOHN, 1735–1826. United States
Interpersonal relationships with Jefferson, Thomas related to their attitudes to political events in United States, 1775–1812
PETERSON, Merrill Daniel
Adams and Jefferson. – Oxford : Oxford University Press, 1978
xiv,146p. – (A galaxy book). – Includes index
0–19–502355–2 X.709/32199
1415158
– [Another copy.] X.709/32871

ADAMS, JOHN, B.1735. United States
related to their interpersonal relationships. Attitudes to political events in United States, 1775–1812
PETERSON, Merrill Daniel
Adams and Jefferson. – Oxford : Oxford University Press, 1978
xiv,146p. – (A galaxy book). – Includes index
0–19–502355–2 X.709/32199
141514x
– [Another copy.] X.709/32871

ADAMS, JOHN BODKIN. Eastbourne. East Sussex
Trial for murder
HOSKINS, Percy
Two men were acquitted. – London : Secker & Warburg, 1984
220p,[8]p of plates : ill
0–436–20161–5 X.800/39396
3189287

ADAMS, JOHN QUINCY. United States – Biographies
ADAMS, John Quincy
Diary of John Quincy Adams. – Cambridge, Mass. ; London : Belknap Press of Harvard University Press. – (The Adams papers. Series 1, diaries)
Vol.1: November 1779–March 1786. – 1981
lxii,415p : ill
B82–35395 10712.p.1/1b
2775271

Vol.2: March 1786–December 1788, index. – 1981
xiv,521p : ill. – Includes index
B82–35396 10712.p.1/1b
2775271

SHEPHERD, Jack, 1937–
Cannibals of the heart. – New York : McGraw-Hill, c1980
xiv,440p
0–07–056730–1 X.800/34435
8446814

ADAMS, LOUISA CATHERINE. United States – Biographies
SHEPHERD, Jack, 1937–
Cannibals of the heart. – New York : McGraw-Hill, c1980
xiv,440p
0–07–056730–1 X.800/34435
8446814

ADAMS-MORGAN. Washington, D.C.
Self-sufficiency. Social aspects
HESS, Karl
Community technology. – New York ; London : Harper and Row, 1979
[4],107p
0–06–011874–1 X.529/34048
1954342

ADAMS, NICK
See Nick Adams

ADAMS, NORMAN, 1927–. English paintings 1971–1981
ADAMS, Norman, 1927–
Norman Adams. – Glasgow : Third Eye Centre, 1981
36p : ill(some col.). – Bibliography: p31
0–906474–20–5 X.429/14375
2895978

ADAMS, PETER. Prisoners. Great Britain
Interpersonal relationships with Adams, Shirley
ADAMS, Peter, 1940–
Knockback. – London : Duckworth, 1982
176p
0–7156–1677–3 X.529/52330
2849755

ADAMS, PETER, 1940–. Prisoners. Great Britain
Interpersonal relationships with Cooklin, Shirley

ADAMS, Peter, 1940–
Knockback. – London : Ariel, 1984, c1982
183p
0–563–20260–2 X.958/28897
2849755

ADAMS REVOLVERS
1851–1881

CHAMBERLAIN, William Henry Jason
Adams' revolvers. – London : Barrie and Jenkins, 1976
240p : ill. – Bibliography: p.224–235. – Includes index
0–214–20089–2 X.622/5608 Woolwich
072260x

ADAMS, RICHARD, 1920–. Fiction in English
– Catalogues

ADAMS, Richard, 1920–
Watership Down. – London : Henry Sotheran Ltd, 1982
[43p] : ill
B84–12203 X.950/28390
3046893

ADAMS, RICHARD, 1920–. Fiction in English
Watership Down

IRONS, John
Watership Down in context. – [Odense] : [Odense University], 1975
19p. – (Pre-publications of the English Institute of Odense University ; no.1)
r6060905 X.0519/744(1)
179258X

ADAMS, SHIRLEY. Acting. Great Britain
Interpersonal relationships with Adams, Peter

ADAMS, Peter, 1940–
Knockback. – London : Duckworth, 1982
176p
0–7156–1677–3 X.529/52330
2849763

ADAMS, STANLEY. Hoffmann–La Roche (Firm)

ADAMS, Stanley
Roche versus Adams. – London : Cape, 1984
xii,236p,[8]p of plates : ill. – Includes index
0–224–02180–x X.200/42452
3058255

Roche versus Adams. – London : Fontana, 1985, c1984
xioi,236p,[8]p of plates : ill. – Includes index
0–00–636880–8 X.529/70368
3058255

ADAMS, THOMAS, 1871–1940
Role in town planning, 1900–1940

SIMPSON, Michael, 19—
Thomas Adams and the modern planning movement. – London : Mansell, 1985
viii,262p : ill. – (Studies in history, planning and the environment) (An Alexandrine Press book). – Includes index
0–7201–1714–3 X.800/41240
3155943

ADAMS, TOM. British illustrations

ADAMS, Tom
Tom Adams' Agatha Christie cover story. – Limpsfield : Paper Tiger, [1981]
144p : chiefly col ill
0–905895–62–2 X.955/1939
9044434

ADAMS, WILLIAM, 1564–1620. Samurai. Japan

TAMES, Richard
Servant of the shogun. – Tenterden : Paul Norbury, 1981
132p : ill. – Bibliography: p126–127
0–904404–39–0 X.809/54496
2871378

ADAMSEZ LTD.
to 1979

ADAMSEZ. – Newcastle upon Tyne : West End Resource Centre, 1980
8p : ill
B84–30284 X.525/9267
2881837

ADAMSEZ. – Newcastle upon Tyne : Benwell Community Project, 1980
71p : ill. – (Benwell Community Project final report series ; no.8)
0–906316–06–5 X.0510/525(8)
2881837

ADAMSKI, GEORGE. Investigation. Unidentified flying objects

ZINSSTAG, Lou
George Adamski. – Beckenham : Ceti, 1983
xii,208p,[12]p of plates : ill. – Includes index
0–9508414–0–4 X.629/20914
294183x

ADAMSON, JOY. Conservation. Wildlife. Kenya
– Biographies

ADAMSON, Joy
The searching spirit. – London : Collins : Harvill Press, 1978
224p,[36]p of plates : ill(some col). – Includes index
0–00–216035–8 X.800/27354
1732102

ADAMSON, ROBERT. Calotypes. Exhibits. Canongate Tolbooth Museum
Hill, David Octavius & Adamson, Robert – Catalogues

SCOTTISH NATIONAL PORTRAIT GALLERY
Retained images. – [Edinburgh] : [City of Edinburgh Museums and Art Galleries], [1977]
Folder ([6]p) : 1 ill
B78–09694 X.615/1987 Woolwich
1353349

ADAMSON, ROBERT. Scottish calotypes. Exhibits. Scottish National Portrait Gallery
Hill, David Octavius & Adamson, Robert – Catalogues

SCOTTISH NATIONAL PORTRAIT GALLERY
David Octavius Hill and Robert Adamson. – Edinburgh : National Galleries of Scotland, 1981
220p : ill
0–903148–37–4 X.425/2368 Woolwich
2585642
– [Another copy] X.425/2400 Woolwich

ADAMSON, ROBERT. Scottish portrait photography. Stock. National Portrait Gallery
Hill, David Octavius & Adamson, Robert

HILL, David Octavius
The Hill/Adamson albums. – London : Times Newspapers, 1973
[15]p,64p of plates : chiefly ill. – Bibliography: p.[4]
0–7230–0086–7 X.619/16749 Woolwich
0652806

ADAN, DE LE HALE
See Adam, de la Halle †

ADÁN, MARTÍN. Peruvian writers. Poetry in Spanish
– Bibliographies

WELLER, Hubert P.
Bibliografía analítica y anotada de y sobre Martín Adán (Rafael de la Fuente Benavides, 1927–1974). – Lima : Instituto Nacional de Cultura, 1975
139p
r6705962 X.958/13370
8361177

ADAPTABILITY. Adults. United States
To transitions. Psychosocial aspects

FISKE, Marjorie
Four stages of life. – San Francisco ; London : Jossey-Bass, 1975
xxv,292p. – (The Jossey-Bass behavioral science series). – Bibliography: p.263–281. – Includes index
0–87589–246–9 X.520/10082
0359254

ADAPTABILITY. Personnel
Psychological aspects

BRANDES, Holger
Flexibilität und Qualifikation. – Darmstadt : Steinkopff, 1980
xi,191p. – (Psychologie und Gesellschaft ; Bd 13). – Bibliography
3–7985–0584–5 X.529/43066
8770115

PSIKHOLOGICHESKIE problemy vzaimnoi adaptatsii cheloveka i mashiny v sistemakh upravleniia / [otvetststvennye redaktory: B.F. Lomov, V.F. Venda, IU.M. Zabrodin]. – Moskva : Nauka, 1980
319p
r6496876 X.529/67033
8770115

ADAPTABILITY. Personnel. Industries. Great Britain
Development. Teaching methods – For industrial training

INDUSTRIAL TRAINING RESEARCH UNIT
Training for versatility. – Cambridge : The Unit, 1980
[4],50p. – (Publications / Industrial Training Research Unit ; TR12)
B80–40949 X.622/8954 Woolwich
2311445

ADAPTATION
See also
Anhydrobiosis

ADAPTATION. Algonquians. Taiga. North America
To environment

BOREAL forest adaptations / edited by A. Theodore Steegmann, Jr. – New York ; London : Plenum, 1983
xii,360p : ill. – Includes bibliographies and index
0–306–41238–1 X.322/12646
2992280

ADAPTATION. Anatomy. Animals
To environment

CLOUDSLEY-THOMPSON, J. L. (John Leonard)
Form and function in animals. – Shildon : Meadowfield Press, 1978
[6],81p : ill. – (Patterns of progress. zoology series ; 10). – Bibliography: p.73–74. – Includes index
0–900541–90–3 X.319/19247 Woolwich
2044951

ADAPTATION. Animals
To environment

JENKINS, Alan C. (Alan Charles)
Mysteries of nature. – Exeter : Webb & Bower
208p : ill(some col.). – Includes index
0–906671–79–5 X.322/14793
2073773

LITTLE, Colin
The colonisation of land. – Cambridge : Cambridge University Press, 1983
vi,290p : ill. – Bibliography: p241–272. – Includes index
0–521–25218–0 SRL
2073773

ADAPTATION. Animals
To environment – Conference proceedings

EUROPEAN SOCIETY FOR COMPARATIVE PHYSIOLOGY AND BIOCHEMISTRY. Conference (1st : 1979 : Liège)
Animals and environmental fitness. – Oxford : Pergamon.
Vol.1: Invited lectures. – 1980
xviii,619p : ill. – Includes bibliographies and index
0–08–024938–8 SRL
2073773

Vol.2: Abstracts. – 1980
[4],186p : ill. – Includes bibliographies and index
0–08–024939–6 SRL
2073773

ADAPTATION. Animals
To high altitudes – Conference proceedings

LIFE, HEAT, AND ALTITUDE CONFERENCE (1979 : University of Nevada)
Environmental physiology. – New York ; Oxford : Elsevier/North-Holland, c1981
468p : ill. – Includes bibliographies and index
0–444–00583–8 SRL
2500329

ADAPTATION. Animals
To high temperatures – Conference proceedings

LIFE, HEAT, AND ALTITUDE CONFERENCE (1979 : University of Nevada)
Environmental physiology. – New York ; Oxford : Elsevier/North-Holland, c1981
468p : ill. – Includes bibliographies and index
0–444–00583–8 SRL
2500337

ADAPTATION. Animals. Arctic

SIMON, Seymour
Life on ice. – New York ; London : F. Watts, 1976
[6],65,[1]p : ill. – Bibliography: p.62. – Includes index
0–531–01133–x X.329/10684 Woolwich
0752452

ADAPTATION. Aquatic insects

The ECOLOGY of aquatic insects / edited by Vincent H. Resh and David M. Rosenberg. – New York ; Eastbourne : Praeger, 1984
x,625p : ill. – Includes bibliographies and index
0–03–059684–x SRL
2519607

ADAPTATION. Areas for use of educational resources. Schools. Great Britain

ATHERTON, Barbara
Adapting spaces for resource-based learning. – London : Council for Educational Technology for the United Kingdom, 1980
96p : ill. – (Guidelines / Council of Educational Technology for the United Kingdom ; 8). – Bibliography: p.96
0–86184–009–7 X.529/38457
2107503

ADAPTATION. Behaviour. Man
To environment

KRUSE, Lenelis
Environment and behavior. - Paris : London : Saur, 1984
2v.(x,1424p)
3-598-10494-4 2725.c.306
3241629

ADAPTATION. Behaviour. Man
To environment - Conference proceedings

HUMAN behaviour and adaptation / edited by N. Blurton Jones and V. Reynolds. - London : Taylor and Francis, 1978
vii,314p : ill. - (Symposia / Society for the Study of Humn Biology ; vol.18). - Includes bibliographies and index
0-85066-137-4 SRL
1774476

ADAPTATION. Behaviour. Prehistoric man
To environment expounded by adaptation to environment ofbehaviour of Western Australian aborgines

GOULD, Richard Allan
Living archaeology. - Cambridge : Cambridge University Press, 1980
xv,270p : ill. - (New studies in archaeology). - Bibliography: p.255-264. - Includes index
0-521-23093-4 X.421/21048 Woolwich
2237849

ADAPTATION. Birds

GOODWIN, Derek
Birds of man's world. - Ithaca : Cornell University Press ; London : British Museum (Natural History), 1978
viii,183p : ill. - Includes index
0-8014-1167-x X.529/41539
2223082

ADAPTATION. Cave organisms
To environment. Mathematical models

CULVER, David C.
Cave life. - Cambridge, Mass. ; London : Harvard University Press, 1982
viii,189p : ill. - Bibliography: p171-185. - Includes index
0-674-10435-8 X.421/23822
2880768

ADAPTATION. Environment
Of man

RETHINKING human adaptation / edited by Rada Dyson-Hudson and Michael A. Little. - Epping : Bowker, c1983
xii,180p : ill. - (A Westview special study). - Bibliography: p149-175. - Includes index
B83-34052 X.322/12171
2938022

ADAPTATION. Grasses. Crops. Tropical regions
To environment

HUMPHREYS, L. R.
Environmental adaptation of tropical pasture plants. - London : Macmillan, 1981
ix,261p : ill. - Bibliography: p221-252. - Includes index
0-333-26820-2 SRL
2045737

ADAPTATION. Hips & thighs. Lorises
To environment

MCARDLE, John E.
Functional morphology of the hip and thigh of the lorisiformes. - Basel ; London : Karger, 1981
132p : ill. - (Contributions to primatology ; v.17). - Bibliography: p126-132
3-8055-1767-x X.329/13608 Woolwich
2376482

ADAPTATION. Intestines. Man
- Conference proceedings

INTERNATIONAL CONFERENCE ON INTESTINAL ADAPTATION (2nd : 1981 : Titisee)
Mechanisms of intestinal adaptation. - Lancaster : MTP Press, c1982
646p : ill. - (Falk symposium ; 30). - Includes index
0-85200-442-7 SRL
274547x

ADAPTATION. Legumes. Crops. Tropical regions
To environment

HUMPHREYS, L. R.
Environmental adaptation of tropical pasture plants. - London : Macmillan, 1981
ix,261p : ill. - Bibliography: p221-252. - Includes index
0-333-26820-2 SRL
2045745

ADAPTATION. Mammals
To stress. Psychobiological aspects - Conference proceedings

NATO INTERNATIONAL WORKSHOP ON COPING AND HEALTH (1979 : Bellagio)
Coping and health. - New York ; London : Plenum Press [for] NATO Scientific Affairs Division, 1980
viii,364p : ill. - (NATO conference series: 3, human factors / North Atlantic Treaty Organization ; vol.12). - Includes bibliographies and index
0-306-40422-2 X.325/652 Woolwich
2291835

ADAPTATION. Man
Medical aspects

DUBOS, René
Man adapting. - [Enl. ed.]. - New Haven ; London : Yale University Press, c1980
xxii,538p : ill. - (Yale University Mrs. Hepsa Ely Silliman memorial lectures). - Bibliography: p467-519. - Includes index
0-300-02580-7 X.329/13427 Woolwich
2352680

ADAPTATION. Man
Psychological aspects

HERBERT, Heidrun
Anpassung. - Hamburg : [H. Herbert], 1976
iii,215p. - Bibliography
r6304570 X.529/48181
8846936

HUMAN adaptation / edited by Rudolf H. Moos. - Lexington, Mass. ; London : D.C. Heath, 1976
xv,447p. - Includes bibliographies and index
0-669-00390-5 X.520/10674
0647268

ADAPTATION. Man
related to cognitive styles

BERRY, John Widdup
Human ecology and cognitive style. - [Beverly Hills] : Sage Publications ; New York ; London : Distributed by Wiley, 1976
xiii,242p : ill. - (Cross-cultural research and methodology series ; [vol.3]). - Bibliography: p.229-242
0-470-15103-x X.0520/500(3)
0799599

ADAPTATION. Man
To environment

ENVIRONMENT and population / edited and integrated by John B. Calhoun. - New York : Praeger, 1983
xviii,486p : ill. - (Praeger special studies)
0-03-063337-0 X.525/8083
2363976

FRISANCHO, A. Roberto
Human adaptation. - St. Louis ; London : Mosby, 1979
xi,209p : ill. - Includes index
0-8016-1693-x SRL
2363976

MORAN, Emilio F.
Human adaptability. - Epping : Bowker, c1982
xii,404p : ill. - Bibliography: p339-389. - Includes index
0-86531-430-6 X.520/29946
2363976

ADAPTATION. Man
To environment. Archaeology - Conference proceedings

SITE, environment and economy / edited by Bruce Proudfoot. - Oxford : B.A.R., 1983
vi,169p : ill. - (Symposia of the Association for Environmental Archaeology ; no.3) (BAR. International series ; 173). - Includes bibliographies
0-86054-222-x X.425/5267
3277828

ADAPTATION. Marine animals
To environment

ADAPTATION to environment / edited by R.C. Newell. - London : Butterworth, 1976
[13],539p : ill. - Includes bibliographies and index
0-408-70778-x SRL
0780073

ADAPTATION. Marine animals. Abyssal zone. Oceans
To pressure. Biochemistry

BIOCHEMISTRY at depth / edited and organized by P.W. Hochachka. - Oxford : Pergamon, 1976
[5],202p,plate : ill. - Includes bibliographies and index
0-08-019960-7 X.622/5532 Woolwich
0555703

ADAPTATION. Marine organisms
To environment

FUNCTIONAL adaptations of marine organisms / edited by F John Vernberg, Winona B Vernberg. - New York : London Academic Press, c1981
xi,347p : ill. - (Physiological ecology). - Includes bibliographies and index
0-12-718280-2 SRL
2551845

ADAPTATION. Marine organisms
To environment. Evolutionary aspects

VERMEIJ, Geerat J.
Biogeography and adaptation. - Cambridge, Mass. ; London : Harvard University Press, 1978
xv,332p : ill. - Bibliography: p.279-321. - Includes index
0-674-07375-4 SRL
1882074

ADAPTATION. Microorganisms
To extreme environments

MICROBIAL life in extreme environments / edited by D J Kushner. - London : Academic Press, 1978
xii,465p,plate : ill(incl 1 col). - Includes bibliographies and index
0-12-430250-5 SRL
1582852

ADAPTATION. Microorganisms
To extreme environments - Conference proceedings

EXTREME environments / edited by Milton R. Heinrich. - New York ; London : Academic Press, 1976
xvii,362p : ill. - (Academic Press rapid manuscript reproductions). - Includes bibliographies and index
0-12-337850-8 SRL
1582852

SOCIETY FOR APPLIED BACTERIOLOGY. Autumn Demonstration Meeting (1978 : Polytechnic of the South Bank)
Microbial growth and survival in extremes of environment. - London : Academic Press, 1980
xv,244p : ill. - (Technical series / Society for Applied Bacteriology ; no.15). - Includes bibliographies and index
0-12-293680-9 SRL
1582852

ADAPTATION. Microorganisms
To interfaces between states of matter

MARSHALL, K. C.
Interfaces in microbial ecology. - Cambridge, Mass. ; London : Harvard University Press, 1976
xiii,156p : ill. - Bibliography: p.129-152. - Includes index
0-674-45822-2 SRL
1032364

ADAPTATION. Organisms

TRIBUTSCH, Helmut
How life learned to live. - Cambridge, Mass. ; London : MIT Press, c1982
ix,218p : ill. - Bibliography: p207-211. - Includes index
0-262-20045-7 X.322/12357
2203294

ADAPTATION. Organisms
- Conference proceedings

INTERNATIONAL SYMPOSIUM ON ADAPTATIONS TO TERRESTRIAL ENVIRONMENT (1982 : Khalkidiki)
Adaptations to terrestrial environments. - New York ; London : Plenum, 1983
viii,247p : ill. - Includes bibliographies and index
0-306-41468-6 SRL
2203294

ORIGINS and development of adaptation. - London : Pitman, 1984
viii,273p : ill. - (Ciba Foundation symposium ; 102). - Includes bibliographies and index
0-272-79749-9 SRL
2203294

ADAPTATION. Organisms
Biochemical adaptation to environmental change - Conference proceedings

SMELLIE, R. M. S. (Robert Martin Stuart)
Biochemical adaptation to environmental change. - London : Biochemical Society, 1976
ix,240p,[3] leaves of plates : ill. - (Symposia / Biochemical Society ; no.41). - Includes bibliographies and index
0-904498-01-8 SRL
114457x

ADAPTATION. Organisms
Genetic aspects
FORD, Edmund Brisco
Genetics and adaptation. - London : Edward Arnold, 1976
v,58p : ill - (Studies in biology / Institute of Biology ; no.69). - Bibliography: p.57-58
0-7131-2562-4 X.0319/54(69)
0780383

ADAPTATION. Organisms
To cold - Conference proceedings
INTERNATIONAL SYMPOSIUM FOR SURVIVAL IN THE COLD (1980 : Prague)
Survival in the cold. - New York ; Oxford : Elsevier/North Holland, c1981
xiii,225p : ill. - Includes bibliographies and index
0-444-00635-4 SRL
2563231

ADAPTATION. Organisms
To effects of man on environment. Genetic aspects
GENETIC consequences of man made change. - London : Academic Press, 1981
xiv,409p : ill. - Bibliography: p343-400. - Includes index
0-12-101620-x SRL
2595761

ADAPTATION. Physiology. Man
To environment - Conference proceedings
ENVIRONMENTAL stress / edited by Lawrence J. Folinsbee ... [et al.]. - New York ; London : Academic Press, 1978
xv,393p : ill. - Includes bibliographies and index
0-12-261350-3 SRL
1755153

ADAPTATION. Physiology. Man
To environment. Research
INTERNATIONAL BIOLOGICAL PROGRAMME
Human adaptability. - London : Taylor and Francis, 1977
ix,356p : ill. - Includes index
0-85066-087-4 SRL
1047205

ADAPTATION. Physiology. Man
To renal failure
RENAL adaptation to nephron loss / editors Georges Peters, Jacques Diezi, Jean-Pierre Guignard. - Basel ; London : Karger ; [Chichester] : [Distributed by Wiley], 1979
[7],182p : ill
B79-35899 X.311/9063 Woolwich
1971727

ADAPTATION. Plants
To high temperature - Conference proceedings
ADAPTATION of plants to water and high temperature stress / edited by Neil C. Turner, Paul J. Kramer. - New York ; Chichester : Wiley, c1980
xiii,482p : ill. - Includes index
0-471-05372-4 SRL
2337835

ADAPTATION. Plants
To water deficiency - Conference proceedings
ADAPTATION of plants to water and high temperature stress / edited by Neil C. Turner, Paul J. Kramer. - New York ; Chichester : Wiley, c1980
xiii,482p : ill. - Includes index
0-471-05372-4 SRL
2337827

ADAPTATION. Prose in English, 1558-1625
Into drama in English, 1558-1625
BLUESTONE, Max
From story to stage. - Hague : Mouton, 1974
341p. - (Studies in English literature ; vol.70). - Bibliography : p. 320-327
r6028693 X.0909/164.(70.)
0386626

ADAPTATION. Residences. Old persons. England
WELSH NURSING AND MIDWIFERY COMMITTEE
The home environment. - [Wales?] : [The Committee?], 1979
8p. - (Ageing and disablement ; 2)
0-904251-30-6 BS.128a/297
2697025

ADAPTATION. Residences. Physically handicapped persons. England
Administration by local authorities
PRESCOTT-CLARKE, Patricia
Organising house adaptations for disabled people. - London : H.M.S.O., 1982
iv,46p,[27]p
0-11-751651-1 BS.414/2110
2888734

ADAPTATION. Residences. Physically handicapped persons. Great Britain
LOCKHART, Terence
Housing adaptations for disabled people. - London : Published for the Disabled Living Foundation by The Architectural Press, 1981
xiv,130p : ill. - Bibliography: p128-130
0-85139-331-4 SRL
2579014

ADAPTATION. Residences. Physically handicapped persons. London
By local authority welfare services. Administration
KEEBLE, Ursula
Aids and adaptations. - [London] : Bedford Square Press, 1979
320p : ill. - (Occasional papers on social administration ; no.62). - Bibliography: p.315-320
0-7199-0963-5 8299.i.7(62)
1952595

ADAPTATION. Students. Higher education institutions. United States
To stress
MECHANIC, David
Students under stress. - Madison ; London : University of Wisconsin Press, 1978
xxxv,231p : ill. - Bibliography: p.222-226. - Includes index
0-299-07470-6 X.529/32726
1728393
- [Another copy: paperback]
ISBN 0-299-07474-9 X.519/28757

ADAPTATION. Students. Secondary schools. Northern Ireland
To secondary schools
SPELMAN, B. J.
Pupil adaptation to secondary school. - [Belfast] : Northern Ireland Council for Educational Research, c1979
x,370p : ill. - (Publications of the Northern Ireland Council for Educational Research ; 18). - Bibliography: p346-359
0-903478-10-2 X.525/4248
2383780

ADAPTATION NURSING. Medicine
RANDELL, Brooke
Adaptation nursing. - St. Louis ; London : Mosby, 1982
xvii,297p : ill. - Includes bibliographies and index
0-8016-4024-5 X.322/11310
2805901

ADAPTATIONS
See also
Dramatisations

ADAPTATIONS. Australian writers. Fiction in English
Cinema films, 1971-1983
MCFARLANE, Brian, 1934-
Words and images. - Richmond, Vic. : Heinemann Publishers Australia in association with Cinema Papers, 1983
v,210p : ill
0-436-27600-3 X.950/42691
3329550

ADAPTATIONS. Böll, Heinrich. Fiction in German
Cinema films
KILBORN, Richard W.
Whose lost honour?. - Glasgow : Scottish Papers in Germanic Studies, 1984
73p : ill. - (Scottish papers in Germanic studies ; v.4). - Bibliography: p71-73
0-907409-03-2 X.529/70564
3298604

ADAPTATIONS. Books
Cinema films - Indexes
ENSER, A. G. S. (Alfred George Sidney)
Filmed books and plays. - Aldershot : Gower, c1982
x,84p. - (A Grafton book)
0-566-03475-1 X.950/19697
0325864
- [Another copy] BB.E.dd.40 Reading Room
Filmed books and plays. - [New ed.]. - Aldershot : Gower, 1985
xii,705p. - (A Grafton book)
0-566-03510-3 X.950/40264
0325864

ENSER, Alfred George Sidney
Filmed books and plays. - Revised cumulated ed. - London : Deutsch, 1975
549p. - (A Grafton book)
0-233-96676-5 X.989/31148
0325864
- Reprinted. - 1976 BB.E.dd.40 Reading Room

ADAPTATIONS. Books. United States
16mm cinema films: Short films - Bibliographies
PARLATO, Salvatore J. (Salvatore Joseph)
Films ex libris. - Jefferson, N.C. : McFarland, 1980
xii,271p : ill. - Includes index
0-89950-006-4 X.950/7795
2561328

ADAPTATIONS. Brecht, Bertolt. Drama in German
Adaptations of drama in English, 1558-1745
YUILL, William Edward
The art of vandalism. - [London] : [Bedford College], [1977]
19p. - (Inaugural lectures / Bedford College)
0-900145-40-4 X.900/25550
1477277

ADAPTATIONS. Brecht, Bertolt. Drama in German
Adaptations of Western European drama
KITCHING, Laurence Patrick Anthony
Der Hofmeister. - Munich : Wilhelm Fink, 1976
182p,. - Bibliography : p.173-178
3-7705-1185-9 X.909/85102
0691585

SUBIOTTO, Arrigo Victor
Bertolt Brecht's adaptations for the Berliner Ensemble. - London : Modern Humanities Research Association, 1975
x,207p. - (Dissertation series / Modern Humanities Research Association ; vol.8). - Bibliography: p.204-207
0-900547-37-5 X.0510/110(8)
0691585

ADAPTATIONS. Drama
American cinema films: Musicals, 1949-1977
DRUXMAN, Michael B.
The musical. - South Brunswick ; New York : Barnes ; London : Yoseloff, 1980
202p : ill
0-498-02282-x X.435/715
2207060

ADAPTATIONS. Drama
Cinema films, to 1980
LEONARD, William Torbert
Theatre. - Metuchen ; London : Scarecrow, 1981
2v.(vii,1804p). - Includes index
0-8108-1374-2 X.950/10230
2614170

ADAPTATIONS. Drama
Cinema films, 1936-1973
MANVELL, Roger
Theater and film. - Rutherford : Farleigh Dickinson University Press ; London : Associated University Presses, 1979
303p : ill. - List of films: p.274-292. - Includes index
0-8386-2077-9 X.989/88372
215711x

ADAPTATIONS. Fiction
Cinema films, 1930-1979
CINEMA / edited by Frank N. Magill. - Epping : Bowker, 1980
xi,504p. - (Magill surveys). - Includes index
0-89356-306-4 X.950/29852
3047695

ADAPTATIONS. Ibsen, Henrik. Drama in Norwegian
Cinema films - Catalogues
HANSEN, Karin Synnøve
Henrik Ibsen, 1828-1978. - Oslo : Norsk filminstitutt, 1978
30p : ill
r6017654 X.909/82811
1986899

ADAPTATIONS. Kleist, Heinrich von. Short stories in German
Cinema films
ERZÄHLSTRUKTUREN - Filmstrukturen / herausgegeben von Kalus Kanzog. - Berlin : Schmidt, 1981
172p,[6]p of plates : ill,charts. - (Jahresgabe der Heinrich-von-Kleist-Gesellschaft ; 1978/79). - "Heinrich von Kleist und der Film : Eine Bibliographie von Klaus Kanzog" : p.142-172
3-503-01669-4 X.0900/63(1978/79)
8759170

ADAPTATIONS. Literature
Cinema films, 1920-1971
BEJA, Morris
Film & + and Literature : an introduction. - New York ; London : Longman, 1979
xv,335p : ill. - Includes bibliographies and index
0-582-28094-x X.900/27640
1969471

ADAPTATIONS. May, Karl. Fiction in German
Cinema films, to date

KARL May im Film / herausgegeben von Christian Unucka. - Dachau : Franke, [1980]
303p : chiefly ill (some col). - Filmographies
3-88626-000-3 X.955/559
8759189

ADAPTATIONS. Mozart, Wolfgang Amadeus. Opera in German
Cinema films, to date

RÉMY, Pierre-Jean
Don Giovanni - Mozart, Losey. - Paris : Albin Michel, 1979
168p : ill
2-226-00880-2 X.439/10780
9005633

ADAPTATIONS. Novels. Forms. Fiction
Use in cinema films

HOLM, Ingvar
Roman blir film. - [Staffanstorp] : Cavefors, 1975
159 p. : ill. - Includes index
91-504-0431-8 X.909/33193
093674X

SPIEGEL, Alan
Fiction and the camera eye. - Charlottesville : University Press of Virginia, 1976
203p
0-8139-0598-2 X.989/54479
093674X

WAGNER, Geoffrey
The novel and the cinema. - Rutherford : Fairleigh Dickinson University Press ; London : Tantivy Press, 1975
394p : ill. - Bibliography: p.358-364. - Includes index
0-904208-80-x X.989/52748
093674x

ADAPTATIONS. Russian literature, 1800-1945
Use in cinema films, stage presentations & television programmes

SAKHAROV, V. I
Obnovlíāíūshchiísíā mir. - Moskva : Sovremennik, 1980
286p
r6178862 X.989/89499
2469847

ADAPTATIONS. Shakespeare, William. Drama in English
By Davenant, Sir William

RADDADI, Mongi
Davenant's adaptations of Shakespeare. - Uppsala ; Stockholm : Distributor : Almqvist & Wiksell International, 1979
181p : ill. - (Studia anglistica upsaliensis ; 36). - Includes bibliographies
91-554-0890-7 Ac.1075/6(13)[vol.36]
2119080

ADAPTATIONS. Shakespeare, William. Drama in English
Cinema films

JORGENS, Jack J.
Shakespeare on film. - Bloomington ; London : Indiana University Press, 1977
xiii,337p : ill. - Includes index
0-253-35196-0 X.950/1050
1249134

Shakespeare on film. - Bloomington ; London : Indiana University Press, 1979
xiii,337p : ill. - (A Midland book)
0-253-20234-5 X.900/27704
1249134

LIPKOV, Aleksandr Iosifovich
Shekspirovskiĭ ėkran. - Moskva : Iskusstvo, 1975
350p : ill
r6093496 X.950/598
1249134

ADAPTATIONS. Soviet literatures
Cinema films, 1945-

GUREVICH, S. D. (Stėlla Davidovna)
Puti vzaimodeĭstviíā literatury i kino (na materiale sovetskikh fil'mov 50-60-kh godov). - Leningrad : Leningradskiĭ gosudarstvennyĭ institut teatra, muzyki i kinematografii, 1980
105p
r6484550 X.958/6710
8786690

ADAPTATIONS. Tolstoĭ, L. N.. Fiction in Russian
Cinema films, to date

ANNINSKIĬ, Lev Aleksandrovich
Lev Tolstoĭ i kinematograf. - Moskva : Iskusstvo, 1980
287p : ill. - List of films: p.283-287
r6402537 X.958/1860
8798214

ADAPTATIONS. Wagner, Richard, 1813-1883. Opera in German
Cinema films, to date

SYBERBERG, Hans-Jürgen
Parsifal. - München : Heyne, 1982
285p : ill. - (Heyne-Bücher ; Nr.6024)
3-453-01626-2 X.439/12947
9005633

ADAPTATIONS. Williams, Tennessee. American writers. Drama in English
Cinema films

PHILLIPS, Gene D.
The films of Tennessee Williams. - Philadelphia : Art Alliance Press ; London : Associated University Presses, c1980
336p : ill. - Bibliography: p321-327. - List of films: p328-331. - Includes index
0-87982-025-x X.950/24972
2962101

ADAPTATIONS IN ENGLISH. Nabokov, Vladimir. Fiction in Russian
By Nabokov, Vladimir compared with originals

GRAYSON, Jane
Nabokov translated. - Oxford : Oxford University Press, 1977
xi,257p. - (Oxford modern languages and literature monographs). - Bibliography: p.232-251. - Includes index
0-19-815527-1 WP.13860/34
1076663

ADAPTATIONS IN MODERN GREEK. Philoctetes. Sophocles. Drama in Greek

MARTINI, Lidia
Fortuna neogreca del Filottete. - Padova : Liviana Editrice, 1974
vi,121p. - (Università di Padova. Studi bizantini e neogreci ; 7)
r6059237 X.900/18998
1387898

ADAPTATIONS IN SERBO-CROATIAN. Molière. Drama in French
Croatian writers, 1600-1800

DUBROVAČKE preradbe Molièreovih komedija / priredio Mirko Deanović. - Zagreb : Jugoslavenska akademija znanosti i umjetnosti, 1972-73
2 knj : ill. - (Stari pisci hrvatski ; knj. 36-37)
r6043830 Ac.741/14[36,37]
1318241

ADAPTIVE-ARRAY SENSOR SYSTEMS

HUDSON, J. E.
Adaptive array principles. - Stevenage : Peregrinus on behalf of the Institution of Electrical Engineers, c1981
xiv,253p : ill. - (IEE electromagnetic waves series ; 11). - Bibliography: p229-234p. - Includes index
0-906048-55-9 SRL
2337428

MONZINGO, Robert A.
Introduction to adaptive arrays. - New York ; Chichester : Wiley, c1980
xii,543p : ill. - Includes index
0-471-05744-4 SRL
2337428

ADAPTIVE BEHAVIOUR
See also
Set. Mental processes

ADAPTIVE BEHAVIOUR. Economic activity
Mathematical models

CROSS, John G.
A theory of adaptive economic behaviour. - Cambridge : Cambridge University Press, 1983
viii,199p : ill. - Bibliography: p189-196. - Includes index
0-521-25110-9 X.520/34524
0625353

ADAPTIVE BEHAVIOUR. Economic activity
Mathematical models - Conference proceedings

SYMPOSIUM ON ADAPTIVE ECONOMICS (1974 : University of Wisconsin-Madison)
Adaptive economic models. - New York ; London : Academic Press, 1975
ix,581p : ill. - (Publications / University of Wisconsin-Madison. Mathematics Research Center ; no.34) (Academic Press rapid manuscript reproduction). - Includes bibliographies and index
0-12-207350-9 X.520/10573
0625353

ADAPTIVE BEHAVIOUR. Foreign relations

SMITH, Steven M. (Steven Murray)
Foreign policy adaptation. - Aldershot : Gower, 1981
vii,152p : ill. - Includes index
0-566-00370-8 X.529/44796
2594978

ADAPTIVE BEHAVIOUR. Foreign students. Higher education institutions. United States

HULL, W. Frank
Foreign students in the United States of America. - New York ; London : Praeger, 1978
xxi,225p. - (Praeger special studies)
0-03-046151-0 X.520/13619
1836587

ADAPTIVE BEHAVIOUR. Man

ADAPTIVE behavior / edited by W. Alan Coulter, Henry W. Morrow. - New York ; London : Grune and Stratton, 1978
xvii,266p : ill. - Bibliography: p.227-252. - Includes index
0-8089-1092-2 X.320/11756 Woolwich
1904434

ADAPTIVE BEHAVIOUR. Politics

MACKENZIE, W. J. M. (William James Millar)
Biological ideas in politics. - Manchester : Manchester University Press, 1979
93p
0-7190-0731-3 X.529/33200
1721356

MACKENZIE, William James Millar
Biological ideas in politics. - Harmondsworth : Penguin, 1978
93p. - (Penguin political science) (A pelican original). - Includes bibliographies
0-14-022162-x X.708/21601
1721356

ADAPTIVE CONTROL SYSTEMS
- Conference proceedings

ADAPTIVE systems in control and signal processing 1983 / edited by I.D. Landau, M. Tomizuka and D.M. Auslander. - Oxford : Published for the International Federation of Automatic Control by Pergamon, 1984
xi,376p : ill. - Includes bibliographies and index
0-08-030565-2 SRL
2603624

INTERNATIONAL WORKSHOP ON APPLICATIONS OF ADAPTIVE CONTROL
(1979 : Yale University)
Applications of adaptive control. - New York ; London : Academic Press, 1980
xiii,554p : ill
0-12-514060-6 SRL
2603624

SELF-TUNING and adaptive control / edited by C.J. Harris and S.A. Billings. - Rev. 2nd ed. - London : Peregrinus on behalf of Institution of Electrical Engineers, 1985
[340]p : ill. - (IEE control engineering series ; 15). - Includes index
0-86341-036-7 X.629/26383
2603624

ADAPTIVE CONTROL SYSTEMS. Complex systems

GOODWIN, Graham C. (Clifford)
Adaptive filtering, prediction and control. - Englewood Cliffs ; London : Prentice-Hall, c1984
xii,540p : ill. - (Prentice Hall information and system sciences series). - Bibliography: p516-534. - Includes index
0-13-004069-x X.622/21586
3172023

ADAPTIVE CONTROL SYSTEMS. Complex systems
- Conference proceedings

NATO ADVANCED RESEARCH INSTITUTE ON ADAPTIVE CONTROL OF ILL DEFINED SYSTEMS (1981 : Moretonhampstead)
Adaptive control of ill-defined systems. - New York ; London : Plenum in cooperation with NATO Scientific Affairs Division, c1984
x,349p : ill. - (NATO conference series. II Systems science ; v.16). - Includes bibliographies and index
0-306-41475-9 SRL
3172023

ADAPTIVE ELECTRIC FILTERS

HAYKIN, S. S. (Simon Saher)
Introduction to adaptive filters. - New York : Macmillan ; London : Collier Macmillan, c1984
xii,217p : ill. - Includes index
0-02-949460-5 SRL
3300587

ADAS
See Agricultural Development and Advisory Service

ADASKIN, HARRY. Violin playing
- Biographies

ADASKIN, Harry
A fiddler's world. - Vancouver : November House, c1977
293p,[16]p of plates
r6807996 X.431/11895
8783152

ADAT LAW. Indonesia
HOOKER, M. B. (Michael Barry)
Adat law in modern Indonesia. - Kuala Lumpur ; Oxford : Oxford University Press, 1978 [i.e. 1979]
viii,181p. - (East Asian historical monographs). - Bibliography: p.154-166. - Includes index
0-19-580394-9 X.200/32095
1911031

'ADAWIYYAH, RABI'AH AL-
See Rabi'ah al-'Adawiyyah

ADBASTON. Staffordshire
Churches: St. Michael and All Angels (Church : Adbaston)
Parish registers - Texts
ST. MICHAEL AND ALL ANGELS (Church : Adbaston)
Registers of St. Michael & All Angels, Adbaston Staffordshire. - [Sedgley] : [Birmingham and Midland Society for Genealogy and Heraldry], [1980].
Pt.2: Baptisms, marriages and burials 1727-1839
i,89,9p. - Includes index
0-905105-43-5 X.805/2573
2728486

ADBASTON. Staffordshire
Parish registers, 1600-1726 - Texts
ADBASTON (Parish)
The parish registers of St Michael and All Angels, Adbaston, Staffordshire. - Birmingham : Birmingham and Midland Society for Genealogy and Heraldry, 1979
[3],58,7p. - Includes index
0-905105-22-2 X.102/2200
1876791

ADC ACCUTRAC (SHIP). Yachts. Whitbread Round-the-World Race (1977-1978)
FRANCIS, Clare
Come wind or weather. - London : Sphere, 1979, c1978
253p : ill
0-7221-3639-0 X.629/16609 Woolwich
1707221

ADC ACCUTRAC (SHIP). Yachts. Whitbread Round-the-World Race (1977-1978)
- Biographies
FRANCIS, Clare
Come wind or weather. - London : Pelham, 1978
224p,[16]p of plates : ill(some col)
0-7207-1104-5 X.620/8789
1707221

ADDED VALUE
Applications in financial analysis of companies
BRITISH PLASTICS FEDERATION
Added value as a management tool. - London : The Federation, 1978
[3],22p : 2 ill
B78-32306 SRL
1576763

WOOD, E. G. (Edward Geoffrey)
Added value and its uses ; [Added value bonus schemes]. - [Sheffield] : [Centre for Innovation and Productivity, Sheffield City Polytechnic], [1984?]
58p.in various pagings. - Bibliography: 8p
0-86339-054-4 X.525/9548
1576763

WOOD, Edward Geoffrey
Added value. - London : Business Books, 1978
xii,148p : ill. - Bibliography: p.135-140. - Includes index
0-220-66349-1 X.520/13753
1576763

ADDED VALUE. Great Britain
Role in improvement of industrial productivity
WOOD, E. G. (Edward Geoffrey)
Added value, wealth creation and productivity. - [Sheffield] : [Centre for Innovation and Productivity, Sheffield City Polytechnic], [1984?]
1v.(various pagings)
0-86339-044-7 X.525/9547
1702157

ADDED VALUE. Productivity. Manufacturing industries. Great Britain
Role in improvement - For accounting
COX, Bernard
Value added. - London : Heinemann [for] the Institute of Cost and Management Accountants, 1979
xii,220p : ill. - Bibliography: p.212-213. - Includes index
0-434-90276-4 X.529/34085
1961233

ADDED VALUE STATEMENTS. Accounts. Companies. Great Britain
GRAY, Sidney
Value added reporting. - London : Association of Certified Accountants, c1980
71p. - Bibliography: p64-71
0-900094-39-7 X.525/5050
1447742

MORLEY, Michael Francis
The Value Added Statement. - London : Gee and Co. for the Institute of Chartered Accounts of Scotland, 1978
x,150p. - Bibliography: p.149-150
0-85258-162-9 X.519/27950
1447742

ADDERBURY REGION. Oxfordshire
Folk songs in English. Singing, 1800-1900. Social aspects
PICKERING, Michael
Village song & culture. - London : Croom Helm, c1982
187p. - Bibliography: p182-184. - Includes index
0-7099-0059-7 X.809/51044
2694956

ADDERS
ASTON, Oliver
Adders. - Oxford : Blackwell, 1977
32p : ill(chiefly col). - (Wild life studies. series 2)
0-631-93430-8 X.319/18147 Woolwich
1484605

ADDICTION
See also
Alcoholism
Drug addiction

ADDICTION
Existential psychotherapy
PATTERSON, Margaret A.
Pharmakon addiction treatment. - London : Pharmakon Clinics, [1978?]
18p. - (Pharmakon publications ; no.2)
B80-15893 X.319/19274 Woolwich
2046695

ADDICTION
Medical aspects
COMMONALITIES in substance abuse and habitual behavior / edited by Peter K. Levison, Dean R. Gerstein, Deborah R. Maloff. - Lexington, Mass. : Lexington Books, 1983
xx,355p : ill. - Includes bibliographies and index
0-669-06293-6 X.520/32269
2100886

ADDICTION
Medical aspects - Serials
ADDICTIVE Behaviors. - Vol.1, no.1- ; July 1975-. - Oxford : Pergamon, 1975-
B75-24113 SRL
0368105

ADDICTION
Medical aspects related to schizophrenia
BIOCHEMISTRY of schizophrenia and addiction / edited by Gwynneth Hemmings. - Lancaster : MTP Press, c1980
xvi,344p : ill. - Includes bibliographies and index
0-85200-310-2 SRL
2046571

ADDICTION
Neuro-electric therapy
PATTERSON, Margaret A.
Pharmakon addiction treatment. - London : Pharmakon Clinics, [1978?]
18p. - (Pharmakon publications ; no.2)
B80-15893 X.319/19274 Woolwich
2046687

ADDICTION
Therapy
ADDICTION - theory and treatment / compiled and edited by George D. Goldman, Donald S. Milman. - Dubuque, Iowa : Kendall / Hunt, [1980]
xv,263p. - (Adelphi University Postdoctoral Program in Psychotherapy Conference Series)
0-8403-2183-x X.322/11069
2085070

The ADDICTIVE behaviors / edited by William R. Miller. - Oxford : Pergamon, 1980
x,353p. - Includes bibliographies and index
0-08-030837-6 X.329/17421
2085070

HATTERER, Lawrence J.
The pleasure addicts. - South Brunswick ; New York : Barnes ; London : Yoseloff, 1980
387p. - Bib.: p.316-387
0-498-02285-4 X.520/21411
2085070

ADDICTION POTENTIAL. Benzodiazepines
MARKS, John, 1924-
The benzodiazepines. - Lancaster : M.T.P. Press, 1978
ix,111p : ill. - Includes index
0-85200-232-7 SRL
152447x

PETURSSON, Hannes
Dependence on tranquilizers. - Oxford : Oxford University Press, 1984
130p : ill. - (Maudsley monographs ; no.28). - Bibliography: p114-128. - Includes index
0-19-712152-7 Ac.2666p./3(28)
152447x

ADDICTION POTENTIAL. Depressants
Evaluation - Conference proceedings
CONFERENCE ON PREDICTION OF ABUSE LIABILITY OF STIMULANT AND DEPRESSANT DRUGS (1976 : Washington, D.C.)
Predicting dependence liability of stimulant and depressant drugs. - Baltimore ; London : University Park Press, 1977
xv,328p : ill. - Includes index
0-8391-1147-9 SRL
1255096

ADDICTION POTENTIAL. Stimulants
Evaluation - Conference proceedings
CONFERENCE ON PREDICTION OF ABUSE LIABILITY OF STIMULANT AND DEPRESSANT DRUGS (1976 : Washington, D.C.)
Predicting dependence liability of stimulant and depressant drugs. - Baltimore ; London : University Park Press, 1977
xv,328p : ill. - Includes index
0-8391-1147-9 SRL
125510x

ADDICTIVE DRUGS
to date
RAUSCH und Realität / [Materialenband zu einer Ausstellung des Rautenstrauch-Joest-Museums für Völkerkunde der Stadt Köln 7. August bis 11. Oktober 1981] ; herausgegeben von Gisela Völger unter Mitarbeit von Karin von Welck und Aldo Legnaro. - Köln : Gesellschaft für Völkerkunde, 1981
2v. : ill. - (Ethnologica ; n.F., Bd.9)
3-923158-00-9 Ac.6257[Bd.9]
8993904

ADDICTIVE DRUGS
Effects on man
LEMBERGER, Louis
Physiologic disposition of drugs of abuse. - New York : Spectrum Publications ; New York ; London : Distributed by Wiley, 1976
[13],401p : ill. - (Monographs in pharmacology and physiology ; vol.1). - Includes bibliographies and index
0-470-15021-1 SRL
1374974

ADDICTIVE DRUGS
Social aspects, to 1974
INGLIS, Brian
The forbidden game. - Sevenoaks : Coronet, 1977
256p. - Bibliography: p.234-243. - Includes index
0-340-21575-5 X.319/17398 Woolwich
0327697

ADDICTIVE DRUGS. Brazil
Law
MENDONÇA, Yolanda
A tragédia das toxicomanias. - Rio de Janeiro : Arsgráfica Editora, 1972
10 p.
r6003399 X.311/4518 Woolwich
0912999

ADDICTIVE DRUGS. Hong Kong
Law
FAULKNER, R. J.
Vanquishing the dragon. - [Hong Kong] : Centre of Asian Studies, University of Hong Kong, 1975
[4], 104p. - (Centre of Asian Studies Occasional papers and monographs ; no. 20)
r6071641 X.100/17416
1378384

ADDICTIVE DRUGS. Italy
Law, to date
MIRENNA, Nicolò
La droga. - Roma : Il pensiero scientifico, 1981
210p : ill
r6510381 X.322/13327
9071016

ADDICTIVE DRUGS. Victoria. Australia
to date
VICTORIA (Australia). Interdepartmental Working Party on the Drug Problem in Victoria
Report. - Melbourne : Government Printer, 1980
2v.. - ([Parliament of Victoria, Parliamentary Paper ; 1980, no.69])
r6805618 CS.G.393/160
8860726

ADDICTIVE NARCOTICS
Effects on human tissues - Conference proceedings
TISSUE RESPONSES TO ADDICTIVE DRUGS (Conference : 1975 : State University of New York)
Tissue responses to addictive drugs. - New York : Spectrum ; New York ; London : Distributed by Wiley, [1977]
[19],704p : ill. - Includes bibliographies and index
0-470-15192-7 SRL
1090836

ADDICTS
See also
Alcoholics
Drug addicts

ADDICTS
See also
Work addicts

ADDING MACHINES. United States
Use in business practices
LOCKE, Flora Mattoon
Electronic calculators for business use. - New York ; Chichester : Wiley, 1978
xv,264p : ill
0-471-03579-3 X.611/8403 Woolwich
1764004

ADDINGTON. Croydon (London Borough)
to 1983
WARREN, Frank
Addington. - Chichester : Phillimore, 1984
xi,104p : ill. - Includes index
0-85033-521-3 X.800/39520
3198871

ADDIS
1780-1980
BEAVER, Patrick
Addis 1780-1980. - Stevenage : Publications for Companies, 1980
v,55p,[4]p of plates : ill(some col.)
0-904928-10-1 X.800/35560
2920867

ADDIS COLLECTION. Chinese porcelain, 1300-1500. Exhibits. British Museum
- Catalogues
BRITISH MUSEUM
Chinese porcelain from the Addis collection. - London : Published for the Trustees of the British Museum by British Museum Publications, c1979
90p : ill(some col.). - Bibliography: p89. - Includes index
0-7141-1410-3 X.425/2148 Woolwich
1106368

ADDISON, CHRISTOPHER, VISCOUNT ADDISON. Politics. Great Britain
MORGAN, Kenneth O. (Kenneth Owen)
Portrait of a progressive. - Oxford : Clarendon Press, 1980
xiii,326p,[4]p of plates : ill. - Bibliography: p.306-317. - Includes index
0-19-822494-x X.809/46087
1926136

ADDISON, JOSEPH. Essays in English
ORTMEIER, Anno
Taste und Imagination. - Frankfurt am Main : Lang, 1982
198p. - (Arbeiten zur Ästhetik, Didaktik, Literatur- und Sprachwissenschaft ; Bd. 8). - Includes bibliography
3-8204-5705-4 X.950/28038
8793352

ADDISON, JOSEPH. Essays in English
Addison, Joseph & Steele, Sir Richard
ADDISON and Steele, the critical heritage / edited by Edward A. Bloom and Lillian D. Bloom. - London : Routledge and Kegan Paul, 1980
xii,480p. - (The critical heritage series). - Bibliography: p.452-457. - Includes index
0-7100-0375-7 X.989/89435
2067676

ADDISONIAN ANAEMIA
See Pernicious anaemia

ADDITION
See also
Sums

ADDITION. Arithmetic
HARPER, Eon
Adding and subtracting whole numbers. - London : Addison-Wesley, 1976
[1],17p : ill(chiefly col). - (Maths minipacks ; 1)
0-201-03122-1 X.0611/1056(1) Woolwich
2309645
HESSE, K. A. (Kenneth Anderson)
Adding & subtracting. - Harlow : Longman, 1980
32p. - (Basic arithmetic practice ; 1)
Subsequent volumes when received will be placed at this shelfmark
0-582-39170-9 X.0622/635
2184184
HOLLANDS, Roy
Addition and subtraction. - St Albans : Hart-Davis Educational. - (Basic maths)
Answer book. - 1980
32p
0-247-13069-9 X.622/7848 Woolwich
2184184
[Pupil's book]. - 1980
32p
0-247-12989-5 X.622/7847 Woolwich
2184184
LAPPING, John
Add up with the number chums. - Ipswich : Studio Publications, 1981
[64]p : chiefly col.ill. - (Number chums books)
0-86215-031-0 X.629/19825
2877066
The OLD fashioned adding-up book. - London : Ward Lock Educational, 1981
11 leaves
0-7062-4086-3 X.625/660 Woolwich
2383284
WROE, Geoffrey
Addition and subtraction problems. - London : Harrap, 1980
32p : ill. - (Maths takes off. Level 3. Number)
0-245-53750-3 X.629/14939 Woolwich
2184184

ADDITION. Arithmetic
- For Caribbean students
SAUNDERS, J. G. (John Graham)
Addition practice for the Caribbean. - [London] : Nelson Caribbean.
[Pupil's book]. - 1976
31 p. : ill
0-17-566201-0 X.611/7332 Woolwich
0688223
Teacher's Book. - 1976
31 p. : ill
0-17-566202-9 X.611/7332a Woolwich
0688223

ADDITION. Arithmetic
- For pre-school children
MARSHALL, Ray, 1957-
Pop-up addition. - [Harmondsworth] : Kestrel, 1984
[8]p : col.ill. - (Maths-pops)
0-7226-5919-9 X.629/24350
2406276

ADDITION. Decimals. Arithmetic
HARPER, Eon
Adding and subtracting. - London : Addison-Wesley, 1976
[1],17p : col ill. - (Maths minipacks ; 6)
0-201-03127-2 X.0611/1056(6) Woolwich
2309688

ADDITION. Fractions. Arithmetic
HARPER, Eon
Adding and subtracting fractions. - London : Addison-Wesley, 1976
[1],17p : ill(chiefly col). - (Maths minipacks ; 4)
0-201-03125-6 X.0611/1056(4) Woolwich
2309661
WROE, Geoffrey
Addition and subtraction of fractions. - London : Harrap, 1980
32p : col.ill. - (Maths takes off. Level 3. Arithmetic)
0-245-53753-8 X.629/14936 Woolwich
2309661

ADDITION POLYMERS
See also
Polyolefins
Polystyrene
Polythene
Polyvinyl acetate
Polyvinyl alcohol
Polyvinyl chloride
Polyvinylidene chloride

ADDITION REACTIONS
See also
Electrophilic addition reactions
Pericyclic reactions

ADDITION REACTIONS. DNA. Organisms
- Conference proceedings
PROTEINS AND OTHER ADDUCTS TO DNA : THEIR SIGNIFICANCE TO AGING, CARCINOGENESIS, AND RADIATION BIOLOGY (Conference : 1975 : Williamsburg)
Aging, carcinogenesis, and radiation biology. - New York ; London : Plenum Press, 1976
xi,561p : ill. - Includes bibliographies and index
0-306-30911-4 X.322/2626 Woolwich
1047302
- [Another copy] X.322/8240 Woolwich

ADDITIVE GROUPS. Rings. Algebra
FEIGELSTOCK, S.
Additive groups of rings. - Boston, Mass. ; London : Pitman Advanced Publishing Program, 1983
113p. - (Research notes in mathematics ; 83). - Bibliography: p105-109. - Includes index
0-273-08591-3 SRL
2953382

ADDITIVE NUMBER THEORY
Hardy-Littlewood method
VAUGHAN, R. C.
The Hardy-Littlewood method. - Cambridge : Cambridge University Press, 1981
vii,172p. - (Cambridge tracts in mathematics). - Bibliography: p152-168. - Includes index
0-521-23439-5 SRL
2220296

ADDITIVES
See also
Admixtures
Anti-knock additives
Fibrous additives
Fillers. Plastics
Powdered additives
Weighting agents

ADDITIVES. Beers. Great Britain
Use. Public health aspects
GREAT BRITAIN. Food Additives and Contaminants Committee
Food Additives and Contaminants Committee report on the review of additives and processing aids used in the production of beer. - London : H.M.S.O., 1978
31,[1]p. - (FAC/REP/26)
0-11-240869-9 BS.3/619
1689665

ADDITIVES. Cheeses & cured meat. Great Britain
Nitrates & nitrites. Public health aspects
GREAT BRITAIN. Food Additives and Contaminants Committee
Food Additives and Contaminants Committee report on the review of nitrites and nitrates in cured meats and cheese. - London : H.M.S.O., 1978
14,[1]p. - (FAC/REP/25)
0-11-240311-5 BS.3/652
1845284

ADDITIVES. Feedingstuffs
See also
Supplements. Feedingstuffs

ADDITIVES. Feedingstuffs. Pigs. Livestock
Copper - Conference proceedings
COPPER in animal wastes and sewage sludge / sponsored by the Commission of the European Communities Directorate-General for Agriculture and Directorate-General for Research, Science and Education ; edited by P. L'Hermite and J. Dehandt. - Dordrecht ; London : Reidel, c1981
xiv,378p : ill. - Includes bibliographies
90-277-1293-x SRL
2527227

ADDITIVES. Food
FARRER, K. T. H. (Keith Thomas Henry)
Fancy eating that!. - Carlton : Melbourne University Press ; London : Eurospan, 1983
xx,194p : ill. - Bibliography: p177-185. - Includes index
0-522-84243-7 X.629/25341
0769207
FOOD additives tables / compiled by Food Law Research Centre, Institute of European Studies, University of Brussels. - Updated ed. - Amsterdam ; Oxford : Elsevier Scientific.
Classes 1-4. - 1980
172p in various pagings
0-444-41937-3 SRL
0340618

ADDITIVES. Food -continued
Classes 5-8. - Updated ed. - 1982
1v.(various pagings)
0-444-42069-x SRL
0340618
- Amsterdam : Oxford : Elsevier.
Classes 9-12. - 1984
1v.(various pagings)
0-444-42286-2 SRL
0340618

HANSSEN, Maurice
E for additives. - Wellingborough : Thorsons, 1984
223p : ill
0-7225-1150-7 X.629/25683
0340618

JONES, Norman Richard
Food additives. - [Leatherhead] : British Food Manufacturing Industries Research Association, 1976
1v.
0-905748-00-x SRL
0769207

TAYLOR, R. J. (Reginald James)
Food additives. - Chichester : Wiley, c1980
xv,126p. - (The Institution of Environment Sciences series). - Includes index
0-471-27684-7 SRL
0769207

WHY additives? / devised and edited by the British Nutrition Foundation. - London : Forbes Publications, 1977
xvi,75p. - Includes index
0-901762-09-1 X.319/17453 Woolwich
0769207

ADDITIVES. Food
Acidulants

ARNOLD, Michael Henry Miller
Acidulants for foods and beverages. - London : Food Trade Press, 1975
vi,137p : ill. - Includes index
0-900379-19-7 SRL
0355119

ADDITIVES. Food
Gums - Conference proceedings

GUMS and stabilisers for the food industry 2 / edited by Glyn O. Phillips, David J. Wedlock and Peter A. Williams. - Oxford : Pergamon, 1984
xii,569p : ill. - Includes bibliographies and index
0-08-029819-2 SRL
3018091

ADDITIVES. Food
Role in hyperactivity of children

CONNERS, C. Keith
Food additives and the hyperactive children. - New York ; London : Plenum Press, 1980
xv,167p : ill. - Bibliography: p.159-163. - Includes index
0-306-40400-1 X.322/9463 Woolwich
2177617

ADDITIVES. Food
Toxic effects

HANDBOOK of international food regulatory toxicology. - Lancaster : M.T.P. Press.
Vol.1: Evaluations. - 1980
[10],161p. - Includes index
0-85200-517-2 SRL
2300206
- Lancaster : MTP.
Vol.2: Profiles. - c1981
191p. - Bibliography: p132-185. - Includes index
0-85200-552-0 SRL
2300206

ADDITIVES. Food. Great Britain
Antioxidants. Public health aspects

GREAT BRITAIN. Food Additives and Contaminants Committee
Report on the antioxidant in Food Regulations 1966 and 1974. - London : H.M.S.O., 1974 [i.e. 1975]
22p. - (FAC-REP ; 18)
0-11-241107-x BS.3/494
0258350

ADDITIVES. Food. Great Britain
Artificial sweeteners

GREAT BRITAIN. Food Additives and Contaminants Committee
Food Additives and Contaminants Committee report on the review of sweeteners in food. - London : H.M.S.O., 1982
iii,61p. - (FAC/REP/34)
0-11-241211-4 SRL
2739798

ADDITIVES. Food. Great Britain
Bulking aids

GREAT BRITAIN. Food Additives and Contaminants Committee
Food Additives and Contaminants Committee review of remaining classes of food additives used as ingredients in food. - London : H.M.S.O., 1980
iii,13p. - (FAC/REP ; 32)
0-11-240870-2 SRL
263841x

ADDITIVES. Food. Great Britain
Enzymes

GREAT BRITAIN. Food Additives and Contaminants Committee
Food Additives and Contaminants Committee review of remaining classes of food additives used as ingredients in food. - London : H.M.S.O., 1982
iv,84p : ill. - (FAC/REP ; 35)
0-11-242302-7 SRL
2821567

ADDITIVES. Food. Great Britain
Flavour modifiers

GREAT BRITAIN. Food Additives and Contaminants Committee
Food Additives and Contaminants Committee review of remaining classes of food additives used as ingredients in food. - London : H.M.S.O., 1978
24p. - (FAC/REP/28). - Bibliography: p.17-19
0-11-240319-0 BS.3/653
1914642

ADDITIVES. Food. Great Britain
Modified starches

GREAT BRITAIN. Food Additives and Contaminants Committee
Food Additives and Contaminants Committee report on modified starches. - London : H.M.S.O., 1980
39p. - (FAC/REP ; 31). - Bibliography: p21-26
0-11-240871-0 SRL
2638401

ADDITIVES. Food. Great Britain
Solvents. Public health aspects

GREAT BRITAIN. Food Additives and Contaminants Committee
Food Additives and Contaminants Committee report on the review of solvents in food. - London : H.M.S.O., 1978
63,[1]p. - (FAC/REP/25)
0-11-240868-0 BS.3/618
1697315

ADDITIVES. Food. Great Britain
Sorbic acid. Use. Public health aspects

GREAT BRITAIN. Food Additives and Contaminants Committee
Food Additives and Contaminants Committee report on the review of the use of sorbic acid in food. - London : H.M.S.O., 1977
20p. - (FAC/REP/24)
0-11-241139-8 BS.3/578
1341626

ADDITIVES. Food. United States
Nitrates & nitrites. Public health aspects

ALTERNATIVES to the current use of nitrite in foods / by the Committee on Nitrite and Alternative Curing Agents in Food. - Washington, D.C. : National Academy Press, 1982
1v(various pagings)
0-309-03277-6 X.622/13878
274130x

The HEALTH effects of nitrate, nitrite and N-nitroso compounds / by the Committee on Nitrite and Alternative Curing Agents in Food, Assembly of Life Sciences. - Washington, D.C. : National Academy Press, c1981
538p(in various pagings) : ill. - Includes bibliographies
0-309-03230-x SRL
274130x

ADDITIVES. Food. United States
Regulation. Role of United States. Food and Drug Administration

The CHEMICAL feast / James S. Turner, project director. - Harmondsworth : Penguin, 1976
xiii,273p. - Includes index
0-14-004376-4 X.319/17220 Woolwich
1074571

ADDITIVES. Plastics

ADDITIVES for plastics / edited by Raymond B. Seymour. - New York ; London : Academic Press.
Vol.1: State of the art. - 1978
viii,279p : ill. - Bibliography: p.217. - Includes index
0-12-637501-1 SRL
1781685

ADDITIVES. Plastics
- Conference proceedings

ADDITIVES for plastics. - New York ; London : Academic Press.
Vol.2: New developments. - 1978
xi,125p : ill. - Includes index
0-12-637502-x SRL
1781685

ADDITIVES. Plastics
Chemical analysis. Laboratory techniques

CROMPTON, T. R. (Thomas Roy)
Chemical analysis of additives in plastics. - 2nd ed. - Oxford : Pergamon, 1977
xii,366p : ill. - (International series in analytical chemistry ; vol.46). - Includes index
0-08-020497-x SRL
111445x

ADDITIVES. Tobacco products. Great Britain
Safety. Testing

INDEPENDENT SCIENTIFIC COMMITTEE ON SMOKING AND HEALTH
Tobacco substitute and additives in tobacco products. - London : H.M.S.O., 1975
iii,29p
0-11-320625-9 BS.385/149
0509345

ADDITIVES. Wines
Use

CHANT, Maurice
Additives for winemaking. - [Bristol] : M. Chant, [c1981]
[9]p. - Includes index
B82-00960 X.629/16657 Woolwich
2581698

ADDLESTONE. Surrey
1860-1960

CHERTSEY & Addlestone in the past / [compiled by] H.J.M. Stratton. - Ottershaw : H.J.M. Stratton, 1980
[2],65p : chiefly ill
B80-38606 X.702/6452
2285673

ADDRESSES
See also
Postal addresses

ADDRESSES. Frequently accessed records. Direct access storage devices
Allocation

HANSON, Owen
Improving the efficiency of randomly organised files by loading in access frequency order. - [London] : [City University Business School], [1979?]
[2],38 leaves : ill. - (Working paper series / City University (London). Business School ; no.7)
B79-37113 X.512/7078
1998242

ADDUCTS
See Inclusion complexes

ADELAIDE. South Australia
to date

LAMSHED, Max
Adelaide sketchbook. - 2nd rev. ed. - Adelaide : Rigby, 1981
54p : ill
0-7270-0806-4 X.809/56336
9008578

ADELAIDE. South Australia
Adelaide, South Australia & Melbourne, Victoria & Sydney, New South Wales. Town planning, 1900-1975

SANDERCOCK, Leonie
Cities for sale. - London : Heinemann, 1976
x,260p. - (Series / Centre for Environmental Studies). - Bibliography: p.251-256. - Includes index
0-435-85980-3 X.0809/505(5)
0640786

ADELAIDE. South Australia
Brighton & Seacliff. Description & travel

BERRY, Peg
Brighton and Seacliff sketchbook. - Adelaide : Rigby, 1975
64p : ill. - (The sketchbook series)
0-85179-903-5 X.809/42895
1353837

ADELAIDE. South Australia
Buildings

EMANUEL, Cedric
Portrait of Adelaide. - Adelaide : Rigby, 1975
[80]p : chiefly ill
0-7270-0026-8 Cup.1285/250
1287060

ADELAIDE. South Australia
Buildings. Architecture, to 1930

BURDEN, Michael
Lost Adelaide. - Melbourne ; Oxford : Oxford University Press, c1983
223p : ill. - Bibliography: p223. - Includes index
0-19-554402-1 X.805/6108
3215075

ADELAIDE. South Australia
Description & travel, to date

ADELAIDE in color. - Repr. - Windsor, N.S.W. : Lloyd O'Neil for Currey O'Neil, 1980
[24]p : chiefly col ill
0-85550-142-1 X.805/3827
8939829

COLWELL, Max
Adelaide. - Melbourne : Lansdowne Press, 1974
ca.128p. : ill
0-7018-0287-1 X.805/4708
8939829

ADELAIDE. South Australia
Kensington & Norwood. Description & travel

PEAKE, A. J.
Kensington and Norwood sketchbook. - Adelaide : Rigby, 1976
63p : ill. - (The sketchbook series)
0-7270-0072-1 X.809/42961
1378805

ADELAIDE. South Australia
Magill. Description & travel

MCCLAREN, Moira
Magill sketchbook. - Adelaide : Rigby, 1976
63p : ill. - (The sketchbook series)
0-7270-0075-6 X.809/42758
1333720

ADELAIDE. South Australia
Unemployed persons. Effects of economic depressions, 1929-1938

BROOMHILL, Ray
Unemployed workers. - St. Lucia : University of Queensland Press, c1978
xiv,220p : ill. - Bibliography: p201-213. - Includes index
0-7022-1235-0 X.529/41705
2495317

ADELAIDE. South Australia
Wines industries, to 1975

WALLS, Bill
Southern vineyards sketchbook. - Adelaide : Rigby, 1976
63p : ill. - (The sketchbook series)
0-7270-0122-1 Cup.1281/753
1341146

ADELBODEN. Switzerland
Social conditions, 1874-1974

DOYON, Josy
Im Schatten des Lohners. - Bern, [1974]
96p
3-85580-036-7 X.808/10201
0886335

ADELBORG, OTTILIA. Swedish illustrations

RÅDSTRÖM, Anne Marie
Fröken Ottil. - Falun : Dalarnas museum, 1980
104p,4leaves of plates : ill(some col.). - (Dalarnas hembygdsbok ; 1980)
91-85378-33-x P.801/561[1980]
8881006

ADÉLIE PENGUINS

BONNERS, Susan
A penguin year. - London : Gollancz, 1981
[46]p : col.ill
0-575-03025-9 L.49/1112
2609908

ADELPHOE. Terence. Drama in Latin

PÖSCHL, Viktor
Das Problem der Adelphen des Terenz. - Heidelberg, 1975
24p.. - (Sitzungsberichte der Heidelberger Akademie der Wissenschaften. Phil.-hist. Klasse ; Jahrg. 1975, Abh.4)
3-533-02426-1 Ac.892/2 [1975.Abh.4]
0873071

ADELSBERG
See Postojna

ADELUNG, JOHANN CHRISTOPH. Lexicography. German language

STROHBACH, Margrit
Johann Christoph Adelung. - Berlin : de Gruyter, 1984
xii,290p,plate. - (Studia linguistica Germanica ; 21)
3-11-009612-9 X.0981/462(21)
8738688

ADEN
See also
Gulf of Aden

ADEN. Yemen (People's Democratic Republic)
1839-1872

KOUR, Z. H. (Zaki Hanna)
The history of Aden 1839-72. - London : Cass, 1981
240p. - Bibliography: p233-234. - Includes index
0-7146-3101-9 X.809/46536
1763296
- [Another copy.] T 45956 IOL

ADEN. Yemen (People's Democratic Republic)
Social conditions, 1952-1958

KNOX-MAWER, June
The sultans came to tea. - Gloucester : Alan Sutton, 1984, c1961
vii,218p,[12]p of plates : ill
0-86299-121-8 X.950/32768
3016501

ADENAUER, KONRAD. West Germany

OSTERHELD, Horst
Konrad Adenauer. - Stuttgart : Verlag Bonn Aktuell, 1981
63p. : ill
3-87959-170-9 X.805/4226
8721661

ADENAUER, KONRAD. West Germany
- Conference proceedings

KONRAD Adenauer e Alcide De Gaspari / a cura di Umberto Corsini e Konrad Repgen. - Bologna : Il Mulino, 1984
399p. - (Annali dell' Istituto storico italo-germanico. Quaderno ; 15)
88-15-00617-6 X.809/64768
8721661

ADENAUER, KONRAD. West Germany
- Correspondence, diaries

ADENAUER, Konrad
[Selections]. Adenauer. - Berlin : Siedler.
Briefe. - 1983-. - Includes bibliography
r6311015 X.0809/2434a
8721661

ADENIUM

ROWLEY, G. D. (Gordon D)
The Adenium and Pachypodium handbook. - Oxford : British Cactus and Succulent Society, [1983?]
95p : ill. - Bibliography: p89-91. - Includes index
0-902099-07-8 SRL
3047032

ADENOIDS. Children. Scotland
Adenotonsillectomy. Incidence

BLOOR, M. J.
An epidemiological and sociological study of variations in the incidence of operations on the tonsils and adenoids. - [Aberdeen] : [University of Aberdeen Institute of Medical Sociology], 1978
80p : ill. - (Occasional paper ; no.2). - Bibliography: p78-80
B84-25268 X.525/9113
3124053

ADENOIDS. In-patients. Hospitals. Scotland
Adenotonsillectomy - Statistics

ADENO-TONSILLECTOMY. - Edinburgh : Information Services Division, Scottish Health Service Common Services Agency, [1982?]
15[i.e.13]leaves : ill. - (Scripps reviews ; no.1) (Occasional papers)
B82-40983 BS.155/255(1)
2801647

ADENOMAS
See also
Microadenomas
Prolactinomas

ADENOMAS. Pituitary gland. Man

The PITUITARY adenoma / edited by Kalmon D. Post, Ivor M.D. Jackson, Seymour Reichlin. - New York ; London : Plenum Medical, 1980
xxiv,507p : ill. - Includes index
0-306-40382-x SRL
2290766

SHELINE, Glenn E.
Pituitary adenomas. - New York ; Oxford : Pergamon, c1981
xvi,SG1-SG25,197,PT1-PT18p : ill. - (Oncologic, multidisciplinary decisions in oncology ; v.12). - Bibliography: p187-193. - Includes index
0-08-027463-3 X.325/1331 Woolwich
2499452

ADENOSINE. Animals
Physiological aspects - Conference proceedings

INTERNATIONAL SYMPOSIUM ON ADENOSINE (1982 : Charlottesville)
Regulatory function of adenosine. - Boston, [Mass.] : Lancaster : Nijhoff, 1983
xxi,558p : ill. - (Developments in pharmacology ; 2). - Includes index
90-247-2779-0 SRL
2977834

ADENOSINE TRIPHOSPHATASE
See ATPase

ADENOSINE TRIPHOSPHATE
See ATP

ADENOSYLMETHIONINE
See also
S-adenosylmethionine

ADENOSYLMETHIONINE. Man
Role in central nervous system - Conference proceedings

INTERNATIONAL ROUND TABLE ON ADENOSYLMETHIONINE AND CENTRAL NERVOUS SYSTEM (1978 : Naples)
Biochemical and pharmacological roles of adenosylmethionine and the central nervous system. - Oxford : Pergamon, 1979
xii,171p : ill. - Includes index
0-08-024929-9 SRL
1875418

ADENOSYLMETHIONINE. Organisms
- Conference proceedings

The BIOCHEMISTRY of adenosylmethionine / Francesco Salvatore ... [et al.], editors. - New York ; Guildford : Columbia University Press, 1977
xv,588p. - Includes index
0-231-03895-x SRL
1376268

ADENOTONSILLECTOMY. Children. Scotland
Incidence

BLOOR, M. J.
An epidemiological and sociological study of variations in the incidence of operations on the tonsils and adenoids. - [Aberdeen] : [University of Aberdeen Institute of Medical Sociology], 1978
80p : ill. - (Occasional paper ; no.2). - Bibliography: p78-80
B84-25268 X.525/9113
3124053

ADENOTONSILLECTOMY. In-patients. Hospitals. Scotland
- Statistics

ADENO-TONSILLECTOMY. - Edinburgh : Information Services Division, Scottish Health Service Common Services Agency, [1982?]
15[i.e.13]leaves : ill. - (Scripps reviews ; no.1) (Occasional papers)
B82-40983 BS.155/255(1)
2801647

ADENOVIRUSES

The ADENOVIRUSES / edited by Harold S. Ginsberg. - New York ; London : Plenum, c1984
xvii,605p : ill. - (The Viruses). - Includes bibliographies and index
0-306-41592-5 SRL
3274241

ADF
See Automatic direction finder

ADHESION

CATHERALL, Ed
Adhesion. - Hove : Wayland, 1983
32p : col.ill. - (Young scientist)
0-85078-290-2 X.990/21532
3043126

WAKE, William C. (William Charles)
Adhesion and the formulation of adhesives. - 2nd ed. - London : Applied Science, 1982
xiii,332p : ill. - Bibliography: p308-320. - Includes index
0-85334-134-6 SRL
0640239

WAKE, William Charles
Adhesion and the formulation of adhesives. - London : Applied Science Publishers, 1976
xii,325p : ill. - Bibliography: p.303-313. - Includes index
0-85334-660-7 SRL
0640239

ADHESION
- Conference proceedings

ADHESION. - London : Applied Science Publishers.
1. - 1977
ix,315p : ill. - Includes index
0-85334-735-2 SRL
0640239

3. - 1979
viii,236p : ill. - Includes index
0-85334-808-1 SRL
0640239

SOCIETE DE CHIMIE PHYSIQUE. International Meeting (34th . 1981 : Paris)
Microscopic aspects of adhesion and lubrication. - Amsterdam ; Oxford : Elsevier Scientific, 1982
xix,812p : ill. - (Tribology series ; 7). - Includes bibliographies and index
0-444-42071-1 SRL
0155845

ADHESION. Cells. Organisms
To surfaces - Conference proceedings

CELL adhesion and motility / edited by A.S.G. Curtis, J.D. Pitts. - Cambridge : Cambridge University Press, 1980
[9],500p,plate : ill(some col). - (Symposia / British Society for Cell Biology ; 3). - Includes bibliographies and index
0-521-22936-7 SRL
2211246

ADHESION. Microorganisms
To surfaces - Conference proceedings

ADHESION and microorganism pathogenicity. - London : Pitman Medical, 1981
x,346p : ill. - (Ciba Foundation symposium ; 80). - Includes bibliographies and index
0-272-79615-8 SRL
2150875

MICROBIAL adhesion to surfaces / editors R.C.W. Berkeley ... [et al.]. - Chichester : Published for the Society of Chemical Industry, London by Horwood, 1980
559p : ill. - Includes bibliographies and index
0-85312-186-9 SRL
2150875

ADHESION. Microorganisms
To surfaces of human teeth - Conference proceedings

BACTERIAL ADHESION TO BIO (SURFACES) : WITH SPECIAL EMPHASIS ON DENTAL PROBLEMS (Workshop : 1983 : Paterswolde)
Bacterial adhesion and preventive dentistry. - Oxford : IRL, Mar.1984
1v.
0-904147-63-0 SRL
3021351

ADHESION. Polymer coatings
- Conference proceedings

SYMPOSIUM ON ADHESION ASPECTS OF POLYMERIC COATINGS (1981 : Minneapolis)
Adhesion aspects of polymeric coatings. - New York ; London : Plenum, c1983
xii,657p : ill. - Includes index
0-306-41250-0 SRL
2969882

ADHESION. Powders

ZIMON, A. D. (Anatoliĭ Davydovich)
Adhesion of dust and powder. - 2nd ed. - New York ; London : Consultants Bureau, c1982
xi,438p : ill
0-306-10962-x SRL
2744392

ADHESION. Solids

DERÎAGIN, B. V. (Boris Vladimirovich)
Adhesion of solids. - New York ; London : Consultants Bureau, 1978
xvi,457p : ill. - (Studies in Soviet science. physical sciences)
0-306-10941-7 SRL
1734563

ADHESIVE-BONDED JOINTS. Structural components

ADAMS, Robert D.
Structural adhesive joints in engineering. - London : Elsevier Applied Science, c1984
ix,309p : ill. - Bibliography: p283-291. - Includes index
0-85334-263-6 SRL
3281485

ADHESIVE-BONDED JOINTS. Structural components
Durability - Conference proceedings

SYMPOSIUM ON DURABILITY OF ADHESIVE BONDED STRUCTURES (1976 : Picatinny Arsenal)
Durability of adhesive bonded structures. - New York ; London : Wiley, 1977
vi,443p : ill. - (Applied polymer symposium ; 32). - Includes index
0-471-04564-0 SRL
1495283

ADHESIVE STAMPS
British adhesive revenue stamps, to 1974 - Catalogues

BOOTH, R. G. (Roger George)
A catalogue of the adhesive revenue stamps of the UK, Isle of Man and Channel Islands, (including pre-1922 Ireland and Free State overprints). - Newcastle upon Tyne : J.M. Booth (Stamps), 1976
xlvi,173p : ill
0-900571-02-0 X.519/25647
0625884

ADHESIVES

DEVELOPMENTS in adhesives. - London : Applied Science Publishers. - ([Developments series])
1. - 1977
xii,318p : ill. - Includes index
Subsequent volumes, when received, will be found at SRL
0-85334-749-2 SRL
0640247

HANDBOOK of adhesives / edited by Irving Skeist. - 2nd ed. - New York ; London : Van Nostrand Reinhold, 1977
xix,921p : ill. - Includes bibliographies and index
0-442-27634-6 SRL
0640247

SHIELDS, J. (John)
Adhesives handbook. - 3rd ed. - London : Butterworths, 1984
360p : ill. - Bibliography: p353-359
0-408-01356-7 SRL
0640247

WAKE, William Charles
Adhesion and the formulation of adhesives. - London : Applied Science Publishers, 1976
xii,325p : ill. - Bibliography: p.303-313. - Includes index
0-85334-660-7 SRL
0640247

ADHESIVES
- Serials

INTERNATIONAL journal of adhesion and adhesives. - Vol.1, no.1- ; July 1980-. - Guildford : IPC Science and Technology Press, 1980- : ill
B80-35244 SRL
227468x

ADHESIVES
Adhesives & sealants

INDUSTRIAL adhesives and sealants / edited by B.S. Jackson. - London : Hutchinson, 1976
303p,[24]p of plates : ill. - Includes index
0-09-126900-8 SRL
1073575

ADHESIVES
Adhesives & sealants - Serials

EUROPEAN adhesives & sealants. - June 1983-. - Redhill : Fuel & Metallurgical Journals, 1983-
v. : ill
B84-14795 SRL
3063127

ADHESIVES
Pressure-sensitive adhesives

HANDBOOK of pressure-sensitive adhesive technology / edited by Donatas Satas. - New York ; London : Van Nostrand Reinhold, c1982
xvi,620p : ill. - Includes bibliographies and index
0-442-25724-4 SRL
2829517

ADHESIVES
Testing

ANDERSON, Garron P.
Analysis and testing of adhesive bonds. - New York ; London : Academic Press, 1977
xviii,255p : ill. - Includes index
0-12-056550-1 SRL
1261193

ADHESIVES. Bonding

SHIELDS, J. (John)
Adhesives handbook. - 2nd ed. - London : Newnes-Butterworths, 1976
[10],348p : ill. - Bibliography: p.345-348
0-408-00210-7 SRL
0143316

ADHESIVES. Building materials

DURABILITY of structural adhesives / edited by A.J. Kinloch. - London : Applied Science, c1983
x,360p : ill. - Includes index
0-85334-214-8 SRL
1038990

ADHESIVES INDUSTRIES. Great Britain
- Statistics

ADHESIVE manufacturers / [Business Ratios]. - 1st ed.: [1978]. - London : Business Ratios, [1978]
[93]p : ill
B80-35801 L.49/518
227048x

ĀDI-GRANTH. Sikhism

DHILLON, N. S. (Nasib Singh)
What is what in the Holy Granth. - [London] : [Nasib Singh Dhillon], [1983?]
140p : ill. - Includes index
0-9505807-2-4 X.200/43344
3122123
- [Another copy.] T 46346 IOL

ĀDI-GRANTH. Sikhism
Characters

DHILLON, N. S.
Pillars of divine philosophy. - London : The author, 1977
[4],xiv,100p
B77-21697 X.108/15861
1140280

ĀDI-GRANTH. Sikhism
Hymns. Words

SHACKLE, C. (Christopher)
A Gurū Nānak glossary. - London : School of Oriental and African Studies, University of London, 1981
xxxi,276p
0-7286-0074-9 X.950/8971
2591855

ADJECTIVAL SUFFIXES. French language, German language & Polish language

GAWEŁKO, Marek
Sufiksy przymiotnikowe w języku polskim, niemieckim i francuskim. - [Cracow] : Nakładem Uniwersytetu Jagiellońskiego, [1976]
79 p.. - (Zeszyty naukowe Uniwersytetu Jagiellońskiego ; 421. Prace językoznawcze ; zesz. 49). - Bibliography : p. 74-79
r6014888 Ac.749/27a[zesz.49]
0888303

ADJECTIVAL SUFFIXES. German language

INGHULT, Göran
Die semantische Struktur desubstantivischer Bildungen auf -mässig. - Stockholm : Almqvist & Wiksell International, 1975
206p. - (Stockholmer germanistische Forschungen ; 18). - Includes bibliographies
r6033897 Ac.2650.m/5[vol.18]
1535153

ADJECTIVE-NOUN COMBINATIONS. German language, 1100-1500

LUTZ, Hans Dieter
Zur Formelhaftigkeit der Adjektiv - Substantiv - Verbindung im Mittelhochdeutschen. - München, 1975
154 p.. - (Münchener Texte und Untersuchungen zur deutschen Literatur des Mittelalters ; Bd 52)
3-406-05182-0 Ac.713d/10[Bd52]
0893455

ADJECTIVES
See also
Adjective-noun combinations

ADJECTIVES. English language

BACHE, Carl
The order of premodifying adjectives in present-day English. - [Odense] : Odense University Press, 1978
99p. - (Odense University studies in English ; vol.3)
87-7492-276-9 X.0900/358(3)
1985892

HERBST, Thomas
Untersuchungen zur Valenz englischer Adjektive und ihrer Nominalisierungen. - Tübingen : Narr, 1983
xix,433p. : ill. - (Tübinger beiträge zur Linguistik ; 233). - Includes bibliography
3-87808-233-9 X.950/26825
1985892

ISITT, David
Crazic, menty and idiotal. - Göteborg : Acta Universitatis Gothoburgensis, 1983
309p. - (Gothenburg studies in English ; 52)
91-7346-122-9 Ac.2651.d/3[no.52]
1985892

ADJECTIVES. English language –continued

WARREN, Beatrice
Classifying adjectives. – Göteborg : Acta Universitatis Gothoburgensis, 1984
318p. – (Gothenburg studies in English ; 56)
91–7346–133–4 Ac.2651.d/3[no.56]
1985892

ADJECTIVES. English language
Comparative adjectives

BRAUN, Albert
Studien zu Syntax und Morphologie der Steigerungsformen im Englischen. – Bern : Francke, [1982]
156p. – (Schweizer anglistische Arbeiten ; Bd.110)
3–7720–1565–4 WP.1693/110
8844801

CAMPBELL, Rod
Great, greater, greatest. – London : Abelard-Schuman, 1980
[24]p : chiefly col ill
0–200–72699–4 X.990/16866 Woolwich
0728799

ADJECTIVES. English language
Comparative adjectives

FREUDBERG, Judy
Some more most. – Glasgow : Collins, 1977
[30]p : col ill. – (A strawberry book)
0–00–138216–0 X.990/8843 Woolwich
0728799

HELLAN, Lars
Towards an integrated analysis of comparatives. – Tübingen : Narr, 1981
iv,258p.. – (Ergebnisse und Methoden moderner Sprachwissenschaft ; Bd.11). – Bibliography
3–87808–986–4 X.950/13624
8844801

MORTON, Gabrielle
The second more and less book. – Andover : Philograph, 1975
24p : col ill
0–85370–028–1 X.990/8184 Woolwich
0728799

ADJECTIVES. English language
Gradable adjectives. Semantics

RUSIECKI, Jan
Adjectives and comparison in English. – London : Longman, 1985
xiv,206p : ill. – (Longman linguistic library ; no.31). – Bibliography: p193–196. – Includes index
0–582–29129–1 X.0989/48(40)
3019543

ADJECTIVES. Estonian language

ERELT, Mati
Eesti adjektiivide süntaks. – Tallinn : ENSV TA, Ühiskonnateaduste Osakond, 1977
62p. – (Preprint KKI ; 8). – Includes bibliography
r6461672 X.0950/608(8)
8956286

Predikatiivne adjektiiv (lausemallid). – Tallinn : Eesti Raamat, 1979
53p
r6461037 X.958/4410
8955328

ADJECTIVES. Finnish language

LEPPÄJÄRVI, Eila
Adjektivit 1960–luvun suomen lehti–, radio– ja yleispuhekielessä. – Oulu : Yliopisto, 1976
32p. – (Oulun yliopiston suomen ja saamen kielen laitoksen tutkimusraportteja ; 7)
951–420–376–3 X.0900/581(7)
1975447

ADJECTIVES. Finnish language
Meaning. Opposition

HAKANEN, Aimo
Adjektiivien vastakohtasuhteet suomen kielessä. – Helsinki : Suomalaisen Kirjallisuuden Seura, 1973
279p.. – (Suomalaisen Kirjallisuuden Seuran toimituksia ; 311). – Bibliography: p.261–271
951–7170–21–1 Ac.9080[no.311]
0857920

ADJECTIVES. French language

BARTNING, Inge
Remarques sur la syntaxe et la sémantique des pseudo-adjectifs dénominaux en français. – 2.éd. – Stockholm : Almqvist & Wiksell International, 1980
174p. – (Romanica Stockholmiensia ; 10). – Includes bibliographies
91–220–0373–8 Ac.2650m/23[vol.10]
1989952

FORSGREN, Mats
La place de l'adjectif épithète en français contemporain. – Uppsala ; Stockholm : Distributed by Almqvist & Wiksell International, 1978
231p. – (Studia romanica upsaliensia ; 20)
91–554–0724–2 Ac.1075/6(19)[vol.20]
1989952

REINER, Erwin
Studie zur Stellung des attributiven Adjektivs im neueren Französischen. – Wien : Braumüller, [1976]
118p.
3–7003–0130–8 X.950/14634
1989952

ADJECTIVES. French language
to date

KOGELSCHATZ, Brigitte
Theorie und Praxis des sprachlichen Feldes. – München : Fink, c1981
290p. – Includes bibliography
3–7705–1853–5 X.950/22519
897070x

ADJECTIVES. German language

BZDĘGA, Andrzej Zdzisław
Das Adjektiv im Polnischen und Deutschen. – Wrocław : Zakład Narodowy im. Ossolińskich – Wydawnictwo PAN, 1980
100p
83–04–00045–8 X.950/3747
878373x

EICHINGER, Ludwig M.
Syntaktische Transposition und semantische Derivation. – Tübingen : Niemeyer, 1982
x,250p. – (Linguistische Arbeiten ; 113). – Bibliography
3–484–30113–9 X.0900/372(113)
8811989

HÄRD, John Evert
Adjektivadverb oder adverbiales Adjektiv?. – Abo : Abo akademi, 1976
75p. – (Acta Academiae Aboensis. ser.A ; v.54, nr 1)
951–648–265–1 Ac.1093/2a[vol.54.no.1]
1473751

HOLMLANDER, Inger O.
Zur Distribution und Leistung des Pronominaladverbs. – Uppsala ; Stockholm : Distributed by Almqvist & Wiksell International, 1979
187p. – (Studia germanistica upsaliensia ; 21). – Includes bibliographies
91–554–0859–1 AC:1075/6(23)[vol.21]
1473751

HUNDSNURSCHER, Franz
Semantik der Adjektive des Deutschen. – Opladen : Westdeutscher Verlag, 1982
112p.. – (Forschungsberichte des Landes Nordrhein–Westfalen ; Nr.3137. Fachgruppe Geisteswissenschaften). – Includes bibliography
3–531–03137–6 X.950/34945
878373x

LAFRENZ, Peter Goerges
Zu den semantischen Strukturen der Dimensionsadjektive in der deutschen Gegenwartssprache. – Göteborg : Acta Universitatis Gothoburgensis, 1983
166p. – (Göteborger germanistische Forschungen ; 24)
91–7346–135–0 Ac.2651.db/12[no.24]
878373x

MOILANEN, Markku
Zum lokalen Gebrauch der Demonstrativadverbien 'da' und 'dort'. – Helsinki : Suomalainen tiedeakatemia, 1973
145p. – (Annales Academiae Scientiarum Fennicae. dissertationes humanarum litterarum ; 2)
951–410–146–4 Ac.1094.c/7b(1)[vol.2]
1473751

ADJECTIVES. Hebrew language

WERNER, Fritz
Die Wortbildung der hebräischen Adjektiva. – Wiesbaden : Harrassowitz, 1983
593p. – Includes bibliography
3–447–02280–9 X.950/37905
9141537

ADJECTIVES. Latin language

KESSISSOGLU, Alexander I.
Das elativische Adjektiv in der frühlateinischen Literatur. – Frankfurt am Main ; Cirencester : Lang, 1982
138p. – (Europäische Hochschulschriften. Reihe 15, Klassische Sprachen und Literaturen ; Bd.19). – Bibliography
3–8204–6792–0 X.950/16544
8884471

ADJECTIVES. Old English language

RÜDEN, Michael von
Wlanc und Derivate im Alt– und Mittelenglischen. – Frankfurt am Main : Lang, 1978
324p. – (Europäische Hochschulschriften. Reihe 14, Angelsächsische Sprache und Literatur ; Bd.61). – Bibliography
3–261–02511–5 X.950/9579
8823235

ADJECTIVES. Old Provençal language
Inflexion

JENSEN, Frede
The Old Provençal noun and adjective declension. – [Odense] : Odense University Press, 1976
177p. – (Études romanes de l'Université d'Odense ; vol.9). – Bibliography : p.150–154 – Includes index
87–7492–193–2 X.0900/295(9)
1690027

ADJECTIVES. Old Provençal language
Verbal adjectives

BETZ, Manfred L.
Aussagegehalt und Syntax deverbaler Adjektive im Altprovenzalischen. – [Lampertheim] : Schäuble, c1975
vi, 129 p.. – (Romanistik ; 4). – Bibliography : p. 117–121
3–87718–704–8 X.0900/328(4)
0935093

ADJECTIVES. Polish language

BZDĘGA, Andrzej Zdzisław
Das Adjektiv im Polnischen und Deutschen. – Wrocław : Zakład Narodowy im. Ossolińskich – Wydawnictwo PAN, 1980
100p
83–04–00045–8 X.950/3747
8783721

ADJECTIVES. Romansch language

WINZAP, Isidor
La posiziun digl adjectiv en romontsch. – Cuera : Ligia romontscha, 1981
221p
r6507720 X.950/22056
8964653

ADJECTIVES. Russian language

NEMCHENKO, V. N. (Vasiliĭ Nikolaevich)
O razgranichenii chasteĭ rechi v sovremennom russkom i͡azyke. – Gor'kiĭ : Gor'kovskiĭ gosudarstvennyĭ universitet im. N.I. Lobachevskogo, Kafedra sovremennogo russkogo i͡azyka i obshchego i͡azykoznanii͡a, 1975
77p. – Bibliography
r6402716 X.958/6308
8790841

Slovoobrazovatel'nai͡a struktura imen prilagatel'nykh v sovremennom russkom i͡azyke. – Gor'kiĭ : Gor'kovskiĭ gosudarstvennyĭ universitet im. N.I. Lobachevskogo, Kafedra russkogo i͡azyka i obshchego i͡azykoznanii͡a, 1973
137p. – Includes bibliography
r6462860 X.950/30071
8790841

ADJECTIVES. Russian language
Predicative adjectives. Use with copulas: 'byt'

GUSTAVSSON, Sven
Predicative adjectives with the copula byt' in modern Russian. – Stockholm : Almvistand Wiksell International, 1976
[4],399p. – (Acta Universitatis Stockholmiensis. Stockholm Slavic studies ; 10). – Bibliography : p.392–399
91–220–0076–3 X.0900/258(10)
1116479

IVERSEN, Malvin
Predikativt adjektiv ved preteritum av kopula byt' i moderne russisk. – Bergen : Universitetet i Bergen, Russisk institutt, 1978
87p. – (Skrifter ; nr 2). – Includes bibliographies
82–902490–0–4 X.0909/858(2)
1116479

ADJECTIVES. Russian language
Transformational–generative theories

BABBY, Leonard Harvey
A transformational grammar of Russian adjectives. – The Hague : Mouton, 1975
242 p. : ill. – (Janua linguarum. series practica ; 235). – Bibliography: p. 237–239. – Includes index
r6010163 X.0981/78c(235)
0782874

ADJECTIVES. Spanish language

AGGERNÆS, Gretter
Lista de adjectivos que se construyen con ser o estar en oraciones atributivas. - København : Københavns Universitet, Romansk Institut, 1982
33p. - (RIDS ; nr 99)
r6610111 X.0902/109(99)
8766622

PELZING, Ernst
Das attributive Adjektiv des Spanischen. - Frankfurt am Main : Lang, 1981
iv,199p. - (Europaische Hochschulschriften. Reihe 24, Ibero-Romanische Sprachen und Literaturen ; Bd.13). - Bibliography
3-8204-6892-7 X.950/9502
8766622

ADJECTIVES. Spanish language
Qualifying adjectives

ZIERER, Ernesto
[El adjetivo calificativo en el idioma español]. The qualifying adjective in Spanish. - The Hague ; Paris : Mouton, 1974
54 p.. - (Janua linguarum. Series practica ; no. 192)
r6003204 X.0981/78c(192)
090306X

ADJECTIVES DENOTING SIZE. Swedish language to 1520

JOSEFSON, Ingela
Die Bezeichnungen für 'gross-klein', 'lang-kurz' im Altschwedischen. - Göteborg : Acta Universitatis Gothoburgensis, 1976
vii,119p. - (Nordistica gothoburgensia ; 9). - Includes bibliographies
91-7346-025-7 Ac.2651.db/14(9)
1228781

ADJUDICATION. Complaints against solicitors by clients. England
By Law Society

SOLICITORS Act 1974, annual report of the Lay Observer / laid before Parliament by the Lord High Chancellor ... - London : H.M.S.O.
1st : 1975-1976. - [1976]
[2],13p. - ([1975-76 H.C.]332)
Later issues, when received, are placed at this shelfmark
0-10-233276-2 BS.24f/111
0673471

ADJUDICATION. Complaints against solicitors by clients. Scotland
By Law Society of Scotland

SOLICITORS (Scotland) Act 1976, annual report of the Scottish lay observer / laid before Parliament by the Secretary of State for Scotland ... - 1st : 1976-1977. - Edinburgh : H.M.S.O, [1977]
[2],9p
Subsequent volumes, when received, will be placed at this shelfmark
0-10-200378-5 BS.179/13
1337513

ADJUDICATION. Festivals. Amateur theatre. Great Britain

LAURIE, Rona
Festivals and adjudication. - London : Pitman, 1975
vii,150p. - Bibliography: p.143-145. - Includes index
0-273-25257-7 X.989/40142
0376469

ADJUDICATION. International disputes. Foreign trade

HUDEC, Robert E.
Adjudication of international trade disputes. - London : Trade Policy Research Centre, 1978
vi,88p. - (Thames essays ; 16)
0-900842-39-3 X.519/28987
1765922

ADJUDICATION. Prisons. Great Britain

GREAT BRITAIN. Working Party on Adjudication Procedures in Prisons
Report of the Working Party on Adjudication Procedures in Prisons. - London : H.M.S.O., 1975
[3],76p
0-11-340578-2 BS.18/443
0520748

SMITH, David, 1946-
Board of visitor adjudications. - London : Home Office, c1981
v,45p. - (Research Unit paper ; 3). - Bibliography: p40
0-86252-004-5 BS.18/684(3)
0520748

ADJUSTABLE HOODS. Local exhaust ventilation equipment. Foundries

JOINT STANDING COMMITTEE ON HEALTH, SAFETY AND WELFARE IN FOUNDRIES. Sub-committee on Dust and Fume
Foundry dust control. - London : H.M.S.O., 1975
25p : ill
0-11-361077-7 BS.385/161
0513512

ADJUSTMENT
See also
Personal adjustment
Social adjustment

ADJUSTMENT
To leisure. Role of education

BRIGHTBILL, Charles Kestner
Educating for leisure-centered living. - 2nd ed.. - New York ; London : Wiley, 1977
xii,128p
0-471-94914-0 X.519/26974
1138138

ADJUSTMENT
To social change related to personal adjustment to bereavement

MARRIS, Peter
Loss and change. - London : Routledge and Kegan Paul, 1978
vi,178p. - (Reports / Institute of Community Studies). - Bibliography: p.172-174. - Includes index
0-7100-8908-2 X.519/28119
0129852

ADJUSTMENT. Balance of payments
Mathematical models

BUITER, Willem H.
International balance of payments. - [London] : [Centre for Labour Economics, London School of Economics], [1983]
33p. - (Discussion paper ; no.159). - Bibliography: p32-33
B84-22241 X.529/60998
3084671

ADJUSTMENT. Claims. Insurance. Property. United States

THOMAS, Paul Irving
Adjustment of property losses. - 4th ed.. - New York ; London : McGraw-Hill, 1977
xiv,577p. - Bibliography: p.558-559. - Includes index
0-07-064215-x X.520/12189
1311441

ADJUSTMENT. Labour market. Great Britain

THOMAS, Barry, 1941-
Labour shortage and economic analysis. - Oxford : Blackwell, 1977
xvi,264p : ill. - (Warwick studies in industrial relations). - Bibliography: p.238-250. - Includes index
0-631-18310-8 X.529/32301
1072439

ADJUSTMENT. Measurement. Surveying

MIKHAIL, Edward M.
Analysis and adjustment of survey measurements. - New York ; London : Van Nostrand Reinhold, c1981
xii,340p : ill. - Bibliography: p333-335. - Includes index
0-442-25369-9 SRL
2364859

ADJUSTMENT. Personal income tax. Organisation for Economic Co-operation and Development countries
Implications of inflation

ORGANISATION FOR ECONOMIC CO-OPERATION AND DEVELOPMENT. Committee on Fiscal Affairs
The adjustment of personal income tax systems for inflation. - Paris : O.E.C.D. ; [London] : [H.M.S.O.], 1976
71p : 1 ill. - Bibliography: p.68-71
B76-13975 UN.P.106/536
0635162

ADJUSTMENT. Prices. Contracts. Construction. Great Britain
Formulae

JONES, Glyn Price
An appraisal of formulae used for fluctuations in construction costs. - Ascot : Institute of Building, 1975
[1],21p. - (Occasional papers / Institute of Building ; 9). - Bibliography: p.21
0-901822-31-0 X.512/5145
0626074

ADJUSTMENT. Rates. Foreign exchange
Theories of monetarists

The MONETARY approach to international adjustment / edited by Bluford H. Putnam and D. Sykes Wilford. - New York ; London : Praeger, 1978
xvi,283p : ill. - (Praeger special studies). - Bibliography: p.251-275. - Includes index
0-03-046711-x X.520/14232
1917919

ADJUSTMENT. Rates. Foreign exchange. Italy
Effects of monetary policies, 1945-1978

TULLIO, Giuseppe
The monetary approach to external adjustment. - London : Macmillan, 1981
xx,127p : ill. - Bibliography: p114-123. - Includes index
0-333-27651-5 X.529/43740
2566516

ADJUSTMENT. Remuneration. Personnel. Civil service. Great Britain
For differences between superannuation benefits of civil service & other personnel

GREAT BRITAIN. Government Actuary
Civil service pay research, the 1980 review of the adjustment for differences in superannuation benefits. - London : H.M.S.O., 1980
[3],24p
0-11-700859-1 BS.121/10
2257270

ADJUSTMENT. United States
To leisure. Role of education

MUNDY, Jean
Leisure education. - New York ; Chichester : Wiley, 1979
xi,244p : ill. - Includes index
0-471-01347-1 X.520/14330
1936972

ADJUVANT THERAPY. Cancer. Man
- Conference proceedings

INTERNATIONAL CONFERENCE ON THE ADJUVANT THERAPY OF CANCER (3rd : 1981 : Tucson)
Adjuvant therapy of cancer III. - New York ; London : Grune & Stratton, c1981
xiv,603p : ill. - Includes bibliographies and index
0-8089-1407-3 SRL
2560496

ADJUVANTS. Drugs

MATERIALS used in pharmaceutical formulation / edited by A.T. Florence. - Oxford : Published for the Society of Chemical Industries by Blackwell Scientific, 1984
viii,161p : ill. - (Critical reports on applied chemistry ; v.6). - Includes index
0-632-01257-9 SRL
3018636

ADLER, ALFRED
Theories of psychology

Die INDIVIDUALPSYCHOLOGIE Alfred Adlers / herausgegeben von Rainer Schmidt ; unter Mitarbeit von Elsa Andriessens ... [et al.]. - Stuttgart : Kohlhammer, 1982
316p : ill. - Includes bibliography
3-17-007537-3 X.520/34584
8779201

ADLER, ALFRED
Theories of social adjustment

CRANDALL, James E.
Theory and measurement of social interest. - New York ; Guildford : Columbia University Press, 1981
xi,181p : ill. - Bibliography: p163-173. - Includes index
0-231-05256-1 X.529/45813
2622564

ADLER, ALFRED. Psychology

MCDERMOT, Violet
The social vision of Alfred Adler. - Ditchling : New Atlantis Foundation, 1980
19p. - (Foundation lecture / New Atlantis Foundation ; 20). - Bibliography: p19
B81-15750 X.529/41001
0619744

ORGLER, Hertha
Alfred Adler, the man and his work. - 4th ed. (enlarged). - London : Sidgwick and Jackson, 1973 [i.e. 1976]
xxi,270p,[2]p of plates. - Bibliography: p.263-265. - Includes index
0-283-98260-8 X.319/16442 Woolwich
0619744

WE knew Alfred Adler. - [London] : Adlerian Society of Great Britain, 1977
[2],12p. - (Individual psychology pamphlets ; no.2)
B78-01921 X.519/27475
0619744

ADLER, ALFRED, 1870-1937. Psychology

RATTNER, Josef
[Individual psychologie. English]. Alfred Adler. - New York : Ungar, c1983
xiv,226p. - (Literature and life series)
0-8044-5988-6 X.629/23013
8710597

ADLER, FELIX. Judaism
- Biographies

KRAUT, Benny
From reform Judaism to ethical culture. - Cincinnati : Hebrew Union College Press, 1979
xv,285p. - (Monographs of the Hebrew Union College ; no.5) (An I. Edward Kiev Library Foundation book). - Bibliography: p.264-277
0-87820-404-0 X.520/24977
8273464

ADLER, FELIX. Moral philosophy
- Biographies

FRIESS, Horace L.
Felix Adler and ethical culture. - New York ; Guildford, Surrey : Columbia University Press, 1981
xi,272p,[13]p of plates. - Bibliography: p264-266. - Includes index
0-231-05184-0 X.200/36837
2563363

ADLER, GUIDO. Criticism. Music

REILLY, Edward R
Gustav Mahler und Guido Adler. - Wien : Universal Edition, c1978
69p. - (Bibliothek der Internationalen Gustav Mahler Gesellschaft)
3-7024-0127-x X.0439/331(2)
8892881

ADLER, GUIDO. Musicology. Austria
Interpersonal relationships with Mahler, Gustav

REILLY, Edward R.
[Mahler und Guido Adler. English]. Gustav Mahler and Guido Adler. - Cambridge : Cambridge University Press, 1982
viii,163p. - Bibliography: p144-155. - Includes index
0-521-23592-8 X.439/11913
2729970

ADLER, HANS, 1904-1979. South African music

HANS Adler 1904-1979 / compiled by Reuben Musiker. - Johannesburg : University of the Witwatersrand Library, 1980
60p,port : ill
0-85494-621-7 X.435/1115
9066578

ADLER, JACOB. Acting. Theatre
- Biographies

ROSENFELD, Lulla
Bright star of exile. - London : Barrie and Jenkins, 1978
xvi,368p[16]p of plates : ill. - Bibliography: p.359-360. - Includes index
0-214-20590-8 X.981/22391
2127857

ADLER, JEREMY. English drawings

ADLER, Jeremy
The Amsterdam quartet. - London : Alphabox Press, 1976
1v. : of ill
0-904504-07-7 Cup.510.dgb.1
1080245

ADLER, LARRY. Harmonica playing
- Biographies

ADLER, Larry
It ain't necessarily so. - London : Collins, 1984
254p,[16]p of plates : ill. - Includes index
0-00-216036-6 X.950/39095
311208x

ADLER, MORTIMER JEROME. American philosophy
- Biographies

ADLER, Mortimer Jerome
Philosopher at large. - London : Weidenfeld and Nicolson, 1977
xiii,349p,[32]p of plates : ill. - Bibliography: p.330-337. - Includes index
0-297-77281-3 X.520/12006
125233x

ADLER, PAUL. Poetry in German

ABICHT, Ludo
Paul Adler. - Wiesbaden : Humanitas, 1972
viii, 317 p.. - (Studien zur Germanistik). - Bibliography : p. 290-307
3-7997-0202-4 X.510/9389
0849839

ADLER PETERSEN, LENE. Danish drawings

ADLER PETERSEN, Lene
Tag en sten op. - [København] : Borgen, [1977]
317p : all ill
Copy no.441
87-418-4120-4 X.415/3048 Woolwich
2138638

ADLERIAN SYSTEM. Psychology

ROM, Paul
Alfred Adler's individual psychology and its history. - London : Adlerian Society of Great Britain, 1976
[14]p. - (Individual psychology pamphlets : no.1)
0-905860-00-4 X.519/26503
1038206

ADLHART, JAKOB. Austrian sculptures

HAHNL, Adolf
Der Bildhauer Jakob Adlhart. - Salzburg : Müller, [1980]
160p. : chiefly ill.(some col.). - Bibliography and lists of Adlhart's works
3-7013-0596-x L.42/1382
8793220

ADLIB

ASHFORD, J. H. (John Hatfield)
ADLIB. - Maidenhead : Lipman Management Resources Limited.
Vol.1: Design. - 1977
[3]p,39 leaves : ill
0-906168-00-7 2719.gg.26(1)
1402021
- 2nd ed. (revised). - Maidenhead : Lipman Management Resources Limited.
Vol.1: System description. - 1979
[3],79 leaves : ill
0-906168-01-5 X.0622/557(1) Woolwich
1402021
- [Another copy] X.525/3152

ROSS, M. R. (Madeleine Rachel)
ADLIB design. - Maidenhead : Lipman Management Resources, c1982
112p. - Includes index
0-906168-03-1 2702.f.71
1402021

ADLINGTON. Cheshire
Country houses: Adlington Hall

LEGH, Charles
Adlington Hall, Cheshire. - [Derby] : English Life Publications, [1979?]
[27]p : ill(some col)
B79-10705 X.708/21755
1822861

ADLINGTON. Cheshire
Country houses: Adlington Hall, to 1978

HANCOX, Joy
Adlington Hall, Cheshire. - [Salford] : [The author], [1979]
31p : ill. - Includes index
B79-35617 X.709/33511
1968106

ADLINGTON HALL

LEGH, Charles
Adlington Hall, Cheshire. - [Derby] : English Life Publications, [1979?]
[27]p : ill(some col)
B79-10705 X.708/21755
1822861

ADLINGTON HALL
to 1978

HANCOX, Joy
Adlington Hall, Cheshire. - [Salford] : [The author], [1979]
31p : ill. - Includes index
B79-35617 X.709/33511
1968106

ADLINGTON HALL
Musical instruments: Pipe organs, to 1977

LEGH, Charles
The organ at Adlington Hall. - Macclesfield : [The author], [1978]
[8]p,[8]p of plates : ill
r6805909 X.709/52103
2304007

ADLUN. Lebanon
Stone Age antiquities. Archaeology

ADLUN in the Stone Age / edited by Derek A. Roe with contributions by L. Copeland ... [et al.]. - Oxford : B.A.R., 1983
2v.(xxiii,455p) : ill. - (BAR. International series ; 159). - Bibliography: p443-455
0-86054-203-3 X.425/5215
3278069

ADMININSTRATION. Christian church

WATKINS, Peter, 1934-
Here are the people. - London : MacRae, c1984
104p : ill. - Includes index
0-86203-160-5 X.200/44473
316697x

ADMINISTRATION
See also
Bureaucracy
Maladministration
Management
Offences relating to administration of justice
Organisation
Postage stamps issued by the British administration of Palestine
Public administration
Secretariats. Governments
Site administration
Social administration

ADMINISTRATION

CHEVALLIER, Jacques
La science administrative. - Paris : Presses Universitaires de France, 1980
127p. - (Que sais-je? ; 1817)
2-13-036330-x X.629/13480 Woolwich
0036374

SINGER, G.
The administrative waltz, or, Ten commandments for the administrator. - Oxford : Pergamon Press, 1976
xiii,75p : ill
0-08-020985-8 X.519/26007
0701602

ADMINISTRATION
- Dictionaries

NÆSHAGEN, Erik
Norsk-engelsk ordbok for utredningsterminologi og administrasjonsspråk. - Oslo : Universitetsforlaget, [1977]
198p
82-00-05040-8 X.989/39841
1236598
- [Another copy.] X.989/72960

ADMINISTRATION
- Serials

GREENHILL journal of administration. - Vol.1, no.1 ([1973])-. - Achimota : Ghana Institute ofManagement and Public Administration, [1973]-
r6202479 P.611/326
0036374

ADMINISTRATION
to 1975

RAJA, Irfan-ur-Rehman
Administration. - Lahore : Catapult, 1976
xiii,235p. - Bibliography : p.218-225. - Includes index
r6072152 X.809/43431
1568272

ADMINISTRATION
Linguistic aspects

CALVO RAMOS, L.
Introducción al estudio del lenguaje administrativo. - Madrid : Editorial Gredos, 1980
290p. - (Biblioteca románica hispánica. II. Estudios y ensayos ; 309)
84-249-0026-x X.958/14226
8890870

ADMINISTRATION
Theories

HODGKINSON, Christopher
Towards a philosophy of administration. - Oxford : Blackwell, 1978
iii-xiii,244p : ill. - Bibliography: p.228-237. - Includes index
0-631-19110-0 X.529/32777
0036366

ADMINISTRATION. 10-year trusts. United States

HANCOCK, William A.
Saving money through 10-year trusts. - New York ; London : McGraw-Hill, 1979
xi,177p. - Includes index
0-07-025983-6 X.520/14158
1913298

ADMINISTRATION. Accommodation. Patients. Hospitals. Great Britain

BURROUGH, Evan Jerome Ridgway
A scandalous impromptu. - Oxford : The author, 1976
23p
0-9505008-0-1 X.319/16555 Woolwich
0658480

ADMINISTRATION. Adaptation of residences of physically handicapped persons by local authority welfare services. London

KEEBLE, Ursula
Aids and adaptations. - [London] : Bedford Square Press, 1979
320p : ill. - (Occasional papers on social administration ; no.62). - Bibliography: p.315-320
0-7199-0963-5 8299.i.7(62)
1952595

ADMINISTRATION. Adaptation. Residences. Physically handicapped persons. England
By local authorities

PRESCOTT-CLARKE, Patricia
Organising house adaptations for disabled people. - London : H.M.S.O., 1982
iv,46p,[27]p
0-11-751651-1 BS.414/2110
2888734

ADMINISTRATION. Agriculture. Developing countries
Role of ministries of agriculture

HOWELL, John, 19—
Ministries of agriculture and the administration of agricultural development. - London : Overseas Development Institute, 1980
14p : . - (Agricultural administration network discussion paper ; 4)
B81-27056 X.329/14219 Woolwich
2506173

ADMINISTRATION. Agriculture. India (Republic)

ROY, Ramashray
Bureaucracy and development. - New Delhi : Manas Publications, 1975
176 p. : ill. - Includes index
r6010710 X.329/10805 Woolwich
1049798

ADMINISTRATION. Albany Prison
Protest by Provisional IRA prisoners, 1976

AMNESTY INTERNATIONAL
A dossier concerning treatment of prisoners during and after a disturbance at Albany Prison in September 1976. - [London] : [Amnesty International] ; [London] : [Howard League for Penal Reform] : [N.C.C.L.], [1977]
[2],23p
0-900058-55-2 X.102/2079
1113240

ADMINISTRATION. Allowances & rebates. Rents. Residences. England

LEGG, Charles
The administration of the rent rebate and rent allowance schemes. - [London] : Department of the Environment : Welsh Office, 1976
[5],198p : ill. - Bibliography: p.198
B77-30012 BS.414/238
1254693

ADMINISTRATION. Ancient Egypt
Officials. Economic status, B.C.2686-B.C.2160. Archaeological sources: Tombs

KANAWATI, Naguib
The Egyptian administration in the Old Kingdom. - Warminster : Aris and Phillips, 1977
[2],iii,165p : ill. - Bibliography: p.157-160. - Includes index
0-85668-102-4 X.802/10560
1407694

ADMINISTRATION. Ancient Roman provinces. Spain

LE ROUX, Patrick
L'armée romaine et l'organisation des provinces ibériques. - [Talence] : Centre Pierre Paris, 1982
493p,plates : ill. - (Publications du Centre Pierre Paris ; vol.8) (Collection de la maison des pays ibériques ; vol.9). - Includes bibliography
2-7018-0002-1 X.800/41548
8596174

ADMINISTRATION. Ancient Roman provinces. Spain
ca A.D.70

ØRSTED, Peter
Vespasian og Spanien. - København : Museum Tusculanum, 1977
117p,[6]p of plates, [2] fold leaves of plates : ill. - (Opuscula graecolatina ; nr 14)
r6054538 X.0900/517(14)
1430718

ADMINISTRATION. Ancient Rome
to 476

TEITLER, H. C.
Notarii en exceptores. - Utrecht : HES, 1983
ix,379p. - Includes bibliography
90-619-4154-7 X.529/60466
9065121

ADMINISTRATION. Ancient Rome
Officials. Lists: Notitia dignitatum - Conference proceedings

ASPECTS of the 'Notitia dignitatum' / by J.C. Mann ... [et al.] ; edited by R. Goodburn and P. Bartholomew. - Oxford : British Archaeological Reports, 1976
xix,224p(2 fold) : ill(incl 3 col). - (British archaeological reports. supplementary series ; 15)
0-904531-58-9 2702.ff.7
1171062

ADMINISTRATION. Ancient Rome
Officials. Lists: Notitia dignitatum. Illustrated manuscripts. Illustrations

BERGER, Pamela C.
The insignia of the Notitia dignitatum. - New York ; London : Garland, 1981
xxvi,300p,[126]p of plates : ill. - (Outstanding dissertations in the fine arts). - Bibliography: p289-300
0-8240-3927-0 X.421/25220
3084272

ADMINISTRATION. Ancient Rome. Aerarium Militare
By senators, 6-ca 250

CORBIER, Mireille
L'aerarium Saturni et l'aerarium militare. - Rome : École Française de Rome, 1974
792 p.. - (Collections de l'École française de Rome ; 24). - Bibliography: p. 721-732
r6001170 Ac.5233(a.) [no. 24.]
0929832

ADMINISTRATION. Ancient Rome. Aerarium Saturni
By senators, B.C.28-ca A.D.360

CORBIER, Mireille
L'aerarium Saturni et l'aerarium militare. - Rome : École Française de Rome, 1974
792 p.. - (Collections de l'École française de Rome ; 24). - Bibliography: p. 721-732
r6001170 Ac.5233(a.) [no. 24.]
0929824

ADMINISTRATION. Antwerp. Belgium
1477-1830

WELLS, Guy Edward
Antwerp and the government of Philip II 1555-1567. - [Wellington, Ohio] : [G.E. Wells], [1982]
2v.(597leaves). - Bibliography: p.555-597
r6806266 X.805/3393
8911010

ADMINISTRATION. Approval plans. Selection. Acquisition. Books. Stock. Libraries. Universities. United States

CARGILL, Jennifer S.
Practical approval plan management. - Phoenix : Oryx Press ; London : Distributed by Mansell Information Publishing, 1979
viii,95p. - Bibliography: p.89-91. - Includes index
0-912700-52-1 X.520/21081
2130319

ADMINISTRATION. Archives

ARCHIVAL choices / edited by Nancy E. Peace. - Lexington, Mass. : Lexington Books, 1984
xi,163p : ill. - Bibliography: p149-153. - Includes index
0-669-05354-6 2719.e.960
2186446

ARKISTONHOIDON opas / [Valtionarkisto]. - Helsinki : [Valtionarkisto], 1978
207p : ill. - (Valtionarkiston julkaisuja ; 10). - Includes bibliographies
951-463-380-6 S.D.76/16
1084305

COOK, Michael, 1931-
Archives administration. - Folkestone : Dawson, 1977
x,258p : ill. - Bibliography: p.228-247. - Includes index
0-7129-0749-1 2719.bb.2
1084305

JENKINSON, Sir Hilary
Selected writings of Sir Hilary Jenkinson. - Gloucester : Alan Sutton, 1980
380p : ill
0-904387-52-6 X629/18284
2186446

NORDAHL-HANSEN, S. J.
Arkivhandboken. - [Oslo] : Kommunalforlaget, c1979
156p : ill
82-7242-038-4 X.529/55284
8964149

PAPRITZ, Johannes
Archivwissenschaft. - Marburg : Archivschule Marburg, Institut für Archivwissenschaft, 1976
3T. in 4 Bde
r6041527 2719.b.42
1084305

RIGSARKIVET. 2. Afdeling
Kassation og bevaring af arkivalier. - [København] : Rigsarkivets 2. afdeling, 1979
83p
r6112569 X.525/3392
1084305

TAYLOR, Hugh A.
The arrangement and description of archival materials. - München ; London : Saur, 1980
181p. - (ICA handbooks series ; v.2). - Bibliography: p147-153. - Includes index
0-89664-260-7 X.421/23824
1084305

ADMINISTRATION. Archives
- Bibliographies

DUCHEIN, Michel
Basic international bibliography of archive administration. - New York ; London : Saur, 1978
250p. - (Archivum ; v.25). - Includes index
0-89664-005-1 2719.e.495
2874849

ADMINISTRATION. Archives
- Dictionaries

RIGSARKIVET. 2. Afdeling
Ordliste til brug for medarbejdere ved Rigsarkivets 2. afdeling. - [Copenhagen] : Rigsarkivets 2. afdeling, 1977
273p
r6060212 S.B.286/46
1537911

ADMINISTRATION. Archives
Applications of automatic data processing systems - Conference proceedings

INTERNATIONAL SEMINAR ON AUTOMATIC DATA PROCESSING IN ARCHIVES (1974 : Chelwood Gate)
Proceedings of an International Seminar on Automatic Data Processing in Archives. - London : H.M.S.O., 1975
[2],viii,349p : ill
0-11-440079-2 BS.33/123
054051x

ADMINISTRATION. Archives
Applications of automatic data processing systems - Serials

ADPA, automation-archives-informatique. - Vol.1, no.1- ; Aug. = aout 1972-. - [London] : [International Council on Archives], [1972]-
r6071614 P.703/577
1378287

ADMINISTRATION. Archives
Applications of digital computer systems

BARTLE, Rachel
Computer applications in archives. - [Liverpool] : Archives Unit University of Liverpool, 1983
i,58p : ill. - (British Library Research and Development Department report ; no.5749)
0-907156-01-0 X.622/16887
2019590

COOK, Michael, 1931-
Archives and the computer. - London : Butterworths, 1980
152p : ill. - Bibliography: p137-145. - Includes index
0-408-10734-0 X.622/9512 Woolwich
2019590

ADMINISTRATION. Archives. Companies

BUSINESS archives / Committee on Business Archives of the International Council on Archives. - München ; London : Saur, 1983
167p. - Bibliography: p165-167
3-598-10516-9 X.520/33829
3130177

ADMINISTRATION. Archives. Developing countries
Planning - Conference proceedings

GENERAL CONFERENCE ON THE PLANNING OF ARCHIVAL DEVELOPMENT IN THE THIRD WORLD (1975 : Dakar)
Proceedings of the General Conference on the Planning of Archival Development in the Third World (Dakar, 28-31 January 1975) = Actes de la Conférence generale sur la planification du développement des Archives dans le Tiers-Monde (Dakar, 28-31 janvier 1975). - München [i.e. Munich] : Verlag Dokumentation ; [London] : [Distributed by Mansell Information Publishing], 1976
117p. - ('Archivum' special volume = 'Archivum' cahier hors série ; no.1)
B77-11033 PP.3554.vcd(2)
1074709

ADMINISTRATION. Archives. France
to date

Les ARCHIVES des administrations / Ministère de l'éducation nationale, Direction de l'administration générale et des personnels administratifs, Service de la formation administrative. - Paris : Ministère de l'éducation nationale, [1982]
97p : ill
r6299172 S.E.400/1164
8976546

ADMINISTRATION. Archives. France
Samaran, Charles - Biographies

SAMARAN, Charles
Enfance et jeunesse d'un centenaire. - Paris : Société d'Édition "Les Belles Lettres", 1980
99p,15leaves of plates,18p of plates : ill
2-251-39000-6 X.429/13138 Woolwich
8793336

ADMINISTRATION. Archives. Great Britain
Bond, Maurice - Biographies

A TRIBUTE to Maurice Bond. - Windsor : Clewer Parochial Church Council, 1984
14p
0-9509490-0-0 X.950/38037
3221091

ADMINISTRATION. Archives. Italy

ITALY
Manuale di legislazione amministrativa sui beni culturali archivistici. - Roma : Centro di ricerca, 1978
xix,1005p. - (Fonti e studi di storia legislazione e tecnica degli archivi moderni ; 15)
r6508851 X.0700/435(15)
8976538

ADMINISTRATION. Archives. Italy
to date

ANTONIELLA, Augusto
L'archivio comunale postunitario. - Firenze : Giunta regionale toscana, 1979
vi,148p. - (Archivi e biblioteche ; 1)
r6502134 X.520/25183
8754411

ADMINISTRATION. Archives. Länder. West Germany

TAUSENDPFUND, Alfred
Die Archivorganisation in den Ländern der Bundesrepublik Deutschland. - Hannover : Schroedel, 1977
64 leaves,[1]fold leaf of plates. - (Akademie für Raumforschung und Landesplanung. Beiträge ; Bd 18) (Behördliche Raumorganisation seit 1800. Grundstudie ; 5). - Includes bibliographies
3-507-91463-8 X:0512/475(18)
1543423

ADMINISTRATION. Archives. Scandinavia
Applications of data processing systems

NORDISKA ARKIVDAGARNA (1974 : Uppsala)
Arkiven i dataaldern. - [Stockholm] : [s.n.], 1975
126 p. : ill
r6003606 X.519/20236
0923435

ADMINISTRATION. Archives. Spain
- Conference proceedings

CONGRESO NACIONAL DE ARCHIVOS (4th : 1972 : Barcelona)
Ponencias, comunicaciones y crónica. - Madrid : Asociación Nacional de Bibliotecarios, Archiveros y Arqueólogos, 1975
305p.,4p. of plates. - (Biblioteca profesional de Anaba. congresos)
84-500-7021-x 2719.b.28
122767X

ADMINISTRATION. Archives. United States

NORTON, Margaret Cross
Norton on archives. - Carbondale : Southern Illinois University Press ; London : Feffer and Simons, 1975
xxi,288p. - Bibliography: p.269-271. - Includes index
0-8093-0738-3 X.800/25300
0561819

ADMINISTRATION. Archives. United States
to 1982

BERNER, Richard C.
Archival theory and practice in the United States. - Seattle ; London : University of Washington Press, c1983
xv,219p. - Bibliography: p199-211. - Includes index
0-295-95992-4 X.800/38000
3074129

ADMINISTRATION. Armagh Prison
Protest by Provisional IRA women prisoners, 1979-1980

D'ARCY, Margaretta
Tell them everything. - London : Pluto, 1981
127p
0-86104-349-9 X.808/36138
2716283

ADMINISTRATION. Arts
- Bibliographies

CITY UNIVERSITY. Library
Arts administration. - 2nd ed. - London : The Library, 1980
102p. - Includes index
0-904683-04-4 2719e.909
192236x

ADMINISTRATION. Arts. Great Britain

PICK, John, 1936-
Arts administration. - London : Spon, 1980
ix,194p : ill. - Includes bibliographies and index
0-419-11540-4 X.529/38663
1958240

ADMINISTRATION. Arts. Great Britain
- Serials

The JOURNAL of arts policy and management. - Vol.1, no.1 (Feb. 1984)-. - London : Department of Arts Administration, City University, 1984-
v. : ill
B85-11082 P.413/297
3231348

ADMINISTRATION. Arts. Great Britain
Organisations: Society of Professional Arts Administrators - Serials

SOCIETY OF PROFESSIONAL ARTS ADMINISTRATORS
SPAA News. - [No.1]- ; [1975]-. - London : The Society, [1975]-
B75-27912 P.423/401
0516724

ADMINISTRATION. Arts. Sweden
Schein, Harry - Biographies

SCHEIN, Harry
Schein. - [Stockholm] : Bonniers, 1980
529p
91-00-44843-5 X.421/24001
891852x

ADMINISTRATION. Audiovisual materials collections. Libraries

CHIBNALL, Bernard
The organisation of media. - London : Bingley [etc.], 1976
80p. - Includes index
0-85157-212-x 2719.aa.21
0659479

HICKS, Warren B. (Warren Braukman)
Managing multimedia libraries. - New York ; London : Bowker, 1977
xv,264p : ill. - Bibliography: p.223-243. - Includes index
0-8352-0628-9 2719.ff.78
0659479

ADMINISTRATION. Audiovisual materials. Teaching aids. United States

AMERICAN ASSOCIATION OF SCHOOL LIBRARIANS
Media programs. - Chicago : American Library Association ; Washington, D.C. : Association for Educational Communications and Technology, 1975
viii, 128p. - Includes index
0-8389-3159-6 X.519/26806
1098640

ADMINISTRATION. Automation. Cataloguing. Stock. Libraries. United States
- Conference proceedings

ACM CONFERENCE ON MANAGEMENT ISSUES IN AUTOMATED CATALOGING
(1977 : Chicago)
Requiem for the card catalog. - London : Aldwych Press [etc.], 1979
xvi,205p : ill. - Includes index
0-86172-002-4 2719.dd.4
1948814

ADMINISTRATION. Baptist Union of Great Britain and Ireland

BACON, Fred
Church administration. - Bristol : Bristol and District Association of Baptist Churches, c1981
187p : ill. - Bibliography: p181-182. - Includes index
0-9507449-0-5 X.208/6437
2497905

ADMINISTRATION. Barcelona. Spain
Officials, 1090-ca 1500

BOSCÀ, Joan Francesc
Memorial històric. - Barcelona : Associació de Bibliòfils de Barcelona, 1977
169p : facsim
84-400-3132-7 Cup.900.ww.18
2430282

ADMINISTRATION. Belize
By Great Britain. Disputes between Great Britain & Guatemala, 1821-1974

SKINNER-KLEE, Jorge
Intervención del licenciado Jorge Skinner-Klee ... ante la IV comisión de la Organización de Naciones Unidas. - [Guatemala] : Ministerio de Relaciones Exteriores. Departamento de Relaciones Públicas, [1975]
9 p.
r6003419 LAS.E.118/15
0919284

ADMINISTRATION. Belize
By Great Britain. Disputes between Great Britain & Guatemala, 1977

GUATEMALA. Ministerio de Relaciones Exteriores
El caso de Belice, después de Washington. - Guatemala : el Ministerio, 1977
22 leaves
r6025347 LAS.E.170/3
1240374

ADMINISTRATION. Bessarabia
By Russia, 1774-1828

JEWSBURY, George Frederick
The Russian annexation of Bessarabia, 1774-1828. - Boulder : East European Quarterly ; New York ; London : Distributed by Columbia University Press, 1976
[6],202p. - (East European monographs ; no.15). - Bibliography: p.185-193. - Includes index
0-914710-09-5 X.809/41101
0696234

ADMINISTRATION. Brazil. Departmento Nacional de Obras Contra as Secas
to date

BRAZIL. Departamento Nacional de Obras Contra as Secas
Pareceres jurídicos. - Fortaleza : DNOCS, Divisão Jurídica, 1978
306p
r6701750 X.529/43733
8774846

ADMINISTRATION. Bremen (Archbishopric) Minden & Verden (Duchy)
By Sweden, 1630-1720

MODÉER, Kjell Å
Gerichtsbarkeiten der schwedischen Krone im deutschen Reichsterritorium. - Stockholm : Nordiska Bokhandeln, in Distribution, 1975-
1-. - (Skrifter utgivna av Institutet för rättshistorisk forskning. ser. 1 ; bd 24)
r6001306 Ac.2140b/2a [bd 24]
0941174

ADMINISTRATION. Broadcasting services. France
to 1974

THOMAS, Ruth, 1944-
Broadcasting and democracy in France. - [Bradford] : Bradford University Press ; London : Crosby Lockwood Staples, 1976
xi,211p. - Bibliography: p.199-204. - Includes index
0-258-97006-5 X.989/50604
0620718

ADMINISTRATION. Broadcasting services. France
1972-1975

CHEVALLIER, Jacques
La radio télévision française entre deux réformes. - Paris : L.G.D.J., 1975 [1976]
342 p.. - Bibliography : p. 335-338
2-275-01393-8 X.908/33917
0938912

ADMINISTRATION. Broadcasting services. West Germany
1945-1974

WILLIAMS, Arthur
Broadcasting and democracy in West Germany. - [Bradford] : Bradford University Press ; London : Crosby Lockwood Staples, 1976
xx,198p. - Bibliography: p.187-189. - Includes index
0-258-96996-2 X.989/50603
0620726

ADMINISTRATION. Building industries. Soviet Union 1945–

ABRAMOVICH, A. M. (Aleksandr Mikhaĭlovich)
Gosudarstvennoe upravlenie stroitel'stvom v SSSR. - Minsk : "Nauka i tekhnika.", 1982
269 p.. - Includes bibliographical references
r6464901 X.629/23285 959 1195
909122X

ADMINISTRATION. Buildings of historical importance. Great Britain
Policies of government

GREAT BRITAIN. Department of the Environment
The way forward. - London : H.M.S.O., 1982
41p
0-11-751624-4 BS.414/2087
2774178

ADMINISTRATION. Buildings. Sweden

STATENS RAD FÖR BYGGNADSFORSKNING
Forvaltning av byggnader och anlaggningar. - Stockholm : Statens rad för byggnadsforskning, 1977
vii,53,[20]p : ill. - (T16. 1977)
91-540-2676-8 X.510/10466
1420941

ADMINISTRATION. Byzantine Empire
Officials

GUILLAND, Rodolphe
Titres et fonctions de l'Empire byzantin. - London : Variorum, 1976
528p in various pagings. - (Collected studies series ; CS50)
0-902089-94-3 X.800/25618
0559466

ADMINISTRATION. Camping & caravanning. England
- For local authorities

ENGLISH TOURIST BOARD
The management of touring caravans and camping. - [London] : The Board, [1979]
iv,50,[2]p,[3]leaves of plates. - (Planning advisory note). - Bibliography: p.49-50
B79-31263 BS.411/271
1950843

ADMINISTRATION. Care. Long-stay inmates. Residential institutions for incurably sick persons. United States

LONG-TERM care administration / edited by Samuel Levey and N. Paul Loomba. - New York : Spectrum ; New York ; London : Distributed by Wiley.
Vol.1. - 1977
[12],300,8p : ill. - (Health systems management ; vol.10). - Includes bibliographies and index
0-470-99111-9 X.520/13202
1494171
- New York : Spectrum Publications ; New York ; London : Distributed by Wiley. - (Health systems management ; vol.10)
Vol.2. - 1977
[14]p,p301-651 : ill. - Bibliography: p.619-644. - Includes index
0-470-99110-0 X.0320/232(2) Woolwich
1494171

ADMINISTRATION. Catholic Church
800-1400

SIEBEN, Hermann Josef
Die Konzilsidee des lateinischen Mittelalters (847-1378). - Paderborn : Schöningh, 1984
xx,484p. - (Konziliengeschichte. Reihe B, Untersuchungen). - Includes bibliography
3-506-74722-3 X.0200/435b(2)
9149368

ADMINISTRATION. Catholic Church. Diocese of Lucca
ca 1050-ca 1350

OSHEIM, Duane J.
An Italian lordship. - Berkeley ; London : University of California Press, 1977
xviii,211p. - (Publications / University of California at Los Angeles. Center for Medieval and Renaissance Studies ; 11). - Includes index
0-520-03005-2 X.200/31491
1569562

ADMINISTRATION. Catholic Church. England
1100-1135

BRETT, M. (Martin)
The English Church under Henry I. - London : Oxford University Press, 1975
xii,282p. - (Oxford historical monographs). - Bibliography: p.247-266. - Includes index
0-19-821861-3 X.200/20154
052641x

ADMINISTRATION. Central Europe
1815- - Conference proceedings

ENTWICKLUNGSFRAGEN der Verwaltung in Mitteleuropa / Redakteur Andor Csizmadia. - Pécs, 1972
298p. - (Studia iuridica auctoritate Universitatis Pécs publicata ; 80)
r6455174 Ac.818/7[80]
9160566

ADMINISTRATION. Charities. England
Legal aspects

LONGLEY, A. R. (Adrian R)
Charity trustees' guide. - London : Bedford Square Press, 1979
viii,32p. - Bibliography: p.26-27. - Includes index
0-7199-0959-7 X.519/42875
0192619
Charity trustees' guide. - 2nd ed.. - London : Bedford Square Press, 1982
viii,51p. - Bibliography: p33-36. - Includes index
0-7199-1062-5 X.529/54004
0192619

ADMINISTRATION. China
Role of European advisers, to date

SCHMIDT, Vera
Aufgabe und Einfluss der europäischen Berater in China. - Wiesbaden : Harrassowitz, 1984
viii,178p. - (Veroffentlichungen des Ostasien-Instituts der Ruhr-Universität Bochum ; Bd.34). - Includes bibliography
3-447-02483-6 X.0700/678(34)
9166041

ADMINISTRATION. Christian church

HOOKER, Richard, 1553 or 4-1600
[Of the lawes of ecclesiasticall politie. Selections]. Of the lawes of ecclesiasticall politie. - Abridged ed.. - London : Sidgwick and Jackson, 1975
413p
0-283-97970-4 X.200/30525
0704032

ADMINISTRATION. Christian church. England
to 1534

RODES, Robert E.
This house I have built. - Notre Dame ; London : University of Notre Dame Press.
Ecclesiastical administration in medieval England. - 1977
xvi,287p. - Bibliography: p.205-211. - Includes index
0-268-00903-1 X.200/31094
1245295

ADMINISTRATION. Christian church. Great Britain
- Serials

The LOOKOUT magazine. - No.1- ; 1978-. - Chipstead : [Lookout Magazine], [1978]- : ill
Holdings: no.7 ; June 1980, etc
B80-38778 P.103/583
2284057

ADMINISTRATION. Christian voluntary organisations. Great Britain
- Serials

The LOOKOUT magazine. - No.1- ; 1978-. - Chipstead : [Lookout Magazine], [1978]- : ill
Holdings: no.7 ; June 1980, etc
B80-38778 P.103/583
2284049

ADMINISTRATION. Church of England

ADAIR, John, 1934-
The becoming Church. - London : S.P.C.K., 1977
x,244p : ill. - Bibliography: p.237-238. - Includes index
0-281-02964-4 X.208/5105
0098965

ADMINISTRATION. Church of England
- For parochial church councils

BOULT, Anthony
Learning about the PCC. - London : Mowbray, 1984
31p. - (Enquirer's library)
0-264-66974-6 X.208/10056
2546051

MACMORRAN, Kenneth M. (Kenneth Mead)
A handbook for churchwardens and parochial church councillors. - New ed.. - London : Mowbray, 1980
178p. - Includes index
0-264-66720-4 X.208/6658
2546051

PITCHFORD, John
An ABC for the PCC. - Completely rev. ed. - London : Mowbray, 1985
x,179p. - (Mowbray's popular Christian paperbacks)
0-264-67050-7 X.208/10885
2546051

ADMINISTRATION. Church of England
- For parochial church councils - Directories

MACMORRAN, Kenneth M. (Kenneth Mead)
A handbook for churchwardens and parochial church councillors. - New ed.. - London : Mowbray, 1983
ix,178p. - (Mowbray parish handbooks). - Includes index
0-264-67004-3 X.208/9900
2546051

ADMINISTRATION. Church of England
- Serials

[RESOURCE paper (Administry)]. Resource paper / Administry. - No.82:1-. - St. Albans : Administry, 1982-
v. : ill
B83-22053 P.103/625
289520x

ADMINISTRATION. Church of England
to 1642

CONTINUITY and change / edited by Rosemary O'Day & Felicity Heal. - [Leicester] : Leicester University Press, 1976
303p. - Includes index
0-7185-1138-7 X.200/20261
0596620

ADMINISTRATION. Church of England
ca 1646

SALTMARSH, John, d.1647
Perfume against the sulphurous stinke of the snuffe of the light for smoak, called Novello-Mastix .... - St Peter Port : Toucan Press, 1977
8p
B78-01964 X.709/31806
1284622

ADMINISTRATION. Church of England. Diocese of Lincoln
1521-1547

BOWKER, Margaret
The Henrician reformation. - Cambridge : Cambridge University Press, 1981
xx,229p : ill. - Bibliography: p188-193. - Includes index
0-521-23639-8 X.200/37111
2581833

ADMINISTRATION. Churches. Church of Scotland

HERRON, Andrew
A guide to congregational affairs. - Edinburgh : St Andrew Press, 1978
viii,132p
0-7152-0390-8 X.108/17033
151749x

ADMINISTRATION. Civil rights agencies. United States
- Conference proceedings

REGIONAL CIVIL RIGHTS CONFERENCE (1st : 1974 : St. Louis)
Towards a more cooperative and productive relationship among civil rights agencies and officials. - Washington : U.S. Commission on Civil Rights, 1975
94 p.
r6000232 AS.413/418
0904481

ADMINISTRATION. Collection. Fees. Radio licences & television licences. France

FRANCE. Ministère de l'Économie et des Finances. Service de la Redevance
La redevance radio-télévision. - [Paris] : le Ministère, 1975
iv,139p
r6038981 S.E.129/127 Woolwich
0397733

ADMINISTRATION. Collections. Audiovisual materials
- Bibliographies

MANN, Margaret
Archival problems of audiovisual materials. - [Sheffield] : Department of Information Studies, University of Sheffield, 1982
ii,48p. - (BL R & D report ; 5746). - Includes index
B83-34507 X.622/17911
2939371

ADMINISTRATION. Collections. Films. Stock. Libraries

REHRAUER, George
The film user's handbook. - New York ; London : Bowker, 1975
xiv,301p : ill. - Bibliography: p.227-263. - Includes index
0-8352-0659-9 X.981/10320
0601152

ADMINISTRATION. Collections. Geological specimens. Stock. Museums. Great Britain

DOUGHTY, P. S.
The state and status of geology in United Kingdom museums. - London : Geological Society, [1982]
118p. - (Miscellaneous paper ; no.13)
B83-18038 X.0319/343
2879573

ADMINISTRATION. Collections. Photographs. Stock. Museum of English Rural Life

MUSEUM OF ENGLISH RURAL LIFE
Museum procedure, photographs. - Reading : [The Museum], 1978
14,[2]p
0-7049-0700-3 2719.gg.80
1871889

ADMINISTRATION. Collections. Sound tape recordings. Orkney. Scotland

MURRAY, J. F.
The Orkney Sound Archive Project. - Glasgow : S.C.E.T., 1982
87p. - (Occasional working paper / Scottish Council for Educational Technology ; 10)
0-86011-059-1 X.629/19112
2819961

ADMINISTRATION. Colleges. England
- Conference proceedings

ASSOCIATION OF COLLEGES FOR FURTHER AND HIGHER EDUCATION.
General Meeting (1980 : Institution of Electrical Engineers)
The government of colleges. - Sheffield : [The Association], [1980]
19p
B83-00796 X.529/50643
280669x

ADMINISTRATION. Colleges of education. India (Republic)
Participation of students

SRIVASTAVA, Ramesh Chandra
Student participation in administration. - New Delhi : Sterling Publishers, 1975
ix, 168 [i.e. 164] p., (4 fold). - Bibliography : p. 120-121. - Includes index
r6031600 X.529/31355
1083910

ADMINISTRATION. Community land scheme. Scotland

PLANNING EXCHANGE
The community land scheme in Scotland. - Glasgow : Planning Exchange, 1976
[4],33p. - (Newsheet ; 5)
0-905011-11-2 X.313/1348 Woolwich
0785318

ADMINISTRATION. Community recreation facilities. United States

MURPHY, James Frederick
Delivery of community leisure services. - Philadelphia : Lea and Febiger ; London : Kimpton, 1977
vii,219p : ill. - (Health, human movement and leisure studies). - Includes index
0-8121-0575-3 X.510/15636
133476x

ADMINISTRATION. Complaints by public about social services. Great Britain
Procedure

PERSONAL SOCIAL SERVICES COUNCIL
Complaints procedures in the personal social services. - London : The Council, 1976
[2],17,[1]p
0-905250-03-6 BS.439/4
1059491

ADMINISTRATION. Complaints by public against local government. England
Organisations: Commission for Local Administration in England

YOUR Local Ombudsman / [prepared by the Commission for Local Administration]. - London : The Commission, [1982?]
8p
B83-00631 X.529/50736
2805553

ADMINISTRATION. Complaints by public against local government. England
Organisations: Commission for Local Administration in England. Operation, to 1977

LEWIS, Norman, 1940-
The Commission for Local Administration. - London : Royal Institute of Public Administration, 1978
68p. - (Studies / Royal Institute of Public Administration)
0-900628-18-9 X.709/32308
1492950

ADMINISTRATION. Complaints by public against local government. England
Organisations: Commission for Local Administration in England. Operation, to 1979

JUSTICE
The local Ombudsmen. - London : JUSTICE, 1980
160p
0-9505247-9-4 X.200/39636
2842645

ADMINISTRATION. Complaints by public against police. England
Procedure

GREAT BRITAIN. Home Office
The establishment of an independent element in the investigation of camplaints against the police. - London : H.M.S.O., [1981]
ii,12p. - (Cmnd. ; 8193)
0-10-181930-7 BS.18/689
2497751

Police and public. - [London] : H.M.S.O., 1977
Folder ([4]p)
B78-13418 BS.18/575
1409514

Police complaints and discipline procedures. - London : H.M.S.O., [1983]
ii,23p. - (Cmnd. ; 9072)
0-10-190720-6 BS.18/798
0081108

ADMINISTRATION. Complaints by public against police. England
Procedure - Translations

GREAT BRITAIN. Home Office
Yr heddlu a'r cyhoedd = Police and public. - [London] : H.M.S.O., 1977
Folder ([6]p)
B78-13419 BS.18/576
1409522

ADMINISTRATION. Complaints by public against police. Great Britain

HEWITT, Patricia
A fair cop. - London : National Council for Civil Liberties, 1982
44p
0-946088-01-2 X.208/8548
2914832

ADMINISTRATION. Complaints by public against police. Great Britain
Great Britain. Parliament. House of Commons. Home Affairs Committee. Fourth report ... session 1981-82

GREAT BRITAIN
The Government reply to the Fourth report from the Home Affairs Committee session 1981-82 HC 98-1. - London : H.M.S.O., [1982]
19p. - (Cmnd. ; 8681)
0-10-186810-3 BS.18/760
2875160

ADMINISTRATION. Complaints by public against police. Great Britain
Organisations: Police Complaints Board

POLICE COMPLAINTS BOARD
Account ... together with the report of the Comptroller and Auditor General thereon. - London : H.M.S.O. 1976-1978 : 8 December-31 March. - [1979]
Folder(3p). - ([1978-79 H.C.]175)
0-10-217579-9 BS.18/631
1569066

POLICE COMPLAINTS BOARD
Report of the Police Complaints Board ... - London : H.M.S.O.
1977. - [1978]
iv,20p. - ([1977-78 H.C.]359)
Subsequent volumes, when received, will be placed at this shelfmark
0-10-235978-4 BS.18/585
1569066

ADMINISTRATION. Complaints by public against police. Northern Ireland
Organisations: Police Complaints Board for Northern Ireland - Serials

POLICE COMPLAINTS BOARD FOR NORTHERN IRELAND
Annual report. - 1977-. - Belfast : H.M.S.O
0-10-244978-3 CS.A.210/58
1569074

ADMINISTRATION. Complaints by public against police. Northern Ireland
Organisations: Police Complaints Board for Northern Ireland - Statistics - Serials

POLICE COMPLAINTS BOARD FOR NORTHERN IRELAND
Statement of accounts for the period 15 June to 31 March, together with the report of the Comptroller and Auditor General thereon .... - 1977-1978-. - Belfast : H.M.S.O, [1979]-
0-10-225479-6 CS.A.210/108
1914197

ADMINISTRATION. Complaints by public against siting & construction of proposed nuclear power stations. United States
Procedure. Legal aspects

COOK, Constance Ewing
Nuclear power and legal advocacy. - Lexington, Mass. Heath : [Farnborough, Hants.] : [Distributed by] Gower Press, 1980
xvii,157p. - Bibliography: p.137-146. - Includes index
0-669-03441-x X.520/22258
2300281

ADMINISTRATION. Complaints by public in Northern Ireland against Royal Ulster Constabulary
Procedure

The HANDLING of complaints against the police, report of the Working Party for Northern Ireland / presented to Parliament by the Secretary of State for Northern Ireland. - Belfast : H.M.S.O., [1976]
[1],iii,56p. - (Cmnd.6475)
0-10-164750-6 BS.186/29
0359688

ADMINISTRATION. Concentration camps. Poland
Stangl, Franz

SERENY, Gitta
Into that darkness. - London : Pan Books, 1977
[2],380p,[8]p of plates : ill. - (Picador). - Bibliography: p.369-372. - Includes index
0-330-25016-7 X.108/16587
0170984

ADMINISTRATION. Contracts. Building construction. Great Britain

AQUA GROUP
Contract administration for architects and quantity surveyors. - 4th ed. - London : Crosby Lockwood Staples, 1979
[1],vii,87p : ill. - Includes index
0-258-97139-8 X.100/18523
0330051

CONTRACT administration / The Aqua Group. - 5th ed. with supplement on the JCT 1980 form. - London : Granada, 1981
121p : ill. - Includes index
0-246-11555-6 X.520/24058
0330051

POWELL-SMITH, Vincent
Contract documentation for contractors. - London : Collins, 1985
x,308p : ill. - Bibliography: p305. - Includes index
0-00-383018-7 X.622/22876
0330051

ADMINISTRATION. Contracts. Civil engineering. Great Britain

ATKINSON, A. V.
Civil engineering contract administration. - London : Hutchinson, 1985
[128]p : ill. - Includes index
0-09-159981-4 X.520/37296
3156303

ADMINISTRATION. Contracts. Data processing. United States

BRANDON, Dick H.
Data processing contracts. - New York ; London : Van Nostrand Reinhold, 1976
xxii,465p : ill. - Includes index
0-442-21031-0 X.622/5569 Woolwich
0697656

Data processing contracts. - 2nd ed.. - New York ; London : Van Nostrand Reinhold, c1984
xxvii,496p : 1ill. - Includes index
0-442-21034-5 X.525/9210
0697656

ADMINISTRATION. Contracts. Installation. Computer systems. Great Britain
- For management

NATIONAL COMPUTING CENTRE
Computer installation contracts. - Manchester : NCC Publications ; [Letchworth] : [Distributed by Dent], 1977
ix,25p. - (Computerguide ; 12)
0-85012-157-4 X.619/18081 Woolwich
1280139

ADMINISTRATION. Contracts. Small works. Building construction. Great Britain

BOWYER, Jack
Small works contract documentation and how to administer it. - London : Architectural Press, 1976
viii,95p
0-85139-558-9 X.203/2083
1032747

ADMINISTRATION. Conveyancing. Real property. England
By local authorities

MANAGEMENT of conveyancing in local government / Department of the Environment Audit Inspectorate. - London : H.M.S.O., 1982
74p in various pagings
0-11-751973-1 X.525/7842
2945908

ADMINISTRATION. Correctional institutions. United States

DUFFEE, David
Correctional management. - Englewood Cliffs ; London : Prentice-Hall, 1980
viii,407p. - (Prentice-Hall series in criminal justice). - Bibliography: p. 381-398. - Includes index
0-13-178400-5 X.520/21698
2110547

ADMINISTRATION. Criminal law. United States
to date

CRITICAL issues in criminal justice / edited by R.G. Iacovetta and Dae H. Chang. - Durham, N.C. : Carolina Academic Press, 1979
xxiii,618p
0-89089-139-7 X.200/45593
9134255

ADMINISTRATION. Cultural institutions. Hungary
Law

HUNGARY
[Laws, statutes, etc]. A kulturális igazgatás kézikönyve. - Budapest : Közgazdasági és Jogi könyvkiadó, 1977
1319p
963-220-410-7 X.200/11381
1472666

ADMINISTRATION. Cultural institutions. Hungary
Law, 1946-

BULGARIA
Sbornik ot normativni aktove za upravlenie na kulturata. - Sofiia : Nauka i izkustvo, 1981
381p
r6427677 X.529/51150
8893454

ADMINISTRATION. Customs duties. China
1868-1907

HART, Sir Robert, bart
The IG in Peking. - Cambridge, Mass. ; London : Belknap Press of Harvard University Press, 1975
2v.(xxix,1625p) : ill. - Includes index
0-674-44320-9 X.802/10557
1406132

ADMINISTRATION. Customs duties. China
Organisations: Chinese Maritime Customs, to 1950

HALL, B. Foster
The Chinese Maritime Customs. - Greenwich : National Maritime Museum, 1977
[2],iv,47p : ill. - (Maritime monographs and reports ; no.26). - Bibliography: p.47
0-905555-03-1 BS.353/19[no.26]
2274159

ADMINISTRATION. Daily newspaper collections. Research libraries

HÖFIG, Willi
Die Behandlung von Tageszeitungen an wissenschaftlichen Bibliotheken. - Pullach bei München : Verlag Dokumentation, 1975
163p : ill. - Includes bibliographies
3-7940-3274-8 X.989/39812
1209825

ADMINISTRATION. Day care centres for children. United States

SCHOOL-AGE child care / Ruth Kramer Baden ... [et al.]. - Boston, Mass. : Auburn House, c1982
xx,486p : ill. - Bibliography: p427-434
0-86569-112-6 X.520/32741
1892746

SCIARRA, Dorothy June
Developing and administering a child care center. - Boston [Mass.] ; London : Houghton Mifflin, 1979
xi,393p : ill. - Bibliography: p.387-389. - Includes index
0-395-26263-1 X.525/2361
1892746

ADMINISTRATION. Denmark
Role of clergy, to date

NØRR, Erik
Præst og administrator. - 2. opl. - [København] : Rigsarkivet : Gad, 1981
559p : ill. - (Administrationshistoriske studier ; [4]). - Bibliography
87-7497-082-8 X.0519/729(4)
8890889

ADMINISTRATION. Denmark. Haeren
1660-1763

KYHL, O.
Den landmilitære centraladministration 1660-1763. - København : Rigsarkivet, 1975-1976
2v
r6001243 X.0800/697
0914185

ADMINISTRATION. Detention institutions. Great Britain

HOWARD LEAGUE FOR PENAL REFORM
Participation in prisons. - [London] : [The League], 1979
[1],15p. - Bibliography: p.13
B79-32409 X.102/2224
1954415

ADMINISTRATION. Development. Agriculture. South-east Asia

GABLE, Richard Walter
Administering agricultural development in Asia. - Boulder : Westview Press : Tonbridge : Distributed by Benn, 1976
xxviii,398p : ill. - (Westview special studies in social, political, and economic development). - Includes bibliographies and index
0-89158-206-1 X.320/11331 Woolwich
1520709

ADMINISTRATION. Development projects. Developing countries

BRIDGER, Gordon A.
Planning development projects. - London : H.M.S.O., 1983
xiv,209p
0-11-580242-8 X.520/34148
3084728

DAVEY, Kenneth
Development administration revisited. - Birmingham : Development Administration Group, 1983
25p. - (Occasional paper ; no.20)
0-7044-0671-3 P.525/395
3084728

ADMINISTRATION. Development projects. Rural regions. Developing countries

BELSHAW, D. G. R. (Deryke Gerald Rooten)
Rethinking monitoring and evaluation systems for rural development institutions. - Norwich : School of Development Studies, University of East Anglia, 1982
15p : ill. - (Discussion paper ; no.107). - Bibliography: p14-15
B84-40805 X.525/9962
3187411

ADMINISTRATION. Dioceses. Christian church. England
Decisions of bishops, 1066-1300

ENGLISH episcopal acta. - London : Oxford University Press for the British Academy.
1: Lincoln, 1067-1185. - 1980
lxiv,257p,iv p of plates : ill. - Bibliography: p.xx-xxvi. - Includes index
0-19-725986-3 X.0200/1082(1)
2316250

ADMINISTRATION. Distance study institutions

ADMINISTRATION of distance-teaching institutions / compiled by Tony Dodds. - Cambridge : International Extension College, 1983
vii,156p
0-903632-25-x X.525/7986
2955105

ADMINISTRATION of distance-teaching institutions. - Cambridge : International Extension College, 1983
v,109p : ill
0-903632-26-8 X.525/7985
2955091

ON course. - Cambridge : International Extension College, [1983]
1portfolio : ill
B83-38908 Cup.918/122
2955083

ADMINISTRATION. Districts. National health services. Great Britain

NICHOL, D. K.
Problems of integration at the district and unit levels. - Manchester : Health Services Management Unit, University of Manchester, 1981
11leaves. - (Working paper series / University of Manchester. Health Services Management Unit ; no.49)
B82-05614 X.322/10465 Woolwich
260146x

ADMINISTRATION. Dodecanese. Greece
1912-1946

MARONGIU BUONAIUTI, Cesare
La politica religiosa del fascismo nel Dodecanneso. - [Napoli] : Giannini, 1979
120p. - (Quaderni della Facoltà di scienze politiche / Università di Napoli ; 8)
r6504073 X.800/33330
8901597

ADMINISTRATION. Donations by deeds of covenant. Charities. Great Britain

CHARITABLE deeds of convenant. - 2nd ed. - Tonbridge : CAF Publications, 1981
39p : ill
0-904757-10-2 X.529/41025
0110418

ADMINISTRATION. Economic development. Middle East

DEVELOPMENT administration in the Middle East / Delwin A. Roy, editor and contributor ; contributors, Muhammad Diab ... [et al.]. - Beirut : American University of Beirut, 1975
vii,344p
r6811523 X.520/32591
9052755

ADMINISTRATION. Economic development. Rural regions. Developing countries

OYUGI, Walter Ouma
Rural development administration. - New Delhi : Vikas, c1981
x,227p. - Bibliography: p215-223
0-7069-1131-8 X.529/52496
2836149
- [Another copy] X:809/60972

REIDY, Angela
Administration as a change agent in rural development. - [Swansea] : Centre for Development Studies, University College of Swansea, 1979
13leaves. - (Occasional paper / Centre for Development Studies, University College of Swansea ; no.10)
B83-08493 X.525/6853
2836149

ADMINISTRATION. Education

HUGHES, Meredydd Glyn
Education administration. - Birmingham : University of Birmingham, [1978]
[2],16p. - (Inaugural lectures / University of Birmingham)
0-7044-0347-1 X.529/37693
0145904

ADMINISTRATION. Education
- Conference proceedings

INTERNATIONAL INTERVISITATION PROGRAMME ON EDUCATIONAL ADMINISTRATION (3rd : 1974 : Bristol, etc.)
Administering education. - London : Athlone Press, 1975
vii,319p : ill. - Includes bibliographies and index
0-485-12026-7 X.519/25817
0145904

ADMINISTRATION. Education
Information sources

DALE, Sheila
Educational administration, using the literature. - Milton Keynes : Open University Press, 1975
29p. - (Educational studies. a second level course, decision-making in British education systems) (E221). - Includes bibliographies
0-335-03231-1 OPL
059900x

ADMINISTRATION. Education. Commonwealth countries
Training courses

EDUCATIONAL administration and supervision. - London : Education Division, Commonwealth Secretariat, c1981
ii,123p : ill
0-85092-195-3 BS.12/167
281532x

ADMINISTRATION. Education. Czechoslovakia
1945-

CIRBES, Milan
Základy vedeckého riadenia školstva. - 2. vyd. - Bratislava : Slovenské pedagogické nakladate lstvo, c1980
418p. - Bibliography
r6440015 X.529/44631
8767459

ADMINISTRATION. Education. England

RAISON, Timothy
The act and the partnership. - London : Bedford Square Press for the Centre for Studies in Social Policy ; London : Distributed by Research Publications Services, 1976
80p. - (Doughty Street paper ; no.3)
0-7199-0917-1 X.519/26147
0730297

ADMINISTRATION. Education. England
Implications of decline in number of students

BRIAULT, Eric William Henry
The management of a contracting education system. - Swansea : University College of Swansea, 1979
24p : ill. - (Charles Gittins memorial lecture ; 1979)
0-86076-017-0 X.529/36077
215580x

ADMINISTRATION. Education. England
Leadership

NEW directions in educational leadership / edited and introduced by Paul Harling. - Lewes : Falmer, 1984
lx,426p. - Includes bibliographies and index
0-905273-60-5 X.520/36988
3277682

ADMINISTRATION. Education. France

CATTEAUX, André
L'inspection académique. - Rennes : imprimé par le C.R.D.P., [1976]
100p : ill
r6012389 S.E.123/615 Woolwich
122543X

ADMINISTRATION. Education. France
to date

BOURSIN, Jean-Louis
L'administration de l'éducation nationale. - Paris : Presses Universitaires de France, 1981
127p. - (Que sais-je? ; 1957)
2-13-037077-2 X.529/61178
897134x

HISTOIRE de l'administration de l'enseignement en France 1789-1981 / Pierre Bousquet [et al.]. - Genève : Librairie Droz, 1983
154p. - (Hautes études médiévales et modernes ; 49)
r6208839 X.0700/48(49)
897134x

ADMINISTRATION. Education. Great Britain

EDUCATIONAL administration / written by members of the Society of Education Officers ; and edited by Kenneth Brooksbank. - London : Councils and Education, 1980
300p : ill. - Includes bibliographies and index
0-900313-65-x X.529/57531
2224496

EDUCATIONAL administration. - 2nd ed.. - Harlow : Councils and Education Press, 1984
353p. - Bibliography: p342-343. - Includes index
0-900313-74-9 X.529/67314
2224496

FENWICK, Keith
The government of education in Britain. - Oxford : Robertson, 1981
xii,260p : ill. - Bibliography: p238-247. - Includes index
0-85520-255-6 X.529/43321
2224496

ADMINISTRATION. Education. Great Britain
- Bibliographies

HOWELL, D. A. (David Antony)
A bibliography of educational administration in the United Kingdom. - Windsor : NFER, [1978]
88p
0-85633-151-1 X.519/28515
1590626
- [Another copy] OPL

ADMINISTRATION. Education. Great Britain
- Conference proceedings

BRITISH EDUCATIONAL ADMINISTRATION SOCIETY. Conference (8th : 1979 : Sheffield)
Current issues in the management of education. - [Sheffield] : Sheffield City Polytechnic Department of Education Management, 1980
viii,280p. - (Sheffield papers in education management ; no.12). - Includes bibliographies
B84-23015 X.529/61153
2224496

ADMINISTRATION. Education. Great Britain
- Serials

EDUCATIONAL administration. - Vol.4, no.1- ; Autumn 1975-. - Bristol : 'Educational administration', [1975]- : ill
B79-27248 P.521/3264
1934627

EDUCATIONAL administration bulletin. - Vol.1, no.1- Vol. 3, no.2 ; Autumn 1972-Summer 1975. - Bristol : 'Educational administration bulletin', 1972-1975 : ill
Holdings: vol.3, no.1 and vol.3, no.2
B79-27247 P.521/3264
1934627

EDUCATIONAL management & administration. - Vol.10, no.1 (Feb. 1982)-. - Harlow : Longman, 1982-
v. : ill
B82-38528 P.521/3264
1934627

ADMINISTRATION. Education. Inner London
Reform

STERN, Michael
Quis custodiet?. - London : Bow Publications, 1980
7p. - (A Bow Group memorandum) (A Bow paper)
B83-15030 X.525/6840
2868709

ADMINISTRATION. Education. Italy
- Conference proceedings

La GESTIONE democratica della scuola / [testi di Silvano Filippelli et al.]. - Firenze : La Nuova Italia, 1975
394 p.. - (Scuola e educazione nel mondo ; n. 18)
r6014159 X.519/20732
0860123

ADMINISTRATION. Education. Italy
to date

SACCO, Piero
L'organizzazione amministrativa della pubblica istruzione. - Nuova ed. riv. e aggiornata. - [Milano] : Giuffrè, 1980
340p
r6507169 X.520/29714
8266603

ADMINISTRATION. Education. Nigeria

EDEM, D. A.
Introduction to educational administration in Nigeria. - Ibadan : Spectrum ; Chichester : Wiley, c1982
xiv,143p. - (Education in Africa). - Bibliography: p136-138. - Includes index
0-471-27984-6 X.529/50425
2778033

NWANKWO, John I.
Educational administration. - New Delhi : Vikas, c1982
ix,117p : ill
0-7069-1576-3 X.529/52385
2778033

ADMINISTRATION. Education. United States

MORPHET, Edgar L. (Edgar Leroy)
Educational organization and administration. - 4th ed. - Englewood Cliffs ; London : Prentice-Hall, c1982
x,422p : ill. - Includes bibliographies and index
0-13-236729-7 X.520/26443
2699486

ADMINISTRATION. Education. United States
Effects of law, 1954-1976

WISE, Arthur Edward
Legislated learning. - Berkeley ; London : University of California Press, 1979
xvii,219p. - Includes index
0-520-03759-6 X.520/21774
2182971

ADMINISTRATION. Education. United States
Participation of public

COMMUNITY participation in education / [edited by] Carl A. Grant. - Boston, Mass. ; London : Allyn and Bacon, 1979
xiv,262p. - Includes bibliographies and index
0-205-06052-8 X.520/13176
1756524

ADMINISTRATION. Education. United States
Political aspects

The POLITY of the school / edited by Frederick M. Wirt. - Lexington, Mass. ; London : D.C. Heath, 1975
xvii,334p. - (Lexington Books politics of education series)
0-669-97618-0 X.520/10350
0563854

ADMINISTRATION. Education. United States
Research - Conference proceedings

PROBLEM-FINDING in educational administration / edited by Glenn L. Immegart, William Lowe Boyd. - Lexington, Mass. : Heath ; [Farnborough, Hants.] : [Distributed by] Gower Press, 1980
xix,300p : ill. - Includes index
0-669-02438-4 X.520/21549
2163799

ADMINISTRATION. Education. Urban regions. United States

ORNSTEIN, Allan C.
Metropolitan schools. - Metuchen, N.J. : Scarecrow Press, 1974
xii, 293 p.. - Bibliography : p. 274-281
0-8108-0653-3 X.529/19922
1022172

ADMINISTRATION. Education. Western Europe

LEARNING from Europe / [Deborah Greig, editor]. - Harlow : Published for AEC Trust by Longman (Councils and Education), 1984
xviii,132p
0-900313-19-6 X.529/66641
3026027

ADMINISTRATION. Educational institutions
See also
Governing boards
Offices. Schools

ADMINISTRATION. Educational institutions

NEWELL, Clarence A.
Human behavior in educational administration. - Englewood Cliffs ; London : Prentice-Hall, 1978
x,261p : ill. - Includes bibliographies and index
0-13-444638-0 X.520/12680
151539x

ADMINISTRATION. Educational institutions
Applications of theories of organisational behaviour

OWENS, Robert G. (Robert Goronwy)
Organizational behaviour in education. - 2nd ed. - Englewood Cliffs ; London : Prentice-Hall, c1981
xii,340p : ill. - Includes bibliographies and index
0-13-641050-2 X.520/24030
2487071

ADMINISTRATION. Educational institutions
Influence of organisational behaviour

HANSON, E. Mark
Educational administration and organizational behavior. - Boston, Mass. ; London : Allyn and Bacon, 1979
x,395p : ill. - Includes index
0-205-06164-8 X.520/13376
1781537

ADMINISTRATION. Elections. Members. Local authorities. England

GREENSMITH, R. Colin
Shaw's local authorities election diary. - London : Shaw
viii,[34]p
0-7219-0950-7 X.529/55892
2928965

ADMINISTRATION. Employment Protection Act 1975
Procedure - For employers

The EMPLOYMENT Protection Act 1975, institutions and administration of the Act. - London : E.E.F., 1976
[2],12p. - (EEF guide ; 1)
0-901700-15-0 X.100/16696
0552410

ADMINISTRATION. Enemy property. Great Britain
1939-1965

MCKENZIE, A. W. (Alexander William)
The treatment of enemy property in the United Kingdom during and after the Second World War. - [Chislehurst] : A.W. McKenzie, c1981
62p. - Bibliography: p52. - Includes index
0-9507837-0-6 X.809/50522
2626675

ADMINISTRATION. Environment planning by local authorities. United States

LEVIN, Melvin Richard
Community and regional planning. - 3rd ed. - New York ; London : Praeger, 1977
xxiii,279p. - (Praeger special studies in US economic, social and political issues). - Bibliography: p.260-274. - Includes index
0-275-23690-0 X.510/15531
1189069

ADMINISTRATION. Estates. Fyn. Denmark
to date

DOMBERNOWSKY, Lotte
Lensbesidderen som amtmand. - [København] : Rigsarkivet, 1983
525p : ill. - (Administrationshistoriske studier ; nr.8). - Includes bibliography
87-7497-065-8 X.0519/729(8)
9035826

ADMINISTRATION. Estates. Hungary
1699-1848

KÁLLAY, István
A magyarországi nagybirtok kormányzata 1711-1848. - Budapest : Akadémiai Kiadó, 1980
337p : ill. - (A Magyar Országos Levéltár kiadványai ; 3. Hatóság és hivataltörténet ; 5). - Bibliography: p317-321
963-05-1742-6 Ac.7308/4(c)[5]
2469367

ADMINISTRATION. Estates. Inherited property. England

TAYLOR, J. N. R. (James Nigel Rodney)
Executorship law and accounts. - 4th ed. - Plymouth : Macdonald and Evans, 1979
xv,176p. - (The M. & E. handbook series). - Includes index
0-7121-0562-x X.208/6593
0090786

TAYLOR, James Nigel Rodney
Executorship law and accounts. - 3rd ed. - Estover : Macdonald and Evans, 1977
xv,173p. - (The M. & E. handbook series). - Includes index
0-7121-0560-3 X.108/16743
0090786

ADMINISTRATION. Estates. Inherited property. England
Law - For accountancy

SHERRING, Tony
Law and accounts for executors and trustees. - London : Gee, 1981
xvi,257p. - Bibliography: p248. - Includes index
0-85258-195-5 X.200/35773
0017205

VICKERY, Bertram George
Law and accounts of executors, administrators and trustees. - 19th ed.. - Reading : Donnington Press ; London : Distributed by Cassell and Collier Macmillan, 1975
xii,496p. - Includes index
0-304-29629-5 X.200/20231
0017205

Law and accounts of executors, administrators, and trustees. - 20th ed. - London : Cassell, 1980
xii,371p. - (A Cassell professional handbook)
0-304-30452-2 X.200/33713
0017205

ADMINISTRATION. Estates. Inherited property. England
Law. Cases: Swinfen v. Swinfen - Biographies

CLAYTON, Howard
The great Swinfen case. - London : Regency Press, c1980
211p,[4]p of plates : ill. - Includes index
0-7212-0630-1 X.809/46455
2353954

ADMINISTRATION. Estates. Inherited property. England
Law. Procedure

NELSON, Bertram
Nelson's tables. - 7th ed.. - London : Oyez Longman, 1983
24p
0-85120-721-9 X.525/7838
0769231

Nelson's tables, administration of estates. - Revised 6th ed.. - London : Oyez Publishing, 1978
23p
0-85120-359-0 X.100/18719
0769231

Nelson's tables of procedure. - 6th ed.. - London : Oyez Publishing, 1975
vi,65p
0-85120-259-4 X.100/16642
0769231

ADMINISTRATION. Estates. Inherited property. South Africa
Law

MEYEROWITZ, David
The law and practice of administration of estates. - 5th ed. - Cape Town : Juta, 1976
xxx,720p. - Includes index
0-7021-0728-x X.200/39914
0779288
- [1982 cumulative supplement] X.200/46546
- [1984 cumulative supplement] X.200/44134

ADMINISTRATION. Exchange visits. Young persons. Great Britain
Reform

COCKERILL, G. F. (Geoffrey Fairfax)
Report on youth exchanges. - [London] : [Foreign and Commonwealth Office], 1983
188p : ill
B83-42062 BS.14/3625
2968363

Report on youth exchanges. - [Great Britain] : [G.F. Cockerill], 1983
188p : ill
B85-14705 BS.14/3726
2968363

ADMINISTRATION. Extension services. Agriculture. Kenya

LEONARD, David K.
Reaching the peasant farmer. - Chicago ; London : University of Chicago Press, 1977
xxi,297p : ill. - Includes index
0-226-47260-4 X.320/11001 Woolwich
1282239
- [Another copy] X.320/11047 Woolwich

ADMINISTRATION. Falkland Islands
By Great Britain. Disputes between Great Britain & Spain, 1766-1811

ZORRAQUIN BECU, Ricardo
Inglaterra prometio abandonar las Malvinas. - Buenos Aires : Platero, 1975
203p
r6084426 X.709/26994
1790188

ADMINISTRATION. Federal courts. United States

MCLAUCHLAN, William P.
Federal court caseloads. - New York ; Eastbourne : Praeger, 1984
xiv,219p : ill. - (Praeger special studies). - Bibliography: p211-214. - Includes index
0-03-057956-2 X.200/44819
3211789

ADMINISTRATION. Federal government assitance for economic development in Georgia (State)
1933-1945

HOLMES, Michael S.
The New Deal in Georgia. - Westport, Conn. ; London : Greenwood Press, 1975
xi,364,[1]p. - (Contributions in American history ; no.36). - Bibliography: p.347-353. - Includes index
0-8371-7375-2 X.809/41931
1112163

ADMINISTRATION. Film collections. Public libraries. Denmark

BIBLIOTEKSTILSYNET
Film og biblioteker. - [Copenhagen] : Bibliotekscentralen, 1977
91p
87-552-0451-1 S.B.286/36
1539922

ADMINISTRATION. Fire brigades. Great Britain
Use of digital computer systems

LOCAL AUTHORITIES MANAGEMENT SERVICES AND COMPUTER COMMITTEE
The administration use of computers by fire brigades. - London : The Committee, 1975
82,[23]p. - (Research reports / Joint Committee on Fire Research ; no.5)
B79-27217 SRL
1925032

ADMINISTRATION. Fire prevention. Great Britain
By fire brigades

A GUIDE to fire prevention planning. - [London] : Home Office. - (Research reports / Joint Committee on Fire Research ; no.2)
Part 1: Principles of approach. - [1974]
[107]leaves : ill
B79-27550 SRL
1925016

Part 2: Detailed notes for guidance. - [1974]
[57]leaves : ill
B79-27551 SRL
1925016

Part 3: Examples and case study. - [1974]
[96]leaves : ill
B79-27552 SRL
1925016

ADMINISTRATION. Fire prevention. Great Britain
By fire brigades. Use of digital computer systems

LOCAL AUTHORITIES MANAGEMENT SERVICES AND COMPUTER COMMITTEE. Computer Panel
Fire prevention systems by computer. - London : LAMSAC, 1975
23p
0-85497-078-9 X.615/1538
0522511

ADMINISTRATION. Fire protection. United States

ROBERTSON, James Craig
Introduction to fire prevention. - Beverly Hills : Glencoe Press ; London : Collier Macmillan, 1975
xix,318p : ill. - (Glencoe Press fire science series). - Includes index
0-02-477080-9 X.622/5387 Woolwich
062330x

ADMINISTRATION. Florida
By Cuba, 1693-1821

ISERN, José
Gobernadores cubanos de la Florida. - Miami : [AIP Publications Center?], 1974
64p : ill. - (Pioneros cubanos en U.S.A. ; III)
r6099608 X.519/41786
2043467

ADMINISTRATION. Free schools. United States

SWIDLER, Anne
Organization without authority. - Cambridge, Mass. ; London : Harvard University Press, 1979
xiii,199p. - Bibliography: p.185-194. - Includes index
0-674-64340-2 X.520/21082
2130831

ADMINISTRATION. Further education institutions. England

COLLEGE administration / editor Ian Waitt. - London : National Association of Teachers in Further and Higher Education, 1980
[2],iii,890p : 1 ill. - Includes bibliographies and index
0-901390-30-5 X.529/37550
2258552

ADMINISTRATION. Further education institutions. Great Britain

PRATT, John
Costs and control in further education. - Windsor : NFER, 1978
242p : ill. - Bibliography: p.233-235
0-85633-159-7 X.529/37921
1199935

ADMINISTRATION. Further education institutions. Great Britain
- Bibliographies

WATSON, Leonard E.
The management of the further education college. - [Sheffield] : Sheffield City Polytechnic, Department of Education Management, 1981
viii,123p. - (Sheffield papers in education management ; no.21). - Includes index
0-903761-36-x X.529/54062
2896516

ADMINISTRATION. Further education institutions. Great Britain
- Serials

[STUDY conference (Further Education Staff College)]. Study conference / Staff College. - [19-]. - [Bristol] : [The College], [19-]-
v.
Holdings: 82/1-
r6813627 P.803/1294
2896516

ADMINISTRATION. German military forces
to date

SICKEN, Bernhard
Die Militärverwaltung. - Hannover : Vincentz, 1982
304p.,12p.of plates,[7]folded leaves of plates. - (Beiträge / Akademie für Raumforschung und Landesplanung ; Bd. 55) (Behördliche Raumorganisation seit 1800. Grundstudie ; 12). - Bibliography
3-87870-701-0 X.0512/475(55)
8868921

ADMINISTRATION. Glyndebourne Festival Opera
Caplat, Moran - Biographies

CAPLAT, Moran
Dinghies to divas, or, Comedy on the bridge. - London : Collins, 1985
281p,[8]p of plates : ill. - Includes index
0-00-217237-2 X.950/42431
3304442

ADMINISTRATION. Government departments. Great Britain
Use of computer systems

LONGER term review of administrative computing in central government / by the Steering Committee. - London : Civil Service Department, 1978
[3],53[i.e.57]p
B78-35952 BS.6/109
1696424

ADMINISTRATION. Government departments. Great Britain
Use of computer systems & telecommunication systems. Advisory services: Central Computer and Telecommunications Agency. Technical Services Organisation. Role

The TECHNICAL Services Organisation of Central Computer and Telecommunications Agency. - London : Civil Service Department, 1981
106p
0-7115-0033-9 BS.6/172
2561662

ADMINISTRATION. Government departments. Great Britain
Use of computer systems. Cost-effectiveness

GREAT BRITAIN. Investment Appraisal Working Party
Review of investment appraisal and monitoring procedures for administrative computer projects. - London : Civil Service Department, 1980
33,[21]p. - Bibliography: p[8]
0-7115-0020-7 X.622/14150
2817381

ADMINISTRATION. Government departments. Latin America
Reform

HAMMERGREN, Linn A.
Development and the politics of administrative reform. - Epping : Bowker, 1983
xvi,213p. - (A Westview replica edition) (Bowker (UK) replica edition). - Bibliography: p195-204. - Includes Index
0-86531-956-1 X.529/59047
3044599

ADMINISTRATION. Government policies on industries. Great Britain
Responsibility of government agencies & government departments

HOGWOOD, Brian W.
In search of accountability. - Glasgow : Centre for the Study of Public Policy, University of Strathclyde, 1981
31p. - (Studies in public policy ; no.82)
B81-27743 X.525/4636
2525550

ADMINISTRATION. Government sponsored research in health services. Great Britain
1973-1977

FIVE years after / edited by Gordon McLachlan. - Oxford : Oxford University Press for the Nuffield Provincial Hospitals Trust, 1978
[7],85p. - (Occasional hundreds ; 9)
0-19-721398-7 X.319/18163 Woolwich
1491628

ADMINISTRATION. Grants from local authorities for improvement of housing. Great Britain

IMPROVEMENT grant administration. - London : The Agency, 1979
94p
B83-00524 X.525/6367
2805588

ADMINISTRATION. Great Britain. Customs and Excise

ST QUINTON, P. E.
Customs attendance. - [London] : [HM Customs and Excise], 1981
[106]p : ill. - (The Rayner scrutiny programme ; 1981)
B84-38197 BS.9/67
3171779

STUTTLE, W. I.
Review of support services for administrative work. - [London] : H.M. Customs and Excise, 1983
1v.(various pagings)
B85-02494 BS.9/70
3171779

ADMINISTRATION. Great Britain. Parliament

GRANADA TELEVISION
'The state of the nation - Parliament'. - London : Granada Television Ltd, 1973
272p,fold leaf : 1 ill
0-904005-01-1 X.808/36085
0013064

ADMINISTRATION. Greater London Council
1965-1981 - Biographies

CUTLER, Horace
The Cutler files. - London : Weidenfeld and Nicolson, 1982
v,183p : ill
0-297-78017-4 X.529/46728
2701286

ADMINISTRATION. Greater London Council
Participation of personnel

INDUSTRIAL democracy in local government. - [London] : Greater London Council, [1977]
[1],44p : ill
0-7168-0873-0 LCC.542
1123327

ADMINISTRATION. Greenland
By Denmark, to 1976

VIEMOSE, Jørgen
Dansk kolonipolitik i Grønland. - [Copenhagen] : Demos, 1977
192p : ill. - Includes bibliographies
87-87337-75-4 X.709/24587
1388274

ADMINISTRATION. Haiti
to dte

FRANCISQUE, Reynold
Pour Haïti. - Port-au-Prince : R. Francisque, 1981
117p
r6208664 X.529/52586
891088x

ADMINISTRATION. Health insurance. United States
Costs

BLAIR, Roger Duncan
The cost of health insurance administration. - Lexington, Mass. ; London : D.C. Heath, 1975
xxi,179p : ill. - Bibliography: p.165-168. - Includes index
0-669-00165-1 X.322/8119 Woolwich
0586323

ADMINISTRATION. Health services
- Bibliographies

BUNCH, Antonia J. (Antonia Janette)
Health care administration. - London : Capital Planning Information, 1979
120p. - (CPI sourcebooks ; no.1). - Includes index
0-906011-05-1 X.329/19363
1824228

ADMINISTRATION. Health services
Evaluation

NUTT, Paul C.
Evaluation concepts & methods. - [Rev. ed.]. - New York : SP Medical & Scientific, c1982
374p : ill. - (Health systems management ; v.14)
0-89335-175-x X.520/27897
2743876

ADMINISTRATION. Health services
Statistical methods

BROYLES, Robert W.
Statistics in health administration. - Germantown, Md ; London : Aspen Systems Corporation ; [London] : [Distributed by Eurospan Ltd].
Vol.1: Basic concepts and applications. - 1979
xvi,570p : ill. - Includes index
0-89443-153-6 X.0322/215(1) Woolwich
2181940

Vol.2: Advanced concepts and applications. - 1980
xxi,557p : ill. - Includes index
0-89443-166-8 X.0322/215(2) Woolwich
2181940

ADMINISTRATION. Health services. Cities. India (Republic)

BANERJEE, Usha
Health administration in a metropolis. - New Delhi : Abhinav Publications, 1976
xv,256p,[10]leaves of plates (1 fold),[2]p of plates : ill. - Bibliography : p.200-207.-Includes index
r6031943 X.329/10952 Woolwich
111591X

ADMINISTRATION. Health services for ambulatory patients

ROSS, Austin
Ambulatory care organization and management. - New York ; Chichester : Wiley, c1984
xvii,453p : ill. - (Wiley series in health services). - Bibliography: p416-443. - Includes index
0-471-08951-6 X.322/14004
3165841

ADMINISTRATION. Health services for pre-school children. Great Britain
Applications of digital computer systems

PRE-SCHOOL health. - London : Child Health Computing Committee, Department of Health and Social Security, c1979
56,9p : ill. - (The Standard child health system)
B83-10326 BS.385/837
2841037

ADMINISTRATION. Health services. Great Britain
Applications of organisation theory

WARMINGTON, Allan
Organisation theory and health service administration. - Manchester : Health Services Management Unit, University of Manchester, 1981
53leaves. - (Working paper series / University of Manchester Health Services Management Unit ; no.48). - Bibliography: leaves 51-53
B82-05613 X.322/10466 Woolwich
2601451

ADMINISTRATION. Health services. Great Britain
Great Britain. Local government

WOOD, Bruce
The organisation of local government for Health Service administrators. - Manchester : Health Services Management Unit Department of Social Administration, University of Manchester, 1981
14,3leaves. - (Working paper series / University of Manchester. Health Services Management Unit)
B81-16696 X.325/1082 Woolwich
2393778

ADMINISTRATION. Health services. India (Republic)

SEAL, S. C.
Health administraion in India. - Calcutta : Dawn Books, 1975
[13],639p : ill. - Includes bibliographies and index
r6071764 X.320/11174 Woolwich
140668X

ADMINISTRATION. Health services. New Zealand
- Conference proceedings

HEALTH administration in New Zealand / edited by R.J.Latimer. - Wellington [N.Z.] : New Zealand Institute of Public Administration, 1969
127p. - (Studies in public administration ; no.15) - Includes index
r6071711 Ac.2273.h/2(15)
1398229

ADMINISTRATION. Health services. Schools. United States

MAYSHARK, Cyrus
Administration of school health programs. - 2nd ed. - Saint Louis : Mosby ; London : Distributed by Kimpton, 1977
xv,509p : ill. - Includes bibliographies and index
0-8016-3166-1 X.311/8779 Woolwich
1481533

ADMINISTRATION. Health services. South Africa

CLOETE, Jacobus Johannes Nicholaas
Administration of health services. - Pretoria : Van Schaik, 1975
[7], 88 p. : ill. - Includes index
0-627-00355-9 X.311/8270 Woolwich
0733466

ADMINISTRATION. Health services. Sweden
1945-

NORDESTEDT, Bo
Att välja framtidens sjukvård. - Lund : Studentlitteratur, 1980
159p
91-441-7671-6 X.329/15649
886750x

ADMINISTRATION. Health services. United States

MECHANIC, David
The growth of bureaucratic medicine. - New York ; London : Wiley-Interscience, 1976
xiii,345p : ill. - (Health, medicine and society). - Includes index
0-471-59021-5 X.320/10343 Woolwich
0646628

ADMINISTRATION. Health services. West Germany

STONE, Deborah A.
The limits of professional power. - Chicago ; London, 1980
xi,212p : ill. - Bibliography: p201-205. - Includes index
0-226-77553-4 X.325/1185 Woolwich
2409232

ADMINISTRATION. Health services. Western world

CHESTER, Theodore Edward
Conflicts in health care policy. - Manchester : Health Services Management Unit, University of Manchester, [1978]
[1],21 leaves. - (Working paper series / University of Manchester. Health Services Management Unit ; no.1)
B79-32392 X.512/6949
1955934

ADMINISTRATION. Health services. Western world
- Conference proceedings

NATO ADVANCED RESEARCH INSTITUTE ON HEALTH SERVICES SYSTEMS
(1982 : The Hague)
Reorienting health services. - New York ; London : Published in cooperation with NATO Scientific Affairs Division [by] Plenum, c1984
xi,374p : ill. - (NATO conference series. II, Systems science ; v.15). - Includes index
0-306-41481-3 X.322/13736
1955934

ADMINISTRATION. Higher education. Brazil
Role of coordinating agencies

CROSS, Jack L
Coordination and control of higher education in the United States and Brazil. - Brasília : Ministério da Educação e Cultura, Departamento de Documentação e Divulgação, 1979
235p
r6084306 LAS.B.275/136 Woolwich
2452472

ADMINISTRATION. Higher education. England
Reorganisation. Great Britain. Department of Education and Science. Higher education in England outside the universities

PRATT, John, 1945-
Agenda for a central body. - London : NELPCO, 1981
20p. - (CIS commentary ; no.21). - Bibliography: p20
0-901987-43-3 X.525/5930
2750090

ADMINISTRATION. Higher education. Great Britain

APPROACHES to post-school management / edited by Oliver Boyd-Barrett ... [et al.]. - London : Harper & Row in association with the Open University, 1983
xv,445p : ill. - (Harper education series). - Includes index
0-06-318263-7 X.529/63220
1969056

ADMINISTRATION. Higher education. Great Britain
By local authorities. Advisory services: National Advisory Body for Local Authority Higher Education

BEVAN, J. S. (John S)
The National Advisory Body for local authority higher education. - [Sheffield] : Association of Colleges for Further and Higher Education, [1982]
4p
B82-37173 X.529/49806
2782251

ADMINISTRATION. Higher education institutions. England

COLLEGE administration / editor Ian Waitt. - London : National Association of Teachers in Further and Higher Education, 1980
[2],iii,890p : 1 ill. - Includes bibliographies and index
0-901390-30-5 X.529/37550
2113597

ADMINISTRATION. Higher education institutions. Great Britain
Participation of personnel & students

TOLLEY, George
Participatory management in institutions of higher education. - [Sheffield] : Sheffield City Polytechnic Department of Education Management, 1979
24leaves. - (Sheffield papers in education management : 6)
B84-23008 X.529/61155
3087352

ADMINISTRATION. Higher education institutions. United States

CORSON, John Jay
The governance of colleges and universities. - Revised ed. - New York ; London : McGraw-Hill, [1976]
xiii,297p. - (The Carnegie series in American education). - Includes index
0-07-013205-4 WP.17184/21
1035274

CRISIS in campus management / edited by George J. Mauer. - New York ; London : Praeger ; [London] : [Distributed by Martin Robertson], 1976
xi,267p. - (Praeger special studies in US economic, social and political issues). - Bibliography: p.250-262. - Includes index
0-275-55710-3 X.520/10924
0700436

DRESSEL, Paul L. (Paul Leroy)
Administrative leadership. - San Francisco ; London : Jossey-Bass, 1981
xviii,243p. - (The Jossey-Bass series in higher education). - Bibliography: p221-233. - Includes index
0-87589-500-x X.520/32356
1035274

HELLER, Jack F.
Increasing faculty and administrative effectiveness. - San Francisco ; London : Jossey-Bass, 1982
xv,126p. - (Jossey-Bass series in higher education) (Jossey-Bass series in social and behavioral science). - Bibliography: p121-122. - Includes index
0-87589-520-4 X.520/28492
1035274

JEDAMUS, Paul
Improving academic management. - San Francisco : London : Jossey-Bass, 1980
xxii,679p : ill. - (The Jossey-Bass series in higher education). - Includes bibliographies and index
0-87589-477-1 X.525/4314
2392739

KAROL, Nathaniel H.
Managing the higher education enterprise. - New York ; Chichester : Wiley, c1980
xiv,269p. - Bibliography: p247-263. - Includes index
0-471-05022-9 X.520/22909
1035274

LEADERSHIP in higher education / edited by Bob W. Miller, Robert W Hotes and Jack D Terry, Jr - Westport, Conn. ; London : Greenwood Press, 1983
xxi,585p : ill. - Bibliography: p555-576. - Includes index
0-313-22263-0 X.520/36062
1035274

LEE, Sang M.
Academic administration - Lincoln, [Neb.] ; London University of Nebraska Press, c1983
xiii,252p : ill. - Includes bibliographies and index
0-8032-2856-2 X.520/34377
1035274

STARK, Joan S.
The many faces of educational consumerism. - Lexington, Mass. : D.C. Heath ; Farnborough, Hants. : [Distributed by Teakfield Ltd], 1978
xv,224p : 1 ill. - Includes index
0-669-01631-4 X.520/12768
1035274

WOLOTKIEWICZ, Rita J.
College administrator's handbook. - Boston, Mass. ; London : Allyn and Bacon, 1980
x,265p : ill. - Bibliography: p.255-258. - Includes index
0-205-06873-1 X.520/21197
1035274

ADMINISTRATION. Higher education institutions. United States
Decision making. Effects of surveys of students

ASTIN, Alexander William
Academic gamesmanship. - New York ; London : Praeger [for] the Higher Education Research Institute ; [London] : [Distributed by Martin Robertson], 1976
xi,209,[1]p. - (Praeger special studies in US economic, social and political issues). - Includes index
0-275-56720-6 X.520/11247
0786918

ADMINISTRATION. Higher education institutions. United States
Protest by students

CAMPUS power struggle / edited by Howard S. Becker. - 2nd ed. - New Brunswick, N.J. : Transaction, c1973
236p. - (Transaction / Society book series ; 1)
0-87855-059-3 X.529/58753
0623946

The POWER of protest / Alexander W. Astin ... [et al.]. - San Francisco ; London : Jossey-Bass : [Distributed by Dent], 1975 [i.e. 1976]
xv,208p : ill. - (The Jossey-Bass series in higher education). - Bibliography: p.199-203 - Includes index
0-87589-266-3 X.800/11568
0623946

ADMINISTRATION. Higher education. Oxford
Brookes, John, b.1891 - Biographies

JOHN Henry Brookes, craftsman, educator, administrator / edited by A. Stuart Addison. - Oxford : Oxford Polytechnic Press, 1979
xi, 76p : ill. - Bibliography: p.x. - Includes index
0-902692-18-6 L.49/636
1899783

ADMINISTRATION. Higher education. United States
Role of coordinating agencies

CROSS, Jack L
Coordination and control of higher education in the United States and Brazil. - Brasília : Ministério da Educação e Cultura, Departamento de Documentação e Divulgação, 1979
235p
r6084306 LAS.B.275/136 Woolwich
0803561

ADMINISTRATION. Hospitals. England
England. Personal injuries. Liability of occupiers of real property. Law

WILLIAMS, Barry, 1940-
Occupiers' Liability Act 1957 and the liability of hospitals. - Beckenham : Ravenswood Publications Ltd, 1977
xi,69p. - (Case studies on health service management law and practice ; vol.6)
0-901812-20-x X.0319/108(6)
1032720

ADMINISTRATION. Hospitals. United States

HOSPITAL organization and management. - 2nd ed.. - New York ; London : Spectrum ; New York ; London : Distributed by Wiley, 1978
[29],473p : ill. - (Health systems management). - Includes bibliographies and index
0-470-99179-8 X.520/13212
1759868

HOSPITAL organization and management. - 3rd ed.. - New York ; London : SP Medical & Scientific Books, c1983
xix,684p : ill. - (Health systems management ; v.16). - Includes bibliographies and index
0-89335-176-8 X.322/13687
1759868

ISSUES in hospital administration / edited by John B. McKinlay. - Cambridge, Mass. ; London : MIT Press, c1982
xiv,444p : ill. - (Milbank reader ; 9). - Includes bibliographies and index
0-262-13184-6 X.322/11347
1759868

ADMINISTRATION. Hospitals. United States
Costs. Control

GRAY, Steven P.
Hospital cost containment through productivity management. - New York ; London : Van Nostrand Reinhold, c1983
ix,233p : ill. - Includes index
0-442-22921-6 X.520/30003
2903210

ADMINISTRATION. Hospitals. United States
Costs. Control. Role of government

JOSKOW, Paul L.
Controlling hospital costs. - Cambridge, Mass. ; London : MIT, c1981
211p : ill. - (MIT Press series in health and public policy ; 2). - Bibliography: p193-205. - Includes index
0-262-10024-x X.322/10808 Woolwich
2717484

ADMINISTRATION. Hospitals. United States
Economic aspects

BAROCCI, Thomas A.
Non-profit hospitals. - Boston, Mass. : Auburn House, c1981
xix,232p : ill. - Bibliography: p209-220. - Includes index
0-86569-054-5 X.322/10614 Woolwich
2634570

ADMINISTRATION. Hungary
Joseph II, Holy Roman Emperor

HAJDU, Lajos
II József igazgatási reformjai Magyarországon. - Budapest : Akadémiai Kiadó, 1982
527p,[1]folded leaf of plates
963-05-2715-4 X.520/26727
8901600

ADMINISTRATION. Identification parades. Scotland
Procedure - For police

GUIDELINES on the conduct of identification parades. - [Edinburgh] : Scottish Home and Health Department, 1982
9p
B83-42061 BS.155/263(1)
2968355

ADMINISTRATION. Ilchester Gaol
1821

ILCHESTER gaol. - St. Peter Port : Toucan, 1980
46p : 2ill. - (Ilchester and district occasional papers ; no.26)
0-85694-224-3 X.0800/1801(76)
2498154

ADMINISTRATION. Incunabula collections
1945- - Conference proceedings

Das HISTORISCHE und wertvolle Buchgut in der Bibliotheksverwaltung / herausgegeben im Auftrag der Vereinigung Österreichischer Bibliothekare von Otto Mazal und Eva Irblich. - Wien : Die Vereinigung, 1980
216p. - (Biblos-Schriften ; Bd 104)
r6302163 Ac.9596.b/3[Bd.104]
8794545

ADMINISTRATION. Industries
Information sources

ANDERER, Jeanne
Selected information sources on industrial administration. - Farnborough, Hants. : Saxon House [for] the International Industrial Administration Centre, 1977
[5],148p
0-566-00177-2 OPL
1191861
- [Another copy] X.529/32675

ADMINISTRATION. Institutions
- Serials

INSTITUTE forum. - No.1 (autumn 1978)-. - London : Chameleon Press, 1978-
v. : ill
Holdings: no 18 (Winter '82)-
B84-24354 P.525/434
3128652

ADMINISTRATION. Institutions for persons with mental disorders. Great Britain
- Conference proceedings

'AFTER Normansfield'. - London : Institute of Health Service Administrators, 1979
79p
B80-00641 X.512/7100
200187x

ADMINISTRATION. Instituto Brasileiro do Sal
1962-1964

ROSADO, Vingt-un
IBS, 1964. - [Mossoró] : [V. Rosado?], 1978
185p. - (Coleção mossoroense ; v.93)
r6159719 X.529/37895
2440814

ADMINISTRATION. Inter-library loans
Applications of microcomputer systems

GADSDEN, S. R.
The administration of interlending by microcomputer. - London : British Library, c 1984
vii,58p : ill. - (Library and information research report ; 30)
0-7123-3044-5 2719.e.951
320023x

ADMINISTRATION. International organisations
- Conference proceedings

REPORTS of the German Section of the International Institute of Administrative Sciences. - Deventer ; London : Kluwer Law and Taxation, c1983
v,152p
90-654-4136-0 X.520/33072
3076547

ADMINISTRATION. Intramural sports & intramural games. Educational institutions. United States
Participation of students

COLGATE, John A.
Administration of intramural and recreational activities. - New York ; Chichester : Wiley, 1978
x,278p : ill. - Includes index
0-471-01728-0 X.620/18210 Woolwich
1779265

ADMINISTRATION. Intramural sports & intramural games. United States

HYATT, Ronald W.
Intramural sports. - Saint Louis : Mosby ; London : Distributed by Kimpton, 1977
ix,299p : ill. - Includes bibliographies and index
0-8016-2320-0 X.611/7703 Woolwich
1166395

MUELLER, Pat
Intramural-recreational sports. - 5th ed.. - New York ; Chichester : Wiley, 1979
xi,366p : ill. - Bibliography: p.334-336. - Includes index
0-471-04911-5 X.520/14320
1166395

ADMINISTRATION. Ireland
By Great Britain, 1800-1830

BRYNN, Edward
Crown & castle. - Dublin : The O'Brien Press, 1978
172p : ill. - Bibliography: p.160-166. - Includes index
0-905140-11-7 X.809/43931
1771906

ADMINISTRATION. Ivory Coast
Law, to date

MESCHERIAKOFF, Alain Serge
Le droit administratif iviorien. - Paris : Economica, 1982
247p. - (La vie du droit en Afrique)
2-7178-0427-7 X.200/41967
901778x

ADMINISTRATION. Jackson State College
Reddix, Jacob L

REDDIX, Jacob L.
A voice crying in the wilderness. - Jackson : University Press of Mississipi, 1974
xi,238p,[4]p of plates : ill. - Includes index
0-87805-024-8 X.800/27232
1687786

ADMINISTRATION. Justice

The POLITICS of informal justice / edited by Richard L. Abel. - New York ; London : Academic Press. - (Studies on law and social control)
Vol.1: The American experience. - c1982
ix,335p. - Includes bibliographies and index
0-12-041501-1 X.520/27588
2758369

The POLITICS of informal justice / edited by Richard L. Abel. - New York ; London : Academic Press. - (Studies on law and social control)
Vol.2: Comparative studies. - c1982
x,338p : ill. - Includes bibliographies and index
0-12-041502-x X.520/27589
2758369

ADMINISTRATION. Justice
to date

DUMON, F. (Frédéric)
Billijkheid en recht. - Brussel : Koninklijke Academie voor Wetenschappen, Letteren en Schone Kunsten van België, 1980
19p. - (Mededelingen / van de Koninklijke Academie voor Wetenschappen, Letteren en Schone Kunsten van België ; jg.42, nr.5)
r6303281 Ac.988.ea/3.[42 no.5]
8273650

JAIN, P. C.
Administrative adjudication. - New Delhi : Sterling Publishers, [1981]
xv,224p.. - Includes bibliography
r6819538 X.200/46178
8273650

ADMINISTRATION. Justice
Evaluation

THIBAUT, John W.
Procedural justice. - Hillsdale : Erlbaum ; New York ; London : Distributed by Wiley, 1975
ix,150p : ill. - Bibliography: p.142-145. - Includes index
0-470-85869-9 X.520/10404
0583294
- [Another copy: paperback] X.519/25469

ADMINISTRATION. Justice
Research. Methodology

CLARK, Robert S. (Robert Segura)
Fundamentals of criminal justice research. - Lexington : D.C. Heath, 1977
xvii,181,[1]p : ill. - Bibliography: p.167-174. - Includes index
0-669-01005-7 X.200/31668
1733842

ADMINISTRATION. Justice. Argentina
1930-1972

BALBÍN, Armando
?Hacia dónde vamos?. - Buenos Aires : [A. Balbin], 1973
xi, 232 p.
r6025087 X.108/14769
1016733

ADMINISTRATION. Justice. Asturias. Spain
to 1865

SANGRADOR Y VITORES, Matias
Historia de la administración de justicia y del antiguo gobierno del principado de Asturias, y colección de sus fueros, cartas pueblas y antiguas ordenanzas. - Oviedo : [s.n.], 1975
xxii,499p
84-500-6738-3 X.200/10984
1158635

ADMINISTRATION. Justice. Austria
1945- - Statistics - Serials

STATISTIK der Rechtspflege / bearbeitet im Österreichischen Statistischen Zentralamt. - 1979-. - Wien : das Zentralamt, 1981-. - (Beiträge zur österreichischen Statistik)
r6303595 S.36/30[Hft.610,644,688,722,etc.]
8734461

ADMINISTRATION. Justice. Barcelona. Spain
Role of batlles reals, 1375-1378

MOREY, Berenguer
Llibre del Batlle Reial de Barcelona .... - Barcelona : Rafael Dalmau, 1976
135p. - (Fundació Salvador Vives Casajuana. Publicacions ; XLV)
84-232-0092-2 X.100/27234
2033658

ADMINISTRATION. Justice. Belgium
to date

MATTHIJS, J. (Jacques)
Openbaar ministerie. - Gent : Story-Scientia, [1983]
xli,392p. - (Algemene practische rechtsverzameling). - Includes bibliography
90-643-9324-9 X.0203/1.(31)
9123962

ADMINISTRATION. Justice. Canada
to date

FINKLER, Harold W.
[Inuit and the administration of criminal justice in the Northwest Territories]. Les Inuit et l'administration de la justice. - Ville La Salle : Les Éditions Hurtubise HMH, 1980
331p. - (Cahiers du Québec. Collection Cultures amérindiennes ; 61). - Includes bibliography
2-89045-449-5 X.0709/692(61)
9015779

ADMINISTRATION. Justice. Canada
1957-

BAUM, Daniel Jay
Discount justice. - [Don Mills] : Burns & MacEachern, [1979]
148p
0-88768-080-1 X.200/42540
9042113

ADMINISTRATION. Justice. Canada
Costs, to date - Conference proceedings

CONFERENCE ON THE COST OF JUSTICE (1979 : Toronto)
Cost of justice. - Toronto : Carswell, 1980
ix,160p
0-459-32940-5 X.200/46066
9154744

ADMINISTRATION. Justice. Catalonia. Spain
1931-1939

RUBIÓ I TUDURÍ, Marian
La justicia a Catalunya, 19 de juliol del 1936 - 19 de febrer del 1937. - Barcelona : Lletra Viva, [1978]
41p. - (Antecedents i documents ; 10)
r6704248 X.808/36640
8861218

ADMINISTRATION. Justice. Civil law. Italy
Discretion

RASELLI, Alessandro
Studi sul potere discrezionale del giudice civile. - Milano : A. Giuffrè, 1975
xxi, 399 p.. - (Università di Firenze. Fondazione Piero Calamandrei. [Pubblicazioni] ; n. 18)
r6024153 X.102/1105
1019252

ADMINISTRATION. Justice. Colombia

ARIZA GÓMEZ, Hilario José
Aportes al sistema judicial colombiano. - Bogotá : Editorial Kelly, 1978
187p
r6700598 X.208/6117
1240137

CASTRO, Jaime
La justicia en Colombia. - [Bogotá] : Instituto Colombiano de Cultura, 1975
104p : ill. - (Serie "Publicaciones especiales")
r6041141 X.100/26071
1240137

DIRECTORIO de organismos del sector justicia, 1977 / Departamento Administrativo Nacional de Estadistica. - Bogotá : el Departamento, 1977
142p
r6700703 LAS.c.230/89
1240137

GALVIS B, Pedro Vicente
Justicia a la colombiana. - Bogotá : Editorial Marca, [1978]
180p. - (Ensayo)
r6157789 X.208/5365
1240137

ADMINISTRATION. Justice. Coroners' courts. England
Expenditure by local authorities - Forecasts - Statistics

ADMINISTRATION of justice statistics, based on estimates / the Society of County Treasurers. - 1980/81. - Reading : The Society, 1980
8p
B81-08227 X.525/4011
1944827

ADMINISTRATION. Justice. Coroners' courts. England
Expenditure by local authorities - Statistics

ADMINISTRATION of justice statistics, based on estimates / the Society of County Treasurers. - 1978-79 / [compiler J. Unsworth]. - Reading : [The Society], 1978
[1],7p : ill
Subsequent volumes, when received, will be placed at this shelfmark. - Holdings: 1978/79, etc
B80-15112 P.203/136
1944827

ADMINISTRATION. Justice. Criminal law

CRIMINAL justice studies / [editor] Gordon E. Misner. - St. Louis ; London : Mosby, 1981
x,381p. - Includes index
0-8016-3457-1 X.200/40595
2912449

CRIMINOLOGICAL diagnosis / edited by Franco Ferracuti, Marvin E. Wolfgang. - Lexington, Mass. : Lexington Books. - Bibliography: p543-613
Vol.2. - 1983
p299-619
0-669-05971-4 X.200/41884
2991039

ADMINISTRATION. Justice. Criminal law. Athens. Ancient Greece ca B.C.400-ca B.C.500

HANSEN, Mogens Herman
Apagoge, endeixis and ephegesis against kakourgoi, atimoi and pheugontes. - [Odense] : Odense University Press, 1976
171 p.. - (Odense University classical studies ; vol. 8). - Bibliography : p. 153-155. - Includes index
87-7492-167-3 X.0902/54(8)
0738085

ADMINISTRATION. Justice. Criminal law. Australia

SALLMANN, Peter
Criminal justice in Australia. - Melbourne ; Oxford : Oxford University Press, 1984
xi,240p. - Bibliography: p227-235. - Includes index
0-19-554459-5 X.200/46248
3280772

ADMINISTRATION. Justice. Criminal law. Austria 1945- - Statistics - Serials

GERICHTLICHE Kriminalstatistik ... / bearbeitet im Österreichischen Statistischen Zentralamt. - 1975/1976-. - Wien : das Zentralamt, 1982-. - (Beiträge zur österreichischen Statistik)
r6306025 S.36/30[Hft.633,636,675,708,719,etc.]
8820694

ADMINISTRATION. Justice. Criminal law. Austria Role of Joseph II, Holy Roman Emperor, 1765-1790

BERNARD, Paul Peter
The limits of enlightenment. - Urbana ; London : University of Illinois Press, 1979
xiii,146p. - Bibliography : p. 138-142. - Includes index
0-252-00735-2 X.200/19030
2048116

ADMINISTRATION. Justice. Criminal law. Canada Plea bargaining

KLEIN, John F.
Let's make a deal. - Lexington, Mass. : D.C. Heath ; Farnborough, Hants. : Distributed by Teakfield Ltd, 1976
ix,161p. - Bibliography: p.151-160
0-669-00796-x X.200/30916
111381x

ADMINISTRATION. Justice. Criminal law. England

CRIMINAL justice / edited by John Baldwin and A. Keith Bottomley. - London : Martin Robertson, 1978
viii,311p : ill. - Includes bibliographies and index
0-85520-234-3 X.100/17865
0271047

FITZGERALD, Mike
System of justice. - Oxford : Basil Blackwell, 1983
216p : ill. - Bibliography: p205-209. - Includes index
0-631-13248-1 X.200/42424
0271047

GABBAY, Edmond
Open justice. - London : The author, 1978
ix,60p. - Includes index
0-9501448-4-3 X.100/17349
0271047

GREAT BRITAIN. Home Office
Evidence to the Royal Commission on Criminal Procedure. - London : Home Office.
Memorandum no.1 : The background to the work of the Commission ; [and], Memorandum no.2 : The accountability of the police. - 1978
v,55,[16]p
0-903727-56-0 BS.18/625(1)
1879634

Memorandum no.3: The powers of the police to arrest or otherwise stop a person, to search him, to stop and search vehicles, and to enter and search premises. - 1978
iii,38,[59]p
0-903727-57-9 BS.18/625(3)
1879634

Memorandum no.4: The law and procedures relating to the detention and treatment of persons in police custody. - 1978
iii,49p
0-903727-61-7 BS.18/625(4)
1879634

Memorandum no.5: The law and procedures relating to the questioning of persons in the investigation of crime. - 1978
iv,74p
0-903727-64-1 BS.18/625(5)
1879634

Memorandum no.6: Cautioning by the police. - 1979
iv,53p : ill
0-903727-72-2 BS.18/625(10)
1879634

Memorandum no.7: Preparation for trial and disclosure of evidence. - 1979
iv,50p
0-903727-70-6 BS.18/625(7)
1879634

Memorandum no.8: The prosecution process. - 1978
iii,52p
0-903727-60-9 B.S.18/625(6)
1879634

Memorandum no.9: The law of evidence in criminal proceedings. - 1978
iii,43p
0-903727-63-3 BS.18/625(9)
1879634

Memorandum no.10: Some aspects of relations between the police and the public. - 1979
iii,21p
0-903727-73-0 BS.18/625(10)
1879634

WALKER, Nigel, 1917-
Curiosities of criminal justice. - Chichester : Rose, [1975]
16p. - (The tenth Frank Newsam memorial lecture)
0-85992-034-8 X.100/16023
0271047

ADMINISTRATION. Justice. Criminal law. England - Statistics

STATISTICS of the criminal justice system, England and Wales / Home Office. - London : H.M.S.O.
1968-78. - 1979
xi,89p : ill. - Bibliography: p.89
Subsequent volumes are also placed at this shelfmark
0-11-340726-2 BS.18/675
2252767

ADMINISTRATION. Justice. Criminal law. England 1550-1800

CRIME in England, 1550-1800 / edited by J.S. Cockburn. - London : Methuen, 1977
xiv,364p : ill. - Bibliography: p.270-298. - Includes index
0-416-83960-6 X.200/31202
1307967

ADMINISTRATION. Justice. Criminal law. England 1926-1977

WOOTTON, Barbara,
Crime and penal policy. - London : Allen and Unwin, 1978
3-261p. - Includes index
0-04-364011-7 X.200/31402
148088x
- [Another copy : paperback]
ISBN 0-04-364013-3 X.200/35586

ADMINISTRATION. Justice. Criminal law. England Plea bargaining

BALDWIN, John, 1945-
Negotiated justice. - London : Martin Robertson, 1977
xvi,128p. - (Law in society series). - Bibliography: p.121-126. - Includes index
0-85520-171-1 X.200/31129
1261746

PLEA-BARGAINING / edited by William F. McDonald, James A. Cramer. - Lexington : Lexington Books, 1980
xi,204p : ill. - Includes bibliographies and index
0-669-02363-9 X.200/35554
2406381

ADMINISTRATION. Justice. Criminal law. England Policies of government, 1966-1976

GREAT BRITAIN. Home Office
A review of criminal justice policy, 1976. - London : H.M.S.O., 1977
[5],49p : ill. - (Home Office working paper)
0-11-340056-x BS.18/556
1332252

ADMINISTRATION. Justice. Criminal law. France

LEFÈVRE, Paul
Les serviteurs de la justice. - Paris : Julliard, 1974
172 p.
r6002846 X.100/14589
0920118

ADMINISTRATION. Justice. Criminal law. France Procedure compared with administration of criminal justice in Scotland

SHEEHAN, Albert V.
Criminal procedure in Scotland and France. - Edinburgh : H.M.S.O., 1975
[2],xvii,230p. - Bibliography: p.229-230
0-11-491329-3 BS.156/5
0766224

ADMINISTRATION. Justice. Criminal law. Great Britain

CAUSES for concern / edited by Phil Scraton and Paul Gordon. - Harmondsworth : Penguin, 1984
287p. - (Pelican books). - Includes bibliographies
0-14-022464-5 X.208/10858
0334154

DECISION making in the legal system / edited by Joanna Shapland. - Leicester : British Psychological Society for the Division of Criminological and Legal Psychology, 1983
73p. - (Issues in criminological and legal psychology ; no.5)
0-901715-24-7 X.525/10529
0334154

GREAT BRITAIN. Central Office of Information. Reference Division
Criminal justice in Britain. - London : H.M.S.O., 1975
[3],56p. - (Central Office of Information reference pamphlet ; 129). - Bibliography: p.56
0-11-700763-3 BS.81/35
0334154

UNFAIR justice / compiled by R.D. Forbes. - [Aberdeen] : [Forbes & Williamson], [1983]
74,[58]p
0-9508953-0-x X.200/42066
0334154

ADMINISTRATION. Justice. Criminal law. India (Republic) Procedure

SARKAR, S. C. (Subodh Chandra)
The law of criminal procedure. - 4th ed.. - Calcutta : S. C. Sarkar and Sons Ltd, 1975
885 p.. - Includes index
r6010239 X.200/30716
0794767

SARKER, U.C
Commentaries on the code of criminal procedure, 1973. - Allahabad : Central Law Agency, 1974
xxxvi,1131p
r6025665 X.200/11531
0794767

ADMINISTRATION. Justice. Criminal law. Languedoc. France 1750-1789

CASTAN, Nicole
Justice et répression en Languedoc à l'époque des lumières. - Paris : Flammarion, 1980
313p : ill,maps,plans. - (Science). - Bibl
2-08-210656-x X.200/33812
2439921

ADMINISTRATION. Justice. Criminal law. Scotland Procedure compared with administration of criminal justice in France

SHEEHAN, Albert V.
Criminal procedure in Scotland and France. - Edinburgh : H.M.S.O., 1975
[2],xvii,230p. - Bibliography: p.229-230
0-11-491329-3 BS.156/5
0766216

ADMINISTRATION. Justice. Criminal law. Scotland Procedure. Law - Texts

GREAT BRITAIN
[Criminal Justice (Scotland) Act 1980]. The Criminal Justice (Scotland) Act 1980. - Edinburgh : W. Green, 1981
xxvii,165p. - Includes index
0-414-00674-7 X.200/35750
1054694

GREAT BRITAIN. Crown Office
Criminal Procedure (Scotland) Act 1975. - Edinburgh : H.M.S.O., 1976, 1976-
2v.(loose-leaf)
0-11-491403-6 BS.39/81
1054694

ADMINISTRATION. Justice. Criminal law. South Africa

JACKSON, John D. (John David), 1940-
Justice in South Africa. - Harmondsworth : Penguin, 1980
239p. - (a Pelican book). - Includes index
0-14-022245-6 X.208/5757
227552x

ADMINISTRATION. Justice. Criminal law. Soviet Union
Role of advocates

STET͡SOVSKIĬ, I͡U. I. (I͡Uriĭ Isaakovich)
Advokat v ugolovnom sudoproizvodstve. - Moskva : I͡Uridicheskai͡a literatura, 1972
157p
r6462729 X.200/42337
9007393

ADMINISTRATION. Justice. Criminal law. United States

An ANATOMY of criminal justice / edited by Cleon H Foust, D. Robert Webster. - Lexington, Mass. : Heath ; [Farnborough, Hants.] : [Distributed by] Gower Press, 1980
xv,325p
0-669-02854-1 X.200/33247
0293350

CONRAD, John P. (John Patrick)
Justice and consequences. - Lexington, Mass. : Lexington Books, c1981
xv,175p. - Includes index
0-669-02190-3 X.200/37264
0293350

CRIMINAL justice planning and development / edited by Alvin W. Cohn. - Beverly Hills ; London : Sage Publications [for] the American Society of Criminology, 1977
134p. - (Sage research progress series in criminology ; vol.4). - Includes bibliographies
0-8039-0918-7 X.100/17564
0293350

The CRIMINAL justice process / edited by William B. Sanders and Howard C. Daudistel. - New York : Praeger, 1976
xi, 291 p. : 1 ill. - Includes bibliographies and index
0-275-89320-0 X.100/16772
1048449

The CRIMINAL justice system / Ronald J. Waldron ... [et al.]. - Boston [Mass.] ; London : Houghton Mifflin, 1976
xiii,478p : ill. - Includes bibliographies and index
0-395-18592-0 X.200/31316
0293350

The CRIMINAL justice system / edited by Vladimir J. Konečni and Ebbe B. Ebbesen. - Oxford : Freeman, c1982
xiv,418p : ill. - Includes bibliographies and index
0-7167-1312-8 X.520/28007
2756013

CRIMINAL law in action / [compiled by] William J. Chambliss. - 2nd ed. - New York ; Chichester : Wiley, c1984
x,461p : ill. - Includes bibliographies
0-471-89678-0 X.200/44467
0293350

CRITICAL issues in criminal justice / edited by Donald O. Schultz. - Springfield, Ill. : Thomas, 1975
xi,289p : ill. - Bibliography : p.285-289
0-398-03331-5 X.200/31326
0293350

DAUDISTEL, Howard C.
Criminal justice. - New York ; London : Holt, Rinehart and Winston, 1979
xiv,401p : ill. - Includes bibliographies and index
0-03-033051-3 X.200/31850
0293350

EDELSTEIN, Charles D.
An introduction to criminal justice. - New York ; London : Gregg Division, McGraw-Hill, 1977
ix,2-296p : ill. - Bibliography: p.245-249. - Includes index
0-07-018980-3 X.100/17049
0293350

The EVOLUTION of criminal justice / edited by John P. Conrad. - Beverly Hills ; London : Sage Publications [for] the American Society of Criminology, 1978
168p : ill. - (Sage research progress series in criminology ; vol.7). - Includes bibliographies
0-8039-1146-7 X.200/31927
0293350

FOSTER, Charles Alfred
Introduction to the administration of justice. - 2nd ed.. - New York ; Chichester : Wiley, 1979
ix,347p,fold plate : ill. - Includes index
0-471-04079-7 X.200/32063
0293350

GOLDSTEIN, Joseph, 1923-
Criminal law. - [New ed.]. - New York : Free Press ; London : Collier Macmillan, 1974
lxxii,1287p : ill. - Includes index
0-02-912310-0 X.203/2013
0293350

GRAY, Virginia
The organizational politics of criminal justice. - Lexington : Lexington Books, 1981, c1980
xiv,175p : ill. - Includes index
0-669-02108-3 X.200/35711
0293350

JOHNSTON, Norman
Legal process and corrections. - New York ; Chichester : Wiley, c1982
viii,352p : ill. - Includes bibliographies
0-471-08337-2 X.200/38039
0293350

KRATCOSKI, Peter C.
Criminal justice in America. - Dallas : London : Scott, Foresman, 1978
xxii,424p : ill. - Includes index
0-673-15051-8 X.200/19072
0293350

LAW and the legal process / edited by Victoria L. Swigert. - Beverly Hills, [Calif.] ; London : Sage, c1982
160p. - (Sage research progress series in criminology ; v.25)
0-8039-1900-x X.200/40243
2506327

LETMAN, Sloan T.
Criminal justice. - Jefferson, N.C. ; London : McFarland, 1983
vi,154p. - Bibliography: p145-152. - Includes index
0-89950-039-0 X.200/41912
0293350

LITTRELL, W. Boyd
Bureaucratic justice. - Beverly Hills ; London : Sage Publications, 1979
284p : 2 ill. - (Sage library of social research ; vol.93). - Includes index
0-8039-1264-1 X.102/2264
0293350

MAYHALL, Pamela D.
Community relations and the administration of justice. - 2nd ed.. - New York ; Chichester : Wiley, 1979
xi,359p : ill. - Includes bibliographies and index
0-471-04135-1 X.520/13782
0293350

NAGEL, Stuart S. (Stuart Samuel)
The legal process. - Beverly Hills ; London : Sage Publications, 1977
215p : ill. - (Sage library of social research ; vol.45). - Includes bibliographies and index
0-8039-0853-9 X.100/17268
0293350

POUND, Roscoe
Criminal justice in America. - New York : Da Capo Press, 1975
[14],226p. - Includes index
0-306-80007-1 X.100/18447
0293350

RICH, Vernon
Law and the administration of justice. - 2nd ed. - New York ; Chichester : Wiley, 1979
ix,362p : ill. - Includes bibliographies and index
0-471-04961-1 X.200/32062
0293350

SEYMOUR, Whitney North
United States attorney. - New York : Morrow, 1975
248p
0-688-02945-0 X.200/12389
0293350

SKOLER, Daniel L.
Organizing the non-system. - Lexington, Mass. : Heath, 1977
xxvi,310p : ill
0-669-00941-5 X.200/31841
0293350

SMITH, Alexander B.
Criminal justice. - [New ed.]. - New York ; London : Holt, Rinehart and Winston, 1980
xi,251p. - Bibliography - Includes index
0-03-046586-9 X.200/32898
0293350

SWATON, J. Norman
Administration of justice. - 2nd ed. - New York ; London : D. Van Nostrand, 1980
xiii,449p : ill. - Bibliography : p.417-419. - Includes index
0-442-25789-9 X.200/32856
0293350

WALDRON, Ronald J.
The criminal justice system. - 2nd ed. - Boston [Mass.] ; London : Houghton Mifflin, 1980
xxii,522p : ill. - Includes bibliographies and index
0-395-28669-7 X.200/32845
0293350

WEINREB, Lloyd L.
Denial of justice. - New York : Free Press ; London : Collier Macmillan, 1977
xiii,177p. - Includes index
0-02-934900-1 X.200/30972
0293350

WRIGHT, Jack
Modern criminal justice. - New York ; London : McGraw-Hill, 1978
xi,338p : ill. - Includes index
0-07-072075-4 X.200/31350
0293350

ADMINISTRATION. Justice. Criminal law. United States - Conference proceedings

IMPROVING management in criminal justice / edited by Alvin W. Cohn and Benjamin Ward. - Beverly Hills ; London : Published in cooperation with the American Society of Criminology [by] Sage, c1980
159p. - (Sage research progress series in criminology ; v.17). - Includes bibliographies
0-8039-1515-2 X.200/35987
2506327

ADMINISTRATION. Justice. Criminal law. United States - Forecasts

CRIME and justice in America : edited by John T. O'Brien, Marvin Marcus. - New York ; Oxford : Pergamon, 1979
xx,361p : ill. - (Pergamon policy studies on crime and justice). - Includes bibliographies and index
0-08-023857-2 X.200/33196
2072513

ADMINISTRATION. Justice. Criminal law. United States to 1975

LAW and order in American history / edited by Joseph M. Hawes. - Port Washington ; London : Kennikat Press, 1979
[6],184p. - (Kennikat Press national university publications. multi-disciplinary studies in the law)
0-8046-9238-6 X.200/33491
2105667

ADMINISTRATION. Justice. Criminal law. United States to 1979

WALKER, Samuel
Popular justice. - New York ; Oxford : Oxford University Press, 1980
xiii,287p. - Bibliography: p.254-256. - Includes index
0-19-502655-1 X.200/33898
2242567

ADMINISTRATION. Justice. Criminal law. United States 1606-1660

CHAPIN, Bradley
Criminal justice in colonial America, 1606-1660. - Athens, Ga. : University of Georgia Press, c1983
203p : ill. - Bibliography: p187-195
0-8203-0624-x X.200/46122
3182061

ADMINISTRATION. Justice. Criminal law. United States
Applications of psychology

FERSCH, Ellsworth A.
Psychology and psychiatry in courts and corrections. - New York ; Chichester : Wiley, 1980
xiii,370p. - (Wiley series on personality processes). - Bibliography: p.323-353. - Includes index
0-471-05604-9 X.325/724 Woolwich
2317419

ADMINISTRATION. Justice. Criminal law. United States
Decision making

NAGEL, Stuart S. (Stuart Samuel)
Decision theory and the legal process. - Lexington, Mass. : Heath ; [Farnborough, Hants.] : Distributed by Teakfield Limited, 1979
xviii,295p : ill. - Bibliography: p.275-281. - Includes index
0-669-02742-1 X.200/19107
2053853

ADMINISTRATION. Justice. Criminal law. United States
Discretion - Conference proceedings

DISCRETION and control / edited by Margaret Evans. - Beverly Hills ; London : Sage Publications [for] the American Society of Criminology, 1978
160p : 2 ill. - (Sage research progress series in criminology ; vol.9). - Includes bibliographies
0-8039-1128-9 X.200/31959
1871676

ADMINISTRATION. Justice. Criminal law. United States
Discretion, to date

DOW, Paul E.
Discretionary justice. - Cambridge, Mass. : Ballinger, [1981]
286p
0-88410-835-x X.200/43507
9118322

ADMINISTRATION. Justice. Criminal law. United States
Effects on negroes

BLACKS and criminal justice / edited by Charles E. Owens, Jimmy Bell. - Lexington, Mass. : Heath ; Farnborough, Hants. : Distributed by Teakfield Ltd, 1977
xi,155p. - Includes bibliographies and index
0-669-01101-0 X.200/32098
1912445

ADMINISTRATION. Justice. Criminal law. United States
Ethical aspects - Conference proceedings

LEGALITY, morality and ethics in criminal justice / edited by Nicholas N. Kittrie, Jackwell Susman. - New York ; London : Praeger [for] the American Society of Criminology, 1979
x,307p : ill. - (Praeger special studies). - Includes bibliographies
0-03-047521-x X.200/32138
1922106

ADMINISTRATION. Justice. Criminal law. United States
Evaluation

HANDBOOK of criminal justices evaluation / edited by Malcolm W. Klein and Katherine S. Teilmann. - Beverly Hills ; London : Sage, c1980
693p. - Includes bibliographies and index
0-8039-1052-5 X.200/35341
2386925

ADMINISTRATION. Justice. Criminal law. United States
Models

MODELING the criminal justice system / Stuart S. Nagel, editor. - Beverly Hills ; London : Sage Publications, 1977
319p : ill. - (Sage criminal justice system annuals ; vol.7). - Includes bibliographies
0-8039-0596-3 X.100/17994
1720929

ADMINISTRATION. Justice. Criminal law. United States
Participation of community action groups

ALPER, Benedict S. (Benedict Solomon)
Beyond the courtroom. - Lexington, Mass. : Lexington Books, 1981
xx,299p : ill. - Bibliography: p273-288. - Includes index
0-669-02724-3 X.200/36987
2574101

NEIGHBORHOOD justice / [edited by] Roman Tomasic, Malcolm M. Feeley. - New York ; London : Longman, c1982
xviii,286p. - Bibliography: p249-275. - Includes index
0-582-28253-5 X.200/38793
2574101

ADMINISTRATION. Justice. Criminal law. United States
Participation of North American Indians, 1878-1900

HAGAN, William T. (William Thomas)
Indian police and judges. - Lincoln [Neb.] ; London : University of Nebraska Press, 1980
xiii,194p,[12]p of plates : ill. - Bibliography: p.177-183. - Includes index
0-8032-2308-0 X.208/5813
2307405

ADMINISTRATION. Justice. Criminal law. United States
Personnel. Interpersonal relationships with clients

NEIL, Thomas C.
Interpersonal communications for criminal justice personnel. - Boston, Mass. ; London : Allyn and Bacon, 1980
xiii,200p : ill. - Bibliography: p.193-195. - Includes index
0-205-06829-4 X.200/33264
2160900

ADMINISTRATION. Justice. Criminal law. United States
Planning

SHANAHAN, Donald T.
The dimensions of criminal justice planning. - Boston, Mass. ; London : Allyn and Bacon, 1980
x,458p : ill. - Bibliography: p.433-434. - Includes index
0-205-06668-2 X.200/33391
2210738

ADMINISTRATION. Justice. Criminal law. United States
Plea bargaining

BUCKLE, Suzann R. Thomas
Bargaining for justice. - New York ; London : Praeger, 1977
v,183p. - (Praeger special studies in US economic, social and political issues). - Bibliography: p.170-177. - Includes index
0-275-56070-8 X.200/31049
120260x

HEUMANN, Milton
Plea bargaining. - Chicago ; London : University of Chicago Press, 1978
viii,220p : ill. - Bibliography: p.207-211. - Includes index
0-226-33187-3 X.200/31473
1551485

JONES, David A. (David Arthur)
Crime without punishment. - Lexington, Mass. : Heath ; [Farnborough, Hants.] : [Distributed by] Teakfield Limited, 1979
xv,277p. - Bibliography: p.251-267. - Includes index
0-669-02512-7 X.200/32410
1551485

MATHER, Lynn M.
Plea bargaining or trial?. - Lexington, Mass. : Heath ; [Farnborough, Hants.] : [Distributed by] Teakfield Limited, 1979
xi,173p : ill. - Bibliography: p.159-164.- Includes index
0-669-00467-7 X.200/33394
2072734

MAYNARD, Douglas W.
Inside plea bargaining. - New York ; London : Plenum, c1984
viii,257p : ill. - Includes index
0-306-41577-1 X.520/36610
1551485

PLEA-BARGAINING / edited by William F. McDonald, James A. Cramer. - Lexington : Lexington Books, 1980
xi,204p : ill. - Includes bibliographies and index
0-669-02363-9 X.200/35554
1551485

ADMINISTRATION. Justice. Criminal law. United States
Policies of government. Implementation

IMPLEMENTING criminal justice policies / edited by Merry Morash. - Beverly Hills, [Calif.] ; London : Sage, c1982
127p. - (Sage research progress series in criminology)
0-8039-1884-4 X.200/40268
2892235

ADMINISTRATION. Justice. Criminal law. United States
Policies of government. Political aspects

RHODES, Robert P.
The insoluble problems of crime. - New York ; London : Wiley, 1977
ix,269p : ill. - Includes index
0-471-71799-1 X.529/36300
2084147

ADMINISTRATION. Justice. Criminal law. United States
Psychological aspects

COHN, Alfred
The criminal justice system and its psychology. - New York ; London : Van Nostrand Reinhold, 1979
xiii,345p : ill. - Bibliography: p.313-232. - Includes index
0-442-28882-4 X.200/32339
1996282

ADMINISTRATION. Justice. Criminal law. United States
Racial factors

RACE, crime, and criminal justice / edited by R.L. McNeely and Carl E. Pope. - Beverly Hills ; London : Sage, c1981
176p : 1ill. - (Perspectives in criminal justice ; 2). - Includes bibliographies
0-8039-1584-5 X.200/36157
2532670

ADMINISTRATION. Justice. Criminal law. United States
related to community mental health services

COMMUNITY mental health and the criminal justice system / editor John Monahan. - New York ; Oxford : Pergamon, 1976
xvi,332p. - (Pergamon international library) (Pergamon general psychology series ; vol.57). - Includes bibliographies and index
0-08-018759-5 X.320/10322 Woolwich
0552941

ADMINISTRATION. Justice. Criminal law. United States
Role in prevention of juvenile delinquency

FALKIN, Gregory P.
Reducing delinquency. - Lexington, Mass. : Heath ; [Farnborough, Hants.] : [Distributed by] Gower Press, 1980
xxiii,213p : ill. - Bibliography: p.197-207. - Includes index
0-669-02318-3 X.520/21288
215157x

ADMINISTRATION. Justice. Criminal law. United States
Role of crime victims

CARRINGTON, Frank G.
The victims. - New Rochelle, N.Y. : Arlington House, 1977
326p
0-87000-302-x X.200/32468
1072528

CRIMINAL justice and the victim / William F. McDonald, editor. - Beverly Hills ; London : Sage Publications, [1977]
288p : 1 ill. - (Sage criminal justice system annuals ; vol.6). - Includes bibliographies and index
0-8039-0508-4 X.100/16850
1072528

ADMINISTRATION. Justice. Criminal law. United States
Systems analysis

LAPATRA, J. W.
Analysing the criminal justice system. - Lexington, Mass. : Heath ; [Farnborough, Hants.] : [Distributed by] Teakfield Ltd, 1978
xv,175p : ill. - Includes index
0-669-01625-x X.200/31651
1720937

ADMINISTRATION. Justice. Customary law. Yorubaland. Nigeria
to date

AJIBOLA, J. O.
Administration of justice in the customary courts of Yorubaland. - Ibadan : University Press, 1982
ix,89p
978-15-4623-9 X.208/9213
9061568

ADMINISTRATION. Justice. Czechoslovakia
1945-

PLUNDR, Otakar
Organizace justice a prokuratury. - Praha : Panorama, 1980
248p. - (Učebnice). - Bibliography: p.220-234
r6439617 X.529/53948
8937788

ADMINISTRATION. Justice. Delhi. India (Republic)
1977-

LAPRISE, Huguette
L'affaire Marie-Andrée Leclerc. - [Montréal] : Les Éditions de la Presse, 1977
214p,24p of plates
0-7777-0180-4 X.200/37087
877448X

ADMINISTRATION. Justice. Ecuador

ITURRALDE VELASCO, Enrique
La administración de justicia en el Ecuador. - Guayaquil : Editorial El Sol, 1978
238p
r6162332 X.200/34243
2463490

ADMINISTRATION. Justice. England
1200-1300

MEEKINGS, C. A. F. (Cecil Anthony Francis)
Studies in 13th century justice and administration. - London : Hambledon, c1981
342p in various pagings. - (History series ; no.3). - Includes index
0-9506882-3-1 X.800/39524
3209350

ADMINISTRATION. Justice. Europe
1100-1517

FOWLER-MAGERL, Linda
Ordo iudiciorum vel ordo iudiciarius. - Frankfurt am Main : Klostermann, 1984
xii,341p. - (Ius commune. Sonderhefte ; 19. Repertorien zur Frühzeit der gelehrten Rechte)
3-465-01623-8 X.0200/99a(19)
9113177

ADMINISTRATION. Justice. Europe
Political aspects - Conference proceedings

INSTITUT D'ÉTUDES POLITIQUES DE STRASBOURG
Justice et politique. - [Colmar] : Presses Universitaires d'Alsace, [1974]
262p. : ill. - (Institut d'études politiques de Strasbourg. Cahier ; n.s. no. 3)
r6012295 X.0709/163(3)
0874892

ADMINISTRATION. Justice. European Community countries
Discretion compared with discretion in administration of justice in United States

DAVIS, Kenneth Culp
Discretionary justice in Europe and America. - Urbana ; London : University of Illinois Press, 1976
xi,203p
0-252-00579-1 X.200/30595
0739871

ADMINISTRATION. Justice. Filipino military forces
1974 - Conference proceedings

ADMINISTRATION of justice by the military, 1974 / edited by Vicente Pascual, Jr. - Quezon City : University of the Philippines, Law Center, 1974
xxv, 390 p.. - Includes index
r6031254 X.200/30793
1050508
- [Another copy] X.200/37367

ADMINISTRATION. Justice. Finland
Political aspects

OIKEUSLAITOKSEN poliitisominen / [toimittanut Oikeustieteen tutkimussäätiö ; kirjoittaneet Eero Backman...et al.]. - Porvoo : W. Söderström, 1972
98p. - (Taskutieto : 89)
951-00-0214-3 X.708/19970
1127446

ADMINISTRATION. Justice. Flensburg. West Germany
to date

RECHTSPFLEGE und Justizverwaltung in Flensburg - Flensburg : Gesellschaft für Flensburger Stadtgeschichte, 1982
95p. : ill. - (Kleine Reihe der Gesellschaft für Flensburger Stadtgeschichte ; Hft.8)
r6307505 L.49/1314(8)
8978115

ADMINISTRATION. Justice. France
to date

DEBRÉ, Jean-Louis
La justice au XIXe siècle. - Paris : Librairie académique Perrin, 1980
223p,16p of plates
2-262-00215-0 X.200/37352
8765790

ADMINISTRATION. Justice. France
1789-1804

WILLETTE, Luc
Le tribunal révolutionnaire. - Paris : Denoël, 1981
189p. - (Les erreurs judiciaires de l'histoire)
r6206751 X.809/51253
8836175

ADMINISTRATION. Justice. France
1815-1870

CONSTANT, Benjamin
[Selections]. L'affaire Regnault. - Grenoble : Université des langues et lettres de Grenoble, 1979
161p : ill
2-902709-07-2 X.200/36677
8734453

ADMINISTRATION. Justice. France
1958-

CASAMAYOR, Louis
L'idole et le citoyen. - Paris : Gallimard, 1982
131p
2-07-022604-2 X.200/42999
8765804

DEBARD, Mireille
Dossiers classés. - [Paris] : Albin Michel, 1982
246p. - (Collection "Illustrations")
2-226-01463-2 X.200/44714
8765804

EMSELLEM, Denise
Pratique et organisation dans l'institution judiciaire. - Paris : La Documentation française, 1982
287p. - (Ministère de la Justice, Service de Coordination de la Recherche. Sociologie Judiciaire)
2-11-001135-1 S.E.400/1345
8765804

KNOBELSPIESS, Roger
L'acharnement, ou, La volonté d'erreur judiciaire. - Paris : Stock, 1981
177p
2-234-01456-5 X.208/7159
8765804

PAUCHET, Catherine
Les prisons de l'insécurité. - Paris : Les Éditions Ouvrières, 1982
201p. - (Comprendre pour agir)
2-7082-2238-4 X.529/63366
8765804

PEYREFITTE, Alain
Les chevaux du Lac Ladoga. - Paris : Plon, 1981
425p
2-259-00782-1 X.200/38016
8765804

ADMINISTRATION. Justice. France
1958- - Serials

FRANCE. Ministère de la justice
Bulletin officiel du Ministère de la justice. - no 1(1980)-. - Paris : [le Ministère], [1980]-
Holdings: no 6, 1982-
r6210121 S.E.400/1158
8765804

ADMINISTRATION. Justice. France
1958- - Statistics - Serials

ANNUAIRE statistique de la justice / Ministère de la justice. - (1978)-. - Paris : Service de l'administration générale et de l'équipement, Division de la statistique de la justice, 1981-
r6206609 SE.400/915
8765804

ADMINISTRATION. Justice. Germany
1933-1945 - Conference proceedings

RECHT, Justiz und Faschismus Helmut D. Fangmann, Norman Paech (Hg.). - Köln : Theurer, 1984
179p.. - (Rechtspolitische Schriften : 6)
3-8161-0106-2 X.809/62447
9170332

ADMINISTRATION. Justice. Germany
Law, 1871-1918

SCHUBERT, Werner, 1936-
Die deutsche Gerichtsverfassung (1869-1877) - Frankfurt am Main : Klostermann, 1981
xxvi,1030p.. - (Ius commune Sonderhefte : 16)
3-465-01442-1 X.0200/99a(16)
8814600

ADMINISTRATION. Justice. Great Britain

STREET, Harry
Justice in the welfare state. - 2nd ed. - London : Stevens and Sons Ltd [for] the Hamlyn Trust, 1975
xi,130p. - (The Hamlyn lectures ; 20th series)
0-420-44450-5 WP.3630/20a
0364649

ADMINISTRATION. Justice. Great Britain
Committees of inquiry. Reports

BURTON, Frank
Official discourse. - London : Routledge and Kegan Paul, 1979
xi, 148p. - (International library of sociology). - Bibliography : p. 140-144. - Includes index
0-7100-0328-5 X.529/34601
2021234

ADMINISTRATION. Justice. Great Britain
Law

GREAT BRITAIN
[The Administration of Justice Act 1982]. The Administration of Justice Act 1982. - London : Sweet & Maxwell, 1983
vii,83p. - (Current law statutes annotated reprints)
0-421-31260-2 X.200/46768
1250280

ADMINISTRATION. Justice. Holy Roman Empire
to 1519

BATTENBERG, Friedrich
Die Gerichtsstandsprivilegien der deutschen Kaiser und Könige bis zum Jahre 1451. - Köln : Böhlau, 1983
2v.. - (Quellen und Forschungen zur höchsten Gerichtsbarkeit im alten Reich ; Bd.12). - Includes bibliography
3-412-04182-3 X.0800/573(12)
8997500

DRÜPPEL, Hubert
Index civitatis. - Köln : Böhlau, 1981
xliii,463p.,plate. - (Forschungen zur deutschen Rechtsgeschichte ; Bd.12). - Includes bibliography
3-412-06580-3 X.800/37672
8997500

ADMINISTRATION. Justice. Holy Roman Empire
1519-1815

HESSISCHES HAUPTSTAATSARCHIV WIESBADEN
Reichskammergericht. - Wiesbaden : Hessisches Hauptstaatsarchiv. - (Repertorien des Hessischen Hauptstaatsarchivs Wiesbaden ; Abt.1)
Tl.2: Prozessakten der Landgrafschaft Hessen-Homburg. - 1984
xiii,87p.. - (Inventar der Akten des Reichskammergerichts ; Nr.7)
3-921930-51-0 X.0700/3071(1)[Tl.2]
872136x

Die KAISERLICHEN privilegia de non appellando / mit einer Abhandlung eingeleitet und in Zusammenarbeit mit Elsbeth Markert regestiert und in einer Auswahl herausgegeben von Ulrich Eisenhardt. - Köln : Böhlau, 1980
xx,351p.. - (Quellen und Forschungen zur höchsten Gerichtsbarkeit im alten Reich ; Bd.7)
3-412-00580-0 X.0800/573(7)
872136x

MENCKE, Klaus
Die Visitationen am Reichskammergericht im 16. Jahrhundert. - Köln : Böhlau, 1984
xxxv,141p.. - (Quellen und Forschungen zur höchsten Gerichtsbarkeit im Alten Reich ; Bd.13). - Includes bibliography
3-412-07683-x X.0800/573(13)
872136x

ADMINISTRATION. Justice. Holy Roman Empire
Historical sources, to date

NIEDERSACHSISCHES STAATSARCHIV IN OLDENBURG
Findbuch zu den Reichskammergerichtsakten 1524-1806. - Göttingen : Vandenhoeck & Ruprecht, 1981
xix,418p.. - (Inventare und kleinere Schriften des Staatsarchivs in Oldenburg ; Hft.15) (Veröffentlichungen der Niedersächsischen Archivverwaltung) (Inventar der Akten des Reichskammergerichts ; Nr.5)
3-525-85961-9 X.0809/2308(15)
910545x

NIEDERSACHSISCHES STAATSARCHIV IN STADE
Findbuch zum Bestand 27. Reichskammergericht (1500-1648) - Göttingen : Vandenhoeck & Ruprecht, 1981
xii,452p.. - (Inventare und Kleinere Schriften des Staatsarchivs in Stade ; Hft.1) (Veröffentlichungen der Niedersächsischen Archivverwaltung) (Inventar der Akten des Reichskammergerichts ; Nr.4)
3-525-85960-0 X.0809/2305(1)
910545x

NIEDERSACHSISCHES STAATSARCHIV IN WOLFENBUTTEL
Findbuch zum Bestand Reichskammergericht und Reichshofrat 1489-1806 (6 Alt) - Göttingen : Vandenhoeck & Ruprecht, 1981
410p.. - (Inventare und Kleinere Schriften des Staatsarchivs in Wolfenbüttel ; Hft.2) (Veröffentlichungen der Niedersächsischen Archivverwaltung) (Inventar der Akten des Reichskammergerichts ; Nr.3)
3-525-85955-4 X.0809/2310(2)
910545X

ADMINISTRATION. Justice. Italy
to date

BRUNI, Domenico
Sul sistema di giustizia amministrativa in Italia dopo i provvedimenti di riforma. - 2a ed. - Bologna : Pàtron, 1978
144p. - (Collana di scienze giuridiche ; 11). - Bibliography
r6505993 X.809/54234
8836159

ADMINISTRATION. Justice. Italy
to date - Conference proceedings

GIUSTIZIA AMMINISTRATIVA COME FUNZIONE DELLO STATO DEMOCRATICO (Conference : 1978 : Venice)
La giustizia amministrativa come funzione dello stato democratico. - Venezia : Marsilio, 1979
318p. - (Ricerche ; n34)
r6502734 X.200/38252
8836159

ADMINISTRATION. Justice. Japan
to date

La JUSTICE au Japon. - [S.l.] : Cour Suprême du Japon, 1980
48p : ill
r6205687 X.205/827
8721513

ADMINISTRATION. Justice. Juvenile courts. Chicago. Illinois

MURPHY, Patrick T.
Our kindly parent, the state. - Harmondsworth : Penguin, 1977
xi,180p : 1 ill
0-14-004230-x X.108/16866
1283758

ADMINISTRATION. Justice. Law
Discretion. Psychosocial aspects

SOCIAL psychology and discretionary law / edited by Lawrence Edwin Abt and Irving R. Stuart. - New York ; London : Van Nostrand Reinhold, 1979
x,370p : ill. - Includes bibliographies and index
0-442-27907-8 X.200/32258
1952196

ADMINISTRATION. Justice. Magistrates' courts. Criminal courts. England

KING, Michael, 1942-
The framework of criminal justice. - London : Croom Helm, c1981
159p. - Bibliography: p153-156. - Includes index
0-7099-0430-4 X.200/34820
2208431

ADMINISTRATION. Justice. Magistrates' courts. England

CARLEN, Pat
Magistrates' justice. - London : Martin Robertson, 1976
xii,134p. - (Law in society series). - Bibliography: p.131-134
0-85520-121-5 X.200/30511
0700541

ADMINISTRATION. Justice. Magistrates' courts. England
Expenditure by local authorities - Forecasts - Statistics

ADMINISTRATION of justice statistics, based on estimates / the Society of County Treasurers. - 1980/81. - Reading : The Society, 1980
8p
B81-08227 X.525/4011
1944819

ADMINISTRATION. Justice. Magistrates' courts. England
Expenditure by local authorities - Statistics

ADMINISTRATION of justice statistics, based on estimates / the Society of County Treasurers. - 1978-79 / [compiler J. Unsworth]. - Reading : [The Society], 1978
[1],7p : ill
Subsequent volumes, when received, will be placed at this shelfmark. - Holdings: 1978/79, etc
B80-15112 P.203/136
1944819

ADMINISTRATION. Justice. Majorca. Spain
1230-1835

La JUSTICIA en el antiguo reino de Mallorca / [documentos y un estudio por] Carlos Álvarez-Novoa. - Palma de Mallorca : [s.n.], 1971
223p. - Bibliography : p. 213-220
r6035554 X.100/15896
1153307

ADMINISTRATION. Justice. Massachusetts
1765-1775

REID, John Phillip
In a defiant stance. - University Park ; London : Pennsylvania State University Press, 1977
[9],225p. - Includes index
0-271-01240-4 X.800/27414
1758519

ADMINISTRATION. Justice. Mendoza. Argentina
to date

ACEVEDO, Edberto Oscar
Orígenes de la organización judicial de la Provincia de Mendoza. - Buenos Aires : Fundación para la Educación, la Ciencia y la Cultura, 1979
212p
r6701452 X.200/36326
8770190

ADMINISTRATION. Justice. Nebraska
to date

HUTTON, Harold
Vigilante days. - Chicago : Swallow Press, c1978
365p
0-8040-0738-1 X.200/37894
8814597

ADMINISTRATION. Justice. Netherlands
to date

STAD en recht. - Arnhem : Gouda Quint, 1982
x,260p
90-600-0263-6 X.200/41567
9023356

ADMINISTRATION. Justice. Northern Ireland
Social aspects

BOYLE, Kevin
Ten years on in Northern Ireland. - London : Cobden Trust, 1980
119p. - Includes bibliographies
0-900137-16-9 X.208/6188
0001503

ADMINISTRATION. Justice. Philippines
to date

ARALAR, Reynaldo B.
Comments on the Judiciary Reorganization Act of 1980 (Batas Pambansa Blg 129). - Manila : Rex Book Store, [1982]
vii,156p
r6814021 X.200/43985
9105468

ADMINISTRATION. Justice. Portugal
- Conference proceedings

ENCONTRO INTERNACIONAL DE MAGISTRADOS (2nd : 1977 : Lisbon)
O estatuto do magistrado e as perspectivas futuras do direito. - Coimbra : Livraria Almedina, 1978
298p
r6086321 X.108/26880
2025043

ADMINISTRATION. Justice. Prussia
to 1871

EBEL, Friedrich
200 Jahre preussischer Zivilprozess. - Berlin : de Gruyter, 1982
28p. - (Schriftenreihe der Juristischen Gesellschaft e.V. Berlin ; Heft 71)
3-11-008905-x X.200/42580
9042121

ADMINISTRATION. Justice. Puerto Rico

TRÍAS MONGE, José
El sistema judicial de Puerto Rico. - [Rio Piedras] : Editorial Universitaria, Universidad de Puerto Rico, 1978
xvii,287p
0-8477-3014-x X.200/34037
2457334

ADMINISTRATION. Justice. Puerto Rico
1800-1898. Documents on administration of justice in Puerto Rico, 1800-1898 in stock of Archivo Histórico Nacional (Spain). Sección de Ultramar - Indexes

ARCHIVO HISTÓRICO NACIONAL. Sección de Ultramar
Inventario de la Serie Gracia y Justicia de Puerto Rico. - Madrid : Servicio de Publicaciones del Ministerio de Educación y Ciencia, 1975
xv,354p.,4 leaves of plates. - (Archivo Histórico Nacional. Sección de Ultramar. [Inventario] ; n.3)
84-369-0433-8 X.700/18535
1321773
- [Another copy] X.800/31134

ADMINISTRATION. Justice. Rural regions. Cambridgeshire
1154-1399

ROYAL justice and the medieval English countryside / edited, translated and annotated with an introduction and biographical register by Anne Reiber DeWindt and Edwin Brezette DeWindt. - Toronto : Pontifical Institute of Mediaeval Studies, 1981
2v.. - (Studies and texts ; 57). - Bibliography
0-88844-057-x Ac.2702.d/4(57)
9023135

ADMINISTRATION. Justice. Santo Domingo. Dominican Republic
1500-1900

MALAGÓN BARCELÓ, Javier
El Distrito de la Audiencia de Santo Domingo en los siglos XV a XIX. - 2a ed. - Santiago, República Dominicana : Universidad Católica Madre y Maestra, 1977
302p. - (Colección "Estudios" ; 27)
r6162085 X.200/33981
245775x

ADMINISTRATION. Justice. Schlitz (County). West Germany
to date

HESSISCHES STAATSARCHIV DARMSTADT
Schlitzer Gerichtsakten. - Darmstadt : Hessisches Staatarchiv, 1984
viii,237p.. - (Repertorien des Hessischen Staatsarchivs Darmstadt ; 17/1) (Abt.F 23, Herrschaft Schlitz ; Bd.1)
3-88443-228-1 X.200/44499
9112065

ADMINISTRATION. Justice. South Pacific islands
- Conference proceedings

PACIFIC courts and justice. - London : Commonwealth Magistrates Association [etc.], 1977
ix,146p,[3] leaves of plates : ill. - Includes index
B79-30574 X.100/18577
1946595

ADMINISTRATION. Justice. Southern Italy

GIANNINI, Mirella
Giustizia e Mezzogiorno. - Venezia : Marsilio, 1978
x,161p. - (Documentazione territorio e società). - Bibliography
r6500640 X.200/35027
2477599

ADMINISTRATION. Justice. Soviet Union
1945-

KONSTITUTSIONNYE osnovy pravosudiia v SSSR / pod redaktsieĭ V.M. Savitskogo. - Moskva : Nauka, 1981
359 p. - Includes bibliographical references
r6465965 X.200/45453
8932220

SUD i pravosudie v SSSR ... / [otvetstvennyĭ redaktor : M.P. Shalamov]. - Moskva : IUridicheskaia literatura, 1974
357p
r6462786 S.N.46/8
8932220

TEPLOV, V. A. (Vladimir Alekseevich)
Nekotorye voprosy primeneniia novogo zakonodatel'stva. - Saratov : Izdatel'stvo Saratovskogo universiteta, 1981
94p
r6463260 X.200/40111
8932220

VO imia pravosudiia / [avtor-sostavitel' N.S. Aref'ev]. - Kuĭbyshev : Kuĭbyshevskoe knizhnoe izdatel'stvo, 1982
133p
r6490183 X.208/10486
8932220

ADMINISTRATION. Justice. Soviet Union
1945- - Serials

VOPROSY osushchestvleniia pravosudiia v SSSR. - Vyp.1-. - Kaliningrad : Kaliningradskiĭ gosudarstvennyĭ universitet, [1971?]-
Holdings: Vyp.7(1979)-
r6463706 P.201/622
8932220

ADMINISTRATION. Justice. Spain

FERNÁNDEZ VIAGAS, Plácido
Qué es la justicia democrática. - Barcelona : Editorial La Gaya Ciencia, 1977
80p : ill. - (Biblioteca de divulgación política)
84-7080-990-3 X.108/15779
1528068

ADMINISTRATION. Justice. Spain
to 1516

VARONA GARCÍA, María Antonia
La Chancillería de Valladolid en el reinado de los Reyes Católicos. - Valladolid : Universidad de Valladolid, 1981
462p,8leaves of plates. - Includes bibliography
84-600-2216-1 X.200/44345
9112057

ADMINISTRATION. Justice. Spain
1975-

MIGUEL GARCILÓPEZ, Adolfo de
Jaque a la justicia. - Barcelona : Planeta, 1980
208p : ill. - (Colección Documento ; 30)
84-320-3570-x X.208/7913
8556016

ADMINISTRATION. Justice. Spain
Law, to 1936

AGÚNDEZ FERNÁNDEZ, Antonio
Historia del poder judicial en España. - Madrid : Editora Nacional, [1974]
196 p.. - (España en 3 tiempos)
84-276-1199-4 X.709/20767
0873934

ADMINISTRATION. Justice. Spanish America
Documents on administration of justice in Spanish America, 1529-1599 in stock of Archivo General de Indias - Catalogues

ARCHIVO GENERAL DE INDIAS
Catálogo de las consultas del Consejo de Indias (1529-1591 [i.e. 1599]). - Madrid : Dirección General de Archivos y Bibliotecas, 1972
2t.
r6066442 X.510/11241
1860208

ADMINISTRATION. Justice. Sweden

BRÅTÉNIUS, Ingvar
Så döms oskyldiga i Sverige. - [Nyköping, 1974]
115 p.
r6004178 X.708/18485
0889431

FIALHO, Sylvio Abreu
O mundo dos olhos. - Rio de Janeiro : Livraria São José, 1975
266 p.
r6011714 X.909/34057
0889431

ADMINISTRATION. Justice. Sweden
to date

SJÖNANDER, Bengt
[Articles. Selections]. Inkast. - [Lund] : Rättsforum, [1981]
175p : ill
91-85826-07-3 X.200/42669
9057935

SWEDEN. Högsta domstolen
Högsta domstolens och Regeringsrättens arbetsformer. - [Stockholm] : Liber förlag, 1980
59p. - (Ds Ju ; 1980 : 12)
91-380-5928-2 X.200/42721
9057935

ADMINISTRATION. Justice. United States
to date

PERRINE, Garrith D.
Administration of justice. - St. Paul : West Publishing, c1980
xvi,311p
0-8299-0345-3 X.205/1068
8867828

ADMINISTRATION. Justice. United States
Discretion compared with discretion in administration of justice in European Community countries
DAVIS, Kenneth Culp
Discretionary justice in Europe and America. - Urbana : London : University of Illinois Press, 1976
xi,203p
0-252-00579-1 X.200/30595
073988x

ADMINISTRATION. Justice. United States
Reform
NAGEL, Stuart S. (Stuart Samuel)
Improving the legal process. - Lexington, Mass. ; London : D.C. Heath, 1975
xx,395p : ill. - Includes bibliographies and index
0-669-97824-8 X.200/20099
0502286

ADMINISTRATION. Justice. Villages. South Korea
1945-
EIKEMEIER, Dieter
Documents from Changjwa-ri. - Wiesbaden : Harrassowitz, 1980
viii,261p,[2]p of plates : ill. - (Veröffentlichungen des Ostasien-Instituts der Ruhr-Universität Bochum). - Bibliography
3-447-02044-x X.0700/678(25)
8753326

ADMINISTRATION. Justice. West Germany
OSTERMEYER, Helmut
Die juristische Zeitbombe. - München : W. Goldmann, 1973
156 p.
3-442-30282-x X.809/22317
0907219

SCHOREIT, Armin
Gerichte und Justizbehörden der Bundesrepublik Deutschland. - Karlsruhe : Müller, 1976
67p : ill. - (Wegweiser Recht, Verwaltung, Politik)
3-8114-4076-4 OPL
0892254

ADMINISTRATION. Justice. West Germany
1945-
WASSERMANN, Rudolf
Ist Bonn doch Weimar?. - Neuwied : Luchterhand, 1983
75p
3-472-05101-9 X.208/10036
8528780

ADMINISTRATION. Justice. West Sumatra (Province). Indonesia
to date
BENDA-BECKMANN, C. E. von
The broken stairways to consensus. - [Nijmegen] : [C.E. von Benda-Beckmann], 1984
xviii,210p : ill. - Includes bibliography
90-900067-5-3 X.520/33879
9093133

ADMINISTRATION. Justice. Western Europe
Participation of public, to date - Catalogues
SCAPARONE, Paolo
La partecipazione popolare all'amministrazione della giustizia. - Milano : Giuffrè, 1980. - (Università di Torino, Memorie dell'Istituto giuridico. Ser.3 ; Memoria 9)
1: Profili storici e comparativi
158p
r6500414 Ac.2308[Ser.3,Memoria9]
8715939

ADMINISTRATION. Justice. Western world
Participation of public, to date
VIGORITI, Vincenzo
Interessi collettivi e processo. - Milano : Giuffrè, 1979
332p. - (Università di Firenze. Pubblicazioni della Facoltà di giurisprudenza ; 28)
r6502018 X.200/37078
8786097

ADMINISTRATION. Justice. Young offenders. England
PARKER, Howard
Receiving juvenile justice. - Oxford : Basil Blackwell, 1981
259p : 1ill. - Bibliography: p253-256. - Includes index
0-631-12727-5 X.200/36307
2537907

PROVIDING criminal justice for children / edited by Allison Morris and Henri Giller. - London : Edward Arnold rc 1983
164p : ill. - Includes index
0-7131-6397-6 X.200/41579
2537907

WALTERS, Graham
Criminal proceedings against juveniles. - London : Oyez Longman, 1984
xxv,188p. - Includes index
0-85120-846-0 X.200/45724
2537907

ADMINISTRATION. Justice. Young offenders. England
Research. Methodology
REDMOND-PYLE, David
Investigating the local juvenile system. - Lancaster : Centre of Youth, Crime and Community. Dept. of Social Administration, University of Lancaster, [1982?]
60p : ill
0-901699-93-4 X.205/1107
2814021

ADMINISTRATION. Justice. Young offenders. Great Britain
- Conference proceedings
JUVENILE justice / Allison Morris ... [et al.]. - [Lancaster] : Centre of Youth, Crime and Community. Department of Social Administration. University of Lancaster, [1980?]
49p. - ([CYCC occasional paper] : [no.2]). - Bibliography: p48-49
B81-25945 X.205/817
2500442

ADMINISTRATION. Justice. Young offenders. Great Britain
to 1977
PARSLOE, Phyllida
Juvenile justice in Britain and the United States. - London : Routledge and Kegan Paul, 1978
x,325p. - (Library of social work). - Bibliography: p.303-325
0-7100-8772-1 X.200/31433
1277480

ADMINISTRATION. Justice. Young offenders. Northampton
REDMOND-PYLE, David
Report on the Northampton juvenile criminal justice system. - Lancaster : Centre of Youth, Crime and Community, Dept. of Social Administration, University of Lancaster, [1982]
49p
0-901699-92-6 X.205/1108
2814013

ADMINISTRATION. Justice. Young offenders. Northern Ireland
The JUVENILE justice system in Northern Ireland / edited by Brian Caul, John Pinkerton, Fred Powell. - [Newtownabbey] : Ulster Polytechnic, 1983
115p. - (The Polytechnic series). - Includes bibliographies
B84-37967 X.525/9755
3164691

ADMINISTRATION. Justice. Young offenders. Scotland
LEGALITY and community / Paul Brown and ... [others]. - Aberdeen : Aberdeen Peoples Press, 1979
[4],104p : ill
0-906074-05-3 X.100/18802
2022974

MARTIN, F. M. (Frederick Morris)
Children out of court. - Edinburgh : Scottish Academic, c1981
xi,331p,4p of plates : ill. - Bibliography: p322-324. - Includes index
0-7073-0287-0 X.520/23981
2022974

ADMINISTRATION. Justice. Young offenders. United States
to 1977
PARSLOE, Phyllida
Juvenile justice in Britain and the United States. - London : Routledge and Kegan Paul, 1978
x,325p. - (Library of social work). - Bibliography: p.303-325
0-7100-8772-1 X.200/31433
1277499

ADMINISTRATION. Justice. Yugoslavia
TRAJKOVIĆ, Josif
Sud. - Beograd : Radnička štampa, 1975
301p. - (Ideje. Kolo 3)
r6097078 X.208/3422
203395X

ADMINISTRATION. Laboratories. Schools. Great Britain
ARCHENHOLD, W. F. (Wolfgang Friedrich)
School science laboratories. - London : J. Murray, 1978
[9],303p : ill. - Includes bibliographies and index
0-7195-3436-4 X.620/17544 Woolwich
1438050

ADMINISTRATION. Large comprehensive schools. England
LYONS, Geoffrey
Heads' tasks. - Windsor : NFER, 1976
[5],254p : ill. - Bibliography: p.5-6
0-85633-068-x LR.421/13
0723193

ADMINISTRATION. Law courts. England
SCOTT, Ian Richard
Court administration. - [Birmingham] : University of Birmingham, 1979
[1],19p. - (Inaugural lectures / University of Birmingham)
0-7044-0334-x X.100/18665
1962566

ADMINISTRATION. Law courts. United States
SAARI, David J.
American court management. - Westport, Conn. ; London Quorum, 1982
xviii,163p : ill. - Bibliography: p147-152. - Includes index
0-89930-006-5 X.520/30046
2906201

ADMINISTRATION. Law. Financial resources. Claimants Supplementary benefits. Great Britain
GREAT BRITAIN. Department of Health and Social Security
Guidance to supplementary benefit officers on the treatment of resources. - [London] : [Department of Health and Social Security], c1981
47p
B83-07535 BS.385/840
2832151

ADMINISTRATION. Law of primary education in England
By Great Britain. Education Department, 1870-1885
ROPER, Henry
Administering the Elementary Education Acts, 1870-1885 - [Leeds] : Museum of [the] History of Education, University of Leeds, 1976
[2],50p. - (Educational administration and history monographs ; no.5)
0-904427-03-x X.108/16605
1078062

ADMINISTRATION. Lending by banks. Ireland (Republic)
SWEENEY, Loughlin
The control of advances. - [Dublin] : Institute of Bankers in Ireland, 1977
64,[1]p. - (Irish banking practice ; 3)
B78-31252 X.519/28421
1578286

ADMINISTRATION. Lérida (Province). Spain
to date
BERTRÁN ROIGÉ, Primo
La procuració reial de Lleida. - Lérida : el autor, [1981]
25p. - (Studia històrica ; 1)
r6707701 X.529/62413
9095357

ADMINISTRATION. Lesotho
By Cape Province, 1871-1884
BURMAN, S. B.
The justice of the Queen's government. - Leiden : Afrika-Studiecentrum ; Cambridge : African Studies Centre, 1976
viii,131p. - (African social research documents ; vol.9). - Bibliography: p.122-125. - Includes index
B77-33085 X.0311/143(9) Woolwich
1267477

ADMINISTRATION. Libraries
EVANS, G. Edward (Gayle Edward)
Management techniques for librarians. - New York ; London : Academic Press, 1976
xii,276p : ill. - (Library and information science). - Includes bibliographies and index
0-12-243850-7 2719.aa.12
0167967

GORE, Daniel
To know a library. - Westport, Conn. ; London : Greenwood Press, 1978
[7],379p : ill. - (New directions in librarianship ; no.1). - Includes index
0-8371-9881-x 2719.cc.20
1804766

HUMBERTUS, de Romanis
[Instructiones de officiis ordinis. capitulum XIII de officio librarii]. Regulations for the operation of a medieval library. - [Minneapolis] : Associates of the James Ford Bell Library, University of Minnesota, 1980
[8p.]
r6108449 2719.e.76
0167967

ADMINISTRATION. Libraries –continued

LUPU, Atanase
Relaţiile bibliotecii cu publicul. – Bucureşti : Editura Litera. – (Probleme de biblioteconomie)
1. – 1973
72p
r6043432 2719.b.29
0021776

PĂPURICĂ, Mihai D.
Completarea colecţiilor din biblioteci. – Bucureşti : Editura Litera, 1973
49p. – (Probleme de biblioteconomie)
r6036285 2719.b.30
0021776

ADMINISTRATION. Libraries
Operations research. Mathematical models

CHEN, Ching–Chih
Applications of operations research models to libraries. – Cambridge, Mass. ; London : MIT Press, 1976
xxi,212p : ill. – Bibliography: p.199–207. – Includes index
0–262–03056–x X.320/10471 Woolwich
0712604

ADMINISTRATION. Libraries
Quantitative methods – Conference proceedings

QUANTITATIVE measurement and dynamic library service / edited with a preface and introduction by Ching–chih Chen. – Phoenix : Oryx Press ; London : Mansell Information Publishing, 1978
xxi,290p : ill. – (A Neal–Schuman professional book). – Bibliography: p.271–284. – Includes index
0–7201–0826–8 2719.cc.43
1499599

ADMINISTRATION. Libraries
Teaching. Use of simulations

ZACHERT, Martha Jane K.
Simulation teaching of library administration. – New York ; London : Bowker, 1975 [i.e. 1976]
xvii,297p : ill. – (Problem–centered approaches to librarianship). – Includes bibliographies and index
0–8352–0612–2 2719.aa.33
0706280

ADMINISTRATION. Libraries. Austria
to date

MAZAL, Otto
[Articles. Selections]. Bibliothekswesen und Forschung. – Graz, Austria : Akademische Druck– u. Verlagsanstalt, 1982
xxii,316p,[1]leaf of plates : ill(1col). – (Studien zur Bibliotheksgeschichte ; Bd.4). – "Schrifttumverzeichnis Otto Mazal (1954–1981)" : p.309–314
3–201–01209–2 2714.c.32/4
8856915

ADMINISTRATION. Libraries. Colleges. Great Britain

COLLEGE librarianship / edited by A. Rennie McElroy. – London : The Library Association, 1984
xix,427p. – (Handbooks on library practice)
0–85365–785–8 2719.e.912
3213935

ADMINISTRATION. Libraries. England

HEWITT, R. G.
An in–tray exercise for librarians and managers. – Birmingham : R.G. Hewitt, 1977
[2]p,[16]leaves
B78–07269 2719.gg.22
1343661

ADMINISTRATION. Libraries. Mass media organisations

WHATMORE, Geoffrey
The modern news library. – London : Library Association, 1978
[11],202p,[4]p of plates : ill. – Includes bibliographies and index
0–85365–530–8 2719.bb.85
120338x
– [Another copy: paperback]
ISBN 0–85365–551–0 2719.dd.58

ADMINISTRATION. Libraries. North America
Role of librarians

REEVES, William Joseph
Librarians as professionals. – Lexington, Mass. : Heath ; [Farnborough, Hants] : [Distributed by] Gower Press, 1980
xxi,169p : ill. – Bibliography: p.151–157. – Includes index
0–669–03163–1 2719i21
2305593

ADMINISTRATION. Libraries. Schools

CARROLL, Frances Laverne
Recent advances in school librarianship. – Oxford : Pergamon, 1981
xiii,249p. – (Recent advances in library and information services ; vol.1). – Includes index
0–08–026084–5 2719.i.83
2548321

HERRING, James E.
School librarianship. – London : Bingley, 1982
116p. – (Outlines of modern librarianship). – Bibliography: p97–101. – Includes index
0–85157–347–9 2719.e.396
2548321

ADMINISTRATION. Libraries. Schools. Great Britain

ROUTINES / [School Library Association]. – Oxford : The Association, 1980
14p : ill. – (School Library Association guidelines ; 1). – Bibliography: p14
0–900641–36–3 2719.e.147
2390205

SCHOOL LIBRARY ASSOCIATION
The way ahead. – Oxford : School Library Association, 1980
28p : ill
0–900641–38–x 2719.e.119
2365669

ADMINISTRATION. Libraries. Schools. United States

NICKEL, Mildred L.
Steps to service. – Chicago : American Library Association, 1975
xii, 124 p. – Bibliography : p. 21–34. – Includes index
0–8389–0161–1 2719.ff.23
0728403

ADMINISTRATION. Libraries. Tuscany. Italy
Law

TUSCANY. Dipartimento istruzione e cultura
La legge toscana per le biblioteche. – Firenze : Giunta regionale toscana, 1977
248p
r6035111 2719.b.75
1316990

ADMINISTRATION. Libraries. United States
Law

LADENSON, Alex
Library law and legislation in the United States. – Metuchen ; London : Scarecrow, 1982
x,191p. – (Scarecrow library administration series ; no.1). – Includes index
0–8108–1513–3 2702.b.8
276394x

ADMINISTRATION. Libraries. Universities

UNIVERSITY library problems. – [Stockholm] : [Almqvist and Wiksell International], 1975
124 p.. – (Acta Bibliothecae R. Universitatis Upsaliensis ; vol. 19)
91–85092–05–3 Ac.1077.e[19]
029974X

ADMINISTRATION. Libraries. Universities. Great Britain

THOMPSON, James, 1932–
An introduction to university library administration. – 3rd ed. – London : Bingley [etc.], [1979]
160p,leaf of plate,[8]p of plates : ill. – Bibliography: p.152–157. – Includes index
0–85157–288–x 2719.cc.91
0108820

UNIVERSITY librarianship / edited by John F. Stirling. – London : Library Association, 1981
xv,229p : ill. – (Handbooks on library practice). – Includes index
0–85365–621–5 2719.e.206
0108820

ADMINISTRATION. Libraries. Universities. India (Republic)

SAINI, B. S.
Library organisation for higher education. – Delhi : Ess Ess Publications, 1976
ix,116p. – Bibliography : p.101–112. – Includes index
r6072214 2719.bb.91
1582151

ADMINISTRATION. Libraries. Universities. United States

MARTELL, Charles R.
The client–centred academic library. – Westport, Conn. ; London : Greenwood Press, 1983
xii,136p : ill. – (Contributions in librarianship and information science ; no.42). – Bibliography: p119–126. – Includes index
0–313–23213–x 2719.e.765
3040666

ADMINISTRATION. Local authorities. Colombia
to date

MEJÍA MAYA, Ignacio
Régimen municipal. – [Medellín] : Editorial Argemiro Salazar, 1979
xiii,345p
r6707702 X.520/34183
9101772

ADMINISTRATION. Local authorities. Great Britain
Ad hoc committees of inquiry. Procedure

AD hoc inquiries in local government. – Lewes : SOLACE, 1978
[2],viii,57p. – (Occasional papers / Society of Local Authority Chief Executives)
0–905566–02–5 X.709/32167
1434063

ADMINISTRATION. Local authorities. Norway

ØSTBERG, Carl S.
Hvordan blir kommunene styrt?. – Oslo : Fabritius, 1977
135p : ill
82–07–00163–1 X.709/25522
1503200

ADMINISTRATION. Local authorities. Spain
to date

CASTRO, Concepción de
La revolución liberal y los municipios españoles (1812–1868). – Madrid : Alianza, 1979
236p. – (Alianza Universidad ; 249)
84–206–2249–4 X.529/43111
8733635

ADMINISTRATION. Local authorities. Strathclyde Region. Scotland
related to social development of Ferguslie Park

PAISLEY COMMUNITY DEVELOPMENT PROJECT
Local authority structures in relation to Ferguslie Park. – Paisley : Community Development Project, Ferguslie Park ; [Glasgow] : [Distributed by Administration Department, Strathclyde Regional Council], 1977
[2],14 leaves : 1 ill
B79–35278 X.512/7031
1968963

ADMINISTRATION. Local councils. England

OUSBY, W. H.
A practical guide to local council administration. – London : C. Knight, 1976
141p. – Includes index
0–85314–264–5 X.709/30930
0791911

ADMINISTRATION. Local councils. Hungary

TANULMÁNYOK a magyar helyi önkormányzat múltjából / [szerkesztette Bónis György, Degré Alajos ; irták : Bácskai Vera et al]. – Budapest : Közgazdasági és Jogi Könyvkiadó, 1971
511p
r6044794 X.800/12915
132473X

ADMINISTRATION. Local education authorities. England

TOMLINSON, J. R. G.
The profession of education officer. – [Sheffield] : Department of Education Management, Sheffield City Polytechnic, 1982
26leaves. – (Sheffield papers in education management ; 25)
0–903761–54–8 X.529/49973
1963260

WINTER, Geoffrey A.
The position of the Education Service following local government reorganisation. – [London] : Society of Education Officers, [1977]
[3],ii,95p
B79–34069 X.512/6996
1963260

ADMINISTRATION. Local government. Great Britain
Organisations: Institute of Local Government Administration

INSTITUTE OF LOCAL GOVERNMENT ADMINISTRATORS
Annual report. – 1975–76–. – Colchester : [The Institute], 1976–
Subsequent volumes, when received, will be placed at this shelfmark
r6072075 P.803/891
1491393

ADMINISTRATION. Local history collections. Public libraries. Great Britain

NICHOLS, Harold
Local studies librarianship. – London : Bingley [etc.], 1979
128p. – (Outlines of modern librarianship). – Bibliography: p.123–127. – Includes index
0–85157–272–3 2719.cc.51
0200840

ADMINISTRATION. Los Monges. Colombia
By Venezuela. Disputes between Colombia & Venezuela, to 1974

CAVELIER, Germán
Memoria histórico-jurídica sobre el asunto de Los Monjes. - Bogotá : Kelly, 1977
543p : ill. - Includes bibliographies
r6084293 X.700/22258
1162780

HOLGUÍN PELÁEZ, Hernando
Los Monjes. - Bogotá : Prosartes, 1975
435p : ill
r6030924 X.702/3255
1162780

ADMINISTRATION. Maintenance. Local authority housing. Great Britain
Applications of digital computer systems

LOCAL AUTHORITIES MANAGEMENT SERVICES AND COMPUTER COMMITTEE. Computer Panel
The benefits of using computer systems for housing maintenance and repairs. - London : The Committee, 1980
30p : ill
0-85497-119-x X.525/3400
0684422

Computers and housing maintenance. - London : L.A.M.S.A.C., 1975
39p : ill
0-85497-083-5 X.615/1647 Woolwich
0684422

PETTITT, Raymond
Computer aids to housing maintenance management. - London : H.M.S.O., 1981
68p : ill
0-11-751547-7 BS.414/2059
0684422

ADMINISTRATION. Map collections. Libraries

The MAP librarian in the modern world / edited by Helen Wallis and Lothar Zogner ; presented by the IFLA Section of Geography and Map Libraries. - München [i.e. Munich] ; London : K.G.Saur ; London : [Distributed by] Bingley, 1979
295p : ill. - Includes bibliographies and index
0-89664-131-7 Map Library
2116006

NICHOLS, Harold
Map librarianship. - London : Bingley [etc.], 1976
298p. - Bibliography: p.281-290. - Includes index
0-85157-204-9 Maps 202.c.44 Map library
0614416

Map librarianship. - 2nd ed. - London : Bingley, 1982
272p. - Bibliography: p258-266. - Includes index
0-85157-327-4 Map Library
0614416

RISTOW, Walter W. (Walter William)
The emergence of maps in libraries. - Hamden : Linnet ; London : Mansell, 1980
358p
0-7201-1620-1 Map Library
0614416

ADMINISTRATION. Mathematics departments. Secondary schools. Hertfordshire
Role of heads of departments

A PROFESSIONAL handbook for heads of mathematics departments. - [Hertford] : Hertfordshire County Council, 1981
20p. - (Hertfordshire guidelines in mathematics, supplementary booklet)
B84-03682 X.625/855
2992493

ADMINISTRATION. Medical libraries. Great Britain

MATTHEWS, David
Medical librarianship. - London : Bingley [etc.], 1979
173p. - Bibliography - Includes index
0-85157-273-1 2719.dd.45
1907220

MEDICAL librarianship / edited by Michael Carmel. - London : Library Association, 1981
xi,359p. - (Handbook on library practice). - Includes index
0-85365-502-2 X.329/13721 Woolwich
1907220
- [Another copy : paperback]
ISBN 0-85365-703-3 X.329/13722 Woolwich

ADMINISTRATION. Mental health services. United States

PSYCHIATRIC administration / edited by John A. Talbott, Seymour R. Kaplan. - New York ; London : Grune & Stratton, c1983
xiv,410p : ill. - Includes index
0-8089-1529-0 X.322/12345
2955741

ADMINISTRATION. Michigan State University 1945- - Biographies

HANNAH, John A.
A memoir. - [East Lansing, Mich.] : Michigan State University Press, 1980
143p,plates
0-87013-214-8 X.529/55067
8972672

ADMINISTRATION. Military forces. France to 1974

DOFING, Jean
Le contrôle général des Armées. - [s.l.] : [s.n.], 1974
viii,111p : ill
r6012600 S.E.65/161
1225405

ADMINISTRATION. Military forces. Soviet Union
Political aspects

DEANE, Michael J.
Political control of the Soviet armed forces. - London : Macdonald and Jane's, 1977
[5],xi,297p. - Includes index
0-354-01129-4 X.800/26571
1293168

ADMINISTRATION. Monasteries. Russia to 1917

VOTCHINNYE khozi͡aĭstvennye knigi XVI v. / pod redakt͡sieĭ A.G. Man'kova. - Moskva ; Leningrad : Akademii͡a nauk SSSR, Institut istorii, Leningradskoe otdelenie, 1979
2v.
r6484571 X.205/939
8773386

ADMINISTRATION. Monastery of Corbie

ADALHARD, Saint, Abbot of Corbie
[CONSUETUDINES Corbeiensis. English]. The customs of Corbie ..... - (In: Horn, Walter : The plan of St. Gall, -Berkeley, 1979. - Vol.3, p.91-128)
r6108557 L.45/448
2416379

ADMINISTRATION. Multinational companies compared with administration of universities in Africa

MAZRUI, Ali Al'Amin
The African university as a multinational corporation. - Brighton : Institute of Development Studies, University of Sussex, [1976]
[3], 24p. - (Discussion paper ; no.96)
r6031828 X.512/5597
1102524

ADMINISTRATION. Museums

INTERNATIONAL COUNCIL OF MUSEUMS
ICOM statutes. - [Paris], 1974
11p.
r6109164 X.419/3892 Woolwich
0392006

MUSEITEKNIK / utgiven av Phebe Fjellström. - Lund : Liber Läromedel, 1976
160p : ill. - Includes bibliographies
91-400-4183-2 X.410/6112 Woolwich
0392006

ADMINISTRATION. Museums. Great Britain

The LOCAL museum. - Milton Keynes : Area Museums Service for Southern England, c1984
72p : ill(some col.)
0-904752-02-x X.425/5127
2106299

ADMINISTRATION. Museums. United States
Valentiner, William R. - Biographies

STERNE, Margaret (Margaret Heiden), 1902-1977
The passionate eye. - Detroit : Wayne State University Press, 1980
407p : ill
0-8143-1631-x X.805/2438
8901813

ADMINISTRATION. Naples (Kingdom). Italy
By Spain, 1494-1796. Documents on the administration of Naples Kingdom by Spain - Catalogues

ARCHIVO GENERAL DE SIMANCAS
Títulos y privilegios de Nápoles siglos XVI-XVIII. - Valladolid : El Archivo, 1980-. - (Catálogo del Archivo de Simancas ; 28)
r6701752 X.0800/1626(1)
8909687

ADMINISTRATION. National health services. England

HELLER, Tom
Restructuring the health service. - London : Croom Helm, 1978
3-114p. - Includes index
0-85664-583-4 X.329/11322 Woolwich
1137867

HUNTER, Robert Brockie, Baron Hunter
Community physicians-clinical administrators. - Birmingham : University of Birmingham, 1979
[1],7p
0-7044-0369-2 X.329/13144 Woolwich
2315564

ADMINISTRATION. National health services. Scotland

SCOTTISH COUNCIL OF SOCIAL SERVICE
NHS re-organisation, Scotland and local health councils. - Edinburgh : The Council, [1973]
[3],13[i.e.14]p. - (Topic ; 1)
0-903589-05-2 X.0512/504(1)
0156930

ADMINISTRATION. National parks. England
Regional committees

GREAT BRITAIN. Countryside Commission
Details of administrative arrangements in national parks. - [Cheltenham] : [Countryside Commission], [1976]
[11]leaves
0-902590-43-x BS.116/125
0731676

ADMINISTRATION. National parks. United States

EVERHART, William C.
The National Park service. - Epping : Bowker, c1982
x,197p,[14]p of plates. - (Westview library of Federal departments, agencies, and systems). - Bibliography: p189-190. - Includes index
0-86531-130-7 X.622/16335
2938200

ADMINISTRATION. Negotiations. Increases. Remuneration. Personnel. Police. Great Britain
Procedure

GREAT BRITAIN. Committee of Inquiry on the Police
Committee of Inquiry on the Police reports on negotiating machinery and pay ... - London : H.M.S.O., [1978]
ix,52,104,[2]p,[2]leaves : ill. - (Cmnd.7283)
0-10-172830-1 BS.18/598
1698966

ADMINISTRATION. Netherlands
By Spain. Alba, Ferdinand Alvarez de Toledo, duque de

GRAPPERHAUS, Ferdinand H. M.
Alva en de tiende penning. - Zutphen : Walburg, [1982]
399p : ill. - Includes bibliography
90-601-1130-3 X.525/7873
9141944

ADMINISTRATION. New Forest. Hampshire
Organisations: Court of Verderers of the New Forest, 1877-1977

PASMORE, Anthony
Verderers of the New Forest. - [Beaulieu] : Pioneer Publications Ltd, [1976]
viii,298p : ill. - Includes index
0-9502786-2-9 X.809/42088
1143042

ADMINISTRATION. Norway
By Denmark, 1618-1629. Juel, Jens, b.1580

RIAN, Øystein
Jens Juels stattholderskap 1618-29. - Oslo : Universitetsforlaget, [1976]
254p
82-00-01499-1 X.709/22077
1027255

ADMINISTRATION. Nursery schools. United States

BUTLER, Annie Louise
Early childhood education. - New York ; London : D. Van Nostrand Company, 1974
xi,240p : ill. - Bibliography: p.207-219. - Includes index
0-442-20897-9 X.519/25321
0529648

ADMINISTRATION. Oral history collections. Local history collections. Libraries. Great Britain

BIRD, Julie
Oral history collections. - London : Polytechnic of North London, School of Librarianship and Information Studies, 1983, c1981
52p : 1ill. - (Occasional publication ; no.5). - Bibliography: p52
0-900639-29-6 2719.k.121
2909340

ADMINISTRATION. Organisations. United States
Policies. Formulation

HODGETTS, Richard Michael
Administrative policy. - 2nd ed. - New York ; Chichester : Wiley, 1980
xvi,712p : ill. - (The Wiley series in management). - Includes bibliographies and index
0-471-03605-6 X.520/21174
0349054

ADMINISTRATION. Patronage. Arts. Great Britain
By government, 1964-1976

JENKINS, Hugh, b.1908
The culture gap. - London : Boyars : [Distributed by Calder and Boyars], 1979
x,271p,[4]p of plates : ill. - Includes index
0-7145-2643-6 X.809/44251
1874160

ADMINISTRATION. Payment. Mandatory benefits. Payments. Great Britain
Legal aspects

GREAT BRITAIN. Civil Service Department
Legal entitlements and administrative practices. - London : H.M.S.O., 1979
[3],20p
0-11-630425-1 BS.6/146
2109077

ADMINISTRATION. Periodicals collections. Libraries

PERIODICALS administration in libraries / edited by Paul Mayes. - London : Bingley [etc.], 1978
[1],156p : ill. - Includes index
0-85157-262-6 2719.bb.86
1568906

PERIODIKAKONTROLL / av Ellen Hagelsteen ... [et al.]. - Oslo : Statens bibliotekskole, 1978
94p. - (Statens bibliotekskole. Publikasjoner ; 2)
r6113477 2719.h.32(2)
1568906

ADMINISTRATION. Physical education. Curriculum subjects. Educational institutions. United States

BUCHER, Charles A. (Charles Augustus)
Administration of physical education & athletic programs. - 8th ed. - St. Louis ; London : Mosby, 1983
vii,632p : ill. - Includes bibliographies and index
0-8016-0852-x X.622/17827
1035959

DAUGHTREY, Greyson
Physical education and intramural programs. - 2nd ed. - Philadelphia ; London : Saunders, 1976
ii-xii,547p : ill. - Includes bibliographies and index
0-7216-2884-2 X.622/5779 Woolwich
1035959

JENSEN, Clayne R.
Administrative management of physical education and athletic programs. - Philadelphia : Lea & Febiger, 1983
xi,385p : ill. - Includes bibliographies and index
0-8121-0817-5 X.520/32959
1035959

The ORGANIZATION and administration of physical education. - 5th ed.. - Englewood Cliffs ; London : Prentice-Hall, 1979
xii,480p : ill. - Includes bibliographies and index
0-13-641100-2 X.320/11835 Woolwich
1035959

RESICK, Matthew C. (Matthew Clarence)
Modern administrative practices in physical education and athletics. - 3rd ed. - Reading, Mass. ; London : Addison-Wesley, c1979
xvi,459p : ill. - Includes bibliographies and index
0-201-06249-6 X.520/30133
1035959

ADMINISTRATION. Physical education. Curriculum subjects. Schools. United States

KELLER, Irvin A.
Administration of high school athletics. - 7th ed.. - Englewood Cliffs ; London : Prentice-Hall, c1984
viii,408p : ill. - Includes bibliographies and index
0-13-005728-2 X.520/35286
0304379

WILLGOOSE, Carl E. (Carl Edward)
The curriculum in physical education. - 4th ed. - Englewood Cliffs ; London : Prentice-Hall, c1984
x,406p : ill. - Includes bibliographies and index
0-13-196072-5 X.520/34506
0304379

ADMINISTRATION. Picture collections. Libraries

EVANS, Hilary
Picture librarianship. - New York ; London : K. G. Saur ; [London] : Bingley, 1980
136p. - (Outlines of modern librarianship). - Bibliography: p.133. - Includes index
0-85157-294-4 2719.i.24
208970x

PICTURE librarianship / edited by Helen P. Harrison. - London : Library Association, 1981
xii,542p : ill. - (Handbooks on library practice). - Bibliography: p494-517. - Includes index
0-85365-912-5 2719.e.275
208970x

ADMINISTRATION. Playgroups

ATTFIELD, Janet
The business side of playgroups. - London : Pre-school Playgroups Association, 1976
50p : ill
0-901755-15-x X.519/26358
012558x

PLAYGROUPS in local authority premises. - [Exeter] : Devon County Council Education Department, c1983
8p
0-86114-403-1 X.529/64807
012558x

PRE-SCHOOL PLAYGROUPS ASSOCIATION
The business side of playgroups. - Revised ed.. - London : The Association, 1980
3-74p : ill
B80-39215 X.529/36878
012558x

ADMINISTRATION. Police. India (Republic)
to 1972

GHOSH, S. K. (Srikanta)
Police administration. - Calcutta : Eastern Law House, 1973
xvii, 307 p.. - Bibliography : p. 297-307
r6025782 X.200/10680
1019554

ADMINISTRATION. Police. New York (City)

REUSS-IANNI, Elizabeth
Two cultures of policing. - New Brunswick, [N.J.] ; London : Transaction Books, c1983
145p. - Bibliography: p137-142. - Includes index
0-87855-469-6 X.200/41121
2940922

ADMINISTRATION. Police. United States

BOUZA, Anthony V.
Police administration. - New York ; Oxford : Pergamon, 1978
xix,328p. - Bibliography: p.310-317. - Includes index
0-08-022220-x X.200/31371
145577x

EASTMAN, George D.
Short of merger. - Levington, Mass. : D.C. Heath, 1976
xiii,161p : ill
0-669-00373-5 X.200/31523
1582739

VANAGUNAS, Stanley
Administration of police organizations. - Boston ; London : Allyn and Bacon, 1980
ix,386p : 1 ill. - Bibliography: p.371-375. - Includes index
0-205-06666-6 X.200/32401
145577x

WILSON, Orlando Winfield
Police administration. - 4th ed. - New York ; London : McGraw-Hill, 1977
xiii,703p : ill. - Bibliography: p.673-680. - Includes index
0-07-070726-x X.200/31485
145577x

ADMINISTRATION. Police. United States
Wilson, O W - Biographies

BOPP, William J.
'O.W.'. - Port Washington ; London : Kennikat Press, 1977
ix,158p. - (Kennikat Press national university publications. interdisciplinary urban series). - Bibliography: p.152-154. - Includes index
0-8046-9179-7 X.200/31442
1520008

ADMINISTRATION. Polish Pomerania
By Sweden, 1630-1720

MODÉER, Kjell Å
Gerichtsbarkeiten der schwedischen Krone im deutschen Reichsterritorium. - Stockholm : Nordiska Bokhandeln, in Distribution, 1975-
1-. - (Skrifter utgivna av Institutet för rättshistorisk forskning. ser. 1 ; bd 24)
r6001306 Ac.2140b/2a [bd 24]
0941174

ADMINISTRATION. Polytechnic of North London
1970-1974

CAMPBELL, Frank Jeffrey
High command. - [London] : 'Time Out', [1974]
[1],132p : ill. - Bibliography: p.5
0-903446-02-2 X.529/52569
0215996

ADMINISTRATION. Polytechnic of North London
Participation of students

VIEWS on the democratisation of the governors. - [London] : North London Poly Students' Union, [1973]
26p
r6865017 X.529/49283
274306x

ADMINISTRATION. Polytechnics. Great Britain

CUTHBERT, Robert E.
The relationship between planning and budgeting in the management of polytechnics. - [Sheffield] : Sheffield City Polytechnic Department of Education Management, 1979
49p. - (Sheffield papers in education management ; no.5). - Bibliography: p45-49
0-903761-18-1 X.529/61152
3087387

ADMINISTRATION. Ponthieu. France
By England, 1361-1369

STOREY-CHALLENGER, Sheila Bredon
[The English administration of the county of Ponthieu, 1360-1369]. L'administration anglaise du Ponthieu après le traité de Brétigny, 1361-1369. - Abbeville : Société d'Émulation Historique et Littéraire d'Abbeville, 1975
321 p., 1 leaf of plates ; 8 p. of plates : ill. - (Études picardes ; 4). - Bibliography : p. 11-16
r6001108 Ac.230/10 [no. 4]
0927449

ADMINISTRATION. Popular movements. Sweden
to date

BEVARA din historia / [redaktör: Bo Heurling] ; [utgiven av Folkrörelsernas arkivförbund och Kommundepartementets Referensgruppen för folkrörelsefrågor]. - Stockholm : Liber förlag, 1981
176p : ill
91-380-6259-3 X.529/50753
8904502

ADMINISTRATION. Ports

GOSS, R. O.
A comparative study of seaport management and administration. - London : Department of Industry, Department of Trade, Department of Prices and Consumer Protection, 1979
2v.([2],396p)
0-85605-298-1 BS.424/174
1921193

ADMINISTRATION. Ports
Economic aspects

JANSSON, Jan Owen
Port economics. - Cambridge, Mass. ; London : MIT Press, c1982
ix,183p : ill. - (MIT Press series in transportation studies ; 8). - Includes index
0-262-10025-8 X.520/30363
2920417

ADMINISTRATION. Portugal. Assembleia Constituinte

PORTUGAL. Assembleia Constituinte
Cenas parlamentares. - [Lisbon] : Editus, 1976
419p
r6048233 X.709/23818
1314580

ADMINISTRATION. Postgraduate education. Great Britain
Administration & finance. Great Britain. Parliament. House of Commons. Expenditure Committee. 'Postgraduate education'

GREAT BRITAIN. Department of Education and Science
Postgraduate education, government observations on third report [of the] Expenditure Committee, session 1973-74. - London : H.M.S.O., [1976]
iii,11p. - (Cmnd.6611)
0-10-166110-x BS.10/487
0743437

ADMINISTRATION. Prayer groups. Christian church

GUNSTONE, John
The charismatic prayer group. - London : Hodder and Stoughton, 1975
159p. - (Hodder Christian paperbacks)
0-340-19958-x X.108/16091
0512834

ADMINISTRATION. Primary schools. England

ASSOCIATION OF ASSISTANT MISTRESSES
Organisation in the primary school. - London : Association of Assistant Mistresses, 1976
24p,[2]leaves
B78-29436 X.519/28376
1571397

ADMINISTRATION. Primary schools. United States
- For head teachers

HUGHES, Larry W.
The elementary principal's handbook. - Boston [Mass.] : London : Allyn and Bacon, 1978
xvii,364p : ill. - Includes bibliographies and index
0-205-06091-9 Cup.1285/310
1497537

ADMINISTRATION. Prisons. England
- Serials

PRISON service news. - Vol.1, no.1 (Dec. 1982)- - [London] : Press and Publicity Unit, Prisons Department, Home Office, 1982-
v. : ill
B83-22046 BS.18/749
2895129

ADMINISTRATION. Prisons. England
1750-

MCCONVILLE, Seán
A history of English prison administration. - London : Routledge & Kegan Paul.
Vol.1: 1750-1877. - 1981
xvii,534p. - Bibliography: p483-505. - Includes index
0-7100-0694-2 X.0809/2076(1)
2506424

ADMINISTRATION. Prisons. Great Britain
Role of ombudsmen

WENER, Geoffrey
A legitimate grievance. - London : Prison Reform Trust, 1983
23p. - Bibliography: p20
0-946209-03-0 X.525/10052
3209784

ADMINISTRATION. Prisons. Great Britain
Secrecy

COHEN, Stanley, 1942-
Prison secrets. - London : National Council for Civil Liberties : Radical Alternatives to Prison, 1978
iv,98p : ill
0-901108-69-3 X.200/33269
2157799

ADMINISTRATION. Prisons. United States
Reform, to 1980

STASTNY, Charles
Who rules the joint. - Lexington, Mass. : Lexington Books, 1983, c1982
xix,229p : ill. - Bibliography: p217-229. - Includes index
0-669-02661-1 X.200/40777
2922568

ADMINISTRATION. Prisons. United States
related to objectives of imprisonement

DUFFEE, David
Correctional policy and prison organization. - [Beverly Hills] : Sage ; New York ; London : Wiley, 1975
xiv,232p : ill. - Bibliography: p.224-228. - Includes index
0-470-22372-3 X.200/20323
0624756

ADMINISTRATION. Prisons. Unted States

MCGEE, Richard A.
Prisons and politics. - Lexington, [Mass.] : Lexington Books, 1981
xiv,161p. - Bibliography: p157. - Includes index
0-669-04527-6 X.200/37708
2627531

ADMINISTRATION. Procuracy. Czechoslovakia
1945-

PLUNDR, Otakar
Organizace justice a prokuratury. - Praha : Panorama, 1980
248p. - (Učebnice). - Bibliography: p.220-234
r6439617 X.529/53948
8937796

ADMINISTRATION. Provinces. Thailand
1892-1915

BUNNAG, Tej
The provincial administration of Siam, 1892-1915. - Kuala Lumpur ; London : Oxford University Press, 1977
xi,322p. - (East Asian historical monographs). - Bibliography: p.288-301. - Includes index
0-19-580343-4 X.809/25311
1346849

ADMINISTRATION. Provision of equipment for physically handicapped persons by local authority welfare services. London

KEEBLE, Ursula
Aids and adaptations. - [London] : Bedford Square Press, 1979
320p : ill. - (Occasional papers on social administration ; no.62). - Bibliography: p.315-320
0-7199-0963-5 8299.i.7(62)
1952587

ADMINISTRATION. Ptolemaic Empire
B.C.322-B.C.40

BAGNALL, Roger S.
The administration of the Ptolemaic possessions outside Egypt. - Leiden : E. J. Brill, 1976
xvi, 286 p., fold plate. - (Columbia studies in the classical tradition ; vol. 4). - Bibliography : p. xiii - xvi. - Includes index
90-04-04490-6 X.800/25893
1041398

ADMINISTRATION. Public bodies. Italy
- Conference proceedings

L'ALTRO potere in economia / saggi e interventi di Peggio ... [et al.]. - Bari : De Donato, 1978
259p. - (Riforme e potere ; 22)
r6107978 X.529/35940
2444321

ADMINISTRATION. Public bodies. Italy
to date - Conference proceedings

CONVEGNO DI STUDI DI SCIENZA DELL'AMMINISTRAZIONE (24th 1978 : Varenna)
Realtà e prospettive dei controlli sugli enti locali territoriali. - Milano : Giuffrè, 1979
412p
r6504718 X.800/32908
8268967

ADMINISTRATION. Public bodies. Race relations. Great Britain
Great Britain. Parliament. House of Commons. Select Committee on Race Relations and Immigration. Organisation of race relations administration

GREAT BRITAIN. Home Office
The organisation of race relations administration, observations on the report of the Select Committee on Race Relations and Immigration .... - London : H.M.S.O., [1976]
21p. - (Cmnd.6603)
0-10-166030-8 BS.18/521
0752479

ADMINISTRATION. Public health services, 1895-1918. Newark. New Jersey

GALISHOFF, Stuart
Safeguarding the public health. - Westport, Conn. ; London : Greenwood Press, 1975
xv,191,[1]p : ill. - Bibliography: p.173-180. - Includes index
0-8371-7956-4 X.329/11244 Woolwich
1308513

ADMINISTRATION. Public health services. Urban regions. United States

ANALYSIS of urban health problems / Irving Leveson and Jeffrey H. Weiss, [editors] ; [for the Health Services Administration of the City of New York]. - New York : SP Books ; New York ; London : Distributed by Wiley, 1976
[10],390p : 2 ill. - (Health systems management ; vol.6). - Includes index
0-470-14983-3 X.320/10349 Woolwich
0648027

ADMINISTRATION. Public lending right. Great Britain

SUMSION, John
Setting up Public Lending Right. - Stockton-on-Tees : Registrar of Public Lending Right, 1984
36p : ill. - Bibliography: p35
0-9509825-0-4 2719.k.304
3277151

ADMINISTRATION. Public libraries

CAMPBELL, H. C.
Developing public library systems and services. - Paris : Unesco, 1983
186p. - (Documentation, libraries and archives. studies and research ; 11). - Includes bibliographies
92-310-1995-3 UNL.244/11
1821962

SINCLAIR, Dorothy
Administration of the small public library. - 2nd ed. - Chicago : American Library Association, 1979
xii,156p. - Includes bibliographies
0-8389-0291-x 2719.e.464
1821962

STOAKLEY, R. J. (Roger John)
Presenting the library service. - London : Bingley, 1982
109p. - (Outlines of modern librarianship). - Includes bibliographies and index
0-85157-320-7 X.529/46904
1821962

ADMINISTRATION. Public libraries. Great Britain

BROWN, Royston
Public library administration. - London : Bingley [etc.], 1979
95p. - (Outlines of modern librarianship). - Includes bibliographies and index
0-85157-276-6 2719.cc.94
0011754

PHILLIPS, Barry
Public libraries. - London : Library Association : Chartered Institute of Public Finance and Accountancy, 1977
[39]p. - Includes bibliographies
0-85365-750-5 2719.gg.27
0011754

ADMINISTRATION. Public libraries. Leicestershire
- For senior librarians

LEICESTERSHIRE LIBRARIES AND INFORMATION SERVICE
Work programming. - [Leicester] : The Service, 1977 [i.e.1978]
[13]leaves
0-85022-028-9 2719.hh.10
1995286

ADMINISTRATION. Public sector. Sweden

ANDERSSON, Simon
Myndigheten i samhället. - Stockholm : Liber förlag, 1978
221p. - (Publica)
91-380-2899-9 X.519/41443
204210X

ADMINISTRATION. Rare book collections
1945- - Conference proceedings

Das HISTORISCHE und wertvolle Buchgut in der Bibliotheksverwaltung / herausgegeben im Auftrag der Vereinigung Österreichischer Bibliothekare von Otto Mazal und Eva Irblich. - Wien : Die Vereinigung, 1980
216p. - (Biblos-Schriften ; Bd 104)
r6302163 Ac.9596.b/3[Bd.104]
8794537

ADMINISTRATION. Rare book collections
1945- - Directories - Serials

RARE books / edited by Alice D. Schreyer. - 1983/84-. - New York : R.R. Bowker Co., 1984-
v.
r6814465 2708.e.524
8794537

ADMINISTRATION. Rare book collections. Libraries
- Bibliographies

SIMPSON, Murray Charles Thomas
Rare book librarianship. - Leeds : Department of Librarianship, Leeds Polytechnic, 1976
16p. - Includes index
0-900738-05-7 2719.aa.34
0712132

ADMINISTRATION. Recreation services. United States

KRAUS, Richard G. (Richard Gordon)
Creative administration in recreation and parks. - 2nd ed. - Saint Louis : Mosby ; London : Distributed by Kimpton, 1977
xiii,331p : ill. - Bibliography: p.322-324. - Includes index
0-8016-2739-7 X.620/16899 Woolwich
004489x

SHIVERS, Jay S. (Jay Sanford)
Essentials of recreational services. - Philadelphia : Lea and Febiger ; London : Kimpton, 1978
ix,348p : ill. - Includes bibliographies and index
0-8121-0638-5 X.622/6740 Woolwich
004489x

ADMINISTRATION. Recruitment. Personnel. Health authorities. Wales
Organisations: National Health Service Staff Commission for Wales, to 1975

NATIONAL HEALTH SERVICE STAFF COMMISSION FOR WALES
National Health Service Staff Commission for Wales (1972-1975). - Cardiff : H.M.S.O., 1975
vii,60p
Later issues are also placed at this shelfmark
0-11-790068-0 BS.128a/75
0599662

ADMINISTRATION. Recruitment. Personnel. Local authorities. England
Organisations: Local Government Staff Commission for England. to 1976

LOCAL GOVERNMENT STAFF COMMISSION FOR ENGLAND
The report of the Local Government Staff Commission for England. - London : H.M.S.O., 1977
v.160p : 1 ill
0-11-751138-2 BS.414/2195
1244167

ADMINISTRATION. Recruitment. Police. United States

HEWITT, William H.
New directions in police personnel administration. - Lexington, Mass. ; London : D.C. Heath, 1975
xiii,121p. - Bibliography: p.105-112. - Includes index
0-669-97832-9 X.200/20238
0590657

ADMINISTRATION. Remuneration. Officers. Great Britain. Army 1714-1763

GUY, Alan J. (Alan James)
Oeconomy and discipline. - Manchester : Manchester University Press, 1985
x,188p. - Bibliography: p173-183. - Includes index
0-7190-1099-3 X.800/40888
3053504

ADMINISTRATION. Residential institutions for old persons. England

PERRY, M.
Professional support for staff in private and voluntary homes for the elderly in England. - [London] : [Department of Health and Social Security], 1981
15p
B83-01585 BS.385/707
2809532

ADMINISTRATION. Resources centres. Libraries. Schools. United States

GILLESPIE, John T. (John Thomas)
Administering the school library media center. - New York ; London : Bowker, 1983
x,381p. - Includes bibliographies and index
0-8352-1514-8 X.520/32461
3042952

HICKS, Warren B. (Warren Braukman)
Managing the building-level school library media program. - Chicago : American Association of School Librarians, 1981
vi,43p. - (School media centers ; no.7). - Bibliography: p41-43
0-8389-3262-2 X.520/33768
3042952

ADMINISTRATION. Rhône (Department). France 1958- - Serials

RHÔNE (Département). Conseil général. Commission départementale
Rapport au Conseil .... - 2ème session ordinaire (1980)-. - Villeurbanne : Imprimerie Chambefort, 1981-
r6209545 S.E.400/1038
8885141

ADMINISTRATION. Royal forests. England 1066-ca 1450

YOUNG, Charles Robert
The royal forests of medieval England. - Leicester : Leicester University Press, 1979
ix,220p : ill. - Bibliography: p.205-212. - Includes index
0-7185-1179-4 X.320/11897 Woolwich
175453x

ADMINISTRATION. Royal Opera House
Webster, Sir David, b.1903 - Biographies

HALTRECHT, Montague
The quiet showman. - London : Collins, 1975
319p, leaf of plate,[16]p of plates : ill. - Bibliography: p.308. - Includes index
0-00-211163-2 X.431/10193
0573914

ADMINISTRATION. Rural regions. France to date

ARCHIVES DÉPARTEMENTALES DU GARD
Répertoire numérique de la série o. - Nîmes : [les Archives], 1981
xix,225p
2-86030-002-3 S.E.400/912
8764603

ADMINISTRATION. Rural regions. Kenya - Conference proceedings

COMPARATIVE ADMINISTRATION IN EAST AFRICA (Conference : 1971 : Arusha)
Rural administration in Kenya. - Nairobi : East African Literature Bureau, 1973
x,165p - (Management and administration series ; no.2). - Includes bibliographies
r6072218 X.709/32499
1580981

ADMINISTRATION. Rural regions. Tanzania to date

FINUCANE, James R.
Rural development and bureaucracy in Tanzania. - Uppsala : Scandinavian Institute of African Studies, 1974
192p
91-7106-076-6 X.519/39681
9119531

ADMINISTRATION. Sacraments. Catholic Church

CATHOLIC CHURCH. Episcopal Conference of Ireland
The sacraments. - Dublin : Veritas Publications, 1976
173p. - Bibliography: p.170-173
0-905092-13-9 X.108/16636
1101188

ADMINISTRATION. Sardinia. Italy

AUTONOMIE locali e riforma della Regione / [rapporto presentato al Consiglio regionale dalla Giunta regionale]. - Cagliari : EDES, 1978
267p. - (Documenti & opinioni ; 10)
r6195644 X.708/48786
2475413

ADMINISTRATION. Schools - Bibliographies

COULSON, Alan Arthur
School administration and management. - [1st ed.] reprinted with additions. - Hull : Flag Publications ; Driffield : Distributed by 'Studies in Education', 1977
[1],36p
B77-31656 X.519/27333
0296910
- [Another copy.] OPL

ADMINISTRATION. Schools
Applications of critical path analysis

VAUGHAN, B. W.
Planning in education. - Cambridge : Cambridge University Press, 1978
vii,120p : ill. - Bibliography: p.119. - Includes index
0-521-21817-9 Cup.1285/415
1834053

ADMINISTRATION. Schools
Applications of organisation development - For counselling

ARENDS, Richard I.
Systems change strategies in educational settings. - New York ; London : Human Sciences Press ; London : Distributed by Eurospan Ltd, 1977
[2],120p. - (New vistas in counseling series ; vol.3). - Bibliography: p.112-120
0-87705-310-3 X.529/33074
1812300

ADMINISTRATION. Schools. Denmark

ANDERSEN, Erik Vitus
En vurdering af skolelederens arbejdsvilkår frem til 1990. - Hellerup : [E.V. Andersen], [1976]
73 leaves
r6060089 X.512/2795
1504320

ADMINISTRATION. Schools. Denmark
Law

DENMARK. Folketinget
Forslag om Ændring of skolestyrelseloven med bemærkninger. 14. februar 1974. - Vejle : Kroghs skolehåndbog, [1974]
40p.. - (Folketinget 1973-74 (2. Samling). Lovforslag Nr.99)
87-7469-441-3 X.512/2465
0887129

FOLKESKOLEN i dag / redigeret af Hans Cornelius. - Totalt revid. og udvidede 2. udg. - København : Suenson, [1976]
336p. - Bibliography : p.312-328
87-87201-14-3 X.519/23034
1228560

ADMINISTRATION. Schools. England
By governing boards

BULLIVANT, Barbara
Governor's guide. - 1979 ed. - Billericay : Home and School Council Publications, [1979]
[2],29p
0-901181-30-7 X.529/37229
0273155

SALLIS, Joan
The effective school governor. - 2nd ed. - London : The Advisory Centre, 1982
31p : ill. - (An ACE handbook)
0-900029-65-x X.529/49196
0273155

School governors. - London : The Advisory Centre, 1979
28p : ill. - (An ACE handbook). - Bibliography: p28
0-900029-49-8 X.529/49193
0273155

ADMINISTRATION. Schools. England
By governing boards & managing boards

BACON, William
Public accountability and the schooling system. - London : Harper and Row, 1978
viii,236p : 1 ill. - Bibliography: p.208-210. - Includes index
0-06-318082-0 X.529/32421
1340018

ADMINISTRATION. Schools. England
Legal aspects

ADAMS, Neil
Law and teachers today. - London : Hutchinson, 1983
285p : ill. - Includes index
0-09-151610-2 X.200/41904
2755742

Law and teachers today. - 2nd ed. - London : Hutchinson, 1984
285p. - Bibliography: p277. - Includes index
0-09-156720-3 X.200/44294
2755742

COUNTY and voluntary schools. - 6th ed., completely rev.. - [Harlow] : Councils and Education Press, c1982
vii,197p. - Includes index
0-900313-40-4 X.529/48452
2755742

ADMINISTRATION. Schools. Germany to date

NEVERMANN, Knut
Der Schulleiter. - Stuttgart : Klett-Cotta, 1982
314p. - (Veröffentlichungen des Max-Planck-Instituts für Bildungsforschung). - Includes bibliography
3-12-936030-1 X.529/59194
9052240

ADMINISTRATION. Schools. Great Britain

BARRY, C. H. (Charles Harold)
Running a school. - 2nd ed. - London : Temple Smith, 1975
247p. - (Teaching in practice). - Bibliography: p239-247
0-85117-190-7 X.529/45360
0108812

BARRY, Charles Harold
Running a school. - 2nd ed. - London : Temple Smith, 1975
247p. - ([Teaching in practice]). - Bibliography: p.239-247
0-85117-069-2 X.529/30157
0108812

JOHN, Denys
Leadership in schools. - London : Heinemann Educational, 1980
x,181p : ill. - (Heinemann organization in schools series). - Bibliography: p173. - Includes index
0-435-80280-1 X.529/64843
0108812

Leadership in schools. - London : Heinemann Educational, 1980
x,181p : ill. - (Heinemann organization in schools series). - Bibliography: p173. - Includes index
0-435-80468-5 X.529/42428
0108812

ADMINISTRATION. Schools. Great Britain
Applications of microcomputer systems

BIRD, Patrick
Microcomputers in school administration. - London : Hutchinson, 1984
208p : ill. - Bibliography: p199-202. - Includes index
0-09-154510-2 X.622/20624
3087395

ADMINISTRATION. Schools. Great Britain
Applications of microcomputer systems - Conference proceedings

MICROCOMPUTERS in school administration / edited by David Lancaster. - [Sheffield] : Sheffield City Polytechnic Department of Education Management, 1983
80p. - (Sheffield papers in education management ; 34)
0-86339-006-4 X.629/22880
3087395

ADMINISTRATION. Schools. Italy
1946–

ARDIGÒ, Achille
Informazione, partecipazione e mondi vitali. – Firenze : Sansoni, 1982
vii,255p. – (Fondazione Angelo Rizzoli ; 7)
r6509408 X.529/66233
8768110

SUMMA, Antonio
Ricorsi in materia scolastica nella vigente legislazione. – Milano : Giuffrè, 1979
iv,124p. – (Biblioteca della Rivista giuridica della scuola. Nuova serie ; 1)
r6501189 X.520/27130
8768110

ADMINISTRATION. Schools. Italy
1946– – Serials

SCUOLA democratica. – n.1 (febbr. / mar.1978)– . – Venezia : Marsilio, 1978–
Holdings: n.1 (febbr. / mar.1978) only
r6598045 X.525/5002
8768110

ADMINISTRATION. Schools. Italy
Law, to date

PAUCIULLO DELLA VALLE, Giuseppina
I consigli di circolo e di istituto. – Milano : Giuffrè, 1978
63p. – (Biblioteca della Rivista giuridica della scuola ; 21)
r6069507 X.520/26860
8834121

ADMINISTRATION. Schools. Nigeria

MUSAAZI, J. C. S.
The theory and practice of educational administration. – London : Macmillan Nigeria, 1982
274p : ill. – (Studies in Nigerian education). – Includes bibliographies and index
0–333–31489–1 X.529/49178
2765489
– [Another copy] X.529/56197

OZIGI, A. O.
A handbook on school administration and management. – London : Macmillan, 1977
viii,144p. – Bibliography: p.139–140. – Includes index
0–333–21325–4 X.519/27281
1247778

ADMINISTRATION. Schools. Sweden
1945–

MARKLUND, Sixten
Educational administration and educational development. – Stockholm : University of Stockholm, Institute of International Education, 1979
63p. – (Studies in comparative and international education ; no.1)
r6610851 X.0520/472(1)
8894361

MUNKNÄS, Stig
Statlig eller kommunal skola?. – [Stockholm] : S. Munknäs, [1981]
237p. – (Stockholm studies in politics ; 19). – Bibliography
r6610104 X.0700/438(19)
8874581

ADMINISTRATION. Schools. United States

CAMPBELL, Roald F. (Roald Fay)
Introduction to educational administration. – 6th ed.. – Boston ; London : Allyn and Bacon, c1983
x,269p. – Includes index
0–205–07983–0 X.520/32337
1106864

CAMPBELL, Roald Fay
Introduction to educational administration. – 5th ed.. – Boston, Mass. ; London : Allyn and Bacon, 1977
xii,396p : ill. – Includes bibliographies and index
0–205–05678–4 X.520/11617
1106864

LANDERS, Thomas J.
Essentials of school management. – Philadelphia ; London : Saunders, 1977
x,445p : ill. – Includes bibliographies and index
0–7216–5619–6 X.525/2186
1106864

A PRINCIPAL'S workbook / edited by Keith Tronc. – St Lucia, Queensland : University of Queensland Press, 1977
[6],264p : ill. – (Educational administration and organization). – Bibliography : p.257–261. – Includes index
0–7022–1382–9 X.510/15841
1551671

STOOPS, Emery
Handbook of educational administration. – 2nd ed. – Boston [Mass.] ; London : Allyn and Bacon, c1981
xiv,499p : ill. – Includes bibliographies and index
0–205–07133–3 X.520/22674
1106864

STRATEGIES for instructional management / Lanore A. Netzer ... [et al.]. – Boston, Mass. ; London : Allyn and Bacon, 1979
xix,288p. – Includes bibliographies and index
0–205–06448–5 X.520/14045
1106864

WILES, Jon
Principles of school administration. – Columbus ; London : Merrill, c1983
x,355p : ill. – Includes bibliographies and index
0–675–20054–7 X.520/36832
1106864

WILSON, L. Craig
School leadership today. – Boston [Mass.] ; London : Allyn and Bacon, 1978
xiv,346p : ill. – Includes index
0–205–06019–6 X.520/12601
1106864

ADMINISTRATION. Schools. United States
– Biographies

WHO'S who biographical record, school district officials / compiled by the editors of 'Who's who in America'. – Chicago : Marquis Who's Who ; London : Distributed by Prior, 1976
xiv,666p
0–8379–3801–5 X.802/10308
1066390

ADMINISTRATION. Schools. United States
1809–1981

TYACK, David Bruce
Managers of virtue. – New York : Basic Books, c1982
vii,312p
0–465–04376–3 X.520/36655
9145982

ADMINISTRATION. Schools. United States
Applications of microcomputer systems

RADIN, Stephen
Computer literacy for school administrators and supervisors. – [Lexington, Mass.] : Lexington Books, 1983
xvi,260p : ill. – Includes index
0–669–06330–4 X.622/19946
3143112

ADMINISTRATION. Schools. United States
Decentralisation

GREENHALGH, John
School site budgeting. – Lanham ; London : University Press of America, c1984
xii,193p. – Includes index
0–8191–3774–x X.529/66101
3215156

ADMINISTRATION. Schools. United States
Political aspects

MANN, Dale
The politics of administrative representation. – Lexington, Mass. ; London : D.C. Heath, 1976
xv,187p : 1 ill. – (Lexington Books politics of education series). – Bibliography: p.167–175. – Includes index
0–669–97188–x X.800/25585
0647918

POLITICAL science and school politics / edited by Samuel K. Gove, Frederick M. Wirt. – Lexington, Mass. : D.C. Heath, 1976
xi,147p. – (Policy studies organization series)
0–669–00739–0 X.800/26739
0647918

WILES, David K.
Practical politics for school administrators. – Boston [Mass.] ; London : Allyn and Bacon, c1981
x,225p : ill. – Includes index
0–205–07132–5 X.520/22955
0647918

ADMINISTRATION. Schools. United States
Role of head teachers

BLUMBERG, Arthur
The effective principal. – Boston, Mass : Allyn and Bacon, 1980
viii,280p. – Bibliography: p.271–273. – Includes index
0–205–06812–x X.520/21426
2181223

ADMINISTRATION. Schools. United States
Role of women

WOMEN and educational leadership / edited by Sari Knopp Biklen, Marilyn B. Brannigan. – Lexington, Mass. : Heath ; [Farnborough, Hants.] : [Distributed by] Gower Press, 1980
xi,269p. – Includes bibliographies and index
0–669–03216–6 X.520/21882
2235773

ADMINISTRATION. Schools. Western world
Political aspects

The POLITICS of school government / edited by George Baron. – Oxford : Pergamon, 1981
viii,298p. – (International studies in education and social change). – Includes bibliographies and index
0–08–025213–3 X.529/42272
2162695

ADMINISTRATION. Secondary education. England
Political aspects, 1944–1970

FENWICK, I. G. K. (Ian Graham Keith)
The comprehensive school, 1944–1970. – London : Methuen, 1976
xi,187p. – Includes index
0–416–65910–1 X.520/10508
0609218

The comprehensive school 1944–1970. – London : Methuen, 1976
187p. – (Education paperbacks). – Bibliography p164–165. – Includes index
0–416–73490–1 X.529/39071
0609218

ADMINISTRATION. Secondary schools. Dundee. Tayside Region Scotland

TEN secondary schools / [edited by] Iain Smith. – Dundee : Department of Education, University of Dundee, 1977
[2],24 leaves
B78–18184 X.515/628
145563x

ADMINISTRATION. Secondary schools. England

GRAY, Henry Lewis
Change and management in schools. – Driffield : Studies in Education Ltd, [1979]
81p. – (Nafferton books)
0–905484–10–x X.519/30170
1948628

ADMINISTRATION. Secondary schools. England
Role of head teachers & senior teachers

MANAGEMENT and headship in the secondary school / editor Arnold Jennings. – London : Ward Lock, 1977
ix,195p : ill. – Includes index
0–7062–3684–x X.529/32142
1377310

ADMINISTRATION. Secondary schools. United States

HAMPTON, Bill R.
Solving problems in secondary school administration. – Boston [Mass.] ; London : Allyn and Bacon, c1981
xii,312p : ill. – Includes index
0–205–06951–7 X.520/22685
2328852

NOTTINGHAM, M. A. (Marvin Aubrey)
Principles for principals. – Rev. ed. – Lanham ; London : University Press of America, c1983
viii,212p
0–8191–3570–4 X.529/64882
2328852

ADMINISTRATION. Serials collections. Libraries
– Serials

The SERIALS librarian. – vol.1, no.1– : Fall 1976–. – New York : Haworth Press, 1976–
r6051909 2719.b.93
1385100

ADMINISTRATION. Services. Local authorities. France
1945–

AUBY, Jean-François
Les services publics locaux. – [Paris] : [Presses universitaires de France], [1982]
127p. – (Que sais-je? ; no.2023)
2–13–037434–4 X.529/65880
9137653

ADMINISTRATION. Sheriff courts. Scotland

GREAT BRITAIN. Committee on the Administration of Sheriffdoms by Full-Time Sheriffs Principal
Administration of sheriffdoms. – Edinburgh : H.M.S.O., [1982]
v,28p. – (Cmnd. ; 8548)
0–10–185480–3 BS.39/110
2767228

ADMINISTRATION. Silesia
By Bohemia (Kingdom), 1327-1339
PUSTEJOVSKY, Otfrid
Schlesiens Übergang an die böhmische Krone. - Köln : Böhlau, 1975
xlviii, 267 p. - (Forschungen und Quellen zur Kirchen- und Kulturgeschichte Ostdeutschlands ; Bd 13). - Bibliography : p. xiii-xlvi
3-412-82573-5 X.0100/45(13)
0863173

ADMINISTRATION. Small churches. United Reformed Church
UNITED REFORMED CHURCH. Church Life Department
Handbook for smaller churches. - London : The Department, 1976
[4],55p
0-902256-24-6 X.102/2044
0721743

ADMINISTRATION. Smallholdings. Agriculture. Gloucestershire
GLOUCESTERSHIRE. Performance Review Sub-committee
Performance review of smallholdings. - Gloucester : [The Council], 1978
[5],41[i.e.45]p : ill
0-904950-15-8 X.512/7199
2052598

ADMINISTRATION. Social security benefits. Great Britain
Great Britain. Parliament. House of Commons. Social Services Committee. First report ... session 1979-80
GREAT BRITAIN. Department of Health and Social Security
Reply to the first report from the Social Services Committee on arrangements for paying social security benefits. - London : H.M.S.O., [1980]
60p : ill. - (CMND ; 8106)
0-10-181060-1 BS.385/675
2643987

ADMINISTRATION. Social security benefits. Great Britain
Reform
DILNOT, A. W.
The reform of social security. - Oxford : Clarendon, 1984
viii,166 : ill. - Bibliography: p158-160. - Includes index
0-19-877226-2 X.529/63424
3023974
- [Another Copy: paperback.]
ISBN 0-19-877225-4 X.529/63425

ADMINISTRATION. Social security. Foreign countries
Applications of digital computer systems
GREAT BRITAIN. Department of Health and Social Security. Management Support and Computers Division
Studies of foreign social security computer systems. - [London] : [The Department].
Report on a visit in December 1978 to SSA, USA. - [1979]
[3],16,2p, fold leaf of plate
B80-37890 BS.385/578(2)
228670x

Report on a visit in December 1978 to Italy. - [1979]
34p in various pagings
B80-37891 BS.385/578(1)
228670x

ADMINISTRATION. Social security. Great Britain
SOCIAL security operational strategy. - [London] : Department of Health and Social Security, 1982
22p : ill(some col.)
B83-33379 BS.385/833
2807750

SOCIAL security operational strategy. - London : H.M.S.O., 1982
i,62p : 2ill
0-11-760861-0 BS.385/725
2807750

TRAFFIC in social security work between DHSS local and regional offices and headquarters / N.R. Warner ... [et al.]. - [London] : [Department of Health and Social Security], 1981
77p : 1ill
B83-01107 BS.385/701
2807750

ADMINISTRATION. Social services. Great Britain
Discretion - Conference proceedings
DISCRETION and welfare / [edited by] Michael Adler and Stewart Asquith. - London : Heinemann Educational, 1981
vii,296p. - ([Studies in social policy and welfare]). - Bibliography: p273-287. - Index
0-435-82009-5 X.529/44200
2574543

ADMINISTRATION. Social services. United States
GILBERT, Neil
Coordinating social services. - New York ; London : Praeger, 1977
xvii,84,[1]p : ill. - (Praeger special studies in US economic, social and political issues)
0-275-24330-3 X.520/11976
1249533

PROGRAMMING effective human services / edited by Walter P Christian, Gerald T Hannah and T J Glahn. - New York ; London : Plenum, c1984
xxiv,514p : ill. - Includes bibliographies and index
0-306-41526-7 X.520/35939
1249533

ADMINISTRATION. Social services. United States
Applications of microcomputer systems
TAYLOR, James B. (James Bentley)
Using microcomputers in social agencies. - Beverly Hills ; London : Sage in cooperation with the Continuing Education Program in the Human Services of the University of Michigan School of Social Work, c1981
119p : ill. - (A Sage human services guide ; 19). - Includes bibliographies
B81-35657 X.529/43391
2548518

ADMINISTRATION. Spanish colonies
Historical sources: Documents, to date - Conference proceedings
SEMANA INTERNACIONAL DE ARCHIVOS (1979 : La Rábida)
Documentación y archivos de la colonización española. - [Madrid] : Ministerio de Cultura Dirección General de Bellas Artes, Archivos y Bibliotecas, Subdirección General de Archivos, 1980
2v.
84-7483-146-6 X.809/61908
9157700

ADMINISTRATION. Special collections. Stock. Libraries - Conference proceedings
INTEGRATED MANAGEMENT OF LIBRARIES AND COLLECTION DEVELOPMENT (Conference : 1973 : Sackville, New Brunswick)
Three papers on collections. - London : Vine Press, 1975
v,47,[4]leaves. - (Dalhousie University Libraries and Dalhousie University School of Library Service occasional papers / Dalhousie University). - Includes bibliographies
B79-03340 X.0512/507(10)
1765124

ADMINISTRATION. Special libraries
ASHWORTH, Wilfred
Special librarianship. - London : Bingley [etc.], 1979
[2],120p. - (Outlines of modern librarianship). - Includes index
0-85157-277-4 2719.cc.63
1878832

ADMINISTRATION. Special schools. Great Britain
MANAGEMENT & the special school / edited by Tony Bowers. - London : Croom Helm, c1984
228p : ill. - Bibliography: p215-223. - Includes index
0-7099-3231-6 X.529/62735
3048039

ADMINISTRATION. Special schools. United States
MAYER, C. Lamar
Educational administration and special education. - Boston [Mass.] ; London : Allyn and Bacon, c1982
xii,383p : ill. - Includes bibliographies and index
0-205-07555-x X.520/25941
261443x

ADMINISTRATION. Sports & games. Activities. Educational institutions. United States
VANDERZWAAG, Harold J.
Sport management in schools and colleges. - New York ; Chichester : Wiley, c1984
xviii,265p : ill. - Includes bibliographies and index
0-471-87135-4 X.622/20961
3186849

ADMINISTRATION. Sports. Curriculum subjects. Training. Youth workers
SPORT and youth leadership / Eastern Council for Sport and Recreation. - Bedford : Eastern Council for Sport and Recreation, 1983
16p. - (Report / Eastern Council for Sport and Recreation)
B85-02889 X.529/65766
3211126

ADMINISTRATION. Sports. United States
BROYLES, J. Frank
Administration of athletic programs. - Englewood Cliffs ; London : Prentice-Hall, 1979
x,405p : ill. - Includes index
0-13-005249-3 X.622/7359 Woolwich
205521x

ADMINISTRATION. St Vincent's Community Home
APPROVED school to community home / [edited by Barbara Kahan]. - London : H.M.S.O., 1976
[3],73p : ill. - (Report series / Great Britain. Department of Health and Social Security. Social Work Service. Development Group)
0-11-320639-9 BS.385/247
1057804

ADMINISTRATION. State courts. United States
GAZELL, James A.
The future of state court management. - Port Washington ; London : Kennikat Press, 1978
[9],90p. - (Kennikat Press national university publications. multi-disciplinary studies in the law). - Includes index
0-8046-9195-9 X.200/31354
0641804

State trial courts as bureaucracies. - New York ; London : Dunellen Publishing Co. ; London : Distributed by Robertson, [1976]
[8],168p : ill. - Bibliography: p.165-168. - Includes index
0-8046-7101-x X.200/20365
0641804

ADMINISTRATION. State. Cuba
Law
CUBA
Ley de Organización de la Administración Central del Estado. - La Habana : Ministerio de Justicia, 1979
62p
r6162153 LAS.D.300/17
2458624

ADMINISTRATION. Sterile equipment - Serials
ASSOCIATION OF STERILE SUPPLY ADMINISTRATORS
ASSA journal. - vol.1, no.1 - : Sept. 1972 -. - Matlock : Newton Mann Ltd., 1972-
r6040789 P.323/63
1329359

ADMINISTRATION. Stock. Libraries. Higher education institutions. United States
CLINE, Hugh F.
Building library collections. - Lexington, Mass. : Lexington Books, 1981
xvi,170p : ill. - Bibliography: p161-163. - Includes index
0-669-04321-4 2719.e.274
2574136

ADMINISTRATION. Stock. Museums. Great Britain
Implications of Wildlife and Countryside Act 1981 - Conference proceedings
The WILDLIFE and Countryside Act (1981) and its implications for museums / compiled and edited by G. Stansfield. - [Great Britain] : Biology Curators Group, 1983
18p. - (Report / Biology Curators Group ; no.2). - Includes bibliographies
0-9507092-1-2 X.322/13478
3126064

ADMINISTRATION. Sweden
Role of aristocracy, to date
ELMROTH, Ingvar
För kung och fosterland. - Lund : Gleerup, 1981
282p. - (Bibliotheca historica lundensis ; 50). - Includes bibliography
91-400-4800-4 WP.2374/50
9127453

ADMINISTRATION. Telecommunication services. Great Britain. Post Office
to 1978
PITT, Douglas
The telecommunications function in the British Post Office. - Farnborough, Hants. : Saaxon House, 1980
xiii,193p. - Includes index
0-566-00273-6 X.529/38433
1934414

ADMINISTRATION. Town planning. Italy
CAMPOS VENUTI, Giuseppe
Amministrare l'urbanistica. - 3. ed. - Torino : Giulio Einaudi, 1974
300 p., 5 leaves of plates, 14 p. of plates : ill. - (Piccola biblioteca Einaudi ; n. 97)
r6014214 X.708/19834
0868914

ADMINISTRATION. Towns
to date

EISFELD, Dieter
Stadt der Zukunft. – Stuttgart : Deutsche Verlags-Anstalt, 1981
142p
3-421-02573-8 X.529/54033
8943184

ADMINISTRATION. Towns
to date – Conference proceedings

Die UNREGIERBARKEIT der Staedte. – Rüschlikon / Zürich : Gottlieb Duttweiler-Institut, [1982]
250p : ill
r6309434 X.529/62488
8943184

ADMINISTRATION. Towns. India (Republic)

MUNICIPAL administration in India / [edited by] S.K. Sharma, V.N. Chawla. – Jullundur : International Book Co., 1975
429[i.e.439],[52]p. – Bibliography : p.i–viii
r6032083 X.809/42170
1169491

ADMINISTRATION. Towns. Soviet Union
1945–

SISTEMA organov gorodskogo upravleniîa / ; pod redaktsieĭ P.N. Lebedeva. – Leningrad : Izdatel'stvo Leningradskogo universiteta, 1980
215p
r6487028 X.529/48669
8854416

TEORIÎA i praktika kompleksnogo razvitiîa goroda / [otvetstvennyĭ redaktor : V.I. Senchenko]. – Kiev : Naukova dumka, 1980
205p. – Bibliography
r6464120 X.529/54082
8854416

ADMINISTRATION. Training. Volunteers. Welfare services. Great Britain

NATIONAL YOUTH BUREAU
Planning your own training (and using a consultant). – Leicester : The Bureau, 1978
[8]p. – (Social action special)
0-902095-29-3 X.512/6071
157549x

ADMINISTRATION. Trusts. England

SLADEN, Michael
Aspects of practical trust administration. – London : Institute of Bankers, [1982?]
29p. – Includes index
0-85297-074-9 X.529/54578
112305x

Practical trust administration. – London : Europa, 1977
xxvi,286p. – Includes index
0-900362-99-5 X.520/11704
112305x

Practical trust administration. – 2nd ed. – London : Europa, c1983
xxxii,312p. – Includes index
0-905118-94-4 X.520/33776
112305x

Practical trust administration. – London : Europa. Supplement to the first edition. – 1980
30p
0-905118-58-8 X.529/47844
112305x

ADMINISTRATION. Ulriksholm (Estate)
1630–1650

JENSEN, Jytte
Organisering og finansiering af byggeriet på Ulriksholm i 1630'rne og 1640'rne. – [Odense] : Odense Universitetsforlag, 1977
98p. – (Odense University studies in history and social sciences ; v.50). – Includes bibliographies
87-7492-233-5 X.0700/318(50)
1848186

ADMINISTRATION. Unemployment benefits. Great Britain
to 1976

FULBROOK, Julian
Administrative justice and the unemployed. – London : Mansell Information Publishing, 1978
xvii,338p. – (Studies in labour and social law ; vol.1). – Bibliography: p.319–335. – Includes index
0-7201-0555-2 X.0520/542 (1)
1114387

ADMINISTRATION. Unemployment insurance. Great Britain
1920–1931

DEACON, Alan, 1946–
In search of the scrounger. – London : Bell [for the Social Administration Research Trust], 1976
110p. – (Occasional papers on social administration ; no.60)
0-7135-1992-4 8299.i.7/60 Woolwich
1050524

ADMINISTRATION. United Reformed Church

UNITED REFORMED CHURCH
The United Reformed Church. – [London] : [United Reformed Church], [1982]
8p
B82-21770 X.200/38328
2723123

UNITED REFORMED CHURCH. Long Range Policy Group
Preparing the way. – London : United Reformed Church, 1983
28p
0-902256-60-2 X.529/58337
2723123

ADMINISTRATION. Universitätsbibliothek Tübingen
By Mohl, Robert von, 1836–1844

EHRLE, Peter Michael
Robert von Mohl als Leiter der Tübinger Universitätsbibliothek (1836–1844). – Tübingen : Mohr, 1975
xviii,183,[16]p of plates : ill. – (Contubernium ; Bd 10). – Includes bibliographies
3-16-937881-3 P.521/1282[10]
1129279

ADMINISTRATION. Universities. Africa
compared with administration of multinational companies

MAZRUI, Ali Al'Amin
The African university as a multinational corporation. – Brighton : Institute of Development Studies, University of Sussex, [1976]
[3], 24p. – (Discussion paper ; no.96)
r6031828 X.512/5597
1102516

ADMINISTRATION. Universities. Great Britain
– Conference proceedings – Serials

CONFERENCE OF UNIVERSITY ADMINISTRATORS
Proceedings. – [No.1]– ; 1974–. – [Aberystwyth] : [The Conference], [1976]– : ill
Holdings: 1976, etc
B80-24357 P.523 / 1352
2105276

ADMINISTRATION. Universities. Great Britain
Organisations: Committee of Vice-Chancellors and Principals of the Universities of the United Kingdom

COMMITTEE OF VICE-CHANCELLORS AND PRINCIPALS OF THE UNIVERSITIES OF THE UNITED KINGDOM
Report on the period 1972–76. – [London] : [The Committee], [1976]
74p
0-85143-051-1 X.519/29913
1100300

ADMINISTRATION. Universities. Great Britain
Participation of students

COMMITTEE OF VICE-CHANCELLORS AND PRINCIPALS OF THE UNIVERSITIES OF THE UNITED KINGDOM
Student participation in university government. – [2nd ed.]. – [London] : [The Committee], 1974
95p
B80-21693 X.529/36117
2075202

ADMINISTRATION. Universities. Italy

CLARK, Burton R.
Academic power in Italy. – Chicago ; London : University of Chicago Press, 1977
xii,205p. – Bibliography: p.191–199. – Includes index
0-226-10847-3 X.529/32432
1524445

IORIO, Giuseppe
Gli organi collegiali dell'università. – Napoli : Società editrice napoletana, 1976
188p
r6067539 X.510/10767
1529668

ADMINISTRATION. Universities. Italy
Legal aspects

AUTONOMIA universitaria. – Roma : Bulzoni, 1978
vii,381p
r6195507 X.520/21449
2418940

ADMINISTRATION. Universities. Ontario
– Serials

COUNCIL OF ONTARIO UNIVERSITIES
Current Review. – Vol. 1, no. 1– ; Winter 1970?–. – Toronto : The Council, [1970?]–
Holdings : Vol. 7, no. 1–
r6005240 P.521/3055
0723584

ADMINISTRATION. Universities. United States

CONFERENCE OF UNIVERSITY ADMINISTRATORS
Study visit to the USA, April 1978. – [Keele] : The Conference, [1978]
iii,40p : ill. – (CUA occasional paper series / Conference of University Administrators ; no.1)
B79-17423 X.512/6620
1879758

WALKER, Donald
The effective administrator. – San Francisco ; London : Jossey-Bass, 1979
iii–xix,208p. – (The Jossey-Bass series in higher education). – Bibliography: p.196–202. – Includes index
0-87589-399-6 X.520/13924
1892606

ADMINISTRATION. Universities. United States
Economic aspects

GARVIN, David A.
The economics of university behavior. – New York ; London : Academic Press, 1980
xv,176p : ill. – Bibliography: p162–168. – Includes index
0-12-276550-8 X.520/23926
2484218

ADMINISTRATION. Universities. United States
Participation of personnel

POWERS, David R.
Making participatory management work. – San Francisco ; London : Jossey-Bass, c1983
xvi,249p. – (The Jossey-Bass higher education series). – Bibliography: p233–241. – Includes index
0-87589-567-0 X.622/17457
2997029

ADMINISTRATION. University of Calcutta
Mukherjee, Sir Asutosh – Biographies

DAS GUPTA, A. P.
Asutosh Mukherjee. – New Delhi : National Book Trust, India, 1973
xii,127,[1]p. – (National biography series). – Bibliography : p.126–127
r6071832 X.708/21148
143506X

ADMINISTRATION. University of London

UNIVERSITY OF LONDON. Committee on Academic Organisation
University of London Committee on Academic Organisation third discussion document. – [London] : [University of London], 1982
32p
B83-39951 X.525/8077
2960222

ADMINISTRATION. University of Reading
Childs, William Macbride – Biographies

CHILDS, Hubert
W.M. Childs. – Reading : The author, 1976
[9],204p,[1],7 leaves of plates : ill. – Includes index
0-9505142-0-9 X.520/10941
0703621

ADMINISTRATION. Value-added tax. Great Britain

GREAT BRITAIN. Customs and Excise
Review of value added tax. – London : H.M.S.O., [1978]
23p. – (Cmnd.7415)
0-10-174150-2 BS.44/515
1801600

SMITH, B. R. (Bryan Richard)
Scrutiny in the VAT Central Unit Customs and Excise. – [London]
1v.(various pagings) : ill
B85-02493 BS.9/69
1801600

ADMINISTRATION. Venerable institutions

SAMUEL, Edwin, Viscount Samuel
See how they run. – London : Woburn Press, 1976
ix,268p : ill
0-7130-0131-3 X.800/11510
0662313

ADMINISTRATION. Voluntary organisations. Great Britain

EDGINTON, John
Legal structures for voluntary organisations. – London : Bedford Square Press/NCVO, 1984
16p. – (An NCVO practical guide)
0-7199-1116-8 X.529/65608
0043885

ADMINISTRATION. Voluntary organisations. Great Britain -continued

SLADEN, Edward
Honorary officers and what they do. - 2nd ed. - London : Bedford Square Press, 1978
[1],14p
B79-00031 X.519/28707
0043885

ADMINISTRATION. Vorpommern & Wismar
By Sweden, 1630-1720

MODÉER, Kjell A
Gerichtsbarkeiten der schwedischen Krone im deutschen Reichsterritorium. - Stockholm : Nordiska Bokhandeln, in Distribution, 1975-
1-. - (Skrifter utgivna av Institutet for rattshistorisk forskning. ser. 1 ; bd 24)
r6001306 Ac.2140b/2a [bd 24]
0941174

ADMINISTRATION. Welfare benefits. United States
Law

COFER, M. Donna Price
Administering public assistance. - Port Washington, N.Y. ; London : National University Publications : Kennikat, 1982
118p. - (Multidisciplinary studies in law and jurisprudence). - Bibliography: p111-112. - Includes index
0-8046-9298-x X.529/49192
2764237

ADMINISTRATION. Welfare services

BILLIS, David
Welfare bureaucracies. - London : Heinemann, 1984
252p : ill. - Bibliography: p238-248. - Includes index
0-435-82058-3 X.529/67074
0503657

WARHAM, Joyce
An introduction to administration for social workers. - Revised ed. - London : Routledge and Kegan Paul, 1975
viii,103p. - (Library of social work). - Bibliography: p.100-103
0-7100-8224-x X.529/30174
0503657

ADMINISTRATION. Welfare services. Developing countries
Use of research in social sciences

RESEARCH for social welfare / edited by L.G. Moseley. - London : Bedford Square Press, 1979
143p : ill
0-7199-0948-1 X.519/29909
1921304

ADMINISTRATION. Welfare services. United States

ABELS, Paul
Administration in the human services. - Englewood Cliffs ; London : Prentice-Hall, c1981
xi,244p : ill. - Includes index
0-13-005850-5 X.520/25009
2556693

ADMINISTRATION. Welfare work
Social services

GIRAUD, Pierre
Questions sociales.... - [Paris] : Imprimerie Nationale, 1975-
vol. 1-
r6012697 S.E.400/8
0871273

ADMINISTRATION. Welfare work
Social services - Serials

ADMINISTRATION in social work. - vol.1,no.1 Spring 1977-. - New York : Haworth Press, 1977-
r6051781 P.521/2029
0871273

ADMINISTRATION. Western Sumatra
By East India Company, 1760-1785

KATHIRITHAMBY-WELLS, J.
The British West Sumatran Presidency, 1760-1785. - Kuala Lumpur : Penerbit University Malaya, 1977
xvi,270p(2fold),[3]fold leaves. - Bibliography : p.229-243.- Includes index
r6072178 X.800/27147
1572407

ADMINISTRATION. William Tyndale Infants School & William Tyndale Junior School. Islington (London Borough) 1973-1975

GRETTON, John
William Tyndale. - London : Allen and Unwin, 1976
126p. - ([A 'Times Educational Supplement' special])
0-04-371047-6 X.529/31079
0762873

ADMINISTRATION. Youth clubs. Scotland
Participation of members. Training

SCOTTISH ASSOCIATION OF YOUTH CLUBS
SMIT, senior member involvement training handbook. - Edinburgh : [S.A.Y.C.], [1978]
[4],71p : ill
B79-04225 X.510/15990
1772392

ADMINISTRATION AREAS. Public services. East Sussex

EAST Sussex boundaries / by Local Government Unit - County Library. - [Lewes] : [East Sussex County Council], 1982
38p
B82-35814 X.805/3084
2777150

EAST Sussex boundaries / by Local Government Unit-County Library. - 2nd ed. - [Lewes] : [East Sussex County Council], 1984
57p
0-86147-041-9 L.49/3032
2777150

ADMINISTRATION LAW. Western European countries
to date

GESCHICHTE der Verwaltungsrechtswissenschaft in Europa / herausgegeben von Erk Volkmar Heyen. - Frankfurt am Main : Klostermann, 1982
x,184p. - (Ius commune. Sonderhefte ; 18)
3-465-01528-2 X.0200/99a(18)
886277x

ADMINISTRATION TRAINEE SCHEME. Recruitment, selection & training. Graduate personnel. Administrative grades. Civil service. Great Britain

GREAT BRITAIN. Administration Trainee Review Committee
Report of the Administration Trainee Review Committee. - London : Civil Service Department, 1978
[2],85[i.e.87]p,fold leaf : ill
B80-37897 BS.6/122
2253771

ADMINISTRATIVE AND CLERICAL OFFICERS ASSOCIATION
to 1979

JUDDERY, Bruce
White collar power. - Sydney ; London : Allen & Unwin, 1980
x,319p. - Includes index
0-86861-138-7 X.529/40952
2390698

ADMINISTRATIVE AREAS
See also
Burghs
Cantons
Districts
Local government areas
Regions
Sheriff courts districts

ADMINISTRATIVE AREAS. Åland Islands. Finland
to date

BERTELL, Erik
Skattedistrikt och förvaltningsområden på Åland under tidigt 1500-tal. - Mariehamn : Ålands kulturstiftelse, 1983
248p([3]folded leaves). - (Skrifter utgivna av Ålands kulturstiftelse ; 12)
951-95089-7-x Ac.9233.lt/3[no.12]
9004602

ADMINISTRATIVE AREAS. Andalusia. Spain
to date

ARJONA CASTRO, Antonio
Andalucia musulmana. - Córdoba : Monte de Piedad y Caja de Ahorros de Córdoba, 1980
164p. - (Colección Medina y Corella ; 10)
84-7231-535-5 X.529/65318
911730x

ADMINISTRATIVE AREAS. Aragón. Spain
to date

UBIETO ARTETA, Antonio
Historia de Aragón. - Zaragoza : Anubar.
Divisiones administrativas. - 1983
392p,plates
84-7013-196-6 X.809/62091
9139370

ADMINISTRATIVE AREAS. Belgorod (oblast'). Russia (RSFSR)
1945-

BELGORODSKAÎA oblast' / [spravochnik podgotovlen k izdaniîu ispolnitel'nym komitetom Belgorodskogo oblastnogo Soveta deputatov trudîashchikhsîa] ; [otvetstvennyĭ za vypusk : Solodovnikov, N.G.]. - Belgorod : Belgorodskiĭ oblastnoĭ Sovet deputatov trudîashchikhsîa, 1972
170p
r6498246 S.NA.28/6
8720657

ADMINISTRATIVE AREAS. Belorussia
1945- - Directories

BELORUSSKAÎA SSR / [pod redaktsieĭ I.V. Likhacha]. - Izd. 6-e. - Minsk : Belarus', 1981
244p
r6484996 S.N.J.11
8840288

ADMINISTRATIVE AREAS. Brazil
to date

DIVISÃO territorial do Brasil. - 9a ed. - Rio de Janeiro : IBGE, 1980
vx[i.e.xv],459p
r6708140 LAS.B.275/343
9073876

ADMINISTRATIVE AREAS. Catalonia. Spain
to date

VILA, Pau
[Selections]. Selecció d'escrits de geografia. - Barcelona : Curial. - (Biblioteca de cultura catalana ; 31)
v.1: La divisió territorial de Catalunya. - 3a ed. - 1984
330p
84-7256-135-6 X.529/62643
9097112

ADMINISTRATIVE AREAS. Catalonia. Spain
1975-

VILA, Marco-Aurelio
Els municipis de Catalunya. - Barcelona : Generalitat de Catalunya, Departament de la Presidència, Servei Central de Publicacions, 1982
136p
84-500-5192-4 X.809/52858
8851964

ADMINISTRATIVE AREAS. Costa Rica
to date

COSTA RICA
[División territorial administrativa]. División territorial administrativa de la República de Costa Rica. - San José, Costa Rica : Imprenta Nacional, 1977
115p
r6701328 LAS.D.62/19 Woolwich
8734119

ADMINISTRATIVE AREAS. Cuba
to date

MATEO DOMINGO, Alfredo
Historia de la división político-administrativa de la Isla de Cuba, 1607-1976. - La Habana : Editorial Arte y Literatura, 1977
135p
r6706986 X.529/55645
8963584

ADMINISTRATIVE AREAS. Dagestan ASSR. Russia (RSFSR)
1945-

DAGESTANSKAÎA ASSR / Prezidium Verkhovnogo Soveta Dagestanskoĭ ASSR. - Makhachkala : Dagestanskoe knizhnoe izdatel'stvo, 1980
201p
r6499257 S.NAE.1/2
8953201

ADMINISTRATIVE AREAS. Ecuador
1895-

ECUADOR. Instituto Nacional de Estadística y Censos
División política-territorial de la República del Ecuador. - [Quito] : El Instituto, 1979
134p
r6706744 LAS.E.70/28
8916888

ADMINISTRATIVE AREAS. Germany
1250-1519

HEINIG, Paul-Joachim
Reichsstädte, Freie Städte und Königtum 1389-1450. - Wiesbaden : Steiner, 1983
viii,439p. - (Veröffentlichungen des Instituts für Europäische Geschichte Mainz, Abt. Universalgeschichte ; Bd.108) (Beiträge zur Sozial- und Verfassungsgeschichte des Alten Reiches ; Nr.3). - Includes bibliography
3-515-03531-1 Ac.6431[Bd.108]
9032991

ADMINISTRATIVE AREAS. Mari ASSR. Russia (RSFSR)

MARI AUTONOMOUS SOVIET SOCIALIST REPUBLIC. Verkhovnyi Sovet. Prezidium
Mariĭskaîa ASSR. - Ioshkar-Ola : Mariĭskoe knizhnoe izdatel'stvo, 1978
149p
r6172576 S.NAJ.2/2
2029332

ADMINISTRATIVE AREAS. North Ossetian ASSR. Russia (RSFSR) 1945-
SEVERO-OSETINSKAÏA ASSR / Prezidium Verkhovnogo Soveta Severo-Osetinskoï ASSR. - Ordzhonikidze : Ir, 1982
43p
r6463703 S.NAL.2/2
8953732

ADMINISTRATIVE AREAS. Northern Italy 774-1122
PAULER, Roland
Das Regnum Italiae in ottonischer Zeit. - Tübingen : Niemeyer, 1983
vii,199p. - (Bibliothek des Deutschen Historischen Instituts in Rom ; Bd.54). - Bibliography
3-484-82054-3 Ac.6545/4[Bd.54]
8953198

ADMINISTRATIVE AREAS. Public services
Location & size
MASSAM, Bryan H.
Location and space in social administration. - London : Edward Arnold, 1975
ix,192p : ill. - Bibliography: p.171-183. - Includes index
0-7131-5776-3 X.510/15033
0377481

ADMINISTRATIVE AREAS. Public services. East Sussex - Atlases
EAST SUSSEX COUNTY LIBRARY. Local Government Unit
Atlas of governmental and public utilities administrative areas in East Sussex. - 2nd ed. - [Lewes] : [East Sussex County Council], 1977
[3],35[i.e.37]p
B78-02826 Maps 56.b.51 Map library
1285858
Atlas of governmental and public utilities administrative areas in East Sussex. - 2nd ed., amended. - [Lewes] : [East Sussex County Council], 1977
[3],35[i.e.37]p
B79-24778 X.902/5472
1285858

ADMINISTRATIVE AREAS. Public services. Zamora (Province). Spain
Reorganisation
NAFRÍA COLLADO, Porfirio
Replanteamiento territorial de la Provincia de Zamora para una concentración demográfica. - Zamora : Caja de Ahorros Provincial, 1975
169p.,2 fold. leaves of plates
84-500-7089-9 X.700/18870
1325191

ADMINISTRATIVE AREAS. Sweden
to date
EKBOM, Carl Axel
Västergötland under sen vikingatid ; Den äldsta Sveariket. - [S.l.] : [s.n.], 1983
63p
91-7260-865-x X.520/36961
9103821
WIJKANDER, Keith
Kungshögar och sockenbildning. - Nyköping : Södermanlands museum, 1983
164p : ill. - (Sörmländska handlingar ; 39). - Includes bibliography
91-85066-49-4 P.801/983[no.39]
9103821

ADMINISTRATIVE AREAS. Tadzhikistan
to date
TADZHIKSKAÏA SSR / [sostavitel': B. Ėshonkulov] ; [otvetstvennyĭ redaktor : A. Babadzhanov]. - Dushanbe : Irfon, 1978
334p
r6462809 S.NO.13
9073868

ADMINISTRATIVE AREAS. Udmurt ASSR. Russia (RSFSR) 1945- - Statistics
UDMURTSKAÏA ASSR / [redaktor B.S. Bukharev]. - Izhevsk : Udmurtiīa, 1972
148p
r6498222 S.NAN.11
8727422

ADMINISTRATIVE AREAS. Vladimir (oblast'). Russia (RSFSR)
to date
VLADIMIR (Oblast'). Vladimirskiĭ oblastnoĭ Sovet narodnykh deputatov. Ispolnitel'nyĭ komitet
Vladimirskaīa oblast'. - [Vladimir] : Ispolnitel'nyĭ komitet Vladimirskogo oblastnogo Soveta narodnykh deputatov, 1981
174p
r6463585 S.NA.28/10
8916896

ADMINISTRATIVE ASPECTS. Arms control. 1945- West Germany
STEINBERG, Rudolf
Abrüstungs- und Rüstungskontrollverwaltung in der Bundesrepublik Deutschland - Berlin : Duncker & Humblot, 1982
128p. - (Schriften zum öffentlichen Recht ; Bd 425). - Includes bibliography
3-428-05176-9 X.520/33552
9068449

ADMINISTRATIVE ASPECTS. Economic development. Latin America
DEVELOPMENT administration in Latin America / edited by Clarence E. Thurber and Lawrence S. Graham - Durham, N.C. : Duke University Press in cooperation with the Comparative Administration Group of the American Society for Public Administration, 1973
xx,453p.. - (Comparative Administration Group series)
0-8223-0292-6 X.800/12603
1242504

ADMINISTRATIVE ASPECTS. Economic development. Rural regions. Karnataka. India (Republic)
RAY, Amal
Organisational aspects of rural development. - Calcutta : World Press, 1976
xi,84p : 1 ill. - (ISEC monograph ; 4). - Includes index
r6091821 X.529/33180
1891669

ADMINISTRATIVE ASPECTS. Legislation. States. United States
BALUTIS, Alan P.
Public administration and the legislative process. - Beverly Hills ; London : Sage Publications, 1974
59p. - (Sage professional papers in administrative and policy studies ; vol.2, no.03-024). - Bibliography: p.56-58
0-8039-0459-2 X.709/30240
0570044

ADMINISTRATIVE ASPECTS. School districts. Italy
PROFILI istituzionali e prospettive del distretto scolastico nel processo di rinnovamento nella scuola. - Milano : Giunta della Regione Lombardia, Assessorato Istruzione, 1978
156p. - (Quaderni della Regione Lombardia ; 52, Istruzione, studi e documenti)
r6195613 X.520/21731
2475243

ADMINISTRATIVE CHANGE
See Organisational change

ADMINISTRATIVE DUTIES. Head teachers & senior teachers. Large secondary schools. England
LYONS, Geoffrey
The administrative tasks of head and senior teachers in large secondary schools. - [Bristol] : University of Bristol, 1974
[5],358p : ill
0-901239-13-5 X.512/5097
0588768

ADMINISTRATIVE DUTIES. Head teachers. Secondary schools. United States
HUGHES, Larry W.
The secondary principal's handbook. - Boston, Mass. ; London : Allyn and Bacon, 1980
xvi,334p : ill. - Includes bibliographies and index
0-205-06875-8 X.520/22039
2180391

ADMINISTRATIVE GRADES. Civil service. Great Britain
Graduate personnel. Recruitment, selection & training. Administration Trainee scheme
GREAT BRITAIN. Administration Trainee Review Committee
Report of the Administration Trainee Review Committee. - London : Civil Service Department, 1978
[2],85[i.e.87]p,fold leaf : ill
B80-37897 BS.6/122
2253771

ADMINISTRATIVE LAW
REAL, Alberto Ramón
Nuevos estudios de derecho administrativo. - Montevideo : Acali, 1976
128p
r6067165 X.108/26893
1786091

ADMINISTRATIVE LAW. Argentina
CASSAGNE, Juan Carlos
El acto administrativo. - Buenos Aires : Abeledo-Perrot, 1974
421p. - Bibliography : p.11-24
r6043177 X.700/17739
0387266

ADMINISTRATIVE LAW. Argentina
MARIENHOFF, Miguel S.
Tratado de derecho administrativo .... - 2. ed. actualizada. - Buenos Aires : Abeledo-Perrot, 1975-78
[t.1, 1977]
4t.in 5. - Includes bibliographies
r6157729 X.200/32596
0387266

ADMINISTRATIVE LAW. Australia
BRETT, Peter
Cases and materials on administrative law. - 3rd ed. - Sydney : Butterworth, 1975
xxxvi,517p. - Includes index
0-409-43404-3 X.100/17309
1332538

ADMINISTRATIVE LAW. Austria
1945-
AUSTRIA
[Laws, etc.]. Österreichische Verfassungs- und Verwaltungsgesetze. - Wien : Manz, 1981-
1v.(loose-leaf)
r6305253 X.0208/122
8842280

ADMINISTRATIVE LAW. Belgium
1945-
MAST, André
Overzicht van het Belgisch administratief recht. - 8e geheel opnieuw bewerkte en aangevulde uitg - Gent : Story-Scientia, 1981
xxxi,653p
90-643-9248-x X.200/38244
8821038

ADMINISTRATIVE LAW. Brazil
CAETANO, Marcello
Principios fundamentais do direito administrativo. - Rio de Janeiro : Forense, 1977
583p
r6067103 X.200/11775
1154354
- [Another copy] X.200/11829
CRETELLA JÚNIOR, J. (José)
Dicionário de direito administrativo. - 3a ed. revista e aumentada. - Rio de Janeiro : Forense, 1978
viii,555p
r6158357 X.200/12266
1154354
FAGUNDES, M. Seabra
O controle dos atos administrativos pelo poder judiciário. - 5a ed. revista e atualizada. - Rio de Janeiro : Forense, 1979
x,449p. - Bibliography
r6163040 X.200/34630
1154354
FALCÃO, Amílcar de Araújo
Introdução ao direito administrativo. - [2a ed]. - Saõ Paulo : Resenha Universitaria, 1977
86p
r6084604 X.100/28851
1154354
FRANCO SOBRINHO, Manoel de Oliveira
Da competência administrativa. - São Paulo : Editora Resenha Universitária, 1977
282p. - Bibliography : p.241-274
r6065915 X.200/11769
1154354
MEDAUAR, Odete
Contrôle administrativo das autarquias. - São Paulo : José Bushatsky Editor, 1976
159p. - (Coleção jurídica ; 23). - Bibliography : p.155-159
r6053766 X.100/27314
1154354
MEIRELLES, Hely Lopes
Direito administrativo brasileiro. - 3. ed. revista, ampliada e atualizada. - São Paulo : Editora Revista dos Tribunais, 1975
xvii,745p
r6037633 X.100/15911
1154354
VALLER, Wladimir
Os funcionários públicos à luz da jurisprudência. - São Paulo : Sugestões Literárias, 1976
586p
r6066053 X.200/11561
1154354

ADMINISTRATIVE LAW. Brazil
1964-
COSTA, José Armando da
Teoria e prática do direito disciplinar. - Rio de Janeiro : Forense, 1981
xi,432p
r6707260 X.529/57860
8559619

ADMINISTRATIVE LAW. Canada
to date – Conference proceedings

ADMINISTRATIVE LAW CONFERENCE (1979 University of British Columbia)
Proceedings of the Administrative Law Conference. – [Vancouver] : University of British Columbia, Faculty of Law, c1981
391p
r6805282 X.200/38310
8818282

ADMINISTRATIVE LAW. Castile. Spain
1200–1500

GARCÍA MARÍN, Jose Maria
El oficio público en Castilla durante la Baja Edad Media. – Sevilla : Universidad de Sevilla, 1974
383p. – (Anales de la Universidad Hispalense. Serie Derecho ; no. 20)
84–600–6202–3 X.700/16661
0924407

ADMINISTRATIVE LAW. Chile
to date

ARGANDOÑA, Manuel Daniel
La organización administrativa en Chile. – Santiago : Editorial Jurídica de Chile, 1982
167p. – (Monografías de derecho público)
r6707614 X.208/9598
8843090

CALDERA DELGADO, Hugo
Manual de derecho administrativo. – Santiago [Chile] : Editorial Jurídica de Chile, 1979
661p. – (Manuales jurídicos ; no.74). – Bibliography
r6700387 X.808/36469
8843090

ADMINISTRATIVE LAW. Colombia

COLOMBIA
Código contencioso administrativo.... – 6.ed.,actualizada. – Bogotá : Editorial Temis, 1976
682p
r6025444 X.200/11267
1421344

Código contencioso administrativo. – 7a ed. actualizada. – Bogotá : Temis, 1979
689p
r6163016 X.200/34575
1421344

Código de régimen político y municipal. – 5a ed., actualizada. – Bogotá : Temis, 1976
893p
r6085084 X.200/11824
1421344

[Código de régimen político y municipal]. Código de régimen político y municipal. – 6.ed. actualizada. – Bogotá : Editorial Temis, 1978
917p
r6157730 LAS.C.230/49 Woolwich
1421344

SARRIA, Eustorgio
Derecho administrativo. – 7. ed. – Bogotá : Publicaciones C.E.I.D.A., 1978
vii,672p
r6101932 X.200/32627
2421054

SIERRA JARAMILLO, Francisco
Derecho contencioso administrativo. – Bogotá : Temis, 1978
xviii,381p
84–8272–063–5 X.200/35005
2421054

ADMINISTRATIVE LAW. East Africa

OLUYEDE, Peter
Administrative law in East Africa. – Nairobi : East African Literature Bureau, 1973
xxi,255p. – Includes index
r6071988 X.200/31358
1451146

ADMINISTRATIVE LAW. East Asia
to date

STATE and law in East Asia / edited by Dieter Eikemeier and Herbert Franke. – Wiesbaden : Harrassowitz, 1981
ix,318p. – "Verzeichnis der Schriften von Karl Bünger": p.313–318
3–447–02164–0 X.200/39885
8889864

ADMINISTRATIVE LAW. England

FOULKES, David, 1924–
Administrative law. – 5th ed. – London : Butterworths, 1982
xxxii,460p. – Includes index
0–406–58407–9 X.205/1083
0115045

Introduction to administrative law. – 4th ed. – London : Butterworth, 1976
xxii,323p. – Includes index
0–406–58405–2 X.200/30562
0115045

GARNER, J. F. (John Francis)
Administrative law. – 5th ed. – London : Butterworth, 1979
xxxiii,525p. – Bibliography: p.500–503. – Includes index
0–406–58796–5 X.100/18507
0115045

WADE, H. W. R. (Henry William Rawson)
Administrative law. – 5th ed. – Oxford : Clarendon, 1982
li,892p. – Includes index
0–19–876138–4 X.200/39851
0115045
– [Another copy: paperback.] ISBN 0–19–876139–2 X.200/39852
– [A re-issue]. – 1983 OPL

WADE, Henry William Rawson
Administrative law. – 4th ed. – Oxford : Clarendon Press, 1977
li,855p. – Includes index
0–19–876070–1 X.200/42953
0115045

Administrative law. – 4th ed. – Oxford : Clarendon Press, 1979
li,855p. – Includes index
B79–30568 X.100/18595
0115045

YARDLEY, D. C. M. (David Charles Miller)
Principles of administrative law. – London : Butterworths, 1981
xx,238p. – Includes index
0–406–68991–1 X.200/37614
0115045

ADMINISTRATIVE LAW. England
compared with administrative law in Western Europe

NEDJATI, Zaim M.
English and continental systems of administrative law. – Amsterdam ; Oxford : North-Holland Publishing Co., 1978
xi,211p. – Includes index
0–444–85108–9 X.200/31713
1762192

ADMINISTRATIVE LAW. England
Remedies. Reform

GREAT BRITAIN. Law Commission
Report on remedies in administrative law. – London : H.M.S.O., [1976]
iv,47p. – (Law Com. ; no.73)
0–10–164070–6 BS.24f/107
0634778

ADMINISTRATIVE LAW. European countries
to date

WISSENSCHAFT und Recht der Verwaltung seit dem Ancien Régime / herausgegeben von Erk Volkmar Heyen. – Frankfurt am Main : Klostermann, 1984
xxv,394p. – (Ius commune. Sonderhefte ; 21)
3–465–01629–7 X.0200/99a(21)
9123644

ADMINISTRATIVE LAW. Finland
to date

ROSAS, Allan
Förvaltningsklagan. – Åbo : Abo akademi, 1980
319p. – (Acta Akademiae Aboensis. Ser.A, Humaniora ; v.59). – Bibliography
951–648–658–4 Ac.1093/2a[vol.59]
8885389

ADMINISTRATIVE LAW. France

BROWN, L. Neville (Lionel Neville)
French administrative law. – 3rd ed.. – London : Butterworths, 1983
xvii,218p : ill. – Bibliography: p205–207. – Includes index
0–406–56152–4 X.200/41849
0087742
– [Another copy] OPL

CODE administratif. – 14e édition. – Paris : Jurisprudence Générale Dalloz, 1977
viii,1434,14p. – (Petits codes Dalloz)
2–247–01363–5 X.108/15661
0087742

ADMINISTRATIVE LAW. Great Britain

BAILEY, S. H.
Cases and materials in administrative law. – London : Sweet and Maxwell, 1977
xxxix,635p. – Bibliography: p.xxxix. – Includes index
0–421–17360–2 X.100/17911
1595652

BEATSON, J.
Administrative law. – Oxford : Clarendon, 1983
xxxiv,674p. – Includes index
0–19–825340–0 X.200/43057
1595652
– [Another Copy: paperback.]
ISBN 0–19–825341–9 X.200/43058

CRAIG, P. P. (Paul P)
Administrative law. – London : Sweet & Maxwell, 1983
xxxiii,636p. – Includes index
0–421–27050–0 X.200/42935
0223689

DE SMITH, Stanley Alexander
Constitutional and administrative law. – 3rd ed.. – Harmondsworth : Penguin, 1977
729p. – (Foundations of law) (Penguin education). – Includes index
B77–24086 X.108/16732
0223689

GRIFFITH, John Aneurin Grey
Administrative law and the judges. – [London] : Haldane Society of Socialist Lawyers, [1978]
[1],23p. – (Pritt memorial lectures ; 1978)
B79–31628 X.100/18617
0223689

HARLOW, Carol
Law and administration. – London : Weidenfeld and Nicolson, c1984
xxvi,718p. – (Law in context). – Bibliography: p686–695. – Includes index
0–297–78239–8 X.200/45328
0223689

PHILLIPS, Owen Hood
O. Hood Phillips' Constitutional and administrative law. – 6th ed.. – London : Sweet and Maxwell, 1978
l,746p. – Includes index
0–421–23970–0 X.100/18081
0223689
– [Paperback edn]
ISBN 0–421–23980–8 OPL

ADMINISTRATIVE LAW. Honduras
to date

FLORES VALERIANO, Enrique
Derecho administrativo. – Tegucigalpa : Universidad Nacional Autónoma de Honduras, 1980
487p
r6708553 X.529/60259
9054286

ADMINISTRATIVE LAW. India (Republic)
1947–

JAIN, M. P.
Principles of administrative law. – 3rd ed. – Bombay : Tripathi, 1979
xl,727p
r6807252 X.200/42606
8952485

KAGZI, Mangal Chandra Jain
The Indian administrative law. – 4th ed. – New Delhi : Metropolitan, 1981
xiv,619p
r6815293 X.200/46771
8952485

MASSEY, I. P.
Administrative law. – Lucknow : Eastern Book Company, 1980
xlvi,352p
r6803936 X.200/40579
8952485

ADMINISTRATIVE LAW. Italy

BARBAGALLO, Renato
La regione Valle d'Aosta. – Milano : Giuffrè, 1978
vii,144p. – (L'ordinamento amministrativo delle regioni ; 2)
r6500650 X.200/35058
0859540

CASSESE, Sabino
La formazione dello stato amministrativo. – Milano : A. Giuffrè, 1974
409 p.
r6014060 X.709/20530
0859540

ADMINISTRATIVE LAW. Italy
– Conference proceedings

PROBLEMI attuali della giustizia amministrativa. – Milano : Giuffrè, 1976
xvi,139p. – (Collana di diritto pubblico ; 1)
r6068754 X.100/28239
0859540

ADMINISTRATIVE LAW. Italy
to date

ITALY
Guistizia costituzionale, giustizia amministrativa. - Torino : UTET, c1981
220p. - (Legislazione sistematica ; 2)
88-02-03618-7 X.208/9509
8814333

PETRONE, Marino
La tutela penale degli ordini amministrativi. - Milano : Giuffrè, 1980
vii,324p
r6507430 X.200/40469
8814333

Il PROCESSO amministrativo. - Milano : Giuffrè, 1979
viii,593p. - (Università di Firenze, Fondazione Piero Calamandrei ; 19)
r6504823 X.200/38283
8814333

TABARRINI, Agostino
I contratti dei Comuni. - 2a edizione. - Milano : Giuffrè, 1979
324p. - (Teoria e pratica del diritto. Sezione 4, diritto amministrativo ; 3)
r6504606 X.200/38011
8814333

VIRGA, Pietro
Diritto amministrativo. - Milano : Giuffrè, 1983-
r6512061 X.0200/1354
8814333

ADMINISTRATIVE LAW. Italy
to date - Serials

DIRITTO processuale amministrativo. - Anno 1, fasc.1 (marzo 1983)-. - Milano : Giuffrè, 1983-
r6510891 P.803/1425
8814333

ADMINISTRATIVE LAW. Kazakhstan

KAZAKHSTAN (Kazakh SSR)
[Laws, statutes, etc]. Zakonodatel'nye i normativnye akty Kazakhskoĭ SSR ob administrativnoĭ otvetstvennosti. - Alma-Ata : Kazakhstan, 1977
317p
r6186071 S.NI.2/5.
203428X

ADMINISTRATIVE LAW. Malaysia
1963-

JAIN, M. P.
Administrative law of Malaysia and Singapore. - Singapore : Malayan Law Journal, 1980
xxviii,473p
I.S.B.N. 9971 70 004 2
9-9717000-3-4 OPL
8933677

ADMINISTRATIVE LAW. Mozambique

MOZAMBIQUE
[Laws, statutes, etc]. Processo revolucionário.. - (In Mozambique. President, 1975- (Machel). Revolução. transformação profunda das estruturas, transformação profunda da nossa vida.-[Maputo]. 1975.-p.37-116)
r6064988 X.519/24956
1473573

[Laws, statutes, etc]. Tarefas e funções que compete a cada ministério.. - (In. Mozambique. Constitution. Constituição da República Popular de Moçambique ; Lei da nacionalidade. - [Maputo?], 1975. - p. 29-57)
r6065108 S.W.290/4
1542567

ADMINISTRATIVE LAW. Netherlands
1945-

WIJMEN, P. C. E. (Peter Cornelis Evert van)
Recht, belang en rechtsbescherming. - 's-Gravenhage : Vuga, 1981
xvi,632p. - Bibliography
r6330121 X.200/38634
8848971

ADMINISTRATIVE LAW. Norway

FRIHAGEN, Arvid
Forvaltningsrett. - 2. revid. utg. - Bergen : Universitetsforlaget, 1977
3v
r6060120 X.200/11429
1503332

ADMINISTRATIVE LAW. Poland

LANG, Jacek
Wnioski obywatelskie w administracji państwowej. - Warszawa : Wydawnictwa Uniwersytetu Warszawskiego, 1976
213p. - (Dissertationes Universitatis Varsoviensis ; 108)
r6075337 Ac.1147/22(108)
1316478

MAISEL, Witold
Wojewódzkie sądy admistracyjne w Drugiej Rzeczypospolitej. - Warszawa : Państwowe Wydawnictwo Naukowe, 1976
258p
r6044299 X.108/15263
1316478

ADMINISTRATIVE LAW. Portugal

FONSECA, Guilherme da
A constituição e a defesa dos adminstrados [sic]. - Lisboa : Diabril, 1977
62p. - (Colecção Temas da constituição ; n.7)
r6064908 X.108/15597
142372X

ADMINISTRATIVE LAW. Russia
Reform. Influence of Swedish administrative law, ca1720

PETERSON, Claes
Peter the Great's administrative and judical reforms. - Stockholm : Nordiska bokhandeln in distribution, 1979
x,448p. - (Skrifter utgivna av Institutet för rättshistorisk forskning. ser.1, Rättshistoriskt bibliotek ; bd 29). - Includes bibliographies
91-85190-11-x Ac.2140.b/2a[bd.29]
2217929

ADMINISTRATIVE LAW. Singapore
1963-

JAIN, M. P.
Administrative law of Malaysia and Singapore. - Singapore : Malayan Law Journal, 1980
xxviii,473p
I.S.B.N. 9971 70 004 2
9-9717000-3-4 OPL
8933685

ADMINISTRATIVE LAW. South Africa
to date

BAXTER, Lawrence
Administrative law. - Cape Town : Juta, 1984
lxi,796p
0-7021-1511-8 X.200/45911
9148833

ADMINISTRATIVE LAW. Soviet Union

LUCHIN, Viktor Osipovich
Profsessual'nye normy v sovetskom gosudarstvennom prave. - Moskva : Iuridicheskaia literatura, 1976
166p
r6055738 X.108/15708
1002619

SHERGIN, Anatoliĭ Pavlovich
Administrativnaia iurisdiktsiia. - Moskva : IUridicheskaia literatura, 1979
141p
r6183162 X.208/5386
1002619

VOPROSY gosudarstvennogo i administrativnogo prava / [redaktsionnaia kollegiia A.V. Rybin (glavnyĭ redaktor) ... [i dr.] otvetstvennyĭ redaktor A.V. Moskalev]. - Perm' : Permskiĭ gosudarstvennyĭ universitet, 1976
202p
r6021180 X:709/22087
1002619

ZAKONODATEL'NYE akty ob administrativnoĭ otvetstvennosti / [sostaviteli : L.L. Dedkov, N.A. Stoliar]. - Minsk : Belarus', 1977
303p
r6087105 X.200/11801
1002619

ADMINISTRATIVE LAW. Soviet Union
- Serials

AKTUAL'NYE voprosy pravovedeniia. - [Vyp.1]-. - Tomsk : Izdatel'stvo Tomskogo universiteta, 1978-
r6077135 P.101/2100
1002619

ADMINISTRATIVE LAW. Soviet Union
to date - Serials

VOPROSY teorii prava i gosudarstvennogo stroitel'stva. - [Vyp.1?]-. - Tomsk : Izdatel'stvo Tomskogo universiteta, [1978?]-
26cm
Holdings: Vyp.[1],3-
r6485581 X.0205/174
8536112

ADMINISTRATIVE LAW. Soviet Union
1945-

KOVAL', L. V. (Leonid Vasil'evich)
Administrativno-deliktnoe otnoshenie. - Kiev : Vyshcha shkola, 1979
229p
r6484974 X.208/7439
8835187

OSNOVY upravleniia narodnym khoziaistvom / [redaktsionnaia kollegiia IU.M. Kozlov ... i dr.]. - Moskva : "IUridicheskaia literatura, 1981
278 p.. - (Sovetskoe administrativnoe pravo). - Includes bibliographical references
r6403669 X.200/45260
8835187

PETROV, G. I. (Georgiĭ Ivanovich)
Sovetskie administrativno-pravovye otnosheniia. - Leningrad : Izdatel'stvo Leningradskogo universiteta, 1972
156p
r6462725 X.200/42236
8835187

PRONINA, V. S. (Vera Sergeevna)
Konstitutsionnyĭ status organov mezhotraslevogo upravleniia. - Moskva : "IUridicheskaia literatura", 1981
156 p.. - Includes bibliographical references
r6482559 X.208/9917
8835187

SOVETSKOE administrativnoe pravo / [uchebnik napisan kollektivom avtorov ... G.P. Emel'ianov ... i dr.] ; [pod obshcheĭ redaktsieĭ V.I. Popovoĭ i M.S. Studenikinoĭ]. - Moskva : IUridicheskaia literatura, 1982
285 p.
r6403537 X.200/43492
8835187

UPRAVLENIE i pravo / Moskovskiĭ gosudarstvennyĭ universitet im. M.V. Lomonosova, IUridicheskiĭ fakul'tet, Laboratoriia pravovykh problem upravleniia. - [Vyp.1] (1974)-. - Moskva : Izdatel'stvo Moskovskogo universiteta, 1974-
r6464021 X.200/39803
8835187

[ZAKONODATEL'STVO o sovetskom gosudarstvennom stroitel'stve i prave]. Osnovnye zakonodatel'nye akty no sovetskomu gosudarstvennomu stroitel'stvu i pravu ... / [sostaviteli : A.P. Kositsyn, A.L. Nedavniĭ, T.A. Pomykaeva]. - Moskva : Mysl', 1972
2v.
r6484939 X.200/39600
8835187

ADMINISTRATIVE LAW. Soviet Union
1945- - Serials

KHOZIAISTVO, pravo, upravlenie / Saratovskiĭ iuridicheskiĭ institut im. D.I. Kurskogo. - [Vyp.1]-. - Saratov : Izdatel'stvo Saratovskogo universiteta, [197-]-
r6464026 X.208/8227
8835187

ADMINISTRATIVE LAW. Spain

SPAIN
[Laws, statutes, etc]. Leyes fundamentales.... - 4.ed. - Madrid : Editorial García Enciso, 1976
64p. - (Colección García Enciso)
r6025443 X.108/15565
1421336

SPAIN
[Laws, statutes, etc]. Leyes constitucionales y administrativas de España. - Pamplona : Universidad de Navarra, 1976
xviii,1240p.,xvip.of fold.plates
84-313-0229-1 X.520/9828
1421336

ADMINISTRATIVE LAW. Sri Lanka
1948- - Texts

PEIRIS, G. L.
Essays on administrative law in Sri Lanka. - Colombo : Lake House Investments Limited, c1980
xii,372p
r6805173 X.200/39947
8940398

ADMINISTRATIVE LAW. Sweden

HELLNERS, Trygve
Förvaltningslagen med kommentarer. - 2.omarb. uppl. - Stockholm : Liber förlag : Jurist- och samhällsvetareförbundet, 1978
268p. - (Publica/Rätt och samhälle)
91-380-1890-x X.100/28582
1711318

SWEDEN. Justitiedepartementet
Myndigheten och människan. - 2. uppl. - [Stockholm] : Justitiedepartementet, 1973
47p
91-380-1552-8 S.B.963/24
1711318

ADMINISTRATIVE LAW. Sweden
1945-
SWEDEN. Statsradsberedningen
Förvaltningslagens tillämpning hos regeringen. - 2. omarb. uppl. - [Stockholm] : Statsradsberedningen, 1979
54p
r6610584 X.200/39781
8504091

ADMINISTRATIVE LAW. Sweden
Influence on reform of administrative law in Russia, ca1720
PETERSON, Claes
Peter the Great's administrative and judical reforms. - Stockholm : Nordiska bokhandeln in distribution, 1979
x,448p. - (Skrifter utgivna av Institutet för rättshistorisk forskning. ser.1, Rättshistoriskt bibliotek ; bd 29). - Includes bibliographies
91-85190-11-x Ac.2140.b/2a[bd.29]
2217929

ADMINISTRATIVE LAW. Switzerland
to date
FLEINER-GERSTER, Thomas
Grundzüge des allgemeinen und schweizerischen Verwaltungsrechts. - 2. überarb. Aufl. - Zürich : Schulthess, [1980]
xxxii,564p.
3-7255-2045-3 X.200/36090
8753598

ADMINISTRATIVE LAW. Territories administered by federal government. Brazil
TEMER, Michel
Território federal nas constituições brasileiras. - São Paulo : Editora da Universidade Católica : Editora Revista dos Tribunais, 1976
103p
r6047574 X.100/26482
1416634

ADMINISTRATIVE LAW. United States
ROBINSON, Glen O.
The administrative process. - St Paul : West Publishing, 1974
928p. - (American casebook series)
r6092456 X.802/10947
0269182

ADMINISTRATIVE LAW. United States
to date
SCHWARTZ, Bernard
Administrative law. - Boston : Little, Brown, 1976
xiv,730p
r6802096 X.200/39107
8876479

ADMINISTRATIVE LAW. Urban regions. Catalonia. Spain
1975-
CATALONIA
Legislació i disposicions d'urbanisme. - Barcelona : Servei Central de Publicacions, Department de la Presidència, Generalitat de Catalunya, 1979
67p
r6702077 X.200/35801
8730059

ADMINISTRATIVE LAW. Urban regions. Spain
to date - Bibliographies
GONZÁLEZ NAVARRO, Francisco
Bibliografía sobre derecho urbanístico. - Madrid : Montecorvo, 1981
272p
84-7111-158-6 X.520/33826
9095403

ADMINISTRATIVE LAW. Val d'Aosta. Italy
BARBAGALLO, Renato
La regione Valle d'Aosta. - Milano : Giuffrè, 1978
vii,144p. - (L'ordinamento amministrativo delle regioni ; 2)
r6500650 X.200/35058
2482916

ADMINISTRATIVE LAW. West Germany
1945-
GERMANY (Federal Republic)
[Laws, etc.]. Verfassungs- und Verwaltungsgesetze der Bundesrepublik Deutschland. - München : Beck, 1981-. - (Sartorius ; Bd.1)
3-406-08145-2 X:0208/112
8745412

MAURER, Hartmut
Die Gesetzgebungskompetenz für das Staatshaftungsrecht. - Konstanz : Universitätsverlag, 1981
65p.. - (Konstanzer Universitätsreden ; 118)
3-87940-199-3 X.200/39178
8745412

WABNITZ, Reinhard Joachim
Politische Fragen, Rechtsfragen und Sachfragen. - Frankfurt am Main : Cirencester : Lang, 1980
328p. - (Europäische Hochschulschriften. Reihe 31, Politikwissenschaft ; Bd.23). - Bibliography: p.296-328
3-8204-6117-5 X.200/35925
8745412

ADMINISTRATIVE LAW. Western Europe compared with administrative law in England
NEDJATI, Zaim M.
English and continental systems of administrative law. - Amsterdam ; Oxford : North-Holland Publishing Co., 1978
xi,211p. - Includes index
0-444-85108-9 X.200/31713
1762206

ADMINISTRATIVE LAW. Yugoslavia
ANDROJNA, Vilko
Upravni postopek in upravni spor. - Ljubljana : Časopisni zavod Uradni list SR Slovenije, 1977
312p
r6178041 X.200/33173
1387758

GATARIĆ, Đuro
Upravno pravo. - Zagreb : Narodne novine, 1978
xxv,474p. - (Udžbenici ; 81). - Bibliography : p.471-474
r6178171 X.200/33045
1387758

HRISTOV, Aleksandar T.
Upravno pravo. - 3 izm. i dop. izd. - Skopje : Univerzitet Kiril i Metodij, 1975
709p
r6057064 X.100/26991
1387758

MARKOVIĆ, Branislav
Položaj i uloga stranke u upravnom postupku .... - Beograd : Privredni pregled, 1977
205p. - Bibliography : p.197-201
r6186146 X.100/28792
1387758

POPOVIĆ, Ljubiša
Zakon o opštem upravnom postupku sa pratećim propisima i registrom pojmova ; i Zakona o upravnim sporovim sa registrom pojmova. - Beograd : Savremena administracija, 1977
320p
r6056901 X.108/15511
1387758

YUGOSLAVIA
[Laws, statutes, etc]. Zakon o spremembah in dopolnitvah zakona o upravnih taksah.
19p. - (In. Yugoslavia. Laws, statutes, etc. Zakon o splošnem upravnem postopku s komentarjem in sodno prakso in zakon o upravnih taksah.- Ljubljana, 1977)
r6096947 X.108/26868
1387758
[Laws, statutes, etc]. Kancelarijsko poslovanje.... - Beograd : Službeni list SFRJ, 1974
514p
r6074208 S.R.195/45
1387758
[Laws, statutes, etc]. Zakon o splošnem upravnem postopku s komentarjem in sodno prakso in Zakon o upravnih taksah. - Ljubljana : Časopisni zavod Uradni list SR Slovenije, 1977
347p
r6096946 X.108/26868
1387758

ZBIRKA obrazaca i primera akata / [priredio Mladen Kosovac]. - Beograd : Službeni list, 1975
341p
r6086877 X.102/1295
1387758

ADMINISTRATIVE MEMORANDA. Health services. England - Indexes
SUBJECT index [to] official memoranda to hospital authorities, NHS Reorganisation Circulars, Health Service Circulars (interim series), Health Circulars, Health Service Whitley Council Circulars, statutory instruments issued under the authority of the National Health Service Acts, 1946-76, England and Wales, current 31 March / [compiled by the Institute of Health Service Administrators]. - 1977. - London : The Institute, 1977
[1],33p
B78-04616 X.311/8640 Woolwich
0749710

SUBJECT index [to] official memoranda to hospital authorities, NHS Reorganisation Circulars, Health Service Circulars (interim series), Health Circulars, Health Service Whitley Council Circulars, statutory instruments issued under the authority of the National Health Service Acts, 1946-77, England and Wales, current 31 March / [compiled by the Institute of Health Service Administrators]. - 1978. - London : The Institute, 1978
[1],32p
B79-29145 X.100/18572
0749710

SUBJECT index to statutory instruments hospital memoranda, NHS reorganisation circulars, interim health service circulars and health circulars. - London : Institute of Health Service Administrators, c1983
29p
B84-02816 X.322/12637
0749710
- [Another copy.] 942.0043 OPL

ADMINISTRATIVE OFFICERS
See also
Intendants

ADMINISTRATIVE PERSONNEL. Cooperative societies. Great Britain
Trade unions: National Association of Co-operative Officials
NATIONAL ASSOCIATION OF CO-OPERATIVE OFFICIALS
Rules of the National Association of Co-operative Officials. - Manchester : [The Association], [1982]
47p
B83-15420 X.208/8251
286973x

ADMINISTRATIVE PERSONNEL. Helsingin yliopisto
1939-1968 - Biographies
HELSINGIN yliopisto : opetlajat ja virkamiehet 1939-1968 = Helsingfors universitet : lärare och tjänstemän 1939-1968 / [toimittaneet Yrjö Blomstedt, Jukka Petänen, Veli-Matti Autio]. - Porvoo : W. Söderström, 1977
452p
951-00-7324-5 X.520/9624
1528963

ADMINISTRATIVE PERSONNEL. Higher education institutions. Denmark
Industrial democracy
ARNOLDUS, Marly
TAP-undersøgelsen. - København : Nyt fra Samfundsvidenskaberne, [1976]
53,7 leaves. - (Institut for Organisation og Arbejdssociologi. Arbejdsnoter ; 76-2)
r6054508 X.0512/366 [1976, no.2]
1464450

ADMINISTRATIVE PERSONNEL. Higher education institutions. United States
Negro administrative personnel
HOSKINS, Robert L.
Black administrators in higher education. - New York ; London : Praeger, 1978
xvi,207p. - (Praeger special studies). - Bibliography: p.203-206
0-03-046611-3 X.520/13781
1879480

ADMINISTRATIVE PERSONNEL. Industries & trades. Hong Kong
Remuneration - Serials
SALARY trends & fringe benefits. - [Hong Kong] : Hong Kong Productivity Centre, [1973?]
r6864521 X.512/6109
1693751

ADMINISTRATIVE PERSONNEL. National health services. Great Britain
Training courses
GUIDE for NHS administrative and clerical staff on qualifications, study, facilities and courses. - London : National Health Service, National Staff Committee for Administrative & Clerical Staff, 1981
54p
B83-10525 BS.385/773
2841436

GUIDE for NHS administrative and clerical staff on qualifications, study facilities and courses. - London : National Staff Committee for Administrative & Clerical Staff, 1984
66p
B85-11816 X.525/10477
2841436

ADMINISTRATIVE PERSONNEL. Secondary schools. Scotland
ANCILLARY staff in secondary schools. - Edinburgh : H.M.S.O., 1976
[6],41p : 1 ill
0-11-491452-4 BS.152/197
1060821

ADMINISTRATIVE PERSONNEL. Welfare services. Local authorities. Great Britain
Relations with fieldworkers
ROWBOTTOM, Ralph
Linking administration to fieldwork services in Croydon Social Services Department. - [Uxbridge] : Brunel Institute of Organisation and Social Studies, 1979
31p in various pagings : 2 ill. - (Case studies / Brunel University. Social Services Unit)
0-902215-46-9 X.622/8839 Woolwich
2317214

ADMINISTRATIVE SERVICES. Great Britain. Parliament. House of Commons

COMPTON, Sir Edmund
Review of the administrative services of the House of Commons. - London : H.M.S.O., [1975]
61p. - ([1974 H.C.]254)
0-10-225475-3 BS.91/152
0333247

ADMINISTRATIVE SERVICES. Great Britain. Parliament. House of Commons
Inquiry reports: Compton, Sir Edmund. Review of the administrative services of the House of Commons

GREAT BRITAIN. Parliament. House of Commons
House of Commons (administration). - London : H.M.S.O., 1975
135p : ill. - ([1974-75 H.C.] 624)
0-10-026245-7 BS.92/121
0533807

ADMINISTRATIVE TRAINEES. Civil service. Great Britain
Recruitment & selection

ATKINSON, Sir Alec
Selection of fast-stream graduate entrants to the Home Civil Service, the Diplomatic Service and the Tax Inspectorate, and of candidates from within the service. - [London] : Management and Personnel Office, 1983
iv,69p
0-7115-0053-3 BS.6/219
2320037

GREAT BRITAIN. Committee on the Selection Procedure for the Recruitment of Administration Trainees
Report of the Committee on the Selection Procedure for the Recruitment of Administration Trainees. - [Basingstoke] : Civil Service Commission, 1979
v,63p. - Bibliography: p.62-63
0-903741-54-7 BS.6/162
2320037

ADMINISTRATORS
See also
Bureaucrats
Course directors
Intendants
Managers
Registrars

ADMINISTRATORS
United States. Higher education institutions. Collective bargaining by teachers - For administrators

ANGELL, George W.
Handbook of faculty bargaining. - San Francisco ; London : Jossey-Bass, 1977
iii-xxix,593p. - (The Jossey-Bass series in higher education). - Includes bibliographies and index
0-87589-320-1 X.520/11908
1192566

ADMINISTRATORS. Arts. Great Britain
Conditions of service

SOCIETY OF PROFESSIONAL ARTS ADMINISTRATORS
Scheme of conditions of service for arts administrators, including house managers appendix. - Romford : [The Society], [1978]
19,[6] leaves
B79-35198 X.512/7028
1969013

ADMINISTRATORS. Districts. National health services. Great Britain
Role

The DISTRICT administrator in the National Health Service / by Rosemary Stewart ... [et al.]. - [Tunbridge Wells] : Pitman Medical for King Edward's Hospital Fund for London, 1980
[12],211p. - Bibliography: p.205-206. - Includes index
0-900889-81-0 X.520/22505
2100363

ADMINISTRATORS. Health services
Senior administrators. Role - Conference proceedings

The HEALTH service administrator, innovator or catalyst? / edited, with commentaries by Leslie Paine. - London : King Edward's Hospital Fund for London ; Tunbridge Wells : Distributed by Pitman Medical, 1978
[12],198p. - Bibliography: p.190-191. - Includes index
0-900889-69-1 X.520/13026
1695282

ADMINISTRATORS. Health services. Great Britain
Development: Great Britain. National Staff Committee for Administrative and Clerical Staff. Recruitment and career development of administration

UNIVERSITY OF MANCHESTER. Health Services Management Unit
The views of the Health Services Management Unit, University of Manchester, on the report of the National Staffs Committee for Administrative and Clerical Staff. 'The recruitment and career development of administration'. - Manchester : The Unit, [1978]
[1],6 leaves. - (Working paper series / University of Manchester. Health Services Management Unit ; no.11)
B79-13029 X.313/1629 Woolwich
1838156

ADMINISTRATORS. Higher education institutions. United States
Evaluation

MUTUAL benefit evaluation of faculty and administrators in higher education / William J. Genova ... [et al.]. - Cambridge, Mass. : Ballinger, 1976
xv, 222 p. : ill. - Includes bibliographies
0-88410-165-7 X.520/10993
0712752

ADMINISTRATORS. Hospitals
Time. Allocation

DOUGLASS, Merrill E.
Successful time management for hospital administrators. - New York : AMACOM, c1980
v,143p : ill
0-8144-5602-2 X.622/19338
3070220

ADMINISTRATORS. National health services. Great Britain
Training schemes: National Administrative Training Scheme

STEWART, Rosemary
Review of the national administrative training scheme. - [Oxford] : [Kennington, Oxford OX1 5NY] : Oxford Centre for Management Studies, [1983?]
1v. : 1ill. - Includes bibliography
B84-03988 X.525/8426
2993635

ADMINISTRATORS. National health services. Scotland
Training schemes: National Administrative Training Scheme. 1956-1981

SCOTTISH Health Service National Administrative Training Scheme / [edited and produced by G.H. France]. - [Edinburgh] : [Management Education and Training Division], [1981]
21p
B82-04943 X.322/10442 Woolwich
2597772

ADMINISTRATORS. Schools. United States
Sex roles

EDUCATIONAL policy and management / edited by Patricia A. Schmuck, W.W. Charters, Jr, Richard O. Carlson. - New York ; London : Academic Press, 1981
xiv,337p. - (Educational psychology). - Includes bibliographies and index
0-12-627350-2 X.520/26420
2693682

ADMINISTRATORS. United States
Professional conduct. Ethics

COOPER, Terry L.
The responsible administrator. - Port Washington ; London : National University Publications : Kennikat, 1982
viii,175p. - (Series in political science). - Bibliography: p171-175
0-8046-9292-0 X.529/47928
2735490

ADMINISTRATORS. Universities. Commonwealth countries
Awards

GRANTS for study visits. - London : Association of Commonwealth Universities, 1979
15p. - Includes index
0-85143-059-7 X.529/42939
2540134

ADMINISTRATORS. Universities. Europe
Training - Conference proceedings

TRAINING university administrators in Europe / edited by D. Porter and J.S. Padley. - Aldershot : Gower, c1982
vi,137p : 1ill
0-566-00522-0 X.529/50784
2810166

ADMINSTRATION. Justice. Criminal law. England
Policies of government, 1945-1980

TAYLOR, Ian, 1944-
Law and order. - London : Macmillan, 1981
xxii,234p. - Bibliography: p224-234
0-333-21442-0 X.529/46584
2696967

ADMIRALS. Great Britain. Royal Navy
Leadership. 1750-1945

HORSFIELD, John
The art of leadership in war. - Westport, Conn. : London : Greenwood Press, 1980
xiv,240p : ill. - (Contributions in military history ; no.21). - Bibliography: p207-231. - Includes index
0-313-20919-7 X.809/47958
2491125

ADMIRALS. Great Britain. Royal Navy
Relations with Churchill, Winston S. (Winston Spencer), 1874-1965, 1939-1945

ROSKILL, S. W. (Stephen Wentworth)
Churchill and the admirals. - London : Collins, 1977
351p,[12]p of plates : ill. - Includes index
0-00-216127-3 X.800/26575
1295810

ADMIRALS. Naval operations
1536-1969 - Biographies

HOUGH, Richard
The great admirals. - London : Weidenfeld and Nicolson, 1977
271p : ill(some col). - Bibliography: p.267. - Includes index
0-297-77308-9 X.802/10385
1168320

ADMIRALS. Naval operations
1733-1966

WARNER, Oliver
Command at sea. - London : Cassell, 1976
xii,196p,[12]p of plates : ill. - Includes index
0-304-29667-8 X.809/41073
0684732

ADMIRALS. Union of Soviet Socialist Republics. Voenno-morskoĭ flot

VYSOKOE zvanie / [sostaviteli : I.A. Kondaurov, S.I. Mokrousov]. - Izd. 2-oe, perer. i dop. - Perm' : Permskoe knizhnoe izdatel'stvo, 1978
261p : ill
r6172569 X.809/27121
2043734

ADMIRALS. United States. Navy
to 1977 - Biographies

REYNOLDS, Clark G.
Famous American admirals. - New York ; London : Van Nostrand Reinhold, 1978
xvii,446p. - Bibliography: p.431-433. - Includes index
0-442-26068-7 X.802/10824
1818562

ADMIRALTY
See Great Britain. Admiralty

ADMIRALTY ACTIONS. Law. Shipping. England
Procedure

MCGUFFIE, Kenneth C.
Admiralty practice. - 3rd impression. - London : Stevens, 1976
lxxvii,773p. - (British shipping laws ; 1)
r6089210 OPL
0649430
- [Latest supplement.] OPL

ADMIRALTY COURT (ENGLAND AND WALES)
See England and Wales. Court of Admiralty

ADMIRALTY LAW
See also
Civilian lawyers

ADMISSION. Baptised non-Catholics
To Catholic Church

CATHOLIC CHURCH
[Ordo initiationis Christianae adultorum. English]. Rite of reception of baptized Christians into full communion with the Catholic Church. - London : Catholic Truth Society, 1977
40p
B77-33607 X.100/17186
1269712

ADMISSION. Cigarette smoking related disease patients. Hospitals. Scotland
- Statistics

The SCOTTISH epidemic / compiled by K.G. Brotherston. - [Scotland] : ASH Scottish Committee of Action on Smoking and Health, 1982
viii,215p : ill
B84-06802 X.329/17209
3037231

ADMISSION. Council for National Academic Awards courses. Polytechnics. Great Britain
Applications
CROSS, Penelope
Choosing a polytechnic. - Cambridge : Advisory Centre for Education, 1976
[3],38p. - (Advisory booklets / Advisory Centre for Education ; no.6). - Bibliography: p.32-38
0-900029-45-5 X.519/26321
0769959

ADMISSION. Ethnic minorities. United States
To higher education institutions. Use of positive discrimination. Law
SINDLER, Allan Paul
Bakke, DeFunis, and minority admissions. - New York ; London : Longman, 1978
xi,358p. - Includes index
0-582-28053-2 X.100/18358
1882961

ADMISSION. Ethnic minorities. United States
To medical schools. Use of positive discrimination. Law. Cases: Bakke case
DREYFUSS, Joel
The Bakke case. - New York ; London : Harcourt Brace Jovanovich, 1979
ix,278p. - Bibliography: p.268. - Includes index
0-15-110536-7 X.200/33597
207155x

ADMISSION. Ethnic minority students. Church schools. England
DUMMETT, Ann
Race and church schools. - London : Runnymede Trust, [1983?]
65p
B83-28520 X.200/40724
2920085

ADMISSION. Ethnic minority students. Higher education institutions. United States
BAILEY, Robert L.
Minority admissions. - Lexington, Mass. : Heath ; Farnborough [Hants.] : [Distributed by] Teakfield Ltd, 1978
xxi,215p : 2 ill. - Bibliography: p.187-205. - Includes index
0-669-02095-8 X.520/13391
1784005

ADMISSION. Ethnic minority students. Higher education institutions. United States
Use of positive discrimination. Law
O'NEIL, Robert M.
Discriminating against discrimination. - Bloomington ; London : Indiana University Press, 1975 [i.e. 1976]
xi,271p. - Includes index
0-253-31800-9 X.200/20423
0666467

ADMISSION. Inmates. Barlavington Manor Children's Home
Racial discrimination
GREAT BRITAIN. Commission for Racial Equality
Barlavington Manor Children's Home. - London : The Commission, 1979
15,[5]leaves
B79-29979 BS.400/68
1941011

ADMISSION. Inmates. Residential institutions for old persons. Great Britain
Role of welfare workers - Conference proceedings
SERVICES for the elderly - the admission process. - [Cardiff?] : Welsh Office, Social Work Service, [1979?]
iii,99p
B83-07613 BS.128A/323
2832704

ADMISSION. Inmates. Residential institutions. Great Britain
BREARLEY, C. Paul (Christopher Paul)
Admission to residential care. - London : Tavistock Publications, 1980
xi,225p. - (Residential social work). - Bibliography: p206-216. - Includes index
0-422-76930-4 X.529/38897
2110199

ADMISSION. Mature students. Joint Matriculation Board universities. England
Applications
21+ returning to learning. - Manchester : Joint Matriculation Board, [1982]
10p
B83-12411 X.529/52544
2849879

ADMISSION. Medical schools. United States
LERNER, Marguerite Rush
Medical school. - 2nd ed. - Woodbury, N.Y. ; London : Barron's Educational Series, c1981
xi,163p : ill. - Bibliography: p163
0-8120-2277-7 X.329/16018
2860570

ADMISSION. Medical schools. United States
Applications
SOLOMON, Donald J.
Getting into medical school. - Philadelphia ; London : Saunders, 1979
xiii,167p : ill. - (Saunders survival series ; 2). - Bibliography: p.165-167
0-7216-8499-8 X.311/8955 Woolwich
1898922

ADMISSION. National health services patients & private patients. Hospitals. Great Britain
Common waiting lists
HEALTH SERVICES BOARD
Common waiting lists for NHS and private patients in NHS hospitals. - London : H.M.S.O., [1977]
15p. - (Cmnd.6828)
0-10-168280-8 BS.385/290
0690899

ADMISSION. Orthopaedic patients. Hospitals. Great Britain
Waiting lists
ORTHOPAEDIC services. - London : H.M.S.O., 1981
x,84p : ill
0-11-320754-9 X.322/10375 Woolwich
2580756

ADMISSION. Patients. Mental illness hospitals. Avon
Procedure
AVON. Mental Health Working Group
Mental health admissions. - Bristol : Avon County Council, Social Services Department, 1977
[5],43[i.e.53]p : ill. - (Info / Avon (County). Social Services Department ; 25)
0-86063-020-x X.313/1593 Woolwich
1765728

ADMISSION. Patients. Mental illness hospitals. California
Compulsory admission. Legal aspects
WARREN, Carol A. B.
The court of last resort. - Chicago ; London : University of Chicago Press, c1982
xii,265p. - Bibliography: p243-257. - Includes index
0-226-87388-9 X.200/39749
2849496

ADMISSION. Patients. Mental illness hospitals. England
Compulsory admission
BEAN, Philip
Compulsory admissions to mental hospitals. - Chichester : Wiley, c1980
xii,278p. - Bibliography: p267-273. - Includes index
0-471-27758-4 X.325/888 Woolwich
2057689

ADMISSION. Patients. Mental illness hospitals. England
Compulsory admission. Procedure
EDWARDS, Arthur Hubert
Shaw's summary of procedural requirements in respect of compulsory admission to hospital and guardianship. - London : Shaw and Sons, 1976
Folder ([3]p)(1 fold))
0-7219-0195-6 X.102/2075
0237671

ADMISSION. Patients. Mental illness hospitals. England
Compulsory admission. Role of welfare workers
DRAFT guidelines for approval of social workers under the proposed Mental Health (Amendment) Act. - [London] : [Department of Health and Social Security], 1981
11leaves
B83-10528 BS.385/776
2841460

ADMISSION. Patients. Mental illness hospitals. Great Britain
Duration. Reduction
SHORTENING hospital stay for psychiatric care / S.R. Hirsch ... [et al.]. - London : Charing Cross Hospital and Medical School, 1980, c1978
1v.(loose-leaf) : ill. - Includes bibliographies
B83-41792 BS.385/787
2966727

ADMISSION. Patients. Mental illness hospitals. San Francisco. California
Compulsory admission, 1870-1930
FOX, Richard W.
So far disordered in mind. - Berkeley ; London : University of California Press, 1978
xvi,204,[1]p. - Includes index
0-520-03653-0 X.529/37311
2078023

ADMISSION. Patients. Mental illness hospitals. Scotland
Compulsory admission. Procedure - For mental health officers
SOCIAL work practice in formal admissions. - [Edinburgh] : Social Work Services Group, [1980]
37p. - (Notes for mental health officers)
B82-32993 BS.39/113
2767775

ADMISSION. Patients. Psychiatric hospitals. United States
Compulsory admission. Legal aspects
SZASZ, Thomas
Psychiatric slavery. - New York : Free Press ; London : Collier Macmillan, 1977
xv,159p. - Includes index
0-02-931600-6 X.329/11058 Woolwich
1172018

ADMISSION. Pollyanna's. Restaurants
Racial discrimination
GREAT BRITAIN. Commission for Racial Equality
Report of a formal investigation into certain activities of Genture Restaurants Ltd and its chairman, Mr J. Weston-Edwards. - London : The Commission, 1978
[1],15 leaves
B80-40930 BS.400/33
2321920

ADMISSION. Students
See also
Selection. Students

ADMISSION. Students, 16-21 years. Open University
Projects: Younger Students Pilot Scheme. Evaluation
WOODLEY, Alan
The door stood open. - Lewes : Falmer, 1980
vii,261p : ill
0-905273-14-1 X.529/39339
2336529

ADMISSION. Students. Distance study institutions
FRIEDMAN, H. Zvi
The admission system in distance teaching institutions. - Milton Keynes : Open University Distance Education Research Group, 1981
35p : ill. - (DERG papers ; no.2)
B82-06690 X.520/25772
2606984

ADMISSION. Students. Higher education institutions. Great Britain
Applications
SUTHERLAND, Francis
Going to university or polytechnic. - Gloucester : Thornhill Press, 1978
57p : ill. - Bibliography: p.49
0-904110-58-3 X.519/28878
1753444

ADMISSION. Students. Higher education institutions. Great Britain
Applications by secondary school students. Results related to academic achievement - Serials
DEGREE course offers. - 1973- / edited by Brian W. Heap. - London : Career Consultants, 1973-
r6000258 P.521/3006
0055557

ADMISSION. Students. Higher education institutions. United States
1870-1970
WECHSLER, Harold S.
The qualified student. - New York ; London : Wiley, 1977
xix,341p. - Bibliography: p.309-331. - Includes index
0-471-92441-5 X.520/12079
1275755

ADMISSION. Students. Higher education institutions. United States
Use of positive discrimination
SWANSON, Kathryn
Affirmative action and preferential admissions in higher education. - Metuchen ; London : Scarecrow, 1981
viii,336p. - Includes index
0-8108-1411-0 X.529/42104
2504553

ADMISSION. Students. Infant schools. Liverpool. Merseyside (Metropolitan County)
Spatial aspects
MASSER, Ian
Spatial issues in school admission policy within Liverpool. - Liverpool : Department of Civic Design, University of Liverpool, 1977
20leaves,[8]leaves of plates : ill. - (Working paper / University of Liverpool. Department of Civic Design ; 5). - Bibliography: leaf 20
B83-20170 X.525/7122
2887800

ADMISSION. Students not fulfilling standard entrance requirements. Colleges of higher education & polytechnics. Great Britain
EVANS, Norman
Access to higher education. - London : Council for National Academic Awards, 1984
20p. - (CNAA Development Services publication ; 6)
B85-26662 X.525/11023
3299066

ADMISSION. Students. Schools. Buckinghamshire - For parents
BUCKINGHAMSHIRE. Education Department
Admissions to first and combined schools at age 5 - 1984/85. - [Aylesbury] : [The Department], [1984?]
26p
B85-02367 X.529/65647
2585677

BUCKINGHAMSHIRE. Education Department
Admissions to schools 1982-83. - [Aylesbury] : [The Department], [1981?]
40p
0-86059-225-1 X.529/44574
2585677

ADMISSION. Students. Secondary schools. Buckinghamshire - For parents
TRANSFER to secondary schools at 12+, September 1984 / Buckinghamshire County Council Education Department. - [Aylesbury] : [The Department], [1984?]
33p
B85-03376 X.529/65761
3213234

ADMISSION. Students. Universities. German speaking countries to date
STUDIENAUFNAHME und Studienzulassung / Paul Kellermann (Hg.). - Klagenfurt : Kärntner Druck- und Verlagsgesellschaft, c1984
303p. - (Klagenfurter Beiträge zur bildungswissenschaftlichen Forschung ; 15)
3-85391-046-7 X.0529/544(15)
9084657

ADMISSION. Students. Universities. Great Britain
Applications
GARNER, John Barrie
A guide to university entrance. - 1980. - London : Ward Lock, 1979
95p : ill
0-7062-3855-9 X.529/35410
0164011
A guide to university entrance. - London : Ward Lock. 1978. - 1977
91p : ill. - Includes index
0-7062-3655-6 X.519/27927
0769967
1981. - 1980
95p : ill. - Includes index
0-7062-4092-8 X.529/37432
0769967

HATTON, Stella
Choosing a university. - 3rd revised ed. - [Cambridge] : Advisory Centre for Education, 1976
[4],43p. - (Advisory booklets / Advisory Centre for Education ; no.5). - Bibliography: p.37-43
0-900029-44-7 X.519/26322
0769967

ADMISSION. University of Cambridge
CAMFORD, Peter
How to get into Oxbridge. - Birmingham : Oxbridge Publications, 1983
14p
0-946839-00-x X.629/21769
2997541

ADMISSION. University of Oxford
CAMFORD, Peter
How to get into Oxbridge. - Birmingham : Oxbridge Publications, 1983
14p
0-946839-00-x X.629/21769
2997533

ADMISSION. Users. Libraries. University of London - Directories
UNIVERSITY OF LONDON. Library Resources Co-ordinating Committee
Guide to admission to libraries in the University of London. - [4th revised ed.]. - [London] : The Committee, 1977
[1],v,31p
B78-22415 2719.bb.73
1269534
- [Another copy] 2719.cc.18

ADMISSION. Women students. University College, London 1878
HARTE, Negley
The admission of women to University College, London. - [London] : University College, London, 1979
29p
B83-19257 X.529/53500
2884801

ADMISSION CHARGES. Public libraries. Great Britain
LIBRARIES and the future. - Welwyn : Libraries Open and Free, 1980
5p
0-907288-00-6 2719.k.92
2888688

ADMISSION CHARGES. Royal Opera House
BLAUG, Mark
Why are Covent Garden seat prices so high?. - London : Royal Opera House Covent Garden Ltd, 1976
13,[2]p : ill
B79-04908 X.415/5628 Woolwich
1773623

ADMISSION STANDARDS. Foreign applicants. Postgraduate courses for nursing specialists. Great Britain
JOINT BOARD OF CLINICAL NURSING STUDIES
Post-basic, postgraduate courses in specialist aspects of clinical nursing. - London : The Board, 1977
[6]p
B77-25647 X.319/17497 Woolwich
1192124

ADMIXTURES. Concrete. Materials
Chemical admixtures
HIGGINS, D. D.
Admixtures for concrete. - Slough : Cement and Concrete Association, 1984
8p : ill(some col.). - Bibliography: p8
0-7210-1303-1 X.622/22435
1492764

RIXOM, M. R.
Chemical admixtures for concrete. - London : Spon, 1978
xii,234p : ill. - Includes index
B78-22061 SRL
1492764

ADMONT. Austria
Abbeys: Stift Admont, to 1974
LIST, Rudolf
Stift Admont 1074-1974. - Ried im Innkreis, 1974
xxiii, 559 p. : ill
r6002373 X.421/8779 Woolwich
0881007

ADNDA
ADNDA user guide. - 3rd ed. - Maidenhead : Lipman Management Resources, 1982
104p in various pagings. - Includes index
0-906168-04-x SRL
2895781

ADNEXA OCULI
See also
Eyelids

ADNEXA OCULI. Man
Eyes & adnexa oculi. Reconstructive surgery - Conference proceedings
INTERNATIONAL SYMPOSIUM OF PLASTIC AND RECONSTRUCTIVE SURGERY OF THE EYE AND ADNEXA (3rd : 1980 : Manhattan Eye, Ear, and Throat Hospital)
Third International Symposium of Plastic and Reconstructive Surgery of the Eye and Adnexa. - Baltimore ; London : Williams & Wilkins, c1982
xviii,362p : ill. - Includes bibliographies and index
0-683-05951-3 SRL
2967340

ADOLESCENCE
L'ADOLESCENZA / Luigi Del Gatto...[et al.]. - Roma : Bulzoni, [1976]
320p : ill. - Includes bibliographies
r6067570 X.510/10859
1708325

COLEMAN, John C. (John Christopher)
The nature of adolescence. - London : Methuen, 1980
ix,214p : ill. - Bibliography: p191-206. - Includes index
0-416-72620-8 X.529/38838
1585126

DEVELOPMENT in adolescence / [edited by] W. Everaerd ... [et al.]. - Boston, [Mass.] ; Lancaster : Nijhoff, 1983
x,255p : ill. - Includes index
0-89838-581-4 X.520/31863
1585126

DOBSON, James C. (James Clayton)
Preparing for adolescence. - Eastbourne : Kingsway, 1982, c1978
155p
0-86065-167-3 X.529/48457
2756323

GIBSON, Sir Ronald
Young people growing up. - London : British Medical Association, [1978]
30p : ill. - (A 'Family doctor' booklet)
B78-37512 7583.n.5/71
170317x

JASPER, Tony
Religion for living. - Edinburgh : Holmes McDougall
Book 1: My scene. - [1976]
64p : ill(some col)
0-7157-1302-7 X.0102/502(1)
0747866

MALMQUIST, Carl P.
Handbook of adolescence. - New York ; London : Aronson ; London : Distributed by Eurospan Ltd, 1978
x,910p : ill. - Includes index
0-87668-270-0 X.520/13364
1780751

MILLER, Derek
Adolescence. - New York : Aronson ; London : Distributed by Eurospan Ltd, 1974
xvii,544p. - Includes bibliographies and index
0-87668-101-1 X.520/13363
1780751

NATIONAL COMMISSION ON YOUTH
The transition of youth to adulthood. - Boulder : Westview Press, 1980
xii,228p
0-89158-675-x X.520/23082
1708325

ODLUM, Doris
Adolescence. - [New ed.]. - Hove : Wayland, 1978
[5],162p. - Includes index
0-85340-491-7 X.529/32519
1585126

PRINCE, Gordon Stewart
Teenagers today. - [Revised ed.]. - London : MIND, 1978
32p : ill
B79-11471 X.519/29395
1827316

RAYNER, Claire
Growing pains. - London : Corgi, 1984
198p : ill
0-552-12220-3 X.520/36985
1585126
Growing pains and how to avoid them. - London : Heinemann, c1984
198p : ill
0-434-98042-0 X.520/37049
1585126

ROGERS, Dorothy, 1914-
Adolescents and youth. - 4th ed. - Englewood Cliffs ; London : Prentice-Hall, c1981
xvii,476p : ill. - Bibliography: p449-470. - Includes index
0-13-008748-3 X.520/23916
1585126

SLEIGH, Julian
Thirteen to nineteen : growing free. - Edinburgh : Floris, 1982
30p
0-903540-58-4 X.529/52347
2756323

SOMMER, Barbara Baker
Puberty and adolescence. - New York : Oxford University Press, 1978
258p
0-19-502376-5 Cup 351/233
1827316

THORNBURG, Hershel D.
Development in adolescence. - 2nd ed. - Monterey : Brooks / Cole, c1982
xvi,656p : ill. - Bibliography: p577-642
0-8185-0448-x X.520/34177
1585126

ADOLESCENCE
- Serials

JOURNAL of adolescence. - Vol.1, issue 1- ; Mar. 1978-. - London : Academic Press for the Association for the Psychiatric Study of Adolescents, 1978- : ill
B78-23210 P.321/741
1498061

ADOLESCENCE. Special subjects. Crane, Stephen. American writers. Fiction in English

HELLER, Arno
Experiments with the novel of maturation. - Innsbruck : Innsbrucker Gesellschaft zur Pflege der Geisteswissenschaften, 1976
32p. - (Innsbrucker Beiträge zur Kulturwissenschaft ; Sonderheft 38)
3-85124-058-8 X.0510/186a (38)
1404172

ADOLESCENCE. Special subjects. Fiction in English. 1700-1970

SPACKS, Patricia Meyer
The adolescent idea. - London : Faber, 1982, c1981
ix,308p. - Includes index
0-571-11915-8 X.520/27480
2756358

ADOLESCENCE. Special subjects. James, Henry, 1843-1916. American writers. Fiction in English

HELLER, Arno
Experiments with the novel of maturation. - Innsbruck : Innsbrucker Gesellschaft zur Pflege der Geisteswissenschaften, 1976
32p. - (Innsbrucker Beiträge zur Kulturwissenschaft ; Sonderheft 38)
3-85124-058-8 X.0510/186a (38)
1404180

ADOLESCENT BOYS
Interpersonal relationships with adolescent girls

EPHRON, Delia
Teenage romance, or, How to die of embarrassment. - London : Gollancz, 1982, c1981
109p : ill
0-575-03146-8 X.958/12357
280378x

ADOLESCENT BOYS. Boston. Massachusetts
Italian immigrant adolescent boys. Development

YOUNG, Harben Boutourline
Puberty to manhood in Italy and America. - New York ; London : Academic Press, 1981
xiii,283p : ill. - (Developmental psychology series). - Bibliography: p267-276. - Includes index
0-12-773150-4 X.520/23941
2485508

ADOLESCENT BOYS. Boston. Massachusetts
Negro adolescent boys. Social behaviour

MANCINI, Janet K.
Strategic styles. - Hanover, N.H. ; London : University Press of New England, 1980, c1981
xv,330p : ill. - Bibliography: p302-326. - Includes index
0-87451-179-8 X.520/31143
295012x

ADOLESCENT BOYS. London
Social behaviour. Effects of violence in television programmes

BELSON, William A. (William Albert)
Television violence and the adolescent boy. - Farnborough, Hants. : Saxon House, 1978
x,529p : ill. - Bibliography: p.52-59
0-566-00211-6 X.529/33394
1564617
- Accompanying microfiche Cup.580.h.24

ADOLESCENT BOYS. Manchester
Social conditions, 1905

RUSSELL, Charles E. B. (Charles Edward Bellyse)
Manchester boys. - Manchester : N. Richardson, 1984
44p : ill
0-907511-62-7 X.805/6816
3287076

ADOLESCENT BOYS. Palermo. Italy
Development

YOUNG, Harben Boutourline
Puberty to manhood in Italy and America. - New York ; London : Academic Press, 1981
xiii,283p : ill. - (Developmental psychology series). - Bibliography: p267-276. - Includes index
0-12-773150-4 X.520/23941
2485524

ADOLESCENT BOYS. Rome Italy
Development

YOUNG, Harben Boutourline
Puberty to manhood in Italy and America. - New York ; London : Academic Press, 1981
xiii,283p : ill. - (Developmental psychology series). - Bibliography: p267-276. - Includes index
0-12-773150-4 X.520/23941
2485516

ADOLESCENT BOYS Sheffield South Yorkshire (Metropolitan County)
Maladjusted adolescent boys. Intermediate treatment. Use of outdoor activities, 1971-1974

PAYNE, Bob
An experience of an intermediate treatment group. - London : Family Service Units, 1975
[3],29p
0-905175-00-x X.512/5429
053028x

ADOLESCENT BOYS Students. Secondary schools. United States
Social adjustment

ADOLESCENT boys in high school / edited by James G. Kelly. - Hillsdale : Erlbaum ; New York ; London : Distributed by Wiley, 1979
xv,267p : ill. - Includes bibliographies and index
0-470-26591-4 X.520/14672
1998072

ADOLESCENT BOYS. Urban regions. Great Britain
Working class adolescent boys. Social conditions, 1973-1978

WILLIAMSON, Howard
Five years. - Leicester : National Youth Bureau, 1981
115p : ill
0-86155-046-3 X.950/10444
2623730

ADOLESCENT FICTION IN ENGLISH
to 1900-

YOUNG adult literature / compiled by Millicent Lenz and Ramona M. Mahood. - Chicago : American Library Association, 1980
viii,516p : ill. - Includes bibliographies
0-8389-0302-9 X.950/12643
2719835

ADOLESCENT GIRLS
See also
Activities for adolescent girls
Correctional institutions for delinquent adolescent girls
Welfare work with adolescent girls

ADOLESCENT GIRLS
Anthologies in English - For adolescent girls

OH boy! annual. - 1978. - London : IPC Magazines, 1977
95p : ill(some col)
Subsequent volumes, when received, will be placed at this shelfmark
0-85037-370-0 P.991/333
0210382

ADOLESCENT GIRLS
Children, to 5 years. Development & care - For adolescent girls

MASON, Honor
The child's world. - Glasgow : Blackie, 1977
[2],37p : ill. - (Home economics topic books)
0-216-90348-3 X.519/27488
1289438

ADOLESCENT GIRLS
Cleanliness & personal appearance

SAUNDERS, Rubie
Good grooming for girls. - New York ; London : F. Watts, 1977
[4],60p : ill. - (A concise guide). - Includes index
0-85166-614-0 X.620/16707 Woolwich
1094246

[Good grooming for girls]. Top to toe. - [London] : Piccolo, 1981, c1977
64p : ill. - Includes index
0-330-25999-7 X.529/39416
1094246

ADOLESCENT GIRLS
Education - For teaching

CHANDLER, E. M.
Educating adolescent girls. - London : Allen & Unwin, 1980
xv,217p. - (Unwin education books). - Bibliography: p205-211. - Includes index
0-04-370096-9 X.529/38915
210041x

ADOLESCENT GIRLS
Great Britain. Society - For adolescent girls

SAINSBURY, Vera
Planning your future. - London : Mills and Boon, 1975
5v. : ill
0-263-05817-4 X.519/25121
0367753

ADOLESCENT GIRLS
Great Britain. Women. Equality of opportunity - For adolescent girls

FEMINISM for girls / edited by Angela McRobbie and Trisha McCabe. - London : Routledge & Kegan Paul, 1981
212p : ill
0-7100-0961-5 X.520/25830
2612348

ADOLESCENT GIRLS
Housework - For adolescent girls

GILLETT, Cynthia
Looking after your home. - Glasgow : Blackie, 1977
[2],36,[1]p : ill. - (Home economics topic books)
0-216-90349-1 X.619/18131 Woolwich
1289535

ADOLESCENT GIRLS
Hygiene

SALTER, Mary
Mainly for girls. - 2nd (revised) ed.. - London : Mills and Boon, 1974
128p : ill
0-263-05464-0 X.319/16381 Woolwich
0601527

ADOLESCENT GIRLS
Interpersonal relationships. Conversation

WINTERS, Catherine
How to talk to boys and other important people. - Toronto ; London : Bantam, 1983
130p. - (Sweet dreams)
0-553-23604-0 X.529/60363
3058433

ADOLESCENT GIRLS
Interpersonal relationships with adolescent boys

EPHRON, Delia
Teenage romance, or, How to die of embarrassment. - London : Gollancz, 1982, c1981
109p : ill
0-575-03146-8 X.958/12357
280378x

ADOLESCENT GIRLS
Man. Health. Role of diet - For adolescent girls

GILLETT, Cynthia
Eating for health and pleasure. - Glasgow : Blackie, 1977
[2],36,[1]p : ill. - (Home economics topic books)
0-216-90181-2 X.319/17465 Woolwich
1174266

ADOLESCENT GIRLS
Man. Sexual intercourse - For adolescent girls

COVENTRY WOMEN'S EDUCATION GROUP
Please yourself. - Coventry : The Group ; London : Distributed by Southern Distribution ; Hebden Bridge : Distributed by Scottish and Northern Distributors, [1980]
[2],34p : ill
0-9506878-0-4 Cup.365.p.12
2131994

ADOLESCENT GIRLS
Medical aspects

ADOLESCENT obstetrics and gynecology / edited by A. Karen Kessler Kreutner and Dorothy Reycroft Hollingsworth ; with 18 contributors. - Chicago ; London : Year Book Medical Publishers, 1978
xvii,658p : ill. - Includes index
0-8151-5200-0 SRL
1523597

ADOLESCENT GIRLS
Participation in Youth Training Scheme

BRELSFORD, Pauline
Widening opportunities for young women within YTS. - Sheffield : Manpower Services Commission, 1983
20p
0-86392-035-7 X.525/10123
3195473

ADOLESCENT GIRLS
Personal finance - For adolescent girls

GILLETT, Cynthia
Shopping and money matters. - Glasgow : Blackie, 1977
[2],36,[1]p : ill. - (Home economics topic books)
0-216-90347-5 X.519/27497
1289543

ADOLESCENT GIRLS
Physical fitness
DAVIS, Julie, 1956–
The sweet dreams body book. - Toronto ; London : Bantam, 1983
122p : ill. - (A Sweet dreams special)
0-553-23376-9 X.329/17150
2986914

ADOLESCENT GIRLS
Pregnancy
RUSSELL, J. K.
Early teenage pregnancy. - Edinburgh : Churchill Livingstone, 1982
x,108p : ill. - (Current reviews in obstetrics and gynaecology ; 3). - Includes bibliographies and index
0-443-02359-x X.322/16281
2892510

ADOLESCENT GIRLS
Pregnancy & abortion. Psychosocial aspects
ALLEN, James E. (James Edward)
Managing teenage pregnancy. - New York : Praeger, 1980
xxvii,300p : ill. - (Praeger special studies). - Bibliography: p290-295. - Includes index
0-03-053991-9 X.322/9480 Woolwich
055376X

ADOLESCENT GIRLS
Psychological development
[TAGEBUCH eines halbwüchsigen Mädchens. 1975. French].
Journal psychanalytique d'une petite fille / Adaptation de l'allemand par Clara Malraux .... - Paris : Denoël, 1975
197 p.. - (Freud et son temps)
r6000984 X.319/8336 Woolwich
0913782

ADOLESCENT GIRLS
Social aspects
WARD, John Powell
Social reality for the adolescent girl. - Swansea : Faculty of Education, University College of Swansea ; Swansea : [Distributed by] Singleton Bookshop, [1976]
94p. - Bibliography: p.88-91
0-900944-14-5 X.510/15159
0642002

ADOLESCENT GIRLS
Women's clothing. Cleaning - For adolescent girls
JEANS, Heather
Yourself and your clothes. - Glasgow : Blackie, 1977
[2],36,[1]p : ill. - (Home economics topic books)
0-216-90182-0 X.619/17751 Woolwich
1169319

ADOLESCENT GIRLS
Women's clothing. Selection - For adolescent girls
JEANS, Heather
Yourself and your clothes. - Glasgow : Blackie, 1977
[2],36,[1]p : ill. - (Home economics topic books)
0-216-90182-0 X.619/17751 Woolwich
1169300

ADOLESCENT GIRLS. Denmark
Social conditions, 1945–
ERNSTSEN, Inger
Barn af fede tider!. - [Odense] : Odense Universitet, Nordisk Institut, [1982]
1v.(various pagings) : ill. - ([Skrifter fra Nordisk Institut, Odense Universitet ; v.17])
r6610847 P.903/409[no.17]
8965269

ADOLESCENT GIRLS. England
Education, 1870-1920. Social aspects
DYHOUSE, Carol
Girls growing up in late Victorian and Edwardian England. - London : Routledge & Kegan Pual, 1981
ix,224p. - (Studies in social history). - Includes index
0-7100-0821-x 9618.1.2/33
259188x

ADOLESCENT GIRLS. England
Social conditions, 1960-1970
DIX, Carol
Say I'm sorry to mother. - London : Pan Books, 1978
158p
0-330-25481-2 X.708/21559
1737368

ADOLESCENT GIRLS. Great Britain
Adolescent Sikh girls. Integration into society
KALRA, Surjit Singh
Daughters of tradition. - Birmingham : Diana Balbir Publications ; Birmingham : [Distributed by] Third World Publications, 1980
A-J,88p. - Bibliography: p.84-86
B80-19624 X.200/33025
2104342

ADOLESCENT GIRLS. Great Britain
Expenditure - Statistics
TEENAGE girls. - London : IPC Magazines.
1977. - 1977
54,[20]p,[10]leaves
Subsequent volumes, when received, will be placed at this shelfmark
B77-33678 Cup.1281/652
1268031

ADOLESCENT GIRLS. Great Britain
Interpersonal relationships with parents. Conflict
WILLANS, Angela
Breakaway. - London : Temple Smith, 1977
201p
0-85117-129-x X.529/31681
1189263

ADOLESCENT GIRLS. Great Britain
Pregnancy & parenthood
WELLS, N. E. J. (Nicholas E J)
Teenage mothers. - Liverpool : European Collaborative Committee for Child Health of the Children's Research Fund, c1983
24p : ill
B84-13930 X.529/59363
3060446

ADOLESCENT GIRLS. Midlands. England
South Asian adolescent girls. Social conditions - For welfare work
JAMDAGNI, Laxmi
Hamari rangily zindagi. - [Leicester] : NAYC, [1980?]
131p. - (Special report series)
B83-35436 X.525/7831
2942348
- [Another copy.] W 7081 IOL

ADOLESCENT GIRLS. Poland
- Biographies
NIELATWO być dziewczyna / wybór i rozmowy z autorkami : Wojciech Wishniewski. - Warszawa : Nasza Księgarnia, 1977
366p : ill
r6016247 X.519/35219
1505564

ADOLESCENT GIRLS. Rural regions. United States
Negro adolescent girls. Social development
DOUGHERTY, Molly Crocker
Becoming a woman in rural black culture. - New York ; London : Holt, Rinehart and Winston, 1978
xi,111p : ill. - (Case studies in cultural anthropology). - Bibliography: p.109-111
0-03-014921-5 X.510/15771
1490257

ADOLESCENT GIRLS. Samoa
FREEMAN, Derek
Margaret Mead and Samoa. - Cambridge, Mass. ; London : Harvard University Press, 1983
xvii,379p. - Includes index
0-674-54830-2 X.520/30874
1267388

Margaret Mead and Samoa. - Harmondsworth : Penguin, 1984
xx,379p : ill. - (Pelican books). - Includes index
0-14-022555-2 X.529/65212
1267388

MEAD, Margaret
Coming of age in Samoa. - Harmondsworth : Penguin, 1977
240p. - (Pelican books). - Includes index
0-14-020127-0 X.709/31709
1267388

ADOLESCENT GIRLS. Special subjects. English photography, 1950-
HAMILTON, David, 1933–
[Bilitis]. Hamilton's movie Bilitis. - London : Columbus, 1982, c1977
109p : chiefly ill(some col.)
0-86287-004-6 Cup.713/44
0629952

Tender cousins. - London : Columbus, 1982, c1981
109p : chiefly ill(some col.)
0-86287-003-8 Cup.713/45
0629952

ADOLESCENT GIRLS. Special subjects. English photography, 1950-
- Collections
HAMILTON, David, 1933–
The best of David Hamilton. - London : Collins, 1977, c1976
143p : chiefly ill(some col.)
B82-14989 Cup.713/10
0629952

Dreams of a young girl. - London : Collins, 1980
143p : chiefly ill(some col)
0-00-211640-5 Cup.821dd19
0629952

Sisters. - London : Collins, 1976
3-144p : of ill(some col)
0-00-216765-4 Cup.815/12
0629952

Sisters. - New York ; London : Collins, 1980
135p : chiefly ill(some col.)
0-00-216810-3 Cup.821.d.33
0629952

HAMILTON, David, 1933–
Souvenirs. - London : Collins, 1978
3-120p : chiefly ill(some col)
0-00-216736-0 Cup.820.pp.35
0629952

A summer in Saint Tropez. - London : Collins, 1982
118p : chiefly col.ill
0-00-216666-6 Cup.713/26
0629952

ADOLESCENT GIRLS. Surinam
Social behaviour, to date
SRIRAHAJU
Ik wil gelukkig zijn. - Paramaribo : Perwakis, [1975]
95p : ill
r6303297 X.529/46388
8795258

ADOLESCENT GIRLS. United States
KONOPKA, Gisela
Young girls. - Englewood Cliffs ; London : Prentice-Hall, 1976
xv,176p. - Includes bibliographies and index
0-13-977215-4 X.519/25782
0654833

ADOLESCENT GIRLS. United States
Moral development
O'BYRNE, Simon
Moral development and logical reasoning of adolescents. - Dublin : Frederick Press, 1976
[1],xii,166p. - Bibliography: p.111-120
B77-20881 X.529/31558
1142186

ADOLESCENT GIRLS. United States
Pregnancy
PREGNANCY in adolescence / edited by Irving R. Stuart and Carl F. Wells. - New York ; London : Van Nostrand Reinhold, 1982
xxvi,441p. - Includes bibliographies and index
0-442-21225-9 X.322/10916 Woolwich
2736705

ADOLESCENT GIRLS. United States
Pregnancy, childbirth & parenthood
TEENAGE parents and their offspring / edited by Keith G. Scott, Tiffany Field, Euan G. Robertson. - New York ; London : Grune & Stratton, 1981
xv,328p : ill. - Includes index
0-8089-1314-x X.520/23799
239913x

ADOLESCENT GIRLS. United States
Pregnancy. Social aspects
BOLTON, Frank G.
The pregnant adolescent. - Beverly Hills ; London : Sage Publications, 1980
247p. - (Sage library of social research ; vol.100). - Bibliography: p.235-246
0-8039-1433-4 X.529/36877
2224070

ADOLESCENT GIRLS. United States
Sexual intercourse & pregnancy. Social aspects
ZELNIK, Melvin
Sex and pregnancy in adolescence. - Beverly Hills ; London : Sage, c1981
272p : ill. - (Sage library of social research ; v.133). - Bibliography: p265-269
0-8039-1733-3 X.529/46751
2700220

ADOLESCENT GIRLS. West Germany
Heroin addiction

F., Christiane
[Wir Kinder vom Bahnhof Zoo. English] H. autobiography of a child prostitute and heroin addict. - [London] : Corgi, 1981, c1980
297p
0-552-11772-2 Cup.365.pp.16
2558769

ADOLESCENT GIRLS. West Germany
Social conditions, 1945-

KOLL, Regine
Wir pfeifen auf den Mist. - Berlin : [R. Koll, 1980]
100p : ill
r6305853 X.529/51172
8880417

ADOLESCENT GIRLS' SERIALS IN ENGLISH
- Texts

The BEST of My guy monthly. - No.1 (May 1984)-. - [London] : [IPC Magazines], 1984-
v. : chiefly ill
B85-07848 P.823/548
0225398

[FAB (Annual)]. Fab. - 1983-. - London : IPC Magazines, 1982-
v. : ill(some col.)
B83-07975 P.991/374
0225398

HEARTBEAT annual. - 1983-. - London : IPC Magazines, 1982-
v. : ill(some col.)
B83-07972 P.991/375
0225398

'MY guy' annual. - '79. - London : IPC Magazines, [1978]- : ill(some col)
B79-09235 P.971/414
0225398

PHOTO-LOVE annual. - 1981-. - London : IPC Magazines, 1980-
v. : ill(some col.)
B82-18723 P.901/3223
0225398

ADOLESCENT GIRLS SERIALS IN ENGLISH
Serials with British imprints: 'Jackie'. Stories. Special themes: Femininity

MCROBBIE, Angela
Jackie. - Birmingham : Centre for Contemporary Cultural Studies, 1978
54p : ill. - (Stencilled occasional paper) (Women series ; no.53)
B83-22246 X.525/7220
2894823

ADOLESCENT GIRLS WITH UNWANTED PREGNANCIES. United States to date

BEDGER, Jean E.
Teenage pregnancy. - Springfield, Ill. : Charles C. Thomas, [1980]
xiii,203p
0-398-03923-2 X.520/26134
9005625

OETTINGER, Katherine B.
"Not my daughter". - 3. [print.]. - Englewood Cliffs : Prentice-Hall, c1979
xviii,184p
0-13-623850-5 X.529/52351
9005625

ADOLESCENT MEDICINE

ADOLESCENT health care / edited by Robert W. Blum. - New York ; London : Academic Press, 1982
xxi,297p : ill. - Includes bibliographies and index
0-12-788080-1 X.322/11457
2833778

ADOLESCENT medicine topics. - New York : Spectrum Publications ; New York ; London : Distributed by Wiley.
Vol.1. - [1976]
[11],260p : ill. - Includes bibliographies and index
0-470-15151-x SRL
2833778

NEINSTEIN, Lawrence S.
Adolescent health care. - Baltimore : Urban & Schwarzenberg, 1984
xv,701p : ill. - Includes bibliographies
0-8067-1351-8 X.329/19347
2833778

ADOLESCENTS
See also
Advisory services for adolescents
Autistic adolescents
Brain damaged adolescents
Cartoons for adolescents in multiracial communities
Catholic adolescents
Cerebral palsied adolescents
Christian adolescents
Compositions by adolescents
COR weekends
Deaf adolescents
Fiction for adolescents
Handicapped adolescents
Homosexual adolescents
Hostels for adolescents
Learning disordered adolescents
Left-wing adolescents
Leukaemic adolescents
Male adolescents
Mentally ill adolescents
Negro adolescents
Newspapers for adolescents
Physically handicapped adolescents
Puberty
Serials for American adolescents in Great Britain
Services for adolescents
Slow learning adolescents
Slow reading adolescents
Socially disadvantaged adolescents
Suicidal adolescents
Unemployed adolescents
Welfare work with adolescents
Working class adolescents

ADOLESCENTS

CRABTREE, Tom
Tom Crabtree on teenagers. - London : Elm Tree Books, 1980
x,198p
0-241-10398-3 X.529/38710
0281948
[Tom Crabtree on teenagers]. Living with teenagers. - London : Macdonald Futura, 1981, c1980
254p
0-7088-2083-2 X.529/42593
0281948

CSIKSZENTMIHALYI, Mihaly
Being adolescent. - New York : Basic, c1984
xvi,332p : ill. - Bibliography: p314-322
0-465-00646-9 X.520/37215
0281948

GRINDER, Robert Eugene
Adolescence. - 2nd ed. - New York ; London : Wiley, 1978
xvii,610p : ill. - Includes index
0-471-32767-0 X.520/13179
0281948

STORR, Catherine
Growing up. - London : Arrow Books, 1975
127p : ill. - (Arrow family handbooks). - Bibliography: p.123-125
0-09-910640-x X.319/10200
0281948

ADOLESCENTS
Aggression. Psychiatric aspects

The AGGRESSIVE adolescent / Charles R. Keith, editor. - New York : Free Press ; London : Collier Macmillan, c1984
xiii,487p : ill. - Includes bibliographies and index
0-02-916720-5 X.520/36730
3257797

ADOLESCENTS
Alcoholic drinks - For adolescents

PICKERING, John, 1949-
What every teenager should know about alcohol. - London : Hope Press Publications [for the UK Band of Hope Union], [1980]
[3],44p : ill
0-9505333-6-x X.329/13129 Woolwich
2315157

ADOLESCENTS
Alcoholic drinks. Consumption. Health aspects - For adolescents

FOSTER, Fiona
All about drinking. - Edinburgh : Macdonald, c1981
46p : ill
0-904265-52-8 X.622/13465
0481297

ADOLESCENTS
Attitudes to family planning. Effects of social environment

ADOLESCENTS and fertility planning / International Planned Parenthood Federation ; adapted from research undertaken by Philip Kreager for IPPF. - London : The Federation, 1979
41,xlip. - (Research for action ; no.5). - Bibliography: pii-xli
B81-05485 X.325/831 Woolwich
2350696

ADOLESCENTS
Behaviour modification

JONES, Vernon F.
Adolescents with behavior problems. - Abridged ed. - Boston, Mass. ; London : Allyn and Bacon, 1980
xv,302p : ill. - Bibliography: p.285-293. - Includes index
0-205-06824-3 X.520/21188
2147823

ADOLESCENTS
Behaviour. Psychosocial aspects

RUTTER, Michael
Changing youth in a changing society. - London : Nuffield Provincial Hospitals Trust, 1979
xii,323p : ill. - Bibliography: p295-323
0-900574-31-3 X.529/44768
2592592

ADOLESCENTS
Behavioural disorders

NIELSON, Gary
Borderline and acting out adolescents. - New York : Human Sciences, c1983
256p. - Bibliography: p237-244
0-89885-109-2 X.529/61563
3070204

ADOLESCENTS
British adolescents. Overseas vacation employment - Directories

WORKING holidays. - 1980 / ...compiled by Hilary Sewell. - London : Central Bureau for Educational Visits and Exchanges, 1979
138p : ill
0-900087-56-0 X.529/36446
0077976

WORKING holidays at home & abroad. - 1975. - [London] : Central Bureau for Educational Visits and Exchanges, 1975
[7],97p : ill
Subsequent volumes, when received, will be placed at this shelfmark
0-900087-33-1 P.801/3003
0077976

ADOLESCENTS
Camping - For adolescents

GREENBANK, Anthony
Camping for young people. - London : Harrap, 1979
143,[1]p : ill. - Includes index
0-245-53260-9 X.629/12325 Woolwich
1819747

Camping for young people. - London : Pan Books, 1979
144p : ill. - (Piccolo). - Includes index
0-330-25745-5 X.619/19603 Woolwich
1844172

ADOLESCENTS
Careers guidance - For parents

HEAP, Brian
Your child's career and how you can help. - London : Careers Consultants Ltd, 1975
177p. - Bibliography: p.176-177
0-85660-013-x X.519/26115
0559482

ADOLESCENTS
Children, to 3 years. Development - For adolescents

BALDWIN, Dorothy
Understanding your baby. - Revised ed. - London : Ebury Press, 1979
144p : ill
0-85223-155-5 X.319/18900 Woolwich
0324698

Understanding your baby. - 2nd ed. - Windsor : NFER-Nelson, 1983
141p : ill
0-7005-0621-7 X.329/17676
0324698

ADOLESCENTS
Children, to 5 years. Home care - For adolescents

CREESE, Angela
Young homemaker's book of parentcraft. - London : Mills and Boon, 1978
221,[1],96p : ill
0-263-06358-5 X.319/18467 Woolwich
1698478

ADOLESCENTS
Christian life

BARCLAY, William, 1907-1978
Marching on. - Evesham : James, 1983, c1974
223p
0-85305-254-9 X.208/9113
009398x

ADOLESCENTS
Christian life–continued

DISCOVERING the Bible. – Redhill : International Bible Reading Association : National Christian Education Council.
Part 1. – 1979
192p
0–7197–0232–1 X.208/5896
009398x
– Redhill : International Bible Reading Association.
Part 3. – c1981
191p
0–7197–0304–2 X.208/6834
009398x

SMITH, Jim, 1947–
More like Jesus. – London : Pathfinders–CYFA, 1981
22p : ill. – (An Operation Breakthrough booklet)
0–907750–02–8 X.208/9306
1544764

YOUNG, John, 1937–
Discipleship. – London : Scripture Union, 1978
24p : ill
0–85421–535–2 X.108/17076
1544764

ADOLESCENTS
Christian life. Prayers

MORE prayers for young people / [compiled by] William Barclay. – [London] : Fountain Books, 1977
127p
0–00–624668–0 X.108/16683
0620688

ADOLESCENTS
Christian life. Prayers – Collections

GREEN, Janet, 1939–
Home-made prayers. – Tring : Lion, 1983
80p : ill. – (A Lion paperback)
0–85648–211–0 X.208/8555
0620688

JASPER, Tony
Thank God. – Tring : Lion Publishing, 1982
76p : ill
0–85648–346–x X.990/20058
0620688

ADOLESCENTS
Christian religious education

MOFFETT, Bob
Bob Moffett's power pack. – London : Scripture Union, 1984, c1983
36[i.e.72]p : ill
0–86201–190–6 X.529/65133
3184595

ADOLESCENTS
Christian religious education – Serials

TEACHING over 13s. – [No.1]– ; Oct. [1975]–. – London : Scripture Union, 1976 : ill
B76–16803 P.103/504
0645699

ADOLESCENTS
Christian religious education. Projects

BLUEPRINT / general editor John Bailey. – London : Galliard.
Book 1. – 1976
lxvp,p1–299,vii p. – Includes index
0–85249–335–5 X.0102/500
0588164
Book 2. – 1976
xii p,p301–644,vii p. – Includes index
0–85249–352–5 X.0102/500
0588164
Book 3. – 1976
xii p,p645–1000,vii p. – Includes index
0–85249–353–3 X.0102/500
0588164
Book 4. – 1976
xii p,p1001–1334,xxxviii p. – Includes index
0–85249–354–1 X.0102/500
0588164

ADOLESCENTS
Christian religious education. Use of drama – For teaching

LOVE, Margaret
Let's dramatise!. – Rev. and repr. – Redhill : National Christian Education Council, 1982
80p
0–7197–0329–8 X.958/19749
3034356

ADOLESCENTS
Christianity – For adolescents – Serials

CLEARWAY. – No.1– ; [Nov. 1975]–. – London : Inter-school Christian Fellowship, Scripture Union, [1975]– : ill
Holdings: no.2–
B76–12336 P.973/301
0624454

ADOLESCENTS
Cigarette smoking. Health aspects – For adolescents

WILLIAMS, Howard, 1919–
To smoke or not to smoke. – London : Chest, Heart and Stroke Association, c1981
11p
B82–08055 X.329/14870 Woolwich
2611236

ADOLESCENTS
Cleanliness & personal appearance

CULLEN, Margaret
Good grooming and clothes care. – 2nd ed. – London : Heinemann Educational, 1979
101p : ill. – (Living today ; book 3)
0–435–42211–1 X.0519/1072(3)
201985x

ADOLESCENTS
Climbing. Techniques – For adolescents

GREENBANK, Anthony
Climbing for young people. – London : Harrap, 1977
128p : ill. – Includes index
0–245–53112–2 X.629/11349 Woolwich
1142917
Climbing for young people. – London : Pan Books, 1977
128p : ill. – (Piccolo). – Includes index
0–330–25139–2 X.619/17839 Woolwich
1142917

ADOLESCENTS
Depression

MCCOY, Kathleen
Coping with teenage depression. – New York : New American Library, 1982
xiv,331p. – (Mosby medical library). – Bibliography: p283–284
0–452–25407–8 X.529/58051
3028453

ADOLESCENTS
Development

JUROVSKÝ, Anton
Mládež a spoločnosť. – Bratislava : Veda, 1974
280p. – Bibliography: p.257–261
r6007263 X.525/1114
1128485

LERNER, Richard Martin
Adolescent development. – New York ; London : McGraw-Hill, 1980
xx,537p : ill(some col). – Bibliography: p.485–509. – Includes index
0–07–037186–5 X.520/22390
2230666

ADOLESCENTS
Development – Conference proceedings

ADOLESCENCE in the life cycle / edited by Sigmund E. Dragastin and Glen H. Elder, Jr. – Washington, D.C. : Hemisphere Publishing Corporation ; New York ; London : Distributed by Wiley, 1975
xi,324p : ill. – Includes bibliographies and index
0–470–22155–0 X.520/10218
2230666

ADOLESCENTS
Development – For parents

BATCHELOR, Mary
Bringing up a family. – Tring : Lion Publishing, c1981. – (A Lion guide)
10–18 years
80p : ill
0–85648–308–7 X.529/42271
2506009

PARENTS & teenagers / The Open University in association with the Health Education Council and the Scottish Health Education Group. – London : Harper & Row, 1982
222p : ill(some col.)
0–06–318243–2 X.525/6849
284866x

ADOLESCENTS
Deviance

KAPLAN, Howard B.
Deviant behaviour in defense of self. – New York ; London : Academic Press, 1980
xi,255p. – Bibliography: p231–241. – Includes index
0–12–396850–x X.520/22876
233612x

ADOLESCENTS
Diabetes. Personal adjustment

FARQUHAR, J. W. (James Watson)
Diabetes in your teens. – Edinburgh : Churchill Livingstone, 1982
156p : ill. – (Churchill Livingstone patient handbook ; 10)
0–443–02220–8 X.329/15859
2818167

ADOLESCENTS
Diseases

CHRONIC disorders in adolescence / edited by Mark Scott Smith. – Boston, [Mass.] ; Bristol : Wright, 1983
xxi,426p : ill. – Includes bibliographies and index
0–7236–7031–5 SRL
0636126

COMERCI, George D.
Adolescent medicine. – London : Kimpton, 1979
560p : ill. – Includes bibliographies and index
0–87488–053–x X.319/18771 Woolwich
1873083

DANIEL, William Andrew
Adolescents in health and disease. – Saint Louis : Mosby ; London : Distributed by Kimpton, 1977
xi,392p : ill. – Includes bibliographies and index
0–8016–1201–2 SRL
0636126

MEDICAL care of the adolescent. – 3rd ed.. – New York : Appleton–Century–Crofts ; London : Prentice-Hall, 1976
xxiii,774p : ill. – Includes bibliographies and index
0–8385–6199–3 SRL
0636126

TOPICS in adolescent medicine / edited by Robert B. Shearin. – Chicago ; London : Year Book Medical, c1983
308p : ill. – Includes bibliographies and index
0–8151–7675–9 SRL
0636126

ADOLESCENTS
Diseases – For general practice

ANYAN, Walter R.
Adolescent medicine in primary care. – New York ; Chichester : Wiley, 1978
xii,387p : ill. – Includes bibliographies and index
0–471–03976–4 SRL
1775073

ADOLESCENTS
Education. Psychological aspects

WALL, William Douglas
Constructive education for adolescents. – Paris : Unesco ; London : Harrap, 1977
iii–xi,333p. – Includes bibliographies and index
0–245–52966–7 X.520/12133
1293303

ADOLESCENTS
Effects of violence in television programmes, to date

BELSON, William A. (William Albert)
Adolescents, violence and television. – Stockholm : Akademilitteratur, 1980
22p. – (Mass ; 3)
91–7410–125–0 X.529/54340
8945144

ADOLESCENTS
Emotional problems

GALLAGHER, James Roswell
Emotional problems of adolescents. – 3rd ed. – New York : Oxford University Press, 1976
xiii, 203 p.. – Bibliography : p. 197–199. – Includes index
0–19–501972–5 X.529/30869
0688126

ADOLESCENTS
Family life. Effects of cannabis smoking

LOVENIUS, Margaretha
Det är ditt problem. – [Stockholm] : Aldus, 1975
121 p.. – (Aldusserien ; 442)
91–00–40440–3 X.519/19929
0905712

ADOLESCENTS
Fear

BAMBER, J. H.
The fears of adolescents. – London : Academic Press, 1979
viii,228p. – (Personality and psychopathology ; 20). – Includes bibliographies and index
0–12–077550–6 X.520/14261
182659x

ADOLESCENTS
Great Britain. Recreations: Long–distance walking – For adolescents

DUERDEN, Frank
Adventure walking for young people. – London : Kaye & Ward, 1980
xv,142p : ill
0–7182–1252–5 X.809/46219
2328941
– [Another copy: paperback]
ISBN 0718212576 X.709/52280

ADOLESCENTS
Great Britain. Social conditions - For adolescents
MAY, John, 1931-
Living in Britain. - Birmingham : Clearway, c1979
128p : ill. - (Britain today series)
0-902336-64-9 X.529/45901
2626969

ADOLESCENTS
Great Britain. Society - For adolescents
TURNER, Dorothy
Leaving home. - Hove : Wayland, 1983
60p : ill. - (In the news). - Bibliography: p57. - Includes index
0-85078-360-7 X.520/33793
3129489

ADOLESCENTS
Hairdressing. Cutting - For adolescents
BRENT, Patricia Jenkins
Haircutting. - New York ; London : F. Watts, 1977
[4],60p : ill. - (A concise guide). - Includes index
0-85166-615-9 X.620/16716 Woolwich
109775x

ADOLESCENTS
Head. Injuries. Therapy
HEAD injury rehabilitation / edited by Mark Ylvisaker. - London : Taylor & Francis, c1985
xxii,463p : ill. - Includes bibliographies and index
0-85066-567-1 SRL
3303861

ADOLESCENTS
Health
KEABLE-ELLIOTT, David
You and your body. - [London] : Beaver, 1984, c1983
188p : ill. - Bibliography: p183-184. - Includes index
0-09-937350-5 X.329/19421
292062x

PHYTOPATHOGENIC prokaryotes / edited by Mark S. Mount, George H. Lacy. - New York ; London : Academic Press.
Vol.1. - 1982
xvii,541p : ill. - Includes index
0-12-509001-3 SRL
292062x

SURVIVAL guide for school-leavers. - [1978]. - [Aberdeen] : [Workers' Educational Association], [1978]
[5],39p : ill
Subsequent volumes, when received, will be placed at this shelfmark
B79-01749 P.523/878
1753053

ADOLESCENTS
Household management - For adolescents
PATERSON, Barbara
Help!. - Harmondsworth : Penguin, 1977
378p : ill. - (Peacock books). - Includes index
0-14-047088-3 X.619/17515 Woolwich
1096524

ADOLESCENTS
Intellectual development
BALKE-AURELL, Gudrun
Changes in ability as related to educational and occupational experience. - Göteborg : Acta Universitatis Gothoburgensis, 1982
203p. - (Göteborg studies in educational sciences ; 40). - Bibliography
91-7346-107-5 X.0519/569(40)
8890897

ADOLESCENTS
Interpersonal relationships
HEMMING, James
Teenage living and loving. - London : British Medical Association, [1976]
30p : ill. - (A 'Family Doctor' booklet)
0-7279-0017-x 7583.n.5/65 Woolwich
0188190

LIMB, Sue
Problem page. - London : Edward Arnold, 1979
63p : ill
0-7131-0256-x X.510/16218
1906429

ADOLESCENTS
Interpersonal relationships - For adolescents
GREEN, Ernest J.
Personal relationships. - New York ; London : McGraw-Hill, 1978
xiii,466p : ill(some col). - Bibliography: p.435-450. - Includes index
0-07-024270-4 X.520/13874
1882341

ADOLESCENTS
Interpersonal relationships with parents
CHUA, Wee Hian
'Dear Mum and Dad -' - Leicester : Inter-Varsity, 1984
159p. - Bibliography: p158-159
0-85110-460-6 X.529/64403
0261718

HEMMING, James
Good Housekeeping you and your adolescent. - London : Ebury Press, 1975
143p. - Bibliography: p.139. - Includes index
0-85223-080-x X.529/30363
0567027

LAWTON, Anthony
Parents and teenagers. - London : Unwin Paperbacks, 1985
xii,155p. - Includes bibliographies
0-04-649038-8 X.529/69982
0567027

The TROUBLE with parents / edited by Tim Stafford ; designed by Joan Nickerson. - London : Pickering and Inglis, 1979
155p : ill
0-7208-0434-5 X.519/30519
0567027

WARREN, Julie
How to handle your parents. - Edinburgh : Macdonald Publishers, c1983
54p : ill
0-904265-99-4 X.520/32848
0567027

ADOLESCENTS
Judaism - For adolescents
RABINOWITZ, E. S.
Ascend the mountain. - London : Anthony Blond.
Vol.1: From Bar/Bat-Mitzvah to pre-marriage. - 1984
xii,353p. - Includes index
0-85634-174-6 X.200/45033
3025411

ADOLESCENTS
Leisure activities - Serials
APOTELESMATICS. - Issue 1-82-. - London : Explorers UNLimited, 1982-
v. : ill
B83-34745 P.901/3381
2947021

ADOLESCENTS
Living alone
INDEPENDENT living / Craig M. Pearson, contributing editor ... [et al.]. - New York ; London : McGraw-Hill, 1980
x,454p : ill(some col). - (The Gregg/McGraw Hill series for independent living). - Includes index
0-07-049061-9 X.520/21542
1571109

ON your own / Craig M. Pearson contributing editor ... [et al.]. - New York ; London : McGraw-Hill, 1978
ix,117p : ill(some col). - (The Gregg/McGraw-Hill series for independent living). - Includes index
0-07-049051-1 X.519/28379
1571109

ADOLESCENTS
Love. Psychosocial aspects - For adolescents
HOLT, Alan
The two of us. - Cambridge : Cambridge University Press, 1976
24p : ill
0-521-21060-7 X.512/5300
0701882

ADOLESCENTS
Man. Contraception - For adolescents
MAYLE, Peter
We're not pregnant. - London : Macmillan, 1981
[56]p : col.ill
0-333-31849-8 Cup.702.tt.15
2566680

ADOLESCENTS
Man. Health - For adolescents
BALDWIN, Dorothy
Healthwise. - London : Harrap, 1979
39p : ill
0-245-53336-2 X.311/5009
1819852

ADOLESCENTS
Man. Hygiene - For adolescents
HOLT, Alan
Come clean. - Cambridge : Cambridge University Press, 1975
24p : col ill
0-521-20900-5 X.313/1227
0506168

ADOLESCENTS
Man. Sex - For adolescents
JOHNSON, Eric Warner
Love and sex in plain language. - Revised ed.. - London : Deutsch, 1975
140p : ill. - Includes index
0-233-96669-2 Cup.351/3
0374628

KAPLAN, Helen Singer
Making sense of sex. - New York ; London : Quartet Books, 1979
xxi,154p : ill. - Includes index
0-7043-2202-1 Cup.351/255
0374628

KAY, Eleanor
Sex and the young teen-ager. - London : F. Watts, 1978
[5],66p : ill. - (A concise guide). - Includes index
0-85166-662-0 X.320/11078 Woolwich
0374628

SEX for modern teenagers / by I. S. Edwards...[et al.] ; art work by M. Lopert. - Adelaide : Rigby, 1975
[2],77p : ill. - (Seal books)
0-85179-013-5 X.519/26915
0374628

STOPPARD, Miriam
Talking sex. - London : Piccolo, 1982
126p : ill. - Includes index
0-330-26752-3 Cup.366.t.55
0374628

ADOLESCENTS
Man. Skin. Acne - For adolescents
HOEHN, Gustave H.
Acne can be cured. - New York : Arco, 1978
144p
0-668-04153-6 X.319/19285 Woolwich
1352482

MURRAY, David, 1943-
The anti-acne book. - London : Arlington, 1981
64p : ill. - (Arlington pocket books)
0-85140-282-8 X.329/13892 Woolwich
1352482

NOURSE, Alan Edward
Clear skin, healthy skin. - [New] ed. - New York ; London : F. Watts, 1978
[7],59p : ill. - (A concise guide). - Includes index
0-85166-661-2 X.320/11119 Woolwich
1352482

ADOLESCENTS
Marriage - For adolescents
GREEN, Ernest J.
Personal relationships. - New York ; London : McGraw-Hill, 1978
xiii,466p : ill(some col). - Bibliography: p.435-450. - Includes index
0-07-024270-4 X.520/13874
1882333

ADOLESCENTS
Medical aspects
ADOLESCENT medicine / editors C. Andrew Rigg, Robert B. Shearin. - Miami : Symposia Specialists, c1980
348,xvp : ill. - (A Home study textbook program). - Includes bibliographies and index
0-8151-7341-5 SRL
2369370

ADOLESCENTS
Medicine - For adolescents
BALDWIN, Dorothy
Know your body. - Harmondsworth : Kestrel Books, 1979
150p : ill
0-7226-5591-6 X.329/12086 Woolwich
2019477

Know your body. - Harmondsworth : Puffin, 1981, c1979
149p : ill. - (Puffin plus)
0-14-031340-0 Cup.358/106
2019477

ADOLESCENTS
Mental disorders
ADOLESCENT psychiatry. - Chicago ; London : University of Chicago Press. - (Annals / American Society for Adolescent Psychiatry)
Vol.6. - 1978
x,521p : ill. - Includes bibliographies and index
0-226-24051-7 X.520/14057
0262617

FRIEDRICH, Max H.
Adoleszentenpsychosen. - Basel ; London : Karger, c1983
xi,141p : ill. - (Bibliotheca psychiatrica ; no.163). - Bibliography: p116-125. - Includes index
3-8055-3640-2 X.322/12799
3039153

ADOLESCENTS
Mental disorders–continued
IRWIN, Edna M.
Growing pains. – Plymouth : Macdonald and Evans, 1977
x,310p. – (Psychiatric topics for community workers). – Includes bibliographies and index
0–7121–0724–x X.319/17701 Woolwich
0262617
Growing pains. – London : Woburn Press, 1978
x,310p : 1 ill. – ([Psychiatric topics for community workers]). – Includes bibliographies and index
0–7130–0166–6 X.329/11404 Woolwich
0262617

ADOLESCENTS
Motorcycles. Riding – For adolescents
OLNEY, Ross R. (Ross Robert)
Motorcycling. – British ed. – London : F. Watts, 1980
[2],46p : ill. – (A concise guide). – Bibliography: p.43. – Includes index
0–85166–749–x X.622/7263 Woolwich
2144794

ADOLESCENTS
Nutrition. Health aspects
ADOLESCENT nutrition / edited by Myron Winick. – New York : Chichester : Wiley, c1982
viii,188p : ill. – (Current concepts in nutrition ; v.11). – Includes index
0–471–86543–5 SRL
2766558
MAHAN, L. Kathleen
Nutrition in adolescence. – St. Louis : Times Mirror / Mosby, 1984
xv,331p : ill
0–8016–3070–3 X.322/12866
2766558

ADOLESCENTS
Nutrition. Physiology – Conference proceedings
NUTRIENT REQUIREMENTS IN ADOLESCENCE (Conference : 1973 : Airlie House, Va.)
Nutrient Requirements in Adolescence. – Cambridge, Mass. ; London : M.I.T. Press, 1976
xviii,365p : ill. – Includes index
0–262–13119–6 SRL
1123610

ADOLESCENTS
Personal conduct. Ethics
BROWN, Brian
Club chats. – London : Chester House, 1976
35p
0–7150–0062–4 X.529/41420
2408899

ADOLESCENTS
Personal evangelism by Christian church
SMITH, Jim, 1947–
Breakthrough. – London : Pathfinders–CYFA, 1981
40p : ill. – (An Operation Breakthrough booklet). – Bibliography: p39–40
0–907750–00–1 X.200/42433
3056031
How to share Jesus with your friends. – Rev. ed. – London : Pathfinders–CYFA, 1981
[20]p : ill. – (An Operation Breakthrough booklet)
0–907750–01–x X.200/42425
3056031

ADOLESCENTS
Personal problems. Decision making
RAINBOW, Claire
Support group. – Cambridge : Cambridge University Press, 1977
1v. : ill. – (Living well)
0–521–21763–6 Cup.911/58(2)
1396730

ADOLESCENTS
Physical development
FISCHBEIN, Siv
Heredity-environment influences on growth and development during adolescence. – Lund : Gleerup, 1979
52,[132]p. – (Studies in education and psychology ; 4). – Bibliography
91–400–4681–8 X.0519/751(4)
8851956

ADOLESCENTS
Political socialisation. Influence of mental development
MERELMAN, Richard M.
Political reasoning in adolescence. – Beverly Hills ; London : Sage Publications, 1976
39p. – (Sage professional papers in American politics ; vol.3, no.04–026). – Bibliography: p.35–37
0–8039–0538–6 X.709/30697
0724084

ADOLESCENTS
Political socialisation. Influence of moral development
IJZENDOORN, Marinus H. van
Moralität und politisches Bewusstsein. – Weinheim : Beltz, 1980
250p. – Bibliography
3–407–54603–3 X.529/42718
8724776

ADOLESCENTS
Psychiatry
ADOLESCENT psychiatry / edited by Sherman C. Feinstein, Peter L. Giovacchini. – New York : Basic Books. – (Annals of the American Society for Adolescent Psychiatry)
Vol.3: Developmental and clinical studies. – [1974]
xvi,456p
0–465–00063–0 X.320/5015
006274X
ADOLESCENT psychiatry. – New York : Aronson : London : Distributed by Eurospan. – (Annals / American Society for Adolescent Psychiatry)
Vol.5. – 1977
xv,515p. – Includes bibliographies and index
0–87668–258–1 X.320/11849 Woolwich
1939343
ADOLESCENT psychiatry. – Chicago ; London : University of Chicago Press. – (Annals / American Society for Adolescent Psychiatry)
Vol.7. – 1979
x,571p. – Includes bibliographies and index
0–226–24052–5 X.320/12064 Woolwich
1939343
– New York : Aronson ; London : Distributed by Eurospan. – (Annals / American Society for Adolescent Psychiatry)
Vol.14. – 1976
xiii,418p. – Includes bibliographies and index
B79–30122 X.320/11848 Woolwich
1939343
CHILD and adolescent psychiatry / edited by Michael Rutter, Lionel Hersov. – 2nd ed. – Oxford : Blackwell Scientific, 1985
xix,960p : ill. – Includes bibliographies and index
0–632–01228–5 X.525/10570
006274x
EVANS, John, 1926–
Adolescent and pre-adolescent psychiatry. – London : Academic Press, 1982
xix,448p. – Includes bibliographies and index
0–12–791168–5 X.322/11795
006274x
MILLER, Derek
The age between. – New York ; London : Aronson, c1983
xiv,441p. – Includes bibliographies and index
0–87668–639–0 X.520/35364
006274x
STEINBERG, Derek
The clinical psychiatry of adolescence. – Chichester : Wiley, c1983
x,412p : ill. – (Wiley series on studies in child psychiatry). – Bibliography: p353–395. – Includes index
0–471–10314–4 X.322/12689
006274x

ADOLESCENTS
Psychiatry – Serials
ADOLESCENT psychiatry. – Vol.8. – Chicago ; London : University of Chicago Press, 1980
x,567p
0–226–24053–3 X.520/24557
1939343

ADOLESCENTS
Psychological development
AUSUBEL, David Paul
Theory and problems of adolescent development. – 2nd ed.. – New York ; London : Grune and Stratton, 1977
xvii,572p : ill. – Bibliography: p.495–541. – Includes index
0–8089–1031–0 X.520/12389
1409735
CAMPHILL working papers on adolescence / compiled and edited by Michael Luxford. – Botton Village : Camphill Press, 1983
49p : ill
0–904145–34–4 X.529/67115
1409735
GROPPELLI, Angela
Materie espressive e maturazione psicologica dell'alunno nella scuola media. – Roma : Bulzoni, [1977]
117p.,[16] fold. leaves of plates. – (Università di Roma, Facoltà di magistero, Cattedra di psicologia scolastica. Quaderni di psicologia scolastica ; 1). – Includes bibliographies
r6100007 X.510/11461
1409735
HAVILAND, Jeannette M.
Adolescent development in contemporary society. – New York ; London : Van Nostrand, c1981
xvii,337p : ill. – Includes bibliographies and index
0–442–25862–3 X.520/24164
1409735
MCCANDLESS, Boyd Rowden
Adolescents. – 2nd ed.. – New York ; London : Holt, Rinehart and Winston, 1979
xvi,494p : ill. – Includes bibliographies and index
0–03–021456–4 X.520/14023
1409735
MEDINNUS, Gene Roland
Child and adolescent psychology. – 2nd ed. – New York ; London : Wiley, 1976
ix,553p : ill. – Includes bibliographies and index
0–471–59022–3 X.520/13195
1409735
The PSYCHOLOGY of adolescence / Aaron H. Esman, editor. – New York : International Universities Press, 1975
x, 564 p.. – Bibliography : p. 535–550. – Includes index
0–8236–5565–2 X.529/30990
0740330

ADOLESCENTS
Psychological development. Social factors
CONGER, John Janeway
Adolescence and youth. – 2nd ed. – New York ; London : Harper and Row, 1977
xvii,670p : ill. – Includes bibliographies and index
0–06–041362–x X.510/15465
1114220
Adolescence and youth. – 3rd ed.. – New York ; London : Harper & Row, c1984
xix,732p : ill. – Includes bibliographies and index
0–06–041357–3 X.520/34278
1114220

ADOLESCENTS
Psychology
ASPIŌTĒS, Ar. A.
Apo ta psychologika provlēmata tou ephēvou. – 3e ekdosis. – Athēnai : Institouton psychologias kai psychikēs hygieinēs, 1976–1977
2v. – (Psychologia kai zōē / Institouton psychologias kai psychikēs hygieinēs). – Includes bibliographies
r6501659 X.529/43083
8724768
Apo ta psychologika provlēmata tou koritsiou. – 3ē ekdosē. – Athēnai : Institouton psychologias kai psychikēs hygieinēs, 1978–1979
2v. – (Psychologia kai zōē / Institouton psychologias kai psychikēs hygieinēs). – Includes bibliographies
r6501657 X.529/43083
8724768
The EXPERIENCE of young adults / [edited] by Stephen J. Goldburgh. – Revised ed. – Cambridge, Mass. : Schenkman, [1976]
iii–xii,108,[1]p
0–87073–431–8 X.519/29686
0041068
FADELY, Jack L.
Confrontation in adolescence. – St. Louis ; London : C.V. Mosby, 1979
xiii,147p : ill. – Includes bibliographies and index
0–8016–1553–4 X.520/27825
0041068
– [Another copy] X.520/26157
GUARDO, Carol Joan
The adolescent as individual. – New York ; London : Harper and Row, 1975
xiii,321p : ill. – Includes bibliographies
0–06–042228–9 X.510/15018
035998x
HANDBOOK of adolescent psychology / edited by Joseph Adelson. – New York ; Chichester : Wiley, 1980
xv,624p : ill. – (Wiley series on personality processes). – Includes bibliographies and index
0–471–03793–1 X.525/3233
2154080
HENDIN, Herbert
The age of sensation. – New York : Norton, 1975
354p
0–393–01122–4 X.529/22129
0041068
HORROCKS, John Edwin
The psychology of adolescence. – 4th ed. – Boston [Mass.] ; London : Houghton Mifflin, 1976
xv,574p : ill. – Includes bibliographies and index
0–395–21918–3 X.520/12371
0041068
JERSILD, Arthur Thomas
The psychology of adolescence. – 3rd ed.. – New York : Macmillan ; London : Collier Macmillan, 1978
xv,616p : ill. – Includes bibliographies and index
0–02–360610–x X.520/12629
0041068
– [Another copy: paperback]
ISBN 0–02–979230–4 X.510/16240

ADOLESCENTS
Psychology–continued

LÓPEZ IBOR, Juan José
Alienación y nenúfares amarillos. – Barcelona : Dopesa, 1976
222p. – (Testimonio de actualidad ; n.25)
84–7235–255–2 X.519/22418
0041068

MANASTER, Guy J.
Adolescent development and the life tasks. – Boston [Mass.] ; London : Allyn and Bacon, 1977
xiii,337p : ill. – Bibliography: p.301–325. – Includes index
0–205–05547–8 X.520/11452
0041068

MITCHELL, John J.
The adolescent predicament. – Toronto : Holt, Rinehart and Winston of Canada, 1975
vii, 263 p.. – Bibliography : p. 247–251. – Includes index
0–03–928029–2 X.519/26699
0041068

Adolescent psychology. – Toronto : Holt, Rinehart and Winston of Canada, 1979
viii,247p : ill. – Bibliography : p.235–241.– Includes index
0–03–920018–3 X.510/16284
0041068

MUUSS, Rolf E.
Grundlagen der Adoleszentenpsychologie. – Lübeck : Hansisches Verlagskontor, 1982
132p : ill. – (Documenta pädiatrica ; Bd 9)
r6308615 X.529/61009
8724768

OFFER, Daniel
The adolescent. – New York : Basic Books, c1981
x,195p. – Bibliography: p175–183
0–465–00054–1 X.529/62985
0041068

READINGS in adolescent psychology / edited by Thomas J. Cottle. – New York ; London : Harper and Row, 1977
[10],338p : ill. – (Contemporary perspectives reader series). – Includes bibliographies
0–06–047057–7 X.512/5683
035998x

UNDERSTANDING adolescence / edited by James F. Adams. – 3rd ed. – Boston, Mass. ; London : Allyn and Bacon, 1976
x,563p. – Includes bibliographies and index
0–205–05410–2 X.520/10415
0041068

UNDERSTANDING adolescence / [edited by] James F. Adams. – 4th ed. – Boston, Mass. ; London : Allyn and Bacon, 1980
xii,451p : ill. – Bibliography – Includes index
0–205–06931–2 X.520/20873
0041068

ADOLESCENTS
Psychology – For teaching

SPICER, Faith
Adolescence and stress. – London : Forbes Publications, 1977
v,97p
0–901762–27–x X.519/27606
1335693

ADOLESCENTS
Psychopathology

DELINQUENCY and adolescent psychopathology / edited by Scott W. Henggeler. – Boston, [Mass.] ; Bristol : John Wright, 1982
xi,259p. – Includes bibliographies and index
0–7236–7041–2 X.520/29134
2864185

ADOLESCENTS
Psychoses. Psychotherapy – Texts

HAYDEN, Torey L.
Murphy's boy. – London : Gollancz, 1983
324p
0–575–03287–1 Nov.49616
2982285

ADOLESCENTS
Psychotherapy

HANDBOOK of treatment of mental disorders in childhood and adolescence / Benjamin B. Wolman, editor ; James Egan, co-editor, Alan O. Ross, co-editor. – Englewood Cliffs ; London : Prentice-Hall, 1978
xv,475p : ill. – Includes bibliographies and index
0–13–382234–6 X.322/8597 Woolwich
0103667

THERAPIES for adolescents / [edited by] Michael D. Stein, J. Kent Davis. – San Francisco ; London : Jossey-Bass, 1982
xxv,393p. – (Guidebooks for therapeutic practice) (The Jossey-Bass social and behavioral science series). – Includes index
0–87589–513–1 X.520/29190
0103667

ADOLESCENTS
Role in society – Conference proceedings

INTERNATIONAL COLLOQUIUM ON ADOLESCENCE IN THE YEAR 2000 (1975 : Amsterdam)
Adolescence and youth in prospect. – Guildford : IPC Science and Technology Press, 1977
viii,217p : ill. – Includes bibliographies
0–902852–66–3 X.529/31417
1103962

INTERNATIONAL RESEARCH CONFERENCE ON ADOLESCENCE (1973 : Oshkosh)
Youth in a changing world. – The Hague : Mouton, 1976
xiii, 346 p. leaf of plate, [4] p. of plates : ill. – (World anthropology). – Includes bibliographies and index
90–279–7779–8 X.520/11344
1038850

ADOLESCENTS
Separation from parents. Psychological aspects

BLOOM, Michael W.
Adolescent–parental separation. – New York : Gardner, c1980
177p. – Bibliography: p169–173. – Includes index
0–470–26739–9 X.520/23127
2366916

ADOLESCENTS
Sex relations

BARNETT, Leonard
Sex and young lovers. – Redhill : National Christian Education Council, 1979
96p
0–7197–0234–8 Cup.366.h.23
0745448

HUGGETT, Joyce
Just good friends?. – Leicester : Inter-Varsity Press, 1985
204p : ill
0–85110–465–7 X.529/69391
0745448

MATTHEWS, Simon
Boy, girl relationships. – St Austell : Good News Crusade, 1976
55p
0–903437–08–2 X.108/16432
0745448

MAYLE, Peter
'Will I like it?'. – London : Star Books, 1978
[78]p : ill(some col)
0–352–30270–4 Cup.351/193
1491210

MAYLE, Peter
'Will I like it?'. – London : W.H. Allen, 1978
[78]p : ill(some col)
0–491–02423–1 Cup.355/17
1491210

RESPONSIBLE SOCIETY
Dear parents .... – Hilton : The Society, [1978?]
12p
B78–24006 X.519/28221
1499068

ADOLESCENTS
Sex relations – For adolescents

CLAXTON, Ernest
Tomorrow's parents. – London : Grosvenor Books, 1977
[1],40,[1]p : ill. – Bibliography: p.40
0–901269–26–3 X.519/26958
1132849

COUSINS, Jane
Make it happy. – 2nd ed. – Harmondsworth : Penguin, 1980
[7],128p : ill. – Bibliography: p.121–123. – Includes index
0–14–005427–8 Cup.358/79
1132849

HAMILTON, Eleanor
Sex with love. – Boston [Mass.] : Beacon Press, 1978
xi,179p. – Bibliography : p.168–173.– Includes index
0–8070–2580–1 Cup.351/197
1132849

KEHLE, Mary
You're nearly there. – London : Pickering and Inglis, [1980?]
79p : ill
0–7208–0453–1 Cup.358/82
1948490

SEX for modern teenagers / by I. S. Edwards...[et al.] ; art work by M. Lopert. – Adelaide : Rigby, 1975
[2],77p : ill. – (Seal books)
0–85179–013–5 X.519/26915
1118390

STAFFORD, Tim
A love story. – London : Lakeland, 1978
127p
0–551–00793–1 Cup.358/56
1948490

The TUG of love. – Wrexham : [National Young Life Campaign], [1984?]
11p : ill
B84–42162 X.205/1581
1948490

VÆRE sammen / [av] Kirsti Bergem...[et al.]. – [Oslo] : Pax, 1977
79p : ill
82–530–0826–0 Cup.364.c.125
1132849

ADOLESCENTS
Sex relations – For parents

PARENTS, teenagers & sex / editor: John Robson. – [Sydney] : Family Life Movement of Australia, 1979
42p : ill(some col.). – (A Family life guide book)
0–909922–92–6 X.529/56248
2945703

ADOLESCENTS
Skin. Diseases

SOLOMON, Lawrence Marvin
Adolescent dermatology. – Philadelphia ; London : Saunders, 1978
xii,485p,xii p of plates : ill(some col). – (Major problems in clinical pediatrics ; vol.19). – Includes index
0–7216–8492–0 SRL
1690167

ADOLESCENTS
Social development

STUDIES in adolescence / selected and edited by Robert E. Grinder. – 3rd ed. – New York : Macmillan ; London : Collier Macmillan, 1975
xiii,576p : ill. – Includes index
0–02–347300–2 X.510/15047
050503x

ADOLESCENTS
Social development – For youth work

COCKRAM, Linda
Rehearsing to be adults. – Leicester : National Youth Bureau, 1978
[7],94p. – Bibliography: p.83–94
0–902095–68–4 X.519/28083
1462660

ADOLESCENTS
Social development. Evaluation. Methodology

BONKE, Paul
Opvoedingsproblemen als waarderingskonflikten. – Lisse : Swets & Zeitlinger, 1984
xii,236p. – (Sociaal-wetenschappelijke proefschriften relevant voor de hulpverlening). – Includes bibliography
90–265–0534–5 X.520/33786
9090495

ADOLESCENTS
Social education

FINNEGAN, Gerry
From school to work–. – Belfast : Standing Conference of Youth Organisations in Northern Ireland, 1980
84p : ill
0–906797–05–5 X.529/39770
2360330

ADOLESCENTS
Social education. Teaching methods

SOCIAL education. – Edinburgh : Scottish Community Education Centre, [1981]
58p : ill. – Includes bibliographies
B81–40006 X.525/5048
2572532

ADOLESCENTS
Social values – For adolescents

THAT'S life – is it? / [editor Brenda Lealman]. – London : Christian Education Movement, c1983
41p : ill
0–905022–89–0 X.205/1517
3139948

ADOLESCENTS
Socialisation

TALLMAN, Irving
Adolescent socialization in cross-cultural perspective. – New York ; London : Academic Press, 1983
xvii,325p : ill. – (Quantitative studies in social relations). – Bibliography: p297–313. – Includes index
0–12–683180–7 X.520/31947
2990342

ADOLESCENTS
Spine. Diseases
KEIM, Hugo Adam
The adolescent spine. - New York ; London : Grune and Stratton, 1976
xv,219p : ill. - (Modern orthopedic monographs). - Includes bibliographies and index
0-8089-0923-1 SRL
0623482

ADOLESCENTS
Suicide, to date
RABKIN, Brenda
Growing up dead. - Toronto : McClelland and Stewart, [1978]
188p
0-7710-7230-9 X.529/56509
876882x

Growing up dead. - Nashville : Abingdon, 1979
188p
0-687-15970-9 X.529/43821
876882x

ADOLESCENTS
United States. Home economics - For adolescents
BRINKLEY, Jeanne Hayden
Teen guide to homemaking. - 4th ed.. - New York ; London : Webster Division, McGraw-Hill, 1977
528p : ill(some col). - Includes index
0-07-007840-8 X.629/11019 Woolwich
0797790

ADOLESCENTS
Visual arts. Techniques - For adolescents
KEIGHTLEY, Moy
Investigating art. - London : Elek, 1976
160p,[4]p of plates : ill(some col). - Includes index
0-236-31143-3 X.429/10079
0601683

ADOLESCENTS
Weight. Control
BENZIGER, Barbara
Controlling your weight. - New York ; London : F. Watts, 1978
[7],61p : ill. - (A concise guide). - Includes index
0-85166-663-9 X.320/11117 Woolwich
1351958

ADOLESCENTS. Audiences. Television programmes. France
Viewing behaviour, 1958-
PIERRE, Evelyne
Les nouveaux téléspectateurs de 9 à 18 ans. - Paris : La Documentation française, [1982]
224p. - (Audiovisuel et communication)
2-11-000814-8 S.E.400/1055
8942579

ADOLESCENTS. Audiences. Television programmes. West Germany
Viewing behaviour, 1945-
HORN, Imme
Jugend und Fernsehen. - 3., erg. Aufl. - Mainz : ZDF, 1978
64p. - (ZDF Schriftenreihe ; Hft.14. Medienforschung)
r6300882 X.0629/565(14) Woolwich
8756090

ADOLESCENTS. Brazil
Development. Influence of socioeconomic status of families
ULHOA, Maria José Carneiro
Características do comportamento do adolescente brasileiro. - Petrópolis : Vozes, 1974
135 p. : ill. - Bibliography : p. 131-135
r6003541 X.519/20411
093593X

ADOLESCENTS. Bristol. Avon
Social conditions
FRED'S people. - [Bristol] : [Bristol Broadsides (Co-op) Ltd], [c1980]
44p : ill. - (A Bristol Broadsides booklet)
0-906944-08-2 X.529/50598
2807718

ADOLESCENTS. Calgary. Alberta
Sex relations
MEIKLE, Stewart
Teenage sexuality. - London : Taylor & Francis, c1985
171p : ill. - Bibliography: p141-146. - Includes index
0-85066-505-1 X.520/37393
3207196

ADOLESCENTS. Canada
Social development
CHILDHOOD and adolescence in Canada / edited and with an introduction by K. Ishwaran. - Toronto ; London : McGraw-Hill, 1979
[9],396p : ill. - (Interdisciplinary studies in family and marriage). - Includes bibliographies
0-07-082935-7 X.529/36390
2065320

ADOLESCENTS. Czechoslovakia
Behaviour
MAJOROVÁ, Milena
Dámy a pánové, život začíná. - Praha : Mladá fronta, 1980
200p : ill. - (Klub mladých 13 ; sv.126)
r6439643 X.529/42917
8749922

ADOLESCENTS. Denmark
Social development, 1965-1973
FABER, Mogens
Ungdomsoprøret i Danmark 1965-1973. - København : Danmarks Biblioteksskole, 1976
87p. - (Studier fra Danmarks Biblioteksskole ; 22)
87-7415-037-5 2714.abc.16/22
1852981

ADOLESCENTS. East & West Sussex
- Serials
MARTLET. - Feb. 1976-. - [Brighton] : National Federation of 18 Plus Groups, Sussex Region, [1976]-
r6031563 P.801/3065
108223X

ADOLESCENTS. England
Residential care
JACKIE, 19—
Jackie's story. - [London] : [Centerprise], [c1984]
126p : ill
0-903738-63-5 X.529/68937
327246x

WOOD, Kay
Towards independence. - London : Church of England Children's Society, [1982?]
10p. - (Child care study paper ; no.4)
B83-19666 X.525/7108
2886537

ADOLESCENTS. England
Sexual intercourse. Age of consent. Law. Reform
GREAT BRITAIN. Policy Advisory Committee on Sexual Offences
Report on the age of consent in relation to sexual offences. - London : H.M.S.O., 1981
v,32p. - (Cmnd. ; 8216)
0-10-182160-3 BS.18/690
2411180
Working paper on the age of consent in relation to sexual offences. - London : H.M.S.O., 1979
[2],v,31p
0-11-340204-x BS.18/646
199705x

ADOLESCENTS. Europe
Sex education
KOZAKIEWICZ, Mikołaj
Sex education and adolescence in Europe. - London : [IPPF Europe], 1981
118p : 1ill. - Bibliography: p107-118
0-904983-07-2 Cup.351/286
2526492

ADOLESCENTS. Europe
Sexuality
KOZAKIEWICZ, Mikołaj
Sex education and adolescence in Europe. - London : [IPPF Europe], 1981
118p : 1ill. - Bibliography: p107-118
0-904983-07-2 Cup.351/286
2526506

ADOLESCENTS. Finland
Behaviour, to date
AALTONEN, Ulla-Maija
Murkun tapatieto. - 2. painos. - Helsingissä : Otava, 1980
140p : ill
951-10-5503-8 X.529/56187
899367x

ADOLESCENTS. Finland
Social development, to date
PULKKINEN, Aimo
Asumistavoitteiden valinnat mahdollisuuksien rajaamina. - Helsinki : Väestöntutkimuslaitos, 1974
138,[10],leaves,[10]p. - (Väestöntutkimuslaitoksen julkaisuja. Sarja B ; n:o 34)
951-90481-3-8 X.0512/156b(34)
8945152

ADOLESCENTS. France
Social conditions
CHILDHOOD and youth in France. - Aldershot : Associated Examining Board for the General Certificate of Education, 1982
55p : ill
B83-13765 X.525/6792
286360x

ADOLESCENTS. France
Suicide. Role of oriental religious cults
IKOR, Roger
Je porte plainte. - Paris : Albin Michel, 1981
113p,2leaves of plates
2-226-01094-7 X.529/43147
872475x

ADOLESCENTS. Gravesham (District). Kent
Leisure activities
GRAVESHAM youth & community service information handbook. - London : Pyramid Press, 1975
7-19p : ill
0-902824-21-x X.519/25307
0524859

ADOLESCENTS. Great Britain
Asian adolescents. Interpersonal relationships with parents
GREAT BRITAIN. Commission for Racial Equality
Between two cultures. - 2nd ed.. - London : The Commission, 1978
73p
B80-20248 BS.400/43
2130092

ADOLESCENTS. Great Britain
Attitudes
FISHER, Susie
Too much too young?. - London : Pan, 1981
216p
0-330-26480-x X.529/41579
2488396

SIMMONS, Cyril
I like to say what I think. - London : Kogan Page, 1984
224p. - Includes index
0-85038-702-7 X.529/69198
2488396

ADOLESCENTS. Great Britain
Counselling
NOONAN, Ellen
Counselling young people. - London : Methuen, 1983
xi,157p. - Bibliography: p153-154. - Includes index
0-416-36210-9 X.529/60648
2206080
- [Another Copy: paperback.] ISBN 0-416-36220-6 X.529/60649

WYLIE, Tom
Counselling young people. - Leicester : National Youth Bureau, 1980
7p : ill. - Bibliography: p.7
0-86155-032-3 X.525/3511
2206080

ADOLESCENTS. Great Britain
Effects of permissiveness
GREENING, G.
Youth in the permissive society. - Waterford : Uni books, 1979
140p. - Bibliography: p.131-134. - Includes index
0-85606-091-7 X.529/37680
0347019
- [Another copy.]
With Epilogue (January 1980) pasted in. - Signed by the author X.529/69655

ADOLESCENTS. Great Britain
Employment. Role of community work
BALL, Colin, 1943-
Community service and the young unemployed. - Leicester : National Youth Bureau [for] the Office of the Manpower Services Commission, 1977
24p : ill
0-902095-28-5 X.512/6048
1548247

ADOLESCENTS. Great Britain
Homosexual adolescents. Social conditions
BREAKING the silence / edited by Michael Burbidge and Jonathan Walters. - London : Joint Council for Gay Teenagers, c1981
58p
0-9506958-1-5 X.529/60531
3077233

ADOLESCENTS. Great Britain
Leisure activities

HENDRY, L. B. (Leo Brough)
Growing up and going out. - Aberdeen : Aberdeen University Press, 1983
xii,212p : ill. - Bibliography: p190-209. - Includes index
0-08-025768-2 X.529/66139
319910x

ADOLESCENTS. Great Britain
Occupations

ASHTON, D. N.
Young workers. - London : Hutchinson, 1976
192p : ill. - (Industry in action). - Bibliography: p.175-184. - Includes index
0-09-127370-6 X.529/31235
1044435

ADOLESCENTS. Great Britain
Participation in playgroups

FAWCETT, Mary
Helpers, students or playmates?. - Cambridge : Pre-school Playgroups Association, Eastern Region, c1984
68p : ill. - Bibliography: p66-67
0-901755-48-6 X.529/68758
3059847

WELCOMING teenagers into PPA groups / Susan Monson. - London : Pre-school Playgroups Association, [1983?]
23p : ill
0-901755-38-9 X.329/17438
3059847

ADOLESCENTS. Great Britain
Protection against religious cults - Serials

PEOPLES ORGANISED WORKGROUP ON ERSATZ RELIGIONS
[Miscellaneous publications not catalogued separately]. - London : P.O.W.E.R., [1976]-
r6009767 X.0102/501
070881X

ADOLESCENTS. Great Britain
Social conditions - Serials

YOUTH and policy. - Vol.1, no.1 (summer 1982)-. - Newcastle-upon-Tyne : Youth and Policy, 1982-
v. : ill
B83-35905 P.523/1873
2946548

ADOLESCENTS. Great Britain
Social education - For teaching

ELLIS, Jane, 19—
Life skills training manual. - Rev., enl. ed. - London : Community Service Volunteers, 1983
1portfolio : ill
0-907829-15-5 Cup.918/119
2951541

ADOLESCENTS. Great Britain
Social education. Organisations: National Youth Bureau

COCKERILL, G. F. (Geoffrey Fairfax)
National youth bureau. - [London] : [H.M.S.O.], 1983
165p
B84-39769 X.525/9899
3174824

ADOLESCENTS. Great Britain
Social values

KITWOOD, T. M. (Thomas Marris)
Disclosures to a stranger. - London : Routledge and Kegan Paul, 1980
[7],291p : ill. - (Social worlds of childhood). - Bibliography:p.285-288. - Includes index
0-7100-0463-x X.529/38460
1195816

ADOLESCENTS. Great Britain
Unemployment - Bibliographies

YOUTH unemployment and special measures. - Rev. and updated. - Sheffield : Manpower Services Commission, Training Division, 1983
56p. - (Special programmes research and development series ; no.1). - Includes index
0-905932-72-2 BS.421/56[no.1]
2941902

ADOLESCENTS. Great Britain
Vacation employment. Organisations

A YEAR off. - Completely rev. - [London] : CRAC, c1983
64p. - Includes index
0-86021-488-5 X.529/56770
0037397

ADOLESCENTS. Harlesden. Brent (London Borough)
Residential care, 1974-1976

HARLESDEN COMMUNITY PROJECT
Caring for children. - Ilkley : Owen Wells Publishing Co., 1979
[2],157p
0-904553-08-6 X.529/38331
2316390

ADOLESCENTS. India (Republic)
Personality

SHANMUGAM, T. E.
Adolescent personality. - Madras : University of Madras, 1975
xii,202p. - Bibliography: p.170-174. - Includes index
r6031872 X.520/11625
1109154

ADOLESCENTS. Ireland (Republic)
Drug abuse

MACKEN, Ultan
Drug abuse in Ireland. - Cork : Mercier Press Ltd, 1975
[1],132p. - Bibliography: p.132
0-85342-343-1 X.319/18258 Woolwich
1524372

ADOLESCENTS. Ireland (Republic)
Further education

NATIONAL YOUTH COUNCIL OF IRELAND
Second chance education. - Dublin : The Council, 1977
[2],32p. - Bibliography: p.30
B78-21002 X.519/28136
1482866

ADOLESCENTS. Israel
Sexual behaviour

ANTONOVSKY, Helen F. (Helen Faigin)
Adolescent sexuality. - Lexington, Mass. : Lexington Books, c1980
xii,162p. - Bibliography: p153-159. - Includes index
0-669-04030-4 X.520/24751
2545136

ADOLESCENTS. Leeds. West Yorkshire (Metropolitan County)
Activity spaces

LEACH, Bridget
Youth and spatial poverty. - Bristol : S.S.R.C. Research Unit on Ethnic Relations, 1978
[4],78p. - (Working papers on ethnic relations / Great Britain. Research Unit on Ethnic Relations ; no.9). - Bibliography: p.76-78
B80-21431 X.525/3215
2068451

ADOLESCENTS. Pacific region
Social development, to date

PACIFIC youth. - Suva, Fiji : Institute of Pacific Studies, c1981
104p : ill
r6012938 X.529/59062
9066527

ADOLESCENTS. Road accident victims
Severe injuries. Recovery

MILLER, Barbara
[Kathy]. All I need is time. - Eastbourne : Kingsway, 1981, c1980
126p
0-86065-106-1 X.958/2848
240589x

ADOLESCENTS. Rural regions. Japan
Peer groups. Role in socialisation

JOHNSON, Thomas Wayne
Shonendan. - Taipei : Chinese Association for Folklore, 1975
[3],x,291p : ill. - (Asian folklore and social life monographs ; vol. 68). - Bibliography : p.278-291
r6031625 X.529/31357
1081365

ADOLESCENTS. Scotland
Education - Conference proceedings

14-19 in and out of school options in education and society / edited by Cameron Slater. - Glasgow : Scottish Television, 1982
68p : ill
B85-02981 X.525/10170
3212335

ADOLESCENTS. Scotland
Rights

BASIC rights / [produced by Citizens Rights Office]. - 3rd ed. - Edinburgh : Citizens Rights Office, 1983
18p : ill
0-907560-07-5 8299.m.4
2780747

BASIC rights information for young people. - Edinburgh : Edinburgh Citizens Rights Office, [1982?]
[20]p : ill
B82-36899 X.529/49771
2780747

ADOLESCENTS. Soviet Union
Social conditions

OUT into the big wide world / [compiled by Tatyana Yakovleva and Alexei Ivkin] ; [translated from the Russian by Inna Medova]. - Moscow : Progress Publishers, c1980
155p
0-7147-1668-5 X.529/44771
2589184

ADOLESCENTS. Spain
Sexual behaviour

MORALEDA, Mariano
Vida sexual de los adolescentes españoles. - Madrid : San Pío X, 1977
155p : ill
84-7221-089-8 X.519/42036
2097508

ADOLESCENTS. Spain
Social behaviour, 1940-1974

MARTÍN VIGIL, José Luis
Carta abierta a un adolescente. - Madrid : Ediciones 99, [1974]
126 p.. - (Colección "Carta abierta")
84-251-0160-3 X.519/20062
091214X

ADOLESCENTS. Special subjects. Humour in English, 1945- - Anthologies

KILGARRIFF, Michael
Best teenage jokes. - London : Wolfe, 1976
64p. - (The Wolfe mini ha-ha books)
0-7234-0701-0 X.907/25082
0623997

ADOLESCENTS. Stockholm. Sweden
Social development, to date

ANDERSON, Maja
Hur går det för 50-talets Stockholmspojkar?. - Stockholm : Stockholms kommunalförvaltning, 1976
172p. - (Monografier utgivna av Stockholms kommunalförvaltning ; nr 38)
r6610552 Ac.2258.b[no.38]
891690x

ADOLESCENTS. Sweden
Attitudes

STATENS UNGDOMSRÅD
Ungdomens fritid och samhällssyn. - Stockholm : Prisma, 1975
166 p.. - (Prismaserien)
91-518-0964-8 X.519/20077
0913979

ADOLESCENTS. United States

COHEN, Stewart
Instructor's manual to accompany The adolescent. - Boston ; London : Allyn and Bacon, c1981
180p : ill
0-205-07304-2 X.520/23876
1493833

EXPERIENCING adolescents / edited by Richard M. Lerner, Nancy L. Galambos. - New York ; London : Garland, 1984
xi,422p : ill. - (Garland reference library of social science ; v.201). - Includes bibliographies and index
0-8240-9051-9 X.529/69435
1493833

RICE, F. Philip
The adolescent. - 2nd ed. - Boston, Mass. ; London : Allyn and Bacon, 1978
xix,714p : ill. - Includes bibliographies and index
0-205-05988-0 X.520/12637
1493833

The adolescent. - 3rd ed. - Boston [Mass.] ; London : Allyn and Bacon, c1981
xii,596p : ill. - Includes bibliographies and index
0-205-07303-4 X.520/23352
1493833

STEVENS-LONG, Judith
Adolescence and early adulthood. - Palo Alto : Mayfield, c1983
xxi,506p : ill. - Bibliography: p477-491
0-87484-507-6 X.520/31563
1493833

ADOLESCENTS. United States
Attitudes to Jews
ADOLESCENT prejudice / Charles Y. Glock ... [et al.]. - New York ; London : Harper and Row, 1975
xviii,29p : ill. - (Patterns of American prejudice series ; vol.7). - Includes index
0-06-011567-x X.520/13068
1706853

ADOLESCENTS. United States
Interpersonal relationships with families. Influence on juvenile delinquency
PEREZ, Joseph F.
The family roots of adolescent delinquency. - New York ; London : Van Nostrand Reinhold, 1978
xix,231p
0-442-26535-2 X.520/13422
1801015

ADOLESCENTS. United States
Marriage
WEEKS, John R.
Teenage marriages. - Westport, Conn. ; London : Greenwood Press, 1976
xxi,3-171,[1]p : ill. - (Studies in population and urban demography ; no.2). - Bibliography: p.149-166. - Includes index
0-8371-8898-9 X.529/31659
1175785

ADOLESCENTS. United States
Murder by adolescents. Psychosocial aspects
GARDINER, Muriel
The deadly innocents. - London : Hogarth Press, 1977
xxv,190p
0-7012-0426-5 X.529/31453
0657123

ADOLESCENTS. United States
Parenthood
ADOLESCENT parenthood / edited by Max Sugar. - Lancaster : MTP, c1984
xiv,237p. - Includes bibliographies and index
0-85200-626-8 X.520/35355
3198480

ADOLESCENTS. United States
Psychosocial aspects
JESSOR, Richard
Problem behavior and psychosocial development. - New York ; London : Academic Press, 1977
xvii,281p : ill. - Bibliography: p.249-258. - Includes index
0-12-384750-8 X.520/12149
1146998

LIPSITZ, Joan
Growing up forgotten. - Lexington, Mass. : D.C. Heath, 1977
xvi,269p : ill. - Bibliography : p.241-254.-Includes index
0-669-00975-x X.520/12784
1146998

SEBALD, Hans
Adolescence. - 2nd ed. - Englewood Cliffs ; London : Prentice-Hall, 1977
xiv,530p. - Includes index
0-13-008599-5 X.519/27025
1146998
Adolescence. - 3rd ed. - Englewood Cliffs ; London : Prentice-Hall, c1984
xi,336p. - Includes index
0-13-008516-2 X.520/34598
1146998

ADOLESCENTS. United States
Psychosocial aspects - Conference proceedings
PREVENTING adolescent alienation / edited by L. Eugene Arnold. - Lexington, Mass. : Lexington Books, c1983
xiii,146p : ill. - Includes bibliographies and index
0-669-06269-3 X.520/31843
1146998

ADOLESCENTS. United States
Reading behaviour
CARLSEN, G. Robert
Books and the teenage reader. - 2nd revised ed. - New York ; London : Harper and Row, 1980
[9],290p. - Includes bibliographies and index
0-06-010626-3 X.989/88383
2157950

ADOLESCENTS. United States
Sex relations. Legal aspects
RODMAN, Hyman
The sexual rights of ad[illegible]ts. - New York ; Guildford : Columbia Univer[illegible]ess, 1984
xii,183p. - Biblio[illegible]y: p159-171. - Includes index
0-231-04916-[illegible] X.520/35089
3168263

ADOLESCENTS. United States
Social aspects
SCHWARZROCK, Shirley
Coping with series. - Circle Pines : American Guidance Service, 1970-73
24v : ill. - Includes bibliographies
r6032541 X.519/27630
1340271

ADOLESCENTS. United States
Social aspects, ca 1960-1974
IRWIN, John
Scenes. - Beverly Hills ; London : Sage Publications, 1977
236p : ill
0-8039-0823-7 X.519/28581
1687875

ADOLESCENTS. United States
Social development
MORIARTY, Alice E. (Alice Ewell)
Adolescent coping. - New York ; London : Grune and Stratton, 1976
xv,220p. - Includes index
0-8089-0942-8 X.520/11250
0788570

SELTZER, Vivian Center
Adolescent social development. - [Lexington, Mass.] : Lexington Books, 1982
xii,286p. - Bibliography: p255-271. - Includes index
0-669-04511-x X.520/27349
0175692

ADOLESCENTS. United States
Special education
SPECIAL education for adolescents / [edited by] Douglas Cullinan, Michael H. Epstein. - Columbus ; London : Merrill, c1979
vii,424p : ill. - (The Merrill personal perspectives in special education series). - Bibliography: p355-389. - Includes index
0-675-08407-5 X.529/39563
2353288

ADOLESCENTS. United States
Unemployment
The YOUTH labor market problem / edited by Richard B. Freeman and David A. Wise. - Chicago ; London : University of Chicago Press, 1982
ix,555p : ill. - (A National Bureau of Economic Research conference report). - Includes bibliographies and index
0-226-26161-1 X.520/27987
2774925

ADOLESCENTS. Urban regions
Social aspects
GROWING up in cities / edited by Kevin Lynch, from the reports of ... [others]. - Cambridge, Mass. ; London : M.I.T. Press [etc.], 1977
ix,177p : ill. - Bibliography: p.165-166. - Includes index
0-262-12078-x X.529/32295
1455966

ADOLESCENTS. Urban regions. India (Republic)
Employment
NAIK, R. D. (Ramesh Devidas)
The profiles of urban youth seeking employment. - Bombay : Tata Institute of Social Sciences, Unit for Study of the Urban Child and Youth, 1971
113 p.
r6028495 X:510/9830
1026283

ADOLESCENTS. [illegible]a. Australia
Attitudes
POOLE, [illegible]incent E.
Yout[illegible] Melbourne ; London : Routledge & Kegan Paul, 19[illegible]
[illegible]9p : ill. - Bibliography: p309-330. - Includes [illegible]ex
[illegible]-7100-9283-0 X.529/55785
2936976

ADOLESCENTS. Villages. Hungary
Social education, to date
CSÁKY, Csaba
Munkamegosztás a falusi ifjúság nevelésében. - [Budapest] : Ifjúsági Lapkiadó Vállalat, 1975
62 p.
r6007965 X.700/16928
086613x

ADOLESCENTS. West Germany
Political socialisation
OPPENHEIM, Abraham Naftali
Civic education and participation in democracy. - London : Sage Publications, 1977
52p. - (Sage professional papers in contemporary political sociology ; vol.2, no.06-023). - Bibliography: p.50-51
0-8039-9874-0 X.519/29565
1878328

ADOLESCENTS. Westminster (London Borough)
Reading of books in stock of Westminster City Libraries. Promotion. Projects: Bookmaster scheme
BIRD, Jean, 1946-
Young teenage reading habits. - [London] : British Library, c1982
x,118p : ill. - (British National Bibliography research report ; 9)
0-7123-3007-0 X.950/18139
2358824

ROBERTS, Lorna
The 1981 Bookmaster scheme. - [London] : Westminster City Libraries, 1982
66p
0-900802-05-7 X.955/2294
2358824
The 1982 bookmaster scheme. - London : Westminster City Libraries, [1982?]
44p
0-900802-06-5 2725.g.163
2358824

WESTMINSTER CITY LIBRARIES
The 1980 Bookmaster scheme. - [London] : Westminster City Libraries, 1980
61p : ill
0-900802-04-9 2719.i.47
2358824

ADOLESCENTS. Wolverhampton. West Midlands (Metropolitan County)
Leisure activities
SANDHU, R. (Resham)
Youth provision in Wolverhampton. - Wolverhampton : Crypt Association, c1983
x,54leaves. - Bibliography: p54
0-946935-00-9 X.525/9244
3138216

ADOLESCENTS, 11-15 YEARS
Christian religious education
CHILDREN no longer / with illustrations by Charles Bannerman and 11 vignettes by Haro. - London : CIO Publishing for the General Synod Board of Education, 1977
[2],64p : ill. - (Seeds for the growing). - Bibliography: p.53-57. - List of sound discs : p.57-58. - List of films : p.59-60
0-7151-0365-2 X.100/18032
1733532

ADOLESCENTS, 12-15 YEARS
Christian religious education
GO teach, young teens. - Leamington Spa : Go Teach Publications.
1- ; Winter [1978]-. - [1978]-
v. : ill
B78-17983 P.103/524
1442325

ADOLESCENTS, 12-15 YEARS. United States
LIPSITZ, Joan
Growing up forgotten. - New Brunswick ; London : Transaction Books ; [Eastbourne] : Distributed by [Holt-Saunders], 1980
xvi,269p : ill. - Bibliography: p.241-254. - Includes index
0-87855-792-x X.529/36492
2210177

ADOLESCENTS, 12-15 YEARS. United States
Drug abuse
LIGHT, Patricia Kahn
Let the children speak. - Lexington, Mass. ; London : D.C. Heath, 1975
xiii,97p. - Bibliography: p.93-95
0-669-92676-0 X.320/10117
0504696

ADOLESCENTS, 13-19 YEARS. United States
Suicide. Psychological aspects
GIOVACCHINI, Peter L. (Peter Louis)
The urge to die. - New York : Macmillan ; London : Collier Macmillan, c1981
viii,216p. - Bibliography: p193-195
0-02-543440-3 X.529/44414
258199x

ADOLESCENTS, 14-17 YEARS. Kent
Foster care, 1975-1980

HAZEL, Nancy
A bridge to independence. - Oxford : Blackwell, 1981
173p : ill. - (The Practice of social work ; 9). -
Bibliography: p170. - Includes index
0-631-12943-x X.529/41200
2396599

ADOLESCENTS, 14-18 YEARS
Sunday schools. Group discussions. Topics - For adolescents, 14-18 years

PRITCHARD, William Hugh
Y ffydd heddiw. - Caernarfon : Llyfrfa'r Methodistiaid Calfinaidd, 1978
106p
B79-04837 X.100/18151
1772228

ADOLESCENTS, 15-19 YEARS. Great Britain
Community work

ENGLISH SPEAKING BOARD
The case for community service and the young adult. - Southport : The Board, [1976]
[10]p
0-9502789-3-9 X.519/25708
0637262

ADOLESCENTS, 16-18 YEARS. Great Britain
Education

ASSOCIATION OF PRINCIPALS OF COLLEGES
From school life to life at 16+. - Broxbourne : A.P.C., [1980]
[1],6p
B80-39222 X.529/38021
2301040

ON to eighteen. - London : Secondary Heads Association, c1979
101p. - (Topics)
0-906916-00-3 X.529/45733
2621231

ADOLESCENTS, 16-18 YEARS. Great Britain
Education. Provision. Role of government - Conference proceedings

BIRD, R. H. (Richard Herries)
Education and training of 16-18s. - Sheffield : Association of Colleges for Further and Higher Education, [1979]
7,[2]p
B80-41647 X.529/38387
2319381

ADOLESCENTS, 16-18 YEARS. Great Britain
Vocational preparation

VOCATIONAL preparation / Further Education Curriculum Review and Development Unit. - [London] : The Unit, 1981
45p : ill
B84-30718 BS.10/1028
2395967

VOCATIONAL preparation / Further Education Curriculum Review and Development Unit. - Stanmore : Department of Education and Science, 1981
45p : ill
0-85522-090-2 BS.10/908
2395967

ADOLESCENTS, 16-18 YEARS. Great Britain
Vocational preparation - Conference proceedings

GETTING ready for work. - [London] : Department of Education and Science, [1984]
19p
B85-14775 BS.10/1056
2395967

ADOLESCENTS, 16-19 YEARS. ACSET. Theories of Great Britain
Vocational preparation. Great Britain. Manpower Services Commission. Courses. Effects on career development of teachers in further education institutions. ACSET. Teacher training and staff development implications of the expansion of MSC-related programmes

FEU response to ACSET advice 'Teacher training and staff development implications of the expansion of MSC-related programmes'. - London : Further Education Curriculum Review and Development Unit, 1982
[7]p
B83-13761 X.525/6791
2863421

ADOLESCENTS, 16-19 YEARS. England
Education

ASSOCIATION OF COLLEGES FOR FURTHER AND HIGHER EDUCATION
16-19, towards a tertiary system. - Sheffield : The Association, [1980]
20p
B81-17923 X.529/41176
2398311

NATIONAL ASSOCIATION OF TEACHERS IN FURTHER AND HIGHER EDUCATION
Education and training for the 16-19s. - [London] : [National Association of Teachers in Further and Higher Education], 1979
15p
B82-08496 X.529/45386
2398311

ADOLESCENTS, 16-19 YEARS. England
Special further education

BRADLEY, Judy
Stretching the system. - London : Further Education Curriculum Review and Development Unit, 1982
vii,94p. - (A Special needs document). - Bibliography: p93-94
0-85522-113-5 BS.10/978
2923084

ADOLESCENTS, 16-19 YEARS. England
Vocational preparation. Courses

GREAT BRITAIN. Department of Education and Science. HM Inspectorate (England)
Report by HM Inspectors on curricula for non-participant 16-19s. - [London] : Department of Education and Science, 1976
[1],55[i.e.64]p : 1 ill
B78-12677 BS.10/838
1400681

ADOLESCENTS, 16-19 YEARS. England
Vocational preparation. Courses organised by Great Britain. Manpower Services Commission. Implementation. Role of further education institutions

O'BRIEN, Sir Richard, 1920-
MSC's new programmes for the 16-19 age group, and their implications for further education. - [London] : The Association, [1978]
[1],2p
B78-28126 X.519/28325
1545329

ADOLESCENTS, 16-19 YEARS. Great Britain
Education

16-19, education and training / Association of Colleges for Further and Higher Education. - Sheffield : ACGHE, 1981
28p. - Bibliography: p27-28
B81-17810 X.529/41178
239796x

ADOLESCENTS, 16-19 YEARS. Great Britain
Employment. Effects of vocational education

GLEESON, Denis
Youth training and the search for work. - London : Routledge & Kegan Paul, 1983
x,360p : ill. - (Routledge education books). - Bibliography: p337-356. - Includes index
0-7100-9513-9 X.529/60522
3079880

ADOLESCENTS, 16-19 YEARS. Great Britain
Employment. Policies of government

YOUTH unemployment and state intervention / [edited by] Teresa L. Rees and Paul Atkinson. - London : Routledge & Kegan Paul, 1982
x,139p. - (Routledge direct editions). - Bibliography: p128-134. - Includes index
0-7100-9263-6 X.520/29516
2816857

ADOLESCENTS, 16-19 YEARS. Northern Ireland
Employment - For youth work

YOUNG people without work / prepared by the Standing Conference of Youth Organisations in Northern Ireland. - Belfast : SCOYO, 1981
16p : ill
0-906797-08-x X.525/5073
2579006

ADOLESCENTS, 16-21 YEARS. United States
Careers guidance

OTTO, Luther B.
Design for a study of entry into careers. - Lexington, Mass. : Lexington Books, c1981
xvi,233p : ill. - (Entry into careers series ; v.1). - Bibliography: p107-120. - Includes index
0-669-03643-9 X.0520/703(1)
2719207

ADOLESCENTS, 16 YEARS. Scotland

YOUNG Scot. - Edinburgh : Scottish Community Education Centre, [1981]
88p : ill. - Includes index
B81-40020 X.529/44243
2572605

ADOLESCENTS, 17-19 YEARS. Great Britain
Vocational preparation. Courses

ABC in action. - Stanmore : Further Education Curriculum Review and Development Unit, 1981
[150]p : ill
0-85522-099-6 BS.10/921
263225x

ADOLESCENTS, 19 YEARS. United States
Young persons, 9 & 19 years. Aggression

GROWING up to be violent / Monroe M. Lefkowitz ... [et al.]. - New York ; Oxford : Pergamon, 1977
ix,236p : ill. - (Pergamon international library) (Pergamon general psychology series ; vol.66). - Bibliography: p.212-224. - Includes index
0-08-019515-6 X.519/27130
1188399

ADOLESCENTS' BOOKS
Books for adolescents, to 16 years: Books with British imprints - Bibliographies

BARLOW, Richard
Attitudes and adventure. - 5th ed. - Chippenham : Library Association, Public Libraries Group, 1978
60p. - (Readers' guides / Library Association. Public Libraries Group ; no.12). - Includes index
0-85365-970-2 X.972/583
0169056

YORKSHIRE LIBRARIES JOINT SERVICES JUNIOR BOOK EXHIBITION COLLECTION
Teenage reading. - [Wakefield] : The Collection, [1975]
[1],13p
B78-07272 X.909/42142
1341928

ADOLESCENTS' BOOKS
Books with British imprints - Bibliographies

WELSH LIBRARY ASSOCIATION. Welsh Youth Libraries Group
Into books. - [Caernarfon] : The Group, 1978
22p. - Includes index
B79-15778 2719.cc.50
1876724

ADOLESCENTS' BOOKS. United States
- For teaching

YOUNG adult literature in the seventies / edited by Jana Varlejs. - Metuchen ; London : Scarecrow Press ; [Folkestone] : [Distributed by Bailey and Swinfen], 1978
ix,452p. - Includes bibliographies
0-8108-1134-0 X.989/52874
1720864

ADOLESCENTS' BOOKS. United States
Books in English

CARLSEN, G. Robert
Books and the teenage reader. - 2nd revised ed. - New York ; London : Harper and Row, 1980
[9],290p. - Includes bibliographies and index
0-06-010626-3 X.989/88383
2157969

ADOLESCENTS' BOOKS IN SPANISH
- Bibliographies

SCHON, Isabel
Books in Spanish for children and young adults. - Metuchen, N.J. ; London : Scarecrow, 1983
x,162p. - Includes index
0-8108-1620-2 X.950/28877
1816861

Books in Spanish for children and young adults = Libros infantiles y juveniles en español. - Metuchen ; London : Scarecrow Press ; [Folkestone] : [Distributed by Bailey and Swinfen], 1978
xi,153p. - Includes index
0-8108-1176-6 2719.cc.25
1816861

ADOLESCENTS' BROADCASTING SERVICES. Great Britain

BROADCASTING and youth / commissioned by the British Broadcasting Corporation ... [et al.]. - London : Calouste Gulbenkian Foundation, UK and Commonwealth Branch, 1979
[3],138p. - Bibliography: p.135-138
0-903319-14-4 X.529/36078
2147386

YOUNG people & broadcasting. - London : Calouste Gulbenkian Foundation UK and Commonwealth Branch, 1981
32p
0-903319-19-5 X.955/439
2147386

ADOLESCENTS' CHOIRS. United States
Christian missional adolescents' choirs. Organisation

DOWNS, Thomas A.
All together now. - Notre Dame, Ind : Ave Maria Press, 1972
112p : ill. - Bibliography:p.23
r6091757 X.519/29598
1884883

ADOLESCENTS' CLOTHING

CULLEN, Margaret
Good grooming and clothes care. - 2nd ed. - London : Heinemann Educational, 1979
101p : ill. - (Living today ; book 3)
0-435-42211-1 X.0519/1072(3)
2019868

ADOLESCENTS DEPRIVED OF PERSONAL CONTACT
Language skills. Acquisition

CURTISS, Susan
Genie. - New York ; London : Academic Press, 1977
xvi,288p : ill. - (Perspectives in neurolinguistics and psycholinguistics). - Bibliography: p.277-283. - Includes index
0-12-196350-0 X.981/20986
1377353

ADOLESCENTS LEAVING CARE
Great Britain. Society - For adolescents leaving care

NELSON, John, 1939-
Out of care - what now?. - London : Church of England Children's Society, 1980
31p : ill. - Bibliography: p30
0-907324-11-8 X.529/51585
2869721

ADOLESCENTS LEAVING CARE. Great Britain
Vocational preparation

BURGESS, Charles
In care and into work. - London : Tavistock, 1981
xiii,146p. - (Residential social work). - Bibliography: p134-140. - Includes index
0-422-77640-8 X.529/67847
3249506

ADOLESCENTS LEAVING RESIDENTIAL CARE. Oxfordshire
Social problems

PORTER, Roger
Teenagers leaving care. - Norwich : University of East Anglia, 1984
34 leaves. - (Social work monographs ; 24)
0-946751-04-8 X.525/10846
3279820

ADOLESCENTS' SERVICES. Eucharist. Catholic Church

A LITURGY of life / David Konstant in collaboration with ... [others]. - Great Wakering : Mayhew-McCrimmon, 1975
133p. - Bibliography: p.133
0-85597-086-3 X.100/16726
1035452

WARD, Shirley
Ten masses for teenagers. - Pinner : Grail Publications, 1978
42p : ill
0-901829-44-7 X.102/2202
1035452

ADOLESCENTS, TO 18 YEARS. Ireland (Republic)
Consumption of alcoholic drinks. Welfare aspects

TEENAGE DRINKING COMMITTEE
Teenage drinking. - Dublin : National Youth Council of Ireland, 1976
[5],27p. - Bibliography: p.26-27
B77-20153 X.519/26976
1137328

ADOLF FREDRIKS KYRKA
1774-1974

ANKER, Bo
Ett litet stycke Stockholm. - Stockholm : Adolf Fredriks Församling, 1974
112 p. : ill
91-7080-261-0 X.800/11378
0932728

ADOLPHE. Constant, Benjamin. Fiction in French

DELBOUILLE, Paul
Genèse, structure et destin d'"Adolphe". - Paris : Société d'Édition "Les Belles Lettres", 1971
643 p.. - (Bibliothèque de la Faculté de philosophie et lettres de l'Université de Liège ; fasc. 195). - Bibliography: p. 603-629
r6012422 Ac.990.b [fasc. 195]
0011312

MERCKEN-SPAAS, Godelieve
Alienation in Constant's Adolphe. - Bern : Lang, [1977]
173p. - Includes bibliographies
3-261-03030-5 X.900/20058
0011312

VERHOEFF, Han
"Adolphe" et Constant. - Paris : Éditions Klincksieck, 1976
134 p.. - (Bibliothèque française et romane ; sér C. no. 56). - Bibliography: p. 129-131
2-252-01779-1 Ac.2633/4 (3) [no. 56]
0011312

ADONIS. Ancient Greek myths

RIBICHINI, Sergio
Adonis. - Roma : CNR, 1981
212p. - (Pubblicazioni del Centro di studio per la civiltà fenicia e punica ; 22) (Studi semitici ; 55)
r6505111 Ac.104e/2[vol.55]
8936420

ADONIS. Byblos. Lebanon
Worship, to ca 400

SERVAIS-SOYEZ, Brigitte
Byblos et la fête des Adonies .... - Leiden : Brill, 1977
94p,15p of plates : ill. - (Études préliminaires aux religions orientales dans l'Empire romain ; tom 60)
90-04-04885-5 3071.bb.5/60
2042320

ADOPTED CHILDREN
See also
Social services for families with adopted children

ADOPTED CHILDREN
See also
Welfare work with adopted children

ADOPTED CHILDREN
Medical aspects

MEDICAL aspects of adoption and foster care / edited by Stephen Wolkind. - London : Heinemann Medical [etc.] [for] Spastics International Medical Publications, 1979
ix,102p. - (Clinics in developmental medicine ; no.74). - Includes bibliographies
0-433-36905-1 X.322/9246 Woolwich
2152037

MEDICAL practice in adoption and fostering / edited by Marie Oxtoby. - London : British Agencies for Adoption and Fostering, 1981
47p. - (Practice series / British Agencies for Adoption & Fostering ; no.4). - Bibliography: p45-47
0-903534-34-7 X.329/14281 Woolwich
2152037

ADOPTED CHILDREN. Great Britain
Social development compared with social development of children in care subsequently restored to natural parents

TIZARD, Barbara
Adoption. - London : Open Books, 1977
xii,251p : ill. - Includes index
0-7291-0191-6 X.529/31799
124485x

ADOPTED CHILDREN. Scotland
Social development compared with social development of children in care

TRISELIOTIS, John
Hard to place. - London : Heinemann Educational, 1984
x,228p. - (Studies in deprivation and disadvantage ; 13). - Bibliography: p215-221. - Includes index
0-435-82892-4 X.529/62267
301584x

ADOPTED CHILDREN. United States
Coloured adopted children. Interpersonal relationships with white parents

SIMON, Rita James
Transracial adoption. - Lexington : Lexington Books, c1981
xi,147p. - Includes index
0-669-04357-5 X.520/26246
2642174

ADOPTED PERSONS
- Serials

[NEWS (NORCAP)]. News / NORCAP. - 1-. - Warwick : NORCAP, 1982-
v.
B85-24398 P.525/660
3283755

ADOPTED PERSONS. England
Discovery of natural parents

SEARCHING for family connections. - Warwick : NORCAP, c1983
14p. - (NORCAP publication ; no.1)
0-9509901-0-8 X.525/10607
3153584

ADOPTED PERSONS. England
Rights of access to birth certificates

DAY, Cyril
Access to birth records. - London : Association of British Adoption & Fostering Agencies, 1980
36p. - (Research series ; 1)
0-903534-29-0 X.529/53326
2881683
- [Another copy] 347.426 OPL

ADOPTED PERSONS. Great Britain
Discovery of natural parents

The OTHER side of adoption / [editor Angela Hamblin ; assistant editor Alan Neale ; for] Jigsaw. - [London] : [Jigsaw Publications], [1977]
[1],98p. - Bibliography: p.98
B78-17294 X.512/5952
1446789

ADOPTED PERSONS. United States
Discovery of natural parents

LIFTON, Betty Jean
Twice born. - Harmondsworth : Penguin, 1977
[5],281p
0-14-004404-3 H.77/613 Woolwich
1100467

A TIME to search / [compiled by] Henry Ehrlich. - New York ; London : Paddington Press, 1977
232p
0-448-22241-8 X.529/32443
1545752

ADOPTION
Children. Genetics - For adoption

CARTER, C. O. (Cedric Oswald)
Genetics in adoption and fostering. - London : British Agencies for Adoption and Fostering, 1982
44p : ill. - Bibliography: p44
0-903534-41-x X 529/53920
2895323

ADOPTION. Children

ALTHEA
David and his sister Carol. - Cambridge : Dinosaur Publications, 1976
[31]p : col ill. - (Althea's dinosaur books)
0-85122-116-5 Cup.1281/206
016335x

BENET, Mary Kathleen
The character of adoption. - London : Cape, 1976
237p. - Bibliography: p.231. - Includes index
0-224-01199-5 X.529/30465
0595322

HASLER, Joy
'One of the family'. - [London] : [Westminster City Council], [1982]
16p : ill
B82-28845 X.809/18840
016335x

KREMENTZ, Jill
How it feels to be adopted. - London : Gollancz, 1984, c1982
xvi,108p : ill
0-575-03425-4 X.529/66154
320801x

LIVINGSTON, Carole
'Why was I adopted?'. - London : Angus & Robertson, 1981, c1978
[45]p : col.ill
0-207-14617-9 L.42/2102
016335x

NEW developments in foster care and adoption / edited by John Triseliotis. - London : Routledge and Kegan Paul, 1980
xi,243p : ill. - Bibliography: p.238-239. - Includes index
0-7100-0368-4 X.529/36872
1956949

SCOTT, Elaine
Adoption. - New York ; London : Watts, 1980
58p : ill. - (A First book). - Bibliography: p55. - Includes index
0-531-02937-9 X.529/39343
016335x

SNELL, Nigel
Steve is adopted. - London : Hamilton, 1985
[32]p : col.ill
0-241-11511-6 X.990/24367
016335x

STANFORD, James
So, you're adopted!. - Edinburgh : Macdonald, c1981
47p : ill
0-904265-53-6 X.520/28012
016335x

WAGSTAFF, Sue
Wayne is adopted. - London : A. & C. Black, [1981]
25p : col.ill. - Bibliography: p25
0-7136-2141-9 X.529/44843
016335x

ADOPTION. Children
- Serials

ADOPTION & Fostering. - [No.1]- ; 1976-. - [London] : The Association, [1976]-
B76-31610 PP.1103.ccu
0739421

ADOPTION. Children, 1-16 years. Great Britain

DONLEY, Kay
Opening new doors. - London : Association of British Adoption Agencies, 1975
56p. - Bibliography: p.56
0-903534-15-0 X.519/25952
0689912

ADOPTION. Children, 2-12 years
Personal adjustment of children - For parents

ASTON, Elaine
Getting to know you. - [London] : British Agencies for Adoption & Fostering, [1981]
44p : ill. - Bibliography: p43-44
0-903534-36-3 X.529/44189
2576279

ADOPTION. Children, 7-11 years. Great Britain
Effects on development compared with effects of illegitimacy

LAMBERT, Lydia
Children in changing families. - London : Macmillan, 1980
xii,196p : ill. - (National Children's Bureau series). - Bibliography: p.183-190. - Includes index
0-333-28696-0 X.529/37802
1954598

ADOPTION. Children, 7-13 years. Great Britain

JEWETT, Claudia L.
For ever and ever -. - [New ed.]. - London : Association of British Adoption Agencies, [1976]
[2],30p : ill
0-903534-19-3 X.519/25542
0601721

ADOPTION. Children. Czechoslovakia
Law, to date

PLANKOVÁ, O lga
Osvojenie die, ta, ta. - Bratislava : Obzor, 1979
259p. - (Právo pre každého)
r6440965 X.208/9231
9032495

ADOPTION. Children. England
Law

JOSLING, J. F. (John Francis)
Adoption of children. - 8th ed. - London : Oyez, 1977
xviii,225p. - (Oyez practice notes ; 3). - Includes index
0-85120-321-3 WP.1357/3g
1191411

Adoption of children. - 9th ed. - London : Oyez, 1980, c1981
xvi,256p. - (Oyez practice notes ; 3). - Includes index
0-85120-541-0 X.529/40786
1191411

PEARCE, Nasreen
Adoption practice and procedure. - London : Fourmat, 1984
xii,133p. - (Lawyers practice and procedure series). - Includes index
0-906840-79-1 X.529/70371
1191411

ADOPTION. Children. England
Law - Collections

GREAT BRITAIN
[Laws, etc.]. Adoption '82. - Durham : Comyn, 1982
xiv,82p
0-907267-02-5 X.529/50025
2786672

ADOPTION. Children. England
Law - For welfare workers

GLOUCESTERSHIRE. Family Services Section
Guide to adoption work. - [Gloucester] : [The Department], 1982
66,[123]p. - Bibliography: p[123]
0-904950-51-4 X.525/8051
1846140

LEEDING, Alfred Ellis
The guardian ad litem and adoption. - [New ed.]. - London : Association of British Adoption and Fostering Agencies, 1978
55p. - Bibliography: p.51-52. - Includes index
B79-13561 X.519/29481
1846140

ADOPTION. Children. England
Organisations: Parents for Children

PARENTS for children / edited by Phillida Sawbridge. - London : British Agencies for Adoption & Fostering, 1983
88p. - (Practice series ; 10)
0-903534-44-4 X.529/65197
3188388

ADOPTION. Children. Great Britain

ADOPTING a child. - 1981 ed. - London : British Agencies for Adoption & Fostering, 1981
26p : ill. - Bibliography: p8. - Includes index
0-903534-32-0 X.529/40606
1106473

ADOPTING a child. - 1982 ed. - London : British Agencies for Adoption & Fostering, 1982
24p. - Includes index
0-903534-39-8 X.529/46517
1106473

ADOPTION. - Rev. - London : British Agencies for Adoption & Fostering, 1982
[7]p : ill
B82-37719 X.529/5002
069066x

ADOPTION / edited by Philip Bean. - London : Tavistock, 1984
xvii,313p. - Includes bibliographies and index
0-422-78410-9 X.520/36134
069066x

ADOPTION in the 70s. - London : Association of British Adoption & Fostering Agencies, 1976
16p
B82-34820 X.525/6132
069066x

ADOPTION RESOURCE EXCHANGE
Working together. - London : Adoption Resource Exchange, 1980
52p. - Bibliography: p52
0-9506807-0-2 X.529/50056
069066x

ASSOCIATION OF BRITISH ADOPTION AGENCIES
Which children and what plan?. - London : The Association, [1975]
13p. - (Ending the waiting ; 1)
0-903534-18-5 X.519/25956
069066x

ASSOCIATION OF BRITISH ADOPTION AND FOSTERING AGENCIES
[Miscellaneous publications not catalogued separately]. - London : The Association, 1976-
r6010775 P.521/3092
1054996

CHILD adoption. - London : Association of British Adoption and Fostering Agencies, [1979]
viii,236p : ill. - Includes bibliographies
B79-10910 X.519/29354
1824643

CHILD from the past. - London : The Association of British Adoption & Fostering Agencies, [1982?]
1folded sheet([5]p)
B82-37722 X.529/50018
069066x

CONSUMERS' ASSOCIATION
How to adopt. - Revised ed. - London : the Association, 1973
144p.. - (A Consumer publication)
0-85202-061-9 X0978/8(106)
1106473

HEIM, A. W. (Alice Winifred)
Thicker than water?. - London : Secker & Warburg, 1983
211p. - Bibliography: p210-211
0-436-19155-5 X.529/55624
069066x

IF you are adopted. - London : British Agencies for Adoption & Fostering, [198-?]
[4]p : ill
B82-37720 X.529/50019
069066x

KORNITZER, Margaret
Adoption. - 5th ed. - London : Putnam, 1976
186p. - Bibliography: p.177-178. - Includes index
0-370-10058-1 X.529/31120
069066x

SUBSTITUTE family care. - London : Association of British Adoption & Fostering Agencies.
1: The use and misuse of resources. - 1978
16p
B79-32397 X.519/30229
1821318
- London : Association of British Adoption and Fostering Agencies.
2: What shall we do with the children. - 1978
28p
B79-10172 X.519/29316
1821318

ADOPTION. Children. Great Britain
- For adoptive parents

ROWE, Jane
Yours by choice. - New ed. - London : Routledge & Kegan Paul, 1982
xi,188p : ill. - Bibliography: p184. - Includes index
0-7100-9035-8 X.529/47657
2729253

TALKING about origins. - Rev. - London : British Agencies for Adoption and Fostering, 1980
1folded sheet([5]p)
B82-37718 X.529/50014
2729253

ADOPTION. Children. Great Britain
- For unmarried mothers

ASSOCIATION OF BRITISH ADOPTION AND FOSTERING AGENCIES
Single and pregnant. - London : The Association, 1978
Folder([5]p). - Bibliography: p.5
B79-14584 X.519/29515
1873334

ADOPTION. Children. Great Britain
Adoption & foster care

MEETING children's needs through adoption and fostering. - Rev. ed. - London : British Agencies for Adoption & Fostering, 1980
[8]p
0-903534-28-2 X.529/50063
2784580

SMITH, Carole R.
Adoption and fostering. - London : Macmillan, 1984
165p. - (Practical social work). - Bibliography: p153-163. - Includes index
0-333-35230-0 X.529/67293
2784580

ADOPTION. Children. Great Britain
Adoption & foster care. Organisations: Association of British Adoption and Fostering Agencies

ASSOCIATION OF BRITISH ADOPTION AND FOSTERING AGENCIES
Annual report. - 1975-76. - London : The Association, [1976]
23p
Subsequent volumes, when received, will be placed at this shelfmark
B77-32543 P.521/3138
1262270

ADOPTION. Children. Great Britain
Adoption & foster care. Organisations: British Agencies for Adoption & Fostering - Serials

ASSOCIATION OF BRITISH ADOPTION AND FOSTERING AGENCIES
ABAFA news. - No. 1- ; Apr. 1979-. - London : Association of British Adoption and Fostering Agencies, [1979]-
B80-26818 P.523/1338
2102927

BRITISH AGENCIES FOR ADOPTION & FOSTERING
BAAF news. - No.1 (Jan.1981)-. - London : BAAF, 1981-
v.
B81-32239 P.523/1529
2102927

ADOPTION. Children. Great Britain
By foster parents

ROWE, Jane
Long term fostering and the Children Act. - London : British Agencies for Adoption & Fostering, 1984
34p. - (Research series ; 3)
0-903534-49-5 P.881/322
3171914

ADOPTION. Children. Great Britain
Inter-country adoption - For welfare workers

ASSOCIATION OF BRITISH ADOPTION AND FOSTERING AGENCIES
Inter-country adoption. - London : The Association, 1979
52p. - (Practice series ; 2). - Bibliography: p.34-35. - Includes index
B80-29426 X.529/37669 Woolwich
2105349

ADOPTION. Children. Great Britain
Organisations: Association of British Adoption and Fostering Agencies

ASSOCIATION OF BRITISH ADOPTION AND FOSTERING AGENCIES
Annual report. - 1975-76. - London : The Association, [1976]
23p
Subsequent volumes, when received, will be placed at this shelfmark
B77-32543 P.521/3138
1262289

ADOPTION. Children. Great Britain
Organisations: British Agencies for Adoption & Fostering - Serials

ASSOCIATION OF BRITISH ADOPTION AND FOSTERING AGENCIES
ABAFA news. - No. 1- ; Apr. 1979-. - London : Association of British Adoption and Fostering Agencies, [1979]-
B80-26818 P.523/1338
2102935

BAAF practice notes. - 1 (Apr. 1982)-. - London : British Agencies for Adoption & Fostering, 1982-
v.
B83-18188 P.523/1758
2102935

ADOPTION. Children. Great Britain
Organisations: British Agencies for Adoption & Fostering - Serials–continued

BRITISH AGENCIES FOR ADOPTION & FOSTERING
BAAF news. - No.1 (Jan.1981)-. - London : BAAF. 1981-
v.
B81-32239 P.523/1529
2102935

Report & accounts. - 1981-82-. - London : The Agencies, 1982-
v.
r6812503 P.881/322
2102935

ADOPTION. Children. Italy
Law, 1946-

DAVÌ, Angelo
L'adozione nel diritto internazionale privato italiano. - Milano : Giuffrè, 1981-. - (Pubblicazioni dell'Istituto di diritto internazionale dell'Università di Roma ; 16)
r6505887 X.200/41800
9017518

ADOPTION. Children. Northern Ireland
Law

CHILDREN AND YOUNG PERSONS REVIEW GROUP
Adoption of children in Northern Ireland. - Belfast : H.M.S.O., 1980
[6],68p
0-337-07204-3 CS.A210/142
2287110

CHILDREN AND YOUNG PERSONS REVIEW GROUP
Adoption of children in Northern Ireland. - Belfast : H.M.S.O., 1982
vii,86p
0-337-07221-3 CS.A.210/211
2287110

ADOPTION. Children. Scotland
Decision making

MACKAY, Sheila J.
Making decisions in adoption. - Edinburgh : Family Care, 1982
x,82p : ill. - Includes bibliographies
0-9508117-0-x X.525/8468
2993457

ADOPTION. Children. Scotland
Law

MCNEILL, Peter G. B.
Adoption of children in Scotland. - Edinburgh : Green, 1982
xx,208p. - Includes index
0-414-00689-5 X.520/28683
2817330

ADOPTION. Children. Scotland
Law. Great Britain. Parliament. House of Lords. Adoption (Scotland) Bill (H.L.). Session 1977-78

GREAT BRITAIN. Scottish Law Commission
Adoption (Scotland) Bill, report on the consolidation of certain enactments relating to adoption in Scotland .... - Edinburgh : H.M.S.O., [1978]
6p. - (Scot. Law Com. ; no.50) (Cmnd.7187)
0-10-171870-5 BS.179/14
1480588

ADOPTION. Children. Strathclyde Region. Scotland - For welfare work

STRATHCLYDE. Regional Council
Adoption policy and practice guidelines. - [Glasgow] : Social Work Department Strathclyde Regional Council, 1980
77p
B84-22536 X.525/9022
3086135

ADOPTION. Children. Sweden

BOHMAN, Michael
Adoptivbarn och deras familjer. - Stockholm : Almqvist & Wiksell, 1973
284 p.. - (Monografier utgivna av Stockholms kommunalförvaltning ; 34). - Bibliography : p. 274-281
91-200-4316-3 Ac.2258b[no. 34]
0909092

ADOPTION. Children. Sweden
Inter-country adoption, to date

GARDELL, Ingegärd
A Swedish study on intercountry adoptions. - [Stockholm] : Swedish Council for Intercountry Adoptions, 1980
66p
r6611402 X.525/9038
9077596

ADOPTION. Children. Sweden
Inter-country adoption, to date - Conference proceedings

INTERNATIONELLA adoptioner / [redigerad av Ia Hillborg] - [Stockholm] : Liber förlag, 1982
197p. - (DSF skriftserie : nr 4)
91-380-7084-7 X.529/64126
9077596

ADOPTION. Children. Switzerland
Law

EICHENBERGER, Rolf
Die materiellen Voraussetzungen der Adoption Unmündiger nach neuem schweizerischem Adoptions recht - Brugg, 1974
xx, 232 p.
r6002303 X.100/14895
0883255

ADOPTION. Children. United States

FEIGELMAN, William
Chosen children. - New York : Praeger, 1983
xxiv,261p. - Bibliography: p243-252
0-03-062343-x X.520/33538
0033111

FISHER, Florence
The search for Anna Fisher. - London : Michael Dempsey : Cassell, 1975
270p
0-86018-102-2 X.809/40394
0530433

LAUFFER, Armand
Resources. - Beverly Hills ; London : Published in cooperation with the North American Center on Adoption, Inc. and the Continuing Education Program in the Human Services of the University of Michigan School of Social Work [by] Sage, c1979
192p : ill. - (A Project CRAFT publication) (Sage human services guides ; v.6). - Includes bibliographies
0-8039-1218-8 X.529/43711
0033111

ADOPTION. Children. United States
Suitability of families. Assessment

HARTMAN, Ann
Finding families. - Beverly Hills ; London : Sage in cooperation with the North American Center on Adoption, Inc. and the Continuing Education Program in the Human Services of the University of Michigan School of Social Work, c1979
108p : ill. - (A Project CRAFT publication) (Sage human services guides ; v.7). - Bibliography: p107
0-8039-1216-1 X.529/47788
2733463

ADOPTION. Coloured children. Great Britain
By white parents

ASSOCIATION OF BRITISH ADOPTION AGENCIES
Adopting a black child. - London : The Association, [1975]
[2],14p. - Bibliography: p.14
0-903534-16-9 X.519/25932
0683051

GILL, Owen
Adoption and race. - London : Batsford Academic and Educational in association with British Agencies for Adoption & Fostering, 1983
151p. - (Child care policy and practice). - Bibliography: p147. - Includes index
0-7134-2023-5 X.529/54663
0683051

JACKSON, Barbara, 1944-
Family experiences of inter-racial adoption. - London : Association of British Adoption and Fostering Agencies, [1976]
32p. - Bibliography: p.30-32
0-903534-20-7 X.519/25965
0683051

Family experiences of inter-racial adoption. - London : The Association of British Adoption & Fostering Agencies, [198-?]
32p. - Bibliography: p30-32
B82-37723 X.529/50022
0683051

ADOPTION. Coloured children. United States
By white parents

GROW, Lucille J.
Black children-white parents. - [New York] : Research Center, Child Welfare League of America, 1975
239p
0-87868-177-9 X.519/30841
1145169

SIMON, Rita James
Transracial adoption. - New York ; London : Wiley, 1977
x,197p. - Includes index
0-471-79208-x X.520/11747
1145169

ADOPTION. Foreign children
Conflict of laws

DELUPIS, Ingrid
International adoptions and the conflict of laws. - Stockholm : Almqvist och Wiksell, 1975
87 p.
91-220-0021-6 X.200/30821
1067923

ADOPTION. Foreign children. Denmark

MADSEN, Monna S.
Urimeligheder omkring adoptioner. - [Nykøbing Mors] : Attika, [1975]
63 p.. - (Attikas dokumentarserie)
87-85189-66-9 X.519/20923
0877069

ADOPTION. Foreign children. Sweden

KATS, Madeleine
Adoption av utländska barn. - [Stockholm] : Bonniers, 1975
198 p.. - (Bonniers ugglebõcker). - Includes bibliographies
91-00-39283-9 X.519/19927
0905844

ADOPTION. Handicapped children. Great Britain

DONLEY, Kay
Opening new doors. - London : Association of British Adoption Agencies, 1975
56p. - Bibliography: p.56
0-903534-15-0 X.519/25952
114488x

ADOPTION. Hydrocephalic children

BAUMGARTNER, Diane
Melissa. - London : Triangle, 1982
103p : ill
0-281-04010-9 X.529/51198
2817950

ADOPTION. Negro children. United States
Effects of racial discrimination

DAY, Dawn
The adoption of black children. - [Lexington, Mass.] : Heath ; [Farnborough] : Distributed by Teakfield Ltd, 1977
xii,157p. - Includes index
0-669-02107-5 X.520/14344
1939203

ADOPTION. Socially disadvantaged children. United States

NO child is unadoptable / a Project CRAFT publication ; edited by Sallie R. Churchill, Bonnie Carlson and Lynn Nybell. - Beverly Hills ; London : Sage, c1979
173p. - (Sage human services guides ; v.8). - Includes bibliographies
0-8039-1215-3 X.529/41023
2391279

ADOPTION. Step-children. England
Law

STEP-CHILDREN and adoption. - London : British Agencies for Adoption and Fostering, 1980
1folded sheet([5]p)
B82-37716 X.529/50020
2784572

ADOPTION. Step-children. Great Britain

MASSON, Judith
Mine, yours or ours. - London : H.M.S.O., 1983
vi,130p : ill. - Bibliography: p123-126. - Includes index
0-11-320835-9 BS.385/863
3184277

ADOPTION. Step-children. Scotland
Law

STEP-CHILDREN and adoption. - London : Association of British Adoption & Fostering Agencies, c1980
1folded sheet([6]p)
B82-37721 X.529/50013
2784564

ADOPTION. Vietnamese children. Norway

SCHJELDERUP-MATHIESEN, Kristin
Vietnambarn i Norge. - Oslo : Grøndahl, 1977
176p : ill
82-504-0258-8 X.519/38708
1542095

ADOPTION. West Indian children. London
Adoption & foster care by West Indian families. Projects: Soul Kids

SOUL KIDS. Steering Group
Report of the steering group of the Soul Kids campaign. - London : Association of British Adoption & Fostering Agencies on behalf of the Soul Kids Campaign, c197?
48p : ill
B82-37713 X.529/50066
2784602

ADOPTION (SCOTLAND) BILL (H.L.) Great Britain. Parliament. House of Lords
Session 1977-78

GREAT BRITAIN. Scottish Law Commission
Adoption (Scotland) Bill. report on the consolidation of certain enactments relating to adoption in Scotland .. - Edinburgh : H.M.S.O., [1978]
6p. - (Scot. Law Com. ; no.50) (Cmnd.7187)
0-10-171870-5 BS.179/14
1480588

ADOPTIVE PARENTS
Great Britain. Children. Adoption - For adoptive parents

ROWE, Jane
Yours by choice. - New ed. - London : Routledge & Kegan Paul, 1982
xi,188p : ill. - Bibliography: p184. - Includes index
0-7100-9035-8 X.529/47657
2729253

TALKING about origins. - Rev. - London : British Agencies for Adoption and Fostering, 1980
1folded sheet([5]p)
B82-37718 X.529/50014
2729253

ADORATION OF THE MAGI. Botticelli, Sandro. Italian paintings

HATFIELD, Rab
Botticelli's Uffizi 'Adoration'. - Princeton ; Guildford : Princeton University Press, 1976
xv,151p,[50]p of plates(2 fold) : ill(incl 1 col). - (Princeton essays on the arts ; 2). - Bibliography: p.133-139. - Includes index
0-691-03912-7 X.421/10301 Woolwich
0756903

ADORAZIONE DEI MAGI. Angelico, fra. Italian paintings

BOSKOVITS, Miklòs
Un'Adorazione dei Magi e gli inizi dell'Angelico. - Bern : Abegg-Stiftung, [1976]
63p : ill. - (Monographien der Abegg-Stiftung Bern ; 11)
r6065201 X.410/6500 Woolwich
1470957

ADORNMENT
See also
Personal adornment

ADORNO, T. W.
See Adorno, Theodor W.

ADORNO, THEODOR W.
Theories of aesthetics

BRÄUTIGAM, Bernd
Reflexion des Schönen - schöne Reflexion. - Bonn, 1975
278 p.. - (Literatur und Wirklichkeit ; Bd 12)
3-416-01016-7 X.0900/144 (12)
0871885

GROHOTOLSKY, Ernst
Ästhetik der Negation. - Meisenheim : Hain, 1984
viii,150p. - (Hochschulschriften Literaturwissenschaft ; 62) (Forum academicum). - Includes bibliography
3-445-02324-7 X.950/36730
8723729

PROHIĆ, Kasim
Prizma i ogledalo. - Beograd : Nolit, 1980
359p
r6435392 X.529/52886
8723729

ADORNO, THEODOR W
Theories of aesthetics

SCHWARZ, Ullrich
Rettende Kritik und antizipierte Utopie. - München : Fink, 1981
308p. - Includes bibliography
3-7705-1987-6 X.529/56729
8723729

ADORNO, THEODOR W.
Theories of music

NIELSEN, Poul
Musik og materialisme. - [København] : Borgen, [1978]
351p,[1]fold leaf of plates : ill. - Bibliography : p.313-343
87-418-4763-6 X.439/10604
150536x

SERRAVEZZA, Antonio
Musica, filosofia e società in Th. W. Adorno. - Bari : Dedalo, 1976
255p. - (La scienza nuova ; 40)
r6078138 X.439/4955
150536x

ADORNO, THEODOR W.
Theories of sociology

GEYER, Carl-Friedrich
Kritische Theorie. - Freiburg : Alber, 1982
204p. - (Kolleg Philosophie). - Includes bibliography
3-495-47501-x X.529/65704
8408076

ADORNO, THEODOR W.
Theories of teaching of music

WIETUSCH, Bernd
Die Zielbestimmung der Musikpädagogik bei Theodor W. Adorno. - Regensburg : Bosse, 1981
245p. - (Perspektiven zur Musikpädagogik und Musikwissenschaft ; Bd.7). - Bibliography
3-7649-2231-1 X.431/11742
8887330

ADORNO, THEODOR W.. Criticism. Music - Conference proceedings

ADORNO und die Kunst / herausgegeben von Otto Kolleritsch. - Graz : Universal Edition, 1979
243p. - (Studien zur Wertungsforschung ; Bd.12)
3-7024-0139-3 X.0439/55(12)
8892881

ADORNO, THEODOR W.. French music - Biographies

COURT, Raymond
Adorno et la nouvelle musique. - Paris : Klincksieck, 1981
155p. - (Collection d'esthétique ; 33)
2-86563-004-8 X.431/11747
8172625

ADORNO, THEODOR W.. German philosophy

BEIER, Christel
Zum Verhältnis von Gesellschaftstheorie und Erkenntnistheorie. - Frankfurt am Main : Suhrkamp, 1977
197p. - Includes bibliographies
3-518-07479-2 X.519/38880
111882x

BRAUN, Carl, of Bonn
Kritische Theorie versus Kritizismus. - Berlin : De Gruyter, 1983
xi,311p. - (Kantstudien. Ergänzungshefte ; 115). - Bibliography
3-11-009541-6 PP.1253.bba(1)[no.115]
8176965

BUCK-MORSS, Susan
The origin of negative dialectics. - Hassocks : Harvester Press, [1978]
xv,335p. - Bibliography: p.307-322. - Includes index
0-85527-960-5 X.520/12402
111882x

JAY, Martin
Adorno. - [London] : Fontana, 1984
192p. - (Fontana modern masters). - Bibliography: p188-192
0-00-636287-7 X.529/65434
111882x

KNAPP, Gerhard P.
Theodor W. Adorno. - Berlin : Colloquium, 1980
94p. - (Köpfe des XX. Jahrhunderts ; Bd.93)
3-7678-0506-5 X.529/45016
8176965

MÖRCHEN, Hermann
Macht und Herrschaft im Denken von Heidegger und Adorno. - Stuttgart : Klett-Cotta, 1980
195p
3-12-915270-9 X.520/23305
111882x

NEBULONI, Roberto
Dialettica e storia in Th. W. Adorno. - Milano : Vita e pensiero, 1978
viii,147p. - (Scienze filosofiche ; 17). - Bibliography, including works of Adorno : p.137-147
r6067391 X.519/24789
111882x

PETTAZZI, Carlo
Th. Wiesengrund Adorno. - Firenze : La nuova Italia, 1979
xvi,295p. - (Pubblicazioni della Facoltà di lettere e filosofia dell'Università di Milano ; 86. Sezione a cura dell'Istituto di storia della filosofia ; 24). - Bibliography: p250-289
r6504555 X.520/26483
8176965

PROTTI, Mauro
Homo theoreticus. - Milano : Franco Angeli, [1978]
251p. - (Sociologia). - Bibliography : p.203-251
r6197067 X.529/35801
111882x

RATH, Norbert
Adornos kritische Theorie. - Paderborn : Schöningh, 1982
266p. - (Konstruktionen der Moderne ; Bd.1). - Includes bibliography
3-506-77005-5 X.529/56338
8176965

ROSE, Gillian
The melancholy science. - London : Macmillan, 1978
x,212p. - Bibliography: p.193-205. - Includes index
0-333-23213-5 X.520/13618
111882x

ADORNO, THEODOR W.. Sociology

ADORNO, Theodor W. (Theodor Wiesengrund)
[Selections. Italian]. Adorno e la teoria critica della società. - Firenze : Sansoni, 1974
123 p.. - (Scuola aperta. scienze umane). - Bibliography : p. 45-49
r6013649 X.519/20534
0940216

ADORNO, THEODOR W. (THEODOR WIESENGRUND)
Theories of aesthetics of arts

GALEAZZI, Umberto
L'estetica di Adorno. - Roma : Città nuova, 1979
275p. - (Idee ; 40)
r6500639 X.429/11859 Woolwich
8704155

ADORNO, THEODOR W. (THEODOR WIESENGRUND). German philosophy

ALLKEMPER, Alo
Rettung und Utopie. - Paderborn : Schöningh, 1981
267p. - (Reihe Sprach- und Literaturwissenschaften ; Bd.1) (Schriften der Gesamthochschule Paderborn). - Bibliography
3-506-78451-x X.0950/442(1)
8176965

DECKER, Peter
Die Methodologie kritischer Sinnsuche. - Erlangen : Palm & Enke, 1982
256p. - (Erlanger Studien ; Bd 37). - Bibliograpy
3-7896-0137-3 X.0909/700(37)
8176965

LANG, Peter Christian
Hermeneutik, Ideologiekritik, Ästhetik. - Königstein / Ts. : Forum Academicum, 1981
207p. - (Monographien zur philosophischen Forschung ; Bd 200). - Bibliography
3-445-02103-1 WP.10999/200
8176965

LÜDKE, W. Martin
Anmerkungen zu einer "Logik des Zerfalls". - Frankfurt am Main : Suhrkamp, 1981
159p. - (Edition Suhrkamp ; 926). - Bibliography
3-518-10926-x X.958/6402
8176965

MÖLLER, Hans-Martin
Adorno, Proust, Beckett. - Frankfurt / Main : Haag + Herchen, 1981
176p. - Bibliography
3-88129-454-6 X.950/18145
8176965

MÖRCHEN, Hermann
Adorno und Heidegger. - Stuttgart : Klett-Cotta, 1981
716p. - Bibliography
3-12-915390-x X.520/28184
8176965

ADP-RIBOSYLATION. Enzymes & proteins. Cells. Organisms - Conference proceedings

FOGARTY INTERNATIONAL CONFERENCE ON NOVEL ADP-RIBOSYLATIONS OF REGULATORY ENZYMES AND PROTEINS (1979 : National Institutes of Health, Bethesda, Md)
Novel ADP-ribosylations of regulatory enzymes and proteins. - New York ; Oxford : Elsevirer/North-Holland, c1980
xi,452p : ill. - (Developments in cell biology ; 6). - Includes index
0-444-00403-3 SRL
2342294

ADRA. Spain
Antiquities. Archaeology

FERNÁNDEZ-MIRANDA, Manuel
Abdera. - [Madrid] : Servicio de Publicaciones del Ministerio de Educación y Ciencia, [1975?]
266p.,2 leaves of plates,16p. of plates : ill. - (Excavaciones arqueológicas en España ; n.85)
84-369-0391-9 X.410/6021 Woolwich
1021249

ADRENAL CORTEX. Children
Physiology related to physiology of hypothalamus
ZURBRÜGG, Rolf Peter
Hypothalamic-pituitary adrenocortical regulation. - Basel ; London : Karger : [Distributed by Wiley], 1976
vii,83p : ill. - (Monographs in paediatrics ; vol.7). - Bibliography: p.74-83
B76-27039 X.319/16766 Woolwich
070914x

ADRENAL CORTEX. Man
Diseases
ADRENAL cortex / edited by David C. Anderson and Jeremy S.D. Winter. - London : Butterworths, 1985
ix,297p,[2]p of plates : ill(some col.). - (Clinical endocrinology) (Butterworths international medical reviews). - Includes bibliographies and index
0-407-02275-9 SRL
2086670

ADRENAL CORTEX. Man
Hormones - Conference proceedings
ENDOCRINE FUNCTION OF THE HUMAN ADRENAL CORTEX (Conference : 1977 : Florence)
The endocrine function of the human adrenal cortex. - London : Academic Press, 1978
ix,628p : ill. - (Proceedings of the Serono symposia ; vol.18). - Includes bibliographies and index
0-12-380160-5 SRL
1821350

ADRENAL CORTEX. Man
Secretion
SUZUKI, Tatuzi
Physiology of adrenocortical secretion. - Basel ; London : Karger, 1983
vi,215p : ill. - (Frontiers of hormone research ; v.11). - Bibliography: p139-191. - Includes index
3-8055-3644-5 SRL
3129691

ADRENAL CORTEX. Vertebrates
Physiology
GENERAL, comparative and clinical endocrinology of the adrenal cortex / edited by I. Chester Jones and I.W. Henderson. - London : Academic Press.
Vol.1. - 1976
xvii,461p : ill. - Includes bibliographies and index
0-12-171501-9 SRL
0644285

Vol.2. - 1978
xv,670p : ill. - Includes bibliographies and index
0-12-171502-7 SRL
0644285

ADRENAL GLANDS
See also
Adrenal cortex

ADRENAL GLANDS. Children
Cancer
ADRENAL and endocrine tumors in children / edited by G. Bennett Humphrey ... [et al.]. - Boston, Mass. ; Lancaster : Nijhoff, 1983
xvi,377p : ill. - (Cancer treatment and research). - Includes index
0-89838-590-3 SRL
3171477

ADRENAL GLANDS. Children
Congenital adrenal hyperplasia
ADRENAL diseases in childhood / volume editors Maria I. New and Lenore S. Levine. - Basel ; London : Karger, c1984
235p : ill. - (Pediatric and adolescent endocrinology ; v. 13). - Includes index
3-8055-3777-8 SRL
1205226

ADRENAL GLANDS. Children
Congenital adrenal hyperplasia - Conference proceedings
TREATMENT OF CONGENITAL ADRENAL HYPERPLASIA : A QUARTER OF A CENTURY LATER (Conference : 1975 : Johns Hopkins Hospital)
Congenital adrenal hyperplasia. - Baltimore ; London : University Park Press, 1977
xxiii,532p : ill. - Includes bibliographies and index
0-8391-0974-1 SRL
1205226

ADRENAL GLANDS. Man
Diagnosis. Tomography. Applications of digital computer systems
COMPUTED tomography of the kidneys and adrenals / edited by Stanley S. Siegelman, Olga M.B. Gatewood, Stanford M. Goldman. - New York ; Edinburgh : Churchill Livingstone, 1984
x,287p : ill. - (Contemporary issues in computed tomography ; v.3). - Includes index
0-443-08267-7 SRL
3195538

ADRENAL GLANDS. Man
Surgery
SURGICAL disorders of the adrenal gland / Timothy S. Harrison ... [et al.]. - New York ; London : Grune and Stratton, 1975
x,150p,plate : ill(incl 1 col). - (Modern surgical monographs). - Includes bibliographies and index
0-8089-0864-2 SRL
0503797

ADRENAL HYPERPLASIA. Adrenal glands. Children
Congenital adrenal hyperplasia - Conference proceedings
TREATMENT OF CONGENITAL ADRENAL HYPERPLASIA : A QUARTER OF A CENTURY LATER (Conference : 1975 : Johns Hopkins Hospital)
Congenital adrenal hyperplasia. - Baltimore ; London : University Park Press, 1977
xxiii,532p : ill. - Includes bibliographies and index
0-8391-0974-1 SRL
1205226

ADRENAL HYPERPLASIA. Children
Congenital adrenal hyperplasia
ADRENAL diseases in childhood / volume editors Maria I. New and Lenore S. Levine. - Basel ; London : Karger, c1984
235p : ill. - (Pediatric and adolescent endocrinology ; v. 13). - Includes index
3-8055-3777-8 SRL
1205226

ADRENALINE. Drug therapy. Glaucoma. Man
Adrenaline & guanethidine
HOYNG, F. J.
Pharmacological denervation and glaucoma. - The Hague ; London : Junk, c1981
ix,150p : ill. - (Monographs in ophthalmology ; 2). - Includes bibliographies
90-619-3802-3 SRL
2408066

ADRENERGIC
See also
Adrenaline

ADRENERGIC NERVOUS SYSTEM. Man
BURNSTOCK, Geoffrey
Adrenergic neurons. - London : Chapman and Hall, 1975
xi,225p, [14]p of plates, leaf of plate : ill. - Bibliography: p.137-217. - Includes index
0-412-14060-8 SRL
0512540

ADRENERGIC NERVOUS SYSTEM. Man
Catecholamines. Release. Mechanism
The RELEASE of catecholamines from adrenergic neurons / edited by David M. Paton. - Oxford : Pergamon, 1979
xi,393p : ill. - Includes bibliographies and index
0-08-021536-x SRL
1587129

ADRENERGIC NEURONS. Central nervous system. Mammals
Role in regulation of cardiovascular system - Conference proceedings
CENTRAL adrenaline neurons / edited by Kjell Fuxe ... [et al.]. - Oxford : Pergamon, 1980
xiv,347p : ill. - (Wenner-Gren Center international symposium series ; v.33). - Includes bibliographies and index
0-08-025927-8 SRL
2083183

ADRENERGIC RECEPTORS. Mammals
Effects of drugs - Conference proceedings
ADRENOCEPTOR SYMPOSIUM (1978 : Manchester)
Recent advances in the pharmacology of adrenoceptors. - Amsterdam ; Oxford : Elsevier : North-Holland Biomedical Press, 1978
xiv,403p : ill. - Includes index
0-444-80083-2 SRL
1809830

ADRENERGIC RECEPTORS. Man
ADRENOCEPTORS and catecholamine action / edited by George Kunos. - New York ; Chichester : Wiley. - (Neurotransmitter receptors ; v.1)
Pt.B. - c1983
vii,327p : ill. - Includes index
0-471-05726-6 SRL
2598167

ADRENERGIC RECEPTORS. Man
- Bibliographies - Serials
ADRENERGIC receptors. - Jan.1981-. - Sheffield : University of Sheffield Biomedical Information Service, 1981-
12v.
B81-38517 SRL
2565668

CATECHOLAMINES and adrenergic receptors. - Vol.1, no.1 (Jan. 1982)-. - Sheffield : University of Sheffield Biomedical Information Service, 1982-
v.
B83-21471 SRL
2565668

ADRENERGIC RECEPTORS. Man
Alpha adrenergic receptors. Blocking agents
ADRENOCEPTER blockers in cariovascular disease / edited by Helge Refsum and Ole D. Mjøs. - Edinburgh : Churchill Livingstone, 1985
328p : ill. - Includes bibliographies and index
0-443-03334-x SRL
3200086

ADRENERGIC RECEPTORS. Man
Beta adrenergic receptors. Blocking agents
FRISHMAN, William H.
Clinical pharmacology of the beta-adrenoceptor blocking drugs. - New York : Appleton-Century-Crofts ; London : Prentice-Hall, 1980
xiv,221p : ill. - Includes index
0-8385-1143-0 SRL
2338785

ADRENERGIC RECEPTORS. Man
Beta adrenergic receptors. Blocking agents - Conference proceedings
ADVANCES in beta-blocker therapy II / editor A. Zanchetti. - Amsterdam ; Oxford : Excerpta Medica, 1982
vii,312p : ill. - (International congress series ; no.593). - Includes index
90-219-0593-0 SRL
2338785

BETA-ADRENOCEPTOR BLOCKING AGENTS (Conference : 1975 : Rotterdam)
Beta-adrenoceptor blocking agents. - Amsterdam ; Oxford : North-Holland Publishing Co. [etc.], 1976
xii,380p : ill. - Includes bibliographies and index
0-7204-0660-9 SRL
2338785

The CLINICAL impact of beta-adrenoceptor blockade / editors, D.M. Burley and G.F.B. Birdwood. - Horsham : CIBA Laboratories, 1980
287p
B81-29514 SRL
2338785

ADRENERGIC RECEPTORS. Man
Beta adrenergic receptors. Disorders. Role in asthma - Conference proceedings
BETA adrenoceptors in asthma. - London : Academic, May 1984
[270]p. - (Perspectives in asthma ; 2)
0-12-506440-3 SRL
3050556

ADRENERGIC TRANSMITTERS. Synapses. Mammals
- Conference proceedings
OHOLO BIOLOGICAL CONFERENCE ON NEUROACTIVE COMPOUNDS AND THEIR CELL RECEPTORS (1979 : Zichron Ya'acov)
Neurobiology of cholinergic and adrenergic transmitters. - Basel ; London : Karger, 1980
xiv,197p : ill. - (Monographs in neural sciences ; v.7). - Includes bibliographies and index
3-8055-0828-x SRL
2526425

ADRENOCORTICOTROPHIC HORMONES. Drug therapy. Medicine
- Bibliographies - Serials
ACTH & related peptides. - Jan.1981-. - Sheffield : University of Sheffield Biomedical Information Service, 1981-
v.
B81-38516 SRL
256565x

ADRETS, FRANÇOIS DE BEAUMONT, BARON DES. France. Armée
- Biographies
DALET, Gilbert
L'étrange figure du baron des Adrets. - Meylan : Les Éditions de L'Aurore, 1982
196p
2-903950-04-0 X.808/40254
8794200

ADRIA
See also
Catholic Church. Diocese of Adria

ADRIAMYCIN
See Doxorubicin

ADRIAN, LEOPOLD. Austrian writers. Fiction in German
RENNER, Ursula
Leopold Andrians 'Garten der Erkenntnis'. - Frankfurt am Main : Lang, 1981
344p. - (Literatur & Psychologie ; Bd.3). - Bibliography: p.308-335
3-8204-6150-7 X.0900/638(3)
8156352

ADRIANOPLE
See Edirne

ADRIATIC COASTAL REGIONS. Italy
- For sailing
DENHAM, Henry Mangles
The Adriatic. - 2nd ed. - London : J. Murray, 1977
xii,259p,[12]p of plates : ill. - (Sea guides / Henry Mangles Denham). - Includes index
0-7195-3359-7 X.800/26359
1203401

ADRIATIC COASTAL REGIONS. Italy
Boats, to ca 1900
BONINO, Marco
Archeologia e tradizione navale tra la Romagna e il Po. - Ravenna : Longo, 1978
240p : ill. - (Antichità, archeologia, storia dell'arte ; 5)
r6069577 X.629/14661 Woolwich
2467453

ADRIATIC COASTAL REGIONS. Italy
Ports
PARIGI, Guido
I porti regione per regione. - Pisa : Nistri-Lischi. - (Il tagliamare ; 21)
Adriatico. - 1974
145p : all ill
r6068278 X.809/26299
1851241

ADRIATIC COASTAL REGIONS. Italy
Salt trades, to date - Conference proceedings
SALE e saline nell'Adriatico / a cura di Antonio Di Vittorio ; saggi di S. Anselmi ... [et al.]. - Napoli : Gianini, 1981
xix,330p : ill. - (Biblioteca di studi meridionali ; 8)
r6505545 X.0700/362(8)
8902860

ADRIATIC COASTAL REGIONS. Yugoslavia
Ancient Greek & Ancient Roman towns
SUIĆ, Mate
Antički grad na istočnom Jadrana. - Zagreb : Liber, 1976
328p,[9]leaves of plates(some fold.) : ill(some col). - (Institut za arheologiju sveučilišta u Zagrebu. Studije iz historije urbanizma ; Knj.1). - Bibliography : p.269-281
r6097944 X.705/694
2188341

ADRIATIC COASTAL REGIONS. Yugoslavia
Maritime aspects - Serials
ADRIATICA maritima .... - Sv.1-. - Zadar : Jugoslavenska akademija znanosti i umjetnosti, 1974- : ill
r6186465 X:0802/393
2139812

ADSORBENTS
See also
Immunoadsorbents

ADSORPTION
See also
Chemisorption
Physical adsorption

ADSORPTION
OŚCIK, J. (Jarosław)
Adsorption. - Chichester : Horwood, 1982
xv,206p : ill. - Includes index
0-85312-166-4 SRL
2926768

ADSORPTION
- Conference proceedings
ADSORPTION at the gas-solid and liquid-solid interface / editors J. Rouquerol and K.S.W. Sing. - Amsterdam ; Oxford : Elsevier Scientific, 1982
xii,512p : ill. - (Studies in surface science and catalysis ; 10). - Includes index
0-444-42087-8 SRL
2786184

ADSORPTION
- Serials
ADSORPTION science & technology. - Vol.1, no.1 (Jan. 1984)-. - Oxford : Blackwell Scientific, 1984-
v. : ill
B84-32925 SRL
2926768
- [Another copy.] P.611/872

ADSORPTION
By solids - Conference proceedings
INTERDISCIPLINARY SURFACE SCIENCE CONFERENCE (2nd : 1975 : University of Warwick)
Adsorption at solid surfaces. - Amsterdam : North-Holland Publishing Co., 1975
iii-x, 744 p. : ill. - Includes index
r6010174 X.620/16484 Woolwich
0782157

ADSORPTION
Statistical mechanics. Simulations. Applications of digital computer systems
NICHOLSON, D. (David)
Computer simulation and the statistical mechanics of adsorption. - London : Academic Press, 1982
xv,398p : ill. - Includes index
0-12-518060-8 SRL
2892162

ADSORPTION. Gases
By surfaces of solids
The CHEMICAL physics of solid surfaces and heterogeneous catalysis / edited by D.A. King and D.P. Woodruff. - Amsterdam ; Oxford : Elsevier.
Vol.2: Adsorption at solid sufaces. - 1983
384p : ill. - Includes index
0-444-42026-6 SRL
276668x

GREGG, S. J. (Sidney John)
Adsorption, surface area and porosity. - 2nd ed. - London : Academic Press, 1982
xi,303p : ill. - Includes index
0-12-300956-1 SRL
276668x

SAXENA, S. C.
Thermal accommodation and adsorption coefficients of gases. - New York ; London : McGraw-Hill, c1981
xxxvi,412p : ill. - (McGraw-Hill/CINDAS data series on material properties ; v.II-1). - Includes index
0-07-065031-4 SRL
2540398

ADSORPTION. Polymers
LIPATOV, †I Uriĭ Sergeevich
Adsorption of polymers. - New York ; Chichester : Wiley ; Jerusalem ; London : Israel Program for Scientific Translations, [1976]
ix,177p : ill. - Includes index
0-470-54040-0 SRL
0724564

ADSORPTION. Solutes. Aqueous solutions
- Conference proceedings
ADSORPTION from aqueous solutions / [proceedings of a symposium on adsorption from aqueous solutions held March 24-27, 1980, as a satellite symposium to the meeting of the American Chemical Society Division of Colloid and Surface Chemistry, Houston, Texas] ; edited by P.H. Tewari. - New York ; London : Plenum, c1981
viii,248p : ill. - Includes index
0-306-40747-7 SRL
2640244

ADSORPTION from solution. - London : Academic Press, Feb.1983
[400]p
0-12-530980-5 SRL
2640244

ADSORPTION. Solutions
By surfaces of solids
ADSORPTION from solution of the solid/liquid interface / edited by G.D. Parfitt and C.H. Rochester. - London : Academic, 1983
xi,416p : ill. - Includes bibliographies and index
0-12-544980-1 SRL
3124622

ADSORPTION. Surface-active agents
- Conference proceedings
COLLOID AND SURFACE SCIENCE SYMPOSIUM (52nd : 1978 : Knoxville)
Solution chemistry of surfactants. - New York ; London : Plenum Press.
Vol.1. - 1979
515,xxvii p : ill. - Includes index
0-306-40174-6 SRL
2053810
Vol.2. - 1979
xvii p,p519-962 : ill. - Includes index
0-306-40175-4 SRL
2053810

INTERNATIONAL SYMPOSIUM ON SOLUTION BEHAVIOR OF SURFACTANTS: THEORETICAL AND APPLIED ASPECTS (1980 : Clarkson College of Technology)
Solution behavior of surfactants. - New York ; London : Plenum Press, 1982
2v. : ill. - Includes index
0-306-41025-7 SRL
2053810

ADULT BAPTISM
See also
Ordo initiationis Christianorum adultorum. Roman Catholic Church. Liturgy and Ritual

ADULT BAPTISM. Catholic Church
CATHOLIC CHURCH
[Ordo initiationis christianae adultorum. English]. Rite of Christian initiation of adults. - Study ed., provisional text. - London : Catholic Truth Society, [1978]
[5],151p
B79-03380 X.100/18089
1765949

ADULT BAPTISM. Christian church
BALL, Peter, 1932-
Journey into faith. - London : SPCK, 1984
vii,52p. - Bibliography: p51-52
0-281-04082-6 X.200/44287
3173216

ADULT CANDIDATES. G.C.E. examinations. Education. Great Britain
Adult external candidates
BUTLER, Linda
Adult external candidates for GCE examinations. - Leicester : ACACE, 1981
iii,101p. - (Enquiry officer's report / Advisory Council for Adult and Continuing Education)
0-906436-06-0 X.525/6265
2790041

ADULT EDUCATION
See also
Chautauquas
Extra-mural departments
Extra-mural education
Further education
University of Nottingham. Department of Adult Education
University of Southampton. Department of Adult Education

ADULT EDUCATION
BROOKFIELD, Stephen
Adult learners, adult education and the community. - Milton Keynes : Open University Press, 1983
229p. - Bibliography: p204-221. - Includes index
0-335-10409-6 X.529/64804
024516x

EDUCATION for adults / edited by Malcolm Tight. - London : Croom Helm, c1983
2v. : ill. - Includes index
B83-23341 X.529/62570-62571
024516x

EDUCATION for adults. - Milton Keynes : Open University Press. - (Education)
Block C: The institutions. - 1984
66p : ill. - (E355 ; block C). - Includes bibliographies
0-335-13267-7 P.523/290
024516x
Block B: Adult learning. - 1984
89p : ill. - (E355 ; block B). - Includes bibliographies
0-335-13266-9 P.523/290
024516x

ADULT EDUCATION -continued

ESSAYS in adult learning and teaching. - Manchester : Dept. of Adult and Higher Education Manchester University, c1984
vi,50p. - Bibliography: p49-50
0-903717-33-6 X.529/68371
024516x

ETAVE, Roberto J.C.
Uma pedagogia para o homem. - Petrópolis : Editora Vozes ; [São Luis] : Universidade do Maranhão, 1971
210p
r6159653 X.529/37290
061968X

FREIRE, Paulo
Ação cultural para a liberdade e outros escritos. - Rio de Janeiro : Paz e Terra, 1976
149p. - (Coleção O mundo hoje ; v.10)
r6066194 X.519/39305
061968X

HARRIS, W. J. A. (William Jack Augustus)
Comparative adult education. - London : Longman, 1980
ix,198p. - Bibliography: p185-194. - Includes index
0-582-29510-6 X.529/39074
2088746

JARVIS, Peter
The sociology of adult & continuing education. - London : Croom Helm, c1985
278p. - Bibliography: p243-259. - Includes index
0-7099-1438-5 X.529/69642
3204030

JONES, R. Kenneth
Sociology of adult education. - Aldershot : Gower, c1984
ix,159p : ill. - Bibliography: p143-159
0-566-00663-4 X.529/66785
2345080

KREITLOW, Burton W.
Examining controversies in adult education. - San Francisco ; London : Jossey-Bass, 1981
xxiv,290p. - (The AEA handbook series in adult education) (The Jossey-Bass series in higher education). - Bibliography: p257-276. - Includes index
0-87589-489-5 X.520/24230
024516x

LAWSON, K. H. (Kenneth Harold)
Analysis and ideology. - Nottingham : Dept. of Adult Education, University of Nottingham, c1982
vii,100p. - (Nottingham studies in the theory and practice of the education of adults). - Includes index
0-902031-77-5 X.0519/573
024516x

LAWSON, Kenneth Harold
Philosophical concepts and values in adult education. - [Nottingham] : Department of Adult Education, University of Nottingham ; [London] : National Institute of Adult Education, 1975
[1],113p. - (Nottingham studies in the theory and practice of the education of adults ; [2]). - Bibliography: p.107-109. - Includes index
0-902031-37-6 X.529/30821
061968x

LAWSON, Kenneth Harold
Philosophical concepts and values in adult education. - Revised ed. - Milton Keynes : Open University Press, 1979
120p. - (The Open University Press series in adult and continuing education). - Bibliography: p.115-117. - Includes index
0-335-00254-4 X.519/30211
061968x

NEW pathways in adult and continuing education / edited by Roger H. Mercer, Michael A. Redmond. - Sheffield : Sheffield City Polytechnic Department of Education Management, 1981
183p. - (Sheffield papers in education management ; no.20)
0-903761-35-1 X.525/8023
024516x

OSNOVNO izobraževanje odraslih / [ureja uredniški odbor : Drago Glogovšek ... et al.]. - Ljubljana : Univerzum, 1978
220p. - (Strokovne publikacije. Priročniki za dopolnilno pedagoško in andragoško izobraževanje)
r6178042 X.529/35917
0835668

PATERSON, Ronald William Keith
Values, education and the adult. - London : Routledge and Kegan Paul, 1979
viii,306p. - (International library of the philosophy of education). - Includes index
0-7100-0102-9 X.529/32896
061968x

THOMAS, Sir Ben Bowen
Syr Ben Bowen Thomas, 1899-1977. - Aberystwyth : Cymdeithas Addysg y Gweithwyr, Rhanbarth Gogledd Cymru, 1978
55p,plate
0-901833-85-1 X.510/15899
1685260

THOMAS, J. E. (James Edward)
Radical adult education. - Nottingham : Dept. of Adult Education, University of Nottingham, c1982
viii,73p. - (Nottingham studies in the theory and practice of the education of adults). - Bibliography: p67-71. - Includes index
0-902031-87-2 X.0519/573
024516x

TURNER, John D. (John Derfel)
The meaning of adult education. - Manchester : John Rylands University Library of Manchester, 1980
p171-193
B81-35542 X.520/24916
024516x

WILTSHIRE, Harold
The spirit and the form. - [Nottingham] : Department of Adult Education, University of Nottingham, 1976
[7],159p,plate. - (Nottingham studies in the theory and practice of the education of adults). - Includes index
0-902031-38-4 X.0519/573(2)
024516x

ADULT EDUCATION
- Bibliographies

HAYWOOD, Peter G. (Peter George)
A bibliography of adult teaching, psychology and research. - Nottingham : Nottingham University Department of Adult Education, [1983]
48p
0-902031-67-8 X.525/8148
2965720

ADULT EDUCATION
- Conference proceedings

POLICY and research in adult education / selected by Brian Harvey ... [et al.]. - [Nottingham] : Department of Adult Education, University of Nottingham, c1981
150p. - (Nottingham studies in the theory and practice of the education of adults)
0-902031-60-0 X.0519/573(8)
024516x

ADULT EDUCATION
Adult literacy education

BERGGREN, Carol
The literacy process. - London : Writers and Readers Publishing Cooperative, 1975
47p. - Bibliography: p.46-47
0-904613-04-6 X.519/25393
0563927

CARTER, Edward, 1902-
Literacy, libraries and liberty. - [London] : Library Association, c1969
15p. - (Arundell Esdaile memorial lectures ; 1968)
0-85365-321-6 2719.e.827
0563927

LITERACY and the adult learner / [editor John Beynon]. - Treforest : Polytechnic of Wales, 1977
[1],ii,200[i.e.219]p : ill. - (Occasional publications / Polytechnic of Wales). - Includes bibliographies
B79-10932 X.512/6498
0563927

ADULT EDUCATION
Adult literacy education - Bibliographies - Serials

LITERACY bibliographies / [issued by] International Institute for Adult Literacy Methods. - 1- ; May 1977-. - Tehran : The Institute, 1977-
r6051843 OPL
131923X

ADULT EDUCATION
Adult literacy education - Conference proceedings

INTERNATIONAL SYMPOSIUM FOR LITERACY (1975 : Persepolis)
A turning point for literacy. - Oxford : Pergamon, 1976
viii,277p : ill. - (Pergamon international library)
0-08-021385-5 X.900/25394
0563927

ADULT EDUCATION
Adult literacy education - For teaching

MACFARLAND, Tom
Literacy in action. - Basingstoke : Macmillan, 1979
ix,77p : ill. - (Adult literacy guides). - Bibliography: p.76
0-333-26358-8 X.519/30659
1703412

SCOTTISH ADULT LITERACY AGENCY
Training for adult literacy. - [Edinburgh] : [The Agency], 1979
1v. : ill. - Includes bibliographies
0-86082-161-7 X.525/3109
1703412
- [Accompanying sound tape.] Cup.580.e.26

ADULT EDUCATION
Adult literacy education - Serials

[NEWSLETTER (National Institute of Adult Education (England and Wales). Adult Literacy & Basic Skills Unit)].
Newsletter / Adult Literacy & Basic Skills Unit. - No 1 (Apr./May 1980)-. - London : The Unit, 1980-
v. : ill
B81-07694 P.903/660
2360012

ADULT EDUCATION
Adult literacy education. British organisations - Directories

A DIRECTORY of British-based resources for adult literacy work overseas / The British Committee on Literacy. - Reading : The Committee, c1983
v,74p
0-9505439-1-8 X.529/64977
3164950

ADULT EDUCATION
Adult literacy education. Curriculum subjects: English language. Composition - For teaching

BBC writing and spelling handbook / edited by Catherine Moorhouse. - London : British Broadcasting Corporation, 1979
96p : ill. - Bibliography: p.89-94. - Includes index
0-563-16292-9 BS.129/172
2225697

ADULT EDUCATION
Adult literacy education. Implications of visual communication

ZIMMER, Anne
Visual literacy in communication. - Amersham : Hulton [for] the International Institute for Adult Literacy Methods, 1978
144p : ill. - (Literacy in development). - Bibliography: p.141-143
0-7175-0806-4 X.519/41531
1757415

ADULT EDUCATION
Adult literacy education. Teaching methods

NEWMAN, Anabel P.
Adult basic education-reading. - Boston [Mass.] : London : Allyn and Bacon, 1980
xii,235p : ill. - Bibliography: p.187-191. Includes index
0-205-06954-1 X.520/21624
1060791

Appendix C.1 to accompany Adult basic education : reading. - Boston ; London : Allyn and Bacon, c1980
38p
0-205-07275-5 X.529/53617
1060791

SINGH, Sohan
Learning to read and reading to learn. - Tehran : Hulton [for] the International Institute for Adult Literacy Methods, 1976
116p. - (Literacy in development. a series of training monographs)
0-7175-0765-3 UNP638/7
1060791

WORKING together / Adult Literacy & Basic Skills Unit. - [2nd ed.]. - London : ALBSU, 1981
75p
0-906509-06-8 X.529/41630
1060791

ADULT EDUCATION
Adult literacy education. Use of programmed texts - For teaching

THIAGARAJAN, Sivasailam
Programmed instruction for literacy workers. - Amersham : Hulton [for] the International Institute for Adult Literacy Methods, 1976
136p : ill. - (Literacy in development)
0-7175-0764-5 X.909/41163
1031368

ADULT EDUCATION
Adult literacy education. Use of radio programmes - For teaching

BURKE, Richard Cullen
The use of radio in adult literacy education. - Amersham : Hulton [for] the International Institute for Adult Literacy Methods, 1976
116p : ill. - (Literacy in development)
0-7175-0760-2 X.510/15274
055944x

ADULT EDUCATION
Adult numeracy education - Bibliographies

NUMERACY resource guide. - London : Adult Literacy Support Services Fund, [1979]
9p
B82-40585 X.525/6325
2799812

ADULT EDUCATION
Audio-visual aids - Bibliographies
OHLIGER, John
Media and adult learning. - New York ; London : Garland, 1975
xxiii,486p. - (Garland reference library of the humanities ; vol.18). - Includes index
0-8240-1088-4 X.529/36829
2225816

ADULT EDUCATION
Audiovisual materials - Bibliographies
CITY OF LONDON POLYTECHNIC Library and Learning Resources Service
Media resources on educational trends for lecturers. - [London] : [The Service], [1978]
[5],14p. - Includes index
B78-37417 X.512/6708
1697633

ADULT EDUCATION
Comparative adult education. Documents on comparative adult education & comparative lifelong education in stock of University of Nottingham. Library - Bibliographies
HAYWOOD, Peter G. (Peter George)
Comparative adult education and lifelong education bibliography. - [Nottingham] : University of Nottingham, [1979]
[3] leaves,90p
B80-38673 X.525/3751
2308002
- [Another copy.] OPL

ADULT EDUCATION
Curriculum subjects: Foreign languages. Self-directed learning
HOLEC, Henri
Autonomy and foreign language learning. - Oxford : Published for and on behalf of the Council of Europe by Pergamon, 1981, c1979
v,51p. - Bibliography: p50-51
0-08-025357-1 X.950/5505
2525070

ADULT EDUCATION
Curriculum subjects: Reading
SCALES, Alice M.
Reading to achieve. - Columbus, Ohio ; London : Merrill, c1983
xvi,256p : ill
0-675-20034-2 X.955/3045
3287637

ADULT EDUCATION
Curriculum subjects: Social sciences
HARRISON, J. (John)
Catching the colour of life. - Leicester : Department of Adult Education, University of Leicester, 1980
35p. - (Vaughan paper ; no.24). - Bibliography: p35
0-901507-15-6 Ac.1354.b[no.24]
2380439

ADULT EDUCATION
Distance study
DISTANCE teaching for higher and adult education / edited by Anthony Kaye and Greville Rumble. - London : Croom Helm in association with the Open University Press, c1981
342p : ill. - Includes bibliographies and index
0-7099-0468-1 X.529/41304
2208393

ADULT EDUCATION
Distance study - Conference proceedings
EDUCATION of adults at a distance / prologue by Michael J. Pentz ; commentary by Michael W. Neil. - London : Kogan Page in association with the Open University Press, 1981
270p : ill
0-85038-415-x X.529/46080
2632748

ADULT EDUCATION
Distance study. Teaching
JENKINS, Janet
Materials for learning. - London : Routledge & Kegan Paul, 1981
xii,209p : ill. - Bibliography: p192-205. - Includes index
0-7100-0808-2 X.520/25380
2578948

ADULT EDUCATION
Implications of sex differences
BERRYMAN, Julia C.
Sex differences in behaviour. - Nottingham : Department of Adult Education, University of Nottingham, c1981
ii,27p. - (Adults ; 6). - Includes bibliographies
0-902031-51-1 X.529/46756
2702258

ADULT EDUCATION
Research. Methodology
CHADWICK, A. F. (Alan F)
Approaches to research in adult education. - [Nottingham] : University of Nottingham School of Education, 1979
23p. - (Rediguides ; 27). - Bibliography: p.20-21. - Includes index
0-85359-127-x X.519/29908
1921282

LONG, Huey B.
Changing approaches to studying adult education. - San Francisco ; London : Jossey-Bass, 1980
xxi,154p. - (Handbook series in adult education). - Bibliography: p.138-151. - Includes index
0-87589-447-x X.520/21574
1921282

ADULT EDUCATION
Role of mass media
MUKHERJEE, Biswa Nath
Mass media and political modernity. - Agra : National Psychological Corporation, 1979
186p. - (Bhargava research monographs series ; no.4)
r6093133 X.520/14956
0438758

ADULT EDUCATION
Role of public libraries - Conference proceedings
ADULT education and public libraries in the 1980s. - London : Library Association, 1980
94p
0-85365-662-2 2725.b.97
1965336

ADULT EDUCATION
Role of Unesco
SUCUPIRA, Newton
A UNESCO e o conceito de educação de adultos. - João Pessoa : Editora Universitária, UFPb, 1978
42p
r6158574 X.519/43783
0839337

ADULT EDUCATION
Serials with West German imprints: 'Hessische Blätter für Volksbildung' & 'Volkshochschule im Westen', to 1975
KURZE, Wilhelm
Zeitschriften der Erwachsenenbildung. - Bochum : Brockmeyer, 1977
288p. - (Bochumer Studien zur Publizistik- und Kommunikationswissenschaft ; Bd13). - Includes bibliographies
3-921543-74-6 X.519/35398
1530542

ADULT EDUCATION
Study techniques - For adult education
APPS, Jerold W.
Study skills for adults returning to school. - 2nd ed. - New York ; London : McGraw-Hill, c1982
xvi,200p : ill. - Includes index
0-07-002165-1 X.529/46618
2697343

ADULT EDUCATION
Study techniques & teaching
MILLARD, Lesley
Adult learners. - Nottingham : Department of Adult Education, University of Nottingham, c1981
ii,26p. - (Adults ; 4). - Includes bibliographies
0-902031-49-x X.529/46754
2702266

ADULT EDUCATION
Teaching
ADULTS in education / edited by Jenny Rogers. - 2nd rev. ed. - London : British Broadcasting Corporation, 1984
88p : ill. - Bibliography: p87
0-563-21112-1 X.525/10593
3257223

ADULT EDUCATION
Teaching methods
BERGEVIN, Paul
Adult education procedures. - New York : Seabury, c1963
245p : ill. - (A Crossroad book)
0-8164-2000-9 X.529/57585
2975696

ADULT EDUCATION
Teaching. Theories
MACKIE, Karl
The application of learning theory to adult teaching. - Nottingham : Department of Adult Education, University of Nottingham, c1981
18p. - (Adults ; 2). - Bibliography: p17-18
0-902031-47-3 X.529/41885
2497859

TOWARDS a developmental theory of andragogy by the Nottingham Andragogy Group. - [Nottingham] : Department of Adult Education, University of Nottingham, 1983, c1981
viii,48p : ill. - (Adults. Psychological and educational perspectives ; 9)
0-902031-88-0 X.529/64290
2497859

ADULT EDUCATION
Theories - Serials
JOURNAL of adult education. - Vol.1, no.1 (Sept. 1982)-. - Lusaka : Centre for Continuing Education, 1982-
v.
r6807362 P.521/3843
9023917

ADULT EDUCATION. Attleborough. Norfolk
AMISON, Paul
The Attleborough experiment. - Cambridge : National Extension College Trust, 1982
36p : 1ill
0-86082-340-7 X.529/51892
2835827

ADULT EDUCATION. Australian aborigines. Deserts. Western Australia
- Serials
WESTERN AUSTRALIA. Adult Aboriginal Education
Annual report. - 1976-. - West Perth : Education Department of Western Australia, Technical Education Division, Adult Aboriginal Education, 1977-
r6051911 CS.G.485/109
1319779

ADULT EDUCATION. Austria
to date
MÖRTH, Ingo
Bildung neben Beruf. - Linz : Trauner, 1983
viii,184p.. - (Sozialwissenschaftliche Materialien ; Bd.1)
3-85320-276-4 X.529/57345
9000887

ADULT EDUCATION. Austria
to date. Social aspects
KUZMICS, Helmut
Zweiter Bildungsweg und Chancengleichheit. - Graz : K. Freisitzer, 1983
115p. : ill. - (Grazer soziologische Schriften ; Bd.9)
r6308208 P.523/67(9)
9032983

ADULT EDUCATION. Avon
ADULT education / County of Avon Education Committee. - Bristol : Avon County Public Relations Department, 1982
6,3p
0-86063-174-5 X.525/7074
2886286

ADULT EDUCATION. Baja. Hungary
1945-
GERGELY, Ferenc
Baja népoktatásának főbb kérdései 1944-1948 között. - Baja, 1980
60p. - (A bajai Türr István Múzeum kiadványai ; 24)
963-01-2444-0 X.0709/645(24)
8712778

ADULT EDUCATION. Belorussia
SOROKA, Zoīā Nikolaevna
Deīātel'nost' obshchestva "Znanie" Belorusskoī SSR po kommunisticheskomu vospitaniīū trudīāshchikhsīā (1959-1975gg). - Minsk : Izdatel'stvo "Nauka i tekhnika", 1976
228p
r6044108 X.519/23369
1312235

ADULT EDUCATION. Bolivia
Adult literacy education
ALFABETIZACIÓN / [documentos reunidos por] Mariano Baptista Gumucio. - La Paz : Editorial "Los Amigos del Libro", 1973
243p : ill
r6100971 X.519/42452
2132923

ADULT EDUCATION. Bradford. West Yorkshire (Metropolitan County)
Advisory, information & counselling services: Education Advice Service for Adults. Information retrieval services using digital computer systems. Machine-readable files

TILLEY, N. V.
Education Advice Service for Adults. - [London] : British Library, c1982
vi,53p : ill. - (Library and information research report ; 8)
0-7123-3011-9 2719.e.625
2935449

ADULT EDUCATION. Brazil
1964-

EDUCAÇÃO de massa e ação comunitária / Arlindo Lopes Corrêa, editor. - Rio de Janeiro : Mobral, 1979
472p
r6700762 X.529/42532
8712905

ADULT EDUCATION. Broadstairs. Kent
Role of public libraries

WALTERS, John, 1936-
The library role in adult education. - [Maidstone] : Kent County Library, 1981
20,[24]p,[3]leaves of plates : ill. - (Research & development report ; no.3)
B84-02739 X.525/8380
2990903

ADULT EDUCATION. Cerebral palsied persons
Adult literacy education - For teaching

WIDENING horizons. - London : Adult Literacy and Basic Skills Unit, 1981
36p : ill
0-906509-10-6 X.529/43514
2558157

ADULT EDUCATION. China

ADULT education in China / edited by Carman St John Hunter & Martha Makee Keehn. - London : Croom Helm, c1985
147p : ill. - (Croom Helm series in international adult education). - Bibliography: p116-117
0-7099-3721-0 X.529/68336
3155056

ADULT EDUCATION. Colombia

TALLER de marginados -? / Ingrid Becker-Ross ... [et al.]. - Bogotá : Punta de Lanza, 1976
96p : ill
r6066209 X.519/39348
1719521

ADULT EDUCATION. Denmark

BUNNAGE, David
Voksenuddannelse. - København : Socialforskningsinstituttet : I kommission hos Teknisk Forlag, 1978
310p. - (Socialforskningsinstituttet. Publikation ; 81)
87-7487-136-6 S.B.210/9 (81)
2137208

ADULT EDUCATION. Denmark
to date

FOLKEOPLYSNING og forskningsformidling i 80'erne / redaktør: Carl August Bernhard. - Århus : Folkeuniversitetet i Århus, 1981
91p : ill
r6611256 X.520/32721
9047603

ADULT EDUCATION. Developing countries

TOWNSEND COLES, Edwin
Adult education in developing countries. - 2nd ed. - Oxford : Pergamon, 1977
xviii,199p : ill. - (Pergamon international library). - Bibliography: p.195-196. - Includes index
0-08-021293-x X.529/31653
140363x

ADULT EDUCATION. Developing countries
- Conference proceedings

INTERNATIONAL CONFERENCE ON ADULT EDUCATION AND DEVELOPMENT
(1976 : Dar es Salaam)
Adult learning, a design for action. - Oxford : Pergamon, 1978
vii,337p. - Includes index
0-08-022245-5 X.529/32651
140363x
- [Another copy: paperback]
ISBN 0-08-023007-5 X.519/28934

ADULT EDUCATION. Developing countries
Adult literacy education. Implications of economic development

CAMPOS, S. N.
Adult literacy in national development. - Norwich : Geo Abstracts [for] Centre for Development Studies, University College of Swansea, 1980
[1],ii,31p. - (Monographs / University College of Swansea. Centre for Development Studies ; no. 10). - Bibliography: p.28-31
0-86094-053-5 X.529/38045
2286564

ADULT EDUCATION. Developing countries
Adult literacy education. Teaching methods: Games

EVANS, David Russell
Games and simulations in literacy training. - Amersham : Hulton [for] the International Institute for Adult Literacy Methods, 1979
xi,136p : ill. - (Literacy in development). - Bibliography: p.134-136
0-7175-0822-6 X.619/22219 Woolwich
1971751

ADULT EDUCATION. Developing countries
Curriculum subjects: Science - For teaching

THOMAS, Frederick J.
Towards scientific literacy. - Amersham : Hulton [for] the International Institute for Adult Literacy Methods, 1978
xi,96p : ill. - (Literacy in development)
0-7175-0805-6 X.619/19966 Woolwich
1923501

ADULT EDUCATION. Developing countries
Role in economic development

COMBATTING poverty through adult education / edited by Chris Duke. - London : Croom Helm, c1985
253p : ill. - (Croom Helm series in international adult education). - Bibliography: p238-247. - Includes index
0-7099-0861-x X.529/68243
3155013

ADULT EDUCATION. Doncaster. South Yorkshire (Metropolitan County)
1800-1900

THRALL, Alan
The history of adult education in 19th century Doncaster. - [Doncaster] : Doncaster Metropolitan Borough Museums and Arts Service, [1977]
[3],102p,[4] leaves of plates : 3 ill
B78-40126 X.512/6182
1738542

ADULT EDUCATION. Dublin (County)
Courses - Directories - Serials

The GUIDE to evening classes and leisure learning in Dublin (City and County). - 1980-1981. - Dublin : Wolfhound Press, c1980
112p
Subsequent volumes, when received, will be placed at this shelf mark
0-905473-54-x P.521/3518
2401959

ADULT EDUCATION. Dunántúl. Hungary
1918 - Conference proceedings

INTERNATIONALES KULTURHISTORISCHES SYMPOSION MOGERSDORF
(8th : 1976 : Kőszeg)
Nemzetközi kultúrtörténeti szimpozion Megersdorf 1976 Kőszegen. - Szombathely : Vas megyei Levéltár, 1978
384p. - (Internationales Kulturhistorisches Symposion Mogersdorf ; Bd.8)
963-01-1349-x X.0709/595(8)
8706646

ADULT EDUCATION. Educationally disadvantaged persons. Great Britain

JONES, H. A. (Henry Arthur)
Education and disadvantage. - [Leicester] : University of Leicester, Department of Adult Education, 1977
[3],18p. - (Vaughan paper ; no.22)
0-901507-10-5 Ac.1354.b[no.22]
1195808

ADULT EDUCATION. England

CANTOR, Leonard M. (Leonard Martin)
Further education today. - 2nd fully rev. ed. - London : Routledge & Kegan Paul, 1983
vi,265p : ill. - Bibliography: p239-240. - Includes index
0-7100-9501-5 X.529/58327
1902504

CANTOR, Leonard Martin
Further education today. - London : Routledge and Kegan Paul, 1979
viii,255p : ill. - Bibliography: p.229. - Includes index
0-7100-0412-5 X.529/34243
1902504

MEE, Graham
Structure and performance in adult education. - London : Longman, 1978
viii,127p : ill. - Includes bibliographies and index
0-582-48944-x X.519/27666
1346385

ADULT EDUCATION. England
- Serials

ADVISORY COUNCIL FOR ADULT AND CONTINUING EDUCATION - 1979-. - London : H.M.S.O., 1980-
v.
B81-08274 B.S.10/884
2363348

ADVISORY COUNCIL FOR ADULT AND CONTINUING EDUCATION
Annual report of the Advisory Council for Adult and Continuing Education for the year ending 31st March ... presented to the Secretaries of State for Education and Science and for Wales. - 1980. - London : H.M.S.O., 1981
35p
0-11-270514-6 BS.10/884
2363348

ADULT EDUCATION. England
1800-1900

NINETEENTH century self-help in education. - [Nottingham] : Department of Adult Education, University of Nottingham. - (Nottingham studies in the history of adult education)
Vol.2: Case study: The Carlisle Working Men's Reading Rooms. - c1983
71p. - Bibliography: p65-69. - Includes index
0-902031-91-0 X.809/59940
2999315

SAMUEL Smiles and nineteenth century self-help in education / [by A.H. Thornton et al.]. - Nottingham : Department of Adult Education, University of Nottingham, c1983
72p. - (Nottingham studies in the history of adult education ; v.1). - Includes index
0-902031-86-4 X.809/57303
2999315

ADULT EDUCATION. England
Accommodation

PRIME use accommodation for adult education / by Keith Percy ... [et al.] ; [for] ACACE. - Leicester : ACACE, [1982]
71p : 2ill. - (Report / Advisory Council for Adult and Continuing Education)
0-906436-15-x X.529/50235
2791137

ADULT EDUCATION. England
Adult & lifelong education. Organisations: National Institute of Adult Continuing Education - Serials

[NEWSLETTER (National Institute of Adult Continuing Education)]. Newsletter / National Institute of Adult Continuing Education (England and Wales). - No.1 (Oct. 1984)-. - Leicester : The Institute, 1984-
v. : ill
B85-28925 P.903/1038
3304868

ADULT EDUCATION. England
Adult literacy education - Conference proceedings

The ADULT illiterate in the community / editor R.J. Kedney. - Bolton : Bolton College of Education (Technical), 1975
[5],167p : ill
0-9504588-0-5 X.519/25289
0147648

ADULT EDUCATION. England
Adult literacy education - Serials

[SPECIAL development projects (National Institute of Adult Education (England and Wales). Adult Literacy and Basic Skills Unit)]. Special development projects / ALBSU. - No.1 (Nov. 1981). - London : The Unit, 1981
1v. : ill
B84-23826 P.525/453
3125165

[SPECIAL development projects report (National Institute of Adult Education (England and Wales). Adult Literacy and Basic Skills Unit)]. Special development projects report / ALBSU. - No.2-. - London : The Unit, 1982-
v. : ill
B84-23825 P.525/437
3125165

ADULT EDUCATION. England
Adult literacy education. Organisations: National Institute of Adult Education (England and Wales). Adult Literacy and Basic Skills Unit. Development projects, 1978-1980

ADULT Literacy Unit development projects 1978-1980 / Adult Literacy & Basic Skills Unit. - London : Adult Literacy & Basic Skills Unit, 1981
68p : ill
0-906509-05-x X.525/4446
2410443

ADULT EDUCATION. England
Adult literacy education. Organisations: National Institute of Adult Education (England and Wales). Adult Literacy Unit - Serials

NATIONAL INSTITUTE OF ADULT EDUCATION (ENGLAND AND WALES).
Adult Literacy Unit
Adult literacy - 1978/79. - London : H.M.S.O., 1980
v.69p
Later reports, when received, will be placed at this shelfmark
0-11-270463-8 BS.10/909
2327376

ADULT EDUCATION. England
Adult literacy education. Teaching aids: Worksheets

FROM wages to Windscale / Friends Centre Brighton. - London : Adult Literacy & Basic Skills Unit, [198]
[20]p : ill
B81-07953 X.525/4013
236221x

ADULT EDUCATION. England
Adult literary education, 1975-1977

JONES, Henry Arthur, b.1917
Adult Literacy. - [Leicester] : National Institute of Adult Education, [1978?]
[2],v,121p. - Bibliography: p.119-121
0-900559-39-x X.519/30809
204949x

ADULT EDUCATION. England
Adult numeracy education. Teaching aids: Worksheets

FROM wages to Windscale / Friends Centre Brighton. - London : Adult Literacy & Basic Skills Unit, [198]
[20]p : ill
B81-07953 X.525/4013
236221x

ADULT EDUCATION. England
Advisory, information & counselling services

ADVISORY COUNCIL FOR ADULT AND CONTINUING EDUCATION
Links to learning. - Leicester : ACACE, c1979
70p. - (Report / Advisory Council for Adult and Continuing Education)
0-906436-02-8 X.529/50277
2791145

ADULT EDUCATION. England
Curriculum subjects: Arts. Provision

ADKINS, Geoffrey
The arts and adult education. - Leicester : ACACE, 1981
xi,164p
0-906436-11-7 X.529/50242
2791161

ADULT EDUCATION. England
Curriculum subjects: Microcomputer systems

MICROTECHNOLOGY and the education of adults / ACACE. - London : ACACE, c1983
44p
0-906436-29-x X.529/62587
3138364

ADULT EDUCATION. England
Curriculum subjects: Science

ADVISORY COUNCIL FOR ADULT AND CONTINUING EDUCATION
Basic science education for adults. - Leicester : ACACE, [1981]
43p. - (Report / Advisory Council for Adult and Continuing Education)
0-906436-12-5 X.529/50273
2791153

ADULT EDUCATION. England
Distance study

DISTANCE learning and adult students. - Leicester : ACACE, c1983
iv,72p. - (Report / ACACE). - Bibliography: p65
0-906436-22-2 X.629/21916
3034992

ADULT EDUCATION. England
Organisations: Guild of St George, to 1983

MORLEY, Catherine W.
John Ruskin. - New York ; London : Garland, 1984
ca 650p in various pagings : ill. - (Outstanding theses from the Courtauld Institute of Art). - Bibliography: p319-327
0-8240-5981-6 X.520/36792
325965x

ADULT EDUCATION. England
Provision

ADVISORY COUNCIL FOR ADULT AND CONTINUING EDUCATION
Protecting the future for adult education. - Leicester : ACACE, [1981]
64p. - (Report / Advisory Council for Adult and Continuing Education)
0-906436-07-9 X.529/50275
2791188

A strategy for the basic education of adultsreport commissioned by the Secretary of State for Education and Science. - Leicester : ACACE, 1979
107p. - (Report / Advisory Council for Adult and Continuing Education)
0-906436-01-x X.529/50276
2791188

POST-INITIAL education in the north west of England / by Keith Percy ... [et al.]. - Leicester : ACACE, 1983
ix,252p : ill - (Report / ACACE). - Bibliography: p243
0-906436-20-6 X.622/17664
3035034

ADULT EDUCATION. England
Provision. Role of universities, 1870-1900

MARRIOTT, Stuart
A backstairs to a degree. - Leeds : Department of Adult Education and Extramural Studies, University of Leeds, 1981
iv,107p. - (Leeds studies in adult and continuing education)
0-907644-00-7 X.529/45074
2603217

ADULT EDUCATION. England
Role of universities

ELLWOOD, Caroline
Adult learning today. - London : Sage Publications, [1977]
[6],265,[1]p : ill. - (Sage studies in social and educational change ; vol.4). - Bibliography: p.261-265
0-8039-9979-8 X.519/26702
1080105

ADULT EDUCATION. England
Role of volunteers

ADVISORY COUNCIL FOR ADULT AND CONTINUING EDUCATION
Volunteers in adult education. - Leicester : ACACE, c1983
viii,132p : ill. - Bibliography: p109-110
0-906436-25-7 X.529/62895
3139972

ADULT EDUCATION. Ethnic minorities. England
Provision

LITTLE, Alan, 1934-
Adult education and the black communities. - Leicester : ACACE, 1982
vi,35p
0-906436-16-8 X.529/50234
2791129

ADULT EDUCATION. Europe
to date - Serials

ADULT education in Europe / European Centre for Leisure and Education. - No.1-. - Prague : European Centre for Leisure and Education, 1977-
r6441201 P.803/1426
9047611

ADULT EDUCATION. Europe
1945-

REDEN ist Leben / [Redaktion : Uwe Meyer, Ernst Tipke]. - Lüneburg : Neubauer, 1981
246p. - (Wissenschaft und Praxis ; Bd 6). - Includes bibliography of Tölke's works
3-88456-014-x X.529/68865
9162062

ADULT EDUCATION. Europe
Adult basic education. Distance study. Teaching aids: Mass media

USING the media for adult basic education / edited by Anthony Kaye and Keith Harry. - London : Croom Helm, c1982
255p : ill. - Includes index
0-7099-1506-3 X.529/46558
2696762

ADULT EDUCATION. Europe
Curriculum subjects: Foreign languages. European unit/credit system

SYSTEMS development in adult language learning / prepared for the Council of Europe by J.L.M. Trim ... [et al.]. - [New] ed. - Oxford : Pergamon for the Council of Europe, 1980
[5],146p
0-08-024587-0 X.900/27909
2146312

TRIM, J. L. M. (John Leslie Melville)
Developing a unit/credit scheme of adult language learning. - Oxford : Published for and on behalf of the Council of Europe by Pergamon, 1980
v,74p : ill. - Bibliography: p73-74
0-08-024596-x X.520/22958
2146312

ADULT EDUCATION. Europe
Distance study

HARRY, Keith
The European experience of the use of mass media and distance methods for adult basic education. - Milton Keynes : Open University Distance Education Research Group. - (DERG papers ; no.3a)
Vol.1: Main report. - 1981
ii,101p : ill. - Bibliography: p95-97
B82-06689 X.520/25774
2606976

- [Milton Keynes] : Open University Distance Education Research Group. - (DERG papers ; no.3b)
Vol.2: Appendices. - 1981
ii,114p. - Includes bibliographies
B82-06691 X.520/25773
2606976

ADULT EDUCATION. Faversham & Sittingbourne. Kent
Role of public libraries

WALTERS, John, 1936-
The library role in adult education. - [Maidstone] : Kent County Library, 1981
20,[24]p,[3]leaves of plates : ill. - (Research & development report ; no.3)
B84-02739 X.525/8380
2990911

ADULT EDUCATION. Finland
to date

BJÖRKSTRAND, Gustav
Den finlandssvenska folkhögskolan och allmogemobilisering. - Åbo : Åbo akademi, Kyrkohistoriska arkivet, 1980
52p. - (Meddelanden / Kyrkohistoriska arkivet vid Åbo akademi ; 10). - Bibliography
951-648-659-2 X.0102/90(10)
8945160

ADULT EDUCATION. France
to date

CACÉRÈS, Bénigno
Les deux rivages. - Paris : François Maspero, 1982
159p. - (Actes et mémoires du peuple)
2-7071-1319-0 X.629/22051
8851948

POUJOL, Geneviève
L'éducation populaire. - Paris : Les Éditions Ouvrières, [1981]
225p. - (Politique sociale). - Bibliography
2-7082-2201-5 X.529/50191
8851948

ADULT EDUCATION. France
Progressive teaching methods

KAEPPELIN, Philippe
Pratique de l'autogestion éducative. - [Paris] : Éditions Resma, 1974
237 p.. - (Le fait humain). - Bibliography : p. 233-234
r6025911 X.519/20725
0861170

ADULT EDUCATION. Germany
to 1945

ARBEITERBILDUNG nach dem Fall des Sozialistengesetzes (1890-1914) / [herausgegeben von] Josef Olbrich. - Braunschweig : Westermann, 1982
410p.. - (Dokumentationen zur Geschichte der Erwachsenenbildung). - Includes bibliography
3-14-160807-5 X.529/68385
8820023

TEWS, Johannes
Geistespflege in der Volksgemeinschaft. - Frankfurt/M. : Pädagogische Arbeitsstelle des Deutschen Volkshochschul-Verbandes, 1981
282p. - (Schriften zur Erwachsenenbildung. Materialien zur Erwachsenenbildung). - Bibliography
3-12-937950-9 X.809/51361
8820023

ADULT EDUCATION. Ghana
- Conference proceedings

UNIVERSITY OF GHANA. Institute of Adult Education
Adult education and the human environment. - Legon : The Institute, 1974
xi, 157 p., [9] leaves of plates (1 fold) : ill
r6010786 X.510/15392
1052187

ADULT EDUCATION. Glasgow. Strathclyde Region. Scotland
Curriculum subjects: Urban studies. Courses

HARVEY, Colin
The cultures of affluence and poverty. - Milngavie : Heatherbank Press for the Museum of Social Work, 1980
22leaves. - (Occasional paper / Museum of Social Work ; no.4)
0-905192-25-7 X.525/4700
252872x

ADULT EDUCATION. Great Britain

ADULT education for a change / edited by Jane L Thompson. - London : Hutchinson, 1980
256p : 1 ill. - Includes index
0-09-141620-5 X.529/36365
0530166

ADVISORY COUNCIL FOR ADULT AND CONTINUING EDUCATION
Continuing education. - Leicester : ACACE, c1983
8p
0-906436-24-9 X.525/9409
0530166

BELL, Judith
Never too late to learn. - London : Longman, c1982
viii,110p. - Bibliography: p107. - Includes index
0-582-49706-x X.529/46247
1875736

COMMUNIST PARTY OF GREAT BRITAIN. Education Sub-Committee
Education for adults. - [London] : CPGB, [1982]
20p
0-86224-018-2 X.529/48934
0530166

CONTINUING education for the post-industrial society / [edited by] Neil Costello and Michael Richardson. - Milton Keynes : Open University Press, 1982
x,146p. - Includes index
0-335-10186-0 X.529/54287
0530166

EDUCATION for adults. - London : National Association of Teachers in Further and Higher Education, [1979]
15p
B83-00877 X.529/50735
0530166

EDUCATION for adults. - Milton Keynes : Open University Press. - (Education)
Block D: Present and future. - 1984
64p : ill. - (E55 ; Block D). - Bibliography: p60-64
0-335-13268-5 X.525/10785
0530166

GROWING persons. - London : British Council of Churches, [1977]
vi,59,[1]p. - (Britain today & tomorrow ; no.2)
B78-15372 X.519/27892
1434772

HOPPER, Earl
Adult students. - London : Frances Pinter (Publishers) Limited, 1975
[4],187p. - Bibliography: p.176-182. - Includes index
0-903804-10-7 X.529/30271
0530166

LEGGE, Derek
The education of adults in Britain. - Milton Keynes : Open University Press, 1982
xi,244p. - (The Open University Press series in adult and continuing education). - Bibliography: p223-225. - Includes index
0-335-00267-6 X.520/26987
0530166

MEE, Graham
Organisation for adult education. - London : Longman, 1980
xiv,114p : ill. - Bibliography: p110-111. - Includes index
0-582-49701-9 X.529/39852
2090481

NEWMAN, Michael
The poor cousin. - London : Allen and Unwin, 1979
iii-xi,249p
0-04-374002-2 X.529/38676
0767697

PATES, Andrew
Second chances for adults. - London : Macmillan, 1979
viii,202p. - Bibliography: p.193-194. - Includes index
0-333-26251-4 X.529/34299
1875736

RICHARDS, E. J. (Elfyn John)
Adult education. - [Cardiff] : University College, Cardiff, [1975?]
18p
0-901426-14-8 X.529/53495
0530166

ROGERS, Jennifer, 1939-
Right to learn. - London : Arrow Books, 1976
202p,[16]p of plates : ill. - Includes index
0-09-914030-6 X.519/26356
0767697

ADULT EDUCATION. Great Britain
- Bibliographies

MORRISON, Charles M. (Charles Murchland)
An annotated bibliography of adult education. - [Edinburgh] : Scottish Council for Research in Education, [1980]
62 leaves
0-901116-28-9 X.525/4425
0231649

A SELECT bibliography of adult education in Great Britain / edited by Thomas Kelly. - 3rd ed. - London : National Institute of Adult Education [for] the Institute and the Universities Council for Adult Education, 1974
xiii,220p. - Includes index
0-900559-26-8 BB.c.f.13 Reading Room
0231649
- [Another copy] OPL
- [Another copy] X.529/30826

ADULT EDUCATION. Great Britain
- Conference proceedings

The OPEN College and alternatives / edited by K.A. Percy and S.M. Lucas. - Lancaster : University of Lancaster, 1980
51p. - (Lancaster series in adult education ; no.2)
B84-33583 X.529/63564
0530166

SOCIAL developments affecting adult education / Educational Centres Association. - [London] : ECA, [1982]
21p
B84-38458 X.525/9762
0530166

WORKSHOP of the adult education world. - London : Educational Centres Association, [1980?]
vii,34p : 1ill
B84-21642 X.525/8991
0530166

ADULT EDUCATION. Great Britain
- For teaching

GOODMAN, Sylvia J.
Mirror images. - Bushey Heath : Fresh Start Centre, [1981?]
[10]p
B82-04109 X.525/5165
2594412

ADULT EDUCATION. Great Britain
- Serials

ECA news. - [19-]-. - London : Educational Centres Association, [19-]-
v
Holdings: 1983-
r6816680 P.523/1724
0530166

The JOURNAL of the W.E.A. South Wales District. - [No.1]-. - Cardiff : The District, [1982]-
v. : ill
Holdings: no 2-
B84-27835 P.861/555
0530166

ADULT EDUCATION. Great Britain
1823-1973

BRIGGS, Asa, Baron Briggs
Communications and culture, 1823-1973. - London : Birkbeck College, [1975]
18p
0-900975-86-5 X.700/25016
0355933

ADULT EDUCATION. Great Britain
1918

The W.E.A. education year book 1918. - [Nottingham] : [Department of Adult Education, University of Nottingham], c1981
26,512p. - ([Nottingham studies in the theory and practice of the education of adults] ; [7]). - Includes index
0-902031-57-0 X.0519/573(7)
3260100

ADULT EDUCATION. Great Britain
1919

GREAT BRITAIN. Ministry of Reconstruction. Adult Education Committee
The 1919 report. - Nottingham : Department of Adult Education, University of Nottingham, 1980
157,vi,409p. - Includes index
0-902031-45-7 X.0519/573(6)
2712385

ADULT EDUCATION. Great Britain
1970-1979

WOODHALL, Maureen
The scope and costs of the education and training of adults in Britain. - Leicester : ACACE, 1980
iv,27p. - (Occasional paper / Advisory Council for Adult and Continuing Education ; 4)
0-906436-05-2 X.529/50158
2790076

ADULT EDUCATION. Great Britain
Adult & lifelong education. Role of universities

CONTINUING education within universities and polytechnics - Leicester : Advisory Council for Adult and Continuing Education, 1983, c1982
37p. - (Occasional paper ; 5). - Bibliography: p35-37
0-906436-21-4 X.0519/250
239930x

ROGERS, Alan, 1933-
Knowledge and the people. - Londonderry : Institute of Continuing Education, 1980
31p. - Bibliography: p31
B81-18303 X.529/41289
239930x

ADULT EDUCATION. Great Britain
Adult basic education - Conference proceedings

PAPERS on adult basic education / editor: Monica Hayes - [Southampton] : Adult Education Training and Research Unit, Department of Adult Education, University of Southampton, [1980]
94p
0-85432-213-2 X.525/4003
2364689

ADULT EDUCATION. Great Britain
Adult basic education - Serials

BASIC education / National Extension College. - [No. 1]- : Sept. 1979-. - Cambridge : NEC Journals, 1979- : ill
B80-28183 P.523/1367
2114372

ADULT EDUCATION. Great Britain
Adult basic education. Applications of microcomputer systems

The USE of microcomputers in basic education / ALBSU. - London : Adult Literacy and Basic Skills Unit, [1982?]
1folded sheet
B83-05300 X.950/17742
2822598

ADULT EDUCATION. Great Britain
Adult Christian religious education

RICHARDSON, Mary K. (Mary Kathleen)
Adult education. - Langley : Published on behalf of the Depart of Catechetics of the Education Commission by Catholic Information Services, [1980]
30p
B83-04704 X.529/51426
2819783

ADULT EDUCATION. Great Britain
Adult literacy education

JEFFREY, Jane
Who needs literacy provision?. - Basingstoke : Macmillan, 1979
iv,44p : ill. - (Adult literacy guides)
0-333-26390-1 X.909/44275
1703404

LIGHT on literacy / edited by Esther Gordon. - Pinner : Grail Publications, 1978
89p : ill. - Bibliography: p.86-88
0-901829-39-0 X.519/29020
1771922

LITERACY. - London : Adult Literacy and Basic Skills Unit, [1982]
8p : ill
B83-14857 Cup.918/79
1703404

MACE, Jane
Working with words. - London : Writers and Readers Publishing Cooperative [for] Chameleon, 1979
[9],118p. - (Chameleon books)
0-906495-14-8 X.908/42930
1703404

SMITH, Delroy
A second chance. - London : Peckham Publishing Project, c1981
11p : ill
0-906464-75-7 X.529/49698
2779269

ADULT EDUCATION. Great Britain
Adult literacy education - Bibliographies

HAY, Wendy
Adult literacy in Britain. - London : Library Association, 1978
104p. - Includes index
0-85365-811-0 X.519/29359
1571540

ADULT EDUCATION. Great Britain
Adult literacy education - For teaching

BBC adult literacy handbook / edited by Chris Longley. - Revised ed. - London : British Broadcasting Corporation, 1977
[2],94p : ill. - Includes bibliographies and index
B78-02180 BS.129/65
0330523

GUIDELINES to literacy teaching. - Leeds : Printed Resource Unit for Continuing Education, Leeds City Council Department of Education, c1982
47p : ill. - Includes index
B83-24545 X.525/7354
0330523

MACFARLANE, Tom
Helping adults to read. - [New ed.]. - Basingstoke : Macmillan, 1979
vii,53p : ill. - (Adult literacy guides). - Bibliography: p.53
0-333-26576-9 X.909/44192
0330523

ROBERTS, Janet
Harrap's handbook for literacy tutors. - London : Harrap, 1976
[2],48p : ill. - Bibliography: p.36-47
0-245-52928-4 X.908/40268
0330523

ADULT EDUCATION. Great Britain
Adult literacy education, 1966-1977

CHARNLEY, A. H. (Alan Huntington)
The concept of success in adult literacy. - London : Adult Literacy & Basic Skills Unit, 1981, c1979
iv,200p
0-906509-07-6 X.529/49015
1840967

CHARNLEY, Alan Huntington
The concept of success in adult literacy. - Cambridge : Huntington Publishers Limited, 1979
iv,200p. - Bibliography: p.186-188. - Includes index
0-906389-02-x X.909/43470
1840967

ADULT EDUCATION. Great Britain
Adult literacy education. Applications of adult educational broadcasting services of British Broadcasting Corporation

HARGREAVES, David
Adult literacy and broadcasting. - London : Pinter, c1980
257p : ill. - Bibliography: p257
0-903804-68-9 X.950/1877
2359847

ADULT EDUCATION. Great Britain
Adult literacy education. Courses. Students. Academic achievement

A SURVEY of attainment and progress of students in adult literacy schemes / ALBSU. - London : Adult Literacy and Basic Skills Unit, [1983?]
11p
B84-06636 X.529/58451
3036782

ADULT EDUCATION. Great Britain
Adult literacy education. Courses. Students. Publications - Bibliographies

A LIST of books by adults learning to write. - [London] : Write First Time, 1982
1folded sheet(7p) : ill
B83-17634 2725.c.141
287864x

ADULT EDUCATION. Great Britain
Adult literacy education. Organisations: Adult Literacy Resource Agency, 1975-1976

ADULT LITERACY RESOURCE AGENCY. Management Committee
Adult literacy. - London : H.M.S.O., 1976
x,58p
0-11-270379-8 BS.152/191
077877x

ADULT EDUCATION. Great Britain
Adult literacy education. Projects: British Association of Settlements. Adult Literacy Project 1975-1977

BRITISH ASSOCIATION OF SETTLEMENTS. Adult Literacy Project 1975-1977
Adult literacy. - [London] : [British Association of Settlements], [1977]
24p
B78-18187 X.902/5508
1455575

ADULT EDUCATION. Great Britain
Advisory, information & counselling services

BUTLER, Linda
Case studies in educational guidance for adults. - Leicester : Published jointly by National Institute of Adult Continuing Education and Advisory Council for Adult and Continuing Education, 1984
x,172p
0-900559-48-9 X.529/66416
2591502

EDUCATIONAL information and guidance for adults / edited by Roger H. Mercer. - [Sheffield] : Sheffield City Polytechnic, Department of Education Management, 1981
106p. - (Sheffield papers in education management ; 16)
0-903761-31-9 X.529/44762
2591502

ADULT EDUCATION. Great Britain
Advisory, information & counselling services - Bibliographies

MARKS, Harold
Bibliography of educational information advisory & counselling services for adults in the United Kingdom. - London : London Association for Countinuing Education, 1979
[1],38p
0-907009-00-x X.529/37345
218429x

ADULT EDUCATION. Great Britain
Applications of adult educational broadcasting services of British Broadcasting Corporation

ROGERS, Jennifer
Using broadcasts with adults. - London : British Broadcasting Corporation, 1980
[1],32p
0-563-16389-5 X.529/36922
0209791

ADULT EDUCATION. Great Britain
Applications of computer systems

GERVER, Elisabeth
Computers and adult learning. - Milton Keynes : Open University Press, 1984
xix,124p : ill. - Bibliography: p113-118. - Includes index
0-335-10577-7 X.529/63717
3014053

ADULT EDUCATION. Great Britain
Curriculum subjects: Arts

ADULT education and the arts / edited D.J. Jones and A.F. Chadwick. - [Nottingham] : Department of Adult Education, University of Nottingham, c1981
82p. - (Nottingham working papers in the education of adults ; 2)
0-902031-54-6 X.529/46740
2700239

ADULT EDUCATION. Great Britain
Curriculum subjects: European studies - For teaching

LIGGETT, Eric
The European studies syllabus. - Glasgow : E. Liggett, 1981
83leaves. - Bibliography: leaves 71-77
0-9507058-3-7 X.525/4726
2538172

ADULT EDUCATION. Great Britain
Curriculum subjects: Man. Health. Teaching

HEALTH EDUCATION COUNCIL
Look after yourself. - [London] : [The Council], [1980]
[152]p in various pagings : ill
B80-39427 X.525/3754
2307812

ADULT EDUCATION. Great Britain
Curriculum subjects: World studies - For teaching

LIGGETT, Eric
The international studies syllabus. - Glasgow : Gressingham Press, 1980
[2],42 leaves : ill. - Bibliography: leaves 40-42
B80-31321 X.702/6388
2175983

ADULT EDUCATION. Great Britain
Debates by Great Britain. Parliament

EDUCATION debates. - London : Educational Centres Association, [1981]
[3]leaves
B82-29224 X.525/5970
2755416

ADULT EDUCATION. Great Britain
Effects of reduction in government expenditure

CRISIS in education / edited by K.H. Lawson. - [Nottingham] : [Save Adult Education Action Group], 1979
[4],39p : ill
B80-19713 X.200/33024
2089025

ADULT EDUCATION. Great Britain
Financial aspects

DRAKE, Keith
Financing adult education and training. - [Manchester] : [Department of Adult and Higher Education, University of Manchester], 1983
vi,137p : 2ill. - (Manchester monographs ; 21). - Bibliography: p127-137
0-903717-31-x X.0519/387(21)
3133451

ADULT EDUCATION. Great Britain
Grants - Serials

ADULT education bursaries / Residential Colleges Committee. - 1984-. - Oxford : The Committee, 1984-
v.
r6863654 P.861/733
3179915

ADULT EDUCATION. Great Britain
Organisations. Cooperation - Conference proceedings

The MANAGEMENT of inter-agency co-operation in adult education / edited by Joyce Machin. - Sheffield : Sheffield City Polytechnic Department of Education Management, 1982
55p. - (Sheffield papers in education management ; no.26). - Bibliography: p29-30
0-903761-52-1 X.525/8024
2972425

ADULT EDUCATION. Great Britain
Organisations: Educational Centres Association

EDUCATIONAL CENTRES ASSOCIATION
The Educational Centres Association. - [London] : Educational Centres Association, [1982?]
1folded sheet(6p). - (ECA advisory leaflet ; no.2)
B82-29227 X.0529/745(2)
2755424

Why the Educational Centres Association?. - [London] : [Educational Centres Association], [1982]
[4]p : ill
B82-29225 X.529/48526
2755424

ADULT EDUCATION. Great Britain
Organisations: Educational Centres Association - Serials

[NEWSLETTER (Educational Centres Association)]. Newsletter / Educational Centres Association. - [19-]-. - London : The Association, [19-]-
v.
Holdings: Feb. 1982-
r6865222 P.523/1724
2794071

ADULT EDUCATION. Great Britain
Organisations: Workers' Educational Association. Curriculum subjects: Arts - Serials

WEA arts newsletter. - No.1 (Spring 1981)-. - London : Workers' Educational Association, 1981-
v. : ill
B81-33240 P.421/768
2544431

ADULT EDUCATION. Great Britain
Organisations: Workers' Educational Associations. Curriculum subjects: Trade unionism - Serials

TRADE union studies journal. - 1- ; Spring 1980-. - London : Workers' Educational Association, 1980-
B80-25270 P.523/1354
2209128

ADULT EDUCATION. Great Britain
Peers, Robert

BROWN, Geoffrey F.
Robert Peers and the Department of Adult Education. - [Nottingham] : Department of Adult Education, University of Nottingham, c1981
20p
0-902031-59-7 X.529/46726
2699869

ADULT EDUCATION. Great Britain
Peers, Robert. Documents - Catalogues

UNIVERSITY OF NOTTINGHAM. Manuscripts Department
Professor Robert Peers (1888-1972) papers. - [Nottingham] : University of Nottingham Manuscripts Department, [c1984]
117p
B85-02848 X.955/2701
3210510

ADULT EDUCATION. Great Britain
Provision. Law. Reform

CHANGING the legislation. - London : Educational Centres Association, 1981
[3]leaves
B82-29223 X.525/5971
2755408

ADULT EDUCATION. Great Britain
Role of public libraries

ALLRED, John
A preliminary study of the involvement of public libraries with adult learners. - Leeds : Public Libraries Management Research Unit, School of Librarianship, Leeds Polytechnic, 1979
59leaves. - Includes index
0-900738-17-0 2719.k.35
2805065

ADULT EDUCATION. Great Britain
Role of universities

JENNINGS, Bernard
New lamps for old?. - Hull : University of Hull, 1976
26p. - (Inaugural lectures / University of Hull)
0-85958-410-0 X.519/26058
0614653

STEPHENS, Michael D. (Michael Dawson)
Regional universities?. - [Nottingham] : University of Nottingham, [1976]
[2],12p
0-900572-56-6 X.519/25603
0614653

TAYLOR, Richard, 19—
University adult education in England and the USA. - London : Croom Helm, c1985
245p. - (Radical forum on adult education series). - Includes index
0-7099-2431-3 X.529/70456
3270572

WILLIAMS, I. M. (Ieuan M)
Continuity and change - the role of the university in the education of adults. - [Swansea] : University College of Swansea, 1981
23p
0-86076-026-x X.529/45551
0614653

ADULT EDUCATION. Great Britain
Role of universities, 1908-1958

BLYTH, John A.
English university adult education 1908-1958. - Manchester : Manchester University Press, c1983
xii,363p. - Includes index
0-7190-0903-0 X.529/56476
2950391

ADULT EDUCATION. Great Britain
Role of volunteers

HALL, Dorothea
Volunteers in the adult education context case studies and volunteer profiles. - Leicester : Advisory Council for Adult and Continuing Education ; Berkhamsted : Volunteer Centre, 1983
152p : ill
0-904647-33-1 X.525/9653
316165x

ADULT EDUCATION. Great Britain
Teaching methods

MCLAGAN, Patricia A.
Helping others learn. - Revised [ed.]. - Reading, Mass. ; London : Addison-Wesley, 1978
[5],101p : ill. - Bibliography: p.98-101
0-201-04617-2 X.512/6027
033328x

ROGERS, Jennifer, 1939-
Adults learning. - 2nd ed. - Milton Keynes : Open University Press, 1977
256p. - Bibliography: p.250-252. - Includes index
0-335-00044-4 X.519/27307
033328x

WILSON, John T. (John Thomas)
Surviving the first lesson. - [London] : Educational Centres Association, [1982]
[12]p : ill. - (An E.C.A. booklet)
B82-29226 X.529/48524
033328x

ADULT EDUCATION. Great Britain
Use of local radio services

BROWN, Jonathan F.
Local radio and the education of adults. - Leicester : ACACE, 1983
81p : ill
0-906436-27-3 X.529/61681
3122204

ADULT EDUCATION. Great Britian
1945-1980

REES, D. Ben (David Ben)
Preparation for crisis. - Ormskirk : Hesketh, 1982
ix,360p. - Bibliography: p309-336. - Includes index
0-905777-15-8 X.529/50170
2791064

ADULT EDUCATION. Guinea-Bissau

FREIRE, Paulo
Cartas à Guiné-Bissau. - Rio de Janeiro : Paz e Terra, 1977
173p : ill. - (Coleção O mundo, hoje ; v.22)
r6084138 X.519/42340
2164353

ADULT EDUCATION. Guinea-Bissau
1975-1976

FREIRE, Paulo
Pedagogy in process. - London : Writers and Readers Publishing Cooperative, 1978
[7e,178p,[6]p of plates : ill
0-904613-86-0 X.809/45722
2060604

ADULT EDUCATION. Hackney (London Borough)
Adult literacy education. Organisations - Directories

DIRECTORY of adult literacy teaching in Hackney. - London : Hackney Library Services.
1977. - [1977]
[24]p
Subsequent volumes, when received, will be placed at this shelfmark
B77-25523 P.523/769
1186973

ADULT EDUCATION. Hajdu-Bihar (megye). Hungary

"ÉHE a kenyérnek, éhe a szónak, éhe a szépnek-" / szerkesztette Gazdag István. - Debrecen, 1978
244p. - (A Hajdú-Bihar megyei Levéltár közleményei ; sz.11)
r6174283 P.701/631(11)
218902X

ADULT EDUCATION. Hajdu-Bihar (megye). Hungary
to date

"HA majd a szellem napvilága - " / szerkesztette Gazdag István ; [a kötet munkatázsai Béres András ... et al.]. - Debrecen, 1979
238p. - (A Hajdú-Bihar megyei Levéltár közleményei ; sz.13). - Bibliography: p.207-222
r6449022 P.701/631(13)
855451x

ADULT EDUCATION. Hungary
1945- - Serials

FELNŐTTNEVELÉS, művelődés = Acta andragogiae et culturae ... - 1980, sz.1-. - Debrecen : KLTE, 1980-
v.
r6455058 X.529/61233
8916918

ADULT EDUCATION. Hungary
1945- - Statistics

HUNGARY, Központi Statisztikai Hivatal
Adult education in Hungary 1970-1980. - Budapest, 1982
81p
r6499242 S.461/216
8916918

ADULT EDUCATION. Immigrants. Western world
- Conference proceedings

EDUCATION for adult immigrants / edited by A.K. Stock and D. Howell. - [Leicester] : National Institute of Adult Education, [1976]
142p. - Bibliography: p.114-115
0-900559-38-1 X.519/26824
1044575

ADULT EDUCATION. India (Republic)
to 1974

MEHTA, Mohan Singh
Progress and development of adult education in India. - Bombay : Indian Council of Basic Education, [1974]
32 p.
r6028519 X.510/9761
101711X

ADULT EDUCATION. India (Republic)
1947- - Conference proceedings

NEW TRENDS IN ADULT EDUCATION FOR WOMEN WITH SPECIAL REFERENCE TO LITERACY (National Seminar : 1980 : New Delhi)
New trends in adult education for women. - New Delhi : Indian Adult Education Association, 1980
95p : ill
r6814415 X.529/67893
8609950

ADULT EDUCATION. India (Republic)
Projects: National Adult Education Programme, to 1980

NON-FORMAL education and the NAEP / edited by A.B. Shah and Susheela Bhan. - Delhi ; Oxford : Oxford University Press, 1980
viii,245p. - Includes index
0-19-561256-6 X.529/42190
2505983

ADULT EDUCATION. Inner London
1870-1980

DEVEREUX, W. A. (William A)
Adult education in Inner London 1870-1980. - London : Shepheard-Walwyn in collaboration with Inner London Education Authority, 1982
x,342p,[16]p of plates : ill. - Bibliography: p332-335. - Includes index
0-85683-059-3 X.520/27362
2749564

ADULT EDUCATION. Ireland (Republic)

IRELAND. Commission on Adult Education
Lifelong learning. - Dublin : Stationery Office, 1984
161p : ill
r6816859 X.529/70271
3329909

ADULT EDUCATION. Ireland (Republic)
- Serials

AONTAS review. - Vol.1, no.1 (1979)-. - Dublin : National Association of Adult Education, 1979-
v. : ill
Holdings: vol 3, no.2-
B84-08752 P.861/356
3042154

ADULT EDUCATION. Islington (London Borough)

NEALE, Brenda
An investigation of the learning needs of adults in Islington. - London : Inner London Education Authority/London Borough of Islington, Library Service, 1983
29p
0-902260-10-3 X.525/9822
3173038

ADULT EDUCATION. Italy

PICCO, Iclea
Educazione popolare e scuola del lavoro. - Roma : Luciano Lucarini, 1977
270p. - (Problemi della scuola di oggi ; 6). - Includes bibliographies
r6085270 X.519/40657
1533010

SIMONE, Raffaele
L'educazione linguistica per gli adulti. - Roma : Istituto della Enciclopedia Italiana, 1976
200p. - (Biblioteca di educazione avanzata ; 3). - Bibliography : p.171-198
r6078284 X.510/10771
1533010

ADULT EDUCATION. Italy
to date

COLELLA, Brizio
L'educazione permanente nella realtà sociale. - Lecce : Milella, 1979
139p. - (Problemi dell'educazione degli adulti)
r6502705 X.520/25324
8576076

DE SANCTIS, Filippo M.
La programmazione dell'educazione degli adulti nella regione e nel distretto scolastico. - Venezia : Marsilio, 1980
142p. - (Ricerche ; n.54)
r6507440 X.529/54390
8576076

VALERI, Mario
Tempo libero ed educazione. - Firenze : La nuova Italia, 1979
156p. - (Strumenti ; 98. Guide. Pedagogia). - Includes bibliography
r6506296 X.529/57329
8576076

ADULT EDUCATION. Italy
1848–1976
DE SANCTIS, Filippo M.
L'educazione degli adulti in Italia, 1848-1976. - Roma : Editori riuniti, 1978
358p. - (Paideia ; 64)
r6107601 X.519/43994
2216132

ADULT EDUCATION. Lambeth (London Borough)
Adult literacy education. Organisations - Directories
LAMBETH AMENITY SERVICES
Help with reading, writing and spelling, help with numbers, help with basic English. - London : Lambeth Amenity Services, [1978?]
60p : ill
B79–04232 X.519/28989
1765582

ADULT EDUCATION. Leigh Park. Havant. Hampshire
Projects: New Communities Project
FORDHAM, Paul
Learning networks in adult education. - London : Routledge and Kegan Paul, 1979
viii,250p : ill. - (Routledge education books). - Includes index
0–7100–0112–6 X.529/33186
158717x

ADULT EDUCATION. London
NEWMAN, Michael, Writer on adult education
Adult education and community action. - [London] : Writers and Readers Publishing Cooperative, 1975
32p
0–904613–08–9 X.519/19135
0361828

ADULT EDUCATION. London
- Serials
CENTRAL Institute news. - [19–]-. - London : The Institute. [19–]-
v : ill
Holdings: 1984–1985–
r6863756 P.2000/1324
0361828

ADULT EDUCATION. London
Adult literacy education
WE made a go of it / by their tutors. - Pinner : Grail Publications, 1975
51p : ill
0–901829–28–5 X.908/40245
0578916

ADULT EDUCATION. London
Curriculum subjects: Irish studies - Directories - Serials
IRISH studies in London. - No.1 (1980/81)-. - London : Addison Press in association with Irish Cultural Activities, London, 1981–
v. : ill
B83–40327 P.801/3413
2964848

ADULT EDUCATION. London
Role of University of London, 1876–1976
BURROWS, John, 1919
University adult education in London. - [London] : University of London, Senate House, c1976
x,122p. - Includes index
0–7187–0428–2 X.529/61441
3069850

ADULT EDUCATION. Mexico
to date
HÖRNEMANN, Maria
Erfahrungen in der lateinamerikanischen Volksbildung .... - Frankfurt / Main : Haag + Herchen, 1981
227p. : ill
3–88129–426–0 X.529/57499
9014659

ADULT EDUCATION. Middlesbrough. Cleveland
to 1982
SAUNDERS, John Whiteside
What happened to a town without a university. - [Middlesbrough] : [J.W. Saunders], [1983]
84p : ill
0–9509004–0–0 X.529/63336
3143848

ADULT EDUCATION. Netherlands
to date
NIJENHUIS, Herman
Volksopvoeding tussen elite en massa. - Meppel : Boom, [1981]
138p
90–600–9512–x X.529/53045
8916926

ADULT EDUCATION. Non-English speaking students. Europe
Curriculum subjects: English language. European unit/credit system. Threshold level
EK, J. A. van (Jan Ate van)
Theshold level English. - Oxford : Published for and on behalf of the Council of Europe by Pergamon, 1980
xi,253p. - Bibliography: p253
0–08–024588–9 X.950/12657
1971239

ADULT EDUCATION. Non-English speaking students. Europe
Curriculum subjects: English language. European unit/credit system. Waystage. Teaching
EK, J. A. van (Jan Ate van)
Waystage English. - Oxford : Published for and on behalf of the Council of Europe by Pergamon, 1980
viii,101p
0–08–024590–0 X.950/10855
1971220

ADULT EDUCATION. Norway
FINSTAD, Nils
Voksenopplæring for hvem? - [Oslo] : Tiden : i samarbeid med Arbeidernes opplysningsforbund, [1976]
214,[47]p. - Includes bibliographies
82–10–01222–3 X.515/326
1331213

ADULT EDUCATION. Nottingham region
Adult literacy education. Role of volunteers
ELSEY, Barry
Voluntary tutors in adult literacy. - Nottingham : University of Nottingham, Department of Adult Education, 1981
67p. - (Nottingham working papers in the education of adults ; 3). - Bibliography: p51
0–902031–56–2 X.529/46718
2702290

ADULT EDUCATION. Organisation for Economic Co-operation and Development countries
ORGANISATION FOR ECONOMIC CO-OPERATION AND DEVELOPMENT
Learning opportunities for adults. - Paris : O.E.C.D. ; [London] : [H.M.S.O.].
Vol.1: General report. - 1977
81p
B77–14166 UN.P.108/108
1098098
Vol.4: Participation in adult education. - 1977
440p : ill. - Includes bibliographies
B78–17314 UNP.108/108
1098098

ADULT EDUCATION. Personnel. Companies. Organisation for Economic Co-operation and Development countries
Provision of paid leave
CENTRE FOR EDUCATIONAL RESEARCH AND INNOVATION
Alternation between work and education. - Paris : Organisation for Economic Co-operation and Development ; [London] : [H.M.S.O.], 1978
97p
B78–24958 UNP.108/152
151900x

ADULT EDUCATION. Personnel. England
Provision of paid leave, 1976–1977
KILLEEN, John, 19—
Education and work. - Leicester : National Institute of Adult Education, 1981
147p : ill. - Bibliography: p146–147
0–900559–43–8 X.529/40545
237711x

ADULT EDUCATION. Personnel. Europe
Provision of paid leave
MATHER, Jim
Growing food for your freezer. - London : Foulsham, 1975
96p : ill. - Includes index
0–572–00914–3 X.329/10036
0128708
MOLTKE, Konrad von
Educational leaves for employees. - San Francisco ; London : Jossey-Bass, 1977
xvii,269p. - (Series / Carnegie Council on Policy Studies in Higher Education). - Bibliography: p.251–60. - Includes index
0–87589–316–3 X.0520/501(9)
0128708
- [Another copy] X.520/9580

ADULT EDUCATION. Personnel. Italy
Provision of paid leave
PAGNONCELLI, Lucio
Le 150 ore. - Firenze : La Nuova Italia, 1978
127p. - (Scuola e cittadini ; 2)
r6106882 X.519/43267
2164019

ADULT EDUCATION. Personnel. Organisation for Economic Co-operation and Development countries
Provision of paid leave
CENTRE FOR EDUCATIONAL RESEARCH AND INNOVATION
Developments in educational leave of absence. - Paris : Organisation for Economic Co-operation and Development ; [London] : [H.M.S.O.], 1976
237p
B76–13960 UN.P.106/537
0635170

ADULT EDUCATION. Poland
1920–1940
KORNIŁOWICZ, Kazimierz
[Selections]. Pomoc społeczno-kulturalna dla młodzieży pracującej i dorosłych. - Wrocław : Zakład Narodowy im Ossolińskich Wydawnictwo Polskiej Akademii Nauk, 1976
xxxv, 157 p., 3 leaves of plates, 4 p. of plates : ill. - (Polska Akademia Nauk. Komitet Nauk Pedagogicznych. Źródła do dziejów myśli pedagogicznej ; tom 14). - Bibliography: p. 153–157
r6014904 X.510/9547
0884634

ADULT EDUCATION. Poland
!
WSPOMNIENIA o Ignacym Solarzu "Chrzestnym" / wybór i opracowanie Zofia Mierzwińska-Szybka / całość opatrzyła przypisami Maria Mioduchowska. - Warszawa : Ludowa Spółdzielnia Wydawnicza, 1983
426p,plates : ill
83–205–3352–x X.529/57474
9000895

ADULT EDUCATION. Rural regions. Great Britain
Adult basic education. Provision
ADULT literacy and basic skills provision in rural areas. - London : Adult Literacy and Basic Skills Unit, [1982?]
7p
B83–05301 X.950/17743
2822601

ADULT EDUCATION. Rural regions. Great Britain
Role of mass media
RURAL adult education workshop report / [written by members of a workshop on rural adult education convened by the National Extension College and the Chelmer Institute of Higher Education]. - Cambridge : National Extension College, [1980]
iv,33p. - (National Extension College report. series 2 ; no.5)
0–86082–145–5 X.525/3987
2360861

ADULT EDUCATION. Rural regions. Wiltshire
REPORT of a Working Party on the Organisation of Adult Education in Rural Areas. - [Trowbridge] : Wiltshire Development Council for Adult Education, 1980
29leaves
0–86080–072–5 X.805/3691
2862999

ADULT EDUCATION. Scandinavia
NOVU - Nordisk voksenundervisningsprosjekt. - København : Sekretariatet, 1977
294p. - (NU–serien ; B1977. 23)
91–7052–315–0 X.510/11982
2138549

ADULT EDUCATION. Scandinavia
- Conference proceedings
NORDISKT FOLKBILDNINGSSEMINARIUM (1974 : Kungälv)
Rapport från Nordiskt folkbildningsseminarium vid Nordens folkliga akademi i Kungälv 14–16 oktober 1974. - [Kungälv : Nordens folkliga akademi, 1975]
135 leaves(various pagings)
r6004240 X.512/2683
1384783

ADULT EDUCATION. Scandinavia
to date
VOKSENUNDERVISNINGEN i Norden / redaktion Johan Engelhardt. - [København] : Nord, c1982
285p
87–7303–009–0 X.520/35587
9128093

ADULT EDUCATION. Scandinavia
1945– - Conference proceedings
ADULT education for social change / Robert Höghielm & Kjell Rubenson (eds.). - Lund : Gleerup, 1980
169,13p. - (Studies in education and psychology ; 8). - Each paper includes a bibliography
91–400–4765–2 X.0519/751(8)
8512485

ADULT EDUCATION. Scandinavia
Organisations: Nordens folkliga akademi

NORDENS folkliga akademi - in i 80-talet / [redaktör : Olav Klonteig]. - Kungälv : Akademien, 1981
106p : ill
r6612007 X.520/36352
916202x

ADULT EDUCATION. Scotland
to date

UNIVERSITY OF EDINBURGH. Extra-Mural Association
When work is done. - Edinburgh : The Association, 1971
48p
r6804990 X.529/45444
8389462

ADULT EDUCATION. Scotland
Adult literacy education

NETWORKS. - Edinburgh : Scottish Adult Literacy Unit. Issue no. 1- ; July 1979-. - [1979]- : ill
B80-26863 P.903/631
2082462

ADULT EDUCATION. Scotland
Adult literacy education - Directories

A DIRECTORY of adult literacy in Scotland / [Scottish Adult Literacy Agency]. - 1978. - Edinburgh : The Agency, [1978]
[1],106p
Subsequent volumes, when received, will be placed at this shelfmark
B78-32866 P.521/3186
1587269

ADULT EDUCATION. Scotland
Adult Literacy education, 1975-1978

MUNRO, Helen
An assessment of the Adult Literacy Campaign in Scotland, 1975-77. - Edinburgh : Scottish Institute of Adult Education, 1979
[1],vi,178p
B80-24352 X.525/3262
2101122
- [Another copy.] X.525/3487

ADULT EDUCATION. Scotland
Adult literacy education. Organisations

SCOTTISH ADULT LITERACY AGENCY. Management Committee
Adult literacy in Scotland. - Edinburgh : H.M.S.O., 1980
ix,73p. - Bibliography: p.60-61
0-11-491647-0 BS.152/228
2256789

ADULT EDUCATION. Scotland
Adult literacy education. Organisations: Scottish Adult Literacy Agency. Mobile Educational Resource Unit. Stock: Documents on adult literacy - Bibliographies - Catalogues

DICKSON, Christine M.
A classified list of the books and resources for organisers, tutors and students displayed in the Mobile Educational Resources Unit. - [Edinburgh] : Scottish Adult Literacy Agency, 1978
[1],53p. - Includes index
B78-40000 X.512/6184
1734504

ADULT EDUCATION. Sheffield. South Yorkshire (Metropolitan County)

ADULT education in Sheffield / edited by L.E. Watson from the work of Mary E. Jackman ... [et al.]. - [Sheffield] : Sheffield City Polytechnic in association with the Sheffield Working Group on Mature Entry to Education, 1980
135p. - (Sheffield papers in education management ; no.10)
0-903761-12-2 X.525/4851
255125x

ADULT education in Sheffield / edited by L.E. Watson from the work of Mary E. Jackman ... [et al.]. - [Sheffield] : Sheffield City Polytechnic in association with the Sheffield Working Group on Mature Entry to Education, 1980
135p. - (Sheffield papers in education management ; no.10)
0-903761-13-0 X.525/4851
255125x

ADULT EDUCATION. Sierra Leone

STEWARD, T. G.
The concept of need in adult education. - [Edinburgh] : Centre of African Studies, [1982?]
9p
B84-32782 X.529/63355
3144771

ADULT EDUCATION. Slovakia
to 1974

PASIAR, Štefan
Dejiny výchovy dospelých na Slovensku. - 1. vyd.. - Bratislava : Obzor, 1975
297 p.. - (Encyklopédia osvetového pracovníka ; 2)
r6008220 X.808/10175
0877301

ADULT EDUCATION. Slovenia. Yugoslavia

KRAJNC, Ana
Motivacija odruslih za izobraževanje. - Ljubljana : Pedagoški inštitut, 1976
137p. - Bibliography : p.129-137
r6186017 X.512/3302
2139618

ADULT EDUCATION. Slovenia. Yugoslavia
to date - Catalogues

IZOBRAŽEVANJE odrasle mladine. 1870-1941 / [katalog je uredila Slavica Pavlič]. - Ljubljana : Slovenski šolski muzej, 1975
27p : ill. - Bibliography
r6435520 X.529/52698
8510377

ADULT EDUCATION. Socially disadvantaged persons. Northern Ireland

PANEL ON THE SOCIALLY DISADVANTAGED
Continuing education in socially disadvantaged areas of Northern Ireland. - Belfast : H.M.S.O., 1978
viii,29p
0-337-04087-7 CSA 210/74
1689916

ADULT EDUCATION. South London
Adult literacy education. Organisations - Directories

LAMBETH AMENITY SERVICES
Help with reading, writing and spelling, help with numbers, help with basic English. - London : Lambeth Amenity Services, [1978?]
60p : ill
B79-04232 X.519/28989
1765590

ADULT EDUCATION. South Wales
Organisations: Workers' Educational Association. South Wales District, to 1982

WORKERS' EDUCATIONAL ASSOCIATION. South Wales District
75 years of service 1907-1982. - [Cardiff] : [WEA - South Wales District], [1982]
85p : ill
B83-16501 X.529/52980
2874393

ADULT EDUCATION. Spain
1939-1969

HIERRO, A.
Explorations into the adult education system of Spain, 1939-1969. - London : Hierro and Grosse, 1977
[1],62p
B78-00188 X.519/27421
1273833

ADULT EDUCATION. Sutton-in-Ashfield. Nottinghamshire
Adult health education

PERKINS, Elizabeth R. (Elizabeth Robina)
Health and community involvement. - [Nottingham] : University of Nottingham, 1977
9leaves. - (Occasional paper / Leverhulme Health Education Project ; no.4)
B81-22821 X.325/1238 Woolwich
2487365

ADULT EDUCATION. Sweden

ABRAHAMSSON, Kenneth
Den svara dialogen - om kontakten mellan högskolan och presumtiva vuxenstuderande. - Lund : Studentlitteratur, 1976
134p
91-441-2831-2 X.519/22946
1156616

BROSTRÖM, Anders
Vuxenutbildning och fördelningspolitik. - Uppsala : [s.n.], 1977
296p. - Includes bibliographies
91-506-0115-6 X.519/24148
1156616

PALM, Olof
Betydelsen av undervisningsgruppens sammansättning vid kompetensinriktade vuxengymnasiala studier. - [Uppsala] : [s.n.], 1976
160p. - Includes bibliographies
91-506-0069-9 X.519/24156
1156616

SUNDSTRÖM, Björn
Education for adults. - Stockholm : Swedish Institute, 1975
7p. - (Current Sweden ; no.99)
r6029119 X.0512/423(99)
0800627

VIKLUND, Birger
Rising with one's class, not above it. - Stockholm : Swedish Institute, 1977
5p. - (Current Sweden ; no 153)
r6029172 X.0512/423(153)
0800627

WENDT, Charlotte
The Swedes and their education. - Stockholm : Swedish Institute, 1979
8p. - (Current Sweden ; no210)
r6029292 X.0512/423(210)
0800627

ADULT EDUCATION. Sweden
- Conference proceedings

TJÄNSTEMÄNNENS CENTRALORGANISATION. Kongress (15th 1976 : Stockholm)
Program för vuxenutbildning. - Stockholm : Tjänstemännens centralorganisation, 1976
62p. - (Kongresstryck ; nr 11)
91-7168-048-9 X.519/24906(11)
1473255

ADULT EDUCATION. Sweden
to ca 1870

LEANDER, Sigfrid
Folk och bildning. - Stockholm : Folkbildningsförbundet, 1974
477 p.. - Bibliography: p. 435-466
91-85294-01-2 X.909/32950
0924687

ADULT EDUCATION. Sweden
to date

ARVIDSON, Lars
Folkbildning i rörelse. - Lund : Gleerup, 1985
417p : ill. - (Studies in education and psychology ; 16). - Includes bibliography
91-400-5089-0 X.0519/751(16)
8504105

ADULT EDUCATION. Sweden
1809-

ÅBERG, Göran
Kyrkligt studiearbete i Växjö stift 1930-1980. - [Växjö] : Växjö stiftshistoriska sällskap, 1980
100p : ill. - (Meddelanden / Växjö stiftshistoriska sällskap ; nr 3)
r6611139 X.0809/2219(3)
8504113

ADULT EDUCATION. Sweden
1870-1977

ANDERSSON, Bo, 1940-
Folkbildningens samhällsroller 1870-1990. - Södertälje : Studieförbundet Vuxenskolan, [1978]
79p : ill. - (Folkbildning och samhälle ; 1)
r6062174 X.529/34796
2251256

ADULT EDUCATION. Sweden
1968-1970

JOHANSSON, Roland
Vuxenstuderande i Sverige 1968-1970. - Uppsala : [s.n.], 1976
84p. - Includes bibliographies
91-506-0067-2 X.519/23157
1232088

ADULT EDUCATION. Sweden
Cederblad, Carl

DAHLLÖF, Tordis
Folkbildning och livsmiljö pa 1920-talet. - Stockholm : LT, 1981
69p. - Includes bibliography
91-360-1842-2 X.529/61224
9082247

ADULT EDUCATION. Sweden
Organisations: Arbetarnas bildningsförbund

ABF-ARE berättar / redaktör : Allan Malmgren. - [Stockholm] : [Arbetarnas bildningsförbund], 1982
242p : ill
r6611588 X.529/64521
9110844

ADULT EDUCATION. Sweden
Role of libraries, to date

BOKEN ater i centrum / redaktör Lars Ljunggren. - Stockholm : Bibliotek i samhälle, 1983
334p : ill. - (Bibliotek i samhälle ; nr 72 (1-2/1983))
91-7712-016-7 2719.f.66/72
9000879

ADULT EDUCATION. Tanzania
- Conference proceedings

ADULT education handbook / [edited by G. Rydstrom ; for the] Institute of Adult Education. - Dar es Salaam : Tanzania Publishing House, 1973
xi, 269 p. : ill. - Includes bibliographies
r6010505 X.519/26509
1041010

ADULT EDUCATION. United States

SOLMON, Lewis C. (Lewis Calvin)
The characteristics and needs of adults in postsecondary education. - Lexington, Mass. : Lexington Books, c1981
xv,155p. - Bibliography: p141-144. - Includes index
0-669-04361-3 X.520/25632
0083437

ADULT EDUCATION. United States
Adult literacy education. Projects: Right to Read project. Role of public libraries

The RIGHT to read and the nation's libraries / edited by the Right to Read Committees of the American Association of School Librarians, Children's Services Division, Public Library Association. - Chicago : American Library Association for the American Association of School Librarians and the Public Library Association, 1975
[1], vi, 109 p. : ill. - Includes bibliographies
0-8389-0193-x 2719.ff.53
1032496

ADULT EDUCATION. United States
Advisory, information & counselling services

HEFFERNAN, James M.
Educational and career services for adults. - Lexington, Mass. : Lexington Books, 1981
xvi,272p. - Bibliography: p255-266. - Includes index
0-669-03440-1 X.520/25386
2579634

ADULT EDUCATION. United States
Costs

ANDERSON, Richard E.
The costs and financing of adult education and training. - Lexington, Mass. : Lexington Books, c1982
xiv,328p. - Includes index
0-669-04570-5 X.520/28976
2837188

ADULT EDUCATION. United States
Organisation

PETERS, John M. (John Marshall)
Building an effective adult education enterprise. - San Francisco ; London : Jossey-Bass, 1980
xix,180p : ill. - (Handbook series in adult education). - Bibliography: p164-170. - Includes index
0-87589-455-0 X.520/22049
2255251

ADULT EDUCATION. United States
Role of universities

TAYLOR, Richard, 19—
University adult education in England and the USA. - London : Croom Helm, c1985
245p. - (Radical forum on adult education series). - Includes index
0-7099-2431-3 X.529/70456
3270580

ADULT EDUCATION. United States
Social aspects

BOONE, J. Edgar
Serving personal and community needs through adult education. - San Francisco ; London : Jossey-Bass, 1980
xxi,338p. - (Handbook series in adult education). - Bibliography: p.294-322. - Includes index
0-87589-451-8 X.520/22228
227924x

ADULT EDUCATION. Victoria. Australia

ADULT education in Victoria / Council of Adult Education, Victoria. - 1st-. - [Melbourne] : The Council, [197-?]- : ill
Holdings: 29th (1975/76)
r6071728 P.521/3159
1404342

ADULT EDUCATION. Wales
Curriculum subjects: Welsh language. Teaching - Translations

COUNCIL FOR THE WELSH LANGUAGE
Dysgu Cymraeg i oedolion = Welsh for adults. - Cardiff : H.M.S.O., 1976
viii,63p
0-11-790086-9 BS.128a/202
1074202

ADULT EDUCATION. Warrington. Cheshire
1800-1900. Social aspects

STEPHENS, William Brewer
Adult education and society in an industrial town. - [Exeter] : University of Exeter, 1980
vii,158p. - Includes index
0-85989-105-4 X.529/37869
2272164

ADULT EDUCATION. West Africa

A HANDBOOK of adult education for West Africa / edited by Lalage Bown, S.H. Olu Tomori. - London : Hutchinson, 1979
296p. - (Hutchinson university library for Africa). - Includes bibliographies and index
0-09-138810-4 X.529/33977
1761366

ADULT EDUCATION. West Germany
1945-

MITARBEITER-INNEN in Familienbildungsstätten / Rolf Jansen ... [et al.] ; [Herausgeber : Siegfried Keil]. - Stuttgart : Kohlhammer, 1981
310p : ill. - (Schriftenreihe des Bundesministers für Jugend, Familie und Gesundheit ; Bd.91)
3-17-007317-6 S.F.476/13[Bd.91]
8766517

MÜLLER, Peter, 1943-
Methoden in der kirchlichen Erwachsenenbildung. - München : Kösel, 1982
247p
3-466-36122-2 X.520/32473
8752087

REALISMUS und Reflexion / herausgegeben von Volker Otto, Wolfgang Schulenberg, Klaus Senzky. - München : Hueber-Holzmann, 1982
252p. - Includes bibliography
3-8096-3841-2 X.629/22015
8752087

ADULT EDUCATION. West Germany
1945- - Conference proceedings

LEBENSLANGES Lernen. - Karlsruhe : Müller, 1981
123p. - (Arbeit und Beruf ; Bd.27)
r6302336 SF.281/1 [Bd.27]
8752087

ADULT EDUCATION. West Germany
Role of sports, 1945-

DUCHROW, Michael
Sport als Bildungsurlaub. - Wuppertal : Putty, 1980
123p. - (Sport & Lernen ; Bd. 2). - Bibliography
3-87650-027-3 X.629/16888 Woolwich
8756872

ADULT EDUCATION. West Germany
Role of television, 1945-

HEJL, Peter M.
Telekolleg II. - Köln : Greven, 1982
250p. - (Materialien zur Mediendidaktik ; 10). - Includes bibliography
r6310041 X.529/63967
9157247

ADULT EDUCATION. Western Europe

TITMUS, Colin
Strategies for adult education. - Milton Keynes : Open University Press, 1981
vi,239p. - Bibliography: p227-231. - Includes index
0-335-10042-2 X.529/44062
2327694

ADULT EDUCATION. Women. Great Britain

NEW futures / [edited by] Mary Hughes and Mary Kennedy. - London : Routledge & Kegan Paul, 1985
vii,183p. - Bibliography: p176-181. - Includes index
0-7100-9988-6 X.529/70454
3207390

ADULT EDUCATION. Women. India (Republic)
Special subjects: Reading - Conference proceedings

NATIONAL SEMINAR ON ADULT EDUCATION OF WOMEN (1968 : New Delhi)
Adult education of women in the changing pattern of society. - New Delhi : Indian Adult Education Association, 1973
[8], 96 p., plate : ill
r6010659 X.519/26557
1048074

ADULT EDUCATION. Women. Scotland

RECENT developments in new opportunities for women in Scotland / edited by Valerie Wilson. - Edinburgh : Scottish Institute of Adult Education, 1981
49p. - Bibliography: p46-49
0-903647-01-x X.529/58276
2997517

ADULT EDUCATION. Women. Wales
- Conference proceedings

NEW opportunities and second chance education for women in Wales. - Manchester : Equal Opportunities Commission, 1980
54p
B83-02081 BS.433/29
2811081

ADULT EDUCATION. Working class women. Liverpool. Merseyside (Metropolitan County)

COUSINS, Liz
I'm a new woman now. - [Liverpool] : [Priority], [1983?]
72p : ill. - Bibliography: p71-72
0-9502785-1-3 X.529/58688
3040941

ADULT EDUCATION. Working classes
Adult literacy education

DEMETRIO, Duccio
Alfabetizzazione degli adulti e classe operaia. - Milano : Franco Angeli, [1976]
202p : ill. - (Collana Condizionamenti educativi ; 23)
r6100720 X.519/41891
2116626

ADULT EDUCATION. Zambia
to date - Serials

JOURNAL of adult education. - Vol.1, no.1 (Sept. 1982)-. - Lusaka : Centre for Continuing Education, 1982-
v.
r6807362 P.521/3843
9023925

ADULT EDUCATION. Zambia
1964-

ALEXANDER, David J.
Origins and development of university extension in Zambia, 1966-1975. - [Lusaka] : University of Zambia, Centre for Continuing Education, [1975]
133p. - (Occasional paper ; no.3 (1975))
r6811292 X.525/8912
9064796

ADULT EDUCATION INSTITUTIONS
See also
Adult literacy education institutions

ADULT EDUCATION INSTITUTIONS
Curriculum. Development

VERDUIN, John R.
Curriculum building for adult learning. - Carbondale, Illinois : Southern Illinois University Press, [1980]
xiii,171p. - Includes bibliographies
0-8093-0960-2 X.529/34934
0827657

ADULT EDUCATION INSTITUTIONS
International adult education institutions: Friedrich-Naumann-Stiftung

PABST, Siegfried
Friedrich-Naumann-Stiftung. - Düsseldorf : Droste, 1982, c1983
113p : ill. - (Ämter und Organisationen der Bundesrepublik Deutschland ; 63)
3-7700-7064-x X.0708/70(63)
9005668

ADULT EDUCATION INSTITUTIONS
Non-English speaking students. Curriculum subjects: English language. Syllabuses. Development

GENERAL English syllabus design / edited by C.J. Brumfit. - Oxford : Published in association with the British Council by Pergamon, 1984
vi,82p. - (ELT documents ; 118). - Includes bibliographies
0-08-031513-5 X.950/39085
238759x

MUNDAY, John
Communicative syllabus design. - Cambridge : Cambridge University Press, 1981, c1978
vi,232p : ill. - Bibliography: p219-228. - Includes index
0-521-28294-2 X.529/40897
238759x

ADULT EDUCATION INSTITUTIONS. Austria
to date

SZERELMES, Richard
Leopold Teufelsbauer und das bäuerliche Volksbildungsheim Hubertendorf 1929–38. – St. Pölten : Niederösterreichisches Pressehaus, c1982
142p,plates
3–85326–660–6 X.520/33960
9084975

ADULT EDUCATION INSTITUTIONS. Austria
Curriculum subjects, 1945–: Literature

JANITSCHEK, Norbert
Der Literarische Fernsehkreis. – Graz : Leykam, [1973]
96p.. – (Schriftenreihe des Verbandes Österreichischer Volkshochschulen ; Bd.3)
3–7011–1177–4 X.529/47884
8820813

ADULT EDUCATION INSTITUTIONS. Birmingham. West Midlands (Metropolitan County)
Residential adult education institutions: Fircroft College

REPORT of a committee of inquiry into the problem at Fircroft College, under the Chairmanship of Andrew Leggatt QC. – London : H.M.S.O., [1976]
[2],33p. – ([1975–76 H.C.] 319)
0–10–231976–6 BS.91/131
0652849

ADULT EDUCATION INSTITUTIONS. Boston. Lincolnshire
Pilgrim College, 1945–1975

CHAMPION, Alan
Pilgrim's progress, Boston, 1945–1975. – Boston, Lincs. : Richard Kay for the Governors of Pilgrim College [and] the Department of Adult Education, University of Nottingham, 1976
60p : ill
0–902662–40–6 X.529/31275
1060163

ADULT EDUCATION INSTITUTIONS. Bucharest (Judeţ). Romania

PROGRAMUL universităţilor populare din municipiul Bucureşti 1972–73. – Bucureşti : Comitetul Culturii şi Educaţei Socialiste al Municipiului Bucureşti, 1973
453p
r6094458 X.519/41014
1989391

ADULT EDUCATION INSTITUTIONS. Camden (London Borough)
Working Men's College, 1954–1979

WORKING MEN'S COLLEGE
The Working Men's College, 1954–1979. – London : The College, 1983
84p : 1ill
0–9503935–1–7 X.525/9414
3126382

ADULT EDUCATION INSTITUTIONS. England
Adult education institutions & further education institutions. Part-time teachers. Professional education

GREAT BRITAIN. Advisory Committee on the Supply and Training of Teachers. Sub-committee on the Training of Teachers for Further Education
The training of adult education and part-time further education teachers. – London : A.C.S.T.T., [1978]
[3],15p
B79–31668 BS.10/734
1950746

The TRAINING of part-time teachers of adults / T.B. Graham ... [et al.]. – Nottingham : Department of Adult Education, University of Nottingham, c1982
x,217p. – (Nottingham studies in the theory and practice of the education of adults). – Bibliography: p197
0–902031–79–1 X.529/53412
1950746

ADULT EDUCATION INSTITUTIONS. England
Students. Political education

POLITICAL education for adults / prepared for the Advisory Council by a committee convened by H.D. Hughes ; with papers by Torsten Böhmer — [et al.]. – Leicester : ACACE, 1983
vi,106p
0–906436–28–1 X.529/62680
3137430

ADULT EDUCATION INSTITUTIONS. Finland
to date – Directories – Serials

SUOMEN KESÄYLIOPISTOT
Kesäyliopistot. – 1977. – [Helsinki], 1977
52p
r660021 X.529/54440
8945241

ADULT EDUCATION INSTITUTIONS. France
Teaching aids: Audiovisual materials, to date

FLAGEUL, Alain
Les organismes de formation d'adultes et l'audiovisuel. – Paris : Centre Inffo, 1980
85p
r6209231 X.520/30694
8971927

ADULT EDUCATION INSTITUTIONS. Germany
to date

BERKE, Silvia
Volkshochschule im Bonn 1904–1984. – Bonn : Röhrscheid, 1984
339p,8p of plates. – (Veröffentlichungen des Stadtarchivs Bonn ; Bd 34). – Includes bibliography
3–7928–0482–4 X.0809/10(34)
9133003

ADULT EDUCATION INSTITUTIONS. Germany
Curriculum subjects, to date: Literature

HAMANN, Wolf-Dieter
Ästhetische Literatur als Gegenstand der Erwachsenenbildung. – Frankfurt am Main : Lang, 1983
195p.. – (Studien zur Germanistik und Anglistik ; Bd.3). – Includes bibliography
3–8204–7726–8 X.0950/313(3)
9018506

ADULT EDUCATION INSTITUTIONS. Great Britain
– Serials

SECOND chances. – 1982–. – Cambridge : Great Ouse Press, 1982–
v. : ill
B83–16029 P.521/3747
2873966

ADULT EDUCATION INSTITUTIONS. Great Britain
Adult education institutions & further education institutions. Part-time students. Teaching

SOUTHERN REGIONAL COUNCIL FOR FURTHER EDUCATION. Course Tutors' Research Group
Case studies involving part-time students for use in technical teacher training. – Reading : The Council, [1979]
[1],ii,27p
B80–19150 X.525/4502
2065886

ADULT EDUCATION INSTITUTIONS. Great Britain
Closure. Prevention

IF the axe threatens you. – [London] : Educational Centres Association, [1982?]
[4]p. – (ECA advisory leaflet ; no.1)
B82–29618 X.0529/745(1)
2756595

ADULT EDUCATION INSTITUTIONS. Great Britain
Curriculum subjects: Classics – Conference proceedings

DIDASKALOS CONFERENCE (14th : 1975 : University of London Institute of Education)
Classics in adult education. – [Sheffield] : [Joint Association of Classical Teachers], [1975]
[1],18p
0–903242–52–4 X.512/5106
0599611

ADULT EDUCATION INSTITUTIONS. Great Britain
Curriculum subjects: Modern languages. Teaching

TEACHING languages to adults / edited by Duncan Sidwell. – London : Centre for Information on Language Teaching and Research, 1984
viii,117p : ill
0–903466–72–4 X.529/62364
0087785

ADULT EDUCATION INSTITUTIONS. Great Britain
Curriculum subjects: Women's studies

THOMPSON, Jane L. (Jane Lindsay)
Learning liberation. – London : Croom Helm, c1983
207p. – (Radical forum on adult education series). – Includes index
0–7099–2414–3 X.529/61247
3085678

ADULT EDUCATION INSTITUTIONS. Great Britain
Curriculum subjects: Women's studies. Courses – Directories

BRADSHAW, Jan
Women's studies courses in the UK. – London : Women's Research and Resources Centre, 1981
71p. – Bibliography: p63–66
0–905969–11–1 X.529/53408
2882965

ADULT EDUCATION INSTITUTIONS. Great Britain
Publicity

REACHING the public. – [London] : Educational Centres Association, [1982?]
[4]p. – (ECA advisory leaflet ; no.6)
B82–29617 X.0529/745(6)
2756587

ADULT EDUCATION INSTITUTIONS. Great Britain
Residential adult education institutions – Conference proceedings

RESIDENTIAL adult education / edited by Bob Houlton. – [Waltham Cross] : Society of Industrial Tutors, [1977]
[4],152p
B80–12787 X.519/30884
205292x

ADULT EDUCATION INSTITUTIONS. Great Britain
Students' associations. Organisation

A MEMBERS association. – London : Educational Centres Association, [1982?]
4p. – (Advisory leaflet ; no.4)
B83–00769 X.0529/745(4)
2806576

A MEMBERS association, partnership in action / Educational Centres Association. – London : ECA, [1982]
1sheet. – (ECA advisory leaflet ; no.5). – Includes bibliography
B82–39515 X.0529/745(5)
2790556

ADULT EDUCATION INSTITUTIONS. Great Britain
Teachers. Professional education

TEACHERS of adults. – Leicester : ACACE, c1983
52p
0–906436–26–5 X.529/63342
0296546

ADULT EDUCATION INSTITUTIONS. Great Malvern. Hereford and Worcester
Private presses: Malvern Hills College Press. Publications – Bibliographies

MALVERN HILLS COLLEGE PRESS
The Malvern Hills College Press, 1974–79. – [Andoversford] : [Whittington Press], 1979
[4]p
B80–32785 2725.g.19
2231301

ADULT EDUCATION INSTITUTIONS. Hampshire
Curriculum subjects: Retirement. Courses. Evaluation

PRE-RETIREMENT education / WEA Southern District. – Southampton : W.E.A. Southern District, 1980
28p
B83–02662 X.529/50973
2813408

ADULT EDUCATION INSTITUTIONS. Harlech. Gwynedd
Coleg Harlech, to 1977

STEAD, Peter, 1943–
Coleg Harlech. – Cardiff : University of Wales Press, 1977
xv,135p,[16]p of plates : ill. – Includes index
0–7083–0657–8 X.510/15584
126253x

ADULT EDUCATION INSTITUTIONS. London
City Lit (Institution). Centre for Adult Studies. Part-time courses: Fresh horizons, to 1977

HUTCHINSON, Enid
Learning later. – London : Routledge and Kegan Paul, 1978
x,200p. – Includes index
0–7100–8952–x X.529/32758
1701576

ADULT EDUCATION INSTITUTIONS. London
Mentally handicapped students. Integration into adult education institutions

BILLIS, Jacquie
From theory to practice. – [London] : [MENCAP Metropolitan Region], c1984
[21]p
B85–03985 X.529/65991
3173437

WILLIS, Penny
A new way evaluated. – London : Thomas Coram Research Unit, c1984
32p. – Bibliography: p32
0–85537–089–0 X.529/64891
3173437

ADULT EDUCATION INSTITUTIONS. Londonderry
New University of Ulster. Institute of Continuing Education. Students. Attitudes

COURTNEY, Sean
The Magee experiment. - Londonderry : Institute of Continuing Education, New University of Ulster, c1981
vi,169p. - (The Magee papers ; 1). - Bibliography: p164-169
0-901229-29-6 X.0525/436(1)
2632624

ADULT EDUCATION INSTITUTIONS. Londonderry
New University of Ulster. Institute of Continuing Education. Students. Attitudes - Serials

The MAGEE papers / Institute of Continuing Education, New University of Ulster. - Vol.1-. - Londonderry : Institute of Continuing Education, 1981-
r6805941 X.0525/436
2632624

ADULT EDUCATION INSTITUTIONS. Marcham. Oxfordshire
Women's adult education institutions: Denman College, to 1978

NATIONAL FEDERATION OF WOMEN'S INSTITUTES
Denman College. - London : The Federation, 1978
32p : ill
0-900556-52-8 X.525/3768
2253801

ADULT EDUCATION INSTITUTIONS. Oxford
Ruskin College, to 1909

YORKE, Paul
Education and the working class. - Oxford : Ruskin College, [1977]
[3],41,[1]p. - (Ruskin students' labour history pamphlets ; no.1). - Bibliography: p.41
B78-16698 X.0519/1044
144624x

ADULT EDUCATION INSTITUTIONS. Oxford
Ruskin College, to 1984

RUSKIN COLLEGE
The history of Ruskin College. - Oxford : Ruskin College Library, 1984
71p,[8]p of plates : ill. - (Ruskin College Library occasional publication ; no.3). - Bibliography: p67-70
B85-11528 X.809/62860
3247473

ADULT EDUCATION INSTITUTIONS. Oxford
Ruskin College. Social conditions, 1919-1920 - Biographies

HODGKINSON, George
A shop steward at Oxford 1919-1920. - Coventry : University of Warwick Library, 1980
29p. - (Occasional publications / University of Warwick Library)
0-903220-08-3 2718.ca.78/8
239071x

ADULT EDUCATION INSTITUTIONS. Portsmouth. Hampshire
New Road Centre (Portsmouth), to 1980

WORKERS' EDUCATIONAL ASSOCIATION. Southern District
The first year. - [Southampton] : [W.E.A. Southern District], [1980]
[22]p
B81-08573 X.529/40006
2363755

ADULT EDUCATION INSTITUTIONS. Salisbury. Wiltshire
Salisbury Mechanics Institution, to 1841

LITTLE, Monte
Salisbury Mechanics Institution 1833-1841. - Trowbridge : Wiltshire Library & Museum Service, 1982
27p. - (Wiltshire monographs ; no.2)
0-86080-098-9 X.805/5774
3151190

ADULT EDUCATION INSTITUTIONS. South-east England

PATES, Andrew
Thames Television's Second thoughts. - London : Thames Television, [1983?]
49p : ill
B84-11550 X.525/8647
3045358

ADULT EDUCATION INSTITUTIONS. Sweden

JAKOBSSON, Ingel
The Swedish folk high school. - Stockholm : Swedish Institute, 1977
9p. - (Current Sweden ; no. 166)
r6029185 X.0512/423(166)
1384813

STRAWN, William
Sweden's study associations. - Stockholm : Swedish Institute, 1977
6p. - (Current Sweden ; no.170)
r6029189 X.0512/423(170)
1384813

UPP till kamp för folkhögskolan! / [av] Alf Ahlberg...[et al.] ; sammanställd av Jan Peterson. - [Nässjö] : Plebs, [1976]
199p
91-85350-01-x X.519/24133
1384813

ADULT EDUCATION INSTITUTIONS. Sweden
Curriculum subjects, to date: Visual arts

ROSÉN, Gull-Mari
Konst- och konsthantverkscirklar under 1970-talet. - Umea : Universitetet i Umea, 1984
236p. - (Umea studies in the humanities ; 65). - Includes bibliographies
91-7174-176-3 X.0510/277(65)
9125892

ADULT EDUCATION INSTITUTIONS. Sweden
Folkuniversitetet

ALEXANDERSSON, Olle
Det svenska Folkuniversitetet. - Stockholm : Studieförbundet Folkuniversitetet, 1982
75p : ill
91-970398-0-2 X.529/66028
9128107

ADULT EDUCATION INSTITUTIONS. West Germany
Curriculum subjects, to date: History

HISTORISCHES Lernen in der Erwachsenenbildung / Ulrich Kröll (Hrsg.). - Münster : Regensberg, 1984
278p : ill. - (Forum Geschichtsdidaktik ; 2)
3-7923-0516-x X.529/65616
9128085

ADULT EDUCATIONAL BROADCASTING SERVICES. British Broadcasting Corporation

BRITISH BROADCASTING CORPORATION
An introduction to further education broadcasts. - Revised [ed.]. - London : B.B.C., 1977
27p
0-563-16182-5 BS.129/32
1132148

ADULT EDUCATIONAL BROADCASTING SERVICES. British Broadcasting Corporation
Applications in adult education

ROGERS, Jennifer
Using broadcasts with adults. - London : British Broadcasting Corporation, 1980
[1],32p
0-563-16389-5 X.529/36922
0209791

ADULT EDUCATIONAL BROADCASTING SERVICES. British Broadcasting Corporation
Applications in adult literacy education

HARGREAVES, David
Adult literacy and broadcasting. - London : Pinter, c1980
257p : ill. - Bibliography: p257
0-903804-68-9 X.950/1877
2359847

ADULT EDUCATIONAL GUIDANCE SERVICES. Great Britain
- Conference proceedings

CONTINUING educational opportunities for adults / edited by A.W. Bacon, M.A. Redmond, A. Sellars. - [Sheffield] : Sheffield City Polytechnic Department of Education Management, 1979
68leaves. - (Sheffield papers in education management ; no.7)
B84-23013 X.529/61154
3087409

ADULT EDUCATIONAL GUIDANCE SERVICES. Great Britain
- Directories

DIRECTORY of educational guidance services for adults / compiled by the Educational Advisory Services Project ; funded by the British Library. - Rev. ed. - Leicester : Advisory Council for Adult and Continuing Education, 1982
[8]p
B82-39326 X.529/50123
2559897

ADULT EDUCATIONAL GUIDANCE SERVICES. Public libraries. United States

REILLY, Jane A.
The public librarian as adult learners' advisor. - Westport, Conn. ; London : Greenwood, 1981
viii,152p. - (Contributions in librarianship and information science ; no.38). - Bibliography: p143-148. - Includes index
0-313-22134-0 2719.e.444
2826763

ADULT EDUCATIONAL RADIO PROGRAMMES. British Broadcasting Corporation
Special subjects: Decision making. Living decisions

STEPHENS, Michael, 1925-1984
Living decisions in family and community. - London : British Broadcasting Corporation, 1976
32p
0-563-16080-2 BS.129/199
1804111

ADULT EDUCATIONAL RADIO PROGRAMMES. Broadcasting services. Great Britain
Adult educational radio & television programmes, to 1982

ROBINSON, John, 1919-
Learning over the air. - London : British Broadcasting Corporation, 1982
256p,[24]p of plates : ill. - Bibliography: p249-250. - Includes index
0-563-20092-8 X.520/29267
2872420

ADULT EDUCATIONAL TELEVISION PROGRAMMES. Broadcasting services. Great Britain
Adult educational radio & television programmes, to 1982

ROBINSON, John, 1919-
Learning over the air. - London : British Broadcasting Corporation, 1982
256p,[24]p of plates : ill. - Bibliography: p249-250. - Includes index
0-563-20092-8 X.520/29267
2872420

ADULT LITERACY AND BASIC SKILLS UNIT
See National Institute of Adult Education (England and Wales). Adult Literacy and Basic Skills Unit

ADULT LITERACY EDUCATION INSTITUTIONS. Great Britain
Students. Assessment

HOWATT, A. P. R. (Anthony Philip Reid)
Making progress. - [Edinburgh] : Scottish Adult Literacy Agency, [1979]
8v.
B80-40642 X.520/22368G
2308940

ADULT LITERACY RESOURCE AGENCY
1975-1976

ADULT LITERACY RESOURCE AGENCY. Management Committee
Adult literacy. - London : H.M.S.O., 1976
x,58p
0-11-270379-8 BS.152/191
077877x

ADULT LITERACY UNIT
See National Institute of Adult Education (England and Wales). Adult Literacy Unit

ADULT MAMMALS
Influence of young mammals

BELL, Richard Q.
Child effects on adults. - Lincoln [Neb.] ; London : University of Nebraska Press, 1980, c1977
xii,253p. - (Bison Books in clinical psychology). - Bibliography: p215-236. - Includes index
0-8032-6058-x X.529/40596
2379155

ADULT OFFENDERS
See also
Day training centres for adult offenders
Hostels for adult offenders

ADULT OFFENDERS. Great Britain
Rehabilitation

BEAN, Philip
Rehabilitation and deviance. - London : Routledge and Kegan Paul, 1976
viii,168p. - (Radical social policy). - Bibliography: p.159-165. - Includes index
0-7100-8270-3 X.200/20291
061243x

ADULT OFFENDERS. Los Angeles (County). California
Prosecution

PROSECUTION of adult felony defendants / Peter W. Greenwood ... [et al.]. - Lexington, Mass. ; London : Lexington Books, 1976
xxx,174p : 1 ill. - Bibliography: p.165-166. - Includes index
0-669-00389-1 X.203/2033
071366x

ADULT STUDENTS
See also
Mature students

ADULT STUDENTS. Courses. Foreign languages. Western Europe
Needs. Assessments

CASE studies in identifying language needs / [edited and] prepared for the Council of Europe by Rene Richterich. - Oxford : Published for and on behalf of the Council of Europe by Pergamon, 1983
vi,149p : ill
0-08-029447-2 X.950/35093
2708795

RICHTERICH, René
Identifying the needs of adults learning a foreign language. - Oxford : Published for and on behalf of the Council of Europe by Pergamon, 1980
lx,103p : ill. - Bibliography: p99-103
0-08-024592-7 X.520/26738
2708795

ADULT T-CELL LEUKAEMIA. Man

ADULT T cell leukemia and related diseases / edited by Masao Hanaoka, Kiyoshi Takatsuki, Masanori Shimoyama. - Tokyo : Japan Scientific Societies Press ; New York ; London : Plenum Press, 1982
viii,243p : ill(some col.). - (GANN monograph on cancer research ; no.28). - Includes index
0-306-41335-3 X.322/12032
2934752

ADULT TRAINING CENTRES. Avon
Clients: Mentally handicapped persons

SURVEY of adult training centre clients aged over 25 living at home / Avon County Council Social Services Department. - Bristol : Director of Social Services, [1980]
49p : 1ill
0-86063-086-2 X.525/4401
2406640

ADULT TRAINING CENTRES FOR MENTALLY HANDICAPPED PERSONS. Wales

The ROLE and function of adult training centres / Welsh Office, Social Work Service. - [Cardiff] : [Welsh Office], [1981]
74p. - Bibliography: p58-74
B81-40342 BS.128A/291
2574942

ADULT TRAINING COURSES. Higher education institutions. Great Britain
Marketing - Conference proceedings

BRITISH ASSOCIATION FOR COMMERCIAL AND INDUSTRIAL EDUCATION. National Conference (1984 : University of Nottingham)
Marketing for colleges, polytechnics, universities. - [Nottingham] : [Published on Behalf of the East Midlands Regional Committee of BACIE by EMFEC], [1984]
81p : ill
0-906794-09-9 X.525/10992
3288005

ADULTERY
See also
Cuckolds

ADULTERY

ADULTERY and the decline of marriage. - New York ; London : Garland, 1984
130p in various pagings. - (Marriage, sex and the family in England 1660-1800)
0-8240-5905-0 X.529/69065
3262693

ESKAPA, Shirley
Woman versus woman. - London : Heinemann, 1984
210p. - Bibliography: p200-205. - Includes index
0-434-24520-8 X.529/63861
1736442

LAKE, Tony
Affairs. - London : Open Books, 1979
vi,205p. - Includes index
0-7291-0162-2 X.529/33300
1736442

PEARCE, William W.
Caught in the act. - New York : Day Books, 1978
153p
r6193722 X.108/17823
1736442

ROTHMAN, Joel
Adultery can be funny!. - London : Robson, 1983
[80]p : ill
0-86051-219-3 X.629/20470
2915782

ADULTERY. Men
to date

BARTUSIS, Mary Ann
Every other man. - New York : Dutton, c1978
xv,207p. - (Thomas Congdon books)
0-525-10064-4 X.529/68410
9152369

ADULTERY. Scotland
Liability of paramours. Law. Reform

GREAT BRITAIN. Scottish Law Commission
Family law, report on liability for adultery and enticement of a spouse. - Edinburgh : H.M.S.O., [1976]
iv,19p. - (Scot. Law Com. : no.42) ([1975-76 H.C.] 497)
0-10-249776-1 BS.179/12
0714364

ADULTERY. Special subjects. Fiction in European languages. 1761-1856

TANNER, Tony
Adultery in the novel. - Baltimore ; London : Johns Hopkins University Press, 1979
xii,383p : ill. - Bibliography: p.378-379. - Includes index
0-8018-2178-9 X.981/22424
2152606

ADULTERY. Special subjects. Fiction in European languages. 1837-1901

ARMSTRONG, Judith
The novel of adultery. - London : Macmillan, 1976
viii,182p. - Bibliography: p.175-178. - Includes index
0-333-19680-5 X.989/51361
1063901

ADULTERY. Special subjects. Koran. Islam

ZAFRULLA KHAN, Muhammad
Punishment of adultery in Islam. - London : London Mosque, [1981?]
12p
B84-12779 X.208/9299
3056236

ADULTHOOD

ADULT development through relationships / edited by Vivian Rogers. - New York ; Eastbourne : Praeger, 1984
xii,116p : ill. - (Praeger scientific studies). - Bibliography: p103-107. - Includes index
0-03-069787-5 X.520/35405
1895311

TURNER, Jeffrey S.
Contemporary adulthood. - Philadelphia ; London : Saunders, 1979
x,259p : ill. - Bibliography: p.237-247. - Includes index
0-7216-8935-3 X.512/6632
1895311

ADULTHOOD
Crises. Psychological aspects

GOLAN, Naomi
Passing through transitions. - New York : Free Press ; London : Collier Macmillan, c1981
xxii,330p. - Bibliography: p301-323. - Includes index
0-02-912070-5 X.520/26831
2714744

MURGATROYD, Stephen
Coping with crisis. - London : Harper & Row, c1982
179p. - Bibliography: p168-172. - Includes index
0-06-318228-9 X.529/52823
2714744

ADULTHOOD
Psychological aspects

ROGERS, Dorothy, 1914-
The adult years. - 2nd ed. - Englewood Cliffs ; London : Prentice-Hall, c1982
xvi,445p : ill. - Bibliography: p420-440. - Includes index
0-13-008961-3 X.520/27106
2730111

ADULTHOOD
Psychological aspects - Conference proceedings

VERMONT CONFERENCE ON THE PRIMARY PREVENTION OF PSYCHOPATHOLOGY (4th : 1978)
Competence and coping during adulthood. - Honover, N.H. ; London : University Press of New England, 1980
viii,388p : ill. - (Primary prevention of psychopathology ; 4). - Includes bibliographies and index
0-87451-159-3 X.520/22052
2255707

ADULTHOOD. India (Republic)
- Conference proceedings

IDENTITY and adulthood / edited by Sudhir Kakar ; with an introductory lecture by Erik H. Erikson. - Delhi ; Oxford : Oxford University Press, 1979
xiii,134p. - Includes index
0-19-561119-5 X.529/34351
2045451

ADULTHOOD. United States

TURNER, Jeffrey S.
Contemporary adulthood. - 2nd ed. - New York ; London : Holt, Rinehart and Winston, c1982
xiv,400p : ill. - Bibliography: p369-387. - Includes index
0-03-060143-6 X.520/27181
2739054

ADULTHOOD. United States
Personal adjustment of young men

HOGAN, Dennis P.
Transitions and social change. - New York ; London : Academic Press, 1981
xiv,232p : ill. - (Studies in population). - Bibliography: p221-225. - Includes index
0-12-352080-0 X.520/26008
2620766

ADULTS
See also
Adulthood
Day care centres for adults
Deaf-blind adults
Homeless adults
Illiterate adults
Mental health services for adult outpatients
Mentally handicapped adults
Mentally ill adults
Non-fiction for adults
Sick adults
Single adults
Slow learning adults
Slow reading adults
Unemployed adults

ADULTS
Abuse of children

ASSAULT against children / edited by John H. Meier. - London : Taylor & Francis, c1985
xxxii,304p : ill. - Bibliography: p271-292. - Includes index
0-85066-507-8 X.529/69972
1736116

ADULTS
Abuse of children - Bibliographies

WELLS, Dorothy P.
Child abuse. - Metuchen ; London : Scarecrow Press ; [Folkestone] : [Distributed by Bailey and Swinfen], 1980
viii,450p. - Includes index
0-8108-1264-9 X.529/36170
2181495

ADULTS
Abuse of children - Conference proceedings

CHILD abuse. - [Dundee] : [Tayside Regional Council Department of Social Work?], [1979?]
36p
B83-32828 X.525/7708
1736116

CHILD abuse. - [Dundee] : [Tayside Regional Council Department of Social Work?], [1982]
29p
B83-32829 X.525/7706
1736116

DECISION making in child abuse. - [Dundee] : [Tayside Regional Council Department of Social Work?], [1981]
31p
B83-32827 X.525/7709
1736116

INTERNATIONAL CONGRESS ON CHILD ABUSE AND NEGLECT (2nd : 1978 : London)
The abused child in the family and in the community. - Oxford : Pergamon, 1980
2v.(1092,xli p) : ill. - Includes bibliographies and index
Subsequent volumes, when received, will be placed at this shelfmark
0-08-023430-5 X.0525/355
1736116

INTERNATIONAL CONGRESS ON CHILD ABUSE AND NEGLECT (2nd : 1978 : London)
Second International Congress on Child Abuse & Neglect, London, 12-15 September 1978. - [Oxford, etc.] : Pergamon for the Organizing Committee of the Congress.
Abstracts. - [1979]
iv[i.e. vi],278,[1],11p. - Includes bibliographies and index
0-08-023438-0 X.519/29453
1736116

ADULTS
Abuse of children - Conference proceedings-continued

NATO ADVANCED STUDY INSTITUTE ON RESEARCH AND INNOVATION IN CHILD ABUSE AND NEGLECT (1982 : Les Arcs)
Child abuse and neglect. - The Hague ; Lancaster : Nijhoff in cooperation with NATO Scientific Affairs Division, 1983
xi,340p : ill. - (NATO advanced study institutes series Series D, Behavioural and social sciences ; no.16). - Includes index
90-247-2862-2 X.520/32624
1736116

ADULTS
Abuse of children. Cultural aspects - Conference proceedings

CHILD abuse and neglect / edited by Jill E. Korbin. - Berkeley ; London : University of California Press, c1981
xiv,217p. - Includes bibliographies and index
0-520-04432-0 X.520/26209
2635178

ADULTS
Ageing

EVANS, Peter, 1940-
Getting on. - London : Granada, 1982
viii,136p. - Bibliography: p130-131. - Includes index
0-246-11136-4 X.529/47789
0130567

The HANDBOOKS of aging / editor-in-chief James E. Birren. - [New York] ; [London etc.] : [Van Nostrand Reinhold]. [Handbook of aging and the social sciences]. - [1976]
xvi,684p : ill. - Includes bibliographies and index
0-442-20798-0 X.525/2143
0130567
- [Another copy: paperback]
ISBN 0-442-20804-9 X.510/16183

[Handbook of the biology of aging]. - [1977]
xvi,771p : ill. - Includes bibliographies and index
0-442-20796-4 SRL
0130567
- [Another copy: paperback]
ISBN 0-442-20806-5

[Handbook of the psychology of aging]. - [1977]
vii,787p : ill. - Includes bibliographies and index
0-442-20797-2 X.322/8383 Woolwich
0130567
- [Another copy: paperback]
ISBN 0-442-20805-7 X.311/8952

STOPPARD, Miriam
50 plus lifeguide. - London : Dorling Kindersley, 1983
320p : ill(some col.). - Includes index
0-86318-016-7 X.529/61992
0130567

WALFORD, Roy L.
Maximum life span. - New York ; London : Norton, c1983
xiv,256p : ill. - Includes index
0-393-01649-8 X.329/16747
0130567

ADULTS
Ageing - Conference proceedings

AGING / [James G. March, editor in chief]. - New York ; London : Academic Press.
Social change. - c1981
xxiv,631p : ill. - Includes bibliographies and index
0-12-040002-2 X.520/29327
0130567
- [Another copy : paperback] X.520/32523

PHILADELPHIA SYMPOSIUM ON AGING (1974)
Explorations in aging. - New York ; London : Plenum Press, 1975
ix,307p : ill. - (Advances in experimental medicine and biology ; vol.61). - Includes bibliographies and index
0-306-39061-2 SRL
0130567

SUCCESSFUL AGING (Conference : 1973 : Duke University)
Successful aging. - Durham, N.C. : Center for the Study of Aging and Human Development, Duke University, 1975
[6],156[i.e.161]p,[8]p of plates. - Includes bibliographies
r6032038 X.311/8481 Woolwich
0130567

ADULTS
Ageing. Effects on diseases

DILMAN, V. M. (Vladimir Mikhaĭlovich)
The law of deviation of homeostasis and diseases of aging. - Boston, [Mass.] ; Bristol : John Wright, c1981
xiii,380p : ill. - Bibliography: p331-372. - Includes index
0-88416-250-8 SRL
2695472

ADULTS
Ageing. Effects on diseases - Conference proceedings

COMPARATIVE pathobiology of major age related diseases / editors Dante G. Scarpelli, George Migaki. - New York : Liss, c1984
xvi,441p : ill. - (Modern aging research ; v. 4). - Includes bibliographies
0-8451-2304-1 X.322/13483
2695472

ADULTS
Ageing. Psychological aspects

FREESE, Arthur S.
The end of senility. - South Yarmouth : John Curley, 1979, c1978
xx,287p. - Bibliography: p282-287
0-89340-189-7 X.329/15033 Woolwich
0173673

KAUSLER, Donald H.
Experimental psychology and human aging. - New York ; Chichester : Wiley, c1982
x,720p : ill. - Bibliography: p637-687. - Includes index
0-471-08163-9 X.520/27916
0173673

ADULTS
Ageing. Social aspects

AGEING and life course transitions / edited by Tamara K. Hareven and Kathleen J. Adams. - London : Tavistock, 1982
xviii,281p. - Includes bibliographies and index
0-422-78360-9 X.520/29109
1133446

AGEING in modern society / edited by Dorothy Jerrome. - London : Croom Helm, c1983
307p : ill. - Includes bibliographies and index
0-7099-1422-9 X.529/59980
1133446

CARLSON, Avis D.
In the fullness of time. - South Yarmouth : Curley, 1979, c1977
xxiii,388p
0-89340-172-2 Nov.45938 Woolwich
1133446

HAMMARSTRÖM TORNSTAM, Gunhild
Det sociala åldrandet. - Uppsala, 1975
176 p.. - Bibliography: p. 165-176
91-506-0025-7 X.519/20987
0884006

The SOCIAL challenge of ageing / edited by David Hobman. - London : Croom Helm, 1978
5-287p : 1 ill. - Includes bibliographies and index
0-85664-428-5 X.529/32145
126642x
- [Another copy: paperback.] X.529/41191

ADULTS
Ageing. Social psychology

TORNSTAM, Lars
Åldrandets socialpsykologi. - [Stockholm] : Rabén & Sjögren, 1978
293p. - (Stora tema nova). - Includes bibliographies
91-295-1241-7 X.519/39528
1785540

ADULTS
Behavioural disorders. Diagnosis

BEHAVIORAL assessment of adult disorders / edited by David H. Barlow. - New York ; London : Guilford, c1981
xii,500p : ill. - Includes bibliographies and index
0-89862-140-2 X.520/25697
2601834

ADULTS
Brain. Minimal brain dysfunction. Psychological aspects - Conference proceedings

ADULT MBD CONFERENCE (1978 : Scottsdale)
Psychiatric aspects of minimal brain dysfunction in adults. - New York ; London : Grune & Stratton, c1979
xi,208p : ill. - Includes index
0-8089-1192-9 X.322/10574 Woolwich
2627264

ADULTS
Christian religious education

BRENNAN, Gemma
Adult religious education. - Great Wakering : Mayhew McCrimmon, 1984
95p : ill
0-85597-364-1 X.200/46758
3299783

ADULTS
Development

ALLMAN, Paula
Adult development. - Nottingham : Department of Adult Education, University of Nottingham, [1981?]
iii,42p. - (Adults ; 1). - Bibliography: p32-42
0-902031-46-5 X.529/41884
2497875

STEVENS-LONG, Judith
Adult life. - 2nd ed. - Palo Alto, Calif. : Mayfield, c1984
xv,636p : ill. - Bibliography: p581-621
0-87484-577-7 X.529/63041
2497875

ADULTS
Development & ageing

HULTSCH, David F.
Adult development and aging. - New York ; London : McGraw-Hill, c1981
xiv,431p : ill. - Bibliography: p377-399. - Includes index
0-07-031156-0 X.520/23833
0069108

KIMMEL, Douglas C. (Douglas Charles)
Adulthood and aging. - 2nd ed. - New York ; Chichester : Wiley, c1980
xiii,574p : ill. - Includes bibliographies and index
0-471-05229-9 X.520/23789
0069108

ADULTS
Development. Role of emotions

EMOTION in adult development / edited by Carol Zander Malatesta, Caroll E. Izard. - Beverly Hills ; London : Sage, c1984
343p : ill. - Includes bibliographies
0-8039-2156-x X.529/67481
3244245

ADULTS
Developmental psychology

ELIAS, Merrill F.
Basic processes in adult developmental psychology. - Saint Louis : Mosby ; London : Distributed by Kimpton, 1977
xi,336p : ill. - Bibliography: p.304-327. - Includes index
0-8016-1519-4 X.525/2152
1200941

READINGS in adult psychology / edited by Lawrence R. Allman and Dennis T. Jaffe. - New York ; London : Harper and Row, 1977
[13],388p : ill. - (Contemporary perspectives reader series). - Includes bibliographies
0-06-047054-2 X.512/5710
1200941

ADULTS
Developmental psychology - Conference proceedings

WEST VIRGINIA UNIVERSITY LIFE-SPAN DEVELOPMENTAL PSYCHOLOGY CONFERENCE (4th : 1974?)
Life-span developmental psychology, normative life crisis. - New York ; London : Academic Press, 1975
xvii,313p : ill. - Includes bibliographies and index
0-12-203550-x X.520/10502
1200941

ADULTS
Dialogues with children - Texts

LAING, R. D. (Ronald David)
Conversations with children. - London : Allen Lane, 1978
[1],x,89p
0-7139-1079-8 X.529/32476
1569791

ADULTS
Effects of childhood cancer. Psychosocial aspects

KOOCHER, Gerald P.
The Damocles syndrome. - New York ; London : McGraw-Hill, c1981
xx,219p : ill. - Bibliography: p191-199. - Includes index
0-07-035340-9 X.520/24311
2506246

ADULTS
Enuresis. Physiotherapy. Exercises

MONTGOMERY, Eileen
Regaining bladder control. - Bristol : J. Wright, 1974
[5],26p : ill
0-7236-0392-8 X.319/10103
0373281

MONTGOMERY, Eileen
Regaining bladder control. - Rev. reprint. - Bristol : Wright PSG, 1983
27p : ill
0-7236-0695-1 X.329/17042
0373281

ADULTS
Foreign language skills. Acquisition. Theories

KRASHEN, Stephen
Second language acquisition and second language learning. - Oxford : Pergamon, 1981
vii,151p : ill. - (Language teaching methodology series) (Pergamon Institute of English (Oxford)). - Bibliography: p139-151
0-08-025338-5 X.950/8784
2152770

ADULTS
Illiteracy

WHERE do we go from here?. - Manchester : Gatehouse Project, c1983
80p : ill
0-906253-20-9 X.529/58048
2964937

ADULTS
Interpersonal relationships. Emotional problems. Self treatment: Behaviour modification. Techniques

PHILIPS, Debora
How to fall out of love. - London : Macdonald Futura, 1981, c1978
191p. - Includes index
0-7088-1986-9 X.529/41142
2396556

SELF-CONTROL and self-modification of emotional behavior / edited by Kirk R. Blankstein and Janet Polivy. - New York ; London : Plenum, c1982
xi,204p. - (Advances in the study of communication and affect ; v.7). - Includes bibliographies and index
0-306-40945-3 X.520/28796
2396556

ADULTS
Interpersonal relationships with children

ASTROP, John
Deeper secrets. - Harmondsworth : Puffin, 1984
[112]p : ill
0-14-031703-1 X.990/23776
321768x

DEAR adults - / selected and edited by Alan Davies ... [et al.]. - London : Published for the General Synod Board of Education by CIO Publishing, 1981
72p
0-7151-0402-0 X.950/11318
2643693

NUPPONEN, Ritva
Vahvistajaroolit aikuisen ja lapsen vuorovaikutuksessa. - Jyväskylä : Jyväskylan yliopisto, 1977
209p. - (Jyväskylä studies in education, psychology and social research ; 34). - Includes bibliographies
951-677-788-0 Ac.9234.pl/2[no.34]
226708X

SCHAFFER, Ulrich
For the love of children. - Tring : Lion Publishing, 1980, c1979
viii,94p : ill. - (An Aslan Lion paperback)
0-85648-317-6 X.529/39495
2350912

ADULTS
Interpersonal relationships with children. Role in development of children

LANGEVELD, M. J.
Personal help for children growing up. - [Exeter] : University of Exeter, 1975
15p. - (The W.B. Curry lecture ; 1974)
0-85989-080-5 X.519/25169
037332x

ADULTS
Learning

ADULT learning / edited by Michael J.A. Howe. - London : Wiley, 1977
xvii,291p : ill. - Includes bibliographies and index
0-471-99458-8 X.520/11743
1144065

GUTMANN, Werner
Selbstgesteuerter Erwerb von Regelwissen mit modularen Lerntexten. - Zürich : [W. Gutmann], 1980
xiii,263p.. - Bibliography
r6300383 X.529/42944
8737894

LOVELL, R. Bernard
Adult learning. - London : Croom Helm [etc.], 1980
[2], 170. - (New patterns of learning). - Bibliography: p.148-156. - Includes index
0-7099-0237-9 X.529/38520
1956973

SMITH, Robert M. (Robet McCaughan)
Learning how to learn. - Milton Keynes : Open University Press, 1983, c1982
201p : ill. - Bibliography: p183-194. - Includes index
0-335-10115-1 X.520/29919
1144065

ADULTS
Learning of foreign languages. Self-assessment

OSKARSSON, Mats
Approaches to self-assessment in foreign language learning. - Oxford : Published for and on behalf of the Council of Europe by Pergamon, 1980
47p : ill. - Bibliography: p35-36
0-08-024594-3 X.950/12946
1953990

ADULTS
Number skills. Teaching

An INTRODUCTION to numeracy teaching. - London : ALBSU, 1982
64p : ill
0-906509-19-x X.529/50156
1878271

RILEY, Terry, 1945-
Teaching numeracy. - Cambridge : National Extension College, 1978
[5],89p : ill
0-86082-113-7 X.619/19669 Woolwich
1878271

ADULTS
Oral communication with children - Conference proceedings

ADULT-CHILD conversation. - London : Croom Helm, c1981
310p : ill. - Includes bibliographies and index
0-7099-0069-4 X.529/39741
2057948

ADULTS
Physical abuse of children

KEMPE, Ruth S.
Child abuse. - [London] : Fontana, 1978
159p. - (The developing child). - Bibliography: p.152-154. - Includes index
0-00-635200-6 X.519/28575
1685988

The MALTREATMENT of children / edited by Selwyn M. Smith. - Lancaster : M.T.P. Press, 1978
xii,452p. - Bibliography: p.416-447. - Includes index
0-85200-141-x X.520/13149
174609x

ADULTS
Physical abuse of children - Serials

CHILD abuse & neglect. - Vol.1, no.1- ; 1977-. - Oxford : Pergamon, 1977- : ill
B78-04965 P.521/3146
1303716

ADULTS
Psychological development

COLARUSSO, Calvin A. (Calvin Anthony)
Adult development. - New York ; London : Plenum Press, c1981
xxx,290p : ill. - (Critical issues in psychiatry). - Includes bibliographies and index
0-306-40619-5 X.520/25071
1913115

READINGS in adult psychology / edited by Lawrence R. Allman, Dennis T. Jaffe. - 2nd ed. - New York ; London : Harper & Row, c1982
xii,407p : ill. - (Harper & Row's contemporary perspectives reader series). - Includes bibliographies
0-06-040234-2 X.525/5935
1913115

STAUDE, John-Raphael
The adult development of C.G. Jung. - London : Routledge & Kegan Paul, 1981
xxiv,134p : 2ill. - Bibliography: p125-130. - Includes index
0-7100-0749-3 X.529/43623
256288x

WHITBOURNE, Susan Krauss
Adult development. - New York ; London : Holt, Rinehart and Winston, 1979
x,329p : ill. - Includes bibliographies and index
0-03-017741-3 X.520/14160
1913115

ADULTS
Psychological development - Conference proceedings

THEMES of work and love in adulthood / edited by Neil J. Smelser and Erik H. Erikson. - London : Grant McIntyre, 1980
xi,297p : ill. - Includes bibliographies and index
0-86216-008-1 X.520/21454
2210908

ADULTS
Reading habits, ca 1930-1975 - Bibliographies

MANN, Margaret Gwendoline
The reading habits of adults. - London : British Library, 1977
72p. - (Reports / British National Bibliography Research Fund ; no.1). - Includes index
0-85350-143-2 BS.10/468
0758531

ADULTS
Reading habits. Research, 1900-1939

KARETZKY, Stephen
Reading research and librarianship. - Westport, Conn ; London : Greenwood, 1982
xxi,385p. - (Contributions in librarianship and information science ; no.36). - Bibliography: p359-370. - Includes index
0-313-22226-6 2719.e.440
2828316

ADULTS
Reading skills. Teaching

JONES, Edward V.
Reading instruction for the adult illiterate. - Chicago : American Library Association, 1981
xi,169p. - Bibliography: p161-166. - Includes index
0-8389-0317-7 X.520/26912
2719770

ADULTS
Sex differences. Social aspects. Role of socialisation of children

GIANINI BELOTTI, Elena
Little girls. - London : Writers and Readers Publishing Cooperative, 1975
158p
0-904613-06-2 X.529/30310
0540021

ADULTS
Sexual abuse of children

CHILD sexual abuse / ... prepared by members of BASPCAN's Sub-Committee on Child Sexual Abuse. - 2nd repr. with minor amendments. - Rochdale : British Association for the Study and Prevention of Child Abuse and Neglect, 1981
13p. - Bibliography: p12-13
0-907778-00-3 X.200/42439
2564343

CHILD sexual abuse within the family / the Ciba Foundation ; edited by Ruth Porter. - London : Tavistock, 1984
xx,156p. - Bibliography: p143-146. - Includes index
0-422-79280-2 Cup.719/98
2564343
- [Another copy: paperback] ISBN 0-422-79290-X Cup.719/99

DE YOUNG, Mary
The sexual victimization of children. - Jefferson ; London : McFarland, 1982
ix,179p. - Bibliography: p165-172. - Includes index
0-89950-063-3 X.520/30064
2564343

I never told anyone / edited by Ellen Bass and Louise Thornton with Jude Brister ... [et al.]. - New York ; London : Harper & Row, c1983
278p. - Bibliography: p272-276
0-06-015149-8 X.529/66643
2564343

KEMPE, Ruth S.
The common secret. - New York : W.H. Freeman, c1984
viii,284p : ill. - Includes bibliographies
0-7167-1624-0 X.529/67016
2564343

KROTH, Jerome A.
Child sexual abuse. - Springfield, IL : Thomas, c1979
x,200p
0-398-03906-2 X.322/10711 Woolwich
2564343

SEXUALLY abused children and their families / edited by Patricia Beezley Mrazek and C. Henry Kempe. - Oxford : Pergamon, 1981
xii,271p : ill. - Includes bibliographies and index
0-08-026796-3 Cup.365.pp.25
2564343
- [Another copy: paperback.] ISBN 0-08-030194-0 Cup.351/309

ADULTS
Sexual abuse of children - Bibliographies

SCHLESINGER, Benjamin
Sexual abuse of children. - Toronto ; London : University of Toronto Press, c1982
xiii,200p. - Includes bibliographies and index
0-8020-6481-7 Cup.366.y.112
2564343

ADULTS
Socialisation of children. Influence of children

BELL, Richard Q.
Child effects on adults. - Hillsdale : Erlbaum : New York : London : Distributed by Wiley, 1977
xiii,253p : 2 ill. - Bibliography: p.215-236. - Includes index
0-470-99267-0 X.520/12578
1481630

ADULTS
Surgery. Preoperative medication

JOHNSTONE, Michael, 1919-
Adult preoperative medication. - Macclesfield : Imperial Chemical Industries. Pharmaceuticals Division, [1976]
[1],30p : col ill. - (Anaesthesia rounds ; no.10)
0-905398-01-7 X.0311/182(10) Woolwich
0640492

ADULTS
Talking to pre-school children. Speech. Grammar & vocabulary

KARPOVA, Sof'ja Nikolaevna
The realization of the verbal composition of speech by preschool children. - The Hague : Mouton, 1977
ix,275p. - (Janua linguarum. Series maior ; no.56). - Bibliography : p.269-275
90-279-3186-0 X.0981/78b(56)
0808547

ADULTS
Teaching. Implications of cognitive styles

SQUIRES, Geoffrey
Cognitive styles and adult learning. - Nottingham : Department of Adult Education, University of Nottingham, c1981
ii,21p. - (Adults ; 3). - Includes bibliographies
0-902031-48-1 X.529/46753
270224x

ADULTS. Caddis flies. Europe

MALICKY, Hans
Atlas of European Trichoptera. - The Hague ; London : Junk, 1983
x,298p : ill. - (Series entomologica ; v.24). - Includes index
90-619-3134-7 SRL
2902273

ADULTS. Canada
Abuse of children, to date

VAN STOLK, Mary
The battered child in Canada. - Rev. ed.. - Toronto : McClelland and Stewart, 1979
xii,178p
0-7710-8712-8 X.529/62247
9088768

ADULTS. Developed countries
Vocational education, to date

BERUFLICHE Weiterbildung - zweiter Bildungsweg / [Herausgeber:] Marburger Forschungsstelle für vergleichende Erziehungswissenschaft. - Marburg : Philipps-Universität, Fachbereich Erziehungswissenschaften, 1978
xi,98p. - (Texte, Dokumente, Berichte ; Heft 22)
r6301249 X.0529/611(22)
8719497

ADULTS. Developing countries
Literacy programmes. Role of British organisations

BRITISH COMMITTEE ON LITERACY
Illiteracy in the developing world. - Reading : British Committee on Literacy, 1976
viii,64p : ill
0-9505439-0-x X.909/41142
0798339

ADULTS. Developing countries
Number skills. Teaching methods: Games

EVANS, David Russell
Games and simulations in literacy training. - Amersham : Hulton [for] the International Institute for Adult Literacy Methods, 1979
xi,136p : ill. - (Literacy in development). - Bibliography: p.134-136
0-7175-0822-6 X.619/22219 Woolwich
197176x

ADULTS. England
Day care - Bibliographies

NATIONAL DAY CARE PROJECT
Adult day care. - London : National Institute for Social Work, 1978
[2],v,34p
B80-38670 X.525/3722
2281643

ADULTS. England
Illiteracy

HAVILAND, R. Michael
Provision for adult illiteracy in England. - Reading : Centre for the Teaching of Reading, University of Reading, School of Education, 1973
68p. - Bibliography: p63-68
0-7049-0294-x X.529/59060
0138657

ADULTS. England
Physical abuse of children

CREIGHTON, Susan J.
Child victims of physical abuse. - London : National Society for the Prevention of Cruelty to Children, [1977]
[1],52p : ill. - Bibliography: p.34-36
B78-18150 X.512/6061
1449605

ADULTS. England
Residential care

LIVING & working in residential homes. - London : Personal Social Services Council, 1975
[2],33p
0-905250-00-1 X.512/5182
0649414

RESIDENTIAL care reviewed. - London : Personal Social Services Council, 1977
[1],59,[1]p
B78-08893 X.512/5822
0649414

ADULTS. England
Residential care. Policies

POLICY issues in residential care / report to the PSSC from an advisory group. - London : Personal Social Services Council, 1978
[1],75p
B79-13585 X.519/29476
1846779

ADULTS. Finland
Religious education, to date

TOTRO, Timo
Uskontokasvatus kansalais- ja kotiopintoliikkeessä. - Helsinki : [Suomalainen teologinen kirjallisuusseura], 1979
413p. - (Suomalaisen teologisen kirjallisuusseuran julkaisuja ; 112). - Bibliography: p.377-404
951-91112-2-0 X.200/40580
8945233

ADULTS. Frankfurt am Main. West Germany
Political education

BÖHMER, Torsten
Political education for adults in Frankfurt. - Leicester : Advisory Council for Adult and Continuing Education, 1983
iii,48p. - (Research report)
0-906436-23-0 X.525/9410
3145824

ADULTS. Great Britain
Higher education

JONES, H. A. (Henry Arthur)
Adult students and higher education. - [New ed.]. - Leicester : ACACE, c1979
vi,64p. - (Occasional paper / Advisory Council for Adult and Continuing Education ; 3)
0-906436-03-6 X.529/50154
2790092

ADULTS. Great Britain
Mathematical skills. Surveys

ADULTS' mathematical ability and performance / ACACE. - Leicester : ACACE, [1982]
59p : ill. - (Report / Advisory Council for Adult and Continuing Education)
0-906436-09-5 X.529/50274
2791102

SEWELL, Bridgid
Use of mathematics by adults in daily life. - Leicester : ACACE, [1981]
ii,83p : ill. - (Enquiry officers report)
0-906436-10-9 X.622/13824
2791102

ADULTS. Great Britain
Numeracy

ADULT numeracy study / conducted by Social Surveys (Gallup Poll) Ltd ; on behalf of the Advisory Council for Adult and Continuing Education in February/March 1981. - Leicester : ACACE, c1981
84p. - (Gallup report)
0-906436-08-7 L.49/1402
279005x

ADULTS. Great Britain
Physical abuse of children, 1970

CASTLE, Raymond Leslie
A study of suspected child abuse. - London : National Society for the Prevention of Cruelty to Children, Battered Child Research Department, 1972
[1],19p
B78-36735 X.512/6126
1700324

ADULTS. Great Britain
Socialisation

MUSGROVE, Frank, b.1922
Margins of the mind. - London : Methuen, 1977
vi,247p. - Includes index
0-416-55040-1 X.529/31902
1278916

ADULTS. Great Britain
Unemployed adults. Further education

EDUCATION for unemployed adults / report commissioned from the Advisory Council for Adult and Continuing Education by the Secretaries of State for Education and Science and for Wales. - Leicester : ACACE, c1982
58p
0-906436-18-4 X.529/53792
277819x

STORRIE, T. C.
Adult unemployed. - Sheffield, c/o Hon. Sec. Sheffield City Polytechnic, Pond St., Sheffield S1 1WB : Association of Colleges for Further and Higher Education, [1982]
10p
B82-36020 X.529/49668
277819x

ADULTS. Great Britain
Unemployed adults. Training

TOWARDS an adult training strategy. - Sheffield : Manpower Services Commission, 1983
12p
0-905932-84-6 BS.421/65
2964600

ADULTS. Hungary
Reading habits

DURKÓ, Mátyás
A felnőttek szépirodalmi műelemző készségének állapota és fejleszthetősége. - Debrecen : Kossuth Lajos Tudományegyetem, 1977
80p,[2]fold.leaves of plates. - (Nevelés, művelődés. Acta Debrecina sorozat ; sz.67 Paedagogica)
r6174420 Ac.829/7a(3)[s2.67]
2249677

ADULTS. India (Republic)
Reading skills. Retention

ROY, Prodipto
The retention of literacy. - Delhi : Macmillan [for the] Council for Social Development, 1975
xiii, 119p : ill. - Bibliography : p. 113-119
0-333-90065-0 X.529/31405
1099744

ADULTS. Northern Ireland
Educational guidance

GREAT BRITAIN. Department of Education for Northern Ireland
Educational guidance for adults in Northern Ireland. - Belfast : H.M.S.O., 1979
34p
0-337-04108-3 CS.A.210/140
2302063

ADULTS. Patients. Hospitals. United States
Nursing

CURRENT practice in nursing care of the adult / [editors] Maureen Shawn Kennedy, Gail Molnar Pfeifer. - St. Louis ; London : Mosby, 1979. - (Mosby current practice and perspectives in nursing series ; v.1) - Includes bibliographies
Vol.1
xiii,314p : ill
0-8016-2646-3 X.322/11612
273639x

ADULTS. Scandinavia
Vocational education

ARBETSMARKNADSUTBILDNING i Norden / [utgiven av Nordiska arbetsmarknadsutskottets underutskott för arbetsmarknadsservice]. - Oslo : Nordiska ministerrådets sekretariat, 1977
ix,113p : charts,forms,maps. - (NU-serien : B ; 1977,27)
91-7052-319-3 X.520/21088
242374X

ADULTS. Scotland
Attitudes to cigarette smoking & consumption of alcoholic drinks by young persons
AITKEN, P. P.
Adults' attitudes towards drinking and smoking among young people in Scotland. - Edinburgh : H.M.S.O., 1981
xxiv,119p
0-11-491742-6 BS.155/270
2902753

ADULTS. United States
Abuse of children, to date
HERBRUCK, Christine Comstock
Breaking the cycle of child abuse. - Minneapolis : Winston Press, c1979
205p
0-03-045691-6 X.529/67723
9164057

ADULTS. United States
Adaptability to transitions. Psychosocial aspects
FISKE, Marjorie
Four stages of life. - San Francisco ; London : Jossey-Bass, 1975
xxv,292p. - (The Jossey-Bass behavioral science series). - Bibliography: p.263-281. - Includes index
0-87589-246-9 X.520/10082
0359254

ADULTS. United States
Ageing. Psychosocial aspects
ROGERS, Dorothy, 1914-
The adult years. - Englewood Cliffs ; London : Prentice-Hall, 1979
xiv,482p : ill. - Bibliography: p.437-463. - Includes index
0-13-008987-7 X.520/14707
2002493

ADULTS. United States
Effects of television programmes compared with effects on children
WINICK, Mariann Pezzella
The television experience[by] Mariann Pezzella Winick, Charles Winick. - Beverly Hills ; London : Sage Publications, 1979
215p : ill. - ISBN 0-8039-1143-2 Pbk : £5.00
0-8039-1142-4 X.519/30807
2050900

ADULTS. United States
Illiteracy
HUNTER, Carman St. John
Adult illiteracy in the United States. - New York ; London : McGraw-Hill, 1979
xiii,206p : ill. - Includes bibliographies and index
0-07-031380-6 X.520/21614
2172712

ADULTS. United States
Physical abuse of children
CHASE, Naomi Feigelson
A child is being beaten. - New York : Holt, Rinehart and Winston, 1975
225p
0-03-013601-6 X.529/34371
0604801

WALTERS, David R.
Physical and sexual abuse of children. - Bloomington ; London : Indiana University Press, 1975 [i.e. 1976]
xii,192p. - Bibliography: p.185-188. - Includes index
0-253-34490-5 Cup.363.M.40
0604801

WALTERS, David R.
Physical and sexual abuse of children. - Bloomington ; London : Indiana University Press, [1978]
xii,192p. - Bibliography: p.185-188. - Includes index
B78-36736 X.519/28662
0604801

ADULTS. United States
Physical abuse of children. Prevention. Law
COSTA, Joseph J.
Child abuse and neglect. - Lexington, Mass. : D.C. Heath ; Farnborough, Hants. : Distributed by Teakfield Ltd, 1978
xxi,417,[1]p. - Bibliography: p.319-402. - Includes index
0-669-01670-5 X.520/12854
1576666

ADULTS. United States
Physical abuse of children. Prevention. Role of welfare services for children
NAGI, Saad Zaghloul
Child maltreatment in the United States. - New York ; Guildford : Columbia University Press, 1977
xii,162p : ill. - Includes index
0-231-04394-5 X.520/12387
1409379

ADULTS. United States
Psychological development
BISCHOF, Ledford Julius
Adult psychology. - 2nd ed. - New York ; London : Harper and Row, 1976
xv,430p. - Bibliography: p.365-414. - Includes index
0-06-040719-0 X.510/15150
0553727

ADULTS. United States
Sexual abuse of children
FINKELHOR, David
Sexually victimized children. - New York : Free Press ; London : Collier Macmillan, 1979
xii,228p. - Bibliography: p.215-221. - Includes index
0-02-910210-3 Cup.351/225
2045095

Sexually victimized children. - New York : Free Press ; London : Collier Macmillan, c1979
xii,228p. - Bibliography: p215-221. - Includes index
0-02-910400-9 X.520/24937
2045095

ADULTS. United States
Sexual abuse of children. Intervention by welfare services
SGROI, Suzanne M. (Suzanne Mary)
Handbook of clinical intervention in child sexual abuse. - Lexington, Mass. : Lexington Books, 1982
xi,387p : ill. - Includes bibliographies and index
0-669-04720-1 Cup.366.y.43
2736721

ADULTS. West Germany
Reading habits, 1945-
BERGINZ-PLANK, Gabriele
Literaturrezeption in einer Kleinstadt. - Stuttgart : Heinz, 1981
420p. - (Stuttgarter Arbeiten zur Germanistik ; Nr.84). - Bibliography: p.394-420
3-88099-088-3 X.0909/1020(84)
8752095

ADUR (DISTRICT). COUNCIL
Trade unions: National and Local Government Officers Association. Adur District Branch - Serials
NATIONAL AND LOCAL GOVERNMENT OFFICERS ASSOCIATION. Adur District Branch
Newsletter. - Jan. 1977-.
r6032206 P.511/5050
1196790

ADVAITA
See also
Visishtadvaita

ADVAITA
COHEN, Sulman Samuel
Advaitic sâdhanâ, or, The yoga of direct liberation. - Delhi : Motilal Banarsidass, 1975
xiii,92p. - Includes index
0-8426-0989-x X.529/32828
1733907

MAHADEVAN, T. M. P.
The philosophy of Advaita. - 4th ed. - New Delhi : Arnold-Heinemann, 1976
iii-xviii,301p. - Bibliography : p.287-289. - Includes index
r6072660 X.529/32772
1733907

ADVAITA
Citsukha. Pratyak-tattva-pradipikâ
SARMA, V. A.
Citsukha's contribution to Advaita. - Mysore : Kavyalaya Publishers, 1974
xv,248p. - Bibliography : p.225-229.- Includes index
r6091825 X.529/33408
1890646

ADVAITA VEDANTA
KLEIN, Jean
Neither this nor that I am. - London : Watkins, c1981
x,138p
0-7224-0189-2 X.200/35595
2408880

ADVAITA VEDANTA
Theories of human consciousness
INDICH, William M.
Consciousness in Advaita Vedanta. - Delhi : South Asia books, 1980
144p
r6801524 X.329/16452
8972052

ADVANCE BOOKING CHARTER FLIGHTS. Air passenger transport services
- Texts
GREAT BRITAIN
[Treaties, etc. United States, 1973 Mar.30. Protocols, etc., 1975 Apr.2]. Exchange of notes between the Government of the United Kingdom of Great Britain and Northern Ireland and the Government of the United States of America further amending the agreement concerning procedures for the implementation and enforcement of rules for advance booking charter flights, signed at Washington on 30 March 1973, London, 2-3 April 1975 ... - London : H.M.S.O., [1975]
4p. - (Treaty series ; no.88(1975)) (Cmnd.6107)
0-10-161070-x BS.14/137
0217123

[Treaties, etc. United States, 1973 Mar.30. Protocols, etc., 1975 June 4]. Exchange of notes between the Government of the United Kingdom of Great Britain and Northern Ireland and the Government of the United States of America further extending the agreement concerning procedures for the implementation and enforcement of rules for advance booking charter flights signed at Washington on 30 March 1973, London 4 June 1975 ... - London : H.M.S.O., [1975]
4p. - (Treaty series ; no.114(1975)) (Cmnd.6211)
0-10-162110-8 BS.14/137
0217123

ADVANCE FACTORIES. Great Britain
Role in regional economic development
SLOWE, Peter M.
The advance factory in regional development. - Aldershot : Gower, c1981
xiii,271p : ill. - Bibliography: p241-271
0-566-00437-2 X.529/46898
2706741

ADVANCE GROUP OF COMPANIES
to 1978
ADVANCE GROUP OF COMPANIES
Advance, 50th anniversary, 1928-1978. - [London] : [Advance Group of Companies], [1978]
16p : ill
B78-23271 X.512/5999
149502x

ADVANCE VARIATION. French Defence. Openings. Chess
HARDING, T. D. (Timothy David)
French, MacCutcheon and Advance lines. - London : Batsford, 1979
3-130p : ill. - (Batsford algebraic chess openings) (Batsford chess books). - Bibliography: p.9-10. - Includes index
0-7134-2026-x X.619/19760 Woolwich
1885839

ADVANCED GAS-COOLED NUCLEAR REACTORS. Great Britain
Safety aspects
The SAFETY of the AGR / edited by G.C. Dale ; written and compiled by J.M. Bowerman. - London : Central Electricity Generating Board ; Glasgow : South of Scotland Electricity Board, 1982
111p : ill(some col.). - Bibliography: p103. - Includes index
0-902543-64-4 SRL
2701758

'ADVANCED GROUND TRANSPORT'. Great Britain. Parliament. House of Commons. Select Committee on Science and Technology
GREAT BRITAIN. Department of Transport
Select Committee on Science and Technology, 'Advanced ground transport'. - London : H.M.S.O., [1977]
12p. - (Cmnd.6825)
0-10-168250-6 BS.43/435
1518909

ADVANCED PASSENGER TRAIN
to 1980
BRITISH RAIL
Advanced Passenger Train. - Weston-super-Mare : Avon-Anglia Publications and Services in association with British Rail, London Midland and Scottish Regions, c1981
48p : ill
0-905466-37-3 X.629/15648 Woolwich
2491486

ADVANCED PHYSICS PROJECT FOR INDEPENDENT LEARNING
ADVANCED PHYSICS PROJECT FOR INDEPENDENT LEARNING
Student's handbook. - London : J. Murray [for the] Inner London Education Authority, Learning Materials Service, 1978
[5],98p : ill
0-7195-3587-5 X.529/35995
2126494

ADVANCED PHYSICS PROJECT FOR INDEPENDENT LEARNING
- For teaching

ADVANCED PHYSICS PROJECT FOR INDEPENDENT LEARNING
Teacher's handbook. - London : J. Murray [for the] Inner London Education Authority, Learning Materials Service, 1978
vi,89p : ill. - Bibliography: p.80-87
0-7195-3589-1 X.529/35994
2126508

ADVENT
Christian life - For Advent

NASSAN, Maurice
Advent. - Dublin : Irish Messenger, 1983
43p : ill
B84-30744 X.208/9821
1273434

SMITH, Delia
A feast for Advent. - London : Bible Reading Fellowship, 1983
88p
0-900164-61-1 X.200/43466
1273434

TERRA, Russell G.
The coming God. - Slough : St Paul Publications, 1977
xii,123p
0-85439-135-5 X.100/17190
1273434

ADVENT
Church of England. Ministry for housebound persons - For advent

FORD, Deirdre
Parish visiting. - Great Wakering : Mayhew McCrimmon, c1982
25p
0-85597-344-7 X.208/8320
2882655

ADVENT. Celebrations. Catholic Church
Prayers

ADVENT prayer book / compiled by Paul James. - Great Wakering : Mayhew McCrimmon, 1983
14p : ill
0-85597-502-4 X.208/9701
3124789

ADVENT. Christian church

GUNSTONE, John
Advent and Christmas. - 2nd ed. - London : Church Literature Association, 1976
15p
0-85191-074-2 X.108/16325
0691607

RYAN, Vincent
Advent to Epiphany. - Dublin : Veritas Publications, 1977
96p. - Bibliography: p.95-96
0-905092-46-5 X.108/16864
0691607

ADVENT. Christian church
to date

BEKH, Wolfgang Johannes
Adventgeschichten. - Pfaffenhofen : Ludwig, [1981]
135p : ill
3-7787-3181-5 X.208/8109
8895201

ADVENT CALENDARS
- Texts

NEDERLANDSCHE ZONDAGSSCHOOL VEREENIGING
Jesus came to us. - London : Church Information Office, 1978
Folder ([6]p) : chiefly col ill
0-7151-0384-9 Cup.1281/1008
1758764

ADVENTURE
- Texts

TESTA, Fulvio
If you seek adventure. - London : Andersen, 1984
26p : col.ill
0-86264-069-5 L.42/2933
3174484

ADVENTURE
to 1963 - Texts

SPY and mystery stories / edited by Kenneth Allen. - [London] : Octopus Books, [1978]
399p : ill
0-7064-0763-6 Nov.24086 Woolwich
1723847

ADVENTURE
1425-1977 - Texts

ADVENTURE stories for girls / edited by John Canning. - [London] : Octopus Books, [1978]
400p : ill
0-7064-0691-5 X.990/10705 Woolwich
1731815

ADVENTURE
1860-1960 - Texts

ADVENTURE stories for boys / edited by John Canning. - London : Octopus Books, 1978
400p : ill
0-7064-0692-3 Nov.37601 Woolwich
1690450

ADVENTURE. Special subjects. Children's literature 1800-
- Bibliographies

SHALASHOVA, Zoia Petrovna
Prikliucheniia i puteshestviia. - Moskva : Kniga, 1979
127p. - Bibliography : p.117-120
r6172965 X.990/6656 Woolwich
2098253

ADVENTURE. Special subjects. Fiction in English, 1700-1979

GREEN, Martin, 1927-
Dreams of adventure, deeds of empire. - London : Routledge and Kegan Paul, 1980
xv,429p. - Bibliography: p.405-410. - Includes index
0-7100-0509-1 X.989/88429
1971263

ADVENTURE. Special subjects. Literature
- Serials

MAGAZIN für Abenteuer-, Reise- und Unterhaltungsliteratur. - Sammelband.1-. - Braunschweig : Graff, [1978-] : ill
r6063574 X.0989/1028
1718363

ADVENTURE. Special subjects. Novels. Forms. Fiction to 1960

BARDAVÍO, José María
La novela de aventuras. - Madrid : Sociedad General Española de Librería, 1977
222p.,[1]folded leaf of plates. - (Colección "Temas" ; no.12)
84-7143-137-8 X.958/453
2472376

ADVENTURE. Special subjects. Poetry in English, 1800-1978
- Anthologies

THRILLS and spills / collected by Zenka and Ian Woodward ; illustrated by Deborah Jane Kartun. - London : Evans Bros, 1979
96p : ill. - Includes index
0-237-44990-0 X.989/54106
1738054

ADVENTURE FILMS. Cinema films
1900-

BERTIERI, Claudio
I film di carta. - Firenze : Vallecchi, 1979
140p,[40]p of plates : ill. - (Il pellicano. Cinema)
r6505858 X.950/18143
8660883

ADVENTURE GAMES

SMITH, Mark, 19—
Avenger!. - [Sevenoaks] : Knight, 1985. - (The way of the tiger ; 1)
Avenger!. - 1985
256p,[1]leaf of plates : ill
0-340-37787-9 X.990/24497
3021173

The WRECKERS' tower game / based on Enid Blyton's Five go down to the sea ; illustrated by Gary Rees. - London : Hodder and Stoughton, 1984
ca.186p : ill. - (Adventure games book)
0-340-35871-8 Cup.920/86
3021173

ADVENTURE GAMES
Electronic adventure games. Applications of Atari microcomputer systems

BRIDGE, Tony
Atari adventures. - London : Sunshine, 1984
xiii,146p : ill. - Includes index
0-946408-18-1 X.622/23745
301391x

ENGEL, C. W. (C William)
Stimulating simulations. - 2nd ed. - Rochelle Park, N.J. : Hayden, c1979
118p : ill. - (The Hayden microcomputer series)
0-8104-5197-2 X.629/22630
301391x

GERRARD, Peter
Exploring adventures on the Atari 48K. - London : Duckworth, 1984
viii,246p : ill. - (Duckworth home computing). - Includes index
0-7156-1924-1 X.629/25218
301391x

ADVENTURE GAMES
Electronic adventure games. Applications of BBC microcomputer systems

WATT, Ian, 1967-
Creating adventure programs on the BBC Micro. - London : Interface, c1983
128p
0-201-14678-9 X.622/22696
3154610

ADVENTURE GAMES
Electronic adventure games. Applications of Commodore 64 microcomputer systems

GRACE, Mike
Commodore 64 adventures. - London : Sunshine, 1983
ii,215p : ill
0-946408-11-4 X.622/23036
3159884

ORKIN, Michael
Survival on Planet X with the Commodore 64. - Reston : Reston Publishing, c1984
vii,163p : ill. - (A Reston computer group book)
0-8359-7413-8 X.622/19844
3071758

RIPPER!. - Harlow : Avalon Hill Game Co., c1984
1program. - (Eclipse software)
0-582-83619-0 Cup.930/70
3159884

ADVENTURE GAMES
Electronic adventure games. Applications of microcomputer systems

GERRARD, Mike
The adventurer's companion. - London : Duckworth, 1984
128p : ill. - (Duckworth home computing)
0-7156-1898-9 X.629/23525
302315x

The adventurer's notebook. - London : Duckworth, 1984
45[i.e.124]p. - (Duckworth home computing)
0-7156-1892-x X.629/23511
302315x

GERRARD, Peter
Exploring adventures on the Electron. - London : Duckworth, 1984
viii,238p : ill. - (Duckworth home computing). - Includes index
0-7156-1820-2 X.629/23523
302315x

MCGATH, Gary
Compute!'s guide to adventure games. - Greensboro : Compute! Publications, c1984
viii,203p : ill. - Bibliography: p199
0-942386-67-1 X.622/23857
302315x

ORKIN, Michael
Random Alley adventure. - Reston : Reston, c1984
v,88p : ill(some col.). - (A Creative pastimes book)
0-8359-6407-8 X.629/26038
3203220

PALMER, Nicholas
Beyond the arcade. - London : Mosaic, 1984
151p : ill
0-946855-03-x X.629/23140
302315x

WILLIAMS, Noel
Intelligent adventures for the Electron and the BBC Micro computers. - London : McGraw-Hill, c1984
x,228p : ill
0-07-084749-5 X.629/25592
302315x

ADVENTURE GAMES
Electronic adventure games. Applications of microcomputer systems - Serials

MICRO adventurer. - Nov. 1983-. - London : Sunshine, 1983-
v. : ill
B84-27049 P.625/368
302315x

ADVENTURE GAMES
Electronic adventure games. Applications of microcomputer systems - Texts

BUCKHOLTZ, Eileen
Space attack. - London : Dragon, 1984
128p. - (Micro adventures ; 1)
0-583-30752-3 X.629/26322
302315x

ADVENTURE GAMES
Electronic adventure games. Applications of microcomputer systems - Texts-continued

GLICK, Ruth, 1942-
Mindbenders. - London : Dragon, 1985
128p. - (Micro adventures ; 5)
0-583-30783-3 X.629/26360
3203220

KRAFT, David Anthony
Robot race. - London : Dragon, 1985
127p. - (Micro adventures ; 6)
0-583-30784-1 X.629/26462
3203220

ADVENTURE GAMES
Electronic adventure games. Applications of microcomputer systems. Programming

CAMPBELL, Keith, 19—
The Computer and video games book of adventure. - Tring : Melbourne House, c1983
138p : ill
0-86161-143-8 X.629/26762
3058565

HORN, Delton T.
Golden flutes and great escapes. - Beaverton : Dilithium, c1984
ix,217p
0-88056-089-4 X.629/23602
3058565

KILLWORTH, Peter
How to write adventure games. - Harmondsworth : Penguin, 1984
214p : ill. - (Penguin Acorn computer library)
0-14-007814-2 X.629/25276
3058565

TYLER, Jenny
Adventure programs. - London : Usborne, 1983
48p : ill(some col.). - (Usborne gamewriters' guides). - Includes index
0-86020-741-2 X.622/18155
3058565

ADVENTURE GAMES
Electronic adventure games. Applications of Sinclair 48K Spectrum microcomputer systems

EWBANK, K. (Kay)
The Spectrum Gamesmaster. - London : Granada, 1984
159p : ill
0-246-12515-2 X.622/22087
3131149

GERRARD, Peter
Exploring adventures on the Spectrum 48K. - London : Duckworth, 1984
viii,245p : ill. - (Duckworth home computing). - Includes index
0-7156-1796-6 X.629/23620
3131149

HURLEY, Richard G.
Graphic adventures for the Spectrum 48K. - Tunbridge Wells : Micro, 1984
194p : ill
0-7447-0013-2 X.622/19616
3131149

ADVENTURE GAMES
Electronic adventure games. Applications of Sinclair ZX Spectrum microcomputer systems

BRIDGE, Tony
Spectrum adventures. - London : Sunshine, 1983
187p : ill
0-946408-07-6 X.622/23033
3159892

NELSON, Andrew
Games of action and excitement for the ZX Spectrum. - London : Corgi, 1984
118p : ill. - Bibliography: p101-105
0-552-99129-5 X.629/26177
3159892

ADVENTURE GAMES
Electronic adventure games. Applications of Tandy TRS-80 microcomputer systems. Programming

DACOSTA, Frank
Writing BASIC adventure programs for the TRS-80. - Blue Ridge Summit : TAB, c1982
vi,222p : ill. - Includes index
0-8306-2422-8 X.629/26042
3278352

ADVENTURE GAMES
Electronic adventure games: Hobbit, The

ELKAN, David
A guide to playing the Hobbit. - Tring : Melbourne House, c1984
76p : ill. - Includes index
0-86161-161-6 X.629/26753
3300595

ADVENTURE GAMES
Electronic adventure games. Use in education

MCLEOD, Rosetta
Learning with adventure programs. - Tring : Melbourne House, c1984
86p : ill
0-86161-169-1 X.629/26755
3300609

ADVENTURE HOLIDAYS. Foreign countries

ADVENTURE holidays. - Oxford : Vacation-Work. 1980 : 3rd ed.. - 1980
231p : ill. - Includes index
Subsequent editions, when received, will be placed at this shelfmark
0-901205-78-8 P.701/1711
1550047

ADVENTURE HOLIDAYS. Foreign countries
- Directories

ADVENTURE holiday guide, abroad. - 1978 / edited by David Stevens. - Oxford : Vacation Work, 1978
viii,118p : ill
Subsequent volume, when received, will be placed at this shelfmark
0-901205-49-4 P.701/1604
1550047

ADVENTURE HOLIDAYS. Foreign countries
- Serials

INTERNATIONAL adventure travelguide. - 1978-. - Seattle : American Adventurers Association, 1978- : ill
r6091789 P.441/870
1550047

ADVENTURE HOLIDAYS. Great Britain
- Directories

ADVENTURE holiday guide, Britain. - 1978 / edited by David Stevens. - Oxford : Vacation Work, 1978
[6],162p : ill
Subsequent volumes, when received, will be placed at this shelfmark
0-901205-47-8 P.701/1605
1550039

ADVENTURE PLAYGROUNDS
- Conference proceedings

INTERNATIONAL PLAYGROUND ASSOCIATION. International Conference (6th : 1975 : Milan)
Adventure playgrounds and children's creativity. - Sheffield : The Association, [1976]
36p : ill
0-904166-03-1 X.611/7188 Woolwich
0611778

ADVENTURE PLAYGROUNDS
Organisation

BALMFORTH, Nick
Jubilee Street. - London : British Broadcasting Corporation, 1978
128p : ill. - Bibliography: p.127-128
0-563-17503-6 BS.129/94
0253022

CALDER, Dave
Adventure playgrounds. - [Liverpool] : [Merseyside Play Action Council], [1979]
[121]p : ill. - Bibliography (5p.)
B80-32187 X.525/3475
0253022

ADVENTURE PLAYGROUNDS
Use by mentally handicapped children

SUTHERLAND, Allan T.
Adventure play with handicapped children. - London : Souvenir, 1984
204p,[16]p of plates : ill. - (Human horizons series). - Bibliography: p185-192. - Includes index
0-285-64999-x X.529/69885
3283631

ADVENTURE PLAYGROUNDS. Cleator Moor. Cumbria
Big Hill Adventure Playground, 1972-1976

DANBY, Simon
The Big Hill Adventure Playground, 1972-1976. - [Cleator Moor] : Cumbria Community Development Project ; York : [Distributed by] Department of Social Administration and Social Work, University of York, [1978]
[2],30p : ill
B79-07468 X.512/6378
1798111

ADVENTURE PLAYGROUNDS. Great Britain
Organisation. Information sources - For play leaders

WILLS, Ray
Contacts for advice and information in the community. - Revised [ed.]. - [London] : National Playing Fields Association, 1977
[1],19p
B78-26249 X.512/6029
1522264

ADVENTURE PLAYGROUNDS FOR CHILDREN, TO 8 YEARS
Organisation

BERRINGTON Walk, Highgate, Birmingham / compiled by Dave Swingle. - [Birmingham] : [East Birmingham Family Service Unit], [1978?]
22p : ill. - (A West Midlands Arts 'self-help' publication ; no.1)
B82-14649 L.50/210
264178x

ADVENTURE STORIES IN ENGLISH
to date - Catalogues

AUSBRUCH und Abenteuer / Kevin Carpenter und Bernd Steinbrink mit Beiträgen von Dennis Butts ... [et al.]. - Oldenburg : Bibliotheks- und Informationssystem der Universität Oldenburg, 1984
136p. - Includes bibliography
3-8142-0103-5 X.990/24105
9159207

ADVENTURE STORIES IN ENGLISH
Children's adventure short stories in English, 1848-1981 - Anthologies

The BEAVER book of adventure stories / compiled by Raymond Wilson ; illustrated by Peter Dennis. - London : Beaver, 1982
159p : ill
0-600-20479-0 X.990/19711
2527308

MORE adventure stories / edited by Hayden McAllister ; illustrated by Sam Thompson. - London : Octopus, 1981
297p : ill
0-7064-1550-7 X.990/17495 Woolwich
2527308

ADVENTURE STORIES IN ENGLISH
Children's adventure short stories in English, 1945- - Anthologies

THRILLING stories / illustrated by Kay Wilson ; cover illustration by Ivan Lapper. - London : Hamlyn, 1982
256p : ill. - (Falcon fiction club)
0-600-36679-0 X.950/15775
2788152

ADVENTURE STORIES IN ENGLISH
Girls' adventure stories in English, 1945- - Anthologies

GIRLS' adventure stories. - Maidenhead : Purnell, 1981
223p : ill
0-361-05058-5 X.990/17354 Woolwich
2485427

ADVENTURE STORIES IN GERMAN
to date - Catalogues

AUSBRUCH und Abenteuer / Kevin Carpenter und Bernd Steinbrink mit Beiträgen von Dennis Butts ... [et al.]. - Oldenburg : Bibliotheks- und Informationssystem der Universität Oldenburg, 1984
136p. - Includes bibliography
3-8142-0103-5 X.990/24105
9159193

ADVENTURE STORIES IN GERMAN
1830-1945

STEINBRINK, Bernd
Abenteuerliteratur des 19. Jahrhunderts in Deutschland. - Tübingen : Niemeyer, 1983
309p : ill. - (Studien zur deutschen Literatur ; Bd.72). - Bibliography
3-484-18072-2 X.0909/187(72)
8953724

ADVENTURES
ca 500-1925

GREAT adventure stories. - [London] : Dean, 1982, c1975
173p : col.ill
0-603-00314-1 X.992/4979
2838168

ADVENTURES
1670-ca 1970 - Texts

BRETT, Bernard
The Hamlyn book of true adventures. - London : Hamlyn, 1978
4-157p : col ill
0-600-30357-8 X.802/10780
1803778

ADVENTURES
1850–1981

GREAT stories of courage & endurance / selected and condensed by the editors of Reader's digest. - London : Reader's Digest Association, [1983?]
2v. : ill(some col.)
B84–06044 X.950/27068
2998904

ADVENTURES
1860–ca 1970

BRETT, Bernard
Daring adventures!. - London : Hamlyn, 1981
110p : ill(some col.)
0–600–36661–8 X.805/1922
2565951

ADVENTURES
1900–1980

CRANFIELD, Ingrid
Skiing down Everest. - London : Severn House, 1983
208p,[16]p of plates : ill
0–7278–2032–x X.629/22469
3073858
- [Another copy.] T 45960 IOL

ADVENTURES
1940–1980

ROBERTS, Emrys
Achub. - [Aberystwyth] : Gwasg Cambria, [1982?]
78p : ill
B83–17083 X.950/19658
2876930

ADVENTURES
1946–1981

BONINGTON, Chris
Quest for adventure. - London : Hodder and Stoughton, 1981
448p : ill(some col.). - Bibliography: p442–443. - Includes index
0–340–25599–4 X.805/2048
2585715

BONINGTON, Chris
Quest for adventure. - [New] ed. - London : Pan, 1983
345p,[16]p of plates : ill(some col.). - Includes index
0–330–28113–5 X.800/37190
2585715

ADVERBIAL CONSTRUCTIONS. Languages

BARTSCH, Renate
The grammar of adverbials. - Amsterdam ; Oxford : North-Holland Publishing Co., 1976
xii,390p : ill. - (North-Holland linguistic series ; 16). - Bibliography: p.387–390
0–7204–6206–1 X.989/50902
0720453

TAYLOR, Barry, 19—
Modes of occurrence. - Oxford : Blackwell, 1985
ix,132p. - (Aristotelian Society series ; v.2). - Bibliography: p126–128. - Includes index
0–631–14026–3 X.950/43028
3155021

ADVERBS
See also
Adverbial constructions

ADVERBS

NØLKE, Henning
Udviklingslinjer inden for nyere adverbialforskning. - København : Københavns Universitet, Romansk Institut, 1981
22p. - (RIDS ; nr 96)
r6610159 X.0902/109(96)
885193x

WELTE, Werner
Bemerkungen zur Problematik der Adverbialsyntax im Lichte verschiedener Modelle. - Wiesbaden : Vieweg, c1971
19leaves. - (Linguistische Berichte ; LB-Papier Nr.17). - Includes bibliography
r6309084 Cup.918/214
885193x

ADVERBS. American usage. English language
Preverbal adverbs. Position related to position of auxiliary verbs

JACOBSON, Sven
Factors influencing the placement of English adverbs in relation to auxiliaries. - Stockholm : Almqvist & Wiksell International, 1975
599 p.. - (Stockholm studies in English ; 33). - Bibliography : p. 582–588
r6001223 Ac.2650m.[no. 33]
0909726

ADVERBS. Czech language
Verbal adverbs

DVOŘÁK, Emil
Přechodníkové konstrukce v nové češtině. - Praha : Univerzita Karlova, 1983
152p. - (Acta Universitatis Carolinae. Philologica ; monographia 72). - Includes bibliography
r6441948 Ac.798/8g[72]
9167080

ADVERBS. English language

FJELKESTAM-NILSSON, Brita
Also and too. - Stockholm : Almqvist & Wiksell International, 1983
viii,149,15p. - (Stockholm studies in English ; 58). - Includes bibliography
91–220–0616–8 Ac.2650.m[no.58]
9006168

ADVERBS. English language
Preverbal adverbs

AHLBERG, Sture
Messianism i staten Israel. - Stockholm : [Almqvist & Wiksell International], 1977
167p : ill. - (Stockholm studies in comparative religion ; 14). - Includes bibliographies
91–220–0093–3 Ac.2650.m/20[vol.14]
1535935

ADVERBS. English language
Preverbal adverbs

JACOBSON, Sven
On the use, meaning, and syntax of English preverbial adverbs. - Stockholm : Almqvist and Wiksell, 1978
160p : ill. - (Stockholm studies in English ; 44). - Bibliography : p.154–157.- Includes index
91–220–0138–7 Ac.2650.m [no.44]
1535935
- [Another copy] X.950/3818

ADVERBS. English language
Stress

TAGLICHT, Josef
Message and emphasis. - London : Longman, 1984
xvii,214p : ill. - (English language series ; no.15) (Longman English language series). - Bibliography: p206–210. - Includes index
0–582–29128–3 X.950/32492
3127605

ADVERBS. Finnish language

PALOLA, Helena
Adverbit, postpositiot, prepositiot ja interjektiot 1960–luvun suomen yleiskielessä. - Oulu : Yliopisto, 1975
60p. - (Oulun yliopiston suomen ja saamen kielen laitoksen tutkimusraportteja ; 2)
951–420–219–8 X.0900/581(2)
1975420

ADVERBS. French language

GRELSSON, Sigvard
Les adverbes en -ment. - Lund : Gleerup, 1981
246picharts. - (Études romanes de Lund ; 34). - Bibliography
91–400–4788–1 WP.3119/34
1792903

HANSÉN, Iah
Les adverbes prédicatifs français en -ment. - Göteborg : Institut d'études romanes, 1979
269p. - Bibliography
r6301295 X.955/542
1792903

Les adverbes prédicatifs français en -ment. - Göteborg : Acta Universitatis Gothoburgensis, 1982
230p. - (Romanica gothoburgensia ; 19). - Bibliography
91–7346–098–2 X.950/16677
1792903

NØLKE, Henning
Problemer ved opstilling af en typologi for de franske adverbialled. - [Copenhagen] : Københavns Universitet, Romansk Institut, 1978
27p. - (RIDS ; nr 57)
r6060894 X.0902/109(57)
1792903

ADVERBS. German language

BARTSCH, Angelika
Zur Stellung der Adverbiale in den Werken von Thomas Mann. - Frankfurt am Main : Lang, 1982
viii,377p. - (Europäische Hochschulschriften. Reihe 1, Deutsche Sprache und Literatur ; Bd 521)
3–8204–6264–3 X.950/25161
9004114

HÄRD, John Evert
Adjektivadverb oder adverbiales Adjektiv?. - Åbo : Åbo akademi, 1976
75p. - (Acta Academiae Aboensis. ser.A ; v.54, nr 1)
951–648–265–1 Ac.1093/2a[vol.54.no.1]
1473751

HOLMLANDER, Inger O.
Zur Distribution und Leistung des Pronominaladverbs. - Uppsala ; Stockholm : Distributed by Almqvist & Wiksell International, 1979
187p. - (Studia germanistica upsaliensia ; 21). - Includes bibliographies
91–554–0859–1 AC:1075/6(23)[vol.21]
1473751

MOILANEN, Markku
Zum lokalen Gebrauch der Demonstrativadverbien 'da' und 'dort'. - Helsinki : Suomalainen tiedeakatemia, 1973
145p. - (Annales Academiae Scientiarum Fennicae. dissertationes humanarum litterarum ; 2)
951–410–146–4 Ac.1094.c/7b(1)[vol.2]
1473751

ADVERBS. Languages

BARTSCH, Renate
The grammar of adverbials. - Amsterdam ; Oxford : North-Holland Publishing Co., 1976
xii,390p : ill. - (North-Holland linguistic series ; 16). - Bibliography: p.387–390
0–7204–6206–1 X.989/50902
0720453

ADVERBS. Latin language

KARLSSON, Keith E.
Syntax and affixation. - Tübingen : Niemeyer, 1981
ix,163p.. - (Beihefte zur Zeitschrift für romanische Philologie ; Bd.182). - Bibliography
3–484–52182–1 PP.5044.ac/3[Bd.182]
8779864

ADVERBS. Norwegian language

HEGGELUND, Kjell Tørres
Setningsadverbial i norsk. - Oslo : Novus, 1981
249p. - (Tromsø-studier i språkvitenskap ; bd.2). - Includes bibliography
82–7099–064–7 X.0950/482(2)
9033335

ADVERBS. Old Greek language

RUSÍNOVÁ, Zdenka
Tvoření staročeských adverbií. - V Brně : Univerzita J.E. Purkyně, 1984
129p. - (Spisy Univerzity J.E. Purkyně v Brně, Filozofická fakulta ; č.253)
r6441998 Ac747/2[253]
9156283

ADVERBS. Romance languages

KARLSSON, Keith E.
Syntax and affixation. - Tübingen : Niemeyer, 1981
ix,163p.. - (Beihefte zur Zeitschrift für romanische Philologie ; Bd.182). - Bibliography
3–484–52182–1 PP.5044.ac/3[Bd.182]
8779872

ADVERBS. Russian language

ĪĀNOVICH, Elena Ivanovna
Narechie v istorii russkogo īāzyka. - Minsk : Izdatel'stvo BGU im. V.I. Lenina, 1978
241p
r6098145 X.908/87428
1865226

STERNBERG, Luise
Die russischen Proadverbien. - Frankfurt am Main : Lang, 1984
138,xp. - (Slavistische Linguistik ; Bd.8). - Includes bibliography
3–8204–5490–x X.0909/1116(8)
1865226

ADVERBS. Spanish language

EGEA, Esteban Rafael
Los adverbios terminados en -mente en el español contemporáneo. - Bogotá : Instituto Caro y Cuervo, 1979
xx,432p. - (Publicaciones / del Instituto Caro y Cuervo ; 49). - Bibliography: p.379–411
r6703241 X.950/8531
8761965

ADVERSE ENVIRONMENTS
Effects on operation of instrumentation systems - Conference proceedings

OPERATION OF INSTRUMENTS IN ADVERSE ENVIRONMENTS (Conference : 1976 : London)
Operation of instruments in adverse environments, 1976. - Bristol : Institute of Physics, 1977
vii,134p : ill. - (Conference series / Institute of Physics ; no.34). - Includes bibliographies
0–85498–125–x SRL
1047582

ADVERSE ENVIRONMENTS
Structural engineering materials: Timber - Conference proceedings
STRUCTURAL use of wood in adverse environments / edited by Robert W. Meyer, Robert M. Kellogg ; [for the] Society of Wood Science and Technology. - New York ; London : Van Nostrand Reinhold, c1982
xi,510p : ill. - Includes bibliographies and index
0-442-28744-5 SRL
2760444

ADVERSE REACTIONS. Children
To vaccination against whooping cough
WHOOPING cough. - London : H.M.S.O., 1981
xiii,184p : ill
0-11-320764-6 SRL
2580691

ADVERSE REACTIONS. Conference proceedings. Man
To psychotropic drugs
STUDY GROUP ON FUNCTIONAL TOXICITY
Adverse effects of environmental chemicals and psychotropic drugs. - Amsterdam ; Oxford : Elsevier.
Vol.2: Neurophysiological and behavioral tests. - 1976
xv,334p : ill. - Includes bibliographies
0-444-41498-3 SRL
0018961

ADVERSE REACTIONS. Man
To chemical pollutants of environment - Conference proceedings
STUDY GROUP ON FUNCTIONAL TOXICITY
Adverse effects of environmental chemicals and psychotropic drugs. - Amsterdam ; Oxford : Elsevier.
Vol.2: Neurophysiological and behavioral tests. - 1976
xv,334p : ill. - Includes bibliographies
0-444-41498-3 SRL
0019720

ADVERSE REACTIONS. Man
To drugs
TEXTBOOK of adverse drug reactions / edited by D.M. Davies. - Oxford : Oxford University Press, 1977
xvi,503p : ill. - (Oxford medical publications). - Includes bibliographies and index
0-19-263206-x SRL
0681946

TEXTBOOK of adverse drug reactions / edited by D.M. Davies. - 2nd ed. - Oxford : Oxford University Press, 1981
693p : ill. - Includes bibliographies and index
0-19-261270-0 SRL
0681946

WADE, Owen Lyndon
Adverse reactions to drugs. - 2nd ed.. - London : Heinemann Medical, 1976
ix,230p : ill. - Bibliography: p.201-207. - Includes index
0-433-34551-9 SRL
0681946

ADVERSE REACTIONS. Man
To drugs - Serials
ADVERSE drug reactions and acute poisoning reviews. - Vol.1, no.1 (Spring 1982)-. - Oxford : Oxford University Press, 1982-
v. : ill
B82-38501 SRL
2797844

ADVERSE REACTIONS. Man
To drugs. Evaluation
ADVERSE drug reaction bulletin. - Hong Kong : Far East Trade Press.
no.1- ; [Jan. 1973]-. - [1973]-
Holdings : no 74- ; Feb. 1979-
r6109315 P.321/572
144106X

ADVERSE REACTIONS. Man
To drugs. Evaluation - Conference proceedings
PHARMACOLOGICAL methods in toxicology / section editors G. Zbinden and F. Gross. - Oxford : Pergamon, 1979
xi,612p : ill. - (International encyclopedia of pharmacology and therapeutics ; section 102). - Includes bibliographies and index
0-08-024900-0 SRL
144106x

WORKSHOP ON STANDARDIZING METHODS OF ASSESSING CAUSALITY OF ADVERSE DRUG REACTIONS (1981 : Morges)
Assessing causes of adverse drug reactions. - London : Academic Press, 1982
xvi,233p,[1]leaf of plates : ill(some col.). - Includes bibliographies
0-12-717350-1 SRL
144106x

ADVERSE REACTIONS. Man
To intravenous drugs
ADVERSE response to intravenous drugs / [edited by] J. Watkins and A. Milford Ward. - London : Academic Press [etc.], 1978
xii,188p,plate : ill(incl 1 col). - Includes bibliographies and index
0-12-737850-2 SRL
1758241

ADVERTISEMENT TICKETS
See Farthings. Tradesmen's tokens

ADVERTISEMENTS
See also
Slogan postmarks
Slogans

ADVERTISEMENTS
British advertisements - Directories - Serials
CAMPAIGN portfolio. - [1982]-. - London : Marketing Publications, 1981-
v. : ill(chiefly col.)
B83-21470 P.515/473
289355x

ADVERTISEMENTS
Complaints by public - Serials
ADVERTISING STANDARDS AUTHORITY
ASA case report. - 1-. - London : [Advertising Standards Authority], [1973]-
Holdings: no.41, etc
B79-25822 P.513/2326
1921932

ADVERTISEMENTS
Copywriting. Techniques
CROMPTON, Alastair
The craft of copywriting. - London : Business Books [for] the CAM Foundation, 1979
xii,164p : ill. - Includes index
0-220-67006-4 X.529/33966
1940600

ADVERTISEMENTS
Layout. Techniques
BORGMAN, Harry
Advertising layout techniques. - New York : Watson-Guptill, 1983
143p : ill(some col.). - Includes index
0-8230-0154-7 X.425/4243
3037983

ADVERTISEMENTS
Psychological aspects
WILLIAMSON, Judith, 1954-
Decoding advertisements. - London : Boyars : Distributed by Calder and Boyars, 1978
180p : ill. - (Ideas in progress). - Bibliography: p.180
0-7145-2614-2 X.529/32160
1382772

ADVERTISEMENTS
Psychological aspects - Conference proceedings
ADVERTISING and consumer psychology / edited by Larry Percy, Arch G. Woodside. - [Lexington, Mass.] : Lexington Books, 1983
xxiv,400p : ill. - Bibliography: p345-385. - Includes index
0-669-05766-5 X.529/63147
1382772

ADVERTISEMENTS
Serials with Swedish imprints, to date
SWEDEN. Utbildningsdepartementet. Dagspresskommittén
Annonsbladen och dagspressen. - [Stockholm] : Liber förlag, 1980
205p. - (Ds U ; 1979 : 10)
91-380-5451-5 X.950/23081
9005676

ADVERTISEMENTS
Special subjects: Cinema films produced by Paramount Pictures Corporation, 1929-1945 expounding the films
HALLIWELL, Leslie
Mountain of dreams. - London : Hart-Davis MacGibbon, 1976
xxii,196p : ill
0-246-10825-8 X.985/1178
1463489

ADVERTISEMENTS
Special subjects: Soap manufactured by Pears (A. & F.) Limited, 1884-1920
A. & F. PEARS LIMITED
Bubbles. - [London] : Fontana, 1978
72p : chiefly ill(chiefly col)
0-00-635420-3 X.415/5590 Woolwich
1746405

ADVERTISEMENTS. American wood engravings 1800-1900
GOODS and merchandise / compiled and arranged by William Rowe. - New York : Dover ; London : Constable, c1982
60p : chiefly ill. - (Dover pictorial archive series)
0-486-24410-5 L.45/2858
3145255

ADVERTISEMENTS. Commercial television services. West Germany
Use of English language, to date
STEINBACH, Horst-Ralf
Englisches im deutschen Werbefernsehen. - Paderborn : Schöningh, 1984
310p. - (Schriften der Gesamthochschule Paderborn. Reihe Sprach- und Literaturwissenschaften ; Bd.2). - Includes bibliography
3-506-78452-8 X.0950/442(2)
9143157

ADVERTISEMENTS. English humorous cartoons - Collections
ROBINSON, W. Heath (William Heath)
The gentle art of advertising. - London : Duckworth, 1979
88p : of ill
0-7156-1428-2 X.423/10931 Woolwich
2049236
- [Another copy : paperback] ISBN 0-7156-1651-X X.421/23112

ADVERTISEMENTS. General serials in Danish to date
LIEBST, Asger
Tresserbilleder. - [København] : Borgen, c1982
94p : ill
87-418-3378-3 X.950/28195
9009655

ADVERTISEMENTS. Great Britain
Heineken (Firm) advertisements, 1973-1983
THIRSTY work / text, Peter Mayle. - London : Macmillan, 1983
[122]p : ill(some col.)
0-333-36052-4 X.525/9726
3165329

ADVERTISEMENTS. Great Britain
Shell UK Limited advertisements - Catalogues
"THAT'S SHELL - THAT IS!" (Exhibition) (1983 : London)
"That's Shell - that is!". - [London] : The Gallery, 1983
47p : ill(some col.). - Includes index
0-946372-03-9 X.421/25858
3165353

ADVERTISEMENTS. London gazette
Special subjects: Maps, 1668-1719 - Collections
LONDON map-sellers, 1660-1720 / [compiled] by Sarah Tyacke. - Tring : Map Collector Publications Ltd, 1978
xxviii,160p,fold plate : ill. - Bibliography: p.149-150. - Includes index
0-906430-00-3 Map Library
1947893

ADVERTISEMENTS. Magazines
Portrayal of women
FLICK, Marjan
Kvinner i ramme. - Bergen : Universitetet i Bergen, Sosiologisk institutt, Senter for mediaforskning, 1977
75,27 leaves. - (Rapport ; nr 1). - Includes bibliographies
r6060168 X.0512/395(1)
1530372

ADVERTISEMENTS. Magazines with American imprints 1920-1978. Special subjects: Families
BROWN, Bruce W.
Images of family life in magazine and advertising: 1920-1978. - New York : Praeger, 1981
xxi,131p : ill. - (Praeger studies on changing issues in the family). - Bibliography: p117-124
0-03-059697-1 X.520/32476
3064344

ADVERTISEMENTS. Newspapers & periodicals with Serbian imprints 1804-1918 - Catalogues
JANC, Zagorka
Oglasi u staroj srpskoj štampi, 1834-1915. - Beograd : Muzej primenjene umetnosti, 1978
48p,[60]p of plates : chiefly ill
r6176856 X.419/4487 Woolwich
2260646

ADVERTISEMENTS. Newspapers & serials with American & British imprints
Small advertisements, to 1978
The WORLD of small ads / [compiled by] Mario Lippa & David Newton. - London : Hamlyn, 1979
3-124p : ill
0-600-36603-0 X.423/10954 Woolwich
208161x

ADVERTISEMENTS. Newspapers with German imprints to date. Linguistic aspects

HOHMEISTER, Karl-Heinz
Veränderungen in der Sprache der Anzeigenwerbung. - Frankfurt am Main : R.G. Fischer, 1981
xv,354p. - Bibliography
3-88323-233-5 X.950/9396
8819300

STOLZE, Peter
Untersuchungen zur Sprache der Anzeigenwerbung in der zweiten Hälfte des 18. Jahrhunderts. - Göppingen : Kümmerle, 1983
vii,260p. - (Göppinger Arbeiten zur Germanistik ; Nr.375). - Includes bibliography
r6307340 X.0989/486(375)
8819300

ADVERTISEMENTS. Radio
See also
Radio advertising

ADVERTISEMENTS. Radio & television. United States
Copywriting

ORLIK, Peter B.
Broadcast copywriting. - 2nd ed. - Boston, Mass. ; London : Allyn and Bacon, c1982
xiii,511p : ill. - Includes index
0-205-07691-2 X.622/12909 Woolwich
2754150

ADVERTISEMENTS. Serials
Copywriting. Techniques

JEFKINS, Frank William
Advertisement writing. - Plymouth : Macdonald and Evans, 1976
vii,248p,[8]p plates : ill. - Includes index
0-7121-0138-1 X.519/26566
1046764

ADVERTISEMENTS. Serials
Small advertisements: Personal advertisements - Collections

RIMMER, Gary
Lonely hearts. - London : Unwin Paperbacks, 1983
125p : ill
0-04-827082-2 X.958/20698
3058719

ADVERTISEMENTS. Serials in German
Personal advertisements, 1945-

FLESCH, Hans Werner
Nur ernstgemeinte Zuschriften erbeten. - Düsseldorf : Erb, 1982
174p
3-88458-035-3 X.329/16838
9005579

ADVERTISEMENTS. Serials with Brazilian imprints. Serials in Portuguese
to date. Linguistic aspects

JOHANNEMANN, Annelie
Hyperbolisches in der Werbesprache brasilianischer Zeitungen und Zeitschriften. - Mettingen : Brasilienkunde-Verlag, 1981
81p : ill. - (Aspekte der Brasilienkunde ; Bd.3)
3-88559-053-0 X.0529/581(3)
8865442

ADVERTISEMENTS. Serials with British imprints
Classified advertisements - Serials

TELEMART print-out. - Issue no.1 (Oct. 1983)-. - London : Telesale Ltd, 1983-
v. : ill
B84-28576 P.823/312
3142175

ADVERTISEMENTS. Serials with British imprints. Women's serials in English
1956-1974. Sex roles

WINSHIP, Janice
Advertising in women's magazines, 1956-74. - Birmingham : Centre for Contemporary Cultural Studies, University of Birmingham, 1980
65p. - (Women series ; SP no.59) (Stencilled occasional paper / Centre for Contemporary Cultural Studies). - Bibliography: p64-65
B83-14484 X.525/6842
2865793

ADVERTISEMENTS. Small advertisements. Serials
Personal advertisements - Collections

RIMMER, Gary
Lonely hearts. - London : Unwin Paperbacks, 1983
125p : ill
0-04-827082-2 X.958/20698
3058719

ADVERTISEMENTS. Television
See also
Television advertising

ADVERTISEMENTS. Television
Acting. Cats: Arthur

MONTGOMERY, John
Arthur. - London : W.H. Allen, 1975
128p : ill
0-491-01705-7 X.329/10164
0524255

ADVERTISEMENTS. Television
Influence on children

The EFFECTS of television advertising on children / Richard P. Adler ... [et al.]. - Lexington, Mass. : Health ; [Farnborough, Hants.] : [Distributed by] Gower Press, 1980
xi,371p : ill
0-669-02814-2 X.520/22141
226918x

ADVERTISEMENTS. Television. Great Britain
Costs

WOOD, John W.
The question of costs. - London : Wood, Brigdale and Company Ltd, 1978
[2],28p. - (Strategic advertising series ; no.5)
B79-25628 X.519/29923
1922785

ADVERTISEMENTS. Television. United States
Costs. Control

BELLAIRE, Arthur
Controlling your TV commercial costs. - Chicago : Crain Books, 1976
166p,[6]p of plates
0-87251-026-3 X.622/2594 Woolwich
172441X

ADVERTISEMENTS. Television. United States
Production

CONRAD, Jon J.
The TV commercial. - New York ; London : Van Nostrand Reinhold, c1983
160p : ill. - Includes index
0-442-21866-4 X.529/57893
2984091

WHITE, Hooper
How to produce an effective TV commercial. - Chicago : Crain, c1981
xix,284p : ill
0-87251-036-0 X.622/21913
2984091

ADVERTISEMENTS. United States
Surveys of advertisements. Use in forecasting social change

FOWLES, Jib
Mass advertising as social forecast. - Westport, Conn. ; London : Greenwood Press, 1976
xi,153,[1]p : 1 ill. - Bibliography: p.127-149. - Includes index
0-8371-8595-5 X.529/31506
0655635

ADVERTISEMENTS. West Germany
Personal advertisements, 1945-

EHEPAAR sucht gleichgesinntes / herausgegeben von Boris Wittich. - München : Deutscher Taschenbuchverlag, 1982
108p : ill. - (DTV ; 1783)
3-423-01783-x X.529/62417
908293x

ADVERTISERS
Boycotts of Göteborgs Handels-och Sjöfarts-Tidning, 1940

JONSSON, Sverker
Annonsbojkotten mot Göteborgs Handels- och Sjöfarts-Tidning 1940. - Göteborg : [Göteborgs universitet, Ekonomisk-historiska institutionen], 1979
33p. - (Meddelanden från Ekonomisk-historiska institutionen vid Göteborgs universitet ; 43)
91-85196-16-9 Ac.2651.db/2[no.43]
2119226

ADVERTISERS. Great Britain
Attitudes to advertising industries, 1972

BRITISH MARKET RESEARCH BUREAU
Advertising and advertising agencies. - London : The Institute, 1972
[2],18 leaves
0-85294-092-0 X.512/5361
0735027

ADVERTISERS. Great Britain
Scottish advertisers. Attitudes to advertising agencies

CROSIER, Keith
The advertising iceberg. - Glasgow : Struthers Advertising Group, c1980
60p
B82-28338 X.529/48402
2752875

ADVERTISING
See also
Prospectuses. Books
Public relations
Publicising

ADVERTISING
See also
Publicity
Window dressing

ADVERTISING

AAKER, David A. (David Allen)
Advertising management. - Englewood Cliffs ; London : Prentice-Hall, 1975
xvii,612p : ill. - (Prentice-Hall international series in management). - Includes index
0-13-015974-3 X.520/10114
1091336
Advertising management. - 2nd ed. - Englewood Cliffs ; London : Prentice-Hall, c1982
xiii,560p : ill. - (Prentice-Hall international series in management). - Includes index
0-13-016006-7 X.520/27011
1091336

ADVERTISING. - 4th ed.. - New York ; London : McGraw-Hill, 1977
xiii,769p,[8]p of plates : ill(some col). - Includes bibliographies and index
0-07-072067-3 X.520/11808
1091336

ADVERTISING ASSOCIATION
Speaking up for advertising. - London : The Association, [1976]
[1],82p
0-902878-21-2 X.519/26193
0739901

BURKE, John D.
Advertising in the market place. - 2nd ed. - New York ; London : McGraw-Hill, 1980
x,454p,16p of plates : ill(some col). - (The Gregg/McGraw-Hill marketing series). - Includes index
0-07-009035-1 X.520/20781
1091336

DYER, Gillian
Advertising as communication. - London : Methuen, 1982
xiv,230p : ill. - (Studies in communication). - Bibliography: p215-226. - Includes index
0-416-74520-2 X.529/52517
1091336
- [Another copy: paperback.] ISBN 0-416-74530-X X.529/52518

FARBEY, David
The business of advertising. - London : Associated Business Press, 1979
x,260p : ill. - Includes index
0-85227-222-7 X.529/34284
1091336

FIND out about advertising / [general editor Peter Rutland] ; [contributors: Rosemary McRobert, Richard Whitburn]. - London : The Advertising Association, c1979
1case : ill(some col.)
B83-19136 Cup.918/68
1091336

FLADER, Dieter
Stategien der Werbung. - Kronberg Ts, 1974
235 p.. - (Skripten. Linguistik und Kommunikationswissenschaft ; 6)
3-589-20019-7 X.909/34429
1004395

FLETCHER, Winston
Advertising. - [New ed.]. - London : Teach Yourself Books, 1978
284p : ill. - (Teach yourself books. business and management studies). - Bibliography: p.274-276. - Includes index
0-340-22239-5 WP.706/697
1091336

FULOP, Christina
Advertising, competition and consumer behaviour. - London : Holt, Rinehart and Winston with the Advertising Association, c1981
x,176p. - Bibliography: p167-171. - Includes index
0-03-910295-5 X.520/26009
1091336

GRAY, Ernest
Profitable methods for small business advertising. - New York ; Chichester : Wiley, c1984
xix,285p : ill. - (Small business management series). - Includes index
0-471-86962-7 X.520/34721
0200859

GREYSER, Stephen A.
Cases in advertising and communications management. - 2nd ed. - Englewood Cliffs ; London : Prentice-Hall, c1981
xviii,644p : ill. - Includes index
0-13-118513-6 X.520/24273
250524x

ADVERTISING –continued

JEFKINS, Frank
Advertising today. – 3rd ed. – Glasgow : International Textbook, 1984
318p : ill. – Bibliography: p303. – Includes index
0–7002–0285–4 X.622/19713
1091336

JEFKINS, Frank William
Advertising today. – 2nd ed. – London : International Textbook Co., 1977
viii,312p : ill. – Bibliography: p.302. – Includes index
0–7002–0263–3 X.620/16633 Woolwich
1091336

JOHNSON, J. Douglas
Advertising today. – Chicago ; Henley-on-Thames : Science Research Associates, c1978
x,395p,[16]p of plates : ill(some col.). – Includes index
0–574–19355–3 X.525/4124
1091336

KAUFMAN, Louis, b.1922
Essentials of advertising. – New York ; London : Harcourt Brace Jovanovich, 1980
xiv,575p : ill(some col). – Includes bibliographies and index
0–15–524100–1 X.520/21646
1091336

KENNEDY, Sherril Helen
The processes of advertising. – Bradford : MCB (Management Decision) Limited ; Bradford : Distributed by MCB Books, 1976
[2],66p : ill. – (Bibliography & review). – Bibliography: p.62–66
0–903763–39–7 X.510/15216
0678856

KLEPPNER, Otto
Advertising procedure. – 7th ed.. – Englewood Cliffs ; London : Prentice-Hall, 1979
[8],631p,[8]p of plates,[2]leaves of plates : ill(some col). – Includes bibliographies and index
0–13–018119–6 X.520/13584
1091336

Otto Kleppner's Advertising procedure. – 8th ed.. – Englewood Cliffs ; London : Prentice-Hall, c1983
xvi,636p,[12]p of plates : ill(some col.). – (The Prentice-Hall series in marketing). – Includes bibliographies and index
0–13–643213–1 X.520/30169
1091336

MANDELL, Maurice I. (Maurice Ira)
Advertising. – 3rd ed. – Englewood Cliffs ; London : Prentice-Hall, c1980
xiv,752p,[4]p of plates : ill(some col.). – Bibliography: p738–744. – Includes index
0–13–014449–5 X.520/25309
1091336

Advertising. – 4th ed. – Englewood Cliffs ; London : Prentice-Hall, c1984
xix,689p,[4]p of plates(1folded) : ill(some col.). – Bibliography: p683–689. – Includes index
0–13–014499–1 X.520/36066
1091336

The NEW world of advertising / compiled by the editors of 'Advertising age' ; edited and with an introduction by Vernon Fryburger. – Chicago : Crain Books, 1975
v,129p : ill
0–87251–021–2 X.512/5924
1091336

NICHOLL, David Shelley
Advertising. – 2nd ed. – Plymouth : Macdonald and Evans, 1978
xi,228p : ill. – Bibliography: p.213–214. – Includes index
0–7121–0166–7 X.519/27952
1091336

PATTI, Charles H.
Cases in advertising and promotion management. – New York ; Chichester : Wiley, c1983
viii,256p : ill. – Includes index
0–471–87507–4 X.520/31713
250524x

PERCY, Larry
Advertising strategy. – New York : Praeger, 1980
vii,301p : ill. – (Praeger special studies). – Bibliography : p274–295. – Includes index
0–03–055906–5 X.520/21742
1091336

ROMAN, Kenneth
How to advertise. – London : Kogan Page, 1979
xiii,159p : ill
0–85038–248–3 X.519/30430
0200859

SEIDEN, Hank
Advertising pure and simple. – New York : Amacom, 1976
ix,198p
0–8144–5427–5 X.529/31601
1091336

SWITKIN, Abraham
Ads. – New York ; London : Van Nostrand Reinhold, c1981
160p : ill. – Bibliography: p157. – Includes index
0–442–24342–1 X.425/2345 Woolwich
0200859

The THREE faces of advertising / edited by Michael Barnes. – London : Advertising Association, 1975
vii,277p : ill. – Includes bibliographies
0–902878–11–5 X.519/25593
0613118

TWENTY advertising case histories / edited and introduced by Simon Broadbent. – London : Holt, Rinehart and Winston, c1984
xv,248p : ill
0–03–910543–1 X.520/33773
250524x

WEILBACHER, William M. (William Manning)
Advertising. – New York : Macmillan ; London : Collier Macmillan, 1979
xxv,754,[16]p of plates : ill(some col). – Includes index
0–02–425100–3 X.520/14043
1091336

WEILBACKER, William M. (William Manning)
Advertising. – 2nd ed. – New York : Macmillan ; London : Collier Macmillan, c1984
xvii,605p,[16]p of plates : ill(some col.). – Includes index
0–02–425250–6 X.520/35414
1091336

WHITE, Roderick
Advertising. – London : McGraw-Hill in association with the Advertising Association, c1980
xi,230p,[16]p of plates : ill. – Bibliography: p219–220. – Includes index
0–07–084538–7 X.520/22602
0200859

WILLIAMS, David, 1926–
Advertising and social conscience. – [New ed.]. – London : Foundation for Business Responsibilities, [1976]
15p
B76–31066 X.519/26156
1091336

WILLIAMSON, Judith
Decoding advertisements. – London : Boyars, 1983, c1978
180p : ill. – (Ideas in progress). – Bibliography: p180
B84–05822 X.529/58326
1091336

WILMSHURST, John
The fundamentals of advertising. – London : Published on behalf of the Institute of Marketing and the CAM Foundation, 1985
xiv,267p,[8]p of plates : ill. – Includes index
0–434–92330–3 X.629/26482
1091336

WRIGHT, John S. (John Sherman)
Advertising. – 5th ed.. – New York ; London : McGraw-Hill, c1982
xiv,545p,[16]p of plates : ill(some col.). – (McGraw-Hill series in marketing). – Includes bibliographies and index
0–07–072069–x X.525/5895
1091336

Advertising. – 5th ed. – New Delhi : Tata McGraw-Hill, 1983, c1982
xiv,545p : ill(some col.). – (McGraw-Hill series in marketing). – Includes bibliographies
0–07–099658–x X.520/37591
1091336

ADVERTISING
– Bibliographies

POLLAY, Richard W.
Information sources in advertising history. – Westport, Conn. ; London : Greenwood Press, 1979
xiv, 330p. – Includes index
0–313–21422–0 X.520/21179
2061236

ADVERTISING
– Serials

BRADAMPS. – Vol.1, no.1 (June 1982)–v.1, no.2 (Dec. 1982). – London : British Rate & Data, 1982–1982
1v. : ill
HOLDINGS: VOL.1.No.2 Dec. 1982 etc
B85–07864 P.843/280
1686259

CREATIVE review. – Vol.1, no.1 (May 1980)–. – London : Marketing Week Communications, 1980–
v. : ill(some col.)
Holdings: vol 2 no 1, (September 1981), etc
B82–09077 P.2000/1026
1686259

INTERNATIONAL journal of advertising. – Vol.2, no.1 (Jan.-Mar. 1983)–. – Eastbourne : Holt, Rinehart and Winston, 1983–
v. : ill
B84–18305 P.861/322
1686259

[JOURNAL of advertising (Eastbourne)]. Journal of advertising. – Vol.1, no.1 (Jan.-Mar. 1982) – v.1, no.4 (Oct.-Dec. 1982). – Eastbourne : Holt, Rinehart and Winston, 1982–1982
v. : ill
B84–18306 P.861/322
1686259

The JOURNAL of advertising history. – No.1– , Dec. 1977–. – London : History of Advertising Trust, [1977]– : ill
B79–27670 P.513/2335
1932810

ADVERTISING
– Statistics – Serials

The ADVERTISING Association's statistical yearbook. – Vol 1 (Oct. 1983)–. – London : The Association, 1983–
v.
B85–11087 P.823/591
1686259

ADVERTISING
to date

MOTTI, Germano
Guida alla pubblicità industriale. – Milano : Angeli, [1980]
120p : ill. – (Azienda moderna ; 158)
88–204–1648–4 X.629/19446
800255x

ADVERTISING
to date – Serials

AD Q / by World Executive's Digest. – Feb. 1983–. – [Hong Kong] : World Executive's Digest, 1983–
v. : ill
r6811553 X.0529/908
800255x

ADVERTISING
to date. Social aspects

GRUNERT, Klaus G.
Werbung. – Frankfurt : Campus, 1981
168p.. – Includes bibliography
3–593–32868–2 X.529/57656
9014667

ADVERTISING
Advertising & public relations

JEFKINS, Frank
Introduction to marketing, advertising and public relations. – London : Macmillan, 1982
290p : ill. – Includes index
0–333–32400–5 X.529/55296
2930021

ADVERTISING
Applications of transactional analysis

BARRETO, Roberto Menna
Análise transacional da propaganda. – São Paulo : Summus, c1981
327p : ill
r6707958 X.629/24910
9139362

ADVERTISING
Attitudes in Great Britain

ATTITUDES towards advertising. – London : Institute of Practitioners in Advertising, 1975
[5],80[i.e.81]p
B79–23670 X.512/6673
1913158

ADVERTISING
Commercial advertising

HOW to produce an advertisement. – London : Benn, 1978
35p : ill
B79–11883 X.519/29398
1834738

ADVERTISING
Communication

RAY, Michael L. (Michael Laurence)
Advertising and communication management. – Englewood Cliffs ; London : Prentice-Hall, c1982
xiii,514p : ill. – Includes index
0–13–015230–7 X.520/29197
2865785

ADVERTISING
Creativity

BAKER, Stephen, b.1921
Systematic approach to advertising creativity. – New York ; London : McGraw-Hill, 1979
viii,280p : ill. – Includes index
0–07–003352–8 X.525/2935
1969307

ADVERTISING
Creativity - Serials
CREATIVE voice. - No.1 (June 1983)- . - London : Creative Communications Pub., 1983-
v. : ill(some col.)
B84-17246 P.823/419
3075370

ADVERTISING
Direct-mail advertising
HODGSON, Richard S.
The Dartnell direct mail and mail order handbook. - 2nd ed. - Chicago ; London : Dartnell ; [Epsom] : [Distributed by Leviathan House], 1974
1575p : ill. - (Dartnell handbooks). - Bibliography: p.1312-1323. - Includes index
0-85013-007-7 X.529/30058
0350168

ADVERTISING
Economic aspects
CHIPLIN, Brian
Economics of advertising. - 2nd ed.. - London : Holt, Rinehart and Winston with the Advertising Association, c1981
xii,145p : ill. - Includes bibliographies and index
0-03-910315-3 X.520/26012
0324027

ADVERTISING
Effectiveness. Evaluation
CORKINDALE, David Reay
Measuring the effect of advertising. - Farnborough, Hants. : Saxon House [etc.], 1975
xvii,242p : ill. - ([Saxon House studies]). - Includes bibliographies and index
0-347-01110-1 X.520/10595
0586463

LOVELL, Mark
Assessing the effectiveness of advertising. - London : Business Books, 1975
xvi,281p : ill. - Bibliography: p.273-275. - Includes index
0-220-69730-2 X.529/30426
0586463

ADVERTISING
Effectiveness. Evaluation, to date
KEITZ, Wolfgang von
Der Saarbrücker Aktivierungs-Test (SAT). - Köln : Heymann, 1981
xii,156p : col.ill. - (Annales Universitatis Saraviensis. Rechts- und Wirtschaftswissenschaftliche Abteilung ; Bd.97). - Bibliography
3-452-18893-0 Ac.2634(2a)[Bd.97]
8799970

ADVERTISING
Effectiveness. Forecasting. Mathematical models
SÖLLNER, Walter Johannes
Modelle zur Werbewirkungsprognose für Konsumgüter-Markenartikel. - Augsburg, 1975
xviii, 180 p.
r6003733 X.519/20763
0863211

ADVERTISING
Effects on economic conditions
ALBION, Mark S.
The advertising controversy. - Boston, Mass. : Auburn House, c1981
xxi,226p : ill. - Bibliography: p201-217. - Includes index
0-86569-057-x X.520/27024
2728796

ADVERTISING
Graphics. Design
BOCKUS, H. William
Advertising graphics. - 3rd ed. - New York : Macmillan ; London : Collier Macmillan, 1979
xi,273p,plate : ill(some col). - Bibliography: p.260. - List of films: p.260. - Includes index
0-02-311490-8 X.415/5745 Woolwich
013774x

DONAHUE, Bud
The language of layout. - Englewood Cliffs ; London : Prentice-Hall, 1978
vi,215p : ill(some col). - (The art and design series)
0-13-522953-7 X.421/20589 Woolwich
013774x

ADVERTISING
Graphics. Designs - Catalogues
INTERNATIONAL GRAPHICAL ALLIANCE
107 grafici dell' AGI (Alliance Graphique Internationale). - [Venice] : Alfieri, [1974?]
ca 208 p. : chiefly ill
r6013664 X.421/8668 Woolwich
0148393

ADVERTISING
Graphics. Designs, to 1981
HOLME, Bryan
Advertising. - London : Heinemann, c1982
324p : ill(some col.). - Includes index
0-434-34540-7 L.45/1723
2807343

ADVERTISING
Great Britain - Statistics - For advertising
INCORPORATED SOCIETY OF BRITISH ADVERTISERS
ISBA area information. - London : Incorporated Society of British Advertisers Ltd.
1978. - [1978]
147p : ill
Subsequent volumes, when received, will be placed at this shelfmark
0-906241-00-6 Cup.1281/833
144655x

ADVERTISING
Great Britain. Commercial radio services - For advertising - Directories - Statistics - Serials
The RADIO advertisers' guide. - Ed.1 (July 1980)- . - Totnes : Hamilton House, 1980-
v.
B81-02336 P.903/677
232914x

ADVERTISING
Great Britain. Mass media - For advertising
BILL GIBBS-PNA GROUP
Editorial media analysis. - London : The Group, [1976]
1v.
0-905873-00-9 X.625/537 Woolwich
1038583

ADVERTISING
Great Britain. Sex discrimination. Law: Sex Discrimination Act 1975 - For advertising
GREAT BRITAIN. Equal Opportunities Commission
The Sex Discrimination Act and advertising. - Manchester : The Commission, 1980
15p : ill
0-905829-31-x BS.433/9
2388502

GREAT BRITAIN. Equal Opportunities Commission. Advertising Unit
So you think you've got it right. - Manchester : Advertising Unit, Equal Opportunities Commission, [1983?]
23p : ill
0-905829-69-7 X.529/61651
2388502

ADVERTISING
Jamaica. Economic conditions - For advertising - Serials
JAMAICA exports. - Vol.1, no.1 (Aug. / Sept. 1979)- . - Kingston, Jamaica : the Gleaner Company, 1979-
r6202206 P.2000/523
0257885

ADVERTISING
Lettering. Design - For advertising
LEACH, Mortimer
Lettering for advertising. - New York ; London : Van Nostrand Reinhold, [1975]
xi,228p : chiefly ill(some col)
0-442-24718-4 X.415/5040
0539376

ADVERTISING
Linguistic aspects
CARDONA, Giorgio Raimondo
La lingua della pubblicità. - Ravenna : Longo, 1974
80p. - (Laboratorio di scrittura ; 2)
r6052489 X.519/24827
1004387

FLADER, Dieter
Stategien der Werbung. - Kronberg Ts, 1974
235 p.. - (Skripten. Linguistik und Kommunikationswissenschaft ; 6)
3-589-20019-7 X.909/34429
1004387

ADVERTISING
Malaysia. Economic conditions - For advertising
REDIFFUSION INTERNATIONAL LIMITED. Advertising and Programmes Division
West Malaysia and Singapore. - [4th ed.]. - London : [The Division], 1975
[13],43 leaves,plate : ill. - (Market profile)
0-904734-14-5 X.515/514
0257818

ADVERTISING
Management
CORKINDALE, David Reay
Managing advertising effectively. - Bradford : MCB Books, 1975
89p : ill
0-903763-15-x X.510/15051
0513954

GILLIGAN, Colin
Advertising management. - Oxford : P. Allan, 1976
viii,215p : ill. - (Philip Allan textbooks in business studies). - Includes index
0-86003-500-x X.529/30885
0513954

PEEBLES, Dean M.
Management of international advertising. - Boston, [Mass.] ; London : Allyn and Bacon, c1984
296p : ill. - Includes index
0-205-07988-1 X.525/9195
0513954

WADEMAN, Victor
Risk-free advertising. - New York ; London : Wiley, 1977
xiii,145p : ill. - (Wiley series on marketing management). - Includes index
0-471-02714-6 X.520/12091
0513954

YOUNG, Robert F. (Robert Freeman)
Managing cooperative advertising. - Lexington, Mass. : Lexington Books, c1983
x,170p : ill. - Bibliography: p169-170
0-669-06301-0 X.622/18592
0513954

ADVERTISING
Middle East. Media. Audiences - For advertising - Statistics
MMEMS. - [London] : McCann-Erickson Advertising Ltd].
79. Bahrain. Decision makers. - [1979]
127 leaves in various pagings
B80-20219 X.702/6322
2148447
79. Bahrain. General public. - [1979]
118 leaves in various pagings
B80-20220 X.702/6315
2148447
79. Egypt. Decision makers. - [1979]
109 leaves in various pagings
B80-20221 X.702/6316
2148447
79. Egypt. General public. - [1979]
105 leaves in various pagings
B80-20222 X.702/6326
2148447
79. Jordan. Decision makers. - [1979]
102 leaves in various pagings
B80-20223 X.702/6323
2148447
79. Jordan. General public. - [1979]
101 leaves in various pagings
B80-20224 X.702/6324
2148447

MMEMS. - [London] : McCann-Erickson Advertising Ltd].
79. Kuwait. Decision makers. - [1979]
115 leaves in various pagings
B80-20225 X.702/6321
2148447
79. Kuwait. General public. - [1979]
114 leaves in various pagings
B80-20226 X.702/6320
2148447
79. Qatar. Decision makers. - [1979]
iii,8,[69] leaves
B80-20227 X.702/6327
2148447
79. Saudi Arabia. Decision makers. - [1979]
111 leaves in various pagings
B80-20228 X.702/6318
2148447
79. Saudi Arabia. General public. - [1979]
102 leaves in various pagings
B80-20229 X.525/3172
2148447
79. UAE/Oman. Decision makers. - [1979]
118 leaves in various pagings
B80-20230 X.702/6317
2148447
79. UAE/Oman. General public. - [1979]
154 leaves in various pagings
B80-20231 X.702/6319
2148447

ADVERTISING
Pictorial advertising

ADVERTISEMENT parade 300. - Worthing : Visual Publications Ltd.
No.1: TSF, TV, disques = Fernsehgeräte, Rundfunkgeräte = TV & radio sets, hi-fi, records, cassettes. - 1975
[38]leaves : chiefly ill. - Includes index
0-903887-27-4 Cup.1281/193(1)
0048089

No.2: Apéritifs, liqueurs = Aperitifs, Liköre = Aperitifs, beers, spirits. - 1975
[39]leaves : chiefly ill. - Includes index
0-903887-28-2 Cup.1281/193(2)
0048089

No.3: Vins, bières, boissons diverses = Weine, Bier, alkoholfreie Getränke = Wines, beers, soft drinks. - 1975
[41]leaves : chiefly ill. - Includes index
0-903887-29-0 Cup.1281/193(3)
0048089

No.4: Biscuits, desserts, chocolats, confiserie = Konfekt, Schokolade, Keks = Biscuits, desserts, chocolates, sweets. - 1975
[40]leaves : chiefly ill. - Includes index
0-903887-30-4 Cup.1281/193(4)
0048089

No.5: Habillement dames, fourrures, tissus = Damen-Konfektion, Pelze = Women's clothing, furs, cloth. - 1975
[39] leaves : chiefly ill. - Includes index
0-903887-31-2 Cup.1281/193(5)
0048089

No.6: Habillement hommes & enfants, imperméables = Herren-und Kinder-Konfektion, Regenbekleidung = Men's & children's clothing, raincoats. - 1975
[39]leaves : chiefly ill. - Includes index
0-903887-32-0 Cup.1281/193(6)
0048089

ADVERTISING
Pictorial advertising

ADVERTISEMENT parade 300. - Worthing : Visual Publications Ltd.
No.7: Lingerie, dames = Damenwäsche, Strumpfwaren = Underwear, stockings. - 1975
[38]leaves : chiefly ill. - Includes index
0-903887-33-9 Cup.1281/193(7)
0048089

No.8: Chaussures = Schuhe = Shoes. - 1975
[38]leaves : chiefly ill. - Includes index
0-903887-34-7 Cup.1281/193(8)
0048089

No.9: Maquillage, rouges à lèvres, vernis ongles, instituts de beauté = Make-up's, Lippenstifte, Nagellack, Kosmetik-Salons = Face make-up, lipsticks, nail polish, beauty institutes. - 1975
[38]leaves : chiefly ill. - Includes index
0-903887-35-5 Cup.1281/193(9)
0048089

No.10: Automobiles = Kraftfahrzeuge = Automobiles. - 1975
[38]leaves : chiefly ill. - Includes index
0-903887-36-3 Cup.1281/193(10)
0048089

No.11: Poids lourds, autocars, tracteurs = LKW's, Autobusse, Traktoren = Trucks, coaches, tractors. - 1975
[38]leaves : chiefly ill. - Includes index
0-903887-37-1 Cup.1281/193(11)
0048089

No.12: Meubles = Möbel = Furniture. - 1975
[39]leaves : chiefly ill. - Includes index
0-903887-38-x Cup.1281/193(12)
0048089

No.13: Automobiles = Kraftfahrzeuge = Automobiles. - 1975
[38]leaves : chiefly ill. - Includes index
0-903887-20-7 Cup.1281/193(13)
0048089

No.14: Maquillages, rouges à lèvres, vernis ongles, instituts de beauté = Make-up, Lippenstifte, Nagellacke, Kosmetiksalons = Face make-ups, lipsticks, nail polishes, beauty institutes. - 1976
[38]leaves : chiefly ill. - Includes index
0-903887-21-5 Cup.1281/193(14)
0048089

No.15: Produits et appareils pour cheveux = Haarpflegemittel = Hair grooming. - 1975
[39] leaves : chiefly ill. - Includes index
0-903887-22-3 Cup.1281/193(15)
0048089

No.16: Conserves, soupes, condiments = Konserven, Suppen, Gewürze = Canned foods, soups, condiments. - 1975
[39]leaves : chiefly ill. - Includes index
0-903887-23-1 Cup.1281/193(16)
0048089

No.17: Automobiles = Kraftfahrzeuge = Automobiles. - 1976
[38]leaves : chiefly ill. - Includes index
0-903887-24-x Cup.1281/193(17)
0048089

No.23: Automobiles = Kraftfahrzeuge = Automobiles. - 1976
[39]leaves : chiefly ill. - Includes index
0-903887-19-3 Cup.1281/193(23)
0048089

BASIC advertising themes and their presentation / [compiled by] Juergens, McCarthy. - Rustington : Visual Publications Ltd, [1979]
[6],297[i.e.340]p : chiefly ill. - Includes index
0-903887-54-1 LR.425/349
1916955

ADVERTISING
Positioning

CROSIER, Keith
Ladders in the mind. - Glasgow : Struthers Advertising Group, c1981
26p : ill
B82-25685 X.529/48143
2335050

Ladders in the mind. - Glasgow : Struthers Advertising Group, c1982
26p. - (Struthers occasional report ; no.2)
B83-19133 X.950/20251
2335050

RIES, Al
Positioning. - New York ; London : McGraw-Hill, c1981
xii,246p. - Includes index
0-07-065263-5 X.529/39531
2335050

ADVERTISING
Printed advertising - For industrial marketing

STOCK, Raymond F.
Sales literature for industry. - London : McGraw-Hill, 1976
148p : ill. - Includes index
0-07-084467-4 X.900/25134
0663751

ADVERTISING
Psychosocial aspects

VESTERGAARD, Torben
The language of advertising. - Oxford : Blackwell, 1985
x,182p : ill. - (Language in society). - Bibliography: p178-180. - Includes index
0-631-10741-x X.520/37478
3115194

ADVERTISING
Public service advertising

PUBLIC communication campaigns / edited by Ronald E. Rice, William J. Paisley. - Beverley Hills ; London : Sage, c1981
328p. - Bibliography: p293-313. - Includes index
0-8039-1578-0 X.529/43982
2567598

ADVERTISING
Sexism

COURTNEY, Alice E.
Sex stereotyping in advertising. - Lexington, Mass. : Lexington Books, c1983
xv,239p. - Bibliography: p209-231. - Includes index
0-669-03955-1 X.520/34565
3083055

ADVERTISING
Singapore. Economic conditions - For advertising

REDIFFUSION INTERNATIONAL LIMITED. Advertising and Programmes Division
West Malaysia and Singapore. - [4th ed.]. - London : [The Division], 1975
[13],43 leaves,plate : ill. - (Market profile)
0-904734-14-5 X.515/514
0257826

ADVERTISING
Social aspects

SMITH, Gillian Crampton
Seeing's believing. - [London] : [Longman], c1981
1portfolio(10parts) : ill. - (Longman thinkstrips)
0-582-22300-8 X.525/4220
2367882

ADVERTISING
Techniques

OGILVY, David, 1911-
Ogilvy on advertising. - London : Pan, 1983
224p : ill(some col.). - Bibliography: p218. - Includes index
0-330-26985-2 X.421/24723
1747894

WADEMAN, Victor
Money-making advertising. - New York ; Chichester : Wiley, c1981
xiii,142p : ill. - Includes index
0-471-06276-6 X.520/23795
1747894

ADVERTISING
Television advertising. Language

GEIS, Michael L.
The language of television advertising. - New York ; London : Academic Press, 1982
xiii,257p : ill. - (Perspectives in neurolinguistics, neuropsychology and psycholinguistics). - Bibliography p245-247. - Includes index
0-12-278980-6 X.950/19944
288013x

ADVERTISING
Word of mouth advertising by consumers

SCHERRER, Anton P.
Das Phaenomen der Mund-zu-Mund-Werbung und seine Bedentung fur das Konsumentenverhalten. - Freiburg, 1975
216 p
r6003815 X.519/20935
0878944

ADVERTISING
World - For advertising - Serials

ADMAP world index. - Vol.1, no.1- Admap. - [London] : [Admark Publications], [1977?]- : ill
r6072628 P.703/599
1699040

ADVERTISING. Agricultural industries & trades. United States

MCGUIRE, James E.
Advertising and display in agribusiness. - New York ; London : McGraw-Hill, 1979
vi,138p : ill. - (Career preparation for agriculture/agribusiness). - Includes index
0-07-045129-x X.510/16346
1946242

ADVERTISING. Air travel
to date

MAIER, Lotte
Aufbruch zum Himmel. - Dortmund : Harenberg, c1982
99p : ill. - (Die bibliophilen Taschenbücher ; 328)
3-88379-328-0 X.429/15404
9047409

ADVERTISING. Argentina
1800-1974

ALONSO PIÑEIRO, Armando
Breve historia de la publicidad argentina. - Buenos Aires : Alzamor Editores, 1974
333 p., 16 p. of plates. - (Colección del Alba ; v. 3)
r6011802 X.708/18388
0885851

ADVERTISING. Austria
Television advertising, to date

STEINER, Karin
Werbung im Fernsehen. - Wien : Österreichischer Rundfunk, [1980]
89p. - (Berichte zur Medienforschung ; 24)
r6302407 X.0950/278(24)
8775974

ADVERTISING. Books. West Germany
to date

TIELEBIER-LANGENSCHEIDT, Florian
Werbung für deutsche Gegenwartsliteratur. - Frankfurt am Main : Buchhändler-Vereinigung, 1983
386columns : ill. - Includes bibliography
3-7657-1203-5 2708.h.77
9121064

ADVERTISING. Brazil
to date. Social aspects

PÔRTO, Sérgio
A nova Opulência das Geraes. - São Paulo : Cortez, 1982
142p : ill. - Includes bibliography
r6709689 X.529/68786
9153144

ADVERTISING. Cars
to ca 1939

ROBERTS, Peter, 1925-
Any color so long as it's black -. - Newton Abbot : David and Charles, 1976
144p : ill(some col). - Includes index
0-7153-7239-4 Cup.1285/154
0792632

ADVERTISING. Cars. Great Britain
Graphics. Design - Serials

AUTO-GRAPH. - Issue 1-. - Bracknell : 3M United Kingdom, 1982-
v. : ill(chiefly col.)
r6865209 P.613/301
2793784

ADVERTISING. Cigarettes. Germany
to date

DICHTERWORT und blauer Dunst / [herausgegeben von Werner Fritsch unter Mitarbeit von Claus-Peter Strempel und Ulrich Hobbeling]. - Memmingen : Visel, [1981?]
96p. : ill
3-922406-11-4 X.950/32107
9077251

ADVERTISING. Cigarettes. West Germany
Effectiveness. Econometric models

REUIJL, Jan C
On the determination of advertising effectiveness. - Boston, Mass. ; London : Kluwer, 1982
226p. - Bibliography: p207-214. - Includes index
0-89838-125-8 X.529/52262
2847949

ADVERTISING. Colman (J and J) Limited
to 1938

YELLOW white & blue - [Norwich] [Colman Foods], 1978
82p : ill(some col)
B79-08083 X.519/29221
1808362

ADVERTISING. Consumer credit. Great Britain
Law

CONSUMER Credit Act 1974 / [prepared by the Office of Fair Trading and the Central Office of Information]. - [London] : [H.M.S.O.], 1980
18p : ill(some col.)
B83-14013 BS.419/45
2864592

ADVERTISING. Consumer credit. Great Britain
Requirements: Truth. Law

KARPINSKI, Jan
Truth in lending. - London : Current Account, 1980
x,80p. - (Current account special report)
0-906840-31-7 X.525/5238
2606852

ADVERTISING. Consumer goods. Japan
Pictorial advertising

JAPAN ADVERTISING PHOTOGRAPHERS' ASSOCIATION
Advertising photography in Japan. - Tokyo : Kodansha International.
1. - 1982
[248]p : chiefly ill(some col.)
0-87011-524-3 L.45/1934
2856344

ADVERTISING. Consumer goods. Western Europe
Economic aspects related to industrial competition. Econometric models

LAMBIN, Jean Jacques
Advertising, competition and market conduct in oligopoly over time. - Amsterdam ; Oxford : North-Holland Publishing Co. [etc.], 1976
x,312p : ill. - (Contributions to economic analysis ; 94). - Bibliography: p.290-308. - Includes index
0-7204-3200-6 WP.C.19/94
0595373

ADVERTISING. Cosmetics, perfumes & toiletries. Germany
1871-1945

MUSEUM FÜR KUNST UND GEWERBE HAMBURG. Pädagogische Abteilung
Kunst-Kommerz und Körperpflege. - Hamburg : Das Museum, 1982
11postcards : col
r6305595 Cup.900.X.102
8893357

ADVERTISING. Denmark
1900-1974

KJÆR-HANSEN, Max
Reklamen i det 20. arhundredes Danmark. - København : Nyt Nordisk Forlag, 1974
346 p. : ill
87-17-01958-3 X.809/22906
0869325

ADVERTISING. Dumbarton (District). Strathclyde Region. Scotland
Outdoor advertising. Regulation

GUIDELINES for display of advertisements / [Dumbarton District Council, Planning & Development Department]. - Dumbarton : Dumbarton District Council, [1982]
6leaves
B83-21985 X.525/7200
2894025

ADVERTISING. England
By barristers. Restrictions. Effects on supply of services of barristers

GREAT BRITAIN. Monopolies and Mergers Commission
Barristers' services. - London : H.M.S.O., [1976]
[4],24,lvi p. - ([1975-76 H.C.]559)
0-10-255976-7 BS.83/80
0736260

ADVERTISING. England
By solicitors. Restrictions. Effects on supply of services of solicitors

GREAT BRITAIN. Monopolies and Mergers Commission
Services of solicitors in England and Wales. - London : H.M.S.O., [1976]
ii-vi,41,lxxx p. - ([1975-1976 H.C.] 557)
0-10-255776-4 BS.83/117
0732400

ADVERTISING. Europe
- Serials

ADVERTISING age's focus. - 1982-. - London : Advertising Age's Focus, 1982-
v. : ill(some col.)
B83-09884 P.521/3736
2836785

ADVERTISING. European Community countries
Law

LAWSON, R. G. (Richard Grenville)
Advertising and labelling laws in the Common Market. - 2nd ed. - Bristol : Jordans, 1981
1v.(loose-leaf). - Includes index
0-85308-063-1 L.42/995
0361151

Advertising and labelling laws in the Common Market. - 3rd ed. - Bristol : Jordans, 1982
1v.(loose-leaf). - Includes index
0-85308-071-2 X.205/1282
0361151

LAWSON, Richard Grenville
Advertising and labelling. - Bristol : Jordan, 1975
xv,393p. - Bibliography: p.361. - Includes index
0-85308-040-2 X.520/10089
0361151
- 1st [etc.] supplement. 1975-.
25 cm X.520/10089 a

ADVERTISING. European Community countries
Misleading advertising. Law. Harmonisation

ADVERTISING ASSOCIATION
Memorandum from the Advertising Association of the United Kingdom to the Director-General, Directorate of the International Market and Approximation of Legislation of the Commission of the European Communities on harmonisation of legislation on misleading advertising. - London : The Association, 1973 [i.e. 1976]
[11]p
0-902878-17-4 X.512/5368
0740012

ADVERTISING. France
to date

CHEVASSU, François
La publicité. - Paris : Edilig, [1982]
125p : ill. - (Les cahiers de l'éducation permanente)
2-85601-023-7 X.629/22978
907385x

ADVERTISING. France
1945-. Linguistic aspects

KUBARTH, Hugo
L'image dans la publicité française moderne. - Rheinfelden : Schäuble, [1981]
vi,70p. - (Reihe Romanistik ; Nr 19)
3-87718-719-6 X.0900/328(19)
8712891

ADVERTISING. France
Law

GREFFE, Pierre
La publicité et la loi. - Troisième édition. - Paris : Librairies techniques, 1977
429p
2-7111-0073-1 X.100/26931
1470590

ADVERTISING. Fund raising. Charities
Direct-mail advertising - Conference proceedings

DIRECT response advertising for fund raising organisations. - St Albans : IBIS Information Services Ltd, [1978]
[2]leaves,i,150p : ill
B79-11640 X.512/6517
1836102

ADVERTISING. Goods. Great Britain
By retailers: Advertising without revealing business interests. Law. Reform

GREAT BRITAIN. Consumer Protection Advisory Committee
Disguised business sales. - London : H.M.S.O., [1976]
iii,11p. - ([1975-76 H.C.] 355)
0-10-235576-2 BS.419/9
0317020

ADVERTISING. Goods. Retail trades. Great Britain

LOWE, Eric
Successful retailing through advertising. - London : McGraw-Hill, c1983
184p : ill. - Bibliography: p172. - Includes index
0-07-084588-3 X.520/31179
2950987

ADVERTISING. Great Britain

ADVERTISING STANDARDS AUTHORITY
What makes advertising tick?. - London : The Authority, [1977]
Folder([6]p)
r6092530 X.512/6933
0202428

ATTWOOD, Tony, 1947-
Careers in advertising. - Totnes : Careerscope, 1980
[14]leaves
0-906888-14-x X.525/6394
1569090

Careers in advertising. - 2nd rev. ed. - Totnes : Hamilton House. - (Careerscope ; [1])
[1]: Careers in advertising
24p
0-906888-18-2 X.529/44779
1569090

CRITCHLEY, Ronald Alfred
U.K. advertising statistics. - London : Advertising Association, [1973?]
50 leaves
r6028766 X.515/298
1027743

INSTITUTE OF PRACTITIONERS IN ADVERTISING
Advertising conditions in the United Kingdom. - 2nd ed. - London : I.P.A., 1976
22p
B77-25061 X.519/29620
0065854

JEFKINS, Frank
Advertising made simple. - Reprinted. - London : W. H. Allen, 1980
xvi,304p. - (Made simple books). - Bibliography: p.293. - Includes index
0-491-00704-3 X.519/28611
0065854
- [A re-issue].-1979 X.529/49997

The PRACTICE of advertising / edited by Norman A. Hart and James O'Connor. - London : Heinemann [for] the CAM Foundation and the Institute of Marketing, 1978
xxiii,196p : ill. - Includes bibliographies and index
0-434-90362-0 X.510/15830
0065854

The PRACTICE of advertising / edited by Norman A. Hart and James O'Connor. - 2nd ed. - London : Heinemann on behalf of the CAM Foundation and the Institute of Marketing, 1983
xx,252p : ill. - Includes bibliographies and index
0-434-90716-2 X.622/18591
0065854

SMITH, Gillian Crampton
Advertising. - York : Longman Resources Unit, 1983
9p : ill. - (Its your life). - Bibliography: p9
0-582-38952-6 X.955/2445
0516864

TAYLOR, Felicity
Careers in marketing, public relations and advertising. - London : Kogan Page, c1983
96p
0-85038-658-6 X.629/24135
1569090

THOM, Vivien
Advertising. - London : Edward Arnold, 1980
16p : ill. - (Checkpoint ; 12). - Bibliography: p16
0-7131-0368-x X.622/14467
0516864

WOOD, John W.
The instrument of advertising. - [London] : [Wood, Brigdale and Company Ltd], 1978
[2],19p. - (Strategic advertising series ; no.4)
B79-04487 X.519/29051
0065854

ADVERTISING. Great Britain
- Directories - Serials

DIRECT mail databook. - 1973. - London : Gower Press, 1973
xiii,266p : ill
0-7161-0157-2 P.513/620
0065854

ADVERTISING. Great Britain
- Serials
MANCHESTER PUBLICITY ASSOCIATION
MPA news. - Vol. 1, no. 1 - Vol. 3, no. 4 ; 1975?- Winter 1977. - [Manchester] : [M.P.A.], [1975?-77] : ill
B80-24647 P.523/1251
2068354

ADVERTISING. Great Britain
to 1981
NEVETT, T. R. (Terry R)
Advertising in Britain. - London : Heinemann on behalf of the History of Advertising Trust, 1982
xiii,231p,[32]p of plates : ill(some col.). - Bibliography: p218-225. - Includes index
0-434-49642-1 X.520/26546
2700603

ADVERTISING. Great Britain
1939-1945
BEGLEY, George
Keep Mum!. - London : Lemon Tree Press, 1975
77p : ill. - (Time remembered)
0-904291-02-2 X.809/40562
0578797

ADVERTISING. Great Britain
1945-. Linguistic aspects
GREVEN, Hubert A.
La langue des slogans publicitaires en anglais contemporain. - Paris : Presses universitaires de France, 1982
386p,plates. - (Publications de l'Université de Rouen. études anglaises ; 76). - Includes bibliography
2-13-037626-6 X.950/23292
8977046

ADVERTISING. Great Britain
Attitudes of society, 1972
BRITISH MARKET RESEARCH BUREAU
Public attitudes to advertising. - London : The Association, [1976]
[4],11p
0-902878-18-2 X.519/26181
0737151

ADVERTISING. Great Britain
Attitudes of society, 1976 - Statistics
ADVERTISING ASSOCIATION
Public attitudes to advertising, 1976. - London : The Association, 1976
47 leaves : ill
0-902878-19-0 X.512/5360
0738778

ADVERTISING. Great Britain
By accountants. Restrictions. Effects on supply of accountancy services
GREAT BRITAIN. Monopolies and Mergers Commission
Accountancy services. - London : H.M.S.O., [1976]
[4],104p. - (Cmnd. 6573). - Includes index
0-10-165730-7 BS.83/93
0736295

ADVERTISING. Great Britain
By stockbrokers. Restrictions. Effects on supply of services of stockbrokers
GREAT BRITAIN. Monopolies and Mergers Commission
Stockbrokers' services. - London : H.M.S.O., [1976]
iv,55,[3]p. - (Cmnd.6571). - Includes index
0-10-165710-2 BS.83/89
0736279

ADVERTISING. Great Britain
By veterinary surgeons. Restrictions. Effects on supply of veterinary medical services
GREAT BRITAIN. Monopolies and Mergers Commission
Veterinary services. - London : H.M.S.O., [1976]
iv,71p. - (Cmnd.6572). - Includes index
0-10-165720-x BS.83/91
0736287

ADVERTISING. Great Britain
Commercial advertising. Statistical models: Manufacturers' Consumer Advertising
ADVERTISING ASSOCIATION
Advertising expenditure, 1960-74 including an index of media rates 1961-74 ; and, MCA - a new tool for economists. - London : The Association, [1976]
11p : ill
0-902878-13-1 X.512/5365
0738824

ADVERTISING. Great Britain
Decision making
BROADBENT, Simon
Spending advertising money. - Completely revised 2nd ed. - London : Business Books, 1975
xi,368p : ill. - Includes index
0-220-66282-7 X.529/30338
0563293

Spending advertising money. - 3rd ed. - London : Business Books [for] CAM (Communication, Advertising and Marketing Education Foundation), 1979
xi,381p : ill. - Includes index
0-220-67024-2 X.529/33995
0563293

BROADBENT, Simon
Spending advertising money. - 4th ed.. - London : Business, 1984
424p. - Includes bibliographies and index
0-09-155970-7 X.529/67346
0563293

ADVERTISING. Great Britain
Direct-mail advertising
A GUIDE to effective direct mail. - London : Post Office, [1982?]
27p : col.ill
B82-19583 BS.31/255
2712075

ADVERTISING. Great Britain
Direct-mail advertising. Role in attraction of new users of public libraries
CRONIN, Blaise
Direct mail advertising and public library use. - [London] : British Library ; Wetherby : British Library Lending Division, 1980
iv,53p : ill. - (Research and development reports / British Library ; 5539)
0-905984-58-7 2719.h.70
2088584

ADVERTISING. Great Britain
Effectiveness
ADVERTISING works / edited and introduced by Simon Broadbent. - London : Holt, Rinehart and Winston, 1981
xvi,207p : ill. - Includes index
0-03-910322-6 X.520/24126
2498006

ADVERTISING. Great Britain
Effects on economic conditions
ADVERTISING and economic behaviour / Keith Cowling ... [et al.]. - London : Macmillan, 1975
viii,202p : ill. - Includes bibliographies and index
0-333-17685-5 X.529/30572
0627534

ADVERTISING. Great Britain
Expenditure - Statistics
ADVERTISING expenditure / [JWT Information Services]. - 1976 : Top 250 companies. - London : JWT Information Services, [1978]
[6],98p. - Includes index
Subsequent issues, when received, will be placed at this shelfmark
B80-38994 P.2000/819
2302578

ADVERTISING. Great Britain
Expenditure, 1960-1974
ADVERTISING ASSOCIATION
Advertising expenditure, 1960-74 including an index of media rates 1961-74 ; and, MCA - a new tool for economists. - London : The Association, [1976]
11p : ill
0-902878-13-1 X.512/5365
0738816

ADVERTISING. Great Britain
Law
LAWSON, R. G.
Advertising law. - Plymouth : Macdonald and Evans, 1978
xxxiv,366p. - Includes index
0-7121-1239-1 X.200/31649
0045144

WOOLLEY, Diana
Advertising law handbook. - 2nd ed. - London : Business Books for the Institute of Practitioners in Advertising, 1976
xiv,106p. - Includes bibliographies and index
0-220-66306-8 X.200/31111
0045144

ADVERTISING. Great Britain
Management
KENNEDY, Sherril Helen
Managing the advertising process. - Farnborough, Hants. : Saxon House [etc.], 1976
x,284p : ill. - ([Saxon House studies]). - Bibliography: p.245-278. - Includes index
0-347-01109-8 X.520/10607
0634042

ADVERTISING. Great Britain
Newspaper advertising. Influence on editorial policies of newspapers
SIMPSON, D. H.
Commercialisation of the regional press. - Aldershot : Gower, c1981
xv,224p : ill. - Bibliography: p217-224
0-566-00441-0 X.529/43429
0330310

ADVERTISING. Great Britain
Organisations: Institute of Practitioners in Advertising - Directories
INSTITUTE OF PRACTITIONERS IN ADVERTISING
Member-agencies. - 1977. - London : I.P.A, 1977
[2],19p
Subsequent volumes, when received, will be placed at this shelfmark
B78-04280 P.511/5064
014178x

ADVERTISING. Great Britain
Outdoor advertising
INCORPORATED SOCIETY OF BRITISH ADVERTISERS
Guide to outdoor advertising. - [New ed.]. - London : ISBA, 1979
28p : 1 ill
B80-38313 X.529/38352
2286122

ADVERTISING. Great Britain
Role in society - Conference proceedings
ADVERTISING in perspective in industry and society. - London : Advertising Association, [1976]
[3],64p : ill
0-902878-10-7 X.512/5130
0613045

ADVERTISING. Great Britain
Role of postal services, 1837-1901
The MULREADY advertisements / [compiled] by Malcolm G. Lowe. - Laguna Hills : Mulready Research Foundation, 1983
xviii,285p
0-911451-00-5 X.955/2037
9064230

ADVERTISING. Great Britain
Self-regulation
The SELF-REGULATORY system of advertising control / [for the] Department of Trade. - London : The Department, 1980
[3],28p
B80-41852 BS.41/1236
2318946

ADVERTISING. Great Britain
Self-regulation, 1974-1975
ADVERTISING ASSOCIATION
The self-disciplinary control system for advertising. - London : [The Association], 1976
Folder([6]p)
0-902878-20-4 X.512/5372
0742058

ADVERTISING. Great Britain
Self-regulation, 1976-1977
GREAT BRITAIN. Office of Fair Trading
Review of the UK self-regulatory system of advertising control. - [London] : Office of Fair Trading, 1978
91p in various pagings
B79-13929 BS.419/27
1843427

ADVERTISING. Great Britain
Television advertising. Costs. Implications of proposed fourth channel
HENRY, Harry
The commercial implications of a second (and complementary) Independent Television channel. - London : ADMAP Publications Limited, 1979
28p : ill. - (Admap monograph). - Bibliography : p.3
B80-14866 X.512/7139
2022427

ADVERTISING. Huntingdon (District). Cambridgeshire
Outdoor advertising. Regulation. Policies of local authorities

HUNTINGDON. District Council
A policy for - advertisement control. - [Huntingdon] : The Council, [1978]
[20]p(8 fold)
B80-27290 X.525/3337
2111241

ADVERTISING. Italy
to date

CESERANI, Gian Paolo
Vetrina della Belle époque. - Roma : Laterza, 1980
267p : chiefly ill. - (Grandi opere)
r6508894 X.421/24344
8799598

FORTE, Francesco
Tattiche del desiderio. - Milano : Sugar Co, 1979
220p. - (Fatti e misfatti ; 73)
r6503077 X.529/45541
8799598

ADVERTISING. Italy
Effects on consumer behaviour

MURA, Antonio
La pubblicità. - Roma : Bulzoni, 1975
214p. - (Pedagogia sociale. Testi e studi ; 4)
r6085314 X.519/40851
0854212

NOVAGA, M.
Pubblicità, comportamento di acquisto e matrici socioculturali. - Roma : Edizioni "C.E.P.I.", 1973
p. [137]-144
r6013758 X.512/2430
0854212

PUBBLICITÀ e consumi in Italia / di Ruggero Paladini...[et al.] ; a cura di Giuseppe Campa. - Milano : Franco Angeli, [1976]
389p. - (Collana "La società" ; 26). - Bibliography : p.371-389
r6052633 X.519/25027
0854212

ADVERTISING. Italy
Law

ITALY
[]. Codice della publicità, della promozione e delle vendite. - Milano : Franco Angeli, [1975]
545p. - (Collana L'azienda moderna ; 114)
r6078104 X.200/11434
1510479

ADVERTISING. Middle East
- Serials

CAMPAIGN Mid East. - [No.1]- ; [Mar./May, 1979]-. - London : Haymarket Publishing, [1979]- : ill(some col)
B80-09313 P.521/3423
2046369
- [Another copy.] SRL

ADVERTISING. North Bedfordshire (District)
Outdoor advertising. Regulation

NORTH BEDFORDSHIRE. Borough Council
Advertisements. - [Bedford] : [North Bedfordshire Borough Council], [1977]
2,[10]leaves : ill
B81-26358 X.425/2080 Woolwich
2503271

ADVERTISING. Prepacked food. Great Britain
Law

INSTITUTE OF PRACTITIONERS IN ADVERTISING
Is it legal?. - 3rd ed. - London : The Institute, 1979
23p
B79-24325 X.102/2207
0037508

ADVERTISING. Prices. Consumer goods. Scandinavia

NORDIC COUNCIL OF MINISTERS. Ämbetsmannakommitté för konsumentfrågor. Utskottet för reklam- och marknadsföringsfrågor. Projektgrupp
Otillbörliga prisjämförelser i reklam och annan marknadsföring. - Stockholm : Nordiska radet, 1978
147,[87]p. - (NU ; B1978. 26)
A slip with the imprint "Fritzes hovbokhandel i distribution" is pasted on the titlepage
91-7052-391-6 X.520/20703
2294575

ADVERTISING. Products. Great Britain
Expenditure - Statistics

ADVERTISING expenditure / [JWT Information Services]. - 1977 : Top 600 brands. - London : JWT Information Services, [1978]
[2],46p
Subsequent issues, when received, will be placed at this shelfmark
B80-38999 P.513/2530
2302586

ADVERTISING. Recruitment of personnel. Great Britain

BRAITHWAITE, Roderick
Communications and the job-seeker. - London : Advertising Association, [1976]
[4],107,[9]p : ill. - (Research studies in advertising ; 9)
0-902878-15-8 X.0519/373(9)
0733822

ADVERTISING. Recruitment. Personnel. Great Britain

RAY, Maurice Edward
Recruitment advertising. - London : Institute of Personnel Management, 1980
xii,228p : ill. - (Management in perspective)
0-85292-259-0 X.529/38720
2074567

ADVERTISING. Scotland
By advocates. Restrictions. Effects on supply of services of advocates

GREAT BRITAIN. Monopolies and Mergers Commission
Advocates' services. - London : H.M.S.O., [1976]
[4],23,liv p. - ([1975-1976 H.C.]560)
0-10-256076-5 BS.83/95
0734926

ADVERTISING. Scotland
By solicitors. Restrictions. Effects on supply of services of solicitors

GREAT BRITAIN. Monopolies and Mergers Commission
Services of solicitors in Scotland. - London : H.M.S.O., [1976]
[5],34,lv p. - ([1975-1976 H.C.]558)
0-10-255876-0 BS.83/78
0732397

ADVERTISING. Scotland
Control

ADVERTISEMENT control. - [Scotland] : [Argyll & Bute District Council], [1983?]
14p
r6815581 X.525/9425
3176789

ADVERTISING. Soviet Union
1945-

ROZENTAL', D. É. (Ditmar Él'îashevich)
ÎAzyk reklamnykh tekstov .... - Moskva : Vysshaîa shkola, 1981
124p
r6486512 X.629/17185 Woolwich
8799601

ADVERTISING. Spain
to date

CARDONA, Domingo
Lingüistica de la publicidad. - Madrid : Júcar, 1979
149p. - (Biblioteca Júcar de ciencias humanas ; 63). - Bibliography
84-334-1063-6 X.958/4385
8724741

ADVERTISING. Supplements. Newspapers with British imprints
By advertising agencies

INCORPORATED SOCIETY OF BRITISH ADVERTISERS
Press supplements and the advertiser. - 2nd ed. - London : ISBA, 1977
8p
B81-18664 X.529/41294
2400863

ADVERTISING. Sweden
to date

JONSSON, Sverker
Pressen, reklamen och konkurrensen 1935-1978. - Göteborg : Göteborgs universitet, Ekonomisk-historiska institutionen, 1982
406p. - (Meddelanden fran Ekonomisk-historiska institutionen vid Göteborgs universitet ; 52). - Includes bibliography
91-85196-24-x Ac.2651.db/2[no.52]
9097104

ADVERTISING. Tobacco
Pictorial advertising

PIPE dreams / edited by Mike Dempsey. - London : Pavilion Books, 1982
94p : chiefly ill(some col.)
0-907516-12-2 CUP.408.N.27
2756714

ADVERTISING. Tobacco products
to date

MOELLER, Magdalena
Plakate für den blauen Dunst. - Dortmund : Harenberg, c1983
156p : chiefly ill. - (Die bibliophilen Taschenbücher ; Nr.387)
3-88379-387-6 X.529/66063
9148353

ADVERTISING. United States

ADVERTISING and society / edited by Yale Brozen. - New York : New York University Press, 1974
189p. - (The key issues lecture series)
r6193715 X.520/14994
1542559

ARONSON, Steven M. L.
Hype. - London : Vermilion, 1984
198p,[8]p of plates : ill
0-09-156251-1 X.529/64380
1542559

BOLEN, William H.
Advertising. - New York ; Chichester : Wiley, c1981
xviii,504p,[8]p of plates : ill(some col.). - (Wiley series in marketing). - Includes index
0-471-03486-x X.520/23671
1542559

Advertising. - 2nd ed. - New York ; Chichester : Wiley, c1984
xvi,649p,[8]p of plates : ill(some col.). - Includes index
0-471-86348-3 X.520/35277
1542559

CONE, Fairfax M.
The Blue Streak. - [Chicago] : Crane Communications, 1973
[8],200p
0-87251-012-3 X520/13172
1542559

DUNN, S. Watson (Samuel Watson)
Advertising. - 5th ed.. - Chicago ; London : Dryden, c1982
xxi,710p,[8]p of plates : ill(some col.). - (Dryden Press series in marketing). - Includes bibliographies and index
0-03-060049-9 X.520/26915
1542559

DUNN, Samuel Watson
Advertising. - 4th ed. - Hinsdale, Ill. : Dryden Press, 1978
xxi,741p,[8]p of plates : ill. - (The Dryden Press series in marketing). - Includes bibliographies and index
0-03-014341-1 X.525/2319
1542559

FAISON, Edmund W. J.
Advertising. - New York ; Chichester : Wiley, 1980
xiii,783p, [4]p of plates : ill. - (Wiley series in marketing). - Includes bibliographies and index
0-471-04956-5 X.520/21154
1542559

KING, Francis S.
Modern advertising skills. - New York ; London : Van Nostrand Reinhold, c1983
iv,282p,[8]p of plates : ill(some col.). - Includes index
0-442-24749-4 X.622/18687
1935569

NORRIS, James S.
Advertising. - 3rd ed. - Reston, Va. : Reston, c1984
xii,484p : ill. - Includes bibliographies
0-8359-0048-7 X.520/37592
1542559

POPE, Daniel
The making of modern advertising. - New York : Basic Books, c1983
x,340p
0-465-04325-9 X.529/57201
1542559

RHETORIC and ideology in advertising / Gunnar Andrén...[et al.]. - [Stockholm] : Liber förlag, 1978
175p,[4]p of plates : ill
91-380-3711-4 X.519/38722
1542559

SCHRANK, Jeffrey
Deception detection. - Boston [mass.] : Beacon Press, 1975
[5],154p : ill
0-8070-3161-5 X.519/30416
1542559

ADVERTISING. United States –continued

WRAY, Ralph D.
Advertising services. – New York ; London : McGraw-Hill, 1979
vi,186p : ill. – (Career competencies in marketing series)
0–07–071900–4 X.512/6782
1935569

ADVERTISING. United States
to date

ATWAN, Robert
Edsels, luckies, & frigidaires. – New York : Dell Publishing Co., 1979
xx,363p : chiefly ill. – (A delta special)
0–440–53487–9 X.425/2182 Woolwich
8724717

EWEN, Stuart
Captains of consciousness. – New York : McGraw-Hill, c1976
x,261p
0–07–019845–4 X.529/55998
8724717

ROHRER, Daniel Morgan
Mass media, freedom of speech, and advertising. – Dubuque : Kendall / Hunt, c1979
xvi,700p
0–8403–1988–6 X.950/13364
8724717

SÉGUÉLA, Jacques
Hollywood lave plus blanc. – Paris : Flammarion, 1982
253p
2–08–064439–4 X.529/56405
8724717

ADVERTISING. United States
By companies

GARBETT, Thomas F.
Corporate advertising. – New York ; London : McGraw-Hill, c1981
xi,252p : ill. – Includes index
0–07–022787–x X.520/24260
2504634

SIEGEL, Gonnie McClung
How to advertise and promote your small business. – New York ; Chichester : Wiley, 1978
viii,128p. – (The small business series). – Bibliography: p.125–126. – Includes index
0–471–04032–0 X.519/28364
1570668
– [Another copy] X.629/17510

SLUTSKY, Jeff
Streetfighting. – Englewood Cliffs ; London : Prentice-Hall, c1984
240p : ill. – (A Reward book). – Includes index
0–13–851550–6 X.520/35353
1570668

ADVERTISING. United States
Effects on competition in industries. Mathematical models. Applications of digital computer systems

BLOOM, Paul N.
Advertising, competition and public policy. – Cambridge, Mass. : Ballinger Publishing Co., 1976
xiii, 205 p. : ill. – Includes index
0–88410–280–7 X.620/16226 Woolwich
0704539

ADVERTISING. United States
Illuminated signs. Designs

DI LEMME, Philip
American streamline. – New York ; Wokingham : Van Nostrand Reinhold, c1984
[160]p : ill
0–442–28104–8 L.49/3198
327778x

ADVERTISING. United States
Industrial advertising, 1900–1982

The 100 greatest corporate and industrial ads / [compiled by] Fred C. Poppe. – New York ; London : Van Nostrand Reinhold, c1983
xii,208p. – Bibliography: p203. – Includes index
0–442–27246–4 X.525/8597
3043924

ADVERTISING. United States
Modelling

HUNT, Cecily
How to get work & make money in commercials & modeling. – New York ; London : Van Nostrand Reinhold, c1982
176p : ill. – Includes index
0–442–23643–3 X.525/6224
2787628

ADVERTISING. United States
Public service television advertising. Political aspects

PALETZ, David L.
Politics in public service advertising on television. – New York ; London : Praeger, 1977
xiii,125p. – (Praeger special studies in US economic, social and political issues). – Includes index
0–275–23880–6 X.520/11924
1194992

ADVERTISING. United States
Radio advertising. Techniques

KEITH, Michael C.
Production in format radio handbook. – Lanham ; London : University Press of America, c1984
xi,206p : ill. – Bibliography: p187–197. – Includes index
0–8191–3886–x X.629/25903
3263800

ADVERTISING. United States
Role in social change

BERMAN, Ronald
Advertising and social change. – Beverly Hills ; London : Sage, c1981
159p. – (The Sage commtext series ; v.8). – Bibliography: p151–154. – Includes index
0–8039–1737–6 X.529/45991
2627426

EWEN, Stuart
Channels of desire. – New York ; London : McGraw-Hill, 1982
viii,312p. – Includes index
0–07–019850–0 X.529/50426
2627426

ADVERTISING. United States
Role in society

PACKARD, Vance
The hidden persuaders. – Reprinted with introduction and epilogue. – Harmondsworth : Penguin, 1981
248p. – Includes index
0–14–020585–3 X.529/42240
2524589

ADVERTISING. United States
Television advertising

BUSCH, H. Ted
The making of a television commercial. – New York : Macmillan ; London : Collier Macmillan, c1980
xi,175p : ill
0–02–518830–5 X.629/16440 Woolwich
2566370

POLTRACK, David
Television marketing. – New York ; London : McGraw-Hill, c1983
xii,395p. – Includes index
0–07–050406–7 X.622/7674
2566370

PRICE, Jonathan
The best thing on TV. – Harmondsworth : Penguin, 1978
[7],184p : ill
0–14–005004–3 X.902/5455
0497703

ADVERTISING. United States
Television advertising by political parties

DIAMOND, Edwin
The spot. – Cambridge, Mass. ; London : MIT, c1984
xiv,416p : ill. – Bibliography: p391–400. – Includes index
0–262–04075–1 X.629/25273
3242315

ADVERTISING. West Germany
to date. Linguistic aspects

REGER, Harald
[Articles. Selections]. Metaphern und Idiome in szenischen Texten, in der Werbe- und Pressesprache. – Hamburg : Buske, 1980
152p. – Bibliography
3–87118–440–3 X.950/6970
8724725

ADVERTISING. West Germany
By armaments industries, 1945–

RÜSTUNGSWERBUNG in der Bundesrepublik. – Stuttgart : ESG-Geschäftsstelle, [1978]
101,24p : ill. – (Militärpolitik ; Hft.9/10). – Bibliography
r6300829 X.0802/1544(9,10)
8821216

ADVERTISING. West Germany
Linguistic aspects

FRIMAN, Kirsti
Zum angloamerikanischen Einfluss auf die heutige deutsche Werbesprache. – Jyväskylä : Universität Jyväskylä, 1977
353p. – (Studia philologica jyväskyläensia ; 9). – Includes bibliographies
951–677–803–8 P.981/127[vol.9]
1474030

ADVERTISING. West Germany
Sexism, 1945–

FRAUENFEINDLICHE Werbung / Christiane Schmerl [Hrsg.]. – Reinbek bei Hamburg : Rowohlt, 1983
297p : ill. – (Rororo aktuell ; 7747)
3–499–17747–1 X.529/66069
9162208

ADVERTISING. West Germany
Television advertising, to date

KRAUSS, Wolfgang
Insertwirkungen im Werbefernsehen. – Bochum : Brockmeyer, 1982
v,302p. : ill. – (Bochumer Studien zur Publizistik- und Kommunikationswissenschaft ; Bd.32). – Includes bibliography
3–88339–223–5 X.950/30032
9057544

ADVERTISING. West Germany
Use of visual arts, to date

KUNST im Rahmen der Werbung / herausgegeben von Uwe Geese und Harald Kimpel ... – Marburg : Jonas, c1982
144p.,xvip.of plates : ill
3–922561–11–x X.421/25851
9110836

ADVERTISING. Western world
Attitudes of society, 1964–1976

PUBLIC attitudes to advertising in the UK, the USA and the EEC. – London : Advertising Association, [1977]
26p : ill
B78–04732 X.519/27551
1298194

ADVERTISING. Yugoslavia
to date

SPAHIĆ, Besim
Aspekti reklame u jugoslovenskom društvu. – Sarajevo : Izdavačka djelatnost, 1983
175p. – Includes bibliography
r6496135 X.958/27164
9147047

ADVERTISING AGENCIES
In-house advertising agencies

HOLTJE, Herbert F.
How to be your own advertising agency. – New York ; London : McGraw-Hill, c1981
viii,215p : ill. – (A James Peter Book). – Bibliography: p206–209. – Includes index
0–07–029665–0 X.520/25156
2569086

ADVERTISING AGENCIES. Advertising in supplements. Newspapers with British imprints

INCORPORATED SOCIETY OF BRITISH ADVERTISERS
Press supplements and the advertiser. – 2nd ed. – London : ISBA, 1977
8p
B81–18664 X.529/41294
2400863

ADVERTISING AGENCIES. Great Britain

The LEO Burnett book of advertising / edited by Simon Broadbent. – London : Business, 1984
199p : ill. – Includes index
0–09–155980–4 X.520/35992
3021254

ADVERTISING AGENCIES. Great Britain
– Directories

UNITED Kingdom advertising agents / [Data Research Group]. – 1975. – Amersham : The Group, [1975]
58 leaves in various pagings
Subsequent volumes, when received, will be placed at this shelfmark
0–905570–28–6 P.515/405
0765023

ADVERTISING AGENCIES. Great Britain
– Directories – Serials

CAMPAIGN portfolio. – [1982]–. – London : Marketing Publications, 1981–
v. : ill(chiefly col.)
B83–21470 P.515/473
2893541

ADVERTISING AGENCIES. Great Britain
to 1939

TREASURE, John
The history of British advertising agencies, 1875–1939. - Edinburgh : Scottish Academic Press, 1977
[2],20p. - (Jubilee lectures / University of Edinburgh. Commerce Graduates' Association ; 1976)
0-7073-0134-3 X.519/27080
1173677

ADVERTISING AGENCIES. Great Britain
Agreements with clients. Standard conditions

INSTITUTE OF PRACTITIONERS IN ADVERTISING
Some suggested provisions for use in agency-client agreements. - 5th ed. - London : The Institute, 1976
[2],27p
0-85294-091-2 X.512/5337
0725048

ADVERTISING AGENCIES. Great Britain
Attitudes of Scottish advertisers

CROSIER, Keith
The advertising iceberg. - Glasgow : Struthers Advertising Group, c1980
60p
B82-28338 X.529/48402
2752875

ADVERTISING AGENCIES. Great Britain
Clients. Changes

TOP thirty agencies, account gains & losses / [JWT Information Services]. - 1977. - London : J.W.T. Information Services, [1978]
[4],ii,91p
Subsequent volumes, when received, will be placed at this shelfmark
B80-16255 P.513/2482
2058693

ADVERTISING AGENCIES. Great Britain
Use in recruitment of personnel

INSTITUTE OF PRACTITIONERS IN ADVERTISING
Why you need a recruitment advertising agency. - London : The Institute, 1977
20p
B78-37805 X.519/28690
1705202

ADVERTISING AGENCIES. United States

GARDNER, Herbert S.
The advertising agency business. - Chicago : Crain Books, 1976
[7],211p. - Includes index
0-87251-025-5 X.520/12677
1513559

ADVERTISING AGENCIES. Western Europe
- Serials

CAMPAIGN Europe annual agency survey. - July 1979–. - London : Campaign Europe, 1979–
v. : ill
Holdings : 1980, etc
B81-04869 P.523/1442
2342677

ADVERTISING AGENCIES. Westminster (London Borough)
Ogilvy Benson and Mather Limited, to 1975

PIGOTT, Stanley
OBM. - London : Ogilvy Benson and Mather Limited, 1975
[11],84p. - Bibliography: p.81–84
0-9504899-0-5 X.529/30535
0613169

ADVERTISING ASSOCIATION
Role

ADVERTISING ASSOCIATION
Action for advertising. - London : [The Association], 1974
[1],10p
0-902878-12-3 X.519/26194
0739987

ADVERTISING FILMS
German advertising films, to date

WESTBROCK, Ingrid
Der Werbefilm. - Hildesheim : Olms, 1983
ii,118p.,plates. - (Studien zur Filmgeschichte ; Bd.1). - Includes bibliography
3-487-07453-2 X.950/37492
9162313

ADVERTISING INDUSTRIES

REEKIE, W. Duncan (William Duncan)
The economics of advertising. - London : Macmillan, 1981
xii,194p : ill. - Includes index
0-333-27204-8 X.529/44210
2126044

ADVERTISING INDUSTRIES
Attitudes of Catholic Church - Conference proceedings

ADVERTISING and the Vatican. - Brussels : European Association of Advertising Agencies, 1976
[1], 10, [1] p.
r6010761 X.100/16782
1053493
- [Another copy] X:519/39592

ADVERTISING INDUSTRIES
Management

PATTI, Charles H.
Advertising management. - New York ; Chichester : Wiley, c1978
xiv,298p : ill. - (Wiley series in marketing). - Includes bibliographies and index
0-471-86999-6 X.520/28130
2780011

ADVERTISING INDUSTRIES. England

MELLORS, John
Memoirs of an advertising man. - London : London Magazine Editions, 1976
220p
0-904388-16-6 X.529/30834
0687227

ADVERTISING INDUSTRIES. England
Ogilvy, David, b.1911 - Biographies

OGILVY, David, 1911-
Blood, brains and beer. - London : Hamilton, 1978
[7],118p. - Includes index
0-241-89874-9 X.809/43073
1203894

ADVERTISING INDUSTRIES. Europe
Direct-mail advertising industries - Directories

EUROPEAN direct mail databook. - Epping : Gower Press.
1976 : [2nd ed.]. - 1976
xi,344p : ill. - Includes index
Subsequent volumes, when received will be placed at this shelfmark
0-7161-0284-6 P.513/620
0645796

ADVERTISING INDUSTRIES. Great Britain

KLEINMAN, Philip
Advertising inside out. - London : W.H. Allen, 1977
viii,248p. - Bibliography: p.239–240. - Includes index
0-491-01678-6 X.529/31323
107539x

ADVERTISING INDUSTRIES. Great Britain
- Directories

The CREATIVE handbook. - 1976. - London : The Creative Handbook Ltd, 1976
316p : ill(some col)
Subsequent volumes, when received, will be placed at this shelfmark
0-903617-03-x P.441/809
0642843

ADVERTISING INDUSTRIES. Great Britain
- Statistics

ADVERTISING. - London : Euromonitor Publications, 1984
88p
0-86338-018-2 X.525/10250
107539x

ADVERTISING INDUSTRIES. Great Britain
Attitudes of advertisers, 1972

BRITISH MARKET RESEARCH BUREAU
Advertising and advertising agencies. - London : The Institute, 1972
[2],18 leaves
0-85294-092-0 X.512/5361
0735027

ADVERTISING INDUSTRIES. Great Britain
Direct-mail advertising industries - Directories

DIRECT mail databook. - 4th ed.. - Aldershot : Gower, 1983
x,296p. - Includes index
0-566-02386-5 X.525/9848
207835x

ADVERTISING INDUSTRIES. Great Britain
Organisations: Advertising Association. Role

ADVERTISING ASSOCIATION
Action for advertising. - London : [The Association], 1974
[1],10p
0-902878-12-3 X.519/26194
0739987

ADVERTISING INDUSTRIES. Great Britain
Proposals: Labour Party (Great Britain). 'Advertising'

ADVERTISING ASSOCIATION
A commentary on the Labour Party 'Green Paper' on advertising of March, 1972. - London : The Association, 1972
16p
0-902878-14-x X.519/26153
073179x

ADVERTISING MEDIA
See also
Advertisements
Cigarette cards
Display boards
Trade cards

ADVERTISING MEDIA

DAVIS, Martyn P. (Martyn Paul)
The effective use of advertising media. - London : Published in association with the Institute of Marketing and the CAM Foundation [by] Business Books, 1981
xxiv,322p. - Includes index
0-09-142970-6 X.520/23172
2174138
- [Another copy : paperback]
ISBN 0-09-142971-4 X.520/23173

ADVERTISING MEDIA
- Directories

OVERSEAS media guide / Overseas Press and Media Association. - 1978 : 12th ed. - [London] : [The Association], [1978]
[1],212p : ill
Subsequent editions, when received, will be placed at this shelfmark
B79-17715 P.901/3140
0025690

ADVERTISING MEDIA
Planning

ADAMS, James Ralph
Media planning. - 2nd ed. - London : Business Books [for the] Communication Advertising and Marketing Education Foundation, 1977
xiv,232p,[4]p of plates : ill. - Includes bibliographies and index
0-220-66335-1 X.981/21327
1516574
- [Another copy: Pbk]
ISBN 0-220-66335-1 X.900/25650

ADVERTISING MEDIA
Portrayal of women - Conference proceedings

The PACKAGING of women / ... report ... produced by members of Women in Media ... - London : W.I.M., 1976
[2],26,v p
0-9505435-0-0 X.519/26403
0788864

ADVERTISING MEDIA
Portrayal of women, to date

PIGNOTTI, Lamberto
Marchio & femmina. - Firenze : Vallecchi, 1978
204p : ill
r6500059 Cup.821.d.38.
8801754

WEISSER, Michael
Die Frau in der Reklame. - Münster : Coppenrath, [1981]
128p. : chiefly ill.. - Includes bibliography
3-88547-125-6 X.425/3911
8801754

ADVERTISING MEDIA. Great Britain
- For marketing - Forecasts

RAVEN, Walter
Promotional media in the next decade. - Bradford : MCB Business Futures Group ; Bradford : Distributed by MCB Books, 1975
99p. - (Delphi studies)
0-903763-09-5 X.510/15124
0611050

ADVERTISING MEDIA. Great Britain
Portrayal of women. Attitudes of women consumers

HAMILTON, Robert, 1937-
Adman and Eve. - Manchester : Equal Opportunities Commission, 1982
ii,41leaves. - Bibliography: p40-41
0-905829-52-2 BS.433/22
2747022

ADVERTISING MEDIA. Western Europe
- Directories - Serials

EUROPEAN media guide / JWT/Campaign Europe. - 1982-. - [London] : Campaign Europe, [1981]-
v. : ill
B82-11814 P.511/5316
2625067

ADVERTISING MEDIA. Western Europe
- Statistics

ADVERTISING in Western Europe. - London : Euromonitor, 1984
1v.(loose-leaf). - (European market surveys)
0-86338-059-x L.45/3367
3286649

ADVERTISING PHOTOGRAPHY
American advertising photography. Stern, Bert

STERN, Bert
The photo illustration. - London : Thames and Hudson, 1974
[1],97p : ill(chiefly col). - (Masters of contemporary photography)
0-500-54028-4 Cup.702.n.4
0217379

ADVERTISING PHOTOGRAPHY
Semiotic aspects

BURGIN, Victor
Two essays on art photography and semiotics. - [London] : Robert Self (Publications), 1976
24p : ill
B78-30774 X.415/5529 Woolwich
1575031

ADVERTISING POSTERS
to 1945. Special subjects: Rail travel

The GOLDEN age of the railway poster / [compiled by] J.T. Shackleton. - London : New English Library, 1976
128p : ill
0-450-02875-5 X.425/1320 Woolwich
1170406

ADVERTISING POSTERS
to ca 1960. Special subjects: North-east England. Coastal regions - Catalogues

POSTERS of the Yorkshire coast. - [Kingston upon Hull] : [Ferens Art Gallery], [1980]
16p : ill
B83-13981 X.425/3707
286455x

ADVERTISING POSTERS
to date. Special subjects: Music - Catalogues

15 ans d'affiches musicales (1964-1979) Musée de Durbuy 5, 6, 7 avril 1980 / Jean-Patrick Duchesne. - [Durbuy] : Musée de Durbuy, 1980
19leaves : ill
r6210394 X.435/1102
9013237

ADVERTISING POSTERS
1880-1900

MASTERPIECES of the poster / selected and edited by Hayward and Blanche Cirker. - New York : Dover : London : Constable, 1983
466p : all col.ill
0-486-24549-7 L.45/3285
3276481

ADVERTISING POSTERS
ca 1890-1937. Special subjects: Travel

TRAVEL posters / [selected by] Bevis Hillier. - Oxford : Phaidon, 1976
[16]p,96p of plates : chiefly ill(chiefly col)
0-7148-1623-x LR.425/125
0726001

ADVERTISING POSTERS
American advertising posters, to ca 1975

MARGOLIN, Victor
The promise and the product. - New York : Macmillan ; London : Collier Macmillan, 1979
viii,152p,[32]p of plates : ill(some col). - Bibliography: p.147-148. - Includes index
0-02-579480-9 X.622/8513 Woolwich
2269112

ADVERTISING POSTERS
American advertising posters, ca 1828-ca 1874. Special subjects: Theatre

[SPECIMENS of theatrical cuts. Selections]. Scenes from the 19th-century stage in advertising woodcuts / selected and edited by Stanley Appelbaum. - New York : Dover Publications [etc.] ; London : Constable, 1977
xvii,152p : chiefly ill. - (Dover pictorial archive series). - Includes index
0-486-23434-7 X.415/5447 Woolwich
1401807

ADVERTISING POSTERS
American advertising posters. Special subjects: Circuses, ca 1870-ca 1950

OLD time circus cuts / edited by Charles Philip Fox. - New York : Dover Publications [etc.] ; London : Constable, 1979
[5],119p : ill. - (Dover pictorial archive series)
0-486-23653-6 X.425/1688 Woolwich
2287587

ADVERTISING POSTERS
American advertising posters. Special subjects: Circuses, ca 1890-ca 1950

AMERICAN circus posters / edited by Charles Philip Fox. - New York : Dover Publications [etc.] ; London : Constable, 1978
[4],44p
0-486-23693-5 L.45/577
2226170

ADVERTISING POSTERS
French advertising posters, 1850-1950. Special subjects: Entertainments - Catalogues

WILD, Nicole
Les arts du spectacle en France. - Paris : Bibliothèque Nationale, 1976
454p,24p. of plates,4 leaves of plates : ill. - (Catalogues de la bibliothèque de l'opéra ; 2). - Includes bibliographies
2-7177-1323-9 X.0410/254(2) Woolwich
121375X

ADVERTISING POSTERS
French advertising posters, 1868-1930. Special subjects: Opera

FRENCH opera posters, 1868-1930 / [compiled by] Lucy Broido. - New York : Dover Publications [etc.] ; London : Constable, 1976
xxv p,52p of plates : chiefly ill(chiefly col). - Bibliography: p.vii. - Includes index
0-486-23306-5 X.415/5350 Woolwich
1197800

ADVERTISING POSTERS
French advertising posters, 1896-1900

MASTERS of the poster, 1896-1900 / preface by Roger Marx ; introduction by Alain Weill ; notes by Jack Rennert ; [translated from the French by Bernard Jacobson]. - London : Academy Editions, 1978
32p,[64]p of plates : chiefly ill(chiefly col)
B78-15117 X.415/5457 Woolwich
1412590

ADVERTISING POSTERS
French advertising posters. Chéret, Jules - Catalogues

BROIDO, Lucy
The posters of Jules Cheret. - New York : Dover ; London : Constable, 1980
xii,60p,46p of plates : ill(some col.). - Bibliography: p55. - Includes index
0-486-24010-x L.45/1006
2401991

ADVERTISING POSTERS
French advertising posters: Parisian advertising posters. Special subjects: Music hall, 1874-1974

WEILL, Alain
100 ans d'affiches des music halls parisiens. - Paris : Chêne, [1977]
112p : chiefly col. ill
2-85108-119-5 LR.430.c.11
153744X

ADVERTISING POSTERS
German advertising posters, to date

MUSEUM FÜR KUNST UND GEWERBE HAMBURG. Pädagogische Abteilung
Kunst-Kommerz und Körperpflege. - Hamburg : Das Museum, 1982
11postcards : col
r6305595 Cup.900.X.102
8765197

WEISSER, Michael
Söhnlein Rheingold. - Frankfurt/M. : Fricke, [1980]
126p : ill(some col.). - (Ästhetik der Alltagswelt ; Bd.[3])
3-88184-037-0 L.42/1359(3)
8765197

ADVERTISING POSTERS
German advertising posters, to date. Special subjects: Drama in German, to date

PLAKATANSCHLAG für Friedrich Schiller / herausgegeben von Frieder Mellinghoff ; mit einer Einführung von Cornelia Naumann. - Dortmund : Harenberg, [1980]
174p : ill. - (Die bibliophilen Taschenbücher ; 165)
3-88379-165-2 X.429/13290 Woolwich
8834512

ADVERTISING POSTERS
German advertising posters. Hohlwein, Ludwig

HOHLWEIN, Ludwig
Hohlwein posters in full color. - New York : Dover Publications [etc.] ; London : Constable, 1976
48p : of col ill
0-486-23408-8 Cup.1285/237
1271660

ADVERTISING POSTERS
London Transport advertising posters, 1908-1976

LONDON Transport posters / [compiled by] Michael F. Levey. - Oxford : Phaidon ; [London] : London Transport Executive, 1976
[16],80p : chiefly ill(chiefly col)
0-7148-1729-5 X.415/5153 Woolwich
0705179

ADVERTISING POSTERS
Outdoor advertising posters

HENDERSON, Sally
Billboard art. - London : Angus & Robertson, 1981, c1980
112p : ill(some col.). - Bibliography: p4
0-207-95973-0 L.49/1075
239779x

ADVERTISING POSTERS
Special subjects: Buffalo Bill's Wild West, 1885-1905

100 posters of Buffalo Bill's Wild West / [compiled] by Jack Rennert. - London : Hart-Davis MacGibbon, 1976
112p,2p of plates(2 fold) : chiefly ill(chiefly col). - (The poster art library)
0-246-10959-9 LR.425/204
1247360

ADVERTISING POSTERS
Special subjects: Cinema films, 1940-1980

WORST movie posters of all time / Gregory J. Edwards and Robin Cross. - London : Sphere, 1984
63p : chiefly col.ill. - (A Charles Herridge book)
0-7221-3229-8 X.425/5398
3284077

ADVERTISING POSTERS
Special subjects: Cinema films: Horror films, 1933-1973

SCIENCE-FICTION and horror movie posters in full color / edited by Alan Adler. - New York : Dover Publications [etc.] ; London : Constable, 1977
viii,40p : chiefly col ill
0-486-23452-5 Cup.1285/239
1271040

ADVERTISING POSTERS
Special subjects: Cinema films produced by Ealing Studios, 1940-1957

PROJECTING Britain / edited by David Wilson. - London : BFI, 1982
xiii,67p : col.ill. - Includes index
0-85170-122-1 X.429/14083
2844915

ADVERTISING POSTERS
Special subjects: Cinema films: Science fiction films, 1933-1973

SCIENCE-FICTION and horror movie posters in full color / edited by Alan Adler. - New York : Dover Publications [etc.] ; London : Constable, 1977
viii,40p : chiefly col ill
0-486-23452-5 Cup.1285/239
1271032

ADVERTISING POSTERS
Special subjects: Conjuring, 1867-1975

100 years of magic posters / [compiled] by Charles and Regina Reynolds. - London : Hart-Davis MacGibbon, 1976
112p : of ill(chiefly col)
0-246-10958-0 Cup.1285/57
064885x

ADVERTISING POSTERS
Special subjects: Dancing, 1879-1974

100 years of dance posters / [compiled] by Walter Terry and Jack Rennert. - London : Dempsey : [Distributed by] Studio Vista, 1975
112p : chiefly ill(chiefly col). - (The poster art library)
0-904518-06-x LR.425/92
062814x

ADVERTISING POSTERS
Special subjects: France. Paris. Music halls, 1870-1977

100 years of the posters of the Folies Bergère and music halls of Paris / [compiled by] Alain Weill. - London : Hart-Davis MacGibbon, 1977
112p : chiefly ill(chiefly col). - Bibliography: p.5
0-246-10960-2 Cup.1285/266
1333186

ADVERTISING POSTERS
Special subjects: Soviet cinema films, 1924–1933

REVOLUTIONARY Soviet film posters / [compiled by] Mildred Constantine and Alan Fern. – Baltimore : London : Johns Hopkins University Press, 1974
xiii,98p : ill(some col). – Bibliography: p.92–93. – Includes index
0–8018–1641–6 X.423/5133
0627399

ADVERTISING POSTERS
Special subjects: Theatre, to 1982

HAILL, Catherine
Theatre posters. – London : H.M.S.O., 1983
48p : col.ill
0–11–290418–1 X.421/25555
3141012

ADVERTISING POSTERS
Western European advertising posters, to date. Special subjects: Spas

Die REISE ins Bad / ausgewählt und mit einem Nachwort von Peter Graf von Wedel. – Dortmund : Harenberg, c1982
139p : chiefly ill. – (Die bibliophilen Taschenbücher ; 320)
3–88379–320–5 X.429/15405
9051554

ADVERTISING POSTERS. Exhibits. Bella C. Landauer Collection. New-York Historical Society
American advertising posters, ca 1842–1897

NEW-YORK HISTORICAL SOCIETY
American advertising posters of the nineteenth century. – New York : Dover Publications [etc.] ; London : Constable, 1976
viii,119p : chiefly ill(some col). – Includes index
0–486–23356–1 LR.425/203
1247859

ADVERTISING POSTERS. Exhibits. Welsh Arts Council
American advertising posters. Special subjects: Cinema films, to 1976 – Catalogues

WELSH ARTS COUNCIL
Selling dreams. – [Cardiff] : Welsh Arts Council, [1977]
[1],84p(6 fold) : chiefly ill(some col)
0–905171–20–9 BS.185/10
1123319

ADVERTISING POSTERS. Exhibits. Welsh Arts Council
British advertising posters. Special subjects: Cinema films, to 1976 – Catalogues

WELSH ARTS COUNCIL
Selling dreams. – [Cardiff] : Welsh Arts Council, [1977]
[1],84p(6 fold) : chiefly ill(some col)
0–905171–20–9 BS.185/10
1123300

ADVERTISING POSTERS. Exhibits. Welsh Arts Council
Special subjects: Circuses, 1923–1976 – Catalogues

The GREATEST show on earth. – [Cardiff] : [Welsh Arts Council], [1980]
[28]p(2 fold) : ill(some col)
B80–32127 X.429/11325 Woolwich
2242982

ADVERTISING POSTERS. Great Britain
Roadside advertising posters. Supply

GREAT BRITAIN. Monopolies and Mergers Commission
Roadside advertising services. – London : H.M.S.O., [1981]
v,130p. – ([H.C.] ; 365). – Includes index
0–10–236581–4 BS.83/136
2545381

ADVERTISING POSTERS. Great Britain
Sites

POSTER advertising site evaluation in Great Britain / prepared for Malcrest Advertising Ltd. – London : Economist Intelligence Unit. – (E.I.U. report)
Vol.2: Findings and conclusions. – 1981
53,xivleaves : ill
B82–02542 X.525/5129
258879x

ADVERTISING SERIALS
See Serials consisting of advertisements

ADVERTISING SIGNS
British enamel advertising signs, to ca 1950

BAGLEE, Christopher
More street jewellery. – London : New Cavendish, 1982
96p : ill(some col.). – Bibliography: p91. – Includes index
0–904568–39–3 L.42/1722
1495771

Street jewellery. – London : New Cavendish, 1978
87p : ill(some col.). – Bibliography: p84. – Includes index
B82–00679 X.425/2344
1495771

Street jewellery. – London : New Cavendish Books, 1978
3–88p : ill(some col). – Bibliography: p.84. – Includes index
0–904568–16–4 X.525/2237
1495771
– Reprinted. – 1979 X.425/2344

ADVERTISING SIGNS
Enamel advertising signs, to date

RIEPENHAUSEN, Axel
Blechplakate. – Münster : Coppenrath, [1979]
72p.,plates
3–920192–73–7 X.425/3912
8976694

ADVERTISING SIGNS
Swedish advertising signs, to date

CONRADSON, Birgitta
Svenska skyltar. – Stockholm : LT, 1980
83,16p of plates : ill(some col.)
91–360–1442–7 X.800/34147
8874263

ADVERTISING SIGNS. Hong Kong
to date

COSTANTINI, Frank
Hong Kong sign book. – [Hong Kong] : [South China Morning Post], [1981]
74p : chiefly ill
r6807281 X.429/15280
9045422

ADVERTISING WITHOUT REVEALING BUSINESS INTERESTS. Advertising of goods by retailers. Great Britain
Law. Reform

GREAT BRITAIN. Consumer Protection Advisory Committee
Disguised business sales. – London : H.M.S.O., [1976]
iii,11p. – ([1975–76 H.C.] 355)
0–10–235576–2 BS.419/9
0317020

ADVICE
See also
Scientific advice
Technical advice

ADVICE. Housing. Homeless persons. London
Provision. Role of SHAC, 1975–1977

RICHARDSON, Paul, 19—
The homeless in London. – London : SHAC (the London Housing Aid Centre), 1980
60p. – (Research report / SHAC ; 2)
B81–10724 X.525/4084
2372363

ADVICE & INFORMATION SERVICES FOR OLD PERSONS. England
Organisation

AGE CONCERN
Advice and information services. – [Mitcham] : Age Concern, 1979
[2],20p. – (Action guides / Age Concern)
0–904502–76–7 W.P.15381
2066084

ADVICE CENTRES
See also
Law centres

ADVICE CENTRES
Great Britain. Immigration. Entry controls. Law – For advice centres

YOU don't have to be a lawyer to help someone being threatened with immediate arrest, detention or expulsion under the Immigration Act. – Manchester : Manchester Law Centre, [1982?]
56p : ill. – (Manchester Law Centre immigration handbook ; no.4)
B82–17760 X.200/38080
270482x

ADVICE CENTRES. Braunstone. Leicestershire
Community advice centres: Braunstone Advice Centre, to 1974

BROWSE, Barbara
Braunstone Advice Centre. – Leicester : Leicester Family Service Unit, 1976
[3],35p
0–905815–00–9 X.512/5477
079693x

ADVICE CENTRES. Castlemilk. Glasgow. Strathclyde Region. Scotland
Community advice centres: Castlemilk Advice Centre

CASTLEMILK LAW CENTRE
Castlemilk Advice Centre. – Glasgow : Castlemilk Law Centre, 1983
58p
B84–29744 X.805/5573
3136353

ADVICE CENTRES. Central Lancashire New Town

ENQUIRE within. – Preston : Lancashire Library, 1981
xi,68p. – Includes index
0–902228–42–0 X.525/5155
2594897

ADVICE CENTRES. Cleator Moor. Cumbria
Community advice centres: Cleator Moor Community Information and Action Centre, to 1974

COMMUNITY information and action centre / [written by Hugh Butcher et al.]. – Heslington : University of York, Department of Social Administration and Social Work, CDP Unit, [1974]
94,[17]p : ill. – (Papers in community studies ; no.1)
B79–07158 X.512/6391
1804308

ADVICE CENTRES. Cleator Moor region. Cumbria
Mobile community advice centres. Provision

BUTCHER, Hugh
Information and action services for rural areas. – [Heslington] : [University of York, Department of Social Administration and Social Work, CDP Unit], 1978
[6],79p : ill. – (Papers in community studies ; no.4)
B79–08624 X.512/6404
1811738

ADVICE CENTRES. England
Family advice centres. Role in intermediate treatment of juvenile delinquency

NATIONAL CHILDREN'S BUREAU
Intermediate treatment. – London : The Bureau, 1977
viii,86p : ill. – Includes bibliographies
0–902817–11–6 X.510/15516
1171275

ADVICE CENTRES. England
Housing advice centres: Shelter. National Housing Aid Trust, to 1982

HOUSING aid facts & figures / [editors Jane Ansell & Roger Matthews]. – [London] : Shelter National Housing Aid Trust, c1983
23p : ill
0–85265–995–4 X.525/8463
2986280
– [Another copy.] X.525/8880

ADVICE CENTRES. Ferguslie Park. Paisley. Strathclyde Region. Scotland
Community advice centres: Ferguslie Park Information and Action Centre

PAISLEY COMMUNITY DEVELOPMENT PROJECT
The Information and Action Centre. – Paisley ; Glasgow : [Distributed by] Administration Department, Strathclyde Regional Council, 1978
[3],30p
B79–08623 X.702/5938
1815245

ADVICE CENTRES. Great Britain

NATIONAL CONSUMER COUNCIL
The fourth right of citizenship. – London : The Council, 1977
[2],iv,80p. – (Discussion paper ; 4)
0–905653–10–6 X.519/27397
1268945

PRICE, Stephanie
Information support for independent advice centres. – London : Lambeth Umbrella Group, 1984
58p. – (British Library research and development report ; 5813)
B85–17987 X.525/10747
1268945

ADVICE CENTRES. Great Britain
– Directories

A DIRECTORY of independent advice centres / compiled by the Community Information Project. – London : National Consumer Council, 1984
72p
0–905653–79–3 X.525/9893
1268945

ADVICE CENTRES. Great Britain
Housing advice centres
HARLOE, Michael
Housing advice centres. - London : Shelter : Centre for Environmental Studies, 1976
28p : 1 ill
B78-02157 X.512/5743
1284584

ADVICE CENTRES. Great Britain
Role in representation of appellants at social security tribunals
LAWRENCE, Roger
Tribunal representation. - London : Published in association with the National Association of Citizens Advice Bureaux by the Bedford Square Press of the National Council for Voluntary Organisations, 1980
vii,98p
0-7199-1044-7 X.525/4764
2528517

ADVICE CENTRES. Italy
Family advice centres, to date
CONSULTORIO familiare e bisogni sociali / a cura di Pierpaolo Donati ; con la collaborazione di Paola Di Nicola. - Milano : Angeli, [1980]
248p. - (Servizi sociali e territorio ; 4). - Bibliography
88-204-1694-8 X.529/47651
8810710

RONFANI, Paola
I consultori familiari pubblici. - Milano : Giuffrè, 1979
viii,204p. - (Studi e ricerche di sociologia del diritto ; n.1)
r6501169 X.520/27678
8810710

ADVICE CENTRES. Italy
Family advice centres, to date - Conference proceedings
ANALISI E PROSPETTIVE DEL CONSULTORIO FAMILIARE PUBBLICO (Conference : 1980 : Venice)
Analisi e prospettive del consultorio familiare pubblico. - Venezia : Marsilio, 1981
189p. - (Ricerche ; n.79)
r6512327 X.529/69345
8810710

ADVICE CENTRES. Kentish Town. Camden (London Borough)
Community advice centres: Camden Neighbourhood Centre
BAKER, John, 1945-
The neighbourhood advice centre. - London : Routledge and Kegan Paul, 1978
ix,310p. - (Reports / Institute of Community Studies). - Bibliography: p.292-296. - Includes index
0-7100-8731-4 WP.4411/27
1482912

ADVICE CENTRES. London
Housing advice centres: SHAC
SHAC
SHAC. - London : SHAC.
75/77. - [1979]
[2],20p : ill
Subsequent volumes, when received, will be placed at this shelfmark
B79-22311 P.523/1213
1903306

ADVICE CENTRES. London
Housing advice centres: SHAC, to 1972
SHAC
SHAC, the first three years. - London : Shelter Housing Aid Centre, [1973?]
[47]p : ill
B79-26284 X.619/19969 Woolwich
1926594

ADVICE CENTRES. London
Housing advice centres: SHAC, to 1980
SHAC
A decade of housing aid. - London : SHAC (The London Housing Aid Centre), [1980]
33p : ill
B81-16308 X.525/7054
2391694

ADVICE CENTRES. London
Housing advice centres: SHAC. Inquiries - Statistics
SHAC
SHAC - statistics. - 1977. - London : SHAC, [1978]
[1],15p : ill
Subsequent volumes, when received, will be placed at this shelfmark
B79-22310 P.523/1212
1903314

ADVICE CENTRES. London
Housing advice centres: SHAC. Role in provision of housing advice for homeless persons, 1975-1977
RICHARDSON, Paul, 19—
The homeless in London. - London : SHAC (the London Housing Aid Centre), 1980
60p. - (Research report / SHAC ; 2)
B81-10724 X.525/4084
2372363

ADVICE CENTRES. Moss Side. Manchester
Community advice centres: Moss Side Family Advice Centre, to 1978
The FAC book. - Manchester : Youth Development Trust, c1981
iv,77p : ill
0-903178-04-4 X.520/25409
2579731

ADVICE CENTRES. North Kenton. Newcastle. Tyne and Wear (Metropolitan County)
Community advice centres: North Kenton Advice Centre, to 1979
PHILLIPS, Hilary
Two years on. - [London] : Dr Barnardo's, c1980
48p : ill. - (Barnardo social work paper ; no.10)
B81-16029 X.529/41007
2390183

ADVICE CENTRES. Terminals. Passenger transport services. London
Welfare advice centres
HELP for the other travellers. - London : London Council of Social Service, [1976]
8p
0-901171-21-2 X.512/5176
0640549

ADVICE COLUMNS
to 1978
KENT, Robin
Aunt Agony advises. - London : W.H. Allen, 1979
[6],265p
0-491-02476-2 X.529/33503
1907786

ADVICE COLUMNS. Annabella
GASPERINI, Brunella
[Correspondence. Selections]. Più botte che risposte. - Milano : Rizzoli, 1981
219p. - (La Scala)
r6509913 X.950/26507
9058109

ADVICE COLUMNS. Girl's own paper
1880-1883
OH, no dear! / selected and compiled by Roy Hindle. - Newton Abbot : David & Charles, c1982
64p
0-7153-8330-2 X.529/51351
282051x

ADVICE COLUMNS. Manchester evening news - Biographies
CRITCHLOW, Dorothy
'Dear Jane Dawson - '. - London : Hale, 1977
192p,[8]p of plates : ill. - Includes index
0-7091-6363-0 X.981/20854
1272071

ADVICE COLUMNS. 'TV Times' - Collections
BOYLE, Katie
Boyle's lore. - London : Weidenfeld and Nicolson, 1982
192p : ill. - Includes index
0-297-78148-0 X.629/19176
2824868
[Boyle's lore]. Dear Katie. - London : Corgi, 1984, c1982
288p : ill. - Includes index
0-552-99078-7 X.629/25203
2824868

ADVICE COLUMNS. 'Woman' - Biographies
MAKINS, Peggy
The Evelyn Home story. - London : Collins, 1975
224p
0-00-211223-x X.989/40422
0573469
The Evelyn Home story. - [London] : Fontana, 1976
224p
0-00-634571-9 X.519/28134
0573469

ADVISERS
See also
Economic advisers
Fiduciaries
Political advisers
Scientific advisers

ADVISERS. China
European advisers. Role in administration, to date
SCHMIDT, Vera
Aufgabe und Einfluss der europäischen Berater in China. - Wiesbaden : Harrassowitz, 1984
viii,178p. - (Veröffentlichungen des Ostasien-Instituts der Ruhr-Universität Bochum ; Bd.34). - Includes bibliography
3-447-02483-6 X.0700/678(34)
9166041

ADVISERS. Companies. Great Britain
Professional advisers. Selection
CHOOSING and using professional advisers / edited by Paul Chaplin. - Hemel Hempstead : Enterprise, 1983
viii,120p : ill. - (Small business success ; bk.1). - Bibliography: p113-115. - Includes index
0-906896-06-1 X.529/64729
2927624

HARRISON, Brian Fraser
How to select your professional advisers. - Chester : B Harrison, c1982
48p
B83-30874 X.529/55203
2927624

ADVISERS. Personnel. Local education authorities. Great Britain
Remuneration - Serials
COMMITTEE ON SALARY SCALES AND SERVICE CONDITIONS OF INSPECTORS, ORGANISERS AND ADVISORY OFFICERS OF LOCAL EDUCATION AUTHORITIES
The Report of the Committee on Salary Scales and Service Conditions of Inspectors, Organisers and Advisory Officers of Local Education Authorities. - [16th (1980)]. - London : [The Committee], 1980
27p
B81-12418 X.529/40560
2377535

ADVISERS. Personnel. Social services departments. Local authorities. England
Role
BRUNEL UNIVERSITY. Social Services Unit
Advisors, development officers and consultants in Social Service Departments. - [Uxbridge] : Brunel Institute of Organisation and Social Studies, 1977
36,[10]p
0-902215-37-x X.512/6588
1871285

ADVISERS. Wilhelm II, Emperor of Germany
Personal advisers
HULL, Isabel V.
The entourage of Kaiser Wilhelm II. - Cambridge : Cambridge University Press, 1982
xii,413p. - Bibliography: p387-402. - Includes index
0-521-23665-7 X.800/33991
2780550

ADVISORY COMMITTEE ON LOCAL GOVERNMENT
See Great Britain. Advisory Committee on Local Government Audit

ADVISORY COMMITTEE ON THE LANDSCAPE TREATMENT OF TRUNK ROADS
See Great Britain. Advisory Committee on the Landscape Treatment of Trunk Roads

ADVISORY COMMITTEES. Conservation areas. England
Role
RAY, Patricia
Conservation area advisory committees. - London : Civic Trust, 1978
31p
0-900849-90-8 X.702/5671
1449850

ADVISORY COMMITTEES. European Community
Socioeconomic advisory committees
COMMUNITY advisory committees for the representation of socio-economic interests / Economic and Social Committee of the European Communities, General Secretariat Directorate-General Studies and Documentation Division. - Farnborough, Hants. : Saxon House, c1980
xvii,215p. - Bibliography: pxii-xiv
0-566-00328-7 X.529/40358
1941208

ADVISORY COMMITTEES. Special education. Handicapped children. Great Britain
A NATIONAL advisory committee for special educational needs?. - London : Published by the National Children's Bureau for the Voluntary Council for Handicapped Children, c1984
v,46p
0-902817-23-x X.529/69989
3284441

ADVISORY COMMITTEES. Vocational education institutions. United States

COCHRAN, Leslie H.
Advisory committees in action. - Boston, Mass. ; London : Allyn and Bacon, 1980
xv,336p : ill. - Bibliography: p.325-333. - Includes index
0-205-06661-5 X.525/3153
211058x

ADVISORY COMMITTEES. Vocational education institutions. United States
Organisation

RIENDEAU, Albert J.
Advisory committees for occupational education. - New York ; London : Gregg Division, McGraw-Hill, 1977
vi,90p. - Bibliography: p.88. - Includes index
0-07-052680-x X.519/27364
1263668

ADVISORY, CONCILIATION AND ARBITRATION SERVICE

ACCOUNTS for the period ended 31 March / Advisory, Conciliation and Arbitration Service [and] Certification Officer [and] Central Arbitration Committee. - [1975]-1976. - London : H.M.S.O, [1976]
Folder (4p)
Subsequent volumes, when received, will be placed at this shelfmark
0-10-009156-3 BS.23/849
1043064

ADVISORY, CONCILIATION AND ARBITRATION SERVICE
Annual report. - 1st : 1975. - London : H.M.S.O, 1976
[5],56p,plate : ill
Subsequent reports, when received, will be placed at this shelfmark
0-11-881105-3 BS.427/1
0648469

ADVISORY, CONCILIATION AND ARBITRATION SERVICE to 1984

ARMSTRONG, E. G. A. (Eric George Abbott)
Improving industrial relations. - London : Croom Helm, c1985
229p : ill. - Includes index
0-7099-0554-8 X.529/70145
3236706

ADVISORY COUNCIL ON THE MISUSE OF DRUGS
Report on a review of the classification of controlled drugs

LEGALISE CANNABIS CAMPAIGN
Trash rehashed. - London : Legalise Cannabis Campaign, 1979
46p,1folded leaf of plates : ill
0-906497-01-9 X.329/14717 Woolwich
2589117

ADVISORY SERVICES
See also
Advice & information services for old persons
Advice centres
Technical units

ADVISORY SERVICES
Administration of higher education by local authorities: National Advisory Body for Local Authority Higher Education

BEVAN, J. S. (John S)
The National Advisory Body for local authority higher education. - [Sheffield] : Association of Colleges for Further and Higher Education, [1982]
4p
B82-37173 X.529/49806
2782251

ADVISORY SERVICES. Adult education. Bradford. West Yorkshire (Metropolitan County)
Advisory, information & counselling services: Education Advice Service for Adults. Information retrieval services using digital computer systems. Machine-readable files

TILLEY, N. V.
Education Advice Service for Adults. - [London] : British Library, c1982
vi,53p : ill. - (Library and information research report ; 8)
0-7123-3011-9 2719.e.625
2935449

ADVISORY SERVICES. Adult education. England
Advisory, information & counselling services

ADVISORY COUNCIL FOR ADULT AND CONTINUING EDUCATION
Links to learning. - Leicester : ACACE, c1979
70p. - (Report / Advisory Council for Adult and Continuing Education)
0-906436-02-8 X.529/50277
2791145

ADVISORY SERVICES. Adult education. Great Britain
Advisory, information & counselling services

BUTLER, Linda
Case studies in educational guidance for adults. - Leicester : Published jointly by National Institute of Adult Continuing Education and Advisory Council for Adult and Continuing Education, 1984
x,172p
0-900559-48-9 X.529/66416
2591502

EDUCATIONAL information and guidance for adults / edited by Roger H. Mercer. - [Sheffield] : Sheffield City Polytechnic, Department of Education Management, 1981
106p. - (Sheffield papers in education management ; 16)
0-903761-31-9 X.529/44762
2591502

ADVISORY SERVICES. Adult education. Great Britain
Advisory, information & counselling services - Bibliographies

MARKS, Harold
Bibliography of educational information advisory & counselling services for adults in the United Kingdom. - London : London Association for Coutinuing Education, 1979
[1],38p
0-907009-00-x X.529/37345
218429x

ADVISORY SERVICES. Adult education. United States
Advisory, information & counselling services

HEFFERNAN, James M.
Educational and career services for adults. - Lexington, Mass. : Lexington Books, 1981
xvi,272p. - Bibliography: p255-266. - Includes index
0-669-03440-1 X.520/25386
2579634

ADVISORY SERVICES. Alyn and Deeside (District). Clwyd
Community advisory services. Provision. Implications of unemployment

ASTBURY, Raymond
A survey of information, advice and counselling providers in North-East Clwyd, February-September 1981. - [Mold] : Clwyd Information Providers Steering Group, c1982
24p
0-904444-77-5 X.529/50492
2798956

ADVISORY SERVICES. Companies. Great Britain

COLLINS, John, 19—
A study paper on sources of advice and information for small businesses in the United Kingdom. - Bristol : Bristol Polytechnic, South West Regional Management Centre, 1979
48,[18]p. - Bibliography: p66
0-904951-03-0 X.525/5906
275035x

ADVISORY SERVICES. Companies. Great Britain - Directories

SOURCES of assistance to small firms in the East Midlands / compiled by the Department of Trade and Industry. - [2nd ed.]. - [London] : [The Department], [1983]
iv,33p
B84-33622 BS.41/1399
275035x

ADVISORY SERVICES. Consumer protection. Rural regions. Scotland

ATHERTON, Graham
A call for advice. - Glasgow : Scottish Consumer Council, 1983
113p : ill
0-907067-14-x X.525/8168
2965437

ADVISORY SERVICES. Government. United States to date

ENCYCLOPEDIA of governmental advisory organizations / Denise Allard Adzigian, editor, Donna Batter, associate editor. - 4th ed. - Detroit : Gale Research Co, c1983
964p
0-8103-0254-3 973.0061 OPL
9169709

ADVISORY SERVICES. Government. United States to date - Encyclopaedias

NEW governmental advisory organizations. - Issue no.1-. - Detroit : Gale Research Co, [1984]-
v.
r6815146 973.0061 OPL
9169709

ADVISORY SERVICES. Great Britain
Community advisory services. Personnel. Professional education

REID, Julia M.
Education and training for community information and advice work. - [Sheffield] : University of Sheffield, Department of Information Studies, 1982
154p. - (Occasional publications series / University of Sheffield Department of Information Studies ; no.1). - Bibliography: p91-97
0-903522-12-8 X.525/6567
2824914

ADVISORY SERVICES. Great Britain
Community advisory services. Provision

INFORMATION and advice services in the United Kingdom. - [London] : [National Consumer Council], [1983]
i,114p
0-905653-75-0 X.525/8065
2866943
- [Another copy] 361.942 OPL

ADVISORY SERVICES. Great Britain. Ministry of Agriculture, Fisheries and Food
Personnel - Biographies

KINLOCH, Ian
Don't forget your wellies. - Ely : Providence, 1983
145p : ill
0-903803-09-7 X.950/31857
3120686

ADVISORY SERVICES. Inner areas. Birmingham. West Midlands (Metropolitan County)
Community advisory services. Provision

LOCAL advice needs in the Birmingham inner city partnership area. - [Birmingham] : [Birmingham Citizens' Advice Bureau], 1981
iv,70p
B81-36912 8299.m.4
2550970

ADVISORY SERVICES. Local education authorities. England
Role in educational innovation

BOLAM, R. (Raymond)
Local education authority advisers and the mechanisms of innovation. - Windsor : NFER Publishing, 1978
viii,247p. - Bibliography: p241-247
0-85633-175-9 X.529/44148
1265318

BOLAM, Raymond
Local education authority advisers and educational innovation. - [Bristol] : University of Bristol, School of Education, 1976
419p in various pagings : ill. - Bibliography: (9p.)
B77-32567 X.525/1331
1265318

ADVISORY SERVICES. National health services. Great Britain

DOLTON, Duncan
Medical advisory machinery. - Manchester : Health Services Management Unit, University of Manchester, 1980
[1],26,2 leaves. - (Working paper series / University of Manchester. Health Services Management Unit ; no.28)
B80-40561 X.325/709 Woolwich
2319934

ADVISORY SERVICES. Paediatrics
Use of telephones

BROWN, Jeffrey L.
Telephone medicine. - St. Louis ; London : Mosby 1980
xi,154p. - Includes index
0-8016-0856-2 X.329/13484 Woolwich
234937x

KATZ, Harvey P.
Telephone manual of pediatric care. - New York ; Chichester : Wiley, c1982
xii,184p : ill. - Includes bibliographies
0-471-09923-6 X.322/11743
234937x

ADVISORY SERVICES. Personnel. Residential institutions for children & day care centres for children. Church of England Children's Society

STONE, Winifred
The nature and function of the advisory service for residential and day care staff. - London : Church of England Children's Society, [1979]
[25]p : ill. - (Child care study paper ; no.2)
B79-35289 X.512/7030
1969005

ADVISORY SERVICES. Public participation in environment planning. England
Town and Country Planning Association. Planning Aid Service
CURTIS, B. (Brian)
Planning aid. - Reading : School of Planning Studies, University of Reading, 1980
x,142p. - (Occasional papers / School of Planning Studies, University of Reading ; OP1). - Bibliography: p139-142
0-7049-0750-x X.525/3871
2322994

ADVISORY SERVICES. Sheffield. South Yorkshire (Metropolitan County)
SHEFFIELD INFORMATION AND ADVICE SERVICES WORKING PARTY
Report on information and advice provision in Sheffield. - [Sheffield] : The Working Party, 1980
205p : ill. - Bibliography: p.195-199
B80-40141 X.622/8840 Woolwich
2275007

ADVISORY SERVICES. Sheffield. South Yorkshire (Metropolitan County)
Community advisory services. Provision
SHEFFIELD INFORMATION AND ADVICE SERVICES WORKING PARTY
Report on information and advice provision in Sheffield. - [London] : [The Working Party], [1980]
[2],19p
B80-40553 X.622/8841 Woolwich
2315122

ADVISORY SERVICES. Use. Concrete
Wigmore, Victor S. - Biographies
WIGMORE, Victor S.
Memoirs of a concrete consultant. - Bognor Regis : New Horizon, c1979
218p. - Includes index
0-86116-127-0 X.629/15586 Woolwich
2484056

ADVISORY SERVICES. Use of computer & telecommunication systems in administration of government departments. Great Britain
Central Computer and Telecommunications Agency. Technical Services Organisation. Role
The TECHNICAL Services Organisation of Central Computer and Telecommunications Agency. - London : Civil Service Department, 1981
106p
0-7115-0033-9 BS.6/172
2561662

ADVISORY SERVICES FOR ADOLESCENTS, 16-18 YEARS. Great Britain
Provision
MILLER, John C. (John Charles)
Towards a personal guidance base. - [London] : [Further Education Unit], 1983
vii,61p. - (A Project report / Further Education Unit). - Bibliography: p60-61
0-946469-01-6 X.525/7918
2949342

ADVISORY SERVICES FOR AGRICULTURAL INDUSTRIES. Great Britain
Policies of European Economic Community
JONES, Gwyn Evans
Socio-economic advisory work. - [Reading] : University of Reading, Agricultural Extension and Rural Development Centre, 1977
98[i.e.99],[19]p : 1 ill
0-7049-0511-6 X.515/619
1812599

ADVISORY SERVICES FOR AGRICULTURAL INDUSTRIES. Ireland (Republic)
Organisation
PROGRAMME development handbook. - Blackrock : An Chomhairle Oiliuna Talmhaiochta, c1982
v,35p : ill
B84-05160 X.525/8477
2987856

ADVISORY SERVICES FOR AGRICULTURAL INDUSTRIES. Ireland (Republic)
Reorganisation
IRISH FARMERS' ASSOCIATION
Proposed operational system for agricultural advisory, education & training services & links with research service. - Dublin : The Association, [1978?]
[2],42p : ill
B80-25320 X.329/12666 Woolwich
2221616

ADVISORY SERVICES FOR CONSTITUENTS. Members. Great Britain. Parliament. House of Commons
MARSH, James W.
The value of the constituency Member of Parliament. - Hull : Department of Politics, University of Hull, 1983
26p. - (Hull papers in politics ; no.30)
B84-06520 X.525/8505
3036685

ADVISORY SERVICES FOR DEBTORS. Citizens' advice bureaux. West Midlands (Metropolitan County)
GIBBONS, Sheila
Debt in the recession. - Birmingham : National Association of Citizens Advice Bureaux, 1983
vi,106p. - (CAB occasional paper series ; no.15)
B83-29953 X.525/7533
2924846

ADVISORY SERVICES FOR ECONOMIC PLANNING. Great Britain
Economist Advisory Group - Serials
ECONOMISTS ADVISORY GROUP
EAG news. - [No.1]- ; 1977-. - London : Economists Advisory Group, [1977]-
Holdings: Nov / Dec. 1978, etc
B80-23086 P.513 / 2517
2161079

ADVISORY SERVICES FOR ENVIRONMENTAL HEALTH. England
Public Health Advisory Service
FARBER, Andrew Seth
A study of the effectiveness of the Public Health Advisory Service, London, England. - [London] : [The author], 1976
[1], iii, 30 p.
r6010112 X.313/1342 Woolwich
0768286

ADVISORY SERVICES FOR INDUSTRIES. London
London Industrial Centre - Serials
LONDON INDUSTRIAL CENTRE
LIC news. - No.1- ; Jan. 1978-. - London : London Industrial Centre, 1978- : ill
Subsequent volumes, when received, will be placed at this shelfmark
B79-32338 P.511/5148
1951777

ADVISORY SERVICES FOR PERSONS WITH PERSONAL PROBLEMS. Great Britain
ADVICE services in welfare rights / editor Rosalind Brooke. - London : Fabian Society, 1976
[2],36,[1]p. - (Fabian research series ; 329)
0-7163-1329-4 WP.10405/329
0754234

ADVISORY SERVICES FOR PREGNANT WOMEN. West Germany
JUNGERMANN, Helmut
Beratung bei Schwangerschaftskonflikten. - Stuttgart : Kohlhammer, 1981
167p. - (Schriftenreihe des Bundesministers für Jugend, Familie und Gesundheit ; Bd 122). - Bibliography
3-17-007254-4 S.F.467/13[Bd.122]
894315x

ADVISORY SERVICES FOR SOCIALLY DISADVANTAGED PERSONS. Stockport (District). Greater Manchester (Metropolitan County)
Voluntary advisory services for socially disadvantaged persons: AID. Organisation
HAMPSON, Fredrick Roy
AID. - [Stockport] : [Elgincourt Office Supplies for AID], [1976]
[3],16 leaves,6 leaves of plates
0-9505259-0-1 X.512/5367
0741272

ADVISORY SERVICES FOR SUBCULTURES. England
Release (Association) - Serials
RELEASE (Association)
Monthly newsletter. - Vol.1, no.1-v.2, no.13 ; May 1974-Nov. 1975. - London : Release, [1974]-[1975]
B78-23306 P.523/831
1495240
News Release. - Vol.4, no.2- ; [July-Sept. 1977]-. - London : Release Publications Ltd, 1977- : ill
B78-23307 P.523/831
1495240
Release newsletter. - Vol.3, no.2- v.4, no.1 ; [1976-1977]. - London : Release, [1976]-[1977] : ill
B78-23308 P.523/831
1495240

ADVISORY SERVICES FOR YOUNG PERSONS. Great Britain
LAWTON, Anthony
Youth counselling matters. - Leicester : National Youth Bureau, 1984
27p. - Bibliography: p27
0-86155-080-3 X.525/10575
1498142

TYLER, Mary, b.1921
Advisory and counselling services for young people. - London : H.M.S.O., 1978
x,93p : ill. - (Research reports / Great Britain. Department of Health and Social Security ; no.1). - Bibliography: p.81-93
0-11-320675-5 BS.385/380(1)
1498142

ADVISORY SERVICES FOR YOUNG PERSONS WITH SEXUAL PROBLEMS
London
- Directories
HELP with sexual problems / by David Heyman ... [et al.]. - Leicester : National Youth Bureau, 1979
16p : ill
0-86155-008-0 Cup.351/169
1892738

ADVOCACY
See also
Self-advocacy

ADVOCACY. Criminal law. England
CROWTHER, Eric
Advocacy for the advocate. - Chichester : Rose, 1984
ix,154p. - Includes index
0-85992-351-7 X.200/46754
330003x

ADVOCACY. Criminal law. Scotland
HARPER, J. Ross
My client My Lord. - Edinburgh : Law Society of Scotland, 1981
61p
0-902023-03-9 X.200/35115
2374110

ADVOCACY. Interpersonal relationships of nurses with patients
KOHNKE, Mary F.
Advocacy. - St Louis ; London : Mosby, 1982
xii,204p : ill. - Bibliography: p190-193. - Includes index
0-8016-2721-4 X.520/28523
2812916

ADVOCACY. Law
DU CANN, Richard
The art of the advocate. - [1st ed.] reprinted with revisions. - Harmondsworth : Penguin, 1980
186p. - (A pelican original)
0-14-020665-5 X.108/17834
0372455

The GREAT power at the Bar and Bench. - Quezon City : [the Author], 1972
xliii,1273p[17] leaves of plates
r6025823 X.200/11102
0372455

NAPLEY, David
The technique of persuasion. - 2nd ed. - London : Sweet and Maxwell, 1975
xiv,160p
0-421-20850-3 X.519/25164
0372455

NAPLEY, Sir David
The technique of persuasion. - 3rd ed. - London : Sweet & Maxwell, 1983
xvi,187p : ill
0-421-25280-4 X.529/58794
0372455

ADVOCACY. Magistrates' courts. England
CHATELIER, Trevor M.
Criminal advocacy in the magistrates' court. - Chichester : Rose, c1981
53p
0-85992-219-7 X.200/35484
2401673

ADVOCATES. India (Republic)
Professional conduct, to 1973
DUTT-MAJUMDAR, Nirmalendu
Conduct of Advocates and legal profession. - Calcutta : Eastern Law House, 1974
96 p.
r6025800 X.200/10675
1018159

ADVOCATES. Scotland
Services. Supply. Effects of restrictions on advertising by advocates
GREAT BRITAIN. Monopolies and Mergers Commission
Advocates' services. - London : H.M.S.O., [1976]
[4],23,liv p. - ([1975-1976 H.C.]560)
0-10-256076-5 BS.83/95
0734926

ADVOCATES. Soviet Union
Role in administration of criminal justice

STEȚSOVSKII, ÎU. I. (ÎUrii Isaakovich)
Advokat v ugolovnom sudoproizvodstve. - Moskva : ÎUridicheskaîa literatura, 1972
157p
r6462729 X.200/42337
9007393

'ADWIYYAH, RABI'AH AL-
See Rabi'ah al-'Adawiyyah

ADY, ENDRE. Hungarian literature

BÖLÖNI MÁRKUS, Ottilia
Ady és Bölöni Párizsában. - Budapest : Petőfi Irodalmi Múzeum . Népművelési Propaganda Iroda, 1974
207 p.. - (Irodalmi Múzeum)
r6023260 X.0909/436(5)
0888443

ADY, ENDRE. Poetry in Hungarian

ADY, Endre
Életem nyitott könyve. - Budapest : Gondolat, 1977
528p
963-280-492-9 X.989/73712
121439X

ADYTÓL Adyról / a kötetet összeállította Fenyvesi Margit. - Budapest : Tankönyvkiadó, 1977
378p : ill
963-17-2563-4 X.989/80725
121439X

BABITS, Mihály
Babits Adyról. - Budapest : Magvető, 1975
280p,[4]p of plates
963-270-125-9 X.989/71417
121439X

BARÓTI, Dezső
Ady Endre. - Budapest : Magyar Helikon, 1977
168p[40]p of plates : ill. - (Kézirattár)
963-207-364-9 X.902/3150
121439X

BÖLÖNI MÁRKUS, Ottilia
[Selections]. Ady Párizsban. - Bukarest : Kriterion Könyvkiadó, 1977
296p. - (Téka)
r6455222 X.958/28807
8356521

DÉNES, Zsófia
Élet helyett órák ; Talán Hellász küldött. - 4. bőv. kiad. - Budapest : Magvető Kiadó, 1980
271p,[24]p of plates : ill
963-271-491-1 X.950/16013
8356521

FÁBRY, Zoltán
[Essays. Selections]. Ady igaza. - Bratislava : Madách, 1977
261p
r6449310 X.958/13235
8356521

HORVÁTH, Béla
Czigány Dezső Ady-képei. - Budapest : Magvető, 1977
85, [1] p. : ill.. - (Gyorsuló idő). - Bibliography: p. 83-[86]
963-270-616-1 X.958/26536
8356521

JÉKEL, Pál
Ady Endre összes költői műveinek fonémastatisztikája. - Budapest : Akadémiai Kiadó, 1974
179p
963-05-0095-7 X.950/21027
8356521

KIRÁLY, István
Intés az őrzőkhöz. - Budapest : Szépirodalmi Könyvkiadó, [1982]
2v.
963-15-2151-6 X.950/19106
8356521

LUKÁCS, György, 1885-1971
Ady Endréről. - Budapest : Magvető, 1977
81 p.. - (Gyorsuló idő)
963-270-676-5 X.958/24021
8356521

SCHWEITZER, Pál
Szépség és totalitás. - Budapest : Akadémiai Kiadó, 1980
296p
963-05-1600-4 X.958/20088
8356521

SZABÓ, Dezső
Ady. - Budapest : Magvető Könyvkiadó, 1982
59p. - (Gondolkodó magyarok)
963-271-772-4 X.958/14422
8356521

VITÁLYOS, László
Ady-bibliográfia 1896-1970. - Budapest : Magyar Tudományos Akadémia Könyvtára, 1972
425p[24]p of plates. - (A Magyar Tudományos Akadémia Könyvtárának kiadványai ; 67)
r6056072 Ac.825.g.[no.67]
1469371

ADY, ENDRE. Poetry in Hungarian
- Bibliographies

VITÁLYOS, László
Ady bibliográfia 1896-1977. - 2. bőv. kiad.. - Budapest, 1980
408p,[12]p of plates : ill. - (A Magyar Tudományos Akadémia Könyvtárának közleményei. Új sorozat ; 9(84))
963-7301-36-4 Ac.825.g.[84]
1469371

ADY, ENDRE. Poetry in Hungarian
- Biographies

BÖLÖNI, György
Az igazi Ady. - 4. kiad.. - Budapest : Magyar Helikon, 1974
593p,[46]p. of plates : ill
963-207-064-x X.989/39220
121439X

HATVANY, Lajos
Ady. - 2., átnézett és bőv. kiad.. - Budapest : Szépirodalmi Könyvkiadó, 1974
846p
963-15-0234-1 X.950/10529
8356521

PÉTER, László
Ady nálunk. - Szeged : Somogyi-könyvtár, 1977
150p : ill. - (A Somogyi-könyvtár kiadványai ; 21)
963-7581-09-x P.1491/3.(21)
121439X

ROBOTOS, Imre
Az igazi Csinszka. - Budapest : Magvető, 1975
207 p.. - Includes bibliographical references
963-270-092-9 X.958/23698
8356521

TEGNAPOK és holnapok árján / szerkesztette : Láng József. - Budapest : Petőfi Irodalmi Múzeum : Népművelési Propaganda Iroda, 1977
457p[42]p of plates : ill
r6083553 X.900/20742
121439X

ADY, ENDRE. Poetry in Hungarian
- Biographies

VEZÉR, Erzsébet
Ady Endre élete és pályája. - 2., átdolg. kiad.. - Budapest : Gondolat, 1977
556p,[48]p of plates : ill. - Bibliography
963-280-507-0 X.958/7498
8356521

VITÁLYOS, László
Ady - Léda - Csinszka. - Budapest : MTAK, 1977
163p : ill. - (A Magyar Tudományos Akadémia Könyvtárának közleményei. Új sorozat ; 4(79))
963-7301-26-7 Ac.825g[4(79)]
121439X

ADY, ENDRE. Poetry in Hungarian
- Conference proceedings

"AKAROM : TISZTÁN LÁSSATOK" (Conference : 1977 : Budapest)
"Akarom : tisztán lássatok". - Budapest : Akadémiai Kiadó, 1980
261p. : ill
963-05-2272-1 X.950/23286
8356521

ADY, ENDRE. Poetry in Hungarian
- Correspondence, diaries

ADY, Endre
[Correspondence]. Ady Endre levelei. - Budapest : Szépirodalmi Könyvkiadó, 1983
3v.. - (Ady Endre művei)
963-15-2257-1 X.950/30037
8356521

ADYGEY
See Adyghe

ADYGHE AUTONOMOUS OBLAST. Russia (RSFSR)
1917-

SBORNIK stateĭ po istorii Adygei / [otvetstvennye redaktory : N.F. Bugai i B.M. Dzhimov]. - Maĭkop, 1977-
v.
r6448452 X.0629/651
9086323

ADYGHE AUTONOMOUS OBLAST. Russia (RSFSR)
Moral education. Role of folklore

SHOROV, Ibragim Askhadovich
Idei nravstvennogo vospitaniîa v adygskom ustnom narodnom tvorchestve. - Rostov-na-Donu : Rostovskiĭ-na-Dony pedagogicheskiĭ institut, 1977
79p. - Bibliography : p.75-77
r6086831 X.519/41421
2118947

ADYGHE AUTONOMOUS OBLAST. Russia (RSFSR)
Place names - Dictionaries

MERETUKOV, K. Kh. (Kasim Khamosovich)
Adygeĭskiĭ toponomicheskiĭ slovar'. - Maĭkop : Adygeĭskoe otdelenie Krasnodarskogo knizhnogo izdatel'stva, 1981
180p
r6432159 X.808/36916
8873593

ADYGHE AUTONOMOUS OBLAST. Russia (RSFSR)
Political events, 1917-1941

USTANOVLENIE Sovetskoĭ vlasti i naîsional'no-gosudarstvennoe stroitel'stvo v Adygee (1917-1923 gg.) / [sostaviteli: L.T. Arutîunova, G.S. Serebrîanyĭ, G.T. Chuchmaĭ (otvetstvennyĭ sostavitel')]. - Maĭkop : Krasnodarskoe knizhnoe izdatel'stvo, Adygeĭskoe otdelenie, 1980
267p
r6402466 X.809/50039
8815100

ADYGHE AUTONOMOUS OBLAST. Russia (RSFSR)
Social conditions, to date

NOGMOV, Sh. B. (Shora Bekmurzovich)
Istoriîa adykheĭskogo naroda. - Nal'chik : El'brus, 1982
165p
r6464691 X.529/61536
9086366

ADYGHE LANGUAGE

AKTUAL'NYE voprosy adygskikh îazykov / [redaktory: P.M. Bagov, A.S. Kishev]. - Nal'chik : Kabardino-Balkarskiĭ institut istorii, filologii i ėkonomiki, 1981
111p
r6483499 X.950/13993
1331191

KUMAKHOVA, Zara ÎUsufovna
Funktsional'naîa stilistika adygskikh îazykov. - Moskva : Nauka, 1979
359p. - Bibliography : p.349-357
r6175811 X.989/82605
1331191

OSOBENNOSTI grammaticheskikh kategoriĭ i ikh primeneniîa v adygeĭskoĭ rechi / [otvetstvennye redaktory : Z.U. Blîagoz, U.S. Zekokh]. - Maĭkop : Adygeĭskiĭ nauchno-issledovatel'skiĭ institut ėkonomiki, îazyka, literatury i istoric, 1978
143p
r6186178 X.989/80574
1331191

SBORNIK stateĭ po adygeĭskomu îazyku / [otvetstvennye redaktory M.A. Kumakhov i K. Kh. Meretukov]. - Maĭkop : Adygeĭskiĭ nauchno-issledovatel'skiĭ institut ėkonomiki, îazyka, literatury i istorii, 1976
422p
r6045245 X.989/71094
1331191

SHAGIROV, Amin Kabîsuevich
Material'nye i strukturnye obshchnosti leksiki abkhazo-adygskikh îazykov. - Moskva : Nauka, 1982
163p
r6496923 X.950/30288
1331191

STANOVLENIE stileĭ i norm adygeĭskogo literaturnogo îazyka. - Maĭkop : Adygeĭskoe otdelenie krasnodarskogo knizhnogo izdatel'stva, 1982
190p
r6448399 X.950/30827
1331191

STROENIE predlozheniîa v adygeĭskom îazyke / [redaktory Kumakhova Z.ÎU., Tkharkakho ÎU.A.]. - Maĭkop : Adygeĭskiĭ nauchno-issledovatel'skiĭ institut ėkonomiki, îazyka, literatury i istorii, 1976
155p
r6055752 X.989/73846
1331191

VOPROSY sopostavitel'nogo izucheniîa îazykov narodov Karachaevo-Cherkesiĭ / [redaktsionnaîa kollegiîa: M.A. Khabichev (otvetstvennyĭ redaktor), Kh. Dzh. Tekeev, S.K. Khalilova]. - Stavropol' : Stavropol'skiĭ gosudarstvennyĭ pedagogicheskiĭ institut, 1978
109p
r6098533 X.908/90078
1331191

ADYGHE LANGUAGE
- Bibliographies

GISHEV, Nukh Turkubievich
Bibliografiîa lingvisticheskoĭ literatury po adygskym îazykam. - Maĭkop : Adygeĭskiĭ nauchno-issledovatel'skiĭ institut ėkonomiki, îazyka, literatury i istorii, 1975
100p
r6185052 X.989/83054
1331191

ADYGHE LANGUAGE
- Dictionaries

ADYGEĬSKO-RUSSKIĬ slovar' ... / pod redakfsieĭ Zh.A. Shaova. - Maĭkop : Adygeĭskiĭ nauchno-issledovatel'skiĭ institut ėkonomiki, îazyka, literatury i istorii, 1975
440p
r6402134 X.950/3810
1331191

SHAGIROV, Amin Kabfsuevich
Ėtimologicheskiĭ slovar' adygskikh (cherkesskikh) îazykov.... - Moskva : Izdatel'stvo "Nauka", 1977
2t
r6055743 X.989/73799
1331191

ADYGHE LANGUAGE
- Indexes

BIBLIOGRAFIÎA lingvisticheskoĭ literatury po adygskim îazykam / sostavil N.T. Gishev. - Maĭkop : Adygeĭskiĭ nauchno-issledovatel'skiĭ institut ėkonomiki, îazyka, literatury i istorii, 1975
100p
r6093459 X.989/77775
1331191

ADYGHE LANGUAGE
compared with Kabardian dialect

TEMIROVA, Raisa Khasanovna
Leksicheskie osobennosti rechi cherkesov. - Cherkessk : Stavropol'skoe knizhnoe izdatel'stvo, Karachaevo-Cherkesskoe otdelenie, 1975
204p. - Bibliography : p.197-203
r6098105 X.989/79098
186601X

ADYGHE LANGUAGE
Dialects

KUMAKHOVA, Zara ÎUsufovna
Abadzekhskiĭ dialekt i ego mesto sredi drugikh adygskikh dialektov. - Nal'chik : Ėl'brus, 1972
116p. - Bibl.: p.113-115
r6178246 X.989/85952
2446987

ADYGHE LITERATURE
See also
Epic poetry in Adyghe
Fiction in Adyghe
Poetry in Adyghe

ADYGHE LITERATURE

ABITOVA, Svetlana Dzhambotovna
Geroĭ literatury Oktîabrîa. - Cherkessk : Stavropol'skoe knizhnoe izdatel'stvo, Karachaevo-Cherkesskoe otdelenie, 1971
181p
r6178114 X.989/85283
2418843

ADYGHE LITERATURE
Kabardian dialect literature, to date

KURASHINOV, B. (Betal)
Sodruzhestvo muz. - Nal'chik : Ėl'brus, 1982
202p
r6487174 X.950/34015
9161902

ADYGHE PEOPLE. Russia (RSFSR)
Cultural processes - Serials

KUL'TURA i byt adygov. - Maĭkop, [1977?]-
Holdings: vyp.2- , 1978-
r6183738 X.809/29009
2462109

ADYGHE TALES
- Anthologies

SKAZKI adygskikh narodov / [sostavlenie, vstupitel'naîa stat'îa, primechaniîa A.I. Alievoĭ]. - Moskva : Nauka, Glavnaîa redakfsiîa vostochnoĭ literatury, 1978
404p. - (Skazki i mify narodov Vostoka). - Bibliography : p.391-[396]
r6094374 X.989/78800
1866788

ADZHAR ASSR. Georgia (Soviet Socialist Republic)
Cultural processes, to date

OCHERKI ėtnografii Adzharii / [redaktor, A. Robakidze]. - Tbilisi : Mefsniereba, 1982
244 p. : ill.. - Includes bibliographical references
r6482806 X.529/64065
9115730

ADZHEMIAN, VARDAN. Directing. Theatre. Soviet Union
- Biographies

RIZAEV, Sabir Alekperovich
Vardan Adzhemîan. - Moskva : Iskusstvo, 1978
255p,[31]p of plates : ill. - (Mastera sovetskogo teatra i kino)
r6184021 X.907/20177
2216388

ADŽIĆ, DRAGOLJUB. Yugoslav sculptures
- Catalogues

ADŽIĆ, Dragoljub
Izložba keramike "Osećaj vremena i prostora".... - Beograd : Muzej primenjene umetnosti, [1975]
[33]p. : chiefly ill
r6036690 X.410/6147 Woolwich
1226908

AEC BUSES
to 1979

HANNAY, R. N. (Robin Neil)
AEC buses in camera. - London : Ian Allan, 1982
112p : ill
0-7110-1160-5 X.622/12389 Woolwich
2705281

AEC BUSES
AEC Bridgemaster buses

AEC Bridgemaster / [compiled by Grahame Wareham]. - Oxford : Oxford Publishing Co. [for the Oxford Bus Preservation Society], 1978
[37]p : chiefly ill(some col)
0-9506739-0-0 X.629/15015 Woolwich
2312646

AEC LORRIES
to 1979

KENNETT, Pat
AEC. - Cambridge : Stephens, 1980. - (World trucks ; no.10)
No.10
87p : ill
0-85059-398-0 X.622/12548 Woolwich
2704323

STEVENS-STRATTEN, S. W. (Seymour Walter)
AEC. - London : Ian Allen, 1984
96p : ill. - (Trucks in camera)
0-7110-1443-4 X.622/21791
2704323

AECLANUM. Italy
Christianity, to 476

GAMBINO, Nicola
Aeclanum cristiana. - Torella dei Lombardi : Associazione F. De Sanctis, 1982
236p : ill. - (Civiltà altirpina ; anno 7, genn./apr. 1982, fasc.1-2. Supplemento). - Includes bibliography
r6507682 P.803/641[anno 7, fasc.1/2, supplemento]
8958084

AEG-TELEFUNKEN (UK) LIMITED
Relations with Chasgift Limited

FREEMAN, David John
AEG-Telefunken (U.K.) Limited, Chasgift Limited (formerly known as Credit Collections Limited). - London : H.M.S.O., 1979
38,[16]p
0-11-512498-5 BS.41/1195
2005328

AEGADIAN ISLES
See Egadi Islands

AEGATES
See Egadi Islands

AEGEAN ARTS
Aegean visual arts, to ca B.C.600. Special subjects: Gods

COLDSTREAM, John Nicolas
Deities in Aegean art. - London : Bedford College, 1977
19p : ill. - (Inaugural lectures / Bedford College)
0-900145-38-2 X.410/10176 Woolwich
1248480

AEGEAN CAMPAIGN. World War 1
Naval operations by Great Britain. Navy. Role of Myres, Sir John Linton

MYRES, J. N. L.
Commander J.L. Myres, R.N.V.R.. - London : Leopard's Head Press Limited, 1980
35p. - (J.L. Myres memorial lecture ; 10th)
0-904920-04-6 X.700/26578
2321904

AEGEAN CIVILIZATION
to B.C.775 - Conference proceedings

GRIECHENLAND, die Agäis und die Levante während der "Dark Ages" vom 12. bis zum 9. Jh. v. Chr. / herausgegeben von Sigrid Deger-Jalkotzy. - Wien : Verlag der Österreichischen Akademie der Wissenschaften, 1983
483p,plates. - (Sitzungsberichte / Österreichische Akademie der Wissenschaften, Philosophisch-historische Klasse ; Bd.418) (Veröffentlichungen der Kommission für Mykenische Forschung ; Bd.10). - Includes bibliographies
3-7001-0596-7 Ac.810/6a[Bd.418]
8992762

AEGEAN ISLANDS. Greece

GREEK Islands of the Aegean / by the staff of Editions Berlitz ; [area specialist Don Larrimore ; photography Daniel Vittet ; cartography Falk-Verlag, Hamburg]. - Lausanne : Editions Berlitz ; [London] : [Distributed by Cassell], 1979
128p : ill(chiefly col). - (Berlitz travel guide). - Includes index
0-304-96972-9 X.708/21971
0340758

AEGEAN ISLANDS. Greece
Description & travel, 1974-

QUATREPOINT, Robert
Terres d'Égée. - Paris : Denoël, 1981
158p : ill
r6206938 X.809/52791
8232830

AEGEAN ISLANDS. Greece
Description & travel, 1975-1978

EBDON, John
Ebdon's odyssey. - London : Peter Davies, 1979
177p : ill
0-432-04020-x X.809/46961
2373033

AEGEAN ISLANDS. Greece
Description & travel, 1975-1978 - Texts

EBDON, John
Ebdon's Odyssey. - London : Unwin, 1984
177p : ill
0-04-910077-7 X.958/24331
2373033

AEGEAN ISLANDS. Greece
Doric civilization

CRAIK, Elizabeth M.
The Dorian Aegean. - London : Routledge and Kegan Paul, 1980
xi,263p : ill. - (States and cities of ancient Greece). - Bibliography: p.230-251. - Includes index
0-7100-0378-1 X.809/45144
1935267

AEGEAN ISLANDS. Greece
Windmills

LIMONA-TREBELA, Eleni
Windmills of the Aegean Sea. - Reading : International Molinological Society, 1983
27p : ill. - (Bibliotheca Molinlogica)
B84-32642 X.425/4716
3145441

AEGEAN REGION
- For sailing

DENHAM, H. M. (Henry Mangles)
The Aegean. - 5th ed. - London : Murray, 1983
xxix,278p,[16]p of plates : ill. - Includes index
0-7195-3980-3 X.800/36009
0638722

DENHAM, Henry Mangles
The Aegean. - 3rd ed. - London : J. Murray, 1975 [i.e.1978]
xxvii,253p,[16]p of plates : ill. - Includes index
0-7195-3173-x X.800/26846
0638722

DENHAM, Henry Mangles
The Aegean. - 3rd ed. - London : J. Murray, 1975
xxvii,253p,[16]p of plates : ill. - Includes index
0-7195-3405-4 X.800/25536
0638722

AEGEAN REGION
- For sailing–continued
The Aegean. - 4th ed. - London : J. Murray, 1979
xxix,305p,[16]p of plates : ill. - (Sea guides / Henry Mangles Denham). - Includes index
0–7195–3675–8 X.800/28367
0638722

AEGEAN REGION
B.C 323–A.D.323
BURASELIS, Kostas
Das hellenistische Makedonien und die Agais. - Munchen : Beck, 1982
xi,207p - (Munchener-Beitrage zur Papyrusforschung und antiken Rechtsgeschichte ; Hft.73). - Bibliography
3–406–07673–4 WP.3042/73
8916349

AEGEAN REGION
Bronze Age antiquities
HOPE SIMPSON, R.
A gazetteer of Aegean civilization in the Bronze Age. - Göteborg : Aström, 1979–
Vol.1–. - (Studies in Mediterranean archaeology ; vol.52 etc). - Includes bibliographies
91–85058–81–5 X.0415/9(52,) Woolwich
2247933

AEGEAN REGION
Bronze Age antiquities. Archaeology, to date
TREUIL, René
Le Néolithique et le Bronze Ancien Égéens. - Athènes : École française d'Athènes, 1983
541p,7p of plates : ill. - ([Publications] / Bibliothèque des écoles françaises d'Athènes et de Rome ; fasc.248). - Includes bibliography
r6208865 Ac.5206/2[248]
8892814

AEGEAN REGION
Bronze Age antiquities. Archaeology, to date - Conference proceedings
SVENSKA INSTITUTET I ATHEN. International Symposium (1st : 1980 : Athens)
Sanctuaries and cults in the Aegean Bronze Age. - Stockholm : The Institute, 1981
226p : ill. - (Skrifter utgivna av Svenska institutet i Athen. 4° ; 28)
91–85086–43–6 Ac.5566.b[no.28]
8892814

AEGEAN REGION
Bronze Age tombs
PELON, Olivier
Tholoi, tumuli et cercles funéraires. - Athènes : École Française d'Athènes, 1976
xvi,537p,3 fold leaves of plates, 182p.of plates : ill. - (Bibliothèque des écoles françaises d'Athènes et de Rome ; fasc. 229)
r6028366 Ac.5206/2[fasc.229]
1236466

AEGEAN REGION
Ionian tombs. Reliefs, B.C.500–B.C.450
HILLER, Hilde
Ionische Grabreliefs der ersten Hälfte des 5. Jahrhunderts v. Chr. - Tübingen, 1975
203p : ill. - (Istanbuler Mitteilungen ; Beihft 12)
r6002339 Ac.5388.e/2a[Beihft 12]
1212303

AEGEAN REGION
Stoneware: Bronze Age faience
FOSTER, Karen Polinger
Aegean faience of the Bronze Age. - New Haven ; London : Yale University Press, 1979
xxi,205p : ill. - Bibliography: p.175–197. - Includes index
0–300–02316–2 X.421/20591 Woolwich
2143216

AEGEAN SEA
See also
Dardanelles

AEGEAN SEA
See also
Straits of Salamis

AEGEAN SEA
Maps, ca 1590: Mehmed Reis ibn Menemenli. Aegean sea-chart
BRICE, William
The Aegean sea-chart of Mehmed Reis Ibn Menemenli, AD 1590–1. - [Manchester] : University of Manchester, 1977
[18]p : ill. - (Monographs / Seminar on Early Islamic Science ; no.2)
B79–08345 X.709/32948
1811622

AEGEAN SEA
Northern Aegean Sea. Voyages by sailing ships, 1444–1445
CIRIACO DE' PIZZICOLLI
Cyriacus of Ancona's journeys in the Propontis and northern Aegean, 1444–1445. - Philadelphia : American Philosophical Society, 1976
viii,90p : ill. - (Memoirs of the American Philosophical Society ; vol. 112). - Includes index
0–87169–112–4 Ac.1830/8(112)
1072064

AEGIDIUS, SAINT
See Giles, Saint †

AEGIDIUS ROMANUS
See Giles of Rome †

AEGIDIUS ROMANUS, ABP OF BOURGES
See Giles of Rome, Abp of Bourges

AEGINA. Greece
Ancient Greek temples: Temple of Aphaea. Sculptures: Exhibits at Glyptothek - Catalogues
OHLY, Dieter
Die Aegineten. - München : Beck, [1976–]
Bd1– : ill
3–406–03333–4 LR.415.c.9
1129228

AEGINA. Greece
Antiquities. Archaeology, to date
WALTER, Hans
Die Leute im alten Ägina 3000–1000 v.Chr.. - Stuttgart : Urachhaus, 1983
147p : ill
3–87838–381–9 X.429/17007
8775877
Die vorgeschichtliche Stadt. - Mainz/Rhein : von Zabern, [1981]
2v.(187p.,128p.of plates,12folded leaves of plates) : ill. - (Alt-Ägina ; Bd.3/1)
3–8053–0007–7 X.0423/103c(1) Woolwich
8775877

AEGINA TREASURE. Stock. British Museum
HIGGINS, R. A. (Reynold Alleyne)
The Aegina Treasure. - London : British Museum Publications, c1979
71p : ill. - (A Colonnade book). - Includes index
0–7141–8006–8 X.421/22624 Woolwich
2629615

AEGNA. Estonia
GUSTAVSON, Heino
[Aegna. Russian]. Ostrov Aegna. - Tallin : Eesti raamat, 1978
50p : ill
r6185749 X.708/47583
2248867

AELFRIC
Lives of the saints. Grammar. Transformational–generative theories
JOHNSON, Judith Anne
A transformational analysis of the syntax of Aelfric's "Lives of saints". - The Hague : Mouton, 1975
111 p. : ill. - (Janua Linguarum. series practica ; 212). - Bibliography: p. 110–111
90–279–3084–8 X.0981/78C (212)
0769525

AELIUS HADRIANUS
See Hadrian, Emperor of Rome

AELMANNIA (DUCHY)
to 750
BEHR, Bruno
Das alemannische Herzogtum bis 750. - Bern, 1975
242 p.. - (Geist und Werk der Zeiten ; no. 41)
3–261–01493–8 X.709/20215
0937401

AELRED, SAINT. Cistercians
SQUIRE, Aelred
Aelred of Rievaulx. - London : S.P.C.K., 1969
xiii,177p. - Bibliography: p153. - Includes index
0–281–03851–1 X.200/37916
2695952
- [Another copy.] X.200/38814

AELRED, SAINT. Cistercians
- Biographies
DANIEL, Walter
[Vita Ailredi Abbatis Rievall'. English]. The life of Ailred of Rievaulx. - Oxford : Clarendon Press, 1978
ciii,81,88p. - (Oxford medieval texts). - Includes index
0–19–822256–4 WP.3189/42
1906348

AEM
See Andes Evangelical Mission

AEMILIUS MACER
See Macer, Aemilius †

AEMTM
See Association of European Machine Tool Merchants

AENEID. Virgil. Epic poetry in Latin
BASSON, W.P
Pivotal catalogues in the Aeneid. - Amsterdam : Hakkert, 1975
xii,208p. - Bibliography : p.193–200
90–256–0767–5 X.909/37753
0311936
JOHNSON, W. R. (Walter Ralph)
Darkness visible. - Berkeley ; London : University of California Press, 1976
xi,179p
0–520–02942–9 X.981/20421
0311936
KRAGELUND, Patrick
Dream and prediction in the Aeneid. - Copenhagen : Museum Tusculanum, 1976
91p. - (Opscula graecolatina ; v.7)
r6028924 X.0900/517(7)
0311936
ZINTZEN, Clemens
Die Laokoonepisode bei Vergil. - Mainz : Akademie der Wissenschaften und der Literatur Wiesbaden : [in Komm. bei] Steiner, 1979
77p. - (Abhandlungen der Geistes– und Sozialwissenschaftlichen Klasse / Akademie der Wissenschaften und der Literatur ; Jahrg. 1979, Nr 10)
3–515–03172–3 Ac:706(1)[1979,Nr.10]
0311936

AENEID. Virgil. Epic poetry in Latin
Half–lines & repetitions
SPARROW, John
Half–lines and repetitions in Virgil. - New York ; London : Garland, 1977
[13],156p. - (The Garland library of Latin poetry)
0–8240–2979–8 X.989/53672
193323x

AENEID. Virgil. Epic poetry in Latin
Influence of scholia on Homer
SCHLUNK, Robin R.
The Homeric scholia and the 'Aeneid'. - Ann Arbor : University of Michigan Press, 1974
xi,144p. - Bibliography : p.139–142
0–472–08780–0 X.989/53377
1885170

AENEID. Virgil. Epic poetry in Latin
Special themes: Cosmology
THORNTON, Agathe
The living universe. - Lugduni Batavorum [i.e.Leiden] : Brill, 1976
xiii,223p. - (Mnemosyne ; supp.46). - Bibliography : p.211–216.–Includes index
90–04–04579–1 PP.4962/3[suppl.26]
1513702

AENEID. Virgil. Epic poetry in Latin
Speeches
LUNDSTRÖM, Sven
Acht Reden in der Aeneis. - Uppsala ; Stockholm : Distributor, Almqvist & Wiksell International, 1977
127p. - (Studia latina upsaliensia ; 10)
91–554–0584–3 Ac.1075/6(17)[vol.10]
142128X

AENEID. Virgil. Poetry in Latin
Translations into Swedish language
KORKKANEN, Irma
Studier i de svenska översättningarna av Vergilius Aeneis I,II,VI,VII,XI,XII och Bucolica. - Helsingfors : Institutionen för nordisk filologi vid Helsingfors universitet, 1974
vi, 192p. - (Meddelanden fran institutionen för nordisk filologi vid Helsingfors universitet. ser. A ; nr 1). - Includes bibliographies
951–450–285–x X.0909/913(1)
0395102

AENEOLOTHIC
See Copper Age

AEOLIC DIALECTS. Greek language
BLÜMEL, Wolfgang
Die aiolischen Dialekte. - Göttingen : Vandenhoeck & Ruprecht, 1982
326p.. - (Ergänzungshefte zur Zeitschrift für vergleichende Sprachforschung ; Nr.30). - Bibliography
3–525–26218–3 PP.5043.b(1)[no.30]
8911118

AEOLIC STYLE. Architecture. Ancient world B.C.1000-B.C.500

BETANCOURT, Philip P.
The Aeolic style in architecture. - Princeton ; Guildford : Princeton University Press, 1977
viii,169p,[24]p of plates : ill. - Bibliography: p.157-162. - Includes index
0-691-03922-4 X.423/5286 Woolwich
125314x

AEPLI, JOHANN MELCHIOR. Medicine

PETER, Claude Henri
Der Briefwechsel zwischen Paulus Usteri und Melchior Aepli. - Zürich : Juris, 1983
161p. - (Zürcher medizingeschichtliche Abhandlungen ; neue Reihe, Nr. 155)
3-260-04976-2 X.329/17943
816746x

AER LINGUS
to 1980

RE-INTRODUCING Aer Lingus. - [Dublin] : Aer Lingus Public Relations and Publicity Departments, 1983
39p : ill
B84-22364 X.629/22793
3085236

AERARIUM MILITARE
See Ancient Rome. Aerarium Militare

AERARIUM SATURNI
See Ancient Rome. Aerarium Saturni

AERATED CONCRETE
Autoclaved aerated concrete

COMITÉ EURO-INTERNATIONAL DU BÉTON
Autoclaved aerated concrete. - Lancaster : Construction Press, 1978
90p : ill. - Bibliography: p.88. - Includes index
0-904406-76-8 SRL
1441655

AERATION. Cereals

MYSON GROUP MARKETING LIMITED. Agricultural Division
Grain drying and aeration. - Colchester : Myson Group Marketing Ltd, Agricultural Division, [1975]
[2],16p : ill(some col)
0-9504706-0-0 X.319/10324 Woolwich
0578886

AERATION. Waste water
- Conference proceedings

SYMPOSIUM ON THE PROFITABLE AERATION OF WASTE WATER (1980 : London)
Papers presented at the Symposium on the Profitable Aeration of Waster Water, held at the Sudbury Conference Hall, London, April 1980, organised and sponsored by BHRA Fluid Engineering. - Cranfield : BHRA Fluid Engineering, 1980
iii,98p : ill
0-906085-44-6 SRL
2086379

AERE
See Atomic Energy Research Establishment

AERIAL APPLICATION. Chemicals
See Aircraft
Application of chemicals

AERIAL PHOTOGRAMMETRY

SEARS, G. M.
First steps in photogrammetry. - [London] : North East London Polytechnic, Department of Land Surveying, 1982
vii,158p : ill. - (Working paper / North East London Polytechnic. Department of Land Surveying ; no.5). - Bibliography: p158
0-907382-04-5 X.525/5657
2712822

AERIAL PHOTOGRAPHS
Interpretation

DIETZ, Klaus R.
Grundlagen und Methoden geographischer Luftbildinterpretation. - München : Institut für Geographie der Universität, 1981
109p : ill. - (Münchener geographische Abhandlungen ; Bd 25). - Bibliography
3-88618-091-3 P.803/473[Bd.25]
8816689

STÖCKLHUBER, Klaus
Erfassung von Ökotopen und ihren zeitlichen Veränderungen am Beispiel des Tegernseer Tales. - München : Institut für Geographie der Universität München, 1982
123p,[4]folded leaves of plates : ill. - (Münchener geographische Abhandlungen ; Bd.26). - Bibliography
3-88618-092-1 P.803/473[Bd.26]
8816689

WIECZOREK, Ulrich
Methodische Untersuchungen zur Analyse der Wattmorphologie aus Luftbildern mit Hilfe eines Verfahrens der digitalen Bildstrukturanalyse. - München : Institut für Geographie der Universität München, 1982
208p,2folded leaves of plates : ill. - (Münchener geographische Abhandlungen ; Bd.27). - Bibliography
3-88618-093-x P.803/473[Bd.27]
8816689

AERIAL PHOTOGRAPHS
Pattern recognition

NAGAO, Makoto
A structural analysis of complex aerial photographs. - New York ; London : Plenum, c1980
xxiii,199p,[10]p of plates : ill(some col.). - (Advanced applications in pattern recognition). - Bibliography: p185-190. - Includes index
0-306-40571-7 SRL
2351145

AERIAL PHOTOGRAPHS. Sources of evidence. Sites. Ancient Roman antiquities. Great Britain

FRERE, Sheppard
Roman Britain from the air. - Cambridge : Cambridge University Press, 1983
xvii,232p : ill. - (Cambridge air surveys). - Includes index
0-521-25088-9 X.805/4452
2967871

AERIAL PHOTOGRAPHS. Sources of evidence. Sites. Antiquities. Buckinghamshire

ALLEN, Denise
Bucks archaeology from the air. - Aylesbury : Buckinghamshire County Museum, 1979
18p,[4]p of plates : ill. - Bibliography: p18
0-86059-197-2 X.429/12074 Woolwich
2398001

AERIAL PHOTOGRAPHS. Sources of evidence. Surface features. North-west Queensland
Multispectral colour aerial photographs. Evaluation

COLE, Monica M.
The use of LANDSAT imagery in relation to air survey imagery for terrain analysis in Northwest Queensland, Australia. - London : Department of Industry, Research and Technology Requirements Division.
Vol.1. - 1977
[3]leaves,157[i.e.160]p. - Includes bibliographies
B79-13727 BS.424/155
1838520
Vol.2. - 1977
[5]leaves,91 leaves of plates : chiefly ill(some col)
B79-13728 BS.424/155
1838520
Vol.3. - 1977
[2]leaves,50[i.e.53]leaves of plates(some fold) : of ill
B79-13729 BS.424/155
1838520

AERIAL PHOTOGRAPHS TAKEN FROM ARTIFICIAL SATELLITES. Surface features. Hungary
to date

MAGYARORSZÁG az űrből / [szerzők: Bak Antal ... et al.]. - Budapest : Zrínyi Katonai Kiadó, 1981
95p : chiefly ill(some col.)
963-326-300-x L.45/1571
8901074

AERIAL PHOTOGRAPHS TAKEN WITH HEAT CAMERAS. Sources of evidence. Surface features. North-west Queensland
Colour aerial photographs taken with heat cameras. Evaluation

COLE, Monica M.
The use of LANDSAT imagery in relation to air survey imagery for terrain analysis in Northwest Queensland, Australia. - London : Department of Industry, Research and Technology Requirements Division.
Vol.1. - 1977
[3]leaves,157[i.e.160]p. - Includes bibliographies
B79-13727 BS.424/155
1838520
Vol.2. - 1977
[5]leaves,91 leaves of plates : chiefly ill(some col)
B79-13728 BS.424/155
1838520
Vol.3. - 1977
[2]leaves,50[i.e.53]leaves of plates(some fold) : of ill
B79-13729 BS.424/155
1838520

AERIAL PHOTOGRAPHY

BRITISH AIR SURVEY ASSOCIATION
Specification for vertical air photography, September 1978. - Boreham Wood, Herts. : The Association, 1978
[16]p
B79-06698 X.519/29148
1778528

DICKINSON, Gordon Cawood
Maps and air photographs. - 2nd ed. - London : Edward Arnold, 1979
xiii,348p : ill. - Bibliography: p.331-338. - Includes index
0-7131-6145-0 Map Library
1734334

GAGNON, Hugues
La photo aérienne. - Montreal ; Toronto : Les Éditions HRW, 1974
ix, 278 p. : ill. - Bibliography : p. 269-274
0-03-928248-1 X.702/2929
0853070

SPECIFICATION for vertical air photography. - 2nd ed.. - London : Surveyors, c1984
13p
0-85406-243-2 Map Library
1778528

AERIAL PHOTOGRAPHY
Applications

The USES of air photography / edited by J.K.S. St Joseph. - New ed.. - London : J. Baker, 1977
196p : ill. - Includes bibliographies and index
0-212-97016-x SRL
1133926

AERIAL PHOTOGRAPHY
Applications in archaeology

AERIAL archaeology / Committee for Archaeological Air Photography, Anglian Region. - Vol.1 : 1977 / [edited by Derek A. Edwards and Clive Partridge]. - Hertford : The Committee, 1978
viii,54p : ill
Subsequent volumes, when received, will be placed at this shelfmark
B78-23847 P.413/203
1498053

AERIAL PHOTOGRAPHY
Applications in archaeology - Conference proceedings

AERIAL reconnaissance for archaeology / edited by D.R. Wilson. - London : Council for British Archaeology, 1975
ix,158p : ill. - (Research reports / Council for British Archaeology ; no.12). - Includes bibliographies
0-900312-29-7 X.0410/35(12)
0612294

AERIAL PHOTOGRAPHY
Applications in archaeology, to date

CHRISTLEIN, Rainer
Das unterirdische Bayern. - Stuttgart : Theiss, 1982
275p : ill
3-8062-0233-8 L.42/2698
9080317

AERIAL PHOTOGRAPHY
Applications in cartography

BURNSIDE, Clifford Donald
Mapping from aerial photographs. - London : Crosby Lockwood Staples, 1979
xiv,304p : ill. - (Aspects of modern land surveying). - Bibliography: p.283-287. - Includes index
0-258-97035-9 SRL
191166x

AERIAL PHOTOGRAPHY
Applications in environment planning - Conference proceedings

AKADEMIE FÜR RAUMFORSCHUNG UND LANDESPLANUNG. Arbeitskreis "Verwendung der Fernerkundung für die Planung". Sitzung (6th : 1981 : Freiburg im Breisgau)
Thermalluftbilder für die Stadt- und Landesplanung. - Hannover : Vincentz, 1982
96p. - (Beiträge / Akademie für Raumforschung und Landesplanung ; Bd.62)
3-87870-853-x X.0512/475(62)
8861854

AERIAL PHOTOGRAPHY
Applications in geography

LO, C. P.
Geographical applications of aerial photography. - New York : Crane, Russak ; Newton Abbot : David and Charles, 1976
330p,[16]p of plates : ill. - Includes index
0-7153-7199-1 SRL
077796x

AERIAL PHOTOGRAPHY
Applications in hydrogeology

NEFEDOV, K. E.
Deciphering of groundwater from aerial photographs. - New Delhi : Amerind, c1972
xii,191p : ill. - Bibliography: p168-191
0-86249-367-6 X.622/17578
3029883

AERIAL PHOTOGRAPHY
Applications in remote sensing

CURRAN, Paul J.
Remote sensing. - Reading : Department of Geography, University of Reading, 1981
39p : ill. - (Geographical papers ; no.75). - Bibliography: p37-39
0-7049-0663-5 X.0519/1008(75)
2562553

AERIAL PHOTOGRAPHY
Applications in research on sites of antiquities - Serials

[ORBIT (London)]. Orbit. - Vol.1, 1980-. - London : Aerial Archaeology Foundation, 1980-
v. : ill
B81-08764 P.411/424
2365189

AERIAL PHOTOGRAPHY
Applications in soil mapping

CARROLL, Douglas Michael
Air photo-interpretation for soil mapping. - Harpenden : Soil Survey, 1977
vi,85p : ill. - (Technical monographs / Soil Survey of Great Britain (England and Wales) ; no.8). - Bibliography: p.82-83. - Includes index
B77-26374 SRL
119156x

AERIAL PHOTOGRAPHY. Austria
Items on applications of aerial photography in archaeology in Austria, to date. Applications in archaeology, to date - Catalogues

FENSTER zur Urzeit / [Austellungsleitung und Redaktion: Helmut Windl]. - [Wien] : [Niederösterreichisches Landesmuseum], [1982]
57p,plates. - ([Ausstellungskatalog des Vorarlberger Landesmuseums] ; [Nr.106]) (Katalog des Niederösterreichischen Landesmuseums ; n.F. 117)
Slip pasted on titlepage: "Vorarlberger Landesmuseum ... Bregenz - 3. Mai bis 12. Juni 1983"
r6307269 X.0429/537(106)
8964386

AERIAL PHOTOGRAPHY. Buckinghamshire
Applications in research of sites of antiquities

ALLEN, Denise
Buckinghamshire archaeology from the air. - Aylesbury : Buckinghamshire County Museum, 1979
18p,[4]p of plates : ill. - Bibliography: p.18
0-86059-054-2 X.429/11150 Woolwich
2162296

AERIAL PHOTOGRAPHY. Construction
Applications in selection of construction sites

WAY, Douglas S.
Terrain analysis. - 2nd ed. - Strondsburg : Dowden, Hutchinson and Ross ; New York ; London : McGraw-Hill, 1978
x,438p,[4] leaves of plates : ill(some col). - (Community development series). - Includes bibliographies and index
0-07-068641-6 SRL
1917315

AERIAL PHOTOGRAPHY. Great Britain
Applications in archaeology

RILEY, D. N.
Aerial archaeology in Britain. - Princes Risborough : Shire, 1982
56p : ill. - (Shire archaeology ; 22). - Bibliography: p54-55. - Includes index
0-85263-592-3 X.429/13668
2790211

AERIAL PHOTOGRAPHY. Netherlands
Applications in research on Celtic fields

BRONGERS, J. A.
Air photography and Celtic field research in the Netherlands. - Amersfoort : Rijksdienst voor het Oudheidkundig Bodemonderzoek, 1976
2v. : ill. - (Nederlandse oudheden ; 6). - Bibliography : p.114-119
90-12-01363-1 Ac.5487/2[no.6]
1091352

AERIAL PHOTOGRAPHY. Soviet Union
Applications in reclamation of land, to date

PRIMENENIE aerofotos"emki pri izuchenii lesnogo i bolotnogo meliorativnogo fondov / [otvetstvennye redaktory G.G. Samoĭlovich, E.A. Galkina]. - Leningrad : Geograficheskoe obshchestvo SSSR, Komissiia aerofotos"emki i fotogrammetrii, 1973
131p. - Includes bibliographies
r6462872 X.629/22522
9058028

AERIAL RECONNAISSANCE. Military operations by Great Britain. Army
to 1945

MEAD, Peter
The eye in the air. - London : H.M.S.O., 1983
xiii,274p,[32]p of plates : ill. - Bibliography: p247-259. - Includes index
0-11-771224-8 X.622/16070
2931907

AERIAL SURFACES. Plants
Microorganisms - Conference proceedings

MICROBIOLOGY of aerial plant surfaces / edited by C.H. Dickinson, T.F. Preece. - London : Academic Press, 1976
ix,669p : ill. - Includes bibliographies and index
0-12-215050-3 SRL
0696900

AERIAL SURVEYING
Applications in economic planning - Conference proceedings

SURVEYS FOR DEVELOPMENT (Conference : 1976 : International Institute for Aerial Survey and Earth Sciences)
Surveys for development. - Amsterdam ; Oxford : Elsevier, 1977
[4],182p : ill
0-444-41603-x X.529/32135
1372149

AERIAL SURVEYING
Techniques

KILFORD, William Kenneth
Elementary air survey. - 4th (metric) ed. - London : Pitman, 1979
xii,345p : ill. - Bibliography : p.321-332. - Includes index
0-273-01078-6 X.619/22391 Woolwich
003150X

AERIALS
See Antennas

AEROACOUSTICS
See also
Vibration induced by fluid flow

AEROACOUSTICS

GOLDSTEIN, Marvin E.
Aeroacoustics. - New York ; London : McGraw-Hill, 1976
xvii,293p : ill. - Includes index
0-07-023685-2 SRL
1175300

AEROBATIC DISPLAY TEAMS
See also
Great Britain. Royal Air Force. Aerobatic Team

AEROBATICS

BERLINER, Don
Aerobatics. - Minneapolis : Lerner, c1980
47p : col.ill. - (Superwheels & thrill sports)
0-8225-0436-7 X.629/18218
2743140
- [Another copy.] X.629/22903

CAMPBELL, R. D.
Basic aerobatics. - London : Granada, 1984
vi,142p : ill. - Includes index
0-246-11705-2 X.622/21621
0608181

MÜLLER, Eric
Flight unlimited. - High Wycombe : E. Müller and A. Carson, c1983
179p,[4]p of plates : ill(some col.). - Bibliography: p178. - Includes index
0-9509252-0-9 X.622/23506
0608181

WILLIAMS, Neil, 1934-
Aerobatics. - Shrewsbury : Airlife, 1975
266p : ill
B81-39359 X.629/16465 Woolwich
0608181

Aerobatics. - Shrewsbury : Airlife Publications, 1975
[4],266p : ill
0-9504543-0-3 X.629/10402 Woolwich
0608181

AEROBATICS
Competitions: Lockheed Aerobatic Trophy, to 1965

LLOYD, Tony
The Lockheed aerobatic trophy, 1955-1965. - Hinckley : Midland Counties Publications, 1976
48p : ill
0-904597-02-4 X.709/30497
0651338

AEROBIC BACTERIA
See also
Escherichia coli
Pseudomonas
Vibrios

AEROBIC BACTERIA
Endospore-forming aerobic bacteria - Conference proceedings

The AEROBIC endospore-forming bacteria / edited by R.C.W. Berkeley and M. Goodfellow. - London : Published for the Society for General Microbiology by Academic Press, 1981
xv,373p : ill. - (Special publications of the Society for General Microbiology ; v.4). - Includes bibliographies and index
0-12-091250-3 SRL
2230356

AEROBICS. Physical fitness

COOPER, Kenneth H. (Kenneth Hardy)
The aerobics program for total well-being. - Toronto ; London : Bantam, 1983, c1982
320p : ill. - Bibliography: p310-317. - Includes index
0-553-34018-2 X.622/20518
2969718

COOPER, Kenneth Hardy
The aerobics way. - London : Corgi, 1978
[7],312p. - Bibliography: p.294-307. - Includes index
0-552-10970-3 X.319/18663 Woolwich
2969718

ZEBROFF, Kareen
Controlling hip and tummy fat through yogactivity. - Wellingborough : Thorsons, 1985
[112]p : ill
0-7225-1130-2 X.629/26562
2969718

AEROBICS. Physical fitness. Women

BARTH, Christina
Bodywork. - London : Foulsham, c1984
119p : ill(some col.). - Includes index
0-572-01260-8 X.622/22981
2969726

SORENSEN, Jacki
Aerobic dancing. - London : Angus & Robertson, [1982], c1979
224p : ill. - Includes index
0-207-14722-1 X.622/23072
2969726

VOX, Valentine
Aerobics stretch and shape workout. - Kingswood : World's Work, c1983
95p : ill
0-437-83555-3 X.629/21141
2969726

AERODROME FLIGHT INFORMATION SERVICE OFFICERS. Great Britain
Role

CIVIL AVIATION AUTHORITY
Aerodrome flight information service and the AFISO licence. - London : C.A.A., 1979
vi,24p. - (CAP. 427)
0-86039-097-7 BS.73/258
2279460

AERODROMES
See also
Airports

AERODROMES
British military aerodromes - Directories

MERCER, Paul S.
British military airfields. - Loughborough : Jackson, c1981
43p : ill
0-9507133-8-4 SRL
2485176

AERODROMES
Pavements. Construction

PROPERTY SERVICES AGENCY. Directorate of Civil Engineering Development
Airfield pavements, construction and maintenance works, series 100, preliminaries. - London : H.M.S.O.
Guide notes. - 1975
[5],9p
0-11-670596-5 BS.414/1803(1)
0600709

AERODROMES
Pavements. Construction–continued
Standard specification clauses. - 1975
[5],16p : 1 ill
0-11-670595-7 BS.414/1803
0600709

AERODROMES. Barnet (London Borough)
Hendon Aerodrome. Aeroplanes. Flights, 1911-1947
FLYING at Hendon / compiled by Clive R. Smith. - London : Routledge and Kegan Paul, 1974
31p : chiefly ill
0-7100-8294-0 Cup.1281/17
0230197

AERODROMES. Beaulieu. Hampshire
Beaulieu Airfield, to 1946
COLES, Robert, 1939–
History of Beaulieu airfield. - Brockenhurst : R. Coles, c1982
64p : ill. - Bibliography: p64
B83-13814 X.805/3639
286388x

AERODROMES. Croydon (London Borough)
Croydon Aerodrome - Serials
[JOURNAL (Croydon Airport Society)]. Journal / Croydon Airport Society. - Issue 1 (1981)-. - [Carshalton] : The Society, 1981–
v. : ill
B82-22693 P.801/3299
2733595

AERODROMES. Croydon (London Borough)
Croydon Aerodrome, to 1928
LEARMONTH, Bob
The first Croydon Airport, 1915-1928. - Sutton, Surrey : Sutton Libraries and Arts Services, 1977
viii,87,[1]p : ill. - Includes index
0-9503224-3-1 X.519/27425
1275534

AERODROMES. Croydon (London Borough)
Croydon Aerodrome, 1928-1939
CLUETT, Douglas
Croydon Airport. - Sutton, Surrey : London Borough of Sutton Libraries and Arts Services, 1980
ii-xii,204p : ill. - (A history of Croydon Airport ; vol. 2). - Bibliography: p.197. - Includes index
0-9503224-8-2 X.629/14123 Woolwich
2282232

AERODROMES. Ealing (London Borough)
Acton Aerodrome, to 1917
LONDON BOROUGH OF EALING LIBRARY SERVICE
The story of Acton Aerodrome and the Alliance Factory, with notes on other early aviation in the Borough. - 2nd ed. - London : The Service, 1978
[7],31p
0-9502264-4-0 X.702/5915
1799185

AERODROMES. East Fortune. Lothian Region. Scotland
East Fortune Airfield, to 1961
BUNYAN, I. T.
East Fortune. - Edinburgh : [Royal Scottish Museum], 1983
68p : ill. - (Royal Scottish Museum information series. Technology ; 2). - Bibliography: p68
B85-21598 X.805/6619
3280640

AERODROMES. Gibraltar
Herring gulls. Control measures
HORTON, N.
Gull control. - [London] : ADAS, c1983
4p
B84-39885 BS.3/861
317543x

AERODROMES. Great Britain
BRITISH Isles airfield guide / edited by P.H. Butler. - 7th ed. - Liverpool : Merseyside Aviation Society Ltd, 1976
[28]p
0-902420-11-9 X.519/25906
0676594

AERODROMES. Great Britain
- Directories
BRITISH Isles airfield guide / edited by P.H. Butler. - 8th ed. - Liverpool : Merseyside Aviation Society Ltd, 1978
30p
0-902420-25-9 X.619/18866 Woolwich
0676594

BRITISH Isles airfield guide / edited by P.H. Butler. - 9th ed.. - Liverpool : Merseyside Aviation Society, 1981
32p : ill. - (A Merseyside Aviation Society publication)
0-902420-37-2 X.809/50665
0676594

AERODROMES. Great Britain
Birds. Control measures
BIRD control on aerodromes. - 2nd ed. - London : Civil Aviation Authority, 1981
iv,24p : ill. - (CAP ; 384)
0-86039-136-1 BS.73/268
1030930

CIVIL AVIATION AUTHORITY. Directorate of Aerodrome Standards
Bird control on aerodromes. - London : C.A.A., 1976
[3],28p : ill. - (CAP ; 384). - Bibliography: p.11
0-86039-017-9 BS.73/172
1030930

AERODROMES. Great Britain
Unlicensed aerodromes. Safety
CIVIL AVIATION AUTHORITY. Directorate of Aerodrome Standards
Safety standards at unlicensed aerodromes. - London : The Authority, 1980
iii,24p : ill. - (CAP 428)
0-86039-106-x BS.73/254
2274515

AERODROMES. Great Britain. Royal Air Force
Brawdy RAF Station - For airmen
ROYAL AIR FORCE, BRAWDY
Arrival information handbook. - Gloucester : British Publishing Co., [1977]
64p : ill
0-7140-1586-5 X.708/20863
1103482

AERODROMES. Great Britain. Royal Air Force
Mildenhall RAF Station, to 1979
DRING, Colin M.
A history of RAF Mildenhall. - Mildenhall : Mildenhall Museum Publications, 1980
64p : ill. - (Local history of the Mildenhall area ; no.3). - Bibliography: p.64
0-905457-03-x X.700/26293
2186020

AERODROMES. Great Britain. Royal Air Force
Odiham RAF Station, to 1975
PITCHER, Anne
'Promise and fulfil'. - [Basingstoke] : The author, [1975]
[1],40p : ill
0-9502641-3-x X.709/30262
0578827

AERODROMES. Great Britain. Royal Air Force
RAF Rivenhall, to 1946
STAIT, B. A. (Bruce A)
Rivenhall. - Cheltenham, 1984
109p : ill
0-9509438-0-0 X.809/61573
3192997

AERODROMES. Great Britain. Royal Air Force
Royal Air Force Ashbourne, to 1945
GIDDINGS, Malcolm L.
Royal Air Force Ashbourne. - Ashbourne : Colerne Debden, c1984
96p : ill
0-9509371-0-x X.809/61584
3194337

AERODROMES. Great Britain. Royal Air Force
Royal Air Force Bawdsey, to 1982 & Royal Air Force Woodbridge, to 1982
KINSEY, Gordon
Bawdsey - birth of the beam. - Lavenham : Dalton, 1983
ix,205p : ill. - Includes index
0-86138-017-7 X.800/37950
3084361

AERODROMES. Great Britain. Royal Air Force
Royal Air Force Station, Kai Tak
ALDERSON, G. L. D.
History of Royal Air Force Kai Tak. - [Hong Kong] : Royal Air Force Kai Tak, [1972?]
104p : ill
r6811417 CS.B.104/226
9051740

AERODROMES. Great Britain. Royal Air Force
Royal Air Force Station, Martlesham Heath, to 1974
KINSEY, Gordon
Martlesham Heath. - Rev. ed. - Lavenham : Dalton, 1983
264p : ill. - Includes index
0-86138-023-1 X.800/36339
0328251

AERODROMES. Great Britain. Royal Air Force
Sealand RAF Station, to 1977
FERGUSON, Aldon P. (Aldon Patrick)
A history of Royal Air Force Sealand. - Liverpool : Merseyside Aviation Society Ltd, 1978
80p : ill. - (Military airfield history ; no. 2)
0-902420-28-3 X.709/51865
2304325

AERODROMES. Great Britain. Royal Air Force
Shawbury RAF Station, to 1976
FERGUSON, Aldon P. (Aldon Patrick)
A history of Royal Air Force, Shawbury. - Liverpool : Merseyside Aviation Society Ltd, 1977
59p : ill. - (Military airfield history ; no.1)
0-902420-20-8 X.619/18032 Woolwich
1270885

AERODROMES. Great Britain. Royal Air Force. Great Britain to 1981
ACTION stations. - Cambridge : Stephens.
1: Wartime military airfields of East Anglia, 1939-1945. - 1979
232p : ill. - Includes index
0-85059-335-2 X.800/27791
1516981
2: Military airfields of Lincolnshire and the East Midlands. - 1981
217p : ill. - Includes index
0-85059-484-7 X.800/30598
1516981
3: Military airfields of Wales and the North-West. - 1981
217p : ill. - Includes index
0-85059-485-5 X.800/31699
1516981
4: Military airfields of Yorkshire. - 1982
216p : ill. - Includes index
0-85059-532-0 X.622/12577 Woolwich
1516981
5: Military airfields of the South West. - 1982
233p : ill. - Includes index
0-85059-510-x X.800/34937
1516981
6: Military airfield of the Cotswolds and the Central Midlands. - 1983
312p : ill. - Includes index
0-85059-529-0 X.800/37200
1516981
- Includes index
8: Military airfields of Greater London. - 1984
248p : ill
0-85059-585-1 X.800/38098
1516981
9: Military airfields of the Central South and South-East. - 1985
[312]p : ill. - Includes index
0-85059-608-4 X.800/41249
1516981

SMITH, David J. (David John), 1943–
Action stations. - Cambridge : Stephens.
7: Military airfields of Scotland, the North-East and Northern Ireland. - 1983
232p : ill
0-85059-563-0 X.800/39093
1516981

AERODROMES. Great Britain. Royal Air Force. Kent to 1980
KENT airfields in the Battle of Britain / the Kent Aviation Historical Research Society. - Repr. (with some revision). - Gillingham : Meresborough, 1982
192p : ill
0-905270-36-3 X.809/54498
287072x

AERODROMES. Great Britain. Royal Air Force. Lincolnshire to 1978
BLAKE, Ron
The airfields of Lincolnshire since 1912. - Leicester : Midland Counties Publications, 1984
296p : ill. - Bibliography: p199-200
0-904597-32-6 X.805/5740
1729276

HANCOCK, T. N.
Bomber country. - Lincoln : Lincolnshire Library Service, 1978
[9],142p : ill. - (Lincolnshire history series ; 2). - Bibliography: p.133-136. - Includes index
0-86111-100-1 X.709/51209
1729276

AERODROMES. Great Britain. Royal Naval Air Service
Felixstowe Air Station, to 1963
KINSEY, Gordon
Seaplanes - Felixstowe. - Lavenham : T. Dalton, 1978
xii,228p : ill. - Bibliography: 218-219. - Includes index
0-900963-89-1 X.800/27499
1774948

AERODROMES. Great Britain. Royal Naval Air Service Yeovilton RNAS Station. Gifts by Great Britain. Ministry of Defence: Gifts to Fleet Air Arm Museum

GREAT BRITAIN. Treasury
Copy of a Treasury minute dated 14th June 1977 concerning the conditional gift of land and buildings to the Trustees of the Fleet Air Arm Museum. - London : H.M.S.O., [1977]
Sheet ([2]p). - (Cmnd.6829)
0-10-168290-5 BS.44/472
1191659

AERODROMES. Norfolk
to 1979

NORFOLK AND SUFFOLK AVIATION MUSEUM
Airfields of Norfolk and Suffolk. - [Norwich] : [The Museum].
[Part 1]. - 1977
[2],22p
B77-32581 X.0705/201
1262335

AERODROMES. Pershore. Hereford and Worcester
Pershore RAF Station, to 1981

WARREN, Glyn
RAF Pershore. - Newport Pagnell : Enthusiasts for G. Warren, c1982
59p : ill
0-907700-05-5 X.809/52950
2774852

AERODROMES. Suffolk
to 1979

NORFOLK AND SUFFOLK AVIATION MUSEUM
Airfields of Norfolk and Suffolk. - [Norwich] : [The Museum].
[Part 1]. - 1977
[2],22p
B77-32581 X.0705/201
1262343

AERODROMES. Sywell. Northamptonshire
Sywell Aerodrome, to 1978

PAUL, Christopher, b.1907
Sywell. - [Sywell] : [Sywell Aerodrome Ltd], [1978]
ix,94p,[36]p of plates : ill. - Includes index
B78-28167 X.700/25740
1547763

AERODROMES. United States. Army Air Force. Air Force, 8th. East Anglia
1942-1945 compared with aerodromes, 1977

AIRFIELDS of the Eighth then and now / [edited by Winston G. Ramsey] ; text Roger A. Freeman ; photographs, 'After the battle' magazine. - London : Battle of Britain Prints International Ltd, 1978
240p : chiefly ill
0-900913-09-6 X.622/6668 Woolwich
1755730

AERODROMES. Vigna di Valle. Italy
Military aerodromes, to date

PESCE, Giuseppe
Vigna di Valle. - Roma : Ateneo & Bizzarri, 1979
179p : ill(some col)
r6505229 X.622/14263
8876711

AERODROMES. White Waltham. Berkshire
White Waltham Airfield. Environment planning

BERKSHIRE. County Council
White Waltham airfield. - Amended. - [Reading?] : [County Planning Department?], 1982
24p
B83-39083 X.805/4347
2956063

AERODYNAMIC DRAG. Motor vehicles
- Conference proceedings

SYMPOSIUM ON AERODYNAMIC DRAG MECHANISMS OF BLUFF BODIES AND ROAD VEHICLES (1976 : Warren, Mich.)
Aerodynamic drag mechanism of bluff bodies and road vehicles. - New York ; London : Plenum Press, 1978
xii,380p : ill. - Includes bibliographies and index
0-306-31119-4 SRL
141531x

AERODYNAMIC FACTORS. Safety. Aeroplanes

ROED, Aage
Flight safety aerodynamics. - Shrewsbury : Airlife Publications, 1977
[2],158p : ill
0-9504543-5-4 X.629/11953 Woolwich
1524321

AERODYNAMICS
See also
Aerothermodynamics
Dynamics. Aircraft
Wind tunnels

AERODYNAMICS

ALLEN, John E. (John Elliston)
Aerodynamics. - [2nd ed.]. - New York : McGraw-Hill ; St Albans : Granada, 1982
205p : ill. - Includes index
0-07-001074-9 X.622/13119
0789585
Aerodynamics. - 2nd ed. - London : Granada, 1982
viii,205p : ill. - Includes index
0-246-11300-6 SRL
0789585

ANDERSON, John D. (John David)
Fundamentals of aerodynamics. - New York ; London : McGraw-Hill, c1984
xi,563p : ill. - Includes index
0-07-001656-9 X.622/20856
0789585

KRASNOV, N. F.
Aerodynamics. - New Delhi : Amerind, c1978
xx,740p : ill
0-86249-265-3 X.622/17632
0789585

KUETHE, Arnold Martin
Foundations of aerodynamics. - 3rd ed.. - New York ; London : Wiley, 1976
xii,527p,fold plate : ill. - Bibliography: p.508-518. - Includes index
0-471-50953-1 X.620/16508 Woolwich
0789585

AERODYNAMICS
Applications of digital computer systems. Programming languages: Basic language

STRIBLING, C. B. (C Barry)
BASIC aerodynamics. - London : Butterworths, 1984
162p : ill. - (Butterworths BASIC series). - Bibliography: p157. - Includes index
0-408-01310-9 X.629/26284
3113566

AERODYNAMICS. Aerofoils. Aeroplanes

GLAUERT, H. (Hermann)
The elements of aerofoil and airscrew theory. - 2nd ed. - Cambridge : Cambridge University Press, 1947
232p : ill. - (Cambridge science classics series). - Bibliography: p230. - Includes index
0-521-27494-x X.629/21293
2962810

AERODYNAMICS. Aerofoils. Turbomachinery

GOSTELOW, J. P.
Cascade aerodynamics. - Oxford : Pergamon, 1984
xvi,270p : ill. - (Thermodynamics and fluid mechanics series) (Pergamon international library). - Includes index
0-08-020428-7 X.629/24106
3165752

AERODYNAMICS. Aeroplanes

BERTIN, John J.
Aerodynamics for engineers. - Englewood Cliffs ; London : Prentice-Hall, 1979
xvii,410p : ill. - Includes index
0-13-018234-6 X.620/19047 Woolwich
1763598

HOUGHTON, E. L. (Edward Lewis)
Aerodynamics for engineering students. - 3rd ed.. - London : Arnold, 1982
viii,696p : ill. - Bibliography: p676. - Includes index
0-7131-3433-x X.622/13866
1763598

HOUGHTON, Edward Lewis
Aerodynamics for engineering students. - 2nd ed., S.I. units. - London : Edward Arnold, 1978
viii,458p : ill. - Bibliography: p.422. - Includes index
0-7131-3395-3 X.619/19284 Woolwich
1763598

AERODYNAMICS. Aircraft

KÜCHEMANN, Dietrich
The aerodynamic design of aircraft. - Oxford : Pergamon, 1978
x,564p : ill. - (Pergamon international library). - Bibliography: p.515-555. - Includes index
0-08-020515-1 SRL
1448382

MORAN, Jack
An introduction to theoretical and computational aerodynamics. - New York ; Chichester : Wiley, c1984
xiii,464p : ill. - Includes index
0-471-87491-4 SRL
1448382

The PRINCIPLES of flight. - London : British Petroleum Educational Service, 1977
1wall chart : ill. (col.)
av00193 Cup.900.NN.15/60
2011778

SHEVELL, Richard S. (Richard Shepherd)
Fundamentals of flight. - Englewood Cliffs ; London : Prentice-Hall, c1983
xxiv,405p : ill. - Includes index
0-13-339093-4 X.322/12269
2011778

AERODYNAMICS. Aircraft
Transonic aerodynamics

MOULDEN, T. H. (Trevor Holmes)
Fundamentals of transonic flow. - New York ; Chichester : Wiley, c1984
xvii,332p : ill. - Includes index
0-471-04661-2 X.622/23918
3287165

AERODYNAMICS. Aircraft. Curriculum subjects. Students, 5-14 years. Schools. Great Britain
- For teaching

SLACK, Derek
Flying starts here. - Basingstoke : Globe Education [for] West Sussex County Council, 1981
36p : ill. - (Science horizons. Level 2b). - Bibliography: p10. - List of films: p10
0-333-28541-7 X.622/10912 Woolwich
2536811

AERODYNAMICS. Aircraft. United States
Research, to 1935

HANLE, Paul A.
Bringing aerodynamics to America. - Cambridge, Mass. ; London : MIT, c1982
xiv,184p : ill. - Includes index
0-262-08114-8 X.629/18482
2783738

AERODYNAMICS. Buildings

AYNSLEY, R. M.
Architectural aerodynamics. - London : Applied Science Publishers, 1977
xii,254p : ill. - (Architectural science series). - Includes index
0-85334-698-4 SRL
1109413

AERODYNAMICS. Cars
Effects on consumption of petrol

MOTOR INDUSTRY RESEARCH ASSOCIATION
Vehicle fuel economy. - [London] : Department of Industry, [1976]
[3],113p : ill
0-85605-222-1 SRL
0709638

AERODYNAMICS. High speed land transport

ADVANCED GROUND TRANSPORT PANEL. Aerodynamics Working Group
The aerodynamic problems of advanced ground transport. - Swindon : Science Research Council, [1978]
[7],106[i.e.117]p(6 fold) : ill
B79-05164 BS.375/78
1778102

AERODYNAMICS. Hovercraft
The PRINCIPLE of the hovercraft. - London : British Petroleum Educational Service, 1977
1wall chart : ill. (col.)
av00194 Cup.900.NN.15/61
2011735

AERODYNAMICS. Model aircraft

SIMONS, Martin, b.1930
Model aircraft aerodynamics. - Watford : Model and Allied Publications, 1978
vi,271p,[10]p of plates : ill. - Includes index
0-85242-441-8 SRL
1589881

AERODYNAMICS. Motor vehicles

SCIBOR-RYLSKI, A. J. (Adam Julius)
Road vehicle aerodynamics. - London : Pentech Press, 1975
[8],213p : ill. - Bibliography: p.202-204. - Includes index
0-7273-1802-0 SRL
0362026
Road vehicle aerodynamics. - London : Pentech Press, 1979
[8],213p : ill. - Bibliography: p.202-204. - Includes index
0-7273-1804-7 X.619/20090 Woolwich
0362026
Road vehicle aerodynamics. - 2nd ed.. - London : Pentech, 1984
xii,244p : ill. - Bibliography: p232-234. - Includes index
0-7273-1805-5 X.629/25486
0362026

AERODYNAMICS. Natural environment
WARD-SMITH, A. J. (Alfred John)
Biophysical aerodynamics and the natural environment. - Chichester : Wiley, c1984
x,172p : ill. - Bibliography: p167-168. - Includes index
0-471-90436-8 SRL
3055256

AERODYNAMICS. Rockets
Glushko, Valentin Petrovich
ROMANOV, Aleksandr Petrovich
Raketam pokorīaetsīa prostranstvo. - Moskva : Izdatel'stvo politicheskoĭ literatury, 1976
109 p., 8 p. of plates. - (Geroi Sovetskoĭ Rodiny)
r6023014 X.708/18604
0892408

AERODYNAMICS. Vehicle tunnels
- Bibliographies
PURSALL, B. R.
The aerodynamics and ventilation of vehicle tunnels. - Bedford : BHRA Fluid Engineering, 1976
vii,237p : ill. - (Series / BHRA Fuild Engineering ; vol.2). - Includes index
0-900983-62-0 SRL
0098108

AERODYNAMICS. Vehicle tunnels
- Conference proceedings
INTERNATIONAL SYMPOSIUM ON THE AERODYNAMICS & VENTILATION OF VEHICLE TUNNELS (4th : 1982 : York)
Papers presented at the Fourth International Symposium on the Aerodynamics & Ventilation of Vehicle Tunnels held at York, England March 23-25, 1982. - Bedford : BHRA Fluid Engineering, [1982]
viii,504p : ill
0-906085-63-2 SRL
0098108

INTERNATIONAL SYMPOSIUM ON THE AERODYNAMICS AND VENTILATION OF VEHICLE TUNNELS (2nd : 1976 : Cambridge)
Proceedings of the Second International Symposium on the Aerodynamics and Ventilation of Vehicle Tunnels, Cambridge, 1976. - [Bedford] : British Hydromechanics Research Association, Fluid Engineering, [1977]
611p in various pagings : ill. - Includes index
0-900983-51-5 SRL
0098108

INTERNATIONAL SYMPOSIUM ON THE AERODYNAMICS AND VENTILATION OF VEHICLE TUNNELS (3rd : 1979 : Sheffield)
Papers presented at the Third International Symposium on the Aerodynamics and Ventilation of Vehicle Tunnels, Sheffield England, March 1979. - Bedford : BHRA Fluid Engineering, 1979
2v.(703p) : ill
0-906085-28-4 SRL
0098108

AERODYNES
See Heavier than air aircraft

AEROELASTICITY. Aircraft
BISPLINGHOFF, Raymond Lewis
Principles of aeroelasticity. - New York : Dover Publications ; London : Constable, 1975
xi,527p : ill. - Includes index
0-486-61349-6 X.619/15388 Woolwich
0565415

AEROEMBOLISM
See Decompression sickness

AEROFOILS. Aeroplanes
Aerodynamics
GLAUERT, H. (Hermann)
The elements of aerofoil and airscrew theory. - 2nd ed. - Cambridge : Cambridge University Press, 1947
232p : ill. - (Cambridge science classics series). - Bibliography: p230. - Includes index
0-521-27494-x X.629/21293
2962810

AEROFOILS. Aircraft
See also
Rotors. Aircraft
Vortella
Wings. Aircraft

AEROFOILS. Flying model aeroplanes
Construction
PRESSNELL, Martyn
Aerofoils for aeromodellers. - London : Pitman, 1977
207p : ill
0-273-01015-8 SRL
1037900

AEROFOILS. Turbomachinery
Aerodynamics
GOSTELOW, J. P.
Cascade aerodynamics. - Oxford : Pergamon, 1984
xvi,270p : ill. - (Thermodynamics and fluid mechanics series) (Pergamon international library). - Includes index
0-08-020428-7 X.629/24106
3165752

AERONAUTICAL COMMUNICATIONS SATELLITES
- Conference proceedings
INTERNATIONAL CONFERENCE ON MARITIME AND AERONAUTICAL SATELLITE COMMUNICATION AND NAVIGATION (1978 London)
International Conference on Maritime and Aeronautical Satellite Communication and Navigation, 7-9 March 1978, organised by the Electronics Division of the Institution of Electrical Engineers in association with ... [others]. - [London] : Institution of Electrical Engineers, 1978
vii,158p : ill. - (Conference publications / Institution of Electrical Engineers ; no.160)
0-85296-188-x SRL
1453149

AERONAUTICAL ENGINEERING
See also
Aircraft

AERONAUTICAL ENGINEERING
Shute, Nevil - Biographies
SHUTE, Nevil
Slide rule. - London : Pan in association with Heinemann, 1968
221p,[4]p of plates : ill
0-330-02251-2 H.84/262
3033090

AERONAUTICAL ENGINEERING
Wallis, Sir Barnes - Biographies
MORPURGO, J. E. (Jack Eric)
Barnes Wallis. - Updated ed. - London : Ian Allan, 1981
xvi,400p,[16]p of plates : ill. - Includes index
0-7110-1119-2 X.629/15449 Woolwich
1180622

AERONAUTICAL ENGINEERING
Wells, Godfrey Nugent - Biographies
WELLS, Godfrey Nugent
Head in the clouds. - Auckland ; London : Hodder and Stoughton, 1973
253p,[32]p of plates : ill. - Includes index
0-340-17737-3 X.809/25406
0170321

AERONAUTICAL ENGINEERING. Great Britain
Examinations
CIVIL AVIATION AUTHORITY
Aeronautical engineering certificate. - 3rd ed. - London : C.A.A., 1980
viii,29p. - (CAP ; 416). - Includes bibliographies
0-86039-104-3 BS.73/257
2285517

AERONAUTICAL ENGINEERING. Great Britain
Organisations: Royal Aeronautical Society. Publications, 1897-1977 - Indexes
ROYAL AERONAUTICAL SOCIETY
Publications index 1897-1977. - London : The Society, 1979
560p
B81-23615 X.629/15675 Woolwich
2493225

AERONAUTICAL ENGINEERING. Great Britain
Projects, 1945-1972. Abandonment
WOOD, Derek
Project cancelled. - London : Macdonald and Jane's, 1975
253p : ill
0-356-08109-5 X.620/15158
0501441

AERONAUTICAL ENGINEERING. Great Britain
Research organisations: Royal Aircraft Establishment
TURNILL, Reginald
Farnborough. - London : Hale, 1980
224p,[40]p of plates : ill. - Includes index
0-7091-8584-7 Nov.48049
0261807

AERONAUTICAL ENGINEERING. Soviet Union
ARLAZOROV, Mikhail Saulovich
Konstruktory. - Moskva : Sovetskaīa Rossiīa, 1975
278p : ill. - Bibliography : p.269-[279]
r6094001 X.629/8089 Woolwich
1982133

AERONAUTICAL ENGINEERING. Soviet Union
T͡Sander, Fridrikh Arturovich - Biographies
T͡SANDER, Fridrikh Arturovich
[Selections]. Sobranie trudov. - Riga : Zinatne, 1977
565p.,[8]p. of plates.
r6051471 X.629/8133 Woolwich
1865242

AERONAUTICAL ENGINEERING INDUSTRIES. Europe
Directories - Directories - Serials
FLIGHT international directory of European aviation. - 1982-. - Sutton : IPC Transport Press, 1982-
v. : ill
B82-31716 P.521/3695
2767864

AERONAUTICAL ENGINEERING INDUSTRIES. Pennsylvania to 1980
TRIMBLE, William F.
High frontier. - Pittsburgh : University of Pittsburgh Press ; London : Feffer and Simons, c1982
xiv,344p : ill. - Bibliography: p319-330. - Includes index
0-8229-5340-4 X.622/17727
3033910

AERONAUTICS
See also
Air sports
Aircraft
Aviation
Flying. Aircraft

AERONAUTICS
BENHAM, Peter Philip
Flight of fancy. - [Belfast] : Queen's University of Belfast, 1978
55p : ill. - (New lecture series / Queen's University of Belfast ; no.109)
0-85389-141-9 Ac.1522/2[no.109]
1767070

VAN SICKLE, Neil D. (Neil David)
Van Sickle's modern airmanship. - 5th ed.. - New York ; London : Van Nostrand Reinhold, c1981
ix,878p : ill. - Includes index
0-442-25793-7 X.622/10272 Woolwich
1767070

AERONAUTICS
- Serials
AEROGRAM. - Vol.1, no.1 (Nov. 1977)-. - Cranfield : The College, 1977-
v. : ill
Holdings: vol 1, no 4 (May '79)-
B84-24310 P.625/284
1912216

AIR international. - Vol.7, no.1- ; July 1974-. - [London] : Fine Scroll Ltd : Distributed by Ducimus Books, 1974- : ill(some col)
B79-23328 P.513/598
1912216

JANE'S aviation annual. - 1981-82. - London : Jane's, 1981
1v. : ill
B83-19460 SRL
1912216

JANE'S aviation review. - 1982-83-. - London : Jane's, 1982-
v. : ill
B83-19461 P.631/312
1912216

AERONAUTICS
to 1980
MAYNARD, Christopher
Aircraft. - London : Kingfisher, 1981
93p : ill(some col.). - (A Kingfisher factbook). - Includes index
0-86272-014-1 X.629/16530 Woolwich
2571544

AERONAUTICS
Aeronautics & astronautics - Dictionaries
AVIATION-SPACE dictionary. - 6th ed.. - Fallbrook : Aero, 1980
272p : ill
0-8168-3002-9 SRL
2490455

AERONAUTICS
Celestial navigation
UNITED STATES. Defense Mapping Agency. Hydrographic Center
Sight reduction tables for air navigation. - London : H.M.S.O.
Vol.1: Selected stars, epoch 1980.0. - 1977
iii-xii,324p. - (A.P.3270)
0-11-772172-7 BS.40/70
1703552

AERONAUTICS
Navigation

CANNER, W. H. P.
Air navigation. - Glasgow : Brown and Ferguson, 1976
[9],334p : ill. - Includes index
0-85174-243-2 X.629/10990 Woolwich
0765910

FARRELL, James L. (James Lawrence)
Integrated aircraft navigation. - New York ; London :
Academic Press, 1976
xv,351p : ill. - Includes index
0-12-249750-3 SRL
0765910

AERONAUTICS
Research. Cryogenic wind tunnels - Conference proceedings

INTERNATIONAL SYMPOSIUM ON CRYOGENIC WIND TUNNELS (1st
1979 : University of Southampton)
First International Symposium on Cryogenic Wind Tunnels.
- [Southampton] : [The Department], [1979?]
330p in various pagings : ill
B81-39547 SRL
2570580

AERONAUTICS
Research. Low speed wind tunnels

RAE, William H.
Low-speed wind tunnel testing. - 2nd ed.. - New York ;
Chichester : Wiley, c1984
xiii,534p : ill. - Includes index
0-471-87402-7 SRL
3287181

AERONAUTICS
Research. Wind tunnels

CATALYTIC combustion studies in a free piston tunnel-final
report / D.R.J. Baxter ... [et al.]. - [Southampton] :
[University of Southampton, Department of Aeronautics and
Astronautics], 1983
28leaves,[24]leaves of plates : ill. - (AASU memo ;
83/14)
r6863691 X.622/21855
3179680

AERONAUTICS. Brazil
ca 1700-ca 1920

AGUIAR, Pinto de
Os precursores brasileiros da aeronáutica. - Rio de
Janeiro : Civilização Brasileira, 1975
212 p. : ill. - (Coleção Retratos do Brasil ; v. 92)
r6011650 X.619/9336 Woolwich
0889334

AERONAUTICS. Cumbria
1825-1914

CONNON, Peter
In the shadow of the eagle's wing. - Penrith : St.
Patrick's, 1982
134p : ill. - Includes index
0-9508287-0-x X.805/3590
2847345

AERONAUTICS. Dumfries and Galloway Region. Scotland
1825-1914

CONNON, Peter
In the shadow of the eagle's wing. - Penrith : St.
Patrick's, 1982
134p : ill. - Includes index
0-9508287-0-x X.805/3590
2847353

AERONAUTICS. Italy
to date

FALESSI, Cesare
Storia fotografica dell'aviazione italiana. - Roma :
Newton Compton, 1984
404p : chiefly ill. - (Quest' Italia ; 73)
r6511313 X.629/25759
8265461

AERONAUTICS. Italy
ca 1600-ca 1900

RARITÀ bibliographiche aeronautiche dei secoli XVII,
XVIII, XIX con riproduzione integrale dei testi originali
/ [a cura di] Luigi Pescasio. - Mantova : Editoriale
Padus, [1975]
606 p.
r6024155 X.625/174 Woolwich
1016555

AERONAUTICS. Italy
1884-1915

LODI, Angelo
Storia delle origini dell'aeronautica militare,
1884-1915. - Roma : Ateneo & Bizzarri, 1976-77
2v. : ill
r6100648 X.0622/202 Woolwich
2038617

AERONAUTICS. Italy
Steffanini, Cirillo

LODI, Angelo
Steffanini story, 1859-1946. - Roma : Ateneo & Bizzarri,
1979
156p : ill
r6502591 X.622/11282 Woolwich
8752796

AERONAUTICS. United States

FOSTER, Timothy R. V.
The aviator's catalog. - New York ; London : Van Nostrand
Reinhold, c1981
255p : ill. - Includes index
0-442-21201-1 X.622/9790
2364891
- [Another copy: paperback] ISBN 0-442-22465-6
X.622/9649

AERONCA AEROPLANES
Aeronca, Citabria & Decathlon aeroplanes, to 1983

DOWNIE, Don
The complete guide to Aeroncas, Citabrias and Decathlons.
- Blue Ridge Summit : Tab, c1984
viii,264p : ill. - Includes index
0-8306-2317-5 X.629/25612
3253570

AERONCA AEROPLANES
Flying, 1935-1937

PARHAM, Jack
Flying for fun. - Cambridge : Belinda Parham, c1983
96p : ill
B84-10981 X.629/22042
3044440

AEROPHONES
See Wind instruments

AEROPLANE AND ARMAMENT EXPERIMENTAL ESTABLISHMENT
1939-1945

JOHNSON, Brian, 1925-
A most secret place. - London : Jane's, 1982
x,193p,[32]p of plates : ill. - Includes index
0-7106-0203-0 X.800/34923
2844982

AEROPLANE INDUSTRIES

BELL, Rivian
Careers in an airplane factory. - Minneapolis : Lerner,
c1983
[36]p : col ill. - (An Early career book)
0-8225-0349-2 X.629/21435
2974363

AEROPLANE INDUSTRIES. England
Airspeed (Firm), to 1963

MIDDLETON, D. H. (Don H)
Airspeed. - Lavenham : Terence Dalton, 1982
ix,206p : ill. - Includes index
0-86138-009-6 X.622/15053
2874660

AEROPLANE INDUSTRIES. Filton. Avon
Bristol Aeroplane Company. Air raids by Germany. Luftwaffe,
1940. Bomber aeroplanes

WAKEFIELD, Kenneth
Luftwaffe encore. - London : Kimber, 1979 [i.e. 1980]
223p : ill. - Bibliography: p.17-19. - Includes index
0-7183-0077-7 X.800/29065
222738x

AEROPLANES
See also
Aeronca aeroplanes
Arado aeroplanes
Armstrong Whitworth aeroplanes
Attack aeroplanes
Avenger aeroplanes
Aviation
Avro aeroplanes
Beaufort aeroplanes
Belfast aeroplanes
Blackbird aeroplanes
Boeing aeroplanes
Bomber aeroplanes
Boston aeroplanes
Breda aeroplanes
Bristol aeroplanes
Britannia aeroplanes
British Aerospace aeroplanes
British Aircraft Corporation aeroplanes
British Airways aeroplanes
Brush aeroplanes
C-46 aeroplanes
C97 aeroplanes
Caravelle aeroplanes
Catalina aeroplanes
Chance Vought aeroplanes
Citabria aeroplanes
Civil aeroplanes
Comet aeroplanes
Commercial aeroplanes
Concorde aeroplanes
Consolidated Aircraft Corporation aeroplanes
Convair aeroplanes
Crusader aeroplanes
Curtiss aeroplanes
Dakota aeroplanes
Dart Herald aeroplanes
Dassault aeroplanes
Dassault-Breguet aeroplanes
DC-3 aeroplanes
DC-4 aeroplanes
DC-6 aeroplanes
DC-8 aeroplanes

AEROPLANES
See also
DC-9 aeroplanes
DC-10 aeroplanes
De Havilland aeroplanes
Decathlon aeroplanes
Derelict aeroplanes
Dornier aeroplanes
Eagle aeroplanes
Electra aeroplanes
English Electric Aeroplanes
Experimental aeroplanes
F-27 aeroplanes
Fairey aeroplanes
Falon aeroplanes
Focke-Wulf aeroplanes
Fokker aeroplanes
Galaxy aeroplanes
General Dynamics aeroplanes
Gloster aeroplanes
Grumman aeroplanes
Halifax aeroplanes
Hampden aeroplanes
Handley Page aeroplanes
Harrier aeroplanes
Hawker aeroplanes
Heinkel aeroplanes
Helldiver aeroplanes
IAI aeroplanes
Inex-Adria aeroplanes
Islander aeroplanes
Jet aeroplanes
Jet Commander aeroplanes
Junkers aeroplanes
Light aircraft
Lockheed aeroplanes
Lysander aeroplanes
Martin aeroplanes
McDonnell Douglas aeroplanes
Messerschmitt aeroplanes
Mew aeroplanes
Microlight aeroplanes
Mig aeroplanes
Military aeroplanes
Mitsubishi aeroplanes
Model aeroplanes
Museums exhibiting aeroplanes
North American aeroplanes
North American Aviation aeroplanes
Panavia aeroplanes
Passenger aeroplanes
PBY aeroplanes
Percival aeroplanes
Pilatus Britten-Norman aeroplanes
Preserved aeroplanes

AEROPLANES
See also
Private aeroplanes
Propeller-driven aeroplanes
Reggiane aeroplanes
Republic Aircraft aeroplanes
Rutan aeroplanes
SAAB aeroplanes
Sabre aeroplanes
Savoia Marchetti aeroplanes
Seaplanes
Short Brothers aeroplanes
S.I.A.I. aeroplanes
Skyraider aeroplanes
Stratocruiser aeroplanes
Stuka aeroplanes
Sud-Aviation aeroplanes
Supermarine aeroplanes
Supersonic aeroplanes
Swordfish aeroplanes
Training aeroplanes
Transport aeroplanes
Trident aeroplanes
Turboprop aeroplanes
VC 10 aeroplanes
Vickers aeroplanes
Vickers Armstrong aeroplanes
Victor aeroplanes
Viking aeroplanes
Viscount aeroplanes
Washington aeroplanes
Westland aeroplanes
Westwind aeroplanes
Wildcat aeroplanes

AEROPLANES

PETTY, Mike, 1947-
The airport. - London : Jackdaw [for] Thames Television, 1977
Portfolio : ill(some col). - (Magpies). - Bibliography: (1p.)
0-305-62135-1 Cup.911/43
1189417

AEROPLANES
to 1910

GIBBS-SMITH, Charles Harvard
Pioneers of the aeroplane. - London : Usborne Publishing Limited, 1975
48p : ill(chiefly col). - ([The museum of discovery]). - Includes index
0-86020-012-4 X.622/5037
0369829

AEROPLANES
to 1945

RILEY, Gordon, 1950-
Vintage aircraft of the world. - London : Ian Allan, 1983
192p : ill. - Includes index
0-7110-1111-7 X.629/21186
2956357

AEROPLANES
to 1975

MUNSON, Kenneth George
Famous aircraft of all time. - Poole : Blandford Press, 1976
vii,189p : ill(chiefly col). - ([Blandford colour series]). - Includes index
0-7137-0785-2 X.629/11017 Woolwich
0660957

RUTLAND, Jonathan
The amazing world of aeroplanes. - London : Angus and Robertson, 1976
32p : ill(chiefly col). - (The amazing world). - Includes index
0-207-95661-8 X.625/520 Woolwich
0662372

AEROPLANES
to 1975 - Biographies

ORMES, Ian
The sky masters. - London : Kimber, 1976
224p,[12]p of plates : ill. - Bibliography: p.215-216. - Includes index
0-7183-0284-2 X.620/16083 Woolwich
0660957

AEROPLANES
to 1980

JERRAM, Michael F.
The world's classic aircraft. - London : Muller, 1981
189p : ill(some col.). - (A Charles Herridge Book). - Bibliography: p189
0-584-97076-5 X.622/11484 Woolwich
2581531

AEROPLANES
to date

PLANEMAKERS. - London : Jane's Publishing Co., 1982-.
r6805961 X.0622/659
8840326

ROEDER, Jean
Giganten am Himmel. - Munchen : Leuthold, 1982
520p : ill
3-8129-0004-1 X.622/22999
8840326

AEROPLANES
to date - Catalogues

MUSEO STORICO DELL'AERONAUTICA MILITARE
Velivoli. - Bologna : Patron, 1980
281p : ill. - Includes bibliography
r6508514 X.622/16051
8840326

AEROPLANES
1900-1945

BERLINER, Don
Yesterday's airplanes. - Minneapolis : Lerner, c1980
47p : col.ill. - (Superwheels & thrill sports)
0-8225-0444-8 X.629/18596
2330792
- [Another copy.] X.629/22906

JERRAM, Michael F.
Antiques of the air. - London : New English Library, 1980
192p : ill(some col.). - Includes index
0-450-04813-6 X.622/9364 Woolwich
2330792

AEROPLANES
Accidents

The BLACK box / edited by Malcolm MacPherson. - London : Panther, c1984
156p
0-586-06362-5 X.629/24460
1371371

PILOT error / by the editors of 'Flying magazine'. - New York ; London : Van Nostrand Reinhold, 1977
160p. - Includes index
0-442-22451-6 X.620/17408
1371371
- [Another copy: paperback] ISBN 0-442-22481-8 X.629/18874

AEROPLANES
Accidents. Survival

BARKER, Ralph
Survival in the sky. - London : Kimber, 1976
186p,[12]p of plates : ill
0-7183-0354-7 X.620/16078 Woolwich
0659134

Survival in the sky. - London : Pan Books, 1977
206p,[8]p of plates : ill
0-330-25109-0 X.708/20947
0659134

AEROPLANES
Accidents. Survival. Techniques

NESBITT, Paul Homer
A pilot's survival manual. - New York ; London : Van Nostrand Reinhold, 1978
v,338p : ill. - Bibliography: p.322-323. - Includes index
0-442-25992-1 X.620/18415 Woolwich
1816268
- [Another copy: paperback]
ISBN 0-442-25994-8 X.619/19462 Woolwich

AEROPLANES
Aerodynamics

BERTIN, John J.
Aerodynamics for engineers. - Englewood Cliffs ; London : Prentice-Hall, 1979
xvii,410p : ill. - Includes index
0-13-018234-6 X.620/19047 Woolwich
1763598

HOUGHTON, E. L. (Edward Lewis)
Aerodynamics for engineering students. - 3rd ed.. - London : Arnold, 1982
viii,696p : ill. - Bibliography: p676. - Includes index
0-7131-3433-x X.622/13866
1763598

HOUGHTON, Edward Lewis
Aerodynamics for engineering students. - 2nd ed., S.I. units. - London : Edward Arnold, 1978
viii,458p : ill. - Bibliography: p.422. - Includes index
0-7131-3395-3 X.619/19284 Woolwich
1763598

AEROPLANES
Aerofoils. Aerodynamics

GLAUERT, H. (Hermann)
The elements of aerofoil and airscrew theory. - 2nd ed. - Cambridge : Cambridge University Press, 1947
232p : ill. - (Cambridge science classics series). - Bibliography: p230. - Includes index
0-521-27494-x X.629/21293
2962810

AEROPLANES
Ageing - Conference proceedings

LONG-LIFE aircraft structures / jointly organised by the Royal Aeronautical Society and The Society of Licensed Aircraft Engineers and Technologists. - London : 4 Hamilton Place, W1V 0BQ : The Royal Aeronautical Society, [1980?]
1v.(various pagings) : ill
B82-07064 SRL
2607875

AEROPLANES
American aeroplanes, to 1978 - Encyclopaedias

The ENCYCLOPEDIA of American aircraft / foreword by Ira C Eaker ; edited by Anthony Robinson. - London : Orbis Books, 1979
416p : ill(chiefly col). - Includes index
0-85613-245-4 L.42/447
2072971

AEROPLANES
British aeroplanes, to 1974

TAPPER, Oliver
Roots in the sky. - London : IPC Transport Press for 'Flight international', 1980
96p,[2]p of plates : ill(some col)
0-617-00323-8 X.622/8756 Woolwich
2286688

AEROPLANES
Canadian aeroplanes, to 1981

MOLSON, K. M. (Kenneth M)
Canadian aircraft since 1909. - London : Putnam, 1982
530p : ill
0-370-30095-5 X.629/17774 Woolwich
273642x

AEROPLANES
Control systems. Safety. Assessment

LLOYD, E.
Systematic safety. - London : Civil Aviation Authority, 1982
xi,159p : ill
0-86039-141-8 SRL
2808900

AEROPLANES
De Havilland aeroplanes, 1919-1939

DE Havilland / edited and compiled by Richard Riding. - Sutton : I.P.C. Transport, c1981
223,4p of plates : ill(some col.). - (A Flight international special)
0-617-00332-7 X.622/9758 Woolwich
2370336

AEROPLANES
Design

HALE, Francis J. (Francis Joseph)
Introduction to aircraft performance, selection, and design. - New York ; Chichester : Wiley, c1984
xii,290p : ill. - Bibliography: p279. - Includes index
0-471-07885-9 X.622/21400
1702262

THURSTON, David B.
Design for flying. - New York ; London : McGraw-Hill, 1978
xiii,274p : ill. - Includes index
0-07-064553-1 SRL
1702262

AEROPLANES
Design. Safety aspects

THURSTON, David B.
Design for safety. - New York ; London : McGraw-Hill, c1980
x,196p : ill. - (McGraw-Hill series in aviation). - Includes index
0-07-064554-x SRL
2326779

AEROPLANES
Digital electronic equipment - Conference proceedings

DIGITAL avionics. - London : Royal Aeronautical Society, [1980?]
[89p] : ill
B82-07065 SRL
2607883

AEROPLANES
Electrical systems

PALLETT, Edwin Henry John
Aircraft electrical systems. - London : Pitman, 1976
vii,159p : ill. - (Introduction to aeronautical engineering series). - Includes index
0-273-36159-7 X.620/16487 Woolwich
078334x

AEROPLANES
Engines

BENT, Ralph Daniel
Aircraft powerplants. - 4th ed. - New York ; London : McGraw-Hill, 1978
v,554p : ill. - (Aviation technology series). - Includes index
0-07-004792-8 SRL
0035726

AEROPLANES
F-9F aeroplanes

SULLIVAN, Jim
F9F Panther Cougar. - Carrollton : Squadron / Signal, c1982
49p : ill(some col.)
0-89747-127-x L.49/1327
2742586

AEROPLANES
Flight

SHEAHAN, Denis
Aeroplanes. - London : Black, c1982
32p : ill. - (Science explorers). - Includes index
0-7136-2149-4 X.990/18528 Woolwich
2709449

AEROPLANES
Flight. Stability

ETKIN, Bernard
Dynamics of flight. - 2nd ed. - New York ; Chichester : Wiley, c1982
xiii,370p : ill. - Includes index
0-471-08936-2 X.622/12679 Woolwich
2733374

GENTRY, Everett
All about stalls & spins. - Blue Ridge Summit, Pa. : Tab, c1983
viii,152p : ill. - Bibliography: p148-149. - Includes index
0-8306-2349-3 X.629/22286
2733374

AEROPLANES
Flights, 1903-1908

GIBBS-SMITH, C. H. (Charles Harvard)
The world's first aeroplane flights, (1903-1908). - 5th impression with amendments. - London : H.M.S.O., 1977
32p : ill. - (A Science Museum booklet)
0-11-290033-x X.629/16522 Woolwich
2571099

AEROPLANES
Flights, 1903-1936

TURNER, John Frayn
Famous flights. - London : A. Barker, 1978
160p
0-213-16688-7 X.809/43733
1721224

AEROPLANES
Flying

BAUER, Harry
The mystique of flying. - London : W.H. Allen, 1980
160p : ill. - Bibliography: p.153-160
0-491-02613-7 X.629/14119 Woolwich
2277433

BIRCH, N. H. (Neville Hamilton)
Flight briefing for pilots. - London : Pitman.
Vol.1: An introductory manual of flying training complete with air instruction. - 4th ed. - 1978
ix,228p : ill
0-273-01164-2 X.629/12270 Woolwich
029568x

Vol.2: An advanced manual of flying training complete with air instruction. - 5th ed. - 1978
vii,283p : ill
0-273-01165-0 X.629/12271 Woolwich
029568x

Vol.3: Radio aids to air navigation. - 4th ed. - 1979
viii,214p,[4]fold p of plates : ill(incl 1 col)
0-273-01166-9 X.629/12567 Woolwich
029568x

Vol.4: Associated ground subjects. - 3rd ed. - 1979
viii,207p,[10]p of plates(2 fold) : ill
0-273-01167-7 X.629/12566 Woolwich
029568x

Vol.6: Examination questions and answers. - 2nd ed. - 1980
viii,106p. - Includes index to vols 1-5
0-273-01409-9 X.629/13661 Woolwich
029568x

The IMC rating manual. - 1981
vii,375p,[10]p of plates(2 folded) : ill. - Includes index
0-273-01732-2 X.0629/151 Woolwich
029568x

ENGEL, Lyle Kenyon
The complete book of flying. - New York : Four Winds, c1976
297p : ill. - Includes index
0-590-17377-4 X.622/8215 Woolwich
029568X

FLYING wisdom / by the editors of 'Flying magazine'. - New York ; London : Van Nostrand Reinhold, 1979
xiii,362p. - Includes index
0-442-22452-4 X.620/18455 Woolwich
1888455

GARRISON, Paul
A funny thing happened on the way to the airport. - Blue Ridge Summit, PA : Tab Books, 1980
148p : ill. - (Modern aviation series)
0-8306-9929-5 X.950/31550
3084647

GILBERT, James
The flier's world. - Walton-on-Thames : M. and J. Hobbs ; London : Joseph, 1977
252p : ill(chiefly col). - Includes index
0-7181-1655-0 X.622/6047 Woolwich
1251465

GRIFFIN, Jeff W.
Foundations of flying. - Blue Ridge Summit : TAB, c1982
viii,136p : ill. - Includes index
0-8306-2345-0 X.629/26044
029568x

JACOB, Fred. E.
Takeoffs and touchdowns. - San Diego : Barnes ; London : Tantivy, c1981
xii,301p : ill
0-498-02540-3 X.950/38089
273172x

JERRAM, Michael F.
To be a pilot. - Shrewsbury : Airlife, 1982
116p,[20]p of plates : ill. - Bibliography: p113-114
0-906393-11-6 X.629/17729 Woolwich
2731711

MORE I learned about flying from that / by the editors of Flying ; compiled by Nigel Moll. - London : Secker & Warburg, 1985, c1982
xii,174p : ill
0-436-16051-x X.629/26406
1251465

MURCHIE, Guy
The world aloft. - Toronto ; London : Bantam, 1983, c1960
xi,241p : ill. - (Aviator's bookshelf). - Includes index
0-553-23947-3 X.629/23951
273172x

PENROSE, Harald
Airymouse. - Shrewsbury : Airlife, 1982
162p : ill
0-906393-16-7 Nov.46454 Woolwich
273172x

SCHWEINITZ, Dagmar von
Aeroplanes. - St Albans : Hart-Davis Educational, 1978
32p : ill(chiefly col). - (Questions answered)
0-247-12809-0 X.629/12300 Woolwich
1808087

UNITED STATES. Federal Aviation Administration
How to become a pilot. - New York : Sterling Publishing Co. ; London : Oak Tree Press, 1980
ix,207p : ill
0-7061-2734-x L.49/484
1251465

AEROPLANES
Flying - For private pilots

BRAMSON, Alan Ellesmere
Be a better pilot. - [London] : Martin Dunitz, 1980
239p : ill(chiefly col). - Includes index
0-906348-12-9 X.622/8638 Woolwich
0319139

CAMPBELL, R. D.
Flying training for the private pilot licence. - London : Aviation Training : Granada.
Instrument flying, radio navigation and instrument approach procedure
Instructor manual. - 1981
250,Q1-Q38p : ill. - Includes index
0-246-11695-1 X.622/14273
0042994

Ground training for the private pilot licence. - London : Aviation Training : Granada.
Manual 4: Specific aircraft type, fire, first aid and safety equipment, aeromedical facts. - 1982
1v.(various pagings) : ill. - Includes index
0-246-11704-4 X.622/14340
0042994

PPL revision. 1200 questions and answers for the private pilot. - London : Granada, 1982
[96]p : ill
0-246-11882-2 X.622/14003
0042994

CONES, Betty
Your private pilot's licence. - London : Pitman, 1978
96p : ill. - Bibliography: p.96
0-273-01190-1 X.620/17804 Woolwich
0319139

DWIGGINS, Don
Welcome to flying. - Blue Ridge Summit : Tab, c1983
ix,197p : ill. - Includes index
0-8306-2362-0 X.629/24472
0319139

EVANS, Julien Spencer
Pilot's manual. - Berkhamsted : Poyser, 1978
367p : ill
0-85661-020-8 X.620/17824 Woolwich
0319139

PHILLIPS, Pearson
I learned to fly. - [London] : Warne, 1982
144p : ill(some col.)
0-7232-2973-2 L.42/1878
0319139

TAYLOR, S. E. T. (Sydney Ernest Thomas)
Private pilot studies. - 3rd ed. - Berkhamsted : Poyser, 1977
192p : ill. - Includes index
0-85661-019-4 X.629/12295 Woolwich
0319139

Private pilot studies. - 4th ed. - Berkhamsted : Poyser, 1979
224p : ill. - Includes index
0-85661-024-0 X.629/16896 Woolwich
0319139

AEROPLANES
Flying - Serials

PILOT yearbook. - No.1 (1982)-. - New Malden : Lernhurst Publications, 1982-
v. : ill(some col.)
B82-30483 P.633/112
2759969

AEROPLANES
Flying, to ca 1975

I learned about flying from that! / by the editors of Flying. - London : Secker & Warburg, 1984
x,310p : ill
0-436-16050-1 X.629/24434
2061163

IN the cockpit / introduction by Jeffrey Quill ; edited by Anthony Robinson. - London : Orbis Books, 1979
[8],304p : ill(chiefly col). - Includes index
0-85613-246-2 L.42/376
2061163

IN the cockpit / introduction by Jeffrey Quill ; edited by Anthony Robinson. - Rev. ed. - London : Orbis Publishing, 1981, c1979
304p : ill(some col.). - Includes index
0-85613-388-4 X.622/16036
2061163

AEROPLANES
Flying, 1928-1945

ROBERTSON, Don, 19—
Those magnificent flying machines. - Poole : Blandford, 1984
160p : ill. - Includes index
0-7137-1402-6 X.622/23396
3280934

AEROPLANES
Flying, ca 1945-1970

WILLIAMS, Neil, 1934-
Airborne. - Shrewsbury : Airlife Publications, 1977
[7],175p : ill
0-9504543-2-x X.809/43054
1415115

AEROPLANES
Flying. Applications of microcomputer systems

GARRISON, Paul
Cockpit computers. - New York ; London : McGraw-Hill, c1982
vi,249p : ill. - (McGraw-Hill series in aviation). - Includes index
0-07-022893-0 SRL
2750384

AEROPLANES
Flying. Batten, Jean - Biographies
BATTEN, Jean
Alone in the sky. - Shrewsbury : Airlife, 1979
190p,[12]p of plates : ill
0-906393-01-9 X.629/16466 Woolwich
2569329

AEROPLANES
Flying. Braunburg, Rudolf - Biographies
BRAUNBURG, Rudolf
Ein Leben auf Flügeln. - München : Kindler, [1981]
351p
3-463-00826-2 X.950/16613
8002746

AEROPLANES
Flying by Australian pilots, to 1942
GWYNN-JONES, Terry
True Australian air stories. - London : Hale [etc.], 1977
[14],176p : ill. - Bibliography: p.176
0-7091-6490-4 X.620/17354 Woolwich
1336754

AEROPLANES
Flying by women pilots, to 1945
MOOLMAN, Valerie
Women aloft. - Alexandria, Va. : Time-Life Books, c1981
176p : ill(some col.). - (The Epic of flight). -
Bibliography: p172-173
0-8094-3289-7 L.42/2175
2740435

AEROPLANES
Flying by women pilots, to 1978
BOASE, Wendy
The sky's the limit. - London : Osprey Publishing, 1979
223p : ill. - Bibliography: p.216. - Includes index
0-85045-318-6 X.622/9123 Woolwich
1889087

AEROPLANES
Flying. Cobham, Sir Alan J - Biographies
COBHAM, Sir Alan J.
A time to fly. - London : Shepheard-Walwyn, 1978
x,214p,leaf of plate,xxiv p of plates : ill
0-85683-037-2 X.620/8798 Woolwich
1706934

AEROPLANES
Flying. De Bernardi, Mario - Biographies
FALORSI DE BERNARDI, Maria Vittoria
Mario de Bernardi. - Roma : Bizzarri, 1977
113p : ill
r6065306 X.622/2544 Woolwich
150892X

AEROPLANES
Flying. Doolittle, Jimmy - Biographies
GLINES, Carroll V. (Carroll Vane)
Jimmy Doolittle. - New York ; London : Van Nostrand
Reinhold, 1980
202p : ill. - Includes index
0-442-23102-4 X.809/46384
1204394

THOMAS, Lowell, b.1892
Bomber commander. - London : Sidgwick and Jackson, 1977
xvi,368p,[32]p of plates : ill. - Bibliography:
p.350-351. - Includes index
0-283-98382-5 X.809/42354
1204394

AEROPLANES
Flying. Emergencies - For pilots
CASS, Martin
The pilot in command. - Shrewsbury : Airlife, 1980
213p : ill
0-906393-05-1 X.629/16470 Woolwich
2569175

AEROPLANES
Flying in adverse weather conditions
NEWTON, Dennis W.
Severe weather flying. - New York ; London : McGraw-Hill,
c1983
xix,149p : ill. - (AOPA/McGraw-Hill series in general
aviation). - Includes index
0-07-046402-2 X.622/15593
2904721

AEROPLANES
Flying. Jackson, Archibald Stewart - Biographies
JACKSON, Archibald Stewart
Both feet in the air. - Lavenham : T. Dalton, 1977
123p : ill
0-900963-73-5 X.620/17194 Woolwich
1261398

AEROPLANES
Flying. Johnson, Amy
SILVERED wings. - [Kingston upon Hull] : [Kingston upon
Hull] City Council], [1980]
[32]p : ill
B80-31827 L.49/448
1164430

AEROPLANES
Flying. Johnson, Amy - Biographies
BABINGTON SMITH, Constance
Amy Johnson. - London : White Lion Publishers, 1977
384p,[24]p of plates,[2] leaves of plates : ill. -
Includes index
0-7274-0264-1 X.620/16922 Woolwich
1164430

SNELL, Gordon
Amy Johnson. - London : Hodder and Stoughton, 1980
128p : ill. - (Twentieth century people). - Includes
index
0-340-25203-0 X.622/9420 Woolwich
2068893

AEROPLANES
Flying. Lindbergh, Charles A.
GILL, Brendan
Lindbergh alone. - New York ; London : Harcourt Brace
Jovanovich, 1977
[7],216p : ill
0-15-152401-7 X.802/10471
074882x

AEROPLANES
Flying. Lindbergh, Charles A. - Biographies
GRIERSON, John, 1909-
I remember Lindbergh. - New York ; London : Harcourt
Brace Jovanovich, 1977
192p : ill. - Includes index
0-15-238895-8 X.809/44411
159124x

LINDBERGH, Charles A. (Charles Augustus)
Autobiography of values. - New York : Harcourt Brace
Jovanovich, c1978
xxi,423p.;[40]p.of plates
0-15-110202-3 X.800/32266
074882X
- [Another copy.] X.622/7221 Woolwich

MOSLEY, Leonard
Lindbergh. - London : Hodder and Stoughton, 1976
xxxi,446p,[24]p of plates : ill. - Includes index
0-340-20117-7 X.800/25702
074882x

AEROPLANES
Flying. Markham, Beryl - Biographies
MARKHAM, Beryl
West with the night. - London : Virago, 1984, c1942
xii,293p. - (Virago travellers)
0-86068-541-1 X.958/26266
3021041

AEROPLANES
Flying. Martin, Lajos
MÉSZÁROS, Vince
Martin Lajos a repülés magyar úttörője. - Budapest :
Közlekedési Múzeum, 1976
63p : ill. - (A Közlekedési Múzeum füzetei ; 5)
r6183640 X.0519/821(5)
2460769

AEROPLANES
Flying. Martínez de Pinillos, Carlos, 1920-1947 -
Biographies
MARTÍNEZ DE PINILLOS, Carlos
Mis veinte mil horas de vuelo. - Lima : Asteria C. de
Martínez de Pinillos, 1947 [i.e. 1973]
506 p., 42 p. of plates
r6014777 X.809/22980
0878960

AEROPLANES
Flying. Negrescu, Gheorghe - Biographies
NEGRESCU, Gheorghe
Din amintirile unui vechi aviator. - Bucureşti : Editura
militarâ, 1977
366p.,[8]p.of plates
r6051519 X.619/10780 Woolwich
1981579

AEROPLANES
Flying. Safety measures
COLLINS, Richard L.
Flying safely. - London : A. and C. Black, 1978
xi,276p. - Includes index
0-7136-1890-6 X.629/12100 Woolwich
1696610

MASON, Sammy
Stalls, spins and safety. - New York ; London :
McGraw-Hill, c1982
xviii,165p : ill. - (McGraw-Hill series in aviation). -
Bibliography: p153-157. - Includes index
0-07-040696-0 X.622/14063
1696610

AEROPLANES
Flying. Saint-Exupéry, Antoine de
SAINT-EXUPÉRY, Antoine de
[Selections]. Écrits de guerre 1939-1944 ; avec la
Lettre à un otage et des témoignages et documents. -
Paris : Éditions Gallimard, 1982
649p,plates
2-07-025893-9 X.629/23123
8002746

AEROPLANES
Flying. Saint-Exupéry, Antoine de - Biographies
SAINT-EXUPÉRY, Antoine de
Carnets. - Édition intégrale. - Paris : Gallimard, 1975
284 p.
r6012124 X.519/20694
0857823

AEROPLANES
Flying. Shuttleworth, Richard - Biographies
DESMOND, Kevin
Richard Shuttleworth. - London : Jane's, 1982
190p : ill
0-7106-0185-9 X.805/3113
0621323

GUTTERY, Thomas Ewart
Richard Shuttleworth. - [Biggleswade] : The author,
[1975]
28p : ill
0-901319-08-2 X.709/30388
0621323

AEROPLANES
Flying. Smith, Elinor - Biographies
SMITH, Elinor
Aviatrix. - New York ; London : Harcourt Brace
Jovanovich, c1981
xiii,304p,[8]p of plates : ill. - Includes index
0-15-110372-0 X.629/19706
2871424

AEROPLANES
Flying. Teaching. Smith Barry, Robert - Biographies
TREDREY, Frank D.
Pioneer pilot. - London : P. Davies, 1976
x,164p,8p of plates : ill. - Bibliography: p.151-152. -
Includes index
0-432-16515-0 X.800/25420
0600733

AEROPLANES
Flying. Use of navigation instruments
FLATAU, Courtney L.
The instrument pilot handbook. - New York ; London : Van
Nostrand Reinhold, 1980
372p,[4]p of plates : ill(some col). - Bibliography:
p.333. - Includes index
0-442-22411-7 X.622/8735
2280922
- [Another copy: paperback]
ISBN 0-442-22412-5 X.622/8736

AEROPLANES
Flying. Vlaicu, Aurel
BĂLAN, Ion Dodu
Copilăria unui Icar. - Bucureşti : Editura Ion
Creangă, 1974
93p : ill
r6036153 X.900/18583
1161695

AEROPLANES
Flying. Welch, Ann - Biographies
WELCH, Ann
Happy to fly. - London : Murray, 1983
viii,248p,[16]p of plates : ill. - Includes index
0-7195-4033-x X.950/28138
3045951

AEROPLANES
Flying. Zumbach, Jean, 1935-1967 - Biographies
ZUMBACH, Jean
On wings of war. - London : Deutsch, 1975
3-252p,[8]p of plates : ill
0-233-96623-4 X.809/40222
0378070
On wings of war. - London : Corgi, 1977
285p
0-552-10521-x X.709/31528
0378070

AEROPLANES
French aeroplanes, 1945-1979

CHILLON, J. (Jacques)
French postwar transport aircraft. - Tonbridge : Air-Britain, c1980
176p,[56]p of plates : ill
0-85130-078-2 X.622/13627
278758x

AEROPLANES
French aeroplanes, 1979 - Directories

FRANCE. - 8th ed. - [Dagenham] : [LAAS International], 1979
144p : ill
0-85075-031-8 X.629/15526 Woolwich
2410281

AEROPLANES
Gas turbines

KERREBROCK, Jack L.
Aircraft engines and gas turbines. - Cambridge, Mass. : London : MIT Press, c1977
x,285p : ill. - Includes index
0-262-11064-4 SRL
2211505

TREAGER, Irwin E.
Aircraft gas turbine engine technology. - 2nd ed. - New York ; London : McGraw-Hill, 1979
vi,586p : ill. - Includes index
0-07-065158-2 SRL
2211505

AEROPLANES
Handley Page Dart Herald aeroplanes, to 1979

COWELL, G. (Graham)
Handley Page Herald. - London : Jane's, 1980
154p : ill. - (An Airline Publications book)
0-7106-0045-3 X.629/14921 Woolwich
2346923

AEROPLANES
Hijacking, to 1973

BLAIR, Ed
Odyssey of Terror. - Nashville : Broadman Press, [1977]
316p
0-8054-7908-2 X.809/45460
006033X

AEROPLANES
Italian pilots: Giorgi, Cantardo - Biographies

PADOVAN, Antonio
Contardo Giorgi. - Milano : Italia letteraria, 1978
95p. - (Collana documentazione)
r6198059 X.629/15116 Woolwich
8785848

AEROPLANES
Landings

BRAMSON, A. E. (Alan Ellesmere)
Make better landings. - London : Dunitz, 1982
252p : ill. - Includes index
0-906348-38-2 X.622/20605
3174522

AEROPLANES
Maintenance & repair

INSPECTING your own plane. - Blue Ridge Summit, Pa. : Tab, c1982
v,106p : ill. - (Modern aviation series)
0-8306-2305-1 X.629/24930
215126x

AEROPLANES
Operation - For private pilots

FOSTER, Timothy R. V.
Aircraft owner's handbook. - New York ; London : Van Nostrand Reinhold, 1978
[6],249p : ill. - Includes index
0-442-22462-1 X.622/6897 Woolwich
1815997

AEROPLANES
Overwater flying

SACCHI, Louise
Ocean flying. - New York ; London : McGraw-Hill, 1979
xix,230p : ill. - (McGraw-Hill series in aviation). - Includes index
0-07-054405-0 X.620/19178 Woolwich
208838x

AEROPLANES
Pilots, to 1975 - Biographies

ORMES, Ian
The sky masters. - London : Kimber, 1976
224p,[12]p of plates : ill. - Bibliography: p.215-216. - Includes index
0-7183-0284-2 X.620/16083 Woolwich
0660949

AEROPLANES
Pilots. Stress. Psychological aspects

NEUMAN, Thomas
Dimensionering och validering av percept-genesens försvarsmekanismer. - Stockholm : Försvarets forskningsanstalt, 1978
255p. - (FOA rapport. C ; 55020-H6). - Includes bibliographies
r6029081 X.512/3074
1795023

AEROPLANES
Polish pilots. Training

BARTNIKOWSKI, Bogdan
Zielone skrzydła. - Warszawa : Wydawnictwo Ministerstwa Obrony Narodowej, 1976
182p
r6045096 X.619/10125 Woolwich
1423452

DALECKI, Tadeusz
Gościnne niebo. - Poznań : Wydawnictwo Poznańskie, 1977
359p,16p of plates : ill
r6044347 X.619/9904 Woolwich
1423452

KRÓL, Wacław
Za sterami odrzutowca. - Warszawa : Wydawnictwo Ministerstwa Obrony Narodowej, 1978
228p
r6088762 X.619/10622 Woolwich
1423452

AEROPLANES
Polish women pilots, to date

DALECKI, Tadeusz
"Orlice". - Warszawa : Wydawnictwo Ministerstwa Obrony Narodowej, 1982
295p,plates
83-11-06694-9 X.629/21698
9042903

AEROPLANES
Precision astrolabes. Development. Role of Portuguese pilots, 1919-1927

ROGERS, Francis Millet
Precision astrolabe. - Lisboa : Academia Internacional da Cultura Portuguesa, 1971
xxxi, 397 p., 6 leaves of plates : ill. - Bibliography : 369-376
r6003544 X.619/9154 Woolwich
0939277

AEROPLANES
Racing aeroplanes. Racing, 1919-1926

FOXWORTH, Thomas Gordon
The speed seekers. - London : Macdonald and Jane's, 1975
xi,560p : ill. - Includes index
0-356-08123-0 Cup.1281/227
0641685

AEROPLANES
Racing. Races, 1909-1936

GWYNN-JONES, Terry
The air racers. - London : Pelham, 1984
224p : ill(some col.). - Bibliography: p219. - Includes index
0-7207-1500-8 X.622/21294
3194760

AEROPLANES
Radio equipment

POWELL, J. (James)
Aircraft radio systems. - London : Pitman, c1981
viii,255p : ill. - Includes index
0-273-08444-5 SRL
2565102

AEROPLANES
Radio frequency navigation equipment. Use

BRAMSON, Alan Ellesmere
Radio navigation for pilots. - Shrewsbury : Airlife Publications, 1979
[7],198p : ill(some col)
0-9504543-8-9 X.629/14078 Woolwich
2239183

AEROPLANES
Rolls-Royce Merlin engines, to 1945

HARVEY-BAILEY, Alec
The Merlin in perspective - the combat years. - [Derby] : Rolls-Royce Heritage Trust, 1983, c1981
119p : ill(some col.). - (Historical series ; no.2)
B84-34773 X.809/60717
3150437

AEROPLANES
Russian pilots

KOROLEVA, Evgeniia Vladimirova
Soperniki orlov. - Izd. 3, perer. i dop.. - Odessa : Maiak, 1976
230 p., 16 p. of plates : ill
r6023104 X.629/7600 Woolwich
0887641

AEROPLANES
Safety. Aerodynamic factors

ROED, Aage
Flight safety aerodynamics. - Shrewsbury : Airlife Publications, 1977
[2],158p : ill
0-9504543-5-4 X.629/11953 Woolwich
1524321

AEROPLANES
Speed & altitude. Measurement

GRACEY, William
Measurement of aircraft speed and altitude. - New York ; Chichester : Wiley, c1981
xiv,262p : ill. - Includes index
0-471-08511-1 SRL
2545691

AEROPLANES
Speed records, to 1976

GUNSTON, Bill
Record breakers, air. - London : Scimitar, [1977]
32p : ill(chiefly col). - (Scatterbox)
0-241-89683-5 X.619/18417 Woolwich
1407090

AEROPLANES
Speed records, to 1982

BERLINER, Don
Victory over the wind. - New York ; London : Van Nostrand Reinhold, c1983
xiii,141p,[16]p of plates : ill. - Includes index
0-442-21301-8 X.622/20401
3161250

AEROPLANES
Test flights. Beaumont, Roland

BEAUMONT, Roland
Testing years. - London : I. Allen, 1980
160p : ill
0-7110-1072-2 X.622/8678 Woolwich
2281198

AEROPLANES. Air passenger transport services

TAYLOR, John, b.1922
Passenger aircraft and airlines. - London : Marshall Cavendish, 1975
[2],152p : ill(chiefly col)
0-85685-142-6 X.622/5370 Woolwich
0618950

AEROPLANES. Air passenger transport services - Serials

AIRLINERS. - No.1-. - Hounslow : Airline Publications and Sales Ltd, [197-?]- : ill(some col)
Holdings: no.8, etc. [1976 etc.]
B76-28894 X.0619/15026 Woolwich
0720739

AEROPLANES. Aircraft carriers. Japan. Kaigun
Naval air operations during Pacific campaigns of World War 2

FRANCILLON, René Jacquet
Japanese carrier air groups, 1941-45. - London : Osprey Publishing, 1979
48p : ill(some col). - (Airwar series ; 21)
0-85045-295-3 X.700/26057
1920065

AEROPLANES. Aircraft carriers. United States. Navy
Naval air operations during Pacific campaigns of World War 2

FRANCILLON, René Jacquet
US Navy carrier air groups, Pacific, 1941-45. - London : Osprey Publishing, 1978
48p : ill(some col). - (Airwar series ; 16)
0-85045-291-0 X.700/25748
1551418

SOWINSKI, Larry
The Pacific War. - London : Conway Maritime, 1981
208p : ill(some col.)
0-85177-217-x L.42/1217
257764x

AEROPLANES. Aircraft carriers. United States. Navy
Naval air operations during World War 2

KILDUFF, Peter
US carriers at war. - London : Ian Allan, 1981
128p : ill. - Bibliography: p128
0-7110-1077-3 X.805/1719
2538512

AEROPLANES. Airlines

BARRETT, Norman
Airliners. – London : Franklin Watts, 1984
32p : ill(some col.). – (Picture library). – Includes index
0–86313–134–4 X.622/23081
3272095

EUROPA. – 78 / compiled by A.B. Eastwood. – Luton : LAAS International, 1978
56p : ill
0–85075–015–6 X.700/26260
0713473

GREEN, William, 1927–
An illustrated guide to the world's civil airliners. – London : Salamander, c1982
279p : ill(some col.)
0–86101–107–4 X.629/22780
2543613

MCALLISTER, Chris
Planes and airports. – London : Batsford, 1981
64p : ill(some col.). – Includes index
0–7134–3911–4 X.629/16144 Woolwich
2543613

WORLD airline fleets. – 1977 / [compiled and edited] by Günter G. Endres. – Hounslow : Airline Publications and Sales Ltd, 1977
[7],436p : ill
Subsequent volumes, when received, will be found at SRL
0–905117–32–8 SRL
0713473

WORLD airline fleets, 'easy reference'. – 1978 / [edited] by Günter G. Endres. – Hounslow : Airline Publications and Sales Ltd, 1978
[2],92p
Subsequent volumes, when received, will be placed at this shelfmark
0–905117–43–3 P.511/5097
0713473

YOUNG, John, 1930–
Airliners. – [London] : Fontana, 1979
1folded sheet([12]p) : col.ill. – (Domino ; 12)
0–00–685455–9 Cup.911/129
240771x

AEROPLANES. Airlines
– Directories

EUROPA. – 79 / compiled by A.B. Eastwood. – [Luton] : LAAS International Publications, [1979]
58p,[6]p of plates : ill
0–85075–028–8 X.629/13640 Woolwich
0713473

JP airline fleets. – 76 / by F.E. Bucher and U. Klee. – Hounslow : Airline Publications and Sales Ltd, 1976
[6],272p,[48]p of plates : ill
Subsequent volumes, when received, will be placed at this shelfmark
0–905117–08–5 P.621/710
0713473

TOMKINS, Nigel Milton
Airliner production list. – Hounslow : Airline Publications and Sales Ltd, 1978
[4],361p : ill
Subsequent volumes, when received, will be placed at this shelfmark
0–905117–42–5 P.621/801
0713473

WORLD airline fleets. – 1981. – Hounslow : Airline Publications & Sales, c1981
592p
0–905117–74–3 X.629/15623 Woolwich
0713473

WORLD airline fleets handbook. – 1978. – Tonbridge : Air-Britain, [1978]
201p : ill
Subsequent volumes, when received, will be placed at this shelfmark
0–85130–063–4 P.611/795
0713473

AEROPLANES. Airlines
– Serials

AIRLINE data news. – Issue no.1 (June 1982)–. – Feltham : Aviation Data Centre, 1982–
v. : ill(some col.)
B83–33964 P.621/878
0713473

AEROPLANES. Allied air forces
Losses, 1939–1945

PHILPOTT, Bryan
In enemy hands. – Cambridge : Stephens, 1981
89p : ill
0–85059–499–5 X.622/11414 Woolwich
2573016

AEROPLANES. Eastern Airlines
Ghosts. Investigation

FULLER, John G. (John Grant)
The ghost of flight 401. – London : Souvenir Press, 1978
319p
0–285–62356–7 X.809/43403
1490419

The ghost of flight 401. – London : Corgi, 1979
300p
0–552–11020–5 H.79/1989 Woolwich
1490419

AEROPLANES. Eastern Airlines
Ghosts. Investigation – Biographies

FULLER, Elizabeth
My search for the ghost of Flight 401. – London : Corgi, 1979
[3],216p
0–552–11220–8 H.80/569 Woolwich
1490419

AEROPLANES. Europe
Flying – For private pilots

ALLAN, James, 1928–
Going foreign – VFR. – Elstree : Pooley, 1983
viii,143p : ill. – Includes index
0–902037–10–2 X.629/26449
3283801

AEROPLANES. Germany. Kriegsmarine
Naval air operations during World War 2

GERMAN maritime aircraft / [compiled by] Bryan Philpott. – Cambridge : Stephens, 1981
94p : chiefly ill. – (World War 2 photo album ; no.18)
0–85059–445–6 X.622/10800
2159651

AEROPLANES. Germany. Luftwaffe
1939–1945. Camouflage & markings

MERRICK, K. A.
German aircraft markings, 1939–1945. – London : I. Allan, 1977
176p,[8]p of plates : ill(some col)
0–7110–0739–x X.622/6147 Woolwich
1277367

AEROPLANES. Germany. Luftwaffe
1939–1945. Camouflage & markings – For modelling

PHILPOTT, Bryan
Luftwaffe camouflage of World War 2. – Cambridge : Stephens [for] Airfix Products Ltd, 1975
64p : ill. – ('Airfix Magazine' guide ; 10)
0–85059–213–5 X.629/10185
0517259

AEROPLANES. Germany. Luftwaffe
Aeroplanes selected for evaluation by Great Britain. Royal Air Force, 1945

BUTLER, P. H. (Philip Henry)
'Air Min'. – Liverpool : Merseyside Aviation Society Ltd, 1977
32p : ill
0–902420–21–6 X.619/18468 Woolwich
1432311

AEROPLANES. Germany. Luftwaffe
Air operations during World War 2

COOPER, Matthew
The German Air Force 1933–1945. – London : Jane's, 1981
v,406p,[32]p of plates : ill. – Bibliography: p396–399. – Includes index
0–7106–0071–2 X.622/10667
0543411

SHEPHERD, Christopher
German aircraft of World War II with colour photographs. – London : Sidgwick and Jackson, 1975
144p : ill(some col). – Includes index
0–283–98179–2 X.622/5132 Woolwich
0543411

AEROPLANES. Great Britain
1910–1954. Preservation

CORNWELL, E. L.
Still flying. – London : I. Allan, 1979
80p : ill. – Includes index
0–7110–0905–8 X.620/18229 Woolwich
180023x

AEROPLANES. Great Britain
Flying – For private pilots – Serials

POOLEY'S pilots information guide / compiled with the assistance of the Civil Aviation Authority and National Air Traffic Services. – 1st ed. (Apr. 1982)–. – [Elstree] : Robert Pooley Ltd., 1982–
v. : ill(some col.),col.maps
B82–31708 P.621/846
2765586

AEROPLANES. Great Britain
Flying. Penrose, Harald – Biographies

PENROSE, Harald
Cloud Cuckooland. – Shrewsbury : Airlife, 1981
155p : ill
0–906393–07–8 X.992/4306 Woolwich
2385643

AEROPLANES. Great Britain
Flying. Statutory regulations – For private pilots

CIVIL AVIATION AUTHORITY
Aviation law for applicants for the private pilot's licence. – 4th ed. – London : C.A.A., 1977
v,50p : ill. – (CAP ; 85)
0–86039–031–4 BS.73/180
0122289

AEROPLANES. Great Britain. Royal Air Force

GETHING, Michael J.
RAF air power today. – London : Arms and Armour, 1984
64p : col.ill. – (Warbirds illustrated ; no.25)
0–85368–634–3 X.622/21600
3199452

AEROPLANES. Great Britain. Royal Air Force
to 1980

TURNER, Michael, 1934–
Royal Air Force. – London : Hamlyn, c1981
208p : ill(some col.). – Bibliography: p205. – Includes index
0–600–34933–0 X.425/3329 Woolwich
2578808

AEROPLANES. Great Britain. Royal Air Force
1939–1945. Camouflage & markings

BOWYER, Michael J. F. (Michael John Frederick)
RAF camouflage of World War 2. – Cambridge : Stephens [for] Airfix Products Ltd, 1975
64p : ill. – ('Airfix Magazine' guide ; 11)
0–85059–215–1 X.629/10275 Woolwich
056009x

BRITISH aviation colours of World War Two. – London : Arms and Armour Press [etc.], 1976
[6],56,[1]p,fold plate : ill(some col). – (RAF Museum series / Royal Air Force Museum ; vol.3)
0–85368–271–2 X.0622/507(3)
056009x

AEROPLANES. Great Britain. Royal Air Force
Air operations during World War 2

TURNER, John Frayn
British aircraft of World War II, with colour photographs. – London : Sidgwick and Jackson, 1975
144p : ill(some col). – Includes index
0–283–98180–6 X.622/5133 Woolwich
0543403

AEROPLANES. Great Britain. Royal Air Force
K serial aeroplanes

HALLEY, James J.
Royal Air Force aircraft K1000 to K9999. – Tonbridge : Air-Britain, 1976
83p,[12]p of plates : ill. – (Monographs / Air-Britain)
B79–13583 X.615/2301 Woolwich
1839128

AEROPLANES. Great Britain. Royal Air Force
L series aeroplanes

HALLEY, James J.
Royal Air Force aircraft L1000 to L9999. – Tonbridge : Air-Britain, c1979
68p : ill. – (An Air-Britain publication)
0–85130–068–5 X.622/13943
280607x

AEROPLANES. Great Britain. Royal Air Force
N serial aeroplanes

HALLEY, James J.
Royal Air Force aircraft N1000 to N9999 (1937 to 1942 production). – Tonbridge : Air-Britain, 1977
62p,[9]p of plates : ill
0–85130–057–x X.615/2302 Woolwich
1839209

AEROPLANES. Great Britain. Royal Air Force
P serial aeroplanes

HALLEY, James J.
Royal Air Force aircraft P1000 to P9999. – Tonbridge : Air-Britain, 1978
58[i.e.60],[12]p of plates : ill. – (Monographs / Air-Britain)
0–85130–064–2 X.615/2295 Woolwich
1839217

AEROPLANES. Great Britain. Royal Air Force
R series aeroplanes

HALLEY, James J.
Royal Air Force aircraft R1000-R9999. - Tonbridge : Air-Britain, c1980
63p : ill. - (An Air-Britain publication)
0-85130-082-0 X.622/13942
2806088

AEROPLANES. Great Britain. Royal Air Force
W series aeroplanes

ROYAL Air Force aircraft WA100-WZ999 / compiled by James J Halley. - Saffron Walden : Air-Britain (Historians), c1983
113p : ill. - Includes index
0-85130-109-6 X.622/19252
3081419

AEROPLANES. Great Britain. Royal Air Force
X & Z series aeroplanes

HALLEY, James J.
Royal Air Force aircraft X1000 to X9999, Z1000 to Z9999 - Tonbridge : Air-Britain (Historians), c1984
96p : ill
0-85130-117-7 X.622/21976
3228851

AEROPLANES. Great Britain. Royal Air Force. Special subjects. Wootton, Frank. English paintings
1915-1977

WOOTTON, Frank
The aviation art of Frank Wootton. - New York ; London : Peacock Press ; [London] : [Distributed by Pan Books], 1976
[16]p,[80]p of plates : chiefly ill(chiefly col)
0-330-25048-5 Cup.1281/921
1586130

AEROPLANES. Hendon Aerodrome
Flights, 1911-1947

FLYING at Hendon / compiled by Clive R. Smith. - London : Routledge and Kegan Paul, 1974
31p : chiefly ill
0-7100-8294-0 Cup.1281/17
0230197

AEROPLANES. Italy. Aeronautica militare
German aeroplanes, 1940-1945

ARENA, Nino
La Luftflotte dell'Aeronautica italiana. - Firenze : Edizioni aeronautiche italiane, 1978
111p : ill
r6195619 X.700/22659
2474662

AEROPLANES. Japanese military forces
Air operations during World War 2

COLLIER, Basil
Japanese aircraft of World War II. - London : Sidgwick and Jackson, 1979
144p : ill(some col). - Includes index
0-283-98399-x X.800/28763
2112035

AEROPLANES. Mount Erebus. Antarctic
Accidents, 1979

HICKSON, Ken
Flight 901 to Erebus. - Christchurch [N.Z.] ; London : Whitcoulls, 1980
278p,[16]p of plates : ill(come col.)
0-7233-0641-9 X.809/47335
2385775

AEROPLANES. Munich. West Germany
Accidents involving Manchester United Football Club players, 1958. Survivors' experiences

TAYLOR, Frank, 1920-
The day a team died. - London : Souvenir, 1983
192p,[16]p of plates : ill
0-285-62563-2 X.629/20038
2889757

AEROPLANES. Salmon River Mountains. Idaho
Accidents, 1979. Survivors experiences

GZOWSKI, Peter
The Sacrament. - London : Hale, 1981, c1980
203p : ill
0-7091-8916-8 X.622/9829 Woolwich
2377837

AEROPLANES. South Africa. South African Air Force
to 1980

POTGIETER, Herman
Aircraft. - London : Jane's, 1981
180p : col.ill. - Includes index
0-7106-0117-4 X.622/10812 Woolwich
2534398

AEROPLANES. Soviet Union

I͡AKOVLEV, Aleksandr Sergeevich
Sovetskie samolety. - Izd. 3-e, perer. i dop. - Moskva : Nauka, 1979
396p : ill
r6184173 X.629/12952 Woolwich
2245671

AEROPLANES. Soviet Union
to date

I͡AKOVLEV, Aleksandr Sergeevich
Sovetskie samolety. - Izd. 4-e, perer. i dop. - Moskva : Nauka, 1982
404p : ill
r6463580 X.629/20170
8536139

AEROPLANES. Soviet Union
Flying - Biographies

LETCHIKI / [sostavitel' : V. Mitroshenkov]. - Moskva : Molodai͡a gvardii͡a, 1978
238p,[32]p of plates. - (Zhizn' zamechatel'nykh li͡udeĭ. serii͡a biografii ; vyp.14)
r6094570 WP16340/182
1989626

AEROPLANES. Stock. Museums. Great Britain
to ca 1930

OGILVY, David, 1929-
Discovering old aeroplanes. - Aylesbury : Shire Publications, 1976
56p : ill. - Includes index
0-85263-332-7 X.619/16730 Woolwich
0647373

AEROPLANES. Stock. Shuttleworth Collection
1909-1939. Flying

FROM Bleriot to Spitfire / compiled and edited by David Ogilvy ; with specialist contributions from Roland Beamont ... [et al.]. - Shrewsbury : Airlife Publications, 1977
[12],180p : ill
0-9504543-4-6 X.629/11803 Woolwich
143408x

AEROPLANES. Tuskar Rock. Wexford (County)
Accidents, 1968

WALSH, Dermot
Tragedy at Tuskar Rock. - Dublin : Mercier, c1983
89p
0-85342-696-1 X.529/59561
3060411

AEROPLANES. United States
Flying - For private pilots

NELSON, John L. (John Lewis)
The beginner's guide to flight instruction. - Blue Ridge Summit : Tab, c1983
vi,298p : ill. - Includes index
0-8306-2324-8 X.629/22006
3041174

TOWER, Merrill E.
Flight facts for private pilots. - 2nd ed.. - Fallbrook [Calif.] : Aero, 1983
212p : ill
0-8168-5804-7 X.622/22860
3041174

AEROPLANES. United States
Flying in adverse weather conditions

BUCK, Robert Nietzel
Weather flying. - London : A. and C. Black, 1977
xxiii,261p : ill. - Bibliography: p.252. - Includes index
0-7136-1793-4 X.629/11648 Woolwich
1293354

GARRISON, Paul
Flying VFR in marginal weather. - Blue Ridge Summit : TAB, c1980
192p : ill. - (Modern aviation series). - Includes index
0-8306-9954-6 X.629/22937
1293354

AEROPLANES. United States
Flying. Teaching

TAYLOR, Albert J.
How to teach flying. - Blue Ridge Summit, Pa. : Tab, c1983
x,294p : ill. - Includes index
0-8306-2343-4 X.629/23449
3137694

AEROPLANES. United States
Supersonic passenger aeroplanes. Development. Policies of government. Influence of public opinion

HORWITCH, Mel
Clipped wings. - Cambridge, Mass. ; London : MIT, c1982
x,473p : ill. - Includes index
0-262-08115-6 X.622/13571
2783746

AEROPLANES. United States. Army Air Force
1937-1945. Camouflage & markings

FREEMAN, Roger Anthony
Camouflage & markings. United States Army Air Force. - London : Ducimus Books Ltd, 1974
[8],240p : ill(some col). - Includes bibliographies
0-903234-01-7 X.620/15150
0500089

AEROPLANES. United States. Army Air Force
1941-1945. Camouflage & markings

SCUTTS, Jerry
USAAF camouflage of World War 2. - Cambridge : Stephens [for] Airfix Products Ltd, 1976
64p : ill. - ('Airfix Magazine' guide ; 18)
0-85059-233-x X.629/10858 Woolwich
0746304

AEROPLANES. Yorkshire
Flights, to 1919

REDMAN, Ronald Nelson
Yorkshire's early flying days. - Clapham : Dalesman, 1981
80p : ill. - (Dalesman heritage)
0-85206-646-5 X.808/35484
2604019

AEROPLANES CAPTURED BY GREAT BRITAIN. Germany. Luftwaffe
1939-1945

WEST, Kenneth S.
The captive Luftwaffe. - London : Putnam, 1978
viii,147p : ill
0-370-30020-3 X.629/11775 Woolwich
1397435

AEROPLANES FITTED WITH LICHTENSTEIN RADAR. Germany. Luftwaffe
Junkers JU88 aeroplanes fitted with Lichtenstein radar. Crews. Surrender of aeroplane to Great Britain. Royal Air Force, 1943

HILL, Robert, 1929-
The great coup. - London : Corgi, 1978
172p
0-552-10959-2 X.708/21602
1289683

The great coup. - London : Arlington Books, 1977
208p
0-85140-263-1 X.629/11643
1289683

AEROPLANES MANUFACTURED BY BRUSH ELECTRICAL ENGINEERING COMPANY
to 1946

JARRAM, A. P.
Brush aircraft production at Loughborough. - Leicester : Midland Counties Publications, 1978
40p : ill. - Bibliography: p.40
0-904597-07-5 X.619/19255 Woolwich
1755315

AEROSKOBING. Denmark
Houses. Architecture

NATIONALMUSEET (Denmark)
Historiske huse i Ærøskøbing. - [Copenhagen] : Nationalmuseet, 1975
118 p., 3 fold. plates : ill
87-480-0074-4 X.415/1964 Woolwich
0942251

AEROSOL CONTAINERS
See also
Consumer goods in aerosol containers

AEROSOL CONTAINERS

BRITISH AEROSOL MANUFACTURERS' ASSOCIATION
Code of practice of the British Aerosol Manufacturers' Association. - 4th ed. - London : The Association, 1980
39p : 1 ill
B80-28529 X.622/9342
2207192

ROBERTS, David J. (David John), 1932-
Aerosol ABC. - London : British Aerosol Manufacturers' Association, 1975
47p : ill. - Bibliography: p.44
0-9504900-0-8 X.611/7192 Woolwich
0089036

AEROSOL CONTAINERS
Aerosols

SANDERS, Paul Amsdon
Handbook of aerosol technology. - 2nd ed. - New York ; London : Van Nostrand Reinhold, 1979
xii,526p : ill. - Bibliography: p.507-511. - Includes index
0-442-27348-7 SRL
1886665

AEROSOL CONTAINERS
Propellants
BRITISH AEROSOL MANUFACTURERS' ASSOCIATION
Report of the British Aerosol Manufacturers' Association on aerosol propellants. - London : The Association, 1978
67p in various pagings. - Bibliography: p.[65]
B79-04489 SRL
1771809

AEROSOL CONTAINERS
Testing
STANDARD test methods / [British Aerosol Manufacturers' Association]. - London : The Association, 1980
68p : ill
B82-13323 SRL
2633019

AEROSOL CONTAINERS MANUFACTURED BY METAL BOX PACKAGING LIMITED.
Great Britain
Open top food & drink cans & aerosol containers manufactured by Metal Box Packaging Limited. Prices
GREAT BRITAIN. Price Commission
Metal Box Ltd - open top food and beverage and aerosol cans .... - London : H.M.S.O., [1978]
vii,34p : 3 ill. - ([1977-78 H.C.]135)
0-10-213578-9 BS.416/16(3)
1451448

AEROSOL CONTAINERS MANUFACTURING INDUSTRIES
Industrial safety
BRITISH AEROSOL MANUFACTURERS' ASSOCIATION
A guide to safety in aerosol manufacture. - London : The Association, 1977
[2]p,62 leaves in various pagings : ill. - Bibliography (2 leaves)
B79-00368 X.615/2162 Woolwich
1740601

AEROSOL VACCINATION. Communicable diseases. Man
LEBEDINSKIĬ, V. A.
The inhalation (aerosol) method of vaccination. - Boston Spa : British Library Lending Division, c1981
vi,182p : ill. - Bibliography: p162-182
0-85350-179-3 SRL
2559862

AEROSOLS
- Conference proceedings
INTERNATIONAL AEROSOL CONFERENCES (1st : 1984 : Minneapolis)
Aerosols. - New York ; Oxford : Elsevier, c1984
xxxiii,1069p : ill. - Includes index
0-444-00947-7 SRL
0065765

SYMPOSIUM ON AEROSOL SCIENCE AND TECHNOLOGY (1976 : Atlantic City)
Fundamentals of aerosol science. - New York ; Chichester : Wiley, 1978
xi,372p : ill. - Includes index
0-471-02949-1 SRL
0065765

Recent developments in aerosol science. - New York ; Chichester : Wiley, 1978
xv,327p : ill. - Includes index
0-471-02950-5 SRL
0065765

SYMPOSIUM ON FINE PARTICLES (1975 : Minneapolis)
Fine particles. - New York ; London : Academic Press, 1976
xiii,837p : ill. - (Academic Press rapid manuscript reproduction). - Includes bibliographies and index
0-12-452950-x SRL
0065765

AEROSOLS
Effects on human lungs
AEROSOLS, airways and asthma / edited by Joseph J. Trautlein. - Lancaster : MTP, c1981
102p : ill. - Includes index
0-85200-569-5 SRL
2543532

AEROSOLS
Fog & smoke. Formation & composition - Conference proceedings
SYMPOSIUM ON FOGS AND SMOKES (1973 : Swansea)
Fogs and smokes. - London : Faraday Division, Chemical Society, 1973
235p,[4] leaves of plates,[8]p of plates : ill. - (Symposia / Faraday Society ; no.7). - Includes index
0-85186-828-2 SRL
0516627

AEROSOLS. Aerosol containers
SANDERS, Paul Amsdon
Handbook of aerosol technology. - 2nd ed. - New York ; London : Van Nostrand Reinhold, 1979
xii,526p : ill. - Bibliography: p.507-511. - Includes index
0-442-27348-7 SRL
1886665

AEROSOLS. Atmosphere
TWOMEY, S.
Atmospheric aerosols. - Amsterdam ; Oxford : Elsevier, 1977
xiv,302p : ill. - (Developments in atmospheric science ; 7). - Includes bibliographies and index
0-444-41527-0 SRL
1144057

AEROSOLS. Pollutants of atmosphere
HINDS, Williams C.
Aerosol technology. - New York ; Chichester : Wiley, c1982
xix,424p : ill. - Includes bibliographies and index
0-471-08726-2 SRL
2823764

AEROSOLS. Pollutants of atmosphere
Measurement
CADLE, Richard Dumbar
The measurement of airborne particles. - New York ; London : Wiley-Interscience, 1975
xiii,342p : ill. - (Environmental science and technology). - Includes bibliographies and index
0-471-12910-0 SRL
0583200

AEROSOLS. Pollutants of atmosphere
Physical properties
FRIEDLANDER, S. K.
Smoke, dust and haze. - New York ; London : Wiley, 1977
xix,317p : ill. - Includes bibliographies and index
0-471-01468-0 X.620/16883 Woolwich
114300x

AEROSPACE COMPONENTS INDUSTRIES. Great Britain
- Statistics
AEROSPACE component manufacturers / [Business Ratios]. - 1st ed.: [1979]. - London : Business Ratios, [1979]
[109]p : ill
0-904540-17-0 X.622/8597 Woolwich
2269325

AEROSPACE ENGINEERING
See also
Aeronautical engineering

AEROSPACE ENGINEERING
ANDERSON, John D. (John David)
Introduction to flight. - New York ; London : McGraw-Hill, 1978
xiii,432p : ill. - Includes bibliographies and index
0-07-001637-2 X.620/17619 Woolwich
1460528

CRANFIELD INSTITUTE OF TECHNOLOGY
[Miscellaneous publications not catalogued separately]. - Cranfield : the Institute, 1980-
r6108700 X.0625/103 Woolwich
1305352

STANGER, Jack
The Armada sky high quiz book. - London : Armada Books, 1975
120p : ill
0-00-690884-5 X.978/205
070461x

AEROSPACE ENGINEERING
- Dictionaries
GUNSTON, Bill
Jane's aerospace dictionary. - London : Jane's, 1980
493p
0-7106-0048-8 S.R.L.
2716097

AEROSPACE ENGINEERING
- Serials
AEROSPACE & science. - Vol.1, no.1- ; Autumn 1977-. - Christchurch, Dorset : Dschubba Scientific Directorate, [1977]- : ill
B78-05220 SRL
1305352

AEROSPACE dynamics. - Issue 1 (1979)-. - Stevenage : The Group, 1979-
v. : ill(some col.)
B82-28120 SRL
1305352

AEROSPACE ENGINEERING
Heat transfer - For aerospace engineering - Conference proceedings
HEAT transfer with thermal control applications / edited by M. Michael Yovanovich. - New York : American Institute of Aeronautics and Astronautics ; Cambridge, Mass. ; London : M.I.T. Press, [1976]
xv,490p : ill. - (Progress in astronautics and aeronautics ; vol.39). - Includes bibliographies and index
0-262-24019-x SRL
0707252

AEROSPACE ENGINEERING
Serials - Catalogues
ROYAL AIR FORCE MUSEUM. Library
Periodicals catalogue. - [London] : The Library, 1975
[154]p. - (Research aids / Royal Air Force Museum) - Includes index
0-9504788-1-4 BS.2a/71
0602000

AEROSPACE ENGINEERING
Testing - Conference proceedings
ADVANCES in test measurement. - Pittsburgh, Pa. : Instrument Society of America.
Vol.15: Proceedings of the 24th International Instrumentation Symposium, May 1-5, 1978, Albuquerque, New Mexico. - c1978
2v(672p) : ill. - (Instrumentation in the aerospace industry ; v.24). - Includes indexes
r6864773 X.622/12632
2740257

AEROSPACE ENGINEERING INDUSTRIES
See also
Aeronautical engineering industries

AEROSPACE ENGINEERING INDUSTRIES. Great Britain
British Aerospace
BESWICK, Frank, Baron Beswick
The role of British aerospace in the British economy. - [Belfast] : [Queen's University of Belfast], 1979
23p : ill. - (Harry Ferguson memorial lecture ; 4th)
0-85389-173-7 X.529/35986
2126745

AEROSPACE ENGINEERING INDUSTRIES. Great Britain
British Aircraft Corporation, to 1977
GARDNER, Charles
British aircraft corporation. - London : Batsford, 1981
320p,[32]p of plates : ill(some col.). - Includes index
0-7134-3815-0 X.622/10746 Woolwich
2530139

AEROSPACE ENGINEERING INDUSTRIES. Great Britain
Lucas Aerospace
LÖW-BEER, Peter
Rüstungskonversion und Produktion sozial nützlicher Güter, dargestellt am Beispiel des 'corporate plan' von Lucas Aerospace .... - [Heidelberg] : [P. Löw-Beer], [1982]
357p.. - Includes bibliography
r6307710 X.629/21734
9026991

AEROSPACE ENGINEERING INDUSTRIES. Great Britain
Lucas Aerospace. Diversification. Proposals: Lucas Aerospace. Combine Committee. Lucas Aerospace alternative corporate plan, to 1982
WAINWRIGHT, Hilary
The Lucas plan. - London : Allison & Busby, 1982
280p : ill. - Includes index
0-85031-429-1 X.529/52300
2849321

AEROSPACE ENGINEERING INDUSTRIES. Great Britain
Lucas Aerospace. Diversification. Proposals: Lucas Aerospace Combine Shop Stewards Committee. Lucas Aerospace alternative corporate plan
ELLIOTT, David, 1943-
The Lucas Aerospace workers' campaign. - London : Fabian Society, 1977
[2],20p. - (Young Fabian pamphlets ; no.46)
0-7163-2046-0 X.519/28754
1727451

AEROSPACE ENGINEERING INDUSTRIES. Great Britain
Lucas Aerospace. Diversification. Proposals: Lucas Aerospace Combine Shop Stewards Committee. Lucas Aerospace alternative corporate plan, to 1978
LUCAS AEROSPACE COMBINE SHOP STEWARDS' COMMITTEE
Democracy versus the Circumlocution Office. - Nottingham : Institute for Workers' Control, [1979]
[2],15p. - (Pamphlet series / Institute for Workers' Control ; no.65)
B79-30564 X.519/30119
1944398

AEROSPACE ENGINEERING INDUSTRIES. Great Britain
Lucas Aerospace. Personnel. Redundancy. Prevention
LUCAS AEROSPACE TRADE UNION COMMITTEE
Turning industrial decline into expansion - a trade union initiative. - Dagenham : Centre for Alternative Industrial and Technological Systems, 1979
461p in various pagings : ill. - Bibliography: 3p
B79-20711 X.512/6650
1904426

AEROSPACE ENGINEERING INDUSTRIES. United States
Relations with government, to 1973
KUCERA, Randolph C.
The aerospace industry and the military. - Beverly Hills ; London : Sage Publications, [1975]
64p. - (Sage professional papers in administrative and policy studies ; vol.2, no.03-023). - Bibliography: p.63-64
0-8039-0458-4 X.519/25942
0682063

AEROSPACE INDUSTRIES
- Forecasts - Conference proceedings
AEROSPACE into the eighties and beyond / sponsors Financial Times, The Royal Aeronautical Society. - London : Financial Times Ltd, [1980?]
202p : ill
B82-07066 SRL
2607891

AEROSPACE INDUSTRIES
Finance - Serials
AIRFINANCE annual. - 1984-. - [Coggeshall] : Airfinance Journal Ltd, 1984-
v. : ill(some col.)
B85-28927 P.625/518
3297500

AEROSPACE INDUSTRIES. France
to date
OFFICE NATIONAL D'ETUDES ET DE RECHERCHES AEROSPATIALES
[Miscellaneous publications not catalogued separately]. - Chatillon : [l'Office], 1980-
r6206907 S.E.400/849
8840067

AEROSPACE INDUSTRIES. Great Britain
Costs - Serials
AEROSPACE and electronics cost indices. - July 1983-. - Newport, Gwent : Business Statistics Office Library, 1983-
v.
B85-11122 P.625/451
3230244

AEROSPACE INDUSTRIES. Great Britain
Machine tools. Requirements - Forecasts
MACHINE TOOL INDUSTRY RESEARCH ASSOCIATION
The future machine tool requirements of the UK automobile and aerospace industries. - [Macclesfield] : [The Association], 1974
45,[17]p,[2] leaves of plates : ill
B77-30029 X.615/1891 Woolwich
124714x

AEROSPACE INDUSTRIES. United States
Instrumentation
INTERNATIONAL INSTRUMENTATION SYMPOSIUM (28th : 1982 : Las Vegas)
Proceedings of the 28th International Instrumentation Symposium May 3-6, 1982 Las Vegas, Nevada. - Research Triangle Park : Instrument Society of America, c1982
513-1027p : ill. - (Instrumentation in the aerospace industry ; v.28) (Advances in test measurement ; v.19, pt.2)
0-87664-690-9 X.622/14092
2794950

AEROSPACE MEDICINE
AVIATION medicine. - London : Tri-med.
[1]: Physiology and human factors. - 1978
xiv,777p : ill. - Bibliography: p748-749. - Includes index
0-905402-04-9 SRL
0461326

AEROSPACE VEHICLES
See also
Aircraft
Rockets
Space vehicles

AEROSPACE VEHICLES
Alloys. Mechanical properties. Effects of forging - Conference proceedings
FORGING and properties of aerospace materials. - London : Metals Society, 1978
vi,456p : ill. - (Books / Metals Society ; no.188). - Includes index
0-904357-12-0 SRL
1527142

AEROSPACE VEHICLES
Boeing aerospace vehicles, to 1983
REDDING, Robert
Boeing. - London : Arms and Armour, 1983
255p : ill(some col.)
0-85368-595-9 L.45/2558
3072843

AEROSPACE VEHICLES
British Aerospace aerospace vehicles, to date - Serials
BRITISH Aerospace, Inc. quarterly. - [1979]-. - Washington, D.C. : British Aerospace, [1979]-
v.
Holdings: Vol.4, no.3 (Fall 1982) ; v.5, no.1 (Jan.-March 1983) ; June 1983-
r6811569 X.0622/906
9043438

AEROSPACE VEHICLES
Instrumentation systems - Conference proceedings
INTERNATIONAL INSTRUMENTATION SYMPOSIUM (22nd : 1976 : San Diego)
Proceedings of the 22nd International Instrumentation Symposium, May 25-27, 1976, San Diego, California, programmed by ISA's Aerospace Industries Division, Test Measurement Division. - Pittsburgh : Instrument Society of America, 1976
[13], 729 p. : ill. - Includes index
0-87664-290-3 X.611/4736 Woolwich
1040154

INTERNATIONAL INSTRUMENTATION SYMPOSIUM (22nd : San Diego). 1976
Tutorial proceedings of the 22nd International ISA Instrumentation Symposium, May 25-27, 1976, San Diego, California, programmed by ISA's Aerospace Industries Division, Test Measurement Division. - Pittsburgh : Instrument Society of America, 1976
[9], 141 p. : ill. - Includes index
0-87664-291-1 X.611/4562 Woolwich
1040154

AEROSPACE VEHICLES
Instruments - Conference proceedings
INTERNATIONAL AEROSPACE INSTRUMENTATION SYMPOSIUM (8th : 1975 : Cranfield)
Proceedings of the 8th International Aerospace Instrumentation Symposium, 24th-27th March 1975, Cranfield, England. - [London] : [Royal Aeronautical Society], [1975]
552p in various pagings : ill. - Includes bibliographies
B75-27291 SRL
0513911

AEROSPACE VEHICLES
Materials. Mechanics
NICA, Alexandru
Mechanics of aerospace materials. - Amsterdam ; Oxford : Elsevier Scientific, 1981
346p : ill. - (Materials science monographs ; 9). - Includes index
0-444-99729-6 SRL
2606550

AEROSPACE VEHICLES
Rocket engines
SUTTON, George Paul
Rocket propulsion elements. - 4th ed.. - New York ; London : Wiley-Interscience, 1976
ix,557p,fold leaf : ill. - Includes index
0-471-83836-5 SRL
0667455

AEROSPACE VEHICLES. Atmosphere
Flight. Trajectories. Optimisation
VINH, Nguyen X.
Optimal trajectories in atmospheric flight. - Amsterdam ; Oxford : Elsevier Scientific, 1981
xviii,402p : ill. - (Studies in astronautics ; v.2). - Includes index
0-444-41961-6 SRL
2486393

AEROSTATS
See Lighter than air aircraft

AEROTHERMODYNAMICS. Gas turbines
WHITTLE, Sir Frank
Gas turbine aero-thermodynamics. - Oxford : Pergamon, 1981
xii,261p,[1]leaf of plates : ill. - (Pergamon international library)
0-08-026719-x SRL
2525089

AERTSEN, PIETER. Dutch paintings
Religious aspects
MOXEY, Keith P. F.
Pieter Aertsen, Joachim Beuckelaer, and the rise of secular painting in the context of the Reformation. - New York ; London : Garland, 1977
[5],xii,285p,72p of plates : ill. - (Outstanding dissertations in the fine arts). - Bibliography: p.269-284
0-8240-2715-9 X.429/11535 Woolwich
2316951

AESCHINES. Public speaking. Ancient Greece
KINDSTRAND, Jan Fredrik
The stylistic evaluation of Aeschines in antiquity. - Uppsala : Uppsala universitet, 1982
104p. - (Studia graeca upsaliensia ; 18). - Bibliography
91-554-1320-x Ac.1075/6(14)[vol.18]
8010234

AESCHLIMANN, RICHARD. Swiss drawings
AESCHLIMANN, Richard
Dix fenêtres ouvertes à dessein sur le gris du temps. - Lausanne : Éditions Clind'Oeil, 1980
[50]p : ill
r6210836 X.421/25764
8437246

AESCHYLUS. Drama in Greek
BECK, Robert Holmes
Aeschylus. - The Hague : Martinus Nijhoff, 1975
xviii,204p. - Bibliography : p.193-204
90-247-1736-1 X.900/18351
1093460

CERRI, Giovanni
Il linguaggio politico nel Prometeo di Eschilo. - Roma : Edizioni dell'Ateneo, 1976
155p. - (Filologia e critica ; 17)
r6078044 X.909/39799
1093460

CHODKOWSKI, Robert
Funkcja obrazów scenicznych w tragediach Ajschylosa. - Wrocław : Zakład Narodowy im. Ossolińskich, Wydawnictwo Polskiej Akademii Nauk, 1975
115p. - (Polska Akademia Nauk. Komitet Nauk o Kulturze Antycznej. Archiwum filologiczne ; 31). - Bibliography : p.108-115
r6015283 Ac.2985.br/12[no.31]
1093460

CONACHER, D. J.
Aeschylus' Prometheus bound. - Toronto : University of Toronto Press, c1980
xii,198p
0-8020-2391-6 X.950/35165
8705690

DI BENEDETTO, Vincenzo
L'ideologia del potere e la tragedia greca. - Torino : Einaudi, 1978
xii,295p. - (Piccola biblioteca Einaudi ; 355)
r6501243 X.958/5797
8705690

GAGARIN, Michael
Aeschylean drama. - Berkeley ; London : University of California Press, 1976
xi,239p. - Bibliography: p.219-228. - Includes index
0-520-02943-7 X.989/51461
1093460

GRUYS, J. A.
The early printed editions (1518-1664) of Aeschylus. - Nieuwkoop : De Graaf, 1981
356p. - (Bibliotheca humanistica & reformatorica ; v.32)
90-600-4368-5 X.520/26954
8705690

LUPAŞ, Liana
Commentaire aux "Sept contre Thèbes" d'Eschyle. - Bucureşti : Editura Academiei ; Paris : Société d'Édition "Les Belles Lettres", 1981
x,301p. - Bibliography
2-251-32605-7 X.950/14406
8705690

PADUANO, Guido
Sui Persiani di Eschilo. - Roma : Ateneo & Bizzarri, 1978
106p. - (Filologia e critica / Università di Urbino, Istituto di filologia classica ; 27)
r6501911 X.950/6572
8705690

AESCHYLUS. Drama in Greek -continued

ROBERTS, Deborah H.
Apollo and his oracle in the Oresteia. - Göttingen : Vandenhoeck & Ruprecht, 1984
136p.. - (Hypomnemata ; Hft.78). - Includes bibliography
3-525-25176-9 11880.m.1/78
8705690

ROSENMEYER, Thomas G. (Thomas Gustav)
The art of Aeschylus. - Berkeley ; London : University of California Press, c1982
393p. - Bibliography: p377-384. - Includes index
0-520-04440-1 X.950/16294
1093460

SCHOLIA Graeca in Aeschylum quae exstant omnia / edidit Ole Langwitz Smith. - Leipzig : Teubner, 1976-
r6309633 2048.f.7 Reading Room
8705690

SEECK, Gustav Adolf
Dramatische Strukturen der griechischen Tragödie. - München : Beck, 1984
85p.. - (Zetemata ; Hft.81)
3-406-09681-6 WP.3194/81
8705690

USCATESCU, Jorge
Tragedia y politica. - Madrid : Fundación Pastor de Estudios Clásicos, 1980
102p. - (Cuadernos / de la "Fundación Pastor" ; 25)
84-300-2613-4 X.950/23191
8705690

WINNINGTON-INGRAM, R. P.
Studies in Aeschylus. - Cambridge : Cambridge University Press, 1983
xiii,225p. - Bibliography: pxii-xiii. - Includes index
0-521-24938-4 X.950/25696
1093460

AESCHYLUS. Drama in Greek
- Conference proceedings

AISCHYLOS und Pindar / herausgegeben von Ernst Günther Schmidt. - Berlin : Akademie-Verlag, 1981
379p.. - (Schriften zur Geschichte und Kultur der Antike ; 19)
r6306273 X.950/17789
8705690

AESCHYLUS. Drama in Greek
- Indexes

INDEX analyticus Graecitatis Aeschyleae / edited by Harry G. Edinger. - Hildesheim : Olms, 1981
ix,474p.. - (Alpha-Omega. Reihe A ; 2)
3-487-06978-4 X.950/22385
8705690

AESCHYLUS. Drama in Greek
Choephori

MARTINA, Antonio
Il riconoscimento di Oreste nelle Coefore e nelle due Elettre. - Roma : Edizioni dell'Ateneo, 1975
xi,195p. - (Bibliotheca athena ; n.16)
r6035106 X.909/37910
1314955

AESCHYLUS. Drama in Greek
Imagery

PETROUNIAS, Evangelos
Funktion und Thematik der Bilder bei Aischylos. - Göttingen : Vandenhoeck & Ruprecht, 1976
xx,439p. - (Hypomnemata ; Heft 48). - Bibliography : p.421-431
r6078910 11880.m.1/48
1004417

AESCHYLUS. Drama in Greek
Influence on composition of Ring des Nibelungen by Wagner, Richard, 1813-1883

EWANS, Michael
Wagner and Aeschylus. - London : Faber, 1982
271p. - Bibliography: p261-265. - Includes index
0-571-11808-9 X.439/12032
2773929

AESCHYLUS. Drama in Greek
Metaphor

SANSONE, David
Aeschylean metaphors for intellectual activity. - Wiesbaden : Franz Steinerc, 1975
xii,100p. - (Hermes, Zeitschrift für klassische Philologie, Einzelschriften ; Hft 35)
3-515-02001-2 PP.5044.b(2)[Hft 35]
1475142

AESCHYLUS. Drama in Greek
Music

SCOTT, William C. (William Clyde)
Musical design in Aeschylean theater. - Hanover, N.H. ; London : For Dartmouth College by University Press of New England, 1984
xxi,228p. - Bibliography: p223-224. - Includes index
0-87451-291-3 X.950/41270
3273342

AESCHYLUS. Drama in Greek
Oresteia

GOLDHILL, Simon
Language, sexuality, narrative. - Cambridge : Cambridge University Press, c1984
ix,315p. - Bibliography: p284-313. - Includes index
0-521-26535-5 X.950/38957
2220865

AESCHYLUS. Drama in Greek
Prometheus bound. Authenticity

GRIFFITH, Mark
The authenticity of 'Prometheus bound'. - Cambridge : Cambridge University Press, 1977
xii,419p. - (Cambridge classical studies). - Bibliography: p.366-379. - Includes index
0-521-21099-2 WP.11639/32
1249371

AESCHYLUS. Drama in Greek
Seven against Thebes

THALMANN, William G.
Dramatic art in Aeschylus's 'Seven against Thebes'. - New Haven ; London : Yale University Press, 1978
ix,193p. - (Yale classical monographs ; 1). - Bibliography: p.183-187. - Includes index
0-300-02219-0 X.981/21312
1441205

AESCHYLUS. Drama in Greek
Stagecraft

TAPLIN, Oliver
The stagecraft of Aeschylus. - Oxford : Clarendon Press, 1977
ix,508p. - Bibliography: p.480-489. - Includes index
0-19-814006-1 X.989/52126
0799920

AESOP. Fables in Greek
Translations into French language, 1500-1600

MOMBELLO, Gianni
Le raccolte francesi di favole esopiane dal 1480 alla fine del secolo XVI. - Genève : Éditions Slatkine, 1981
182p. - (Textes et études - domaine français / Universités de Turin et de Savoie, Centre d'études franco-italien ; no.5)
2-05-100353-x X.950/22248
8957258

AESTHETIC ASPECTS. Architectural design

VIL vi være det foruden = Can't some of it be pulled down = Sans tout ceci - / [introduktion : B.J. Rambøll] ; [billedteksterne : N.P. Christiansen, Jørgen Leisner] ; [tegningerne : Bodil Sørensen]. - [Copenhagen] : I kommission hos Schultz Boghandel, 1977
112p : chiefly ill
87-7338-000-8 X.415/2402 Woolwich
1469592

AESTHETIC ASPECTS. Design, 1910-1924. Bugatti cars

HALLUMS, E.
The quality of work and the quality of art. - London : Mithra Press, 1979
90,[1]p, plate : ill(some col). - Includes index
0-906573-00-9 X.425/1570 Woolwich
2070375

AESTHETIC ASPECTS. Design. Cities

ART and life and the building and decoration of cities / T.J. Cobden-Sanderson et al. - New York ; London : Garland, 1978
[11], 260p. - (The Aesthetic movement and the arts and crafts movement)
0-8240-2484-2 X.429/11093 Woolwich
2076357

AESTHETIC ASPECTS. Design. Roads. Ireland (Republic)

HYDE, N. (Niall)
Guidelines on road design aesthetics. - Dublin : An Foras Forbartha, 1981
56p : ill. - (RT ; 182)
0-906120-42-x X.525/5515
2693771

AESTHETIC ASPECTS. Environment. Urban regions. Great Britain
Projects: Art and the Built Environment Working Parties Project - Serials

ART AND THE BUILT ENVIRONMENT WORKING PARTIES PROJECT
Bulletin. - 1-. - [London] : The Project, [1981]-
v.
B82-19860 P.423/523
271518x
Newsletter. - 1-20. - London : The Project, [1980]-[1982]
20v. : ill
B84-06839 P.423/580
271518x

AESTHETIC ASPECTS. Exteriors. Buildings

WESTERMAN, Ann
Estetisk värdering av byggnadsexteriorer = On aesthetic judgement of building exteriors. - Stockholm : Avdelningen för Arkitektur, KTB, 1976
211 leaves : ill. - (Skrift 1976 ; 16)
91-85212-15-6 X.415/2101 Woolwich
0387592

AESTHETIC ASPECTS. Geometry
to 1975

PEDOE, Daniel
Geometry and the liberal arts. - Harmondsworth : Penguin, 1976
296p,24p of plates : ill. - (Peregrine books). - Bibliography: p.287. - Includes index
0-14-055119-0 X.419/3136 Woolwich
0663603

AESTHETIC ASPECTS. Landscape. England
Evaluation by public

PREECE, R. A.
An evaluation by the general public of scenic quality in the Cotswolds area of outstanding natural beauty. - [Oxford] : Oxford Polytechnic, [1980]
84p : ill. - (Working paper / Oxford Polytechnic Department of Town Planning ; no.48). - Bibliography: p80-84
B81-06146 X.622/9542 Woolwich
2351811

AESTHETIC ASPECTS. Maintenance, repair & improvement. Exteriors. Houses. Great Grimsby (District). Humberside

GRIMSBY. Borough Council
Grimsby design guide, house improvements and the street scene. - [Grimsby] : [The Council], [1975]
[1],21 leaves : ill
B79-06663 X.519/29159
1799150

AESTHETIC ASPECTS. Maintenance, repair & improvement. Exteriors. Terraced houses. Great Britain

PYKE, Beverley
The good looking house. - Bristol : Redcliffe, 1980
64p : ill
0-905459-23-7 X.622/12597 Woolwich
2722577

AESTHETIC ASPECTS. Methodology. Sociology

BROWN, Richard Harvey
A poetic for sociology. - Cambridge : Cambridge University Press, 1977
x,302p : ill. - Bibliography: p.266-294. - Includes index
0-521-21121-2 X.520/11799
116922x

AESTHETIC ASPECTS. Plants

BAGER, Bertel
Nature as designer. - London : F. Warne, 1976
3-176p : ill
0-7232-0035-1 X.320/10879 Woolwich
1188526

AESTHETIC ASPECTS. Plastics

BLACK, Sir Misha
The aesthetics of plastics. - [Bradford] : [University of Bradford], [1977]
12p. - (Professor Moore memorial lecture ; 1977)
0-901945-27-7 X.0519/1012(2)
1408550

AESTHETIC ASPECTS. Play. Special themes. German literature, 1750-1945

KOWATZKI, Irmgard
Der Begriff des Spiels als ästhetisches Phänomen. - Bern : Lang, [1973]
165p. - (Stanford German studies ; vol.4). - Bibl
3-261-00772-9 X:0909/675(4)
2462680

AESTHETIC ASPECTS. Proportions

DOCZI, György
The power of limits. - Boulder ; London : Shambhala, 1981
150p : ill. - Includes index
0-87773-194-2 L.42/1258
2591804

AESTHETIC ASPECTS. Proportions, to date. Visual arts

FALUS, Róbert
Az aranymetszés legendája. - Budapest : Magvető, c1982
271 p. : ill.. - (Gyorsuló idő). - Includes bibliographical references
963-271-695-7 X.429/16133
915700x

AESTHETIC ASPECTS. Science

ON aesthetics in science / edited by Judith Wechsler - Cambridge, Mass. ; London : M.I.T. Press, 1978
[11].180p : ill. - Bibliography: p 165-169. - Includes index
0-262-23088-7 X.620/17851 Woolwich
1549014

AESTHETIC ASPECTS. Skylines. Cities

ATTOE, Wayne
Skylines. - Chichester : Wiley, c1981
xiv,128p : ill. - Bibliography: p121-123. - Includes index
0-471-27940-4 L.49/964
255240x

AESTHETIC ASPECTS. Spirals

COOK, Theodore Andrea
The curves of life. - New York : Dover ; London : Constable, 1979
xxx,479p : ill. - Includes index
0-486-23701-x X.429/12429 Woolwich
2545608

AESTHETIC ASPECTS. Sports

LOWE, Benjamin
The beauty of sport. - Englewood Cliffs ; London : Prentice-Hall, 1977
xix,327p : ill. - Bibliography: p.304-320. - Includes index
0-13-066589-4 X.620/16805 Woolwich
0293407

AESTHETIC ASPECTS. Television, to date

BROGLIE, Gabriel de
Une image vaut dix mille mots. - Paris : Plon, 1982
222p
2-259-00949-2 X.529/60924
8787816

SCHWEIKHARDT, Josef
Ästhetik des Fernsehens. - Wien : Österreichischer Rundfunk, [1980]
91p : ill(some col). - (Berichte zur Medienforschung ; 23). - Bibliography
r6302406 X.0950/278(23)
8787816

AESTHETIC ASPECTS. Town planning. Great Britain

BURKE, Gerald
An approach to aesthetics of town planning. - Reading : College of Estate Management, c1975
32p : ill
0-902132-21-0 L.49/1592
0682322

Townscapes. - [Harmondsworth] : Penguin, 1976
xviii,246p : ill. - (Pelican geography and environmental studies). - Includes bibliographies and index
0-14-021821-1 X.419/3158 Woolwich
0682322

Townscapes. - Harmondsworth : Penguin, 1980
xviii,246p : ill. - (Pelican geography and environmental studies). - Bibliography : p.237. - Includes index
B80-16350 X.708/48339
0682322

AESTHETIC ASPECTS. Town planning. Sweden

HESSELGREN, Sven
Vad vacker är. - [Stockholm] : Byggforskningsradet, 1977
216p : ill(some col). - (T32:77)
91-540-2777-2 X.619/10426 Woolwich
1390082

KONST i ett bostadsomrade / utgiven av Sven Sandström...[et al.]. - Lund : Liber läromedel, 1977
132p : ill. - (Konstnärliga element i publik miljö ; 1)
91-400-4411-4 X.0510/311(1)
1390082

AESTHETIC DESIGNS. Computer assisted graphic displays

PRUEITT, Melvin L.
Computer graphics. - New York : Dover Publications ; London : Constable, 1975
[1],iii-ix,71p : chiefly ill(some col). - (Dover pictorial archive series)
0-486-23178-x X.611/7269 Woolwich
0652814

AESTHETIC EDUCATION. Young persons. Soviet Union 1917-

ISKUSSTVO i esteticheskoe vospitanie molodezhi / [redakt͡sionnai͡a kollegii͡a: V.G. Kisun'ko (otvetstvennyi redaktor) ... i dr.]. - Moskva : Nauka, 1981
334p
r6464117 X.529/54292
8965471

AESTHETIC FACTORS. Natural selection. Evolution. Man

GOULSTONE, J
Human evolution through aesthetic selection. - Bexleyheath : J. Goulstone, c1981
[14]p,[8]leaves of plates : ill
B81-25417 X.325/1316 Woolwich
2498936

GOULSTONE, J. (John)
The role of aesthetic rejection and selection in human evolution. - [Bexleyheath] : J. Goulstone, c1981
[16]p,[7]leaves of plates : ill
B82-06457 X.520/25767
2498936

AESTHETIC JUDGMENT. Children
Theories, to date

MENZEN, Karl-Heinz
Eine Erziehungsgeschichte der Ästhetik. - Weinheim : Beltz, 1982
228p. - (Beltz Forschungsberichte)
3-407-58138-6 X.950/29347
9079424

AESTHETIC MOVEMENT. English arts
- Texts

The AESTHETES / edited by Ian Small. - London : Routledge and Kegan Paul, 1979
xxxii,204p,8p of plates : ill. - Bibliography: p.203-204
0-7100-0145-2 X.989/53413
1811835

AESTHETIC MOVEMENT. English visual arts
Items associated with Aesthetic Movement exhibited at Julian Hartnoll (Gallery) - Catalogues

ASPECTS of the aesthetic movement. - [London] : [Dan Klein], [1978]
48p : ill(some col.)
B81-19777 X.429/12111 Woolwich
2404702

AESTHETIC MOVEMENT WRITERS. English literature 1837-1900 - Anthologies

The AESTHETES / edited by Ian Small. - London : Routledge and Kegan Paul, 1979
xxxii,204p,8p of plates : ill. - Bibliography: p.203-204
0-7100-0145-2 X.989/53413
1811827

AESTHETICS
See also
Aesthetic education
Beauty
Kitsch

AESTHETICS

ADORNO, Theodor W. (Theodor Wiesengrund)
Aesthetic theory. - London : Routledge & Kegan Paul, 1984
x,526p. - (The International library of phenomenology and moral sciences). - Includes index
0-7100-9204-0 X.529/64752
0201537

ÆSTETISKE teorier / ved Jørgen Dehs. - [Odense] : Odense Universitetsforlag, 1984
278p. - (Odense University studies in literature ; v.18)
87-7492-493-1 P.901/555[vol.18]
8821380

ÄSTHETISCHE Erfahrung und das Wesen der Kunst / Redactores : Helmut Holzhey et Jean-Pierre Leyvraz. - Bern : Haupt, 1984
253p,plates. - (Studia philosophica ; v.43)
3-258-03398-6 Ac.631[vol.43]
8821380

BÄCHLER, Hagen
Ästhetik, Mensch, gestaltete Umwelt. - Berlin : Deutscher Verlag der Wissenschaften, 1982
125p,[16]p of plates. - (Weltanschauung heute ; Bd.37)
r6305486 X.529/50962
8821380

BARABASH, I͡Uriĭ I͡Akovlevich
Voprosy ėstetiki i poėtiki. - Izd.3. - Moskva : Sovetskai͡a Rossii͡a, 1978
382p
r6056783 X.989/78225
0201537

BAUMGARTEN, Alexander Gottlieb
[Selections]. Texte zur Grundlegung der Ästhetik. - Hamburg : Meiner, 1983
xxiv,110p. - (Philosophische Bibliothek ; Bd 351)
3-7873-0573-4 X.429/17108
8821380

BOSELIE, Frans Antonius Johan Marie
Over visuele schoonheidservaring .... - Nijmegen : [F.A.J.M. Boselie], 1982
235p. - Includes bibliography
r6306572 X.429/14223
8821380

BROŽÍK, Vladimir
O užitočnosti krásy. - Bratislava : Pravda, 1978
175p. - (Človek a umenie)
r6096487 X.529/21694
0201537

BUKIN, Viktor Rodionovich
Voskhozhdenie k prekrasnomu. - Leningrad : Lenizdat, 1979
174p
r6183423 X.529/36612
0201537

CARRERA, Mario Alberto
Cuando el arte muera. - Guatemala : José de PinedaIbarra, 1973
229p. - (Colección Festival)
r6099255 X.908/86769
0336718

CONTEMPORARY aesthetics in Scandinavia / edited by Lars Aagaard-Mogensen and Göran Hermerén. - Lund : Doxa, 1980
224p : ill
91-578-0063-4 X.429/13677
8821380

DAMNJANOVIĆ, Milan
Suština i povest. - Beograd : Univerzitet umetnosti, 1976
316p
r6186492 X.512/3432
0201537

ECO, Umberto
La definizione dell'arte. - Milano : Garzanti, 1978
303p. - (I Garzanti. Argomenti ; 43)
r6197520 X.439/11010
0336718

ERMASH, Gela Leonidovna
Tvorcheskai͡a priroda iskusstva. - Moskva : "Iskusstvo", 1977
317p
r6076993 X.809/25900
107153X

ESTETIIKAN kenttä / [toimittaneet Markku Envall, Aarne Kinnunen ja Yrjö Sepänmaa]. - [Porvoo] : W. Söderström, 1972
230p. - (Universitas ; 12)
951-00-0525-8 X.808/10595
0336718

FOCHT, Ivan
Uvod u estetiku. - 3 izd. - Sarajevo : Svjetlost, 1984
259p
r6496357 X.520/37402
8821380

GARAI, László
Szabadságszükséglet és esztétikum. - Budapest : Akadémiai Kiadó, 1980
159p. - (Muszeion könyvtár ; 3)
963-05-2274-8 X.0421/248(3)
8821380

GOMBRICH, E. H. (Ernst Hans)
Nature and art as needs of the mind. - Liverpool : Liverpool University Press, 1981
24p
0-85323-354-3 X.529/44192
0201537

GORSEN, Peter
[Selections]. Ausgewählte Schriften. - Frankfurt am Main : Europäische Verlagsanstalt, 1980- : ill
r6303579 X.0529/634
8821380

HEPBURN, Ronald W. (Ronald William)
'Wonder'. - Edinburgh : Edinburgh University Press, c1984
viii,192p. - Includes index
0-85224-488-6 X.958/25399
0201537

JORN, Asger
Held & hasard. - 3. opl. - [København] : Borgen, [1980]
176p : ill. - (Meddelelse / Skandinavisk Institut for Sammenlignende Vandalisme ; nr.3)
87-418-3079-2 X.529/55288
8821380

KOVACH, Francis Joseph
Philosophy of beauty. - Norman, Okl : University of Oklahoma Press, 1974
x,350p. - Bibliography : p.318-337
0-8061-1186-0 X.520/9354
0201537

AESTHETICS -continued

KUNST und Kunstforschung / herausgegeben von Gerd Wolandt.
- Bonn : Bouvier, 1983
98p.. - (Aachener Abhandlungen zur Philosophie ; Bd.3)
3-416-01749-8 X.520/32869
8821380

KUPFER, Joseph H.
Experience as art. - Albany : State University of New York Press, c1983
vii,216p. - (SUNY series in philosophy)
0-87395-693-1 X.429/15357
0201537

LEONE DE CASTRIS, Arcangelo
Estetica e marxismo. - Roma : Editori riuniti, 1976
229p. - (Argomenti ; 58)
r6500739 X.529/40662
1739247

LICHTENTHAL, Pietro
Estetica, ossia dottrina del bello e delle arti belle. - Bologna : A.M.I.S., 1971
xx,435p. - (Biblioteca storico giuridica e artistico letteraria. Arti ; 21)
r6041931 X.519/23745
0201537

LØGSTRUP, K. E.
Kunst og erkendelse. - [København] : Gyldendal, 1983
110p. - (Metafysik ; 2)
87-00-50252-9 X.529/69852
8821380

LOSEV, Alekseĭ Fedorovich
[Dialektika khudozhestvennoĭ formy]. Dialektika chudoẑestvennoj formy. - München : Sagner, 1983
xxviii,250p. - (Specimina philologiae Slavicae ; Bd 55)
3-87690-236-3 X.0909/738(55)
8821380

MARCUSE, Herbert
The aesthetic dimension. - London : Macmillan, 1979
xv,88p. - (Communications and culture). - Bibliography: p.81-82. - Includes index
0-333-26674-9 X.519/30417
1739247

[MARKSISTSKO-LENINSKAÎA ėstetika i zhizn'. English].
Marxist-Leninist aesthetics and life / [editors I. Kulikova and A. Zis ; translated from the Russian]. - Moscow : Progress Publishers, 1976
269 p.
r6031411 X.529/31309
107153X

MÁRQUEZ, Pedro José
[Selections. 1972]. Sobre lo bello en general, y Dos monumentos de arquitectura mexicana, Tajín y Xochicalco.
- México : Universidad Nacional Autónoma de México, Instituto de Investigaciones Estéticas, 1972
[286] p., 5 fold. leaves of plates. - (Estudios de arte y estética ; no. 13)
r6009179 X.429/6815 Woolwich
0924709

MATERIALISTISK æstetik / redigeret af Peter Nyord og Kurt Dahl Christiansen. - Odense : Odense Universitetsforlag, 1978
388p. - (Odense University studies in literature ; vol.7). - Includes bibliographies
87-7492-269-6 P.901/555[vol.7]
107153X

A MODERN book of esthetics / edited with introduction and notes by Melvin Rader. - New York ; London : Holt, Rinehart and Winston, 1979
[10],563p : ill. - Bibliography: p.523-552. - Includes index
0-03-019331-1 X.981/13142
0201537

MOTHERSILL, Mary
Beauty restored. - London : Clarendon, 1984
vii,438p. - Bibliography: p429-434. - Includes index
0-19-824729-x X.950/39474
0201537

NEW perspectives in German literary criticism / edited by Ricard E. Amacher and Victor Lange ; translated [from the German] by David Henry Wilson and others. - Princeton ; Guildford : Princeton University Press, 1979
x,480,[1]p. - Includes index
0-691-06380-x X.981/22370
0201537

OBERTI, Elisa
Estetica. - 1a ristampa anastatica. - Milano : Celuc Libri, 1974
371p. - (Scienze umane ; 35)
r6052826 X.519/23566
0201537

PLATĒS, E.N
Hē aisthētikē koinōnia. - Athēna : [E.N. Platēs], 1976
163p
r6107458 X.519/44031
0336718

PLEASURE, preference and value / edited by Eva Schaper. - Cambridge : Cambridge University Press, 1983
xi,172p. - Includes index
0-521-25101-x X.520/30896
0201537

POLE, David
Aesthetics, form and emotion. - London : Duckworth, 1983
viii,248p. - Includes index
0-7156-1709-5 X.520/30376
0201537

POSPELOV, Gennadiĭ Nikolaevich
Iskusstvo i ėstetika. - Moskva : Iskusstvo, 1984
325 p. - Includes bibliographical references
r6492637 X.950/41211
8821380

AESTHETICS

ROSSI, Lino
Studi di estetica. - Bologna : CLUEB, 1979
225p
r6509612 X.529/57016
8821380

SCRUTON, Roger
The aesthetic understanding. - London : Methuen, 1983
259p : ill. - Includes index
0-416-36160-9 X.529/62585
0201537
The aesthetic understanding. - Manchester : Carcanet, 1983
259p : ill. - Includes index
0-85635-487-2 X.429/16057
0201537
Art and imagination. - London : Rougledge & Kegan Paul, 1982, c1974
viii,259p. - Bibliography: p250-253. - Includes index
0-7100-9014-5 X.529/46636
0201537

ŠIMUNEK, Eugen
Filozofické problémy estetiky. - Bratislava : nakladatelstvo Pravda, 1976
124p,[8]p of plates : ill. - (Filozofické aktuality)
r6016706 X.519/24167
0201537

SITUAZIONE dell'estetica in Italia / a cura di Lino Rossi.
- Torino : Paravia, 1976
ccxcii,236p. - (La tradizione del nuovo. Antologie ; n.4). - Includes bibliographies
r6067243 X.510/10487
1849735

SMOL'ÎANINOV, I. F (Ivan Fedorovich)
Sushchnost' cheloveka i gumanizm iskusstva. - Leningrad : "Khudozhnik RSFSR", 1983
368 p, [72] p. of plates : ill.. - Includes bibliography: p. 362-[367]
r6465778 X.950/38152
8821380

STUDIA o współczesnej estetyce polskiej / [praca wykonana w Zespole Estetyki pod redakcją Sławą Krzemienia Ojaka].
- Warszawa : Państwove Wydawnictwo Naukowe, 1977
253p
r6075467 X.520/9725
0336718

SUASSUNA, Ariano
Iniciação à estética. - Recife : Universidade Federal de Pernambuco, Editora Universitária, 1975
287p
r6052199 X.519/23598
0201537

TATARKIEWICZ, Władysław
A history of six ideas. - The Hague ; London : Nijhoff, 1980
xiii,383p : ill. - (Melbourne international philosophy series ; v.5). - Bibliography: p349-374. - Includes index
90-247-2233-0 X.520/28367
0201537

TEORIÎA ėsteticheskogo vospitaniîa / [otvetstvennye redaktory N.I. Kiîashchenko ; N.L. Leizerov]. - Moskva : Iskusstvo, 1979
254p
r6183084 X.529/35209
1739247

TRÍAS, Eugenio
El artista y la ciudad. - Barcelona : Anagrama, [1976?]
236p. - (Colección Argumentos ; n.40)
84-339-0040-4 X.519/23065
0336718

VARIETIES of aesthetic experience / edited by Earle J. Coleman. - Lanham ; London : University Press of America, c1983
xii,256p. - Bibliography: p255-256
0-8191-3276-4 X.950/34090
0201537

ZIS', A. (Avner)
Foundations of Marxist aesthetics. - Moscow : Progress Publishers, 1977
298p
r6032104 X.529/31657
107153X

AESTHETICS
- Conference proceedings

INTERNATIONAL CONGRESS OF AESTHETICS (7th : 1972 : Bucharest)
Proceedings of the VIIth International Congress of Aesthetics = Actes du VIIe Congrès International d'Esthétique Bucarest 28 Août - 2 Septembre, 1972. - Bucureşti : Editura Academiei Republicii Socialiste România, 1976-77
2v.
r6015826 X.520/9477
0201537

KOLLOQUIUM Kunst und Philosophie / herausgegeben von Willi Oelmüller. - Paderborn : Schöningh, 1981-1983
3v.. - (Uni-Taschenbücher ; 1105, 1178, 1276)
r6308352 X.429/12888
8821380

AESTHETICS
- Dictionaries

GIRALDI, Giovanni
[Dizionario di estetica e di linguistica generale]. - Milano : Pergamena, 1975
946p.,15 leaves of plates : ill
r6068287 X.975/30
1854917

AESTHETICS
- Serials

AESTHETICA upsaliensia. - 1-. - Uppsala : Universitetet, 1981-. - (Acta Universitatis Upsaliensis)
r6610136 X.0429/558
8821380

AESTHETICS. - Vol.1, no.1 (Oct. 1983)-. - Tokyo : Japanese Society for Aesthetics, 1983-
v.
r6818524 X.0525/697
8821380

ĖSTETIKA i zhizn' ... / [Akademiîa nauk SSSR, Institut filosofii]. - Vyp.1-. - Moskva : Iskusstvo, [1972?]-
Holdings: vyp.2 (1973)-
r6020115 X.0529/600
115740X

HELSINGIN yliopiston Yleisen kirjallisuustieteen, teatteritieteen ja estetiikan laitoksen monistesarja. - N:o 1-. - Helsinki : [Helsingin yliopisto], [1976?]-
r6600264 X.0950/873
8821380
- Holdings: N:o 3, 7- X.0950/873

SUOMEN ESTETIIKAN SEURA
Suomen estetiikan seuran vuosikirja. - no. 1, 1974- / toimituskunta : Irma Rantavaara...[et al.]. - Porvoo : W. Söderström, 1974-. - (Taskutieto ; 117, etc)
r6001402 P.411/141
115740X

UNIVERSITY OF TOKYO. faculty of letters. Institute of Aesthtics
Journal of the faculty of letters, the University of Tokyo. - Vol.1- ; 1976-. - Tokyo : the University, 1976-
r6109478 P.521/2257
0201537

AESTHETICS
Ancient Roman theories

LOSEV, Alekseĭ Fedorovich
Ėllinisticheski-rimskaîa ėstetika I-II vv. n.ė. - Moskva : Izdatel'stvo Moskovskogo universiteta, 1979
416p,plates. - Includes bibliography
r6465120 X.529/67369
9147055

AESTHETICS
Belorussian theories, 1600-1800

KONON, Vladimir Mihaĭlovich
Ot Renessansa k klassitsizmu. - Minsk : Izdatel'stvo "Nauka i tekhnika", 1978
156p. - Bibliography : p.152-[157]
r6074271 X.808/11479
1716867

AESTHETICS
Bulgarian theories, to date

FILIPOV, Dimitŭr
Krasotata na zhivota. - Sofiîa : Nauka i izkustvo, 1980
378p
r6427335 X.950/12081
882004x

AESTHETICS
Documents on aesthetics, 1946-: Documents with Bulgarian imprints - Bibliographies

MILADINOVA, Milka Khristova
Za vsenarodno estetichesko vŭzpitanie. - Sofiîa : Narodna biblioteka "Kiril i Metodiĭ" ; Varna : Okrŭzhna biblioteka P. Slaveĭkov, 1979
554p
r6184895 X.520/20928
2425416

AESTHETICS
French theories, to date

IKNAYAN, Marguerite
The concave mirror. - Saratoga, Calif. : Anma Libri, 1983
227p. - (Stanford French and Italian studies ; v.30). -
Includes bibliography
0-915838-07-9 X.421/26278
9147195

AESTHETICS
Hellenistic theories

LOSEV, Aleksei Fedorovich
Ellinisticheski-rimskai͡a estetika I-II vv. n.e. -
Moskva : Izdatel'stvo Moskovskogo universiteta, 1979
416p.plates. - Includes bibliography
r6465120 X.529/67369
9147063

AESTHETICS
Hungarian theories, 1700-1900

NAGY, Endre, 1923-
A magyar esztétika történetéből. - Budapest :
Kossuth Könyvkiadó, 1983
534p
963-09-2221-5 X.429/15587
9039554

AESTHETICS
Linguistic aspects

STRUBE, Werner
Sprachanalytische Ästhetik. - München : Fink, 1981
275p. - Includes bibliography
3-7705-1904-3 X.950/22607
8964394

THOBO-CARLSEN, John
Æstetik og kommunikation. - [Odense] : Odense
Universitetsforlag, 1984
441p. - (Odense University studies in literature ; v.19).
- Includes bibliography
87-7492-502-4 P.901/555[no.19]
8964394

AESTHETICS
Marxist theories, to date

DEMIROVIĆ, Alex
Jenseits der Ästhetik. - Frankfurt am Main : Materialis,
1982
333p. - (Materialis Ideen ; MI 1. Kollektion Philosophie,
Ökonomie, Politik). - Includes bibliography
3-88535-051-3 X.329/17777
9073833

PETROVIĆ, Sreten
Marksistička kritika estetike. - Beograd : Prosveta,
1982
277p. - Includes bibliography
r6494980 X.529/66081
9073833

AESTHETICS
related to information theory

COELHO NETTO, J. Teixeira
Introdução a teoria da informação estética. -
Petrópolis : Vozes, [1973]
141 p.. - (Coleção Textos introdutórios ; no. 1)
r6011489 X.519/20463
0939250

AESTHETICS
Role in theology. Theories of Balthasar, Hans Urs von

KAY, Jeffrey Ames
Theological aesthetics. - Bern : H. Lang [etc.], 1975
[1], viii, 105 p.. - Bibliography : p. 97-105
r6017398 X.100/16795
1058290

AESTHETICS
Role in theology. Theories of Balthasar, Hans Urs von

SCHMID, Johannes
Im Ausstrahl der Schönheit Gottes. - Münsterschwarzach
: Vier-Türme-Verlag, 1982
xvi,264p. - (Münsterschwarzacher Studien ; Bd.35). -
Bibliography
3-87868-150-x X.200/40211
894234x

AESTHETICS
Russian theories

PLEKHANOV, G. V. (Georgiĭ Valentinovich)
Estetika i sot͡siologii͡a iskusstva .... - Moskva :
Iskusstvo, 1978
2v. - (Istorii͡a estetiki v pami͡atnikakh i
dokumentakh)
r6175025 X.529/21712
2132044

AESTHETICS
Soviet theories, 1945-1978

SWIDERSKI, Edward M.
The philosophical foundations of Soviet aesthetics. -
Dordrecht ; London : Reidel, c1979
xviii,225p. - (Sovietica ; v.42). - Bibliography:
p210-219. - Includes index
90-277-0980-7 Ac.607.e/2[vol.42]
2391465
- [Another copy.] X.520/23630

AESTHETICS
Theories, to date

ASSUNTO, Rosario
Filosofia del giardino e filosofia nel giardino. - Roma :
Bulzoni, c1981
203p. - (Biblioteca di cultura ; 188)
r6510499 X.950/29736
8820031

DUFRENNE, Mikel
Trattato di estetica. - Milano : Mondadori, 1981
2v.. - (Gli Oscar studio ; 78). - Includes bibliography
r6509703 X.529/57211
8820031

FANIZZA, Franco
Sulla coscienza estetica contemporanea. - Bari :
Adriatica, 1979
129p
r6510138 X.529/58989
8820031

HESSE, Bettina
Religionskritik und Ästhetik. - Frankfurt am Main :
Lang, 1981
176p.. - (Europäische Hochschulschriften. Reihe 20,
Philosophie ; Bd.76). - Bibliography
3-8204-5911-1 X.200/38281
8820031

JHANJI, Rekha
Aesthetic meaning. - Delhi : Ajanta Publications, 1980
199p. - ([Ajauta's series on aesthetics ; 3])
r6808061 X.329/16222
8820031

SHESTAKOV, V. P. (Vi͡acheslav Pavlovich)
Esteticheskie kategorii. - Moskva : Iskusstvo, 1983
358 p. : ill.. - Bibliography: p. 349-[357]
r6489806 X.529/66668
8820031

AESTHETICS
Theories, to date - Conference proceedings

INTERNATIONAL WITTGENSTEIN SYMPOSIUM (8th : 1983 :
Kirchberg am Wechsel)
Akten des 8. Internationalen Wittgenstein Symposiums, 15.
bis 21. August 1983, Kirchberg am Wechsel (Österreich).
- Wien : Hölder-Pichler-Tempsky, 1984
2v.. - (Schriftenreihe der Wittgenstein-Gesellschaft ;
Bd.10)
3-209-00549-4 X.800/39071
8820031

AESTHETICS
Theories, 1700-1800

CHOUILLET, Jacques
L'esthétique des lumières. - [Paris] : Presses
Universitaires de France, 1974
230 p.. - (SUP Littératures modernes ; 4)
r6000972 X.907/14381
0908541

AESTHETICS
Theories of Adorno, Theodor W.

BRÄUTIGAM, Bernd
Reflexion des Schönen - schöne Reflexion. - Bonn, 1975
278 p.. - (Literatur und Wirklichkeit ; Bd 12)
3-416-01016-7 X.0900/144 (12)
0871885

GROHOTOLSKY, Ernst
Ästhetik der Negation. - Meisenheim : Hain, 1984
viii,150p. - (Hochschulschriften Literaturwissenschaft ;
62) (Forum academicum). - Includes bibliography
3-445-02324-7 X.950/36730
8723729

PROHIĆ, Kasim
Prizma i ogledalo. - Beograd : Nolit, 1980
359p
r6435392 X.529/52886
8723729

AESTHETICS
Theories of Adorno, Theodor W

SCHWARZ, Ullrich
Rettende Kritik und antizipierte Utopie. - München :
Fink, 1981
308p. - Includes bibliography
3-7705-1987-6 X.529/56729
8723729

AESTHETICS
Theories of Augustine, Saint, Bishop of Hippo

O'CONNELL, Robert J.
Art and the Christian intelligence in St Augustine. -
Oxford : Blackwell, 1978
xv,251p. - Includes index
0-631-19170-4 X.200/31853
1370782

AESTHETICS
Theories of Beattie, James

KLOTH, Karen
James Beatties ästhetische Theorien. - München : Fink,
1974
291 p.. - Bibliography : p. 264-286
r6019217 X.519/20580
0942332

AESTHETICS
Theories of Belinskiĭ, V. G.

SOBOLEV, Petr Vasil,evich
Estetika Belinskogo. - Moskva : Iskusstvo, 1978
239p. - Bibliography : p.234-[240]
r6097737 X.529/21390
2024365

AESTHETICS
Theories of Belyĭ, Andreĭ

DEPPERMANN, Maria
Andrej Belyjs ästhetische Theorie des schöpferischen
Bewusstseins. - München : Sagner, 1982
256p. : ill. - (Slavistische Beiträge ; Bd.150). -
Bibliography
3-87690-215-0 11879.aa.2/150
8723729

AESTHETICS
Theories of Benjamin, Walter

SCHWARZ, Ullrich
Rettende Kritik und antizipierte Utopie. - München :
Fink, 1981
308p. - Includes bibliography
3-7705-1987-6 X.529/56729
8723729

AESTHETICS
Theories of Bloch, Ernst

PROHIĆ, Kasim
Prizma i ogledalo. - Beograd : Nolit, 1980
359p
r6435392 X.529/52886
8723729

STROHSCHEIN, Barbara
Tagträume hinter Schulmauern. - Frankfurt [am Main] :
Päd. Extra Buchverlag, 1982
223p. - (Forschung ; Bd.1). - Includes bibliography
3-88704-025-2 X.429/16231
8723729

AESTHETICS
Theories of Caudwell, Christopher

SZILI, József
A művészi visszatükrözés szerkezete. - Budapest :
Akadémiai Kiadó, 1981
179p
963-05-2707-3 X.520/29914
8723729

AESTHETICS
Theories of Church Fathers, to 300

BYCHKOV, V. V. (Viktor Vasil'evich)
Estetika pozdneĭ antichnosti II-III veka. - Moskva :
Nauka, 1981
323p. - Bibliography
r6464111 X.208/8430
8950172

AESTHETICS
Theories of Cicero, Marcus Tullius

DESMOULIEZ, André
Cicéron et son goût. - Bruxelles : Latomus, 1976
637p. - (Collection Latomus ; vol.150). - Bibliography :
p.597-621
2-87031-000-5 PP.5044.an(2) [vol.150]
1149814

AESTHETICS
Theories of Croce, Benedetto

D'ANGELO, Paolo
L'estetica di Benedetto Croce. - Roma : Laterza, 1982
xiii,131p. - (Biblioteca di cultura moderna ; 857)
r6506865 X.429/14515
8723729

AESTHETICS
Theories of Czechoslovak writers, 1900–
DOSTÁL, Vladimír
V tomto znamení. – Praha : Melantrich, 1975
658p
r6016726 X.429/7744 Woolwich
2094630

AESTHETICS
Theories of Dewey, John
ZELTNER, Philip M.
John Dewey's aesthetic philosophy. – Amsterdam : B. R. Grüner, 1975
vii, 125 p.. – (Philosophical currents ; vol. 12). – Bibliography : p. 121-123
90-603-2029-8 X.0510/162(12)
1034375

AESTHETICS
Theories of Diderot, Denis
MECCHIA, Renata
Le teorie linguistiche e l'estetica di Diderot. – Roma : Carucci, 1980
241p. – (Biblioteca Carucci ; 2)
88-85027-36-9 X.958/18286
8723729

AESTHETICS
Theories of Dragomirescu, Mihail
MATEI, Dumitru
Mihail Dragomirescu. – Bucureşti : Editura Ştiinţifică, 1974
238 p.
r6007755 X.529/19637
0873810

AESTHETICS
Theories of Edwards, Jonathan
ERDT, Terrence
Jonathan Edwards. – Amherst, Mass. : University of Massachusetts Press, c1980
xiv,123p. – Includes bibliography
0-87023-304-1 X.529/58937
8723729

AESTHETICS
Theories of Engels, Friedrich
KOCH, Hans
Marx, Engels und die Ästhetik. – Berlin : Dietz, 1983
205p. – Includes bibliography
r6309294 X.208/9669
8723729

LIFSHITS, Mikh.
Karl Marks. – Izd. 2-e. – Moskva : Khudozhestvennaia literatura, 1979
469p
r6484714 X.529/46412
8723729

AESTHETICS
Theories of Flaubert, Gustave
TONDEUR, Claire-Lise
Gustave Flaubert, critique. – Amsterdam : Benjamins, 1984
xiv,119p. – (Purdue University monographs in Romance languages ; v.15). – Includes bibliography
90-272-1725-4 X.950/43527
8723729

AESTHETICS
Theories of Hamann, Johann Georg
BRÄUTIGAM, Bernd
Reflexion des Schönen – schöne Reflexion. – Bonn, 1975
278 p.. – (Literatur und Wirklichkeit ; Bd 12)
3-416-01016-7 X.0900/144 (12)
0871869

AESTHETICS
Theories of Hegel, Georg Wilhelm Friedrich
VOSS, Dietmar
Wahrheit und Erfahrung im ästhetischen Diskurs. – Frankfurt am Main : Lang, 1983
vii,369p. – (Europäische Hochschulschriften. Reihe 1, Deutsche Sprache und Literatur ; Bd 572). – Includes bibliography
3-8204-7163-4 X.529/63084
8723729

AESTHETICS
Theories of Kant, Immanuel
AFASIZHEV, M. N. (Marat Nurbievich)
Ėstetika Kanta. – Moskva : Izdatel'stvo "Nauka", 1975
133 p.. – (Akademiia nauk SSSR. Seriia "Iz istorii mirovoĭ kul'tury")
r6021310 Ac.1125/434(22)
0595284

COLEMAN, Francis Xavier Jerome
The harmony of reason. – [Pittsburgh] : University of Pittsburgh Press ; London : Feffer and Simons : [Distributed by Transatlantic Book Service], 1974
xv,221p. – Bibliography: p.205-216. – Includes index
0-8229-3282-2 X.529/30486
0595284

ESSAYS in Kant's aesthetics / edited and introduced by Ted Cohen & Paul Guyer. – Chicago ; London : University of Chicago Press, 1982
x,323p. – Bibliography: p307-323
0-226-11226-8 X.950/15179
0595284

GUYER, Paul, 1948–
Kant and the claims of taste. – Cambridge, Mass. ; London : Harvard University Press, 1979
xiv,447p. – Includes index
0-674-50020-2 X.520/13823
0595284

MÜLLER, Gerald
Philosophische Ästhetik versus ästhetisches Manifest. – Konstanz : Hartung-Gorre, 1983
v,342p. – (Konstanzer Dissertationen ; Bd.21). – Includes bibliography
3-923200-27-7 X.429/17241
8723729

SCHAPER, Eva
Studies in Kant's aesthetics. – Edinburgh : Edinburgh University Press, 1979
viii,141p. – Includes index
0-85224-359-6 X.529/34007
0595284

AESTHETICS
Theories of Kant, Immanuel; Hegel, Georg Wilhelm Friedrich & Schopenhauer, Arthur
KNOX, Israel
The aesthetic theories of Kant, Hegel, and Schopenhauer. – [Atlantic Highlands] : Humanities Press ; [Brighton] : Harvester Press, 1978, c1958
xi,219p. – Bibliography: p195-207. – Includes index
0-85527-207-4 X.520/24377
2526107

AESTHETICS
Theories of Kierkegaard, Søren
ADORNO, Theodor W. (Theodor Wiesengrund)
Kierkegaard. – Frankfurt am Main, 1974
294 p.
3-518-07674-4 X.519/21698
0868353

SCIACCA, Michele Federico
L'estetismo, Kierkegaard, Pirandello. – Milano : Marzorati, 1974
425 p., 1 leaf of plates : ill. – (Opere complete di Michele F. Sciacca ; n. 39)
r6014326 X.519/20815
0868353

AESTHETICS
Theories of Lenin, V. I.
MAZEPA, V. I. (Volodymyr Ivanovych)
Ėsteticheskoe nasledie V.I. Lenina i sovremennye problemy khudozhestvennoĭ kul'tury. – Kiev : Naukova dumka, 1980
162 p.. – Includes bibliographical references
r6482988 X.529/65910
8723729

AESTHETICS
Theories of Lévi-Strauss, Claude
MERQUIOR, José Guilherme
L'esthétique de Lévi-Strauss. – [Paris] : PUF, 1977
155p. – (Croisées)
r6059080 X.519/35483
1537466

AESTHETICS
Theories of Lukács, Georg
PROHIĆ, Kasim
Prizma i ogledalo. – Beograd : Nolit, 1980
359p
r6435392 X.529/52886
8723729

AESTHETICS
Theories of Lukács, György
DAHL CHRISTIANSEN, Kurt
Kulturstrategi og litteraturteori. – [Odense] : Odense Universitet, Udgivelsesudvalget : I kommission hos Centerboghandelen, 1978
163p. – (Udgivelsesudvalgets samling af studenterafhandlinger ; 13). – Includes bibliographies
87-87766-02-7 X.909/86839
0351482

EBADIAN, Mahmoud
Die Problematik der Kunstauffassung Georg Lukacs'. – Hamburg : Buske, 1977
viii,154p. – (Hamburger Studien zur Philosophie ; Bd 6). – Includes bibliographies
3-87118-316-4 X.0519/558(6)
0351482

KIRÁLYFALVI, Béla
The aesthetics of György Lukács. – Princeton ; London : Princeton University Press, 1975
xi,166p. – (Princeton essays in literature). – Bibliography: p.149-155. – Includes index
0-691-07205-1 X.529/30063
0351482

AESTHETICS
Theories of Lukács, György, 1885-1971
SZILI, József
A művészi visszatükrözés szerkezete. – Budapest : Akadémiai Kiadó, 1981
179p
963-05-2707-3 X.520/29914
8723729

AESTHETICS
Theories of Lunacharskii, A. V.
BARANENKOVA, A. F. (Askaniia Fedorovna)
A.V. Lunacharskiĭ o problemakh sotsialisticheskoĭ ėstetiki. – Riga : Zinatne, 1981
158p
r6464238 X.950/26865
8723729

AESTHETICS
Theories of Marx, Karl, 1818-1883
KOCH, Hans
Marx, Engels und die Ästhetik. – Berlin : Dietz, 1983
205p. – Includes bibliography
r6309294 X.208/9669
8723729

LIFSHITS, Mikh.
Karl Marks. – Izd. 2-e. – Moskva : Khudozhestvennaia literatura, 1979
469p
r6484714 X.529/46412
8723729

AESTHETICS
Theories of Moritz, Karl Philipp
RAU, Peter
Identitätserinnerung und ästhetische Rekonstruktion. – Frankfurt (Main) : R.G. Fischer, 1983
486p. – (Literatur und Kommunikation ; Bd 1). – Includes bibliography
3-88323-451-6 X.529/65179
8723729

AESTHETICS
Theories of Mukařovský, Jan
SCHWARZ, Ullrich
Rettende Kritik und antizipierte Utopie. – München : Fink, 1981
308p. – Includes bibliography
3-7705-1987-6 X.529/56729
8723729

AESTHETICS
Theories of Natorp, Paul
KREBS, Inge
Paul Natorps Ästhetik. – Berlin : de Gruyter, 1976
xii,327p. – (Kantstudien. Ergänzungshefte ; 109). – Includes bibliographies
3-11-006587-8 PP.1253.baa(1)[no.109]
1153110

AESTHETICS
Theories of Nietzsche, Friedrich
BRÄUTIGAM, Bernd
Reflexion des Schönen – schöne Reflexion. – Bonn, 1975
278 p.. – (Literatur und Wirklichkeit ; Bd 12)
3-416-01016-7 X.0900/144 (12)
0871877

AESTHETICS
Theories of Osborne, Harold
ENVALL, Markku
Harold Osbornen varhaisestetiikka. – Helsinki : [Helsingin yliopisto], 1980
vi,180p. – (Helsingin yliopiston Yleisen kirjallisuustieteen, teatteritieteen ja estetiikan laitoksen monistesarja ; n:o 7). – Includes bibliography
951-452-049-1 X.0950/873(7)
8723729

AESTHETICS
Theories of Pirandello, Luigi

SCIACCA, Michele Federico
L'estetismo. Kierkegaard. Pirandello. - Milano : Marzorati, 1974
425 p., 1 leaf of plates : ill. - (Opere complete di Michele F. Sciacca ; n. 39)
r6014326 X.519/20815
0868361

AESTHETICS
Theories of Ralea, Mihai

POPESCU, Titu
Idei estetice in scrierile lui Mihai Ralea. - Cluj : Editura Dacia, 1974
182
r6022163 X.519/22612
1129449

AESTHETICS
Theories of Schiller, Friedrich

POTT, Hans-Georg
Die schöne Freiheit. - München : Fink, 1980
160p - (Theorie und Geschichte der Literatur und der schönen Künste ; Bd 56)
3-7705-1894-2 X.529/43141
8723729

WESSELL, Leonard P.
The philosophical background to Friedrich Schiller's aesthetics of living form. - Frankfurt am Main : Lang, 1982
174p. - (European university studies. Ser.1, German language and literature ; v.578). - Includes bibliography
3-8204-7195-2 X.529/63083
8723729

AESTHETICS
Theories of Schopenhauer, Arthur

POTHAST, Ulrich
Die eigentlich metaphysische Tätigkeit. - Frankfurt am Main : Suhrkamp, 1982
412p.
3-518-57619-4 X.958/24115
8723729

AESTHETICS
Theories of Stasov, V. V.

OBRAZTSOV, G. A.
Estetika V.V. Stasova i razvitie russkogo natsional'no-realisticheskogo iskusstva. - Leningrad : Izdatel'stvo Leningradskogo universiteta, 1975
190p
r6196110 X.519/41075
1991949

AESTHETICS
Theories of Vischer, Friedrich Theodor

GÖBEL, Wendelin
Friedrich Theodor Vischer. - Würzburg : Königshausen + Neumann, 1983
246p. - (Epistemata. Reihe Philosophie ; Bd 15). - Includes bibliography
3-88479-119-2 X.529/68560
8723729

AESTHETICS. African literatures, to date
- Conference proceedings

COLLOQUE SUR LITTÉRATURE ET ESTHÉTIQUE NÉGRO-AFRICAINES (1974 : Abidjan)
Colloque sur littérature et esthétique négro-africaines. - Abidjan : Les nouvelles éditions africaines, 1979
358p
2-7236-0544-2 X.950/5562
8719500

AESTHETICS. Architectural design

ARNAU AMO, Joaquín
Arquitectura estética empírica. - [Valencia] : Escuela Técnica Superior de Arquitectura, Universidad Politécnica de Valencia, [1975]
183p : ill
84-600-6627-4 X.410/6062 Woolwich
0621749

CHING, Francis D. K. (Francis Dai-Kam)
Architecture. - New York ; London : Van Nostrand Reinhold, 1979
395p : ill. - Bibliography: p.388. - Includes index
0-442-21534-7 L.49/295
0621749

GRILLO, Paul Jacques
Form, function and design. - New York : Dover Publications ; London : Constable, 1975
238p : ill
0-486-20182-1 X.415/5075
0621749

AESTHETICS. Architectural design
Psychological aspects

ARNHEIM, Rudolf
The dynamics of architectural form. - Berkeley ; London : University of California Press, 1977
vi,289p : ill. - Bibliography: p.279-283. - Includes index
0-520-03305-1 X.421/10599 Woolwich
1379380

SMITH, Peter Frederick
Architecture and the human dimension. - London : Godwin, 1979
vii,232p : ill. - Bibliography : p.226-228. - Includes index
0-7114-3103-5 X.421/20677 Woolwich
1379380

AESTHETICS. Architecture

GREENE, Herb
Mind & image. - Lexington, Ky : University Press of Kentucky, 1976
210p
0-8131-1323-7 L.49/170
1964291

SCRUTON, Roger
The aesthetics of architecture. - London : Methuen, 1979
x,302p : ill. - Bibliography: p.291-292. - Includes index
0-416-85970-4 X.421/11039 Woolwich
1964291

AESTHETICS. Armenia (Soviet Socialist Republic)

APRESIAN, Grant Zakharovich
Iz istorii armianskoi esteticheskoi mysli. - Erevan : Sovetakan grokh.
kn. 11. - 1977
341p
r6094902 X.809/26391
1013564

OCHERK razvitiia esteticheskoi mysli v Armenii / [pod obshchei nauchnoi redaktsiei Ia. I. Khachikiana]. - Moskva : Iskusstvo, 1976
278 p.
r6022949 X.529/19845
1013564

AESTHETICS. Arts

AESTHETICS and politics / Ernst Bloch ... [et al.] ; afterword by Frederic Jameson ; [translated from the German] ; translation editor Ronald Taylor. - London : NLB, 1977
220p. - Includes index
0-902308-38-6 X.809/42983
1399330

AESTHETICS in twentieth-century Poland / edited by Jean G. Harrell and Alina Wierzbiańska [translated from the Polish]. - Lewisburg : Bucknell University Press, 1973
285p
0-8387-1100-6 X.529/31896
0281689

BEHNE, Adolf
Zur neuen Kunst. - Nendeln : Kraus, 1974
32p. - (Sturm-Bücher ; 7)
r6063457 X.981/12165(7)
0242608

CHERNYSHEVSKII, N. G. (Nikolai Gavrilovich)
[Selections]. Izbrannye esteticheskie proizvedeniia. - Izd. 2-oe. - Moskva : Iskusstvo, 1978
558p,[1] leaf of plates
r6172599 X.529/21519
0105090

CHIARI, Joseph
Art and knowledge. - London : Elek, 1977
x,132p. - Bibliography: p.125-127. - Includes index
0-236-40017-7 X.429/10250
0242608

DELLA VOLPE, Galvano
Critique of taste. - London : NLB, 1978
272p. - Includes index
0-902308-48-3 X.989/52567
1399330

DORFLES, Gillo
Le oscillazioni del gusto. - 2. ed. - Torino : Giulio Einaudi, 1971
131 p.. - (Piccola biblioteca Einaudi ; n. 137)
r6006533 X.519/20309
0311340

FISCHER, Ernst, 1899-1972
The necessity of art. - Harmondsworth : Penguin, 1978
234p. - (Peregrine books). - Includes index
0-14-055151-4 X.519/28548
1399330

GAISER, Gerd
Umgang mit Kunst. - Tübingen : Katzmann, 1974
136 p. : ill. - (Jugend, Bildung, Eoziehung)
3-7805-0323-9 X.419/1551 Woolwich
0242608

GARCÍA MARTÍNEZ, J.A.
Dimensiones de la creación estética. - Buenos Aires : Hachette, 1977
135p. - (Colección Las ideas y las formas)
r6158180 X.419/4216 Woolwich
0242608

GERMAN aesthetic and literary criticism / edited and introduced by David Simpson. - Cambridge : Cambridge University Press, 1984
x,294p. - Bibliography: p284-290. - Includes index
0-521-23630-4 X.950/31883
0242608

GHOSH, Ranjan K.
Aesthetic theory and art. - Delhi : Ajanta Publications, 1979
vi,152p.. - (Ajanta's series on aethetics ; 1)
r6109364 X.429/7865 Woolwich
0242608

GOŁASZEWSKA, Maria
Estetyka. - Warszawa : Państwowe Wydawnictwo Naukowe, 1975
41 p.. - (Nauka dla wszystkich ; nr 246)
r6008349 P971/165(276)
0242608

HARBISON, Robert
Eccentric spaces. - London : Deutsch, 1977
xii,177p. - Includes index
0-233-96878-4 X.620/16886 Woolwich
0242608

HARTH-TERRÉ, Emilio
Formas estéticas. - Lima : Edit. Mejia Baca, 1976
240p
r6065149 X.419/7957 Woolwich
0105090

HEGEL, Georg Wilhelm Friedrich
[Vorlesungen über die Ästhetik. Einleitung. English]. Hegel's introduction to 'Aesthetics'. - Oxford : Clarendon Press, 1979
lxxvi,94p. - Includes index
0-19-824377-4 X.529/34502
0242608

HESTER, Marcus B.
Sensibility and criticism. - Lanham ; London : University Press of America, c1983
x,168p : ill. - Bibliography: p161-165. - Includes index
0-8191-3457-0 X.950/35362
0242608

IASTREBOVA, Natal'ia Aleksandrovna
Formirovanie esteticheskogo ideala i iskusstvo. - Moskva : Izdatel'stvo ""Nauka", 1976
293 p.
r6023100 X.519/20963
0242608

INNOVATION/RENOVATION / edited by Ihab Hassan & Sally Hassan. - Madison ; London : University of Wisconsin Press, 1983
viii,365p : 2ill. - (Theories of contemporary culture). - Includes index
0-299-09390-5 X.529/62238
0242608

JOHNSON, Pauline
Marxist aesthetics. - London : Routledge & Kegan Paul, 1984
viii,168p. - Bibliography: p163-166. - Includes index
0-7100-9927-4 X.529/63553
1399330

AESTHETICS. Arts

KOSTYRKO, Teresa
O wyjaśnianiu faktów i procesów kultury artystycznej. - Poznań : Wydawnictwo Naukowe Uniwersytetu im. Adama Mickiewicza, 1975
144 p.. - (Seria. Filozofia i logika ; nr. 15). - Bibliography : p. 135-138
r6015240 X.410/5893 Woolwich
0242608

LANG, Berel
Art & inquiry. - Detroit : Wayne State University Press, 1975
227, [1] p. : 1 ill. - Bibliography : p. 221-224. - Includes index
0-8143-1531-3 X.429/10233 Woolwich
0242608

LESSING, Gotthold Ephraim, 1729-1781
Laocoön. - Baltimore ; London : Johns Hopkins University Press, 1984
xxx,259p
0-8018-3139-3 X.958/27116
0242608

LOMAZZO, Giovan Paolo
Idea del tempio della pittura. - Firenze : Istituto Nazionale di Studisul Rinascimento, 1974
2 v. (792 p.) ; 8 leaves of plates
r6019626 X.415/1922
0242608

AESTHETICS. Arts -continued

MARGOLIS, Joseph
Art and philosophy. - Brighton : Harvester, 1980
vi,350p. - Bibliography: p307-333. - Includes index
0-85527-837-4 X.520/22747
0242608

MARKSIZAM i umjetnost / izbor tekstova, predgovor, bilješke o autorima : Vjekoslav Mikecin. - Beograd : Komunist, 1972
xvi,345p. - (Antologije)
r6086884 X.520/9850
1399330

O niektorých problémoch ideologického boja v umeni. - Bratislava : Pravda, 1977
141p
r6183356 X.429/11321 Woolwich
0242608

PAPU, Edgar
Arta şi umanul. - Bucureşti : Editura Meridiane, 1974
199 p., 36 p. of plates : ill. - (Biblioteca de artă. biografii, memorii, eseuri)
r6008178 X.419/1560 Woolwich
0242608

PLEKHANOV, G. V. (Georgiĭ Valentinovich)
Ėstetika i sot͡siologii͡a iskusstva. - Moskva : Iskusstvo, 1978
2t.
r6088046 X.429/7602 Woolwich
0311340

REDFERN, Betty
Dance, art and aesthetics. - London : Dance Books, 1983
131p. - Bibliography: p120-124. - Includes index
0-903102-73-0 X.421/25066
0242608

ROSS, Malcolm, 1932-
The aesthetic impulse. - Oxford : Pergamon, 1984
xvi,149p : ill. - (Pergamon international library). - Bibliography: p148. - Includes index
0-08-030234-3 X.950/33618
0242608

RUBERT DE VENTÓS, Xavier
La estética y sus herejías. - Barcelona : Editorial Anagrama, [1974]
394 p. : ill. - (Colección Argumentos ; no. 28)
84-339-0028-5 X.908/33313
0242608
Heresies of modern art. - New York ; Guildford : Columbia University Press, 1980
ix,262p : ill. - Includes index
0-231-04458-5 X.421/20935 Woolwich
0242608

AESTHETICS. Arts

SAVILE, Anthony
The test of time. - Oxford : Clarendon, 1982
xiii,319p. - Includes index
0-19-824590-4 X.529/49411
0242608

"SCHÖN" / Hrsg. Siegfried J. Schmidt. - München : Fink, [1976]
218p. - (Grundfragen der Literaturwissenschaft ; n.F., Bd 1)
3-7705-1321-5 P:901/749a[Bd.1]
0242608

SCHREYER, Lothar
Die neue Kunst. - Nendeln : Kraus, 1974
60p. - (Sturm-Bücher)
r6063465 X:981/12165(15)
0242608

SHARPE, R. A.
Contemporary aesthetics. - Brighton : Harvester, 1983
199p. - Includes index
0-7108-0456-3 X.329/16824
0242608

ŠIMUNEK, Eugen
Estetika umeleckej tvorby. - v Bratislave : Tatran, 1980
160p. - (Okno ; zv.30)
r6439061 X.529/40676
0242608

SZABÓ, László, Sz.
A művészeti ízlés fogalomköre.... - Debrecen : [s.n.], 1974
61p. - (Nevelés, művelődés. Acta paedagogica Debrecina sorozat ; sz.60)
r6007659 Ac.829/7a.(3)[sz.60]
0242608

TAYLOR, Roger, 1940-
Art, an enemy of the people. - Hassocks : Harvester Press, 1978
[7],155p. - (Philosophy now ; vol.3)
0-85527-941-9 X.429/10422 Woolwich
0242608

TEORII, shkoly, kont͡sept͡sii / [redkollegii͡a I͡U.V. Borev (otvetstvennyĭ redaktor) ... i dr.]. - Moskva : Nauka, 1975
229p. - (Ideologicheskai͡a bor'ba v sovremennom mire)
r6186589 X.429/7886 Woolwich
0242608

USCATESCU, Jorge
Estructuras de la imaginación. - Madrid : Reus, 1976
162 p.
84-290-1038-6 X.519/21054
0105090
Fundamentos de estética y estética de la imagen. - Madrid : Reus, 1979
216p
84-290-1256-7 X.529/40230
0242608

WALKER, John A.
Art in the age of mass media. - London : Pluto, 1983
124p : ill. - Bibliography: p121-124
0-86104-718-4 X.429/15912
1399330

WOLFF, Janet
Aesthetics and the sociology of art. - London : Allen & Unwin, 1983
120p. - (Controversies in sociology ; 14). - Bibliography: p109-116. - Includes index
0-04-301152-7 X.529/55268
0311340
Hermeneutic philosophy and the sociology of art. - London : Routledge and Kegan Paul, 1975
vii,149p. - (International library of sociology). - Bibliography: p.139-146. - Includes index
0-7100-8048-4 WP.8084/394
0311340
- [Another copy: paperback] WP.8084/445
- [Another copy: paperback]

WOLLEN, Peter
Readings and writings. - London : NLB, c1982
ix,228p. - Includes index
0-86091-055-5 X.950/19304
1399330

WOLLHEIM, Richard
Art and its objects. - Harmondsworth : Penguin, 1975
189p. - Bibliography: p.171-189
0-14-055108-5 X.419/3040
0242608
Art and its objects. - Harmondsworth : Penguin, 1978
189p. - (Peregrine books). - Bibliography: p.171-189
B78-33908 X.519/28540
0242608
Art and its objects. - 2nd ed.. - Cambridge : Cambridge University Press, c1980
xv,270p. - Bibliography: p241-270
0-521-22898-0 X.429/13673
0242608

AESTHETICS. Arts
- For teaching

The AESTHETIC in education / edited by Malcolm Ross. - Oxford : Pergamon, 1985
xxii,235p : ill. - (Curriculum issues in arts education ; v.5). - Bibliography: p52-53. - Includes index
0-08-031843-6 X.0529/590(5)
2777010

The ARTS / edited by Malcolm Ross. - Oxford : Pergamon, 1983
xii,244p : ill. - (Curriculum issues in arts education ; v.4). - Includes index
0-08-030180-0 X.0529/590(4)
2777010

AESTHETICS. Arts
- For teaching - Conference proceedings

The DEVELOPMENT of aesthetic experience / edited by Malcolm Ross. - Oxford : Pergamon, 1982
xiii,210p. - (Curriculum issues in arts education ; v.3). - Includes bibliographies and index
0-08-028908-8 X.0529/590(3)
2777010

AESTHETICS. Arts
- Texts

READ, Sir Herbert
Aristotles mother. - Harrow : Philip Ward, 1961
14p. - (Herbert Read reprints ; no.1)
r6814654 X.950/33694
0242608

AESTHETICS. Arts
to date

AESTHETICA / collana del Centro internazionale studi di estetica. - Bologna : il Mulino, 1982-
r6509432 X.0429/610
8713197

AKTUAL'NYE voprosy metodologii sovremennogo iskusstvoznanii͡a / [redkollegii͡a, M.N. Afasizhev, E.I͡A. Basin, N.A. I͡Astrebova] ; [otvetstvennyĭ redaktor A.I͡A. Zis']. - Moskva : Nauka, 1983
366 p.. - Includes bibliographical references
r6403224 X.429/15939
8713197

BATTEUX, Charles
[Les beaux arts réduits en un même principe. Italian]. Le belle arti ricondotte ad unico principio. - Bologna : Il mulino, 1983
196p. - (Aesthetica ; 5. Classici)
88-15-00036-4 X.0429/610(5)
8713197

BOREV, I͡Uriĭ Borisovich
Ėstetika. - Izd 3-e. - Moskva : Izdatel'stvo politicheskoĭ literatury, 1981
399p : ill
r6496642 X.429/15503
8713197

BOZONES, Georgios A.
Aisthetikoi provlematismoi. - Athena : [s.n.], 1982-. - Includes bibliography
r6598265 X.429/15842
8715572

AESTHETICS. Arts
to date

CAPONE, Vittorio Ugo
Arte e originalità. - Napoli : SEN, [1979]
vii,247p
r6503494 X.421/22702 Woolwich
8713197

DELLA VOLPE, Galvano
[Critica del gusto. Spanish]. Crítica del gusto. - 2a ed. revisada y aumentada por el autor. - La Habana : Editorial Arte y Literatura, 1978
lxvii,335p. - (Arte y sociedad)
r6706994 X.529/55805
8713197

FORMAGGIO, Dino
L'arte. - [3a. ed. ampliata]. - Milano : Arnoldo Mondadori, 1981
267p : ill. - (Enciclopedia Filosofica Oscar Studio ; 7). - Includes bibliography
r6511041 X.429/16643
8713197
La "morte dell'arte" e l'estetica. - Bologna : Il mulino, 1983
305p. - (Aesthetica ; 7. Saggi)
88-15-00290-1 X.0429/610(7)
8713197

GRIMALDI, Nicolas
L'art, ou, La feinte possion. - [Paris] : Presses Universitaires de France, [1983]
300p. - (Épiméthée). - Includes bibliography
2-13-037953-2 X.950/39908
8713197

IL'IN, I. A. (Igor' Arkad'evich)
Istorii͡a iskusstva i estetika. - Moskva : Iskusstvo, 1983
287 p.. - Includes bibliographical references
r649003 X.950/36934
8713197

ISKUSSTVO i tvorcheskai͡a dei͡atel'nost' / [otvetstvennyĭ redaktor : V.I. Mazepa]. - Kiev : Naukova dumka, 1979
311p
r6432199 X.429/13550
8713197

KATENIN, P. A. (Pavel Aleksandrovich)
[Selections]. Razmyshlenii͡a i razbory. - Moskva : Iskusstvo, 1981
373p. - (Istorii͡a ėstetiki v pami͡atnikakh i dokumentakh)
r6432175 X.429/13517
8713197

KLÉBANER, Daniel
L'art du peu. - [Paris] : Gallimard, 1983
162p. - (Le chemin)
2-07-024646-9 X.958/27970
8713197

KUNCHEVA, L. I. (Li͡udmila Ivanovna)
Ėsteticheskie vzgli͡ady obshchestva i khudozhestvennai͡a kul'tura. - Moskva : Nauka, 1979
246p
r6484766 X.429/13098 Woolwich
8713197

LUKÁCS, György, 1885-1971
Die Eigenart des Ästhetischen. - Berlin : Aufbau-Verlag, 1981
2v.
r6304650 X.529/50174
8713197

MARX, Wolfgang, 1940-
Ästhetische Ideen. - Bonn : Bouvier, 1981
134p. - (Abhandlungen zur Philosophie, Psychologie und Pädagogik ; Bd.155)
3-416-01581-9 X.529/45284
8713197

AESTHETICS. Arts
to date-continued

MICHEL, Alain
La parole et la beauté. - Paris : Les Belles lettres, 1982
458p. - (Collection d'études anciennes)
2-251-32846-7 X.421/23346
8713197

AESTHETICS. Arts
to date

PAŠTEKA, Július
Estetické paralely umenia - V Bratislave : Veda, 1976
425p,[90]p of plates : ill
r6440194 X.421/22760
8713197

PETROVIĆ, Sreten
Savremena sociologija umetnosti. - 2. izm. i dop. izd. - Beograd : Privredni pregled, 1979
346p
r6435081 X.520/25322
8713197

REIJNDERS, Frank
Kunst - geschiedenis. - Amsterdam : SUA, [1984]
168p : ill. - (Kunstreeks ; 1)
90-622-2116-5 X.429/17033
8713197

REUCHER, Theo
Von der Logik des Sinns zum Sinn der Kunst. - Sankt Augustin : Richarz, 1981
vii,200p. - (Duisburger Studien ; Bd 4)
3-88345-603-9 X.421/24159
8713197

ROOKMAAKER, H. R. (Hans R)
Art needs no justification. - 2nd printing. - Downers Grove : Inter-Varsity Press, 1979
61p
r6801800 X.429/12604 Woolwich
8713197

ROSENKRANZ, Karl
[Aesthetik des Hässlichen. Italian]. Estetica del brutto. - Bologna : Il mulino, 1984
346p. - (Aesthetica ; 9. Classici). - Includes bibliography
88-15-00539-0 X.0429/610(9)
8713197

SANTORO, Liberato
Presupposti filosofici dell'arte moderna. - Urbino : Argalia, 1978
175p. - (Studi filosofici)
r6500904 X.429/12204 Woolwich
8715572

ŠIMUNEK, Eugen
Estetika a všeobecná teória umenia. - Bratislava : Obzor, 1976
192p,[16]p of plates : ill. - (Malá moderná encyklopédia)
r6440650 X.429/13607
8713197

STELLA, Vittorio
L'apparizione sensibile. - Roma : Bulzoni, c1979
283p. - (Biblioteca di cultura ; 143)
r6501851 X.429/12354 Woolwich
8713197

VEĬDLE, V. (Vladimir)
Gestalt und Sprache des Kunstwerks. - Mittenwald : Mäander, [1981]
120p
3-88219-000-0 X.421/22556 Woolwich
8713197

VOPROSY istorii i teorii ėstetiki / pod redakt͡sieĭ M.N. Afasizheva. - Moskva : Izdatel'stvo Moskovskogo universiteta, 1975
213p
r6484684 X.429/13068 Woolwich
8713197

VUK-PAVLOVIĆ, Pavao
Duševnost i umjetnost. - Zagreb : Liber, 1976
257p. - (Biblioteka znanstvenih radova). - Bibliography
r6435501 X.421/23728
8713197

ZIS', A. (Avner)
Konfrontat͡sii v ėstetike. - Moskva : Iskusstvo, 1980
237p
r6432057 X.950/14678
8713197

AESTHETICS. Arts
to date - Conference proceedings

WITTGENSTEIN - Ästhetik und transzendentale Philosophie / herausgegeben von Kjell S. Johannessen und Tore Nordenstam. - Wien : Hölder-Pichler-Tempsky, 1981
193p : ill. - (Schriftenreihe der Wittgenstein-Gesellschaft ; Bd.6)
3-209-00279-7 X.520/26425
8713197

AESTHETICS. Arts
French theories, to date

MORTIER, Roland
L'originalité. - Genève : Droz, 1982
218p. - (Histoire des idées et critique littéraire ; v.207)
r6298052 X.429/16737
9160213

AESTHETICS. Arts
Hegel, Georg Wilhelm Friedrich

BUNGAY, Stephen
Beauty and truth. - Oxford : Oxford University Press, 1984
x,235p,111p of plates. - (Oxford modern languages and literature monographs). - Bibliography: p220-227. - Includes index
0-19-815540-9 WP.13860/46
3026485

AESTHETICS. Arts
Implications of forgeries

The FORGER'S art / edited by Denis Dutton. - Berkeley ; London : University of California Press, c1983
x,276p : ill. - Bibliography: p265-272. - Includes index
0-520-04341-3 X.421/25436
3133192

AESTHETICS. Arts
Marxist theories, to date

ERMASH, Gela Leonidovna
Iskusstvo kak myshlenie. - Moskva : Iskusstvo, 1982
276 p.. - Bibliography: p. 268-[275]
r6403518 X.809/59797 CAI 010 83-119489
9005757

ISKUSSTVO i ideologii͡a / [otvetstvennyĭ redaktor i sostavitel' N.M. Fed']. - Moskva : Iskusstvo, 1973
501p
r6487407 X.529/62103
9005757

MARKSISTSKO-LENINSKAI͡A ėstetika v bor'be za progressivnoe iskusstvo / [otvetstvennye redaktory: M.F. Ovsi͡annikov, I.S. Kulikova, V.N. Ermolaeva]. - Moskva : Nauka, 1980
285p. - (Ėstetika za rubezhom)
r6403080 X.429/14992
9005757

AESTHETICS. Arts
Marxist viewpoints - Bibliographies

BAXANDALL, Lee
Marxism and aesthetics. - [Atlantic Highlands] : Humanities Press ; [Hassocks] : Harvester Press, 1978
xxiii,261p. - (AIMS bibliographical series / American Institute for Marxist Studies ; no.4)
0-85527-257-0 X.525/2363
1895494

AESTHETICS. Arts
Mathematical models. Applications of digital computer systems. Algorithms

STINY, George
Algorithmic aesthetics. - Berkeley ; London : University of California Press, 1978
xi,220p : ill. - Bibliography: p.211-215. - Includes index
0-520-03467-8 X.620/18971 Woolwich
1996940

AESTHETICS. Arts
Psychological aspects

ADVANCES in intrinsic motivation and aesthetics / edited by Hy I. Day. - New York ; London : Plenum, c1981
xii,503p : ill. - Includes bibliographies and index
0-306-40606-3 X.520/24634
0341002

PSYCHOLOGY and the arts / edited by David O'Hare. - Brighton : Harvester, 1981
335p : ill. - Bibliography: p305-332. - Includes index
0-85527-958-3 X.529/44850
0341002

AESTHETICS. Arts
related to physiology of human sensory perception

ALLEN, Grant
Physiological aesthetics. - New York ; London : Garland, 1977
[4],xii,283p. - (The decadent consciousness)
0-8240-2752-3 X.0989/2044(28)
2273489

AESTHETICS. Arts
Structuralism

MUKAŘOVSKÝ, Jan
Structure, sign, and function. - New Haven ; London : Yale University Press, 1977
xxxix,269p. - (Yale Russian and East European studies ; 14). - Bibliography: p.251-266. - Includes index
0-300-02108-9 X.981/20972
1355678

The PRAGUE School / edited by Peter Steiner ; translated by John Burbank ... [et al.]. - Austin : University of Texas Press, 1982
xii,219p. - (University of Texas Press Slavic series ; no.6)
0-292-78043-5 X.520/28458
1355678

TEORII, shkoly, kont͡sept͡sii / [redkollegii͡a ͡IU.V. Borev (otvetstvennyĭ redaktor) ... i dr.]. - Moskva : Nauka, 1975
229p. - (Ideologicheskai͡a bor'ba v sovremennom mire)
r6186589 X.429/7886 Woolwich
1355678

AESTHETICS. Arts
Theories

HIGGINS, Dick
[An exemplativist manifesto. Swedish]. Ett exemplativistiskt manifest. - [Lund] : Kalejdoskop, 1977
p.52-59
91-85552-07-0 X.410/6965 Woolwich
0854670

TRIONE, Aldo
Struttura e istituzioni dell'arte. - Lecce : Edizioni Milella, 1974
179 p.. - (Collezione di studi e testi diretta da M. Marti e A. Vallone ; n. 21)
r6025885 X.519/20638
0854670

AESTHETICS. Arts
Theories, 1700-

FREY, Peter
Die Philosophie der Kunst denkt sich zu ende. - [Frankfurt am Main] : [P. Frey], 1977
190p. - Bibliography
r6305302 X.429/13460
877224x

LACOSTE, Jean
La philosophie de l'art. - Paris : Presses universitaires de France, 1981
127p. - (Que sais-je? ; 1887)
2-13-036704-6 X.429/12625 Woolwich
877224x

RAVERA, Marco
Estetica posthegeliana. - Milano : Mursia, [1978?]
155p. - (Saggi di estetica e di poetica ; n.21)
r6504586 X.429/13119 Woolwich
877224x

AESTHETICS. Arts
Theories, 1945-

BATTCOCK, Gregory
Why art. - New York : Dutton, [1977]
viii,134p : ill. - (A Dutton paperback)
0-525-47286-x X.429/12607 Woolwich
8629579

BONVILLE, W. J.
Footnotes to a fairytale. - St. Louis : Warren H. Green, c1979
xvi,174p. - Bibliography
0-87527-192-8 X.950/7460
8629579

BRAUN, Jerzy
Krytyka rozumu twórczego. - Roma : Towarzystwo Przyjaciół Teologii i Filozofii Chrześcijańskiej, 1975
192p
r6444024 X.429/12287 Woolwich
8629579

FOWKES, William I. (William Ivor)
A Hegelian account of contemporary art. - Ann Arbor : UMI Research Press, c1981
xi,104p. - (Studies in the fine arts : art theory)
0-8357-1187-0 X.421/22346 Woolwich
8629579

LASCAULT, Gilbert
Boucles et nœuds. - [Paris] : Balland, 1981
102p. - (Le commerce des idées)
2-7158-0295-1 X.958/5404
8629579

ŠABOUK, Sáva
Tři polemické studie. - Praha : Československý spisovatel, 1976
261p,plates
r6440988 X.429/15321
8629579

AESTHETICS. Arts
Theories, 1945—continued

TEMME, J. E. V.
Over smaak valt te twisten. - [Utrecht] : [J.E.V. Temme]. 1983
192p. - Includes bibliography
r6330765 X.421/25004
8629579

AESTHETICS. Arts
Theories of Adorno, Theodor W. (Theodor Wiesengrund)

GALEAZZI, Umberto
L'estetica di Adorno. - Roma : Città nuova, 1979
275p. - (Idee ; 40)
r6500639 X.429/11859 Woolwich
8704155

AESTHETICS. Arts
Theories of Benjamin, Walter

STOESSEL, Marleen
Aura. - München : Hanser, c1983
255p. - (Edition Akzente). - Includes bibliography
3-446-13766-1 X.429/16343
8704155

AESTHETICS. Arts
Theories of Burckhardt, Jacob

REIJNDERS, Frank
Marx, Burckhardt en het probleem van de kunsthistoriese wetenschap. - Amsterdam : Talsma & Hekking, [1979]
52p,[12]p of plates : ill. - (Kunsthistoriese schriften ; 1)
90-647-4003-8 X.429/12212 Woolwich
8704155

AESTHETICS. Arts
Theories of Eco, Umberto

DE LAURETIS, Teresa
Umberto Eco. - Firenze : La nuova Italia, 1981
111p. - (Il castoro ; 179). - Bibliography
r6504610 X.0907/36(179)
8704155

AESTHETICS. Arts
Theories of Hegel, Georg Wilhelm Friedrich

GETHMANN-SIEFERT, Annemarie
Die Funktion der Kunst in der Geschichte. - Bonn : Bouvier, 1984
iii,442p. - (Hegel-Studien. Beiheft ; 25). - Includes bibliography
3-416-01795-1 8458.e.1a/25
8704155

AESTHETICS. Arts
Theories of Hegel, Georg Wilhelm Friedrich - Conference proceedings

HEGEL SOCIETY OF AMERICA. Conference (4th : 1974 : Georgetown University)
Art and logic in Hegel's philosophy. - [Atlantic Highlands, N.J.] : Humanities Press ; [Hassocks] : Harvester Press, 1980
[11], 274p. - Bibliography : p. 239-270. - Includes index
0-85527-317-8 X.421/21326 Woolwich
1271237

AESTHETICS. Arts
Theories of Indian writers, 1861-1974

NANDI, S.K
Studies in modern Indian aesthetics. - Simla : Indian Institute of Advanced Studies, 1975
xii,307p. - Bibliography : p.295-300
r6038848 X.529/20327
132134X

AESTHETICS. Arts
Theories of Katenin, P. A. - Correspondence, diaries

KATENIN, P. A. (Pavel Aleksandrovich)
[Selections]. Razmyshleniīa i razbory. - Moskva : Iskusstvo, 1981
373p. - (Istoriīa ėstetiki v pamīatnikakh i dokumentakh)
r6432175 X.429/13517
8704155

AESTHETICS. Arts
Theories of Lukács, György

LUKÁCS, György
[Frühe Schriften zur Ästhetik 1, Heidelberger Philosophie der Kunst. French]. Philosophie de l'art (1912-1914). - Paris : Éditions Klincksieck, 1981
262p. - (L'esprit et les formes)
2-252-02325-2 X.429/13219 Woolwich
8704155

AESTHETICS. Arts
Theories of Marx, Karl, 1818-1883

REIJNDERS, Frank
Marx, Burckhardt en het probleem van de kunsthistoriese wetenschap. - Amsterdam : Talsma & Hekking, [1979]
52p,[12]p of plates : ill. - (Kunsthistoriese schriften ; 1)
90-647-4003-8 X.429/12212 Woolwich
8704155

ROSE, Margaret A. (Margaret Anne)
Marx's lost aesthetic. - Cambridge : Cambridge University Press, 1984
x,216p : ill. - Bibliography: p199-207. - Includes index
0-521-25666-6 X.421/26122
0230472

AESTHETICS. Arts
Theories of Marxists

ERMASH, Gela Leonidovna
Tvorcheskaīa priroda iskusstva. - Moskva : "Iskusstvo", 1977
317p
r6076993 X.809/25900
0858889

NEKOTORYE problemy marksistsko-Leninskoi ėstetiki / [otvetstvennyi redaktor : S. Shermukhamedov]. - Tashkent : Uzbekistan, 1975
214p
r6095237 X.989/78031
0858889

RESZLER, André
Le marxisme devant la culture. - Paris : Presses Universitaires de France, 1975
146 p.. - (SUP. Le philosophe ; 119)
r6019730 X.709/20524
0858889

RĪŪRIKOV, Boris Sergeevich
Ideologicheskie osnovy ėstetiki. - Moskva : "Iskusstvo", 1978
214p
r6088827 X.527/859
0858889

VISHNĪAKOV, Boris Vladimirovich
Vysokoe naznachenie iskusstva. - Leningrad : Khudozhnik RSFSR, 1977
281p,[64]p of plates : ill
r6076980 X.429/7515 Woolwich
0858889

AESTHETICS. Arts
Theories of Mengs, Anton Raphael

CIOFFI MARTINELLI, Rosanna
La ragione dell'arte. - Napoli : Liguori, 1981
170p,16p of plates
88-207-1035-8 X.429/15537
8704155

AESTHETICS. Arts
Theories of Pisarev, D. I.

KONKIN, S. S. (Semen Semenovich)
Ėsteticheskie i literaturno-kriticheskie vzglīady Pisareva. - Saransk : Mordovskiĭ gosudarstvennyĭ universitet im. N.P. Ogareva, 1973
138p
r6462861 X.950/30070
8704155

AESTHETICS. Arts
Theories of Plato

GAWLIK, Ladislav
Umělecká tvorba a estetické vnímání u Platóna. - Praha : Universita Karlova, 1974
127p. - (Acta Univeritatis Carolinae Philosophica et Historica. Monographia ; 1974 ; 56)
r6015405 Ac.798/8e [Monog.56]
1157205

AESTHETICS. Arts
Theories of Read, Sir Herbert

THISTLEWOOD, David
Herbert Read. - London : Routledge & Kegan Paul, 1984
xiv,209p. - Includes index
0-7102-0147-8 X.950/31183
3017729

AESTHETICS. Arts
Theories of Romanian writers, 1860-

ROMANIAN aestheticians = esthéticiens roumains .... - Bucureşti : Editura Minerva, 1972
529 p.
r6008169 X.519/20619
0853038

AESTHETICS. Arts
Theories of Winckelmann, Johann

CIOFFI MARTINELLI, Rosanna
La ragione dell'arte. - Napoli : Liguori, 1981
170p,16p of plates
88-207-1035-8 X.429/15537
8704155

AESTHETICS. Arts
Theories of Winckelmann, Johann Joachim

PAPAÏOANNOU, Angelos A
Ho Winckelmann kai he genesis tes historias tes technes. - Athenai : [s.n.], 1976
124p
r6194550 X.421/20337 Woolwich
2418525

Winckelmann kai Platon. - Athenai : [s.n.], 1977
244p. - Bibl
r6194835 X.421/20336 Woolwich
2418525

AESTHETICS. Arts
Ukrainian theories, to date

IVAN'O, I. V. (Ivan Vasil'ovich)
Ocherk razvitiīa ėsteticheskoi mysli Ukrainy. - Moskva : Iskusstvo, 1981
422p. - (Ėsteticheskaīa mysl' narodov SSSR). - Bibliography
r6486710 X.950/14759
8908591

AESTHETICS. Arts. Great Britain
1695-1810

MALEK, James S.
The arts compared. - Detroit : Wayne State University Press, 1974
175,[1]p. - Includes index
0-8143-1519-4 X.981/12436
1845888

AESTHETICS. Ballet
to 1974

AUSTIN, Richard
Images of the dance. - London : Vision, 1975
183p,[8]p of plates : ill. - Bibliography: p.177-178. - Includes index
0-85478-293-1 X.439/5042
0516872

AESTHETICS. Belorussian arts

SALEEV, Vadim Alekseevich
Sovremennaīa ėstetika Belorussii. - Minsk : Vyshėĭshaīa shkola, 1979
218p. - Bibl: p.203-[219]
r6185379 X.429/7953 Woolwich
2450577

AESTHETICS. Bohemia. Czechoslovakia
1800-1900

JŮZL, Miloš
Otakar Hostinský. - Praha : Melantrich, 1980
378p,[32]p of plates : ill. - (Odkazy pokrokových osobností naší minulosti ; sv.57). - Bibliography: p.347-370
r6440185 X.439/11849
8822883

AESTHETICS. Cinema films

BORDWELL, David
Film art. - Reading, Mass. ; London : Addison-Wesley, c1979
xi,339p,[2]p of plates : ill(some col.). - (Addison-Wesley series in speech, drama, and film). - Bibliography: p322-326. - Includes index
0-201-00566-2 X.950/8093
0584614

CAVELL, Stanley
The world viewed. - Enl. ed. - Cambridge, Mass. ; London : Harvard University Press, 1979
xxv,253p. - (Harvard film studies). - Includes index
0-674-96197-8 X.950/13504
0584614

GIANNETTI, Louis Daniel
Godard and others. - Rutherford : Fairleigh Dickinson University Press ; London : Tantivy Press, [1976]
184p : ill. - Includes index
0-8386-1291-1 X.981/20478
079130x

MAST, Gerald
Film/cinema/movie. - Chicago ; London : University of Chicago Press, 1983
xvii,299p,[4]p of plates : ill(some col.). - Bibliography: p287-290. - Includes index
0-226-50974-5 X.950/36163
0584614

AESTHETICS. Cinema films –continued

SCOTT, James Frazier
Film. – New York ; London : Holt, Rinehart and Winston, 1975
xii,340p : ill. – Bibliography: p.319–323. – List of films. – Includes index
0–03–079445–5 X.900/25091
0584614

STEPHENSON, Ralph
The cinema as art. – 2nd ed. – Harmondsworth : Penguin, 1976
318p,[32]p of plates : ill. – (Pelican book). – Bibliography: p.280–281. – Includes index
0–14–020677–9 X.908/40273
0584614

USCATESCU, Jorge
Fundamentos de estética y estética de la imagen. – Madrid : Reus, 1979
216p
84–290–1256–7 X.529/40230
0584614

WEISS, Paul
Cinematics. – Carbondale : Southern Illinois University Press ; London : Feffer and Simons, 1975
xii,227p. – Bibliography: p.211–213. – Includes index
0–8093–0671–9 X.981/21027
0584614

AESTHETICS. Cinema films
to 1973

MCCONNELL, Frank DeMay
The spoken seen. – Baltimore ; London : Johns Hopkins University Press, 1975
xv,199p : ill. – Includes index
0–8018–1725–0 X.981/20158
0195219

XAVIER, Ismail
O discurso cinematográfico. – Rio de Janeiro : Paz e Terra, 1977
151p : ill. – (Coleção cinema ; v.4)
r6158278 X.419/4246 Woolwich
0195219

AESTHETICS. Cinema films
Theories of Mitry, Jean – Biographies

LEWIS, Brian, 1948–
Jean Mitry and the aesthetics of the cinema. – Ann Arbor : UMI Research, c1984
140p : ill. – (Studies in cinema ; no.25). – Bibliography: p133–135. – Includes index
0–8357–1553–1 X.950/40888
3263258

AESTHETICS. Classical antiquity

ANTIIKIN estetiikka / toimittaneet Aarne Kinnunen ja Teivas Oksala. – Porvoo : W. Söderström, 1977
207p. – (Suomen estetiikan seuran vuosikirja ; 2) (Taskutieto ; 136)
951–00–6817–9 X.519/23001
1157426

AESTHETICS. Curriculum subjects. Education. Bulgaria

GORANOV, Krŭst'o Ankov
Esteticheskoto vŭzpitanie. – Sofiia : Narodna prosveta, 1977
181p
r6093875 X.529/21324
2030675

AESTHETICS. Curriculum subjects. Education. Bulgaria
– Bibliographies

MILADINOVA, Milka Khristova
Za vsenarodno estetichesko vŭzpitanie. – Sofiia : Narodna biblioteka "Kiril i Metodiĭ" ; Varna : Okrŭzhna biblioteka P. Slaveĭkov, 1979
554p
r6184895 X.520/20928
2030675

AESTHETICS. Curriculum subjects. Education. Bulgaria
– Conference proceedings

BULGARIA. Komitet za izkustvo i kultura. Plenum (8th : 1975)
Kompleksna programa za vsenarodno estetichesko vŭzpitanie. – 2. izd. – Sofiia : Nauka i izkustvo, 1980
137p
r6427051 X.529/40586
2030675

AESTHETICS. Czechoslovak arts
1800–1900

ČÍTANKA z dějin české a slovenské estetiky XIX. století / vybral a uvádí Vladimír Dostál. – Praha : Československý spisovatel, 1972
215p
r6187259 X.529/21706
2123991

AESTHETICS. Dancing

LABAN, Rudolf
A vision of dynamic space. – London : Laban Archives in association with Falmer, 1984
80p : ill(some col.)
1–85000–008–5 X.425/4893
0255602

SHEETS-JOHNSTONE, Maxine
The phenomenology of dance. – 2nd ed. – London : Dance Books, 1979
xv,158,[1]p. – Bibliography: p.153–154. – Includes index
0–903102–52–8 X.439/10991
0255602

AESTHETICS. Dancing
– Conference proceedings

PHILOSOPHICAL essays on dance / edited by Gordon Fancher and Gerald Myers. – New York : Dance Horizons, c1981
166p
0–87127–126–5 X.529/53240
0255602

AESTHETICS. Design

BAYLEY, Stephen, 1951–
Taste. – London : Boilerhouse Project, Victoria and Albert Museum, c1983
110p,[24]p of plates : ill
0–946410–04–6 X.529/59524
3061906

AESTHETICS. Design. Great Britain
Social aspects

KLEIN, Bernat
Design matters. – London : Secker and Warburg, 1976
[10],245p : ill
0–436–23470–x X.429/10093
0627569

AESTHETICS. Dinesen, Isak. Fiction in Danish

WHISSEN, Thomas R.
Isak Dinesen's aesthetics. – Port Washington ; London : Kennikat Press, 1973
[7],130p. – (Kennikat Press national university publications. series in literary criticism). – Bibliography: p.124–128. – Includes index
0–8046–9059–6 X.989/40318
0529176

AESTHETICS. Drama

FÖLDÉNYI, László
A drámától a színpadig. – [Budapest] : Kossuth Könyvkiadó. – (Esztétikai kiskönyvtár)
a színjátékművészet esztétikája. – 1976
311 p.
963–09–0601–5 X.908/35012
0872040

AESTHETICS. Drama
– Texts

MATOŠ, Antun Gustav
[Works]. Sabrana djela Antuna Gustava Matoša (1873–1914–1973)....
sv.10: Theatralia/uredio Nikola Batušić. O glazbi/uredio Lovro Županović. – Zagreb : Jugoslavenska akademija znanosti i umjetnosti, Liber, Mladost, 1973
451p
r6007279 X.0981/699(10)
1231294

AESTHETICS. Drama
to date

BLOK, Vladimir
Dialektika teatra. – Moskva : Iskusstvo, 1983
296p
r6484413 X.950/23092
8500258

VODNIK, France
Misli in presoje. – Ljubljana : Slovenska matica, 1983
188p
r6494996 X.950/37368
8500258

AESTHETICS. European architecture
1900–

CASTEELS, Maurice
Die Sachlichkeit in der modernen Kunst. – München : Kraus Reprint, 1981
123p,144p of plates : ill. – (Dokumente der modernen Architektur. Documents of modern architecture)
3–601–00352–x X.425/3079 Woolwich
8800340

AESTHETICS. European arts
to 1500

WACKERS, P.
Met ogen van toen. – Nijmegen : ALFA, 1980
90p,8p of plates : ill. – (Tekst en tijd ; 2). – Bibliography
90–7040–710–8 X.0950/389(2)
8820627

AESTHETICS. European decorative arts
1900–

CASTEELS, Maurice
Die Sachlichkeit in der modernen Kunst. – München : Kraus Reprint, 1981
123p,144p of plates : ill. – (Dokumente der modernen Architektur. Documents of modern architecture)
3–601–00352–x X.425/3079 Woolwich
8800359

AESTHETICS. European literatures
1840–1910

STRANGENESS and beauty / edited by Eric Warner and Graham Hough. – Cambridge : Cambridge University Press, 1983
2v. : ill. – Includes bibliographies
B83–23682 X.950/23172–23173
2925621

AESTHETICS. European music
to date

TARASTI, Eero
Myth and music. – Helsinki : Suomen musiikkitieteellinen seura, 1978
364p. – (Acta musicologica fennica ; 11). – Bibliography
951–95335–3–2 P.431/193[vol.11]
893634x

AESTHETICS. European music, 1700–1800

POLÁK, Pavol
Hudobnoestetické náhlady v 18. storočí. – Bratislava : Veda, vydavatel'stvo Slovenskej akadémie vied, 1974
299 p., 32 p. of plates : ill
r6007369 X.431/2223
1011294

AESTHETICS. European visual arts
1840–1910

STRANGENESS and beauty / edited by Eric Warner and Graham Hough. – Cambridge : Cambridge University Press, 1983
2v. : ill. – Includes bibliographies
B83–23682 X.950/23172–23173
292563x

AESTHETICS. Finland
to date

HUUHTANEN, Päivi
Tunteesta henkeen. – Helsinki : Suomalaisen kirjallisuuden seura, 1978
323p. – (Suomalaisen kirjallisuuden seuran toimituksia ; 341). – Bibliography
951–7171–57–9 Ac.9080[no.341]
9000860

AESTHETICS. Flaubert, Gustave. Fiction in French

DIAMOND, Marie J.
Flaubert. – Port Washington ; London : Kennikat Press, 1975
[5],141p. – (Kennikat Press national university publications. literary criticism series). – Includes index
0–8046–9075–8 X.989/40301
0526339

NADEAU, Maurice
Gustave Flaubert écrivain. – Nouvelle édition revue. – Paris : Les lettres nouvelles, Maurice Nadeau, [1980]
282p
2–86231–024–7 X:950/3604
0526339

AESTHETICS. French literature
to date

DENS, Jean-Pierre
L'honnête homme et la critique du goût. – Lexington : French Forum, 1981
157p. – (French Forum monographs ; 28). – Bibliography: p151–157
0–917058–27–5 X.950/12699
8901236

AESTHETICS. French literature
to date

VACHON, G.-André
Esthétique pour Patricia. suivi d'un écrit de Patricia B.. – Montréal : Les Presses de l'Université de Montréal, 1980
144p. – (Lectures)
2–7606–0491–8 X.958/20897
8901236

AESTHETICS. Graphic arts

GIARDELLI, Arthur
The delight of painting. - [Swansea] : University College of Swansea, 1976
25p,11p of plates : ill. - (W.D. Thomas memorial lectures). - Bibliography: p.23
0-901626-98-8 X.419/3127 Woolwich
0652016

AESTHETICS. Holography

WALTON, Paul
Space light. - London : Routledge & Kegan Paul, 1982
96p : ill(some col.). - Bibliography: p91-92
0-7100-9229-6 X.421/25586
3129365

AESTHETICS. Human emotions. Special themes. Literature
Theories of English writers, 1500-1640

PLETT, Heinrich F.
Rhetorik der Affekte. - Tübingen, 1975
xi, 222 p. : ill. - (Studien zur englischen Philologie ; Neue Folge ; Bd 18)
3-484-45017-7 12981.ff.1/18
0893420
- [Another copy.] X.900/19054

AESTHETICS. Hungarian arts
1600-

KLANICZAY, Tibor
Hagyományok ébresztése. - Budapest : Szépirodalmi Könyvkiadó, 1976
579 p.
963-15-0684-3 X.529/19687
0885797

AESTHETICS. Hungarian music

ZOLTAI, Dénes
Bartók nem alkuszik-avagy az úgynevezett esztétikai kompromisszumról. - Budapest : Magvető Könyvkiadó, 1976
283p. - (Elvek és utak)
963-270-358-8 X:439/4525
1153005

AESTHETICS. Hungary

JÓZSA, Péter
Esztétikai alkotások társadalmi hatása. - Budapest : Népművelési Propaganda Iroda, [1976]
184p
r6056038 X.611/4730 Woolwich
1001809

SZERDAHELYI, István
A magyar esztétika története (1945-1975). - Budapest : Kossuth Könyvkiadó, 1976
563p
963-09-0660-0 X.809/23873
1001809

AESTHETICS. Hungary
to date

TERTULIAN, Nicolas
Georges Lukács. - Paris : Le sycomore, [1980]
295p. - (Arguments critiques)
2-86262-095-5 X.529/42465
855269x

AESTHETICS. Hungary
1800-1900

ERDÉLYI, János
Filozófiai és esztétikai írások. - Budapest : Akadémiai Kiadó, 1981
1115p. - (A magyar irodalomtörténetírás forrásai ; 10)
963-05-1973-9 Ac.825/287(10)
8759626

AESTHETICS. Italy
1945- - Conference proceedings

SITUAZIONI E INTENZIONI DELLA RICERCA ESTETICA OGGI IN ITALIA (Conference : 1979 : Reggio Emilia)
Orizzonte e progetti dell'estetica. - Parma : Pratiche, 1980
253p. - (Archivi ; 3)
r6510682 X.429/15635
906480x

AESTHETICS. Japanese arts

IZUTSU, Toshihiko
The theory of beauty in the classical aesthetics of Japan. - The Hague ; London : Nijhoff, 1981
x,167p. - (Philosophy and world community)
90-247-2381-7 X.950/3942
2013495

AESTHETICS. Landscape

NATURE over again after Poussin. - Glasgow : Collins Exhibition Hall, University of Strathclyde, [1980?]
26p : ill
0-907114-00-8 X.439/12345
2894882

AESTHETICS. Landscape
- Conference proceedings

AESTHETICS OF LANDSCAPE (Conference 1976 University of Hull)
The aesthetics of landscape. - Didcot : Rural Planning Services Ltd, 1980
95p. - (Publications / Rural Planning Services ; no. 7). - Includes bibliographies
0-906269-03-2 X.529/38003
2301059

AESTHETICS. Latvia
to date

ZEILE, Pēteris
Ocherk razvitiia ėsteticheskoĭ mysli v Latvii. - Moskva : Iskusstvo, 1980
323p. - (Ocherki istorii ėsteticheskoĭ mysli narodov SSSR). - Bibliography
r6484973 X.529/47983
8820791

AESTHETICS. Literature

BARABASH, IUriĭ IAkovlevich
Aesthetics and poetics. - Moscow : Progress Publishers ; [London] : Distributed by Central Books, 1977
292p,plate
B78-03612 X.989/52061
1296922

BARTHES, Roland
The pleasure of the text. - London : Cape, 1976
xi,67p
0-224-01187-1 X.989/40462
0583014

CHETUNOVA, Natal'ia Ivanovna
V sporakh o prekrasnom. - Izd.2.,dop. - Moskva : Sovetskiĭ pisatel', 1977
338p
r6055889 X.989/77543
1296922

GORDIENKO, Anatoliĭ Terent'evich
V bor'be za cheloveka. - Kiev : "Naukova Dumka", 1977
265p
r6086694 X.989/76790
1296922

JEAN PAUL
[Vorschule der Aesthetik. English]. Horn of Oberon. - Detroit : Wayne State University Press, 1973
lx, 368 p.
0-8143-1500-3 X.520/9278
0583014

NEKOTORYE problemy marksistsko-Leninskoĭ ėstetiki / [otvetstvennyĭ redaktor : S. Shermukhamedov]. - Tashkent : Uzbekistan, 1975
214p
r6095237 X.989/78031
1296922

OLSEN, Stein Haugom
The structure of literary understanding. - Cambridge : Cambridge University Press, 1978
xi,235p. - Bibliography: p.224-228. - Includes index
0-521-21731-8 X.989/52371
0583014

PHILOSOPHY and fiction / edited by Peter Lamarque. - [Aberdeen] : Aberdeen University Press, 1983
ix,111p. - Bibliography: p109-111
0-08-030353-6 X.950/41147
0583014

PROBLEMY literatury i ėstetiki / [redaktsionnaia kollegiia ... Z.M. Salagaeva (otvetstvennyĭ redaktor) ... i dr.]. - Ordzhonikidze : Severo-Osetinskiĭ gosudarstvennyĭ universitet, 1976
187p
r6076014 X.519/39314
1296922

ŠKAMLA, Ján
Determinanty literatúry a umenia. - Bratislava : Slovenský spisovatel, 1973
213p. - (Knižnica Slovenského spisovatela ; 47)
r6016639 X.908/37445
0583014

AESTHETICS. Literature
to date

CHETUNOVA, Natal'ia Ivanovna
A zhizn' idet—. - Moskva : Sovetskiĭ pisatel', 1982
366 p., [1] leaf of plates : port.. - Includes bibliographical references
r6490386 X.950/41298
8847843

CROCE, Benedetto
[La poesia. English]. Benedetto Croce's Poetry and literature. - Carbondale : Southern Illinois University Press, [1981]
lxxiv,210p
0-8093-0982-3 L.42/1606
8847843

FRAGMENTS / guest editor Lawrence D. Kritzman. - New York : New York Literary Forum, 1981
x,308p : ill. - (New York Literary Forum ; v.8-9)
0-931196-05-1 X.0950/585(8)
8847843

IENGO, Francesco
Le muse conservatrici. - Lanciano : Rocco Carabba, 1981
181p. - (Collana di studi e ricerche ; 7)
r6512702 X.950/41343
8847843

JAUSS, Hans Robert
Asthetische Erfahrung und literarische Hermeneutik. - Frankfurt am Main : Suhrkamp, 1982
876p. - Includes bibliography
3-518-57617-8 X.950/38528
8847843

KHMARA, Valentin
Ėsteticheskiĭ ideal i tvorchestvo. - Moskva : Sovetskiĭ pisatel', 1983
269 p.. - Includes bibliographical references
r6489793 X.950/36899
8847843

KOPPE, Franz
Grundbegriffe der Ästhetik. - Frankfurt am Main : Suhrkamp, 1983
251p. - (Edition Suhrkamp ; 1160 = n.F., Bd.160). - Includes bibliography
3-518-11160-4 X.958/29554
8847843

LITTERÆR værdi og vurdering / Jørgen Dines Johansen og Erik Nielsen (eds.). - [Odense] : Odense Universitetsforlag, 1984
330p : ill. - (Odense University studies in literature ; v.16)
87-7492-472-9 P.901/555[vol.16]
8847843

LUKÁCS, György, 1885-1971
Esztétikai írások 1930-1945. - Budapest : Kossuth Könyukiadó, 1982
837p
r6455068 X.950/18938
8847843

MUKAŘOVSKÝ, Jan
[Estetická funkce, norma a hodnota jako sociální fakty. English]. Aesthetic function, norm and value as social facts. - Ann Arbor : University of Michigan, Dept. of Slavic Languages and Literature, 1979
102p. - (Michigan Slavic contributions ; no.3)
r6439400 X.0909/589(3)
8847843

PEREKHODNYE ėsteticheskie iavleniia v literaturnom protsesse XVIII-XX vekov / pod redaktsieĭ N.P. Mikhal'skoĭ. - Moskva : Moskovskiĭ gosudarstvennyĭ pedagogicheskiĭ institut imeni V.I. Leiuna, 1981
165p
r6492223 X.950/43397
8847843

AESTHETICS. Literature
1800-1900

KIREEVSKIĬ, I. V. (Ivan Vasil'evich)
[Works]. Polnoe sobranie sochineniĭ Ivana Vasil'evicha Kireevskago. - Ann Arbor : Ardis, 1983
2v
r6403194 X.950/32140
9078061

AESTHETICS. Literature
1945-

PONIŽ, Denis
Pot. - Ljubljana : ŠKUC, 1979
168p. - (Študentski kulturni center ; 18)
r6434298 X.958/4071
8720851

AESTHETICS. Literature
Swedish theories

OLSSON, Thomas
Idealism och klassicism. - Göteborg : Göteborgs universitet, Litteraturvetenskapliga institutionen, 1981
152p. - (Skrifter utgivna av Litteraturvetenskapliga institutionen vid Göteborgs universitet ; 4)
r6610609 X.0909/1113(4)
8914400

AESTHETICS. Literature
Theories of Amirkhan, Fatikh

SAĬGANOV, A. (Abdulla)
U istokov estetiki realizma. - Kazan' : Tatarskoe knizhnoe izdatel'stvo, 1982
165p
r6462268 X.958/15452
8864985

AESTHETICS. Literature
Theories of Bakhtin, M.

SALSANO, Roberto
Estetica e "cronotopo" romanzesco. - Roma : Lucarini, 1981
56p. - (Saggi)
r6510675 X.958/22062
8864985

AESTHETICS. Literature
Theories of Brecht, Bertolt

GOLDHAHN, Johannes
Vergnügungen unseres Zeitalters. - Berlin : Brecht-Zentrum der DDR, [1980], c1981
243p. - (Brecht-Studien ; 1)
r6305628 X.0958/172(1)
8864985

AESTHETICS. Literature
Theories of Freud, Sigmund

RUSSO, Luigi
La nascita dell'estetica di Freud. - Bologna : Il Mulino, 1983
275p. - (Aesthetica ; 4. Saggi). - Includes bibliography
88-15-00007-0 X.0429/610(4)
8864985

AESTHETICS. Literature
Theories of Hungarian writers

NÉMETH, G. Béla
Tragikum és történetfelfogás. - Budapest : Akadémiai Kiadó, 1971
126p. - (Irodalomtörténeti füzetek ; 71)
r6043415 W.P.4992.a/71
1240641

AESTHETICS. Literature
Theories of Ingarden, Roman

FALK, Eugene H. (Eugene Hannes)
The poetics of Roman Ingarden. - Chapel Hill : University of North Carolina Press, c1981
xxi,213p
0-8078-1436-9 X.950/12403
2727730

AESTHETICS. Literature
Theories of Jean Paul

MÜLLER, Götz
Jean Pauls Ästhetik und Naturphilosophie. - Tübingen : Niemeyer, 1983
x,305p. - (Studien zur deutschen Literatur ; Bd 73). - Includes bibliography
3-484-18073-0 X.0909/187(73)
8864985

AESTHETICS. Literature
Theories of Nyblom, Carl Rupert

OLSSON, Thomas
Idealism och klassicism. - Göteborg : Göteborgs universitet, Litteraturvetenskapliga institutionen, 1981
152p. - (Skrifter utgivna av Litteraturvetenskapliga institutionen vid Göteborgs universitet ; 4)
r6610609 X.0909/1113(4)
8864985

AESTHETICS. Literature
Theories of Whitman, Walt

AHLUWALIA, Harsharan Singh
Whitman,s idea of the poet in his ,preface 1855,and ,Democratic Vistas,. - Amritsar : Guru Nanuk University, 1974
[27]p
r6025779 X.900/18096
1023691

AESTHETICS. Movement. Man
- For teaching

FOSTER, Ruth
Knowing in my bones. - London : A. and C. Black, 1976
ix,131p : ill. - Bibliography: p.131
0-7136-1653-9 X.439/8236
0773123

AESTHETICS. Music

BUSONI, Ferruccio
Entwurf einer neuen Ästhetik der Tonkunst. - Frankfurt am Main, 1974
84 p.
3-518-01397-1 X.439/4190
088488X

DAHLHAUS, Carl
Esthetics of music. - Cambridge : Cambridge University Press, 1982
xii,115p. - Bibliography: p101-112. - Includes index
0-521-23508-1 X.439/11874
0798908

FARBSHTEĬN, Aleksandr Aleksandrovich
Muzyka i estetika. - Leningrad : Izdatel'stvo "Muzyka" Leningradskoe otdelenie, 1976
206p. - Bibliography : p.188-202
r6079069 X.439/4853
1430025

FUBINI, Enrico
L'estetica musicale dall'antichità al Settecento. - Torino : Einaudi, 1976
xii,167p. - (Piccola biblioteca Einaudi ; 285)
r6052830 X.439/4663
088488X
Musica e linguaggio nell' estetica contemporanea. - Torino : Giulio Einaudi, 1972
xiii, 170 p.. - (Piccola biblioteca Einaudi ; n. 201)
r6006575 X.908/33610
0932000

UJFALUSSY, József
Zenéről, esztétikáról. - Budapest : Zeneműkiadó, 1980
206p
963-330-332-x X.431/11041
088488X

WOLFF, Ernst Georg
Grundlagen der autonomen Musikästhetik. - 2. Aufl. - Baden-Baden : Koerner, 1976
388p.. - (Collection d'études musicologiques ; Bd 15/27)
3-87320-527-0 WP.9947/15a
0798908

ZICH, Jaroslav
Kapitoly a studie z hudební estetiky. - Praha : Supraphon, 1975
238p : ill
r6096509 X.431/2977
088488X

ZUCKERKANDL, Victor
Man the musician. - Princeton, N. J. : Princeton University Press, 1976
XV, 370 p.. - (Sound and Symbol, Vol. 2) (Bollingen Series ; XLIV, 2). - Bibliography : p. 357-359. - Includes index
0-691-01812-x X.439/8250
0798908

AESTHETICS. Music
- Texts

MATOŠ, Antun Gustav
[Works]. Sabrana djela Antuna Gustava Matoša (1873-1914-1973)....
sv.10: Theatralia/uredio Nikola Batušić. O glazbi/uredio Lovro Županović. - Zagreb : Jugoslavenska akademija znanosti i umjetnosti, Liber, Mladost, 1973
451p
r6007279 X.0981/699(10)
0932000

AESTHETICS. Music
to 1900

HENNEBERG, Gudrun
Idee und Begriff des musikalischen Kunstwerks. - Tutzing : Schneider, 1983
vi,324p. - (Mainzer Studien zur Musikwissenschaft ; Bd 17). - Includes bibliography
3-7952-0384-8 X.0431/88(17)
9039376

AESTHETICS. Music
to date

BIMBERG, Siegfried
Kontrast als musikästhetische Kategorie. - Berlin : Neue Musik, [1981]
266p. - Bibliography
r6304618 X.439/12000
876364x

BROECKX, Jan L
Contemporary views on musical style and aesthetics. - Antwerp : Metropolis, [1979]
163p
r6303240 X.439/11577
8640033

KREMLEV, ÎÛliĭ Anatol'evich
Ocherki po ėstetike muzyki. - Izd. 2-e, dop. - Moskva : Sovetskiĭ kompozitor, 1972
270p
r6463404 X.439/12395
8640033

LIESS, Andreas
Kritische Erwägungen zu einer Ganzheitsphilosophie der Musik. - Wien : Böhlau, 1980
148p
3-205-07151-4 X.439/11604
8640033

MÂCHE, François-Bernard
Musique, mythe, nature, ou, Les dauphins d'Arion. - Paris : Klincksieck, 1983
136p. - (Collection d'Esthétique)
r6212031 X.431/13547
8640033

PHRANKOU-PSYCHOPAIDĒ, Olympia
Hē aisthetikē hermēneia tou mousikou ergou. - Athēna : Pylē, 1979
91p
r6598097 X.439/12728
8640033

PSYCHOPEDIS-FRANGOU, Olympia
Einführung in Probleme der Musikästhetik. - Wilhelmshaven : Heinrichshofen, 1981
112p.. - (Veröffentlichungen zur Musikforschung ; 5)
3-7959-0265-7 X.439/12631
8640033

SOKHOR, Arnol'd Naumovich
Voprosy sotŝiologii i ėstetiki muzyki .... - Leningrad : Sovetskiĭ kompozitor, 1980-
v.
r6402592 X.0439/325
876364X

SONNTAG, Brunhilde
Ästhetische Theorie und ästhetische Erziehung. - Wolfenbüttel : Möseler, c1981
179p : ill. - (Schriften zur Musikpädagogik ; 8). - Includes bibliography
r6307424 X.431/12085
8640033

ZICH, Otakar
Estetické vnímání hudby ; Estetika hudby. - Praha : Supraphon, 1981
451p. - (Klasikové hudební vědy a kritiky. 1. řada ; sv.6)
r6440808 X.439/12371
8640033

AESTHETICS. Music
Theories of Hostinský, Otakar

JŮZL, Miloš
Otakar Hostinský. - Praha : Melantrich, 1980
378p,[32]p of plates : ill. - (Odkazy pokrokových osobností naší minulosti ; sv.57). - Bibliography: p.347-370
r6440185 X.439/11849
8822875

AESTHETICS. Music
Theories of Tinctoris, Jean

ZANONCELLI, Luisa
Sulla estetica di Johannes Tinctoris. - [Bologna] : Forni, 1979
126p. - (Forni università. Storia della musica. Studi di teoria e estetica della musica)
r6506987 X.439/12542
8822875

AESTHETICS. Norris, Frank. American writers. Fiction in English

GRAHAM, Don
The fiction of Frank Norris. - Columbia ; London : University of Missouri Press, 1978
[8],172p : 1 ill. - Bibliography: p.159-168. - Includes index
0-8262-0252-7 X.989/53159
1811630

AESTHETICS. Oriental visual arts

COOMARASWAMY, Ananda Kentish
On the traditional doctrine of art. - Ipswich : Golgonooza Press, 1977
24p
B77-33332 Cup.407.g.9
1269216

AESTHETICS. Paintings
to date

CARDUCHO, Vicente
Diálogos de la pintura. - Madrid : Turner, [1979]
cxxxiii,483p : ill
84-85137-94-9 X.421/25791
9116273

AESTHETICS. Photography

PETERSON, Norman
Photographic art. - Epping : Bowker, c1984
139p : ill. - (Studies in photography ; no.4). - Bibliography: p133-135. - Includes index
0-8357-1529-9 X.421/26025
0305324

SONTAG, Susan
Susan Sontag on photography. - London : Allen Lane, 1978
[7],207p
0-7139-1128-x X.429/10410 Woolwich
0305324

AESTHETICS. Photography
to date

ADAMS, Robert, 1937-
Beauty in photography. - New York : Aperture, [1981]
108p : ill
0-89381-080-0 X.950/37552
8709351

SCHMOLL, J. A. (Josef Adolf), gen. Eisenwerth
Vom Sinn der Photographie. - München : Prestel, 1980
247p. : ill
3-7913-0533-6 X.421/21927 Woolwich
8709351

TURRONI, Giuseppe
Guida alla critica fotografica. - 3a ed. - [Milano] : Il castello, 1980
171p : ill
r6506562 X.429/14250
8709351

AESTHETICS. Photography
1945-

FIRINU, Piergiorgio
La gabbia sui Pirenei. - Torino : Studio 46, 1979
191p : ill
r6505381 X.429/14105
8621349

KELLY, John
La photo de charme. - [Paris] : Filipucchi, [1982]
175p : ill
2-85018-413-6 Cup.713/91
8621349

AESTHETICS. Poetry

COLERIDGE, Samuel Taylor
Biographia literaria, or, Biographical sketches of my literary life and opinions. - [New ed.]. - London : Dent [etc.], 1975
xxvi,303p. - ([Everyman's university library]). - Bibliography: p.xxiv. - Includes index
0-460-10011-4 X.989/50992
0160385

LOURENÇO, Eduardo
Tempo e poesia. - Porto : Editorial Inova, 1974
303 p.. - (Colecção Civilização portuguesa ; v. 20)
r6011369 X.908/33565
0160385

RAMUZ, C. F. (Charles Ferdinand)
Poésie. - Lausanne : Fondation C. F. Ramuz, [1975]
19 p.
r6000826 X.909/32145
0160385

RASMUSSEN, Dennis
Poetry and truth. - The Hague : Mouton, 1974
123p. - (De proprietatibus litterarum, series minor ; no.20). - Bibliography : p.116-118
r6025752 X.909/37705
0160385

SKELTON, Robin
Poetic truth. - London : Heinemann [etc.], 1978
x,131p. - Includes index
0-435-18821-6 X.989/52796
0160385

AESTHETICS. Poetry
to date

BAUMGARTEN, Alexander Gottlieb
[Meditationes philosophicae de nonnullis ad poema pertinentibus. German & Latin]. Meditationes philosophicae de nonnullis ad poema pertinentibus. - Hamburg : Meiner, 1983
lx,94p. - (Philosophische Bibliothek ; Bd 352)
3-7873-0574-2 X.958/28708
8500487

MATOS PAOLI, Francisco
Diario de un poeta. - Santurce, P.R. : Ediciones Puerto, 1973
357p
r6705905 X.950/17270
8500487

PAGINE di poetica dal Sei al Novecento / [a cura di] Ada Ruschioni. - Ristampa. - Milano : Vita e pensiero, 1980
375p. - (I corsi universitari)
r6509292 X.950/27989
8500487

AESTHETICS. Poetry
Coleridge, Samuel Taylor. Biographia literaria

WALLACE, Catherine Miles
The design of Biographia literaria. - London : Allen & Unwin, 1983
x,169p. - Includes index
0-04-800016-7 X.950/24188
2044714

WHEELER, Kathleen M.
Sources, processes and methods in Coleridge's Biographia literaria. - Cambridge : Cambridge University Press, 1980
xiii,229p. - Bibliography: p212-221. - Includes index
0-521-22690-2 X.950/1350
2044714

AESTHETICS. Poetry in German
1750-1856

TODOROW, Almut
Gedankenlyrik. - Stuttgart : Metzler, 1980
140p.. - (Germanistische Abhandlungen ; 50). - Bibliography
3-476-00447-3 X.950/5051
8704678

AESTHETICS. Poetry in Russian
to date

DONET͡SKIKH, L. I. (Li͡udmila Ivanovna)
Ėsteticheskie funkt͡sii slova. - Kishinev : Shtiint͡sa, 1982
152p
r6464523 X.958/19087
9038124

AESTHETICS. Poetry in Slovak

KRČMÉRY, Štefan
Estetické reflexie. - Bratislava : Slovenský spisovateľ, 1975
166 p.. - (Knižnica Slovenského spisovateľa ; 68)
r6008216 X.908/36566
0385522

AESTHETICS. Poetry in Spanish
to date

BALLESTERO, Manuel
Poesía y reflexión. - Madrid : Taurus, 1980
153p. - (Persiles ; 125)
84-306-2125-3 X.950/16541
8905223

AESTHETICS. Role of artificial light in architectural design

GABRIEL, Hans
Das künstliche Licht in der Architektur. - Stuttgart : Deutsche Verlags-Anstalt, c1974
155 p. : ill. - Includes bibliographies
3-421-02350-6 X.423/2338 Woolwich
0932752

AESTHETICS. Russia
1800-1900

RUSSKIE ėsteticheskie traktaty pervoĭ treti XIX veka ... / [sostavlenie, vstupitel'nai͡a stat'i͡a i primechanii͡a Z.A. Kamenskogo]. - Moskva : Iskusstvo, 1974
2t. - (Istorii͡a ėstetiki v pami͡atnikakh i dokumentakh)
r6186194 X:529/21748
2121360

AESTHETICS. Russia
1800-1917

EGOROV, B. F.
Bor'ba ėsteticheskikh ideĭ Rossii serediny XIX reka. - Leningrad : Iskusstvo, Leningradskoe otdelenie, 1982
266p
r6463733 X.629/20563
8524742

VALIT͡SKAI͡A, A. P. (Alisa Petrovna)
Russkai͡a ėstetika XVIII veka. - Moskva : Iskusstvo, 1983
236p. - Includes bibliography
r6463722 X.629/20569
8524742

AESTHETICS. Russian arts

KANTOR, V. K.
Russkai͡a ėstetika vtoroĭ poloviny XIX stoletii͡a i obshchestvennai͡a bor'ba. - Moskva : Iskusstvo, 1978
172p
r6187290 X.419/3723 Woolwich
1219324

KORNILOV, Evgeniĭ Alekseevich
Na fronte iskusstva (pechat' Dona 20-kh godov v bor'be za vysokui͡u ideĭnost' i khudozhestvennost' sot͡sialisticheskogo iskusstva. - Rostov-na-Donu : Izdatel'stvo Rostovskogo universiteta, 1977
62p
r6075578 X.419/8115 Woolwich
1219324

VOLKOVA, Elena Vasil'evna
Proizvedenie iskusstva ' predmet ėsteticheskogo analiza.... - Moskva : Izdatel'stvo Moskovskogo universiteta, 1976
282p
r6046458 X.429/7122
1219324

AESTHETICS. Russian arts
to date

Le AVANGUARDIE artistiche in Russia / antologia critica a cura e con introduzione di Michaela Bohmig. - Bari : De Donato, 1979
293p : ill. - (Rapporti). - Bibliography
r6504923 X.809/52012
8511926

AESTHETICS. Russian literature
1800-1917

CHERNYSHEVSKIĬ, N. G. (Nikolaĭ Gavrilovich)
[Selections]. Izbrannye ėsteticheskie proizvedenii͡a. - Moskva : Iskusstvo, 1974
549p. - (Istorii͡a ėstetiki v pami͡atnikakh i dokumentakh)
r6185564 X.529/47978
2190583

KIREEVSKIĬ, I. V. (Ivan Vasil'evich)
Kritika i estetika. - Moskva : Iskusstvo, 1979
438p. - (Istorii͡a ėstetiki v pami͡atnikakh i dokumentakh). - Bibliography: p.393-427
r6498122 X.950/8461
2190583

AESTHETICS. Russian literature
Russian theories, 1917-

TVORCHESKIE vzgli͡ady sovetskikh pisateleĭ / [otvetstvennyĭ redaktor : V.A. Kovalev]. - Leningrad : Nauka, Leningradskoe otdelenie, 1981
284p
r6486675 X.950/14248
8908605

AESTHETICS. Soviet arts

ADAMI͡AN, A. A. (Arshak Abgarovich)
Voprosy ėstetiki i teorii iskusstva. - Moskva : Iskusstvo, 1978
299p. - (Iz istorii sovetskoĭ ėstetiki i teorii iskusstva)
r6173571 X:429/7767 Woolwich
1853740

EGOROV, Anatoliĭ Grigor'evich
Problemy ėstetiki. - 2. izd. - Moskva : Sovetskiĭ pisatel'.
A. Egorov. - 1977
461p
r6076757 X.529/21016
1853740

AESTHETICS. Soviet arts
1917-1945

IZ istorii sovetskoĭ ėsteticheskoĭ mysli 1917-1932 / [sostavitel' G.A. Belai͡a]. - Moskva : Iskusstvo, 1980
454p
r6498465 X.429/12440 Woolwich
8728356

AESTHETICS. Soviet arts
1917-

BAĬGUSHEV, Aleksandr
Tochka zrenii͡a. - Moskva : Sovremennik, 1979
430p
r6484454 X.429/14699
8962804

AESTHETICS. Soviet literatures
to date

SUROVT͡SEV, I͡Uriĭ
Neobkhodimost' dialektiki. - Moskva : Khudozhestvennai͡a literatura, 1982
534p
r6403141 X.950/27328
9029036

AESTHETICS. Soviet Union
to date

KALANTAR, Aleksandr
Problema prekrasnogo. - Erevan : Sovetakan grokh, 1981
306p
r6432211 X.529/49603
8858462

AESTHETICS. Special themes. Dante Alighieri. Poetry in Italian

TOOK, J. F.
'L'etterno piacer'. - Oxford : Clarendon, 1984
x,128p,[1]leaf of plates : 1ill. - Bibliography: p115-122. - Includes index
0-19-815793-2 X.950/40130
3012441

AESTHETICS. Special themes. Forster, E. M.. Fiction in English

DOWLING, David
Bloomsbury aesthetics and the novels of Forster. - London : Macmillan, 1985
xii,249p,[12]p of plates : ill. - Includes index
0-333-35474-5 X.950/40168
3254275

AESTHETICS. Special themes. Ulysses. Joyce, James, 1882-1941. Fiction in English

MULTHAUP, Uwe
Das künstlerische Bewusstsein und seine Gestaltung in James Joyces "A portrait of the artist as a young man" and "Ulysses". - Bern, 1973
253 p.. - (Europäische Hochschulschriften. Reihe 14, Angelsachsische Sprache und Literatur ; Bd 11)
3-261-01024-x X.909/33776
0878227

AESTHETICS. Special themes. Woolf, Virginia. Fiction in English

DOWLING, David
Bloomsbury aesthetics and the novels of Forster. - London : Macmillan, 1985
xii,249p.[12]p of plates : ill. - Includes index
0-333-35474-5 X.950/40168
3254283

AESTHETICS. Tadzhik arts

TURSUN-ZADE, Firīūza Mirzoevna
Khudozhestvennye tradifsii i progress iskusstva. - Dushanbe : Izdatel'stvo "Irfon", 1976
124 p.
r6023049 X.908/36338
0896616

AESTHETICS. Tales & legends

NAGY, Olga
A táltos törvénye. - Bukarest : Kriterion, 1978
358 p.. - Bibliography: p. 329-336
r6455186 X.950/37110
9157018

AESTHETICS. Tragedy. Drama
to date

CARCHIA, Gianni
Orfismo e tragedia. - Milano : Celuc libri, 1979
89p. - (Produzione e interpretazione ; 5)
r6504501 X.958/8148
8806837

AESTHETICS. Ukrainian literature, 1798-1917

ÎĀTSENKO, M. T. (Mykhaïlo Trokhymovych)
Pytannīā realizmu i pozytyvnyĭ heroĭ v ukrains'koï literaturno-estetychniĭ dumtsi pershoï polovyny XIX st.. - Kyiv : Naukova dumka, 1979
333p
r6463382 X.950/21271
8961220

AESTHETICS. Visual arts

ANAND, Mulk Raj
Seven little-known birds of the inner eye. - Rutland, Vt : Tuttle, 1978
157p,[1]leaf of plates : ill. - Bibliography: p151-152. - Includes index
0-8048-0936-4 X.421/21879 Woolwich
0196355

BEARDSLEY, Monroe C. (Monroe Curtis)
The aesthetic point of view. - Ithaca, [N.Y.] ; London : Cornell University Press, 1982
385p. - Includes index
0-8014-1250-1 X.520/30890
0196355

KANDINSKY, Wassily
Concerning the spiritual in art. - New York : Dover Publications [etc.] ; London : Constable, 1977
iii-xxiii,57p,[10]p of plates,[2] leaves of plates : ill
0-486-23411-8 X.419/34565 Woolwich
0196355

KNOBLER, Nathan
The visual dialogue. - 3rd ed. - New York ; London : Holt, Rinehart and Winston, 1980
x,324p : ill(some col). - Bibliography: p.301-305. - Includes index
0-03-049316-1 X.421/20796 Woolwich
0196355

LANGMUIR, Erika
Art history and quality in art. - Milton Keynes : Open University, 1984, c1983
24p,[4]p of plates : ill
0-335-10601-3 X.425/4815
0196355

SCHORR, Justin
Toward the transformation of art. - Rutherford : Fairleigh Dickinson University Press ; London : Associated University Press, 1974
118p. - Bibliography: p.114-116. - Includes index
0-8386-1382-9 X.429/10284 Woolwich
0196355

WEIMARCK, Torsten
Konst och varuform. - Ahus : Kalejdoskop, 1979
212p,[8]p of plates : ill. - Includes bibliographies
91-85552-14-3 X.419/3668 Woolwich
0196355

AESTHETICS. Visual arts
- Texts

MATOŠ, Antun Gustav
[Works]. Sabrana djela Antuna Gustava Matoša (1873-1914-1973)....
sv.11. O likovnim umjetnostima/uredio Slavko Batušić. Putopisi/uredio Dubravko Jelčić. - Zagreb : Jugoslavenska akademija znanosti i umjetnosti. Liber. Mladost, 1973
411p
r6007280 X.0981/699(11)
1221086

AESTHETICS. Visual arts
1945-

BLOCKER, H. Gene
Philosophy of art. - New York : Charles Scribner's Sons, c1979
viii,278p[8]p of plates
0-684-16049-8 X.429/13130 Woolwich
8653720

CLAIR, Jean
Considérations sur l'état des beaux-arts. - [Paris] : Gallimard, 1983
193p. - (Les essais ; no.224)
2-07-024719-8 X.429/16930
8653720

HABIG, Inge
Sich ein Bild machen. - Frankfurt am Main : Lang, 1982
139p. - (Europäisches Forum ; Bd 5). - Bibliography
3-8204-5709-7 X.0529/667(5)
8653720

LANGER, Michael
Ästhetik des Sichtbaren. - Mittenwald : Mäander, c1980
iv,209p. : ill. - (Schriften des Lehrstuhls für Kunsterziehung der Ludwig-Maximilians-Universität München ; Bd.2)
3-88219-083-3 X.429/15863
8653720

LE BOT, Marc
L'œil du peintre. - Paris : Gallimard, 1982
158p. - (Le chemin)
2-07-020130-9 X.429/15504
8653720

AESTHETICS. Visual arts
1945- - Texts

LEDUC, Fernand
[Selections]. Vers les îles de lumière. - Québec : Hurtubise HMH, 1981
xxxvi,288p. - (Cahiers du Québec ; 65)
2-89045-497-5 X.0709/692(65)
8653720

AESTHETICS. Visual arts
Geometric aspects

GHYKA, Matila
The geometry of art and life. - New York : Dover Publications, 1977
iii-xvii,174p,lxxxp of plates : ill
0-486-23542-4 X.619/18978 Woolwich
1690922

AESTHETICS. Visual arts
Theories of Cochin, Charles Nicolas, 1715-1790

TAVERNIER, Ludwig
Das Problem der Naturnachahmung in den kunstkritischen Schriften Charles Nicolas Cochins d.J.. - Hildesheim : Olms, 1983
139p. - (Studien zur Kunstgeschichte ; Bd.20). - Includes bibliography
3-487-07398-6 X.0419/45(20)
8840210

AESTHETICS. Visual arts
Theories of Montel, Jean-Claude - Texts

MONTEL, Jean-Claude
Partage et lisières. - Paris : Flammarion, 1981
95p : ill. - (Textes / Flammarion)
2-08-064374-6 X.429/14300
8840210

AESTHETICS. Visual arts
Theories of Ruskin, John

HEWISON, Robert
John Ruskin. - London : Thames and Hudson, 1976
228p : ill(some col). - Bibliography: p.219-223. - Includes index
0-500-01148-6 X.421/10199 Woolwich
0669954

MORLEY, Catherine W.
Ruskin's philosophy. - Penzance : Triton Press, 1975
[3],28p
B77-33930 X.908/41324
0669954

AESTHETICS. Visual arts
Theories of Wölfflin, Heinrich

LURZ, Meinhold
Heinrich Wölfflin. - Worms : Werner, 1981
xii,388p.. - (Heidelberger kunstgeschichtliche Abhandlungen ; n.F., Bd.14). - Bibliography
3-88462-003-7 7811.t.23a/14
8840210

AESTHETICS. Visual arts, 1900-1979

HEYER, George S.
Signs of our times. - Edinburgh : Handsel Press Limited, 1980
[7],73p,leaf of plate,[28]p of plates : ill(incl 1 col)
0-905312-04-x X.200/33683
2235447

AESTHETICS. Visual arts. Australia
1770-1914

DOCUMENTS on art and taste in Australia / edited by Bernard Smith. - Melbourne ; London : Oxford University Press, 1975 [i.e. 1976]
xi,299p. - Includes index
0-19-550476-3 X.419/3131 Woolwich
055135x

AESTHETICS. Western architecture
to date

NAREDI-RAINER, Paul von
Architektur und Harmonie. - Köln : DuMont, 1982
303p : ill. - (DuMont Dokumente. Reihe Kunstgeschichte, Wissenschaft). - Includes bibliography
3-7701-1196-6 X.429/15908
908326x

AESTHETICS. Western arts
1945-

DAVYDOV, ÎŪriĭ Nikolaevich
Ėstetika nigilizma. - Moskva : Iskusstvo, 1975
270p
r6403703 X.429/15944
9083219

AESTHETICS. Western Europe
1700-1900

OVSĪĀNNIKOV, M. F. (Mikhail Fedotovich)
Iskusstvo i kapitalizm. - Moskva : Iskusstvo, 1979
341p
r6484979 X.529/47978
8820775

AESTHETICS. Western European music
Theories, 1700-1900

SCHMIDT, Marlene
Zur Theorie des musikalischen Charakters. - München : Katzbichler, 1981
154p. - (Beiträge zur Musikforschung ; Bd.9). - Bibliography
3-87397-258-1 X.0431/123(9)
8860122

AESTHETICS. Western music

KIVY, Peter
The corded shell. - Princeton ; Guildford : Princeton University Press, c1980
xiv,167p : ill. - (The Princeton essays on the arts ; 9). - Bibliography: p159-162. - Includes index
0-691-07258-2 X.431/11286
2370859

AETIOLOGY
See also
Pathogenesis

AETIOLOGY. Acquired immune deficiency syndrome. Man
Nutritional factors

FOOD imprint as cause of AIDS. - Dover : Collation Research, 1984
[14]p
0-906470-05-6 X.329/18352
3167860

AETIOLOGY. Bacterial diseases. Animals
- Conference proceedings

The DETERMINANTS of bacterial and viral pathogenicity / organised and edited by Harry Smith, J.P. Arbuthnott and C.A. Mims. - London : Royal Society, 1983
viii,163p,[2]p of plates : ill. - Includes bibliographies
0-85403-216-9 X.322/13280
308762x

AETIOLOGY. Cancer. Breasts. Women
Role of barrier contraceptives

GJORGOV, Arne N.
Barrier contraception and breast cancer. - Basel ; London : Karger ; [Chichester] : [Distributed by Wiley], 1980
x,162p : ill. - (Contributions to gynecology and obstetrics ; 8). - Bibliography: p.133-153. - Includes index
B80-39481 SRL
2281082

AETIOLOGY. Cancer. Man

WHELAN, Elizabeth M.
Preventing cancer. - London : Sphere, 1980
255p : ill. - Includes index
0-7221-9055-7 X.319/19347 Woolwich
2081458

AETIOLOGY. Cancer. Man
- Conference proceedings

PERSONS at high risk of cancer / edited by Joseph F. Fraumeni, Jr. - New York ; London : Academic Press, 1975
xix,544p : ill. - Includes index
0-12-265950-3 SRL
2081458

AETIOLOGY. Cancer. Women

STOLL, Basil Arnold
But why cancer, Sally?. - London : Heinemann Health, 1976
[6],122p : ill. - Includes index
0-433-31736-1 X.319/16787 Woolwich
0727512

AETIOLOGY. Deafness. Children

FRASER, George Robert
The causes of profound deafness in childhood. - London : Baillière Tindall, 1976
xv,410p. - Bibliography: p.357-390. - Includes index
0-7020-0640-8 SRL
1055275

AETIOLOGY. Diseases. Man

HOY, Ronald
Nature and causation of disease. - London : McGraw-Hill, c1981
188p : ill. - (McGraw-Hill nursing studies series). - Bibliography: p180-181. - Includes index
0-07-084617-0 X.322/10262 Woolwich
1925229

TAYLOR, Frederick Kräupl
The concepts of illness, disease and morbus. - Cambridge : Cambridge University Press, 1979
ix,131p. - Bibliography: p.120-124. - Includes index
0-521-22433-0 X.329/11949 Woolwich
1925229

AETIOLOGY. Diseases. Man
Parapsychological aspects

REYNER, J. H. (John Hereward)
Psionic medicine. - 2nd ed. - London : Routledge & Kegan Paul, 1982
151p. - Bibliography: p147-148. - Includes index
0-7100-9088-9 X.329/15306 Woolwich
0217557

AETIOLOGY. Diseases. Man
Parapsychological aspects - Serials

INSTITUTE OF PSIONIC MEDICINE
Newsletter. - 1978-. - Hindhead : [The Institute], [1978]-
B80-39455 P.323/377
2291282

AETIOLOGY. Diseases. Man
Theories

MURDOCK, George Peter
Theories of illness. - Pittsburgh : University of Pittsburgh Press ; London : Feffer and Simons, c1980
xv,127p. - Includes bibliographies and index
0-8229-3428-0 X.329/15337
2732181

AETIOLOGY. Diseases. Man. Great Britain

NATIONAL health and social sickness - what are the real causes of illness? How can we be healthy / [prepared by the Sheffield Branch of the Socialist Health Association]. - London : Socialist Health Association, [1982]
24p : ill. - (A charter for health publication)
B82-37449 X.529/49879
2783606

AETIOLOGY. Diseases of man
Nutritional factors - Conference proceedings

AMERICAN COLLEGE OF NUTRITION. Annual Meeting (16th : 1975 : New York)
Nutritional imbalances in infant and adult disease. - New York : Spectrum Publications ; London : Distributed by Eurospan Ltd, 1977
[19],170p : ill. - Includes bibliographies and index
0-89335-007-9 X.320/12044 Woolwich
2075466

AETIOLOGY. Epilepsy. Man
Role of limbic system - Conference proceedings

INTERNATIONAL SYMPOSIUM ON LIMBIC EPILEPSY AND THE DYSCONTROL SYNDROME (1st : 1980 : Sydney)
Limbic epilepsy and the dyscontrol syndrome. - Amsterdam ; Oxford : Elsevier/North-Holland Biomedical Press, 1980
x,281p : ill. - (Developments in psychiatry ; vol.4). - Includes index
0-444-80251-7 SRL
2288842

AETIOLOGY. Leukaemia. Man

RAYMOND, James
The possible cause of leukaemia. - Burnley : F.H. Brown Ltd, 1978
[9]p
B79-00375 X.319/18530 Woolwich
1741225

AETIOLOGY. Mental disorders. Man
Longitudinal studies. Implications for privacy

ORIGINS of psychopathology / edited by David F. Ricks, Barbara S. Dohrenwend. - Cambridge : Cambridge University Press, 1983
xiii,281p : ill. - Includes bibliographies and index
0-521-25298-9 X.520/32000
2971895

AETIOLOGY. Mental handicaps. Man

DUTTON, Gordon
Mental handicap. - London : Butterworth, 1975
ix,176p : ill. - (Postgraduate psychiatry series). - Includes bibliographies and index
0-407-00035-6 X.329/10199
0536318

AETIOLOGY. Periodontal diseases. Man

WORLD HEALTH ORGANIZATION. Scientific Group on Epidemiology, Etiology and Prevention of Periodontal Diseases
Epidemiology, etiology and prevention of periodontal diseases. - Geneva : W.H.O. ; [London] : [H.M.S.O.], 1978
60p : ill. - (Technical report series / World Health Organization ; 621)
B78-35334 SRL
1686836

AETIOLOGY. Petit-mal epilepsy. Man
Role of wave-spike activity

MYSLOBODSKY, Michael
Petit mal epilepsy. - New York ; London : Academic Press, 1976 [i.e. 1977]
xi,218p : ill. - Bibliography: p.191-208. - Includes index
0-12-511950-x SRL
1088556

AETIOLOGY. Sudden deafness. Man

SUDDEN loss of cochlear and vestibular function / volume editors M. Hoke and M.E. Wigand. - Basel ; London : Karger, 1981
viii,197p : ill. - (Advances in oto-rhino-laryngology ; v.27). - Includes bibliographies
3-8055-2630-x X.325/1365 Woolwich
2633175

AETIOLOGY. Virus diseases. Animals
- Conference proceedings

The DETERMINANTS of bacterial and viral pathogenicity / organised and edited by Harry Smith, J.P. Arbuthnott and C.A. Mims. - London : Royal Society, 1983
viii,163p,[2]p of plates : ill. - Includes bibliographies
0-85403-216-9 X.322/13280
3087638

AEZKOA. Navarre. Spain
Land. Ownership, 1784-1978

GOÑI, Fermín
Aezkoa : 200 años de lucha. - Pamplona : Euskal Bidea, 1978
142p : ill,facsims
84-85490-02-9 X.708/48956
2448920

AFAN
See Thomas, Afan

AFAN. West Glamorgan
Upper Afan. Agriculture. Farms. Social conditions, 1570-1851

BROWN, Roger Lee
The farms of the upper Afan. - [Tongwynlais] : R.L. Brown, 1982
iv,60p
B83-16477 X.809/54578
2874326

AFAN (DISTRICT). West Glamorgan

BOROUGH of Afan official guide. - Carshalton : Home Publishing Co., [1977]
32,[1]p : ill
B77-32967 X.709/31686
1264699

AFAN (DISTRICT). West Glamorgan
- Serials

AFAN uchaf. - Vol.1 (1978)-. - Port Talbot : The Society, 1978-
v. : ill
B82-33858 P.801/3307
2772973

AFAN ARGOED COUNTRY PARK

AFAN Argoed Country Park, Cynonville, Port Talbot, West Glamorgan. - Swansea : West Glamorgan County Council, Planning Department in conjunction with the Forestry Commission, 1979
30p : ill(some col.)
B82-17502 X.329/15110 Woolwich
2702975

WEST GLAMORGAN. County Council
Afan Argoed Country Park. - [Swansea] : West Glamorgan County Council, Planning Department in conjunction with the Forestry Commission, [1979]
32p : ill(some col.)
0-907599-03-6 L.49/3178
2702975

AFÁN DE RIBERA, PERO. Spain
- Biographies

FERNÁNDEZ PERALTA, Ricardo
Reinado de Felipe II. - San José, Costa Rica : Ministerio de Obras Públicas y Transportes, Instituto Geográfico Nacional, 1974
53p.,[1] fold.leaf of plates
r6047669 X.700/20035
1716832

AFAN VALLEY. West Glamorgan
Railways, 1853-1971

SMITH, Clive, 1934-
Bygone railways of the Afan. - Port Talbot : Alun Books, c1982
70p : ill. - Bibliography: p70
0-907117-17-1 X.629/20037
2889897

AFAN VALLEY. West Glamorgan
Upper Afan Valley. Church in Wales, to 1981

BROWN, Roger Lee
Through cloud and sunshine. - Port Talbot : Alun, c1982
159p : ill. - Bibliography: p158-159
0-907117-13-9 X.200/40823
2926555

AFEWERK, TEKLÉ. Ethiopian graphic arts
- Catalogues

AFEWERK, Teklé
Afewerk Teklé. - [Bonn?] : [s.n.], [1981]
[12]p. : chiefly ill.
r6306294 X.429/13572
8863989

AFFECT
See also
Emotions
Feelings

AFFECT. Children
Development

The DEVELOPMENT of affect / edited by Michael Lewis and Leonard A. Rosenblum. - New York ; London : Plenum Press, 1978
xii,426p : ill. - (Genesis of behavior ; vol.1). - Includes bibliographies and index
0-306-34341-x X.520/12792
1567268

AFFECT. Children, to 5 years
Development related to cognitive development

TURNER, Johanna
Made for life. - London : Methuen, 1980
xiii,159p. - Bibliography: p133-152. - Includes index
0-416-72690-9 X.529/63249
2100649

AFFECT. Man
related to memory

DUTTA, Satrajit
Affect and memory. - Oxford : Pergamon, 1975
vii,138p : ill. - (International series in experimental psychology ; vol.20). - Bibliography: p.129-135. - Includes index
0-08-018270-4 X.0520/18(20)
0686891

AFFECTIVE DISORDERS. Man

AFFECTIVE disorders / [edited by] Eduardo R. Val, F. Moises Gaviria, Joseph A. Flaherty. - Chicago ; London : Year Book Medical, c1982
xxi,487p : ill. - Includes bibliographies and index
0-8151-8952-4 X.520/28422
2639319

HANDBOOK of affective disorders / edited by E.S. Paykel. - Edinburgh : Churchill Livingstone, 1982
viii,457p : ill. - Includes bibliographies and index
0-443-02036-1 X.322/10644 Woolwich
2639319

NEUROBIOLOGY of mood disorders / edited by Robert M. Post, James C. Ballenger. - Baltimore ; London : Williams & Wilkins, c1984
xvii,887p : ill. - (Frontiers of clinical neuroscience ; v.1). - Includes index
0-683-06949-7 SRL
2639319

WHITLOCK, F. A. (Francis Antony)
Symptomatic affective disorders. - Sydney ; London : Academic Press, 1982
xv,218p. - (Personality and psychopathology). - Bibliography: p171-211. - Includes index
0-12-747580-x X.322/11959
2639319

WHYBROW, Peter C.
Mood disorders. - New York ; London : Plenum, c1984
xv,228p : ill. - (Critical issues in psychiatry). - Includes index
0-306-41568-2 X.525/10158
2639319

AFFECTIVE DISORDERS. Man
- Conference proceedings

ORIGIN, prevention and treatment of affective disorders / edited by M. Schou, E. Strömgren. - London : Academic Press, 1979
xiii,307p : ill. - Includes bibliographies and index
0-12-628950-6 SRL
1901281

AFFECTIVE DISORDERS. Man
Drug therapy

PHARMACOTHERAPY of affective disorder / edited by W.G. Dewhurst and G.B. Baker. - London : Croom Helm, c1985
628p : ill. - Includes bibliographies and index
0-7099-0857-1 SRL
3233960

AFFECTIVE DISORDERS. Man
Effects of drugs

PSYCHOPHARMACOLOGY of affective disorders / edited by E.S. Paykel and A. Coppen. - Oxford : Oxford University Press, 1979
x,261p : ill. - (Oxford medical publications) (Monographs / British Association for Psychopharmacology). - Includes bibliographies and index
0-19-261140-2 SRL
1490427

PSYCHOPHARMACOLOGY of affective disorders / edited by E.S. Paykel and A. Coppen. - [1st ed.] reprinted with corrections. - Oxford : Oxford University Press, 1980
x,261p : ill. - (Oxford medical publications) (Monographs / British Association for Psychopharmacology). - Includes bibliographies and index
0-19-261278-6 X.325/782 Woolwich
1490427

AFFECTIVE DISORDERS. Man
Psychoanalysis

AFFECT / edited by Morton B. Cantor and Myron L. Glucksman. - New York ; Chichester : Wiley, c1983
viii,320p : ill. - Includes bibliographies and index
0-471-88071-x X.520/34726
3169103

AFFECTIVE DISORDERS. Young persons

AFFECTIVE disorders in childhood and adolescence / edited by Dennis P. Cantwell, Gabrielle A. Carlson. - Lancaster : MTP, c1983
xi,484p : ill. - (Child behavior and development). - Includes bibliographies and index
0-85200-630-6 X.322/13293
308714x

AFFECTIVE MEANING. Words. Languages

SOYDAN, Haluk
Social structure and subjective culture. - Uppsala : Uppsala University, 1975
[5], 180 p. : ill. - Bibliography: p. 170-178. - Includes index
91-506-0035-4 X.519/26387
0779121

AFFECTIVE METHOD. Teaching methods. Schools

MILLER, John P. (John Pearse)
Humanizing the classroom. - New York : Praeger, 1976
vii, 183 p. : ill. - Includes bibliographies and index
0-275-49940-5 X.510/15349
0795291

AFFECTIVE METHOD. Teaching methods. Schools
Role in emotional development of students

STEINBERG, John M.
Emotional growth in the classroom. - Stockholm : Almquist & Wiksell International, 1976
245 p.. - Bibliography : p. 237-245
91-220-0041-0 X.519/20222
0924385

AFFILIATES. American multinational companies. Manufacturing industries. Great Britain

DUNNING, John H. (John Harry)
US industry in Britain. - London : Wilton House Publications, 1976
v,122p : ill. - (Business research studies / Economists Advisory Group)
0-904655-14-8 X.512/5489
1031414

AFFILIATES. American multinational companies. Manufacturing industries. Scotland

HOOD, Neil
European development strategies of US owned manufacturing companies located in Scotland. - Edinburgh : H.M.S.O., 1980
100p : 1 ill
0-11-491636-5 BS.564
2291681

AFFILIATES. Commercial banks

WHO owns what in world banking. - London : Financial Times Business Publishing, c1983
ix,186p. - Includes index
0-902101-32-3 X.525/9257
0778338

AFFILIATES. Companies
See also
Linkages

AFFILIATION. Illegitimate children. England
Law

WILKINSON, George Stephen
Affiliation law and practice. - 4th ed.. - London : Oyez Publishing, 1977
xx,163p. - (Oyez practice notes ; 41). - Includes index
B78-15470 WP.1357/41
143229x

AFFILIATION. Interpersonal relationships

MEHRABIAN, Albert
A theory of affiliation. - Lexington, Mass. ; London : D.C. Heath, 1974
xvi,213p : ill. - Bibliography: p.189-201. - Includes index
B75-26357 X.520/10260
0510971

AFFILIATION. Trade unions. Great Britain
Of nuclear disarmament movements

TRADE union powers and the anti-nuclear campaign. - [London] : Lawyers Versus the Bomb, [1982?]
19p. - (Pamphlet / Lawyers Versus the Bomb ; 2)
B83-22896 X.529/54045
2898241

AFFINITY CHROMATOGRAPHY

AFFINITY chromatography / edited by P.D.G. Dean, W.S. Johnson, F.A. Middle. - Oxford : IRL Press, c1985
xv,215p : ill. - (Practical approach series). - Includes index
0-904147-71-1 X.329/19553
0184896

SCOUTEN, William H.
Affinity chromatography. - New York ; Chichester : Wiley, c1981
xiii,348p : ill. - (Chemical analysis ; v.59). - Includes index
0-471-02649-2 SRL
0184896

TURKOVÁ, Jaroslava
Affinity chromatography. - Amsterdam ; Oxford : Elsevier, 1978
ix,405p : ill. - ('Journal of chromatography' library ; vol.12). - Includes bibliographies and index
0-444-41605-6 SRL
0184896

AFFINITY CHROMATOGRAPHY. Biochemistry

THEORY and practice in affinity techniques / edited by P.V. Sundaram and F. Eckstein. - London : Academic Press, 1978
x,246p : ill. - Includes index
0-12-677150-2 SRL
1873709

AFFINITY CHROMATOGRAPHY. Biochemistry
- Conference proceedings

AFFINITY chromatography / edited by O. Hoffmann-Ostenhof ... [et al.]. - Oxford : Pergamon, 1978
xvi,374p : ill. - Includes index
0-08-022632-9 SRL
1873709

AFFINITY chromatography and related techniques / edited by T.C.J. Gribnau, J. Visser and R.J.F. Nivard. - Amsterdam ; Oxford : Elsevier Scientific, 1982
xviii,584p : ill. - (Analytical chemistry symposia series ; v.9). - Includes index
0-444-42031-2 SRL
1873709

AFFINITY LABELLING
Use in research into active sites of biopolymers

THEORY and practice in affinity techniques / edited by P.V. Sundaram and F. Eckstein. - London : Academic Press, 1978
x,246p : ill. - Includes index
0-12-677150-2 SRL
1873717

AFFIXES
See also
Prefixes
Suffixes

AFFIXES. Proto-Chinese language

SCHÜSSLER, Axel
Affixes in Proto-Chinese. - Wiesbaden : Franz Steiner, 1976
[4], 3-292p. - (Münchener ostasiatische Studien ; Bd 18). - Bibliography : p.131-136. - Includes index
3-515-02565-0 X.0900/230(18)
1592602

AFFOR
See All Faiths for One Race

AFFOR
- Serials

AFFOR news. - No.1 (Spring 1978)-. - Birmingham : AFFOR, 1978-
v. : ill
Holdings: Spring 1981, etc
B81-33291 P523/1530
2544873

AFFORESTATION
Project management

FRASER, A. I.
A manual on the management of plantation forests. - Penicuik : International Forest Science Consultancy, [1980]
127p : ill. - Bibliography: p127
B81-03577 SRL
2333945

AFFORESTATION. Central Lowlands. Scotland

CENTRAL SCOTLAND WOODLANDS PROJECT. Steering Group
Central Scotland Woodlands Project. - Edinburgh : Lothian Regional Council, Department of Physical Planning, 1978
43p
B82-20918 SRL
2717085

AFFORESTATION. Central Lowlands. Scotland
- Serials

CENTRAL SCOTLAND WOODLANDS PROJECT
Interim report. - 1st (Feb. 1979-Jan. 1980)-. - [Falkirk] : [The Project], 1980-
v. : ill
Holdings: 2nd (Feb. 1980-Jan. 1981)-
B82-14787 P.313/538
2641992

AFFORESTATION. Dumfries and Galloway Region. Scotland
Effects on antiquities

JACKSON, Angela M.
Forestry & archaeology. - Hertford : RESCUE, 1978
viii,52p : ill. - Bibliography: p.47-48
0-903789-06-x X.319/18574 Woolwich
176182x

AFFORESTATION. Farms. Agriculture. Upland regions. Scotland

MUTCH, W. E. S.
The interaction of forestry and farming. - [Edinburgh] : Department of Forestry and Natural Resources, University of Edinburgh, 1980
113p : ill. - (Economics and management series ; no.2). - Bibliography: p107
B81-14288 X.325/1009 Woolwich
2383500

AFFORESTATION. Great Britain
Organisations: Men of the Trees. Archives

BASSETT, Philippa
A list of the historical records of the Men of the Trees. - [Birmingham] : Centre for Urban and Regional Studies, University of Birmingham, 1980
iv,4leaves. - (Lists of historical records)
B81-31791 X.325/1338 Woolwich
2524805

AFFORESTATION. Hungary

TÓTH, Béla
Szikesek fásítása. - Budapest : Akadémiai Kiadó, 1972
266p,[5]fold.leaves of plates : ill. - Bibliography : p.243-248
r6174430 X.620/8945 Woolwich
2212757

AFFORESTATION. India (Republic)

SONI, R. C.
Afforestation in India's economy. - New Delhi : Economic and Scientific Research Foundation, 1975
[7], 119 p.
r6031347 X.311/8406 Woolwich
106990X

AFFREIGHTMENT
See also
Contracts of affreightment

AFFRY, ADÈLE D'
See Marcello †

AFGHAN HOUNDS

AFGHAN hounds / [edited] by Beverly Pisano. - Neptune ; Reigate : T.F.H., 1980
4-125p : ill(some col)
0-87666-682-9 X.629/13077 Woolwich
0165581

GIE, Daphne
Afghan hounds. - Newton Abbot : David and Charles, 1978
208p : ill. - Bibliography: p.202. - Includes index
0-7153-7423-0 X.329/11290 Woolwich
0165581

HARRISSON, Charles
The Afghan hound. - 4th ed., revised. - London : Popular Dogs Publishing Co., 1975
248p,leaf of plate,[16]p of plates : ill. - (Popular Dogs' breed series). - Bibliography: p.239-240. - Includes index
0-09-124880-9 SRL
0165581

HARRISSON, Charles
The Afghan hound. - 5th ed. - London : Popular Dogs, 1979
250p,[16]p of plates : ill. - (Popular Dogs' breed series). - Bibliography: p.241-242. - Includes index
0-09-136140-0 SRL
0165581

MCCARTHY, Dennis
The Afghan hound. - Edinburgh : J. Bartholomew, 1977
96p : ill(chiefly col). - Bibliography: p.94. - Includes index
0-7028-1054-1 X.319/9450 Woolwich
0165581

AFGHAN JEWELLERY
to date

JANATA, Alfred
Schmuck in Afghanistan. - Graz / Austria : Akademische Druck- u. Verlagsanstalt, [1981]
211p,[1]folded leaf of plates : ill(some col). - Bibliography
3-201-01150-9 X.425/2106 Woolwich
8721491

SCHMUCK und Silberschmiedearbeiten in Afghanistan und Zentralasien / ... herausgegeben von Paul Bucherer-Dietschi. - Liestal : Bibliotheca Afghanica, 1981
79p. : ill. - (Schriftenreihe / Bibliotheca Afghanica ; 2)
r6303823 X.0529/648(2)
8814473

AFGHAN LORRIES
Decorations: Paintings

BLANC, Jean Charles
Afghan trucks. - London : Mathews Miller Dunbar, 1976
[83]p : of col ill
0-903811-19-7 X.619/19104 Woolwich
1728520

AFGHAN WAR, 1839-1842
- Biographies

EYRE, Sir Vincent
[The military operations at Cabul]. Journal of an Afghanistan prisoner. - [1st ed. reprinted]. - London : Routledge and Kegan Paul, 1976
xxvi,330p,[3] leaves of plates : ill
0-7100-8349-1 X.809/40887
0640689

AFGHAN WAR, 1839-1842
Army operations by Great Britain. Army. 41st (Welsh) Regiment of Foot - Biographies

MOORE, Geoffrey, 1914-
Vincent of the 41st. - Huntingdon : The author, [1979]
60p : ill
0-9506360-4-5 X.709/33742
2052016

AFGHAN WAR, 1839-1842
Causes

IQBAL, Afzal
Circumstances leading to the First Afghan War. - Lahore : Research Society of Pakistan, 1975
viii,94p. - (Publication ; no.33). - Bibliography : p.85-87. - Includes index
r6042996 X.800/25915
1048295

AFGHAN WARS

HEATHCOTE, T. A. (Thomas Anthony)
The Afghan Wars 1839-1919. - London : Osprey, 1980
224p : ill. - Bibliography: p218-219. - Includes index
0-85045-354-2 X.800/29305
2087693

AFGHĀNĪ, JAMĀL AL-DĪN. Islam

KEDDIE, Nikki R. (Nikki Ragozin)
An Islamic response to imperialism. - Berkeley ; London : University of California Press, 1983
xxiv,212p. - Bibliography: p191-200. - Includes index
0-520-04766-4 X.529/70005
3284255

AFGHANISTAN

MICHAUD, Roland
Afghanistan. - London : Thames and Hudson, 1980
23p,98p of plates : chiefly col.ill
0-500-54067-5 L.42/785
2344653

AFGHANISTAN
- Serials

AFGHAN studies. - Vol.1 (1978)-. - London : Society for Afghan Studies, 1978-
v. : ill
Holdings: vol 3&4 (1982)-
B84-23665 P.705/325
3120155

AFGHANISTAN journal / in Zusammenarbeit mit der Arbeitsgemeinschaft Afghanistan. - Jahrg. 1, Hft.1-Jahrg. 9, Hft.4. - Graz : Akademische Druck- u. Verlagsanstalt, 1974-1982
9v.
r6000087 P.803/596
0904201

AFGHANISTAN
to 1950. Documents on Afghanistan, to 1950 in stock of India Office Library and Records. Collections: India Office Records - Catalogues

INDIA OFFICE LIBRARY AND RECORDS
A brief guide to sources for the study of Afghanistan in the India Office Records. - London : The Library, 1981
v,60p : 1ill
0-903359-20-0 X.525/5496
253651x

AFGHANISTAN
to 1980

AFGHANISTAN. - Moscow : Social Sciences Today Editorial Board USSR Academy of Sciences, 1981
270p. - (Oriental studies in the USSR ; no.3). - Bibliography: p254-261
0-7147-1682-0 X.809/50165
2502526

GILFOND, Henry
Afghanistan. - New York ; London : Watts, c1980
64p : col.ill. - (A First book). - Bibliography: p61. - Includes index
0-531-04157-3 X.809/46308
2344327

GRIFFITHS, John C. (John Charles)
Afghanistan. - London : Deutsch, 1981
225p,[8]p of plates : ill. - Includes index
0-233-97350-8 X.809/48137
2502526

SHAH, Sirdar Ikbal Ali
Afghanistan of the Afghans. - London : Octagon, c1982
272p
B82-32899 X.950/14820
2502526
- [Another copy.] T44112 IOL

AFGHANISTAN
to 1980 - Bibliographies

MCLACHLAN, K. S. (Keith S)
A bibliography of Afghanistan. - Wisbech : Middle East & North African Studies Press, 1983
xiii,671p. - Includes index
0-906559-12-x X.809/56153
2951630
- [Another copy.] T 45287 IOL CH/ZABJD
- [Another copy.] 958.0016 OPL

AFGHANISTAN
to date

AFGHANISTAN / Aktion Afghanistan. - Liestal : Aktion Afghanistan, [1980 or 1981]
[16]p : ill
r6302442 X.805/2143
8002835

BALDEV, Prof.
Global impact of Afghan crisis. - New Delhi : International Reporter Publications, [1980]
210p
r6804428 X.809/49166
8002835

DAOUD, Zemaray
L'État monarchique dans la formation sociale afghane. - Berne : Peter Lang, [1982]
400p. - (European university studies. Ser. xxxi, sciences politiques ; v.37) (Publications Universitaires Européennes ; 37). - Includes bibliography
3-261-05050-0 X.809/63550
8002835
- [Another copy] T 45046 IOL

DEMOKRATICHESKAI͡A Respublika Afganistan / [redakt͡sionnai͡a kollegii͡a : M.R. Arunova (otvetstvennyĭ redaktor) ... i dr.]. - Moskva : Nauka, Glavnai͡a redakt͡sii͡a vostochnoĭ literatury, 1981
172p
r6402697 X.808/35232
8002835

ISTORII͡A Afganistana / [otvetstvennyĭ redaktor I͡U. V. Gankovskiĭ]. - Moskva : Mysl', 1982
366p
r6463420 X.809/55409
8002835

NAYAR, Kuldip
Report on Afghanistan. - New Delhi : R.N. Sachdev for Allied Publishers, 1981
ix,212p,8p.of plates : ill. - Bibliography
r6804714 X.809/49591
8002835

ROMODIN, V. A. (Vadim Aleksandrovich)
Ocherki po istorii i istorii kul'tury Afganistana. - Moskva : Nauka, Glavnai͡a redakt͡sii͡a vostochnoĭ literatury, 1983
189, [3] p. : ill.. - Bibliography: p. 183-[190]
r6482957 X.529/67370
8002835

SHRAR, S. A.
Afghanistan. - Hamburg : Hanseatische Edition, c1981
217p
3-921554-07-1 X.809/57444
8002835
- [Another copy.] T 47071 IOL

SRIVASTAVA, M. P.
Soviet intervention in Afghanistan. - New Delhi : Ess Ess Publications, 1980
xii,128p
r6804821 X.529/43964
8002835

AFGHANISTAN
to date – Conference proceedings
ARBEITSGEMEINSCHAFT AFGHANISTAN Arbeitstagung (5th 1979 : Mannheim)
Neue Forschungen in Afghanistan. – Opladen : Leske & Budrich, 1981
245p,[2]folded leaves of plates : ill. – (Schriften des Deutschen Orient-Instituts)
3-8100-0326-3 X.800/31769
8002835
– [Another copy.] V 25572 IOL

AFGHANISTAN
to date – Serials
SCHRIFTENREIHE / Bibliotheca Afghanica. – 1- – Liestal : Bibliotheca Afghanica, 1980-
r6301709 X.0529/648
8002835

AFGHANISTAN
1709-1919
SCHINASI, May
Afghanistan at the beginning of the twentieth century .... – Naples : Ist. universitario orientale, 1979
302p,plates. – (Series minor / Seminario di studi asiatici, Istituto universitario orientale ; 3). – Includes bibliography
r6507640 X.800/35528
8572828

AFGHANISTAN
1800-1900
AHMAD, N. D.
The survival of Afghanistan. – Lahore : People's Publishing House, 1973
92 p.
r6000424 X.809/23312
1008757

AFGHANISTAN
1919-
AFGHANISTAN seit dem Sturz der Monarchie / [Herausgeber: Deutsches Orient-Institut im Verbund der Stiftung Deutsches Übersee-Institut. Dokumentations-Leitstelle Moderner Orient]. – Hamburg : [Das Institut], 1981
298p. – (Aktueller Informationsdienst Moderner Orient. Sondernummer ; 9). – Bibliography: p247-274
3-88693-009-2 X.805/3486
8877211
– [Another copy.] X.805/3486

AFGHANISTAN
1973-1979
VERCELLIN, Giorgio
Afghanistan 1973-1978. – Venezia : [Università degli studi di Venezia], 1979
157p. – (Quaderni del seminario di iranistica, uralo-altaistica e caucasologia dell'Università degli studi di Venezia ; 4). – Bibliography
r6503809 X.800/33597
8840040
– [Another copy] 14773.e.501 OMPB

AFGHANISTAN
1979-
AFGHANISTAN. – Varese : Aurora, 1980
145p
88-85039-03-0 X.808/37476
8752117

ARNOLD, Anthony
Afghanistan. – Stanford : Hoover Institution Press, c1981
xvi,126p. – (Hoover Press publication ; 251) (Hoover international studies)
0-8179-7512-8 X.809/50965
8752117

BUCHHORN, Martin
40 Tage in Kabul. – Weinheim : Beltz, 1982
187p : ill
3-407-85011-5 X.809/58199
8752117

RATNAM, Perala
Afghanistan's uncertain future. – New Delhi : Tulsi Publishing House, [1981]
100p. – Bibliography: p.93-100
r6801851 X.809/49537
8752117

AFGHANISTAN
'Abd al-Rahmān Khān – Biographies
'ABD AL-RAHMĀN KHĀN
The life of Abdur Rahman. – Karachi ; Oxford : Oxford University Press, 1980
2v. : ill. – (Oxford in Asia historical reprints). – Includes index
0-19-577258-x X.809/49745
2579448

AFGHANISTAN
Antiquities, to 1500. Archaeology
The ARCHAEOLOGY of Afghanistan / edited by F.R. Allchin and Norman Hammond. – London : Academic Press, 1978
xxiii,451p : ill. – (Studies in archeology). – Bibliography: p.415-433. – Includes index
0-12-050440-5 X.421/10889 Woolwich
1869175

AFGHANISTAN
Army operations by Great Britain. Army. South Afghanistan Field Force, 1878-1880
MAXWELL, Leigh
My God – Maiwand. – London : Cooper, 1979
x,277p,[4]p of plates : ill. – Bibliography: p.268. – Includes index
0-85052-247-1 X.809/44655
1939009

AFGHANISTAN
Bactrian antiquities
PUGACHENKOVA, G. A. (Galina Anatol'evna)
Iskusstvo Baktrii épokhi Kushan. – Moskva : Iskusstvo, 1979
247p : ill. – (Iz istorii mirovogo iskusstva)
r6174303 X.421/20108 Woolwich
2168170

AFGHANISTAN
Bactrian antiquities
[ZOLOTO Baktrii. English]. Bactrian gold / [text and selection by Victor Sarianidi]. – Leningrad : Aurora Art Publishers, 1985
258p : ill
r6490524 L.45/3360
2168170

AFGHANISTAN
Badakhshan. Social conditions, to date
KUŠKAKĪ, Burhān ad-Dīn
[Rāhnamā-yi Qataġan va Badahšan. Persian and French]. Qataghan et Badakhshân. – Paris : Éditions du CNRS, 1979
3v.. – (Publications de l'URA ; no.10. mémoire ; no.3)
2-222-02603-2 X.525/9070
8724695

MOOS, Iren von
Die wirtschaftlichen Verhältnisse im Munjan-Tal und der Opiumgebrauch der Bevölkerung. – Liestal : Bibliotheca Afghanica, 1980
63p,[4]p of plates : ill. – (Schriftenreihe / Bibliotheca Afghanica ; 1)
r6301710 X.0529/648(1)
8724695

AFGHANISTAN
Baghlan & Pul-i-Khumri (Province). Village communities
TOEPFER, Helmuth
Untersuchungen zur Wirtschafts- und Sozialstruktur der Dorfbevölkerung der Provinz Baghlan (Afghanistan). – Meisenheim am Glan : Hain, 1976
156 p.. – (Afghanische Studien ; Bd 15). – Bibliography : p. 143-154
3-445-01341-1 X.0700/383(15)
1007386
– [Another copy.] V 25537 IOL

AFGHANISTAN
Balkh, to date
MUKHTAROV, A. (Akhror)
Pozdnesrednevekovyĭ Balkh. – Dushanbe : Donish, 1980
123p
r6465093 X.808/40811
9128115

AFGHANISTAN
Baluchis. Social conditions, to date
ORYWAL, Erwin
Die Balūč in Afghanisch-Sīstān. – Berlin : Reimer, 1982
vii,299p,plates. – (Kölner ethnologische Studien ; Bd.4). – Includes bibliography
3-496-00170-4 X.529/54705
8948461
– [Another copy.] T 47075 IOL

AFGHANISTAN
Constitution, to date
MOLTMANN, Gerhard
Die Verfassungsentwicklung Afghanistans 1901-1981. – Hamburg : Deutsches Orient-Institut, 1982
184p. – (Mitteilungen des Deutschen Orient-Instituts ; Nr.18)
r6305434 X.0512/459(18)
8851921
– [Another copy.] W 7040 IOL

AFGHANISTAN
Description & travel
AFGHANISTAN / herausgegeben von...Willy Kraus im Auftrag der ""Arbeitsgemeinschaft Afghanistan". – 2., revid. Aufl. – Tübingen, 1974
427 p. : ill
3-7711-0150-6 X.809/23021
0883964

HISTORICAL and political gazetteer of Afghanistan / edited by Ludwig W.Adamec.
Vol.1-
r6002255 Maps 57.bb.21 Map Library
0883964

AFGHANISTAN
Description & travel, to date – Serials
SCHRIFTENREIHE / Bibliotheca Afghanica. – 1-. – Liestal : Bibliotheca Afghanica, 1980-
r6301709 X.0529/648
800286x

AFGHANISTAN
Description & travel, 1967-1972
MICHAUD, Roland
Caravans to Tartary. – London : Thames and Hudson, 1978
[106]p : chiefly ill(chiefly col)
0-500-54053-5 Cup.1281/939
1693271

AFGHANISTAN
Description & travel, 1970-1972
LEVI, Peter
The light garden of the angel king. – [New] ed. – Harmondsworth : Penguin, 1984
256p. – (Penguin travel library). – Includes index
0-14-009525-x X.808/40467
3163164

AFGHANISTAN
Economic development
FRY, Maxwell J.
The Afghan economy. – Leiden : E.J. Brill, 1974
x,332p[2] fold leaves. – (Social, economic and political studies of the Middle East ; Vol. 15). – Bibliography : p. 301-316
90-04-03986-4 X.520/9473
1236318

AFGHANISTAN
Economic development. Financial assistance by Great Britain
GREAT BRITAIN
[Treaties, etc. Afghanistan, 1974 Aug.24]. Exchange of notes concerning financial assistance by the Government of the United Kingdom of Great Britain and Northern Ireland to the Government of the Republic of Afghanistan, (United Kingdom-Afghanistan Financial Agreement, 1974), Kabul, 24 August-30 September 1974 ... – London : H.M.S.O., 1975
18p. – (Treaty series ; no.91(1975)) (Cmnd.6087)
0-10-160870-5 BS.14/137
0516708

AFGHANISTAN
Education, to date
SCHARIFI, Nadjibullah
Probleme und Erfordernisse des Schulwesens in Afghanistan. – Frankfurt am Main : Lang, 1981
iii,187p. – (Übergange. Reihe B, Monographien ; Bd.2). – Bibliography
3-8204-7085-9 X.0529/587(2)
8746427

AFGHANISTAN
Food supply. Foreign assistance. Grants from European Economic Community: Grants of powdered skimmed milk
AFGHANISTAN
[Treaties, etc. European Economic Community, 1975 Feb.20]. Agreement between the European Economic Community and the Republic of Afghanistan on the supply of skimmed milk powder as food aid, Brussels, 20 February 1975 ... – London : H.M.S.O., [1975]
7p. – (European Communities ; no. 102 (1975)) (Cmnd. 6156)
0-10-161560-4 BS.14/1821
0533955

AFGHANISTAN
Foreign relations, 1900-1950
ADAMEC, Ludwig W.
Afghanistan's foreign affairs to the mid-twentieth century. – Tucson : University of Arizona Press, [1974]
ix,324p : ill. – Bibliography : p.311-315
0-8165-0388-5 X.800/12726
1316664

AFGHANISTAN
Foreign relations with India, to date

SAREEN, Anuradha
India and Afghanistan. - Delhi : Seema Publications, 1981
vi,238p. - Includes bibliography
r6814428 X.809/63070
9168125

AFGHANISTAN
Foreign relations with Soviet Union, 1917-1980

BRADSHER, Henry S. (Henry St. Amant)
Afghanistan and the Soviet Union. - Durham, N.C. : Duke University Press, 1983
viii,324p. - (Duke Press policy studies). - Bibliography: p297-311
0-8223-0563-1 X.800/36790
2739658
- [Another issue] V 25206 IOL

MAPRAYIL, Cyriac
The Soviets and Afghanistan. - London : Cosmic, 1982
xii,165p. - Includes index
B82-25650 X.809/52080
2739658

AFGHANISTAN
Foreign relations with Soviet Union, 1919-

POLVEKA mirnogo sotrudnichestva / [redaktsionnaia kollegiia : A.L. Voronin, N.A. Kuznetsova (otv. red.) ... i dr]. - Moskva : Nauka, Glavnaia redaktsiia vostochnoĭ literatury, 1973
239p
r6498276 X.808/34514
8713731

TEPLINSKIĬ, L. B. (Leonid Borisovich)
SSSR i Afganistan, 1919-1981. - Moskva : Nauka, Glavnaia redaktsiia vostochnoĭ literatury, 1982
293p. - (SSSR i strany Vostoka). - Bibliography
r6464053 X.529/53041
8713731

AFGHANISTAN
Guerilla wars, 1979-1983

FULLERTON, John
The Soviet occupation of Afghanistan. - London : Methuen, 1984
205p,16p of plates : ill. - Bibliography: p203-205. - Includes index
0-413-55780-4 X.808/40219
3146049

AFGHANISTAN
Guerrilla wars, 1982

GALL, Sandy
Behind Russian lines. - London : Sidgwick & Jackson, 1983
194p,[24]p of plates : ill
0-283-99039-2 X.800/37045
3042332

RYAN, Nigel, 19—
A hitch or two in Afghanistan. - London : Weidenfeld and Nicolson, c1983
xiv,210p,[8]p of plates : ill
0-297-78363-7 X.809/57580
3042332
- [Another copy.] T 45457 IOL

AFGHANISTAN
Herat, to date

ALLEN, Terry, Archaeologist
Timurid Herat. - Wiesbaden : Reichert, 1983
95p,plate. - (Beihefte zum Tübinger Atlas des Vorderen Orients. Reihe B, Geisteswissenschaften ; Nr 56). - Includes bibliography
r6309816 X.0700/644b(56)
9064818

AFGHANISTAN
Hindu Kush. Description & travel, 1956

NEWBY, Eric
A short walk in the Hindu Kush. - London : Pan, 1974, c1958
247p,[8]p of plates : ill. - (Picador)
0-330-26623-3 X.808/35452
2612321

NEWBY, Eric
A short walk in the Hindu Kush. - London : Harvill, 1984, c1958
247p
0-00-272772-2 X.809/60607
2612321

AFGHANISTAN
Invasion by Soviet Union, 1979

BEHRENS, Henning
Die Afghanistan-Intervention der UdSSR. - München : Tuduv, 1982
227p. - (Tuduv-Buch)
3-88073-109-8 X.809/55922
8990832
- [Another copy.] T 47076 IOL

HAMMOND, Thomas T. (Thomas Taylor)
Red flag over Afghanistan. - Epping : Bowker, [1984]
xvii,261p : ill. - Bibliography: p241-252. - Includes index
0-86531-444-6 X.520/35287
2253755

HELLER, Mark, 19—
The Soviet invasion of Afghanistan. - [Tel Aviv] : Center for Strategic Studies, Tel Aviv University, 1980
40p. - (CSS memorandum ; no.2)
r6807675 X.805/4823
9007040

HUSSAIN, Syed Shabbir
Afghanistan under Soviet occupation. - Islamabad : World Affairs Publications, 1980
208p : ill
r6819736 X.809/62853
9007040

VERSTAKOV, Viktor Glebovich
Afganskiĭ dnevnik. - Moskva : Voennoe izdatel'stvo, 1983
139p
r6496746 X.808/39268
8990832

AFGHANISTAN
Invasion by Soviet Union. Implications for policies on foreign relations of Great Britain government. Great Britain. Parliament. House of Commons. Foreign Affairs Committee. Fifth report ... session 1979-80

GREAT BRITAIN. Foreign and Commonwealth Office
Fifth report from the Foreign Affairs Committee, sesion 1979/80. - London : H.M.S.O., [1980]
5p. - (Miscellaneous ; no.23 (1980)) (Cmnd. ; 8068)
0-10-180680-9 BS.14/3512
2641860

AFGHANISTAN
Journalism, to date

SCHINASI, May
Afghanistan at the beginning of the twentieth century .... - Naples : Ist. universitario orientale, 1979
302p,plates. - (Series minor / Seminario di studi asiatici, Istituto universitario orientale ; 3). - Includes bibliography
r6507640 X.800/35528
8966478

AFGHANISTAN
Kabul. Civilians. Evacuation by Great Britain. Royal Air Force. No.70 Squadron, 1928-1929

BAKER, Anne
Wings over Kabul. - London : Kimber, 1975
191p : ill
0-7183-0184-6 X.800/25301
0561797

AFGHANISTAN
Kirgiz & Wakhi. Cultural processes, to 1978

SHAHRANI, M. Nazif Mohib
The Kirghiz and Wakhi of Afghanistan. - Seattle ; London : University of Washington Press, 1979
xxiii,264p : ill. - (Publications on ethnicity and nationality / University of Washington. School of International Studies ; no.1). - Bibliography: p.243-258. - Includes index
0-295-95669-0 X.520/15057
2052490

AFGHANISTAN
Military operations by British military forces, 1839-1879 - Biographies

The GORDON Creeds in Afghanistan, 1839 and 1878-79 / edited by William Trousdale. - London : BACSA, 1984
x,189p : ill. - Includes index
0-907799-08-6 X.809/64647
3287807

AFGHANISTAN
Mundigak. Prehistoric antiquities. Archaeology, 1955-1957

MATHESON, Sylvia A. (Sylvia Anne)
Time off to dig. - Karachi ; Oxford : Oxford University Press, 1982
286p,[24]p of plates : ill. - Includes index
0-19-577276-8 X.950/32325
3123855
- [Another copy.] T 45396 IOL

AFGHANISTAN
Nuristan. Buildings, to date

EDELBERG, Lennart
Nuristani buildings. - Aarhus : Jutland Archaeological Society, 1984
xxvii,223p,plate : ill. - (Jutland Archaeological Society publications ; 18)
87-88415-28-7 Ac.5538.c/2[no.18]
9162291

AFGHANISTAN
Nuristan. Religions, to date

FREMBGEN, Jürgen
Religiöse Funktionsträger in Nuristan. - Sankt Augustin : VGH Wissenschaftsverlag, 1983
xx,195p. - (Beiträge zur Zentralasienforschung ; Bd.3). - Includes bibliography
3-88280-017-8 X.0809/666(3)
9021515

AFGHANISTAN
Pashai. Social conditions, to date

WUTT, Karl
Pashai. - Graz : Akademische Druck- u. Verlagsanstalt, 1981
140p.,[64]p.of plates : ill(some col.)
3-201-01170-3 X.805/2262
8872627

AFGHANISTAN
Pathans, to 1957

CAROE, Sir Olaf
The Pathans, 550 BC-AD 1957. - Karachi ; London : Oxford University Press, 1976 [i.e. 1977]
xxii,521p,[16]leaves of plates(4 fold) : ill. - ([Oxford in Asia historical reprints from Pakistan]). - Includes index
0-19-577221-0 X.809/41684
1060872
- [Another issue, reprinted, with an Epilogue on Russia, 1983]. - xxiv, 533p., [16] leaves of plates [4 fold.] : ill., geneal. tables, ports T 45660 IOL

AFGHANISTAN
Pathans. Codes of conduct, to date

STEUL, Willi
Paschtunwali. - Wiesbaden : Steiner, 1981
xiv,313p,[4]folded leaves of plates. - (Beiträge zur Südasienforschung / Südasien-Institut, Universität Heidelberg ; Bd.54). - Bibliography
3-515-03167-7 X.800/32816
8833648

AFGHANISTAN
Political events

MARTIN, Mike, 1951-
Afghanistan. - Poole : Blandford, 1984
256p : ill. - Includes index
0-7137-1388-7 X.520/36415
3050041
- [Another copy.] V 25283 IOL

AFGHANISTAN
Political events - Serials

AFGHAN voice. - Vol.1 no.1 (May 1980)-. - London : Afghan Muslims, 1980-
v. : ill
B81-13092 P.2000/859
237997x

AFGHANISTAN
Political events, to date - Conference proceedings

REVOLUTIONS & rebellions in Afghanistan / M. Nazif Shahrani & Robert L. Canfield, editors. - Berkeley, Calif. : University of California, Institute of International Studies, [1984]
xiv,394p. - (Research series / University of California, Berkeley. Institute of International Studies ; no.57). - Includes bibliography
0-87725-157-6 X.0809/130(57)
9138625

AFGHANISTAN
Political events, 1838-1844

MOHAN LAL
Life of the Amir Dost Mohammed Khan of Kabul. - Karachi ; Oxford : Oxford University Press. - (Oxford in Asia historical reprints from Pakistan)
Vol.2. - [1st ed. reprinted]. - 1978
xiii,498p,[8]leaves of plates
0-19-577237-7 X.809/44882
1999176

AFGHANISTAN
Political events, 1973-

MITRA, Samar
Curtain unveiled. - Calcutta : Pagoda House, [1981?]
139p : ill
r6815151 X.809/63197
8840059

PEOPLES' PERMANENT TRIBUNAL
Afghanistan ockuperat. - Stockholm : Ordfront, 1982
175p
91-7324-157-1 X.808/39712
8840059

AFGHANISTAN
Political events, 1973—continued

POLÍAKOV, G. A (Genrikh Anatol'evich)
Afganistan revolîûtsionnyi. - Moskva : Mezhdunarodnye otnosheniîa, 1981
68p
r6486661 X.808/36716
8840059

AFGHANISTAN
Political events, ca 1975-1980

NEWELL, Nancy Peabody
The struggle for Afghanistan. - Ithaca : London Cornell University Press, 1981
236p : ill. - Includes index
0-8014-1389-3 X.809/49222
2554879

AFGHANISTAN
Political events, 1978-1980

The TRUTH about Afghanistan. - 2nd ed. - Moscow : Novosti Press, 1981
201p,[32]p.of plates : ill
r6864844 X.808/36670
2712792

The UNDECLARED war / [compiled by A.S. Grachev] ; [translated from the Russian by Dmitry Belyavsky]. - Moscow : Progress, c1980
156p,[16]p of plates : ill. - Includes index
0-7147-1708-8 X.808/36102
2712792

AFGHANISTAN
Political events, 1978-1981

AFGHANISTAN in crisis / edited by K.P. Misra. - London : Croom Helm, 1981
150p. - Bibliography: p120-150
0-7099-1727-9 X.800/31695
2577550
- [Another copy] 14163.f.53 OMPB

ENGIN, Emine
The revolution in Afghanistan. - [London] : İşçinin Sesi, 1982
174p. - (English series ; 16)
B84-02398 X.808/38800
2577550

The TRUE face of Afghan counter-revolution. - Kabul : Editorial Office of the newspaper Haqiqat Enqelab Sawer, 1982
63p : ill
r6865345 X.808/37401
2577550

AFGHANISTAN
Political events, 1979-

FRANCESCHI, Patrice
Ils ont choisi la liberté. - [Paris] : Arthaud, 1981
267p,16p of plates
r6207492 X.800/33796
8713502

ILYINSKY, Mikhail
Afghanistan, onward march of the revolution. - New Delhi : Sterling Publishers, [1982]
88p
r6807904 X.809/58243
8713502

MISSEN, François
La nuit afghane. - [Paris] : Le Pré aux Clercs, c1983
313p
2-7144-1581-4 X.809/63264
8713502

VOÍNA v Afganistane. - Frankfurt / Main : Posev, 1980
200p : ill
r6498201 X.808/34380
8713502

AFGHANISTAN
Political events. Role of tribal societies, 1800-1980 - Conference proceedings

The CONFLICT of tribe and state in Iran and Afghanistan / edited by Richard Tapper. - London : Croom Helm, c1983
463p. - Includes index
0-7099-2440-2 X.809/56536
2964058

AFGHANISTAN
Political prisoners. Prison life, 1979-

MISSEN, François
La nuit afghane. - [Paris] : Le Pré aux Clercs, c1983
313p
2-7144-1581-4 X.809/63264
9145745

AFGHANISTAN
Politics, 1965-1980

MALE, Beverley
Revolutionary Afghanistan. - London : Croom Helm, c1982
229p. - Bibliography: p216-220. - Includes index
0-7099-1716-3 X.809/52196
2748304

AFGHANISTAN
Politics. Role of Germany, 1919-1939

FLEURY, Antoine
La penetration allemande au Moyen-Orient, 1919-1939. - Leiden : A.W. Sijthoff ; Genève : Institut Universitaire de Hautes Études Internationales, 1977
xiii,432p. - (Collection de relations internationales ; 5). - Bibliography : p. 399-422
90-286-0297-6 X.800/13269
150956X
- [Another copy.] X.800/13541
- [Another copy.] X.800/13557

AFGHANISTAN
Pottinger, Eldred - Biographies

POTTINGER, George
The Afghan connection. - Edinburgh : Scottish Academic, c1983
xix,239p : ill. - Bibliography: p218-219. - Includes index
0-7073-0286-2 X.809/56415
2960435
- [Another copy.] T 45456 IOL
- [Another copy.] T 46001 IOL

AFGHANISTAN
Qataghan. Social conditions, to date

KUŠKAKĪ, Burhān ad-Dīn
[Rāhnamā-yi Qaṭaġan va Badaḫšan. Persian and French]. Qataghan et Badakhshân. - Paris : Éditions du CNRS, 1979
3v.. - (Publications de l'URA ; no.10. mémoire ; no.3)
2-222-02603-2 X.525/9070
9073825

AFGHANISTAN
Resistance movements, 1979

MISSEN, François
Le syndrome de Kaboul. - Aix-en-Provence : Edisud, 1980
185p,16p of plates : ill,maps
2-85744-062-6 X.709/51363
2418711

AFGHANISTAN
Resistance movements, 1979-

CHALIAND, Gérard
Rapport sur la résistance afghane. - Paris : Berger-Levrault, 1981
163p. - (Stratégies). - Bibliography
2-7013-0436-9 X.529/49922
8572933

DURAN, Ragip
Afghanistan - ryssarnas Vietnam. - Stockholm : Oktoberförlaget, 1980
96p,plates
91-7242-199-1 X.529/63687
8572933

AFGHANISTAN
Seistan. Antiquities. Archaeology, to date

PREHISTORIC Sistān / edited by Maurizio Tosi. - Roma : IsMEO, 1983- : ill. - (Reports and memoirs / Istituto italiano per il Medio ed Estremo Oriente, Centro studi e scavi archeologici in Asia ; v.19)
r6510245 LR.408.wl[vol.19]
9047263

AFGHANISTAN
Seistan. Buildings. Architecture, to date

KLINKOTT, Manfred
Islamische Baukunst in Afghanisch-Sistān. - Berlin : Reimer, 1982
295p.,[7]folded leaves of plates : ill. - (Archaeologische Mitteilungen aus Iran. Ergänzungsband ; 8)
3-496-00107-0 P.413/10a[no.8]
8843732
- [Another copy.] V 25543 IOL

AFGHANISTAN
Social conditions, to 1972

SARWARI, Mohammad Sadiq
Afghanistan zwischen Tradition und Modernisierung. - Bern : Lang, 1974
312 p.. - (Europäische Hochschulschriften. Reihe 31 ; Bd 2). - Bibliography : p. 297-312
3-261-01445-8 X.709/21367
1007319

AFGHANISTAN
Social conditions, to date - Conference proceedings

REVOLUTIONS & rebellions in Afghanistan / M. Nazif Shahrani & Robert L. Canfield, editors. - Berkeley, Calif. : University of California, Institute of International Studies, [1984]
xiv,394p. - (Research series / University of California, Berkeley. Institute of International Studies ; no.57). - Includes bibliography
0-87725-157-6 X.0809/130(57)
8572909

AFGHANISTAN
Social conditions, 1880-1901

KAKAR, Hasan Kawun
Government and society in Afghanistan. - Austin ; London : University of Texas Press, 1979
xxiv,328p. - (Modern Middle East series ; no.5). - Bibliography: p.307-312. - Includes index
0-292-72718-6 X.800/28905
2149486

AFGHANISTAN
Social conditions, 1973- - Conference proceedings

ARBEITSGEMEINSCHAFT AFGHANISTAN. Arbeitstagung (6th : 1982 : Bielefeld)
Forschungen in und über Afghanistan. - Hamburg : Deutsches Orient-Institut, 1983
230p : ill. - (Mitteilungen des Deutschen Orient-Instituts ; Nr.22). - Includes bibliographies
r6308139 X.0512/459(22)
8995559
- [Another copy.] X.0512/459(22)

AFGHANISTAN
Social conditions, 1980

CHALIAND, Gérard
[Rapport sur la résistance afghane. English]. Report from Afghanistan. - New York : Viking ; Harmondsworth : Penguin, 1982
112p. - Includes index
0-14-006516-4 X.808/37704
2865637
- [Another copy.] T 44812 IOL

AFGHANISTAN
Society. Role of women, to date

DELLOYE, Isabelle
Des femmes d'Afghanistan. - Paris : Des Femmes, 1980
158p. - (Pour chacune ; 32)
r6299083 X.529/43373
8724709

AFGHANISTAN
Soviet imperialism, 1964-1981

HYMAN, Anthony
Afghanistan under Soviet domination 1964-81. - London : Macmillan, 1982
xi,223p,[8]p of plates : ill. - Bibliography: p201-202. - Includes index
0-333-30579-5 X.809/54724
2879840
- [Another copy.] T 44941 IOL

AFGHANISTAN
Soviet imperialism, 1964-1983

HYMAN, Anthony
Afghanistan under Soviet domination, 1964-1983. - 2nd ed. - London : Macmillan, 1984
xi,247p,[8]p of plates : ill. - Bibliography: p221-222. - Includes index
0-333-36352-3 X.529/67517
3240371
- [Another copy.] T 46002 IOL

AFGHANISTAN
Uzbek folk music

THALHAMMER, Ingeborg
Die Liedkategorien der Özbeken Nordwestafghanistans. - Wien : Verband der wissenschaftlichen Gesellschaften Österreichs, 1984
iii,258p. - (Beihefte zur Wiener Zeitschrift für die Kunde des Morgenlandes ; Bd.12)
3-85369-560-4 P.P.5057(2)[Bd.12]
9073310

AFGHANISTAN
Waigal Valley. Kalashum. Social structure

JONES, Schuyler
Men of influence in Nuristan. - London : Seminar Press, 1974
xiv,299p : ill. - (Seminar studies in anthropology ; 3). - Bibliography: p.283-288. - Includes index
0-12-785393-6 X.800/25491
0619086

AFGHANISTANI TALES
- Anthologies
SHAH, Amina
Tales of Afghanistan. - London : Octagon, 1982
115p
0-900860-94-4 X.958/14336
2868741

AFGHANS
See also
Pashai
Pathans

AFGHANS
ADAMEC, Ludwig W.
Historical and political who's who of Afghanistan. - Graz : Akademische Druk-u Verlagsanstalt, 1975-
ix,385,[120]p
3-201-00921-0 958.10092 OPL
1241087

AFGHANS. Coverlets
Crocheting
FELDMAN, Annette
The big book of afghans. - New York ; Van Nostrand Reinhold, c1981
160p : ill(some col.). - (A Genie book). - Includes index
0-442-22528-8 X.622/12236 Woolwich
2241676

WEISS, Rita
Crocheting afghans. - New York : Dover Publications [etc.] ; London : Constable, 1979
[2],45p : ill. - (Dover needlework series)
0-486-23883-0 X.622/8237 Woolwich
2241676

AFGHANS. Coverlets
Crocheting & knitting
KURITA, Valerie
The second big book of afhgans. - New York ; London : Van Nostrand Reinhold, c1982
168p : ill(some col.). - (A Genie book). - Includes index
0-442-24863-6 X.622/15558
2903458

AFHANISTAN
1973-
BRÖNNER, Wolfram
Afghanistan. - Frankfurt am Main : Marxistische Blätter, 1980
277p : ill. - (Marxismus aktuell ; 148)
3-88012-612-7 X.808/34446
8706654

AFIKPO. Nigeria
Rituals. Use of carved masks
OTTENBERG, Simon
Masked rituals of Afikpo. - Seattle ; London : University of Washington Press for the Henry Art Gallery, 1975
[15],229p,16p of plates : ill(some col). - (Index of art in the Pacific northwest ; no.9). - Bibliography: p.223-225. - Includes index
0-295-95391-8 X.423/5076
0567299

AFIKPO MASKS. Nigeria
Carved Afikpo masks. Use in rituals
OTTENBERG, Simon
Masked rituals of Afikpo. - Seattle ; London : University of Washington Press for the Henry Art Gallery, 1975
[15],229p,16p of plates : ill(some col). - (Index of art in the Pacific northwest ; no.9). - Bibliography: p.223-225. - Includes index
0-295-95391-8 X.423/5076
0567299

AFINOGENOV, ALEKSANDR NIKOLAEVICH. Drama in Russian
AFINOGENOV, Aleksandr Nikolaevich
[Selections]. Izbrannoe .... - Moskva : Iskusstvo, 1977
2t.
r6094149 X.989/80117
2034808

AFJORD. Norway
Antiquities
HUMSTAD, Ragnvald
Fortidsminner i Afjord. - [A i Afjord] : [s.n.], 1976
161p : ill
r6034192 X.421/9094 Woolwich
1213741

AFLÁJ. Irrigation. Oman
Sociopolitical aspects, to 1974
WILKINSON, John Craven
Water and tribal settlement in South-east Arabia. - Oxford : Clarendon Press, 1977
xvi,276p : ill. - (Oxford research studies in geography). - Bibliography: p.249-257. - Includes index
0-19-823217-9 X.800/26443
1056808

AFLATOXINS
HEATHCOTE, J. G.
Aflatoxins. - Amsterdam ; Oxford : Elsevier, 1978
ix,212p : ill. - (Developments in food science ; 1). - Includes index
0-444-41686-2 SRL
1774417

AFN LTD
to 1983
JENKINSON, Denis
From chain drive to turbocharger. - Wellingborough : Stephens, 1984
320p : ill. - Includes index
0-85059-631-9 X.622/21427
3015653

AFP. Diagnosis. Neural tube. Embryos & foetuses. Man
- Conference proceedings
PREVENTION of neural tube defects / edited by Barbara F. Crandall, Mary A.B. Brazier. - New York ; London : Academic Press, 1978
xvii,264p : ill. - (Forums in medical sciences / University of California at Los Angeles ; no.20). - Includes bibliographies and index
0-12-195350-5 SRL
1583964

AFP. Man
NORGAARD-PEDERSEN, Bent
Human alpha-fetoprotein. - Oslo : Universitetsforlaget, 1976
45 p. : ill. - ('Scandinavian Journal of Immunology' supplement ; no. 4). - Bibliography : p. 37-45
82-00-37183-2 X.311/8368 Woolwich
1036718

AFRAGOLA. Italy
to 1860
CAPASSO, Gaetano
Afragola. - Napoli : Athena Mediterranea, 1974
415 p.. - (Nuova collana di storia napoletana ; n. 10)
r6024221 X.700/17500
101689X

AFRANIUS, LUCIUS. Latin literature
PASQUAZI BAGNOLINI, Anna
Note sulla lingua di Afranio. - Firenze : Felice Le Monnier, 1977
viii,88p. - (Bibliotechina del saggiatore ; 41)
r6078721 X.908/85733
1849034

AFRICA
See also
Books with African imprints
Documents with African imprints
Horn of Africa
Maghreb
Organization of African Unity
Pan-Africanism
Records on Commonwealth countries in Southern Africa
Research papers on environment of Africa
Serials with African imprints
Theses on Africa

AFRICA
AFRICAN dimensions / edited by Mark Karp. - Boston, Mass. : African Studies Center, Boston University, 1975
213p
0-915118-05-x X.809/28024
0438464

BRADT, Hilary
Backpacker's Africa. - 2nd ed. - Chalfont St. Peter : Bradt Enterprises, c1983
192p : ill
0-9505797-9-3 X.809/61090
0595055

CROWTHER, Geoff
Africa on the cheap. - South Yarra : Lonely Planet : Lascelles, 1980
362p : ill
0-908086-19-9 X.808/34408
0595055
- [Another copy] X.809/55958

GRIFFITHS, Ieuan Ll.
An atlas of African affairs. - London : Methuen, 1984
200p : ill. - Bibliography: p193-194. - Includes index
0-416-30940-2 MAP LIBRARY
0438464

GUIDES touristiques de l'Afrique... / [édités] sous la direction d'Arlette Eyraud, avec la collaboration d'Air Afrique. - Paris : Hatier, 1975-
tom.1-
r6019957 X.0702/398
0595055

NEW African yearbook - 1978 / edited and compiled by Alan Rake. - London : I.C. Magazines Ltd, [1978]
3-258p : ill(some col)
Subsequent volumes, when received, will be placed at this shelfmark
0-905268-02-4 P.701/1626
0118400

ODETOLA, T. O. (Theophilus Olatunde)
Man and society in Africa. - London : Longman, 1983
171p : ill. - Includes bibliographies and index
0-582-64313-9 X.529/59696
3062449

TRAVELLER'S guide to Africa. - London : 'African Development' (African Buyer and Trader Co. Ltd) : 'African Contemporary Record'.
1974. - 1974
432p : ill(some col)
Subsequent volumes, when received will be placed at this shelfmark
0-9502755-1-4 X.0702/1006
0595055

AFRICA
- Atlases
FIRST atlas for Sierra Leone schools / [T.A. Lucan, editorial consultant] ; [E.A. Pabs-Garnon and C. Roy-Macauley, editorial advisers]. - [Freetown?] : Macmillan Education, 1982
49p : col.ill
0-333-33954-1 X.805/4702
218530x

AFRICA
- Bibliographies
SCHEVEN, Yvette
Bibliographies for African studies, 1980-1983. - [Oxford] : Zell, c1984
xiii,300p. - Includes index
0-905450-13-2 X.809/62489
2872366

AFRICA
- Catalogues
STADT- UND UNIVERSITÄTSBIBLIOTHEK FRANKFURT AM MAIN
Fachkatalog Afrika. - München ; London : Saur.
Bd.10: Ostafrika. - 1984
2v(ix,1343,21p). - Includes index
3-598-20929-0 2725.g.306
2872366

AFRICA
- Directories
AFRICA guide. - 1977. - Saffron Walden : Africa Guide Company, 1976
311p : ill
Subsequent volumes, when received, will be placed at this shelfmark
B77-28318 P.701/1565
0118400
- [Another copy of the latest edition] OPL

AFRICA year book and who's who. - 1977 / [editor-in-chief Raph Uwechue]. - London : Africa Journal Ltd, 1976
xlvii,1364p,[8]p of plates : ill(some col)
Subsequent volumes, when received, will be placed at this shelfmark
0-903274-05-1 P.701/1533
0118400
Le GUIDE du voyageur en Afrique. - [London] : [International Communications].
[1978?]. - [1978?]
424p : ill
0-905268-06-7 X.708/22593
2268264

OWEN'S commerce and travel and international register. - 1977 : 24th ed. - London : Owen's Commerce and Travel Ltd, 1977
[3],1211[i.e. 1229]p : ill
0-900576-08-1 SRL
0118400

AFRICA
- For British businessmen
AFRICA / regional editor Toby Milner. - London : Joseph, c1981
511p. - (The Business traveller's handbook ; v.1). - Includes index
0-7181-1981-9 X.809/48199
2500043

AFRICA
- For motoring

MCELDUFF, Colin
Trans-Africa motoring. - London : Wilton House Gentry Ltd, 1975
248p : ill. - Bibliography: p.29-30. - Includes index
0-905064-00-3 X.800/11670
065289x

AFRICA
- Serials

AFRICA contemporary record. - 1968-1969- / [edited by] Colin Legum and John Drysdale. - London : Rex Collings, 1969 (1975 printing-)
Wanting the issue for 1970-1971
r6050968 OPL
0118400

AFRICA International Perspective. - No. 1- , Dec. 1975-Jan. 1976-. - Monrovia : African Publishing House, 1976- : ill
r6004492 P.701/1515
0673358

[AFRICA now (London : 1981)]. Africa now. - Feb. 1981 ; 1 (Apr. 1981)-. - [London] : [Pan-African Publishers], 1981-
v. : ill
B81-31568 P.701/1743
0087963

The AFRICAN review. - vol.1, no.1- ; 1971-. - Nairobi : East African Literature Bureau, 1971-
Holdings : vol.2, no.1-4 ; vol.3, no.1-4 ; vol.5, no.1,3 and 4, 1975-
r6077633 P.701/1077
0673358

AFRICAN STUDY GROUP
African Study Group. - Vol.1, no.1- ; [July 1978]-. - London : The Group, [1978]- : ill
B79-15161 P.703/605
1846019

AFRIQUE. - No.1- ; Mai-juin 1977-. - Londres [i.e. London] : Africa Journal Ltd, 1977- : ill(some col)
B78-02903 P.901/1573
1287362

BASLER AFRIKA BIBLIOGRAPHIEN
Nachrichten = Newsletter. - 1976, Nr.11- 15. Sept. 1976-. - Basel : Basler Afrika Bibliographien, 1976-
r6002510 P.703/403
1709925

IKENGA : journal of African studies / [issued by] University of Nigeria, Institute of African Studies. - Vol.1, no.1 (Jan. 1972)-. - Nsukka : The University, 1972-
r6077598 P.971/246
0673358

ISTORIÎA Afriki. - 1979, no.2-. - Moskva : Nauka, Glavnaîa redaktsiîa vostochnoĭ literatury, 1979-
r6183579 X:809/28459
1846019

MARBURGER Studien zur Afrika- und Asienkunde / Herausgeber : H.-J. Greschat ... W. Haenisch ... H. Jungraithmayr. - Bd. 1-. - Marburg an der Lahn : Selbstverlag, 1973-
r6000146 X.0100/683
0087963

POSITIVE review. - Vol.1, no.1- ; [1979?]-. - Ibadan : Oxford University Press Nigeria, [1979?]-
Holdings : Vol.1, no.2-
r6108373 P.521/2293
0673358

PULA / [issued by] University College of Botswana and Swaziland. - vol.1, no.1- ; June, 1978-. - Gaborone : The University College, 1978-
r6095862 P.701/1192
0673358

AFRICA
to 1964

DU BOIS, W. E. B. (William Edward Burghardt)
The world and Africa. - [New enlarged ed. reprinted]. - Millwood, N.Y. : Kraus-Thomson, 1976
20,xll,352p : ill. - Includes index
0-527-25340-5 X.800/26049
1094653

AFRICA
to 1970

HALL, Richard, 1925-
Discovery of Africa. - Feltham, 1982, c1970
159p : col.ill. - Bibliography: p156. - Includes index
B83-01380 X.808/37284
030459x

AFRICA
to 1974

DAVIDSON, Basil
Africa in history. - Revised and enlarged ed. - London : Paladin, 1978
366p,[16]p of plates : ill. - Bibliography: p.327-337. - Includes index
B78-31128 X.708/21363
0565555

OLIVER, Roland
A short history of Africa. - 5th ed. - Harmondsworth : Penguin, 1975
304p. - (Penguin African library). - Bibliography: p.275-289. - Includes index
B76-02778 010058.m.69/47
0565555
- Reprinted. - 1977 X.808/39070

AFRICA
to 1974 - Bibliographies

HARTWIG, Gerald W.
The student Africanist's handbook. - Cambridge, Mass. : Schenckman ; New York ; London : Wiley, 1974
viii,152p
0-470-35773-8 X.809/40130
036374x

PANOFSKY, Hans E.
A bibliography of Africana. - Westport, Conn. ; London : Greenwood Press, 1975
xi,350p. - (Contributions in librarianship and information science ; no.11). - Includes index
0-8371-6391-9 BB.R.e.12 Reading Room
036374x
- [Another copy.] X.809/41989

AFRICA
to 1974 - Catalogues

INTERNATIONAL AFRICAN INSTITUTE. Library
Cumulative bibliography of African studies.... - Boston : G. K. Hall & Co, 1973
2 v.
0-8161-1045-x LR.419.b.6
036374X

Cumulative bibliography of African studies.... - Boston : G. K. Hall & Co, 1973
3 v.
0-8161-1076-x LR.419.b.5
036374X

AFRICA
to 1975

The CAMBRIDGE history of Africa. - Cambridge : Cambridge University Press.
Vol.1: From the earliest times to c.500 BC. - 1982
xxiii,1157p,[20]p of plates : ill. - Includes bibliographies and index
0-521-22215-x X.0800/1504
1035606

Vol.2. - 1978
xvii,840p,[24]p of plates : ill. - Bibliography: p.719-770. - Includes index
0-521-21592-7 2090.g. Reading Room
1035606
- [Another copy.] X.0800/1504(2)

Vol.3. - 1977
xiii,803p. - Bibliography: p.702-747. - Includes index
0-521-20981-1 2090g. Reading Room
1035606
- [Another copy.] X.0800/1504(3)

Vol.4. - 1975
xiv,738p. - Bibliography: p.652-693. - Includes index
0-521-20413-5 2090g. Reading Room
1035606
- [Another Copy.] X.0800/1504.(4.)

Vol.5. - 1976
xv,617p. - Bibliography: p.539-580. - Includes index
0-521-20701-0 2090g. Reading Room
1035606
- [Another copy.] X.0800/1504(5)

LOFTUS, Ernest
A visual history of Africa. - 2nd ed. - London : Evans Bros, 1976
80p : ill
0-237-50025-6 X.700/25244
0736619

AFRICA
to 1975 - Encyclopaedias

MACDONALD'S encyclopedia of Africa. - London : Macdonald Educational, 1976
223p : ill(some col). - Bibliography: p.216-217. - Includes index
0-356-05053-x X.802/10163
1035606

AFRICA
to 1976 - Atlases

FAGE, J. D. (John Donnelly)
An atlas of African history. - 2nd ed. - London : Edward Arnold, 1978
[84]p. - Includes index
0-7131-5963-4 2060.eee Reading Room
1548212
- [Another copy] OPL
- [Another copy] X.702/5740

AFRICA
to 1977

AFRICA / edited by Phyllis M. Martin and Patrick O'Meara ; maps and charts by Cathryn L. Lombardi ; photographic selection by Mary Joy Pigozzi. - Bloomington ; London : Indiana University Press, 1977
xxiii,482p : ill. - Bibliography: p.415-466. - Includes index
0-253-30210-2 X.800/27855
141383x

AFRICAN history / Philip Curtin ... [et al.]. - London : Longman, 1978
xv,612,[4]p,[8]p of plates : ill. - Includes bibliographies and index
0-582-64665-0 X.709/32856
141383x

DAVIDSON, Basil
Discovering Africa's past. - London : Longman, 1978
vi,218p : ill. - Bibliography: p.210-212. - Includes index
0-582-22048-3 X.800/26839
1434802
- [Another copy: paperback]
ISBN 0-582-22049-1 X.700/25658

FAGE, J. D. (John Donnelly)
A history of Africa. - London : Hutchinson, 1978
x,534p,[8]p of plates : ill. - (The history of human society). - Bibliography: p.487-514. - Includes index
0-09-132850-0 2090g Reading Room
141383x
- [Another copy] X.0700/74(12)

MILLUM, Trevor
Exercises in African history. - [London] : Longman, [1978]
ii,70p
0-582-65006-2 X.700/25797
159530x

OLIVER, Roland
A short history of Africa. - 5th ed. - Harmondsworth : Penguin, 1975
304p. - (Pelican books). - Includes index
0-14-022467-x X.808/40586
141383x

AFRICA
to 1978 - Atlases

AFRICAN history in maps / Michael Kwamena-Poh ... [et al.]. - Harlow : Longman, 1982
76p. - Includes index
0-582-60331-5 Map Library
2205807

MCEVEDY, Colin
The Penguin atlas of African history. - Harmondsworth : Penguin, 1980
142p. - Includes index
0-14-051083-4 Map Library
2205807

The Penguin atlas of African history. - London : Allen Lane, 1980
142p. - Includes index
0-7139-1127-1 Map Library
2205807

AFRICA
to 1979

GIGLIO, C. (Carlo)
L'Africa. - Torino : UTET, 1980
xviii,698p,[87]p of plates : ill. - (Nuova storia universale dei popoli e delle civiltà ; vol.14'). - Bibliography
88-02-03516-4 X.0802/473.(14)
2238500

ROBERTS, George O.
Afro-Arab fraternity. - Beverly Hills ; London : Sage Publications, 1980
251p : ill. - (Sage library of social research ; vol.95). - Bibliography: p.237-243. - Includes index
0-8039-1334-6 X.700/26381
2238500

AFRICA
to 1979 - For East African students

SINGH, Malkiat
African history. - London : Evans, 1980
202p
0-237-50441-3 X.709/52212
2323338

AFRICA
to 1979. Historiology, 1953–1979

HOPKINS, Antony Gerald
Africa's age of improvement. – [Birmingham] : University of Birmingham, 1980
[2],20p. – (Inaugural lectures / University of Birmingham)
0–7044–0380–3 X.529/38141
2302195

AFRICA
to 1980

AFRICAN history and culture / edited by Richard Olaniyan. – Lagos : Longman Nigeria ; Harlow : Longman, 1982
ix,259p : ill. – Includes bibliographies and index
0–582–64369–4 X.809/52528
2607794

An AFRICAN miscellany for John Hargreaves / edited by Roy Bridges. – [Aberdeen] : Aberdeen University African Studies, 1983
x,122p : ill
B84–35722 X.525/9553
2607794

CULTURAL atlas of Africa / edited by Jocelyn Murray. – Oxford : Phaidon, c1981
240p : ill(some col.). – Bibliography: p227–229. – Includes index
0–7148–2045–8 Map Library
2607794

DAVIDSON, Basil
Africa. – [London] : Commonwealth Institute, c1984
73p : ill(some col.). – Bibliography: p70–71
0–946140–12–x X.800/40063
2607794

GENERAL history of Africa. – Edinburgh : Heinemann.
1: Methodology and African prehistory. – 1981
xxvii,819p : ill. – Bibliography: p753–800. – Includes index
0–435–94807–5 X.800/35182
2607794
2: Ancient civilizations of Africa. – 1981
xvii,804p : ill. – Bibliography: p742–785. – Includes index
0–435–94805–9 X.800/35181
2607794
– London : Heinemann, 1984. – Bibliography: p692–733. – Includes index
4: Africa from the twelfth to the sixteenth century
xxvii,751p : ill
0–435–94810–5 X.800/40037
2607794

AFRICA
to 1980 – Encyclopaedias

The CAMBRIDGE encyclopedia of Africa / general editors Roland Oliver, Michael Crowder. – Cambridge : Cambridge University Press, 1981
492p : ill(some col.). – Bibliography: p485–491. – Includes index
0–521–23096–9 X.805/2141
2607794
– [Another copy] OPL

AFRICA
to date

AFRICA past and present. – Lesotho : Mazenod Book Centre, 1979
119p : ill
r6818540 X.800/40783
8003009

The AFRICAN heritage. – Harare : Zimbabwe Educational Books, [1982]-
v.
r6813235 X.0809/2367
8003009

AFRIKANSKIĬ sbornik / [otvetstvennyĭ redaktor D.A. Ol'derogge]. – Moskva : Nauka, Glavnai͡a redakt͡sii͡a vostochnoĭ literatury, 1983
133 p. – Includes bibliographical references
r6465766 X.529/66961
8724679

AFRICA
to date

COLLANA di studi africani / Istituto italo-africano. – Roma : l'Istituto, [197–?]-
v.
r6512158 X.0800/2020
8003009

GENERAL history of Africa / [prepared by] UNESCO, International Scientific Committee for the Drafting of a General History of Africa. – London ; Paris ; Berkeley : Heinemann Educational Books : UNESCO : University of California Press, 1981-
v.
r6801429 UN.L.331/185
8003009

GUBER, Aleksandr Andreevich
Novai͡a istorii͡a stran Azii i Afriki. – Izd. 3–e perer. i dop. – Moskva : Nauka, Glavnai͡a redakt͡sii͡a vostochnoĭ literatury, 1982
559p
r6448741 X.809/53514
8003009

HISTÓRIA da África. – [Maputo?] : Frelimo, Departamento de Trabalho Ideológico, 1978
60p : ill. – (Colecção Conhecer ; 2)
r6704879 X.809/51616
8003009

KAKÉ, Ibrahima Baba
Combats pour l'histoire africaine. – Paris : Présence africaine, 1982
300p
2–7087–0400–1 X.809/56251
8003009

KANDERT, Josef
Afrika. – Praha : Mladá fronta, 1984
374p,plates : ill. – Includes bibliographies
r6442536 X.800/41500
8724679

NUSCHELER, Franz
Politische Herrschaft in Schwarzafrika. – München : Beck, 1980
199p. – (Beck'sche Schwarze Reihe ; Bd 219). – Bibliography
3–406–06019–6 X.808/34321
8003009

OSNOVNYE problemy afrikanistiki / [redkollegii͡a : I͡U.V. Bromleĭ (otvetstvennyĭ redaktor) ... i dr.]. – Moskva : Nauka, Glavnai͡a redakt͡sii͡a vostuchnoĭ liteatury, 1973
445p,[1]leaf of plates
r6403215 X.529/60893
8003009

Le SOL, la parole et l'écrit. – Paris : Société française d'histoire d'Outre–mer, 1981
2v : ill. – (Bibliothèque d'histoire d'outre–mer. Nouvelle série études 5–6). – Bibliography: t.1, p5–20
2–85970–005–6 Ac.6835m/10[5,6]
8003009

STORIA dell'Africa e del Vicino Oriente / a cura di Alessandro Triulzi ; con la collaborazione di Guido Valabrega e Anna Bozzo. – Firenze : La nuova Italia, 1979
xvii,521p. – (Il mondo contemporaneo ; 4). – Includes bibliographies
r6505045 X.800/34403
8003009

AFRICA
to date – Bibliographies

AFRIKALITERATUR in Basler Bibliotheken / Bibliothek und Dokumentation COOP Schweiz .... – Basel : Basler Afrika Bibliographien, 1982
28p. – (Mitteilungen der Basler Afrika Bibliographien = Communications from the Basel Africa Bibliography ; vol 28)
r6304695 X.809/53478
8003009

BIBLIOGRAFII͡A stran Afriki i Arabskogo Vostoka. – Moskva : Nauka, Glavnai͡a redakt͡sii͡a vostochnoĭ literatury, 1979–1980
2v
r6498211 X.809/48397
8003009

AFRICA
to date – Bibliographies – Serials

AFRIKA–STUDIECENTRUM (Leiden)
Documentatieblad. – Leiden : Afrika–Studiecentrum, 1980–.
Vol.1, no.1 (1968)-
Holdings: Vol.11, no.1 (1980)-
r6303376 P.523/931
8003009

SCHEVEN, Yvette
Bibliographies for African studies. – 1970–1975–. – Waltham, Mass. : Crossroads Press, [1977]-. – (Archival and bibliographic series)
r6803835 X.0805/59
8003009

AFRICA
to date – Catalogues – Serials

AFRICANA i nordiska afrikainstitutets bibliotek / [Nordiska afrikainstitutet]. – [Nr 1]–. – Uppsala : Institutet, [1965?]-
Holdings: Nr 42 (1981)-
r6610415 X.0805/69
8003009

AFRICA
to date – Conference proceedings

INTERNATIONALES GESPRÄCH DER GESCHICHTSDIDAKTIKER (3rd 1980 : Tutzing)
Afrika im Geschichtsunterricht europäischer Länder. – München : Minerva, 1982
233p.. – (Politik, Recht, Gesellschaft ; Bd.4)
3–597–10257–3 X.529/59723
8003009

AFRICA
to date – Serials

AKTUELLER Informationsdienst Afrika / Institut für Afrika–Kunde ... Dokumentations–Leitstelle Afrika = Institute of African Studies, Africa Documentation Center = Institut d'études africaines, Centre de documentation africaine. Sondernummer. – Hamburg : das Institut, 1979–
r6300853 X.0802/1549
8003009

[BULLETIN (African Studies Association of the West Indies)]. Bulletin / African Studies Association of the West Indies. – [No.1 (Dec. 1968)?]- – Kingston [Jamaica] : African Studies Association of the West Indies, [1968?]-
v.
Holdings: no.4 (Dec. 1971)-
r6807622 X.0809/2323
8003009

DENGA. – June 1977 – [June 1978?]. – Zomba : Department of English, Chancellor College, 1977–[1978?]
2v.
Holdings: (June 1977)-(June 1978)-
r6810144 X.0525/601
8003009

ESTUDOS afro–asiáticos / do Centro de Estudos Afro–Asiáticos do Conjunto Universitário Candido Mendes. – Ano 1, no.1 (jan–abr de 1978)-. – Rio de Janeiro : O Centro, 1978–. – (Cadernos Candido Mendes)
r6707062 P.701/1835
8003009

MAINZER Afrika Studien / Herausgeber ... Institut für Ethnologie und Afrika–Studien, Mainz. – 1(1976)-. – Wiesbaden : B. Heymann, [1976]-
r6301614 X.0950/198
8003009

OBJECTIF. – no.1 (janvier 1980)-. – [Yaoundé] : [s.n.], 1980– : ill
Holdings: no.12 (déc. janvier 1980–1981)-
r6209942 P.703/938
8003009

OUTLOOKLOOKOUT. – 1(75)–2(75). – Zomba : Department of English, Chancellor College, 1975
2v.
r6810103 X.0525/602
8003009

[RESEARCH reports (Helsinki)]. Research reports / Department of Asian and African Studies, University of Helsinki. – No.1–. – Helsinki : [The University], 1979–
Holdings: no.2 (1980)-
r6301604 X.0955/126
8724679

TRANSAFRICAN journal of history. – Vol.1, no.1 (1971)-. – Nairobi : East African Literature Bureau, 1971–
v.
Holdings: vol.5, no.1 (1976)-
r6807377 P.701/1830
8003009

VERÖFFENTLICHUNGEN aus dem Übersee–Museum Bremen. Reihe F, Bremer Afrika–Archiv. – [Bd.1]-. – Bremen : Das Museum, 1977–
r6301624 X.0705/111f
8003009

AFRICA
to date. Information sources, to date

AMARAL, Ilídio do
África. – Lisboa : Centro de Estudos Geográficos, 1983
48leaves. – (Estudos de geografia das regiões tropicais ; 14). – Includes bibliography
r6706307 X.805/5412
9073817

AFRICA
ca B.C.500–A.D.1400

OLIVER, Roland
Africa in the Iron Age, c500 BC to AD 1400. – Cambridge : Cambridge University Press, 1975
xi,228p : ill. – Bibliography: p.215–224. – Includes index
0–521–20598–0 X.429/10052
0538388

AFRICA
1000–1974 – Atlases

WILSON, Derek
A student's atlas of African history. – 2nd ed. – London : University of London Press, 1975
64p
0–340–19302–6 X.702/5012
0360171

AFRICA
1400–1800

OLIVER, Roland
The African Middle Ages 1400–1800. – Cambridge : Cambridge University Press, 1981
viii,216p. – Includes index
0–521–23301–1 X.809/48157
2086387

AFRICA
1450–

AFRICA in the nineteenth and twentieth centuries / edited by Joseph C. Anene and Godfrey N. Brown. – Repr. – Ibadan : Ibadan University Press, 1981
xvi,555p,[13]p of plates : ill
r6805442 X.809/55404
8953716

OKOTH, Assa
A history of Africa 1855–1914. – Repr. – Nairobi : Bookwise, 1980
viii,260p. – (Standard textbooks)
r6815008 X.809/60928
8953716

AFRICA
1800–1971

OLIVER, Roland
Africa since 1800. – 2nd ed, 1st paperback ed.. – Cambridge : Cambridge University Press, 1977
343p
0–521–29240–9 2090.f. Reading Room
1319787

AFRICA
1800–1972

AYANDELE, Emmanuel Ayankanmi
African historical studies. – London : Cass, 1979
xii,314p. – Includes index
0–7146–2942–1 X.809/45071
1583476

AFRICA
1800–1975 – For African students

SAGAY, J. O.
Africa, a modern history (1800–1975). – London : Evans Bros, 1978
[5],425p : ill. – Includes index
0–237–50417–0 X.709/32972
181723x

AFRICA
1800–1980

OLIVER, Roland
Africa since 1800. – 3rd ed. – Cambridge : Cambridge University Press, 1981
xii,372p. – Bibliography: p354–357. – Includes index
0–521–23485–9 X.809/48957
2502224

THATCHER, Paul
Students' notes on the history of Africa in the 19th and 20th centuries. – Harlow : Longman, 1981
138p
0–582–60362–5 X.809/49716
2502224

AFRICA
1840–1914 – For East African students

TIDY, Michael
A history of Africa 1840–1914. – London : Hodder and Stoughton.
Vol.1: 1840–1880. – 1980
188p. – Bibliography: p178–180. – Includes index
0–340–24419–4 X.0809/2164(1)
2086417
Vol.2: 1880–1914. – 1981
220p. – Bibliography: p211. – Includes index
0–340–24411–9 X.809/50080
2086417

AFRICA
1885–1960

SEARS, Mason
Years of high purpose. – Washington, D.C. : University Press of America, [1980]
xii,183p,plates
0–8191–1052–3 X.809/56089
8579423

AFRICA
1885–

GROMYKO, Anatolii
[Afrika segodnia: progress, trudnosti, perspektivy. English]. Africa today: progress difficulties, perspectives. – Moscow : Social Sciences Today, 1983
198p. – (African Studies by Soviet Scholars ; no 4). – Includes bibliography
r6465995 X.529/67818
8579431

AFRICA
1885– – Conference proceedings

AFRIKA zwischen Tradition und Fortschritt / herausgegeben von Heinz-Dietrich Ortlieb und Jürgen Zwernemann. – Hamburg : Weltarchiv, 1980
119p. – (Veröffentlichungen des HWWA-Institut für Wirtschaftsforschung Hamburg)
3–87895–198–1 X.800/31988
8579431

AFRICA
1900–1980

CAMPLING, Elizabeth
Africa in the twentieth century. – London : Batsford Academic and Educational, 1980
96p : ill. – (Twentieth century world history). – Includes index
0–7134–2492–3 X.800/15530
232329x

AFRICA
1914–1982

DAVIDSON, Basil
Modern Africa. – London : Longman, 1983
vi,234p : ill. – Bibliography: p228–229. – Includes index
0–582–65525–0 X.809/55429
2920379

AFRICA
1945–1980

COOK, Chris
African political facts since 1945. – London : Macmillan, 1983
263p. – Bibliography: p257. – Includes index
0–333–25682–4 X.800/35663
2927381
– [Another copy.] OPL

MAZRUI, Ali A. (Ali Al'Amin)
Nationalism and new states in Africa. – Nairobi ; London : Heinemann, 1984
xxix,402p. – Bibliography: p385–390. – Includes index
0–435–94145–3 X.520/35727
2927381

AFRICA
1950–1976

CLARE, Roger
The new Africa. – London : Macdonald Educational, 1977
[2],24,[1]p : ill(chiefly col). – (Macdonald Educational colour units. world topics). – Bibliography(1p.)
0–356–05420–9 X.702/5382
108948x

AFRICA
1960–

GANN, L. H. (Lewis Henry)
Africa between East and West. – Cape Town : Tafelberg, 1983
132p
0–624–01813–x X.809/60071
8003076

GROMYKO, Anatoliĭ
Afrika, progress, trudnosti, perspektivy. – Moskva : Mezhdunarodnye otnosheniia, 1981
271 p.. – Includes bibliographical references
r6403983 X.809/60860
8003076

KIRCHHERR, Eugene C.
Abyssinia to Zimbabwe: a guide to the political units of Africa in the period 1947–1978. – 3rd ed. – Athens, Ohio : Ohio University Center for International Studies, c1979
x,80p. – (Papers in international studies. Africa series ; no.25)
0–89680–100–4 X.805/3324
8003076

Abyssinia to Zona Al Sur Del Draa : a guide to the political units of Africa in the period 1950–1974. – 2nd ed. – Athens, Ohio, : Ohio University Center for International Studies, c1975
x,40p. – (Papers in international studies. Africa series ; no.25)
r6802102 X.622/14070
8003076

Abyssinia to Zona Al Sur Del Draa : a guide to the political units of Africa in the period 1950–1974. – 2nd ed. – Athens, Ohio, : Ohio University Center for International Studies, c1975
x,40p. – (Papers in international studies. Africa series ; no.25)
r6865475 X.622/14070
8003076

MESTIRI, Ezzedine
Les Cubains et L'Afrique. – Paris : Éditions Karthala, [1980]
239p : ill. – (Méridiens)
2–86537–012–7 X.809/48981
8003076

RÖSNER, Dieter
Die afrikanische Herausforderung. – München : Langen-Müller / Herbig, c1982
295p
3–7844–7103–x X.809/62738
8003076

AFRICA
1960–

TIMAMO, Chrétien
Réflexions sur la condition humaine et sur quelques problèmes actuels de l'Afrique. – Yaoundé : Éditions Le Flambeau, 1981
60p
r6211121 X.808/38966
8003076

AFRICA
1960– – Bibliographies

MITNIK, T.
Bibliography of the works by V.G. Solodovnikov, Corresponding Member of the USSR Academy of Sciences. – Moscow : USSR Academy of Sciences, Institute of Africa, 1973
122p
r6462855 X.809/58959
8003076

AFRICA
1960– – Bibliographies – Serials

DOKUMENTATIONSDIENST Afrika = Africa documentation service. Kurzbibliographie / Institut für Afrika-Kunde ... Dokumentations-Leitstelle Afrika. – Hamburg : die Dokumentations-Leitstelle, 1975–
Holdings: 1981–
r6310662 X.0805/184
8003076

AFRICA
1960– – Serials

AKTUELLER Informationsdienst Afrika. Beiheft / Institut für Afrika-Kunde ... Dokumentations-Leitstelle Afrika. – Beiheft 1–. – Hamburg : Die Leitstelle, 1981–
r6304747 X.0802/1617
8003076

JONCTION. – no 1 (juin 1980) – no 6 (janvier 1982). – Paris : Jonction, 1980–1982
Holdings: no 2, sept 1980 – no 6, janvier 1982
r6209080 X.0809/2229
8003076

JOURNAL of contemporary African studies. – Vol.1, no.1 (Oct. 1981)–. – Pretoria : Africa Institute of South Africa, 1981–
v.
r6807364 X.0529/898
8003076

POLITIQUE africaine. – v.1 (janvier 1981)–. – Paris : Éditions Karthala, 1981–
r6205782 X.0800/1635
8003076

AFRICA
1970–1979. Description & travel

L'AFRIQUE du Sahel / rédaction : Suzanne Agnely et Jean Barraud. – Paris : Librairie Larousse, [1980]
[84] : ill,maps. – (Beautés du monde)
2–03–013482–1 L.42/656
1157035

BLUM, Dieter
Africa. – Montrouge : Draeger, 1976
206p : col ill
2–85119–005–9 LR.416.s.14
1157035

ŠIRCELJ, Jože
Živa Afrika. – Ljubljana : Prešernova družba, 1977
198p : ill
r6098564 X.808/11892
1157035

AFRICA
Africa & Asia. Communism
UL'ĨANOVSKIĬ, Rostislav Aleksandrovich
[OCHERKI naťsional'no-osvoboditel'noi bor'by. English].
National liberation. - Moscow : Progress, 1978
396p
0-7147-1427-5 X.809/28123
2418614

AFRICA
African elephants. Hunting, ca 1900-ca 1920
BELL, W. D. M. (Walter Dalrymple Maitland)
The wanderings of an elephant hunter. - [New] ed. - London : Spearman, 1976
vii[i.e.viii],188p,[2]leaves of plates,[50]p of plates : ill. - Includes index
0-85435-063-2 X.320/11265 Woolwich
1479652

AFRICA
African historiography
TEMU, A. J. (Arnold James)
Historians and Africanist history. - London : Zed, c1981
xiv,187p. - (Africa series). - Bibliography: p177-183. - Includes index
0-905762-78-9 X.809/51795
2726653

AFRICA
Afro-American Christian missions, 1877-1900
WILLIAMS, Walter L.
Black Americans and the evangelization of Africa 1877-1900. - Madison : London : University of Wisconsin Press, 1982
xviii,259p : ill. - Bibliography: p237-250. - Includes index
0-299-08920-7 X.200/39733
2849771

AFRICA
Agricultural machinery. Use. Economic factors
FARM equipment innovations in eastern and central southern Africa / edited by Iftikhar Ahmed and Bill H. Kinsey. - Aldershot : Gower, c1984
xvii,345p : ill. - Includes bibliographies and index
0-566-00697-9 X.629/25852
3055353

KINSEY, B. H. (Bill H)
Economic research and farm machinery design in eastern Africa. - Norwich : School of Development Studies, University of East Anglia, 1983
20leaves. - (Discussion paper ; no.10). - Bibliography: leaves 19-20
B84-39704 X.322/14040
3055353

AFRICA
Agriculture, 1960-
AGRARNYE preobrazovaniĩa v stranakh Afriki na sovremennom étape / [redkollegiĩa : L.V. Goncharov (otvetstvennyĭ redaktor) ... dr.]. - Moskva : Nauka, Glavnaĩa redaktšiĩa vostochnoĭ literatury, 1982
227p
r6490171 X.329/19090
9125604

AFRICA
Agriculture. Economic development
BADOUIN, Robert
Les agricultures de subsistance et le développement économique. - Paris : Éditions A. Pedone, 1975
204 p.. - (Collection du Centre de Recherche, d'Étude et de Documentation sur les Instutions et les Législations Africaines ; 12). - Bibliography : p. 199-202
2-233-00011-0 X.311/4492 Woolwich
0905127
- [Another copy.] X.311/4598

JEROME, W.A.
National agrarianism. - Kampala : East African Literature Bureau, 1973-1976
3v.
r6108552 X.0808/2003
0905127

LEVI, John
Economics of African agriculture. - Harlow : Longman, 1982
vii,175p : ill. - Includes index
0-582-64147-0 X.529/47366
2718707

AFRICA
Agriculture. Economic development - Conference proceedings
AFRICAN RURAL SECTOR INTERDISCIPLINARY PERSPECTIVES (Conference : 1978 : Los Angeles)
Agricultural development in Africa. - New York : Praeger, 1980
453p
0-03-056173-6 X.322/9504 Woolwich
0905127

AFRICA
Agriculture. Productivity. Improvement - Conference proceedings
COMMONWEALTH AGRICULTURAL BUREAUX. Scientific Conference (1st : 1984 : Arusha)
Advancing agricultural production in Africa. - Slough : Commonwealth Agricultural Bureau, c1984
x,454p : ill. - Includes bibliographies
0-85198-537-8 SRL
3306836

AFRICA
Agriculture. Small farms. Stored maize. Losses. Evaluation
ADAMS, John Mervyn
The evaluation of losses in maize stored on a selection of small farms in Zambia with particular reference to the development of methodology. - London : Tropical Products Institute, 1977
xi,149p : ill. - (G109). - Bibliography: p.126-128
0-85954-058-8 BS.38X
0368644

AFRICA
Air services, 1914-1953
JONES, David, b.1926
The time shrinkers. Africa. - London : Beaumont Aviation Literature, 1977
xii,273p,leaf of plate,[24]p of plates : ill. - Bibliography: p.266-268. - Includes index
0-902230-03-4 X.529/32545
158734x

AFRICA
Air transport - Serials
AFRICAN air transport. - Vol.1, no.1 (Jan./Feb.1980)-. - Winchester : Kingswood Publications, 1980-
v. : ill
B81-33284 SRL
2544539

AFRICA
Alternative technology - Serials
POPULAR technology. - Vol.1, no.1-. - London : Wenca, 1980-
v. : ill
B81-05930 P.621/821
2349000

AFRICA
American photojournalism. Kaplan, Marion - Biographies
KAPLAN, Marion
Focus Africa. - London : Elm Tree Books, 1983, c1982
x,468p : ill. - Includes index
0-241-10602-8 X.800/35355
2901811

AFRICA
American studies, 1945- - Conference proceedings
SYMPOSIUM ON AMERICAN STUDIES IN AFRICA (1st : 1983 : Roma)
American studies in Africa. - Roma : University Press of Lesotho, 1984
133p
r6814463 X.809/63076
9160329

AFRICA
Animals
FELIX, Jiří
Animals of Africa. - London : Hamlyn, 1981
314p : col.ill. - Includes index
0-600-36488-7 SRL
2198452

GALARDI, Fernando
Savana. - [Florence] : Olimpia, 1976
183p : ill
r6065354 X.322/2606 Woolwich
1280341

HUXLEY, Elspeth
Last days in Eden. - London : Harvill, 1984
192p : col.ill. - Bibliography: p188. - Includes index
0-00-272447-2 X.950/35277
2198452

MANDLEY, Gwen
Wild creatures of Africa. - Cobham : Starfish Books, [1975]
176p : ill. - Includes index
0-900708-84-0 X.319/10294 Woolwich
0569682
Wild creatures of Africa. - London : White Lion Publishers, 1976
176p : ill. - Includes index
0-85686-164-2 X.329/10708 Woolwich
0569682

PAYSAN, Klaus
Wild animals of Africa. - Minneapolis : Lerner Publications Co., 1974
108p. - (A nature and man book)
0-8225-0563-0 L.49/428
0569682

PIA, Maria
Africa. - Bristol : Purnell, 1984
77p : col.ill. - (Purnell's animals of the world). - Includes index
0-361-06597-3 X.322/14763
0569682

AFRICA
Animals - Texts
WRIGHT, Kit
Professor Potts meets the animals in Africa. - London : Watts, c1981
[32]p : col.ill
0-85166-895-x X.992/4153 Woolwich
2374099

AFRICA
Animals. Hunting
MELLON, James
African hunter. - London : Cassell, 1975
xx,522p : ill. - Bibliography: p.505-506. - Includes index
0-304-29666-x SRL
0756342

AFRICA
Animals. Trypanosomiasis. Vectors: Tsetse flies. Control - Serials
TSETSE and trypanosomiasis information quarterly / [Tsetse and Trypanosomiasis Information and News Service]. - Vol.1, parts 1-4- ; 1978-. - London : Centre for Overseas Pest Research, 1978-
B79-17186 SRL
1884646

AFRICA
Antiquities
PRÉHISTOIRE africaine / mélanges offerts au doyen Lionel Balout ; réunis par Colette Roubet, Henri-Jean Hugot et Georges Souville ... - Paris : Éditions A.D.P.F., 1981
376p : ill. - (Recherche sur les grandes civilisations. synthèse : no.6). - Includes bibliography
2-86538-022-x X.425/4074
8726183

AFRICA
Antiquities - Conference proceedings
PHILLIPSON, D. W. (David Walter)
African archaeology. - Cambridge : Cambridge University Press, 1985
ix,234p : ill. - (Cambridge world archaeology). - Bibliography: p212-227. - Includes index
0-521-25234-2 X.421/26437
3158101

AFRICA
Antiquities - Serials
AFRIQUE ancienne / Société d'ethnographie [Paris]. - no.1 (1981)-. - [Paris] : Société d'ethnographie, 1981- : ill
r6205836 X.0622/619 Woolwich
8726183

AFRICA
Antiquities. Archaeology - Serials
The AFRICAN archaeological review. - Vol.1 (1983)-. - Cambridge : Cambridge University Press, 1983-
v. : ill
B84-38371 P.413/286
3170993

AFRICA
Archaeology - Serials
SANKOFA. - Vol.1- ; 1975-. - Legon : Legon Archaeological Society, [1975]-
ill,maps
r6072601 P.411/309
1699318

SOUTHERN AFRICAN MUSEUMS ASSOCIATION
Anthropological newsletter. - No.10- ; 1976-. - Pretoria : the Association, 1976-
r6108102 X.0415/303 Woolwich
1699318

AFRICA
Arid regions. Natural resources: Water. Evaporation
LAUER, Wilhelm
Untersuchungen zur Humidität und Andität von Afrika. - Bonn : In Kommission bei Dümmler, 1981
127p,[4]folded leaves of plates. - (Bonner geographische Abhandlungen ; Hft.66) (Dümmlerbuch ; 7616). - Bibliography
3-427-76161-4 Ac.9234.la.[Hft.66]
8796688

AFRICA
Attitudes of American negroes - Bibliographies
HELMREICH, William B.
Afro-Americans and Africa. - Westport, Conn. ; London : Greenwood Press, 1977
xxxv,74,[1]p. - (Special bibliographic series : new series / African Bibliographic Center ; no.3). - Includes index
0-8371-9439-3 X.800/26631
1333224
- [Another copy] X.800/37701

AFRICA
Attitudes of French, 1870-1900
SCHNEIDER, William H. (William Howard)
An empire for the masses. - Westport, Conn. ; London : Greenwood, 1982
xxi,222p : ill. - (Contributions in comparative colonial studies ; no.11). - Includes index
0-313-23043-9 X.800/35396
2905256

AFRICA
Babies. Home care
KAINE, Winifred
Baby care. - London : Macmillan, 1984
vii,136p,4p of plates : ill(some col.). - Includes index
0-333-36002-8 X.322/14214
3208702

AFRICA
Banking
ONOH, J. K.
Money and banking in Africa. - London : Longman, 1982
ix,214p : 1ill. - Includes bibliographies and index
0-582-64439-9 X.529/53393
2884089

AFRICA
Bantu, to date
CHIGWEDERE, A. S. (Aeneas S.)
Birth of Bantu Africa. - [Bulawayo] : Books for Africa, 1982
iii,141p
0-949933-02-3 X.809/63218
914630x

AFRICA
Bantu. Religious beliefs, to date
MULAGO GWA CIKALA M.
La religion traditionnelle des Bantu et leur vision du monde. - 2. éd, revue et corrigée. - Kinshasa : Faculté de théologie catholique, 1980
211p. - (Bibliothèque du Centre d'études des religions africaines ; 5). - Includes bibliography
r6211107 X.200/42084
9024662

AFRICA
Bibliography. Organisations - Serials
LAGOS SEMINAR ON BIBLIOGRAPHIC CONTROL AND STANDARDS
Afribiblios. - Lagos : the Seminar, 1978-
r6108893 P.701/1309
2445085

AFRICA
Big game
ROWLAND WARD (PUBLICATIONS)
Rowland Ward's records of big game. - 16th ed. (Africa). - London : Rowland Ward (Publications) Ltd, 1975
[4],537p,fold leaf, plate : ill(incl 1 col). - Includes index
0-902463-05-5 X.320/10324 Woolwich
0097535

AFRICA
Big game. Hunting
CAPSTICK, Peter Hathaway
Death in the long grass. - London : Cassell, 1978
[5],xv,297p : ill
0-304-30012-8 X.320/11440 Woolwich
1688588

ROWLAND WARD (PUBLICATIONS)
Rowland Ward's records of big game. - 16th ed. (Africa). - London : Rowland Ward (Publications) Ltd, 1975
[4],537p,fold leaf, plate : ill(incl 1 col). - Includes index
0-902463-05-5 X.320/10324 Woolwich
0097543

AFRICA
Big game. Hunting, to date
WYNNE-JONES, Aubrey
Hunting. - Johannesburg : Macmillan South Africa, 1980
180p : ill
0-86954-080-7 X.622/17966
9040307

AFRICA
Birds
AMUCHASTEGUI, Axel
Some birds and mammals of Africa. - London : Tryon Gallery, c1979
[41]p,xiv leaves of plates : col.ill
B81-36406 C.180.K.3
2554763

BROWN, Leslie, 1917-1980
The birds of Africa. - London : Academic Press. Vol.1. - 1982
xiii,521p,32p of plates : ill(some col.). - Bibliography: p479-507. - Includes index
0-12-137301-0 SRL
2803445

AFRICA
Birds. Migration
CURRY-LINDAHL, Kai
Bird migration in Africa. - London : Academic Press, 1981
2v.(xxiii,695p,[16]p of plates) : ill. - Bibliography: p619-652. - Includes index
B82-12626 SRL
2086425

AFRICA
Book industries & trades - Directories
The AFRICAN book world & press, a directory = Répertoire du livre et de la presse en Afrique / compiled by the 'African book publishing record' : Hans M. Zell, editor. - Oxford : Hans Zell (Publishers) Ltd, 1977
xxvi,[1],296,[1]p
0-905450-00-0 X.955/307
1169351
- [Another copy] 2719.ff.92

The AFRICAN book world and press / edited by Hans M. Zell = Répertoire du livre et de la presse en Afriqüe / editée par Hans M. Zell. - 2nd ed.. - London : Zell, 1980
xxiv,244p : ill
0-905450-06-x X.525/5661
1169351

The AFRICAN book world and press = Répertoire du livre et de la presse en Afrique. - 3rd ed.. - Oxford : Zell, 1983
xx,285p. - Includes index
0-905450-10-8 2725.g.293
1169351

AFRICA
Book trades, to date - Serials
PAN African book world. - Vol.1 no.1-. - Enugu : Pan African Book World, 1981-
v.
Holdings: Vol.2, no.1 (Feb. 1982)-
r6813204 P.903/950
909993x

AFRICA
Books printed before 1870. Prints - Catalogues
AFRICANA MUSEUM
Catalogue of prints in the Africana Museum and in books in the Strange Collection of Africana in the Johannesburg Public Library up to 1870. - Johannesburg : The Museum, 1975
2v.([729]p) : ill. - Includes index
0-620-01927-1 LR.425/209
1256122

AFRICA
Books with Scandinavian imprints, to date - Bibliographies
NORBERG, Viveca Halldin
Litteratur om Afrika utgiven i Norden 1975-78. - Uppsala : Nordiska afrikainstitutet, 1982
81p
r6610878 2719.k.120
8964114

AFRICA
British colonies. Colonial administration, 1870-1914
GANN, L. H. (Lewis Henry)
The rulers of British Africa, 1870-1914. - London : Croom Helm, 1978
xiii,406p : ill. - (Publications / Hoover Institution on War, Revolution and Peace). - Includes index
0-85664-771-3 X.800/28296
1341464

AFRICA
British colonies. Colonial administration, 1938-1948
PEARCE, Robert D. (Robert Desmond)
The turning point in Africa. - London : Cass, 1982
223p. - Bibliography: p211-217. - Includes index
0-7146-3160-4 X.809/52277
2750767

AFRICA
British colonies. Colonial administration. Baring, Sir Evelyn - Biographies
DOUGLAS-HOME, Charles
Evelyn Baring. - London : Collins, 1978
344p,8p of plates : ill. - Bibliography: p.336-337. - Includes index
0-00-216457-4 X.809/43393
1545647

AFRICA
British colonies. Social conditions, 1924-1945
ROSS, M. A.
The old third world. - Bishop's Castle : The author ; Shrewsbury : Distributed by Wildings, 1977
v,71p,[16]p of plates : ill
0-9506082-0-3 X.708/21091
1396323

AFRICA
British colonies. Social development. Attitudes of Britons, 1920-1940
HETHERINGTON, Penelope
British paternalism and Africa, 1920-1940. - London : Cass, 1978
xvi,196p. - Bibliography: p.159-192. - Includes index
0-7146-3051-9 X.809/42914
1353071

AFRICA
British colonies. Society. Role of European women, 1850-1950
OLIVER, Caroline, 19—
Western women in colonial Africa. - Westport, Conn. ; London : Greenwood, 1982
xv,201p. - (Contributions in comparative colonial studies ; no.12). - Includes index
0-313-23388-8 X.529/57602
2979470

AFRICA
British imperialism, 1815-1901
ROBINSON, Ronald, 1920-
Africa and the Victorians. - 2nd ed. - London : Macmillan, 1981
xxv,519p,[3]folded leaves of plates. - Includes index
0-333-31006-3 X.809/58850
3079902

AFRICA
Britons. Social conditions, 1900-1960
TALES from the Dark Continent / edited by Charles Allen ; edited in association with Helen Fry. - London : Macdonald Futura, 1980, c1979
xix,200p,[8]p of plates : ill
0-7088-1930-3 X.808/37877
2000822

AFRICA
Britons. Social conditions, 1900-1960 - Collections
TALES from the Dark Continent / edited by Charles Allen in association with Helen Fry. - London : Deutsch : British Broadcasting Corporation, 1979
xvii,166p,[16]p of plates : ill
0-233-97171-8 X.800/28246
2000822

AFRICA
Business enterprise - Serials
AFRICAN business. - [No.1]- ; Sept. 1978-. - London : IC Magazines Ltd, 1978- : ill
B78-30491 P.511/5101
1573608

MODERN Africa. - Vol.1, no.1- ; Jan./Feb. 1977-. - London : Johnston International Publishing Corp, 1977- : ill
Holdings: vol.3, no.1 ; Jan / Feb. 1979, etc
B80-20348 P.523 / 1342
1573608

AFRICA
Butterflies
CARCASSON, R. H. (Robert Herbert)
Collins handguide to the butterflies of Africa. - London : Collins, 1981, c1980
xix,188p : ill(some col.). - Includes index
0-00-219783-9 X.329/14581 Woolwich
2570920
- [Another copy] SRL

D'ABRERA, Bernard
Butterflies of the afrotropical region. - East Melbourne : Lansdowne, 1980
xx,593p : chiefly col.ill. - ([Butterflies of the world] ; [v.2]). - Bibliography: p577-582. - Includes index
0-7018-1029-7 SRL
2570920

AFRICA
Cannabis, to date

DU TOIT, Brian M
Cannabis in Africa. - Rotterdam : Balkema, 1980
xv,512p : ill. - Includes bibliography
90-619-1030-7 X.520/27617
8843414

AFRICA
Capitalism, ca 1890-ca 1980

ILIFFE, John, 1939-
The emergence of African capitalism. - London : Macmillan, 1983
ix,113p. - Bibliography: p107-109. - Includes index
0-333-33156-7 X.529/62675
3138313

AFRICA
Catholic Church. Catechists. Duties

MCGRATH, Michael
Africa. - [London] : Chapman, c1979
vi,183p : ill
0-225-66246-9 X.208/7271
1836668

AFRICA
Catholic Church. Eucharist

CATHOLIC CHURCH
[Mass]. Worship!. - London : Collins, 1981
63p : col.ill. - Includes index
0-00-599686-4 X.208/6905
2588749

AFRICA
Catholic Church. Eucharist. Assistants. Training

LOBINGER, F.
Training assistant ministers of the Eucharist. - London : Collins Liturgical, 1978
32p : ill. - (Training for community ministries ; no.2)
0-00-599617-1 X.208/8211
2860775

NORDKAMP, B.
Training sacristans. - London : Collins Liturgical, 1978
31p : ill. - (Training for community ministries ; no.3)
0-00-599618-x X.200/39753
2860775

AFRICA
Catholic Church. Funeral leaders. Training

TRAINING funeral leaders / by H. Rickert ... [et al.]. - London : Collins Liturgical, 1978
v.52p : illill. - (Training for community ministries ; no.8)
0-00-599616-3 X.200/39778
2860767

AFRICA
Catholic Church. Lay preachers. Training

LOBINGER, F.
Leading the community service. - London : Collins Liturgical, 1978
92p : ill. - (Training for community ministries ; no.7)
0-00-599615-5 X.200/39752
2860783

NORDKAMP, B.
Training for preaching. - London : Collins Liturgical, 1978
84p : ill. - (Training for community ministries ; no.6). - Bibliography: p81-83
0-00-599614-7 X.208/8212
2860783

AFRICA
Catholic Church. Liberation theology

ELA, Jean-Marc
From charity to liberation. - London : Catholic Institute for International Relations, c1984
20p. - (Third World theology)
0-946848-15-7 X.208/10138
3174999

AFRICA
Catholic Church. Married auxiliary clergy

HICKEY, Raymond
Africa. - London : Geoffrey Chapman, c1980
143p. - Bibliography: p133-143
0-225-66275-2 X.200/35774
196528x

AFRICA
Catholic Church. Public worship. Hymn leaders. Training

HIRMER, O.
The training of hymn leaders. - London : Collins Liturgical, 1978
26p : ill. - (Training for community ministries ; no.4)
0-00-599612-0 X.208/8218
2860759

AFRICA
Catholic Church. Public worship. Lay readers. Training

NORDKAMP, B.
Training readers. - London : Collins Liturgical, 1978
51p : ill. - (Training for community ministries ; no.5)
0-00-599613-9 X.529/52542
2860791

AFRICA
Catholic missions

BONZANINO, Giovanni
Cittadini d'Africa. - [Bologna] : Editrice Missionaria Italiana, 1974
208p. - (Collana "Incontri" ; 13)
r6052616 X.108/15755
1500503

Ó FIANNACHTA, Pádraig
Ón bhFuacht go dtí an Teas. - Maigh Nuad : An Sagart, 1982
76p,[8]p of plates : ill
B83-11311 X.950/18590
2846519

AFRICA
Catholic missions, to date

MEESTER DE RAVESTEIN, Paul de
Où va l'Église d'Afrique?. - Paris : Les Éditions du cerf, 1980
230p. - (Rencontres / Nouvelle serie ; 18)
2-204-01559-8 X.208/6475
800322x

AFRICA
Catholic missions. Pfanner, Franz - Biographies

BALLING, Adalbert Ludwig
Der Trommler Gottes. - Freiburg im Breisgau : Herder, 1981
350p
3-451-19185-7 X.200/37321
8791945

AFRICA
Catholics. Christian life

GREGOIRE, Nicole
Africa. - London : Chapman, 1981
196p. - Includes index
0-225-66289-2 X.208/7027
2608723

AFRICA
Cattle industries

BAKER, Randall
Perceptions of pastoralism. - Norwich : School of Development Studies, University of East Anglia, 1983
24p. - (Discussion paper ; no 3). - Bibliography: p22-24
B84-39709 X.322/14043
3174662

AFRICA
Cattle industries. Socioeconomic aspects

BAKER, Randall
"Sociological factors" in the commercialisation of cattle in Africa. - Norwich : School of Development Studies, University of East Anglia, 1983
19p. - (Discussion paper ; no.61). - Bibliography: p18-19
B85-03256 X.322/14310
321303x

AFRICA
Central & East Africa. Geographical features

CLAYTON, Robert
Central and East Africa. - Revised ed. - London : Hart-Davis, 1974
48p : ill(some col). - (Finding out about geography ; 9)
0-298-12062-3 X.809/19816
0366846

AFRICA
Central & East Africa. Insects

SWEENEY, Charles
Insect life in East and Central Africa. - London : Longman, 1976
65p : ill. - (Biology background books). - Bibliography: p.62-63. - Includes index
0-582-60635-7 X.319/16824 Woolwich
0733660

AFRICA
Central & East Africa. Luo. Cultural processes

OCHOLLA-AYAYO, A. B. C.
Traditional ideology and ethics among the southern Luo. - Uppsala : Scandinavian Institute of African Studies, 1976
248 p. : ill. - Bibliography : p. 244-248
91-7106-100-2 X.709/30754
0734918

AFRICA
Central & East Africa. Prehistoric settlements. Archaeology

HODDER, Ian
Symbols in action. - Cambridge : Cambridge University Press, 1982
x,244p : ill. - (New studies in archaeology). - Bibliography: p230-238. - Includes index
0-521-24176-6 X.421/22656 Woolwich
2635356

AFRICA
Central & East Africa. Rural regions. Social conditions compared with social conditions in urban regions - Conference proceedings

INTERNATIONAL AFRICAN SEMINAR (12th . 1972 . Lusaka)
Town and country in Central and Eastern Africa. - London . Oxford University Press [for the] International African Institute, 1975 [i.e. 1976]
ix,362p : ill. - Bibliography: p.337-358. Includes index
0-19-724199-9 X.529/30676
0645192

AFRICA
Central & Southern Africa - Serials

TRAVELLER'S guide to Central and Southern Africa. - [No.1]-. - London : IC Magazines, 1981-
v. : ill
B81-29436 X.808/34644
2530082

AFRICA
Central & Southern Africa. Commonwealth countries. Land. Surveying - Conference proceedings

The DEVELOPMENT of African land resources. - London : Commonwealth Association of Surveying and Land Economy, 1984
285p
0-903577-26-7 X.805/5816
3171728

AFRICA
Central & West Africa. Livestock. Production

SOCIÉTÉ D'ÉTUDES POUR LE DÉVELOPPEMENT ÉCONOMIQUE ET SOCIAL
Recueil statistique de la production animale. - [Paris] : Éditions du Ministère de la Coopération, [1976]
xxv,1201p.28 fold leaves of plates : ill
r6027186 SE.400/158
0393886

AFRICA
Central Africa, to 1870

BIRMINGHAM, David
Central Africa to 1870. - Cambridge : Cambridge University Press, 1981
vii,177p. - Bibliography: p159-164. - Includes index
0-521-24116-2 X.800/32525
2693054

AFRICA
Central Africa, to 1980

HISTORY of Central Africa / edited by David Birmingham and Phyllis Martin. - London : Longman, 1983
2v. : ill. - Includes bibliographies and index
B84-18986 X.809/58538
3076237

NEEDHAM, D. E. (David Edward)
From Iron Age to Independence. - New ed.. - Harlow : Longman, 1984
208p : ill. - Bibliography: p205. - Includes index
0-582-65111-5 X.800/40820
3263509

WILLS, A. J. (Alfred John)
An introduction to the history of central Africa. - 4th ed. - Oxford : Oxford University Press, 1985
xiii,556p : ill. - Bibliography: p521-533. - Includes index
0-19-873075-6 X.809/64566
3076237
- [Another copy: paperback] ISBN 0-19-873076-4 X.809/64567

AFRICA
Central Africa, to date

Les MILIEUX tropicaux d'altitude. - Talence : C.N.R.S. Centre d'Études de Géographie Tropicale, 1981
307p,plates. - (Travaux et documents de géographie tropicale ; 42)
2-222-02858-2 X.805/5191
8579377

PRIOUL, Christian
Entre Oubangui et Chari vers 1890. - Nanterre : Université de Nanterre, 1982
199p : ill. - (Recherches oubanguiennes ; 6). - Includes bibliography
2-901161-19-7 X.800/37057
8579377

AFRICA
Central Africa, to date–continued

ZOCTIZOUM, Yarisse
Histoire de la Centrafrique ... - Paris : L'Harmattan, 1983–. - (Bibliothèque du Développement)
r6211629 X.0809/2378
8579377

AFRICA
Central Africa, to date - Bibliographies

WEISS, Marianne
Nicht-konventionelle Literatur aus Zentralafrika. - Hamburg : Institut für Afrika-Kunde, 1982
63p.. - (Dokumentationsdienst Afrika. Reihe A : 19)
3-922887-23-6 X.0525/312a(19)
8579377

AFRICA
Central Africa & East Africa. Exploration by Europeans. Role of Africans, 1840–1890

SIMPSON, Donald Herbert
Dark companions. - London : Elek, 1975
xi,228p,[8]p of plates : ill. - Bibliography: p.201-207. - Includes index
0-236-40006-1 X.800/11573
0622591

AFRICA
Central Africa & East Africa. Mammals

BERE, Rennie
Mammals of East and Central Africa. - London : Longman, 1975
[5],76p : ill. - (Biology background books). - Includes index
0-582-60636-5 X.319/16346 Woolwich
0588644

AFRICA
Central Africa. Anthropology, to date

[NEW perspectives in Belgian anthropology]. New perspektives [sic] in Belgian anthropology, or, The postcolonial awakening / Rik Pinxten (ed.). - Göttingen : Edition Herodot, 1984
272p. : ill. - (Forum ; 6). - Includes bibliographies
3-88694-032-2 X.520/35060
9160280

AFRICA
Central Africa. Babinga. Social conditions

DEMESSE, Lucien
Changements techno-économiques et sociaux chez les pugmées Babinga (Nord Congo et Sud Centrafrique). - Paris : SELAF, 1978
2v. - (Langues et civilisations à tradition orale ; 26) (Études pygmées ; 1)
2-85297-012-0 P.701/1326
2436191

AFRICA
Central Africa. Butterflies

PINHEY, Elliot Charles Gordon
A guide to the butterflies of Central and Southern Africa. - London : Sir Joseph Causton, 1977
3-106p : ill(chiefly col). - Bibliography: p.104. - Includes index
0-340-22605-6 SRL
1476696

AFRICA
Central Africa. Central & Southern Africa. Rural regions. Economic conditions, 1840–1970

The ROOTS of rural poverty in Central and Southern Africa / edited by Robin Palmer, Neil Parsons. - London : Heinemann Educational, 1977
xviii,430p : ill. - Bibliography: p.422-423. - Includes index
0-435-94752-4 X.329/11236 Woolwich
1289578

AFRICA
Central Africa. Christian missions. Arnot, Frederick Stanley - Biographies

TATFORD, Frederick A. (Frederick Albert)
Frederick Stanley Arnot. - [Bath] : [Echoes of Service], 1981
[18]p : ill
B81-27571 X.208/6525
2507102

AFRICA
Central Africa. Companies, to date - Directories - Serials

The DIRECTORY of Central Africa. - 1981–. - Bulawayo : B & T Directories, 1981–
r6804693 P.P.2579Cqa
8730067

AFRICA
Central Africa. Correspondence courses

WAKATAMA, M. A. (Matthew A)
Correspondence education in central Africa. - Lanham, Md. : London : University Press of America, c1983
xiv,535p : ill. - Bibliography: p506-520
0-8191-3097-4 X.529/61805
3124266

AFRICA
Central Africa. Description & travel

SLEEN, Marc
Safari. - [London] : New English Library, 1980
224p : chiefly ill(some col.)
0-450-04445-9 X.322/9732 Woolwich
2329425

AFRICA
Central Africa. Description & travel, 1834–1836

SMITH, Andrew, 1797–1872
Andrew Smith's journal of his expedition into the interior of South Africa, 1834–36. - Cape Town : A. A. Balkema for the South African Museum, 1975
xii, 323 p., [16] p. of plates : ill. - Includes index
0-86961-069-4 X.802/10282
1033050

AFRICA
Central Africa. Description & travel, 1960–

LEVÝ, Miroslav
Od pramenu Nilu. - V Curychu : Demos, 1977
249p,[18]p of plates : ill
3-85633-001-1 X.808/35763
8798818

AFRICA
Central Africa. Description & travel, 1973–1974

EMERSON, P. J.
Inflation? try a bicycle. - Belfast : Northern Whig Limited ; Belfast : [Distributed by] the author, 1978
xv,147,[1]p,[2]leaves of plates(1 fold) : ill
0-9506028-0-9 X.709/32131
140914x

AFRICA
Central Africa. Economic conditions, 1960– - Serials

COMMUNAUTÉ ÉCONOMIQUE DES PAYS DES GRANDS LACS
Journal officiel. - no.1 (1980)–. - Gisenyi : [la Communauté], 1980–
Holdings: no.3 (avril 1981)–
r6207463 S.V.890/6
8860149

ÉTUDES et statistiques / Banque des états de l'Afrique centrale. - no 1 (1974)–. - Yaoundé : Banque des états del'Afrique centrale, [1974]–
Holdings: no.55, octobre 1978–
r6206858 X.0802/1605
8860149

GRANDS lacs. - no.1 (1980)–. - Gisenyi : C.E.P.G.L., 1980– : ill
Holdings: no.3 (mars 1981)–
r6207464 S.V.890/5
8860149

AFRICA
Central Africa. Geographical features

MICHIE, W. D.
The lands and peoples of Central Africa. - New ed. - London : Longman, 1981
iv,202p : ill. - Includes index
0-582-60380-3 X.800/30195
2388057

Lands and peoples of Central Africa. - Complete rev. ed.. - Harlow : Longman, 1983
vi,204p : ill. - Includes index
0-582-65548-x X.800/38282
2388057

AFRICA
Central Africa. Gestures

HOCHEGGER, Hermann
Le langage gestuel en Afrique centrale. - Bandundu : Ceeba Publications, [1978]
x,226p : ill. - (CEEBA Publications. série 2, mémoires et monographies ; v.47). - Bibliography
r6207287 X.520/27966
8863873

AFRICA
Central Africa. Wildlife. Conservation

MARKS, Stuart A. (Stuart Alexander)
The imperial lion. - Epping : Bowker, 1984
xv,196p : ill. - Bibliography: p167-177. - Includes index
0-86531-818-2 X.329/19035
322046x

AFRICA
Central Africa. Yao, to 1919 - Translations

ABDALLAH, Yohanna Barnaba
The Yaos = Chiikala cha Wayao. - 2nd ed. [i.e. 1st ed. reprinted]. - London : Cass, 1973
xvi,3-60p,3-60p. - (Cass library of African studies. missionary researches and travels ; no.25)
0-7146-2462-4 X.809/40439
0535362

AFRICA
Cerambycidae. Larvae

DUFFY, E. A. J. (Evelyn Arthur Joseph)
A monograph of the immature stages of African timber beetles (cerambycidae).
Supplement. - Farnham Royal : Commonwealth Institute of Entomology, 1980
186p,[4]p of plates : ill. - Bibliography: p151-163. - Includes index
0-85198-473-8 SRL
2822520

AFRICA
Chemical industries - Directories

CHEMICAL DATA SERVICES
Chemical company profiles Africa, Asia, Australasia. - London : Chemical Data Services, IPC Industrial Press, 1978
166p. - Includes index
B78-18082 SRL
0667226

ECN CHEMICAL DATA SERVICES
The chemical industry of Africa & Asia. - [1975]. - [London] : I.P.C. Business Press, 1975
viii,204p,[6]leaves
Subsequent volumes, when received, will be placed at this shelfmark
0-617-00156-1 P.623/818
0667226

AFRICA
Chemical technology - For teaching

STEWARD, J. W. (John William)
Chemical technology in Africa. - Cambridge : Cambridge University Press, 1984
vii,176p : ill. - Bibliography: p170-173. - Includes index
0-521-31927-7 X.622/22778
3050297

AFRICA
Children. Health - Conference proceedings

DAG HAMMARSKJÖLD SEMINAR ON THE DILEMMA OF QUALITY, QUANTITY AND COST IN AFRICAN CHILD CARE (1973 : Addis Ababa)
Action for children. - Uppsala : Dag Hammarskjöld Foundation, 1975
238p : ill. - Includes bibliographies
91-85214-04-3 X.329/11162 Woolwich
1258192

AFRICA
Children. Influence of environment

The AFRICAN child and his environment / editors R. Ogbonna Ohuche, Barnabas Otaala for the Science Education Programme for Africa (SEPA), with the assistance of the United Nations Environment Programme (UNEP). - Oxford : Published for the United Nations Environment Programme by Pergamon, 1981
vi,97p,[4]p of plates : ill. - (UNEP studies ; v.3). - Bibliography: p88-97
0-08-025671-6 X.529/40834
1958224

AFRICA
Christian church

ANDREW, Brother
Battle for Africa. - London : Lakeland, 1978
5-156p
0-551-00796-6 WPD.404/356
1338773

ELA, Jean Marc
Le cri de l'homme africain. - Paris : Librairie-Éditions l'Harmattan, [1980]
173p
2-85802-145-7 X.200/34747
1338773

AFRICA
Christian church - Texts

CHAPMAN, Lois
Kidnapped by the dancers. - Gerrards Cross : W.E.C. Youth Crusade, 1965 [i.e. 1978]
15p : ill. - (Camp fire series ; no.2)
B79-18082 X.108/26999
1891286

AFRICA
Christian church, to 1975

HELLESKOV, Børge
Punktlys pa Afrika. - Christiansfeld : Savanne, 1976
120p : ill
87-85190-50-0 X.108/15484
1384228

AFRICA
Christian church, to date

HILDEBRANDT, Jonathan
History of the church in Africa. - Achimota, Ghana : Africa Christian Press, 1981
xi,269p,plates. - Includes bibliography
0-85352-320-7 X.20[illegible]
8579474

PHEKO, S. E. M.
The early church in Africa (1-7 century) and today. - Lusaka : Multimedia, 1982
141p
r6813352 X.200/43493
8579474

AFRICA
Christian church, 1960- - Conference proceedings

The CHURCH in Africa 1977 / Charles R. Taber, editor. - South Pasadena : William Carey Library, c1978
xii,212p. - Bibliography
0-87808-161-5 X.200/40578
8579466

AFRICA
Christian church. Preaching, to date

REISACH, Christian
Das Wort und seine Macht in Afrika. - Münsterschwarzach : Vier-Türme-Verlag, 1981
xx,450p. : ill. - (Münsterschwarzacher Studien ; Bd.34)
3-87868-138-0 X.200/39334
8891680

AFRICA
Christian church. Relations with Islam - Serials

BULLETIN on Islam and Christian-Muslim relations in Africa. - Vol.1, no.1 (Jan. 1983)-. - Birmingham : Centre for the Study of Islam and Christian-Muslim Relations, 1983-
v.
B83-42558 P.201/638
2977400

AFRICA
Christian church. Sermons, to date - Texts

M'PASSOU, Denis
Stories with a sting. - Blantyre, Malaŵi : Claius, 1973
25p. - (Voices from Africa series)
r6202079 X.208/8711
8982023

AFRICA
Christian mission schools, to date

ERNY, Pierre
Écoles d'église en Afrique noire. - Immensee : Nouvelle revue de science missionnaire, 1982
191p. - (Nouvelle revue de science missionnaire. Supplementa ; v.30). - Includes bibliography
3-85824-059-1 P.101/449a[vol.30]
8934215

AFRICA
Christian missions - Forecasts

BÜHLMANN, Walbert
The missions on trial. - Slough : St Paul Publications, 1978
160p. - Bibliography: p.159-160
0-85439-152-5 X.100/17964
1695142

AFRICA
Christian missions, 1913-1931

STUDD, Charles Thomas
Fool and fanatic?. - Gerrards Cross : Worldwide Evangelization Crusade, 1980
128p : ill
B80-26486 X.208/5538
211593x

AFRICA
Christian missions, 1960-

PERRIN JASSY, Marie-France
Face au changement. - Bandundu : Ceeba, [1981]
164p. - (Ceeba publications. série 1 ; v.8)
r6206766 X.200/38155
8003254

AFRICA
Christian missions, 1960- - Serials

SOCIÉTÉ INTERNATIONALE MISSIONAIRE
SIM Actualités. - No.1 (1982)-. - [Lausanne] : [La Société], [1982-]
v.
r6212629 P.105/209
8003254

AFRICA
Christian missions. Comboni, Daniele

CHIOCCHETTA, Pietro
Daniele Comboni. - Bologna : Editrice missionaria italiana, 1978
335p. - (Studi comboniani ; 2)
r6504686 X.200/38103
890183x

AFRICA
Christian missions. Missionaries

OKWUOSA, Vincent Esimai Akubueze
In the name of Christianity. - Philadelphia : Dorrance, c1977
50p
0-8059-2420-5 X.200/32935
2041715

UOMINI per l'Africa / Bertulli Cesare ... [et al.] ; a cura di Antonio Scarin. - Milano : Istituto propaganda libraria, 1976
112p.,[8]p.of plates : ill. - (Biblioteca di pastorale missionaria)
r6100434 X.100/28295
2041715

AFRICA
Christian religious education

MEGILL, Esther L.
Education in the African church. - Rev. ed. - London : Geoffrey Chapman, 1981
222p. - Bibliography: p213-222
0-225-66280-9 X.200/35023
1934392

AFRICA
Christian sects. Durojaiye, J. O.. Influence of Cerullo, Morris

DUROJAIYE, J.O.
Morris, Joshua and the Holy Spirit in Nigeria. - Ibadan : 20th Century World Evangelism, [1979?]
63p.
r6108410 X.208/5244
2418622

AFRICA
Christian theology

DICKSON, Kwesi A. (Kwesi Abotsiq)
Theology in Africa. - London : Darton, Longman and Todd, 1984
ix,243p. - Bibliography: p229-237. - Includes index
0-232-51551-4 X.529/70033
3273822

AFRICA
Christianity

DONDERS, Joseph G.
Non-bourgeois theology. - Maryknoll, N.Y. : Orbis, c1985
vii,200p
0-88344-352-x X.200/46810
1730940

AFRICA
Christianity, to date

NTETEM, Marc
Die negro-afrikanische Stammesinitiation. - Münsterschwarzach : Vier-Türme-Verlag, 1983
xlvi,381p. - (Münsterschwarzacher Studien ; Bd 36). - Includes bibliography
3-87868-161-5 X.200/41798
9017666

AFRICA
Christianity, 1960-

COMBATS pour un Christianisme africain / édité par A. Ngindu Mushete [et al.]. - Kinshasa : Faculté de théologie Catholique, 1981
324p. - (Bibliothèque du Centre d'études des religions africaines ; 6). - Includes bibliography
r6210641 X.200/42290
8579490

AFRICA
Christianity. Attitudes of society, to date

EBOUSSI BOULAGA, F.
Christianisme sans fétiche. - Paris : Présence Africaine, 1981
219p. - (Situations et perspectives)
r6210304 X.200/39258
8925062

AFRICA
Civil service. Professional personnel & executives. Remuneration. Effects of remuneration of British professional personnel & executives

BENNELL, Paul
African pay structures in a transnational context. - Brighton : Institute of Development Studies, 1980
xvi,309columns : ill. - (IDS research reports)
0-903354-70-5 L.49/855
249602x

AFRICA
Civil wars, 1960-

WIRZ, Albert
Krieg in Afrika. - Wiesbaden : Steiner, 1982
x,640p. - (Beitrage zur Kolonial- und Uberseegeschichte ; Bd 23). - Bibliography: p.602-634
3-515-03752-7 X.0700/119(23)
8977038

AFRICA
Coastal regions, to 1984

AFRICA and the sea / edited by Jeffrey C. Stone. - Aberdeen : Aberdeen University African Studies Group, 1985
412 columns : ill
0-9509989-0-7 L.49/3302
3236641

AFRICA
Coastal waters. Big game fish. Angling

GODDARD, John
Big fish from salt water. - London : E. Benn, 1977
xi,244p : ill. - Bibliography: p.238. - Includes index
0-510-22504-7 X.320/10795 Woolwich
1144987

AFRICA
Colonial administration, 1830-1962

COLONIAL rule in Africa / edited with an introduction by Bruce Fetter. - Madison ; London : University of Wisconsin Press, 1979
xiii,223,[1]p. - Bibliography: p.iv-vi. - Includes index
0-299-07780-2 X.809/45715
2160250
- [Another copy: paperback]
ISBN 0-299-07784-5 X.700/26542

AFRICA
Colonial administration, 1870-1960

COLONIALISM in Africa 1870-1960 / edited by L.H. Gann and Peter Duignan. - Cambridge : Cambridge University Press, 1969
2v. : ill. - Includes index
B81-36238 X.0809/2176
0039845

AFRICA
Colonial administration, 1885-1960

MEKER, Maurice
Le temps colonial. - Dakar : Les Nouvelles Éditions Africaines, 1980
263p,plates
2-7236-0207-9 X.520/33448
8579512

AFRICA
Colonial administration, ca 1945

NKRUMAH, Kwame
Towards colonial freedom. - London : Panaf, 1973, c1962
xviii,45p
0-901787-20-5 X.808/38075
2902621

AFRICA
Colonial administration by Germany. Role of Dernbury, Bernhard, 1906-1918

SCHIEFEL, Werner
Bernhard Dernburg, 1865-1937. - Zürich & Freiburg im Breisgau : Atlantis, 1974
277 p.. - (Beiträge zur Kolonial- und Überseegeschichte ; Bd. 11). - Bibliography : p. 254-271
3-7611-0445-6 X.0700/119(11)
0900923
- [Another copy] X:700/18122

AFRICA
Colonial administration. Effects of Arab nationalist movements, to 1973 - Conference proceedings

SETTLER regimes in Africa and the Arab world / edited by Ibrahim Abu-Lughod, Baha Abu-Laban. - Wilmette : Medina University Press International, 1974
x,255p. - (AAUG monograph series ; no.4). - Includes index
0-914456-06-7 X.800/27800
1893092

AFRICA
Colonial administration. Governors, to 1976 - Biographies

AFRICAN proconsuls / L.H. Gann and Peter Duignan, editors - New York : Free Press ; London : Collier Macmillan [etc.], 1978
xi,548p.[8]p of plates : ill. - Includes index
0-02-911190-0 X.802/10892
1902156

AFRICA
Colonies Portuguese. Independence movements, 1880-1975

LIBERATION movements in Lusophone Africa / [produced by the] Center for Research Libraries. Cooperative Africana Microforms Project. - Chicago : CAMP, 1976-1977
3microfilms
r6703040 Mic.A.6566-6568
8721165

AFRICA
Colonies. Alcoholic drinks trades. Control, to 1974

PAN, Lynn
Alcohol in colonial Africa. - Helsinki : Finnish Foundation for Alcohol Studies ; Uppsala : Scandinavian Institute of African Studies, 1975
121 p.. - (Finnish Foundation for Alcohol Studies. [Publications] ; vol. 22). - Bibliography : p. 112-119
951-91912-1-6 X.311/4526 Woolwich
0920649

AFRICA
Colonies. Independence, 1938-1968

BOATENG, Ernest Amano
Independence and nation building in Africa. - Tema : Ghana Publishing, 1973
xii,56p,plate. - Bibliography : p.51-52
r6091799 X.519/29707
1891707

AFRICA
Colonisation

FANON, Frantz
Toward the African revolution. - London : Writers and Readers, 1980, c1967
x,197p
0-904613-97-6 X.808/35206
2583399

AFRICA
Colonisation, 1450-1960 - Conference proceedings

SCIENCES de l'homme et conquête coloniale / communications [d'un colloque consacré aux élaborations et applications politiques des sciences humaines en Afrique au XIX siècle, tenu à l'École normale supérieure, 1977] réunies et présentées par Daniel Nordman et Jean-Pierre Raison. - Paris : Presses de l'École normale supérieure, 1980
238p : ill
2-7288-0066-9 X.520/25057
8781575

AFRICA
Colonisation, 1870-1900

STRAGE, Mark
Cape to Cairo. - Harmondsworth : Penguin, 1977
397p. - Includes index
0-14-004182-6 X.708/20852
109856x

AFRICA
Colonisation by Ancient Romans, to 200 A.D.

KOLENDO, Jerzy
Le colonat en Afrique sous le Haut-Empire. - Paris : Les Belles Lettres, 1976
116p. - (Annales littéraires de l'Université de Besançon ; 177) (Centre de recherches d'histoire ancienne ; vol.17)
r6012585 Ac.282/6[no.177]
1025279

AFRICA
Colonisation by Europe, ca 1850- ca 1910

MACKENZIE, John M. (John MacDonald)
The partition of Africa. - London : Methuen, 1983
x,48p. - (Lancaster pamphlets). - Bibliography: p47-48
0-416-35050-x X.809/55487
0153079

UZOIGWE, G.N
Britain and the conquest of Africa. - Ann Arbor : University of Michigan Press, [1974]
xii,403p.. - Bibliography : p.371-391
0-472-08920-x X.802/3914
0153079

AFRICA
Colonisation by European countries, 1875-1900. International political aspects

GANIAGE, Jean
L'expansion coloniale et les rivalités internationales. - Paris : Centre de Documentation Universitaire, 1975-
tom. 1-. - ("Les cours de Sorbonne")
r6012505 X.0709/817
0945609

AFRICA
Colonisation by European countries, 1880-1920

JACOBS, Sylvia M.
The African nexus - Westport, Conn. ; London : Greenwood Press, 1981
xiv,311p : ill. - (Contributions in Afro-American and African studies : no.55). - Bibliography: p275-299. - Includes index
0-313-22312-2 X.529/44550
2585782

AFRICA
Commercial law. Harmonisation - Translations - Conference proceedings

L'HARMONISATION du droit en Afrique = The harmonisation of African law = L'armonizzazione del diritto in Africa. - Milano : Giuffè, 1974
296p
r6029617 X.100/15578
0391441

AFRICA
Commonwealth countries. Foreign assistance. Organisations - Directories

COMMONWEALTH African directory of aid agencies / consultant editor Michael Collins. - London : Commonwealth Foundation, 1979
[1],125p. - Bibliography: p.115-116. - Includes index
0-903850-16-8 X.529/37489
2255464

AFRICA
Commonwealth countries. Government. Role of judiciary

NWABUEZE, B. O. (Benjamin Obi)
Judicialism in Commonwealth Africa. - London : C. Hurst [etc.], 1977
xii,324p. - Includes index
0-903983-56-7 X.200/31033
1098470

AFRICA
Community health services

ADEGOROYE, Anu
Community health care. - London : Macmillan, 1984, c1983
vi,140p : ill. - (Macmillan tropical nursing and health sciences series). - Bibliography: p135-137. - Includes index
0-333-28623-5 X.322/14581
3226204

AFRICA
Companies. Taxation

CONFEDERATION OF BRITISH INDUSTRY
Taxation in the Middle East, Africa & Asia .... - London : C.B.I., 1977
135p. - Bibliography: p.122-123. - Includes index
B78-00768 X.102/2100
1277332

AFRICA
Computer systems. Applications

The COMPUTER and Africa / edited by D.R.F. Taylor, R.A. Obudho. - New York ; London : Praeger, 1977
xxiii,341p : ill. - (Praeger special studies in international economics and development). - Bibliography: p.309-332. - Includes index
0-275-56820-2 X.620/17083 Woolwich
1193465

AFRICA
Conference proceedings, to date - Indexes

The AFRICANA conference paper index / compiled and published by the Melville J. Herskovits Library of African Studies, Northwestern University. - Boston, Mass. : Hall, 1982
2v.
0-8161-1309-2 X.525/8004
8995540

AFRICA
Correspondence courses. Students

HARRIS, W. J. A. (William Jack Augustus)
The home study adult in Africa. - Manchester : Department of Adult and Higher Education, University of Manchester, 1978
[1],iii,44p. - (Manchester monographs ; 11). - Bibliography: p.43-44
B79-03513 X.0519/387(11)
1766899

AFRICA
Crimes by young persons, 1960-

LEUNMI, Benjamin
Der Kulturkonflikt und die Jugendkriminalität in Afrika. - Bremen : Übersee-Museum, 1978
vi,123p. - (Veröffentlichungen aus dem Übersee-Museum Bremen. Reihe F, Bremer Afrika-Archiv ; Bd.5)
3-88299-013-9 X.0705/111f(5)
8753741

AFRICA
Crops, to 1975

ORIGINS of African plant domestication / editors Jack R. Harlan, Jan M.J. De Wet, Ann B.L. Stemler. - The Hague : Mouton, 1976
xiii,498p,leaf of plate,[4]p of plates : ill. - (World anthropology). - Bibliography- Includes index
r6051227 X.320/11710 Woolwich
1890735

AFRICA
Crops: Cassava. Bacterial blight - Conference proceedings

La BACTÉRIOSE du manioc en Afrique / éditeurs Eugene R. Terry, Gabrielle J. Persley et Sheila C.A. Cook. - London : Centre for Overseas Pest Research, 1979
vii,72p : ill. - Includes bibliographies
0-85135-093-3 BS.251/241
1945351

CASSAVA bacterial blight in Africa, past, present and future : report of an Interdisciplinary Workshop held at IITA, Ibadan, Nigeria, 26-30 June 1978 cosponsored by the International Development Research Centre, Ottawa, Canada and the International Institute of Tropical Agriculture, Ibadan, Nigeria / editors Eugene R. Terry, Gabrielle J. Persley and Sheila C.A. Cook. - London : Centre for Overseas Pest Research, 1979
vii,71p : ill. - Includes bibliographies
0-85135-092-5 SRL
1945343

AFRICA
Cultural processes

BESHIR, Mohamed Omer
Diversity, regionalism and national unity. - Uppsala : Scandinavian Institute of African Studies, 1979
50p. - (Nordiska afrikainstitutet. Research report ; no 54)
91-7106-166-5 P.803/212(54)
0608084

ONWUEJEOGWU, M. Angulu
The social anthropology of Africa. - London : Heinemann Educational, 1975 [i.e. 1976]
xxiv,296p : ill. - Includes bibliographies and index
0-435-89700-4 X.700/25091
0608084

TURNBULL, Colin Macmillan
Man in Africa. - Newton Abbot : David and Charles, 1976
ii-xxi,313p,[8]p of plates : ill. - Includes index
0-7153-7348-x X.809/41551
0608084

Man in Africa. - Harmondsworth : Penguin, 1978
272p,[8]p of plates : ill. - (Pelican books). - Includes index
0-14-022035-6 X.519/28427
0608084

AFRICA
Cultural processes - Serials

The PAN-AFRICANIST. - no.1- ; [1971?]-. - Evanston, Ill. : Northwestern University, [1971?]
Holdings : no.8- ; July 1979-
r6109278 P.701/1249
0608084

UNIVERSITY OF NIGERIA. Institute of African Studies
Ikoro. - Vol.1, no.1 (1974?)-. - Nsukka : The Institute, [1974?]-
Holdings: Vol.3, no.1 (Jan. 1976)-
r6202387 P.521/2394
0608084

AFRICA
Cultural processes, to date

COMOÉ KROU, Barthélemy
Les morts qui ne veulent pas mourir. - [Abidjan] : Nouvelles Éditions Africaines, 1982
79p
2-7236-0553-1 X.809/64324
8719519

DIOP, Anta, Cheikh
Civilisation ou barbarie. - Paris : Présence Africaire, 1981
526p : ill. - Bibliography
2-7087-0394-3 X.529/53443
8719519

AFRICA
Cultural processes, to date–continued

ZWERNEMANN, Jürgen
Culture history and African anthropology. – Uppsala : Uppsala University, 1983
170p. – (Uppsala studies in cultural anthropology ; 6). – Includes bibliography
91–554–1493–1 X.0520/566(6)
8719519

AFRICA
Cultural processes, to date – Serials

MAINZER ethnologische Arbeiten. – Bd 1–. – Berlin : Reimer, 1980–
v.
r6302082 X:0529/624
8719519

AFRICA
Decolonisation

NKRUMAH, Kwame
Consciencism. – London : Panaf, 1964
vi,122p. – Includes index
0–901787–11–6 X.529/54381
2906171

AFRICA
Decolonisation, to 1975

BONNET, Jean
Les drames de la décolonisation, 1900–1975. – Paris : Éditions Roblot, 1975
141 p.. – (Histoire du monde). – Bibliography : p. 135–141
r6012048 X.709/20157
0933201

AFRICA
Decolonisation, 1919–1963

GRIMAL, Henri
Decolonization. – London : Routledge and Kegan Paul, 1978
xii,443p. – Includes bibliographies and index
0–7100–8709–8 X.809/43381
1122274

AFRICA
Decolonisation, 1940–1960 – Conference proceedings

The TRANSFER of power in Africa / edited by Prosser Gifford and Wm. Roger Louis. – New Haven ; London : Yale University Press, c1982
xi,654p. – Bibliography: p603–635. – Includes index
0–300–02568–8 X.800/35421
2898632

AFRICA
Decolonisation, 1945–1979

DECOLONIZATION and dependency / edited by Aguibou Y. Yansané. – Westport ; London : Greenwood Press, 1980
xix,321p. – (Contributions in Afro-American and African studies ; no.48). – Bibliography: p291–304. – Includes index
0–313–20873–5 X.520/26349
2643464

AFRICA
Decolonisation. Attitudes of Christian Church, to date

LAY, Michael van
Kirche im Entkolonisierungskonflikt. – München : Kaiser, 1981
399p.. – (Entwicklung und Frieden ; 23). – Includes bibliography
3–459–01358–3 X.208/9097
9017755

AFRICA
Description & travel

GORDON, René
Africa. – Richmond upon Thames : Country Life, 1980
280p : col.ill. – Bibliography: p280
0–600–36782–7 L.45/813
2335689

HILLS, Denis
The rock of the wind. – London : Deutsch, 1984
205p
0–233–97598–5 X.809/60550
2335689

NICHOLSON, Michael, 1937–
Across the Limpopo. – London : Robson, c1985
219p,[16]p of plates : ill
0–86051–304–1 X.809/64289
3112810

SMITH, Anthony, 1926–
Smith & son. – London : Hodder and Stoughton, 1984
271p : ill
0–340–28733–0 X.800/39942
3112810

AFRICA
Description & travel, 1325–1354

IBN BAṬŪṬAH, Muḥammad ibn ʻAbd Allāh
[Tuḥfat an-nuẓẓār fī gharāʼib al-amṣār wa-ʻajāʼib al-asfār. English]. Travels in Asia and Africa, 1325–1354. – London : Routledge & Kegan Paul, 1983
vii,398p,[8]p of plates : ill. – Includes index
0–7100–9568–6 X.809/56310
2959119

AFRICA
Description & travel, 1885–1960

LEIRIS, Michel
L'Afrique fantôme. – Paris : Gallimard, 1981
536p,plates. – (Bibliothèque des Sciences Humaines)
r6211608 X.950/31057
8754624

SABATÉS, Fabien
1924–1925, la croisière noire Citroën. – [Paris] : Eric Baschet, 1980
146p : ill
2–86309–011–9 X.805/1953
8754624

AFRICA
Description & travel, 1940

FARSON, Negley
Behind God's back. – Feltham : Zenith, 1983, c1940
400p,[16]p of plates : ill
0–600–20739–0 X.808/38782
2989867

AFRICA
Description & travel, 1960–

BELTRANI, Pietro
Parlammo di Cézanne sul Nilo bianco. – Milano : L'erba voglio, 1980
101p
r6506568 X.808/37858
8916373

MORAVIA, Alberto
Lettere dal Sahara. – Milano : Bompiani, 1981
204p. – (Opere di Alberto Moravia)
r6509713 X.809/56534
8916373

AFRICA
Description & travel, 1970–1979

CUNHA, Breno Duarte da
Do Tâmisa ao Vale do Pó. – Recife : Editora Universitária, Universidade Federal de Pernambuco, 1977
111p
r6099817 X.709/28732
203560X

OLIVER, Dorothy
Four wheels across Africa. – St. Ives, Cornwall : United Writers, c1980
156p,[9]p of plates : ill
0–901976–59–8 X.809/51477
203560x

AFRICA
Description & travel, 1971–1972

TOPOLSKI, Daniel
Muzungu. – London : Arlington Books, 1976
240p,[32]p of plates : ill
0–85140–250–x X.800/25758
0766380

AFRICA
Description & travel, 1976

FINLAY, Iain
Africa overland. – London : Angus and Robertson, 1977
x,264p,[16]p of plates : col ill
B78–24726 X.800/26955
1492861

Africa overland. – London : Angus and Robertson, [1978]
x,264p,[16]p of plates : col ill
0–207–13510–x X.800/26860
1492861

AFRICA
Desertification, to date

CALDWELL, John C. (John Charles)
Desertification. – [Canberra] : Australian National University. Development Studies Centre, 1984
50p. – (Occasional paper / Australian National University ; no.37). – Includes bibliography
0–86784–489–2 X.805/6659
9097090

AFRICA
Desertification, to date – Bibliographies

LENG, Gunter
Desertification. – Bremen : Schwerpunkt Geographie, Fachbereich 1, Universität Bremen, 1982
xviii,177p.. – (Bremer Beiträge zur Geographie und Raumplanung ; Hft.4)
3–88722–075–7 X.0802/1548(4)
9097090

AFRICA
Developing countries. Capital. Accumulation

POL'SHIKOV, P. I. (Petr Ivanovich)
Capital accumulation and economic growth in developing Africa. – Moscow : Progress, c1981
206p. – (Problems of the developing countries)
0–7147–1738–x X.529/48520
2755645

AFRICA
Developing countries. Economic conditions, to date

GURA, V. K. (Viktor Konstantinovich)
Problemy i puti razvitii͡a osvobodivshikhsi͡a stran Afriki. – Kiev : Naukova dumka, 1981
190p
r6485880 X.529/48498
8840482

AFRICA
Developing countries. Economic conditions, to date – Bibliographies – Serials

BIBLIOGRAPHIES occasionnelles / Centre africain de formation et de recherche administratives pour le développement. – no 1(1974)–. – Tanger : Centre africain de formation et de recherche administratives pour le développement, 1974–
r6206866 X.0525/549
8840482

AFRICA
Developing countries. Economic development

PAN AFRICAN INSTITUTE FOR DEVELOPMENT
Working papers. – No.1–. – Buea : the Institute, 1978–
r6105912 P.513/1238
086434X

AFRICA
Developing countries. Economic development – Bibliographies – Serials

[INFORMATION bulletin (Centre africain de formation et de recherche administratives pour le développement, Centre de documentation)]. Information bulletin / Centre africain de formation et de recherche administratives pour le développement, Centre de documentation. – Vol.1 (1973)–. – Tangier : Centre africain de formation et de recherche administratives pour le développement. Centre de Documentation, [1973]–
Holdings: vol.1, no.2. [1973]–
r6208889 X.0525/535
086434x

AFRICA
Developing countries. Economic development – Conference proceedings

COLLOQUE INTERNATIONAL "ENTREPRISES ET ENTREPRENEURS EN AFRIQUE XIXE ET XXE SIÈCLES (1981)
Actes du Colloque Entreprises et entrepreneurs en Afrique (XIXe et XXe siècles). – Paris : Éditions L'Harmattan, 1983–. – (Racines du présent)
r6211693 X.0529/959
086434x

CONFERENCE ON THE IMPLEMENTATION PROBLEMS OF ECONOMIC DEVELOPMENT PLANS AND GOVERNMENT DECISIONS IN THE COUNTRIES OF BLACK AFRICA (1969 : Budapest)
Proceedings of the Conference on the Implementation Problems of Economic Development Plans and Government Decisions in the Countries of Black Africa. – Budapest, 1971
3v. – (Center for Afro-Asian Research of the Hungarian Academy of Sciences. Studies on Developing Countries ; no.50)
r6033243 X.510/9481
086434X

AFRICA
Developing countries. Economic development – Serials

AFRICA development ; Afrique et développement. – Vol.1 (1976)–. – Dakar : CODESRIA, [1976]–
Holdings: v.3, no.4, 1978–
r6211814 X.0520/887
086434x

AFRICA
Developing countries. Economic development. Research projects sponsored by IFC-Institut für Wirtschaftsforschung. München. Afrika-Studienstelle: Africa research programme

IFO-INSTITUT FÜR WIRTSCHAFTSFORSCHUNG
Developing countries : research and policies. - München : Ifo - Institut : Weltforum, 1974
vii, 117 p.. - (Ifo - Institut für Wirtschaftsforschung. Forschungsberichte der Afrika - Studienstelle ; 41). - Bibliography : p. 105-117
r6002180 X.512/2427
0850128

AFRICA
Developing countries. Food supply. Foreign assistance by developed countries

BLOGG, Keith
Family London. - London : London Transport, 1982
63p : col.ill. - (A Visitor's London guide)
0-85329-112-8 X.808/37154
2225522

AFRICA
Developing countries. Food supply. Foreign assistance by developed countries

STEVENS, Christopher, 1948-
L'aide alimentaire et le monde en developpement. - London : Overseas Development Institute, [1980]
245p : ill. - Bibliography: p.238-245
B80-27983 X.529/36814
2225522

Food and the developing world. - London : Croom Helm [for] the Overseas Development Institute, 1979
224p : ill. - Bibliography: p.212-219. - Includes index
0-85664-622-9 X.529/33276
1757741

AFRICA
Developing countries. Mental health services - Conference proceedings

CONFERENCE ON THE FUTURE OF MENTAL HEALTH SERVICES (1979 : Nairobi)
The future of mental health services. - Amsterdam ; Oxford : Excerpta Medica, 1980
xxii,257p : ill. - (International congress series ; no. 504). - Bib. - Includes index
0-444-90127-2 X.322/9530 Woolwich
2231956

AFRICA
Developing countries. Monetary systems

FURNESS, Eric Longfellow
Money and credit in developing Africa. - London : Heinemann Educational, 1975
xi,308p : ill. - (Studies in the economics of Africa). - Includes bibliographies and index
0-435-84360-5 X.529/30364
056723x

AFRICA
Developing countries. Political development, 1947-1974

NEW states in the modern world / edited by Martin Kilson. - Cambridge, Mass. ; London : Harvard University Press, 1975
xi,254p. - Bibliography: p.245-247. - Includes index
0-674-62261-8 X.800/25361
0584363

AFRICA
Developing countries. Society. Role of middle classes, 1960-

INTELLIGENTSIIA i sotsial'nyĭ progress v razvivaiushchikhsia stranakh Azii i Afriki / [otvetstvennyĭ redaktor V.F. Li]. - Moskva : Nauka, Glavnaia redaktsiia vostochnoĭ literatury, 1981
327p. - (Puti razvitiia osvobodivshikhsia stran Vostoka). - Bibliography
r6486899 X.529/51259
887669x

AFRICA
Developing countries. Trade, to 1980

OROSZ, Árpád
Trade of African developing countries up to 1970 and prognosis to 1980 with special reference to trade with the European Socialist countries. - Budapest : Institute for World Economics of the Hungarian Academy of Sciences, 1975
98 p.. - (Studies on developing countries ; [no. 70])
r6007849 X.700/16834
0852805

AFRICA
Developing countries. Women personnel

WOMEN and work in Africa / edited by Edna G. Bay. - Boulder : Westview, 1982
xiv,310p. - (Westview special studies on Africa). - Includes bibliographies
0-86531-312-1 X.529/63038
3071820

AFRICA
Docks. Personnel, to date

WATERMAN, Peter
Aristocrats and plebeians in African unions? .... - [Nijmegen] : [P. Waterman], 1983
xii,426p. - Includes bibliography
r6330516 X.520/30766
8978271

AFRICA
Documents on Africa in stock of British libraries - Conference proceedings

UK library resources for southern African studies / Standing Conference on Library Materials on Africa. - London : SCOLMA, 1984
36p
0-904090-04-3 2719.k.220
0129739

AFRICA
Documents on Africa in stock of libraries in Chicago, Illinois - Catalogues

CHICAGO PUBLIC LIBRARY. George Cleveland Hall Branch
The Chicago Afro-American union analytic catalog. - Boston : G. K. Hall & Co., 1972
5 v.
0-8161-0978-8 LR.417.a.15
0923699

AFRICA
Documents on Africa in stock of University of Cape Town Libraries. Accessions - Catalogues - Serials

UNIVERSITY OF CAPE TOWN LIBRARIES. Africana and Manuscripts Divisions
List of titles added to the catalogue. - Cape Town : University of Cape Town Libraries, 1978-
Holdings: July-September 1977, May-June, September-October 1978-
r6108598 2725.g.7
2451212

AFRICA
Dravidians, to date

ARAVAANAN, K. P.
Anthropological studies on the Dravido-Africans. - Madras : Tamil Koottam, 1980
126p
r6805042 X.809/50091
8756546

AFRICA
Drought

AMERICAN UNIVERSITIES FIELD STAFF
The drought in Africa. - Hanover, N.H. : American Universities Field Staff, [1978?]
1v.. - (Fieldstaff portfolio)
r6189841 L.45/366
1842579

DROUGHT in Africa = Sécheresse en afrique. - Revised and expanded ed.. - London : International African Institute [for] the Environment Training Programme, UNEP-IDEP-SIDA, 1977
viii,200p : ill. - (African environment special reports ; 6). - Includes bibliographies
0-85302-056-6 P.701/1503
1842579

AFRICA
Drought - Conference proceedings

SYMPOSIUM ON DROUGHT IN BOTSWANA (1978 : Gaborone)
Proceedings of the Symposium on Drought in Botswana, National Museum, Gaborone, Botswana June 5th to 8th, 1978. - [Botswana] : Published by the Botswana Society in collaboration with Clark University Press, c1979
305p : ill,charts. - Bibliography: p294-305
0-87451-171-2 X.700/22891
0024546

AFRICA
Drought. Causes. Role of public administration

BAKER, Randall
The administrative trap. - Norwich : School of Development Studies, University of East Anglia, 1975
13p : 2ill. - (Development studies discussion paper ; no.5). - Bibliography: p13
B84-36447 X.525/9629
3161048

AFRICA
East & North-east Africa. Birds

MACKWORTH-PRAED, Cyril Winthrop
Birds of Eastern and North Eastern Africa. - 2nd ed. - London : Longman. - (African handbook of birds. Series 1)
Vol.1. - 1980
xxxiv,836p,53p of plates,vip of plates : ill(chiefly col). - Includes index
0-582-46082-4 X.329/13863
1875817
- (African handbook of birds, series 1)
Vol.2. - 1980
xiii,1113,[62]p,96p of plates,xixp of plates : ill(chiefly col). - Includes index
0-582-46083-2 X.329/13864
1875817

AFRICA
East Africa

MAGARY, Alan
East Africa. - New York ; London : Harper and Row, 1975
xxii,714p,[8]p of plates : ill. - Includes bibliographies and index
0-06-012792-9 X.809/40934
0647802

RIEFENSTAHL, Leni
Leni Riefenstahl's Africa. - London : Collins, 1982
231p : chiefly col.ill
0-00-262509-1 L.45/1781
2816946

TRAVELLERS' guide to East Africa. - 5th ed., fully rev.. - London : Thornton Cox, 1980
164p : ill(some col.). - Includes index
0-902726-33-1 X.809/50030
0647802
- [Another copy.] X.809/55007

AFRICA
East Africa - Serials

MAKERERE historical journal. - vol.1, no. 1- ; 1975-. - Nairobi : East African Literature Bureau, 1975
Holdings : vol.2, no.1 1976-
r6051850 P.701/1049
1536419

TRAVELLER'S guide to East Africa and the Indian Ocean. - [No.1]-. - London : IC Magazines, 1981-
v. : ill
B81-29433 X.808/34643
253004x

AFRICA
East Africa, to 1970 - Bibliographies

COMMONWEALTH INSTITUTE
Kenya, Tanzania, Uganda. - [London] : [The Institute], 1971[i.e.1979]
[1],27p. - (Commonwealth bibliographies ; no.12)
B79-30963 X.0709/737(12)
1948911

AFRICA
East Africa, to 1975

SSEKAMWA, J. C. (John Crysostom)
A sketch map history of East Africa. - New enlarged ed. - Amersham : Hulton, 1976
276p : 1 ill. - Bibliography: p.261-263. - Includes index
0-7175-0757-2 X.700/25265
0759546

AFRICA
East Africa, to 1976 - For East African students

ODHIAMBO, E. S. Atieno
A history of East Africa. - London : Longman, 1977
ix,192p : ill. - Includes index
0-582-60886-4 X.700/25640
1402951

AFRICA
East Africa, to 1983

SSEKAMWA, J. C. (John Crysostom)
A sketch map history of East Africa. - New enlarged ed. - Amersham : Hulton Educational, 1984
277p. - Includes index
0-7175-1135-9 X.800/39358
3175766

AFRICA
East Africa, 1000-1970 - For East African students

WERE, Gideon S.
East Africa through a thousand years. - 2nd ed. - London : Evans, 1972
ix,372p : ill. - Includes index
0-237-28873-7 X.809/53950
1059890

WILSON, Derek Alan
East African history, 1000-1970. - [New] ed. - London : Evans Bros, 1976
64p. - (Evans pocket facts)
0-237-49970-3 X.708/20729
1059890

AFRICA
East Africa, 1000–1975 – For Kenyan students
WILSON, Derek
A new primary history course for Kenya. – London : Hodder and Stoughton.
Teacher's book 1. – 1976
16p
0–340–17978–3 X.0700/1009a(1)
0555819

Book 1: East Africa, AD1000–1885. – 1976
64p : ill
0–340–17977–5 X.0700/1009(1)
0555819

AFRICA
East Africa, 1900–1980
NABUDERE, D. Wadada (Dan Wadada)
Imperialism in East Africa. – London : Zed. – (Africa series)
Vol.1: Imperialism and exploitation. – 1981
132p
0–905762–99–1 X.0529/733(1)
2726661

Vol.2: Imperialism and integration. – 1982
187p
0–905762–05–3 X.0529/733(2)
2726661

AFRICA
East Africa. Administrative law
OLUYEDE, Peter
Administrative law in East Africa. – Nairobi : East African Literature Bureau, 1973
xxi,255p. – Includes index
r6071988 X.200/31358
1451146

AFRICA
East Africa. African elephants. Hunting, 1897–1903
BELL, W. D. M. (Walter Dalrymple Maitland)
Karamojo safari. – Sudbury, Suffolk : Spearman, 1984, c1949
288p
0–85435–165–5 Nov.53610
3280438

AFRICA
East Africa. Age-group systems
AGE, generation and time / edited by P.T.W. Baxter and Uri Almagor. – London : C. Hurst, 1978
viii,276p : ill. – Includes bibliographies and index
0–903983–80–x X.529/32764
173685x

AFRICA
East Africa. Agriculture
NGUGI, D.
East African agriculture. – London : Macmillan, 1978
vi,266p : ill. – (Macmillan books for Africa). – Includes index
0–333–21271–1 X.311/8728 Woolwich
1413376

AFRICA
East Africa. Agriculture – For East African students
AGRICULTURE in East Africa / A.D.R. Ker ... [et al.]. – London : Edward Arnold, 1978
vi,218p : ill. – Includes bibliographies and index
0–7131–0160–1 X.311/8849 Woolwich
1565753

NGUGI, David
East African agriculture. – Rev. ed. – London : Macmillan, 1982
vi,282p : ill. – Bibliography: p276. – Includes index
0–333–32380–7 X.322/11967
1565753

East African agriculture. – [London] : English Language Book Society, 1982
vi,282p : ill. – Bibliography: p276
0–333–34154–6 X.800/38121
1565753

AFRICA
East Africa. Agriculture. Economic development. Effects of colonialism, to 1980
IMPERIALISM, colonialism and hunger / edited by Robert I. Rotberg. – Lexington, Mass. : Lexington, 1983
xiv,270p : ill. – Includes index
0–669–05871–8 X.800/36424
2971429

AFRICA
East Africa. Amaryllidaceae
NORDAL, Inger
Amaryllidaceae. – Rotterdam : published on behalf of the East African Governments by Balkema, 1982
31p : ill. – (Flora of tropical East Africa). – Includes index
90–619–1305–5 X.322/12047
2880601

AFRICA
East Africa. Animals
COTT, Hugh Bamford
Looking at animals. – London : Collins, 1975
221p,[64]p of plates : ill. – Bibliography: p.211–214. – Includes index
0–00–219093–1 SRL
001575x

HART, Susanne
Vet in the wild. – [London] : Fontana, 1977
189p,[4]p of plates : ill. – Includes index
0–00–634154–3 X.319/17773 Woolwich
1298364

AFRICA
East Africa. Animals. Ecology
BURTON, Jane
Animals of the African year. – [London] : Peter Lowe, 1979
141p : col.ill. – Bibliography: p137. – Includes index
0–85654–626–7 X.325/1141 Woolwich
2398427

AFRICA
East Africa. Balsaminaceae
GREY-WILSON, Christopher
Balsaminaceae. – Rotterdam : published on behalf of the East African Governments by Balkema, 1982
77p : ill. – (Flora of tropical East Africa). – Includes index
90–619–1304–7 X.322/12046
2880636

AFRICA
East Africa. Big game. Hunting. Finch Hatton, Denys – Biographies
TRZEBINSKI, Errol
Silence will speak. – London : Heinemann, 1977
xx,348p,[16]p of plates : ill. – Includes index
0–434–79500–3 X.800/26403
1245902

AFRICA
East Africa. Birds
BROWN, Leslie, 1917–
Birds of the African bush. – London : Collins, 1975
[111]p : ill(chiefly col)
0–00–216069–2 LR.425/64
0561185

KARMALI, John
Birds of Africa. – London : Collins, 1980
191p : ill(some col.). – Bibliography: p187. – Includes index
0–00–219031–1 L.45/1284.
0561185
– [Another issue: – Viking Press] L.45/1284

WILLIAMS, John G. (John George)
A field guide to the birds of East Africa. – [Completely rev. ed.]. – London : Collins, c1980
415p : col.ill. – Bibliography: p400–401. – Includes index
0–00–219179–2 SRL
2334461

AFRICA
East Africa. British government publications, 1897–1963 – Collections
GOVERNMENT publications relating to Kenya and the East Africa High Commission, 1897–1963. – East Ardsley : EP Microform, 1976
134 microfilm reels. – (Government publications relating to Africa in microform)
r6050925 SPR.Mic.B.25/7
1539655
– Introduction by H.F. Morris. – [32]p ; 21cm OPL

AFRICA
East Africa. British military postal services, 1939–1945
ROSSITER, Stuart
History of the East African army postal service. – [Heathfield] : Proud-Bailey, [1984?]
111p : ill. – Bibliography: p111
B85–02089 X.800/39593
3199509

AFRICA
East Africa. Christian broadcasting services, to date
The LIVELY word. – Limbe : Popular Publications, [1975?]
68p : ill
r6804706 X.200/36593
8777144

AFRICA
East Africa. Christian church. Revivals, 1927–1965 – Correspondence, diaries
CHURCH, J. E. (John Edward)
Quest for the highest. – Exeter : Paternoster Press, c1981
284p : ill. – Includes index
0–85364–328–8 X.200/37430
2613441

AFRICA
East Africa. Christian missions: Africa Inland Mission, to 1976
RICHARDSON, Kenneth
Garden of miracles. – [1st ed. reprinted]. – London : Africa Inland Mission, 1976
270p,[8]p of plates : ill
B78–24880 X.108/17049
1519727

AFRICA
East Africa. Christian missions: Africa Inland Mission International – Serials
AIM international. -- Mar. 1984–. – London : The Mission, 1984–
v. : ill
B85–14351 H.85/548
3252434

AFRICA
East Africa. Christian missions. New, Charles – Biographies
KENDALL, R. Elliott
Charles New and the East Africa mission. – Nairobi : Kenya Literature Bureau, 1978
xi,194p. – Includes bibliography
r6814068 X:208/9935
9165088

AFRICA
East Africa. Christianity. Books. Publishing, to date
The LIVELY word. – Limbe : Popular Publications, [1975?]
68p : ill
r6804706 X.200/36593
8777136

AFRICA
East Africa. Coastal regions, to date
FROM Zinj to Zanzibar / edited by J. de V. Allen and Thomas H. Wilson. – Wiesbaden : Steiner, 1982
231p : ill. – (Paideuma ; 28)
r6304634 Ac.659/2[vol.28]
8099286

AFRICA
East Africa. Coastal regions, ca 100– ca 1900
The EAST African coast / [edited by] G.S.P. Freeman-Grenville. – 2nd ed. – London : Collings, 1975
[15],314p,fold plate. – Includes index
0–901720–85–2 X.809/40473
0538035

AFRICA
East Africa. Coastal waters. Big game fish. Angling
CULLEN, Anthony
Crash strike!. – Nairobi : East African Publishing House, 1971
xiv,175p,[16]p of plates : ill
r6031697 X.319/17240 Woolwich
1091441

AFRICA
East Africa. Coastal waters. Voyages by dhows, 1972–1977: Mir-El-Lah (Ship)
RICCIARDI, Lorenzo
The voyage of the 'Mir-El-Lah'. – London : Collins, 1980
256p ill(some col), maps, plans, ports(some col). – Includes index
0–00–211887–4 X.800/29114
2207982

AFRICA
East Africa. Colonisation by Italy, to 1919
DEL BOCA, Angelo
Gli italiani in Africa Orientale. – Roma : Laterza, 1976. – (Storia e società)
[Vol 1]: Dall'unità alla marcia su Roma. – Roma : Laterza, 1976
909p. – (Storia e società)
r6078213 X.809/25567
1529110

AFRICA
East Africa. Companies
PRIVATE enterprise and the East African company / edited by Philip Aneurin Thomas. - Dar es Salaam : Tanzania Publishing House ; Birmingham : [Distributed by] Third World Publications, 1969[i.e.1979]
xiv,283p
B80-00620 X.519/30432
2000431

AFRICA
East Africa. Conches. Shells
PINN, Frederick M. J.
Spider and wing shells. - London : Collings, 1977
[8],41p : ill(chiefly col). - (Seashells of East Africa)
0-901720-95-x X.319/17993 Woolwich
1407147

AFRICA
East Africa. Cruciferae
JONSELL, Bengt
Cruciferae. - Rotterdam : published on behalf of the East African Governments by Balkema, 1982
73p,[1]leaf of plate : ill. - (Flora of tropical East Africa). - Includes index
90-619-1301-2 X.322/12045
2880644

AFRICA
East Africa. Demography
MØNSTED, Mette
A demographic analysis of East Africa. - Uppsala : Scandinavian Institute of African Studies, 1978
211p : charts
91-7106-126-6 X.709/51076
2444542

AFRICA
East Africa. Description & travel
MOSER, Giorgio
Le montagne della luce. - Torino : ERI, [1976]
156p : ill
r6100625 LR.430.ss.10
2041707

AFRICA
East Africa. Description & travel, to date
BARDEY, Alfred
Barr-Adjam. précédé de "le Patron de Rimbaud". - Paris : Éditions du Centre national de la recherche scientifique, 1981
386p
2-222-02878-7 X.809/56417
8099294

RIEFENSTAHL, Leni
Mein Afrika. - München : List, 1982
232p : chiefly ill
3-471-78531-0 L.45/2555
8099294

AFRICA
East Africa. Description & travel, 1960-1970
MATTHIESSEN, Peter
The tree where man was born. - London : Picador in association with Collins, 1984, c1972
256p. - Bibliography: p242-245. - Includes index
0-330-28196-8 X.958/26558
3216330

AFRICA
East Africa. Description & travel, 1972
WHITTINGHAM, John
Shoe-string safari. - Oxford : Hannon, [c1978]
vii,191p : ill
0-904233-23-5 X.809/46380
2348624

AFRICA
East Africa. Ecology
BERGLUND, Björn
Noah's ark is stranded. - London : Macdonald and Jane's, 1977
xiv,210p : ill. - Bibliography: p.207-208. - Includes index
0-354-04051-0 X.322/8387 Woolwich
1165682

AFRICA
East Africa. Economic conditions
ASOÎAN, Nadezhda Samuilovna
Vostochnaia Afrika. - Moskva : Izdatel'stvo ""Mysl"", 1976
250 p.. - Bibliography: p. 240-245
r6023173 X.329/8530 Woolwich
0883018

LIVINGSTONE, I. (Ian)
Economics for Eastern Africa. - Revised and enlarged ed. - Nairobi ; London : Heinemann Educational, 1980
xx,572p : col ill. - (Studies in the economics of Africa). - Includes index
0-435-97400-9 X.520/21483
2181460

AFRICA
East Africa. Economic conditions - Bibliographies
KILLICK, Tony
The economies of East Africa. - Boston, Mass. : G.K. Hall, 1976
xxiv,150p. - (Bibliographies and guides in African studies). - Includes index
0-8161-7916-6 X.520/11927
1194135

AFRICA
East Africa. Economic development
KJEKSHUS, Helge
Ecology control and economic development in East African history. - London : Heinemann Educational, 1977
viii,215p : ill. - Bibliography: p.190-209. - Includes index
0-435-94526-2 X.329/15364
1096834

AFRICA
East Africa. Economic development, 1950-1960
SEIDMAN, Ann
Comparative development strategies in East Africa. - Nairobi : East African Publishing House ; Birmingham : [Distributed by] Third World Publications, 1972
vii,299p. - Includes index
B79-34713 X.510/10851
1965824
- [Another copy.] X.510/16397

AFRICA
East Africa. Economic development. Effects of exporting, 1959-1971
STEIN, Leslie
The growth of East African exports and their effect on economic development. - London : Croom Helm, 1979
3-272p. - Bibliography: p.254-268. - Includes index
0-85664-948-1 X.529/33620
1777416

AFRICA
East Africa. Education. Role of Catholic missions, 1885-
MIRTSCHINK, Bernhard
Zur Rolle christlicher Mission in kolonialen Gesell-schaften. - Frankfurt/Main : Haag+Herchen, 1980
148p. - (Berliner Studien zur Erziehung und Internationalität ; Bd 2). - Bibliography
3-88129-344-2 X.200/36158
8726620

AFRICA
East Africa. Ethnic groups. Social interactions related to ethnocentrism
BREWER, Marilynn Bolt
Ethnocentrism and intergroup attitudes. - [Beverly Hills] : Sage Publications ; New York ; London : Wiley, 1976
[5],218p : ill. - (Cross-cultural research and methodology series ; [Vol.2]). - Bibliography: p.147-155. - Includes index
0-470-10330-2 X.0520/500(2)
0648558

AFRICA
East Africa. Ethnocentrism related to social interactions of ethnic groups
BREWER, Marilynn Bolt
Ethnocentrism and intergroup attitudes. - [Beverly Hills] : Sage Publications ; New York ; London : Wiley, 1976
[5],218p : ill. - (Cross-cultural research and methodology series ; [Vol.2]). - Bibliography: p.147-155. - Includes index
0-470-10330-2 X.0520/500(2)
064854x

AFRICA
East Africa. Field crops. Improvement
CROP improvement in East Africa / edited by C.L.A. Leakey. - Farnham Royal : Commonwealth Agricultural Bureaux, 1970
280p : ill. - (Technical communication ; 19). - Includes bibliographies and index
0-85198-498-3 X.329/19304
325531x

AFRICA
East Africa. Flacourtiaceae
SLEUMER, Hermann
Flacourtiaceae. - Nairobi ; London : Crown Agents for Oversea Governments and Administrations [for] the East African Community, 1975
iv,68p : ill. - (Flora of tropical East Africa). - Includes index
0-85592-035-1 BS. 7c/8
0502146

AFRICA
East Africa. Geographical features
BRITTAIN, Herbert, b.1922
East Africa atlas notebook. - 5th ed. - London : University of London Press, 1975
63p : ill
0-340-16790-4 X.702/5301
1029924

EVANS, F. C. (Frederick Colin)
Geographical photographs for East Africa. - Cambridge : Cambridge University Press, 1983
48p : ill
0-521-28615-8 L.49/1939
1029924

MORGAN, William Thomas Wilson
East Africa. - London : Longman, 1973
xx,410p : ill. - (Geographies for advanced study). - Includes bibliographies and index
0-582-48465-0 2060.ff Reading Room
0028614

East Africa. - London : Longman, 1973 [i.e. 1976]
xx,410p : ill. - (Geographies for advanced study). - Includes bibliographies and index
0-582-48573-8 WPB.375/51
0028614

AFRICA
East Africa. Geographical features - For East African students
CARTER, Charles Eric
Case studies in East African geography. - Edinburgh : Johnston and Bacon, 1972
96p : ill. - Bibliography: p.94-95
0-7179-1577-8 X.700/25099
0614068

MINNS, W. J.
African geography. - London : Evans Bros, 1979
64p : ill
0-237-50196-1 X.700/25952
182239x

SENIOR, M. W. (Michael Wraith)
Certificate notes. - Harlow : Longman, 1982
96p : ill. - (Study for success) (Study for success)
0-582-60383-8 X.809/51842
0614068

AFRICA
East Africa. Geographical features - For Kenyan students
MINNS, W. J.
East Africa. - London : Evans Bros, 1976
65 p. : ill. - (Kenya primary geography)
0-237-50077-9 X.702/5195
0688002

AFRICA
East Africa. Goodeniaceae
DAVIES, F. G.
Goodeniaceae. - Nairobi ; London : Crown Agents for Oversea Governments and Administrations [for] the East African governments, 1978
4,ivp : 1 ill. - (Flora of tropical East Africa). - Includes index
0-85592-050-5 SRL
1703560

AFRICA
East Africa. Grasses
GRAMINEAE. - Rotterdam : published on behalf of the East African Governments by Balkema. - (Flora of tropical East Africa)
Pt.3. - 1982
p451-898 : ill. - Includes index
90-619-1303-9 X.322/12044
288061x

AFRICA
East Africa. Gregory Rift Valley. Fossil vertebrates
GEOLOGICAL background to fossil man / edited by Walter W. Bishop. - Edinburgh : Scottish Academic Press [etc.] for the Geological Society of London, 1978
xii,585p,[5]leaves of plates(3 fold),[6]p of plates(4 fold) : ill. - Includes bibliographies and index
0-7073-0143-2 X.622/23877
156661x
- [Another copy.] SRL

AFRICA
East Africa. Gregory Rift Valley. Geological features

GEOLOGICAL background to fossil man / edited by Walter W. Bishop. - Edinburgh : Scottish Academic Press [etc.] for the Geological Society of London, 1978
xii,585p,[5]leaves of plates(3 fold),[6]p of plates(4 fold) : ill. - Includes bibliographies and index
0-7073-0143-2 X.622/23877
1566601
- [Another copy.] SRL

AFRICA
East Africa. Guttiferae

BAMPS, P.
Guttiferae. - Nairobi ; London : Crown Agents for Overseas Governments and Administrations [for] the East African governments, 1978
35,iv p : ill. - (Flora of tropical East Africa). - Includes index
0-85592-051-3 SRL
1703579

AFRICA
East Africa. Historical events. Chronology. Historical sources: Droughts

CHRONOLOGY, migration and drought in interlacustrine Africa / edited by J.B. Webster. - London : Longman [etc.], 1979
xvi,345p : ill. - (Dalhousie African studies series). - Includes index
0-582-64624-3 X.322/9470 Woolwich
2229536

AFRICA
East Africa. Historiology

ROBERTS, Andrew
Recording East Africa's past. - 2nd ed. - Nairobi : East African Publishing House for the History Department, University of Dar es Salaam, 1973
24p. - Bibliography : p.20-24
r6072086 X.709/32347
1513729

AFRICA
East Africa. Human settlements - Bibliographies

CURUTCHET, Mirina
Bibliography on human settlements in developing countries. - Stockholm : Swedish Council for Building Research, 1982
224p. - (Document / Swedish Council for Building Research ; D12:1982)
91-540-3727-1 X.800/40472
9142274

AFRICA
East Africa. Indians, to date

RAI, Kauleshwar
Indians and British colonialism in East Africa 1883-1939. - Patna : Associated Book Agency, 1979
ix,159p
r6053040 X.809/50506
8795401

AFRICA
East Africa. Influence of Orient, to ca 1500 - Conference proceedings

EAST Africa and the Orient / edited by H. Neville Chittick and Robert I. Rotberg. - New York : Africana ; London : Holmes and Meier Publishers, Ltd, 1975
viii,343p. - Bibliography: p.291-305. - Includes index
0-8419-0142-2 X.809/41152
0710113

AFRICA
East Africa. Italian prisoners of war, to 1945

CORSINI, Giovanni
Lunga fuga verso il sud. - Milano : Mursia, c1979
207p,[16]p of plates. - (Testimonianze fra cronaca e storia ; 104. Resistenza e campi di prigionia)
r6502631 X.809/49633
8770492

AFRICA
East Africa. Ivory trade. Stokes, Charles - Biographies

LUCK, Anne
Charles Stokes in Africa. - Nairobi : East African Publishing House, 1972
211 p., [4] of plates. - (Historical studies ; no. 6). - Bibliography : p. 211. - Includes index
r6031224 X.708/20722
1056484

AFRICA
East Africa. Lake ecosystems

BROWN, Leslie, 1917-
East African mountains and lakes. - Nairobi : East African Publishing House, 1971
[5], 122, [2] p., plate : ill
r6031226 X.800/25934
1055518

AFRICA
East Africa. Lake Region, to date

AYOT, Henry Okello
Historical texts of the Lake Region of East Africa. - Nairobi : Kenya Literature Bureau, 1977
320p
r6803619 X.809/49092
8727848

La CIVILISATION ancienne des peuples des Grands Lacs / organisé par le Centre de civilisation burundaise (Ministère de la jeunesse, des sports et de la culture du Burundi). - Paris : Karthala, [1981]
497p
2-86537-026-7 X.800/41535
8727848

AFRICA
East Africa. Livestock. Production - For East African students

BARBER, John, 1920-
Livestock management for East Africa. - London : Edward Arnold, 1976
[6],126p : ill. - Bibliography: p.117. - Includes index
0-7131-0063-x X.311/8272 Woolwich
0735434

AFRICA
East Africa. Livestock production industries, to date

RAIKES, Philip L.
Livestock development and policy in East Africa. - Uppsala : Scandinavian Institute of African Studies, 1981
x,254p. - (Publications from the Centre for Development Research, Copenhagen ; no.6)
91-7106-182-7 X.0519/814(6)
8947821

AFRICA
East Africa. Mammals

KINGDON, Jonathan
East African mammals. - London : Academic Press.
Vol.3. Part A: (Carnivores). - 1977
viii,476p,[3]leaves of plates : ill(some col). - Bibliography: p.423-451. - Includes index
0-12-408303-x SRL
0273694

Vol.3 Part C: Bovids. - 1982
393p,[1]p of plates : ill(some col.)
0-12-408344-7 SRL
0273694

Vol.3 Part D: Bovids. - 1982. - Bibliography: p643-704. - Includes index
Part D: (Bovids)
395-746p : ill
0-12-408345-5 SRL
0273694

AFRICA
East Africa. Mammals. Behaviour

MOSS, Cynthia
Portraits in the wild. - London : Hamilton, 1976
xviii,363p,[36]p of plates : ill(some col). - Bibliography: p.336-348. - Includes index
0-241-02453-6 SRL
0583472

AFRICA
East Africa. Masai, to date

KALTER, Johannes
Die materielle Kultur der Massai und ihr Wandel. - Bremen : Übersee-Museum, 1978
258p : ill. - (Veröffentlichungen aus dem Übersee-Museum Bremen. Reihe F, Bremer Afrika-Archiv ; Bd.4)
3-88299-012-0 X.0705/111f(4)
8773998

AFRICA
East Africa. Masai. Social conditions

OLE SAIBULL, Soloman
Herd and spear. - London : Collins and Harvill, 1981
141p,[16]p of plates : ill
0-00-262303-x X.809/48170
2502143

AFRICA
East Africa. Medical services: African Medical and Research Foundation. Wood, Michael, b.1919 - Biographies

WOOD, Michael, b.1919
Go an extra mile. - London : Collins, 1978
160p,[16]p of plates : ill(some col)
0-00-216267-9 X.320/11328 Woolwich
1517791

Go an extra mile. - London : Fount Paperbacks, 1980
160p
0-00-625918-9 H.80/503 Woolwich
1517791

AFRICA
East Africa. Mountain ecosystems

BROWN, Leslie, 1917-
East African mountains and lakes. - Nairobi : East African Publishing House, 1971
[5], 122, [2] p., plate : ill
r6031226 X.800/25934
1055526

AFRICA
East Africa. Nilotes. Religious beliefs compared with religious beliefs of Proto-Indo-Iranians

LINCOLN, Bruce
Priests, warriors, and cattle. - Berkeley ; London : University of California Press, c1981
xiii,242p : ill. - (Hermeneutics ; 10). - Bibliography p185-215. - Includes index
0-520-03880-0 X.200/35981
2507757

AFRICA
East Africa. Oliniaceae

VERDCOURT, B. (Bernard)
Oliniaceae. - [London] : Crown Agents for Oversea Governments and Administrations [for] the East African Community, 1975
4, iv p : 1 ill. - (Flora of tropical East Africa). - Includes index
0-85592-037-8 BS.7c/8
052493x

AFRICA
East Africa. Passifloraceae

WILDE, W. J. J. O. de
Passifloraceae. - London : Crown Agents for Oversea Governments and Administrations [for the] East African Community, 1975
71,iv p : ill. - (Flora of tropical East Africa). - Includes index
0-85592-036-x BS.7c/8
0516481

AFRICA
East Africa. Pastoral communities

BRANDSTRÖM, Per
Aspects of agro-pastoralism in East Africa. - Uppsala : Scandinavian Institute of African Studies, 1979
60p. - (Nordiska afrikainstitutet. Research report ; no 51). - Includes bibliographies
91-7106-155-x P.803/212[no.51]
2261898

AFRICA
East Africa. Plants

BYWATER, Marie
Myrothamnaceae. - Rotterdam : Published on behalf of the East African Governments by Balkema, 1984
iv,5p : ill. - (Flora of tropical East Africa). - Includes index
90-619-1309-8 X.322/14490
3222608

CRIBB, Phillip
Orchidaceae. - Rotterdam : Published on behalf of the East African Governments by Balkema, 1984
p237-411 : ill. - (Flora of tropical East Africa). - Includes index
90-619-1312-8 X.322/14495
3222608

HALLIDAY, P.
Myrsinaceae. - Rotterdam : Published on behalf of the East African Governments by Balkema, 1984
iv,21p : ill. - (Flora of tropical East Africa). - Includes index
90-619-1310-1 X.322/14493
3222608

MARAIS, W.
Cyclocheilaceae. - Rotterdam : Published on behalf of the East African Governments by Balkema, 1984
iv,5p : ill. - (Flora of tropical East Africa). - Includes index
90-619-1306-3 X.322/14488
3222608

Nesogenaceae. - Rotterdam : Published on behalf of the East African Governments by Balkema, 1984
iv,2p : ill. - (Flora of tropical East Africa). - Includes index
90-619-1311-x X.322/14494
3222608

THULIN, Mats
Lobeliaceae. - Rotterdam : Published on behalf of the East African Governments by Balkema, 1984
iv,59p : ill. - (Flora of tropical East Africa). - Includes index
90-619-1314-4 X.322/14489
3222608

AFRICA
East Africa. Plants–continued

VERDCOURT, B. (Bernard)
Erythroxylaceae. – Rotterdam : Published on behalf of the East African Governments by Balkema, 1984
iv,11p : ill. – (Flora of tropical East Africa). – Includes index
90–619–1307–1 X.322/14491
3222608

Ixonanthaceae. – Rotterdam : Published on behalf of the East African Governments by Balkema, 1984
iv,11p : 1ill. – (Flora of tropical East Africa). – Includes index
90–619–1308–x X.322/14492
3222608

AFRICA
East Africa. Politics, ca 1950–1975

SAUL, John S.
The state and revolution in Eastern Africa. – London : Heinemann Educational, 1979
[6],454p. – Includes index
0–435–96541–7 X.708/22153
1970410

AFRICA
East Africa. Primary schools. Teaching

MARY JACINTA, Sister
Primary methods handbook. – London : Hodder and Stoughton, 1981
128p. – Bibliography: p128
0–340–25952–3 X.529/40321
2372878

AFRICA
East Africa. Quaternary era. Environmental aspects

HAMILTON, A. C. (Alan Charles)
Environmental history of East Africa. – London : Academic Press, 1982
xiv,328p : ill. – Bibliography: p294–311. – Includes index
0–12–321880–2 SRL
2760673

AFRICA
East Africa. Rangeland

RANGELAND management and ecology in East Africa / edited by D.J. Pratt and M.D. Gwynne ; from contributions by J.R. Blackie ... [et al.]. – London : Hodder and Stoughton, 1977
x,310p,[8]leaves of plates,[16]p of plates : ill. – Bibliography: p.289–303. – Includes index
0–340–19766–8 X.320/10833 Woolwich
1165232

AFRICA
East Africa. Rift Valley. Description & travel

EDBERG, Rolf
Dalens ande. – Stockholm : Norstedts, 1976
229p
91–17–63052–5 X.809/24365
122333X

[DALENS ande. English]. The dream of Kilimanjaro. – London : Heinemann, 1979
[8],179p. – Bibliography: p.177–179
0–434–22197–x X.809/44304
1882287

AFRICA
East Africa. Rural industries, to date

KYEWALYANGA, Francis-Xavier S. (Francis-Xavier Sserufusa)
Einige Aspekte des Kulturwandels in Ostafrika. – Hohenschäftlarn : Renner, 1982
iii,372p. – Bibliography: p.270–348
r6305922 X.529/53204
8940576

AFRICA
East Africa. Rural industries & trades. Development

BERG, Liv
Towards village industry. – London : Intermediate Technology Publications, 1978
87p : ill
0–903031–52–3 X.611/8855 Woolwich
1886770

AFRICA
East Africa. Rural regions. Social inequality

KITCHING, G. N. (Gavin N)
Economic and social inequality in rural East Africa. – Swansea : Centre for Development Studies, University of Swansea, 1977
[1],ii,65,[1] leaves : 1 ill. – (Monographs / University College of Swansea. Centre for Development Studies ; no.1)
B77–23140 X.512/5653
1171070

AFRICA
East Africa. Rural regions. Society. Economic aspects

SCHNEIDER, Harold K.
Livestock and equality in East Africa. – Bloomington ; London : Indiana University Press, 1979
xi,291p : ill. – Bibliography: p.265–285. – Includes index
0–253–19565–9 X.529/37152
2231034

AFRICA
East Africa. Rutaceae

KOKWARO, J. O
Rutaceae. – Rotterdam : published on behalf of the East African Governments by Balkema, 1982
52p : ill. – (Flora of tropical East Africa). – Includes index
90–619–1302–0 X.322/12048
2880628

AFRICA
East Africa. Slavery, 1800–1890

COOPER, Frederick
Plantation slavery on the east coast of Africa. – New Haven ; London : Yale University Press, 1977
xviii,314p. – (Yale historical publications. miscellany ; 113). – Bibliography: p.283–304. – Includes index
0–300–02041–4 Ac.2692.md/3(113)
118766x

NWULIA, Moses D. E.
Britain and slavery in East Africa. – Washington, D.C. : Three Continents Press, 1975
224p
0–914478–12–5 X.709/50191
118766X

AFRICA
East Africa. Social change, 1876–1976 – Conference proceedings

INTERNATIONAL CONGRESS OF ANTHROPOLOGICAL AND ETHNOLOGICAL SCIENCES (9th : 1973 : Oshkosh and Chicago)
A century of change in Eastern Africa. – The Hague : Mouton, 1976
xii,309p : 1 ill. – (World anthropology). – Includes bibliographies and index
90–279–7879–4 X.800/26749
1378392

AFRICA
East Africa. Social conditions

TOWNSEND, Derek
Let's visit Kenya, Uganda and Tanzania. – 3rd rev. ed. – London : Burke, 1983
95p : ill(some col.). – Includes index
0–222–00924–1 X.809/63413
1957082

AFRICA
East Africa. Social conditions, 1929–1936

READ, David, b.1922
Barefoot over the Serengeti. – London : Cassell, 1979
[11],187p,[8]p of plates : ill
0–304–30057–8 X.809/44545
1920251

AFRICA
East Africa. Social structure, 1900–1979

SIVONEN, Seppo
Yhteiskunnalliset rakennemuutokset Itä-Afrikassa 1900-luvulla. – Joensuu : Joensuun korkeakoulu, 1980
80p. – (Joensuun korkeakoulu. Historian, maantieteen ja muiden aluetieteiden osaston julkaisuja ; n:o 15). – Bibl
951–696–251–3 P.703/811.(15)
2437937

AFRICA
East Africa. Students. Political behaviour

EDUCATION and political values / edited by Kenneth Prewitt. – Nairobi : East African Publishing House ; Birmingham : [Distributed by] Third World Publications, 1971
[4],xv,249p : ill. – (EAPH political studies ; 12)
B80–00585 X.519/30441
2001624

AFRICA
East Africa. Students. Political socialisation

EDUCATION and political values / edited by Kenneth Prewitt. – Nairobi : East African Publishing House ; Birmingham : [Distributed by] Third World Publications, 1971
[4],xv,249p : ill. – (EAPH political studies ; 12)
B80–00585 X.519/30441
2001616

AFRICA
East Africa. Swahili speaking persons. Cultural processes, ca 1900

MTORO BIN MWINYI BAKARI
The customs of the Swahili people. – Berkeley ; London : University of California Press, 1981
xv,342p. – Bibliography: p313–327. – Includes index
0–520–04122–4 X.800/33619
276086x

AFRICA
East Africa. Swallowtail butterflies

CARCASSON, R. H. (Robert Herbert)
The swallowtail butterflies of East Africa (Lepidoptera, Papilionidae). – Faringdon : Classey, 1975
[3],31p,11p of plates : ill. – Includes index
0–900848–86–3 SRL
0358614

AFRICA
East Africa. Thymelaeaceae

PETERSON, B.
Thymelaeaceae. – Nairobi ; London : Crown Agents for Oversea Governments and Administrations [for] the East African governments, 1978
38,iv p : ill. – (Flora of tropical East Africa). – Includes index
0–85592–052–1 SRL
1703587

AFRICA
East Africa. Tribal societies. Cultural processes, 1890–1976

KESBY, John D.
The cultural regions of East Africa. – London : Academic Press, 1977
xiv,320p : ill. – Bibliography: p.297–302. – Includes index
0–12–405450–1 X.800/26663
1343378

AFRICA
East Africa. Tribal societies. Cultural processes. Effects of education, to date

KYEWALYANGA, Francis-Xavier S. (Francis-Xavier Sserufusa)
Einige Aspekte des Kulturwandels in Ostafrika. – Hohenschäftlarn : Renner, 1982
iii,372p. – Bibliography: p.270–348
r6305922 X.529/53204
8940568

AFRICA
East Africa. Urbanisation. Legal aspects

URBAN legal problems in Eastern Africa / edited by G.W. Kanyeihamba and J.P.W.B. McAuslan. – Uppsala : Scandinavian Institute for African Studies, 1978
298p. – (Studies of law in social change and development ; no.2)
91–7106–136–3 X.0100/805(2)
1990675

AFRICA
East Africa. Vahliaceae

BRIDSON, D. M.
Vahliaceae. – [London] : Crown Agents for Oversea Governments and Administrations [for] the East African Community, 1975
6,iv p : 1 ill. – (Flora of tropical East Africa). – Includes index
0–85592–038–6 BS.7c/8
0524948

AFRICA
East Africa. Wildlife reserves: National parks

WILLIAMS, John G. (John George)
A field guide to the national parks of East Africa. – New ed.. – London : Collins, c1981
336p : ill(some col.). – Includes index
0–00–219215–2 SRL
2608529

AFRICA
East Africa. Wildlife reserves. Vertebrates

THOMPSON, Ralph, b.1913
An artist's safari. – [Revised ed.]. – London : Collins [for] the Tryon Gallery, 1979
[160]p : ill(some col)
0–00–211033–4 X.423/10969 Woolwich
2045931

AFRICA
East Africa. Witchcraft

WITCHCRAFT and sorcery in East Africa / John Beattie ... [et al.] ; edited by John Middleton and E.H. Winter. – London : Routledge and Kegan Paul, 1978
ix,302p : 2 ill. – Bibliography: p.25–26. – Includes index
0–7100–1838–x X.529/32614
1690221

AFRICA
East Africa. Women. Economic conditions
OBBO, Christine
African women. - London : Zed Press, 1980
x,166p. - Bibliography: p.165-166
0-905762-33-9 X.529/37812
2281430
African women. - London : Hutchinson University Library for Africa in association with Zed Press, 1982, c1980
x,166p : ill. - Bibliography: p165-166
0-905762-48-7 X.529/50124
2281430

AFRICA
East Central Africa. Ivory trade, 1498-ca 1860
ALPERS, Edward Alter
Ivory & slaves in east central Africa. - London : Heinemann Educational, 1975
xx,296p : ill. - Bibliography: p.269-286. - Includes index
0-435-32005-x X.529/30842
0688916

AFRICA
East Central Africa. Slave trade, 1770-ca 1860
ALPERS, Edward Alter
Ivory & slaves in east central Africa. - London : Heinemann Educational, 1975
xx,296p : ill. - Bibliography: p.269-286. - Includes index
0-435-32005-x X.529/30842
0688924

AFRICA
Eastern Africa - For backpacking
BRADT, Hilary
Backpacker's Africa. - Chalfont St Peter : Bradt Enterprises, 1977
100p : ill. - Bibliography: p.95
0-9505797-1-8 X.709/32080
1396625

AFRICA
Eastern Africa. Agriculture. Mechanisation
KINSEY, B. H. (Bill H)
Economic research and farm machinery designs in East Africa. - Norwich : School of Development Studies, University of East Anglia, 1976
20leaves. - (Development studies discussion paper ; no.10). - Bibliography: leaves 19-20
B84-36446 X.525/9651
3161021

AFRICA
Eastern Africa. Slave trade, to 1922
BEACHEY, Raymond Wendell
The slave trade of Eastern Africa. - London : Collings, 1976
vii,324p. - Bibliography: p.298-307. - Includes index
0-901720-72-0 X.809/45616
2076608

AFRICA
Eastern Africa. Slave trade, 1800-1900
A COLLECTION of documents on the slave trade of Eastern Africa / [compiled by] R.W. Beachey. - London : Collings, 1976
xii,140p. - Includes index
0-86036-000-8 X.809/28721
2064499

AFRICA
Ecology - Bibliographies
ROGERS, Dilwyn J.
A bibliography of African ecology. - Westport ; London : Greenwood Press, 1979
xxi,499,[1]p. - (Special bibliographic series : new series / African Bibliographic Center ; no.6)
0-313-20552-3 SRL
1892231
- [Another copy] X:322/13144

AFRICA
Economic conditions
AFRICA projected / edited by Timothy M. Shaw and Olajide Aluko. - London : Macmillan, 1985
xv,217 : ill. - Bibliography: p210-212. - Includes index
0-333-34064-7 X.529/68633
3015793
ANNALS of African Economy. - Calcutta : K.K. Roy. 1978-. - 1978-
r6109125 P.513/1147
0534390
RELATIONS of production / edited by David Seddon ; translated [from the French] by Helen Lackner. - London : Cass, 1978
xiv,414p : ill. - Bibliography: p.413-414
0-7146-3000-4 X.529/32112
1354248

AFRICA
Economic conditions, to ca 1930
WICKINS, P. L. (Peter L)
An economic history of Africa. - Cape Town ; Oxford : Oxford University Press, 1981, c1980
323p : ill. - Includes bibliographies and index
0-19-570192-5 X.809/50230
2613131

AFRICA
Economic conditions, 1885-
AFRIKA zwischen Subsistenzökonomie und Imperialismus / Georg Elwert, Roland Fett (Hg.). - Frankfurt : Campus, 1982
295p
3-593-32842-9 X.529/60256
9043470

AFRICA
Economic conditions, 1950-1980
NEW acquisitions in the UNECA Library = nouvelles acquisitions de la bibliothèque de la CEA. - Vol.1, nos.1-2[(1963)]-. - Addis-Ababa : the Commission, [1963]-
Holdings: Vol.14, nos.5-6 ; Vol.15, nos.1-2, 5-6 - 9-10 ; vol.16, nos.3-4-
r6202462 OPL
2076225
POLITICAL economy of Africa / edited and introduced by Dennis L. Cohen and John Daniel. - London : Longman, 1981
vi,289p. - Includes bibliographies
0-582-64284-1 X.520/25313
2076225

AFRICA
Economic conditions, 1950-1980 - Serials
UNIVERSITY OF LAGOS
[Faculty of Social Sciences]. The journal of business and social studies. New series (N.S.). - Vol.1, no.1 (Dec. 1977)-. - Lagos : The University, 1977-
r6108375 P.511/1645
2076225

AFRICA
Economic conditions. Effects of British imperialism, ca 1780-ca 1940
The IMPERIAL impact / edited by Clive Dewey and A.G. Hopkins. - London : Athlone Press for the Institute of Commonwealth Studies, 1978
x,409p : 1 ill. - (Commonwealth papers ; 21). - Includes index
0-485-17621-1 Ac.2666.da/2[no.21]
1341553

AFRICA
Economic conditions. Effects of British imperialism, 1870-1895
HYNES, William G.
The economics of Empire. - London : Longman, 1979
x,162p. - Bibliography: p.143-151. - Includes index
0-582-64234-5 X.529/36580
2181355

AFRICA
Economic conditions. Effects of British imperialism, 1880-1960
MUNRO, J. Forbes (John Forbes)
Britain in tropical Africa 1880-1960. - London : Macmillan, 1984
89p. - (Studies in economic and social history). - Bibliography: p73-85. - Includes index
0-333-32330-0 X.809/60944
3184625

AFRICA
Economic conditions. Information sources
SOURCES of African and Middle-Eastern economic information / edited by Euan Blauvelt and Jennifer Durlacher. - Aldershot : Gower, c1982
2v.. - Includes index
0-566-02321-0 SRL
2338971
- [Another copy] OPL

AFRICA
Economic conditions. Political aspects
SHAW, Timothy M.
Towards a political economy for Africa. - London : Macmillan, 1985
xiii,134p. - Includes index
0-333-36195-4 X.809/63786
3015807

AFRICA
Economic development
ABOYADE, Oketunji
Integrated economics. - London : Addison-Wesley, c1983
viii,568p : ill. - Includes bibliographies and index
0-201-14068-3 X.520/37113
0697869

BANINI, L. Y.
Some problems of economic development in Africa. - [Kumasi] : [Department of Economics and Industrial Management, University of Science and Technology], [1976]
[4],26p
r6072699 X519/28886
0697869

GLADKIĬ, ĬUriĭ Nikiforovich
Afrika. - Leningrad : Nauka, Leningradskoe otdelenie, 1979
188p,[1]fold.leaf of plates. - Bibliography : p.180-[189]
r6183038 X.529/34878
0697869

HYDEN, Goran
No shortcuts to progress. - London : Heinemann, 1983
xv,223p. - Includes index
0-435-96303-1 X.529/57766
0697869

INDIGENIZATION of African economics / edited by Adelbayo Adedeji. - London : Hutchinson University Library for Africa in association with the African Association for Public Administration and Management, 1981
413p. - (Hutchinson university library for Africa). - Bibliography: p395-396. - Includes index
0-09-143730-x X.529/42129
2324334
- [Another copy: paperback]
ISBN 0-09-143731-8 X.529/42130

LIVINGSTONE, I. (Ian)
Choices for rural and urban development in Africa in the 1980's and beyond. - Norwich : School of Development Studies, University of East Anglia, 1983
40p. - (Discussion paper ; no.128)
B84-42810 X.322/14166
0697869

OKWUOSA, Emanuel A.
New direction for economic development in Africa. - London : Africa Books, 1976
3-222p
0-903274-06-x X.520/10918
0697869

RODNEY, Walter
How Europe underdeveloped Africa. - Washington : Howard University Press, 1974
x,288p
0-88258-013-2 X.800/13597
0697869

AFRICA
Economic development - Serials
AFRICA journal. - Vol.1, no.1- ; [1976?]-. - Monrovia : Chester's Publishing Company, [1976?]-
Holdings: Vol.4, no.5-
r6108905 P.701/1329
0697869

AFRICA
Economic development, to 1969
ARRIGHI, Giovanni
[The political economy of Rhodesia. Italian]. Sviluppo economico e sovrastrutture in Africa. - 2. ed. - Torino : Giulio Einaudi, 1974
358 p.. - (Serie politica ; n. 14)
r6013846 X.519/20605
0852333

AFRICA
Economic development, to 1975
DEVELOPMENT paths in Africa and China / edited by Ukandi G. Damachi, Guy Routh, Abel-Rahman E. Ali Taha. - London : Macmillan [for the] International Institute for Labour Studies, 1976
xv,251p : 1 ill. - Bibliography: p.240-248. - Includes index
0-333-18987-6 X.529/30748
0657409

MAFEJE, Archie
Science, ideology and development. - Uppsala : Scandinavian Institute of African Studies, 1978
86p
91-7106-134-7 X.629/12981 Woolwich
0657409

AFRICA
Economic development, 1960-
AFRIKA v 70-80-e gody / [redaktsionnaia kollegiia : An.A. Gromyko (otvetstvennyi redaktor), L.V. Goncharov (rukovoditel' avtorskogo kollektiva) ... i dr.]. - Moskva : Nauke, Glavnaia redaktsiia vostochnoi literatury, 1980
324p
r6464595 X.529/59924
857961x

AFRICA
Economic development, 1960—continued

AMIN, Samir
L'avenir industriel de l'Afrique. - Paris : Éditions l'Harmattan : Agence de coopération culturelle et technique, [1980]
227p
2-85802-171-6 X.529/45704
857961x

GLOBAL'NYE problemy sovremennosti i Afrika / [otvetstvennyi redaktor G.S. Khozin]. - Moskva : Nauka, Glavnaia redaktsiia vostochnoi literatury, 1983
234, [4] p.. - Bibliography: p. 229-[235]
r6482861 X.529/66675
857961x

GRUHN, Isebill V.
Regionalism reconsidered. - Boulder, Col. : Westview Press, 1979
xiv,154p. - (Westview special studies on Africa)
0-89158-576-1 X.520/24974
857961x

IKOKU, Emman U.
Self-reliance. - Enugu, Nig. : Fourth Dimension Publishers, 1980
399p
978-15-6028-2 X:529/63823
857961x

INTÉGRATION africaine / revue trimestrielle de la C.E.A.O.. - no.1 (1977)-. - Ouagadougou : Communauté économique de l'Afrique de l'Ouest, 1977- : ill
Holdings: no.2 (1977)-
r6207148 P.523/926
857961x

AFRICA
Economic development, 1960- - Conference proceedings

SYMPOSIUM LEO FROBENIUS (2nd : 1979 : Dakar)
Le rôle des traditions dans le développement de l'Afrique. - Bonn : Deutsche UNESCO-Kommission ; München, etc : K.G. Saur, 1980
417p : ill. - (Seminarbericht der Deutschen UNESCO-Kommission ; Nr 33)
3-598-20423-x X.529/48628
857961x

AFRICA
Economic development, 1960- - Serials

BANQUE ARABE POUR LE DÉVELOPPEMENT ÉCONOMIQUE EN AFRIQUE
Rapport annuel. - (1977)-. - Khartoum : [la Banque], 1977- : ill
r6209507 X.0525/489
857961x

RESEARCH for development. - [Vol.1, no.1 (Jan. 1981)?]-. - [Ibadan?] : The Institute, [1981?]-
v.
Holdings: vol.1, no.2 (July 1981)-
r6806864 P.511/5396
857961x

AFRICA
Economic development. Effects of European colonisation, to ca 1970

RODNEY, Walter
How Europe underdeveloped Africa. - London : Bogle-L'Ouverhure Publications, 1978
316p
B80-30903 X.529/37155
2208156

AFRICA
Economic development. Effects of imperialism, to date

OFFIONG, Daniel A.
Imperialism and dependency. - Enugu : Fourth Dimension, 1980
xii,304p
978-15-6111-4 X.529/58890
900680x

AFRICA
Economic development. Effects of mining industries

LANNING, Greg
Africa undermined. - Harmondsworth : Penguin, 1979
592p : ill. - (African affairs) (A pelican original). - Includes index
0-14-021964-1 X.708/22326
2057751

AFRICA
Economic development. Environmental aspects

ENVIRONMENT and development in Africa / prepared by the Environmental Development Action (ENDA) for the United Nations Environment Programme. - Oxford : Published for the United Nations Environment Programme by Pergamon, 1981
x,76p : ill. - (UNEP studies ; v.2). - Bibliography: p64-75
0-08-025667-8 X.529/41862
196206x

AFRICA
Economic development. Financial assistance by developed countries

ARNOLD, Guy
Aid in Africa. - London : Kogan Page [etc.], 1979
240p. - Bibliography: p.223-225. - Includes index
0-85038-218-1 X.520/21910
2080982

AFRICA
Economic development. Influence of mass media - Conference proceedings

SEMINAR ON MOTIVATION, INFORMATION AND COMMUNICATION FOR DEVELOPMENT IN AFRICAN AND ASIAN COUNTRIES (1974 : Ibadan)
Seminar on Motivation, Information and Communication for Development in African and Asian Countries, Ibadan, Nigeria, 1-9 July 1974. - London : International Broadcast Institute, 1975
[3],73p : 2 ill
0-904776-00-x X.902/5057
064837x

AFRICA
Economic development. Political aspects

AKE, Claude
A political economy of Africa. - Harlow : Longman, 1981
viii,196p. - Includes bibliographies and index
0-582-64374-0 X.520/24726
1581996

DUMONT, René
Stranglehold on Africa. - London : Deutsch, 1983
xvi,255p. - Bibliography: p255
0-233-97471-7 X.520/31822
1581996

The POLITICAL economy of contemporary Africa / Peter C.W. Gutkind and Immanuel Wallerstein, editors. - Beverly Hills ; London : Sage Publications, [1977]
318p. - (Sage series on African modernization and development ; vol.1). - Includes bibliographies
0-8039-0506-8 X.519/26666
1072307

AFRICA
Economic development. Political aspects - Conference proceedings

CONFERENCE OF DIRECTORS OF SOCIAL AND ECONOMIC RESEARCH INSTITUTES IN AFRICA (1971 : University of Nairobi)
Economic independence in Africa. - Nairobi : East African Literature Bureau, 1973
xxii,235p. - Bibliography : p.148-150
r6072185 X.520/12879
1581996

AFRICA
Economic development. Role of Sweden, 1960-

JINADU, L. Adele
Idealism and pragmatism as aspects of Sweden's development policy in Africa. - Lagos : Nigerian Institute of International Affairs, [1982?]
107p. - (Monograph series)
978-227-698-7 X.529/60044
9090517

AFRICA
Economic development. Role of women. Social aspects

AFRICAN women in the development process / edited by Nici Nelson. - London : Cass, 1981
viii,136p. - Includes bibliographies
0-7146-3175-2 X.529/45956
2629178

AFRICA
Economic growth. Role of governments, 1955-1975

GOSUDARSTVENNYĬ sektor v stranakh Afriki / [redaktsionnaia kollegiia L. I. Aleksandrovskaia (otvetstvennyĭ redaktor), I. B. Matsenko, Iu. N. Cherkasov]. - Moskva : Izdatel'stvo "Nauka", Glavnaia redaktsiia vostochnoĭ literatury, 1976
295p
r6037209 X.709/22532
0395889

AFRICA
Economic planning

HELLEINER, Gerald Karl
Beyond growth rates and plan volumes. - [Brighton] : [Institute of Development Studies], [1975]
18p. - (Communications / University of Sussex. Institute of Development Studies ; 112)
0-903354-22-5 X.0512/206(112)
0659665

AFRICA
Economic relations with Europe - Conference proceedings

EUROPE and Africa. - London : The Society, 1978
49p
B78-20059 X.709/32233
1461915

AFRICA
Economic relations with India (Republic), to date - Conference proceedings

SEMINAR ON TRADING WITH AFRICA (1979 : Bombay)
Seminar on Trading with Africa. - New Delhi : Indian Institute of Foreign Trade, [1979?]
1v.
r6803983 X.525/8683
9041699

AFRICA
Economic relations with Middle East petroleum exporting countries, 1973-1977

CHIBWE, Ephraim Chipampe
Afro-Arab relations. - London : Friedmann, 1977
x,150p. - Includes index
0-904014-34-7 X.809/42741
1311484

AFRICA
Education

EDUCATION in Africa / edited by A. Babs Fafunwa, J.U. Aisiku. - London : Allen & Unwin, c1982
270p. - Includes index
0-04-370113-2 X.529/47751
0749281

MURPHREE, Marshall Warne
Education, development and good change in Africa. - [Johannesburg] : South African Institute of Race Relations, [1976]
[8], 25 p.. - (The Alfred and Winifred Hoernlé memorial lecture ; 1976)
0-86982-129-6 X.519/26225
0749281

AFRICA
Education, 1960-

KAAYK, Jan
Search and research for educational reform in Africa. - 2e oplage. - Leiden : CA/SNWV, 1984
v,227,6p. - (ICA publication ; no.40). - Includes bibliography
r6330986 X.529/66266
8579628

AFRICA
Education, 1960- - Bibliographies

COURTOIS, N.
Pour une éducation et un enseignement africains. - Dakar : Association d'études linguistiques interculturelles africaines, 1980
263p. - (Les langues parlées en Afrique. études documents et bibliographies)
2-222-02801-9 X.520/30518
8579628

FARWER, Christine
Bildungswesen in Afrika. - Hamburg : Institut für Afrikakunde, Dokumentations-Leitstelle Afrika, 1984
2v.. - (Dokumentationsdienst Afrika. Reihe A ; 18)
r6309489 X.0525/312a(18)
8579628

AFRICA
Education, 1960- - Conference proceedings

ATELIER SUR "TECHNIQUES PÉDAGOGIQUES ET MÉTHODES D'ÉVALUATION DES TRAVAUX PRATIQUES" (2e : 1975 : Douala)
Techniques d'enseignement et méthodes d'évaluation de travaux pratiques. - Addis Abeba : Association pour l'enseignement social en Afrique, 1976
121p
r6205713 X.529/42827
8579628

CONFERENCE OF THE INSTITUTES/CENTRES OF AFRICAN/CULTURAL STUDIES IN AFRICA (1978 : Khartoum)
African studies in Africa. - [Khartoum] : Institute of African and Asian Studies, University of Khartoum, 1981
viii,152p. - (Sudan library series ; 10)
r6807399 X.520/32911
8579628

AFRICA
Education. Psychological aspects - For teaching

DUROJAIYE, Michael Oladele Akanbi
A new introduction to educational psychology. - London : Evans Bros, 1976
ix,265p : ill. - Includes bibliographies and index
0-237-49923-1 X.519/25703
0635006

AFRICA
Educational institutions. Non-English speaking students. Curriculum subjects: English language. Teaching. Projects

LANGUAGE teaching projects for the Third World / edited by C.J Brumfit. - Oxford : Published in association with the British Council by Pergamon, 1983
viii,174p : ill. - (ELT documents ; 116). - Includes bibliographies
0-08-030342-0 X.950/30638
3078736

AFRICA
Educational institutions. Students. Assessment - For teaching

NWANA, Obioha Chima
Educational measurement for teachers. - Lagos ; Sunbury-on-Thames : Nelson, 1979
vi,234p : ill. - (Nelson teacher education). - Includes index
0-17-511552-4 X.529/35696
2087901

AFRICA
Employment related to distribution of income

FRY, James, 1946-
Employment and income distribution in the African economy. - London : Croom Helm, 1979
3-186p : ill. - Includes index
0-85664-715-2 X.529/33267
1737589

AFRICA
Employment. Labour

HINCHLIFFE, Keith
An introduction to labour issues in Africa. - Norwich : School of Development Studies, University of East Anglia, 1975
42,3p. - (Development studies discussion papers ; no.6). - Bibliography: p.1-3
r6863857 X.525/10714
318269x

AFRICA
Energy resources

PROIZVODITEL'NYE sily stran Afriki / redkollegiīa L.V. Goncharov (otvetstvennyĭ redaktor). - Moskva : Nauka, Glavnaīa redakt͡sii͡a vostochnoĭ literatury, 1982
196p
r6463258 X.629/19943
8927774

AFRICA
Energy resources - Serials

ENERGY, environment and development in Africa / Beijer Institute ... Scandinavian Institute of African Studies. - 1-. - Stockholm : The Institutes, 1984-
r6612243 X.0529/1056
8927774

QUARTERLY energy review. Africa. - Ist quarter 1981-. - London : Economist Intelligence Unit, 1981-
v.
B81-32798 P.513/2616
2541572

AFRICA
Environment

AFRICAN environment, problems and perspectives / editor Paul Richards assisted by Nicola Harris. - London : International African Institute, 1975
[2],xii,117p : ill. - (African environment special report ; 1). - Includes bibliographies
0-85302-046-9 P.701/1503(1)
0576255

ENVIRONNEMENT Africain / rédacteur Paul Richards, assistante de rédaction Nicola Harris ; [traduit en francais par Alice Bordat et al.]. - London : Institut Africain International : Programme de[de] Formation pour l'Environment, 1976
[2],xiv,126p,4p of plates : ill. - (Environnement africain dossier spécialisé ; 1). - Includes bibliographies
0-85302-048-5 X.311/8214 Woolwich
0694495

AFRICA
Ethnic groups

ETHNICITY in modern Africa / edited by Brian M. du Toit. - Epping : Bowker, c1978
x,319p. - (Westview special studies on Africa). - Bibliography: p303-315
0-89158-314-9 X.800/36365
2197715
- [Another copy] X.520/33211

AFRICA
Ethnic groups. Political aspects

STATE versus ethnic claims / edited by Donald Rothchild and Victor A. Olorunsola. - Epping : Bowker, [1983?]
x,356p. - (Westview special studies on Africa). - Includes index
0-86531-503-5 X.520/30881
294006x
- [Another copy: paperback.] ISBN 0-86531-504-3 X.520/30882

AFRICA
European colonies. Colonial administration, ca 1850-1970

CHRISTOPHER, A. J. (Anthony John)
Colonial Africa. - London : Croom Helm, c1984
232p. - (Croom Helm historical geography series). - Bibliography: p211-224. - Includes index
0-7099-0368-5 X.809/60560
3165140

AFRICA
Exploration, to 1877

HUXLEY, Elspeth
The challenge of Africa. - London : Reader's Digest Association, 1980, c1971
191p : ill(some col.). - (Discovery and exploration). - Includes index
B82-31531 X.805/2953
0775770

MOUNTFIELD, David
A history of African exploration. - London : Hamlyn, 1976
160p : ill(some col). - Bibliography: p.158. - Includes index
0-600-01131-3 X.802/10254
0775770

AFRICA
Exploration, to 1885

BARTH, Heinrich, 1821-1865
[Reisen und Entdeckungen in Nord- und Central-Afrika. Selections]. Im Sattel durch Nord- und Zentralafrika. - 2. Aufl. - Wiesbaden : Brockhaus, 1980
403p.,[24]p.of plates : ill. - (Klassiker der Entdeckung)
3-7653-0159-0 X.800/31172
8579679

VAN ORMAN, Richard A.
The explorers. - Albuquerque : University of New Mexico Press, c1984
xiii,243p : ill
0-8263-0711-6 X.622/20472
8579679

AFRICA
Exploration, to 1980

CAMERON, Ian, 1924-
Exploring Africa. - Harlow : Longman, 1984
61p : ill(some col.). - (Royal Geographical Society exploring series). - Includes index
0-582-39285-3 X.800/39694
3012786

AFRICA
Exploration, 1450-1885

PIAGGIA, Carlo
[Viaggi di Carlo Piaggia nell'Africa centrale. Selections]. Nella terra dei Niam-Niam (1863-1865) .... - [Lucca] : Pacini Fazzi, 1978
lxiv,215p,[1]folded leaf of plates[32]p of plates : ill. - (Movimento operaio ; 52). - Bibliography
r6500452 X.800/30646
8710317

AFRICA
Exploration by Alexander, Boyd, b.1873 - Biographies

ALEXANDER, Joan
Whom the gods love. - London : Heinemann, 1977
[10],310p,[16]p of plates : ill. - Bibliography: p.299-301. - Includes index
0-434-01814-7 X.809/41553
1031392

AFRICA
Exploration by Baker, Sir Samuel White & Baker, Florence, Lady

HALL, Richard, b.1925
Lovers on the Nile. - London : Collins, 1980
254p,[16]p of plates : ill. - Bibliography: p.233-237. - Includes index
0-00-216471-x X.800/28775
2069423

HALL, Richard, 1925-
Lovers on the Nile. - London : Quartet, 1981, c1980
254p,[8]p : ill. - Bibliography: p233-237. - Includes index
0-7043-3365-1 X.808/34466
2069423

AFRICA
Exploration by Europeans, 1769-1889

HIBBERT, Christopher
Africa explored. - London : Allen Lane, 1982
335p,[24]p of plates : ill. - Bibliography: p321-325. - Includes index
0-7139-1371-1 X.800/34641
2836971

Africa explored. - Harmondsworth : Penguin, 1984, c1982
336p,[12]p of plates : ill. - Bibliography: p321-325. - Includes index
0-14-005692-0 X.808/40292
2836971

AFRICA
Exploration by Magyar, László

KRIZSÁN, László
Magyar László. - Budapest : Akadémiai Kiadó, 1983
227p. - (A múlt magyar tudósai). - Includes bibliographies
963-05-3046-5 X.808/38043
8003475

AFRICA
Exploration by Piaggia, Carlo - Biographies

PIAGGIA, Carlo
Due anni tra i cannibali. - Torino : SEI, 1981
207p : ill. - (Collana reporter)
88-05-03726-5 X.808/39363
8003475

AFRICA
Exploration by Piaggia, Carlo - Catalogues

CARLO PIAGGIA E L'AFRICA (1979 : Nairobi)
Carlo Piaggia e l'Africa. - Lucca : Pacini Fazzi, 1979
173p : ill. - (La balestra ; n.5). - Bibliography
r6503261 X.800/32166
8003475

AFRICA
Exploration by Poles, to date

GOŁĄBEK, Stefan
Związki Polski i Polaków z Afryką do roku 1945. - Warszawa : Państwowe Wydawnictwo Naukowe, 1978
100p
r6483562 X.800/33280
8840466

AFRICA
Exploration by Teleki, Sámuel

ERDÉLYI, Lajos
Teleki Samu Afrikában. - Bukarest : Kriterion Könyvkiadó, 1977
82p,[48]p of plates : ill
r6449459 X.800/36386
8003475

TELEKI Sámuel gróf felfedező útja Kelet-Afrika egyenlítői vidékein 1887-1888-ban ... / [leírta kísérője Höhnel Lajos] ; [a művet sajtó alá rendezte, az utószót írta Véber Károly]. - Budapest : Magvető Könyvkiadó, 1984
2v.. - (Magyar hírmondó). - Includes Bibliography
963-14-0181-2 X.808/39997
8003475

AFRICA
Exploration. Grogan, Ewart Scott - Biographies

FARRANT, Leda
The legendary Grogan. - London : Hamilton, 1981
xii,260p,[8]p of plates : ill. - Bibliography: p250-252. - Includes index
0-241-10592-7 X.809/48944
2541734

AFRICA
Famines - Conference proceedings

La FAMINE en Afrique / editeur James P. Carter. - Oxford : Pergamon, 1982
xvi,42p
0-08-028885-5 X.950/15451
2776987

FAMINE in Africa / editor James P. Carter. - Oxford : Pergamon, 1982
xvi,35p
0-08-027998-8 X.529/47899
2736780

AFRICA
Foreign assistance by capitalist countries, 1945-

BASKIN, V. S. (Vladimir Sergeevich)
Kapitalisticheskaīa "pomoshch" razvivaīushchimsīa stranam. - Moskva : Nauka, Glavnaīa redakt͡sii͡a vostochnoĭ literatury, 1982
231p. - Bibliography
r6462074 X.529/52832
890264x

AFRICA
Foreign assistance by China, 1960–
SOVIET and Chinese aid to African nations / edited by Warren Weinstein, Thomas H. Henriksen. – New York : Praeger, 1980
xv,184p
0–03–052756–2 X:800/38118
9147020

AFRICA
Foreign assistance by Soviet Union, 1960–
SOVIET and Chinese aid to African nations / edited by Warren Weinstein, Thomas H. Henriksen. – New York : Praeger, 1980
xv,184p
0–03–052756–2 X:800/38118
9147012

AFRICA
Foreign investment by investors from Middle Eastern petroleum exporting countries
CHIBWE, Ephraim Chipampe
Arab dollars for Africa. – London : Croom Helm, 1976
[5],147p,[16]p of plates. – Includes index
0–85664–174–x X.529/30816
0678082

AFRICA
Foreign relations
CALVOCORESSI, Peter
Independent Africa and the world. – London : Longman, 1985
151p : ill. – Includes index
0–582–29654–4 X.529/70353
3015777

The POLITICAL economy of African foreign policy / edited by Timothy M. Shaw, Olajide Aluko. – Aldershot : Gower, c1984
xiii,397p. – Bibliography: p380–392. – Includes index
0–566–00380–5 X.809/62506
3016854

AFRICA
Foreign relations, to 1976
The FOREIGN policies of African states / edited by Olajide Aluko. – London : Hodder and Stoughton, 1977
vi,243p
0–340–21030–3 X.700/25475
1172344

KONTCHOU KOUOMEGNI, Augustin
Le système diplomatique africain. – Paris : Éditions A. Pédone, 1977
ix,279p : ill. – Includes bibliographies
2–233–00037–4 X.700/19674
1172344

MAZRUI, Ali Al'Amin
Africa's international relations. – London : Heinemann Educational [etc.], 1977
ix,310p. – Includes index
0–435–96520–4 X.809/43173
1172344

Africa's international relations. – London : Heinemann [etc.], 1979
ix,310p. – Includes index
0–435–96521–2 X.709/33786
1172344

AFRICA
Foreign relations, to 1976 – Serials
READINGS in African international relations / compiled and edited with commentary by Yashpal Tandon. – Nairobi : East African Literature Bureau, 1972–
r6071352 X.0809/625
1172344

AFRICA
Foreign relations, 1960–
EJIOFOR, Lambert U.
Africa in world politics. – Onitsha : Africana Educational Publishers, 1981
154p
r6810747 X.808/39231
904759x

AFRICA
Foreign relations. United States. Policies of United States government
COHEN, Barry
The black and white minstrel show. – Nottingham : Bertrand Russell Peace Foundation for 'Spokesman', 1977
[1],21p. – ('Spokesman' pamphlet ; no.58)
B79–13396 X.0702/139(58)
1842609

AFRICA
Foreign relations with Arab countries
BESHIR, Mohamed Omer
Terramedia. – London : Ithaca, 1982
205p. – Bibliography: p173–205
B82–33285 X.529/49341
2768518

AFRICA
Foreign relations with China, 1955–1974
HUTCHISON, Alan
China's African revolution. – Boulder : Westview Press, 1976
313p
0–89158–025–5 X.809/44974
0283975

AFRICA
Foreign relations with China & Soviet Union, 1926–1970 – Bibliographies
LINIGER–GOUMAZ, Max
Bibliographies Africaines. – Genève : Les Éditions du Temps, 1971–
tom.1–
r6012347 X.0909/953
1010204

AFRICA
Foreign relations with communist countries
AFRICA and international communism / edited by David E. Albright. – London : Macmillan, 1980
ix,278p. – (Contemporary African issues)
0–333–29121–2 X.802/11300
2019655
– [Another copy: paperback] X.700/26583

AFRICA
Foreign relations with European Community countries
EUROPE & Africa / edited by Margaret Cornell. – London : Overseas Development Institute, c1981
vi,106p : ill
0–85003–075–7 X.809/48364
2505932

AFRICA
Foreign relations with European Community countries – Conference proceedings
AFRICA and the EEC in the aftermath of Lomé and UNCTAD IV / edited by A.G. Kemp. – Aberdeen : Aberdeen University African Studies Group, 1977
v,101p,[3]leaves of plates : ill. – Bibliography: p.16–17
B79–00111 X.512/6191
1739646

AFRICA
Foreign relations with Israel. Policies of Israeli government, 1948–1974
PONOMAREV, D. K. (Dmitriĭ Konstantinovich)
Politika Izraili͡a v Tropicheskoĭ Afrike. – Moskva : Nauka, Glavnai͡a redakt͡sii͡a vostochnoĭ literatury, 1981
162p
r6432240 X.808/37050
8861366

AFRICA
Foreign relations with Soviet Union, 1917–1977
USSR and countries of Africa / general editor E.A. Tarabrin ; [translated from the Russian by David Fidlon and Stanislav Ponomarenko]. – Moscow : Progress, c1980
319p. – Includes index
0–7147–1684–7 X.809/51499
2712768

AFRICA
Foreign relations with United States
HOWE, Russell Warren
Along the Afric shore. – New York : Barnes and Noble, 1975
197p
0–06–493021–1 X.809/45251
0679895

US policy toward Africa / edited by Frederick S. Arkhurst. – New York ; London : Praeger [for] the Phelps-Stokes Fund ; [London] : [Distributed by Martin Robertson], 1975
xii,259p. – (Praeger special studies in international politics and government). – Includes index
0–275–05330–x X.700/25179
0679895

AFRICA
Foreign relations with United States, 1960–
JACKSON, Henry F.
From the Congo to Soweto. – New York : William Morrow, 1982
324p
0–688–00772–4 X.529/68575
8965617

JINADU, L. Adele
Human rights and US African policy under President Carter. – [Lagos] : Nigerian Institute of International Affairs, 1980
iii,120p. – (NIIA monograph series ; no.5)
r6807689 X.529/54751
8965617

AFRICA
Foreign relations with Western world, to date
CHINWEIZU
The West and the rest of us. – London : Nok Publishers, [1978]
xxxiv,544p. – Includes bibliography
0–88357–015–7 X.520/36526
9162097

AFRICA
Foreign trade, to date. Legal aspects
LEGAL aspects of doing business with Black Africa / Donald R. Allen, chairman and editor. – New York : Practising Law Institute, c1981
2v.. – (Commercial law and practice, course handbook series ; no.252)
r6813286 OPL
9072489

AFRICA
Foreign trade with Cadiz, 1400–1600
RUMEU DE ARMAS, Antonio
Cádiz, metrópoli del comercio con África en los siglos XV y XVI. – [Cádiz] : Caja de Ahorros de Cádiz, 1976
56p.,4p. of plates
84–500–7565–3 X.519/23496
1326309

AFRICA
Former French colonies. Role of French language, to date
OUMAR BA
La langue française après la décolonisation. – Paris : La Pensée Universelle, 1980
96p
r6211718 X.808/39912
9090614

AFRICA
Former German colonies. Influence of German political movements, to date
The NAZIS in Africa / edited by L. Smythe Barron. – Salisbury, N.C. : Documentary Publications, [1978]
vii,207p. – (Lost Documents on the Third Reich ; volume 3)
0–89712–076–0 X.805/1956
8784744

AFRICA
Franciscan missions. Kevin, Mother – Biographies
O'HARA, S.
Dare to live!. – Dublin : Veritas, 1979
22p
0–905092–79–1 X.208/9127
2989093

AFRICA
French colonies. Colonial administration, 1885–1960
BRUNSCHWIG, Henri
Noirs et Blancs dans l'Afrique noire française, ou, Comment le colonisé devient colonisateur (1870–1914). – [Paris] : Flammarion, 1982
243p : ill. – (Nouvelle bibliothèque scientifique). – Includes bibliography
r6298171 X.809/63185
8619999

AFRICA
French colonies. Colonial administration, 1900–1945
SURET–CANALE, Jean
French colonialism in tropical Africa, 1900–1945. – London : Heinemann, 1976
xvii,521p. – Bibliography: p.491–507. – Includes index
0–435–89220–7 X.709/30335
0607851

AFRICA
French colonies. Colonial administration. Deschamps, Hubert – Biographies
DESCHAMPS, Hubert
Roi de la brousse. – Nancy : Berger-Levrault, 1975
359 p. : ill
2–7013–0040–1 X.709/20474
0853445

AFRICA
Freshwater Gastropoda - For medicine
BROWN, D. S. (David Seymour)
Freshwater snails of Africa and their medical importance. - London : Taylor & Francis, 1980
x,487p : ill. - Includes bibliographies and index
0-85066-145-5 SRL
2070669

AFRICA
Freshwater regions. Water birds
BROWN, Leslie, 1917-
Birds of the African waterside. - London : Collins, 1979
[111]p : ill(chiefly col)
0-00-216079-x SRL
1778161

AFRICA
Frontiers - Encyclopaedias
BROWNLIE, Ian
African boundaries. - London : C. Hurst [etc.] for the Royal Institute of International Affairs, 1979
xxxvi,1355p. - Includes bibliographies and index
0-903983-87-7 X.800/27875
1567365
- [Another copy] OPL

AFRICA
Frontiers. Disputes
BOUTROS-GHALI, Boutros
Les conflits de frontières en Afrique. - Paris : Éditions Techniques et Économiques, [1973]
158p
r6059018 X.709/25569
1504312

AFRICA
Game wardens
KINLOCH, Bruce, 19—
Game wardens in Africa. - St. Albans : Hart-Davis, c1981
32p : col.ill. - (Natural science series). - Includes index
0-247-13155-5 X.322/10308 Woolwich
2569876

AFRICA
Garnier, Christine - Biographies
GARNIER, Christine
Jusqu'où voient mes yeux. - Paris : Éditions Robert Laffont, 1975
365 p.: 8 p. of plates. - (Vécu)
r6000106 X.700/16431
0904430

AFRICA
Geographical features
An ADVANCED geography of Africa / John I. Clarke (editor) ... [et al.]. - Amersham : Hulton, 1975
528p : ill. - Includes index
0-7175-0600-2 2060.ff Reading Room
0021253
- [Another copy] X.520/8671

AFRICA / H.A. Curtis ... [et al.]. - London : Evans, 1979
91p : ill(some col.)
0-237-50663-7 X.805/4261
0090026

AFRICA and the islands / by R.J. Harrison Church ... [et al.]. - 4th ed. - London : Longman, 1977
xvii,542p : ill. - (Geographies, an intermediate series). - Includes bibliographies and index
0-582-35197-9 9999.aa.2/12
0090026

AFRICAN geography for schools / [edited by B. Brouillette, Norman J. Graves, Geoffrey Last]. - London : Longman [for] Unesco, 1974 [i.e.1975]
ix,309p : ill. - (Source books / Unesco). - Bibliography: p.286-293. - Includes index
0-582-60307-2 X.802/10010
0359068

BEST, Alan C. G.
African survey. - New York ; London : Wiley, 1977
xiii,626p : ill. - Includes bibliographies and index
0-471-20063-8 X.802/10353
0021253

GROVE, Alfred Thomas
Africa. - 3rd ed. - Oxford : Oxford University Press, 1978
xi,337p : ill. - Bibliography: p.320-326. - Includes index
0-19-913244-5 X.802/10656
0021253

HICKMAN, Gladys
The new Africa. - 3rd ed.. - London : Hodder and Stoughton, 1980
192p,[4]p of plates : ill. - Includes index
0-340-25343-6 X.800/33043
0090026

HICKMAN, Gladys Minnie
The new Africa. - 2nd ed. - London : Hodder and Stoughton, 1976
179p,[4]p of plates : ill. - Includes index
0-340-20959-3 X.700/25331
0090026

JARRETT, Harold Reginald
Africa. - 5th ed. - Plymouth : Macdonald and Evans, 1979
xxi,598p : ill. - (The new Certificate geography series. Advanced Level). - Bibliography: p.580. - Includes index
0-7121-0153-5 9999.d.1/28
0090026

PRITCHARD, John Malcolm
Africa. - Revised metric ed. - London : Longman, 1979
vii,282p : ill. - Includes index
0-582-64636-7 X.702/6111
0021253

RICHARDSON, Geoffrey Alan
A geographical notebook of Africa. - London : Hodder and Stoughton, 1977
[4],60p : 1 ill. - (Basic geographical notebooks ; 4)
0-340-21276-4 X.700/25608
0090026

SALL, Mamadou
La géographie de l'Afrique. - Dakar : Les Nouvelles éditions africaines, 1979
224p : ill
2-7236-0436-5 X.800/36877
0021253

SOURCE book on African geography / edited by John Renner, John Macaulay, Colin Gunn. - Wellington, N.Z. : Hicks Smith, 1977
212p : ill. - Includes bibliographies and index
0-456-02230-9 X.709/33044
0021253
- [Another copy] X.809/59139

STEEL, Robert W. (Robert Walter)
Africa. - 2nd ed. - London : Longman, 1981
328p : ill. - (Geographies). - Includes index
0-582-33079-3 9999.b.2/16
0090026

WHITE, Richard, 1943-
Africa. - London : Heinemann Educational, 1978
viii,208p : ill. - Bibliography: p.203. - Includes index
0-435-34916-3 X.700/25674
0090026

Africa. - 2nd ed. - London : Heinemann Educational, 1984
vii,216p : ill. - Bibliography: p211. - Includes index
0-435-34917-1 X.800/37948
0090026

AFRICA
Geographical features - For African students
PRITCHARD, J. M. (John Malcolm)
Africa. - London : Arnold, 1984
58p : ill
0-7131-8071-4 X.800/41370
3079155

TURNER, Howard
Africa. - Harlow : Longman, 1983
204p : ill. - Includes index
0-582-60381-1 X.800/37796
3079155

AFRICA
Geographical features - For East African students
MINNS, W. J.
A geography of Africa. - London : Macmillan, 1978
[4],188p : ill
0-333-21928-7 X.700/25846
1735152

A geography of Africa. - New ed. - London : Macmillan, 1984
iii,195p : ill. - Includes index
0-333-37170-4 X.800/40676
1735152

AFRICA
Geological features
CAHEN, L. (Lucien)
The geochronology and evolution of Africa. - Oxford : Clarendon, 1984
xiii,512p : ill. - Bibliography: p455-496. - Includes index
0-19-857544-0 SRL
3168484

AFRICA
Geological features - Serials
JOURNAL of African earth sciences. - Vol.1, no.1 (1983)-. - Oxford : Pergamon, 1983-
v. : ill
B83-40353 SRL
2959224

AFRICA
Gestures
HOCHEGGER, Hermann
Le langage des gestes rituels. - Bandundu : Ceeba Publications, [1981]- : ill. - (CEEBA Publications. série 2 ; v.66, etc.). - Bibliography
r6207281 X.0950/411
8856141

AFRICA
Government departments. Names - For cataloguing
AFRICAN legislative and ministerial bodies / compiled by the IFLA International Office for UBC in association with the African Standing Conference on Bibliographic Control - London : The Office, 1980
viii,37p. - Includes index
0-903043-25-4 2714.abc.38
2146460
- [Another copy.] OPL

AFRICA
Government departments responsible for cooperatives - Directories
INTERNATIONAL CO-OPERATIVE ALLIANCE
Directory of addresses of government ministries/departments. - London : I.C.A., [1978]
ii,28,[7]p
B78-22553 Cup.1285/307
1493310

AFRICA
Governments, 1960-
OBSHCHESTVO i gosudarstvo v Tropicheskoi Afrike / [otvetstvennyi redaktor Anat. A. Gromyko]. - Moskva : Nauka, Glavnaia redaktsiia vostochnoi literatury, 1980
275p
r6483036 X.529/43927
8752125

PAMBOU TCHIVOUNDA, Guillaume
Essai sur l'État africain postcolonial. - Paris : Librairie Générale de Droit et de Jurisprudence, 1982
165p. - (Bibliothèque africaine et malgache ; 37). - Includes bibliography
2-275-01055-6 X.800/39157
8752125

AFRICA
Governments. Bureaucracy
PRICE, Robert M. (Robert Morris)
Society and bureaucracy in contemporary Ghana. - Berkeley ; London : University of California Press, 1975
xiv,261p : ill. - Bibliography: p.245-255. - Includes index
0-520-02811-2 X.800/25394
0592382

AFRICA
Governments. Decentralisation. Political aspects
LOCAL government in the Third World / edited by Philip Mawhood. - Chichester : Wiley, c1983
xiii,261p. - (Public administration in developing countries) (A Wiley series on public administration in developing countries). - Includes bibliographies and index
0-471-10510-4 X.800/36113
2948664

AFRICA
Governments. Ministers, 1950-1976
BIDWELL, Robin
Guide to African ministers. - London : Collings, 1978
[4],79p
0-86036-064-4 X.802/10659
1570366
- [Another copy] OPL

AFRICA
Guerrilla warfare, to 1980
DAVIDSON, Basil
The people's cause. - London : Longman, 1981
xi,210p : ill. - (Longman studies in African history). - Includes index
0-582-64680-4 X.800/31936
2597616
- [Another copy: paperback]
ISBN 0-582-64681-2 X.800/31937

AFRICA
Hawk moths
CARCASSON, R. H. (Robert Herbert)
Revised catalogue of the African Sphingidae (Lepidoptera) with descriptions of the East African species. - 2nd ed. - Faringdon : Classey, 1976
[2],148,[7]p,[2],xvii p of plates : ill. - Bibliography: p.135. - Includes index
0-900848-85-5 SRL
0754374

AFRICA
Heads of state - Biographies
VAN RENSBURG, A. P. J.
Contemporary leaders of Africa. - Cape Town : Haum, 1975
x, 529 p.
0-7986-0156-6 X.800/25808
0778907

AFRICA
Heads of state, to date
BAUM, Edward
Chief executives in black Africa and Southeast Asia. - Athens, Ohio : Ohio University Center for International Studies, c1976
viii,35p. - (Papers in International Studies. Southeast Asia series ; no.39) (Papers in International studies. Africa series ; no.29)
0-8214-0310-9 X.805/3337
8003572

AFRICA
Heads of state, to date - Biographies
ONADIPE, Kola
Builders of Africa. - Ijebu-Ode : Natona, 1980
91p : ill
978-17-8004-5 X.808/38081
8003572

AFRICA
Higher education
PORTER, Arthur T.
The concept of the African university. - [Edinburgh] : [Centre of African Studies, University of Edinburgh], [1984?]
9p. - (The Africanus Horton memorial lecture)
r6815714 X.529/61108
3069338

AFRICA
Higher education institutions. Curriculum subjects, to date: African studies - Conference proceedings
CONFERENCE OF THE INSTITUTES/CENTRES OF AFRICAN/CULTURAL STUDIES IN AFRICA (1978 : Khartoum)
African studies in Africa. - [Khartoum] : Institute of African and Asian Studies, University of Khartoum, 1981
viii,152p. - (Sudan library series ; 10)
r6807399 X.520/32911
9061185

AFRICA
Historical demography
AFRICAN historical demography. - Edinburgh : Centre of African Studies, University of Edinburgh.
Vol.2: Proceedings of a seminar held in the Centre of African Studies, University of Edinburgh, 24th and 25th April 1981. - c1981
802p : ill. - Includes bibliographies
B84-38772 X.529/64648
3167879

AFRICA
Historiography
ERIKSEN, Tore Linné
Modern African history. - Uppsala : Scandinavian Institute of African Studies, 1979
27p. - (Nordiska afrikainstitutet. Research report ; no 55)
91-7106-167-3 P.803/212[no.55]
226188X

FILESI, Teobaldo
Realtà e prospettive della storiografia africana. - [Napoli] : Giannini, 1978
317p. - (Quaderni della Facolta di Scienze politiche / Università di Napoli ; vol.5). - Bibliography
r6195933 X.800/29582
226188X

AFRICA
Human rights, 1960-
JINADU, L. Adele
Human rights and US African policy under President Carter. - [Lagos] : Nigerian Institute of International Affairs, 1980
iii,120p. - (NIIA monograph series ; no.5)
r6807689 X.529/54751
8965595

AFRICA
Income. Distribution related to employment
FRY, James, 1946-
Employment and income distribution in the African economy. - London : Croom Helm, 1979
3-186p : ill. - Includes index
0-85664-715-2 X.529/33267
1737570

AFRICA
Income. Distribution. Inequalities
A WORLD of differentials / Ragia Abdin ... [et al.]. - London : Hodder and Stoughton in association with the International Development Research Centre, 1983
152p : ill. - Bibliography: p145-148. - Includes index
0-340-33415-0 X.520/32046
2986213

AFRICA
Independence movements, 1885-
BEFREIUNGSBEWEGUNGEN in Afrika / Rainer Falk, Peter Wahl (Hrsg.). - Koln : Pahl-Rugenstein, 1980
532p. - (Kleine Bibliothek ; 204)
3-7609-0551-x X.808/34959
8747717

DAVIDSON, Basil
[Africa in modern history]. Let freedom come. - Boston : Little, Brown and Co., c1978
431p. - (An Atlantic Monthly Press book)
0-316-17435-1 X.809/58119
8747717

LEADERSHIP and national liberation movement in Africa / edited by Thea Büttner. - Berlin : Akademie-Verlag, 1980
144p. - (Asia, Africa, Latin America. Special issue ; 7)
r6309225 X.800/37924
8747717

AFRICA
Independent Christian churches, to date
KORTE, Werner
Organisation und gesellschaftliche Funktion unabhängiger Kirchen in Afrika. - Wiesbaden : Heymann, 1977
253p. - (Studien und Materialien der anthropologischen Forschung ; Bd.2, Nr.1) (Edition Ethnos). - Bibliography
3-88055-025-5 X.0520/618(2)[pt.1]
875649x

AFRICA
Industrialisation, to 1980
INDUSTRY and accumulation in Africa / edited by Martin Fransman. - London : Heinemann, 1982
438p : ill. - (Studies in the economics of Africa). - Bibliography: p416-431. - Includes index
0-435-97139-5 X.529/49667
2778793

AFRICA
Industries - Serials
MODERN Africa. - Vol.1, no.1- ; Jan./Feb. 1977-. - London : Johnston International Publishing Corp, 1977- : ill
Holdings: vol.3, no.1 ; Jan / Feb. 1979, etc
B80-20348 P.523 / 1342
2142708

AFRICA
Industries, to 1900
MODES of production in Africa / Donald Crummey and C.C. Stewart, editors. - Beverly Hills ; London : Sage, c1981
256p. - (Sage series on African modernization and development ; v.5). - Includes bibliographies and index
0-8039-1133-5 X.529/54019
289758x

AFRICA
Industries & trades - Directories - Serials
OWEN'S. - 29th ed.-. - Bulle, Switzerland ; Alperton : Owen's Publications, 1982-
v. : ill
B82-27658 PP2579cdy
2750058

AFRICA
Influence of Great Britain, 1780-1850
CURTIN, Philip D. (Philip DeArmond)
The image of Africa. - Madison ; London : University of Wisconsin Press.
Vol.1. - 1973
xvi,286p,[8]p of plates : ill
0-299-83025-x X.0709/1011(1)
0570338

AFRICA
Influence of Islam
TRIMINGHAM, John Spencer
The influence of Islam upon Africa. - 2nd ed.. - London : Longman [etc.], 1980
xi,182[2]p. - (Arab background series). - Bibliography: p.165-168. - Includes index
0-582-78499-9 X.200/34376
1935526

AFRICA
Influence of Scotsmen, 1600-1980
SCOTLAND & Africa. - Edinburgh : National Library of Scotland, 1982
63p : ill. - Includes index
0-902220-50-0 X.800/33721
2761017

AFRICA
Influence on Caribbean writers of Western European literatures, 1900-1970
BANGOU, Henri
Influencia del África en las literaturas antillanas. - Montevideo : I.L.A.C., 1972
54p. - (ILAC ; 1)
r6101151 X.909/84551
2095866

AFRICA
Initiation rites
NTETEM, Marc
Die negro-afrikanische Stammesinitiation. - Münsterschwarzach : Vier-Türme-Verlag, 1983
xlvi,381p. - (Münsterschwarzacher Studien ; Bd 36). - Includes bibliography
3-87868-161-5 X.200/41798
9017674

AFRICA
Insects
SKAIFE, Sydney Harold
African insect life. - New ed.. - London : Country Life Books, 1979
279p,[72]p of plates : ill(some col). - Includes bibliographies and index
0-600-34920-9 SRL
0178438

AFRICA
Intergovernmental organisations
ĖTINGER, Īākov Īākovlevich
Mezhgosudarstvennye organizat͡sii stran Azii i Afriki. - Moskva : Izdatel'stvo "Nauka", Glavnai͡a redakt͡sii͡a vostochnoĭ lite ratury, 1976
189 p.
r6020428 X.519/21149
0892912

AFRICA
Intergovernmental organisations, 1958-1969
BASIC documents of African regional organizations / edited by Louis B. Sohn. - Dobbs Ferry : Oceana Publications for the Inter-American Institute of International Legal Studies, 1971-
4v.(x,1854p). - Includes bibliographies
0-379-00361-9 X.802/10504
1340379

AFRICA
Internal migration. Effects on African culture, to date
BINET, Jacques
Migrations et appauvrissement culturel en Afrique. - Paris : Centre de Hautes Études sur l'Afrique et l'Asie Modernes, 1980
82leaves. - (Cahiers du C.H.E.A.M. ; no.9)
2-903182-03-5 X.525/5821
8901805

AFRICA
International relations with West Germany, 1960- - Serials
AFRIKA und die Deutschen. - 1981-. - Pfullingen : Neske, 1981-
r6304718 X.0809/2227
8876770

AFRICA
Intervention by great powers, to 1982
GAVSHON, Arthur
Crisis in Africa. - Epping : Bowker, [1984?]
320p. - Includes index
0-86531-680-5 X.520/35946
322368x

AFRICA
Intervention by great powers, 1960-1979
CHALIAND, Gérard
The struggle for Africa. - London : Macmillan, 1982
xiii,121p. - (Contemporary African issues series). - Bibliography: p115-120. - Includes index
0-333-32395-5 X.529/52786
2871459

AFRICA
Islam, 1960–
WESTERLUND, David
From socialism to Islam?. – Uppsala : Scandinavian Institute of African Studies, 1982
62p. – (Research report / Scandinavian Institute of African Studies ; no.61). – Bibliography
91–7106–203–3 P.803/212[no.61]
8882606

AFRICA
Islam. Relations with Christian church – Serials
BULLETIN on Islam and Christian–Muslim relations in Africa. – Vol.1, no.1 (Jan. 1983)–. – Birmingham : Centre for the Study of Islam and Christian–Muslim Relations, 1983–
v.
B83–42558 P.201/638
2977419

AFRICA
Islands
AFRICAN islands and enclaves / Robin Cohen, editor. – Beverley Hills ; London : Sage, 1983
279p. – (Sage series on African modernization and development ; v.7). – Includes bibliographies
0–8039–1966–2 X.809/57048
2985802

AFRICA
Italian colonies, 1885–1960
ROSSI, Gianluigi
L'Africa italiana verso l'indipendenza (1941–1949). – [Milano] : Giuffrè, 1980
xviii,626p. – ([Pubblicazioni] / Facoltà di scienze politiche, Università di Roma ; 33). – Includes bibliography
r6507326 X.800/35276
8916365

AFRICA
Italian colonies. Guerrilla wars, to 1946
BANDINI, Franco
Gli italiani in Africa. – Milano : Mondadori, 1980
425p. – (Gli Oscar ; 1218. Documenti ; 51)
r6506757 X.808/38036
8950180

AFRICA
Italian exploration. Political aspects, 1867–1900
SACCONE, Salvatore
Aspetti politici ed economici nell'esplorazione italiana dell'Africa, 1867–1900. – Bologna : Pàtron, 1976
113p.,[1]fold. leaf of plates
r6078203 X.709/25671
1511351

AFRICA
Items associated with Africa in stock of Ipswich (County Borough). Department of Recreation and Amenities
JONES, D. L. (David Lloyd), 1948–
Africa. – Ipswich : Department of Recreation and Amenities, [1979?]
[15]p : ill
0–906688–01–9 L.49/338
2065657

AFRICA
Journeys by cars, 1933
GILG, A. Cameron (Alan Cameron)
Turn left – the Riffs have risen. – London : Royal Automobile Club, 1981
191p : ill. – Includes index
0–86211–016–5 X.950/2327
2367157

AFRICA
Land. Surveying – Conference proceedings
ROLE OF SURVEYING AND LAND ECONOMY IN NATIONAL DEVELOPMENT IN AFRICA (Conference : 1978 : Nairobi)
The role of surveying and land economy in national development in Africa. – London : Commonwealth Association of Surveying and Land Economy, 1978
[4],150p : ill
B79–13437 X.702/5994
184380x

AFRICA
Land use
LANDUSE and development / editors Phil O'Keefe and Ben Wisner. – London : International African Institute [for] the Environment Training Programme, 1977
[6],232p. – (African environment special report ; 5). – Includes bibliographies
0–85302–058–2 P.701/1503
1734814

AFRICA
Land use. Management – Conference proceedings
ROLE OF SURVEYING AND LAND ECONOMY IN NATIONAL DEVELOPMENT IN AFRICA (Conference : 1978 : Nairobi)
The role of surveying and land economy in national development in Africa. – London : Commonwealth Association of Surveying and Land Economy, 1978
[4],150p : ill
B79–13437 X.702/5994
1843796

AFRICA
Landforms
BUCKLE, Colin
Landforms in Africa. – London : Longman, 1978
[5],249p : ill. – Includes bibliographies and index
0–582–60334–x X.700/25666
1444697

PRITCHARD, John Malcolm
Landform and landscape in Africa. – London : Edward Arnold, 1979
160p : ill. – Bibliography: p.151–152 – Includes index
0–7131–0204–7 X.702/6215
1193910

AFRICA
Languages – Conference proceedings
AFRICA LANGUAGES CONGRESS OF UNISA (3rd : 1979 : University of South Africa)
Third Africa Languages Congress of Unisa. – Pretoria : University of South Africa, [1980]
vii,389p : ill. – (Miscellanea Congregalia ; 12). – Includes bibliographies
0–86981–206–8 X.950/21894
8003637

AFRICA
Languages – Serials
AFRICAN languages = Langues africaines. – Vol.1– ; 1975–. – London : International African Institute, 1975– : ill
B78–02223 P.983/106
0032972

AFRICA
Languages, to date. Political aspects
REH, Mechthild
Sprachpolitik in Africa. – Hamburg : Buske, 1982
332p.
3–87118–510–8 X.950/35953
9122346

AFRICA
Languages, 1885– – Bibliographies
REH, Mechthild
Problems of linguistic communication in Africa. – Hamburg : Buske, 1981
230p.. – (African linguistic bibliographies ; v.1)
3–87118–509–4 X.950/24518
8993629

AFRICA
Legislatures. Names – For cataloguing
AFRICAN legislative and ministerial bodies / compiled by the IFLA International Office for UBC in association with the African Standing Conference on Bibliographic Control. – London : The Office, 1980
viii,37p. – Includes index
0–903043–25–4 2714.abc.38
2146479
– [Another copy.] OPL

AFRICA
Libraries – Directories
The AFRICAN book world & press, a directory = Répertoire du livre et de la presse en Afrique / compiled by the 'African book publishing record' ; Hans M. Zell, editor. – Oxford : Hans Zell (Publishers) Ltd, 1977
xxvi,[1],296,[1]p
0–905450–00–0 X.955/307
1169343
– [Another copy] 2719.ff.92

The AFRICAN book world and press / edited by Hans M. Zell = Répertoire du livre et de la presse en Afrique / editée par Hans M. Zell. – 2nd ed.. – London : Zell, 1980
xxiv,244p : ill
0–905450–06–x X.525/5661
1169343

The AFRICAN book world and press = Répertoire du livre et de la presse en Afrique. – 3rd ed.. – Oxford : Zell, 1983
xx,285p. – Includes index
0–905450–10–8 2725.g.293
1169343

AFRICA
Livestock. Vectors: Tsetse flies. Control measures, to 1972
MCKELVEY, John J.
Man against tsetse. – Ithaca ; London : Cornell University Press, 1975
xvii,306p : ill. – Bibliography: p.239–292. – Includes index
B77–27114 X.329/11101 Woolwich
1133160

AFRICA
Mammals
AMUCHÁSTEGUI, Axel
Some birds and mammals of Africa. – London : Tryon Gallery, c1979
[41]p.xivleaves of plates : col.ill
B81–36406 C.180.K.3
2554771

HALTENORTH, Theodor
A field guide to the mammals of Africa. – London : Collins, c1980
400p : col.ill. – Bibliography: p391–392. – Includes index
0–00–219778–2 SRL
233741x

KINGDON, Jonathan
African mammal drawings. – Islip : Pangolin, c1983
16,254p : chiefly ill
0–9508837–0–0 Tab.800/17
2554771

AFRICA
Mammals. Evolution
EVOLUTION of African mammals / edited by Vincent J. Maglio and H.B.S. Cooke. – Cambridge, Mass. ; London : Harvard University Press, 1978
xiii,641p : ill. – Includes bibliographies and index
0–674–27075–4 SRL
1890034

AFRICA
Man. Cholera. Epidemiological aspects, 1970–1975
STOCK, Robert F.
Cholera in Africa. – London : International African Institute [for] the Environment Training Programme, 1976
vii,127p : ill. – (African environment special report ; 3). – Bibliography: p.119–127
0–85302–050–7 P.701/1503(3)
1065726

AFRICA
Man. Degenerative diseases. Epidemiology – Conference proceedings
EAST AFRICAN MEDICAL RESEARCH COUNCIL. Annual Scientific Conference (1975 : Dar es Salaam)
Degenerative disorders in the African environment. – Nairobi : East African Literature Bureau, 1976
xv,390p : ill. – Includes bibliographies and index
r6072141 X.311/8843 Woolwich
1551701

AFRICA
Man. Diseases, to 1981
RANSFORD, Oliver
'Bid the sickness cease'. – London : Murray, 1983
x,235p,[8]p of plates : ill. – Bibliography: p223–226. – Includes index
0–7195–3986–2 X.329/16655
2937247

AFRICA
Man. Eyes. Diseases
EVANS, Stanley C.
Prevention of blindness in Nigeria and other African states. – Oulton : Teecoll, 1981
x,235p : ill
0–7109–0008–2 X.329/15628
2780798

AFRICA
Man. Trypanosomiasis. Vectors: Tsetse flies. Control – Serials
TSETSE and trypanosomiasis information quarterly / [Tsetse and Trypanosomiasis Information and News Service]. – Vol.1, parts 1–4– ; 1978–. – London : Centre for Overseas Pest Research, 1978–
B79–17186 SRL
1884638

AFRICA
Man. Vectors: Tsetse flies. Control measures, to 1972
MCKELVEY, John J.
Man against tsetse. – Ithaca ; London : Cornell University Press, 1975
xvii,306p : ill. – Bibliography: p.239–292. – Includes index
B77–27114 X.329/11101 Woolwich
1133152

AFRICA
Manpower
HUMAN resources and African development / edited by Ukandi G. Damachi, Victor P. Diejomaoh. - New York ; London : Praeger, 1978
xix,378p : ill. - (Praeger special studies). - Bibliography: p.349-375
0-03-022826-3 X.520/14622
1969439

AFRICA
Manpower, 1960- - Bibliographies
MIKKELSEN, Britha
Industrial labour in Africa. - Uppsala : Scandinavian Institute of African Studies, 1979
59p
91-7106-158-4 X.529/53036
8933723

AFRICA
Maps, 1486-1886
NORWICH, Oscar I.
Maps of Africa. - Johannesburg : Donker, 1983
444p : ill(some col.)
0-86852-028-4 X.805/5620
3071707

AFRICA
Marriage
CHURCH and marriage in modern Africa / edited by T.D. Verryn. - Groenkloaf : Ecumenical Research Unit, 1975
[3],ii,497p. - Includes bibliographies
r6079113 X.100/17622
1120301

HASTINGS, Adrian
Christian marriage in Africa. - London : S.P.C.K., 1974
vi,185p. - Includes index
0-281-02814-1 X.100/16344
0651451

KISEMBO, Benezeri
African Christian marriage. - London : G. Chapman, 1977
xvii,242p. - Bibliography: p.217-225. - Includes index
0-225-66181-0 X.100/17091
0651451

MCGRATH, Michael
Africa, our way to love and marriage. - London : Cassell and Collier Macmillan, 1977
ix,201p,[8]p of plates : ill. - Bibliography: p.197-198. - Includes index
0-225-66163-2 Cup.358/36
1120301

AFRICA
Marxism, to date
The DEBATE / edited by Yash Tandon. - Dar es Salaam : Tanzania Publishing House, 1982
310p
9-9761001-1-6 X.520/34074
9102159

AFRICA
Mass communication
MYTTON, Graham
Mass communication in Africa. - London : Edward Arnold, 1983
v,159p,[8]p of plates : ill. - Bibliography: p150-152. - Includes index
0-7131-8140-0 X.629/23528
3127532

AFRICA
Medical services: African Medical & Research Foundation International
AFRICAN MEDICAL & RESEARCH FOUNDATION INTERNATIONAL
[Miscellaneous publications not catalogued separately]. - London : the Foundation, 1980-
r6108495 X.0325/100
2298414

AFRICA
Medicine - Serials
AFRICA health. - Vol.1, no.1 (Oct.1978)-. - Sutton, Surrey : IPC Middle East Publishing Company, c1978-
v. : ill(some col.)
B81-28382 SRL
1266780

AFRICA
Medicine. Natural remedies
SOFOWORA, Abayomi
Medicinal plants and traditional medicine in Africa. - Chichester : Wiley, c1982
xviii,256p : ill. - Includes bibliographies and index
0-471-10367-5 SRL
2874253

AFRICA
Metals mining industries. Prain, Sir Ronald - Biographies
PRAIN, Sir Ronald
Reflections on an era. - [Worcester Park] : Metal Bulletin, c1981
262p. - Bibliography: p246. - Includes index
0-900542-52-7 X.800/30985
2528274

AFRICA
Midwifery. Kéita, Diallo Aoua
KÉITA, Diallo Aoua
Femme d'Afrique. - Paris : Présence Africaine, 1975
397 p.
r6012208 X.709/20699
0871060

AFRICA
Military governments, to date
Os MILITARES e o poder. - Lisboa : Publicações Dom Quixote, 1975
138p. - (Colecção Novos cadernos D. Quixote ; 27)
r6702372 X.0708/491(27)
800370x

AFRICA
Military operations by British military forces, 1887-1920
MAGOR, R. B.
African general service medals. - [Wadhurst] : [The author], [1979]
[3],vi,193p,[9]leaves of plates(3 fold) : ill. - Bibliography: p.192-193
B79-29821 X.800/28068
1939076

AFRICA
Mineral deposits - Bibliographies
RIDGE, John Drew
Annotated bibliographies of mineral deposits in Africa, Asia (exclusive of the USSR) and Australasia. - Oxford : Pergamon, 1976
viii,547p. - Includes index
0-08-020459-7 SRL
0609161

AFRICA
Monarchy. Effects of social change. Political aspects
AFRICAN kingships in perspective / edited by René Lemarchand. - London : Cass, 1977
xiv,325p : ill. - Bibliography: p.305-314. - Includes index
0-7146-3027-6 X.800/26024
1083457

AFRICA
Multinational companies - Conference proceedings
MULTINATIONAL CORPORATIONS IN AFRICA (Conference : 1974 : Dakar)
Multinational firms in Africa. - Uppsala : Scandinavian Institute of African Studies, 1975
vii,425p : ill. - Includes bibliographies
91-7106-093-6 X.519/26911
1118404

AFRICA
Muslims, to 1974
CUOQ, Joseph M.
Les musulmans en Afrique. - Paris : Éditions G.-P. Maisonneuve et Larose, 1975
522 p.. - Bibliography : p. 517-519
2-7068-0603-6 X.102/1068
0853615

AFRICA
National accounting, to date
ANSON-MEYER, Monique
La nouvelle comptabilité des Nations-Unies en Afrique. - Paris : L'Harmattan, 1982
421p. - (Bibliothèque du développement). - Includes bibliography
2-85802-253-4 X.520/35699
9139400

AFRICA
Nationalism
AUMA-OSOLO, Agola
Cause-effects of modern African nationalism on the world market. - Lanham ; London : University Press of America, c1983
xiii,218p : ill. - Bibliography: p199-210. - Includes index
0-8191-3475-9 X.529/64231
3161943

AFRICA
Nationalism, 1960-
NATSIONALIZM v sovremennoi Afrike / [G.B. Starushenko (otvetstvennyi redaktor)]. - Moskva : Nauka, Glavnaia redaktsiia vostochnoi literatury, 1983
261 p.. - Includes indexes. - Includes bibliography: p. 246-[256]
r6403986 X.809/60862
9119892

AFRICA
Nationalism. Role of military forces
OGUERI, Eze
African nationalism and military ascendancy. - Owerri ; London : Conch Magazine Limited (Publishers), 1976
vi,85p. - (Contemporary African politics series). - Includes index
0-914970-13-5 X.809/41620
1044583

AFRICA
Nationalist movements, to 1977. Socioeconomic aspects
UL'IANOVSKII, R. A. (Rostislav Aleksandrovich)
Present-day problems in Asia and Africa. - Moscow : Progress Publishers, c1980
239p
0-7147-1627-8 X.808/34738
2542528

AFRICA
Nationalist movements, 1640-1918
ISTORIIA natsional'no-osvoboditel'noi bor'by narodov Afriki v novoe vremia / [rukovoditel' izdaniia ... V. G. Solodovnikov ; redaktsionnaia kollegiia ... M. Iu. Frenkel' (otv. redaktor) i dr.]. - Moskva : Izdatel'stvo "Nauka", Glavnaia redaktsiia vostochnoi literatury, 1976
634p,2p of plates. - Bibliography : p.590-610
r6021738 X.809/23907
0389994

AFRICA
Nationalist movements, 1885-
AKTUAL'NYE problemy ideologii natsional'no-osvoboditel'nogo dvizheniia v stranakh Azii i Afriki / [redaktsionnaia kollegiia: G.F. Kim ... i dr.]. - Moskva : Nauka, 1982
445p
r6448868 X.529/54843
8964076

GEISTIGE Profile Asiens und Afrikas / Herausgeberkollegium: G.F. Kim ... [et al.]. - Berlin : Akademie-Verlag, 1982
ix,401p.
r6309781 X.800/38676
8964076

AFRICA
Nationalist movements. Foreign assistance by Denmark, 1971-1977
LA COUR, Peter
Danmarks hjælp til såkaldte "frihedsbevægelser". - København : Pluralisme, 1978
123p : ill. - Includes bibliographies
87-980677-0-2 X.709/50285
2195909

AFRICA
Negroes. Immigration from United States, 1787-1863
MILLER, Floyd John
The search for a black nationality. - Urbana ; London : University of Illinois Press, 1975
xv,295p. - (Blacks in the New World). - Bibliography: p.275-282. - Includes index
0-252-00263-6 X.800/25288
0542733

AFRICA
Neocolonialism
NKRUMAH, Kwame
Neo-Colonialism. - London : Nelson, 1965
xx,280p : ill. - Bibliography: p260-262
r6863137 X.800/35173
285810x

AFRICA
Nile River region, 1800-1900
MOOREHEAD, Alan
The Blue Nile. - Rev. ed. - Harmondsworth : Penguin, 1983, c1972
304p,[16]p of plates : ill(some col.). - Bibliography: p294-295. - Includes index
0-14-006673-x X.800/36644
2985187
The White Nile. - Rev. ed. - Harmondsworth : Penguin, 1973, c1971
320p,[24]p of plates : ill(some col.). - Bibliography: p312-314. - Includes index
B84-03348 X.800/36643
2985187

AFRICA
Nile River region. Description & travel

BOLWELL, L. H. (Laurence Henry)
A journey up the Nile. - Hove : Wayland, 1984
64p : ill(some col.). - (River journeys). - Includes index
0-85078-403-4 X.809/61535
3184935

AFRICA
Nomadic communities, to date - Conference proceedings

The NOMADIC alternative / editor: Wolfgang Weissleder. - The Hague : Mouton, [1978]
xviii,423p,[4]p of plates : ill,maps. - (World anthropology). - Includes bibliography
90-279-7520-5 X.800/35727
8974810

AFRICA
Non-passerine birds. Distribution - Atlases

An ATLAS of speciation in African non-passerine birds. - London : British Museum (Natural History), 1978
vii,390p. - (Publications / British Museum (Natural History) ; no.787). - Bibliography: p.375-378. - Includes index
0-565-00787-4 SRL
1590766

AFRICA
North Africa - Bibliographies

CHRISTENSEN, Peter
Litteratur om Mellemøsten og Nordafrika. - Nyt udvalg. - København : Mellemfolkeligt Samvirke, 1979
132p
87-7028-267-6 X.909/86834
2266342

AFRICA
North Africa - Serials

FODOR'S North Africa. - 1980-. - London : Hodder and Stoughton, 1980-
v. : ill
Holdings: 1981, etc
B82-04924 P.701/1757
0219169

TRAVELLER'S guide to North Africa. - [No.1]-. - London : IC Magazines, 1981-
v. : ill
B81-29435 X.808/34646
2530074

AFRICA
North Africa, to 640

DECRET, François
L'Afrique du Nord dans l'antiquité. - Paris : Payot, 1981
391p. - (Bibliothèque historique). - Bibliography
2-228-12900-3 X.809/53057
8853282

GSELL, Stéphane
Études sur l'Afrique antique. - [Lill] : Université de Lille 3, [1981]
310p : ill. - (Travaux et recherches / Université de Lille 3). - Bibliography: p19-35
2-86531-007-8 X.800/34682
8853282

RAVEN, Susan
Rome in Africa. - New ed. - London : Longman, 1984
xxx,268p : ill. - Bibliography: p254-255. - Includes index
0-582-78355-0 X.800/40132
3017443

AFRICA
North Africa, to 1983

AROIAN, Lois A.
The modern Middle East and North Africa. - New York : Macmillan ; London : Collier Macmillan, c1984
xxiv,455p. - Bibliography: p422-431. - Includes index
0-02-304200-1 X.525/9781
3175243

AFRICA
North Africa, 1500-1980

MATTHEWS, Noel
A guide to manuscripts and documents in the British Isles relating to the Middle East and North Africa. - Oxford : Oxford University Press, 1980
xvii,482p. - Includes index
0-19-713598-6 X.950/39652
3249212

AFRICA
North Africa, 1800-1900

MORSY, Magali
North Africa 1800-1900. - London : Longman, 1984
356p : ill. - Includes bibliographies and index
0-582-78376-3 X.800/40015
3218058
- [Another copy: paperback] ISBN 0582783771 X.800/40016

AFRICA
North Africa, 1960-

DUIGNAN, Peter
The Middle East and North Africa. - Stanford : Hoover Institution Press, c1981
xi,141p. - (Hoover Press publication ; 239)
0-8179-7392-3 X.809/51127
8318220

The GOVERNMENT and politics of the Middle East and North Africa / edited by David E. Long and Bernard Reich. - Boulder : Westview Press, c1980
xiv,480p
0-89158-593-1 X.520/25727
8318220

AFRICA
North Africa, 1968-1983. Politico-economic aspects

NORTH Africa / edited by Richard Lawless and Allan Findlay. - London : Croom Helm, 1984
xv,282p : ill. - Includes index
0-7099-1609-4 X.809/59794
3128911

AFRICA
North Africa & Middle East. Western European women, 1763-1904 - Biographies

BLANCH, Lesley
The wilder shores of love. - London : Abacus, 1984
323p. - Bibliography: p312-316. - Includes index
0-349-10329-1 X.808/40779
0144266

AFRICA
North Africa. Ancient Roman ceramics

GARBSCH, Jochen
Zur spätantiken Keramik aus Nordafrika. - Augst-Kaiseraugst : Rei Cretariae Romanae Fautores, 1980
70p,plates. - (Acta RCRF. Supplementa ; v.5)
r6307262 X.0415/233(5)
8962294

AFRICA
North Africa. Ancient Roman civilization

BERTHIER, André
La Numidie, Rome et le Maghreb. - Paris : Picard, 1981
224p,plates. - Includes bibliography
2-7084-0063-0 X.421/25370
8983224

AFRICA
North Africa. Ancient Roman civilization - Conference proceedings

150-JAHR-FEIER Deutsches Archäologisches Institut Rom. - Mainz : von Zabern, 1982
209p,plates. - (Mitteilungen des Deutschen Archaeologischen Instituts, Roemische Abteilung. Ergänzungsheft ; 25)
3-8053-0558-3 07705.m.26/25
8983224

AFRICA
North Africa. Ancient Roman mosaics

DUNBABIN, Katherine M. D.
The mosaics of Roman North Africa. - Oxford : Clarendon Press, 1978 [i.e. 1979]
xx,303p,[8]p of plates,lxxx p of plates : ill(some col). - (Oxford monographs on classical archaeology). - Bibliography: p.280-289. - Includes index
0-19-813217-4 WP.8912/16
117049x

MOSAÏQUE romaine tardive / édité par Yvette Duval. - [Paris] : Institut de Recherche Universitaire, d'histoire de la Connaissance, des Idées et des Mentalités Université de Paris-Val de Marne, [1982]
92p,plates : ill
r6298392 X.425/5419
117049x

AFRICA
North Africa. Antiquities - Bibliographies

REBUFFAT, René
Bibliographie de l'Afrique du Nord antique. - Paris : Presses de l'École normale supérieure, 1980
94p. - (Guides et inventaires bibliographiques / Bibliothèque de l'École normale supérieure ; 1)
2-7288-0071-5 X.425/3498
8855250

AFRICA
North Africa. Byzantine fortifications, 500-800

PRINGLE, Denys
The defence of Byzantine Africa from Justinian to the Arab Conquest. - Oxford : B.A.R., 1981
2v.(xxi,699p) : ill. - (BAR. International series ; 99) - Bibliography: p449-514. - Index
0-86054-119-3 X.425/2250 Woolwich
2556006

AFRICA
North Africa. Catholic missions. Foucauld, Charles de, vicomte - Biographies

DU PERRON, Marguerite Castillon
Charles de Foucauld. - Paris : Grassef, 1982
521p
2-246-27361-7 X.200/44346
9077642

AFRICA
North Africa. Catholic missions. Lavigerie, Charles Martial Allemand - Biographies

GSTREIN, Heinz
Der Karawanenkardinal. - Mödling-Wien : St. Gabriel, c1982
110p : ill. - (Missionare, die Geschichte machten)
3-85264-202-7 X.208/9666
9077642

AFRICA
North Africa. Christian holy places, to date

DUVAL, Yvette
Loca sanctorum Africae. - Rome : École française de Rome, 1982
2v.((xvi,818p)) : ill. - (Collection de l'École française de Rome ; 58). - Bibliography
2-7283-0021-6 Ac.5233.a/58
8932891

AFRICA
North Africa. Christianity. Conversion of non-Christians

BUTCHER, Muriel
By faith -. - 2nd ed - Loughborough : North Africa Mission, 1975
84p
0-904971-00-7 X.108/16112
0527629

AFRICA
North Africa. Christians: Martyrs. Veneration, to 700

DUVAL, Yvette
Loca sanctorum Africae. - Rome : École française de Rome, 1982
2v.((xvi,818p)) : ill. - (Collection de l'École française de Rome ; 58). - Bibliography
2-7283-0021-6 Ac.5233.a/58
8932905

AFRICA
North Africa. Description & travel, 1960-

UHLIG, Eleonore
Der Melonenkönig, und andere Reiseanekdoten. - Darmstadt : Garuda, [1981]
107p. : ill
3-921019-30-3 X.800/34912
8318263

AFRICA
North Africa. Developing countries. Social change. Role of political elites

HERMASSI, Elbaki
Leadership and national development in North Africa. - Berkeley ; London : University of California Press, 1975 [i.e. 1976]
xi,241p : ill. - Bibliography: p.220-237. - Includes index
0-520-02894-5 X.708/20399
0667773

AFRICA
North Africa. Economic conditions - For British businessmen

HARRISON, Roger, 1947-
North Africa. - [London] : Metra Consulting, c1981
xiii,361p
0-902231-29-4 X.525/4811
2549360

AFRICA
North Africa. Economic conditions - For British businessmen - Serials

EIU annual regional review. The Middle East & North Africa / the Economist Intelligence Unit. - 1984-. - London : The Unit, 1984-
v. : ill
B85-28991 P.905/247
2549360

AFRICA
North Africa. Economic development, 1800–1980
ISSAWI, Charles
An economic history of the Middle East and North Africa. - London : Methuen, 1982
xiii,304p : 1ill. - Bibliography: p265-297. - Includes index
0-416-34330-9 X.520/28418
2803569

AFRICA
North Africa. English women. Interpersonal relationships with Arab men
LEGH-JONES, Alison
English woman, Arab man. - London : Elek, 1975
275p
0-236-40003-7 X.809/40490
0561649

AFRICA
North Africa. Exploration. Hanno. Periplus - Translations
RAMIN, Jacques
Le Périple d'Hannon = The periplus of Hanno. - Oxford : British Archaeological Reports, 1976
[6],121,4p. - (British archaeological reports. supplementary series ; 3)
0-904531-35-x X.415/5200 Woolwich
0768278

AFRICA
North Africa. Exploration. Hanno. Periplus. Translations into Greek language
BLOMQVIST, Jerker
The date and origin of the Greek version of Hanno's Periplus. - Lund : Gleerup, 1979
66p. - (Scripta minora Regiae Societatis Humaniorum Litterarum Lundensis ; 1979-1980. 3)
91-400-4696-6 Ac.1068/3[1979/80,no.3]
2261030

AFRICA
North Africa. Foreign relations, 1535–1975
The MIDDLE East and North Africa in world politics / compiled, translated, and edited by J.C. Hurewitz. - 2nd ed. revised and enlarged. - New Haven ; London : Yale University Press.
Vol.1: European expansion, 1535-1914. - 1975
xxiv,616p. - Bibliography: p.587-616
0-300-01294-2 X.0802/1500(1)
037184x
- [Another copy.] OPL
The MIDDLE East and North Africa in world politics / compiled, translated, and edited by J.C. Hurewitz. - 2nd ed., revised and enlarged. - New Haven ; London : Yale University Press.
Vol.2: British-French supremacy, 1914-1945. - 1979
xxii,861p. - Bibliography: p.825-861
0-300-02203-4 X.0802/1500(2)
037184x
- [Another copy.] OPL

AFRICA
North Africa. German prisoners of war, 1939–1945
PASCHEN, Konrad
Auf unsicherem Terrain 1940 bis 47. - Frankfurt / Main : Haag + Herchen, 1981
144p
r6308235 X.809/55630
8973911

AFRICA
North Africa. Historical sources
SOURCES of the history of North Africa, Asia and Oceania in Finland, Norway and Sweden / the National Archives of Finland, the National Archives of Norway, the National Archives of Sweden ; [compiled by Berndt Federley ... et al.] ; [translated into English by Mogens Møller ... et al.]. - München ; London : Saur, 1981
233p. - (Guides to the sources for the history of the nations. 3rd series : North Africa, Asia and Oceania = Guides des sources de l'histoire des nations. 3ème série : Afrique du Nord, Asie et Océanie ; v.3). - Includes bibliographies and index
3-598-21475-8 X.800/34895
2872250

AFRICA
North Africa. Historical sources - Bibliographies
SOURCES de l'histoire du Proche-Orient et de l'Afrique de Nord dans les archives et bibliothèques françaises. - München ; London : Saur.
2: Bibliothèque Nationale. - 1984
ix,480p. - (Guides to the sources for the history of the nations. 3rd series: North Africa, Asia and Oceania = Guides des sources de l'histoire des nations. 3 ème Série: Afrique du Nord, Asie et Océanie ; v.5, pt.2). - Includes index
3-598-21479-0 X.800/40133
3240177

AFRICA
North Africa. Human settlements, to BC 2000
The SAHARA and the Nile / edited by Martin A.J. Williams, Hugues Faure. - Rotterdam : Balkema, 1980
xvi,607p;[12]p. of plates
90-619-1012-9 X.800/14730
219452X

AFRICA
North Africa. Ibadites, to 1517
SCHWARTZ, Werner
Die Anfange der Ibaditen in Nordafrika. - Wiesbaden : in Komm. bei Harrassowitz, 1983
349p.,plates. - (Studien zum Minderheitenproblem im Islam ; 8) (Bonner orientalistische Studien ; neue Serie, Bd.27). - Includes bibliography
3-447-02353-8 X.0200/455(8)
9070109

AFRICA
North Africa. Islamic architecture
HILL, Derek
Islamic architecture in North Africa. - London : Faber, 1976
167p,[272]p of plates : ill(some col). - Bibliography: p.155-156. - Includes index
0-571-10460-6 X.421/10193 Woolwich
0666815

HUTT, Antony
North Africa. - London : Scorpion Publications Ltd, 1977
192p : chiefly ill(some col). - (Islamic architecture). - Bibliography: p.27. - Includes index
0-905906-01-2 X.0419/218(2) Woolwich
1779818

AFRICA
North Africa : Islamic pilgrimages to Mecca. Routes, to ca 1975
BIRKS, J. S.
Across the Savannas to Mecca. - London : C. Hurst, 1978
xiii,161p : ill. - Bibliography: p.153-157. - Includes index
0-903983-93-1 X.200/31543
1581945

AFRICA
North Africa. Medicine, 1924–1943
DENTI DI PIRAJNO, Alberto, duca
A cure for serpents. - London : White Lion Publishers, 1977
[7],277p,[16]p of plates : ill. - Includes index
0-85617-808-x X.329/11083 Woolwich
1187066

AFRICA
North Africa. Moors. Social conditions, to ca 1800
HAWTING, Gerald
The Moors. - Maidenhead : Sampson Low, 1978
48p : ill(chiefly col). - (Sampson Low library of the past)
0-562-00092-5 X.421/10733 Woolwich
1579347

AFRICA
North Africa. Mosques. Architecture, to date
EWERT, Christian
Forschungen zur almohadischen Moschee. - Mainz am Rhein : von Zabern, 1981-
v. : ill(some col.). - (Madrider Beiträge). - Lfg.1 includes a bibliography
r6305585 X.0423/99(9,etc.)
8831831

AFRICA
North Africa. Muslims. Relations with Jews, 1800–1980 - Conference proceedings
Les RELATIONS entre Juifs et Musulmans en Afrique du nord, XIXe-XX siècles. - Paris : Éditions du Centre national de la recherche scientifique, 1980
228p
2-222-02690-3 X.205/694
2462052

AFRICA
North Africa. Palaeozoic strata
LOWER Palaeozoic of north-western and west-central Africa / edited by C.H. Holland. - Chichester : Wiley, c1985
xii,512p,[16]p of plates : ill. - (Lower Palaeozoic rocks of the world ; v. 4). - Includes bibliographies and index
0-471-10357-8 SRL
3266036

AFRICA
North Africa. Personnel. Migration
MANPOWER and international labor migration in the Middle East and North Africa / Ismail Serageldin ... [et al]. - Oxford : Published for The World Bank [by] Oxford University Press, 1983
xi,138p : ill. - Includes index
0-19-520406-9 L.49/2386
308583x

AFRICA
North Africa. Portuguese exploration. Economic aspects, 1415–1471
COELHO, António Borges
Raízes da expansão portuguesa. - 3.ed. - Lisboa : Prelo, 1976
93p. - (Cadernos de hoje ; n.1)
r6066397 X.519/40656
1984187

AFRICA
North Africa. Prehistoric rock paintings. Archaeology, to date
STRIEDTER, Karl Heinz
Felsbilder Nordafrikas und der Sahara. - Wiesbaden : Steiner, 1983
viii,287p : ill. - (Studien zur Kulturkunde ; Bd.64). - Bibliography
3-515-03397-1 Ac.659[Bd.64]
8949298

AFRICA
North Africa. Public administration, to 640
PULIATTI, Salvatore
Ricerche sulla legislazione "regionale" di Giustiniano. - Milano : Giuffrè, 1980
130p. - (Seminario giuridico della Università di Bologna ; 84)
r6507166 X.520/29713
8920761

AFRICA
North Africa. Quaternary strata
The SAHARA and the Nile / edited by Martin A.J. Williams, Hugues Faure. - Rotterdam : Balkema, 1980
xvi,607p;[12]p. of plates
90-619-1012-9 X.800/14730
2194511

AFRICA
North Africa. Rābih Fadl Allāh - Biographies
HALLAM, W. K. R.
The life and times of Rabih Fadh Allah. - Ilfracombe : Stockwell, 1977
367p,[16]p of plates : ill. - Bibliography: p.330-339. - Includes index
0-7223-0959-7 X.809/42217
1184474

AFRICA
North Africa. Relations with Spain, ca 1450–ca 1600
HESS, Andrew C.
The forgotten frontier. - Chicago ; London : University of Chicago Press, 1978
xiv,278p : ill. - (Publications / University of Chicago. Center for Middle Eastern Studies ; 10). - Bibliography: p257-265. - Includes index
0-226-33028-1 X.809/43917
1766708

AFRICA
North Africa. Slavery, to 1900
ELBASHIR, Ahmed E.
The United States, slavery and the slave trade in the Nile Valley. - Lanham ; London : University Press of America, c1983
x,189p. - Bibliography: p175-183. - Includes index
0-8191-3490-2 X.809/60880
3160459

AFRICA
North Africa. Social conditions
WILKINS, Frances
Let's visit North Africa. - 2nd rev. ed. - London : Burke, 1983
96p : ill(some col.). - Includes index
0-222-00917-9 X.809/61581
3197034

AFRICA
North Africa. Social conditions, to 640
SCHULTEN, P. G. G. M.
De circumcellionen. - [Leiden] : [P. Schulten], 1984
194p. - Includes bibliography
r6330931 X.529/65528
9139478

AFRICA
North Africa. Social stratification - Bibliographies
BANUAZIZI, Ali
Social stratification in the Middle East and North Africa. - London : Mansell, 1984
xviii,248p. - Includes index
0-7201-1711-9 X.520/34866
305229x

AFRICA
North Africa. Society. Role of women - Bibliographies
AL-QAZZAZ, Ayad
Women in the Middle East and North Africa. - Austin : Center for Middle Eastern Studies, University of Texas at Austin, 1977
xii,178p. - (Middle East monographs ; no. 2). - Includes index
0-292-79009-0 X.519/27951
1447297
- [Another copy.] X.529/48397

AFRICA
North Africa. South African prisoners of war, 1941-1943
OGILVIE, Peter
In the bag. - Johannesburg : Macmillan South Africa ; London : [Macmillan], 1975
101p : ill
0-333-19448-9 X.802/10101
0589470

AFRICA
North Africa. Spanish slaves, 1500-1700
FRIEDMAN, Ellen G.
Spanish captives in North America in the early modern age. - Madison, [Wis.] ; London : University of Wisconsin Press, 1983
xxvii,215p. - Bibliography: p195-202. - Includes index
0-299-09380-8 X.809/60488
3144178

AFRICA
North Africa. Women, to date - Bibliographies
OTTO, Ingeborg
Frauenfragen in modernen Orient. - Hamburg : Deutsches Orient-Institut, Dokumentations-Leitstelle Moderner Orient, 1982
xv,247p.. - (Dokumentationsdienst moderner Orient. Reihe A ; 12)
3-88693-010-6 X.0702/370(12)
8838917

AFRICA
North-east Africa - Serials
N.E.A.. - Vol.1, no.1 (Summer '81)-. - Oxford : N.E.A., 1981-
v.
B82-12464 P.701/1761
2629399

AFRICA
North-east Africa. Ethnic groups
GROTTANELLI, Vinigi L.
Gerarchie etniche e conflitto culturale. - Firenze : Franco Angeli, [1976]
346p : ill. - (Collana di antropologia culturale e sociale)
r6041900 X:709/24122
1387952

AFRICA
Norwegian Christian missions, to date
NORSK misjon og afrikanske samfunn / redaktør Jarle Simensen. - [Trondheim] : Tapir, 1984-
r6612055 X.0200/1376
915602x

AFRICA
One-party states. Human rights - Conference proceedings
HUMAN rights in a one-party state. - London : Search Press [for] the Commission, 1978
133p
0-85532-389-2 X.100/17678
1482432

AFRICA
Orchids
POTTINGER, Mollie
African orchids. - Wokingham : H.G.H., c1983
61p : ill(some col.)
0-946445-00-1 X.329/16775
2585499

STEWART, J. (Joyce)
Orchids of Africa. - London : Macmillan, 1981
ix,159p : ill(some col.). - Bibliography: p155-157. - Includes index
0-333-32536-2 SRL
2585499

AFRICA
Participation in United Nations, to date
WILKINS, Gregory L.
African influence in the United Nations 1967-1975. - Washington, D.C. : University Press of America, c1981
xii,251p : ill. - Includes bibliography
0-8191-1424-3 X.800/39985
9167846

AFRICA
Partition by European countries, ca 1800-ca 1900
EUROPEAN imperialism and the partition of Africa / edited by E.F. Penrose. - London : Cass, 1975
vii,171p. - (New orientations series ; no.2)
0-7146-3058-6 X.800/11663
0555592

AFRICA
Penis sheaths
LAGERCRANTZ, Sture
Penis sheats [i.e. sheaths] and their distribution in Africa. - Uppsala : [Institutionen för allmän och jämförande etnografi vid Uppsala Universitet], 1976
63p : ill. - (Institutionen för allmän och jämförande etnografi vid Uppsala Universitet. Occasional papers ; 7. 2)
r6031807 X.0700/758(7,pt.2)
1103458

AFRICA
Persons, to 1960 - Biographies - Dictionaries
LIPSCHUTZ, Mark R.
Dictionary of African historical biography. - London : Heinemann Educational, 1978
xi,292p. - Bibliography: p.259-280. - Includes index
0-435-94710-9 AA.oo.g Reading Room
1376497
- [Another copy] OPL
- [Another copy] X.800/26737

AFRICA
Persons. Reproduction system. Mutilation, to date
OMBOLO, Jean-Pierre
Les mutilations sexuelles en Afrique noire. - Yaoundé : [J.-P. Ombolo], 1981
180p
r6209066 X.322/11292
8908214

AFRICA
Polish immigrants, to date
GOŁĄBEK, Stefan
Związki Polski i Polaków z Afryką do roku 1945. - Warszawa : Państwowe Wydawnictwo Naukowe, 1978
100p
r6483562 X.800/33280
8840474

AFRICA
Polish refugees, 1939-1945
KORABIEWICZ, Wacław
Gdzie słoń a gdzie Polska. - Warszawa : Wydawnictwo Ministerstwa Obrony Narodowej, 1980
269p,[40]p of plates : ill
83-11-06494-6 X.808/34621
8759154

AFRICA
Political development
DAVIDSON, Basil
Crossroads in Africa. - Nottingham : Spokesman Books, 1980
97p
0-85124-245-6 X.708/22617
2281899

AFRICA
Political development. Role of military governments
ODETOLA, T. O. (T Olatunde)
Military regimes and development. - London : Allen & Unwin, 1982
200p. - Bibliography: p190-195. - Includes index
0-04-301154-3 X.529/52341
2860430

AFRICA
Political events - Serials
AFRICA events. - Oct. 1984-. - London : [Dar est Salaam Ltd.]
v. : ill
B85-28997 P.823/698
0209732

AFRICA in struggle. - No.1- ; Oct. 1975-. - London : International Marxist Group 'Africa Commission', [1975]-
: ill
B77-01952 P.703/524
0776823

AFRICAN freedom annual / [issued by] Southern African Freedom Foundation ; editor-in-chief : F.R. Metrowich. - 1977-. - Sandton : The Foundation, 1977-
r6079320 P.885/6
0209732

AFRIQUE informations politiques et économiques. - Année 6, no.283 (4 juillet 1978)-. - Paris : Édiafric, La Documentation Africaine, 1978-
r6038016 P.703/362
0209732

AFRO-ASIAN affairs. - No.1- ; 1975-. - [London] : [International Currency Review], 1975-
Holdings: no.52, etc. : 28 July 1977, etc
B79-26790 P.803/930
0209732

ÉDIAFRIC-SÉLECTION. - Année 6, no.213 (jan. 1977) - année 6, no.282 (27 juin 1978). - Paris : Édiafric, 1977-1978
70pts
r6802131 P.703/362
0209732

THIRD world first. - July 1978-1981?. - Lagos : Third World First Publications, 1978-1981
r6108376 P.901/1855
0209732

AFRICA
Political events, 1880-1977
DAVIDSON, Basil
Africa in modern history. - Harmondsworth : Penguin, 1978
431p. - (Pelican books). - Bibliography: p.401-402. - Includes index
0-14-022105-0 X.708/22492
1527444

Africa in modern history. - London : Allen Lane, 1978
431p. - Includes index
0-7139-0874-2 X.809/43853
1527444

AFRICA
Political events, 1900-1975
AKINTOYE, Stephen Adebanji
Emergent African states. - London : Longman, 1976
[5],250p. - Includes index
0-582-60127-4 X.709/31167
1095218

AFRICA
Political events, 1950-1974
FODERARO, Salvatore
Independent Africa. - Gerrards Cross : Smythe, 1976
183p. - Includes index
0-901072-31-1 X.709/30330
0607444

AFRICA
Political events, 1960-
KOTECHA, Ken C.
African politics. - Washington, D.C. : University Press of America, c1981
vi,473p
0-8191-1565-7 X.809/60263
8003742

AFRICA
Political events, 1960- - Conference proceedings
COLLOQUE IDÉOLOGIQUE INTERNATIONAL (1978 : Conakry)
Colloque idéologique international de Conakry du 13 au 16 novembre 1978 sur le thème "L'Afrique en marche". - Conakry : Bureau de presse de la présidence de la République, [1978]
373p. - ([Publications] / Parti-État de Guinée ; tome 24)
r6211227 X.800/37507
8003742

AFRICA
Political events, 1960- - Serials
L'ANNÉE politique et économique africaine. - 1981-. - Dakar : Société africaine d'édition, 1981-
Holdings: 1981-
r6211242 X.0805/134
8003742

SERVIR. - no 1 [(197?)]-. - Ivry : Comité international culturel N'Krumah, [197?]-
Holdings: no 9, 1980-
r6206853 X.0802/1607
8003742

AFRICA
Political events. Policies of government of United States, 1960-1975

AFRICA / edited by Helen Kitchen. - Lexington, Mass. : D.C Heath [for the Commission on Critical Choices for Americans], 1976
xxiii,415p. - (Critical choices for Americans ; vol 11). - Includes index
0-669-00425-1 X.800/26656
1340093

AFRICA
Political events. Role of communist countries, 1960-1977

GREIG, Ian
The Communist challenge to Africa. - Richmond, Surrey : Foreign Affairs Publishing Co., 1977
[10],306p. - Bibliography: p.287-288. - Includes index
0-900380-21-7 X.809/42646
1292331

AFRICA
Political events. Role of communist countries & Western bloc countries, 1945-1980

GAVSHON, Arthur
Crisis in Africa. - Harmondsworth : Penguin, 1981
320p : ill. - (Pelican books. African affairs). - Includes index
0-14-022239-1 X.529/45231
2608294

AFRICA
Political science, to date

THEMES in African social and political thought / edited by Onigu Otite. - Enugu : Fourth Dimension, 1978
xii,347p
r6813423 X.529/58569
9034293

AFRICA
Political systems, 1960-1981. Socioeconomic aspects

YOUNG, Crawford
Ideology and development in Africa. - New Haven ; London : Yale University, c1982
xviii,376p : ill. - (A Council on Foreign Relations book). - Includes index
0-300-02744-3 X.809/52400
2745410

AFRICA
Politicians, to 1974

The MAKING of politicians / edited by W.H. Morris-Jones. - London : Athlone Press for the Institute of Commonwealth Studies, 1976
x,249p. - (Commonwealth papers ; 20). - Includes index
0-485-17620-3 Ac.2666da/2[20]
1040685

AFRICA
Politics

AUSTIN, Dennis
Politics in Africa. - Manchester : Manchester University Press, 1978
[10],202p. - Bibliography: p.165-170. - Includes index
0-7190-0721-6 X.800/14004
1370693

CHALIAND, Gérard
L'enjeu africain. - Paris : Éditions du Seuil, [1980]
155p. - (L'histoire immédiate). - Bibl
2-02-005505-8 X.709/51934
1370693

GONIDEC, Pierre François
African politics. - The Hague ; London : Nijhoff, 1981
367p. - Includes index
90-247-2391-4 X.800/29667
2363747

HODDER, B. W. (Bramwell William)
Africa today. - London : Methuen, 1978
ix,166p. - Bibliography: p.159-160. - Includes index
0-416-71010-7 X.708/21103
1370693

JACKSON, Robert H.
Personal rule in black Africa. - Berkeley ; London : University of California Press, c1982
xi,316p. - Includes index
0-520-04185-2 X.520/27440
1370693

LIYONG, Taban
Thirteen offensives against our enemies. - Nairobi : East African Literature Bureau, 1973
[8],124p
r6073372 X.708/21701
1802909

MARKOVITZ, Irving Leonard
Power and class in Africa. - Englewood Cliffs ; London : Prentice-Hall, 1977
xiv,398p. - Bibliography: p.350-377. - Includes index
0-13-686691-3 X.800/26122
0612960

MAZRUI, Ali Al'Amin
The African condition. - London : Heinemann Educational, 1980
xviii,142p. - (The Reith lectures ; 1979). - Includes index
0-435-96524-7 X.709/52076
0176176

POTHOLM, Christian P. (Christian Peter)
The theory and practice of African politics. - Englewood Cliffs ; London : Prentice-Hall, 1979
xii,307p : ill. - Bibliography: p.248-298. - Includes index
0-13-913533-2 X.709/33171
1370693

WILMOT, Patrick F.
In search of nationhood. - Lagos : Lantern Books, 1979
xi,196p.
978-228-101-8 X.709/50276
2049090

AFRICA
Politics - Conference proceedings

The POLITICS of Africa / edited by Timothy M. Shaw and Kenneth A. Heard. - London : Longman [etc.], 1979
xii,400p : ill. - (Dalhousie African studies series). - Includes index
0-582-64229-9 X.800/27985
1928767

The POLITICS of Africa / edited by Timothy M. Shaw and Kenneth A. Heard. - London : Longman, 1979
xii,400p : ill. - (Dalhousie African studies series). - Includes index
0-582-64378-3 X.520/28774
1928767

AFRICA
Politics - Serials

AFRICA Currents. - No.1- ; Spring 1975-. - London : Africa Publications Trust, 1975-
B75-20083 P.701/1773
0350079

AFRICA journal. - Vol.1, no.1- ; [1976?]-. - Monrovia : Chester's Publishing Company, [1976?]-
Holdings: Vol.4, no.5-
r6108905 P.701/1329
1370693

JOURNAL of African Marxists = Journal des Marxistes africains. - Issue 1 (Nov. 1981)-. - London : Zed Press, 1981-
v. : ill
B83-03103 P.701/1774
280543x

AFRICA
Politics, to 1975 - Bibliographies

GORDON DRABEK, Anne
The politics of African and Middle Eastern states. - Oxford : Pergamon, 1976
x,192p. - (Pergamon international library)
0-08-020584-4 OPL
1074245
- [Another copy] X.709/31105

AFRICA
Politics, 1885-1976

RUBIN, Leslie
Introduction to African politics. - 2nd ed. - New York : Praeger, 1977
x,338p. - (Praeger university series). - Bibliography : p.317-329. - Includes index
0-275-64700-5 X.700/25589
130741X

AFRICA
Politics, 1960-

MENGES, Dietrich Wilhelm von
Afrika morgen. - Düsseldorf : Econ, 1980
140p
3-430-16492-3 X.809/48079
8579814

AFRICA
Politics, 1960- - Conference proceedings

NEW perspectives and conceptions of international law / edited by K. Ginther and W. Benedek. - Wien : Springer, c1983
ix,261p.. - (Österreichische Zeitschrift für öffentliches Recht und Völkerrecht. Supplementum ; 6)
3-211-81780-8 X.200/45132
8579814

AFRICA
Politics, 1960- - Serials

The AFRICAN interpreter. - [No 1]-. - Köln : African Interpreter European Bureau, [1976?]-
Holdings: No.55 (June 1981)-
r6307392 P.703/931
8579814

ORGANISATION COMMUNE AFRICAINE ET MAURICIENNE. Secrétariat Général
Chronique mensuelle. - No.1 (1972)-. - Bangui : Organisation commune africaine et mauricienne. Secrétariat Général, [1972]-
Holdings: no.67 (1977)-
r6211298 X.0805/141
8579814

AFRICA
Politics. Geographical factors

BOATENG, Ernest Amano
A political geography of Africa. - Cambridge : Cambridge University Press, 1978
xix,292p. - Bibliography: p.272-279. - Includes index
0-521-21764-4 X.709/32590
1695398

AFRICA
Politics. Intervention by China, 1965-

SINO-SOVIET intervention in Africa / edited by Roger Pearson. - Washington, D.C. : Council on American Affairs, [1977]
103p
r6805207 X.800/32887
8820066

AFRICA
Politics. Intervention by Soviet Union, 1945-

LINT, George J. de
The United Nations giving countenance to Soviet strategic penetration of Africa. - Zwolle : W.E.J. Tjeenk Willink, 1977
14p
90-271-1333-5 X.800/35757
8820074

SINO-SOVIET intervention in Africa / edited by Roger Pearson. - Washington, D.C. : Council on American Affairs, [1977]
103p
r6805207 X.800/32887
8820074

AFRICA
Politics. Leadership, 1945-1981

CARTWRIGHT, J. (John)
Political leadership in Africa. - London : Croom Helm, c1983
310p. - Bibliography: p301-303. - Includes index
0-7099-0751-6 X.529/54791
2914026

AFRICA
Politics. Leadership. Role of Nigeria

WAYAS, Joseph
Nigeria's leadership role in Africa. - London : Macmillan, 1979
vii,132p. - Includes index
0-333-26295-6 X.809/44539
1686402

AFRICA
Politics. Role of international organisations

AFRICA and international organization / edited by Yassin El-Ayouty, Hugh C. Brooks. - The Hague : Martinus Nijhoff, 1974
xii,250p : 1 ill. - Includes bibliographies and index
90-247-1615-2 X.510/15900
1688200

AFRICA
Politics. Role of Liberia

HOLLOWAY, Joseph E.
Liberian diplomacy in Africa. - Washington, D.C. : University Press of America, c1981
xxiv,230p : ill. - Includes bibliography
0-8191-1790-0 X.529/66999
9128638

AFRICA
Politics. Role of military forces

Os MILITARES e o poder na América Latina e na África / [colaboração de Denis Martin et al.]. - Lisboa : Publicações Dom Quixote, 1975
138 p.. - (Colecção Novos cadernos D. Quixote ; n. 27)
r6013342 X.708/18299
0874191

UKPABI, Sam C.
Military involvement in African politics. - New Paltz, N.Y. : Conch Magazine, 1972
ii,27p.. - (Contemporary African politics ; no.4)
r6028460 X.709/26068
0874191

AFRICA
Politics. Women leaders, to 1980
SWEETMAN, David
Women leaders in African history. - London : Heinemann, 1984
100p : ill. - (African historical biographies). - Includes bibliographies and index
0-435-94480-0 X.809/59882
3131009
- [Another copy.] X.809/61251

AFRICA
Population, to date
Les STRUCTURES par sexe et âge en Afrique / par Gilles Roger, Dominique Waltisperger, Christine Corbille-Guitton [pour le] Groupe de démographie africaine. - Paris : Institut de démographie de Paris, 1981
556p. - Bibliography: p521-556
r6209576 X.525/6504
8003769

AFRICA
Population. Censuses, 1960-
GROUPE DE TRAVAIL DE DÉMOGRAPHIE AFRICAINE
Recensements africains. - Paris : Groupe de travail de démographie africaine, 1980-1981
2v
r6209175 X.525/6543
8877300

AFRICA
Population. Distribution
REDISTRIBUTION of population in Africa / edited by John I. Clarke and Leszek A. Kosiński. - London : Heinemann, 1982
x,212p : ill. - Includes bibliographies and index
0-435-95030-4 X.805/2399
2693224

AFRICA
Population. Growth - Conference proceedings
POPULATION growth and economic development in Africa / edited by S.H. Ominde and C.N. Ejiogu. - London : Heinemann in association with the Population Council, New York, 1972
xxii,421p : ill. - Includes bibliographies and index
0-435-97470-x X.525/5120
258543x

AFRICA
Portugeuse speaking countries. Political events - Serials
PEOPLE'S power in Mozambique, Angola and Guinea-Bissau. - No.5 (1976)-. - London : Mozambique, Angola and Guine Information Centre, 1976-
v. : ill
Holdings: no 14 (Summer 1979)-
B82-17260 P.701/1763
2704064

AFRICA
Portuguese colonies
COSTA, Eduardo Freitas da
Para uma nova fundação de Portugal. - Lisboa : Edições do Templo, 1978
122p. - (Colecção País real)
r6701099 X.809/46851
2477882

AFRICA
Portuguese colonies, to 1976
ANTUNES, José Freire
O império com pés de barro. - Lisboa : Dom Quixote, 1980
128p. - (Colecção Participar ; 17)
r6701088 X.809/46852
2477890

AFRICA
Portuguese colonies, 1961-1975
CUNHA, J. M. da Silva (Joaquim Moreira da Silva)
Assim construímos o futuro. - Lisboa : Agência-Geral do Ultramar, 1972
215p
r6703066 S.L. 276/288 Woolwich
8724687

Lutando pelo destino comum. - Lisboa : Agência-Geral do Ultramar, 1973
xxvii,173p
r6703073 S.L. 39/80 Woolwich
8724687

Unidos no mesmo ideal. - Lisboa : Agência-Geral do Ultramar, 1971
xxiv,303p
r6703067 S.L. 276/289 Woolwich
8724687

AFRICA
Portuguese colonies. Attitudes of Portugal, 1974
PORTUGAL. Presidência do Conselho
Reflexão sobre o Ultramar. - Lisboa : Secretaria de Estado da Informação e Turismo, 1974
34p
r6086183 S.L.276/245 Woolwich
0935468

SANTOS, António de Almeida
Preparar o futuro. - Lisboa : Agência-Geral do Ultramar, 1974
14 p.
r6011403 S.L.201/152
0935468

AFRICA
Portuguese colonies. Colonial administration, to 1974
CAPELA, José
As ideias coloniais de Marcelo Caetano.. - (In Capela, José. O imposto de palhota.-Porto, 1977.-p.197-255) (Colecção As armas e os varões ; n.6)
r6064919 X.708/42790
0519537

JAFFE, Hosea
El neo-imperialismo portugués. - Madrid : Zero, 1976
74p. - (Colección "Lee y discute". serie 5 ; n.61)
84-317-0331-8 X.708/45403
0519537

NEWITT, Malyn
Portugal in Africa. - Harlow : Longman, 1981
viii,278p. - Bibliography: p250-263. - Includes index
0-582-64379-1 X.809/48933
0519537

PONTE, Bruno da
The last to leave. - London : International Defence and Aid Fund, 1974
[1],77p. - Bibliography: p.74
0-901500-36-4 X.709/30155
0519537

AFRICA
Portuguese colonies. Colonial administration. Carvalho, Henrique Augusto Dias de - Biographies
CARVALHO, João Augusto de Noronha Dias de
Henrique de Carvalho. - Lisboa : [s.n.], 1975
351 p., 2 fold. leaves of plates : ill
r6013382 X.709/20875
0880884

AFRICA
Portuguese colonies. Colonial administration. Economic aspects, 1930-1970
FERREIRA, Eduardo de Sousa
Aspectos do colonialismo português. - Lisboa : Seara Nova, 1974
363 p.. - (Colecção De leste a oeste)
r6008835 X.708/17455
0911585

AFRICA
Portuguese colonies. Decolonisation
A DESCOLONIZAÇÃO portuguesa / [autores : Mário António Fernandes de Oliveira et al.]. - Lisboa : Instituto Democracia e Liberdade, 1979-. - (Colecção Estudos e ensaios)
r6701102 X.0800/1563
2117371

MARQUES, Silvino Silvério
Portugal : e agora?. - Lisboa : Edições do Templo, 1978
250p. - (Colecção País real)
r6086171 X.709/28532
2117371

AFRICA
Portuguese colonies. Decolonisation, to 1975
ARRIAGA, Kaúlza de
Guerras coloniais e descolonização. - [Portugal?] : [s.n], [1973?]
[20]p
r6709741 X.809/64213
8626855

AFRICA
Portuguese colonies. Economic development. Role of Portuguese companies
FERREIRA, Eduardo de Sousa
Portugal e o neocolonialismo. - Lisboa : Livraria Sá da Costa, 1975
36 p.. - (Cadernos livres ; n. 5)
r6013340 X.709/20848
0880043

AFRICA
Portuguese colonies. Guerrilla wars, 1961-1974. Political aspects
CRESPO, Manuel Pereira
Porque perdemos a guerra. - [Lisboa?] : Ed. Abril, 1977
168p
r6066354 X.709/27766
1988042

AFRICA
Portuguese colonies. Guerrilla wars, 1961-1975
CALVÃO, Alpoim
De Conakry ao M.D.L.P.. - Lisboa : Intervenção, c.1976
237p
r6704653 X.809/51083
8003777

RODRIGUES, Paulo Madeira
4 países libertados. - Lisboa : Livraria Bertrand, 1975
140p : chiefly ill
r6064969 X.809/25400
1473964

AFRICA
Portuguese colonies. Habeas corpus. Applications by political prisoners
VALENTIM, Eduardo Artur Santana
Angolanos no Tarrafal. - Porto : Afrontamento, 1974
139 p.. - (Colecção Bezerro d'ouro ; n. 8)
r6013391 X.708/18372
0882429

AFRICA
Portuguese colonies. Imports from Portugal: Wines
O VINHO para o preto / [edição de] José Capela. - Porto : Afrontamento, 1973
170 p.
r6011285 X.519/20343
0930946

AFRICA
Portuguese colonies. Military aspects, 1973-1974
ÁFRICA : a vitória traída / Joaquim da Luz Cunha, Kaúlza de Arriaga, Bethencourt Rodrigues, Silvino Silvério Marques. - Braga : Intervenção, 1977
276p.,[3]fold. leaves of plates : ill
r6064928 X.708/42800
1423460

AFRICA
Portuguese colonies. Negro personnel. Employment. Law, 1962-1966
CAPELA, José
Legislação do trabalho nas colónias nos anos 60.. - (In Capela, José. O imposto de palhota. - Porto. 1977. - p.257-273) (Colecção As armas e os varões ; n.6)
r6064918 X.708/42790
1428136

AFRICA
Portuguese colonies. Revolutionary movements
NEVES, Fernando
Negritude, independência, revolução. - Paris : Edições Etc, 1975
450 p.
r6027397 X.709/22020
1026380

AFRICA
Portuguese colonies. Revolutionary movements, 1960-1970
COLONIALISMO e lutas de libertação / [editores : António Melo ... et al.]. - Reedição. - Porto : Afrontamento, 1974
300 p.. - (Arquivo ; no. 1). - Includes bibliographies
r6011559 X.709/20617
0865001

AFRICA
Portuguese colonies. Revolutionary movements, 1961-1975
MACHEL, Samora Moisés
Mensagem do camarada Samora Moisés Machel...à XXIV sessão do Comité de Libertação da Organização da Unidade Africana, Dar-es-Salaam, 8 de Janeiro de 1975. - Lourenço Marques : Imprensa Nacional de Moçambique, 1975
29p
r6065101 X.709/25976
1541307

AFRICA
Portuguese colonies. Revolutionary movements, 1961-1975
OVER de bevrijdingsstrijd in de Portugese kolonies / [uitgegeven door de Derde Wereld Beweging]. - Leuven : de Beweging, [1971?]
55p : ill
r6330947 X.809/61736
8003807

AFRICA
Portuguese colonies. Revolutionary movements. Political aspects

ARRIAGA, Kaúlza de
A solução portuguesa. - [Porto?] : [s.n.], 1976
31p : ill
r6047253 X.708/40024
0910147

NEVES, Fernando
As colónias portuguesas e o seu futuro. - Paris : Edições Etc., 1974
335 p.
r6008844 X.709/19712
0910147

QUINTINHA, Santana
O terrorismo e os partidos subversivos no Ultramar português. - Lisboa : [s.n.], 1973
472p
r6086184 X.700/20965
0910147

VALLES, Edgar
Africa : colonialismo e socialismo. - Lisboa : Seara Nova, 1974
166 p.. - (Cadernos Seara Nova. Série "de leste a oeste")
r6009078 X.708/17628
092038X

AFRICA
Portuguese colonies. Slavery, 1436-1959

ALMEIDA, Pedro Ramos de
Portugal e a escravatura em África. - Lisboa : Estampa, 1978
151p. - (Polémica ; n.18). - Includes bibliographies
r6086136 X.708/46055
2039257

AFRICA
Portuguese colonies. Social conditions, to 1961

PORTUGAL no século XIX. - Lisboa : Sá da Costa, 1979.
Vol.3: Origens do colonialismo português moderno, 1822-1891
vii,219p
r6703148 X.0809/2110(3)
862688x

AFRICA
Portuguese colonies. Social conditions, 1830-1920

CAPELA, José
O imposto de palhota e a introdução do modo de produção capitalista nas colónias ; As ideias coloniais de Marcelo Caetano ; Legislação do trabalho nas colónias nos anos 60. - Porto : Afrontamento, 1977
273p.,6 leaves. - (Colecção As armas e os varões ; n.6)
r6064920 X.708/42790
1420194

AFRICA
Portuguese colonies. Social conditions, 1956-1959

CASTRO, Armando
O sistema colonial português em África. - Lisboa : Caminho, 1978
434p. - (Temas políticos e sociais ; n.39)
r6086138 X.708/46048
204062X

AFRICA
Portuguese colonies. Social conditions compared with social conditions of Zaire

AMARO, Eugénio
Colónias portuguesas, seu futuro. - [Lisboa?] : [s.n.], [1974]
39 p. : ill
r6014719 X.709/20673
0869473

AFRICA
Portuguese ex-colonies. Lighthouse, to date

LISTA de faróis / Instituto Hidrográfico, Portugal. - Lisboa : O Instituto, 1978
2v.
r6159830 S.L.276/269
9098305

AFRICA
Portuguese exploration, 1435-1900

SANTOS, Maria Emília Madeira
Viagens de exploração terrestre dos portugueses em África. - Lisboa : Junta de Investigações Científicas do Ultramar, Centro de Estudos de Cartografia Antiga, 1978
414p.,[3]leaves of plates,[10]p.of plates(some col.) : ill,facsims,maps. - Bibl: p.369-387
r6161104 X.709/51723
2459868

AFRICA
Press

FEUEREISEN, Fritz
Die Presse in Afrika. - 2 Aufl. - Pullach/München : Verlag Dokumentation, 1973
280p
3-7940-3058-3 X.950/18846
2854783

AFRICA
Press, to 1978

BARTON, Frank
The press of Africa. - London : Macmillan, 1979
xi,304p. - Bibliography: p.286-290. - Includes index
0-333-24521-0 X.989/53313
1377299

AFRICA
Press, to date

EPULE, Kome
Challenge of responsibility. - Buea : The Author, [1978]
165p
r6804384 X.955/708
8579822

AFRICA
Primary schools. Curriculum subjects: English language

AFOLAYAN, Adebisi
Teaching primary English. - London : Longman, 1980
103p : ill. - Bibliography: p100. - Includes index
0-582-60196-7 X.950/19493
2873540

AFRICA
Protestant churches. Public worship

BÜRKI, Bruno
L'assemblée dominicale. - Immensee : Nouvelle Revue de Science Missionnaire, 1976
200p. - (Neue Zeitschrift für Missionswissenschaft. Supplementa ; vol. 25)
r6028179 P.101/449a.[vol.25]
1149598

AFRICA
Psychology, to 1974

WOBER, Mallory
Psychology in Africa. - London : International African Institute, 1975
xi,247p : ill. - Bibliography: p.217-239. - Includes index
0-85302-045-0 X.320/10366 Woolwich
0551643
- [Another copy: paperback] X.520/21172

AFRICA
Public bodies - Directories

DUIC, Walter Z.
African administration. - New York : Saur.
Vol.1. - 1978
xxxii,1285p
0-89664-018-3 X.529/52962
2874563

AFRICA
Publishing - Conference proceedings

INTERNATIONAL CONFERENCE ON PUBLISHING AND BOOK DEVELOPMENT
(1973 : University of Ife)
Publishing in Africa in the seventies. - Ile-Ife : University of Ife Press, 1975
ix,377p. - Includes index
r6032453 2708.gg.86
1299735

AFRICA
Publishing, to date

KOTEI, S. I. A.
The book today in Africa. - Paris : Unesco, 1981
210p. - (Books about books)
92-310-1876-0 UN.L.331/196
8003831

AFRICA
Refugees

AIBONI, Sam Amaize
Protection of refugees in Africa. - Uppsala : Svenska institutet för internationell rätt, 1978
xii,163p. - (Studies in international law ; vol.3). - Includes bibliographies
91-506-0171-7 X.0809/2077(3)
198148X

AFRICA
Refugees, 1960-

An ANALYSING account of the Conference on the African refugee problem / edited by L.G Eriksson, G Melander & P Nobel. - Uppsala : Scandinavian Institute of African studies, 1981
233p : ill
91-7106-185-1 X.529/50686
8579849

AFRICA
Refugees. Law

INTERNATIONAL legal instruments on refugees in Africa = Instruments legaux internationaux sur les réfugiés en Afrique / edited by Göran Melander & Peter Nobel. - Uppsala : Scandinavian Institute of African Studies, 1979
xix,413p
91-7106-154-1 X.200/33055
244478X

AFRICA
Regional geography - For Nigerian students

REGIONAL geography. - London : Macmillan, 1980
iv,91p. - (Macmillan certificate model answers)
0-333-28524-7 X.700/26539
2281678

AFRICA
Relations with Aquileia, Italy, B.C.241-A.D.1300 - Conference proceedings

SETTIMANA DI STUDI AQUILEIESI (1973 : Aquileia)
Aquileia e l'Africa. - Udine : Arti Grafiche Friulane, 1974
399 p., 4 leaves of plates, 46 p. of plates : ill. - (Antichità altoadriatiche ; n. 5)
r6024232 X.410/5946 Woolwich
1015591

AFRICA
Relations with European Economic Community

DJAMSON, Eric Christopher
The dynamics of Euro-African co-operation. - Hagne : Nijhoff, 1975
519p. - Bibliography : p.511-519
90-247-1841-4 X.510/10161
1239074

MORINO, Lina
La Comunità europea e l'Africa. - Bari : Dedalo, 1975
367p. - (Problemi del giorno ; 2)
r6052652 X.709/25403
1239074

AFRICA
Relations with United States, to 1914

MCCARTHY, Michael
Dark continent. - Westport, Conn. ; London : Greenwood Press, 1983
xxviii,192p : 1ill. - (Contributions in Afro-American and African studies ; no.75). - Bibliography: p155-184. - Includes index
0-313-23828-6 X.529/66632
3220222

AFRICA
Religions - Conference proceedings

TOWARDS a global congress of world religions / edited by Warren Lewis. - Barrytown, N.Y. : Unification Theological Seminary, 1978
297p. - (Conference series ; no.2)
0-932894-01-1 X.100/29303
1488260

AFRICA
Religions, to 1975

PARRINDER, Geoffrey
Africa's three religions. - 2nd ed. - London : Sheldon Press, 1976
253p. - Bibliography: p.239-242. - Includes index
0-85969-096-2 X.108/16418
073893x

AFRICA
Religions, to date

AFRICAN religious groups and beliefs / Simon Ottenberg, editor. - Meerut : Folklore Institute, [1982]
xii,355p,plates
r6813262 X.200/42976
8579857

SHPAZHNIKOV, G. A. (Genrikh Alekseevich)
Religii stran Afriki. - 2-e izd. - Moskva : Nauka Glavnaîa redakt͡siîa vostochnoĭ literatury, 1981
364p. - Bibliography: p.319-344
r6487089 X.200/38783
8579857

AFRICA
Religions, to date-continued

TRINCAZ, Jacqueline
Colonisations et religions en Afrique noire. - Paris : Éditions l'Harmattan, 1981
viii,357p. - Bibliography: p340-357
2-85802-181-3 X.200/39406
8579857

AFRICA
Religions. Attitudes of Western world, to date

P'BITEK, Okot
African religions in Western scholarship. - Kampala : East African Literature Bureau, [1971?]
ix,139p. - Includes bibliography
r6812067 X.208/1987
9063951

AFRICA
Residences. Inexpensive building materials - Conference proceedings

APPROPRIATE building materials for low cost housing / hosted by the Government of the Republic of Kenya, Ministry of Works and Housing & University of Nairobi and organized by CIB and RILEM in co-operation with UNCHS ... [et al.]. - London : Spon, 1983
ix,524p : ill
0-419-13280-5 SRL
3075885

APPROPRIATE building materials for low cost housing / hosted by the Government of the Republic of Kenya, Ministry of Works and Housing & University of Nairobi and organized by CIB and RILEM in co-operation with UNCHS ... [et al.]. - London : Spon.
Vol.2. - 1985
[500]p : ill
0-419-13430-1 X.622/24104
3075885

AFRICA
Revolutionary movements

ISTORIÎA natsional'no-osvoboditel'noĭ bor'by narodov Afriki v noveĭshee vremia / [otvetstvennyĭ redaktor : Anat. A. Gromyko]. - Moskva : Glavnaia redaktsiia vostochnoĭ literatury, 1978
614p. - Bibliography : p.570-594
r6172823 X.809/27255
1988549

AFRICA
Revolutionary movements, to 1978

JAFFE, Hosea
Africa. - Milano : Arnoldo Mondadori, 1978
343p : ill,maps
r6160104 X.700/22380
2418827

AFRICA
Revolutionary movements, to date

IDEOLOGIÎA revoliutsionnykh demokratov Afriki / [otvetstvennyĭ redaktor G.B. Starushenko]. - Moskva : Nauka, 1981
288 p.. - Includes bibliographies
r6403984 X.529/63984
8579865

SÉKOU TOURÉ, Ahmed
L'Afrique et la révolution. - [Conakry] : [Parti-État de Guinée], [1978?]
398p
r6211588 SV.600/17
8579865

AFRICA
Revolutionary movements, 1945-1981

The STRUGGLE for Africa / Mai Palmberg (editor) ; translated by E.M.K. Andrée, Mai Palmberg and Howard Simson. - London : Zed, 1983
iii,286p : ill
0-86232-100-x X.809/61454
312469x

AFRICA
Revolutionary movements. Role of communist countries

POSADAS, J.
The tour of Fidel Castro and Podgorny in Africa, the Permanent Revolution and the world objective necessity of socialism. - London : Revolutionary Workers Party (Trotskyist), 1977
[1],17,[1]p. - (A 'European Marxist review' publication)
B77-29986 X.709/31618
1252755

AFRICA
Rituals: Sacrifice

HENSCH, Luc de
Sacrifice in Africa. - Manchester : Manchester University Press, c1985
232p : ill. - (Themes in social anthropology). - Bibliography: p217-225. - Includes index
0-7190-1716-5 X.529/69807
3018059

AFRICA
Rock carvings

WILLCOX, A. R. (Alexander Robert)
The rock art of Africa. - London : Croom Helm, c1984
xvi,287p,[24]p of plates : ill(some col.). - Bibliography: p267-279. - Includes index
0-7099-2743-6 X.421/25420
3130452

AFRICA
Rock paintings

RITCHIE, Carson I. A. (Carson Irving Alexander)
Rock art of Africa. - South Brunswick : Barnes ; London : Yoseloff, c1979
157p,[8]p of plates : ill(some col.). - Includes index
0-498-01753-2 L.45/1234
2559544

WILLCOX, A. R. (Alexander Robert)
The rock art of Africa. - London : Croom Helm, c1984
xvi,287p,[24]p of plates : ill(some col.). - Bibliography: p267-279. - Includes index
0-7099-2743-6 X.421/25420
2559544

AFRICA
Rural regions - Serials

MICHIGAN STATE UNIVERSITY. African Studies Center
Rural Africana. - New series : no.1 (Spring 1978)-. - Michigan : The Center, 1978-
r6202129 P.801/1690
2437422

AFRICA
Rural regions. Agriculture

GROVE, Alfred Thomas
Rural Africa. - Cambridge : Cambridge University Press, 1979
[6],122p : ill. - (Cambridge topics in geography series). - Bibliography: p.120. - Includes index
0-521-21825-x X.800/27854
190230x

AFRICA
Rural regions. Community health services. Organisation

MCCUSKER, Jane
How to measure and evaluate community health. - [Rev. ed.]. - London : Macmillan, 1982, c1978
[228]p : ill. - (Macmillan tropical community health manuals)
0-333-31680-0 X.322/12106
2929538

AFRICA
Rural regions. Economic development. Political aspects, to 1982

BATES, Robert H. (Robert Hinrichs)
Essays on the political economy of rural Africa. - Cambridge : Cambridge University Press, 1983
vii,178p : ill. - (African studies series ; 38). - Includes index
0-521-24563-x X.520/31415
2960052

AFRICA
Rural regions. Health services

BELLONCLE, Guy
Santé et développement en milieu rural africain. - Paris : Éditions Économie et Humanisme ; Les Éditions Ouvrières, 1975
236 p. : ill. - (Développement et civilisations). - Bibliography : p. 233-236
r6002961 X.319/8450 Woolwich
0859648

AFRICA
Rural regions. Poverty

ELLIOTT, Charles, 1939-
Rural poverty in Africa. - [Norwich] : [Geo Abstracts for] the Centre for Development Studies, University College of Swansea, 1980
[3],43p. - (Occasional papers / University College of Swansea. Centre for Development Studies ; no.12)
0-86094-050-0 X.529/38005
2303086

AFRICA
Rural regions. Printing presses, to date

RURAL press for village post-literacy literature / edited by Charles T. Hein and Keith K. Kanyogonya. - Nairobi : Afrolit Society, 1979
96p. - (Afrolit paper ; no.5)
r6813274 2708.e.436
9085440

AFRICA
Rural regions. Small scale industries

LIVINGSTONE, I. (Ian)
Alternative approaches to small industry promotion. - Norwich : School of Development Studies, University of East Anglia, 1983
35p. - (Discussion paper : no.62). - Bibliography: p35
B85-01220 X.322/14225
3197468

AFRICA
Rural regions. Transport - Texts

LEWIN, Hugh
Jafta -. - London : Evans Brothers, 1983
24p : col.ill. - (Jafta's family series)
0-237-45676-1 X.990/24048
3230945

AFRICA
Rural regions. Water supply

NISSEN-PETERSEN, Erik
Rain catchment and water supply in rural Africa. - London : Hodder and Stoughton in association with the Danish National Fund for Developing Countries, 1982
x,83p : ill. - Bibliography: p82-83
0-340-28429-3 X.808/37734
2866978

AFRICA
Sahara desert. Prehistoric rock carvings

LHOTE, Henri
Les gravures de l'oued Mammanet (Nord-Ouest du massif de l'Aïr). - Dakar : Les Nouvelles éditions africaines, 1979
430p : ill
2-7236-0423-3 X.421/24989
8722994

AFRICA
Sahara desert. Prehistoric rock carvings - Catalogues

MUSEUM FÜR VÖLKERKUNDE UND SCHWEIZERISCHES MUSEUM FÜR VOLKSKUNDE BASEL
Als die Sahara grün war. - Basel : Das Museum, [1977]
11p,[17]p of plates : 9ill
r6302108 X.429/12214 Woolwich
8722994

AFRICA
Savanna regions. Animals

CUISIN, Michel
Animals of the African plains. - London : Kingfisher, 1980
44p : ill(some col.). - (Nature's hidden world)
0-7063-6034-6 X.322/9819 Woolwich
2327821

AFRICA
Schools. Curriculum subjects: African languages. Teaching

AWONIYI, Timothy Adedeji
The teaching of African languages. - London : Hodder and Stoughton, 1982
165p : 3ill. - (Foundations of education in Africa). - Bibliography: p160-165
0-340-28171-5 X.950/18396
2840316

AFRICA
Schools. Curriculum subjects: African literature. Teaching, to date

PRETORIUS, W. J.
Teaching African literature. - 2nd. impression. - Pretoria : University of South Africa, 1982
iv,68p. - Includes bibliographies
0-86981-277-7 X.529/63688
9001611

TEACHING of African literature in schools / edited by Eddah Gachukia and S. Kichamu Akivaga. - Nairobi : Kenya Literature Bureau, 1978-
v.
r6807883 X.529/56841
9001611

AFRICA
Science, to 1980

BLACKS in science / editor Ivan van Sertima. - New Brunswick, N.J. ; London : Transaction, c1983
302p : ill. - Bibliography: p295-297
0-87855-941-8 X.629/22072
3046125

AFRICA
Scientific relations with Soviet Union. Asia & Africa

KORNEEV, S. G. (Stepan Gavrilovich)
Nauchnye svîazi Akademii nauk SSSR so stranami Azii i Afriki. - Moskva : Nauka, Glavnaîa redaktsiîa vostochnoĭ literatury, 1976
382p : ill. - Bibliography : p.376-381
r6186088 X.629/8204 Woolwich
2123614

AFRICA
Secondary schools. Curriculum subjects: Economics. Teaching

HENDERSON, Willie
Teaching economics in African secondary schools. - London : Heinemann, 1980
118p : ill. - Bibliography: p.117-118
0-435-97220-0 X.529/37557
2140977

AFRICA
Serial articles - Indexes

ASAMANI, J.O
Index Africanus. - Stanford, Calif : Hoover Institution Press, Stanford University, [1975]
xiii,659p. - (Hoover Institution bibliographies ; 53)
0-8179-2531-7 Ac.2692.nf/3(53)
1239473

AFRICA
Serial articles - Indexes - Serials

AFRICA index / [issued by] Africa Bibliographic Centre. - no.1- ; 1976-. - Dar es Salaam : The Centre, 1977-
r6082567 P.523/559
1239473

AFRICA
Serials - Bibliographies - Serials

CURRENT contents Africa. - Vol.1, no.1- ; 1976-. - Oxford : Hans Zell (Publishers) Ltd [for] the Stadt- und Universitätsbibliothek Frankfurt am Main, [1976]-
Holdings: vol.4, no.1, etc: 1978, etc
B79-19807 P.513/2316
1894714

AFRICA
Serials, to date - Catalogues - Serials

TIDNINGAR och tidskrifter i nordiska afrikainstitutets bibliotek = Periodicals in the library of the Scandinavian Institute of African Studies / [Nordiska afrikainstitutet]. - 1982-. - [Uppsala] : Institutet, 1982-
r6610416 X.0805/70
8877319

AFRICA
Slave trade, to date

LOTH, Heinrich
Das Sklavenschiff. - Berlin : Union, 1981
346p,[48]p of plates : ill. - Bibliography
r6301931 X.808/35425
8756880

AFRICA
Slave trade, 1600-1900

REYNOLDS, Edward
Stand the storm. - London : Allison & Busby, 1985
182p. - Bibliography: p150-172. - Includes index
0-85031-575-1 X.809/64073
302203x

AFRICA
Slave trade, 1820-1850

CONNEAU, Theophilus
A slaver's log book, or, 20 years' residence in Africa. - London : Hale, 1977
xiv,370p. - Includes index
0-7091-6401-7 X.520/11992
1245260

AFRICA
Slave trade. Policies of British government, 1809-1919

MIERS, Suzanne
Britain and the ending of the slave trade. - London : Longman, 1975
xvi,405p,[8]p of plates : ill. - Bibliography: p.320-345. - Includes index
0-582-64079-2 X.520/10574
0624969

AFRICA
Slave trade with America, to ca 1850

DAVIDSON, Basil
Black mother. - Revised and expanded ed. - Harmondsworth : Penguin, 1980
304p. - (Pelican books). - Bibliography: p.289-293. - Includes index
0-14-022278-2 X.708/22586
2240300

RAWLEY, James A.
The transatlantic slave trade. - New York ; London : Norton, c1981
xiv,452p. - Includes index
0-393-01471-1 X.800/39382
2240300

AFRICA
Slave trade with America, 1450-1885

FRESE, Jens
Den atlantiske slavehandel og England. - Copenhagen : Gyldendal, 1981
172p : ill
87-01-98271-0 X.520/28656
8894019

AFRICA
Slave trade with America, 1518-1865

MANNIX, Daniel Pratt
Black cargoes. - New York : Penguin, 1976
xiv, 306 p., [16] p. of plates : ill. - Bibliography : p. 288-295. - Includes index
0-14-004253-9 X.519/26134
0725773

AFRICA
Slave trade with America, ca 1640-ca 1890. Economic aspects - Conference proceedings

ECONOMIC HISTORY OF THE TRANSATLANTIC SLAVE TRADE (Conference : 1975 : Colby College)
The uncommon market. - New York ; London : Academic Press, 1979
xvi,448p : ill. - (Studies in social discontinuity). - Bibliography: p.72-76. - Includes index
0-12-279850-3 X.520/14680
1998536

AFRICA
Slave trade with America, ca 1700-1850

HOGG, Peter C. (Peter Carlquist)
Slavery. - London : British Library, c1979
51p : ill(some col.). - Bibliography: 2p
0-904654-28-1 X.800/31045
1874608

AFRICA
Slave trade with Brazil, 1793. Medical aspects

MENDES, Luís António de Oliveira
Memória a respeito dos escravos e tráfico da escravatura entre a costa d'África e o Brazil apresentada à Real Academia das Ciências de Lisboa, 1793. - Porto : Escorpião, 1977
89p. - (Cadernos "O Homem e a sociedade" ; n.26)
r6064910 X.319/9260 Woolwich
1428128

AFRICA
Slave trade with Great Britain, to ca 1900

KAY, F. George (Frederick George)
The shameful trade. - London : White Lion Publishers, 1976
[6],218p,[8]p of plates : ill. - Bibliography: p.208-212. - Includes index
0-7274-0037-1 X.529/30520
0612847

AFRICA
Slave trade with Spanish America, 1595-1640

VILA VILAR, Enriqueta
Hispanoamérica y el comercio de esclavos. - Sevilla : Escuela de Estudios Hispano-Americanos de Sevilla, 1977
306p.,10p.of plates,2 leaves of plates : ill. - (Publicaciones de la Escuela de Estudios Hispano-Americanos de Sevilla ; n.239). - Includes bibliographies
84-00-03665-4 X.700/21245
2134241

AFRICA
Slavery, to 1920

LOVEJOY, Paul E.
Transformations in slavery. - Cambridge : Cambridge University Press, 1983
xvi,349p. - (African studies series ; 36). - Bibliography: p309-336. - Includes index
0-521-24369-6 X.520/31345
2700204

AFRICA
Slavery. Role of women

WOMEN and slavery in Africa / edited by Claire C. Robertson, Martin A. Klein. - Madison, Wis. ; London : University of Wisconsin Press, 1983
x,380p : ill. - Includes bibliographies and index
0-299-09460-x X.520/35435
3196135

AFRICA
Social change, to date

ASANTE, Molefi K. (Molefi Kete)
Afrocentricity. - Buffalo : Amulefi, [1980]
iii,125p
r6807240 X.529/58594
8902119

GENTILI, Anna Maria
Africa come storia. - Milano : Angeli, [1980]
437p. - (La società ; 77). - Bibliography
88-204-1777-4 X.529/51880
8902119

AFRICA
Social change. Effects of technological development

GOODY, Jack
Technology, tradition and the state in Africa. - London [etc.? : Hutchinson, 1980
viii,88p : ill. - (Hutchinson university library for Africa). - Bibliography: p.77-84. - Includes index
0-09-141321-4 X.529/38545
1958089

AFRICA
Social change. Influence on African languages, to date

SOZIALER Wandel in Afrika und die Entwicklung von Formen und Funktionen afrikanischer Sprachen / herausgegeben von Dmitrij Alekseevič Olderogge und Siegmund Brauner. - Berlin : Zentralinstitut für Sprachwissenschaft, 1980
272p.. - (Linguistische Studien. Reihe A, Arbeitsberichte ; 64)
r6300299 X.958/3987
8822255

AFRICA
Social class. Political aspects

STATE and class in Africa / edited by Nelson Kasfir. - London : Cass, 1984
125p
0-7146-3239-2 X.529/65987
3022293

AFRICA
Social conditions

AFRICAN Spark. - Lagos : African Spark Publications. Vol.1, no.1- ; [1973]-. - [1973]-
Holdings : Vol.1, nos. 3, 7, 8, 10 ; Vol.4, no.9 : Vol.5, no.7- ; April, Sept, Oct., Dec., 1973 ; Sept., 1976 ; May, 1977-
r6109119 P.511/1600
0628093

AFRIKA - kontinent i krise / ved Gunilla Andrea ... [et al.] ; redigeret af Ellen Buch-Hansen. - [København] : Mellemfolkeligt Samvirke, 1980
219p : ill. - Includes bibliographies
87-7028-281-1 X.529/37592
0628093

AKE, Claude
Revolutionary pressures in Africa. - London : Zed Press, 1978
3-109p. - (Africa series)
0-905762-14-2 X.809/43306
151380x

CONTEMPORARY Africa / edited by C. Gregory Knight, James L. Newman. - Englewood Cliffs ; London : Prentice-Hall, 1976
xiv,546p : ill. - Bibliography: p.507-530. - Includes index
0-13-170035-9 X.800/25812
0628093

KANZA, Thomas
Evolution & revolution in Africa. - London : Collings, 1971
141p
B77-03967 X.519/26426
0628093

LAURENT, André
Life in Africa. - St. Albans : Hart-Davis, 1981
62p : ill(some col.). - (Signposts series). - Includes index
0-247-13038-9 X.805/1943
2569906

TIMOTHY, Bankolé
Africa. - Blandford : Davison Publishing Ltd, 1976
X,170p : ill. - Includes index
0-904130-08-8 X.809/40814
0628093

AFRICA
Social conditions - Conference proceedings

UNION GÉNÉRALE DES TRAVAILLEURS SÉNÉGALAIS EN FRANCE
Qui est responsable du sous-développement?. - Paris : François Maspero, 1975
82 p.
2-7071-0766-2 X.519/20652
0857211

AFRICA
Social conditions, to 1976
AFRICAN social studies / edited by Peter C.W. Gutkind and Peter Waterman. - London : Heinemann Educational, 1977
xi,481p : 1 ill. - Bibliography: p.436-462. - Includes index
0-435-89042-5 X.809/42648
1291858

QUILICI, Folco
Io Africa. - Bari : De Donato, 1977
348p.,[78]p. of plates : ill
r6068042 X.809/25969
1784099

AFRICA
Social conditions, to date
Les FAUX prophètes de l'Afrique, ou, L'Afr(eu)canisme / sous la direction de Alf Schwarz. - Québec : Les Presses de l'Université Laval, 1980
244p
2-7637-6905-5 X.520/27910
857989X

AFRICA
Social conditions, to date - Serials
INTER-AFRICA now. - [No.1]-. - Mbabane : African Unity Press, [1981]-
v.
r6813230 X.0525/592
857989x

AFRICA
Social conditions, 1934-1947
SEYMOUR, John, 1914-
On my own terms. - Revised ed. - London : Faber, 1980
3-225p
0-571-18016-7 X.329/13188 Woolwich
1948148

AFRICA
Social conditions, 1960-
AFRIKA im Aufbruch / [Herausgeberkollektiv: Klaus Ulrich et al.]. - Leipzig : Brockhaus, 1980
[334]p,[1]folded leaf : col.ill
r6307126 L.45/1959
8579881

DOGBÉ, Yves-Emmanuel
Lettre ouverte aux pauvres d'Afrique. - [Le Mée-sur-Seine] : Éditions Akpagnon, 1981
77p
2-86427-012-9 X.529/55698
8579881

OCTAVE
Demain l'Afrique. - Paris : Presses Universitaires de France, 1981
xi,195p. - (Perspectives internationales). - Bibliography: p192-194
2-13-037095-0 X.529/47742
8579881

AFRICA
Social conditions, 1960- - Serials
AFRIQUE message. - no.1 (1979)-. - Paris : C.A.C.E.F., [1979]- : ill
Holdings: no.5, avril-mai 1980-
r6207204 P.2000/556
8579881

BAOBAB / herausgegeben vom Arbeitskreis der Freunde Afrikanischer Literatur. - 1-. - Göttingen : Edition Herodot, [1981]-
r6304653 X.0950/468
8579881

TRIBUNE africaine. - no 1 (1983)-. - Paris : Tribune africaine, 1983-
r6210193 P.521/3809
8579881

AFRICA
Social conditions. Effects of slave trade, 1400-1900
FORCED migration / edited by J.E. Inikori. - London : Hutchinson University Library, 1982
349p. - (Hutchinson university library for Africa). - Bibliography: p341-343. - Includes index
0-09-145900-1 X.809/52771
276492x
- [Another copy: paperback.] ISBN 0-09-145901-X X.809/52772

AFRICA
Social conditions. Statistics
KPEDEKPO, G. M. K.
Social and economic statistics for Africa. - London : Allen & Unwin, 1981
x,259p. - Includes bibliographies and index
0-04-310011-2 X.529/43014
2543753
- [Another copy] OPL

AFRICA
Social development. Policies of political leaders, ca 1950-1975
DAMACHI, Ukandi Godwin
Leadership ideology in Africa. - New York ; London : Praeger ; [London] : [Distributed by Martin Robertson], 1976
xiii,113p. - (Praeger special studies in international economics and development). - Bibliography: p.107-112
0-275-56760-5 X.520/11417
1052616

AFRICA
Social development. Role of education
THOMPSON, A. R.
Education and development in Africa. - London : Macmillan, 1981
viii,358p : ill. - Bibliography: p339-353. - Includes index
0-333-30018-1 X.529/44325
2579375

AFRICA
Social sciences, 1800- - Conference proceedings
SCIENCES de l'homme et conquête coloniale / communications [d'un colloque consacré aux élaborations et applications politiques des sciences humaines en Afrique au XIX siècle, tenu à l'École normale supérieure, 1977] réunies et présentées par Daniel Nordman et Jean-Pierre Raison. - Paris : Presses de l'École normale supérieure, 1980
238p : ill
2-7288-0066-9 X.520/25057
8781567

AFRICA
Social structure. Influence of technology, to 1980
GOODY, Jack
Technology, tradition and the state in Africa. - Cambridge : Cambridge University Press, 1980, c1971
88p : ill. - Bibliography: p77-84. - Includes index
0-521-29892-x X.529/64595
3165280

AFRICA
Socialism
BABU, Abdul Rahman Mohamed
African socialism or socialist Africa?. - London : Zed, 1981
xv,174p. - (Africa series)
0-905762-19-3 X.809/51798
2726688

AFRICA
Socialism, to date
[IDEOLOGIÎA afrikanskoĭ revoliūtsionnoĭ demokratii. English]. The ideology of African revolutionary democracy / [editorial board An. Gromyko, consulting editor ... et al.]. - Moscow : Social Sciences Today, 1984
202p. - (African Studies by Soviet Scholars ; no.5). - Includes bibliography
r6465996 X.529/67819
8003971

KONTER, Jan Herman
Ujamaa. - Assen : Gorcum, 1978
139p : ill. - (Terreinverkenningen in de culturele antropologie ; 11). - Includes bibliography
90-232-1562-1 X.529/48428
8003971

AFRICA
Socialism, 1960-1977
MOHIDDIN, Ahmed
African socialism in two countries. - London : Croom Helm, c1981
231p. - Includes index
0-7099-1702-3 X.529/46218
2208245

African socialism in two countries. - London : Croom Helm, c1981
231p. - Includes index
0-7099-1702-3 X.529/46218
2208253

AFRICA
Socialism related to economic development - Conference proceedings
PROBLEMS of Socialist orientation in Africa / contributors: Mohamed Aden ... [et al.] ; edited by Mai Palmberg. - Uppsala : Scandinavian Institute of African Studies ; Stockholm : Distributed by Almqvist & Wiksell International, 1978
243p
91-7106-141-x X.529/35514
2425319

AFRICA
Socialist parties, 1960-
INTERAFRICAINE SOCIALISTE. Comité préparatoire [Réunion du] Comité préparatoire de l'Interafricaine socialiste, Tunis 13-14 juillet 1980. - Tunis : Division des relations extérieures du Parti socialiste destourien, 1980
20p
r6210420 X.520/31872
902302x

AFRICA
Society. Role of graduates
BARKAN, Joel D.
An African dilemma. - Nairobi ; London : Oxford University Press, 1975 [i.e. 1976]
xvii,259p : ill. - Bibliography: p.247-254. - Includes index
0-19-572374-0 X.809/40944
0651796
- [Another copy: paperback]
ISBN 0-19-572373-2 X.709/31401

AFRICA
Society. Role of intellectuals, 1945-1977. Political aspects
MAZRUI, Ali Al'Amin
Political values and the educated class in Africa. - London : Heinemann Educational, 1978
xiv,392p. - Includes index
0-435-96522-0 X.809/43448
1569589

AFRICA
Society. Role of literature. Political aspects
NGŨGĨ WA THIONG'O
Writers in politics. - London : Heinemann, 1981
142p. - (Studies in African literature). - Includes index
0-435-91751-x X.529/41683
2404532

AFRICA
Society. Role of women
CUTRUFELLI, Maria Rosa
Donna perché piangi?. - Milano : Gabriele Mazzotta, 1976
253p. - (Nuova informazione ; 47). - Bibliography : p.246-253
r6052727 X.519/35063
1531433
Women of Africa. - London : Zed Press, 1983
186p. - Bibliography: p179-186
0-86232-083-6 X.529/58177
1531433

WOMEN in Africa and development assistance. - Uppsala : Scandinavian Institute of African Studies, 1978
55p. - Includes bibliographies
r6113179 X.529/34994
1531433

AFRICA
Society. Role of women - Conference proceedings
SOCIÉTÉ AFRICAINE DE CULTURE
La civilisation de la femme dans la tradition africaine = The civilization of the woman in African tradition. - Paris : Présence africaine, 1975
606 p., 16 p. of plates : ill
r6012230 X.519/20838
1531433

AFRICA
Society. Role of women - Serials
AFRICA Woman. - No.1- ; Oct./Nov.1975-. - London : Africa Journal Limited, 1975- : ill(some col)
B76-21622 P.521/3021
0676985

FEMAIL. - Vol.1, no.1- ; June 1977-. - [Freetown] : Sierra Leone Association of University Women, 1977-
r6108226 P.525/159
1531433

AFRICA
South-eastern Africa. Expeditions, 1888-1891
WIESE, Carl
Expedition in East-Central Africa, 1888-1891. - London : Collings, 1983
xvi,383p : ill. - Bibliography: p371-375. - Includes index
0-86036-206-x X.805/5546
3135454

AFRICA
South-eastern Africa. Pioneers, 1552-1878 - Biographies - Dictionaries
TABLER, Edward Corban
Pioneers of Natal and Southeastern Africa, 1552-1878. - Cape Town : A.A. Balkema, 1977
ix,117p. - (South African biographical and historical studies). - Bibliography : p.109-114. - Includes index
0-86961-090-2 X.0700/226(19)
1565583

AFRICA
South-eastern Africa. Plants

FLORA Zambesiaca. - [Richmond] : Flora Zambesiaca Managing Committee : [Distributed by British Museum (Natural History)].
Vol.4. - 1978
658p,plate : ill(incl 1 col). - Includes index
0-85592-044-0 SRL
1491237
- London : Published by the Managing Committee on behalf of the contributors to Flora Zambesiaca.
Vol.7
Pt.1. - 1983
394p : ill(some col.). - Includes index
0-9507682-0-0 SRL
1491237

AFRICA
Southern Africa

ARNOLD, Guy
The last bunker. - London : Quartet Books, 1976
[13],286p. - Bibliography: p.270-275. - Includes index
0-7043-2099-1 X.809/40746
0614548

The last bunker. - London : Quartet Books, 1976 [i.e. 1977]
[13],286p. - Bibliography: p.270-275. - Includes index
0-7043-3190-x X.709/31778
0614548

TRAVELLERS' guide to Southern Africa / [orginally compiled by Richard Cox]. - 3rd ed.. - London : Geographia, 1981
176p,[12]p of plates : ill(some col.). - (Thornton Cox travellers' guides). - Includes index
0-09-208200-9 X.809/50823
2638320

AFRICA
Southern Africa - Serials

JOURNAL of Southern African Affairs : an interdisciplinary research quarterly / [issued by] Southern African Research Association and Afro-American Studies, University of Maryland. - vol.1-; October, 1976-. - College Park, Md. : The Association and the University, 1976-
r6089457 X.0510/344
0326178

MBIRE : historical research south of the Zambezi. - no.1- ; 1971-. - Salisbury, Rhod. : D.N. Beach, 1971-
r6098905 P.705/148
0326178

AFRICA
Southern Africa, to 1974 - For East African students

PARKER, Graham, 1943-
History of Southern Africa. - London : Bell, 1975
xii,308p : ill. - Includes index
0-7135-1783-2 X.709/30202
0539430

AFRICA
Southern Africa, to 1975

WILLCOX, A. R. (Alexander Robert)
Southern land. - Cape Town ; London : Purnell, 1976
xiii,274[i.e.282]p,[34]p of plates : ill(some col). - Bibliography: p.258-260. - Includes index
0-360-00324-9 X.421/10508 Woolwich
1264494

AFRICA
Southern Africa, to 1975 - Conference proceedings - Serials

SOUTHERN African research in progress / [University of York, Centre for Southern African Studies]. - 1- ; 1975-. - [York] : [The Centre], [1975]-
0-905055-00-4 P.701/1640
1264494

AFRICA
Southern Africa, to 1976. Ethnographic aspects

BERNARDI, Bernardo
Africa meridionale. - Novara : Istituto Geografico De Agostini, 1977
120p : ill(some col),maps. - (Popoli nel mondo)
r6195093 L.42/454
2425920

AFRICA
Southern Africa, to 1977

SOUTHERN Africa in crisis. - London : Relgocrest Ltd, [1977?]
48p : ill. - ('Red weekly' pamphlet). - Bibliography: p.48
0-85612-017-0 X.809/45790
222948x

AFRICA
Southern Africa, to 1978 - For East African students

ODHIAMBO, E. S. Atieno
A history of Southern Africa. - New ed.. - London : Longman, 1978
vii,200p : ill. - Bibliography: p.195. - Includes index
0-582-60349-8 X.700/25904
1776363

AFRICA
Southern Africa, to 1980

HULL, Richard W. (Richard William)
Southern Africa. - New York ; London : New York University Press, c1981
vii,168p : ill. - Includes bibliographies and index
0-8147-3410-3 X.520/28233
2786311
- [Another copy: paperback.] ISBN 0-8147-3411-1 X.520/28234

LEEMING, Donald
'O' level Southern African history. - Harlow : Longman, 1981
92p. - (Study for success) (Study for success)
0-582-60392-7 X.529/48606
2759047

AFRICA
Southern Africa, to 1980 - For Southern African students

JUNIOR Certificate history of Southern Africa. - London : Heinemann.
Bk.1: Southern African societies before the Scramble. - 1979
144p : ill. - Bibliography: p141. - Includes index
0-435-94160-7 X.0809/2181(1)
1929852

JUNIOR certificate history of southern Africa. - London : Heinemann Educational.
Bk. 2: Southern Africa in the late nineteenth and twentieth centuries. - 1981
xviii,126p : ill. - Bibliography: p120. - Includes index
0-435-94161-5 X.809/49691
1929852

JUNIOR Certificate history of Southern Africa. - London : Heinemann Educational.
Bk.3: The modern world. - 1982
xiv,130p : ill. - Bibliography: p127. - Includes index
0-435-94760-5 X.0809/2181(3)
1929852

PARSONS, Neil, 1944-
A new history of southern Africa. - London : Macmillan Education, 1982
vi,330p : ill. - Bibliography: p321-325. - Includes index
0-333-26220-4 X.800/34040
1929852

AFRICA
Southern Africa, to 1982

DANZIGER, Christopher
A history of southern Africa. - Oxford : Oxford University Press, 1983
125p : ill. - Includes index
0-19-913284-4 X.800/37660
3074587

AFRICA
Southern Africa, to date

CHANGING realities in Southern Africa. - Berkeley : Institute of International Studies, University of California, c1982
xi,318p. - (Research series ; no.47)
0-87725-147-9 X.0809/130.(47)
8611483

HOEK, P. W.
Southern Africa. - Pretoria : Haum, 1980
123p
0-7986-0520-0 X.809/50967
8611483

MUNGAZI, Dickson A.
To honor the sacred trust of civilization. - Cambridge, Mass. : Schenkman Publishing Co., [1983]
ix,318p
0-87073-454-7 X.809/64765
8611483

ROTBERG, Robert I.
Suffer the future. - Cambridge, MA : Harvard University Press, 1980
xvi,311p
0-674-85401-2 X.520/26522
8611483

SOUTHERN Africa since the Portuguese coup / edited by John Seiler. - Boulder : Westview Press, 1980
xix,252p. - (Westview special studies on Africa)
0-89158-767-5 X.520/25358
8611483

AFRICA
Southern Africa, to date - Catalogues

A CATALOGUE of books on Africa, mainly south of the equator, in the library of Ida and F.W. Hosken, Copperhouse, Honeydew, Transvaal. - Northcliff : F.W. Hosken, 1981
ix,224p,plates
0-620-04973-1 2725.c.217
8611483

AFRICA
Southern Africa, to date - Conference proceedings

CONFERENCE ON CONFLICT AND CHANGE IN SOUTHERN AFRICA : SCANDINAVIAN AND CANADIAN PERSPECTIVES AND POLICY OPTIONS
(1978 : Ottawa)
Conflict and change in Southern Africa. - Washington, D.C. : University Press of America for the Scandinavian Institute of African Studies in Uppsala ; the Norman Pearson School of International Affairs, Carleton University ; and the Centre for African Studies of Dalhousie University, [1978]
vi,269p
r6864600 X.809/49215
8611483

AFRICA
Southern Africa, to date - Conference proceedings - Serials

WORKING papers in Southern African studies / [edited by] P. Bonner. - Vol.2-. - Johannesburg : Ravan Press, 1981-
v.
r6812134 X.0809/2387
8611483

AFRICA
Southern Africa, ca 1800- ca 1970

UNIVERSITY OF LONDON. Institute of Commonwealth Studies
Collected seminar papers on the societies of Southern Africa in the 19th and 20th centuries. - London : The Institute.
Vol.9: October 1977 - June 1978. - [1980]
iii,160p. - (Collected seminar papers ; no.24)
0-902499-22-x X.0700/225
022393X

AFRICA
Southern Africa, ca 1800- ca 1970 - Conference proceedings

UNIVERSITY OF LONDON. Institute of Commonwealth Studies
Collected seminar papers on the societies of Southern Africa in the 19th and 20th centuries. - London : The Institute.
[Vol.7]: October 1975-June 1976. - [1977]
[4],211p. - (Collected seminar papers ; no.21)
0-902499-19-x X.0700/225
0196711

AFRICA
Southern Africa, 1800-1983

DENOON, Donald
Southern Africa since 1800. - New ed. - London : Longman, 1984
x,246p : ill. - Bibliography: p236-238. - Includes index
0-582-72707-3 X.809/64667
3324486

AFRICA
Southern Africa, 1960-

AERSCHOT, Jan van
De slag om het palmengordijn. - Antwerpen : Nederlanden, [1982]
160p
90-658-3023-5 X.809/54830
8916381

KRIZIS na i͡uge Afriki / [otvetstvennyĭ redaktor : An.A. Gromyko]. - Moskva : Nauka, Glavnai͡a redakt͡sii͡a vostochnoĭ literatury, 1984
268p
r6490280 X.809/62667
8916381

AFRICA
Southern Africa. Aloes

JEPPE, Barbara
South African Aloes. - 2nd ed. - Cape Town ; London : Purnell, 1977
xxvii,150[i.e.287]p : ill(chiefly col). - Bibliography: p.viii. - Includes index
0-360-00018-5 X.320/11967 Woolwich
2002507

AFRICA
Southern Africa. American multinational companies

SEIDMAN, Ann
US multinationals in Southern Africa. - Dar es Salaam : Tanzania Publishing House ; Birmingham : [Distributed by] Third World Publications, 1977
xii,251p : ill. - Includes index
B80-07707 X.519/30569 Woolwich
2022044

AFRICA
Southern Africa. Animals

HART, Susanne
Back in the wild. - London : Collins ; London : Harvill Press, 1977
158p : ill
0-00-262049-9 X.320/10866 Woolwich
1186795

AFRICA
Southern Africa. Apartheid - Serials

BULLETIN of the Continuation Committee of the World Conference against Apartheid, Racism and Colonialism in Southern Africa. - No.1 (Aug.1978)- No.2 (Nov.[1978]). - London : The Committee, 1978-1978
2v. : ill
B81-05680 P.703/730
2348543

ICSA bulletin. - [No.1]-. - London : International Committee against Apartheid, Racism and Colonialism in Southern Africa, 1979-
v. : ill
Holdings: no.4 ; June 1979, etc
B81-05683 P.703/717
2348543

AFRICA
Southern Africa. Bantu, to 1964

MUTWA, Vusa'mazulu Credo
My people. - Harmondsworth : Penguin, 1977
332p : ill
B78-01858 X.709/31799
1282662

AFRICA
Southern Africa. Bantu, to date - Bibliographies

ANTHROPOLOGY of southern Africa in periodicals to 1950 / compiled under the direction of N.J. van Warmelo. - Johannesburg : Witwatersrand University Press, 1977
1484p
0-85494-378-1 X.805/4243
8981744

AFRICA
Southern Africa. Bantu. Cultural processes

The BANTU-SPEAKING peoples of Southern Africa / edited by W.D. Hammond-Tooke. - 2nd ed. - London : Routledge & Kegan Paul, 1974
xxii,525p,[24]p of plates : ill. - Bibliography: p473-510. - Includes index
B81-11214 X.529/40552
0171042

BOURDILLON, M. F.
The Shona peoples. - Gwelo : Mambo Press, 1976
399p. - (Shona heritage series ; vol.1)
0-86922-020-9 X.809/45270
0171042
- [Another copy] X.529/60883

KRIGE, Eileen Jensen
The realm of a rain-queen. - Cape Town : Juba & Company, 1980
xv,335p,16p. of plates
0-7021-1060-4 X.520/20610
0171042

TYRRELL, Barbara
African heritage. - Johannesburg : Macmillan South Africa, 1983
276p : ill(some col.). - Bibliography: p271-272
0-86954-116-1 X.805/4754
0171042

AFRICA
Southern Africa. Bantu. Cultural processes, to date

TEMPELS, Placide
Plaidoyer pour la philosophie bantu et quelques autres textes. - Kinshasa : Département de Philosophie et Religions Africaines, Faculté de théologie Catholique, 1982
vi,99p : ill. - (Cours et documents ; 6)
r6210770 X.529/61351
8611467

AFRICA
Southern Africa. Bantu. Religious beliefs

MATUNGULU OTENE
Être avec pour vivre vrai. - Lubumbashi, 1981
107p
r6211649 X.200/43230
9067507

AFRICA
Southern Africa. Birds

BIRDS of Southern Africa. - Johannesburg : Macmillan South Africa ; London : Macmillan. - Includes index
1: Kruger National Park. - c1980
xii,242p : ill(some col.)
0-86954-103-x X.0329/239(1) Woolwich
2365995

MACKWORTH-PRAED, C. W. (Cyril Winthrop)
Birds of the southern third of Africa. - London : Longman.
Vol.1. - 1962
xxvi,688p,[49]p of plates : ill(some col.). - (African handbook of birds. Series 2). - Includes index
0-582-46084-0 X.329/14982 Woolwich
2628570

MACKWORTH-PRAED, C. W. (Cyril Winthrop)
Birds of the southern third of Africa. - London : Longman.
Vol.2. - 1963
x,747p,[47]p of plates : ill(some col.). - (African handbook of birds. Series 2). - Includes index
0-582-46085-9 X.329/14981 Woolwich
2628570

AFRICA
Southern Africa. Birds of prey

STEYN, Peter
Birds of prey of southern Africa. - Cape Town : David Philip ; Beckenham : Croom Helm, 1983, c1982
xxiii,309p,24leaves of plates : ill(some col.). - Bibliography: p289-303. - Includes index
0-7099-2382-1 SRL
2964260

AFRICA
Southern Africa. Butterflies

PINHEY, Elliot Charles Gordon
A guide to the butterflies of Central and Southern Africa. - London : Sir Joseph Causton, 1977
3-106p : ill(chiefly col). - Bibliography: p.104. - Includes index
0-340-22605-6 SRL
147670x

AFRICA
Southern Africa. Catholic Church. Effects of nationalist movements, 1959-1979

LINDEN, Ian
The Catholic Church and the struggle for Zimbabwe. - London : Longman, 1980
x,310p. - Bibliography: p.306-307. - Includes index
0-582-64302-3 X.200/33974 Woolwich
2254123

AFRICA
Southern Africa. Cheetahs

WROGEMANN, Nan
Cheetah under the sun. - Johannesburg ; London : McGraw-Hill, 1975
[15],159p,[8]p of plates : ill(some col). - Bibliography: p.149-156. - Includes index
0-07-091260-2 X.320/11506 Woolwich
1755765

AFRICA
Southern Africa. Christian church, to date - Conference proceedings

INSTITUTE FOR THEOLOGICAL RESEARCH
Church unity and diversity in the southern African context. - Pretoria : University of South Africa, [1979?]
146p. - (Miscellanea Congregalia 17)
0-86981-214-9 X.200/36922
8821402

AFRICA
Southern Africa. Coastal waters. Echinoderms

CLARK, Ailsa McGown
The echinoderms of Southern Africa. - London : British Museum (Natural History), 1976
[6],277p : ill. - (Publications / British Museum (Natural History) ; no.776). - Bibliography: p.254-259. - Includes index
0-565-00776-9 SRL
0155381

AFRICA
Southern Africa. Cuckoos

ROWAN, M. K. (Mary Kathrine)
The doves, parrots, louries and cuckoos of southern Africa. - London : Published for the John Voelcker Bird Book Fund [by] Croom Helm, 1983
xx,429p,8leaves of plates : ill(some col.). - Bibliography: p412-421. - Includes index
0-7099-2247-7 SRL
3072215

AFRICA
Southern Africa. Cultural processes

SOCIAL system and tradition in Southern Africa / edited by John Argyle, Eleanor Preston-Whyte. - Cape Town ; Oxford : Oxford University Press, 1978
xxi,251p,plate : ill. - Includes bibliographies
0-19-570089-9 X.520/13204
1762923

AFRICA
Southern Africa. Description & travel, to date

BULPIN, T.V. (Thomas Victor)
Reader's Digest illustrated guide to Southern Africa. - 2nd ed.. - Cape Town : The Reader's Digest, [1980]
544p : ill,all col
0-620-04650-3 X.805/1906
8415080

AFRICA
Southern Africa. Description & travel, 1951 - Correspondence, diaries

LISTER, Douglas S. (Douglas Spence)
Africa calls. - Bognor Regis : New Horizon, c1980
317p,[39]p of plates : ill
0-86116-042-8 X.809/47855
2484064

AFRICA
Southern Africa. Description & travel, 1972

WHITTINGHAM, John
Shoe-string safari. - Oxford : Hannon, [c1978]
vii,191p : ill
0-904233-23-5 X.809/46380
2348632

AFRICA
Southern Africa. Developing countries. Foreign assistance by Great Britain

JONES, David, 1942-
Aid and development in Southern Africa. - London : Croom Helm [for] the Overseas Development Institute, 1977
313p. - Includes index
0-85664-511-7 X.529/31547
1058428

AFRICA
Southern Africa. Doves

ROWAN, M. K. (Mary Kathrine)
The doves, parrots, louries and cuckoos of southern Africa. - London : Published for the John Voelcker Bird Book Fund [by] Croom Helm, 1983
xx,429p,8leaves of plates : ill(some col.). - Bibliography: p412-421. - Includes index
0-7099-2247-7 SRL
3072193

AFRICA
Southern Africa. Economic conditions - Serials

QUARTERLY economic review of Namibia, Botswana, Lesotho, Swaziland. - 1st quarter 1982-. - London : Economist Intelligence Unit, 1982-
v.
B82-36707 P.523/1703
2780372

AFRICA
Southern Africa. Economic conditions. Effects of multinational companies in South Africa, ca 1960-1980

MAKGETLA, Neva
Outposts of monopoly capitalism. - Westport : Lawrence Hill ; London : Zed, c1980
xiii,370p : ill. - Includes index
0-86232-020-8 X.529/39316
2333538

AFRICA
Southern Africa. Economic development - Conference proceedings

SADCC2 (Conference : 1980 : Maputo)
SADCC2-Maputo. - London : SADCC Liaison Committee, 1981
287p : ill
0-9507150-2-6 X.809/49361
0118893

AFRICA
Southern Africa. Economic development, to date - Serials

DEVELOPMENT Southern Africa. - Vol.1, no.1-. - Sandton, S.A. : Development Bank of Southern Africa, 1984-
v.
r6812352 P.881/464
9023836

GROEI = Growth. - [197-]-. - [Pretoria] : Corporation for Economic Development, [197-]-
v.
r6810100 P.513/3227
9023836

AFRICA
Southern Africa. Economic development. Conferences

GLASER, Kurt
Accelerated development and racial peace in South Africa. - New York : American African Affairs Association, [1973]
10 p.
r6025696 X.510/9838
1023527

AFRICA
Southern Africa. Economic policies - Conference proceedings
SOUTHERN Africa / edited by Amon J. Nsekela. - London : Collings, 1981
xix,274p
0-86036-154-3 X.520/24682
254055x

AFRICA
Southern Africa. Education
RUPERTI, R. M.
The education system in Southern Africa [by] R.M. Ruperti. - Pretoria : Van Schaik, 1976
[10],180p,plate. - Includes index
0-627-00386-9 X.519/27068
1040316

AFRICA
Southern Africa. Education - Conference proceedings
EDUCATION for development in Southern Africa / edited by David Hirschmann, Brian Rose. - Braamfontein : South African Institute of International Affairs for the Foundation for Foreign Affairs [and] the Rand Afrikaans University, 1974
vi [i.e. vii], 195 p. : 1 ill
0-909239-02-9 X.512/5501
1040316

AFRICA
Southern Africa. Education. Grant, Jack - Biographies
GRANT, Jack
Jack Grant's story. - Guildford : Lutterworth Press, 1980
x,198p. - Includes index
0-7188-2452-0 X.429/11218 Woolwich
220486x

AFRICA
Southern Africa. Education. Socioeconomic aspects - Conference proceedings
INDEPENDENCE without freedom / edited and with an introduction by Agrippah T. Mugomba and Mougo Nyaggah. - Santa Barbara ; Oxford : ABC-Clio, c1980
x,289p. - (Studies in international and comparative politics ; 13). - Bibliography: p263-276. - Includes index
0-87436-293-8 X.520/23469
2381419

AFRICA
Southern Africa. Exploration, to 1856
BECKER, Peter
The pathfinder. - Harmondsworth : Viking, 1985
281p,[16]p of plates : ill. - Bibliography: p267-271. - Includes index
0-670-80126-7 X.800/41312
327991x

AFRICA
Southern Africa. Exploration. Baines, Thomas
LUCKETT, Helen
Thomas Baines, 1820-1875. - [Norwich] : Norfolk Museums Service ; [Southampton] : Southampton Art Gallery and Museums, [1975]
[13]p : ill
B79-04724 X.708/21630
0782114

WALLIS, John Peter Richard
[Thomas Baines of King's Lynn]. Thomas Baines. - 2nd ed. - Cape Town : A. A. Balkema, 1976
xx, 235 p., [16] p. of plates : ill. - Bibliography: p. 225-228. - Includes index
0-86961-059-7 LR.425/152
0782114

AFRICA
Southern Africa. Food, to date. Cultural aspects
COETZEE, Renata
Funa. - Durban : Butterworths, [1982]
182p : ill
0-409-10300-4 X.622/19671
9100229

AFRICA
Southern Africa. Foreign investment by American investors
ROGERS, Barbara
White wealth and black poverty. - Westport, Conn. ; London : Greenwood Press, 1976
xv,333p. - (Studies in human rights / University of Denver. Center on International Race Relations ; no.2). - Includes index
0-8371-8277-8 X.529/31730
120128x

AFRICA
Southern Africa. Foreign relations
SOUTHERN Africa in the 1980s / edited by Olajide Aluko and Timothy M. Shaw. - London : Allen & Unwin, 1985
[320]p. - Includes index
0-04-320169-5 X.800/40951
3152650

AFRICA
Southern Africa. Foreign relations with South Africa
SOUTH Africa in southern Africa / edited by Thomas M. Calloghy. - New York : Praeger, 1983
viii,420p. - Bibliography: p405-412
0-03-060306-4 X.800/37304
0795399

SPENCE, John Edward
South African foreign policy in today's world. - Braamfontein : South African Institute of International Affairs, 1975
[2], 16 p.
0-909239-07-x X.702/5294
0795399

AFRICA
Southern Africa. Foreign relations with South Africa - Conference proceedings
INTERNATIONAL RELATIONS IN SOUTHERN AFRICA (Conference : 1973 : Pretoria)
International relations in Southern Africa. - Braamfontein : South African Institute of International Affairs, 1974
vii, 44 p.. - Bibliography : p. 37
0-909239-09-6 X.702/5368
108240X

AFRICA
Southern Africa. Foreign relations with United States - Bibliographies
AMERICAN-SOUTHERN African relations / edited by Mohamed A. El-Khawas, Francis A. Kornegay, Jr. - Westport, Conn. ; London : Greenwood Press, 1975
xix,188p. - (Special bibliographic series : new series / African Bibliographic Center ; no.1)
0-8371-8398-7 X.809/41990
0526428
- [Another copy] X:809/58514

AFRICA
Southern Africa. Frontier life, to 1900 compared with frontier life in North America, to 1900
The FRONTIER in history / edited by Howard Lamar and Leonard Thompson. - New Haven ; London : Yale University Press, c1981
xii,360p. - Bibliography: p317-333. - Includes index
0-300-02624-2 X.800/32108
261667x

AFRICA
Southern Africa. Geographical features, ca 1800-1975
CHRISTOPHER, A. J.
Southern Africa. - Folkestone : Dawson [etc.], 1976
292p : ill. - (Studies in historical geography). - Bibliography: p.270-285. - Includes index
0-7129-0694-0 X.809/41605
0707457

AFRICA
Southern Africa. Hairdressing, to date
FRISUREN, Kopfbedeckungen & Schmuck = Hair-styles, head-dresses & ornaments ... = Haarstyle, hooftooisels & ornamente ... / A. Scherz ... [et al.]. - Windhoek : Gamsberg, 1981
110p : ill(some col.)
r6330193 L.42/1707
8881502

AFRICA
Southern Africa. Higher education institutions. Students. Christian life - Serials
The CHRISTIAN Student. - Vol. 1. no. 1- ; Feb. 1975-. - Johannesburg : Students' Christian Association, 1975- : ill
Holdings : Vol. 1, no. 3- ; Aug. 1975-
r6004525 P.105/104
1055771

AFRICA
Southern Africa. Human rights, to date - Conference proceedings
CONFERENZA NAZIONALE DI SOLIDARIETÀ CON I POPOLI DELL'AFRICA AUSTRALE (1978 : Reggio Emilia)
Indipendenza e sovranità dei popoli dell'Africa australe. - Roma : Edizioni delle autonomie, 1979
232p : ill. - (Atti ; 3)
r6509832 X.529/57687
9015116

AFRICA
Southern Africa. Industrial development
SELWYN, Percy
Industries in the Southern African periphery. - London : Croom Helm [for] the Institute of Development Studies, Sussex, 1975
156p. - Bibliography: p.151-155. - Includes index
0-85664-277-0 X.529/30410
0581054

AFRICA
Southern Africa. Journalism. Weiss, Ruth - Biographies
WEISS, Ruth
Ein Lied ohne Musik. - Gelnhausen : Burckhardthaus-Laetare, 1981
336p. - (Laetare-Frauenprogramm)
3-7664-0105-x X.958/18777
9015795

AFRICA
Southern Africa. Kalahari Desert. Description & travel
LUARD, Nicholas
The last wilderness. - London : Elm Tree, 1981
222p,[32]p of plates : col.ill
0-241-10299-5 X.800/30777
2339110

The last wilderness. - Feltham : Hamlyn Paperbacks, 1982, c1981
222p
0-600-20635-1 H.82/1225
2339110

AFRICA
Southern Africa. Kalahari Desert. San, to date
TANAKA, Jiro
The San. - Tokyo : University of Tokyo Press, c1980
xvi,200p,plates. - Includes bibliography
0-86008-276-8 X.800/36735
9037624

AFRICA
Southern Africa. Kalahari Desert. San. Cultural processes
SOUTH AFRICAN MUSEUM. Ethnology Department
The Bushmen. - Cape Town : Trustees of the South African Museum, 1976
[2],38p : ill. - Bibliography on inside back cover
0-949940-82-8 X.700/25715
1493507

AFRICA
Southern Africa. Kalahari Desert. San. Ritual dancing
SCHADEBERG, Jurgen
The Kalahari Bushmen dance. - [S.l.] : J. Schadeberg, 1982
[58]p : chiefly ill
0-7045-0472-3 X.629/21086
2950588

AFRICA
Southern Africa. Kalahari Desert. San. Social conditions
BARNARD, Alan
Bushmen. - London : British Museum Publications Ltd [for] Museum of Mankind, 1978
[2],16p : ill. - (Discovering other cultures). - Bibliography: p.16
0-7141-0092-7 BS.4/269(2)
2003708

TYLER, J. S.
The Bosjesmans or bush people. - Reprinted. - Pretoria : The State Library, 1974
8p. - (State Library reprints ; 70)
r6800041 X.800/31117
2003708

VAN DER POST, Laurens
Testament to the Bushmen. - [Harmondsworth] : Viking, 1984
176p,[48]p of plates : ill(some col.). - Bibliography: p171-172. - Includes index
0-670-80065-1 X.805/6224
2003708

WANNENBURGH, Alf
The Bushmen. - London : Country Life Books, 1980
192p : ill(chiefly col)
0-600-31575-4 L.42/268
2003708

AFRICA
Southern Africa. Kavango. Cultural processes, to date
GIBSON, Gordon D.
The Kavango peoples. - Wiesbaden : Steiner, 1981
viii,275p.,[13]p of plates,[1]folded leaf of plates : ill. - (Studien zur Kulturkunde ; Bd.56). - Bibliography
3-515-03343-2 Ac.659[Bd.56]
8761264

AFRICA
Southern Africa. Lories & parrots

ROWAN, M. K. (Mary Kathrine)
The doves, parrots, louries and cuckoos of southern Africa. - London : Published for the John Voelcker Bird Book Fund [by] Croom Helm, 1983
xx,429p,8leaves of plates : ill(some col.). - Bibliography: p412-421. - Includes index
0-7099-2247-7 SRL
3072207

AFRICA
Southern Africa. Lutheran missions, 1957-1961

FOSSEUS, Helge
Mission blir kyrka. - [Stockholm] : Verbum, [1974]
155 p., 12 p. of plates : ill
r6003610 X.100/14716
0932035

AFRICA
Southern Africa. Man. Snake bites. Therapy

VISSER, John
Snakes and snakebite. - Cape Town ; London : Purnell, 1978
[5],152p : ill(some col). - Bibliography: p.137-138. - Includes index
0-86843-011-0 X.320/11218 Woolwich
1446444

AFRICA
Southern Africa. Methodist Church

BROOKS, Mike
Wrestling for God. - Belfast : Christian Journals Limited, 1976
127p
0-904302-19-9 X.108/16228
0627674

AFRICA
Southern Africa. Military relations with United States, to date

U.S. military involvement in southern Africa / edited by Western Massachusetts Association of Concerned African Scholars. - Boston [Mass.] : South End Press, c1978
xii,276p
0-89608-041-2 X.809/56207
9062424

AFRICA
Southern Africa. Mineral resources. Exploitation. International economic aspects - Conference proceedings

SOUTHERN AFRICAN METALS & MINERALS CONFERENCE (2nd : 1983 : Johannesburg)
Proceedings of the Second Southern African Metals & Minerals Conference. - Worcester Park : Metal Bulletin Conferences Limited, c1983
207p in various pagings : ill
0-900542-84-5 SRL
2699168

SOUTHERN AFRICAN METALS & MINERALS IN A WORLD CONTEXT (Conference : 1981 : Johannesburg)
Proceedings of the Southern African Metals & Minerals in a World Context conference. - Worcester Park : Metal Bulletin Congresses Limited, c1981
207p in various pagings : ill
0-900542-61-6 SRL
2699168

AFRICA
Southern Africa. Monomotapa, 1500-1900

RANDLES, W. G. L.
L'empire du Monomotapa du XVe au XIXe siècle. - Paris : Mouton, 1975
167 p., 3 leaves of plates, 4 p. of plates. - (Civilisations et sociétés ; 46). - Bibliography : p. 131-157
r6000998 X.0520/40(46)
0915203

AFRICA
Southern Africa. Nationalist movements

ESSAYS on the liberation of southern Africa / edited by N.M. Shamuyarira. - Dar es Salaam : Tanzania Publishing House, 1975
xiii,95p. - (University of Dar es Salaam. Studies in political science ; no. 3)
r6031778 X.709/31184
1099809

AFRICA
Southern Africa. Nationalist movements, 1963-1969

ESSAYS on the liberation of Southern Africa / edited by N.M. Shamuyarira. - Dar es Salaam : Tanzania Publishing House ; Birmingham : [Distributed by] Third World Publications, 1972[i.e.1979]
xiii,95p. - (Studies in political science / University of Dar es Salaam ; no.3)
B79-36193 X.709/33524
1993038

AFRICA
Southern Africa. Negro personnel. Industrial relations. Strikes, 1980-1981

ROGERS, John, 1952-
Striking against apartheid. - London : Produced and distributed for the SWP by Socialists Unlimited, 1982
30p : ill. - (A Socialist Workers Party pamphlet)
0-905998-30-8 X.529/47846
2734478

AFRICA
Southern Africa. Negroes. Cultural processes, to 1973

BECKER, Peter
Tribe to township. - St Albans : Panther, 1974
256p,16p of plates : ill
0-586-04007-2 X.708/20583
0759090

AFRICA
Southern Africa. News. Reporting, 1960-

HARRIS, Phil
Reporting southern Africa. - [Paris] : Unesco, [1981]
168p
92-310-1700-4 UN.L.331/188
8964564

AFRICA
Southern Africa. Pelargoniums

VAN DER WALT, J. J. A.
Pelargoniums of Southern Africa. - Cape Town : Purnell, 1977-
v : ill(chiefly col). - Includes bibliographies
0-86843-006-4 X.0322/280
1491768

AFRICA
Southern Africa. Policies of Canadian government

ADAIR, Dennis G.
Canada and Southern Africa. - Don Mills, Ont. : Fulcrum Press, 1973
38p. - (Canada and the world ; no.1)
r6028439 X.709/24339
1388479

AFRICA
Southern Africa. Policies of Canadian government - Conference proceedings

CANADA, Scandinavia and southern Africa / edited by Douglas Anglin, Timothy Shaw and Carl Widstrand. - Uppsala : Scandinavian Institute of African Studies, 1978
190p
91-7106-143-6 X.629/19899
1388479

AFRICA
Southern Africa. Policies of Soviet government

WALKER, Sir Walter
The bear at the back door. - Richmond, Surrey : Foreign Affairs Publishing Co., 1978
xvii,246p : ill
0-900380-23-3 X.809/43603
1685759

AFRICA
Southern Africa. Policies of United States government

UNITED STATES. National Security Council. Interdepartmental Group for Africa
The Kissinger study on southern Africa. - Nottingham : Spokesman Books, 1975
134p
0-85124-122-0 X.708/41155
0504254

AFRICA
Southern Africa. Political events

AFRICAN crisis. - London : Tory Reform Group, 1977
36p
B79-10732 X.709/32997
1818643

DAVIDSON, Basil
Southern Africa. - London : British Defence and Aid Fund for Southern Africa, 1984
34p. - (Canon Collins memorial lecture)
0-9509576-0-7 X.809/60097
3055116

AFRICA
Southern Africa. Political events, 1877-1895

SCHREUDER, Deryck Marshall
The scramble for Southern Africa, 1877-1895. - Cambridge : Cambridge University Press, 1980
xiv, 384p. - (Cambridge Commonwealth series). - Bibliography: p.372-379. - Includes index
0-521-20279-5 X.800/29168
2273861

AFRICA
Southern Africa. Political events, 1945-1978

SOUTHERN Africa / edited by Gwendolen M. Carter and Patrick O'Meara. - London : Macmillan, 1979
xii,404 : 1ill. - (Contemporary African issues series). - Bibliography: p363-381. - Includes Index
0-333-27495-4 X.700/26630
1834533

AFRICA
Southern Africa. Political events, ca 1955-1974

WALL, Patrick
Prelude to détente. - London : Stacey International : Distributed by Bayard Books, 1975
[1],108p,[8]p of plates : ill. - Includes index
0-9503304-5-0 X.708/41567
0507067

AFRICA
Southern Africa. Political events, ca 1960-1975

DAVIDSON, Basil
Southern Africa. - Harmondsworth : Penguin, 1976
374p. - (A pelican original). - Includes index
0-14-021963-3 X.708/20472
0707856

AFRICA
Southern Africa. Political events, 1960-1976

SOUTHERN Africa in crisis / edited by Gwendolen M. Carter and Patrick O'Meara. - Bloomington ; London : Indiana University Press, 1977
vii,279p : 1 ill. - Bibliography: p.268-271. - Includes index
0-253-35399-8 X.800/26208
1167324

AFRICA
Southern Africa. Political events, 1974

LEGUM, Colin
Southern Africa, the secret diplomacy of detente ; [and], South Africa at the crossroads. - London : Collings, 1975
[4],91p. - (Current affairs series)
0-901720-97-6 X.700/25061
0242527

AFRICA
Southern Africa. Political events, 1974-1976

CALLINICOS, Alex
Southern Africa after Soweto. - London : Pluto Press, 1977
viii,229p. - Bibliography: p.214-220. - Includes index
0-904383-42-3 X.709/31379
073103x

AFRICA
Southern Africa. Political events, 1974-1977

CALLINICOS, Alex
Southern Africa after Soweto. - 2nd ed. - London : Pluto Press, 1978
[8],246p. - Bibliography: p.230-236. - Includes index
B78-32640 X.709/32513
1585029

AFRICA
Southern Africa. Political events, 1976-1977

LEGUM, Colin
Southern Africa, the year of the whirlwind. - London : Collings, 1977
72p. - (Current affairs series)
0-86036-066-0 X.709/31387
117245x

AFRICA
Southern Africa. Political events, 1976-1980

CALLINICOS, Alex
Southern Africa after Zimbabwe. - London : Pluto, 1981
186p. - Includes index
0-86104-336-7 X.808/34032
2392070

AFRICA
Southern Africa. Political events. Role of Tanzania & United States, 1976

NYERERE, Julius K. (Julius Kambarage)
Crusade for liberation. - Dar es Salaam ; Oxford : Oxford University Press, 1978
94p,[8]p of plates : ill
0-19-572462-3 X.529/46219
2640929

AFRICA
Southern Africa. Politics

CLUTTON-BROCK, Guy
Southern Africa. - London : Christian Socialist Movement. [1979]
[2],19p
B79-36192 X.519/30374
1993704

LEGUM, Colin
The western crisis over southern Africa. - New York ; London : Africana, 1979
xi,260p : ill. - (Current affairs series)
0-8419-0492-8 X.809/48508
1993704
- [Another copy: paperback] ISBN 0-8419-0496-0 X.809/48509

AFRICA
Southern Africa. Politics, to 1978

SOUTHERN Africa / edited by Gwendolen M. Carter and Patrick O'Meara. - Bloomington ; London : Indiana University Press, 1979
xii,404p : 1 ill. - Bibliography: p.363-381. - Includes index
0-253-31181-0 X.800/28888
2064154

AFRICA
Southern Africa. Politics, 1960-1975

STOCKHOLM INTERNATIONAL PEACE RESEARCH INSTITUTE
Southern Africa. - New York ; London : Praeger, 1976
x,235p,2p of plates : ill. - (Praeger special studies in international politics and government). - Bibliography: p.224-228. - Includes index
0-275-56840-7 X.800/25889
1040014

AFRICA
Southern Africa. Politics, 1960-

RITSCHER, Hans
Fernsehen - genau betrachtet. - Opladen : Leske, 1981
101p.. - Includes bibliography
3-8100-0278-x X.950/24852
9008144

AFRICA
Southern Africa. Predatory animals

GROBLER, Hans
Predators of Southern Africa. - Johannesburg : Macmillan South Africa, 1984
134p : ill
0-86954-194-3 X.520/36847
9153217

AFRICA
Southern Africa. Prehistoric antiquities. Archaeology

FRONTIERS / edited by M. Hall ... [et al.]. - Oxford : B.A.R., 1984
379p : ill. - (Cambridge monographs in African archaeology ; 10) (BAR. International series ; 207). - Includes bibliographies
0-86054-268-8 X.425/5220
3277925

AFRICA
Southern Africa. Protea

ROURKE, John P.
The proteas of Southern Africa. - Capetown ; London : Purnell, 1980
236,xxp : ill(chiefly col). - Bibliography: p.xiii-xiv. - Includes index
0-86843-027-7 L.45/636
222982x

AFRICA
Southern Africa. Race relations. Political aspects - Serials

BRIEFING paper on Southern Africa / International Defence & Aid Fund. - No.1 (Mar. 1981)-. - London : The Fund, 1981- v.
B82-03426 P.703/753
0760706

FACT Paper on Southern Africa. - No.1- ; May 1976-. - [London] : International Defence and Aid Fund, 1976-
B76-35083 P.703/523
0760706

AFRICA
Southern Africa. Racism. Attitudes of Commonwealth

COMMONWEALTH SECRETARIAT
Racism in Southern Africa. - London : The Secretariat, 1979
12p : ill
B80-41355 BS.12/152
2147149

AFRICA
Southern Africa. Religion related to social change

RELIGION and social change in Southern Africa / edited by Michael G. Whisson and Martin West. - Cape Town : David Philip ; London : Collings, 1975
[1],xii,223p. - Bibliography: p.207-218. - Includes index
0-901720-99-2 X.800/25344
0577448

AFRICA
Southern Africa. Resistance movements, 1950-1971 - Bibliographies

ANSARI, S.
Liberation struggle in Southern Africa. - Gurgaon : Indian Documentation Service, 1972
118p. - Includes index
r6032093 X.809/42196
1173561

AFRICA
Southern Africa. Rock carvings

LEWIS-WILLIAMS, J. David
The rock art of Southern Africa. - Cambridge : Cambridge University Press, 1983
68p,[64]p of plates : ill(some col.). - (The Imprint of man)
0-521-24460-9 X.421/25107
3076431

AFRICA
Southern Africa. Rock paintings

LEWIS-WILLIAMS, J. David
The rock art of Southern Africa. - Cambridge : Cambridge University Press, 1983
68p,[64]p of plates : ill(some col.). - (The Imprint of man)
0-521-24460-9 X.421/25107
0152803

AFRICA
Southern Africa. Rock paintings. Special subjects: Bees

PAGER, H.
Rock paintings in Southern Africa showing bees and honey hunting. - Gerrards Cross : Bee Research Association, 1973
8 p. : ill. - (Bee Research Association Reprint ; M68)
r6000551 X.419/2317 Woolwich
1021079

AFRICA
Southern Africa. Rulers, 1800-1900 - Biographies

BLACK leaders in southern African history / edited by Christopher Saunders. - London : Heinemann Educational, 1979
160 : ill. - ([African historical biographies]). - Includes bibliographies and index
0-435-94477-0 OPL
1896180
- [Another copy] X.809/44396

AFRICA
Southern Africa. Rural regions. Economic conditions, 1840-1970

The ROOTS of rural poverty in Central and Southern Africa / edited by Robin Palmer, Neil Parsons. - London : Heinemann Educational, 1977
xviii,430p : ill. - Bibliography: p.422-423. - Includes index
0-435-94752-4 X.329/11236 Woolwich
1289586

AFRICA
Southern Africa. Rural regions. Houses. Architecture, to date

FRESCURA, Franco
Rural shelter in Southern Africa. - Johannesburg : Ravan, 1981
x,208p : ill. - Includes bibliography
0-86975-205-7 X.429/15296
9044418

AFRICA
Southern Africa. San. Social conditions

STEPHEN, David, 1942-
The San of the Kalahari. - London : Minority Rights Group, 1982
16p : 1ill. - (Report / Minority Rights Group ; no.56). - Bibliography: p16
B83-35294 P.803/299(56)
2941929

AFRICA
Southern Africa. Social change related to religion

RELIGION and social change in Southern Africa / edited by Michael G. Whisson and Martin West. - Cape Town : David Philip ; London : Collings, 1975
[1],xii,223p. - Bibliography: p.207-218. - Includes index
0-901720-99-2 X.800/25344
0577456

AFRICA
Southern Africa. Social conditions

SOUTHERN Africa under attack. - London : Catholic Institute for International Relations, c1983
21p. - (Comment). - Bibliography: p21
0-904393-97-6 X.529/64414
3164756

AFRICA
Southern Africa. Social conditions, to ca 1000. Archaeological sources

INSKEEP, R. R.
The peopling of Southern Africa. - Cape Town ; London : David Philip ; London : Distributed by Global Book Resources Ltd, 1978
160p : ill(some col). - (The peoples of Southern Africa). - Bibliography: p.155-157. - Includes index
0-949968-69-2 X.800/27889
190602x

AFRICA
Southern Africa. Social conditions, ca 1970

WÄSTBERG, Per
Afrika - ett uppdrag. - Stockholm : Wahlström & Widstrand, 1976
286p
91-461-2714-3 X.809/24333
1221809

AFRICA
Southern Africa. Social conditions. Political aspects

HANNON, Peter
Southern Africa. - Johannesburg : Grosvenor Books, 1977
136p
0-620-02330-9 X.708/21812
1843710

AFRICA
Southern Africa. Social conditions. Political aspects - Serials

FOCUS on Political Repression in Southern Africa. - No.1- ; November 1975-. - London : I.D.A.F, 1975- : ill
B76-09796 P.701/1504
0611859

AFRICA
Southern Africa. Social stratification

ROSS, Robert, 1949-
Adam Kok's Griquas. - Cambridge : Cambridge University Press, 1976
xiv,194p. - (African studies series ; 21). - Bibliography: p.175-189. - Includes index
0-521-21199-9 X.800/25916
1048651

AFRICA
Southern Africa. Sotho-Tswana. Christianity

SETILOANE, Gabriel M.
The image of God among the Sotho-Tswana. - Rotterdam : A.A. Balkema, 1976
x,298p. - Bibliography : p.296-298
90-619-1007-2 X.100/16881
1087843

AFRICA
Southern Africa. Stone Age man

SCHWEITZER, F. R.
The stone age = Die steentydperk. - Cape Town : South African Museum, 1975
[25] p. : ill. - Bibliography : p. [25]
0-949940-60-7 X.415/5259 Woolwich
1056042

AFRICA
Southern Africa. Trees

PALMER, Eve
A field guide to the trees of Southern Africa. - London : Collins, 1977
352p,32p of plates : ill(some col). - Bibliography: p.321-322. - Includes index
0-00-219339-6 SRL
155008x

AFRICA
Southern Africa. Tribes

BECKER, Peter
Inland tribes of southern Africa. - London : Granada, 1979
183p : ill(some col). - Includes index
0-246-10926-2 L.R.421/68
0508098

Trails & tribes in Southern Africa. - London : Hart-Davis MacGibbon, 1975
211p : ill(some col). - Includes index
0-246-10768-5 X.802/10040
0508098

AFRICA
Southern Africa. Tribes–continued
WEST, Martin
Abantu. – Cape Town : Struik, 1976
184p : ill(chiefly col). – Bibliography : p.182. – Includes index
0–86977–057–8 X.525/2900
0508098

AFRICA
Southern Africa. Tswana, to 1980 & Sotho, to 1980
LYE, William F. (William Frank)
Transformations on the highveld. – Cape Town : London : David Philip, 1980
160p : ill(some col.). – (The People of Southern Africa ; 2). – Bibliography: p153–154. – Includes index
0–908396–28–7 X.800/36598
2990520

AFRICA
Southern Africa. Tswana. Cultural identity
ALVERSON, Hoyt
Mind in the heart of darkness. – New Haven ; London : Yale University Press, 1978
xiv,299p,[12]p of plates : ill. – Bibliography: p.281–284. – Includes index
0–300–02244–1 X.529/32948
1774190

AFRICA
Southern Africa. Tswana. Disputes
COMAROFF, John L. (John Lionel)
Rules and processes. – Chicago ; London : University of Chicago, 1981
ix,293p : ill. – Bibliography: p277–288. – Includes index
0–226–11424–4 X.529/44801
2594234

AFRICA
Southern Africa. Venomous snakes
VISSER, John
Snakes and snakebite. – Cape Town ; London : Purnell, 1978
[5],152p : ill(some col). – Bibliography: p.137–138. – Includes index
0–86843–011–0 X.320/11218 Woolwich
1446452

AFRICA
Southern Africa. Visits by British royal families, 1860–1947
ARONSON, Theo
Royal ambassadors. – Cape Town : David Philip, 1975
144 p. : ill. – Bibliography : p. 139–140. – Includes index
0–949968–46–3 X.802/10279
1029746

AFRICA
Southern Africa. White persons. Emigration to South America, 1976
UPPSALA UNIVERSITET. Avdelningen för freds- och konfliktforskning. Gruppen för beroendestudier
[DOCUMENTS on colonialist export from South Africa to South America. Swedish]. Dokument om exporten av kolonialister från Södra Afrika till Latinamerika. – [Uppsala] : [Uppsala universitet, Avdelningen för freds- och konfliktforskning], 1978
56,iiip : ill,facsims,map. – (Analys och debatt ; 11)
91–506–0102–4 X.0512/450(11)
2431319

AFRICA
Southern Central Africa. Orchids
WILLIAMSON, Graham
The orchids of South Central Africa. – London : Dent, 1977
237p,[48]p of plates : ill(some col). – Bibliography: p.16. – Includes index
0–460–04295–5 SRL
1262424

AFRICA
Sports. Ganga, Jean-Claude – Biographies
GANGA, Jean Claude
Combats pour un sport africain. – Paris : Librairie-Éditions L'Harmattan, [1979]
271p,8p of plates : ill
2–85802–109–0 X.629/14230 Woolwich
2461870

AFRICA
Statistics – Bibliographies
HARVEY, Joan Margaret
Statistics Africa. – Ed. 2, revised and enlarged. – Beckenham : C.B.D. Research Ltd, 1978
ii–xii,374p. – Includes index
0–900246–26–x X.525/2335
177915x
– [Another copy] OPL

AFRICA
Steam locomotives, to 1980
DURRANT, A. E. (Anthony Edward)
Steam in Africa. – London : Hamlyn, [c1981]
207p : ill(some col.). – Bibliography: p205. – Includes index
0–600–34946–2 L.45/1084
2499444

AFRICA
Stone age cutting tools
ROCHE, Hélène
Premiers outils taillés d'Afrique. – [Paris] : Société d'ethnographie, 1981
263p : ill. – (Afrique ancienne ; 1). – Bibliography
2–901161–14–6 X.0622/619(1) Woolwich
8753911

AFRICA
Sufism, 1800–1900
MARTIN, B. G.
Muslim brotherhoods in nineteenth-century Africa. – Cambridge : Cambridge University Press, 1976 [i.e. 1977]
xiii,267p. – (African studies series ; 18). – Bibliography: p. 238–246. – Includes index
0–521–21062–3 X.200/30780
1049356

AFRICA
Syndicalism, to date
SIBIRI, Sieba
Syndicalisme et politique, vrais et faux problèmes. – [Ouagadougou] : [s.n.], [1972?]
14leaves
r6210657 X.525/8626
9039546

AFRICA
Taxation
PLATT, C. J.
Tax systems of Africa, Asia and the Middle East. – Aldershot : Gower, c1982
xxii,240p
0–566–02335–0 X.529/49578
2706768
– [Another copy.] X.200/38042

AFRICA
Taxation, 1960–
AFRICAN tax systems / prepared ... by Robert C. Hammond ... Marc J. van den Abeelen. Current editors : Elizabeth A. de Braun-Hay ... Françoise Butzelaar-Mohr. – Amsterdam : International Bureau of Fiscal Documentation, [1980–]
2v(loose-leaf)
r6303182 OPL
8741824

AFRICA
Technology – Conference proceedings
TECHNOLOGY & information / by Africa Technology Discussion Group. – London : Africa Technology Discussion Group, c1982
43p
B82–23046 X.525/5747
2729121

AFRICA
Textile industries – Serials
AFRICAN textiles. – [1]– ; June 1980–. – London : Alain Charles Publishing Ltd, [1980]– : ill
B80–37820 SRL
2280205
– [Another copy]
Holdings: October 1980– P.625/116

AFRICA
Towns. Social conditions – Texts
LEWIN, Hugh
Jafta – the town. – London : Evans, 1983
[24]p : ill. – (Jafta's family series)
0–237–45677–x X.990/23427
3174948

AFRICA
Trade unionism, to 1977
ANANABA, Wogu
The trade union movement in Africa. – London : C. Hurst, 1979
xiii,248p. – Includes index
0–905838–13–0 X.529/33101
1736094
– [Another copy: paperback]
ISBN 0–905838–27–0 X.519/29912

AFRICA
Trade unionism, to date
WATERMAN, Peter
Aristocrats and plebeians in African unions? .... – [Nijmegen] : [P. Waterman], 1983
xii,426p. – Includes bibliography
r6330516 X.520/30766
857992x

AFRICA
Trade unionism, to date – Bibliographies
WESSELS, Elizabeth
Trade unions in Africa. – [Pretoria] : [Africa Institute], [1981]
35p. – (Bibliography series ; no.3)
r6813261 X.529/61160
9090231

AFRICA
Transport, to date – Conference proceedings
SYMPOSIUM AFRICAIN DES TRANSPORTS PUBLICS (1st 1978 Abidjan)
Textes des rapports ... [et compte rendu du] premier Symposium africain des transports publics, Abidjan, 12–15 novembre 1978. – Bruxelles : Union internationale des transports publics, 1978–1980
2v : ill
r6209140 X.805/3374
8004021

AFRICA
Transport, 1861–1970 – Bibliographies
LINIGER-GOUMAZ, Max
Bibliographies Africaines. – Genève : Les Éditions du Temps, 1971–
tom.1–
r6012347 X.0909/953
1010182

AFRICA
Transport networks. Political aspects
ARNOLD, Guy
Strategic highways of Africa. – London : Friedmann, 1977
[1],178p. – Bibliography: p.166–167. – Includes index
0–904014–12–6 X.529/31610
1165437

AFRICA
Tribal societies. Cultural identity
CERULLI, Ernesta
Popoli e culture dell'Africa. – 2a ed. interamente riveduta e aggiornata. – Genova : Tilgher, 1977
208p.,[1]leaf of plates
r6100026 X.510/11469
2093197

AFRICA
Tribal societies. Cultural processes
PEOPLES of Africa / [editors Lynette Singer, Robin Wood]. – London : Marshall Cavendish, 1978
200p : ill(chiefly col). – Includes index
0–85685–494–8 X.802/10739
1732595

AFRICA
Tribes. Personal adornment
FISHER, Angela
Africa adorned. – London : Collins, 1984
304p : ill(some col.). – Bibliography: p300–301. – Includes index
0–00–216622–4 L.45/2890
3022587

AFRICA
Tropical rain forests. Animals in danger of extinction
DISAPPEARING animals of the forests of Africa / compiled by Gill Gould ; wildlife adviser Michael M. Scott ; illustrations by John Butler, Sheila Smith and Alan R. Thomson. – Edinburgh : Chambers, c1980
[28]p : ill(some col.). – (Animals in danger)
0–550–32001–6 X.329/13510 Woolwich
2341565

AFRICA
Tropical regions. Agriculture
LIVINGSTONE, I. (Ian)
Agricultural economics for Tropical Africa. – London : Heinemann, 1981
x,294p : ill. – (Studies in the economics of Africa). – Includes index
0–435–97431–9 X.520/24051
2140721

AFRICA
Tropical regions. Agriculture. Development

RURAL development in tropical Africa / edited by Judith Heyer, Pepe Roberts and Gavin Williams. - London : Macmillan, 1981
x,375p : ill. - Includes bibliographies and index
0-333-28448-8 X.809/49468
2566494

AFRICA
Tropical regions. Agriculture. Development. Socioeconomic aspects

AGRICULTURAL change in tropical Africa / Kenneth R.M Anthony ... [et al.]. - Ithaca ; London : Cornell University Press, 1979
327p : ill
0-8014-1159-9 X.320/11790 Woolwich
1914375

AFRICA
Tropical regions. Agriculture. Policies of government

BATES, Robert H. (Robert Hinrichs)
Markets and states in tropical Africa. - Berkeley ; London : University of California Press, c1981
xi,178p : ill. - (California series on social choice and political economy). - Bibliography: p147-166. - Includes index
0-520-04253-0 X.329/14836 Woolwich
2607646

The POLITICS of agriculture in tropical Africa / Jonathan Barker editor. - Beverly Hills ; London : Sage, c1984
320p. - (Sage series on African modernization and development ; v.9). - Includes bibliographies
0-8039-2295-7 X.329/19316
2607646

AFRICA
Tropical regions. Animals. Ecology

OWEN, Denis Frank
Animal ecology in tropical Africa. - 2nd ed. - London : Longman, 1976
132p,[8]p of plates : ill(some col). - (Tropical ecology series). - Bibliography: p.124-126. - Includes index
0-582-44363-6 SRL
0771538
- [Another copy: paperback]
ISBN 0-582-44362-8 X.311/8316 Woolwich

READER, John
Pyramids of life. - London : Collins, 1977
222p : ill. - Bibliography: p.219. - Includes index
0-00-219004-4 X.322/8364 Woolwich
0771538

AFRICA
Tropical regions. Christianity. Effects of independence - Conference proceedings

CHRISTIANITY in independent Africa / edited by Edward Fasholé-Luke ... [et al.]. - London : Collings, 1978
x,630p. - Includes index
0-86036-058-x X.200/31557
1594869

AFRICA
Tropical regions. Economic conditions

An ECONOMIC history of tropical Africa / selected and edited by Z.A. Konczacki and J.M. Konczacki. - London : Cass.
Vol.1: The pre-colonial period. - 1977
xii,310p. - Includes bibliographies and index
0-7146-2919-7 X.0529/513(1)
1188992
Vol.2: The colonial period. - 1977
xv,260p : ill. - Includes index
0-7146-2915-4 X.0529/513(2)
1188992

AFRICA
Tropical regions. Economic development. Geographical factors

O'CONNOR, A. M. (Anthony Michael)
The geography of tropical African development. - 2nd ed. - Oxford : Pergamon, 1978
xviii,229p,[12]p of plates : ill. - (Pergamon Oxford geography series) (Pergamon international library). - Includes bibliographies and index
0-08-021847-4 X.700/25808
1200135

AFRICA
Tropical regions. Education. Social aspects - Conference proceedings

CONFLICT AND HARMONY BETWEEN TRADITIONAL AND WESTERN EDUCATION IN AFRICA (Conference : 1973 : School of Oriental and African Studies, University of London)
Conflict and harmony in education in tropical Africa. - London : Allen and Unwin, 1975
3-496p. - (Studies on modern Asia and Africa ; no.10). - Includes index
0-04-370057-8 010127.dd.17/10
0364843

AFRICA
Tropical regions. Foreign trade with Great Britain

AID, credit and trade in tropical Africa / report by a sub-group of the Tropical Africa Advisory Group. - London : British Overseas Trade Board, 1983
123p
B85-02549 BS.41/1403
3209407

AFRICA
Tropical regions. Fresh waters

BALEK, Jaroslav
Hydrology and water resources in tropical Africa. - Amsterdam ; Oxford : Elsevier, 1977
208p,[2]p of plates : ill(some col). - (Developments in water science ; 8). - Includes bibliographies and index
0-444-99814-4 SRL
1334972

AFRICA
Tropical regions. Freshwater ecosystems

BEADLE, L. C. (Leonard Clayton)
The inland waters of tropical Africa. - 2nd ed. - London : Longman, 1981
x,475p : ill. - Bibliography: p417-462. - Includes index
0-582-46341-6 SRL
0667447

BEADLE, Leonard Clayton
The inland waters of tropical Africa. - London : Longman, 1974
viii,365p : ill. - Bibliography: p.324-346. - Includes index
0-582-44852-2 SRL
0667447

AFRICA
Tropical regions. Health services, ca 1880-ca 1960 - Conference proceedings

HEALTH AND MEDICINE IN TROPICAL AFRICA DURING THE COLONIAL PERIOD (Conference : 1977 : Oxford)
Health in tropical Africa during the colonial period. - Oxford : Clarendon Press, 1980
ix,276p,plate : ill. - ([Oxford science publications]). - Includes bibliographies and index
0-19-858165-3 X.322/9374 Woolwich
1926683

AFRICA
Tropical regions. Human geographical features - For African students

UDO, Reuben K. (Reuben Kenrick)
The human geography of tropical Africa. - Idadan : Heinemann Educational (Nigeria) ; London : Heinemann Educational, 1982
xii,244p : ill. - Bibliography: p234-237. - Includes index
0-435-95919-0 X.800/32501
2639122

AFRICA
Tropical regions. Internal migration

MASSER, Ian
Inter-regional migration in tropical Africa. - London : Institute of British Geographers, 1975
iv,107p : ill. - (Special publications / Institute of British Geographers ; no.8). - Bibliography: p.101-105. - Includes index
0-901989-19-3 P.801/666
0527556

AFRICA
Tropical regions. Mammals. Ecology

DELANY, M. J.
Ecology of African mammals. - London : Longman, 1979
viii,434p : ill. - (Tropical ecology series). - Bibliography: p.413-415. - Includes index
0-582-44176-5 SRL
1893939

AFRICA
Tropical regions. Muslim communities, to 1979 - Conference proceedings

INTERNATIONAL AFRICAN SEMINAR (5th : 1964 : Zaria)
Islam in Tropical Africa. - 2nd ed. - London : Hutchinson [for the] International African Institute, 1980
xii,310p. - (Hutchinson university library for Africa). - Includes bibliographies and index
0-09-141311-7 X.200/34395
1958097

AFRICA
Tropical regions. Nomadic pastoral communities. Social conditions - Conference proceedings

INTERNATIONAL AFRICAN SEMINAR (13th : 1972 : Niamey)
Pastoralism in tropical Africa. - London : Oxford University Press for the International African Institute, 1975
xv,502p : ill. - Includes bibliographies and index
0-19-724196-4 X.329/10213
0540463

AFRICA
Tropical regions. Organisms. Distribution - For West African students

BISHOP, O. N. (Owen Neville)
The distribution of organisms. - London : Macmillan Education, 1977
[1],iv,28,[1]p : ill. - (Advanced readers in tropical biology)
0-333-21396-3 X.319/17244 Woolwich
1091727

AFRICA
Tropical regions. Political systems, 1945-1975

COLLIER, Ruth Berins
Regimes in tropical Africa. - Berkeley ; London : University of California Press, c1982
xi,221p : ill. - Bibliography: p181-207. - Includes index
0-520-04313-8 X.809/52638
2761076

AFRICA
Tropical regions. Politics

AFRIKA v novoe i noveishee vremia / [otvetstvennye redaktony G.A. Nersesov, V.A. Subbotin]. - Moskva : Izdatel'stvo "Nauka", Glavnaia͡ redakt͡sii͡a vostochnoĭ literatury, 1976
211 p.
r6023127 X.709/21023
0887595

HODDER-WILLIAMS, Richard
An introduction to the politics of tropical Africa. - London : Allen & Unwin, 1984
xxv,262p. - Bibliography: p239-251. - Includes index
0-04-320162-8 X.529/66621
0887595

AFRICA
Tropical regions. Population. Mobility. Small area sampling

PROTHERO, Ralph Mansell
Micro-scale studies of population mobility in tropical Africa. - [Liverpool] : [University of Liverpool, Department of Geography], [1975]
[2]leaves,15p. - (Working papers / African Population Mobility Project ; no.21). - Bibliography : p.14-15
B80-14443 X.702/6297
2056569

AFRICA
Tropical regions. Women. Emancipation, to date

SHVET͡SOVA, S. I. (Svetlana Ivanovna)
Tropicheskaia͡ Afrika. - Moskva : Nauka Glavnaia͡ redakt͡sii͡a vostochnoĭ literatury, 1982
173p. - Includes bibliography
r6499319 X.529/58137
9031669

AFRICA
Unity

NKRUMAH, Kwame
Africa must unite. - London : Panaf, 1963
xvii,229p
0-901787-13-2 X.809/54913
2858096

AFRICA
Unity. International legal aspects

UDECHUKU, E. C.
African unity and international law. - [London] : Africa Press, [1974]
iii-xiii[i.e.xiv],228p
0-904437-00-0 X.108/16891
129735x

AFRICA
Universities. Administration compared with administration of multinational companies

MAZRUI, Ali Al'Amin
The African university as a multinational corporation. - Brighton : Institute of Development Studies, University of Sussex, [1976]
[3], 24p. - (Discussion paper ; no.96)
r6031828 X.512/5597
1102516

AFRICA
Universities. Curriculum subjects, to date: History

OBENGA, Théophile
La dissertation historique en Afrique. - Dakar : Les Nouvelles Éditions Africaines, 1981
111p
2-7236-0229-x X.809/57800
9056033

AFRICA
Universities. Curriculum subjects: Political science

POLITICAL science in Africa / edited by Yolamu Barongo. - London : Zed, 1983
xii,257p. - Includes index
0-86232-033-x X.529/58481
3037584

AFRICA
Urban development

AFRICAN urban development / by Treufried Grau...[et al.] ; Michael Koll (editor). - Düsseldorf, c 1972
215p. - (Materialien des Arnold-Bergstraesser-Instituts für kulturwissenschaftliche Forschung ; Bd 31)
3-571-09139-6 X.709/20935
0885169

MABOGUNJE, Akin Ladipo
Cities and African development. - Ibadan : Oxford University Press, 1976
52p.8p.of plates : maps. - (Studies in the development of African Resources ; 3)
0-19-575255-4 X.529/35725
2447053

AFRICA
Urban regions. Poverty. Political aspects

SANDBROOK, Richard
The politics of basic needs. - London : Heinemann, 1982
vi,250p. - Includes index
0-435-96537-9 X.529/47392
270644x

AFRICA
Urban regions. Working classes. Political development, to 1973

GUTKIND, Peter C. W.
The emergent African urban proletariat. - Montreal : Centre for Developing Area Studies, McGill University, 1974
[2],vii,79p. - (Occasinal paper series ; no.8). - Bibliography : p.67-79
r6071874 X.709/32189
144042X

AFRICA
Vernacular architecture

DENYER, Susan
African traditional architecture. - London : Heinemann, 1978
xiv,210p : ill. - Bibliography: p.202-204. - Includes index
0-435-89059-x X.421/10594 Woolwich
0620785
- [Another copy: paperback]
ISBN 0-435-89058-1 X.410/10219 Woolwich

AFRICA
Vernacular architecture. Social aspects

SHELTER in Africa / edited by Paul Oliver. - London : Barrie and Jenkins, 1976
240p : ill. - Includes bibliographies
0-214-20205-4 X.700/25072
0586838

AFRICA
Vertebrates. Cinematography

PLAGE, Dieter
Wild horizons. - London : Collins, 1980
216p : ill(some col.). - Includes index
0-00-216029-3 X.800/29374
2328968

AFRICA
Villages. Children. Social conditions - Texts

LEWIN, Hugh
Jafta. - London : Evans, c1981
[24]p : ill. - (Jafta's family series)
0-237-45543-9 X.990/17424 Woolwich
2182327

Jafta. - Cambridge : Dinosaur, c1982
[24]p : ill. - (Dinosaur's Althea books)
0-85122-267-6 X.990/18900
2182327

AFRICA
Villages. Fathers. Social conditions - Texts

LEWIN, Hugh
Jaffa - my father. - London : Evans, 1981
[24]p : ill. - (Jaffa's family series)
0-237-45545-5 X.990/17466 Woolwich
2182351

Jafta - my father. - Cambridge : Dinosaur, c1983
[24]p : ill. - (Dinosaur's Althea books)
0-85122-397-4 X.990/21049
2182351

AFRICA
Villages. Mothers. Social conditions - Texts

LEWIN, Hugh
Jafta - my mother. - London : Evans, 1981
[24]p : ill. - (Jafta's family series)
0-237-45544-7 X.990/17425 Woolwich
2182343

Jafta - my mother. - Cambridge : Dinosaur, c1982
[24]p : ill. - (Dinosaur's Althea books)
0-85122-268-4 X.990/18907
2182343

AFRICA
Villages. Social conditions - Texts

LEWIN, Hugh
Jaffa - the wedding. - London : Evans, 1981
[24]p : ill. - (Jaffa's family series)
0-237-45546-3 X.990/17467 Woolwich
2182335

AFRICA
Villages. Social conditions, 1960-

ELA, Jean-Marc
L'Afrique des villages. - Paris : Éditions Karthala, [1982]
228p. - (Les Afriques)
2-86537-052-6 X.529/60661
8004153

AFRICA
Villages. Weddings - Texts

LEWIN, Hugh
Jafta - the wedding. - Cambridge : Dinosaur, c1983
[24]p : ill. - (Dinosaur's Althea books)
0-85122-398-2 X.990/21139
2971763

AFRICA
Volcanoes

KRAFFT, Katia
Dans l'antre du diable. - Paris : Presses de la Cité, [1981]
124p : chiefly ill. - (Connaissance du monde)
2-258-00904-9 X.622/13975
8877327

AFRICA
Volta River basin. Man. Onchocerciasis. Control measures. Financial assistance

ONCHOCERCIASIS 1974 Fund Agreement, Washington, 14 February 1974 ... - London : H.M.S.O., 1975
25p. - (Treaty series ; no.72(1975)) (Cmnd.6048)
0-10-160480-7 BS.14/137
0362689

AFRICA
War. International law, to 1980

BELLO, Emmanuel G.
African customary humanitarian law. - [Northampton] : Oyez Publishing [for the] International Committee of the Red Cross, Geneva, 1980
xi,157p. - Bibliography: p.131-154. - Includes index
0-85120-464-3 X.200/33363
2210134

AFRICA
Water industries - Serials

AFRICAN water and sewage. - Vol.1, no.1 (Mar. 1982)-. - Redhill : Fuel & Metallurgical Journals, 1982-
v. : ill(some col.)
B82-28123 P.623/1192
2753286

AFRICA
Water supply & waste water - Conference proceedings

WEDC CONFERENCE (6th : 1980 : Zaria)
Water and waste engineering in Africa. - Loughborough : Department of Civil Engineering, University of Technology, 1981
203p : ill
0-906055-11-3 SRL
2410958

AFRICA
West & Central Africa. Birds

MACKWORTH-PRAED, C. W. (Cyril Winthrop)
Birds of West Central and Western Africa. - London : Longman.
Vol.1. - 1970
xxxii,671p,[45]p of plates : ill(some col.). - (African handbook of birds. Series 3). - Includes index
0-582-46086-7 X.329/14984 Woolwich
2525003
Vol.2. - 1973
x,818p,[47]p of plates : ill(some col.). - (African handbook of birds. Series 3). - Includes index
0-582-46087-5 X.329/14983 Woolwich
2525003

AFRICA
West Africa

HISTORY of West Africa / edited by J.F.A. Ajayi and Michael Crowder. - 2nd ed. - London : Longman.
Vol.1. - 1976
xiv,649p : ill. - Includes index
0-582-64187-x 2090.g. Reading Room
0707597
- [Another copy.] X.0700/1015(1)

AFRICA
West Africa - For British businessmen

TRADE contacts in West African countries / [editor Ethel de Keyser]. - London : Kogan Page [for] 'Export times', 1977
189p : ill. - Includes index
0-85038-063-4 X.525/2326
1815253
- [Another copy] OPL

AFRICA
West Africa - Serials

TRAVELLER'S guide to West Africa. - [No.1]-. - London : IC Magazines, 1981-
v. : ill
B81-29434 X.808/34645
2530066

AFRICA
West Africa, to 1976 - Bibliographies

KEANE, Christiane
Ghana, Sierra Leone & the Gambia. - London : Commonwealth Institute, 1977
[3],40p. - (Commonwealth bibliographies ; no.10)
0-900906-53-7 X.709/32264
147894x

AFRICA
West Africa, to 1977

CROWDER, Michael
West Africa. - London : Longman, 1977
[9],214p,[8]p of plates : ill(some col). - Bibliography: p.203-204. - Includes index
0-582-60003-0 X.700/25467
1165224

AFRICA
West Africa, to 1980 - For West African students

LUCAN, Talabi A. (Talabi Aisie)
A visual history of West Africa. - London : Evans, 1981
76p : ill
0-237-50145-7 X.800/31914
2592843

SINGH, Malkiat
Model answers in West African history. - Ibadan ; London : Evans, 1980
51p
0-237-50457-x X.809/54790
2592843

AFRICA
West Africa, to date

STRIDE, G. T.
Peoples and empires of West Africa. - Repr. - Walton-on-Thames : Nelson, 1981
373p : ill
r6805443 X.809/53073
8469202

TRUELOVE, Roger
A history of West Africa 1000-1965. - Nairobi : East African Publishing House, 1975
274p : ill. - Includes bibliography
r6810161 X.809/56081
8469202

AFRICA
West Africa, to date - Serials

AFRICA. - no 1 (1962)-. - Dakar : Africa, [1962]- : ill
Holdings: année 16, no.101, 1978-
r6206935 P.821/219
8469202

AFRICA
West Africa, to date - Serials-continued
ENTENTE africaine. - [no.1] (1969)-. - Abidjan : Inter Afrique Presse, [1969]- : ill
Holdings: no.22, Octobre 1975-
r6206468 P.701/1441
8469202

AFRICA
West Africa, 830-1350
CORPUS of early Arabic sources for West African history edited and annotated by N. Levtzion & J.F.P. Hopkins. - Cambridge : Cambridge University Press, 1981
xx,492p : 2ill. - (Fontes historiae Africanae. Series Arabica ; 4). - Bibliography: p434-440. - Includes index
0-521-22422-5 X.800/29853
2368595

AFRICA
West Africa, 1000-1800 - For West African students
DAVIDSON, Basil
A history of West Africa, 1000-1800. - New revised ed. - London : Longman, 1977
xvi,318p : ill. - (The growth of African civilisation). - Bibliography: p.295-296. - Includes index
0-582-60340-4 X.709/31591
124468x

OGINI, F. G.
An outline history of West Africa, 1000-1800. - Lagos : Macmillan Nigeria, 1977
x,214p : ill. - Bibliography : p.199-201.- Includes index
r6032626 X.709/31986
124468X

OGINI, FG
An outline history of West Africa 1000 - 1800. - Reprinted. - Lagos : Macmillan Nigeria, 1978
214p. : ill
r6096265 X.709/28272
124468X

AFRICA
West Africa, 1000-1960
WILSON, Derek
West African history, 1000-1960. - London : Evans Bros, 1976
64p. - (Evans pocket facts)
0-237-49975-4 X.708/20418
0679631

AFRICA
West Africa, 1000-1972 - For West African students
BUAH, Francis Kwamina
West Africa since AD 1000. - London : Macmillan.
Book 1: The people. - 1974
viii,263p : 2 ill. - Bibliography: p.247-250. - Includes index
0-333-15578-5 X.0709/1005
0373796
Book 2: The people and outsiders. - 1977
vii,231p. - Bibliography: p.218-220. - Includes index
B77-28319 X.0709/1005
0373796

AFRICA
West Africa, 1000-1977 - For West African students
LATHAM, Norah
A sketch-map history of West Africa. - Revised [ed.]. - Amersham : Hulton, 1978
96p : 1 ill
0-7175-0803-x X.700/25771
1577182

AFRICA
West Africa, 1000-1978
PARRY, K. J.
Revision guide, essay writing in West African history. - London : Longman, 1979
viii,120p. - (Study for success)
0-582-60348-x X.709/33496
1962949

AFRICA
West Africa, 1000-1979 - For East African students
CROWDER, Michael
A history of West Africa, A.D. 1000 to the present. - London : Longman, 1979
222p : ill. - Bibliography: p.214-215. - Includes index
0-582-60368-4 X.700/26398
2158299

AFRICA
West Africa, 1000-1980
ISICHEI, Elizabeth
West African history. - Oxford : Oxford University Press, 1983
80p
0-19-913269-0 X.800/35631
2402467

THATCHER, Paul
West African history. - New ed.. - Harlow : Longman, 1980
119p. - (Study for success) (Certificate notes)
0-582-60398-6 X.809/47653
2402467

AFRICA
West Africa, 1450-1885
JONES, Adam
German sources for West African history 1599-1669. - Wiesbaden : Steiner, 1983
xi,417p : ill. - (Studien zur Kulturkunde ; Bd 66). - Includes bibliography
3-515-03728-4 Ac.659[Bd.66]
9017763

AFRICA
West Africa, 1800-1979
ISICHEI, Elizabeth
History of West Africa since 1800. - New York : Africana Publishing Co., 1977
ix,380p : ill
0-8419-0311-5 X.809/27941
1847554
History of West Africa since 1800. - New York : Africana Publishing Company, 1977
xii,380p. - Bibliography : p.356-375. - Includes index
0-8419-0312-3 X.709/33697
1847554

AFRICA
West Africa, 1900-1978
FROM colony to sovereign state / edited by A. Fajana and A.O. Anjorin. - Sunbury-on-Thames : Nelson, 1979
viii,280p : ill. - Bibliography: p.271-276. - Includes index
0-17-511551-6 X.709/32981
1819763

AFRICA
West Africa, 1940-1978
PEDLER, Sir Frederick
Main currents of West African history, 1940-1978. - London : Macmillan, 1979
xii,301p. - Bibliography: p.271-281. - Includes index
0-333-25870-3 X.809/44831
1739042

AFRICA
West Africa. Adult education
A HANDBOOK of adult education for West Africa / edited by Lalage Bown, S.H. Olu Tomori. - London : Hutchinson, 1979
296p. - (Hutchinson university library for Africa). - Includes bibliographies and index
0-09-138810-4 X.529/33977
1761366

AFRICA
West Africa. Agriculture
ADUAYI, E. A.
General agriculture and soils. - London : Cassell, 1981
102p : ill. - (Cassell's tropical agriculture series ; book 1). - Includes index
0-304-30207-4 X.325/1195 Woolwich
2408767

AFRICA
West Africa. Agriculture. Role of entrepreneurship
LAWSON, Rowena M. (Rowena Mary)
The agricultural entrepreneurship of upper-income Africans. - [Hull] : [The author], 1977
[9],232p : 1 ill. - Bibliography: p.227-232
B78-00777 X.313/1457 Woolwich
1275151
- [Another copy] BS.251/129
The agricultural entrepreneurship of upper-income Africans. - [Hull] : [The author], 1977
[9],232p : 1 ill. - Bibliography: p.227-232
B78-00777 X.313/1457 Woolwich
127516x
- [Another copy] BS.251/129

AFRICA
West Africa. Akan states. Gold trades, to ca 1900
GARRARD, Timothy F.
Akan weights and the gold trade. - London : Longman, 1980
xxii,393p,fold leaf : ill. - (Legon history series). - Bibliography : p.364-374. - Includes index
0-582-64631-6 X.800/29262
1887424

AFRICA
West Africa. Big game. Hunting
GÖRTZ, Heinrich
Beyond the river bend. - London : Cassell, 1979
x,159p : ill
0-304-30071-3 X.989/53564
1916742

AFRICA
West Africa. Birds
SERLE, William
A field guide to the birds of West Africa. - London : Collins, 1977
351p,48p of plates : ill(chiefly col). - Includes index
0-00-219204-7 X.329/10927 Woolwich
1101382

AFRICA
West Africa. Black magic, to date
ITALIAANDER, Rolf
Schwarze Magie - Magie der Schwarzen. - Frieburg im Breisgau : Aurum, 1983
176p. : ill
3-591-08186-8 X.529/67550
9132813

AFRICA
West Africa. Buildings. Construction
GARDI, René
[Auch im Lehmhaus lässt sich's leben]. Maisons africaines. - Paris ; Bruxelles : Elsevier Séquoia, [1974]
248p : ill
2-8003-0046-9 Cup.1264.ll.26
1538578

AFRICA
West Africa. Children. Foster care
GOODY, Esther N. (Esther Newcomb)
Parenthood and social reproduction. - Cambridge : Cambridge University Press, 1982
xiii,348p : ill. - (Cambridge studies in social anthropology ; 35). - Bibliography: p333-340. - Includes index
0-521-22721-6 X.520/26318
2693046

AFRICA
West Africa. Christian movements, 1910-1960
GRESCHAT, Hans Jürgen
Westafrikanische Propheten. - Marburg an der Lahn : Selbstverlag, 1974
113 p.. - (Marburger Studien zur Afrika- und Asienkunde. Serie A ; Bd. 4). - Bibliography : p. 105-107
r6000150 X.0100/683(4.)
0900966

AFRICA
West Africa. Christianity, to 1978
The HISTORY of Christianity in West Africa / edited and introduced by O.U. Kalu. - London : Longman, 1980
378p. - Includes index
0-582-64693-6 X.200/34894
1847589

AFRICA
West Africa. Christianity, to 1982
SANNEH, Lamin O.
West African Christianity. - London : Allen & Unwin, 1983
xviii,286p. - Bibliography: p253-267. - Includes index
0-04-276001-1 X.200/42459
3058697

AFRICA
West Africa. Church of England. Crowther, Samuel - Biographies
SWEETMAN, David
Bishop Crowther. - London : Longman, 1981
52p : ill. - (Makers of African history)
0-582-60378-1 X.808/35129
2573695

AFRICA
West Africa. Civics - For Nigerian students
KEAY, Elliot Alexander
West African government for Nigerian students. - 3rd ed. - London : Hutchinson, 1977
271p,[4]p of plates : ill. - Includes index
0-09-129181-x X.709/31448
1189522

AFRICA
West Africa. Climate
OJO, Oyediran
The climates of West Africa. - London : Heinemann Educational, 1977
xvii,219p : ill. - Bibliography: p.210-214. - Includes index
0-435-95700-7 X.800/26909
1476033

AFRICA
West Africa. Coinage, 1684–1958
VICE, David
The coinage of British West Africa and St. Helena 1684–1958. – Birmingham : Peter Ireland (Format), 1983
vii,237p : ill. – Bibliography: pvii
0–9508812–0–1 X.800/37489
3073823

AFRICA
West Africa. Colonial administration, ca 1850–1971
CROWDER, Michael
Colonial West Africa. – London : Cass, 1978
[9],341p. – Includes index
0–7146–2943–x X.809/42875
1350919

AFRICA
West Africa. Colonisation by European countries, 1885–
HARGREAVES, John D. (John Desmond)
West Africa partitioned. – London : Macmillan.
Vol.1: The loaded pause, 1885–1889. – 1974
xiv,273p. – Includes index
0–333–14485–6 X.0809/2003(1)
0506702

AFRICA
West Africa. Colonisation by European countries. Resistance by Africans, 1850–1903
WEST African resistance / edited by Michael Crowder. – New ed. with corrections. – London : Hutchinson, 1978
xii,314p,[8]p of plates : ill. – Includes bibliographies and index
0–09–134031–4 X.700/25757
1568752

AFRICA
West Africa. Colonisation by Western world, 1454–1578
BLAKE, John William
West Africa. – 2nd ed., revised and enlarged. – London : Curzon Press [etc.], 1977
xxi,246p,leaves of plates. – Bibliography: p.225–237. – Includes index
0–7007–0098–6 X.809/42488
1074490

AFRICA
West Africa. Commerce – For West African students
LAWSON, Rowena M. (Rowena Mary)
Introduction to commerce in West Africa. – 3rd ed.. – [Harlow] : Longman, 1979
[4],124p : ill. – Includes index
0–582–65022–4 X.510/16522
207818x

AFRICA
West Africa. Commonwealth countries. Commercial banks: Bank of West Africa, to 1965
FRY, Richard
Bankers in West Africa. – London : Hutchinson, 1976
xviii,270p,[32]p of plates : ill. – Bibliography: p.xvii–xviii. – Includes index
0–09–126910–5 X.520/11084
0740748

AFRICA
West Africa. Construction – Serials
WAC – West African Construction. – Vol.1–, no.1– ; 1978–. – Sutton, Surrey : IPC Building and Contract Journals Ltd, 1978– : ill(some col)
Holdings: vol.1, no.5, etc. : 1978, etc
B79–21038 SRL
1902040

AFRICA
West Africa. Construction equipment – Serials
WEST African materials and equipment. – Vol.1, no.1– ; Mar. 1978–. – London : BWS Publishing Ltd, 1978– : ill(some col)
B78–29662 SRL
1571516

AFRICA
West Africa. Construction materials – Serials
WEST African materials and equipment. – Vol.1, no.1– ; Mar. 1978–. – London : BWS Publishing Ltd, 1978– : ill(some col)
B78–29662 SRL
1571524

AFRICA
West Africa. Consumer goods. Consumption
CONSUMER markets in West Africa. – London : Euromonitor, 1984
xi,421p
0–86338–040–9 X.525/10209
3051382

AFRICA
West Africa. Decolonisation
HARGREAVES, John D. (John Desmond)
The end of colonial rule in West Africa. – London : Historical Association, 1976
38p : ill. – (Pamphlets : general series / Historical Association ; 88). – Bibliography: p.37–38
B77–04508 WP.3175/88
0797820

AFRICA
West Africa. Decolonisation, 1930–1978
HARGREAVES, John D. (John Desmond)
The end of colonial rule in West Africa. – London : Macmillan, 1979
xvi,141p. – Includes index
0–333–24843–0 X.809/44623
1699881

AFRICA
West Africa. Dependency, to date
DEPENDENCY and underdevelopment in West Africa / edited by Victor C. Uchendu. – Leiden : Brill, 1980
136p. – (International studies in sociology and social anthropology ; v29)
90–04–06170–3 X.520/24914
8730601

AFRICA
West Africa. Description & travel
PERN, Stephen
The beach of morning. – London : Hodder and Stoughton, 1983
224p,[8]p of plates : col.ill. – Bibliography: p223–224
0–340–26236–2 X.809/56892
2989182

AFRICA
West Africa. Description & travel, to date
MAGE, Eugène
Voyage au Soudan occidental (1863–1866). – Paris : Éditions Karthala, [1980]
xiv,308p
2–86537–009–7 X.808/36840
8469377

AFRICA
West Africa. Description & travel, 1795–1797
PARK, Mungo
[Travels in the interior districts of Africa]. Travels in the interior of Africa. – London : Folio Society, 1984
xxii,203p,[16]leaves of plates : ill
B84–39746 X.800/39327
3175057

AFRICA
West Africa. Description & travel, 1893–1900
KINGSLEY, Mary H. (Mary Henrietta)
Travels in West Africa. – 4th ed.. – London : Virago, 1982
xxiv,741p,[16]p of plates : ill. – Includes index
0–86068–266–8 X.958/11107
2768216

AFRICA
West Africa. Description & travel, 1931–1932 – Correspondence, diaries
PERHAM, Margery
West African passage. – London : Peter Owen, 1983
245p,[8]pof plates : ill. – Includes index
0–7206–0609–8 X.809/57348
303772x

AFRICA
West Africa. Description & travel, 1934
GORER, Geoffrey
Africa dances. – Harmondsworth : Penguin, 1945
218p
0–14–009502–0 X.629/20053
2891263

AFRICA
West Africa. Description & travel, 1941 – Correspondence, diaries
GREENE, Graham
In search of a character. – Harmondsworth : Penguin in association with Bodley Head, 1968, c1961
105p
0–14–002822–6 X.958/680
2354748

AFRICA
West Africa. Diamond mining industries, 1919–1983
GREENHALGH, Peter, 1945–
West African diamonds 1919–1983. – Manchester : Manchester University Press, c1985
xiii,306p : ill. – Bibliography: p291–297. – Includes index
0–7190–1763–7 X.520/37448
3156400

AFRICA
West Africa. Divorce
LOSPINOSO, Mariannita
Il divorzio nell'Africa Occidentale e nel Camerun. – Genova : Tilgher, 1975
174p. – Bibliography : p.165–170
r6052641 X.100/27055
1500023

AFRICA
West Africa. Domestic trade, ca 1850–ca 1980
HARRISS, Barbara
Some lessons from the history of internal trade in the interior of West Africa. – Norwich : School of Development Studies, University of East Anglia, 1981
36p. – (Discussion paper ; no.105). – Bibliography: p33–36
B84–40803 X.525/9960
318739x

AFRICA
West Africa. Economic conditions
CONFEDERATION OF BRITISH INDUSTRY
West African visit report. – [London] : C.B.I., 1978
[1],47p. – Bibliography: p.47
B78–16636 X.0515/29
1446541

MEILLASSOUX, Claude
Maidens, meal and money. – Cambridge : Cambridge University Press, 1981
xiv,196p. – (Themes in the social sciences). – Bibliography: p162–188. – Includes index
0–521–22902–2 X.520/23522
2066165

RIMMER, Douglas
The economies of West Africa. – London : Weidenfeld and Nicolson, c1984
xii,308p : ill. – (The International economies series). – Includes index
0–297–78095–6 X.529/62947
1446541

AFRICA
West Africa. Economic conditions – Serials
BULLETIN de l'Afrique Noire. – Paris : Édiafric, La Documentation Africaine, 1977–
Holdings année 19. no.895, jan 1977–
r6038014 P.703/363
0296805

AFRICA
West Africa. Economic conditions, to 1968 – For West African students
ORD, Henry Wilson
An introduction to West African economics. – London : Heinemann Educational, 1978
x,457p : ill. – Includes index
0–435–97430–0 X.519/29170
1801031

AFRICA
West Africa. Economic conditions, 1960–
ORGANISATION DES ÉTATS RIVERAINS DU SÉNÉGAL. Conseil
Résolutions adoptées par le 5e Conseil des ministres de l'O.E.R.S. tenu à Bamako du 13 au 15 janvier 1971. – Bamako : L'Informateur, 1971
36p : ill
r6209941 X.525/7979
8853266

AFRICA
West Africa. Economic conditions, 1960– – Serials
NOTES d'information et statistiques / Banque centrale des États de l'Afrique de l'ouest. – no 1 (1955)–. – Dakar : [La Banque], [1955]–
Holdings: no 301, 1982–
r6209028 X.0525/480
8853266

AFRICA
West Africa. Economic conditions. Geographical aspects
ONYEMELUKWE, J. O. C. (Joe Okezie C)
Economic geography of West Africa. – London : Longman, 1983
xii,175p : ill. – Includes bibliographies and index
0–582–64671–5 X.529/54151
2901056

AFRICA
West Africa. Economic development, 1960–

The ECONOMIC integration of West Africa / edited by E.C Edozien, E. Osagie. - Ibadan : Ibadan University Press, 1982
xvii,192p
978-12-1125-3 X.529/56503
8994668

AFRICA
West Africa. Economic development. Organisations. Office du Niger

SCHREYGER, Emil
L'Office du Niger au Mali 1932 à 1982. - Wiesbaden : Steiner, 1984
xvi,394p.,plates. - (Beitrage zur Kolonial- und Uberseegeschichte ; Bd.27). - Includes bibliography
3-515-04196-6 X.0700/119(27)
9128123

AFRICA
West Africa. Economic development. Role of international cooperation

RENNINGER, John P.
Multinational cooperation for development in West Africa. - New York ; Oxford : Pergamon for UNITAR, 1979
xiv,162p. - (Pergamon policy studies). - Bibliography: p.147-156. - Includes index
0-08-022490-3 X.520/9950
192057x

AFRICA
West Africa. Economic integration

EZENWE, Uka
ECOWAS and the economic integration of West Africa. - London : Hurst, c1983
xii,210p. - Bibliography: p199-202. - Includes index
0-905838-76-9 X.529/66288
299030x

ROBSON, Peter, 1926–
Integration, development and equity. - London : Allen & Unwin, 1983
viii,181p : ill. - Bibliography: p171-177. - Includes index
0-04-338109-x X.529/57792
299030x

AFRICA
West Africa. Economic organisations: Economic Community of West African States - Conference proceedings

ECONOMIC COMMUNITY OF WEST AFRICAN STATES CONFERENCE (1976)
- Lagos : Nigerian Institute of International Affairs, 1983 [i.e. 1984]
xxxv,779p
978-227-633-2 X.520/37470
9103309

WEST AFRICAN ECONOMIC ASSOCIATION. Biennial Conference (3rd : 1982 : Freetown)
Trade & development in Economic Community of West African States (ECOWAS). - New Delhi : Vikas Publishing House, [1984]
xi,589p
0-7069-2557-2 X.529/67594
9103309

AFRICA
West Africa. Education - For trainee teachers

AWÓNÍYÌ, Timothy A.
Principles and practice of education. - London : Hodder and Stoughton, 1979
xii,111p. - (Nigerian education series). - Bibliography: p.109-111
0-340-24409-7 X.519/30804
202084x

AFRICA
West Africa. Education. Use of African languages

MOTHER tongue education / edited by Ayo Bamgbose. - London : Hodder and Stoughton [etc.], 1976
153p
0-340-20120-7 X.529/30981
0730270

AFRICA
West Africa. Emigration to Jamaica, 1840-1865

THOMAS, Mary Elizabeth
Jamaica and voluntary laborers from Africa. - Gainesville : University Presses of Florida, [1974]
v,211p.. - (a Florida State University book)
0-8130-0438-1 X.800/12805
1511475
- [Another copy.] X.800/32516

AFRICA
West Africa. English speaking countries. Politics, ca 1900-1971

PRICE, Joseph Henry
Political institutions of West Africa. - 2nd ed. - London : Hutchinson, 1975
288p,fold plate : ill. - Bibliography: p.279-281. - Includes index
0-09-123111-6 X.709/30126
0503029

Political institutions of West Africa. - 3rd ed. - London : Hutchinson, 1977
297p : ill. - Bibliography: p.287-289. - Includes index
0-09-131461-5 X.709/31604
0503029

AFRICA
West Africa. Ewe, to date

MAMATTAH, Charles M. K.
The Eves of West Africa (oral traditions). - Keta : Research Publications, [1976?]. - Bibliography
Vol.1: The Anglo-Eves and their immediate neighbours
768p,1plate : ill
r6806942 X.808/38430
9066810

VERDON, Michel
The Abutia Ewe of West Africa. - Amsterdam : Mouton, [1983]
xix,319p. - (Studies in the social sciences. anthropology ; 38). - Includes bibliography
90-279-3410-x X.800/38751
9066810

AFRICA
West Africa. Ewe. Geomancy, to date

SURGY, Albert de
La géomancie et le culte d'Afa chez les Evhé du littoral. - Paris : Publications orientalistes de France, [1981]
444p,16p of plates : ill. - Bibliography
2-7169-0132-5 X.809/50585
8810702

AFRICA
West Africa. Experimental theatre. Performances by companies of International Centre of Theatre Research

HEILPERN, John
Conference of the birds. - London : Faber, 1977
3-317p
0-571-10372-3 X.809/42340
1059971

Conference of the birds. - Harmondsworth : Penguin, 1979
302p
0-14-004825-1 X.619/22093 Woolwich
1059971

AFRICA
West Africa. Exploration, 1893-1900

KINGSLEY, Mary
Travels in West Africa. - London : Folio Society, 1976
276p,[13]leaves of plates : ill
0-85067-100-0 X.802/10234
0759635

AFRICA
West Africa. Exploration. Kingsley, Mary, b.1862 - Biographies

HÖJER, Signe
Mary Kingsley, forskningsresande i Västafrika. - [Ny uppl.]. - Stockholm : LT, 1975
250 p.
91-360-0655-6 X.809/23428
1014692

AFRICA
West Africa. Exploration. Park, Mungo

BRENT, Peter
Black Nile. - London : Gordon and Cremonesi, 1977
[2],200p,[16]p of plates : ill. - Bibliography: p.193-194. - Includes index
0-86033-017-6 X.802/10375
0072850

AFRICA
West Africa. Exploration. Park, Mungo - Biographies

LUPTON, Kenneth
Mungo Park. - Oxford : Oxford University Press, 1979
xxii,272p,12p of plates : ill. - Bibliography: p.241-244. - Includes index
0-19-211749-1 X.809/44163
1397133

AFRICA
West Africa. Family life

OPPONG, Christine
[Marriage among a matrilineal elite]. Middle class African marriage. - London : Allen & Unwin, 1981
xx,190p : 2ill. - Bibliography: p166-177. - Includes index
0-04-301138-1 X.529/46271
0168491

AFRICA
West Africa. Finance, 1960- - Serials

FONDS D'ENTRAIDE ET DE GARANTIE DES EMPRUNTS DU CONSEIL DE L'ENTENTE
Rapport d'activité. - (1966)-. - Abidjan : [le Fonds], [1967]- : ill
Holdings: 1976-
r6206346 L.49/1304
8840504

AFRICA
West Africa. Foreign investment by German companies, to date

SCHMIDT, Christian A.
Betriebswirtschaftliche Probleme beim Aufbau klein- und mittelindustrieller Joint Ventures in westafrikanischen Entwicklungsländern. - Frankfurt/Main : R.G. Fischer, 1980
xi,337p.. - Bibliography: p.313-337
3-88323-142-8 X.529/49739
8866201

AFRICA
West Africa. Foreign relations. Diplomacy, to ca 1900

SMITH, Robert S.
Warfare and diplomacy in pre-colonial West Africa. - [London] : Methuen, 1976
vii,240p. - (Studies in African history ; 15). - Bibliography: p.225-235. - Includes index
0-416-55060-6 X.808/33109
0797898

AFRICA
West Africa. Foreign trade with Great Britain, ca 1860-ca 1920 - Biographies

TRADING in West Africa / [compiled by] P.N. Davies. - London : Croom Helm, 1976
3-209p. - Includes bibliographies and index
0-85664-129-4 X.529/30604
0635367

AFRICA
West Africa. Freight transport. Shipping by liners

H.P. DREWRY (SHIPPING CONSULTANTS) LIMITED. Research Division
West African liner shipping. - London : [Drewry], 1979
[4],67p : ill(some col). - (An economic study ; no.75)
B79-36274 SRL
199493x

AFRICA
West Africa. French colonies. Colonial administration, to 1960

BENOIST, Joseph-Roger de
L'Afrique Occidentale Française de la conférence de Brazzaville (1944) à l'Indépendance (1960). - [Dakar] : Nouvelles Éditions Africaines, 1983
617p. - Includes bibliography
2-7236-0860-3 X.800/39909
8877335

DADIÉ, Bernard B.
Commandant Taureault et ses nègres. - [Abidjan] : [CEDA], [1980]
126p
2-86394-021-x X.809/41190
8877335

SPITTLER, Gerd
Verwaltung in einem afrikanischen Bauernstaat. - Wiesbaden : Steiner, [1981]
208p.. - (Beiträge zur Kolonial- und Überseegeschichte ; Bd.21). - Bibliography
3-515-03792-6 X.0700/119(21)
8877335

AFRICA
West Africa. French colonies. Islam, to date

KABA, Lansiné
The Wahhabiyya. - Evanston : Northwestern University Press, 1974
xv,285p. - (Studies in African religion)
0-8101-0427-x X.800/35148
8931224

AFRICA
West Africa. French speaking countries. Economic conditions - For British businessmen

TROPICAL AFRICA ADVISORY GROUP
Francophone West Africa. - [London] : British Overseas Trade Board, [1983?]
33p : ill
B84-29161 BS.41/1391
2779021

AFRICA
West Africa. Fulani, to date

L'HISTOIRE des Peuls Férôbé du Diamaré, Maroua et Pétté / traditions transmises par Modibbo Bakari Yana [et al.] ; éditées par Eldridge Mohammadou. - Tokyo : Institute for the Study of Languages and Cultures of Asia and Africa, 1976
iii,409p. - (African languages and ethnography ; 3). - Bibliography: p377-382
r6207098 X.808/35754
8808317

NILSSON, Christine
Le Sahara des Peuls. - Paris : Éditions René Moser, [1982]
85p
2-902906-08-0 L.42/2641
8808317

AFRICA
West Africa. Fulani. Cultural processes

AZARYA, Victor
Aristocrats facing change. - Chicago ; London : University of Chicago Press, 1978
xv,293p. - Bibliography: p.273-281. - Includes index
0-226-03356-2 X.809/43478
1578316

AFRICA
West Africa. Fuloni. Christian missions, 1960-1980

DENFFER, Ahmad von
The Fulani evangelisation-scheme in West-Africa. - Leicester : The Islamic Foundation, c1980
27p,[3]p of plates : ill. - (Situation report / Islamic Foundation ; no.1)
0-86037-074-7 X.205/761
2387603

AFRICA
West Africa. Geographical features

CHURCH, R. J. Harrison (Ronald James Harrison)
West Africa. - 8th ed. - London : Longman, 1980
xxxi,526p : ill. - (Geographies for advanced study). - Bibliography: p485-516. - Includes index
0-582-30020-7 WPB.375/60
0102121

CLAYTON, Robert
Western Africa. - 2nd ed. - St Albans : Hart-Davis, 1975
48p : ill(some col). - (Finding out about geography ; [13])
0-298-12090-9 X.809/40269
0506761

EVANS, F. C. (Frederick Colin)
Geographical photographs for West Africa. - Cambridge : Cambridge University Press, 1983
48p : ill
0-521-28616-6 L.49/1938
2960559

ILOEJE, N. P. (Nwadilibe P.)
A new geography of West Africa. - New rev. ed. - London : Longman, 1980
201p,[8]p of plates : ill(some col.). - Includes index
0-582-60379-x X.800/30474
0506761

ILOEJE, Nwadilibe P.
A new geography of West Africa. - Metricated ed. - London : Longman, 1977
iv,172p : ill. - Bibliography: p.169. - Includes index
0-582-60344-7 X.700/25513
0506761

UDO, Reuben Kenrick
A comprehensive geography of West Africa. - Ibadan ; London : Heinemann Educational, 1978
xvi,304p : ill. - Bibliography: p.294-295. - Includes index
0-435-95914-x X.800/27027
0506761

AFRICA
West Africa. Geographical features - For West African students

GILBERT, D. A. E.
Regional geography of West Africa. - London : Macmillan, 1982
217p : ill. - Includes index
0-333-34245-3 X.800/35592
1565990

JARRETT, Harold Reginald
A geography of West Africa. - New revised ed. - London : Evans, 1980
[8],184p : ill. - Includes index
0-237-50151-1 X.700/26243
1565990

OBOLI, Herbert Oguejiofo Nkemka
A new outline geography of West Africa. - 8th ed., greatly enlarged and revised. - London : Harrap [etc.], 1978
143p : ill. - Bibliography: p.140. - Includes index
0-245-53213-7 X.700/25752
1565990

AFRICA
West Africa. Government, to 1980 - For West African students

OYENEYE, Ibiyemi
O-level West African government. - Rev. ed. - Harlow : Longman, 1982
iv,92p : 1ill. - (Study for success) (Study for success)
0-582-65098-4 X.809/55132
2543761

WARD, W. E. F. (William Ernest Frank)
Government in West Africa. - 6th rev. ed.. - London : Allen & Unwin, 1981
383p. - Bibliography: p10. - Includes index
0-04-351059-0 X.809/48927
2543761

AFRICA
West Africa. Government, 1765-1973

ADIGWE, Francis
Essentials of government for West Africa. - Ibadan ; London : Oxford University Press, 1974 [i.e. 1976]
viii,389p. - Bibliography: p.383-384. - Includes index
0-19-575242-2 X.809/41757
1074156

AFRICA
West Africa. Governments - For West African students

KELLY, Robert, 1946-
Multiple choice tests in government. - London : Macmillan, 1982
ix,91p
0-333-33165-6 X.520/35493
0726702

OYENEYE, Ibeyimi
West African government for O-level. - Harlow : Longman, 1981
xiv,191p : ill. - Includes index
0-582-60958-5 X.520/26149
0726702

OYENEYE, Ibiyemi
West African government. - London : Longman, 1976
[5],96p : ill. - (Longman certificate notes)
0-582-60316-1 X.708/20524
0726702

AFRICA
West Africa. Handicrafts: Weaving, to 1974

LAMB, Venice
West African weaving. - London : Duckworth, 1975
228p : ill(some col). - Bibliography: p.222-224. - Includes index
0-7156-0905-x X.423/5016
0360457

AFRICA
West Africa. Hausa, to date

ADAMU, Mahdi
The Hausa factor in West African history. - Zaira : Ahmadu Bello University Press, c1978
xii,224p. - (Ahmadu Bello University history series). - Includes bibliography
978-12-5006-2 X.800/36171
8997799
- [Another copy] X.520/32985

AFRICA
West Africa. Industrialisation

ONYEMELUKWE, J. O. C. (Joe Okezie C)
Industrialization in West Africa. - London : Croom Helm, c1984
226p. - (Croom Helm industrial geography series). - Bibliography: p209-214. - Includes index
0-7099-1936-0 X.529/67687
3243206

AFRICA
West Africa. Industries, to date

SCHMIDT, Christian A.
Betriebswirtschaftliche Probleme beim Aufbau klein- und mittelindustrieller Joint Ventures in westafrikanischen Entwicklungsländern. - Frankfurt/Main : R.G. Fischer, 1980
xi,337p.. - Bibliography: p.313-337
3-88323-142-8 X.529/49739
886621x

AFRICA
West Africa. Initiation rites

NGOMA NGAMBU
Initiation dans les sociétés traditionnelles africaines (le cas kongo). - Kinshasa : Presses universitaires du Zaïre, 1981
227p
r6210306 X.529/56749
8861633

Une RÉCITATION du Bagré / publiée par Jack Goody et S.W.D.K. Gandah. - Paris : Armand Colin, 1980
405p. - (Classiques africains ; 20)
r6207344 X.0981/53(20.)
8861633

AFRICA
West Africa. Insects

BOORMAN, John, 19—
West African insects. - Harlow : Longman, 1981
viii,88p : ill(some col.). - (West African nature handbooks). - Bibliography: p82-83. - Includes index
0-582-60626-8 7010.f.2.
2718650

AFRICA
West Africa. Islam

STUDIES in West African Islamic history. - London : Cass.
Vol.1: The cultivators of Islam. - 1979
xi,325p,[3]fold plates : ill. - Includes index
0-7146-1737-7 X.800/27987
1272853

AFRICA
West Africa. Islam, to 1980

CLARKE, Peter B.
West Africa and Islam. - London : Edward Arnold, 1982
viii,275p : ill. - Bibliography: p268-270. - Includes index
0-7131-8029-3 X.200/38488
2734133

HISKETT, Mervyn
The development of Islam in West Africa. - London : Longman, 1984
xi,353p : ill. - (Longman studies in African history). - Bibliography: p331-336. - Includes index
0-582-64692-8 X.200/45352
2734133

AFRICA
West Africa. Islam. Al-Hadj Omar Tal Al-Futi - Biographies

DUMONT, Fernand
L'anti-sultan, ou AC-Hajj Omar Tal du Fouta, combattant de la Foi, 1794-1864. - Dakar-Abidjan : Les Nouvelles Éditions Africaines, 1974
xviii. 244 p.. - Bibliography : p. 241-244
r6001162 X.700/16920
0864897

AFRICA
West Africa. Leaders, 1885-1924

REACTION and protest in the West African press / edited with introduction and notes by Georgia McGarry. - Leiden : Afrika-Studiecentrum ; Cambridge : African Studies Centre, 1978
[16],194p. - (African social research documents ; vol.10). - Bibliography: p.185-186. - Includes index
B80-34370 X.0311/143
223372x

AFRICA
West Africa. Librarianship - Bibliographies

DAVIES, Helen, 1936-
Libraries in West Africa. - [Oxford] : Zell, 1982
xix,170p. - Includes index
B83-17224 2719.e.520
2877392

AFRICA
West Africa. Livestock

OSITELU, G. S.
Animal science. - London : Cassell, 1981
105p : ill. - (Cassell's tropical agriculture series ; Bk.3). - Bibliography: p95. - Includes index
0-304-30209-0 X.322/10014 Woolwich
2525178

AFRICA
West Africa. Migration, 1960-1975. Demographic aspects

ZACHARIAH, K. C. (Kunniparampil Curien)
Migration in West Africa. - New York ; Oxford : Published for the World Bank [by] Oxford University Press, c1981
xiv,130p,22leaves of plates : ill. - Bibliography: p124-126. - Includes index
0-19-520186-8 X.525/4760
2540932

AFRICA
West Africa. Monetary policies. Effects of international monetary system, 1960–
NKUETE, Jean
Le franc CFA face aux mutations des grandes unités de compte : dollar-DTS-ECU. - Yaounde : Editions CLE, 1981
174p
2-7235-0086-1 X.529/53711
8929831

AFRICA
West Africa. Multinational cooperatives, 1960– - Directories
KOOPERATION in Westafrika / [herausgegeben von] Harald Voss, Frank Sandvoss = Co-operation in West Africa = Coopération en Afrique de l'Ouest. - Hamburg : Institut fur Afrika-Kunde, Dokumentations-Leitstelle Afrika, 1982
345p. - (Aktueller Informationsdienst Afrika. Sondernummer ; 4). - Bibliography: p.289-327
3-922887-26-0 X.0802/1549(4)
8842434

AFRICA
West Africa. Nationalist movements. Influence of invasion of Ethiopia by Italy, 1934–1941
ASANTE, S. K. B.
Pan-African protest. - London : Longman, 1977
xii,243p,[8]p of plates : ill. - (Legon history series). - Bibliography: p.218-233. - Includes index
0-582-64194-2 X.800/26665
1343327

AFRICA
West Africa. Niger River, to ca 1900
DE GRAMONT, Sanche
The strong brown god. - London : Hart-Davis MacGibbon, 1975
350p,[32]p of plates : ill. - Bibliography: p.332-339. - Includes index
0-246-10759-6 X.800/25232
0525472

AFRICA
West Africa. Niger River region, to 1975
FINLEY, Diana
The Niger. - London : Macdonald Educational, 1975
[1],44p : col ill. - ([Toppers rivers ; 3]). - Includes index
0-356-05160-9 X.800/25268
0534889

AFRICA
West Africa. Niger River region. Economic development. Role of British communities, to 1960
NZEMEKE, Alexander D.
British imperialism and African response. - Paderborn : Schöningh, 1982
379p. - (Sammlung Schöningh zur Geschichte und Gegenwart). - Includes bibliography
3-506-77453-0 X.800/36974
9037241

AFRICA
West Africa. Niger River region. Exploration, 1976
ODILI, Herbert A.
My endurance - my conquest. - Bognor Regis : New Horizon, c1979
vi,148p,[16]p of plates : ill
0-86116-392-3 X.809/47836
2484099

AFRICA
West Africa. Niger River region. Social conditions, to date
CLAIR, Andrée
Niger. - Paris : Messidor/La Farandole, 1982
235p,plates. - (Document)
2-7047-0275-6 X.800/39718
9162267

AFRICA
West Africa. Non-Christian religions. Attitudes of Crowther, Samuel
MCKENZIE, Peter Rutherford
Inter-religious encounters in West Africa. - [Leicester] : Study of Religion Sub-department, University of Leicester, 1976
115p,4p of plates : ill. - (Leicester studies in religion ; 1). - Bibliography: p.107-110. - Includes index
0-905510-00-3 X.100/16764
1046039

AFRICA
West Africa. Palaeozoic strata
LOWER Palaeozoic of north-western and west-central Africa / edited by C.H. Holland. - Chichester : Wiley, c1985
xii,512p,[16]p of plates : ill. - (Lower Palaeozoic rocks of the world ; v. 4). - Includes bibliographies and index
0-471-10357-8 SRL
3266044

AFRICA
West Africa. Plants, 1783–1803
HEPPER, Frank Nigel
The west African herbaria of Isert and Thonning. - Kew : Bentham-Moxon Trust, 1976
227p,7p of plates : ill. - Bibliography: p.165-166. - Includes index
0-9504876-0-0 SRL
060092x

AFRICA
West Africa. Political events. Role of British military forces, 1807–1874
MBAEYI, Paul Mmegha
British military and naval forces in West African history, 1807-1874. - New York ; London : NOK Publishers, 1978
xii,263p. - ([NOK library of African affairs]). - Bibliography: p.241-253. - Includes index
0-88357-029-7 X.800/28353
2023598
- [Another copy] X.800/15054

AFRICA
West Africa. Politics, to 1982 - For West African students
OKWODU, R. N.
'O' Level government for West Africa. - [New ed.]. - London : Macmillan, 1983
viii,229p. - Includes index
0-333-35010-3 X.529/56288
1896350

School certificate government for West Africa. - London : Macmillan, 1979
vii,198p. - Includes index
0-333-23447-2 X.709/33209
1896350

AFRICA
West Africa. Politics, 1945–1977
WEST African states / edited by John Dunn. - Cambridge : Cambridge University Press, 1978
[8],259p. - (African studies series ; 23). - Includes index
0-521-21801-2 X.800/27025
1520261

AFRICA
West Africa. Population, to date - Conference proceedings
WEST AFRICAN SEMINAR ON POPULATION STUDIES (1972 : University of Ghana)
Interdisciplinary approaches to population studies. - Legon : Population Dynamics Programme, University of Ghana, 1975
ix,333p
r6806855 X.809/54693
891639x

AFRICA
West Africa. Postal services, 1792–1898
MCCAIG, Colin
The postal history and handstamps of British West Africa. - [London] : [R. Lowe], [1978]
[1],64p
B79-20018 X.510/16190
1898590

AFRICA
West Africa. Primary schools. Curriculum - For teaching
GRANT, Margaret
School methods with younger children. - 4th ed. - London : Evans Bros, 1978
vi,250p : ill. - Bibliography: p.vi. - Includes index
0-237-50001-9 X.519/28812
1736655

AFRICA
West Africa. Primary schools. Students. Intelligence tests - Texts
ARMAR, Theophilus Quancoo
Common Entrance and scholarship tests. - London : Macmillan, 1978
vi,169p : ill
0-333-24750-7 X.512/6431
1816136

AFRICA
West Africa. Religion
FORTES, Meyer
Oedipus and Job in West African religion. - Reissued. - Cambridge : Cambridge University Press, 1983
ix,92p. - (Cambridge studies in social anthropology ; 48). - Bibliography: p88-89
0-521-26208-9 X.200/44081
3150453

AFRICA
West Africa. Rice production industries - Serials
WEST AFRICA RICE DEVELOPMENT ASSOCIATION
Annual report. - Jan-Dec. [1972?]-. - Monrovia : The Association, [1973?]-
Holdings : Annual report, Jan-Dec, 1975–
r6055483 P.611/299
1718134

AFRICA
West Africa. Rice. Production. Organisations: West Africa Rice Development Association
WEST AFRICA RICE DEVELOPMENT ASSOCIATION
Warda, what it is, what it does, how it works. - Revised ed. - Monrovia : Warda, 1976
29,[2]p
r6072415 X.311/8888 Woolwich
1692658

AFRICA
West Africa. Rural communities. Assistance by Secours Catholique Français
MERIAN, Pierre
Toute la nuit on danse, autour d'un puits au Sahel. - Paris : Éditions S.O.S., 1975
127 p., 8 p. of plates : ill
2-7185-0788-8 X.709/19705
0909750

AFRICA
West Africa. Rural regions. Social conditions, to date
KLINGSHIRN, Agnes
Frauen und ländliche Entwicklung in Afrika. - Köln : Weltforum, 1982
xiv,334p. : ill. - (Forschungsberichte des Bundesministeriums für wirtschaftliche Zusammenarbeit ; Bd.32). - Includes bibliographies
3-8039-0240-1 X.529/65947
8469555

AFRICA
West Africa. Sahel. Agriculture, to date
PERETTI, M.
Projet mise en place de l'ORD du Sahel. - [s.l.] : Ministère du developpement rural, 1976
2v((maps in a folder))
r6210519 SY81/29
9008403

AFRICA
West Africa. Sahel. Description & travel, to date
GIBOUDEAU, Gilbert
Les scories. - Chamblay : G.G. "Le Cèdre", 1982
178p : ill
2-904000-00-3 L.45/2594
9073981

AFRICA
West Africa. Sahel. Economic conditions, to date - Serials
UNITED NATIONS. Sahelian Office
DP/SO/PROJECTS. - 1 (26 Oct. 1978)-. - [New York] : UNDP, 1978–
v
r6802260 UN.A.356/148
8886105

AFRICA
West Africa. Sahel. Employment, to date
INFORMATION sur le marché de l'emploi dans les pays de l'Afrique sahélienne / Bureau international du travail. - Genève : Le Bureau, 1981
viii,117p
92-220-2861-9 UN.H.421/154
8889503

AFRICA
West Africa. Sahel. Famines caused by drought, to date
FRANKE, Richard W.
Seeds of famine. - Montclair, N.J. : Allanheld, Osmun, 1980
xi,266p
0-916672-26-3 X.322/11754
8928525

AFRICA
West Africa. Sahel. Famines caused by drought, 1968–1975
The POLITICS of natural disaster / edited by Michael H. Glantz. - New York ; London : Praeger ; [London] : [Distributed by Martin Robertson], 1976
xix,340p : ill. - (Praeger special studies in international economics and development). - Includes bibliographies and index
0-275-01180-1 X.800/25552
0915513

AFRICA
West Africa. Sahel. Famines caused by drought, 1968-1975–continued

SÉCHERESSES et famines du Sahel ... / sous la direction de Jean Copans. – Paris : François Maspero, 1975
2 tom. : ill. – (Dossiers africains). – Bibliography : tom. 2. p. 118–140
r6000982 X.519/20106
0915513

AFRICA
West Africa. Sahel. Famines caused by drought, 1969-1973. Responsibility of developed countries

COMITÉ INFORMATION SAHEL
Qui se nourrit de la famine en Afrique?. – Paris : François Maspero, 1974
279 p.. – (Cahiers libres ; 292-293)
r6012703 X.709/20788
087616X

AFRICA
West Africa. Sahel. Nomadic tribes. Social conditions, to date

KRINGS, Thomas F.
Kulturgeographischer Wandel in der Kontaktzone von Nomaden und Bauern im Sahel von Obervolta. – Hamburg : Hirt in Verbindung mit dem Institut für Geographie und Wirtschaftsgeographie der Universität Hamburg, 1980
x,116p,[15]p of plates[4]folded leaves of plates : ill(some col). – (Hamburger geographische Studien ; Hft.36). – Bibliography
r6302443 X.800/32079
8579199

AFRICA
West Africa. Sahel. Peasants. Social conditions, to date

BILLAZ, R.
Enquêtes en milieu rural sahélien. – Paris : Presses universitaires de France, 1981
195p : ill. – (Techniques vivantes). – Includes bibliography
2–85319–098–6 X.529/56545
8785376

KRINGS, Thomas F.
Kulturgeographischer Wandel in der Kontaktzone von Nomaden und Bauern im Sahel von Obervolta. – Hamburg : Hirt in Verbindung mit dem Institut für Geographie und Wirtschaftsgeographie der Universität Hamburg, 1980
x,116p,[15]p of plates[4]folded leaves of plates : ill(some col). – (Hamburger geographische Studien ; Hft.36). – Bibliography
r6302443 X.800/32079
8785376

LATEEF, Noel V.
Crisis in the Sahel. – Boulder : Westview Press, 1980
xxix,287p. – (Westview special studies in social, political and economic development)
0–89158–991–0 X.520/23062
8785376

PAROLES de brousse / présentées par le GRAAP [Groupe de recherche et d'appui pour l'autopromotion paysanne]. – Paris : Éditions Karthala, [1982]
115p : ill
r6210458 X.529/57461
8785376

AFRICA
West Africa. Sahel. Social conditions

DALY, Bridget
Tarlift, Tuareg boy. – London : Macdonald, 1984
47p : ill(some col.)
0–356–11153–9 L.42/3135
3023362

AFRICA
West Africa. Savanna regions. Agriculture

KOWAL, J. M.
Agricultural ecology of Savanna. – Oxford : Clarendon Press, 1978
xi,403p : ill. – Bibliography: p.349–379. – Includes index
0–19–859462–3 SRL
1188887

AFRICA
West Africa. Savanna regions. Soils. Cultivation

JONES, M. J. (Michael John), 19––
Soils of the West African savanna. – Slough : Commonwealth Agricultural Bureaux, 1975
[15],246p : ill. – (Technical communications / Commonwealth Bureau of Soils ; no.55). – Bibliography: p.221–241. – Includes index
0–85198–348–0 Ac.3484.db[no.55]
0693162

AFRICA
West Africa. Savanna regions. Wildlife. Conservation – Conference proceedings

IBADAN/GAROUA INTERNATIONAL SYMPOSIUM ON WILDLIFE MANAGEMENT (Ibadan)
Wildlife management in savannah woodland. – London : Taylor and Francis, 1979
xv,237p : ill. – Includes bibliographies and index
0–85066–175–7 SRL
1810715

AFRICA
West Africa. Sex roles

FEMALE and male in West Africa / edited by Christine Oppong. – London : Allen & Unwin, 1983
xix,402p : ill. – Bibliography: p374–391. – Includes index
0–04–301158–6 X.329/16801
2953595

AFRICA
West Africa. Slave trade, to date

LOTH, Heinrich
Das Sklavenschiff. – Berlin : Union, 1981
346p,[48]p of plates : ill. – Bibliography
r6301931 X.808/35425
8579180

AFRICA
West Africa. Slave trade. Bosman, Willem – Biographies

HEUVEL, Aad van den
Willem Bosman in goud en slaven. – Amsterdam : Meulenhoff, [1981]
159p : ill
90–290–1467–9 X.525/7135
8926387

AFRICA
West Africa. Slave–trade with America, to 1885

SAVIGNY, Carl Werner von
Schwarze Handelsware. – Straubing : Donau-Verlag, 1980
259p. – (Münchener Hochschulschriften. Reihe : Philosophie und Geisteswissenschaften ; Bd.2). – Bibliography: p.235–259
3–921570–11–5 X.808/35289
8787255

AFRICA
West Africa. Slave trade with America. Role of Portugal, 1500–1900

CARREIRA, António, Writer on Portuguese-speaking Africa
Notas sobre o tráfico português de escravos. – Lisboa : Universidade Nova de Lisboa, 1978
82p.,7p.of plates : map. – (Ciências humanas e sociais ; n.6 : Série Investigação). – Bibl
r6101776 X.709/50158
2431386

AFRICA
West Africa. Slave trade with Brazil, 1786–1832

CARREIRA, António, Writer on Portuguese-speaking Africa
O tráfico português de escravos na costa oriental africana nos começos do século XIX. – Lisboa : Junta de Investigações Científicas do Ultramar, Centro de Estudos de Antropologia Cultural, 1979
135p.,1 fold.leaf. – (Estudos de antropologia cultural ; n.12)
r6101891 X.0700/608(12)
2215845

AFRICA
West Africa. Slavery

GRACE, John, 1936–
Domestic slavery in West Africa, with particular reference to the Sierre Leone Protectorate, 1896–1927. – London : Muller, 1975
ix,294p. – Bibliography: p.263–275. – Includes index
0–584–10146–5 X.520/10166
0377384

AFRICA
West Africa. Slavery – For Caribbean students

HALCROW, Elizabeth M.
Canes and chains. – Kingston, [Jamaica] ; London : Heinemann, 1982
viii,88p : ill. – (Heinemann CXC history). – Bibliography: p87–88
0–435–98223–0 X.809/52529
2757729

AFRICA
West Africa. Social conditions

GREEN, Malcolm
Through the year in West Africa. – London : Batsford Academic and Educational, 1982
72p : ill. – Bibliography: p70. – Includes index
0–7134–3964–5 X.800/33800
1061089

PHILLIPS, Howard
Progress and change in West Africa. – London : Longman, 1976
49p : ill(chiefly col). – (Longman revised colour geographies). – Bibliography: p.49. – Lists of filmstrips: p.49
0–582–20126–8 X.709/31057
1061089

AFRICA
West Africa. Social conditions, to 1967

LIFE before the drought / edited by Earl Scott. – Boston ; London : Allen & Unwin, 1984
xiii,196p : ill. – Includes bibliographies and index
0–04–910076–9 X.322/14007
3053687

AFRICA
West Africa. Social conditions, 1900–1970

KHRISTOV, Angel
Sot͡sialna struktura i sotsialno razvitie na zapadnoafrikanskite strani 1900–1970. – Sofii͡a : Izdatelstvo na Bŭlgarskata akademii͡a na naukite, 1972
107p. – (Afrikano-aziatski problemi ; 3)
r6044088 P.511/1200[3]
1430548

AFRICA
West Africa. Social conditions, 1960–

MARIKO, Kélétigui A.
Souvenirs de la boucle du Niger 1953–1976. – Dakar : Les Nouvelles éditions africaines, 1980
178p
2–7236–0214–1 X.529/48395
8840032

AFRICA
West Africa. Society – Serials

The WEST African Journal of Sociology and Political Science. – Vol.1, no.1– ; Oct. 1975–. – Exeter : 'West African Journal of Sociology and Political Science', 1975–
B76–18766 P.521/3017
0657387

AFRICA
West Africa. Student unions: West African Students Union

OLUSANYA, G. O.
The West African Students Union and the politics of decolonisation, 1925–1958. – Ibadan : Daystar Press, 1982
127p
978–12–2164–x X.529/62244
9095489

AFRICA
West Africa. Sufism. Cerno Bokar Saalif Taal – Biographies

BRENNER, Louis
West African Sufi. – London : C. Hurst, 1984
viii,215p : ill. – Bibliography: p203–207. – Includes index
0–905838–71–8 X.200/46529
3257320

AFRICA
West Africa. Teachers. Professional education. Teaching practice

OLAITAN, Samson O.
Principles of practice teaching. – Chichester : Wiley, c1981
165p : ill. – (Education in Africa). – Includes index
0–471–27805–x X.520/24235
2502828

AFRICA
West Africa. Teaching

ROBINSON, Adjai
Principles and practice of teaching. – London : Allen & Unwin, 1980
160p : ill. – Includes index
0–04–370098–5 X.529/47064
2130505

AFRICA
West Africa. Tropical regions. Population. Growth. Social aspects

POPULATION growth and socioeconomic change in West Africa / edited by John C. Caldwell with the collaboration of N.O. Addo ... [et al.]. – New York ; London : Columbia University Press for the Population Council, New York, 1975
xv,763p : ill. – Includes index
0–231–03732–5 X.525/2033
0583154

AFRICA
West Africa. Urbanisation. Social aspects

GUGLER, Josef
Urbanization and social change in West Africa. - Cambridge : Cambridge University Press, 1978
xvii,237p : ill. - (Urbanization in developing countries). - Bibliography: p.202-220. - Includes index
0-521-21348-7 X.800/27149
1579754

AFRICA
West Africa. Uthman dan Fodio - Biographies

BALLARD, Martin
Uthman dan Fodio. Commander of the Faithful. - London : Longman, 1977
viii,44p : ill. - (Makers of African history)
0-582-60317-x X.709/31854
1292358

AFRICA
West Africa. Warfare, to ca 1900

SMITH, Robert S.
Warfare and diplomacy in pre-colonial West Africa. - [London] : Methuen, 1976
vii,240p. - (Studies in African history ; 15). - Bibliography: p.225-235. - Includes index
0-416-55060-6 X.808/33109
079788x

AFRICA
West Africa. Women. Circumcision, to date

DINSLAGE, Sabine
Mädchenbeschneidung in Westafrika. - Hohenschäftlarn bei München : Renner, 1981
185p.,[1]leaf of plates. - (Kulturanthropologische Studien ; Bd.5). - Bibliography
3-87673-079-1 X.329/16121
8924090

AFRICA
West Africa. Women. Social conditions, to date

KLINGSHIRN, Agnes
Frauen und ländliche Entwicklung in Afrika. - Köln : Weltforum, 1982
xiv,334p. : ill. - (Forschungsberichte des Bundesministeriums für wirtschaftliche Zusammenarbeit ; Bd.32). - Includes bibliographies
3-8039-0240-1 X.529/65947
9085300

MARCINOWSKI, Bettina
Die Frau in Afrika. - Frankfurt am Main : Lang, 1982
246p.. - (Europäische Hochschulschriften. Reihe 13, Französische Sprache und Literatur ; Bd.77). - Includes bibliography
3-8204-7237-1 X.529/61212
9085300

AFRICA
West Africa. Yams. Production - Conference proceedings

YAMS = Ignames / edited by J. Miège and S.N. Lyonga. - Oxford : Clarendon, 1982
xviii,411p : ill. - Includes bibliographies and index
0-19-854557-6 SRL
2842211

AFRICA
West Africa. Yoruba, to 1893

SMITH, Robert, 1919-
Kingdoms of the Yoruba. - 2nd ed. - [London] : Methuen, 1976
xvi,266p. - (Studies in African history ; 2). - Bibliography: p.245-256. - Includes index
0-416-84710-2 X.808/33110
1031732

AFRICA
West Africa. Yoruba, to date

ATANDA, J. A. (Joseph Abedowale)
An introduction to Yoruba history. - Ibadan : Ibadan University Press, 1980
vii,77p,14p of plates,1leaf of plates
978-12-1061-3 X.809/49553
8579164

AFRICA
West Africa. Yoruba. Cultural processes

EADES, J. S.
The Yoruba today. - Cambridge : Cambridge University Press, 1980
xix,188p : ill. - (Changing cultures). - Bibliography: p.169-181. - Includes index
0-521-22656-2 X.809/45717
2180499

AFRICA
West Africa. Yoruba. Religious beliefs

BURT, Ben
The Yoruba and their gods. - London : British Museum, c1977
16p : ill. - (Discovering other cultures)
0-7141-0057-9 X.200/42489
3066576

AFRICA
West Africa. Yoruba. Religious beliefs & sacrifices

AWOLALU, J. Omosade
Yoruba beliefs and sacrificial rites. - London : Longman, 1979
xvi,203p,[8]p of plates : ill. - Includes index
0-582-64203-5 X.100/18765
2007568

AFRICA
White persons. Interpersonal relationships with negroes, to date

GÖRÖG-KARADY, Veronika
Noirs et Blancs. - Paris : Société d'Études Linguistiques et Anthropologiques de France, 1976
427p. - (Langues et civilisations à tradition orale ; 23). - Bibliography: p415-418
r6207133 X.520/26576
8925836

AFRICA
Women

MURRAY, Maggie
Our own freedom. - London : Sheba, 1981
112p : ill
0-907179-09-6 X.429/13213 Woolwich
2708396

AFRICA
Women, to date

BONNET, Doris
Africaines. - Paris : L'Harmattan, [1983]
[72]p : ill. - (Collection Cairn)
2-85802-271-2 X.520/37358
9085319

AFRICA
Women, to date - Bibliographies

ZANZIG, Heike
Die Rolle der Frau in Afrika. - Hamburg : Institut für Afrika-Kunde, 1984
131p.. - (Dokumentationsdienst Afrika. Reihe A ; 21)
3-922887-33-3 X.0525/312a(21)
9085319

AFRICA
Women. Fertility. Effects of breast feeding. Physiological aspects

MONDOT-BERNARD, Jacqueline M.
Relationships between fertility, child mortality and nutrition in Africa. - Paris : Development Centre of the Organisation for Economic Co-operation and Development ; [London] : [H.M.S.O.], 1977
105p : ill. - (Technical papers / Organisation for Economic Co-operation and Development. Development Centre). - Bibliography: p.97-105
B79-00361 SRL
1739778

AFRICA
Women. Mental disorders, to date

DORÈS, Maurice
La femme village. - Paris : Éditions l'Harmattan, [1981]
215p
2-85802-176-7 X.329/15198
8838968

AFRICA
Women. Reproductive system. Deliberate disfigurement - Bibliographies

CHARPENTIER, C-J
An annotated bibliography on female genital mutilations in Africa. - Lidköping : C-J Charpentier, 1977
14p
91-970174-1-8 X.319/9339
1501135

AFRICA
Women. Reproductive system. Mutilation

FEMALE circumcision, excision and infibulation / edited by Scilla McLean ; with contributions from Marie Assaad ... [et al.]. - London : Minority Rights Group, 1980
20p : ill. - (Report / Minority Rights Group ; no.47)
B81-05033 P.803/299(47)
2343959

SANDERSON, Lilian Passmore
Against the mutilation of women. - London : Ithaca, 1981
117p
0-903729-66-0 X.529/40925
2343959

AFRICA
Women. Reproductive system. Mutilation, to date

SAUREL, Renée
L'enterrée vive ; suivi de la Conférence de la mi-décennie de la femme à Copenhague. - Genève ; Paris : Éditions Slatkine, 1981
x,310p
2-05-100338-6 X.529/53205
8616027

AFRICA
Women. Social conditions, 1885-

DIX femmes d'Afrique en Europe / [entretiens recueillis par Bishikwabo N. Mukala]. - Paris : Silex éditions, 1981
69p
r6209989 X.529/56984
8616019

WOMEN in Africa / edited by Nancy J. Hafkin and Edna G. Bay. - Stanford : Stanford University Press, 1976
x,306p
0-8047-0906-8 X.809/50496
8616019

AFRICA
Women. Social conditions, ca 1900-1975

NJOKU, John E. Eberegbulam
The world of the African woman. - Metuchen ; London : Scarecrow, 1980
v,124p
0-8108-1350-5 X.529/39402
2344246

AFRICA
Women. Vulva. Circumcision

ABDALLA, Raqiya Haji Dualeh
Sisters in affliction. - London : Zed Press, 1982
ii,122p : 2ill. - (Women in the Third World). - Bibliography: p119-122
0-86232-093-3 X.529/56867
2957027

AFRICA
Working classes, 1960-

RABOCHIĬ klass stran Afriki / [M.I. Braginskiĭ (otvetstvennyĭ redaktor)]. - Moskva : Nauka, Glavnai͡a redakt͡sii͡a vostochnoĭ literatury, 1983
219 p.. - Includes bibliographical references
r6482608 X.529/63242
9112243

AFRICA
Working classes. Trade unionism - Conference proceedings

WORKERS, UNIONS AND DEVELOPMENT (Conference : 1973 : University of Toronto)
The development of an African working class. - London : Longman, 1975
X,330p. - Bibliography: p.317-323. - Includes index
0-582-64179-9 X.520/10589
0629839

AFRICA. Curriculum subjects. Middle schools & secondary schools. England
Teaching

AFRICAN studies / edited by Margaret Killingray. - London : Extramural Division, University of London School of Oriental and African Studies, 1979
[2],ii,81p. - Bibliography: p.50-68. - List of films: p.73-81
0-7286-0061-7 X.702/5963
1821075
- [Another copy] OPL

AFRICAN studies / edited by Margaret Killingray. - 2nd ed.. - London : Extramural Division, School of Oriental and African Studies, 1983
iii,87p
0-7286-0110-9 X.525/8719
1821075

AFRICA. Special subjects. African writers. Literatures
Sociological perspectives of role of women in African urban communities

LITTLE, Kenneth
The sociology of urban women's image in African literature. - London : Macmillan, 1980
x,174p. - Bibliography: p159-161. - Includes index
0-333-28845-9 X.989/89984
2018667

AFRICA. Special subjects. Children's stories in English

LASSAM, Christine
Stories and settings, Africa. - London : National Book League, 1977
20p
0-85353-270-2 X.909/42292
1400991

AFRICA. Special subjects. English & West African writers, 1900–. Fiction in English
West Africa. European & African stereotypes

MILBURY-STEEN, Sarah L.
European and African stereotypes in twentieth-century fiction. – London : Macmillan, 1980
xiii,188p. – Bibliography: p178–183. – Includes index
0–333–29143–3 X.981/22632
2079968

AFRICA. Special subjects. English literature – For African students

AFRICAN literary appreciation / [compiled by] K.M. Lillis. – London : University of London Press, 1975
[12],132p. – Bibliography: p.132
0–340–17259–2 X.909/40062
0360686

AFRICA. Special subjects. English literature, 1837–1945

ZINS, Henryk
Joseph Conrad and Africa. – Nairobi : Kenya Literature Bureau, 1982
176p
r6811446 X.958/21813
9053506

AFRICA. Special subjects. Epic poetry in Mandinka
West Africa. Kaabu & Fuladu – Anthologies – Translations

KAABU and Fuladu / [transcribed by Bakari Sidibe] ; [translated and edited by Gordon Innes]. – London : University of London School of Oriental and African Studies, 1976
[7],310p. – Bibliography: p.309–310
0–7286–0016–1 X.900/25152
0670901

AFRICA. Special subjects. Fiction in English, to date

GRANQVIST, Raoul
Stereotypes in Western fiction on Africa. – Umea : Umea University, Department of English, 1984
39p. – (Umea papers in English ; no.7)
r6611490 X.0950/590(7)
9097414

AFRICA. Special subjects. French & African writers, 1900–. Fiction in French
European & African stereotypes

MILBURY-STEEN, Sarah L.
European and African stereotypes in twentieth-century fiction. – London : Macmillan, 1980
xiii,188p. – Bibliography: p178–183. – Includes index
0–333–29143–3 X.981/22632
2079976

AFRICA. Special subjects. Lessing, Doris. Fiction in English

THORPE, Michael
Doris Lessing's Africa. – London : Evans Bros, 1978
x,117p. – (Modern African writers). – Bibliography: p.105–111. – Includes index
0–237–49976–2 X.909/42739
1565710

AFRICA. Special subjects. Shepherd, David. English paintings
Animals

SHEPHERD, David, 1931–
Paintings of Africa and India. – London : Tryon Gallery, c1978
[75]p : col.ill
0–902189–08–5 C.135.K.12
2556359

AFRICA. Special themes. African writers, 1945–. Fiction in English
Society

EJRNÆS, Anne Marie
Afrika – litteratur og samfund. – København : Medusa, 1977
260p. – (Kritiske studier). – Includes bibliographies
87–7332–013–7 X.908/82215
1430807

AFRICA. Special themes. African writers, 1945–. Fiction in French
Society

EJRNÆS, Anne Marie
Afrika – litteratur og samfund. – København : Medusa, 1977
260p. – (Kritiske studier). – Includes bibliographies
87–7332–013–7 X.908/82215
1430815

AFRICA. Special themes. American negro writers, to 1976. English literature

BERGHAHN, Marion
Images of Africa in Black American literature. – London : Macmillan, 1977
x,230p. – Bibliography: p.224–228. – Includes index
0–333–22326–8 X.989/51968
126981x

AFRICA. Special themes. Classical literatures

HANSBERRY, William Leo
Africa and Africans as seen by classical writers. – Washington, D.C. : Howard University Press, 1977
xxii,162p. – (The William Leo Hansberry African history notebook ; vol.2)
r6812194 X.950/37482
9129332

AFRICA INLAND MISSION
See also
Africa Inland Mission International

AFRICA INLAND MISSION
to 1976

RICHARDSON, Kenneth
Garden of miracles. – [1st ed. reprinted]. – London : Africa Inland Mission, 1976
270p,[8]p of plates : ill
B78–24880 X.108/17049
1519727

AFRICA INLAND MISSION INTERNATIONAL
– Serials

AIM international. – Mar. 1984–. – London : The Mission, 1984–
v. : ill
B85–14351 H.85/548
3252434

AFRICA RESEARCH PROGRAMME

IFO-INSTITUT FÜR WIRTSCHAFTSFORSCHUNG
Developing countries : research and policies. – München : Ifo – Institut : Weltforum, 1974
vii, 117 p.. – (Ifo – Institut für Wirtschaftsforschung. Jorschungsberichte der Afrika – Studienstelle ; 41). – Bibliography : p. 105–117
r6002180 X.512/2427
0850128

AFRICA SOUTH OF THE SAHARA
See also
Books with Africa south of the Sahara imprints
Documentary films on animals in Africa South of the Sahara
Sub-Saharan

AFRICA SOUTH OF THE SAHARA
– Bibliographies

ERIKSEN, Tore Linné
Litteratur om Afrika söder om Sahara. – Uppsala : Nordiska afrikainstitutet, 1979
116p
91–7106–148–7 X.709/28812
2096579

AFRICA SOUTH OF THE SAHARA
– Directories

AFRICA south of the Sahara. – 1979–80: 9th ed. – London : Europa, 1979
xxiv, 1325p
0–905118–37–5 X.800/28856
0072060

AFRICA SOUTH OF THE SAHARA
– Serials

MOHLOMI. – vol. 1– ; 1976–. – Roma : The University, 1976–
r6051854 P.701/1047
1428837

AFRICA SOUTH OF THE SAHARA
to ca 1600

WILSON, Derek Alan
Early Africa. – London : Evans Bros, 1978
48p : ill. – (Knowing world history). – Bibliography: p.48
0–237–29188–6 X.700/25659
1435434

AFRICA SOUTH OF THE SAHARA
to ca 1900 – Serials

AFRIKA zamani. – no. 1–. – Yaoundé : [s.n.], 1973–
r6012491 P.701/940
0851248

AFRICA SOUTH OF THE SAHARA
to date

ADIKO, Assoi
Histoire des peuples noirs. – éd. revue, corrigée et augmentée. – Abidjan : Centre d'édition et de diffusion africaines, [1979]
192p : ill. – (Collection d'histoire à l'usage des écoles d'Afrique)
2–218–00910–2 X.809/56194
8995516

AFRICA SOUTH OF THE SAHARA
1450–

GREINDL, Léopold
L'Afrique noire au XIX siècle. – Kinshasa : Centre de recherches pédagogiques, [c1980]
46p : ill
r6210270 X.809/55957
8652120

MUTIBWA, Phares Mukasa
African heritage and the new Africa. – Nairobi : East African Literature Bureau, 1977
xi,221p
r6806307 X.809/56093
8652120

UMOZURIKE, U. O.
International law and colonialism in Africa. – Enugu : Nwamife, [1979]
x,173p. – Bibliography
978–12–4034–2 X.200/36116
8652120

AFRICA SOUTH OF THE SAHARA
ca 1800–ca 1900

MURRAY-BROWN, Jeremy
Faith and the flag. – London : Allen and Unwin, 1977
3–238p : ill. – Includes index
0–04–960008–7 X.800/27073
1551108

AFRICA SOUTH OF THE SAHARA
1800–1900. Historical sources

ROBINSON, David, 1938–
Sources of the African past. – London : Heinemann, 1979
xiv,203p : ill. – Includes bibliographies and index
0–435–94248–4 X.800/31676
2575574

AFRICA SOUTH OF THE SAHARA
1800–1979

HULL, Richard William
Modern Africa. – Englewood Cliffs ; London : Prentice-Hall, 1980
xii,274p : ill. – Bibliography: p.261–270. – Includes index
0–13–586305–8 X.709/51307
2089440

AFRICA SOUTH OF THE SAHARA
1885–1960

ÉTUDES africaines offertes à Henri Brunschwig. – Paris : École des Hautes Études en Sciences sociales, 1982
xxvii,426p. – Includes bibliography of the works of Henri Brunschwig
2–7132–0788–6 X.800/40353
8652112

AFRICA SOUTH OF THE SAHARA
1960–

DECRAENE, Philippe
Vieille Afrique, jeunes nations. – Paris : Presses universitaires de France, 1982
301p. – (Perspectives internationales). – Includes bibliography
2–13–037189–2 X.809/56148
800420x

GLÉLÉ, Maurice Ahanhanzo
Religion, culture et politique en Afrique noire. – Paris : Economica : Présence africaine, [1981]
206p. – (Politique comparée). – Bibliography: p195–204
2–7178–0291–6 X.520/27030
800420X

AFRICA SOUTH OF THE SAHARA
1960– – Serials

TOGO-DIALOGUE. – no.1 (jan. 1975)–. – Lomé, 1975–
Holdings: no.3 (mars 1975)–
r6209682 P.803/1159
800420x

AFRICA SOUTH OF THE SAHARA
Agriculture. Role of multinational companies

DINHAM, Barbara
Agribusiness in Africa. – London : Earth Resources Research, [1983]
224p b ill. – Includes index
0–946281–00–9 X.529/69516
328087x

AFRICA SOUTH OF THE SAHARA
American negro Protestant missions, to 1960

BLACK Americans and the missionary movement in Africa / edited by Sylvia M. Jacobs. – Westport, Conn. ; London : Greenwood Press, 1982
xii,255p. – (Contributions in Afro-American and African studies ; no.66). – Includes index
0–313–23280–6 X.809/55236
2907186

AFRICA SOUTH OF THE SAHARA
Christian church, 1885–
CHRISTIANITY in independent Africa / edited by Edward Fasholé-Luke [et al.]. – Ibadan : Ibadan University Press, 1978
630p
978-12-1052-4 X.200/36710
8652139

DI GENNARO, Joseph
A la rencontre des Azande. – [Bologna] : [E.M.I.], [1980]
95p : ill. – Includes bibliography
r6298116 X.208/10475
8652139

POBEE, J. S. (John S.)
Toward an African theology. – Nashville : Abingdon, c1979
174p
0-687-42420-8 X.200/38618
8652139

AFRICA SOUTH OF THE SAHARA
Christian church, 1950–1975
HASTINGS, Adrian
A history of African Christianity, 1950–1975. – Cambridge : Cambridge University Press, 1979
xi,336p. – (African studies series ; 26). – Bibliography: p.303–321. – Includes index
0-521-22212-5 X.200/32051
1899333

AFRICA SOUTH OF THE SAHARA
Christian life. Marriage
MARRIAGE before and after. – Achimota : African Christian Press, 1982
xiv,79p : ill
r6812176 X.208/9749
9088512

AFRICA SOUTH OF THE SAHARA
Christian mission schools, ca 1950–ca 1960
AFRICAN reactions to missionary education / Edward H. Berman [editor]. – New York ; London : Teachers College Press, [1976]
xix,231p,[3]p of plates. – (Publications / Columbia University. Center for Education in Africa). – Includes index
0-8077-2445-9 X.100/16351
0655546

AFRICA SOUTH OF THE SAHARA
Christian missions
KENDALL, R. Elliott
The end of an era. – London : S.P.C.K., 1978
vii,197p. – Bibliography: p.187–192. – Includes index
0-281-02989-x X.100/17889
1586572

AFRICA SOUTH OF THE SAHARA
Christian missions. Moffat, Mary – Biographies
DICKSON, Mora
Beloved partner. – London : Dobson, 1976
[8],244,[2]p : ill. – Bibliography: p.225 – Includes index
0-234-72015-8 X.809/41501
0105228

AFRICA SOUTH OF THE SAHARA
Christian sects
AFRICAN Christianity / edited by George Bond, Walton Johnson, Sheila S. Walker. – New York ; London : Academic Press, 1979
xvi,175p. – (Studies in anthropology). – Includes bibliographies and index
0-12-113450-4 X.200/33037
214316x

AFRICA SOUTH OF THE SAHARA
Coffee industries, to date
DUSABEYEZU
La place du café dans l'économie africaine. – 2e éd. – Kinshasa : Presses universitaires du Zaire, 1978
198p. – Bibliography: p185–189
r6207138 X.520/26413
8807086

AFRICA SOUTH OF THE SAHARA
Colonies. Negroes. Economic conditions, ca 1880–ca 1930
NZULA, A. T.
Forced labour in colonial Africa. – London : Zed Press, 1979
xi,218p. – Includes bibliographies
0-905762-30-4 X.529/33508
1908758

AFRICA SOUTH OF THE SAHARA
Commonwealth countries. Young persons. Employment
MBITHI, Philip M.
Youth employment problems. – London : Commonwealth Secretariat, 1975
[4],27p. – (Occasional papers : employment series / Commonwealth Youth Programme ; 2)
0-85092-101-5 BS.12/102
0674567

AFRICA SOUTH OF THE SAHARA
Communication systems, to date
THIAM, Thierno-Djibi
Les flux de l'information sud-sud en Afrique noire. – Fribourg : Éditions Universitaires, 1982
178p. – (Communication Sociale ; v.17). – Includes bibliography
2-8271-0227-7 X.622/21833
9165312

AFRICA SOUTH OF THE SAHARA
Constitutions, 1960–
REYNTJENS, Filip
Off the horseback?. – Baden-Baden : Nomos, 1980
vi,85p. – (Verfassung und Recht in Übersee. Beihefte ; Heft 10)
3-7890-0606-8 X.800/30792
8751897

AFRICA SOUTH OF THE SAHARA
Copper, to ca 1880
HERBERT, Eugenia W.
Red gold of Africa. – Madison ; London : University of Wisconsin Press, 1984
xiii,413p,[4]p of plates : ill(some col.). – Bibliography: p373–391. – Includes index
0-299-09600-9 X.622/22662
3255697

AFRICA SOUTH OF THE SAHARA
Criminal law, 1885–
BRILLON, Yves
Ethnocriminologie de l'Afrique noire. – Paris : Librarie philosophique J. Vrin ; Montréal : Les Presses de l'Université de Montréal, 1980
368p : ill. – (Bibliothèque criminologique). – Bibliography: p346–362
2-7606-0508-6 X.208/6772
8768846

AFRICA SOUTH OF THE SAHARA
Cultural processes
AYISI, Eric O.
An introduction to the study of African culture. – 2nd ed. – London : Heinemann Educational, 1979
xxv,125p. – Bibliography: p.115. – Includes index
0-435-89051-4 X.519/29189
1803204

SCHNEIDER, Harold K.
The Africans. – Englewood Cliffs ; London : Prentice-Hall, c1981
x,278p : ill. – (Prentice-Hall series in anthropology). – Bibliography: p255–264. – Includes index
0-13-018648-1 X.809/47360
1803204

AFRICA SOUTH OF THE SAHARA
Cultural processes, to date
GLÉLÉ, Maurice Ahanhanzo
Religion, culture et politique en Afrique noire. – Paris : Economica : Présence africaine, [1981]
206p. – (Politique comparée). – Bibliography: p195–204
2-7178-0291-6 X.520/27030
8652155

JUNGRAITHMAYR, Herrmann
Gedächtniskultur und Schriftlichkeit in Afrika. – Wiesbaden : Steiner, 1981
25p. – (Sitzungsberichte der Wissenschaftlichen Gesellschaft an der Johann Wolfgang Goethe-Universität ; Bd.17, Nr.5)
3-515-03569-9 Ac.548c/3[Bd.17,no.5]
8652155

AFRICA SOUTH OF THE SAHARA
Description & travel
COLE, Mary
Dirtroads. – Dublin : Gill and Macmillan, 1975
[7],246p,[8]p of plates : ill
0-7171-0770-1 X.809/40472
0147508

AFRICA SOUTH OF THE SAHARA
Description & travel, 1331–1354
IBN BAṬŪṬAH, Muḥammad ibn 'Abd Allāh
[Tuḥfat al-nuẓẓār fī gharā'ib al-amṣār wa-ajā'ib al-asfār. English. Selections]. Ibn Battuta in Black Africa. – London : Collings, 1975
[4],99p. – Bibliography: p.90–94. – Includes index
0-901720-57-7 X.708/41188
0504882

AFRICA SOUTH OF THE SAHARA
Description & travel, 1960–
BALEINE, Philippe de
Le petit train de la brousse. – Paris : Plon, 1982
232p
2-259-00879-8 X.809/55884
8004293

DÉCOUVERTE de l'Afrique noire / [Bernard Nantet ... et al.]. – Paris : Librairie Larousse, 1981
239p : ill. – (Monde et voyages)
2-03-513183-9 X.805/2653
8004293

THIEULOY, Jack
Le continent maudit. – Paris : Presses de la Renaissance, 1982
352p
2-85616-225-8 X.800/38056
8004293

AFRICA SOUTH OF THE SAHARA
Description & travel, 1963–1972
MORAVIA, Alberto
Which tribe do you belong to?. – St Albans : Panther, 1976
222p
0-586-04162-1 X.709/30788
0144681

AFRICA SOUTH OF THE SAHARA
Description & travel, 1975–1978
DODWELL, Christina
Travels with fortune. – London : W.H. Allen, 1979
x,316p,[8]p of plates : ill
0-491-02407-x X.809/44498
1913093

Travels with fortune. – London : Arrow, 1981, c1979
315p
0-09-924640-6 X.808/33941
1913093

AFRICA SOUTH OF THE SAHARA
Détente with South Africa
BALLINGER, Ronald
Détente in Southern Africa, two views. – Braamfontein : South African Institute of International Affairs, 1976
[3],29p
0-909239-18-5 X.702/5781
1721445

AFRICA SOUTH OF THE SAHARA
Developing countries. Agriculture. Policies of governments
BADU-DANQUAH, M. E.
Agricultural policy in a developing country. – Tema : Ghana Publishing, 1973
64p
r6091860 X.319/18857
1891995

AFRICA SOUTH OF THE SAHARA
Economic conditions, 1800–1960
MUNRO, John Forbes, 1940–
Africa and the international economy, 1800–1960. – London : Dent [etc.], 1976
230p : ill. – Bibliography: p.223–225. – Includes index
0-460-10088-2 X.529/31078
0761907

AFRICA SOUTH OF THE SAHARA
Economic conditions, 1960–
GRELLET, Gérard
Les structures économiques de l'Afrique noire. – Paris : I.E.D.E.S., 1982
179p. – (Tiers-monde)
2-13-037586-3 X.520/33408
8771464

HANCE, William A.
Black Africa develops. – Waltham, Mass. : Crossroads Press, c1977
xiii,158p,[24]p of plates. – (The Basic Africana library). – Includes bibliography
r6808156 X.800/35888
8771464

AFRICA SOUTH OF THE SAHARA
Economic conditions, 1960—continued

KANINDA, Kasonga Citenga
Regroupements régionaux et développement économique en Afrique noire. - Lubumbashi : Éditions du Centre d'Études Socio-politiques pour l'Afrique Centrale, 1979
55p. - (Travaux et recherches / Université nationale du Zaïre)
r6210773 X.529/61491
8771464

AFRICA SOUTH OF THE SAHARA
Economic conditions, 1960- - Conference proceedings

CONGRÈS INTERNATIONAL DES ÉTUDES AFRICAINES (4th : 1978 : Kinshasa)
La dépendance de l'Afrique et les moyens d'y remédier. - Paris : Agence de coopération culturelle et technique : Berger-Levrault, 1980
792p : ill
2-7013-0347-8 X.520/25202
8771464

AFRICA SOUTH OF THE SAHARA
Economic conditions, 1960- - Statistics

SÉRIES statistiques sur les économies africaines et Malgache [sic]. - [Paris] : Ministère des relations extérieurs, 1982
2v.
2-11-084480-9 S.E.400/1187
8771464

AFRICA SOUTH OF THE SAHARA
Economic conditions. Geographical aspects

AFRICAN perspectives / assembled and organized by Harm de Blij, Esmond Martin ; assisted by Ali Memon. - New York ; London : Methuen, c1981
xxiii,264p : ill. - Includes bibliographies and index
0-416-60231-2 X.520/26592
270336x

AFRICA SOUTH OF THE SAHARA
Economic conditions. Political aspects

GANNON, Edmund J.
Sub-Saharan Africa. - Washington, D.C. : Council on American Affairs, 1978
189p
r6189050 X.709/33781
0586706

The POLITICAL economy of Africa / edited by Richard Harris. - Cambridge, Mass. : Schenkman ; New York ; London : Distributed by Wiley, 1975
ix,270p
0-470-35420-8 X.520/10412
0586706

AFRICA SOUTH OF THE SAHARA
Economic cooperation with Arab countries, ca 1975-1980

SYLVESTER, Anthony
Arabs and Africans. - London : Bodley Head, 1981
251p,[8]p of plates : ill. - Includes index
0-370-30332-6 X.809/48492
2527138

AFRICA SOUTH OF THE SAHARA
Economic development

ACCELERATED development in sub-Saharan Africa / [report written by the African Strategy Review Group]. - Washington, D.C. : World Bank, 1981
viii,198p
r6865089 X.525/6200
2792370

AFRICA SOUTH OF THE SAHARA
Economic development. Role of peasants. Political aspects

HYDEN, Goran
Beyond Ujamaa in Tanzania. - London : Heinemann Educational, 1980
x,270p : 1 ill. - Includes index
0-435-96300-7 X.529/38537
2002264

AFRICA SOUTH OF THE SAHARA
Economic policies. Decision making

ROTHCHILD, Donald
Scarcity, choice and public policy in middle Africa. - Berkeley ; London : University of California Press, 1978
viii,357p : ill. - Includes index
0-520-03378-7 X.520/12848
1577123

AFRICA SOUTH OF THE SAHARA
Education

BLAKEMORE, Kenneth
A sociology of education for Africa. - London : Allen & Unwin, 1980 c1981
274p. - Bibliography: p262-270. - Includes index
0-04-370105-1 X.529/40922
2390418

ERNY, Pierre
The child and his environment in black Africa. - Nairobi ; Oxford : Oxford University Press, 1981
xxii,230p. - Bibliography: p206-223. - Includes index
0-19-572507-7 X.529/59368
2390418

AFRICA SOUTH OF THE SAHARA
Education related to politics

MORRISON, David Robert
Education and politics in Africa. - London : C. Hurst, 1976
352p. - Bibliography: 326-339. - Includes index
0-903983-21-4 X.520/10466
0598062

AFRICA SOUTH OF THE SAHARA
Effects of European imperialism, 1870-1976

WILSON, Henry Summerville
The imperial experience in sub-Saharan Africa since 1870. - Minneapolis : University of Minnesota Press[etc.], 1977
xv,415p. - (Europe and the world in the age of expansion ; vol.8). - Bibliography : p.369-393.-Includes index
0-19-690423-4 X.800/27785
1883038

AFRICA SOUTH OF THE SAHARA
Elites - Conference proceedings

INTERNATIONAL AFRICAN SEMINAR (6th : 1964 : Ibadan)
The new elites of tropical Africa. - London : Oxford University Press for the International African Institute, 1977
x,390p. - Bibliography: p.62-65. - Includes index
0-19-724203-0 X.519/29273
1808001

AFRICA SOUTH OF THE SAHARA
Employment. Labour, 1897-1972

AFRICAN labor history / Peter C.W. Gutkind, Robin Cohen, and Jean Copans, editors. - Beverly Hills ; London : Sage Publications, 1978
280p. - (Sage series on African modernization and development ; vol.2). - Includes bibliographies
0-8039-1063-0 X.519/29357
1824899

AFRICA SOUTH OF THE SAHARA
English speaking countries. Primary schools. Curriculum subjects: Social studies - For teaching

THOMPSON, Fred, 1923-
Social studies. - London : Longman, 1977
vi,198p : ill. - Bibliography: p.163-166
0-582-60761-2 X.519/26893
1114948

AFRICA SOUTH OF THE SAHARA
English speaking countries. Primary schools. Teachers. In-service training

The IN-SERVICE training of primary school teachers in English-speaking Africa / edited by Jeremy Greenland ; from country profiles and case studies written by Ray Molomo ... [et al.]. - London : Macmillan, 1983
x,324p : ill. - Includes index
0-333-36298-5 X.529/62682
3138534

AFRICA SOUTH OF THE SAHARA
Ethnic groups. National integration

The SEARCH for national integration in Africa / edited by David R. Smock and Kwamena Bentsi-Enchill. - New York : Free Press ; London : Collier Macmillan, 1976
xii,340p : 1 ill. - Includes index
0-02-929560-2 X.800/25608
0721301

AFRICA SOUTH OF THE SAHARA
Europeans. Social conditions, 1972-1977

WEST, Richard, 1930-
The white tribes revisited. - London : Private Eye Publications Ltd : Deutsch, 1978
[8],199p : ill
0-233-97045-2 X.709/32716
1757792

AFRICA SOUTH OF THE SAHARA
Exploration, 1866-1873

QUEST / edited by Sue Newson-Smith. - London : Arlington Books, 1978
267p : ill. - Bibliography: p.266-267
0-85140-277-1 X.800/27488
1689576

AFRICA SOUTH OF THE SAHARA
Exploration. Livingstone, David, 1813-1873

CESERANI, Gian Paolo
The travels of Livingstone. - Harmondsworth : Kestrel Books, 1979
[33]p : col ill
0-7226-5567-3 Cup.1285/409
0109606

DACHS, Anthony J.
Livingstone, missionary explorer. - Salisbury, Rhodesia : Central Africa Historical Association, 1974
[3],11p. - (Local series ; no.28)
r6072296 X.0705/92(28)
0621897

AFRICA SOUTH OF THE SAHARA
Exploration. Livingstone, David, 1813-1873 - Biographies

DAVID Livingstone. - Basingstoke : Marshalls, 1982
92p. - (Heroes of the Cross)
0-551-00942-x X.990/23312
0109606

JEAL, Tim
Livingstone. - London : Futura Publications, 1975
511p : ill. - (An omega book). - Bibliography: p.472-476. - Includes index
0-86007-721-7 X.108/16216
0621897

LIVINGSTONE / edited by B.W. Lloyd. - Cape Town : Struik, 1973
99p.,13p. of plates
0-86977-027-6 X.200/11116
0021229

RANSFORD, Oliver
David Livingstone. - London : J. Murray, 1978
xi,332p,12p of plates : ill. - Bibliography: p.320-321. - Includes index
0-7195-3489-5 X.800/27111
0621897

AFRICA SOUTH OF THE SAHARA
Exploration. Livingstone, David, 1813-1873 & Stanley, H. M. - Bibliographies

CASADA, James A.
Dr David Livingstone and Sir Henry Morton Stanley. - New York ; London : Garland, 1976
[4],224p. - (Garland reference library of social science ; vol.21). - Includes index
0-8240-9967-2 X.809/44555
1921428

AFRICA SOUTH OF THE SAHARA
Exploration. Livingstone, David, 1813-1873. Documents - Catalogues

CLENDENNEN, G. W.
David Livingstone. - Edinburgh : National Library of Scotland for the David Livingstone Documentation Project, 1979
iii-xxiv,348p
0-902220-22-5 X.702/6439
2275694

AFRICA SOUTH OF THE SAHARA
Exploration. Stanley, H. M.

TURNBULL, Patrick
Search for Livingstone. - London : Macdonald Educational, 1978
31p : ill(chiefly col). - (Macdonald adventures)
0-356-05929-4 X.990/10264 Woolwich
1572008

AFRICA SOUTH OF THE SAHARA
Families

KAYONGO-MALE, Diane
The sociology of the African family. - London : Longman, 1984
124p : ill. - Bibliography: p110-120. - Includes index
0-582-64703-7 X.529/68144
3253988

AFRICA SOUTH OF THE SAHARA
Federalism

DIOP, Anta, Cheikh
Les fondements économiques et culturels d'un état fédéral d'Afrique Noire. - Édition revue et corrigée. - Paris : Présence africaine, c1974
124 p.
r6019323 X.519/20498
094050X

AFRICA SOUTH OF THE SAHARA
Foreign policies, 1960-

NGANGA-MUNGWA
Essai de politique extérieure états africains (1960-1970). - [Kinshasa] : [Nganga-Mungwa], [1975?]
184leaves
Includes MS. emendations
r6210292 X.805/4418
8995532

AFRICA SOUTH OF THE SAHARA
Foreign relations with Arab countries, 1973–1975 – Bibliographies

ZOGHBY, Samir M.
Arab–African relations, 1973–1975 : a guide. – Washington, D.C. : Govt Print. Off. [for] Library of Congress, 1976
iii,26p. – (Maktaba Afrikana series)
0–8444–0197–8 OPL
1539604
– [Another copy] AS.287/90

AFRICA SOUTH OF THE SAHARA
Foreign relations with Egypt, 1885–

SAWANT, Ankush B
Egypt's Africa policy. – New Delhi : National, 1981
xiv,285p. – Bibliography: p.263–278
r6804967 X.529/56167
8977011

AFRICA SOUTH OF THE SAHARA
Foreign relations with France, 1885–

La POLITIQUE africaine du général de Gaulle (1958–1969). – Paris : Éditions A. Pedone, [1981]
xiv,421p. – (Bibliothèque / Institut d'études politiques de Bordeaux Centre d'étude d'Afrique novre ; série Afrique noire ; 10)
2–233–00096–x X.800/34965
8929378

AFRICA SOUTH OF THE SAHARA
Foreign relations with Israel, 1973–1975 – Bibliographies

ZOGHBY, Samir M.
Arab–African relations, 1973–1975 : a guide. – Washington, D.C. : Govt Print. Off. [for] Library of Congress, 1976
iii,26p. – (Maktaba Afrikana series)
0–8444–0197–8 OPL
1539612
– [Another copy] AS.287/90

AFRICA SOUTH OF THE SAHARA
Foreign relations with South Africa, ca 1965–1975

TOWARDS dialogue and détente / edited by F. R. Metrowich. – Sandton : Valiant, 1975
vi, 139 p.
0–86884–003–3 X.809/41773
1080725

AFRICA SOUTH OF THE SAHARA
Foreign relations with Soviet Union, to 1939

WILSON, Edward Thomas
Russia and Black Africa before World War II. – New York ; London : Holmes and Meier Publishers Ltd, [1976]
xvi,397p. – Bibliography: p.375–388. – Includes index
0–8419–0109–0 X.800/25770
0767433

AFRICA SOUTH OF THE SAHARA
Former French colonies. Families. Law, 1960–1973. Effects of French colonial administration

KOUASSIGAN, Guy A.
Quelle est ma loi?. – Paris : Éditions A. Pedone, [1974]
311 p.. – (Collection du Centre de recherche, d'étude et de documentation sur les institutions et les législations africaines ; 11). – Bibliography: p. 296–306
2–233–00000–5 X.100/14852
0875635

AFRICA SOUTH OF THE SAHARA
French colonies. Decolonisation. Role of Gaulle, Charles de

WHITE, Dorothy Shipley
Black Africa and De Gaulle. – University Park, Pa. ; London : Pennsylvania State University Press, c1979
314p : ill. – Bibliography: p283–301. – Includes index
0–271–00214–x X.800/35325
2901528

AFRICA SOUTH OF THE SAHARA
French colonies. Negroes. Attitudes of French, 1450–1885

M'BOKOLO, Elikia
Noirs et Blancs en Afrique équatoriale. – Paris : Éditions de l'École des hautes études en sciences sociales, [1981]
302p,. – (Civilisations et sociétés ; 69)
2–7132–0776–2 X.0520/40(69)
865221x

AFRICA SOUTH OF THE SAHARA
French colonies. Negroes. Attitudes of French, 1530–1880

COHEN, William B.
The French encounter with Africans. – Bloomington ; London : Indiana University Press, 1980
xxii,360p : ill. – Includes index
0–253–34922–2 X.800/29108
2243768

AFRICA SOUTH OF THE SAHARA
French ex-colonies, to 1980

PASCAL, Roger
L'opium du tiers monde. – Paris : Nouvelles Éditions Latines, [1979]
185p
2–7233–0088–9 X.708/48998
2457008

AFRICA SOUTH OF THE SAHARA
French-speaking countries, 1960–

BENOIST, Joseph-Roger de
La balkanisation de l'Afrique occidentale française. – Dakar, [etc] : Les Nouvelles Éditions africaines, 1979
283p
2–7236–0457–8 X.800/33630
8756902

BIARNÈS, Pierre
L'Afrique aux Africains. – Paris : Armand Colin, [1980]
479p : ill
r6206547 X.809/49824
8756902

DE VILLIERS, C. F.
Francophone Africa. – Pretoria : Africa Institute of South Africa, 1974. – (Countries of Africa ; no.2)
0–7983–0041–8 X.805/2769
8756902

AFRICA SOUTH OF THE SAHARA
French-speaking countries. Christian church. Relations with state

SARTORIUS, Joachim
Staat und Kirchen im francophonen Schwarzafrika und auf Madagaskar. – München : Claudius, 1973
220 p.. – (Jus ecclesiasticum ; Bd 19). – Bibliography: p. 212–217
r6000482 X.100/14692
0934208

AFRICA SOUTH OF THE SAHARA
French-speaking countries. Cinema industries

POMMIER, Pierre
Cinéma et développement en Afrique Noire francophone. – Paris : Éditions A. Pedone, 1974
184p. – (Institut d'études politiques de Bordeaux. Centre d'étude d'Afrique noire. série Afrique noire ; 4)
2–233–00001–3 X.0700/855(4)
1241729

AFRICA SOUTH OF THE SAHARA
French-speaking countries. Employment. Labour. Law

KIRSCH, Martin
Le droit du travail africain. – éd. nouvelle. – Paris : Travail et Profession d'Outre-Mer, 1975–1976
2v.. – Includes bibliographies
r6041605 X.0100/731
1218964
– [Another copy.] X.510/10941

AFRICA SOUTH OF THE SAHARA
French-speaking countries. Monetary policies, to date

BOURDIN, Joël
Monnaie et politique monétaire dans les pays africains de la zone franc. – Dakar : Les Nouvelles éditions africaines, [1980]
64p. – (Collection du Centre international de formation de la profession bancaire)
2–7236–0202–8 X.525/7968
8999694

AFRICA SOUTH OF THE SAHARA
French-speaking countries. Publishing, to date

SIMON, Günter, 1929–
Der Buchmarkt im frankophonen Schwarzafrika. – Frankfurt / M. : Ausstellungs- und Messe GmbH des Börsenvereins des deutschen Buchhandels, [1982?]
50p. – (Buch international ; 1)
3–923352–00–x 2708.e.363
8968322

AFRICA SOUTH OF THE SAHARA
Historical sources: Oral traditions

The AFRICAN past speaks / edited by Joseph C. Miller. – Folkestone : Dawson, 1980
xii,284p : ill. – Includes index
0–7129–0951–6 X.989/89855
1967932

AFRICA SOUTH OF THE SAHARA
Holy Ghost Fathers missions – Serials

MISSIONWIDE. – [No.1]– ; Spring 1978–. – Bickley : Holy Ghost Fathers, [1978]– : ill
B78–30391 P.101/863
1569058

AFRICA SOUTH OF THE SAHARA
Industrial relations

INDUSTRIAL relations in Africa / edited by Ukandi G. Damachi, H. Dieter Seibel, Lester Trachtman. – London : Macmillan [for] International Institute for Labour Studies, 1979
xiii,373p : ill. – Includes bibliographies and index
0–333–26079–1 X.529/34212
1739220

AFRICA SOUTH OF THE SAHARA
Influence of developed countries

MARNHAM, Patrick
Fantastic invasion. – London : Cape, 1980
xiv,271p
0–224–01829–9 X.809/45726
2070782

[Fantastic invasion]. Dispatches from Africa. – [London] : Abacus, 1981, c1980
xii,241p
0–349–12280–6 X.529/44038
2070782

AFRICA SOUTH OF THE SAHARA
Influence of Scotsmen, ca 1700–ca 1980

HARGREAVES, John D. (John Desmond)
Aberdeenshire to Africa. – Aberdeen : Aberdeen University Press, 1981
viii,95p,[4]p of plates
0–08–025764–x X.809/60919
3174433

AFRICA SOUTH OF THE SAHARA
Initiation rites

OBELITALA, Alphonse
L'initiation en Afrique noire et en Grèce. – Brazzaville : P. Kivouvou Verlag, 1982
121p. – Includes bibliography
3–88827–005–7 X.809/62204
9141014

AFRICA SOUTH OF THE SAHARA
Internal migration, to date

DAVID, Philippe
Les Navétanes. – Dakar : Les Nouvelles Éditions Africaines, 1980
525p,plates
2–7236–0194–3 X.520/32842
9042520

AFRICA SOUTH OF THE SAHARA
Intervention by great powers – Conference proceedings

The INTERVENTIONS in Africa. – [Cheltenham] : British Atlantic Publications, 1978
15p
B80–40457 X.709/52035
2310325

AFRICA SOUTH OF THE SAHARA
Investment by foreign investors, 1960– – Bibliographies

HILBERT, Roger
Foreign direct investments and multinational corporations in Sub-Saharan Africa. – Frankfurt : Campus, 1980
xxxvi,699p. – (Berliner Studien zur internationalen Politik)
3–593–32793–7 X.520/24302
8722390

AFRICA SOUTH OF THE SAHARA
Islam, to date

NICOLAS, Guy
Dynamique de l'Islam au sud du Sahara. – Paris : Publications orientalistes de France, [1981]
335p. – Bibliography: p287–310
2–7169–0156–2 X.200/39666
8912653

AFRICA SOUTH OF THE SAHARA
Ivory trade, to 1973

WILSON, Derek
White gold. – London : Heinemann, 1976
[8],184p,[12]p of plates : ill
0–434–87227–x X.329/10495 Woolwich
0671495

AFRICA SOUTH OF THE SAHARA
Kingdoms, to ca 1970

MAIR, Lucy
African kingdoms. – Oxford : Clarendon Press, 1977
[6],151p. – Bibliography: p.141–143. – Includes index
0–19–821698–x X.809/42442
0799963

AFRICA SOUTH OF THE SAHARA
Kinship

Les DOMAINES de la parenté / sous la direction de Marc Augé. - Paris : François Maspero, 1975
139 p. : ill. - (Dossiers africains). - Bibliography : p. 119-132
2-7071-0757-3 X.709/28433
0912301

AFRICA SOUTH OF THE SAHARA
Land resources. Development - Conference proceedings

The DEVELOPMENT of land resources in East, Central and Southern Africa. - London : Commonwealth Association of Surveying and Land Economy, 1976
[2],134p
0-903577-11-9 X.313/1297 Woolwich
0697842

AFRICA SOUTH OF THE SAHARA
Land resources. Development, 1960- - Conference proceedings

ENJEUX fonciers en Afrique noire / études réunies et présentées par E. Le Bris, E. Le Roy, et F. Leimdorfer. - Bondy : ORSTOM, [1982]
425p. - (Hommes et sociétés)
2-86537-057-7 X.800/40046
8004498

AFRICA SOUTH OF THE SAHARA
Law

HOLLEMAN, J.F
Issues in African law. - Hague : Mouton, [1974]
vi,179p. - (Change and continuity in Africa)
90-279-7571-x X.200/10971
0242691

AFRICA SOUTH OF THE SAHARA
Law - Serials

PENANT. - Paris : Édiafric, La Documentation Africaine, 1976-
Holdings : année 85. no. 751, jan/mars 1976-
r6038011 P.201/308
0543721

AFRICA SOUTH OF THE SAHARA
Law, 1960-

OCRAN, Tawia Modibo
Law in aid of development. - Tema : Ghana Publishing Corporation, 1978
xii,247p. - Bibliography: p.235-247
r6804922 X.200/38292
8652279

AFRICA SOUTH OF THE SAHARA
Legal systems, to date - Conference proceedings

CONFERENCE ON THE INTEGRATION OF CUSTOMARY AND MODERN LEGAL SYSTEMS IN AFRICA (1964 : Ibadan)
Integration of customary and modern legal systems in Africa. - Ile-Ife : University of Ife Press, c1971
xxiv,461p
r6805050 X.200/37487
890278x

AFRICA SOUTH OF THE SAHARA
Librarianship. Cultural factors

AMADI, Adolphe O.
African libraries. - Metuchen, N.J. ; London : Scarecrow Press, 1981
xii,265p. - Bibliography: p247-254. - Includes index
0-8108-1409-9 2719.e.201
2529246

AFRICA SOUTH OF THE SAHARA
Libraries, to 1978

KAUNGAMNO, E. E.
Books build nations. - London : Transafrica Book Distributors [etc.].
Vol.1: Library services in West and East Africa. - 1979
xii,169p. - Includes index
0-86036-111-x 2719.dd.35
2048175

Vol.2: Library services in Tanzania. - 1979
xii,273p,[8]p of plates : ill. - Includes index
0-86036-112-8 2719.dd.34
2048175

AFRICA SOUTH OF THE SAHARA
Livestock production industries

LIVESTOCK development in Subsaharan Africa / edited by James R. Simpson and Phylo Evangelou. - Epping : Bowker, 1984
xvii,407p,[4]p of plates : ill. - Includes bibliographies and index
0-86531-803-4 X.329/19027
3220451

AFRICA SOUTH OF THE SAHARA
Livestock production industries, ca 1960-1977

KONCZACKI, Z. A.
The economics of pastoralism. - London : Cass, 1978
xi,185p : ill. - Includes index
0-7146-3086-1 X.529/21343
1398792

AFRICA SOUTH OF THE SAHARA
Man. Infertility. Social aspects

RETEL-LAURENTIN, A.
Infécondité en Afrique noire. - Paris : Masson & Cie, 1974
vi, 188 p. : ill. - Bibliography : p. 141-188
2-225-39314-1 X.320/4991 Woolwich
0920754

AFRICA SOUTH OF THE SAHARA
Marriage, to date

MULAZO GUA CIKALA MUSHARHAMINA
Mariage traditionnel africain et mariage chrétien. - Kinshasa : Éditions Saint Paul Afrique, [1981]
92p
r6209698 X.529/53266
8933901

AFRICA SOUTH OF THE SAHARA
Mass media. Influence of governments

WILCOX, Dennis L.
Mass media in black Africa. - New York ; London : Praeger ; [London] : [Distributed by Martin Robertson], 1975
xvi,171p. - (Praeger special studies in international politics and government). - Bibliography: p.145-164. - Includes index
0-275-05990-1 X.520/10818
068175x

AFRICA SOUTH OF THE SAHARA
Matabele. Christian missions, 1829-1860

MOFFAT, Robert, b.1795
The Matabele journals of Robert Moffat, 1829-1860. - Salisbury, Rhodesia : National Archives of Rhodesia. - (Oppenheimer series ; no.1, vol.1)
Vol 1. - 1976
[6],xv,382p,[4]leaves of plate(1 fold)
0-86923-103-0 X.802/3167(1a)
1699180

Vol.2. - 1976
[11],295p,[4]leaves of plates (1 fold). - (Oppenheimer series ; no.1 vol.2). - Bibliography : p.281-282. - Includes index
0-86923-104-9 X.802/3167(1a)
1699180

AFRICA SOUTH OF THE SAHARA
Mental health, to date

MENTAL health in Africa. - [Ibadan] : Ibadan University Press, 1982
xv,266p. - Includes bibliography
978-12-1132-6 X.329/17981
9102205

AFRICA SOUTH OF THE SAHARA
Middle schools & primary schools

DEL CORNÒ, Lucio
Esperienze pedagogiche in Africa. - Milano : Emme, 1976
247p : ill. - (Il puntoemme ; 51)
r6100146 X.519/41358
1863525

AFRICA SOUTH OF THE SAHARA
Military equipment trades: Weapons trades. International political aspects

ARMS for Africa / edited by Bruce E. Arlinghaus. - Lexington, Mass. : Lexington Books, 1983
xiv,233p : ill
0-669-05527-1 X.800/35844
2940612

AFRICA SOUTH OF THE SAHARA
Military governments, 1958-1978

KIRK-GREENE, A. H. M. (Anthony Hamilton Millard)
'Stay by your radios'. - Leiden : Afrika-Studiecentrum ; Cambridge : African Studies Centre, 1981
156p. - (African social research documents ; v.12)
90-7011-029-6 X.0311/143(12) Woolwich
2496216

AFRICA SOUTH OF THE SAHARA
Military governments, 1959-1975

DECALO, Samuel
Coups and army rule in Africa. - New Haven ; London : Yale University Press, 1976
x,284p. - Bibliography: p.255-268. - Includes index
0-300-01942-4 X.809/41303
0748846

AFRICA SOUTH OF THE SAHARA
Military operations by British military forces, 1885-1960

AFRICAN military accounts. - East Ardsley : E. P. Microform, [1978]
136microfiche
Wanting reports 7, 11 (fiche 1), 13, 20, 38
r6802066 SPRMic.E.295
8739242

AFRICA SOUTH OF THE SAHARA
Muslims. Education

SANTERRE, Renaud
Pédagogie musulmane d'Afrique Noire. - Montréal : Les Presses de l'Université de Montréal, 1973
174 p. ; 4 p. of plates
0-8405-0218-4 X.519/20082
0913367

AFRICA SOUTH OF THE SAHARA
Nationalism

NANGOLI, Petero
The African cause. - [Sheffield : [The author], 1980
ix,22p.
B80-38877 X.529/38039
2255928

AFRICA SOUTH OF THE SAHARA
Ophioninae

GAULD, I. D.
The taxonomy, distribution and host preferences of African parasitic wasps of the subfamily Ophioninae. - Slough : Commonwealth Agricultural Bureaux [for the] Commonwealth Institute of Entomology, 1978
[3],287p : ill. - Bibliography: p.162-165. - Includes index
0-85198-409-6 SRL
1824775

AFRICA SOUTH OF THE SAHARA
Peasant communities. Socioeconomic aspects

PEASANTS in Africa / Martin A. Klein, editor. - Beverly Hills ; London : Sage Publications, 1980
319p. - (Sage series on African modernization and development ; vol.4). - Includes bibliographies
0-8039-1406-7 X.700/26497
2283441

AFRICA SOUTH OF THE SAHARA
Personnel. Remuneration, 1960-

DUBOIS, Daniel
Les salaires en Afrique francophone au sud du Sahara. - Yaoundé : CRADAT, 1979
162p. - (Les manuels du Centre régional africain d'administration du travail)
r6206355 X.529/43145
8754012

AFRICA SOUTH OF THE SAHARA
Political development. Participation of ethnic minorities

KASFIR, Nelson
The shrinking political arena. - Berkeley ; London : University of California Press, 1976
xvi,323p. - Bibliography: p.291-309. - Includes index
0-520-02576-8 X.800/25691
0745669

AFRICA SOUTH OF THE SAHARA
Political events, to date

ANALYZING political change in Africa / edited by James R. Scarritt. - Boulder : Westview Press, 1980
xv,360p. - (Westview replica editions)
0-89158-275-4 X.800/31723
8652333

IBINGIRA, Grace Stuart
African upheavals since independence. - Boulder : Westview Press, 1980
xvii,349p. - (Westview special studies on Africa)
0-89158-585-0 X.800/33653
8652333

AFRICA SOUTH OF THE SAHARA
Political events, 1950-1979

SCOTT, David, 1919-
Ambassador in black and white. - London : Weidenfeld and Nicolson, 1981
xii,258p,[8]p of plates : ill. - Bibliography: p250-251. - Includes index
0-297-77865-x X.800/30522
2486202

AFRICA SOUTH OF THE SAHARA
Political events, 1976-1980

LAMB, David, 1940-
The Africans. - London : Bodley Head, 1983, c1982
xv,363p. - Bibliography: p349-351. - Includes index
0-370-30968-5 X.800/35850
2927306

AFRICA SOUTH OF THE SAHARA
Political events. Role of United States. Central Intelligence Agency, 1960–1979

DIRTY work 2 / edited by Ellen Ray [et al.]. - London Zed Press, 1980
xvii, 523p. - (Africa series). - Bibliography: p. 281–284
0–905762–81–9 X.700/26504
2277220

AFRICA SOUTH OF THE SAHARA
Politics

AUSTIN, Dennis
Politics in Africa. - 2nd ed. - Hanover, N.H. ; London : University Press of New England for University of Rhode Island, c1984
187p. - Bibliography: p177–182. - Includes index
0–87451–272–7 X.520/35980
0315109

SILVEIRA, Onésimo
Africa south of the Sahara. - Stockholm : Distributed by Rabén och Sjögren, 1976
212 p.. - (Publications of the Political Science Association in Uppsala ; 71). - Bibliography : p. 205–212
91–294–6797–7 Ac.1077.r [no.71]
0315109

AFRICA SOUTH OF THE SAHARA
Politics - Forecasts

AFRICA in the 1980s / Colin Legum ... [et al.]. - New York ; London : McGraw-Hill [for the] 1980s Project, Council on Foreign Relations, 1979
xv,232p : 1 ill. - (Studies / 1980s Project). - Bibliography: p.213–218. - Includes index
0–07–037081–8 X.800/28251
2002043

AFRICA SOUTH OF THE SAHARA
Politics - Serials

MAJI Maji. - no.1- : [1972?]-. - Dar es Salaam : University of Dar es Salaam, Tanu Youth League, [1972?]-
Holdings : no.13, Jan. 1974-
r6033301 P.705/143
0315109

SOUTH AFRICAN FREEDOM FOUNDATION
SAFF position paper. - Vol.1, no.1 (1977)-. - Sandton : the Foundation, 1977-
r6108907 P.521/2336
0315109

AFRICA SOUTH OF THE SAHARA
Politics, 1856–1970

IDEOLOGIES of liberation in black Africa, 1856–1970 / [compiled and introduced by] J. Ayo Langley. - London : Collings, 1979
[11],858 p
0–86036–039–3 X.800/28994
2078767

AFRICA SOUTH OF THE SAHARA
Politics, 1960-

DOSUMU-JOHNSON, T. O.
Reflections of an African nationalist. - New York : Vantage, c1980
301p
0–533–04144–9 X.809/57319
8652341

GANN, L. H. (Lewis Henry)
Africa south of the Sahara. - Stanford : Hoover Institution Press, c1981
xiv,114p. - (Hoover Press publication ; 238)
0–8179–7382–6 X.529/45057
8652341

KABONGO MAKANDA MWADIAMVITA
Armées et politique en Afrique au sud du Sahara. - Kinshasa : Presses Universitaires du Zaïre, 1979
209p
r6211646 X.800/38020
8652341

AFRICA SOUTH OF THE SAHARA
Politics related to education

MORRISON, David Robert
Education and politics in Africa. - London : C. Hurst, 1976
352p. - Bibliography: 326–339. - Includes index
0–903983–21–4 X.520/10466
0598070

AFRICA SOUTH OF THE SAHARA
Politics. Decision making

ROTHCHILD, Donald
Scarcity, choice and public policy in middle Africa. - Berkeley ; London : University of California Press, 1978
viii,357p : ill. - Includes index
0–520–03378–7 X.520/12848
1577131

AFRICA SOUTH OF THE SAHARA
Politics. Participation of university students

HANNA, William John
University students and African politics. - New York : Africana Publishing Co. ; London : Holmes and Meier Publisher, Ltd, 1975
xix,296p. - Includes index
0–8419–0145–7 X.800/11630
0708216

AFRICA SOUTH OF THE SAHARA
Population

POPULATION education source book for sub-saharan Africa / edited by Reuben K. Udo ... [et al.] ; commissioned by the African Social Studies Programme. - Nairobi ; London : Heinemann Educational, 1979[i.e.1980]
[6],311p : ill. - Includes index
0–435–95917–4 X.525/3250
218463x

AFRICA SOUTH OF THE SAHARA
Population, 1960- - Conference proceedings

SEMINAR ON MORAL AND RELIGIOUS ISSUES IN POPULATION DYNAMICS AND DEVELOPMENT (1974 : University of Ghana)
Religion, morality, and population dynamics. - Legon : Population Dynamics Programme, University of Ghana, 1977
xii,270p. - (University of Ghana population studies ; no.8)
r6807268 X.200/42253
865235x

AFRICA SOUTH OF THE SAHARA
Pre-school children. Education, 1960-

SCHULTZ, Magdalene
Frühkindliche Erziehung in Afrika südlich der Sahara. - Saarbrücken : Breitenbach, 1980
viii,307p. - (Sozialwissenschaftliche Studien zu internationalen Problemen ; Heft 54). - Bibliography: p273–307
3–88156–152–8 X.529/43072
8751811

AFRICA SOUTH OF THE SAHARA
Primary schools. Curriculum. Planning

HAWES, Hugh
Curriculum and reality in African primary schools. - [Harlow] : Longman, 1979
iv,220p : ill. - Bibliography: p.201–216. - Includes index
0–582–60769–8 X.510/16280
1918915

AFRICA SOUTH OF THE SAHARA
Prostitution, to date

STUDIES in prostitution (East, West and South Africa, Zaire and Nevada) / edited by Erasto Muga. - Nairobi : Kenya Literature Bureau, 1980
279p
r6804497 X.529/43309
8774269

AFRICA SOUTH OF THE SAHARA
Puppet theatre, to date

DARKOWSKA-NIDZGÓRSKA, Oleńka
Théâtre populaire de marionnettes en Afrique sud-saharienne. - Bandundu : Publication du Centre d'études ethnologiques, [1980]
259p : ill. - (Ceeba publications. serie 2, mémoires et monographies ; v.60). - Bibliography
r6205843 X.950/6738
8768919

AFRICA SOUTH OF THE SAHARA
Relations with Soviet Union

STEVENS, Christopher, 1948-
The Soviet Union and black Africa. - London : Macmillan, 1976
xiii,236p. - Bibliography: p.221–229. - Includes index
0–333–18052–6 X.809/40983
0664391

AFRICA SOUTH OF THE SAHARA
Religions - Conference proceedings

WORKSHOP IN RELIGIOUS RESEARCH (1967 : Nairobi)
African initiatives in religion. - Nairobi : East African Publishing House, 1971
[4], xviii, 288 p., [4] p. of plates : ill
r6031551 X.200/30847
1081578

AFRICA SOUTH OF THE SAHARA
Religions, to date

OMBOLO, Jean-Pierre
La genèse du fait religieux en Afrique noire. - Yaoundé : [s.n.], 1980
361p
r6205718 X.205/834
8004544

AFRICA SOUTH OF THE SAHARA
Rice. Production - Conference proceedings

RICE IN AFRICA (Conference : 1977 : International Institute of Tropical Agriculture)
Rice in Africa. - London : Academic Press, 1978
xvi,356p : ill. - Includes bibliographies and index
0–12–139350–x SRL
1796933

AFRICA SOUTH OF THE SAHARA
Rituals. Use of carved masks

TIÉROU, Alphonse
La vérité première du second visage africain .... - Paris : G.-P. Maisonneuve & Larose, 1975
135 p. ; 8 p. of plates
2–7068–0602–8 X.709/19860
0914967

AFRICA SOUTH OF THE SAHARA
Rural regions. Economic planning, ca 1960–1980

AGRARIAN policies and rural poverty in Africa / edited by Dharam Ghai and Samir Radwan. - Geneva : International Labour Office, 1983
ix,311p : ill. - (A WEP study)
92–210–3100–4 X.322/15140
3330400

AFRICA SOUTH OF THE SAHARA
Schools. Curriculum subjects: Visual arts. Teaching

TALABI, George
Art teaching in African schools. - Ibadan ; London : Heinemann Educational, 1979
xiv,109p : ill. - Bibliography: p.106. - Includes index
0–435–89850–7 X.419/34837 Woolwich
180863x

AFRICA SOUTH OF THE SAHARA
Serials, to date - Indexes

AFRICA south of the Sahara / compiled in the African Section, African and Middle Eastern Division, Library of Congress. - Boston, Mass. : G.K. Hall. - 1982-.
0–8161–0293–7 OPL
8926123

AFRICA SOUTH OF THE SAHARA
Slave trade, to ca 1870

HAIR, Paul Edward Hedley
The Atlantic slave trade and Black Africa. - London : Historical Association, 1978
36p. - (General series / Historical Association ; 93). - Bibliography: p.35–36
0–85278–196–2 WP.3175/93
1771116

AFRICA SOUTH OF THE SAHARA
Slavery, to 1976

SLAVERY in Africa / edited by Suzanne Miers and Igor Kopytoff. - Madison ; London : University of Wisconsin Press, 1977
xvii,474,[1]p. - Includes bibliographies and index
0–299–07330–0 X.520/12144
1297414
- [Another copy.] X.529/36701

AFRICA SOUTH OF THE SAHARA
Slavery. Economic aspects

ASIAN and African systems of slavery / edited by James L. Watson. - Oxford : Blackwell, 1980
vii,348p. - Bibliography: p.318–339. - Includes index
0–631–11011–9 X.809/45619
207656x

AFRICA SOUTH OF THE SAHARA
Social change. Effects of colonialism

COLONIALISM and change / edited (with introduction) by Maxwell Owusu. - The Hague : Mouton, 1975
264p,plate : 2 ill. - (Studies in anthropology ; 4). - Includes bibliographies and index
r6072327 X.0700/822(4)
1684965

AFRICA SOUTH OF THE SAHARA
Social change. Political aspects

SMITH, Anthony D. (Anthony David)
State and nation in the Third World. - Brighton : Wheatsheaf, 1983
171p. - Bibliography: p147–162. - Includes index
0–7108–0199–8 X.809/55227
2904829

AFRICA SOUTH OF THE SAHARA
Social conditions

SUB-SAHARAN Africa / edited by Chris Allen and Gavin Williams. - London : Macmillan, 1982
xxi,217p. - (Sociology of 'developing societies'). - Bibliography: p203–213. - Includes index
0–333–27674–4 X.529/53621
2890119

AFRICA SOUTH OF THE SAHARA
Social conditions - Conference proceedings

SYMPOSIUM 77 / Lumumba Society, National University of Lesotho. - Lesotho : Lumumba Society, National University of Lesotho, 1977
[3],ii,53p
r6032664 Cup.1281/763
1349929

AFRICA SOUTH OF THE SAHARA
Social conditions, to date

CROCE-SPINELLI, Michel
Les enfants de Poto-Poto. - Paris : Éditions L'Harmattan, 1982
363p
2-85802-220-8 X.529/57291
8652368

MATTHEY, Piero
Afrique tropicale. - Paris : Éditions Atlas, [1981]
120p : ill
2-7312-0087-1 X.805/4239
8652368

AFRICA SOUTH OF THE SAHARA
Social conditions, 1800-1983

FREUND, Bill
The making of contemporary Africa. - London : Macmillan, 1984
xv,357p,[8]p of plates : ill. - Bibliography: p289-338. - Includes index
0-333-29499-8 X.809/61185
3164837

AFRICA SOUTH OF THE SAHARA
Social conditions, 1960-1983

AFRICA in the post-decolonization era / edited by Richard E. Bissell, Michael S. Radu. - New Brunswick ; London : Transaction, c1984
x,263p : ill. - Includes index
0-87855-496-3 X.800/40649
3256693

AFRICA SOUTH OF THE SAHARA
Social conditions, 1960-

MUDIMBÉ, Valentin Yves
L'odeur du père. - Paris : Présence africaine, 1982
203p
2-7087-0398-6 X.529/56954
8995524

AFRICA SOUTH OF THE SAHARA
Social conditions, 1960- - Bibliographies

ANDOR, L. E.
Psychological and sociological studies of the black people of Africa, South of the Sahara 1960-1975. - Braamfontein : National Institute for Personnel Research, 1983
vi,397p
0-7988-2329-1 X.525/9383
8995524

AFRICA SOUTH OF THE SAHARA
Social conflict

BOZEMAN, Adda Bruemmer
Conflict in Africa. - Princeton ; Guildford : Princeton University Press, 1976
xiv,430p. - Bibliography: p.383-416. - Includes index
0-691-03104-5 X.800/11624
070590x

AFRICA SOUTH OF THE SAHARA
Social minorities - Conference proceedings

STRANGERS in African societies / edited by William A. Shack and Elliott P. Skinner ... - Berkeley ; London : University of California Press, 1979
x,325p : 1 ill. - Bibliography: p.289-302. - Includes index
0-520-03458-9 X.520/14510
1955241

AFRICA SOUTH OF THE SAHARA
Social security, 1960-1973

MOUTON, Pierre
La sécurité sociale en Afrique au sud du Sahara. - Genève : Bureau International du Travail, 1974
ix, 173 p.
92-220-1192-9 X.700/16630
092170X

AFRICA SOUTH OF THE SAHARA
Social structure. Analysis

MANGHEZI, Alpheus
Class, elite, and community in African development. - Uppsala : Scandinavian Institute of African Studies, 1976
118 p.. - Bibliography : p. 117-118
91-7106-099-5 X.519/26160
0734292

AFRICA SOUTH OF THE SAHARA
Society - Serials

POSITIVE review. - No.1- ; 1978-. - Ile-Ife : 'Positive Review' ; Oxford : Distributed by Hans Zell (Publishers) Ltd, 1978- : ill
B80-30839 P.2000/803
220424x

AFRICA SOUTH OF THE SAHARA
Society. Role of women

AFRICAN women south of the Sahara / edited by Margaret Jean Hay and Sharon Stichter. - London : Longman, 1984
xiv,225p : ill. - Bibliography: p195-211. - Includes index
0-582-64373-2 X.520/36484
3250156

AFRICA SOUTH OF THE SAHARA
Steam locomotives, 1968-1974

TALBOT, Edward
Steam from Kenya to the Cape. - [Harrow] : Continental Railway Circle ; Stafford : Halcyon Books, 1975
[6],86p : ill. - Bibliography: p.82-83
0-9503469-2-6 X.620/16030 Woolwich
0642614

AFRICA SOUTH OF THE SAHARA
Technological innovation. Effects of foreign investment

THOMAS, D. Babatunde
Importing technology into Africa. - New York ; London : Praeger ; [London] : [Distributed by Martin Robertson], 1976
xv,202,[1]p : ill. - (Praeger special studies in international economics and development). - Bibliography: p.188-191. - Includes index
0-275-05740-2 X.620/16304
0730416

AFRICA SOUTH OF THE SAHARA
Theatre, to date

VAZ, Carlos
Para um conhecimento do teatro africano. - Lisboa : Ulmeiro, 1978
204p : ill
r6704651 X.950/11790
8815666

AFRICA SOUTH OF THE SAHARA
Towns, to 1885

HULL, Richard William
African cities and towns before the European conquest. - New York : Norton, c1976
xxi,138p
0-393-05581-7 X.529/45061
8787751

AFRICA SOUTH OF THE SAHARA
Tribal societies. Communication

LOHISSE, Jean
La communication tribale. - Paris : Éditions Universitaires, 1974
215 p.. - (Encyclopédie universitaire). - Bibliography : p. 201-210
r6011034 X.510/9461
0857718

AFRICA SOUTH OF THE SAHARA
Tribal societies. Cultural processes

PEOPLES of Africa / edited by James Lowell Gibbs, Jr. - Abridged [ed.]. - New York ; London : Holt, Rinehart and Winston, 1978
xv,362p : ill. - Includes bibliographies and index
0-03-039371-x X.510/15829
1544950

AFRICA SOUTH OF THE SAHARA
Tribal societies. Cultural processes, to date

GHONEIM-GRAF, Ingrid
Möglichkeiten und Grenzen archäologischer Interpretation. - Frankfurt am Main : Lang, 1978
105p : ill. - (Arbeiten zur Urgeschichte des Menschen ; Bd.3)
3-261-02611-1 X.425/3462
8652422

IORDANSKIĬ, Vladimir Borisovich
Khaos i garmoniîa. - Moskva : Nauka, Glavnaîa redakfsiîa vostochnoĭ literatury, 1982
341, [2] p.. - Bibliography: p. 336-[342]
r6403520 X.529/62020
8652422

AFRICA SOUTH OF THE SAHARA
Tropical regions. Cities

O'CONNOR, Anthony
The African city. - London : Hutchinson, 1983
359p : ill. - (Hutchinson university library for Africa). - Bibliography: p327-351. - Includes index
0-09-152980-8 X.809/57368
3037444
- [Another Copy: paperback.] ISBN 0-09-152981-6 X.809/57369

AFRICA SOUTH OF THE SAHARA
Tropical regions. Flies. Species - Catalogues

CATALOGUE of the Diptera of the Afrotropical region / editor R.W. Crosskey ; assistant editors ... [others]. - London : British Museum (Natural History), 1980
437p. - (Publications / British Museum (Natural History) ; no.821). - Bibliography: p.889-1234. - Includes index
0-565-00821-8 SRL
225204x

AFRICA SOUTH OF THE SAHARA
Tropical regions. Population. Fertility

CHILD-SPACING in tropical Africa / edited by Hilary J Page, Ron Lesthaeghe. - London : Academic Press, 1981
xxiii,332p : ill. - (Studies in population). - Includes bibliographies and index
0-12-543620-3 X.520/23862
2402866

AFRICA SOUTH OF THE SAHARA
Universities

IKE, Vincent Chukwuemeka
University development in Africa. - Ibadan ; Oxford : Oxford University Press, 1976
xi,232p
0-19-575324-0 X.529/31959
1297406

AFRICA SOUTH OF THE SAHARA
Urban regions, to data - Bibliographies

O'CONNOR, Anthony M. (Anthony Michael)
Urbanization in tropical Africa. - Boston, Mass : G.K. Hall, [1981]
xxx,381p. - (Bibliographies and guides in African Studies)
0-8161-8262-0 X.800/30713
8715459

AFRICA SOUTH OF THE SAHARA
Urban regions. Social change

PEIL, Margaret
African urban society. - Chichester : Wiley, c1984
xi,392p,[8]p of plates : ill. - Bibliography: p351-377. - Includes index
0-471-90062-1 X.520/36162
3013154

AFRICA SOUTH OF THE SAHARA
Urbanisation, to date

HUSSON, B.
Éléments d'analyse urbaine. - [s.l.] : Secrétariat d'État à la promotion humaine, Centre national de Mbour, 1979
[64]leaves : ill
r6210579 SV.850/30
8652430

AFRICA SOUTH OF THE SAHARA
Watch Tower Bible and Tract Society. Persecution

HODGES, Tony
Jehovah's Witnesses in Central Africa. - London : Minority Rights Group, 1976
16p. - (Reports / Minority Rights Group ; no.29). - Bibliography: p.16
0-903114-31-3 P.803/299
0688797

AFRICA SOUTH OF THE SAHARA
Witchcraft. Rituals

RETEL-LAURENTIN, Anne
Sorcellerie et ordalies. - Paris : Anthropos, 1974
364 p.. - Bibliography : p. 239-317 ; index to bibl. : p. 319-364
r6006093 X.709/19908
0917389

AFRICA SOUTH OF THE SAHARA
Witchcraft. Rituals, to date

BUAKASA, Tulu Kia Mpansu
L'impensé du discours. - 2e éd. revue et corrigée. - Kinshasa : Faculté de Théologie catholgique, 1980
317p. - (Bibliothèque du Centre d'étude des religions africaines ; 4). - Includes bibliography
r6209929 X.622/16713
8652449

AFRICA SOUTH OF THE SAHARA
Women. Equality of opportunity
SACKS, Karen
Sisters and wives. - Westport, Conn. ; London : Greenwood Press, 1979
ix,274,[1]p. - (Contributions in women's studies ; no. 10). - Bibliography: p.253-268. - Includes index
0-313-20983-9 X.529/36918
2114070

AFRICA SOUTH OF THE SAHARA. Special themes. African writers. 1945-. English literature
Cultural awareness of negroes of Africa south of the Sahara
SOYINKA, Wole
Myth, literature and the African world. - Cambridge : Cambridge University Press, 1976
xii,168p. - Includes index
0-521-21190-5 X.989/51009
0746525

AFRICAN
See also
Ancient African
Negro

AFRICAN ARMIES
to date
BENCHENANE, Mustapha
Les armees africaines. - [Paris] : Publisud, [1983]
221p
2-86600-019-6 X.809/64815
8890463

KAKÉ, Ibrahima Baba
Les armées traditionnelles de l'Afrique. - Libreville : Lion, 1980
192p : ill
r6209236 L.45/1923
8890463

AFRICAN ARTS
- Serials
NEW culture. - Vol.1, no.1- ; November 1978-. - Ibadan : New Culture Studios, 1978-
r6108519 P.421/460 Woolwich
1488163

AFRICAN ARTS
1945- - Serials
ISALA. - No.1 (1982)-. - Ile-Ife : Ife Studies in African Literature and the Arts, 1982-
v.
r6806290 X.0950/542
8616035

AFRICAN ARTS
African religious arts, to date
COMMEMORATIVE figures. - Leiden : Brill, 1982
ix,174p : ill. - (Visible religion ; vol.1)
90-04-06779-5 X.205/1344
9000631

AFRICAN ARTS
African textile arts
PICTON, John
African textiles. - London : Published for the Trustees of the British Museum by British Museum Publications, c1979
208p : ill(some col.). - Bibliography: p204-205. - Includes index
0-7141-1552-5 X.425/2127 Woolwich
1812831

AFRICAN ARTS
African visual arts
AFRICAN images / edited by Daniel F. McCall, Edna G. Bay. - New York : Africana Publishing Co. ; London : Holmes and Meier Publishers Ltd for the African Center, Boston University, 1975 [i.e. 1976]
xv,327p,[64]p of plates : ill. - (Papers on Africa / Boston University ; vol.6)
0-8419-0147-3 X.421/10243 Woolwich
0500186

DUERDEN, Dennis
African art. - London : Hamlyn, 1974
96p : ill(some col). - Bibliography: p.91. - Includes index
0-600-34853-9 X.423/5036
0500186

GILLON, Werner
Collecting African art. - London : Studio Vista, 1979
x,183p,[24]p of plates : ill(some col.). - Includes bibliographies and index
0-289-70897-4 X.425/3227 Woolwich
271034x

VANSINA, J. (Jan)
Art history in Africa. - London : Longman, 1984
xiv,233p : ill. - Bibliography: p214-223. - Includes index
0-582-64367-8 X.429/16815
0500186

AFRICAN ARTS
African visual arts - Catalogues
GILFOY, Peggy Stoltz
African art from the Harrison Eiteljorg Collection. - [Indianapolis] : Indianapolis Museum of Art, [1976]
83p
r6189015 X.410/10471 Woolwich
0500186

AFRICAN ARTS
African visual arts, to ca 1900
GILLON, Werner
A short story of African art. - Harmondsworth : Viking, 1984
405p : ill. - Bibliography: p375-394. - Includes index
0-670-80055-4 X.421/26561
3274667

AFRICAN ARTS
African visual arts, to date
BAMERT, Arnold
[Afrika. French]. Africana. - [Paris] : Herscher, c1980
xiii,329p : ill
2-7335-0003-1 L.45/2294
8004765

MVENG, Englebert
L'art et l'artisanat africains. - Yaoundé : Éditions CLE, 1980
163p : ill
2-7235-0050-0 X.429/15500
8004765

AFRICAN ARTS
African visual arts, to date - Catalogues
ARTE AFRICANO EN COLECCIONES VENEZOLANAS (1976 : Caracas)
Arte africano en colecciones venezolanas, 15-29 de febrero de 1976. - Caracas : Biblioteca Nacional, 1976
80p : ill. - Bibliography
r6701545 X.429/12236 Woolwich
8004765

AFRICAN ARTS
African visual arts, to date - Serials
SOKROVISHCHA iskusstva stran Azii i Afriki / Vsesoiuznyĭ nauchno-issledovatel'skiĭ institut iskusstvoznaniīā Ministerstva kul'tury SSSR. - [Vyp 1]-. - Moskva : Izobrazitel'noe iskusstvo, [197-?]- : ill(some col). - Bibliographies
Holdings: Vyp.3 (1979)-
r6402600 X.421/22047 Woolwich
8004765

AFRICAN ARTS
African visual arts. Influence in Western visual arts, to 1900
L'IMAGE du Noir dans l'art occidental... / documentation et coordination par Ladislas Bugner. - Paris : Bibliothèque des Arts, 1976
tom1- : ill
r6028301 LR.415.c.7
112613X

AFRICAN ARTS
African visual arts. Influence of tourism. Semiotic aspects
JULES-ROSETTE, Bennetta
The messages of tourist art. - New York ; London : Plenum, c1984
xviii,266p : ill. - (Topics in contemporary semiotics). - Bibliography: p239-246. - Includes index
0-306-41598-4 X.421/26610
3274233

AFRICAN ARTS
East African visual arts
MILLER, Judith von D. (Judith von Daler)
Art in East Africa. - London : Muller [etc.], 1975
125p,leaf of plate,[8]p of plates : ill(some col). - Bibliography: p.111-113. - Includes index
0-584-10154-6 X.421/10156
0636150
- [Another copy] X.421/25297

AFRICAN ARTS
East African visual arts, to date - Bibliographies
BURT, Eugene C.
An annotated bibliography of the visual arts of East Africa. - Bloomington : Indiana University Press, [1980]
xv,371p. - (Traditional arts of Africa)
0-253-17225-x X.421/24515
9002316

AFRICAN ARTS
Social aspects
BRAIN, Robert
Art and society in Africa. - London : Longman, 1980
304p : ill. - Bibliography: p285-290. - Includes index
0-582-64578-6 X.429/12223
2399091

AFRICAN ARTS
Southern African visual arts, 1961-
LIP (Collective)
LIP from Southern African women. - Johannesburg : Raven Press, [1983]
vii,232p : ill
0-86975-153-0 X:950/34215
916653x

AFRICAN ARTS
West African visual arts
AFRICAN art as philosophy / edited by Douglas Fraser. - New York : Interbook, 1974
[2],xii,140p : ill. - Bibliography : p.127-140
0-913456-22-5 X.410/10340 Woolwich
1881906

ARMSTRONG, Robert Plant
The powers of presence. - Philadelphia : University of Pennsylvania Press, 1981
xiii,130p,[88]p of plates : ill. - Bibliography: p 127-128
0-8122-7804-6 X.520/29453
1881906

PRICE, Christine
Made in West Africa. - London : Studio Vista, 1976
x,150p : ill. - Bibliography: p.147. - Includes index
0-289-70742-0 X.421/10378 Woolwich
1067362

AFRICAN ARTS
West African visual arts, to date
HOLAS, B.
L'art sacré sénoufo, ses différentes expressions dans la vie sociale. - Abidjan : Nouvelles éditions africaines, 1978
332p : ill
r6209828 X.429/14662
8579121

AFRICAN ARTS. Private collections. West Germany
African visual arts
SCHÄDLER, Karl Ferdinand
Afrikanische Kunst in deutschen Privatsammlungen = African art in private German collections = L'art africain daus les collections privées allemandes. - München : Münchner Buchgewerbehaus, [1973]
363p : ill
r6019027 LR.416.pp.14
1160753

AFRICAN ASSEMBLY OF GOD. East London. South Africa
DUBB, Allie A.
Community of the saved. - Johannesburg : Witwatersrand University Press for African Studies Institute, 1976
xvii, 175 p., 10 p. of plates : ill. - Bibliography : p. 173-175
0-85494-292-0 X.200/30767
1041878

AFRICAN BUFFALOES. Tanzania
Population. Effects of availability of natural resources
SINCLAIR, A. R. E.
The African buffalo. - Chicago ; London : University of Chicago Press, 1977
xii,355p,[22]p of plates : ill. - (Wildlife behavior and ecology). - Bibliography: p.327-345. - Includes index
0-226-76030-8 SRL
1521233

AFRICAN CAMPAIGNS
See also
East African campaign
North African campaign

AFRICAN, CARIBBEAN AND PACIFIC STATES
See ACP countries

AFRICAN CARVINGS
African wood carvings
SEATTLE ART MUSEUM
African masks and muses. - Seattle : The Museum, 1977
[50]p : ill. - Bibliography: p50
r6863365 X.429/14966
2974665

AFRICAN CARVINGS. Exhibits. Royal Pavilion, Art Gallery and Museums, Brighton

ROYAL PAVILION, ART GALLERY AND MUSEUMS, BRIGHTON
African carvings. - [Brighton] : [Royal Pavilion, Art Gallery and Museums], 1975
63p : chiefly ill. - Bibliography: p.13
0-9502372-2-1 X.415/5068
0607258

AFRICAN CICHLIDAE
Colour. Patterns. Behavioural aspects

VOSS, J.
Color patterns of African cichlids. - Neptune, N.J. ; Reigate : T.F.H., c1980
123p : ill(some col.)
0-87666-503-2 SRL
2348055

AFRICAN CINEMA FILMS
to 1973

VIEYRA, Paulin Soumanou
Le cinéma africain. - Paris : Présence Africaine. - Bibliography: p. 426-440
tom. 1: Des origines à 1973. - [1975]
444 p., 28 p. of plates : ill
r6012711 X.909/33677
0873268

AFRICAN CINEMA FILMS
to date

CHERIAA, Tahar
Écrans d'abondance - ou, Cinémas de libération en Afrique. - Tunis : Société d'impression et d'édition Laplume, 1978
312p
r6206353 X.950/5217
8616043

HAFFNER, Pierre
Essai sur les fondements du cinéma africain. - Abidjan ; Dakar : Les Nouvelles Éditions africaines, 1978
274p,8p of plates
2-7236-0524-8 X.950/15185
8616043

AFRICAN CIVILIZATION
- Serials

SOAS AFRICA SOCIETY
SOAS Africa Society Magazine. - Vol.1, no.1- ; Spring 1976-. - London : [The Society], 1976- : ill
B76-16743 P.703/514
0641855

AFRICAN CIVILIZATION
to ca 1600

GARLAKE, Peter
The kingdoms of Africa. - Oxford : Elsevier-Phaidon, 1978
152p : ill(chiefly col). - (The making of the past). - Bibliography: p.137. - Includes index
0-7290-0049-4 X.423/10684 Woolwich
1704729

AFRICAN CIVILIZATION
to 1973

PAULME, Denise
Les civilisations africaines. - 6e édition mise à jour. - Paris : Presses Universitaires de France, 1974
126 p. : ill. - (Que sais-je? no. 606)
r6001002 X.708/17546
0915335

AFRICAN CIVILIZATION
to 1983

DAVIDSON, Basil
Africa in history. - Rev. and enl. ed.. - London : Paladin, 1984
384p,[16]p of plates : ill. - Bibliography: p342-352. - Includes index
B85-02343 X.808/40675
3164888

AFRICAN CIVILIZATION
to date

ALEXANDRE, Pierre
Les Africains .... - Paris : Éditions Lidis, 1981
607p,plates. - (Histoire ancienne des peuples ; 4)
2-85032-062-5 X.805/4062
8616051

DIKA-AKWA NYA BONAMBELA, Prince
Problèmes de l'anthropologie et de l'histoire africaines. - Yaoundé : Éditions Clé, 1982
372p,plate. - (Etudes et documents africains)
r6299200 X.529/62308
8616051

HERMANN, Hans H.
Europa kam aus Afrika. - Hamburg : Hoffmann und Campe, 1980
436p,[24]p of plates : ill(chiefly col.)
3-455-08859-7 X.529/44623
8616051

MOZHEĬKO, Igor'
7 u 37 chudes. - Moskva : Nauka, Glavnaīa redakfsiīa vostochnoĭ literatury, 1980
358p : ill
r6402296 X.808/34974
8616051

AFRICAN CIVILIZATION
to date compared with Antillean civilization, to date

DUFRENOT, Max-Auguste
Des Antilles à l'Afrique. - Abidjan : Les Nouvelles éditions africaines, 1981
134p
2-7236-0803-4 X.809/54957
8943516

AFRICAN CIVILIZATION
ca 1800-1976

SALMON, Pierre
L'Afrique noire. - Bruxelles : Meddens, [1976]
220p : ill(some col)
2-87013-007-4 Cup.1264.f.14
1128558

AFRICAN CIVILIZATION
Central African civilization, 1600-1910

OBENGA, Théophile
Afrique centrale précoloniale. - Paris : Présence Africaine, 1974
187 p.
r6012839 X.709/20366
094257X

AFRICAN CIVILIZATION
Influence of Western civilization

AFRICA & the West / edited by Philip D. Curtin. - Madison ; London : University of Wisconsin Press, 1972
x,259p
r6039083 X.709/24539
0254592

THAIRU, Kihumbu
The African civilization = Utamaduni wa Kiafrika. - Kampala : East African Literature Bureau, 1975
xii,223p,[15]leaves of plates,[4]p of plates : ill. - Includes index
r6031740 X.709/31162
0254592

AFRICAN CIVILIZATION
Prehistoric African civilization. Archaeological sources

The ARCHAEOLOGICAL and linguistic reconstruction of African history / edited by Christopher Ehret and Merrick Posnansky. - Berkeley ; London : University of California Press, c1982
vii,299p : ill. - Bibliography: p269-295. - Includes index
0-520-04593-9 X.950/25923
2982269

AFRICAN CIVILIZATION
Prehistoric Eastern African civilization

PHILLIPSON, D. W. (David Walter)
The later prehistory of eastern and southern Africa. - London : Heinemann Educational, 1977
xii,323p : ill. - Bibliography:p.294-316. - Includes index
0-435-94750-8 X.421/10507 Woolwich
1267264

AFRICAN CIVILIZATION
Prehistoric Southern African civilization

DERRICOURT, R. M. (Robin M.)
Prehistoric man in the Ciskei & Transkei. - Cape Town : Struik, 1977
xviii,284p : ill
0-86977-077-2 X.421/22757 Woolwich
1267272

PHILLIPSON, D. W. (David Walter)
The later prehistory of eastern and southern Africa. - London : Heinemann Educational, 1977
xii,323p : ill. - Bibliography:p.294-316. - Includes index
0-435-94750-8 X.421/10507 Woolwich
1267272

AFRICAN CIVILIZATION
Research, 1945-

RECENT German research on Africa / edited by Bernd Heine = Rapport sur la recherche africanistique allemande. - Bonn : die Forschungsgemeinschaft, [1982]-
r6306497 X.0950/532
8890935

AFRICAN CIVILIZATION
West African civilization, to date - Conference proceedings

WEST African culture dynamics / editors: B.K. Swartz, Raymond E. Dumett. - The Hague : Mouton, 1980
xiv,636p : ill. - (World anthropology)
90-279-7920-0
8579156

AFRICAN CIVILIZATION
West African civilization, 1000-1600

NIANE, Djibril Tamsir
Le Soudan occidental au temps des grands empires, XIe-XVIe siècle. - Paris : Présence Africaine, 1975
271 p., 30 p. of plates : ill
r6000535 X.709/20303
0940763

AFRICAN CIVILIZATION Curriculum subjects, to date. Schools. Nigeria

ANDAH, Bassey W.
African development in cultural perspective. - [Ibadan] : Dept. of Archaeology and Anthrology, University of Ibadan, 1982
92p
978-300-210-4 X.529/59381
9062750

AFRICAN COMMUNITIES. Namakgale. Phalaborwa region. South Africa
Poverty related to social conditions

SCHLEMMER, Lawrence
Poverty, family patterns and material aspirations among Africans in a border industry township. - [Durban] : [University of Natal, Institute for Social Research], 1974
[1],vii,89 leaves
r6091775 X.519/29678
1893025

AFRICAN COMPETITORS
Participation in Olympic Games, to 1972

ALI, Ramadhan
Africa at the Olympics. - London : Africa Books, 1976
5-162p : ill
0-903274-07-8 X.611/7355 Woolwich
0705616

AFRICAN COSTUME
North African folk costume

TILKE, Max
Folk costumes from East Europe, Africa and Asia. - London : Zwemmer, 1978
39p,128p of plates : chiefly ill(chiefly col)
0-302-02782-3 LR.425/385
1938665

AFRICAN COUNTRIES
Agriculture. Research, to date

INTERNATIONAL INSTITUTE OF TROPICAL AGRICULTURE
Tasks for the eighties. - Ibadan : The Institute, 1981
iii,60p
r6810511 X.322/12906
9063943

AFRICAN COUNTRIES
Commonwealth African countries. Economic relations with Great Britain, 1951-1975

BANGURA, Yusuf
Britain and Commonwealth Africa. - Manchester : Manchester University Press, c1983
242p. - Bibliography: p214-236. - Includes index
0-7190-0854-9 X.809/56821
2962756

AFRICAN COUNTRIES
Defence, 1960-

NKOA ATENGA, Camille
Réflexion sur la stratégie de la défense en Afrique. - Yaoundé : Éditions Clé, 1980
131p
r6211237 X.809/56138
900047x
- [Another copy] X.809/56138

AFRICAN COUNTRIES
Economic cooperation, 1960-

SEKOU TOURÉ, Ahmed
Pour la paix, la fraternité, la concorde et la coopération entre les Nations africaines. - Conakry : Bureau de presse de la présidence de la République, [1978]
24p
r6209298 X.520/29524
8916403

AFRICAN COUNTRIES
Economic relations with European Economic Community, to 1975. International legal aspects

DJAMSON, Eric C.
The dynamics of Euro-African co-operation. - The Hague : Nijhoff, 1976
xxiii,370p. - Bibliography : p.355-360. - Includes index
r6083475 X.520/12980
1684523

AFRICAN COUNTRIES
Economic relations with European Economic Community, 1957-1978

TWITCHETT, Carol Cosgrove
Europe and Africa. - Farnborough, Hants. : Saxon House, 1979
xv,195p. - Bibliography: p.179-190. - Includes index
0-566-00182-9 X.529/33726
1377159

AFRICAN COUNTRIES
Employment. Role of education, to date

NDAYAMBAJE, Jean Damascène
Rapports entre l'éducation et l'emploi en Afrique noire. - [Fribourg] : [Université de Fribourg], 1983
278p. - Includes bibliography
r6211477 X.529/67410
9131418

AFRICAN COUNTRIES
Foreign relations, to date

VNESHNÎAÎ a politika stran Afriki / otvetstvennyĭ redaktor Anat. A. Gromyko. - Moskva : "Mezhdunarodnye otnosheniîa", 1981
333 p.. - (Vneshnîaî a politika razvivaîushchikhsîa stran). - Bibliography: p. 316-[322]
r6403670 X.809/60084
8003483
- Russian title transliterated X.809/60084

AFRICAN COUNTRIES
Foreign relations, 1960-

SAXENA, S. C. (Suresh Chandra)
Foreign policy of African states. - New Delhi : Deep and Deep, [1982]
227p
r6807100 X.529/54652
8579660

AFRICAN COUNTRIES
Foreign relations, 1960-. Soviet Union

RUSSIA in Asia & Africa / edited by J.A. Naik. - Kolhapur : Avinash, [1982?]
260p
r6807987 X.529/56179
8976422

AFRICAN COUNTRIES
Foreign relations with Arab countries, to date

LE VINE, Victor Theodore
The Arab-African connection. - Boulder, Colo : Westview Press, 1979
xvii,155p. - (Westview Special Studies on Africa The Middle East). - Bibliography
0-89158-398-x X.520/25863
883735x

AFRICAN COUNTRIES
Foreign relations with Arab countries, 1960-

AFRIKA und die arabischen Staaten = Africa and the Arab states / [zusammengestellt von] Hartmut Neitzel, Renate Nötzel. - Hamburg : Institut für Afrika-Kunde, 1979
458p. - (Aktueller Informationsdienst Afrika. Sondernummer ; 1). - Bibliography
r6300854 X.0802/1549(1)
8724954

AFRICAN COUNTRIES
Foreign relations with China, 1960-

CHANG, Ya-chun
Actividades de los comunistas chinos en Africa. - [Taiwan?] : La Liga Mundial Anticomunista, Capitulo Chino, 1981
61p
r6708289 X.808/40986
9129995

AFRICAN COUNTRIES
Foreign relations with communist countries, to date

COMMUNIST powers and Sub-Saharan Africa / Thomas H. Henriksen, editor. - Stanford, Calif. : Hoover Institution Press, [1981]
xii,137p. - (Hoover international studies)
0-8179-7522-5 X.800/35143
8916411

AFRICAN COUNTRIES
Foreign relations with East Germany, 1885-

POST, Ulrich
Die Afrikapolitik der DDR. - Hamburg : Institut für Afrika-Kunde, 1982
ix,158p. - (Arbeiten aus dem Institut für Afrika-Kunde ; 43). - Includes bibliography
3-923519-53-2 X.809/56737
900145x

AFRICAN COUNTRIES
Foreign relations with France, to date

La FRANCE contre l'Afrique. - Paris : Maspero, 1981
271p. - (Tricontinental. nouvelle série, numéro spécial) (Petite collection Maspero ; 251)
2-7071-1229-1 X.808/35608
8792763

FRANKREICH und Afrika = La France et l'Afrique. - Hamburg : Institut für Afrika-Kunde, Dokumentations-Leitstelle Afrika, 1981
125p. - (Aktueller Informationsdienst Afrika. Beiheft ; 3)
3-922887-17-1 X.0802/1617(3)
8792763

AFRICAN COUNTRIES
Foreign relations with United States, to date

CHESTER, Edward W.
Clash of titans. - Maryknoll : Orbis Books, c1974
ix,316p. - Bibliography: p.281-298
0-88344-065-2 X.805/2711
8860955

OGENE, F. Chidozie
Interest groups and the shaping of foreign policy. - Lagos : Nigerian Institute of International Affairs, 1981
224p. - Includes bibliography
978-227-657-x X.520/31774
8860955

AFRICAN COUNTRIES
Government, to date

TABUWE, Aletum
The one-party system and the African traditional institutions. - Bilingual ed. - Yaoundé : A. Tabuwe, 1980
63p
r6208524 X.529/49682
8909695

AFRICAN COUNTRIES
International relations, to date - Bibliographies

DELANCEY, Mark W. (Mark Wakeman)
African international relations. - Boulder : Westview Press, 1981
xviii,365p. - (Westview special studies on Africa)
0-89158-680-6 X.520/28781
8820082

AFRICAN COUNTRIES
International relations with Soviet Union, 1885-

[SSSR i Afrika. English]. The USSR and Africa / [editorial board ... An. Gromyko, consulting editor ... et al.]. - Moscow : Social Sciences Today, 1983
205p. - (African Studies by Soviet Scholars ; no.3). - Includes bibliography
r6465994 X.809/62927
916670x

AFRICAN COUNTRIES
International relations with Soviet Union, 1960-

SOVETSKO-AFRIKANSKIE otnosheniîa / [L.V. Goncharov (otv. redaktor.) .. P.ÎA. Koshelev (otv. redaktor) ... E.A. Tarabrin (otv. redaktor)]. - Moskva : Nauka, Glavnaîa redaktsîa vostochnoĭ literatury, 1982
237p
r6448088 X.809/59168
9064834

AFRICAN COUNTRIES
Law, to date - Bibliographies

BENNETT, T. W.
A bibliography of African law. - Salisbury : University of Rhodesia, 1975
xxxiv,291p. - (Bibliographical series ; no.4)
0-86924-005-6 OPL
9044531

AFRICAN COUNTRIES
Law, to date - Serials

JAHRBUCH für afrikanisches Recht = Annuaire de droit africain = Yearbook of African law / herausgegeben ... im Auftrag der Gesellschaft für afrikanisches Recht. - Bd.1 (1980)-. - Heidelberg : Müller, [1981]-
r6305886 P.203/153
8890900

AFRICAN COUNTRIES
Legal systems, to date

NDING, Dyatlm
Civilisation et science juridique en Afrique et dans le monde. - Yaoundé : Éditions Clé, 1982
132p. - Includes bibliography
r6211396 X.200/44404
8890919

AFRICAN COUNTRIES
Legal systems, to date - Serials

JAHRBUCH für afrikanisches Recht = Annuaire de droit africain = Yearbook of African law / herausgegeben ... im Auftrag der Gesellschaft für afrikanisches Recht. - Bd.1 (1980)-. - Heidelberg : Müller, [1981]-
r6305886 P.203/153
8890919

AFRICAN COUNTRIES
Military equipment: Weapons. Acquisition. International political aspects

ARLINGHAUS, Bruce E.
Military development in Africa. - Epping : Bowker, c1984
xiv,152p. - (Westview special studies on Africa). - Includes index
0-86531-434-9 X.520/37477
3286959

AFRICAN CULTURE

ROSENFELD, Erwin M.
Afro-Asian culture studies. - 2nd rev. ed. - Woodbury : Barron's Educational, c1976
469p : ill. - Bibliography: p455-462
0-8120-0648-8 X.525/6638
2853515

AFRICAN CULTURE
to date

Das AFRIKA der Afrikaner / herausgegeben von Rüdiger Jestel. - Frankfurt am Main : Suhrkamp, c1982
366p. - (Edition Suhrkamp ; 1039 ; n.F., Bd.39)
3-518-11039-x X.529/58923
8788073

FUGLESANG, Andreas
About understanding. - Uppsala : Dag Hammarskjöld Foundation, 1982
231p
91-85214-09-4 X.950/29667
8788073

GULDAGER, Reinhardt
Afrikanische Impressionen. - Düsseldorf : Droste, 1980
255p. : chiefly ill.
3-7700-0559-7 L.42/2347
8788073

MARTÍNEZ FURÉ, Rogelio
Diálogos imaginarios. - La Habana : Editorial Arte y Literatura, 1979
283p. - (Cuadernos de arte y sociedad). - Includes bibliographies
r6706974 X.808/38304
8788073

PATRIMOINE culturel et creation contemporaine en Afrique et dans le monde arabe / sous la direction de Mohamed Aziza avec la collaboration de: Mohamed Boughali [et al.]. - Dakar ; Abidjan : Les Nouvelles Éditions Africaines, 1978
247p,1leaf of plates,16p of plates : ill. - Bibliography
2-7236-0185-4 X.800/32431
8788073

AFRICAN CULTURE
African negro culture - Conference proceedings

COLLOQUIUM ON BLACK AESTHETICS (1971 : University of Nairobi)
Black aesthetics. - Nairobi : East African Literature Bureau, 1973
vii,217p : 1 ill
r6072637 X.908/42062
1733869

AFRICAN CULTURE
African negro culture. Influence on race awareness of American negroes, 1816-1970

WEISBORD, Robert G.
Ebony kinship. - Westport, Conn. ; London : Greenwood Press, 1974
xi,256p. - (Contributions in Afro-American and African studies ; no.14). - Bibliography: p.227-240. - Includes index
0-8371-7340-x X.709/31305
0043729

AFRICAN CULTURE
African negro culture. Influence on race awareness of American negroes, 1960–1975

DISCOVERING Afro-America / edited by Roger D. Abrahams and John F. Szwed. – Leiden : Brill, 1975
[5], 94 p.. – (International studies in sociology and social anthropology ; vol. 28). – Includes bibliographies and index
90-04-04264-4 X.800/25749
0763454

AFRICAN CULTURE
Effects of internal migration, to date

BINET, Jacques
Migrations et appauvrissement culturel en Afrique. – Paris : Centre de Hautes Études sur l'Afrique et l'Asie Modernes, 1980
82leaves. – (Cahiers du C.H.E.A.M. ; no.9)
2-903182-03-5 X.525/5821
8901805

AFRICAN CULTURE
Influence of Islamic culture, to date

ANE, Mohamed Mustapha
L'Islam et la culture dans la république du Sénégal. – Dakar : Librairie Dar-Senegalia, 1973
39p
r6209813 X.809/55604
8825904

DIA, Mamadou
Islam et civilisation négro-africaines. – Dakar, etc. : Les Nouvelles Éditions Africanes, [1980]
154p. – (Sociologie et civilisation) (Essai sur l'Islam ; t.3). – Bibliography
2-7236-0212-5 X.529/47815
8825904

AFRICAN CULTURE
Influence of Western culture, to date

TOURÉ, Abdou
La civilisation quotidienne en Côte-d'Ivoire. – Paris : Éditions Karthala, [1981]
279p. – (Les Afriques). – Includes bibliography
2-86537-039-9 X.809/55977
8982244

AFRICAN CULTURE
Influence on Brazilian culture

BASTIDE, Roger
Estudos afro-brasileiros. – São Paulo : Editora Perspectiva, 1973
xxiv, 384 p., 2 fold. p. of plates : ill. – (Coleção Estudos ; no. 18)
r6011384 X.519/20420
093514X

BRAZIL. Ministério das Relações Exteriores
The impact of African culture on Brazil = L'impact de la culture africaine au Brésil = O impacto da cultura africana no Brasil .... – [s.l.] : o Ministério, [1978?]
296p : ill(some col.)
r6157747 X.972/1021
093514X

ORTIZ ODERIGO, Néstor
Macumba, culturas africanas en el Brasil. – Buenos Aires : Editorial Plus Ultra, 1976
237p : ill. – (Colección Ensayos ; no 16). – Includes bibliographies
r6065849 X.708/44436
093514X

AFRICAN CULTURE
Influence on Caribbean culture, to 1978

AFRICA and the Caribbean / edited by Margaret E. Crahan and Franklin W. Knight. – Baltimore ; London : Johns Hopkins University Press, 1979
xiii,159,[1]p : 1 ill. – (Johns Hopkins studies in Atlantic history and culture). – Bibliography: p.146–157
0-8018-2186-x X.800/28278
2000970

AFRICAN CULTURE
Influence on cultural processes of Plate River Basin countries

ORTIZ ODERIGO, Néstor
Aspectos de la cultura africana en el Río de la Plata. – Buenos Aires : Plus Ultra, 1974
200 p., 16 p. of plates. – (Colección Ensayos ; v. 9). – Bibliography: p. 189–193
r6011661 X.708/18295
0874000

AFRICAN CULTURE
Influence on Latin America culture, to date

ÁFRICA en América Latina / relator: Manuel Moreno Fraginals. – México : Siglo Veintiuno Editores ; Paris : UNESCO, 1977
xvi,436p. – (Serie El mundo en América Latina). – Bibliography
92-330-1436-3 X.520/28328
8754209

GRANDA, Germán de
Materiales complementarios para el estudio sociohistórico de los elementos lingüísticos afroamericanos en el área hispánica. – Bogotá : Instituto Caro y Cuervo, 1971
16p
r6702321 X.529/44334
8754209

AFRICAN CULTURE
Influence on Venezuelan culture – Bibliographies

POLLAK-ELTZ, Angelina
Bibliografía afrovenezolana. – Caracas : Universidad Católica "Andrés Bello", Instituto de Investigaciones Históricas, 1976
25p
r6065703 X.510/11428
1989138

AFRICAN CULTURE
Influence on Venezuelan culture, to date

POLLAK-ELTZ, Angelina
Vestigios africanos en la cultura del pueblo venezolano. – Caracas : Universidad Católica "Andrés Bello", Instituto de Investigaciones Históricas, 1972
171p. – Bibliography
r6705674 X.529/48119
8845468

AFRICAN CULTURE
North African culture, to date – Serials

QUADERNI dell'Instituto italiano di cultura di Tripoli. – Nuova serie, 1-. – Roma : L'erma di Bretschneider, 1979– : ill
r6508958 P.523/1919
8993262

AFRICAN CULTURE
North African culture, B.C. 146–A.D. 640

MACKENDRICK, Paul
The North African stones speak. – London : Croom Helm, c1980
xxi,434p : ill. – Bibliography: p334–409. – Includes index
0-7099-0394-4 X.421/21511 Woolwich
205969x

AFRICAN CULTURE
Role of religion, to date

DICKSON, Kwesi A. (Kwesi Abotsia)
Aspects of religion and life in Africa. – Accra : Ghana Academy of Arts and Sciences, 1977
viii,41p. – (The J. B. Danguah memorial lectures ; series 10 (1977))
r6811323 X.200/42863
9068961

AFRICAN CULTURE
West African culture, to 1978

ATMORE, Anthony
Black kingdoms, black peoples. – London : Orbis Books, 1979
128p : col ill. – Bibliography: p.126. – Includes index
0-85613-303-5 X.805/1074
2003279

AFRICAN CULTURE. Brazil
African negro culture, to date

NASCIMENTO, Abdias do
O quilombismo. – Petrópolis : Vozes, 1980
281p. – Includes bibliographies
r6704105 X.809/51632
8861188

AFRICAN CULTURE. Latin America
African negro culture, to date – Conference proceedings

HOMENAJE a Lydia Cabrera / Reinaldo Sánchez, editor, José Antonio Madrigal, editor ; Ricardo Viera, editor de arte, José Sánchez-Boudy, editor asesor. – Miami, Florida : Ediciones Universal, 1978
xiv,349p,plates : ill. – (Colección "Ebano y canela" ; no.3)
r6708266 X.950/36840
9156682

AFRICAN CUSTOMS
African death customs, to date

THOMAS, Louis-Vincent
La mort africaine. – Paris : Payot, 1982
272p. – (Bibliothèque Scientifique)
2-228-13010-9 X.529/61695
9086846

AFRICAN CUSTOMS
Southern African marriage customs: Bridewealth

KUPER, Adam
Wives for cattle. – London : Routledge & Kegan Paul, 1982
xiii,202,14p : ill. – (International library of anthropology). – Bibliography: p187–197. – Includes index
0-7100-0989-5 X.529/47945
2736675

AFRICAN DAILY NEWS
Editing, to 1964

WASON, Eugene
Banned. – London : Hamilton, 1976
vii,161p,[4]p of plates
0-241-89473-5 X.989/50899
0720054

AFRICAN DANCING

AKESSON, Birgit
Källvattnets mask. – Stockholm : Atlantis, 1983
323p : ill
91-7486-297-9 X.622/23390
8883017

HUET, Michel
Danses d'Afrique. – Paris : Éditions du Chêne, 1978
241p : chiefly ill(some col.)
2-85108-189-6 L.45/1706
8883017

AFRICAN DESIGNS. Canvas embroidery

TURNER, Diana Oliver
African needlepoint designs. – New York : Dover Publications [etc.] ; London : Constable, 1976
v,41p : chiefly ill(some col). – (Dover needlework series)
0-486-23244-1 X.415/5173 Woolwich
0731072

AFRICAN DEVELOPMENT BANK
1964–1974

EBONG, Ime Ekop
Development financing under constraints. – Bonn-Bad Godesberg, 1974
172 p.. – ([Research Institute of the Friedrich-Ebert-Foundation.] Vierteljahresberichte ; special issue no. 8)
3-87831-190-7 P.703/101.a [no. 8]
0850063

AFRICAN DEVELOPMENT BANK
Establishment

AGREEMENT establishing the African Development Bank. – London : H.M.S.O., [1981]
51p. – (Cmnd. ; 8284) (Miscellaneous ; no.15)
0-10-182840-3 BS.14/3425
2549875

AFRICAN DEVELOPMENT BANK
Privileges. Accordance by Great Britain

GREAT BRITAIN
[Treaties, etc. African Development Bank, 1978 July 17]. Exchange of notes between the Government of the United Kingdom of Great Britain and Northern Ireland and the African Development Bank concerning the status and privileges of the office of the bank in London, London, 17 July 1978 ... – London : H.M.S.O., 1978
Folder (4p). – (Miscellaneous ; no.25(1978)) (Cmnd.7345)
0-10-173450-6 BS.14/3197
1845411

AFRICAN DISHES. Food

OLAORE, Ola
African cooking. – London : Foulsham, c1980
96p,[8]p of plates : ill(some col.). – Includes index
0-572-01131-8 X.629/14557 Woolwich
1591568

OMINDE, Mary
African cookery book. – Nairobi ; London : Heinemann Educational, 1975
[4],iv,152p : ill. – Includes index
B78-33877 X.619/18914 Woolwich
1591568
– [Another copy] X619/10004 Woolwich

VAN DER POST, Laurens
First catch your eland. – London : Hogarth Press, 1977
x,211p
0-7012-0413-3 X.809/42460
1118544

AFRICAN DISHES. Food
North African dishes

DER HAROUTUNIAN, Arto
North African cookery. - London : Century, 1985
336p : ill. - Bibliography: p323-324. - Includes index
0-7126-0925-3 X.622/22659
3154432

AFRICAN DISHES. Food
Southern African dishes

COETZEE, Renata
Funa. - Durban : Butterworths, [1982]
182p : ill
0-409-10300-4 X.622/19671
9100237

AFRICAN DISHES. Food. Brazil

LODY, Raul
Santo também come. - Recife : Instituto Joaquim Nabuco de Pesquisas Sociais, 1979
127p : ill. - (Série Estudos e pesquisas ; 11)
r6701656 X.0950/208(15)
8718156

AFRICAN DOCTORS
Medicine - Serials

MEDICINE digest. - Vol.1, no.1- ; May 1974-. - London : Medical Digest Ltd, 1974- : ill
B78-14987 SRL
1433784

AFRICAN ELEPHANTS

ANTHONY, John, b.1925
The elephant. - London : Ravenwill Disco Books Ltd, 1975
13p : chiefly ill(chiefly col). - (Animals of the world)
0-905129-00-8 Cup.575.n.3
0627178

COLEMAN, Jill
The elephant. - London : Pan Books, 1979
[24]p : col ill. - (Piccolo explorer books. wildlife) (A Piccolo original)
0-330-25770-6 X.319/19113 Woolwich
0627178

AFRICAN ELEPHANTS
Conservation

PARKER, Ian, 1936-
Ivory crisis. - London : Chatto & Windus, 1983
184p : ill(some col.). - Includes index
0-7011-2633-7 L.42/2191
2958139

AFRICAN ELEPHANTS. Africa
Hunting, ca 1900-ca 1920

BELL, W. D. M. (Walter Dalrymple Maitland)
The wanderings of an elephant hunter. - [New] ed. - London : Spearman, 1976
vii[i.e.viii],188p,[2]leaves of plates,[50]p of plates : ill. - Includes index
0-85435-063-2 X.320/11265 Woolwich
1479652

AFRICAN ELEPHANTS. East Africa
Hunting, 1897-1903

BELL, W. D. M. (Walter Dalrymple Maitland)
Karamojo safari. - Sudbury, Suffolk : Spearman, 1984, c1949
288p
0-85435-165-5 Nov.53610
3280438

AFRICAN ELEPHANTS. Lake Manyara National Park
Social behaviour

DOUGLAS-HAMILTON, Iain
Among the elephants. - [London] : Fontana, 1978
320p,[32]p of plates : ill(some col). - Bibliography: p.311-314. - Includes index
0-00-634993-5 X.319/18631 Woolwich
0260800

Among the elephants. - Harmondsworth : Penguin, 1978
285p,[72]p of plates : ill(some col). - Bibliography: p.277-280. - Includes index
0-14-004837-5 SRL
0260800

AFRICAN ELEPHANTS. Tsavo National Park

SHELDRICK, Daphne
An elephant called Eleanor. - London : Dent, 1980
128p : ill
0-460-06900-4 X.322/9809 Woolwich
1926152

AFRICAN EMBROIDERY
North African embroidery

STONE, Caroline
The embroideries of North Africa. - Harlow : Longman, 1985
218p : ill(some col.). - Bibliography: p211-215. - Includes index
0-582-78371-2 X.622/23692
3117456

AFRICAN EQUIPMENT. Exhibits. Museums. United States
African household equipment - Catalogues

SIEBER, Roy
African furniture & household objects. - New York : American Federation of Arts ; Bloomington ; London : Indiana University Press, 1980
279p : ill(some col). - Bibliography: p.269-276
0-253-11927-8 X.425/1683
2288230
- [Another copy : paperback]ISBN 0-253-28242-X X.425/1684

AFRICAN FISH EAGLES

BROWN, Leslie, 1917-
The African fish eagle. - Folk[e]stone : Bailey Bros and Swinfen, 1980
[9],168p, [6]p of plates : ill(some col). - Bibliography: p.167-168
0-561-00304-1 X.325/731 Woolwich
2316099

AFRICAN FOLKLORE

AFRICAN folklore / [edited by] Richard M. Dorson. - Bloomington ; London : Indiana University Press, 1979
ix,587p. - Includes index
0-253-20220-5 X.709/33254
1911384

ELLE ne mangea que des oiseaux! / [transcrits et traduits par] Ritiy Mutz. - Bandundu : Ceeba publications, 1979
130p : ill. - (Ceeba publications. Sér. 2 ; v.51)
r6205154 X.955/758
1335898

FESTSCHRIFT zum 60. Geburtstag von P. Anton Vorbichler / herausgegeben von I. Hofmann. - Wien : Afro-Pub, 1981
2v. : ill. - (Beiträge zur Afrikanistik ; Bd.11, 12) (Veröffentlichungen der Institute für Afrikanistik und Ägyptologie der Universität Wien ; Nr.14, 15). - List of Anton Vorbichler's works
r6300615 X.0909/1053(11,12)
1335898

FORMS of folklore in Africa / edited by Bernth Lindfors ; illustrated by Adebisi Akanji. - Austin ; London : University of Texas Press, 1977
viii,281p : ill. - Includes bibliographies and index
0-292-72418-7 X.709/31946
1335898

AFRICAN FOLKLORE
Influence on North American negro folklore

AFRICAN folklore in the New World / edited by Daniel J. Crowley. - Austin ; London : University of Texas Press, 1977
x,98p. - Includes bibliographies and index
0-292-70326-0 X.909/42101
1333275

AFRICAN FOLKLORE
Influence on South American negro folklore

AFRICAN folklore in the New World / edited by Daniel J. Crowley. - Austin ; London : University of Texas Press, 1977
x,98p. - Includes bibliographies and index
0-292-70326-0 X.909/42101
1333283

AFRICAN FURNITURE. Exhibits. Museums. United States - Catalogues

SIEBER, Roy
African furniture & household objects. - New York : American Federation of Arts ; Bloomington ; London : Indiana University Press, 1980
279p : ill(some col). - Bibliography: p.269-276
0-253-11927-8 X.425/1683
2288249
- [Another copy : paperback]ISBN 0-253-28242-X X.425/1684

AFRICAN GENERAL SERVICE MEDAL. Medals
Awards to British military forces, 1887-1920

MAGOR, R. B.
African general service medals. - [Wadhurst] : [The author], [1979]
[3],vi,193p,[9]leaves of plates(3 fold) : ill. - Bibliography: p.192-193
B79-29821 X.800/28068
1939084

AFRICAN GOVERNMENT PUBLICATIONS
East African government publications - Bibliographies

HOWELL, John Bruce
East African Community. - Washington [D.C.] : Library of Congress, 1976
xvi,269p. - Includes index
0-8444-0208-7 AS.287/29
1291599

AFRICAN GREY PARROTS. Pets
Care

PARADISE, Paul R.
African grey parrots. - Neptune, N.J. ; Reigate : T.F.H., c1979
93p : ill(some col.)
0-87666-977-1 SRL
2367858

AFRICAN GREY PARROTS. Pets
Training

TEITLER, Risa
Training African grey parrots. - Hong Kong : T.F.H. Publications, c1979
93p : ill(some col.). - Bibliography: p93
0-87666-994-1 SRL
2400081

AFRICAN HANDSTAMPS
West African handstamps, 1827-1905

MCCAIG, Colin
The postal history and handstamps of British West Africa. - [London] : [R. Lowe], [1978]
[1],64p
B79-20018 X.510/16190
1898604

AFRICAN HISTORIOGRAPHY. Africa

TEMU, A. J. (Arnold James)
Historians and Africanist history. - London : Zed, c1981
xiv,187p. - (Africa series). - Bibliography: p177-183. - Includes index
0-905762-78-9 X.809/51795
2726653

AFRICAN IMMIGRANT PERSONNEL. Jamaica
African immigrant indentured personnel, 1841-1865

SCHULER, Monica
'Alas, alas, Kongo'. - Baltimore ; London : Johns Hopkins University Press, c1980
x,186p : 1ill. - (Johns Hopkins studies in Atlantic history and culture). - Bibliography: p165-177. - Includes index
0-8018-2308-0 X.800/29407
2331187

AFRICAN IMMIGRANTS. Buenos Aires. Argentina to date

ANDREWS, George Reid
The Afro-Argentines of Buenos Aires, 1800-1900. - Madison : University of Wisconsin Press, 1980
xiv,286p : ill. - Bibliography
0-299-08290-3 X.520/32866
9054502

AFRICAN IMMIGRANTS. Spanish America
Social conditions, 1502-1975

ROUT, Leslie Brennan
The African experience in Spanish America, 1502 to the present day. - Cambridge : Cambridge University Press, 1976
xv,404p. - (Cambridge Latin American studies ; 23). - Bibliography: p.391-392. - Includes index
0-521-20805-x X.809/41162
0712086

AFRICAN JEWS. Israel
North African Jews. Ethnicity. Political aspects

SHOKEID, Moshe
Distant relations. - New York : Praeger, 1982
x,179p : ill. - Bibliography: p165-174
0-03-059856-7 X.800/33844
2743787

AFRICAN LANGUAGE
Dialects. Sounds

BARBOT, M
Évolution de l'arabe contemporain. - Paris : Librairie d'Amérique et d'Orient A. Maisonneuve, [1981]
2v((1008p)). - (Publications de la Sorbonne. recherches ; no.47). - Bibliographie d'arabe moderne et du Levant: v.1 p9-196
r6207103 Ac.442/5f[47]
8955395

AFRICAN LANGUAGES
See also
Fala Kpelle language
Kordofanian languages
Laal language
Niger-Congo languages
Yoruba language

AFRICAN LANGUAGES

FESTSCHRIFT zum 60. Geburtstag von P. Anton Vorbichler / herausgegeben von I. Hofmann. - Wien : Afro-Pub, 1981
2v. : ill. - (Beiträge zur Afrikanistik ; Bd.11, 12) (Veröffentlichungen der Institute für Afrikanistik und Ägyptologie der Universität Wien ; Nr.14, 15). - List of Anton Vorbichler's works
r6300615 X.0909/1053(11,12)
0164720

GHANA INSTITUTE OF LINGUISTICS
[Miscellaneous publications in languages not elsewhere specified]. - [Accra] : this Institute, [1980?]-
r6202260 P.901/1961
1112465

GREGERSEN, Edgar A.
Language in Africa. - New York ; London : Gordon and Breach, 1977
xvii,237p : ill. - (Library of anthropology). - Bibliography: p.215-226. - Includes index
0-677-04380-5 X.981/20635
1112465

ÎAZYKI Azii i Afriki ... - Moskva : Nauka, Glavnaîa redakîsiîa vostochnoĭ literatury, 1976-
v.. - Includes bibliographies
r6463940 X.0955/206
0164720

Les LANGUES dans le monde ancien et moderne / ouvrage publié sous la direction de Jean Perrot ... - Paris : Éditions du Centre national de la recherche scientifique, 1981
2v(xii,691p,12leaves of plates in separate folder). - Bibliography
2-222-01720-3 X.955/1306
0164720

Die SPRACHEN Afrikas / herausgegeben von Bernd Heine, Thilo C. Schadeberg, Ekkehard Wolff. - Hamburg : Buske, 1981
665p : ill
3-87118-433-0 X.950/21673
0164720

WELMERS, William Everett
African language structures. - Berkeley ; London : University of California Press, [1973]
xi,488p
0-520-02210-6 X.981/11545
0164720

AFRICAN LANGUAGES
- Bibliographies

MEIER, Wilma
Bibliography of African languages .... - Wiesbaden : Harrassowitz, 1984
lxxi,888p.
3-447-02415-1 X.950/37237
0164720

AFRICAN LANGUAGES
- Conference proceedings

DEUTSCHER ORIENTALISTENTAG (21st : 1980 : Berlin)
Berliner afrikanistische Vorträge (XXI. Deutscher Orientalistentag, Berlin 24.-29.3.1980). - Berlin : Reimer, 1981
276p. - (Marburger Studien zur Afrika- und Asienkunde. Serie A, Afrika ; Bd.28)
3-496-00505-x X.0100/683(28)
0164720

AFRICAN LANGUAGES
- Serials

BINNDI e jande. - no.1(1980)-. - Paris : Binndi e jande, 1980-
Holdings: no.6, 1981-
r6206833 P.903/445
0164720

BINNDI e jang de. - no.1 (1980)-. - [Paris?] : NUBIA, [1980]-
Holdings: no.2, 2e Trimestre 1980-
r6205694 P.903/435
0164720

CAHIER du Département des langues africaines et linguistique / Faculté des lettres et sciences humaines, Université de Yaoundé. - No.1 (1981)-. - Yaoundé : Université de Yaoundé, [1981]-
Holdings: No.2 (1982)-
r6210764 X.0955/239
0164720

RESEARCH notes / [issued by University of Ibadan, Department of Linguistics and Nigerian Languages] ; editor : Isaac George. - vol.1, no.1- ; [1968?]-. - Ibadan : The University, [1968?]-
Holdings : vol.6, nos.1-3, Jan-June, 1973 ; vol.7, nos.1-2- ; Jan. March, 1976-
r6077769 P.901/1631
1112465

SPRACHE und Geschichte in Afrika. Beihefte. - 1-. - Hamburg : Buske, 1981-
r6301405 X.0909/1133a
0164720

AFRICAN LANGUAGES
Classification

FODOR, István
A fallacy of contemporary linguistics. - [4th ed.]. - Hamburg : Buske, 1982
114p. - Includes bibliography
3-87118-421-7 X.958/21296
9049460

AFRICAN LANGUAGES
Dictionaries - Bibliographies

HENDRIX, Melvin K.
An international bibliography of African lexicons. - Metuchen ; London : Scarecrow, 1982
xxi,348p. - Includes index
0-8108-1478-1 X.950/15171
2774127
- [Another copy] 2050.d. Reading Room

AFRICAN LANGUAGES
Documents on African languages & literatures in stock of Stadtbibliothek Winterthur - Catalogues

STADTBIBLIOTHEK WINTERTHUR
Der Katalog der Africana-Sammlung in der Stadtbibliothek Winterthur. - Basel : Basler Afrika Bibliographien, 1977-. - (Mitteilungen der Basler Afrika Bibliographien ; v.17, 27, etc.)
r6063473 2725.c.88
1712705

AFRICAN LANGUAGES
Influence of Portuguese language, to date

BAL, Willy
Afro-românica studia. - Albufeira : Poseidon, 1979
161p
r6701482 X.950/6526
8724660

AFRICAN LANGUAGES
Influence of social change, to date

SOZIALER Wandel in Afrika und die Entwicklung von Formen und Funktionen afrikanischer Sprachen / herausgegeben von Dmitrij Alekseevič Olderogge und Siegmund Brauner. - Berlin : Zentralinstitut für Sprachwissenschaft, 1980
272p.. - (Linguistische Studien. Reihe A, Arbeitsberichte ; 64)
r6300299 X.958/3987
8822255

AFRICAN LANGUAGES
Influence on Colombian dialect of Spanish language, to date

CASTILLO MATHIEU, Nicolás del
Esclavos negros en Cartagena y sus aportes léxicos. - Bogotá : Instituto Caro y Cuervo, 1982
xvi,247p,plates. - (Publicaciones / del Instituto Caro y Cuervo ; 62). - Includes bibliographies
r6706843 X.520/30303
8975205

LEZAMA Y URRUTIA, Yosu de
Apuntes de lingüística aborigen. - Bogotá : Universidad Santo Tomás, Centro de Enseñanza Desescolarizada, 1979
271p. - Includes bibliography
r6707611 X.950/33522
8975205

AFRICAN LANGUAGES
Influence on Gullah dialect of English language

TURNER, Lorenzo Dow
Africanisms in the Gullah dialect. - [1st ed. reprinted]. - Ann Arbor : University of Michigan Press, 1974
xviii,317p : 2 ill. - Bibliography : p.293-299. - Includes index
0-472-08915-3 X.909/41837
1259458

AFRICAN LANGUAGES
Intonation

TONS et accents dans les langues africaines / Gladys Guarisma (éditeur). - Paris : SELAF, 1981
143p. - (Documents / LACITO. Afrique ; 7)
2-85297-109-7 X.950/32165
9073809

AFRICAN LANGUAGES
Linguistics, to date - Conference proceedings

CENTRES DE LINGUISTIQUE APPLIQUÉE D'AFRIQUE FRANCOPHONE. Table ronde (2e : 1975 : Lomé)
Deuxième Table ronde des Centres de linguistique comparée d'Afrique francophone, Lomé, 13-15 décembre 1975. - Dakar : Centre de Linguistique appliquée de Dakar, 1975
59leaves. - (Réalités africaines et langue française ; numéro spécial)
r6207231 X.955/1050
8955387

AFRICAN LANGUAGES
Linguistics, to date - Serials

CURRENT approaches to African linguistics. - Vol.1- - Dordrecht : Foris, 1983-. - (Publications in African languages and linguistics ; 1, 5 etc.)
90-7017-668-8 X.0950/898
8955387

AFRICAN LANGUAGES
Loanwords - Dictionaries

CABRAL, António Carlos Pereira
Empréstimos linguísticos nas línguas moçambicanas. - Lourenço Marques : [s.n.], 1975
78p
r6030914 X.909/37228
115480X

AFRICAN LANGUAGES
Prehistoric African languages

The ARCHAEOLOGICAL and linguistic reconstruction of African history / edited by Christopher Ehret and Merrick Posnansky. - Berkeley ; London : University of California Press, c1982
vii,299p : ill. - Bibliography: p269-295. - Includes index
0-520-04593-9 X.950/25923
2982250

AFRICAN LANGUAGES
Research, 1945-

RECENT German research on Africa / edited by Bernd Heine = Rapport sur la recherche africanistique allemande. - Bonn : die Forschungsgemeinschaft, [1982]-
r6306497 X.0950/532
8890927

AFRICAN LANGUAGES
Research, 1945- - Conference proceedings

ATELIER INTERNATIONAL DE DIALECTOLOGIE AFRICAINE (1980 : Ivry)
Dialectologie et comparatisme en Afrique noire. - Paris : Société d'études linguistiques et anthropologiques de France, 1980
306p. - (Oralité-documents ; 2)
2-85297-064-3 X.950/24032
8890927

AFRICAN LANGUAGES
Syntax

HEINE, Bernd
A typology of African languages based on the order of meaningful elements. - Berlin : Reimer, [1976]
89p,[2]folded leaves of plates. - (Kölner Beiträge zur Afrikanistik ; Bd 4). - Bibliography
r6302065 X.0950/105(4)
8706662

AFRICAN LANGUAGES
Translation of Bible into African languages

DAMMANN, Ernst
Die Übersetzung der Bibel in afrikanische Sprachen. - München, 1975
167p.. - (Abhandlungen der Marburger Gelehrten Gesellschaft ; Jahrg. 1972, Nr 3)
3-7705-1122-0 P.971/203[1972.no.3]
0886785

AFRICAN LANGUAGES
Vocabulary

SKINNER, A. Neil
Afroasiatic vocabulary. - Berlin : Reimer, 1984
65p. - (Africana marburgensia. Special issue ; 7)
r6310389 P.801/731a[no.7]
9143351

AFRICAN LANGUAGES
West African languages - Serials

The JOURNAL of West African languages. - Vol.1, no.1 (1965)-. - Ibadan : Published by the West African linguistic society at the Ibadan University Press, [1965]-
Holdings: Vol.9, no.1 (1974)-
r6801841 P.901/2086
8972443

AFRICAN LANGUAGES
Written African languages

GUREVICH, N. M. (Naum Manuilovich)
Mladopis'mennye îazyki Afriki. - Moskva : Nauka. Glavnaîa redakfsiîa vostochnoĭ literatury, 1981
153 p.
r6482548 X.950/32696
9084355

AFRICAN LANGUAGES Curriculum subjects. Schools. Africa
Teaching

AWONIYI, Timothy Adedeji
The teaching of African languages. - London : Hodder and Stoughton, 1982
165p : 3ill. - (Foundations of education in Africa). - Bibliography: p160-165
0-340-28171-5 X.950/18396
2840316

AFRICAN LANGUAGES. West Africa
Use in education

MOTHER tongue education / edited by Ayo Bamgbose. - London : Hodder and Stoughton [etc.], 1976
153p
0-340-20120-7 X.529/30981
0730270

AFRICAN LEGENDS
African tales & legends

CAREY, Margaret
Myths and legends of Africa. - Feltham : Hamlyn, c1970
159p : col.ill. - (Hamlyn all-colour paperbacks). - Bibliography: p157. - Includes index
B82-38338 X.958/11826
1443046

KOULIBALY, Isaïe Biton
La légende de Sadjo. - Abidjan : Éditions CEDA, [1981]
63p : ill
r6211072 X.958/18910
8004722

AFRICAN LEGENDS
African tales & legends - Anthologies

ELLIOT, Geraldine
The hunter's cave. - London : Routledge & Kegan Paul, 1951
174p : ill
0-7100-0861-9 X.990/17796 Woolwich
2561034

The long grass whispers. - London : Routledge & Kegan Paul, 1949
132p : ill
0-7100-0860-0 X.990/17797 Woolwich
2561034

The singing chameleon. - London : Routledge & Kegan Paul, 1957
168p : ill
0-7100-0862-7 X.950/9131
2561034

Where the leopard passes. - London : Routledge & Kegan Paul, 1949
133p : ill
0-7100-0863-5 X.990/17798 Woolwich
2561034

FABLES from Africa / collected by Jan Knappert ; illustrated by Jeroo Roy. - London : Evans, 1980
64p : ill
0-237-44985-4 X.950/12814
1888218

PITCHER, Diana
Tokoloshi. - London : Skilton and Shaw, 1980
64p : ill
0-7050-0081-8 X.981/22444
1888218

AFRICAN LITERATURE. Curriculum subjects. Schools. Africa
Teaching, to date

PRETORIUS, W. J.
Teaching African literature. - 2nd. impression. - Pretoria : University of South Africa, 1982
iv,68p. - Includes bibliographies
0-86981-277-7 X.529/63688
9001611

TEACHING of African literature in schools / edited by Eddah Gachukia and S. Kichamu Akivaga. - Nairobi : Kenya Literature Bureau, 1978-
v.
r6807883 X.529/56841
9001611

AFRICAN LITERATURES
See also
Ballads in Asante-Twi
Drama in Efik
Drama in Ewe
Drama in Idoma
Drama in Kikuyu
Drama in Ndebele
Drama in Pedi
Drama in Shona
Drama in Sotho
Drama in Tsonga
Drama in Venda
Drama in Wolof
Drama in Xhosa
Drama in Zulu
Epic poetry in Mandinka
Epic poetry in Nyanga

AFRICAN LITERATURES
See also
Essays in Tsonga
Essays in Xhosa
Fiction in African languages
Fiction in Bemba
Fiction in Kikuyu
Fiction in Lozi
Fiction in Ndebele
Fiction in Nyanga
Fiction in Nyanja
Fiction in Pedi
Fiction in Shona
Fiction in Sotho
Fiction in Tsonga
Fiction in Twi
Fiction in Venda
Fiction in Xhosa
Fiction in Zulu
Igbo literature
Orally transmitted literature in African languages
Orally transmitted literature in Fiba
Orally transmitted literature in Mende
Orally transmitted literature in Plateau Tonga
Orally transmitted poetry in Bantu languages
Orally transmitted poetry in Nyanga
Poetry in Acholi
Poetry in African languages
Poetry in Bambara
Poetry in Basa
Poetry in Bemba
Poetry in Kikuyu
Poetry in Lozi
Poetry in Ndebele
Poetry in Nupe
Poetry in Pedi
Poetry in Rwanda
Poetry in Shona
Poetry in Sotho
Poetry in Swazi
Poetry in Tsonga
Poetry in Venda
Poetry in Xhosa
Poetry in Zulu
Songs in Tiv
Swahili literature
Tswana literature
Yoruba literature

AFRICAN LITERATURES
to 1980

GÉRARD, Albert S. (Albert Stanislas)
African language literatures. - Harlow : Longman, 1981
xv,398p. - Includes index
0-582-64352-x X.950/11109
2641453

NKOSI, Lewis
Tasks and masks. - Harlow : Longman, 1981
202p. - (Longman studies in African literature). - Includes index
0-582-64145-4 X.950/9730
2608839

AFRICAN LITERATURES
to 1982. Myth

OKPEWHO, Isidore
Myth in Africa. - Cambridge : Cambridge University Press, 1983
x,305p : ill. - Bibliography: p289-298. - Includes index
0-521-24554-0 X.950/27560
3040224

AFRICAN LITERATURES
to date

COMPARATIVE approaches to modern African literature / edited by S.O. Asein. - Ibadan : University of Ibadan. Department of English, [1982]
x,153p
r6814279 X.950/37483
8724970

EGEJURU, Phanuel Akubueze
Towards African literary independence. - Westport, Conn. : Greenwood Press, 1980
173p. - (Contributions in Afro-American and African studies ; no.53)
0-313-22310-6 X.950/23462
8724970

RAZVITIE literatury v nezavisimykh stranakh Afriki (60-70-e gody XX veka) / [redkollegiîa: I.D. Nikiforova (otv. redaktor) ... i dr.]. - Moskva : Nauka, Glavnaîa redakfsiîa vostochnoĭ literatury, 1980
253p
r64299292 X.950/29117
8724970

SCHIPPER-DE LEEUW, Mineke
Afrikaanse letterkunde. - Utrecht : Spectrum, [1983]
336p. - (Aula pockets ; 696). - Includes bibliography
90-274-5463-9 X.958/23593
8724970

TOWARD the decolonization of African literature. - Enugu : Fourth Dimension, 1980-
v.
978-15-6102-5 X.0950/767
8724970

WHEN the drumbeat changes / edited by Carolyn A. Parker, Stephen H. Arnold ... - Washington, D.C. : Three Continents, 1981
293p
0-89410-262-1 X.950/32780
8724970

AFRICAN LITERATURES
to date - Anthologies

MUTLOATSE, Mothobi
Reconstruction. - Johannesburg : Ravan Press, 1981
320p. - (Staffrider series ; no.8)
0-86975-207-3 X.958/23682
8724970

AFRICAN LITERATURES
to date - Conference proceedings

DUKH Tashkenta torzhestvuet / [otvetstvennyĭ redaktor : L. Kaĭumov] ; [sostavitel' : E. Kalina]. - Tashkent : Izdatel'stvo literatury i iskusstva im. Gafura Guliama, 1980
158p,[4]p of plates : ill
r6484765 X.950/11734
8724970

LITERATUREN Asiens und Afrikas / herausgegeben von Fritz Gruner. - Berlin : Akademie, 1981
x,396p. - (Studien über Asien, Afrika und Lateinamerika ; Bd 26)
r6301284 X.0709/574(26)
8724970

NEO-AFRICAN literature and culture / editors : Bernth Lindfors and Ulla Schild. - Wiesbaden : B. Heymann, [1976]
352p. - (Mainzer Afrika-Studien ; 1) (Edition Ethnos). - "A bibliography of the works of Janheinz Jahn" : p.24-31
3-88055-500-1 X.0950/198.(1)
8724970

AFRICAN LITERATURES
to date - Serials

BAOBAB / herausgegeben vom Arbeitskreis der Freunde Afrikanischer Literatur. - 1-. - Göttingen : Edition Herodot, [1981]-
r6304653 X.0950/468
8724970

GANGA. - Vol.1 [1979]-. - [Maiduguri] : University of Maiduguri. Department of English, [1979]-
v.
Holdings: vol.2, no.3, 1980-
r6814284 X.0950/901
8724970

AFRICAN LITERATURES
to date. Aesthetics - Conference proceedings

COLLOQUE SUR LITTÉRATURE ET ESTHÉTIQUE NÉGRO-AFRICAINES (1974 : Abidjan)
Colloque sur littérature et esthétique négro-africaines. - Abidjan : Les nouvelles éditions africaines, 1979
358p
2-7236-0544-2 X.950/5562
8719500

AFRICAN LITERATURES
1945-

SOVREMENNYE literatury Azii i Afriki i ideologicheskaîa bor'ba / [otvetstvennye redaktory I.D. Nikiforova, N.I. Nikulin]. - Moskva : Nauka, Glavnaîa redakfsiîa vostochnoĭ literatury, 1982
270 p.. - Includes bibliographies
r6403847 X.950/38132
9129790

AFRICAN LITERATURES
African folk literatures

AFRIKANSKAÎÂ skazka / [otvetstvennyĭ redaktor : V.A. Vinogradov]. - Moskva : Nauka, Glavnaîa redakîsiîa vostochnoĭ literatury, 1984
346p
r6490290 X.950/39031
8746214

ATLANTIS / herausgegeben von Leo Frobenius. - Nendeln : Kraus Reprint, 1978
10v.in 5 : ill
3-262-00024-8 X.950/7543
8746214

MÉLANGES de culture et de linguistique africaines / édités par C. Faïk-Nzuji Madiya et Erika Sulzmann. - Berlin : Reimer, 1983
xii,611p,plate : ill. - (Mainzer Afrika-Studien ; Bd.5). - Includes bibliographies
3-496-00160-7 X.0950/198(5)
8746214

AFRICAN LITERATURES
African folk literatures - Bibliographies

GÖRÖG, Veronika
Littérature orale d'Afrique noire. - Paris : Maisonneuve et Larose, 1981
394p
2-7068-0819-5 X.950/23575
8746214

AFRICAN LITERATURES
Documents on African languages & literatures in stock of Stadtbibliothek Winterthur - Catalogues

STADTBIBLIOTHEK WINTERTHUR
Der Katalog der Africana-Sammlung in der Stadtbibliothek Winterthur. - Basel : Basler Afrika Bibliographien, 1977-. - (Mitteilungen der Basler Afrika Bibliographien ; v.17, 27, etc.)
r6063473 2725.c.88
1712705

AFRICAN LITERATURES
Influence of religions, to date

UMEZINWA, Wilberforce A.
La religion dans la littérature africaine. - Kinshasa : Presses universitaires du Zaïre, 1978
185p. - (Bibliothèque du Centre d'études des religions africaines ; 2)
r6206604 X.200/37384
8949042

AFRICAN LITERATURES
Orally transmitted literature - For Kenyan students

LUSWETI, B. M.
The hyena and the rock. - London : Macmillan, 1984
viii,125p
0-333-34628-9 X.950/35767
3175200

AFRICAN LITERATURES
Orally transmitted literature, to 1969

FINNEGAN, Ruth
Oral literature in Africa. - Nairobi ; Oxford : Oxford University Press, 1976
xix,558p,fold plate. - (Oxford library of African literature). - Bibliography: p.522-536. - Includes index
0-19-572413-5 X.0909/44(22)
1174487

AFRICAN LITERATURES
Orally transmitted literature, to 1979

OKPEWHO, Isidore
The epic in Africa. - New York ; Guildford : Columbia University Press, 1979
xx,288p : 2 ill. - Bibliography: p.271-279. - Includes index
0-231-04400-3 X.989/53586
1919970

AFRICAN LITERATURES
Orally transmitted literature, to 1979 - Conference proceedings

GENRES, forms, meanings / edited by Veronika Görög-Karady. - Oxford : JASO, 1982
vi,122p : ill. - (JASO occasional papers ; no.1). - Includes bibliographies
B82-39561 X.950/16078
1919970

AFRICAN LITERATURES
Research - Conference proceedings

RESEARCH priorities in African literatures / edited by Bernth Lindfors. - München ; London : Zell, 1984
222p. - Includes bibliographies
0-905450-17-5 2719.e.988
328123x

AFRICAN LITERATURES
Translations into Russian language, to date - Bibliographies

ĖL'VOVA, V. A. (Valeriîa Abramovna)
Khudozhestvennaîa literatura stran Afriki v sovetskoi pechati 1965-1974. - Moskva : Kniga, 1976
303p
r6463700 X.950/21923
8953708

AFRICAN MEDICAL & RESEARCH FOUNDATION INTERNATIONAL

AFRICAN MEDICAL & RESEARCH FOUNDATION INTERNATIONAL
[Miscellaneous publications not catalogued separately]. - London : the Foundation, 1980-
r6108495 X.0325/100
2298414

AFRICAN MEDICAL AND RESEARCH FOUNDATION
Wood, Michael, b.1919 - Biographies

WOOD, Michael, b.1919
Go an extra mile. - London : Collins, 1978
160p,[16]p of plates : ill(some col)
0-00-216267-9 X.320/11328 Woolwich
1517791

WOOD, Michael, b.1919
Go an extra mile. - London : Fount Paperbacks, 1980
160p
0-00-625918-9 H.80/503 Woolwich
1517791

AFRICAN METHODIST EPISCOPAL CHURCH. Zambia

JOHNSON, Walton R.
Worship and freedom. - London : International African Institute, 1977
xx,152p : 1 ill. - Bibliography: p.137-146. - Includes index
0-85302-053-1 X.200/31209
1301330

AFRICAN MIGRANTS
to ca 1970

The AFRICAN diaspora / edited by Martin L. Kilson, Robert I. Rotberg. - Cambridge, Mass. ; London : Harvard University Press, 1976
xv,510p. - Includes index
0-674-00779-4 X.800/25874
1032542

AFRICAN MUSIC

CHERNOFF, John Miller
African rhythm and African sensibility. - Chicago ; London : University of Chicago Press, c1979
xv,261p,[20]p of plates : ill. - (A Phoenix book). - Bibliography: p224-247. - Includes index
B85-01592 X.439/13139
2877457
African rhythm and African sensibility. - Chicago ; London : University of Chicago Press, 1979
xvii,261p,[20]p of plates : ill. - Bibliography: p.223-247. - Includes index
0-226-10344-7 X.431/11116
2101351

KEBEDE, Ashenafi
Roots of black music. - Englewood-Cliffs ; London : Prentice-Hall, c1982
xiv,162p : ill. - Includes index
0-13-783167-6 X.431/12079
2877457

NKETIA, J. H. Kwabena (Joseph Hanson Kwabena)
The music of Africa. - London : Gollancz, 1975, c1974
x,278p : ill. - Bibliography: p263-271. - List of sound discs: p247-249. - Includes index
0-575-03234-0 X.439/12279
2877457

AFRICAN MUSIC
to 1974

NKETIA, J.H. Kwabena
History and the organisation of music in West Africa. - Legon : University of Ghana, Institute of African Studies, [1975?]
24p.
r6028633 X.435/418
0263206

AFRICAN MUSIC
to date

AUSSEREUROPÄISCHE Musik in Einzeldarstellungen. - Gemeinschaftliche Ausg. - München : Deutscher Taschenbuch Verlag ; Kassel : Bärenreiter, 1980
479p. - (Edition MGG). - Bibliography: p.447-477
3-423-04330-x X.439/11414
8616124

KAUFMAN, Fredrick
The African roots of jazz. - [Sherman Oaks, Calif.] : Alfred Publishing, c1979
ix,147p : ill
0-88284-065-7 Cup.575.N.44
8616124

MERRIAM, Alan P.
African music in perspective. - New York : Garland Publishing, 1982
xvi,506p : ill. - (Critical studies on black life and culture ; vol.6). - Includes bibliography
0-8240-9461-1 X.439/13010
8616124

MUSIK in Afrika / herausgegeben von Artur Simon : mit 20 Beiträgen ... von A. M. Dauer ... [et al.]. - Berlin : Museum für Völkerkunde, c1983
432p : ill. - (Veröffentlichungen des Museums für Völkerkunde Berlin ; n.f., 40. Abteilung Musikethnologie ; 4). - Includes bibliography
3-88609-121-x Ac.9233.tl(5)[vol.4]
8616124

AFRICAN MUSIC
to date - Conference proceedings

SYMPOSIUM ON ETHNOMUSICOLOGY (1980 : Grahamstown)
Papers presented at the Symposium on Ethnomusicology. - Grahamstown : International Library of African Music, 1981
35p
r6814168 X.439/13098
8616124

AFRICAN MUSIC
to date. Sound recordings, to date - Catalogues - Serials

UNIVERSITÄT BAYREUTH. Archiv moderner afrikanischer Musik
Kommentierte Kataloge zur afrikanischen Musik. - Nr.1-. - Bayreuth : Archiv moderner afrikanischer Musik, Iwalewa, Universität Bayreuth, 1982-
r6308131 X.0435/126
8995508

AFRICAN MUSIC
African folk music

BEBEY, Francis
African music. - London : Harrap, 1975
viii,184p : ill. - List of records: p.150-176
0-245-52735-4 X.431/5010
0351407

African music. - New York : Lawrence Hill, 1975
184p
0-88208-051-2 X.431/10968
0351407

AFRICAN MUSIC
African guitar music, to date

LOW, John
Shaba diary. - Wien-Föhrenau : E. Stiglmayr, 1982
123p. - (Acta ethnologica et linguistica ; Nr 54. Series musicologica ; 4. Series Africana ; 16)
r6308748 X.0802/102(54)
909248x

AFRICAN MUSIC
East African music, to date

KUBIK, Gerhard
Ostafrika. - Leipzig : Deutscher Verlag für Musik, 1982
250p : ill. - (Musikgeschichte in Bildern ; Bd.1, Musikethnologie ; Lfg.10). - Includes bibliography
r6308085 L.20.p.8(1)[Lfg.10]
8973946

AFRICAN MUSIC
North African folk music

LACHMANN, Robert
Zwei Aufsätze ; Orientalische Musik und Antike. - Jerusalem : Magnes Press, 1974
59p : ill. - (Yuval monograph series ; 2) (Posthumous works / Robert Lachmann ; 1)
r6302189 X.431/11428 Woolwich
8750033

AFRICAN MUSIC
Sound discs, 1902-1975. Collections - Catalogues

STONE, Ruth M.
African music and oral data. - Bloomington ; London : Indiana University Press, 1976
xix,412p. - Includes index
0-253-30262-5 X.431/10260
0559849

AFRICAN MUSICAL INSTRUMENTS
East African musical instruments

HYSLOP, Graham
Musical instruments of East Africa. - Nairobi : Nelson Africa.
Vol.1: Kenya. - 1975
viii,64p : ill
0-17-511250-9 X.0439/119
0505323

AFRICAN MYTHS

PARRINDER, Geoffrey
African mythology. - New rev. ed. - London : Hamlyn, 1982
144p : ill(some col.). - (Library of the world's myths and legends). - Bibliography. - Includes index
0-600-34279-4 X.955/1285
2806363

AFRICAN MYTHS
Special subjects: Creation

The ORIGIN of life and death / edited by Ulli Beier. - London : Heinemann, 1966
x,65p. - (African writers series ; 23)
0-435-90023-4 012212.e.1/23
3282554

AFRICAN NATIONALIST MOVEMENTS. South Africa to 1973

WALSHE, Peter
Black nationalism in South Africa. - Johannesburg : Spro-cas [etc.], 1973
[3],40p,[8]p of plates : ill
0-86975-028-3 X.709/32393
1523511

AFRICAN NATIONALIST MOVEMENTS. Zimbabwe to 1982

ASTROW, André
Zimbabwe. - London : Zed, 1983
xv,254p. - Bibliography: p225-242. - Includes index
0-86232-140-9 X.529/63144
3120953

AFRICAN NATIONALIST MOVEMENTS. Zimbabwe 1960-1977

MAALØE, Erik
Zimbabwe. - Arhus : Internationalt Forum, 1978
244p : ill. - Includes bibliographies
87-87780-00-3 X.709/50365
2266601

AFRICAN NATIONALIST MOVEMENTS. Zimbabwe 1966-1974

MAXEY, Kees
The fight for Zimbabwe. - [New ed.]. - London : Collings, 1975
[4],196p
0-901720-81-x X.708/41356
052803x

AFRICAN PARTY FOR THE INDEPENDENCE OF CAPE VERDE
See PAICV

AFRICAN PARTY FOR THE INDEPENDENCE OF GUINÉ AND CAPE VERDE
See PAIGC

AFRICAN PHILOSOPHY

AFRICAN philosophy / edited by Richard A. Wright. - 3rd ed. - Lanham ; London : University Press of America, c1984
xvi,317p : ill. - Bibliography: p265-306. - Includes index
0-8191-3784-7 X.529/65040
2964716

HOUNTONDJI, Paulin J.
African philosophy. - London : Hutchinson University Library for Africa, 1983
221p. - Includes index
0-09-151940-3 X.529/57005
2964716
- [Another Copy : paperback.] ISBN 0-09-151941-1 X.529/59583

AFRICAN PHILOSOPHY
to date

APOSTEL, Leo
African philosophy. - Gent : Story-Scientia, 1981
428p. - (Philosophy and anthropology ; vol.2). - Bibliography
90-643-9183-1 X.520/28237
8833486

FILOSOFSKOE nasledie narodov Vostoka i sovremennost' / [redaktsionnaia kollegiia V.G. Burov, D.V. Dzhokhadze, E.A. Frolova]. - Moskva : Nauka, Glavnaia redaktsiia vostochnoi literatury, 1983
243 p.. - Includes bibliographical references
r6489832 X.529/68031
8833486

MBARGANE GUISSÉ, Youssouph
Philosophie, culture et devenir social en Afrique noire. - Dakar : Nouvelles éditions africaines, 1979
186p
2-7236-0435-7 X.529/59748
8833486

AFRICAN PHILOSOPHY
to date - Conference proceedings

SEMINAR ON AFRICAN PHILOSOPHY (1976 : Addis Ababa)
Proceedings of the Seminar on African Philosophy, Addis Ababa, 1-3 December 1976. - Addis Ababa : Addis Ababa University, 1980
xxii,473p : ill
r6209178 X.520/28767
8833486

SYMPOSIUM on philosophy in the present situation of Africa / edited by Alwin Diemer. - Wiesbaden : Steiner, 1981
58p
3-515-03273-8 X.520/27203
8833486

AFRICAN PHILOSOPHY
to date - Serials

AFRIQUE et philosophie. - no 1(1979)- - Kinshasa / Limete : Département de philosophie et religions africaines, Faculté de théologie catholique de Kinshasa, [1979]-
Holdings: no 6, 1982-
r6210281 X.0529/863
8833486

CONSÉQUENCE / revue du Conseil Inter-Africain de Philosophie = Journal of the Inter-African Council for Philosophy. - 1 (Jan. June 1974)-. - Cotonou : Inter-African Council for Philosophy, 1974-
r6207388 X.0529/865
8833486

AFRICAN POSTAGE STAMPS
- Catalogues - Serials

STANLEY Gibbons stamp catalogue. Part 13, Africa since independence, F-M. - 1st ed. (1981)-. - London : Stanley Gibbons, 1981-
v. : ill
B81-35118 Cup.711/32
2550172

STANLEY Gibbons stamp catalogue. Part 14, Africa since independence, N-Z. - 1st ed. (1981)-. - London : Stanley Gibbons, 1981-
v. : ill
B82-09680 X.429/12928
2550172

AFRICAN POSTAGE STAMPS
East African postage stamps

MINNS, John
British East Africa. - London : Royal Philatelic Society, 1981
viii,134p,[4]p of plates : ill(some col.). - Bibliography: p132. - Includes index
B81-34791 X.520/24850
2549689

AFRICAN POSTMARKS
African paquebot postmarks, to 1979

DRECHSEL, Edwin
The paquebot marks of Africa, the Mediterranean countries, and their islands. - London : Pall Mall Stamp Co. on behalf of Robson Lowe, c1980
72p : ill
0-85397-185-4 L.49/794
2387786

AFRICAN PROVERBS

MBOUBI, Pierre Claver
Recueil de proverbes koukouya. - [Brazzaville?] : P.C. Mboubi, [1974?]
40leaves
r6208551 X.955/1268
8004676

NDZOMO MALLA, Jean-Marie
Proverbes, vrais ou faux?. - Bandundu : CEEBA publications, 1982
86p. - (CEEBA publications. Série II ; v.81)
r6209815 X.955/1687
8004676

AFRICAN PROVERBS
- Anthologies

A POCKET book of proverbs and other original fragments / [compiled by] Dau Aldulpha. - London : The compiler, 1976
[4]p,[28]leaves. - (The African heritage series. philosophy ; P2)
0-9505106-0-2 X.907/25109
0677760

AFRICAN REFORMED CHURCH IN ZIMBABWE

VAN DER MERWE, W. J.
From mission field to autonomous church in Zimbabwe. - [Transvaal] : N.G. Kerkboekhandel Transvaal, [1981]
252p : ill. - Bibliography
0-7987-0278-8 X.200/42875
9087109

AFRICAN RELIGIONS
African negro religions

MITCHELL, Robert Cameron
African primal religions. - Niles, Ill. : Argus Communications, 1977
xii,100p : col ill. - (Major world religions). - Bibliography : p.95
0-913592-97-8 X.100/17324
0154547

MUZUNGU, Bernardin
Une théologie anthropologique. - Bujumbura : Presses Lavigerie, 1981
xlv,156p. - (Le Dieu de nos pères ; 3)
r6211563 X.200/42955
0154547

RAY, Benjamin Caleb
African religions. - Englewood Cliffs ; London : Prentice-Hall, 1976
xii,239p : ill. - (Prentice-Hall studies in religion). - Bibliography: p.218-231. - Includes index
0-13-018630-9 X.100/16226
0154547

SHORTER, Aylward
African Christian theology - adaptation or incarnation?. - London : G. Chapman, 1975
viii,167p. - Includes index
0-225-66112-8 X.100/16265
0626767

THEORETICAL explorations in African religion / edited by Wim van Binsbergen and Matthew Schoffeleers. - London : KPI, 1985
x,389p. - (Monographs from the African Studies Centre, Leiden). - Includes bibliographies and index
0-7103-0049-2 X.200/46249
0154547

WEST African religion. - No.1 (May 1963)-. - Nsukka : Department of Religion, University of Nigeria, 1963-
Holdings: No.1, 4, 10-
r6202384 P.101/2029
0154547

AFRICAN RELIGIONS
African negro religions. Doctrines. Eschatology compared with Christian doctrine of eschatology

MBITI, John Samuel
New Testament eschatology in an African background. - London : S.P.C.K., 1978
xii,216p. - Bibliography: p.192-199. - Includes index
0-281-03659-4 X.100/17863
1580000

AFRICAN RELIGIONS
African negro religions. Doctrines. God

MBITI, John Samuel
Concepts of god in Africa. - London : S.P.C.K., 1975
xv,348p. - Bibliography: p.317-323. - Includes index
0-281-02902-4 X.100/16461
0692921

AFRICAN RELIGIONS
Influence on Brazilian culture

VALENTE, Waldemar
Sincretismo religioso afro-brasileiro. - 2a ed. - São Paulo : Companhia Editora Nacional, 1976
xxiv,117p. - (Brasiliana ; v. 280). - Bibliography : p. 111-117
r6040862 X.100/15788
1130439

AFRICAN RELIGIONS. Bahia. Brazil
African negro religions

BASTIDE, Roger
[Le CANDOMBLÉ de Bahia, rito Nagô. Portuguese]. O candomblé da Bahia, rito nagô. - 2a ed. - São Paulo : Companhia Editora Nacional, 1978
298p. - (Brasiliana ; v.313). - Bibl
r6159485 X.200/33095
2443724

AFRICAN RELIGIONS. Bahia. Brazil
African negro religions, to date

AMADO, Jorge
Iconografia dos deuses africanos no Candomblé da Bahia. - [São Paulo] : Raízes, 1980
266,[32]p : col.ill
r6704043 Cup.408.rr.7
8817979

AFRICAN RELIGIONS. Brazil
African negro religions

BASTIDE, Roger
The African religions in Brazil. - Baltimore ; London : Johns Hopkins University Press, 1978
xxix,494,[1]p. - (Johns Hopkins studies in Atlantic history and culture). - Includes index
0-8018-2056-1 X.200/31847
1809997

AFRICAN RELIGIONS. Brazil
African negro religions–continued

FARELLI, Maria Helena
Os rituais secretos da magia negra e do candomblé. - Rio de Janeiro : Editora Eco, 1976
159p : ill
r6066058 X.519/38875
090077X

FLASCHE, Rainer
Geschichte und Typologie afrikanischer Religiosität in Brasilien. - Marburg an der Lahn : Selbstverlag, 1973
302 p. : ill. - (Marburger Studien zur Afrika- und Asienkunde. Serie A ; Bd. 1). - Bibliography : p. 271-291
r6000147 X.0100/683(1)
090077X

MIRANDA, Roberto dos Santos
Candomblé, umbanda e suas obrigações. - Rio de Janeiro : Editora Espiritualista, [1975?]
167p : ill
r6035538 X.908/37678
090077X

O'GORMAN, Frances
Aluanda. - Rio de Janeiro : Livraria Francisco Alves Editora, 1977
108p
r6084271 X.709/51463
090077X

RIBEIRO, José
Culto malê. - [Rio de Janeiro] : Editora Espiritualista, [1973?]
203p
r6066129 X.519/39103
090077X

Pomba-Gira Mirongueira. - 2.ed. - Rio de Janeiro : Editora Espiritualista, [1974?]
147p
r6066128 X.519/39104
090077X

VELHO, Yvonne Maggie Alves
Guerra de Orixá. - Rio de Janeiro : Zahar, 1975
170p.. - (Biblioteca de antropologia social)
r6025157 X.907/22808
090077X

AFRICAN RELIGIONS. Brazil
African negro religions - Dictionaries

CACCIATORE, Olga Gudolle
Dicionário de cultos afro-brasileiros com origem das palavras. - Rio de Janeiro : Forense Universitária, 1977
279p. - Bibliography : p.271-279
r6066625 X.208/3413
2117290

AFRICAN RELIGIONS. Brazil
African negro religions, to date

LODY, Raul
Santo também come. - Recife : Instituto Joaquim Nabuco de Pesquisas Sociais, 1979
127p : ill. - (Série Estudos e pesquisas ; 11)
r6701656 X.0950/208(15)
8718148

AFRICAN RELIGIONS. Cuba
African negro religions

CROS SANDOVAL, Mercedes
La religión afrocubana. - Madrid : Playor, [1975]
285p : ill. - (Colección Plaza Mayor libre). - Includes bibliographies
84-359-0123-8 X.100/27791
1858157

AFRICAN RELIGIONS. Cuba
African negro religions. Rituals

CABRERA, Lydia
La laguna sagrada de San Joaquín. - Madrid : Ediciones Erre, 1973
105p : ill. - (Los de hoy)
84-399-0714-1 X.802/4224
1500546

La Regla Kimbisa del Santo Cristo del Buen Viaje. - Miami : [s.n.], 1977
85p : ill. - (Colección del Chichereku en el exilio)
r6160533 X.208/10265
1500546

Yemayá y Ochún. - Madrid : [s.n.], 1974
359p. - (Colección del Chicherukú en el exilio)
84-400-7153-1 X.200/11833
1500546

AFRICAN RELIGIONS. Cuba
African negro religions. Rituals, to date

CABRERA, Lydia
Koeko Iyawó. - [S.l.] : CR, c1980
xv,231p. - (Colección del Chicherekú en el exilio)
r6708692 X.200/43489
8625700

AFRICAN RELIGIONS. Latin America
African negro religions, to date

DIOSES COMO LOS VEMOS (Exhibition) (1977 : Caracas)
Africa América .... - Caracas : Instituto Autónomo Biblioteca Nacional, 1977
91p : ill. - (Serie "Juan Pablo Sojo" ; v.2)
r6701676 L.49/932
8718164

VERGER, Pierre Fatumbi
[Orixas, deuses iorubás na Africa e no Novo Mundo. French]. Orisha. - Paris : Métailié, 1982
293p : ill. - Includes bibliography
2-86424-017-3 X.205/1625
8718164

AFRICAN RELIGIONS. Latin America
African negro religions, to date - Conference proceedings

ENCUENTRO SOBRE PASTORAL CON GRUPOS AFROAMERICANOS (Cartagena, Colombia : 1980)
Los grupos afroamericanos. - [Bogotá] : CELAM, [1981?]
251p. - (Documentos / CELAM ; 45)
r6708738 X.200/44306
8718164

AFRICAN RELIGIONS. Maranhão. Brazil
African negro religions, to date

PEREIRA, Nunes
A casa das minas. - 2a ed. - Petrópolis : Vozes, 1979
245p.,plates. - Includes bibliography
r6707744 X.200/43553
9102477

AFRICAN RELIGIONS. Western Zimbabwe
African negro religions. Influence of Christian missions, to 1923

BHEBE, Ngwabi
Christianity and traditional religion in Western Zimbabwe, 1859-1923. - London : Longman, 1979
xvi,190p. - Bibliography: p.174-183. - Includes index
0-582-64237-x X.200/19070
1884719

AFRICAN RELIGIOUS LIFE
African negro religious life. Prayer

MBITI, John Samuel
The prayers of African religion. - London : S.P.C.K., 1975
xii,193p. - Bibliography: p.179-181. - Includes index
0-281-02872-9 X.100/16454
0691631

SHORTER, Aylward
Prayer in the religious traditions of Africa. - Nairobi ; London : Oxford University Press, 1975 [i.e. 1976]
xvii,146p : ill. - Bibliography: p.144-146
0-19-572350-3 X.200/20386
065275x
- [Another copy : paperback]
ISBN 0-19-572349-X X.100/18687

AFRICAN SCULPTURES
to date

GALLOIS DUQUETTE, Danielle
Dynamique de l'art bidjogo (Guinée-Bissau). - Lisbôa : Instituto de Investigação Científica Tropical, [1983]
261p,plates : ill. - Includes bibliography
r6210844
881578x
- [Another copy] X.421/25869

MUSÉE BARBIER-MÜLLER
Art ancien du Mali. - Genève : [le Musée], [1983]
[12]p : ill
r6210833 X.421/25722
881578x

AFRICAN SCULPTURES
to date - Catalogues

BASSANI, Ezio
Scultura africana nei musei italiani. - Bologna : Calderini, 1977
xx,131p : ill. - (Musei d'Italia-Meraviglie d'Italia ; 12)
r6504004 X.421/22670 Woolwich
881578x

AFRICAN SCULPTURES
African wood sculptures - Catalogues

ADAMS, Monni
Designs for living. - Cambridge, Mass. : Carpenter Center for the Visual Arts, c1982
143p : ill
0-674-19969-3 X.429/14564
2917203

AFRICAN SCULPTURES
West African sculptures

BAMERT, Arnold
[Afrika. English]. Africa. - London : Thames and Hudson, c1980
xii,332p : col.ill. - Bibliography: p327-329. - Includes index
0-500-23318-7 L.45/824
2344556

AFRICAN SCULPTURES
West African sculptures, to ca 1900

SEGY, Ladislas
African sculpture speaks. - 4th ed., enlarged. - New York : Da Capo Press, 1975
[12],346p : ill. - Bibliography : p. 281-292.-Includes index
0-306-80018-7 X.415/5295 Woolwich
1116541

AFRICAN SCULPTURES. Stock. Ewing Museum of Nations - Catalogues

EWING MUSEUM OF NATIONS
African art. - [Normal] : Ewing Museum of Nations : Illinois State University Foundation, [1975]
[126] p. : of ill
r6009619 X.419/3173 Woolwich
0691291

AFRICAN SECRET SERVICES
to date

FALIGOT, Roger
Services secrets en Afrique. - Paris : Le Sycomore, [1982]
133p
2-86262-146-3 X.809/58087
9052267

AFRICAN SPORTSMEN
- Serials

AFROSPORT. - Vol.1, no.1 (1978)-. - London : PDI Publications, 1978-
v. : ill
Holdings : vol.2, no.11 ; Aug. 1980, etc
B81-03280 P.443/406
2329506

AFRICAN STEREOTYPES. Africa. Special subjects. French & African writers, 1900-. Fiction in French
European & African stereotypes

MILBURY-STEEN, Sarah L.
European and African stereotypes in twentieth-century fiction. - London : Macmillan, 1980
xiii,188p. - Bibliography: p178-183. - Includes index
0-333-29143-3 X.981/22632
2079976

AFRICAN STEREOTYPES. West Africa. Special subjects. English & West African writers, 1900-. Fiction in English
European & African stereotypes

MILBURY-STEEN, Sarah L.
European and African stereotypes in twentieth-century fiction. - London : Macmillan, 1980
xiii,188p. - Bibliography: p178-183. - Includes index
0-333-29143-3 X.981/22632
2079968

AFRICAN STUDENTS
Accounting - For West African students

OKOLO, M. N. (Michael N.)
Key to book-keeping and accounts. - London : Longman, 1980
204p. - (Longman commercial studies)
0-582-65029-1 X.520/23418
237983x

OKOLO, Michael N.
Questions and answers, O-Level principles of accounts. - London : Longman, 1979
vii,216p. - (Study for success)
0-582-60692-6 X.529/36196
2154226

WOOD, Frank
Business accounting 1. - West African ed.. - Harlow : Longman, 1982
424p. - Includes index
0-582-65602-8 X.0529/729(1)
237983x

Business accounting 2. - West African ed.. - Harlow : Longman, 1983
498p. - Includes index
0-582-65603-6 X.0529/729(2)
237983x

AFRICAN STUDENTS
Africa, to 1979 - For East African students

SINGH, Malkiat
African history. - London : Evans, 1980
202p
0-237-50441-3 X.709/52212
2323338

AFRICAN STUDENTS
Africa, 1800-1975 - For African students

SAGAY, J. O.
Africa, a modern history (1800-1975). - London : Evans Bros, 1978
[5],425p : ill. - Includes index
0-237-50417-0 X.709/32972
181723x

AFRICAN STUDENTS
Africa, 1840-1914 - For East African students

TIDY, Michael
A history of Africa 1840-1914. - London : Hodder and Stoughton.
Vol.1: 1840-1880. - 1980
188p. - Bibliography: p178-180. - Includes index
0-340-24419-4 X.0809/2164(1)
2086417

Vol.2: 1880-1914. - 1981
220p. - Bibliography: p211. - Includes index
0-340-24411-9 X.809/50080
2086417

AFRICAN STUDENTS
Africa. Geographical features - For African students

PRITCHARD, J. M. (John Malcolm)
Africa. - London : Arnold, 1984
58p : ill
0-7131-8071-4 X.800/41370
3079155

TURNER, Howard
Africa. - Harlow : Longman, 1983
204p : ill. - Includes index
0-582-60381-1 X.800/37796
3079155

AFRICAN STUDENTS
Africa. Geographical features - For East African students

MINNS, W. J.
A geography of Africa. - London : Macmillan, 1978
[4],188p : ill
0-333-21928-7 X.700/25846
1735152

A geography of Africa. - New ed. - London : Macmillan, 1984
iii,195p : ill. - Includes index
0-333-37170-4 X.800/40676
1735152

AFRICAN STUDENTS
Africa. Tropical regions. Human geographical features - For African students

UDO, Reuben K. (Reuben Kenrick)
The human geography of tropical Africa. - Idadan : Heinemann Educational (Nigeria) ; London : Heinemann Educational, 1982
xii,244p : ill. - Bibliography: p234-237. - Includes index
0-435-95919-0 X.800/32501
2639122

AFRICAN STUDENTS
Africa. Tropical regions. Organisms. Distribution - For West African students

BISHOP, O. N. (Owen Neville)
The distribution of organisms. - London : Macmillan Education, 1977
[1],iv,28,[1]p : ill. - (Advanced readers in tropical biology)
0-333-21396-3 X.319/17244 Woolwich
1091727

AFRICAN STUDENTS
Agriculture - For African students

AYERST, Peter William
Agricultural science. - London : Evans Bros, 1978
63p : ill. - (Evans pocket facts)
0-237-50297-6 X.319/18537 Woolwich
1742639

OWEN, G. H.
O-level agriculture for Central Africa. - Harlow : Longman, 1984
236p : ill. - Includes index
0-582-65103-4 X.322/14001
1742639

AFRICAN STUDENTS
Agriculture - For East African students

NGUMY, James
Agriculture in practice. - [Nairobi] : Macmillan Kenya ; London : Macmillan, 1983
v,85p : ill. - Includes index
0-333-36083-4 X.322/13544
3131556

AFRICAN STUDENTS
Agriculture - For West African students

AKINSANMI, Oluyemi
Objective tests. - Harlow : Longman, 1980
vii,55p : ill. - (Study for success) (Study for success)
0-582-60688-8 X.329/14513 Woolwich
2559102

AKINYOSOYE, Vincent Oladoyin
Senior tropical agriculture. - London : Macmillan, 1976
[6],199p : ill. - Includes index
0-333-19239-7 X.311/8225
0028193

AKUBUILO, C. J. C.
Certificate practical agricultural science. - Harlow : Longman, 1984
iv,76p : ill
0-582-65118-2 X.322/14522
0028193

An introduction to practical agriculture for West Africa. - London : Macmillan, 1978
[6],154p : ill. - Includes index
0-333-21978-3 X.311/8889 Woolwich
0028193

ANTHONIO, Q. B. Olatunji (Quirino Bandele Olatunji)
General agriculture for West Africa. - London : Allen & Unwin, 1979
250p : ill. - Includes index
0-04-631012-6 X.329/19249
0028193

FAKOREDE, M. A. B.
Agricultural science. - London : Macmillan, 1977
[8],104p : ill. - (School Certificate revision course)
0-333-21229-0 X.319/17250 Woolwich
0028193

OKORIE, J. U.
Certificate notes. - Harlow : Longman, 1983
viii,119p : ill. - (Study for success)
0-582-65535-8 X.329/16792
0028193

AFRICAN STUDENTS
Association football - For African students

FOOTBALL and great footballers. - Harlow : Longman, 1981
48p : ill
0-582-78522-7 X.622/11823 Woolwich
2608952

AFRICAN STUDENTS
Bible. N.T. - For West African students

COUTTS, John
Revision guides, Bible knowledge, New Testament. - Harlow : Longman, 1982
iv,140p. - (Study for success) (Study for success)
0-582-65069-0 X.200/40004
2876841

AFRICAN STUDENTS
Bible. N.T. Acts - For West African students

TARRANT, C. J. C.
Mark and the Acts of the Apostles. - Harlow : Longman, 1983
iv,156p : ill. - (Certificate Bible knowledge)
0-582-64396-1 X.200/41127
0052604

AFRICAN STUDENTS
Bible. N.T. Matthew - For West African students

TARRANT, C. J. C.
Matthew and the Acts of the Apostles 1-15. - Harlow : Longman, 1982
v,153p : ill. - (Certificate Bible knowledge)
0-582-65096-8 X.200/38354
2727935

WILLIAMS, David, 19—
The gospel of Matthew. - [Glasgow] : Collins, 1982
192p. - (Bible knowledge for the West African School Certificate)
0-00-326135-2 X.200/38729
2727935

AFRICAN STUDENTS
Bible. N.T. Matthew - For West African students - Texts

[BIBLE. N.T. Matthew. English. Revised Standard. 1982]. St Matthew's Gospel / R.H. Horton, E.O.N. Aghaegbuna. - London : Edward Arnold in association with the National Association of Bible Knowledge Teachers of Nigeria and African Universities Press, 1982
xv,226p : ill. - (Bible Knowledge)
0-7131-8085-4 X.200/39978
2875888

AFRICAN STUDENTS
Bible. O.T. - For West African students

OKAFOR, S. N. O.
Bible knowledge for School Certificate. - London : Macmillan Education, 1981
vi,90p. - Includes index
0-333-31976-1 X.200/38562
2727951

TARRANT, C. J. C.
From the united monarchy to the fall of Samaria. - Harlow : Longman, 1981
vi,138p : ill. - (Certificate Bible knowledge)
0-582-65082-8 X.200/38353
2727951

AFRICAN STUDENTS
Bible. O.T. Kings - For West African students

WILLIAMS, David, 19—
From the institution of the Monarchy to the fall of the Northern Kingdom. - [London] : Collins, 1981
224p : ill. - (Bible knowledge for the West African school certificate)
0-00-326134-4 X.809/50449
2620510

AFRICAN STUDENTS
Bible. O.T. Samuel - For West African students

WILLIAMS, David, 19—
From the institution of the Monarchy to the fall of the Northern Kingdom. - [London] : Collins, 1981
224p : ill. - (Bible knowledge for the West African school certificate)
0-00-326134-4 X.809/50449
2620502

AFRICAN STUDENTS
Biology - For African students

STONE, R. H. (Robert Henry)
Revision biology for tropical schools. - Oxford : Oxford University Press, 1981
190p : ill. - Includes index
0-19-914060-x X.325/1132 Woolwich
2399261

AFRICAN STUDENTS
Biology - For West African students

SAVORY, Henry Jarvis
Junior biology. - Sunbury-on-Thames : Nelson. - (Science for tropical secondary schools)
Part 1. - 1979
viii,176p. - Includes index
Subsequent volumes, when received, will be placed at this shelfmark
0-17-511075-1 X.0322/213
2111713

AFRICAN STUDENTS
Biology. Graphs. Interpretation - For African students

HUNT, J. B.
Graph questions for Advanced Level tropical biology. - London : Macmillan, 1977
[1],iv,77p : ill
0-333-21150-2 X.319/17731 Woolwich
1295683

AFRICAN STUDENTS
Book-keeping - For East African students

MUTABIIRWA, D. E. R.
Book-keeping and accounts for East Africa. - Walton-on-Thames : Nelson, 1981
186p. - (Nelson Africa business studies). - Includes index
0-17-511491-9 X.520/25597
2591774

AFRICAN STUDENTS
Business arithmetic - For African students

NEWCOMB, V. N.
Practical calculations for business studies. - Chichester : Wiley, c1981
vii,156p : ill. - Includes index
0-471-27966-8 X.520/25383
2366258

AFRICAN STUDENTS
Business arithmetic - For East African students

BISHOP, K. R. (Kenneth Richard)
Business calculations for East Africa. - London : Cassell, 1981
106p : ill
0-304-30804-8 X.529/44705
2591545

AFRICAN STUDENTS
Business practices - For West African students

OMOTOSHO, Jacob 'Niyi
Multiple choice tests in business methods. - London : Macmillan Education, 1981
121p : ill
0-333-30507-8 X.520/25782
2607328

AFRICAN STUDENTS
Catholic Church, to 1980 - For African students

MCGRATH, Michael
Africa. - London : Mission Book Service, c1981
viii,227p. - Bibliography: p225
0-9507888-0-5 X.200/37754
2635143

AFRICAN STUDENTS
Chemistry - For African students

GRIMSHAW, Geoffrey
Chemistry. - London : Evans Bros, 1978
64p : ill. - (Evans pocket facts)
0-237-50074-4 X.619/18504 Woolwich
1444972

AFRICAN STUDENTS
Chemistry - For West African students

AKUSOBA, E. U.
Certificate notes. - Harlow : Longman, 1982
150p : ill. - (Study for success) (Study for success)
0-582-65081-x X.629/17713 Woolwich
0054925

BAJAH, Sam 'Tunde
Chemistry. - S.I. ed. - London : Longman, 1974
[4],274p : ill(some col). - Includes index
B76-12581 X.611/7207 Woolwich
0054925

PILLAI, G. C. (Geevarghese C)
Chemistry. - New ed. - London : Macmillan, [1984?]
viii,120p : ill. - (School certificate/GCE 'O' Level revision course)
0-333-32600-8 X.629/25354
0054925

AFRICAN STUDENTS
Christian life - For East African students

CHRISTIAN living today. - London : G. Chapman.
Book 1. - 1974
[1],192,C1-C27,B1-B18p : ill
0-225-66032-6 X.0102/121(1)
0525227
- London : Chapman.
Teacher's handbook. - 1981
xv,326p : ill
0-225-66307-4 X.200/37471
0525227

AFRICAN STUDENTS
Christianity - For African students

TEACHER'S handbook for Developing in Christ. Course one : Christ and my humanity (units 1 to 13) / prepared by teachers throughout Africa and written at The AMECEA Pastoral Institute. - London : Chapman, 1981
198p : ill
0-225-66298-1 X.200/36669
2551578

TEACHER'S handbook for Developing in Christ. Course two : Christian life in community (units 14 to 25) / prepared by teachers throughout Africa and written at the AMECEA Pastoral Institute. - London : Cassell, 1981.
Course 2: Teacher's handbook
223p : ill
0-225-66299-x X.200/44667
2551578

AFRICAN STUDENTS
Commerce - For East African students

ANDERSON, David J. (David John)
Multiple choice tests in commerce for East Africa. - London : Macmillan, 1981
v,98p
0-333-30975-8 X.520/26894
2626616

BUTT, Safdar Ali
Essentials of commerce in East Africa. - London : Cassell, 1976
x,225p : ill. - Includes index
0-304-29668-6 X.519/25529
0595829

Essentials of commerce in East Africa. - 2nd ed. - London : Cassell, 1977
x,225p : ill. - Includes index
0-304-29888-3 X.519/27925
0595829

MUKIIBI, P.
O-level commerce for East Africa. - London : Longman, 1981
105p. - (Study for success) (Study for success)
0-582-65529-3 X.529/45899
2626616

MUTABIIRWA, D. E. R.
Revision Commerce for East Africa. - Sunbury-on-Thames : Nelson, 1979
[46]p. - (Nelson Africa business studies). - Includes index
0-17-511555-9 X.520/20914
0595829

AFRICAN STUDENTS
Commerce - For West African students

ANDERSON, David J. (David John)
Commerce. - London : Macmillan, 1981
vii,96p : ill. - (School certificate revision course)
0-333-29166-2 X.529/44203
116872x

ANDERSON, David J. (David John)
Commerce for West Africa. - London : Macmillan, 1977
[4],252p : ill. - Includes index
0-333-21418-8 X.519/27045
116872x
Multiple choice tests in commerce. - 1st ed. for West Africa. - London : Macmillan, 1982
v,98p
0-333-32678-4 X.520/36006
2575108

COMMERCE. - London : Macmillan, 1981
124p. - (Macmillan certificate model answers)
0-333-29165-4 X.529/44204
2575108

LAWAL, O. A.
Commerce today for West Africa. - London : Hodder and Stoughton, 1982
178p : ill
0-340-26437-3 X.520/29166
116872x

LOBLEY, Derek
Success in commerce. - West African ed.. - London : African University Press in association with Murray, 1982
xii,240p : ill. - Includes index
0-7195-3901-3 X.529/46351
2693011

MBA, R. O.
Commercial calculations. - London : Edward Arnold in association with African Universities Press, 1982
v,200p : ill. - Includes index
0-7131-8025-0 X.629/19835
2575108

OKOLO, M. N. (Michael N)
Objective tests. - Harlow : Longman, 1980
iv,124p : ill. - (Study of success) (Study for success)
0-582-65052-6 X.529/44551
2575108

AFRICAN STUDENTS
Cookery - For African students

NDUNGI, Harriet Karura
Food and nutrition for schools and colleges. - London : Evans, 1982
vi,218p : ill. - Includes index
0-237-50460-x X.629/17136 Woolwich
2623404

AFRICAN STUDENTS
Cookery - For West African students

O'REILLY-WRIGHT, Enid
The student's cookery book. - 2nd ed. (metric). - Oxford : Oxford University Press, 1979
vi,298p : ill. - Includes index
0-19-832711-0 X.619/19924 Woolwich
1917579

AFRICAN STUDENTS
Correspondence in English - For African students

TAIWO, Oladele
Writing essays and letters. - London : Nelson, 1975 [i.e. 1976]
139p
0-17-511426-9 X.909/40493
0089869

AFRICAN STUDENTS
Crops. Production - For West African students

ONWUEME, I. C.
Crop science. - London : Cassell, 1979
106p : ill. - (Cassell's tropical agriculture series ; book 2). - Includes index
0-304-30208-2 X.311/8967 Woolwich
1456466

AFRICAN STUDENTS
Demography. Techniques - For African students

KPEDEKPO, G. M. K.
Essentials of demographic analysis for Africa. - London : Heinemann, 1982
xii,210p : ill. - (Studies in the economics of Africa). - Includes bibliographies and index
0-435-97390-8 X.520/27489
2756986

AFRICAN STUDENTS
Drama - For African students

CROW, Brian
Studying drama. - Harlow : Longman, 1983
ix,171p : ill. - Bibliography: p166-167. - Includes index
0-582-64425-9 X.950/25519
2976099

AFRICAN STUDENTS
East Africa, to 1976 - For East African students

ODHIAMBO, E. S. Atieno
A history of East Africa. - London : Longman, 1977
ix,192p : ill. - Includes index
0-582-60886-4 X.700/25640
1402951

AFRICAN STUDENTS
East Africa, 1000-1970 - For East African students

WERE, Gideon S.
East Africa through a thousand years. - 2nd ed. - London : Evans, 1972
ix,372p : ill. - Includes index
0-237-28873-7 X.809/53950
1059890

WILSON, Derek Alan
East African history, 1000-1970. - [New] ed. - London : Evans Bros, 1976
64p. - (Evans pocket facts)
0-237-49970-3 X.708/20729
1059890

AFRICAN STUDENTS
East Africa. Geographical features - For East African students

CARTER, Charles Eric
Case studies in East African geography. - Edinburgh : Johnston and Bacon, 1972
96p : ill. - Bibliography: p.94-95
0-7179-1577-8 X.700/25099
0614068

MINNS, W. J.
African geography. - London : Evans Bros, 1979
64p : ill
0-237-50196-1 X.700/25952
182239x

SENIOR, M. W. (Michael Wraith)
Certificate notes. - Harlow : Longman, 1982
96p : ill. - (Study for success) (Study for success)
0-582-60383-8 X.809/51842
0614068

AFRICAN STUDENTS
East Africa. Livestock. Production - For East African students

BARBER, John, 1920-
Livestock management for East Africa. - London : Edward Arnold, 1976
[6],126p : ill. - Bibliography: p.117. - Includes index
0-7131-0063-x X.311/8272 Woolwich
0735434

AFRICAN STUDENTS
Economics - For African students

OBONE, A. E.
Economics. - London : Evans Bros, 1977
x,290p : ill. - Bibliography: p.281-282. - Includes index
0-237-49912-6 X.519/26710
208712x

SEIDMAN, Ann
An economics textbook for Africa. - 3rd ed. - London : Methuen, 1980
xxii,346p : ill. - Bibliography: p331-339. - Includes index
0-416-74060-x X.529/67937
208712x

AFRICAN STUDENTS
Economics - For East African students

BRADLEY, G. R.
Economics for East Africa. - London : Cassell, 1979
vi,244p : ill
0-304-30288-0 X.519/30599
202201x

AFRICAN STUDENTS
Economics - For West African students

AJAYI, Simeon Ibi. (Simeon Ibidayo)
Economics in a West African setting. - Harlow : Longman, c1984
194p : ill. - Includes index
0-582-65539-0 X.520/34432
1078267

ANDERSON, David J. (David John)
Intermediate economics for West Africa. - London : Macmillan, 1980
442p : ill. - Includes index
0-333-26193-3 X.529/40588
1207032

BENEMY, Frank William Georgeson
Economics. - London : Evans Bros, 1977
64p : ill. - (Evans pocket facts)
0-237-50072-8 X.519/26834
1078267

ECONOMICS. - London : Macmillan, 1979
[5],112p : ill. - (Macmillan certificate model answers)
0-333-27318-4 X.519/30823
2078228

NWANKWO, G. O.
Basic economics. - Cambridge : Cambridge University Press, 1977
vii,263p : ill
0-521-21122-0 X.510/15556
1207032
Basic economics. - 2nd ed. - Cambridge : Cambridge University Press, 1982
viii,263p : ill
0-521-28625-5 X.529/48689
1078267

OLAYIDE, S. O. (Samson Olajuwon)
Certificate economics for West Africa. - Oxford : Oxford University Press, 1977
vii,216p : ill. - Includes index
0-19-575416-6 X.520/21618
1078267

TERIBA, O.
Certificate economics for West Africa. - London : Longman, 1976
[6],250p : ill. - Includes index
0-582-60931-3 X.510/15425
1078267
Certificate economics for West Africa. - New ed.. - London : Longman, 1980
250p : ill. - Includes index
0-582-60649-7 X.520/22823
1078267

UDU, Ewa
West African economics for School Certificate and G.C.E.. - London : Evans Bros, 1978
iv,172p : ill. - Includes index
0-237-50175-9 X.519/28911
1078267

UMO, Joe U.
Objective tests : essential economics. - Harlow : Longman, 1983
40p : ill. - (Study for success) (Study for success)
0-582-65112-3 X.529/55778
2078228

AFRICAN STUDENTS
Ecosystems - For East African students

MOSS, Stanley
Beginning ecology. - London : Macmillan. - (Macmillan ecology project)
Student's workbook. - 1976
[2],29p : ill
0-333-19656-2 X.0311/1010 Woolwich
1119680
- (The Macmillan ecology project)
Teachers' guide. - 1976
v,26p : ill
0-333-21131-6 X.0311/1010a Woolwich
1119680

AFRICAN STUDENTS
English language - For African students

ALLEY, Elisabeth
Food from plants. - London : Macmillan, 1983
14p : ill. - (Macmillan primary English readers)
0-333-36770-7 X.950/35908
0054542
Homes of Africa. - London : Macmillan, 1983
14p : ill. - (Macmillan primary English readers)
0-333-36769-3 X.950/35909
0054542
The praying mantis and others. - London : Macmillan, 1983
14p : ill. - (Macmillan primary English readers)
0-333-36768-5 X.950/35910
0054542

ANORUE, J. C.
Teach yourself good English, better sentences, good letters and compositions. - Englarged ed. - [Onitsha] : [J.C. Brothers Bookshop], [1979?]
80p.
r6108165 X.0909/588(200)
173900X

BAILEY, D.
Ants. - London : Macmillan, 1983
14p : ill. - (Macmillan primary English readers)
0-333-36766-9 X.950/35903
0054542
The bus ride. - London : Macmillan, 1983
14p : ill. - (Macmillan primary English readers)
0-333-36772-3 X.950/35905
0054542
Modise's ball. - London : Macmillan, 1983
14p : ill. - (Macmillan primary English readers)
0-333-36771-5 X.950/35904
0054542
Rain. - London : Macmillan, 1983
14p : ill. - (Macmillan primary English readers)
0-333-36767-7 X.950/35907
0054542
What do they do?. - London : Macmillan, 1983
14p : ill. - (Macmillan primary English readers)
0-333-36773-1 X.950/35906
0054542

CHISUNGO, Jefurio
The big radio mystery. - Harlow : Longman, 1981
39p : col.ill. - (Action books. Level two)
0-582-59593-2 X.990/17342 Woolwich
0054542
Call Mr Africa. - Harlow : Longman, 1981
48p : col.ill. - (Action book / Level three)
0-582-59591-6 X.990/17343 Woolwich
0054542

CREW, C. C.
Seek and find English. - Teacher's ed. - Exeter : Wheaton, 1960
80,15p
0-08-022875-5 X.950/11353
0099260

GRANT, Neville J. H.
English in use. - [London] : Longman.
Book 2. - 1979
202p
0-582-60175-4 X.0909/2096(2)
173900X

MACKENZIE, Malcolm Douglas Munro
English language. - London : Evans Bros, 1978
64p. - (Evans pocket facts)
0-237-50140-6 X.908/41595
0099260

MACMILLAN reading scheme. - London : Macmillan.
Book 5: In the market. - 1977
[2],16p : chiefly col ill
0-333-22721-2 X.0519/1071
0054542
Book 6: The dirty goat. - 1977
[2],16p : chiefly col ill
0-333-22722-0 X.0519/1071
0054542

MACMILLAN reading scheme. - London : Macmillan.
Book 7: The new football. - 1978
[2],16p : col ill
B79-27348 X.0519/1071
0054542
Book 10: The snake. - 1978
[2],16p : chiefly col ill
B79-08684 X.0519/1071
0054542

MODERN method English / D.W. Grieve [and others]. - London : Nelson.
Teachers' guide for books 1 to 4. - 1978
[7],120p : 2 ill
0-17-511508-7 X.909/43047
0099260

MONTGOMERY, Michael George Patrick
Effective English. - 2nd ed.. - London : Evans Bros.
1. Students' book. - 1976
x,238p : ill
Subsequent volumes, when received, will be placed at this shelfmark
0-237-50081-7 X.0909/2040
173900x
- [2nd ed.]. - London : Evans Bros.
3. Teacher's book. - 1978
x,158p
0-237-50302-6 X.0909/2040
173900x

AFRICAN STUDENTS
English language - For African students - Texts

DIBBA, Ebou
Olu and the smugglers. - [Harlow] : Longman, 1980
40p : col.ill. - (Action books. Level 2)
0-582-59592-4 X.990/17053 Woolwich
0054542

SWEETMAN, David
The amulet. - [London] : Longman, 1980
47p : col.ill. - (Action books. Level 2)
0-582-59584-3 X.992/4084 Woolwich
0054542
The Moyo Kids. - [Harlow] : Longman, 1980
32p : col.ill. - (Action books. Level 1)
0-582-59553-3 X.990/17148 Woolwich
0054542
Skyjack over Africa. - [Harlow] : Longman, 1980
30p : col.ill. - (Action books. Level 1)
0-582-59585-1 X.990/17149 Woolwich
0054542

AFRICAN STUDENTS
English language - For East African students

GRANT, Neville J. H.
English in use. - [London] : Longman.
[Students'] bk.3. - 1980
v,187p : ill
0-582-60176-2 X.0909/2096(3)
1779664
- Harlow : Longman.
Bk.4. - 1981
170p : ill
0-582-60177-0 X.0909/2096(5)
1779664
- [Harlow] : Longman.
Teacher's book for year 1. - 1978
[3],109p : 1 ill
0-582-60191-6 X.0909/2096 (1)
1779664
Teacher's book for year 2. - 1979
[4],84p : 1 ill
0-582-60384-6 X.0909/2096(2)
1779664
- Harlow : Longman, 1981.
Teacher's book for year four
51p
0-582-60386-2 X.0909/2096(6)
1779664

GRANT, Neville J. H.
English in use. - [Harlow] : Longman.
Teacher's book for year three. - 1980
76p
0-582-60385-4 X.0909/2096(4)
1779664

AFRICAN STUDENTS
English language - For Southern African students

BEVAN, R. B. (Roy B)
Secondary English project for Botswana, Lesotho and Swaziland. - Harlow : Longman.
Students bk. 2. - 1983
186p : ill
0-582-65532-3 X.950/33778
2920387

BEVAN, Roy B.
Secondary English project. - Harlow : Longman.
Students' bk. 3. - 1984
vi,170p : ill
0-582-65531-5 X.950/40902
2920387

SECONDARY English project. - [London] : Longman.
Teacher's guide for year one. - 1982
154p : ill
0-582-60975-5 X.950/22009
2920387
- Harlow : Longman.
Teacher's guide for year two. - 1983
140p : ill
0-582-60976-3 X.950/33887
2920387

AFRICAN STUDENTS
English language - For West African students

BERRY, Ann
Evans primary English course. - Ibadan : Evans Brothers (Nigeria Publishers) ; London : Evans Bros.
Pupils' transition book. - New ed.. - 1976
128p : ill
0-237-50087-6 X.0908/2012(1)
0376132

GRANT, Neville J. H.
Secondary English project. - Nigeria : Longman.
Book 1. [Students' book]. - 1976
vii, 185 p. : ill
0-582-60153-3 X.0900/1013
0214531
- [Nigeria] : Longman.
Year 1. Teacher's book. - 1976
[5], 161 p. : ill
0-582-60155-x X.0900/1013
0376132
- [London] : Longman.
Book 2. - 1977
[6],184p : ill
0-582-60154-1 X.0900/1013(2)
0214531

AFRICAN STUDENTS
English language - For West African students–continued
[Book 2]. Teacher's book for year two. - 1977
[4],137p
0-582-60168-1 X.0900/1013.a
0214531
- [Nigeria] : Longman.
Book 3: [Students' book]. - [Nigeria] : Longman
vii,184p : ill
0-582-60159-2 X.0900/1013(3)
0376132
- [London] : Longman.
Book 4. - 1979
v,181p : ill
0-582-60160-6 X.0900/1013(4)
0214531

HYDE, Elizabeth, 1924-
New Evans primary English. - London : Evans
48p : col.ill
Subsequent volumes when received will be placed at this shelfmark
0-237-50591-6 X.0950/164
0376132

NWOSU, A. I.
English language. - London : Macmillan, 1980
vi,121p. - (School certificate revision course)
0-333-26997-7 X.909/44862
0376132

OGUNDIPE, P. A. (Phebean Ajibola)
Practical English. - [New ed.]. - London : Longman.
[Book 1]. Workbook. - 1979
[1],60p : ill
Subsequent volumes, when received, will be placed at this shelfmark
0-582-60150-9 X.0900/1067
0376132

PARKER, T. H. (Thomas Henry)
Practice papers in English language. - London : Longman, 1979
vii,151p : 1 ill
0-582-60172-x X.900/27629
0214531

PARRY, Helen
Use of English course for West African students. - London : Macmillan, 1979
277p : ill
0-333-25344-2 X.529/45547
0376132

SECONDARY English project. - [Harlow] : Longman.
Cassette tape year 1. - 1976
1 sound cassette(ca 30min.)
0-582-60956-9 Cup.580.ff.13
0214531

STODDART, John
English language for the Certificate year. - Revised ed. - London : Allen and Unwin, 1977
140p
0-04-370081-0 X.909/41622
0376132

AFRICAN STUDENTS
English language. Comprehension - For African students

BERRY, John Pearson
English practice for school certificate. - London : Evans Bros, 1978
ix,94p
0-237-50259-3 X.909/43060
0609676

BROWN, Nigel, 1939-
Reading, talking, writing. - London : Heinemann, 1975
viii,120p : ill
0-435-92060-x X.909/40454
0609676

AFRICAN STUDENTS
English language. Comprehension - For East African students

BRACEY, N. W.
Comprehension and summary for O-level. - Harlow : Longman, 1981
viii,87p
0-582-60194-0 X.950/7911
2559072

AFRICAN STUDENTS
English language. Comprehension - For West African students

BISONG, J. O. (Joseph Obi)
Comprehension and summary for the new syllabus. - London : Evans, 1982
x,245p
0-237-50665-3 X.529/52904
2622629

MORAKINYO, Rosemary E.
English language tests for school certificate. - London : Macmillan, 1981
ix,125p
0-333-32611-3 X.950/10442
2622629

NOBLE, R. W. (Robert William)
English comprehension and summary. - New ed. rev. and expanded. - Harlow : Longman, 1981
124p. - Bibliography: p121-124
0-582-65091-7 X.529/46017
2630303

AFRICAN STUDENTS
English language. Grammar - For African students

OGUNDIPE, P. A. (Phebean Ajibola)
Brighter grammar. - New ed.. - Harlow : Longman.
Bk.1. - 1983
80p : ill
0-582-60971-2 X.529/60100
3073386

OGUNDIPE, P. A. (Phebean Ajibola)
Brighter grammar. - Harlow : Longman.
Bk.2. - New ed. - 1983
82p : ill
0-582-60972-0 X.629/21996
1913700
- New ed.. - Harlow : Longman.
Bk.4. - 1983
87p : ill
0-582-60974-7 X.958/22899
1913700

TREGIDGO, Philip Sillince
English grammar in practice. - London : Longman, 1979
xi,238p : ill. - Includes index
0-582-60186-x X.909/43672
1913700

AFRICAN STUDENTS
English language. Précis - For African students

BYRAM, R. S.
Guided summary & note-making. - London : Heinemann Educational, 1977
ix,58p : ill
0-435-92065-0 X.519/27694
1351729

AFRICAN STUDENTS
English language. Reading books - For African students

GRANT, N. J. H.
Reading for a purpose. - [Nigeria] : Longman.
Book 3. - 1978
[3],77p : ill
0-582-60169-x X.0900/1023(3)
0328022

AFRICAN STUDENTS
English language. Reading books - For West African students

AGWU, Ada K.
The boy in the ark. - Lagos : African Universities Press, 1977
24p : ill. - (African junior library ; 15)
0-410-80180-1 X.990/9056 Woolwich
0502553
The dreamer boy. - Lagos : African Universities Press, 1977
32p : ill. - (African junior library ; 16)
0-410-80179-8 X.990/9059 Woolwich
0502553
Spider's land. - Lagos : African Universities Press, 1977
47p : ill. - (African junior library ; 11)
0-410-80100-3 X.990/9057 Woolwich
0502553

AFRICAN STUDENTS
English language. Reading books - For West African students - Texts

CHUKWUKA, J. I. N.
Zandi and the wonderful pillow. - Lagos : African Universities Press, 1977
48p : ill. - (African junior library ; 13)
0-410-80099-6 X.990/9054 Woolwich
0502553

IROAGANACHI, John
A fight for honey. - Lagos : African Universities Press, 1977
30,[1]p : ill. - (African junior library ; 14)
0-410-80181-x X.990/9055 Woolwich
0502553

UWEMEDIMO, Rosemary
Boma and his friends ; and, Okon and Nko in trouble. - Lagos : African Universities Press, 1977
32p : ill. - (African junior library ; 12)
0-410-80177-1 X.990/9060 Woolwich
0502553

AFRICAN STUDENTS
English language. Writing skills - For African students

TETLOW, J. G.
Developing writing skills. - Harlow : Longman, 1981
vi,72p : ill
0-582-65055-0 X.950/10816
2630338

AFRICAN STUDENTS
English languages. Composition - For African students

BROWN, Nigel, 1939-
Reading, talking, writing. - London : Heinemann, 1975
viii,120p : ill
0-435-92060-x X.909/40454
0205176

SHEAL, Peter
Advanced level writing skills. - Harlow : Longman, 1981
viii,135p : ill
0-582-64307-4 X.950/15689
0205176

AFRICAN STUDENTS
English literature, 1900-1976. Comprehension - For West African students

PROSE and poetry unseens for certificate literature in English / [compiled by] Rosina Umelo. - London : Macmillan.
[Pupils' ed.]. - 1976
vi,90p
0-333-19824-7 X.909/41514
1121790
Teachers' ed. - 1976
vi,138p
0-333-19825-5 X.909/41514a
1121790

AFRICAN STUDENTS
English literature. Comprehension - For West African students

MOODY, H. L. B. (Harold Leonard Birch)
Unseen prose and poetry. - London : Longman, 1980
v,121p : ill
0-582-60166-5 X.950/4401
2410931

AFRICAN STUDENTS
English literature. Special subjects: Africa - For African students

AFRICAN literary appreciation / [compiled by] K.M. Lillis. - London : University of London Press, 1975
[12],132p. - Bibliography: p.132
0-340-17259-2 X.909/40062
0360686

AFRICAN STUDENTS
Essays in English. Composition - For African students

TAIWO, Oladele
Writing essays and letters. - London : Nelson, 1975 [i.e. 1976]
139p
0-17-511426-9 X.909/40493
0614114

AFRICAN STUDENTS
Foreign relations - For African students

OFOEGBU, Ray
A foundation course in international relations for African universities. - London : Allen & Unwin, 1980
x,243p. - Bibliography: p233-238. - Includes index
0-04-327058-1 X.529/67661
2106078

AFRICAN STUDENTS
French language. Grammar - For African students

GRANDSAIGNE, Jean de
France Afrique. - London : Macmillan.
Teacher's book for years 1-4. - 1976
[7],150p : ill
0-333-18977-9 X.0989/503
0090638

AFRICAN STUDENTS
Geographical features - For East African students

TAYLOR, P. F.
Geography. - [New ed.]. - London : Evans Bros, 1976
64p : ill. - (Evans pocket facts)
0-237-49972-x X.708/20733
1060287

AFRICAN STUDENTS
Geography - For West African students

ADELEKE, B. O.
Certificate physical and human geography. - West African ed. - Ibadan ; Oxford : Oxford University Press, 1978
xi,353p : ill(some col). - Includes index
0-19-575317-8 X.702/6131
1444670

IMOGIE, A. Inanoya
Geography. - London : Macmillan, 1977
iv,124p : ill. - (School Certificate revision course)
B78-16483 X.700/25668
1444670

OKUNROTIFA, P. Olatunde
Revision guide, O Level geography. - London : Longman, 1979
vi,106p : ill. - (Study of success)
0-582-60345-5 X.709/33323
1927299

AFRICAN STUDENTS
Geography - For West African students–continued

PHYSICAL and human geography. - London : Macmillan, 1980
[4],92p : ill. - (Macmillan certificate model answers)
0–333–28525–5 X.700/26351
1444670

AFRICAN STUDENTS
Geography. Field studies. Methodology - For African students

PRITCHARD, J. M. (John Malcolm)
Practical geography for Africa. - Harlow : Longman, 1984
234p,16p of plates : ill(some col.). - Includes index
0–582–60366–8 X.800/40680
326145x

AFRICAN STUDENTS
Ghana. Geographical features - For West African students

DICKSON, Kwamina Busumafi
A new geography of Ghana. - Metricated ed. - London : Longman, 1977
173p : ill. - Includes index
0–582–60343–9 X.700/25445
005755x

AFRICAN STUDENTS
Human geography - For West African students

LEWIS, G. E. D. (Gerwyn Elidor David)
Human geography for West Africa. - London : Hodder and Stoughton, 1982
vi,154p : ill(some col.). - Includes index
0–340–27596–0 X.800/35048
2881330

AFRICAN STUDENTS
Islam. Hadith - For West African students

DOI, A. Raḥman I. ('Abdur Raḥmân I)
Introduction to the Hadith. - Sevenoaks : Arewa, 1981
130p. - Bibliography: p130
0–340–26706–2 X.200/37518
2605570

AFRICAN STUDENTS
Islam. Koran - For West African students

DOI, A. Raḥmân I. ('Abdur Raḥmân I)
Introduction to the Qur'an. - Sevenoaks : Arewa, 1981
v,90p. - Bibliography: p90
0–340–26705–4 X.200/37520
2605546

AFRICAN STUDENTS
Jews, to 132. Historical sources: Bible - For West African students

RATTEY, Beatrice Katharine
A short history of the Hebrews from the patriarchs to Herod the Great. - 3rd ed.. - London : Oxford University Press, 1976
144p : ill. - Bibliography: p.135. - Includes index
0–19–832121–x X.709/30417
063204x

AFRICAN STUDENTS
Macroeconomics - For African students

HARVEY, Charles, 1937-
Macroeconomics for Africa. - London : Heinemann Educational, 1977[i.e.1978]
xvi,240p : ill. - (Studies in the economics of Africa). - Includes index
0–435–97201–4 X.519/27942
1446177

AFRICAN STUDENTS
Malikite Islamic doctrine - For West African students

DOI, A. Raḥmân I. ('Abdur Raḥmân I)
The cardinal principles of Islam. - Sevenoaks : Arewa, 1981
ix,144p. - Bibliography: p143-144
0–340–26704–6 X.200/37519
2605554

AFRICAN STUDENTS
Man. Biology - For African students

GODMAN, A. (Arthur)
Human and social biology for Junior Certificate. - Harlow : Longman, 1982
vii,224p : ill
0–582–58821–9 X.322/11598
2863626

AFRICAN STUDENTS
Man. Diagnosis. Physical examination - For East African students

ESSEX, B. J.
Diagnostic pathways in clinical medicine. - Edinburgh : Churchill Livingstone, 1977
[11],173p : ill. - (Medicine in the tropics). - Includes index
0–443–01514–7 X.311/8401 Woolwich
1066293

AFRICAN STUDENTS
Man. Diseases - For African students

PRINCIPLES of medicine in Africa / edited by E.H.O. Parry. - Oxford : Oxford University Press, 1976
xi,604p : ill. - (Oxford medical publications). - Includes bibliographies and index
0–19–264223–5 SRL
1198467

AFRICAN STUDENTS
Man. Health - For West African students

ADEMUWAGUN, Z. A.
Certificate health science. - Cambridge : Cambridge University Press, 1976
iv,156p : ill
0–521–21148–4 X.311/8361 Woolwich
1031376

GODMAN, A. (Arthur)
Certificate notes. - Harlow : Longman, 1981
vi,122p : ill. - (Study for success) (Study for success)
0–582–60621–7 X.329/14606 Woolwich
1031376

GODMAN, Arthur
A new health science for Africa. - [New ed.]. - London : Longman, 1979
[10],258p : ill. - Bibliography: p.[10]. - Includes index
0–582–65011–9 X.311/9066 Woolwich
1031376

HEALTH science. - London : Macmillan, 1979
v,80p : ill. - (Macmillan certificate model answers)
0–333–26806–7 X.322/11403
2573687

KURIEN, A. V.
Objective tests in health science for WASC/GCE O Level. - London : Hodder and Stoughton, 1982
140p : ill
0–340–27595–2 X322/11713
2573687

MACKEAN, D. G. (Donald Gordon)
Certificate health science for West Africa. - London : John Murray in association with African Universities Press, 1983
vi,298p : ill. - Bibliography: p288. - Includes index
0–7195–4099–2 X.322/13560
1031376

USUA, E. J. (Efana James)
Questions and answers. - Harlow : Longman, 1981, c1980
91p : ill. - (Study for success) (Study for success)
0–582–60652–7 X.329/14614 Woolwich
2573687

AFRICAN STUDENTS
Map reading - For Central African students

MAP reading for central Africa. - Southerton : Longman.
Book 2. - Metric ed. - 1978
53[i.e.63]p : ill(some col). - Includes index
0–582–65004–6 X.702/6100
1572237

AFRICAN STUDENTS
Map reading - For East African students

MCMASTER, David Newcombe
Map reading for East Africa. - New metric ed.. - Nairobi : Longman, 1978
96p : ill
0–582–60341–2 X.619/19847 Woolwich
0085820

PRITCHARD, John Malcolm
First lessons in map reading for East Africa. - New ed. - London : Evans Bros, 1976
[2],37p : ill(chiefly col)
0–237–49999–1 X.700/25146
0085820

AFRICAN STUDENTS
Map reading - For West African students

NIMAKO, D. Annor
Map reading for West Africa. - New ed. - London : Longman, 1977
iv,60p : ill(some col)
0–582–60347–1 Maps 57.bb.7 Map Library
1267965

Map reading for West Africa. - Rev. ed. with metric examples. - Harlow : Longman, 1982
76p : ill
0–582–60391–9 Map Library
0745235

Map reading for West Africa practice book. - Rev. ed. - Harlow, 1982
62p : ill
0–582–60395–1 X.805/4703
0745235

WAREHAM, Albert Kenneth
Map-reading for West African schools. - 2nd ed.. - Oxford : Oxford University Press [etc.], 1980
vii,72p : ill
0–19–913212–7 Map Library
0745235

AFRICAN STUDENTS
Mathematics - For African students

JARMAN, E. A. (Eric Arthur)
Certificate notes. - London : Longman, 1980
153p : ill. - (Study for success)
0–582–60654–3 X.629/14896 Woolwich
1824759

MACRAE, Murray
Modern mathematics. - London : Evans Bros, 1979
64p : ill. - (Evans pocket facts)
0–237–50206–2 X.619/22293 Woolwich
0029122

WAKANO, Katambo
Solved revision problems for Ordinary Level mathematics. - Sunbury-on-Thames : Nelson, 1979
94p : ill. - Includes index
0–17–511547–8 X.611/8713 Woolwich
1824759

AFRICAN STUDENTS
Mathematics - For East African students

DALRYMPLE, P. J.
C.P.E. mathematics. - Nairobi : Macmillan Kenya, 1982
94p : ill
0–333–31652–5 X.622/15031
2550865

MODERN mathematics for East Africa / [edited by E.A. Jarman]. - London : Longman.
Teacher's guide 2. - 1980
131p : ill
0–582–60642–x X.622/14423
2550865

MODERN mathematics for East Africa. - London : Longman.
Workbook 3. - 1978
61,[2]p : ill
Subsequent volumes, when received, will be placed at this shelfmark
0–582–65002–x X.0611/1094
1992279

AFRICAN STUDENTS
Mathematics - For economics - For African students

ONIMODE, Bade
Basic mathematics for economists. - London : Allen & Unwin, 1980
x,247p : ill. - Includes bibliographies and index
0–04–330304–8 X.629/14706 Woolwich
2102366

AFRICAN STUDENTS
Mathematics - For Southern African students

BICKNELL, E. D.
Revision mathematics for Junior Certificate. - Harlow : Longman, 1980
160p : ill
0–582–60695–0 X.622/11238 Woolwich
2559110

NEW general mathematics / J.B. Channon ... [et al.] ; advisory panel D. Samudzimu ... [et al.]. - Harlow : Longman.
1. - 1983
250p : ill
Subsequent volumes when received will be placed at this shelfmark
0–582–58847–2 X.0622/933
2559110

AFRICAN STUDENTS
Mathematics - For West African students

CHANNON, John Bredin
New general mathematics for West Africa. - [New ed.]. - London : Longman.
5: Revision. - 1978
vi,378p : ill
0–582–65023–2 X.0619/825
0089524

ELVIN, R.
Mathematics. - Ibadan ; Oxford : Oxford University Press, 1978
[4],274p : ill
0–19–832529–0 X.611/8285 Woolwich
0374431

NEW general mathematics for West Africa. - New ed.. - Harlow : Longman.
1. - 1981
188p : ill. - Includes index
Subsequent volumes when received will be placed at this shelfmark
0–582–60644–6 X.0622/602
0089524

2. - 1981
202p : ill. - Includes index
0–582–65056–9 X.0622/602 Woolwich
0374431

AFRICAN STUDENTS
Mathematics - For West African students-continued
4. - 1983
268p : ill. - Includes index
Subsequent volumes when received will be placed at this shelfmark
0-582-58824-3 X.0622/938
0059846

OKEKE, C. E.
Questions and answers. - Lagos : Longman Nigeria ; Harlow : Longman, 1983
v,98p : ill. - (Study for success)
0-582-65536-6 X.629/20585
0374431

OKEKE, Pius N.
Objective tests. - Harlow : Longman, 1982
87p : ill. - (Study for success) (Study for success)
0-582-65079-8 X.629/17470 Woolwich
0374431

UNIVERSAL primary mathematics. - [Revised ed.]. - Ibadan ; London : Evans Bros.
Book 1. [Pupils' book]. - [1976]
[1],64p : chiefly col ill
Later volumes are also placed at this shelfmark
0-237-50059-0 X.0615/508 Woolwich
0598755

AFRICAN STUDENTS
Mechanical engineering - For West African students
TIMINGS, R. L. (Roger Leslie)
Mechanical engineering craft theory and related subjects. - London : Longman. - (Longman industrial crafts series)
Vol.1. - 1983
viii,327p : ill. - Includes index
0-582-65801-2 L.49/2280
3081192

AFRICAN STUDENTS
Metalworking - For African students
EDWARDS, R.
Metalwork projects. - London : Cassell, 1981
99p : ill
0-304-30372-0 L.49/911
2530007

AFRICAN STUDENTS
Nigeria, to 1975 - For West African students
CROWDER, Michael
Nigeria. - London : Longman, 1979
[8],216p,[4]p of plates : ill(some col). - Bibliography: p.209-211. - Includes index
0-582-60346-3 X.700/26109
196853x

ISICHEI, Elizabeth
Junior history of Nigeria. - Yaba : Macmillan Nigeria, 1981
iii,140p : ill
978-13-2150-4 X.800/33934
196853x

AFRICAN STUDENTS
Office practices - For African students
BURTON, Mabel
Secretarial and office practice. - London : Longman, 1977
[7],127p : ill. - (Longman commercial studies)
0-582-60898-8 X.519/27158
1193058

AFRICAN STUDENTS
Office practices - For West African students
HARRISON, John, 1931-
Secretarial duties. - West African ed.. - London : Pitman Education, 1979
v,305p : ill. - (Pitmans African business series). - Includes index
0-273-01273-8 X.529/48556
1949764

STANWELL, Sheila Toft
West African office practice. - West African ed.. - Lagos : African Universities Press ; London : Edward Arnold, 1978
[6],266p : ill. - (Basic business studies). - Bibliography: p.247-251. - Includes index
B79-31268 X.519/30166
1949764

AFRICAN STUDENTS
Organic compounds - For West African students
JOHN, M. O.
A concise School Certificate organic chemistry. - 2nd ed.. - London : Evans Bros, 1975
vii,52p : ill. - Includes index
0-237-49948-7 X.619/15138
0373745

AFRICAN STUDENTS
Organisational behaviour - For African students
BLUNT, Peter
Organizational theory and behaviour. - London : Longman, 1983
xii,195p : ill. - (Modern management in Africa). - Bibliography: p181-192. - Includes index
0-582-64404-6 X.529/58910
3043878

AFRICAN STUDENTS
Organisms
AYERST, Peter William
Biology. - [New ed.]. - London : Evans Bros, 1976
64p : ill. - (Evans pocket facts)
0-237-49971-1 X.319/17134 Woolwich
1059734

AFRICAN STUDENTS
Organisms - For West African students
BECKETT, Brian Samuel
Biology for the West African Certificate. - Oxford : Oxford University Press, 1979
[11],356p : ill. - Includes index
0-19-914059-6 X.311/9004 Woolwich
0090735

ILOEJE, S. O. (Sylvester Onuabeke)
Certificate practical biology. - 2nd ed. - Harlow : Longman, 1981
iv,108p : ill
0-582-65068-2 X 322/10335 Woolwich
0090735

MACKEAN, D. G. (Donald Gordon)
Introduction to biology. - 2nd West African ed. - London : Heinemann Educational (Nigeria) in association with John Murray, 1982
256p : ill. - Bibliography: p240. - Includes index
0-7195-3887-4 X.322/10634 Woolwich
0090735

MACKEAN, Donald Gordon
Introduction to biology. - West African ed. - [Ibadan] : Heinemann Educational Books (Nigeria) ; [London] : J. Murray, 1977
224p : ill. - Includes index
0-7195-3315-5 X.0313/516 Woolwich
0090735

NEAL, Ernest G. (Ernest Gordon)
Biology for today. - Tropical ed. - Poole : Blandford Press, 1983, c1978
vi,417p : ill(some col.). - Includes index
0-7137-1345-3 X.322/12113
0090735

STONE, Robert Henry
New biology for West African schools. - [New ed.]. - London : Longman, 1975
vi,378p : ill. - Includes index
0-582-60637-3 X.311/8194 Woolwich
0090735

AFRICAN STUDENTS
Physical geography - For West African students
BRYANT, R. V. (Raymond Victor)
Elements of physical geography. - London : Collins, 1981
96p : ill. - Includes index
0-00-326510-2 X.800/31278
2547201

WATERS, Grahame H.C. (Grahame Hugh Clement)
First lessons in physical geography. - West African ed. - Harlow : Longman, 1982
iv,140p : ill(some col.). - Includes index
0-582-60399-4 X.809/52532
2547201

AFRICAN STUDENTS
Physical sciences - For East African students
JOHNSON, Harvey
Energy and matter. - London : Edward Arnold, 1977
180p : ill. - Includes index
0-7131-0066-4 X.311/8516 Woolwich
116466x

JOHNSON, Harvey
Study and revision course in junior physical science. - London : Longman, 1979
iv,91p : ill
0-582-60500-8 X.622/8063 Woolwich
2181274

USHER, N.
Physical science. - London : Longman, 1977
[2],174p : ill. - Includes index
0-582-60676-4 X.611/7833 Woolwich
116466x

Physical science. - London : Longman.
Teacher's guide. - 1977
[3],299p : ill. - Bibliography: p.297
0-582-60677-2 X.615/1921 Woolwich
116466x

AFRICAN STUDENTS
Physics - For African students
ATHERTON, Michael Anthony
Beginning physical science. - London : J. Murray, 1976
[4],124p : ill. - Includes index
0-7195-3414-3 X.611/7421 Woolwich
0748404

Beginning physical science. - London : J. Murray.
Teachers' guide. - 1976
[1],141p : ill
0-7195-3317-1 X.611/7421a Woolwich
0748404

VINALL, N. P.
Physics. - London : Evans Bros, 1977
64p : ill. - (Evans pocket facts)
0-237-50073-6 X.619/17645 Woolwich
0748404

WARD, A. H.
Essential senior physics in SI units. - Sunbury-on-Thames : Nelson.
[1]. - 1977
vi,137p : ill. - Includes index
0-17-511497-8 X.0611/1050 Woolwich
1199463

[2]: Energy, heat and the structure of matter. - 1977
iv,124p : ill. - Includes index
0-17-511498-6 X.0611/1048 Woolwich
1199463

[3]: Light and sound. - 1977
iv,124p : ill. - Includes index
0-17-511499-4 X.0611/1047 Woolwich
1199463

[4]: Magnetism, electricity and atomic physics. - 1977
iv,124p : ill. - Includes index
0-17-511500-1 X.0611/1049 Woolwich
1199463

WARD, Elizabeth Honor
Essential senior physics in SI units. - [New ed.]. - Sunbury-on-Thames : Nelson, 1980
[5],506p : ill. - Includes index
0-17-511496-x X.622/7423 Woolwich
0748404

AFRICAN STUDENTS
Physics - For East African students
JOHNSON, Harvey
Modern basic physics. - London : Evan Bros, 1977
[5],215p : ill. - Includes index
0-237-28934-2 X.611/7980 Woolwich
1403303

AFRICAN STUDENTS
Physics - For West African students
ASHWORTH, A. E.
Physics for WAEC. - London : Cassell. - (Cassell's 'O' level science for West Africa)
Bk.1. - 1982
166p : ill. - Includes index
Subsequent volumes when received will be placed at this shelfmark
0-304-30422-0 X.0622/647
2616351

AFRICAN STUDENTS
Physics. Experiments - For African students
OKEKE, P. N.
Preliminary practical physics. - Chichester : Wiley, c1981
x,244p : ill. - Includes index
0-471-27852-1 X.622/11054 Woolwich
2547325

AFRICAN STUDENTS
Poetry, to ca 1975 - For African students - Anthologies - Translations
HEYWOOD, Annemarie
Senior poetry anthology. - London : Macmillan, 1983
xxi,266p : ill
0-333-36638-7 X.950/36756
2046229

WOOLGER, David
Rhymes and rhythms. - Oxford : Oxford University Press, 1978
155p. - Includes index
0-19-911096-4 X.909/44158
2046229

AFRICAN STUDENTS
Poetry, to ca 1980 - For African students
FINNEGAN, Eddie
Essential poetry. - London : Heinemann, 1981
viii,116p : ill
0-435-92256-4 X.950/9337
2598264

AFRICAN STUDENTS
Poetry in African languages - For African students - Anthologies - Translations

AFRICAN poetry for schools. - London : Longman.
Book 1. - 1978
viii,87p : ill
0-582-60163-0 X.0909/2086(1)
1579967

Book 2. - 1978
ix,101p : ill
0-582-60164-9 X.0909/2086(2)
1579967

AFRICAN STUDENTS
Poetry in English, 1558- - For African students - Anthologies

An ANTHOLOGY of poetry for secondary schools / selected by B.S. Kwakwa. - Tema : Ghana Publishing Corporation, 1974
160p. - Includes index
r6032597 X.909/42143
1341162

AFRICAN STUDENTS
Poetry in English, 1800-1980 - For African students - Anthologies

A POETRY anthology for junior secondary schools / Rosina Umelo. - London : Macmillan, 1978
x,118p. - Includes index
0-333-22693-3 X.950/1397
235098x

AFRICAN STUDENTS
Poetry in English, 1900-1977 - For West African students - Texts

IMAGES and impressions / [selected and introduced by] Kolawole Ogungbesan and David Woolger. - Oxford : Oxford University Press, 1978
153p. - Bibliography: p.151-153. - Includes index
0-19-911097-2 X.909/43481
184718x

AFRICAN STUDENTS
Poetry in English. African writers, 1945- - For African students - Anthologies

FRASER, Robert, 1947-
Reading African poetry. - London : Collins, 1975
96p : ill
0-00-321866-x X.909/40794
0711896

An INTRODUCTION to East African poetry / [edited by] Jonathan Kariara, Ellen Kitonga. - Nairobi : Oxford University Press, 1976
126p
0-19-572064-4 X.909/44397
0711896

AFRICAN STUDENTS
Politics - For West African students

GOVERNMENT. - London : Macmillan, 1981
92p. - (Macmillan certificate model answers)
0-333-30493-4 X.529/41639
2488582

PRICE, Joseph Henry
Comparative government. - 2nd ed. - London : Hutchinson, 1975
384p : ill. - Includes index
0-09-123121-3 X.700/25031
0505056

AFRICAN STUDENTS
Primary schools. Curriculum subjects: Mathematics. Teaching - For West African students

ASHWORTH, A. E.
The teaching of mathematics. - London : Hodder and Stoughton, 1981
ix,118p : ill. - (Nigerian education series)
0-340-25953-1 X.629/15072 Woolwich
2223848

AFRICAN STUDENTS
Reading skills. Self-development - For West African students

GRANT, N. J. H.
Reading for a purpose. - [Nigeria] : Longman.
Book 1. - 1976
[3], 44p : ill. - Includes index
0-582-60113-4 X.0900/1023(1)
1102567

Book 2. - 1977
[3],61p : ill. - Includes index
0-582-60112-6 X.0900/1023
1102567

AFRICAN STUDENTS
Reference books. Use - For African students

ALLAN, Alastair
Look it up. - London : Heinemann, 1980
88p. - Includes index
0-435-92010-3 X.950/11960
2695324

AFRICAN STUDENTS
Science - For Southern African students

BOLESWA integrated science. - London : Heinemann Educational.
Book 1. - 1978
[8],151p : ill(some col). - Includes index
0-435-57035-8 X.0611/1067(1) Woolwich
1463314

Book 2. - 1979
[8],136p : ill(some col). - Includes index
0-435-57036-6 X.611/8714 Woolwich
1463314

AFRICAN STUDENTS
Science - For West African students

HAGGIS, Sheila Mary
Finding out in science. - 2nd ed.. - London : Cambridge University Press.
Book 1. - 1975
80p : ill
0-521-20641-3 X.611/7152 Woolwich
0586188

HAGGIS, Sheila Mary
Finding out in science. - 2nd ed.. - London : Cambridge University Press.
Teacher's handbook. - 1975
144p : ill. - Bibliography: p.137-138
0-521-20640-5 X.619/15425 Woolwich
0586188

UNIVERSAL primary science. - Ibadan ; London : Evans Bros.
Year 4: Looking at our earth. - 1978
[2],45p : ill(chiefly col)
0-237-49936-3 X.0611/1068 Woolwich
1481185

Year 4: Looking at our earth: Teacher's guide. - 1978
45p : ill
0-237-49938-x X.0611/1068(a) Woolwich
1481185

AFRICAN STUDENTS
Sierra Leone, to 1978 - For West African students

FYFE, Christopher
A short history of Sierra Leone. - New ed. - London : Longman, 1979
ix,179p : ill. - Bibliography: p.158-159. - Includes index
0-582-60358-7 X.709/33782
2081466

AFRICAN STUDENTS
Sierra Leone. Geographical features - For West African students

A NEW geography of Sierra Leone / D.R.G. Gwynne-Jones ... [et al.]. - London : Longman, 1978
vii,212p : ill. - (School Certificate geography). - Includes index
0-582-60270-x X.700/25809
1691899

AFRICAN STUDENTS
Social sciences. Research. Methodology - For African students

PEIL, Margaret
Social science research methods. - London : Hodder and Stoughton, 1982
vi,234p : ill. - Bibliography: p226-231. - Includes index
0-340-26205-2 X.520/27041
2725525

AFRICAN STUDENTS
Social sciences. Statistical methods - For African students

LERCHE, C. O. (Charles Olsen)
Social and economic statistics. - London : Longman, 1983
141p : ill. - Bibliography: p135-136. - Includes index
0-582-64332-5 X.529/53077
2876663

AFRICAN STUDENTS
Sociology - For African students

PEIL, Margaret
Consensus and conflict in African societies. - London : Longman, 1977
xv,396p : ill. - Bibliography: p.362-387. - Includes index
0-582-64173-x X.709/31610
1252054

AFRICAN STUDENTS
Southern Africa, to 1974 - For East African students

PARKER, Graham, 1943-
History of Southern Africa. - London : Bell, 1975
xii,308p : ill. - Includes index
0-7135-1783-2 X.709/30202
0539430

AFRICAN STUDENTS
Southern Africa, to 1978 - For East African students

ODHIAMBO, E. S. Atieno
A history of Southern Africa. - New ed.. - London : Longman, 1978
vii,200p : ill. - Bibliography: p.195. - Includes index
0-582-60349-8 X.700/25904
1776363

AFRICAN STUDENTS
Southern Africa, to 1980 - For Southern African students

JUNIOR Certificate history of Southern Africa. - London : Heinemann.
Bk.1: Southern African societies before the Scramble. - 1979
144p : ill. - Bibliography: p141. - Includes index
0-435-94160-7 X.0809/2181(1)
1929852

JUNIOR certificate history of southern Africa. - London : Heinemann Educational.
Bk. 2: Southern Africa in the late nineteenth and twentieth centuries. - 1981
xviii,126p : ill. - Bibliography: p120. - Includes index
0-435-94161-5 X.809/49691
1929852

JUNIOR Certificate history of Southern Africa. - London : Heinemann Educational.
Bk.3: The modern world. - 1982
xiv,130p : ill. - Bibliography: p127. - Includes index
0-435-94760-5 X.0809/2181(3)
1929852

PARSONS, Neil, 1944-
A new history of southern Africa. - London : Macmillan Education, 1982
vi,330p : ill. - Bibliography: p321-325. - Includes index
0-333-26220-4 X.800/34040
1929852

AFRICAN STUDENTS
Tropical medicine - For West African students

SAMUEL, Perumkunnil Skaria
A new tropical health science for West Africa. - London : Macmillan, 1979
vi,233p : ill. - Includes index
0-333-26614-5 X.311/9129 Woolwich
2078244

AFRICAN STUDENTS
Tropical regions. Agriculture - For West African students

AIYEGBAYO, J. T.
An introduction to tropical agricultural science. - Ibadan ; London : Evans Bros, 1980
iv,107p : ill
0-237-50390-5 X.322/9283
2151898

AKINSANMI, Oluyemi
Certificate agricultural science. - London : Longman, 1975
ix,196p : ill. - Includes index
0-582-60884-8 X.311/8075
0541702

JOY, Derek
Tropical agriculture handbook. - London : Cassell, 1979
219p : ill. - (Cassell's tropical agriculture series). - Includes index
0-304-30016-0 X.311/9067 Woolwich
1992686

OKORIE, J. U.
Junior tropical agriculture for West Africa. - London : Macmillan.
[Bk.1]. - 1981
108p : ill(some col.)
0-333-31873-0 X.322/14097
0541702

UGURU, O. O.
An introduction to agricultural science for tropical areas. - Walton-on-Thames : Nelson, 1981
384p : ill. - Bibliography: p378-379. - Includes index
0-17-511085-9 X.322/10424 Woolwich
0541702

AFRICAN STUDENTS
Tropical regions. Organisms - For West African students

MAXWELL-OJO, Bola
Modern tropical biology. - London : Evans, 1982
108p. - Bibliography: p104-108
0-237-50548-7 X.329/15134 Woolwich
2707764

MAXWELL-OJO, Bola
Modern tropical biology. - London : Evans.
Student's book. - 1982
295p : ill(some col.). - Includes index
0-237-50509-6 X.322/10906 Woolwich
2707764

AFRICAN STUDENTS
Typing - For West African students

KERRY, Iris
Longman office typing for West Africa. - Harlow : Longman, 1983
282p : ill. - (Longman commercial studies). - Includes index
0-582-65050-x L.49/2442
3131572

AFRICAN STUDENTS
Urban regions. Ecology. Sanitary engineering aspects - For West African students

GADD, P.
The ecology of urbanisation and industrialisation. - London : Macmillan, 1976
[3],30,[1]p : ill. - (Advanced readers in tropical biology)
B77-09074 X.319/17138 Woolwich
1062352

AFRICAN STUDENTS
West Africa, to 1977

CROWDER, Michael
West Africa. - London : Longman, 1977
[9],214p,[8]p of plates : ill(some col). - Bibliography: p.203-204. - Includes index
0-582-60003-0 X.700/25467
1165224

AFRICAN STUDENTS
West Africa, to 1980 - For West African students

LUCAN, Talabi A. (Talabi Aisie)
A visual history of West Africa. - London : Evans, 1981
76p : ill
0-237-50145-7 X.800/31914
2592843

SINGH, Malkiat
Model answers in West African history. - Ibadan ; London : Evans, 1980
51p
0-237-50457-x X.809/54790
2592843

AFRICAN STUDENTS
West Africa, 1000-1800 - For West African students

DAVIDSON, Basil
A history of West Africa, 1000-1800. - New revised ed. - London : Longman, 1977
xvi,318p : ill. - (The growth of African civilisation). - Bibliography: p.295-296. - Includes index
0-582-60340-4 X.709/31591
124468x

OGINI, F. G.
An outline history of West Africa, 1000-1800. - Lagos : Macmillan Nigeria, 1977
x,214p : ill. - Bibliography : p.199-201.- Includes index
r6032626 X.709/31986
124468X

OGINI, FG
An outline history of West Africa 1000 - 1800. - Reprinted. - Lagos : Macmillan Nigeria, 1978
214p. : ill
r6096265 X.709/28272
124468X

AFRICAN STUDENTS
West Africa, 1000-1960

WILSON, Derek
West African history, 1000-1960. - London : Evans Bros, 1976
64p. - (Evans pocket facts)
0-237-49975-4 X.708/20418
0679631

AFRICAN STUDENTS
West Africa, 1000-1972 - For West African students

BUAH, Francis Kwamina
West Africa since AD 1000. - London : Macmillan.
Book 1: The people. - 1974
viii,263p : 2 ill. - Bibliography: p.247-250. - Includes index
0-333-15578-5 X.0709/1005
0373796

Book 2: The people and outsiders. - 1977
vii,231p. - Bibliography: p.218-220. - Includes index
B77-28319 X.0709/1005
0373796

AFRICAN STUDENTS
West Africa, 1000-1977 - For West African students

LATHAM, Norah
A sketch-map history of West Africa. - Revised [ed.]. - Amersham : Hulton, 1978
96p : 1 ill
0-7175-0803-x X.700/25771
1577182

AFRICAN STUDENTS
West Africa, 1000-1978

PARRY, K. J.
Revision guide, essay writing in West African history. - London : Longman, 1979
viii,120p. - (Study for success)
0-582-60348-x X.709/33496
1962949

AFRICAN STUDENTS
West Africa. Commerce - For West African students

LAWSON, Rowena M. (Rowena Mary)
Introduction to commerce in West Africa. - 3rd ed. - [Harlow] : Longman, 1979
[4],124p : ill. - Includes index
0-582-65022-4 X.510/16522
207818x

AFRICAN STUDENTS
West Africa. Economic conditions - For West African students

ANDERSON, David J. (David John)
Economics of West Africa. - London : Macmillan, 1977
ix,342p : ill. - Includes index
0-333-21575-3 X.519/26988
1142275

Economics of West Africa. - Rev. ed. - London : Macmillan, 1981
ix,342p : ill. - Includes index
0-333-32333-5 X.529/45177
1142275

EKUNDARE, R. Olufemi
An introduction to economics for West African students. - London : Allen & Unwin, 1980
203p : ill. - Includes index
0-04-330301-3 X.529/39073
1142275

AFRICAN STUDENTS
West Africa. Economic conditions, to 1968 - For West African students

ORD, Henry Wilson
An introduction to West African economics. - London : Heinemann Educational, 1978
x,457p : ill. - Includes index
0-435-97430-0 X.519/29170
1801031

AFRICAN STUDENTS
West Africa. Government, to 1980 - For West African students

OYENEYE, Ibiyemi
O-level West African government. - Rev. ed. - Harlow : Longman, 1982
iv,92p : 1ill. - (Study for success) (Study for success)
0-582-65098-4 X.809/55132
2543761

WARD, W. E. F. (William Ernest Frank)
Government in West Africa. - 6th rev. ed.. - London : Allen & Unwin, 1981
383p. - Bibliography: p10. - Includes index
0-04-351059-0 X.809/48927
2543761

AFRICAN STUDENTS
West Africa. Governments - For West African students

KELLY, Robert, 1946-
Multiple choice tests in government. - London : Macmillan, 1982
ix,91p
0-333-33165-6 X.520/35493
0726702

OYENEYE, Ibeyimi
West African government for O-level. - Harlow : Longman, 1981
xiv,191p : ill. - Includes index
0-582-60958-5 X.520/26149
0726702

OYENEYE, Ibiyemi
West African government. - London : Longman, 1976
[5],96p : ill. - (Longman certificate notes)
0-582-60316-1 X.708/20524
0726702

AFRICAN STUDENTS
West Africa. Politics, to 1982 - For West African students

OKWODU, R. N.
'O' Level government for West Africa. - [New ed.]. - London : Macmillan, 1983
viii,229p. - Includes index
0-333-35010-3 X.529/56288
1896350

School certificate government for West Africa. - London : Macmillan, 1979
vii,198p. - Includes index
0-333-23447-2 X.709/33209
1896350

AFRICAN STUDENTS
Woodworking - For African students

GRIMSHAW, Sidney Allan
Cassell's woodwork notebook. - African ed. - [London] : [Cassell], [1976]
52p : ill
0-304-29689-9 X.611/7322 Woolwich
0676721

AFRICAN STUDENTS
World, 1750-1974 - For African students

BUAH, Francis Kwamina
The world since 1750. - 2nd ed. - London : Macmillan, 1975
vi,346p : ill. - Includes index
0-333-19078-5 X.708/41359
0368830

AFRICAN STUDENTS. Secondary schools
Curriculum subjects: Spoken English language - For teaching

DODD, Robert
Practising spoken English. - London : Evans Bros, 1975
[3],90p. - Bibliography: p.84-85. - Includes index
0-237-50035-3 X.909/40195
0535443

AFRICAN STUDIES
See also
University of London. School of Oriental and African Studies

AFRICAN STUDIES

ÉTUDES africaines en Europe. - Paris : Agence de coopération culturelle et technique, 1981
2v.
r6210179 X.800/35776
8004692

AFRICAN STUDIES
- Serials

IFE African studies. - vol. 1, no. 1- , March 1974-. - Ile-Ife : The University, 1974- : ill
r6004454 P.801/1541
0925357

PAPER. - No.1-. - [Nairobi] : [The Institute], [197?]-
v.
Many parts wanting
r6818506 X.0525/702
8004692

PAPERS in African studies. - 1- ; [1975?]-. - Legon : Institute of African Studies [1975?]- : ill
Holdings : No.3-
r6010553 P.701/1534
1039369

SOUTHERN ASSOCIATION OF AFRICANISTS
The bulletin of the Southern Association of Africanists. - Vol.1, no.1- ; [1973?]-. - Gainesville : Center for African Studies, University of Florida, [1973?]-
Holdings : Vol.3, no.3 ; Vol.4, nos.1,2 ; Vol.5, nos.2,3 ; Vol.6, nos.1,2,3 ; Vol.7, no.1-
r6109388 P.801/1631
0925357

AFRICAN STUDIES
to 1973 - Bibliographies

BESTERMAN, Theodore
A world bibliography of African bibliographies. - Oxford : Blackwell, 1975
[9]p,21-241 columns. - Includes index
0-631-16900-8 BB.R.d.19a Reading Room
0583707
- [Another copy] OPL
- [Another copy] X.802/10094

AFRICAN STUDIES
1945-1975 - Conference proceedings

AFRICAN studies since 1945 / edited by Christopher Fyfe ; [for the University of Edinburgh, Centre of African Studies]. - London : Longman, 1976
vii,255p. - Includes index
0-582-64207-8 X.520/11294
0797510
- [Another copy : paperback]
ISBN 0-582-64208-6 X.700/25317

AFRICAN STUDIES
Brunschwig, Henri

ÉTUDES africaines offertes à Henri Brunschwig. - Paris : École des Hautes Études en Sciences sociales, 1982
xxvii,426p. - Includes bibliography of the works of Henri Brunschwig
2-7132-0788-6 X.800/40353
8735719

AFRICAN STUDIES
Jahn, Janheinz

UNIVERSITAT BAYREUTH. Lehrstuhl für Romanische Literaturwissenschaft und Komparatistik
Léopold Sédar Senghor. – Bayreuth : die Universitat, [1979]
33p
r6300710 X.950/8448
8735719

AFRICAN STUDIES
Research – Directories

INTERNATIONAL AFRICAN INSTITUTE
African studies = Etudes africaines. – Enlarged and revised ed. – [London] : The Institute, 1975
x,185p. – Includes index
0–85302–042–6 BB.R.e.13 Reading Room
0316970

AFRICAN STUDIES. Curriculum subjects, to date. Higher education institutions. Africa
– Conference proceedings

CONFERENCE OF THE INSTITUTES/CENTRES OF AFRICAN/CULTURAL STUDIES IN AFRICA (1978 : Khartoum)
African studies in Africa. – [Khartoum] : Institute of African and Asian Studies, University of Khartoum, 1981
viii,152p. – (Sudan library series ; 10)
r6807399 X.520/32911
9061185

AFRICAN TALES

ABRAHAMS, Roger D. (Roger David)
African folktales. – New York : Pantheon, c1983
xviii,354p. – (The Pantheon fairy tale and folklore library). – Bibliography: p345–348
0–394–50236–1 X.950/33476
0892874

AFRICAN TALES
– Anthologies

AUTA the giant killer. Two brothers and two eggs. – [London] : Save the Children Fund, 1981
32p : ill(some col.). – (Round the world folk tales)
0–333–30763–1 X.990/17826 Woolwich
2566664

COURLANDER, Harold
The Crest and the hide. – London : Gollancz, 1984, c1982
137p : ill
0–575–03360–6 Nov.50765
0892874

FABLES from Africa / collected by Jan Knappert ; illustrated by Jeroo Ray. – [London] : Evans Brothers, 1981, c1980
64p : ill. – (Evans Africa library)
0–237–50670–x X.990/17603 Woolwich
0892874

SKAZKI narodov Afriki ; perevod s afrikanskikh i zapadnoevropeĭskikh i͡azykov / [sostaviteli A. A. Zhukov i E. S. Kotli͡ar]. – Moskva : Glavnai͡a redakt͡sii͡a vosbochnoĭ literatury, 1976
685 p. : ill. – (Skazki i mify narodov Vostoka)
r6020476 X.809/23113
0892874

AFRICAN TALES
African tales & legends

CAREY, Margaret
Myths and legends of Africa. – Feltham : Hamlyn, c1970
159p : col.ill. – (Hamlyn all-colour paperbacks). – Bibliography: p157. – Includes index
B82–38338 X.958/11826
1443046

KOULIBALY, Isaïe Biton
La légende de Sadjo. – Abidjan : Éditions CEDA, [1981]
63p : ill
r6211072 X.958/18910
8004722

AFRICAN TALES
African tales & legends – Anthologies

ELLIOT, Geraldine
The hunter's cave. – London : Routledge & Kegan Paul, 1951
174p : ill
0–7100–0861–9 X.990/17796 Woolwich
2561034

The long grass whispers. – London : Routledge & Kegan Paul, 1949
132p : ill
0–7100–0860–0 X.990/17797 Woolwich
2561034

The singing chameleon. – London : Routledge & Kegan Paul, 1957
168p : ill
0–7100–0862–7 X.950/9131
2561034

ELLIOT, Geraldine
Where the leopard passes. – London : Routledge & Kegan Paul, 1949
133p : ill
0–7100–0863–5 X.990/17798 Woolwich
2561034

FABLES from Africa / collected by Jan Knappert ; illustrated by Jeroo Roy. – London : Evans, 1980
64p : ill
0–237–44985–4 X.950/12814
1888218

PITCHER, Diana
Tokoloshi. – London : Skilton and Shaw, 1980
64p : ill
0–7050–0081–8 X.981/22444
1888218

AFRICAN TALES
West African tales. Characters: Tricksters

PELTON, Robert D.
The trickster in West Africa. – Berkeley : University of California Press, 1980
[8],312p. – (Hermeneutics, studies in the history of religions ; 8). – Bibliography: p. 291–304. – Includes index
0–520–03477–5 X.200/34143
229155x

AFRICAN TERRACOTTAS. Exhibits. Museum of Mankind to 1969

FAGG, William
The potter's art in Africa. – 2nd ed. – London : Published for the Trustees of the British Museum by British Museum Publications, 1978
47p : ill
0–7141–1545–2 X.421/22539 Woolwich
2616025

AFRICAN TRYPANOSOMIASIS
See Sleeping–sickness

AFRICAN USAGE. English language
– Serials

ENGLISH in Africa. – Vol. 1, no. 1– ; Mar. 1974–. – Grahamstown : Institute for the Study of English in Africa, Rhodes University, 1974– : ill
Holdings : Vol. 1, no. 2–
r6004479 P.901/3026
0727687

AFRICAN USAGE. English language
West African usage – Serials

FOURAH Bay studies in language and literature. – Vol.1(1980)–. – Freetown : Fourah Bay College, University of Sierra Leone, 1980–
v.
r6811369 X.0955/227
9051260

AFRICAN USAGE. Portuguese language

MISCELÂNEA luso-africana / coligidos por Marius F. Valkhoff com a colaboração de Willy Bal... [et al.]. – Lisboa : Junta de Investigações Científicas do Ultramar, 1975
319p
r6065191 X.900/20482
154389X

AFRICAN VIOLETS. Pot plants
Cultivation

RECTOR, Carolyn K.
How to grow African violets. – 2nd ed. – Poole : Blandford, 1969, c1971
95p : ill. – Includes index
0–7137–1254–6 SRL
2550776

ROBEY, Melvin J.
African violets. – San Diego : Barnes ; London : Tantivy, c1980
199p,[4]p of plates : ill(some col.). – Includes index
0–498–02349–4 SRL
2550776

ROCHFORD, Thomas
Success with African violets. – Cambridge : Martin Books in association with Thomas Rochford & Sons, 1983
32p : ill(some col.). – (Rochfords' guides to plant care). – Includes index
0–85941–240–7 X.329/17810
2550776

AFRICAN WOMEN
Hair styles

SAGAY, Esi
African hairstyles. – London : Heinemann, 1983
108p : ill. – Bibliography: p108
0–435–89830–2 X.629/19655
2867435

AFRICAN WRITERS. Christian life
– Anthologies

AFRICAN Christian spirituality / edited with an introduction by Aylward Shorter. – London : G. Chapman, 1978 [i.e.1979]
xii,160p
0–225–66238–8 X.100/18293
1695797

AFRICAN WRITERS. Drama

BANHAM, Martin
African theatre today. – London : Pitman, 1976
vii,103p. – (Theatre today). – Bibliography: p.93–99. – Includes index
0–273–00348–8 X.909/40575
0552380

ETHERTON, Michael
The development of African drama. – London : Hutchinson, 1982
368p : ill. – (Hutchinson university library for Africa). – Bibliography: p358–363. – Includes index
0–09–146420–x X.950/15627
0552380

GRAHAM-WHITE, Anthony
The drama of black Africa. – New York : French, [1974]
xll,220p,[8]p of plates : ill. – Bibliography : p.183.– Includes index
r6092164 X.529/33676
0552380

AFRICAN WRITERS. Drama in English
1900– – Anthologies

TEATRO africano / [curatori : G. Bellini, M. Grampa, C. Muschio ; traduzione : Graziella Bellini ...et al.]. – Milano : Jaca Book, 1796 [i.e. 1976]
541p. – (Di fronte e attraverso ; 2)
r6078105 X.989/73636
1506862

AFRICAN WRITERS. Drama in English
1945–

SCHIPPER–DE LEEUW, Mineke
Toneel en maatschappij in Afrika. – Assen : Gorcum, 1977
123p. – (Terreinverkenningen in de culturele antropologie ; 8). – Bibliography
90–232–1506–0 X.950/12316
8091803

AFRICAN WRITERS. Drama in English
1945– – Anthologies

AFRICAN plays for playing / selected and edited by Michael Etherton. – London : Heinemann Educational.
1. – 1975
[3],136p : ill. – (African writers series ; 165)
0–435–90165–6 012212.e.1/165
068984x
2. – 1975
[3],107p. – (African writers series ; 179)
0–435–90179–6 012212.e.1/179
068984x
KUNTU drama / collected and with an introduction by Paul Carter Harrison. – New York : Grove Press, 1974
ii–xi,352p : 1,ill. – (An evergreen book ; E615)
0–394–17806–8 X.908/42699
068984X

TWO African radio plays from the BBC / [edited by] Michael & Annemarie Morris. – Nykøbing Mors : Futurum, 1984
59p : ill. – (Futurum scripts)
87–88497–08–9 X.950/33671
8091803

AFRICAN WRITERS. Drama in French
1960–

SALIEN, François
Panorama du théâtre africain d'expression française. – Bandundu : CEEBA Publications, 1983
218p. – (CEEBA publications. série II ; v.86)
r6210817 X.955/2511
8826757

SCHIPPER–DE LEEUW, Mineke
Toneel en maatschappij in Afrika. – Assen : Gorcum, 1977
123p. – (Terreinverkenningen in de culturele antropologie ; 8). – Bibliography
90–232–1506–0 X.950/12316
8826757

AFRICAN WRITERS. English literature
to 1977

The CRITICAL evaluation of African literature / edited by Edgar Wright. – Washington, D.C. : INSCAPE, 1976
179p
0–87953–404–4 X.989/53713
135468X

ROSCOE, A. A. (Adrian Alan)
Uhuru's fire. – Cambridge : Cambridge University Press, 1977
xi,281p. – Includes index
0–521–21295–2 X.981/20965
135468x

AFRICAN WRITERS. English literature
to date

COMPARATIVE approaches to modern African literature / edited by S.O. Asein. – Ibadan : University of Ibadan. Department of English, [1982]
x,153p
r6814279 X.950/37483
8730369

EGEJURU, Phanuel Akubueze
Black writers: white audience. – Hicksville, N.Y. : Exposition Press, c1978
255p. – Includes bibliography
0–682–48977–8 X.950/35243
8730369

IMFELD, Al
Vision und Waffe. – Zürich : Unionsverlag, 1981
355p. : ill. – Includes bibliographies
3–293–00025–8 X.950/11881
8730369

OGUDE, S. E.
Genius in bondage. – Ile-Ife : University of Ife Press, [1983]
166p. – Includes bibliography
978–13–6048–8 X.950/34399
8730369

VINCENT, Theo
Seventeen black and African writers on literature and life. – 2nd printing. – Lagos : Centre for Black and African Arts and Civilization, 1982
x,108p
978–13–4053–3 X.958/27231
8730369

AFRICAN WRITERS. English literature
to date – Conference proceedings

AFRICAN STUDIES ASSOCIATION. Meeting (23rd : 1980)
African literature in English. – [Philadelphia] : African Studies Association, Literature in English Group, [1981]
67p
r6806220 X.950/19454
8730369

JANHEINZ JAHN-SYMPOSIUM (2nd : 1977 : Mainz)
The East African experience. – Berlin : Reimer, 1980
134p. – (Mainzer Afrika-Studien ; Bd.4)
3–496–00125–9 X.0950/198.(4).
8730369

AFRICAN WRITERS. English literature
to date – Serials

ISALA. – No.1 (1982)–. – Ile-Ife : Ife Studies in African Literature and the Arts, 1982–
v.
r6806290 X.0950/542
8730369

AFRICAN WRITERS. English literature
1945–1977

COOK, David, 1929–
African literature. – Harlow : Longman, 1977
xiv,240p. – Includes index
0–582–64210–8 X.909/42207
1355554

GRIFFITHS, Gareth
A double exile. – London : Boyars : Distributed by Calder and Boyars, 1978
205p. – (Critical appraisals series). – Bibliography: p.194–202. – Includes index
0–7145–2622–3 X.989/52506
1355554

AFRICAN WRITERS. English literature
1945–

The BENIN Review. – Benin City : Ethiope Publishing Corporation.
no.1– ; June 1974–. – 1974–
r6096247 P.901/1760
0727679

DARLITE : a magazine of original writing from University College, Dar es Salaam, vols. 1–4, 1966–1970. – Reprinted. – Nendeln : Kraus Reprint, 1976
r6079447 P.521/704
0727679

IRELE, Abiola
The African experience in literature and ideology. – London : Heinemann, 1981
216p. – (Studies in African literature). – Includes index
0–435–91630–0 X.950/6390
2337185
– [Another Copy: paperback.] ISBN 0–435–91631–9 X.950/30871

MOORE, Gerald, 1924–
Twelve African writers. – London : Hutchinson University Library for Africa, 1980
327p. – Bibliography: p301–317. – Includes index
0–09–141850–x X.950/1357
2337185

AFRICAN WRITERS. English literature
1945– – Serials

ENGLISH in Africa. – Vol. 1, no. 1– ; Mar. 1974–. – Grahamstown : Institute for the Study of English in Africa, Rhodes University, 1974– : ill
Holdings : Vol. 1, no. 2–
r6004479 P.901/3026
0727679

NSUKKA studies in African literature. – Vol.1, no.1 (1978)–. – Nsukka : Department of English and Department of Modern Languages, University of Nigeria, 1978–
r6202385 X.0900/668
0727679

AFRICAN WRITERS. English literature
1945–. Social aspects

SANKAWULO, Wilton
The role of the black and African writer in the changing African society. – [Liberia] : [Wilton Sankawulo], [1977]
17 leaves
r6079737 X.902/3193
0121258

AFRICAN WRITERS. English literature
1945–. Special themes: Cultural awareness of negroes of Africa south of the Sahara

SOYINKA, 'Wole
Myth, literature and the African world. – Cambridge : Cambridge University Press, 1976
xii,168p. – Includes index
0–521–21190–5 X.989/51009
0746525

AFRICAN WRITERS. English literature
1960– – Anthologies

An ANTHOLOGY of African and Caribbean writing in English / an anthology edited by John J. Figueroa for the Third World studies course at the Open University. – London : Heinemann Educational in association with Open University, 1982
xix,297p. – Bibliography: p293–295. – Includes index
0–435–91297–6 X.950/19346
2871920

AFRICAN WRITERS. English literature
1960–. Political aspects

BECKER, Jörg
Afrikanische Literatur in der entwicklungspolitischen Bildungsarbeit. – Köln : Weltforum, 1981
ix,312p. – (Forschungsberichte des Bundesministeriums für wirtschaftliche Zusammenarbeit ; Bd.21). – Bibliography
3–8039–0220–7 X.950/21622
8950563

AFRICAN WRITERS. English literature
Southern African women writers, to date

LIP (Collective)
LIP from Southern African women. – Johannesburg : Raven Press, [1983]
vii,232p : ill
0–86975–153–0 X:950/34215
9166521

AFRICAN WRITERS. English literature
West African writers, 1960– – Serials

FOURAH Bay studies in language and literature. – Vol.1(1980)–. – Freetown : Fourah Bay College, University of Sierra Leone, 1980–
v.
r6811369 X.0955/227
9051252

AFRICAN WRITERS. English literature. Curriculum subjects. Schools. Great Britain
Teaching – Conference proceedings

ASSOCIATION FOR THE TEACHING OF CARIBBEAN, AFRICAN AND ASSOCIATED LITERATURE. National Conference (5th : 1982 : University of Exeter)
Proceedings of the Fifth National Conference of the Association for the Teaching of Caribbean, African and Associated Literature (ATCAL), University of Exeter 17–19 September 1982. – [London] : ATCAL, [1982]
13p
B83–41662 X.525/8191
2966212

AFRICAN WRITERS. European literatures
to date

ORTOVÁ, Jarmila
Modern literatures of Subsaharan Africa. – v Praze : Universita 17. listopadu, 1971
228leaves. – Bibliography
r6439578 X.955/623
8819114

AFRICAN WRITERS. European literatures
1945–1980. Implications of structuralism

ANOZIE, Sunday O.
Structural models and African poetics. – London : Routledge & Kegan Paul, 1981
xi,338p : ill. – Bibliography: p289–323. – Includes index
0–7100–0467–2 X.950/1479
2353350

AFRICAN WRITERS. Fiction in English
to date

MCEWAN, Neil
Africa and the novel. – Atlantic Highlands : Humanities Press, 1983
179p. – Includes bibliography
0–391–02922–3 X.950/32864
9088210

VIVAN, Itala
Tessere per un mosaico africano. – Verona : Morelli, 1984
235p. – (Nuove letterature). – Includes bibliography
r6512758 X.950/41733
9088210

AFRICAN WRITERS. Fiction in English
1945–

FRITSCHI, Gerhard
Africa and Gutenberg. – Berne : Lang, 1983
194p. – (European university studies. Ser.27, Asian and African studies ; v.9). – Includes bibliography
3–261–03322–3 X.950/41548
8153183

GAKWANDI, Shatto Arthur
The novel and contemporary experience in Africa. – London : Heinemann Educational, 1977
[6],136p. – Bibliography: p.131–132. – Includes index
0–435–91320–4 X.989/51966
1269798

MCEWAN, Neil
Africa and the novel. – London : Macmillan, 1983
x,179p. – Bibliography: p172–176. – Includes index
0–333–32477–3 X.950/32776
1269798

NGARA, Emmanuel
Art and ideology in the African novel. – London : Heinemann Educational, 1985
[160]p. – (Studies in African literature). – Includes bibliography and index
0–435–91721–8 X.950/41554
1269798
Stylistic criticism and the African novel. – London : Heinemann, 1982
viii,150p. – Bibliography: p143–146. – Includes index
0–435–91720–x X.950/11211
1269798

RECENT trends in the novel / editor Eldred Durosimi Jones ; associate editor Eustace Palmer ; editorial assisstant Marjorie Evans. – London : Heinemann, 1983
x,245p. – (African literature today ; 13). – Includes index
0–435–91646–7 X.950/21556
1269798

AFRICAN WRITERS. Fiction in English
1945–. Special themes: Africa. Society

EJRNÆS, Anne Marie
Afrika – litteratur og samfund. – København : Medusa, 1977
260p. – (Kritiske studier). – Includes bibliographies
87–7332–013–7 X.908/82215
1430807

AFRICAN WRITERS. Fiction in English
1960–

JANSEN, Karl Heinz
Literatur und Geschichte in Afrika. – Berlin : Reimer, [1981]
353p.. – (Kölner Beiträge zur Afrikanistik ; Bd.8). – Bibliography
3–496–00176–3 X.0950/105(8)
8792941

AFRICAN WRITERS. Fiction in English
1960– – Texts

FAKUNLE, Funmilayo
Chasing the shadow. – Oshogbo : Fakunle Major Press, 1980
viii,151p
r6810746 X.958/20663
8792941

AFRICAN WRITERS. Fiction in English
English & West African writers, 1900–. Special subjects: West Africa. European & African stereotypes

MILBURY-STEEN, Sarah L.
European and African stereotypes in twentieth-century fiction. – London : Macmillan, 1980
xiii,188p. – Bibliography: p178–183. – Includes index
0–333–29143–3 X.981/22632
2079968

AFRICAN WRITERS. Fiction in English
West African writers, 1945–. Social aspects

OBIECHINA, Emmanuel Nwanonye
Culture, tradition and society in the West African novel. – Cambridge : Cambridge University Press, 1975
ix,296p. – (African studies series ; 14). – Bibliography: p.279–284. – Includes index
0–521–20525–5 X.981/20026
0369179

TRENZ, Günter
Die Funktion englischsprachiger westafrikanischer Literatur. – Berlin : Reimer, 1980
iii,590p. – (Mainzer ethnologische Arbeiten ; Bd 1). – Bibliography: p.538–590
3–496–00143–7 X:0529/624(1)
8154635

AFRICAN WRITERS. Fiction in European languages 1945–

RUUTH-BÄCKER, Karin
En litterär resa genom det svarta Afrika. – Uppsala : Nordiska Afrikainstitutet, 1981
64p. – Includes bibliography
91–7106–194–0 X.958/22500
9079645

AFRICAN WRITERS. Fiction in French to date

MCEWAN, Neil
Africa and the novel. – Atlantic Highlands : Humanities Press, 1983
179p. – Includes bibliography
0–391–02922–3 X.950/32864
8957223

RICHTER-NGOGANG, Barbara
Der moderne schwarzafrikanische Roman in französischer Sprache (1921–1979). – Rheinfelden : Schäuble, [1983]
iii,396,vip.. – (Reihe Romanistik ; Nr.33). – Includes bibliography
3–87718–733–1 X.0900/328(33)
8957223

AFRICAN WRITERS. Fiction in French to date. Characters: Women

CHEMAIN-DEGRANGE, Arlette
Émancipation féminine et roman africain. – Dakar : Les Nouvelles Éditions Africaines, 1980
359p. – Bibliography
2–7236–0208–7 X.950/18890
8888353

AFRICAN WRITERS. Fiction in French 1945–

GAKWANDI, Shatto Arthur
The novel and contemporary experience in Africa. – London : Heinemann Educational, 1977
[6],136p. – Bibliography: p.131–132. – Includes index
0–435–91320–4 X.989/51966
1269801

AFRICAN WRITERS. Fiction in French 1945–. Special themes: Africa. Society

EJRNÆS, Anne Marie
Afrika – litteratur og samfund. – København : Medusa, 1977
260p. – (Kritiske studier). – Includes bibliographies
87–7332–013–7 X.908/82215
1430815

AFRICAN WRITERS. Fiction in French 1960–

JANSEN, Karl Heinz
Literatur und Geschichte in Afrika. – Berlin : Reimer, [1981]
353p.. – (Kölner Beiträge zur Afrikanistik ; Bd.8). – Bibliography
3–496–00176–3 X.0950/105(8)
879295x

AFRICAN WRITERS. Fiction in French 1960–. Special themes: Towns

CHEMAIN, Roger
La ville dans le roman africain. – Paris : Éditions l'Harmattan, [1981]
280p
2–85802–201–1 X.950/23013
8970971

AFRICAN WRITERS. Fiction in French French & African writers, 1900–. Special subjects: Africa. European & African stereotypes

MILBURY-STEEN, Sarah L.
European and African stereotypes in twentieth-century fiction. – London : Macmillan, 1980
xiii,188p. – Bibliography: p178–183. – Includes index
0–333–29143–3 X.981/22632
2079976

AFRICAN WRITERS. French literature – Bibliographies

BIBLIOGRAPHIE des auteurs africains de langue française / établie par Thérese Baratte-Eno Belinga, Jacqueline Chauveau-Rabut, Mukala Kadima-Nzuji. – 4e éd. – [Paris] : Fernand Nathan, 1979
245p. – (Classiques du monde)
2–09–160101–2 X.989/89590
2468433

AFRICAN WRITERS. French literature to 1975

BLAIR, Dorothy Sara
African literature in French. – Cambridge : Cambridge University Press, 1976
xx,348p. – Bibliography: p.324–339. – Includes index
0–521–21195–6 X.981/20458
0776351

AFRICAN WRITERS. French literature to date

ANTHOLOGIE africaine d'expression française / [choisie par] Jacques Chevrier. – Paris : Hatier, [1981]–. – (Monde noir poche ; 9)
r6206554 X.0958/136
8172307

EGEJURU, Phanuel Akubueze
Black writers: white audience. – Hicksville, N.Y. : Exposition Press, c1978
255p. – Includes bibliography
0–682–48977–8 X.950/35243
8172307

IMFELD, Al
Vision und Waffe. – Zürich : Unionsverlag, 1981
355p. : ill. – Includes bibliographies
3–293–00025–8 X.950/11881
8172307

SIX conférences sur la littérature africaine de langue française / présentées et éditées par Wolfgang Leiner. – Tübingen : Attempto, [1981]
102p. – (Werkhefte der Universität Tübingen. Reihe B, Geisteswissenschaften ; Nr 1)
3–921552–18–4 X.950/24643
8172307

AFRICAN WRITERS. French literature 1900–

CHEVRIER, Jacques
Littérature nègre. – Paris : Armand Colin, 1974
287 p.. – (U prisme ; 36)
r6006103 X.907/14417
0914479

AFRICAN WRITERS. French literature 1945–

IRELE, Abiola
The African experience in literature and ideology. – London : Heinemann, 1981
216p. – (Studies in African literature). – Includes index
0–435–91630–0 X.950/6390
2337193
– [Another Copy: paperback.] ISBN 0–435–91631–9 X.950/30871

MOORE, Gerald, 1924–
Twelve African writers. – London : Hutchinson University Library for Africa, 1980
327p. – Bibliography: p301–317. – Includes index
0–09–141850–x X.950/1357
2337193

AFRICAN WRITERS. French literature 1945– – Anthologies

LITTÉRATURE de l'Afrique du Nord / ved Marie-Alice Séférian. – København : Nyt Nordisk Forlag, 1976
119p : ill
87–17–02165–0 X.909/37339
1223828

AFRICAN WRITERS. Literature to 1978

INTRODUCTION to African literature / edited by Ulli Beier. – New ed. – London : Longman, 1979
xvii,302p. – Bibliography: p.289–298
0–582–64228–0 X.909/44586
2104105

AFRICAN WRITERS. Literature to date

MANDALA, Yamaina
Propos sur la littérature et la culture. – Kinshasa : Éditions la Grue Couronnée, [1979]
114p. – (Collection Mosaïque)
r6210460 X.958/19862
8284733

AFRICAN WRITERS. Literatures

HALLETT, George
African writers portfolio. – London : Heinemann Educational, [1981?]
1portfolio. – (African writers series)
0–435–90502–3 L.50/202
2598744

AFRICAN WRITERS. Literatures to 1975 – Conference proceedings

PAPERS on African literature / edited by Christopher Heywood. – Sheffield : Department of English Literature, The University ; London : Africa Educational Trust, 1976
vii,127p. – (Sheffield papers on literature and society ; 1)
B78–33979 P.901/3107
1594540

AFRICAN WRITERS. Literatures to 1983

A NEW reader's guide to African literature. – 2nd, completely rev. and exp. ed.. – London : Heinemann, 1983
xvi,553p. – (Studies in African literature). – Includes index
0–435–91999–7 X.950/28613
3059057
– [Another copy.] 2045.e. Reading Room

AFRICAN WRITERS. Literatures to date

EGBULEFU, John Okoro
Die schwarzafrikanische Literatur. – Darmstadt : Thesen, 1981
304p. – (Vergleichende Literaturwissenschaft ; 2). – Bibliography
3–7677–0029–8 X.950/8032
8285241

AFRICAN WRITERS. Literatures 1900–

CONVERSATIONS with African writers / interviews conducted and edited by Lee Nichols. – Washington, D.C. : Voice of America, 1981
xii,302p : ill
r6807257 X.950/23996
8285268

A NEW reader's guide to African literature. – 2nd, completely rev. and exp. ed.. – London : Heinemann, 1983
xvi,553p. – (Studies in African literature). – Includes index
0–435–91999–7 X.950/28613
3059065
– [Another copy.] 2045.e. Reading Room

AFRICAN WRITERS. Literatures 1950–1975 – Conference proceedings

DALHOUSIE/MOUNT SAINT VINCENT CONFERENCE ON AFRICAN WRITING (1973 : Halifax, N.S.)
Exile and tradition. – London : Longman [etc.], 1976
xiv,190p. – (Dalhousie African studies series)
0–582–64188–8 X.900/17621
0663980
– [Another copy ; paperback]
0 582 64189 6 X.900/40452

AFRICAN WRITERS. Literatures African women writers, 1964–1979

BROWN, Lloyd W. (Lloyd Wesley)
Women writers in Black Africa. – Westport, Conn. ; London : Greenwood Press, 1981
204p. – (Contributions in women's studies ; no.21). – Bibliography: p191–196. – Includes index
0–313–22540–0 X.950/8788
2585758

AFRICAN WRITERS. Literatures East African writers, 1945– – Anthologies

EAST African literature / edited by Arne Zettersten. – London : Longman, 1983
222p
0–582–64388–0 X.950/27729
3042715

AFRICAN WRITERS. Literatures Special subjects: Sociological perspectives of role of women in African urban communities

LITTLE, Kenneth
The sociology of urban women's image in African literature. – London : Macmillan, 1980
x,174p. – Bibliography: p159–161. – Includes index
0–333–28845–9 X.989/89984
2018667

AFRICAN WRITERS. Literatures
West African writers, 1930–1970

TAIWO, Oladele
An introduction to West African literature. – Walton-on-Thames : Nelson, 1967
191p. – Bibliography: p183–185. – Includes index
0–17–511277–0 X.950/9475
2603233

AFRICAN WRITERS. Literatures
West African writers, ca 1960–1978

NEW West African literature / edited by Kolawole Ogungbesan. – London : Heinemann Educational, 1979
ix,118p. – Includes index
0–435–91760–9 X.989/88999
2055635

AFRICAN WRITERS. Literatures. Curriculum subjects. Secondary schools. Great Britain
Teaching

GUNNER, Elizabeth
A handbook for teaching African literature. – London : Heinemann in association with the Extramural Division of the School of Oriental and African Studies, 1984
xvi,160p : ill. – Bibliography: p146–155. – Includes index
0–435–92260–2 X.950/30153
3015963

AFRICAN WRITERS. Poetry
– Anthologies – Translations

A CHILD'S book of African poetry / collected and edited by Wendy Ijioma. – London : Macmillan, 1978
[2],46p : col ill
0–333–22662–3 X.990/16810 Woolwich
2253879

AFRICAN WRITERS. Poetry
to 1975 – Anthologies – Translations

A SELECTION of African poetry / introduced and annotated by K.E. Senanu and T. Vincent. – [Harlow] : Longman, 1976
[6],224p
0–582–60141–x X.909/41074
0782246

AFRICAN WRITERS. Poetry
to 1980 – Anthologies – Translations

The PENGUIN book of modern African poetry / edited by Gerald Moore and Ulli Beier. – 3rd ed. – Harmondsworth : Penguin, 1984
315p. – (The Penguin poets)
0–14–042311–7 X.958/22042
3079015

AFRICAN WRITERS. Poetry
1945–1977. Political aspects

EQUDU, Romanus
Modern African poetry and the African predicament. – London : Macmillan, 1978
viii,154p. – Includes index
0–333–23784–6 X.989/52917
1193813

AFRICAN WRITERS. Poetry in English
to 1980 – Anthologies

A NEW book of African verse / compiled and edited by John Reed & Clive Wake. – Rev. ed. – London : Heinemann, 1984
xii,116p. – (African writers series ; v.8). – Bibliography: p116
0–435–90308–x 012212.e.1./8
304842x

AFRICAN WRITERS. Poetry in English
1945– – Anthologies

POESIA africana de hoje / selecção, prefácio, notas e tradução de Manuel de Seabra. – Lisboa : Editorial Futura, 1974
198 p.. – Bibliography : p. 191–198
r6009047 X.907/14462
0922048

AFRICAN WRITERS. Poetry in English
1945– – For African students – Anthologies

FRASER, Robert, 1947–
Reading African poetry. – London : Collins, 1975
96p : ill
0–00–321866–x X.909/40794
0711896

An INTRODUCTION to East African poetry / [edited by] Jonathan Kariara, Ellen Kitonga. – Nairobi : Oxford University Press, 1976
126p
0–19–572064–4 X.909/44397
0711896

AFRICAN WRITERS. Poetry in English
1945–. Practical criticism

BURTON, S. H. (Samuel Holroyd)
African poetry in English. – London : Macmillan, 1979
xiv,154p : ill. – Bibliography: p153–154
0–333–27050–9 X.958/7864
1811266

AFRICAN WRITERS. Poetry in English
1960–

EGUDU, Romanus N.
Modern African poetry and the African predicament. – New York : Barnes & Noble, 1978
viii,154p
0–06–491916–1 X.950/35235
8351805

GOODWIN, K. L. (Kenneth L)
Understanding African poetry. – London : Heinemann, 1982
xix,204p. – Includes index
0–435–91325–5 X.950/17839
2827247

AFRICAN WRITERS. Poetry in English
1960– – Anthologies

AFRICAN voices / selected by Howard Sergeant. – New ed. – London : Evans Bros, 1978
xvii,137p
B78–33203 X.909/42816
027710x

ATTACHMENTS to the sun / edited by Dougal Blackburn, Alfred Horsfall and Chris. L. Wonjala ; illustrated by Hosea Nganga. – London : Edward Arnold, 1978
xii,100p : ill
0–7131–0247–0 X.909/43199
027710x

AFRICAN WRITERS. Poetry in English
1960– – Anthologies

DALSGAARD, Tage C.
Selections from modern African poetry in English. – København : Nyt Nordisk Forlag Arnold Busck, 1974
112 p. : ill
87–17–01855–2 X.909/31968
027710x

POEMS of black Africa / edited and introduced by Wole Soyinka. – London : Heinemann Educational, 1975
378p. – (African writers series ; 171)
0–435–90171–0 012212.e.1/171
027710x

SUMMER fires / edited by Angus Calder, Jack Mapanje & Cosmo Pieterse. – London : Heinemann, 1983
xii,116p. – (African writers series ; 257)
0–435–90257–1 012212.e.1/257
027710x

AFRICAN WRITERS. Poetry in English
West African writers, to date

EGUDU, Romanus N.
Four modern West African poets. – New York : NOK Publishers, [1977]
138p
0–88357–021–1 X.950/13345
8837082

AFRICAN WRITERS. Poetry in English
West African writers, 1745–1800 – Texts

[The DYING negro]. A poetical epistle of the dying negro. – [3rd ed., corrected and enlarged, reprinted]. – London : Helix Press, 1978
43p. – (Africans (Blacks) in history)
B79–13088 X.909/43456
1838385

AFRICAN WRITERS. Poetry in French
to date. Surrealism

MICHEL, Jean-Claude
Les écrivains noirs et le surréalisme. – Sherbrooke (Que.) : Éditions Naaman, 1982
181p. – Includes bibliography
2–89040–212–6 X.950/35167
911596x

AFRICAN WRITERS. Poetry in French
1945– – Texts

CONFIDENCES sur les ondes / Samuel Nkamgnia. – Yaoundé : Éditions Semences africaines, 1979
70p
r6208906 X.950/24068
8355495

LINZE, Georges
Poème science du coeur et du monde. – Liège : Éditions "Anthologie", 1981
58p
r6208859 X.950/23171
8355495

POESIA africana de hoje / selecção, prefácio, notas e tradução de Manuel de Seabra. – Lisboa : Editorial Futura, 1974
198 p.. – Bibliography : p. 191–198
r6009047 X.907/14462
092203X

AFRICAN WRITERS. Poetry in French
1945– – Translations

CHONANG, Antoine de Padoue
L'arc-en-ciel de minuit (poésie) = The midnight rainbow (poems). – Yaoundé : [L'Auteur], 1975
97p
r6038275 X.902/2861
0190411

L'arc-en-ciel de minuit = The midnight rainbow. – Yaoundé : [L'Auteur?], 1975
97p
r6038151 X.902/2794
0190411

AFRICAN WRITERS. Poetry in Portuguese
1900– – Anthologies

ANTOLOGIA temática de poesia africana / [editada por] Mário de Andrade. – Lisboa : Livraria Sá da Costa Editora. – (Colecção Vozes do mundo ; n.5)
1: Na noite grávida de punhais. – 1975
281p : ill
r6066202 X.909/81469
0903302

NO reino de Caliban / organização, selecção, prefácio e notas de Manuel Ferreira. – Lisboa : Seara Nova, 1975–
vol. 1– : ill
r6004834 X.0909/824
0903302

RESISTÊNCIA africana / organizada por Serafim Ferreira. – Lisboa : Diabril, 1975
131 p.. – (Colecção Universidade do povo ; n. 2)
r6014564 X.909/33573
0903302

AFRICAN WRITERS. Poetry in Portuguese
1945– – Anthologies

ANTOLOGIA temática de poesia africana / [editado por] Mário de Andrade. – Lisboa : Livraria Sá da Costa Editora. – (Colecção Vozes do mundo)
2: O canto armado. – 1979
192p
r6703198 X.950/12484
8731810

AFRICAN WRITERS. Portuguese literature
to 1977

FERREIRA, Manuel
Literaturas africanas de expressão portuguesa. – [Lisboa] : Instituto de Cultura Portuguesa, 1977
2v.. – (Biblioteca breve. série literatura ; 6,7). – Includes bibliographies
r6066418 X.908/84787
1715674

HAMILTON, Russell G.
Voices from an empire. – Minneapolis : University of Minnesota Press, 1975
450p. – (Minnesota monographs in the humanities ; vol.8)
0–8166–0745–1 X.989/54299
1715674

AFRICAN WRITERS. Portuguese literature
to date

MARGARIDO, Alfredo
Estudos sobre literaturas das nações africanas de língua portuguesa. – Lisboa : A Regra do Jogo, 1980
558p. – (Ensaios ; 3)
r6704748 X.950/12169
8368694

AFRICAN WRITERS. Portuguese literature
to date – Serials

AFRICA. – Vol.I, ano I, no.I (Julho 1978)–. – Lisboa : Estampa, 1978– : ill
r6706044 X.0950/425
8368694

AFRICAN WRITERS. Short stories in English
African women writers, 1960– – Anthologies

UNWINDING threads / writing by women in Africa ; selected and edited by Charlotte H. Bruner. – London : Heinemann, 1983
xv,208p. – (African writers series ; 256)
0–435–90256–3 012212.e.1/256
3075583

AFRICAN WRITERS. Short stories in English
Southern African writers – Anthologies

STORIES from Central & Southern Africa / edited and introduced by Paul A. Scanlon. – London : Heinemann, 1983
xii,207p. – (African writers series ; 254)
0–435–90254–7 012212.e.1/254
3075591

AFRICAN WRITERS, 1945-. Fiction in French
- Texts

OUSSOU-ESSUI, Denis
Les saisons sèches. - Paris : Éditions l'Harmattan, [1979]
188p
2-85802-133-3 X.989/87472
2435047

AFRICANA MUSEUM
Stock: Prints, to 1870 - Catalogues

AFRICANA MUSEUM
Catalogue of prints in the Africana Museum and in books in the Strange Collection of Africana in the Johannesburg Public Library up to 1870. - Johannesburg : The Museum, 1975
2v.([729]p) : ill. - Includes index
0-620-01927-1 LR.425/209
1256130

AFRICANA MUSEUM JAMES HALL MUSEUM OF TRANSPORT
See James Hall Museum of Transport

AFRICANS
- Biographies - Directories

AFRICA year book and who's who. - 1977 / [editor-in-chief Raph Uwechue]. - London : Africa Journal Ltd, 1976
xlvii,1364p,[8]p of plates : ill(some col)
Subsequent volumes, when received, will be placed at this shelfmark
0-903274-05-1 P.701/1533
0042196

AFRICANS
- Dictionaries - Serials

DICTIONARY of African biography / editor-in-chief : L. H. Ofosu-Appiah. - New York : Reference Publications, [1977]-. - (Encyclopaedia Africana)
r6049014 AA.OO.g Reading Room.
0042196

AFRICANS
1800-1900

LEADERSHIP in 19th century Africa / editor Obaro Ikime. - London : Longman [etc.] for the Historical Society of Nigeria, 1974
xvi,179p : ill. - Bibliography: p.177-179
0-582-60893-7 X.709/30143
0515345

AFRICANS
Bible - For Africans - Serials

SHARING scripture. - 1-. - London : Geoffrey Chapman, 1981-
v. : ill
B84-07264 P.103/642
3033775

AFRICANS
Bible. Special subjects: Christian doctrine. Justification - For Africans

WHITE, Ron
Right with God. - Achimota ; London : Africa Christian Press, 1977
30,[1]p : ill. - (What the Bible teaches booklets ; no.7)
0-85352-166-2 X.0108/515(7)
1480723

AFRICANS
Bible. Special subjects: Spiritual healing - For Africans

BOWEN, Roger, 1934-
Healing. - Achimota ; London : Africa Christian Press, 1977
31,[1]p : ill. - (What the Bible teaches booklets ; no.6). - Bibliography: p.31
0-85352-165-4 X.0108/515(6)
1723618

AFRICANS
Christian doctrine - For West Africans

MAIRE, C. D.
Learning to live for God. - Welwyn : Evangelical Press, 1979
vii,119p
0-85234-132-6 X.100/18778
2007495

AFRICANS
Christian life - For Africans

ASHTON, M. L.
God's plan for your life. - Achimota ; London : Africa Christian Press, 1975
31p : ill
0-85352-031-3 X.108/17245
1462520

OSBORNE, Denis
Dear Mudlick. - Achimota ; London : Africa Christian Press, 1973
30p. - (Letters from Hell series / Denis Osborne)
0-85352-106-9 X.907/25398
1462520

Dear Mudlick, series 2. - Achimota ; London : Africa Christian Press, 1976
37p. - (Letters from Hell series / Denis Osborne ; 2)
0-85352-109-3 X.907/25399
1462520

AFRICANS
Families. Christian life - For Africans

COON, Joyce
Biblical foundations for a Christian home. - Lagos ; London : Africa Christian Press, 1977
118p
0-85352-316-9 X.100/17692
1485857

AFRICANS
Overseas study, to date - Conference proceedings

The IMPACT of overseas training on development / edited by T.L. Maliyomkono and the EAURP Editorial Committee. - Arusha : Eastern Africa Publications, [1981?]
159p
r6804710 X.808/35070
8773939

AFRICANS
Personality - Conference proceedings

COLLOQUE INTERNATIONAL SUR LA NOTION DE PERSONNE EN AFRIQUE NOIRE (1971 : Paris)
La notion de personne en Afrique Noire.... - Paris : Éditions du Centre National de la Recherche Scientifique, 1973
596p., 1 leaf of plates, 6p. of plates : ill. - (Colloques internationaux du Centre national de la recherche scientifique ; no.544)
r6043483 S:E.74/217
1157159

AFRICANS
Role in exploration of Central Africa & East Africa by Europeans, 1840-1890

SIMPSON, Donald Herbert
Dark companions. - London : Elek, 1975
xi,228p,[8]p of plates : ill. - Bibliography: p.201-207. - Includes index
0-236-40006-1 X.800/11573
0622591

AFRICANS. United States
Interpersonal relationships with negroes

MOIKOBU, Josephine Moraa
Blood and flesh. - Westport, Conn. ; London : Greenwood Press, 1981
xiii,226p. - (Contributions in Afro-American and African studies ; no.59). - Bibliography: p205-217. - Includes index
0-313-22549-4 X.529/47056
2707624

AFRICANS. West Africa
Resistance to colonisation by European countries, 1850-1903

WEST African resistance / edited by Michael Crowder. - New ed. with corrections. - London : Hutchinson, 1978
xii,314p,[8]p of plates : ill. - Includes bibliographies and index
0-09-134031-4 X.700/25757
1568752

AFRICANS. West Indies
to date - Serials

[BULLETIN (African Studies Association of the West Indies)]. Bulletin / African Studies Association of the West Indies. - [No.1 (Dec. 1968)?]-. - Kingston [Jamaica] : African Studies Association of the West Indies, [1968?]-
v.
Holdings: no.4 (Dec. 1971)-
r6807622 X.0809/2323
9017771

AFRICO NUOVO. Italy
Social conditions, to date

STAJANO, Corrado
Africo. - 2a ed. - Torino : Einaudi, 1979
203p,[16]p of plates : ill. - (Gli struzzi ; 184)
r6702516 X.808/34565
8722684

AFRIHILI LANGUAGE

ATTOBRAH, K. A. Kumi
Ni Afrihili oluga. - [Akrokerri] : [K.A. Kumi Attobrah], [1971].
No.2
16p
r6804838 X.958/16380
8953694

AFRIKAANS- EN SUID-AFRIKAANS-DUITSE KULTUURVERENIGING

VAN DER MERWE, W. J.
Die geskiedenis van die Afrikaans- en Suid-Afrikaans-Duitse Kultuurvereniging. - Pretoria : Universiteit van Zuid-Afrika, 1982
xi,217p. - (Miscellanea / Unisa ; 39)
0-86981-279-3 X.529/68029
9142533

AFRIKAANS LANGUAGE
See also
Encyclopaedias in Afrikaans

AFRIKAANS LANGUAGE

INLEIDING tot die Afrikaanse taalkunde / hoofredakteur: T.J.R. Botha ... - Pretoria : Academica, 1984
311p
0-86874-184-1 X.950/36271
8977003

RAIDT, Edith H.
Einführung in Geschichte und Struktur des Afrikaans. - Darmstadt : Wissenschaftliche Buchgesellschaft, 1983
viii,199p. - (Germanistische Einführungen in Gegenstand, Methoden und Ergebnisse der Disziplinen und Teilgebiete). - Includes bibliography
3-534-08607-4 X.950/41698
8977003

SCHOOR, J. L. van
Die grammatika van Standaard-Afrikaans. - Kaapstad : Lex Patria, 1983
459p
0-628-02389-8 X.950/30111
8977003

AFRIKAANS LANGUAGE
- Dictionaries

DU PLESSIS, J. A.
Bukantswe ya maleme-Pedi. - Kaapstad : Via Afrika, [1974]
269p
0-7994-0352-0 X.950/32663
8977003

AFRIKAANS LANGUAGE
- Serials

T.N.A.. - Jg.1, no.1 (Maart 1983)-. - Köln : Lukassen, 1983- : ill
r6306717 P.901/3385
8977003

AFRIKAANS LANGUAGE
Comprehension

KURSUS in Afrikaans. - London : Linguaphone Institute, c1983
246p : ill
B84-24875 WP.13705/1
312360x

AFRIKAANS LANGUAGE. Namibia
to date

KLEINZ, Norbert
Deutsche Sprache im Kontakt in Südwestafrika. - Stuttgart : Steiner, 1984
xix,316p.. - (Deutsche Sprache in Europa und Übersee ; Bd.9). - Includes bibliography
3-515-04049-8 X.0900/505(9)
9139184

AFRIKAANS LITERATURE
See also
Fiction in Afrikaans
Poetry in Afrikaans

AFRIKAANS LITERATURE
1904-. Political aspects

COPE, Jack
The adversary within. - Cape Town : David Philip ; London : Collings, c1982
xi,208p. - Bibliography: p199-201. - Includes index
0-86036-198-5 X.950/30352
3076016

AFRIKAANS LITERATURE
1924-1961 - Texts

MARAIS, Eugène N. (Eugène Nielen)
[Works]. Versamelde werke. - Pretoria : J.L. van Schaik, 1984
2v.(1296p) : ill
r6330941 X.950/36584
912814x

AFRIKAANS LITERATURE
1961- - Serials

SA literature. - 1 (1980)-. - Johannesburg : AD. Donker, 1980-
v.. - (Censal publications ; no.1)
r6810678 X.0950/655
9001441

AFRIKAANS LITERATURE
Leipoldt, Christian Frederick Louis

LEIPOLDT 100 / 'n Bundel opstelle saamgestel deur Merwe Scholtz vir die Leipoldt-gedenkjaar 1980. - Kaapstad : Tafelberg, 1980
140p
0-624-01404-5 X.950/32651
9082913

AFRIKAANS LITERATURE
M.E.R. - Correspondence, diaries

M.E.R. (Maria Elizabeth Rothmann)
'n Kosbare erfenis. - Kaapstad : Tafelberg, 1977
239p
0-624-01056-2 X.950/32880
9082913

AFRIKAANS LITERATURE
Special subjects: South Africa. Coloured persons

FEBRUARY, Venie A.
Mind your colour. - London : Kegan Paul International, 1981
viii,248p. - (Monographs from the African Studies Centre, Leyden). - Bibliography: p222-235. - Includes index
0-7103-0002-6 X.950/3977
2395487

AFRIKAANS- UND SÜDAFRIKANISCH-DEUTSCHEN KULTURVEREINIGUNG
See Afrikaans- en Suid-Afrikaans-Duitse Kultuurvereniging †

AFRIKAKORPS
See Germany. Heer. Afrikakorps

AFRIKANERS
See also
Rehoboth Basters

AFRIKANERS. Kalahari Desert. Botswana
Social aspects

RUSSELL, Margo
Afrikaners of the Kalahari. - Cambridge : Cambridge University Press, 1979
xiv,167p : ill. - (African studies series ; 24). - Bibliography: p.157-161. - Includes index
0-521-21897-7 X.800/27755
1880047
- [Another copy] X.800/28141

AFRIKANERS. South Africa
to date

SCHOLTZ, G. D.
Afrikaner, waarheen?. - Johannesburg : Perskor, 1978
178p
0-628-01285-3 X.809/59871
9082778

AFRIKANERS. South Africa
1780-1975 - Correspondence, diaries

DU TOIT, André
Afrikaner political thought. - Berkeley ; London : University of California Press. - (Perspectives on Southern Africa ; 22)
Vol.1: 1780-1850. - c1983
xxx,309p. - Includes index
0-520-04319-7 X.800/37319
306221x

AFRIKANERS. South Africa
1838-1980

HARRISON, David, 19—
The white tribe of Africa. - London : British Broadcasting Corporation, 1981
vii,307p : ill. - Bibliography: p286-290. - Includes index
0-563-17838-8 X.800/31864
2588366
- [Another copy]. - Berkeley: University of California Press, 1981 X.800/39641

AFRIKANERS. South Africa
Attitudes of society - Conference proceedings

LOOKING at the Afrikaner today / compiled by Hendrik W. van der Merwe ; in association with the Centre for Intergroup Studies. - Cape Town : Tafelberg, 1975
[10], 124 p.. - Includes bibliographies
0-624-00702-2 X.809/41580
1036386

AFRIKANERS. South Africa
Nationalism, to 1974

MOODIE, T. Dunbar
The rise of Afrikanerdom. - Berkeley ; London : University of California Press, 1975
xii,328p. - (Perspectives on southern Africa ; 11). - Bibliography: p.301-316. - Includes index
0-520-02310-2 X.809/40211
0376221

AFRIKANERS. South Africa
Nationalism, to 1978

ADAM, Heribert
Ethnic power mobilized. - New Haven ; London : Yale University Press, 1979
xii,308p. - Includes index
0-300-02377-4 X.700/26241
2055856
- [Another copy.] X.809/45599

AFRIKANERS. South Africa
Nationalism, to date

The AFRIKANERS. - Cape Town : Tafelberg, 1979
183p
0-624-01340-5 X.529/54718
8611394

AFRIKANERS. South Africa
Nationalism, 1934-1948

O'MEARA, Dan
Volkskapitalisme. - Cambridge : Cambridge University Press, 1983
xvi,281p. - (African studies series ; 34). - Bibliography: p267-273. - Includes index
0-521-24285-1 X.520/30921
294085x

AFRIKANERS. South Africa
Nationalism. Organisations: Broederbond, to 1978

SERFONTEIN, J. H. P.
Brotherhood of power. - London : Collings, 1979
278p : ill. - Includes index
0-86036-093-8 X.520/21936
2060132

WILKINS, Ivor
The Broederbond. - New York ; London : Paddington Press, 1979
[16]p,458, A 139p : ill. - Bibliography: p.451-452. - Includes index
0-7092-0734-4 X.800/28956
2060132

Broederbond. - London : Corgi, 1980, c1979
450,A299p : ill. - Bibliography: p449-450
0-552-11512-6 X.808/33544
2060132

AFRIKANERS. South Africa
Nationalism. Support by Britons, to date

DAVEY, Arthur
The British pro-Boers 1877-1902. - Cape Town : Tafelberg, [1978]
220p
0-624-01200-x X.809/54312
8959560

AFRIKANERS. South Africa
Puritanism, to 1974

DE KLERK, Willem Abraham
The Puritans in Africa. - London : Collings, 1975
xvi,376p. - Includes index
0-901720-71-2 X.800/25178
0507849

The Puritans in Africa. - Harmondsworth : Penguin, 1976
xvi,376p. - (Pelican books). - Includes index
0-14-021965-x X.708/20526
0507849

AFRO. Italian paintings

AFRO
Afro. - Roma : Editalia, [1977]
225p : ill(some col),port. - (Maestri del XX secolo)
r6195280 L.45/485
2450747

QUARGNAL, Elettra
Afro, 1912-1976. - Udine : Edizioni della Galleria d'arte moderna, 1978
59p : ill(some col.). - (Quaderno / Galleria d'arte moderna ; 1)
r6197502 X.429/11614 Woolwich
826449x

AFRO-AMERICAN CHRISTIAN MISSIONS. Africa
1877-1900

WILLIAMS, Walter L.
Black Americans and the evangelization of Africa 1877-1900. - Madison ; London : University of Wisconsin Press, 1982
xviii,259p : ill. - Bibliography: p237-250. - Includes index
0-299-08920-7 X.200/39733
2849771

AFRO-BRAZILIAN THOUGHT

SILVA, Edson Nunes da
Estrutura do pensamento afrobrasileiro (Eminismo). - Salvador : Prefeitura Municipal da Cidadedo Salvador, Secretaria Municipal de Educação e Cultura, Departamento de Cultura, 1975
96p.
r6025146 X.510/10132
1210025

AFROASIATIC LANGUAGES
See Hamito-Semitic languages

AFROASIATIC LITERATURES
See Hamito-Semitic literatures

AFROBRAZILIAN RELIGIOUS CULTS. Brazil

BASTOS, Wilson de Lima
Figuras presentes em cultos afro-brasileiros. - Juiz de Fora : Centro de Estudos Sociológicos de Juiz de Fora, 1977
42p. - (Arquivos de folclore ; 5)
r6101452 X.510/11951
2190427

AFROBRAZILIAN RELIGIOUS CULTS. Brazil
to date

CISNEIROS, Israel
Fundamentos de umbanda. - Rio de Janeiro : Equipe, [1978?]
155p : ill
r6703427 X.529/45927
8559635

MACUMBA / exposição introdutória e coordenação de D. Cirilo Folch Gomes. - 2.ed., revista e ampliada. - São Paulo : Edições Paulinas, 1976
113p. - (CNBB Leste ; 1)
r6703432 X.529/45928
8559635

AFROCARIBBEAN RELIGIOUS CULTS. America
to date

FICHTE, Hubert
Petersilie. - Frankfurt am Main : S. Fischer, 1980
401p.. - (Die afroamerikanischen Religionen ; 4)
3-10-020708-4 X.200/36108
8774722

AFROCARIBBEAN RELIGIOUS CULTS. Puerto Rico
to date

SÁNCHEZ CÁRDENAS, Julio A.
La religión de los orichas. - [s.l.] : [s.n.], 1978
ix,149p : ill. - (Colección Estudios afrocaribeños). - Bibliography
r6704262 X.200/39054
8873399

AFTER LIFE
See Future life

AFTER SALES SERVICE. Consumer goods. European Community countries

BUREAU EUROPÉEN DES UNIONS DE CONSOMMATEURS
After sales service in the EEC. - London : Graham and Trotman for the Commission of the European Communities, 1976
vii,271p
0-86010-056-1 X.512/5709
1254979

AFTER SCHOOL CARE. Children of working mothers. Great Britain
Provision

WHITE, Paul, 1947-
Starting from scratch!. - London : National Out of School Alliance, 1983
20p : ill. - Includes bibliographies
0-9509139-0-1 X.525/10272
2756722

AFTERCARE
See also
Residential aftercare

AFTERCARE. Children leaving residential care. Scotland
GODEK, Stan
Leaving care. - [Barkingside] : Dr Barnardo's, 1977
54p. - (Barnardo social work papers ; no.2)
B80-41588 X.529/38405
2319551

AFTERCARE. Discharged patients. Hospitals. England
Role of volunteers
GAY, Pat
When I went home - - London King Edward's Hospital Fund for London ; [Tunbridge Wells] : Distributed by Pitman Medical, 1979
[7],82p. - Bibliography: p.79-80. - Includes index
0-900889-72-1 X.320/11570 Woolwich
1781529

AFTERCARE. Discharged patients. Hospitals. Great Britain
Effectiveness
ROBERTS, Ida
Discharged from hospital. - London : Royal College of Nursing, 1975
157p,fold leaf : ill. - (The study of nursing care project reports. series 2 ; no.6). - Bibliography: p.153-157
0-902606-42-5 X.0319/249b(6) Woolwich
0681989

AFTERCARE. Discharged patients. Psychiatric hospitals. Great Britain
Role of volunteers
DARTINGTON, Tim
Volunteers and psychiatric aftercare. - Berkhamsted : Volunteer Centre : MIND, 1978
24,[4]p. - (Occasional papers series / Volunteer Centre)
B78-36732 X.529/49783
169636x

AFTERCARE. Old persons. Discharged patients. Hospitals. Great Britain
CONTINUING CARE PROJECT
Organising aftercare. - London : National Corporation for the Care of Old People, [1979]
[6],34,4p. - Bibliography: p.34
0-904139-17-4 X.325/658 Woolwich
2287897

AFTERCARE SERVICES. Discharged patients. Hospitals
Planning
CONTINUING care / edited by Kathleen M. McKeehan ; instructional design by Barbara R. Pollack. - St. Louis ; London : Mosby, 1981
xviii,254p : ill. - Bibliography: p213-227. - Includes index
0-8016-3285-4 X.325/1138 Woolwich
2402505

AFTERCARE SERVICES. Discharged prisoners. Durham (County) 1882-1982
CRANFIELD, Ruth E. G.
One hundred years of prisoners' aid in County Durham. - Durham : North Eastern Prison After Care Society, 1982
34p
B82-20096 X.809/51487
2713330

AFTERCARE SERVICES. Discharged prisoners. England
GREAT BRITAIN. Central Office of Information
The probation and after-care service in a changing society. - [London] : [Home Office], 1976
[2],30p(2 fold) : ill. - Bibliography: p.30
0-903723-15-8 BS.18/468
109386x

HAXBY, David
Probation. - London : Constable, 1978
351p. - Bibliography: p.333-341. - Includes index
0-09-460240-9 X.200/31342
109386x

AFTERCARE SERVICES. Discharged prisoners. England
- Statistics
REPORT on the probation and after-care statistics (England and Wales). - 1975 & 1976. - [Surbiton] : Home Office, Statistical Department, 1977
[10],23p : ill
Subsequent volumes, when received, will be placed at this shelfmark
B78-28106 BS.18/579
1549650

AFTERCARE SERVICES. Discharged prisoners. England to 1975
BOCHEL, Dorothy
Probation and after-care. - Edinburgh : Scottish Academic Press ; London : Distributed by Chatto and Windus, 1976
viii,289p. - Bibliography: p.241-248. - Includes index
0-7011-2179-3 X.200/30708
0792519

AFTERCARE SERVICES. Discharged prisoners. England
Role of voluntary personnel
CLARKE, A. T.
Volunteers accredited to the Probation Service. - [Hertford] : Hertfordshire Probation and After-Care Service, [1977?]
41p : ill. - Bibliography: p39-40
B83-11349 X.525/6874
1207636

HERTFORDSHIRE PROBATION AND AFTER-CARE SERVICE
Volunteers accredited to the probation service. - [Hertford] : [The Service], 1977
[7],23p : ill
B79-01796 X.515/598
1207636

VOLUNTEER CENTRE
A survey of the involvement of volunteers with probation and after-care services in England and Wales. - Berkhamsted : The Centre, [1977?]
[2],19,[1]p
B77-28526 X.525/6174
1207636

AFTERCARE SERVICES. Discharged prisoners. Great Britain
Dodd, John - Biographies
NORMAN, Diana
Road from Singapore. - Revised ed. - London : Hodder and Stoughton, 1979
238p,[4]p of plates : 2 ill
0-340-23483-0 X.709/33129
0202770

AFTERCARE SERVICES. Old persons. Discharged patients. Hospitals. England
Voluntary aftercare services. Organisation
AGE CONCERN
Hospital after-care schemes. - Mitcham : Age Concern, 1980
[2],20p. - (Action guides / Age Concern). - Bibliography: p.20
0-904502-90-2 X.329/13131 Woolwich
230919x

AFTERCARE SERVICES. Old persons. Discharged patients. Hospitals. London
AGE CONCERN GREATER LONDON
Discharge from hospital. - [London] : [Age Concern Greater London], 1980
[2],21leaves
B80-31254 X.525/3491
2210274

AFTERNOON TEAS
The NATIONAL Trust & the West Country Tourist Board's book of afternoon tea / edited by Marika Hanbury Tenison ; with research by Jackie Gurney and Warren Davis. - Newton Abbot : David and Charles, 1980
96p : ill. - Includes index
0-7153-7929-1 X.629/13931 Woolwich
2002426

AFTONBLADET
AFTONBLADET / Gunnar Fredriksson ... [et al.]. - Stockholm : Tiden, 1980
428p : ill. - Includes bibliographies
91-550-2456-4 X.950/22315
8964505

AFTONBLADET
Reporting, 1876-1877
WIRÉN, Karl Hugo
Från skilda håll. - Stockholm : Almqvist & Wiksell International, 1979
255p. - (Stockholm studies in history ; 25). - Includes bibliographies
91-220-0288-x Ac.2650.m/13[no.25]
2093707

AGA. Norway
Buildings of historical importance, to 1972
STIFTELSEN AGATUNET
Aga i Ullensvang. - [Bergen, 1973]
27p : ill
r6004379 X.419/2419 Woolwich
0390593

AGA KHAN (FAMILY). Ismāilism
to 1980 - Biographies
BOSE, Mihir
The Aga Khans. - Kingswood, Surrey : World's Work, c1984
399p. - Bibliography: p384-390. - Includes index
0-437-01555-6 X.800/40883
3274349

AGA KHAN FAMILY. Islam
BOURRE, Jean-Paul
Aga Khan. - Paris : Encre, 1982
283p
2-86418-134-7 X.200/43871
8261547

AGA LANGUAGE
See Kanuri language

AGAPE. Christian ethics
OUTKA, Gene Harold
Agape. - New Haven ; London : Yale University Press, 1976
ix,321p. - (Yale publications in religion ; 17). - Includes index
0-300-02122-4 Ac.1805.C/3(17)
1199447

AGAPEMONE
Prince, Henry James
MANDER, Charles
The Reverend Prince and his Abode of Love. - Wakefield : EP Publishing, 1976
vii,143p : ill
0-7158-1198-3 X.200/31185
1289268

AGAR, EILEEN. English paintings
- Catalogues
AGAR, Eileen
Eileen Agar. - London : New Art Centre, c1981
[17]p : ill(some col.)
B84-32161 X.429/16148
3142299

AGÂRBICEANU, ION. Fiction in Romanian
- Biographies
CEASURI de seară cu Ion Agârbiceanu / o carte gîndită şi alcătuită de Mircea Zaciu. - Cluj-Napoca : Editura Dacia, 1982
374 p., [24] p. of plates : ill.. - Includes bibliographical references
r6403297 X.950/32601
815774x

AGÂRBICEANU, ION. Romanian literature
VATAMANIUC, Dimitrie
Ion Agârbiceanu. - Bucureşti : Editura Enciclopedică Română, 1974
xlii,618p, 16p of plates
r6043272 X.809/24189
1000454

AGARICACEAE
See Gill fungi

AGARICALES
See also
Mushrooms

AGARICALES
INGOLD, Cecil Terence
The nature of toadstools. - London : Edward Arnold, 1979
iv,57p : ill. - (Studies in biology / Institute of Biology ; no.113). - Bibliography: p.57
0-7131-2748-1 SRL
0150045
MUSHROOMS and toadstools / [written and] with an introduction by Uberto Tosco and Annalaura Fanelli ; [translated from the Italian]. - London : Orbis Publishing, 1975
64p : chiefly ill(chiefly col). - Includes index
0-85613-199-7 X.322/8220 Woolwich
0150045

AGARICALES. Europe
HVASS, Else
Mushrooms and toadstools, in colour. - Poole : Blandford, 1961
156p : col.ill. - Includes index
0-7137-0146-3 X.329/18498
2612798

NONIS, U.
Mushrooms & toadstools. - Newton Abbot : David & Charles, 1982
229p : ill(some col.). - Includes index
0-7153-8282-9 SRL
2612798

PEGLER, David
The Mitchell Beazley pocket guide to mushrooms and toadstools. - London : Mitchell Beazley, 1981
168p : ill(some col.). - Includes index
0-85533-347-2 X.329/14846 Woolwich
2612798

WILKINSON, John, 1934-
Mushrooms & toadstools. - [London] : Fontana, 1979
1folded sheet([12]p) : col.ill. - (Domino ; 3)
0-00-685443-5 Cup.911/133
2407728

AGARICALES. Great Britain
HAHNEWALD, Edgar
Mushrooms and toadstools. - Poole : Blandford, c1981
128p : col. ill. - (A Blandford mini-guide). - Includes index
0-7137-1211-2 X.329/14329 Woolwich
2536234

WILKINSON, John, 1934-
Mushrooms & toadstools. - London : Collins, 1982
240p : col.ill. - (Collins gem guides). - Includes index
0-00-458812-6 X.329/15438 Woolwich
2536234

AGARICALES. Northern hemisphere
JORDAN, Michael
A guide to mushrooms. - London : Millington, 1975
240p : ill(chiefly col). - Bibliography: p.237-238. - Includes index
0-86000-040-0 X.320/10301
0631752
- [Another copy: paperback]
ISBN 0-86000-042-7 X.319/16487 Woolwich

AGASSIZ, LOUIS. Natural history
BALMER, Heinz
Louis Agassiz, 1807-1873. - [Zurich], [1975]
142p,fold leaf of plates : ill. - (Denkschriften der Schweizerischen Naturforschenden Gesellschaft : Bd 89). - Includes bibliographies
r6034505 X.313/740 Woolwich
1224557

AGATE, JAMES. Criticism. Theatre. Great Britain - Biographies
AGATE, James
The selective 'Ego'. - London : Harrap, 1976
xii,269p. - Includes index
0-245-52849-0 X.989/50670
0644153

AGATES. Great Britain
Collecting
RODGERS, Peter Rowland
Agate collecting in Britain. - London : Batsford, 1975
96p,[4]p of plates : ill(some col). - Bibliography: p.92-93. - Includes index
0-7134-3056-7 X.320/10173
0535346

AGAU LANGUAGES
See Central Cushitic languages

AGAVE SISALANA
See Sisal hemp

AGBEBI, MOJOLA. Baptist churches. Nigeria - Biographies
AYANDELE, Emmanuel Ayankanmi
A visionary of the African church. - Nairobi : East African Publishing House, 1971
32 p.
r6031583 X.100/16877
1068040

AGDER. Norway
to date - Bibliographies
ANDREASSEN, Berit
Agder-bibliografien. - Kristiansand : Historielaget, 1983-
r6611595 X.0809/2422
8897832

AGDER. Norway
to date - Serials
KILDER til Agders historie / utgitt av Agder historielag. - 1-. - Kristiansand : Agder historielag, 1979-
r6600077 X.0809/2249
8897832

AGDER. Norway
ca 900-1300
RIKSSAMLINGSTID pa Agder / redigert av Hans Try. - Kristiansand : Studentbokhandelen, 1976
112p : ill
82-7214-000-4 X.709/23415
1229958

AGDER. Norway
Description & travel
AGDER / under redaksjon av Alv Kristiansen. - Oslo : Gyldendal, 1977
509p : ill. - (Bygd og by i Norge)
82-05-09406-3 X.0800/731(9)
1475894

AGDER. Norway
Social conditions, ca 1800-ca 1930
HOVET, Svein
Fra kvardag og fest pa Agder. - Trondheim : Rune, 1976
44p : ill
82-523-0106-1 X.700/19309
1429930

AGDER. Norway
Social conditions, ca 1900
AGDER i manns minne / [redaksjon Bjørn Slettan og Hans Try]. - Oslo : Det Norske samlaget, 1974
222p. - (I manns minne. distriktsband)
82-521-0398-7 X.520/9835
1853910

AGDLUITSUP-PÅ. Greenland
to date
AGDLUITSUP-PÅ = Sydprøven / årqigssuissoq Ove Bak. - Sydprøven : Bygderadet i Sydprøven, 1980
56p : ill
87-7561-002-7 X.809/55779
897537x

AGE
See also
Coming of age
Gestational age
Middle age
Old age

AGE. Earth
Theories of Kelvin, William Thomson, Baron
BURCHFIELD, Joe Donald
Lord Kelvin and the age of the earth. - London : Macmillan, 1975
xii,260p. - Bibliography: p.219-253. - Includes index
0-333-18109-3 X.320/10039
0359475

AGE. Man
related to learning
EXPERIMENTAL studies in adult learning and memory / edited by D. Barry Lumsden, Ronald H. Sherron. - Washington, D.C. : Hemisphere Publishing ; New York ; London : Distributed by Wiley, 1975
xv,208p : ill. - Includes bibliographies and index
0-470-55370-7 X.520/10110
0367265

AGE. Man
related to memory
EXPERIMENTAL studies in adult learning and memory / edited by D. Barry Lumsden, Ronald H. Sherron. - Washington, D.C. : Hemisphere Publishing ; New York ; London : Distributed by Wiley, 1975
xv,208p : ill. - Includes bibliographies and index
0-470-55370-7 X.520/10110
0367257

AGE. Members. Communities
related to social structure - Conference proceedings
ASSOCIATION OF SOCIAL ANTHROPOLOGISTS OF THE COMMONWEALTH.
Conference (1977 : Swansea)
Sex and age as principles of social differentiation. - London : Academic Press, 1978
vii,188p : ill. - (Monographs / Association of Social Anthropologists of the Commonwealth ; 17). - Includes bibliographies and index
0-12-433050-9 X.0529/47(17)
1762265

AGE. Objects
HUNT, Val
Old and new. - London : Abelard, 1979
[20]p : of col ill. - (Bright little books)
0-200-72619-6 X.990/15958 Woolwich
0591319

MOUSDALE, Peter
Old and new. - London : Macdonald Educational, 1976
[29]p : of col ill. - (Opposites ; [4])
0-356-05114-5 X.990/7470 Woolwich
0591319

AGE. Sirenia
Determination - Conference proceedings
INTERNATIONAL CONFERENCE ON DETERMINING AGE OF ODONTOCETE CETACEANS AND SIRENIANS (1978 : La Jolla)
Age determination of toothed whales and sirenians. - Cambridge : International Whaling Commission, 1980
viii,229p : ill. - (Reports / International Whaling Commission ; special issue 3). - Includes bibliographies and index
B81-29171 SRL
2528835

AGE. Toothed whales
Determination - Conference proceedings
INTERNATIONAL CONFERENCE ON DETERMINING AGE OF ODONTOCETE CETACEANS AND SIRENIANS (1978 : La Jolla)
Age determination of toothed whales and sirenians. - Cambridge : International Whaling Commission, 1980
viii,229p : ill. - (Reports / International Whaling Commission ; special issue 3). - Includes bibliographies and index
B81-29171 SRL
2528827

AGE CONCERN
DICKSON, Niall
Age Concern at work. - Mitcham : Age Concern England, c1981
30p : ill
0-86242-001-6 X.525/4465
0786187

AGE CONCERN ENGLAND
Policies
AGE CONCERN
The national policy. - Mitcham : Age Concern, [1978]
[2],19p
B79-02544 X.515/613
176165x

The national policy. - [London?] : [Age Concern?], [1982?]
[17]p
B82-32989 X.525/6085
176165x

AGE CONCERN SCOTLAND
- Serials
ADAGE. - No.1 (Apr./May 1982)-. - Edinburgh : Age Concern Scotland, 1982-
v. : ill
B82-33856 P.523/1694
2772930

AGE CONCERN SCOTLAND
Annual report. - [19-]-. - Edinburgh : Age Concern Scotland, [19-]-
v. : ill
r6864922 P.521/3689
2741822

ÂGE D'HOMME. Leiris, Michel. French literature
LEJEUNE, Philippe
Lire Leiris. - Paris : Éditions Klincksieck, 1975
192 p.. - (Bibliothèque du XX Siècle). - Bibliography : p. 187-190
2-252-01753-8 X.900/17130
0940100

AGE DIFFERENCES
See also
Generation gap

AGE DIFFERENCES. Man
Effects on memory
FOULDS, Alison M.
Age differences in aspects of memory performance. - Nottingham : Department of Adult Education, University of Nottingham, c1981
ii,28p. - (Adults ; 5). - Includes bibliographies
0-902031-50-3 X.529/46755
2702274

AGE DIFFERENCES. Man
Effects on memory - Conference proceedings
NEW directions in memory and aging / edited by Leonard W. Poon ... [et al.]. - Hillsdale, N.J. : Lawrence Erlbaum, 1980
xix,572p : ill. - Includes bibliographies
0-89859-035-3 X.520/32072
2702274

AGE-GROUP SYSTEMS
STEWART, Frank Henderson
Fundamentals of age-group systems. - New York ; London : Academic Press, 1977
xv,381p : ill. - (Studies in anthropology). - Bibliography: p.347-369. - Includes index
0-12-670150-4 X.520/11833
1170910

AGE-GROUP SYSTEMS. East Africa
AGE, generation and time / edited by P.T.W. Baxter and Uri Almagor. - London : C. Hurst, 1978
viii,276p : ill. - Includes bibliographies and index
0-903983-80-x X.529/32764
173685x

AGE-GROUP SYSTEMS. Labwor. Uganda
1972

CHRISTENSEN, Palle Rolf
Generationsmodsætningerne sat i system. samfundsmæssig organisation pa basis av alder. - Moesgard : Etnografisk Afdeling, Arhus Universitet, 1974
70 leaves. - (Studenterfeltrapport ; nr 1)
r6028944 X.702/3472
1417614

AGE GROUPS. Readership. Serials with British imprints

RESEARCH SERVICES LIMITED
The 'Sunday times' supplement to the national readership survey. - London : Times Newspapers.
Fieldwork period July 1978- June 1979. - [1979]
[104]p
B80-36661 L.49/558
1735853

AGE GROUPS. Readership. Serials with British imprints
- Statistics

RESEARCH SERVICES LIMITED
The 'Sunday times' supplement to the national readership survey. - London : Times Newspapers.
Fieldwork period January-December 1978. - [1979]
[1],96p
B80-27847 L.49/415
1735853

RESEARCH SERVICES LIMITED
The 'Sunday times' supplement to the national readership survey. - London : Times Newspapers.
Fieldwork period July 1976-June 1977. - 1977
[1],6,96p
B79-00071 Cup.1281/990
1735853

The SUNDAY times supplement to the national readership survey / prepared by Research Services Ltd. - London : Times Newspapers.
Fieldwork period January-December 1979. - [1980]
6,96p
B83-08903 L.49/1559
1735853

AGE OF CONSENT. Sexual intercourse. Adolescents. England
Law. Reform

GREAT BRITAIN. Policy Advisory Committee on Sexual Offences
Report on the age of consent in relation to sexual offences. - London : H.M.S.O., 1981
v,32p. - (Cmnd. ; 8216)
0-10-182160-3 BS.18/690
2411180

Working paper on the age of consent in relation to sexual offences. - London : H.M.S.O., 1979
[2],v,31p
0-11-340204-x BS.18/646
199705x

AGE OF CONSENT. Sexual intercourse. Homosexual adolescents. England
Law. Reform

JOINT COUNCIL FOR GAY TEENAGERS
I know what I am. - Liverpool : JCGT, 1980
[5],18p
0-9506958-0-7 X.200/34283
2312476

AGE OF CONSENT. Sexual intercourse. Homosexual male adolescents

The AGE taboo / Daniel Tsang, editor. - Boston : Alyson ; London : Gay Men's Press, 1981
178p. - Bibliography: p167-173. - Includes index
0-907040-11-x Cup.351/298
2643707

.AGE OF ELIGIBILITY. State superannuation schemes. Denmark

DENMARK. Socialministeriet. Pensionsreformarbejdsgruppen
Rapport om udvidet adgang til førtidig folkepension og fleksibel pensionsalder. - København : [Socialministeriet], 1976
176p. - (Betænkning ; nr 755)
87-503-1861-6 S.B.9/4(755)
1329057

HILLESTRØM, Karsten
Ønsker om fleksibel pensionsalder og nedsat arbejdstid. - København : Socialforskningsinstituttet i Kommission hos Teknisk Forlag, 1977
59p. - (Socialforskningsinstituttet. Meddelelse ; 23)
87-7487-128-5 X.0702/300(23)
1329057

AGE OF ELIGIBILITY. State superannuation schemes. Great Britain

GREAT BRITAIN. Department of Health and Social Security
Pension age. - London : D.H.S.S., 1976
18p
B78-08904 BS.385/174
1351249

AGE OF ELIGIBILITY. State superannuation schemes. Great Britain
Sex discrimination. Abolition

GREAT BRITAIN. Equal Opportunities Commission
Equalising the pension age. - [London] : [The Commission], 1978
[3],43p : 1 ill
B78-19049 BS.433/49
1103172

AGE OF ELIGIBILITY. State superannuation schemes. Great Britain
Sex discrimination. Abolition

GREAT BRITAIN. Equal Opportunities Commission
Sex equality and the pension age. - Manchester : The Commission, [1977]
23p : 1 ill
0-905829-00-x BS.433
1103172

NATIONAL ASSOCIATION OF PENSION FUNDS
Towards equality in retirement ages. - Croydon : The Association, [1977]
[1],12p. - (Notes on pensions ; no.5)
B78-10813 X.0100/1015(5)
1103172

AGE STRUCTURE. Population

PIROZHKOV, Sergeĭ Ivanovich
Demograficheskie prot͡sessy i vozrastnai͡a struktura naselenii͡a. - Moskva : Statistika, 1976
135p
r6037203 X.709/22694
1124609

AGE STRUCTURE. Population. Organisms
Implications for evolutionary aspects of population genetics

CHARLESWORTH, Brian
Evolution in age-structured populations. - Cambridge : Cambridge University Press, 1980
xiii,300p : ill. - Bibliography: p.275-287. - Includes index
0-521-23045-4 SRL
1994093

AGE STRUCTURE. Population. Portugal

NAZARETH, J. Manuel
O envelhecimento da população portuguesa. - Lisboa : Editorial Presença : Gabinete de Investigações Sociais, 1979
235p : charts,maps. - (Colecção Analise social : Série A, Estudos e investigações ; n.6). - Bibl
r6101818 X.519/43773
2448939

AGE STRUCTURE. Population. Soviet Union

KALINI͡UK, Iraida Vladimirovna
Vozrastnai͡a struktura naselenii͡a SSSR. - Moskva : Statistika, 1975
110p : ill
r6057878 X.519/24679
1430467

AGED PERSONS
See Old persons

AGEE, JAMES. American writers. English literature

DOTY, Mark A.
Tell me who I am. - Baton Rouge ; London : Louisiana State University Press, c1981
xiii,144p. - Bibliography: p135-140. - Includes index
0-8071-0758-1 X.950/8573
1316737

REMEMBERING James Agee / edited by David Madden. - Baton Rouge : Louisiana State University Press, [1974]
172p : ill
0-8071-0086-2 X.981/11655
1316737

AGEE, JAMES. American writers. English literature
- Biographies

BARSON, Alfred T.
A way of seeing. - [Amherst] : University of Massachusetts Press, c1972
217p
r6808175 X.950/31379
8131120

AGEE, JAMES. American writers. English literature
Criticism - Bibliographies

HUSE, Nancy Lyman
John Hersey and James Agee. - Boston, Mass. : Hall, 1978
122p. - (A reference publication in literature)
0-8161-8019-9 X.800/14608
1895877

HUSE, Nancy Lyman
John Hersey and James Agee. - London : Prior [etc.], 1978
ix,122p. - (A reference publication in literature). - Includes index
0-86043-242-4 X.981/12431
1895877

AGEE, JAMES. Criticism. Cinema films

SNYDER, John J.
James Agee. - New York : Arno Press, 1977
229p. - Bibliography
0-405-09895-2 X.950/5622
871651x

AGEEV, NIKOLAĬ VLADIMIROVICH. Metallurgy
- Biographies

NIKOLAĬ Vladimirovich Ageev / vstupitel'nai͡a stat'i͡a L.N. Gusevoĭ ; bibliografii͡a sostavlena G.N. Finashinoĭ. - Moskva : Izdatel'stvo "Nauka", 1976
72p,[1] leaf of plates. - (Akademii͡a nauk SSSR. Materialy k biobibliografii uchenykh SSSR. serii͡a khimicheskikh nauk ; vyp.57)
r6046449 Ac.1125/211(185)
1241095

AGEING
See also
Gerontology

AGEING
Religious aspects

AGING / Opera Pia International ; Ettore Di Filippo, Director ; Francis V. Tiso, Editor. - Lake Worth : Sunday Publications, 1982
ix,256p
0-941850-03-x X.200/42804
9075062

AGEING
Religious aspects - Conference proceedings

ERANOS CONFERENCE (40th : 1971 : Ascona)
The stages of life in creative process. - Leiden : Brill, 1973
vi,442p. - (Eranos yearbook ; v.40 (1971))
90-04-03601-6 PP.1253.bkc[vol.40]
9075062

AGEING. Adults

EVANS, Peter, 1940-
Getting on. - London : Granada, 1982
viii,136p. - Bibliography: p130-131. - Includes index
0-246-11136-4 X.529/47789
0130567

The HANDBOOKS of aging / editor-in-chief James E. Birren. - [New York] ; [London etc.] : [Van Nostrand Reinhold].
[Handbook of aging and the social sciences]. - [1976]
xvi,684p : ill. - Includes bibliographies and index
0-442-20798-0 X.525/2143
0130567
- [Another copy: paperback]
ISBN 0-442-20804-9 X.510/16183

[Handbook of the biology of aging]. - [1977]
xvi,771p : ill. - Includes bibliographies and index
0-442-20796-4 SRL
0130567
- [Another copy: paperback]
ISBN 0-442-20806-5

[Handbook of the psychology of aging]. - [1977]
vii,787p : ill. - Includes bibliographies and index
0-442-20797-2 X.322/8383 Woolwich
0130567
- [Another copy: paperback]
ISBN 0-442-20805-7 X.311/8952

STOPPARD, Miriam
50 plus lifeguide. - London : Dorling Kindersley, 1983
320p : ill(some col.). - Includes index
0-86318-016-7 X.529/61992
0130567

WALFORD, Roy L.
Maximum life span. - New York ; London : Norton, c1983
xiv,256p : ill. - Includes index
0-393-01649-8 X.329/16747
0130567

AGEING. Adults
- Conference proceedings

AGING / [James G. March, editor in chief]. - New York ; London : Academic Press.
Social change. - c1981
xxiv,631p : ill. - Includes bibliographies and index
0-12-040002-2 X.520/29327
0130567
- [Another copy : paperback] X.520/32523

AGEING. Adults
- Conference proceedings-continued

PHILADELPHIA SYMPOSIUM ON AGING (1974)
Explorations in aging. - New York ; London : Plenum Press, 1975
ix,307p : ill. - (Advances in experimental medicine and biology ; vol.61). - Includes bibliographies and index
0-306-39061-2 SRL
0130567

SUCCESSFUL AGING (Conference : 1973 : Duke University)
Successful aging. - Durham, N.C. : Center for the Study of Aging and Human Development, Duke University, 1975
[6],156[i.e.161]p,[8]p of plates. - Includes bibliographies
r6032038 X.311/8481 Woolwich
0130567

AGEING. Adults
Development & ageing

HULTSCH, David F.
Adult development and aging. - New York ; London : McGraw-Hill, c1981
xiv,431p : ill. - Bibliography: p377-399. - Includes index
0-07-031156-0 X.520/23833
0069108

KIMMEL, Douglas C. (Douglas Charles)
Adulthood and aging. - 2nd ed. - New York ; Chichester : Wiley, c1980
xiii,574p : ill. - Includes bibliographies and index
0-471-05229-9 X.520/23789
0069108

AGEING. Adults
Effects on diseases

DILMAN, V. M. (Vladimir Mikhaĭlovich)
The law of deviation of homeostasis and diseases of aging. - Boston, [Mass.] ; Bristol : John Wright, c1981
xiii,380p : ill. - Bibliography: p331-372. - Includes index
0-88416-250-8 SRL
2695472

AGEING. Adults
Effects on diseases - Conference proceedings

COMPARATIVE pathobiology of major age related diseases / editors Dante G. Scarpelli, George Migaki. - New York : Liss, c1984
xvi,441p : ill. - (Modern aging research ; v. 4). - Includes bibliographies
0-8451-2304-1 X.322/13483
2695472

AGEING. Adults
Psychological aspects

FREESE, Arthur S.
The end of senility. - South Yarmouth : John Curley, 1979, c1978
xx,287p. - Bibliography: p282-287
0-89340-189-7 X.329/15033 Woolwich
0173673

KAUSLER, Donald H.
Experimental psychology and human aging. - New York ; Chichester : Wiley, c1982
x,720p : ill. - Bibliography: p637-687. - Includes index
0-471-08163-9 X.520/27916
0173673

AGEING. Adults
Social aspects

AGEING and life course transitions / edited by Tamara K. Hareven and Kathleen J. Adams. - London : Tavistock, 1982
xviii,281p. - Includes bibliographies and index
0-422-78360-9 X.520/29109
1133446

AGEING in modern society / edited by Dorothy Jerrome. - London : Croom Helm, c1983
307p : ill. - Includes bibliographies and index
0-7099-1422-9 X.529/59980
1133446

CARLSON, Avis D.
In the fullness of time. - South Yarmouth : Curley, 1979, c1977
xxiii,388p
0-89340-172-2 Nov.45938 Woolwich
1133446

HAMMARSTRÖM TORNSTAM, Gunhild
Det sociala aldrandet. - Uppsala, 1975
176 p.. - Bibliography: p. 165-176
91-506-0025-7 X.519/20987
0884006

The SOCIAL challenge of ageing / edited by David Hobman. - London : Croom Helm, 1978
5-287p : 1 ill. - Includes bibliographies and index
0-85664-428-5 X.529/32145
126642x
- [Another copy: paperback.] X.529/41191

AGEING. Adults
Social psychology

TORNSTAM, Lars
Aldrandets socialpsykologi. - [Stockholm] : Rabén & Sjögren, 1978
293p. - (Stora tema nova). - Includes bibliographies
91-295-1241-7 X.519/39528
1785540

AGEING. Adults. United States
Psychosocial aspects

ROGERS, Dorothy, 1914-
The adult years. - Englewood Cliffs ; London : Prentice-Hall, 1979
xiv,482p : ill. - Bibliography: p.437-463. - Includes index
0-13-008987-7 X.520/14707
2002493

AGEING. Aeroplanes
- Conference proceedings

LONG-LIFE aircraft structures / jointly organised by the Royal Aeronautical Society and The Society of Licensed Aircraft Engineers and Technologists. - London : 4 Hamilton Place, W1V 0BQ : The Royal Aeronautical Society, [1980?]
1v.(various pagings) : ill
B82-07064 SRL
2607875

AGEING. Amorphous polymers
Physical ageing

STRUIK, L. C. E.
Physical aging in amorphous polymers and other materials. - Amsterdam ; Oxford : Elsevier, 1978
xiv,229p : ill. - Includes index
0-444-41655-2 SRL
1411527

AGEING. Animals

COMFORT, Alex
The biology of senescence. - 3rd ed. - Edinburgh : Churchill Livingstone, 1979
414p : ill. - Includes bibliographies and index
0-443-01928-2 SRL
1279416

LAMB, Marion J.
Biology of ageing. - Glasgow : Blackie, 1977
viii,184p : ill. - (Tertiary level biology). - Bibliography: p.173-177. - Includes index
0-216-90327-0 SRL
1279416

AGEING. Animals
Effects on biological rhythms - Conference proceedings

AGING and biological rhythms / edited by Harvey V. Samis, Jr and Salvatore Capobianco. - New York ; London : Plenum Press, 1978
x,342p : ill. - (Advances in experimental medicine and biology ; vol.108). - Includes bibliographies and index
0-306-40031-6 SRL
177607x

AGEING. Brain. Man
- Conference proceedings

INTERNATIONAL SCHOOL OF PHYSIOPATHOLOGY AND CLINIC OF THE THIRD AGE (2nd : 1980 : Ettore Majorana Center for Scientific Culture)
The aging brain. - New York ; London : Plenum, c1980
xiii,393p : ill. - (Ettore Majorana international science series. Life sciences ; vol.5). - Includes index
0-306-40625-x SRL
2381303

AGEING. Cells. Mammals

CELLULAR ageing of. - Basel ; London : Karger ; [London] : [Distributed by Wiley]. - (Interdisciplinary topics in gerontology ; vol.9)
Part 1 : general concepts ; Mechanisms 1 : fidelity of information flow. - 1976
[6],218p : ill. - Includes bibliographies
B76-24375 SRL
0674621

PHYSIOLOGY of cell aging / volume editor Vladimir V. Frolkis ; translated from the Russian by Alexander Lipinsky. - Basel ; London : Karger, c1984
206p : ill. - (Interdisciplinary topics in gerontology ; v.18). - Bibliography: p175-203. - Includes index
3-8055-3866-9 SRL
0674621

AGEING. Cells. Man

CELLULAR ageing. - Basel ; London : Karger : [Chichester] [Distributed by Wiley].
Part 2: Mechanisms II, translation, transcription and structural properties. - 1976
vi,129p : ill. - (Interdisciplinary topics in gerontology ; vol.10). - Includes bibliographies
B76-29607 SRL
0716723

AGEING. Cells. Man
- Conference proceedings

INSTITUTE FOR MEDICAL RESEARCH. Annual Symposium (2nd 1976)
Senescence. - New York ; London : Plenum Press, 1977
x,123p : ill. - (Cellular senescence and somatic cell genetics ; vol.2). - Includes bibliographies and index
0-306-32602-7 SRL
0716723

SYMPOSIUM ON IMPAIRMENT OF CELLULAR FUNCTIONS DURING AGING AND DEVELOPMENT IN VIVO AND IN VITRO (1974 : Castle of Hrubá Skála, Czechoslovakia)
Cell impairment in aging and development. - New York ; London : Plenum Press, 1975
xv,557p : ill. - (Advances in experimental medicine and biology ; vol.53). - Includes bibliographies and index
0-306-39053-1 SRL
0638595

AGEING. Cells. Organisms

AGING and cell function / edited by John E. Johnson, Jr. - New York ; London : Plenum, c1984
xiii,283p : ill. - Includes bibliographies and index
0-306-41420-1 SRL
1376640

AGING and cell structure. - New York ; London : Plenum.
Vol.1. - c1981
xvi,385p : ill. - Includes bibliographies and index
0-306-40695-0 S.R.L.
1376640

CELL ageing and cell death / edited by I. Davies and D.C. Sigee. - Cambridge : Cambridge University Press, 1984
x,361p : ill. - (Society for experimental biology seminar series ; 25). - Includes bibliographies and index
0-521-26172-4 SRL
3301761

STREHLER, Bernard Louis
Time, cells and aging. - 2nd ed. - New York ; London : Academic Press, 1977
xiv,456p : ill. - Includes index
0-12-673260-4 SRL
1376640

AGEING. Central nervous system. Mammals

AGING and recovery of function in the central nervous system / edited by Stephen W. Scheff. - New York ; London : Plenum, c1984
xiii,221p : ill. - Includes bibliographies and index
0-306-41525-9 SRL
2081970

AGEING. Central nervous system. Old persons

CNS aging and its neuropharmacology / volume editor W. Meier-Ruge. - Basel ; London : Karger ; [Chichester] : [Distributed by Wiley], 1979
viii,218p : ill. - (Interdisciplinary topics in gerontology ; vol.15). - Includes bibliographies and index
B79-34919 SRL
1969250

AGEING. Connective tissues. Man

HALL, David Alan
The ageing of connective tissue. - London : Academic Press, 1976
vii,204p : ill. - Bibliography: p.181-199. - Includes index
0-12-319150-5 SRL
0695580

AGEING. Cytoplasm. Cells. Man

AGE pigments / editor R.S. Sohal. - Amsterdam ; Oxford : Elservier/North-Holland Biomedical, 1981
xvi,394p : ill. - Includes bibliographies and index
0-444-80277-0 SRL
2542013

AGEING. Ethnic minorities. United States
Psychosocial aspects

MINORITY aging / edited by Ron C. Manuel. - Westport, Conn. ; London : Greenwood, 1982
xvii,285p. - (Contributions in ethnic studies ; no.8). - Bibliography: p249-272. - Includes index
0-313-22541-9 X.520/30079
2906945

AGEING. Mammals

KOHN, Robert R.
Principles of mammalian aging. - 2nd ed. - Englewood Cliffs ; London : Prentice-Hall, 1978
xv,240p. - (Foundations of developmental biology). - Includes bibliographies and index
0-13-709352-7 SRL
1551604

AGEING. Mammals
- Conference proceedings

NAITO SYMPOSIUM ON AGING (1978 : Tokyo)
Aging phenomena. - New York ; London : Plenum Press, 1980
xiv,317p : ill. - (Advances in experimental medicine and biology ; vol.129). - Includes index
0-306-40460-5 SRL
1551604

SYMPOSIUM ON THE THEORETICAL ASPECTS OF AGING (1974 : Miami)
Theoretical aspects of aging. - New York ; London : Academic Press, 1974
xi,192p : ill. - (Academic Press rapid manuscript reproduction). - Includes bibliographies and index
0-12-591655-8 SRL
1551604

AGEING. Mammals
Effects on drug metabolism of liver - Conference proceedings

TOKYO SYMPOSIUM ON LIVER AND AGING - LIVER AND DRUGS (2nd : 1982)
Liver and aging - 1982. - Amsterdam ; Oxford : Elsevier Biomedical, 1982
xiii,410p : ill. - Includes index
0-444-80471-4 SRL
2888785

AGEING. Mammals
Effects on eye lenses - Conference proceedings

WORKSHOP ON AGEING OF THE LENS (1977 : Bonn)
Lens ageing and development of senile cataracts. - Basel ; London : Karger ; [Chichester] : [Distributed by Wiley], 1978
xi,294p : ill. - (Interdisciplinary topics in gerontology ; vol.12). - Includes bibliographies and index
B79-04349 SRL
1773372

AGEING. Mammals
Effects on eyes - Conference proceedings

ASSOCIATION FOR EYE RESEARCH. Meeting (18th : 1977 : Bonn)
Gerontological aspects of eye research. - Basel ; London : Karger ; [Chichester] : [Distributed by Wiley], 1978
xi,259p : ill. - (Interdisciplinary topics in gerontology ; vol.13). - Includes bibliographies and index
B79-04348 SRL
1773364

AGEING. Mammals
Effects on liver - Conference proceedings

LIVER AND AGING (Conference : 1978 : Tokyo)
Liver and aging, 1978. - Amsterdam ; Oxford : Elsevier/North-Holland Biomedical Press, 1978
xi,388p : ill. - Includes index
0-444-80088-3 SRL
1804413

AGEING. Mammals
Effects on neuroendocrine system

NEUROENDOCRINOLOGY of aging / edited by Joseph Meites. - New York ; London : Plenum, 1983
xvi,383p : ill. - Includes bibliographies and index
0-306-41310-8 SRL
1907751

AGEING. Mammals
Effects on neuroendocrine system - Conference proceedings

TARBOX PARKINSON'S DISEASE SYMPOSIUM (2nd : 1978 : Lubbock)
Parkinson's disease, II. - New York ; London : Plenum Press, 1978
xii,270p,[2]leaves of plates : ill(some col). - (Advances in experimental medicine and biology ; vol.113). - Includes bibliographies and index
0-306-40156-8 SRL
1907751

AGEING. Mammals
Genetic factors

BURNET, Sir Macfarlane
Intrinsic mutagenesis. - Lancaster : Medical and Technical Publishing Co., 1974
ix,244p : ill. - Includes index
0-85200-099-5 SRL
0602922

AGEING. Man

AGING. - Oxford : Published on behalf of l'Institut de la Vie and the World Health Organization Regional Office for Europe by Oxford University Press. - (Oxford medical publications)
Vol.1: Biology. - 1981
xv,346p : ill. - Includes bibliographies and index
0-19-261254-9 SRL
1685864
- Oxford : Published on behalf of l'Institut de la vie and the World Health Organization Regional Office for Europe by Oxford University Press. - (Oxford medical publications)
Vol.2. - 1981
xvi,403p : ill. - Includes bibliographies and index
0-19-261255-7 X.520/25555
1685864

AGING. - Oxford : Published on behalf of l'Institut de la vie and the World Health Organization Regional Office for Europe by Oxford University Press. - (Oxford medical publications)
Vol.3: [Behavioural sciences and conclusions]. - 1983
xvi,472p : ill. - Includes bibliographies and index
0-19-261256-5 X.520/30917
1685864

BEAVER, Marion L.
Human service practice with the elderly. - Englewood Cliffs, N.J. ; London : Prentice-Hall, c1983
xii,291p : ill. - Bibliography: p269-282. - Includes index
0-13-447482-1 X.520/31691
1685864

CARTER, A. Barham (Alan Barham)
The art of ageing. - London : British Medical Association, [1983?]
30p. - (A Family doctor booklet)
0-7279-0175-3 7583.n.5/81
1685864

DANGOTT, Lillian R.
A time to enjoy. - Englewood Cliffs ; London : Prentice-Hall, 1979
x,182p. - Bibliography: p.169-177. - Includes index
0-13-921684-7 X.619/19681 Woolwich
1685864

DEVELOPMENT, growth and ageing / edited by Nicholas Carter. - London : Croom Helm, 1980
169p : ill. - (Croom Helm biology in medicine series). - Bibliography : p. 160. - Includes index
0-85664-861-2 X.529/38580
1685864

FRIES, James F.
Vitality and aging. - Oxford : W.H. Freeman, c1981
xii,172p : ill. - Bibliography: p153-165. - Includes index
0-7167-1308-x X.322/10444 Woolwich
1685864

FULDER, Stephen
An end to ageing?. - Wellingborough : Thorsons, 1983
112p. - Bibliography: p107-109. - Includes index
0-7225-0769-0 X.529/55185
1685864

HERSHEY, Daniel
A new age-scale for humans. - Lexington, Mass. : Heath ; [Farnborough, Hants.] : [Distributed by] Gower Press, 1980
xiii,157p : ill. - Bibliography: p.149-153. - Includes index
0-669-03374-x SRL
1685864

KART, Cary S. (Cary Steven)
Aging and health. - Menlo Park ; London : Addison-Wesley, 1978
[1],xiv,341p : ill. - Includes bibliographies and index
0-201-03600-2 X.320/11435 Woolwich
1685864

KENNEY, Richard A.
Physiology of aging. - Chicago ; London : Year Book Medical, c1982
xii,137p : ill. - Includes bibliographies and index
0-8151-5016-4 SRL
1685864

LAMBREV, Zhivko
Zhivot, stareene, smŭrt. - Sofiīa : Nauka i izkustvo, 1979
187p : ill. - (Chovekŭt i prirodata ; no.7). - Bibliography : p.184-185
r6173934 X.529/21716
1685864

LOETHER, Herman John
Problems of aging. - 2nd ed. - Encino, Calif. : Dickenson, 1975
170p
0-8221-0152-1 X.519/30165
1685864

AGEING. Man

NEUHAUS, Ruby Hart
Successful aging. - New York ; Chichester : Wiley, c1982
xviii,285p. - Bibliography: p250-274. - Includes index
0-471-08448-4 X.520/27238
1685864

The OXFORD book of ages / chosen by Anthony and Sally Sampson. - Oxford : Oxford University Press, 1985
[v,288]p. - Includes index
0-19-214134-1 X.950/43469
1685864

THRESHOLDS in aging / edited by M. Bergener, M. Ermini, H.B. Stähelin. - London : Academic, 1985
xxii,301p : ill. - (The 1984 Sandoz lectures in gerontology). - Includes bibliographies
0-12-090160-9 X.520/37524
1685864

AGEING. Man
- Conference proceedings

ACTION ON AGEING (Conference : 1976 : Basle)
Action on ageing. - Tunbridge Wells : MCS Consultants for Sandoz, 1976
60p : 1 ill. - Includes bibliographies
0-9505367-1-7 X.311/8319 Woolwich
0564680

AGING / edited by William C. Bier. - New York : Fordham University Press, 1974
xi,292p : ill. - (The pastoral psychology series ; no.8). - Includes bibliographies
0-8232-0980-6 X.520/14270
1685864

INTERNATIONAL CONFERENCE OF SOCIAL GERONTOLOGY (IXth : 1980 : Quebec)
Adaptability and ageing. - Paris : International Center of Social Gerontology, 1981
2v.
r6299196 X.529/62578
8867062

KAN vi fördröja åldrandet? / redaktörer : Carl Ström, Yngve Zotterman ; Arbetsgruppen för studiet av de äldres problem. - [Stockholm] : Liber förlag, 1981
259p : ill. - Each contribution includes a bibliography
91-380-5773-5 X.529/50329
8867062

SYMPOSIUM ON THE PHYSIOLOGY AND PATHOLOGY OF HUMAN AGING (1975 : Miami)
The physiology and pathology of human aging. - New York ; London : Academic Press, 1975
xi,232p : ill. - (Academic Press rapid manuscript reproduction). - Includes bibliographies
0-12-288150-8 SRL
1685864

AGEING. Man
Biochemical aspects

HALL, David A.
The biomedical basis of gerontology. - Bristol : Wright PSG, Jan.1984
[288]p
0-7236-0697-8 SRL
1996711

KANUNGO, M. S.
Biochemistry of ageing. - London : Academic Press, 1980
x,281p : ill. - Includes bibliographies and index
0-12-396450-4 SRL
1996711

AGEING. Man
Development & ageing. Psychosocial aspects

MORRIS, Desmond
The book of ages. - London : Cape, 1983
[207]p : ill. - Includes index
0-224-02166-4 X.805/5422
2303531

NICHOLSON, John, 1945-
Seven ages. - [London] : Fontana, 1980
283p. - Bibliography: p.269-273. - Includes index
0-00-635858-6 X.529/38061
2303531

AGEING. Man
Effects of drugs - Conference proceedings

PHILADELPHIA SYMPOSIUM ON AGING (2nd : 1977)
Pharmacological intervention in the aging process. - New York ; London : Plenum Press, 1978
xii,335p : ill. - (Advances in experimental medicine and biology ; vol.97). - Includes bibliographies and index
0-306-32697-3 SRL
1568973

AGEING. Man
Effects on action of psychotropic drugs

AGE and the pharmacology of psychoactive drugs / editors-in-chief Allen Raskin, Donald S. Robinson, Jerome Levine. - New York ; Oxford : Elsevier, c1981
xviii,212p : ill. - Bibliography: p202-204. - Includes index
0-444-00586-2 SRL
2344300

PSYCHOPHARMACOLOGY of aging / edited by Carl Eisdorfer and William E. Fann. - Lancaster : MTP, c1980
271p : ill. - (Aging). - Includes index
0-85200-549-0 SRL
2344300

AGEING. Man
Effects on drug action - Conference proceedings

PHILADELPHIA SYMPOSIUM ON AGING (2nd : 1977)
Pharmacological intervention in the aging process. - New York ; London : Plenum Press, 1978
xii,335p : ill. - (Advances in experimental medicine and biology ; vol.97). - Includes bibliographies and index
0-306-32697-3 SRL
1568965

AGEING. Man
Effects on endocrine system

ENDOCRINE aspects of aging / edited by Stanley G. Korenman. - New York ; Oxford : Elsevier Biomedical, c1982
xii,275p : ill. - (Current endocrinology). - Includes index
0-444-00681-8 SRL
2841614

AGEING. Man
Effects on eyes

AGING and human visual function / editors: Robert Sekuler, Donald Kline, Key Dismukes. - New York : Liss, c1982
xv,350p : ill. - (Modern aging research ; v.2)
0-8451-2301-7 X.322/11489
2852470

AGEING. Man
Effects on intellectual skills

AGE, learning ability, and intelligence / edited by Richard L. Sprott. - New York ; London : Van Nostrand Reinhold, c1980
xi,170p : ill. - Includes bibliographies and index
0-442-27895-0 X.520/23232
236333x

AGEING. Man
Effects on interpersonal communication

TAMIR, Lois M.
Communication and the aging process. - New York ; Oxford : Pergamon, 1979
xix,195p. - (Pergamon general psychology series ; 86). - Bibliography: p.155-177. - Includes index
0-08-024621-4 X.520/21003
2087960

AGEING. Man
Genetic factors

BURNET, Sir Macfarlane
Endurance of life. - Cambridge : Cambridge University Press [etc.], 1978
[5],230p : ill. - Includes index
0-521-22114-5 X.320/11377 Woolwich
156692x

The GENETICS of aging / edited by Edward L. Schneider. - New York ; London : Plenum Press, 1978
xvi,424p : ill. - Includes bibliographies and index
0-306-31100-3 SRL
156692x

LINTS, F. A.
Genetics and ageing. - Basel ; London : Karger ; [Chichester] : [Distributed by Wiley], 1978
ix,130p. - (Interdisciplinary topics in gerontology ; vol.14). - Bibliography: p.104-118. - Includes index
B79-07310 SRL
156692x

AGEING. Man
Immunological aspects

DEVELOPMENTAL immunology / edited by Edwin L. Cooper, Mary A.B. Brazier. - New York ; London : Academic Press, 1982
xviii,321p : ill. - (UCLA forum in medical sciences ; no.25). - Includes bibliographies and index
0-12-188040-0 SRL
2108488

AGEING. Man
Psychobiological aspects

LUXEMBOURG CONFERENCE ON THE PSYCHOBIOLOGY OF AGING (1st : 1979 : Walferdange)
The psychobiology of aging. - New York ; Oxford : Elsevier/North-Holland, c1980
446p : ill. - Includes bibliographies and index
0-444-00391-6 SRL
2357909

AGEING. Man
Psychosocial aspects

HULICKA, Irene M.
Empirical studies in the psychology and sociology of aging. - New York : Crowell, 1977
xiv,289,[17]p : ill. - Includes index
0-690-01509-7 X.519/27729
1372890

AGEING. Man
related to physical activities. Psychosocial aspects

ROOS, Kjell
Aktivitet och aldrande. - Uppsala : [s.n.], 1975
139p. - Includes bibliographies
91-506-0045-1 X.319/8861 Woolwich
1235575

AGEING. Man
Research

BUTLER, Robert N. (Robert Neil)
The graying of nations. - Mitcham : Age Concern, 1979
12,[3]p. - (Occasional papers / Age Concern ; 6)
B79-25199 X.313/1654 Woolwich
1920030

MORGAN, Robert F.
Growing younger. - Toronto ; London : Methuen, 1982
xviii,245p. - Bibliography: p237-242. - Includes index
0-458-95410-1 X.529/59217
1920030

AGEING. Man
Research, 1945-

ALTERNSFORSCHUNG / Stiftung Volkswagenwerk (Herausgeber) : Beiträge von Mechthild Horn ... [et al.]. - Göttingen : Vandenhoeck & Ruprecht, 1981
209p. - (Schriftenreihe der Stiftung Volkswagenwerk ; Bd.20)
3-525-85361-0 X.0529/680(20)
8796343

AGEING. Man
Research, 1945- - Conference proceedings

Le VIEILLISSEMENT / [sous la direction de A.E. Imhof]. - [Lyon] : Presses Universitaires de Lyon, 1982
224p
2-7297-0158-3 X.329/18520
8796343

AGEING. Man
Role of diet - Conference proceedings

NUTRITION, LONGEVITY AND AGING (Conference : 1976 : University of Miami School of Medicine)
Nutrition, longevity and aging. - New York ; London : Academic Press, 1976
xi,284p : ill. - Includes bibliographies
0-12-591656-6 SRL
1042610

AGEING. Man
Social aspects

AGE and anthropological theory / edited by David I. Kertzer and Jennie Keith. - Ithaca ; London : Cornell University Press, 1984
344p : ill. - Includes bibliographies and index
0-8014-1567-5 X.520/36291
2398575

BARASH, David P.
Aging. - Seattle ; London : University of Washington Press, c1983
xi,240p. - Includes index
0-295-95993-2 X.520/31939
2398575

WARD, Russell A.
The aging experience. - 2nd ed. - New York ; London : Harper & Row, c1984
xiii,444p : ill. - Bibliography: p375-435. - Includes index
0-06-046901-3 X.322/14006
2398575

AGEING. Man
Social aspects - Conference proceedings

TRANSITIONS of aging / edited by Nancy Datan, Nancy Lohmann. - New York ; London : Academic Press, 1980
xv,221p : ill. - Includes bibliographies and index
0-12-203580-1 X.520/22760
2331659

AGEING. Man
Social aspects. Research

INTERNATIONAL handbook on ageing / edited by Erdman Palmore. - London : Macmillan, 1980
xviii,529p. - Includes bibliographies and index
0-333-27828-3 SRL
2336782

AGEING. Man. Western Europe
Attitudes of society, to date - Conference proceedings

Le VIEILLISSEMENT / [sous la direction de A.E. Imhof]. - [Lyon] : Presses Universitaires de Lyon, 1982
224p
2-7297-0158-3 X.329/18520
9124144

AGEING. Nervous system. Old persons

The NEUROLOGY of aging / [edited by] Robert Katzman, Robert Terry. - Philadelphia : F.A. Davis, c1983
ix,249p : ill. - (Contemporary neurology series ; 22). - Includes index
0-8036-5231-3 X.322/12075
2926725

AGEING. Old persons
Psychological aspects

PATTERNS of aging / volume editor H. Thomae. - Basel ; London : Karger ; [London] : [Distributed by Wiley], 1976
vii,179p : ill. - (Contributions to human development ; vol.3). - Bibliography: p.162-179
B76-29402 X.319/16751 Woolwich
0719137

AGEING. Organisms

The BIOLOGY of aging / edited by John A. Behnke, Caleb E. Finch and Gairdner B. Moment. - New York ; London : Plenum Press [for the] American Institute of Biological Sciences, 1978
xi,388p : ill. - (Publications / American Institute of Biological Sciences). - Includes bibliographies and index
0-306-31139-9 SRL
1773224

LECTURES on gerontology. - London ; New York : Academic Press.
Vol.1: On biology of ageing. - 1982
2pt.(516p) : ill. - Includes bibliographies and index
B82-07493 SRL
1773224

AGEING. Plants
Effects of growth hormones - Conference proceedings

GROWTH regulators in plant senescence / edited by M.B. Jackson, B. Grout, I.A. Mackenzie. - Wantage : British Plant Growth Regulator Group, 1982
vii,132p : ill. - (Monograph ; no.8)
0-906673-05-4 SRL
3033546

AGEING. Women
Effects on physiology of reproductive system

AGING and reproductive physiology / edited by E. S. E. Hafez. - Ann Arbor : Ann Arbor Science, 1976
ix, 102 p. : ill. - (Perspectives in human reproduction ; vol. 2). - Includes bibliographies and index
0-250-40108-8 X.320/10567 Woolwich
0765414

AGEING. Women
Role in osteoporosis

The OSTEOPORATIC syndrome / edited by Louis V. Avioli. - New York ; London : Academic, c1983
x,164p : ill. - Includes bibliographies and index
0-8089-1548-7 SRL
3186288

AGEING. Women, 40 years -
Effects. Prevention

LLOYD-JONES, Vernon
Why grow old?. - Swansea : Christopher Davies, 1980
100p
0-7154-0554-3 X.329/14486 Woolwich
2556944

AGEING. Women. Special subjects. American cinema films, 1930-1981

STODDARD, Karen M.
Saints and shrews. - Westport, Conn. ; London : Greenwood Press, 1983
174p. - (Contributions in women's studies ; no.39). - Bibliography: p156-160. - Includes index
0-313-23391-8 X.529/62588
3137155

AGEING. Women. Special subjects. English humorous cartoons - Collections

MARKHAM, Marcella
Old is - great!. - Watford : Exley Publications Ltd, 1978
[63]p : ill
0-905521-17-x X.419/34800 Woolwich
1764381

AGEN. France
Learned societies: Société académique d'Agen

LAUZUN, Philippe
La Société académique d'Agen - Toulouse : Eche, 1981
xviii,409p
r6211357 X.529/62509
9101705

AGEN. France
Public records. Repositories: Archives départementales de Lot-et-Garonne. Stock

ARCHIVES DÉPARTEMENTALES DE LOT-ET-GARONNE
Répertoire numérique sous-séries E. État civil. - Agen : [Archives départementales], 1977
299p
2-86047-000-x S.E.400/194
1508660

AGENCIES
See also
Commercial agencies
Credit reference agencies
Development agencies
Employment agencies
European Community agencies
Export agencies
Government agencies
Job centres
Specialised agencies
Underwriting agencies

AGENCIES
International agencies

TAYLOR, Phillip
Nonstate actors in international politics. - Boulder ; London : Westview, [1984]
xvii,247p : ill. - (Westview special studies in international relations). - Bibliography: p233-238. - Includes index
0-86531-344-x X.800/40744
0566713

WEISS, Thomas George
International bureaucracy. - Lexington, Mass. ; London : D.C. Heath, 1975
xxiv,189p. - Bibliography: p.169-180. - Includes index
0-669-99341-7 X.800/25319
0566713

AGENCIES. British diplomatic service. Persian Gulf countries Records. Collections: India Office Library and Records - Catalogues

INDIA OFFICE LIBRARY AND RECORDS
The records of the British Residency and agencies in the Persian Gulf. - London : India Office Library and Records, 1979
xxi,188,[7]p,leaf of plate,[4]p of plates : ill(some col). - (Guides to archive groups) (IOR R/15). - Bibliography: p.187-188. - Includes index
0-903359-16-2 BS.21/98
1832662
- [Another copy] OPL

AGENCY
See also
Estate agency

AGENCY
International law - Conference proceedings

HAGUE CONFERENCE ON PRIVATE INTERNATIONAL LAW (13th : 1976)
Protocol of closing session of the Thirteenth Session of the Hague Conference on Private International Law, special commission on the law applicable to agency, including draft convention, the Hague, 16 June 1977 ... - London : H.M.S.O., [1977]
10p. - (Miscellaneous ; no.29(1977)) (Cmnd.7020)
0-10-170200-0 BS.14/3098
1337564

AGENCY. England
Law

BOWSTEAD, William
Bowstead on agency. - 14th ed.. - London : Sweet and Maxwell, 1976
xcxx,461p. - (The common law library ; no.7). - Includes index
0-421-17590-7 X.200/30632
058181x

FRIDMAN, G. H. L. (Gerald Henry Louis)
The law of agency. - 5th ed. - London : Butterworths, 1983
xxxvii,386p. - Includes index
0-406-58437-0 X.200/40870
058181x

FRIDMAN, Gerald Henry Louis
The law of agency. - 4th ed. - London : Butterworth, 1976
xxi,341p. - Includes index
0-406-58553-9 X.200/20223
058181x
- [Another copy: paperback.]
ISBN 0 406 58554 7 X.100/16173

KAY, Maurice
Agency in commerce. - London : Sweet and Maxwell, 1979
vi,90p. - (Essential business law). - Bibliography: p.88. - Includes index
0-421-24700-2 X.108/17783
058181x

KOBRIN, D. L. (David Lawrence)
Agency. - London : Anderson Keenan, 1981
viii,68p. - (Law unit). - Includes index
0-906501-23-7 X.200/36465
058181x

MARKESINIS, B. S.
An outline of the law of agency. - London : Butterworth, 1979
xviii,252p. - Includes index
0-406-62269-8 X.100/18328
058181x

AGENCY. Foreign trade

ANTHONY WADE (CONSULTANTS)
The postal course of export-import (London). - [Sutton] : Anthony Wade (Consultants) Ltd, [1976]
[4], ii, 149, [3] leaves : ill. - Includes index
0-9505308-0-8 X.515/524
079905X

AGENCY. France
Law

GUYÉNOT, Jean
The French law of agency and distributorship agreements. - London : Oyez, 1976
xiii,283p. - (European commercial law library ; no.4). - Includes index
0-85120-244-6 X.100/16590
0747424
- [Another copy] OPL

AGENCY. Italy
Law

LA VILLA, G.
The Italian law of agency and distributorship agreements. - London : Oyez Publishing, 1977
xvi,217p. - (European commercial law library ; no.8). - Includes index
0-85120-301-9 X.100/17384
1370278
- [Another copy] OPL

AGENCY. Italy
Law, to date

BALDI, Roberto
Il contratto di agenzia. - 3a ed. completamente aggiornata rielaborata ed ampliata. - Milano : Giuffrè, 1981
433p. - Includes bibliography
r6512621 X.200/46045
9153934

AGENCY. Scotland
Law

MARSHALL, Enid Ann
Scottish cases on agency. - Edinburgh : W. Green, 1980
xiv,193p
0-414-00657-7 X.529/36873
2207141

AGENCY. South Africa
Law, to date

JOUBERT, D. J.
Die Suid-Afrikaanse verteenwoordigingsreg. - Kaapstad : Juta, 1979
xxxviii,264p
0-7021-1014-0 X.200/36492
8729948

AGENCY. West Germany
Law

STAUBACH, Fritz
The German law of agency and distributorship agreements. - London : Oyez Publishing, 1977
xiii,267p. - (European commercial law library ; no.7). - Includes index
0-85120-314-0 X.100/17385
137026x
- [Another copy] OPL

AGENCY NURSES. Great Britain

FEDERATION OF PERSONNEL SERVICES OF GREAT BRITAIN
Agency nurses. - London : The Federation, [1975]
[3],iv,27p
0-9504114-2-6 X.311/8080 Woolwich
056334x

AGENT ORANGE
See Tetrachlorodibenzodioxin

AGENTS
See also
Commercial agents
Estate agents
Fiduciaries

AGENTS
See also
Forwarding agents
Insurance agents
Manufacturers' agents
Parliamentary agents
Record agents
Serial agents
Serials subscription agents
Shipping agents

AGENTS. Dutch companies. Great Britain - Directories

NETHERLANDS-BRITISH CHAMBER OF COMMERCE
Dutch Companies with their UK agents, representations. - [New ed.]. - London : The Chamber of Commerce, 1975
[3],71p. - Includes index
0-900990-15-5 X.512/5117
0257974

AGENTS. Entertainers
- Directories

ARTISTES and their agents. - [1980]. - Eastbourne : John Offord Publications, [1980] : ill
Subsequent issues, when received, will be placed at this shelfmark
0-903931-25-7 P.901/3226
2259214

AGENTS. Governments
Agents in Great Britain: Crown Agents for Oversea Governments and Administrations

GREAT BRITAIN. Ministry of Overseas Development
Future of the Crown Agents. - London : H.M.S.O., [1976]
19p. - (Cmnd.6445)
0-10-164450-7 BS.251/115
0667498

AGENTS. Governments
Agents in Great Britain: Crown Agents for Oversea Governments and Administrations. Financial assistance by British government, 1974

FAY, Edgar Stewart
Report by the Committee of Inquiry appointed by the Minister of Overseas Development into the circumstances which led to the Crown Agents requesting financial assistance from the Government in 1974. - London : H.M.S.O., 1977
[1],vi,203p,[3]fold leaves : ill. - ([1977-78 H.C.]48)
0-10-204878-9 BS.92/168
1306154

AGENTS. Governments
Agents in Great Britain: Crown Agents for Overseas Governments and Administrations

GREAT BRITAIN. Advisory Committee on the Crown Agents
Report by the Advisory Committee on the Crown Agents (the Stevenson Committee), 24 March 1974 [i.e. 1972]. - London : H.M.S.O., [1977]
76p : ill. - ([1977-78 H.C.]50)
0-10-205078-3 BS.92/180
1477676

REPORT of the tribunal appointed to inquire into certain issues arising out of the operations of the Crown Agents as financiers on own account in the years 1967-74. - London : H.M.S.O., [1982?]
xi,604p. - (HL ; 149) (HC ; 364). - Includes index
0-10-414982-5 BS.18/724
1477676

AGENTS. Governments
Agents in Great Britain: Crown Agents for Overseas Governments and Administrations. Financial assistance by British government, 1974. Theories of Fay, Edgar

GREAT BRITAIN
Statement by Her Majesty's Government on the report of the Fay Committee of Inquiry on the Crown Agents (House of Commons return number 48). - London : H.M.S.O., [1977]
[2],41p. - ([1977-78 H.C.]49)
0-10-204978-5 BS.92/166
1306146

AGENTS. Musicians
- Directories

BRITISH ASSOCIATION OF CONCERT AGENTS
Composite list of artists showing sole representation, by members of the British Association of Concert Agents, as at January 1. - 1976. - London : [The Association], [1976]
[1],75p
0-904892-01-8 X.435/518
0555045

Composite list of artists showing sole representation by members of the British Association of Concert Agents, as at January 1. - 1977. - London : [The Association], [1977]
[1],79p
0-904892-02-6 X.435/560
0555045

Composite list of artists showing sole representation by members of the British Association of Concert Agents, as at January 1. - 1978. - London : [The Association], [1978]
[1],77p
B78-13360 X.435/614
0555045

AGENTS OF PRODUCTION
See Factors of production

AGERSO. Denmark
Environment planning

ENGEL, John Dyssel
Agersø. - [Copenhagen] : Dansk bygningskonstruktørforening, 1974
100p : ill
r6028811 Cup.1264.g.16
090287X

AGGADAH. Judaism
Theories of Isaac ben Jedaiah

SAPERSTEIN, Marc
Decoding the rabbis. - Cambridge, Mass. ; London : Harvard University Press, 1980
ix,289p. - (Harvard Judaic monographs ; 3). - Includes index
0-674-19445-4 X.200/35169
2377306

AGGARWALS. India (Republic)
to 1974

GUPTA, Badlu Ram
The Aggarwals. - New Delhi : S. Chand & Co., 1975
viii,146p : ill
r6083395 X.520/12875
1581104

AGGER. Denmark
1910-1920

MADSEN, Kr
Man kom hinanden ved -. - Esbjerg : Fiskeri- og Søfartsmuseet, 1975
153 p. : ill. - (Fiskeri- og Søfartsmuseet. Årsskrift ; 1975)
87-87453-18-5 X.809/22999
0882828

AGGER. Denmark
Social conditions, 1920-1930

MADSEN, Kr
Ung i tyverne. - Fredericia : Lohse, 1977
161p
87-564-0409-3 X.708/44713
178580X

AGGESEN, SVEN. Historiography. Denmark

CHRISTENSEN, Karsten
Om overleveringen af Sven Aggesens værker. - København : Den Danske Historiske Forening, 1978
93p. - (Skrifter udgivet af Det Historiske Institut ved Københavns Universitet ; bd 10). - Bibl
87-87462-14-1 Ac.1025.i.[bd.10]
2451026

AGGLOMERATION PROCESSES
See also
Sintering

AGGLUTININS
See also
Concanavalin A

AGGREGATE DEMAND. Great Britain
Control. Policies of government, 1954-1973

HOPKIN, Sir Bryan
The control of demand. - [Cardiff] : University College, Cardiff, [1974]
15p. - (Inaugural lectures / University College (Cardiff))
B76-02995 X.519/20473
0568759

AGGREGATE DEMAND. Great Britain
Microeconomic factors

BHATTACHARYYA, D. K.
Demand for financial assets. - Farnborough, Hants. : Saxon House, 1978
ix,197p. - Bibliography: p.186-194. - Includes index
0-566-00228-0 X.529/33322
1512439

AGGREGATE INVESTMENT
Aggregate capital investment. Econometric models

AGGREGATE investment / edited by J.F. Helliwell. - Harmondsworth : Penguin Education, 1976
344p : ill. - (Penguin education) (Penguin modern economics readings). - Bibliography: p.331-332. - Includes index
0-14-080871-x X.519/25972
0695475

AGGREGATED DATA. Research. Social sciences
Statistical analysis

AGGREGATE data / Edgar F Borgatta, David J. Jackson, editors. - Beverly Hills ; London : Sage Publications, 1980
192p : ill. - Includes bibliographies
0-8039-1428-8 X.629/13948 Woolwich
2257327

AGGREGATES
See also
Mineral deposits yielding aggregates

AGGREGATES. Concrete
Crushed rock aggregates. Properties

TEYCHENNÉ, D. C. (David Clayton)
The use of crushed rock aggregates in concrete. - Watford : Building Research Establishment, 1978
74p : ill
B81-14332 SRL
2383977

AGGREGATES. Concrete
Mixes. Design. Applications of digital computer systems

LEWIS, P. A. (Peter Alan)
Mixing aggregates to achieve a grading curve. - Birmingham : Department of Transportation and Environmental Planning, University of Birmingham, 1982
37p. - (Departmental publication ; no.59)
0-7044-0616-0 SRL
2750082

AGGREGATES. East Midlands. England
Naturally occurring aggregates. Supply & demand - Forecasts

AGGREGATES in the East Midlands. - [Matlock] : [Derbyshire County Council], 1978
37p
B84-27486 X.622/19528
3130649

AGGREGATES. England
Naturally occurring aggregates. Supply & demand - Forecasts

REGIONAL aggregate working parties. - [London] : Department of the Environment : Welsh Office, 1978
77[i.e.88]p : ill. - Bibliography: p.69-70
B80-36803 BS.414/1932
2253518

AGGREGATES. Grampian Region. Scotland
Extraction. Policies of Grampian (Region). Council

GRAMPIAN (Region). Council
A policy for aggregate extraction. - [Aberdeen] : [The Council], 1979
61p,[15]p of plates (9 fold) : ill
B80-38959 X.325/673 Woolwich
2306298

AGGREGATES. North Wales
Naturally occurring aggregates. Supply & demand - Forecasts

NORTH WALES WORKING PARTY ON AGGREGATES
Interim report of the North Wales Working Party on Aggregates. - [Mold] : [The Working Party], 1976
[5],35p : ill
0-904695-03-4 SRL
1045814

REGIONAL commentary / North Wales Working Party on Aggregates. - Mold : Clwyd County Council on behalf of the Working Party.
Part 1. - 1981
ii,65p,[5]leaves of plates
Subsequent volumes, when received will be placed at this shelf mark
0-904444-55-4 X.0622/630 Woolwich
1045814

AGGREGATES. Northern England
Naturally occurring aggregates. Demand

NORTHERN REGION WORKING PARTY ON AGGREGATES
Report. - Durham : The Working Party.
2nd : 1979. - [1979]
[6],65p,[3] fold leaves of plates : ill
B80-37841 X.325/743 Woolwich
199252x

AGGREGATES. Northern England
Naturally occurring aggregates. Demand - Forecasts

NORTHERN REGION WORKING PARTY ON AGGREGATES
Report. - Durham : The Working Party.
1st : 1978. - [1978]
[10],112p,[11]fold leaves of plates : ill
Subsequent volumes, when received, will be placed at this shelfmark
B79-36224 P.513/2356
199252x

AGGREGATES. South-east England
Naturally occurring aggregates. Supply & demand

STANDING CONFERENCE ON LONDON AND SOUTH EAST REGIONAL PLANNING
Policy guidelines to meet the South East region's need for aggregates in the 1980's. - London : The Conference, 1979
48p : ill. - (SC 1151R)
B80-40447 X.622/8821 Woolwich
2314649

AGGREGATES INDUSTRIES. Great Britain

ADVISORY COMMITTEE ON AGGREGATES
Aggregates, the way ahead. - [London] : H.M.S.O., 1976
xii,118p(6 fold),[8]p of plates,[2]fold leaves of plates : ill. - Bibliography: p.115-118
0-11-750917-5 SRL
0677957

GREAT BRITAIN. Regional Aggregate Working Parties
Collation of the results of the 1977 survey. - [London] : Department of the Environment, 1980
[6],93p : ill
B80-34648 BS.414/2017
0024082

AGGREGATES INDUSTRIES. Great Britain
- Statistics

PRODUCTION of aggregates in Great Britain / Department of the Environment. - 1972 and 1973. - London : H.M.S.O, 1975
[2],20p
Subsequent issues are also placed at this shelfmark
0-11-670586-8 SRL
0024082

READY mixed concrete & aggregates / [Business Ratios]. - 7th ed.: [1978]. - London : Business Ratios, [1978]
[75]p : ill
B80-35799 L.49/520
2270471

AGGREGATES PROCESSING PLANTS
Mobile crushed rock concrete aggregates processing plants

LESTER, D. (David)
Quarrying and rockbreaking. - London : Intermediate Techology, c1981
116p : ill
0-903031-80-9 SRL
2884623

AGGREGATION. Alkylammonium salts
Equilibrium constants - For solvent extraction

KERTES, Aviezer Steven
Critical evaluation of some equilibrium constants involving alkylammonium extractants. - Oxford : Pergamon, 1977
v,30p
0-08-021591-2 SRL
1412108

AGGREGATION. Physical processes
See Accretion

AGGREGATION. Solutions

AGGREGATION processes in solution / edited by E. Wyn-Jones, J. Gormally. - Amsterdam ; Oxford : Elsevier Scientific, 1983
x,632p : ill. - (Studies in physical and theoretical chemistry ; 26). - Includes index
0-444-42187-4 SRL
2962896

AGGREGATION. Surface-active agents
- Conference proceedings

COLLOID AND SURFACE SCIENCE SYMPOSIUM (52nd 1978
Knoxville)
Solution chemistry of surfactants. - New York ; London :
Plenum Press.
Vol.1. - 1979
515,xxvii p : ill. - Includes index
0-306-40174-6 SRL
2053810

Vol.2. - 1979
xvii p,p519-962 : ill. - Includes index
0-306-40175-4 SRL
2053810

INTERNATIONAL SYMPOSIUM ON SOLUTION BEHAVIOR OF
SURFACTANTS THEORETICAL AND APPLIED ASPECTS (1980
Clarkson College of Technology)
Solution behavior of surfactants. - New York ; London
Plenum Press, 1982
2v. : ill. - Includes index
0-306-41025-7 SRL
2053810

AGGREGATORS. Concrete
Lightweight aggregators. Use in construction

COMITÉ EURO-INTERNATIONAL DU BÉTON
Lightweight aggregate concrete. - Lancaster :
Construction Press, 1977
[8],169p : ill. - Includes bibliographies and index
0-904406-24-5 SRL
1286099

FIP manual of lightweight aggregate concrete. - 2nd ed.. -
Glasgow : Surrey University Press, 1983
xii,259p : ill. - Includes index
0-903384-43-4 SRL
1286099

AGGRESSION
See also
Bullying
Non-aggression

AGGRESSION
See also
Violence

AGGRESSION

EIBL-EIBESFELDT, Irenäus
The biology of peace and war. - [London] : Thames and
Hudson, 1979
ix,294p : ill. - Bibliography: p.265-285. - Includes
index
0-500-01188-5 X.520/14638
1971832

AGGRESSION
International aggression. Definition

BLANCO GASPAR, Vicente
La agresión internacional. - Madrid : Consejo Superior
de Investigaciones Científicas, Instituto Francisco de
Vitoria, 1973
viii,89p. - (Colección de estudios internacionales.
Serie 1, Problemas actuales ; no.4). - Bibliography
84-00-03879-7 X.529/38234
2462613

AGGRESSION
Social aspects

MARSH, Peter
Aggro. - London : Dent, 1978
165p. - Includes index
0-460-12026-3 X.200/31460
0159077

AGGRESSION. Adolescents
Psychiatric aspects

The AGGRESSIVE adolescent / Charles R. Keith, editor. - New
York : Free Press ; London : Collier Macmillan, c1984
xiii,487p : ill. - Includes bibliographies and index
0-02-916720-5 X.520/36730
3257797

AGGRESSION. Animals
compared with aggression in man

SCOTT, John Paul
Aggression. - 2nd ed., revised and expanded. - Chicago ;
London : University of Chicago Press, 1975 [i.e. 1976]
xiii,233p : ill. - (The scientist's library). -
Bibliography: p.211-222. - Includes index
0-226-74294-6 Ac.2691d/72(13)
067088x

AGGRESSION. Animals
compared with aggression in man - Conference proceedings

SYMPOSIUM ON COMMUNICATION AND AFFECT (6th : 1976 :
University of Toronto)
Aggression, dominance and individual spacing. - New York
; London : Plenum Press, 1978
x,173p : ill. - (Advances in the study of communication
and affect ; vol.4). - Includes bibliographies and index
0-306-35904-9 X.520/12409
067088x

AGGRESSION. Animals
Effects of drugs - Conference proceedings

ETHOPHARMACOLOGICAL aggression research / editors Klaus A
Miczek, Menno R Kruk, Berend Olivier. - New York : Liss,
c1984
xi,275p : ill. - (Progress in clinical and biological
research ; v.167). - Includes bibliographies
0-8451-5017-0 X.622/22768
3181537

AGGRESSION. Animals
Role of hormones

HORMONES and aggression. - Vol.1 : 1977 / Paul F. Brain. -
Edinburgh : Churchill Livingstone [etc.], 1977
[8],126p. - (Annual research reviews)
Subsequent volumes, when received, will be found at SRL
0-443-01734-4 SRL
1452126

AGGRESSION. Animals
Theories

ZILLMANN, Dolf
Hostility and aggression. - Hillsdale : Erlbaum ; London
: [Distributed by Wiley], 1979
x,422p. - Bibliography: p.377-408. - Includes index
0-470-26832-8 X.520/21075
2127717

AGGRESSION. Children. United States
Effects of portrayal of violence by television programmes

TELEVISION and aggression / J. Ronald Milavsky ... [et
al.]. - New York ; London : Academic Press, 1982
xvi,505p : ill. - (Quantitative studies in social
relations). - Bibliography: p491-499. - Includes index
0-12-495980-6 X.520/30689
2932717

AGGRESSION. Foreign relations
United Nations. General Assembly. Definition of aggression

STONE, Julius
Conflict through consensus. - Baltimore ; London : Johns
Hopkins University Press, 1977
x,234p. - Includes index
0-8018-1989-x X.809/43532
1584413

AGGRESSION. Mammals
Effects of drugs

PSYCHOPHARMACOLOGY of aggression / volume editor L.
Valzelli. - Basel ; London : Karger, 1978
viii,180p : ill. - (Modern problems of pharmacopsychiatry
; vol.13). - Includes bibliographies and index
B78-18351 SRL
1456652

AGGRESSION. Mammals
Nonverbal communication - Conference proceedings

SYMPOSIUM ON COMMUNICATION AND AFFECT (4th : 1974 :
University of Toronto)
Nonverbal communication of aggression. - New York ;
London : Plenum Press, 1975 [i.e. 1976]
ix,196p : ill. - (Advances in the study of communication
and affect ; vol.2). - Includes bibliographies and index
0-306-35902-2 X.320/10610 Woolwich
0784486

AGGRESSION. Man

AGGRESSION / edited by Russell G. Geen, Edward I
Donnerstein. - New York ; London : Academic.
Vol.1: Theoretical and methodological issues. - 1983
xviii,269p : ill. - Includes bibliographies and index
0-12-278801-x X.520/33712
0146870

AGGRESSION / edited by Russell G. Geen, Edward I.
Donnerstein. - New York ; London : Academic.
Vol.2: Issues in research. - 1983
xviii,209p : ill. - Includes bibliographies and index
0-12-278802-8 X.520/33711
0146870

AGGRESSION and violence / edited by Peter Marsh and Anne
Campbell. - Oxford : Basil Blackwell, 1982
vi,242p : ill. - Includes bibliographies and index
0-631-12742-9 X.520/27100
0146870

BACH, George R. (George Robert)
Creative aggression. - London : Coventure, [1977]
416p. - Includes index
0-904576-13-2 X.519/27422
0146870

BARON, Robert A.
Human aggression. - New York ; London : Plenum Press,
1977
xvii,298p : ill. - (Perspectives in social psychology). -
Bibliography: p.275-289. - Includes index
0-306-31050-3 X.520/12363
0146870

CROCE, Maria Angela
Motivazione e aggressività nelle ricerche sulle
comunicazioni di massa: un approccio psicologico. -
Bologna : Pàtron, 1978
66p. - (Quaderni di psicologia ; 2). - Bibliography
r6501881 X.520/24562
8716641

FORBES, Jack D.
A world ruled by cannibals. - Davis, Calif. : D-Q
University Press, [1979]
93p. - (DQU pre-print series). - Includes bibliography
r6814095 X.805/5744
8716641

FROMM, Erich
The anatomy of human destructiveness. - Harmondsworth :
Penguin, 1977
679p. - Bibliography: p.633-660. - Includes index
0-14-004258-x X.519/27187
0146870

GUGEL, Günther
Erziehung und Gewalt. - Waldkirch : Waldkircher Verlag,
1983
118p. - (Tübinger Beiträge zur Friedensforschung und
Friedenserziehung ; Bd.9). - Includes bibliography
3-87885-081-6 X.0709/898(9)
8716641

JARVIS, Maurice
Why war?. - Coleford : Jarvis, Lucy and Co., [1977]
[2],20,[1]p
B78-12485 X.519/27867
1400835

Why war?. - Ross-on-Wye : Hydatum, 1978
20,[3]p
0-905682-11-4 X.519/27868
1400835

KEMPF, Wilhelm F.
Zur Neuorientierung der Aggressionsforschung. - Konstanz
: Universitätsverlag, 1979
35p.. - (Konstanzer Universitätsreden ; 119)
3-87940-145-4 X.529/50463
8716641

KEVIN
Aggression. - Bangor : The Teachers, 1978
[3],11p : 1 ill. - Bibliography: p.11
0-86085-004-8 X.529/37880
0146870

KUDRNOFSKY, Wolfgang
Der Mensch in seinem Zorn. - Wien : Jugend und Volk, 1974
373 p.
3-7141-7419-2 X.519/20856
0146870

The MAKING of human aggression / edited by Herbert Selg
; [translated from the German by Arnold Pomerans]. -
London : Quartet Books, 1975 [i.e. 1978]
[5],202p : ill. - Bibliography: p.181-202
0-7043-3193-4 X.319/18061 Woolwich
0146870

MULTIDISCIPLINARY approaches to aggression research /
edited by Paul F. Brain and David Benton. - Amsterdam ;
Oxford : Elsevier/North-Holland Biomedical, 1981
xvi,549p : ill. - Includes bibliographies and index
0-444-80317-3 SRL
0146870

PERSPECTIVES on aggression / edited by Russell G. Geen,
Edgar C. O'Neal. - New York ; London : Academic Press,
1976
xiv,273p : ill. - Includes bibliographies and index
0-12-278850-8 X.320/10427 Woolwich
0146870

PSYCHOANALYTIC perspectives on aggression / edited by
George D. Goldman, Donald S. Milman. - Dubuque : Kendall
Hunt, c1978
xvii,212p
0-8403-1800-6 X.520/25462
0188050

WILHELM, Christian
Schritte zur Konfliktbewältigung. - Waldkirch i.Br. :
Waldkircher Verlagsgesellschaft, 1981
135p
r6305177 X.529/47450
8716641

AGGRESSION. Man
- Conference proceedings

AGGRESSION and behaviour change / edited by Seymour Feshbach, Adam Frączek. - New York : Praeger, 1979
xi,299p : ill. - (Praeger special studies). - Includes bibliographies and index
0-03-052446-6 X.520/21733
1115936

DETERMINANTS AND ORIGINS OF AGGRESSIVE BEHAVIOR (Conference : 1973 : Monte Carlo)
Determinants and origins of aggressive behavior. - The Hague : Mouton, 1974
xiv,623p : ill. - (New Babylon. studies in the social sciences ; 22). - Includes bibliographies and index
90-279-7671-6 X.320/10735 Woolwich
1115936

AGGRESSION. Man
compared with aggression in animals

SCOTT, John Paul
Aggression. - 2nd ed., revised and expanded. - Chicago ; London : University of Chicago Press, 1975 [i.e. 1976]
xiii,233p : ill. - (The scientist's library). - Bibliography: p.211-222. - Includes index
0-226-74294-6 Ac.2691d/72(13)
067088x

AGGRESSION. Man
compared with aggression in animals - Conference proceedings

SYMPOSIUM ON COMMUNICATION AND AFFECT (6th : 1976 : University of Toronto)
Aggression, dominance and individual spacing. - New York ; London : Plenum Press, 1978
x,173p : ill. - (Advances in the study of communication and affect ; vol.4). - Includes bibliographies and index
0-306-35904-9 X.520/12409
067088x

AGGRESSION. Man
Neurophysiological aspects - Conference proceedings

ISSUES in brain/behavior control / edited by W. Lynn Smith and Arthur Kling. - New York : Spectrum Publications ; New York ; London : Distributed by Wiley, 1976
[13],157p. - Includes bibliographies and index
0-470-15038-6 X.329/10602 Woolwich
0725137

AGGRESSION. Man
Neuropsychological aspects - Conference proceedings

WORKSHOP ON THE NEUROPSYCHOLOGY OF AGGRESSION (1974 : Newport Beach)
The neuropsychology of aggression. - New York ; London : Plenum Press, 1974
viii,214p : ill. - (Advances in behavioral biology ; vol.12). - Includes bibliographies and index
0-306-37912-0 SRL
0370789

AGGRESSION. Man
Physiological aspects

MOYER, K. E.
The psychobiology of aggression. - New York ; London : Harper and Row, 1976
xix,402p : ill. - (Harper's physiological psychology series). - Bibliography: p.301-371. - Includes index
0-06-044641-2 SRL
1056891

AGGRESSION. Man
Projective tests: Rosenzweig Picture-Frustration Study, to 1978

ROSENZWEIG, Saul
Aggressive behavior and the Rosenzweig picture-frustration study. - New York ; London : Praeger, 1978
ix,164p. - (Praeger special studies). - Bibliography: p.99-154. - Includes index
0-03-045656-8 X.520/14245
1922246

AGGRESSION. Man
related to altruism

HORNSTEIN, Harvey Arnold
Cruelty and kindness. - Englewood Cliffs ; London : Prentice-Hall, 1976
vi,154p. - (The patterns of social behavior series). - Includes index
0-13-194928-4 X.519/26036
0706639

AGGRESSION. Man
Research - Conference proceedings

SYMPOSIUM ON BEHAVIOR MODIFICATION (3rd : 1973)
Analysis of delinquency and aggression. - Hillsdale : Erlbaum ; New York ; London : Distributed by Wiley, 1976
xi,237p : ill. - Includes bibliographies and index
0-470-15227-3 X.200/30493
0693529

AGGRESSION. Man
Theories

MACKAL, P. Karl
Psychological theories of aggression. - Amsterdam : Oxford : North-Holland Publishing Co., 1979
xvii,216p : ill. - Bibliography : p.205-210. - Includes index
0-444-85350-2 X.519/30769
0715824

MONTAGU, Ashley
The nature of human aggression. - Oxford : Oxford University Press, 1978 [i.e. 1979]
xi,381p : ill. - Includes index
0-19-285078-4 X.519/29281
0715824

MONTAGUE, Ashley
The nature of human aggression. - New York : Oxford University Press, 1976
xi, 381 p.. - Includes index
0-19-501822-2 X.529/30919
0715824

ZILLMANN, Dolf
Hostility and aggression. - Hillsdale : Erlbaum ; London : [Distributed by Wiley], 1979
x,422p. - Bibliography: p.377-408. - Includes index
0-470-26832-8 X.520/21075
0715824

AGGRESSION. Man
Theories of Seneca, Lucius Annaeus, ca.4 B.C-65

BÄUMER, Änne
Die Bestie Mensch. - Frankfurt am Main : Lang, 1982
239p. - (Studien zur klassischen Philologie ; Bd 4). - Bibliography
3-8204-7211-8 X.0909/1127(4)
8891168

AGGRESSION. Man
Verbal aggression - Serials

MALEDICTA. - Vol.1, no.1 ([Summer, 1977])-. - Waukesha : Maledicta Press, [1977]-
Holdings: vol.4, no.1 (Summer, 1980)-
r6804039 P.901/2033
877532x

AGGRESSION. Offenders
Psychological aspects

AGGRESSION and dangerousness / edited by David P. Farrington and John Gunn. - Chichester : Wiley, c1985
271p : ill. - (Current research in forensic psychiatry and psychology). - Includes bibliographies and index
0-471-90556-9 X.0200/1192(3)
3027848

AGGRESSION. Participation of public in politics
Psychosocial aspects

MULLER, Edward N.
Aggressive political participation. - Princeton ; Guildford : Princeton University Press, 1979
x,306p : ill. - Includes index
0-691-07605-7 X.809/44901
2002035

AGGRESSION. Primates

PRIMATE aggression, territoriality and xenophobia / edited by Ralph L. Holloway. - New York ; London : Academic Press, 1974
xiv,513p : ill. - Includes bibliographies and index
0-12-352850-x SRL
1133144

AGGRESSION. Rhesus monkeys
Role in play

SYMONS, Donald
Play and aggression. - New York ; Guildford : Columbia University Press, 1978
x,246p : ill. - Bibliography: p.221-237. - Includes index
0-231-04334-1 SRL
1457152

AGGRESSION. Vertebrates
Role of hormones

HORMONES and aggressive behavior / edited by Bruce B. Svare. - New York ; London : Plenum, c1983
xiii,595p : ill. - Includes bibliographies and index
0-306-41055-9 SRL
2992310

AGGRESSION. Vertebrates. Laboratory animals

MULTIDISCIPLINARY approaches to aggression research / edited by Paul F. Brain and David Benton. - Amsterdam ; Oxford : Elsevier/North-Holland Biomedical, 1981
xvi,549p : ill. - Includes bibliographies and index
0-444-80317-3 SRL
2584514

AGGRESSION. Young persons
- Conference proceedings

NATO ADVANCED STUDY INSTITUTE ON AGGRESSION IN CHILDREN AND YOUTH (1981 : Maratea)
Aggression in children and youth. - The Hague ; Lancaster : Nijhoff in cooperation with NATO Scientific Affairs Division, 1984
viii,367p. - (NATO advanced study institutes series. Series D, Behavioural and social sciences ; no.17). - Includes bibliographies and index
90-247-2903-3 X.520/34829
3171450

AGGRESSION. Young persons, 9 & 19 years. United States

GROWING up to be violent / Monroe M. Lefkowitz ... [et al.]. - New York ; Oxford : Pergamon, 1977
ix,236p : ill. - (Pergamon international library) (Pergamon general psychology series ; vol.66). - Bibliography: p.212-224. - Includes index
0-08-019515-6 X.519/27130
1188399

AGGRESSIVENESS. Man
Measurement

EDMUNDS, G. (George)
The measurement of human aggressiveness. - Chichester : Ellis Horwood ; New York ; Chichester : [Distributed by] Wiley, 1980
225,[2]p : ill. - Bibliography: p.205-223. - Includes index
0-85312-146-x X.520/22518
2085968

AGGREY, JAMES EAMMAN KWEGYIR. Education. Ghana
- Biographies

OFOSU-APPIAH, L. H.
The life of Dr. J.E.K. Aggrey. - Accra : Waterville Publishing House, 1975
103p : 1 ill. - Bibliography : p.100. - Includes index
r6073371 X.709/32910
180197X

AGGSBACH. Austria
Priories: Kartause Aggsbach, to 1782

ROSSMANN, Heribert
Die Geschichte der Kartause Aggsbach bei Melk in Niederösterreich. - Salzburg : Institut für englische Sprache und Literatur, Universität Salzburg, 1976
2 Bde. - (Analecta cartusiana ; 29, 30)
r6019132 P.101/445[29,30]
1129260

AGGTELEK CAVE. Baradla Caves. Hungary

AGGTELEKI-KARSZTVIDÉK útikalauz / [szerkesztette Jakucs László]. - Budapest : Sport, 1975
404 p., 40 p. of plates : ill
963-253-221-x X.808/10044
0940976

AGHALEE. Lisburn (District)
1677-1976

BRENNAN, S. J.
A brief history of Aghalee Parish to commemorate the three hundredth anniversary of the consecration of Aghalee Parish Church, 1st May, 1677-1st May, 1977. - Newcastle, Co. Down : Mourne Observer Press, 1977
[5],30p : ill. - Bibliography: p.30
B77-23894 X.709/31356
1170015

AGHARTI

MACLELLAN, Alec
The lost world of Agharti. - London : Souvenir Press, 1982
231p,[16]p of plates : ill. - Bibliography: p229-231
0-285-62521-7 X.809/51479
2713942

The lost world of Agharti. - London : Corgi, 1983, c1982
240p. - Bibliography: p238-240
0-552-12143-6 X.529/55299
2713942

AGHIA. Greece
Monasteries: Monastery of Saint Panteleimon. Architecture

KOUMOULIDES, John Thomas Anastasios
Byzantine and post-Byzantine monuments at Aghia in Thessaly, Greece. - London : Zeno, 1975
xviii,77p : ill
0-7228-0004-5 X.410/10052
0642967

AGHNÔN, SAMUEL JOSEPH
See Agnon, Shmuel Yosef

ÄGIDIUS ROMANUS
See Giles of Rome †

AGILITY TESTS. Dog shows. Great Britain

LEWIS, Peter, 1935-
The agility dog. - Portsmouth : Canine Publications, c1981
126p.,[16]p of plates : ill
0-906422-04-3 X.329/14953 Woolwich
262317x

AGIN, ALEKSANDR ALEKSEEVICH. Russian graphic arts - Biographies

KURBATOV, Valentin ÎAkovlevich
A.A. Agin. - Leningrad : Khudozhnik RSFSR, 1979
47p.,[24]p of plates : ill. - (Massovaîa biblioteka po iskusstvu)
r6185471 X.419/4417 Woolwich
224876X

AGINCOURT
See also
Battle of Agincourt

AGING
See Ageing

AGING. Materials
See Ageing. Materials

AGÎRBICEANU, ION. Fiction in Romanian

ZACIU, Mircea
Ion Agârbiceanu 1882-1982. - Bucharest : Cartea românească, [1983?]
103p. - Includes bibliography and "Translations from Agârbiceanu's works into other languages"
r6457278 X.950/37631
815774x

AGISTERS. New Forest. Hampshire

STAGG, David J.
Agisters of the New Forest. - [Lyndhurst?] : New Forest Association, 1983
20p
0-9508428-2-6 X.329/18374
3185419

AGITATED VESSELS
Use in mixing liquids

NAGATA, Shinji
Mixing. - Tokyo : Kodansha ; New York ; London : Wiley, 1975
xviii,458p : ill. - (Kodansha scientific books). - Includes bibliographies and index
0-470-62863-4 SRL
0353523

AGNATHA
See also
Lampreys

AGNELLI, GIOVANNI
Influence on political events in Italy, 1966-1976

ROCCATI, Cesare
Umberto & C.. - Firenze : Vallecchi, 1977
214p. - (Collana diversa)
r6085559 X.708/45841
2030551

AGNELLI, GIOVANNI. Fiat (Firm). Italy

BIAGI, Enzo
Il signor Fiat. - 3. ed. - Milano : Rizzoli, 1976
163p
r6048618 X.629/7748 Woolwich
1226487

CASTRONOVO, Valerio
Giovanni Agnelli. - [enlarged ed.]. - Torino : Einaudi, 1977
xxxii,565p.,12p. of plates. - (Gli struzzi ; 122)
r6052863 X.519/23730
1226487

AGNELLI, UMBERTO
Influence on political events in Italy, 1966-1976

ROCCATI, Cesare
Umberto & C.. - Firenze : Vallecchi, 1977
214p. - (Collana diversa)
r6085559 X.708/45841
2030551

AGNELLO, ARCHBISHOP OF RAVENNA
See Agnellus, Archbishop of Ravenna †

AGNELLUS, ARCHBISHOP OF RAVENNA. Christian church - Biographies

AGNELLO, arcivescovo di Ravenna. - Faenza : Lega, 1971
vii,102p.,xxleaves of plates : ill. - (Saggi e repertori / Società di Studi romagnoli ; 14)
r6505610 X.200/39315
8059918

AGNELLUS OF PISA, BLESSED. Franciscans - Biographies

HARDING, John, 1956-
Agnellus of Pisa, 1194-1236. - Canterbury : Franciscan Study Centre, [1979]
28,[20]leaves
B79-18079 X.102/2203
1888552

AGNES, OF BOHEMIA. Nuns. Catholic Church. Bohemia. Czechoslovakia

BERAN, Josef
Blahoslavena Anežka Česká. - Řim : Křes tanská akademie, 1974
511p.,[4]p of plates : ill. - (Sul země ; sv.7)
r6440333 X.329/15529
8859558

AGNEW (FAMILY)
Role in army operations by Great Britain. Army, 1639-1915

AGNEW, Douglas
The whites of their eyes. - Edinburgh : Agnew Association, 1982
248p : ill
0-9508502-0-9 X.809/56579
296662x

AGNEW, SPIRO T.. Politics. United States

AGREW, Spiro T.
Go quietly - or else. - New York : William Morrow, 1980
288p
0-688-03668-6 X.800/35376
8793506

AGNEW, THOS. & SONS LTD.
See Thos. Agnew & Sons Ltd.

AGNI
See Anyi

AGNI LANGUAGE
See Anyi language

AGNIHOTRA. Hinduism

SARASWATI, Satya Prakash
Agnihotra. - 2nd revised ed. - New Delhi : Jan Gyan Prakashan, 1974
vi,99p : ill
r6028481 X.329/8616 Woolwich
0393118

AGNON, SAMUEL JOSEPH
See Agnon, Shmuel Yosef

AGNON, SHMUEL YOSEF. Fiction in Hebrew

ABERBACH, David
At the handles of the lock. - Oxford : Published for the Littman Library by Oxford University Press, 1984
221p. - (The Littman library of Jewish civilization). - Bibliography: p203-214. - Includes index
0-19-710040-6 X.950/40906
3023184

AGNON, SHMUEL YOSEF. Fiction in Hebrew - Bibliographies

MARTIN, Werner
Samuel Josef Agnon. - Hildesheim : Olms, 1980
xi,118p
3-487-06972-5 X.950/6225
871522x

AGNOSIC ATAXIA. Children

GUBBAY, Sasson S.
The clumsy child. - London : Saunders, 1975
xiii,194p : ill. - (Major problems in neurology ; vol.5). - Bibliography: p.177-189. - Includes index
0-7216-4340-x SRL
1310690

AGNOSTICISM

COX, Edwin
What it means to be agnostic. - Ely : EARO, c1977
15p : ill
0-904463-19-2 X.208/7985
2816237

AGORA. Athens. Greece

AMERICAN SCHOOL OF CLASSICAL STUDIES AT ATHENS
The Athenian Agora. - Princeton, N.J. : The school, 1976
[33] p. : ill. - (Excavations of the Athenian Agora, picture book ; no. 16)
r6010430 X.419/3292 Woolwich
1034251

AGORAPHOBIA. Man

HAYDEN, Eric W. (Eric William)
Learning to cope with agoraphobia. - London : Mowbray, 1983
32p. - (Enquirer's library)
0-264-66947-9 X.529/56721
2703351

MATHEWS, Andrew M.
Agoraphobia. - London : Tavistock, 1981
xiii,233p : ill. - Bibliography: p214-223. - Includes index
0-422-78060-x X.322/10718 Woolwich
2703351

ROWAN, Joan E.
Agoraphobia phobias and anxiety. - Cleadon : J.E. Rowan, 1984
24p. - Bibliography: p24
0-9509314-0-3 X.529/62999
2703351

VOSE, Ruth Hurst
Agoraphobia. - London : Faber, 1981
204p. - Bibliography: p189-191. - Includes index
0-571-11752-x X.529/44476
0270830

AGORAPHOBIA. Man
Behaviour therapy

THORPE, Geoffrey L.
The agoraphobic syndrome. - Chichester : Wiley, c1983
xviii,237p. - Bibliography: p207-224. - Includes index
0-471-10495-7 X.322/12099
2928353

AGORAPHOBIA. Man
Self-treatment

FRAMPTON, Muriel
Agoraphobia. - Completely rev. and reset ed. - Wellingborough : Turnstone, 1984
96p. - (A Life crisis book). - Bibliography: p93. - Includes index
0-85500-213-1 X.529/65781
0204277

LYLE, Ronald C.
Manual for the treatment of agoraphobia by behavioural methods. - Glasgow : The author, 1980
[2],24 leaves
B80-36118 X.525/3626
0204277

PEEL, R. (Russell)
Agoraphobia. - Halifax : R. Peel, c1981
36p
B81-40883 X.329/14636
0204277

AGORAPHOBIA. Man
Therapy

AGORAPHOBIA / edited by Dianne L. Chambless, Alan J. Goldstein. - New York ; Chichester : Wiley, c1982
x,227p : ill. - (Wiley series on personality processes). - Includes bibliographies and index
0-471-07947-2 X.520/29036
1104713

WEEKES, Claire
Agoraphobia. - London : Angus and Robertson, 1977
x,178p. - Includes index
0-207-95723-1 X.329/10933 Woolwich
1104713

AGOSTINHO, DA PIEDADE, FREI. Brazilian sculptures

SILVA-NIGRA, Clemente Maria da
Os dois escultores Frei Agostinho da Piedade, Frei Agostinho de Jesus e o arquiteto Frei Macário de São João. - [Salvador, Bahia] : Universidade Federal da Bahia, 1971
160p : ill
r6707026 X.425/3982
8709564

AGOSTINHO, DE JESUS, FREI. Brazilian sculptures

SILVA-NIGRA, Clemente Maria da
Os dois escultores Frei Agostinho da Piedade, Frei Agostinho de Jesus e o arquiteto Frei Macário de São João. - [Salvador, Bahia] : Universidade Federal da Bahia, 1971
160p : ill
r6707026 X.425/3982
8709564

AGOSTINI, NICOLÒ DEGLI
See degli Agostini, Nicolò †

AGOSTINO, BATTISTA DE
See De Agostino, Battista

AGOURAI. Morocco
Social conditions, to date

HØIRIS, Ole
Agourai. - [København] : Nationalmuseet, 1980
155p : ill
87-480-0279-8 X.529/49583
8903190

AGRA. India
Memorial buildings: Taj Mahal, to 1972

CARROLL, David, 1942-
The Taj Mahal. - New York : Newsweek ; London : Reader's Digest Association, [1975]
3-174p : ill(some col). - (Wonders of man). - Bibliography: p.168. - Includes index
0-88225-024-8 X.423/5026
0373613
- [Another copy.] X.423/5052

OAK, P.N.
The Taj Mahal is Tejo Mahalaya. - [New Delhi] : the author, [1978]
27p.
r6079962 X.100/28026
0373613

AGRA. India (Republic)
Memorial buildings: Taj Mahal

AURORA, Kamal
Story of the Taj. - Ghaziabad : Vikas, [1980?]
24p. - (Pathfinders)
0-86144-500-7 X.800/35055
2009617

KANWAR, H.I.S
[Offprints not catalogued separately]. - New Delhi, 1973-
r6028658 X.0702/399
0398969

OAK, P.N
The Taj Mahal is a Temple Palace. - Third revised edition. - New Delhi : [the Author], [1974]
272p.,[1] fold leaf. - Bibliography : p.269-272
r6086517 X.808/11560
0398969

AGRAMONTE, IGNACIO
See Agramonte y Loynaz

AGRAMONTE PIÑA, EDUARDO. Medicine. Cuba
- Biographies

GODÍNEZ SOSA, Emilio
Eduardo Agramonte Piña. - La Habana : Editorial Arte y Literatura, Instituto Cubano del Libro, 1975
325p.,[1]leaf of plates,[14]p.of plates : facsims,ports
r6157733 X.329/12256 Woolwich
2414015

AGRAMONTE Y LOYNAZ, IGNACIO. Revolutionary movements. Cuba

11 de Mayo 1873-1973. - La Habana : Museo de la Ciudad de La Habana, 1976
118p : ill
r6037950 X.802/3961
1162268

AGRAMONTE Y LOYNAZ, Ignacio
Ignacio Agramonte. - La Habana : Instituto Cubano del Libro, Editorial de Ciencias Sociales, 1974
415p : ill
r6037931 X.520/9442
1162268

AGRAMONTE Y LOYNAZ, IGNACIO. Revolutionary movements. Cuba
- Biographies

CRUZ, Mary
El Mayor. - [Havana] : UNEAC, 1972
263p. - (Contemporáneos). - Includes bibliographies
r6066234 X.709/27070
179082X

AGRARIAN CAPITALISM
Study regions: India (Republic). Tamil Nadu, 1975-1980

HARRISS, John
Capitalism and peasant farming. - Bombay ; Oxford : Oxford University Press, 1982
xviii,358p : ill. - Bibliography: p340-352. - Includes index
0-19-561340-6 X.529/54132
2900424
- [Another copy.] T 46329 IOL

AGRARIAN ECONOMICS
Soviet theories, 1917-1945

FIGUROVSKAÎÂ, Nadezhda Konstantinovna
Agrarnye problemy v sovetskoĭ ėkonomicheskoĭ literature 20-kh godov. - Moskva : Nauka, 1978
257p. - (Problemy sovetskoĭ ėkonomiki)
r6077188 X.519/40319
1858661

AGRARIAN ECONOMICS
Theories, to ca 1900

TRIBE, Keith
Land, labour and economic discourse. - London : Routledge and Kegan Paul, 1978
[8],182p : ill. - Bibliography: p.169-177. - Includes index
0-7100-0002-2 X.529/32812
151251x

AGRARIAN REFORM
See Land tenure. Agricultural land
Reform

AGREED SYLLABUS. England

BRINE, Alan
The agreed syllabus for religious education in the middle year. - London : Schools Council, 1984
161p : ill. - Bibliography: p141
B85-04159 X.205/1602
0784850

AGREED SYLLABUS. Religion. Curriculum subjects. Schools. Cambridgeshire

A FRAMEWORK for religious education in Cambridgeshire. - Ely : EARO, c1983
42p
0-904463-98-2 X.200/43518
3130460

AGREED SYLLABUS. Religion. Curriculum subjects. Schools. West Sussex

WILSON-VOKE, Maurice
The debate on the agreed syllabus of religious education in West Sussex 1979-1980. - Chichester : M. Wilson-Voke, 1981
42p : 2ill
0-9507413-0-2 X.200/35064
2372622

AGREEMENT. Predicates. Russian language

CORBETT, Greville G.
Predicate agreement in Russian. - Birmingham : Department of Russian Language & Literature, University of Birmingham, c1979
xii,111p : ill. - (Birmingham Slavonic monographs ; no.7). - Bibliography: p105-111
0-7044-0349-8 X.950/2395
2370514

AGREEMENT. Predicates. Slavonic languages

CORBETT, Greville G.
Hierarchies, targets and controllers. - London : Croom Helm, c1983
260p. - (Croom Helm linguistics series). - Bibliography: p247-254. - Includes index
0-7099-0744-3 X.950/26467
299450x

AGREEMENT FOR MINOR BUILDING WORKS. Joint Contracts Tribunal
1980

WOOD, R. D. (Reginald Douglas)
The J.C.T. agreement for minor building works. - London : Estates Gazette, 1983
lx,270p : ill. - Bibliography: p254-256. - Includes index
B84-10179 X.629/21999
304257x

AGREEMENTS
See also
Anglo-German Naval Agreement
Arbitration agreements
Contracts
Covenants
Foreign agreements
International agreements
Mixed agreements
National agreements
Subject to contract agreements

AGREEMENTS
See also
Tenancy agreements
Treaties
Written agreements

AGREEMENTS. Advertising agencies. Great Britain
With clients. Standard conditions

INSTITUTE OF PRACTITIONERS IN ADVERTISING
Some suggested provisions for use in agency-client agreements. - 5th ed. - London : The Institute, 1976
[2],27p
0-85294-091-2 X.512/5337
0725048

AGREEMENTS. Clients. Advertising agencies. Great Britain
With advertising agencies. Standard conditions

INSTITUTE OF PRACTITIONERS IN ADVERTISING
Some suggested provisions for use in agency-client agreements. - 5th ed. - London : The Institute, 1976
[2],27p
0-85294-091-2 X.512/5337
0725048

AGREEMENTS. Compensation. Hit & run victims. Great Britain
- Texts

GREAT BRITAIN. Department of Transport
Motor Insurers' Bureau (compensation of victims of untraced drivers). - London : H.M.S.O., 1977
5p
0-11-550459-1 BS.43/428
1311255

AGREEMENTS. Cooperation. Research libraries. Scandinavia
Scandiaplanen

SKÆRBAK, Helli
Scandiaplanen. - København : BIFF'en, 1974
48p. - (BIFF'en publikation ; no.2). - Includes bibliographies
r6001390 2719.b.131
1423177

AGREEMENTS. Educational institutions. Great Britain
With personnel: Agreements on copyright

CRABB, Geoffrey
Copyright agreements between employers and staff in education. - London : Council for Educational Technology for the United Kingdom, 1978
27p
0-902204-80-7 X.519/28064
146048x

AGREEMENTS. Exploitation. Petroleum deposits. United Arab Emirates
1939-1981 - Texts

The PETROLEUM concession agreements of the United Arab Emirates, 1939-1981 (Abu Dhabi) / [compiled by] Mana Saeed al-Otaiba. - London : Croom Helm, c1982
2v.
0-7099-1915-8 X.525/9638
3173178

AGREEMENTS. Foreign investment
Renegotiation

STOEVER, William A.
Renegotiations in international business transactions. - Lexington, Mass. : Lexington Books, 1981
xiii,380p : ill. - Bibliography: p351-365. - Includes index
0-669-03057-0 X.520/25096
2565897

AGREEMENTS. Government. Catalonia. Spain
1975-

CATALONIA
Convenis signats per la Generalitat de Catalunya. - Barcelona : Servei Central de Publicacions de la Generalitat de Catalunya, 1980
196p. - ([Publicacions] / [de la Generalitat] ; 5)
r6702441 X.529/45836
8807353

AGREEMENTS. Government. Great Britain
With industries. Agreements on industrial planning: Planning Agreements

GREAT BRITAIN. Department of Industry
The contents of a planning agreement. - London : The Department, 1975
[1],ii,17p
0-85605-177-2 BS.424/37
0173789

AGREEMENTS. Guarantee payments. Remuneration. Personnel. Great Britain
- Collections

GUARANTEED week agreements. - London : Incomes Data Services, 1981
62p. - (IDS handbook series ; no.20)
B82-02873 X.529/44760
2590395

AGREEMENTS. Labour Party (Great Britain)
With Liberal Party, 1977-1978

MICHIE, Alistair
The pact. - London : Quartet Books, 1978
183p : ill
0-7043-2193-9 X.808/33262
1698281
- [Another copy : paperback]
ISBN 0-7043-3236-1 X.709/32584

AGREEMENTS. Landowners. England
With local authorities: Agreements on landscape conservation

GREAT BRITAIN. Countryside Commission
Management agreements. - Cheltenham : Countryside Commission, 1983
27p. - (CCP ; 156)
0-86170-077-5 BS.116/200
191314x

WAGER, Jonathan F.
Management agreements in principle and practice. - [Bakewell] : The Board, 1976
36p : ill
0-901428-30-2 Bs.116/163
205051x

AGREEMENTS. Landowners. Great Britain
With local authorities: Agreements on landscape conservation in rural regions

MANAGEMENT agreements in the countryside / Country Landowners Association, National Farmers Union, the Royal Institution of Chartered Surveyors. - London : Surveyors, c1984
60p
0-85406-248-3 X.329/19385
2732165

AGREEMENTS. Landowners. Scotland
With local authorities: Agreements on landscape conservation

MANAGEMENT agreements. - Redgorton : Countryside Commission for Scotland, 1982
9p
0-902226-60-6 BS.182/42
2767929

AGREEMENTS. Liberal Party. Great Britain
With Labour Party (Great Britain), 1977-1978

MICHIE, Alistair
The pact. - London : Quartet Books, 1978
183p : ill
0-7043-2193-9 X.808/33262
1698281
- [Another copy : paperback]
ISBN 0-7043-3236-1 X.709/32584

AGREEMENTS. Local authorities. England
With landowners: Agreements on landscape conservation

GREAT BRITAIN. Countryside Commission
Management agreements. - Cheltenham : Countryside Commission, 1983
27p. - (CCP ; 156)
0-86170-077-5 BS.116/200
191314x

WAGER, Jonathan F.
Management agreements in principle and practice. - [Bakewell] : The Board, 1976
36p : ill
0-901428-30-2 Bs.116/163
205051x

AGREEMENTS. Local authorities. Great Britain
With landowners: Agreements on landscape conservation in rural regions

MANAGEMENT agreements in the countryside / Country Landowners Association, National Farmers Union, the Royal Institution of Chartered Surveyors. - London : Surveyors, c1984
60p
0-85406-248-3 X.329/19385
2732165

AGREEMENTS. Local authorities. Great Britain
With public utilities: Agreements on reinstatement of roads

GREAT BRITAIN. Department of the Environment
Reinstatements under the Public Utilities Street Works Act 1950. - London : H.M.S.O., 1976
ix,19p
B76-25172 BS.414/351
0209937

AGREEMENTS. Local authorities. Scotland
With landowners: Agreements on landscape conservation

MANAGEMENT agreements. - Redgorton : Countryside Commission for Scotland, 1982
9p
0-902226-60-6 BS.182/42
2767929

AGREEMENTS. Marketing
Negotiation

MCCALL, J. B.
Marketing by agreement. - Chichester : Wiley, c1984
x,274p : ill. - Bibliography: p261-265. - Includes index
0-471-90461-9 X.622/22190
3015998

AGREEMENTS. Personnel. Educational institutions. Great Britain
With educational institutions: Agreements on copyright

CRABB, Geoffrey
Copyright agreements between employers and staff in education. - London : Council for Educational Technology for the United Kingdom, 1978
27p
0-902204-80-7 X.519/28064
146048x

AGREEMENTS. Politics. Spain
Pactos de la Moncloa - Texts

Los PACTOS de la Moncloa. - 2.ed. - Madrid : Servicio Central de Publicaciones, Secretaría General Técnica, Presidencia del Gobierno, 1977
96p. - (Colección Informe ; 17)
84-500-2323-8 X.708/49294
2470241

AGREEMENTS. Procedure. Industrial relations. Great Britain
- For management

MUIR, John, 19—
Industrial relations procedures and agreements. - Aldershot : Gower, c1981
xv,184p. - Includes index
0-566-02275-3 X.529/43672
255917x

AGREEMENTS. Public access to private land in rural regions of Scotland

ACCESS agreements. - Redgorton : Countryside Commission for Scotland, 1982
11p
0-902226-61-4 BS.182/45
0710091

COUNTRYSIDE COMMISSION FOR SCOTLAND
Access agreements. - Redgorton : The Commission, [1971?]
i,10p. - (Information sheet ; no.2)
0-902226-06-1 BS.116/115
0710091

AGREEMENTS. Public utilities. Great Britain
With local authorities: Agreements on reinstatement of roads

GREAT BRITAIN. Department of the Environment
Reinstatements under the Public Utilities Street Works Act 1950. - London : H.M.S.O., 1976
ix,19p
B76-25172 BS.414/351
0209937

AGREEMENTS. Recognition. Closed shops. Companies. Great Britain
- For management

SWEENEY, Shelagh
Closed shop agreements. - London : Institute of Personnel Management, 1976
[6],41p
0-85292-147-0 X.512/5543
1062174

AGREEMENTS. Reinsurance. War risks. British shipping
- Texts

AGREEMENTS. - London : H.M.S.O., 1982
6p
0-11-513647-9 BS.41/1328
2811103

AGREEMENTS. Remuneration. Personnel. Great Britain
- Serials

PAY & benefits bulletin. - No.1- ; Oct. 1979-. - London : Eclipse Publications Ltd [for] Industrial Relations Services, [1979]-
B80-19642 P.511/5210
211142x

AGREEMENTS. Tenancies. Local authority housing. Brent (London Borough)
Improvement. Role of Brent Federation of Tenants' and Residents' Associations

BRENT FEDERATION OF TENANTS' AND RESIDENTS' ASSOCIATIONS
'How to gain a better Council Tenancy Agreement'. - [London] : The Federation, 1978
11,[5]p
B80-36734 X.529/37920
2239221

AGREEMENTS. Tenancies. Local authority housing. England
Conditions

GREAT BRITAIN. Department of the Environment. Housing Services Advisory Group
Tenancy agreements. - London : Department of the Environment, [1977]
i,22p
r6071526 BS.414/1781
075613X

HOUSING SERVICES ADVISORY GROUP
Tenancy agreements. - [London] : Department of the Environment, 1977
[1],i,22p
B77-26871 BS.414/766
075613x

NATIONAL CONSUMER COUNCIL
Tenancy agreements between councils and their tenants. - London : The Council, 1976
44p. - (Discussion paper)
0-905653-01-7 BS.364/15
075613x

AGREEMENTS. Tenancies. Local authority housing. Newcastle upon Tyne. Tyne and Wear (Metrpolitan County)

FRAMEWORK for a tenants' charter / by Newcastle Tenants' Federation and Project. - London : Community Projects Foundation, c1983
35p
0-902406-28-0 X.529/56252
294815x

AGREEMENTS. Tenancies. Local authority housing. Scotland
Conditions

ATHERTON, Graham
Terms of tenancy. - Glasgow : Scottish Consumer Council, 1983
vi,95p
0-907067-16-6 X.525/8478
2995395

AGRELL, ALFHILD. Fiction in Swedish
- Biographies

NORDIN-HENNEL, Ingeborg
Dömd och glömd. - Umeå : Universitetet i Umeå, 1981
183p. - (Umeå studies in the humanities ; 38). - Bibliography
91-7174-072-4 X.0510/277(38)
8158967

AGRÉMENT BOARD

GREAT BRITAIN. Agrément Study Group
The future of agrément in the United Kingdom. - London : H.M.S.O., 1978
[2],25p
0-11-751380-6 BS.414/1933
1878174

AGRICOLA, GNAEUS JULIUS
Role in army operations by Ancient Roman armies in Scotland - Conference proceedings

SCOTTISH ARCHAEOLOGICAL FORUM (12th : 1980 : University of Edinburgh)
Agricola's campaigns in Scotland. - Edinburgh : Edinburgh University Press, c1981
114p : ill. - Includes bibliographies
0-85224-418-5 X.800/32299
2628872

AGRICOLA, JOHANN. Lutheran churches

KJELDGAARD-PEDERSEN, Steffen
Gesetz, Evangelium und Busse. - Leiden : Brill, 1983
x,394p. - (Acta theologica danica ; v.16)
90-04-06966-6 W.P.16199/16
8737460

AGRICOLA, MIKAEL. Reformation. Christian church. Finland

HEININEN, Simo
Nuori Mikael Agricola. - Helsinki Suomalaisen Kirjallisuuden Seura, 1976
110 p.. - (Suomi ; 120. 3). - Includes bibliographies
951-7170-92-0 PP.4852.d[vol.120.no.3]
1007823

NIKKILÄ, Osmo
Mikael Agricolan teosten paikallissijojen loppuheitto. - Tampere : [Tampereen yliopisto], 1980
97p. - (Folia fennistica & linguistica ; 1). - Includes bibliography
951-440-949-3 X.0950/872(1)
9112715

AGRICOLA, MIKAEL. Reformation. Christian church. Finland
Works. Indefinite pronouns

SUOJANEN, MK
Mikael Agricolan teosten indefiniittipronominit : totalitiivit. - Helsinki : Suomalaisen kirjallisuuden seura, 1977
207p. - (Suomalaisen kirjallisuuden seuran toimituksia ; 334). - Includes bibliographies
951-7171-18-8 Ac.9080[no.334]
1710257

AGRICULTURAL ADMINISTRATION NETWORK
- Serials

[NEWSLETTER (Agricultural Administration Network)].
Newsletter / Agricultural Administration Network. - No.1 (Nov.1979)-. - London : Overseas Development Institute, 1979-
v.
Holdings: no.2; March 1980, etc.
B81-06903 X.0322/258 Woolwich
2354276

AGRICULTURAL AND HORTICULTURAL COLLEGE. PILTOWN
College farm. Soils

COLLINS, J. F.
Soils of Agricultural and Horticultural College Farm, Kildalton Piltown, Co. Kilkenny. - [Dublin] : Soil Science Department, University College, Dublin, 1976
[2], 39 leaves. - (Soil bulletin ; no.4). - Bibliography : leaf 39
r6072089 X.311/8820 Woolwich
1513672

AGRICULTURAL BUILDINGS
See Buildings. Farms

AGRICULTURAL CHEMICALS
See also
Fertilisers
Herbicides
Horticultural chemicals

AGRICULTURAL CHEMICALS
Antimicrobials - Conference proceedings

INTERNATIONAL SYMPOSIUM ON ANTIBIOTICS IN AGRICULTURE : BENEFITS AND MALEFITS (4th : : University of Nottingham)
Antimicrobials and agriculture. - London : Butterworths, 1984
xiv,583p : ill. - (Studies in the agricultural and food sciences). - Includes bibliographies and index
0-408-11155-0 SRL
3013847

AGRICULTURAL CHEMICALS
Names

MARLER, E. E. J.
Pharmacological and chemical synonyms. - 6th ed. - Amsterdam ; Oxford : Excerpta Medica, 1976
[8],472p
B77-04840 SRL
1030841

AGRICULTURAL CHEMICALS
Names - Dictionaries

MARLER, E. E. J.
Pharmacological and chemical synonyms. - 6th ed.. - Amsterdam ; Oxford : Exerpta Medica, 1978
viii,506p
B79-04371 SRL
1030841

AGRICULTURAL CHEMICALS
Toxic effects

AGRICULTURAL chemicals and pesticides / edited by Edward J. Fairchild ; for the U.S. Department of Health, Education and Welfare, Public Health Service, Centre for Disease Control, National Institute for Occupational Safety and Health. - Tunbridge Wells : Castle House Publications, 1978
xxi,229,[1]p
0-7194-0002-3 SRL
1775278

AGRICULTURAL CHEMICALS
Use

FARRINGTON, William S.
Fertilizer, chemicals and seed. - New York ; London : McGraw-Hill, 1980
viii,136p : ill. - (Career preparation for agriculture/agribusiness). - Includes index
0-07-019965-5 X.322/9595 Woolwich
2256886

AGRICULTURAL CHEMICALS
Use of recombinant DNA

LEE, A. M. T.
Report on the congress on standardization and control of biologicals produced by recombinant DNA technology. - [Pinner] : ADAS, c1984
6p
B84-43523 BS.3/883
3193292

AGRICULTURAL CHEMICALS. Bracknell. Berkshire
Research organisations: Jealott's Hill Research Station, to 1978

JEALOTT'S Hill / edited by F.C. Peacock. - Bracknell : Imperial Chemical Industries, Plant Protection Division, 1978
viii,160p,[28]p of plates : ill(some col). - Includes index
0-901747-01-7 X.320/11450 Woolwich
1586351

AGRICULTURAL CHEMICALS. Contaminants of food
Chemical analysis

ANALYTICAL methods for pesticides and plant growth regulators / edited by Gunter Zweig. - New York ; London : Academic Press.
Vol.9: Spectroscopic methods of analysis. - 1977
xi,299p : ill. - Includes bibliographies and index
0-12-784309-4 SRL
0149772
Vol.10: New and updated methods. - 1978
xviii,593p : ill. - Includes bibliographies and index
0-12-784310-8 SRL
0149772

AGRICULTURAL CHEMICALS. Great Britain
Toxic agricultural chemicals. Use. Safety measures. Law

DRAFT proposals for poisonous substances in agriculture regulations. - London : H.M.S.O., 1980
18p. - (Consultative document / Health and Safety Commission)
0-11-883412-6 BS.438/130
271597x

AGRICULTURAL CHEMICALS APPROVAL SCHEME
See also
Pesticides approved under Agricultural Chemicals Approval Scheme

AGRICULTURAL CHEMICALS EQUIPMENT INDUSTRIES
See also
Fertiliser equipment industries

AGRICULTURAL CHEMICALS INDUSTRIES
See also
Fertiliser industries

AGRICULTURAL CHEMICALS INDUSTRIES. Great Britain
- Serials

MAY AND BAKER. Agrochemicals Division
News-sheet. - no. 1- : April, 1975-. - Dagenham : May and Baker, 1975-
r6040723 P.311/470
139410X

AGRICULTURAL CHEMICALS INDUSTRIES. Great Britain
Organisations: British Agrochemicals Association

BRITISH AGROCHEMICALS ASSOCIATION
Manual. - London : The Association, 1978
1v. : ill
B79-09379 SRL
1818295

AGRICULTURAL CHEMICALS MANUFACTURED BY FISONS LIMITED. AGROCHEMICAL DIVISION. Great Britain
Prices

GREAT BRITAIN. Price Commission
Fisons Ltd Agrochemical Division - agrochemical and horticultural products .... - London : H.M.S.O., [1978]
v,18p. - ([1977-78 H.C.]151)
0-10-215178-4 BS.416/16(4)
1453440

AGRICULTURAL CHEMICALS MANUFACTURING EQUIPMENT
See also
Fertiliser manufacturing equipment

AGRICULTURAL CHEMICALS MANUFACTURING PLANTS
See also
Fertiliser manufacturing plants

AGRICULTURAL CHEMISTRY
Research. Use of radioisotopes

L'ANNUNZIATA, Michael F.
Radiotracers in agricultural chemistry. - London : Academic Press, 1979
xxi,536p : ill. - Includes bibliographies and index
0-12-436250-8 SRL
1825534

AGRICULTURAL COLLEGES. Aargau. Switzerland
to date - Serials

BERICHT ... über die landwirtschaftliche Berufsbildung / Kanton Aargau, Finanzdepartement, Abteilung Landwirtschaft. - 1979/81-. - [Aarau] : [die Abteilung], [1981]-
r6303923 S.P.2/6
8805989

LANDWIRTSCHAFTLICHE Berufsbildung / Kanton Aargau, Finanzdepartement, Abteilung Landwirtschaft. - 1977/79. - Aarau : die Abteilung, [1979]
1v.
r6303922 S.P.2/6
8805989

AGRICULTURAL COLLEGES. Australia
to 1975

BLACK, Alan W.
Organizational genesis and development. - St Lucia : University of Queensland Press, 1976
xii,316p : 1 ill. - (Educational administration and organization). - Bibliography : p.291-304. - Includes index
0-7022-1191-5 X.529/21344
1729586

AGRICULTURAL COLLEGES. Aylesbury Vale (District). Buckinghamshire
Colleges of further education & of agriculture: Aylesbury College of Further Education and Agriculture

AYLESBURY COLLEGE OF FURTHER EDUCATION AND AGRICULTURE
Full-time courses. - Session 1963-64-. - [Aylesbury] : Buckinghamshire County Council Education Department, [1963]-
0-902805-68-1 P.521/3229
1839624

AGRICULTURAL COLLEGES. Aylesbury Vale (District). Buckinghamshire
Colleges of further education and agriculture: Aylesbury College of Further Education and Agriculture. Department of Agriculture and Horticulture. Day release and evening courses

AYLESBURY COLLEGE OF FURTHER EDUCATION AND AGRICULTURE.
Department of Agriculture and Horticulture
Part-time day and evening classes. - 1976-77-. - [Stoke Mandeville] : [The Department of Agriculture and Horticulture], [1976]-
[1], 16, [2] p.
0-902805-41-x X.319/17061 Woolwich
0794678

AGRICULTURAL COLLEGES. Aylesbury Vale (District). Buckinghamshire
Colleges of further education and agriculture: Aylesbury College of Further Education and Agriculture. Department of Agriculture and Horticulture. Short courses

AYLESBURY COLLEGE OF FURTHER EDUCATION AND AGRICULTURE.
Department of Agriculture and Horticulture
Short courses. - 1976-77 -. - [Stoke Mandeville] : [The Department of Agriculture and Horticulture], [1976]-
[1], 24, [2] p.
0-902805-40-1 X.319/17060 Woolwich
079466X

AGRICULTURAL COLLEGES. Denmark
- Serials

FORTEGNELSE over godkendte landbrugsskoler, gartnerskoler og mejeriskoler / Undervisningsministeriet, Direktoratet for Folkeskolen, Folkeoplysning, Seminarier m.v.. - København : Undervisningsministeriet, [1975-]
r6112651 SB.278/43
246067x

AGRICULTURAL COLLEGES. Great Britain
Libraries. Users. Education

BROADBENT, Leonard
Survey of user education in horticultural education and training curricula. - [London] : [British Library, Research and Development Dept.], 1982
49p. - (BL R & D report ; 5679)
B83-01081 X.525/6914
2808242

AGRICULTURAL COLLEGES. Keszthely. Hungary
Keszthelyi Agrártudományi Egyetem, 1797-1972

GEORGIKON 175 / szerkesztette Sági Károly. - Budapest : Mezőgazdasági Kiadó, 1972
278 p.
r6007873 X.320/5017 Woolwich
0859400

AGRICULTURAL COLLEGES. Kirkmichael. Strathclyde Region. Scotland
Barony Agricultural College, to 1977

MUIR, Jean, 1954-
The Barony Agricultural College. - [Kirkmichael?] : [J. Muir?], [1983?]
22p : ill. - Bibliography: p22
B84-03406 X.809/57043
2985683

AGRICULTURAL COLLEGES. Lacock. Wiltshire
Libraries: Lackham College of Agriculture. Library. Stock - Bibliographies - Serials

LACKHAM COLLEGE OF AGRICULTURE
Lackham College of Agriculture Library Bulletin. - [No.] 1- - Chippenham : Wiltshire Library and Museum Service. 1975?-
Holdings : no 10/75 etc. 25 Nov. 1975 etc
r6010739 2719.ff.68
1048082

AGRICULTURAL COLLEGES. Piltown. Kilkenny (County)
Agricultural and Horticultural College. Piltown. College farm. Soils

COLLINS, J. F.
Soils of Agricultural and Horticultural College Farm. Kildalton Piltown, Co. Kilkenny. - [Dublin] : Soil Science Department, University College. Dublin. 1976
[2]. 39 leaves. - (Soil bulletin ; no.4). - Bibliography : leaf 39
r6072089 X.311/8820 Woolwich
1513672

AGRICULTURAL COLLEGES. Poland

DWUDZIESTOPIĘCIOLECIE Akademii Rolniczo-Technicznej w Olsztynie, 1950-1975 / [komitet redakcyjny : przewodniczący : Stefan Tarczyński]. - Olsztyn : Wydawnictwo Akademii Rolniczo-Technicznej w Olsztynie. 1975
2v : ill
r6182812 X.0322/233 Woolwich
2466325

AGRICULTURAL COLLEGES. St Albans. Hertfordshire
Hertfordshire College of Agriculture and Horticulture, to 1971

PELHAM, Eric Claude
The Oaklands story, 1921-1971. - [St Albans] : Hertfordshire College of Agriculture and Horticulture, 1971
[1],136p : ill
0-903883-02-3 X.319/10056
0358266

AGRICULTURAL COLLEGES. Sweden
Sveriges lantbruksuniversitet - Serials

SVERIGES LANTBRUKSUNIVERSITET
Katalog. - 1977-78-. - [Ultuna] : [Sveriges lantbruksuniversitet], 1977-
r6029025 X.0519/732
1789295

AGRICULTURAL COLLEGES. United States
to date

EDMOND, Joseph Bailey
The magnificent charter. - Hicksville : Exposition Press, [1978]
xvi,204p,1plate. - (An Exposition-University book)
0-682-49081-4 X.329/14477 Woolwich
8768765

AGRICULTURAL COLLEGES. Universities. Wye. Kent
Libraries: Wye College. Library. Stock: Books printed between 1543 & 1918 - Bibliographies

WYE COLLEGE. Library
A catalogue of agricultural & horticultural books, 1543-1918, in Wye College Library. - Ashford, Kent : The Library, [1977]
[1],ii,100p
0-901859-63-x X.313/1407 Woolwich
1137654

AGRICULTURAL COLLEGES. West Germany
to date

WITZENHAUSEN - 85 Jahre im Dienste der Agrarentwicklung in den Tropen und Subtropen / zusammengestellt von Peter Wolff ; Herausgeber: Gesamthochschule Kassel, Fachbereich Internationale Agrarwirtschaft in Witzenhausen ... [et al.]. - Witzenhausen : Selbstverlag des Verbandes der Tropenlandwirte aus Witzenhausen, 1983
253p : ill. - (Der Tropenlandwirt. Beiheft ; Nr.18)
r6308724 X.329/17844
9082824

AGRICULTURAL COMMUNITIES
See also
Crofting communities

AGRICULTURAL COMMUNITIES
Industrialisation. Role of upper classes & peasants, to 1965

MOORE, Barrington
Social origins of dictatorship and democracy. - Boston [Mass.] : Beacon Press, 1967
v-xix,559p. - Bibliography : p.524-546. - Includes index
0-8070-5075-x X.708/20812
1090518

AGRICULTURAL COMMUNITIES
Marriage

GOODY, Jack
Production and reproduction. - Cambridge : Cambridge University Press, 1976
xiii,157p : ill. - (Cambridge studies in social anthropology ; 17). - Bibliography: p.145-152. - Includes index
0-521-21294-4 X.800/26092
1111108

AGRICULTURAL COMMUNITIES. Alto. Portugal
Socioeconomic aspects

JENKINS, Robin, 1942-
The road to Alto. - London : Pluto Press, 1979
158p : ill. - Bibliography: p.158
0-86104-077-5 X.519/29952
1926977

AGRICULTURAL COMMUNITIES. Andhra Pradesh. India (Republic)
Displaced agricultural communities. Socioeconomic aspects

MUTHAYYA, B. C.
Rehabilitation of displaced villagers. - Hyderabad : National Institute of Community Development, 1975
vii,183p. - Includes index
r6091763 X.510/16155
1884832

AGRICULTURAL COMMUNITIES. Argentina
Economic conditions

ARCHETTI, Eduardo P.
Explotación familiar y acumulación de capital en el campo argentino. - Buenos Aires : Siglo Veintiuno Editores, 1975
229p. - (Sociología y política). - Bibliography : p. 225-229
r6032890 X.519/22191
039131X

AGRICULTURAL COMMUNITIES. Argentina
Social conditions

VIALE, Jorge O.
Ensayos sociológicos del agro argentino. - [Rosario] : [s.n.], 1972
223p
r6084032 X.510/11378
1986856

AGRICULTURAL COMMUNITIES. Bologna (Province). Italy
Politics, ca 1900-1930

CARDOZA, Anthony L.
Agrarian elites and Italian fascism. - Princeton : Princeton University Press, c1982
xvi,477p. - Bibliography: p455-462. - Includes index
0-691-05360-x X.809/55620
2929953

AGRICULTURAL COMMUNITIES. By region. Sweden
Peasants, 1680-1860. Socioeconomic aspects

ISACSON, Maths
Ekonomisk tillväxt och social differentiering 1680-1860. - Uppsala ; Stockholm : Distributor : Almqvist & Wiksell International, 1979
284p. - (Uppsala studies in economic history ; 18). - Includes bibliographies
91-554-0900-8 X.0510/38(18)
209472X

AGRICULTURAL COMMUNITIES. Central America
Social conditions

TENENCIA de la tierra y desarrollo rural en Centroamérica / estudio elaborado por expertos de CEPAL ... [et al.]. - 2.ed. - San José, Costa Rica : Editorial Universitaria Centroamericana, 1976
199p. - (Colección Seis. serie mayor)
r6047904 X.319/9660 Woolwich
1713469

AGRICULTURAL COMMUNITIES. China
Communes

BANGLADESH ACADEMY FOR RURAL DEVELOPMENT
Rural development in China. - Comilla : Bangladesh Academy for Rural Development, 1977
vi,200p : ill
r6110471 IS.BA.119/54
1769936

AGRICULTURAL COMMUNITIES. China
Communes, 1949-

ROMICH, Manfred F.
Chinas Volkskommunen. - Frankfurt (Main) : R.G. Fischer, 1981
190p.. - (Aachener Beiträge zur China-Forschung ; Bd.1)
3-88323-292-0 X.329/15625
8542767

AGRICULTURAL COMMUNITIES. China
Communes. Economic development. Effects of economic policies, 1949-1979

DOMES, Jürgen
Socialism in the Chinese countryside. - London : C. Hurst [etc.], 1980
x,189p. - Bibliography: p.182-186. - Includes index
0-905838-33-5 X.529/36464
2110148

AGRICULTURAL COMMUNITIES. Chuvash ASSR. Russia (RSFSR)
Russian families, 1917-

BUSYGIN, Evgeniĭ Prokop'evich
Russkaia sel'skaia sem'ia Chuvashskoĭ ASSR. - Kazan : Izdatel'stvo Kazanskogo universiteta, 1980
190p : ill
r6497634 X.329/13991 Woolwich
8706670

AGRICULTURAL COMMUNITIES. Cochabamba. Bolivia
Peasants. Political movements, 1952-1953

DANDLER, Jorge
Campesinado y reforma agraria en Cochabamba, 1952-1953 : dinámica de un movimiento campesino en Bolivia. - La Paz : Centro de Investigación y Promoción del Campesinado, 1976
40p.,[3]leaves of plates. - (Cuadernos de investigación CIPCA ; 9)
r6158927 X.322/2689 Woolwich
226059X

AGRICULTURAL COMMUNITIES. Colombia
Education

INSTITUTO COLOMBIANO DE PEDAGOGÍA
Experiencias educativas en el medio rural colombiano. - Bogotá : Centro Nacional de Documentación e Información Pedagógica, 1972
51 p., 10 leaves of plates (some fold.). - (Serie Divulgación ; n. 2)
r6003579 LAS.C.203/16(2)
0937185

AGRICULTURAL COMMUNITIES. Denmark
1905-1974

FUGLSANG, Mads
Bondedreng på valsen. - [Copenhagen] : Gyldendal, 1975
259 p.
87-00-54813-8 X.900/17979
101904X

AGRICULTURAL COMMUNITIES. Denmark
Role of newspapers, 1880-1890

LEHRMANN, Ulrik
Avisen og agrarsmåborgerskabet. - [Odense] : Odense Universitetsforlag, 1978
125p : ill. - (Dansk kulturhistorie og bevidsthedsdannelse 1880-1920 ; bd 5). - Includes bibliographies
87-7492-255-6 X.0709/929(5)
1793942

AGRICULTURAL COMMUNITIES. Denmark
Social conditions, to 1900

NATIONALMUSEET (Denmark)
Dansk bondekultur før 1900. - København : Nationalmuseet, [1979]
92p : ill(some col)
87-480-0188-0 X.319/20266
2192659

AGRICULTURAL COMMUNITIES. Denmark
Social conditions, 1500-1800

SKRUBBELTRANG, Fridlev
Det danske landbosamfund 1500-1800. - [København] : Den danske historiske Forening, 1978
480p. - Includes bibliographies
87-87462-09-5 X.320/5408 Woolwich
226613X

AGRICULTURAL COMMUNITIES. Developing countries
Poverty related to land tenure

GRIFFIN, Keith
Land concentration and rural poverty. - London : Macmillan, 1976
ix,303p : ill. - Includes index
0-333-18398-3 X.329/10362 Woolwich
0619892
Land concentration and rural poverty. - 2nd ed. - [London] : Macmillan, 1981
vi,345p. - Includes index
0-333-28153-5 X.322/19130
0619892

WHITTEMORE, Claire
Land for people. - Oxford : Oxfam Public Affairs Unit, 1981
55p,[8]p of ill
0-85598-046-x X.529/42180
0619892

AGRICULTURAL COMMUNITIES. Developing countries
Social change, ca 1945–1974

PAIGE, Jeffery M.
Agrarian revolution. - New York : Free Press ; London : Collier Macmillan, 1975
xiv,435p : ill. - Bibliography: p.403–430. - Includes index
0–02–923580–4 X.320/10430 Woolwich
0692670

Agrarian revolution. - New York : Free Press ; London : Collier Macmillan, 1978
xii,435p : ill. - Bibliography: p.403–430. - Includes index
0–02–923550–2 X.319/18568 Woolwich
0692670

AGRICULTURAL COMMUNITIES. Developing countries
Social structure

STAVENHAGEN, Rodolfo
Las clases sociales en las sociedades agrarias. - 7. ed. - México : Siglo Veintiuno Editores, 1975
viii, 292p. - (Sociología y política). - Bibliography : p.275–292
r6053992 X.319/9398 Woolwich
1506935

AGRICULTURAL COMMUNITIES. Doab region. India (Republic)
Effects of irrigation canals, 1830–1930

STONE, Ian
Canal irrigation in British India. - Cambridge : Cambridge University Press, 1984
xiv,174p : ill. - (Cambridge South Asian studies ; 29). - Bibliography: p359–368. - Includes index
0–521–25023–4 X.529/67447
3016412

AGRICULTURAL COMMUNITIES. East Anglia
Social conditions. Influence of land owners

PROPERTY, paternalism and power / Howard Newby ... [et al.]. - London : Hutchinson, 1978
432p : ill. - Bibliography: p.387–402. - Includes index
0–09–133670–8 X.529/32756
1334905

AGRICULTURAL COMMUNITIES. East central Wisconsin
Social conditions, 1866–1936

OLESON, Thurine
Wisconsin, my home. - Madison ; London : University of Wisconsin Press, 1975 [i.e. 1976]
[9],230p : ill
0–299–00711–1 X.809/40913
0645567

AGRICULTURAL COMMUNITIES. Emilia-Romagna. Italy
Social conditions, to date

BROCCOLI, Armide
Chiamavano pane il pane. - Bologna : Edagricole, 1979
xi,371p. - (Testimonianze letterarie del mondo rurale)
88–206–2065–0 X.529/51015
8877351

AGRICULTURAL COMMUNITIES. Europe
to date

Das BILD vom Bauern / Texte : Theodor Kohlmann et al.]. - Berlin : Museum für Deutsche Volkskunde, 1978
160p : ill. - (Schriften des Museums für Deutsche Volkskunde ; Bd.3)
r6305160 X.0322/270(3) Woolwich
8820090

AGRICULTURAL COMMUNITIES. France
Communes, to date

SHAFFER, John W.
Family and farm. - Albany : State University of New York Press, 1982
258p : ill. - (SUNY series on European social history). - Includes bibliography
0–87395–563–3 X.439/13232
9073566

THOLIN, G.
Abrégé de l'histoire des communes du département de Lot-et-Garonne, arrondissement d'Agen. - Toulouse : Eché, 1982
xiv,154p
r6212015 X.809/59250
9073566

AGRICULTURAL COMMUNITIES. Grampian Region. Scotland
Social conditions, 1840–1914

CARTER, Ian
Farmlife in northeast Scotland, 1840–1914. - Edinburgh : Donald, 1979
xiv,258p,[4]p of plates : ill. - Bibliography: p.225–248. - Includes index
0–85976–029–4 X.800/27615
1819143

AGRICULTURAL COMMUNITIES. Great Britain
Use of plants useful to man, 1600–1800

PALLISER, Susan M.
Plants and their uses in the agricultural community. - Leeds : Leeds Folklore Group, 1982
13leaves. - (Leeds Folklore Group papers in folk life studies ; 1)
B82–30881 X.0525/470
2760088

AGRICULTURAL COMMUNITIES. Guatemala
Peasants

FLORES ALVARADO, Humberto
[Proletarisation des paysans du Guatemala. Spanish]. Proletarización del campesino de Guatemala. - [Guatemala] : Rumbos Nuevos, 1971
323p. - Includes bibliographies
r6159124 X.329/12196 Woolwich
2245698

AGRICULTURAL COMMUNITIES. Guatemala
Peasants, to date

FIGUEROA IBARRA, Carlos
El proletariado rural en el agro guatemalteco. - Guatemala : Editorial Universitaria de Guatemala, 1980
475p. - (Colección "Realidad nuestra" ; no.9). - Includes bibliography
r6709225 X.429/16771
8548951

FLORES ALVARADO, Humberto
[Proletarisation des paysans du Guatemala. Spanish]. Proletarización del campesino de Guatemala. - 2a ed. - [Guatemala] : Editorial Piedra Santa, 1977
288p. - (Biblioteca centroamericana de las ciencias sociales). - Bibliography
r6702096 X.329/13957 Woolwich
8548951

AGRICULTURAL COMMUNITIES. Hohenlohe-Franken. West Germany
Catholic agricultural communities compared with Protestant agricultural communities

GOLDE, Günter
Catholics and Protestants. - New York ; London : Academic Press, 1975
xiii,198p : ill. - (Studies in anthropology). - Bibliography: p.187–194. - Includes index
0–12–288250–4 X.320/10241 Woolwich
059993x

AGRICULTURAL COMMUNITIES. Hohenlohe-Franken. West Germany
Protestant agricultural communities compared with Catholic agricultural communities

GOLDE, Günter
Catholics and Protestants. - New York ; London : Academic Press, 1975
xiii,198p : ill. - (Studies in anthropology). - Bibliography: p.187–194. - Includes index
0–12–288250–4 X.320/10241 Woolwich
059993x

AGRICULTURAL COMMUNITIES. India
Effects of British colonial administration, to 1947

STOKES, Eric
The peasant and the Raj. - Cambridge : Cambridge University Press, 1978
viii,308p. - (Cambridge South Asian studies ; [no.23]). - Includes index
0–521–21684–2 X.809/43128
1450387

AGRICULTURAL COMMUNITIES. India (Republic)
Social structure

DASGUPTA, Satadal
Community and agriculture in two Indian villages. - Calcutta : Editions Indian, 1978
188p
r6093089 X.529/34445
0256153

AGRICULTURAL COMMUNITIES. Italy
Social conditions, to date

BIANCHI, Luisito
I contadini salariati nella Bassa Padana. - Roma : Nuove edizioni operaie, 1979
95p,[16]p of plates. - (Quaderni Sapere ; 35)
r6503085 X.329/14907 Woolwich
879880x

BOSIO, Gianni
Il trattore ad Acquanegra. - Bari : De Donato, 1981
lii,281p. - (Opere di Gianni Bosio ; 1)
r6505105 X.329/15826
879880x

AGRICULTURAL COMMUNITIES. Japan
Protest against proposals for use of agricultural land as sites for airports

APTER, David E. (David Ernest)
Against the state. - Cambridge, Mass. ; London : Harvard University Press, 1984
ix,271p, [12]p of plates : ill. - Includes index
0–674–00920–7 X.520/35943
0480304

AGRICULTURAL COMMUNITIES. Jungles. South-east Asia

BOULBET, Jean
Paysans de la forêt. - Paris : École Française d'Extrême-Orient, 1975
146 p. ; 6 fold. leaves of plates, 32 p. of plates : ill. - (Publications de l'École Française d'Extrême-Orient ; 105). - Bibliography : p. 145–146
r6011014 Ac.8814.e[vol.105]
0857246

AGRICULTURAL COMMUNITIES. Lower Saxony. West Germany
Prehistoric agricultural communities

FRÜHE Bauernkulturen in Niedersachsen. - Oldenburg : Staatliches Museum für Naturkunde und Vorgeschichte, 1983
263p. : ill. - (Archäologische Mitteilungen aus Nordwestdeutschland. Beiheft ; 1). - Includes bibliographies
r6309801 X.800/39837
9162305

AGRICULTURAL COMMUNITIES. Mexico
1920–

Los CAMPESINOS de la tierra de Zapata. - Tlalpan : Centro de Investigaciones Superiores, Instituto Nacional de Antropología e Historia.
1: Adaptación, cambio y rebelión. - 1974
221p
r6700191 X.0329/223(1) Woolwich
8726590
- Includes bibliographies
2: Subsistencia y explotación. - 1974
260p : ill
r6700193 X.0329/223(2) Woolwich
8726590
- [Another copy] X.0329/223(2) Woolwich
3: Política y conflicto. - 1976
255p
r6700192 X.0329/223(3) Woolwich
8726590

AGRICULTURAL COMMUNITIES. Northern Ireland
Social structure

BUCKLEY, Anthony D.
A gentle people. - [Holywood] : Ulster Folk and Transport Museum, 1982
184p. - Bibliography: p181–184
0–902588–08–7 X.805/4572
2982153

AGRICULTURAL COMMUNITIES. Perak. Malaysia
Peasants. Social conditions

WAN HASHIM
A Malay peasant community in Upper Perak. - Bangi : Universiti Kebangsaan Malaysia, 1978
viii,192p,[2]folded leaves of plates : ill. - (Monograph series / Department of Anthropology and Sociology, National University of Malaysia ; 4). - Bibliography: p189–192
r6813809 X.529/65715
3178986

AGRICULTURAL COMMUNITIES. Peru
Social change, ca 1960–ca 1970

MACDONALD, Alphonse Leonie
Social factors related to the use of some modern techniques in agriculture by comuneros in two highland regions of Peru. Proefschrift... - Rotterdam : Universitaire Pers, 1975
xi,236p. - Bibliography : p.225–231
r6028681 X.311/4777 Woolwich
1049224

WHYTE, William Foote
Power, politics and progress. - New York ; Oxford : Elsevier, 1976
xiii,307p : ill. - Bibliography: p.301–307
0–444–99028–3 X.800/25917
1049224

AGRICULTURAL COMMUNITIES. Poland
Social conditions

ROLA przeorana, dom piękny / wybór, opracowanie i wstęp Eugenia Jagiełło-Łysiowa. - Warszawa : Książka i Wiedza, 1977
669p
r6044328 X.319/8971 Woolwich
1323229

AGRICULTURAL COMMUNITIES. Poland
Socil conditions–continued

STRUKTURALNE przemiany wsi i rolnictwa w Polsce Ludowej / [Andrzej Stelmachowski redaktor waczelny]. - Warszawa : Państwowe Wydawnictwo Naukowe, 1978
283p. - (Problemy rejonów uprzemysławianych)
r6088687 X.311/4980 Woolwich
1323229

AGRICULTURAL COMMUNITIES. Punjab. India (Republic)
Mer:hants

LONDON, Paul A.
Merchants as promoters of rural development. - New York : London : Praeger : [London] : [Distributed by Martin Robertson], 1975
xiii,160p. - (Praeger special studies in international economics and development). - Bibliography: p.145–154. - Includes index
0-275-05430-6 X.520/10898
0696048

AGRICULTURAL COMMUNITIES. Rural regions. Brazil

IANNI, Octavio
A classe operária vai ao campo. - São Paulo : Brasiliense, 1976
84p. - (Cadernos Cebrap ; n.24)
r6067197 X.311/4947 Woolwich
78885X

AGRICULTURAL COMMUNITIES. Rural regions. Brazil
Social conditions

MARTINS, José de Souza
Capitalismo e tradicionalismo. - São Paulo : Livraria Pioneira Editora, 1975
161p. - (Biblioteca Pioneira de ciências sociais. Sociologia)
r6037818 X.319/8998 Woolwich
326392

AGRICULTURAL COMMUNITIES. Rural regions. Brazil
Working conditions, to date

GNACCARINI, José César
Latifúndio e proletariado. - São Paulo : Polis, 1980
185p. - (Coleção Teoria e história ; 7). - Bibliography
r6704399 X.809/53937
1877343

AGRICULTURAL COMMUNITIES. Rural regions. Colombia
Socia conditions

ACEROS CÁCERES, Hugo
La Negra y la Asociación de Usuarios Campesinos de El Carmen. - Bogotá : Ediciones Tercer Mundo, 1974
150 p.
r6003236 X.519/19966
0907375

AGRICULTURAL COMMUNITIES. Rural regions. Galicia. Spain
Socia structure, 1300–1600

RODRÍGUEZ GALDO, María Xosé
Señores y campesinos en Galicia. - Santiago de Compostela : Editorial Pico Sacro, 1976
36p : ill. - (Biblioteca de bolsillo ; no 9). - Includes bibliographies
84-85170-12-1 X.708/46134
2040174

AGRICULTURAL COMMUNITIES. Rural regions. Mexico
Political aspects, 1910–1970

CACIQUISMO y poder político en el México rural / por Roger Bartra [et al. ; edición al cuidado de Eugenia Huerta]. - México : Siglo Vientiuno Editores, 1975
vii, 203 p., 1 fold. leaf of plates. - (Sociología y política)
r6011677 X.709/20879
0381651

AGRICULTURAL COMMUNITIES. Rural regions. Peru
Organisation

HURTADO GUTIÉRREZ, Hugo
Formación de las comunidades campesinas en el Perú. - Lima : Editorial Tercer Mundo, 1974
121 p.. - Bibliography : p. 113–120
r6005126 X.709/19662
0007553

MARTÍNEZ ALIER, Juan
Los huacchilleros del Perú. - [Paris] : Ruedo Ibérico, [1973]
vii,100p
r6085060 X.519/40172
0007553

AGRICULTURAL COMMUNITIES. Rural regions. Peru
Peasants

ESCALANTE ALVAREZ, Elsa C.
Personalización del campesino y trabajo social. - Lima : Pontificia Universidad Católica del Perú, 1974
187 p., 14 leaves of plates. - (Colección Servicio social)
r6008825 X.510/9303
0910740

AGRICULTURAL COMMUNITIES. Rural regions. Peru
Social conditions

MARTÍNEZ ALIER, Juan
Los huacchilleros del Perú. - Lima : Instituto de Estudios Peruanos ; Paris : Ruedo Ibérico, [1973]
viii, 100 p.. - (Colección Mínima ; no. 3)
r6011693 X.319/8515 Woolwich
0881619

AGRICULTURAL COMMUNITIES. Rural regions. Savelinia. Italy
Peasants. Socioeconomic aspects

ANGIONI, Giulio
Rapporti di produzione e cultura subalterna. - Cagliari : Editrice Democratica Sarda, 1974
326 p.. - (Saggi ; n. 3). - bibl. : p. 313–317
r6014216 X.709/20676
0869287

AGRICULTURAL COMMUNITIES. Russia
Communes, to 1917

PUSHKAREV, Sergeĭ Germanovich
Krest'ianskaia pozemel'no–peredel'naia obshchina v Rossii. - Newtonville, Mass., : Oriental Research Partners, 1976
161p in various pagings. - Bibliography
0-89250-101-4 X.622/14099
887736x

AGRICULTURAL COMMUNITIES. Sardinia. Italy
Social conditions

ANGIONI, Giulio
Sa laurera. - Cagliari : Editrice democratica sarda, 1976
292p : ill. - (Saggi ; 6)
r6100423 X.319/20067 Woolwich
2117002

AGRICULTURAL COMMUNITIES. Sardinia. Italy
Social conditions, ca 1944–1974

LEDDA, Gavino
Padre padrone. - 6. ed.. - Milano : Giangiacomo Feltrinelli, 1975
226 p.. - (Franchnarratori ; n. 19)
r6006512 X.908/33544
0929077

AGRICULTURAL COMMUNITIES. Siberia. Russia RSFSR
Communes

KRESTʹĪANSKAĪA obshchina v Sibiri XVII–nachala XXv. / [redaktsionnaīa kollegiīa L.M. Goriushkin (otv. redaktor)...i dr.]. - Novosibirsk : Izdatel'stvo "Nauka", Sibirskoe otdelenie, 1977
284p
r6076091 X.329/8854 Woolwich
1789775

AGRICULTURAL COMMUNITIES. Sicily. Italy
Social conditions

BATTAGLIA, Aristide
L'evoluzione sociale in rapporto alla proprietà fondiaria in Sicilia. - [Palermo] : Edizioni della Regione siciliana, 1974
312p. - (Edizioni della Regione siciliana ; 22)
r6100160 X.311/5045 Woolwich
2036509

AGRICULTURAL COMMUNITIES. South Australia
Settlement, 1869–1884

MEINIG, Donald William
On the margins of the good earth. - Adelaide : Rigby, 1976
xiv,231p,[16]p of plates : ill. - (Seal books). - Bibliography : p.223–224. - Includes index
0-85179-046-1 X.708/21835
1843664

AGRICULTURAL COMMUNITIES. South-east Asia
Socioeconomic aspects

SCOTT, James Cameron
The moral economy of the peasant. - New Haven ; London : Yale University Press, 1976 [i.e. 1977]
ix,246p : ill. - Includes index
0-300-01862-2 X.520/11641
1091794

The moral economy of the peasant. - New Haven ; London : Yale University Press, 1976 [i.e. 1977]
ix,246p : ill. - Includes index
0-300-01862-2 X.520/11641
1091808

AGRICULTURAL COMMUNITIES. South-western Europe to ca B.C.2000. Archaeological sources

PHILLIPS, Patricia
Early farmers of West Mediterranean Europe. - London : Hutchinson, 1975
176p,[4]p of plates : ill. - (Hutchinson university library). - Bibliography: p.156–172. - Includes index
0-09-123971-0 WP.1413/432
0534978

AGRICULTURAL COMMUNITIES. South-western Wisconsin
Social conditions, ca 1920–1936

LOGAN, Ben
The land remembers. - Large print [ed.]. - Boston, Mass. : Hall ; London : George Prior Publishers, 1975
[12],527p : ill
0-904000-27-3 X.800/25202
0516007

AGRICULTURAL COMMUNITIES. Southern Sweden
Poverty, 1800–1900

SÖDERBERG, Johan
Agrar fattigdom i Sydsverige under 1800–talet. - Stockholm : Distr. Almqvist & Wiksell International, 1978
217p : charts,maps. - (Stockholm studies in economic history; 4). - Bibl
91-220-0190-5 X.0510/262(4)
2427893

AGRICULTURAL COMMUNITIES. Spain
Social conditions

PIÑERO, José Antonio
El campo. - Madrid : Emiliano Escolar, editor, [1977]
135p. - (Tribuna popular. Serie Democracia para el pueblo ; 4)
84-7393-045-2 X.519/42312
2164485

AGRICULTURAL COMMUNITIES. Swale (District). Kent
Social conditions, ca 1930–ca 1945

KENT, Joan
Binder twine & rabbit stew. - Folkestone : Bailey and Swinfen, 1976
[10],149p : ill
0-561-00294-0 X.529/30912
0714313

Binder twine and rabbit stew. - London : Coronet, 1978
[10],149p : ill
0-340-22334-0 X.708/21192
0714313

KENT, Joan
Hay wains & cherry ale. - Folkestone : Bailey and Swinfen, 1980
[6],153p : ill
0-561-00302-5 X.989/88373
0714313

AGRICULTURAL COMMUNITIES. Sweden
Unmarried mothers, 1800–1900

FRYKMAN, Jonas
Horan i bondesamhället. - Lund : Liber Läromedel, 1977
237p. - Includes bibliographies
91-40-04432-7 X.519/24252
1394789

AGRICULTURAL COMMUNITIES. Tamil Nadu (State). India (Republic)
Social conflict, 1967–1968

ALEXANDER, K. C
Agrarian tension in Thanjavur. - Hyderabad : National Institute of Community Development, 1975
192p.
r6028599 X.320/5248 Woolwich
1020358

AGRICULTURAL COMMUNITIES. Texas
German immigrant agricultural communities, ca 1830–ca 1885

JORDAN, Terry Gilbert
German seed in Texas soil. - Austin ; London : University of Texas Press, 1975 [i.e. 1976]
xiv,237p,[8]p of plates : ill. - Bibliography: p.209–227. - Includes index
0-292-72707-0 X.519/15294
0744905

AGRICULTURAL COMMUNITIES. Thailand
to date

COOPER, Robert G.
Patterns of work organization in a situation of agricultural transition. - Singapore : Institute of Southeast Asian studies, [1981?]
vii,28p. - (ISEAS Occasional Paper ; no.62). - Bibliography
r6804545 X.0702/265(62)
8773904

AGRICULTURAL COMMUNITIES. Trondelag. Norway
Social conditions, 1854–1876

BRUNSTAD, Ingebrigt
Dagbok 1854–1876. – [Steinkjer] : Nord-Trøndelag historielag, 1978
271p : ill. – (Kulturminne i Nord-Trøndelag ; 1)
r6061194 X.0329/105 Woolwich
2190532

AGRICULTURAL COMMUNITIES. Tropical regions
Poverty, 1947-1974 related to economic development, 1947-1974

HASWELL, Margaret Rosary
The nature of poverty. – London : Macmillan, 1975
xvii,234p : ill. – Bibliography: p.221–224. – Includes index
0–333–17432–1 X.529/30325
0560936

AGRICULTURAL COMMUNITIES. Valle di Senales. Italy
Social conditions

KEIM, Maike
Schnals. – Bozen : Athesia, 1975
234 p. : ill. – (Schriftenreihe des Südtiroler Kulturinstitutes ; Bd 2). – Bibliography : p. 199–205
r6002034 X:0708/591(2)
0916765

AGRICULTURAL COMMUNITIES. Vietnam
Communes. Social conditions, 1975–1983

HOUTART, François
Hai Van. – London : Zed, 1984
207p : ill. – (Third World studies) (Asia. Political economy). – Includes index
0–86232–234–0 X.529/68250
3023931

AGRICULTURAL COMMUNITIES. Western Ireland
Cultural processes – Conference proceedings

RURAL development in the west of Ireland / editor Proinnsias Breathnach. – Maynooth : Geography Department, St. Patrick's College, 1983
iii,87p. – (Occasional paper series ; no.3)
0–901519–25–1 X.322/13725
2315742

AGRICULTURAL COMMUNITIES. Western Uttar Pradesh. India (Republic)
Merchants

LONDON, Paul A.
Merchants as promoters of rural development. – New York ; London : Praeger ; [London] : [Distributed by Martin Robertson], 1975
xiii,160p. – (Praeger special studies in international economics and development). – Bibliography: p.145–154. – Includes index
0–275–05430–6 X.520/10898
069603x

AGRICULTURAL COOPERATIVE

DESHPANDE, S. H.
Some problems of co-operative farming. – Bombay : Himalaya Publishing House, 1977
196p
r6093149 X.329/12122 Woolwich
1081608

AGRICULTURAL COOPERATIVE
– Conference proceedings

ASIAN TOP-LEVEL COOPERATIVE LEADERS' CONFERENCE (1973 : Tokyo)
Report of the Asian Top-Level Cooperative Leaders' Conference [held in] Tokyo (Japan), October 25–27, 1973. – New Delhi : International Cooperative Alliance, [1974]
[6], 425 p., [8] p. of plates : ill
r6031557 X.529/31346
1081608

CO-OPERATIVE RESEARCH SEMINAR (2nd : 1977 : Wolfson College)
Second Co-operative Research Seminar, Wolfson College, Oxford, 22nd-24th September 1977. – Oxford : Plunkett Foundation for Co-operative Studies, [1978]
101p in various pagings : 2 ill. – (Research reports / Plunkett Foundation for Co-operative Studies). – Includes bibliographies
B79–31591 X.512/6915
1081608

AGRICULTURAL COOPERATIVE
Role in economic development

GARCÍA, Antonio
Cooperación agraria y estrategias de desarrollo. – México : Siglo Veintiuno, 1976
291p. – (Economía y demografía)
r6085786 X.519/40493
1975234

AGRICULTURAL COOPERATIVE MOVEMENTS. Ireland
to 1975

BOLGER, Patrick
The Irish co-operative movement. – Dublin : Institute of Public Administration, 1977
iii–xiv,434p : ill. – Includes index
0–902173–75–8 X.700/25493
1192329

AGRICULTURAL COOPERATIVES
See also
Moshavs

AGRICULTURAL COOPERATIVES

SARGENT, M. J. (Malcolm James)
Agricultural co-operation. – Aldershot : Gower, c1982
ix,156p : ill. – Bibliography: p135–149. – Includes index
0–566–00460–7 X.329/15876
2822164

AGRICULTURAL COOPERATIVES
Communication

SMITH, Dudley
Communications in co-operatives. – [London] : [Central Council for Agricultural and Horticultural Co-operation], [1976]
35p : ill
B84–11839 X.529/59131
3046346

AGRICULTURAL COOPERATIVES
Financial assistance by governments, to 1972 – Conference proceedings

DAVIDOVIC, George-Krikor
State financing of co-operatives leading to self-financing. – [Oxford] : [Plunkett Foundation for Co-operative Studies], [1973]
54p. – Bibliography: p.53–54
B76–28808 X.319/16749 Woolwich
0720755

AGRICULTURAL COOPERATIVES
Role in economic development

SUDARSKY, John
Cooperativas rurales, política clientelista y cambio social. – Bogotá : Andina, 1977
xvii,92p. – Includes bibliographies
r6099221 X.519/40883
1973975

AGRICULTURAL COOPERATIVES. Alentejo. Portugal

FISCHER, Regina
New life in Portugal. – London : Kay Beauchamp, 1976
36p : ill
0–9504283–2–9 X.709/30575
0681636

MORAIS, Jorge
Alentejo saqueado. – Lisboa : Perspectivas & Realidades, 1976
182p
r6025168 X.708/20070
0681636

NAVARRO, António Modesto
Das árvores mortas à reforma agrária : reportagem. – Lisboa : Prelo, 1976
79 p., 4 p. of plates. – (Documentos)
r6027438 X.319/8584 Woolwich
0681636

SILVA, Antunes da
Terras velhas semeadas de novo. – Amadora : Livraria Bertrand, 1976
232p.,8p. of plates. – (Documentos de todos os tempos)
r6047223 X.519/22618
0681636

AGRICULTURAL COOPERATIVES. Aragon. Spain
1936–1937

SOUCHY, Agustín
With the peasants of Aragon. – Orkney : Cienfuegos, 1982
145p
B83–25723 X.809/55226
2893738

AGRICULTURAL COOPERATIVES. Balmazújváros. Hungary
Balmazújvárosi Vörös Csillag Termelöszövetkezet, 1949–1974

NEGYEDSZÁZAD a balmazújvárosi Vörös Csillagban / [szerkesztette Csige József]. – Debrecen : [s.n.], 1975
147p : ill
r6000640 X.311/4592 Woolwich
0863831

AGRICULTURAL COOPERATIVES. Bangladesh
– Conference proceedings

WORKSHOP ON CO-OPERATIVE FARMING (1972 : Bangladesh Academy for Rural Development)
Co-operative farming in Bangladesh. – Comilla : Bangladesh Academy for Rural Development, 1974
60p in various pagings (2 fold)
r6072414 X.311/8887 Woolwich
169264X

AGRICULTURAL COOPERATIVES. Bangladesh
– Serials

BANGLADESH PALLI UNNAYAN SANGSTHA
Monthly report. – [no.1]–. – Dacca : Integrated Rural Development Programme, Statistics, Research and Evaluation Division, [1973?]-
r6072358 P.523/844
1684728

AGRICULTURAL COOPERATIVES. Barcouço. Portugal

SALVADOR, José A.
Camponeses de Barcouço. – Coimbra : Centelha, 1977
178p. – (Colecção O que faz falta ; n.2)
r6064940 X.319/9249 Woolwich
1424335

AGRICULTURAL COOPERATIVES. Bolivia
to date

GARCÍA GARCÍA, Facundo
Problemas del cooperativismo agropecuario ejemplarizado en las cooperativas del Altiplano Central. – La Paz : Universidad Católica Boliviana, Departamento de Estudios Cooperativos, [1979]
xv,192p. – (Serie Estudios ; no.5)
r6704010 X.322/10607 Woolwich
8807019

AGRICULTURAL COOPERATIVES. Bolivia
1952–

IRIARTE, Gregorio
El cooperativismo y la comunidad indígena. – La Paz : Ediciones Puerta del Sol, 1979
97p. – (Colección Luces y sombras ; 2)
r6702305 X.329/14565 Woolwich
8757623

AGRICULTURAL COOPERATIVES. Budapest. Hungary
"Sasad" Kertészeti és Mezögazdasági Termelöszövetkezet

25 év a sasadi dombokon / [szerkesztés, fotók : Kő Tamás]. – [Budapest] : [s.n.], [1975]
40 p. : ill
r6007963 X.319/8474 Woolwich
0863858

AGRICULTURAL COOPERATIVES. Catalonia. Spain
Buildings. Architecture

MARTINELL BRUNET, Cèsar
Construcciones agrarias en Cataluña. – Barcelona : La Gaya Ciencia, [1975]
124p : ill. – (Publicaciones del Colegio de Arquitectos de Cataluña y Baleares. Estudios históricos)
84–7080–419–7 X.319/8661 Woolwich
0387614

AGRICULTURAL COOPERATIVES. Cered. Hungary
1950–1975

SÓTÉR, István
Huszonöt év a közösben. – [Cered] : [s.n.], 1975
51 p. : ill
r6007910 X.319/8473 Woolwich
086384X

AGRICULTURAL COOPERATIVES. Colombia
1969–1974

Las EMPRESAS comunitarias campesinas. – Bogotá : Centro de Investigación y Acción Social, 1975
177p. – Bibliography : p. 169–177
r6037717 X.809/24401
1154427

AGRICULTURAL COOPERATIVES. Commonwealth countries
Law

SURRIDGE, Brewster Joseph
A manual of co-operative law and practice. – 3rd ed. – Oxford : Plunkett Foundation for Co-operative Studies, 1972
x,274p. – Bibliography: p.264–266. – Includes index
B79–31635 X.100/18612
1951416

AGRICULTURAL COOPERATIVES. Czechoslovakia Law, 1945-

HUŇÁČEK, Zdeněk
Rechtliche Aspekte der Kooperation in der Landwirtschaft der ČSSR und der DDR. - Berlin : In Kommission bei Duncker & Humblot, 1977
136p. - (Osteuropastudien der Hochschulen des Landes Hessen. Reihe 1, Giessener Abhandlungen zur Agrar- und Wirtschaftsforschung des europäischen Ostens ; Bd 83). - Includes bibliographies
3-428-03915-7 X.100/26717
8803307
- [Another copy.] X.100/26717

AGRICULTURAL COOPERATIVES. Danszentmiklos. Hungary

GALI, Sándor
Negyedszázad a homokon. - [Vác] : Micsurin Tsz vezetősége és pártszervezete, [1972]
200p : ill
r6044797 X.320/5120 Woolwich
133039X

AGRICULTURAL COOPERATIVES. Denmark

BJERRE, Poul
Andelssamfundet. - [København] : Gyldendal, 1979
187p
87-01-77311-9 X.319/20287 Woolwich
2136856

AGRICULTURAL COOPERATIVES. Developing countries

SURRIDGE, Brewster Joseph
Co-operative thrift, credit, marketing and supply in developing countries. - Revised [ed.]. - [Oxford] : [Plunkett Foundation for Co-operative Studies], 1978
[5],ii,88p. - (Study series / Plunkett Foundation for Co-operative Studies ; 1)
0-85042-018-0 X.519/29684
189126x

AGRICULTURAL COOPERATIVES. Developing countries
Proposed agricultural cooperatives. Members. Education

INTERNATIONAL CO-OPERATIVE ALLIANCE. Co-operative Educational Materials Advisory Service
Member education, a campaign plan for Tinsa Valley. - London : The Alliance, 1980
[6],55[i.e. 109]p
0-904380-46-7 X.525/4187
2314681

AGRICULTURAL COOPERATIVES. East Germany Law, 1945-

HUŇÁČEK, Zdeněk
Rechtliche Aspekte der Kooperation in der Landwirtschaft der ČSSR und der DDR. - Berlin : In Kommission bei Duncker & Humblot, 1977
136p. - (Osteuropastudien der Hochschulen des Landes Hessen. Reihe 1, Giessener Abhandlungen zur Agrar- und Wirtschaftsforschung des europäischen Ostens ; Bd 83). - Includes bibliographies
3-428-03915-7 X.100/26717
8803315
- [Another copy.] X.100/26717

AGRICULTURAL COOPERATIVES. Eastern Europe & Soviet Union
Regional agricultural cooperatives

BONDARS, Arvid Rikhardovich
Raionnye agrarno-promyshlennye ob'edineniia. - Riga : Izdatel'stvo "Zinatne", 1976
177p
r6044825 X.329/8699 Woolwich
1322923

AGRICULTURAL COOPERATIVES. Estonia
Ekbaum, Artur - Biographies

ÜHISTEGEVUS on päike... / [koostanud ja toimetanud Salme Ekbaum]. - [Toronto] : Estoprint, 1977
205p : ill
r6021988 X.320/5128 Woolwich
1416715

AGRICULTURAL COOPERATIVES. Europe
1965-1975 - Bibliographies

COMMONWEALTH AGRICULTURAL BUREAUX
Agricultural co-operatives in Europe. - Slough : C.A.B. ; Oxford : The Foundation, 1976
[4],150,6p. - (Annotated bibliography ; no. CP1). - Includes index
0-85042-012-1 UNP.266/74
1754556

AGRICULTURAL COOPERATIVES. France
Law, to date

BASILE, Eva Rook
La coltivazione dei terreni in società nell'esperienza guiridica italiana e francese. - Milano : Giuffrè, [1981]
174p
r6512609 X.200/46126
9166130

AGRICULTURAL COOPERATIVES. Füzesgyarmat. Hungary

25 éves a Füzesgyarmati Vörös Csillag Termelőszövetkezet 1949-1974. - [Füzesgyarmat] : [s.n.], [1974]
[15] p. : ill
r6007929 X.319/8557 Woolwich
0892955

AGRICULTURAL COOPERATIVES. Great Britain

CENTRAL COUNCIL FOR AGRICULTURAL AND HORTICULTURAL CO-OPERATION
Benefits of agricultural co-operation. - London : CCAHC, [1979?]
[2]leaves,28p
B80-38962 X.329/13106 Woolwich
2303647

MORLEY, J. A. E. (John Arthur Elwell)
British agricultural cooperatives. - London : Hutchinson, 1975
168p. - Includes index
0-09-123960-5 X.329/10040
0355801

AGRICULTURAL COOPERATIVES. Great Britain
- Directories

DIRECTORY of agricultural co-operatives in the United Kingdom / [Plunkett Foundation for Co-operative Studies]. - 1978. - Oxford : The Foundation, 1978
[1],114,xiv p
Subsequent volumes, when received, will be placed at this shelfmark
0-85042-011-3 P.311/743
1522752

AGRICULTURAL COOPERATIVES. Great Britain
- Directories - Serials

DIRECTORY of agricultural, horticultural and fishery co-operatives in the United Kingdom. - 1982-. - Oxford : Plunkett Foundation for Co-operative Studies, 1982-
v.
Holdings: 1983-
B84-18275 P.311/743
1522752

AGRICULTURAL COOPERATIVES. Great Britain
Finance

ABSALOM, W. L. G.
A directors' guide to financing marketing co-operatives. - London : Central Council for Agricultural and Horticultural Co-operation, [1981]
48p : ill
B81-37344 X.529/43685
2558955

AGRICULTURAL COOPERATIVES. Great Britain
Law

CENTRAL COUNCIL FOR AGRICULTURAL AND HORTICULTURAL CO-OPERATION
Guide to agricultural co-operative legislation. - 2nd ed. - [London] : [The Council], [1979]
1v.. - Includes index
B79-20776 LR.425/319
1738488

Guide to agricultural cooperative legislation. - [London] : [The Council], [1977]
1v.. - Includes index
B78-40096 X.205/512
1738488

CENTRAL COUNCIL FOR AGRICULTURAL AND HORTICULTURAL COOPERATION
A directors' manual of cooperative law. - London : The Council, 1977
56,[1]p : ill. - Bibliography: p.50. - Includes index
B78-08889 X.100/17361
1350420

GUIDE to agricultural co-operative legislation. - 3rd ed. - [London] : [Central Council for Agricultural and Horticultural Co-operation], [1982]
1v.(various pagings). - Includes index
B83-15233 L.45/1916
1738488

AGRICULTURAL COOPERATIVES. Great Britain
Membership

KIMBLE, Helen
Effective membership of agricultural co-operatives. - [Oxford] : Plunkett Foundation for Co-operative Studies, [1977]
[2],24 leaves. - (Research reports / Plunkett Foundation for Co-operative Studies). - Bibliography: leaf 17
B80-02343 X.512/7113
2007320

AGRICULTURAL COOPERATIVES. Greece
to date

AVDELIDES, Parmenion S.
Synetairistika provlemata. - Athena : Nea synora, 1981
179p. - (Politike, oikonomia, koinoniologia)
r6503912 X.329/16676
8982546

AGRICULTURAL COOPERATIVES. Hajdú-Bihar (megye). Hungary

DOKUMENTUMOK a termelőszövetkezeti mozgalom Hajdú-Bihar megyei történetéhez ... / szerkeszti Gazdag István ; [a kötet munkatársai Gazdag István et al.]. - Debrecen : Hajdú-Bihar megyei Levéltár, 1977
199p. - (A Hajdú-Bihar megyei Levéltár Közleményei ; 10)
r6094686 P701/631[sz.10]
2038153

AGRICULTURAL COOPERATIVES. Honduras

ASTORGA LIRA, Enrique
Evaluación de los asentamientos y cooperativas campesinas en Honduras. - 2.ed. revisada. - Colonia Alameda, [Honduras] : [s.n.], 1975
76,[15p] : ill
r6064976 X.702/3555
150200X

MENCIA, Miguel Angel
Organización y funcionamiento de las empresas campesinas comunitarias de Honduras. - Tegucigalpa : Instituto Nacional Agrario, Departamento de Promoción y Organización Campesina, 1977
25leaves
r6160779 X.525/3863
150200X

AGRICULTURAL COOPERATIVES. Hungary

BALATINÁCZ, Jeromos
Hat falu a közös úton.... - [Belvárdgyula] : Belvárdgyulai Közös Út Termelőszövetkezet, 1972
208p
r6056279 X.320/5150 Woolwich
1000217

ERDEI, Ferenc
Szövetkezeti írások .... - Budapest : Akadémiai Kiadó, 1979
2v.. - (Erdei Ferenc összegűjtött művei)
963-05-0028-0 X.0320/87(10)
1000217

FAZEKAS, Béla
A mezőgazdasági termelőszövetkezeti mozgalom Magyarországon. - Budapest : Kossuth Könyvkiadó, 1976
314p
963-09-0643-0 X.319/8858 Woolwich
1000217

RIDEG, Margit
A mezőgazdasági termelőszövetkezetek és a cukoripari vállalatok kapcsolatai. - Budapest : [s.n.], 1974
50p. - (Szövetkezeti Kutató Intézet. Közlemények ; 106)
r6056312 X.0512/379(106)
1000217

SZAKÁCS, Sándor
A földmüvesszövetkezeti mozgalom története 1945-1949. - [Budapest] : Szövetkezeti Kutató Intézet, 1976
127p. - (Szövetkezeti Kutató Intézet ; közlemények ; 115)
r6075011 X.0512/379(115)
1000217

SZÖVETKEZETPOLITIKAI kérdések, 1976 / [sajtó alá rendezte és szerkesztette Szlameniczky István, Zsuffa Ervin...Szaszkó László]. - Budapest : Kossuth Könyvkiadó, 1976
135p
963-09-0809-3 X.708/19731
1000217

AGRICULTURAL COOPERATIVES. Hungary
Dunarecsei Béke Mezőgazdasági Termelőszövetkezet, 1950-1975

NASZÁLY, Sándor
A dunavecsei Béke Mezőgazdasági Termelőszövetkezet (a volt Virágzó és Új Élet) története. 1950-1975. - [Dunavecse] : [s.n.], [1975]
80 p., 11 p. of plates
963-02-0149-6 X.319/8475 Woolwich
0863823

AGRICULTURAL COOPERATIVES. Hungary
Finance

SZAKÁL, Pál
A hitel szerepe a mezőgazdasági termelőszövetkezetek fejlődésében 1968-1974 között. - Budapest : [s.n.], 1976
74p. - (Szövetkezeti Kutató Intézet. Közlemények ; 109)
r6056310 X.0512/379(109)
1541668

AGRICULTURAL COOPERATIVES. Hungary
Law

HUNGARY
A mezőgazdasági termelőszövetkezeti jogszabályok. - Budapest : Közgazdasági és Jogi Könyvkiadó, 1978
471p
963-220-586-3 S.183/6
2471221

AGRICULTURAL COOPERATIVES. Hungary
Leadership

LACZKÓ, István
Az elnöki munkakör tartalma és struktúrája a mezőgazdasági termelőszövetkezetekben. - Budapest, 1976
113p. - (Közlemények / Szövetkezeti Kutató Intézet ; 118)
r6183778 X.0512/379(118)
2469774

AGRICULTURAL COOPERATIVES. Hungary
Mencshelyi "Balatonfelvidék" Mezogazdasági Termeloszövetkezet, 1959-1974

KAZAL, Gábor
Tizenöt év. - [Mencshely] : [s.n.], [1974]
103 p., 12 p. of plates : ill
r6007674 X.708/17922
0943185

AGRICULTURAL COOPERATIVES. Hungary
Names

SOLYMÁR, Imre
Mezőgazdasági termelőszövetkezetek nevei Magyarországon. - Budapest : ELTE Magyar Nyelvészeti Tanszékcsoport Névkutató Munkaközössége, 1981
52p. - (Magyar névtani dolgozatok ; 10)
r6449100 X.0529/650(10)
8753628

AGRICULTURAL COOPERATIVES. Hunya. Hungary

TERMELŐSZÖVETKEZETÜNK 25 éve. - [Hunya] : [s.n.], [1974]
36 p. : ill
r6007927 X.319/8558 Woolwich
0892939

AGRICULTURAL COOPERATIVES. India

DADHICH, C. L.
Overdues in farm co-operative credit. - Bombay : Popular Prakashan, 1977
258p
r6093084 X.529/34462
0478121

RESERVE BANK OF INDIA. Committee on Integration of Co-operative Credit Institutions
Report of the Committee on Integration of Co-operative Credit Institutions. - Bombay : Agricultural Refinance and Development Corporation, 1976
260p
r6201089 X.520/15135
0478121

AGRICULTURAL COOPERATIVES. India (Republic)
Management

KRISHNASWAMI, O. R.
Co-operative democracy in action. - Bombay : Somaiya, 1976
xi,271p. - Bibliography : p.257-267. - Includes index
r6071959 X.529/32267
1447807

AGRICULTURAL COOPERATIVES. Ireland
Role in economic development of sparsely populated regions - Conference proceedings

CO-OPERATION and community development / edited by John Sewel and Diarmuid Ó Cearbhaill. - [Galway] : Social Sciences Research Centre, University College Galway in association with the Institute for the Study of Sparsely Populated Areas, University of Aberdeen, 1982
167p
0-903299-02-x X.525/7057
2886332

AGRICULTURAL COOPERATIVES. Ireland (Republic)
Lough Egish Cooperative Agricultural and Dairy Society - Serials

LOUGH Egish news / [Lough Egish Co-operative Agricultural and Dairy Society]. - [No.1]- ; June 1973-. - [Eire] : Farmlink, [1973]- : ill
r6032523 P.313/430
1332376

AGRICULTURAL COOPERATIVES. Italy
Cooperativa agricola Il Forteto

SAM, Maria ... il giardino della verità / a cura di Luigi Goffredi. - [Barberino di Mugello] : Cooperativa agricola Il Forteto, 1981
229p
r6511169 X.329/19139
9166122

AGRICULTURAL COOPERATIVES. Italy
Law, to date

BASILE, Eva Rook
La coltivazione dei terreni in società nell'esperienza guiridica italiana e francese. - Milano : Giuffrè, [1981]
174p
r6512609 X.200/46126
9166149

AGRICULTURAL COOPERATIVES. Kalasin. Thailand
Role in development of agricultural industries

DEMAINE, Harvey
The role of farmers' associations in agricultural development. - London : University of London School of Oriental and African Studies, Department of Geography, 1976
[2]p,21 leaves. - (Occasional papers / University of London. School of Oriental and African Studies. Department of Geography)
0-7286-0021-8 X.0702/1007(1)
0608122

AGRICULTURAL COOPERATIVES. Kenya
to date

BAGER, Torben
Marketing cooperatives and peasants in Kenya. - Uppsala : Scandinavian Institute of African Studies, 1980
116p : ill. - (Publications from the Centre for Development Research, Copenhagen ; 5)
91-7106-174-6 X.0519/814(5)
8857598

AGRICULTURAL COOPERATIVES. Kiskunfélegyháza. Hungary

TÓTH, Miklós
A kiskunfélegyházi Dózsa Mg. Termelőszövetkezet 25 éves krónikája. - Kiskunfélegyháza : kiskunfélegyházi Dózsa Mg. Termelőszövetkezet, 1974
190p : ill
r6499045 X.311/5203 Woolwich
2472201

AGRICULTURAL COOPERATIVES. Latin America

FALS BORDA, Orlando
[Cooperatives and rural development in Latin America. Spanish]. El reformismo por dentro en América Latina. - 2a ed. - México : Siglo Veintiuno, 1974
215p. - (Colección mínima ; n.48)
r6085135 X.319/9855 Woolwich
1863266

AGRICULTURAL COOPERATIVES. Latin America
- Conference proceedings

REUNIÓN INTERAMERICANA DE EXPERTOS SOBRE EMPRESAS COMUNITARIAS CAMPESINAS (1973 : Santiago de Chile)
Reunión interamericana de expertos sobre empresas comunitarias campesinas, Abril 30- Mayo 5 de 1973, Santiago, Chile. - [Santiago de Chile?] : IICA ; CIRA, [1973?]
52p. - (Serie de reuniones, cursos y conferencias ; n.10)
r6065032 X.702/3585
1509217

AGRICULTURAL COOPERATIVES. Latin America
1945-

La EMPRESA comunitaria / José Emilio G. Araujo, editor. - San José, Costa Rica : IICA, Dirección General, División de Información Pública de la Subdirección General Adjunta para la Coordinación Externa, 1975
xix,387p. - (Serie Desarrollo institucional ; n.4)
r6701219 X.329/13945 Woolwich
8563357

AGRICULTURAL COOPERATIVES. Latin America
Role in economic development

GARCÍA, Antonio
Las cooperativas agrarias en el desarrollo de América Latina. - Bogotá : Ediciones Colatina, 1976
276p
r6041180 X.519/23398
1315145

AGRICULTURAL COOPERATIVES. Lökösháza. Hungary
1950-1975

BÉLA, Ottó
A sikerek munkával Születnek. - [Lökösháza] : [s.n], [1975]
130 p. : ill
963-02-0007-4 X.319/8476
0863432

AGRICULTURAL COOPERATIVES. Makó. Hungary

FORGÓ, István
Tagosítás Makón (1949-1952). - Makó : Makó város Tanácsa V.B. Művelődésügyi Osztálya, 1976
41p. - (A makói múzeum füzetei ; 17)
r6098768 X.0700/426(17)
2038412

AGRICULTURAL COOPERATIVES. Mexico
Policies of government, ca 1910-1975

AGUIRRE AVELLANEDA, Jerjes
La política ejidal en México. - México : Instituto Mexicano de Sociología, 1976
200p. - Includes bibliographies
r6066801 X.709/26713
1718932

AGRICULTURAL COOPERATIVES. Mitchelstown. Cork (County)
Mitchelstown Co-operative Agricultural Society - Serials

YOUR business. - [No.1] - ; Dec.1973-. - Mitchelstown : Mitchelstown Co-operative Agricultural Society, [1973]- : ill
r6072507 P.513/2264
1701541

AGRICULTURAL COOPERATIVES. Mongolia (People's Republic)
1945-

GRAĬVORONSKIĬ, Vladimir Viktorovich
Kooperirovannoe aratstvo MNR. - Moskva : Nauka, Glavnaia redaktsiia vostochnoi literatury, 1982
213p. - Includes bibliography
r6448486 X.329/17790
9072659

AGRICULTURAL COOPERATIVES. Mor. Hungary

BARÓTI, Dezső
Kortárs útlevelére. - Budapest : Szépirodalmi Könyvkiadó, 1977
504p : ill
963-15-0795-5 X.989/80968
1001582

GALI, Sándor
Szövetkezet az Ezerjó hazájában. - Mór : Kossuth Tsz vezetősége, 1975
169p,22p of plates
r6023861 X.709/22610
1001582

NEMES, István, 1915-
Radnóti Miklós költői nyelve .... - Budapest : Akadémiai Kiadó, 1979
310p. - (Irodalomtörténeti füzetek ; sz.94)
963-05-1679-9 WP.4992.a/94
1001582

POMOGÁTS, Béla
Radnóti Miklós. - Budapest : Gondolat Kiadó, 1977
228p[24]p of plates : ill. - (Nagy magyar írók)
963-280-495-3 X.908/83304
1001582

AGRICULTURAL COOPERATIVES. Nagykovacsi. Hungary
Nagykovacsi Vörös Hajnal Mezögazdasági Termelöszövetkezet, 1949-1974

SZATMÁRI, Lajos
Régi és új barázdák nyomában. - Budapest : Nagykovácsi Vörös Hajnal Mezőgazdasági Termelőszövetkezet, 1974
180 p. : ill
r6007846 X.700/16877
0858552

AGRICULTURAL COOPERATIVES. Nemesnádudvar. Hungary

A NEMESNÁDUDVARI "Kossuth" Termelőszövetkezet története (1949-1974). - [Nemesnádudvar] : A Nemesnádudvari "Kossuth" Termelőszövetkezet Vezetősege és Nőbizottsága, [1974]
58 p. : ill
r6007925 X.708/18581
0892920

AGRICULTURAL COOPERATIVES. North Korea
1948-1974

HISTORICAL experience of agricultural cooperation in our country. - Pyongyang : Foreign Languages Publishing House, 1975
[2] leaves, 172 p.
r6031357 X.319/17173 Woolwich
1070797

AGRICULTURAL COOPERATIVES. Orosháza. Hungary

ARY, Róza
Küzdelmes negyedszázad az orosházi Petőfi Tsz.-ben. - Orosháza : Petőfi Tsz., 1974
120p : ill
r6044768 X.319/9097 Woolwich
1416189

AGRICULTURAL COOPERATIVES. Peru

EGUREN LÓPEZ, Fernando
Reforma agraria, cooperativización y lucha campesina. - Lima : DESCO. Centro de Estudios y Promoción del Desarrollo, 1975
239p.,[1] fold. leaf of plates
r6053993 X.319/9399 Woolwich
1529153

AGRICULTURAL COOPERATIVES. Peru
Law

PERU
[Laws, statutes, etc]. Estatuto de comunidades campesinas (Decreto supremo n.37-70-A). - Lima : Mercurio, 1977
119p. - (Legislación peruana)
r6084530 X.108/26447
1791540

AGRICULTURAL COOPERATIVES. Peru
Management. Participation of peasants, 1968-1975. Political aspects

MCCLINTOCK, Cynthia
Peasant cooperatives and political change in Peru. - Princeton ; Guildford : Princeton University Press, c1981
xvii,418p. - Bibliography: p381-410. - Includes index
0-691-07627-8 X.520/24081
2494787

AGRICULTURAL COOPERATIVES. Peru
Members. Participation in politics, 1969-1975. Implications of control of management of agricultural cooperatives by members

MCCLINTOCK, Cynthia
Self-management and political participation in Peru, 1969-1975. - London : Sage Publications, 1977
64p. - (Sage professional papers in contemporary political sociology ; vol.2, no.06-022). - Bibliography: p.59-63
0-8039-9873-2 X.519/29206
1808974

AGRICULTURAL COOPERATIVES. Philippines

SACAY, Orlando J.
Samahang Nayon. - [Quezon City] : [National Publishing Co-operative], [1974]
142 p. : ill
r6025806 X.529/19926
102339X

AGRICULTURAL COOPERATIVES. Poland
1945-

JARECKA-KIMLOWSKA, Stanisława
Z problemow spółdzielczości wiejskiej w Polsce w latach 1944-1957. - Warszawa : Ludowa Spoldzielnia Wydawnicza 1977
175p
r6044399 X.319/8983 Woolwich
1233211

KOOPERACJA w polskim rolnictwie / pod redakcją naukową ... Czesława Farkowskiego. - Warszawa : Państwowe Wydawnictwo Rolnicze i Leśne, 1978
235p. - Bibliography : p232-236
r6097381 X.529/21336
1233211

AGRICULTURAL COOPERATIVES. Poland
Brzostowski, Karol - Biographies

BARTYŚ, Julian
Czerwony hrabia Karol Brzostowski. - Warszawa : Iskry, 1978
354p,[8]leaves of plates : ill
r6097473 X329/8917 Woolwich
2039184

AGRICULTURAL COOPERATIVES. Portugal

CAVACO, Carminda
Considerações gerais acerca de alguns tipos de cooperativas de empresários agrícolas, no Continente. - Lisboa : Centro de Estudos Geográficos, 1977
73 leaves.,25 leaves of plates
r6158923 X.313/964 Woolwich
1980130

A cooperação agrícola em Portugal. - Lisboa : [s.n.], 1977
79 leaves., 30 leaves of maps (1 fold.). - Includes bibliographies
r6066371 X.512/3146
1980130

AGRICULTURAL COOPERATIVES. Raheen. Clare (County)
Ralahine Agricultural and Manufacturing Co-operative Association, 1831-1833

CRAIG, Edward Thomas
An Irish commune. - Blackrock : Irish Academic, c1983
xv,208p. - (Cooperative studies ; 4)
0-7165-2349-3 X.809/56180
2937417

AGRICULTURAL COOPERATIVES. Rice production industries. Northern Peru

SCOTT, C. D.
Agrarian reform and agricultural labour markets. - Norwich : School of Development Studies, University of East Anglia, 1979
96,[8]p : ill. - (Development studies occasional paper ; no.2). - Bibliography: p90-92
B84-39936 X.322/14077
318403x

AGRICULTURAL COOPERATIVES. Romania

FULEA, Maria
Stimulente materiale şi morale în cooperativele agricole de producţie. - Bucureşti : Editura Academiei Republicii Socialiste România, 1977
199p
r6039194 X.319/9142 Woolwich
1330098

AGRICULTURAL COOPERATIVES. Rural regions. Developing countries
Role in economic development - Conference proceedings

CO-OPERATION AS AN INSTRUMENT FOR RURAL DEVELOPMENT (Conference : 1976 : University of Ghent)
Co-operation as an instrument for rural development. - London : International Co-operative Alliance, 1978
160p. - Includes bibliographies
0-904380-36-x X.319/18248 Woolwich
1523058

AGRICULTURAL COOPERATIVES. Serbia. Yugoslavia
1945-

ŠULJMANAC, Nedeljko
Razvoj zadruga, kooperacije u socijalistička transformacija sela. - Novi Sad : Matica srpska, Odeljenje za društvene nauke, 1975
317p. - Includes bibliography
r6496023 X.322/14777
915132x

AGRICULTURAL COOPERATIVES. South-west England
West of England Farmers Ltd, to 1977

SARGENT, M. J. (Malcolm James)
Agricultural cooperation in practice. - [Bath] : University of Bath, 1977
vi,211p : ill. - Bibliography: p.183-192
0-900843-85-3 X.519/29872
1918567

AGRICULTURAL COOPERATIVES. Southern Italy
1970-1975

CESARINI, Giuliano, of Bologna University
Rural production cooperatives in South Italy. - Langholm : Arkleton Trust, 1979
[2],iii,80p,6 leaves of plates
B79-24278 X.512/6687
1917943

AGRICULTURAL COOPERATIVES. Soviet Union
to 1975

ARKHIPOV, A.
How farmers' cooperatives were organized. - Moscow : Novosti Press Agency Publishing House, 1977
64p. - (The Soviet experience)
r6072346 X.708/21462
1687379

AGRICULTURAL COOPERATIVES. Swaziland
Attitudes of farmers, to date

GUMA, Xolile P.
Small former credit and small farmer attitudes towards co-operatives. - [Kwaluseni] : University of Swaziland, Social Science Research Unit, 1982
94p. - (Research paper / University of Swaziland, Social Science Research Unit ; no.8)
r6818738 X.322/14756
9166157

AGRICULTURAL COOPERATIVES. Sweden

SANCHEZ, Carlos C.
Agricultural co-operation in Sweden. - Oxford : Plunkett Foundation for Co-operative Studies, c1984
viii,132p : ill. - Bibliography: p121-123. - Includes index
0-85042-064-4 X.322/14587
3231631

AGRICULTURAL COOPERATIVES. Sweden
to date

MARKEN - ett arbetskooperativ / Christina Rahmn, Kenneth Karmaeus och medlemmar ur kooperativet Marken. - Stockholm : Liber förlag, [1981]
159p : ill
91-380-6107-4 X.629/18981
8883440

AGRICULTURAL COOPERATIVES. Tiszapöldvár. Hungary

BALÁZS, Árpád
A Lenin Tsz. Tiszaföldvárott. - Szolnok : Damjanich János Múzeum, 1970
251p,[16]p of plates : ill. - (A Damjanich János Muzeum közleményei, 24)
r6094630 X.0709/753(24)
203087X

AGRICULTURAL COOPERATIVES. Western Europe
1975-1980 - Bibliographies

AGRICULTURAL co-operatives in Western Europe. - Slough : Commonwealth Agricultural Bureaux, c1980
[36]p. - (Commonwealth Agricultural Bureaux annotated bibliographies. General series). - Includes index
B82-27995 SRL
2751941

AGRICULTURAL COOPERATIVES. Western Europe
Law - Conference proceedings

ISSERT, J.
The legal status of co-operatives in European countries. - [Oxford] : Plunkett Foundation for Co-operative Studies, 1978
[5],ii,112p. - (Occasional papers / Plunkett Foundation for Co-operative Studies ; no.44)
0-85042-017-2 X.319/18571 Woolwich
175839x

AGRICULTURAL COOPERATIVES. Wielkopolska. Poland

LAWNICZAK, Irena
Rolnicza spółdzielczość produkcyjna w Wielkopolsce w latach 1949-1974. - Warszawa : Państwowe Wydawnictwo Naukowe, 1978
146p : ill. - (Badania z dziejów społecznych i gospodargych ; nr.54)
r6075335 X.0200/68(54)
1533029

AGRICULTURAL COOPERATIVES. Zala (megye). Hungary

VASKA, Miklós
Húsz küzdelmes év. - Zalaegerszeg : Zala megyei Könyvtár, 1973
115p,[5]leaves of plates : ill. - (Göcseji Helikon. Honismereti közlemények ; 5. Mezőgazdasági üzemtörténet ; 1)
r6174247 P.801/1325(no.5)
2189038

AGRICULTURAL COOPERATIVES. Zselic. Hungary

PAKUTS, Károly
25 esztendő munkája a szocialista mezőgazdaságért a Zselicben. - Pécs : A szentlászlói "Búzakalász" mgtsz, 1975
113 p. : ill
r6023367 X.311/4638 Woolwich
0893684

AGRICULTURAL CREDIT. Agriculture. European Community countries

COMMISSION OF THE EUROPEAN COMMUNITIES
Credit to agriculture in the E.E.C. member states. - Luxembourg : Office for Official Publications of the European Communities ; [London] : [H.M.S.O.], 1977
[2],104p. - (Information on agriculture ; no.28)
B78-35902 SRL
1695746

CREDIT to agriculture. - Luxembourg : Office for Official Publications of the European Communities ; [London] : [H.M.S.O.].
4: The Netherlands. - 1976
[4],116p. - (Information on agriculture ; no.4)
B79-02434 UNP.131/181
1695746

AGRICULTURAL DEVELOPMENT AND ADVISORY SERVICE

AGRICULTURAL DEVELOPMENT AND ADVISORY SERVICE
Annual report. - 1977. - London : H.M.S.O, 1978
[4],101p,[8]p of plates : ill(some col)
Subsequent volumes, when received, will be found at SRL
0-11-240302-6 SRL
0728667

AGRICULTURAL DEVELOPMENT AND ADVISORY SERVICE
Policies

POLICY review of the Agricultural Development and Advisory Service. - Pinner : Ministry of Agriculture, Fisheries and Food, 1979
[3],35p
B80-31846 BS.3/660
2235366

AGRICULTURAL DEVELOPMENT AND ADVISORY SERVICE. Agricultural development projects. East Midlands. England

AGRICULTURAL DEVELOPMENT AND ADVISORY SERVICE. East Midland Region
Research and development, agriculture, summary report. - [Nottingham] : [A.D.A.S.].
1978. - [1979]
[1],xv,84p : 1 ill
Subsequent volumes, when received, will be found at SRL
B79-37362 SRL
1996053

AGRICULTURAL DEVELOPMENT AND ADVISORY SERVICE. Agricultural development projects. South-west England

AGRICULTURAL DEVELOPMENT AND ADVISORY SERVICE
Review of development work. - 1976. - Bristol : H.M.S.O. [1977]
[1],xv,149p
Subsequent volumes, when received, will be found at SRL
B79-20188 SRL
1896547

AGRICULTURAL DEVELOPMENT AND ADVISORY SERVICE. AGRICULTURE SERVICE

AGRICULTURE service annual report / Agricultural Development and Advisory Service. - 1975-76. - London : H.M.S.O, 1977
117p,[4]p of plates : ill
Later issues are placed at this shelfmark
0-11-241132-0 SRL
210766x
- [Another copy]

AGRICULTURAL DEVELOPMENT AND ADVISORY SERVICE. NATIONAL COLLECTION OF PLANT PATHOGENIC BACTERIA
See National Collection of Plant Pathogenic Bacteria

AGRICULTURAL DEVELOPMENT AND ADVISORY SERVICE. PLANT PATHOLOGY LABORATORY
See Plant Pathology Laboratory

AGRICULTURAL DEVELOPMENT AND ADVISORY SERVICE. PLANT PATHOLOGY LABORATORY. NATIONAL COLLECTION OF PLANT PATHOGENIC BACTERIA
See National Collection of Plant Pathogenic Bacteria

AGRICULTURAL DEVELOPMENT AND ADVISORY SERVICE. SCIENCE ARM

AGRICULTURAL DEVELOPMENT AND ADVISORY SERVICE. Science Service
Agricultural Development and Advisory Service Science Service annual report. - 1975. - London : H.M.S.O, 1976
ii-xxi,308p,iv p of plates : ill
Subsequent volumes, when received, will be placed at this shelfmark
0-11-241094-4 BS.3/599
0093645

REGIONAL AGRICULTURAL SCIENCE SERVICE
Regional Agricultural Science Service annual report. - London : H.M.S.O.
1976. - [1977]
xviii,238p,iv p of plates : ill. - Includes bibliographies
Subsequent volumes, when received, will be placed at this shelfmark
0-11-241124-x SRL
0093645

AGRICULTURAL DEVELOPMENT AND ADVISORY SERVICE. STOCKBRIDGE HOUSE EXPERIMENTAL HORTICULTURE STATION
See Stockbridge House Experimental Horticulture Station

AGRICULTURAL DEVELOPMENT AND ADVISORY SERVICE (WALES)
Research projects

AGRICULTURAL DEVELOPMENT AND ADVISORY SERVICE (WALES)
Development work. - [Pinner] : Ministry of Agriculture Fisheries and Food ; [Aberystwyth] : Welsh Office, Agriculture Department.
1978. - [1979]
[1],xiv,125p
B80-00775 BS.3/655(1)
1999222

AGRICULTURAL DEVELOPMENT PROJECTS. Developing countries
Management - Conference proceedings

INSTITUTIONS management and agricultural development. - London : Overseas Development Institute, 1979
[3],61p. - (Occasional papers / Overseas Development Institute. Agricultural Administration Unit ; 3). - Includes bibliographies
B80-27289 X.329/12720 Woolwich
2154064

AGRICULTURAL DEVELOPMENT PROJECTS. Developing countries
Role of foreign multinational companies, 1920-1979

VOLL, Sarah Potts
A plough in field arable. - Hanover, N.H. ; London : Published for University of New Hampshire by University Press of New England, 1980
x,213p. - Includes bibliographies and index
0-87451-186-0 X.325/909 Woolwich
2374781

AGRICULTURAL DEVELOPMENT PROJECTS. East Midlands. England
By Agricultural Development and Advisory Service

AGRICULTURAL DEVELOPMENT AND ADVISORY SERVICE. East Midland Region
Research and development, agriculture, summary report. - [Nottingham] : [A.D.A.S.].
1978. - [1979]
[1],xv,84p : 1 ill
Subsequent volumes, when received, will be found at SRL
B79-37362 SRL
1996053

AGRICULTURAL DEVELOPMENT PROJECTS. South-west England
By Agricultural Development and Advisory Service

AGRICULTURAL DEVELOPMENT AND ADVISORY SERVICE
Review of development work. - 1976. - Bristol : H.M.S.O. [1977]
[1],xv,149p
Subsequent volumes, when received, will be found at SRL
B79-20188 SRL
1896547

AGRICULTURAL ECONOMICS

DOLL, John P.
Production economics. - 2nd ed. - New York ; Chichester : Wiley, c1984
ix,470p : ill. - Includes bibliographies and index
0-471-87470-1 X.520/35081
3147320

AGRICULTURAL ECONOMICS
to date

DOLL, John P.
Production economics. - Columbus, Ohio : Grid. Inc., c1978
x,406p.. - (Grid series in agricultural economics)
0-88244-118-3 X.520/36865
9154620

AGRICULTURAL ECONOMICS. Great Britain
1900-1940

WHETHAM, Edith H.
Agricultural economists in Britain 1900-1940. - [Oxford] : Institute of Agricultural Economics, University of Oxford, c1981
vi,97p,[1]leaf of plates. - Includes index
0-900034-11-4 X.329/14372 Woolwich
2541785

AGRICULTURAL ECONOMICS. United States

SJO, John
Economics for agriculturalists. - New York ; Chichester : Wiley, c1976
xii,232p : ill. - Includes bibliographies and index
0-471-87003-x X.520/27570
2758709

WESSEL, James
Trading the future. - San Francisco : Institute for Food and Development Policy, c1983
xii,250p : ill. - Bibliography: p219-226. - Includes index
0-935028-13-7 X.529/66756
2758709

AGRICULTURAL ENGINEERING
See also
Field engineering. Agriculture

AGRICULTURAL ENGINEERING

SHINN, Glen C.
Working in agricultural mechanics. - New York ; London : McGraw-Hill, 1978
viii,280p : ill(chiefly col). - (Career preparation for agriculture/agribusiness). - Includes index
0-07-000843-4 X.311/8851 Woolwich
1565419

AGRICULTURAL ENGINEERING
- Serials

NATIONAL COLLEGE OF AGRICULTURAL ENGINEERING
NCAE news. - No.1- ; [Dec. 1978]-. - Silsoe : National College of Agricultural Engineering, 1978- : ill
B80-27202 P.313/490
2209802

AGRICULTURAL ENGINEERING. Great Britain
Organisations: Agricultural Engineers Association - Directories

AGRICULTURAL ENGINEERS ASSOCIATION
List of members. - [London] : [The Association].
1975. - 1975
8,2p,leaf
Subsequent issues, when received, will be placed at this shelfmark
0-901852-01-5 X.0313/503 Woolwich
0584649

AGRICULTURAL ENGINEERING. Great Britain
Organisations: Agricultural Engineers Association, to 1975

TROW-SMITH, Robert
Power on the land. - London : Agripress Publicity Ltd for the Association, 1975
3-93p : ill
0-903458-01-2 X.320/10297 Woolwich
0633666

AGRICULTURAL ENGINEERING. Scotland
Research organisations: Scottish Institute of Agricultural Engineering - Serials

SCOTTISH INSTITUTE OF AGRICULTURAL ENGINEERING
Biennial report. - 1974-1976-. - Penicuik : SIAE, 1976- v. : ill
B82-18530 SRL
2710935

AGRICULTURAL ENGINEERING INDUSTRIES. Great Britain
ca 1964-1974

ROTHWELL, Roy
Technical change and competitiveness in agricultural engineering. - Brighton : Science Policy Research Unit, University of Sussex, 1979
[5],124p : ill. - (Occasional paper series / University of Sussex. Science Policy Research Unit ; no.9)
0-903622-10-6 SRL
2086700
- [Another copy.] X.622/8346 Woolwich

AGRICULTURAL ENGINEERS ASSOCIATION
- Directories

AGRICULTURAL ENGINEERS ASSOCIATION
List of members. - [London] : [The Association].
1975. - 1975
8,2p,leaf
Subsequent issues, when received, will be placed at this shelfmark
0-901852-01-5 X.0313/503 Woolwich
0584649

AGRICULTURAL ENGINEERS ASSOCIATION
to 1975

TROW-SMITH, Robert
Power on the land. - London : Agripress Publicity Ltd for the Association, 1975
3-93p : ill
0-903458-01-2 X.320/10297 Woolwich
0633666

AGRICULTURAL EQUIPMENT
See also
Bee-keeping equipment
Crop equipment
Dairy equipment
Horticultural equipment
Scythes
Shepherding tools
Shepherds' crooks
Sprinkler systems. Irrigation equipment

AGRICULTURAL EQUIPMENT

HUWS, Emily
Ffermydd. - Llandysul : Gwasg Gomer, 1983
[20]p : col.ill. - (Nawr gellwch chi ddarllen a dysgu)
0-85088-958-8 X.622/16898
2962594

AGRICULTURAL EQUIPMENT
- Directories

AL-MA'RAḌ az-zirā'ī s-su'ūdī '83 = Saudi agriculture 83. - Riyadh : Al-Dhiafa Exhibition Services, [1983]
[83]p : ill(some col.)
r6863318 X.622/15848
291767x

AGRICULTURAL EQUIPMENT
- For developing countries

BOYD, John, 1946-
Tools for agriculture. - 2nd ed. - London : Intermediate Technology Publications Ltd, 1976
173p : ill
0-903031-22-1 X.313/1372 Woolwich
1058894

AGRICULTURAL EQUIPMENT
- Serials

FARM equipment international. - Vol.1, no.1 (1980)-. - Horley : Agraria Press, 1980-
v. : ill
B83-34770 P.623/1243
2204681

WHAT'S new in farming. - Vol.1, no.1- ; Oct. 1977-. - London : Morgan-Grampian, 1977- . ill
Holdings: Nov.1979, etc
B80-27203 P.2000/748
2204681

AGRICULTURAL EQUIPMENT
to 1975

BLANDFORD, Percy William
Old farm tools and machinery. - Newton Abbot : David and Charles, 1976
188p. - Bibliography: p.183-184. - Includes index
0-7153-7151-7 X.329/10515 Woolwich
0679070

AGRICULTURAL EQUIPMENT
Bellows operated oil drum forges & fan operated oil drum forges

OIL drum forges. - London : Intermediate Technology Publications Ltd, [1975]
[1],40p : ill. - (Agricultural equipment and tools for farmers designed for local construction)
0-903031-17-5 X.615/1539
0514314

AGRICULTURAL EQUIPMENT
Electric pumping systems. Design

FARM-ELECTRIC pumping and irrigation. - Stoneleigh : Electricity Council, Farm-Electric Centre, [1980?]
61p : ill(some col.). - (A farmelectric handbook)
B82-40739 SRL
2800438

AGRICULTURAL EQUIPMENT
Ploughs, B.C.100-A.D.500

KRASNOV, ÎUriĭ Alekseevich
Drevneĭshie upriazhnye pakhotnye orudiia. - Moskva : Nauka, 1975
180p : ill
r6097014 X.319/20165 Woolwich
1991701

AGRICULTURAL EQUIPMENT
Research. Organizations: International Secretariat for Research on the History of Agricultural Implements

LERCHE, Grith
Agricultural tools and field shapes. - Copenhagen : National Museum of Denmark, 1980
106p : ill. - (Publication / International Secretariat for Research on the History of Agricultural Implements ; no.3). - Bibliography
87-480-0309-3 X.322/10047 Woolwich
8712646

AGRICULTURAL EQUIPMENT. Ancient Rome

WHITE, K. D. (Kenneth Douglas)
Farm equipment of the Roman world. - Cambridge : Cambridge University Press, 1975
xvii,258p,16p of plates : ill. - Bibliography: p.241-242. - Includes index
0-521-20333-3 X.320/10187
0541567

AGRICULTURAL EQUIPMENT. Central Europe
Winnowing machines, to date

MEINERS, Uwe
Die Kornfege in Mitteleuropa. - Münster : Coppenrath, 1983
ii,481p.,plates : ill. - (Beiträge zur Volkskultur in Nordwestdeutschland ; Hft.28). - Includes bibliography
3-88547-216-3 X.0709/743(28)
9047581

MOSER, Oskar
Materialien zur Geschichte und Typologie der Getreidewinde (Kornfege). - Wien : Verlag der Österreichischen Akademie der Wissenschaften, 1984
70p,15p of plates. - (Sitzungsberichte / Österreichische Akademie der Wissenschaften. Philosophisch-historische Klasse ; Bd 434) (Mitteilungen des Instituts für Gegenwartsvolkskunde ; Nr 13)
3-7001-0633-5 Ac.810/6a[Bd.434]
9047581

AGRICULTURAL EQUIPMENT. Cotswolds. England
ca 1800-1900

COTSWOLD COUNTRYSIDE COLLECTION
Cotswold Countryside Collection. - Northleach : Cotswold District Council for the Cotswold Countryside Collection, 1981
[12]p : ill
B83-14689 X.800/34876
2867087

AGRICULTURAL EQUIPMENT. Developing countries

GIBBON, David, 19—
The hashasha and atulba toolbar. - Norwich : School of Development Studies, University of East Anglia, 1983
23[i.e.27]p : ill. - (Discussion paper ; no.21)
B85-00829 X.322/14227
3196410

AGRICULTURAL EQUIPMENT. England
to 1974. Social aspects

GOODLAND, Norman
Down on the farm. - Southampton : Paul Cave Publications, 1982
96p : ill
0-86146-029-4 X.329/16994
297116x

AGRICULTURAL EQUIPMENT. Great Britain

HUDSON, Kenneth
Farm furniture. - London : Bodley Head, 1980
40p : ill. - Includes index
0-370-30169-2 X.320/12065 Woolwich
1813501

AGRICULTURAL EQUIPMENT. Great Britain
to ca 400

REES, Sian E.
Ancient agricultural implements. - Princes Risborough : Shire, 1981
72p. - (Shire archaeology series ; 15). - Bibliography: p31. - Includes index
0-85263-535-4 X.329/14630 Woolwich
2577305

AGRICULTURAL EQUIPMENT. Great Britain
to 410

REES, Sian E.
Agricultural implements in prehistoric and Roman Britain. - Oxford : British Archaeological Reports, 1979
2v.([24],772p) : ill. - (British archaeological reports. British series ; 69). - Bibliography: p.744-769
0-86054-064-2 X.425/1358 Woolwich
2156121

AGRICULTURAL EQUIPMENT. Great Britain
to 1900

VINCE, John
Old farms. - London : Murray, 1982
160p : ill. - Includes index
0-7195-3978-1 X.800/34759
284737x

AGRICULTURAL EQUIPMENT. Great Britain
1500-1900

FUSSELL, G. E. (George Edwin)
The farmer's tools. - London : Orbis Publishing, 1981
246p,111p of plates : ill. - Bibliography: p225-230. - Includes index
0-85613-359-0 X.622/10619 Woolwich
2504715

WRIGHT, Philip, 1908-
Old farm implements. - Gloucester : Alan Sutton, 1983
91p,[23]p of plates : ill
0-86299-083-1 X.629/24450
2504715

AGRICULTURAL EQUIPMENT. Great Britain
ca 1880-1940

LAWSON, Douglas
Hand to the plough. - Sevenoaks : Ashgrove, c1982
[120]p : chiefly ill
0-906798-05-1 X.622/13602
2779552

AGRICULTURAL EQUIPMENT. Great Britain
Design Council approved designs

DESIGN COUNCIL
The Design Council's catalogue of farm buildings. - [London] : The Council, [1977]
132p : ill(some col)
0-85072-078-8 BS.80/30
1519239

AGRICULTURAL EQUIPMENT. Great Britain
Hand tools, to ca 1940

BRIGDEN, Roy
Agricultural hand tools. - Princes Risborough : Shire, 1983
32p : ill. - (Shire album ; 100). - Bibliography: p32
0-85263-630-x X.0972/143(100)
2999277

AGRICULTURAL EQUIPMENT. Guernsey
Horse ploughs, ca 1750-1978

JOHNSTON, Peter
The Guernsey plough. - [Guernsey?] : [The author?], [1979?]
[8]p : ill
B80-03206 X.319/19222 Woolwich
2020300

AGRICULTURAL EQUIPMENT. Norfolk
ca 1500-ca 1920

BARRY Lady, 1903-
Lady Margaret Barry & Mrs Gertrude Smith. - Ipswich : Boydell Press, 1977
31p : ill. - (Bygones)
0-85115-078-0 X.709/31918
130528x

EBBAGE, Sheridan
Barns and granaries in Norfolk. - Ipswich : Boydell Press, 1977
48p : ill. - (Bygones). - Bibliography: p.48
0-85115-079-9 X.709/31917
130528x

JOICE, Dick
Arthur Randell, fenman, talking to Dick Joice. - Ipswich : Boydell Press, 1977
46p : ill. - (Bygones)
0-85115-076-4 X.709/31920
130528x

Talking about bygones. - Ipswich : Boydell Press, 1977
45p : ill. - (Bygones)
0-85115-077-2 X.709/31919
130528x

AGRICULTURAL EQUIPMENT. Suffolk
Use, 1800-1950

The FARMING year in Suffolk / [compiled by S. Bowell et al. ; edited by J.G. Fairclough]. - Ely : EARO, The Resource and Technology Centre, [1977]
vi,54p : ill. - Bibliography: p.54
0-904463-16-8 X.311/8537 Woolwich
1190601

AGRICULTURAL EQUIPMENT. Wales
Ploughs, to ca 1930

PAYNE, Ffransis G. (Ffransis George)
Yr aradr Gymreig. - Caerdydd [Cardiff] : Gwasg Prifysgol Cymru, 1975
205p,[2] leaves of plates,[20]p of plates : ill. - Includes index
0-7083-0583-0 X.329/10141
0512885

AGRICULTURAL EQUIPMENT. Wales
Use, to ca 1945

WELSH FOLK MUSEUM
The Agricultural Gallery at the Welsh Folk Museum. - Cardiff : The Museum, 1976
43p : ill
0-85485-037-6 X.708/20508
0723029

Yr Oriel Amaethyddiaeth, Amgueddfa Werin Cymru. - [Cardiff] : Amgueddfa Werin Cymru, 1976
[1],42p : ill
0-85485-038-4 X.709/30937
0793515

AGRICULTURAL EQUIPMENT INDUSTRIES
See also
Potato production equipment industries

AGRICULTURAL EQUIPMENT INDUSTRIES
Agricultural equipment industries & trades - Serials

MIDDLE East agribusiness buyers guide = Dalīl al-mushtarin lil-zirā'ah fī al-Sharq al-Awsaṭ. - '84-. - Redhill : International Trade Publications, 1983-
v. : ill(some col.)
B84-24953 X.322/13864
3128474

AGRICULTURAL EQUIPMENT INDUSTRIES. Colombia
to date - Statistics

INSUMOS agropecuarios 1976 / Ministerio de Agricultura, O.P.S.A.. - Bogotá : Oficina de Divulgación del Ministerio de Agricultura, [1977?]
215p
r6701770 LAS.c.230/142 Woolwich
8768757

AGRICULTURAL EQUIPMENT INDUSTRIES. East Anglia
Agricultural equipment & materials industries & trades - Serials

EAST Anglian farmers guide. - [Issue no. 1]- ; [Sept. '79]-. - [Ipswich] : [Champaul Ltd], [1979]- : ill
B80-28200 P.313/502
2106795

AGRICULTURAL EQUIPMENT INDUSTRIES. Great Britain
AGRICULTURAL business manufacturers / ICC Business Ratios. - 4th ed. - London : ICC Business Ratios, [1981]
[72]p. - (ICC business ratio report)
0-86261-076-1 X.525/5044
2269309

AGRICULTURAL EQUIPMENT INDUSTRIES. Great Britain - Statistics
AGRICULTURAL equipment manufacturers / [Business Ratios]. - 3rd ed.: [1979]. - London : Business Ratios, [1979]
[81]p : ill
0-904540-30-8 X.622/8596 Woolwich
2269309

AGRICULTURAL EQUIPMENT INDUSTRIES. Wales
John Williams & Son & Powell Brothers & Whitaker
SCOURFIELD, Elfyn
Agricultural implement makers. - [Cardiff] : National Museum of Wales : Welsh Folk Museum, 1983
33p : ill
0-85485-067-8 X.329/17202
3037819

AGRICULTURAL EQUIPMENT TRADES
See also
Horticultural equipment trades

AGRICULTURAL EQUIPMENT TRADES
Agricultural equipment industries & trades - Serials
MIDDLE East agribusiness buyers guide = Dalil al-mushtarin lil-zirā'ah fi al-Sharq al-Awsat. - '84-. - Redhill : International Trade Publications, 1983-
v. : ill(some col.)
B84-24953 X.322/13864
3128474

AGRICULTURAL EQUIPMENT TRADES. East Anglia
Agricultural equipment & materials industries & trades - Serials
EAST Anglian farmers guide. - [Issue no. 1]- ; [Sept. '79]-. - [Ipswich] : [Champaul Ltd], [1979]- : ill
B80-28200 P.313/502
2106795

AGRICULTURAL EQUIPMENT TRADES. Great Britain
Agricultural equipment & materials trades - Directories
UNITED Kingdom agricultural merchants / [Data Research Group]. - [1976]. - Amersham : The Group, 1976
186 leaves in various pagings
Subsequent volumes, when received, will be placed at this shelfmark
0-905570-30-8 X.0313/509 Woolwich
0769061

AGRICULTURAL FAMILIES. North Wales
Social conditions
GWYNDAF, Robin
A farming family from Wales. - London : A. and C. Black, 1979
[1],25p : chiefly ill. - (The 'Strands' series ; 7)
0-7136-1919-8 X.329/11896 Woolwich
1763334

AGRICULTURAL HOLDINGS. England
Law
DENSHAM, H. A. C.
Scammell and Densham's Law of agricultural holdings. - London : Butterworths, 1980
x,49p
0-406-36811-2 X.200/38355
0254142

SCAMMELL, Wilfrid Stanley
Scammell and Densham's law of agricultural holdings. - 6th ed.. - London : Butterworth, 1978
xxxix,646p. - Includes index
0-406-36807-4 X.200/31805
0254142

AGRICULTURAL HOLDINGS. England
Law. Reform
ROYAL INSTITUTION OF CHARTERED SURVEYORS
Agricultural holdings legislation. - [London] : The Institution, 1978
19 leaves in various pagings
B79-01023 X.102/2162
1745980

AGRICULTURAL HOLDINGS. England
Tenancies. Law
DONELL, R. A. (Richard Antony)
Business and agricultural tenancies. - [Guildford] : The College of Law, 1977
[1],28p. - (Crash course lectures / College of Law ; 1977)
B78-01430 X.319/17692 Woolwich
1282115

AGRICULTURAL HOLDINGS. England
Tenancies. Policies of National Farmers' Union
NATIONAL FARMERS' UNION
The tenant farmers and the NFU. - [London] : National Farmers' Union, 1983
[8]p
B84-18318 X.322/13091
3074986

AGRICULTURAL HOLDINGS. England
Tenancies. Succession. Law. Decisions of agricultural land tribunals
HAMILTON, Richard Neville Dalton
Agricultural land tribunals succession decisions. - London : Royal Institution of Chartered Surveyors, 1978
1v.
B80-21607 X.0329 / 211
215174x
- Supplement. First, etc X.0329/211a Woolwich

AGRICULTURAL HOLDINGS. Farnham Manor
Tenancies, 1674
The ANCIENT and true customs of the manor of Farnham, as presented by the jury, at the court baron, held in the said manor, on the eighth day of September, 1674, in the 26th year of the reign of King Charles II. - [Farnham] : Farnham Museum Society, 1980
9p
B83-08402 X.955/1395
2835304

AGRICULTURAL HOLDINGS. Italy
Collective tenancies. Role of Catholics, 1890-1920
GUCCIONE, Eugenio
Le affittanze collettive nel pensiero politico ed economico dei Cattolici tra Otto e Novecento. - Palermo : Editrice ila Palma, 1978
99p. - (Athena)
r6069510 X.329/13137 Woolwich
2471744

AGRICULTURAL HOLDINGS. Madhya Pradesh. India (Republic)
Size. Restrictions. Law - Texts
MADHYA PRADESH
[Madhya Pradesh Ceiling on Agricultural Holdings Act, 1960]. Madhya Pradesh Ceiling on Agricultural Holdings Act, 1960. - 3rd ed. - Indore : Lawyers Home, 1975
xvi, 203 p.. - Includes index
r6031582 X.200/30851
1082485

AGRICULTURAL HOLDINGS. Maryland
Tenancies, 1700-1800
STIVERSON, Gregory A.
Poverty in a land of plenty. - Baltimore ; London : Johns Hopkins University Press, 1977
xvii,187,[1]p. - (Maryland bicentennial studies). - Includes index
0-8018-1966-0 X.800/26815
1411969

AGRICULTURAL HOLDINGS. Scotland
Law
GILL, Brian
The law of agricultural holdings in Scotland. - Edinburgh : Green, 1982
xxxiv,451p. - Includes index
0-414-00697-6 X.200/39985
288044x

AGRICULTURAL HOLDINGS. Wales
Classification
JENKINS, T. N.
The classification of agricultural holdings in Wales. - Aberystwyth : University College of Wales, Department of Agricultural Economics, 1982
v,258p : ill. - Bibliography: p251-258
B83-37692 X.329/16799
2950812

AGRICULTURAL, HORTICULTURAL AND FORESTRY INDUSTRY TRAINING BOARD
- Serials
AGRICULTURAL TRAINING BOARD
[Miscellaneous publications not catalogued separately.]. - [S.l.] : The Board, [1979-]
r6801996 X.0525/692
0041564

AGRICULTURAL INDUSTRIES
See also
Manila hemp production industries
United States. Federal Extension Service

AGRICULTURAL INDUSTRIES
Great Britain. Capital transfer tax - For agricultural industries
CAPITAL transfer tax / the Scottish agricultural colleges. - Revised [ed.]. - [Edinburgh] : [East of Scotland College of Agriculture etc.], 1978
[2],14p. - (Scottish agricultural colleges publications, no.27)
B79-02456 X.519/28931
1373277

AGRICULTURAL INDUSTRIES
Management. Leadership - For agricultural industries
STEWART, Bob R.
Leadership for agricultural industry. - New York : London : McGraw-Hill, 1978
viii,152p : ill(some col). - (Career preparation for agriculture/agribusiness). - Includes index
0-07-000847-7 X.311/8857 Woolwich
1571249

AGRICULTURAL INDUSTRIES
Norway. Accountancy - For agricultural industries - Serials
MODELLBRUKSBEREGNINGER / Budsjettnemnda for jordbruket - 1978-. - Oslo : Budsjettnemnda, 1980-
r6112771 SB.795/32
2465388

AGRICULTURAL INDUSTRIES. Great Britain
Small scale agricultural industries - Conference proceedings
SMALL farming and the nation / edited by R.B. Tranter. - Reading : Centre for Agricultural Strategy, University of Reading, 1981
40p. - (CAS paper ; 9)
0-7049-0615-5 X.329/13724 Woolwich
2385031

AGRICULTURAL INSTITUTE
- Serials
HORTICULTURE research report / An Foras Talúntais (the Agricultural Institute). - [19-]-. - Dublin : The Institute, [19-]-
v.
Holdings: 1982-
r6813520 P.313/564
3066045

AGRICULTURAL INSTITUTE
Libraries. Stock: Serials
AGRICULTURAL INSTITUTE
Liosta na dtréimseachán i gcuid leabharlann An Fhorais Talúntais 1980. - [Dublin] : [An Foras Talúntais], 1980
62p
B81-24774 2725.g.55
2496313

AGRICULTURAL INSTITUTE. KINSEALY RESEARCH CENTRE
See Kinsealy Research Centre

AGRICULTURAL LAND
See also
Agricultural holdings
Enclosures
Fields
Marginal land
Pastures
Rangeland
Vineyards

AGRICULTURAL LAND
Arterial drainage
FIELD DRAINAGE SERVICE
Arterial drainage and agriculture. - [Revised ed.]. - Pinner : Ministry of Agriculture, Fisheries and Food, 1979
21,[3]p : ill(chiefly col). - (Getting down to drainage ; no.16) (Drainage leaflets ; 16)
B79-36364 SRL
1971794

AGRICULTURAL LAND
Drainage
FIELD drainage / edited by D.A. Castle, J. McCunnall and I.M. Tring. - London : Batsford Academic and Educational, 1984
250p,[8]p of plates : ill. - Bibliography: p235-246. - Includes index
0-7134-3833-9 X.322/14119
2825880

KENNEDY, M. N.
A handbook of irrigation and drainage terms, English-French. - Bishop's Waltham : J.W.A. Newhouse, c1981
44p
0-9502591-3-6 SRL
2550059

AGRICULTURAL LAND
Drainage–continued

SMEDEMA, Lambert K.
Land drainage. – London : Batsford Academic & Educational, 1983
376p : ill. – Includes bibliographies and index
0–7134–3500–3 SRL
2825880

AGRICULTURAL LAND
Drainage – Abstracts – Serials

IRRIGATION and Drainage Abstracts. – Vol.1, no.1– ; May 1975–. – Slough : Commonwealth Agricultural Bureaux, 1975–
B76–30625 SRL
0734624

AGRICULTURAL LAND
Drainage – Dictionaries

REES, Tom, b.1925
'There's a word for it'. – Stroud : Critchley Bros Ltd. [1977]
[3],42,[1]p
B77–27190 SRL
1201182

AGRICULTURAL LAND
Drainage systems. Filters

DENNIS, C. W.
Field drainage pipes and filters. – [London] : ADAS, 1982
15p : 1ill. – (ADAS overseas study tour programme ; 1981/82)
B82–40444 BS.3/737
2799014

AGRICULTURAL LAND
Fertilisers. Environmental aspects

FERTILISERS and our environment. – London : Fertiliser Manufacturers Association, c1981
22p
B82–21418 X.329/15544
2721562

HANDBOOK on environmental aspects of fertilizer use / Centre d'étude de l'azote, International Fertilizer Industry Association, International Potash Institute. – The Hague ; London : Nijhoff for the CEA, IFA, IPI, 1983
103p : ill. – Includes bibliographies
90–247–2801–0 X.322/12619
2721562

AGRICULTURAL LAND
Irrigation

JAMES, D. W. (David Winston)
Modern irrigated soils. – New York ; Chichester : Wiley, c1982
x,235p : ill. – (A Wiley-Interscience publication). – Includes bibliographies and index
0–471–06351–7 SRL
2818078

MICHAEL, A. M.
Irrigation. – New Dehli : Vikas, 1978
vi,801p : ill
0–7069–1513–5 X.322/11532
2818078

AGRICULTURAL LAND
Irrigation – Abstracts – Serials

IRRICAB. – Vol.1 (1976)–. – Bet Dagan : International Irrigation Information Center, 1976–
v.
B81–03613 SRL
0734632

IRRIGATION and Drainage Abstracts. – Vol.1, no.1– ; May 1975–. – Slough : Commonwealth Agricultural Bureaux, 1975–
B76–30625 SRL
0734632

AGRICULTURAL LAND
Irrigation – Serials

ADVANCES in irrigation. – Vol.1–. – New York ; London : Academic Press, 1982–
v. : ill
B83–13371 SRL
2863391

IRRIGATION news. – No.1 (summer 1981)–. – Silsoe : UK Irrigation Association, 1981–
v. : ill
B84–06855 P.843/126
2863391

AGRICULTURAL LAND
Irrigation. Economic aspects

CLARK, Colin
The economics of irrigation. – [revision and updating]. – Liverpool : Liverpool University Press, 1981
300p : ill. – Bibliography: p269–283. – Includes index
0–85323–464–7 X.322/13491
2090376

CLARK, Colin, 1905–
The economics of irrigation. – [revision and updating]. – Liverpool : Liverpool University Press, 1981
xviii,300p : ill. – Bibliography: p269–283. – Includes index
0–85323–254–7 X.622/10306 Woolwich
2090376

AGRICULTURAL LAND
Irrigation. Management. Applications of action research

BOTTRALL, Anthony
The action research approach to problem solving, with illustrations from irrigation management. – London : Overseas Development Institute, 1982
20p. – (ODI working paper ; no.9). – Bibliography: p19–20
0–85003–083–8 X.525/6155
2776774

Action research towards improved water distribution. – London : Overseas Development Institute, Agricultural Administration Unit, [1981]
20p. – (Network paper ; 1/81/2)
B84–30331 X.329/17995
2776774

AGRICULTURAL LAND
Irrigation. Management. Applications of operations research – Conference proceedings

ORAGWA INTERNATIONAL CONFERENCE (1979 : Jerusalem)
Operations research in agriculture and water resources. – Amsterdam ; Oxford : North-Holland, 1980
xviii,586p : ill. – Includes bibliographies and index
0–444–86044–4 SRL
2348292

AGRICULTURAL LAND
Irrigation systems

BAILEY, R. J.
To study and evaluate various techniques used for irrigation investigations. – [London] : ADAS, [1983?]
24p : ill
B84–43958 BS.3/906
3016447

AGRICULTURAL LAND
Irrigation systems. Modernisation & replacement

INTERNATIONAL COMMISSION ON IRRIGATION AND DRAINAGE. International Congress (11th : 1981 : Grenoble)
Report on 11th International Congress of the International Commission on Irrigation and Drainage, Grenoble, France, August-September 1981. – [London] : ADAS, [1983?]
13p in various pagings
B84–30427 BS.3/815
2935503

AGRICULTURAL LAND
Land tenure

KAUTSKY, Karl
[Die Agrarfrage. Spanish]. La cuestión agraria. – [Bogotá?] : Editorial Latina, [ca.1975?]
501p
r6065051 X.709/25582
1508296

AGRICULTURAL LAND
Land tenure. Economic aspects

CURRIE, J. M. (John Martin)
The economic theory of agricultural land tenure. – Cambridge : Cambridge University Press, 1981
vii,194p : ill. – Bibliography: p188–191. – Includes index
0–521–23634–7 X.322/10019 Woolwich
2183501

AGRICULTURAL LAND
Land tenure. Law. Terminology – Dictionaries

PORRAS DEL CORRAL, Manuel
Léxico de equivalencias inglés–español de términos legales relacionados con los arrendamientos rústicos. – Madrid : Instituto Nacional de Investigaciones Agrarias, 1976
38p. – (Comunicaciones I.N.I.A. ; serie general ; n.1)
84–500–1728–9 S.O.157/8 Woolwich
1318748

AGRICULTURAL LAND
Land tenure. Reform related to political violence

HARRISON, Fred, 1944–
Land reform or red revolution. – [Ilford] : Economic and Social Science Research Association, 1980
vi,37p. – (Centenary essays / Economic and Social Science Research Association ; no.1)
B80–21519 X.525/3208
2178788

AGRICULTURAL LAND
Land tenure. Terminology

MATERIALIEN zur Terminologie der Agrarlandschaft / Internationale Arbeitsgruppe für die geographische Terminologie der Agrarlandschaft = Basic material for the terminology of the agricultural landscape ... – Giessen : Lenz. – Includes bibliographies
Vol.1: Flur und Flurformen. – 2., stark überarb. Fassung. – 1978
136p : ill
r6308987 X.0525/680(1)
9117296

AGRICULTURAL LAND
Law. Theories, 1945–

AKSENENOK, G. A. (Georgiĭ Aleksandrovich)
Kritika sovremennykh burzhuaznykh agrarnopravovykh teoriĭ. – Moskva : Nauka, 1972
319p. – (Ideologicheskaiâ bor'ba v sovremennom mire)
Imperfect: p.293–294 is torn.
r6462708 X.200/42141
9026983

AGRICULTURAL LAND
Management. Implications of conservation of environment – Conference proceedings

MAN AND HIS ENVIRONMENT (Conference : 1975 : University of Birmingham)
Conservation and agriculture. – London : Duckworth, 1978
xvi,284p : ill. – Includes bibliographies and index
0–7156–1032–5 SRL
1568868

AGRICULTURAL LAND
Sharecropping

SHARECROPPING and sharecroppers / edited by T.J. Byres. – London : Cass, 1983
283p : ill. – Includes bibliographies and index
0–7146–3223–6 X.809/56716
2977133

AGRICULTURAL LAND
Slow-release ureaform fertilisers. Use

SCHNEIDER, H.
Practical experience with ureaform slow-release nitrogen fertiliser during the past 20 years, and outlook for the future. – Preliminary ed. – London : Fertiliser Society, [1979]
43,[3]p : ill. – (Proceedings / Fertiliser Society ; no.180)
B79–27581 SRL
1933604

AGRICULTURAL LAND
Soils. Conservation. Engineering aspects

SOIL and water conservation engineering. – 3rd ed.. – New York ; Chichester : Wiley, c1981
xv,525p : ill. – Includes bibliographies and index
0–471–03078–3 SRL
2595435

SPOOR, Gordon
Soil management or reclamation?. – Bedford : The College, 1983
16p : ill. – (Occasional paper ; no.11). – Bibliography: p15–16
B84–28201 X.325/1393
2595435

AGRICULTURAL LAND
Soils. Nutrient cycle – Conference proceedings

CYCLING OF MINERAL NUTRIENTS IN AGRICULTURAL ECOSYSTEMS (Conference : 1976 : Amsterdam)
Cycling of mineral nutrients in agricultural ecosystems. – Amsterdam ; Oxford : Elsevier, 1978
vii,356p : ill. – (Developments in agricultural and managed-forest ecology ; 3) (Mededelingen / Nederlandsche Heide Maatschappij). – Bibliography: p.317–346. – Includes index
0–444–41660–9 SRL
1570749
– [Another copy] X.620/18019 Woolwich

AGRICULTURAL LAND
Water. Conservation. Engineering aspects

SOIL and water conservation engineering. - 3rd ed.. - New York ; Chichester : Wiley, c1981
xv,525p : ill. - Includes bibliographies and index
0-471-03078-3 SRL
2595443

AGRICULTURAL LAND. Alentejo. Portugal
Land tenure. Reform, 1975-1977

CARVALHO, António Aleixo Pais Vacas de
O fracasso de um processo. - [Lisboa?] : o Autor, 1977
342p
r6066178 X.709/26678
1718320

AGRICULTURAL LAND. Ancient Rome
Land tenure

SETÄLÄ, Päivi
Private domini in Roman brick stamps of the empire. - Helsinki : Suomalainen tiedeakatemia, 1977
316p : geneal tables. - (Annales Academiae Scientiarum Fennicae : Dissertationes humanarum litterarum : 10). - Bibl
r6698004 Ac.1094.c/7b(1)[vol.10]
2447800

AGRICULTURAL LAND. Andes. South America
Land tenure. Reform, to date

DERECHOS humanos en las zonas rurales / seminario organizado por la Comisión Internacional de Juristas y el Consejo Latinoamericano de Derecho y Desarrollo, Bogotá, septiembre 1979. - Bogotá : Sociedad Ediciones Internacionales, 1979
306p
r6708390 X.429/16945
9143270

AGRICULTURAL LAND. Argentina
Land tenure, to 1975

FLICHMAN, Guillermo
La renta del suelo y el desarrollo agrario argentino. - México : Siglo Veintiuno, 1977
241p : ill
r6084821 X.319/9863
1976966

ZEMBORAIN, Saturnino M.
La verdad sobre la propiedad de la tierra en la Argentina. - [Buenos Aires?] : Instituto de Estudios Económicos de la Sociedad Rural Argentina, 1973
65p.,32leaves
r6101984 X.702/15439
1976966

AGRICULTURAL LAND. Argentina
Land tenure. Policies of Sarmiento, Domingo Faustino

PISANO, Natalio J.
La política agraria de Sarmiento. - Buenos Aires : Ediciones Depalma, 1980
xxiv,449p
r6705605 X.200/38464
8855765

AGRICULTURAL LAND. Asket Hill. Leeds. West Yorkshire
Transfer by Great Britain. Department of the Environment to Leeds. City Council

GREAT BRITAIN. Treasury
Treasury Minute dated 11 November 1983 relating to the gift of 7.62 Acres of grazing land to the Leeds City Council. - London : H.M.S.O., [1984]
1sheet. - (Cmnd. ; 9086)
0-10-190860-1 BS.44/685
3257622

AGRICULTURAL LAND. Asturias & León (Kingdom). Spain
Land tenure, ca 700-ca 1000

SÁNCHEZ-ALBORNOZ, Claudio
El régimen de la tierra en el reino asturleonés hace mil años. - Buenos Aires : Universidad de Buenos Aires, Instituto de Historia de España, 1978
xiv,248p
r6162438 X.709/52189
2465841

AGRICULTURAL LAND. Atencingo. Mexico
Land tenure. Reform, 1910-1970

RONFELDT, David
[Atencingo : the politics of agrarian struggle in a Mexican ejido. Spanish]. Atencingo. - México : Fondo de Cultura Económica, 1975
278p. - (Tierra firme)
r6037895 X.319/8996 Woolwich
1326201

AGRICULTURAL LAND. Békés (megye). Hungary
Land tenure. Reform

SZABÓ, István
Agrárforradalom a Viharsarokban (1945-1947). - Budapest : Akadémiai Kiadó, 1976
193p
963-05-1110-x X.520/9463
1209221

AGRICULTURAL LAND. Belorussia
Land tenure. Reform, 1900-1917

LIPINSKIĬ, Leonid Pavlovich
Stolypinskaia agrarnaia reforma v Belorussii. - Minsk : Izdatel'stvo Belorusskogo gosudarstvennogo universiteta im. V.I. Lenina, 1978
221p
r6087114 X.529/20981
1848755

AGRICULTURAL LAND. Bengal & Bihar. India (Republic)
Land tenure, to date

CHAUDHARY, R. B.
The British agrarian policy in Eastern India. - Patna : Ram Brit Singh [for] Janaki Prakashan, 1980
viii,222p. - Bibliography: p.209-218
r6804728 X.329/14571 Woolwich
8773947

AGRICULTURAL LAND. Bengal. India
Land tenure. Reform, 1769-1784

GUHA, Ranajit
A rule of property for Bengal. - 2nd ed. - New Delhi : Orient Longman, 1982, c1981
222p. - Bibliography: p211-216
0-86131-289-9 X.529/49286
2742977

AGRICULTURAL LAND. Bolivia
Land tenure, 1975

CONSEJO NACIONAL DE REFORMA AGRARIA. Departamento de Estadística
El proceso de reforma agraria en cifras 1975. - La Paz : el Consejo, 1975
163p : ill
r6052342 X.329/8720 Woolwich
1393499

AGRICULTURAL LAND. Bolivia
Land tenure. Reform, 1930-1970

IRIARTE, Gregorio
Sindicalismo campesino. - La Paz : Centro de Investigación y Promoción del Campesinado, 1974
100p. - (Cuadernos de investigación CIPCA ; n.5)
r6085907 X.525/1243
1985132

AGRICULTURAL LAND. Bolivia
Land tenure. Reform, 1952-

ALBÓ, Javier
?Bodas de plata? o requiem por una reforma agraria. - La Paz : CIPCA, 1979
105p. - (Cuaderno de investigación ; 17). - Bibliography
r6703639 X.322/10304 Woolwich
8562768

ANTEZANA E., Luis
Proceso y sentencia a la reforma agraria en Bolivia. - La Paz : Ediciones Puerta del Sol, 1979
171p : ill. - (Colección Luces y sombras ; 1)
r6703636 X.329/14570 Woolwich
8562768

AGRICULTURAL LAND. Bolivia
Land tenure. Reform, 1953-1975

URQUIDI, Arturo
Temas de reforma agraria. - La Paz : Juventud, 1976
120p
r6085096 X.709/27435
1857762

AGRICULTURAL LAND. Brazil
Land tenure, 1500-1970

GUIMARÃES, Alberto Passos
Quatro séculos de latifúndio. - 3a ed. - Rio de Janeiro : Paz e Terra, [1975]
xv,255p. - (Série Estudos brasileiros ; v. 3)
r6032888 X.709/22419
0392456

AGRICULTURAL LAND. Brazil
Land tenure. Law, to 1977

BORGES, Paulo Torminn
Institutos básicos do direito agrário. - 3a ed. revista e ampliada. - São Paulo : Pró-Livro, 1978
306p. - Bibliography
r6162387 X.529/38788
2469332

AGRICULTURAL LAND. Brazil
Land tenure. Law, 1964-1971

BORGES, Paulo Torminn
Institutos básicos do direito agrario. - São Paulo : Editôra Juriscredi, [1974]
301 p.
r6009123 X.100/14577
0919578

AGRICULTURAL LAND. Brazil
Land tenure. Law, 1964-

DEMÉTRIO, Nelson
Doutrina e prática do direito agrário. - São Paulo : Pró-Livro, 1980
383p
r6707596 X.329/17974
8559651

RIZZARDO, Arnaldo
O uso da terra no direito agrário. - Porto Alegre : Sintese, 1981
268p
r6707591 X.200/44068
8559651

AGRICULTURAL LAND. Brazil
Land tenure. Reform

LIMA, Rafael Augusto de Mendonça
Direito agrário, reforma agrária e colonização. - Rio de Janeiro : Livraria Francisco Alves Editora, 1975
147 p.
r6013579 X.319/8565 Woolwich
0876348

RIOS, Arthur Edmundo de Souza
Direito agrário. - Goiânia : Universidade Federal de Goiás, 1974
91 p.
r6014723 X.100/14868
0876348

AGRICULTURAL LAND. Brazil
Land tenure. Reform, 1960-1964

GÓMEZ GARCÍA, Zoila
Amadeo Roldán. - La Habana : Arte y Literatura, 1977, 1978 printing
243p : ill,facsims,music,ports
r6160633 X.439/11019
2457121

PRADO JÚNIOR, Caio
A questão agrária no Brasil. - São Paulo : Brasiliense, 1979
188p
r6160683 X.329/13059 Woolwich
2457121

AGRICULTURAL LAND. Brazil
Ownership. Law

BORGES, Paulo Torminn
O imóvel rural e seus problemas jurídicos. - São Paulo : Pró-livro, 1976
205p
r6066085 X.329/8821 Woolwich
1711199

AGRICULTURAL LAND. Brinsbury region. West Sussex
Agricultural land use, landscape conservation & nature conservation. Interrelationships

FARMING and wildlife study 1980 / Sussex Farming and Wildlife Advisory Group and West Sussex College of Agriculture and Horticulture, Brinsbury. - Chichester : West Sussex County Council, [1981]
18p : ill
B82-04948 SRL
2597829

AGRICULTURAL LAND. British occupation zone. West Germany
Land tenure. Reform, 1945-1949

TRITTEL, Günter J.
Die Bodenreform in der Britischen Zone 1945-1949. - Stuttgart, [1975]
184p. - (Schriftenreihe der Vierteljahrshefte für Zeitgeschichte ; Nr 31)
3-421-01730-1 PP.3481.iad(2)[no.31]
088149X

AGRICULTURAL LAND. Buckinghamshire
Amenities: Trees. Planting

BUCKINGHAMSHIRE. County Council
Tree planting on farms. - Aylesbury : The Council, 1976
[2], 12 p. : ill
0-902805-49-5 X.319/17033 Woolwich
0788023

AGRICULTURAL LAND. Burnt Fen. East Anglia
Drainage, to 1981

BECKETT, John
The urgent hour. - Updated. - [Ely] : Ely Local History Publications Board, 1983, c1974
53p : ill
0-904463-88-5 X.629/21114
2938081

AGRICULTURAL LAND. Calabria. Italy
Land tenure, to date

CHIODO, Gabriela
Lotte per la terra e movimento cooperativo in Cosenza. - Napoli : Guida, 1981
254p. - (Studio sud ; 17)
88-7042-050-7 X.329/17972
8890986

DI BELLA, Saverio
Strutture agrarie e lotte per la terra nel Mezzogiorno contemporaneo. - Soveria Mannelli : Rubbettino, 1979
xv,769p,[3]leaves of plates : ill. - (Collana "quaderni di scienze umane" ; 1)
r6506075 X.800/34717
8890986

MOTTURA, Giovanni
Il diritto alla terra. - Milano : Feltrinelli, 1981
208p. - (Feltrinelli economica) (I nuovi testi ; 238. Le classi subalterne)
r6508692 X.329/16672
8890986

AGRICULTURAL LAND. Calabria. Italy
Land tenure. Reform, 1950-1970

PEZZINO, Paolo
La riforma agraria in Calabria. - Milano : Feltrinelli, 1977
253p. - (I fatti e le idee ; 394. Saggi e biografie: storia) (Istituto nazionale per la storia del movimento di liberazione in Italia. [Pubblicazioni])
r6160154 X.329/12379 Woolwich
2425092

AGRICULTURAL LAND. Canada
Land use, to date

EHLERS, Eckart
Die agraren Siedlungsgrenzen der Erde. - Wiesbaden : Steiner, 1984
82.,plates. - (Erdkundliches Wissen ; Hft.69). - Includes bibliography
3-515-04211-3 WP.934/69
9127984

AGRICULTURAL LAND. Capitalist countries
Land tenure, 1875-1975 related to industrialisation

PUHLE, Hans Jürgen
Politische Agrarbewegungen in kapitalistischen Industriegesellschaften. - Göttingen : Vandenhoeck und Ruprecht, 1975
496 p., 1 fold leaf of plates : ill. - (Kritische Studien zur Geschichtswissenschaft ; 16). - Bibliography : p.426-476
3-525-35967-5 X.0700/530(16)
1010379

AGRICULTURAL LAND. Central America
Land tenure

TENENCIA de la tierra y desarrollo rural en Centroamérica / estudio elaborado por expertos de CEPAL ... [et al.]. - 2.ed. - San José, Costa Rica : Editorial Universitaria Centroamericana, 1976
199p. - (Colección Seis. serie mayor)
r6047904 X.319/9660 Woolwich
1713450

AGRICULTURAL LAND. Central America
Land tenure. Reform, to date

GARCÍA, Antonio
El nuevo problema agrario de la América Latina. - México : Universidad Nacional Autónoma de México, 1981
86p
r6709837 X.809/63554
9152881

AGRICULTURAL LAND. Chile
Land tenure

GÓNGORA, Mario
Origen de los inquilinos de Chile central. - 2.ed. - [Santiago de Chile] : Instituto de Capacitación e Investigación en Reforma Agraria, 1974
160p
r6013508 X.708/18413
0886963

AGRICULTURAL LAND. Chile
Land tenure. Reform, 1965-1970

REFORMA agraria chilena / David Alaluf [et al.]. - 2a ed. - Santiago de Chile : ICIRA, 1972
125p
r6032961 X.319/8708 Woolwich
0397326

AGRICULTURAL LAND. Chile
Land tenure. Reform, 1965-1972

CHILE : reforma agraria y gobierno popular / Solon Barraclough ... [et al.]. - Buenos Aires : Ediciones Periferia, 1973
244p. - (Colección Estados Unidos y América Latina). - Includes bibliographies
r6065803 X.519/39048
1712241

AGRICULTURAL LAND. Chile
Land tenure. Reform. Organisations: Sociedad Nacional de Agricultura, 1919-1940

WRIGHT, Thomas C.
Landowners and reform in Chile. - Urbana ; London : University of Illinois Press, c1982
xix,249p : ill. - Bibliography: p224-238. - Includes index
0-252-00853-7 X.520/31200
2953633

AGRICULTURAL LAND. Chile
Land tenure. Reform. Policies of government, 1967-1973

KAY, Cristóbal
El reformismo agrario y la transición al socialismo en América Latina. - Medellín : La Oveja Negra, 1976
71p. - (Cuaderno no.8). - Includes bibliographies
r6158726 X.319/20362 Woolwich
0290513

AGRICULTURAL LAND. China
Irrigation

WATER management organization in the People's Republic of China / edited with an introduction by James E. Nickum. - Armonk, N.Y. : Sharpe, c1981
xvi,269p : ill. - (The China book project)
0-87332-140-5 X.520/26553
2701081

AGRICULTURAL LAND. China
Irrigation. Conservation of water. Socioeconomic aspects

VERMEER, Eduard Boudewijn
Water conservancy and irrigation in China. - The Hague : Leiden University Press, 1977
xviii,350,[2]p : ill. - Bibliography : p.332-345. - Includes index
90-602-1409-9 X.619/19694 Woolwich
188302X

AGRICULTURAL LAND. China
Land tenure, 1644-1912

VOGEL, Hans Ulrich
Lokale Administration und Bodenpolitik der Himmlischen Dynastie des Grossen Friedens (Taiping Tianguo, 1850-1864). - Hamburg : Gesellschaft für Natur- und Völkerkunde Ostasiens, 1981
239p. - (Mitteilungen / Gesellschaft für Natur- und Völkerkunde Ostasiens ; 85). - Bibliography
r6303617 Ac.1944[vol.85]
8803293

AGRICULTURAL LAND. China
Land tenure. Reform, 1946-1953

MOISE, Edwin E. (Edwin Evariste), 1946-
Land reform in China and North Vietnam. - Chapel Hill ; London : University of North Carolina Press, c1983
xiv,305p : 2ill. - Bibliography: p289-296. - Includes index
0-8078-1547-0 X.520/31208
2951436

AGRICULTURAL LAND. Coastal regions. Ecuador
Land tenure. Reform, 1970-1975

REDCLIFT, M. R.
Agrarian reform and peasant organization on the Ecuadorian Coast. - London : Athlone Press for the Institute of Latin American Studies, 1978
xii,186p. - (Monographs / University of London. Institute of Latin American Studies ; 8). - Bibliography: p.168-180. - Includes index
0-485-17708-0 X.0809/331(8)
1113445

AGRICULTURAL LAND. Colombia
to 1974 - Conference proceedings

SIMPOSIO PROBLEMAS DE TIERRA, TRADICIÓN Y PODER EN COLOMBIA (1974 : Mexico City)
Tierra, tradición y poder en Colombia : enfoques antropológicos. - [Bogotá] : Instituto Colombiano de Cultura, Subdirección de Comunicaciones Culturales, 1976
171p.,1 fold leaf : ill. - (Biblioteca básica colombiana ; n.12). - Includes bibliographies
r6065128 X.519/38641
1539485

AGRICULTURAL LAND. Colombia
Land tenure, to 1975

GARCÍA, Antonio
Colombia. - Bogotá : Cruz del Sur, 1977
121p
r6085610 X.519/40426
1390589

LÓPEZ G., Fernando
Evolución de la tenencia de la tierra en una zona minifundista. - Bogotá : Centro de Estudios sobre Desarrollo Económico, Facultad de Economía, Universidad de los Andes, 1975
vii,78p.,[2]leaves of plates. - (Documentos CEDE ; no.029)
r6158932 X.311/5198 Woolwich
1390589

REYES POSADA, Alejandro
Aparcería y capitalismo agrario. - Bogotá : Centro de Investigación y Educación Popular, 1975
67p. - (Controversia ; n.38)
r6064810 X.319/9079 Woolwich
1390589

TÉLLEZ A., Neftalí
Aparcería, arriendo y colonato en el desarrollo del capitalismo en Colombia. - Bogotá : Editorial Latina, [1977]
124p. - Includes bibliographies
r6158818 X.708/47842
1390589

AGRICULTURAL LAND. Colombia
Land tenure, 1500-1970

ZULETA, Estanislao
La tierra en Colombia. - Medellín : Editorial La Oveja Negra, 1973
151 p.. - (Cuadernos La oveja negra ; n. 6)
r6027508 X.808/10398
1023772

AGRICULTURAL LAND. Colombia
Land tenure, 1970-1975

La TIERRA para el que la trabaja / [presentado por la] Asociación Colombiana de Ingenieros Agrónomos. - Bogotá : Asociación Colombiana de Ingenieros Agrónomos : Editorial Punto y Coma, 1975
401p
r6037956 X.108/15161
1234102

AGRICULTURAL LAND. Colombia
Land tenure. Law, 1819-

COLOMBIA
Legislación agraria. - Bogotá : Temis, 1980
347p
r6707342 X.200/42260
856809x

MORALES BENÍTEZ, Otto
Derecho agrario y otros temas de la tierra. - [Bogotá] : Universidad Externado de Colombia, 1981
277p,[1]leaf of plates. - Includes bibliography
r6709117 X.322/14020
856809x

AGRICULTURAL LAND. Colombia
Land tenure. Law, 1821-1974

PÉREZ SALAZAR, Honorio
Derecho agrario colombiano. - Bogotá : Temis, 1975-
v.1-
r6037813 X.0100/747
1232258

AGRICULTURAL LAND. Colombia
Land tenure. Reform, 1850-1970

TOBÓN, Alonso
La tierra y la reforma agraria en Colombia. - 2a ed. - Bogotá : Ediciones Cáncer, [1976]
162p. - (Colección Tierra y revolución ; no. 2)
r6052046 X.319/8926 Woolwich
1314602

AGRICULTURAL LAND. Colombia
Land tenure. Reform, 1917-1970

GILHODÉS, Pierre
Las luchas agrarias en Colombia. - 3a ed. - Medellín : La Carreta, 1976
92p
r6085190 X.708/45312
1865854

AGRICULTURAL LAND. Colombia
Land tenure. Reform, 1937-1966

CASTRO MONSALVO, Pedro
Un campesino previno al país. - Bogotá : Ediciones Tercer Mundo, 1972
xxvii,263p
r6027564 X.708/19490
0391239

AGRICULTURAL LAND. Colombia
Land tenure. Reform, 1944-1973

INSTITUTO COLOMBIANO DE LA REFORMA AGRARIA
Reforma agraria colombiana. - Bogotá : INCORA, 1973
19p : ill
r6035492 X.319/8778 Woolwich
1150693

AGRICULTURAL LAND. Colombia
Land tenure. Reform. Role of workers' organisations, 1930-1940

GAITÁN, Gloria
Colombia : la lucha por la tierra en la década del treinta. - Bogotá : Ediciones Tercer Mundo, 1976
102p. - Includes bibliographies
r6065840 X.708/44433
1718231

AGRICULTURAL LAND. Costa Rica
Law

SALAS MARRERO, Oscar A.
Derecho agrario. - San José : Publicaciones de la Universidad de Costa Rica, 1973
xxvi,897p.,[1]fold.leaf of plates. - (Serie Ciencias jurídicas y sociales ; n.22)
r6085989 X.102/1387
1974246

AGRICULTURAL LAND. Csongrád (megye). Hungary
Land tenure. Reform

SZABÓ, István
Agrárforradalom a Viharsarokban (1945-1947). - Budapest : Akadémiai Kiadó, 1976
193p
963-05-1110-x X.520/9463
120923X

AGRICULTURAL LAND. Cuba
Land tenure. Reform, 1930-1976

MARTINEZ-ALIER, Juan
Haciendas, plantations and collective farms. - London : Cass, 1977
viii,185p. - (Library of peasant studies ; no.2). - Bibliography: p.171-179. - Includes index
0-7146-3048-9 X329/15586
1083481

AGRICULTURAL LAND. Czech Socialist Republic
Settlement

SLEZÁK, Lubomír
Zemědělské osídlování pohraničí českých zemí po druhé světové válce. - v Brně : Blok, 1978
189p : ill
r6015696 X.320/5316 Woolwich
2095955

AGRICULTURAL LAND. Denmark
Land tenure, to date

PORSMOSE, Erland
Bønder, bryder og tjenere. - [København] : Landbohistorisk Selskab, c1983
109p : ill. - (Odense University studies in history and social sciences ; v.81)
87-7492-413-3 X.0700/318(81)
9107711

AGRICULTURAL LAND. Denmark
Land tenure. Reform, to date

HANSGAARD, Torben
Landboreformerne i Danmark i det 18. arhundrede. - København : Landbohistorisk Selskab, 1981
222p,plate : ill
87-7526-081-6 X.322/12613
9017747

AGRICULTURAL LAND. Developing countries
Irrigation. Economic aspects

HAZLEWOOD, Arthur
Irrigation economics in poor countries. - Oxford : Pergamon, 1982
viii,144p : ill. - Includes index
0-08-027451-x X.529/52515
2851229

LIVINGSTONE, I. (Ian)
The economics of water storage for irrigation. - Norwich : School of Development Studies, University of East Anglia, 1983
33p : ill. - (Discussion paper ; no.87)
B84-39702 X.322/14042
2851229

AGRICULTURAL LAND. Developing countries
Land tenure. Reform

AGRARIAN reform in contemporary developing countries / edited by Ajit Kumar Ghose. - London : Croom Helm, c1983
364p. - Includes bibliographies and index
0-7099-1312-5 X.329/17506
0232211

PEASANTS, landlords and governments / edited by David Lehmann. - New York : Holmes & Meier, 1974
320p.
0-8419-0162-7 X.329/8635 Woolwich
0232211

AGRICULTURAL LAND. Developing countries
Land tenure. Reform, to date

CRAIG, Jane
Land, people and power. - Toronto : Oxfam-Canada, [1977]
80p : ill
r6811471 X.322/12833
9006699

AGRICULTURAL LAND. Developing countries
Land tenure. Reform, 1945-

FONES-SUNDELL, Melinda
Agrarian reform and rural development. - Uppsala : Swedish University of Agricultural Sciences, 1980
66p. - (Rural development studies ; no.6)
r6611223 X.0313/56(6)
8601755

AGRICULTURAL LAND. Developing countries
Land tenure. Reform, ca 1950-1976

KING, Russell
Land reform. - London : Bell, 1977
xvi,446p. - (Bell's advanced economic geographies). - Includes index
0-7135-1995-9 X.329/11179 Woolwich
1268880

AGRICULTURAL LAND. Developing countries
Land tenure. Reform. International aspects

INTERNATIONAL dimensions of land reform / edited by John D. Montgomery. - Epping : Bowker, c1984
ix,239p. - (A Westview special study). - Includes index
0-86531-781-x X.800/39927
3220664

AGRICULTURAL LAND. Developing countries
Soils. Degradation. Role of capitalism

BLAIKIE, Piers M. (Piers Macleod)
Capitalism, the peasantry and environmental degradation. - Norwich : School of Development Studies, University of East Anglia, 1983
26p. - (Discussion paper ; no.81). - Bibliography: p25-26
B85-03255 X.322/14311
3213021

AGRICULTURAL LAND. Developing countries
Soils. Erosion. Mathematical models

STOCKING, Michael A.
A working model for the estimation of soil loss suitable for underdeveloped areas. - Norwich : School of Development Studies, University of East Anglia, 1981
[60]p : ill. - (Development studies occasional paper ; no. 15). - Bibliography: p55-60
B84-39960 X.322/14071
3184226

AGRICULTURAL LAND. Developing countries
Soils. Erosion. Politico-economic aspects

BLAIKIE, Piers M. (Piers Macleod)
The political economy of soil erosion in developing countries. - London : Longman, 1985
118p : ill. - (Longman development studies). - Bibliography: p158-182. - Includes index
0-582-30089-4 X.529/69082
301665x

AGRICULTURAL LAND. Developing countries. Asia
Land tenure. Reform, 1934-1975

LADEJINSKY, Wolf
Agrarian reform as unfinished business. - New York ; Oxford : Oxford University Press for the World Bank, 1977
xi,603p. - Bibliography: p.580-585. - Includes index
0-19-920095-5 X.322/8645 Woolwich
1550349

AGRICULTURAL LAND. Dominican Republic
Land tenure. Reform, to date

DORE Y CABRAL, Carlos
Problemas de la estructura agraria dominicana. - Santo Domingo : Taller, 1979
122p. - (Biblioteca Taller ; 112). - Includes bibliography
r6708059 X.329/17669
9064842

Reforma agraria y luchas sociales en la República Dominicana 1966-1978. - 2a ed. - Santo Domingo : Taller, 1981
101p. - (Biblioteca Taller ; 121). - Includes bibliography
r6799103 X.329/17651
9064842

AGRICULTURAL LAND. East Asia
Land tenure. Reforms

AHMED, Zahir
Land reforms in South-east Asia. - New Delhi : Orient Longman, 1975
xii,196p. - Includes index
r6083258 X.809/43742
1724347

AGRICULTURAL LAND. Eastern Sumatra. Sumatra. Indonesia
Land tenure, 1863-1947

PELZER, Karl J.
Planter and peasant. - 's-Gravenhage : Nijhoff, 1978
vii,163p.;[35]p of plates. - (Verhandelingen van het Koninklijk Instituut voor Taal-, Land- en Volkenkunde ; 84)
90-247-2174-1 Ac.7519/32[84]
219077X

AGRICULTURAL LAND. Ecuador
Land tenure

GALARZA ZAVALA, Jaime
Los campesinos de Loja y Zamora. - Quito : Ediciones Solitierra, 1976
255p
r6047906 X.319/9659 Woolwich
1713477

AGRICULTURAL LAND. Ecuador
Land tenure, to 1974

GALARZA ZAVALA, Jaime
El yugo feudal. - 4.ed. - Quito : Solitierra, 1975
197p
r6047880 X.319/9622 Woolwich
1709224

AGRICULTURAL LAND. Ecuador
Land tenure. Law

CHECA COBO, Marco Antonio
Régimen de la propiedad de la tierra en el Ecuador. - Quito : Lexigrama, 1973
vi,363,22p.
r6047403 X.319/8875 Woolwich
1239813

ORTEGA JARAMILLO, Rubén
Manual de reforma agraria. - Guayaquil : Imprenta Editorial Didáctica, 1978
176p
r6159133 X.909/87102
1239813

AGRICULTURAL LAND. Ecuador
Land tenure. Law, to date

JURISPRUDENCIA agraria del Ecuador / [recopilación de] Jorge Duque G.. - Quito : Voluntad, 1981
166p
r6707992 X.200/45329
9125701

AGRICULTURAL LAND. El Priorat. Spain
Land use, 1960-1963

IGLÉSIES, Josep
Les minves dels cultius i de la població a la comarca del Priorat. - Barcelona : Fundació Salvador Vives Casajuana, 1973
88 p.. - (Publicacions de la Fundació Salvador Vives Casajuana ; no. 34)
84-232-0322-0 X.700/17085
0888702

AGRICULTURAL LAND. England
AGRICULTURAL DEVELOPMENT AND ADVISORY SERVICE
Agricultural land classification of England and Wales. - Pinner : The Ministry.
Report to accompany Sheet 85. Durham. 1971. - [1977]
[3]leaves,15p,fold plate. - (ALC 85). - Bibliography: p.15
0-85521-202-0 BS.3/485
1082906
Report to accompany Sheet 168. Winchester. 1975. - [1977]
[5],14,[4],fold plate. - (ALC 168). - Bibliography: (1p.)
B77-18606 BS.3/485
1082906
Report to accompany Sheet 183. Eastbourne. 1972. - [1977]
[4],12,[4]p,fold plate. - (ALC 183). - Bibliography: (1p.)
B77-18607 BS.3/485
1082906

AGRICULTURAL LAND. England
Compulsory purchase. Compensation. Law
HAMILTON, Richard Neville Delton
Compensation for compulsory acquisition of agricultural land and of rights in and over such land. - 3rd ed. - London : Royal Institution of Chartered Surveyors, 1980
[4],82p
B80-24254 X.329/12609 Woolwich
0209120

AGRICULTURAL LAND. England
Inheritance. Capital transfer tax. Law - For tax avoidance
THORNHILL, A. R. (Andrew Robert)
Passing down the family farm. - London : Oyez Longman, c1983
vii,53leaves. - (Oyez Longman intelligence reports)
0-85120-736-7 X.525/7830
2942100

AGRICULTURAL LAND. England
Irrigation
AGRICULTURAL DEVELOPMENT AND ADVISORY SERVICE
Thinking irrigation. - Revised [ed.]. - Pinner : The Ministry, 1976
[2],18p : 1 ill. - (Farm water supply ; leaflet no.2)
0-85521-188-1 BS.3/426
1083082

AGRICULTURAL LAND. England
Irrigation. Finance
AGRICULTURAL DEVELOPMENT AND ADVISORY SERVICE
Economics of irrigation. - Pinner : Ministry of Agriculture, Fisheries and Food, [1978]
[3]leaves,35p
B78-26623 BS.3/593
1527894

AGRICULTURAL LAND. England
Landscape conservation. Projects
HAMILTON, Peter, 1947-
Agricultural landscapes. - Cheltenham : The Commission, c1984
72p : ill(some col.). - (CCP ; 169)
0-86170-101-1 L.49/3070
3228770

AGRICULTURAL LAND. England
Law
GREGORY, Michael, 1925-
Essential law for landowners and farmers. - London : Granada, 1980
xxii,265. - Includes index
0-246-11213-1 X.329/13353 Woolwich
2125803

AGRICULTURAL LAND. England
Rights of way. Ploughing. Legal aspects
UNIVERSITY OF READING. Joint Centre for Land Development Studies
Ploughing footpaths and bridleways. - Cheltenham : Countryside Commission, c1985
104p. - (CCP ; 190)
0-86170-127-5 X.622/23622
3307123

AGRICULTURAL LAND. Europe
Birds
FLEGG, James John Maitland
Birds of lake, river, marsh and field. - Harmondsworth : Penguin, 1978
132p : col ill. - (Penguin nature guides). - Bibliography: p.128-129. - Includes index
0-14-063009-0 SRL
1745573

AGRICULTURAL LAND. European Community countries
Land tenure & transfer. Economic aspects
KELLY, P. W.
Agricultural land, tenure and transfer. - Dublin : An Foras Talúntais, 1982
xiv,100p. - (Socio-economic research series ; no.1). - Bibliography: p100
0-905442-64-4 X.329/16524
2926385

AGRICULTURAL LAND. Exmoor National Park
Heathland. Improvement. Effects on landscape
PORCHESTER, Henry George Reginald Molyneux Herbert, Baron
A study of Exmoor. - London : H.M.S.O., 1977
[3],93p. - Bibliography: p.89-93
0-11-751255-9 BS.414/1763
1335375

AGRICULTURAL LAND. Fens. East Anglia
Drainage, to 1982
DARBY, H. C. (Henry Clifford)
The changing Fenland. - Cambridge : Cambridge University Press, 1983
xv,267p : ill. - Includes index
0-521-24606-7 X.800/35947
2946874

AGRICULTURAL LAND. Fens. East Anglia
Drainage, 1319-1324
The RECORDS of a commission of sewers for Wiggenhall 1319-1324 / A.E.B. Owen [editor]. William Assheboume's book / Dorothy M. Owen [editor]. - [Norwich] : Norfolk Record Society, 1981
122p. - (Norfolk Record Society ; v.48). - Includes index
B84-41609 X.805/5955
3187543

AGRICULTURAL LAND. Fens. East Anglia
Drainage, ca 1600-1975
SUMMERS, Dorothy
The great level. - Newton Abbot : David and Charles, 1976
295p : ill. - Includes index
0-7153-7041-3 X.329/10539 Woolwich
0701394

AGRICULTURAL LAND. Fens. East Anglia
Drainage. Political aspects, 1626-1719
LINDLEY, Keith
Fenland riots and the English Revolution. - London : Heinemann Educational, 1982
ix,276p : ill. - Includes index
0-435-32535-3 X.809/51198
2706466

AGRICULTURAL LAND. Fiji
Land tenure, to date
MUNRO-FAURE, P. W.
The Agricultural Landlord and Tenant Act. - [Suva?] : Institute of Valuation and Estate Management of Fiji, [1980?]
xix,169p
r6807263 X.200/42193
9031022

AGRICULTURAL LAND. Finland
Land tenure, to date
JUTIKKALA, Eino
Talonpoika, aatelismies, kruunu. - [Helsinki] : Suomen historiallinen seura, 1983
117p. - (Tietolipas ; 92)
951-7173-29-6 Ac.9080/9[no.92]
9103880

AGRICULTURAL LAND. France
Irrigation, to date
INSTITUT NATIONAL DE LA RECHERCHE AGRONOMIQUE
L'INRA, l'eau et la production agricole. - Paris : [l'Institut], [1979]
xiii,269p,12p,13leaves of plates : ill
2-85340-196-0 S.E.400/869
8786461

AGRICULTURAL LAND. France
Land tenure, 1589-1815
RESENDE, Hernâni A
[Socialisme utopique et question agraire dans la transition du féodalisme au capitalisme : sur le concept d'égalitarisme agraire dans la Révolution française. Portuguese]. Igualitarismo agrário e socialismo utópico na transição do feudalismo para o capitalismo em França no século XVIII. - Lisboa : Livros Horizonte, 1979
187p. - (Colecção Dialéctica ; 1)
r6703214 X.329/14564 Woolwich
8828253

AGRICULTURAL LAND. France
Land tenure, 1958-
REY, Violette
Besoin de terre des agriculteurs. - Paris : Economica, 1982
389p. - (Geographie économique). - Includes bibliography
2-7178-0471-4 X.622/17353
9026975

AGRICULTURAL LAND. France
Land use
HUDSON, D. J. A.
Planning for agriculture in the urban fringe. - [London] : ADAS, c1983
26p : ill. - (ADAS overseas study tour programme ; 1982/83. Project ; no.19)
B84-39500 BS.3/865
3174018

AGRICULTURAL LAND. France
Land use, ca 1050-ca 1800
BLOCH, Marc
French rural history. - London : Routledge and Kegan Paul, 1978
xxxiii,258p,[20]p of plates. - Bibliography: p.xxxii-xxxiii. - Includes index
0-7100-8902-3 X.709/32128
1407422

AGRICULTURAL LAND. Gambia
Soils. Water
WILLIAMS, J. B. (James Benjamin)
Soil water investigation in the Gambia. - Surbiton : Land Resources Development Centre, Ministry of Overseas Development, 1979
xiv,180[i.e.188]p(36 fold),[13]leaves,[8]leaves of plates : ill. - (Technical bulletins / Land Resources Development Centre ; 3). - Bibliography: p.133-136
0-902409-79-4 BS.251/242(3)
2257696

AGRICULTURAL LAND. Germany
Land use, to date
HILDEBRANDT, Helmut
Studien zum Zelgenproblem. - Mainz : Geographisches Institut der Johannes Gutenberg-Universität, 1980
268p,[2]folded leaves of plates. - (Mainzer geographische Studien ; Hft.14)
3-88250-014-x X.322/10477 Woolwich
8756910

AGRICULTURAL LAND. Great Britain
- For historical geography
ADAMS, Ian Hugh
Agrarian landscape terms. - London : Institute of British Geographers, 1976
xii,314p : 1 ill. - (Special publications / Institute of British Geographers ; no.9). - Bibliography: p.207-276. - Includes index
0-901989-20-7 2020.c Reading Room
1049194
- [Another copy.] P.801/666[no.9]

AGRICULTURAL LAND. Great Britain
Acquisition & occupancy
GREAT BRITAIN. Committee of Inquiry into the Acquisition and Occupancy of Agricultural Land
Report of the Committee of Inquiry into the Acquisition and Occupancy of Agricultural Land. - London : H.M.S.O., [1979]
xiii,378p : ill. - (Cmnd.7599). - Bibliography: p.370-373
0-10-175990-8 BS.3/665
196514x

ROYAL INSTITUTION OF CHARTERED SURVEYORS
The acquisition and occupancy of agricultural land. - London : The Institution, 1977
[3],37 leaves
B78-14143 X.313/1527 Woolwich
1408275

AGRICULTURAL LAND. Great Britain
Land use
CONSERVATION SOCIETY. Land Use and Planning Working Party
Priority for agriculture. - Chertsey : The Society, 1976
[2],13p : ill
0-903127-05-9 2719.ff.45
0766070

ENVIRONMENTAL EDUCATION PROJECT (DEVON SCHOOLS)
Land for agriculture. - Exeter : Education Department, Devon County Council, 1978
[3]p,15 leaves. - Bibliography: leaf 9
0-86114-089-3 X.512/6583
1840258

AGRICULTURAL LAND. Great Britain
Land use-continued

NATIONAL ECONOMIC DEVELOPMENT COUNCIL. Economic Development Committee for Agriculture
Agriculture into the 1980s. - London : National Economic Development Office, 1977
iv,32p
0-7292-0191-0 BS.352/325
0766070

AGRICULTURAL LAND. Great Britain
Management. Implications of conservation of environment

CONSERVATION and agriculture / edited by Joan Davidson and Richard Lloyd. - Chichester : Wiley, 1977
xii,250p,[16]p of plates : ill. - Includes index
0-471-99502-9 SRL
1345761

AGRICULTURAL LAND. Great Britain
Nature conservation

BOVINGDON Hall / Countryside Commission. - [Cheltenham] : Countryside Commission, [1984]
32p : ill
0-86170-107-0 L.49/3206
1113917

NATURE CONSERVANCY COUNCIL
Nature conservation and agriculture. - London : The Council, 1977
[2],40p : ill. - Bibliography: p.36
B77-16467 SRL
1113917

AGRICULTURAL LAND. Great Britain
Ownership

ROYAL INSTITUTION OF CHARTERED SURVEYORS. Land Agency and Agriculture Division
The agricultural resources of the United Kingdom. - [London] : [R.I.C.S.], 1977
[6],69p. - Includes bibliographies
0-85406-079-0 X.313/1383 Woolwich
1086146

AGRICULTURAL LAND. Great Langdale. Cumbria
Landscape conservation. Projects: Lake District Upland Management Experiment. Trails: Upland Management Trail, Great Langdale - Walkers' guides

GREAT BRITAIN. Countryside Commission
The Upland Management Trail, Great Langdale. - [Cheltenham] : [The Commission], [1976]
[16]p(2 fold) : ill
B77-17649 BS.116/114
1122029

AGRICULTURAL LAND. Guatemala
Irrigation

MATHEWSON, Kent, 1946-
Irrigation horticulture in highland Guatemala. - Epping : Bowker, 1984
xxvi,183p,[8]p of plates : ill. - (Bowker (UK) replica edition). - Bibliography: p119-138. - Includes index
0-86531-973-1 X.329/18456
3175944

AGRICULTURAL LAND. Guatemala
Land tenure related to economic conditions

INTER-AMERICAN COMMITTEE FOR AGRICULTURAL DEVELOPMENT
Tenencia de la tierra y desarrollo socio-económico del sector agrícola en Guatemala ... - 2.ed. - Guatemala : Editorial Universitaria, 1971
395p,[14]p of plates : maps. - (Colección Realidad nuestra ; n.2). - Bibl
r6159767 X.329/13118 Woolwich
2456214

AGRICULTURAL LAND. Hartsop region. Cumbria
Agricultural land use, landscape conservation & recreational land use. Interrelationships

FEIST, Michael James
A study of the Hartsop Valley. - [Cheltenham] : [Countryside Commission], 1976
[1],xii,146p : ill. - (CCP. 92). - Bibliography: p.145-146
0-902590-49-9 BS.116/51
077314x

AGRICULTURAL LAND. Hereford and Worcester
Landscape conservation. Projects

The HEREFORD and Worcester new agricultural landscapes project / Richard Jackson, project officer. - [Worcester] : [The Council], [1983?]
35p : ill
B84-34240 L.49/2830
3148483

AGRICULTURAL LAND. Hesse. West Germany
Land tenure. Reform, to date

FLECK, Peter
Agrarreformen in Hessen-Darmstadt. - Darmstadt : Hessische Historische Kommission, 1982
xii,384p. - (Quellen und Forschungen zur hessischen Geschichte ; 43). - Includes bibliography
3-88443-132-3 Ac.7354[no.43]
8976953

AGRICULTURAL LAND. Highland Region. Scotland
Land tenure. Reform. Highland Clearances

MCKICHAN, Finlay
The Highland Clearances. - London : Longman, 1977
96p : ill. - (Then and there series)
0-582-20541-7 WP.c.9/133
1376373

AGRICULTURAL LAND. Highlands & Islands. Scotland
Land use. Policies of Highlands and Islands Development Board

HIGHLANDS AND ISLANDS DEVELOPMENT BOARD
Highland agriculture and land use. - Inverness : The Board, 1977
28,[1]p : ill. - (Occasional bulletin ; 7). - Bibliography: p.28
0-902347-52-7 BS.178/6
1295020

AGRICULTURAL LAND. Hilly regions. Great Britain
Fertilisers. Use

SANDFORD, H.
The use of fertilisers for hill and upland improvement. - [London] : Fertiliser Society, [1979]
42p. - (Proceedings / Fertiliser Society ; no.181)
B79-27580 SRL
1933582

AGRICULTURAL LAND. Honduras
Land tenure. Reform

CID, Rafael del
Reforma agraria y capitalismo dependiente. - Tegucigalpa : Editorial Universitaria, 1977
xviii,162p
r6158808 X.311/5190 Woolwich
2195224

ESTRATEGIA de desarrollo y reforma agraria / [por C. Santos de Morais ... et al.]. - Tegucigalpa : [Instituto Nacional Agrario], [1975]
129p. - (Serie didáctica / PROCCARA ; no.7)
r6157947 X.529/38188
2192667

HONDURAS / COHEP ; ANEC ; FENAGH ; INA. - Tegucigalpa : Instituto Nacional Agrario, [1975?]
81p. - (PROCCARA: Programa de capacitación campesina para la reforma agraria ; n.77)
r6101935 X.329/12259 Woolwich
2192667

HONDURAS, debates sobre reforma agraria / [compilación de textos por el Consejo Hondureño de la Empresa Privada] COHEP... [et al.]. - Tegucigalpa : Instituto Nacional Agrario, [1976?]
81p. - (PROCCARA. Programa de ediciones y divulgación ; 77)
r6099994 X.319/20268
2192667

AGRICULTURAL LAND. Honduras
Land tenure. Reform, to date

HONDURAS / PROCCARA - INA. - Tegucigalpa : [s.n.], 1975
13p
r6703307 X.525/5147
8549192

AGRICULTURAL LAND. Hungary
Land tenure. Reform - Conference proceedings

ORSZÁGOS FÖLDREFORM-EMLÉKÜNNEPSÉG (1970 : Békéscsaba)
A földreform történelmi jelentősége. - Budapest : Magyar Mezőgazdasági Múzeum, 1972
69p. - (Mezőgazdaságtörténeti tanulmányok ; 5)
r6075015 Ac.817/3[no.5.]
1542869

AGRICULTURAL LAND. Hungary
Land tenure. Reform, 1945-

ILLYÉS, Gyula
Honfoglalók között. Az Ideiglenes Nemzeti Kormány 600/1945. M.E. számú rendelete a nagybirtokrendszer megszüntetéséről és a földmíves nép földhözjuttatásáról. - Budapest : Mezőgazdasági Kiadó, 1980
77p
963-231-010-1 X.329/15836
8877246

AGRICULTURAL LAND. India (Republic)
Land tenure. Reform. Effects on economic growth, 1951-1971

SIDHU, B. S.
Land reform, welfare and economic growth. - Bombay : Vora, 1976
xvi,271p : ill. - Bibliography : p.257-271
r6091732 X.529/33164
1891839

AGRICULTURAL LAND. Indonesia
Conversion of forests

ROSS, M. S.
A method for selecting agricultural land from production and conversion forests in Indonesia. - [Oxford] : Commonwealth Forestry Institute, University of Oxford, 1983
x,72p. - (CFI occasional papers ; no.22). - Bibliography p68-72
B85-09166 X.322/14575
3229025

AGRICULTURAL LAND. Investments. Great Britain
Unit trusts: Agricultural Property Unit Trust for Pension Funds and Charities - Serials

AGRICULTURAL PROPERTY UNIT TRUST FOR PENSION FUNDS AND CHARITIES
Report and accounts 31st December .... - 197?-. - [London] : The Trust, [197-]-
v.
Holdings: 1981-
r6864792 P.513/2858
2741164

AGRICULTURAL LAND. Ireland
Land tenure. Political aspects, 1879-1882

CLARK, Samuel
Social origins of the Irish land war. - Princeton ; Guildford : Princeton University Press, 1979[i.e.1980]
xiii,418p : 1 ill. - Bibliography: p.383-404. - Includes index
0-691-05272-7 X.520/22108
2151448

AGRICULTURAL LAND. Ireland (Republic)
Drainage. Economic aspects

BRUTON, Richard
Land drainage policy in Ireland. - Dublin : Economic and Social Research Institute, 1982
93p : ill. - (Policy research series ; no.4)
0-7070-0049-1 X.809/53323
2788462

AGRICULTURAL LAND. Ireland (Republic)
Economic aspects

KELLY, Paul W.
The price of agricultural land in the Republic of Ireland 1978-1981. - Dublin : Economics & Rural Welfare Research Centre An Forás Talúntais, [1983]
v,25p : ill. - (Situation and outlook bulletin ; no.5 1983). - Bibliography: p25
0-905442-74-1 X.329/18090
2880903

AGRICULTURAL LAND. Ireland (Republic)
Economic aspects - Conference proceedings

The LAND question / An Foras Taluntais, Economics and Rural Welfare Research Centre. - Dublin : The Centre, [1980]
170p : ill. - Includes bibliographies
B83-18159 X.329/16192
2880903

AGRICULTURAL LAND. Italy
Land tenure. Reform, to date

BOGLIARI, Francesco
Il movimento contadino in Italia. - Torino : Loescher, [1980]
346p. - (Documenti della storia ; 28)
r6507093 X.329/16239
8805830

La RIFORMA fondiaria / saggi e ricerche di Carlo Barberis ... [et al.]. - Milano : Angeli, [1979]
2v. - (Collana Insor ; 6). - Bibliography
r6599066 X.329/15082 Woolwich
8805830

AGRICULTURAL LAND. Italy
Settlement by Ancient Roman army ex-servicemen, B.C.47-B.C.14

KEPPIE, L. J. F. (Lawrence J F)
Colonisation and veteran settlement in Italy. - London : British School At Rome, 1983
xv,233p,viiip of plates : ill. - Bibliography: pxiii-xv. - Includes index
0-904152-06-5 X.800/38961
3151379

AGRICULTURAL LAND. Itatiba & Morungaba. Brazil
Land use

NAVARRA, Wanda Silveira
O uso da terra em Itatiba e Morungaba. - São Paulo : Universidade de São Paulo, Instituto de Geografia, 1977
216p.,[2]fold.leaves of plates : ill. - (Serie Teses e monografias ; no.29). - Includes bibliographies
r6158406 X.700/21723
2188287

AGRICULTURAL LAND. Japan
Land tenure. Reform, 1946-1956

DORE, Ronald
Land reform in Japan. - London : Athlone, 1984
xxxi,510p.[4]p of plates : ill. - Bibliography: p492-500. - Includes index
0-485-11241-8 X.809/62936
3013790

AGRICULTURAL LAND. Kápolna. Hungary
Land tenure. Reform, to date

KRÚDY, Gyula
A kápolnai földosztás. - Budapest : Magvető Kiadó, [1982]
55p. - (Gondolkodó magyarok)
963-271-713-9 X.629/19614
8890978

AGRICULTURAL LAND. Kazakhstan
Land use

BREZHNEV, L. I. (Leonid Il'ich)
[Speeches. Selections]. Voprosy agrarnoĭ politiki KPSS i osvoenie tselinnykh zemel' Kazakhstana. - Moskva : Izdatel'stvo politicheskoĭ literatury, 1974
357p.,[1]leaf of plates
r6402063 X.329/13582 Woolwich
2482436

AGRICULTURAL LAND. Kenya
Land tenure. Reform, 1950-1983

LEO, Christopher
Land and class in Kenya. - Toronto ; London : University of Toronto Press, c1984
xii,244p. - (The Political economy of world poverty ; 3). - Bibliography: p231-236. - Includes index
0-8020-2532-3 X.520/36998
3273350

AGRICULTURAL LAND. Kenya
Settlement schemes

CLAYTON, Eric
A comparative study of settlement schemes in Kenya. - [Ashford] : Wye College, University of London, 1978
iii,60p. - (Occasional paper / Wye College, Agrarian Development Unit ; no.3)
B83-05415 X.0525/446(3)
2823624

AGRICULTURAL LAND. Kenya
Soils. Degradation

BAKER, Randall
Land degradation in Kenya. - Norwich : School of Development Studies, University of East Anglia, 1983
28p. - (Discussion paper ; no.82). - Bibliography: p28
B85-03254 X.322/14312
3213013

AGRICULTURAL LAND. La Laguna. Mexico
Land tenure. Policies of Cárdenas, Lázaro

ANDA, Gustavo de
El cardenismo. - México : el Autor, 1974
136 p.
r6011776 X.709/20932
0886009

AGRICULTURAL LAND. Lake District National Park
Landscape conservation. Projects: Lake District Upland Management Experiment

The LAKE District Upland Management Experiment. - [Cheltenham] : [The Countryside Commission], 1976
[1],vi,41p : ill(chiefly col). - (CCP ; 93)
0-902590-48-0 BS.116/51
0777595

AGRICULTURAL LAND. Lapland. Sweden
Land tenure, to date

STENMAN, Lennart
Avvittringen i Västerbottens läns lappmarker. - Uppsala : Uppsala universitet, Kulturgeografiska institutionen, 1983
1v.(various pagings). - (Forskningsrapporter fran Kulturgeografiska institutionen, Uppsala universitet ; nr 83). - Includes bibliography
91-506-0421-x P.803/199[no.83]
9063617

AGRICULTURAL LAND. Latin America
Land tenure, to date

HACIENDAS and plantations in Latin American history / edited by Robert G. Keith. - New York ; London : Holmes & Meier Publishers, 1977
vi,200p
0-8419-0319-0 X.322/12607
9016538

AGRICULTURAL LAND. Latin America
Land tenure related to economic conditions

FEUDALISMO, capitalismo, subdesarrollo / Luis Vitale ... [et al.]. - Bogotá : Latina, 1977
281p
r6085162 X.519/40423
1863177

AGRICULTURAL LAND. Latin America
Land tenure. Reform

BELL, W. S.
Issues in Latin American agrarian reform. - Liverpool : Centre for Latin American Studies, University of Liverpool, c1983
42p. - (Working paper ; 1). - Includes bibliographies
0-902806-10-6 P.801/943
2245981

GARCÍA, Antonio
Sociología de la reforma agraria en América Latina. - Buenos Aires : Amorrortu, 1973
239p. - (Biblioteca de sociología : Serie América Latina)
r6162067 X.329/12995 Woolwich
2245981

La REFORMA agraria y el desarrollo / [por Jacques Chonchol ... et al.]. - Tegucigalpa : [Instituto Nacional Agrario], 1975
173p : ill. - (PROCCARA: Serie didáctica ; no 4)
r6157946 X.529/38178
2245981

AGRICULTURAL LAND. Latin America
Land tenure. Reform - Dictionaries

SANTOS DE MORAIS, Clodomir
Diccionario de reforma agraria. - Ciudad Universitaria Rodrigo Facio, San José, Costa Rica : Editorial Universitaria Centroamericana, 1973
533p. - (Colección Seis. Serie mayor). - Includes bibliographies
r6159123 X.329/12197 Woolwich
2245981

AGRICULTURAL LAND. Latin America
Land tenure. Reform, 1945-

SHAW, R. Paul
Land tenure and the rural exodus in Chile, Colombia, Costa Rica, and Peru. - Gainesville : University Presses of Florida, 1976
180p. - (Center for Latin American studies, University of Florida : Latin American monographs, second series ; 19)
0-8130-0528-0 AC.2687.hb/2.[19.]
8563365

AGRICULTURAL LAND. Latin America
Land tenure. Reform, 1967-1972

LINDQVIST, Sven
Land and power in South America. - Shortened ed. - Harmondsworth : Penguin, 1979
333p : ill. - (A pelican original). - Bibliography: p.311-322. - Includes index
0-14-022129-8 X.519/29915
1922351

AGRICULTURAL LAND. Latin America
Settlement

MAOS, Jacob O.
The spatial organization of new land settlement in Latin America. - Epping : Bowker, 1984
xix,179p : ill. - (Dellplain Latin American studies ; no.15). - Bibliography: p167-176. - Includes index
0-86531-626-0 X.529/65244
3194205

AGRICULTURAL LAND. León region. Mexico
Land tenure, 1700-1860

BRADING, David Anthony
Haciendas and ranchos in the Mexican Bajío, León, 1700-1860. - Cambridge : Cambridge University Press, 1978
xviii,255p : ill. - (Cambridge Latin American studies ; 32). - Bibliography: p.243-248. - Includes index
0-521-22200-1 X.809/44256
1875604

AGRICULTURAL LAND. Locknevi. Sweden
Land tenure, to date

FOGELVIK, Stefan
Man, landscape and society. - Stockholm : Almqvist & Wiksell International, 1981
84p : ill. - (Stockholm studies in human geography ; 2)
91-220-0393-2 X.0525/528(2)
897697x

AGRICULTURAL LAND. Lowland regions. England
Landscape conservation

GREAT BRITAIN. Countryside Commission
New agricultural landscapes, a discussion paper. - [Cheltenham] : [The Commission], 1974
[1],ii,12p. - (CCP 76a)
0-902590-19-7 X.319/10097
0251380

New agricultural landscapes, issues, objectives and action. - Cheltenham : [The Commission], 1977
[3],12p. - (CCP102)
0-902590-51-0 BS.116/51
0251380
- 2nd.ed. - 1980. - ISBN 0-902590-84-7 BS.116/51

AGRICULTURAL LAND. Matapuquio. Peru
Land tenure. Reform, 1969

SKAR, Harald O. (Harald Olav)
The warm valley people. - Oslo : Universitetsforlaget, c1982
xi,350p : ill. - (Oslo studies in social anthropology ; no.2). - Bibliography: p332-344. - Includes index
82-00-05682-1 X.529/54504
2909960

AGRICULTURAL LAND. Mexico
Land tenure, to 1895

OROZCO, Wistano Luis
Los ejidos de los pueblos. - México : Ediciones "El Caballito", 1975
260p. - (Colección Clásicos de fragua mexicana)
r6041153 X.319/8909 Woolwich
1313924

AGRICULTURAL LAND. Mexico
Land tenure, 1900-1970

GUTELMAN, Michel
[Réforme et mystification agraires en Amérique Latine. Le cas du Mexique. Spanish]. Capitalismo y reforma agraria en México. - México : Ediciones Era, 1974
290 p.. - (Colección Problemas de México)
r6011816 X.319/8542 Woolwich
0888672

AGRICULTURAL LAND. Mexico
Land tenure. Reform, 1325-1963

SILVA HERZOG, Jesús
El agrarismo mexicano y la reforma agraria. - 2. ed. aumentada, 1. reimpresión. - México : Fondo de Cultura Económica, 1974
627 p.. - (Vida y pensamiento de México). - Bibliography: p. 596-613
r6014707 X.319/8491 Woolwich
0870285

AGRICULTURAL LAND. Mexico
Land tenure. Reform, 1917-1976

SANDERSON, Steven E.
Agrarian populism and the Mexican state. - Berkeley ; London : University of California Press, c1981
xx,290p : 1ill. - Bibliography: p263-277. - Includes index
0-520-04056-2 X.520/24309
2507374

AGRICULTURAL LAND. Mexico
Land tenure. Reform. Law, to 1971

MENDIETA Y NÚÑEZ, Lucio
El problema agrario de México y la Ley Federal de Reforma Agraria. - 13a ed. Corregida, aumentada y puesta al día. - México : Editorial Porrúa, 1975
589p.,2 fold. leaves of plates
r6052087 X.320/5177 Woolwich
1315889

AGRICULTURAL LAND. Morocco
Irrigation, to date

La QUESTION hydraulique / N. Bouderbala [et al.]. - [Rabat] : [Presses de Graphitec], 1984-
v. : ill
r6298257 X.0800/2034
9103872

AGRICULTURAL LAND. Nepal
Land tenure. Reform, 1951-1975

NEPAL RASTRA BANK
Agricultural credit survey Nepal. - Kathmandu : Nepal Rastra Bank, Research Department, 1972
4v.
r6005340 X.311/4879 Woolwich
0763314

REGMI, Mahesh C. (Mahesh Chandra)
Landownership in Nepal. - Berkeley : London : University of California Press, 1976
xvii,252p : 1 ill. - Bibliography: p.237-247. - Includes index
0-520-02750-7 X.320/10566 Woolwich
0763314

AGRICULTURAL LAND. Nepal
Land tenure. Reform. Policies of government, 1951-ca 1985

NEPAL. Ministry of Land Reforms
Evaluation of land reform in Nepal. - Kathmandu, 1973
124p
r6028688 X.320/5169 Woolwich
1027808

AGRICULTURAL LAND. Netherlands
Conservation. Role of environment planning

MARTYN, E. B. H. (Eldred Bridgenan H)
Planning control to protect agricultural land. - [London] : ADAS, [1983?]
16,7p : ill
B84-36765 BS.3/871
2862565

AGRICULTURAL LAND. Nigeria
Land tenure, to date

FAMORIYO, Segun
Land tenure and agricultural development in Nigeria. - [Ibadan] : Nigerian Institute of Social and Economic Research, 1979
xi,146p. - Includes bibliography
r6811141 X.800/37239
9047557

AGRICULTURAL LAND. Norfolk
Trees - For landscape conservation

HARDY, P. N. G.
Farmland tree survey of Norfolk. - [Norwich] : Norfolk County Council, [1977]
[2],41p : ill
B78-00362 SRL
1273906

AGRICULTURAL LAND. Norfolk Broads
Drainage. Windmills, ca 1750-ca 1920

SMITH, Arthur Carlton
Drainage windmills of the Norfolk marshes. - Stevenage, Herts. : Stevenage Museum, 1978
[1],46p : ill. - Bibliography: p.45
0-9504239-5-5 X.313/1595 Woolwich
1768042

AGRICULTURAL LAND. North American Indians. Western states. United States
Land tenure. Law: General Allotment Act of 1887. Operation, to 1934

CARLSON, Leonard A.
Indians, bureaucrats, and land. - Westport, Conn. ; London : Greenwood, 1981
xii,219p : ill. - (Contributions in economics and economic history ; no.36). - Bibliography: p207-212. - Includes index
0-313-22533-8 X.322/10403 Woolwich
2588951

AGRICULTURAL LAND. North-east Brazil
Land tenure, to date

ANDRADE, Manuel Correia de
Nordeste. - Recife : Editora Guararapes, 1981
119p. - (Cadernos Guararapes ; v.2)
r6709260 X.529/67490
9130292

AGRICULTURAL LAND. North-east Brazil
Land use, 1945-

MELO, Mário Lacerda de
Regionalização agrária do Nordeste. - Recife : SUDENE, Divisão de Política Espacial, 1978
225p,8folded leaves of plates. - (Estudos regionais ; 3)
r6701529 LAS.B.275/33 Woolwich
8713553

AGRICULTURAL LAND. North Vietnam
Land tenure. Reform, 1953-1956

MOISE, Edwin E. (Edwin Evariste), 1946-
Land reform in China and North Vietnam. - Chapel Hill ; London : University of North Carolina Press, c1983
xiv,305p : 2ill. - Bibliography: p289-296. - Includes index
0-8078-1547-0 X.520/31208
2951444

AGRICULTURAL LAND. Northern Ireland
Land tenure

MCBURNEY, S. A.
Land leasing in Northern Ireland. - Belfast : Economics & Statistics Division, Department of Agriculture, N.I., [1983?]
33p. - (Studies in agricultural economics)
B84-03234 X.525/8405
2985020

AGRICULTURAL LAND. Northern Ireland
Nature conservation - Conference proceedings

NATURE conservation and agriculture. - [Londonderry] : [The Institute], 1980
[51]p : ill
B81-21762 SRL
2411067

AGRICULTURAL LAND. Oettingen. West Germany
Land tenure, to date

HOPFENSITZ, Josef
Studien zur oberdeutschen Agrarstruktur und Grundherrschaft. - München : Beck, 1982
178p,plates. - (Schriftenreihe zur bayerischen Landesgeschichte ; Bd.75). - Bibliography
3-406-10475-4 X.800/35418
8945225

AGRICULTURAL LAND. Östergötland. Sweden
Land use, to date

WIDGREN, Mats
Settlement and farming systems in the early Iron Age. - Stockholm : Almqvist & Wiksell International, 1983
132p : ill. - (Stockholm studies in human geography ; 3). - Includes bibliography
91-220-0602-8 X.0525/528(3)
8976961

AGRICULTURAL LAND. Pakistan
Small scale solar energy irrigation systems

HOWES, Michael, 1947-
The potential for small-scale solar-powered irrigation in Pakistan. - Brighton : University of Sussex, Institute of Development Studies, 1982
vi,41p. - (Commissioned study / IDS). - Bibliography: p32
B83-35067 P.623/1254
2941341

AGRICULTURAL LAND. Paraguay
Land tenure, 1500-1970

PASTORE, Carlos
La lucha por la tierra en el Paraguay. - Montevideo : Editorial Antequera, 1972
xvi, 526 p.
r6011782 X.708/18385
088507X

AGRICULTURAL LAND. Paraguay
Land tenure. Reform, 1963-1973

INSTITUTO DE BIENESTAR RURAL (Paraguay)
La reforma agraria en el Paraguay. - Asunción : I.B.R., [1975]
xxvi,256p : ill
r6048331 X.311/4809 Woolwich
1317776

AGRICULTURAL LAND. Paraíba. Brazil
Dispossession, to date

GRUPO DE ASSESORIA E EDUCAÇÃO POPULAR
Pela reforma agrária. - [João Pessoa?] : O Grupo, [1980]
[1]leaf : ill
r6159526 Cup.21.g.2(62)
875344X

AGRICULTURAL LAND. Pécs region. Hungary
Drainage

LOVÁSZ, György
Mezőgazdasági célú felszíni víztárolási lehetőségek Pécs közelében. - Pecs : [s.n.], 1976
30p : ill. - (Magyar Tudományos Akadémia. Dunántúli Tudományos Intézet. Közlemények ; 14)
r6007713 P.803/207(14)
1500279

AGRICULTURAL LAND. Peru
Land tenure. Reform

ATUSPARIA, Pedro
La izquierda y la reforma agraria peruana. - [Lima] : Labor, [1977]
58p
r6085819 X.519/40611
198232X

EGUREN LÓPEZ, Fernando
Reforma agraria, cooperativización y lucha campesina - Lima : DESCO, Centro de Estudios y Promoción del Desarrollo, 1975
239p.,[1] fold. leaf of plates
r6053993 X.319/9399 Woolwich
1529145

AGRICULTURAL LAND. Peru
Land tenure. Reform, 1895-

ALBERTS, Tom
Agrarian reform and rural poverty. - Lund : University of Lund, Research Policy Institute ; Stockholm : Institute of Latin American Studies, 1981
306p. - (Lund economic studies ; no 23). - Bibliography
91-86002-10-4 X.0700/366(23)
860116x

DOLZER, Hermann
Die bäuerliche Gesellschaft. - Frankfurt am Main ; Cirencester : Lang, 1979
246p. - (Europäische Hochschulschriften. Reihe 22. Soziologie ; Bd.35). - Bibliography: p.213-246
3-8204-6531-6 X.529/45119
860116x

MONHEIM, Felix
Die Entwicklung der peruanischen Agrarreform 1969-1979 und ihre Durchführung im Departement Puno. - Wiesbaden : Steiner, 1981
37p. - (Erdkundliches Wissen ; Hft.55)
3-515-03629-6 WP.934/55
860116x

AGRICULTURAL LAND. Peru
Land tenure. Reform, 1964-1973

HARDING, Colin, 1945-
Agrarian reform and agrarian struggles in Peru. - Cambridge : Centre of Latin American Studies, University of Cambridge, 1974
[1],26 leaves. - (Working papers / University of Cambridge. Centre of Latin American Studies)
0-904927-04-0 X.313/1367 Woolwich
069407x

AGRICULTURAL LAND. Peru
Land tenure. Reform, 1968-1977

CLEAVES, Peter S. (Peter Shurtleff)
Agriculture, bureaucracy, and military government in Peru. - Ithaca ; London : Cornell University Press, 1980
329p : ill. - Bibliography: p299-321. - Includes index
0-8014-1300-1 X.329/13589 Woolwich
2373726

GUILLET, David
Agrarian reform and peasant economy in southern Peru. - Columbia ; London : University of Missouri Press, 1979
xv,227p : ill. - Bibliography: p.213-222. - Includes index
0-8262-0263-2 X.529/34143
1964631

AGRICULTURAL LAND. Peru
Land tenure. Reform, 1969-1976

VALDERRAMA LEÓN, Mariano
7 años de reforma agraria peruana, 1969-1976. - Lima : Pontificia Universidad Católica del Perú, 1976
632p
r6065143 X.100/27410
1534831

AGRICULTURAL LAND. Philippines
Land tenure. Law - Texts

PHILIPPINES
[Code of Agrarian Reforms annotated (as amended by presidential decrees)]. Code of Agrarian Reforms annotated (as amended by presidential decrees). - 1973 revised ed. - Quezon City : Central Lawbook Publishing Co., 1973
vii,434p
r6072540 X.200/31605
1704249

AGRICULTURAL LAND. Philippines
Land tenure. Law, to 1974 - Conference proceedings

INSTITUTE ON AGRARIAN LAWS (1st : 1974 : UP Law Center)
Agrarian reform and law. - Quezon City : U. P. Law Center, 1975
xl, 233 p.. - (Law institute series ; 87). - Includes index
r6010649 X.200/30777
1047159

AGRICULTURAL LAND. Poland
Land tenure

POLICIŃSKI, Alojzy
Darowizna gospodarstwa rolnego. - Warszawa : Panstwowe Wydawnictwo Naukowe, 1976
108p. - (Polska Akademia Nauk. Problemy rejonów uprzemyslawianych)
r6044260 X.311/4788 Woolwich
1241753

AGRICULTURAL LAND. Poland
Land tenure, 1945

TKACZYK, Tadeusz Paweł
Panstwowy Fundusz Ziemi. - Warszawa : Ludowa Spółdzielnia Wydawnicza, 1979
382p
83-205-3079-2 X.319/20155 Woolwich
2033429

AGRICULTURAL LAND. Poland
Land tenure. Reform. Policies of political parties, 1918-

WOJTAS, Andrzej
Problematyka agrarna w polskiej myśli politycznej 1918-1948. - Warszawa : Ludowa Spółdzielnia Wydawnicza, 1983
483p. - Includes bibliography
83-205-3362-7 X.329/19015
9162240

AGRICULTURAL LAND. Portugal
Land tenure, to date

DRAIN, Michel
Occupations de terres et expropriations dans les campagnes portugaises. - Paris : Éditions du Centre National de la Recherche Scientifique, 1982
227p : ill. - (Travaux de la R.C.P. ; 479)
2-222-03052-8 X.322/12993
9047956

AGRICULTURAL LAND. Portugal
Land tenure, 1926-1974

TRIBUNAL CÍVICO SOBRE A REFORMA AGRARIA
A Reforma Agrária acusa. - Lisboa : Caminho, 1980
431p
r6701405 X.329/14305 Woolwich
8571422

AGRICULTURAL LAND. Portugal
Land tenure, 1950-1970

FREITAS, Eduardo de
Modalidades de penetração do capitalismo na agricultura. - Lisboa : Presença, 1976
179p.,8p. of plates. - (Biblioteca de textos universitários ; n. 16)
r6047449 X.319/8953 Woolwich
1315749
- [Another copy.] X.319/8727 Woolwich

AGRICULTURAL LAND. Portugal
Land tenure. Reform

EM defesa da reforma agrária. - Lisboa : Avante!, 1977
136p
r6066157 X.519/39452
1784196

GERVÁSIO, António
Pela reforma agrária : a terra a quem a trabalha. - Lisboa : Avante!, 1975
14p
r6066325 X.709/27519
1784196

O PCP e a luta pela reforma agrária. - Lisboa : Avante!, 1975
203p. - (Cadernos do PCP ; n.7)
r6066341 X.709/27531
1784196

AGRICULTURAL LAND. Portugal
Land tenure. Reform, 1974-1976

CARDOSO, António Lopes
Luta pela reforma agrária. - Lisboa : Diabril, 1976
243p.. - (Colecção Teoria e prática ; 18)
r6025518 X.319/9357 Woolwich
1473522

GONÇALVES, Manuela
Uma legalidade revolucionária. - [Lisboa?] : Comissão Revolucionária de Apoio à Reforma Agrária, 1976
21p : ill. - (Série História da reforma agrária ; n.2)
r6101656 X.709/29643
1473522

AGRICULTURAL LAND. Portugal
Land tenure. Reform, 1974-1977

FERNANDES, Blasco Hugo
Reforma agrária. - Lisboa : Seara Nova, 1978
163p.,4 fold.leaves. - (Colecção Que país? ; n.12)
r6101807 X.708/47692
1863630

GARIN, João
Reforma agrária. - Lisboa : Edições do Templo, 1977
695p : ill. - (Colecção Pais real)
r6066339 X.102/1283
1863630

AGRICULTURAL LAND. Portugal
Land tenure. Reform, 1974-

BARRETO, António
Memória da reforma agrária. - [Mem Martins] : Publicações Europa-América, [198-?]
361p. - (Estudos e documentos ; 184) (A reforma agrária ; v.1)
r6706308 X.809/59629
8571430

AGRICULTURAL LAND. Portugal
Land tenure. Reform, 1975

BANDARRA, Álvaro
A estrutura agrária portuguesa transformada?. - Lisboa : Iniciativas Editoriais, 1976
139p. - (Colecção Século XX-XXI)
r6047313 X.319/8783 Woolwich
1222678

REBELLO, José Pequito
Boas e mas reformas agrárias : o decreto-lei 406-A/75. - Porto : o Autor, 1975
16p
r6027561 X.319/8735 Woolwich
1125265

AGRICULTURAL LAND. Portugal
Land tenure. Reform, 1975-1976

FERREIRA, Vitor Matias
Da reconquista da terra à reforma agrária. - Lisboa : A Regra do Jogo, 1977
180p
r6064974 X.709/25396
1500767

ROSA, Eugénio
Nikolaï Kochin. - Lisboa : Caminho, 1977
218p. - (Temas políticos e sociais ; n.17)
r6086017 X.708/45698
1500767

AGRICULTURAL LAND. Portugal
Land tenure. Reform, 1977

CARDOSO, António Lopes
A nova lei da reforma agrária. - Lisboa : Livros Horizonte, 1977, 1978 printing
184p. - (Colecção Movimento ; n.22)
r6086296 X.319/20107 Woolwich
2033542

AGRICULTURAL LAND. Portugal
Land tenure. Reform. Implications of constitutional law

REBELLO, José Pequito
O problema da Constituição. - [Lisboa] : o Autor, 1977
8p
r6065187 X.100/27465
1541196

AGRICULTURAL LAND. Pre-Colombian Latin America
Irrigation - Conference proceedings

DRAINED field agriculture in Central and South America / edited by J.P. Darch. - Oxford : B.A.R., 1983
xiv,263p : ill. - (Proceedings) (BAR. International series ; 189). - Includes bibliographies
0-86054-241-6 X.425/5281
3278158

AGRICULTURAL LAND. Proposed sites for airports. Japan
Protest by agricultural communities

APTER, David E. (David Ernest)
Against the state. - Cambridge, Mass. ; London : Harvard University Press, 1984
ix,271p, [12]p of plates : ill. - Includes index
0-674-00920-7 X.520/35943
0480304

AGRICULTURAL LAND. Queensland
Land use, to date

FAUTZ, Bruno
Agrarlandschaften in Queensland. - Wiesbaden : Steiner, 1984
195p.. - (Erdkundliches Wissen ; Hft.65). - Includes bibliography
3-515-03890-6 WP.934/65
9127992

AGRICULTURAL LAND. Romania
Land tenure. Reform, to 1944

ŞANDRU, D.
Reforma agrară din 1921 în România. - Bucureşti : Academia Republicii Socialiste România, 1975
357p. - (Biblioteca istorică ; 43)
r6457267 Ac.743/84[43]
9091300

AGRICULTURAL LAND. Rural regions. England
Land use. Planning

ADVISORY COUNCIL FOR AGRICULTURE AND HORTICULTURE IN ENGLAND AND WALES
Agriculture and the countryside. - [London] : [The Council], 1978
[1],vii,65p
B79-33984 BS.3/614
1964992

AGRICULTURAL LAND. Russia
Land tenure

DUBROVSKIĬ, Sergeĭ Mitrofanovich
Sel'skoe khozīaĭstvo i krest'īanstvo Rossii v period imperializma. - Moskva : Nauka, 1975
397p,[1]leaf of plates
r6175876 X.329/9036 Woolwich
2216256

AGRICULTURAL LAND. Russia
Land tenure, to 1917 - Statistics

MATERIALY po istorii agrarnykh otnoshenii v Rossii v kontse XIX-nachale XXvv. / [sostavitel' N.A. Proskurīakova] ; [otvetstvennyĭ redaktor A.P. Korelin]. - Moskva : Akademiīa nauk SSSR, Institut istorii SSSR, 1980
79p
r6485839 L.45/1616
8524173

AGRICULTURAL LAND. Russia
Land tenure. Reform

DRUZHININ, N. M. (Nikolaĭ Mikhaĭlovich)
Russkaīa derevnīa na perelome 1861-1880 gg.. - Moskva : Nauka, 1978
286p
r6097188 X.800/13983
1233181

PETROV, Aleksandr Petrovich
Kritika fal'sifikatsii agrarno-krest'īanskogo voprosa v trekh russkikh revolīutsiīakh. - Moskva : Izdatel'stvo "Mysl'", 1977
398p
r6044013 X.329/8665 Woolwich
1233181

WOJNA, Romuald
Walka o ziemię w Rosji w 1917 roku. - Wroclaw : Zakład Narodowy im. Ossolińskich-Wydawnictwo Polskiej Akademii Nauk, 1977
290p
r6016063 X.313/865 Woolwich
1233181

AGRICULTURAL LAND. Salisbury region. Wiltshire
Irrigation systems: Floated water meadows, to ca 1968

COWAN, Michael
Floated water meadows in the Salisbury area. - Salisbury : South Wilts Industrial Archaeology Society, 1982
[12]p : ill. - (Historical monograph / SWIAS ; 9). - Bibliography: p[12]
0-906195-08-x X.805/3612
2847965

AGRICULTURAL LAND. Scotland
Land capability. Classification

LAND capability classification for agriculture / J.S. Bibby ... [et al.] ; edited by J.S. Bibby. - Aberdeen : Macaulay Institute for Soil Research, 1982
viii,75p : ill. - (Monograph / Soil Survey of Scotland). - Bibliography: p73-75
0-7084-0251-8 X.329/16512
2924110

AGRICULTURAL LAND. Scotland
Settlement. Role of government, to 1976

MATHER, Alexander Smith
State aided land settlement in Scotland. - Aberdeen : Department of Geography, University of Aberdeen, 1978
34,[12]p,[4]p of plates,[3] leaves of plates : ill. - (O'Dell memorial monographs / University of Aberdeen. Department of Geography ; no.6)
B78-20905 X.313/1560 Woolwich
1484931

AGRICULTURAL LAND. Semi-arid regions
Irrigation

IRRIGATION of field and orchard crops under semi-arid conditions / compiled and edited by J. Shalhevet... [et al.]. - Revised ed. - Bet Dagan : International Irrigation Information Center ; Oxford : Distributed by Pergamon, 1979
[7],124p : ill. - (IIIC publications / International Irrigation Information Center ; no.1). - Includes bibliographies
Subsequent volumes, when received, will be placed at this shelfmark
0-08-025511-6 X.0322/232 Woolwich
1944002

AGRICULTURAL LAND. Semi-arid regions. North-east Brazil
Irrigation

BRAZIL. Departamento Nacional de Obras Contra as Secas
O trabalho do DNOCS e o programa de irrigação no Nordeste semi-árido. - [Brasília?] : DNOCS, 1978
53p. - (Publicação ; no.07/78)
r6159613 LAS.B.275/180 Woolwich
2440466

AGRICULTURAL LAND. Sicily. Italy
Land tenure. Reform, to date

CORLEO, Simone
Storia della enfiteusi dei terreni ecclesiastici di Sicilia. - Caltanissetta : Sciascia, 1977
lxxxiii,598p,[1]leaf of plate,[1]folded leaf. - (Collezione "viaggi e studi" ; 18)
r6505583 X.429/14958
8890994

AGRICULTURAL LAND. Sierra Leone
Land tenure. Reform, to date

EBERT, Kurt Hanns
Gewohnheitsrecht und ländliche Entwicklung in Afrika. - München ; London : Weltforum, [1982]
iv,208p. - (Afrika-Studien ; Nr.108). - Bibliography
3-8039-0235-5 X.529/50028
8853258

AGRICULTURAL LAND. Snowdonia National Park
Landscape conservation. Projects: Snowdonia Upland Management Experiment

The SNOWDONIA Upland Management Experiment. - Cheltenham : Countryside Commission, 1979
[1],iii,34p : ill. - (CCP ; 122)
0-902590-85-5 BS.116/51
214428x

AGRICULTURAL LAND. Somalia
Land tenure. Law, to date

GUADAGNI, Marco
Xeerka beeraha. - Milano : Giuffrè, 1981
x,354p. - (Università di Trieste, Facoltà di scienze politiche ; 21). - Includes bibliography
r6510378 X.322/13052
9043462

AGRICULTURAL LAND. South Asia
Irrigation. Role in alleviation of poverty

WHO gets a last resource? / by IDS Study Seminar 88. - Brighton : Institute of Development Studies, 1980
13p. - (Discussion paper ; 156)
B82-35054 X.529/49562
277349x

AGRICULTURAL LAND. South Asia
Land tenure. Reform, 1950-1975

HERRING, Ronald J.
Land to the tiller. - New Haven ; London : Yale University Press, c1983
x,314p. - Bibliography: p289-307. - Includes index
0-300-02725-7 X.520/31479
2965577
- [Another copy.] V 24446 IOL

AGRICULTURAL LAND. South Norfolk (District)
Drainage, 1800-1899

BURGESS, Ben
Nineteenth century tile drainage. - Norwich : B.B. & Co., [1982?]
12p : ill
B82-38944 X.629/18642
2788411

AGRICULTURAL LAND. South Norfolk (District)
Underground drainage systems, 1836-1900

BURGESS, Ben
Nineteenth century tile drainage. - Norwich : B.B. & Co., 1981
120p : ill
B83-00455 X.629/18829
2804905

AGRICULTURAL LAND. Southern Italy
Land tenure, 1918-

CINANNI, Paolo
Lotte per la terra nel Mezzogiorno, 1943-1953. - Venezia : Marsilio, 1979
154p,[2]p of plates. - (Materialimarsilio ; 22)
r6505938 X.809/54130
8902100

AGRICULTURAL LAND. Southern states. United States
Ownership by negro. Decline

The BLACK rural landowner - endangered species / edited by Leo McGee and Robert Boone. - Westport, Conn. ; London : Greenwood Press, 1979
xxi,200,[1]p : ill. - (Contributions in Afro-American and African studies ; no.44). - Includes index
0-313-20609-0 X.529/34385
2045710

AGRICULTURAL LAND. Soviet Union
Land tenure, 1900-1917

KOSTRIKIN, Vasiliĭ Ivanovich
Zemel'nye komitety v 1917 godu. - Moskva : Nauka, 1975
335p
r6098430 X.329/8980 Woolwich
2193043

AGRICULTURAL LAND. Soviet Union
Land tenure. Reform, 1861-1930

YANEY, George
The urge to mobilize. - Urbana ; London : University of Illinois Press, c1982
viii,599p : ill. - Bibliography: p571-590. - Includes index
0-252-00910-x X.800/35054
2882361

AGRICULTURAL LAND. Soviet Union
Land use

ĖKONOMICHESKIE problemy ispol'zovaniia zemel'nykh i vodnykh resursov v sel'skom khoziaistve / [otvetstvennyĭ redaktor : A.M. Onishchenko]. - Kiev : Naukova dumka, 1978
198p. - Bibliography : p.194-197
r6093246 X.329/8877 Woolwich
1856286

AGRICULTURAL LAND. Spain
Ancient Roman boundaries

ESTUDIOS sobre centuriaciones romanas en España / [por V. M. Rosselló Verger...et al.]. - Cantoblanco : Universidad Autónoma de Madrid, 1974
155 p., 3 leaves of plates, 2 p. of plates. - Includes bibliographies
84-600-6453-0 X.700/17430
1012673

AGRICULTURAL LAND. Spain
Land tenure

CARRIÓN, Pascual
Los latifundios en España.... -- 2.ed. - Barcelona : Ariel, 1975
393p.,6 leaves of plates,18p. of plates(some fold.) : ill. - (Horas de España)
84-344-2484-3 X.319/8802 Woolwich
1154869

SANZ-PASTOR MELLADO, Fernando
Es urgente la reforma agraria. - Madrid : E.M.E.S.A., [1977]
277p : ill. - (Ensayos Aldaba)
84-265-2504-0 X.329/8950 Woolwich
1154869

AGRICULTURAL LAND. Spain
Land tenure, to date

BRINES BLASCO, Joan
La desamortización eclesiástica en el País Valenciano durante el Trienio Constitucional. - Valencia : Universidad de Valencia, 1978
xii,252p,2leaves of plates. - Bibliography
84-370-0065-3 X.800/31140
8551979

AGRICULTURAL LAND. Spain
Land tenure. Law - Conference proceedings

JORNADAS ITALO-ESPAÑOLAS DE DERECHO AGRARIO (1972 : Salamanca and Valladolid)
Jornadas italo-españolas de derecho agrario, Salamanca-Valladolid, noviembre 1972. - Valladolid : Universidades de Salamanca y Valladolid, Cátedras de Derecho Civil, 1976
591p
84-600-6937-0 X.100/26448
1389475

AGRICULTURAL LAND. Spain
Land tenure. Political aspects, ca 1760-1855

TOMÁS Y VALIENTE, Francisco
El marco político de la desamortización en España. - 3. ed. - Barcelona : Editorial Ariel, 1977
172p. - (Ariel quincenal ; no 54)
84-344-0704-3 X.319/9679 Woolwich
1717286

AGRICULTURAL LAND. Spain
Land tenure. Political aspects, 1765-1970 - Conference proceedings

COLOQUIO DE PAU (6th : 1975)
La cuestión agraria en la España contemporánea. - Madrid : Editorial Cuadernos para el Diálogo, 1976
565p : ill. - (Colección ITS)
84-229-4027-2 X.319/9114 Woolwich
1386786

AGRICULTURAL LAND. Spain
Land tenure. Reform, to date

ESTEPA GIMÉNEZ, Jesús
'Gimeñez!Giménez!. - Puente-Genil (Córdoba) : [s.n], 1980
250p. - (Colección Anzur ; 10). - Includes bibliography
84-500-3849-9 X.329/17910
8551987

AGRICULTURAL LAND. Spain
Land tenure. Reform, 1900-1936

MAURICE, Jacques
La reforma agraria en España en el siglo XX. - Madrid : Siglo XXI de España, 1975
159 p. : ill. - (Estudios de historia contemporánea siglo XXI). - Bibliography: p. 73-79
84-323-0178-7 X.319/8511 Woolwich
0878332

AGRICULTURAL LAND. Strabane (District)
Settlement, 1600-1641

The PLANTATION in Ulster in Strabane Barony, Co. Tyrone c.1600-41 / Strabane Local History Group ; organised R.J. Hunter. - [Londonderry] : New University of Ulster, Institute of Continuing Education, 1982
vii,61p. - (Local history series / New University of Ulster. Institute of Continuing Education ; no.1). - Bibliography: p61
B82-40443 X.805/3231
2799006

AGRICULTURAL LAND. Sucre (Department). Colombia
Land tenure. Political aspects

REYES POSADA, Alejandro
Latifundio y poder político. - Bogotá : CINEP, 1978
iv,183p.,[4]leaves of plates(some fold.). - (Serie Colombia agraria ; 2). - Includes bibliographies
r6158741 X.319/20360 Woolwich
2248980

AGRICULTURAL LAND. Sumer. Ancient Mesopotamia
Land tenure, to 637

FOSTER, Benjamin R.
Administration and use of institutional land in Sargonic Sumer. - Copenhagen : Akademisk Forlag, 1982
132p : ill. - (Mesopotamia ; v 9)
87-500-2393-4 P.411/189[no.9]
8981604

AGRICULTURAL LAND. Sweden
Land tenure, 1945-1965

IVARSSON, Rune
Jordförvärvslagen, socialdemokratin och bönderna. - Lund : Gleerup, 1977
ix,350p. - (Bibliotheca historica lundensis ; 42). - Includes bibliographies
91-400-4468-8 WP.2374/42
1395335

AGRICULTURAL LAND. Sweden
Land tenure. Attitudes of political parties, to date

HOLMSTRÖM, Barry
Äganderätt och brukningsansvar. - Uppsala : [Uppsala universitet], 1983
271p. - (Skrifter utgivna av Statsvetenskapliga föreningen i Uppsala ; 94) (Politik son rationellt handlande ; 7)
91-554-1379-x Ac.1077.r[no.94]
9004580

AGRICULTURAL LAND. Szentes. Hungary
Land tenure. Reform

BARTA, László
A szentesi örökváltság. - Szeged : Somogyi-könyvtár, 1979
101p. - (Csongrád megyei könyvtári füzetek ; 10)
963-7581-16-2 2719b.135(10)
2038110

AGRICULTURAL LAND. Tanzania
Irrigation. Mathematical models

HAZLEWOOD, A.
Complementarity and competitiveness of large and small scale irrigated farming. - Norwich : School of Development Studies, University of East Anglia, 1978
25p : ill. - (Development studies discussion paper ; no.28)
B84-40874 X.525/9932
3175707

AGRICULTURAL LAND. Tornedalen. Sweden
Land tenure

MUOTKA, Astrid
Dödsbona. - Lulea : Distribution : Tornedalica, 1978
117p. - (Tornedalica ; nr 24)
91-7538-009-9 X.0700/436(24)
198845X

AGRICULTURAL LAND. Tropical regions
Land use. Effects of management of water resources - Conference proceedings

TROPICAL agricultural hydrology / edited by R. Lal and E.W. Russell. - Chichester : Wiley, c1981
xiv,482p.[22]p of plates : ill. - Includes bibliographies and index
0-471-27931-5 SRL
2573660

AGRICULTURAL LAND. United States
Availability

The CROPLAND crisis / Pierre R. Crosson, editor. - Baltimore ; London : Published for Resources for the Future by the Johns Hopkins University Press, c1982
xxiii,250p : ill. - Includes index
0-8018-2816-3 SRL
2791374

AGRICULTURAL LAND. United States
Drainage systems

UNITED STATES. Soil Conservation Service
[National engineering handbook. Section 16]. Drainage of agricultural land. - Port Washington : Water Information Center, 1973
[9], 430 p. : ill
0-912394-06-4 X.311/8335 Woolwich
0788619

AGRICULTURAL LAND. United States
Foreign investment

LABAND, David N.
Foreign ownership of U.S. farmland. - Lexington : Lexington Books, c1984
xiii,125p : ill. - Bibliography: p117-121
0-669-07305-9 X.322/14770
3181421

AGRICULTURAL LAND. United States
Purchase

OSTER, Merrill
Farmland buying strategies. - New York ; Chichester : Wiley, c1984
xiii,230p : ill. - Includes index
0-471-80454-1 X.322/14340
3214788

AGRICULTURAL LAND. Venezuela
Land tenure, 1950-1972

LOSADA ALDANA, Ramón
La tierra venezolana en la dialéctica del subdesarrollo. - Caracas : Universidad Central de Venezuela, 1976
2t. - Includes bibliographies
r6084802 X.700/20482
1848615

AGRICULTURAL LAND. Venezuela
Land tenure. Reform, to 1970

DREYER LANDAETA, Armando
Reforma agraria y desarrollo económico. - Caracas : Monte Ávila Editores, 1971
185p. - (Colección Temas venezolanos). - Bibliography : p.181-185
r6053720 X.319/9179 Woolwich
1390783

AGRICULTURAL LAND. Venezuela
Land tenure. Reform, 1900-

GARCÍA, Antonio
El nuevo problema agrario de la América Latina. - México : Universidad Nacional Autónoma de México, 1981
86p
r6709837 X.809/63554
8570272

AGRICULTURAL LAND. Wales
Ploughing. Cooperation. Law, ca 1250

JENKINS, Dafydd
Agricultural co-operation in Welsh medieval law. - [Cardiff] : Amgueddfa Werin Cymru, 1982
18p
0-85485-056-2 X.329/15971
2840820

AGRICULTURAL LAND. West Germany
Land tenure, 1945- - Statistics - Serials

LAND- und Forstwirtschaft, Fischerei. Reihe 2.1.6, Besitzverhältnisse und Grundstücksverkehr / Herausgeber: Statistisches Bundesamt Wiesbaden. - 1977. - Stuttgart : Kohlhammer, 1980
1v.. - (Fachserie 3 / Statistisches Bundesamt)
r6308707 S.F.555/99(7a)
9047549

AGRICULTURAL LAND. West Germany
Land use, 1945- - Statistics - Serials

LAND- und Forstwirtschaft, Fischerei. Reihe 2.1.2, Bodennutzung der Betriebe / Herausgeber : Statistisches Bundesamt Wiesbaden. - 1977-. - Stuttgart : Kohlhammer, 1979-. - (Fachserie 3 / Statistisches Bundesamt)
r6308699 S.F.555/99(4)
9047573

AGRICULTURAL LAND. West Germany
Reclamation, 1945-

ARBEITSGEMEINSCHAFT FLURBEREINIGUNG
Flurbereinigung - Naturschutz und Landschaftspflege. - Münster-Hiltrup : Landwirtschaftsverlag, 1980
78p,[5]leaves of plates : ill. - (Schriftenreihe des Bundesministers für Ernährung, Landwirtschaft und Forsten. Reihe B, Flurbereinigung ; Sonderhft.) (Schriftenreihe der Arbeitsgemeinschaft Flurbereinigung ; Hft.5)
3-7843-1119-9 SF.583/309
8756929

AGRICULTURAL LAND. Western Scotland
Fertilisers. Use

WEST OF SCOTLAND AGRICULTURAL COLLEGE
Farm fertiliser recommendations. - [Auchincruive] : The College, [1975?]
[1],11p. - (Advisory publication ; no.1)
B77-18235 X.319/17345 Woolwich
1122207

AGRICULTURAL LAND. Western States. United States
Irrigation

FREDERICK, Kenneth D. (Kenneth Dyer)
Water for western agriculture. - Washington : Resources for the Future, 1982
xiv,241p
0-8018-2832-5 X.322/11561
2854600

AGRICULTURAL LAND. Wicken. Cambridgeshire
Reclamation, 1938-1943

BLOOM, Alan
The farm in the fen. - [1st ed. reprinted]. - Wakefield : EP Publishimg, 1977
195p,[20]p of plates,[2]fold leaves of plates : ill. - Includes index
0-7158-1221-1 X.320/11103 Woolwich
1051474

AGRICULTURAL LAND. Widdington region. Essex
Agricultural land use, landscape conservation & nature conservation. Interrelationships - Conference proceedings

ESSEX EXERCISE (1975 : Widdington)
Essex Exercise. - [Chelmsford] : Essex County Council, 1976
[7],58p : ill
0-901355-70-4 X.313/1377 Woolwich
1065289

AGRICULTURAL LAND. Wielkopolska. Poland
Land tenure

JAKÓBCZYK, Witold
Pruska Komisja Osadnicza 1886-1919. - Poznań : Wydawnictwo Poznańskie, 1976
199p,1 fold leaf of plates
r6057217 X.319/9161 Woolwich
138788X

AGRICULTURAL LAND. Zala (megye). Hungary
Land tenure. Reform

BÉLI, József
Az 1945-ös földreform végrehajtása Zala megyében. - Zalaegerszeg : Zala megyei Levéltár, 1977
239p. - (Zalai gyűjtemény ; sz.4)
963-01-0847-x X.0700/770(4)
2029049

AGRICULTURAL LAND CLASSIFICATION. Great Britain. Ministry of Agriculture, Fisheries and Food

BODDINGTON, Michael A. B.
The classification of agricultural land in England and Wales. - Ipsden : Rural Planning Services Ltd, 1978
[2],vi,48p. - (Publications / Rural Planning Services ; no.4). - Bibliography: p.46-48
0-906269-00-8 X.319/18051 Woolwich
1447467

WORTHINGTON, Tom
Agricultural land. - Wallingford : Reading Agricultural Consultants, 1981
42p
0-9507931-0-8 X.809/52466
1447467

AGRICULTURAL LAND DISTURBED BY SAND & GRAVEL EXTRACTION. Great Britain
Reclamation. Feasibility studies

GREAT BRITAIN. Department of the Environment
Joint agricultural land restoration experiments. - [London] : [The Department].
Progress report no.1 : 1974-1977. - [1978]
[4],141p : ill(some col)
B78-34659 SRL
1584006
Progress report no.1 : 1974-1977. - [1978]
[4],141p : ill(some col)
B78-34659 SRL
1584014

AGRICULTURAL LAND RESOURCES. Jos Plateau. Nigeria

The JOS plateau / M.T. Alford ... [et al.] ; edited by W.T.W. Morgan. - Durham : Department of Geography, University of Durham, 1979
v,45p : ill. - (Occasional publications. New series / Department of Geography, University of Durham ; no.14). - Bibliography: p44-45
B83-01456 X.0700/584
280946x

AGRICULTURAL LAND TRIBUNALS. England
Decisions on law of succession to tenancies of agricultural holdings

HAMILTON, Richard Neville Dalton
Agricultural land tribunals succession decisions. - London : Royal Institution of Chartered Surveyors, 1978
1v.
B80-21607 X.0329 / 211
215174x
- Supplement. First, etc X.0329/211a Woolwich

AGRICULTURAL LIBRARIES. Bonn. West Germany
Zentralbibliothek der Landbauwissenschaft Bonn. Stock: Newspapers - Catalogues

ZENTRALBIBLIOTHEK DER LANDBAUWISSENSCHAFT
Zeitschriftenverzeichnis der Zentralbibliothek der Landbauwissenschaft Bonn. - Bonn : Universitätsbibliothek, 1977
v,942p
r6063400 2719.c.37
1712985

AGRICULTURAL MACHINERY
See also
Harvesters
Milking machines
Seed cleaning machines
Seed treating machines
Shellers
Silage production machinery

AGRICULTURAL MACHINERY

ALTHEA
Machines on a farm. - London : Evans Bros, 1979
[25]p : chiefly col ill
0-237-44951-x LR.421/113
0537322
Machines on a farm. - Over : Dinosaur Publications, 1979
[24]p : chiefly col ill. - (Althea series. blue label) (Dinosaur's Althea books)
0-85122-167-x X.990/15206 Woolwich
0537322
Machines on a farm. - Cambridge : Produced by Dinosaur for Cambridge University Press, 1982, c1979
[25]p : chiefly col.ill
0-521-25045-5 X.329/17865
0537322

CULPIN, Claude
Farm machinery. - 9th ed. - London : Crosby Lockwood Staples, 1976
x,410p : ill. - Includes bibliographies and index
0-258-97048-0 SRL
0110744
Farm machinery. - 9th ed. - [London] : English Language Book Society [etc], 1978
x,410p : ill. - Includes bibliographies and index
0-258-97110-x X.319/18916 Woolwich
0110744
Farm machinery. - 10th ed. - London : Granada, 1981
450p : ill. - Includes bibliographies and index
0-246-11539-4 SRL
0110744

DENTON, John
Machines on the farm. - Harmondsworth : Puffin Books, 1975
32p : ill(chiefly col). - (A new puffin picture book)
0-14-049125-2 X.319/10251 Woolwich
0537322

AGRICULTURAL MACHINERY –continued

JACOBS, Clinton O.
Agricultural power and machinery. – New York ; London : McGraw-Hill, c1983
472p : ill. – Includes index
0–07–032210–4 SRL
0110744

LADYMAN, Phyllis
About farm machines. – Revised ed. – London : Hodder and Stoughton, 1980
32p : chiefly col ill. – (A picture science book)
0–340–25279–0 X.322/8987 Woolwich
0537322

MILSOME, John
Machines and farms. – London : Burke, 1977
32p : col ill. – (The world of machines)
0–222–00414–2 X.329/11154 Woolwich
0537322

SHIPPEN, John Matthew
Basic farm machinery. – 3rd ed. – Oxford : Pergamon, 1980
x,288p : ill(chiefly col). – (Pergamon international library). – Includes index
0–08–024912–4 SRL
0110744

STONE, Archie Augustus
Machines for power farming. – 3rd ed. – New York ; London : Wiley, 1977
ix,533p : ill. – Includes index
0–471–82556–5 SRL
0110744

WYKEHAM, Nicholas
Farm machines. – Hove : Wayland, 1979
32p : chiefly col ill. – (Machines of today and tomorrow). – Includes index
0–85340–720–7 X.322/8843 Woolwich
0537322

AGRICULTURAL MACHINERY
– Directories

Al-MA'RAḌ az-zirā'i s-su'ūdi '83 = Saudi agriculture 83. – Riyadh : Al-Dhiafa Exhibition Services, [1983]
[83]p : ill(some col.)
r6863318 X.622/15848
291767x

AGRICULTURAL MACHINERY
British horse-drawn agricultural machinery

SMITH, D. J. (Donald John)
Discovering horse-drawn farm machinery. – Princes Risborough : Shire, 1979
96p : ill. – Bibliography: p94. – Includes index
0–85263–464–1 X.629/16394 Woolwich
2562839

Horse-drawn farm machinery. – 2nd ed. – Aylesbury : Shire Publications, 1984
96p : ill. – (Discovery ; 245). – Bibliography: p94. – Includes index
0–85263–664–4 X.629/26010
2562839

AGRICULTURAL MACHINERY
British horse-drawn agricultural machinery, ca 1800–ca 1900

HORSE-DRAWN farm implements. – Fleet : J. Thompson.
Part 2: Preparing the soil. – 1979
82p : ill
0–9505775–7–x X.322/10576 Woolwich
2622181

Part 3: Sowing and haymaking. – 1978
78p : ill
0–9505775–8–8 X.322/10577 Woolwich
2622181

Part 4: Harvesting. – 1979
68p : ill
0–9505775–9–6 X.322/10578 Woolwich
2622181

AGRICULTURAL MACHINERY
Cattle-drawn agricultural machinery

BARTON, D.
An animal drawn toolcarrier for small farm systems. – Norwich : School of Development Studies, University of East Anglia, 1982
8,[32]p : ill. – (Discussion paper ; no.110). – Bibliography: p8
B84–40090 X.622/20803
3185540

AGRICULTURAL MACHINERY
Design

KRUTZ, Gary
Design of agricultural machinery. – New York ; Chichester : Wiley, c1984
vii,472p : ill. – Includes index
0–471–08672–x X.622/20894
3188523

AGRICULTURAL MACHINERY
Maintenance & repair

FARM workshop & maintenence. – 3rd ed.. – London : Granada, 1984
248p : ill. – Includes index
0–246–12019–3 X.622/21218
1889621

FARM workshop and maintenance. – 2nd ed.. – London : Crosby Lockwood Staples, 1979
vi,183p : ill. – Includes index
0–258–97133–9 P.621/760
1889621

AGRICULTURAL MACHINERY
Manufacture. Savage, Frederick

BRAITHWAITE, David
Savage of King's Lynn. – Cambridge : Stephens, 1978
136p : ill. – Bibliography: p.133. – Includes index
0–85059–333–6 X.611/8320 Woolwich
0311111

AGRICULTURAL MACHINERY. Africa
Use. Economic factors

FARM equipment innovations in eastern and central southern Africa / edited by Iftikhar Ahmed and Bill H. Kinsey. – Aldershot : Gower, c1984
xvii,345p : ill. – Includes bibliographies and index
0–566–00697–9 X.629/25852
3055353

KINSEY, B. H. (Bill H)
Economic research and farm machinery design in eastern Africa. – Norwich : School of Development Studies, University of East Anglia, 1983
20leaves. – (Discussion paper ; no.10). – Bibliography: leaves 19–20
B84–39704 X.322/14040
3055353

AGRICULTURAL MACHINERY. Burgenland. Austria
to date

SIMON, Franz
Bäuerliche Bauten und Geräte. – Oberschützen : F. Simon, 1981
539p. : chiefly ill.
r6306307 L.45/1677
8908079

AGRICULTURAL MACHINERY. England
ca 1850–1972

VINCE, John
Vintage farm machines. – 2nd ed. – Aylesbury : Shire Publications, 1978
33p : ill. – (Shire album ; 1). – Bibliography: p.33
0–85263–440–4 X.0972/143(1)
1813684

AGRICULTURAL MACHINERY. Exports. Great Britain

ADVISORY COUNCIL FOR AGRICULTURE AND HORTICULTURE IN ENGLAND AND WALES
Report of an inquiry into agricultural exports. – [London] : [The Council], 1976
[2],61p : 1 ill
0–905551–00–1 BS.3/654
069780x

AGRICULTURAL MACHINERY. Great Britain
to 1900

VINCE, John
Old farms. – London : Murray, 1982
160p : ill. – Includes index
0–7195–3978–1 X.800/34759
284737x

AGRICULTURAL MACHINERY. Great Britain
1500–1900

FUSSELL, G. E. (George Edwin)
The farmer's tools. – London : Orbis Publishing, 1981
246p,111p of plates : ill. – Bibliography: p225–230. – Includes index
0–85613–359–0 X.622/10619 Woolwich
2504715

WRIGHT, Philip, 1908–
Old farm implements. – Gloucester : Alan Sutton, 1983
91p,[23]p of plates : ill
0–86299–083–1 X.629/24450
2504715

AGRICULTURAL MACHINERY. Great Britain
ca 1920–1977

LEE, Richard, b.1930
A source book of tractors and farm machinery. – London : Ward Lock, 1979
128p : ill. – Includes index
0–7063–5774–4 X.629/12449 Woolwich
1692089

AGRICULTURAL MACHINERY. Great Britain
Depreciation. Calculation – For inflation accounting

HILL, G. P.
Calculating machinery depreciation on farms during a period of inflation. – Ashford, Kent : Wye College, School of Rural Economics and Related Studies, Farm Business Unit, 1978
[4],28p. – (FBU occasional papers / Wye College, Farm Business Unit ; no.2)
0–901859–79–6 X.322/9448 Woolwich
2059304

AGRICULTURAL MACHINERY. Great Britain
Materials: Polymers

The USE of polymers in agricultural machinery / for the Polymer Engineering Directorate of the Science Research Council by the National College of Agricultural Engineering, Cranfield Institute of Technology. – Bedford : [The College], 1980
82p
B81–35671 SRL
2552639

AGRICULTURAL MACHINERY. Great Britain
Steam diggers, to ca 1900

TYLER, Colin
Digging by steam. – Watford : Model and Allied Publications, 1977
173p : ill. – ([MAP technical publication]). – Bibliography: p.169. – Includes index
0–85242–522–8 SRL
1191500

AGRICULTURAL MACHINERY. Great Britain
Use. Economic factors

CULPIN, Claude
Profitable farm mechanization. – 3rd ed. – London : Crosby Lockwood Staples, 1975
xi,307p : ill. – Bibliography: p.283–284. – Includes index
0–258–96984–9 SRL
0370851

AGRICULTURAL MACHINERY. United States

SMITH, Harris Pearson
Farm machinery and equipment. – 6th ed.. – New York ; London : McGraw-Hill, [1977]
viii,488p : ill. – (McGraw-Hill publications in the agricultural sciences). – Includes bibliographies and index
0–07–058957–7 SRL
1088505

AGRICULTURAL MACHINERY INDUSTRIES
See also
Harvester industries
Soil preparation machinery industries

AGRICULTURAL MACHINERY INDUSTRIES. Great Britain

GREAT BRITAIN. Department of Industry
The agricultural engineering industry. – London : The Department, 1978
20,[10]p : 1 ill
B78–14830 SRL
1414976

AGRICULTURAL MACHINERY INDUSTRIES. Great Britain
– Serials

OEM newsletter. – No.1 (1982)–. – Ilford : Wordsworth Trade Press, 1982–
v. : ill
B85–32413 P.625/504
3311481

AGRICULTURAL MACHINERY INDUSTRIES. Great Britain
Howard Machinery Limited – Serials

HOW. – Issue no.1 – ; Aug. 1976–. – [Ipswich] : Howard Machinery Limited, 1976– : ill(some col)
r6032715 P.515/437
1354949

AGRICULTURAL MACHINERY INDUSTRIES. Ireland (Republic)

A REVIEW of the Irish agricultural machinery industry / [National Board for Science and Technology, IDA Ireland]. – [Dublin] : [The Board], 1982
v,27p : 1ill
0–86282–016–2 X.622/15025
2872552

AGRICULTURAL MACHINERY INDUSTRIES. Islington (London Borough)
International Harvester Company of Great Britain – Serials

INTERNATIONAL HARVESTER COMPANY OF GREAT BRITAIN
IH Reporter. – No. 1– ; May 1976–. – London : The Company, 1976– : ill
r6009686 P.623/821
0698172

AGRICULTURAL MACHINERY INDUSTRIES. Rostov-na-Donu. Russia (RSFSR)
Factories
ARSENAL stepnykh korablei / [avtor-sostavitel' : ÎÛ P Krymskii]. - Rostov-na-Donu : Rostovskoe knizhnoe izdatel'stvo, 1979
190p,[18]p of plates : ill(some col.)
r6178667 X.809/28959
2458578

AGRICULTURAL MACHINERY INDUSTRIES. Salford Priors. Warwickshire
Bomford and Evershed Ltd. to 1977
SHERWEN, Theo
The Bomford story. - Evesham : Bomford and Evershed Ltd, [1979]
[7],104p : ill(some col)
B79-11587 X.529/21537
1827235

AGRICULTURAL MACHINERY INDUSTRIES. Wales
1800-1900 - Directories
WELSH FOLK MUSEUM
A directory of agricultural machinery and implement makers in Wales. - [Cardiff] : The Museum, 1980
[41]p. - Bibliography: p.[39]-[41]
B80-37821 X.329/13078 Woolwich
2286165

AGRICULTURAL MACHINERY POWERED BY STEAM ENGINES
to 1976
WILLIAMS, Michael, 1935- (Nov.)
Steam power in agriculture. - Poole : Blandford Press, 1977
184p : ill(chiefly col). - ([Blandford colour series]). - Includes index
0-7137-0819-0 X.329/11039 Woolwich
1166891

AGRICULTURAL MACHINERY TRADES
See also
Soil preparation machinery trades

AGRICULTURAL MACHINERY TRADES
- Serials
TRACTOR & farm machinery trader. - No.1 (Mar./Apr. 1982)-. - Ilford : Wordsworth Trade Press, 1982-
v. : ill
B83-07947 P.621/854
2832062

AGRICULTURAL MACHINERY TRADES. Great Britain
Apprentices. Training
TRAINING of apprentices in agricultural machinery. - Gloucester : Food, Drink and Tobacco Industry Training Board, 1978
7p
B82-26132 BS.23/1145
2745259

AGRICULTURAL MARKETING BOARDS
IZRAELI, Dov
Societal marketing boards. - New York ; Chichester : Wiley [etc.], 1977
xviii,265p : ill
0-470-99308-1 X.520/12335
1400339

AGRICULTURAL MATERIALS
Thermal properties
MOHSENIN, Nuri Nurdin
Thermal properties of foods and agricultural materials. - New York ; London : Gordon and Breach, 1980
[10],407p : ill. - Bibliography: p.376-398. - Includes index
0-677-05450-5 SRL
2252058

AGRICULTURAL MATERIALS INDUSTRIES. East Anglia
Agricultural equipment & materials industries & trades - Serials
EAST Anglian farmers guide. - [Issue no. 1]- ; [Sept. '79]-. - [Ipswich] : [Champaul Ltd], [1979]- : ill
B80-28200 P.313/502
2106795

AGRICULTURAL MATERIALS INDUSTRIES. Great Britain
- Serials
AGRITRADE. - Vol.58, no.3- ; Mar. 1978-. - London : Agricultural Supply Trade Association, 1978- : ill(some col)
B78-14825 SRL
1433709

AGRICULTURAL MATERIALS TRADES. East Anglia
Agricultural equipment & materials industries & trades - Serials
EAST Anglian farmers guide. - [Issue no. 1]- ; [Sept. '79]-. - [Ipswich] : [Champaul Ltd], [1979]- : ill
B80-28200 P.313/502
2106795

AGRICULTURAL MATERIALS TRADES. Great Britain
Agricultural equipment & materials trades - Directories
UNITED Kingdom agricultural merchants / [Data Research Group]. - [1976]. - Amersham : The Group, 1976
186 leaves in various pagings
Subsequent volumes, when received, will be placed at this shelfmark
0-905570-30-8 X.0313/509 Woolwich
0769061

AGRICULTURAL PRODUCTS
See also
Cereals products
Crops
Hive products
Horticultural products

AGRICULTURAL PRODUCTS
1945-
INTERNATIONAL trade and agriculture / edited by Jimmye S. Hillman and Andrew Schmitz. - Boulder : Westview Press, 1979
xviii,333p. - (Westview Special Studies in International Economics and Business). - Bibliography: p299-321
0-89158-498-6 X.520/25191
8004978

AGRICULTURAL PRODUCTS
Chemical analysis. Near infrared spectroscopy - Conference proceedings
BISHOP, C. K.
Report on Third Annual Users Conference for NIR Researchers. - [London] : ADAS, c1984
13p
B85-11819 BS.3/971
3248755

AGRICULTURAL PRODUCTS
Cooperative marketing - Conference proceedings
ALLEN, George, 1923-
Agricultural co-operatives in the marketing of agricultural produce and inputs. - [Oxford] : [Plunkett Foundation for Co-operative Studies], [1973]
21p
B76-28806 X.319/16752 Woolwich
0721131

PEDERSEN, Anders
Agricultural co-operative marketing and production with special reference to the changes in Denmark. - [Oxford] : [Plunkett Foundation for Co-operative Studies], [1973]
17p
B76-28807 X.319/16753 Woolwich
0720887

AGRICULTURAL PRODUCTS
Foreign trade between Japan & United States
U.S.-JAPANESE agricultural trade relations / edited by Emery N. Castle and Kenzo Hemmi with Sally A. Skillings. - Washington, D.C. : Resources for the Future, 1982
xvii,436p : 3ill. - Includes bibliographies and index
0-8018-2815-5 X.322/11455
2833336

AGRICULTURAL PRODUCTS
Foreign trade. Negotiations
JOSLING, Timothy Edward
Agriculture in the Tokyo Round negotiations. - London : Trade Policy Research Centre, 1977
vi,42p. - (Thames essays ; no.10)
0-900842-32-6 X.319/17978 Woolwich
1405276

AGRICULTURAL PRODUCTS
Foreign trade. Negotiations between European Economic Community & United States, 1973-1976
HARRIS, Simon
EEC trade relations with the USA in agricultural products. - Ashford, Kent : Centre for European Agricultural Studies, Wye College, 1977
57,[17]p. - (Occasional papers / Wye College. Centre for European Agricultural Studies ; no.3)
0-905378-00-8 X.0313/51(3) Woolwich
1166166

AGRICULTURAL PRODUCTS
Foreign trade. Non-tariff controls
HILLMAN, Jimmye S.
Nontariff agricultural trade barriers. - Lincoln [Neb.] ; London : University of Nebraska Press, 1978
xvii,236p : 3 ill. - Bibliography: p.221-226. - Includes index
0-8032-2301-3 X.529/33484
1901141

AGRICULTURAL PRODUCTS
Free-flowing agricultural products. Packaging in sacks. Mechanisation
NEW, J. H.
Sack filling. - London : The Institute, 1975
iii,8p. - (G98)
0-85954-044-8 BS.7F/4
0521973

AGRICULTURAL PRODUCTS
Marketing - Bibliographies
DELAGNEAU, Bernard
Marketing in an agricultural environment, a selected bibliography = La commercialisation dans un environment agricole, une bibliographie selectionnée. - [Winchester] : Business Information Press, 1978
28p
B80-32801 X.525/4522
2206641
- [Another copy.] X.325/1258 Woolwich

AGRICULTURAL PRODUCTS
Marketing, to date
WAHL, Andrea A. von
Multinationale Unternehmen in der Agroindustrie. - Kiel : Vauk, 1982
2,xiv,345p : ill. - (Schriften zum Agribusiness ; Bd.1). - Includes bibliography
3-922553-13-3 X.329/17433
9047530

AGRICULTURAL PRODUCTS
Marketing. Econometric models
ADAMS, Francis Gerard
Econometric models of world agricultural commodity markets. - Cambridge, Mass. : Ballinger, 1976
xi, 161 p. : ill. - Bibliography : p. 157-160
0-88410-290-4 X.520/11351
1041037

AGRICULTURAL PRODUCTS
Supply & demand
ORGANISATION FOR ECONOMIC CO-OPERATION AND DEVELOPMENT
Study of trends in world supply and demand of major agricultural commodities. - Paris : O.E.C.D. ; [London] : [H.M.S.O.], 1976
349p : 1 ill
B77-00667 U.N.P.108/79
0772607

AGRICULTURAL PRODUCTS. Belgium
Prices, to date
DEMOULIN, Bruno
La principauté de Liège et l'Europe. - Liège ; Louvain : Centre belge d'histoire rurale, 1982
33p,plates. - (Publication / Centre belge d'histoire rurale ; no 74)
r6210350 X.0311/197(74)
9001433

AGRICULTURAL PRODUCTS. Brazil
Marketing, to date
BRANDT, Sergio Alberto
Mercado agrícola brasileiro. - São Paulo : Livraria Nobel, 1979
145p : ill
r6702221 X.329/14419 Woolwich
8726582

AGRICULTURAL PRODUCTS. Cape Verde Islands
to date
CARREIRA, António, Writer on Portuguese-speaking Africa
Estudos de economia caboverdiana. - [Lisboa?] : Imprensa Nacional / Casa de Moeda, 1982
342p : ill. - (Estudos de história de Portugal e dos portugueses)
r6709506 X.322/14526
9147802

AGRICULTURAL PRODUCTS. Colombia
Marketing - Bibliographies
CENTRO DE ESTUDIOS E INVESTIGACIONES SOBRE MERCADEO AGROPECUARIO
Bibliografía sobre mercadeo agropecuario [1973-]. - Bogotá : Cámara de Comercio, 1973-
r6024836 LASc 230/2
0385883

AGRICULTURAL PRODUCTS. Denmark
Prices. Control, to date

DENMARK. Monopoltilsynet
Kontrollen med priserne for landbrugsvarer i relation til EF-retten. - København : Tilsynet, 1981
1v.(various pagings)
r6610714 S.B.286/92
8916551

AGRICULTURAL PRODUCTS. Developing countries
Cooperative marketing

RAHMAN, Khandker Mahmudur
Development of agricultural marketing in Bangladesh. - Comilla : Bangladesh Academy for Rural Development, 1973
[7],iv,135p. - Bibliography : p.127-135
r6091846 X.529/21898
1893793

AGRICULTURAL PRODUCTS. Developing countries
Prices related to supply. Econometric models

ASKARI, Hossein
Agricultural supply response. - New York ; London : Praeger ; [London] : [Distributed by Martin Robertson], 1976
xvii,444p : ill. - (Praeger special studies in international economics and development). - Bibliography: p.414-435. - Includes index
0-275-23260-3 X.320/10673 Woolwich
1037242

AGRICULTURAL PRODUCTS. Developing countries
Prices. Policies of governments

TOLLEY, George S.
Agricultural price policies and the developing countries. - Baltimore ; London : Published for the World Bank [by] Johns Hopkins University Press, c1982
xii,242p : ill. - (A World Bank publication). - Includes index
0-8018-2704-3 X.520/27725
2761092

AGRICULTURAL PRODUCTS. European Community countries
Cooperative marketing

FOXALL, Gordon R.
Co-operative marketing in European agriculture. - Aldershot : Gower, c1982
xi,101p. - Bibliography: p96-101
0-566-00512-3 X.329/15335
2726084

AGRICULTURAL PRODUCTS. European Community countries
Prices. Policies of European Economic Community

COMMON prices and Europe's farm policy / by Theodor Heidhues ... [et al.]. - London : Trade Policy Research Centre, 1978
viii,74p : ill. - (Thames essays ; no.14)
0-900842-36-9 X.319/18004 Woolwich
1410075

AGRICULTURAL PRODUCTS. Exports. Eastern states. Nigeria to ca 1850

NORTHRUP, David
Trade without rulers. - Oxford : Clarendon Press, 1978
x,269p : ill. - (Oxford studies in African affairs). - Bibliography: p.241-258. - Includes index
0-19-822712-4 X.529/32867
1338692

AGRICULTURAL PRODUCTS. Exports. Great Britain

ADVISORY COUNCIL FOR AGRICULTURE AND HORTICULTURE IN ENGLAND AND WALES
Report of an inquiry into agricultural exports. - [London] : [The Council], 1976
[2],61p : 1 ill
0-905551-00-1 BS.3/654
0697796

AGRICULTURAL PRODUCTS. Exports. Great Britain - Serials

AGRICULTURAL exporter. - [No.1]- ; July/Aug. 1977-. - Wilmslow : Jalna Publicity Services Limited, [1977]- : ill
B78-25975 P.313/449
1522728

AGRICULTURAL PRODUCTS. Exports. Sudan
Role in economic development, 1900-1967

BESHAI, Adel Amin
Export performance & economic development in Sudan, 1900-1967. - London : Ithaca Press for the Middle East Centre, St Anthony's College, 1976
[11],358p : ill. - (Monographs / St Antony's College. Middle East Centre ; no.3). - Bibliography: p.343-350. - Includes index
0-903729-15-6 X.0809/544(3)
0789402

AGRICULTURAL PRODUCTS. Exports. United States
Policies of government, 1953-1961

PETERSON, Trudy Huskamp
Agricultural exports, farm income, and the Eisenhower administration. - Lincoln [Neb.] ; London : University of Nebraska Press, 1979
xiii,222p. - Bibliography: p.197-204. - Includes index
0-8032-3659-x X.329/12860 Woolwich
2242834

AGRICULTURAL PRODUCTS. Great Britain - Statistics

OUTPUT and utilisation of farm produce in the United Kingdom / Ministry of Agriculture, Fisheries and Food. - 1972 to 1978. - [Pinner] : [The Ministry], 1979
[1],36p
B80-14843 BS.3/676
0027464

AGRICULTURAL PRODUCTS. Great Britain - Statistics - Serials

OUTPUT and utilisation of farm produce in the United Kingdom / Ministry of Agriculture, Fisheries and Food. - 1973 to 1979. - Pinner : The Ministry, 1980
35p
B81-06676 SRL
0027464

AGRICULTURAL PRODUCTS. Great Britain
Cooperative marketing

FOOD FROM BRITAIN. Co-operative Development Division
Outline operating strategy. - [London] : [The Division], [1984
16p
B85-09415 X.525/10445
3228738

AGRICULTURAL PRODUCTS. Great Britain
Cooperative marketing. Attitudes of farmers

LE VAY, Clare
Cooperative theory and farmers' attitudes. - Aberystwyth : University College of Wales, Department of Agricultural Economics, [1975]
[7],83p. - Bibliography: p.80-83
0-902124-13-7 X.313/1264 Woolwich
062117x

Farmers' attitudes to cooperation. - Aberystwyth : University College of Wales, Department of Agricultural Economics, [1976]
[5],83p,[2]leaves of plates. - Bibliography: p.83
0-902124-16-1 X.313/1306 Woolwich
062117x

AGRICULTURAL PRODUCTS. Great Britain
Cooperative marketing. Participation of farmers

LE VAY, Clare
Farmer participation in cooperatives. - Aberystwyth : University College of Wales, Department of Agricultural Economics, 1977
[7],117,[12]p,[5]leaves of plates : ill. - Bibliography: p.[9]-[12]
B78-02067 Ac.1375.b/7
128505x

AGRICULTURAL PRODUCTS. Great Britain
Direct sale

PEACHEY, P. R.
Direct farm sales to the public. - [London?] : Ministry of Agriculture, Fisheries and Food, 1977
20p. - (Socio-economics paper ; 8)
B83-08414 BS.3/832(8)
2835479

AGRICULTURAL PRODUCTS. Great Britain
Exporting. Finance

ADVISORY COUNCIL FOR AGRICULTURE AND HORTICULTURE IN ENGLAND AND WALES
Agricultural exports. - [London] : [The Council], 1978
[1],iii,22p
B80-21709 X.322/9256 Woolwich
2163268

AGRICULTURAL PRODUCTS. Great Britain
Marketing

BARKER, J. W.
Agricultural marketing. - Oxford : Oxford University Press, 1981
vi,226p : ill. - (Oxford science publications). - Bibliography: p219-221. - Includes index
0-19-859468-2 X.329/13939 Woolwich
2324415
- [Another copy: paperback] ISBN 0-19-859469-0 X.329/13940 Woolwich

CANNON, Tom
Marketing problems of the food chain. - [Stirling] : University of Stirling Department of Business Studies, 1984
23p. - (Working paper : 8409). - Bibliography: p23
B85-26724 X.525/11004
2324415

A DIRECTORS' guide to practical marketing. - [London] CCAHC, [1982]
48p : ill
B82-34561 X.529/49523
2324415

AGRICULTURAL PRODUCTS. Great Britain
Marketing - Serials

FARM & food marketing. - No 1- ; 1976-. - London : The Marketing Unit, Central Council for Agricultural and Horticultural Co-operation, [1976]- : ill
Holdings: no.7, etc. : 1978, etc
B79-01070 P.613/220
1751506

AGRICULTURAL PRODUCTS. Great Britain
Marketing. Economic aspects

CARPENTER, E. M.
Marketing and the producer. - [Belfast] : Queen's University of Belfast, 1977
19p. - (The George Scott Robertson memorial lecture ; 27th). - Bibliography: p.19
0-85389-133-8 WPA.219/19
1403370

AGRICULTURAL PRODUCTS. Great Britain
Prices - Forecasts

The OUTLOOK for agriculture / School of Agriculture, Aberdeen. - [Aberdeen] : [The School].
1978. - 1977
[3],10p. - (Outlook series ; no.5)
B78-37558 X.319/18484 Woolwich
1372009

AGRICULTURAL PRODUCTS. Great Britain
Prices, 1978-1979

NATIONAL FARMERS' UNION OF SCOTLAND
UK farm prices for 1978/79. - [Edinburgh] : [National Farmers' Union of Scotland, [1978]
8p,3 leaves
Subsequent volumes, when received, will be placed at this shelfmark
B80-20352 P.313 / 495
2109263

AGRICULTURAL PRODUCTS. India (Republic)
Cooperative marketing

KAMAT, Gopal Sadashiv
Marketing. - Poona : Harshad Prakashan, 1974
176 p.
r6028515 X.529/19853
1015079

AGRICULTURAL PRODUCTS. India (Republic)
Marketing

GUPTA, A. P.
Marketing of agricultural produce in India. - Bombay : Vora, 1975
xv, 264 p.. - Bibliography : p. 257-260. - Includes index
r6010423 X.329/10785 Woolwich
1034278

AGRICULTURAL PRODUCTS. Ireland (Republic)
Marketing

MCKINSEY AND COMPANY
A marketing opportunity for agricultural products. - [London] : [McKinsey and Co.], [1977]
viii,70p : ill(chiefly col)
B78-03200 X.512/5762
1295950

AGRICULTURAL PRODUCTS. Malawi
Marketing

KINSEY, B. H. (Bill H)
Overcoming impediments to the effective marketing of staples. - Norwich : School of Development Studies, University of East Anglia, 1983
24p,3p of plates : 1ill. - (Discussion paper ; no.9)
B84-39708 X.322/14039
3174654

AGRICULTURAL PRODUCTS. Mexico
to date

NETTEL ROSS, Margarita
Geografía agrícola estatal de México en el siglo XIX. - México : INAH, Departamento de Investigaciones Históricas, [1977?]
iv,121p. - (Cuaderno de trabajo del Departamento de Investigaciones Históricas ; 23). - Bibliography
r6703310 X.322/10428 Woolwich
8756937

AGRICULTURAL PRODUCTS. Spain
Prices, to date

SÁNCHEZ-ALBORNOZ, Nicolás
Los precios agrícolas durante la segunda mitad del siglo XIX. - Madrid : Tecnos, Servicio de Estudios del Banco de España. - (Materiales para la historia económica de España)
vol.2: Vino y aceite. - 1981
xii,216p,1fold.leaf
84-309-0879-x X.322/10886 Woolwich
841856x

AGRICULTURAL PRODUCTS. Spain
Prices, 1856-1890

SÁNCHEZ-ALBORNOZ, Nicolás
Los precios agrícolas durante la segunda mitad del siglo XIX. - Madrid : Servicio de Estudios del Banco de España. - (Materiales para la historia económica de España)
vol.1: Trigo y cebada. - 1975
xii,283p
84-500-6793-6 X.311/4756 Woolwich
1154591

AGRICULTURAL PRODUCTS. Tanzania
Cooperative marketing

BRYCESON, Deborah
Second thoughts on marketing co-operatives in Tanzania. - Oxford : Plunkett Foundation for Co-operative Studies, [1983?]
45p : ill. - (Plunkett development series ; 5)
0-85042-056-3 X.629/23921
3169588

AGRICULTURAL PRODUCTS. United States
Marketing. Economic aspects

DAVIS, John, 1922-
Developments in food marketing in the USA. - Belfast : Economics & Statistics Division, Dept. of Agriculture, 1984
xviii,172p : ill
B84-41487 P.843/279
2156814

KOHLS, Richard L. (Richard Louis)
Marketing of agricultural products. - 5th ed.. - New York : Macmillan ; London : Collier Macmillan, 1980
xi,612p : ill. - Includes bibliographies and index
0-02-365640-9 X.520/22033
2156814

RHODES, V. James
The agricultural marketing system. - 2nd ed. - New York ; Chichester : Wiley, c1983
xvii,557p : ill. - Includes index
0-471-87466-3 X.520/30703
2156814

AGRICULTURAL PRODUCTS. United States
Prices

TOMEK, William G. (William Goodrich)
Agricultural product prices. - 2nd ed. - Ithaca ; London : Cornell University Press, 1981
367p : ill. - Includes bibliographies and index
0-8014-1337-0 X.520/23265
2373696

AGRICULTURAL PRODUCTS. West Germany
Marketing

BLAKEY, J. P.
Report of a visit to the Federal Republic of Germany. - [London] : ADAS, c1984
22p : ill
B84-43952 BS.3/904
3193160

AGRICULTURAL PRODUCTS. Western Europe
Cooperative marketing. Effects of law of European Economic Community - Conference proceedings

HULLE, A. van
The effect of EEC regulations on marketing co-operatives. - [Oxford] : [Plunkett Foundation for Co-operative Studies], [1973]
45p
B76-28838 X.319/16750 Woolwich
0720747

AGRICULTURAL PRODUCTS TRADES
See Agricultural trades

AGRICULTURAL PROPERTY. Britons
In Morocco. Expropriation. Compensation

GREAT BRITAIN
[Treaties, etc. Morocco, 1981 Feb.12]. Agreement between the Government of the United Kingdom of Great Britain and Northern Ireland and the Government of the Kingdom of Morocco concerning compensation for properties transferred to the state under the Dahir of 2 March 1973. - London : H.M.S.O., 1981
3p. - (Cmnd. ; 8249) (Morocco ; no.1 (1981))
0-10-182490-4 BS.14/3406
2526514

AGRICULTURAL PROPERTY UNIT TRUST FOR PENSION FUNDS AND CHARITIES
- Serials

AGRICULTURAL PROPERTY UNIT TRUST FOR PENSION FUNDS AND CHARITIES
Report and accounts 31st December .... - 197?-. - [London] : The Trust, [197-]-
v.
Holdings: 1981-
r6864792 P.513/2858
2741164

AGRICULTURAL RESEARCH COUNCIL. JOINT CONSULTATIVE ORGANISATION FOR RESEARCH AND DEVELOPMENT IN AGRICULTURE AND FOOD
See Joint Consultative Organisation for Research and Development in Agriculture and Food

AGRICULTURAL RESEARCH COUNCIL. POULTRY RESEARCH CENTRE
See Poultry Research Centre

AGRICULTURAL SCIENTIFIC SERVICES
See Great Britain. Department of Agriculture and Fisheries for Scotland. Agricultural Scientific Services

AGRICULTURAL SETTLEMENTS
See also
Settlement schemes

AGRICULTURAL SETTLEMENTS
Prehistoric agricultural settlements

HUNTERS, gatherers and first farmers beyond Europe / edited by J.V.S. Megaw. - [Leicester] : Leicester University Press, 1977
256p : ill. - Includes bibliographies and index
0-7185-1136-0 X.311/8575 Woolwich
1207059
- [Another copy] X.410/10211 Woolwich

AGRICULTURAL SHOWS. Bedwellty. Gwent
Bedwellty Show, to 1972

JONES, Reginald Thomas
Bedwellty Show. - 2nd ed. - Risca : Starling Press, 1976
91p,[20]p of plates : ill
0-903434-20-2 X.319/16948 Woolwich
0764507

AGRICULTURAL SHOWS. Shepton Mallet. Somerset
Bath & West Show, to 1975

HUDSON, Kenneth
The Bath & West. - Bradford-on-Avon : Moonraker Press, 1976
xiii,251p : ill. - Includes index
0-239-00156-7 X.800/25596
0550922

AGRICULTURAL SHOWS. Skipton. North Yorkshire
Skipton Show, ca 1900

MASON, Roger, 1940-
The great Skipton Show. - London : P. Davies, 1979
[10],190p,[4]p of plates : ill
0-432-09211-0 X.809/45951
2278464

AGRICULTURAL STUDIES. Curriculum subjects. Great Britain
Role in curriculum of schools - For teaching

ILLSLEY, Terence Walker
Farms and schools. - London : Association of Agriculture, 1975
19p
0-903313-24-3 X.319/16526 Woolwich
0643459

AGRICULTURAL SYSTEMS

SPEDDING, Colin Raymond William
An introduction to agricultural systems. - London : Applied Science Publishers, 1979
ix,169p : ill. - Bibliography: p.162-166. - Includes index
0-85334-823-5 X.529/36650
177459x

AGRICULTURAL SYSTEMS
Research. Methodology

MARTIN, Adrienne
Concepts and approaches to farm systems research in Syria 1977-1980. - Norwich : School of Development Studies, University of East Anglia, 1983
18p. - (Discussion paper ; no.120). - Bibliography: p17-18
B84-41080 X.525/9935
0515027

AGRICULTURAL SYSTEMS
Research. Methodology - Conference proceedings

STUDY of agricultural systems / edited by G.E. Dalton. - London : Applied Science Publishers, 1975
xiv,441p : ill. - Includes index
0-85334-640-2 SRL
0515027

AGRICULTURAL SYSTEMS. Mediterranean region

GIBBON, David, 1939-
Rainfed farming systems in the Mediterranean region. - Norwich : School of Development Studies, University of East Anglia, 1983
20[i.e.31]p : ill. - (Discussion paper ; no.70). - Bibliography: p17-20
B84-41073 X.525/9951
318885x

AGRICULTURAL TRACTOR INDUSTRIES. Western Europe
Agricultural tractor industries & trades - Statistics

A MARKET studies report on agricultural tractors in Europe. - London : Inter Company Comparisons, [1981]
89p : ill(some col.). - (Market studies)
0-86191-905-x SRL
2610248

AGRICULTURAL TRACTOR TRADES. Western Europe
Agricultural tractor industries & trades - Statistics

A MARKET studies report on agricultural tractors in Europe. - London : Inter Company Comparisons, [1981]
89p : ill(some col.). - (Market studies)
0-86191-905-x SRL
2610248

AGRICULTURAL TRACTORS

BALDWIN, Nick
Farm tractors. - London : F. Warne, 1977
64p : ill. - (Warne's transport library). - Includes index
0-7232-2060-3 SRL
1289128

Farm tractors. - New ed. - London : Warne, c1983
64p : ill. - (Warne's transport library)
0-7232-3092-7 L.49/2082
1289128

BUTTERWORTH, Bill
Farm tractors. - London : Spon, 1984
149p : ill. - Includes index
0-419-13230-9 X.629/23301
1289128

AGRICULTURAL TRACTORS
- Serials

VINTAGE tractor magazine. - 1981-. - Perth : Vintage Tractor Publications, 1981-
v. : ill
B82-32134 P.621/848
276945x

AGRICULTURAL TRACTORS
to 1946

VIGZOL OIL REFINING COMPANY (LONDON)
Farm tractors. - Hinckley : Traction Engine Enterprises, 1976
63p : ill
0-905100-07-7 X.319/17266 Woolwich
1096354

AGRICULTURAL TRACTORS
to 1950 - Serials

VINTAGE tractor extra. - No.1-. - Perth : Vintage Tractor Publications, 1982-
v. : ill
B84-23834 P.625/343
3125203

AGRICULTURAL TRACTORS
to 1959

CAWOOD, Charles L.
Vintage tractors. - Aylesbury : Shire, 1980
32p : ill. - (Shire album ; 48)
0-85263-499-4 X.0972/143(48)
1342541

AGRICULTURAL TRACTORS
to 1959–continued

An 'OLD motor' kaleidoscope of farm tractors / [compiled] by Nick Baldwin. – London : 'Old motor' magazine, 1977
[96]p : chiefly ill. – ([Kaleidoscopes ; 2]). – Includes index
0–906116–01–5 X.322/8528 Woolwich
1342541

AGRICULTURAL TRACTORS
to 1973

WILLIAMS, Michael, 1935 Nov.4–
Great tractors. – Poole : Blandford, 1982
vi,160p : ill(some col.). – Includes index
0–7137–1205–8 X.622/13245
0232416

AGRICULTURAL TRACTORS
to 1980

VINTAGE tractor album / compiled & edited by Nick Baldwin. – London : Marshall Harris & Baldwin, [1983?]
95p : ill. – Includes index
0–906116–24–4 X.622/16356
293924x

AGRICULTURAL TRACTORS
to 1983

WILLIAMS, Michael, 1935– Nov. 4–
Classic farm tractors. – Poole : Blandford, 1984
144p,[16]p : ill(some col.). – Includes index
0–7137–1421–2 X.622/22420
3055590

AGRICULTURAL TRACTORS
1950–1960

MELLOY, John
Forty tractors of the fifties. – Perth : Vintage Tractor, c1984
[40]p : ill. – (Milestones series ; no.4)
0–907742–30–0 X.629/26083
3273741

AGRICULTURAL TRACTORS
American agricultural tractors

MORLAND, Andrew
Classic American form tractors. – London : Osprey, 1985
120p : col.ill. – (Osprey colour series)
0–85045–617–7 X.622/23581
3204006

AGRICULTURAL TRACTORS
American agricultural tractors, to date – Encyclopaedias

WENDEL, C. H.
Encyclopedia of American farm tractors. – Sarasota : Crestline Publishing, 1979
352p : chiefly ill
0–912612–15–0 X.622/12228 Woolwich
8815151

AGRICULTURAL TRACTORS
British agricultural tractors, to 1979

WILLIAMS, Michael, 1935– (Nov.)
British tractors for world farming. – Poole : Blandford Press, 1980
112p : ill. – Includes index
0–7137–0959–6 X.622/9006
1995731
– [Another copy: paperback]
ISBN.0–71371–407–7 X.622/21226

AGRICULTURAL TRACTORS
David Brown agricultural tractors, to 1965

CONDIE, Allan T.
David Brown album. – Perth : Vintage Tractor & Machinery Publications, [1983]
[76]p : chiefly ill
0–907742–16–5 X.629/21422
296564x

AGRICULTURAL TRACTORS
Field Marshall series 3 diesel agricultural tractors. Maintenance & repair

MARSHALL SONS AND COMPANY
Field Marshall diesel tractor series 3. – Hinckley : Traction Engine Enterprises, 1976
[1],48p : ill. – Includes index
0–905100–08–5 X.619/17533 Woolwich
1097466

AGRICULTURAL TRACTORS
Oliver agricultural tractors, 1930–1948

CONDIE, Allan T.
Oliver album. – Perth : Allan T. Condie Vintage Tractor, [c1984]
[63]p : ill. – (Vintage tractor & machinery publications)
0–907742–14–9 X.629/24814
3216543

AGRICULTURAL TRACTORS. Great Britain
Accidents: Overturning

HUNTER, A. G. M.
A discussion of tractor overturning accidents on slopes due to stability or control loss. – Penicuik : [Scottish Institute of Agricultural Engineering], 1981
25p : ill. – (Technical report / Scottish Institute of Agricultural Engineering ; no.3)
0–7084–0187–2 SRL
2905329

Tractor accidents on slopes. – Penicuik : [Scottish Institute of Agricultural Engineering], 1981
30p in various pagings : ill. – (Technical report / Scottish Institute of Agricultural Engineering ; no.4)
0–7084–0188–0 SRL
2905329

AGRICULTURAL TRADES
See also
Cattle trades
Feedingstuffs trades
Horticultural trades
Potato trades

AGRICULTURAL TRADES
Agricultural industries & trades

The STATE of food and agriculture / [Food and Agriculture Organization]. – 1977 : World review ; Some factors affecting progress in food and agriculture in developing countries ; [and], The state of natural resources and the human environment for food and agriculture. – Rome : F.A.O. ; [London] : [H.M.S.O.], 1978
234p in various pagings(2 fold) : ill
B79–24284 SRL
0066958

AGRICULTURAL TRADES. Grupo Andino countries

JUNTA DEL ACUERDO DE CARTAGENA
Grupo Andino guia agropecuaria. – Lima : la Junta, 1972
300 p.
r6024879 X.319/8615 Woolwich
1016881

AGRICULTURAL TRADES. South–west England
to 1900 – Conference proceedings

POPULATION and marketing / edited by Walter Minchinton. – [Exeter] : University of Exeter, 1976
139p : ill. – (Exeter papers in economic history ; no.11)
0–85989–035–x X.512/5645
1141511

AGRICULTURAL TRADES. Tamil Nadu (State). India (Republic)

HARRISS, Barbara
Transitional trade and rural development. – New Delhi : Vikas, c1981
xiii,263p : ill. – Bibliography: p247–253
0–7069–1036–2 X.520/29239
2854376

AGRICULTURAL TRADES. Tanzania

ELLIS, Frank
Agricultural marketing and peasant–state transfers in Tanzania. – Norwich : School of Development Studies, University of East Anglia, 1983
40p. – (Discussion paper ; no.116). – Bibliography: p38–40
B84–41078 X.525/9938
3188906

AGRICULTURAL TRADES. United States

KOHLS, Richard L. (Richard Louis)
Marketing of agricultural products. – 5th ed.. – New York : Macmillan ; London : Collier Macmillan, c1980
xi,612p : ill. – Includes bibliographies and index
0–02–979360–2 X.322/9679 Woolwich
2323591

AGRICULTURAL TRADES. United States
Agricultural industries & trades

GOLDBERG, Ray Allan
Agribusiness coordination. – Boston [Mass.] : Harvard University, Graduate School of Business Administration, Division of Research, 1968
xix, 256 p. : ill. – Bibliography : p. 243–252. – Includes index
r6009871 X.322/8250 Woolwich
0730378

AGRICULTURAL TRADES. United States
Agricultural industries & trades. Advertising

MCGUIRE, James E.
Advertising and display in agribusiness. – New York ; London : McGraw–Hill, 1979
vi,138p : ill. – (Career preparation for agriculture/agribusiness). – Includes index
0–07–045129–x X.510/16346
1946242

AGRICULTURAL VEHICLES
See also
Agricultural tractors
Model agricultural vehicles

AGRICULTURAL VEHICLES
British horse–drawn agricultural vehicles, 1802–ca 1940

HORSE–DRAWN farm vehicles / [selected and with additional material] by John Thompson. – Fleet : John Thompson, 1980
78p : ill
0–906922–03–8 X.622/8379 Woolwich
2258463

AGRICULTURAL VEHICLES
Tyres

RUTHERFORD, Ian
Tyres and traction. – Alnwick : MAFF Publications, 1983
11p : col.ill
B85–22529 X.629/26570
328431x

AGRICULTURAL VEHICLES. Aker. Norway
1650–1800

WINGE, Harald
Føringsredskaper i Aker fogderi ca. 1650–1800. – Oslo : Universitetet i Oslo, Institutt for folkelivsgransking, 1976
102p,[3] leaves of plates. – (Undersøkelser for magistergraden i etnologi ; 1) (Universitetet i Oslo. Institutt for folkelivsgransking. Avhandling ; nr 5)
r6061042 X.0319/398(1) Woolwich
1858319

AGRICULTURAL VEHICLES. Belgium
Diesel agricultural vehicles. Marketing

ECONOMIST INTELLIGENCE UNIT
A study of the agricultural vehicles market in Belgium. – London : E.I.U., 1976
[3]59 leaves. – (EIU multi–client project)
0–900351–51–9 X.313/1332 Woolwich
0753750

AGRICULTURAL VEHICLES. Estonia
to date

VIIRES, Ants
Talurahva veovahendid. – Tallinn : Valgus, 1980
267p,[90]p of plates : ill
r6461015 X.329/14384 Woolwich
8726574

AGRICULTURAL VEHICLES. France
Diesel agricultural vehicles. Marketing

ECONOMIST INTELLIGENCE UNIT
A study of the agricultural vehicles market in France. – London : E.I.U., 1975
67,[4] leaves. – (An EIU multi–client project)
0–900351–41–1 SRL
0580759

AGRICULTURAL VEHICLES. Great Britain
Diesel agricultural vehicles. Marketing

ECONOMIST INTELLIGENCE UNIT
A study of the agricultural vehicles market in the United Kingdom. – London : E.I.U., 1975
[3],83 leaves. – (An EIU multi–client project)
0–900351–35–7 X.313/1302 Woolwich
0677582

AGRICULTURAL VEHICLES. Great Britain
Steam powered agricultural vehicles, to 1975. Steam powered agricultural vehicles & steam powered road vehicles, to 1975

STEAM on the road. – London : I. Allan, 1979
96p : chiefly ill. – Includes index
0–7110–0894–9 LR.421/163
1940996

AGRICULTURAL VEHICLES. Italy
Diesel agricultural vehicles. Marketing

ECONOMIST INTELLIGENCE UNIT
A study of the agricultural vehicles market in Italy. – London : E.I.U., 1976
[2],79 leaves. – (An EIU multi–client project)
0–900351–52–7 X.313/1331 Woolwich
0753769

AGRICULTURAL VEHICLES. Netherlands
Diesel agricultural vehicles. Marketing

ECONOMIST INTELLIGENCE UNIT
A study of the agricultural vehicles market in the Netherlands. – London : E.I.U., 1976
[3],58 leaves. – (An EIU multi–client project)
0–900351–53–5 X.313/1329 Woolwich
0753785

AGRICULTURAL VEHICLES. Spain
Diesel agricultural vehicles. Marketing
ECONOMIST INTELLIGENCE UNIT
A study on the agricultural vehicles market in Spain. - London : E.I.U., 1977
[4],94 leaves. - (An EIU multi-client project)
B77-33728 Cup.702.n.36
1270729

AGRICULTURAL VEHICLES. West Germany
Diesel agricultural vehicles. Marketing
ECONOMIST INTELLIGENCE UNIT
A study of the agricultural vehicles market in West Germany. - London : E.I.U., 1976
[4],69 leaves. - (An EIU multi-client project)
0-900351-54-3 X.313/1330 Woolwich
0753777

AGRICULTURAL WASTE MATERIALS
See also
Waste materials. Livestock

AGRICULTURAL WASTE MATERIALS
- Serials
AGRICULTURAL wastes. - Vol.1, no.1- ; Feb. 1979-. - [London] : Applied Science Publishers, 1979- : ill
B79-13858 SRL
1845063

AGRICULTURAL WASTE MATERIALS
Use in manufacture of food - Conference proceedings
FOOD from waste / edited by G.G. Birch, K.J. Parker and J.T. Worgan. - London : Applied Science Publishers, 1976
xii,301p : ill. - Includes bibliographies and index
0-85334-659-3 SRL
0585041

AGRICULTURAL WASTE MATERIALS
Use in production of methane for fuel
HOBSON, P. N.
Methane production from agricultural and domestic wastes. - London : Applied Science, c1981
xi,269p : ill. - (Energy from wastes series). - Includes bibliographies and index
0-85334-924-x SRL
2491710

AGRICULTURAL WASTE MATERIALS. Energy sources. United States
- Conference proceedings
CORNELL AGRICULTURAL WASTE MANAGEMENT CONFERENCE (1975 : 7th)
Energy, agriculture and waste management. - Ann Arbor : Ann Arbor Science, 1975
ix, 540 p. : ill. - Includes index
0-250-40113-4 X.320/10447 Woolwich
0699403

AGRICULTURAL WASTE MATERIALS. Great Britain
Treatment
HOBSON, P. N.
Waste treatment in agriculture. - London : Applied Science Publishers, 1977
x,257p : ill. - Bibliography: p.243-249. - Includes index
0-85334-736-0 SRL
1257234

AGRICULTURAL WASTE MATERIALS. Great Britain
Treatment & disposal
WELLER, John B. (John Brian)
Farm wastes management. - London : Crosby Lockwood Staples, 1977
235p : ill. - Bibliography: p.221-227. - Includes index
0-258-96928-8 SRL
1251481

AGRICULTURAL WASTE MATERIALS. United States
Treatment
LOEHR, Raymond C.
Pollution control for agriculture. - New York ; London : Academic Press, 1977
xiii,383p : ill. - Includes index
0-12-455260-9 SRL
1120476

AGRICULTURE
See also
Agronomy
Alternative agriculture
Aquaculture

AGRICULTURE
See also
Arable farming
Bee-keeping
Brazil. Ministério da Fazenda
Dairy farming
Farm museums
Farms
Food and Agriculture Organization
Forestry
Great Britain. Department of Agriculture for Northern Ireland
Great Britain. Ministry of Agriculture, Fisheries and Food
Harvesting
Horticulture
Irrigation
Mixed farming
Norway. Landbruksdepartementet
Planting times
Ploughing
Sharecropping
Subsoiling
Theses on agriculture

AGRICULTURE
APPLIED biology. - London : Academic Press.
Vol.1. - 1976
x,358p : ill. - Includes bibliographies and index
0-12-040901-1 SRL
0146285
Vol.2. - 1977
x,272p : ill. - Includes bibliographies and index
0-12-040902-x SRL
0146285
Vol.3. - 1978
x,418p : ill. - Includes bibliographies and index
0-12-040903-8 SRL
0146285

BAKER, Susan
Farms. - [London] : Macdonald Educational, [1977]
3-46p : ill(chiefly col). - (Topic books). - Includes index
0-356-05448-9 X.322/8421 Woolwich
0325481

BARFIELD, Arnold
Farming. - Brighton : Wayland, 1984
48p : col.ill. - (Endeavour books). - Includes index
0-85078-346-1 X.322/14450
0325481

BLANCO, Josette
On the farm. - Purton : Child's Play (International) Ltd, [1977]
[20]p : chiefly col ill. - (Child's Play moments ; book 2)
0-85953-040-x X.329/10891 Woolwich
1091891

BREWIS, Henry
Don't laugh till he's out of sight. - Ipswich : Farming Press, 1984
152p : ill
0-85236-153-x X.958/29904
3284980

CARDOSO, Ciro F.S. (Ciro Flamarion Santana)
Agricultura, escravidão e capitalismo. - Petrópolis : Vozes, 1979
210p
r6161837 X.529/36969
0308145

CASTELLIZ, K.
Life to the land. - East Grinstead : Lanthorn Press, 1980
72p. - Bibliography: p.72
B80-41811 X.329/1311
2306336

CHERRINGTON, John
A farming year. - London : Hodder and Stoughton, c1983
136p : ill(some col.)
0-340-28209-6 X.322/12356
2912406
Lifting my nose from the grindstone. - London : Northwood, 1981
183p : ill
0-7198-2548-2 X.329/16443
2912406

DEJARDIN, Elizabeth
Illustrated rural studies. - Slough : University Tutorial Press, c1982
128p : ill
0-7231-0846-3 X.322/11661
0031089

DUNLOP, Stewart
Farming and the countryside. - London : Heinemann Educational, 1976
77p : ill(some col). - (Place and people ; 4)
0-435-34700-4 X.311/8221 Woolwich
0672378

DUNMAN, Jack
Agriculture, capitalist and socialist. - London : Lawrence and Wishart, 1975
256p. - Includes index
0-85315-330-2 X.329/10261 Woolwich
0577553

FOOD and agriculture. - San Francisco : W.H.Freeman, 1976
viii,154p : ill. - Bibliography : p.151-152.-Includes index
0-7167-0382-3 WP.5721/15
0132713

FREAM, William
Fream's agriculture. - 16th ed.. - London : Murray, 1983
xxxi,816p,[24]p of plates : ill. - Bibliography: p759-768. - Includes index
0-7195-4034-8 SRL
0132713
- [Another copy.] 338.1842 OPL

GRAHAM-CAMERON, M. G. (Malcolm Gordon)
The farmer. - Cambridge : Dinosaur Publications, 1975
[32]p : col ill. - (Althea's dinosaur books)
0-85122-106-8 Cup.1281/106
0325481

GREASLEY, Brian
Agriculture and rural development. - London : Harrap, 1983
63p : ill(some col.). - (Harrap's world topic geography)
0-245-53708-2 X.322/12806
0672378

GRIERSON, Ian
Farming. - London : Edward Arnold, 1983
64p : ill. - (Foundation geography)
0-7131-0666-2 X.322/12329
0672378

HOLLYER, Belinda
Farms and farming. - London : Macdonald Educational, 1982
45p : ill(some col.). - (Macdonald new reference library). - Bibliography: p44. - Includes index
0-356-05829-8 X.622/13441
0325481

INTERNATIONAL CONFERENCE OF AGRICULTURAL ECONOMISTS (17th : 1979 : Banff, Alberta)
Rural change. - Oxford : International Association of Agricultural Economists, Institute of Agricultural Economics : Gower, 1981
xv,738p : ill. - Includes index
0-566-00401-1 X.329/14497 Woolwich
0020036

JONES, Barbara, 19—
Food, farming and famine. - Walton-on-Thames : Nelson for the Schools Council, 1982
48p : ill(some col.). - (Geography and change)
0-17-434183-0 X.322/11047 Woolwich
0672378

LADEJINSKY, Wolf
Ladejinsky papers. - [Ottawa] : [International Development Research Centre], 1977
25 microfiches
r6086462 Mic.F.118
0020036

ON the farm. - Maidenhead : Purnell, 1979
[32]p : ill. - (Colour and learn)
B79-04443 X.311/3942 Woolwich
1771973

PAARLBERG, Don
Farmers of five continents. - Lincoln [Neb.] ; London : University of Nebraska Press, 1984
ix,114p : 2ill
0-8032-3670-0 X.329/19193
3245942

PADGET, Sheila
World farming and fishing. - London : Macdonald Educational, 1976
24p : ill(chiefly col). - (Macdonald Educational colour units. world topics)
0-356-04876-4 X.311/8193 Woolwich
0672378

PERSPECTIVES in world agriculture / compiled by CAB. - Farnham Royal : Commonwealth Agricultural Bureaux, c1980
xvii,532p,1folded leaf of plate : ill. - Includes bibliographies
0-85198-458-4 SRL
0132713

PFEIFFER, Ehrenfried
Soil fertility, renewal & preservation. - East Grinstead : Lanthorn, c1983
199p,[8]p of plates : ill. - Bibliography: p187-191. - Includes index
0-906155-12-6 X.329/18126
2306336

RAEBURN, J. R. (John Ross)
Agriculture. - Chichester : Wiley, c1984
xv,329 : ill. - Includes bibliographies and index
0-471-10308-x X.322/14452
0132713

AGRICULTURE -continued

RICHARDSON, William Norman
Plants, agriculture and human society. - Reading, Mass. ; London : Benjamin, 1978
xii,353p : ill. - Includes bibliographies and index
0-8053-8215-1 X.311/8789 Woolwich
0132713

RITSON, Christopher
Agricultural economics. - London : Crosby Lockwood Staples, 1977
xiii,409p : ill. - Includes index
0-258-96938-5 X.520/12066
0020036

Agricultural economics. - London : Crosby Lockwood Staples, 1978
xiii,409p : ill. - Includes index
0-258-97119-3 X.510/15865
0020036

SCARRY, Richard
On the farm. - London : Hamlyn, 1977
[24]p : col ill. - (A golden look-look book)
0-600-33629-8 X.990/9150 Woolwich
0325481

SNODGRASS, Milton Moore
Agriculture, economics and resource management. - 2nd ed. - Englewood Cliffs ; London : Prentice-Hall, 1980
xiii,546p : ill. - Includes index
0-13-018820-4 X.520/20842
0020036

SUSTAINABLE food systems / edited by Dietrich Knorr. - Chichester : Ellis Horwood, 1983
416p : ill. - (Ellis Horwood series in food science and technology). - Includes bibliographies and index
0-85312-623-2 X.520/31294
0020036

SWAYNE, Dick
I am a farmer. - London : Dent, 1978
[28]p : chiefly col ill
0-460-06791-5 X.320/11249 Woolwich
1091891

UPTON, Martin
Farming as a business. - [2nd ed.]. - Oxford : English Language Book Society [etc.], 1979
135p
0-19-442358-1 X.319/19119 Woolwich
0020036

WRIGHT, Jill Ann
Problems in world farming. - London : Hodder and Stoughton, 1975
64p : ill(some col). - (Living in the modern world). - Includes index
0-340-16135-3 X.319/16642 Woolwich
0672378

AGRICULTURE
- Abstracts - Serials

ADVANCES in applied biology. - Vol.6-. - London : Academic Press, 1981-
v. : ill
B82-07634 SRL
0146285

AGRICULTURE
- Bibliographies

LIBRARY ASSOCIATION. Public Libraries Group
Agriculture. - 5th ed.. - [Bury St Edmunds] : [The Group], 1975
58p. - (Readers' guides / Library Association. Public Libraries Group ; no.2)
0-85365-158-2 2725.bb.47
0202975

MULLAY, Marilyn
Agriculture. - Penzance : Library Association, Public Libraries Group, 1983
47p. - (Readers' guide ; no.52). - Includes index
0-86378-005-9 2725.c.309
0202975

AGRICULTURE
- Conference proceedings

AGRICULTURE in the twenty-first century / edited by John W. Rosenblum. - New York ; Chichester : Wiley, c1983
xx,415p : ill. - Bibliography: p369-392. - Includes index
0-471-88538-x X.322/13528
0132713

INTERNATIONAL CONFERENCE OF AGRICULTURAL ECONOMISTS (16th : 1976 : Nairobi)
Contributed papers read at the 16th International Conference of Agricultural Economists. - Oxford : University of Oxford Institute of Agricultural Economics ; [Oxford] : International Association of Agricultural Economists, 1977
[1],iv,286p : ill. - (Papers on current agricultural economic issues) (Occasional papers / International Association of Agricultural Economists). - Includes bibliographies
B78-14824 X.519/27866
0020036
- [Another copy] X.519/30932

INTERNATIONAL CONFERENCE OF AGRICULTURAL ECONOMISTS (17th : 1979 : Banff, Alberta)
The rural challenge. - Aldershot : Gower, c1981
xii,329p : ill. - (IAAE occasional paper ; no.2). - Includes bibliographies and index
0-566-00472-0 X.329/15119 Woolwich
0020036

INTERNATIONAL CONFERENCE OF AGRICULTURAL ECONOMISTS (18th : 1982 : Jakarta)
Growth and equity in agricultural development. - Aldershot : Gower, c1983
xiv,619p : ill. - Includes bibliographies and index
0-566-00636-7 X.329/17814
0020036

AGRICULTURE
- Dictionaries

DALAL-CLAYTON, D. B. (D. Barry)
Black's agricultural dictionary. - London : Black, 1981
xii,499p : ill
0-7136-2130-3 SRL
2530686

HAENSCH, Günther
Dictionary of agriculture. - 4th completely revised and enlarged ed.. - Amsterdam ; Oxford : Elsevier, 1975
xxiv,1000p
0-444-99849-7 SRL
0565466

NEWHOUSE, John Walter Albert
A handbook of agricultural terms, English-French = Guide pratique des termes agricoles, français-anglais. - 2nd ed. revised and enlarged. - Bishop's Waltham : J.W.A. Newhouse, 1977
39p
B78-10291 SRL
1372254

NORSK landbruksordbok = [Norwegian dictionary of agriculture ...] / utgitt av Nemnda for Norsk landbruksordbok ; redaktør : Magne Rommetveit ... [et al.]. - Oslo : Det Norske samlaget, 1979
2v
82-521-0870-9 X.322/9570 Woolwich
0565466

AGRICULTURE
- For African students

AYERST, Peter William
Agricultural science. - London : Evans Bros, 1978
63p : ill. - (Evans pocket facts)
0-237-50297-6 X.319/18537 Woolwich
1742639

OWEN, G. H.
O-level agriculture for Central Africa. - Harlow : Longman, 1984
236p : ill. - Includes index
0-582-65103-4 X.322/14001
1742639

AGRICULTURE
- For Caribbean students

HENRY, A. I.
Caribbean agricultural science. - [Port-of-Spain] : Macmillan Caribbean ; London : Macmillan.
Book 2: Practices. - 1979
[5],216p : ill. - Includes index
0-333-23390-5 X.311/9084 Woolwich
0681164

PERSAD, Ralph Sakaldip
Agricultural science for the Caribbean.
1. - London : Nelson, 1975
136p : ill. - Includes index
0-17-566166-9 X.0311/1007(1) Woolwich
0681164
- [Another copy: paperback]
ISBN 0-17-566309-2 X.322/10972
- Sunbury-on-Thames : Nelson.
2. - 1977
176p : ill. - Includes index
0-17-566167-7 X.0311/1013 Woolwich
0681164
3. - 1978
[4],188p : ill. - Includes index
0-17-566168-5 X.0311/1013(3) Woolwich
0681164

AGRICULTURE
- For East African students

NGUMY, James
Agriculture in practice. - [Nairobi] : Macmillan Kenya ; London : Macmillan, 1983
v,85p : ill. - Includes index
0-333-36083-4 X.322/13544
3131556

AGRICULTURE
- For Iranian readers - Serials

IRANIAN journal of agricultural research. - Vol.5, no.1- , 1977-. - Oxford : Pergamon, 1977- : ill
B78-02456 SRL
1290002

AGRICULTURE
- For Kenyan students

MIGWI, James
Beginning agriculture. - London : Macmillan. - (The Macmillan primary agriculture project)
Pupil's book 1. - 1979
[3],69p : ill
Subsequent volumes, when received, will be placed at this shelfmark
0-333-23287-9 X.0322/209 Woolwich
2144816
- Nairobi : Macmillan Kenya. - (Macmillan primary agriculture project)
Pupil's book 3. - 1982
122p : ill
0-333-33138-9 X.322/11286
2142430
- London : Macmillan. - (The Macmillan primary agriculture project) (Macmillan books for Africa) - Bibliography: p.65
Teacher's book 2. - 1979
[5],65p : ill
0-333-26643-9 X.322/9596 Woolwich
2142430

AGRICULTURE
- For West African students

AKINSANMI, Oluyemi
Objective tests. - Harlow : Longman, 1980
vii,55p : ill. - (Study for success) (Study for success)
0-582-60688-8 X.329/14513 Woolwich
2559102

AKINYOSOYE, Vincent Oladoyin
Senior tropical agriculture. - London : Macmillan, 1976
[6],199p : ill. - Includes index
0-333-19239-7 X.311/8225
0028193

AKUBUILO, C. J. C.
Certificate practical agricultural science. - Harlow : Longman, 1984
iv,76p : ill
0-582-65118-2 X.322/14522
0028193
An introduction to practical agriculture for West Africa. - London : Macmillan, 1978
[6],154p : ill. - Includes index
0-333-21978-3 X.311/8889 Woolwich
0028193

ANTHONIO, Q. B. Olatunji (Quirino Bandele Olatunji)
General agriculture for West Africa. - London : Allen & Unwin, 1979
250p : ill. - Includes index
0-04-631012-6 X.329/19249
0028193

FAKOREDE, M. A. B.
Agricultural science. - London : Macmillan, 1977
[8],104p : ill. - (School Certificate revision course)
0-333-21229-0 X.319/17250 Woolwich
0028193

OKORIE, J. U.
Certificate notes. - Harlow : Longman, 1983
viii,119p : ill. - (Study for success)
0-582-65535-8 X.329/16792
0028193

AGRICULTURE
- Serials

CENTRE FOR AGRICULTURAL STRATEGY
CAS report. - 1- ; Oct. 1976-. - Reading : The Centre, [1976]-
r6010778 P.311/716
105497X

INTERNATIONAL agricultural development. - Nov./Dec.1980-. - Gowborough : Pharos Pub. Services, 1980-
v. : ill(some col.)
B81-32702 SRL
0146285

AGRICULTURE
- Serials–continued

INTERNATIONAL ASSOCIATION OF AGRICULTURAL ECONOMISTS
IAAE members bulletin. - No.1- ; July 1977-. - Oxford : [International Association of Agricultural Economists]. [1977]- : ill
B78-16651 P.311/738
1447076

AGRICULTURE
to 1000

PERSPECTIVES in agricultural geography / edited by Noor Mohammad. - New Delhi : Concept Publishing Company, 1981
5v.(plates) : ill
r6804984 X.329/14758 Woolwich
8005001

AGRICULTURE
to 1974

LET'S look at - farming. - Double Bay : McGregor ; London : Schlesinger, 1975
23p : col ill. - (A school project book)
0-85921-112-6 X.319/10216
0523674

AGRICULTURE
to 1977

SUTTON, A.
Food, farms and shops. - Edinburgh : Holmes McDougall, 1978
[2],32p : col ill. - (Then & now)
0-7157-1311-6 X.708/21513
1702432

WILKES, Peter
An illustrated history of farming. - Bourne End : Spurbooks, 1978
159p : ill. - Includes index
0-904978-84-2 X.329/11530 Woolwich
1439227

AGRICULTURE
to 1980

HEISER, Charles B. (Charles Bixler)
Seed to civilization. - 2nd ed. - Oxford : W.H. Freeman, c1981
xii,254p : ill. - Bibliography: p239-245. - Includes index
0-7167-1264-4 X.322/10208 Woolwich
255481x

AGRICULTURE
to 1981

CASE, S. L. (Stephen Leslie)
Farming through the ages. - London : Evans Brothers, 1982
48p : ill
0-237-29213-0 X.800/34462
2817748

GRIGG, David, 1934-
The dynamics of agricultural change. - London : Hutchinson, 1982
260p : ill. - Includes index
0-09-147790-5 X.329/15717
2797011
- [Another copy: paperback] ISBN 0-09-147791-3 X.329/15718

AGRICULTURE
to date

AGRARPOLITIK, Landentwicklung und Umweltschutz / herausgegeben von Adolf H. Malinsky. - Wien : Springer, 1982
270p.. - (Linzer Universitätsschriften. Festschriften ; Bd.2). - Includes a list of Bach's publications
3-211-81709-3 X.0972/219a(2)
800501x

A MEZŐGAZDASÁG története / [szerkesztette Gunst Péter, Lőkös László]. - Budapest : Mezőgazdasági Kiadó, 1982
269p. - Bibliography
963-231-427-1 X.322/11732
800501x

AGRICULTURE
to date - Bibliographies

ACTUALIZACIÓN del Catálogo colectivo de publicaciones periódicas / [compilación del] Centro de Documentación e Información Agrícola (CEDIA), Dirección de Planificacion Sectorial, Secretaría de Recursos Naturales. - Tegucigalpa : El Centro, 1982
105p
r6708987 LAS.E.356/13
800501x

AGRICULTURE
to date - Encyclopaedias

MEZŐGAZDASÁGI lexikon / [szerkesztették Barna József et al.] ... - 2., teljesen átdolg. kiad.. - Budapest : Mezőgazdasági Kiadó, 1982
2v. : ill. - Includes a list of current Hungarian and foreign publications on subjects related to agriculture
963-230-030-0 X.322/10691 Woolwich
800501x

AGRICULTURE
to date. Geographical aspects

ANDREAE, Bernd
[Agrargeographie. English]. Farming, development and space. - Berlin : de Gruyter, 1981
345p.,[1]folded leaf of plates. - Bibliography
r6306465 X.800/34209
8853193

AGRICULTURE
1700-1800

BECKER, Rudolph Zacharias
Noth- und Hülfsbüchlein für Bauersleute .... - Dortmund : Harenberg, [1980]
499p. : ill. - (Die bibliophilen Taschenbücher ; 207). - Contains a list of Becker's publications
3-88379-207-1 X.329/14644 Woolwich
876218x

AGRICULTURE
1945-

AGROSELEKT / Herausgeber : Akademie der Landwirtschaftswissenschaften der Deutschen Demokratischen Republik, Institut für Landwirtschaftliche Information und Dokumentation (ILID). - 4., überarb. Aufl. - Berlin : Akademie-Verlag, 1985
v,95p
r6310745 P.1211/10a
8916470

AGRICULTURE
1945- - Bibliographies - Serials

AGROSELEKT. Reihe 1, Landtechnik / Akademie der Landwirtschaftswissenschaften der Deutschen Demokratischen Republik, Institut für Landwirtschaftliche Information und Dokumentation. - Bd 31, Heft 1-. - Berlin : Akademie-Verlag, 1985-
Issues under the earlier title are entered in the General Catalogue
r6310726 P.1211/10
8916470

AGRICULTURE
1945- - Forecasts

AGRICULTURE. - Rome : Food and Agriculture Organization, 1981
xx,134p
92-510-1080-3 UN.F.131/100
8916470

AGRICULTURE
Agricultural industries & trades

The STATE of food and agriculture / [Food and Agriculture Organization]. - 1977 : World review ; Some factors affecting progress in food and agriculture in developing countries ; [and], The state of natural resources and the human environment for food and agriculture. - Rome : F.A.O. ; [London] : [H.M.S.O.], 1978
234p in various pagings(2 fold) : ill
B79-24284 SRL
0066958

AGRICULTURE
Anthroposophical viewpoints. Steiner, Rudolf. Landwirtschaftlicher Kursus

SOPER, John
Studying the Agriculture Course. - London : Bio-Dynamic Agricultural Association, 1976
88p
0-9503780-2-x X.319/16561 Woolwich
0659967

AGRICULTURE
Applications of digital computer systems

BARKER, M.
Report on the Association of the European Chambers of Agriculture convened to discuss development in the videotex and associated technology by the advisory services. - [Pinner] : ADAS, c1984
7p : ill
B85-09125 BS.3/955
2593882

REHMAN, T. (Tahir)
Computers in farming. - Reading : Farm Management Unit, University of Reading, 1981
vii,70p : ill. - (Study ; no.1). - Includes bibliographies
0-7049-0244-3 X.625/936
2593882
- [Another copy.] SRL

AGRICULTURE
Applications of digital computer systems - Bibliographies

RAMIREZ, G. G.
Agriculture. - London : Heyden on behalf of the British Computer Society, c1982
viii,42p. - (Use of computers for national development)
0-85501-694-9 SRL
279084x

AGRICULTURE
Applications of digital computer systems - Conference proceedings

EUROPEAN SEMINAR ON EXTENSION EDUCATION (6th : 1983 : Italy)
Sixth European Seminar on Extension Education. - [London] : ADAS, c1984
8p
B84-43953 BS.3/911
2593882

IFIP TC5 WORKING CONFERENCE ON FOOD PRODUCTION AND AGRICULTURAL ENGINEERING (1981 : Havana)
Computer applications in food production and agricultural engineering. - Amsterdam ; Oxford : North-Holland, 1982
xi,334p : ill. - Includes bibliographies
0-444-86382-6 SRL
2593882

AGRICULTURE
Applications of genetics - Conference proceedings

GENETIC engineering : applications to agriculture / Lowell D. Owens, editor ; organized by the BARC Symposium VII Committee ; sponsored by the Beltsville Agricultural Research Center, Northeastern Region, Agricultural Research Service, United States Department of Agriculture. - Totowa : Rowman & Allanheld ; London : Granada, 1983
x,327p : ill. - (Beltsville symposia in agricultural research ; 7). - Includes bibliographies and index
0-246-11947-0 X.322/13505
0103896

AGRICULTURE
Applications of meteorology

SMITH, Lionel P. (Lionel Percy)
Methods in agricultural meteorology. - Amsterdam ; Oxford : Elsevier, 1975
xii,210p : ill. - (Developments in atmospheric science ; 3). - Includes bibliographies and index
0-444-41286-7 SRL
0507547

AGRICULTURE
Applications of microcomputer systems

BUTTERWORTH, K. (Keith)
Microcomputer developments. - [London] : ADAS, [1982]
21p
B82-40879 BS.3/747
2801191

JARRETT, Dennis
Business systems handbook for farmers. - [London] : New Technology, [1982]
viii,152p : ill. - Bibliography: p147-151
0-86330-205-x X.525/6709
2801191
- [Another copy.] X.525/7035

AGRICULTURE
Applications of microelectronic devices

COX, S. W. R. (Sidney Walter Reginald)
Microelectronics in agriculture and horticulture. - London : Granada, 1982
230p : ill. - Includes bibliographies and index
0-246-11717-6 SRL
2811812

AGRICULTURE
Applications of neutron activation analysis

CAMPBELL, J. A. (Jacqueline Ann)
Neutron activation analysis. - Slough : Commonwealth Agricultural Bureaux, 1978
36p. - (Special publication / Commonwealth Bureau of Soils ; no.7). - Bibliography: p20-36
0-85198-438-x X.322/11903
2923211

AGRICULTURE
Applications of parasitology - Conference proceedings

The RELEVANCE of parasitology to human welfare today / edited by Angela E.R. Taylor and Ralph Muller. - Oxford : Blackwell Scientific, 1978
viii,135p. - (Symposia / British Society for Parasitology ; vol.16). - Includes bibliographies and index
0-632-00422-3 SRL
1337408

AGRICULTURE
Applications of solar energy - Conference proceedings

INTERNATIONAL SEMINAR ON ENERGY CONSERVATION AND THE USE OF SOLAR AND OTHER RENEWABLE ENERGIES IN AGRICULTURE. HORTICULTURE AND FISHCULTURE (1980 : Polytechnic of Central London)
Energy conservation and use of renewable energies in the bio-industries. - Oxford : Pergamon, 1981
xi,574p : ill
0-08-026866-8 SRL
0756350

SOLAR energy in agriculture. - [London] : [UK Section, I.S.E.S.], [1976]
[3],87p : ill. - Includes bibliographies
0-904963-13-6 SRL
0756350

AGRICULTURE
Applications of taxonomy - Conference proceedings

BELTSVILLE SYMPOSIUM IN AGRICULTURAL RESEARCH (2nd : 1977)
Biosystematics in agriculture. - Montclair : Allanheld, Osmun ; New York ; Chichester : Wiley, 1978
xii,340p : ill. - Includes bibliographies and index
0-470-26416-0 SRL
176599x

AGRICULTURE
Applied research projects. Organisation

ANDREW, Chris O.
Planning and conducting applied agricultural research. - Epping : Bowker, c1982
xii,94p : ill. - (A Westview special study). - Bibliography: p73-79. - Includes index
0-86531-461-6 X.329/16343
2900602

AGRICULTURE
Arithmetic - For agriculture

O'LOAN, Arthur
Farm calculations. - [Blackrock] : ACOT, c1981
93p : ill(some col.)
B82-05192 L.49/1080
2599376

AGRICULTURE
Automated planning, to date

TÓTH, József, Dr.
Mezőgazdasági vállalatok automatizált tervezése. - Budapest : Mezőgazdasági Kiadó, 1981
245p. - Bibliography
963-231-099-3 X.322/11848
8949867

AGRICULTURE
Biological aspects - Serials

BIOLOGICAL agriculture & horticulture. - Vol.1, no.1 (1982)-. - Berkhamsted : A B Academic Publishers, 1982-
v. : ill
B83-40346 P.621/891
2963264

AGRICULTURE
Bonded employment

ENNEW, Judith
Debt bondage - a survey. - London : Anti-Slavery Society, 1981
85p. - (Human rights series report ; no.1). - Bibliography: p72-85
0-900918-14-4 X.529/45416
2617358

AGRICULTURE
Book-keeping - For agriculture

BRUNI, Franco
La contabilità agraria nei centri di gestione. - Milano : Franco Angeli, [1977]
241p. - (Università di Napoli, Cassa per il Mezzogiorno, Centro di specializzazione e ricerche economico-agrarie per il Mezzogiorno. Collana di studi e richerche ; 10)
r6100287 X.319/20040 Woolwich
1179918

AGRICULTURE
Business records

LEE, Delene W.
Agribusiness procedures and records. - New York : London : McGraw-Hill, 1980
viii,135p : ill. - (Career preparation for agriculture/agribusiness). - Includes index
0-07-036737-x X.520/22406
2317516

AGRICULTURE
Change. Research. Applications of multivariate analysis

SHORT, Brian
The application of multivariate statistical techniques to the study of agricultural change through time. - [London] : [Geography Department, King's College], 1975
[2],25[i.e. 30]leaves,[8]leaves of plates : ill. - (Occasional papers / King's College (London). Geography Department)
0-904547-01-9 X.313/1284 Woolwich
0681539

AGRICULTURE
Chemistry - For agriculture

HAY, R. K. M. (Robert K. M.)
Chemistry for agriculture and ecology. - Oxford : Blackwell Scientific, 1981
xi,243p : ill. - Includes bibliographies and index
0-632-00699-4 X.629/15421 Woolwich
2402149

AGRICULTURE
Collectivisation. Theories of Marx, Karl, 1818-1883

CLIFF, Tony
Marxism and the collectivisation of agriculture. - London : Socialists Unlimited, 1980
36p. - (International socialism reprint ; no.1)
B83-05048 X.0700/3113(1)
2315335
- [Another copy.] X.529/51492

AGRICULTURE
Conferences - Serials

CONFERENCE reports, agriculture, food and biology. - Vol.1, [no.1]- ; 1977-. - London : Applied Science Publishers, 1977-
B77-15924 SRL
1108433

AGRICULTURE
Decision making - Conference proceedings

INTERNATIONAL CONFERENCE OF AGRICULTURAL ECONOMISTS (16th : 1976 : Nairobi)
Decision-making and agriculture. - Oxford : Agricultural Economics Institute for International Association of Agricultural Economists ; [London] : [Distributed by University of Nebraska Press], 1977
xiii,603p : ill. - Includes bibliographies and index
0-8032-7201-4 X.319/18209 Woolwich
1519131

AGRICULTURE
Development related to economic development

AHMAD, Syed
Induced adjustment and the role of agriculture in economic development. - Hamilton, Ont. : Department of Economics, McMaster University, 1977
49 leaves. - (McMaster University. Department of Economics. Working paper ; no.77-20)
r6189004 P.515/190 [no.77-20]
0370312

GROWTH and adjustment in national agricultures / edited by J.P. O'Hagan. - London : Macmillan [for] the Food and Agriculture Organization of the United Nations, 1978
viii,242p. - Bibliography: p.237-238. - Includes index
0-333-22352-7 X.329/11427 Woolwich
111784x

JOHNSTON, Bruce Foster
Agriculture and structural transformation. - New York ; London : Oxford University Press, 1975
xxi,474p : ill. - (Economic development series). - Includes index
0-19-501869-9 X.329/10075
0370312

Agriculture and structural transformation. - New York ; London : Oxford University Press, 1975 [i.e. 1978]
xxi,474p : ill. - (Economic development series). - Includes index
0-19-501870-2 X.319/18224 Woolwich
0370312

AGRICULTURE
Development. Social factors

OSUNTOGUN, Adeniyi
Institutional determinants and constraints on agricultural development. - [London] : Joint Programme on Agricultural Development Overseas, [1976?]
[1],iv,57p
0-85003-061-7 X.311/8316 Woolwich
0772526

AGRICULTURE
Ecological aspects

BAYLISS-SMITH, Timothy P.
The ecology of agricultural systems. - Cambridge : Cambridge University Press, 1982
112p : ill. - (Cambridge topics in geography. Second series). - Bibliography: p110-112
0-521-23125-6 X.322/11422
078589x

DUCHET, René
Perspectiva ecológica da agricultura: China, Estados Unidos, Terceiro Mundo. - Porto : Afrontamento, 1975
144 p.. - (Viver é preciso ; n. 2)
r6013389 X.319/8507 Woolwich
0877549

RADICAL agriculture / edited by Richard Merrill. - New York ; London : Harper and Row, 1976
xix,459p : ill. - (Harper colophon books). - Includes bibliographies and index
0-06-090337-6 X.319/17030 Woolwich
078589x

RESPONSIBLE husbandry as a basic educational principle. - London : 4 Willifield Way, NW11 7XT : Farm and Food Society, [1984]
[8]p : 2ill. - (Discussion paper)
B85-04787 X.322/14386
078589x

AGRICULTURE
Ecological aspects - Conference proceedings

REES, W. A.
Report on the 3rd European Ecological Symposium. - [London] : ADAS, c1984
12p
B85-05687 BS.3/928
0877549

AGRICULTURE
Economic aspects. Archaeology - Conference proceedings

INTEGRATING the subsistence economy / edited by Martin Jones. - Oxford : B.A.R., 1983
300p : ill. - (Symposia of the Association for Environmental Archaeology ; no.4) (BAR. International series ; 181). - Includes bibliographies
0-86054-232-7 X.425/5208
3278263

AGRICULTURE
Economics - For agriculture

HILL, Berkeley
An introduction to economics for students of agriculture. - Oxford : Pergamon, 1980
ix,346p : ill. - (Pergamon international library). - Bibliography: p300-301. - Includes index
0-08-020510-0 X.529/38895
1807250

AGRICULTURE
Economics - For agriculture - Serials

WEST African journal of agricultural economics. - Vol.1, no.1- ; 1972-. - Ibadan : Progress Publishers, 1972-
r6108891 P.311/521
1807250

AGRICULTURE
Effects of changes of climate, to 1977

PARRY, M. L. (Martin Lewis)
Climatic change, agriculture and settlement. - Folkestone : Dawson [etc.], 1978
214p : ill. - (Studies in historical geography). - Bibliography: p.194-207. - Includes index
0-7129-0794-7 X.329/11435 Woolwich
1260839

AGRICULTURE
Electric equipment. Safety

SAFE use of electricity. - [Stoneleigh] : [Farm-electric Centre], [1980]
[2],60p : col ill. - Bibliography: p.55
B80-38266 X.629/14162 Woolwich
2281902

AGRICULTURE
Energy. Conservation

HUGHES, D. F.
Report of conference "Rational use of electric thermal consumers in agriculture" (sic) Balatonfured Hungary 16–20 May, 1983. – [London] : ADAS, c1983
17p
B84–43633 BS.3/903
2842165

AGRICULTURE
Energy. Conservation – Conference proceedings

INTERNATIONAL SEMINAR ON ENERGY CONSERVATION AND THE USE OF RENEWABLE ENERGIES IN THE BIO-INDUSTRIES (2nd 1982 Oxford)
Energy conservation and use of renewable energies in the bio-industries 2. – Oxford : Pergamon, 1982
xvi,750p : ill
0–08–029781–1 SRL
2842165

AGRICULTURE
English language – For agriculture – For non-English speaking students

MOUNTFORD, Alan
English in agriculture. – Oxford : Oxford University Press, 1977
ix,113p : ill. – (English in focus)
0–19–437514–5 X.900/25481
1286366

English in agriculture. – Oxford : Oxford University Press. – (English in focus)
Teacher's ed. – 1977
xxi,156p : ill
0–19–437506–4 X.900/25481a
1286366

AGRICULTURE
English languages – For non-English speaking students – For agriculture

AGRICULTURE / English Language Services, Inc. ; consultants Frank J. Bowman, Marla A. Jackson. – New York : Macmillan ; London : Collier Macmillan, c1984
104p : ill. – (Macmillan career English)
0–02–973620–x X.329/17132
2992639

AGRICULTURE / English Language Services, Inc. ; consultant Marla A. Jackson. – New York : Macmillan ; London : Collier Macmillan, c1984
115p : ill. – (Macmillan career English)
0–02–973630–7 X.329/17133
2992639

AGRICULTURE / English Language Services Inc. ; consultants Frank J. Bowman, James G. Doherty, Marla A. Jackson. – New York : Macmillan ; London : Collier Macmillan, c1984
149p : ill. – (Macmillan career English)
0–02–973640–4 X.329/17134
2992639

AGRICULTURE
Environmental aspects

WALLER, Robert
The agricultural balance sheet. – London : Green Alliance and the Conservation Society, c1982
27p
B83–27663 X.329/16426
2913437

AGRICULTURE
Extension services. Information services. Requirements

DEY, Jennie
Information needs of staff of agricultural extension services. – Reading : University of Reading, Agricultural Extension and Rural Development Centre, 1977
[4],ii,44p
0–7049–0512–4 X.515/616
1800256

AGRICULTURE
Extension services. Management

CHAMBERS, Robert, 1932–
Two frontiers in rural management. – [Brighton] : [Institute of Development Studies], [1975]
12p. – (Communications / University of Sussex. Institute of Development Studies ; 113). – Bibliography: p.12
0–903354–17–9 X.0512/206(113)
0659827

AGRICULTURE
Extension services. Role in social development of rural regions

EXTENSION education and rural development / edited by Bruce R. Crouch, Shankariah Chamala. – Chichester : Wiley, c1981.
Vol.1: International experience in communication & innovation
xxvi,371p : ill. – Bibliography: p331–361. – Includes index
0–471–27829–7 X.520/23222
1942352
– Bibliography: p283–313. – Includes index
Vol.2: Experience in strategies for planned change
xxvi,325p : ill
0–471–27675–8 X.520/23223
1942352

AGRICULTURE
Factors of production

UPTON, Martin
Agricultural production economics and resource-use. – London : Oxford University Press, 1976
x,357p : ill. – Includes bibliographies and index
0–19–859452–6 X.329/10470 Woolwich
0656054

AGRICULTURE
Farms

AMERY, Heather
On the farm. – London : Usborne, 1984
[16]p : col.ill. – (What's happening?)
0–86020–853–2 X.990/24224
175033x

BOLWELL, L. H. (Laurence Henry)
Farms and farming. – Hove : Wayland, 1984
32p : ill(some col.). – (Young explorer). – Includes index
0–85078–418–2 X.322/14563
175033x

HEEKS, Peggy
Buildings on the farm. – Hove : Wayland, 1984
33p : col.ill. – Includes index
0–85078–475–1 X.629/26135
175033x

HURT, Mary
On the farm. – Loughborough : Ladybird, c1983
[44]p : col.ill. – (Ladybird toddler books)
0–7214–0786–2 X.990/21833
175033x

AGRICULTURE
Farms. Buildings. Concrete foundations. Construction

BARNES, Maurice M.
Concrete foundations for farm buildings. – 2nd ed. – Slough : Cement and Concrete Association, 1981
1folded sheet : ill. – (Farm note ; 9)
0–7210–1230–2 X.622/10472 Woolwich
2494361

AGRICULTURE
Farms. Buildings. Design & construction

WATTS, John W.
The supervision and inspection of field and building works on farms. – London : Batsford Academic and Educational, 1984
219p : ill. – (Mitchell's professional library). – Bibliography: p215–216. – Includes index
0–7134–3534–8 X.622/22284
296466x

AGRICULTURE
Farms. Buildings. Design & construction. Safety aspects

REPORT on the Commission internationale du genie rural (CIGR) Section 2 / by P.A. Mason ... [et al.]. – [London] : ADAS, c1983
18p
B84–39495 BS.3/858
3174034

AGRICULTURE
Farms. Concrete pavements. Construction

BARNES, Maurice M. (Maurice Malcolm)
Farm construction: laying concrete. – Slough : Cement and Concrete Association, 1982
[8]p : col.ill. – (Pocket guide ; 1)
0–7210–1268–x X.808/37252
2798034

AGRICULTURE
Farms. Fences. Construction

HART, Edward
Hedge laying and fencing. – Wellingborough : Thorsons, 1981
128p : ill. – Bibliography: p126. – Includes index
0–7225–0701–1 SRL
2320517

AGRICULTURE
Farms. Financial management

STURROCK, Ford Gibson
Farm accounting and management. – 6th ed. – London : Pitman, 1976
ix,262p : ill. – Includes index
0–273–00230–9 X.319/16926 Woolwich
0760196

AGRICULTURE
Farms. Intensive animal housing

SAINSBURY, David, 1928–
Livestock health and housing. – 2nd ed. – London : Baillière Tindall, 1979
xi,388p : ill. – Includes bibliographies and index
0–7020–0617–3 SRL
1404814

AGRICULTURE
Farms. Management

BARNARD, Christopher Stephen
Farm planning and control. – Cambridge : Cambridge University Press, 1979
x,600p : ill. – Bibliography: p.577–581. – Includes index
0–521–22658–9 X.322/9186 Woolwich
2127849

BOEHLJE, Michael D.
Farm management. – New York ; Chichester : Wiley, c1984
xv,806p : ill. – Includes bibliographies and index
0–471–04688–4 X.322/14246
2127849

BUCKETT, Maurice
An introduction to farm organisation and management. – Oxford : Pergamon, 1981
x,313p : ill. – (Pergamon international library). – Includes index
0–08–024433–5 X.322/10150 Woolwich
2127849

AGRICULTURE
Farms. Management. Applications of small computer systems. Programs written in Basic language

PUGH, C. L.
Farm planning systems for small computers. – [Ashford, Kent] : Farm Business Unit, School of Rural Economics and Related Studies, 1979
[4],82[i.e. 85]p : 1 ill. – (FBU occasional papers / Wye College. Farm Business Unit ; no. 3)
0–901859–83–4 BS.475/5(3)
2085496

AGRICULTURE
Farms. Self-built buildings. Design & construction

EAST OF SCOTLAND COLLEGE OF AGRICULTURE. Farm Buildings Advisory and Development Department
DIY farm buildings. – Edinburgh : The College, 1975
[4],10p : 1 ill. – (Farm buildings topics. DIY series ; no.1)
0–902164–17–1 P.313/400
0571288

AGRICULTURE
Farms. Shire horses. Use

COCKCROFT, Barry
Princes of the plough. – London : Dent, 1978
128p : ill
0–460–04320–x X.329/11366 Woolwich
1110527

AGRICULTURE
Farms. Surveys

CLARK, G. (Gordon)
The agricultural census. – [Norwich] : [Geo Books], c1982
47p : ill. – (CATMOG ; 35). – Bibliography: p32–36
0–86094–115–9 X.329/18247
3162362

AGRICULTURE
Farms. Tiered cages & flat deck pens for early weaned pigs

MEAT AND LIVESTOCK COMMISSION. Pig Improvement Services
The early weaned pig. – [Bletchley] : [M.L.C.], [1976]
[2],28p
0–904650–03–0 SRL
0651109

AGRICULTURE
Farms. Wildlife. Conservation

CHESHIRE College of Agriculture wildlife and landscape management plan review 1976–1981 / joint editors Michael J. Willis, Simon W. Young. – [Nantwich] : Cheshire County Education Committee, 1983
viii,127p : ill. – Bibliography: p94–96
B83–39523 X.322/12390
2958309

AGRICULTURE
Farms. Workshops

BELL, Brian, 1935–
Farm workshop. – Ipswich : Farming Press, 1981
223p : ill. – (Farming book series). – Bibliography: p217. – Includes index
0–85236–112–2 X.629/19556
2851261

AGRICULTURE
Field experiments. Design & analysis – Conference proceedings

DEVELOPMENTS IN THE DESIGN, ANALYSIS AND INTERPRETATION OF FIELD EXPERIMENTS (Conference : 1973 : Armidale)
Developments in field experiment design and analysis. – Slough : Commonwealth Agricultural Bureaux, 1975
vii,196p : ill. – (Bulletins / Commonwealth Bureau of Pastures and Field Crops ; 50). – Includes bibliographies
0–85198–333–2 SRL
0518573

AGRICULTURE
Financial incentives. Effects of government policies on agricultural industries – Conference proceedings

DISTORTIONS of agricultural incentives / edited by Theodore W. Schultz. – Bloomington ; London : Indiana University Press, 1978
viii,343p : ill. – Includes bibliographies and index
0–253–31806–8 X.320/11637 Woolwich
1838261

AGRICULTURE
Geographical aspects

DINKELE, Geoff
Farming. – London : Harrap, 1976
68p : ill. – (Harrap's course in reformed geography)
0–245–52379–0 X.311/8094 Woolwich
058276x

GRIGG, David, 1934–
An introduction to agricultural geography. – London : Hutchinson, 1984
204p : ill. – Includes bibliographies and index
0–09–156710–6 X.322/14477
0036684

HAYES, John
Food and farming. – London : Hutchinson, 1980
32p : ill. – (Down to earth series)
0–09–138591–1 L.49/78
058276x

HORSFALL, David
Agriculture. – Oxford : Blackwell, 1983
64p : ill. – (Geography applied). – Includes index
0–631–12683–x X.322/13165
1477021

HUGGETT, Richard, 1929–
Agriculture. – London : Harper & Row, 1980
vii,136p : ill. – (Geography ; bk.2)
0–06–318166–5 X.0800/1714(2)
0036684

MEWBURY, Paul Augustus
A geography of agriculture. – Plymouth : Macdonald and Evans, 1980
x,326p : ill. – (Aspect geographies). – Bibliography: p.313–314. – Includes index
0–7121–0733–9 X.329/12676 Woolwich
2222876

POLKINGHORNE, K.
Agriculture. – London : Hutchinson. – (Hutchinson units in geography)
1. – 1978
[7],64p : ill(some col)
0–09–132441–6 Cup.1281/859
1477021

SINGH, Jasbir
Agricultural geography. – New Delhi : Tata McGraw-Hill ; London : McGraw-Hill, c1984
xiv,412p : ill. – Includes bibliographies and index
0–07–451915–8 X.322/14706
2222876

SYMONS, Leslie
Agricultural geography. – Rev. ed. – London : Bell & Hyman, 1978
xiii,285p : ill. – (Bell's advanced economic geographies). – Bibliography: p259–269. – Includes index
0–7135–0001–8 X.809/53453
2222876

WALKER, Michael J.
Agricultural location. – Oxford : Blackwell, 1977
80p : ill(chiefly col). – Bibliography: p.78
0–631–93660–2 X.311/8526 Woolwich
058276x

AGRICULTURE
Great Britain. Biomass fuels. Production – For agriculture

CARRUTHERS, S. P.
Biofuel production strategies for U.K. agriculture. – Reading : Centre for Agricultural Strategy, 1983
143p : ill. – (CAS paper ; 13). – Bibliography: p126–137. – Includes index
0–7049–0619–8 X.329/17820
312455x

AGRICULTURE
Information sources

INFORMATION sources in agriculture and food science / editor G.P. Lilley. – London : Butterworths, 1981
xiv,603p : 1ill. – (Butterworths guides to information sources). – Includes index
0–408–10612–3 SRL
2325888

AGRICULTURE
Information systems

INFORMATION systems for agriculture / edited by M. J. Blackie and J. B. Dent. – London : Applied Science Publishers, 1979
xii, 176p : ill. – Bibliography – Includes index
0–85334–829–4 X.329/12460 Woolwich
1824252

AGRICULTURE
Librarianship & information science

INTERNATIONAL ASSOCIATION OF AGRICULTURAL LIBRARIANS AND DOCUMENTALISTS. Regional European Symposium (1973 : Wageningen)
Progress and prospects in agricultural librarianship and documentation.
155p : ill. – Bibliography : p.39
90–220–0460–0 2719.aa.65
109033x

AGRICULTURE
Librarianship & information science – Conference proceedings

EUROPEAN REGIONAL CONGRESS OF AGRICULTURAL LIBRARIANS AND DOCUMENTALISTS (1978 : Hamburg)
Europäischer Regionalkongreß des Bibliothekswesens und der Dokumentation der Agrarwissenschaften. – München ; London : K.G. Saur, 1978
347p : ill
3–598–07096–9 X.329/16128
109033x

INTERNATIONAL agricultural librarianship / edited by Alan Fusonie and Leila Moran. – Westport, Conn. ; London : Greenwood Press, 1979
ix,127,[1]p : ill. – Includes index
0–313–20640–6 2719.dd.13
109033x

AGRICULTURE
Management

DALTON, G. E. (Graham Eyre)
Managing agricultural systems. – London : Applied Science, c1982
xii,161p : ill. – Includes index
0–85334–165–6 X.520/29631
2317508

LONG, Don L.
Introduction to agribusiness management. – New York ; London : McGraw-Hill, 1979
viii,136p : ill. – (Career preparation for agriculture/agribusiness). – Includes index
0–07–038665–x X.520/22394
2317508

NIX, John
Farm management pocketbook. – 8th ed. – [Wye] : Wye College, School of Rural Economics and Related Studies, Farm Business Unit, 1977
vii,154p. – Includes index
0–901859–67–2 X.319/17915
0228095

NIX, John
Farm management pocketbook. – 9th ed.. – Ashford, Kent : Wye College School of Rural Economics and Related Studies, Farm Business Unit, 1978
viii,156p. – Includes index
0–901859–77–x SRL
0228095

Farm management pocketbook. – 10th ed.. – Ashford, Kent : Wye College, School of Rural Economics and Related Studies, Farm Business Unit, 1979
vii,157p. – Includes index
0–901859–85–0 SRL
0228095

AGRICULTURE
Management – Serials

RESEARCH in domestic and international agribusiness management. – Vol.1 (1980)–. – Greenwich, Conn. : JAI Press, 1980–
v.
B82–02356 P.511/5300
2587823

AGRICULTURE
Management. Simulation games – Collections

MINSHULL, Roger Michael
Factors in farming. – Sunbury-on-Thames : Nelson, 1976
30p : ill
0–17–434202–0 X.311/8313 Woolwich
0769886

Factors in farming. – Sunbury-on-Thames : Nelson. Teachers' book, including printed spirit duplicating masters. – 1976
38p : ill. – Bibliography: p.12
0–17–434201–2 X.311/8313a Woolwich
0769886

AGRICULTURE
Materials handling. Costs. Control

BUTTERWORTH, Bill
Materials handling in farm production. – London : Crosby Lockwood Staples, 1979
[6],122p : ill. – Includes index
0–258–97100–2 X.311/8922 Woolwich
1872516

AGRICULTURE
Mathematical models

FOOD FOR A DOUBLING WORLD POPULATION (Project)
MOIRA. – Amsterdam : Oxford : North-Holland Publishing Co., 1979
xv,379p : ill. – (Contributions to economic analysis ; 124). – Includes bibliographies and index
0–444–85169–0 WPC.19/124
1993577

AGRICULTURE
Mechanisation

SHIPWAY, G. P.
27th Session of the FAO–ECE Working Party on Mechanization of Agriculture. – [London] : ADAS, 1982
10p
B82–24281 BS.3/744
273477x

AGRICULTURE
No-tillage cropping systems

NO–TILLAGE agriculture / edited by Ronald E. Phillips, Shirley H. Phillips. – New York ; London : Van Nostrand Reinhold, c1984
xii,306p : ill. – Includes bibliographies and index
0–442–27731–8 X.322/14480
3220974

AGRICULTURE
Office practices

HOSKEN, Michael John
The farm office. – Ipswich : Farming Press, 1976
5–200p : ill. – Includes index
0–85236–063–0 X.329/10397 Woolwich
062862x

HOSKEN, Michael John
The farm office. – 2nd (revised) ed. – Ipswich : Farming Press, 1978
5–200p : ill. – Includes index
0–85236–085–1 X.529/32530
062862x
– [Another copy] SRL

AGRICULTURE
On-line bibliographic information retrieval services. Machine-readable files

HARVEY, Susan, 19––
CAB/CAIN evaluation project. – Wageningen : Centre for Agricultural Publishing and Documentation ; London : British Library Board, 1979
93p. – (British Library research and development reports ; no.5483)
90–220–0704–9 X.629/22692
2896052

AGRICULTURE
Part-time agriculture – Conference proceedings

RURAL GEOGRAPHY SYMPOSIUM (1st : 1975 : University of Guelph)
Part-time farming. – Norwich : Geo Abstracts Ltd, 1976
vi,291p : ill
0–902246–57–7 X.311/8233 Woolwich
070637x

AGRICULTURE
Personnel-related costs
HAMILTON, Carl
Project analysis in the rural sector with special reference to the evaluation of labour cost. - Stockholm : University of Stockholm, Institute for International Economic Studies, 1974
173 leaves. - ([Universitetet i Stockholm. Institute for International Economic Studies. Monograph series ; 3]). - Includes bibliographies
r6029297 X.0512/449(3)
2031906

AGRICULTURE
Policies of communist parties, to 1945
RIZZI, Franco
Contadini e comunismo. - Milano : Angeli, c1981
251p. - (Storia delle ideologie e delle istituzioni contemporanee ; 1)
88-204-1903-3 X.329/16589
8963606

AGRICULTURE
Policies of governments, ca 1950-1975
NAGLE, John Charles
Agricultural trade policies. - Farnborough : Saxon House [etc. for] the Agricultural Adjustment Unit, University of Newcastle-upon-Tyne, 1976
xviii,171p. - Includes index
0-347-01139-x X.320/10379 Woolwich
066667x

AGRICULTURE
Policies of governments. Economic aspects
HALLETT, Graham
The economics of agricultural policy. - 2nd ed. - Oxford : Blackwell, 1981
xi,365p : ill. - Includes bibliographies and index
0-631-12493-4 X.329/14224
2140918

AGRICULTURE
Prehistoric agriculture
BENDER, Barbara
Farming in prehistory. - London : J. Baker, 1975
xi,268p : ill. - Bibliography: p.239-254. - Includes index
0-212-97003-8 X.311/8050
0511463

ORIGINS of agriculture / editor Charles A. Reed. - The Hague : Mouton, 1977
xvi,1013,[2]p(2 fold.),[10]p of plates,fold leaf of plate : ill. - (World anthropology). - Includes bibliographies and index
90-279-7919-7 X.320/11961 Woolwich
1060406

AGRICULTURE
Prehistoric agriculture - Conference proceedings
The EARLY history of agriculture. - Oxford : Oxford University Press for the British Academy, 1977
[4],213p,5p of plates : ill. - Includes bibliographies
0-19-725967-7 X.322/8239 Woolwich
0511463

AGRICULTURE
Prehistoric agriculture. Origins. Role of increases in population
COHEN, Mark Nathan
The food crisis in prehistory. - New Haven ; London : Yale University Press, 1977
x,341p. - Bibliography: p.287-330. - Includes index
0-300-02016-3 X.329/11088 Woolwich
1187759
The food crisis in prehistory. - New Haven ; London : Yale University Press, 1978
x,341p. - Bibliography: p.287-330. - Includes index
0-300-02351-0 X.410/10328 Woolwich
1187759

AGRICULTURE
Productivity
CROXALL, Harold E.
The fight for food. - London : Allen & Unwin, 1984
xvi,218p. - Bibliography: p208-216. - Includes index
0-04-630011-2 X.329/18442
3048624

AGRICULTURE
Productivity. Effects of agricultural research
ROSSITER, Margaret Walsh
The emergence of agricultural science. - New Haven ; London : Yale University Press, 1975
xiv,275p,[4]p of plates : ill. - (Yale studies in the history of science and medicine ; 9). - Bibliography: p.233-263. - Includes index
0-300-01721-9 X.320/10195
0533092

AGRICULTURE
Project management. Teaching aids
POTTS, D. J.
The tomato concentrate project. - [Bradford] : The Centre, 1981
1portfolio : ill. - (Teaching materials series ; no.2)
0-901945-43-9 X.0525/458(2)
273060x

AGRICULTURE
Renewable resources - Conference proceedings
RENEWABLE resource management for forestry and agriculture / edited by James S. Bethel and Martin A. Massengale. - Seattle ; London : University of Washington Press, 1978
x,126p : ill. - (Publications series / Geo. S.Long Fund). - Includes bibliographies and index
0-295-95624-0 SRL
1933353

AGRICULTURE
Research
RICE RESEARCH STATION (Rokupr)
Work and development programme for rice research. - Rokupr : the Station, [1978]
17p.
r6108463 P.313/231
2265508

RILEY, Ralph
The case for agricultural research. - [London?] : [The College?], [1983?]
10p. - (The fourteenth J. D. Bernal lecture)
B84-34172 X.322/13745
2265508

AGRICULTURE
Research - Serials
[RESEARCH and development note (Scottish Agricultural Colleges)]. Research and development note / the Scottish Agricultural Colleges. - No.1 (Sept. 1981)-. - Aberdeen : The Colleges, 1981-
v.
Holdings: no.3 (Jan. 1982)-
r6865240 P.313/238
1137417

AGRICULTURE
Research & development. Effects on food supply
MCCALLA, A. F.
International agricultural research. - Reading : Centre for Agricultural Strategy, 1978
37p : ill. - (CAS papers / Centre for Agricultural Strategy ; 5). - Bibliography: p.37
0-7049-0607-4 X.319/18760 Woolwich
1846515

AGRICULTURE
Research. Applications of nuclear technology
VOSE, Peter B.
Introduction to nuclear techniques in agronomy and plant biology. - Oxford : Pergamon, 1980
xiii,391p : ill. - (Pergamon international library of science, technology, engineering and social studies). - Includes bibliographies and index
0-08-024924-8 SRL
1896474

AGRICULTURE
Research. Cooperation between Great Britain & Soviet Union
GREAT BRITAIN
[Treaties, etc. Union of Soviet Socialist Republics, 1965 Jan.6. Protocols, etc., 1974 Oct.17]. Exchange of notes between the Government of the United Kingdom of Great Britain and Northern Ireland and the Government of the Union of Soviet Socialist Republics revising the Schedule appended to the Agreed Minute on Co-operation in the Field of Agricultural Research signed on 6 January 1965, Moscow, 17 October 1974 ... - London : H.M.S.O., [1975]
12p. - (Treaty series ; no.64(1975)) (Cmnd.6044)
0-10-160440-8 BS.14/1708
0356360

AGRICULTURE
Research. Effects on productivity of agricultural industries
ROSSITER, Margaret Walsh
The emergence of agricultural science. - New Haven ; London : Yale University Press, 1975
xiv,275p,[4]p of plates : ill. - (Yale studies in the history of science and medicine ; 9). - Bibliography: p.233-263. - Includes index
0-300-01721-9 X.320/10195
0533092

AGRICULTURE
Research. Experiments. Design & statistical analysis
LITTLE, Thomas M.
Agricultural experimentation. - New York ; Chichester : Wiley, 1978
xiii,350p : ill. - Includes index
0-471-02352-3 SRL
1740563

AGRICULTURE
Research. Great Britain. Department of Agriculture for Northern Ireland - Serials
GREAT BRITAIN. Department of Agriculture for Northern Ireland
Annual report on research and technical work of the Department of Agriculture for Northern Ireland. - 1975-. - [Belfast] : [The Department], [1977]-
B77-13714 SRL
1093215

AGRICULTURE
Research in British universities: Research sponsored by Wye College. School of Rural Economics and Related Studies - Bibliographies
WYE COLLEGE. School of Rural Economics and Related Studies
Recent publications and current research. - [Ashford, Kent] : [Wye College], [1978]
[1],v,52p
B78-38153 BS.475/3
172388x

AGRICULTURE
Research organisations - Directories
AGRICULTURAL research centres / consultant editor Nigel Harvey. - 7th ed. - Harlow : Longman, 1983
2v.. - (Reference on research). - Includes index
0-582-90014-x SRL
1496948

ILLES, Doris
Resources directory for agricultural research in semi-arid regions. - Boston, Mass. : G.K. Hall, 1979
757p. - (Bibliographies and guides in African studies)
0-8161-8302-3 X.320/12072 Woolwich
1496948

AGRICULTURE
Research organisations - Indexes
AGRICULTURAL research index / editor J. Burkett. - 6th ed. - [Guernsey] : Hodgson ; Harlow : Longman, 1978
2v.(1020,[8]p)
0-582-90000-x SRL
1496948

AGRICULTURE
Research organisations: Commonwealth Agricultural Bureaux. Information retrieval services using digital computer systems. Machine-readable files: CAB ABSTRACTS. Searching. Use of DIALOG
COMMONWEALTH AGRICULTURAL BUREAUX
CAB abstracts on Dialog. - Farnham Royal : C.A.B., 1977
1v. : ill
B79-09968 UNP.266/75
1816349

AGRICULTURE
Research. Pattern analysis
PATTERN analysis in agricultural science / edited by W.T. Williams. - Melbourne : CSIRO ; Amsterdam ; Oxford : Elsevier, 1976
xii,331p : ill. - Includes bibliographies and index
0-444-99844-6 SRL
1171933
- [Another copy.] X.320/10843 Woolwich

AGRICULTURE
Research. Questionnaires. Design & interpretation - Conference proceedings
SYMPOSIUM ON THE DESIGNING AND INTERPRETING OF QUESTIONNAIRES (1975 : University College of Wales)
Symposium on the Designing and Interpreting of Questionnaires. - Aberystwyth : University College of Wales, Department of Agricultural Economics, 1975
171,[53]p,leaf of plate,[2]p of plates : ill
0-902124-14-5 X.313/1255 Woolwich
0606308

AGRICULTURE
Research. Role in economic growth of developing countries
PINSTRUP-ANDERSEN, Per
Agricultural research and technology in economic development. - London : Longman, 1982
ix,261p : ill. - Bibliography: p237-255. - Includes index
0-582-46048-4 X.329/15877
2824329

AGRICULTURE
Research. Statistical methods

GOMEZ, Kwanchai A.
Statistical procedures for agricultural research. - 2nd ed. - New York ; Chichester : Wiley, c1984
xvi,680p : ill. - (An International Rice Research Institute book) (A Wiley-interscience publication). - Includes index
0-471-87092-7 SRL
296094x

MEAD, R. (Roger)
Statistical methods in agriculture and experimental biology. - London : Chapman and Hall, 1983
xi,335p : ill. - Bibliography: p327. - Includes index
0-412-24230-3 X.322/12446
296094x
- [Another copy: paperback.] ISBN 0-412-24240-0 X.322/12447

AGRICULTURE
Research. Use of transducers

SENSORS in agriculture. - [Roslin] : Agricultural Research Council, [1983?]
[22]p
B84-19173 X.322/13111
3076997

AGRICULTURE
Role in economic development - Conference proceedings

AGRICULTURE in development theory / edited by Lloyd G. Reynolds. - New Haven ; London : Yale University Press, 1975 [i.e. 1976]
x,510p : ill. - Includes index
0-300-01805-3 X.320/10466 Woolwich
0712124

INTERNATIONAL FARM MANAGEMENT CONGRESS (4th : 1980)
The role of agriculture in society. - Slough : Commonwealth Agricultural Bureaux, c1980
256p : ill. - Includes bibliographies
0-85198-480-0 SRL
0712124

AGRICULTURE
Role in economic development of capitalist countries

GOODMAN, David, 1938-
From peasant to proletarian. - Oxford : Basil Blackwell, 1981
244p : ill. - Bibliography: p218-238. - Includes index
0-631-10361-9 X.520/25329
2339250

AGRICULTURE
Role in food supply

LINDSTRÖM, U B
Världens matförsörjning och vår. - Helsingfors : [U. Lindström & P. Tigerstedt], 1978
66p : ill,charts
r6112715 X.329/13121 Woolwich
2461595

AGRICULTURE
Role of plant regulation - Conference proceedings

NATO ADVANCED STUDY INSTITUTE ON PLANT REGULATION AND WORLD AGRICULTURE (1978 : Izmir)
Plant regulation and world agriculture. - New York ; London : Plenum Press [for the] NATO Scientific Affairs Division, 1979
xi,575p : ill. - (NATO Advanced Study Institutes. series A, life sciences ; vol.22). - Includes bibliographies and index
0-306-40180-0 SRL
1941348

AGRICULTURE
Role of Shell International Chemical Company - Serials

SHELL INTERNATIONAL CHEMICAL COMPANY
Shell in agriculture. - Nov. 1976-. - [London] : [Shell International Chemical Company], 1976- : ill
r6071772 P.313/437
1405411

AGRICULTURE
Serial articles - Indexes - Serials

AGRICULTURE and horticulture index / Hertis. - Jan. 1982-. - St. Albans : Hertfordshire College of Agriculture and Horticulture, 1982-
v.
B83-34744 SRL
294703x

AGRICULTURE
Serials

AGRICULTURAL journal titles and abbreviations. - Phoenix : Oryx, 1982
154p
0-89774-028-9 X.322/11696
2878046

AGRICULTURE
Serials in stock of libraries in Spain - Catalogues

CATÁLOGO colectivo de publicaciones periódicas en bibliotecas españolas / [compilado por el Instituto Bibliográfico Hispánico]. - Madrid : Dirección General de Archivos y Bibliotecas, 1973.
Vol.3: Agricultura y veterinaria
xxii,496p
r6700014 X.0515/122(3)
2482142

AGRICULTURE
Serials in stock of Science Reference Library - Catalogues

SCIENCE REFERENCE LIBRARY
Periodicals on agriculture held by the Science Reference Library. - London : Science Reference Library.
Pt.1: Agricultural research and industry. - [1981?]
v,[78p]p. - Includes index
0-902914-66-9 L.49/1474
2702800

AGRICULTURE
Simulations

DENT, John Barry
Systems simulation in agriculture. - London : Applied Science Publishers, 1979
x,180p : ill. - Includes bibliographies and index
0-85334-827-8 SRL
178191x

AGRICULTURE
Simulations: Models - Serials

AGRICULTURAL Systems. - Vol.1-, no.1- ; Jan. 1976-. - Barking : Applied Science Publishers, 1976- : ill
B76-18835 SRL
0657557

AGRICULTURE
Smallholdings. Agriculture

MUNROE, C. J.
The smallholder's guide. - Newton Abbot : David and Charles, 1979
160p. - Includes index
0-7153-7652-7 X.529/33316
1203061

SMITH, Joanna
Farm your garden. - London : Sidgwick and Jackson, 1977
176p,[12]p of plates : ill. - Includes bibliographies and index
0-283-98311-6 X.329/11111 Woolwich
1203061
Farm your garden. - Harmondsworth : Penguin, 1979
205p,[8]p of plates : ill. - (Penguin handbooks). - Includes bibliographies and index
0-14-046348-8 WP.4003/262
1203061

AGRICULTURE
Smallholdings. Crops. Weeds. Control measures - Bibliographies

COMPTON, J. A. F.
Small farm weed control. - London : Intermediate Technology Development Group, 1982
iv,170p
B82-29323 SRL
2755815

AGRICULTURE
Smallholdings. Livestock. Care & management

TREVISICK, Charles Huxtable
Keep your own livestock. - London : Paul, 1978
144p,[8]p of plates : ill. - Includes index
0-09-129460-6 X.329/11386 Woolwich
1440837

AGRICULTURE
Smallholdings. Livestock: Pigs. Care & management

DOWNING, Elisabeth
Keeping pigs. - London : Pelham, 1978, c1985
95p : ill. - (Garden farming series). - Bibliography: p90-91. - Includes index
0-7207-1613-6 X.329/19630
3233138

AGRICULTURE
Smallholdings. Livestock: Sheep. Care & management

DOWNING, Elisabeth
Keeping sheep. - London : Pelham, 1979, c1985
142p : ill. - (Garden farming series). - Bibliography: p129-130. - Includes index
0-7207-1514-8 X.329/19631
3233146

AGRICULTURE
Smallholdings. Self-sufficiency
LIVING on a few acres. - New York : Sterling, 1982
432p : ill(some col)
0-8069-7594-6 X.629/20298
0774413

SAVAGE, Derek Stanley
The cottager's companion. - London : Mayflower, 1980
224p : ill. - Includes index
0-583-12850-5 X.329/12408 Woolwich
0774413

SELF-SUFFICIENT small holding. - 2nd ed. - Stowmarket : Soil Association, 1977
14p : ill
0-905200-10-1 X.329/16104
0774413

SOIL ASSOCIATION
Self-sufficient small holding. - Stowmarket : The Association, 1976
[2],14p(6 fold)
0-905200-05-5 X.319/16973 Woolwich
0774413

THEAR, Katie
The family smallholding. - London : Batsford, 1983
143p : ill. - Bibliography: p140. - Includes index
0-7134-1935-0 X.322/12389
0774413

AGRICULTURE
Smallholdings. Self-sufficiency. Social aspects

SEYMOUR, John, 1914-
Bring me my bow. - London : Turnstone Books, 1977
159p
0-85500-078-3 X.709/32165
1433172

AGRICULTURE
Socioeconomic aspects

AGRICULTURAL sustainability in a changing world order / edited by Gordon K. Douglass. - Epping : Bowker, 1984
xiv,282p : ill. - (Westview special studies in agricultural science and policy). - Includes bibliographies and index
0-86531-669-4 X.322/14201
3196194

AGRICULTURE
Southern Africa. Veld. Plants - For agriculture

HOBSON, Nigel Kayser
Veld plants of southern Africa. - Johannesburg : Macmillan South Africa (Publishers) ; London : Macmillan, 1975
310p : ill(some col). - Includes index
0-86954-013-0 L.R.425/97
0640158

AGRICULTURE
Soviet theories, 1917-1945

FIGUROVSKAÎÂ, Nadezhda Konstantinovna
Razvitie agrarnoĭ teorii v SSSR. - Moskva : Nauka, 1983
383p
r6499261 X.329/16514
8958661

AGRICULTURE
Statistics, 1966-1974

FOOD AND AGRICULTURE ORGANIZATION
Report on the 1970 world census of agriculture. - Rome : F.A.O. ; [London] : H.M.S.O.], 1977
v,289p,fold plate : ill. - (FAO statistics series / Food and Agriculture Organization ; 10)
B79-35259 UNF.131/15
0630705

AGRICULTURE
Techniques. Effects on quality of food - Conference proceedings

AGRICULTURAL practices and food quality. - Dublin : Royal Irish Academy, 1980
viii,120p : ill. - Includes bibliographies
0-901714-16-x SRL
2737922

AGRICULTURE
Technological development. Ecological aspects

DICKSON, Gordon R. (Gordon Ross)
The problems of technical advance in agriculture. - Newcastle upon Tyne : University of Newcastle upon Tyne, 1975
11p. - (Inaugural lectures / University of Newcastle upon Tyne)
0-7017-0002-5 X.319/17517 Woolwich
1199129

AGRICULTURE
Theses, to date - Bibliographies - Serials

LISTA de tesis presentadas en las escuelas de agricultura, ganadería y medicina veterinaria de la República Mexicana / compilado por Margarita Sandoval Guerrero. - 1968-1970-. - [México] : Secretaría de Agricultura y Ganadería, Instituto Nacional de Investigaciones Agrícolas, 1976-
r6700273 X.0520/674
8847894

AGRICULTURE
Use of aircraft - Conference proceedings
TECHNO-ECONOMIC TRENDS IN AIRBORNE EQUIPMENT FOR AGRICULTURE AND OTHER SELECTED AREAS OF THE NATIONAL ECONOMY (Conference : 1978 : Warsaw)
Trends in airborne equipment for agriculture and other areas. - Oxford : Pergamon for the United Nations, 1980
xvi,358p : ill
0-08-022425-3 SRL
1943987

AGRICULTURE
Use of microorganisms
ADVANCES in agricultural microbiology / editor: N.S. Subba Rao. - London : Butterworth Scientific, 1982
xxii,704p,[36]p of plates : ill. - (Studies in the agricultural and food sciences). - Includes bibliographies and index
0-408-10848-7 SRL
2848031

AGRICULTURE
Use of mobile seed cleaning machines & mobile seed treating machines
SLY, J. M. A. (John Michael Alfred)
Mobile seed-cleaning/treating machines, 1972-73. - Pinner : Ministry of Agriculture, Fisheries and Food, Agricultural Development and Advisory Service, 1975
[1],ii,30p : ill. - (Pesticide usage survey report ; 7)
B78-11077 SRL
138029x

AGRICULTURE
Use of plastics
DUBOIS, P.
Plastics in agriculture. - London : Applied Science Publishers, 1978
xiii,176p : ill. - Bibliography: p.169. - Includes index
0-85334-776-x X.629/12089 Woolwich
1412892

AGRICULTURE
Use of zeolites
ZEO-AGRICULTURE / edited by Wilson G. Pond and Frederick A. Mumpton. - Epping : Bowker, 1984
xii,296p : ill. - (Westview special studies in agriculture/aquaculture science and policy). - Includes bibliographies and index
0-86531-602-3 X.322/13915
3196216

AGRICULTURE
Waste materials. Composting - Conference proceedings
COMPOSTING of agricultural and other wastes / [proceedings of a seminar organized by the Commission of the European Communities, Directorate-General [for] Science, Research and Development, Environment Research Programme held at Brasenose College, Oxford, UK, 19-20 March 1984] ; edited by J.K.R. Gasser. - London : Elsevier Applied Science, c1985
x,320p : ill. - (EUR ; 9419). - Includes bibliographies
0-85334-357-8 SRL
3205770

AGRICULTURE
Water supply engineering - For agriculture
WATERHOUSE, James
Water engineering for agriculture. - London : Batsford Academic and Educational, 1982
395p : ill. - Includes bibliographies and index
0-7134-1409-x SRL
2768410

AGRICULTURE. Aceh. Indonesia
Development. Effects on rural economic conditions
GIBBONS, David S.
Agricultural modernization, poverty and inequality. - Farnborough, Hants. : Saxon House, 1980
xi,225p : ill. - Bibliography: p.213-219. -Includes index
0-566-00331-7 X.329/13187 Woolwich
1933558

AGRICULTURE. Africa
1960-
AGRARNYE preobrazovaniîa v stranakh Afriki na sovremennom ètape / [redkollegiîa : L.V. Goncharov (otvetstvennyĭ redaktor) ... dr.]. - Moskva : Nauka, Glavnaîa redaktsiîa vostochnoĭ literatury, 1982
227p
r6490171 X.329/19090
9125604

AGRICULTURE. Africa
Economic development
BADOUIN, Robert
Les agricultures de subsistance et le développement économique. - Paris : Éditions A. Pedone, 1975
204 p. - (Collection du Centre de Recherche, d'Étude et de Documentation sur les Instutions et les Législations Africaines : 12). - Bibliography : p. 199-202
2-233-00011-0 X.311/4492 Woolwich
0905127
- [Another copy.] X.311/4598

JEROME, W.A.
National agrarianism. - Kampala : East African Literature Bureau, 1973-1976
3v.
r6108552 X.0808/2003
0905127

LEVI, John
Economics of African agriculture. - Harlow : Longman, 1982
vii,175p : ill. - Includes index
0-582-64147-0 X.529/47366
2718707

AGRICULTURE. Africa
Economic development - Conference proceedings
AFRICAN RURAL SECTOR : INTERDISCIPLINARY PERSPECTIVES (Conference : 1978 : Los Angeles)
Agricultural development in Africa. - New York : Praeger, 1980
453p
0-03-056173-6 X.322/9504 Woolwich
0905127

AGRICULTURE. Africa
Productivity. Improvement - Conference proceedings
COMMONWEALTH AGRICULTURAL BUREAUX. Scientific Conference (1st : 1984 : Arusha)
Advancing agricultural production in Africa. - Slough : Commonwealth Agricultural Bureau, c1984
x,454p : ill. - Includes bibliographies
0-85198-537-8 SRL
3306836

AGRICULTURE. Africa
Small farms. Stored maize. Losses. Evaluation
ADAMS, John Mervyn
The evaluation of losses in maize stored on a selection of small farms in Zambia with particular reference to the development of methodology. - London : Tropical Products Institute, 1977
xi,149p : ill. - (G109). - Bibliography: p.126-128
0-85954-058-8 BS.38X
0368644

AGRICULTURE. Africa south of the Sahara
Role of multinational companies
DINHAM, Barbara
Agribusiness in Africa. - London : Earth Resources Research, [1983]
224p b ill. - Includes index
0-946281-00-9 X.529/69516
328087x

AGRICULTURE. African countries
Research, to date
INTERNATIONAL INSTITUTE OF TROPICAL AGRICULTURE
Tasks for the eighties. - Ibadan : The Institute, 1981
iii,60p
r6810511 X.322/12906
9063943

AGRICULTURE. Albania
Policies of government, 1944-
LANGE, Klaus
Die Agrarfrage in der Politik der Partei der Arbeit Albaniens. - München : Trofenik, 1981
153p. - (Albanische Forschungen ; Bd.21)
r6399169 X.0800/173(21)
875697x

AGRICULTURE. Alentejo. Portugal
to 1977
RAPOSO, José Hipólito
Alentejo. - Lisboa : Editorial O Século, [1978]
407,[44]p
r6101722 X.319/20329 Woolwich
2218208

AGRICULTURE. Alentejo. Portugal
to date
CORTEZ, João Maria Parreira
[Pandémonios. Selections]. Senhores da terra. - [Lisboa?] : Imprensa Nacional-Casa da Moeda, 1982
410p. - (Temas portugueses). - Includes bibliography
r6710536 X.800/41524
8367086

AGRICULTURE. Alentejo. Portugal
Economic development, 1974-1975
CAUTELA, Afonso
O Alentejo na reforma agrária. - Lisboa : Diabril, 1975
141p : ill. - (Colecção Universidade do povo. política ; n.8)
r6000685 X.319/8521 Woolwich
0885738

AGRICULTURE. Algeria
1962-
ÉTUDE sur le secteur privé agricole / Association Algérienne pour la recherche démographique économique et sociale. - Alger : Association algérienne pour la recherche démographique économique et sociale, 1975.
3: Configuration socio-économique du secteur privé agricole
1v((various pagings))
r6207347 S.V.58/28
8839999

JÖNSSON, Lars
La révolution agraire en Algérie. - Uppsala : Scandinavian Institute of African Studies, 1978
84p. - (Nordiska afrikainstitutet. Research report ; no.47)
91-7106-145-2 P.803/212[no.47]
8839999

AGRICULTURE. Algeria
Economic development, 1962-
BENAMRANE, Djilali
Agriculture et développement en Algérie. - Alger : Société nationale d'édition et de diffusion, 1980
269p.1folded leaf of plates. - Bibliography: p251-260
r6206846 X.520/27387
8863040

HENNI, Ahmed
La colonisation agraire et le sous-développement en Algérie. - [Alger] : [S.N.E.D.], [1982]
246p. - Includes bibliography
r6212588 X.520/37009
8863040

AGRICULTURE. Amazon River Basin. Brazil
1964-1978
IANNI, Octavio
Ditadura e agricultura. - Rio de Janeiro : Civilização Brasileira, 1979
249p.,[1]fold leaf. - (Coleção Retratos do Brasil ; v.131)
r6161854 X.329/12792 Woolwich
2475278

AGRICULTURE. America
Multinational companies
BURBACH, Roger
Agribusiness in the Americas. - New York ; London : Monthly Review Press, c1980
314p : ill. - Includes index
0-85345-535-x X.809/46945
2364174

AGRICULTURE. Ancient Rome
to 476
GABBA, Emilio
Strutture agrarie e allevamento transumante nell' Italia romana, III-I sec.a.C.. - Pisa : Giardini, 1979
197p,[11]folded leaves[40]p of plates. - (Biblioteca di studi antichi ; 18)
r6504621 X.622/12367 Woolwich
8706735

RODBERTUS-JAGETZOW, Carl
Per la storia dell'evoluzione agraria di Roma sotto gl'imperatori. - [Bologna] : Forni, 1977
p458-508
r6500932 X.329/13989
8706735

AGRICULTURE. Ancient Rome
Cato, Marcus Porcius, b.B.C.234. De agri cultura
RICHTER, Will
Gegenständliches Denken, archaisches Ordnen. - Heidelberg : Winter, 1978
191p. - (Bibliothek der klassischen Altertumswissenschaften ; n.F., Reihe 2, Bd 62)
3-533-02677-9 011313.cc1a/62
2030381

AGRICULTURE. Ancient Rome
Law, to 476
BUCK, Robert J
Agriculture and agricultural practice in Roman law. - Wiesbaden : Steiner, 1983
59p. - (Historia. Einzelschriften ; Hft.45)
3-515-04040-4 PP.3481.iae(2)[Hft.45]
8995494

AGRICULTURE. Ancient Rome
Law, to 476–continued

NEEVE, Pieter Willem de
[COLONUS; privé grondpacht in Romeins Italië. English]. Colonus. – Amsterdam : Gieben, 1984
273p. – Includes bibliography
90–7026–515–x X.200/44570
8995494

AGRICULTURE. Ancient Rome
Palladius Rutilius Taurus Aemilianus. Documents

RODGERS, Robert Howard
An introduction to Palladius. – London : Institute of Classical Studies, 1975
xv,181p,viii leaves of plates : ill. – (Bulletin. Supplements / University of London. Institute of Classical Studies ; no.35). – Bibliography: p.163–171. – Includes index
0–900587–32–6 Ac.2666qa[Suppl.35]
0574066

AGRICULTURE. Andalusia. Spain
711–1516

LADERO QUESADA, Miguel Ángel
Diezmo eclesiástico y producción de cereales en el reino de Sevilla. – [Sevilla] : Universidad de Sevilla, Departamento de Historia Medieval, 1979
128p,[5]folded leaves of plates : ill
84–7405–133–9 X.322/10050 Woolwich
8715130

AGRICULTURE. Andalusia. Spain
1808–

HÉRAN HAEN, François
Tierra y parentesco en el campo sevillano. – [Madrid?] : Servicio de Publicaciones Agrarias, [1980]
268p. – (Serie Estudios)
84–7479–093–x X.329/17818
9073795

AGRICULTURE. Andes. South America
Role of women

DEERE, Carmen Diana
Women in Andean agriculture. – Geneva : International Labour Office, 1982
xii,172p : ill. – (Women, work and development ; 4)
92–210–3106–3 X.520/37410
3330389

AGRICULTURE. Andover region. Hampshire
Labourers, 1846–1875

GREAT BRITAIN. Parliament. House of Commons. Select Committee on the Andover Union
[Report]. The north-west Hampshire agricultural labourer 1846, or, 'The new Poor Law' / edited by J.E.H. Spaul ; and, The north-west Hampshire agricultural labourer 1867–1875 / by George E. Brickell. – Andover : Andover Local Archives Committee, 1975
[1],30p
0–903755–03–3 X.702/5102
0607061

AGRICULTURE. Angeln. West Germany
Farms. Buildings. Architecture, to date

HAGEMEIER-KOTTWITZ, Anita
Angeliter Dreiseithofanlagen. – Neumünster : Wachholtz, 1982
436p,[3]folded leaves of plates : ill. – (Studien zur Volkskunde und Kulturgeschichte Schleswig-Holsteins ; Bd.9)
3–529–02458–9 X.0702/468(9)
8953643

AGRICULTURE. Arab countries

HATEM, M. Abdel-Kader
Land of the Arabs. – London : Longman, 1977
323p : ill(some col). – Includes bibliographies and index
0–582–78037–3 X.802/10462
1278894

AGRICULTURE. Arabia
– For British exporters

BOWEN-JONES, Howard
Agriculture in the Arabian peninsula. – London : Economist Intelligence Unit, 1983
174p. – (Special report ; no.145). – Bibliography: p171–174
B84–05096 PP.1423.aeh/68
2995255

AGRICULTURE. Argentina

COSCIA, Adolfo A.
Economía agraria. – Buenos Aires : Editorial Hemisferio Sur, 1976
xv,270p : ill. – (Colección de textos de agronomía y veterinaria ; no 22)
r6065854 X.519/39284
1718983

IVARS, Luis
Campo argentino. – Buenos Aires : Ediciones Colombo, 1977
138p
r6078628 X.619/10529 Woolwich
1848690

AGRICULTURE. Argentina
to 1950

SAN ESTEBAN, Ricardo
El agro argentino. – Buenos Aires : Editorial Cartago, 1975
188p
r6048147 X.319/8829 Woolwich
1229443

AGRICULTURE. Argentina
to date

BALÁN, Jorge
Urbanización regional y producción agraria en Argentina : un análisis comparativo. – Buenos Aires : CEDES, 1979
42p. – (Estudios CEDES ; v.2, no.2)
r6705935 X.629/19070
8876843

CARRACEDO, Orlando
Economía agraria y progreso social. – Buenos Aires : Depalma, 1980
xix,173p
r6708548 X.322/13226
8876843

RANDLE, Patricio H.
Atlas del desarrollo territorial de la Argentina. – Buenos Aires : OIKOS, 1981
3v.
r6708354 Tab.800/23
8876843

AGRICULTURE. Argentina
1973–

MOLINA, Jorge S.
Una nueva conquista del desierto. – 2a impresión. – Buenos Aires : Emecé Editores, 1980
252p,plates. – Includes bibliographies
r6706438 X.329/17988
801583x

SAN ESTEBAN, Ricardo
El agro argentino. – Buenos Aires : Quipo, 1979
206p
r6701248 X.329/13960 Woolwich
801583x

AGRICULTURE. Argentina
Economic development, to 1975

FLICHMAN, Guillermo
La renta del suelo y el desarrollo agrario argentino. – México : Siglo Veintiuno, 1977
241p : ill
r6084821 X.319/9863
1976974

AGRICULTURE. Argentina
Economic development, 1536–1976

RAS, Norberto
Una interpretación sobre el desarrollo agropecuario de la Argentina. – 2a ed. – Buenos Aires : Hemisferio Sur, 1977
278p : ill. – (Colección de textos de agronomía y veterinaria ; 26). – Includes bibliographies
r6084763 X.319/9818 Woolwich
1855360

AGRICULTURE. Argentina
Economic development. Role of Jesuits, 1650–1767

CUSHNER, Nicholas P.
Jesuit ranches and the agrarian development of colonial Argentina, 1650–1767. – Albany : State University of New York Press, c1983
xi,206p : ill. – Bibliography: p195–202
0–87395–706–7 X.329/17367
3066967

AGRICULTURE. Argentina
Law

TABORDA CARO, María Susana
Derecho agrario. – [4a ed.]. – Buenos Aires : Plus Ultra, 1977
588p
r6084973 X.100/27757
1853988

AGRICULTURE. Argentina
Law, 1973–

ARGENTINA
[Laws, etc]. Digesto de legislación agraria. – Rosario : Zeus, 1979
648p
r6701479 X.329/14444 Woolwich
8015864

[Legislación agraria]. Legislación agraria. – 2. ed corregida y actualizada. – Buenos Aires : Editorial Astrea, 1979
xxiii,718p
r6704076 X.200/38093
8820104

AGRICULTURE. Arid regions

SETTLING the desert / edited by L. Berkofsky, D. Faiman, J. Gale. – Sede Boqer : Jacob Blaustein Institute for Desert Research ; London : Gordon and Breach Science, c1981
xvi,274p : ill. – Includes index
0–677–16280–4 X.520/24894
2550393

AGRICULTURE. Arid regions
Role of nitrogen in soils

SOIL water and nitrogen / edited by John Monteith and Colin Webb. – The Hague ; London : Nijhoff, 1981
xviii,338p : ill. – (Developments in plant and soil sciences ; v.1). – Includes bibliographies and index
90–247–2406–6 X.325/1326 Woolwich
2501570

AGRICULTURE. Arid regions
Role of water in soils

SOIL water and nitrogen / edited by John Monteith and Colin Webb. – The Hague ; London : Nijhoff, 1981
xviii,338p : ill. – (Developments in plant and soil sciences ; v.1). – Includes bibliographies and index
90–247–2406–6 X.325/1326 Woolwich
2501562

AGRICULTURE. Arid regions. North-east Brazil
to date

DUQUE, José Guimarães
O Nordeste e as lavouras xerófilas. – 2a ed. – [Fortaleza] : Banco do Nordeste do Brasil, 1973
238p,[65]p of plates : ill
r6705870 X.800/34290
8876886

AGRICULTURE. Armenia (Soviet Socialist Republic)
Production

SHAKHBAZĪAN, Minas Sarkisovich
Vosproizvodstvo v sel'skom khozīaĭstve Armīanskoĭ SSR v period razvitogo sofsializma. – Erevan : Izdatel'stvo AN Armīanskoĭ SSR, 1978
337p
r6097827 X.629/8143 Woolwich
2429551

AGRICULTURE. Asia
Development. Political aspects – Conference proceedings

The POLITICS of agrarian change in Asia and Latin America / edited by Howard Handelman. – Bloomington : Published in association with the American Universities Field Staff [by] Indiana University Press, c1981
viii,136p : ill
0–253–34548–0 X.520/29702
2858460

AGRICULTURE. Asia
Employment. Effects of technological development

BARTSCH, William H.
Employment and technology choice in Asian agriculture. – New York ; London : Praeger, 1977
xxi,127,[1]p. – (Praeger special studies in international economics and development). – Bibliography: p.108–122. – Includes index
0–275–24280–3 X.320/10974 Woolwich
1269399

AGRICULTURE. Asia
Employment. Labour

CONTRACTUAL arrangements, employment and wages in rural labor markets in Asia / edited by Hans P. Binswanger and Mark R. Rosenzweig. – New Haven ; London : Yale Universtiy Press, c1984
xvii,330p
0–300–03214–5 X.520/36179
3021947

AGRICULTURE. Aurskog-Holand. Norway
Part-time agriculture

ARNLJOT, Harald A.
Deltidsjordbruk. – Oslo : Norsk institutt for by- og regionforskning, 1972
92p. – (By- og regionforskning ; rapport 20)
r6033793 X.0802/100(20)
1505947

AGRICULTURE. Australia

AGRICULTURE in the Australian economy / edited by D.B. Williams. – 2nd ed., rev. and reset. – [Sydney] : Sydney University Press ; London : Eurospan, 1982
xviii,422p : ill. – Includes index
0–424–00092–x X.322/14545
323097x

AGRICULTURE. Australia –continued

JONES, William, 19—
o Benygroes i ben draw'r byd. – [Caernarfon] : Cyngor Sir Gwynedd. Gwasanaeth Llyfrgell, 1983
49p : ill. – (Darlith flynyddol Llyfrgell Penygroes ; 1982)
0–904852–36–9 X.958/25666
3175642

AGRICULTURE. Australia
to 1980

DAVIDSON, B. R. (Bruce Robinson)
European farming in Australia. – Amsterdam ; Oxford : Elsevier Scientific, 1981
x,437p. – Includes bibliographies and index
0–444–41993–4 X.322/10142 Woolwich
253472x

AGRICULTURE. Austria
to date

KLEIN-BRUCKSCHWAIGER, Franz
[Essays. Selections]. Beiträge zur österreichischen Agrargeschichte. – Wien : Hochschülerschaft an der Universität für Bodenkultur, 1979
ix,131p. – (Schriftenreihe für Agrarwirtschaft ; Hft 16). – Bibliography: p.97–131
r6301693 X.322/10182 Woolwich
8726485

AGRICULTURE. Austria
1918–1938

MATTL, Siegfried
Agrarstruktur, Bauernbewegung und Agrarpolitik in Österreich 1919–1929. – Wien : Geyer, 1981
iv,471p. – (Veröffentlichungen zur Zeitgeschichte ; Bd.1). – Bibliography
r6305692 X.809/52753
884190x

AGRICULTURE. Austria
1945– – Statistics – Serials

ERGEBNISSE der landwirtschaftlichen Statistik / bearbeitet im Österreichischen Statistischen Zentralamt. – 1980–. – Wien : das Zentralamt, 1981–. – (Beiträge zur österreichischen Statistik)
r6303571 S.36/30[Hft.604,etc.]
8726477

AGRICULTURE. Austria
Credit institutions, to date

KREINECKER, Günther
Die Anfänge der Raiffeisenkassen in Oberösterreich (1889–1914). – Linz : Trauner, c1982
xvi,276p. – (Linzer Schriften zur Sozial- und Wirtschaftsgeschichte ; Bd.6). – Includes bibliography
3–85320–279–9 X.0525/376(6)
8993866

AGRICULTURE. Austria
Farms. Buildings. Architecture, to date

TOMASI, Elisabeth
Die traditionellen Gehöftformen in Niederösterreich. – St. Pölten : Niederösterreichisches Pressehaus, c1984
64p : ill. – (Wissenschaftliche Schriftenreihe Niederösterreich ; 75/76)
3–85326–554–5 X.0709/722(75/76)
912800x

AGRICULTURE. Austria
Personnel, 1945– – Statistics – Serials

LAND- und forstwirtschaftliche Arbeitskräfte / bearbeitet im Österreichischen Statistischen Zentralamt. – 1979–. – Wien : das Zentralamt, 1981–. – (Beiträge zur österreichischen Statistik)
r6303804 S.36/30[Hft.615, etc.]
8766134

AGRICULTURE. Aveyron. France

LEA, K. J.
Agriculture in Aveyron. – [Glasgow] : Department of Geography, University of Strathclyde, 1982
19leaves : ill. – (Research paper ; no.13)
B84–00654 X.805/4533
2981068

AGRICULTURE. Badenoch. Highland Region. Scotland
Farms. Social conditions, 1769–1782

GRANT, I. F. (Isabel Frances)
Every-day life on an old Highland farm 1769–1782. – London : Shepheard-Walwyn, c1981
ix,292p,[16]p of plates : ill. – Includes index
0–85683–058–5 X.322/10606 Woolwich
2632098

AGRICULTURE. Balkan countries
ca 1400–ca 1600

ASDRACHAS, Spyros I
Mechanismoi tes agrotikes oikonomias sten Tourkokratia, 15–16 ai.. – Athena : Themelio, 1978
300p. – (Historike vivliotheke). – Bibl
r6194626 X.519/44128
2415828

AGRICULTURE. Balkan countries
Socioeconomic aspects

KARAVIDAS, Kostas D.
Agrotika. – [Athens] : Ekdoseis Papazese, 1978
xxii,32p.,698p
r6187888 X.319/20233 Woolwich
2166259

AGRICULTURE. Ballylongford. Kerry (County)
Effects of pollution of environment by proposed electrolytic zinc smelters

COULTER, B. S.
An electrolytic smelter at Ballylongford, Co. Kerry. – Dublin : An Foras Talúntais, 1979
[4],88p : ill. – Includes bibliographies
0–905442–39–3 SRL
2319322

AGRICULTURE. Baltic countries
Socioeconomic aspects, 1400– – Conference proceedings

PROBLEMY razvitiia sotsial'no-ėkonomicheskikh formatsii v stranakh Baltiki. – Tartu : Tartuskii gosudarstvennyi universitet, 1978
156p
r6097026 X.519/41413
1867172

AGRICULTURE. Bangladesh

ALIM, A.
An introduction to Bangladesh agriculture. – Dacca : M. Alim, 1974
[4],xxiv,436p,[2]fold leaves. – Bibliography : p.431–432. – Includes index
r6032341 X.319/17602 Woolwich
125863X

JANNUZI, F. Tomasson
The agrarian structure of Bangladesh. – New Delhi : Sangam, 1982
xii,150p. – Bibliography: p139–144
0–86131–434–4 X.529/57848
3028151

AGRICULTURE. Bangladesh
– Bibliographies

GOODLAND, Robert James A.
Agriculture and ecology of Bangladesh. – [New York] : the author, [1979?]
14p.
r6109219 X.315/56 Woolwich
125863X

AGRICULTURE. Bangladesh
– Serials

FARM economy. – Vol.1, no.1–. – Dacca : the Association, 1978–
r6202055 X.0322/229 Woolwich
125863X

AGRICULTURE. Bangladesh
1971–

AHMAD, Kamal Uddin
Bangladesh agriculture and field crops. – Dacca : Mumtaj Kamal, 1980
viii,137p
r6814390 X.322/14640
8610509

JANNUZI, F. Tomasson
The agrarian structure of Bangladesh. – Boulder : Westview Press, 1980
xv,150p. – (Westview special studies on South and Southeast Asia). – Bibliography
0–89158–682–2 X.520/26027
8610509

AGRICULTURE. Bangladesh
1971– – Serials

MENNONITE CENTRAL COMMITTEE. Agriculture Program, Bangladesh
Report. – [No.1 (1973)]–. – Dacca : The Committee, [1973]–
Holdings: report no.7 (July 1980)
r6801849 P.311/528
8610509

AGRICULTURE. Bangladesh
Economic development

WENNERGREN, E. Boyd (Emil Boyd)
Agricultural development in Bangladesh. – Epping : Bowker, [1984?]
xxii,373p : 1ill. – Bibliography: p363–373
0–86531–806–9 X.329/19063
3223655
– [Another issue.] – Boulder, Colo.: Westview Press, 1984 V 25030 IOL

AGRICULTURE. Bangladesh
Mechanisation

FARM power in Bangladesh. – [Reading] : Department of Agricultural Economics and Management, University of Reading.
Vol.1: A comparative analysis of animal and mechanical farm power in Bangladesh. – 1981
iv,248p : ill. – (Development study / University of Reading Department of Agricultural Economics & Management ; no.19). – Bibliography: p245–248
0–7049–0695–3 X.322/10591 Woolwich
2625075
– (Development study / Department of Agricultural Economics and Management, University of Reading ; no.20)
Vol.2. – 1981
145p : ill. – Includes bibliographies
0–7049–0696–1 X.322/10847 Woolwich
2625075

AGRICULTURE. Bangladesh
Technological development

BIGGS, Stephen D. (Stephen Devon)
Appropriate agricultural technology in Bangladesh. – Brighton : Institute of Development Studies, 1976
[1],37[i.e.44]p. – (Discussion papers / University of Sussex. Institute of Development Studies ; no.90). – Bibliography: p.34–37
0–903354–27–6 X.313/1334 Woolwich
0761052

AGRICULTURE. Bangladesh
Technological development – Serials

BANGLADESH ACADEMY FOR RURAL DEVELOPMENT
Evaluation of Thana irrigation programme in Bangladesh. – 1968–69–. – Comilla : Bangladesh Academy for Rural Development, [1969?]–
Holdings: 1973/74–
r6202328 P.311/521
0761052

AGRICULTURE. Barry. South Glamorgan
Ancient Roman agriculture. Farms. Buildings. Archaeology, 1965–1970

JARRETT, Michael G. (Michael Grierson)
Whitton. – Cardiff : Published on behalf of the Board of Celtic Studies of the University of Wales [by] University of Wales Press, 1981
xiii,262p,[21]p of plates, 10 folded leaves of plates : ill(some col.). – Bibliography: p255–262
0–7083–0765–5 L.42/1296
2206277

AGRICULTURE. Bashkir ASSR. Russia (RSFSR)

IZ istorii sel'skogo khoziaistva Bashkirii / [pod redaktsiei V.P. Ivankova i Kh. F. Usmanova]. – Ufa : Akademiia nauk SSSR, Bashkirskii filial, Institut istorii, iazyka i literatury, 1976
156p
r6046226 X.319/8817 Woolwich
1224190

KACHESTVENNAIA i ėkonomicheskaia otsenka zemel' Bashkirskoi ASSR / [redaktsionnaia kollegiia : K.Sh. Bagamanov (otv. redaktor) ... i dr.]. – Ufa : Akademiia nauk SSSR, Bashkirskii filial, 1977
142p
r6187229 X.319/20084 Woolwich
1224190

ZONAL'NYE osobennosti intensifikatsii sel'skokhoziaistvennogo proizvodstva / pod redaktsiei M.S. Gubaidullina i K.Sh. Bagamanova. – Ufa : Akademiia nauk SSSR, Bashkirskii filial, 1974
225p
r6177478 X.329/12427 Woolwich
1224190

AGRICULTURE. Bashkir ASSR. Russia (RSFSR)
to date

ZEMEL'NYI fond Bashkirskoi ASSR i ego ispol'zovanie / [IA.I. Vaniukov (otv. redaktor)]. – Ufa : Bashkirskii filial Akademii nauk SSSR, Otdel ėkonomicheskikh issledovanii, 1979
165p
r6465246 X.429/16956
8382506

AGRICULTURE. Bashkir ASSR. Russia (RSFSR)
Finance
VOPROSY vyravnivaniia ėkonomicheskikh uslovii povysheniia dokhodov v kolkhozakh / [redaktsionnaia kollegiia : A.I. Grekova (otv. redaktor), G.A. Urazbakhtin, R.A. Galin]. - Ufa : Akademiia nauk SSSR, Bashkirskii filial, Otdel ėkonomicheskikh issledovanii, 1972
227p
r6178443 X.809/28714
2438259

AGRICULTURE. Bavaria. West Germany
1918-1933
RATJEN, Wolfgang
Die bayerischen Bauernkammern von 1920 bis 1933. - München : Kommissionsbuchhandlung Wölfle, 1981
vi,204p. - (Neue Schriftenreihe des Stadtarchivs München ; Bd.115) (Miscellanea Bavarica Monacensia ; Hft.94). - Bibliography
3-87913-108-2 WP.8420/115
8806063

AGRICULTURE. Belgium
to date
HAEGEN, Herman van der
Bronnen voor de reconstructie van de agrarische struktuur. - Leuven : Belgisch Centrum voor Landelijke Geschiedenis, 1980
[16]p[2]p of plates,[1]leaf of plates : ill. - (Publikatie / Belgisch Centrum voor Landelijke Geschiedenis ; nr.58)
r6303436 X.0311/197.(58)
8026963

AGRICULTURE. Belgium
1945-. Economic aspects
AGRICOM-AGRIGES
Assemblée générale ordinaire. - Exercise (1980)-. - Bruxelles : AGRICOM-AGRIGES, 1981-
r6206831 X.0425/171 Woolwich
8840008

AGRICULTURE. Belgium
1970
BELGIUM. Institut National de Statistique
Recensement général de l'agriculture et des forêts de 1970. - Bruxelles : Institut National de Statistique, 1976-
tom.1-
r6012384 S.A.19/83
1152092

AGRICULTURE. Belgium
Farms. Buildings. Architecture, to date
TREFOIS, Clemens V.
[Selections]. Het boerendak ; De plattegrondsvormen van zg. Frankische boederijtypen (heruitgave) ; De bouw der boerenhoeven (heruitgave). - Sint-Niklaas : Danthe, 1980
312p : ill
r6330165 L.45/1750
8876835

AGRICULTURE. Belorussia
KOPYSSKII, Zinovii IUr'evich
Istochnikovedenie agrarnoi istorii Belorussii ... - Minsk : Nauka i tekhnika, 1978
134p
r6174152 X.329/12922 Woolwich
2428652

MARTINKEVICH, Feliks Stanislavovich
Soft sial,no-ėkonomicheskie problemy sel,skogo khozti aistva Belorussii na ėtape razvitogo soft sializma. - Minsk : Nauka i tehnika, 1975
470p
r6096850 X.329/8905 Woolwich
2025515

AGRICULTURE. Belorussia
to 1240
KOROBUSHKINA, T. N. (Tat'iana Nikolaevna)
Zemledelie na territorii Belorussii v X-XIII vv.. - Minsk : Nauka i tekhnika, 1979
115p : ill
r6462119 X.329/16146
8916497

AGRICULTURE. Belorussia
1917-1945
KUKHAREV, B. E. (Boris Efimovich)
Sel'skoe khoziaistvo Zapadnoi Belorussii (1919-1939 gg.). - Minsk : Vysheishaia shokola, 1975
110p
r6402681 X.329/14645 Woolwich
8753520

AGRICULTURE. Belorussia
1917-
SEL'SKOE khoziaistvo Belorussii / [pod redaktsiei N.I. Dementeia, F.P. Sen'ko]. - Minsk : Uradzhai, 1980
263p : ill(some col)
r6498395 X.322/10087 Woolwich
8582300

AGRICULTURE. Belorussia
1945-
POLUIAN, I. V. (Ivan Vasil'evich)
Tekhnicheskie kadry belorusskoi derevni i razvitie sel'skogo khoziaistva respubliki. - Minsk : Nauka i tekhnika, 1978
220p. - Bibliography: p.210-[221]
r6498296 X.329/14204 Woolwich
8715173

AGRICULTURE. Belorussia
Collectivisation
PODGOTOVKA sploshnoi kollektivizatsii sel'skogo khoziaistva Belorusskoi SSR (noiabr 1927 g. - noiabr 1929 g.) / pod redaktsiei A. L. Azarova [i dr.] ; [L. N. Iashenko (otvetstvennyi sostavitel')]. - Minsk : Izdatel'stvo "Belarus'", 1976
269 p.. - (Istoriia kollektivizatsii sel'skogo khoziaistva SSSR. Dokumenty i materialy)
r6021063 X.809/23653
1000799

SOROKIN, Anatolii Nikolaevich
Sovkhozy Belorusskoi SSR (1917-1941 g.g.) ... - Minsk : Nauka i tekhnika, 1979
246p
r6185609 X.329/8989 Woolwich
1000799

AGRICULTURE. Belorussia
Collectivisation, 1917-1941
POBEDA kolkhoznogo stroia v Belorusskoi SSR / [kollektiv avtorov: M.P. Kostiuk ... i dr.]. - Minsk : Nauka i tekhnika, 1981
447p
r6448823 X.629/19520
802751x

PROVEDENIE sploshnoi kollektivizatsii sel'skogo khoziaistva Belorusskoi SSR (noiabr' 1929g.-1932g.) / pod redaktsiei A.I. Azarova ... [i dr.] ; [L.N. IAshenko (otvetstvennyi sostavitel')]. - Minsk : Belarus', 1973
422p. - (Istoriia kollektivizatsii sel'skogo khoziaistva SSSR)
r6462804 X.329/17543
802751x

AGRICULTURE. Bengal. India
1920-1946
ISLAM, M. Mufakharul
Bengal agriculture, 1920-1946. - Cambridge : Cambridge University Press, 1978
xiv,283p : ill. - (Cambridge South Asian studies ; 22). - Bibliography: p.270-280. - Includes index
0-521-21579-x X.529/33027
1800132

AGRICULTURE. Bessarabia
1800-1900
ANTSUPOV, Ivan Antonovich
Agrarnye otnosheniia na iuge Bessarabii. - Kishinev : Shtiintsa, 1978
234p
r6187299 X.319/20144 Woolwich
2118955

AGRICULTURE. Binbrook. Lincolnshire
Farms. Social conditions, 1871-1875 - Correspondence, diaries
STOVIN, Cornelius
Journals of a Methodist farmer 1871-1875. - London : Croom Helm, c1982
251p. - Bibliography: p243-244. - Includes index
0-7099-2324-4 X.809/51045
2694751

AGRICULTURE. Birmingham. West Midlands (Metropolitan County)
Farms: Chapman's Hill School Farm Limited. Visits by students of Chivenor County Primary Junior and Infants School
PALMER, Joy Annette
The Chivenor farmers. - London : Association of Agriculture, 1976
16p,[4]p of plates : col ill. - Bibliography: p.15-16
0-903313-27-8 X.319/16706 Woolwich
0701319

AGRICULTURE. Bolivia
OSSIO SIVILA, Hugo
La problemática agropecuaria y agroindustrial en la zona del oriente boliviano. - La Paz : Universidad Católica Boliviana, Instituto de Investigaciones Socioeconómicas, 1977
iv,224,90p.,[3]leaves of plates : maps. - (Documento de trabajo; no.03/77)
r6158919 X.322/2688 Woolwich
2415518

AGRICULTURE. Bolivia
- Forecasts
BOLIVIA. Ministerio de Asuntos Campesinos y Agropecuarios
Plan Quinquenal Agropecuario. - La Paz : [el Ministerio], 1976
189p.,[7]leaves of plates : ill
r6157517 LAS.B.65/30
218866X

AGRICULTURE. Botswana
Policies of government
WILY, Liz
Land allocation and hunter-gatherer land rights in Botswana. - London : Anti-Slavery Society, 1980
117p,[7]leaves of plates. - (Human rights and development working papers ; no.4). - Bibliography: p111-117
B81-16585 X.525/4331
2392461

AGRICULTURE. Bournemouth region. Dorset
Farms: Muscliff Farm. Social conditions, 1800-1804
FARMER West and Muscliff Farm, 1800-04 / by his son. - [Dorchester] : Dorset County Council Education Committee, 1975
[3], 22 p.. - (Booklet ; no. 612)
r6009772 X.319/16722 Woolwich
0711403

AGRICULTURE. Brazil
- For British businessmen
FRITH, Dennis E.
The agriculture of Brazil and Venezuela. - London : Graham & Trotman, 1978
148p
0-86010-124-x X.325/884 Woolwich
2361590

AGRICULTURE. Brazil
- Serials
BRAZILIAN agriculture & commodities / produced by the Centre for Technical & Economic Cooperation of the Institute of Brazilian Studies. - Vol.1, no.1 (July-Aug.1980)-. - Portsmouth : Hambrook, 1980-
v. : ill
B81-06776 P.313/519
2353903

EXPERIENTIAE / orgão da Universidade Federal de Viçosa. - [Vol.1, no.1 (maio 1961)]-. - Viçosa : a Universidade, [1961-]
Holdings: Vol.23, no.8 (agosto 1977) ; Vol.23, no.11 (nov. 1977)-
r6101725 P.311/515 Woolwich
2449536

AGRICULTURE. Brazil
to ca 1900
SCHMIDT, Carlos Borges
Técnicas agrícolas primitivas e tradicionais. - [Brasília?] : Conselho Federal de Cultura e Departamento de Assuntos Culturais, 1976
188p
r6031173 X.319/9477 Woolwich
1539051

AGRICULTURE. Brazil
1960-1970
CONTADOR, Claudio Roberto
Tecnologia e rentabilidade na agricultura brasileira. - Rio de Janeiro : Instituto de Planejamento Econômico e Social, 1975
xvi, 257 p.. - (Coleção Relatórios de pesquisa ; no. 28). - Bibliography: p. 249-257
r6011703 X.319/8518 Woolwich
0881627

AGRICULTURE. Brazil
1963-1973
NICHOLLS, William H.
Mudanças na estrutura e produtividade da agricultura brasileira, 1963/73. - Rio de Janeiro : Instituto de Planejamento Econômico e Social, Instituto de Pesquisas, 1979-. - (IPEA/INPES. Relatório de pesquisa ; no 45). - Includes bibliographies
r6099915 Cup900.w.80
2118688

AGRICULTURE. Brazil
1964– – Statistics
[RECENSEAMENTO geral (1980)]. IX Recenseamento geral do Brasil, 1980 / Fundação Instituto Brasileiro de Geografia e Estatística. – Rio de Janeiro : IBGE.
Vol.2: [Censo agropecuário]
tom.2: Tabulações avançadas do censo agropecuário. – 1982
xxx,228p
8-524-00091-0 LAS.B.275/355
8559686
IX Recenseamento geral do Brasil, 1980 / IBGE. – Rio de Janeiro : IBGE.
Vol.2: Censo agropecuário
tom.3. – 1983–1984
26v.in 37
8-524-00167-4 LAS.B.275/356
8559686

AGRICULTURE. Brazil
1964– – Statistics
[RECENSEAMENTO geral, 1970]. VIII Recenseamento geral, 1970 / Fundação Instituto Brasileiro de Geografia e Estatística, Superintendência de Estatísticas Primárias, Decen. – Rio de Janeiro : Fundação IBGE.
Vol.III: Censo agropecuário. – [1974–1975]
25v.in 27
Wanting Série regional : t.3, 4, 7, 9, 13, 14 pt.2, 18pt.2, 22 ; Série nacional : Brasil
r6701153 LAS.B.118/25 Woolwich
8559686

AGRICULTURE. Brazil
Development. Economic aspects, ca 1970–1977
PAIVA, Ruy Miller
A agricultura no desenvolvimento econômico. – Rio de Janeiro : Instituto de Planejamento Econômico e Social, Instituto de Pesquisas, 1979
204p. – (IPEA/INPES. Monografia ; no 30)
r6099787 X.319/20064 Woolwich
2043440

AGRICULTURE. Brazil
Economic aspects
CAVINA, Romolo
Introdução à economia rural brasileira. – 1a ed. – São Paulo : Atlas, 1979
244p
r6163066 X.329/13392 Woolwich
2476460

AGRICULTURE. Brazil
Economic aspects – Serials
COLEÇÃO Análise e pesquisa / [editada pela] Comissão de Financiamento da Produção, Ministério da Agricultura. – Vol.1 (1977)–. – Brasilia : a Comissão, 1977–
r6101757 LAS.B.275/109 Woolwich
2448874
– Holdings: vols.1–5, 8, 9, 13 onwards
LAS.B.275/109 Woolwich

AGRICULTURE. Brazil
Economic development, 1945–
SORJ, Bernardo
Estado e classes sociais na agricultura brasileira. – Rio de Janeiro : Zahar Editores, 1980
152p. – (Coleção Agricultura e sociedade)
r6707322 X.529/58288
9026959

AGRICULTURE. Brazil
Financing, to date – Serials
PESQUISA / Ministério da Agricultura, Comissão de Financiamento da Produção, Serviço de Pesquisas Econômicas. – [No.1 (1968?)] – no.4 (1971). – Rio de Janeiro : a Comissão, [1969?]–[1972?]
Holdings: no.3 (1970) ; 4 (1971)
r6706072 LAS.B.122/49 Woolwich
8842930

AGRICULTURE. Brazil
Law
BRAZIL
[Laws, statutes, etc]. Legislação agrária. – 8.ed. – São Paulo : Editora Atlas, 1977
600p. – (Biblioteca de legislação trabalhista)
r6157724 LAS.B.275/152 Woolwich
2043548

[Laws, statutes, etc]. Vade-mecum agrário. – Brasília : INCRA, 1978
7v
r6086179 LAS.B.275/100 Woolwich
2043548

AGRICULTURE. Brazil
Personnel. Employment. Law
BRAZIL
[Laws, statutes, etc]. Legislação do trabalho rural. – São Paulo : Editora Atlas, 1974
429 p.
r6011725 X.709/20881
0881767

AGRICULTURE. Brazil
Research organisations: Centro de Energia Nuclear na Agricultura
CERVELLINI, Admar
The development and work of the Centro de Energia Nuclear na Agricultura - CENA, Piracicaba, São Paulo, Brasil. – Piracicaba : Centro de Energia Nuclear na Agricultura, 1976
11 leaves. – (Boletim de divulgação)
r6047540 X.311/4829 Woolwich
186131X

AGRICULTURE. Brazil
Technological development, ca 1960–1975
TECNOLOGIA e desenvolvimento agrícola / editado por Claudio Roberto Contador. – Rio de Janeiro : Instituto de Planejamento Econômico e Social, Instituto de Pesquisas, 1975
xiv,308p : ill. – (IPEA/INPES Monografia ; no 17)
r6065735 X.319/9509 Woolwich
1543814

AGRICULTURE. Brightwell-cum-Sotwell. Oxfordshire
ca 1900
EVEREX, Tom
Farming memories. – [Wallingford] : [K. Owen], c1981
8p
B81-35504 X.809/49208
2552027

AGRICULTURE. Brittany. France
Trade unions: Office central des associations agricoles de Landernau
MÉVELLEC, François
Le combat du paysan breton à travers les siècles. – [Landernau] : [Office Central des Associations Agricoles de [Landernau], 1973–
tom.1–
r6030415 X.0709/857
1126962

AGRICULTURE. Buenos Aires (Province). Argentina
to date
ESTRUCTURA, representatividad y evolución del complejo agro-industrial. – La Plata : Gobernación de la provincia de Buenos Aires, Secretaría de planeamiento y Desarrollo, 1977
154p. – (17-SP DS. ; 7/77)
r6708427 X.322/14774
914532x

AGRICULTURE. Bulgaria
PRŬMOV, Ivan
Bulgarian agriculture today. – 2nd rev. ed.. – [Sofia] : Sofia Press, 1976
107p
r6173824 X.519/43139
1131796

SHOPOV, Vŭlkan
Socialist development of the Bulgarian countryside. – [Sofia] : Sofia-press, 1972
68p : chiefly ill(some col.)
r6044089 X.319/8978 Woolwich
1131796

AGRICULTURE. Bulgaria
to 1878
RAZVITIE na zemedelieto po bŭlgarskite zemi / sŭstavitel : Ivan Venedikov. – Sofiīā : Zemizdat, 1981
308p[56]p of plates : ill
r6427212 X.421/22493 Woolwich
8823898

AGRICULTURE. Bulgaria
Economic planning
IKONOMICHESKIĪAT mekhanizŭm v selskoto stopanstvo / V. Mishev ... [i dr.] ; pod otgovornata redaktsiīā Kr. Brusarski, Sl. Atich. – Sofiīā : Zemizdat, 1979
142p
r6184754 X.529/36229
2438429

AGRICULTURE. Bulgaria
Finance
PALIĪSKI, Ivan Atanasov
Finansirane i kreditirane na NAPK. – Sofiīā : Zemizdat, 1977
150p
r6044403 X.519/23239
1235729

AGRICULTURE. Bulgaria
Grŭncharov, Dimitŭr
GEORGIEV, Nikolaĭ
Dmitŭr Grŭncharov. – 2. prerab. izd. – Sofiīā : Izdatelstvo na BZNS, 1977
80p. – (Sdruzheni zemedeltsi, revolīūtsionni bortsi ; 3)
r6173907 X.319/20124 Woolwich
2028247

AGRICULTURE. Bulgaria
Organisation
SOTSIALISTICHESKA organizatsiīā na truda v selskoto stopanstvo / [sŭstavitel i otgovoren redaktor : Ivan Grozev]. – Sofiīā : Partizdat, 1979
349p
r6184858 X.329/12204 Woolwich
224442X

AGRICULTURE. Bulgaria
Personnel. Remuneration, 1946–
USŬVŬRSHENSTVUVANE organizatsiīata na rabotnata zaplata v selskoto stopanstvo / I. Grozev ... [i dr.]. – Sofiīā : Zemizdat, 1982
156p,[1]folded leaf of plates
r6427873 X.329/16122
889101x

AGRICULTURE. Bulgaria
Planning
LUTSOV, Ivan
Problemi na planiraneto na selskoto stopanstvo v natsionalniīā agrarno-promishlen kompleks. – Sofiīā : Partizdat, 1976
232p
r6056724 X.329/8749 Woolwich
1426982

AGRICULTURE. Bulgaria
Policies of Bŭlgarska komunisticheska partiīā, 1946–
AGRARNATA politika na BKP na sŭvremenniīā etap / sŭstavitel Atanas Tanov ; pod otgovornata redaktsiīā na Atanas Tanov i Vasil Mishev. – Sofiīā : Zemizdat, 1981
211p
r6427245 X.329/14893
8795061

AGRICULTURE. Bulgaria
Trade unions
OBSHTESTVENO-IKONOMICHESKI funktsii na profsŭīuzite v selskoto stopanstvo / sŭstavitel Miladin Shaterov. – Sofiīa : Profizdat, 1978
142p. – Bibliography : p.139–140
r6093861 X.319/20007 Woolwich
198960X

PŬRVICHNATA profsŭīuzna organizatsiīā v agrarnite trudovi kolektivi. – Sofiīā : Profizdat, 1979
145p. – (Biblioteka za profsŭīuzniīā aktiv ; 8)
r6173943 X.319/20170 Woolwich
2132737

AGRICULTURE. Burgenland. Austria
Farms. Buildings. Architecture, to date
SIMON, Franz
Bäuerliche Bauten und Geräte. – Oberschützen : F. Simon, 1981
539p. : chiefly ill.
r6306307 L.45/1677
8908052

AGRICULTURE. Burton Dassett. Warwickshire
Temple, Peter, ca 1517–1578, 1543–1553
ALCOCK, N. W. (Nathaniel Warren)
Warwickshire grazier and London skinner 1532–1555. – London : Published for the British Academy by the Oxford University Press, c1981
xix,281p,6p of plates : ill. – (Records of social and economic history. New series ; 4). – Bibliography: pxix. – Includes index
0-19-726008-x Ac.1186/5(4)
2725193

AGRICULTURE. Buryat ASSR. Russia (RSFSR)
BURUEV, Matveĭ Trofimovich
Spetsializatsiīā khozīaĭstv skotovodcheskoĭ zony Burīatskoĭ ASSR. – Ulan-Udė : Burīatskoe knizhnoe izdatel'stvo, 1974
104p
r6174817 X.319/20187 Woolwich
2123649

AGRICULTURE. Buryat ASSR. Russia (RSFSR)
Butunaev, Alekseĭ - Biographies

BUTUNAEV, Bulat
Krasnyĭ bator Alekseĭ Butunaev. - Ulan-udė : Buri͡atskoe knizhnoe izdatel'stvo, 1982
207p
r6448225 X.629/22712
9076174

AGRICULTURE. Butleigh. Somerset
Hodges, John - Biographies

HEELEY, Ann
Victorian Somerset. - Glastonbury : Friends of the Abbey Barn, 1978
[26]p : ill. - Bibliography: p.[25]
B79-11047 X.708/21771
1827324

AGRICULTURE. California

A GUIDEBOOK to California agriculture / by faculty and staff of the University of California ; edited by Ann Foley Scheuring. - Berkeley [Calif.] ; London : University of California Press, c1983
xv,413p : ill. - Includes bibliographies and index
0-520-04709-5 SRL
3146618

AGRICULTURE. California
1900-1982

SCHEURING, Ann Foley
Tillers. - [New York] : Praeger, 1983
xvii,278p : ill. - (Praeger special studies)
0-03-063796-1 X.800/37127
3065197

AGRICULTURE. California
Filipino personnel. Trade unions, to date

DE WITT, Howard
Violence in the fields. - Saratoga, Calif. : Century Twenty One Publishing, 1980
xii,139p
0-86548-008-7 X.322/11777
8950156

AGRICULTURE. California
Personnel. Social conditions, 1870-1941

DANIEL, Cletus E.
Bitter harvest. - Ithaca ; London : Cornell University Press, c1981
348p : ill. - Includes index
0-8014-1284-6 X.800/31360
2554674

AGRICULTURE. California
Trade unions, 1947-1960

GALARZA, Ernesto
Farm workers and agri-business in California, 1947-1960. - Notre Dame ; London : University of Notre Dame Press, 1977
xvii,405p. - Bibliography: p.389-399. - Includes index
0-268-00941-4 X.322/9404 Woolwich
2101246

AGRICULTURE. Canada
to date

TROUGHTON, M. J. (Michael J.)
Canadian agriculture. - Budapest : Akadémiai Kiadó, 1982
355 p., [1] folded leaf : ill., maps. - (Geography of world agriculture ; 10). - Bibliography: p. 243-249
963-05-2653-0 X.0322/92(10)
8674787

WARLEY, T. K.
Agriculture in an interdependent world. - Washington : Canadian-American Committee, 1977
x,93p
0-89068-040-x X.329/15749
8674787

AGRICULTURE. Canada
Policies of government, 1600-1930

FOWKE, Vernon C.
Canadian agricultural policy. - Toronto ; London : University of Toronto Press ; London : [Distributed by Canada Books International], 1978
xii,304p. - Bibliography: p. 283-288. - Includes index
0-8020-6352-7 X.329/13066 Woolwich
2274485

AGRICULTURE. Canada
Policies of government, 1973-1977

ORGANISATION FOR ECONOMIC CO-OPERATION AND DEVELOPMENT
Recent developments in Canadian agricultural policy. - Paris : O.E.C.D. ; [London] : [H.M.S.O.], 1978
45p : ill. - (Agricultural policy reports / Organisation for Economic Co-operation and Development)
B79-07830 SRL
1809253

AGRICULTURE. Capitalist countries
Family farms

FAMILY and work in rural societies / edited by Norman Long. - London : Tavistock, 1984
viii,222p : 2ill. - Bibliography: p199-214. - Includes index
0-422-78480-x X.529/67865
3025780

AGRICULTURE. Caribbean region
- For Caribbean students

HENRY, A. I.
Caribbean agricultural science. - [Port of Spain] : Macmillan Caribbean ; London : Macmillan.
Book 1: Principles. - 1979
[5],202p : ill. - Includes index
0-333-23389-1 X.311/9060 Woolwich
1967266

PERSAD, Ralph S. (Ralph Sakaldip)
Agricultural science for Caribbean primary schools. - Walton-on-Thames : Nelson Caribbean.
Workbook 1. - 1981
43p
Subsequent volumes when received will be placed at this shelfmark
0-17-566311-4 X.0322/263
2597314

RAMHARACKSINGH, R.
Caribbean primary agriculture. - London : Cassell.
Workbook 1. - 1981
47p : ill
0-304-30765-3 X.0322/238
1967266
Bk.4. - 1982
95p : ill(some col.)
0-304-30867-6 X.0322/238
1967266
Workbook 4. - 1982
64p : ill
0-304-30868-4 X.0322/238
1967266
Workbook for infants. - 1981
31p : ill
0-304-30768-8 X.0322/238
1967266

AGRICULTURE. Carnmoney region. Newtownabbey (District)
Farms: Sentry Hill. Social conditions, to 1981

WALKER, Brian Mercer
Sentry Hill. - Dundonald : Blackstaff, c1981
xii,167p : ill
0-85640-254-0 X.322/10596 Woolwich
2630087

AGRICULTURE. Carpathian Mountains. Ukraine
to 1945

MANDYBURA, M. D. (Mar'i͡an Danylovych)
Polonyns'ke hospodarstvo Hutsul'schyny druhoï polovyny XIX - 30-kh rokiv XX st.. - Kyïv : Naukova dumka, 1978
188p : ill
r6485760 X.329/15177 Woolwich
8819777

AGRICULTURE. Catalonia. Spain
to date

CHECCHI I LANG, Alexandre
L'explotació pagesa a Catalunya. - Barcelona : Edicions Vicens-Vives, 1979
viii,230p : ill
84-316-1699-7 X.329/14476 Woolwich
8753016

AGRICULTURE. Catalonia. Spain
1868-1931

GALÍ, Alexandre
Història de les institucions i del moviment cultural a Catalunya, 1900 a 1936. - Barcelona : Fundació Alexandre Galí. - (Obra completa)
Llibre 6: Ensenyaments i serveis agrícoles. - 1982
221p
84-300-7655-7 X.0708/656
8995451

AGRICULTURE. Catalonia. Spain
1939-1975

BENELBAS I TAPIERO, León A.
Economia agrària de Catalunya. - Barcelona : Ketres Editora, 1981
239p : ill(some col.)
84-85256-21-2 X.322/10366 Woolwich
8753512

AGRICULTURE. Catalonia. Spain
1975-

L'AGRICULTURA catalana / [per] J.M. Brunet ... [et al.] ; [realitzat sota la direcció de] J. Grifoll. - Barcelona : Fundació Jaume Bofill : Banca Catalana, [1980?]
275p
84-85344-07-3 X.322/10459 Woolwich
8755973

AGRICULTURE. Central African Republic
Personnel. Employment

INTERNATIONAL LABOUR ORGANISATION
La formation et l'emploi des artisans ruraux et des jeunes agriculteurs en République Centrafricaine. - Genève : [L'Organisation], 1973
xi,98p. - (République Centrafricaine. Programme de formation professionnelle d'artisans ruraux (CAF/71/503) Rapport technique ; no.1)
r6038015 UNH 419/23
121344X

AGRICULTURE. Central America
to 1535

PALERM, Ángel
Agricultura y sociedad en Mesoamérica. - México : Sep Diana, 1980
195p. - (Sep Diana ; 55)
9-681-30949-9 X.329/15340
8819068

AGRICULTURE. Central Chernozem'e. Russia (RSFSR)
1917-

PARTIĬNYE organizat͡sii T͡Sentral'nogo Chernozem'i͡a v bor'be za sot͡sialisticheskoe preobrazovanie i razvitie sel'skogo khozi͡aĭstva / [nauchnye redaktory A.V. Losev, V.I. Logunov]. - Voronezh : Izdatel'stvo Voronezhskogo universiteta, 1977
310p.
r6087944 X.320/5296 Woolwich
1860712

AGRICULTURE. Central Sweden
Bonded employment, ca 1750-1900. Demographic aspects

ERIKSSON, Ingrid
Rural labor and population change. - Uppsala ; Stockholm : Distributor : Almqvist & Wiksell International, 1978
280p. - (Studia historica upsaliensia ; 100). - Includes bibliographies
91-554-0740-4 Ac.1077.ng[vol.100]
1543113

AGRICULTURE. Central Volga River region. Russia (RSFSR)
to 1917

KABYTOV, Petr Serafimovich
Agrarnye otnoshenii͡a v Povolzh'e v period imperializma (1900-1917 gg.). - Saratov : Izdatel'stvo Saratovskogo universiteta, 1982
199p
r6490170 X.329/19592
8382573

VOPROSY agrarnoĭ istorii Srednego Povolzh'i͡a / [redakt͡sionnai͡a kollegii͡a : M.K. Mukhari͡amov (otvetsvennyĭ redaktor) ... i dr.]. - Ĭoshkar-Ola : Mariĭskiĭ nauchno-issledovatel'skiĭ institut, 1978
192p
r6094281 X.319/20012 Woolwich
199140X

AGRICULTURE. Central Volga River region. Russia (RSFSR)
1900-1917

KABYTOV, Petr Serafimovich
Povolzhskai͡a derevni͡a nakanune fevral'skoĭ burzhuazno-demokraticheskoĭ revoli͡ut͡sii. - Kuĭbyshev : Kuĭbyshevskiĭ gosudarstvennyĭ universitet, 1977
72p
r6097200 X.708/46173
1976109

AGRICULTURE. Chancay River Valley. Lima (Department). Peru
Socioeconomic aspects

MATOS MAR, José
Yanaconaje y reforma agraria en el Perú. - Lima : Instituto de Estudios Peruanos, 1976
268p. - (Serie Perú problema ; n.15). - Includes bibliographies
r6083330 X.519/38646
1541293

AGRICULTURE. Channel Islands
Influence on Kropotkin, Petr Alekseevich, kniaz'

FORD, James Glyn
Prince Peter Kropotkin and the Channel Islands. - St Peter Port : Toucan Press, 1979
17p. - (Guernsey historical monograph ; no.19)
0-85694-208-1 X.708/22649
2144328

AGRICULTURE. Chapel-en-le-Frith region. Derbyshire
1914-1969

VIRTUE, Alan
Memoirs of a Derbyshire hill farmer. - London : Virtue and Co., 1976
ix,129p : ill
0-900778-09-1 X.329/10722 Woolwich
077152x

AGRICULTURE. Chechen-Ingush ASSR. Russia (RSFSR)
Collectivisation

POPOV, Andreĭ Georgievich
Sot͡sialisticheskoe preobrazovanie sel'skogo khozi͡aĭstva v Checheno-Ingushetii (istoricheskiĭ ocherk). - Groznyĭ : Checheno-Ingushskoe knizhnoe izdatel'stvo, 1976
292 p.. - Bibliography : p. 284-291
r6021161 X.329/8591 Woolwich
1001434

AGRICULTURE. Chechen-Ingush ASSR. Russia (RSFSR)
Collectivisation, 1917-1945

EFANOV, K. I. (Konstantin Ivanovich)
Klassovai͡a bor'ba v checheno-ingushskom aule v period sot͡sialisticheskogo stroitel'stva. - Groznyĭ : Checheno-Ingushskoe knizhnoe izdatel'stvo, 1979
116p. - Includes bibliography
r6462919 X.329/18006
838259x

AGRICULTURE. Chernozem'e. Russia (RSFSR)
Collectivisation

PEREVERZEV, Aleksandr Īakovlevich
Sot͡sialisticheskai͡a revoli͡ut͡sii͡a Rossii (t͡sentra - oktīabr'1917-1918). - Voronezh : Izdatel'stvo Voronezhskogo universiteta, 1976
188p
r6037321 X.519/23173
1209507

AGRICULTURE. Chesterfield. Derbyshire
1485-1714 - Dictionaries

MILWARD, Rosemary
A glossary of household, farming and trade terms from probate inventories. - 2nd ed., greatly extended and fully rev. and corrected. - Chesterfield : Derbyshire Record Society, 1982
52p. - (Occasional paper ; no.1)
r6811618 X.0809/2324(1)
9145079

AGRICULTURE. Chilalo (District). Ethiopia
Productivity, to date

LINDER, Harald
Crop production improvement. - [Uppsala] : Swedish University of Agriculture, Forestry and Veterinary Medicine, 1976
ix,104p : ill. - (Rural development studies ; no 5)
r6611140 X.0313/56(5)
9034862

AGRICULTURE. Chile

HERIVEL, Ethel Beatrice
[We farmed a desert. Spanish]. Cultivando el desierto. - Santiago de Chile : Instituto de Capacitacíon e Investigación en Reforma Agraria, 1973
259p
r6013534 X.708/18412
0887188

AGRICULTURE. Chile
- Serials

AGRO Sur. - v. 1, no. 2- ; 1973-. - Valdivia : Facultad de Ciencias Agrarias, Universidad Austral de Chile, 1973-
r6003244 P.311/460
090340X

AGRICULTURE. Chile
Policies of government

DIAGNÓSTICO de la reforma agraria chilena / estudio coordinado por Solon Barraclough y José Antonio Fernández. - México : Siglo Veintiuno Editores, 1974
viii, 323 p.. - (Sociología y política)
r6005044 X.319/8269 Woolwich
0903590

AGRICULTURE. Chile
Policies of government, 1925-

CABALLERO URSÚA, Lorenzo
The reformed units - asentamientos, CERA and U.A.F. - in the Chilean agriculture 1965-1975. - Uppsala : Swedish University of Agriculture, Forestry and Veterinary Medicine, 1975
iii,41p. - (Rural development studies ; no.4)
r6610154 X.0313/56(4)
8877378

AGRICULTURE. Chile
Policies of government, 1973-1978

COLLINS, Joseph
Agrarian reform and counter-reform in Chile. - San Francisco : Institute for Food and Development Policy ; Birmingham : Distributed by Third World Publications, 1979
17p : ill
B80-40411 X.700/26555
2285959

AGRICULTURE. Chile
Policies of government, 1973-

JARVIS, Lovell S.
Chilean agriculture under military rule. - Berkeley : Institute of International Studies, University of California, c1985
xiv,210p. - (Research series / Institute of International Studies, University of California, Berkeley ; no.59)
0-87725-159-2 X.0809/130
8056838

AGRICULTURE. China

The CHINESE agricultural economy / edited by Randolph Barker and Radha Sinha with Beth Rose. - Boulder, Colo. : Westview ; London : Croom Helm, 1982
xiii,266p : ill. - (Westview's special studies on China and East Asia). - Includes bibliographies and index
0-7099-0682-x X.520/30275
2913038

LARDY, Nicholas R.
Agriculture in China's modern economic development. - Cambridge : Cambridge University Press, 1983
xiii,285p : ill. - Bibliography: p251-274. - Includes index
0-521-25246-6 X.322/14029
2913038

AGRICULTURE. China
to 1644

LIEW, Foon Ming
Tuntian farming of the Ming dynasty (1368-1644). - Hamburg : Gesellschaft für Natur- und Völkerkunde Ostasiens, 1984
xiii,329p. - (Mitteilungen der Gesellschaft für Natur- und Völkerkunde ; Bd.97). - Includes bibliography
r6309455 Ac.1944[Bd.97]
9097066

AGRICULTURE. China
to date

LAMB, Jefferson D. H.
The development of the agrarian movement and the agrarian legislation in China. - New York : Garlong Publishing Inc, 1980
xv,228p
0-8240-4261-1 X.329/15750
8057818

AGRICULTURE. China
B.C.206-A.D.220

HSU, Cho-yun
Han agriculture. - Seattle ; London : University of Washington Press, c1980
xxv,377p : ill. - (Han Dynasty China ; v.2). - Bibliography: p337-357. - Includes index
0-295-95676-3 X.0800/506
2331454

AGRICULTURE. China
1949-1975

JACOBY, Erich H.
Agricultural development in China. - Stockholm : EFI, 1974
89p. - (China's developmental strategy ; 4)
91-7258-038-0 X.0702/496(4)
1086669

KUO, Leslie Tse-chiu
Agriculture in the People's Republic of China. - New York ; London : Praeger, [1977]
xx,289p. - (Praeger special studies in international economics and development). - Includes index
0-275-05870-0 X.320/5275 Woolwich
1086669

SIGURDSON, Jon
The role of small scale and rural industry and its interaction with agriculture and large scale industry in China. - Stockholm : EFI, 1974
xi,169p. - (China's developmental strategy ; 5)
91-7258-036-4 X.0702/496(5)
1086669

STAVIS, Benedict
Making green revolution. - Ithaca, N.Y. : Rural Development Committee, Cornell University, 1974 [i.e. 1975]
287p. - (Rural development monograph ; no.1)
r6188680 X.512/7182
1086669

AGRICULTURE. China
Collectivisation, 1949-1956

SHUE, Vivienne
Peasant China in transition. - Berkeley ; London : University of California Press, c1980
xvi,394p : ill. - Bibliography: p347-385. - Includes index
0-520-03734-0 X.800/29438
2336995

AGRICULTURE. China
Collectivisation, 1949-

HSU, Robert C.
Food for one billion. - Boulder, Colo. : Westview, 1982
xvi,156p : ill. - (Westview special studies on China and East Asia)
0-86531-062-9 X.322/14388
8057826

AGRICULTURE. China
Development compared with development of agricultural industries in India

BANDYOPADHYAYA, Kalyani
Agricultural development in China and India. - New Delhi : Wiley Eastern, 1976
xi,204p. - Bibliography : p.185-199. - Includes index
0-85226-054-7 X.329/11644 Woolwich
1693662

AGRICULTURE. China
Mechanisation, 1949-

HSU, Robert C.
Food for one billion. - Boulder, Colo. : Westview, 1982
xvi,156p : ill. - (Westview special studies on China and East Asia)
0-86531-062-9 X.322/14388
8057842

AGRICULTURE. China
Mechanisation, 1956-1976

SINO-BRITISH TRADE COUNCIL
Mechanisation of China's agriculture. - London : The Council, [1977?]
[2],69 leaves : ill. - Bibliography: p.39-40
B79-04166 X.512/6297
1765264

AGRICULTURE. China
Mechanisation. Political aspects, 1950-1975

STAVIS, Benedict
The politics of agricultural mechanization in China. - Ithaca ; London : Cornell University Press, 1978
288p : ill. - Bibliography: p.270-283. - Includes index
0-8014-1087-8 X.329/11580 Woolwich
1574353

AGRICULTURE. Ciscaucasia. Russia (RSFSR)
to 1917

RATUSHNĪAK, V. N. (Valeriĭ Nikolaevich)
Agrarnye otnoshenii͡a na Severnom Kavkaze v kont͡se XIX-nachale XXvv.. - Krasnodar : Kubanskiĭ gosudarstvennyĭ universitet, 1982
83p
r6490172 X.329/19229
9160485

AGRICULTURE. Ciscaucasia. Russia (RSFSR)
1796-

KALOEV, Boris Aleksandrovich
Zemledelie narodov Severnogo Kavkaza. - Moskva : Nauka, 1981
246p
r6464677 X.322/13117
9083987

AGRICULTURE. Ciscaucasia. Russia (RSFSR)
1917-

AGRARNAĪA istorii͡a Dona i Severnogo Kavkaza / Otvetstvennyĭ redaktor : P.G. Chernopit͡skiĭ. - Rostov-na-Donu : Izdatel'stvo Rostovskogo universiteta, 1980
151p
r6402217 X.329/14416 Woolwich
8726515

KPSS - organizator agrarnykh preobrazovaniĭ na Severnom Kavkaze / [redakt͡sionnai͡a kollegii͡a : M.I. Gioev (otv. red.) ... i dr.]. - Ordzhonikidze : Severo-Osetinskiĭ gosudarstvennyĭ universitet im. K.L. Khetagurova, 1981
141p
r6499253 X.329/16245
8726515

AGRICULTURE. Ciskei. South Africa
to date

NOTZKE, Claudia
Die Ciskei. - Köln : Geographisches Institut der Universität, 1980
iv,122p. - (Kölner geographische Arbeiten. Sonderfolge. Beiträge zur Länderkunde Afrikas ; Hft.9)
r6301840 X.329/14655 Woolwich
8773890

AGRICULTURE. Clwyd
Farms. Vernacular buildings, to 1900

WILIAM, Eurwyn
Traditional farm buildings in north-east Wales 1550-1900. - [Cardiff] : National Museum of Wales, Welsh Folk Museum, 1982
334p : ill. - Bibliography: p296-301. - Includes index
0-85485-049-x X.421/22939 Woolwich
2733684

AGRICULTURE. Coleshill. Oxfordshire
Farms. Buildings. Architecture

COLESHILL Model Farm Oxfordshire / report of working seminar April 1980 ; editor John Weller. - [Birmingham] : [Architects in Agriculture Group, [1981]
30p : ill. - (Architects in Agriculture Group occasional paper ; no.1)
B82-39542 X.322/11288
2790890

AGRICULTURE. Coleshill. Oxfordshire
Farms: Coleshill Farm. Buildings. Architecture - Conference proceedings

COLESHILL model farm Oxfordshire. - Birmingham : Architects in Agriculuture Group, 1981
30p : ill. - (Architects in Agriculture Group occasional paper ; no.1)
B83-20208 X.805/3927
2888297

AGRICULTURE. Colombia

VÉLEZ MELGUIZO, Hugo E.
Dos ensayos acerca del desarrollo capitalista en la agricultura colombiana. - Medellin : Centro de Investigaciones Económicas, Universidad de Antioquia, 1975
139p
r6047990 X.311/5050 Woolwich
2039982

VÉLEZ TRUJILLO, Luis Guillermo
Barreras del hambre. - [Bogotá?] : Fondo Editorial ANIF, [1978?]
224p
r6101928 X.529/35074
2039982

AGRICULTURE. Colombia
- For British exporters

AGRICULTURE, and prospects for associated business in Colombia. - London : British Agricultural Export Council, 1982
iv,85p : ill
B83-17130 SRL
2877252

AGRICULTURE. Colombia
- Statistics

CENSO nacional agropecuario 1970-1971 / [Departamento Administrativo Nacional de Estadística]. - Bogotá : DANE, 1974
p.8-48
r6700710 LAS.c.230/108
1329081

COLOMBIA. Oficina de Planeamiento del Sector Agropecuario
Programas agrícolas. - evaluación 1974, programación 1975, proyecciones 1976-. - [Bogotá] : el Ministerio, Oficina de Divulgación, 1974-
r6048379 LAS.C.230/22
1329081

AGRICULTURE. Colombia
to date

MORALES BENÍTEZ, Otto
Derecho agrario y otros temas de la tierra. - [Bogotá] : Universidad Externado de Colombia, 1981
277p,[1]leaf of plates. - Includes bibliography
r6709117 X.322/14020
8067236

AGRICULTURE. Colombia
1920-1975

La AGRICULTURA colombiana en el siglo XX / compilación y prólogo : Mario Arrubla. - [Bogotá] : Instituto Colombiano de Cultura, Subdirección de Comunicaciones Culturales, 1976
600p. - (Biblioteca básica colombiana ; n.17)
r6065127 X.519/38642
1539493

AGRICULTURE. Colombia
1930-

TAUSSIG, Michael T.
Destrucción y resistencia campesina. - Bogotá : Punta de Lanza, 1978 [i.e. 1979]
142p. - Includes bibliographies
r6708397 X.429/16868
8726493

VÉLEZ MELGUIZO, Hugo E.
[Dos ensayos acerca del desarrollo capitalista en la agricultura colombiana]. Dos ensayos en torno al debate sobre el carácter de la agricultura colombiana. - 3a ed. - Bogotá : La Carreta, 1979
290p
r6709104 X.329/18363
8726493

AGRICULTURE. Colombia
1930- - Bibliographies

HERNÁNDEZ DE CALDAS, Ángela
Bibliografía agroindustrial de Colombia. - Bogotá : Fundación Mariano Ospina Pérez, 1978
222p
r6703512 X.322/10185 Woolwich
8726493

AGRICULTURE. Colombia
1950-1976

KALMANOVITZ, Salomón
Desarrollo de la agricultura en Colombia. - Bogotá : La Carreta, 1978
360p
r6158552 X.319/20346 Woolwich
2249944

AGRICULTURE. Colombia
Influence of capitalism, 1800-1900

ARANGO JARAMILLO, Mario
El proceso del capitalismo en Colombia. - Medellín : Aurora, 1978. - (Colección Claves polémicas) - Bibl
[Vol.4]: Evolución del campo hacia el capitalismo, siglo 19
215p
r6157839 X.0519/764(4)
2444399

AGRICULTURE. Colombia
Law

COLOMBIA
Legislación agraria colombiana. - Bogotá : Editorial Presencia, 1978
xv,938p
r6161619 X.200/33198
2450844

AGRICULTURE. Colombia
Policies of government, 1950-1975

La POLÍTICA agraria en Colombia 1950-1975 / por Juan Enrique Araya A....[et al.]. - Bogotá : Fundación para la Educación Superior y el Desarrollo, 1975
v,149p. - Bibliography : p.145-149
r6031033 X.313/764 Woolwich
1391739

AGRICULTURE. Colombia
Role in economic development, 1850-1970 - Conference proceedings

SEMINARIO NACIONAL DE DESARROLLO RURAL (1st : 1976 : Bogotá)
El agro en el desarrollo histórico colombiano. - Bogotá : Punta de Lanza, 1977
379p
r6085075 X.519/40375
1865730

AGRICULTURE. Colombia
Women personnel

LEÓN DE LEAL, Magdalena
Mujer y capitalismo agrario. - Bogotá : Asociación Colombiana para el Estudio de la Población, 1980
ii,295p.. - Bibl
r6159586 X.322/9355 Woolwich
2420600

AGRICULTURE. Columbia Plateau. Washington (State)
to 1982

MCGREGOR, Alexander Campbell
Counting sheep. - Seattle ; London : University of Washington Press, c1982
xii,482p,[24]p of plates : ill. - Bibliography: p439-469. - Includes index
0-295-95894-4 X.322/12133
293230x

AGRICULTURE. Commonwealth countries
Organisations: Royal Agricultural Society of the Commonwealth

PHILIP, Prince, consort of Elizabeth II, Queen of Great Britain
The Royal Agricultural Society of the Commonwealth. - Shepton Mallet : Royal Bath and West and Southern Counties Society, 1980
11p
B83-04741 X.809/53939
282003x

AGRICULTURE. Commonwealth countries
Research organisations: Commonwealth Agricultural Bureaux - Conference proceedings

COMMONWEALTH AGRICULTURAL BUREAUX Review Conference (1980 : London)
Commonwealth Agricultural Bureaux Review Conference, London 1980. - London : H.M.S.O., [1981]
viii,64p : ill. - (Cmnd. ; 8411)
0-10-184110-8 BS.3/720
0755028

COMMONWEALTH AGRICULTURAL BUREAUX REVIEW CONFERENCE (1975 : London)
Report of proceedings [of the] Commonwealth Agricultural Bureaux Review Conference, London, 1975. - London : H.M.S.O., [1976]
viii,56p. - (Cmnd.6619)
0-10-166190-8 BS.3/589
0755028

AGRICULTURE. Commonwealth developing countries
Economic development. Financial assistance by Commonwealth Development Corporation

COMMONWEALTH DEVELOPMENT CORPORATION
CDC in agriculture. - London : Commonwealth Development Corporation, [1977?]
[2],16p : col ill
B79-06323 BS.113/10
179809x

AGRICULTURE. Commonwealth developing countries
Policies of European Economic Community. Common Agricultural Policy. Implications for sugar industries of Commonwealth developing countries - Conference proceedings

SUGAR, the European Community and the developing Commonwealth / rapporteur Carol Cosgrove Twitchett. - London : Public Affairs Department, Royal Commonwealth Society, [1981]
[58]p. - Bibliography: p[45]-[46]
0-905067-95-9 SRL
2556995

AGRICULTURE. Communist countries
Efficiency

ÉFFEKTIVNOST' sot͡sialisticheskogo zemlepol'zovanii͡a / [otvetstvennyĭ redaktor : A.M. Onishchenko]. - Kiev : Naukova dumka, 1975
287p
r6177954 X.329/12898 Woolwich
2441977

AGRICULTURE. Communist countries
Policies of governments, 1945-
[AGRARNAI͡A politika sot͡sialisticheskikh stran na sovremennom ėtape. Bulgarian]. Agrarnata politika na sotsialisticheskite strani na sŭvremennii͡a etap / [pod obshtata redaktsii͡a Dimo Vladov, A.M. Emeli͡anov] ; [prev. ot ruski]. - Sofii͡a : Partizdat, 1981
351p. - (Sotsializŭm)
r6427447 X.329/15492
8840350

AGRICULTURE. Corfu. Greece
to date

CHYTĒRĒS, Ger.
To Kerkyraïko agrotiko provlēma tēn epomenē tēs Enōseōs kai hoi anaphores tou Anglou proxenou. - Kerkyra : Hetaireia Kerkyraïkōn Spoudōn, 1981
93p. - (Dēmosieumata Hetaireias Kerkyraïkōn Spoudōn)
r6598382 X.622/19597
9082875

AGRICULTURE. Cornwall

CREATION, David
Beasts go West. - London : Hodder and Stoughton, 1979
251p
0-340-24527-1 X.329/12002 Woolwich
1942204

CREATON, David
Beasts go west. - Large print ed. - Long Preston : Magna, 1984, c1979
517p
0-86009-567-3 Nov.52975
1942204

ROSENTHALL, Pamela
Agricultural production and marketing in Cornwall and Devon. - [Exeter] : University of Exeter Agricultural Economics Unit, 1981
72p : ill. - (Report ; no.212)
B82-01507 X.322/10382 Woolwich
2584018

AGRICULTURE. Costa Rica
to date - Statistics

CENSO agropecuario 1973 / Dirección General de Estadística y Censos. Costa Rica. - San José, Costa Rica : La Dirección, 1974
61,187,286p. - (Censos nacionales de 1973 ; 3)
r6707396 LAS.D.62/29
9016953

AGRICULTURE. Cuba
Research. Acuña Gale, Julián

MONCADA FERRERA, Milagros
Recordando a Julián Acuña. - La Habana : Academia de Ciencias de Cuba, Centro de Estudios de Historia y Organización de la Ciencia Carlos J. Finlay, 1980
10leaves. - (Conferencias y estudios de historia y organización de la ciencia ; 22)
r6701145 LAS.D.300/15 Woolwich
8747482

AGRICULTURE. Cuba
Research. Acuña Galé, Julián - Biographies

MARTÍNEZ VIERA, Rafael
Julián Acuña Galé en el 80 aniversario de su nacimiento. - La Habana : Academia de Ciencias de Cuba, Centro de Estudios de Historia y Organización de la Ciencia Carlos J. Finlay, 1980
13 leaves. - (Conferencias y estudios de historia y organización de la ciencia ; 21)
r6701147 LAS.D.300/15 Woolwich
8747482

AGRICULTURE. Cuba
Research. Noda, Tranquilino Sandalio de - Biographies

MARTÍNEZ VIERA, Rafael
Tranquilino Sandalio de Noda en el 170 aniversario de su nacimiento. - La Habana : Academia de Ciencias de Cuba, Centro de Estudios de Historia y Organización de la Ciencia Carlos J. Finlay, 1980
11leaves. - (Conferencias y estudios de historia y organización de la ciencia ; 16)
r6701136 LAS.D.300/15 Woolwich
8747482

AGRICULTURE. Cuba
Research. Research. Reynoso, Alvaro

DÍAZ BARREIRO, Francisco
El legado útil de Alvaro Reynoso. - La Habana : Academia de Ciencias de Cuba, Centro de Estudios de Historia y Organización de la Ciencia Carlos J. Finlay, 1979
13leaves. - (Conferencias y estudios de historia y organización de la ciencia ; 15)
r6160801 LAS.D.300/15 Woolwich
2479516

AGRICULTURE. Cuba
Research. Research. Reynoso, Alvaro - Biographies

SÁENZ, Tirso W
Alvaro Reynoso, 1829-1888, en el sesquicentenario de su nacimiento. - La Habana : Academia de Ciencias de Cuba, Centro de Estudios de Historia y Organización de la Ciencia Carlos J. Finlay, 1979
11leaves. - (Conferencias y estudios de historia y organización de la ciencia ; 14)
r6799019 LAS.D.300/15 Woolwich
2479516

AGRICULTURE. Cuba
Research. Reynoso, Alvaro - Biographies

OTERO HERRERA, Eugenio
Alvaro Reynoso. - La Habana : Academia de Ciencias de Cuba, Centro de Estudios de Historia y Organización de la Ciencia Carlos J. Finlay, 1980
16leaves. - (Conferencias y estudios de historia y organización de la ciencia ; 20)
r6701148 LAS.D.300/15 Woolwich
8747482

AGRICULTURE. Cuba
Research stations: Estación Experimental Agronómica de Santiago de las Vegas

MARTÍNEZ VIERA, Rafael
70 años de la Estación Experimental Agronómica de Santiago de las Vegas. - La Habana : Academia de Ciencias de Cuba, 1977
128p : ill
r6703452 X.329/14989 Woolwich
8809097

70 años de la Estación Experimental Agronómica de Santiago de las Vegas. - La Habana : Academia de Ciencias de Cuba, 1977
128p : ill
r6704816 X.329/15169
8839107

AGRICULTURE. Cumbria

CUMBRIA farming year book. - 1977-78. - Carlisle : 'Cumbria Weekly Digest' ; Carlisle : 'Lakescene', 1977
120p : ill
Subsequent volumes, when received, will be placed at this shelfmark
B77-17356 P.511/5042
1120441

AGRICULTURE. Curriculum subjects. Education. Developing countries

TRAINING for agriculture and rural development / Food and Agriculture Organization of the United Nations, United Nations Educational, Scientific and Cultural Organization, International Labour Organisation. - 1977. - Rome : Food and Agriculture Organization of the United Nations ; [London] : [H.M.S.O.], 1977
vii,122p : ill. - (Economic and social development series / Food and Agriculture Organization ; no.7)
B80-11061 UN.F.34/69
2045176

AGRICULTURE. Curriculum subjects. Education. European Community countries
- Conference proceedings

EDUCATING for the year 2000 / [prepared by David Moore]. - Langholm : Arkleton Trust, 1978
viii,28,[3]p
B79-06605 X.313/1601 Woolwich
1783866

AGRICULTURE. Curriculum subjects. Education. Ireland (Republic)

ACOT
General guide to agricultural education. - Blackrock : ACOT, [1981?]
12p : ill
B82-05195 X.329/14795
2599414

AGRICULTURE. Curriculum subjects. Education. Nigeria

OKORIE, J. U.
Agricultural science for teacher training colleges. - Harlow : Longman, 1983
168p : ill. - Includes index
0-582-77911-1 X.622/17906
3043606

AGRICULTURE. Curriculum subjects. Education. Peasants
Teaching

FREIRE, Paulo
Education, the practice of freedom. - London : Writers and Readers Publishing Cooperative, 1976
xiv,162p : ill
0-904613-16-x X.519/25883
0143006

AGRICULTURE. Curriculum subjects. Educational institutions. European Community Countries
- Conference proceedings

L'ÉDUCATION pour l'an 2000 / [préparé par David Moore]. - Langholm : Arkleton Trust, 1978
[2],v,22p
B79-17662 X.315/110 Woolwich
187795x

AGRICULTURE. Curriculum subjects. Educational institutions. Western Europe

AGRICULTURAL education in Europe. - London : British Council ; Slough : Distributed [by] Commonwealth Agricultural Bureaux, 1977
iii,135p : ill
0-900229-28-4 BS.130/20
1143611

AGRICULTURE. Curriculum subjects. Further education institutions & higher education institutions. Commonwealth countries
Courses - Directories

COMMONWEALTH SECRETARIAT. Food Production and Rural Development Division
Training for agricultural development. - London : The Division, 1976
x,360p. - Bibliography: p.340. - Includes index
0-85092-112-0 OPL
0780774
- [Another copy] BS.12/110

AGRICULTURE. Curriculum subjects. Further education institutions & higher education institutions. England
Courses

AGRICULTURAL education / [prepared by the Department of Education and Science and the Central Office of Information]. - 1976-77. - [London] : [The Department], 1976
27p
Subsequent volumes, when received, will be placed at this shelfmark
0-85522-080-5 BS.10/471
0730327

COURSES in agriculture and allied subjects in England and Wales / edited by J.G. Cooke. - York : Agricultural Education Association, c1981
70p
0-901409-02-2 X.322/10124 Woolwich
0730327

COURSES in agriculture and allied subjects in the United Kingdom. - 2nd rev. ed.. - York : Agricultural Education Association, c1984
85p
0-901409-04-9 X.325/1408
0730327

AGRICULTURE. Curriculum subjects. Further education institutions & higher education institutions. England
Courses - Serials

COURSES in agriculture, horticulture and forestry / National Consultative Committee for Agricultural Education. - [198-]-. - Bury St. Edmunds : The Committee, [198-]-
v.
Holdings: 1983-84-
r6813545 P.311/860
0730327

AGRICULTURE. Curriculum subjects. Further education institutions & higher education institutions. Ireland (Republic)

REPORT of the ACOT expert group on agricultural education and training. - Blackrock : ACOT, 1981
xi,171p : ill
0-907816-00-2 X.329/14896 Woolwich
2614928

AGRICULTURE. Curriculum subjects. Secondary schools. Developing countries
Teaching

OLAITAN, Samson O.
Agricultural education in the Tropics. - London : Macmillan, 1984
xii,194p : ill. - (Macmillan intermediate agriculture series). - Includes index
0-333-35946-1 X.329/18555
3208222

AGRICULTURE. Cyprus
Panaretos, Anthimos - Biographies

PANARETOS, Anthimos
Skorpies selides. - Leukōsia : Pankyprios Henōsis Geōponōn, 1973
47p : ill
r6503887 X.322/12395
8995486

AGRICULTURE. Cyprus
Part-time agriculture

PEARCE, Richard, 1944-
Part-time farming in Cyprus. - [Reading] : University of Reading, Department of Agricultural Economics & Management, 1981
93p. - (Development study ; no.21)
0-7049-0697-x X.0311/123(21) Woolwich
2589834

AGRICULTURE. Czech Socialist Republic

LACINA, Vlastislav
Zadlužení kapitalistického zemědělství v českých zemích v letech 1918-1938. - Praha : Academia, 1978
209p
r6015682 X.319/9885 Woolwich
2024551

MAUR, Eduard
Český komorní velkostatek v 17. století. - Praha : Univerzita Karlova, 1976
156p. - (Acta Universitatis Carolinae. Philosophica et historica. Monographia ; 1975 ; 59)
r6015459 Ac.798/8e [Monog 59]
1318527

AGRICULTURE. Czechoslovakia
to 1526

BERANOVÁ, Magdalena
Zemědělství starých Slovanů. - Praha : Academia, 1980
395p,[56]p of plates : ill. - Bibliography: p.328-369
r6440001 X.322/10647 Woolwich
8822301

AGRICULTURE. Czechoslovakia
1945–

ČESKOSLOVENSKÉ zemědělství = Chekhoslovatskoe sel'skoe khoziaĭstvo ... = Czechoslovak agriculture ... / [autor textu a libreta Jaroslav Kabrhel]. – Druhé, přeprac. vyd.. – Praha : Státní zemědělské nakladatelství, 1979
237p : chiefly ill(some col). – (Ekonomické a encyklopedické publikace)
r6439981 X.322/10675 Woolwich
8778892

KABRHEL, Jaroslav
30 let vývoje čs. zemědělství a jeho další perspektivy. – Praha : Horizont, 1975
143p. – (Na pomoc národnímu hospodářství)
r6440588 X.329/15528
8778892
Základy zemědělské politiky KSČ. – Vyd.2.. – Praha : Svoboda, 1976
291p
r6439076 X.329/13692 Woolwich
8792496

KAPITOLY z dejín socialistického po lnohospodárstva v Československu / [vedecký redaktor a zostavovateľ: Samuel Cambel]. – Bratislava : Ústav marxizmu-leninizmu ÚV KSS, 1982
310p. – (Zborník Ústavu marxizmu-leninizmu ÚV KSS ; roč. 21, čís. 1)
r6441696 X.329/19409
8778892

KLÍMA, Jaroslav
Chléb zraje i na hranici. – České Budějovice : Jihočeské nakladatelství, 1980
114p,[16]p of plates : ill,port. – (Hrdinové práce a vědy ; sv.2)
r6177833 X.329/12753 Woolwich
2451867

KUTIL, Jaroslav
Rozvoj socialistického zemědělství v ČSSR. – Praha : Svoboda, 1979
103p. – (Malá politická knihovna)
r6440359 X.529/48022
8778892

AGRICULTURE. Czechoslovakia
Collectivisation

SUCHÁNEK, Jaroslav
Právní regulace združstevňování československého zemědělství. – Praha : Academia, 1977
434p
r6187003 X.319/20210 Woolwich
2120577

AGRICULTURE. Czechoslovakia
Mechanisation, to 1948

PÁTEK, Jaroslav
Racionalizace zemědělské výroby mechanizací v Českých zemích v první polovině 20. století. – Praha : Československá akademie zemědělská, Ústav vědeckotechnických informací, Československé zemědělské muzeum, 1971–72
209p. – (Prameny historie zemědělství a lesnictví ; 10). – Bibliography : p. 201–209
r6015638 X.619/10179 Woolwich
1541064

AGRICULTURE. Czechoslovakia
Organisation, 1945–

SUCHÁNEK, Jaroslav
K obecné platnosti Leninovy agrární teorie.
74p. – (Rozpravy Československé akademie věd. Řada společenských věd ; roč.84 seš.2)
r6440399 Ac798ha/20(2)[R.84s.2]
8840458

AGRICULTURE. Czechoslovakia
Productivity, 1945–

KRILEK, Jozef
Ekonomická efektivnost zemědělství. – Praha : Svoboda, 1979
267p. – (Ekonomie a společnost). – Bibliography
r6440096 X.629/17303
8799415

AGRICULTURE. Czechoslovakia
Teaching, to date

ČERNOHORSKÝ, Zdeněk
Dějiny zemědělského školství v Československu. – Praha : Státní pedagogické nakladatelství, 1980
307p. – (Dějiny školy a výchovy)
r6440368 X.529/48494
8840490

AGRICULTURE. Dagestan ASSR. Russia (RSFSR)
INTENSIFIKATSIIA v promyshlennosti i sel'skom khoziaĭstve DASSR. – Makhachkala : Akademiia nauk SSSR, Dagestanskiĭ filial, Otdel ėkonomiki, 1978
160p
r6177503 X.329/12289 Woolwich
2455439

AGRICULTURE. Dagestan ASSR. Russia (RSFSR)
1800–1900

RAMAZANOV, Khidir Khidirovich
Sel'skoe khoziaĭstvo i promyshlennost' Dagestana v poreformennyi period (1861–1900g.g.). – Makhachkala : Dagestanskoe knizhnoe izdatel'stvo, 1972
198p
r6178231 X.329/12464. Woolwich
2447150

AGRICULTURE. Dagestan ASSR. Russia (RSFSR)
Collectivisation

KOLLEKTIVIZATSIIA sel'skogo khoziaĭstva Dagestanskoĭ ASSR 1927–1940 gg. / [redaktsionnaia kollegiia : A.S. Gadzhiev (otvet. redaktor) ... i dr.]. – Makhachkala : Dagestanskoe knizhnoe izdatel'stvo, 1976
2ch
r6088422 X.529/20954
1794906

KOLLEKTIVIZATSIIA sel'skogo khoziaĭstva Dagestanskoĭ ASSR 1927–1940gg ... / [redaktsionnaia kollegiia : Gadzhiev A.S. (otvet. redaktor) ... i dr.]. – Makhachkala : Dagestanskoe knizhnoe izdatel'stvo, 1976
2ch
r6094668 X.329/8919 Woolwich
2039192

AGRICULTURE. Denmark

HAMMARLUND, Lars
Den kultiverede natur. – [København] : Gyldendal, 1979
78p
87–01–81811–2 X.319/20398 Woolwich
1198351

HOLDEN, S. C. M.
Visit by the NAC Dairy Club to Denmark to study the financial situation in farming 13–15 June 1983. – [London] : ADAS, c1983
12p
B84–39328 BS.3/912
1198351

[LANDBRUGET i Danmark. English]. Agriculture in Denmark / [edited by P.H. Knudsen] ; published by the Agricultural Council of Denmark. – Copenhagen : Agricultural Council of Denmark, 1977
227p : ill(some col)
87–85065–43–9 X.329/8844
1198351

MURRAY, James William
Growth and change in Danish agriculture. – London : Hutchinson, 1977
64p : ill(chiefly col). – Includes index
0–09–127860–0 X.329/11102 Woolwich
1198351

AGRICULTURE. Denmark
– Forecasts

JUUL JØRGENSEN, Erik
Mulige udviklingslinier for dansk jordbrug til ar 2000. – København : Instituttet for Fremtidsforskning, 1977
157p. – Includes bibliographies
r6054492 X.313/811 Woolwich
1423487

AGRICULTURE. Denmark
– Serials

LANDBRUGETS økonomi / Jordbrugsøkonomisk Institut. – Efteråret 1978–. – København : Jordbrugsøkonomisk Institut : I kommission hos Landhusholdningsselskabets Forlag, 1978–
r6112648 SB.286/77
2473143

REGNSKABSRESULTATER fra danske landbrук i aret ... / Jordbrugsøkonomisk Institut. – 1976–77–. – København : Jordbrugsøkonomisk Institut : I kommission hos Landhusholdningsselskabets Forlag, 1978–. – (Jordbrugsøkonomisk Institut. Beretning ; nr 61:2, etc)
r6112646 SB.286/74
2473143

AGRICULTURE. Denmark
to date

CHRISTENSEN, Jens
Rural Denmark 1750–1980. – Copenhagen : Central Co-operative Committee of Denmark, 1983
187p : ill
87–7526–005–0 X.808/41764
8079501

CHRISTIANSEN, Palle Ove
En livsform pa tvangsauktion?. – 3.opl. – [København] : Gyldendal, 1983
199p
87–00–74281–3 X.329/18512
8079501

DANØ, Ivar
Landbruget. – København : Europa, 1982
115p : ill
87–87733–40–4 X.322/14334
8079501

JENSEN, Einar
Dansk landbrug skal overleve – men hvordan?. – København : JPbureau's Forlag, 1981
95p
87–981144–6–8 X.329/16707
8079501

LANDBRUGETS placering i samfundsøkonomien / red. Alex Dubgaard. – København : Statens Jordbrugsøkonomiske Institut, 1983
216p : ill
r6611474 X.520/33929
8079501

NANCKE-KROGH, Søren
Af jord er du kommet. – København : Nyt Nordisk Forlag, 1982
155p : ill
r6611656 X.329/18510
8079501

AGRICULTURE. Denmark
to date – Conference proceedings

BERETNING fra et symposium om dansk landbrug i oldtid og middelalder / udgivet af Henrik Thrane. – Odense : Odense Universitet, Historisk Institut, 1984
82p : ill. – (Skrifter fra historisk institut, Odense universitet ; nr.32)
r6612149 X.0702/305(32)
8079501

AGRICULTURE. Denmark
to date – Serials

MEDDELELSE / Jordbrugsøkonomisk Institut. – 1–. – København : Institutet, 1967–
Holdings: nr.26 (1978)–
r6610329 SB.286/75
8079501

RAPPORT / Jordbrugsøkononisk Institut. – Nr.1–. – Valby : Institutet, 1981–
r6611156 SB.286/94
8079501

UNDERSØGELSE / Jordbrugsøkonomisk Institut. – 1–. – København : Institutet, 1924–
Holdings: 33 (1978)–
r6610328 S.B.286/76
8079501

AGRICULTURE. Denmark
to date – Statistics – Serials

DENMARK. Danmarks Statistik
Statistiske efterretninger. Landbrug. – Årg.1, nr 1–. – København : Danmarks Statistik, 1983–
r6611243 SB.248/6(36)
8079501

AGRICULTURE. Denmark
ca 1680–1800. Information sources

KILDER til belysning af landbrugets økonomiske kar i 1700-tallet / udgivet af Finn Stendal Pedersen. – Odense : [Odense Universitet, Institut for Historie og Samfundsvidenskab], 1975
309p. – ([Skrifter fra Institut for Historie og Samfundsvidenskab] ; nr 15)
r6054441 X.0702/305(15)
1424742

AGRICULTURE. Denmark
Andersen, Ejner

ANDERSEN, Ejner
En bondes erindringer. – Hjørring : [E.Andersen], 1976
59p : ill
r6034376 X.319/9193 Woolwich
1395157

AGRICULTURE. Denmark
Effects of industrialisation, 1860–1914

Det AGRARE Danmark og det industrielle gennembrud 1860–1914. – [København] : Landbohistorisk Selskab, 1977
139p. – (Bol og by : 2. række ; 1)
87–7526–100–6 9341.c.1.
2469553

AGRICULTURE. Denmark
Farms. Social conditions, 1800–1900

KNUDSEN, Gunnar
Dansk bondeliv i forrige århundrede. – København : Erichsen, 1975
156 p. : ill. – (Fortidsreportage)
87–555–0434–5 Cup.1264.cc.18
1014463

AGRICULTURE. Denmark
Olsen, Levin – Biographies

OLSEN, Levin
Også en tilværelse. – [Århus] : Aros, [1979]
109p : ill
87–7003–342–0 X.329/12310 Woolwich
2263068

AGRICULTURE. Denmark
Organisations: Jordbrugsokonomisk Institut - Serials
JORDBRUGSØKONOMISK INSTITUT
Arsberetning. - 1978-. - Valby : Institutet : København
· I kommission hos Landhusholdningsselskabets Forlag. 1979-
r6061133 SB.286/71
2137216

AGRICULTURE. Denmark
Organisations: Landbrugsradet, to 1920
HANSGAARD, Torben
Landbrugsradets tilblivelse. - København : Landbohistorisk Selskab, 1976
287p : ill. - (Landbohistoriske skrifter ; 5). - Includes bibliographies
87-7526-078-6 WP.16884/5
1533355

AGRICULTURE. Denmark
Organisations: Landbrugsradet, 1919-1933
HELMER PEDERSEN, Erik
Landbrugsraadet som erhvervspolitisk toporgan 1919-33. - København : Landbohistorisk Selskab, 1979
667p : ill. - (Landbohistoriske skrifter ; 6). - Bibl
87-7526-084-0 WP.16884/6
2459620

AGRICULTURE. Denmark
Policies of social democratic movements, 1814-1901
LAHME, Hans-Norbert
Sozialdemokratie und Landarbeiter in Dänemark (1871-1900). - [Odense] : Odense University Press, [1982]
415p. - (Odense University studies in history and social sciences ; v.76). - Bibliography
87-7492-391-9 X.0700/318(76)
8891028

AGRICULTURE. Denmark
Polish labourers, to date
NELLEMANN, George
Polske landarbejdere i Danmark og deres efterkommere. - [København] : Nationalmuseet, 1981
413p : ill. - (Folkelivs studier ; 7). - Bibliography
87-480-0324-7 X.322/11335
8877424

AGRICULTURE. Denmark
Smallholdings. Land tenure. Reform, 1800-1920
CHRISTIANSEN, Palle Ove
Husmandsbevægelse og jordreform i Danmark. - [Copenhagen] : Etnologisk arbejdsgruppe for nordiske og europæiske studier, 1975
30p : ill
r6033886 X.319/9206 Woolwich
1418610

AGRICULTURE. Developed countries
BARNAS, Tibor
Agriculture towards the year 2000. - [Brighton] : Sussex European Research Centre, 1979
xiii,133p. - (Sussex European papers ; no.6)
B80-20349 X.0708/1507(6)
2130807

HILL, Brian E.
An economic analysis of agriculture. - London : Heinemann Educational, 1977
viii,296p : ill. - Includes index
0-435-84407-5 X.320/10981 Woolwich
0529702

An economic analysis of agriculture. - 2nd ed. - London : Heinemann, 1982
viii,355p : ill. - Includes index
0-435-84409-1 X.329/10905 Woolwich
0529702

PATTERNS of development. - London : Edward Arnold. Settlement and agriculture in the developed world. - 1981
64p : ill
0-7131-0452-x X.520/24141
1999362

STABLER, Michael John
Agricultural economics and rural land-use. - London : Macmillan, 1975
95p : ill. - (Macmillan studies in economics). - Bibliography: p.85-95
0-333-12700-5 X.519/25320
0529702

AGRICULTURE. Developed countries
Policies of governments
BOWLER, I. R.
Government and agriculture. - London : Longman, 1979
xiv,127p : ill. - (Topics in applied geography). - Bibliography: p.108-117. - Includes index
0-582-48578-9 X.311/9077 Woolwich
1775774

AGRICULTURE. Developing countries
PRICE, Gillian
Population and food resources in the developing world. - London : Edward Arnold, 1983
63p : ill. - (Patterns of development ; [5])
0-7131-0454-6 X.520/30385
2920026

AGRICULTURE. Developing countries
- Conference proceedings
INTERNATIONAL SEMINAR ON CHANGE IN AGRICULTURE (2nd : 1974
: University of Reading)
Policy and practice in rural development. - London : Croom Helm [for] the Overseas Development Institute, [1980]
520p : ill. - Bibliography: p.114. - Includes index
0-85664-958-9 X.529/36044
2144514

RURAL poverty and agribusiness / editor Michael Lipton. - Brighton : Institute of Development Studies, 1977
[6],xiii,86p. - (Discussion papers / University of Sussex. Institute of Development Studies ; 104)
B77-18615 X.313/1404 Woolwich
1123173

AGRICULTURE. Developing countries
to date
LUNDBORG, Per
Trade policy and development. - Göteborg : Universitetet, 1981
xi,189p. - (Ekonomiska studier utgivna av Nationalekonomiska institutionen vid Göteborgs universitet ; 9). - Bibliography
r6610018 X.329/15880
887686x

AGRICULTURE. Developing countries
Administration. Role of ministries of agriculture
HOWELL, John, 19—
Ministries of agriculture and the administration of agricultural development. - London : Overseas Development Institute, 1980
14p : . - (Agricultural administration network discussion paper ; 4)
B81-27056 X.329/14219 Woolwich
2506173

AGRICULTURE. Developing countries
Analysis. Methodology
BIRGEGÅRD, Lars-Erik
Manual for the analysis of rural underdevelopment. - Uppsala : Swedish University of Agricultural Sciences, 1980
150p. - (Rural development studies ; no.7)
91-576-0702-8 X.0313/56(7)
9032967

AGRICULTURE. Developing countries
Consumption of energy supply
MAKHIJANI, Arjun
Energy and agriculture in the Third World. - Cambridge, Mass : Ballinger, 1975
xvii,168p : ill. - Bibliography : p.166-161.- Includes index
0-88410-341-2 X.320/11703 Woolwich
1888722

AGRICULTURE. Developing countries
Economic development
AGRARIAN systems and rural development / [edited by] Dharam Ghai ... [et al.]. - London : Macmillan, 1979
xv,375p : ill. - Includes index
0-333-27343-5 X.529/34554
1836730

AGRICULTURAL development in the Third World / edited by Carl K. Eicher & John M. Staatz. - Baltimore ; London : Johns Hopkins University Press, c1984
x,491p : ill. - (The Johns Hopkins studies in development). - Includes bibliographies and index
0-8018-3014-1 X.322/14697
1094866

ARNON, I. (Isaac)
Modernisation of agriculture in developing countries. - Chichester : Wiley, c1981
xxiii,565p : ill. - (Environmental monographs and symposia). - Includes bibliographies and index
0-471-27928-5 SRL
1094866

CLAYTON, Eric
Agriculture, poverty and freedom in developing countries. - London : Macmillan, 1983
x,277p : ill. - Includes index
0-333-36100-8 X.529/62708
1094866

GHATAK, Subrata
Agriculture and economic development. - Brighton : Wheatsheaf, 1984
xii,380p : ill. - Includes bibliographies and index
0-7108-0142-4 X.329/18226
1094866

GROWING out of poverty / edited by Elizabeth Stamp. - Oxford : Oxford University Press, 1977
x,165p : 1 ill
0-19-857529-7 X.529/31833
1094866

JIGGINS, Janice
Extension, planning and the poor. - London : Overseas Development Institute, 1977
vi,57p. - (Occasional papers / Overseas Development Institute. Agricultural Administration Unit ; 2)
B79-20742 X.519/29786
1094866

SCHULTZ, Theodore W. (Theodore William)
Transforming traditional agriculture. - Chicago ; London : University of Chicago Press, 1983, c1964
xiv,212p. - Includes index
0-226-74075-7 X.329/19733
1094866

AGRICULTURE. Developing countries
Economic development - Conference proceedings
SCIENCE AND AGRIBUSINESS IN THE SEVENTIES (Conference : 1974 : London)
Agricultural initiative in the Third World. - Lexington, Mass. ; London : D.C. Heath, 1975
xvii,210p : 1 ill. - Includes bibliographies
0-669-97253-3 X.320/10094
1094866

AGRICULTURE. Developing countries
Economic development, 1945-
CRECIMIENTO de la población y cambio agrario / compiladores : Víctor L. Urquidi y José B. Morelos. - México : El Colegio de México, 1979
viii,391p : ill. - (Colección Centro de Estudios Económicos y Demográficos ; VIII)
9-681-20040-3 X.529/47339
8083630

AGRICULTURE. Developing countries
Economic development. Financial assistance by United Nations. International Fund for Agricultural Development. Establishment
AGREEMENT establishing the International Fund for Agricultural Development adopted by the United Nations conference at Rome on 13 June 1976 and opened for signature at New York on 20 December 1976 ... - London : H.M.S.O., [1977]
27p. - (Miscellaneous ; no.7(1977)) (Cmnd.6787)
0-10-167870-3 BS.14/2071
1105701

AGRICULTURE. Developing countries
Economic development. Planning
DANIEL, A. V.
Strategy for agricultural development. - Bombay : Vora, 1976
xii,171p. - Bibliography: p.161-170. - Includes index
r6083454 X.329/11606 Woolwich
1587005

AGRICULTURE. Developing countries
Economic development. Projects
BENJAMIN, McDonald P.
Investment projects in agriculture. - Harlow : Longman, 1981
xxiii,297p : ill. - Includes index
0-582-64306-6 X.322/10554 Woolwich
2608901

AGRICULTURE. Developing countries
Economic planning
MOLLETT, J. A.
Planning for agricultural development. - London : Croom Helm, c1984
355p : ill. - Bibliography: p348-252. - Includes index
0-7099-1784-8 X.529/61997
0215562

AGRICULTURE. Developing countries
Effects of taxation, to date
MANIG, Winfried
Steuern und rurale Entwicklung. - Saarbrücken : Breitenbach, 1981
517p. - (Sozialökonomische Schriften zur Agrarentwicklung ; Nr.38). - Bibliography: p.445-486
3-88156-162-5 X.529/46945
8823103

AGRICULTURE. Developing countries
Effects of technological change. Ecological aspects

DAHLBERG, Kenneth A.
Beyond the green revolution. - New York ; London : Plenum Press, 1979
xiii,256p : ill. - Bibliography: p.229-238. - Includes index
0-306-40120-7 SRL
1917366

AGRICULTURE. Developing countries
Extension services

ADAMS, M. E.
Agricultural extension in developing countries. - Harlow : Longman, 1982
xii,108p : ill. - (Intermediate tropical agriculture series). - Includes bibliographies and index
0-582-65025-9 X.809/53446
2338947

AGRICULTURE. Developing countries
Farms. Management. Decision making

AGRICULTURAL decision making / edited by Peggy F. Barlett. - New York ; London : Academic Press, c1980
xv,378p : ill. - (Studies in anthropology). - Includes bibliographies and index
0-12-078880-2 X.325/819 Woolwich
2345218

AGRICULTURE. Developing countries
Farms. Productivity related to size

BERRY, R. Albert
Agrarian structure and productivity in developing countries. - Baltimore ; London : Johns Hopkins University Press, 1979
xiii,248,[1]p : ill. - Bibliography: p.186-188. - Includes index
0-8018-2190-8 X.520/14767
2007029

AGRICULTURE. Developing countries
Innovation, 1967-1972

MAC DONALD, A. L.
Agricultural technology in developing countries. - [Rotterdam] : Rotterdam University Press, 1976
xv,236,[1]p : ill. - Bibliography : p.225-231. - Includes index
90-237-6259-2 X.329/11611 Woolwich
1589598

AGRICULTURE. Developing countries
Insurance

RAY, P. K. (Parimal Kumar)
Agricultural insurance. - 2nd ed. - Oxford : Pergamon, 1981
xxviii,419p,[1]folded leaf of plates : ill. - Bibliography: p407-411. - Includes index
0-08-025787-9 X.520/24317
195959x

AGRICULTURE. Developing countries
Management. British organisations: Agricultural Administration Network - Serials

[NEWSLETTER (Agricultural Administration Network)].
Newsletter / Agricultural Administration Network. - No.1 (Nov.1979)-. - London : Overseas Development Institute, 1979-
v.
Holdings: no.2; March 1980, etc.
B81-06903 X.0322/258 Woolwich
2354276

AGRICULTURE. Developing countries
Mathematical models

BHADURI, Amit
The economic structure of backward agriculture. - London : Academic Press, 1983
ix,151p. - (Studies in political economy). - Bibliography: p142-145. - Includes index
0-12-095420-6 X.322/14689
3243338

AGRICULTURE. Developing countries
Mechanisation

CROSSLEY, Peter
Small farm mechanization. - Chichester : Wiley, c1983
xvi,253p : ill. - Includes bibliographies and index
0-471-90101-6 SRL
1518402

AGRICULTURE. Developing countries
Mechanisation - Conference proceedings

RECENT stage and prospects of agricultural mechanization in developing countries. - [East] Berlin : Akademie-Verlag, 1976
xvii,261p : ill. - (Studien über Asien, Afrika und Lateinamerika ; Bd 22). - Includes bibliographies
r6072103 X.0709/574
1518402

AGRICULTURE. Developing countries
Planning. Spatial models

BISHAY, Fahmi K.
Models for spatial agricultural development planning. - [Rotterdam] : Rotterdam University Press, 1974
xv,172p : ill. - (Studies in development and planning ; vol 3). - Bibliography : p.165-169. - Includes index
90-237-2803-3 X.629/12084 Woolwich
1684531

AGRICULTURE. Developing countries
Project management

SMITH, Peter, 19—
Agricultural project management. - London : Elsevier Applied Science, c1984
ix,190p : ill. - Includes index
0-85334-282-2 X.329/19201
3017796

AGRICULTURE. Developing countries
Projects. Appraisal

BRIDGER, Gordon A.
Guidelines for the appraisal of agricultural projects. - London : Overseas Development Institute, 1980
43p. - (Agricultural administration network papers ; no.10)
B81-27055 X.329/14220 Woolwich
0655430

HARDIE, John David Morrison
A guide to basic agricultural project appraisal in developing countries. - Aberdeen : School of Agriculture, 1976
[3],78,[51]p : 3 ill. - (Miscellaneous publications / School of Agriculture, Aberdeen ; no.12). - Bibliography: p.77-78
0-85281-006-7 X.512/5193
0655430

AGRICULTURE. Developing countries
Projects. Evaluation. Cost-benefit analysis

GITTINGER, J. Price (James Price)
Economic analysis of agricultural projects. - 2nd ed., completely rev. and expanded. - Baltimore ; London : Johns Hopkins University Press [for] the Economic Development Institute of the World Bank, c1982
xxi,505p : ill. - (EDI series in economic development). - Bibliography: p445-455. - Includes index
0-8018-2912-7 X.525/6822
2863316

AGRICULTURE. Developing countries
Spatial analysis

MORGAN, William Basil
Agriculture in the Third World. - London : G. Bell, 1977 [i.e. 1978]
xiii,290p : ill. - (Bell's advanced economic geographies). - Bibliography: p.259-278. - Includes index
0-7135-0082-4 X.329/11565 Woolwich
1565680

AGRICULTURE. Developing countries
Statistics

IDAIKKADAR, N. M.
Agricultural statistics. - Oxford : Pergamon, 1979
xii,139p,[2]fold leaves. - (Pergamon international library). - Bibliography: p.134-135. - Includes index
0-08-023388-0 SRL
1736558

AGRICULTURE. Developing countries. Africa south of the Sahara
Policies of governments

BADU-DANQUAH, M. E.
Agricultural policy in a developing country. - Tema : Ghana Publishing, 1973
64p
r6091860 X.319/18857
1891995

AGRICULTURE. Devon

ROSENTHALL, Pamela
Agricultural production and marketing in Cornwall and Devon. - [Exeter] : University of Exeter Agricultural Economics Unit, 1981
72p : ill. - (Report ; no.212)
B82-01507 X.322/10382 Woolwich
258400x

AGRICULTURE. Devon
Ley, John Carter - Biographies

LEY, John Carter
The story of my life. - Ilfracombe : Stockwell, 1978
164p,[32]p of plates : ill
0-7223-1128-1 X.329/11541 Woolwich
1547690

AGRICULTURE. Devon
Luxton (Family)

CORNWELL, John, 1940-
Earth to earth. - London : Allen Lane, 1982
174p,[16]p of plates : ill
0-7139-1045-3 X.809/54057
2832674

CORNWELL, John, 1940-
Earth to earth. - Harmondsworth : Penguin, 1984, c1982
174p,[16]p of plates : ill
0-14-006438-9 X.808/40293
2832674

AGRICULTURE. Devon
Smallholdings

A STUDY of minor holdings in Devon. - Exeter : University of Exeter, Agricultural Economics Unit, 1983
27leaves. - (Report ; no.219)
B84-27908 X.525/9196
313279x

AGRICULTURE. Dial Post. West Sussex
Farms: Dial Post Farm, 1909-1910 - Biographies

SWAN, William Carter
The diary of a farm apprentice. - Gloucester : Sutton, 1984
141p : ill. - (Regional letters and diaries of the British Isles)
0-86299-109-9 X.809/64027
3113841

AGRICULTURE. Donetsk (oblast'). Ukraine
1917-

LÎAKH, R. D. (Roman Danylovych)
Ahrarni peretvorenni͡a na Donechchyni (1917-1923 rr.). - Donets'k : Donets'kyi derz͡havnyi universytet, 1974
62p
r6485687 X.629/17438 Woolwich
8805679

AGRICULTURE. Dordogne (Department). France
to date

MÉMENTO agricole et rural / Département de la Dordogne, Direction départmentale de l'agriculture, Service statistique. - (1976)-. - Périgueux : Département de la Dordogne, 1976
r6209512 S.E.400/1042
8876878

AGRICULTURE. Dorset
Labourers. Social conditions, 1883

HARDY, Thomas, 1840-1928
The Dorsetshire labourer, 1883. - St. Peter Port : Toucan, 1980
p253-268
0-85694-225-1 X.809/54983
2891913

AGRICULTURE. Draved. Denmark
Stone Age agriculture. Simulations, 1953-1955

STEENSBERG, Axel
Draved. - Copenhagen : National Museum of Denmark, [1979]
116p,[4] fold leaves of plates : ill
87-480-0161-9 X.410/7445 Woolwich
2195267

AGRICULTURE. Dunántúl. Hungary

TAKÁCS, Lajos
Egy irtásfalu földművelése. - Budapest : Akadémiai Kiadó, 1976
287p : ill. - Bibliography : p.243-248
963-05-0880-x X.320/5402 Woolwich
2213451

AGRICULTURE. Dunlop. Strathclyde Region. Scotland
Robert Howie & Sons, to 1978

HOWIE, James M. (James Meldrum)
ROBERT HOWIE & SONS
Things I remember. - [Kilmarnock] : [Howie and Sons], [1978?]
52p : ill
B81-21045 X.529/41477
2409003

AGRICULTURE. Dyfed

CRAGOE, Elizabeth
Buttercups and daisy. - London : Hamilton, 1976
[5],167p
0-241-89444-1 X.329/10549 Woolwich
0705330

CRAGOE, Elizabeth
Buttercups and daisy. - London : Sphere, 1978
189p
0-7221-2650-6 X.319/18201 Woolwich
0705330

AGRICULTURE. Dyfed –continued

Cowslips and clover. - London : Hamilton, 1977
[4],172p
0-241-89761-0 X.809/42421
0705330

AGRICULTURE. Dyfed
- Texts

CRAGOE, Elizabeth
Cowslips and clover. - London : Sphere, 1978
[4],200p
0-7221-2649-2 X.319/18547 Woolwich
0705330

AGRICULTURE. East Africa

NGUGI, D.
East African agriculture. - London : Macmillan, 1978
vi,266p : ill. - (Macmillan books for Africa). - Includes index
0-333-21271-1 X.311/8728 Woolwich
1413376

AGRICULTURE. East Africa
- For East African students

AGRICULTURE in East Africa / A.D.R. Ker ... [et al.]. - London : Edward Arnold, 1978
vi,218p : ill. - Includes bibliographies and index
0-7131-0160-1 X.311/8849 Woolwich
1565753

NGUGI, David
East African agriculture. - Rev. ed. - London : Macmillan, 1982
vi,282p : ill. - Bibliography: p276. - Includes index
0-333-32380-7 X.322/11967
1565753

East African agriculture. - [London] : English Language Book Society, 1982
vi,282p : ill. - Bibliography: p276
0-333-34154-6 X.800/38121
1565753

AGRICULTURE. East Africa
Economic development. Effects of colonialism, to 1980

IMPERIALISM, colonialism and hunger / edited by Robert I. Rotberg. - Lexington, Mass. : Lexington, 1983
xiv,270p : ill. - Includes index
0-669-05871-8 X.800/36424
2971429

AGRICULTURE. East Anglia
- Serials

REPORT on farming in the Eastern counties of England. - 1979-80. - Cambridge : Agricultural Economics Unit, Department of Land Economy, [1980]
iii,134p
Subsequent reports, when received, will be placed at this shelfmark
B81-09082 P.315/28
0256285

AGRICULTURE. East Anglia
ca 1905-1974

LING, W. E.
More time to mardle. - Ipswich : East Anglian Magazine Ltd, [1980]
104p : ill
0-900227-45-1 X.700/26526
2278421

AGRICULTURE. East Anglia
Personnel. Social conditions, 1850-1976

NEWBY, Howard, 1947-
The deferential worker. - London : Allen Lane, 1977
462p
0-7139-0892-0 X.329/11122 Woolwich
1207164

NEWBY, Howard, 1947-
The deferential worker. - Harmondsworth : Penguin, 1979
462p. - (Penguin education)
0-14-080318-1 X.519/29142
1207164

The deferential worker. - Madison : University of Wisconsin Press, 1979
486p
0-299-07890-6 X.529/34562
1207164

AGRICULTURE. East Anglia
Trade unionism. Role of Methodists, 1872-1896

SCOTLAND, Nigel
Methodism and the revolt of the field. - Gloucester : Sutton, 1981
296p : ill. - Bibliography: p267-289. - Includes index
0-904387-46-1 X.200/38408
2738961

AGRICULTURE. East Germany
to date

GRÜNEBERG, Gerhard
Agrarpolitik der Arbeiterklasse zum Wohle des Volkes. - Berlin : Dietz, 1981
615p,[33]p of plates
r6305003 X.629/17154 Woolwich
8726442

SOZIALISTISCHE Betriebswirtschaft in der Landwirtschaft / unter Federführung von Helmut Plesske. - Berlin : Deutscher Landwirtschaftsverlag, [1979]
304p
r6301659 X.322/10132 Woolwich
8726442

AGRICULTURE. East Germany
1945-

Die DURCHFÜHRUNG der Beschlüsse des X. Parteitages der SED in der sozialistischen Landwirtschaft / [Autorenkollektiv unter Leitung von Werner Eckler]. - Berlin : Dietz, 1983
96p.. - (Der Parteiarbeiter)
r6306456 P.971/276(58)
8916489

ECKART, Karl
Die Landwirtschaft der DDR im Wandel. - Paderborn : Schöningh, c1981
31p. - (Fragenkreise ; 23557)
3-506-23557-5 X.329/17358
8916489

AGRICULTURE. East Germany
1945- - Conference proceedings

WISSENSCHAFTLICHER RAT FÜR DIE WIRTSCHAFTSWISSENSCHAFTLICHE FORSCHUNG. Tagung (25th : 1978 : Berlin)
Intensivierung der landwirtschaftlichen Produktion. - Berlin : Akademie-Verlag, 1979
118p.. - (Abhandlungen der Akademie der Wissenschaften der DDR. Abteilung Veröffentlichungen der Wissenschaftlichen Räte ; Jahrg. 1979, Nr.W2)
r6307565 X.0510/288(1979,no.2)
8916489

AGRICULTURE. East Germany
Consumption of energy, 1945-

HOHMANN, Karl
Zur energetischen Effizienz der Agrarproduktion in der DDR. - Berlin : Forschungsstelle für Gesamtdeutsche Wirtschaftliche und Soziale Fragen, 1982
49p. - (FS-Analysen ; 4-1982)
r6305679 SF.583/219[1982,no.4]
8844771

AGRICULTURE. East Germany
Independent smallholdings, 1945-

HOHMANN, Karl
Entwicklung und Bedeutung der privaten Agrarproduktion in der DDR. - Berlin : Forschungsstelle für gesamtdeutsche wirtschaftliche und soziale Fragen, 1984
45p. - (FS-Analysen ; 3-1984)
r6310219 SF.583/219[1984, no.3]
9136916

AGRICULTURE. East Jordan
Economic development. Effects of migration of personnel

SECCOMBE, Ian J.
Manpower and migration. - [Durham] : University of Durham, Centre for Middle Eastern and Islamic Studies, c1981
viii,110p : ill. - (Occasional papers series ; no.11). - Bibliography: p105-109
B82-14475 X.0700/524(11)
2640163

AGRICULTURE. East Midlands. England
- Serials

[SUMMARY report (University of Nottingham. Department of Agriculture and Horticulture)]. Summary report / University of Nottingham, Department of Agriculture and Horticulture. - No.7 (Apr.1981). - [Loughborough] : The Department, 1981
14p
B81-27126 X.322/10012 Woolwich
0527793

AGRICULTURE. East Midlands. England
- Statistics

FARMING in the East Midlands / [University of Nottingham, Department of Agriculture and Horticulture]. - 1978-79 : (twenty-eighth annual report). - Sutton Bonington : The Department, 1980
63p
B80-20351 X.322/9221 Woolwich
0527793

AGRICULTURE. East Midlands. England
- Statistics - Serials

FARMING in the East Midlands, financial results / University of Nottingham Department of Agriculture and Horticulture. - 1979-80. - Sutton Bonington : The Department, 1981
66p
B81-15158 X.325/1032 Woolwich
0527793

AGRICULTURE. East Midlands. England
Farms. Accounts, 1708-1950 - Catalogues

UNIVERSITY OF NOTTINGHAM. Manuscripts Department
Farm records c.1708-1950. - [Nottingham] : University of Nottingham, Manuscripts Department, [c1984]
7p
B85-02916 X.805/6068
321172x

AGRICULTURE. East Prussia
Personnel. Training, to 1945

RICHTER, Gottfried
Beiträge und Quellen zum landwirtschaftlichen Schulwesen in der Provinz Ostpreussen. - Osnabrück : [G. Richter], 1979
vii,405p.. - Bibliography
r6303660 X.329/14937 Woolwich
8902607

AGRICULTURE. Eastern Africa
Mechanisation

KINSEY, B. H. (Bill H)
Economic research and farm machinery designs in East Africa. - Norwich : School of Development Studies, University of East Anglia, 1976
20leaves. - (Development studies discussion paper ; no.10). - Bibliography: leaves 19-20
B84-36446 X.525/9651
3161021

AGRICULTURE. Eastern England
to 1975

AGRICULTURAL DEVELOPMENT AND ADVISORY SERVICE
Agriculture in Eastern Region. - [4th ed.]. - [Pinner] : Ministry of Agriculture, Fisheries and Food, [1976]
[2],iii,63p : 1 ill
B79-14512 BS.3/596
1871412

AGRICULTURE. Eastern England
Personnel. Mobility

GASSON, Ruth
Mobility of farm workers. - Cambridge : University of Cambridge Department of Land Economy, 1974
[7],96p. - (Occasional papers / University of Cambridge. Department of Land Economy ; no.2). - Bibliography: p.91-96
0-901978-53-1 X.0329/97(2)
0227625

AGRICULTURE. Eastern Europe
- Conference proceedings

MEZHRESPUBLIKANSKIĬ SIMPOZIUM PO AGRARNOĬ ISTORII VOSTOCHNOĬ EVROPY (15th : Vologda : 1974)
Tezisy dokladov i soobshchenii XV sessii Mezhrespublikanskogo simpoziuma po agrarnoĭ istorii Vostochnoĭ Evropy, Vologda, 7-13 sentiabria 1974 g.. - Moskva : [b.i], 1974
2v.
r6177611 X.329/12577 Woolwich
2446502

AGRICULTURE. Eastern Europe
Collectivisation, 1945-

KOZLOWSKI, Zygmunt
The integration of peasant agriculture into the socialist economies of Eastern Europe since 1948. - Oxford : Centre for Soviet and East European Studies, St. Anthony's College, 1978
136p. - (Papers in East European economics)
r6813467 X.0515/69(54)
8102082

AGRICULTURE. Eastern Europe & Soviet Union
- Serials

EAST Europe agriculture. - No.1 (Oct. 1982)-. - Tunbridge Wells : Agra Europe (London), 1982-
v.
B83-25101 SRL
2908182

AGRICULTURE. Eastern Europe & Soviet Union
Policies of governments, 1960–1980

WÄDEKIN, Karl-Eugen
Agrarian policies in communist Europe. - Totowa : Allanheld, Osmun ; The Hague ; London : Nijhoff, 1982
x,324p. - (Studies in East European and Soviet Russian Agrarian policy ; v.1). - Bibliography: p266–318. - Includes index
90–247–2572–0 X.800/36578
2966549

AGRICULTURE. Eastern India (Republic)
Effects of increases in crop yields

MANDAL, G. C.
Economics of the Green Revolution. - London : Asia Publishing House, 1976
xvi,115p. - Includes index
0–210–40571–6 X.329/10847 Woolwich
107041x

AGRICULTURE. Eastern states. United States
Migrant personnel. Social conditions, to 1980

THOMAS-LYCKLAMA À NIJEHOLT, G. (Geertje)
On the road for work. - Boston, [Mass.] ; London : Nijhoff, c1980
xii,211p. - (Series on the development of societies ; v.7). - Bibliography: p203–211
0–89838–043–x X.520/29598
2885026

AGRICULTURE. Ecuador
1895- - Statistics

[CENSO agropecuario. 1974]. II Censo agropecuario 1974 / Instituto Nacional de Estadística y Censos. - Quito : INEC, 1977-
r6707173 LAS.E.70/31
9001409

AGRICULTURE. Ecuador
Economic development

SAAD, Pedro
La realidad agropecuaria del Ecuador. - Guayaquil : Libros Claridad, 1972
89p
r6053874 X.519/24609
1427075

AGRICULTURE. Ecuador
Economic development, to date

CUSHNER, Nicholas P.
Farm and factory. - Albany : State University of New York Press, c1982
ix,231p. - Includes bibliography
0–87395–570–6 X.322/14929
8103607

PROAÑO P., Fabián
Ecuador: dinámica del sector agrario. - Quito : La Junta, Oficina de Publicaciones, 1978
238p. - Bibliography
r6703785 LAS.E.70/25 Woolwich
8103607

AGRICULTURE. Egilsay. Orkney. Scotland
Smallholdings. Self-sufficiency, 1974–1980

COLEMAN, Vicki
Living on an island. - Findhorn : Thule Press, 1981
127p
0–906191–55–6 X.809/49988
2594471

AGRICULTURE. Egypt
- For British exporters

DAVIES, Piers
A report on agriculture in Egypt. - London : BAEC, c1981
iv,51p
B82–11356 SRL
2625768

AGRICULTURE. Egypt
Economic development, 1800–1980

RICHARDS, Alan
Egypt's agricultural development 1800–1980. - Boulder : Westview, 1982
xvi,296p : ill. - Bibliography: p266–289
0–86531–099–8 X.322/10775
2732793

AGRICULTURE. Egypt
Economic development, 1952–1970

ABDEL-FADIL, Mahmoud
Development, income distribution and social change in rural Egypt, 1952–1970. - Cambridge : Cambridge University Press, 1975
xiii,157p : ill. - (Occasional papers / University of Cambridge. Department of Applied Economics ; 45). - Bibliography: p.153–157
0–521–21000–3 PP.8004.lh(45)
0662631

AGRICULTURE. El Salvador
to date

EL SALVADOR. Dirección General de Estadística y Censos
[Censo Nacional Agropecuario 1971]. Tercer Censo Nacional Agropecuario 1971. - San Salvador : La Dirección.
Vol.2: Características a nivel nacional, departamental y tamaño de la explotación. - 1975
xxxiv,509p
r6706874 LAS.G.462/38
895366x

MONTES, Segundo
El agro salvadoreño, 1973–1980. - San Salvador : Universidad Centroamericana "José Simeon Cañas", 1980
v,395p
r6709056 X.622/20538
895366x

AGRICULTURE. El Salvador
Economic development, ca 1940–1976

DOWNING, T. J.
Agricultural modernisation in El Salvador, Central America. - Cambridge : Centre of Latin American Studies, University of Cambridge, 1978
[2],ii leaves,54[i.e.70]p. - (Working papers / University of Cambridge. Centre of Latin American Studies ; no.32)
B79–04165 X.313/948. Woolwich
1768050

AGRICULTURE. Emilia-Ramagna. Italy
1946-

PIEPOLI, Giovanni
Una politica per lo sviluppo agricolo. - Venezia : Marsilio, 1980
142p. - (Documentazione territorio e società)
r6507436 X.329/16377
8945195

AGRICULTURE. Emilia-Romagna. Italy
Theories, to 1700

BIGNARDI, Agostino
Le campagne emiliane nel Rinascimento e nell'età barocca. - [Bologna] : Arnaldo Forni, 1978
viii,401p.,[8]leaves of plates
r6160067 X.329/12321 Woolwich
2262681

AGRICULTURE. England

BREWIS, Henry
Anyfoolcanbeafarmer. - Whitehaven : West Cumberland Farmers Ltd, 1974
48p : ill
0–9504367–0–4 X.319/10081
0366056

A load of rural rubbish. - Whitehaven : West Cumberland Farmers, 1972
48p : ill
0–9504367–1–2 X.909/40092
0366056

FARM management survey. - 1975–6 : farm liabilities and assets. - London : Ministry of Agriculture, Fisheries and Food, Economics Division 1, 1977
53p in various pagings
Subsequent volumes, when received, will be placed at this shelfmark
B78–10751 BS.3/570
1381504

HARVEY, Nigel
Farms and farming. - Aylesbury : Shire Publications, 1977
33p : ill. - (Shire album ; 26). - Bibliography: p.33
0–85263–382–3 X.0972/143(26)
1259954

PRACTICAL proficiency tests for agriculture / National Proficiency Tests Council for Agriculture and Horticulture. - Kenilworth : The Council, [1980]
1v.(loose-leaf)
B81–12533 X.329/13680 Woolwich
2379430

RICE, W. Farleigh
Patterns in farming. - Harlow : Common Ground, 1973
1filmstrip(36fr.7title fr.) : col.
32p.. - (Patterns in geography)
av00001 Cup.580.i.2(1)
1259954

AGRICULTURE. England
- Texts

TUSSER, Thomas
Five hundred points of good husbandry. - Oxford : Oxford University Press, 1984
xx,344p. - Includes index
0–19–286040–2 X.808/39507
3084639

AGRICULTURE. England
to 1939

The AGRARIAN history of England and Wales / general editor Joan Thirsk. - Cambridge : Cambridge University Press.
Vol.1. - 1981. - Includes index
1: Prehistory
xxi,51p.12p of plates : ill
0–521–08741–4 X.0311/87(1) Woolwich
2394391

AGRICULTURE. England
to 1980

WHITLOCK, Ralph
The English farm. - London : Dent, 1983
256p,[16]p of plates : ill(some col.). - Bibliography: p248–251. - Includes index
0–460–04584–9 X.800/37180
3056473

AGRICULTURE. England
ca 1350–1870

AGRICULTURAL improvement / edited by Walter Minchinton - [Exeter] : University of Exeter, 1981
vii,137p : ill. - (Exeter papers in economic history ; no.14)
0–85989–142–9 X.329/14929 Woolwich
2620022

AGRICULTURE. England
1837–1936

VICTORIAN and Edwardian farming from old photographs / [compiled by] Edward Hart. - London : Batsford, 1981
120p : chiefly ill
0–7134–3799–5 X.322/10457 Woolwich
2599112

AGRICULTURE. England
Ancient Roman agriculture. Farms. Buildings

MORRIS, Pat
Agricultural buildings in Roman Britain. - Oxford : British Archaeological Reports, 1979
[10],225p : ill. - (British archaeological reports. British series ; 70). - Bibliography: p.203–222. - Includes index
0–86054–065–0 X.322/9219 Woolwich
2111268

AGRICULTURE. England
Cherrington, John - Biographies

CHERRINGTON, John
On the smell of an oily rag. - London : Northwood, 1979
v,178p : ill. - (A Big farm management book)
0–7198–2528–8 X.329/13373
233237x

AGRICULTURE. England
Farms, to ca 1950

SEEBOHM, Mabel Elizabeth
The evolution of the English farm. - Revised 2nd ed. - Wakefield : EP Publishing, 1976
356p,[3] leaves of plates : ill. - Bibliography: p.344–350. - Includes index
0–7158–1173–8 X.329/10778 Woolwich
062506x

AGRICULTURE. England
Farms, to 1980

FOWLER, Peter
Farms in England. - London : H.M.S.O., 1983
vi,[72]p : chiefly ill. - (National Monuments Record photographic archives). - Bibliography: 4th prelim. page. - Includes index
0–11–701130–4 BS.77/123
2907267

AGRICULTURE. England
Farms. Buildings

DARLEY, Gillian
The National Trust book of the farm. - London : National Trust, c1981
256p : ill(some col.). - Bibliography: p250–252. - Includes index
0–297–78006–9 X.325/1344 Woolwich
2559129

AGRICULTURE. England
Farms. Buildings, to 1983

HARVEY, Nigel
A history of farm buildings. - New ed. - Newton Abbot : David & Charles, 1984
279p : ill. - Bibliography: p267–271. - Includes index
0–7153–8383–3 X.421/26091
3013529

AGRICULTURE. England
Farms. Buildings. Architecture

WOODFORDE, John
Farm buildings in England and Wales. - London : Routledge & Kegan Paul, 1983
150p : ill. - Bibliography: p145-147. - Includes index
0-7100-9275-x X.629/20647
2930811

AGRICULTURE. England
Farms. Buildings. Conversion to self-catering vacation accommodation

ENGLISH TOURIST BOARD
Self-catering on the farm. - London : The Board, 1979
48p : ill. - Bibliography: p.40
B79-21115 BS.411/268
1901443

AGRICULTURE. England
Farms. Grassland

GREEN, J. O. (Joe O)
A sample survey of grassland in England and Wales 1970-72. - Maidenhead : Grassland Research Institute, c1982
39p : ill(some col.). - Bibliography: p33
0-7084-0227-5 SRL
2761920

AGRICULTURE. England
Farms. Industrial antiquities

HARVEY, Nigel
The industrial archaeology of farming in England and Wales. - London : Batsford, 1980
232p : ill. - Bibliography: p.225-227
0-7134-1845-1 X.322/9548 Woolwich
2232286

AGRICULTURE. England
Farms. Permanent grassland. Crops: Grasses. Production

JOINT PERMANENT PASTURE GROUP
Factors affecting the productivity of permanent grassland. - Maidenhead : The [Joint] Permanent Pasture Group, 1980
[8],141p : ill. - Bibliography: p.75-76
0-7084-0132-5 SRL
2284332
- [Another copy] BS.3/687

AGRICULTURE. England
Farms. Productivity related to size

BRITTON, D. K. (Denis King)
Farm size, tenure and efficiency. - [Ashford, Kent] : Wye College, Agricultural Economics Unit, School of Rural Economics and Related Studies, 1978
[5],36p
0-901859-78-8 X.512/6394
1810812

AGRICULTURE. England
Farms. Recreation facilities. Provision

DARTINGTON AMENITY RESEARCH TRUST
Farm recreation and tourism in England and Wales. - [Cheltenham] : [Countryside Commission], 1974
[4],62p : ill. - (CCP ; 83) (DART publication ; no.14)
0-9500410-4-1 BS.116/51
0536202

AGRICULTURE. England
Farms. Servants. Social conditions, ca 1550-1850

KUSSMAUL, Ann
Servants in husbandry in early modern England. - Cambridge : Cambridge University Press, 1981
xii,233p : ill. - (Interdisciplinary perspectives on modern history). - Bibliography: p205-229. - Includes index
0-521-23566-9 X.800/31941
2597128

AGRICULTURE. England
Farms. Social conditions, 1900-1930

STREET, A. G. (Arthur George)
Farmer's glory. - Oxford : Oxford University Press, 1959, c1932
292p
0-19-281384-6 X.808/38306
2935104

AGRICULTURE. England
Farms. Tenancies. Law. Reform

CONTRACTUAL relationships in farming. - London : Royal Institution of Chartered Surveyors, 1983
21p. - Bibliography: p21
B84-04123 X.622/17425
2993953

AGRICULTURE. England
Farms. Types. Distribution

GREAT BRITAIN. Ministry of Agriculture, Fisheries and Food. Statistics Division II
Type of farming maps for England and Wales. - Guildford : The Division, [1977]
[1],viii leaves,13 leaves of plates
0-904243-02-8 BS.3/560
1059505

AGRICULTURE. England
Farms. Wildlife. Habitats. Conservation. Attitudes of farmers

AGRICULTURAL DEVELOPMENT AND ADVISORY SERVICE
Wildlife conservation in semi-natural habitats on farms. - London : H.M.S.O., 1976
vii,69p : ill. - Bibliography: p.28
0-11-240539-8 SRL
0617040

AGRICULTURE. England
Geographical aspects

COPPOCK, J. T. (John Terence)
An agricultural atlas of England and Wales. - New and revised ed.. - London : Faber, 1976
3-267p : 1 ill. - Bibliography: p.256-261. - Includes index
0-571-04829-3 2020.c. Reading Room
0584819
- [Another copy.] X.320/10218 Woolwich

AGRICULTURE. England
Grants, 1940-1982

BINGHAM, S. P.
Capital grants for farmers. - [London] : [Ministry of Agriculture, Fisheries and Food], [1983]
15leaves : 1ill
B84-01289 Bs.3/785
2982595
A guide to the development of grants for agriculture and horticulture for England and Wales 1940-1982. - [London] : [Ministry of Agriculture, Fisheries and Food], [1983]
81leaves
B84-01288 Bs.3/784
2982595

AGRICULTURE. England
Implications of capital transfer tax

TAXATION of farmland. - Guildford : College of Law, c1983
v,61p. - (College of Law lectures)
B83-35071 X.200/41142
2941392

AGRICULTURE. England
Labourers. Low pay

WINYARD, Steve
Cold comfort farm. - London : Low Pay Unit, 1982
38p. - (Pamphlet / Low Pay Unit ; no.21). - Bibliography: p37-38
B83-39319 X.529/56767
2957418

AGRICULTURE. England
Labourers. Social conditions, 1850-1890

KEBBEL, T. E. (Thomas Edward)
The agricultural labourer. - New York ; London : Garland, 1984
271p. - (The World of labour)
0-8240-5714-7 X.808/41915
3285804

AGRICULTURE. England
Law

The FARMER, the landowner and the law. - [Guildford] : College of Law, 1978
ix,155p. - (Lectures / College of Law ; 1978)
B78-18112 X.100/17559
1452541

The FARMER, the landowner and the law. - [Guildford] : College of Law. - (Lectures / College of Law ; 1978 supplement)
Supplement to 1st October 1978. - [1979]
[26]p
B80-00625 X.100/18734
0287431

AGRICULTURE. England
Organic methods

BALFOUR, Lady Eve
Towards a sustainable agriculture the living soil. - [Stowmarket] : [Soil Association], [1982]
3p
B83-16273 X.322/11659
2725622

SOIL ASSOCIATION
Soil Association organic husbandry. - [Stowmarket] : [Soil Association], [1982]
11p
B83-16272 X.322/11662
2725622

VINE, Anne
Organic farming systems in England and Wales. - Aberystwyth : Department of Agricultural Economics, University College of Wales, 1981
210p : ill. - Bibliography: p209-210
B82-22443 X.329/10878 Woolwich
2725622

AGRICULTURE. England
Organic methods - Directories

WWOOF directory of organic organisations in the UK and other relevant bodies / [compiled by Susan Coppard and Chris Mager]. - Purley : Working Weekends on Organic Farms, 1984
22p
0-9509502-0-3 X.329/18578
2725622

AGRICULTURE. England
Organisations: National Farmers' Union. Policies on tenancies of agricultural holdings in England

NATIONAL FARMERS' UNION
The tenant farmers and the NFU. - [London] : National Farmers' Union, 1983
[8]p
B84-18318 X.322/13091
3074986

AGRICULTURE. England
Organisations: National Federation of Young Farmers' Clubs - Serials

The NEW young farmer. - Vol.1, no.1 (Apr. 1982)-. - Kenilworth : The Federation, 1982- : ill
B82-30484 P.313/544
2759985

AGRICULTURE. England
Organisations: National Federation of Young Farmers' Clubs, to 1970

NATIONAL FEDERATION OF YOUNG FARMERS' CLUBS
50 not out!. - [Kenilworth] : [National Federation of Young Farmers' Clubs], [1971]
[1],27p : 1 ill. - (Essentially for clubs ; 1)
B78-38499 P.521/3200
1707949

AGRICULTURE. England
Organisations: National Federation of Young Farmers' Clubs, to 1977

SHIELDS, F. E.
The Young Farmers Club movement. - [Kenilworth] : [Young Farmers Club Centre], 1978
[5],40p : ill
B79-12996 X.313/1630 Woolwich
1835637

AGRICULTURE. England
Organisations: National Federation of Young Farmers' Clubs, to 1980

FIFTY years young. - Kenilworth : National Federation of Young Farmers' Clubs, 1982
99p : ill
0-906863-01-5 X.529/49107
2746107

SHIELDS, Tanner
Youthful forward thinking. - 2nd ed. - [Kenilworth] : National Federation of Young Farmers' Clubs, [198-?]
iv,51p : ill
0-906863-00-7 X.529/48105
2746107

AGRICULTURE. England
Organisations: National Federation of Young Farmers' Clubs. Clubs. Leadership

NATIONAL FEDERATION OF YOUNG FARMERS' CLUBS
Leadership in YFC. - Kenilworth : National Federation of Young Farmers' Clubs, [1978]
[1],48p : ill. - (Essentially for clubs ; 4)
B78-38543 X.519/28740
1707957

AGRICULTURE. England
Organisations: National Federation of Young Farmers' Clubs. Clubs. Management

NATIONAL FEDERATION OF YOUNG FARMERS' CLUBS
Effective management. - [Kenilworth] : National Federation of Young Farmers' Clubs, 1972
43p : ill. - (Essentially for clubs ; 3)
B78-38553 X.519/28739
1707930

AGRICULTURE. England
Organisations: National Federation of Young Farmers' Clubs. Clubs. Officers. Duties

NATIONAL FEDERATION OF YOUNG FARMERS' CLUBS
Club officers handbook. - Kenilworth : The Federation, [1978]
11p : ill
B78-38554 X.519/28738
1707914

AGRICULTURE. England
Organisations: National Federation of Young Farmers' Clubs. Clubs. Programmes. Design

NATIONAL FEDERATION OF YOUNG FARMERS' CLUBS
Club programming. - Kenilworth : National Federation of Young Farmers' Clubs, 1971
31p : ill. - (Essentially for clubs ; 2). - Includes bibliographies
B78-38555 P.521/3199
1707922

AGRICULTURE. England
Personnel. Employment, 1960-1980

WAGES and employment in agriculture / by P.J. Lund ... [et al.]. - [London] : Ministry of Agriculture, Fisheries and Food, 1982
60p : ill. - (Government Economic Service working paper ; no.52). - Bibliography: p37
B83-04527 BS.44/505(52)
2819686

AGRICULTURE. England
Personnel: Harvesters. Social conditions, 1840-1900

MORGAN, David Hoseason
Harvesters and harvesting 1840-1900. - London : Croom Helm, c1982
224p : ill. - Bibliography: p206-220. - Includes index
0-7099-1735-x X.329/15185 Woolwich
2713187

AGRICULTURE. England
Personnel. Proficiency. Testing, 1938-1980

SHEPPY, Phillip C.
Proficiency testing for agriculture and horticulture in England and Wales. - [Kenilworth] : [National Proficiency Tests Council for Agriculture and Horticulture], 1980
58p : ill
B81-12531 X.325/959 Woolwich
2379414

AGRICULTURE. England
Personnel. Remuneration

BROWN, Marie
Low pay on the farm. - London : Low Pay Unit, 1975
[2],39p. - (Low pay pamphlet ; no.3)
0-905211-00-6 X.319/17143 Woolwich
0589691

REPORT on wages in agriculture / Ministry of Agriculture, Fisheries and Food. - London : H.M.S.O..
1 Jan.-31 Dec. 1976-. - 1977-
r6050952 BS.3/616
0589691

AGRICULTURE. England
Personnel. Rented residences. Tenancies. Law. Rent (Agriculture) Act 1976

ROSSI, Hugh
Shaw's guide to the Rent (Agriculture) Act 1976. - London : Shaw and Sons, 1977
xxi,268p. - Includes index
0-7219-0750-4 X.100/17264
1296086

AGRICULTURE. England
Protection. Policies of government. Influence of farmers, 1815-1852. Political aspects

CROSBY, Travis L.
English farmers and the politics of protection, 1815-1852. - Hassocks : Harvester Press, 1977
[8],224p. - Bibliography: p.191-198. - Includes index
0-85527-116-7 X.329/11105 Woolwich
0766011

AGRICULTURE. England
Read, Clare Sewell, 1865-1905

FISHER, John Richard
Clare Sewell Read, 1826-1905. - [Hull] : University of Hull, 1975
vi,39p. - (Occasional papers in economic and social history / University of Hull ; no.8). - Bibliography: p.36-37. - Includes index
0-900480-38-6 X.0529/203(8)
0687995

AGRICULTURE. England
Research organisations: Agricultural Development and Advisory Service

AGRICULTURAL DEVELOPMENT AND ADVISORY SERVICE
Annual report. - 1977. - London : H.M.S.O, 1978
[4],101p,[8]p of plates : ill(some col)
Subsequent volumes, when received, will be found at SRL
0-11-240302-6 SRL
0728667

AGRICULTURE. England
Research organisations: Agricultural Development and Advisory Service. Agriculture Service

AGRICULTURE service annual report / Agricultural Development and Advisory Service. - 1975-76. - London : H.M.S.O, 1977
117p,[4]p of plates : ill
Later issues are placed at this shelfmark
0-11-241132-0 SRL
210766x
- [Another copy]

AGRICULTURE. England
Research organisations: Agricultural Development and Advisory Service. Science Arm

AGRICULTURAL DEVELOPMENT AND ADVISORY SERVICE. Science Service
Agricultural Development and Advisory Service Science Service annual report. - 1975. - London : H.M.S.O, 1976
ii-xxi,308p,iv p of plates : ill
Subsequent volumes, when received, will be placed at this shelfmark
0-11-241094-4 BS.3/599
0093645

REGIONAL AGRICULTURAL SCIENCE SERVICE
Regional Agricultural Science Service annual report. - London : H.M.S.O.
1976. - [1977]
xviii,238p,iv p of plates : ill. - Includes bibliographies
Subsequent volumes, when received, will be placed at this shelfmark
0-11-241124-x SRL
0093645

AGRICULTURE. England
Research organisations: Agricultural Development and Advisory Service. Policies

POLICY review of the Agricultural Development and Advisory Service. - Pinner : Ministry of Agriculture, Fisheries and Food, 1979
[3],35p
B80-31846 BS.3/660
2235366

AGRICULTURE. England
Smallholdings, 1200-1500

The PEASANT land market in medieval England / edited by P.D.A. Harvey. - Oxford : Clarendon, 1984
xvi,375p. - Includes index
0-19-822661-6 X.800/40644
3263010

AGRICULTURE. England
Street, A. G. - Biographies

STREET, Pamela
My Father, A.G. Street. - 2nd ed. - London : Hale, 1984
110p,[12]p of plates : ill
0-7090-1441-4 X.958/25882
3195570

My father, A.G. Street. - Large print ed. - Bath : Chivers, 1985, c1984
xvi,112p. - (A Lythway book)
0-7451-0148-8 Nov.54198
3157636

AGRICULTURE. England
Women personnel, to date

KITTERINGHAM, Jennie
Country girls in 19th century England. - Oxford : History Workshop, Ruskin College, [1973]
75p. - (History Workshop pamphlets ; no.11)
r6801180 X.0802/1619(11)
890295x

AGRICULTURE. England
Working conditions, 1500-1900

FUSSELL, G. E. (George Edwin)
The English countryman. - London : Orbis Publishing, 1981
172p,[64]p of plates : ill. - Bibliography: p157-163. - Includes index
0-85613-335-3 X.520/24259
2504723

AGRICULTURE. Enstone. Oxfordshire
ca 1930-1943

HENDERSON, George, b.1904
The farming ladder. - [2nd ed. revised and reset]. - London : Faber, 1978
246[i.e.253]p,plate : ill. - Includes index
0-571-11059-2 X.319/18567 Woolwich
1479776

AGRICULTURE. Eritrea. Ethiopia
Policies of EPLF

BARNETT, Tony, 19—
Agriculture in the Eritean revolution. - Norwich : School of Development Studies, University of East Anglia, 1983
25p. - (Discussion paper ; no.55). - Bibliography: p25
B85-03258 X.322/14308
3213064

AGRICULTURE. Estates. Somogy (megye). Hungary

TÓTH, Tibor
Nagybirtoktól a nagyüzemig. - Budapest : Közgazdasági és Jogi Könyvkiadó, 1977
373p
963-220-412-3 X.320/5415 Woolwich
247221X

AGRICULTURE. Estonia

KAHK, J. (Juhan)
K voprosu o tipologii krest'i͡anskikh khozi͡aĭstv v Estonii v nachale XIX veka. - Tallin : Akademii͡a nauk Ėstonskoĭ SSR, Institut istorii, 1975
78p. - (Talurahva töö-ja võitlusradadelt ; 6)
r6036121 X.519/22802
115091X

Ob ėkonomicheskom potent͡siale krest'i͡anskogo dvora v Ėstonii v nachale XIX veka. - Tallin : Akademii͡a nauk Ėstonskoĭ SSR, Institut istorii, 1975
78p. - (Talurahva töö-ja võitlusradadelt ; 5)
r6036117 X.519/22801
115091X

TÕNURIST, Edgar
Agriculture in Soviet Estonia. - Tallinn : Perioodika, 1978 [i.e. 1979]
147p,[52]p of plates : col.ill
r6461006 X.329/13459 Woolwich
115091X

AGRICULTURE. Estonia
1561-1918

AKADEMIE DER WISSENSCHAFTEN DER DDR. Institut für Wirtschaftsgeschichte
Bäuerliche Wirtschaft und landwirtschaftliche Produktion in Deutschland und Estland (16. bis 19. Jahrhundert). - Berlin : Akademie-Verlag, 1982
360p. - (Jahrbuch für Wirtschaftsgeschichte. Sonderband ; 1981)
r6309298 X.520/33554
8710333

BÄUERLICHE Wirtschaft und landwirtschaftliche Produktion in Deutschland und Estland (16. bis 19. Jahrhundert). - Berlin : Akademie-Verlag, 1982
360p. : ill. - (Jahrbuch für Wirtschaftsgeschichte. Sonderband ; 1981)
r6308274 Ac.855/49a(4)
8710333

FEOKTISTOVA, L. Kh. (Livii͡a Khugovna)
Zemledelie u ėstont͡sev. - Moskva : Nauka, 1980
189p : ill
r6497570 X.329/13962 Woolwich
8710333

AGRICULTURE. Estonia
1918-

RUUSMANN, A.
Nachalo sot͡sialisticheskikh preobrazovaniĭ v ėstonskoĭ derevne. - Tallin : Ėesti raamat, 1982
230p
r6463191 X.329/16082
8891036

AGRICULTURE. Estonia
1945-

PROBLEMY ėkonomicheskogo razvitii͡a sel'skogo khozi͡aĭstva Ėstonskoĭ SSR. - Tallin : Valgus, 1980
169p
r6463513 X.329/16179
8139237

AGRICULTURE. Estonia
Collectivisation

EESTI NSV põllumajanduse kollektiviseerimine / [toimetuse kolleegium ... E. Tõnurist (peatoimetaja) ... et al.]. - Tallinn : Eesti Raamat, 1978
664p
r6020107 S.C.4/4
2475588

AGRICULTURE. Estonia
Farms, 1721-1940

KÄBIN, Abel
Vaim ja muld. - Toronto : [Estoprint], 1975
189p.
r6461512 X.529/45867
8799423

AGRICULTURE. Estonia
Policies of Eesti Töörahva Kommuun

TIDO, Genrikh
Agrarnaĭa politika Bol'shevikov v Estonii. - Tallin : Eesti raamat, 1978
157p
r6098212 X.619/10980 Woolwich
1865293

AGRICULTURE. Estonia
Research organisations: Eesti Põllumajanduse Akadeemia

170 aastat kõrgemat põllumajanduslikku haridust Eestis. - Tallinn : Valgus.
Vol.2: [Eesti Põllumajanduse Akadeemia, 1951-1976]. - 1978
221p : ill
r6461520 X.329/15329 Woolwich
8856796

AGRICULTURE. Ethiopia
to 1974. Political aspects

COHEN, John M.
Land and peasants in imperial Ethiopia. - Assen : Van Gorcum, 1975
x, 115 p. : 1 ill
90-232-1291-6 X.311/8275 Woolwich
0737488

AGRICULTURE. Ethiopia
Development

GREEN, David Alfred George
Ethiopia, an economic analysis of technological change in four agricultural production systems. - [East Lansing] : [African Studies Center, Michigan State University] [etc.], 1974
xv,117p : ill. - (Committee on Ethiopian Studies occasional papers series. monograph ; no. 2). - Bibliography : p.113-117
r6073442 X.519/29268
1814591

AGRICULTURE. Europe

CLAY, Roger C.
Farming and food. - Harlow : Common Ground, 1978
1 filmstrip(36 single fr.,9title fr.)
33p.. - (Europe today). - Includes bibliographies
0-7056-2214-2 Cup.580.ff.25
2008483

AGRICULTURE. Europe
to 1880 - Conference proceedings

EUROPEAN peasants and their markets / edited by William N. Parker and Eric L. Jones. - Princeton ; London : Princeton University Press, 1975
viii,367p : ill. - Bibliography: p.67-71. - Includes index
0-691-05230-1 X.320/10255 Woolwich
0611999

AGRICULTURE. Europe
to date

STORIA dell'agricoltura europea. - Milano : Etaslibri, 1980
398p : ill. - Includes bibliography
r6507925 L.45/2132
8140421

AGRICULTURE. Europe
to date - Serials

AGRARPOLITISCHER Jahresbericht. - 1977-. - Bonn : Bundesministerium für Ernährung, Landwirtschaft und Forsten, 1978-. - (Schriftenreihe des Bundesministers für Ernährung, Landwirtschaft und Forsten. Reihe C, Agrarpolitische Berichte der Organisation für Wirtschaftliche Zusammenarbeit und Entwicklung (OECD))
Holdings: 1978 (1979)-
r6301795 S.F.583/65[Hft.2,etc]
8140421

AGRICULTURE. Europe
ca 1100-1982

WHITLOCK, Ralph
Farming in history. - Hove : Wayland, 1983
63p : col.ill. - (Picture history)
0-85078-357-7 X.809/57880
3047180

AGRICULTURE. Europe
1200-1978

ABEL, Wilhelm
Agricultural fluctuations in Europe. - London : Methuen, 1980
xii,363p : ill. - Bibliography: p329-356. - Includes index
0-416-72110-9 X.322/9687 Woolwich
1999443

AGRICULTURE. Europe
1975-1980. Geographical factors - Bibliographies

PROBLEMS and policies of agriculture in the less-favoured area of Europe (1975-1980). - Slough : Commonwealth Agricultural Bureaux, c1981
iii,111p : ill. - (Commonwealth Agricultural Bureaux annotated bibliographies. General series ; annotated bibliography no.R45)
B82-27996 SRL
275195x

AGRICULTURE. Europe
Applications of solar energy - Conference proceedings

SOLAR energy in agriculture / editor J. Coombs. - London : UK Section of the International Solar Energy Society, 1983
i,89p : ill. - (Conference proceedings ; C33). - Includes bibliographies
0-904963-34-9 SRL
3036669

AGRICULTURE. Europe
Economic development

EVANS, J. M. (John M)
Report on FAO/ECE Working Party on Agrarian Structure and Farm Rationalisation - 6th Session. - [Pinner] : ADAS, c1984
17p
B85-09128 BS.3/946
3230147

AGRICULTURE. Europe
Economic development related to population growth, 1300-1911

GRIGG, David, 1934-
Population growth and agrarian change. - Cambridge : Cambridge University Press, 1980
xi,340p : ill. - (Cambridge geographical studies ; 13). - Includes index
0-521-22760-7 X.520/23238
2363143

AGRICULTURE. Europe
Family farms - Conference proceedings

FUTURE OF THE FAMILY FARM (Conference : 1974 : Exeter)
The future of the family farm in Europe. - [Ashford] : [The Centre], [1977?]
112p. - (Seminar papers / Wye College. Centre for European Agricultural Studies)
0-901859-51-6 X.311/8678 Woolwich
1349678

AGRICULTURE. Europe
Policies of governments, 1945-1975

L'ÉTAT et la politique agraire en Europe. - Paris : Institut de Sciences Mathématiques et Économiques Appliquées, 1976
1343-1652p.. - (Économies et sociétés ; tom.10.no.7-8) (Cahiers de l'I.S.M.E.A. ; série AG, no.14)
2-85898-015-2 P.521/500d.[tom.10.no.7-8]
1124501

AGRICULTURE. Europe
Prehistoric agriculture

BARKER, Graeme
Prehistoric farming in Europe. - Cambridge : Cambridge University Press, 1985
[327]p : ill. - (New studies in archaeology). - Includes bibliography and index
0-521-22810-7 X.421/26842
2843692

EARLY European agriculture / being the third volume of papers in economic prehistory by members and associates of the British Academy Major Research Project in the Early History of Agriculture ; written in honour of Eric Higgs ; edited by M.R. Jarman, G.N. Bailey and H.N. Jarman. - Cambridge : Cambridge University Press, 1982
x,283p : ill. - Bibliography: p259-271. - Includes index
0-521-24359-9 X.322/11507
2843692

AGRICULTURE. Europe
Serials, 1900- - Catalogues - Serials

UNIVERSITÄT GIESSEN. Zentrum für kontinentale Agrar- und Wirtschaftsforschung. Bibliothek
ZZZ. - Stand: 1.12.1982-. - Giessen : Bibliothek des Zentrums, [1983?]-
r6306499 X.0329/242
8891052

AGRICULTURE. Europe
Smallholdings - Conference proceedings

SMALL farming and the rural community / edited by B.J. Marshall and R.B. Tranter. - Reading : Centre for Agricultural Strategy, University of Reading, 1982
48p. - (CAS paper ; 11). - Bibliography: p27
B82-29620 X.529/48521
2756625

AGRICULTURE. Europe
Social aspects, to ca 1970

HUGGETT, Frank E. (Frank Edward)
The land question and European society. - London : Thames and Hudson, 1975
179p : ill. - ([Library of European civilization]). - Bibliography: p.167-171. - Includes index
0-500-32034-9 X.329/10074
0370320

AGRICULTURE. European Community countries

ZARDI, Alfonso
L'agricoltura tra Comunità economica europea, Stato e Regioni. - Udine : Camera di commercio, industria, artigianato e agricoltura, 1976
211p. - (Collana "Tesi di laurea" ; n.8). - Bibliography : p.205-211
r6042565 S.H.420/18
1323881

AGRICULTURE. European Community countries
- Statistics

STATISTICAL OFFICE OF THE EUROPEAN COMMUNITIES
Statistics of animal production : livestock, breeding herd size, meat production, civil year balance sheets, 1965-1975. - Luxembourg : Office for Official Publications of the European Communities, [1977]
xxii,82p
r6077713 UN.P.131/248
0233404

AGRICULTURE. European Community countries
- Statistics - Serials

AGRICULTURAL accounts : agricultural and forestry accounts, unit values / [issued by] Statistical Office of the European Communities. - 1976-. - Luxembourg : Office for Official Publications of the European Communities, [1977]-
r6077716 UN.P.131/246
0233404

AGRICULTURAL price statistics / [issued by] Statistical Office of the European Communities. - 1969-75-. - Luxembourg : Office for Official Publications of the European Communities, 1976-
r6055393 UN.P.131/263
0233404

EARNINGS in agriculture / [issued by] Statistical Office of the European Communities. - [1974]-. - Luxembourg : Office for Official Publications of the European Communities, [1976?]-
Holdings : Earnings in agriculture, 1975-
r6077706 UN.P.131/290
0233404

LAND use and production / [issued by] Statistical Office of the European Communities. - 1976-. - Luxembourg : Office for Official Publications of the European Communities, 1977-
r6055392 UN.P.131/264
0233404

MILK and milk products / [issued by] Statistical Office of the European Communities. - 1977-. - Luxenbourg : Office for Official Publications of the European Communities, 1977-
r6079207 UN.P.131/326
0233404

AGRICULTURE. European Community countries
to date - Statistics

STATISTICAL OFFICE OF THE EUROPEAN COMMUNITIES. Agricultural Statistics Committee
EC surveys on the structure of agricultural holdings. - [Luxembourg] : [The Office], 1981
[116]p
r6208654 UNP131/554
890443x

AGRICULTURE. European Community countries
Agricultural credit

COMMISSION OF THE EUROPEAN COMMUNITIES
Credit to agriculture in the E.E.C. member states. - Luxembourg : Office for Official Publications of the European Communities ; [London] : [H.M.S.O.], 1977
[2],104p. - (Information on agriculture ; no.28)
B78-35902 SRL
1695746

AGRICULTURE. European Community countries
Agricultural credit–continued
CREDIT to agriculture. – Luxembourg : Office for Official Publications of the European Communities ; [London] : [H.M.S.O.].
4: The Netherlands. – 1976
[4],116p. – (Information on agriculture ; no.4)
B79–02434 UNP.131/181
1695746

AGRICULTURE. European Community countries
Effects of enlargement of European Economic Community
The AGRICULTURAL implications of EEC enlargement. – Tunbridge Wells : Agra Europe (London) Ltd. – ('Agra Europe' special report ; no.3)
Part 1: Greece. – 1979
[7],124p
B79–33122 X.0512/585(3)
1956620

AGRICULTURE. European Community countries
Effects of political development in the European Economic Community
POLITICAL change in the European Community. – Tunbridge Wells : Agra Europe (London) Ltd, 1978
[1],v,96p : ill. – ('Agra Europe' special report ; no. 2)
B80–33130 X.0512/585(2)
2238187

AGRICULTURE. European Community countries
Farms. Accounting systems
KERR, Hugh William Thomas
Farm costing in the European Economic Community. – [Loughborough] : University of Nottingham, Department of Agriculture and Horticulture, 1974
[3],33,[7]leaves(2fold) : ill
0–900573–39–2 X.313/1273 Woolwich
0621137

AGRICULTURE. European Community countries
Finance – Conference proceedings
FARM financing and farm indebtedness in the Community / translated and edited by Claude Rosenfeld, Bridget Girling, Ian G. Reid. – Ashford, Kent : Wye College, c1980
iv,169p : ill. – (Seminar paper / Centre for European Agricultural Studies (Agricultural Finance Study Group) ; n.9)
0–905378–35–0 X.325/830 Woolwich
2360187

Le FINANCEMENT de l'agriculture et l'endettement dans les pays membres de la communauté / traduction et présentation par Claude Rosenfeld, Bridget Girling, Ian G. Reid. – Ashford : Wye College (Université de Londres), 1981
164p
B83–12913 X.322/11583
2360187

AGRICULTURE. European Community countries
Financial assistance
SCHEMES of assistance to farmers in less favoured areas of the EEC. – Langholm : Arkleton Trust, c1982
68p. – Includes bibliographies
B82–37959 X.322/11251
278503x

AGRICULTURE. European Community countries
Non-governmental organisations – Directories
COMMISSION OF THE EUROPEAN COMMUNITIES. Directorate General Agriculture
Directory of non-governmental agricultural organizations set up at Community level. – Luxembourg : Office for Official Publications of the European Communities, 1974
854 p.
r6000227 UN.P.130/491
0906395

AGRICULTURE. European Community countries
Organisations, to date
BURKHARDT-REICH, Barbara
Agrarverbände in der EG. – Kehl am Rhein : Engel, 1983
xx,410p. – (Schriftenreihe Europa-Forschung ; Bd.5). – Includes bibliography
3–88357–023–0 X.0809/2199(5)
8965080

AGRICULTURE. European Community countries
Organisations, to date – Directories
VERZEICHNIS der land- und ernährungswirtschaftlichen Verbände / Kommission der Europäischen Gemeinschaften = Directory of non-governmental agricultural organizations / Répertoire des organisations agricoles et alimentaires non gouvernementales / Commission des Communautés européennes. – 6 Ausg.. – München ; London : Saur, 1980
402p. – Includes index
3–598–10147–3 2725.c.133
8965080

AGRICULTURE. European Community countries
Part-time agriculture
ALDEN, Jeremy
Dual activity in the agriculture sector of the European Community. – Cardiff : Department of Town Planning, University of Wales Institute of Science and Technology, 1982
xviii,106p,[3]p of plates : ill. – (Papers in planning research ; 52)
B84–09388 X.805/4775
3040321

AGRICULTURE. European Community countries
Policies of European Community, to date
Die AGRARWIRTSCHAFTLICHE Integration Europas / Leitung der FAG Landwirtschaft: Hermann Priebe : Berichterstatter: Hans-Eberhard Buchholz ... [et al.]. – Baden-Baden : Nomos, 1979
262p. – (Möglichkeiten und Grenzen einer Europäischen Union ; Bd 6)
3–7890–0496–0 X.329/17670
8904340

CLERC, François
Le Marché commun agricole. – Sixième édition refondue. – Paris : Presses Universitaires de France, 1981
127p. – (Que sais-je? ; 1115)
2–13–037088–8 X.329/17691
8904340

MARIADÈS, Stauros
Hē koinē agrotikē politikē tēs E.O.K.. – Thessalonikē : Paratērētēs, 1981
77p. – (Hellada kai E.O.K.)
r6503624 X.329/16972
8904340

MARSH, John S. (John Stanley)
[Agriculture and the European Community. Italian]. La politica agricola comunitaria. – Roma : Laterza, 1982
xix,193p. – (Libri del tempo ; 186). – Includes bibliography
r6509969 X.529/57904
8904340

TØRSLEV, Niels
Dansk landbrug i EF. – [Copenhagen] : Dansk Udenrigspolitisk Institut : Samfundsvidenskabelight Forlag, 1981
98p. – (DUPI hæfte ; 1981/4)
r6610006 X.322/11257
8904340

AGRICULTURE. European Community countries
Policies of European Economic Community
EUROPE and our food. – Brussels ; London : Commission of the European Communities, Directorate General for Press and Information, 1978
[8]p : 1 ill
B79–06325 UNP.131/382
015346x

MARSH, John S. (John Stanley)
European agriculture in an uncertain world. – Paris : Atlantic Institute for International Affairs, 1975
69p. – (The Atlantic papers ; 1975. 1)
r6073586 X.0709/1009.(24.)
0112151

AGRICULTURE. European Community countries
Policies of European Economic Community, to date
Il MERCATO comune agricolo / di C. Bernini Carri ... [et al.]. – Firenze : La nuova Italia, 1979
128p. – (Europa ; 2)
r6502307 X.629/17061 Woolwich
8833176

MORVANT, Michel
Le Parlement européen et le financement de la politique agricole commune. – Rennes : Université de Rennes I, 1980
98p
r6208594 X.322/11382
8833176

AGRICULTURE. European Community countries
Policies of European Economic Community, to date – Conference proceedings
AGRICULTURAL data and economic analysis / A. Dubgaard ... [et al.], editors. – Maastricht : European Institute of Public Administration, 1984
xii,266p
90–677–9006–0 S.B.286/118.
8833176

AGRICULTURE. European Community countries
Policies of European Economic Community: Common Agricultural Policy
AFANAS'EV, L.
Contradictions of agrarian integration in the Common Market. – Moscow : Progress Publishers, 1976
178 p.
r6010142 X.319/16996 Woolwich
0779148

BOWLER, I. R. (Ian Robert)
Agriculture under the Common Agricultural Policy. – Manchester : Manchester University Press, c1985
xii,255p : ill. – Includes index
0–7190–1095–0 X.322/15090
1731483

C.A.P. monitor / Agra Europe. – Tunbridge Wells : Agra Europe (London), c1983
1v.(looseleaf) : ill
B84–21587 L.45/2639
1731483

COMMISSION OF THE EUROPEAN COMMUNITIES
The common agricultural policy. – Revised [ed.]. – Brussels : Commission of the European Communities, Directorate General for Press and Information : London : The Commission Information Office, [1977]
32,[3]p
B79–02463 UNP.131/331
1731483

CONSTANTINIDES-MEGRET, C.
La politique agricole commune en question. – Paris : Éditions A. Pedone, 1982
253p
2–233–00097–8 X.322/12209
8976996

The COSTS of the common agricultural policy / Allan E. Buckwell ... [et al.]. – London : Croom Helm, c1982
184p : ill. – Includes index
0–7099–0671–4 X.329/15325 Woolwich
1731483

FENNELL, Rosemary
The Common Agricultural Policy of the European Community – London : Granada, 1979
xi,243,[1]p : ill. – Bibliography: p.229–235. – Includes index
0–246–11197–6 X.329/12106 Woolwich
1731483

HARRIS, Simon
The food and farm policies of the European Community. – Chichester : Wiley, c1983
xii,354p : ill. – Includes bibliographies and index
0–471–10261–x X.520/32826
1731483

HILL, Brian E.
The Common Agricultural Policy. – London : Methuen, 1984
xvi,168p : ill. – (The Methuen EEC series). – Bibliography: p160–163. – Includes index
0–416–32180–1 X.329/18260
1759450
– [Another copy : paperback] ISBN 0–416–32190–9 X.329/18261

KOESTER, U.
The CAP and our food. – Ashford : Wye College, 1978
15,[3]leaves : 3 ill. – (Occasional papers / Wye College, Centre for European Agricultural Studies ; no.7). – Bibliography(1p.)
0–905378–14–8 X.0313/51(7) Woolwich
1731483

MANSHOLT, Sicco
The Common Agricultural Policy. – Stowmarket : Soil Association, 1979
18p
B83–16006 X.950/19337
1731483

MARSH, John S. (John Stanley)
Agriculture and the European Community. – London : Allen & Unwin, 1980
98p. – (Studies on contemporary Europe ; 2). – Bibliography: p91–96. – Includes index
0–04–338092–1 X.329/19067
1731483

O'CONNOR, R. (Robert)
A review of the Common Agricultural Policy and the implications of modified systems for Ireland. – Dublin : Economic and Social Research Institute, 1983
xxi,151p : ill. – (Broadsheet / Economic and Social Research Institute ; no. 21). – Bibliography: p141–145
0–7070–0060–2 X.0319/107(21)
1731483

PAGGI, Guido
Europa verde. – Bologna : Edagricole, 1976
xii,147p
r6085330 X.311/4996 Woolwich
0246700

PETERSON, Martin
International interest organizations and the transmutation of postwar society. – Stockholm : Almqvist & Wiksell International, 1979
xx,720p. – (Interest organizations and internationalism Report ; 2) (Bulletin of the Department of History, University of Gothenburg ; no.17). – Includes bibliographies
91–220–0255–3 X.0709/587(18)
1759450

AGRICULTURE. European Community countries
Policies of European Economic Community. Common Agricultural Policy-continued

ROLLO, J. M. C.
The CAP and resource flows among EEC member states. - London : Ministry of Agriculture, Fisheries and Food, 1979
[2],19p : ill. - (Government Economic Service working paper ; no. 27). - Bibliography: p.16
B80-37790 BS.44/505(27)
1731483

SCOTT-HOPKINS, Jim
Food for thought. - London : European Conservative Group, [1978]
51p
B78-24064 X.319/18189 Woolwich
0246700

AGRICULTURE. European Community countries
Policies of European Economic Community: Common Agricultural Policy - Serials

AGRA-BRIEFING. - No.1 (Feb. 1984)-. - Tunbridge Wells : Agra Europe (London), 1984-
v. : ill
B85-14364 P.803/1546
1731483

AGRICULTURE. European Community countries
Policies of European Economic Community: Common Agricultural Policy. Effects on agricultural industries in Great Britain, 1973-1976

FRANKLIN, M. D. M.
Four years' experience of the common agricultural policy in a changed world. - [Belfast] : [Queen's University of Belfast], [1977]
18p. - (The George Scott Robertson memorial lecture ; 26th)
0-85389-118-4 WP.A.219/18
1299867

AGRICULTURE. European Community countries
Policies of European Economic Community: Common Agricultural Policy. Effects on agricultural industries in Italy, 1956-

GALLI, Rosemary
La partecipazione italiana alla politica agricola comunitaria. - Roma : Instituto affari internazionali, 1976
260p. - (Collana dello Spettatore Internazionale ; 37)
r6085328 X.519/40857
1981110

AGRICULTURE. European Community countries
Policies of European Economic Community: Common Agricultural Policy. Effects on cereals trades

DEBATISSE, Michel Louis
EEC organisation of the cereals markets. - Ashford : Centre for European Agricultural Studies, Wye College, c1981
43,xxxviip : ill. - (Occasional paper / Wye College. Centre for European Agricultural Studies ; no.10)
0-905378-27-x X.325/1209 Woolwich
2485494

AGRICULTURE. European Community countries
Policies of European Economic Community. Common Agricultural Policy. Implications for oilseeds industries of European Community countries

PARRIS, Kevin
EEC oilseed products sector and the common agricultural policy. - Ashford, Kent : Wye College, Centre for European Agricultural Studies, 1978
[1],vi,103[i.e.109]p. - (Occasional papers / Wye College. Centre for European Agricultural Studies ; no.4). - Bibliography(6p.)
0-905378-04-0 X.0313/51(4)
1481657

AGRICULTURE. European Community countries
Policies of European Economic Community: Common Agricultural Policy. Reform - Forecasts

JOSLING, T. E. (Timothy Edward)
Options for farm policy in the European Community. - London : Trade Policy Research Centre, 1981
xii,84p. - (Thames essay ; no.27)
0-900842-52-0 X.329/14849 Woolwich
2396866

MARSH, John S. (John Stanley)
The CAP in the 1980s. - Ashford : Wye College, 1981
ii,66p. - (Occasional paper / Centre for European Agricultural Studies ; no.13)
0-905378-32-6 X.0313/51.(13) Woolwich
2396866

AGRICULTURE. European Community countries
Policies of European Economic Community. Common Agricultural Policy. Reform - Forecasts - Conference proceedings

REFORM of the Common Agricultural Policy and re-structuring of the EEC budget / organised by Geoffrey Denton. - London : UACES Secretariat, c1983
i,116p. - Includes bibliographies
0-906384-09-5 X.529/56251
2396866

AGRICULTURE. European Community countries
Research projects - Directories - Serials

AGREP / Commission of the European Communities. Vol.1, Main list, research projects. - 1980. - The Hague ; London : Nijhoff, 1980
x,841p
90-247-2466-x X.529/41856
2497484

AGRICULTURE. European Community countries
Research projects - Indexes - Serials

AGREP / Commission of the European Communities. Vol.2, Indexes. - 1980. - The Hague ; London : Nijhoff, 1980
ix,231p
90-247-2467-8 X.529/41855
2497484

AGRICULTURE. European Community countries
Statistics issued by Statistical Office of the European Communities. Agricultural Division. Use - Conference proceedings

EEC agricultural statistics. - Ashford : The College, 1977
[5],37p : ill
0-905378-07-5 X.313/1571 Woolwich
1524577

AGRICULTURE. European Soviet Union
Productivity, 1945-

IVANOV, Valentin Aleksandrovich
Ėffektivnost' intensifikat͡sii proizvodstva v sovkhozakh Evropeĭskogo Severa. - Moskva : Nauka, 1974
164p
r6462779 X.329/17480
9053425

AGRICULTURE. Ewyas Harold. Hereford and Worcester
Farms, 1750-1980

FIELD and furrow / by Ewyas Harold and District W.E.A. Research Group. - [Hereford] : West Midlands W.E.A., [1983?]
41p,[3]folded leaves of plates : ill
B84-24111 X.322/13415
3121976

AGRICULTURE. Far Eastern Territory. Russia (RSFSR)

NESTERENKO, A. D
Ėkonomicheskie problemy sel'skogo khozi͡aĭstva Dal'nego Vostoka. - Vladivostok : Dal'nevostochnoe knizhnoe izdatel'stvo, 1972
370p. - Bibl.: p.357-370
r6178298 X.329/12574 Woolwich
2446138

AGRICULTURE. Far Eastern Territory. Russia (RSFSR)
Collectivisation, 1922-1941

IZ istorii kollektivizat͡sii sel'skogo khozi͡aĭstva Dal'nego Vostoka (1927-1937gg.) / pod redakt͡sieĭ F.F. Tatalina (otvetstvennyĭ redaktor) ... [i dr.]. - Khabarovsk, 1979
237p. - (Istorii͡a kollevtivizat͡sii sel'skogo khozi͡aĭstva SSSR)
r6400024 X.325/939 Woolwich
8783896

AGRICULTURE. Färnäs. Sweden
Farms. Buildings. Architecture

INGELSTRÖM, Ann-Christine
Gammal jordbruksbebyggelse i byn Färnäs. - Stockholm : Byggforskningen, 1977
156p : ill
91-540-2748-9 P.703/360[1977.no.57]
153582X

AGRICULTURE. Farnborough. Hampshire
Smith, John, d.1915 & Smith, Ann, b.1831 - Biographies

STURT, George
A farmer's life. - Firle : Caliban Books, 1979
208p : ill
0-904573-15-x X.809/44258
1876260

AGRICULTURE. Fiji
Indian smallholdings, to 1973

ANDERSON, Allan Grant
Indo-Fijian smallfarming. - [Auckland] : Auckland University Press ; [London] : Oxford University Press, 1974 [i.e. 1975]
xii,199p : ill. - Bibliography: p.191-194. - Includes index
0-19-647805-7 X.320/10337 Woolwich
064224x

AGRICULTURE. Finland

KETTUNEN, Lauri
Finnish agriculture in 1978. - Helsinki : Agricultural Economics Research Institute, 1979
22p. - (Maatalouden taloudellisen tutkimuslaitoksen. Tiedonantoja = The Agricultural Economics Research Institute, Finland. Research reports ; no.55a)
951-91994-8-9 X.325/716 Woolwich
2471809

AGRICULTURE. Finland
1720-1870

SOININEN, Arvo M.
Vanha maataloutemme. - Helsinki : Suomen Historiallinen Seura, 1975
459p.. - (Historiallisia tutkimuksia ; 96). - Includes bibliographies
951-92540-4-8 Ac.7814/8[no.96]
0878138

AGRICULTURE. Finland
ca 1880-ca 1910

ANTTILA, Veikko
Talonpojasta tuottajaksi. - Helsinki : Kirjayhtymä, 1974
212 p. : ill. - Includes bibliographies
951-260-856-1 X.319/8643 Woolwich
1026666

AGRICULTURE. Finland
1900-1975

LAUKKONEN, Ilmari
Alkion miehet. - Vaasa : Vaasa Oy, [1976]
242p : ill
951-99083-0-7 X.329/8949 Woolwich
2099829

AGRICULTURE. Finland
Economic aspects

JOHANSSON, Eira
Utvecklingslinjer och problemer i finländskt jordbruk. - [Helsingfors] : [Helsingfors universitet, Institutionen för politisk historia], 1977
93p. - (Helsingin yliopisto. Poliittisen historian laitos. Julkaisuja = Research reports ; 2/1977)
951-451-215-4 X.0519/819(2/77)
2249766

AGRICULTURE. Finland
Farms, to date

MÄKELÄ, Anneli
Hattulan kihlakunnan ja Porvoon läänin autioituminen myöhäiskeskiajalla ja uuden ajan alussa. - Helsinki : Suomen historiallinen seura, 1979
276p. - (Historiallisia tutkimuksia ; 109). - Bibliography
951-92542-1-8 Ac.7814/8[no.109]
889647x

AGRICULTURE. Finland
Farms, to date - Statistics - Serials

MAATILAREKISTERI = Farm register / Maatilahallitus. - 1 (1980)-. - Helsinki : Maatilahallitus, 1983-. - (Soumen virallinen tilasto. 43)
r6612014 S.D.49/100
889647x

MAATILATALOUS = Farm economy. - 1 (1973)-. - Helsinki : Tilastokeskus, 1977-. - (Suomen virallinen tilasto. 39, Maatilatalous)
r6600015 S.D.49/96
889647x

AGRICULTURE. Finland
Ihamuotila, Veikko

IHAMUOTILA, Veikko
Veikko Ihamuotila. - Porvoo : W. Söderström, 1979
62p. - (Itse asiassa kuultuna)
951-00-9431-5 X.329/16804
899370x

AGRICULTURE. Finland
Looperation, to date

KÖPPÄ, Tapani
Viljelijäperhe, yhteisö ja yhteistoiminta. - Helsinki : [Pellervo], 1979
188p. - (Pellervon taloudellisen tutkimuslaitoksen julkaisuja ; no 1). - Bibliography
951-90830-0-6 X.322/12492
9001425

AGRICULTURE. Finland
Methods, to date

AHLBÄCK, Ragna
Bonden i svenska Finland. - Helsingfors : Svenska litteratursällskapet i Finland, 1983
429p : ill. - (Skrifter utgivna av Svenska litteratursällskapet i Finland ; nr 509) (Folklivsstudier ; 14)
951-90178-5-2 Ac.9082[no.509]
9026940

AGRICULTURE. Finland
Prehistoric agriculture. Origins. Archaeology

ZVELEBIL, Marek
From forager to farmer in the Boreal Zone. - Oxford : British Archaeological Reports, 1981
2a(xiii,472p) : ill. - (BAR. International series ; 115). - Bibliography: p443-472
0-86054-140-1 X.425/5324
3277267

AGRICULTURE. Finland
Role of middle classes, 1700-1800

NIKULA, Oscar
Borgarna som lantbrukare i Finland under 1700-talet. - Åbo : Åbo akademi, 1977
24p. - (Acta Academiae Aboensis. ser.A ; vol.54, nr 4)
951-648-304-6 Ac.1093/2a[vol.54,no.4]
1973738

AGRICULTURE. Flood plains. Danube River. Sárköz. Hungary to 1800

ANDRÁSFALVY, Bertalan
A Sárköz és a környező Duna-menti területek ősi ártéri gazdálkodása és vízhasználatai a szabályozás előtt.... - Budapest : [s.n.], 1973
74p. : ill. - (Vízügyi történeti Süzetek ; 6)
r6007683 P.621/521.(6.)
0386367

AGRICULTURE. France
to date

LE ROY, Pierre
Le problème agricole français. - Paris : Economica, 1982
151p
2-7178-0485-4 X.322/13192
8165408

AGRICULTURE. France
1587-1789

GOYON DE LA PLOMBANIE, Henri de
La France agricole et marchande. - München : Kraus Reprint, 1980
2v(11fold.leaves of plates) : ill
r6206518 X.809/49783
8754306

AGRICULTURE. France
1800-1860

CLOUT, Hugh
Agriculture in France on the eve of the railway age. - London : Croom Helm, c1980
239p. - (Croom Helm historical geography series). - Bibliography: p231-233. - Includes index
0-85664-919-8 X.322/9694 Woolwich
1938959

AGRICULTURE. France
Farms. Financial management

NAISH, R. W.
A study of aspects of supplementing farm income in two areas of France. - [London] : ADAS, c1984
34p
B85-05938 BS.3/933
3219410

AGRICULTURE. France
Farms. Social conditions, to date

LE ROY LADURIE, Emmanuel
Inventaire des campagnes. - [Paris] : J.-C. Lattès, 1980
243p,24p of plates : ill
r6205847 X.322/10160 Woolwich
8165416

AGRICULTURE. France
Farms. Social conditions, 1800-1900

SEGALEN, Martine
Mari et femme dans la société paysanne. - Paris : Flammarion, [1980]
211p : ill,geneal table,plans. - (Bibliothèque d'ethnologie historique). - Bibl
2-08-210957-7 X.329/12992 Woolwich
2442531

AGRICULTURE. France
Finance, 1958-

CARLES, Roland
La situation financière de l'entreprise agricole. - Grignon : Institut National de la Recherche Agronomique. Laboratoire d'Économie Rurale de Grignon, 1981
105p
2-85340-405-6 X.322/10996 Woolwich
886022x

LIFRAN, R.
Mythes scientifiques sur la terre et les paysans. - Montpellier : Institut national recherche agronomique, 1981
50leaves
2-85340-385-8 X.322/11847
886022x

AGRICULTURE. France
Policies of government, 1945-

ILLE-ET-VILAINE (Département). Direction départementale de l'agriculture
Plan d'aménagement rural du Centre-Ouest. - [Rennes] : Direction départementale de l'agriculture, [1982]
134p : ill
r6211012 S.E.400/1268
8820112

ORGANISATION FOR ECONOMIC CO-OPERATION AND DEVELOPMENT
[La politique agricole en France. German]. Die Agrarpolitik in Frankreich. - [Bonn] : Bundesministerium für Ernährung, Landwirtschaft und Forsten, 1975
79p.. - (Berichte über die Agrarpolitik in den OECD-Mitgliedsländern) (Agrarpolitische Berichte)
r6303932 UN.P.108/383
8820112

AGRICULTURE. France
Research. Voisin, André - Biographies

CAÑAS ABRIL, Pedro
La vida y la obra del prof. André Voisin. - La Habana : Academia de Ciencias de Cuba, 1978
16leaves. - (Conferencias y estudios de historia y organización de la ciencia ; 1)
r6160787 LAS.D.300/15 Woolwich
2479834

AGRICULTURE. France
Technological innovation. Attitudes of peasants

BODIGUEL, Maryvonne
Les paysans face au progrès. - Paris : Presses de la Fondation Nationale des Sciences Politiques, 1975
177 p.. - (Travaux et recherches de science politique ; no. 37)
2-7246-0334-6 X.0700/263(37)
0860581

AGRICULTURE. Franconia. West Germany
1945- - Serials

BERICHT über die Durchführung der Unfallverhütung und die Massnahmen für die Erste Hilfe / Landwirtschaftliche Berufsgenossenschaft Unterfranken. - 1980-. - [Würzburg] : [Die Genussenschaft], [1981]-
r6302353 SF.583/295
8778426

AGRICULTURE. Fuenterrabia. Spain
1920-1969

GREENWOOD, Davydd James
Unrewarding wealth. - Cambridge : Cambridge University Press, 1976
x,223p : ill. - Bibliography: p.217-218. - Includes index
0-521-21021-6 X.320/10461 Woolwich
0709085

AGRICULTURE. Gabon
- For British exporters

A REPORT on the current state of the agricultural industry and the developing market for farm inputs in Gabon. - London : BAEC, 1981
ii,23p
B82-11357 SRL
2625776

AGRICULTURE. Galicia. Spain
to date

RODRÍGUEZ GALDO, María Xosé
Crisis agrarias y crecimiento económico en Galicia en el siglo XIX. - Sada, A Coruña : Ediciós do Castro, 1981
188p. - (Serie liminar. Economía)
84-7492-056-6 X.529/63965
8420580

AGRICULTURE. Galicia. Spain
1977

CAJA RURAL PROVINCIAL DE ORENSE. Servicio de Estudios
Estudio económico. - Orense : la Caja, 1977
496p
84-500-2729-2 X.329/12220 Woolwich
2244209

AGRICULTURE. Galicia. Spain
Organisation, 1875-1912

DURÁN, J.A
Agrarismo y movilización campesina en el país gallego (1875-1912). - Madrid : Siglo Veintiuno Editores, 1977
458p : ill. - (Historia de los movimientos sociales). - Bibliography : p.431-444
84-323-0244-9 X.519/38512
1534777

AGRICULTURE. Geita. Tanzania
Change

ANTHONY, Kenneth R. M.
Agricultural change in Geita district, Tanzania. - Nairobi : East African Literature Bureau, 1976
xii,79p. - Bibliography : p. 74-77.-Includes index
r6072200 X.319/18353 Woolwich
1580779

AGRICULTURE. Georgia (Soviet Socialist Republic)

ADAMIÎA, V.I
Sofŝial'no-ėkonomicheskoe razvitie gruzinskoi derevni v poreformennyĭ period (1870-1900 gg.). - Tbilisi : Sabchota Sakartvelo, 1976
272p
r6037468 X.529/20192
1155350

AGRICULTURE. Georgia (Soviet Socialist Republic)
Collectivisation

ZHGENTI, Tinatin Ivanovna
Istoriîa kolzhoznogo stroitel'stva v Gruzii (1921-1950 gg.). - Tbilisi : Mefŝniereba, 1977
180p
r6055818 X.329/8785 Woolwich
150519X

AGRICULTURE. Germany
to 1648

MAĬER, V. E. (Vil'gel'm Evgen'evich)
Derevnîa i gorod Germanii v XIV-XVI vv.. - Leningrad : Izdatel'stvo Leningradskogo universiteta, 1979
165p
r6487450 X.439/12981
9097082

AGRICULTURE. Germany
1519-1918

AKADEMIE DER WISSENSCHAFTEN DER DDR. Institut für Wirtschaftsgeschichte
Bäuerliche Wirtschaft und landwirtschaftliche Produktion in Deutschland und Estland (16. bis 19. Jahrhundert). - Berlin : Akademie-Verlag, 1982
360p. - (Jahrbuch für Wirtschaftsgeschichte. Sonderband ; 1981)
r6309298 X.520/33554
9073744

BÄUERLICHE Wirtschaft und landwirtschaftliche Produktion in Deutschland und Estland (16. bis 19. Jahrhundert). - Berlin : Akademie-Verlag, 1982
360p. : ill. - (Jahrbuch für Wirtschaftsgeschichte. Sonderband ; 1981)
r6308274 Ac.855/49a(4)
9073744

AGRICULTURE. Germany
1848-1918

ELSNER, Michael
Sozialdemokratie und Agrarfrage 1863-1895. - Hamburg : Ergebnisse-Verlag, [1981]
130p. - (Ergebnisse ; 14). - Bibliography
r6307112 X.329/16240
8916454

AGRICULTURE. Germany
1933-1945

ANIKEEV, A. A. (Alekseĭ Alekseevich)
Germanskiĭ fashizm i krest'îanstvo (1933-1945 gg.). - Rostov-na-Donu : Izdatel'stvo Rostovskogo universiteta, 1979
219p
r6485949 X.629/18196
8982368

AGRICULTURE. Germany
Influence of Nationalsozialistische Deutsche Arbeiter-Partei, 1928-1945

FARQUHARSON, John
The plough and the swastika. - London : Sage Publications, 1976
viii,312p. - (Sage studies in 20th century history ; vol.5). - Bibliography: p.272-279. - Includes index
0-8039-9949-6 X.319/17112 Woolwich
1049011

AGRICULTURE. Germany
Koppe, Johann Gottlieb

TUČEK, Helga
Für den Fortschritt der Agrarwissenschaft. - Luckau : Das Rat, 1982
64p : ill
r6308118 X.329/16826
8995478

AGRICULTURE. Germany
Law, 1945-1975 - Bibliographies

WINKLER, Wolfgang
Bibliographie des deutschen Agrarrechts 1966-1975. - Köln : Heymann, [1977]
xxxix,551p. - (Schriftenreihe des Instituts für Landwirtschaftsrecht der Universität Göttingen ; Bd 16)
3-452-18367-x 2727.g.10/16
186324X

AGRICULTURE. Germany
Policies of Marxists, 1871-1918

MARXISMO e questione agraria / a cura di Antonino Criscione. - Catania : Pellicanolibri, 1979
142p. - (Pensiero militante. Testi e documenti ; 4)
r6505587 X.629/19549
8891044

AGRICULTURE. Germany
Policies of Sozialdemokratische Partei Deutschlands, 1890-1907

HUSSAIN, Athar
Marxism and the agrarian question. - London : Macmillan.
Vol.2: Russian Marxism and the peasantry, 1861-1930. - Dec. 1980
[192]p
0-333-28675-8 X.529/44548
209178x

AGRICULTURE. Germany
Roemer, Theodor

FREIE DEUTSCHE JUGEND. Grundorganisation "Edwin Hoernle"
Theodor Roemer - Persönlichkeit und Werk. - Halle (Saale) : Martin-Luther-Universität Halle-Wittenberg, 1984
151p : ill. - (Wissenschaftliche Beiträge / Martin-Luther-Universität Halle-Wittenberg ; 1984/3 = T54). - Includes bibliographies
r6309389 X.329/17987
8995478

AGRICULTURE. Gloucestershire
1796

MARSHALL, Mr
The rural economy of Gloucestershire. - Gloucester : A. Sutton, 1979
2v.(xxxii,332p,fold plate;xxiii,367,[16]p). - Includes index
0-904387-22-4 X.529/34156
1992228

AGRICULTURE. Gloucestershire
1800-1914

FARMING in Gloucestershire 1800-1914. - [Gloucester] : [Gloucester Folk Museum], [1984]
75p : ill. - Bibliography: p73-74
B85-13634 X.805/6383
3252663

AGRICULTURE. Gloucestershire
Organisations: National Farmers' Union. Gloucestershire and North Avon Branch

NATIONAL FARMERS' UNION. Gloucestershire and North Avon Branch
The Gloucestershire and North Avon NFU year book and desk diary. - [1975]-. - Chippenham : Gloucestershire and North Avon Branch of the NFU, [1975-] : ill
r6009997 P.311/711
0745030

AGRICULTURE. Gloucestershire
Organisations: National Farmers' Union. Gloucestershire Branch

NATIONAL FARMERS' UNION. Gloucestershire Branch
The Gloucestershire NFU year book and desk diary. - [197-?-1974]. - Gloucester : Gloucestershire Branch of the N.F.U., [197-?-1974] : ill
Holdings : 1973, 1974
r6010000 P.311/710
0745057

AGRICULTURE. Gloucestershire
Smallholdings. Administration

GLOUCESTERSHIRE. Performance Review Sub-committee
Performance review of smallholdings. - Gloucester : [The Council], 1978
[5],41[i.e.45]p : ill
0-904950-15-8 X.512/7199
2052598

AGRICULTURE. Golodnaya step'. Uzbekistan

TĬURIKOV, Vladimir
U starykh znakomykh. - Tashkent : Izdatel'stvo literatury i iskusstva im. Gafura Gulīama, 1977
156p
r6046181 X.319/8835 Woolwich
1232282

AGRICULTURE. Gotland. Sweden
Farms: Petesgard, to 1975

SVAHNSTRÖM, Karin
Petes i Hablingbo. - Visby : B. Press, 1976
16p : ill
91-7400-053-5 X.319/9196 Woolwich
1395408

AGRICULTURE. Gower. West Glamorgan
Farms - Walkers' guides

HOWELLS, Roscoe
Gower farm trail. - [Swansea] : Planning Department, West Glamorgan County Council, [1980]
20p : ill. - Bibliography: p20
B82-16039 X.329/15055 Woolwich
2695359

AGRICULTURE. Grampian Region. Scotland
Farms. Buildings, to 1979

WALKER, Bruce, 1936-
Farm buildings in the Grampian Region. - [Aberdeen] : Grampian Regional Council, 1979
72p : ill
B81-04868 X.425/1834 Woolwich
2342685

AGRICULTURE. Grassland

SPEDDING, Colin Raymond William
Grassland ecology. - Oxford : Clarendon Press, 1976
xii,221p : ill. - Includes bibliographies and index
0-19-854133-3 X.319/16612 Woolwich
0676012

AGRICULTURE. Grassland
Research. Methodology

MEASUREMENT of grassland vegetation and animal production / edited by L. 't Mannetje. - Farnham Royal : Commonwealth Agricultural Bureaux, 1978
xv,260p : ill. - (Bulletins / Commonwealth Bureau of Pastures and Field Crops ; 52). - Includes bibliographies and index
0-85198-404-5 SRL
1783610

AGRICULTURE. Grassland. Great Britain

SOPER, M. H. R. (Michael Henry Ray)
Grassland. - London : Association of Agriculture, 1982
43p,[4]p of plates : ill. - (Modern agriculture series). - Bibliography: p43
B82-33665 X.809/52840
2769492

AGRICULTURE. Grassland. Great Britain
Research organisations: Grassland Research Institute

GRASSLAND RESEARCH INSTITUTE
The Grassland Research Institute. - Hurley : G.R.I., 1975
[2],25p : ill
0-9500670-5-9 SRL
0585343

AGRICULTURE. Grassland. Hungary

PALÁDI-KOVÁCS, Attila
A magyar parasztság rétgazdálkodása = Praxis rusticorum Hungariae in oeconomia pratorum. - Budapest : Akadémiai Kiadó, 1979
541p : ill
963-05-1770-1 X.320/5360 Woolwich
2170345

AGRICULTURE. Great Britain

AGRICULTURE in Britain / [prepared by Reference Division, Central Office of Information]. - 2nd ed. - London : H.M.S.O., 1980
59p. - (Central Office of Information reference pamphlet ; 43). - Bibliography: p58-59
0-11-701007-3 SRL
0058211

COX, Sarah
Farmworker. - Harmondsworth : Kestrel Books, 1975
[32]p : chiefly ill. - (People working)
0-7226-5396-4 X.322/8009
0268437

CRABTREE, Vickie
Farming today. - London : Muller, 1982
155p : ill. - Includes index
0-584-10412-x X.329/14986 Woolwich
0171441

ELTRINGHAM, Denham
Agriculture. - Revised ed. - Aylesbury : Ginn, 1977
49p : ill. - (Satellite books / Manchester Economics Project ; 5)
0-602-22333-4 X.319/17379 Woolwich
0058211

FARM planning data / University of Cambridge, Department of Land Economy, Agricultural Economics Unit. - Cambridge : [The Unit].
1975. - 1975
[4],97p : ill
0-901978-62-0 P.511/5003
0216925

FARMING and the nation / presented to Parliament by the Secretary of State for Northern Ireland ... [et al.]. - London : H.M.S.O., [1979]
[2],36p. - (Cmnd.7458)
0-10-174580-x BS.3/658
1905465

FREAM, William
Fream's elements of agriculture. - 15th ed. revised and metricated. - London : J. Murray, 1977
xli,735p,lvii p of plates : ill. - Bibliography: p.714-725. - Includes index
B78-00967 SRL
0526177

FRENCH, Mary, 1908-
Farms. - London : Mills and Boon, 1974
96p : ill. - (Mills and Boon on location ; book no.10)
0-263-05582-5 X.329/10613 Woolwich
0171441

FURTH, Charles
Old farms and new farming. - London : Allen and Unwin, 1975
140p,ii leaves of plates : ill(incl 2 col). - Includes index
0-04-630008-2 X.329/10169
0526177

GRAHAM-CAMERON1931-, M. G. (Malcolm Gordon)
The farmer. - 2nd ed. - Cambridge : Dinosaur, 1980
[32]p : col.ill. - (Dinosaur's Althea books)
0-85122-227-7 X.990/19483
0171441

HAINES, Michael
Introduction to farming systems. - London : Longman, 1982
viii,214p : ill. - Bibliography: p210-212. - Includes index
0-582-45081-0 SRL
0058211

HILL, Brian E.
Britain's agricultural industry. - London : Heinemann Educational, 1975
[7],100p : ill. - (Studies in the British economy ; 22). - Bibliography: p.97. - Includes index
0-435-84565-9 X.319/10245
0058211

HUGGETT, Frank E. (Frank Edward)
Farming in Britain. - 3rd ed. - London : A. and C. Black, 1975
64p : ill. - (Black's junior reference books). - Bibliography: p.63. - Includes index
0-7136-1527-3 X.320/10157
0171441

LENGA, Rosalind
Let's go to the farm. - London : F. Watts, 1980
32p : col ill. - (Let's go series)
0-85166-789-9 X.329/12180 Woolwich
0171441

LIBERAL PARTY
A new deal for British farming. - London : Liberal Publication Department, 1979
[4],23p
B79-21622 X.319/18905 Woolwich
1904450

LODGE, Tony
The farm. - London : Collins, 1983
240p : col.ill. - (Collins gem guides). - Includes index
0-00-458813-4 X.329/17159
0526177

MCCONNELL, Primrose
Primrose McConnell's The agricultural notebook. - 16th ed.. - London : Newnes-Butterworths for 'British Farmer and Stockbreeder', 1976
ix,965p : ill. - Includes index
0-408-00208-5 SRL
0553913
Primrose McConnell's The agricultural notebook. - 17th ed.. - London : Butterworth Scientific, 1982
xviii,683p : ill. - Includes bibliographies and index
0-408-10701-4 SRL
0553913

MISTER, Alan A.
Working on a farm. - London : Batsford Academic and Educational, 1982
105p,[8]p of plates : ill. - Includes index
0-7134-3962-9 X.329/15271 Woolwich
0505064

AGRICULTURE. Great Britain –continued

NATIONAL FEDERATION OF YOUNG FARMERS' CLUBS
A 'farming' future for young farmers?. – Kenilworth : The Federation, [1978]
[20]p
B79–32340 X.519/30249
1956744

RESILIENT, resourceful – and at risk. – London : National Farmers' Union, [1981?]
[4]p : ill
B81–13291 X.325/991 Woolwich
0058211

ROBERTS, Frank, 1939–
The countryside. – Basingstoke : Macmillan Education, 1981
30p : ill. – (Looking at Britain ; 2)
0–333–28411–9 X.805/1497
0021822

SALT, Bernard
Rural science. – London : Cassell. – (Cassell's rural science series) – Includes index
1. – 1980
152p : ill
Subsequent volumes, when received, will be placed at this shelfmark
0–304–30410–7 X.0322/216
2211203

STEWART, Anne
The farmer. – London : Hamilton, 1983
24p : col.ill. – (Cherrystones). – Includes index
0–241–10937–x X.329/16737
0171441

THOMAS, Leslie, 1912–
Farming in Britain. – London : Association of Agriculture, 1975
28p,[4]p of plates : ill
0–903313–23–5 X.311/8058
0526177

The U.K. farming industry. – London : Euromonitor, c1983
134p
0–86338–020–4 X.322/13622
0058211

WHITLOCK, Ralph
Working on a farm. – Hove : Wayland, 1983
70p : ill. – (People at work). – Bibliography: p68. – Includes index
0–85340–983–8 X.322/12259
0505064

WINDASS, Stanley
Agriculture. – [Kidlington] : [Alternative Society], 1976
[1],25p : 1 ill. – (Where do we go from here?). – Bibliography: p.25
B80–26664 X.322/9406 Woolwich
1905465

WOODWARD, John, 1954–
The field guide. – Poole : Blandford, 1983
160p : ill(some col.). – Bibliography: p158. – Includes index
0–7137–1234–1 X.329/16612
0526177

AGRICULTURE. Great Britain
– Bibliographies

FUSSELL, G. E. (George Edwin)
The old English farming books. – London : Pindar.
Vol.3: 1793–1839. – 1983
285p. – Bibliography: p266–269. – Includes index
0–907132–03–0 X.329/18102
3026213
– [Another copy] BB.E.d.11 Reading Room

AGRICULTURE. Great Britain
– Conference proceedings

FORESTRY & farming in upland Britain. – Edinburgh : Forestry Commission, c1980
244p : ill. – (Occasional paper ; no.6). – Includes bibliographies
0–85538–083–7 SRL
0058211

OXFORD FARMING CONFERENCE (29th : 1975)
The twenty-ninth Oxford Farming Conference [held at] the Town Hall, Oxford, January 6–8, 1975. – [Oxford] : [M.H.R. Soper], [1976]
152p : 1 ill
0–900034–09–2 X.319/16465 Woolwich
0058211

AGRICULTURE. Great Britain
– For national planning

BRITAIN'S future in farming / edited by Sir Frank Engledow and Leonard Amey. – Berkhamsted : Geographical Publications, 1980
x,165p. – (Studies in land economy). – Bibliography : p.159–160. – Includes index
0–900394–21–8 X.322/9608 Woolwich
2088762

AGRICULTURE. Great Britain
– Forecasts

GREAT BRITAIN. Ministry of Agriculture, Fisheries and Food
Possible patterns of agricultural production in the United Kingdom by 1983. – London : H.M.S.O., 1979
[2],20p : ill
0–11–240326–3 BS.3/656
1967819

AGRICULTURE. Great Britain
– Forecasts – Conference proceedings

The OUTLOOK for agriculture, 1978 and beyond. – Aberdeen : The School, 1977
[4],168p,[4]leaves : ill
B79–07108 X.313/1611 Woolwich
0663700

The OUTLOOK for agriculture, 1979 and beyond / edited by B.J. Revell. – Aberdeen : School of Agriculture, [1979]
[9],170p : ill
B79–16443 X.313/1647 Woolwich
0663700

The OUTLOOK for agriculture 1981 and beyond / edited by B.J. Revell. – Aberdeen : School of Agriculture, Aberdeen, 1981
v,126p : ill
B81–13599 X.325/992 Woolwich
2381915

The OUTLOOK for agriculture 1982 and beyond / Chairmen James Davidson, Sylvester Campbell ; edited by B.J. Revell. – Aberdeen : School of Agriculture, Aberdeen, 1982
iv,111p : ill
B82–17457 SRL
2702312

AGRICULTURE. Great Britain
– Serials

[ADVISOR (Cambridge)]. Advisor. – No.1 (May 1981)–. – Cambridge : Ministry of Agriculture, Fisheries and Food, 1981–
v. : ill
B84–44240 P.311/907
0034991

AGRICULTURAL RESEARCH COUNCIL
ARC Research Review. – Vol.1, no.1– ; Mar. 1975–. – London : The Council, 1975– : ill
B76–07469 BS.3/557
0526177

CARDIGANSHIRE farmer. – Vol.1, no.1 (June 1980)–. – Dorchester : N.F.U. County Pubications, 1980–
v. : ill
B81–06757 P.311/817
2353938

CEREDIGION farmer = Amaethwr Ceredigion. – Vol.4, no.3 (Mar. 1983)–. – Dorchester : N.F.U. County Publications, 1983–
v. : ill
B84–05582 P.511/5476
2353938

FARM diary. – [Mar. 1975]–. – London : Veratbrite Ltd, [1975]–
r6071934 P.313/444
1447181

The FARMER. – Vol.1, no.1– ; Autumn 1975–. – Lincoln : 'The Farmer', 1975– : ill(some col)
B76–01615 SRL
0564656

[The FARMERS advertiser (Manchester)]. The Farmers advertiser. – [19–]–. – Manchester : [Paramount Publicity], [19–]–
v. : ill
r6865180 P.2000/984
2353938

FARMING Industry. – Vol. 1– , no. 1– ; Mar. 13th 1976–. – London : Veratbrite Ltd, [1976]– : ill
B76–13978 SRL
0564656

FARMING progress. – Jan. 1982–. – Penrith : Ministry of Agriculture, Fisheries and Food, 1982–
v. : ill
B84–44239 P.311/903
0034991

The STATE of British agriculture. – 1953–54–. – Oxford : Institute of Agricultural Economics, 1954–
v.
Holdings: 1981–
r6864861 P.313/543
2353938

AGRICULTURE. Great Britain
– Statistics

BURRELL, A. M. (Alison M)
Statistical handbook of U.K. agriculture. – London : Macmillan, 1984
176p. – Bibliography: p173–176
0–333–37910–1 X.322/14405
0216925

STATISTICAL handbook of U.K. agriculture / by Denis K. Britton ... [et al.]. – Ashford, Kent : Wye College, Agricultural Economics Unit, 1980
vii,135p. – Bibliography: p133–135
0–901859–91–5 SRL
0216925
– [Another copy.] OPL

AGRICULTURE. Great Britain
to 55 B.C.

FOWLER, P. J. (Peter Jon)
The farming of prehistoric Britain. – Cambridge : Cambridge University Press, 1983
x,246p,[12]p of plates : ill. – Bibliography: p219–231. – Includes index
0–521–27369–2 X.421/24468
2979292

AGRICULTURE. Great Britain
to 1934

HENNELL, Thomas
Change in the farm. – Wakefield : EP Publishing, 1977
x,201p,[8]p of plates : ill
0–7158–1220–3 X.329/11003 Woolwich
1056840

AGRICULTURE. Great Britain
to 1945

FARMING through the ages in pictures / [compiled by] Robert Trow-Smith. – Ipswich : Farming Press, 1978
111p : chiefly ill. – Includes index
0–85236–091–6 X.320/11516 Woolwich
1763024

AGRICULTURE. Great Britain
to 1964

WHITLOCK, Ralph
A short history of farming in Britain. – Wakefield : EP Publishing, 1977
[1],246p,leaf of plate,[22]p of plates : ill. – Bibliography: p.225–227. – Includes index
0–7158–1208–4 X.329/11123 Woolwich
1248642

AGRICULTURE. Great Britain
220–1977

STRATTON, John M.
Agricultural records, A.D.220–1977. – 2nd ed.. – London : J. Baker, 1978
259p
0–212–97022–4 X.329/11691 Woolwich
1480022
– [Another copy] OPL

AGRICULTURE. Great Britain
ca B.C.700–ca A.D.1

REYNOLDS, Peter J. (Peter John)
Farming in the Iron Age. – Cambridge : Cambridge University Press, 1976
48p : ill. – (Cambridge introduction to the history of mankind. topic book)
0–521–21084–4 X.519/25774
0653810

AGRICULTURE. Great Britain
1400–1979

BURRELL, R. E. C. (Roy Eric Charles)
Food from the fields. – Exeter : Wheaton, 1980
96p : ill. – (The Making of the industrial revolution series ; bk 1)
0–08–020545–3 X.809/46726
2348039

AGRICULTURE. Great Britain
1523–1793

FUSSELL, George Edwin
Old English farming books, 1523–1793. – Collieston : Aberdeen Rare Books, 1978
[7],141,v,186p,leaf of plate,4p of plates. – Includes bibliographies and index
B79–06190 BB.E.d.11 Reading Room
1797069
– [Another copy] X.809/44012

AGRICULTURE. Great Britain
1650–1880

The AGRICULTURAL revolution / edited and with an introduction by G.E. Mingay. – London : A. and C. Black, 1977
ix,322p : ill. – (Documents in economic history). – Includes index
0–7136–1703–9 X.329/10932 Woolwich
1104845

AGRICULTURE. Great Britain
1700–1980

MASON, Simon
Agrarian Britain 1700–1980. - [Oxford] : Blackwell, 1984
64p : ill. - (Blackwell history project). - Includes index
0–631–91420–x X.805/6230
3229483

AGRICULTURE. Great Britain
ca 1750–ca 1950

NIALL, Ian
To speed the plough. - London : Heinemann, 1977
[7],231p : ill
0–434–51024–6 X.329/11052 Woolwich
1146874

AGRICULTURE. Great Britain
1770–1813

YOUNG, Arthur, 1741–1820
Arthur Young and his times. - London : Macmillan, 1975
vii,264p. - Bibliography: p.251–255. - Includes index
0–333–15366–9 X.809/40474
0541451

AGRICULTURE. Great Britain
1800–1977

WORMELL, Peter
Anatomy of agriculture. - London : Harrap : Kluwer, 1978
571p,[24]p of plates : ill. - Includes index
0–245–53302–8 X.320/11493 Woolwich
1739859

AGRICULTURE. Great Britain
ca 1860

PATTERSON, Geoffrey
Chesnut Farm 1860. - London : Deutsch, 1980
[32]p : col.ill
0–233–97208–0 X.992/5523
1995618

AGRICULTURE. Great Britain
1870

STEPHENS, Henry
[The book of the farm. 3rd ed. Selections]. Victorian farming. - Winchester : Shurlock, 1975
xx,138p : ill. - Bibliography: p.135–136. - Includes index
0–903330–12–1 X.322/8007
0350397

AGRICULTURE. Great Britain
1918–1940

CREASEY, John, 19—
The countryside between the wars 1918–1940. - London : Batsford, 1984
144p : ill
0–7134–1186–4 X.322/14582
3240940

AGRICULTURE. Great Britain
1920–1977

BEYNON, Victor Howell
Agriculture and economics. - [Exeter] : University of Exeter, 1979
23,[10]p. - (Inaugural lectures / University of Exeter)
0–85989–135–6 X.329/12626 Woolwich
2102587

AGRICULTURE. Great Britain
1945–1974

BERESFORD, Tristram
We plough the fields. - Harmondsworth : Penguin, 1975
xxvi,257p,[16]p of plates : ill. - (A pelican original). - Bibliography: p.236–243. - Includes index
0–14–021872–6 X.319/10228
0526304

AGRICULTURE. Great Britain
1945–1983

HOLDERNESS, B. A.
British agriculture since 1945. - Manchester : Manchester University Press, c1985
185p. - Bibliography: p176–180. - Includes index
0–7190–1722–x X.329/19673
3050521

AGRICULTURE. Great Britain
1968–1975

GREAT BRITAIN. Ministry of Agriculture, Fisheries and Food
The changing structure of agriculture, 1968–1975. - London : H.M.S.O., 1977
78p : 1 ill
0–11–241103–7 BS.3/592
1257862

AGRICULTURE. Great Britain
Application of microprocessors - Conference proceedings

READING UNIVERSITY AGRICULTURAL CLUB. Conference (14th : 1980)
Microprocessors in agriculture, will farming have its' [sic] chips?. - Reading : The Club, 1980
39p : 1ill. - Includes bibliographies
0–7049–0308–3 X.622/13423
2773503

AGRICULTURE. Great Britain
Consumption of energy related to food production industries

LEACH, Gerald
Energy and food production. - London : International Institute for Environment and Development, [1975]
[4],151p : ill
0–905347–00–5 SRL
0601098

Energy and food production. - [Revised ed.]. - Guildford : IPC Science and Technology Press for the International Institute for Environment and Development, 1976
[6],137p : ill
0–902852–55–8 X.329/10556 Woolwich
0601098

AGRICULTURE. Great Britain
Dairy farms. Management

SLATER, Ken
Dairy farm business management. - Ipswich : Farming Press, 1981
217p : ill. - Includes index
0–85236–114–9 X.529/52800
2872048

AGRICULTURE. Great Britain
Development compared with agricultural development in West Germany, 1870–1975

The DEVELOPMENT of agriculture in Germany and the UK. - Ashford, Kent : Centre for European Agricultural Studies. Wye College.
1: German agriculture, 1870–1970. - 1979
59,[31]p. - (Miscellaneous studies / Wye College. Centre for European Agricultural Studies ; no.2). - Bibliography(2p.)
0–905378–15–6 X.512/6718
1921339

2: UK agricultural policy, 1870–1970. - 1979
110,[28]p. - (Miscellaneous studies / Wye College. Centre for European Agricultural Studies ; no.3). - Bibliography: p.110
0–905378–16–4 X.512/6719
1921339

AGRICULTURE. Great Britain
Development compared with agricultural development in West Germany, 1870–1975 - Statistics

The DEVELOPMENT of agriculture in Germany and the UK. - Ashford, Kent : Centre for European Agricultural Studies, Wye College.
3: Comparative time series, 1870–1975. - 1979
97,[23]p : ill. - (Miscellaneous studies / Wye College. Centre for European Agricultural Studies ; no.4)
0–905378–17–2 X.512/6720
1921339

AGRICULTURE. Great Britain
Development compared with agricultural development in West Germany, 1967–1980 - Conference proceedings

The DEVELOPMENT of agriculture in Germany and the UK. - Ashford, Kent : Wye College. - (Seminar paper ; no.15)
5: Policy implications of their comparative output, structure and productivity. - c1984
iv,71p : ill
0–905378–19–9 X.322/14743
259501x

AGRICULTURE. Great Britain
Effects of Common Agricultural Policy of the European Economic Community, 1973–1976

FRANKLIN, M. D. M.
Four years' experience of the common agricultural policy in a changed world. - [Belfast] : [Queen's University of Belfast], [1977]
18p. - (The George Scott Robertson memorial lecture ; 26th)
0–85389–118–4 WP.A.219/18
1299867

AGRICULTURE. Great Britain
Effects on landscape

SHOARD, Marion
The theft of the countryside. - London : Temple Smith, 1980
269p,[16]p of plates : ill. - Includes index
0–85117–200–8 SRL
2556863

AGRICULTURE. Great Britain
Effects on nature conservation - Bibliographies

NATURE CONSERVANCY COUNCIL
Nature conservation and agriculture. - Banbury : The Council, 1979
[12]p. - (Bibliography series / Nature Conservancy Council. Information and Library Services ; no.1)
B80–28988 BS.32/180(1)
220407x

AGRICULTURE. Great Britain
Effects on organisms

MELLANBY, Kenneth
Farming and wildlife. - London : Collins, 1981
178p,24p of plates : ill. - (The New naturalist ; 67). - Bibliography: p169–174. - Includes index
0–00–219239–x SRL
2490951

AGRICULTURE. Great Britain
Efficiency - Conference proceedings

AGRICULTURAL efficiency. - London : The Society, 1977
[6],227p. - Includes bibliographies
0–85403–095–6 X.322/8816 Woolwich
1901486

AGRICULTURE. Great Britain
Extension services. Evaluation

SHEPPARD, D. (David)
Research on agricultural advice. - [York] : Agricultural Education Association, [1977]
[5],78p
B79–07384 X.313/1614 Woolwich
1804316

AGRICULTURE. Great Britain
Family farms - Conference proceedings

STRATEGIES for family-worked farms in the UK / edited by R.B. Tranter. - Reading : The Centre, 1983
280p : ill. - (CAS paper ; 15). - Includes bibliographies and index
0–7049–0622–8 P.311/716
3215385

AGRICULTURE. Great Britain
Farms. Accounting

CRITCHLEY, R.
Understanding your farm accounts. - [Aberdeen] : [North of Scotland College of Agriculture Agriculture Economics Division School of Agriculture], 1982
i,45p : ill
B83–17097 X.525/7007
2877112

AGRICULTURE. Great Britain
Farms. Accounting. Applications of digital computer systems - Conference proceedings

EFFECTIVE use of computing for farm accounting / edited by Ann Foster. - Aldershot : Gower, c1981
72p : ill. - (Computing in business report)
0–566–03028–4 SRL
2727420

AGRICULTURE. Great Britain
Farms. Book-keeping

MADDOX, J. M. (Jack M)
Farm cash book recording - an example. - Rev. [ed]. - Aberdeen : North of Scotland College of Agriculture, Agricultural Economics Division, School of Agriculture, 1983
19p
B83–31825 X.525/7645
0297755

AGRICULTURE. Great Britain
Farms. Buildings, to ca 1900

HARVEY, Nigel
Old farm buildings. - Aylesbury : Shire Publications, 1975
[1],33p : ill. - (Shire album ; 10). - Bibliography: p.33
0–85263–313–0 X.0972/143(10)
0528986

PETERS, J. E. C. (John Edward Colclough)
Discovering traditional farm buildings. - Princes Risborough : Shire, 1981
80p : ill. - Includes index
0–85263–556–7 X.808/36775
0528986

AGRICULTURE. Great Britain
Farms. Buildings. Architecture

ALLEN, Eleanor
Farms. - London : A. and C. Black, 1977
32p : ill. - (Buildings as history). - Bibliography: p.2. - Includes index
0–7136–1757–8 X.322/8448 Woolwich
1260359

AGRICULTURE. Great Britain
Farms. Buildings. Architecture, 1700–1846

ROBINSON, John Martin
Georgian model farms. – Oxford : Clarendon, 1983
xiv,190p,[64]p of plates : ill. – Bibliography: p173–177. – Includes index
0–19–817366–0 X.800/38700
3141233

AGRICULTURE. Great Britain
Farms. Buildings. Conservation

KAMM, Jonathan
Saving old farm buildings. – Winchester : Hampshire County Planning Dept. in association with the District Councils of Hampshire, 1982
20p : ill. – (Hampshire's heritage). – Bibliography: p20
0–900908–71–8 X.622/12939 Woolwich
2754681

AGRICULTURE. Great Britain
Farms. Buildings. Design Council approved designs

DESIGN COUNCIL
The Design Council's catalogue of farm buildings. – [London] : The Council, [1977]
132p : ill(some col)
0–85072–078–8 BS.80/30
1519220

AGRICULTURE. Great Britain
Farms. Buildings. Environment planning. Law

WELLER, John B. (John Brian)
Agricultural buildings planning and allied controls. – Edinburgh : Capital Planning Information, 1981
59p. – (Planning reviews ; no.2). – Bibliography: p57–59
0–906011–12–4 L.49/1119
2611821

AGRICULTURE. Great Britain
Farms. Camping & caravanning. Sites. Provision

VAUGHN, J. S. (Jack S)
Caravan and camping sites on the farm. – [London?] : Ministry of Agriculture, Fisheries and Food, Agricultural Development and Advisory Service, 1976
18p. – (Socio-economics paper ; 6)
B83–08412 BS.3/832(6)
2835452

AGRICULTURE. Great Britain
Farms. Carthorses. Use

EVANS, George Ewart
Horse power and magic. – London : Faber, 1979
iii,xiv,222p,[12]p of plates : ill. – Bibliography: p.215–216. – Includes index
0–571–11348–6 X.329/11833 Woolwich
1743767

AGRICULTURE. Great Britain
Farms. Dairies. Design & construction

AGRICULTURAL DEVELOPMENT AND ADVISORY SERVICE
Farm dairies. – London : H.M.S.O., 1975
[1],12p. – (Fixed equipment of the farm ; leaflet 3)
0–11–241108–8 SRL
0359963

AGRICULTURE. Great Britain
Farms. Design & construction

NOTON, Nicholas H.
Farm buildings. – Reading : College of Estate Management, 1982
359p : ill. – Bibliography: p347. – Includes index
0–902132–73–3 X.425/4061
2920042

AGRICULTURE. Great Britain
Farms. Electric equipment. Installation

ELECTRICITY COUNCIL
Power. – London : The Council, 1970
[6], 73 p. : ill. – (Farm electrification handbook ; no. 16)
r6031597 X.619/17461 Woolwich
1083767

AGRICULTURE. Great Britain
Farms. Energy supply systems, to 1980

WELLER, John B. (John Brian)
History of the farmstead. – London : Faber, 1982
248p : ill. – Bibliography: p221–228. – Includes index
0–571–11804–6 X.809/51281
2707713

AGRICULTURE. Great Britain
Farms. Equestrian recreation facilities. Provision

EQUESTRIAN enterprises on the farm. – [London?] : ADAS, Ministry of Agriculture, Fisheries and Food, 1978
ii,18p. – (Socio-economics paper ; 9). – Bibliography: p18
B83–08415 BS.3/832(9)
2835487

AGRICULTURE. Great Britain
Farms. Financial management

AGRICULTURE in the 1980's. – London : Peat, Marwick, Mitchell & Co., 1984
v,61p
1–85061–013–4 X.329/18535
002581x

BROWN, Maxwell L.
Farm budgets. – Baltimore ; London : Johns Hopkins University Press for the World Bank, 1979
xviii,136p : ill. – (Staff occasional papers / World Bank ; no.29). – Bibliography: p.133–136
0–8018–2386–2 X.520/20696
002581x

STURROCK, F. G. (Ford Gibson)
Farm accounting and management. – 7th ed. – London : Pitman, 1982
xii,300p : ill. – Includes index
0–273–01788–8 X.529/48925
002581x
– [Another copy: paperback.] ISBN 0–273–01765–9 X.529/48926

WARREN, Martyn F.
Financial management for farmers. – London : Hutchinson, 1982
xi,314p : ill. – Includes index
0–09–148930–x X.520/29389
002581x
– [Another copy: paperback.] ISBN 0–09–148931 8 X.520/30733

AGRICULTURE. Great Britain
Farms. Labourers. Use

WALFORD, N. S.
Labour and machinery use on the larger, mainly arable farm. – Ashford, Kent : Wye College, School of Rural Economics and Related Studies, 1979
[4],42p. – (FBU occasional papers / Wye College. Farm Business Unit ; no. 4). – Bibliography: p.39–40
0–901859–86–9 X.622/8958 Woolwich
214140x

AGRICULTURE. Great Britain
Farms. Livestock. Breeding, to 1982

URQUHART, Judy
Animals on the farm. – London : Macdonald, 1983
x,182p,[32]p of plates : ill. – Bibliography: p175–177. – Includes index
0–356–07897–3 X.322/12033
2923017

AGRICULTURE. Great Britain
Farms. Livestock. Care – Conference proceedings

The WELFARE of the food animals. – South Mimms : U.F.A.W., 1979
68p. – Includes bibliographies
B79–11052 SRL
1827227

AGRICULTURE. Great Britain
Farms. Livestock. Mutilation. Control

FARM ANIMAL WELFARE COUNCIL
Advice to the agriculture ministers of Great Britain on the need to control certain mutilations on farm animals. – Surbiton : The Council, 1981
[8]p
0–9507656–0–0 BS.3/769
2902761

AGRICULTURE. Great Britain
Farms. Livestock. Preventive veterinary medicine

CLARKSON, M. J.
Preventive medicine. – Liverpool : Liverpool University Press, 1978
18p : ill. – (Inaugural lecture series / University of Liverpool)
0–85323–094–3 X.319/18761 Woolwich
1846590

AGRICULTURE. Great Britain
Farms. Livestock. Production

BOATFIELD, Graham
Farm livestock. – Ipswich : Farming Press, 1979
141p : ill. – (Farming book series). – Includes index
0–85236–097–5 X.619/19026 Woolwich
1936913

AGRICULTURE. Great Britain
Farms. Livestock. Production. Profitability. Improvement – Conference proceedings

PROFITS from livestock and ways to achieve them / edited by the Conference Committee. – Reading : The Club, 1979
32p : ill. – Includes bibliographies
0–7049–0307–5 X.525/6938
2629216

AGRICULTURE. Great Britain
Farms. Machinery. Use. Economic factors

WALFORD, N. S.
Labour and machinery use on the larger, mainly arable farm. – Ashford, Kent : Wye College, School of Rural Economics and Related Studies, 1979
[4],42p. – (FBU occasional papers / Wye College. Farm Business Unit ; no. 4). – Bibliography: p.39–40
0–901859–86–9 X.622/8958 Woolwich
2141418

AGRICULTURE. Great Britain
Farms. Management

ASHTON, Peter
Farm business management and land ownership. – [Leicester] : [The author], 1979
128p : ill
B80–29992 X.329/13659 Woolwich
1926721

GILES, A. K. (Anthony Kent)
The farmer as manager. – London : Allen and Unwin, 1980
[7],199p : ill. – Includes bibliographies and index
0–04–658228–2 X.329/13226 Woolwich
1926721

NIX, John
Farm management pocketbook. – 11th ed.. – Ashford : School of Rural Economics, Wye College, 1980
vii,160p. – Includes index
0–901859–88–5 SRL
2873222

Farm management pocketbook. – 12th ed. – Ashford : Wye College (University of London), School of Rural Economics, Farm Business Unit, 1981
viii,160. – Includes index
0–901859–95–8 SRL
2581736

AGRICULTURE. Great Britain
Farms. Management. Applications of digital computer systems

GREAT BRITAIN. Ministry of Agriculture, Fisheries and Food
Farm planning by computer. – 2nd ed. – London : H.M.S.O., 1979
[2],140p : ill. – (GFM ; 2). – Bibliography: p.108–112
0–11–240314–x SRL
1993453

AGRICULTURE. Great Britain
Farms. Management. Role of farmers' wives

BUCHANAN, W. I.
The farmer's wife. – Reading : Farm Management Unit, University of Reading, 1982
ix,46p. – (Study / University of Reading. Farm Management Unit ; no.2)
0–7049–0805–0 X.525/6558
2822083

AGRICULTURE. Great Britain
Farms. Management. Simulation games: WYEFARM

YOUNGMAN, James
Wyefarm management game, listing of programs BONZO and ERROR. – Ashford, Kent : Farm Business Unit, Wye College, [1975]
42p in various pagings
0–901859–44–3 Cup.1281/226
0648221

AGRICULTURE. Great Britain
Farms. Manpower. Labour sharing

GASSON, Ruth
Labour sharing in agriculture. – [Ashford, Kent] : Wye College [School of Rural Economics and Related Studies], 1979
[7],117p. – Bibliography: p.112–117
0–901859–82–6 X.512/6895
195024x

AGRICULTURE. Great Britain
Farms. Mechanisation. Management

BUTTERWORTH, Bill
Farm mechanisation for profit. – London : Granada, 1983
269p : ill. – Includes index
0–246–11562–9 X.520/30429
2922800

AGRICULTURE. Great Britain
Farms. Personnel. Remuneration. Determination

WINYARD, Steve
Poor harvest. – London : Low Pay Unit, 1980
20p. – (Low pay report ; 3)
B83–02625 X.529/53561
0615714

AGRICULTURE. Great Britain
Farms. Productivity related to size

BRITTON, D. K. (Denis King)
Size and efficiency in farming. - Farnborough, Hants. : Saxon House [etc.], 1975
xi,184p : ill. - ([Saxon House studies]). - Includes index
0-347-01043-1 X.320/10047
0362379

AGRICULTURE. Great Britain
Farms. Severance by roads. Cost-benefit analysis

BODDINGTON, Michael A. B.
Motorways and agriculture. - [Ipsden] : Rural Planning Services [Ltd], [1978]
[3],vi,33p. - (Publications / Rural Planning Services : no.5). - Bibliography: p.32-33
0-906269-01-6 X.319/18852 Woolwich
1893289

AGRICULTURE. Great Britain
Farms. Stockmen. Training & personnel management - Conference proceedings

STOCKMANSHIP on the farm. - Potters Bar : Universities Federation for Animal Welfare, c1983
87p : ill
0-900767-32-4 X.329/17162
2996111

AGRICULTURE. Great Britain
Farms. Surveys - Serials

REGISTER of farm surveys in the United Kingdom and Eire. - [No.1] / compiled by a working party formed following a conference on farm surveys in July 1975. - [Exeter] : [The Working Party], 1976-
Subsequent volumes, when received, will be placed at this shelfmark
0-901165-61-1 P.313/419
1074385

AGRICULTURE. Great Britain
Farms. Taxation

CAMAMILE, G. H.
Taxation of farms and woodlands. - 2nd ed. - London : Institute of Chartered Accountants in England and Wales, 1982
xxii,171p. - Bibliography: pxxi-xxii. - Includes index
0-85291-316-8 X.529/51358
2144115

CAMAMILE, G. H. (Gerrard Hugh)
Taxation of farms and woodlands. - London : Institute of Chartered Accountants in England and Wales, 1981
xxii,170p. - (Chartac taxation guides). - Includes index
0-85291-290-0 X.529/41354
2144115

NATIONAL FARMERS' UNION
Tax on the farm. - 2nd ed. - London : The Union ; [Edinburgh] : N.F.U. of Scotland ; [Belfast] : Ulster Farmers' Union, 1979
70p
B80-26653 X.529/36672
2144115

AGRICULTURE. Great Britain
Farms. Taxation - Serials

FARM tax and finance. - Issue no.1 (Nov. 1982)-. - London : Sweet & Maxwell, 1982-
v.
B83-21472 P.513/3073
2893525

AGRICULTURE. Great Britain
Farms. Vernacular buildings

BRUNSKILL, R. W. (Ronald William)
Traditional farm buildings of Britain. - London : Gollancz in association with Peter Crawley, 1982
160p : ill. - Bibliography: p154-156. - Includes index
0-575-03117-4 X.421/23184
2779129

AGRICULTURE. Great Britain
Finance - For banking

CHECKLEY, Keith
Finance for farming. - London : Institute of Bankers, 1982
xi,144p,[8]p of plates : ill(some col.). - Includes index
0-85297-063-3 X.520/30134
2909693

AGRICULTURE. Great Britain
Financial management

CLERY, Peter
Farming finance. - Ipswich : Farming Press, 1975
145p : ill. - Includes index
0-85236-060-6 X.329/10259 Woolwich
0577936

AGRICULTURE. Great Britain
Forecasts: Food from our own resources

BODY, Richard
No way to feed a nation. - [London] : Selsdon Group, [1975?]
16p. - (A selsdon Group viewpoint series ; no.1)
B82-07143 X.529/45254
2608103

WALTON, Brian
Food from our own resources. - London : Economics Association, 1976
[2],24,[1]p : ill. - Bibliography: p.24
B77-17354 X.319/17333 Woolwich
1119109

AGRICULTURE. Great Britain
Government financed research & development. Coordination. Organisations: Joint Consultative Organisation for Research and Development in Agriculture and Food

JOINT CONSULTATIVE ORGANISATION FOR RESEARCH AND DEVELOPMENT IN AGRICULTURE AND FOOD
Reports of the Boards of the Joint Consultative Organisation for Research and Development in Agriculture and Food, and the responses from Agricultural Research Council [and] Department of Agriculture and Fisheries for Scotland [and] Ministry of Agriculture, Fisheries and Food. - London : H.M.S.O.
1976-77. - 1977
[3],158p
Subsequent volumes, when received, will be found at SRL
0-11-241045-6 SRL
0787833

AGRICULTURE. Great Britain
Implications of green money

SWINBANK, A.
The British interest and the green pound. - Reading : Centre for Agricultural Strategy, University of Reading, 1978
48p : ill. - (CAS papers / Centre for Agricultural Strategy ; 6). - Bibliography: p.45
0-7049-0608-2 X.319/18572 Woolwich
1759426

AGRICULTURE. Great Britain
Implications of law of conservation of environment - Conference proceedings

The IMPACT of environmental legislation upon agriculture / sponsored by the Anglo-German Foundation for the Study of Industrial Society held at Wye on March 14-16, 1984 ; edited by Bernarde Duesenberg, Bridget Girling. - Ashford : Wye College, 1984
viii,117p : ill. - (Seminar papers ; no.16)
0-905378-30-x X.322/14946
3280608

AGRICULTURE. Great Britain
Industrial health

AGRICULTURE / [Agricultural Health and Safety Inspectorate] ; [for the] Health and Safety Executive. - 1976. - London : H.M.S.O, 1978
20p,[2]p of plates : ill
Subsequent volumes, when received, will be found at SRL
0-11-883021-x SRL
1491385

AGRICULTURE. Great Britain
Industrial safety

AGRICULTURE / [Agricultural Health and Safety Inspectorate] ; [for the] Health and Safety Executive. - 1976. - London : H.M.S.O, 1978
20p,[2]p of plates : ill
Subsequent volumes, when received, will be found at SRL
0-11-883021-x SRL
1491377

AGRICULTURE. Great Britain
Industrial training. Organisations: Agricultural, Horticultural and Forestry Industry Training Board - Serials

AGRICULTURAL TRAINING BOARD
[Miscellaneous publications not catalogued separately.]. - [S.l.] : The Board, [1979-]
r6801996 X.0525/692
0041564

AGRICULTURE. Great Britain
Information systems

AGRICULTURE INFORMATION REVIEW COMMITTEE
Information systems in UK agriculture. - London : British Library Research and Development Department, 1979
v,74p. - (Researh and development reports / British Library ; no.5469)
0-905984-33-1 X.325/769 Woolwich
1834681

AGRICULTURE. Great Britain
Involvement of royal families, 1760-1979

WHITLOCK, Ralph
Royal farmers. - London : Joseph, 1980
256p : ill. - Bibliography: p.249-251. - Includes index
0-7181-1752-2 X.802/11165
2126273

AGRICULTURE. Great Britain
Manpower - Serials

AGRICULTURAL manpower. - No.1 (Autumn 1980)-. - Haslemere : Agricultural Manpower Society, 1980-
v. : ill
B81-13215 P.513/2583
2380528

AGRICULTURE. Great Britain
Marshall, Mr. - Biographies

HORN, Pamela
William Marshall (1745-1818) and the Georgian countryside. - [Abingdon] : Beacon Publications, 1982
88p,[1] leaf of plates : 1ill. - Includes index
B85-26820 X.809/64716
3297993

AGRICULTURE. Great Britain
Personnel. Careers guidance

BLAKE, R. E.
Retraining, alternative employment and career advice. - [London?] : Agricultural Development and Advisory Service, Ministry of Agriculture, Fisheries and Food, [1974?]
8p. - (Socio-economics paper ; 2)
B83-08410 BS.3/832(2)
283541x

AGRICULTURE. Great Britain
Personnel. Retraining

BLAKE, R. E.
Retraining, alternative employment and career advice. - [London?] : Agricultural Development and Advisory Service, Ministry of Agriculture, Fisheries and Food, [1974?]
8p. - (Socio-economics paper ; 2)
B83-08410 BS.3/832(2)
2151073

AGRICULTURE. Great Britain
Policies of government. Implications of Common Agricultural Policy of European Economic Community

MARSH, John S. (John Stanley)
UK agricultural policy within the European Community. - Reading : Centre for Agricultural Strategy, University of Reading, 1977
63p. - (CAS papers / Centre for Agricultural Strategy ; 1). - Bibliography: p.59
0-7049-0601-5 X.319/18770 Woolwich
1868470

AGRICULTURE. Great Britain
Policies of government. Influence of interest groups, 1956-1970

WILSON, Graham K. (Graham Keith)
Special interests and policymaking. - London : Wiley, 1977
[7],205p. - Bibliography: p.183-200. - Includes index
0-471-99495-2 X.320/11135 Woolwich
1355635

AGRICULTURE. Great Britain
Policies of government. International economic aspects

AGRICULTURE and the state / edited by Brian Davey, T.E. Josling and Alister McFarquhar. - London : Macmillan for the Trade Policy Research Centre, 1976
xxiii,280p : ill. - Bibliography: p.270-273. - Includes index
0-333-21194-4 X.329/10770 Woolwich
1028391

AGRICULTURE. Great Britain
Prehistoric agriculture - Conference proceedings

FARMING practice in British prehistory / edited by Roger Mercer. - Edinburgh : Edinburgh University Press, c1981
xxvi,245p : ill. - Includes bibliographies and index
0-85224-414-2 X.322/10293 Woolwich
2568551

AGRICULTURE. Great Britain
Productivity. Effects of land tenure

GASSON, Ruth
Farm tenure and performance. - [Ashford, Kent] : School of Rural Economics, Wye College, 1984
iv,124p. - Bibliography: p117-124
0-86266-008-4 X.322/14874
327232x

AGRICULTURE. Great Britain
Productivity. Effects of multiple land use
LUKEHURST, Clare T. (Clare Therese)
The effects of multiple land use policies upon the use and productivity of farmland. - Brighton : Brighton Polytechnic, 1981
42p : ill
0-907558-00-3 X.322/10103 Woolwich
2538229

AGRICULTURE. Great Britain
Protection, 1955-1981
BODY, Richard
Agriculture. - London : Temple Smith, 1982
139p : ill
0-85117-228-8 X329/16171
2881365

AGRICULTURE. Great Britain
Protection, 1964-1979
STRAK, John
Measurement of agricultural protection. - London : Macmillan for the Trade Policy Research Centre, 1982
xvi,120p : ill. - (Commercial policy issues ; no.6). - Bibliography: p117-120
0-333-32480-3 X.329/16212
2881357

AGRICULTURE. Great Britain
Protection. Theories of Body, Richard - Conference proceedings
AGRICULTURE - the triumph and the shame. - [Reading] : Centre for Agricultural Strategy, University of Reading, 1983
79p : 2ill. - Includes bibliographies
0-7049-0620-1 X.329/17012
9068473

AGRICULTURE. Great Britain
Research, 1931-1981
AGRICULTURAL research, 1931-1981 / edited by G.W. Cooke. - London : The Council, 1981
ix,367p : ill. - Includes bibliographies and index
0-7084-0180-5 SRL
2841568

AGRICULTURE. Great Britain
Research projects - Directories
REFERENCE information on agricultural economic research within the United Kingdom. - [London] : Ministry of Agriculture, Fisheries & Food, 1980
111p
B83-23639 BS.3/833
2900289

AGRICULTURE. Great Britain
Small farms. Draught animals. Use
BARTLETT, Jane
Nutrition and working efficiency of draught bovines on a Norfolk small holding. - Norwich : School of Development Studies, University of East Anglia, 1983
12p. - (Discussion paper ; no.127). - Bibliography: p11-12
B84-42808 X.322/14167
3192016

AGRICULTURE. Great Britain
Smallholdings
ALLABY, Michael
Making and managing a smallholding. - Newton Abbot : David and Charles, 1979
192p : ill. - ([The making and managing series]). - Includes index
0-7153-7803-1 X.529/34389 Woolwich
1869469
SEYMOUR, John, 1914-
The lore of the land. - Weybridge : Whittet, 1982
159p : ill. - Includes index
0-905483-23-5 X.322/10884 Woolwich
1869469
The lore of the land. - [London] : Corgi, 1984, c1982
159p : ill. - Includes index
0-552-99059-0 X.322/13955
1869469
The smallholder. - London : Sidgwick & Jackson, 1983
127p : ill. - (The Country treasury)
0-283-98923-8 X.629/20839
2937093
SWEENY, Sedley
The challenge of smallholding. - Oxford : Oxford University Press, 1985
x,222p : ill. - Bibliography: p203-206. - Includes index
0-19-286018-6 X.629/25696
1869469
THEAR, Katie
Part-time farming. - London : Ward Lock, 1982
192p : ill. - Includes index
0-7063-5932-1 X.322/11327
1869469

AGRICULTURE. Great Britain
Smallholdings. Cereals. Harvesting & drying
SMALLHOLDER'S harvest. - Stowmarket : Soil Association, 1977
16p : ill
0-905200-08-x X.329/16143
2878410

AGRICULTURE. Great Britain
Smallholdings. Feedingstuffs: Hay. Harvesting & drying
SMALLHOLDER'S harvest. - Stowmarket : Soil Association, 1977
16p : ill
0-905200-08-x X.329/16143
2878429

AGRICULTURE. Great Britain
Smallholdings. Livestock. Management
ISAAC, Peter, 1932-
Taking stock. - London : Jill Norman, 1981
168p : ill
0-906908-42-6 X.529/41320
2404796

AGRICULTURE. Great Britain
Subsidies by government, 1945-1980
BOWERS, J. K. (John Kenneth)
Agriculture, the countryside and land use. - London : Methuen, 1983
x,170p : ill. - Includes index
0-416-31830-4 X.329/17385
3047482

AGRICULTURE. Great Britain
Theses accepted by universities, 1876-1978 - Bibliographies
MORGAN, Raine
Dissertations on British agrarian history. - [Reading] : University of Reading, 1981
xxvi,170p. - (Institute of Agricultural History bibliographies in agricultural history ; no.2). - Bibliography: p137-138. - Includes index
0-7049-0705-4 X.529/42423
2533545
- [Another copy] OPL

AGRICULTURE. Great Britain
Tied cottages
GASSON, Ruth
Provision of tied cottages. - Cambridge : University of Cambridge, Department of Land Economy, 1975
[8],147p. - (Occasional papers / University of Cambridge. Department of Land Economy ; no.4). - Bibliography: p.146-147
0-901978-61-2 X.0329/97(4)
0502219
IRVING, Barrie Leslie
Tied cottages in British agriculture. - London : Tavistock Institute of Human Relations ; London : Distributed by Research Publications Services Ltd. Working paper no.1: Basic statistics. - 1975
[6],104p : ill
B77-00621 X.0319/1010 Woolwich
0502219

AGRICULTURE. Great Britain
Tied cottages. Abolition
GREAT BRITAIN. Department of the Environment
Abolition of the tied cottage system in agriculture. - [London] : [The Department], [1976]
33p
0-7184-0060-7 BS.414/273
0701262

AGRICULTURE. Great Britain
Trade unions. Arch, Joseph - Biographies
ARCH, Joseph
Joseph Arch. - New York ; London : Garland, 1984
412p. - (The World of labour)
0-8240-5700-7 X.958/29852
3285812

AGRICULTURE. Great Britain
Trade unions. Tolpuddle Martyrs, 1834-1836
HUMPHRIES, Barbara
The Tolpuddle Martyrs. - London : World Books, [1976]
[14]p
0-9504915-1-9 X.702/5242
0743798

AGRICULTURE. Great Britain
Use of pesticides. Safety measures
GREAT BRITAIN. Health and Safety Executive
Poisonous chemicals on the farm. - London : H.M.S.O., 1978
iii,38p : ill. - (HS(G) ; 2)
0-11-883215-8 SRL
1594567

AGRICULTURE. Great Britain
Use of petroleum products. Organisations: Shell UK Oil (Firm). Heating and Agricultural Division - Serials
LINK. - [No.1]- ; Winter 1975-76-. - London : Shell UK Oil, [1975]- : ill(some col)
r6079019 P.313/439
1405012

AGRICULTURE. Great Britain
Women personnel
GASSON, Ruth
The role of women in British agriculture. - [Ashford, Kent] : Wye College, 1980
72p. - Bibliography: p57-58
B82-27668 X.329/15401 Woolwich
275102x

AGRICULTURE. Great Britain
Women personnel. Organisations: Women's Land Army, 1939-1945 - Biographies
ABBOTT, Dorothea
Librarian in the Land Army. - Stratford-upon-Avon : D. Abbott, c1984
69p,[1]p of plates : ill
0-9509396-0-9 X.809/61378
317333x

AGRICULTURE. Great Plains. North America
to 1976 - Conference proceedings
CULTURAL HERITAGE OF THE PLAINS SYMPOSIUM (1977 : University of Nebraska-Lincoln)
The Great Plains. - Lincoln [Neb.] ; London : University of Nebraska Press for the Center for Great Plains Studies, University of Nebraska-Lincoln, 1979
xxviii,246p : ill
0-8032-1155-4 X.520/15042
2050099

AGRICULTURE. Greece
The MARKET for farm inputs, consultancy and agro-industrial projects in Greece. - London : BAEC, 1981
iv,48p : ill
B82-11353 SRL
2625733

AGRICULTURE. Greece
to date
STAUROPOULOS, Theodōros
Historikē analysē tou agrotikou zētēmatos stēn Hellada. - Athena : Nea synora, 1979-
v. - (Politikē, oikonomia, koinōniologia). - Includes bibliographies
r6501794 X.0329/229 Woolwich
8766126

AGRICULTURE. Greece
1935-. Economic aspects
AVDELIDĒS, Parmeniōn S.
Themata agrotikēs oikonomias. - Athēna : Nea synora, 1981
122p. - (Politikē, oikonomia, koinōniologia)
r6503914 X.329/16677
8755566
KAMARINOU, Lilē
[L'accumulation du surplus agricole grec. Modern Greek]. Geōrgia kai anaptyxiakē diadikasia stēn Hellada. - Athēna : Nea synora, [1977?]
277p. - (Synchrona hellēnika provlēmata ; 3)
r6509024 X.329/16735
8755566
PAPAIOANNOU, Anastasios
Der Agrarsektor Griechenlands und die Problematik seiner Integration in die Europäische Gemeinschaft. - Hannover : Strothe, 1980
xvi,389p. - (Agrarwirtschaft. Sonderheft ; 85). - Bibliography
r6301308 X.329/14455 Woolwich
8755566

AGRICULTURE. Greece
Social aspects, 1949-1974
WEINTRAUB, Dov
Rural reconstruction in Greece. - Beverly Hills ; London : Sage Publications, 1975
80p. - (Sage research papers in the social sciences. studies in comparative modernization series ; vol.3, no.90-017). - Bibliography: p.79-80
0-8039-0487-8 X.519/25613
061579x

AGRICULTURE. Guam
FAIRBAIRN, Te'o Ian
Agriculture in Guam. - Noumea : South Pacific Commission, 1977
[1],i,19p. - (South Pacific Commission. Occasional paper ; no.1). - Bibliography : p.19
r6072102 UNP.38/18(1)
1516493

AGRICULTURE. Guatemala
1500–1800

SOLANO, Francisco de
Tierra y sociedad en el Reino de Guatemala. - Guatemala : Editorial Universitaria, 1977
462p. - (Colección "Realidad nuestra" ; v.4)
r6162354 X.329/13197 Woolwich
2454270

AGRICULTURE. Guatemala
Economic development, to 1970

VILLACORTA ESCOBAR, Manuel
Apuntes de economia agricola. - Guatemala : Editorial Universitaria, 1973
146 p.. - (Colección Aula ; v. 12)
r6011746 X.311/4626 Woolwich
0888427

AGRICULTURE. Guiting Power. Gloucestershire
Farms: Cotswold Farm Park. Livestock. Rare breeds

COTSWOLD FARM PARK
Cotswold Farm Park, rare breeds survival centre, Guiting Power. - [Cheltenham] : [The Park], [1978?]
[12]p : ill(chiefly col)
B79-32619 X.319/19107 Woolwich
1954326

AGRICULTURE. Gwynedd
- Serials

AMAETHWR Meirion. - Vol.1, no.1 (Oct.1980)-. - Dorchester : N.F.U. Pubications, 1980–
v. : ill
Holdings: Vol.1 No.3 ; Dec 1980, etc
B81-13214 P.511/5260
238051x

AMAETHWR môn. - Vol.1, no.1 (Oct. 1980) - v.4, no.2 (Nov. 1983). - Dorchester : N.F.U. County Publications, 1980–1983
4v.
Holdings: vol 3, no 4–
B84–18304 P.511/5260
238051x

AGRICULTURE. Haciendas. Colombia
to date

TOVAR, Hermes
Grandes empresas agrícolas y ganaderas. - Bogotá : CIEC, 1980
210p
r6709551 X.329/19330
9168451

AGRICULTURE. Haryana. India (Republic)
Geographical aspects

SINGH, Jasbir
An agricultural geography of Haryana. - Kurukshetra : Vishal Publications, 1976
[1],ix,457p : ill. - Bibliography-Includes index
r6031931 X.320/10736 Woolwich
1116525

AGRICULTURE. Herefordshire
Effects of drought, 1976

COX, Evelyn
The great drought of 1976. - London : Hutchinson, 1978
153p : 1 ill
0–09–133200–1 X.329/11473 Woolwich
1280651

AGRICULTURE. Hertfordshire
Smallholdings

BRANDER, Laurence
Four acres of our own. - North Walsham : Warren House Press, 1979
[9],62p
B80–29891 X.909/44966
208094x

AGRICULTURE. Highlands & Islands. Scotland
1965–1975

BRYDEN, John
Agrarian change in the Scottish Highlands. - London : Martin Robertson [for] the Board, 1976
xiv,152p : ill. - (Glasgow social & economic research studies ; 14). - Bibliography: p.143–145. - Includes index
0–85520–151–7 X.320/10690 Woolwich
1057693

AGRICULTURE. Highlands & Islands. Scotland
Crofting, to 1979

GRANT, James Shaw
The part-time holding. - [Langholm] : Arkleton Trust, c1984
28p
B84–35405 X.529/63907
2354861

STEWART, Katharine
Crofts and crofting. - Edinburgh : William Blackwood, 1980
[8],64p : ill. - (Scottish connection). - Bibliography: p[6]
0–85158–137–4 X.329/13447 Woolwich
2354861

AGRICULTURE. Highlands & Islands. Scotland
Crofting. Law

GREAT BRITAIN. Crofters Commission
Guide to the Crofting Acts. - Edinburgh : H.M.S.O., 1976
46p
0–11–491437–0 BS.151/89
1040073

AGRICULTURE. Highlands. Bolivia
Effects of towns

PRESTON, David A. (David Anthony)
Farmers and towns. - Norwich : Geo Abstracts Ltd, 1978
xi,196p : ill. - Bibliography: p.186–194. - Includes index
0–86094–009–8 X.800/27966
1705830
- [Another copy: paperback]
ISBN 0–86094–008–x X.700/25837

AGRICULTURE. Highlands. Morelos (State). Mexico
1800–1980. Political aspects

PEÑA, Guillermo de la
A legacy of promises. - Manchester : Manchester University Press, c1982
ix,289p : ill. - (Manchester Latin American studies). - Bibliography: p271–283. - Includes index
0–7190–0766–6 X.800/33221
2113678

AGRICULTURE. Highlands. Scotland

ORR, Willie
Deer forests, landlords and crofters. - Edinburgh : John Donald, c1982
viii,226p : 2ill. - Bibliography: p220–223. - Includes index
0–85976–081–2 X.800/34590
2831368

AGRICULTURE. Highlands. Scotland
1700–1850

FAIRWEATHER, Barbara
Old Highland farming. - [Glencoe] : Glencoe and North Lorn Folk Museum, c1978
12p
B84–00013 X.329/17040
2978873

AGRICULTURE. Hjortlund. Denmark
Farms: Ovringgard. Social conditions, 1909–1925

JØRGENSEN, Knud Martin
Fra min barndom på Ovringgard, Hedegård i Hjortlund sogn. - Ringsted : Jørn Jørgensen, 1975
123 p. : ill
87–980260–3–8 X.709/20990
0888583

AGRICULTURE. Honduras
Development, to 1974

SANTOS DE MORAIS, Clodomir
El modelo hondureño de desarrollo agrario. - Tegucigalpa : Instituto Nacional Agrario, 1975
141p.(some fold.). - (Serie didáctica / PROCCARA ; no.3)
r6053775 X.319/9251 Woolwich
1424157

AGRICULTURE. Hong Kong
Effects on natural environment, to date

WONG, C. T.
Environmental implications of agricultural development in Hong Kong. - Hong Kong : Department of Geography, Chinese University of Hong Kong, 1983
22leaves. - (Occasional paper ; no.47)
r6813247 X.322/13302
9078177

AGRICULTURE. Horsley. Northumberland
Farms: Slackfields Farm, 1872–1896 - Correspondence, diaries

HILL, Henry
Diaries of Henry Hill of Slackfields Farm, 1872–1896. - Nottingham : University of Nottingham, Department of Adult Education, c1982
[180]p : ill. - (Centre for Local History record series ; 4)
0–902031–70–8 X.809/56213
2949725

AGRICULTURE. Huelva (Province). Spain
1960–1970

ROUX, Bernard
[La crise du capitalisme latifundaire dans les sierras de l'Andalousie Occidentale. Spanish]. Crisis agraria en la sierra andaluza. - Sevilla : Universidad de Sevilla, 1975
274 p., 2 fold. leaves of plates. - (Ediciones del Instituto de Desarrollo Regional ; no. 3)
84–85057–60–0 X.319/8587 Woolwich
1006495

AGRICULTURE. Humberside
Farms. Intensive animal housing. Location - For environment planning

HUMBERSIDE. County Council
Intensive livestock units subject plan. - [Beverley] : The Council, 1979
13,[28]p,[3]leaves of plates(2 fold)
B80–35292 X.325/1259
1478346

HUMBERSIDE. Planning Department
Intensive livestock units. - [Beverley] : [The Department], 1978
12,[24]p,[2]leaves of plates(1 fold)
B78–20400 X.313/1551 Woolwich
1478346

AGRICULTURE. Hungary

BENET, Iván
Az agrár-ipari komplexum elméleti kérdései. - Budapest, 1978
98p : ill. - (A Magyar Tudományos Akadémia Közgazdaságtudományi Intézetének közleményei. Új sorozat ; 8)
r6183527 X.0510/399(8)
221895X

CSIZMADIA, Ernő
Socialist agriculture in Hungary. - Budapest : Akadémiai Kiadó, 1977
178p
963–05–1184–3 X.520/20328
221895X

DONÁTH, Ferenc
Reform and revolution. - Budapest : Corvina Kiadó, 1980
489p
963–13–0911–8 X.329/13406 Woolwich
2249960

MERRE tart a magyar mezőgazdaság? / Bíró Ferenc [et al.]. - Budapest : Kossuth Könyvkiadó, 1980
264p
963–09–1549–9 X.329/12831 Woolwich
243394X

ROSENFELD, Claude H.
Hungarian agriculture. - Ashford, Kent : Wye College, 1979
[1],iv[i.e. v],57[i.e.58]p : 1 ill. - (Miscellaneous studies / Wye College. Centre for European Agricultural Studies ; no.6). - Bibliography: p.48–49
0–905378–20–2 BS.475/8
2272245

AGRICULTURE. Hungary
to date

ECONOMIC studies on Hungary's agriculture / edited by Iván Benet and János Gyenis. - Budapest : Akadémiai Kiadó, 1977
193p
963–05–1112–6 X.320/5382 Woolwich
8247447

A MEZŐGAZDASÁG története / [szerkesztette Gunst Péter, Lőkös László]. - Budapest : Mezőgazdasági Kiadó, 1982
269p. - Bibliography
963–231–427–1 X.322/11732
8247447

AGRICULTURE. Hungary
to date - Biographies

AGRÁRTÖRTÉNETI életrajzok / szerkesztette : Für Lajos és Pintér János. - Budapest : Magyar Mezőgazdasági Múzeum, 1985
711p
963–01–5556–7 X.800/41568
8247447

AGRICULTURE. Hungary
to date - Conference proceedings

UNGARISCH-ÖSTERREICHISCHE HISTORIKERKOMMISSION. Wissenschaftliche Tagung (1st : 1978 : Budapest, Hungary) Demographie, Bevölkerungs- und Agrarstatistik. - Budapest : Akadémiai Kiadó, 1982
129 p. : ill.. - Includes bibliographical references
963–05–2830–4 X.800/38428
8247447

AGRICULTURE. Hungary
1700–1800

WELLMANN, Imre
A magyar mezőgazdaság a XVIII. században. - Budapest : Akadémiai Kiadó, 1979
194p. - (Agrártörténeti tanulmányok ; 6)
963-05-1910-0 X.0311/235(6)
219340X

AGRICULTURE. Hungary
1800–

A MAGYAR mezőgazdaság a XIX - XX században (1849-1949). - Budapest : Akadémiai Kiadó, 1976
471p. - (Agrártörténeti tanulmányok ; 4)
963-05-1036-7 X.0311/235(4) Woolwich
1973061

AGRICULTURE. Hungary
1800– - Conference proceedings

The MODERNIZATION of agriculture / [edited by] Joseph Held. - Boulder : East European Monographs, 1980
508p. - (East European monographs ; no.63)
0-914710-60-5 X.329/12732 Woolwich
1973061

AGRICULTURE. Hungary
1919–

REINERT-TÁRNOKY, Ilona
Radikale Bauernpolitik in Ungarn. - München : Trofenik, 1985
168p. - (Studia hungarica ; 26). - Includes bibliography
r6311075 X.0800/33(26)
8552762

AGRICULTURE. Hungary
1935–1970

ERDEI, Ferenc
[Selections]. Emberül élni. - Budapest : Gondolat, 1974
670 p.
963-280-073-7 X.808/10104
0863815

Agrárgazdasági tanulmányok. - Budapest : Akadémiai Kiadó, 1976
3v.. - (Erdei Ferenc összegyűjtött művei)
963-05-0625-4 X.0320/87(5) Woolwich
0863815

AGRICULTURE. Hungary
1945– - Statistics - Serials

HUNGARY. Központi Statisztikai Hivatal. Mezőgazdasági Statisztikai Főosztály
Mezőgazdasági adattár .... - 1-. - Budapest : Statisztikai Kiadó Vállalat, 1978–
v.
963-340-184-4 S.460/76
8945209

AGRICULTURE. Hungary
1945– - Statistics - Serials

HUNGARY. Központi Statisztikai Hivatal. Mezőgazdasági Statisztikai főosztály
Mezőgazdasági statisztikai évkönyv .... - 1980-. - Budapest, 1981–
r6499254 S.460/94
8945209

AGRICULTURE. Hungary
Collectivisation. Social aspects, 1945–1980

BELL, Peter D.
Peasants in socialist transition. - Berkeley ; London : University of California Press, c1984
ix,322p : ill. - Bibliography: p307-312. - Includes index
0-520-04157-7 X.520/36072
3068072

NEW Hungarian peasants / Marida Hollos and Bella C. Maday, editors. - New York : Social Science Monographs, 1983
ix,41p : ill. - (East European monographs ; no.134) (Atlantic studies) (Brooklyn College studies on society in change ; no.26). - Bibliography: p331-340
0-88033-024-4 X.529/59440
3068072

AGRICULTURE. Hungary
Farms. Social conditions, to date

A MAGYAR tanyarendszer múltja / szerkesztette Pölöskei Ferenc, Szabad György. - Budapest : Akadémiai Kiadó, 1980
449p,[24]p.of plates : ill. - Bibliography
963-05-1893-7 X.322/10330 Woolwich
8828865

AGRICULTURE. Hungary
Law, 1945–

HUNGARY
[Laws, etc]. Mezőgazdasági és élelmezésügyi jogszabályok gyűjteménye. - Budapest : Közgazdasági és Jogi Könyvkiadó, 1983
2v.(1959p.)
963-221-254-1 S.181/2
8957185

AGRICULTURE. Hungary
Personnel. Left-wing political movements. Latinca, Sándor

KÁVÁSSY, Sándor
Latinca Sándor. - Budapest : Akadémiai Kiadó, 1973
163 p., 1 leaf of plates. - (Értekezések a történeti tudományok köréből. Új Sorozat ; 71)
r6007890 Ac.825.b/2 [71]
0859109

AGRICULTURE. Hungary
Personnel. Remuneration related to productivity, 1945–

NAGY, Lajos, 1926–
Munkatermelékenység és személyi jövedelmek a mezőgazdaságban. - Budapest : Közgazdasági és Jogi Könyvkiadó, 1976
407p. - Includes bibliography
963-220-394-1 X.629/24869
9137823

AGRICULTURE. Hungary
Policies of government

PRUGBERGER, Tamás
Az állami felügyelet kulcskérdései a mezőgazdasági termelőszövetkezetekben. - Budapest, 1979
174p. - (-Közlemények / Szövetkezeti Kutató Intézet ; 138)
r6183776 X.0512/379(138)
246246x

AGRICULTURE. Hungary
Policies of government, 1945–

AGRÁRPOLITIKA és agrárátalakulás Magyarországon / szerkesztette : Balogh Sándor és Pölöskei Ferenc. - Budapest : Akadémiai Kiadó, 1979
240p
963-05-1836-8 X.322/11755
8916446

AGRICULTURE. Hungary
Smallholdings

JUHÁSZ, János
A háztáji gazdaságok léte és kilátásai. - Budapest : Szövetkezeti Kutató Intézet, 1975
159p. - (Közlemények / Szövetkezeti Kutató Intézet ; 99)
r6015837 X.0512/379(99)
1238841

KENÉZ, Győzőné
Adalékok a háztartások önellátó és árutermelésének kérdéséhez. - Budapest, 1976
92p. - (Közlemények / Szövetkezeti Kutató Intézet ; 113)
r6183777 X.0512/379(113)
1238841

VITA a mezőgazdasági kistermelésről / szerkesztette Vargha László. - Budapest, 1977
120p. - (A Magyar Tudományos Akadémia Köngazdaságtudományi Intézetének közleményei. Új sorozat ; 7)
r6183526 X.0510/399(7)
1238841

AGRICULTURE. Hungary
Tessedik, Sámuel - Biographies

PENYIGEY, Dénes
Tessedik Sámuel. - Budapest : Akadémiai Kiadó, 1980
273p : ill. - (Agrártörténeti tanulmányok ; 9)
963-05-2378-7 X.0311/235(9)
8706743

AGRICULTURE. Huron (County). Ontario
Farms. Linkages. Spatial analysis

HOGGART, Keith
The determinants of farm linkage patterns. - London : University of London, King's College, Department of Geography, 1979
[1],25p : ill. - (Occasional papers / King's College (London). Geography Department ; no.10). - Bibliography: p.23-25
B79-35258 X.512/7026
1969218

AGRICULTURE. Hvidovre. Denmark
Farms: Beringgard, to 1975

SØRENSEN, Sv Aa
Beringgard. - [Hvidovre] : Lokalhistorisk Arkiv, Hvidovre Kommune, 1976
33p : ill
r6034378 X.319/9188 Woolwich
1395351

AGRICULTURE. Iceland
Economic development - Forecasts

RANNSÓKNARÁD RÍKISINS
Þróun landbúnadar. - Reykjavík : Rannsóknaráð ríkisins, 1976
178p
r6060144 X.319/9426 Woolwich
1511955

AGRICULTURE. Iceland
Farms. Social conditions, 1262–1944

BJÖRN TEITSSON
Eignarhald og ábúð á jördum i Sudur-Pingeyjarsýslu 1703-1930. - Reykjavík : Menningarsjódur, 1973
183p. - (Sagnfrædirannsóknir = Studia historica ; bd.2). - Bibliography
r6600101 X.322/10215 Woolwich
876221x

AGRICULTURE. Iceland
Farms. Social conditions, 1850–1930

GÍSLASON, Magnús
Kvällsvaka. - Uppsala ; Stockholm : Distributor : Almqvist & Wiksell International, 1977
218p : ill. - (Studia ethnologica upsaliensia ; 2)
91-554-0553-3 X.0700/869(2)
1324667

AGRICULTURE. Iddesleigh. Devon
1976–1977

MORPURGO, Michael
All around the year. - London : J. Murray, 1979
208p : ill. - Includes index
0-7195-3577-8 X.320/11684 Woolwich
1884204

AGRICULTURE. India

A REGIONAL profile of Indian agriculture / edited by Vadilal Dagli. - Bombay : Vora & Co. on behalf of Commerce, 1974
iv, 311 p.. - (Commerce economic studies ; no. 11)
r6025796 X.320/5306 Woolwich
1026216

AGRICULTURE. India
to date

GOPAL, Lallanji
Aspects of history of agriculture in ancient India. - Varanasi : Sri Prakash Pandey [for] Bharati Prakashan, 1980
viii,185p
r6804761 X.329/14417 Woolwich
8612005

LAPING, Johannes
Die landwirtschaftliche Produktion in Indien. - Wiesbaden : Steiner, 1982
x,155p. - (Beiträge zur Südasien-Forschung ; Bd.62). - Bibliography
3-515-03521-4 X.520/29768
8612005

SAMPATH, R. K.
Economic efficiency in Indian agriculture. - Delhi : Macmillan, 1979
viii,191p
0-333-90285-8 X.329/14569 Woolwich
8612005

AGRICULTURE. India
1793–1975

SEN, Chiranjib
Essays on the transformation of India's agrarian economy. - New York ; London : Garland, 1984
285p : ill. - (Outstanding dissertations in economics). - Bibliography: p278-285
0-8240-4155-0 X.520/36859
3261638

AGRICULTURE. India
Bonded employment - Conference proceedings

CULTURAL ACTION FOR SOCIAL CHANGE (Conference : 1974 : New Delhi)
Bonded labour in India. - Calcutta : India Book Exchange, 1976
[5],x,136,lxxxxvip. - Bibliography : p.lxxxix-lxxxxix.-Includes index
r6032179 X.529/31703
1195360
- [Another copy] X.108/15211
- [Another copy.] T 46170 IOL

AGRICULTURE. India (Republic)

AGRAWAL, A. N.
Indian agriculture. – 2nd ed. – New Delhi : Vikas Publishing, 1981
xi,556p. – Bibliography: p554–556
0–7069–1378–7 X.329/18108
3177165

BASU, Asok
Technological possibilities of Indian agriculture. – Calcutta : Firma KLM, 1978
185p
r6093140 X.629/12814 Woolwich
1805428

DR Punjabrao Deshmukh and his team of workers. – Madras : Bharat Krishak Samaj, Tamilnadu Branch, 1978
153p.
r6108137 X.320/5379 Woolwich
1805428

AGRICULTURE. India (Republic)
to date

AGRAWAL, A. N.
Indian agriculture. – Sahibabad : Vikas Publishing House, 1980
xi,559p
0–7069–0688–8 X.329/14421 Woolwich
8726450

DASGUPTA, Biplab
The new agrarian technology and India. – Delhi : S.G. Wasani for Macmillan Company of India, 1980
xxxi,405p. – Bibliography: p.383–405
r6804556 X.329/14390 Woolwich
8726450

AGRICULTURE. India (Republic)
to date – Bibliographies

SATYAPRAKASH, 1927–
Agriculture. – Gurgaon : Indian Documentation Service, 1977
384p. – (Subject bibliography series ; no.3)
r6810170 X.322/11990
8726450

AGRICULTURE. India (Republic)
ca 1900–1980

HARRISS, Barbara
The process of the commercialisation of agricultural commodities in India. – Norwich : School of Development Studies, University of East Anglia, 1982
18p. – (Discussion paper ; no.108). – Bibliography: p17–18
B84–40807 X.525/9952
3187438
– [Another copy.] W 6967(e) IOL

AGRICULTURE. India (Republic)
1947–

SEN, Sudhir
A richer harvest. – New Delhi : Tata McGraw-Hill, [1974]
xxiii,573p. – Includes bibliography
r6807217 X.322/13026
9049452

AGRICULTURE. India (Republic)
Administration

ROY, Ramashray
Bureaucracy and development. – New Delhi : Manas Publications, 1975
176 p. : ill. – Includes index
r6010710 X.329/10805 Woolwich
1049798

AGRICULTURE. India (Republic)
Development

ARORA, R. C.
Development of agriculture and allied sectors. – New Delhi : Chand, 1976
xii,212p, fold plate : ill. – Bibliography : p. 207–208.–Includes index
r6032127 X.329/11091 Woolwich
0386642

MENAMKAT, Alexander
Indian agriculture. – Montreal : Centre for Developing-Area Studies, McGill University, 1976
10 leaves. – (Working papers ; no.17)
r6189698 X.311/5195 Woolwich
0386642

NATIONAL COUNCIL OF APPLIED ECONOMIC RESEARCH
Credit requirements for agriculture. – New Delhi : The Council, 1974
xii,121p.
r6028607 X.320/5218 Woolwich
0386642
Demand for tractors. – New Delhi : The Council, 1974
xi,76p.
r6028606 X.529/20686
0386642

SWAMINATHAN, M.S
Agricultural evolution – Mysore : University of Mysore, 1974
48p. – (Princess Leelavathi memorial lectures. Special lectures ; no. 64)
r6028698 X.319/8659 Woolwich
0386642

AGRICULTURE. India (Republic)
Development compared with development of agricultural industries in China

BANDYOPADHYAYA, Kalyani
Agricultural development in China and India. – New Delhi : Wiley Eastern, 1976
xi,204p. – Bibliography : p.185–199. – Includes index
0–85226–054–7 X.329/11644 Woolwich
1693670

AGRICULTURE. India (Republic)
Development. Management aspects

PANDEY, Shiv Mohan
Development of marginal farmers and agricultural labourers. – New Delhi : Shri Ram Centre for Industrial Relations and Human Resources, [1974]
viii,116p
r6025653 X.319/8719 Woolwich
0322059

AGRICULTURE. India (Republic)
Development. Planning by commercial banks

GEORGE, P. S.
Planning and implementation of rural development projects by a banking organization. – Ahmedabad : Centre for Management in Agriculture, Indian Institute of Management, [1975]
[2]p,ii–v leaves, 140 [i.e. 147]p (1 fold) : ill. – (Monograph ; no.55)
r6032359 X.515/550
1256890

AGRICULTURE. India (Republic)
Finance, to date

MOHANAN, N.
Development finance for small farmers. – Coimbatore : Rainbow, 1981
298p. – Includes bibliography
r6810513 X.529/59122
8860637

ROLE of co-operative credit in increasing farm production / [prepared by] Reserve Bank of India. – Bombay : The Bank, 1974
x,214p
r6112397 IS.488/242
8860637
– [Another copy.] IS.488/41

SHARMA, R. D.
Agricultural finance in India. – Bhagalpur : Bharat Book Depot, [1981?]
ii,ii,viii,237p. – Bibliography
r6805673 X.529/46952
8860637

AGRICULTURE. India (Republic)
Finance. Planning by commercial banks

DESAI, D. K.
Planning and implementation of financing agriculture through area approach. – Ahmedabad : Centre for Management in Agriculture, Indian Institute of Management, 1973
[6]p,iii leaves, 148 [i.e.153]p(1 fold), leaf of plate : ill
r6032355 X.313/1439 Woolwich
1256904

AGRICULTURE. India (Republic)
Geographical aspects

MAMORIA, C. B. (Chatur Bhuj)
Geography of India. – Agra : Shiva Lal Agarwala and Co., 1975
xxii,703p,[2]fold leaves,[3]fold leaves of plates : ill. – Bibliography : p. 701–703
r6031873 X.800/26083
110912X

AGRICULTURE. India (Republic)
Income. Distribution. Effects of technological change

HANUMANTHA RAO, Chennamaneni
Technological change and distribution of gains in Indian agriculture. – Delhi : Macmillan, 1975
xv,249p : ill. – (Studies in economic growth ; no.17). – Includes index
0–333–90098–7 Ac:9233.mc/3(17)
180216X
– [Another copy] X.529/33035

AGRICULTURE. India (Republic)
Legislation, 1793–1966 – Bibliographies

WADHWA, D. C.
Agrarian legislation in India (1793–1966). – Poona : Gokhale Institute of Politics and Economics [etc.], 1973–
Vol. 1–. – (Gokhale Institute studies ; no. 61)
r6004522 Ac.2366 [no.61]
0695114

AGRICULTURE. India (Republic)
Modernisation

SINGH, Shrinath
Modernisation of agriculture. – New Delhi : Heritage, 1976
xiii,238p,fold leaf. – Bibliography: p.155–161
r6072393 X.320/11438 Woolwich
1688189

AGRICULTURE. India (Republic)
Policies of Communist Party of India

SOME problems concerning the agrarian movement and our tasks. – New Delhi : Tarun Sengupta for the Communist Party of India, [1975]
[2], 32p. – (Communist Party publication ; no.10. April 1975 (C134))
r6071972 X.709/32207
1451006

AGRICULTURE. India (Republic)
Policies of government, 1947–1974

BANSIL, P. C. (Puran Chand)
Agricultural problems of India. – 2nd, revised and enlarged ed. – Delhi : Vikas, 1975
xi,608p
0–7069–0363–3 X.319/8700 Woolwich
0393932

AGRICULTURE. India (Republic)
Policies of government, 1947–

KRISHNA, Daya
Indian farmer. – New Delhi : Swan Publishers, 1980
viii,207p
r6803661 X.329/14566 Woolwich
8250960

AGRICULTURE. India (Republic)
Policies of government, 1947– – Conference proceedings

ALL-INDIA SEMINAR ON INTENSIVE AGRICULTURAL DISTRICT PROGRAMME (1967 : Madras University)
Agricultural development of India. – Madras : University of Madras, 1972
302p. – (Madras University publications ; no.20). – Bibliography: p.283–300
r6805229 X.622/12147 Woolwich
8250960

AGRICULTURE. India (Republic)
Research organisations: Indian Council of Agricultural Research. Organisation structure

CHOWDHRY, Kamla
An organisation study of the Indian Council of Agricultural Research. – Ahmedabad : Indian Institute of Management, Centre for Management in Agriculture, 1972
iii,71p.
r6028506 X.311/4709 Woolwich
0388491

AGRICULTURE. India (Republic)
Role in economic development

GEORGE, P. S.
The oilseeds economy of India. – New Delhi : Macmillan Company of India Ltd, 1978
274p
0–333–90193–2 X.529/34469
1073109

RASTĪANNIKOV, V. G. (Viktor Georgievich)
Agrarian evolution in a multiform structure society. – London : Routledge & Kegan Paul, 1981
x,373p. – Bibliography: p347–363. – Includes index
0–7100–0755–8 X.809/48612
1073109

SINHA, Sachchidanand
The bitter harvest. – New Delhi : Manas, 1975
viii, 134 p.
r6031433 X.529/31312
1073109

AGRICULTURE. India (Republic)
Role in economic development of rural regions, 1963–1979

ETIENNE, Gilbert
India's changing rural scene. – Delhi ; Oxford : Oxford University Press, 1982
x,231p. – Bibliography: p217–220. – Includes index
0–19–561429–1 X.529/59135
3056740
– [Another copy.] T 45619 IOL

AGRICULTURE. India (Republic)
Role of capitalism

HARRISS, John
The mode of production controversy. - Norwich : School of Development Studies, University of East Anglia, 1983
58p. - (Discussion paper ; no.60). - Bibliography: p54-58
B85-03251 X.322/14316
321298x
- [Another copy.] W 7034 IOL

AGRICULTURE. India (Republic)
Role of irrigation, to date - Conference proceedings

SEMINAR ON ROLE OF IRRIGATION IN THE DEVELOPMENT OF INDIA'S AGRICULTURE (1974 : Bangalore)
[Papers and proceedings]. - Bombay : Indian Society of Agricultural Economics in collaboration with Institute for Social and Economic Change, 1976
xx,201p. - (Seminar series ; no.13)
r6803596 X.800/31219
8749485

AGRICULTURE. India (Republic)
Socioeconomic aspects

NAIR, Kusum
In defense of the irrational peasant. - Chicago ; London : University of Chicago Press, 1979
xxi,154p. - Bibliography: p.145-152.- Includes index
0-226-56798-2 X.329/12481 Woolwich
2064197

AGRICULTURE. Indonesia
- For British exporters

A REPORT on the agricultural industry in Indonesia. - London : British Agricultural Export Council, 1982
iv,59p : 1ill. - Bibliography: p59
B83-17129 SRL
2877244

AGRICULTURE. Indonesia
to date

AGRICULTURAL and rural development in Indonesia / edited by Gary E. Hansen. - Boulder : Westview Press, 1981
xiv,312p. - (Westview special studies in social, political, and economic development)
0-86531-124-2 X.322/10792 Woolwich
882021x

AGRICULTURE. Iowa
Farms. Tenancies, 1850-1900

WINTERS, Donald Lee
Farmers without farms. - Westport, Conn. ; London : Greenwood Press, 1978
xvi,145,[1]p. - (Contributions in American history ; no.79). - Includes index
0-313-20408-x X.809/44441
1902369

AGRICULTURE. Iran
1900-1975

ARESVIK, Oddvar
The agricultural development of Iran. - New York ; London : Praeger, 1976
xxv,271,[1]p : ill. - (Praeger special studies in international economics and development). - Bibliography: p.262-271
0-275-00170-9 SRL
1033743

CHIPASHVILI, Georgiĭ Severianovich
Istorii͡a vozniknovenii͡a i razvitii͡a sel'skokhozi͡aĭstvennykh kooperativov v Irane i ikh mesto v sisteme agrarnykh preobrazovaniĭ strany. - Tbilisi : Mets̆niereba, 1979
264p
r6183438 X.529/36756
1033743

TADJBAKHSCH, Jussef
Entwicklungspolitik in Iran. - Frankfurt / Main : R.G. Fischer, 1981
261p. - Bibliography
3-88323-229-7 X.322/10244 Woolwich
825687x

AGRICULTURE. Iraq
Policies of Iraqi Communist Party, to 1976

GABBAY, Rony
Communism and agrarian reform in Iraq. - London : Croom Helm, 1978
3-240p. - Bibliography: p.223-233. - Includes index
0-85664-567-2 X.329/11317 Woolwich
1137778

AGRICULTURE. Ireland
Cooperation, to 1900

O'DOWD, Anne
Meitheal. - Dublin : Comhairle Bhéaloideas Éireann, 1981
181p : ill. - Bibliography: p178-181
0-906426-06-5 X.529/47553
272121x

AGRICULTURE. Ireland (Republic)

CONNOLLY, L. (Liam)
Appraisal of investment and intensification in farming. - Dublin : Economics & Rural Welfare Research Centre, 1983
v,51p. - (Situation and outlook bulletin ; no.4)
0-905442-71-7 P.311/873
1074598

EVANS, J. B., 19—
Bro Ddyfi Grassland Society summer tour May 1983. - [London] : ADAS, c1984
8p
B84-42549 BS.3/888
3192156

LLEWELLYN, Tom
'Farmers have it easy'. - Dublin : Irish Farmers' Association, [1977]
[2],24p : chiefly col ill
B77-20984 X.512/5643
114149x

The PROGRESSIVE farmers' annual. - [1977]. - Dublin : Farming Publications, [1977]
[1],92p : ill
Subsequent volumes, when received, will be found at SRL
0-906062-00-4 SRL
1106465

SOCIALISM and the farmer. - Cork : Labour comment, 1984
28p
B85-23877 X.525/10962
3285308

AGRICULTURE. Ireland (Republic)
- Conference proceedings

AGRICULTURAL DEVELOPMENT PROSPECTS AND POSSIBILITIES (Conference : 1976 : Dublin)
Proceedings of conference, Agricultural Development Prospects and Possibilities, [organized by] An Foras Talúntais [Economics & Rural Welfare Research Centre]. - Dublin : The Centre, [1976]
[9],145[i.e.151]p : ill
B77-11659 X.319/17218 Woolwich
1074598

AGRICULTURE. Ireland (Republic)
- Serials

[The FARMER (Dublin)]. The Farmer. - Vol.1, no.1 (Mar. 1979)-v.4 (1982). - Dublin : Baggot Pub., 1979-1982
4v. : ill(some col.)
B82-28861 P.313/542
1106465

IRISH agriculture in figures. - [19-]-. - [Dublin] : An Foras Talúntais, [19-]-
v.
Holdings: 1981-
r6865290 P.311/532
2794594

IRISH farmers monthly. - [No.1]- ; 1977?-. - Dun Laoire : 'Irish farmers monthly', 1977?-
Holdings: Aug. 1977, etc
B79-00147 P.313/456
1738186

The NORTH Midland farmer. - Vol.1, no.1 ([198-])-. - [Meath] : [Meath Chronicle], [198-]-
v. : ill
Holdings: vol 1, no 7 (Feb' 83)-
B84-27861 P.2000/1239
1106465

The PRACTICAL farmer. - Vol.1, no.1 (May 1982)-. - Dublin : Baggot Publishing, 1982-
v. : ill(some col.)
B82-28858 SRL
1106465

AGRICULTURE. Ireland (Republic)
Development. Organisations: Council for Development in Agriculture

ACOT
An Chomhairle Oiliúna Talmhaíochta. - Blackrock : [ACOT], [1981?]
20p : col.ill
B82-05191 X.329/14789 Woolwich
2599449

AGRICULTURE. Ireland (Republic)
Energy. Conservation

ACOT
Farmers. - [Blackrock] : [ACOT], [1981?]
[8]p : col.ill
B82-05190 X.329/14797 Woolwich
2599422

AGRICULTURE. Ireland (Republic)
Farms. Management

CONNOLLY, L.
Farm management manual, 1979. - Dublin : Foras Talúntais, 1979
[8],85p. - (Handbook series / Agricultural Institute (Republic of Ireland) ; no. 14)
Subsequent volumes, when received, will be placed at this shelfmark
0-905442-36-9 P.311/525
2302179

AGRICULTURE. Ireland (Republic)
Farms. Partnerships. Legal aspects

FARM family partnerships / sponsored by the Agricultural Credit Corporation. - New rev. ed. - [Dublin] : Macra na Feirme, 1982
95p : ill
B83-15871 X.529/52855
2871785

AGRICULTURE. Ireland (Republic)
Farms. Production. Planning - Serials

MANAGEMENT data for farm planning / ACOT. - 1981/82- - Blackrock : ACOT National Office, 1981-
v.
B83-41493 P.513/3244
297097x

AGRICULTURE. Ireland (Republic)
Industrial training

REPORT of the ACOT expert group on agricultural education and training. - Blackrock : ACOT, 1981
xi,171p : ill
0-907816-00-2 X.329/14896 Woolwich
2614936

AGRICULTURE. Ireland (Republic)
Modernisation. Attitudes of low-income farmers

KELLEHER, Carmel
Adjustment problems of low income farmers. - Dublin : An Foras Talúntais, 1978
[4],81p. - Bibliography: p.79-81
B79-32341 X.519/30253
1956566

AGRICULTURE. Ireland (Republic)
Policies of government

FOUR-YEAR plan for agriculture. - Dublin : Stationery Office, 1984
138p
r6816838 X.322/15142
0153400

AGRICULTURE. Ireland (Republic)
Research & development

DOWNEY, William Kevin
Technological support for the food industry. - Dublin : Stationery Office, [1976]
[3],60p : ill. - Bibliography: p.59-60
0-7076-0017-0 SRL
0773816

AGRICULTURE. Ireland (Republic)
Research organisations: Agricultural Institute - Serials

HORTICULTURE research report / An Foras Talúntais (the Agricultural Institute). - [19-]-. - Dublin : The Institute, [19-]-
v.
Holdings: 1982-
r6813520 P.313/564
3066045

AGRICULTURE. Ireland (Republic)
Research organisations: Agricultural Institute. Libraries. Stock: Serials

AGRICULTURAL INSTITUTE
Liosta na dtréimseachán i gcuid leabharlann An Fhorais Talúntais 1980. - [Dublin] : [An Foras Talúntais], 1980
62p
B81-24774 2725.g.55
2496313

AGRICULTURE. Ireland (Republic)
Research organisations: Institute of Agricultural Research. Research into pigs as livestock

RILEY, J. E.
Report on a visit to the Institute of Agricultural Research, Moore Park, County Fermoy, Eire, March, 1981. - [Pinner] : ADAS, [1981]?
6p : ill
B81-39000 SRL
2567997

AGRICULTURE. Ireland (Republic)
Research organisations: Kinsealy Research Centre. Strawberries. Pick-your-own marketing

BOWBRICK, P.
Pick-your-own fruit marketing. - [Dublin] : Foras Taluntais, [1977]
[5],27p. - (Economic research series / Agricultural Institute (Republic of Ireland) ; no.22). - Bibliography p.27
0-905442-13-x X.319/17904 Woolwich
1373323

AGRICULTURE. Ireland (Republic). Special subjects. Humorous prose in English, 1945-
Farms. Social conditions - Anthologies

O'FARRELL, Padraic
The book of Irish farmers' jokes. - Cork : Mercier Press, 1977
48p. - (Mercier minibooks)
0-85342-493-4 X.908/41953
1688553

AGRICULTURE. Ireland (Repulic)
- Serials

The PROGRESSIVE farmer. - Vol.1, no.1- ; Apr. 1977-. - Dublin : Tara Publishing Company Ltd, 1977- : ill
Holdings: March 1980, etc
B80-25819 P.313 / 487
2209101

AGRICULTURE. Islamic countries
Innovation, 700-1100

WATSON, Andrew M.
Agricultural innovation in the early Islamic world. - Cambridge : Cambridge University Press, 1983
x,260p : ill. - (Cambridge studies in Islamic civilization). - Bibliography: p215-254. - Includes index
0-521-24711-x X.322/12615
298847x

AGRICULTURE. Isle of Wight
1900-1930

RURAL Wight in bygone days / [compiled by] Hilary Scammell. - Newport, Isle of Wight : County Council, Cultural Services, 1982
[52]p : chiefly ill,ports
0-906328-19-5 L.49/1678
2880792

AGRICULTURE. Israel

CORMACK, W. F.
UK/Israel exchange of advisory staff. - [London] : ADAS, c1984
22p
B85-08116 BS.3/936
3227766

AGRICULTURE. Israel
Extension services

FLETCHER, J. T.
Report on a visit to Israel in connection with Agricultural Development and Advisory Service and Israeli Extension Service exchange. - [London] : ADAS, c1984
24p : ill. - Bibliography: p21-23
B85-08117 BS.3/945
3227707

AGRICULTURE. Italy

AVOLIO, Giuseppe
Agricoltura e sviluppo. - Venezia : Marsilio, 1976
364p. - (Socialismo oggi ; 5). - Includes bibliographies
r6048669 X.319/8840 Woolwich
1230999

AGRICULTURE. Italy
- Serials
RIVISTA di storia dell'agricoltura / [a cura dell'Accademia economico-agraria dei Georgofili]. - Anno 1, no.1 (ott. 1961)-. - Firenze : l'Accademia, 1961-
Holdings: Anno 12, no 1/2 (gen / giugno 1972)
r6069367 P.311/520 Woolwich
243928X

AGRICULTURE. Italy
to 1300

FUMAGALLI, Vito
Coloni e signori nell'Italia settentrionale, secoli VI-XI. - Bologna : Patròn, 1978
119p. - (Il mondo medievale. Sezione di storia della società, dell' economia e della politica ; 2)
r6500970 X.329/13994 Woolwich
8706689

AGRICULTURE. Italy
to 1975

ZANGHERI, Renato
Agricoltura e contadini nella storia d'Italia. - 2.ed. - Torino : Giulio Einaudi, 1977
xiii,289p. - (Piccola biblioteca Einaudi ; 311)
r6068604 X.319/20071 Woolwich
2094762

AGRICULTURE. Italy
to date

AMBROSOLI, Mauro
Problemi del lavoro contadino in Italia tra '500 e '700. - Torino : Tirrenia-Stampatori, [1977?]
i,116p : ill
r6504829 X.322/10867 Woolwich
826547x

BRUSCO, Sebastiano
Agricoltura ricca e classi sociali. - Milano : Feltrinelli, 1979
194p. - (I fatti e le idee. Saggi e biografie ; 451) (Biblioteca di economia)
r6504687 X.329/15138 Woolwich
826547x

BUTERA, Maria Maddalena
Le campagne italiane nell'età napoleonica. - Milano : Angeli, c1981
295p. - (Studi e ricerche storiche ; 6)
88-204-1929-7 X.329/16671
826547x

CAVAZZANI, Ada
Il part-time agricolo .... - 2. ed. - Venezia : Marsilio, 1980
166p. - (Biblioteca Marsilio. Sociologia)
r6507438 X.329/16378
826547x

DANEO, Camillo
Breve storia dell'agricoltura italiana, 1860-1970. - Milano : Mondadori, 1980
237p. - (Saggi ; 138). - Includes bibliography
r6507749 X.329/16465
826547x

DESIDERI, Carlo
L'amministrazione dell'agricoltura (1910-1980). - Roma : Officina, 1981
319p. - (Paperback ; 9)
r6510222 X.322/12905
826547x

FABIANI, Guido
L'agricoltura in Italia tra sviluppo e crisi (1945-1977). - Bologna : Il mulino, 1979
271p. - (Universale paperbacks Il mulino ; 95)
r6510583 X.329/17536
826547x

MÅNSSON, Peder
[Bondakonst]. Peder Manssons Bondakonst. - Uppsala : Svenska fornskriftsällskapet, 1983
324p. - (Samlingar utgivna av Svenska fornskriftsällskapet ; hft.254)
r6611529 Ac.9068[no.254]
826547x

La POLITICA agraria in Italia / a cura di Carlo Catena. - Roma : Editori riuniti, 1979
xii,363p
r6508750 X.329/16674
826547x

SALTINI, Antonio
Storia delle scienze agrarie. - Bologna : Edagricole, 1979
viii,339p : ill
r6599037 X.322/10545 Woolwich
826547x

STORIA dell'agricoltura italiana / [Alvise Comel et al.]. - Milano : Etas libri, 1976
403p. - Includes bibliography
r6510166 L.45/2470
826547x

AGRICULTURE. Italy
to date - Conference proceedings

CONGRESSO NAZIONALE DI STORIA DELL'AGRICOLTURA (Milan : 1971)
Atti del Congresso nazionale di storia dell'agricoltura, Milano ... 1971. - Firenze : Accademia economico-agraria dei Georgofili, 1972
2v.,[30]leaves of plates. - (Rivista di storia dell'agricoltura ; anno 12)
r6069368 P.311/520
826547x

Gli ENTI di sviluppo / Gian Giacomo Dell'Angelo ... [et al.] ... - Roma : Monteverde, 1978
96p. - (Quaderni dell'Istituto di legislazione agraria "Alessandro de Feo" ; 1)
r6500127 X.325/952 Woolwich
826547x

AGRICULTURE. Italy
to date. Economic aspects

La VERTENZA agro-industria / a cura di Brunello Cipriani e Alberto Mattei. - Roma : Editrice Sindacale Italiana, 1981
198p. - (Dossier ; 14)
r6511018 X.629/24780
9128018

AGRICULTURE. Italy
to date. Political aspects

AMADEI, Giorgio
La terra non può aspettare. - Bologna : Edagricole, 1980
vii,197p
88-206-2096-0 X.329/16273
8916500

AGRICULTURE. Italy
1494-1918 - Conference proceedings

AGRICOLTURA e aziende agrarie nell'Italia centro-settentrionale (Secoli XVI-XIX) / a cura di Gauro Coppola. - Milano : Angeli, c1983
724p. - (Studi e ricerche storiche ; 20)
r6512687 X.329/19467
9162054

AGRICULTURE. Italy
1700-1800. Effects of capitalism, 1700-1800

GIORGETTI, Giorgio
[Selections]. Capitalismo e agricoltura in Italia. - Roma : Editori Riuniti, 1977
xxvi,591p. - (Biblioteca di storia ; n.66)
r6050475 X.319/9120 Woolwich
1324721

AGRICULTURE. Italy
1860-1900

SERENI, Emilio
Il capitalismo nelle campagne, 1860-1900. - 2. ed., con un nuovo saggio introduttivo. - Torino : Giulio Einaudi, 1975
xxvii, 369 p.. - (Piccola biblioteca Einaudi ; n. 112)
r6014212 X.519/20774
0864471

AGRICULTURE. Italy
1900-1942

SERENI, Emilio
La questione agraria nella rinascita nazionale italiana. - 2. ed. - Torino : Giulio Einaudi, 1975
viii, 450 p.. - (Piccola biblioteca Einaudi ; n. 245)
r6014226 X.319/8483 Woolwich
0867276

AGRICULTURE. Italy
1947-1967

STEFANELLI, Renzo
Lotte agrarie e modello di sviluppo 1947-1967. - Bari : De Donato, 1975
384p. - (Movimento operaio ; 29)
r6048633 X.319/8790
1222627

AGRICULTURE. Italy
1951-1960

DANEO, Camillo
Agricoltura e sviluppo capitalistico in Italia. - 3. ed. aumentata. - Torino : Giulio Einaudi, 1972
244 p.. - (Serie politica ; n. 8)
r6013847 X.319/8425 Woolwich
0852449

AGRICULTURE. Italy
Ancient Roman agriculture. Farms. Buildings

ROSSITER, J. J.
Roman farm buildings in Italy. - Oxford : British Archaeological Reports, 1978
[4],viii,81p : ill. - (British archaeological reports. international series ; 52). - Bibliography: p.76-81
0-86054-040-5 X.415/5786 Woolwich
1931962

AGRICULTURE. Italy
Census data, to date

3O Censimento generale dell'agricoltura / Istituto centrale di statistica. - Roma : l'Istituto, 1983-
r6512146 SH.420/116
9146326

AGRICULTURE. Italy
Corio, Giacinto - Correspondence, diaries

CORIO, Giacinto
[Correspondence. Selections]. Lettere di Giacinto Corio a Camillo Cavour (1843-1855). - Santena : Fondazione Cavour, 1980
lvi,474p. - (Studi e documenti ; 9)
r6507655 X.322/11965
8953686

AGRICULTURE. Italy
Effects of Common Agricultural Policy of European Economic Community, 1956–
GALLI, Rosemary
La partecipazione italiana alla politica agricola comunitaria. - Roma : Instituto affari internazionali, 1976
260p. - (Collana dello Spettatore Internazionale ; 37)
r6085328 X.519/40857
1981110

AGRICULTURE. Italy
Implications of value-added tax, to date
COZZIO, Enzo
IVA agricola. - Bologna : Edagricole, 1981
171p. - (Collana teorico-practica di legislazione agraria)
88-206-2308-0 X.322/12892
9047522

AGRICULTURE. Italy
Law, to date
BIVONA, Giuseppe
I contratti d'integrazione verticale in agricoltura. - Milano : Giuffrè, 1979
138p. - (Università di Catania. Pubblicazioni della Facoltà di giurisprudenza ; 87)
r6502498 X.322/10693 Woolwich
877949x

DIRITTO agrario italiano / Natalino Irti ; con la collaborazione di Tito Bellisario ... [et al.] ; ha raccolto i saggi Carlo Alberto Graziani. - Torino : UTET, [1978]
xxvii,754p
88-02-02491-x X.200/39280
877949x

AGRICULTURE. Italy
Law, 1946- - Abstracts - Serials
BIBLIOGRAFIA IDG. E, Diritto agrario. - 1980-. - Firenze : Istituto per la documentazione giuridica, 1981-. - (Collana di documentazione giuridica)
r6509581 X.0200/1269.e(1979)
8877386

AGRICULTURE. Italy
Law, 1946- - Conference proceedings
PRAVO i sel'skoe khoziaĭstvo v SSSR i Italii / [redaktsionnaia kollegiia : M.I. Kozyr' (otv. redaktor) ... i dr.]. - Moskva : AN SSSR, Institut gosudarstva i prava, 1980
158p
r6432433 X.629/19201
8877386

AGRICULTURE. Italy
Organisations: Istituto nazionale di economia agraria
MAGNARELLI, Paola
L'agricoltura italiana fra politica e cultura. - Milano : Edizioni di Comunità, 1981
128p. - (Studi e ricerche di scienze sociali ; 84)
r6512198 X.329/19465
9162348

AGRICULTURE. Italy
Personnel. Employment
BARBERIS, Corrado
Famiglie senza giovani e agricoltura a mezzo tempo in Italia. - Milano : Angeli, [1979]
3vin5. - (Collana Insor ; 4)
r6500021 X.0329/217 Woolwich
2479672

AGRICULTURE. Italy
Policies. Economic aspects
ORLANDO, Giuseppe
Piano alimentare o politica agraria alternativa?. - Bologna : Il mulino, 1977
180p. - (Studi e ricerche ; 71). - Includes bibliographies
r6107785 X.329/12193 Woolwich
2251248

AGRICULTURE. Italy
Policies of fascist movements, 1921-1945
SITTI, Renato
La politica agraria del fascismo. - Firenze : La nuova Italia, 1978
16p.,20 leaves of plates in portfolio : ill,facsims. - (Le fonti della storia ; 45)
r6107502 Cup.900.ww.31
2415151

AGRICULTURE. Italy
Policies of government, 1946–
MARCORA, Giovanni
La questione agraria e l'Europa. - Bologna : Edagricole, 1979
vi,304p
88-206-2125-8 X.529/50982
8877408

AGRICULTURE. Italy
Political aspects, 1880-1900
DE BERNARDI, Alberto
Questione agraria e protezionismo nella crisi economica di fine secolo. - Milano : Franco Angeli, 1977
229p. - (La società italiana moderna e contemporanea : 3)
r6078524 X.519/39469
1784188

AGRICULTURE. Italy
Regional planning
LECHI, Francesco
L'agricoltura nella pianificazione delle Comunità montane. - Milano : Vita e Pensiero, Pubblicazioni della Università Cattolica, 1974
122 p.. - (Problemi economici d'oggi ; no. 1)
r6005062 X.319/8311 Woolwich
090824X

AGRICULTURE. Italy
Smallholdings. Agriculture, B.C.500-A.D.200
FRAYN, Joan M.
Subsistence farming in Roman Italy. - Fontwell : Centaur Press, 1979
168p,leaf of plate,[2]p of plates : ill. - Bibliography: p.154-161. - Includes index
0-900000-92-9 X.329/11937 Woolwich
192124x

AGRICULTURE. Italy
Trade unions
AGRICOLTURA e movimento operaio / a cura di Giovanni Mottura e Enrico Pugliese. - Roma : Savelli, 1977
158p. - (Cultura politica ; 179)
r6085462 X.319/20169
2025523

AGRICULTURE. Italy
Trade unions, to date - Conference proceedings
SETTIMANA DI STUDIO DELLA FISBA CISL (12th : 1978 : Rome)
Il sindacalismo agricolo in una economia e in una società italiana in crisi. - Roma : Romana, [1979?]
210p. - (Quaderni lotte contadine)
r6509726 X.329/17950
9084363

AGRICULTURE. Ivory Coast
1960–
L'AGRICULTURE ivoirienne. - Paris : Édiafric, 1980
186p
r6206567 X.322/10385 Woolwich
8755981

AGRICULTURE. Ivory Coast
1960- - Serials
Le LIEN. - Abidjan : S.A.T.M.A.C.I., 1977-.
r6209651 X.0625/132
8755981

AGRICULTURE. Jämtland. Sweden
Crofting. Ek, Elsa Christina
EK, Elsa Christina
Odelstorpet Ottslätta. - Stockholm : LT, 1977
138p,[16]p of plates : ill
91-360-0960-1 X.629/7862 Woolwich
1395831

AGRICULTURE. Japan
Policies of government, to date
OGURA, Takekazu
Can Japanese agriculture survive?. - 3rd ed. - Tokyo : Agricultural Policy Research Center, 1982
xi,880p,plates
r6811527 X.322/12837
9032940

AGRICULTURE. Jordan
Economic development, 1946-1975
ARESVIK, Oddvar
The agricultural development of Jordan. - New York ; London : Praeger ; [London] : [Distributed by Martin Robertson], 1976
xxx,377p. - (Praeger special studies in international economics and development). - Bibliography: p.366-375
0-275-00450-3 X.320/10317 Woolwich
055278x

AGRICULTURE. Junin (Department). Peru
1972 - Statistics
[CENSO Nacional Agropecuario]. II Censo Nacional Agropecuario, 4 al 24 de setiembre 1972 / [preparado por el] Instituto Nacional de Estadística [de la] Presidencia de la República [del Perú]. - Lima : I.N.E..
Departamento de Junin. - 1976
2v.
r6157513 LAS.G.381/7 Woolwich
2188678

AGRICULTURE. Kalasin. Thailand
Development. Role of agricultural cooperatives
DEMAINE, Harvey
The role of farmers' associations in agricultural development. - London : University of London School of Oriental and African Studies, Department of Geography, 1976
[2]p,21 leaves. - (Occasional papers / University of London. School of Oriental and African Studies. Department of Geography)
0-7286-0021-8 X.0702/1007(1)
0608122

AGRICULTURE. Kalmyk ASSR. Russia (RSFSR)
1945–
BUDYL'SKIĬ, P. P. (Petr Petrovich)
Agrarnaia politika partii i razvitie sel'skogo khoziaĭstva Kalmykii. - Elista : Kalmytskoe knizhnoe izdatel'stvo, 1979
123p : ill
r6462154 X.329/16206
8916527

AGRICULTURE. Karakalpak ASSR. Uzbekistan
to 1924
SARYBAEV, K
Agrarnyĭ vopros v Karakalpakii (konets XIX - nachalo XX vekov). - Nukus : Karakalpakstan, 1972
197p. - Bibl.: p.183-[198]
r6178287 X.329/12467 Woolwich
2447177

AGRICULTURE. Karcag. Hungary
BELLON, Tibor
Karcag város gazdálkodása. - Szolnok : Damjanich János Muzeum, 1973
233p. - (A Damjanich János Muzeum közleményei ; 34-35). - Bibliography : p.217-224
r6089035 X.0709/753(34,35)
1854453

AGRICULTURE. Karnataka (State). India (Republic)
Economic development
DESAI, Vasant
Agricultural development. - Bombay : Popular Prakashan, 1976
xv,156p,fold leaf. - Includes index
r6091838 X.529/21553
1887777

AGRICULTURE. Kazakhstan
DOVZHIK, Mikhail Egorovich
Vroven' s vekom. - Alma-Ata : Kazakhstan, 1979
285p : ill
r6185467 X.329/9029 Woolwich
1794949

KAZAKHSTANSKIĬ Milliard-76 / [sostaviteli : Kh. Sh. Abdrashitov, V.V. Vladimirov]. - Alma-Ata : Izdatel'stvo "Kazakhstan", 1977
296p,[40]p of plates : ill(chiefly col.)
r6087064 X.629/8034 Woolwich
1794949

KAZAKHSTANSKIĬ milliard-79 ... / [sostaviteli : Kh. Sh. Abdrashitov i V.V. Vladimirov]. - Alma-Ata : Kazakhstan, 1980
422p,[45]leaves of plates : ill(some col.)
r6179080 X.329/13306 Woolwich
1794949

PRODOLZHENIE podviga / [sostavitel' V.V. Vladimirov]. - Moskva : Izdatel'stvo politicheskoĭ literatury, 1979
165p,[16]p of plates : ill
r6185287 X.319/20351 Woolwich
1794949

AGRICULTURE. Kazakhstan
to 1920
KHOZIAĬSTVO kazakhov na rubezhe XIX-XX vekov / [otvetstvennyĭ redaktor : G.F. Dakhshleiger]. - Alma-Ata : Izdatel'stvo "Nauka" Kazakhskoĭ SSR, 1980
252p : ill
r6402234 Cup.900tt138
8726507

AGRICULTURE. Kent
Smallholdings

HALL, Daphne E. E.
Memories of 'Beechcroft'. - Ilfracombe : Stockwell, 1983
24p : ill
0-7223-1731-x X.990/21357
2991144

AGRICULTURE. Kenya

AGRICULTURAL development in Kenya / editors Judith Heyer, J.K. Maitha, W.M. Senga - Nairobi ; London : Oxford University Press, 1976 [i.e. 1977]
xv,372p,[8]p of plates : ill(some col). - Includes index
0-19-572377-5 X.320/10707 Woolwich
1061178

AGRICULTURE. Kenya
to 1939

VAN ZWANENBERG, R. M. A. (Roger Maurice Arthur)
The agricultural history of Kenya to 1939. - Nairobi : East African Publishing House ; Birmingham : [Distributed by] Third World Publications, 1972
56p. - (Papers / Historical Association of Kenya : no.1). - Bibliography: p.56
B79-35733 X.519/30359
1992295

AGRICULTURE. Kenya
1963-

TIBAIJUKA, Anna Kajumulo
Kenya - a study of the agricultural sector. - Uppsala : Swedish University of Agricultural Sciences, 1981
172,7,xxxvp. - (Rural development studies ; no.11). - Includes bibliography
91-576-0790-7 X.0313/56(11)
8275386

AGRICULTURE. Kenya
Extension services. Administration

LEONARD, David K.
Reaching the peasant farmer. - Chicago ; London : University of Chicago Press, 1977
xxi,297p : ill. - Includes index
0-226-47260-4 X.320/11001 Woolwich
1282239
- [Another copy] X.320/11047 Woolwich

AGRICULTURE. Kheiran. Sudan
to date

MANGER, Leif O.
The sand swallows our land. - Bergen : University of Bergen, Department of Social Anthropology, 1981
v,137p. - (Bergen occasional papers in social anthropology ; no.24) (African savannah studies). - Bibliography
r6610353 X.329/15620
8853185

AGRICULTURE. Kirgizia

DUĬSHEMALIEV, Tashtan Duĭshemalievich
Pobeda Leninskogo plana sot͡sialisticheskogo preobrazovanii͡a sel'skogo khozi͡aĭstva v Kirgizii. - Frunze : Kyrgyzstan, 1971
253p. - Bibliography : p.246-252
r6176363 X.329/8997 Woolwich
1973533

MUSAKOZHOEV, Shaĭlobek
Problemy povyshenii͡a ėffektivnosti osnovnykh fondov. - Frunze : Kyrgyzstan, 1975
148p
r6093473 X.329/8887 Woolwich
1973533

AGRICULTURE. Kirgizia
to 1924

PETROV, K. I. (Konstantin Ivanovich)
Ocherki sot͡sial'no-ėkonomicheskoĭ istorii Kirgizii VI-nachala XIII vv.. - Frunze : Ilim, 1981
232p
r6496513 X.329/17316
9010033

AGRICULTURE. Kirgizia
Collectivisation, 1917-

BAKTYGULOV, Dzhumadil Sapalovich
Iz istorii perekhoda kirgizskogo krest'i͡anstva k osedlosti i sot͡sialisticheskim sposobam khozi͡aĭstvovanii͡a. - Frunze : Kyrgyzstan, 1971
112p
r6184569 X.319/20382 Woolwich
2266245

AGRICULTURE. Kirgizia
Farms. 1900-1917

SAPELKIN, Alekseĭ Aleksandrovich
Agrarnye otnoshenii͡a v Kirgizii v nachale XX veka (1900-1917gg.). - Frunze : Izdatel'stvo "Ilim", 1977
279p,[1] fold. leaf of plates. - Bibliography : p.266-275
r6055849 X.529/20630
153114X

AGRICULTURE. Kirov (oblast). Russia (RSFSR)

FOKIN, Gennadiĭ Timofeevich
Slovo o vi͡atskom khlebe. - Kirov : Volgo-Vi͡atskoe knizhnoe izdatel'stvo, Kirovskoe otdelenie, 1976
106 p. : ill
r6020478 X.329/8558 Woolwich
0897272

AGRICULTURE. Kisii. Kenya
Change

UCHENDU, Victor C.
Agricultural change in Teso District, Uganda. - Nairobi : East African Literature Bureau, 1975
vii,115p. - Bibliography : p.107-108. - Includes index
r6072213 X.519/28446
1581902

AGRICULTURE. Komi ASSR. Russia (RSFSR)
1945-

GABOV, L. A. (Leonard Aleksandrovich)
Sel'skoe khozi͡aĭstvo Komi ASSR v period razvitogo sot͡sializma. - Moskva : Nauka, 1981
135p
r6432221 X.329/15688
8853215

AGRICULTURE. Konkan Region. India (Republic)
to date

SHINDE, Suresh D.
Agriculture in an underdeveloped region. - Bombay : Himalaya Publishing House, 1980
143p,32p of plates
r6800396 X.322/10051 Woolwich
8784507

AGRICULTURE. Kuban. Russia (RSFSR)
to 1917

RATUSHNI͡AK, V. N. (Valeriĭ Nikolaevich)
Vkhozhdenie Severo-Zapadnogo Kavkaza v sostav Rossii i ego kapitalisticheskoe razvitie. - Krasnodar : Kubanskiĭ gosudarstvennyĭ universitet, 1978
94p. - (Kubanskiĭ gosudarstvennyĭ universitet ; vyp. 264)
r6484511 X.808/35335
8790701

AGRICULTURE. Kumaon region. India (Republic)
to date

AKHTAR, Rais
Environment, agriculture and nutrition in Kumaon region. - New Delhi : Marwah, 1980
xxiii,196p. - Bibliography: p184-196
r6803652 X.322/10305 Woolwich
8773882

AGRICULTURE. La Laguna. Mexico
Collectivisation

RESTREPO, Iván
La agricultura colectiva en México. - México : Siglo Veintiuno, 1975
320p. - (Economía y demografía)
r6158053 X.319/20285 Woolwich
2136376

AGRICULTURE. La Libertad (Department). Peru
1972 - Statistics

[CENSO Nacional Agropecuario]. II Censo Nacional Agropecuario, 4 al 24 de setiembre 1972 / [preparado por el] Instituto Nacional de Estadística [de la] Presidencia de la República [del Perú] ... - Lima : I.N.E..
Departamento de La Libertad. - Lima : I.N.E.
2t
r6157514 LAS.G.381/8 Woolwich
2188686

AGRICULTURE. Lanarkshire. Strathclyde Region. Scotland
1760-1840

AGRICULTURE in Lanarkshire, 1760-1840 / [compiled by Jacqueline Rosie, Linda Kelly]. - [Glasgow] : [Jordanhill College of Education], [1978]
[6],86p,[2]leaves of plates : ill
B79-33123 X.313/1686 Woolwich
1960024

AGRICULTURE. Land. Docks. Port of London
- Biographies

PETERS, Hilary
Docklandscape. - London : Watkins, 1979
[7],88p,16p of plates : ill
0-7224-0173-6 X.629/14216 Woolwich
2305097

AGRICULTURE. Landes. France
to ca 1900

TOULGOUAT, Pierre
La vie rurale dans l'ancienne Lande. - Pau : Éditions Marrimpouey jeune, 1975
107 p., 6 leaves of plates, 2 p. of plates : ill
r6012404 X.311/4546 Woolwich
0931969

AGRICULTURE. Lapland. Sweden
to date

LAESTADIUS, Lars Levi, 1800-1861
Om möjligheten och fördelen af allmänna uppodlingar i Lappmarken. - Luleå : Tornedalica, 1981
143,[13]p. - (Tornedalica ; nr 33)
r6610371 X.0700/436(33)
9043454

AGRICULTURE. Latin America

LEHMANN, David
A theory of agrarian structure. - Cambridge : [University of Cambridge, Centre of Latin American Studies], [1976]
[2],iv leaves,104[i.e.105]p : 1 ill. - (Working papers / University of Cambridge. Centre of Latin American Studies ; no.25)
0-904927-16-4 X.313/1368 Woolwich
0770558

SOARES, Gláucio Ary Dillon
A questão agrária na América Latina. - Rio de Janeiro : Zahar, 1976
177p
r6084672 X.319/9769 Woolwich
0770558

AGRICULTURE. Latin America
1700-1973 - Conference proceedings

LANDLORD AND PEASANT IN LATIN AMERICA AND THE CARIBBEAN
(Conference : 1972 : King's College, Cambridge)
Land and labour in Latin America. - Cambridge : Cambridge University Press, 1977
xiii,535p : ill. - (Cambridge Latin American studies ; 26). - Includes index
0-521-21206-5 X.329/11336 Woolwich
138189x

AGRICULTURE. Latin America
1945-

DÉVELOPPEMENT agricole dépendant et mouvements paysans en Amérique latine / sous la direction de Lawrence R. Alschuler. - Ottawa : Éditions de l'Université d'Ottawa, 1981
218p. - (Cahiers des sciences sociales ; no.11) (Livres et monographies de l'Institut de coopération internationale ; no.1)
2-7603-3011-7 X.0519/137(11)
8278954

KOVALEV, E. V. (Evgeniĭ Vladimirovich)
Latinskai͡a Amerika. - Moskva : Nauka, 1982
267p
r6490939 X.529/70282
8278954

NUEVAS perspectivas de la integración latinoamericana. - Santiago de Chile : Editorial Universitaria. - (Estudios internacionales)
Vol.2: La agricultura en la integración latinoamericana. - [1977]
183p
r6700326 X.0529/706(2)
8278954

PRODOVOL'STVENNAI͡A problema i sel'skoe khozi͡aĭstvo stran Latinskoĭ Ameriki / [otvetstvennyĭ redaktor : I͡U. G. Onufriev]. - Moskva : Nauka, 1981
311p
r6432314 X.329/15818
8278954

AGRICULTURE. Latin America
1945- - Abstracts

RESÚMENES analíticos en economía agrícola latinoamericana ... / [compilado por la Biblioteca de CIAT, Centro Internacional de Agricultura Tropical]. - [Cali?] : Centro International de Agricultura Tropical : [Centro de Documentación Económica para América Latina], 1976-1978
3v.. - (Serie HS ; 30, 33)
r6702268 X.520/24903
8278954

AGRICULTURE. Latin America
1945- - Catalogues

FUNDACIÓN MARIANO OSPINA PÉREZ. Servicio de Documentación e Información
Fondo bibliográfico Armando Samper Gnecco. - Bogotá : La Fundación, 1980
lxii,313p
r6708163 L.42/2758
8278954

AGRICULTURE. Latin America
1945- - Conference proceedings

SIMPOSIO SOBRE EL SECTOR AGRARIO EN AMÉRICA LATINA (1977 : Ultuna)
El sector agrario en América Latina,. - Stockholm : Instituto de Estudios Latinoamericanos, 1979
187p. - (Publicaciones / del Instituto de Estudios Latinoamericanos, Universidad de Estocolmo. serie A, monografías ; no.4 = Skrifter utgivna av Latinaamerika-institutet i Stockholm. serie A, monografier ; 4)
r6701493 X.0700/366(1)[No.4]
8278954
El sector agrario en América Latina. - Stockholm : Instituto de Estudios Latinoamericanos, Latinamerika-institutet, 1980
36p. - (Separata)
r6703032 X.329/14391 Woolwich
8278954

AGRICULTURE. Latin America
Development, 1940-1980

DE JANVRY, Alain
The agrarian question and reformism in Latin America. - Baltimore, Md. ; London : Johns Hopkins University Press, c1981
xvi,311p : ill. - (Johns Hopkins studies in development). - Bibliography: p303-306. - Includes index
0-8018-2531-8 X.520/27416
275231x

AGRICULTURE. Latin America
Development. Political aspects - Conference proceedings

The POLITICS of agrarian change in Asia and Latin America / edited by Howard Handelman. - Bloomington : Published in association with the American Universities Field Staff [by] Indiana University Press, c1981
viii,136p : ill
0-253-34548-0 X.520/29702
2858452

AGRICULTURE. Latin America
Effects of multinational companies, 1945- - Conference proceedings

Les FIRMES transnationales et l'agriculture en Amérique latine / sous la direction de Gonzalo Arroyo. - Paris : Éditions Anthropos, 1980
256p
2-7157-1006-2 X.329/14838 Woolwich
8785929

AGRICULTURE. Latin America
Law

ALVARENGA, Ivo P
Temas de derecho agrario y reforma agraria. - Ciudad Universitaria Rodrigo Facio : EDUCA, 1977
294p. - (Aula)
r6161972 X.329/12923 Woolwich
2439883

AGRICULTURE. Latin America
Law - Conference proceedings

REUNIÓN IBEROAMERICANA DE ESPECIALISTAS EN DERECHO AGRARIO
(1974 : Mérida, Venezuela)
Memoria [de la] Reunión Iberoamericana de Especialistas en Derecho Agrario. - Mérida : Universidad de los Andes, 1974
350 p.
r6003384 X.100/14504
0910996

AGRICULTURE. Latin America
Workers' organisations, to date

SANTOS DE MORAIS, Clodomir
Organizaciones campesinas en América Latina. - Tegucigalpa : [Instituto Nacional Agrario], 1976
167p. - (Serie didáctica / PROCCARA ; no.12)
r6702053 X.529/51146
8877416

AGRICULTURE. Latin American countries
Technological innovation

TECHNICAL change and social conflict in agriculture / edited by Martín Piñeiro and Eduardo Trigo. - Epping : Bowker, c1983
xvi,248p : ill. - (A Bowker (U.K.) replica edition). - Bibliography: p241-248
0-86531-802-6 X.322/13998
3165442

AGRICULTURE. Latvia

GODY bol'shikh peremen / [sostaviteli : V.A. Risman, V.ÎA. Dimante]. - Riga : Liesma, 1975
268p
r6176598 X.329/8999 Woolwich
0898546

KOLLEKTIVIZAT͡SIÎA krest'îanskikh khozîaistv Litovskoĭ SSR / redkollegiîa : F. Belîaukas, ... A. Efremenko (otvetstvennyĭ redaktor) ... [i dr.]. - Vil'nîus : Mokslas, 1977
508p : ill. - (Istoriîa kollektivizat͡sii sel'skogo khozîaĭstva SSSR. dokumenty i materialy)
r6176186 S.c.105/2
0898546

KOZIN, Mikhail Ivanovich
Latyshskaîa derevnîa v 50-70. gody XIX veka. - Riga : Izdatel'stvo "Zinatne", 1976
372 p.
r6020916 X.809/23386
0898546

AGRICULTURE. Latvia
1945- - Statistics

SEL'SKOE khozîaĭstvo Latviĭskoĭ SSR / [rukovodstvo sostavleniem ... Vanusha, D.A.]. - Riga : T͡Sentral'noe statisticheskoe upravlenie pri Sovete ministrov Latviĭskoĭ SSR, 1973
592p
r6498223 S.C.53/25
8280134

AGRICULTURE. Latvia
Aizsilnieks, Arnolds - Biographies

AIZSILNIEKS, Arnolds
Dzives šūpotnės. - [Stockholm] : Daugava, [1979]
167p : ill
r6487537 X.329/16661
8976945

AGRICULTURE. Latvia
Economic planning. Mathematical models

SPROGIS, Andris
Avtomatizirovannaîa sistema razrabotki planov sel'skokhozîaĭstvennogo proizvodstva. - Riga : Zinatne, 1977
129p,[1]fold.leaf of plates. - Bibliography : p.127-129
r6172803 X.329/8946 Woolwich
2098156

AGRICULTURE. Latvia
Farms

KUNDZIŅŠ, Pauls
Latvju sēta. - [Stockholm] : Daugava, 1974
449p,[1] fold. leaf of plates : ill
r6015701 X.322/2582 Woolwich
1243128

AGRICULTURE. Latvia
Policies of Latvijas Komunistiskā partija, to date

BONDAREV, P. (Petr)
Agrarnaîa politika Kommunisticheskoĭ partii Latvii, 1904-1940gg.. - Riga : Avots, 1982
322p
r6403719 X.529/62774
9101632

AGRICULTURE. Leicestershire
Farms. Social conditions, 1931-1945

SMITH, Ellen, 1907-
Seven pennies in my hand. - Loughborough : Ellen Smith, c1984
108p,[10] leaves of plates : ill
0-9509092-1-1 X.950/43174
3287777

AGRICULTURE. Leinster. Ireland (Republic)
Development. Organisations: Council for Development in Agriculture

ACOT
Midland and East Region. - Newcastle, [Co. Dublin] : [ACOT], [1981?]
[16]p : ill
B82-05196 X.329/14794 Woolwich
2599457

AGRICULTURE. Lérida (Province). Spain
to date

MAJORAL I MOLINÉ, Roser
Anàlisi de l'agricultura a la Vall d'Aran. - [Barcelona] : Generalitat de Catalunya, Departament d'Agricultura, Ramaderia i Pesca, [1983]
197p,[1]leaf of plates : ill
84-393-0206-1 X.322/13462
9082905

AGRICULTURE. Lincolnshire

LINCOLNSHIRE. Planning Department
Agriculture. - [Lincoln] : [The Department], 1976
36p : ill
0-902092-06-5 X.702/5241
073585x

AGRICULTURE. Lincolnshire
1700-1875

BEASTALL, T. W. (Tom William)
The agricultural revolution in Lincolnshire. - Lincoln : History of Lincolnshire Committee for the Society of Lincolnshire History and Archaeology, 1978
xii,256p,leaf of plate,[16]p of plates : ill. - (History of Lincolnshire ; 8). - Bibliography: p.240-244. - Includes index
0-902668-07-2 AA.KK.c. Reading Room
1948970
- [Another copy] X.0800/385(8)

AGRICULTURE. Lindsey. Lincolnshire
1900-1914

FENTON, Michael
Farmers & farming in Lindsey 1900-1914. - Lincoln : Centre for the Study of Rural Society, 1978
33p,[13]p of plates : ill. - Bibliography: p31-33
B83-26231 X.322/11914
2907771

AGRICULTURE. Lithuania
Collectivisation

BUBLYS, Vladas
Agricultural production cooperatives in independant Lithuania 1920-1940 (summary) = Das landwirtschaftliche Produktions-Genossenschaftswesen im unabhängigen Litauen 1920-1940 (Zusammenfassung). - London : Lithuanian House Ltd, 1974
80p
r6015734 X.319/8914 Woolwich
1313118

MANZUROVAS, Grigorijus
Sel'skie partiĭnye organizat͡sii Litvy v period kollektivizat͡sii. - Vil'nîus : Izdatel'stvo "Mintis", 1976
155p. - Bibliography : p.148-[156]
r6074428 X.809/25409
1313118

AGRICULTURE. Llanblethian. South Glamorgan
1660-1750

RIDEN, Philip
Farming in Llanblethian, 1660-1750. - [Cardiff] : University College Cardiff, Department of Extra-Mural Studies, 1980
viii,62p,4p of plates : ill. - (Park Place papers ; no.9)
B81-04900 X.0708/1537(9)
2342944

AGRICULTURE. Lombardy. Italy
to date

CARATTERISTICHE e problemi del lavoro agricolo in Lombardia / contributi di D. Casati ... [et al.]. - Milano : Istituto nazionale di economia agraria, Osservatorio di economia agraria per la Lombardia, 1983
330p
r6509348 X.329/17925
8976937

CARATTERISTICHE e problemi del lavoro agricolo in Lombardia / contributi di D. Casati ... [et al.]. - Milano : Istituto nazionale di economia agraria, Osservatorio di economia agraria per la Lombardia, 1983
330p
r6509364 X.329/17924
8976937

I SERVIZZI per lo sviluppo dell'agricoltura / IReR. - Milano : Franco Angeli, c1981
279p. - (Collana IReR ; 7)
88-204-2127-5 X.329/18627
8976937

Il SISTEMA agricolo-industriale lombardo / IReR. - Milano : Angeli, c1981
490p. - (Collana IReR ; 4)
88-204-1948-3 X.329/16734
8976937

AGRICULTURE. Lombardy. Italy
to date - Conference proceedings

AGRICOLTURA e forze sociali in Lombardia nella crisi degli anni trenta / P. Bertolini ... [et al.]. - Milano : Angeli, c1983
444p. - (Studi e ricerche storiche ; 19)
r6511444 X.329/19468
8976937

AGRICULTURE. Lombardy. Italy
1700–1900
QUESTIONI di storia agricola lombarda nei secoli XVIII–XIX / autori vari ; a cura di Sergio Zaninelli. - Milano : Vita e pensiero, 1979
343p. - (Contributi dell'Istituto di storia economica e sociale [dell'Università cattolica del Sacro Cuore] ; vol.4) (Scienze storiche ; 25)
r6195536 X.0319/396(4)
2454890

AGRICULTURE. Loup Valley region. Nebraska
1910–1970
BREMER, Richard G.
Agricultural change in an urban age. - Lincoln [Neb.] : University of Nebraska, 1976
xii, 239 p. : ill. - (University of Nebraska studies. new series ; no. 51). - Bibliography : p. 228–239
r6031374 Ac.2692.l/11[no.51]
107167X

AGRICULTURE. Lower Austria
to date
GREIF, Franz
Land- und Forstwirtschaft in Niederösterreich. - St. Pölten : Niederösterreichisches Pressehaus, 1983
64p.. - (Wissenschaftliche Schriftenreihe Niederösterreich ; 73/74)
3–85326–552–9 X.0709/722(73/74)
8841837

AGRICULTURE. Lower Austria
to date - Conference proceedings
NIEDERÖSTERREICHISCHES INSTITUT FÜR LANDESKUNDE. Symposium (1st : 1980 : Geras)
Die Auswirkungen der theresianisch–josephinischen Reformen auf die Landwirtschaft und die ländliche Sozialstruktur Niederösterreichs. - Wien : Selbstverlag des Institutes, 1982
ii,267p.. - (Studien und Forschungen aus dem Niederösterreichischen Institut für Landeskunde ; Bd.3)
r6306195 X.0800/1610(3)
8841837

AGRICULTURE. Lower Silesia. Poland
KAPUSTA, Franciszek
Zmiany struktury agrarnej i kierunków produkcji rolniczej w Legnicko–głogowskim Okręgu Miedziowym. - Warszawa : Państwowe Wydawnictwo Naukowe, 1976
139p. - (Problemy rejonów Uprzemysławianych). - Bibliography : p.129–134
r6075418 X.311/4884 Woolwich
1542265

AGRICULTURE. Lowland regions. England
Effects on landscape conservation
COBHAM, Ralph
Agricultural landscapes. - Cheltenham : The Commission, c1984
135p : ill. - (CCP ; 170). - Bibliography: p132–135
0–86170–102–x X.322/14567
0250406

WESTMACOTT, Richard
Agricultural landscapes. - Cheltenham : The Commission, c1984
80p : ill(some col.). - (CCP ; 168)
0–86170–100–3 L.49/3071
0250406

AGRICULTURE. Lowland regions. Scotland
Farms. Labourers & servants. Social conditions, 1770–1914
FARM servants and labour in lowland Scotland, 1770–1914 / edited by T.M. Devine. - Edinburgh : Donald, c1984
ix,262p : ill. - Includes index
0–85976–105–3 X.800/41476
3296369

AGRICULTURE. Lucerne (Canton). Switzerland
1798–1918. Economic aspects
LEMMENMEIER, Max
Luzerns Landwirtschaft im Umbruch. - Luzern : Rex, 1983
463p.,[16]p.of plates : ill. - (Luzerner historische Veröffentlichungen ; Bd.18). - Includes bibliography
3–7252–0438–1 X.0800/625(18)
904745x

AGRICULTURE. Macedonia. Yugoslavia
STOPANSKA KOMORA NA MAKEDONIJA
Majski nagradi '72. - [Skopje] : Zaednica za izdavačka dejnost, [1972]
[36]p : ill(some col.)
r6057074 Cup.1264.kk.16
1502603

AGRICULTURE. Madura. Indonesia
to date
LEUNISSEN, Jef (Joseph A. C. M.)
Landbouwen op Madura .... - Nijmegen : KU Nijmegen, Vakgroep Sociale Antropologie, 1982
215p. - (Sociaal antropologische cahiers ; 5). - Includes Bibliography
r6330419 X.322/11989
8953651

AGRICULTURE. Magdeburg (Bezirk). East Germany
Labourers. Social conditions, to date
BAUER und Landarbeiter im Kapitalismus in der Magdeburger Börde / herausgegeben von Hans-Jürgen Rach und Bernhard Weissel. - Berlin : Akademie–Verlag, 1982
x,438p,[32]p of plates : ill. - (Veröffentlichungen zur Volkskunde und Kulturgeschichte / Akademie der Wissenschaften der DDR. Zentralinstitut für Geschichte ; Bd.66/3) ([Untersuchungen zur Lebensweise und Kultur der werktätigen Dorfbevölkerung in der Magdeburger Börde ; T.3]). - Bibliography
r6305588 Ac.855/43[Bd.66,Tl.3]
8911479

AGRICULTURE. Magdeburg. East Germany
Personnel. Housing, 1789–1900
RACH, Hans–Jürgen
Bauernhaus, Landarbeiterkaten und Schnitterkaserne. - Berlin : Akademie–Verlag, 1974
xiv, 142 p. : ill. - (Akademie der Wissenschaften der DDR. Zentralinstitut für Geschichte. Wissenschaftsbereich Kulturgeschichte/Volkskunde. Veröffentlichungen zur Volkskunde und Kulturgeschichte ; Bd 58). - Bibliography : p. 113–120
r6002176 Ac.855/43[Bd. 58]
0932868

AGRICULTURE. Maharashtra. India (Republic)
Effects of drought
VINCENT, Linden
Dry spells, drought risk and agricultural production in Maharashtra State. - Norwich : School of Development Studies, University of East Anglia, 1981
168p in various pagings : ill. - (Monographs in development studies ; no.9). - Bibliography: p158–168
B84–41283 X.0322/303
3189740
- [Another copy.] W 6939 IOL

AGRICULTURE. Maharashtra. India (Republic)
Smallholdings, to date
MUKHOPADHYAY, Arun K. (Arun Kumar)
Small farms resource use and technology. - Hyderabad : National Institute of Rural Development, 1980
63p
r6818719 X.322/14781
915129x

AGRICULTURE. Malawi
Farms. Surveys
FARRINGTON, John
Farm surveys in Malawi. - [Reading] : University of Reading, Department of Agricultural Economics and Management, 1975
[4],xii,111,[70]p : ill. - (Development studies / University of Reading. Department of Agricultural Economics and Management ; no.16)
0–7049–0191–9 X.0311/123(16)
0631167

AGRICULTURE. Malawi
Information services. Role of mass media
PERRATON, Hilary
Mass media for agricultural extension in Malawi. - Cambridge : International Extension College, 1982
viii,51p : 2ill. - Bibliography: p50–51
0–903632–21–7 X.525/6148
2777886

AGRICULTURE. Malaysia
HILL, R. D.
Agriculture in the Malaysian region. - Budapest : Akadémiai Kiadó, 1982
233p : ill. - (Geography of world agriculture ; 11). - Bibliography: p223–229
963–05–2825–8 X.322/13512
3131017

AGRICULTURE. Malaysia
to date
KELLER, Roger
Agrarpolitik im unabhängigen Malaya, 1957–1982. - Wiesbaden : Steiner, 1984
xiv,375p,plates. - (Beiträge zur Kolonial- und Überseegeschichte ; Bd.30). - Includes bibliography
3–515–04300–4 X.0700/119(30)
9097074

WONG, Diana
The social organization of peasant reproduction. - Bielefeld : Universität Bielefeld, Fakultät für Soziologie, Forschungsschwerpunkt Entwicklungssoziologie, 1984
358leaves. - Includes bibliography
r6309852 X.525/9681
9112413

AGRICULTURE. Malaysia
Development. Effects on rural economic conditions
GIBBONS, David S.
Agricultural modernization, poverty and inequality. - Farnborough, Hants. : Saxon House, 1980
xi,225p : ill. - Bibliography: p.213–219. -Includes index
0–566–00331–7 X.329/13187 Woolwich
193354x

AGRICULTURE. Mali
MALI. Service de la Statistique Générale, de la Comptabilité Nationale et de la Mécanographie
Rapport de l'enquête agricole. - [Bamako] : [La Service], 1973–
r6019938 S:V.864/2
1148583

AGRICULTURE. Mali
to date
SCHREYGER, Emil
L'Office du Niger au Mali 1932 à 1982. - Wiesbaden : Steiner, 1984
xvi,394p.,plates. - (Beiträge zur Kolonial- und Überseegeschichte ; Bd.27). - Includes bibliography
3–515–04196–6 X.0700/119(27)
9128131

AGRICULTURE. Mali
1960–
Le MALI / Moussa Cola Cissé ... [et al] ; textes réunis et presentés par Pierre Jacquemot. - Paris : Éditions L'Harmattan, [1981]
197p. - (Bibliothèque du développement). - Includes bibliography
2–85802–200–3 X.329/16634
8295220

AGRICULTURE. Mali
1960– - Statistics
MALI. Ministère du plan. Direction nationale de la statistique et de l'informatique
Rapport de l'enquête agricole, 1974–1975. - [S.l.] : [le Ministère], 1979
101p
r6205849 SV.870(24
8295220

AGRICULTURE. Mantua (Province). Italy
DRAGONI, Aloisia Nenè
L'Oltrepò mantovano. - Firenze : La nuova Italia, 1977
x,198p : maps. - (Pubblicazioni della Facoltà di lettere e filosofia dell'Università di Milano ; 82) (Sezione a cura dell'Istituto di geografia umana ; 5)
r6160128 X.322/9148. Woolwich
2424541

AGRICULTURE. Marandellas region. Zimbabwe
Political aspects, 1890–1965
HODDER-WILLIAMS, Richard
White farmers in Rhodesia, 1890–1965. - London : Macmillan, 1983
xii,256p. - Includes index
0–333–27237–4 X.809/60492
314447x

AGRICULTURE. Maranhão. Brazil
EMATER. Nidoc
Pacote de informações para a programação da Emater, Ma, 1980. - [São Luis] : Nidoc, 1979
128leaves
r6159602 LAS.B.275/203 Woolwich
2459469

AGRICULTURE. Marches. Italy
to date
ANSELMI, Sergio
Mezzadri e terre nelle Marche. - Bologna : Pàtron, 1978
166p. - (Fonti, ricerche, testi / Istituto di storia economica e sociale dell'Università di Bologna ; 15)
r6500239 X.322/10042 Woolwich
8715149

AGRICULTURE. Mari ASSR. Russia (RSFSR)
PO planu partiĭ / [otvetstvennyĭ redaktor I.S. Gusev ; sostaviteli Boris Nikolaevich Verein...i dr]. - Ĭoshkar–Ola : Mariĭskoe knizhnoe izdatel'stvo, 1976
297 p., 88 p. of plates : ill
r6023155 X.809/23008
0896055

AGRICULTURE. Marshes. Sweden
to 1976

MYRMARKER / [redaktionskommitté : Rune Engström, Sigfrid Leander, Birgitta von Malmborg]. - [Stockholm] : Riksförbundet för hembygdrvard, 1976
118p : ill
91-85124-14-1 X.319/8709 Woolwich
0396400

AGRICULTURE. Mauritius
Development, to date

PARTI SOCIALISTE MAURICIEN
Dossier sur la diversification agricole. - Port Louis : Parti socialiste mauricien, 1980
24p
r6207436 X.809/53044
8907722

AGRICULTURE. Mediterranean region
Research

GIBBON, David, 1939-
A systems approach to resource management and development in the ECWA region. - Norwich : School of Development Studies, University of East Anglia, 1981
30p : ill. - (Discussion paper ; no.104). - Bibliography: p29-30
B84-40804 X.525/9959
3187403

AGRICULTURE. Mehrauli. India (Republic)
Trade unions: Dehat Mazdoor Union, 1973-1975

LEWIS, Primila
Reason wounded. - London : Allen and Unwin, 1979
vii,207p. - Includes index
0-04-301096-2 X.809/45596
2156938

AGRICULTURE. Melin-y-wig. Clwyd
Owen, Huw Parry - Biographies

OWEN, Ellis Parry
Huw Foelgrachen. - Dinbych [i.e. Denbigh] : Gwasg Gee, 1978
87p
B78-28014 X.319/18288 Woolwich
1550365

AGRICULTURE. Meopham. Kent
Smallholdings. Averill, Edward. Personal property. Probate, 1674 - Biographies

CARLEY, James
Edward Averill 1619-1674. - Meopham : Meopham Publications Committee, [1981?]
12p. - (Meopham biographies ; 2)
B81-33684 X.809/49045
2546450

AGRICULTURE. Mernye. Hungary

TÓTH, Tibor
A mernyei uradalom a feudális rend utolsó szakaszában. - Budapest : Akadémiai Kiadó, 1978
177p,[1]fold leaf of plates. - (Agrártörténeti tanulmányok ; 5)
963-05-1294-7 X:0311/235(5) Woolwich
2429217

AGRICULTURE. Messina (Province). Italy
to date

CLASSI sociali e agricoltura meridionale / a cura di Enzo Mingione. - Milano : Giuffrè, 1981
xiv,142p,[1]leaf of plates. - (Studi sociologici / Facoltà di scienze politiche, Università degli studi di Messina ; 2)
r6507943 X.322/12095
8963622

AGRICULTURE. Metropolitan counties. England

THOMSON, Kenneth J.
Farming in the fringe. - Cheltenham : Countryside Commission, c1981
vi,105p : ill. - Bibliography: p102-105
0-86170-034-1 BS.116/177
2531682

AGRICULTURE. Mexico

MAUNDER, W. F. (Wynne Frederick)
University of Guanajuato (Agricultural Engineering Department). - [London] : ADAS, 1983
20p
B84-24855 BS.3/845
3123715

AGRICULTURE. Mexico
- Texts

AGUIRRE, Norberto
Cuestiones agrarias. - México : Editorial Joaquín Mortiz, 1977
230p. - (Cuadernos de Joaquín Mortiz ; 47/48)
r6099793 X.319/20086 Woolwich
1983288

AGRICULTURE. Mexico
1500-1821

FLORESCANO, Enrique
[Estructuras y problemas agrarios de México, 1500-1821]. Origen y desarrollo de los problemas agrarios de México (1500-1821). - México : Era, 1976
158p. - (Colección Problemas de México)
r6084603 X.319/9814 Woolwich
185366X

AGRICULTURE. Mexico
1910-

ESTEVA, Gustavo
La batalla en el México rural. - México : Siglo Veintiuno Editores, 1980
243p. - (Sociología y política)
9-682-31004-0 X.629/18757
8853223

AGRICULTURE. Mexico
1920-

CASTAÑOS, Carlos Manuel
Asistencia técnica programada en base a perfiles de desarrollo vegetativo. - [México?] : FIRA, 1980
89p,[4]p.of plates : ill(some col.)
r6702343 LAS.F.210/13 Woolwich
8302782

COLL-HURTADO, Atlántida
?Es México un país agrícola?. - México : Siglo Veintiuno, 1982
213p. - (Economía y demografía). - Includes bibliography
9-682-31149-7 X.329/19712
8302782

SCHNEIDER, Hans, Geographer
Die Landwirtschaft im Valsequillo. - München : Fink, 1979
214p : ill. - (Lateinamerika-Studien ; Bd.5). - Includes bibliography
3-7705-1510-2 X.0800/1815(5)
8302782

AGRICULTURE. Mexico
1946-1970 - Bibliographies

VELÁSQUEZ, Pablo
Bibliografía agrícola nacional 1946-1970. - México : Instituto Nacional de Investigaciones Agrícolas, 1973
2v.
r6024972 X.322/2508 Woolwich
1013270

AGRICULTURE. Mexico
Economic development, 1521-1630

FRANK, Andre Gunder
Mexican agriculture, 1521-1630. - Cambridge : Cambridge University Press [etc.], 1979
xiv,91p. - (Studies in modern capitalism). - Bibliography: p.85-87. - Includes index
0-521-22209-5 X.329/11824 Woolwich
1872419

AGRICULTURE. Mexico
Economic development, 1921-1970

LOMBARDO TOLEDANO, Vicente
[Selections. 1974]. En torno al problema agrario. - México : Coedición de la Confederación Nacional Campesina y del Partido Popular Socialista, 1974
516 p. : ill
r6011590 X.319/8496 Woolwich
0872334

AGRICULTURE. Mexico
Economic development. Effects of land tenure, to 1970

CENTRO DE INVESTIGACIONES AGRARIAS (Mexico)
Estructura agraria y desarrollo agrícola en México. - México : Fondo de Cultura Económica, 1974
xvi, 1174 p.
r6009011 X.320/4988 Woolwich
0918806

AGRICULTURE. Mexico
Employment

CENTRO DE INVESTIGACIONES AGRARIAS (Mexico)
Empleo, desempleo y subempleo en el sector agropecuario. - México : el Centro, 1976
367p. - Includes bibliographies
r6066825 X.510/11049
1785214

AGRICULTURE. Mexico
Law

IBARROLA, Antonio de
Derecho agrario. - México : Porrúa, 1975
xxxv,862p. - Bibliography : p.825-832
r6037940 X.200/11027
1211277

LUNA ARROYO, Antonio
Derecho agrario mexicano. - México : Editorial Porrúa, 1975
c,827p
r6079145 X.200/11438
1211277

AGRICULTURE. Mexico
Mathematical models

The BOOK of CHAC edited by Roger D Norton and Leopoldo Solis M.. - Baltimore . London : Published for the World Bank [by] the Johns Hopkins University Press, c1983
xx,602p : ill. - (A World Bank research publication). - Includes bibliographies and index
0-8018-2585-7 X.622/16822
2959666

AGRICULTURE. Mexico
Policies of govermment, 1980

REDCLIFT, Michael
Development policymaking in Mexico. - London : University of London, Institute of Latin American Studies, [1982?]
12,vp. - (University of London, Institute of Latin American Studies working papers ; 6)
0-901145-45-9 X.520/27091
2728664

AGRICULTURE. Mexico
Reform, to 1974

AGUILERA G., Manuel
Reforma agraria. - México : Secretaría de la Presidencia, Departamento Editorial, 1975
38p. - (Temas mexicanos)
r6065617 X.319/9432 Woolwich
1507281

AGRICULTURE. Mexico
Working conditions, 1876-1910

La SERVIDUMBRE agraria en México en la época porfiriana / [textos de] Friedrich Katz ... [et al.] ; introducción y selección de Friedrich Katz ; traducción de Antonieta Sánchez Mejorada. - México : Secretaría de Educación Pública, 1976
183p. - (SepSetentas ; 303)
r6084610 X.319/9813 Woolwich
1854038

AGRICULTURE. Mid-West region. North America
1840-1974

BROWN, Ron
The agricultural Mid-West. - London : Heinemann Educational, 1975
[4],64p : ill. - (Contemporary geographies. North American section). - Includes index
0-435-34065-4 X.708/43826
0330922

AGRICULTURE. Mid-West region. United States
Labourers, 1815-1860

SCHOB, David Eugene
Hired hands and plowboys. - Urbana ; London : University of Illinois Press, [1976]
x,329p. - Bibliography: p.273-308. - Includes index
0-252-00509-0 X.320/10215 Woolwich
0579718

AGRICULTURE. Middle East
- Serials

MIDDLE East agribusiness. - Vol.1 no.1 (Oct./Nov. 1981)-. - Redhill : International Trade Publications, 1981-
v. : ill(some col.)
B82-17259 SRL
2704080

AGRICULTURE. Middle East
Development, ca 1960-1975

The DEVELOPING agriculture of the Middle East / R.M. Burrell ... [et al.] ; [maps by Ray Martin]. - London : Graham and Trotman, 1976
[7],74p
0-86010-046-4 X.313/1300 Woolwich
0701750

AGRICULTURE. Midlands. England
Farms. Social conditions, ca 1905-ca 1925

WAINE, M. E. (Muriel Edna)
All our own horses. - Wolverhampton : Mooley, 1981
68p : ill
B81-20795 X.990/17320 Woolwich
240785x

AGRICULTURE. Mindoro. Philippines
Development. Mindco Agro-Industrial Rural Community Development

The MINDORO strategy / edited by Angelito L. Santos. - Metro Manila : Development Academy of the Philippines. [1982]
xix,115p : ill
9-7104-0006- X.329/19353
9151281

AGRICULTURE. Moldavia (Soviet Socialist Republic)

DZHESMEDZHIAN, Artem Arshakovich
Sovershenstvovanie mekhanizma khoziaistvovaniia v partiinaia rabota. - Kishinev : Kartia Moldoveniaske, 1975
278p. - Bibliography : p.267-277
r6174615 X.629/8180 Woolwich
2027275

VIZER, Boris Konstantinovich
Razvitie sel'skogo khoziaistva Moldavskoi SSR 1951-1970. - Kishinev : Shtiintsa, 1975
336p
r6096830 X.329/8903 Woolwich
1987275

AGRICULTURE. Moldavia (Soviet Socialist Republic) 1700-1800

DRAGNEV, Denir Mironovich
Sel'skoe khoziaistvo feodal'noi Moldavii (konets XVII-nachalo XXV.). - Kishinev : Shtiintsa, 1975
285p
r6094527 X.329/8906 Woolwich
2025507

AGRICULTURE. Moldavia (Soviet Socialist Republic) 1940-

TOPUZLU, G. N. (Grigorii Nikolaevich)
Pomoshch' bratskikh soiuznykh respublik sel'skomu khoziaistvu Moldavskoi SSR, 1940-1958 gg.. - Kishinev : "Shtiintsa,, 1981
196 p.. - Includes bibliographical references
r6482631 X.329/17928
9082867

AGRICULTURE. Moldavia (Soviet Socialist Republic) 1945-

AGRIKULTURA Moldovei = Moldavian agriculture / [autor al tekstului shi alkătuitor N. Sosnin]. - Kishinëu : Kartia Moldoveniaskė, 1974
[ca.112]p : chiefly ill
r6487420 X.322/13464
8307261

STRATU, B. N. Boris Nikiforovich)
Spechializaria agrikulturii yn konditsiile khozraschotului..... - Kishinëu : Kartia Moldoveniaskė, 1972
150p
r6484262 X.329/16350
8307261

AGRICULTURE. Moldavia (Soviet Socialist Republic)
Collectivisation

SYTNIK, Mikhai Kondrat'evich
Kollektivizatsiia sel'skogo khoziaistva i formirovanie klassa kolkhoznogo krest'ianstva v Moldavii. - Kishinev : Izdatel'stvo "Shtiintsa", 1976
352p. - Bibliography : p.338-345
r6037486 X.329/8657 Woolwich
1154559

AGRICULTURE. Moldavia (Soviet Socialist Republic)
Cooperation, 1945-

ERKHAN, L. S. (Leonid Semenovich)
Vozniknovenie razvitie mezhkhoziaistvennoi kooperatsii v sel'skom khoziaistve Moldavskoi SSR. - Kishinev : Shtiintso, 1983
153p
r6490080 X.329/19093
9125590

AGRICULTURE. Moldavia (Soviet Socialist Republic)
Finance

USTIIAN, I.
Agrarno-promyshlennoe kooperirovanie i khozraschetnye otnosheniia. - Kishinev : Izdatel'stvo "Kartia Moldoveniaske", 1975
217p
r6036920 X.529/20207
1158260

AGRICULTURE. Moldavia (Soviet Socialist Republic)
Personnel

TIMUSH, Andreĭ Ivanovich
Sotsial'nye protsessy na sele. - Kishinev : Izdatel'stvo "Kartia Moldoveniaske", 1975
138p
r6036466 X.529/20249
1209345

AGRICULTURE. Moldavia (Soviet Socialist Republic)
Productivity, 1945-

RAZVITIE material'nogo proizvodstva i povyshenie ego effektivnosti / [pod redaktsiei G.N. Singura]. - Kishinev : Shtiintsa, 1973
182p,plate
r6462840 X.329/19095
9053417

AGRICULTURE. Moldavia (Soviet Socialist Republic)
Role of Partid Komunist al Moldovei, 1945-

BOBICHEV, M. I. (Moisei Il'ich)
Deiatel'nost' Kompartii Moldavii po razvitiiu sel'skogo khoziaistva (1951-1958). - Kishinev : Shtiintsa, 1978
158p
r6485596 X.329/14973 Woolwich
8795088

AGRICULTURE. Monasteries. Emilia-Romagna. Italy to date

BASINI, Gian Luigi
Le terre di un monastero. - Bologna : Pàtron, 1979
126p.[3]leaves of plates. - (Quaderni di studi e ricerche / Facoltà di economia e commercio dell'Università di Parma ; 3)
r6505627 X.322/11377
889180x

AGRICULTURE. Montenegro. Yugoslavia 1918-

KALEZIĆ, Žarko
Strukturne promjene na crnogorskom selu u XX vijeku. - Titograd : Pobjeda, 1976
198p. - Bibliography
r6435616 X.322/11886
8945217

AGRICULTURE. Mordvinian ASSR. Russia (RSFSR) 1917-

AGEEV, M. V.
Sotsialisticheskoe preobrazovanie sel'skogo khoziaistva Mordovskoi ASSR. - Saransk : Mordovskoe knizhnoe izdatel'stvo, 1979
179p
r6485933 X.329/15502
8840016

AGRICULTURE. Morocco
1956- - Conference proceedings

PARTI DU PROGRÈS ET DU SOCIALISME. Colloque agraire (1980 : Rabat)
La question agraire du Maroc. - [Casablanca] : Édition Al-Bayane, [1981]
661p
r6211142 X.322/12831
9006575

AGRICULTURE. Morocco
Development

PONIKIEWSKI, Augustyn
Problemy rozwojowe produkcji rolnej Maroka na tle warunków przyrodniczych i struktur agrarnych. - Warszawa : Wydawnictwa Uniwersytetu Warszawskiego, 1976
297p,3 fold. leaves of plates. - (Rozprawy Uniwersytetu Warszawskiego ; 98). - Bibliography : p.273-286
r6016413 Ac.1147/22[98]
1125648

AGRICULTURE. Mozambique
1964-

WUYTS, Marc
Camponeses e economia rural em Moçambique. - [Maputo] : Universidade Eduardo Mondlane, Centro de Estudos Africanos, [1978?]
31p. - (CEA relatório ; no.81/8)
r6707366 X.322/12808
9031049

AGRICULTURE. Munster. Ireland (Republic)
Development. Organisations: Council for Development in Agriculture

ACOT
South West Region. - Fermoy : [ACOT], [1981?]
[16]p : ill
B82-05197 X.329/14793 Woolwich
9049568

AGRICULTURE. Namulonge. Uganda
Research organisations: Namulonge Cotton Research Station, to 1972

COTTON RESEARCH CORPORATION
Agricultural research for development. - Cambridge : Cambridge University Press, 1976
xi,343p,8p of plates,leaf of plate : ill(some col). - Bibliography: p.325-343. - Includes index
0-521-21051-8 X.320/10514 Woolwich
0732346

AGRICULTURE. Närke. Sweden
Farms: Kettstaka (Farm). Social conditions, 1867-1970

ADOLFSSON, Axel
Dagbok från en liten gård i Närke. - [Västra Frölunda] : [A. Adolfsson], 1977
126p : ill
91-970250-0-3 X.419/7911
170835X

AGRICULTURE. Natal. South Africa
Baynes, Joseph - Biographies

PEARSE, R. O. (Reg O.)
Joseph Baynes pioneer. - Pietermaritzburg : Shuter & Shooter [for] The Baynesfield Board of Administration, 1983
331p,plates
0-86985-708-8 X.322/14780
9151303

AGRICULTURE. Neath region. West Glamorgan
Farms. Social conditions

ST CLAIR, Isla
The farm on the hill. - London : Pelham, 1980
91p : ill(some col). - Includes index
0-7207-1241-6 X.325/738 Woolwich
2074028

AGRICULTURE. Nebraska
Geographical aspects

AGRICULTURAL atlas of Nebraska / project director Merlin P. Lawson ; edited by James H. Williams, Doug Murfield ; foreword by James H. Zumberge ; cartographer John D. Magill. - Lincoln [Neb.] ; London : University of Nebraska Press, 1977
xvi,110p : ill(some col)
0-8032-0894-4 Cup.1281/903
1570382

AGRICULTURE. Negev. Israel

EVENARI, Michael
The Negev. - 2nd ed. - Cambridge, Mass. ; London : Harvard University Press, 1982
ix,437p : ill. - Bibliography: p418-426. - Includes index
0-674-60672-8 X.421/26659
3277100

AGRICULTURE. Nepal
to date

PANT, Y. P.
Agricultural development in Nepal. - 2nd revised and enlarged edition. - Bombay : Vora & Co., 1979
xii,162p
r6804552 X.329/14567 Woolwich
8762201

AGRICULTURE. Nepal
Finance

JHA, Kumar Kant
Agricultural finance in Nepal. - New Delhi : Heritage, c1978
xvi,241p. - Bibliography: p225-238. - Includes index
B81-09299 X.329/13633 Woolwich
2366630

AGRICULTURE. Netherlands
to 1901

LANDBOUWKUNDIGE beschrijvingen uit de 19e eeuw / een heruitgave door G.H. Kocks en J.M.G. van der Poel. - Wageningen : Vereniging voor Landbouwgeschiedenis, 1981
2v. : ill. - (Agronomisch-historische bijdragen ; dl.9, 10)
r6330822 X.0311/222.(9,10.)
9082891

AGRICULTURE. Netherlands
Farms. Buildings. Architecture, to date

TREFOIS, Clemens V.
[Selections]. Het boerendak ; De plattegrondsvormen van zg. Frankische boederijtypen (heruitgave) ; De bouw der boerenhoeven (heruitgave). - Sint-Niklaas : Danthe, 1980
312p : ill
r6330165 L.45/1750
8876827

AGRICULTURE. Netherlands
Mechanisation, to date

POEL, J. M. G. van der
Honderd jaar landbouwmechanisatie in Nederland. - Wageningen : Vereniging voor Landbouwgeschiedenis, 1983
307p : ill. - (Agronomisch-historische bijdragen ; dl.11). - Includes bibliography
r6330825 X.0311/222.(11.)
9082832

AGRICULTURE. New Castile. Spain
Collectivisation, 1936–1939

GUTIÉRREZ MOLINA, José Luis
Colectividades libertarias en Castilla. - Madrid : Campo Abierto, 1977
113p. - (Debate libertario ; 3. Serie Divulgación)
84-7446-005-0 X.319/9816 Woolwich
1854046

AGRICULTURE. New Zealand

JONES, Derrick A. P.
Final report. - London : ADAS, [1984?]
ii,14p
B85-01977 BS.3/925
3199363

SAVAGE, M. J. (Martin J)
Report of an exchange year in New Zealand 1980-1981. - [London] : ADAS, [1981]
10p : 1ill
B82-11662 BS.3/721
2627299

STANDAGE, M. R. M.
Report on an exchange visit to New Zealand, 1981/82. - [London] : ADAS, c1983
15p
B84-42139 BS.3/924
1805401

AGRICULTURE. New Zealand
to date

YORKE, A. Tremenheere (Amy Tremenheere)
The animals came first. - Auckland : London : Heinemann, 1980
145p : ill
0-908592-06-x X.329/14890 Woolwich
8837341

AGRICULTURE. New Zealand
Farms. Financial management

LATTA, C. R. (C Ralph)
The NZ farmer's guide to financial planning. - Rev ed. - Christchurch, N.Z. ; London : Whitcoulls, 1983
156p. - Includes index
0-7233-0690-7 X.529/56892
2960869

AGRICULTURE. New Zealand
Farms. Management

SQUIRE, J.
Farm business management. - Auckland : Longman Paul, 1982
viii,156p
0-582-71790-6 X.629/21900
9026932

AGRICULTURE. New Zealand
McLeod, David - Biographies

MCLEOD, David
Down from the tussock ranges. - Christchurch [N.Z.] ; London : Whitcoulls, 1980
254p
0-7233-0640-0 X.981/22688
2333376

AGRICULTURE. Nigeria

AGBOOLA, S. A.
An agricultural atlas of Nigeria. - Oxford : Oxford University Press, 1979
xiv,248p : ill. - Bibliography: p.211-230. - Includes index
0-19-575408-5 X.802/11236
2101270

NORMAN, David W.
Farming systems in the Nigerian savanna. - Epping : Bowker, c1982
xxiv,275p : ill. - (A Westview replica edition). - Bibliography: p247-266. - Includes index
0-86531-925-1 X.329/19064
2101270

PHILLIPS, Tom Alan
An agricultural notebook. - Metric ed. - London : Longman, 1977
[8],312p : ill. - Includes index
0-582-60768-x SRL
119139x

AGRICULTURE. Nigeria
- For British exporters

A REPORT on the agricultural industry and the market for UK farm inputs in Nigeria. - London : BAEC, 1981
63p
B82-11354 SRL
2625741

AGRICULTURE. Nigeria
Economic development related to social change

IGBOZURIKE, Uzo M. (Uzo Martin)
Problem-generating structures in Nigeria's rural development. - Uppsala : Scandinavian Institute of African Studies, 1976
140 p. : ill. - Bibliography : p. 134-140
91-7106-102-9 X.519/26200
0740209

AGRICULTURE. Nigeria
Employment

COLEMAN, Gilroy
Labour data collection in African traditional agricultural systems. - Norwich : School of Development Studies, University of East Anglia, 1982
48p : ill. - (Development studies occasional paper ; no. 18). - Bibliography: p48
B84-39962 X.322/14081
3184242

AGRICULTURE. Nigeria
Policies of government, 1960-

JAJA, Joseph S.
Staatliche Förderung der Landwirtschaft in Nigeria. - Tübingen : Erdmann, 1981
x,210p. - (Bochumer Materialien zur Entwicklungsforschung und Entwicklungspolitik ; Bd.17). - Bibliography
3-7711-0351-7 X.809/52288
8840024

AGRICULTURE. Nigeria
Policies of government related to economic development, 1962-1968

WELLS, Jerome C.
Agricultural policy and economic growth in Nigeria, 1962-1968. - Ibadan ; London : Oxford University Press for the Nigerian Institute of Social and Economic Research, 1974 [i.e. 1976]
xviii,490p. - Bibliography: p.459-481. - Includes index
0-19-575230-9 X.320/10334 Woolwich
0642312

AGRICULTURE. Norfolk
1776-1842

PARKER, Robert Alexander Clarke
Coke of Norfolk. - Oxford : Clarendon Press, 1975
xiii,222p,plate : 1 ill. - Includes index
0-19-822403-6 X.329/10238 Woolwich
056785x

AGRICULTURE. Norfolk
Research organisations: Norfolk Agricultural Station - Serials

MORLEY bulletin / Norfolk Agricultural Station. - No.1 (Mar. 1979)-. - Morley, Norfolk : The Station, 1979-
v.
B83-41487 P.621/896
296810x

AGRICULTURE. Norfolk
Research organisations: Norfolk Agricultural Station, to 1978

HUTCHINSON, Sir Joseph, b.1902
Change and innovation in Norfolk farming. - Chichester : Packard Publishing Ltd for the Norfolk Agricultural Station, 1980
72p : ill. - Bibliography: p.67-68. - Includes index
0-906527-04-x X.325/625 Woolwich
2282038

AGRICULTURE. Norra Fjöle, Skasas & Skofterud. Sweden
Farms. Buildings. Architecture, ca 1830-1976

NYSTRÖM, Bengt, 1937-
Skasas, Skofterud, Norra Fjöle. - Stockholm : Byggforskningen, 1977
152p : ill. - (Rapport ; R64. 1977) (Rapport / Kulturhistoriska undersökningen vid Nordiska museet ; 4)
91-540-2733-0 P.703/360[1977.no.64]
1535978
- [Another copy] P.443/219 [no.4]

AGRICULTURE. Norrbotten. Sweden
1809-

BERGSTRÖM, Sten Ove
Kronotorparna. - Lulea : Tornedalica, 1981
205p : ill. - (Tornedalica ; nr 32). - Bibliography
91-7538-030-7 X.0700/436(32)
8928339

AGRICULTURE. North Bedfordshire (District)
Farms. Buildings. Architectural design

FARM buildings / NBBC. - [Bedford] : [North Bedfordshire Borough Council], [1977]
18p : ill. - Bibliography: p18
B81-26360 X.525/4592
2503298

AGRICULTURE. North-east Brazil

DIAS, Gentil Martins
Depois do latifúndio. - Rio de Janeiro : Tempo Brasileiro, 1978
248p : ill. - (Biblioteca Tempo universitário ; 48). - Includes bibliographies
r6158137 X.319/20294 Woolwich
2137321

FIGUEROA, Manuel
O problema agrário no Nordeste do Brasil. - São Paulo : HUCITEC : Recife : SUDENE, 1977
xi,272p
r6066608 X.319/20081 Woolwich
2092743

KUTCHER, Gary P.
The agricultural economy of northeast Brazil. - Baltimore ; London : Published for the World Bank [by] Johns Hopkins University Press, c1981
xiv,217p : ill. - (A World Bank research publication). - Bibliography: p259-264. - Includes index
0-8018-2581-4 X.322/11096
2761084

AGRICULTURE. North-east Brazil
- Catalogues

BRAZIL. Superintendência do Desenvolvimento do Nordeste Departamento de Agricultura e Abastecimento. Biblioteca
Catálogo de publicações. - Recife : SUDENE, 1977
40leaves
r6159622 LAS.B.275/170 Woolwich
2440393

AGRICULTURE. North-east Brazil
to date

MELO, Mário Lacerda de
O açúcar e o homem. - Recife : Instituto Joaquim Nabuco de Pesquisas Sociais, 1975
304p. - (Série Estudos e pesquisas ; 4). - Bibliography
r6701662 X.0950/208(4)
8726469

AGRICULTURE. North-east Brazil
Development

FIGUEROA, Manuel
O problema agrário no Nordeste do Brasil. - São Paulo : HUCITEC : Recife : SUDENE, 1977
xi,272p
r6066608 X.319/20081 Woolwich
2092751

AGRICULTURE. North-east Brazil
Development - Serials

[RELATÓRIO anual (Brazil. Superintendência do Desenvolvimento do Nordeste. Departamento de Agricultura e Abastecimento)]. Relatório anual. - [196-?]-. - Recife : DAA, [196-?]-
Holdings: 1973
r6159791 LAS.B.275/219 Woolwich
2464241

AGRICULTURE. North-east Brazil
Economic development, 1948-1969

PATRICK, George F.
Desenvolvimento agrícola do nordeste. - Rio de Janeiro : Instituto de Planejamento Econômico e Social : Instituto de Pesquisas, 1972
319p. - (Coleção Relatórios de pesquisa ; n.11). - Includes bibliographies
r6067193 X.319/9735 Woolwich
1790455

AGRICULTURE. North-east Brazil
Effects of irrigation projects

HALL, Anthony L.
Drought and irrigation in North-East Brazil. - Cambridge : Cambridge University Press, 1978
xi, 152p. - (Cambridge Latin American studies ; 29). - Bibliography: p.142-149.- Includes index
0-521-21811-x X.809/28688
2064421

AGRICULTURE. North-east Morocco
Effects of irrigation, 1970-1980

SEDDON, David
Changing agriculture in the Sebra Plain, northeast Morocco. - Norwich : School of Development Studies, University of East Anglia, 1980
100p. - (Development studies occasional paper ; no.7). - Bibliography: p100
B84-39938 X.322/14073
3184056

AGRICULTURE. North-eastern France
Farms. Social conditions, to 1328 - Conference proceedings

DEUTSCH-FRANZOSISCHES HISTORIKERKOLLOQUIUM (16th : 1980 : Xanten)
Villa - curtis - grangia. - München : Artemis, 1983
xi,281p : ill. - (Beihefte der Francia ; Bd.11). - Some contributions include a bibliography
3-7608-4661-0 X.0800/635a(11)
8916438

AGRICULTURE. North European Soviet Union. European Soviet Union

KANEV, Genii Vasil'evich
Ėkonomicheskaīa reforma v sovkhozakh Evropeĭskogo Severa.... - Syktyvkar : Akademiīa nauk SSSR, Komi filial, 1975
51p. - (Seriīa "Nauchnye rekomendat͡sii-narodnomu khozīaĭstvu ; vyp.7)
r6036805 X.519/22506
1000845

KOLESNIKOV, Petr Andreevich
Severnaīa derevnīa v XV-pervoi polovine XIX veka. - Vologda : Severo-Zapadnoe knizhnoe izdatel'stvo, 1976
415p
r6078960 X.529/19931
1000845

REZERVY sokrashcheniīa zatrat v sel'skom khozīaĭstve Evropeĭskogo Severa / [otvetstvennyi redaktor V. P. Podoplelov]. - Moskva : Izdatel'stvo "Nauka", 1976
158 p.. - (Problemy sovetskoi ekonomiki)
r6021030 X.519/21710
1000845

AGRICULTURE. North Ossetian ASSR. Russia (RSFSR) to 1917

ORTABAEV, Batyrbek Khadzhumarovich
Zemel'nye otnosheniīa i razvitie kapitalizma v sel'skom khozīaĭstve Severnoi Osetii v kont͡se XIX-nachale XX vekov (1890-1917gg.). - Ordzhonikidze : Severo-Osetinskii gosudarstvennyi universitet im. K.L. Khetagurova, 1979
77p
r6462129 X.329/16145
8916535

AGRICULTURE. North Saskatchewan
Johnson, Jazzer - Biographies

JOHNSON, Jazzer
My god! What a life. - Irchester : Saunders, 1983, c1982
260p
0-907877-95-8 X.809/56681
2976269

AGRICULTURE. North Wales
- Serials

FARMING express. - No.1 (Apr. 1982)-. - Wrexham : St Regis Newspapers, 1982-
v. : ill
B83-03098 P.2000/1088
2805499

AGRICULTURE. North-west Brazil
Industrial relations, 1550-1972

ANDRADE, Manuel Correia de
A terra e o homem no nordeste. - 3. ed., revista e atualizada. - São Paulo : Editora Brasiliense, 1973
251 p. : ill. - Bibliography : p. 241-249
r6003324 X.709/19720
0910198

AGRICULTURE. North-west Russia (RSFSR)
1917-1945

SELEZNEV, Valeriĭ Aleksandrovich
Pervye shagi sot͡sialisticheskikh preobrazovaniĭ v sel'skom khozīaĭstve Severo-Zapada RSFSR (1917-1929gg.) .... - Leningrad : Izdatel'stvo Leningradskogo universiteta, 1983
198p
r6463727 X.329/16515
8962499

AGRICULTURE. North-western Ireland (Republic)
Development. Organisations: Council for Development in Agriculture

ACOT
North West Region. - Sligo : [ACOT], [1981?]
12p : ill
B82-05198 X.329/14791 Woolwich
2599473

AGRICULTURE. North York Moors. North Yorkshire

FUSSEY, Joyce
Cows in the corn. - London : Elek, 1978
144p
0-236-40149-1 X.989/53152
0622648

Cows in the corn. - London : Magnum, 1981, c1978
144p
0-417-05100-x X.329/13431 Woolwich
0622648

'Milk my ewes and weep'. - London : Elek, 1976
131p
0-236-40031-2 X.329/10374 Woolwich
0622648

Milk my ewes and weep. - London : Magnum Books, 1980
5-131p : ill
0-417-05110-7 X.329/12371 Woolwich
0622648

AGRICULTURE. North York Moors. North Yorkshire
Farms. Social conditions, ca 1980

DRYSDALE, Ann
Faint heart never kissed a pig. - London : Routledge & Kegan Paul, 1982
170p
0-7100-0972-0 X.809/51116
2702177

AGRICULTURE. North Zealand. Denmark
Hansen Risegaard, Niels - Biographies

NIELS Hansen Risegaard / ved Marie Risegaard Jakobsen og Børge Bay. - [Riis] : Risegaards Legatstiftelse, [1979]
144p : ill
87-980783-0-5 X.329/12299. Woolwich
2294648

AGRICULTURE. Northavon (District). Avon
Organisations: National Farmers' Union. Gloucestershire and North Avon Branch

NATIONAL FARMERS' UNION. Gloucestershire and North Avon Branch
The Gloucestershire and North Avon NFU year book and desk diary. - [1975]-. - Chippenham : Gloucestershire and North Avon Branch of the NFU, [1975-] : ill
r6009997 P.311/711
0745049

AGRICULTURE. Northern India (Republic)
Development, 1961-1972

CHAUDHRI, D. P.
Education, innovations and agricultural development. - London : Croom Helm, 1979
vii,127p : 1 ill. - Bibliography: p.82-105. - Includes index
0-85664-826-4 X.329/11827 Woolwich
1497952

AGRICULTURE. Northern Ireland
- Statistics - Serials

GREAT BRITAIN. Department of Agriculture for Northern Ireland. Economics and Statistics Division
Northern Ireland farm business planning handbook. - 1977-1978. - Belfast : Economics and Statistics Division, Department of Agriculture, N.I., 1977-1978
2v.
Holdings: 1978, etc
B82-12456 P.513/2726
0241008

STATISTICAL review of Northern Ireland agriculture / Department of Agriculture for Northern Ireland, Economics and Statistics Division. - 1976-1977-. - Belfast : The Division, 1978-
B79-16444 CS.A.210/107
187571x

AGRICULTURE. Northern Ireland
Smallholdings. Livestock: Beef cattle. Profitability. Mathematical models

MOSS, Joan E.
Growth possibilities for small scale beef cattle and sheep farms. - Belfast : Economics & Statistics Division, Dept. of Agriculture, N.I., [1983?]
56p. - (Studies in agricultural economics)
B84-03232 X.322/12634
2985004

AGRICULTURE. Northern Ireland
Smallholdings. Livestock: Sheep. Profitability. Mathematical models

MOSS, Joan E.
Growth possibilities for small scale beef cattle and sheep farms. - Belfast : Economics & Statistics Division, Dept. of Agriculture, N.I., [1983?]
56p. - (Studies in agricultural economics)
B84-03232 X.322/12634
2985012

AGRICULTURE. Northern Scotland
- Forecasts

The OUTLOOK for agriculture / [School of Agriculture, Aberdeen]. - 1976. - Aberdeen : The School, 1975
9p
0-85281-005-9 P.311/706
0693200

AGRICULTURE. Norway
to date

VEVSTAD, Andreas
Aust-Agder landbruksselskap gjennom 150 ar. - [Arendal] : Selskapet, 1981
264p : ill
r6611409 X.809/59426
9073779

AGRICULTURE. Norway
1905-

LIDTVEIT, Aslak
Jordbruket i Noreg 1914-1974. - Oslo : [Landbruksdepartementet], 1979
749p : ill
82-583-0010-5 X.322/11714
8916462

AGRICULTURE. Norway
1905- - Conference proceedings

HVORFOR bør vi drive jordbruk i Norge?. - [Oslo] : Landbruksforlaget, 1980
207p. - (Skrifter fra Norsk samfunnsgeografisk forening ; [7])
82-529-0429-7 X.0702/442(7)
8916462

AGRICULTURE. Norway
Development

ALMAS, Reidar
Norsk jordbruk. - Oslo : Gyldendal, 1977
175p
82-05-09382-2 X.319/9351 Woolwich
1500422

AGRICULTURE. Norway
Farms. Livestock: Goats, to date

NORDANGER, Dag
Livet på Bretningen. - [Bergen] : Nordanger, 1982
185p,plates
82-7051-065-3 X.329/19018
9160493

AGRICULTURE. Norway
Organisations: Nordland landbruksselskap, 1876-1976

MOE, Knut
Bygdefolket i Nordland. - Bodø : [Nordland landbruksselskap], 1976
365p : ill
82-900304-6-0 X.311/5133 Woolwich
2170442

AGRICULTURE. Norway
Subsidies, 1945-1970

BONDEN, Karl
Jordbruksavtalene 1945-70. - Oslo : Landbruksforlaget, 1973
253p
r6112764 X.329/13267 Woolwich
2468379

AGRICULTURE. Nova. Hungary
1919-1945

VASKA, Miklós
Paraszti gazdálkodás Nován a két világháború között. - Zalaegerszeg : Zala megyei Levéltár, 1979
71p : ill. - (Zalai gyűjtemény ; sz.11)
r6499044 X.0700/770(11)
246229x

AGRICULTURE. Nova Scotia
Farms. Social conditions

UNGERER, Tomi
Far out isn't far enough. - London : Methuen, 1984, c1983
170p : col.ill
0-413-55630-1 X.950/31007
3013030

AGRICULTURE. Ohio
Farms. Social conditions, 1900-1915

MCMILLEN, Wheeler
Ohio farm. - Columbus : Ohio State University Press, [1974]
x, 220 p.
0-8142-0217-9 X.320/3069 Woolwich
1015834

AGRICULTURE. Oise. France
to date

FRANCE. Direction départementale de l'agriculture de l'Oise. Service technique de statistique
[Miscellaneous publications not catalogued separately]. - Beavais : [la Direction], 1981-
r6209865 S.E.400/317[1981]
8963614

AGRICULTURE. Öland. Sweden
Farms. Social conditions, to date

EKLÖF, Einar
Mitt liv i Sverige och Amerika och i det stora kriget. - Stockholm : LT, 1981
205p : ill
91-360-1722-1 X.520/29932
8930619

AGRICULTURE. Olsztyn. Poland

SUCHTA, Jerzy
Czynniki wyznaczające rozwój rolnictwa na obszarze województwa olsztyńskiego. - Olsztyn : Ośrodek Badań Naukowych im. Wojciecha Kętrzyńskiego
187 p., 1 fold. leaf of plates. - (Rozprawy i materiały Ośrodka Badań Naukowych im. Wojciecha Kętrzyńskiego w Olsztynie ; nr. 47). - Bibliography : p. 177-183
r6022972 X.311/4646 Woolwich
1006878

AGRICULTURE. Oppland. Norway
to date

CHRISTENSEN, Arne Lie
Et seterbrukslandskap i Nord-Gudbrandsdalen. - Øvre Ervik : Alvheim & Eide, 1982
85p : ill. - (Riksantikvarens rapporter ; 6)
82-903591-3-6 X.0425/212(6)
9032975

AGRICULTURE. Organisation for Economic Co-operation and Development countries
Part-time agriculture

ORGANISATION FOR ECONOMIC CO-OPERATION AND DEVELOPMENT
Part-time farming. - Paris : O.E.C.D. ; [London] : [H.M.S.O.]. - (Agricultural policy reports / Organisation for Economic Co-operation and Development)
Austria, France, Belgium, Canada, Finland, Ireland, Italy, Netherlands, New Zealand, Switzerland. - 1978
175p in various pagings
B79-31053 UN.P.107/51(36)
109811x

ORGANISATION FOR ECONOMIC CO-OPERATION AND DEVELOPMENT
Part-time farming in OECD countries. - Paris : O.E.C.D. ; [London] : [H.M.S.O.], 1978
[4],63p. - (Agricultural policy reports / Organisation for Economic Co-operation and Development)
B78-24979 SRL
109811x

AGRICULTURE. Organisation for Economic Co-operation and Development countries
Policies of governments

ORGANISATION FOR ECONOMIC CO-OPERATION AND DEVELOPMENT
Review of agricultural policies. - 1977. - Paris : O.E.C.D. ; [London] : [H.M.S.O.], 1978
126p. - (Agricultural policy reports / Organisation for Economic Co-operation and Development)
Subsequent volumes, when received, will be placed at this shelfmark
B78-35916 SRL
0618578

Review of agricultural policies in OECD member countries. - 1974-76. - Paris : Organisation for Economic Co-operation and Development ; [London] : [H.M.S.O.], 1977
115p
Subsequent volumes, when received, will be placed at this shelfmark
B79-00150 UNP.108/141
1742426

AGRICULTURE. Ostlandet. Norway
Effects on landscape, to 1974

ASHEIM, Vidar
Kulturlandskapets historie i Østlandets jord- og skogbruksbygder. - Ås : Landbruksbokhandelen, 1975
206 leaves : ill
82-557-0037-4 X.313/808 Woolwich
1421689

AGRICULTURE. Otaki Gorge. New Zealand

MARRIOTT, Les
Life in the Gorge. - Christchurch, N.Z. ; London : Whitcoulls, 1981
148p : ill
0-7233-0654-0 X.809/49113
255075x

AGRICULTURE. Oued R'hir. Algeria

PERENNES, Jean Jacques
Structures agraires et décolonisation. - Alger : Office des Publications universitaires, [1979]
372p : ill. - Bibliography: p.347-364
r6127906 X.325/699 Woolwich
2461498

AGRICULTURE. Outseats. Derbyshire
Farms, 1200-1800

MEREDITH, Rosamond
Farms and families of Hathersage Outseats. - Sheffield : [c/o Sheffield City Libraries].
Pt.1: Nether Hirst, North Lees, Brookfield, Green's House and Cowclose Farm. - 1981
28p : ill
0-9507815-0-9 X.0800/1659(1)
2606682

MEREDITH, Rosamond
Farms and families of Hathersage Outseats Derbyshire. - Sheffield : R. Meredith.
Pt.2: Gatehouse, Upper Hirst, Thorp, Birley, The Hill, The Old Road from Sheffield to Hope. - c1983
p30-56 : ill
0-9507815-1-7 X.0800/1659(2)
2606682

AGRICULTURE. Pakistan

EMPIRICAL studies on Pakistan agriculture / edited by Sayed Mushtaq Hussain, Mohammad Irshad Khan. - Karachi : Pakistan Institute of Development Economics, 1970
xix,1-204p,p.247-373,fold plate : ill. - (Readings in development economics ; no. 3). - Includes bibliographies
r6032252 X.320/10934 Woolwich
1245597

AGRICULTURE. Pamir region. Tadzhikistan
1800-1900

MUKHIDDINOV, Ikromiddin
Zemledelie pamirskikh tadzhikov Vakhana i Ishkashima v XIX - nacale XXv.. - Moskva : Nauka, Glavnaîa redakšiîa vostochnoĭ literatury, 1975
123p : ill
r6095221 X.311/5107 Woolwich
2120526

AGRICULTURE. Panama
1904-

GANDÁSEGUI, Marco A
Situación actual en el agro. - Panamá : Centro de Estudios Latinoamericanos "Justo Arosemena", 1980
24p : ill. - (Cuadernos populares ; 12)
r6700945 X.329/14296 Woolwich
8726523

AGRICULTURE. Panama
Law

PANAMA
[Laws, statutes, etc]. Código agrario. - [Edición especial]. - [Panamá] : [s.n.], [1976]
176p
r6099942 X.100/28694
2122782

AGRICULTURE. Panama
Role in economic development, 1970-1973

PANAMA'S economic development / William C. Merrill ... [et al.]. - Ames : Iowa State University Press, 1975
xii,219p : ill. - Bibliography : p.213-216. - Includes index
0-8138-1205-4 X.319/18837 Woolwich
1891952

AGRICULTURE. Papa Stour. Shetland. Scotland
Smallholdings. Self-sufficiency, 1972-1980

COLEMAN, Vicki
Living on an island. - Findhorn : Thule Press, 1981
127p
0-906191-55-6 X.809/49988
2594463

AGRICULTURE. Papua New Guinea
to date

A TIME to plant and a time to uproot / edited by Donald Denoon and Catherine Snowden for the Department of Primary Industry. - [Papua New Guinea] : Institute of Papua New Guinea Studies, 1981
vi,348p
r6815032 X.322/14019
911727x

AGRICULTURE. Paraiba (State). Brazil
Personnel. Alleged brutality by security forces

A LUTA de Alagamar. - [João Pessoa] : [s.n.], [1980]
68p : ill
r6159524 X.325/571 Woolwich
2422182

AGRICULTURE. Parmana region. Venezuela
Prehistoric agriculture

ROOSEVELT, Anna Curtenius
Parmana. - New York ; London : Academic Press, c1980
xv,320p : ill. - (Studies in archaeology). - Bibliography: p273-314. - Includes index
0-12-595350-x X.421/21856 Woolwich
248420x

AGRICULTURE. Pasco (Department). Peru
1972 - Statistics

[CENSO Nacional Agropecuario]. II Censo Nacional Agropecuario, 4 al 24 de setiembre 1972 / [preparado por el] Instituto Nacional de Estadística [de la] Presidencia de la República [del Perú] ... - Lima : I.N.E..
Departamento de Pasco. - Lima : I.N.E.
xxiii,379p
r6157515 LAS.G.381/9 Woolwich
2188694

AGRICULTURE. Pastures. Tropical regions
Research - For developing countries

TROPICAL pasture research / edited by N.H. Shaw and W.W. Bryan. - Farnham Royal : Commonwealth Agricultural Bureaux, 1976
xiv,454p : ill. - (Bulletins / Commonwealth Bureau of Pastures and Field Crops ; 51). - Bibliography: p.393-442. - Includes index
0-85198-358-8 UNP.266/65
0793183

AGRICULTURE. Peasant communities. Peninsular Malaysia
1874-1941

LIM TECK GHEE
Peasants and their agricultural economy in colonial Malaya, 1874-1941. - Kuala Lumpur ; London : Oxford University Press, 1977
xiv,291p. - (East Asian historical monographs). - Bibliography: p.264-278. - Includes index
0-19-580338-8 X.320/10986 Woolwich
1277251

AGRICULTURE. Peninsular Malaysia
to 1975

TAN, Yee Kew
The land and the agricultural organisation of Peninsula Malaysia. - Norwich : Published for the Centre for Development Studies by Geo Books, 1981
58p. - (Monograph / University College of Swansea. Centre for Development Studies ; no.13). - Bibliography: p56-58
0-86094-075-6 X.329/14221 Woolwich
2506580

AGRICULTURE. Pernambuco. Brazil
Economic aspects, 1950-1970

ANDRADE, Manuel Correia de
A problemática agro-pecuária de Pernambuco. - Recife : Universidade Federal de Pernambuco, Editora Universitária, 1975
74 p.
r6011775 X.319/8522 Woolwich
0885207

AGRICULTURE. Peru

ARRÚS RASCHIO, Manuel
Efecto de la política de uso de mano de obra en el resultado de la gestión empresarial de las actuales empresas agrarias : el caso Caqui. - Lima : Centro de Estudios y Promoción del Desarrollo, 1974
x,103p. - (Cuadernos Desco)
r6159007 X.311/5205 Woolwich
2261324

AGRICULTURE. Peru
to date

KÜCHEMANN, Johann-Christof
Soziale Kreativität und soziale Entwicklung in der Dritten Welt. - Frankfurt : Campus, 1980
497p(one folded). - (Campus. Forschung ; Bd. 180)
r6399111 X.329/14599 Woolwich
8344000

AGRICULTURE. Peru
1877

MARTINET, Jean Baptiste Henri
[L'agriculture au Pérou. Spanish]. La agricultura en el Perú. - Lima : Centro Peruano de Historia Económica, 1977
vi,165p.,[1]fold.leaf of plates
r6099853 X.319/20182 Woolwich
2123657

AGRICULTURE. Peru
1972

PERU. Oficina Nacional de Estadística y Censos
II censo nacional agropecuario, 4 al 24 de setiembre 1972. - Lima : la oficina, 1975-
v. 1-
r6024818 LAS.G. 345/15
0881880

AGRICULTURE. Peru
1972 - Statistics

PERU. Oficina Nacional de Estadistica y Censos
II censo nacional agropecuario 4 al 24 de setiembre. 1972. - Lima : a Oficina. 1972 -
v. 1 -
r6024817 LAS.G. 345/15
0881880

AGRICULTURE. Peru
Law

ROBLES RECAVARREN, Alejandro
La administracion de justicia en la nueva estructura agraria peruana - Lima : Ital Peru, 1977
205p
r6084904 X.108/26512
1852590

AGRICULTURE. Peru
Social aspects, to 1969

HACIENDA. Comunidad y campesinado en el Peru compilador : José Matos Mar. - 2. ed. - Lima : Instituto de Estudios Peruanos, 1976
390p. - (Serie Perú problema ; n.3). - Includes bibliographies
r6065151 X.619/10154 Woolwich
153503X

AGRICULTURE. Peru
Trade union

NEIRA SAMANEZ, Hugo
Huillca: habla un campesino peruano. - La Habana : Casa de las Américas 1974
217 p.
r6013270 X.708/18270
0874825

AGRICULTURE. Piedmont. Italy
1250-1450

ROTELLI, Claudio
Una campagna medievale. - Torino : Einaudi, 1973
xii,378p.,[1]fold leaf. - (Biblioteca di cultura storica ; 120)
r6100020 X.320/5311 Woolwich
2033739

AGRICULTURE. Poland

KOLBUSZ, Franciszek
Polityka rolna lat 1971-1980. - Warszawa : Krajowa Agencja Wydawnicza, 1978
146p. - (ABC gospodarki)
r6075478 X.319/9837 Woolwich
1419986

OLESIAK, Kazimierz
Obrót i gospodarka ziemią w Polsce Ludowej. - Warszawa : Książka i Wiedza, 1977
244p
r6044954 X.319/9227 Woolwich
1419986

PRZEMIANY strukturalne w rolnictwie polskim / praca zbiorowa pod redakcją prof. dr hab. Zdrisława Grochowskiego. - Warszawa : Państwowe Wydawnictwo Rolnicze i Leśne, 1975
241p
r6044268 X.319/8930 Woolwich
1231944

STUDZIŃSKI, Bogumił
Wieś Polska w latach 1944-1978. - Londyn : Polski, 1981
34p. - (Biblioteka Ruchu Chłopskiej Samoobrony Społecznej ; zesz. 2)
B81-40388 X.958/5894
2574500

TOMASZEWSKI, Zenon
O rolnictwie bez szablonów i schematów. - Warszawa : Ludowa Spółdzielnia Wydawnicza, 1979
186p
r6173394 X.319/20151 Woolwich
1419986

ŻABKO-POTOPOWICZ, Antoni
Pionierzy postępu w rolnictwie polskim. - Warszawa : Ludowa Spółdzielnia Wydawnicza, 1977
307p
r6016161 X.329/8773 Woolwich
1231944

ŻMIDZIŃSKI, Franciszek
Realizacja reformy rolnej na Pomorzu 1920-1938. - Warszawa : Państwowe Wydawnictwo Naukowe, 1978
184p. - (Bydgoskie Towarzystwo Naukowe) (Prace Wydziału Nauk Humanistycznych ; ser. C. nr. 21)
r6097219 AC9234ui/3c[ur.21]
1419986

AGRICULTURE. Poland
to date

WAWRZYNIAK, Bogdan
Służba rolna w Polsce. - Warszawa : Ludowa Spółdzielnia Wydawnicza, 1980
250p
83-205-3191-8 X.329/14842 Woolwich
8664390

AGRICULTURE. Poland
1506-1697

ÎAKUBSKIĬ, V. A. (Vladimir Aleksandrovich)
Problemy agrarnoi istorii pozdne-srednevekovoi Pol'shi .... - Leningrad : Izdatel'stvo Leningradskogo universiteta, 1975
116p
r6178865 X.329/13258 Woolwich
2470829

AGRICULTURE. Poland
1945-

MANTEUFFEL, Ryszard
Rolnictwo polskie w reformie. - Warszawa : Ludowa Spółdzielnia Wydawnicza, 1983
222p
83-205-3547-6 X.329/19014
9162321

AGRICULTURE. Poland
Cooperation

OZDOWSKI, Jerzy
Prosta kooperacja w gospodarce żywnościowej PRL. - Warszawa : Państwowe Wydawnictwo Rolnicze i Leśne, 1978
109p
r6077048 X319/9763 Woolwich
179146X

AGRICULTURE. Poland
Effects of technological development

MOSKAL, Stanisław
Ludność dwuzawodowa w procesie społeczno-gospodarczych przeobrażeń wsi. - Warszawa : Państwowe Wydawnictwo Naukowe, 1978
189p. - (Problemy rejonów uprzemysławianych). - Bibliography : p.182-[186]
r6088659 X.311/4986 Woolwich
1975358

AGRICULTURE. Poland
Farms, 1795-1863

KOCHANOWICZ, Jacek
Pańszczyźniane gospodarstwo chłopskie w Królestwie Polskim w I połowie XIXw.. - Warszawa : Wydawnictwa Uniwersytetu Warszawskiego, 1981
201p. - Includes bibliography
83-00-01085-8 X.622/17426
9026924

AGRICULTURE. Poland
Farms. Management

JASSEM, Grażyna
Majątek smogulecki w latach 1918-1937. - Poznań : Poznańskie Towarzystwo Przyjaciół Nauk, 1976
185p,5 fold leaves of plates : ill. - (Badania z dziejów społecznych i gospodarczych ; nr.51)
r6045758 X.0200/68(51)
1503723

AGRICULTURE. Poland
Labourers

GRONIOWSKI, Krzysztof
Robotnicy rolni w Królestwie Polskim, 1871-1914. - Warszawa : Ludowa Spółdzielnia Wydawnicza, 1977
270p
r6057530 X.319/9226 Woolwich
1418882

AGRICULTURE. Poland
Organisations

BORKOWSKI, Jan
Kółka rolnicze w II Rzeczypospolitej. - Warszaw : Ludowa Spółdzielnia Wydownicza : 1978
611p
r6077006 X.320/5250 Woolwich
1473794

KOSTKIEWICZ, Wojciech
Kółka rolnicze. - Warszawa : Państwowe Wydawnictwo Rolnicze i Leśne, 1977
180p
r6016081 X.319/9325 Woolwich
1473794

AGRICULTURE. Poland
Policies of government, 1645-

BAJAN, Konrad
Polityka rolna. - Wyd. 2-gie zmien. - Warszawa : Państwowe Wydawnictwo Ekonomiczne, 1978
290p. - Bibliography: p.288-290
r6444016 X.329/14200 Woolwich
8715165

AGRICULTURE. Poland
Policies of government, 1939-1945

GORZELAK, Eugeniusz
Polityka agrarna PRL. - Warszawa : Państwowe Wydawnictwo Naukowe, 1980
400p
83-01-01476-8 X.329/14347 Woolwich
8795169

AGRICULTURE. Poland
Policies of government, 1945-

KARAŚ, Romuald
Podeptany chleb. - Warszawa : Czytelnik, 1981
287p
83-07-00509-4 X.329/17107
9016341

KUŚMIEREK, Józef
Credo. - [West Berlin] : Pogląd, [1984]
90p
r6483835 X.329/19802
9016341

AGRICULTURE. Poland
Productivity. Forecasting, 1945-

MANTEUFFEL, Ryszard
Przyszłość rolnictwa. - Warszawa : Państwowe Wydawnictwo Naukowe, 1979
139p
83-01-00516-5 X.322/12265
898994x

AGRICULTURE. Poland
Sikorski, Stanisław - Biographies

BORZYSZKOWSKI, Józef
Z dziejów pracy organicznej na Pomorzu. - Gdańsk : Zakład Narodowy im. Ossolińskich-Wydawnictwo, 1979
189, [3] p., [4] leaves of plates : ill.. - (Seria monografii / Gdańskie Towarzystwo Naukowe, Wydział I Nauk Społecznych i Humanistycznych). - Includes index. - Bibliography: p. 183
83-04-00281-7 X.329/17941
9082859

AGRICULTURE. Poland
Smallholdings, 1945-

ROLNICTWO indywidualne w latach siedemdziesiątych / praca zbiorowa pod redakcją Anny Szemberg. - Warszawa : Książka i Wiedza, 1980
218p
r6483515 X.329/14928 Woolwich
879958x

AGRICULTURE. Poland
Statistics

BERGER, Jan
Inwentarz archiwalny spisów rolniczych 1946-1965. - Warszawa : Główny Urząd Statystyczny, 1974
63p. - (Statystyka w dokumencie archiwalnym ; nr.13)
r6016098 X.319/9332 Woolwich
170933X

AGRICULTURE. Portugal
1800-1970

MATERIAIS para a história da questão agrária em Portugal - séc. XIX e XX / selecção, prefácio e notas de Manuel Villarverde [Villaverde] Cabral. - Porto : Editorial Inova, 1974
571 p. : ill. - (Colecção Civilização portuguesa ; v. 19)
r6008993 X.708/17592
0919071

AGRICULTURE. Portugal
1930-1960

CUNHAL, Álvaro
[A questão agrária em Portugal]. Contribuição para o estudo da questão agrária. - Lisboa : Avante!, 1976
2v. - (Colecção Reforma agrária ; n. 1-2)
r6047295 X.319/8767 Woolwich
1218212

AGRICULTURE. Portugal
Mathematical models

EGBERT, Alvin Charles
A development model for the agricultural sector of Portugal. - Baltimore : London : Johns Hopkins University Press for the World Bank, [1976]
xii,97p : ill. - (Staff occasional papers / World Bank ; no.20). - Bibliography: p.95-97
0-8018-1793-5 U.N.Q.121/3
0667722

AGRICULTURE. Potosi (Department). Bolivia
Policies of government, 1973-1974

La AGRICULTURA en Potosi, año agricola 1973-1974 / [escriben : Wilson Mendieta Pacheco et al.]. - Potosi : Universidad Boliviana Mayor "Tomás Frias", 1973
viii,55p. - (Cultura potosina. Serie C. Desarrollo y economia)
r6066408 X.708/44246
1713825

AGRICULTURE. Prairie provinces. Canada
to date

FAULKENOR, C. V. (Cliff V.)
Pen and plow. - reprinted. - Winnipeg : Public Press, 1977
vii,152p,[16]p of plates : ill
r6053114 L.45/1459
8805695

GISCARD, Gaston
Dans la prairie canadienne. - Regina : Canadian Plains Research Center, 1982
xvii,82,82p,[8]p of plates : ill. - (Canadian plains studies ; no.11)
0-88977-024-7 X.529/52305
8805695

GRAY, James H., 1906-
The winter years. - reprinted. - Toronto : MacMillan, 1972
xi,220p.,[24]p.of plates
0-7705-0122-2 X.950/15422
8805695

AGRICULTURE. Prato region. Italy
Political aspects, 1918-1922

CAPONI, Claudio
Leghe bianche e lotte agrarie nel Pratese, 1918-1922. - [Prato?] : Edizioni del Palazzo, 1974
272p
r6042477 X.510/10098
1239783

AGRICULTURE. Preseli (District) & South Pembrokeshire (District). Dyfed
Organisations: Pembrokeshire Agricultural Society, to 1977

REES, Derek
Rings and rosettes. - [Llandysul] : [Gomer Press], 1977
xv,234p,leaf of plate,[16]p of plates : ill. - Bibliography: p.224. - Includes index
0-85088-466-7 X.329/11507 Woolwich
1515713

AGRICULTURE. Prussia
to 1945

JAMIN, Rainer
Aufbau, Tätigkeit und Verfahren der Auseinandersetzungsbehörden bei der Durchführung der preussischen Agrarreformen. - Frankfurt am Main : Lang, 1985
205p. - (Rechtshistorische Reihe ; Bd.39). - Includes bibliography
3-8204-8181-8 X.0102/159(39)
9124853

AGRICULTURE. Pskov region. Russia
1500-1600

AGRARNAĨA istoriĩa severo-zapada Rossii XVI veka / rukovoditel' avtorskogo kollektiva A.L. Shapiro. - Leningrad : Nauka, Leningradskoe otdelenie, 1978
220p
r6173268 X.322/2666 Woolwich
2122022

AGRICULTURE. Puerto Rico
to date - Statistics - Serials

ANUARIO de estadisticas agricolas de Puerto Rico / [compilado por el] ... Departamento de Agricultura, Oficina de Estadisticas Agricolas. - 1971/72-. - Santurce, Puerto Rico : El Departamento, 1972-
r6708131 AS.P.137/12
9073760

AGRICULTURE. Punjab. India (Republic)
to date

CHADHA, G. K.
Production gains of new agricultural technology. - Chandigarh : Punjab University, 1979
144p.,2folded leaves. - Bibliography: p.131-144
r6802002 X.329/13645 Woolwich
8776083

AGRICULTURE. Punjab. India (Republic)
Development. Effects on economic conditions, 1950-1980

CHAUDHRI, D. P.
Agriculture and the development process. - London : Croom Helm, c1985
216p : ill. - Bibliography: p193-207. - Includes index
0-7099-3408-4 X.329/19722
3201279

AGRICULTURE. Punjab. India (Republic)
Economic development. Mathematical models

DAY, Richard Hollis
Economic development as an adaptive process. - Cambridge : Cambridge University Press, 1977
x,326p : ill. - Bibliography: p.310-321. - Includes index
0-521-21114-x X.520/12457
1445111

AGRICULTURE. Puno (Department). Peru
1972 - Statistics

[CENSO Nacional Agropecuario]. II Censo Nacional Agropecuario, 4 al 24 de setiembre 1972 / [preparado por el] Instituto Nacional de Estadística [de la] Presidencia de la República [del Perú] ... - Lima : I.N.E..
Departamento de Puno. - 1976
2t
r6157516 LAS.G.381/10 Woolwich
2188708

AGRICULTURE. Quebec (Province)
to date

MIGNER, Robert
Quand gronde la révolte verte. - Montréal : La Presse, 1980
263p. - (Collection 'Jadis et naguère')
2-89043-055-3 X.329/15704
8868794

AGRICULTURE. Quebec (Province)
1841-1911

JEAN, Bruno
Les idéologies éducatives agricoles (1860-1890) et l'origine de l'agronomie québécoise. - Québec : Université Laval, 1977
ix,237p. - (Cahiers de l'ISSH ; no 7) (Études sur le Québec). - Bibliography: p233-237
r6206599 X.322/10525 Woolwich
8790094

AGRICULTURE. Rogaland. Norway
to date

CHRISTENSEN, Arne Lie
Spadens, plogens og traktorens landskap. - Øvre Ervik : Alvheim & Eide, 1982
96p : ill. - (Riksantikvarens rapporter ; 1)
82-903590-4-7 X.0425/212(1)
9031030

AGRICULTURE. Romania

COLUMBEANU, Sergiu
Grande exploitations domaniales en Valachie au XVIIIe siècle. - Bucarest : Editura Academiei Republicii Socialiste România, 1974
199p. - (Bibliotheca historica Romaniae ; 47(5))
r6015818 Ac.743.ce/4[47]
1216856

CONDUCEREA proceselor economice din agricultură / Constantin Anderca [et.al.]. - Bucureşti : Editura Academiei Republicii Socialiste România, 1972
417p,[5] fold. leaves of plates. - (Ex terra aurum Seria A. Studii şi cercetări de economie agrară ; 2)
r6036228 X.319/8749 Woolwich
1216856

MARIAN, Teodor
Roumanie : agriculture. - Bucarest : Rédaction des revues agricoles, 1971
148p : ill(chiefly col.)
r6174468 X.322/2683 Woolwich
1216856

AGRICULTURE. Romania
to 1944

CONSTANTINESCU, Olga
Critica teoriei "România - ţară eminamente agricolă". - Bucureşti : Editura Academiei Republicii Socialiste România, 1973
323p,[1]p of plates. - (Bibliotheca oeconomica ; 26) - Bibliography
r6457188 X.0519/204
8841853

CORFUS, Ilie
Agricultura in Ţările Române, 1848-1864. - Bucureşti : Editura ştiinţifică şi enciclopedică, 1982
434p
r6457704 X.329/16320
8841853

PĂTRĂŞCANU, Lucreţiu
Studii economice şi social-politice. - Bucureşti : Editura politică, 1978
231 p.. - Bibliography: p. 127-[129]
r6460519 X.329/18435
8841853

AGRICULTURE. Romania
1878-1940

VALOTA CAVALLOTTI, Bianca
Questione agraria e vita politica in Romania (1907-1922) - Milano : Cisalpino-La Goliardica, 1979
xxii,273p
88-205-0214-3 X.329/17017
9001387

AGRICULTURE. Romania
1918-1939

BOZGA, Vasile
Criza agrară in România dintre cele două războaie mondiale. - Bucureşti : Editura Academiei Republicii Socialiste România, 1975
257 p., 2 p. of plates. - Bibliography : p. 225-231
r6008152 X.329/8475 Woolwich
0853623

AGRICULTURE. Romania
1944-

FULEA, Maria
Modernizare şi structură socială în comunitatea rurală. - Bucureşti : Editura Academiei Republicii Socialiste România, 1982
287 p. : ill.. - Includes bibliographical references
r6460515 X.329/18516
9119736

AGRICULTURE. Romania
Law

CRISTEA, Gheorghe
Contribuţii la istoria problemei agrare în România. - Bucureşti : Editura Academili Republicii Socialiste România, 1977
196p. - (Biblioteca istorică ; 47)
r6056169 Ac.743/84[nr47]
1468790

AGRICULTURE. Ross-on-Wye region. Hereford and Worcester
Farms. Social conditions, 1796-1797

HUGHES, Anne
Anne Hughes, her boke. - London : Folio Society, 1981
176p : ill
B81-31968 X.529/42993
2160404

The diary of a farmer's wife, 1796-1797. - London : Allen Lane, 1980
3-162p : ill
0-7139-1327-4 X.809/45494
2160404

AGRICULTURE. Ross-on-Wye region. Hereford and Worcester
Farms. Social conditions, 1796-1797 - Correspondence, diaries

HUGHES, Anne
The diary of a farmer's wife 1796-1797. - Harmondsworth Penguin, 1981, c1980
162p : ill
0-14-005457-x X.809/50183
2160404

AGRICULTURE. Rural regions. Africa

GROVE, Alfred Thomas
Rural Africa. - Cambridge : Cambridge University Press, 1979
[6],122p : ill. - (Cambridge topics in geography series) - Bibliography: p.120. - Includes index
0-521-21825-x X.800/27854
190230x

AGRICULTURE. Rural regions. Developing countries
Development. Role of urban food markets

EPSTEIN, T. Scarlett (Trude Scarlett)
Urban food marketing and third world rural development. -
London : Published in association with the Research
Center for South West Pacific Studies, La Trobe
University [by] Croom Helm, c1982
260p : ill. - Bibliography p255-260
0-7099-0911-x X.529/48148
2747707

AGRICULTURE. Rural regions. Developing countries
Policies of governments. Effects on poverty, 1950-1970 -
Conference proceedings

RURAL poverty and agrarian reform : edited by Steve Jones,
P.C. Joshi, Miguel Murmis. - New Delhi : Published on
behalf of ENDA by Allied Publishers Private, 1982
xxi,384p. - Includes bibliographies and index
0-907108-45-8 X.520/31837
2988895

AGRICULTURE. Rural regions. England
Labourers. Riots, 1830-1831. Spatial aspects

CHARLESWORTH, Andrew
Social protest in a rural society. - Norwich : Geo
Abstracts [for] the Historial Geography Research Group,
[1979]
[7],78p : ill. - (Historial geography research series ;
no. 1)
0-86094-035-7 X.709/51652
214400x

AGRICULTURE. Rural regions. England
Labourers. Social conditions. Effects of industrialisation,
1760-1830

HAMMOND, John Lawrence
The village labourer. - [4th ed. reprinted]. - London :
Longman, 1978
xliii,301p. - Bibliography: p.273-286. - Includes index
0-582-48518-5 2021.e Reading Room
1495011
- [Another copy] X.319/18202 Woolwich

AGRICULTURE. Rural regions. Peru
Effects of workers' organisations

FIORAVANTI, Eduardo
Latifundio y sindicalismo agrario en el Perú. - Lima :
Instituto de Estudios Peruanos, 1974
242 p., 6 p. of plates. - Includes bibliographies
r6005187 X.519/19984
0909602
Latifundio y sindicalismo agrario en el Perú. - 2a ed. -
Lima : Instituto de Estudios Peruanos, 1976
217p. - (Estudios de la sociedad rural ; no. 3). -
Includes bibliographies
r6040935 X.319/8755 Woolwich
0909602

AGRICULTURE. Rural regions. Punjab. India (Republic)

GAMBLE, Paul
The Green Revolution in five Punjabi villages. - [London]
: Commonwealth Institute, 1975
[1],12p : ill. - (Commonwealth topic ; no.13)
0-900906-25-1 X.313/1294 Woolwich
0692336

AGRICULTURE. Russia

DUBROVSKIĬ, Sergeĭ Mitrofanovich
Sel'skoe khozi͡aĭstvo i krest'i͡anstvo Rossii v period
imperializma. - Moskva : Nauka, 1975
397p,[1]leaf of plates
r6175876 X.329/9036 Woolwich
1851837

PROBLEMY agrarnoĭ istorii ... / [redakt͡sionnai͡a
kollegii͡a : V.L. I͡Anin (otv. redaktor) ... i dr.]. -
Minsk : Nauka i tekhnika", 1978
2 ch.
r6088179 X.320/5284 Woolwich
1851837

AGRICULTURE. Russia
to 1917

SOVETSKAI͡A istoriografii͡a agrarnoĭ istorii SSSR (do
1917g.) / [V.L. I͡Anin (otv. redaktor) ... i dr.]. -
Kishinev : Shtiint͡sa, 1978
261p
r6187009 X.629/8153 Woolwich
1972138

AGRICULTURE. Russia
to 1917. Historiography, to date

ISTORIOGRAFII͡A agrarnoĭ istorii dorevoli͡ut͡sionnoĭ
Rossii ... / [redakt͡sionnai͡a kollegii͡a : G.P.
Zhidkov- otv. redaktor ... i dr.]. - Kaliningrad :
Izdanie KGU, 1982
147p
r6463620 X.329/16319
8950091

AGRICULTURE. Russia
ca 1000-ca 1700

SMITH, R. E. F. (Robert Ernest Frederick)
Peasant farming in Muscovy. - Cambridge : Cambridge
University Press, 1977
xii,289p,[12]p of plates : ill. - Bibliography:
p.266-279. - Includes index
0-521-20912-9 X.322/8304 Woolwich
1104497

AGRICULTURE. Russia
1500-1600

IOSIFO-VOLOKOLAMSKIĬ MONASTYR'
Votchinnye khozi͡aĭstvennye knigi XVI v.. - Moskva ;
Leningrad : Akademii͡a nauk SSSR, Institut istorii SSSR,
Leningradskoe otdelenie, 1976
3v
r6183237 X.322/9307 Woolwich
2446197

AGRICULTURE. Russia
1725-1796

MATERIALY po istorii sel'skogo khozi͡aĭstva Rossii
vtoroi poloviny XVIII veka / [sostavite' : K.V. Sivkov].
- Moskva : Akademii͡a nauk SSSR, Institut istorii SSSR,
1982
218p
r6448091 X.429/15740
9063609

AGRICULTURE. Russia
Policies of government, 1800-1917

SIDEL'NIKOV, Stepan Mikhaĭlovich
Agrarnai͡a politika samoderzhavii͡a v period
imperializma. - Moskva : Izdatel'stvo Moskovskogo
universiteta, 1980
286p
r6177701 X.809/28717
2439476

AGRICULTURE. Russia
Policies of Partii͡a sot͡sialistov-revoli͡ut͡sionerov, to
1907

PERRIE, Maureen
The agrarian policy of the Russian
Socialist-Revolutionary Party, from its origins through
the revolution of 1905-1907. - Cambridge : Cambridge
University Press, 1976
xii,216p. - (Soviet and East European studies). -
Bibliography: p.206-209. - Includes index
0-521-21213-8 X.329/10818 Woolwich
1057472

AGRICULTURE. Russia
Theories of Lenin, V. I.

TRAPEZNIKOV, Sergeĭ Pavlovich
Leninizm i agrarno-krest'i͡anskiĭ vopros.... - 2-e.,
dop. izd. - Moskva : Izdatel'stvo "Mysl'", 1976
2t. : ill
r6046547 X.329/8661 Woolwich
1231405

AGRICULTURE. Russia (RSFSR)

GORIZONTY Nechernozem'i͡a / [sostavitel' : L. I͡A.
Florent'ev]. - Moskva : Izdatel'stvo politicheskoĭ
literatury, 1977
253p,[14]p of plates : ill
r6093270 X.329/8879 Woolwich
1856464

MONOV, Aleksandr Ivanovich
Sel'skoe khozi͡aĭstvo Nechernozemnoĭ zony RSFSR. -
Moskva : Kolos, 1978
271p.
r6094828 X.329/8883 Woolwich
1866931

PROIZVODITEL'NYE sily nechernozemnoĭ zony RSFSR /
[otvetstvennye redaktory : A.N. Gladyshev, V.P. Mozhin].
- Moskva : Mysl', 1977
254p
r6087568 X.329/13432 Woolwich
1866931

AGRICULTURE. Russia (RSFSR)
- Conference proceedings

VOPROSY agrarnoĭ istorii T͡Sentra i Severo-Zapada RSFSR /
[redkollegii͡a A.A. Kondrashenkov (otvetstvennyĭ
redaktor) ... i dr.]. - Smolensk, 1972
465p
r6400580 X.329/13686 Woolwich
1856464

AGRICULTURE. Russia (RSFSR)
1945-. Economic aspects

ĖKONOMICHESKIE i sot͡sial'nye problemy razvitii͡a
sel'skogo khozi͡aĭstva na sovremennom ėtape /
[otvetstvennyĭ redaktor : N.A. Sti͡azhkov]. - Ri͡azan'
: Ri͡azanskiĭ gosudarstvennyĭ pedagogicheskiĭ
institut, 1973
165p
r6490573 X.529/69718
9110542

AGRICULTURE. Russia (RSFSR)
Policies of Vsesoi͡uznai͡a Kommunisticheskai͡a partii͡a
(bol'shevikov), 1860-1930

HUSSAIN, Athar
Marxism and the agrarian question. - London : Macmillan.
Vol.2: Russian Marxism and the peasantry, 1861-1930. -
Dec. 1980
[192]p
0-333-28675-8 X.529/44548
2091798

AGRICULTURE. Ryazan (oblast). Russia (RSFSR)
Collectivisation

POBEDA kolkhoznogo stroi͡a na ri͡azanskoĭ zemle /
[redakt͡sionnai͡a kollegii͡a : I.S. Stepichev
(otvetstvennyĭ redaktor), P.P. Kiri͡anov, I͡U.F.
Fulin]. - Ri͡azan : Ri͡azanskiĭ gosudarstvennyĭ
pedagogicheskiĭ institut, 1976
218p
r6098038 X.319/9881 Woolwich
2024624

AGRICULTURE. Sabana de Bogotá. Colombia
1590-1820

VILLAMARÍN, Juan A.
Los factores que afectaron la producción agropecuaria en
la Sabana de Bogotá en la época colonial. - [Tunja?] :
Pato Marino, 1975
27p. - (Colección Lecturas de historia ; n.6). -
Bibliography : p.23-27
r6037798 X.311/4823 Woolwich
1323989

AGRICULTURE. Sahel. West Africa
to date

PERETTI, M.
Projet mise en place de l'ORD du Sahel. - [s.l.] :
Ministère du developpement rural, 1976
2v((maps in a folder))
r6210519 SY81/29
9008403

AGRICULTURE. Sakhalin. Russia (RSFSR)

KOZYNI͡UK, Viktor Mikhaĭlovich
Sel'skoe khozi͡aĭstvo Sakhalinskoĭ oblasti. -
I͡Uzhno-Sakhalinsk : Sakalinskoe otdelenie
Dal'nevostochnogo knizhnogo izdatel'stva, 1975
190p. - Bibliography : p.188-[191]
r6036474 X.619/9618 Woolwich
1209566

AGRICULTURE. San Jose. Palawan. Philippines
Migrant personnel. Social inequality

EDER, James F.
Who shall succeed?. - Cambridge : Cambridge University
Press, 1982
xv,264p : ill. - Bibliography: p253-258. - Includes index
0-521-24218-5 X.322/11090 Woolwich
2760800

AGRICULTURE. São Paulo (State). Brazil
to date

MÜLLER, Geraldo
Estado, estrutura agrária e população. - Petrópolis :
Vozes : CEBRAP, 1980
141p. - (Cadernos CEBRAP ; n.32)
r6703373 X.329/14970 Woolwich
8799431

AGRICULTURE. São Paulo (State). Brazil
Immigrant labourers, to date

HOLLOWAY, Thomas H.
Immigrants on the land. - Chapel Hill : University of
North Carolina Press, c1980
xviii,218p : ill. - Bibliography
0-8078-1430-x X.520/32937
9066551

AGRICULTURE. São Paulo (State). Brazil
Labourers. Social conditions

MELLO, Maria Conceição D'Incao e
O bóia-fria : acumulação e miséria. - 2.ed. -
Petrópolis : Vozes, 1975
154p
r6035531 X.519/22620
1130323

AGRICULTURE. Saskatchewan

DAVIES, Evan, 1882?-1958
[Beyond the Old Bone Trail. Welsh]. Arswyd yr unigeddau. - Llandysul : Gwasg Gomer, 1980
167p
0-85088-992-8 X.909/45287
2335077

AGRICULTURE. Saudi Arabia

DAVIES, Piers
Prospects for agricultural and project business in the kingdom of Saudi Arabia. - London : BAEC, 1981
iv,51p
B82-11355 SRL
262575x

AGRICULTURE. Saudi Arabia
to date

PHILIPP, Hans-Jürgen
Die landwirtschaftliche Modernisierung Saudi-Arabiens im 20. Jahrhundert. - Hamburg : Deutsches Orient-Institut, c1984
132p. - (Mitteilungen des Deutschen Orient-Instituts ; Nr.25). - Includes bibliography
r6310290 X.0512/459(25)
9147071

AGRICULTURE. Savanna regions. Sudan

GIBBON, David, 19—
Subsistence farming in the dry savanna of Western Sudan. - Norwich : School of Development Studies, University of East Anglia, 1977
21p. - (Discussion paper / [School of Development Studies, University of East Anglia] ; no.18). - Bibliography: p21
r6819984
3181901

AGRICULTURE. Savanna regions. West Africa

KOWAL, J. M.
Agricultural ecology of Savanna. - Oxford : Clarendon Press, 1978
xi,403p : ill. - Bibliography: p.349-379. - Includes index
0-19-859462-3 SRL
1188887

AGRICULTURE. Savo region. Finland
to date

TIRKKONEN, Kalevi
Etelä-Savon maaperä- ja tilussuhteet maataloustuotannon kannalta. - Mikkeli : [Itä-Suomen instituutti], 1978
[10,]134leaves. - (Itä-Suomen instituutin julkaisusarja. B ; 26)
951-90071-9-9 X.0702/472(26)
8945187

AGRICULTURE. Scandinavia
Prehistoric agriculture - Conference proceedings

INTRODUKSJONEN av jordbruk i Norden / [arrangert av Det Norske videnskaps-akademi] ; redaktør : Thorleif Sjøvold. - Oslo : Universitetsforlaget, 1982
282p
82-00-09534-7 X.425/3764
8891095

AGRICULTURE. Scotland

THOMSON, J. K. (John K)
The structure of Scottish agriculture and food industries. - Edinburgh : Scottish Council Research Institute, 1978
5,[6]leaves
B81-34342 X.622/11061 Woolwich
2547678

AGRICULTURE. Scotland
- For regional planning

GREAT BRITAIN. Department of Agriculture and Fisheries for Scotland
Agriculture in Scotland. - [Edinburgh] : [Scottish Development Department], 1975
[5],12p. - (Planning advice notes / Great Britain. Scottish Development Department ; no.1) (Regional reports advice)
B78-19947 BS.176/257
1479695

AGRICULTURE. Scotland
- Statistics - Serials

ECONOMIC report on Scottish agriculture / Department of Agriculture and Fisheries for Scotland. - 1980-. - Edinburgh : H.M.S.O., 1981-
v.
B82-12463 BS.151/98
2629380

AGRICULTURE. Scotland
1600-1700

WHYTE, Ian
Agriculture and society in seventeenth-century Scotland. - Edinburgh : Donald, 1979
ix,301p : ill. - Bibliography: p.272-285. - Includes index
0-85976-033-2 X.320/11976 Woolwich
2001594

AGRICULTURE. Scotland
ca 1600-ca 1900

FENTON, Alexander
Scottish country life. - Edinburgh : Donald, 1976
ix,255p : ill. - Bibliography: p.238-244. - Includes index
0-85976-011-1 X.809/40958
0653780

AGRICULTURE. Scotland
1700-1900

SIMPSON, Eric
Farming and the countryside, 1700-1900. - Edinburgh : Holmes McDougall, 1976
48p : ill. - (Scottish search series). - Bibliography: p.47-48
0-7157-1476-7 X.311/8585 Woolwich
1252275

AGRICULTURE. Scotland
Development. Organisations: Edinburgh Centre of Rural Economy - Serials

EDINBURGH CENTRE OF RURAL ECONOMY
Annual report. - 1st- : [1948?]-. - Edinburgh : the Centre, [1948?]-
Holdings : 32nd- ; 1978-79-
r6108476 P.311/782
2298287

AGRICULTURE. Scotland
Farms. Buildings. Architecture, 1750-1970

FENTON, Alexander
The rural architecture of Scotland. - Edinburgh : Donald, 1981
xi,242p : ill. - Bibliography: p225-234. - Includes index
0-85976-020-0 X.425/2018 Woolwich
2404338

AGRICULTURE. Scotland
Farms. Buildings. Architecture, ca 1770- ca 1870

MORTON, Robert Scott
Traditional farm architecture in Scotland. - Edinburgh : Ramsay Head Press, 1976
96p : ill. - Includes index
0-902859-32-3 X.421/10168
0643130

AGRICULTURE. Scotland
Farms. Buildings. Construction. Standard costs

FARM building cost guide. - 1976 / H.J. Wight. - Aberdeen : Scottish Farm Buildings Investigation Unit, 1976
viii,75p
0-902433-10-5 P.311/707
0291331

AGRICULTURE. Scotland
Farms. Buildings. Design - Serials

FARM buildings topics. - No.1-. - Edinburgh : East of Scotland College of Agriculture, Farm Buildings Advisory and Development Department, [197-?]- : ill
B77-30459 SRL
1255851

AGRICULTURE. Scotland
Farms. Livestock

HART, Edward
Scottish farm animals. - Gartocharn, Alexandria : Famedram, [1983?]
[58]p : ill
B83-42303 X.329/16956
2969122

AGRICULTURE. Scotland
Farms. Open days. Attitudes of visitors

LEE, T. R.
The educational effectiveness of the farm open day. - Perth : The Commission, c1980
53p : ill. - Bibliography: p48
0-902226-49-5 X.525/6059
2765160

AGRICULTURE. Scotland
Farms. Open days. Projects: Family Days on the Farm (Project)

FAMILY Days on the Farm / Countryside Commission for Scotland. - Perth : The Commission, c1982
42p : ill
0-902226-63-0 BS.182/47
2842572

AGRICULTURE. Scotland
Farms. Social conditions, ca 1850-1920

CAMERON, David Kerr
The ballad and the plough. - London : Gollancz, 1978
253p,[16]p of plates : ill
0-575-02468-2 X.809/43356
1525085

AGRICULTURE. Scotland
Farms. Social conditions, ca 1850-ca 1920

CAMERON, David Kerr
The cornkister days. - London : Gollancz, 1984
263p,16p of plates : ill
0-575-03492-0 X.809/61145
3053393

AGRICULTURE. Scotland
Geographical aspects

COPPOCK, J. T. (John Terence)
An agricultural atlas of Scotland. - Edinburgh : Donald, 1976
xv,242p : ill. - Bibliography: p.240-242
0-85976-016-2 SRL
0722820

AGRICULTURE. Scotland
Organisations: National Farmers' Union of Scotland

BASSETT, Philippa
A list of the historical records of the National Farmers' Union of Scotland. - [Birmingham] : Centre for Urban and Regional Studies, University of Birmingham, 1980
v,3leaves. - (Lists of historical records)
B81-31792 X.325/1339 Woolwich
2524791

AGRICULTURE. Scotland
Policies of British government

SCOTTISH CONSERVATIVE AND UNIONIST ASSOCIATION
'Food for thought'. - [Edinburgh] : [S.C.U.A.], [1978?]
7p
B79-11576 X.313/1626 Woolwich
1833626

AGRICULTURE. Scotland
Research - Serials

SCHOOL OF AGRICULTURE, ABERDEEN
Research investigations and field trials. - 1978-79-. - Aberdeen : The School, [1980]-
v. : ill
B83-03104 SRL
0128686

AGRICULTURE. Scotland
Role in conservation of environment

LANDSCAPE and wildlife conservation on farms / Cobham Resource Consultants. - [Redgorton] : [The Commission], 1984
[124]p : ill
B85-04751 X.322/14384
3216667

AGRICULTURE. Senegal
to date

SENEGAL. Institut de recherches agronomiques tropicales et des cultures vivrières
Esquisse d'une unité de production en culture irriguée pour le centre-nord Sénégal. - [S.l.] : Centre national de la recherche agronomique, 1973
19leaves
r6209938 X.322/12238
899546x

AGRICULTURE. Serbia
to 1456

BLAGOJEVIĆ, Miloš
Zemljoradnja u srednjovekovnoj Srbiji. - Beograd : Istorijski institut, 1973
463p,22p of plates : ill. - (Posebna izdanja ; knj.15). - "Izvori i literatura" : p433-444
r6057003 P.701/541(15)
1475568

AGRICULTURE. Shropshire
Wyke, Edward - Biographies

ALLWOOD, Arthur
Wyke of Ruckley 1826-1916. - Shrewsbury : A. Allwood, [1983]
5p
B83-39182 X.809/56275
2956608

AGRICULTURE. Siberia. Russia (RSFSR)

ANTOSHIN, ÎÛriĭ Grigor'evich
Sovkhozy Zapadnoĭ Sibiri v gody pervoĭ pîatiletki .... - Novosibirsk : Nauka, Sibirskoe otdelenie, 1971
221p
Imperfect : p.104,101,100,97 are duplicated in place of p.[218],[219],[223],[224]
r6178103 X.329/12323 Woolwich
0899143

GUSHCHIN, N. ÎA. (Nikolaĭ ÎÂkovlevich)
Krest'îanstvo Zapadnoĭ Sibiri v dovoennye gody 1935-1941. - Novosibirsk : Nauka Sibirskoe otdelenie, 1975
284p
r6093533 X.329/8901 Woolwich
0899143

IVANOV, Boris Vasil'evich
Osushchestvlenie leninskogo kooperativnogo plana v Sibiri (1920-1927 gg.). - Tomsk : Izdatel'stvo Tomskogo universteta, 1977
339p
r6172575 X.329/8923 Woolwich
0899143

IVANOV, Leonid Ivanovich
Korpus direktorskiĭ. - Moskva : Sovetskiĭ pisatel', 1979
365p : ill
r6177206 X.329/9042 Woolwich
0899143

MOGIL'NITSKAÎÂ, Kaleriîa Ivanovna
Razvitie sofsialisticheskikh proizvodstvennykh otnosheniĭ v sibirskoĭ derevne. - Tomsk : Izdatel'stvo Tomskogo universiteta, 1976
123p
r6086978 X319/9779 Woolwich
0899143

RAZVITE i razmeshchenie sel'skokhozîaĭstvennogo proizvodstva na Severo-Vostoke SSSR / [otvetstvennyĭ redaktor...F. I. Golovnykh]. - Îâkutsk : Izdanie Îâkutskogo filiala SO AN SSSR, 1976
130 p.
r6020749 X.519/21466
0899143

RAZVITE narodnogo khozîaĭstva Sibiri / [redakfsonnaîa kollegiîa ... M.K. Bandman, B.P. Orlov, Z.P. TSimdina (otv. redaktory)]. - Novosibirsk : Nauka, Sibirskoe otdelenie, 1978
371p
r6173436 X.322/2663 Woolwich
0899143

RED'KIN Pavel Kuzmich
Preobrazhennaîa derevnîa. - Kemerovo : Kemerovskoe knizhnoe izdatel'stvo, 1977
174p.
r6088188 X.319/9803 Woolwich
0899143

SHOTSKIĬ, V. P.
Agro-industrial complexes and types of agriculture in Eastern Siberia. - Budapest : Akadémiai Kiadó, 1979
130p. - (Geography of world agriculture ; 8)
963-05-1845-7 X.0322/92.(8.) Woolwich
0899143

ZABORTIEV, Alekseĭ Ivanovich
Razvitie khozîaĭstvennogo rascheta i vosproizvodstvo v sovkhozakh Zapadnoĭ Sibiri. - Novosibirsk : Nauka, Sibirskoe otdelenie, 1975
296p
r6177583 X.529/22067
0899143

AGRICULTURE. Siberia. Russia (RSFSR)
1800-1900

ASALKHANOV, Innokentiĭ Arsen'evich
Sel'skoe khozîaĭstvo Sibiri kontsa XIX - nachala XXv... - Novosibirsk : Nauka, Sibirskoe otdelenie, 1975
264p
r6098845 X.329/8981 Woolwich
1856448

SOLOV'EVA, E. I.
Krest'îanskaîa promyshlennost' Sibiri vo vtoroĭ polovine XIXv.. - Novosibirsk : Novosibirskiĭ gosudarstvennyĭ pedagogicheskiĭ institut, 1975
196p
r6099329 X.319/9831 Woolwich
1856448

AGRICULTURE. Sicily. Italy
1946-

GEROLD Gerhard
Agrarwirtschaftliche Inwertsetzung Südost-Siziliens. - Hannover : Selbstverlag der Geographischen Gesellschaft Hannover, 1982
xiv,338p,9]folded leaves of plates : ill. - (Jahrbuch der Geographischen Gesellschaft zu Hannover ; 1980). - Bibliography
r6300584 P.801/490[1980]
8876924

AGRICULTURE. Sierra Leone

GLEAVE, M. B. (Michael Barrie)
Agricultural development in Sierra Leone. - Salford : University of Salford, Department of Geography, 1978
26p,[6]p of plates. - (Discussion papers in geography / University of Salford Department of Geography ; no.6). - Bibliography: p26
B82-36204 X.525/6153
1049038

LEVI, John
African agriculture. - Slough : Commonwealth Agricultural Bureaux, 1976
[5],xviii,428p : ill. - (Miscellaneous publications / Commonwealth Bureau of Agricultural Economics ; no.2). - Bibliography: p.409-428
0-85198-374-x UNP.266/66
1049038

AGRICULTURE. Sierra Leone
Smallholdings. Mechanisation

GLEAVE, M. B. (Michael Barrie)
Mechanisation of peasant farming. - Salford : University of Salford, Department of Geography, 1977
27p. - (Discussion paper in geography / Department of Geography, University of Salford ; no.3). - Bibliography: p25-26
B83-02752 X.805/3365
2813734

AGRICULTURE. Silesia. Poland

OLSZEWSKI, Bogusław
"Osthilfe". - Wrocław : Zakład Narodowy im. Ossolińskich, Wydawnictwo, 1974
142p. - (Monografie śląskie Ossolineum ; 27). - Bibliography : p.135-140
r6078897 Ac.7215/107[no.27]
1003453

AGRICULTURE. Silesia. Poland
to 1945

HAASE, A.
Schlesiens Landwirtschaft. - Wolfenbüttel : Grenzland-Verlag, 1981
397p : ill. - Includes bibliography
r6307496 X.329/17301
9032959

AGRICULTURE. Sinaloa. Mexico
Economic development, 1900-1970

CECEÑA CERVANTES, José Luis
Sinaloa, crecimiento agrícola y desperdicio. - México : Universidad Nacional Autónoma de México, Instituto de Investigaciones Económicas, 1973
243 p., 1 fold. leaf of plates
r6011592 X.319/8497 Woolwich
0872318

AGRICULTURE. Skien. Norway
Farms, ca 1700-1960

JENSEN, Jens Per
Gårder og landsteder i Skien og omegn. - Skien : Selskapet for Skien bys vel, 1976
92p : ill. - (Byminner ; 12)
r6033942 X.0802/322(7)
1714597

AGRICULTURE. Slovakia
to 1945 - Bibliographies

KÁŇAVOVÁ, Anna
Po lnohospodárstvo na Slovensku 1918-1928. - Martin : Matica slovenská. - (Slovenská národná retrospektívna bibliografia. Séria C. Články III. Bibliografia článkov zo slovenských movín a časopisov 1918-1945. Odbor XXIV. Po lnohospodárstvo ; zv.1)
Diel 1.: Po lnohospodárstvo všeobecne. - 1982
2v
r6440899 Ac8986/17c
9031057

AGRICULTURE. Slovenia. Yogoslavia
Productivity, 1945-

POSPEŠEVANJE kmetijske proizvodnje / pripravili Anton Robič ... [et al.]. - Ljubljana : Zadružna zveza SR Slovenije, Republiški center za pospeševanje kmetijstva, 1978
68p
r6435534 X.622/15371
8916543

AGRICULTURE. Slovenia. Yugoslavia

MAKAROVIČ, Marija
Kmečko gospodarstvo na slovenskem. - Ljubljana : Mladinska knjiga, 1978
296p : ill. - Bibliography : p.293-296
r6176843 X.329/12318 Woolwich
2262150

AGRICULTURE. Smaland. Sweden
Socioeconomic aspects, 1800-1850

KARLSSON, Folke
Mark och försörjning. - Göteborg : [Göteborgs universitet, Ekonomisk-historiska institutionen], 1978
172p. - (Meddelanden från Ekonomisk-historiska institutionen vid Göteborgs universitet ; 41)
91-85196-14-2 Ac.2651.db/2[no.41]
0392634
Mark och människor. - Göteborg, 1976
686p. - Includes bibliographies
r6028901 X.702/3139
0392634

AGRICULTURE. Smallholdings. Agriculture

MUNROE, C. J.
The smallholder's guide. - Newton Abbot : David and Charles, 1979
160p. - Includes index
0-7153-7652-7 X.529/33316
1203061

SMITH, Joanna
Farm your garden. - London : Sidgwick and Jackson, 1977
176p,[12]p of plates : ill. - Includes bibliographies and index
0-283-98311-6 X.329/11111 Woolwich
1203061
Farm your garden. - Harmondsworth : Penguin, 1979
205p,[8]p of plates : ill. - (Penguin handbooks). - Includes bibliographies and index
0-14-046348-8 WP.4003/262
1203061

AGRICULTURE. Smallholdings. Agriculture. Italy
B.C.500-A.D.200

FRAYN, Joan M.
Subsistence farming in Roman Italy. - Fontwell : Centaur Press, 1979
168p,leaf of plate,[2]p of plates : ill. - Bibliography: p.154-161. - Includes index
0-900000-92-9 X.329/11937 Woolwich
192124x

AGRICULTURE. Socialist countries
Legal aspects

PRUGBERGER, Tamás
A gazdasági szervezetek szabályozásának jogrendszertagozódási kérdései. - [Budapest], 1978
246p. - (Közlemények / Szövetkezeti Kutató Intézet ; 133)
r6499041 X.0512/379(133)
2469782

AGRICULTURE. Socialist countries
Policies of governments, to date

RYDENFELT, Sven
Bönder, mat, socialism. - Malmö : Liber förlag, 1983
159p
91-386-1052-3 X.329/19464
9162410

AGRICULTURE. Somerset
1770-1979

HEELEY, Ann
Victorian Somerset, farming. - Glastonbury : Friends of the Abbey Barn, 1979
52p : ill. - Bibliography: p.52
B80-20350 X.708/48585
2127652

AGRICULTURE. Somogy (megye). Hungary
1918-1945

KANYAR, József
Somogy megye agrártársadalma az első földreformtól a szocialista mezőgazdaságig (1920-1949). - Budapest : Akadémiai Kiadó, 1980
377p
963-05-2063-x X.322/10163 Woolwich
8748780

AGRICULTURE. Sous region. Morocco
to date

BENCHERIFA, Abdellatif
Chtouka et Massa. - [Rabat] : Faculté des lettres et des sciences humaines, Université Mohamed V, 1980
226p,1folded leaf of plates : ill. - (Publications de la Faculté des lettres et des sciences humaines, Université Mohamed V. thèses et mémoires ; 5)
r6209065 X.622/13834
8853207

AGRICULTURE. South Africa
1945-1970

LIPTON, Merle
White farming. - Johannesburg : S.A. Institute of Race Relations, 1975
[3],25p
0-86982-105-9 X.319/17242 Woolwich
1090550

AGRICULTURE. South Asia
Economic development, 1945- - Bibliographies

PATHAK, Angelika
Landwirtschaftliche Entwicklung in Südasien. - Hamburg : Institut für Asienkunde, Dokumentations-Leitstelle Asien, 1978
xxviii,208p. - (Dokumentationsdienst Asien. Reihe A ; 11)
r6308578 X.0802/1563(11)
9016511

AGRICULTURE. South Asia
Technological development

BRADNOCK, Robert W.
Agricultural change in South Asia. - London : Murray, 1984
vi,64p : ill(some col.). - (Case studies in the developing world). - Bibliography: p63. - Includes index
0-7195-3911-0 X.322/13242
2940094
- [Another copy.] V 25018 IOL

SCIENCE, politics, and the agricultural revolution in Asia / edited by Robert S. Anderson ... [et al.]. - Epping : Bowker, [1983?]
xix,512p : 1ill. - (AAAS selected symposium ; 70)
0-86531-320-2 X.800/35849
2940094
- [Another issue]. - Boulder, Colo. : Westview Press, for the American Association for the Advancement of Science, 1982 V 24822 IOL

UNDERSTANDING green revolutions / edited by Tim P. Bayliss-Smith and Sudhir Wanmali. - Cambridge : Cambridge University Press, 1984
x,384p : ill. - Includes bibliographies and index
0-521-24942-2 X.322/13944
2940094
- [Another copy.] V 25285 IOL

AGRICULTURE. South-east Asia
1950-1970

SOUTHEAST Asian agribusiness / [editor] J.D. Drilon, Jr ; [contributors] ... [others]. - Tokyo : Asian Productivity Organization, 1975
iv,iv,326p : ill. - (Agribusiness management resource materials ; vol. 3)
92-8331-034-9 X.529/44511
184766X

AGRICULTURE. South-east Asia
Development. Administration

GABLE, Richard Walter
Administering agricultural development in Asia. - Boulder : Westview Press ; Tonbridge : Distributed by Benn, 1976
xxviii,398p : ill. - (Westview special studies in social, political, and economic development). - Includes bibliographies and index
0-89158-206-1 X.320/11331 Woolwich
1520709

AGRICULTURE. South-east Asia
Management. Applications of statistical mathematics

OÑATE, Burton T.
Statistics in Southeast Asia agriculture. - [Manila?] : Southeast Asian Regional Centre for Graduate Study and Research in Agriculture, 1976
xxxvi,744p : ill. - Includes bibliographies and index
r6072488 X.620/18054 Woolwich
1704052

AGRICULTURE. South-eastern Ireland (Republic)
Development. Organisations: Council for Development in Agriculture

ACOT
ACOT. - Piltown : [ACOT], [1981?]
11p : ill
B82-05199 X.329/14790 Woolwich
2599430

AGRICULTURE. South-eastern United Sates

ANDERSON, James R.
A geography of agriculture in the United States' Southeast. - Budapest : Akadémiai Kiadó, 1973
135p : ill. - (Geography of world agriculture ; 2)
r6022259 X.0322/92(2) Woolwich
1152726

AGRICULTURE. South Tyrol. Italy

FISCHER, Klaus, 1937-
Agrargeographie des westlichen Südtirol. - Wien : Braumüller, c1974
xiv, 365 p., 12 p. of plates, 1 leaf of plates : ill. - Bibliography : p. 343-365
3-7003-0061-1 X.320/5003 Woolwich
0929905

AGRICULTURE. South Vietnam
to date

CALLISON, C. Stuart
The land-to-the-tiller program and rural resource mobilization in the Mekong delta of South Vietnam. - Athens, Ohio : Ohio University Center for International Studies, c1974
vii,41p. - (Papers in International Studies. Southeast Asia series ; no.34)
r6802106 X.622/14067
8881790

AGRICULTURE. South-west England
- Serials

SCIENCE-FARMING link Somerset-Dorset - ADAS - Vol.1, no.1 ([197-])-. - Taunton : Agricultural Development and Advisory Service, [197-]-
v. : ill
Holdings: Vol.10, no.4 (Jan. 1983)
r6813767 P.311/904
3177602

AGRICULTURE. South-west England
to 1900 - Conference proceedings

POPULATION and marketing / edited by Walter Minchinton. - [Exeter] : University of Exeter, 1976
139p : ill. - (Exeter papers in economic history ; no.11)
0-85989-035-x X.512/5645
1141511

AGRICULTURE. South-west England
Farms

FARM management survey. - 1977/78. Financial results on 293 farms in the Exeter province / V.H. Beynon, G.D.D. Davies. - [Exeter] : University of Exeter, Agricultural Economic Unit, 1978
[4],47p : ill
Subsequent volumes, when received, will be placed at this shelfmark
B79-10109 P.313/101
0069876

AGRICULTURE. South-west England
Farms - Statistics

FARM management handbook / University of Exeter Agricultural Economics Unit. - [1980]. - Exeter : The Unit, 1980
vii,85p : ill
B81-12636 X.329/13733 Woolwich
0069876

AGRICULTURE. South-west England
Farms - Statistics - Serials

FARM management handbook / University of Exeter, Agricultural Economics Unit. - [1981]. - Exeter : The Unit, 1981
vii,89p
B81-31004 X.329/14339 Woolwich
0069876

AGRICULTURE. South-west England
Organisations: Royal Bath and West and Southern Countries Society, to 1981

ROYAL BATH AND WEST AND SOUTHERN COUNTIES SOCIETY
A short history of the Royal Bath and West and Southern Counties Society. - Shepton Mallet : Royal Bath & West & Southern Counties Society, 1982
8p : ill
B83-39470 X.809/56327
2957795

AGRICULTURE. South-west Hertfordshire
1600-1850

LONGMAN, Grant
A corner of England's garden. - [Bushey] : [E.G. Longman], [1977]
2v.(iv,96p,[4]p of plates;74p) : ill. - Includes index
B77-17355 X.0700/1035
1121405

AGRICULTURE. Southern Bulgaria
B.C.6000-B.C.2000

DENNELL, Robin
Early farming in South Bulgaria from the VI to the III millennia B.C.. - Oxford : British Archaeological Reports, 1978
[21],363p : ill. - (BAR international series (supplementary) ; 45). - Bibliography: p.169-183
0-86054-021-9 X.512/6883
1946536

AGRICULTURE. Southern England
- Serials

[PROGRESS (Agricultural Development and Advisory Service)]
Progress / ADAS. - Vol.1, no.1 ([19-])-. - [S.l.] : ADAS, [19-]-
v.
Holdings: Vol.22, no.1 (Jan. 1984)-
r6813765 P.311/911
2200503

AGRICULTURE. Southern England
Part-time agriculture

GASSON, Ruth
Gainful occupations of farm families. - [Ashford, Kent] Wye College, School of Rural Economics, 1983
vi,204p. - Bibliography: p197-204
B83-38858 X.322/12366
2954982

AGRICULTURE. Southern France
Neolithic agriculture

PHILLIPS, Patricia
The middle Neolithic in southern France. - Oxford : B.A.R., 1982
viii,204p : ill. - (BAR. International series ; 142). - Bibliography: p189-201. - Includes index
0-86054-173-8 X.425/5242
3277321

AGRICULTURE. Southern Iran
1962-1979. Social aspects

SALMANZADEH, Cyrus
Agricultural change and rural society in Southern Iran. - Wisbech : Middle East & North African Studies Press, 1980
xvi,275p : ill. - Includes index
0-906559-02-2 X.329/13624 Woolwich
2369931

AGRICULTURE. Southern Italy
to 1850

SCIACCA, Marisa
Le terre del sud. - Cosenza : Lerici, 1977
112p
r6052425 X.319/9315 Woolwich
1465465

AGRICULTURE. Southern Italy
to 1970

MOTTURA, Giovanni
Agricoltura, Mezzogiorno e mercato del lavoro. - Bologna : Società Editrice Il Mulino, 1975
327 p.. - (Studi e ricerche ; n. 46)
r6014250 X.319/8487 Woolwich
0868337

AGRICULTURE. Southern Italy
1918-

RIFORMA agraria e questione meridionale / [a cura di] Pasquale Villani, Nunzia Marrone. - Bari : De Donato, 1981
309p. - (Movimento operaio ; 65)
r6509855 X.329/17066
9016295

AGRICULTURE. Southern Italy
1945-1975

MEZZOGIORNO e agricoltura / a cura di E. Zagari. - [Milan] : Giuffrè, 1977
584p. - (SVIMEZ. La questione meridionale dal dopoguerra ad oggi)
r6068244 X.319/9804 Woolwich
1851527

ZAGARI, Eugenio
Il problema agrario in trent'anni di meridionalismo. - Napoli : Cooperativa editrice Economia e commercio, 1976
117p
r6078446 X.311/4919 Woolwich
1718029

AGRICULTURE. Southern Italy
1961-1971

PAGLIANI, Carlo
La spesa pubblica per l'agricoltura nel decennio precedente l'ordinamento regionale. - Milano : Giuffrè, 1977
viii,183p. - (SVIMEZ. Collana Francesco Giordani)
r6100142 X.311/5032 Woolwich
1863533

AGRICULTURE. Southern Moravia. Czechoslovakia
Collectivisation, 1945- - Conference proceedings

MIKULOVSKÉ SYMPOZIUM (9th : 1979)
Mikulovská sympozia, IX ... 23.-24.10.1979. - Praha : TEBS, 1980
237p,[8]p of plates
r6439776 X.322/10342 Woolwich
8762198

AGRICULTURE. Southern states. United States to 1976

HART, John Fraser
The South. - 2nd ed. - New York ; London : D. Van Nostrand, 1976
vi,166p. - (New searchlight series). - Bibliography: p.155-157. - List of maps: p.158-160. - Includes index
0-442-29754-8 X.708/21014
1344439

AGRICULTURE. Southern states. United States 1865-1914

DECANIO, Stephen J.
Agriculture in the postbellum South. - Cambridge, Mass. ; London : M.I.T. Press, 1974
xiii,335p : ill. - (Monographs in economics / Massachusetts Institute of Technology ; 12). - Bibliography: p.312-322. - Includes index
0-262-04047-6 X.320/10577
0345229

AGRICULTURE. Southern states. United States Organisations: Southern Farmers' Alliance, 1880-1890

SCHWARTZ, Michael
Radical protest and social structure. - New York ; London : Academic Press, 1976
xi,302p : ill. - (Studies in social discontinuity). - Includes index
0-12-632850-1 X.320/5171 Woolwich
1105132

AGRICULTURE. Sovet ekonomicheskoi vzaimopomoshchi countries

AGRARNO - promyshlennaia integratsiia stran SEV / [otvetstvennyi redaktor...Iu. F. Kormnov]. - Moskva : Izdatel'stvo "Nauka", 1976
283 p.
r6021154 X.529/19927
1025910

BUKH, Mikhail Efimovich
Problemy effektivnosti sel'skogo khoziaistva v evropeiskikh stranakh SEV. - Moskva : Nauka, 1978
167p
r6077192 X.319/9858 Woolwich
1025910

SHALAMANOV, Stoian
Sotsialisticheskata integratsiia i razvitieto na APK v stranite-chlenki na SIV. - Sofiia : Zemizdat, 1977
193p
r6173870 X.529/21737
1025910

SHMELEV, Gelii Ivanovich
Sotrudnichestvo stran-chlenov SEV v sel'skom khoziaistve. - Moskva : Izdatel'stvo "Ekonomika", 1976
139p
r6036002 X.519/22358
1025910

SHMELEV, G.I
Sotsial'no-ekonomicheskie problemy razvitiia sel'skogo khoziaistva evropeiskikh sotsialisticheskikh stran. - Moskva : Izdatel'stvo "Nauka", 1977
222p
r6055793 X.319/9419 Woolwich
1025910

SUSLOV, Ivan Fedorovich
Agrarnyi sektor ekonomiki stran sotsializma - Moskva : Mysl', 1978
253p.
r6094965 X.319/9843 Woolwich
1025910

AGRICULTURE. Sovet ekonomicheskoi vzaimopomoshchi countries to date

AGRARNAIA politika sotsialisticheskikh stran na sovremennom etape / [pod obshchei redaktsiei Dimo Vladova (NPB), Emel'ianova A.M. (SSSR)]. - Moskva : Izdatel'stvo politicheskoi literatury, 1981
366p. - (Sotsializm)
r6400989 X.329/14818 Woolwich
8416044

AGRICULTURE. Sovet ekonomicheskoi vzaimopomoshchi countries. Europe 1976-1980

UNIVERSITY OF LANCASTER. Comenius Centre
East Europe agriculture and food industry. - Lancaster : Comenius Centre, University of Lancaster.
Report 1: Beer and soft drink. - 1979
[1],vi,65p
B80-41453 SRL
2313537

The commodities : grains and sugar. - 1980
[5],48p : ill
B80-41454 SRL
2313537

AGRICULTURE. Soviet Union

AFANASENKO, I. D. (Ivan Dmitrievich)
Ekonomicheskie sviazi sotsialisticheskoi promyshlennosti i sel'skogo khoziaistva. - Leningrad : Izdatel'stvo Leningradskogo universiteta, 1978
151p
r6176281 X.319/20300 Woolwich
1207962

AGRARNO-PROMYSHLENNYE ob'edineniia / pod obshchei redaktsiei K.D. Karpova. - Ulan-Ude : Buriatskoe knizhnoe izdatel'stvo, 1975
390p.[2]fold leaves
r6176044 X.329/8983 Woolwich
1156675

BOGUSH, Grigorii Mikhailovich
Sel'skoe khoziaistvo SSSR. - Moskva : Kolos, 1977
235p : ill(some col)
r6074574 X.322/2612 Woolwich
1160397

DAVYDOV, Iurii Stepanovich
Agrarno-promyshlennyi sintez v SSSR. - Rostov-na-Donu : Izdatel'stvo Rostovskogo universiteta, 1978
159p. - Bibliography: p.151-158
r6187028 X.319/20050 Woolwich
1156675

EKONOMICHESKIE problemy novogo etapa osushchestvleniia leninskogo kooperativnogo plana / [otvetstvennyi redaktor E.S. Karnaukhova]. - Moskva : Nauka, 1979
365p. - (Problemy sovetskoi ekonomiki)
r6098152 X.329/8911 Woolwich
1207962

HEDLUND, Stefan
Crisis in Soviet agriculture. - London : Croom Helm, c1984
228p : ill. - Bibliography: p215-225. - Includes index
0-7099-1281-1 X.329/18167
1156675

IGNATOVSKII, Pavel Artem'evich
Obshchestvennoe proizvodstvo sovetskoi derevni. - Moskva : Mysl', 1975
342p
r6094053 X.329/8900 Woolwich
1207962

IL'IN, S. S. (Sergei Sergeevich)
Industrializatsiia sotsialisticheskogo sel'skogo khoziaistva. - Moskva : Izdatel'stvo Moskovskogo universiteta, 1975
175p
r6036450 X.519/22949
1156675

KALASOV, Aigum Kalasovich
Agrarno-promyshlennaia integratsiia i sovershenstvovanie sotsial'no- ekonomicheskikh otnoshenii v sel'skom khoziaistve. - Moskva : Nauka, 1978
109p. - (Problemy sovetskoi ekonomiki)
r6172833 X.319/20106 Woolwich
1156675

LATOV, V.
[SSSR : sel'skoe khoziaistvo. English]. Agriculture. - Moscow : Novosti, 1976
44p,[32]p. of plates : ill. - (USSR. Yesterday, today, tomorrow)
r6087990 X.319/9847 Woolwich
1182129

AGRICULTURE. Soviet Union

MODELIROVANIE razvitiia agrarno-promyshlennogo kompleksa / nauchnyi redaktor...A.I. Tianutov. - Novosibirsk : Institut ekonomiki i organizatsii promyshlennogo proizvodstva SO AN SSSR, 1975
183p
r6037515 X.319/8830 Woolwich
1160397

MOROZOV, V. A.
Soviet agriculture. - Moscow : Progress Publishers, 1977
220p. - (Progress socialism today)
r6043499 X.329/11124 Woolwich
1207962

NOVYI etap v razvitii sel'skogo khoziaistva SSSR / [sostavitel' Nikolai Anisimovich Bannikov]. - Moskva : Kolos, 1975
502p
r6036429 X329/8832 Woolwich
1156675

PIVNEV, Nikolai Pavlovich
Kommunisticheskaia partiia - organizator tekhnicheskogo progressa v sel'skom khoziaistve (1959-1970). - Voronezh : Izdatel'stvo Voronezhskogo universiteta, 1977
209p
r6046228 X.319/8811 Woolwich
1160397

REGIONAL'NYE agropromyshlennye kompleksy SSSR / pod redaktsiei N.N. Nekrasova. - Moskva : Nauka, 1979
323p. - (Problemy sovetskoi ekonomiki)
r6186186 X.329/8973 Woolwich
1156675

RUKOVODSTVO KPSS sotsialisticheskim preobrazovaniem v sel'skom khoziaistve / [redaktsionnaia kollegiia : Kosmachev K.N. (otv. redaktor), Budaev D.I., Krylov E.F.]. - Smolensk : Smolenskii gosudarstvennyi pedagogicheskii institut, 1978
107p
r6094384 X.319/9878 Woolwich
1160397

SEMIN, Sergei Ivanovich
Razvitie sotsialisticheskikh proizvodstvennykh otnoshenii v sel'skom khoziaistve. - Moskva : Izdatel'stvo "Nauka", 1977
276p
r6086947 X.329/8866 Woolwich
1156675

AGRICULTURE. Soviet Union - Conference proceedings

OSNOVNYE problemy sotsial'no-ekonomicheskogo razvitiia sovetskoi derevni v usloviiakh sozdaniia razvitogo sotsialisticheskogo obshchestva i stroitel'stva kommunizma / [redaktsionnaia kollegiia: I.M. Volkov ... i dr.]. - Moskva : Akademiia nauk SSSR, Institut istorii SSSR, 1973
203p
r6176578 X.319/20303 Woolwich
1156675

PROBLEMY agrarnoi politiki KPSS na sovremennom etape / [redaktsionnaia kollegiia Golikov V.A., rukovoditel', ... i dr.]. - Moskva : Izdatel'stvo politicheskoi literatury, 1975
2t
r6036833 X.329/12585
1156675

SIMPOZIUM PO IZUCHENIIU PROBLEM AGRARNOI ISTORII (16th : 1976 : Kishinev)
Sel'skoe khoziaistvo i krest'ianstvo SSSR v sovremennoi sovetskoi istoriografii. - Kishinev : Shtiintsa, 1977
213p
r6098034 X.629/8122 Woolwich
1156675

AGRICULTURE. Soviet Union - Statistics

GOZULOV, Avdei Il'ich
Statistika sel'skogo khoziaistva. - Izd. 3-e, perer. i dop.. - Moskva : Statistika, 1975
366p
r6177251 X.329/12194 Woolwich
1156675

AGRICULTURE. Soviet Union to 1976

SOVIET agriculture / edited by Harry G. Shaffer. - New York ; London : Praeger, 1977
xvii,167p. - (Praeger special studies in international economics and development)
0-03-021976-0 X.520/12022
1256513

AGRICULTURE. Soviet Union to date - Conference proceedings

SIMPOZIUM PO IZUCHENIIU PROBLEM AGRARNOI ISTORII (18th : Voronezh : 1980)
Sotsial'no-politicheskii i kul'turnyi oblik derevni v ego istoricheskom razvitii. - Moskva : AN SSSR, Institut istorii SSSR, 1980
2v.
r6402200 X.529/43121
8536155

AGRICULTURE. Soviet Union 1800-1922

SHLIKHTER, O. H. (Oleksandr Hryhorovych)
Agrarnyi vopros i prodovol'stvennaia politika v pervye gody Sovetskoi vlasti. - Moskva : Nauka, 1975
445p
r6098437 X.329/8979 Woolwich
2193027

AGRICULTURE. Soviet Union 1900-1917

KOSTRIKIN, Vasilii Ivanovich
Zemel'nye komitety v 1917 godu. - Moskva : Nauka, 1975
335p
r6098430 X.329/8980 Woolwich
2193035

AGRICULTURE. Soviet Union 1917-1922

DANILOV, Viktor Petrovich
Sovetskaia dokolkhoznaia derevnia. - Moskva : Nauka, 1979
357p
r6402092 X.329/13685 Woolwich
8536163

AGRICULTURE. Soviet Union
1917-1939. Soviet historiography

CHINCHIKOV, Aleksandr Mikhaĭlovich
Sovetskaîa istoriografiîa pervykh agrarnykh preobrazovaniĭ i nachal'nogo ėtapa kolkhoznogo stroitel'stva (1917-1973gg.). - [Saratov] : Izdatel'stvo Saratovskogo universiteta, 1974
235p. - Bibliography : p.230-[236]
r6007396 X.319/20307 Woolwich
2189011

AGRICULTURE. Soviet Union
1917-1941

LENINSKIĬ kooperativnyĭ plan - nauchnaîa osnova sotsialisticheskogo preobrazovaniîa sel'skogo khozîaĭstva ... / [redaktsionnaîa kollegiîa : A.P. Petrov ... i dr.]. - Moskva : Akademiîa nauk SSSR, Institut nauchnoĭ informatsii po obshchestvennym naukam, 1980
2v.
r6484582 X.329/14840 Woolwich
8756015

AGRICULTURE. Soviet Union
1917-

AGRARNAÎA politika kommunisticheskoĭ partii i sovetskogo pravitel'stva v period sotsialisticheskogo stroitel'stva / [otvetstvennyĭ redaktor : V.S. Flerov]. - ÎAroslavl' : ÎAroslavskiĭ gosudarstvennyĭ universitet, 1978
229p
r6486858 X.529/50941
8876908

AGRICULTURE. Soviet Union
1917-. Economic aspects - Bibliographies

DOLGOSHEĬ, G. A. (Gavriil Artemovich)
Ėkonomika sel'skogo khozîaĭstva. - Moskva : Kolos, 1981
395p
r6448835 X.329/16080
8891087

AGRICULTURE. Soviet Union
1917-. Role of Kommunisticheskaîa partiîa Sovetskogo Soîuza

NEKOTORYE voprosy deîatel'nosti KPSS po razvitiîu sel'skogo khozîaĭstva / [redaktsionnaîa kollegiîa Mikhalev G.M. (otv. redaktor), Laktionov I.D., Pesikov M.I.]. - Orel : Kurskiĭ gosudarstvennyĭ pedagogicheskiĭ institut, 1973
261p
r6400969 X.329/14927
879507X

AGRICULTURE. Soviet Union
ca 1920-1975

STANIS, Vladimir Frantsevich
The socialist transformation of agriculture. - Moscow : Progress, 1976
238 p.. - (Socialism today)
r6009849 X.329/10615 Woolwich
0728098

AGRICULTURE. Soviet Union
1922-1941

DANILOV, Viktor Petrovich
Sovetskaîa dokolkhoznaîa derevnîa. - Moskva : Nauka, 1979
357p
r6402092 X.329/13685 Woolwich
8416494

HARRIS, Lement
Harold M. Ware (1890-1935). - New York : The American Institute for Marxist Studies, 1978
72p. - (Occasional paper ; no.30 (1978))
r6800028 X.322/10086 Woolwich
8416494

AGRICULTURE. Soviet Union
1945-

AGRARNYE problemy razvitogo sotsializma / [redaktsionnaîa kollegiîa V.A. Tikhonov (otvetstvennyĭ redaktor) ... i dr.]. - Moskva : Nauka, 1980
551p. - (Problemy sovetskoĭ ėkonomiki / Akademiîa nauk SSSR, Institut ėkonomiki)
r6400809 X.329/14442 Woolwich
8416516

ĖKONOMIKA sel'skogo khozîaĭstva na sovremennom ėtape / pod redaktsieĭ A.M. Emel'îanova, S.I. Polovenko. - Moskva : Izdatel'stvo Moskovskogo universiteta, 1980
267p
r6402277 X.329/14458 Woolwich
8416516

IGNATOV, V. G.
Industrializatsiîa sotsialisticheskogo sel'skogo khozîaĭstva. - Rostov-na-Donu : Izdatel'stvo Rostovskogo universiteta, 1979
151p
r6485587 X.329/14971 Woolwich
8416516

VOPROSY formirovaniîa i razvitiîa agrarno-promyshlennogo kompleksa / pod redaktsieĭ S.E. Il'îushonka. - Novosibirsk : Akademiîa nauk SSSR, Sibirskoe otdelenie, Institut ėkonomiki i organizatsii promyshlennogo proizvodstva, 1984
126p. - Includes bibliographies
r6492379 X.808/41969
8416516

AGRICULTURE. Soviet Union
1945- - Conference proceedings

OSNOVNYE PROBLEMY SOTSIAL'NO-ĖKONOMICHESKOGO RAZVITIÎA SOVETSKOĬ DEREVNI V USLOVIÎAKH SOZDANIÎA RAZVITOGO SOTSIALISTICHESKOGO OBSHCHESTVA I STROITEL'STVA KOMMUNIZMA (Nauchnaîa konferentsiîa : 1973 : Moscow)
Problemy istorii sovremennoĭ sovetskoĭ derevni 1946-1973gg.. - Moskva : Nauka, 1975
507p
r6186263 X.329/8964 Woolwich
2121379

AGRICULTURE. Soviet Union
1945-. Economic aspects

PROTSENKO, V. A. (Vîacheslav Aleksandrovich)
Mezhkhozîaĭstvennaîa kooperatsiîa i vyravnivanie ėkonomicheskikh usloviĭ khozîaĭstvovaniîa. - Kishinev : Kartîa Moldoveniaske, 1982
158p. - Includes bibliography
r6492150 X.529/70514
9159266

AGRICULTURE. Soviet Union
1945-. Legal aspects

KONSTITUTSIÎA SSSR i pravovye problemy sovershenstvovaniîa rukovodstva narodnym khozîaĭstvom / [redaktsionnaîa kollegiîa : K.F. Sheremet (predsedatel') ... i dr.]. - Moskva : Akademiîa nauk SSSR, Institut gosudarstva i prava, 1979
230p
r6487007 X.200/39601
8927588

AGRICULTURE. Soviet Union
Capital investment, 1945-

KLEMYSHEV, P. A. (Petr Aleksandrovich)
Ėffektivnost' kapital'nykh vlozheniĭ v sel'skoe khozîaĭstvo. - Moskva : Nauka, 1974
214p
r6462783 X.529/59754
9047484

AGRICULTURE. Soviet Union
Collectivisation

AGRARIAN relations in the USSR / [by P.A. Ignatovsky et al.] ; [translated from the Russian by Galina Sdobnikova]. - Moscow : Progress, 1980
199p. - (Progress socialism today)
B81-05586 X.329/13426 Woolwich
235246x

BOGDENKO, Mariîa Lukinichna
Sovkhozy SSSR. - Moskva : Izdatel'stvo politicheskoĭ literatury, 1976
277p
r6021840 X.809/24147
1125575

BOGOLÎUBOV, Klavdiĭ
Development of the agrarian sector in the USSR. - Moscow : Novasti, 1983
119p
r6810921 X.329/17158
235246x

LIZOGUB, ÎU. M. (ÎUriĭ Markovich)
Rabochiĭ klass SSSR v bor'be za pobedu kolkhoznogo stroîa (1928-1937). - L'vov : Vishcha shkola, 1978
173p
r6022628 X.329/13257 Woolwich
1125575

TRAPEZNIKOV, Sergeĭ Pavlovich
Leninizm i agrarno-krest'îanskiĭ vopros.... - 2., dop. izd. - Moskva : Izdatel'stvo "Mysl'", 1974
2t
r6057669 X.329/8827 Woolwich
1125575

VARENOV, Vasiliĭ Il'ich
Pomoshch' Krasnoĭ Armii v razvitii kolkhoznogo stroitel'stva 1929-1933gg.. - Moskva : Nauka, 1978
175p
r6087829 X.319/9865
1125575

VESELOV, Anatoliĭ Pavlovich
Bor'ba Kommunisticheskoĭ partii za provedenie kul'turnoĭ revoliutsii v derevne v gody kollektivizatsii. - Leningrad : Izdatel'stvo Leningradskogo universiteta, 1978
127p
r6177155 X.709/50835
1125575

VYLTSAN, Mikhail Avgustovich
Zavershaîushchiĭ ėtap sozdaniîa kolkhoznogo stroîa (1935-1937gg.). - Moskva : Nauka, 1978
261p
r6176265 X.329/8996 Woolwich
1125575

AGRICULTURE. Soviet Union
Collectivisation - Statistics

KREST'ÎANSKIE khozîaistva, kolkhozy i sovkhozy SSSR v 1924/25 - 1927/28 gg. / sostaviteli V.P. Danilov, T.I. Slavko. - Moskva : Akademiîa nauk SSSR, Institut istorii SSSR, 1977
3t
r6093337 X.319/9869
1125575

AGRICULTURE. Soviet Union
Collectivisation, to date

KOLLEKTIVIZATSIÎA sel'skogo khozîaĭstva v SSSR / M.A. Vyltsan ... [i dr.]. - Moskva : Kolos, 1982
398p
r6463008 X.629/18636
8416559

TRAPEZNIKOV, Sergeĭ Pavlovich
Leninizm i agrarno-krest'îanskiĭ vopros. - 3.e.dop.izd. - Moskva : Mysl', 1983
2v
r6490255 X.429/16850
8416559

AGRICULTURE. Soviet Union
Collectivisation, to date - Conference proceedings

SIMPOZIUM PO IZUCHENIÎU PROBLEM AGRARNOĬ ISTORII (17th. 1978 : Rostov-na-Donu)
Leninskiĭ kooperativnyĭ plan i sovremennaîa derevnîa. - Rostov-na-Donu : Izdatel'stvo Rostovskogo universiteta, 1980
221 p. - Includes bibliographical references
r6465857 X.439/13235
8416559

AGRICULTURE. Soviet Union
Collectivisation, 1922-1941

BETTANIN, Fabio
La collettivizzazione delle campagne nell'URSS. - Roma : Editori riuniti, 1978
269p. - (Biblioteca di storia ; 75)
r6505964 X.629/19306
8416567

CHUDESNAÎA sila / [sostaviteli: G.I. Kopanev, V.A. Smyshlîaev]. - Leningrad : Lenizdat, 1978
246p : ill
r6498175 X.809/48290
8416567

AGRICULTURE. Soviet Union
Collectivisation, 1929-1930

DAVIES, Robert William
The socialist offensive. - London : Macmillan, 1980
xxi,491p. - (The industrialisation of Soviet Russia). - Bibliography: p.455-465. - Includes index
Subsequent volumes, when received, will be placed at this shelfmark
0-333-26171-2 X.0802/464
2220733

AGRICULTURE. Soviet Union
Development - Conference proceedings

ZEMLÎA tselinnogo podviga / [otvetstvennye za vypusk A.I. Egorov, N.N. Koroleva]. - Alma-Ata : Kazakhstan, 1979
113p
r6177336 X.329/12241
225056X

AGRICULTURE. Soviet Union
Econometric models, 1945-

KISELEV, Viktor Ivanovich
Organizatsiîa i planirovanie narodno-khozîaĭstvennogo agropromyshlennogo kompleksa. - Moskva : Nauka, 1979
223p. - (Problemy sovetskoĭ ėkonomiki)
r6465136 X.329/19091
8756945

AGRICULTURE. Soviet Union
Econometric models. 1945—continued

PROBLEMY ėkonomicheskoĭ kibernetiki v sel'skom khozíaĭstve / nauchnyĭ redaktor V.A. Kardash. - Novosibirsk : Akademiía nauk SSSR, Sibirskoe otdelenie, Institut ėkonomiki i organizafsii promyshlennogo proizvodstva, 1977
158p
r6483113 X.329/14692 Woolwich
8756945

AGRICULTURE. Soviet Union
Employment. Law, 1945-

CHUBUKOV, G. V. (Georgii Vasil'evich)
Pravovoe regulirovanie truda rabotnikov sel'sko-khozíaĭstvennykh predpriíatiĭ. - Moskva : Nauka, 1980
243p
r6400889 X.329/14677 Woolwich
8781893

AGRICULTURE. Soviet Union
Finance

OSNOVNYE napravleniía sovershenstvovaniía oplaty truda i material'nogo stimulirovaniía v sel'skom khozíaĭstve / [nauchnye redaktory Georgiĭ Kuz'mich Rusakov, Nikolaĭ Ivanovich Popov, ÍUliía Nikolaevna Listova]. - Moskva : Akademiía nauk SSSR, TSentral'nyĭ ėkonomiko-matematicheskiĭ institut, Nauchnyĭ sovet po kompleksnoĭ probleme "Nauchnye osnovy khozíaĭstvennogo rascheta", 1977
220p
r6175702 X.319/20298 Woolwich
1472348

SEMCHENKO, Andreĭ Petrovich
Problemy povysheniía ėffektivnosti kapital'nykh vlozheniĭ i osnovnykh fondov sel'skogo khozíaĭstva SSSR. - Leningrad : Izdatel'stvo Leningradskogo universiteta, 1976
125p
r6055696 X.519/24889
1472348

VALOVOĬ dokhod kolkhozov i sovkhozov i ėffektivnost' ego ispol'zovaniía / [redakfsionnaía kollegiía : M.I. Takumbetov (otv. redaktor), U.G. Gusmanov, R.K. Araslanova]. - Ufa : Akademiía nauk, Bashkirskiĭ filial, 1977
162p
r6187094 X.319/20046 Woolwich
1472348

VNUTRIKHOZÍAĬSTVENNYĬ raschet i material'noe stimulirovanie khozraschetnykh podrazdeleniĭ za rezul'taty truda / [nauchnyĭ redaktor G. Rusakov]. - Akademiía nauk SSSR, TSentral'nyĭ ėkonomiko-matematicheskiĭ institut, Uchenyĭ sovet po kompleksnoĭ probleme "Nauchnye osnovy khozíaĭstvennogo rascheta", Sekfsiía "Khozraschet v sel'skom khozíaĭstve", 1977
187p
r6175714 X.319/20299 Woolwich
1472348

ZABORTSEV, Alekseĭ Ivanovich
Razvitie khozíaĭstvennogo rascheta i vosproizvodstvo v sovkhozakh Zapadnoĭ Sibiri. - Novosibirsk : Nauka, Sibirskoe otdelenie, 1975
296p
r6175831 X.529/22067
1472348

AGRICULTURE. Soviet Union
Finance, to date

BOKAREV, ÍU. P. (ÍUriĭ Pavlovich)
Bíudzhetnye obsledovaniía krest'íanskikh khozíaĭstv 20-kh godov kak istoricheskiĭ istochnik. - Moskva : Nauka, 1981
308p
r6485909 X.529/48995
8416583

AGRICULTURE. Soviet Union
Law, 1945-

BYCHKOVA, TS. V. (TSetsiliía Volodymyrivna)
Dogovor v proizvodstvenno-ėkonomicheskikh svíazíakh mezhdu sel'skim khozíaĭstvom i promyshlennost'íu. - Kiev : Naukova dumka, 1980
227 p.. - Includes bibliographical references
r6482927 X.208/10365
8877394

RAZVITIE agrarno-pravovykh nauk / [M.I. Kozyr' (otvetstvennyĭ redaktor)]. - Moskva : Akademiía nauk SSSR, Institut gosudarstva i prava, 1980
174 p.. - Includes bibliographical references
r6482779 X.200/44202
8877394

AGRICULTURE. Soviet Union
Law, 1945- - Conference proceedings

PRAVO i sel'skoe khozíaĭstvo v SSSR i Italii / [redakfsionnaía kollegiía : M.I. Kozyr' (otv. redaktor) ... i dr]. - Moskva : AN SSSR, Institut gosudarstva i prava, 1980
158p
r6432433 X.629/19201
8877394

AGRICULTURE. Soviet Union
Mechanisation, 1945- - Bibliographies

USPENSKAÍA, G. V. (Galina Vasil'evna)
Segodnía i zavtra sel'skokhozíaĭstvennoĭ tekhniki. - Moskva : Kniga, 1979
15p. - (Novoe v nauke i tekhnike)
r6402313 X.329/14495 Woolwich
8753539

AGRICULTURE. Soviet Union
Personnel. Pensions

MIKHALKEVICH, Vitaliĭ Nikolaevich
Pensii i posobiía chlenam kolkhozov. - Moskva : Profizdat, 1978
126p
r6174773 X.319/20163 Woolwich
2096986

AGRICULTURE. Soviet Union
Planning, 1945-

PROBLEMY formirovaniía i upravleniía agrarno-promyshlennym kompleksom / [redaktor:] A.I. Panchenko. - Novosibirsk : Akademiía nauk SSSR, Sibirskoe otdelenie, Institut ėkonomiki i organizafsii promyshlennogo proizvodstva, 1979
157p
r6483240 X.329/14865 Woolwich
8781885

ZUFAROV, R. M. (Rais Magsumovich)
Problemy sovershenstvovaniía razmeshcheniía sel'skogo khozíaĭstva regiona. - Moskva : Nauka, 1982
193p. - (Problemy sovetskoĭ ėkonomiki)
r6464640 X.329/19042
8781885

AGRICULTURE. Soviet Union
Policies of government

BREZHNEV, L. I. (Leonid Il'ich)
O dal'neĭshem razvitii sel'skogo khozíaĭstva SSSR. - Moskva : Pravda, 1978
62p
r6093218 X.319/5010 Woolwich
0261173

RESHENIÍA partii i pravitel'stva po sel'skomu khozíaĭstvu 1965-1974gg. / [sostaviteli : K.M. Bogolíubov, M.S. Smirtíukov]. - Moskva : Kolos, 1975
926p
r6093297 X.809/50910
0261173

AGRICULTURE. Soviet Union
Policies of government, 1917-1976

TRAPEZNIKOV, S. P.
Leninism and the agrarian and peasant question. - Moscow : Progress, c1981
2v. : ill. - Includes index
0-7147-1852-1 X.809/54319
2846012

AGRICULTURE. Soviet Union
Policies of government, 1917-

ALEKSASHKIN, E. K. (Evgeniĭ Karpovich)
Gosudarstvennoe rukovodstvo sel'skim khozíaĭstvom v raĭone. - Kazan' : Izdatel'stvo Kazanskogo universiteta, 1982
118p
r6490671 X.809/64268
884044x

MERL, Stephan
Der Agrarmarkt und die neue ökonomische Politik. - München : Oldenbourg, 1981
530p.. - (Studien zur modernen Geschichte ; Bd.25). - Bibliography: p.487-516
3-486-50021-x X.0709/559(25)
884044x

AGRICULTURE. Soviet Union
Policies of government, 1953-1964

MCCAULEY, Martin
Khrushchev and the development of Soviet agriculture. - London : Macmillan [for] the School of Slavonic and East European Studies, University of London, 1976
xiii,232p. - (Studies in Russian and East European history). - Bibliography: p.223-228. - Includes index
0-333-19834-4 X.329/10815 Woolwich
1052446

VELIKIĬ podvig partii i naroda - massovoe osvoenie fselinnykh i zalezhnykh zemel' / [sostaviteli : S.R. Levina, O.M. Slonim, Z.A. Troifskaía]. - Moskva : Nauka, 1979
438p
r6185409 X.329/9032 Woolwich
1052446

AGRICULTURE. Soviet Union
Policies of government, 1982-1990

FOR the well-being of all / [translated from the Russian by Lenina Ilyitskaya] ; [translation edited by Peter Tempest]. - [Rev. ed.]. - Moscow : Novosti, 1985
87p,[8]p of plates : ill
r6816739 X.529/69961
3328511

AGRICULTURE. Soviet Union
Policies of government. Implementation, 1973-1980. Political aspects

GUSTAFSON, Thane
Reform in Soviet politics. - Cambridge : Cambridge University Press, 1981
xii,218p : 1ill. - Includes index
0-521-23377-1 X.800/31458
2563045

AGRICULTURE. Soviet Union
Policies of Kommunisticheskaía partiía Sovetskogo Soíuza

AGRARNAÍA politika KPSS / pod redakfsieĭ ÍU. V. Sedykh. - Moskva : Kolos, 1977
302p
r6174682 X.329/8943 Woolwich
2098148

AGRARNAÍA politika KPSS v deĭstvii / pod redakfsieĭ R.É. Prausta. - Leningrad : Lenizdat, 1979
291p,[16]p of plates : ill
r6177175 X.329/9040 Woolwich
2098148

BREZHNEV, L. I. (Leonid Il'ich)
[Speeches. Selections]. Voprosy agrarnoĭ politiki KPSS i osvoenie fselinnykh zemel' Kazakhstana. - Moskva : Izdatel'stvo politicheskoĭ literatury, 1974
357p,[1]leaf of plates
r6402063 X.329/13582 Woolwich
2098148

KOMMUNISTICHESKAÍA PARTIÍA SOVETSKOGO SOÍUZA
Leninskaía agrarnaía politika KPSS. - Moskva : Izdatel'stvo politicheskoĭ literatury, 1978
679p
r6187143 X.809/50910
2098148

AGRICULTURE. Soviet Union
Policies of political parties, to 1922

ILLERIfSKAÍA, E. V. (Ekaterina Vladimirovna)
Agrarnyĭ vopros. - Moskva : Nauka, 1981
164p
r6432237 X.629/18607
885324x

AGRICULTURE. Soviet Union
Productivity

BODRENKOV, K.A. (Kirill Aleksandrovich)
Bor'ba partiĭnykh organizafsiĭ Rossiĭskoĭ Federafsii za pod"em kolkhoznogo proizvodstva respubliki v usloviíakh razvitogo sofsializma 1965-1970gg.. - Kursk : Kurskiĭ gosudarstvennyĭ pedagogicheskiĭ institut, 1975
315p
r6093443 X.329/13351 Woolwich
2219204

AGRICULTURE. Soviet Union
Productivity

SOfSIAL'NYE faktory povysheniía proizvoditel'nosti truda v sel'skom khozíaĭstve ... / [otvetstvennyĭ redaktor : V.I. Staroverov]. - Moskva : Akademiía nauk SSSR, Institut sofsiologicheskikh issledovaniĭ, 1973
2v
r6178089 X.329/12339 Woolwich
2219204

AGRICULTURE. Soviet Union
Productivity, 1945-

ÉFFEKTIVNOST' obshchestvennogo proizvodstva / [redakfsionnaía kollegiía : Purifs V.N. (nauch. redaktor) ... idr.]. - Alma-Ata : Kazakhskiĭ gosudarstvennyĭ universitet im. S.M. Kirova, 1982
226p
r6464808 X.429/16849
8710325

AGRICULTURE. Soviet Union
Productivity, 1945—continued

OKHRIMCHUK, A. I. (Alekseĭ Ivanovich)
Razvitie proizvoditel'nykh sil sel'skogo khozīaĭstva. - Moskva : Ėkonomika, 1981
143p
r6484658 X.329/14925 Woolwich
8710325

POPOV, V. M. (Vladimir Mikhaĭlovich)
Sotsial'no-ėkonomicheskie problemy proizvoditel'nosti truda i vosproizvodstva rabocheĭ sily v sel'skom khozīaĭstve. - Moskva : Nauka, 1979
193p. - (Problemy sovetskoĭ ėkonomiki)
r6498306 X.329/13986
8710325

POVYSHENIE ėffektivnosti sel'skokhozīaĭstvennogo proizvodstva / [redaktor : T. Blink]. - Kishinev : Kartīa Moldoveniaske, 1974
111p
r6432417 X.629/19092
8710325

ZEL'DNER, A. G. (Alekseĭ Grigor'evich)
Regional'nye aspekty vyravnivaniīa ėkonomicheskikh usloviĭ vosproizvodstva v sel'skom khozīaĭstve. - Moskva : Nauka, 1982
158p. - (Problemy sovetskoĭ ėkonomiki)
r6462269 X.629/20113
8710325

AGRICULTURE. Soviet Union
Role of Kommunisticheskaīa partiīa Sovetskogo Soīuza, 1917-1941

KPSS - organizator agrarnykh preobrazovaniĭ na Severnom Kavkaze / [redaktsionnaīa kollegiīa : M.I. Gioev (otv. red.) ... i dr.]. - Ordzhonikidze : Severo-Osetinskiĭ gosudarstvennyĭ universitet im. K.L. Khetagurova, 1981
141p
r6499253 X.329/16245
8941416

AGRICULTURE. Soviet Union
Role of Kommunisticheskaīa partiīa Sovetskogo Soīuza, 1945-

BAZAROV, V. B. (Vandan Badmaevich)
Partiĭnye organizatsii v bor'be za razvitie sel'skogo khozīaĭstva, 1965-1970gg.. - Ulan-Udė : Buriātskoe knizhnoe izdatel'stvo, 1974
194p
r6487357 X.329/17846
8963630

KOMMUNISTICHESKAĪA PARTIĪA SOVETSKOGO SOĪUZA
Leninskaīa agrarnaīa politika KPSS. - Moskva : Izdatel'stvo politicheskoĭ literatury, 1983
735p
r6484415 X.329/16626
8963630

AGRICULTURE. Soviet Union
Role of political parties, 1917

TOCHENYĬ, D. S.
Bor'ba politicheskikh partiĭ v Rossii po agrarnomu voprosu nakanune Oktīabrīa. - Kuĭbyshev : Kuĭbyshevskiĭ pedinstitut, 1981
83p
r6448891 X.329/16954
9001417

AGRICULTURE. Soviet Union
Theories of Lenin, V. I.

LENINSKIĬ Dekret o zemle v deĭstvii / [redkollegiīa : I.I. Mints (otvetstvennyĭ redaktor) ... i dr.]. - Moskva : Nauka, 1979
309p
r6174537 X.329/8952 Woolwich
1231413

TRAPEZNIKOV, Sergeĭ Pavlovich
Leninizm i agrarno-krest'īanskiĭ vopros.... - 2-e., dop. izd. - Moskva : Izdatel'stvo "Mysl'", 1976
2t. : ill
r6046547 X.329/8661 Woolwich
1231413

AGRICULTURE. Soviet Union
Vavilov, Nikolaĭ Ivanovich - Bibliographies

NIKOLAĬ Ivanovich Vavilov / otvetstvennyĭ redaktor D.D. Brezhnev. - Leningrad : Vsesoīuznyĭ NII rastenievodstva im. N.I. Vavilova, 1978
182p
r6098088 X.319/9887 Woolwich
1866842

AGRICULTURE. Soviet Union
Vavilov, Nikolaĭ Ivanovich - Biographies

KOROTKOVA, Tamara Ivanovna
N.I. Vavilov v Saratove (1917-1921). - Saratov : Privolzhskoe knizhnoe izdatel'stvo, 1978
118p,[16]p of plates : ill
r6094166 X.329/8902 Woolwich
1991426

AGRICULTURE. Soviet Union & Italy
Law - Conference proceedings

SOVETSKO-ITAL'ĪANSKIĬ "KRUGLYĬ STOL" PO VOPROSAM SEL'SKOKHOZĪAĬSTVENNOGO PRAVA (2nd : 1975 : Moscow)
Sel'skoe khozīaĭstvo i pravo v SSSR i Italii. - Moskva : Akademiīa nauk SSSR, Institut gosudarstva i prava, 1977
107p
r6186156 X.709/28932
2164949

AGRICULTURE. Spain

MAESTRE ALFONSO, Juan
Crisis y cambio en el campo español. - Madrid : Ediciones de la Torre, [1978]
127p. - (Libro de bolsillo. Vida). - Includes bibliographies
84-85277-40-6 X.319/20380 Woolwich
1004786

SANZ-PASTOR MELLADO, Fernando
Es urgente la reforma agraria. - Madrid : E.M.E.S.A., [1977]
277p : ill. - (Ensayos Aldaba)
84-265-2504-0 X.329/8950 Woolwich
1004786

SOLDEVILLA Y VILLAR, Antonio D.
Sobre la agricultura, la política agraria y el desarrollo rural. - Valladolid : el Autor, 1975
xxxi, 389 p.
84-400-8436-6 X.319/8583 Woolwich
1004786

AGRICULTURE. Spain
to ca 1975

NAREDO PÉREZ, José Manuel
La evolución de la agricultura en España. - 3.ed. - Barcelona : Editorial Laia, 1977
177p : ill. - (Colección Economía, 4. Papel ; 451)
84-7222-475-9 X.319/20090 Woolwich
1984888

AGRICULTURE. Spain
to date

MAAS, Johannes Henricus Maria
Spaanse latifundios. - Nijmegen : [J.H.M. Maas], 1981
416p : ill. - Includes bibliography
r6303447 X.329/15441 Woolwich
8418586

AGRICULTURE. Spain
711-1479

PROPIEDADES del cabildo segoviano, sistemas de cultivo y modos de explotación de la tierra a fines del siglo XIII / estudio realizado por Ángel García Sanz ... [et al.]. - Salamanca : Ediciones Universidad de Salamanca, 1981
184p. - (Acta salmanticensia. Filosofía y letras ; 118) (Documentos y estudios para la historia del occidente peninsular durante la Edad Media ; 2)
84-7481-141-4 Ac.153/3(4)[vol.118]
8726531

AGRICULTURE. Spain
ca 1200

IBN AL-'AWWĀM, Yaḥya ibn Muḥammad
[Kitāb al-filāhah. English. Selections]. A Moorish calendar. - Wantage : Black Swan, 1979, c1978
viii,54p : ill. - Includes index
0-905475-05-4 X.329/13601 Woolwich
2374137

AGRICULTURE. Spain
1814-1868

MORAL RUIZ, Joaquín del
La agricultura española a mediados del s. XIX (1850-1870). - [Madrid] : Servicio de Publicaciones Agrarias, [1979]
226p. - (Serie Estudios). - Bibliography
r6705085 X.629/18515
8853231

AGRICULTURE. Spain
1960-1970

JUANI FENOLLAR, Rafael
La formación de la Agroindustria en España. - [Madrid] : Servicio de Publicaciones Agrarias, [1978]
283p. - (Serie Estudios)
84-7479-017-4 X.329/13403
2475723

AGRICULTURE. Spain
ca 900- - Serials

ESTUDIS d'història agrària / Redacció : Centre d'Estudis Històrics Internacionals. - n.1-. - Barcelona : Curial, [1978]-
r6099298 P.311/516
2267136

AGRICULTURE. Spain
Columella, Lucius Junius Moderatus, ca 22-ca 60 - Conference proceedings

HOMENAJE NACIONAL A LUCIO JUNIO MODERATO COLUMELA (1975 : Cadiz)
Homenaje nacional a Lucio Junio Moderato Columela, Cádiz, junio, 1975. - Cádiz : A.P.A.E., 1975
144p : ill
84-500-6869-x X.700/17865
0392367

AGRICULTURE. Spain
Conflict with government, 1970-1976

CRISIS agrarias y luchas campesinas, 1970-1976 / V.L. Alonso...[et al.]. - Madrid : Editorial Ayuso, 1976
322p : ill. - (Colección Temas actuales, 4. Serie Bolsillo ; no 9)
84-336-0126-1 X.319/9390 Woolwich
1506870

AGRICULTURE. Spain
Economic aspects - Statistics

INDICADORES socio-económicos del campo español. - Madrid : Fondo para la Investigación Económica y Social de la Confederación Española de Cajas de Ahorros. - (Fondo para la Investigación Económica y Social Publicaciones ; 67)
año 1972. - 1975
2t. : ill
84-7231-209-7 X.320/5253
1238507

AGRICULTURE. Spain
Economic development, 1700-1970 - Conference proceedings

COLOQUIO DE HISTORIA ECONÓMICA DE ESPAÑA (1st : 1972 : Barcelona)
Agricultura, comercio colonial y crecimiento económico en la España contemporánea. - Esplugues de Llobregat : Editorial Ariel, 1974
372 p.. - (Ariel Historia ; no. 3)
84-344-6504-3 X.319/8393 Woolwich
0866091

AGRICULTURE. Spain
Economic development, 1939-1972 - Conference proceedings

CENTRO DE ESTUDIOS SOCIALES DEL VALLE DE LOS CAÍDOS. Mesa Redonda (34th : 1973 : Madrid)
La crisis de la agricultura tradicional en España. - Madrid : el Centro, 1974
xxx, 372 p.. - (Anales de moral social y económica ; v. 34)
84-400-7340-2 X.329/8394 Woolwich
0914290

AGRICULTURE. Spain
Effects of Spanish entry into European Economic Community - Conference proceedings

ASOCIACIÓN ESPAÑOLA DE ECONOMÍA Y SOCIOLOGÍA AGRARIAS. Reunión de Estudios (9th : 1977 : Madrid)
La incorporación de la agricultura española al Mercado Común Europeo. - Madrid : Instituto de Relaciones Agrarias, 1977
2v
84-500-2415-3 X.520/20925
2425408

AGRICULTURE. Spain
Finance, 1939-1975

TARRAFETA PUYAL, Luis
La capitalización de la agricultura española 1962-75. - Madrid : Banco de Crédito Agrícola, 1979
418p
84-85482-10-7 X.329/14432 Woolwich
872654x

AGRICULTURE. Spain
Implications of socialism

PARTIDO SOCIALISTA OBRERO ESPAÑOL. Equipo Agrario
Agricultura y socialismo. - Madrid : EDICUSA, 1977
ix,122p
84-229-6013-3 X.519/39190
1715755

AGRICULTURE. Spain
Law

BALLARÍN MARCIAL, Alberto
Estudios de derecho agrario y política agraria. - Madrid : [s.n.], 1975
1026p
84-400-8706-3 X.200/10832
0395803

SANZ JARQUE, Juan José
Derecho agrario. - Madrid : Fundacion Juan March, 1975
xxiv, 717 p. - (Coleccion Compendios)
r6000390 X.100/14658
0928852

AGRICULTURE. Spain
Law - Conference proceedings

REUNIÓN IBEROAMERICANA DE ESPECIALISTAS EN DERECHO AGRARIO
(1974 : Mérida, Venezuela)
Memoria [de la] Reunion Iberoamericana de Especialistas en Derecho Agrario. - Merida : Universidad de los Andes, 1974
350 p.
r6003384 X.100/14504
0911003

AGRICULTURE. Spain
Policies of government, 1939-1975

ORTEGA, Nicolás
Politica agraria y dominación del espacio. - Madrid : Ayuso, 1979
258p : maps. - (Ciudad y sociedad ; 5)
84-336-0155-5 X.329/13028 Woolwich
2458411

AGRICULTURE. Spain
Policies of Partido Comunista de España

ÁLVAREZ, Santiago
El Partido Comunista y el campo. - Madrid : Ediciones de la Torre, [1977]
258p. - (Libro Cícero. Política. Problemas del movimiento obrero ; n.3)
84-85277-24-4 X.708/44996
1793748

AGRICULTURE. Spain
Policies of Partido Socialista Obrero Español

BALLESTERO, Enrique
La política agraria del socialismo español. - [Madrid] : Mañana Editorial, [1979]
150p
84-7421-049-6 X.329/13393 Woolwich
2476479

AGRICULTURE. Spain
Policies of political parties

BALDRICH CABALLE, Juan
Programas agrarios de partidos politicos españoles. - Madrid : Editora Agricola Española, 1977
235p
84-400-2559-9 X.519/39771
1789945

AGRICULTURE. Spain
Role in economic development

La AGRICULTURA en el desarrollo capitalista español, 1940-1970. - Madrid : Siglo Veintiuno Editores, 1975
248p. - (Economía y demografía)
84-323-0198-1 X.319/8681 Woolwich
0390658

AGRICULTURE. Spain
Socioeconomic aspects - Serials

AGRICULTURA y sociedad / [editada por la Secretaria General Técnica del Ministerio de Agricultura]. - no 1- ; oct./dic. 1976-. - Madrid : el Ministerio, 1976-
r6047641 P.311/492
1540416

AGRICULTURE. Special subjects. Alphabets. English language
Farms

AZARIAN, Mary
A farmer's alphabet. - Boston, Mass. : Godine, 1981
61p : all ill
0-87923-394-x X.425/3311
222383X

MILLER, Jane, 1925-
Farm alphabet book. - London : Dent, 1981
[32]p : of col.ill
0-460-06979-9 X.990/17409 Woolwich
222383x

AGRICULTURE. Special subjects. Drawings
Farms. Techniques

PAINE, Brian
You can draw the farmyard. - Henley-on-Thames : Cressrelles Publishing Co. Ltd. 1975
[29]p : chiefly col ill
0-85956-013-9 X.990/7421 Woolwich
0577707

AGRICULTURE. Special subjects. Hungarian visual arts - Catalogues

MAGYAR MEZŐGAZDASÁGI MÚZEUM
Mezőgazdaság a képzőművészetben 1975. - Budapest : Magyar Mezőgazdasági Múzeum, 1975
[46]p : ill
r6000723 X.419/1894 Woolwich
0892580

AGRICULTURE. Special subjects. Vocabulary. Emilia-Romagna dialect. Italian language

MALERBA, Luigi
Le parole abbandonate. - Milano : Bompiani, 1977
xii,138p. - (Nuovi saggi italiani : 22)
r6100402 X.909/83716
2041723

AGRICULTURE. Sri Lanka
to date

The AGRARIAN situation relating to paddy cultivation in five selected districts of Sri Lanka. - Colombo : The Institute, 1974-1975
v.. - (Research study series ; no.s6-11)
wanting part 1
r6108993 X.322/10473 Woolwich
8584168

AGRICULTURE in the peasant sector of Sri Lanka / edited by S.W.R. de A. Samarasinghe. - Peradeniya : Ceylon Studies Seminar, 1977
xiv,202p
r6805695 X.322/10546 Woolwich
8584168

VERHAGEN, Koenraad
Co-operation for survival. - Amsterdam : Royal Tropical Institute, 1984
xix,253p : ill. - Includes bibliography
90-900065-6-7 X.322/14294
8584168

AGRICULTURE. Sri Lanka
Economic development

FARRINGTON, John
Small farm capital in Sri Lanka. - Reading : Department of Agricultural Economics and Management, University of Reading, 1983
viii,117p. - (Development study / Department of Agricultural Economics and Management, University of Reading ; no.25). - Bibliography: p113-117
0-7049-0808-5 X.322/13776
3147479

AGRICULTURE. St Agnes. Cornwall
Farms: Jennings Farm, 1976-1980

INGLIS, Nigel
Nigel Inglis. - [Penzance] : Newlyn Orion, Newlyn Art Gallery, [1980?]
[4]leaves : ill
B82-05836 X.425/2412 Woolwich
2602709

AGRICULTURE. Sudan
to date - Conference proceedings

PROBLEMS of agricultural development in the Sudan / Günter Heinritz (ed.), M. E. Abu Sin ... [et al.]. - Göttingen : Herodot, 1982
123p. - (Forum ; 2)
3-88694-009-8 X.322/12143
8963592

AGRICULTURE. Sukumaland. Tanzania

LARSEN, Arne
Agriculture in Sukumaland, Tanzania. - [Copenhagen] : Kgl. Veterinær- og Landbohøjskole, Økonomisk Institut, 1974
93 leaves. - (Landbrugsøkonomiske studier ; nr. 5). - Bibliography : p.87-89
r6001366 S.B.111/17(5)
0945897

AGRICULTURE. Surrey
Grover, Frederick, to ca 1895

STURT, George
The Bettesworth book. - Firle : Caliban Books, 1978
xv,325p
0-904573-09-5 X.808/33245
176022x

AGRICULTURE. Surrey
Grover, Frederick, 1896-1905

BOURNE, George, 1863-1927
Memoirs of a Surrey labourer. - London : Breslich & Foss, 1983
318p. - (Country classics)
1-85004-006-0 X.950/35758
1742698

AGRICULTURE. Surrey
Grover, Frederick, 1896-1905 - Biographies

STURT, George
A memoir of a Surrey labourer. - Firle : Caliban Books, 1978
vii,318p
0-904573-10-9 X.808/33239
1742698

AGRICULTURE. Surrey
Grover, Lucy, ca 1890-ca 1905 - Biographies

STURT, George
Lucy Bettesworth. - Firle : Caliban Books, 1978
[7],280p
0-904573-11-7 X.808/33246
1760211

AGRICULTURE. Sverdlovsk (oblast'). Russia (RSFSR)

DUDKIN, Nikolaĭ Matveevich
Rubezhi Ural'skogo sela. - Sverdlovsk : Sredne-Ural'skoe knizhnoe izdatel'stvo, 1976
97 p., 16 p. of plates : ill
r6020567 X.329/8546 Woolwich
0898015

AGRICULTURE. Swaziland
Economic development. Financial assistance by Commonwealth Development Corporation, 1950-1974

COMMONWEALTH DEVELOPMENT CORPORATION
CDC in the Swaziland Lowveld. - London : C.D.C., [1975]
[2],16p : col ill
0-903799-05-7 X.311/8107 Woolwich
0595659

AGRICULTURE. Swaziland
Smallholdings. Productivity

LOW, Allan
Farm-household theory and rural development in Swaziland. - [Reading] : [University of Reading], [1982]
103p : ill. - (Development study / University of Reading ; no.23). - Bibliography: p93-103
0-7049-0802-6 X.525/6557
2823217

AGRICULTURE. Sweden

SVERIGES bönder / [utgiven av Lantbrukarnas riksförbund, Informationssektionen]. - [Stockholm : Lantbrukarnas riksförbund, Informationssektionen, 1976]
96p : ill
91-360-0870-2 X.319/8704 Woolwich
0396818

AGRICULTURE. Sweden
to date

BRUKET av vätängarna vid Nötmyran, Västerfärnebo socken / [fältarbete, sammanställning: Lena A-son Palmqvist]. - Stockholm : Nordiska museet, 1980
60p : ill. - (Rapport / Kulturhistoriska undersökningen vid Nordiska museet ; 6)
91-7108-188-7 P.443/219[no.6]
850413x

AGRICULTURE. Sweden
1800-1850

STATTIN, Jan
Hushållningssällskapen och agrarsamhällets förändring. - Uppsala ; Stockholm : Distributor : Almqvist & Wiksell International, 1980
244p. - (Studia historica upsaliensia ; 113). - Bibl
91-554-1006-5 Ac.1077.ng[vol.113.]
2427001

AGRICULTURE. Sweden
1939-1945

EKHOLM, Gunnar
Det svenska jordbruket och folkförsörjningen under andra världskriget. - Lund : [Ekonomisk-historiska föreningen], 1976
192p. - (Skrifter utgivna av Ekonomisk-historiska föreningen i Lund ; v.18). - Includes bibliographies
r6033936 X.313/813 Woolwich
1421700

AGRICULTURE. Sweden
Bonded employment, ca 1900-1945

JERSTEDT, Sven
Pa statarnas tid. - [Stockholm] : Bonniers, 1975
142 p. : ill
91-00-40291-5 X.809/22424
0915351

AGRICULTURE. Sweden
Capital formation, to date

MAGNUSSON, Lars
Kapitalbildning i Sverige 1750–1860. – Uppsala : [Uppsala universitet], 1983
95p. – (Uppsala studies in economic history ; 25)
91–554–1443–5 X.0510/38(25)
9047492

AGRICULTURE. Sweden
Cooperation, to 1979

JOHANSSON, Tore
Samverkan gav styrkan. – Stockholm : LT, 1979
258p : ill,facsims
91–360–1344–7 X.322/9158 Woolwich
244867x

AGRICULTURE. Sweden
Geographical aspects

SÖDERMAN, Inga
Sveriges jordbruk i ekonomisk-geografisk belysning. – Stockholm : Esselte Studium, 1976
108p : ill. – (Scandinavian university books)
91–242–6041–x X.519/22459
1125338

AGRICULTURE. Sweden
Part-time agriculture

BOLIN, Olof
Deltidslantbruket. – Uppsala : Lantbrukshögskolan, 1977
29p : ill. – (Aktuellt från Lantbrukshögskolan ; nr 246) (Ekonomi ; 25)
91–7088–713–6 X.319/9373 Woolwich
1505955

AGRICULTURE. Sweden
Personnel. Health, 1945– – Serials

JORDBRUKETS socialmedicin / [projektledning : Institutionen för socialmedicin, Lunds universitet]. – Nr.1–. – Lund : Institutionen för socialmedicin, 1972–
Holdings: Nr.14 (1975)–
r6039645 P.313/235
8729255

AGRICULTURE. Sweden
Policies of government

GRANATH, Olle
Swedish agricultural policy. – Stockholm : Swedish Institute, 1978
5p. – (Current Sweden ; no. 181)
r6029199 X.0512/423(181)
1858920

AGRICULTURE. Sweden
Policies of government, 1945–

SWEDBORG, Erik
Lantbrukspolitik för 80-talet. – Stockholm : LT, 1980
229p
91–360–1544–x X.329/16415
8432856

AGRICULTURE. Sweden
Policies of government, 1950–1975

TIDERIUS, Gert
Effekten av den statliga rationaliseringspolitiken. – Lund : Distribution : Studentlitteratur, 1977
158p. – (Skrifter utgivna av Ekonomisk-historiska föreningen i Lund ; v.22)
91–441–4381–8 X.319/9346 Woolwich
1476009

AGRICULTURE. Sweden
related to growth in population, to date

YDBORN, Lars
Befolkningstillväxt och jordbruk. – Göteborg : Göteborgs universitet, Historiska institutionen, 1984
xi,215p. – (Meddelanden fran Historiska institutionen i Göteborg ; nr 25). – Includes bibliography
91–7222–705–2 X.0709/587(25)
9151141

AGRICULTURE. Sweden
Research, to date

FRAMTIDSPERSPEKTIV inom svensk skogs- och jordbruksforskning. – [Stockholm] : Liber förlag, 1981
192p. – (Ds Jo ; 1980:9)
91–380–5996–7 X.629/18788
888532x

AGRICULTURE. Syria
1945–

FILONIK, Aleksandr Oskarovich
Agrarnye problemy sovremennoĭ Sirii. – Moskva : Nauka, Glavnaia redaktsiia vostochnoĭ literatury, 1981
193 p.. – Bibliography: p. 186–[190]
r6482938 X.329/18540
9117288

AGRICULTURE. Tanzania
Economic development

GLAESER, Bernhard
Ecodevelopment in Tanzania. – Berlin : Mouton, c1984
xix,230p. – Bibliography: p227–230
3–11–009719–2 X.520/33535
1072838

RURAL cooperation in Tanzania / edited by Lionel Cliffe...[et al.] ; [for the University of Dar es Salaam, Rural Development Research Committee]. – Dar es Salaam : Tanzania Publishing House, 1975
xl, 554 p.
r6031388 X.709/31098
1072838

AGRICULTURE. Tanzania
Künzler, August – Biographies

HÄSLER, Alfred A.
Der Weizenkönig von Tanganjika. – Frauenfeld : Huber, [1980]
192p.,[8]leaves of plates
3–7193–0690–9 X.529/42721
8729492

AGRICULTURE. Tanzania
Policies of government

ELLIS, Frank
Agricultural pricing policy in Tanzania, 1973–79. – Norwich : School of Development Studies, University of Norwich, 1983
55p. – (Discussion paper ; no.84). – Bibliography: p53–55
B84–39707 X.322/14036
2057263

AGRICULTURE. Tatar ASSR. Russia (RSFSR)

MURTAZIN, Il'ias Safinovich
Spetsializatsiia i kontsentratsiia sel'skogo khoziaistva Tatarskoĭ ASSR. – Kazan' : Tatarskoe knizhnoe izdatel'stvo, 1976
175p : ill
r6037525 X.519/23051
1160400

AGRICULTURE. Tázlár. Hungary
Collectivisation. Effects on social conditions, to 1977

HANN, C. M.
Tázlár. – Cambridge : Cambridge University Press, 1980
xii,206p : ill. – (Changing cultures). – Bibliography: p.197–203. – Includes index
0–521–22591–4 X.809/45939
2157985

AGRICULTURE. Temperate regions
Environmental aspects

BRIGGS, David J. (David John)
Agriculture and environment. – London : Longman, 1985
xiv,442p : ill. – Bibliography: p395–434. – Includes index
0–582–30000–2 X.322/15218
3054187

AGRICULTURE. Teso. Uganda
Change

UCHENDU, Victor C.
Agricultural change in Kisii district, Kenya. – Nairobi : East African Literature Bureau, 1975
ix,102p. – Includes index
r6072199 X.319/18354 Woolwich
1580787

AGRICULTURE. Texas
Farms, to date

CAFFEY, David L.
The old home place. – Burnet, Tex : Eakin Press, [1981]
x,213p : ill
0–89015–283–7 X.329/19466
9162356

AGRICULTURE. Thailand
to date

DURRENBERGER, E. Paul
Agricultural production and household budgets in a Shan peasant village in Northwestern Thailand. – Athens, Ohio : Ohio University Center for International Studies, c1978
xii,142p. – (Papers in international studies. Southeast Asia series ; no.49)
0–89680–071–7 X.322/11353
8876894

O'REILLY, F. D.
Thailand's agriculture. – Budapest : Akadémiai Kiadó, 1983
97p : ill. – (Geography of world agriculture ; 12)
963–05–3360–x X.0322/92(12)
8876894

VERHAGEN, Koenraad
Co-operation for survival. – Amsterdam : Royal Tropical Institute, 1984
xix,253p : ill. – Includes bibliography
90–900065–6–7 X.322/14294
8876894

AGRICULTURE. Transcaucasia. Soviet Union
Collectivisation

LOMASHVILI, P. N
Velikiĭ perevorot. – Tbilisi : Sabchota Sakartvelo, 1972
457p
r6178533 X.329/12738 Woolwich
2441349

AGRICULTURE. Transylvania. Romania
Kozán, Imre – Biographies

KOZÁN, Imre
Fekete ugar. – Bukarest : Kriterion, 1978
148 p.
r6403371 X.329/17935
9082840

AGRICULTURE. Trawsgoed. Dyfed
Experimental farms: Trawsgoed Experimental Husbandry Farm

TRAWSGOED Experimental Husbandry Farm, annual review / [ADAS, Ministry of Agriculture, Fisheries and Food]. – 1977. – Trawsgoed : Trawsgoed E.H.F, [1978]
[2],i,57p : ill
Subsequent volumes, when received, will be found at SRL
B79–01240 SRL
1751352

AGRICULTURE. Trinidad
Smallholdings

FLOYD, B. N.
Small-scale agriculture in Trinidad. – [Durham] : Department of Geography, University of Durham, 1977
vi,69p : ill. – (Occasional publications : new series / University of Durham. Department of Geography ; no.10). – Bibliography: p.62–69
B77–22434 X.0700/584(10)
1164627

AGRICULTURE. Tropical regions

ABBOTT, John Cave
Agricultural economics and marketing in the tropics. – London : Longman, 1979
vii,168p : ill. – (Intermediate tropical agriculture series). – Includes bibliographies and index
0–582–60304–8 X.519/41534
0075248

ANDREAE, Bernd
The economics of tropical agriculture. – English ed.. – Farnham Royal : Commonwealth Agricultural Bureaux, c1980
178p : ill. – Includes bibliographies and index
0–85198–453–3 SRL
0075248

MANSHARD, Walther
Tropical agriculture. – London : Longman, 1979
x,226p : ill. – (Geographies for advanced study). – Bibliography: p.201–222. – Includes index
0–582–48989–x WPB.375/58
0075248

RUTHENBERG, Hans
Farming systems in the tropics. – 2nd ed. – Oxford : Clarendon Press, 1976
xviii,366p,[20]p of plates : ill(some col). – Bibliography: p.336–356. – Includes index
0–19–859458–5 SRL
1122444
Farming systems in the tropics. – 3rd ed. – Oxford : Clarendon Press, 1980
xxii,424p,[20]p of plates : ill(some col). – Bibliography: p.397–414. – Includes index
0–19–859481–x SRL
1122444

WEBSTER, Cyril Charles
Agriculture in the tropics. – 2nd ed. – London : Longman, 1980
[8],640p : ill. – (Tropical agriculture series). – Bibliography : p.582–622. – Includes index
0–582–46814–0 SRL
1122444

WRIGLEY, Gordon
Tropical agriculture. – 4th ed. – London : Longman, 1982, c1981
xv,496p : ill. – Bibliography: p460–480. – Includes index
0–582–46037–9 SRL
1122444

AGRICULTURE. Tropical regions
– For West African students

AIYEGBAYO, J. T.
An introduction to tropical agricultural science. – Ibadan ; London : Evans Bros, 1980
iv,107p : ill
0–237–50390–5 X.322/9283
2151898

AGRICULTURE. Tropical regions
- For West African students-continued

AKINSANMI, Oluyemi
Certificate agricultural science. - London : Longman, 1975
ix,196p : ill. - Includes index
0-582-60884-8 X.311/8075
0541702

JOY, Derek
Tropical agriculture handbook. - London : Cassell, 1979
219p : ill. - (Cassell's tropical agriculture series). - Includes index
0-304-30016-0 X.311/9067 Woolwich
1992686

OKORIE, J. U.
Junior tropical agriculture for West Africa. - London Macmillan.
[Bk.1]. - 1981
108p : ill(some col.)
0-333-31873-0 X.322/14097
0541702

UGURU, O. O.
An introduction to agricultural science for tropical areas. - Walton-on-Thames : Nelson, 1981
384p : ill. - Bibliography: p378-379. - Includes index
0-17-511085-9 X.322/10424 Woolwich
0541702

AGRICULTURE. Tropical regions
- Programmed texts - For cooperative officers

YEO, Peter
An initial course in tropical agriculture for the staff of co-operatives. - London : Intermediate Technology Development Group [for] the Co-operative Educational Materials Advisory Service of the International Co-operative Alliance, 1976
[5],54p
0-903031-39-6 X.319/17048 Woolwich
0790680

AGRICULTURE. Tropical regions
to date

WITZENHAUSEN - 85 Jahre im Dienste der Agrarentwicklung in den Tropen und Subtropen / zusammengestellt von Peter Wolff ; Herausgeber: Gesamthochschule Kassel, Fachbereich Internationale Agrarwirtschaft in Witzenhausen ... [et al.]. - Witzenhausen : Selbstverlag des Verbandes der Tropenlandwirte aus Witzenhausen, 1983
253p : ill. - (Der Tropenlandwirt. Beiheft ; Nr.18)
r6308724 X.329/17844
9082816

AGRICULTURE. Tropical regions
Environmental aspects

GOODLAND, Robert J. A. (Robert James A)
Environmental management in tropical agriculture. - Epping : Bowker, 1984
x,237p : 1ill. - (Bowker (UK) replica edition). - Includes bibliographies and index
0-86531-715-1 X.329/18458
3175952

AGRICULTURE. Tropical regions
Financial management

COY, David V.
Accounting and finance for managers in tropical agriculture. - London : Longman, 1982
xii,118p : ill. - (Intermediate tropical agriculture series). - Bibliography: p94. - Includes index
0-582-77502-7 X.329/16044
2836947

AGRICULTURE. Tropical regions
Management

AKUBUILO, C. J. C.
Agricultural economics and farm management. - London : Macmillan, 1977
[5],59p : ill. - (Tropical agricultural handbooks). - Bibliography(1p.)
0-333-21710-1 X.319/18221 Woolwich
152030x

JOHNSON, David T. (David Terrence)
The business of farming. - London : Macmillan, 1982
viii,362p : ill. - (Macmillan tropical agriculture, horticulture and applied ecology series). - Includes bibliographies and index
0-333-25392-2 X.520/28689
152030x

AGRICULTURE. Tropical regions
Multiple cropping systems

BEETS, Willem C.
Multiple cropping and tropical farming systems. - Aldershot : Gower, c1982
xiv,156p : ill. - Bibliography: p146-152. - Includes index
0-566-00567-0 SRL
2877643

AGRICULTURE. Tropical regions. Africa

LIVINGSTONE, I. (Ian)
Agricultural economics for Tropical Africa. - London : Heinemann, 1981
x,294p : ill. - (Studies in the economics of Africa). - Includes index
0-435-97431-9 X.520/24051
2140721

AGRICULTURE. Tropical regions. Africa
Development

RURAL development in tropical Africa / edited by Judith Heyer, Pepe Roberts and Gavin Williams. - London : Macmillan, 1981
x,375p : ill. - Includes bibliographies and index
0-333-28448-8 X.809/49468
2566494

AGRICULTURE. Tropical regions. Africa
Development. Socioeconomic aspects

AGRICULTURAL change in tropical Africa / Kenneth R.M. Anthony ... [et al.]. - Ithaca ; London : Cornell University Press, 1979
327p : ill
0-8014-1159-9 X.320/11790 Woolwich
1914375

AGRICULTURE. Tropical regions. Africa
Policies of government

BATES, Robert H. (Robert Hinrichs)
Markets and states in tropical Africa. - Berkeley ; London : University of California Press, c1981
xi,178p : ill. - (California series on social choice and political economy). - Bibliography: p147-166. - Includes index
0-520-04253-0 X.329/14836 Woolwich
2607646

The POLITICS of agriculture in tropical Africa / Jonathan Barker editor. - Beverly Hills ; London : Sage, c1984
320p. - (Sage series on African modernization and development ; v.9). - Includes bibliographies
0-8039-2295-7 X.329/19316
2607646

AGRICULTURE. Tropical regions. Developing countries
British research organisations: Tropical Development and Research Institute - Serials

[NEWSLETTER (Tropical Development and Research Institute)].
Newsletter / Tropical Development and Research Institute.
- 1 (Sept. 1983)-. - London : The Institute, 1983-
v. : ill
B84-28594 P.313/615
3142396

AGRICULTURE. Tropical regions. Developing countries
British research organisations: Tropical Products Institute - Serials

POST-HARVEST leaflet / Tropical Products Institute. - 1- ; June 1976-. - London : The Institute, [1976]- : ill
B78-36175 BS.38x/18
0680478

AGRICULTURE. Tunisia
1956-1971

ZUDINA, Larisa Petrovna
Agrarnye otnosheniia v Tunise (1956-1971). - Moskva : Izdatel'stvo "Nauka", Glavnaia redaktsiia vostochnoĭ literatury, 1976
180 p.
r6023128 X.319/8586 Woolwich
1006754

AGRICULTURE. Tunisia
1956- - Serials

ANNUAIRE agricole de la Tunisie. - (1978-1979)-. - Tunis : AGEP, 1978- : ill
r6205697 S.W.392.b/4
8444390

AGRICULTURE. Turkey
- For British exporters

A REPORT on the agricultural and agro-industrial sectors in Turkey. - London : British Agricultural Export Council, c1981
iv,67p : ill. - Bibliography: p65
B83-17131 SRL
2877260

AGRICULTURE. Turkey
1923-1974

ARESVIK, Oddvar
The agricultural development of Turkey. - New York ; London : Praeger ; [London] : [Distributed by Pall Mall Press], 1975
xxiv,221p : 2 ill. - (Praeger special studies in international economics and development). - Bibliography: p.213-220
0-275-28851-x SRL
0566322

AGRICULTURE. Turkey
Law, 1945-

AKSOY, Suat
Das türkische Agrarrecht im Vergleich mit der Agrargesetzgebung einiger europäischer Länder. - Köln : Heymann, 1980
l,603p.. - (Schriftenreihe des Instituts für Landwirtschaftsrecht der Universität Göttingen ; Bd.23). - Bibliography: p.xxiii-xlvii
Pages xxxvii-xxxviii and 573-574 reversed
3-452-18820-5 X.200/37122
8756953

AGRICULTURE. Turkey
Sociopolitical aspects

MARGULIES, Ronnie
The politics of the agrarian question in Turkey. - Norwich : School of Development Studies, University of East Anglia, 1983
46p. - (Development studies occasional paper ; no. 23). - Bibliography: p46
B84-39961 X.322/14072
3184234

AGRICULTURE. Turkmenistan
to 1924

ANNANEPESOV, Murad
Khoziaistvo turkmen v XVIII-XIX vv. .... - Ashkhabad : Ylym, 1972
282p. - Bibl.: p.269-276
r6178442 X.529/36989
2483831

AGRICULTURE. Turkmenistan
Collectivisation

KAKABAEV, Sapar Durdy
Kollektivizasiia sel'skogo khoziaistva Turkmenskoĭ SSSR (1927-1937gg.). - Ashkhabad : Ylym, 1978
203p
r6174574 X.319/20145 Woolwich
2034727

AGRICULTURE. Turriff. Grampian Region. Scotland
Organisations: Turriff and District Junior Agricultural Club, to 1982

TURRIFF AND DISTRICT JUNIOR AGRICULTURAL CLUB
Turriff and District Junior Agricultural club 50th anniversary 1932-1982. - [Turriff] : [Turriff and District Junior Agricultural Club], [1982?]
71p : ill
r6816129 X.622/19152
306946x

AGRICULTURE. Tuscany. Italy

TORNAR, Federico
Toscana agricola. - Pisa : Pacini, 1976
228p.,[10]leaves of plates : ill
r6068532 X.313/935 Woolwich
2030918

AGRICULTURE. Tuscany. Italy
to date

PAZZAGLI, Carlo
Per la storia dell'agricoltura toscana nei secoli XIX e XX. - Torino : Einaudi, 1979
146p. - (Studi / Fondazione Luigi Einaudi ; 25)
r6506184 X.322/11560
8270236

PESTELLINI, Tito
La Mezzeria e le sue consuetudini nelle province di Siena, Firenze e Pisa. - Firenze : Accademia economico-agraria dei georgofili, 1980
80,[64]p
r6501949 P.311/520
8270236

AGRICULTURE. Tuscany. Italy
to date - Conference proceedings

CONFERENZA REGIONALE DELL'AGRICOLTURA (2nd : 1977 : Florence)
Atti della II conferenza regionale dell'agricoltura, Firenze ... 1977 [convocata dalla] Regione Toscana, Giunta regionale. - Firenze : Olschki, 1978
2v(785p)
r6500384 X.322/10025 Woolwich
8270236

AGRICULTURE. Tuva ASSR. Russia (RSFSR)
1945-

KONGAR, N. M. (Nikolaĭ Monzuelovich)
Aktual'nye problemy razvitiia sel'skogo khoziaistva v Tuve. - Kyzyl : Tuvinskoe knizhnoe izdatel'stvo, 1974
109p
r6462770 X.329/17364
9047468

AGRICULTURE. Tveit. Norway
Farms, to ca 1970

TVEITE, Johan
Tveits historie. - Kristiansand : Kristiansand kommune, 1973-
v.1-
r6033970 X.0322/94 Woolwich
133140X

AGRICULTURE. Udmurt ASSR. Russia (RSFSR)
to 1917

AGRARNYE otnosheni͡i͡a v Udmurtii vo vtoroi polovine XIX-nachale XX VV. / [redakt͡sionnai͡a kollegi͡i͡a M.M. Martynova (otvetstvennyi redaktor) ... i dr.]. - Izhevsk : Nauchno-issledovatel'skiĭ institut pri Sovete Ministrov Udmurtskoi ASSR, 1981
157p
r6464731 X.329/17678
9076123

AGRICULTURE. Udmurt ASSR. Russia (RSFSR)
to date

OTRASLEVAI͡A ėkonomika narodnogo khozi͡aĭstva Udmurtskoĭ ASSR / [otvetstvennyĭ redaktor S.A. Efremov]. - Izhevsk : Udmurtskiĭ nauchno-issledovatel'skiĭ institut istorii, ėkonomiki, literatury i i͡azyka, 1974
128p
r6465045 X.329/18630
9128026

AGRICULTURE. Ukraine

PUTI povyshenii͡a ėffektivnosti sel'skokhozi͡aĭstvennogo proizvodstva / [nauchnye redaktory L.A. Shepot'ko, P.F. Vedenichev]. - Kiev : Institut ėkonomiki AN USSR, 1975
150p
r6036102 X.519/22805
1150928

RAZVITIE proizvodstvennykh otnosheniĭ v sel'skom khozi͡aĭstve / [otvetstvennyĭ redaktor L.A. Shepot'ko]. - Kiev : Naukova dumka, 1977
282p
r6055677 X.329/8756 Woolwich
1150928

TUPIKOV, Ivan Grigor'evich
Partiĭnye organizat͡sii Ukrainy v bor'be za povyshenie ėffektivnosti kolkhoznogo proizvodstva (1965-1975gg.). - Kharkov : Vishcha shkola, Izdatel'stvo pri Kharkovskom gosudarstvennom universitete, 1976
150p. - Bibliography : p.141-149
r6095168 X.319/4988 Woolwich
1150928

AGRICULTURE. Ukraine
1917-1945

SHLIKHTER, O. H. (Oleksandr Hryhorovych)
Ėkonomicheskie problemy stroitel'stva fundamenta sot͡sializma. - Moskva : "Nauka,", 1982
431 p.. - Includes bibliographical references
r6403606 X.329/17890
9082883

AGRICULTURE. Ukraine
1917-

AKTUAL'NYE problemy agrarnoï istorii Ukrainy / [redakt͡sionnai͡a kollegii͡a : D.P. Poïda (otvetstvennyĭ redaktor) ... i dr.]. - Dnepropetrovsk : DGU, 1980
184p
r6402202 X.329/14350 Woolwich
8445958

AGRICULTURE. Ukraine
1945-

PANCHENKO, P. P. (Petr Panteleĭmonovich)
Razvitie sel'skogo khozi͡aĭstva Ukrainskoĭ SSR, 1959-1980. - Kiev : Naukova dumka, 1980
285p
r6402174 X.329/14258 Woolwich
8715157

VOPROSY razviti͡i͡a sel'skokhozi͡aĭs tvennogo proizvodstva / [redakt͡sionnai͡a kollegi͡i͡a L.A. Shepot'ko, P.F. Vedenichev, A.V. Zharinov]. - Kiev : Institut ėkonomiki AN USSR, 1979
134 p. - Includes bibliographical references
r6465872 X.329/19088
8715157

AGRICULTURE. Ukraine
Collectivisation

CHMYGA, Anna Fedorovna
Kolkhoznoe dvizhenie na Ukraine (1917-1929g.). - Moskva : Izdatel'stvo Moskovskogo universiteta, 1974
319p : ill
r6175882 X.809/28171
1330381

LI͡AKH, R. D. (Roman Danylovych)
Rozv'i͡azanni͡a ahrarnoho pytanni͡a na Ukraïni (1917-1923 rr.). - Kyïv : Vyshcha shkola, 1975
116p. - Bibliography
r6179190 X.329/13445 Woolwich
1330381

PERSHYN, Pavlo Mykolaïovych
Narysy ahrarnykh problem budivnytstva sotsializmu. - Kyïv : Vydavnytstvo "Naukova dumka", 1973
338p[1] leaf of plates
r6055623 X.329/8723 Woolwich
1330381

Z istoriï kolektyvizatsiï sil's'koho hospodarstva zakhidnykh oblasteĭ Ukraïns'koï RSR / [redaktsiina kolehii͡a ... M.K. Ivasi͡uta (vidpovidal'nyĭ redaktor) ... ta in. ; upori͡adnyky Kh. I. Horozhankina ... ta in.]. - Kyïv : Naukova dumka, 1976
509p
r6179020 X.329/13348 Woolwich
1330381

AGRICULTURE. Ukraine
Collectivisation, 1917-

Z istoriï kolektyvizatsiï sil's'koho hospodarstva zakhidnykh oblasteĭ Ukraïns'koï RSR / [redaktsiina kolehii͡a: P.S. Honcharuk ... ta in.]. - Kyïv : Naukova dumka, 1976
509p
r6463980 X.529/57759
8445966

AGRICULTURE. Ukraine
Manpower, 1945-

RADCHENKO, A. I͡A. Andriĭ I͡Akymovych
Problemy znyzhenni͡a sukupnykh zatrat pratsi u sil's'komu hospodarstvi. - Kyïv : Naukova dumka, 1974
189p
r6485679 X.329/15078 Woolwich
8805822

AGRICULTURE. Under River. Kent
Smallholdings

JACKSON, John, b.1929
A bucket of nuts and a herring net. - London : Collins : Harvill Press, 1979
191p : ill
0-00-262045-6 X.329/12058 Woolwich
1996630

AGRICULTURE. United States

The COMPLEAT farmer / compiled by the Main Street Press. - London : J. Murray, 1976
173p : ill. - Includes index
0-7195-3412-7 Cup.1285/99
0709352

CRAMER, Gail L.
Agricultural economics and agribusiness. - New York ; Chichester : Wiley, 1979
xiv,440p : ill. - Bibliography: p.424-425. - Includes index
0-471-04429-6 X.620/18517 Woolwich
1896245
Agricultural economics and agribusiness. - 2nd ed. - New York ; Chichester : Wiley, c1982
xiii,465p : ill. - Bibliography: p448-449. - Includes index
0-471-09393-9 X.322/10789 Woolwich
1896245

DONAHUE, Roy L. (Roy Luther)
Exploring agriculture. - 6th ed.. - Englewood Cliffs ; London : Prentice-Hall, c1984
431p : ill. - (Prentice-Hall agriculture series). - Bibliography: p403-407. - Includes index
0-13-295899-6 X.322/13902
1896245

EPP, Donald J.
Introduction to agricultural economics. - New York : Macmillan ; London : Collier Macmillan, c1981
xi,354p : ill. - Includes index
0-02-333940-3 X.322/10211 Woolwich
1896245
- [Another copy : paperback.]
ISBN 0-02-977560-4 X.520/28181

GREGOR, Howard F. (Howard Frank)
Industrialization of U.S. agriculture. - Epping : Bowker, [1982]
xii,259p. - (Westview special studies in agriculture/aquaculture science and policy). - Includes index
0-86531-236-2 SRL
1896245

JACOBS, Herbert
A practical guide for the beginning farmer. - [1st ed. reprinted]. - New York : Dover Publications [etc.] ; London : Constable, 1978
ix,237p. - Bibliography: p.vii-ix. - Includes index
0-486-23675-7 X.319/18671 Woolwich
1810634

LEE, Jasper S.
Working in agricultural industry. - New York : London : McGraw-Hill, 1978
viii,152p : ill(some col). - (Career preparation for agriculture/agribusiness). - Includes index
0-07-000831-0 X.311/8852 Woolwich
1567594

RAWLINS, N. Omri
Introduction to agribusiness. - Englewood Cliffs . London : Prentice-Hall, 1980
viii,248p : ill. - Bibliography - Includes index
0-13-477703-4 X.322/9123 Woolwich
1896245

RUSSELL, Howard S.
A long, deep furrow. - Hanover, N.H : University Press of New England, 1976
xvi,672p : ill. - Bibliography : p581-620. - Includes index
0-87451-093-7 X.322/9046 Woolwich
1896245

SHOVER, John L.
First majority, last minority. - De Kalb : Northern Illinois University Press, 1976
338p. - (Minorities in American history)
0-87580-056-4 X.809/45224
129931X

VIVIAN, John
The manual of practical homesteading. - Emmaus, Pa : Rodale Press, 1976
ix,340p : ill. - Includes index
0-87857-092-6 X.329/11128 Woolwich
1246798
- [Another copy : Paperback]
ISBN 0-87857-154-X X.319/17556 Woolwich

VOGELER, Ingolf
The myth of the family farm. - Epping : Bowker, 1982, c1981
xix,352p : ill. - (A Westview special study). - Includes index
0-89158-910-4 X.322/12348
1896245

AGRICULTURE. United States
to 1975 - Encyclopaedias

SCHAPSMEIER, Edward L.
Encyclopedia of American agricultural history. - Westport, Conn. ; London : Greenwood Press, 1975
xii,467p. - Includes bibliographies and index
0-8371-7958-0 X.320/10746 Woolwich
1120522

AGRICULTURE. United States
to 1978

EBELING, Walter
The fruited plain. - Berkeley ; London : University of California Press, 1979
xiii,433p : ill. - Bibliography: p.395-413. - Includes index
0-520-03751-0 X.322/9435 Woolwich
2126982

AGRICULTURE. United States
to date

COCHRANE, Willard W. (Willard Wesley)
The development of American agriculture. - Minneapolis : University of Minnesota Press, c1979
xii,464p
0-8166-0926-8 X.322/12212
8447225

DANBOM, David B
The resisted revolution. - Ames : The Iowa State University Press, 1979
ix,195p. - Includes bibliography
0-8138-0945-2 X.520/24640
8447225

HARRIS, Lement
Harold M. Ware (1890-1935). - New York : The American Institute for Marxist Studies, 1978
72p. - (Occasional paper ; no.30 (1978))
r6800028 X.322/10086 Woolwich
8447225

PRIMACK, Martin Leonard
Farm formed capital in American agriculture 1850-1910. - New York : Arno Press, 1977
274p. - (Dissertations in American economic history). - Includes bibliography
0-405-09920-7 X.520/32148
8447225

TWEETEN, Luther
Foundations of farm policy. - 2nd ed., revised. - Lincoln, Neb. ; London : University of Nebraska Press, 1978
xi,567
0-8032-0972-x X.322/10532 Woolwich
8447225

AGRICULTURE. United States
to date-continued

WARLEY, T. K.
Agriculture in an interdependent world. - Washington : Canadian-American Committee, 1977
x,93p
0-89068-040-x X.329/15749
8447225

AGRICULTURE. United States
to date - Texts

TAYLOR, John, b.1753
Arator. - Indianapolis : Liberty Classics, [1977]
394p. : port
0-913966-26-6 X.950/23975
8447225

AGRICULTURE. United States
ca 1620 to 1975. Documents on agricultural industries of United States, ca 1620 to 1975. Documents with American imprints - Conference proceedings

AGRICULTURAL literature / edited by Alan Fusonie, Leila Moran. - Washington, D.C. : Associates of the National Agricultural Library, Inc. [etc.], 1977
[5],371p : ill
0-87771-010-4 X.620/18502 Woolwich
1893750

AGRICULTURE. United States
ca 1860-1973

The AMERICAN farm / [compiled] by Maisie Conrat and Richard Conrat. - London : Scolar Press, 1977
3-256p : chiefly ill
0-85967-404-5 Cup.1285/254
1289306

AGRICULTURE. United States
1933-

ANDREEVA, N. M. (Nina Mikhaĭlovna)
SShA. - Moskva : Nauka, 1974
146p,[2]folded leaves of plates
r6462773 X.329/17363
9047476

SALOUTOS, Theodore
The American farmer and the New Deal. - Ames : Iowa State University Press, 1982
xviii,327p
0-8138-1076-0 X.800/39847
9047476

AGRICULTURE. United States
Agricultural industries & trades

GOLDBERG, Ray Allan
Agribusiness coordination. - Boston [Mass.] : Harvard University, Graduate School of Business Administration, Division of Research, 1968
xix, 256 p. : ill. - Bibliography : p. 243-252. - Includes index
r6009871 X.322/8250 Woolwich
0730378

AGRICULTURE. United States
Agricultural industries & trades. Advertising

MCGUIRE, James E.
Advertising and display in agribusiness. - New York ; London : McGraw-Hill, 1979
vi,138p : ill. - (Career preparation for agriculture/agribusiness). - Includes index
0-07-045129-x X.510/16346
1946242

AGRICULTURE. United States
Alternative energy sources

RITCHIE, James D.
Sourcebook for farm energy alternatives. - New York ; London : McGraw-Hill, c1983
1v.(various pagings) : ill. - Includes index
0-07-052951-5 SRL
2927004

AGRICULTURE. United States
Conservation of energy resources - Conference proceedings

CORNELL AGRICULTURAL WASTE MANAGEMENT CONFERENCE (1975 : 7th)
Energy, agriculture and waste management. - Ann Arbor : Ann Arbor Science, 1975
ix, 540 p. : ill. - Includes index
0-250-40113-4 X.320/10447 Woolwich
069939X

AGRICULTURE. United States
Consumption of energy - Conference proceedings

AGRICULTURE and energy / edited by William Lockeretz. - New York ; London : Academic Press, 1977
xv,750p : ill. - Includes bibliographies and index
0-12-454250-6 SRL
1311492

AGRICULTURE. United States
Ecological aspects

BERRY, Wendell
The unsettling of America. - San Francisco : Sierra Club, c1977
ix,228p
0-87156-194-8 X.520/21689
1747754

AGRICULTURE. United States
Econometric models - Conference proceedings

NEW directions in econometric modeling and forecasting in U.S. agriculture / editor Gordon C. Rausser. - New York ; Oxford : North-Holland, c1982
xiii,830p : ill. - Includes bibliographies
0-444-00736-9 X.322/12165
2937425

AGRICULTURE. United States
Farms. Management

HARSH, Steven B.
Managing the farm business. - Englewood Cliffs ; London : Prentice-Hall, c1981
xv,384p : ill. - Includes bibliographies and index
0-13-550376-0 X.322/10221 Woolwich
239667x

KAY, Ronald D.
Farm management. - New York ; London : McGraw-Hill, c1981
xiii,370p : ill. - Includes bibliographies and index
0-07-033462-5 X.325/1107 Woolwich
239667x

AGRICULTURE. United States
Farms. Social conditions

LEIMBACH, Patricia Penton
All my meadows. - Englewood Cliffs : Prentice-Hall, c1977
ix,235p
0-13-022525-8 X.520/20458
2251507

AGRICULTURE. United States
Farms. Social conditions, to date

HONG, Edna H. (Edna Hatlestad)
A nostalgic almanac. - Minneapolis, Minn. : Augsburg Publishing House, [1980]
173p
0-8066-1790-x X.329/19460
9162364

AGRICULTURE. United States
Financial management - Conference proceedings

MODELING farm decisions for policy analysis / edited by Kenneth H. Baum and Lyle P. Schertz. - Epping : Bowker, c1983
xvii,418p : ill. - (Bowker (UK) special studies in agricultural science and policy). - Includes bibliographies and index
0-86531-589-2 X.322/14012
3165388

AGRICULTURE. United States
Influence of Communist Party of the United States, 1920-1960

DYSON, Lowell K.
Red harvest. - Lincoln [Neb.] ; London : University of Nebraska Press, c1982
xii,259p. - Bibliography: p243-247. - Includes index
0-8032-1659-9 X.529/50399
2798174

AGRICULTURE. United States
Influence on government policies on foreign relations - Conference proceedings

The ROLE of US agriculture in foreign policy / edited by Richard M. Fraenkel, Don F. Hadwiger, William P. Browne. - New York ; London : Praeger [for] the Policy Studies Organization, 1979
xiii,255p : ill. - (Praeger special studies). - Bibliography: p.242-244. - Includes index
0-03-043101-8 X.520/14279
1915061

AGRICULTURE. United States
Management

DOWNEY, W. David (Walter David)
Agribusiness management. - New York ; London : McGraw-Hill, c1981
x,459p : ill. - Includes index
0-07-017645-0 X.520/24078
2493772

AGRICULTURE. United States
Mechanisation. Political aspects

PRICE, Barry L.
The political economy of mechanization in U.S. agriculture. - Epping : Bowker, 1983
xiv,108p : ill. - (A Westview replica edition) (Bowker (UK) replica edition)
0-86531-976-6 X.329/17336
3044572

AGRICULTURE. United States
Mexican immigrant personnel. Social conditions, to date

HERRERA-SOBEK, María
The bracero experience. - Los Angeles : University of California, Latin American Center Publications, [1979]
xv,142p : ill. - (UCLA Latin American studies ; vol.43). - Bibliography
0-87903-043-7 X.322/10184 Woolwich
8795185

AGRICULTURE. United States
Migrant personnel

DUNBAR, Tony
Hard traveling. - Cambridge, Mass. : Ballinger, 1976
xiii,159p. - Includes index
r6078935 X.520/11478
1076590

MCQUARRIE, Jack
Wildcrafting. - Santa Barbara : Capra Press, 1975
96p
0-88496-042-0 X.311/9108 Woolwich
1076590

AGRICULTURE. United States
Migrant personnel - Bibliographies

FODELL, Beverly
Cesar Chavez and the United Farm Workers. - Detroit : Wayne State University Press, 1974
103p.
0-8143-1502-x X.700/19808
1076590

AGRICULTURE. United States
Policies, 1980-1989

PAARLBERG, Don
Farm and food policy. - Lincoln, [Neb.] ; London : University of Nebraska Press, 1980
ix,338p : ill. - Includes bibliographies and index
0-8032-3656-5 X.322/9576 Woolwich
2237296

AGRICULTURE. United States
Policies of government

COCHRANE, Willard W. (Willard Wesley)
American farm policy, 1948-1973. - Minneapolis : University of Minnesota Press, c1976
xiv,431p : ill. - Includes index
0-8166-0783-4 X.800/15189
0079278

KNUTSON, Ronald D.
Agricultural and food policy. - Englewood Cliffs ; London : Prentice-Hall, c1983
x,387p : ill. - Includes index
0-13-018911-1 X.322/12482
0079278

SCHAPSMEIER, Edward L.
Ezra Taft Benson and the politics of agriculture. - Danville : Interstate Printers and Publishers, 1975
374p
r6188681 X.319/19302 Woolwich
0079278

AGRICULTURE. United States
Policies of government. Influence of interest groups, 1956-1970

WILSON, Graham K. (Graham Keith)
Special interests and policymaking. - London : Wiley, 1977
[7],205p. - Bibliography: p.183-200. - Includes index
0-471-99495-2 X.320/11135 Woolwich
1355643

AGRICULTURE. United States
Policies of government. Influence of pressure groups, 1975-1979

GUITHER, Harold D. (Harold Daniel)
The food lobbyists. - Lexington : Lexington Books, 1981, c1980
xiv,358p. - Includes index
0-669-03539-4 X.325/799 Woolwich
235067x

AGRICULTURE. United States
Political aspects - Conference proceedings

The NEW politics of food / edited by Don F. Hadwiger, William P. Browne. - Lexington, Mass. : Heath ; [Farnborough, Hants.] : [Distributed by Teakfield Ltd], [1979]
ix,267,[5]p : ill. - (Series / Policy Studies Organization ; 19). - Includes index
0-669-01986-0 X.520/14403
1943685

AGRICULTURE. United States
Provision of work experience for secondary school students

AMBERSON, Max L.
Learning through experience in agricultural industry. - New York ; London : McGraw-Hill, 1978
viii,152p : ill(some col). - (Career preparation for agriculture/agribusiness). - Includes index
0-07-000851-5 X.311/8861 Woolwich
157132x

AGRICULTURE. United States
Research

BUSCH, Lawrence
Science, agriculture and the politics of research. - Epping : Bowker, [1983]
xvi,303p : ill. - (Rural studies series) (Westview special studies in agriculture science and policy). - Bibliography: p247-271. - Includes index
0-86531-225-7 X.622/20900
3189910

AGRICULTURE. United States
Research, 1840-1880

EVENSON, Robert Eugene
Agricultural research and productivity. - New Haven ; London : Yale University Press [for] the Economic Growth Center, Yale University [and] the World Bank, 1975
xi,204p : ill. - Bibliography: p.195-201. - Includes index
0-300-01815-0 SRL
0533084

AGRICULTURE. United States
Research, 1945-1981

HADWIGER, Don F.
The politics of agricultural research. - Lincoln, [Neb.] ; London : University of Nebraska Press, 1982
x,230p : ill. - Bibliography: p217-222. - Includes index
0-8032-2322-6 X.329/18089
3144135

AGRICULTURE. United States
Resources. Allocation. Economic aspects

BUSE, Reuben C.
Applied economics. - Ames : Iowa State University Press, 1975
x,623p : ill. - Includes bibliographies and index
0-8138-0115-x X.520/12589
1484575

AGRICULTURE. United States
Salesmanship

MILLER, Larry E.
Selling in agribusiness. - New York ; London : McGraw-Hill, 1979
viii,136p : ill. - (Career preparation for agriculture/agribusiness). - Includes index
0-07-041962-0 X.325/718 Woolwich
2317494

AGRICULTURE. United States
Sayre, Ruth Buxton - Biographies

MCDONALD, Julie
Ruth Buxton Sayre. - Ames : Iowa State University Press, 1980
xii,206p : ill
0-8138-0420-5 X.322/13255
9066632

AGRICULTURE. United States
Trade unions, to date

MITCHELL, H. L. (Harry Leland)
Mean things happening in this land. - Montclair, N.J. : Allanheld, Osmun, c1979
xiv,358p
0-916672-25-5 X.529/4915!
8894841

AGRICULTURE. United States
Trade unions: National Nonpartisan League

GASTON, Herbert E. (Herbert Earle)
The Nonpartisan League. - Repr. ed. - Westport, Conn. : Hyperion, 1975
vii,325p. - (The Radical tradition in America)
0-88355-219-1 X.529/63088
916233x

AGRICULTURE. United States
Trade unions: United Farm Workers Union. Role of Chavez, Cesar

CHILD, Christopher
The California Crusade of Cesar Chavez. - London : Quaker Peace and Service, 1980
[6],31p : 1 ill. - (Non-violence in action series)
B80-33074 X.700/26447
2147327

AGRICULTURE. United States
Women labourers, to date

JONES, Calvin
American farm women. - Chicago : National Opinion Research Center, 1981
x,238,[64]p. - (NORC report ; no.130)
0-932132-26-x X.322/11964
8953678

AGRICULTURE. Upland regions. Great Britain
1750-1950

COLLINS, Edward John T.
The economy of upland Britain, 1750-1950. - Reading : Centre for Agricultural Strategy, University of Reading, 1978
116p : ill. - (CAS papers / Centre for Agricultural Strategy ; 4)
0-7049-0606-6 X.319/18242 Woolwich
1523627

AGRICULTURE. Upland regions. Scotland
Farms. Afforestation

MUTCH, W. E. S.
The interaction of forestry and farming. - [Edinburgh] : Department of Forestry and Natural Resources, University of Edinburgh, 1980
113p : ill. - (Economics and management series ; no.2). - Bibliography: p107
B81-14288 X.325/1009 Woolwich
2383500

AGRICULTURE. Upper Afan. West Glamorgan
Farms. Social conditions, 1570-1851

BROWN, Roger Lee
The farms of the upper Afan. - [Tongwynlais] : R.L. Brown, 1982
iv,60p
B83-16477 X.809/54578
2874326

AGRICULTURE. Urals. Russia (RSFSR)

CHISTOPLI͡ASOV, Stepan Ivanovich
Na puti intensifikat͡sii. - Perm' : Permskoe knizhnoe izdatel'stvo, 1975
127p,[30]p of plates : ill
r6178235 X.329/9031 Woolwich
0395919

RUSAKOVA, Lidii͡a Mikhaĭlovna
Sel'skoe khozi͡aĭstvo Srednego Zaural'i͡a na rubezhe XVIII-XIX vv.. - Novosibirsk : Izdatel'stvo "Nauka", Sibirskoe otdelenie, 1976
182p
r6037212 X.709/22524
0395919

AGRICULTURE. Urals. Russia (RSFSR)
1700-

VOPROSY agrarnoĭ istorii Urala / [otvetstvennyĭ redaktor G.V. I͡Arovoĭ]. - Sverdlovsk : Ural'skiĭ gosudarstvennyĭ universitet im. A.M. Gor'kogo, 1975
141p
r6057881 X.319/20038 Woolwich
2040867

AGRICULTURE. Urals. Russia (RSFSR)
1917-

OSUSHCHESTVLENIE agrarnoĭ politiki KPSS na Urale i v I͡Uzhnom Zaural'e / [redakt͡sionnai͡a kollegii͡a : R.N. Ponomareva (otv. redaktor) ... i dr.]. - Cheli͡abinsk : Cheli͡abinskiĭ gosudarstvennyĭ pedagogicheskiĭ institut, 1978
79p
r6484570 X.329/14823 Woolwich
8766150

AGRICULTURE. Urals. Russia (RSFSR)
Collectivisation

PROVEDENIE agrarnoĭ politiki sovetskoĭ vlasti v Zaural'e / [redakt͡sionnai͡a kollegii͡a : R.N. Ponomareva (otvetstvennyĭ redaktor) S.A. Aref'eva, Z.M. Filippova]. - Cheli͡abinsk : Cheli͡abinskiĭ gosudarstvennyĭ pedagogicheskiĭ institut, 1976
84p. - Includes bibliographies
r6046272 X.319/8823 Woolwich
1224417

AGRICULTURE. Urals. Russia (RSFSR)
Collectivisation, 1922-1941

ISTORII͡A kollektivizat͡sii sel'skogo khozi͡aĭstva Urala (1927-1937) / [otvetstvennyĭ sostavitel' L.A. Trefilova]. - Perm' : Permskoe knizhnoe izdatel'stvo, 1983
220p
r6492762 X.329/19589
8383391

AGRICULTURE. Urals. Russia (RSFSR)
Productivity, 1945-

REGIONAL'NYE osobennosti razvitii͡a obshchestvennogo proizvodstva / [otvetstvennye redaktory : A.I. Demenev, I͡U.A. Nagaev]. - Sverdlovsk : AN SSSR, Ural'skiĭ nauchnyĭ t͡sentr, 1978
103p
r6462063 X.329/16079
8891079

AGRICULTURE. Urban regions. Great Britain
- Serials

CITY farms news. - No.1 ([197-])- . - Bristol : National Federation of City Farms, [197-]-
v. : ill
Holdings: Winter 1982/3-
B84-28648 P.313/576
3142493

AGRICULTURE. Uruguay

MÉTODOS de gestion para la explotacion agropecuaria uruguaya / [por] Estefanell, Antonio J. ... [et al.]. - [Montevideo] : Fundación de Cultura Universitaria, 1978
305p. - (Temas de economia ; 2)
r6161699 X.329/12575 Woolwich
2446154

AGRICULTURE. Uruguay
to 1850

BERRO, Mariano B.
La agricultura colonial. - Montevideo : Ministerio de Educación y Cultura, Biblioteca Artigas, 1975
xlvi,356p. - (Colección de clásicos uruguayos ; v.148)
r6041021 X.329/8664 Woolwich
123319X

AGRICULTURE. Uruguay
to date

PÉREZ ARRARTE, Carlos
El agro uruguayo y el mercado internacional. - Montevideo : Fundación de Cultura Universitaria, 1982
74p. - (Colección Temas nacionales ; 4)
r6709335 X.429/16997
856051x

AGRICULTURE. Uruguay
Organisations: Asociación Rural del Uruguay, to 1971

ASOCIACIÓN RURAL DEL URUGUAY
Cien años de la Asociación Rural del Uruguay. - [Montevideo?] : la Asociación, 1971
640p : ill
r6084662 X313/913 Woolwich
1795430

PENSAMIENTO y acción de la Asociación Rural del Uruguay en el siglo XIX / [edición de Carlos Frick Davie]. - [Montevideo] : la Asociación, 1971
257,166p : ill
r6700691 X.325/979 Woolwich
1795430

AGRICULTURE. Uruguay
Role in economic development

PAGÉS, Walter Hugo
Rol de la agropecuaria en el Uruguay que busca su desarrollo. - Montevideo : Editorial Hemisferio Sur, 1978
200p
r6161064 X.329/12737 Woolwich
2441357

AGRICULTURE. Uzbekistan

ALIMOV, Ibragim Abdugapparovich
Sel'skoe khozi͡aĭstvo Uzbekistana v devi͡atoĭ pi͡atiletke. - Tashkent : Uzbekistan, 1976
58p
r6185677 X.319/20297 Woolwich
2188023

Uzbekskoe dekhkanstvo na puti k sot͡sializmu. - Tashkent : Uzbekistan, 1974
236p. - (Bibl. p.231-[237])
r6174821 X.529/21689
2132699

I͡USUPOV, Kamil
Agrarnye otnoshenii͡a v Uzbekistane v period stroitel'stva sot͡sializma. - Tashkent : Uzbekistan, 1977
173p
r6177575 X.329/12280 Woolwich
2024632

RIZAEV, Guli͡am Rizaevich
Sot͡sialisticheskoe sel'skoe khozi͡aĭstvo Uzbekistana. - Tashkent : Uzbekistan, 1978
269p
r6055945 X.329/8910 Woolwich
2024632

AGRICULTURE. Uzbekistan -continued

SAZONOVA, Mariia Vasil'evna
Tradifsionnoe khoziaistvo uzbekov IUzhnogo Khorezma. - Leningrad : Nauka. Leningradskoe otdelenie. 1978
95p : ill
r6176034 X.510/11990
2136112

AGRICULTURE. Uzbekistan
1945-

RUZIEV, I. R. (Ikram Ruzievich)
Agropromyshlennaia integratsiia na sele. - Tashkent : Uzbekistan, 1982
198p
r6448322 X.629/22984
9073787

AGRICULTURE. Uzbekistan
Collectivisation

AMINOVA, Rakhima Khadievna
Osushchestvlenie kollektivizatsii v Uzbekistane (1929-1932gg.). - Tashkent : Izdatel.stvo "Fan" Uzbekskoi SSR, 1977
311p. - Bibliography : p.306-311
r6187292 X.329/8933 Woolwich
1990772

AGRICULTURE. Uzbekistan
Role of Kommunisticheskaia partiia Sovetskogo Soiuza

URUNOV, S. U.
Vozrastanie rukovodiashchei roli KPSS v povyshenii effektivnosti sel'skogo khoziaistva razvitogo sotsialisticheskogo obshchestva (na primere deiatel'nosti Kompartii Uzbekistana). - Tashkent : Izdatel'stvo "Uzbekistan", 1977
234p
r6044837 X.319/8974 Woolwich
1317709

AGRICULTURE. Uzbekistan
Role of Kommunisticheskaia partiia Uzbekistana, 1945-

RUKOVODSTVO Kommunisticheskoi partii Uzbekistana intensifikatsiei sel'skogo khoziaistva v usloviiakh razvitogo sotsializma / [redaktsionnaia kollegiia : I.A. Alimov ... E. IU. IUsupov (otv. redaktor) ... i dr.]. - Tashkent : Uzbekistan, 1979
190p
r6498312 X.629/16356 Woolwich
873920x

AGRICULTURE. Val d'Aosta. Italy
compared with tourist industries

VINCENT, John, 1947-
The political economy of Alpine development. - [Exeter] : [Department of Sociology, University of Exeter], 1978
[1],31p. - (Discussion Papers / Exeter Research Group ; 3)
B80-34646 X.700/26474
227308x

AGRICULTURE. Val Padana. Italy
ca 1940-1965

ZANIBELLI, Amos
Miglioli-Grieco. - 2. ed.. - Roma : Edizioni Cinque Lune, 1975
398 p.. - (Collana di storia del movimento cattolico ; n.36)
r6014286 X.709/20668
0868825

AGRICULTURE. Vale of Evesham. Hereford and Worcester
1875-1900

ROBINSON, G. M. (Guy Martin)
Late Victorian agriculture in the Vale of Evesham. - Oxford : [University of Oxford] School of Geography, 1976
32p. - (Research papers / University of Oxford. School of Geography ; 16). - Bibliography: p.29-32
B77-11660 X.702/5465
1078178

AGRICULTURE. Valencia (Region). Spain

MIRA, Joan Francesc
Els valencians i la terra. - València : Eliseu Climent, 1978
194p. - (Sèrie "La unitat" ; no.41)
84-85211-79-0 X.519/43636
2135698

La QÜESTIÓ agrària al País Valencià / [per] María Josepa Cucó Giner ... [et al.]. - Barcelona : Aedos, 1978
135p. - (Biblioteca Iskra universitària)
84-7003-176-7 X.319/20284 Woolwich
2135698

AGRICULTURE. Västmanland. Sweden
B.C.4000-A.D.1000

WELINDER, Stig
Prehistoric agriculture in eastern middle Sweden. - Lund : Gleerup : Bonn : Habelt, 1975
102p : charts,maps. - (Acta archaeologica lundensia : series in 8° minore ; no.4). - Bibl
91-400-3941-2 Ac.1067.b/2(3)[no.4]
2439247

AGRICULTURE. Venezuela

FRITH, Dennis E.
The agriculture of Brazil and Venezuela. - London : Graham & Trotman, 1978
148p
0-86010-124-x X.325/884 Woolwich
2361604

RANGEL LAMUS, Amenodoro
Los problemas de la tierra y otros temas. - [Caracas] : [s.n.], 1975
288p. - (Biblioteca de autores y temas tachirenses ; no.65)
r6052081 X.319/8939 Woolwich
1321919

AGRICULTURE. Venezuela
1900-

SEQUERA DE SEGNINI, Isbelia
Dinámica de la agricultura y su expresión en Venezuela. - Caracas : Ariel-Seix Barral Venezolana, 1978
331p.,[1]leaf of plates. - Bibliography
84-344-7445-x X.329/14249 Woolwich
8462003

AGRICULTURE. Venezuela
Economic development, to 1970

SOTO, Oscar David
La empresa y la reforma agraria en la agricultura venezolana. - Mérida, Venezuela : Instituto Iberoamericano de Derecho Agrario y Reforma Agraria, Universidad de los Andes, 1973
307 p.. - (Serie Investigaciones ; no. 1). - Bibliography : p. 297-302
r6008818 X.311/4509 Woolwich
0910139

AGRICULTURE. Venezuela
Law

VENEZUELA. President (1974-1979 : Pérez)
Exposición del Presidente de la República Carlos Andrés Pérez acerca de las recientes medidas acordadas por el Consejo de Ministros y sobre la política agraria nacional. - [Caracas] : Oficina Central de Información, [1976?]
70p
r6157550 X.319/20301 Woolwich
2171996

AGRICULTURE. Venezuela
Research stations, to date

20.O aniversario de la Estación Experimental de Occidente. - Caracas : Oficina de Divulgación Agrícola, 1972
151p : ill
r6703053 LAS.H.242/5 Woolwich
8726566

AGRICULTURE. Venezuela
Trade unions, to date

SOTO, Oscar David
El sindicalismo agrario en Venezuela. - [Caracas?] : CIARA, 1977
186p
r6702103 X.325/1303 Woolwich
8711348

AGRICULTURE. Vestfold. Norway
Organisations: Vestfold landbruksselskap - Serials

VESTFOLD LANDBRUKSSELSKAP
Årsmelding. - 1976-. - [s.l] : Vestfold landbruksselskap, 1976-
r6033739 X.0319/394 Woolwich
146826X

AGRICULTURE. Villages. Fyn. Denmark
1600-1700

FRANDSEN, Karl-Erik
Harndrup og Skårup. - Odense : [Odense Universitet], 1976
47p. - (Odense Universitet. Institut for Historie og Samfundsvidenskab. Skrifter ; nr 19). - Includes bibliographies
r6060837 X.0702/305(19)
1792873

AGRICULTURE. Voss. Norway
to date

GJERDAKER, Johannes
Arbeidet i bygdene. - Voss : Vestanbok forlag, 1980
120p : ill. - Bibliography
82-7101-072-7 X.629/20367
8945179

AGRICULTURE. Wales

WELSH COUNCIL
Agriculture in the Welsh economy. - [Cardiff] : The Council, 1977
9p
B78-32002 BS.205/10
1580868

AGRICULTURE. Wales
- Serials

WALES development / ADAS. - 1975-. - [Cardiff?] : ADAS, 1976-
r6086508 BS.3/655
1580868

WELSH studies in agricultural economics. - July 1982-. - Aberystwyth : Department of Agricultural Economics, University College of Wales, 1982-
v.
B83-16039 P.511/5383
2873869

AGRICULTURE. Wales
- Statistics

WELSH agricultural statistics = Ystadegau amaethyddol Cymru / Welsh Office = Y Swyddfa Gymreig. - No.1 : 1979. - Cardiff : H.M.S.O, 1979
v,102p
Subsequent volumes, when received, will be placed at this shelfmark
0-11-790127-x BS.128A/268
0288462

AGRICULTURE. Wales
Farms. Finance - Serials

FARM accounts in Wales / Welsh Office = Cyfrifon ffermydd yng nghymru / y Swyddfa Gymreig. - 1981/82-. - [Cardiff] : Welsh Office, 1983-
v. : ill
B84-44540 P.823/541
2571919

AGRICULTURE. Wales
Farms. Income

FARM management survey / edited by D.A.G. Green. - [Aberystwyth] : [Dept. of Agricultural Economics, University College of Wales, Aberystwyth], 1981
vii,100p
B83-34721 X.525/7788
2939851

AGRICULTURE. Wales
Research organisations: Agricultural Development and Advisory Service (Wales). Research projects

AGRICULTURAL DEVELOPMENT AND ADVISORY SERVICE (WALES)
Development work. - [Pinner] : Ministry of Agriculture Fisheries and Food ; [Aberystwyth] : Welsh Office, Agriculture Department.
1978. - [1979]
[1],xiv,125p
B80-00775 BS.3/655(1)
1999222

AGRICULTURE. Wales
Roberts, Robert Alun - Biographies

WILLIAMS, Melfyn R.
Doctor Alun. - Talybont [Dyfed] : Talybont, [Dyfed SY24 5ER] : Y Lolfa, 1977
175p : ill. - Bibliography: p.165-168. - Includes index
0-904864-29-4 X.319/17681 Woolwich
128004x

AGRICULTURE. Wales
Smallholdings - Translations

GREAT BRITAIN. Welsh Office
Adroddiad blynyddol i'r Senedd ar fandaliadau yng Nghymru = Annual report to Parliament on smallholdings in Wales. - 1977-1978. - [Cardiff] : [Welsh Office], [1978]-
30cm.
B80-26707 BS.128a/259
2082640

AGRICULTURE. Wales
Smallholdings. Self-sufficiency

MCMULLEN, Jeanine
My small country living. - London : Allen & Unwin, 1984
219p : ill
0-04-630010-4 X.950/31647
3018679

AGRICULTURE. Wallachia. Romania
1600–1821

CONSTANTINIU, Florin
Relaţiile agrare din Ţara Românească în secolul al XVIII-lea. – Bucureşti : Editura Academiei Republicii Socialiste România, 1972
210p. – (Biblioteca istorică ; 30). – Includes bibliography
r6457266 Ac.743/84[30]
9117954

AGRICULTURE. Warwickshire
1349–1520

DYER, Christopher
Warwickshire farming 1349–c.1520. – Oxford : Dugdale Society, 1981
41p. – (Dugdale Society occasional papers ; no.27)
0–85220–057–9 Ac.8024/2
2593963

AGRICULTURE. Weald. Kent

CREATON, David
Beasts and babies. – London : Hodder and Stoughton, 1978
240p
0–340–22453–3 X.989/52702
1114026

Beasts and babies. – Sevenoaks : Coronet, 1979
240p
0–340–23833–x X.329/13166 Woolwich
1114026

Beasts and babies. – Large print ed. – Long Preston : Magna Print, 1983, c1978
491p
0–86009–465–0 Nov.49122
1114026

The beasts of my fields. – London : Hodder and Stoughton, 1977
3–254p
0–340–21267–5 X.529/31465
1114026

The beasts of my fields. – Large print [ed.]. – Bolton-by-Bowland : Magna Print Books, [1979]
3–498p
0–86009–129–5 X.329/11989 Woolwich
1114026

AGRICULTURE. Welbeck Colliery Village. Nottinghamshire
Experimental farms: Gleadthorpe Experimental Husbandry Farm. Research on poultry

GLEADTHORPE / ADAS. – [London] : Ministry of Agriculture, Fisheries and Food, 1981
iii,43p : ill. – (Poultry booklet ; no.8)
B82–32225 SRL
2765225

AGRICULTURE. Welsh Marches

HOLGATE, John, 1924–
Make a cow laugh. – London : P. Davies, 1977
[6],234p
0–432–06740–x X.329/11149 Woolwich
1252127

Make a cow laugh. – London : Pan in association with Peter Davies, 1979, c1977
220p
B82–00393 X.329/14637 Woolwich
1252127

Make a cow laugh. – London : Pan Books, 1979
221p
0–330–25780–3 X.319/18993 Woolwich
1252127

On a pig's back. – London : P. Davies, 1978
[6],184p
0–432–06741–8 X.329/11438 Woolwich
1252127

On a pig's back. – London : Pan Books, 1979
173p
0–330–25781–1 X.319/18994 Woolwich
1252127

A sheep's eye view. – London : P. Davies, 1979
[7],199p
0–432–06742–6 X.329/12019 Woolwich
1252127

HOLGATE, John, 1924–
A sheep's eye view. – London : Pan in association with Peter Davies, 1981, c1979
202p
0–330–26408–7 H.81/1312 Woolwich
1252127

AGRICULTURE. West Africa

ADUAYI, E. A.
General agriculture and soils. – London : Cassell, 1981
102p : ill. – (Cassell's tropical agriculture series ; book 1). – Includes index
0–304–30207–4 X.325/1195 Woolwich
2408767

AGRICULTURE. West Africa
Economic development

HARRISS, Barbara
Agricultural marketing in the semi-arid tropics of West Africa. – Norwich : School of Development Studies, University of East Anglia, 1982
iv,133p : ill. – (Development studies occasional paper ; no. 17). – Bibliography: p122–133
B84–39955 X.322/14080
2772884

HART, Keith
The political economy of West African agriculture. – Cambridge : Cambridge University Press, 1982
ix,226p. – (Cambridge studies in social anthropology ; 43). – Bibliography: p174–207. – Includes index
0–521–24073–5 X.520/27988
2772884

AGRICULTURE. West Africa
Role of entrepreneurship

LAWSON, Rowena M. (Rowena Mary)
The agricultural entrepreneurship of upper-income Africans. – [Hull] : [The author], 1977
[9],232p : 1 ill. – Bibliography: p.227–232
B78–00777 X.313/1457 Woolwich
1275151
– [Another copy] BS.251/129
The agricultural entrepreneurship of upper-income Africans. – [Hull] : [The author], 1977
[9],232p : 1 ill. – Bibliography: p.227–232
B78–00777 X.313/1457 Woolwich
127516x
– [Another copy] BS.251/129

AGRICULTURE. West Germany
to date

HERRMANN, Klaus, Writer on agriculture
Kölner Beiträge zur Modernisierung der deutschen Landwirtschaft. – Köln : Forschungsinstitut für Sozial- und Wirtschaftsgeschichte an der Universität, 1981
38p.. – (Kölner Vorträge und Abhandlungen zur Sozial- und Wirtschaftsgeschichte ; Hft.35)
r6306097 X.0519/352(35)
8820201

AGRICULTURE. West Germany
1945–

FISCHER, Erich
Grundlagen des Strahlenschutzes in der Land- und Ernährungswirtschaft. – 3., Aufl. – Bonn : [AID], [1979]
85p : 38ill
r6302329 SF.583/284
8785392

AGRICULTURE. West Germany
1945– – Conference proceedings

GERMANY (Federal Republic). Auswertungs- und Informationsdienst für Ernährung, Landwirtschaft und Forsten. Conference (1982 : Darmstadt-Kranichstein)
Energieverwendung in landwirtschaftlichen Haushalten. – Bad Godesberg : Auswertungs- und Informationsdienst für Ernährung, Landwirtschaft und Forsten, 1983
77p : ill
r6308037 S.F.583/384
8785392

AGRICULTURE. West Germany
1945– – Statistics

LAND- und Forstwirtschaft, Fischerei.
Landwirtschaftszählung 1971 / Herausgeber : Statistisches Bundesamt Wiesbaden. – Stuttgart : Kohlhammer, 1972–1978
17v.. – (Fachserie 3 / Statistisches Bundesamt)
r6308754 S.F.555/72
8785392

AGRICULTURE. West Germany
1945– – Statistics – Serials

LAND- und Forstwirtschaft, Fischerei. Reihe 2.1.4, Betriebssysteme und Betriebseinkommen / Herausgeber: Statistisches Bundesamt Wiesbaden. – 1975–1977. – Stuttgart : Kohlhammer, 1979–1980
2v.. – (Fachserie 3 / Statistisches Bundesamt)
r6308705 S.F.555/99(6)
8785392

LAND- und Forstwirtschaft, Fischerei. Reihe 2.1.4, Betriebssysteme und Standardbetriebs-einkommen / Herausgeber : Statistisches Bundesamt Wiesbaden. – 1979–. – Stuttgart : Kohlhammer, 1981–. – (Fachserie 3 / Statistisches Bundesamt)
r6308706 S.F.555/99(6)
8785392

LAND- und Forstwirtschaft, Fischerei. Reihe 2.1.5, Sozialökonomische Verhältnisse / Herausgeber : Statistisches Bundesamt Wiesbaden. – 1975–. – Stuttgart : Kohlhammer, 1979–. – (Fachserie 3 / Statistisches Bundesamt)
r6308701 S.F.555/99(7)
8785392

LAND- und Forstwirtschaft. Fischerei. Reihe 6. Ausgewählte Zahlen für die Agrarwirtschaft / Statistisches Bundesamt Wiesbaden. – 1972–1975. – Stuttgart : Kohlhammer, 1973–1976
4v.. – (Fachserie B / Statistisches Bundesamt)
Holdings: 1973–1975
r6308750 S.F.555/84
8785392

AGRICULTURE. West Germany
1945–. Reports – Statistics

LAND- und Forstwirtschaft. Fischerei. Reihe 2.S.1. Methodische Grundlagen der Agrarberichterstattung / Herausgeber: Statistisches Bundesamt Wiesbaden – Ausg 1979. – Stuttgart . Kohlhammer, 1980
107p. – (Fachserie 3 / Statistisches Bundesamt)
r6308703 S.F.555/99(29)
9047441

AGRICULTURE. West Germany
Development compared with agricultural development in Great Britain, 1870–1975

The DEVELOPMENT of agriculture in Germany and the UK. – Ashford, Kent . Centre for European Agricultural Studies, Wye College.
1: German agriculture, 1870–1970. – 1979
59,[31]p. – (Miscellaneous studies / Wye College. Centre for European Agricultural Studies ; no.2). – Bibliography(2p.)
0–905378–15–6 X.512/6718
1921347
2: UK agricultural policy, 1870–1970. – 1979
110,[28]p. – (Miscellaneous studies / Wye College. Centre for European Agricultural Studies ; no.3). – Bibliography: p.110
0–905378–16–4 X.512/6719
1921347

AGRICULTURE. West Germany
Development compared with agricultural development in Great Britain, 1870–1975 – Statistics

The DEVELOPMENT of agriculture in Germany and the UK. – Ashford, Kent : Centre for European Agricultural Studies, Wye College.
3: Comparative time series, 1870–1975. – 1979
97,[23]p : ill. – (Miscellaneous studies / Wye College. Centre for European Agricultural Studies ; no.4)
0–905378–17–2 X.512/6720
1921347

AGRICULTURE. West Germany
Development compared with agricultural development in Great Britain, 1967–1980 – Conference proceedings

The DEVELOPMENT of agriculture in Germany and the UK. – Ashford, Kent : Wye College. – (Seminar paper ; no.15)
5: Policy implications of their comparative output, structure and productivity. – c1984
iv,71p : ill
0–905378–19–9 X.322/14743
2595028

AGRICULTURE. West Germany
Farms. Livestock, 1945– – Statistics – Serials

LAND- und Forstwirtschaft, Fischerei. Reihe 2.1.3, Viehhaltung der Betriebe / Herausgeber: Statistisches Bundesamt Wiesbaden. – 1977–. – Stuttgart : Kohlhammer, 1979–. – (Fachserie 3 / Statistisches Bundesamt)
r6308700 S.F.555/99(5)
9047433

LAND- und Forstwirtschaft. Fischerei. Reihe 4, Viehbestand und tierische Erzeugung / Herausgeber : Statistisches Bundesamt Wiesbaden. – 1981–. – Stuttgart : Kohlhammer, 1982–. – (Fachserie 3 / Statistisches Bundesamt)
r6310607 S.F.555/99(20)
9047433

AGRICULTURE. West Germany
Finance, 1945– – Statistics – Serials

LAND- und Forstwirtschaft, Fischerei. Reihe 2.1.7, Ausserbetriebliche Einkommen und Arbeitsverhältnisse für ausgewählte Betriebsgruppen / Herausgeber : Statistisches Bundesamt Wiesbaden. – 1975–. – Stuttgart : Kohlhammer, 1979–. – (Fachserie 3 / Statistisches Bundesamt)
r6308702 S.F.555/99(28)
9047425

AGRICULTURE. West Germany
Implications of law of conservation of environment – Conference proceedings

The IMPACT of environmental legislation upon agriculture / sponsored by the Anglo-German Foundation for the Study of Industrial Society held at Wye on March 14–16, 1984 ; edited by Bernarde Duesenberg, Bridget Girling. – Ashford : Wye College, 1984
viii,117p : ill. – (Seminar papers ; no.16)
0–905378–30–x X.322/14946
3280616

AGRICULTURE. West Germany
Research organisations. 1945-

FORSCHUNG im Geschaftsbereich des Bundesministers fur Ernahrung, Landwirtschaft und Forsten - [Bonn-Duisdorf] : Der Bundesminister, 1981
51p : ill
r604840 S.F.583/341
8590684

AGRICULTURE. West Germany
Safety measures. 1945- - Serials

BERICHT uber die Durchfuhrung der Unfallverhutung und die Massnahmen fur die Erste Hilfe / Landwirtschaftliche Berufsgenossenschaft Unterfranken - 1980- - [Wurzburg] : [Die Genussenschaft], [1981]-
r602353 SF.583/295
878434

AGRICULTURE. West Germany
Superannuation schemes. 1945-

HAGEDORN, Konrad
Agrarsozialpolitik in der Bundesrepublik Deutschland - Berlin : E. Schmidt, c1982
xvi,371p : ill. - (Beiträge zur Sozialpolitik und zum Sozialrecht ; Bd 1). - Includes bibliography
3-503-02145-0 X.322/12558
9011212

AGRICULTURE. West Indies
Economic development - Conference proceedings

WEST INDIAN AGRICULTURAL ECONOMICS CONFERENCE (7th : 1972 : Grand Anse, Grenada)
Proceedings of the Seventh West Indian Agricultural Economics Conference [held at Grand Anse, Grenada, April 9th-15th, 1972]. - St Augustine : University of the West Indies, [Department of Agricultural Economics and Farm Management], 1972
iii-vii, 175p : ill. - Includes bibliographies
r6032073 X.0311/1012(7) Woolwich
1147730

WEST INDIES AGRICULTURAL ECONOMICS CONFERENCE (8th : 1973 : Port of Spain)
Proceedings of the Eighth West Indies Agricultural Economics Conference held at Hilton Hotel, Trinidad, April 1st - 7th, 1973. - St Augustine : University of the West Indies, Department of Agricultural Economics and Farm Management, 1973
iii-vii, 155p : ill. - Bibliography : p.83
r6032074 X.0311/1012(8) Woolwich
1147730

WEST INDIES AGRICULTURAL ECONOMICS CONFERENCE (9th : 1974 : New Kingston, Jamaica and Twickenham Park, Jamaica)
Proceedings of the Ninth West Indies Agricultural Economics Conference held at the Pegasus Hotel, New Kingston, Jamaica, April 3rd-6th, 1974, and at The Jamaica School of Agriculture, Twickenham Park, April 7th-10th, 1974. - St Augustine : University of the West Indies, Department of Agricultural Economics and Farm Management, 1974
iii-xvi, 222p : ill. - Includes bibliographies
r6032075 X.0311/1012(9) Woolwich
1147730

AGRICULTURE. West Indies
Economic development, 1950-1976

FLOTO, Edgardo
Agrarian dualism in a non agricultural economy. - Cambridge : Centre of Latin American Studies, University of Cambridge, 1977
[3],24p,[2]leaves(1 fold). - (Working papers / University of Cambridge. Centre of Latin American Studies ; no.29)
B78-04858 X.313/1467 Woolwich
1303953

AGRICULTURE. West Jutland. Denmark
Farms: Drongstrupgård. Social conditions, to 1974

RISBJERG THOMSEN, S.Kr
En gammel gard og dens beboere. - [Aaskov] : Aaskov kommune, 1975
37p
r6028813 X.702/3185
1125974

AGRICULTURE. West Midlands. England

GREAT BRITAIN. Ministry of Agriculture, Fisheries and Food
Agriculture in the West Midlands. - [Pinner] : [The Ministry], 1979
[2],102p
B80-38986 SRL
2301881

AGRICULTURE. West Riding. Yorkshire
Farms. Social conditions, ca 1905-1933

KITCHEN, Fred
Brother to the ox. - Horsham : Caliban Books, 1981
viii,243p,[8]leaves of plates : ill
0-904573-54-0 Nov.45041 Woolwich
2628791

Brother to the ox. - Harmondsworth : Penguin, 1983, c1940
xvii,243p,[4]p of plates : ill. - (Penguin country library)
0-14-059001-3 X.329/18599
2628791

Brother to the ox. - Large print ed. - Bath : Chivers, 1984
241p. - (A New Portway large print book)
0-85119-251-3 X.950/34031
2628791

AGRICULTURE. West Sussex
Organisations: West Sussex Growers' Association - Serials

WEST SUSSEX GROWERS' ASSOCIATION
Newsletter. - [No.1]- ; Jan. 1977-. - Arundel : The Association, [1977]-
B77-15743 P.513/2105
1108743

AGRICULTURE. Western Europe
to 1914 - Bibliographies

FUSSELL, G. E. (George Edwin)
Agricultural history in Great Britain and Western Europe before 1914. - London : Pindar, 1983
157p. - Includes index
0-907132-04-9 Cup.510.d.mi.1.
3081737

AGRICULTURE. Western Europe
1880-1980

TRACY, Michael
Agriculture in Western Europe. - 2nd ed. - London : Granada, 1982
viii,419p. - Includes bibliographies and index
0-246-11446-0 X.322/11389
2809907

AGRICULTURE. Western Europe
Farms. Mergers

HIRSCH, Gunther Paul
Farm amalgamation in Western Europe. - Farnborough, Hants. : Saxon House, 1978
xi,120p. - Includes bibliographies and index
0-566-00253-1 X.529/33368
1564765

AGRICULTURE. Western Europe
Part-time agriculture - Conference proceedings

PART-TIME farming. - Ashford, Kent : Centre for European Agricultural Studies, Wye College, 1977
[6],42,[6]p : 2 ill. - (Seminar papers / Wye College. Centre for European Agricultural Studies ; no.2)
0-905378-02-4 X.313/1413 Woolwich
1170090

AGRICULTURE. Western Europe
Part-time agriculture. Role in economic development of rural regions - Conference proceedings

The PLACE of part time farming in rural and regional development / edited by Ruth Gasson. - Ashford, Kent : Wye College, 1977
[4],145p. - (Seminar papers / Wye College. Centre for European Agricultural Studies ; no.3)
0-905378-05-9 X.313/1528 Woolwich
1408291

AGRICULTURE. Western Europe
Role of women

ANDERSSON, Rosmarie
Kvinnan i lantbruket. - Stockholm : LT, 1977
120p. - (LT pocket)
91-360-0564-9 X.319/9130 Woolwich
1325493

AGRICULTURE. Western European countries
Law, 1945-

AKSOY, Suat
Das türkische Agrarrecht im Vergleich mit der Agrargesetzgebung einiger europäischer Länder. - Köln : Heymann, 1980
l,603p.. - (Schriftenreihe des Instituts für Landwirtschaftsrecht der Universität Göttingen ; Bd.23). - Bibliography: p.xiii-xlvii
Pages xxxvii-xxxviii and 573-574 reversed
3-452-18820-5 X.200/37122
8756961

AGRICULTURE. Western Scotland
Dairy farms. Productivity

SARGENT, E. D. (Edward David)
Managing the dairy herd in difficult times. - [Auchincruive] : West of Scotland Agricultural College, Economics Division, 1980
33p. - (F.P.I. ; no.3)
B81-27586 X.322/10011 Woolwich
2524481

AGRICULTURE. Western Scotland
Dairy farms. Profitability

GROVES, C. R. (Chris R.)
Trends in profitability of dairy farming. - [Auchincruive] : West of Scotland Agricultural College, Economics Division, 1980
59p : ill. - (FPI ; no.4)
B81-16692 X.325/1071 Woolwich
2393735

AGRICULTURE. Western Scotland
Dairy farms. Profitability, 1980-1981

TWEDDLE, J. M.
Profitability of dairy farms 1980/81. - Auchincruive : West of Scotland Agricultural College, Economics Division, 1982
38p. - (Cooperators' report ; no.9)
B83-05127 X.525/6521
2822229

AGRICULTURE. Western Scotland
Farms. Profitability, 1980-1981

TWEDDLE, J. M.
Profitability of livestock with arable farms 1980-81 (1980 crop year). - Auchincruive : West of Scotland Agricultural College, Economics Division, 1982
14p. - (Cooperators' report ; no.10)
B83-05128 X.525/6522
2822237

Profitability of lowground stock farms 1980-81. - Auchincruive : West of Scotland Agricultural College, Economics Division, 1982
14p. - (Cooperators' report ; no.11)
B83-05129 X.525/6523
2822237

AGRICULTURE. Western Siberia. Russia (RSFSR)

LADENKOV, Vasiliĭ Nikolaevich
Problemy razvitiiā sel'skogo khozīāistva Zapadnoĭ Sibiri. - Novosibirsk : Nauka, Sibirskoe otdelenie, 1977
202p.
r6087909 X.319/9830 Woolwich
1856189

AGRICULTURE. Western states. Nigeria
to date

BEER, Christopher
The politics of peasant groups in Western Nigeria. - Ibadan : Ibadan University Press, 1976
xvii,276p. - (Ibadan social science series ; no.7)
978-12-1014-1 X.520/29689
8916519

AGRICULTURE. Western Yorkshire
- Statistics

AGRICULTURAL DEVELOPMENT AND ADVISORY SERVICE. Harrogate Division
Harrogate Division farming at a glance, 1976. - Harrogate : [The Division], 1977
39p : ill
B78-14159 BS.3/580
1412795

AGRICULTURE. Westphalia. West Germany
Farms. Buildings. Architecture

SCHEPERS, Josef
[Haus und Hof deutscher Bauern. Bd 2]. Haus und Hof westfälischer Bauern. - 3., verb. und erw. Aufl. - Münster : Aschendorff, 1976
x,544p (2 fold. leaves),64,[32]p of plates : ill. - Includes bibliographies
3-402-05253-9 X.322/2570 Woolwich
1161970

AGRICULTURE. Westphalia. West Germany
Farms. Buildings. Architecture, to date

BAUMEIER, Stefan
Das Westfälische Freilichtmuseum Detmold. - Gummersbach : Gronenberg, [1981]
123p. : chiefly ill.
3-88265-069-9 X.425/3154 Woolwich
8811385

WALTER, Friedrich
Das westfälische Bauernhaus. - Hildesheim : Gerstenberg, 1981
128p : ill. - (Gerstenberg Reprint)
3-8067-0863-0 X.322/12010
8811385

WESTFÄLISCHES FREILICHTMUSEUM BÄUERLICHER KULTURDENKMALE
Museumsführer, Westfälisches Freilichtmuseum Detmold, Landesmuseum für Volkskunde. - Detmold : das Museum, 1982
152p. : ill
3-923432-80-1 X.429/14635
8811385

AGRICULTURE. Wilcote. Oxfordshire
Farms: Shakenoak Farm. Antiquities, to 750. Archaeology

BRODRIBB, A. C. C. (Arthur Charles Conant)
Excavations at Shakenoak Farm, near Wilcote, Oxfordshire. - Oxford : Bar.
Pt.5: Sites K and E. - 1978
210p,[11]p of plates : ill
0-86054-022-7 X.429/14298
2887673

AGRICULTURE. Wiltshire
1870-1880

JEFFERIES, Richard
Landscape & labour. - Bradford-on-Avon : Moonraker Press, 1979
199p : ill. - Bibliography: p.182-199
0-239-00188-5 X.529/34032
1951564

AGRICULTURE. Wimborne (District)
Farms: Chilbridge Farm

TROUGHTON, Madge P.
Chilbridge Farm, Wimborne, Dorset. - [Dorchester] : Dorset County Council Education Committee, 1977
[1],16[i.e.32]p : ill
0-85216-156-5 X.319/17895 Woolwich
1371533

AGRICULTURE. Winkton region. Dorset
1855-1875

LAVENDER, Ruth
The farmers of Winkton Tithing, 1855-1875. - [Dorchester] : Dorset County Council Education Committee, [1977]
[3],13[i.e.15]p
0-85216-144-1 X.319/17358 Woolwich
066071x

AGRICULTURE. Worcestershire. Hereford and Worcester
1854

DODD, J. Phillip
Worcestershire agriculture in the mid-nineteenth century. - Worcester : Worcestershire Historical Society, 1979
24p : ill. - (Occasional publications / Worcestershire Historical Society ; no.2)
B79-33124 X.0709/1042(2)
1961772

AGRICULTURE. Wye Valley. England
Smallholdings. Self-sufficiency

RIVERS, Patrick
Living on a little land. - London : Turnstone Books, 1978
159p,[4]p of plates : ill. - Bibliography: p.159
0-85500-086-4 X.319/18133 Woolwich
1483358

AGRICULTURE. Yakut ASSR. Russia (RSFSR)

SEL'SKAÎÂ molodezh' ÎÂkutii / [otvetstvennyĭ redaktor V.I. Boĭko]. - ÎÂkutsk : ÎÂkutskoe knizhnoe izdatel'stvo, 1977
73p
r6175789 X.519/43617
2248840

AGRICULTURE. Yakut ASSR. Russia (RSFSR)
Collectivisation

KOLLEKTIVIZAT͡SIÎÂ sel'skogo khozîaĭstva ÎÂkutskoĭ ASSR (1928-1940) gg. / [redkollegiîa V.V. Nikolaeva (glavnyĭ redaktor) ... i dr.]. - ÎÂkutsk : ÎÂkutskoe knizhnoe izdatel'stvo, 1978
263p. - (Istoriîa kollektivizat͡sii sel'skogo khozîaĭstva SSSR)
r6183226 7086.c.2/5 Woolwich
2454882

AGRICULTURE. Yorkshire

FARMING in Yorkshire / [Askham Bryan College of Agriculture and Horticulture, Department of Management, Farm Management Survey Unit]. - 1975-76. - [York] : The Department, 1976
76p in various pagings
Subsequent volumes, when received, will be placed at this shelfmark
B77-16771 Cup.1281/527
067057x

FARMING in Yorkshire / [Askham Bryan College of Agriculture and Horticulture, Department of Management, Farm Management Survey Unit]. - 1977-78. - [York] : The Department, 1978
52p in various pagings
B80-24205 L.49/352
067057x

AGRICULTURE. Yorkshire
- Serials

The YORK farmers' journal. - Vol.23, no.2 (Feb. 1982)-. - Dorchester : N.F.U. County Publications, 1982-
v. : ill
B82-22656 P.311/110
2725665

AGRICULTURE. Yugoslavia

RADOMIROVIĆ, Vojin
Reprodukcija seljačkog gazdinstva. - Novi Sad : Radivoj Ćirpanov, 1976
203p. - (Nauka i univerzitet). - Bibliography : p.189-199
r6097076 X.311/5095 Woolwich
213358X

YUGOSLAV agriculture / [text by a group of authors : Petar Maksimović ... et al. ; editor Aleksandar Pavlović ... et al.]. - Beograd : Yugoslaviapublic, [1977]
69p,[24]p of plates : ill(some col.)
r6182759 X.329/13098 Woolwich
213358X

AGRICULTURE. Yugoslavia
1945-

POLJOPRIVREDA Jugoslavije / [redakcioni odbor: Milovan Zidar, predsednik ... et al.]. - Beograd : Eksport-pres, 1982
444p : ill
r6496056 X.322/14445
8766118

STIPETIĆ, Vladimir
[Razvoj poljoprivrede u socijalističkoj Jugoslaviji (1945-1975). German]. Die jugoslawische Landwirtschaft, 1945-1975. - Beograd : Komunist, 1975
189p. - (Sozialistische Theorie und Praxis)
r6435194 X.329/14755 Woolwich
8766118

AGRICULTURE. Yugoslavia
Law, 1918-1941

ŽIVKOV, Sava
Agrarno zakonodavstvo Jugoslavije, 1918-1941. - [Novi Sad] : Matica srpska, 1976
152p. - Bibliography
r6462553 X.200/39799
8902844

AGRICULTURE. Yugoslavia
Policies of government

ČUKANOVIĆ, Savo
Selo i seljaštvo u agrarnoj politici Jugoslavije .... - Beograd : Privredni pregled, 1977
263p. - Bibliography: p.247-263
r6178641 X.329/13266 Woolwich
0028096

KONTETZKI, Heinz
Agrarpolitischer Wandel und Modernisierung in Jugoslawien. - Nürnberg : Nürnberger Forschungsvereinigung, 1976
viii, 563 p.. - (Nürnberger Forschungsberichte ; Bd 7). - Bibliography : p. 521-563
3-921453-06-2 P.901/1433[Bd. 7]
0028096

AGRICULTURE. Yugoslavia
Self-management, to date

TOMIN, Miloš
Položaj zemljoradnika i udruživanje u poljoprivredi. - Beograd : Fakultet političkih nauka, Institut za političke studije, 1976
51p
r6434008 X.529/43097
8726558

AGRICULTURE. Zaïre
Economic development, 1960-

NSUMBU TSHAHE DJANGA
Le problème du remplacement des importations alimentaires par une production nationale au Zaïre. - Fribourg : [Université de Fribourg], 1983
179p. - Includes bibliography
r6211476 X.529/67409
913140x

AGRICULTURE. Zala (megye). Hungary

NÉMETH, István
Gépállomások Zalában. - Zalaegerszeg : Zala megyei Könyvtár, 1975
111p,[18]leaves of plates : ill. - (Göcseji Helikon : Honismereti közlemények ; 9 : Mezőgazdasági üzemtörténet ; 2)
r6499006 P.801/1325[no.9]
2434474

AGRICULTURE. Zambia

BAKER, Randall
An evaluation of policy toward agricultural technology choice of power in Zambia. - Norwich : School of Development Studies, University of East Anglia, 1983
24p. - (Discussion paper ; no.121)
B84-41076 X.525/9945
3188884

AGRICULTURE. Zambia
- For Zambian students

OWEN, G. H.
Agricultural Science. - Lusaka : Longman Zambia.
Book 3. - 1976
[3],113p : ill
0-582-60613-6 X.0311/1016(3)
125717X

AGRICULTURE. Zambia
to 1970

RANGER, Terence Osborn
The agricultural history of Zambia. - Lusaka : National Educational Company of Zambia [for the Historical Association of Zambia], 1971 [i.e.1973]
[3],28p. - (HAZ ; 1)
r6072039 X.319/18145 Woolwich
1484893

AGRICULTURE. Zambia
1964-1971

DODGE, Doris Jansen
Agricultural policy and performance in Zambia. - Berkeley : Institute of International Studies, University of California, 1977
285p. - (University of California. Institute of International Studies. Research series ; no.32)
0-87725-132-0 X.0809/130(32)
1970402

LOMBARD, C. Stephen
Agriculture in Zambia since independence. - [Lusaka] : National Educational Company of Zambia for the Institute of [i.e. for] African Studies, University of Zambia ; Birmingham : [Distributed by] Third World Publications, 1974 [i.e.1979]
viii,105p : ill. - (Zambia - past and present). - Bibliography: p.93-102. - Includes index
B79-35260 X.519/30346
1970402

AGRICULTURE. Zambia
1964- - Serials

HARVEST. - Vol.1 (Oct. 1981)-. - Lusaka : Flash Publications, 1981-
v.
r6806720 P.313/553
8486131

AGRICULTURE. Zambia
Policies of government, 1964-1971

DIXON, Carole L.
The development of agricultural policy in Zambia, 1964-1971. - [Aberdeen] : Department of Geography, University of Aberdeen, 1977
[7],44p. - (O'Dell memorial monographs / University of Aberdeen. Department of Geography ; no.5). - Bibliography: p.37-44
B78-14161 X.313/1529 Woolwich
1410407

AGRICULTURE. Zambia
Small scale agriculture. Management

BESSELL, James Edward
Farmer operating efficiency and credit-worthiness. - Sutton Bonington : University of Nottingham, Department of Agriculture and Horticulture, 1976
ix,123p : 1 ill. - (Bulletins / University of Nottingham's Farm Management Investigations for Zambia ; no.2)
B77-29563 X.319/17557 Woolwich
1251139

AGRICULTURE AND POLLUTION. Report. 7th. Great Britain. Royal Commission on Environmental Pollution

FARM AND FOOD SOCIETY
Agriculture and pollution. - London : The Society, [1980]
[3],52p
B80-34517 X.329/12994 Woolwich
2227312

AGRIGENTO. Italy
to ca 1500

BOSCO, Giuseppe
Il Comune di Agrigento nel Medio evo. - Agrigento : Edizioni dioscuri, 1973
199p.,xxxii leaves of plates : ill
Author's presentation copy
r6068859 X.708/46309
2092387

AGRIOTES
See Click beetles

AGROBIOLOGY
Fehér, Dániel

KECSKÉS, Mihály
Fehér Dániel. - Budapest : Akadémiai Kiadó, 1983
215p. - (A múlt magyar tudósai). - Includes bibliographies
963-05-3047-3 X.808/38040
8949948

AGROCHEMICALS
See Agricultural chemicals

AGRONOMY
See also
Field crops

AGRONOMY. Spain
to 1479

BOLENS, Lucie
Agronomes andalous du Moyen-Âge. - Genève : Librarie Droz, 1981
xii,305p : ill,facsims,map. - (Départment d'histoire générale de la Faculté des lettres de l'Université de Genève ; 13). - Bibliography: p245-258
r6206173 X:0700/95(13)
8825319

AGRS
See Advanced gas-cooled nuclear reactors

AGRUPACI ON DE JÓVENES UNIDOS
1970-1973

ALVAREZ CALDERÓN, Jorge
Así comenzamos. - Lima : Centro de Estudios y Publicaciones, 1973
147 p. : ill. - (CEP ; no. 5)
r6011585 X.519/20885
0872873

AGT, A. A. M. VAN. Netherlands

SCHWIETERT, Charl
De formatie van een tijdbom. - Haarlem : De Haan, [1981]
136p,[8]p of plates : ill
90-228-3581-2 X.809/53255
8736928

AGT, A. A. M. VAN. Politics. Netherlands

PUCHINGER, G. (George)
In gesprek met het kabinet-Van Agt. - Kampen : Kok, 1980
377p
90-242-0376-7 X.809/49615
8312915

AGUALONGO, AGUSTÍN. Revolutionary movements. Latin America
- Biographies

IBARRA REVELO, Alfonso
Agualongo. - Pasto : Imprenta Departamental, 1975, 1976 printing
iv,382p
r6162269 X.709/51914
1151703

ORTIZ, Sergio Elías
Agustín Agualongo y su tiempo. - Bogotá : Banco Popular, 1974
549p. - (Biblioteca Banco Popular ; v.61)
r6037745 X.709/23072
1151703

AGUARUNA LANGUAGE

FAST, Gerhard
Introducción al idioma aguaruna. - Edición provisional. - [Lima?] : Instituto Lingüístico de Verano, 1974
111p. - (Instituto Lingüístico de Verano. documentos de trabajo ; n.3)
r6051466 X.902/3365
1865803

PAYNE, David Lawrence
Nasalidad en Aguaruna. - [Yarinacocha, Peru] : Instituto Lingüístico de Verano, 1976
83p. - (Serie lingüística peruana ; n.15)
r6084758 X.902/3332
1853767

AGUARUNA LANGUAGE
- Dictionaries

CORBERA MORI, Angel
Glosario aguaruna-castellano. - Lima : Universidad Nacional Mayor de San Marcos, Centro de Investigación de Lingüística Aplicada, 1981
v,77leaves. - (Documento de trabajo ; no.44)
r6708448 X.955/2882
1853767

AGUARUNA LITERATURE
See also
Fiction in Aguaruna

AGUARUNAN MYTHS
"DUIK múun ... " / Aurelio Chumap-Lucía, Manuel García-Rendueles. - Lima : Centro Amazónico de Antropología y Aplicación Práctica, 1979
2v.(866p) : ill. - (Serie antropológica ; 2)
r6704590 X.622/12277 Woolwich
8814090

["DUIK muun - " Selections. Spanish & Aguaruna]. "Duik múun - " ... / revisión y corrección de textos : Jorge J. Juep Nampin. - Lima : Centro Amazónico de Antropología y Aplicación Práctica, 1979-1980
3v. : ill
r6707038 X.950/23096
8814090

AGUDELO, MIGUEL SAMPER
See Samper, Miguel †

AGUILAR, CÁNDIDO. Politics. Mexico
- Biographies

MANZUR OCAÑA, Justo
La revolución permanente. - México : B.Costa-Amic, 1972
364p., 29p. of plates
r6027545 X.709/22184
038707X

AGUILERA MALTA, DEMETRIO. Ecuadorian writers. Fiction in Spanish

FAMA, Antonio
Realismo mágico en la narrativa de Aguilera-Malta. - Madrid : Playor, [1977]
169p. - (Colección Nova scholar). - Includes bibliographies
84-359-0035-5 X.908/85769
1795465

RABASSA, Clementine C.
En torno a Aguilera Malta. - [Guayaquil] : Casa de la Cultura Ecuatoriana, Núcleo del Guayas, 1981
269p. - (Colección Nuestramérica ; 5)
r6706124 X.958/21978
815855x

AGUILERA, PEDRO. Education. Paraguay
- Biographies

BRUGADA GUANES, Alejandro
Pedro Aguilera. - Asunción : Casa América, 1976
141p : ill(some col.)
r6053867 X.519/24552
1425595

AGUINALDO, EMILIO. Philippines

TUROT, Henri
[Aguinaldo et les philippins. English]. Emilio Aguinaldo. - 2nd ed. - Manila : Foreign Service Institute, 1981
vi,194p
r6811354 X.808/38897
8345260

AGUINALDOS
Puerto Rican aguinaldos, to ca 1960

LÓPEZ CRUZ, Francisco
El aguinaldo en Puerto Rico. - San Juan de Puerto Rico : instituto de Cultura Puertorriqueña, 1972
46p
r6048342 X.439/4667
1326775

AGUIRRE, ANDRÉS POEY Y
See Poey y Aguirre, Andrés †

AGUIRRE, LOPE DE. Conquest. Latin America
- Biographies

LÓPEZ, Casto Fulgencio
Lope de Aguirre el peregrino. - [3a ed.]. - Salsadella, Castellón : Los Libros de Plon, [1977]
288p.,[1] leaf of plates,[6]p. of plates. - (Colección Majos y bajos)
84-85056-06-x X.708/44979
1848860

AGUIRRE, LOPE DE. Conquistadores
- Biographies

LOPE de Aguirre / [textos seleccionados por] Elena Mampel González, Neus Escandell Tur. - Barcelona : Editorial 7 1/2, 1981
xviii,306p,plates. - Includes bibliography
84-85411-51-x X.800/39223
8711046

AGUIRRE, NATANIEL. Bolivian writers. Fiction in Spanish

BARNADAS, Josep M.
Nataniel Aguirre. - Cochabamba : Los Amigos del Libro, 1977
30p. - (Colección Estudios críticos ; n.1) (Realidad sociohistórica y expresión literaria en Bolivia ; n.1)
r6084953 X.907/18541
1859196

DÍAZ MACHICAO, Porfirio
Nataniel Aguirre. - 2a ed. - Cochabamba : Editorial "Los Amigos del Libro", 1972
348p : ill. - (Biblioteca IV centenario ; 12)
r6706239 X.958/22783
8158347

AGULS. Dagestan ASSR. Russia (RSFSR)
AGULY / [otvetstvennyĭ redaktor A. Islammagomedov]. - Makhachkala : Akademiîa nauk SSSR, Dagestanskiĭ filial, Institut istorii, îazyka i literatury im. G. TSadasy, 1975
184p : ill
r6183072 X.520/20660
2299372

AGUNG, GUNUNG
See Gunung Agung

AGUSTÍN, ANTONIO. Catholic Church. Spain
- Correspondence, diaries

AGUSTÍN, Antonio
[Correspondence]. Epistolario. - Salamanca : Universidad de Salamanca, 1980
335p. - (Acta salmanticensia. Filosofía y letras ; 115)
84-7481-109-0 Ac.153/3(4)[v.115]
8419825

AGUSTINI, DELMIRA. Uruguayan writers. Poetry in Spanish

BONADA AMIGO, Roberto
Delmira Agustini en la vida y en la poesía. - 2a ed. ampliada. - Montevideo : Librería Selecta Editorial, 1974
74 p.
r6009054 X.909/32808
0919527

STEPHENS, Doris T.
Delmira Agustini and the quest for transcendence. - Montevideo : Ediciones Geminis, 1975
220p. - (Colección Novus Orbis). - Bibliography : p.207-220
r6041014 X.908/39290
0919527

AGUSTINI, DELMIRA. Uruguayan writers. Poetry in Spanish
- Biographies

AGUSTINI, Delmira
[Correspondence. Selections]. Correspondencia íntima de Delmira Agustini y Tres versiones de Lo inefable. - Montevideo : Biblioteca Nacional, 1978
88p : facsims
r6161506 X.989/85992
2420546

AH VOCABULARY SCALE

HEIM, Alice Winifred
Manual for AH vocabulary scale. - [Windsor] : NFER, [1978]
[1],v,41p : ill
0-7005-0200-9 X.902/5663
2108879

AH2. Intelligence tests

HEIM, Alice Winifred
AH2/AH3 manual. - Windsor : NFER, 1974
47p : ill
0-7005-0013-8 X.512/5036
0364452

AH3. Intelligence tests

HEIM, Alice Winifred
AH2/AH3 manual. - Windsor : NFER, 1974
47p : ill
0-7005-0013-8 X.512/5036
0364452

'AHA'. Suffixes. Place names. Switzerland
to ca 1400

WERNLI, Fritz
Die Frage der Kontinuität vom Altertum ins Mittelalter und die-aha-Namen in der deutschsprachigen Schweiz und in andern der Sprachgrenze nahen Gebieten. Geschichtsforschung und Namenforschung im Zusammenhang. - Degersheim (Schweiz) : [F. Wernli], 1977
x,108p. - (Studien zur mittelalterlichen Verfassungsgeschichte ; Hft 8)
r6058408 08074.r.5/8
1852108

AHAGGAR. Algeria
Tuareg

KEENAN, Jeremy
The Tuareg. - London : Allen Lane, 1977
iii-xiii,385p,[16]p of plates : ill. - Bibliography: p.367-371. - Includes index
0-7139-0636-7 X.800/26670
1344218

AHAGGAR MOUNTAINS. Algeria
Prehistoric rock paintings
TROST, Franz
Die Felsbilder des Zentralen Ahaggar (Algerische Sahara). - Graz : Akademische Druck- und Verlagsanstalt, 1981
251p : ill(some col.). - (Monographien und Dokumentationen. Die afrikanischen Felsbilder). - Bibliography
3-201-01176-2 LR.411.i.6(9)
8802998

AHASCRAGH. Galway (County)
to 1978
MACLOCHLAINN, Tadhg
A historical summary on the Parish of Ahascragh, Caltra and Castleblakeney. - Ballinasloe : The author, [1979]
[4],92p : ill
Signed by the author
B79-28309 X.709/33349
1933035
- [Another copy.]
Signed by the author X.809/63891

AHAUS. West Germany
Community associations
SCHWERING, Burkhard
Nachbarschaften und Vereine in Ahaus. - Münster : Coppenrath, 1979
671p. - (Beiträge zur Volkskultur in Nordwestdeutschland ; Hft 18). - Bibl
3-920192-91-5 X:0709/743(18)
2430266

AHIMSA
ALTMAN, Nathaniel
Ahimsa. - Wheaton, Ill. ; London : Theosophical Publishing House, c1980
xvii,125p. - (A Quest book)
0-8356-0537-x X.529/51682
2828227

AHLBERG, ALF. Criticism. Swedish literature
- Bibliographies
LILLIESTAM, Ake
Alf Ahlbergs bibliografi. - Stockholm : [Kungliga biblioteket], 1982
261p. - (Acta Bibliothecæ Regiæ Stockholmiensis ; 39)
91-7000-088-3 Ac.9233.zk[vol.39]
8436770

AHLÉN & HOLM (FIRM)
to 1978
LARSSON, Valle
Ringvägen 100. - [Stockholm] : Askild & Kärnekull, 1979
161p : ill
91-582-0080-0 X.529/35287
226465X

AHLERS, FRIEDRICH. Conducting. Music
- Biographies
TOECHE-MITTLER, Joachim
Musikmeister Ahlers. - Stuttgart : Spemann, 1981
145p : ill
3-440-81049-6 X.431/12000
8309728

AHLERS-HESTERMANN, FRIEDRICH. German paintings
- Bibliographies
KAYSER, Werner
Friedrich Ahlers-Hestermann. - Hamburg, [1975]
83p. - (Hamburger Bibliographien ; Bd18)
3-7672-0359-6 2782.kt.25/18
1211668

AHLFORS, BENGT. Directing. Theatre. Finland
- Biographies
AHLFORS, Bengt
Fragment av en barndom. - [Helsingfors] : Söderströms, 1978
113p
951-520-521-2 X.989/85440
241760X

AHLGREN, ERNST
See Benedictsson, Victoria

AHLMANN, LIS. Danish textile arts
MOGENSEN, Thomas
Lis Ahlmann. - København : Christian Ejlers, 1974
71 p. : ill
87-7241-707-2 X.421/8798 Woolwich
0889083

AHLQVIST, PEHR. Fliseryd. Sweden
- Biographies
SVENSSON, Estrid
Om Per Ahlqvist och Fliseryd. - [Falun] : [E. Svensson], 1976
64p : ill
r6034132 X.319/8737 Woolwich
1125605

AHLSTEN, JOHAN. Politics. Sweden
- Biographies
AHLSTEN, Johan
Ett försök att minnas. - Visby : Wessman & Pettersson, 1976
176p : ill
91-85122-01-7 X.809/24490
1234021

AHMAD, BEY DE CONSTANTINE. Algeria
TEMIMI, Abdeljelil
Le Beylik de Constantine et Hadj Ahmed Bey (1830-1837). - Tunis : Presses de la Société tunisienne des arts graphiques, 1978
303p[22]p of plates : ill. - (Publications de la Revue d'histoire maghrebine ; v.1). - Bibliography
r6209693 X.0800/1793(1)
8738351

AHMADIYYAT
AHMAD, Ghulam, Hazrat Mirza
The essence of Islam. - London : London Mosque.
Vol.1. - 1979, c1978
xviii,328p
B85-01241 X.200/44837
1745735
Vol.2. - 1981
355p
B85-01240 X.200/44811
1745735
The philosophy of the teachings of Islam. - London : London Mosque, 1979
vii,115p
B85-01234 X.200/44823
1745735
CHAUDHRI, Rashid Ahmad
A book of religious knowledge. - 2nd ed. - London : London Mosque, 1982
79p. - (Children's series ; no.4)
B84-13692 X.208/9340
3058905
RAFIQ, B. A. (Bashir Ahmad)
Truth about Ahmadiyyat. - London : London Mosque, 1978
98p
0-85525-013-5 X.208/9320
1745735

AHMADIYYAT
RASUL, Faizur
Commonsense about Ahmadiyyat. - 2nd ed. - London : London Mosque, [1981]
35p,[1]p of plates
B84-13691 X.208/9339
1745735

AHMADIYYAT. Pakistan
The QADIYANIES. - London : London Mosque, [1975?]
46p
r6810881 X.200/41930
3028380

AHMADIYYAT. Pakistan
Persecution
FROM the world press / compiled by B.A. Rafiq. - London : London Mosque.
[Pt.1]: [English language]. - [1975?]
330p : ill
r6816133 X.809/59457
3069869

AHMADU BELLO UNIVERSITY
- Bibliographies
AMUNE, Stephen A.
A.B.U in perspective. - Zaria : Kashim Ibrahim Library, Ahmadu Bello University, 1982
xvi,105p
r6813455 X.525/9466
9114025

AHMEDABAD. India (Republic)
Economic development. Role of companies
PAPOLA, T. S.
Urban informal sector in a developing economy. - Sahibabad : Vikas, c1981
vi,128p
0-7069-1133-4 X.529/52457
2854988
- [Another copy] X.529/47200

AHMEDI, TACEDDIN AHMED. Poetry in Turkish
KORTANTAMER, Tunca
Leben und Weltbild des altosmanischen Dichters Ahmedi unter besonderer Berücksichtigung seines 'Diwans'. - Freiburg im Breisgau : Schwarz, 1973
509p.. - (Islamkundliche Untersuchungen ; Bd.22). - Includes bibliography
3-87997-022-x P.101/417[Bd 22]
8871779

AHMOSE-NOFRETARI, QUEEN, CONSORT OF AHMOSE I, KING OF EGYPT
Ancient Egypt
Historical sources
GITTON, Michel
L'épouse du Dieu, Ahmes Nefertary. - Paris : Les Belles Lettres, 1975
103 p.; 1 leaf of plates : ill. - (Annales littéraires de l'Université de Besançon ; 172) (Centre de recherches d'histoire ancienne ; vol. 15)
r6012518 Ac.282/6 [no. 172]
0852481

AHNLUND, NILS. Historiography. Sweden
BJÖRK, Ragnar
Den historiska argumenteringen. - Uppsala : [Uppsala universitet], 1983
340p. - (Studia historica upsaliensia ; 132). - Includes bibliography
91-554-1432-x Ac 1077.ng[vol.132]
8433933

AHNOFF, TORE. Swedish drawings. Exhibits. Konsthallen, Gothenburg
- Catalogues
KONSTHALLEN (Gothenburg)
Isabella Laurell, Bengt Kristenson och Tore Ahnoff visar teckningar och malningar. - Göteborg : Konsthallen, 1975
[32]p : ill
r6033858 X.419/7648 Woolwich
1510673

AHO, JUHANI. Fiction in Finnish
KOSKIMIES, Rafael
Kymmenen tutkielmaa Juhani Ahosta. - [Helsinki] : Suomalaisen Kirjallisuuden Seura, 1975
122 p.. - (Tietolipas ; 77)
951-7170-54-8 Ac.9080/9[no.77]
0895660
REMES, Hannu
Juhani Aho Virossa. - Joensuu : Korkeakoulu, 1979
80p. - (Kielten osaston julkaisuja / Joensuun korkeakoulu = Publications of the Department of Linguistics / University of Joensuu ; n:o 4). - Bibliography
951-696-207-6 X.0955/162(4)
8155402

AHRENSBURG. West Germany
Castles: Schloss Ahrensburg
LÜHNING, Frauke
Schloss Ahrensburg. - Neumünster : Wachholtz, 1982
70p,[8]leaves of plates : ill(some col). - (Führer zu schleswig-holsteinischen Museen ; Bd 1)
r6304810 X.429/14800
8983437

AHRENSBURG REGION. West Germany
Religious ceremonies, B.C.8850-B.C.8300. Archaeological sources: Ahrensburgian antiquities
RUST, Alfred
Urreligiöses Verhalten und Opferbrauchtum des eiszeitlichen Homo sapiens. - Neumünster, 1974
152 p. : ill
3-529-01819-8 X.100/15416
102079X

AHRENSBURGIAN ANTIQUITIES. Archaeological sources. Religious ceremonies, B.C.8850-B.C.8300. Ahrensburg region. West Germany
RUST, Alfred
Urreligiöses Verhalten und Opferbrauchtum des eiszeitlichen Homo sapiens. - Neumünster, 1974
152 p. : ill
3-529-01819-8 X.100/15416
102079X

ÄHTÄRI. Finland
Social conditions, to ca 1900
NYYSSÖLÄ, Eino
Entisestä Ähtäristä. - [2. painos]. - [Ähtäri] : [Ähtäri-seura], 1976
157p : ill
951-99098-8-5 X.529/20634
1532316

AHVENANMAA
See Aland Islands

AI
See Artificial insemination

AĬBEK, MUSA TASHMUKHAMEDOV. Uzbek literatures - Biographies
BAT, Lidiīa Grigor'evna
Aĭbek. - Moskva : Sovetskiĭ pisatel', 1976
221p, 1 leaf of plates : ill
r6021526 X.989/38550
0398179

AICHINGER, CARL FRIEDRICH. Linguistics. German language
POPPE, Erich
C.F. Aichingers 'Versuch einer teutschen Sprachlehre'. - Hildesheim : Olms, 1982
xii,539p. - (Documenta linguistica. Studienreihe ; 1) - Bibliography: p.514-539
3-487-07260-2 X.950/20596
8930015

AICHINGER, ILSE. Austrian writers. Fiction in German
LORENZ, Dagmar C. G.
Ilse Aichinger. - Königstein / Ts. : Athenäum, 1981
259p. - Bibliography
3-7610-8119-7 X.950/10066
8156352

AICHL, JAN BLAŽEJ SANTINI
See Santini-Aichl, Jan Blažej

AICKMAN (FAMILY). King's Lynn. Norfolk 1827-1896
AICKMAN, Dorothy Jean
John Aickman's Foundery MDCCC XXVII, King's Lynn. - [Cambridge] : [D.J. Aickman], 1980
ix,52p,[9]p of plates : ill. - Includes index
0-9507068-0-9 X.809/47590
2395525

AID
See Artificial insemination of donor sperm

AID
Organisation
HAMPSON, Fredrick Roy
AID. - [Stockport] : [Elgincourt Office Supplies for AID], [1976]
[3],16 leaves,6 leaves of plates
0-9505259-0-1 X.512/5367
0741272

AID TO FAMILIES WITH DEPENDENT CHILDREN
See also
Families receiving assistance from Aid to Families with Dependent Children

AID TO THE CHURCH IN NEED
to 1980
STRAATEN, Werenfried van
They call me the bacon priest. - 3rd ed. - Chulmleigh : Augustine in association with Aid to the Church in Need (UK), 1981
viii,160p,[8]p of plates : ill
0-85172-732-8 X.208/8054
2826801

AIDA. Verdi, Giuseppe. Opera in Italian
AIDA / [composer] Giuseppe Verdi. - London : J. Calder [etc.] : [Distributed by Calder and Boyars], 1980
96p : ill. - (English National Opera guides ; 2). - Bibliography: p.93. - List of sound discs: p.94-96
0-7145-3770-5 X.439/11096
230094x

AIDES. Prime ministers. Great Britain
JONES, G. W. (George William)
The Prime Minister's aides. - [Hull] : Department of Politics, University of Hull, 1980
[2],23 leaves. - (Hull papers in politics ; no.6). - Bibliography: leaves 22-23
B80-31214 X.702/6408
2242656

AIDOO, AMA ATA. Ghanaian writers. Drama in English
Dilemma of a Ghost
GRANT, Jane W.
Ama Ata Aidoo. - Harlow : Longman, 1980
ix,49p. - (Longman guides to literature)
0-582-65075-5 X.958/669
2364360

AIDS
See Acquired immune deficiency syndrome

AIGION. Greece
Mycenaean tombs, ca B.C.1500-ca B.C.1100. Archaeology, 1970
PAPADOPOULOS, Athanasios J.
Excavations at Aigion, 1970. - Göteborg : Aström, 1976
xvi, 50 p., 98 p. of plates : ill. - (Studies in Mediterranean archaeology ; vol. 46). - Bibliography : p. xiii-xiv
91-85058-68-8 X.0415/9(46)
0735019

AIH
See Homologous artificial insemination

AIKEN, CONRAD. American writers. English literature
ZIMMERMANN-MARTHALER, Rosmarie
Reisen ins Bewusstsein. - Zürich : [s.n.], 1981
ii,218p. - Bibliography
r6302272 X.950/10663
8131120

AIKEN, CONRAD. American writers. English literature - Biographies
AIKEN, Conrad
Selected letters of Conrad Aiken. - New Haven ; London : Yale University Press, 1978
xxv,350p,[20]p of plates : ill. - Includes index
0-300-02180-1 X.981/21082
1477870

LORENZ, Clarissa M.
Lorelei two. - Athens, Ga. : University of Georgia Press, c1983
231p,[8]p of plates : ill
0-8203-0661-4 X.800/40850
3182053

AIKEN, CONRAD. American writers. English literature Criticism - Bibliographies
HARRIS, Catherine Kirk
Conrad Aiken. - New York ; London : Garland, 1983
xxii,324p. - (Garland reference library of the humanities ; v.391). - Includes index
0-8240-9187-6 X.950/28936
3061566

AIKENS, TOM. Politics. Queensland - Biographies
MOLES, Ian
A majority of one. - St Lucia : University of Queensland Press ; Hemel Hempstead : Distributed by Prentice-Hall, 1979
xv,258p : ill. - Bibliography: p.245-251. - Includes index
0-7022-1285-7 X.809/45407 Woolwich
2057786

AIKIDO
LOI, Lee Ah
Tomiki aikido. - London : Crompton.
Bk.2. - 1979
74p : ill
0-901764-46-9 X.629/17403 Woolwich
2698056

STEVENS, John, 1947-
Aikido. - Boulder, Colo. ; London : Shambhala, 1984
ix,198p : chiefly ill
0-87773-229-9 X.525/10342
2698056

UESHIBA, Kisshōmaru
The spirit of aikidō. - Tokyo : Kodansha, 1984
126p : ill
4-7700-1100-8 X.629/25314
3180271

AIKIDO
Ki
TOHEI, Koichi
Book of ki. - Tokyo : Japan Publications, 1976
102p : ill. - Includes index
0-87040-379-6 X.611/8221 Woolwich
1569686

AILRED, SAINT
See Aelred, Saint

AILUROPODA MELANOLEUCA
See Giant pandas

AIM MICROCOMPUTER SYSTEMS
Rockwell AIM 65 microcomputer systems. Programming
SCANLON, Leo J.
AIM 65. - New York ; Chichester : Wiley, c1981
179p : ill
0-471-06488-2 X.622/10156 Woolwich
2400790

AĬMANOV, SHAKEN. Theatre. Kazakhstan - Biographies
BOGATENKOVA, L. (Līūdmila)
Narodnyĭ artist SSSR Shaken Aimanov. - Alma-Ata : Izdatel'stvo "Nauka" Kazakhskoĭ SSR, 1973
113p,[26]p of plates,[1]leaf of plates : ill,ports
r6178420 X.907/20629
2451255

AIMARA
See Aymará

AIMARA LANGUAGE
See Aymara language

AIMARA LITERATURE
See Aymara literature

AIMONE, NINO. Italian visual arts
NINO Aimone. - [Bologna] : Grafis, [1979?]
106p : chiefly ill. - (Rilevamenti nella terza generazione del '900). - Includes bibliography
r6507922 X.425/3936
8265070

AIMS
See Aims of Industry

AIMS FOR FREEDOM AND ENTERPRISE
See also
Aims of Industry

AIMS OF INDUSTRY
See also
Aims for Freedom and Enterprise

AIN (DEPARTMENT). France
to date
HISTOIRE des communes de l'Ain / Guy Brunet [et al.]. - [Le Coteau] : Horvath, 1983-
v. : ill. - (Grande encyclopedie de l'Ain ; vol.I). - Includes bibliography
r6298431 X.0805/211
871598x

AIN (DEPARTMENT). France
to date - Dictionaries
POMMEROL, Jacques
Dictionnaire du département de l'Ain. - Marseille : Laffitte reprints, 1980
xxxi,630p
r6205270 X.800/30815
871598x

AIN (DEPARTMENT). France
Local archives, to date - Serials
ARCHIVES DÉPARTEMENTALES DE L'AIN
Répertoire numérique de la sous-série 3E. - Bourg-en-Bresse : [les Archives], 1980- : ill
r6205945 S.E.400/884
8856354

AĬNI, SADRIDDIN. Tadzhik literature
MULLODZHANOVA, Zakiīa Alimovna
Stil'originala i perevod. - Dushanbe : Izdatel'stvo "Donish", 1976
134p
r6057856 X.900/19357
1430904

AĬNI, SADRIDDIN. Tadzhik literature - Biographies
BRAGINSKIĬ, Iosif Samuilovich
Sadriddin Aĭni. - 2-e dop. izd. - Moskva : Sovetskiĭ pisatel', 1978
244p
r6185476 X.989/82971
1430904

KNIGA zhizni Sadriddina Aĭni / podgotovka izdaniīa i issledovanie K.S. Aĭni ; pod obshcheĭ redakt͡sieĭ M.S. Asimova. - Dushanbe : Irfon, 1978
39,[ca.300]p : chieflyill(some col.),facsims,ports
r6178782 X.981/13665
1430904

AINOLA. Local associations. Sibelius, Jean. Finnish music
SALMENHAARA, Erkki
Sibeliuksen Ainola .... - Helsinki : W. Söderström, 1976
42p,[21]p of plates : ill
951-00-7533-7 X.439/4478
1714813

AINSWORTH, JOHN DAWSON. Kenya

MAXON, R. M.
John Ainsworth and the making of Kenya. – Washington, D.C. : University Press of America, c1980
xvi,456p. – Includes bibliography
0–8191–1155–4 X.809/57392
8275343

AINTREE. Merseyside (Metropolitan County)
Racehorses. Racing. Races: Grand National

HUGHES, John, 1926–
Long live the National. – London : Joseph, 1983
208p,[8]p of plates : ill(some col.). – Includes index
0–7181–2231–3 X.622/15651
2909839

AINU
to date – Bibliographies

ADAMI, Norbert Richard
Verzeichnis der europäischsprachigen Literatur über die Ainu. – Wiesbaden : Harrassowitz, 1981
vii,81p. – (Veröffentlichungen des Ostasien-Instituts der Ruhr-Universität Bochum ; Bd.27)
3–447–02207–8 X.0700/678(27)
8781982

AINU
to date – Catalogues

TAGUCHI, Kirsten Yumiko
An annotated catalogue of Ainu material in the East Asian Institute of Aarhus University. – Lund : Studentlitteratur, 1974
136p. – (Scandinavian Institute of Asian Studies monograph series ; no.20)
91–441–0571–1 X.0700/317(20)
8781982

AINU. Japan
Sakhalin Ainu. Medicine

OHNUKI-TIERNEY, Emiko
Illness and healing among the Sakhalin Ainu. – Cambridge : Cambridge University Press, 1981
xvi,245p : ill. – Bibliography: p226–238. – Includes index
0–521–23636–3 X.200/36846
2567385

AIR
See also
Compressed air
Dry air
Protective air enclosures
Winds

AIR

ARDLEY, Neil
Air and flight. – London : Watts, 1984
31p : col.ill. – (Action science). – Includes index
0–86313–158–1 X.322/14458
0308161

CRYSTAL, David
Air. – London : Edward Arnold, 1982
24p : ill. – (Databank). – Includes index
0–7131–0633–6 X.629/19516
0675628

DAFYDD, Marged
Aer. – Blaenau Ffestiniog : Elfen, 1983
51p : col.ill. – (Gwyddoniaeth yr ifanc). – Includes index
B84–38629 X.629/24151
3166619

FITZPATRICK, Julie
In the air. – London : Hamilton, 1984
29p : col.ill. – (Science spirals). – Includes index
0–241–11205–2 X.629/23202
0308161

JENNINGS, Terry
Air. – Oxford : Oxford University Press, 1982
32p : ill(chiefly col.). – (The Young scientist investigates)
0–19–918044–x X.322/11049 Woolwich
0308161

KERROD, Robin
Into the air. – London : Macdonald Educational, 1983
30p : ill(some col.). – (My first encyclopaedia ; 9). – Includes index
0–356–07825–6 X.622/16443
0308161

KING, Stanley Frederick
Science and air. – St Albans : Hart-Davis Educational, 1976
46p : ill. – (Science action book ; 1). – Bibliography: p.46
0–247–12705–1 Cup.1281/263
0675628

LLOYD, David, 1945–
Air. – London : Methuen, 1982
[26]p : col.ill. – (The Elements)
0–416–06470–1 X.622/13908
0308161

LUCAS, Angela
Focus on air. – London : Methuen, 1977
48p : col ill. – Includes index
0–416–82410–2 X.329/15361
0308161

RAMSDEN, E. N.
Air. – Cheltenham : Thornes, 1983
55p : ill. – (Extending science ; 1). – Includes index
0–85950–388–7 X.622/17449
0675628

SANDAY, A. P. (Alan Philip)
Air. – Loughborough : Ladybird Books, 1975
52p : col ill. – (Ladybird leaders)
0–7214–0389–1 X.629/10097
0308161

TECHNICAL guide to air and heat / unit written by Michael Thompson. – London : Addison-Wesley, for the Inner London Education Authority, 1978
24cards : ill.(col.)
2v. : ill.. – (Insight to science)
0–201–13760–7 Cup.900.tt.74.
174979X

AIR
Compressible flow

CHRISTOPHER, P. A. T.
Bodies in compressible flow. – [Cranfield] : [College of Aeronautics, Cranfield Institute of Technology], 1983
38leaves,[16]leaves of plates : ill. – (Cranfield College of Aeronautics report ; 8329)
0–902937–95–2 X.622/23122
3182681

AIR
Experiments

AIR. – London : Harrap, 1981
48p : ill. – (Access to science)
0–245–53663–9 X.622/11253 Woolwich
255870x
– [Teacher's notes.] ISBN 0–245–53704–X X.622/11254 Woolwich

AIR / general editors: Colin Terry and Terry Gibson. – St. Albans : Hart-Davis Educational, 1982
31p : ill. – (Centre science)
0–247–13135–0 X.629/18899
255870x

KINCAID, Doug
In the air. – [Amersham] : Hulton, 1981
64p : ill(some col.). – (Science in a topic). – Bibliography: 1p
0–7175–0877–3 X.622/9937 Woolwich
0219924

SMITH, Henry, 19—
Amazing air. – London : Methuen/Walker Books, 1982
46p : col.ill. – (Science club). – Includes index
0–416–24250–2 X.622/13595
0219924

AIR
Flow

MARKLAND, Eric
A first course in air flow. – Nottingham : Tecquipment Ltd, 1976
[1],89p : ill
0–904031–04–7 X.619/16988 Woolwich
0711179

AIR
Flow. Measurement

OWER, Ernest
The measurement of air flow. – 5th ed. (in SI units). – Oxford : Pergamon, 1977
xi,362p,[4] leaves of plates,[2]p of plates : ill. – Includes index
0–08–021282–4 SRL
1254138

AIR
Quality control. Mathematical models – Conference proceedings

MATHEMATICAL models for planning and controlling air quality / Giorgio Fronza and Piero Melli editors. – Oxford : Pergamon, 1982
viii,247p : ill. – (IIASA proceedings series ; v.17). – Includes bibliographies and index
0–08–029950–4 SRL
2820668

AIR. Buildings
Infiltration – Conference proceedings

AIC CONFERENCE (4th : 1983 : Elm)
Air infiltration reduction in existing buildings. – Bracknell : Air Infiltration Centre, c1983
1v.(various pagings) : ill. – Includes bibliographies
0–946075–04–2 X.622/19589
3143694
– Supplement to proceedings X.622/19590

AIR. Buildings
Infiltration. Design – Conference proceedings

BUILDING design for minimum air infiltration – Bracknell : Air Infiltration Centre, c1982
iv,215p : ill
0–946075–01–8 X.622/13636
2784882

AIR. Buildings
Infiltration. Measurement – Conference proceedings

AIC CONFERENCE (1st : 1980 : Windsor)
Air infiltration instrumentation and measuring techniques. – Bracknell : Air Infiltration Centre, c1981
xii,371p : ill
B81–26275 SRL
2502453
– [Another copy.] X.525/10619

AIR. Buildings
Pollution

MEYER, Beat
Indoor air quality. – Reading, Mass. ; London : Addison-Wesley, 1983
xiii,434p : ill. – Bibliography: p337–397. – Includes index
0–201–05094–3 SRL
2910411

WADDEN, Richard A.
Indoor air pollution. – New York ; Chichester : Wiley, c1983
xi,213p : ill. – (Environmental science and technology). – Includes bibliographies and index
0–471–87673–9 SRL
2910411

AIR. Chemical engineering plants. Great Britain
Pollutants: Toxic materials. Control measures

A GUIDE to the evaluation and control of toxic substances in the work environment. – London : Health Committee of the Chemical Industry Safety and Health Council of the Chemical Industries Association, c1980
48p
B81–11855 SRL
2376695

AIR. Coal mines
Pollutants: Gases – Programmed texts

NATIONAL COAL BOARD. Industrial Training Branch
More pit gases. – [London] : The Branch, 1977
[5],61p. – (Programmed text ; 11)
B77–23493 BS.82/97
1147668

AIR. Coal mines
Pollutants: Methane. Emission. Prediction

CURL, S. J.
Methane prediction in coal mines. – London : IEA Coal Research, 1978
79p(1 fold) : ill. – (Report / IEA Coal Research : no. ICTIS-TR04). – Bibliography: p.68–77
B80–35121 BS.82/164(4)
2240238

AIR. Coal mines. European Community countries
Pollutants: Methane. Control measures – Conference proceedings

METHANE, climate, ventilation in the coalmines of the European communities, Luxembourg, 4 to 6 November 1980 / [organization of the information symposium by the Commission of the European Communities Directorate-General Energy and Directorate-General for Information Market and Innovation]. – English language ed. – Redhill : Colliery Guardian.
Vol.1. – c1980
v,533p : ill
0–86108–080–7 SRL
2789639

AIR. Curriculum subjects. Primary schools
– For teaching

HINE, Pamela
What is air?. – Basingstoke : Globe Education for West Sussex County Council, 1981
39p : ill. – (Science horizons. Level 2a). – Bibliography: p8
0–333–28534–4 X.325/1254 Woolwich
2487594

AIR. Factories
Pollutants. Assessment & control - Conference proceedings
AIRBORNE hazards at work. - [London] : [QMC Industrial Research Ltd], [1977]
206leaves in various pagings : ill
B77-26367 SRL
1196081

AIR. Factories
Pollutants: Particles. Pollution control equipment
THEODORE, Louis
Industrial air pollution control equipment for particulates. - Cleveland, Ohio : CRC Press, 1976
[10],308p : ill. - Includes index
0-8493-5132-4 X.622/6635 Woolwich
1737783

AIR. Factories
Pollution by asbestos dust. Safety measures
GREAT BRITAIN. Health and Safety Executive
Asbestos. - 2nd ed. - [London] : H.M.S.O., 1975
[2],30p : 1 ill. - (Health and safety at work ; 44). - Bibliography: p.19-21
0-11-360860-8 BS.23/123
0126101

LAJHANS, Gyan S.
Engineering aspects of asbestos dust control. - Ann Arbor : Ann ArborScience, 1978
191p
0-250-40227-0 X.620/19085 Woolwich
0126101

AIR. Factories. Manufacture. Plastics & rubber. Great Britain
Pollution. Monitoring - Conference proceedings
AIR pollution monitoring. - Shawbury : Rubber and Plastics Research Association of Great Britain, 1978
[3],ii,39p : ill
0-902348-17-5 SRL
1940201

AIR. Hospitals
Contamination. Prevention. Engineering aspects
LUCIANO, Joseph R.
Air contamination control in hospitals. - New York ; London : Plenum Press, 1977
xvi,479p : ill. - Bibliography: p.451-466. - Includes index
0-306-30984-x SRL
1410474

THEODORE, Louis
Air pollution control for hospitals and other medical facilities. - New York ; London : Garland STPM, c1981
xii,324p : ill. - Includes index
0-8240-7132-8 SRL
1410474

AIR. Hospitals
Contamination. Prevention. Use of protective air enclosures
WHITE, P. A. F. (Percival Albert Frederick)
Protective air enclosures in health buildings. - London : Macmillan, 1981
viii,182p : ill. - Includes index
0-333-27678-7 SRL
241127x

AIR. Industrial buildings. United States
Pollution. Control measures: Ventilation
MCDERMOTT, Henry J.
Handbook of ventilation for contaminant control. - Ann Arbor : Ann Arbor Science, 1976
viii,368p : ill. - Includes index
0-250-40139-8 X.620/16720 Woolwich
109968X

AIR. Industries. Great Britain
Pollutants: Toxic gases. Control - Conference proceedings
The CONTROL of toxic atmospheres / edited by D.J. Hughes. - Manchester : The Branch, 1980
87p in various pagings : ill. - (Symposium papers / Institution of Chemical Engineers. North Western Branch ; no.2)
0-906636-05-1 SRL
2304163

AIR. Mines
Contaminants: Radon. Control
BYKHOVSKIĬ, A. V.
Experience with radon control in mining. - New Delhi : Amerind, c1974
192p : ill. - Bibliography: p183-192
0-86249-360-9 X.629/21833
3029875

AIR. Storage buildings. Horticultural products. Great Britain
Quality control
AGRICULTURAL DEVELOPMENT AND ADVISORY SERVICE
Control of air conditions in horticultural stores. - [London] : H.M.S.O., 1975
15p : ill(some col). - (Mechanization leaflet ; 12)
0-11-240669-6 SRL
0359947

AIR. Subsonic wind tunnels
Turbulence & laminar flow. Measurement. Applications of hot-wire anemometry
PERRY, A. E. (Anthony Edward)
Hot-wire anemometry. - Oxford : Clarendon, 1982
xix,184p : ill. - (Oxford science publications). - Bibliography: p177-180. - Includes index
0-19-856327-2 SRL
2714450

'AIR'. Words. French language
Semantics
MUDIMBE, Valentin Yves
Air. - Wien : E. Stiglmayr, 1979
454p. - (Acta ethnologica et linguistica ; Nr 46. Series generalis ; 5). - Bibl: p.265-299
r6146171 X.0802/102(46)
2478250

AIR. Workplaces
Pollutants: Heavier than air gases. Dispersion
DENSE gas dispersion / edited by R.E. Britter and R.F. Griffiths. - Amsterdam ; Oxford : Elsevier Scientific, 1982
vii,247p : ill. - (Chemical engineering monographs ; 16)
0-444-42095-9 X.622/14023
2797070

AIR. Workplaces. Aluminium industries
Pollutants. Sampling
The MEASUREMENT of employee exposures in aluminium reduction plants / International Primary Aluminium Institute, Health Committee review. - London : International Primary Aluminium Institute, c1982
52p : ill
B83-38295 SRL
2953110

AIR. Workplaces. Great Britain
Pollutants: Asbestos dust. Control. Standards
ACHESON, E. D. (Ernest Donald)
Asbestos. - London : H.M.S.O. [for the] Health & Safety Commission, 1983
26p : ill. - Bibliography: p19-21
0-11-883700-1 X.525/8720
3061116

AIR. Workplaces. Great Britain
Pollutants: Dangerous gases. Detection & measurement
DETECTION and measurement of hazardous gases / edited by C.F. Cullis and J.G. Firth. - London : Heinemann, 1981
226p : ill. - Includes index
0-435-71030-3 SRL
253262x

DETERMINATION of formaldehyde in the workplace atmosphere. - London : Chemical Industries Association, c1983
8p
B84-31753 X.622/19989
253262x

AIR. Workplaces. Great Britain
Pollutants: Lead. Control
GREAT BRITAIN. Health and Safety Commission
Control of lead at work. - London : H.M.S.O., 1980
vi,27p
0-11-883411-8 SRL
2535491

AIR. Workplaces. Great Britain
Pollutants: Lead. Control. Standards
GREAT BRITAIN. Health and Safety Commission
Control of lead at work. - London : H.M.S.O., 1978
vi,36p : 1 ill
0-11-883206-9 BS.438/72
1696084

AIR. Workplaces. Great Britain
Pollutants: Lead. Control. Standards. Law: Control of Lead at Work Regulations 1980
CONTROL of lead at work. - London : Lead Development Association, [1981]
39p : ill
B81-40570 SRL
2575345

AIR. Workplaces. Great Britain
Pollution. Safety levels. Legal aspects
FARMER, David, 19—
Safe to breathe?. - Horsham : Kingwood, 1984
80p. - (Kingwood focus on occupational health and safety ; 1)
0-946933-00-6 X.950/37452
3026876

AIR (DEPARTMENT). France
Proton synchrotrons: CERN 400 Ge V Super Proton Synchrotron. Design & construction
GOLDSMITH, Maurice
Europe's giant accelerator. - London : Taylor and Francis, 1977
x,261p,xvi p of plates : ill(some col). - Includes index
0-85066-121-8 X.622/5984 Woolwich
1175955

AIR AMERICA
to 1978
ROBBINS, Christopher
The invisible air force. - London : Macmillan, 1979
319p. - Bibliography: p.313-314. - Includes index
0-333-21847-7 X.809/44687
1945874

ROBBINS, Christopher
The invisible air force. - London : Pan in association with Macmillan, 1981, c1979
318p. - Bibliography: p311-312. - Includes index
0-330-26256-4 H.81/401 Woolwich
1945874

AIR BASES. Great Britain
American air bases. Establishment, 1947-1950
THOMAS, Andy
How Britain was sold. - Nottingham : Peace News, [1984]
31p : 2ill. - (A Peace News/Housmans pamphlet)
0-946409-04-8 X.809/61602
3194183

AIR BASES. Hawkinge. Kent
Hawkinge RAF Station, to 1968
HUMPHREYS, Roy S.
Hawkinge, 1912-1961. - Gillingham : Meresborough, 1981
192p : ill
0-905270-35-5 X.629/19643
2866498

AIR BASES. Woodvale. Merseyside (Metropolitan County)
Woodvale RAF Station, to 1979
FERGUSON, Aldon P. (Aldon Patrick)
A history of Royal Air Force Woodvale. - 2nd ed. - Liverpool : Merseyside Aviation Society, 1980
67p : ill. - (Military airfield history ; no.3) (A Merseyside Aviation Society publication)
0-902420-36-4 X.629/14815 Woolwich
2356961

AIR BASES. Yorkshire
1939-1945
RAPIER, Brian J. (Brian James)
White Rose Base. - [2nd ed]. - [York] : [Air Museum York Publications], 1980
94p : ill. - Includes index
0-9507326-0-5 X.622/9555 Woolwich
2357518

AIR BATTLES
to 1967
JACKSON, Robert, 1941-
Aerial combat. - London : Weidenfeld and Nicolson, 1976
160p : ill(some col). - Bibliography: p.156. - Includes index
0-297-77235-x X.622/5689 Woolwich
0756040

AIR BATTLES. Belgrade. Yugoslavia
1941-1945
KOSTIĆ, Božidar S
Borbe u vasduhu za Beograd 1941. - [West Germany] : [b.i.], 1978
xxvi,147p[3]fold.leaves of plates : ill,col maps,ports. - (Poezija i proza). - Bibl.: p.135
r6186799 X.709/51852
2443171

AIR-BORNE FORCES
to 1981
WEEKS, John
The airborne soldier. - Poole : Blandford, 1982
192p : ill(some col.). - Includes index
0-7137-0918-9 X.808/34706
2843765

AIR-BORNE FORCES. Germany. Heer
1939-1945

QUARRIE, Bruce
German airborne troops 1939-45. - London : Osprey, 1983
40p,A-Hp of plates : ill(some col.). - (Men-at-arms series ; 139)
0-85045-480-8 X.622/16002
2930560

AIR COMPRESSORS
See also
Shops with air compressors

AIR COMPRESSORS

HOW to Choose and use an air compressor / prepared for Campbell Hausfeld by Robert Scharff & Associates Ltd. - Harrison, Ohio : Campbell Hausfeld, c1979
154p : ill
0-8359-2943-4 X.622/18403
0544353

AIR CONDITIONING
See also
Ventilation

AIR CONDITIONING

AIR-CONDITIONING AND REFRIGERATION INSTITUTE
Refrigeration and air-conditioning. - Englewood Cliffs ; London : Prentice-Hall, 1979
xii,863p : ill. - Includes index
0-13-770164-0 X.622/6789 Woolwich
0053015

ARORA, C. P. (Chandra Prakash)
Refrigeration and air conditioning. - New Delhi ; London : Tata McGraw-Hill, c1981
xxiii,728p,[12]folded p of plates : ill. - Includes index
0-07-451515-2 X.622/16061
0053015

HARRIS, Norman C. (Norman Clifton)
Modern air conditioning practice. - 3rd ed.. - New York ; London : McGraw-Hill, c1983
xii,433p : ill. - Includes index
0-07-026833-9 X.622/20498
0053015

JONES, William Peter
Air conditioning applications and design. - London : Edward Arnold, 1980
x,330p : ill. - Includes index
0-7131-3423-2 SRL
0053015

LANG, V. Paul
Principles of air conditioning. - 3rd ed. - Albany : Delmar, 1979
326p
0-8273-1001-3 X.622/7124 Woolwich
0053015

LANGLEY, Billy C.
Refrigeration and air conditioning. - Reston : Reston Pub., 1978
xiii,689p : ill. - Includes index
0-8359-6612-7 X.620/18075 Woolwich
0053015

Refrigeration and air conditioning. - 2nd ed. - Reston : Reston Publishing, c1982
xiii,687p : ill
0-8359-6617-8 SRL
0053015

AIR CONDITIONING
Scientific aspects

KIMURA, Ken-Ichi
Scientific basis of air conditioning. - London : Applied Science Publishers, 1977
viii,269p : ill. - (Architectural science series). - Bibliography: p.259-263. - Includes index
0-85334-732-8 SRL
1265059

AIR CONDITIONING. Buildings
Air conditioning & ventilation

CROOME, Derek J. (Derek John)
Airconditioning and ventilation of buildings. - 2nd ed. - Oxford : Pergamon. - (International series in heating, ventilation and refrigeration ; v.14)
Vol.1. - 1981
xx,575p : ill. - Includes bibliographies and index
0-08-024779-2 SRL
0321699

PITA, Edward G. (Edward Gerald)
Air conditioning principles and systems. - New York ; Chichester : Wiley, c1981
xiii,467p : ill. - Includes index
0-471-04214-5 X.622/10161 Woolwich
0321699

TROTT, A. R.
Refrigeration and air-conditioning. - London : McGraw-Hill, c1981
310p : ill. - Includes index
0-07-084543-3 X.622/10460 Woolwich
0321699

AIR CONDITIONING. Buildings
Heating & air conditioning. Use of solar energy

ANDERSON, Bruce
The solar home book. - Harrisville, N.H. : Cheshire Books, 1976
297,[1]p : ill. - Bibliography- Includes index
0-917352-01-7 L.R.421/72
0786683

HARRELL, Joe J.
Solar heating and cooling of buildings. - New York ; London : Van Nostrand Reinhold, c1982
xii,365p : ill. - Includes bibliographies and index
0-442-21658-0 SRL
2052997

KREIDER, Jan F.
Solar heating and cooling. - Washington, D.C. : Scripta Book Company ; New York ; London : McGraw-Hill, [1976]
ix,342p : ill. - Bibliography: p.337-338. - Includes index
0-07-035473-1 SRL
0786683

Solar heating and cooling. - Washington, [D.C] ; London : Hemisphere, c1982
xiii,479p : ill. - Includes index
0-07-035486-3 SRL
0786683

MAZRIA, Edward
The passive solar energy book. - Emmaus, Pa : Rodale Press, 1979
435p
0-87857-237-6 X.611/9189 Woolwich
0786683

SOLAR energy application in buildings / edited by A.A.M. Sayigh. - New York ; London : Academic Press, 1979
xvi,444p : ill. - Bibliography: p.423-433. - Includes index
0-12-620860-3 X.622/7990 Woolwich
2052997

AIR CONDITIONING. Buildings
Heating & air conditioning. Use of solar energy - Conference proceedings

INTERNATIONAL CONFERENCE ON SOLAR BUILDING TECHNOLOGY (1st : 1977 : Royal Institute of British Architects, London)
Solar technology for building. - London : RIBA Publications, 1978
2v.(846p) : ill. - Includes bibliographies and index
0-900630-66-3 SRL
2052997

AIR CONDITIONING. Buildings
Heating, ventilation & air conditioning

HAVRELLA, Raymond A.
Heating, ventilating, and air conditioning fundamentals. - New York ; London : McGraw-Hill, c1981
viii,280p : ill. - (Contemporary construction series). - Includes index
0-07-027281-6 X.622/11696 Woolwich
2596210

AIR CONDITIONING. Buildings
Heating, ventilation & air conditioning - Conference proceedings

HEATING, VENTILATING AND AIR CONDITIONING SEMINAR (1975 : Imperial College of Science and Technology)
Heating, ventilating and air conditioning. - Southampton : Blair Bowden Associates Ltd [for] Lennox Industries, 1975
[57]p : ill
0-903329-01-8 SRL
2596210

AIR CONDITIONING. Buildings
Heating, ventilation & air conditioning. Control systems

FISK, D. J.
Thermal control of buildings. - London : Applied Science, c1981
xvii,246p : ill. - Includes bibliographies and index
0-85334-950-9 SRL
2490021

SCHNEIDER, Raymond K.
HVAC control systems. - New York ; Chichester : Wiley, c1981
x,358p : ill. - Includes index
0-471-05180-2 X.622/10393 Woolwich
2490021

AIR CONDITIONING. Buildings. Special subjects. Comprehension. English language
- For non-English speaking students

COOK, Norman, 1941-
Air-conditioning. - London : Macmillan, 1979
iv,68p : ill. - (Base books)
0-333-25430-9 X.619/22299 Woolwich
2005212

AIR CONDITIONING. Buildings. United States
Heating & air conditioning. Use of solar energy

BRADEN, Spruille
Successful solar energy solutions. - New York ; London : Van Nostrand Reinhold, 1980
xi,302p : ill. - Bibliography: p.297. - Includes index
0-442-20738-7 SRL
2207354

AIR CONDITIONING. Mines
Air conditioning & ventilation

MINE ventilation and air conditioning / [principle author] Howard L. Hartman. - 2nd ed. - New York ; Chichester : Wiley, 1982
xviii,791p : ill. - Includes index
0-471-05690-1 SRL
2789809

AIR CONDITIONING. Residences
Heating & air conditioning

PRICE, Bill
Central heating & air conditioning repair guide. - 2nd ed. - Blue Ridge Summit : Tab, c1983
314p : ill. - Includes index
0-8306-0120-1 X.629/22193
0067261

AIR CONDITIONING. Residences
Heating & air conditioning. Use of solar energy, 1950-1976 - Bibliographies

EGGERS-LURA, A.
Solar energy for domestic heating and cooling. - Oxford : Pergamon, 1979
vi,229p. - (The Pergamon European heliostudies ; vol.2). - Includes index
0-08-022152-1 SRL
1595687

AIR CONDITIONING. Residences. United States
Heating & air conditioning. Use of renewable energy sources - For architectural design

CROWTHER, Richard L.
Sun/earth. - Updated and enl. ed. - New York ; London : Van Nostrand Reinhold, c1983
255p : ill. - Bibliography: p243-251. - Includes index
0-442-21499-5 L.49/2118
3045447

AIR CONDITIONING. Workplaces. Great Britain
Ventilation & air conditioning

EVERSON, I.
Industrial ventilation and air-conditioning. - [Camberley] : Industrial and Commercial Techniques, [1979]
247p in various pagings : ill
B79-23955 SRL
1915169

AIR CONDITIONING EQUIPMENT
See also
Compressors

AIR CONDITIONING EQUIPMENT

KISSELL, Thomas E.
Motors, controls and circuits for air conditioning and refrigeration systems. - Reston, Va. : Reston Publishing, c1984
xv,253p : ill
0-8359-4665-7 X.622/19796
2817675

LANGLEY, Billy C.
Estimating air conditioning systems. - Reston, Va. : Reston Publishing Co., c1983
ix,261p : ill
0-8359-1790-8 X.622/16655
2817675

MAHONEY, Edward F.
Reading and interpreting diagrams in air conditioning and refrigeration. - Reston, Va. : Reston, c1983
xi,176p : ill. - Bibliography: p167
r6811444 X.622/17059
2817675

STOECKER, W. F. (Wilbert Frederick)
Refrigeration and air conditioning. - 2nd ed. - New York ; London : McGraw-Hill, 1982
xvi,443p : ill. - (McGraw-Hill series in mechanical engineering). - Includes index
0-07-061619-1 SRL
2817675

AIR CONDITIONING EQUIPMENT
Air conditioning, heating & ventilation equipment

PORGES, John
Handbook of heating, ventilating and air conditioning. - 7th ed.. - London : Newnes-Butterworths, 1976
308p in various pagings, 6p of plates(6 fold).fold leaf of plate : ill. - Bibliography: p.xiv 1-xiv 5. - Includes index
0-408-00233-6 X.629/10806 Woolwich
0719331

AIR CONDITIONING EQUIPMENT
Air conditioning, heating & ventilation equipment - Directories

AIR conditioning, ventilating and heating equipment. - 1978-79. - Croydon : Heating and Ventilating Publications (Developments) Ltd, [1978]
422p : ill
Subsequent volumes, when received, will be found at SRL
B78-21356 SRL
0227498

AIR CONDITIONING EQUIPMENT
Air conditioning, heating & ventilation equipment. Analysis & design

MCQUISTON, Faye C.
Heating, ventilating, and air conditioning. - New York ; London : Wiley, 1977
xx,651p : ill. - Includes index
0-471-01722-1 X.620/17317 Woolwich
130318x

MCQUISTON, Faye C. (Faye Clem)
Heating, ventilating, and air conditioning. - 2nd ed. - New York ; Chichester : Wiley, c1982
xviii,666p : ill. - Includes index
0-471-08259-7 X.622/12190 Woolwich
130318x

AIR CONDITIONING EQUIPMENT
Air conditioning, heating & ventilation equipment. Design. Effects of conservation of energy resources

CHARTERED INSTITUTION OF BUILDING SERVICES
Building energy code. - [London] : C.I.B.S.
Part 1: Guidance towards energy conserving design of buildings and services. - 1977
24p : ill
B78-12948 SRL
1403192

AIR CONDITIONING EQUIPMENT
Design & installation

UNIT air conditioning / [prepared by the Refrigeration and Unit Air Conditioning Group of the Heating and Ventilating Contractors' Association and the British Refrigeration and Air Conditioning Association, in collaboration with the Chartered Institution of Building Services]. - London : Heating and Ventilating Contractors' Association, 1980
19p : ill
B81-37085 X.622/11193 Woolwich
2558416

AIR CONDITIONING EQUIPMENT. Buildings
Air conditioning, heating & ventilation equipment

FABER, Oscar
Heating and air-conditioning of buildings. - 6th revised ed. (in SI units). - London : Architectural Press, 1979
ix,645p,leaf plate,[22]p of plates : ill. - Includes index
0-85139-288-1 SRL
1909339

HALL, F. (Fred)
Heating, ventilating and air conditioning. - Lancaster : Construction Press, 1980
141p : ill. - Bibliography: p137. - Includes index
0-86095-884-1 SRL
1909339

JENNINGS, Burgess Hill
The thermal environment. - New York ; London : Harper and Row, 1978
xix,648p : ill. - (Series in mechanical engineering). - Includes bibliographies and index
0-06-043311-6 SRL
1909339

PORGES, F. (Fred)
Handbook of heating, ventilating and air conditioning. - 8th ed. - London : Butterworths, 1982
267p[4]p of plates : ill. - Includes index
0-408-00519-x SRL
1909339

AIR CONDITIONING EQUIPMENT. Buildings
Air conditioning, heating & ventilation equipment. Automatic control systems

LETHERMAN, K. M. (Kenneth Murray)
Automatic controls for heating and air conditioning. - Oxford : Pergamon, 1981
xiv,235p : ill. - (International series in heating, ventilation and refrigeration ; v.15). - Bibliography: p199-205. - Includes index
0-08-023222-1 SRL
2587238

AIR CONDITIONING EQUIPMENT. Buildings
Air conditioning, heating & ventilation equipment. Control systems

HAINES, Roger W. (Roger William)
Control systems for heating, ventilating and air conditioning. - 3rd ed. - New York ; London : Van Nostrand Reinhold, c1983
xiv,306p : ill. - (Van Nostrand Reinhold environmental engineering series). - Includes index
0-442-23649-2 SRL
1171895

HAINES, Roger William
Control systems for heating, ventilating and air conditioning. - 2nd ed. - New York ; London : Van Nostrand Reinhold, 1977
xvii,233p : ill. - (Van Nostrand Reinhold environmental engineering series). - Includes index
0-442-23031-1 X.620/17049 Woolwich
1171895

AIR CONDITIONING EQUIPMENT. Buildings
Air conditioning, heating & ventilation equipment. Installation. Estimating

KHASHAB, A. M.
Heating, ventilating, and air-conditioning systems estimating manual. - New York ; London : McGraw-Hill, 1977
xvii,311p : ill. - Bibliography: p.303-304. - Includes index
0-07-034535-x SRL
1290673

KHASHAB, A. M.
Heating, ventilation, and air-conditioning systems estimating manual. - 2nd ed. - New York ; London : McGraw-Hill, 1984
xv,320p : ill. - Bibliography: p313-314. - Includes index
0-07-034536-8 SRL
1290673

AIR CONDITIONING EQUIPMENT. Buildings
Air conditioning, heating & ventilation equipment. Pneumatic automatic control systems

EDWARDS, Harry J.
Automatic controls for heating and air conditioning. - New York ; London : McGraw-Hill, 1980
xix,150p : ill. - Includes index
0-07-019046-1 SRL
2309149

AIR CONDITIONING EQUIPMENT. Cars
Maintenance & repair

CROUSE, William H. (William Harry)
Automotive air conditioning. - New York ; London : McGraw-Hill, 1977
xi,292p. - Includes index
0-07-014591-1 X.615/2012 Woolwich
011801x

Automotive air conditioning. - 2nd ed. - New York ; London : Gregg, c1983
275p : ill. - Includes index
0-07-014857-0 SRL
1455117

DWIGGINS, Boyce H.
Automotive air conditioning. - 4th ed. - New York ; London : Van Nostrand Reinhold, 1978
vii,408p : ill(some col). - Includes index
0-442-22216-5 SRL
1455117

LANGLEY, Billy C.
Principles and service of automotive air conditioning. - Reston, Va. : Reston, c1984
261p : ill
0-8359-5615-6 X.622/19891
1455117

SAMUELS, Clifford L.
Automotive air conditioning. - Englewood Cliffs ; London : Prentice-Hall, c1981
xiv,226p : ill. - Includes index
0-13-054205-9 X.622/14881
1455117

Automotive air conditioning. - Englewood Cliffs ; London : Prentice-Hall, c1981
xiv,226p : ill. - Includes index
0-13-054213-x X.622/11891
1455117

AIR CONDITIONING EQUIPMENT. Great Britain
Heating & air conditioning equipment. Installation. Standard costs - Serials

SPON'S quarterly price book update. - 1 (1981)-. - London : E. & F. N. Spon, 1981-
v.
B83-10265 P.421/786
0028959

AIR CONDITIONING EQUIPMENT. Residences
Maintenance & repair

FISCHER, Roger A.
Successful air conditioning & refrigeration repair. - Blue Ridge Summit : TAB, c1982
224p : ill. - Includes index
0-8306-0015-9 X.629/23155
0118001

AIR CONDITIONING EQUIPMENT INDUSTRIES. Dartford. Kent
Air conditioning, heating & ventilation equipment industries: Hall-Thermotank Limited, to 1984

MILLER, Harry, 1900 June 19—
Halls of Dartford 1785-1985. - London : Hutchinson Benham, 1985
231p[4]p of plates : ill. - Bibliography: p221-222. - Includes index
0-09-158490-6 X.622/22585
3152189

AIR CONDITIONING EQUIPMENT INDUSTRIES. Great Britain
Air conditioning equipment industries & trades - Statistics

REFRIGERATION & air conditioning / [Business Ratios]. - 3rd ed.: [1979]. - London : Business Ratios, [1979]
[101]p : ill
B80-35767 X.622/8603 Woolwich
2269341

AIR CONDITIONING EQUIPMENT INDUSTRIES. Great Britain
Air conditioning, heating & ventilation equipment industries - Directories

UNITED Kingdom heating, refrigeration & ventilating equipment manufacturers / [Data Research Group]. - [1976]. - Amersham : The Group, [1976]
86 leaves in various pagings
Subsequent volumes, when received, will be placed at this shelfmark
0-905570-55-3 X.0615/522 Woolwich
0246891

AIR CONDITIONING EQUIPMENT INDUSTRIES. Great Britain
Air conditioning, heating & ventilation equipment industries - Serials

H & V Index Desk Reference Book. - [No.1]- ; Apr.-Sept. '73-. - Warlingham : Heating and Ventilating Index, Industrial Indexes (International) Ltd, [1973]-
B75-27092 SRL
0512346

AIR CONDITIONING EQUIPMENT INDUSTRIES. Kirkby. Merseyside (Metropolitan County)
KME, to 1979

ECCLES, Tony
Under new managment. - London : Pan, 1981
416p. - (Pan business / industrial relations) (Pan original). - Bibliography: p407-410. - Includes index
0-330-26285-8 X.529/44478
2584212

AIR CONDITIONING EQUIPMENT INDUSTRIES. Western Europe
Air conditioning, heating & ventilation equipment industries - Serials

EXPOCLIMA Journal. - [No.1]- ; Mar. 1976-. - [Karlsruhe] : [Promoter Verlags- und Förderungsgesellschaft etc.] ; [Croydon] : [Heating and Ventilating Publications Ltd], 1976- : ill
B76-17370 SRL
0650196

AIR CONDITIONING EQUIPMENT INDUSTRIES. Western Europe
Air conditioning, heating & ventilation equipment industries & trades - Statistics

A MARKET studies report on heating and ventilation equipment in Europe. - London : Inter Company Comparisons, c1981
83p : ill. - (Market studies)
0-86191-910-6 SRL
2610191

AIR CONDITIONING EQUIPMENT TRADES. Great Britain
Air conditioning equipment industries & trades - Statistics

REFRIGERATION & air conditioning / [Business Ratios]. - 3rd ed.: [1979]. - London : Business Ratios, [1979]
[101]p : ill
B80-35767 X.622/8603 Woolwich
2269341

AIR CONDITIONING EQUIPMENT TRADES. Western Europe
Air conditioning, heating & ventilation equipment industries & trades - Statistics

A MARKET studies report on heating and ventilation equipment in Europe. - London : Inter Company Comparisons, c1981
83p : ill. - (Market studies)
0-86191-910-6 SRL
2610191

AIR CUSHION VEHICLES
See Hovercraft

AIR CYLINDERS. Respiratory protective equipment. Fire brigades. Great Britain
Compressed air cylinders. Recharging

ROYAL INSTITUTE OF PUBLIC ADMINISTRATION. Local Government Operational Research Unit
The recharging of firemen's breathing apparatus cylinders. - Reading : The Unit. - (Research reports / Joint Committee on Fire Research ; no.3)
[Main report]. - 1975
[8],108[i.e.109]p(2 fold) : ill. - (Reports / Royal Institute of Public Administration. Local Government Operational Research Unit ; no.C.214)
B79-27553 SRL
1925040

Ready reckoner tables. - 1975
3,[62]p. - (Reports / Royal Institute of Public Administration. Local Government Operational Research Unit ; no.214a)
B79-27554 SRL
1925040

AIR DEFENCE SYSTEMS
to 1973

HOGG, Ian V. (Ian Vernon)
Anti-aircraft. - London : Macdonald and Jane's, 1978
189p,[48]p of plates : ill. - Includes index
0-354-01163-4 X.620/18218 Woolwich
1781618

AIR DISPLAYS. England
International Air Tattoo, to 1983

AIR tattoo. - Leicester, 1983
80p : ill
0-904597-48-2 X.622/18649
3076636

AIR DISPLAYS. Farnborough. Hampshire
Farnborough Air Display, 1948-1979

BAXTER, Raymond
Raymond Baxter's Farnborough commentary. - Cambridge : Stephens, 1980
112p : chiefly ill
0-85059-434-0 X.622/10777 Woolwich
2109700

AIR DISPLAYS. Reno. Nevada
Reno Air Races (1982)

MOLL, Nigel
Reno. - London : Osprey, 1983
123p,[5]p of plates : col.ill
0-85045-531-6 X.629/22224
3059316

AIR DISPLAYS. Western Europe

AIRSHOWS 1978 / compiled by David Sargent. - Leicester : Midland Counties Publications, 1979
43p : ill
Subsequent volumes, when received, will be placed at this shelfmark
0-904597-19-9 P.623/954
194312x

AIRSHOWS 1979 / compiled by David Sargent. - Leicester : Midland Counties Publications, 1980
43p : ill
0-904597-27-x X.622/8670 Woolwich
2282305

BUSHELL, Susan
Airshows 1982. - Leicester : Midland Counties Publications, c1983
43p : ill
0-904597-44-x X.622/17298
2984040

AIR DISPLAYS. Western Europe
- Serials

AIRSHOWS. - 1980. - Leicester : Midland Counties Publications, c1981
43p
0-904597-35-0 X.622/11152 Woolwich
194312x

AIR DISTRIBUTION SYSTEMS. Refrigerated freight containers

LAWTON, A. R.
Pressure drops through loaded containers. - Cambridge : Shipowners Refrigerated Cargo Research Association, 1983
13leaves,[8]leaves of plates : ill(some col.). - Bibliography: leaf 13
B84-28995 SRL
3135020

AIR FORCES
See also
Australia. Royal Australian Air Force
Bulgaria. Voennovŭzdushni Sili
Canada. Royal Canadian Air Force
Finland. Ilmavoimat
France. Armée de l'Air
Germany. Luftwaffe
Glider forces
Great Britain. Royal Air Force
Great Britain. Royal Flying Corps
Great Britain. Royal Flying Corps Canada
Honduras. Fuerza Aérea
India. Indian Air Force
Israel. Ḥel âvir
Italy. Aeronautica militare
Italy. Regia Aeronautica
Japan. Nihon Kaigun Kōkū-bu
Japan. Nihon Rikugun Kōkū-butai
Netherlands. Koninklijke Luchtmacht
Night fighter forces
Pakistan. Pakistan Air Force
Poland. Lotnicze Siły Zbrojne
Polskie siły powietrzne
Sweden. Flygvapnet
Union of Soviet Socialist Republics. Voenno-vozdushnye sily
United States. Air Force
United States. Army Air Force
Uruguay. Fuerza Aérea

AIR FORCES

AIR forces of the world / compiled by David C. Mondey and Kenneth Munson ; and edited by John W.R. Taylor. - London : Jane's, 1981
256p : ill. - (Jane's pocket book)
0-7106-0012-7 X.629/15482 Woolwich
2404958

AIR forces of the world / Mark Hewish ... [et al.]. - London : Peerage, 1984, c1979
264p : ill(some col.). - Includes index
0-907408-93-1 X.625/934
1767283

AIR power / foreword by Stewart Menaul ; edited by Anthony Robinson. - London : Orbis Books, 1980
viii,304p : ill(chiefly col). - Includes index
0-85613-289-6 L.42/465
2046318

WHEELER, Barry C.
Air forces of the world. - London : I. Allan, 1979
112p,[4]p of plates : ill(some col). - Includes index
0-7110-0979-1 X.809/44861
1997459

AIR FORCES
- Directories

AIR forces of the world / Mark Hewish ... [et al.]. - London : Salamander Books, 1979
264[i.e.260]p(2 fold) : ill(chiefly col). - Includes index
0-86101-039-6 X.805/1075
2046318

AIR FORCES
- Encyclopaedias

The WORLD'S air forces / general editor Chris Chant. - Newton Abbot : David and Charles, 1979
255p : ill. - Includes index
0-7153-7690-x X.802/4614 Woolwich
1767283

AIR FORCES
to 1945

CHANT, Christopher
The illustrated history of the air forces of World War I & World War II. - London : Hamlyn, 1979
287p : ill(some col). - (A quarto book). - Bibliography: p.286. - Includes index
0-600-30326-8 X.802/11081
2048809

AIR FORCES
Allied air forces. Aeroplanes. Losses, 1939-1945

PHILPOTT, Bryan
In enemy hands. - Cambridge : Stephens, 1981
89p : ill
0-85059-499-5 X.622/11414 Woolwich
2573016

AIR FORCES
Allied air forces. Air operations during Pacific campaigns of World War 2

CAIDIN, Martin
The ragged, rugged warriors. - Toronto : London : Bantam, 1979
xiv,366,[1]p,fold plate : ill(incl 1 col). - (The Bantam war book series). - Includes index
0-553-13109-5 H.80/218 Woolwich
208533x

AIR FORCES
Allied air forces. Air operations during Pacific campaigns of World War 2. 1941-1942

CAIDIN, Martin
The ragged, rugged warriors. - London : Severn House, 1980, c1966
xiv,366p : ill. - Includes index
0-7278-0642-4 X.808/33393
232721x

AIR FORCES
Allied air forces. Air operations during Sino-Japanese War, 1937-1945

CAIDIN, Martin
The ragged, rugged warriors. - London : Severn House, 1980, c1966
xiv,366p : ill. - Includes index
0-7278-0642-4 X.808/33393
2327228

AIR FORCES
Allied air forces. Air operations during World War 2

BONACINA, Giorgio
Comando bombardieri operazione Europa. - Milano : Longanesi, 1975
44p.,16p. of plates : ill. - (Il cammeo ; n. 310)
r6000509 X.700/16755
0938327

FOSS, Joe
Joe Foss. - Reprint. - Washington, D.C. : Zenger Publishing Co, 1979
160p,14pages of plates. : ill
0-89201-080-0 X.809/50280
0938327

LICHTENSTEIN, Heiner
Warum Auschwitz nicht bombardiert wurde. - Köln : Bund-Verlag, 1980
183p,[8]p of plates : ill
3-7663-0428-3 X.809/48195
0938327

ONDERWATER, Hans
En toen was het stil-. - Baarn : Hollandia, 1981
340p : ill
90-604-5463-4 X.809/53917
0938327

AIR FORCES
Allied air forces. Air photographic operations during World War 2

POWYS-LYBBE, Ursula
The eye of intelligence. - London : Kimber, 1983
223p : ill. - Includes index
0-7183-0468-3 X.622/17156
0138207

AIR FORCES
Allied air forces. Air raids on Germany, to 1945

HOLMSTEN, Georg
Deutschland Juli 1944. - Düsseldorf : Droste, 1982
160p : ill. - (Fotografierte Zeitgeschichte)
3-7700-0598-8 X.805/5205
8530831

KÖSTERS, Hans G.
Essen Stunde Null. - Düsseldorf : Droste, 1982
117p : ill. - (Fotografierte Zeitgeschichte)
3-7700-0617-8 X.805/6087
8530831

AIR FORCES
Allied air forces. Air raids on Germany, 1944-1945. Strategy

ROSTOW, W. W. (Walt Whitman)
Pre-invasion bombing strategy. - Aldershot : Gower, c1981
xiii,166p. - Includes index
0-566-00482-8 X.809/51303
2706091
- [ANOTHER COPY: PAPERBACK] I.S.B.N. 0-566-00483-6 X.809/51304

AIR FORCES
Allied air forces. Air raids on Koblenz region, to 1945

SCHNATZ, Helmut
Der Luftkrieg im Raum Koblenz 1944/45. - Boppard am Rhein : Boldt, 1981
xiv,592p. : ill. - (Veröffentlichungen der Kommission des Landtages für die Geschichte des Landes Rheinland-Pfalz ; Bd.4). - Bibliography
3-7646-1774-8 X.0802/446(4)
8789711

AIR FORCES
Allied air forces. Boeing B-17 aeroplanes. Air operations during World War 2
WILLMOTT, H. P. (Hedley Paul)
B-17 flying fortress. - London : Arms and Armour Press, 1980
64p : ill(some col.). - ([War planes in colour] ; 4) (A bison book)
0-85368-444-8 X.622/12617 Woolwich
2061988

AIR FORCES
Allied air forces. Bomber aeroplanes. Gunners. Air operations during World War 2
BOWYER, Chaz
Guns in the sky. - London : Dent, 1979
x,182p : ill. - Bibliography: p.179-180. - Includes index
0-460-04429-x X.800/28057
1723561
Guns in the sky. - London : Corgi, 1981, c1979
ix,182p : ill. - Bibliography: p179-180. - Includes index
0-552-98121-4 X.800/30830
1723561
DAVIS, Larry
Gunships. - Carrollton : Squadron / Signal, c1982
64p : ill(some col.)
0-89747-123-7 X.622/13144
1723561

AIR FORCES
Allied air forces. Bombing of Germany, 1939-1945
MORRISON, Wilbur H.
Fortress without a roof. - London : W.H. Allen, 1982
ix,322p,[16]p of plates : ill. - Bibliography: p311-312. - Includes index
0-491-02878-4 X.629/20962
0328804

AIR FORCES
Allied air forces. Bombing of Hamburg, 1943 (July)
MIDDLEBROOK, Martin
The Battle of Hamburg. - London : Allen Lane, 1980
424p,[24]p of plates : ill. - Bibliography: p415. - Includes index
0-7139-1074-7 X.809/46216
2322412
The Battle of Hamburg. - Harmondsworth : Penguin, 1984, c1980
424p,[24]p of plates : ill. - Bibliography: p415. - Includes index
0-14-006745-0 X.808/40365
2322412
MUSGROVE, Gordon
Operation Gomorrah. - London : Jane's, 1981
xiv,197p,xxxiip of plates : ill. - Includes index
0-7106-0079-8 X.800/31564
2322412

AIR FORCES
Allied air forces. Bristol Blenheim aeroplanes. Air operations during World War 2
BOWYER, Chaz
Bristol Blenheim. - London : Ian Allan, 1984
128p : ill. - Bibliography: p128
0-7110-1351-9 X.622/21185
3198162

AIR FORCES
Allied air forces. Consolidated Aircraft Corporation B-24 aeroplanes. Air operations during World War 2
BOWMAN, Martin W.
The B-24 Liberator 1939-1945. - Norwich : Wensum, c1979
128p : ill(some col.)
0-903619-27-x X.622/9901 Woolwich
2381133
FREEMAN, Roger A. (Roger Anthony)
B-24 Liberator at war. - London : Ian Allan, 1983
128p : ill
0-7110-1264-4 X.805/4369
2381133

AIR FORCES
Allied air forces. Curtiss P-40 aeroplanes. Air operations during World War 2
CHRISTY, Joe
P-40 Hawks at war. - London : I. Allan, 1979
128p : ill
0-7110-0983-x X.622/7157 Woolwich
2079097

AIR FORCES
Allied air forces. Czechoslovak airmen. Air operations during Western European campaigns of World War 2
ČEJKA, Eduard
Zlomená křídla. - Vyd. 2. - Praha : Naše vojsko, 1970
275p. : ill. - (Máj ; sv. 146)
r6000744 X.809/22951
0882054

ŠIKL, Václav
Nebe nad Seinou. - Curych : Konfrontace, 1977
193p,[6]leaves of plates : ill
3-85770-036-x X.808/36871
0882054

AIR FORCES
Allied air forces. Douglas A-20 Havoc aeroplanes. Air operations during World War 2
MESKO, Jim
A-20 Havoc. - Carrollton, Tex. : Squadron / Signal, c1983
49p : ill(some col.). - (Aircraft ; no.56)
0-89747-131-8 L.49/1807
2917718

AIR FORCES
Allied air forces. Douglas Boston aeroplanes. Air operations during World War 2
HESS, William N.
A-20 Boston at war. - London : I. Allan, 1979
128p : ill
0-7110-0965-1 X.802/11084
2052520

AIR FORCES
Allied air forces. Naval air operations during World War 2
OLDS, Robert
Helldiver squadron. - Reprint. - Washington, D.C. : Zenger, 1980
xi,263p,36p of plates : ill
0-89201-054-1 X.809/50309
0764914
WINTON, John
Air power at sea, 1939-45. - London : Sidgwick and Jackson, 1976
185,[3]p : ill. - Bibliography: p.184. - Includes index
0-283-98313-2 X.620/16414 Woolwich
0764914

AIR FORCES
Allied air forces. North American B-25 Mitchell aeroplanes. Air operations during World War 2
SCUTTS, Jerry
B-25 Mitchell at war. - London : Ian Allan, 1983
144p : ill
0-7110-1219-9 X.622/16438
2943859

AIR FORCES
Axis air forces. Fighter aeroplanes, 1939-1945
GUNSTON, Bill
An illustrated guide to German, Italian and Japanese fighters of World War II. - London : Salamander, c1980
159p : ill(some col.)
0-86101-064-7 X.629/17467 Woolwich
2704544
MONDEY, David
Concise guide to Axis aircraft of World War II. - [Feltham] : Temple, 1984
256p : ill(some col.). - Includes index
0-600-35027-4 X.622/23798
2704544

AIR FORCES
British air forces. Air operations during World War 1
LEWIS, Cecil
Sagittarius rising. - London : Heinemann, 1982
viii,331p
0-432-08600-5 X.958/14335
2868520

AIR FORCES
British air forces. Air operations during World War 1. Interviews. Sound tape recordings - Catalogues
IMPERIAL WAR MUSEUM. Department of Sound Records
Military and naval aviation 1914-18. - [London] : Imperial War Museum, Dept. of Sound Records, [1982?]
26p. - (Oral history recordings)
B82-26785 X.809/52267
274757x

AIR FORCES
British air forces. Awards of medals, to 1983: Meritorious Service Medal
MCINNES, Ian, 1937-
The Meritorious Service Medal to aerial forces. - Chippenham : Picton, 1984
63p : ill
0-902633-92-9 X.809/62483
323004x

AIR FORCES
Commonwealth air forces. Air operations during World War 1. Decorations: Victoria Cross
BOWYER, Chaz
For valour. - London : Kimber, 1978
548p : ill. - Bibliography: p.519-522. - Includes index
0-7183-0425-x X.800/27029
151802x

AIR FORCES
Commonwealth air forces. Air operations during World War 2. Decorations: Victoria Cross
BOWYER, Chaz
For valour. - London : Kimber, 1978
548p : ill. - Bibliography: p.519-522. - Includes index
0-7183-0425-x X.800/27029
1518038

AIR FORCES
Commonwealth air forces. Hawker Typhoon aeroplanes. Pilots. Air operations during World War 2
SCOTT, Desmond
Typhoon pilot. - London : Secker & Warburg, 1982
xvi,168p,[8]p of plates : ill. - (A Leo Cooper book). - Includes index
0-436-44428-3 X.809/53896
2820919

AIR FORCES
German air forces. Naval air operations during Mediterranean campaign of World War 2
SANTONI, Alberto
La partecipazione tedesca alla guerra aeronavale nel Mediterraneo (1940-1945). - Roma : Ateneo & Bizzarri, 1980
639p,plates. - (Protagonisti ; 5). - Includes bibliography
r6507275 X.800/35258
9057757

AIR FORCES
Insignia, 1939-1945
ROSIGNOLI, Guido
Badges and insignia of World War II. - London : Peerage, 1983
363p : ill(some col.)
0-907408-34-6 X.809/63958
0797529

AIR FORCES
Insignia & uniforms, 1914-1918
MOLLO, Andrew
Army uniforms of World War I. - Poole : Blandford Press, 1977
5-219p : col ill. - ([Blandford colour series]). - Bibliography: p.214-218. - Includes index
0-7137-0821-2 X.429/10334 Woolwich
128262x

AIR FORCES
North Atlantic Treaty Organization air forces. Military aircraft
HORSEMAN, Martin
NATO air power album. - London : Ian Allan, 1980
111p : ill
0-7110-1061-7 X.622/9393 Woolwich
2331896
NATO air power in the 1980s. - London : Arms and Armour, 1982
65p : chiefly ill(some col.). - (Warbirds illustrated ; no.7)
0-85368-545-2 X.622/14256
2331896

AIR FORCES
Personnel. Insignia, 1939-1945
ROSIGNOLI, Guido
Air Force badges and insignia of World War 2. - Poole : Blandford Press, 1976
200p : ill(chiefly col). - ([Blandford colour series]). - Includes index
0-7137-0777-1 X.629/11020 Woolwich
0797529

AIR FORCES
Uniforms, 1939-1945
MOLLO, Andrew
Naval, marine and air force uniforms of World War 2. - Poole : Blandford Press, 1975
231p : col ill. - ([Blandford colour series]). - Bibliography: p.227-229. - Includes index
0-7137-0725-9 X.429/10054
0541613

AIR FORCES
Warsaw Pact air forces. Military aircraft
GETHING, Michael J.
Warsaw Pact air power in the 1980s. - London : Arms and Armour Press, [1982]
67p : chiefly ill(some col.). - (Warbirds illustrated ; no.8)
0-85368-546-0 X.805/3424
2818116

SWEETMAN, Bill
The Hamlyn concise guide to Soviet military aircraft. - London : Hamlyn, 1981
207p : ill(some col.). - Includes index
0-600-34947-0 X.622/10769 Woolwich
159432x

AIR FORCES
Warsaw Pact air forces. Military aircraft - Encyclopaedias
SWEETMAN, Bill
Soviet air power. - London : Salamander Books, 1978
192p : ill(chiefly col). - Includes index
0-86101-015-9 X.625/584 Woolwich
159432x

AIR FORCES
Warsaw Pact air forces. Military equipment: Missiles - Encyclopaedias
SWEETMAN, Bill
Soviet air power. - London : Salamander Books, 1978
192p : ill(chiefly col). - Includes index
0-86101-015-9 X.625/584 Woolwich
1594338

AIR FORCES
Western bloc air forces. Fighter aeroplanes, ca 1950-1979
MONDEY, David
Fighter aircraft of the West. - London : I. Allan, 1979
96p : ill(some col). - ([Ian Allan colour series])
0-7110-0762-4 X.620/19096 Woolwich
2046989

AIR FORCES. World War I
Allied air forces
AIRMEN of World War I / [compiled by] Chaz Bowyer. - London : Arms and Armour Press [etc.], 1975
128p : of ill. - Bibliography: p.128
0-85368-320-4 X.800/25323
056575x

CUTLACK, F. M. (Frederick Morley)
The Australian flying corps in the Western and Eastern theatres of war 1914-1918. - St Lucia, Qld : University of Queensland Press in assn. with the Australian War Memorial, 1983
li,493p,plates : ill. - (The official history of Australia in the war of 1914-1918 ; vol. 8)
0-7022-1728-x X.0800/1645(8)
8481806

AIR GUN SHOOTING
Target air gun shooting
CHURCHILL, Bob
Modern airweapon shooting. - Newton Abbot : David & Charles, c1981
196p : ill. - Bibliography: p189. - Includes index
0-7153-8123-7 X.629/17018 Woolwich
261474x

AIR GUNS
WALTER, John
The airgun book. - [3rd ed.]. - London : Arms & Armour, 1984
176p : ill
0-85368-675-0 X.622/21424
3051242

AIR GUNS
Air guns & gas guns
WALTER, John
The airgun book. - London : Arms and Armour, c1981
146p : ill
0-85368-257-7 X.622/10379 Woolwich
2342960
The airgun book. - 2nd ed. rev. - London : Arms and Armour, 1982
146p : ill
0-85368-518-5 SRL
2342960

WESLEY, L. (Leslie)
Air-guns and air-pistols. - New ed.. - London : Cassell, 1979
xvi,208p,28p of plates : ill. - Includes index
0-304-30375-5 X.629/14916 Woolwich
2342960

AIR HOSTESSES
See Women stewards. Air passenger transport services

AIR MAIL
See Airmail

AIR MANOEUVRES
By Great Britain. Royal Air Force, 1950-1953. B Mk.1 Washington aeroplanes
FOPP, Michael A.
The Washington file. - Saffron Walden : Air-Britain (Historians), c1983
36p : ill
B84-20241 X.622/18772
3081427

AIR OPERATIONS
See also
Air battles
Battle of Arnhem
Battle of Britain
Battle of Crete
Battle of France
Ground attacks
Offshore operations. Aircraft

AIR OPERATIONS
to 1984
JOHNSON, J. E. (James Edgar)
The story of air fighting. - Rev. and enlarged ed. - London : Hutchinson, 1985
306p,[20]p of plates : ill. - Bibliography: p297. - Includes index
0-09-159820-6 X.622/23541
3203867

AIR OPERATIONS
1945-1982
ARMITAGE, M. J.
Air power in the nuclear age, 1945-82. - London : Macmillan, 1983
xv,296p. - Bibliography: p279-283. - Includes index
0-333-25320-5 X.629/21447
2969394

AIR OPERATIONS
By Australia. Royal Australian Air Force, to 1975
PARNELL, N. M.
Australian air force since 1911. - Sydney ; London : A.H. and A.W. Reed, 1976
216p : ill. - Bibliography: p.210. - Includes index
0-589-07153-x X.622/5989 Woolwich
1172352

AIR OPERATIONS
By Austria-Hungary. Heer, to 1918
PETER, Ernst
Die k.u.k. Luftschiffer- und Fliegertruppe Österreich-Ungarns 1794-1919. - Stuttgart : Motorbuch, 1981
300p : ill
3-87943-743-2 X.800/32162
8830584

AIR OPERATIONS
By Avro Lincoln aeroplanes, 1944-1966
GARBETT, Mike
Lincoln at war, 1944-1966. - London : I. Allan, 1979
176p : ill
0-7110-0847-7 X.802/10820
1821865

AIR OPERATIONS
By Boeing B-29 aeroplanes, to 1954
ANDERTON, David A.
B-29 Superfortress at war. - London : I. Allan, [1978]
176p,[4]p of plates : ill(some col). - Bibliography: p.176
0-7110-0881-7 X.622/6601 Woolwich
1728989

AIR OPERATIONS
By bomber aeroplanes, to 1976
BROOKS, Stephen
Bomber. - [London] : Imperial War Museum, 1983
64p : ill. - Bibliography: p63-64
0-901627-27-5 X.629/21069
2113260

CLARK, Ronald W. (Ronald William)
The role of the bomber. - Richmond, Vic. ; London : Hutchinson ; London : Sidgwick and Jackson, 1977
160p : ill(some col). - Bibliography: p.158. - Includes index
0-283-98394-9 X.800/28767
2113260

AIR OPERATIONS
By Breda 65 aeroplanes, to date
GARELLO, Giancarlo
Il Breda 65 e l'aviazione d'assalto. - Roma : Ateneo & Bizzarri, 1980
255p,[4]folded leaves : ill
r6507125 X.622/15403
9058184

AIR OPERATIONS
By Chance Vought Corsair aeroplanes, to 1963
ABRAMS, Richard
F4U Corsair at war. - London : I. Allan, 1977
160p,[4]p of plates : ill(some col)
0-7110-0766-7 X.622/6130 Woolwich
1275283

JOHNSEN, Frederick A.
F4U Corsair. - London : Jane's, 1983
56p(16folded) : ill(some col.)
0-7106-0288-x L.45/2497
1275283

MUSCIANO, Walter A.
Corsair aces. - New York : Arco, 1979
136p
0-668-04600-7 X.611/9228 Woolwich
1275283

AIR OPERATIONS
By Chance Vought Corsair aeroplanes, to 1978
TILLMAN, Barrett
Corsair. - Cambridge : Stephens, [1980]
xii,219p : ill. - Bibliography: p.213-214. - Includes index
0-85059-427-8 X.622/7229 Woolwich
2141566

AIR OPERATIONS
By Chance Vought Crusader aeroplanes, to 1979
TILLMAN, Barrett
MiG master. - Cambridge : Stephens, 1981, c1980
xii,224p : ill. - Bibliography: p221-222. - Includes index
0-85059-525-8 X.629/15245 Woolwich
238793x

AIR OPERATIONS
By Cuba. Fuerza Aérea Revolucionaria, 1961
PRENDES QUINTANA, Álvaro
En el punto rojo de mi kolimador. - La Habana : Ediciones Huracán, Editorial de Arte y Literatura, 1976
214p : maps
r6157622 X.619/11334 Woolwich
2448785

AIR OPERATIONS
By Douglas C-47 Dakota aeroplanes, to 1981
PEARCY, Arthur
Dakota at war. - London : Ian Allan, 1982
127p : ill
0-7110-1217-2 X.622/14921
2869446

AIR OPERATIONS
By fighter aeroplanes, to 1945
GUNSTON, Bill
Fighters, 1914-1945. - London : Hamlyn : Phoebus, 1978
128p : ill(chiefly col). - Bibliography: p.128
0-600-33636-0 X.622/6300 Woolwich
1441337

AIR OPERATIONS
By fighter aeroplanes, to 1964
JOHNSON, J. E. (James Edgar)
Full circle. - New York ; London : Bantam, 1980
280p : ill. - (A Bantam war book). - Includes index
0-553-13568-6 X.808/33860
2377152

AIR OPERATIONS
By fighter aeroplanes, to 1978
HALLEY, James J.
The role of the fighter in air warfare. - London : Barrie and Jenkins, 1979
152p : chiefly ill(some col). - Includes index
0-214-20601-7 X.620/18490 Woolwich
1892460

AIR OPERATIONS
By fighter aeroplanes. Aces, to 1975
SHORES, Christopher
Fighter aces. - London : Hamlyn, 1975
160p : ill(some col). - Bibliography: p.158. - Includes index
0-600-30230-x X.802/10100
058861x

AIR OPERATIONS
By Germany. Luftwaffe, 1930-1945 - Encyclopaedias
WOOD, Tony, 1940-
Hitler's Luftwaffe. - London : Salamander Books, 1977
248p : ill(chiefly col). - Includes index
0-86101-005-1 X.622/6113 Woolwich
1269968

AIR OPERATIONS
By Gloster Javelin aeroplanes, to 1968
ALLWARD, Maurice
Gloster Javelin. - London : Ian Allan, 1983
112p : ill. - (Postwar military aircraft ; 1)
0-7110-1323-3 X.800/38432
3131483

AIR OPERATIONS
By Great Britain. Royal Air Force, 1914-1939
POWERS, Barry Douglas
Strategy without slide-rule. - London : Croom Helm, 1976
3-295p. - Bibliography: p.267-288. - Includes index
0-85664-219-3 X.629/10855 Woolwich
0747580

AIR OPERATIONS
By Great Britain. Royal Air Force, 1972-1978. McDonnell Douglas F-4 aeroplanes
PREST, Robert
F4 Phantom. - London : Cassell, 1979
[7],209p,[16]p of plates : ill
0-304-30311-9 Nov.38937 Woolwich
191717x

PREST, Robert
F4 Phantom. - London : Corgi, 1981, c1979
254p
0-552-11615-7 X.629/16377 Woolwich
191717x

AIR OPERATIONS
By Great Britain. Royal Air Force. Fighter Command, 1936-1968
BOWYER, Chaz
Fighter Command, 1936-1968. - London : Dent, 1980
205p,[32]p of plates : ill. - Bibliography: p.196-198. - Includes index
0-460-04388-9 X.800/28967
1941992

Fighter Command 1936-1968. - London : Sphere, 1981, c1980
242p,[8]p of plates : ill. - Bibliography: p235-237. - Includes index
0-7221-1808-2 X.629/16726 Woolwich
1941992

AIR OPERATIONS
By Great Britain. Royal Air Force. Royal Auxiliary Air Force. Squadron 500, 1931-1957
BROOKS, Robin J.
Kent's own. - Gillingham : Meresborough, 1982
158p : ill. - Bibliography: p158
0-905270-54-1 X.809/56632
2970708

AIR OPERATIONS
By Italy. Aeronautica militare, to 1943. Reggiane 2000 aeroplanes
MALIZIA, Nicola
Il Reggiane 2000. - Roma : Ateneo & Bizzarri, 1978
61p,[2]leaves of plates : ill
r6500547 X.622/9590 Woolwich
2478617

AIR OPERATIONS
By Italy. Aeronnautica militare, to 1943. Torpedo bomber aeroplanes
UNIA, Carlo
Storia degli aerosiluranti italiani. - Roma : Bizzarri, 1974
426p : ill
r6042775 X.802/3924
1225081

AIR OPERATIONS
By Messerschmitt Bf 109 aeroplanes, to 1945
BEAMAN, John R.
Messerschmitt BF 109 in action. - Carrollton, Tex. : Squadron/Signal. - (Aircraft ; no.57)
Pt.2. - c1983
58p : ill(some col.)
0-89747-138-5 L.49/1808
129668x

FEIST, Uwe
The fighting 109. - Newton Abbot : David and Charles, 1978
x,198p,[4]p of plates : ill(some col). - Bibliography: p.192. - Includes index
0-7153-7627-6 X.802/10644
129668x

VAN ISHOVEN, Armand
Messerschmitt Bf 109 at war. - London : I. Allan, 1977
160p : ill
0-7110-0770-5 X.622/6169 Woolwich
129668x

AIR OPERATIONS
By Messerschmitt Bf 109 aeroplanes, to date
ARENA
Bf-109. - Roma : Ateneo & Bizzarri, 1980
238p,[8]p of plates,[1]folded leaf : ill(some col.)
r6507126 X.622/15402
8619573

AIR OPERATIONS
By military helicopters, to 1982
WRAGG, David W. (David William)
Helicopters at war. - London : Hale, 1983
283p : ill. - Includes index
0-7090-0858-9 X.622/18048
3058840

AIR OPERATIONS
By Supermarine Spitfire aeroplanes, to 1954
PRICE, Alfred
Spitfire. - London : Macdonald and Jane's, 1977
159p : ill. - Bibliography: p.158
0-354-01077-8 X.622/5990 Woolwich
1188429

AIR OPERATIONS
By torpedo bomber aeroplanes, to date
LAUCK, Friedrich
Der Lufttorpedo. - München : Bernard & Graefe, [1981]
304p : ill. - (Wehrtechnische Handbücher). - Bibliography
3-7637-5230-7 X.629/17588 Woolwich
8619565

AIR OPERATIONS
By United States. Air Force, to 1953. Boeing B-29 aeroplanes
BIRDSALL, Steve
Superfortress. - Carrollton : Squadron / Signal, c1980
79p : ill(some col.)
0-89747-104-0 X.622/22739
2061953

AIR OPERATIONS
By United States. Air Force, to 1974. Boeing B-52 aeroplanes
DRENDEL, Lou
B-52 Stratofortress in action. - Carrollton : Squadron/Signal, c1975
49p. : ill(some col.). - (Aircraft ; no.23)
0-89747-022-2 L.49/1317
2742195
- [Another copy] L.49/2184

AIR OPERATIONS
By United States. Army Air Force, to 1953. Boeing B-29 aeroplanes
PIMLOTT, John, 1948-
B-29 superfortress. - London : Arms and Armour, 1980
64p : ill(some col.). - (A bison book) ([War planes in colour] ; 2). - Bibliography: p64
0-85368-095-7 X.622/12616
2061953

AIR OPERATIONS
By United States. Army Air Force. Air Force, 5th. Fighter Squadron, 39th, 1940-1980
STANAWAY, John
Cobra in the clouds. - Temple City, Ca. : Historical Aviation Album, 1982
48p : ill
0-911852-92-1 X.805/4438
2973162

AIR OPERATIONS
By United States. Marine Corps, 1912-1980
MERSKY, Peter B.
U.S. Marine Corps aviation. - Annapolis : Nautical & Aviation, c1983
x,310p : ill. - Bibliography: p300-301
0-933852-39-8 X.800/38096
3069389

AIR OPERATIONS
By West German military forces. Military aircraft, 1956-1976
JACKSON, Paul A. (Paul Andrew)
German military aviation, 1956-1976. - Hinckley : Midland Counties Publications, 1976
150p : ill
0-904597-04-0 X.629/10845 Woolwich
0740756

AIR OPERATIONS
Naval air operations by aircraft carriers, 1945-1982
CARRIER air operations since 1945 / [compiled by] Paul Beaver. - London : Arms and Armour, [1983]
72p : chiefly ill(some col.). - (Warbirds illustrated ; no.19)
0-85368-598-3 X.0800/1646(19)
3063216

AIR OPERATIONS
Tactics. Training courses
SKINNER, Michael, 1953-
Red flag. - London : Arms and Armour, c1984
ix,134p : ill(some col.)
0-85368-642-4 X.809/60588
3148521

AIR OPERATIONS. Arctic campaign. World War 2
Naval air operations by Union of Soviet Socialist Republics. Voenno-morskoĭ flot
BOĬKO, Viktor Semenovich
Kryl'ía Severnogo flota. - Murmansk : Murmanskoe knizhnoe izdatel'stvo, 1976
317p : ill
r6045247 X.809/25013
1328999

AIR OPERATIONS. Battle of Britain
By Germany. Luftwaffe
VAN ISHOVEN, Armand
The Luftwaffe in the Battle of Britain. - London : I. Allen, 1980
128p : ill
0-7110-0996-1 X.622/8232 Woolwich
2232936

AIR OPERATIONS. Battle of Britain
By Germany. Luftwaffe. Fighter aeroplanes
GERMANY (Federal Republic). Bundesarchiv
German fighters over England. - Cambridge : Stephens, 1979
94p : chiefly ill. - (World War 2 photo album ; no.10)
0-85059-355-7 X.700/26104
1763121

AIR OPERATIONS. Battle of Britain
By Great Britain. Royal Air Force. Fighter Command. Group, 11
COOKSLEY, Peter G. (Peter George)
1940. - London : Hale, 1983
224p,[16]p of plates : ill. - Includes index
0-7090-0907-0 X.809/56622
2970384

AIR OPERATIONS. Battle of Britain
By Great Britain. Royal Air Force. Fighter Command. Group, no.12
TURNER, John Frayn
The Bader Wing. - Speldhurst : Midas Books, 1981
153p,[16]p of plates : ill. - Includes index
0-85936-279-5 X.809/49674
2575191

AIR OPERATIONS. Battle of Dunkirk
By Great Britain. Royal Air Force. Fighter Command. Group, 11
FRANKS, Norman L. R. (Norman Leslie Robert)
The air battle of Dunkirk. - London : Kimber, 1983
236p : ill. - Bibliography: p229. - Includes index
0-7183-0349-0 X.800/36319
2964317

AIR OPERATIONS. Battle of France. World War 2
By Great Britain. Royal Air Force. No.1 Squadron, 1939-1940 - Biographies
RICHEY, Paul
Fighter pilot. - Revised ed.. - London : Pan Books, 1977
xii,147p,[8]p of plates : ill
B77-26024 X.709/31442
1188054
Fighter pilot. - New ed.. - London : Macdonald and Jane's, 1980
142,[4]p,[44]p of plates : ill
0-7106-0000-3 X.800/28770
1188054

AIR OPERATIONS. Burma campaign. World War 2
By Great Britain. Royal Air Force. Squadron, No.67, 1941-1944
THOMAS, J. Helsdon
Wings over Burma. - Bognor Regis : New Horizon, c1984
152p,[10]p of plates : ill
0-86116-966-2 X.809/63012
3246507

AIR OPERATIONS. Chaco War
By Paraguay
ALVAREZ ALBERT, Abdón
Con llamas en el aire. - Asunción : Dirección de Publicaciones de las FF. AA. NN., 1979
200p
r6706582 X.809/54434
889504x

AIR OPERATIONS. Clwyd
By Germany. Luftwaffe, 1939-1945
JONES, Ivor Wynne
Luftwaffe over Clwyd. - Llandudno : Pegasus Press, [1977]
[2],46p,[2]p of plates : ill
B78-09926 X.709/32018
1353101

AIR OPERATIONS. Cretan campaign. World War 2
By Great Britain. Royal Air Force, 1941 – Biographies

COMEAU, Marcel Gerard
Operation Mercury. – London : Severn House : [Distributed by Hutchinson], 1976
160p
0–7278–0170–8 X.809/41150
070959x

AIR OPERATIONS. Cuba
By Cuban exiles, 1959–1961

FERRER, Eduardo
Operación Puma. – 2a ed. – Miami : International Aviation Consultants, 1976
254p : ill
r6085679 X.709/27521
1976869

AIR OPERATIONS. East African campaign. World War 2

RICCI, Corrado
La guerra aerea in Africa orientale, 1940–41. – Modena : STEM Mucchi, 1980
145p : ill
88–7000–030–3 X.805/4040
8971641

AIR OPERATIONS. Eastern European campaigns. World War 2
By Germany. Luftwaffe – Biographies

RUDEL, Hans Ulrich
Stuka pilot. – New York ; London : Bantam, 1979
xii,290p,col fold leaf of plate : ill(some col). – (Bantam war book series)
0–553–12304–1 H.79/2541 Woolwich
012916x

AIR OPERATIONS. Eastern European campaigns. World War 2
By Union of Soviet Socialist Republics. Voenno-vozdushnye sily

INOZEMTSEV, Ivan Grigor'evich
Pod krylom – Leningrad. – Moskva : Voennoe izdatel'stvo Ministerstva oborony SSSR, 1978
272p,[16]p of plates : ill
r6175523 X.809/27938
0135933

AIR OPERATIONS. Eastern European campaigns. World War 2
By United States. Army Air Force. Cooperation between United States & Soviet Union, 1943–1944

INFIELD, Glenn B. (Glenn Berton)
The Poltava affair. – New York : Macmillan, 1973
265p
r6189469 X.809/28018
0130974

AIR OPERATIONS. European campaigns. World War 2

BOWYER, Chaz
Air war over Europe. – London : Kimber, 1981
235p,[30]p of plates : ill. – Bibliography: p223–224. – Includes index
0–7183–0238–9 X.800/31891
258851x

AIR OPERATIONS. European campaigns. World War 2
By Germany. Luftwaffe, 1939–1941. Fighter aeroplanes

SCUTTS, Jerry
Luftwaffe fighter units, Europe, September 1939–June 1941. – London : Osprey Publishing, 1977
48p : ill(some col). – (Aircam/airwar series ; 6)
0–85045–204–x X.611/8294 Woolwich
1700464

AIR OPERATIONS. European campaigns. World War 2
By Germany. Luftwaffe, 1942–1945. Fighter aeroplanes

SHORES, Christopher
Luftwaffe fighter units, Europe, 1942–45. – London : Osprey Publishing, 1979
48p : ill(some col). – (Airwar series ; 24)
0–85045–298–8 X.700/26049
1918273

AIR OPERATIONS. European campaigns. World War 2
By Great Britain. Royal Air Force, 1942–1945. Fighter aeroplanes

PHILPOTT, Bryan
RAF fighter units, Europe, April 1942–45. – London : Osprey Publishing, 1978
48p : ill(some col). – (Aircam/airwar ; 10)
0–85045–233–3 X.700/25702
1489232

AIR OPERATIONS. European campaigns. World War 2
By United States. Air Force, 1942–1945. Bomber aeroplanes

ETHELL, Jeffrey
Air war over Germany. – London : Arms and Armour, [c1985]
72p : ill(some col.). – (Warbirds illustrated ; no. 31)
0–85368–700–5 X.0800/1646(31)
1296590

SCUTTS, Jerry
USAAF heavy bomber units, ETO and MTO, 1942–45. – London : Osprey Publishing, 1977
48p : ill(some col). – (Aircam/airwar series ; 2)
0–85045–131–0 X.611/7879 Woolwich
1296590

UNITED STATES. Strategic Bombing Survey
The United States Strategic Bombing Survey. – New York ; London : Garland, 1976
10v
r6112491 X.800/14744
1296590

AIR OPERATIONS. European campaigns. World War 2
By United States. Air Force, 1944 (February)

INFIELD, Glenn B. (Glenn Berton)
Big week!. – London : New English Library, 1976
188p. – Bibliography: p.186–188
0–450–02804–6 X.708/20312
0631191

AIR OPERATIONS. European campaigns. World War 2
By United States. Army Air Force, 1942–1945. Fighter aeroplanes

FRANÇILLON, René Jacquet
USAAF fighter units, Europe, 1942–45. – London : Osprey Publishing, 1977
48p : ill(some col). – (Aircam/airwar series ; 8)
0–85045–231–7 X.700/25712
1494716
– [Another copy.] X.622/15418

AIR OPERATIONS. European campaigns. World War 2
By United States. Army Air Force, 1942–1945. Medium bomber aeroplanes

FRANÇILLON, René Jacquet
USAAF medium bomber units, ETO and MTO, 1942–45. – London : Osprey Publishing, 1977
48p : ill(some col). – (Aircam/airwar series ; 7)
0–85045–205–8 X.700/25783
1584723
– [Another copy] X.611/8268 Woolwich

AIR OPERATIONS. Falkland Islands War

ETHELL, Jeffrey
Air war. – London : Sidgwick & Jackson, 1983
260p,[32]p of plates : ill. – Includes index
0–283–99142–9 X.622/22430
3046494

ETHELL, Jeffrey
Air war South Atlantic. – London : Sidgwick & Jackson, 1983
260p,[32]p of plates : ill. – Includes index
0–283–99035–x X.800/37481
3046494

HARRIER / edited by John Godden. – Oxford : Brassey's Defence, 1983
xi,132p,6p of plates : ill(some col.)
0–08–031166–0 X.800/38783
3046494

AIR OPERATIONS. Falkland Islands War
By Argentina. Fuerza Aérea Argentina

ANDRADA, Benigno Héctor
Guerra aérea en las Malvinas. – Buenos Aires : Emecé Editores, 1983
239p,[16]p of plates
9–500–40191–6 X.808/38617
9003517

AIR OPERATIONS. Far Eastern campaigns. World War 2
By Great Britain. Royal Air Force. Military aircraft

PHILPOTT, Bryan
RAF combat units, SEAC, 1941–45. – London : Osprey Publishing, 1979
48p : ill(some col). – (Air series ; 23)
0–85045–297–x X.700/26051
1917838

AIR OPERATIONS. Far Eastern campaigns. World War 2
By Great Britain. Royal Air Force. No.258 Squadron, 1941–1942 – Biographies

KELLY, Terence, b.1920
Hurricane over the jungle. – London : Kimber, 1977
234p : ill. – Includes index
0–7183–0405–5 X.800/26641
133431x

AIR OPERATIONS. Germany
By Germany. Luftwaffe, 1939–1945

FIGHTERS defending the Reich / [selected and with commentaries by] Bryan Philpott. – Cambridge : Stephens, 1978
94p : of ill. – (World War 2 photo albums ; no.4)
0–85059–341–7 X.611/8353 Woolwich
1345427

AIR OPERATIONS. Germany. Luftwaffe
Effects on social life in Great Britain, 1940–1941

HARRISON, Tom
Living through the Blitz. – Harmondsworth : Penguin Books, 1979
3–370p. – Bibliography: p.342. – Includes index
0–14–004612–7 X.708/22194
1994565

HARRISSON, Tom
Living through the Blitz. – London : Collins, 1976
[2],372p. – Bibliography: p.345. – Includes index
0–00–216009–9 X.529/30916
1994565

AIR OPERATIONS. Guerrilla warfare. Himalayas. Asia
By Himalayan Pilots' Association

SINHA, Satyanarayan
Operation Himalayas. – New Delhi : S. Chand, 1975
xi, 200 p.
r6031279 Nov.33658 Woolwich
1052330

AIR OPERATIONS. Indo-Pakistan War, 1965
By Pakistan. Pakistan Air Force

FRICKER, John
Battle for Pakistan. – London : I. Allan, 1979
192p : ill. – Includes index
0–7110–0929–5 X.800/28184
1993895

AIR OPERATIONS. Kent
By Great Britain. Royal Naval Air Services, 1914–1918

FLYING / compiled by David Collyer. – Rochester : North Kent, c1982
[64]p : chiefly ill
0–9505733–6–1 X.629/18039 Woolwich
2757222

AIR OPERATIONS. Korean War

DAVIS, Larry
Air war over Korea. – Carrollton, Tex. : Squadron/Signal, c1982
96p : ill(some col.)
0–89747–137–7 X.622/15407
2858681

AIR OPERATIONS. Korean War
By Australia. Royal Australian Air Force. Mustang aeroplanes

ANDERSON, Peter N.
Mustangs of the RAAF and RNZAF. – Sydney ; London : A.H. and A.W. Reed, 1975
[4],103,[5]p : ill(some col). – Bibliography: p.[2] (first sequence). – Includes index
0–589–07130–0 X.622/5921 Woolwich
1136526

AIR OPERATIONS. Korean War
By United States. Air Force. Fighter aeroplanes

DAVIS, Larry
Mig Alley. – Carrollton : Squadron / Signal, [c1978]
89p : ill(some col.)
0–89747–081–8 X.622/22735
3181227

AIR OPERATIONS. Malta
By Great Britain. Royal Air Force, 1942

BARNHAM, Denis
One man's window. – London : New English Library, 1975
208p
0–450–02548–9 H.75/623
0366188

DOUGLAS-HAMILTON, Lord James
The air battle for Malta. – Edinburgh : Mainstream, 1981
208p,[24]p of plates : ill. – Bibliography: p195–202. – Includes index
0–906391–20–2 X.629/18107
0366188

The air battle for Malta. – Feltham : Hamlyn Paperbacks, 1982, c1981
208p,[8]p of plates : ill. – Bibliography: p195–202. – Includes index
0–600–20659–9 X.808/37677
0366188

AIR OPERATIONS. Mediterranean campaign. World War 2
By Germany. Luftwaffe. Bomber aeroplanes

GERMAN bombers over the Med / [compiled by] Bryan Philpott. – Cambridge : Stephens, 1980
94p : chiefly ill. – (World War 2 photo album ; no.13)
0–85059–393–x X.700/26618
1903454

AIR OPERATIONS. Mediterranean campaign. World War 2
By Germany. Luftwaffe. Fighter aeroplanes

GERMAN fighters over the Med / [compiled by] Bryan Philpott. - Cambridge : Stephens, 1979
95p : chiefly ill. - (World War 2 photo album ; no.6)
0-85059-343-3 X.950/12945
2721899

SHORES, Christopher
Luftwaffe fighter units. Mediterranean, 1941-44. - London : Osprey Publishing, 1978
48p : ill(some col). - (Airwar series ; 20)
0-85045-294-5 X.611/8369 Woolwich
1744526

AIR OPERATIONS. Mediterranean campaign. World War 2
By Great Britain. Royal Air Force. Bristol Blenheim aeroplanes - Biographies

GILLMAN, Ronald Edward
The shiphunters. - London : J. Murray, 1976
xiii,241p,[12]p of plates : ill
0-7195-3299-x X.809/41041
0676470

AIR OPERATIONS. Mediterranean campaign. World War 2
Naval air operations by German air forces

SANTONI, Alberto
La partecipazione tedesca alla guerra aeronavale nel Mediterraneo (1940-1945). - Roma : Ateneo & Bizzarri, 1980
639p,plates. - (Protagonisti ; 5). - Includes bibliography
r6507275 X.800/35258
9057757

AIR OPERATIONS. Mediterranean campaigns. World War 2
By United States. Army Air Force, 1942-1945. Fighter aeroplanes

SHORES, Christopher
USAAF fighter units MTO. 1942-45. - London : Osprey Publishing, 1978
48p : ill(some col). - (Aircam/airwar series ; 12)
0-85045-244-9 X.611/8429 Woolwich
1734091

AIR OPERATIONS. Middle East
By Great Britain. Royal Air Force, 1945-1972

LEE, Sir David, 1912-
Flight from the Middle East. - London : H.M.S.O., 1980
xvi,339p,[46]p of plates(some folded) : ill(some col.). - Includes index
0-11-772356-8 X.809/47336
2385686
- [Another copy.] T 45259 IOL

AIR OPERATIONS. Military aeroplanes. Special subjects. Poetry in English, 1900-1945
- Anthologies

ECHOES in the sky / edited and narrated by Ronald Dixon. - Poole : Blandford, 1982
128p. - Includes index
0-7137-1271-6 X.958/14030
2847914

AIR OPERATIONS. Mongolia (People's Republic)
By Soviet airmen, 1939

SMIRNOV, Boris Aleksandrovich
Ot Madrida do Khalkhin-Gola. - Izd. 2., ispr. i dop.. - Kuĭbyshev : Kuĭbyshevskoe knizhnoe izdatel'stvo
341 p. : ill
r6022909 X.809/23338
1009737

AIR OPERATIONS. Netherlands campaigns. World War 2

KORTHALS ALTES, A. (Alexander)
Luchtgevaar. - Amsterdam : Sijthoff, [1984]
348p : ill
90-218-2603-8 X.800/39546
9128034

AIR OPERATIONS. North African campaign. World War 2
By Great Britain. Royal Air Force, 1942-1943. Polish units

KRÓL, Wacław
Pod niebem Tunezji. - Warszawa : Krajowa Agencja Wydawnicza RSW ""Prasa-Książka-Ruch", 1975
138 p. : ill. - (Miniatury lotnicze)
r6008368 X.708/17918
0943266

AIR OPERATIONS. North African campaign. World War 2
By Great Britain. Royal Air Force. Bomber Command. No 205 Group, 1942-1943 - Biographies

CHAPPELL, F. R.
Wellington wings. - London : Kimber, 1980
282p,[16]p of plates : ill. - Includes index
0-7183-0177-3 X.800/28852
2182629

AIR OPERATIONS. North Atlantic campaign. World War 2
By Australia. Royal Australian Air Force. No. 461 Squadron - Biographies

SOUTHALL, Ivan
Fly west. - Harmondsworth : Penguin, 1978
192p : 2 ill. - (Peacock books)
0-14-047118-9 H.78/467 Woolwich
0244678

AIR OPERATIONS. North Sea campaign. World War 1
Naval air operations by Great Britain. Royal Naval Air Service. Flying boats

HALLAM, T. D.
The spider web. - London : Arms and Armour Press, 1979
[7],278p,[19]p of plates : ill
0-85368-452-9 X.809/44378
1690612

AIR OPERATIONS. North-West Frontier. Pakistan
By Great Britain. Royal Air Force, 1933-1936 - Biographies

LEE, Sir David, 1912-
Never stop the engine when it's hot. - London : Thomas Harmsworth, c1983
330p : ill
0-9506012-4-1 X.809/57050
2992434
- [Another copy.] T 45659 IOL

AIR OPERATIONS. Pacific campaigns. World War 2
By Allied air forces

CAIDIN, Martin
The ragged, rugged warriors. - Toronto ; London : Bantam, 1979
xiv,366,[1]p,fold plate : ill(incl 1 col). - (The Bantam war book series). - Includes index
0-553-13109-5 H.80/218 Woolwich
208533x

AIR OPERATIONS. Pacific campaigns. World War 2
By Allied air forces, 1941-1942

CAIDIN, Martin
The ragged, rugged warriors. - London : Severn House, 1980, c1966
xiv,366p : ill. - Includes index
0-7278-0642-4 X.808/33393
232721x

AIR OPERATIONS. Pacific campaigns. World War 2
By Japan. Navy. Air Service. Shimpu Tokubetsu-Kogekitai, 1944-1945

INOGUCHI, Rikihei
The divine wind. - Toronto ; London : Bantam, 1978
xxii,262,[1]p,fold plate : ill(1 col). - (A Bantam war book). - Includes index
0-553-12578-8 X.708/21829
1841181

AIR OPERATIONS. Pacific campaigns. World War 2
By Japan. Nihon Kaigun Kōkū-bu. Shimpu Tokubetsu-Kōgekitai & Japan. Nihon Rikugun Kōkū-butai. Shimbu Tokubetsu-Kōgekitai, 1944-1945

HOYT, Edwin P. (Edwin Palmer)
The kamikazes. - London : Hale, 1984, c1983
333p : ill. - Bibliography: p304-306. - Includes index
0-7090-1790-1 X.809/62253
0143642

AIR OPERATIONS. Pacific campaigns. World War 2
By United States. Air Force. Bomb Wing, 73rd

The STORY of the 73rd. - Reprinted. - Nashville, Tenn. : Battery Press, 1980
1v.(unpaged (chiefly ill.)). - (Aviation series ; 3)
0-89839-032-x X.805/3321
8483094

AIR OPERATIONS. Pacific campaigns. World War 2
By United States. Army Air Force

MONDEY, David
USAAF at war in the Pacific. - London : Ian Allan, 1980
160p : ill
0-7110-1026-9 X.805/1187
2342235

AIR OPERATIONS. Pacific campaigns. World War 2
By United States. Army Air Force. Bomber aeroplanes

MUNDAY, Eric A.
USAAF bomber units, Pacific, 1941-45. - London : Osprey Publishing, 1979
48p : ill(some col). - (Airwar series ; 22)
0-85045-296-1 X.700/26050
1918265

AIR OPERATIONS. Pacific campaigns. World War 2
By United States. Army Air Force. Air Force, 5th

BIRDSALL, Steve
Flying buccaneers. - Newton Abbot : David and Charles, 1978
viii,312p,[4]p of plates : ill(some col). - Bibliography: p.303-304. - Includes index
0-7153-7750-7 X.622/6801 Woolwich
1494406

AIR OPERATIONS. Pacific campaigns. World War 2
By United States. Army Air Force. Air Force, Seventh, 1942-1945

RUST, Kenn C.
Seventh Air Force. - Temple City : Historical Aviation Album, 1979
64p : ill. - (A Historical Aviation Album publication)
0-911852-84-0 X.622/11395 Woolwich
2575957

AIR OPERATIONS. Pacific campaigns. World War 2
By United States. Army Air Force. Air Force, Thirteenth

RUST, Kenn C.
Thirteenth Air Force. - Temple City : Historical Aviation Album, c1981
64p : ill. - (A Historical Aviation Album publication)
0-911852-90-5 X.622/11896 Woolwich
2610612

AIR OPERATIONS. Pacific campaigns. World War 2
By United States. Army Air Force. Gopher Squadron

PARK, Edwards
Nanette. - London : Deutsch, 1978
186p : ill
0-233-96977-2 X.809/43318
1514075

AIR OPERATIONS. Pacific campaigns. World War 2
Naval air operations by aircraft carriers

BELOTE, James H.
Titans of the seas. - New York ; London : Harper and Row, 1975
x,336p,[16]p of plates : ill. - Bibliography: p.305-322. - Includes index
0-06-010278-0 X.620/17740 Woolwich
1494600

AIR OPERATIONS. Pacific campaigns. World War 2
Naval air operations by Japan. Kaigun. Aircraft carriers. Aeroplanes

FRANCILLON, René Jacquet
Japanese carrier air groups, 1941-45. - London : Osprey Publishing, 1979
48p : ill(some col). - (Airwar series ; 21)
0-85045-295-3 X.700/26057
1920065

AIR OPERATIONS. Pacific campaigns. World War 2
Naval air operations by United States. Army Air Force & United States. Navy, 1942: Tokyo Raid

GLINES, Carroll V. (Carroll Vane)
Doolittle's Tokyo raiders. - New York ; London : Van Nostrand Reinhold, 1981
449p : ill. - Includes index
0-442-21925-3 X.809/48239
2503867

AIR OPERATIONS. Pacific campaigns. World War 2
Naval air operations by United States. Navy, 1941-1942

LUNDSTROM, John B.
The first team. - Annapolis : Naval Institute Press, c1984
xxi,547p : ill. - Bibliography: p524-530
0-87021-189-7 X.805/6772
3329852

AIR OPERATIONS. Pacific campaigns. World War 2
Naval air operations by United States. Navy. Aircraft carriers. Aeroplanes

FRANCILLON, René Jacquet
US Navy carrier air groups, Pacific, 1941-45. - London : Osprey Publishing, 1978
48p : ill(some col). - (Airwar series ; 16)
0-85045-291-0 X.700/25748
1551418

SOWINSKI, Larry
The Pacific War. - London : Conway Maritime, 1981
208p : ill(some col.)
0-85177-217-x L.42/1217
257764x

AIR OPERATIONS. Pacific campaigns. World War 2
Naval air operations by United States. Navy. Grumman F-6F aeroplanes

ANDERTON, David A.
Hellcat. - London : Jane's, 1981
56p : ill(some col.)
0-7106-0036-4 L.45/1101
2506475

AIR OPERATIONS. Pacific campaigns. World War 2
Naval air operations by United States. Navy. Air Group Fifteen, 1944

HOYT, Edwin P. (Edwin Palmer)
McCampbell's heroes. - New York ; London : Van Nostrand Reinhold, c1983
xiii,189p,[8]p of plates : ill. - Includes index
0-442-26289-2 X.800/37168
3056856

AIR OPERATIONS. Pacific campaigns. World War 2
Night air operations by United States. Navy. PBY Catalina aeroplanes

KNOTT, Richard C.
[Black Cat raiders of World War II]. The Black Cats. - Cambridge : Stephens, 1981
x,198p : ill. - Includes index
0-85059-568-1 X.800/32836
2707985

AIR OPERATIONS. Reconnaissance squadrons. Great Britain. Royal Air Force
to 1981

RAWLINGS, John D. R. (John Dunstan Richard)
Coastal support and special squadrons of the RAF and their aircraft. - London : Jane's, 1982
270p : ill. - Includes index
0-7106-0187-5 X.805/3192
2799820

AIR OPERATIONS. Russian campaign. World War 2

FEDOROV, Alekseĭ Grigor'evich
Aviat͡sii͡a v bitve pod Moskvoĭ. - 2. izd., dop. i ispr.. - Moskva : Nauka, 1975
342p. - (Vtorai͡a mirovai͡a voĭna v issledovanii͡akh, vospominanii͡akh, dokumentakh). - Bibliography : p.328-334
r6174961 X.809/27552
212307X

PEMLER, Georg
Der Flug zum Don. - Leoni am Starnberger See : Druffel, 1981
331p,[20]p of plates : ill
3-8061-1011-5 X.809/51504
212307x

POKRYSHKIN, Oleksandr
Nebo viĭny. - Kyïv : Vydavnytstvo politychnoï literatury Ukraïny, 1982
492 p.. - (Memuary)
r6492685 X.809/63714
212307x

AIR OPERATIONS. Russian campaign. World War 2
By Germany. Luftwaffe

NOWARRA, Heinz J. (Heinz Joachim)
Luftwaffeneinsatz "Barbarossa", Russland 1941. - Friedberg : Podzun-Pallas, [1982]
159p. : ill. - ([Die Pallas-Bücher] ; [26])
3-7909-0168-7 X.809/61528
8974217

AIR OPERATIONS. Russian campaign. World War 2
By Germany. Luftwaffe. Bomber aeroplanes

GERMAN bombers over Russia / [compiled by] Bryan Philpott. - Cambridge : Stephens, 1979
94p : chiefly ill. - (World War 2 photo album ; no.8)
0-85059-346-8 X.700/25973
1517023

AIR OPERATIONS. Russian campaign. World War 2
By Germany. Luftwaffe. Fighter aeroplanes

SHORES, Christopher
Luftwaffe fighter units, Russia, June 1941-45. - London : Osprey Publishing, 1978
48p : ill(some col). - (Aircam/airwar ; 11)
0-85045-234-1 X.700/25714
1491369

AIR OPERATIONS. Russian campaign. World War 2
By Union of Soviet Socialist Republics. Voenno-vozdushnye sily

INOZEMT͡SEV, Ivan Grigor'evich
Tarany v severnom nebe. - Moskva : Voennoe izdatel'stvo Ministerstva oborony SSSR, 1981
143p : ill. - (Geroicheskoe proshloe nasheĭ rodiny)
r6462334 X.808/38165
8974217

AIR OPERATIONS. Sino-Japanese War, 1937-1945
By Allied air forces

CAIDIN, Martin
The ragged, rugged warriors. - London : Severn House, 1980, c1966
xiv,366p : ill. - Includes index
0-7278-0642-4 X.808/33393
2327228

AIR OPERATIONS. Sino-Japanese War, 1937-1945
By American military forces

CORNELIUS, Wanda
Ding hao. - Gretna, La. : Pelican, [1980]
x,502p : ill
0-88289-253-3 X.800/36798
9044272

AIR OPERATIONS. Spanish Civil War

MALUQUER I WAHL, Joan J.
L'aviació de Catalunya els primers mesos de la guerra civil. - Barcelona : Pòrtic, 1978
423p.,[16]p.of plates : ill. - (Col—lecció Pòrtic 71 ; no.14)
84-7306-092-x X.708/47992
0271217

SHORES, Christopher
Spanish Civil War air forces. - London : Osprey Publishing, 1977
48p : ill(some col). - (Airwar series ; 3) (Aircam). - Bibliography: p.46
0-85045-135-3 X.611/7764 Woolwich
0271217

AIR OPERATIONS. Spanish Civil War
By fighter aeroplanes

GARCÍA LACALLE, Andrés
Mitos y verdades. - México : Ediciones Oasis, 1973
590 p. : ill
r6004943 X.709/19542
0902918

AIR OPERATIONS. Spanish Civil War
By Italian fighter aeroplanes

ALCOFAR NASSAES, José Luis
La aviación legionaria en la guerra española. - Barcelona : Editorial Euros, 1975
397 p., 64 p. of plates. - (Colección Historia y tiempo). - Bibliography: p. 389-394
84-7364-026-8 X.809/22979
0878952

AIR OPERATIONS. Spanish Civil War
By Soviet airmen

SMIRNOV, Boris Aleksandrovich
Ot Madrida do Khalkhin-Gola. - Izd. 2., ispr. i dop.. - Kuĭbyshev : Kuĭbyshevskoe knizhnoe izdatel'stvo
341 p. : ill
r6022909 X.809/23338
1009745

AIR OPERATIONS. Vietnamese wars
By United States. Air Force

DRENDEL, Lou
Air war over South east Asia. - Carrollton, Tex. : Squadron / Signal.
Vol.1: 1962-1966. - c1982
80p : ill(some col.)
0-89747-134-2 X.622/14243
2795264

AIR OPERATIONS. Vietnamese wars
By United States. Air Force, 1961-1973 - Bibliographies

SMITH, Myron J.
Air war, Southeast Asia, 1961-1973. - Metuchen ; London : Scarecrow Press, 1979
xviii,298p, Plate : 1 ill
0-8108-1261-4 X.809/45891
2108747
- [Another copy] OPL

AIR OPERATIONS. Vietnamese wars
By United States. Air Force, 1964-1975

AIR war - Vietnam. - London : Arms and Armour Press, [1979]
xiii,361p : ill. - Includes index
0-85368-283-6 X.800/28110
1810308

BELL, Dana
Air war over Vietnam. - London : Arms and Armour Press.
Vol.2. - 1983
68p : ill(some col.). - (Warbirds illustrated ; no.14)
0-85368-573-8 X.622/16468
1810308
- London : Arms and Armour.
Vol.3. - 1983
67p : ill(some col.). - (Warbirds illustrated ; no.21)
0-85368-607-6 X.0800/1646(21)
1810308
- (Warbirds illustrated ; no.26)
Vol.4. - 1984
64p : col.ill
0-85368-635-1 X.622/21601
1810308

DRENDEL, Lou
The air war in Vietnam. - New York : Arco, 1978
96p. - (Arco aviation books)
0-668-01817-8 X.702/6270
1810308

AIR OPERATIONS. Vietnamese wars
By United States. Air Force, 1972

ANDERSON, William C.
Bat-21. - Englewood Cliffs, N.J. ; London : Prentice-Hall, c1980
188p
0-13-069500-9 X.800/34310
2812223

AIR OPERATIONS. Western European campaigns. World War 2

KRZEMIŃSKI, Czesław
Wojna powietrzna w Europie 1939-1945. - Warszawa : Wydawnictwo Ministerstwa Obrony Narodowej, 1983
456p : ill. - (Rodzaje wojsk i sił zbrojnych). - Includes bibliography
83-11-06862-3 X.809/57679
9047751

AIR OPERATIONS. Western European campaigns. World War 2
By Allied air forces. Czechoslovak airmen

ČEJKA, Eduard
Zlomená křidla. - Vyd. 2. - Praha : Naše vojsko, 1970
275p. : ill. - (Máj ; sv. 146)
r6000744 X.809/22951
0882054

ŠIKL, Václav
Nebe nad Seinou. - Curych : Konfrontace, 1977
193p,[6]leaves of plates : ill
3-85770-036-x X.808/36871
0882054

AIR OPERATIONS. Western European campaigns. World War 2
By Canada. Royal Canadian Air Force. Squadron, 418. 1943-1944. De Havilland Mosquito aeroplanes

MCINTOSH, Dave
[Terror in the starboard seat]. Mosquito intruder. - London : John Murray, 1982, c1980
184p,[8]p of plates : ill
0-7195-3918-8 X.809/51377
2710803

AIR OPERATIONS. Western European campaigns. World War 2
By Great Britain. Royal Air Force

DEMOULIN, Charles
Mes oiseaux de feux. - Paris : Julliard, 1982
346p
2-260-00310-9 X.809/60338
8483310

AIR OPERATIONS. Western European campaigns. World War 2
By Great Britain. Royal Air Force, 1939-1942. Bomber aeroplanes

PHILPOTT, Bryan
RAF bomber units, September 1939 to June 1942. - London : Osprey Publishing, 1977
48p : ill(some col). - (Aircam/airwar series ; 5)
0-85045-139-6 X.611/8295 Woolwich
1700626

AIR OPERATIONS. Western European campaigns. World War 2
By Great Britain. Royal Air Force, 1939-1942. Fighter aeroplanes

PHILPOTT, Bryan
RAF fighter units, Europe, September 1939-March 1942. - London : Osprey Publishing, 1977
48p : ill(some col). - (Aircam/airwar series ; 1)
0-85045-091-8 X.611/7887 Woolwich
1300768

AIR OPERATIONS. Western European campaigns. World War 2
By Great Britain. Royal Air Force, 1942: Dieppe Raid

FRANKS, Norman L. R. (Norman Leslie Robert)
The greatest air battle. - London : Kimber, 1979
256p : ill. - Bibliography: p.248. - Includes index
0-7183-0396-2 X.800/27942
1918443

AIR OPERATIONS. Western European campaigns. World War 2
By Great Britain. Royal Air Force. Bomber Command. No 2(B) Group

BOWYER, Michael J. F. (Michael John Frederick)
2 Group RAF. - London : Faber, 1979
532p,[48]p of plates : ill. - Includes index
0-571-11460-1 X.700/26114
014097x

AIR OPERATIONS. Western European campaigns. World War 2
By Great Britain. Royal Air Force. Bomber Command. No 8 Group

BOWYER, Chaz
Path Finders at war. - London : I. Allan, 1977
160p : ill
0-7110-0757-8 X.622/5915 Woolwich
0792454

MUSGROVE, Gordon
Pathfinder force. - London : Macdonald and Jane's, 1976
x,302p,[32]p of plates : ill. - Includes index
0-354-01017-4 X.800/25844
0792454

AIR OPERATIONS. Western European campaigns. World War 2
By Great Britain. Royal Air Force. Squadron, No.617, 1943: Operation Chastise

SWEETMAN, John
Operation chastise. - London : Jane's, 1982
xiv,218p,[32]p of plates : ill. - Bibliography: p208-215. - Includes index
0-7106-0124-7 X.800/33237
2744635

AIR OPERATIONS. Western European campaigns. World War 2
By United States. Air Force. Fighter Group, 359th

359TH Fighter Group, 1943-1945 ... / [edited by Thomas H. Raines]. - Reprinted. - Nashville, Tenn. : Battery Press, [1980?]
67p : ill. - (Aviation series ; 1)
0-89839-014-1 X.805/3322
8483310

AIR OPERATIONS. Western European campaigns. World War 2
By United States. Air Force. Troop Carrier Group, 437th

GUILD, Frank
Action of the tiger. - Reprinted. - Nashville, Tenn : Battery Press, 1980
177p. - (Aviation series ; 2)
0-89839-028-1 X.800/34328
8483310

AIR OPERATIONS. Western European campaigns. World War 2
By United States. Army Air Force. 8th Air Force, 1942-1945

FREEMAN, Roger Anthony
The mighty Eighth. - London : Macdonald and Jane's, 1978
[5],311p,[16]p of plates : ill(some col). - Includes index
B79-12683 X.622/6796 Woolwich
1836420

HOSEASON, James
The 1000 day battle. - Lowestoft : Gillingham Publications, 1979
3-256p : ill. - Bibliography: p.256
B80-42018 X.702/6440
1836420

AIR OPERATIONS. Western Front. World War 1

SIMKINS, Peter
Air fighting 1914-18. - [London] : Imperial War Museum, 1978
80p : ill. - Bibliography: p78-79
B82-03982 X.629/16802 Woolwich
019641x

AIR OPERATIONS. Western Front. World War 1
By Baracca, Francesco

MANCA, Vincenzo
L'idea meravigliosa di Francesco Baracca. - Roma : Ateneo [& Bizzarri], 1980
293p : ill
r6508601 X.800/35769
8482020

AIR OPERATIONS. Western Front. World War 1
By Great Britain. Royal Air Force, 1917-1918. Fighter aeroplanes

REVELL, Alex
British fighter units, Western Front, 1917-18. - London : Osprey Publishing, 1978
48p : ill(some col). - (Airwar series ; 18)
0-85045-292-9 X.611/8368 Woolwich
1742566

AIR OPERATIONS. Western Front. World War 1
By Great Britain. Royal Air Force. Bristol F-2B aeroplanes - Biographies

NOBLE, Walter
With a Bristol fighter squadron. - Large print ed. - Bath : Chivers, 1977
viii,198p,vii p of plates : ill
0-85997-230-5 Nov.34777 Woolwich
1170619

AIR OPERATIONS. Western Front. World War 1
By Great Britain. Royal Air Force. No.48 Squadron & Great Britain. Royal Air Force. No.88 Squadron, 1918 - Biographies

VEE, Roger
Flying minnows. - [1st ed. reprinted]. - London : Arms and Armour Press [etc.], 1977
318p,[12]p of plates : ill
0-85368-121-x X.629/11266 Woolwich
1104993

AIR OPERATIONS. Western Front. World War 1
By Great Britain. Royal Air Force. No.100 Squadron, 1917-1918

The ANNALS of 100 Squadron / [compiled by] C. Gordon Burge. - London : Bivouac Books Ltd, [1975]
[12],211p : ill
0-85680-007-4 X.802/10070
0535990

AIR OPERATIONS. Western Front. World War 1
By Great Britain. Royal Air Force. No.203 Squadron, 1917-1918 - Biographies

ROCHFORD, Leonard H.
I chose the sky. - London : Kimber, 1977
224p : ill. - Includes index
0-7183-0085-8 X.800/26573
1293214

AIR OPERATIONS. Western Front. World War 1
By Great Britain. Royal Air Force. No.210 Squadron & Great Britain. Royal Naval Air Service. No.8 Squadron, 1917-1918

CRUNDALL, Eric Duncan
Fighter pilot on the Western front. - London : Kimber, 1975
192p : ill. - Includes index
0-7183-0144-7 X.800/25337
0571911

AIR OPERATIONS. Western Front. World War 1
By Great Britain. Royal Flying Corps - Biographies

BISHOP, William Avery
Winged warfare. - [1st ed. reprinted]. - London : Pan Books, 1978
281p,[8]p of plates : ill
0-330-25443-x X.708/21333
0206202

LEWIS, Cecil
Sagittarius rising. - Harmondsworth : Penguin, 1977
266p
0-14-004367-5 X.708/20896
0206202

AIR OPERATIONS. Western Front. World War 1
By Great Britain. Royal Flying Corps, 1914-1916. Fighter aeroplanes

REVELL, Alex
British fighter units, Western Front, 1914-16. - London : Osprey Publishing, 1978
48p : ill(some col). - (Airwar series ; 14)
0-85045-285-6 X.700/25746
1551426

RIMELL, Raymond Laurence
The Royal Flying Corps. - London : Arms and Armour, 1985
64p : chiefly ill. - (Vintage warbirds ; no.1)
0-85368-693-9 X.800/41264
1551426

AIR OPERATIONS. World War 1

BATCHELOR, John, 1936-
Fighting aircraft of World War One and Two. - London : Phoebus, 1976
256p : ill(some col). - Includes index
0-7026-0007-5 X.625/552 Woolwich
0170372

CAMPBELL, Christopher
Aces and aircraft of World War 1. - London : Treasure, 1984, c1981
144p : ill(some col.). - Includes index
0-907812-62-7 L.42/2846
0170372
Aces and aircraft of World War I. - Poole : Blandford, 1981
144p : ill(some col.). - Includes index
0-7137-0954-5 L.42/866
0170372

MCKEE, Alexander
The friendless sky. - London : Panther, 1984, c1962
256p
0-586-05823-0 X.808/40572
0170372

AIR OPERATIONS. World War 1 - Bibliographies

SMITH, Myron J.
World War I in the air. - Metuchen : Scarecrow, 1977
xix,271p : ill. - Includes index
0-8108-0990-7 X.809/28394
0170372

AIR OPERATIONS. World War 1
By British air forces

LEWIS, Cecil
Sagittarius rising. - London : Heinemann, 1982
viii,331p
0-432-08600-5 X.958/14335
2868520

AIR OPERATIONS. World War 1
By British air forces. Interviews. Sound tape recordings - Catalogues

IMPERIAL WAR MUSEUM. Department of Sound Records
Military and naval aviation 1914-18. - [London] : Imperial War Museum, Dept. of Sound Records, [1982?]
26p. - (Oral history recordings)
B82-26785 X.809/52267
274757x

AIR OPERATIONS. World War 1
By Commonwealth air forces. Decorations: Victoria Cross

BOWYER, Chaz
For valour. - London : Kimber, 1978
548p : ill. - Bibliography: p.519-522. - Includes index
0-7183-0425-x X.800/27029
151802x

AIR OPERATIONS. World War 1
By fighter aeroplanes. Pilots

REYNOLDS, Quentin
They fought for the sky. - London : Pan Books, 1977
285p,[16]p of plates : ill. - Bibliography: p.276-279. - Includes index
B77-20807 X.708/20936
0168637

SMITHERS, Alan Jack
Wonder aces of the air. - London : Gordon and Cremonesi, 1980
212p,[8]p of plates : ill. - Bibliography: p.207. - Includes index
0-86033-077-x X.622/8033 Woolwich
1737929

WALLACE, Melville
Pilots of World War 1. - London : Macdonald Educational, 1979
361,[1]p : ill(chiefly col). - (Macdonald living history). - Includes index
0-356-06300-3 X.802/10760
1783777

WINTER, Denis
The first of the few. - London : Allen Lane, 1982
223p,[24]p of plates : ill. - Includes index
0-7139-1278-2 X.809/51212
0168637
The first of the few. - Harmondsworth : Penguin, 1983, c1982
223p,[24]p of plates : ill. - Bibliography: p217-218. - Includes index
0-14-005256-9 X.629/20637
0168637

AIR OPERATIONS. World War 1
By fighter aeroplanes. Pilots - Biographies

JACKSON, Robert, 1941-
Fighter pilots of World War I. - London : A. Barker, 1977
vii,152p
0-213-16645-3 X.809/42055
0168637

AIR OPERATIONS. World War 1
By Germany. Kriegsmarine. Marine-Luftschiffabteilung

ROBINSON, Douglas H. (Douglas Hill)
The Zeppelin in combat. - 3rd ed. - Seattle ; London : University of Washington Press, 1971
xiv,417p,[34]p of plates (2fold.) : ill. - Bibliography: p370-377. - Includes index
0-295-95752-2 X.809/47631
239880x

AIR OPERATIONS. World War 1
By Germany. Luftstreitkräfte - Biographies

RICHTHOFEN, Manfred, Freiherr von
The Red Baron. - London : Sphere, 1976
xiv,225p,[16]p of plates : ill
0-7221-8776-9 X.708/20696
0170933

AIR OPERATIONS. World War 1
By Germany. Luftstreitkräfte, 1914-1917. Fighter aeroplanes

The GERMAN Army Air Service / [compiled by] Raymond Laurence Rimell. - London : Arms and Armour, 1985
64p : chiefly ill. - (Vintage warbirds ; no.2)
0-85368-694-7 X.800/41265
1551434

IMRIE, Alex
German fighter units, 1914-May 1917. - London : Osprey Publishing, 1978
48p : ill(some col). - (Airwar series ; 13)
0-85045-290-2 X.700/25747
1551434

AIR OPERATIONS. World War 1
By Germany. Luftstreitkräfte, 1917-1918. Fighter aeroplanes

IMRIE, Alex
German fighter units, June 1917-1918. - London : Osprey Publishing, 1978
48p : ill(some col). - (Airwar series ; 17)
0-85045-289-9 X.700/25853
1742736

AIR OPERATIONS. World War 1
By Great Britain. Royal Air Force - Biographies

BARING, Maurice
[R.F.C., H.Q., 1914-1918]. Flying Corps headquarters 1914-1918. - London : Buchan & Enright, 1985
313p. - (Echoes of war). - Includes index
0-907675-44-1 X.809/64772
2798557

AIR OPERATIONS. World War 1
By Great Britain. Royal Air Force - Biographies-continued

LEWIS, Cecil
Sagittarius rising. - 2nd ed. - Harmondsworth : Penguin, 1966, c1936
265p
B83-24224 X.808/38078
2891484

AIR OPERATIONS. World War 1
By Great Britain. Royal Air Force - Correspondence, diaries

WORTLEY, Rothesay Stuart
Letters from a flying officer. - Gloucester : Sutton, 1982
vii,[8]p of plates : ill
0-86299-017-3 X.808/37197
2798557

AIR OPERATIONS. World War 1
By Great Britain. Royal Air Force, 1916-1918 - Biographies

LEWIS, Gwilym Hugh
Wings over the Somme, 1916-1918. - London : Kimber, 1976
205p : ill. - Includes index
0-7183-0324-5 X.800/11552
0679615

AIR OPERATIONS. World War 1
By Great Britain. Royal Air Force. Sopwith Camel aeroplanes

BOWYER, Chaz
Sopwith Camel. - Falmouth : Glasney Press, 1978
192p : ill. - Includes index
0-9502825-7-x X.622/6372 Woolwich
1489534

AIR OPERATIONS. World War 1
By theories of Germany. Luftstreitkräfte

KNIGHT, Dennis
Baron Richthofen. - London : Royal Sovereign Group Ltd, 1976
[2],16,[2]p,[2]leaves of plates : ill(chiefly col). - (Activity books / Dennis Knight)
0-900862-43-2 X.619/17333 Woolwich
1038893

AIR OPERATIONS. World War 1
By theories of Germany. Luftstreitkräfte - Biographies

DEGELOW, Carl
Germany's last knight of the air. - London : Kimber, 1979
218p : ill. - Bibliography: p.207. - Includes index
0-7183-0146-3 X.800/27899
1908782

WRIGHT, Nicolas
The Red Baron. - London : Sidgwick and Jackson, 1976
120p,[16]p of plates : ill. - Includes index
0-283-98298-5 X.809/40633
0590851

AIR OPERATIONS. World War 1
By theories of Great Britain. Royal Air Force - Biographies

REVELL, Alex
The vivid air. - London : Kimber, 1978
255p,[16]p of plates : ill. - Includes index
0-7183-0385-7 X.800/27053
1544284

AIR OPERATIONS. World War 1
By theories of Great Britain. Royal Flying Corps - Biographies

BOWYER, Chaz
Albert Ball, VC. - London : Kimber, 1977
208p,[24]p of plates : ill. - Bibliography: p.200-201. - Includes index
0-7183-0045-9 X.800/26209
1166883

AIR OPERATIONS. World War 2

ARENA, Nino
Assalto dal cielo. - Milano : Mursia, 1975
305 p., 24 p. of plates : ill. - (Biblioteca del cielo ; n. 17)
r6024301 X.629/7645 Woolwich
017372X

BATCHELOR, John, 1936-
Fighting aircraft of World War One and Two. - London : Phoebus, 1976
256p : ill(some col). - Includes index
0-7026-0007-5 X.625/552 Woolwich
017372x

BERTHOLD, Will
Der Sieg, der vor die Hunde ging. - Bayreuth : Hestia, [1981]
316p.,[18]p.of plates : ill
3-7770-0216-x X.809/50421
017372x

BOWYER, Chaz
Images of air war 1939-45. - London : Batsford, 1983
120p : ill. - Bibliography: p120
0-7134-3784-7 X.622/16721
2931877

LOWE, James Trapier
A philosophy of air power. - Lanham , London : University Press of America, c1984
vii,463p. - Includes index
0-8191-3953-x X.529/69283
3279057

NANI, Augusto
Gli idrovolanti della Seconda Guerra Mondiale - Parma Albertelli, 1978
235p : ill
r6599050 L.42/851
017372x

OVERY, R. J. (Richard James)
The air war 1939-1945. - London : Europa, c1980
xii,263p,[12]p of plates : ill. - Bibliography: p240-256. - Includes index
0-905118-53-7 X.800/29388
017372x

WINGS of war / edited by Laddie Lucas. - London : Hutchinson, 1983
409p,[40]p of plates : ill. - Includes index
0-09-154280-4 X.800/41303
3082695

WORLD War II in the air / edited by James F. Sunderman. - New York ; London : Van Nostrand Reinhold, 1981, c1963
2v. : ill. - Includes index
B81-28465 X.800/30918-30919
017372x

AIR OPERATIONS. World War 2
1943-1944. Great Britain. Royal Air Force. Bomber Command. Avro Lancaster aeroplanes

CURRIE, Jack
Lancaster target. - London : Goodall, 1981, c1977
175p
0-907579-00-0 X.808/37192
118458x

AIR OPERATIONS. World War 2
1943-1944. Great Britain. Royal Air Force. Bomber Command. Avro Lancaster aeroplanes - Biographies

CURRIE, Jack
Lancaster target. - London : New English Library, 1977
175p
0-450-03280-9 H.77/942 Woolwich
118458x

AIR OPERATIONS. World War 2
Air operations over Benelux countries, 1945

FRANKS, Norman L. R. (Norman Leslie Robert)
The battle of the airfields. - London : Kimber, 1982
224p : ill. - Bibliography: p216. - Includes index
0-7183-0448-9 X.800/33622
2759136

AIR OPERATIONS. World War 2
Air operations over Maidstone, 1940

KENT BATTLE OF BRITAIN MUSEUM
Battle over Kent 1940. - Chilham : The Museum, 1977
[19]p : ill
B78-04563 X.709/31885
1307185

AIR OPERATIONS. World War 2
Air photographic operations by Allied air forces

POWYS-LYBBE, Ursula
The eye of intelligence. - London : Kimber, 1983
223p : ill. - Includes index
0-7183-0468-3 X.622/17156
0138207

AIR OPERATIONS. World War 2
Allied air operations. Goodson, James A. - Biographies

GOODSON, James A.
Tumult in the clouds. - London : Kimber, 1983
238p,[16]p of plates : ill. - Includes index
0-7183-0329-6 X.800/36325
2964295

AIR OPERATIONS. World War 2
By Allied air forces

BONACINA, Giorgio
Comando bombardieri operazione Europa. - Milano : Longanesi, 1975
44p.,16p. of plates : ill. - (Il cammeo ; n. 310)
r6000509 X.700/16755
0938327

FOSS, Joe
Joe Foss. - Reprint. - Washington, D.C. : Zenger Publishing Co, 1979
160p,14pages of plates. : ill
0-89201-080-0 X.809/50280
0938327

LICHTENSTEIN, Heiner
Warum Auschwitz nicht bombardiert wurde. - Köln : Bund-Verlag, 1980
183p,[8]p of plates : ill
3-7663-0428-3 X.809/48195
0938327

ONDERWATER, Hans
En toen was het stil-. - Baarn : Hollandia, 1981
340p : ill
90-604-5463-4 X.809/53917
0938327

AIR OPERATIONS. World War 2
By Allied air forces. Boeing B-17 aeroplanes

WILLMOTT, H. P. (Hedley Paul)
B-17 flying fortress. - London : Arms and Armour Press, 1980
64p : ill(some col.). - ([War planes in colour] ; 4) (A bison book)
0-85368-444-8 X.622/12617 Woolwich
2061988

AIR OPERATIONS. World War 2
By allied air forces. Bomber aeroplanes. Gunners

BOWYER, Chaz
Guns in the sky. - London : Dent, 1979
182p : ill. - Bibliography: p.179-180. - Includes index
0-460-04429-x X.800/28057
1723561

Guns in the sky. - London : Corgi, 1981, c1979
ix,182p : ill. - Bibliography: p179-180. - Includes index
0-552-98121-4 X.800/30830
1723561

DAVIS, Larry
Gunships. - Carrollton : Squadron / Signal, c1982
64p : ill(some col.)
0-89747-123-7 X.622/13144
1723561

AIR OPERATIONS. World War 2
By Allied air forces. Bristol Blenheim aeroplanes

BOWYER, Chaz
Bristol Blenheim. - London : Ian Allan, 1984
128p : ill. - Bibliography: p128
0-7110-1351-9 X.622/21185
3198162

AIR OPERATIONS. World War 2
By Allied air forces. Consolidated Aircraft Corporation B-24 aeroplanes

BOWMAN, Martin W.
The B-24 Liberator 1939-1945. - Norwich : Wensum, c1979
128p : ill(some col.)
0-903619-27-x X.622/9901 Woolwich
2381133

FREEMAN, Roger A. (Roger Anthony)
B-24 Liberator at war. - London : Ian Allan, 1983
128p : ill
0-7110-1264-4 X.805/4369
2381133

AIR OPERATIONS. World War 2
By Allied air forces. Curtiss P-40 aeroplanes

CHRISTY, Joe
P-40 Hawks at war. - London : I. Allan, 1979
128p : ill
0-7110-0983-x X.622/7157 Woolwich
2079097

AIR OPERATIONS. World War 2
By Allied air forces. Douglas A-20 Havoc aeroplanes

MESKO, Jim
A-20 Havoc. - Carrollton, Tex. : Squadron / Signal, c1983
49p : ill(some col.). - (Aircraft ; no.56)
0-89747-131-8 L.49/1807
2917718

AIR OPERATIONS. World War 2
By Allied air forces. Douglas Boston aeroplanes

HESS, William N.
A-20 Boston at war. - London : I. Allan, 1979
128p : ill
0-7110-0965-1 X.802/11084
2052520

AIR OPERATIONS. World War 2
By Allied air forces. North American B-25 Mitchell aeroplanes

SCUTTS, Jerry
B-25 Mitchell at war. - London : Ian Allan, 1983
144p : ill
0-7110-1219-9 X.622/16438
2943859

AIR OPERATIONS. World War 2
By Allied glider forces

DANK, Milton
The glider gang. - London : Cassell, 1978
3-273p : ill. - Bibliography: p.263-266. - Includes index
0-304-30014-4 X.800/27143
1577727

AIR OPERATIONS. World War 2
By Australia. Royal Australian Air Force. Mustang aeroplanes

ANDERSON, Peter N.
Mustangs of the RAAF and RNZAF - Sydney ; London : A.H and A.W. Reed, 1975
[4],103,[5]p : ill(some col). - Bibliography: p.[2] (first sequence). - Includes index
0-589-07130-0 X.622/5921 Woolwich
1136518

AIR OPERATIONS. World War 2
By bomber aeroplanes

JACKSON, Robert, 1941-
Bomber!. - London : A. Barker, 1980
157p
0-213-16743-3 X.809/45316
2080532

AIR OPERATIONS. World War 2
By Canada. Royal Canadian Air Force

ALCORN, Douglas
From hell to breakfast. - Toronto : Intruder, [1980]
339p,plates
r6812143 X.809/59735
8763518

PEDEN, Murray
A thousand shall fall. - 2nd. ed.. - Stittsville : Canada's Wings, 1982
xiii,490p : ill
0-920002-07-2 X.800/37509
8763518

AIR OPERATIONS. World War 2
By Canada. Royal Canadian Air Force. Bomber aeroplanes. Pilots: Morris, Jerrold - Biographies

MORRIS, Jerrold
Canadian artists and airmen, 1940-45. - Toronto : The Morris Gallery, [1974]
207p : ill
r6072196 X.421/10728 Woolwich
1581937

AIR OPERATIONS. World War 2
By Commonwealth air forces. Decorations: Victoria Cross

BOWYER, Chaz
For valour. - London : Kimber, 1978
548p : ill. - Bibliography: p.519-522. - Includes index
0-7183-0425-x X.800/27029
1518038

AIR OPERATIONS. World War 2
By Commonwealth air forces. Hawker Typhoon aeroplanes. Pilots

SCOTT, Desmond
Typhoon pilot. - London : Secker & Warburg, 1982
xvi,168p,[8]p of plates : ill. - (A Leo Cooper book). - Includes index
0-436-44428-3 X.809/53896
2820919

AIR OPERATIONS. World War 2
By fighter aeroplanes

JACKSON, Robert, 1941-
Fighter!. - London : A. Barker, 1979
168p
0-213-16717-4 X.809/27492
1956590

AIR OPERATIONS. World War 2
By fighter aeroplanes. Pilots

FIGHTERS in the sky / compiled by Aidan Chambers. - Basingstoke : Macmillan, 1976
125p : ill. - (Topliners)
0-333-19451-9 X.708/20345
0640131

JACKSON, Robert, 1941-
Fighter pilots of World War II. - London : Barker, 1976
176p
0-213-16578-3 X.809/40819
0630519

MEN at war / compiled by Aidan Chambers. - Basingstoke : Macmillan, 1977
[6],106p. - (Topliners)
0-333-22748-4 X.709/31804
0640131

AIR OPERATIONS. World War 2
By fighter aeroplanes. Pilots - Biographies

JACKSON, Robert, 1941-
Fighter aces of World War II. - London : Corgi, 1978
143p
0-552-10783-2 X.708/21360
0630519

AIR OPERATIONS. World War 2
By Finland. Ilmavoimat. Hävittäjälaivue 34

KARHUNEN, Joppe
Ässien iskulaivue. - Helsingissä : Otava, 1976
239p
951-10-2295-4 X.809/54113
8763518

AIR OPERATIONS. World War 2
By Germany. Luftwaffe

NAUROTH, Holger
Die deutsche Luftwaffe vom Nordkap bis Tobruk 1939-1945. - Friedberg : Podzun-Pallas, [1981]
234p : chiefly ill
3-7909-0160-1 X.800/35010
8763518

AIR OPERATIONS. World War 2
By Germany. Luftwaffe, to 1945. Messerschmitt Me 163 aeroplanes

ETHELL, Jeffrey
Komet. - London : I. Allan, 1978
160p,[4]p of plates : chiefly ill(some col). - Bibliography: p.159-160
0-7110-0827-2 X.620/17611 Woolwich
1455419

AIR OPERATIONS. World War 2
By Germany. Luftwaffe, 1939-1941. Bomber aeroplanes

SCUTTS, Jerry
Luftwaffe bomber units, 1939-41. - London : Osprey Publishing, 1978
48p : ill(some col). - (Airwar series ; 15)
0-85045-279-1 X.700/25742
1550098

AIR OPERATIONS. World War 2
By Germany. Luftwaffe, 1940-1943

'Der ADLER' / edited by S.L. Mayer and Masami Tokoi. - London : Arms and Armour Press, 1977
[190]p : ill(some col)
0-85368-112-0 X.802/10534
1374281

AIR OPERATIONS. World War 2
By Germany. Luftwaffe, 1940-1944. Fighter units - Biographies

KNOKE, Heinz
I flew for the Führer. - London : Evans Bros, 1979
187p,leaf of plate,[24]p of plates : ill
0-237-45504-8 X.809/45560
2065304

AIR OPERATIONS. World War 2
By Germany. Luftwaffe, 1940-1945. Night fighter forces - Biographies

JOHNEN, Wilhelm
Duel under the stars. - London : New English Library, 1975
157p
0-450-02193-9 X.708/20284
0614866

AIR OPERATIONS. World War 2
By Germany. Luftwaffe, 1944-1945

GIRBIG, Werner
Six months to oblivion. - London : Allan, 1975
140p,[32]p of plates : ill
0-7110-0610-5 X.800/25119
0374261

AIR OPERATIONS. World War 2
By Germany. Luftwaffe. Aeroplanes

COOPER, Matthew
The German Air Force 1933-1945. - London : Jane's, 1981
v,406p,[32]p of plates : ill. - Bibliography: p396-399. - Includes index
0-7106-0071-2 X.622/10667
0543411

SHEPHERD, Christopher
German aircraft of World War II with colour photographs. - London : Sidgwick and Jackson, 1975
144p : ill(some col). - Includes index
0-283-98179-2 X.622/5132 Woolwich
0543411

AIR OPERATIONS. World War 2
By Germany. Luftwaffe. Arado Ar 234 aeroplanes, Messerschmitt Me 163 aeroplanes & Messerschmitt Me 262 aeroplanes

ETHELL, Jeffrey L.
The German jets in combat. - London : Jane's Publishing Company, 1979
143p : ill. - Bibliography: p.139. - Includes index
0-354-01252-5 X.802/11141
2101904

AIR OPERATIONS. World War 2
By Germany. Luftwaffe. Fighter aeroplanes

HELD, Werner
Fighter!. - London : Arms and Armour Press, 1979
224p : chiefly ill
0-85368-133-3 X.800/28201
1995197

AIR OPERATIONS. World War 2
By Germany. Luftwaffe. Fighter aeroplanes - Biographies

KNOKE, Heinz
I flew for the Führer. - Large print ed. - Bath : Chivers, 1984, c1953
viii,275p. - (A Lythway autobiography)
0-7451-0067-8 X.809/62593
3024377

AIR OPERATIONS. World War 2
By Germany. Luftwaffe. Fighter aeroplanes. Pilots

CONSTABLE, Trevor James
Horrido!. - [New ed.]. - Toronto ; London : Bantam, 1979
xvi,411p,fold plate : ill(incl 1 col). - (The Bantam war book series). - Includes index
0-553-12663-6 X.708/22655
2319349

AIR OPERATIONS. World War 2
By Germany. Luftwaffe. Focke-Wulf Fw 190 aeroplanes

PRICE, Alfred
Focke Wulf 190 at war. - London : I. Allan, 1977
160p : ill
0-7110-0768-3 X.625/557 Woolwich
1272039

AIR OPERATIONS. World War 2
By Germany. Luftwaffe. Focke-Wulf Kondor aeroplanes

POOLMAN, Kenneth
Scourge of the Atlantic. - London : Macdonald and Jane's, 1978
192p,[24]p of plates : ill. - Bibliography: p.185. - Includes index
0-354-01164-2 X.809/43877
1755072

AIR OPERATIONS. World War 2
By Germany. Luftwaffe. Ground attack units

PEGG, Martin
Luftwaffe ground attack units, 1939-45. - London : Osprey Publishing, 1977
48p : ill(some col). - (Airwar series ; 4) (Aircam)
0-85045-137-x X.611/7765 Woolwich
1199102

AIR OPERATIONS. World War 2
By Germany. Luftwaffe. Junkers JU87 Stuka aeroplanes

JUNKERS Ju 87 Stuka / G. Borelli ... [et al.]. - Modena : STEM Mucchi, 1974
155p.,[3]fold leaves,[8]p.of plates : ill,chart. - (Le macchine e la storia ; 3)
r6197542 L.49/420
2452081

AIR OPERATIONS. World War 2
By Germany. Luftwaffe. Junkers JU88 aeroplanes - Biographies

STAHL, P. W. (Peter Wilhelm)
The diving eagle. - London : Kimber, 1984
203p : ill. - Includes index
0-7183-0509-4 X.800/39601
3210685

AIR OPERATIONS. World War 2
By Germany. Luftwaffe. Junkers Stuka aeroplanes

BARKER, A. J. (Arthur J.)
Stuka Ju-87. - London : Arms and Armour, 1980
64p : ill(some col.). - ([War planes in colour] ; 5) (A Bison book). - Bibliography: p64
0-85368-474-x X.622/9391 Woolwich
1167197

ROBERTSON, Bruce
JU87 Stuka. - Cambridge : Stephens [for] Airfix Products Ltd, 1977
96p : ill. - (Classic aircraft ; no.5)
0-85059-193-7 X.620/16976 Woolwich
1167197

AIR OPERATIONS. World War 2
By Germany. Luftwaffe. Junkers Stuka aeroplanes–continued

SMITH, Peter C. (Peter Charles), 1940–
Stuka at war. – [Revised ed.]. – London : I. Allan, 1980
128p : ill
0–7110–1017–x X.802/11221
205860x

VANAGS–BAGINSKIS, Alex
Stuka Ju87. – London : Jane's, 1982
55p,8folded : ill(some col.). – (Jane's aircraft spectaculars)
0–7106–0191–3 L.45/1699
1167197

AIR OPERATIONS. World War 2
By Germany. Luftwaffe. Messerschmitt Me 163 aeroplanes

ZIEGLER, Mano
Rocket fighter. – London : Arms and Armour Press, 1976
xii,161p,[34]p of plates : ill(some col)
0–85368–161–9 X.620/16368 Woolwich
0746371

AIR OPERATIONS. World War 2
By Germany. Luftwaffe. Night fighter forces

ADERS, Gebhard
History of the German night fighter force, 1917–1945. – London : Jane's Publishing Company, 1979
[4],284p,[64]p of plates : ill. – Bibliography: p.283–284
0–354–01247–9 X.800/28749
1490974

The DEFENCE of the Reich / [compiled by] Werner Held and Holger Nauroth ; translated by David Roberts. – London : Arms and Armour, 1982
230p : ill
0–85368–414–6 X.622/12572 Woolwich
2722429

SCUTTS, Jerry
Luftwaffe night fighter units, 1939–45. – London : Osprey Publishing, 1978
48p : ill(some col). – (Aircam/airwar ; 9)
0–85045–232–5 X.611/8101 Woolwich
1490974

AIR OPERATIONS. World War 2
By Germany. Luftwaffe. 2. Fallschirmjäger–Division, 1944–1945

RAMCKE, Hermann Bernhard
Fallschirmjäger. – Preussisch Oldendorf, c1973
267 p. : ill
r6016876 X.809/22960
0876542

AIR OPERATIONS. World War 2
By Germany. Luftwaffe. Kampfgeschwader 100 'Wiking'

BALKE, Ulf
Kampfgeschwader 100 'Wiking'. – Stuttgart : Motorbuch, 1981
479p. : ill
3–87943–772–6 X.800/32633
8763518

AIR OPERATIONS. World War 2
By Germany. Luftwaffe. Kampfgeschwader 200

STAHL, P. W. (Peter Wilhelm)
KG 200. – London : Jane's, 1981
206p,[32]p : ill
0–7106–0096–8 X.622/15462
2898438

AIR OPERATIONS. World War 2
By Germany. Luftwaffe. Kampfgruppe 100, 1939–1941

WAKEFIELD, Kenneth
The first pathfinders. – London : Kimber, 1981
256p : ill. – Includes index
0–7183–0318–0 X.622/11788 Woolwich
2606038

AIR OPERATIONS. World War 2
By Germany. Wehrmacht. V–1 flying bombs & V–2 rockets

GARLIŃSKI, Józef
Ostatnia broń Hitlera. – Londyn [i.e. London] : Odnowa, 1977
220p,[32]p of plates : ill. – Bibliography: p.205–209. – Includes index
0–903705–16–8 X.709/32847
1774379

AIR OPERATIONS. World War 2
By gliders

MRAZEK, James Edward
The glider war. – London : Hale [etc.], 1975
304p : ill. – Bibliography: p.295. – Includes index
0–7091–4495–4 X.800/25093
0365084

AIR OPERATIONS. World War 2
By Great Britain. Army. Glider Pilot Regiment

LLOYD, Alan
The gliders. – [London] : Cooper in association with Secker & Warburg, 1982
vi,208p[32]p of plates : ill. – Bibliography: p201–203. – Includes index
0–436–25470–0 Nov.47862
2865289

The gliders. – [London] : Corgi, 1984, c1982
224p. – Bibliography: p216–218. – Includes index
0–552–12167–3 X.629/24020
2865289

AIR OPERATIONS. World War 2
By Great Britain. Army. Glider Pilot Regiment, 1942–1945

CHATTERTON, George
The wings of Pegasus. – Nashville : Battery, 1982
xi,282p,[16]p of plates : ill. – (Airborne series ; 14). – Includes index
0–89839–060–5 X.809/54202
284205x

AIR OPERATIONS. World War 2
By Great Britain. Royal Air Force

BOWYER, Chaz
Royal Air Force handbook 1939–1945. – London : Ian Allan, 1984
128p : ill. – Bibliography: p127–128
0–7110–1318–7 X.800/39078
0290025

CURRIE, Jack
Mosquito victory. – London : Goodall, c1983
174p
0–907579–03–5 X.958/20695
0165212

OXSPRING, Bobby
Spitfire command. – London : Kimber, 1984
196p : ill. – Includes index
0–7183–0537–x X.800/41317
0165212

RICHARDS, Denis
Royal Air Force, 1939–1945. – [Revised ed.]. – London : H.M.S.O.
Vol.1: The fight at odds. – 1974
xii,430p,[9] leaves of plates(3fold),[16]p of plates : ill. – Includes index
0–11–771592–1 BS.88/115
0290025

Vol.2: The fight avails. – 1975
x,415p,[14]leaves of plates(3 fold),[16]p of plates : ill. – Includes index
0–11–771593–x BS.88/115
0290025

RICHARDS, Denis
Royal Air Force, 1939–1945. – [Revised ed.]. – London : H.M.S.O.
Vol.3: The fight is won. – 1975
x,441p,[10]leaves of plates(4 fold),[24]p of plates : ill. – Includes index
0–11–771594–8 BS.88/115
0290025

ROYAL Air Force at war / edited by Sir Christopher Foxley–Norris ; photo research and captioning by Chaz Bowyer ; produced in conjunction with and on behalf of the Royal Air Force Benevolent Fund. – London : Ian Allan, 1983
144p : ill
0–7110–1328–4 X.805/5547
0290025

TERRAINE, John
The right of the line. – London : Hodder & Stoughton, 1985
[640]p : ill
0–340–26644–9 X.800/41286
0290025

AIR OPERATIONS. World War 2
By Great Britain. Royal Air Force – Biographies

FAIRHEAD, R. (Ralph)
An airman's diary. – Bognor Regis : New Horizon, c1982
103p, [11]p of plates : ill
0–86116–599–3 X.809/53997
0165212

GODEFROY, Hugh Constant
Lucky thirteen. – London : Croom Helm, c1983
274p,[16]p of plates : ill
0–7099–1131–9 X.809/56559
0165212

HOUGH, Richard
One boy's war. – London : Heinemann, 1975
viii,151p,[16]p of plates : ill
0–434–34860–0 X.809/40105
0165212

POWELL, Josh
Gone but not forgotten. – [Merthyr Tydfil] : [J. Powell], [1984]
52p : ill
B85–19232 X.809/63765
0165212

AIR OPERATIONS. World War 2
By Great Britain. Royal Air Force – Texts

MCHARRIE, Dennis
Out of the blue. – London : Edward Arnold, 1977
64p : ill
0–7131–0153–9 X.709/31654
1260367

AIR OPERATIONS. World War 2
By Great Britain. Royal Air Force, 1939–1942 – Biographies

HALL, Roger
Clouds of fear. – Folkestone : Bailey and Swinfen, 1975
187p
0–561–00215–0 X.809/40325
0515280

Clouds of fear. – Sevenoaks : Coronet, 1977
156p
0–340–21835–5 X.709/31805
0515280

AIR OPERATIONS. World War 2
By Great Britain. Royal Air Force, 1940. Supermarine Spitfire aeroplanes

SPITFIRE squadron / edited by David Guthrie. – London : Corgi, 1976
142p
0–552–10326–8 X.708/20695
1042998

SPITFIRE squadron / edited by David Guthrie. – London : Hale, 1978
160p
0–7091–6733–4 X.809/43289
1042998

AIR OPERATIONS. World War 2
By Great Britain. Royal Air Force, 1940. Supermarine Spitfire aeroplanes – Biographies

JOHNSTONE, Sandy
Enemy in the sky. – London : Kimber, 1976
191p,[12]p of plates : ill. – Includes index
0–7183–0474–8 X.800/25814
0780340

AIR OPERATIONS. World War 2
By Great Britain. Royal Air Force, 1940–1943. Avro Lancaster aeroplanes – Biographies

RAYMOND, Robert S.
A Yank in Bomber Command. – Newton Abbot : David and Charles [etc.], 1977
159p,[4]p of plates
0–7153–7317–x X.809/41957
1115626

AIR OPERATIONS. World War 2
By Great Britain. Royal Air Force, 1940–1945. Night fighter forces

CHISHOLM, Roderick
Cover of darkness. – Morley : Elmfield Press, 1976
222p. – ([Morley war classics])
0–7057–0069–0 X.808/23093
0762407

AIR OPERATIONS. World War 2
By Great Britain. Royal Air Force, 1943–1944. Fighter aeroplanes – Biographies

WHITE, Roger Henshaw
Spitfire saga. – London : Kimber, 1981
221p : ill. – Includes index
0–7183–0058–0 X.800/30992
2530155

AIR OPERATIONS. World War 2
By Great Britain. Royal Air Force, 1943–1945. Halifax aeroplanes

SMITH, Arthur C. (Arthur Carlton)
Halifax crew. – Stevenage : Carlton, c1983
60p : ill. – Bibliography: p60
0–9508546–0–3 X.805/4610
2606046

AIR OPERATIONS. World War 2
By Great Britain. Royal Air Force. Aeroplanes

TURNER, John Frayn
British aircraft of World War II, with colour photographs. – London : Sidgwick and Jackson, 1975
144p : ill(some col). – Includes index
0–283–98180–6 X.622/5133 Woolwich
0543403

AIR OPERATIONS. World War 2
By Great Britain. Royal Air Force. Armstrong Whitworth Albemarle aeroplanes, Supermarine Spitfire XII aeroplanes & Short Shetland flying boats

BOWYER, Michael J. F. (Michael John Frederick)
Aircraft for the Royal Air Force. - London : Faber, 1980
170p : ill. - Includes index
0-571-11515-2 X.622/9217 Woolwich
2091283

AIR OPERATIONS. World War 2
By Great Britain. Royal Air Force. Avro Lancaster aeroplanes

GARBETT, Mike
The Lancaster at war. - London : I. Allan.
2. - 1979
160p : ill
0-7110-0966-x X.802/11027
1546805

Lancaster at war. - London : Ian Allan.
3. - 1984
152p : ill
0-7110-1377-2 X.805/6244
0338923

AIR OPERATIONS. World War 2
By Great Britain. Royal Air Force. Avro Lancaster aeroplanes - Biographies

WAINWRIGHT, John, 1921-
Tail-end Charlie. - London : Macmillan, 1978
187p : ill
0-333-23047-7 X.809/43387
1546805

AIR OPERATIONS. World War 2
By Great Britain. Royal Air Force. Bomber aeroplanes. Groundcrew

RALEY, Norman
Destination unknown. - Bognor Regis : New Horizon, c1983
175p
0-86116-753-8 X.809/56621
2970848

AIR OPERATIONS. World War 2
By Great Britain. Royal Air Force. Bristol Beaufighter aeroplanes

BOWYER, Chaz
Beaufighter at war. - London : Allan, 1976
160p : ill. - Bibliography: p.160
0-7110-0704-7 X.802/10258
0779830

AIR OPERATIONS. World War 2
By Great Britain. Royal Air Force. Bristol Beaufort aeroplanes

NESBIT, Roy Conyers
Torpedo airmen. - London : Kimber, 1983
234p : ill. - Includes index
0-7183-0369-5 X.622/16754
0554421

ROBERTSON, Bruce
Beaufort special. - London : Allan, 1976
80p,[4]p of plates : ill(some col)
0-7110-0667-9 X.622/5482 Woolwich
0554421

AIR OPERATIONS. World War 2
By Great Britain. Royal Air Force. De Havilland Mosquito aeroplanes

BISHOP, Edward
The wooden wonder. - 2nd rev. ed. - Shrewsbury : Airlife, 1980
189p : ill. - Includes index
0-906393-04-3 X.629/16455 Woolwich
0039772

AIR OPERATIONS. World War 2
By Great Britain. Royal Air Force. Fighter aeroplanes. Pilots

BARTLEY, Anthony
Smoke trails in the sky. - London : Kimber, 1984
208p,[16]p of plates : ill. - Includes index
0-7183-0517-5 X.800/39740
2547910

BOWYER, Chaz
Fighter pilots of the RAF. 1939-1945. - London : Kimber, 1984
223p,[24]p of plates : ill. - Includes index
0-7183-0519-1 X.800/40087
3226859

AIR OPERATIONS. World War 2
By Great Britain. Royal Air Force. Fighter aeroplanes. Pilots - Biographies

PAGE, Geoffrey
Tale of a Guinea Pig. - London : Pelham, 1981
218p,[8]p of plates : ill
0-7207-1354-4 X.629/16230 Woolwich
2547910

Tale of a guinea pig. - London : Corgi, 1983, c1981
218p,[8]p of plates : ill
0-552-12277-7 H.84/414
2547910

AIR OPERATIONS. World War 2
By Great Britain. Royal Air Force. Fighter units. Johnson, James Edgar - Biographies

JOHNSON, J. E. (James Edgar)
Wing leader. - Feltham : Hamlyn, 1979
320p. - Includes index
0-600-39472-7 X.708/22539
2147203

AIR OPERATIONS. World War 2
By Great Britain. Royal Air Force. Fighter units. Lacey, Ginger - Biographies

BICKERS, Richard Townshend
Ginger Lacy [i.e. Lacey] - fighter pilot. - London : Corgi, 1978
[2],220p. - Includes index
0-552-10875-8 X.708/21550
1734881

AIR OPERATIONS. World War 2
By Great Britain. Royal Air Force. Handley Page Halifax aeroplanes

JONES, Geoffrey Patrick
Raider. - London : Kimber, 1978
240p : ill. - Includes index
0-7183-0066-1 X.620/18025 Woolwich
1694359

MERRICK, K. A.
Halifax. - London : I. Allan, 1980
224p : ill
0-7110-0767-5 X.620/19218 Woolwich
1694359

SCUTTS, Jerry
Halifax in action. - Carrollton : Squadron / signal, c1984
50p : ill(some col.). - (Aircraft ; no.66)
0-89747-158-x L.49/3279
1694359

AIR OPERATIONS. World War 2
By Great Britain. Royal Air Force. Handley Page Hampden aeroplanes

BOWYER, Chaz
Hampden special. - London : I. Allan, 1976
64p,[4]p of plates : ill(some col). - Bibliography: p.64
0-7110-0683-0 X.622/6201 Woolwich
1299859

AIR OPERATIONS. World War 2
By Great Britain. Royal Air Force. Hawker Hurricane aeroplanes

ALLWARD, Maurice
Hurricane special. - London : Allan, 1975
64p,[4]p of plates : ill(some col)
0-7110-0541-9 X.622/5057
0502111

STEWART, Adrian
Hurricane. - London : Kimber, 1982
336p,[16]p of plates : ill. - Bibliography: p315-320. - Includes index
0-7183-0009-2 X.800/33859
0502111

AIR OPERATIONS. World War 2
By Great Britain. Royal Air Force. Hawker Hurricane aeroplanes - Biographies

BARCLAY, George
Angels 22. - Revised ed. - London : Arrow Books, 1977
224p,[12]p of plates : ill. - Bibliography: p.212. - Includes index
0-09-916130-3 X.709/31698
0678619

Fighter pilot. - London : Kimber, 1976
224p : ill. - Includes index
0-7183-0294-x X.800/11555
0678619

AIR OPERATIONS. World War 2
By Great Britain. Royal Air Force. Hawker Typhoon aeroplanes

FRANKS, Norman L. R. (Norman Leslie Robert)
Typhoon attack. - London : Kimber, 1984
253p : ill. - Includes index
0-7183-0513-2 X.800/39741
3215555

AIR OPERATIONS. World War 2
By Great Britain. Royal Air Force. Hawker Typhoon aeroplanes - Biographies

KYLE, James, 1922-
A typhoon tale. - Bognor Regis : New Horizon, c1984
175p : ill
0-7125-0058-8 X.809/61763
3211290

AIR OPERATIONS. World War 2
By Great Britain. Royal Air Force. Military aeroplanes. Crews. Stress

AIRCREW stress in wartime operations / edited by E.J. Dearnaley, P.B. Warr. - London : Academic Press, 1979
vii,223p : ill. - Includes bibliographies
0-12-207750-4 X.325/626 Woolwich
1825615

AIR OPERATIONS. World War 2
By Great Britain. Royal Air Force. Polish units

KRÓL, Wacław
Polskie dywizjony lotnicze w Wielkiej Brytanii 1940-1945. - Wyd.2-gie, popr. i uzup.. - Warszawa : Wydawnictwo Ministerstwa Obrony Narodowej, 1982
374p,[24]p of plates : ill
83-11-06745-7 X.809/54013
8925992

AIR OPERATIONS. World War 2
By Great Britain. Royal Air Force. Short Stirling aeroplanes

BOWYER, Michael J. F. (Michael John Frederick)
The Stirling bomber. - London : Faber, 1980
3-225p : ill. - Includes index
0-571-11101-7 X.800/29295
1887351

GOMERSALL, Bryce
The Stirling file. - Tonbridge : Air-Britain, c1979
95p : ill
0-85130-072-3 X.622/13944
1887351

AIR OPERATIONS. World War 2
By Great Britain. Royal Air Force. Short Sunderland flying boats

BOWYER, Chaz
Sunderland at war. - London : Allan, 1976
160p : ill
0-7110-0665-2 X.622/5609 Woolwich
0721859

AIR OPERATIONS. World War 2
By Great Britain. Royal Air Force. Special duty squadrons

JACKSON, Robert, 1941-
The secret squadrons. - London : Robson Books, 1983
199p,[8]p of plates : ill. - Bibliography: p195-196. - Includes index
0-86051-207-x X.629/20540
2924889

AIR OPERATIONS. World War 2
By Great Britain. Royal Air Force. Supermarine Spitfire aeroplanes

BOWYER, Chaz
Supermarine Spitfire. - London : Arms and Armour, 1980
64p : ill(some col.). - (A bison book) ([War planes in colour] ; 1). - Bibliography: p64
0-85368-464-2 X.622/12618 Woolwich
0223662

SMITH, W. G. G. Duncan
Spitfire into battle. - Feltham : Hamlyn, 1982, c1981
235p,[8]p of plates : ill. - Includes index
0-600-20606-8 X.808/37136
255979x

AIR OPERATIONS. World War 2
By Great Britain. Royal Air Force. Supermarine Spitfire aeroplanes & Hawker Hurricane aeroplanes

BADER, Sir Douglas
Fight for the sky. - [London] : Fontana, 1975
168p : ill(some col). - Includes index
0-00-633807-0 X.611/7204 Woolwich
0011002

BADER, Sir Douglas
Fight for the sky. - [London] : Fontana, 1978
157p,[8]p of plates : ill. - Includes index
0-00-635237-5 X.708/21358
0011002

AIR OPERATIONS. World War 2
By Great Britain. Royal Air Force. Vickers Armstrong Wellington aeroplanes

BOWYER, Chaz
Wellington at war. - London : Ian Allan, 1982
128p : ill. - Bibliography: p128
0-7110-1220-2 X.622/14918
0244112

AIR OPERATIONS. World War 2
By Great Britain. Royal Air Force. Westland Lysander aeroplanes

ROBERTSON, Bruce
Lysander special. - London : I. Allan, 1977
64p : ill
0-7110-0764-0 X.622/6338 Woolwich
1458027

AIR OPERATIONS. World War 2
By Great Britain. Royal Air Force. 100 Group

STREETLY, Martin
Confound and destroy. - London : Macdonald and Jane's, 1978
279p : ill. - Bibliography: p.271-274. - Includes index
0-354-01180-4 X.802/10761
1780697

AIR OPERATIONS. World War 2
By Great Britain. Royal Air Force. Bomber Command

GIBSON, Guy
Enemy coast ahead. - London : Pan Books, 1977
xvi,317p,[4]p of plates : 2 ill
0-330-02452-3 X.709/31329
0683167

HALPENNY, Bruce Barrymore
To shatter the sky. - Cambridge : Stephens, 1984
233p : ill. - Includes index
0-85059-678-5 X.800/38099
3020207

HASTINGS, Max
Bomber Command. - London : Joseph, 1979
399p,[16]p of plates : ill. - Bibliography: p.375-377. - Includes index
0-7181-1603-8 X.800/28332
1963708

Bomber Command. - London : Pan, 1981, c1979
489p,[8]p of plates : ill. - Bibliography: p454-458. - Includes index
0-330-26236-x X.808/34134
1963708

JOHNSON, Brian, 1925-
Bombers. - London : Thames Methuen, 1984
301p,[24]p of plates : ill. - Includes index
0-423-00630-4 X.622/21685
1963708

MIDDLEBROOK, Martin
The Bomber Command war diaries. - Harmondsworth : Viking, 1985
804p : ill. - Bibliography: p78-790. - Includes index
0-670-80137-2 X.800/41147
1963708

SAWYER, Tom
Only owls and bloody fools fly at night. - London : Kimber, 1982
191p,[12]p of plates : ill. - Includes index
0-7183-0119-6 X.800/34740
0039934

AIR OPERATIONS. World War 2
By Great Britain. Royal Air Force. Bomber Command - Biographies

CHESHIRE, Leonard
Bomber pilot. - St Albans : Mayflower, 1975
159p,[16]p of plates : ill
0-583-12541-7 X.708/43834
0039934

GIBSON, Guy
Enemy coast ahead. - London : Joseph, 1976
302p
0-7181-1519-8 X.809/41070
0683167

RENAUT, Michael
Terror by night. - London : Kimber, 1982
192p,[12]p of plates : ill. - Includes index
0-7183-0139-0 X.950/18628
0039934

WINFIELD, Roland
The sky belongs to them. - London : Kimber, 1976
188p,[16]p of plates : ill
0-7183-0414-4 X.800/25704
0039934

AIR OPERATIONS. World War 2
By Great Britain. Royal Air Force. Bomber Command, 1940-1944

BARKER, Ralph
Strike hard, strike sure. - London : Pan Books, 1977
208p,[8]p of plates : ill. - Includes index
0-330-02513-9 X.708/20944
1142550

AIR OPERATIONS. World War 2
By Great Britain. Royal Air Force. Bomber Command, 1941-1945. Bomber aeroplanes

GOULDING, James
RAF Bomber Command and its aircraft, 1941-1945. - London : I. Allan, 1978
128p,[8]p of plates : ill(some col)
0-7110-0788-8 X.622/6583 Woolwich
1727672

AIR OPERATIONS. World War 2
By Great Britain. Royal Air Force. Bomber Command, 1942-1945

BARKER, Ralph
Strike hard, strike sure. - London : White Lion Publishers, 1975
x,210p,[16]p of plates : ill. - Includes index
0-7274-0089-4 X.808/23025
0524700

PHILPOTT, Bryan
RAF bomber units, July 1942-1945. - London : Osprey Publishing, 1978
48p : ill(some col). - (Airwar series ; 19)
0-85045-293-7 X.700/25849
0524700

AIR OPERATIONS. World War 2
By Great Britain. Royal Air Force. Bomber Command, 1944-1945 Avro Lancaster aeroplanes

TRIPP, Miles
The eighth passenger. - London : Macmillan, 1978
[11],188p. - Includes index
0-333-23420-0 X.809/43126
1450425
The eighth passenger. - Milton Keynes : Robin Clark, 1979, c1969
188p. - Includes index
0-86072-023-3 X.629/16529 Woolwich
1450425

AIR OPERATIONS. World War 2
By Great Britain. Royal Air Force. Bomber Command. Bomber aeroplanes, to 1940

GOULDING, James
RAF Bomber Command and its aircraft, 1936-1940. - London : Allan, 1975
144p,[8]p of plates : ill(some col)
0-7110-0627-x X.622/5121
0535141

AIR OPERATIONS. World War 2
By Great Britain. Royal Air Force. Bomber Command. Pilots - Biographies

BOWYER, Chaz
Bomber barons. - London : Kimber, 1983
222p,[32]p of plates : ill. - Bibliography: p211. - Includes index
0-7183-0339-3 X.622/16500
294488x

AIR OPERATIONS. World War 2
By Great Britain. Royal Air Force. Bomber Command. Group, No.2, 1938-1940 - Biographies

PASSMORE, Richard
Blenheim boy. - London : Thomas Harmsworth, 1981
254p,[16]p of plates : ill
0-9506012-2-5 Nov.43810 Woolwich
2527650

AIR OPERATIONS. World War 2
By Great Britain. Royal Air Force. Bomber Command. Group, No.5

BOWYER, Chaz
Bomber Group at war. - London : Ian Allan, 1981
160p : ill. - Bibliography: p160
0-7110-1087-0 X.625/685 Woolwich
2487551

AIR OPERATIONS. World War 2
By Great Britain. Royal Air Force. Coastal Command

BOWYER, Chaz
Coastal Command at war. - London : I. Allan, 1979
160p : chiefly ill. - Bibliography: p.159
0-7110-0980-5 X.802/11016
1992856

NESBIT, Roy Conyers
The strike wings. - London : Kimber, 1984
288p : ill. - Bibliography: p274-278. - Includes index
0-7183-0520-5 X.800/40086
1992856

AIR OPERATIONS. World War 2
By Great Britain. Royal Air Force. Coastal Command - Biographies

JAWORZYN, J. F. (Josef F)
No place to land. - London : Kimber, 1984
205p,[16]p of plates : ill. - Includes index
0-7183-0510-8 X.800/39604
1992856

AIR OPERATIONS. World War 2
By Great Britain. Royal Air Force. Desert Air Force, 1940-1943. Fighter aeroplanes. Pilots - Biographies

BOWYER, Chaz
Men of the Desert Air Force, 1940-1943. - London : Kimber, 1984
256p : ill. - Includes index
0-7183-0539-6 X.800/41318
3284867

AIR OPERATIONS. World War 2
By Great Britain. Royal Air Force. Desert Air Force, 1940-1945

BOWYER, Chaz
Desert Air Force. - London : Ian Allan, 1981
128p : ill. - Bibliography: p128
0-7110-1154-0 X.622/12095 Woolwich
2625407

AIR OPERATIONS. World War 2
By Great Britain. Royal Air Force. Fighter Interception Development Squadron

HOWARD-WILLIAMS, Jeremy
Night intruder. - Newton Abbot : David and Charles, 1976
184p : ill. - Bibliography: p.180. - Includes index
0-7153-7054-5 X.809/40752
0613746

AIR OPERATIONS. World War 2
By Great Britain. Royal Air Force. No.71, No.121 & No.133 Squadrons, 1940-1942

HAUGLAND, Vern
The Eagle Squadrons. - Newton Abbot : David and Charles, 1980
xvi,206p,[16]p of plates : ill. - Includes index
0-7153-7940-2 X.800/29185
2147262

AIR OPERATIONS. World War 2
By Great Britain. Royal Air Force. No.158 Squadron

CHORLEY, W. R.
In brave company. - [Woking] : The authors, 1978
[5],172p,[20]p of plates : ill
B78-15935 X.709/32171
1434446

AIR OPERATIONS. World War 2
By Great Britain. Royal Air Force. No.617 Squadron

KNIGHT, Dennis
The Dam Busters. - London : Royal Sovereign Group, 1976
16,[4]p,[2]leaves of plates : ill(chiefly col). - (Activity books / Dennis Knight)
0-900862-48-3 X.709/21908
0267597

AIR OPERATIONS. World War 2
By Great Britain. Royal Air Force. Squadron, 10

RAPIER, Brian J. (Brian James)
Melbourne ten. - York : Air Museum Publications (York), c1982
164p : ill
0-9507326-1-3 X.622/15975
2929813

AIR OPERATIONS. World War 2
By Great Britain. Royal Air Force Squadron, 85

FIGHTER squadron at war / [compiled by] A.J. Brookes. - London : Ian Allan, 1980
128p : ill
0-7110-1083-8 X.622/9518 Woolwich
2338025

AIR OPERATIONS. World War 2
By Great Britain. Royal Air Force. Squadron, No. 217, 1941 - Biographies

NESBIT, Roy Conyers
Woe to the unwary. - London : Kimber, 1981
192p : ill. - Includes index
0-7183-0348-2 X.622/11789 Woolwich
260602x

AIR OPERATIONS. World War 2
By Great Britain. Royal Air Force. Squadron, No.66, 1940-1942 - Biographies

ALLEN, H. R. (Hubert Raymonds)
Fighter Squadron. - London : Granada, 1982, c1979
206p,[8]p of plates : ill. - (A Mayflower book)
0-583-13416-5 H.84/427
2053527

ALLEN, Herbert Raymond
Fighter Squadron. - London : Kimber, 1979
192p : ill
0-7183-0087-4 X.800/28738
2053527

AIR OPERATIONS. World War 2
By Great Britain. Royal Air Force. Squadron, No.76, 1941-1945

CHORLEY, W. R.
To see the dawn breaking. - [Ottery St. Mary] : W.R. Chorley, 1981
272p : ill. - Bibliography: p188. - Includes index
0-9507467-0-3 X.629/17867 Woolwich
2748940

AIR OPERATIONS. World War 2
By Great Britain. Royal Air Force. Squadron, No.101
BROOKES, Andrew
Bomber squadron at war. - London : Ian Allan, 1983
144p : ill. - Bibliography: p4
0-7110-1279-2 X.622/16734
2956365

AIR OPERATIONS. World War 2
By Great Britain. Royal Air Force. Squadron, No.161, 1940-1944 - Biographies
VERITY, Hugh
We landed by moonlight. - London : I. Allan, 1978
256p,[16]p of plates : ill. - Includes index
0-7110-0886-8 X.809/44112
182001x

AIR OPERATIONS. World War 2
By Great Britain. Royal Air Force. Squadron, No.161, 1942-1945
MCCALL, Gibb
Flight most secret. - London : Kimber, 1981
270p,[12]p of plates : ill. - Bibliography: p261-262. - Includes index
0-7183-0038-6 X.800/32473
2641178

AIR OPERATIONS. World War 2
By Great Britain. Royal Air Force. Squadron, No.242
HALLIDAY, Hugh
No.242 Squadron, the Canadian years. - Leicester : Midland Counties, 1982, c1981
177p,[44]p of plates : ill. - Bibliography: p169-171. - Includes index
0-904597-39-3 X.622/14329
2832313

AIR OPERATIONS. World War 2
By Great Britain. Royal Air Force. Squadron, No.617
BRICKHILL, Paul
The dam busters. - Revised ed. - London : Evans Bros, 1977
x,238p,leaf of plate,[12]p of plates : ill. - Includes index
0-237-44886-6 X.809/42507
1261339
The dam busters. - [Revised ed.]. - London : Pan Books, 1978
268p,[8]p of plates : ill
0-330-23618-0 X.708/21178
1261339
The dam busters. - Rev. ed.. - London : Pan, 1983
272p,[8]p of plates : ill. - (Grand strategy). - Includes index
0-330-28083-x X.808/38586
1261339
COOPER, Alan W.
The men who breached the dams. - London : Kimber, 1982
223p : ill. - Includes index
0-7183-0308-3 X.800/33902
1261339

AIR OPERATIONS. World War 2
By Great Britain. Royal Air Force. Squadron, No.617, 1943-1945
COOPER, Alan W.
Beyond the dams to the Tirpitz. - London : Kimber, 1983
253p : ill. - Includes index
0-7183-0359-8 X.800/38294
3127001

AIR OPERATIONS. World War 2
By Italy. Aeronautica militare
BORGIOTTI, Alberto
97° Gruppo antonomo bombardamento a tuffo, 1940-1941. - Roma : Ateneo & Bizzarri, 1980
72p : ill
r6507127 X.622/15404
9057765
LICHERI, Sebastiano
Il Raggruppamento Caccia nella Guerra di Liberazione, 8 settembre 1943-8 maggio 1945. - Roma : Ateneo & Bizzarri, 1978
218p : ill
r6501946 X.805/1788
8763518

AIR OPERATIONS. World War 2
By Italy. Aeronautica militare, to 1943
LICHERI, Sebastiano
L'arma aerea italiana nella seconda guerra mondiale, 10 giugno 1940-8 settembre 1943. - Milano : Mursia, 1976
329p.,24p of plates : ill. - (Biblioteca del cielo ; n. 18). - Bibliography : p.320-322
r6035019 X.809/24402
1225103

AIR OPERATIONS. World War 2
By Japan. Navy. Air Service - Biographies
SAKAI, Saburo
Samurai!. - Toronto ; London : Bantam, 1978
xvi,316p,[2]p of plates(2 fold) : ill(incl 1 col)
0-552-61035-6 X.708/21596
0228516

AIR OPERATIONS. World War 2
By Japan. Navy. Air Services. Mitsubishi Zero aeroplanes
MIKESH, Robert C.
Zero fighter. - London : Jane's, 1981
56p : ill(some col.)
0-7106-0037-2 L.45/1100
2062100
NOHARA, Shigeru
A6M Zero in action. - Carrollton : Squadron / Signal, c1983
49p : ill(some col.). - (Aircraft ; no.59)
0-89747-141-5 L.49/3322
2062100
WILLMOTT, H. P. (Hedley Paul)
Zero A6M. - London : Arms and Armour, 1980
64p : ill(some col.). - ([War planes in colour] ; 6) (A Bison book)
0-85368-085-x X.622/9392 Woolwich
2062100

AIR OPERATIONS. World War 2
By Japanese military forces. Aeroplanes
COLLIER, Basil
Japanese aircraft of World War II. - London : Sidgwick and Jackson, 1979
144p : ill(some col). - Includes index
0-283-98399-x X.800/28763
2112035

AIR OPERATIONS. World War 2
By Lockheed Hudson aeroplanes
HENDRIE, Andrew
Seek and strike. - London : Kimber, 1983
320p,[16]p of plates : ill. - Includes index
0-7183-0489-6 X.800/38293
2243997
JONES, Geoffrey Patrick
Attacker. - London : Kimber, 1980
238p : ill. - Bibliography: p.227. - Includes index
0-7183-0237-0 X.622/8233 Woolwich
2243997

AIR OPERATIONS. World War 2
By military aeroplanes. Pilots - Biographies
JACKSON, Robert, 1941-
Air heroes of World War II. - London : A. Barker, 1978
175p,[4]p of plates : ill
0-213-16669-0 X.809/43061
1433644

AIR OPERATIONS. World War 2
By New Zealand. Royal New Zealand Air Force. Supermarine Spitfire aeroplanes - Biographies
HOULTON, Johnnie
Spitfire strikes. - London : Murray, 1985
xii,257p,[16]p of plates : ill. - Bibliography: p250-251. - Includes index
0-7195-4178-6 X.809/64345
3155692

AIR OPERATIONS. World War 2
By North American Mustang aeroplanes
NORTH-AMERICAN P-51 Mustang 1940-45 / G. Borelli ... [et al.]. - Modena : STEM-Mucchi, 1973
107p.,[6]leaves of plates,[4]p. of plates : ill. - (Le macchine e la storia ; 1)
r6188434 X.709/50011
0199990

AIR OPERATIONS. World War 2
By Poland. Lotnicze Sily Zbrojne
KARPIŃSKI, Stanislaw
Na skrzydlach huraganu. - London : Veritas Foundation Publication Centre.
Tom 1 i 2. - 1976
653p,[4]leaves of plates : 4 ill. - (Biblioteka polska. seria czerwona ; tom 81)
B78-03849 X.989/52060
1295071
Tom 3 i 4. - 1977
478,14p,[4]leaves of plates : 4 ill. - (Biblioteka polska. seria czerwona ; tom 82). - Includes index
B78-03850 X.808/33171
1295071
SZEWCZYK, Witold
Samoloty na których walczyli Polacy. - Warszawa : Wydawnictwa Komunikacji i Łączności, 1978
177p : ill. - (Biblioteczka Skrzydlatej Polski)
r6097216 X.619/10900 Woolwich
1295071

SZUBAŃSKI, Rajmund
Wobronie polskiego nieba. - Warszawa : Książka i Wiedza, 1978
185p,[16]p of plates : ill. - (Biblioteka pamięci pokoleń)
r6095468 X.708/45672
1295071

AIR OPERATIONS. World War 2
By Polish pilots
GNYŚ, Wladek
First kill. - London : Kimber, 1981
207p : ill. - Includes index
0-7183-0397-0 X.800/30982
1393278
KRÓL, Wacław
Walczyłem pod niebem Londynu. - Warszawa : Ludowa Spółdzielnia Wydawnicza, 1982
266p,plates
83-205-3349-x X.808/39132
8482284
WASILEWSKI, Antoni
W szkocką kratę. - Kraków : Wydawnictwo Literackie, 1978
339p,[1]leaf of plates : ill
r6185957 X.708/48713
1393278
WESELI, Jan
Wielkie safari. - Warszawa : Czytelnik, 1976
384p
r6045070 X.708/42282
1393278

AIR OPERATIONS. World War 2
By Polskie Sily Powietrzne
KRÓL, Wacław
Polskie dywizjony lotnicze w Wielkiej Brytanii 1940-1945. - Warszawa : Wydawnictwo Ministerstwa Obrony Narodowej, 1976
428p,32p of plates : ill
r6008548 X.808/10528
0398837
PAWLAK, Jerzy
Brygada Pościgowa - alarm. - Warszawa : Wydawnictwo Ministerstwa Obrony Narodowej, 1977
260p,16p of plates : ill
r6044333 X.708/42842
0398837

AIR OPERATIONS. World War 2
By Polskie Sily Powietrzne
SZEWCZYK, Witold
Samoloty na których walczyli Polacy. - Warszawa : Wydawnictwa Komunikacji i Łączności, 1978
177p : ill. - (Biblioteczka Skrzydlatej Polski)
r6097216 X.619/10900 Woolwich
0398837

AIR OPERATIONS. World War 2
By Polskie Sily Powietrzne - Texts - Biographies
JANCZAK, Jan H.
Podcięte skrzydła. - Warszawa : Wydawnictwo Ministerstwa Obrony Narodowej, 1976
442p,12p of plates
r6057207 X.708/42843
0398837

AIR OPERATIONS. World War 2
By Soviet military forces. Kuibyshev airmen
OGNENNYE tarany / [sostavitel' B.M. Kosenkov]. - Kuĭbyshev : Kuĭbyshevskoe knizhnoe izdatel'stvo, 1978
100p
r6096777 X.708/46289
2094878

AIR OPERATIONS. World War 2
By theories of Great Britain. Royal Air Force - Biographies
FRANKS, Norman L. R. (Norman Leslie Robert)
Fighter leader. - London : Kimber, 1978
207p,[12]p of plates : ill. - Bibliography: p.199. - Includes index
0-7183-0106-4 X.800/27303
1721267
HAMILTON, Alexander, b.1923
Wings of night. - London : Kimber, 1977
206p,[16]p of plates : ill. - Includes index
0-7183-0415-2 X.800/26541
1284940
REVELL, Alex
The vivid air. - London : Kimber, 1978
255p,[16]p of plates : ill. - Includes index
0-7183-0385-7 X.800/27053
1544292

AIR OPERATIONS. World War 2
By theories of Great Britain. Royal Air Force - Biographies-continued

SANDERS, James
Venturer courageous. - Auckland : Hutchinson of New Zealand ; London : Hutchinson, 1983
266p : ill. - Includes index
0-09-154600-1 X.950/34598
3168883

AIR OPERATIONS. World War 2
By theories of Great Britain. Royal Air Force, 1940-1944. Logs - Texts

GIBSON, Guy
Pilot's flying log book No.2. - London : After the Battle Magazine, [1976]
[206]p : 1 ill
B79-06150 MS.Facs.869
1779605

AIR OPERATIONS. World War 2
By theories of Great Britain. Royal Air Force. Fighter Command - Biographies

FORRESTER, Larry
Fly for your life. - [Abridged ed.]. - London : Mayflower, 1979
222p
0-583-12787-8 X.708/21769
1821024

FRANKS, Norman L. R. (Norman Leslie Robert)
Double mission. - London : Kimber, 1976
192p,[12]p of plates : ill. - Includes index
0-7183-0254-0 X.800/11503
0659118

AIR OPERATIONS. World War 2
By theories of Great Britain. Royal Air Force, Fighter Command - Biographies

STOKES, Doug
Paddy Finucane. - London : Kimber, 1983
219p,[16]p of plates : ill. - Includes index
0-7183-0279-6 X.622/16452
2942224

AIR OPERATIONS. World War 2
By theories of Union of Soviet Socialist Republics. Voenno-vozdushnye sily

LITVINOVA, Larisa Nikolaevna
Ulit͡sa Tat'i͡any Makarovoĭ. - Moskva : Moskovskiĭ rabochiĭ, 1976
149p : ill
r6022175 X.708/19694
0397598

AIR OPERATIONS. World War 2
By Union of Soviet Socialist Republics

ANISHCHENKOV, P. S. (Panteleĭmon Stepanovich)
Tret'i͡a vozdushnai͡a. - Moskva : Voennoe izdatel'stvo, 1984
190 p, [24] p. of plates : ill. ports.. - Includes bibliographical references
r6465835 X.809/62205
8763518

AIR OPERATIONS. World War 2
By Union of Soviet Socialist Republics. Voenno-morskoĭ Flot

AVIAT͡SII͡A VMF v Velikoĭ Otechestvennoĭ voĭne / [avtorskiĭ kollektiv : N.M. Lavrent'ev (rukovoditel') ... i dr]. - Moskva : Voennoe izdatel'stvo, 1983
183p
r6448072 X.809/59482
8763518

AIR OPERATIONS. World War 2
By Union of Soviet Socialist Republics. Voenno-vozdushnye sily

17-I͡A vozdushnai͡a armii͡a v boi͡akh ot Stalingrada do Veny / [N.M. Skomorokhov (rukovoditel' avtorskogo kollektiva)]. - Moskva : Voennoe izdatel'stvo Ministerstva oborony SSSR, 1977
261p,[32]p of plates : ill
r6046240 X.809/24446
089771X

BOGDANOV, Nikolaĭ
V nebe - gvardeĭskiĭ Gatchinskiĭ. - Leningrad : Lenizdat, 1980
298p,[16]p of plates : ill
r6483191 X.629/17048 Woolwich
8763518

CHECHNEVA, Marina Pavlovna
Nebo ostaetsi͡a nashim. - Izd. 2, ispr. - Moskva : Voennoe izdatel'stvo Ministerstva oborony SSSR, 1976
236p.,1 leaf of plates,34p. of plates : ill. - (Voennye memuary)
r6020444 X.809/23959
089771X

EFIMOV, Aleksandr Nikolaevich
Nad polem boi͡a. - Moskva : Voennoe izdatel'stvo Ministerstva oborny SSSR, 1976
315p, 1 leaf of plates, 16p of plates : ill. - (Voennye memuary)
r6021887 X.809/23809
089771X

EMEL'I͡ANENKO, Vasiliĭ Borisovich
V voennom vozdukhe surovom-. - Izd.2., dop. i perer. - Moskva : Molodai͡a gvardii͡a, 1976
476p,[1] leaf of plates,[32]p of plates : ill
r6044008 X.809/24481
089771X

FEDOROV, Alekseĭ Grigor'evich
Idu v pike. - 2. izd., dop. i perer.. - Moskva : "Molodai͡a gvardii͡a", 1976
174p, 16p of plates : ill
r6021273 X.708/19572
089771X

FEDOROV, Alekseĭ Grigor'evich
Letchiki na zashchite Moskvy. - Moskva : Nauka, 1979
190p. - (Stranit͡sy istorii nasheĭ Rodiny)
r6098148 X.708/46024
089771X

GOLUBEV, Georgiĭ Gordeevich
V pare s "Sotym". - Izd. 2., dop. - Moskva : Isdatel'stvo DOSAAF SSSR, 1978
251p.,[16]p of plates : ill
r6095598 X.629/8101 Woolwich
089771X

KABEROV, Igor' Aleksandrovich
V prit͡sele svastika. - Leningrad : Lenizdat, 1975
309p,[1]leaf of plates,[32]p of plates : ill
r6185821 X.808/12249
089771X

KOĪANDER, Evgeniĭ Valer'i͡anovich
I͡A - "Rubin", prikazyvai͡u -. - Moskva : Voennoe izdatel'stvo Ministerstva oborony SSSR, 1978
275p,[1]leaf of plates,[10]p of plates : ill. - (Voennye memuary)
r6172975 X.809/27309
089771X

KRYLOV, Alekseĭ Ivanovich
Po prikazu Stavki. - Moskva : Voennoe izdatel'stvo Ministerstva oborony SSSR, 1977
300p,[1] leaf of plates,[16]p of plates : ill. - (Voennye memuary)
r6044524 X.809/24690
089771X

KUBAREV, Vasiliĭ Nikolaevich
Atakui͡u gvardeit͡sy. - Tallin : Ėesti raamat, 1975
229p,[1] leaf of plates,[16]p of plates : ill
r6093343 X.809/26456
089771X

LAVRINENKOV, Vladimir Dmitrievich
Vozvrashchenie v nebo. - Moskva : Voennoe izdatel'stvo Ministerstva oborony SSSR, 1974
238p,plates. - (Voennye memuary)
r6465432 X.629/26188
8763518

LOKSHIN, Vladimir Semenovich
Shest' zolotykh zvezd. - Moskva : Izdatel'stvo DOSAAF SSSR, 1976
127p : ill
r6037386 X.708/19596
089771X

SEMENOV, Aleksandr Fedorovich
Ėskadril'i͡a "Mongol'skiĭ arat". - Izd. 2-oe, dop. - Moskva : Voennoe izdatel'stvo Ministerstva oborony SSSR, 1978
159p,[16]p of plates : ill
r6098637 X.809/27084
089771X

SHVET͡S, Stepan Ivanovich
Pod kryl'i͡ami - noch'. - Izd.2., dop. - Dnepropetrovsk : Izdatel'stvo "Promin'", 1976
293p : ill
r6037459 X.989/38216
089771X

TIMOKHOVICH, Ivan Vasil'evich
Operativnoe iskusstvo sovetskikh VVS v Velikoĭ Otechestvennoĭ voine. - Moskva : Voennoe izdatel'stvo Ministerstva oborony SSSR, 1976
342 p.
r6020809 X.809/23667
089771X

TREKHONIN, Nikolaĭ Filippovich
Tysi͡acha dneĭ nad polem boi͡a. - Volgograd : Nizhne-Volzhskoe knizhnoe izdatel'stvo, 1976
157 p. : ill
r6022090 X.908/36731
089771X

AIR OPERATIONS. World War 2
By Union of Soviet Socialist Republics. Voenno-vozdushnye sily

V nebe - letchiki Baltiki / [sostavitel' A.G. Solov'ev]. - Tallin : Ėesti raamat, 1974
293p,[24]p of plates : ill
r6177146 X.809/28244
089771X

VERSHININ, Konstantin Andreevich
Chetvertai͡a vozdushnai͡a. - Moskva : Voennoe izdatel'stvo Ministerstva oborony SSSR, 1975
346p,[1] leaf of plates,[18]p of plates : ill. - (Voennye memuary)
r6036404 X.808/10540
089771X

AIR OPERATIONS. World War 2
By Union of Soviet Socialist Republics. Voenno-vozdushnye sily - Biographies

RUDENKO, Sergeĭ Ignat'evich
Kryl'i͡a Pobedy. - Moskva : Voennoe izdatel'stvo Ministerstva oborony SSSR, 1976
411p.,16p. of plates, 1 leaf of plates : ill. - (Voennye memuary)
r6021016 X.809/23936
089771X

SHEPELEV, Alekseĭ Lavrent'evich
V nebe i na zemle. - Moskva : Voennoe izdatel'stvo Ministerstva oborony SSSR, 1974
238 p., 1 leaf of plates, 6 p. of plates : ill. - (Voennye memuary)
r6015171 X.809/23366
089771X

AIR OPERATIONS. World War 2
By Union of Soviet Socialist Republics. Voenno-vozdushnye sily - Texts

NALIVAĬKO, Boris Mikhaĭlovich
Borozdy v nebe. - Minsk : Izdatel'stvo "Belarus'", 1976
141p : ill
r6023536 X.708/19571
089771X

AIR OPERATIONS. World War 2
By Union of Soviet Socialist Republics. Voenno-vozdushnye sily. Istrebitel' nyĭ aviat͡sionnyĭ polk, 586-ĭ, Union of Soviet Socialist Republics. Voenno-vozdushnye sily. Bombardirovochnyĭ aviat͡sionnyĭ polk, 587-ĭ & Union of Soviet Socialist Republics. Voenno-vozdushnye sily. Nochnoĭ bombardirovochnyĭ aviat͡sionnyĭ polk, 588-ĭ, 1941-1945

MYLES, Bruce
Night witches. - Edinburgh : Mainstream, c1981
viii,278p,[18]p of plates : ill
0-906391-22-9 X.629/21032
2947161

AIR OPERATIONS. World War 2
By Union of Soviet Socialist Republics. Voenno-vozdushnye sily. Istrebitel'nyĭ aviat͡sionnyĭ polk, 568-ĭ, Union of Soviet Socialist Republics. Voenno-vozdushnye sily. Bombardirovochnyĭ aviat͡sionnyĭ polk, 587-ĭ & Union of Soviet Socialist Republics. Voenno-vozdushnye sily. Nochnoĭ bombardirovochnyĭ aviat͡sionnyĭ polk, 588-ĭ, 1941-1945

MYLES, Bruce
Night witches. - London : Panther, 1982, c1981
271p,[8]p of plates : ill
0-586-05812-5 X.958/21777
2947161

AIR OPERATIONS. World War 2
By United States. Air Force

SIEFRING, Thomas A.
US air force in World War II. - London : Hamlyn, 1977
192p : ill(some col). - Bibliography: p.192. - Includes index
0-600-30349-7 X.622/6222 Woolwich
1342118

US air force in World War II. - Secaucus, N.J. : Chartwell Books, 1977
192p : ill(some col)facsims,maps,ports. - Bibliography : p.192.- Includes index
0-89009-182-x X.802/10961
1342118

AIR OPERATIONS. World War 2
By United States. Air Force (Fighter Group), 4th

FRY, Garry L.
Escort to Berlin. - New York : Arco, c1980
xiv,226p,plates
0-668-04768-2 X.622/21179
8763518

AIR OPERATIONS. World War 2
By United States. Army Air Force

HUSTON, James A. (James Alvin)
Out of the blue. - Reprinted. - Nashville : Battery Press, 1981
xi,327p : ill
0-89839-055-9 X.800/32796
8763518
- [Another copy.] X.800/36130

AIR OPERATIONS. World War 2
By United States. Army Air Force, 1942-1945. Republic Aircraft Thunderbolt aeroplanes

FREEMAN, Roger Anthony
Thunderbolt. - London : Macdonald and Jane's, 1978
152p,[8]p of plates : ill(some col). - Includes index
0-354-01166-9 X.802/10740
0762490

HESS, William N.
P-47 Thunderbolt at war. - London : Allan, 1976
160p,4p of plates : ill(some col)
0-7110-0705-5 X.622/5698 Woolwich
0762490

AIR OPERATIONS. World War 2
By United States. Army Air Force, 1942-1945. Republic Aircraft Thunderbolt aeroplanes - Biographies

BLEDSOE, Marvin
Thunderbolt. - New York ; London : Van Nostrand Reinhold, c1982
282p,[16]p of plates : ill. - Includes index
0-442-21355-7 X.622/16066
0762490

AIR OPERATIONS. World War 2
By United States. Army Air Force, 1942-1945. Special duty squadrons

JACKSON, Robert, 1941-
The secret squadrons. - London : Robson Books, 1983
199p,[8]p of plates : ill. - Bibliography: p195-196. - Includes index
0-86051-207-x X.629/20540
2924897

AIR OPERATIONS. World War 2
By United States. Army Air Force. Boeing B-17 aeroplanes

FREEMAN, Roger Anthony
B-17 Fortress at war. - London : Allan, 1977
192p : ill(some col)
0-7110-0686-5 X.622/5833 Woolwich
020000x

AIR OPERATIONS. World War 2
By United States. Army Air Force. Lockheed Lightning aeroplanes

CHRISTY, Joe
P-38 Lightning at war. - London : I. Allan, 1978
144p,[4]p of plates : ill(some col)
0-7110-0772-1 X.622/6523 Woolwich
1685155

ETHELL, Jeffrey L.
P.38 Lightning. - London : Jane's, 1983
52p : ill(some col.)
0-7106-0289-8 L.45/2905
1685155

STAFFORD, Gene B.
P.-38 lightning in action. - Carrollton : Squadron / Signal, c1976
48p : ill(some col.). - (Aircraft ; no.25)
r6805994 L.49/1324
1685155

AIR OPERATIONS. World War 2
By United States. Army Air Force. Martin B-26 aeroplanes

BIRDSALL, Steve
B-26 Marauder in action. - Carrollton, Tex. : Squadron / Signal, c1981
49p : ill(some col.). - (Aircraft ; no.50)
0-89747-119-9 L.49/1822
1801104

FREEMAN, Roger Anthony
B-26 Marauder at war. - London : I. Allan, 1978
192p : ill. - Includes index
0-7110-0823-x X.622/6736 Woolwich
1801104

AIR OPERATIONS. World War 2
By United States. Army Air Force. Mustang P-51 aeroplanes

GRANT, William Newby
P-51 Mustang. - London : Arms and Armour, 1980
64p : ill(some col.). - ([War planes in colour] ; 3) (A Bison book). - Bibliography: p64
0-85368-454-5 X.622/9865 Woolwich
2061325

AIR OPERATIONS. World War 2
By United States. Army Air Force. Air Division, Second, 1942-1945

BOWMAN, Martin W.
Fields of little America. - Norwich : Wensum Books (Norwich) Ltd, 1977
[8],120p : ill. - Includes index
0-903619-20-2 X.802/10584
1452509

Fields of Little America. - Cambridge : Stephen, 1983
115p : ill. - Includes index
0-85059-644-0 X.805/4465
1452509

AIR OPERATIONS. World War 2
By United States. Army Air Force. Air Force, 8th

BOWMAN, Martin W.
Castles in the air. - Cambridge : Stephens, 1984
209p : ill. - Includes index
0-85059-675-0 X.800/39133
2406322

FREEMAN, Roger A. (Roger Anthony)
Mighty Eighth war diary. - London : Jane's, 1981
508p : ill. - Includes index
0-7106-0038-0 X.805/1993
1481231

Mighty Eighth war manual. - London : Jane's, 1984
320p : ill. - Includes index
0-7106-0325-8 X.622/23211
1481231

MCCRARY, John R. Tex (John Reagan Tex)
First of the many. - London : Robson, 1981, c1944
xvi,241p,[129]p of plates : ill
0-86051-129-4 X.622/10203 Woolwich
2406322

RUST, Kenn C.
Eighth Air Force story. - Temple City : Historical Aviation Album ; London : [Distributed by] Hersant, 1978
72p : ill
0-911852-81-6 X.702/5689
1481231

AIR OPERATIONS. World War 2
By United States. Army Air Force. Air Force, 15th

RUST, Kenn C.
Fifteenth Air Force story. - Temple City : Historical Aviation Album ; London : [Distributed by] Hersant, 1976
64p : ill
0-911852-79-4 X.611/4733 Woolwich
1052268

AIR OPERATIONS. World War 2
By United States. Army Air Force. Air Force, Twentieth. Boeing B-29 aeroplanes

BIRDSALL, Steve
Saga of the Superfortress. - London : Sidgwick & Jackson, 1981, c1980
346p,[4]p of plates : ill(some col.). - Bibliography: p340. - Includes index
0-283-98786-3 X.622/10943 Woolwich
2542900

AIR OPERATIONS. World War 2
By United States. Army Air Force. Airborne Division, 17th

PAY, D. R. (Don Raymond)
Thunder from heaven. - Nashville : Battery Press, [1980]
xii,179p : ill
0-89839-037-0 X.805/1212
2370158

AIR OPERATIONS. World War 2
By United States. Army Air Force. Bombardment Group, 306th, 1942-1946

BOVE, Arthur P. (Arthur Philip)
First over Germany. - [S.l.] : [s.n.], [1981?]
[148]p : chiefly ill
B82-00102 X.805/1994
2578670

AIR OPERATIONS. World War 2
By United States. Army Air Force. Bombardment Group, 460th, 1943-1944

NEWBY, Leroy W.
Target Ploesti. - London : Arms and Armour Press, 1983
xv,253p : ill. - Bibliography: p214. - Includes index
0-85368-610-6 X.800/37305
3062236

AIR OPERATIONS. World War 2
By United States. Army Air Force. Bomber Command

FREEMAN, Roger Anthony
The US strategic bomber. - London : Macdonald and Jane's, 1975
160p : ill. - (Macdonald illustrated war studies). - Bibliography: p.7
0-356-08096-x X.629/10302 Woolwich
0568007

MACISAAC, David
Strategic bombing in World War Two. - New York ; London : Garland, 1976
[4],xi,231p. - Bibliography: p.217-226. - Includes index
0-8240-2025-1 X.800/27401
0568007

AIR OPERATIONS. World War 2
By United States. Army Air Force. Bomber Command. Documents

A GUIDE to the reports of the United States strategic bombing survey / edited by Gordon Daniels. - London : Royal Historical Society, 1981
xxvi,115p. - (Royal Historical Society guides and handbooks ; 12. Supplementary series ; 2)
0-901050-71-7 X.622/22728
2176068

AIR OPERATIONS. World War 2
Naval air operations

MILLER, Nathan
The naval air war, 1939-1945. - London : Conway Maritime Press, 1980
xi,224p : ill. - Bibliography: p.213-214. - Includes index
0-85177-201-3 X.800/29201
2281929

AIR OPERATIONS. World War 2
Naval air operations by allied air forces

OLDS, Robert
Helldiver squadron. - Reprint. - Washington, D.C. : Zenger, 1980
xi,263p,36p of plates : ill
0-89201-054-1 X.809/50309
0764914

WINTON, John
Air power at sea, 1939-45. - London : Sidgwick and Jackson, 1976
185,[3]p : ill. - Bibliography: p.184. - Includes index
0-283-98313-2 X.620/16414 Woolwich
0764914

AIR OPERATIONS. World War 2
Naval air operations by allied forces, 1943-1944. Grumman F-6F aeroplanes

TILLMAN, Barrett
Hellcat. - Cambridge : Stephens, 1979
xiii,265p : ill. - Bibliography: p.257-258. - Includes index
0-85059-383-2 X.800/28373
181026x

AIR OPERATIONS. World War 2
Naval air operations by allied forces. Grumman Wildcat aeroplanes

TILLMAN, Barrett
The Wildcat in WWII. - Annapolis, Md. : Nautical & Aviation Publishing Company of America, c1983
xiv,270p : ill. - Bibliography: p263-264. - Includes index
0-933852-32-0 X.800/36830
3029662

AIR OPERATIONS. World War 2
Naval air operations by Germany. Kriegsmarine. Aeroplanes

GERMAN maritime aircraft / [compiled by] Bryan Philpott. - Cambridge : Stephens, 1981
94p : chiefly ill. - (World War 2 photo album ; no.18)
0-85059-445-6 X.622/10800
2159651

AIR OPERATIONS. World War 2
Naval air operations by Great Britain. Royal Air Force. Fighter aeroplanes catapulted from ships

BARKER, Ralph
The hurricats. - London : Pelham, 1978
xi,207p,[12]p of plates : ill. - Includes index
0-7207-0994-6 X.629/11858 Woolwich
1457675

The hurricats. - London : Sphere, 1979
222p,[8]p of plates : ill. - Includes index
0-7221-1424-9 X.708/22231
1457675

AIR OPERATIONS. World War 2
Naval air operations by Great Britain. Royal Navy. Aircraft carriers: Ark Royal (Ship), 1939-1941

POOLMAN, Kenneth
'Ark Royal'. - London : New English Library, 1978
173p
0-450-03953-6 X.708/21228
0122327

AIR OPERATIONS. World War 2
Naval air operations by Great Britain. Royal Navy. Aircraft carriers: Illustrious (Ship) - Biographies
HANSON, Norman
Carrier pilot. - Cambridge : Stephens, 1979
255p : ill
0-85059-349-2 X.800/29289
1734326

AIR OPERATIONS. World War 2
Naval air operations by Great Britain. Royal Navy. Aircraft carriers: Vindex (Ship)
POOLMAN, Kenneth
Escort carrier. - London : Leo Cooper in association with Secker & Warburg, c1983
xx,216p : ill. - Includes index
0-436-37705-5 X.805/4379
2958600

AIR OPERATIONS. World War 2
Naval air operations by Great Britain. Royal Navy. Fleet Air Arm
HOARE, John
Tumult in the clouds. - London : Joseph, 1976
208p. - Includes index
0-7181-1410-8 X.809/40579
0222232
STURTIVANT, Ray
Fleet air arm at war. - London : Ian Allan, 1982
144p : ill
0-7110-1084-6 X.525/2749
2340690
WINTON, John
Find, fix and strike!. - London : Batsford, 1980
152p,[32]p of plates : ill. - Includes index
0-7134-3488-0 X.800/29463
2340690

AIR OPERATIONS. World War 2
Naval air operations by Great Britain. Royal Navy. Fleet Air Arm. Fairey Swordfish aeroplanes
HARRISON, William Alfred
Swordfish special. - London : I. Allan, 1977
80p : ill
0-7110-0742-x X.622/5950 Woolwich
1166190
KILBRACKEN, John Godley, Baron
Bring back my Stringbag. - London : P. Davies, 1979
x,227p,viiip of plates : ill. - Includes index
0-432-08160-7 X.809/45077
1188410
LAMB, Charles, 1914-
War in a stringbag. - London : Cassell and Collier Macmillan, 1977
[12],340p,[16]p of plates : ill. - Includes index
0-304-29778-x X.809/42248
1188410
War in a stringbag. - London : Arrow Books, 1978
344p,[16]p of plates : ill
0-09-918250-5 X.708/21613
1188410

AIR OPERATIONS. World War 2
Naval air operations by Great Britain. Royal Navy. Fleet Air Arm. Fairey Swordfish aeroplanes - Biographies
KILBRACKEN, John Godley, Baron
Bring back my Stringbag. - London : Pan in association with Peter Davis, 1980, c1979
221p,[4]p of plates : ill
0-330-26172-x X.808/34147
1188410

AIR OPERATIONS. World War 2
Naval air operations by United States. Navy. Aircraft carriers. Aeroplanes
KILDUFF, Peter
US carriers at war. - London : Ian Allan, 1981
128p : ill. - Bibliography: p128
0-7110-1077-3 X.805/1719
2538512

AIR OPERATIONS. World War 2
Naval air operations by United States. Navy. Fighter aeroplanes
O'LEARY, Michael
United States naval fighters of World War II in action. - Poole : Blandford Press, 1980
160p : ill. - Bibliography: p.158. - Includes index
0-7137-0956-1 X.622/9039 Woolwich
2002302

AIR OPERATIONS. World War 2
Naval air operations by United States. Navy. Grumman Avenger aeroplanes
TILLMAN, Barrett
Avenger at war. - Shepperton : I. Allan, 1979
128p : ill
0-7110-0957-0 X.802/11080
2050005

AIR OPERATIONS. World War 2
Night air operations by Great Britain. Royal Air Force. Fighter aeroplanes
RAWNSLEY, C. F. (Cecil Frederick)
Night fighter. - London : Corgi, 1957
359p
0-552-11597-5 X.808/34146
032695x

RAWNSLEY, Cecil Frederick
Night fighter. - Morley : Elmfield Press, 1976
3-382p
0-7057-0077-1 X.809/42188
032695x

AIR OPERATIONS. World War 2
Night air operations by Great Britain. Royal Air Force. Fighter aeroplanes - Biographies
RAWNSLEY, C. F. (Cecil Frederick)
Night fighter. - London : Corgi, 1959
359p
0-552-07967-7 H.84/639
032695x

AIR OPERATIONS. World War 2
Use of radar
ARENA, Nino
Il Radar. - Modena : STEM Mucchi, 1976-1977
3v. : ill. - (Le macchine e la storia)
r6197545 X.702/15529
1168150
PRICE, Alfred
Instruments of darkness. - New expanded and updated ed. - London : Macdonald and Jane's, 1977
284p,[32]p of plates : ill. - Includes index
0-354-01062-x X.620/16948 Woolwich
1168150
Instruments of darkness. - Revised ed. - London : Panther, 1979
333p,[32]p of plates : ill. - Includes index
0-586-04834-0 X.619/19720 Woolwich
1168150

AIR OPERATIONS. World War I
By French fighter aeroplanes. Aces
PORRET, D.
Les "AS" français de la Grande Guerre. - Paris : CEDOCAR, [1983]
2v. : ill
2-7170-0741-5 X.800/36686
9032851

AIR POLICIES. Great Britain
1918-1939
HYDE, Harford Montgomery
British air policy between the wars, 1918-1939. - London : Heinemann, 1976
xx,539p,leaf of plate,[32]p of plates : ill. - Bibliography: p.520-526. - Includes index
0-434-47983-7 X.620/16416 Woolwich
0763047

AIR-POWER
Development
AIR power in the next generation / [edited by] E.J. Feuchtwanger and R.A. Mason. - London : Macmillan, 1979
xvi,151p,[8]p of plates : ill. - Bibliography: p.147-148. - Includes index
0-333-23609-2 X.629/12496
145000x

AIR-POWER. Developing countries
Development
KULLAR, Amarjeet
Effective airpower for developing countries. - Delhi : Macmillan Co. of India, 1975
x,146p,[6]p of plates : ill. - Bibliography : p.137-139. - Includes index
0-333-90072-3 X.629/11125 Woolwich
1082426

AIR PRODUCTS LTD
See also
Industrial gases manufactured by Air Products Ltd

AIR PUMPS
See Compressors

AIR QUALITY MONITORING STATIONS
Networks. Design & siting
MUNN, R. E. (Robert Edward)
The design of air quality monitoring networks. - London : Macmillan, 1981
xiii,109p : ill. - (Air pollution problems ; 2). - Bibliography: p101-107. - Includes index
0-333-30460-8 SRL
2572230

AIR RAID SHELTERS
HULTAKER, Örjan E.
The family and the shelters. - Uppsala : Uppsala University, Department of Sociology, [1976]
15p. - (Disaster studies ; 1)
91-506-0077-x X.0519/709(1)
1535099

AIR RAID SHELTERS
Social aspects. Research
HAMMARSTRÖM-TORNSTAM, Gunhild
Liten handledning om beläggningsförsök. - Uppsala : Uppsala University, Department of Sociology, 1978
31p. - (Disaster studies ; 7)
91-506-0160-1 X.0519/709(7)
1793985

AIR RAID SHELTERS. Great Britain
Design. Role of Great Britain. Ministry of Home Security. Design & Development Section, 1939-1945
BAKER, John Baker, Baron
Enterprise versus bureaucracy. - Oxford : Pergamon, 1978
x,123p : ill. - Includes index
0-08-022149-1 X.800/27005
1184431

AIR RAIDS
See also
Air raid shelters

AIR RAIDS. American ships. Bari. Italy
By Germany. Luftwaffe, 1943
INFIELD, Glenn B. (Glenn Berton)
Disaster at Bari. - London : New English Library, 1976
256p
0-450-02659-0 X.708/20702
0070823

AIR RAIDS. Barrow-in-Furness
By Germany. Luftwaffe, 1940-1941
BARROW at war / [compiled by] B. Trescatheric & D.J. Hughes. - Chorley : Countryside Publications Limited, 1979
48p : chiefly ill
0-86157-027-8 X.700/26383
2223740

AIR RAIDS. Belfast
By Germany. Luftwaffe, 1941
BOMBS on Belfast. - Belfast : Pretani, 1984
xiv,[54]p : ill. - Bibliography: p54
0-9503461-6-0 X.800/40464
3256456

AIR RAIDS. Berlin. Germany
By United States. Army Air Force. Air Force, 8th, 1944 (March)
ETHELL, Jeffrey
Target Berlin. - London : Jane's, 1981
xii,212p,[32]p of plates : ill. - Bibliography: p164. - Includes index
B81-37402 X.800/31482
2559617

AIR RAIDS. Bristol Aeroplane Company
By Germany. Luftwaffe, 1940. Bomber aeroplanes
WAKEFIELD, Kenneth
Luftwaffe encore. - London : Kimber, 1979 [i.e. 1980]
223p : ill. - Bibliography: p.17-19. - Includes index
0-7183-0077-7 X.800/29065
222738x

AIR RAIDS. Bromley (London Borough)
By Germany. Luftwaffe, 1940-1945
BLAKE, Lewis
Bromley in the front-line. - [Whitstable] : L. Blake, c1980
[88]p : ill
B81-06952 X.809/46601
2355140
Bromley in the front-line. - Rev. print. - [Bromley] : Lewis Blake, 1983
96p : ill. - Includes index
B84-19542 X.809/58839
2355140

AIR RAIDS. Cologne. West Germany
By Great Britain. Royal Air Force. Bomber Command, 1942 (May)
BARKER, Ralph
The thousand plan. - London : Pan Books, 1978
271p,[8]p of plates : ill. - Bibliography: p.9-10. - Includes index
0-330-20205-7 X.708/21145
1433512

AIR RAIDS. Convoys. Axis navies. Mediterranean campaign. World War 2
By Great Britain. Royal Navy. 830 Squadron, 1940–1943
POOLMAN, Kenneth
Night strike from Malta. - London : Jane's Publishing Company, 1980
192p : ill. - Bibliography: p.186. - Includes index
0-7106-0003-8 X.800/28764
2113201

AIR RAIDS. Coventry
By Germany. Luftwaffe, 1940 (November)
LONGMATE, Norman
Air raid. - London : Hutchinson, 1976
[2],302p,[16]p of plates : ill. - Bibliography: p.272-276. - Includes index
0-09-127900-3 X.809/41505
0788783
Air raid. - London : Arrow Books, 1979
[2],302p,[8]p of plates : ill. - Bibliography: p.272-276. - Includes index
0-09-920850-4 X.708/22281
0788783

AIR RAIDS. Denmark
By Great Britain. Royal Air Force. Bomber Command, 1940-1945
HERMANSEN, Ove
Gruppen der bombede Shellhuset. - [København] : [Frihedsmuseet], [1979]
72p : ill. - (Arsskrift for Frihedsmuseets Venner ; 1979)
87-480-0186-4 X.709/50390
2260212

AIR RAIDS. Dresden. East Germany
By Great Britain. Royal Air Force. Bomber Command, 1945
MCKEE, Alexander
Dresden 1945. - London : Souvenir, 1982
334p,[24]p of plates : ill. - Bibliography: p327-329. - Includes index
0-285-62515-2 X.809/52410
0135550
Dresden 1945. - London : Granada, 1983, c1982
xxxi,393p,[16]p of plates : ill. - Bibliography: p383-386. - Includes index
0-583-13686-9 X.808/41580
0135550

AIR RAIDS. East End
By Germany. Luftwaffe, ca 1940 - Biographies
NIXON, Barbara
Raiders overhead. - London : Scolar/Gulliver, 1980
176p : ill
0-906428-06-8 X.809/46542
2140780

AIR RAIDS. England
By Germany. Luftwaffe, 1939-1944
PRICE, Alfred
Blitz on Britain. - London : I. Allan, 1977
192p : ill. - Includes index
0-7110-0723-3 X.800/26160
1139959

AIR RAIDS. England
By Germany. Luftwaffe, 1940-1944
GERMAN bombers over England / [selected and with commentaries by] Bryan Philpott. - Cambridge : Stephens, 1978
94p : of ill. - (World War 2 photo albums ; no.2)
0-85059-339-5 X.611/8345 Woolwich
1345435

AIR RAIDS. England
By Germany. Wehrmacht, 1914-1918
CASTLE, H. G.
Fire over England. - London : Secker & Warburg, 1982
254p,[8]p of plates : ill. - (A Leo Cooper book). - Includes index
0-436-08900-9 X.809/53955
2826062

AIR RAIDS. England
By Germany. Wehrmacht, 1944-1945. V-1 flying bombs
LONGMATE, Norman
The doodlebugs. - London : Hutchinson, 1981
548p,[16]p of plates : ill. - Bibliography: p488-513. - Includes index
0-09-144750-x X.809/48001
1576550
YOUNG, Richard Anthony
The flying bomb. - London : I. Allan, 1978
160p : ill. - Bibliography: p.160
0-7110-0842-6 X.802/10669
1576550

AIR RAIDS. England
By Germany. Wehrmacht. Flying bombs & V-2 rockets, 1944-1945
COOKSLEY, Peter George
Flying bomb. - London : Hale, 1979
208p,[16]p of plates : ill. - Bibliography: p.200-201. - Includes index
0-7091-7399-7 X.809/44475
1907441

AIR RAIDS. England
By Germany. Wehrmacht. V-1 flying bombs & V-2 rockets, 1944-1945
COLLIER, Basil
The battle of the V-weapons, 1944-45. - Morley : Elmfield Press, 1976
192p,[8]p of plates : ill. - ([Morley war classics]). - Bibliography: p.184-185. - Includes index
0-7057-0070-4 X.629/10930 Woolwich
0762385

AIR RAIDS. Essen. West Germany
By Great Britain. Royal Air Force. Bomber Command, 1943 (March)
SEARBY, J.
Essen. - Chippenham : Nutshell Press, 1978
[6],94p : ill. - (The great raids ; vol.2)
Subsequent volumes, when received, will be placed at this shelfmark
0-902633-50-3 X.0705/206
2282623

AIR RAIDS. Germany
By Allied air forces, to 1945
HOLMSTEN, Georg
Deutschland Juli 1944. - Düsseldorf : Droste, 1982
160p : ill. - (Fotografierte Zeitgeschichte)
3-7700-0598-8 X.805/5205
8530831

KÖSTERS, Hans G.
Essen Stunde Null. - Düsseldorf : Droste, 1982
117p : ill. - (Fotografierte Zeitgeschichte)
3-7700-0617-8 X.805/6087
8530831

AIR RAIDS. Germany
By Allied air forces, 1944-1945. Strategy
ROSTOW, W. W. (Walt Whitman)
Pre-invasion bombing strategy. - Aldershot : Gower, c1981
xiii,166p. - Includes index
0-566-00482-8 X.809/51303
2706091
- [ANOTHER COPY: PAPERBACK] I.S.B.N. 0-566-00483-6 X.809/51304

AIR RAIDS. Guernica. Spain
1937. Responsibility
URIARTE AGUIRREAMALLOA, Castor
Bombas y mentiras sobre Guernica. - Bilbao : [s.n.], [1976]
221p : ill
84-400-1462-7 X.708/42450
1422073

AIR RAIDS. Guernica. Spain
By Germany. Luftwaffe. Legion Kondor, 1937
SOUTHWORTH, Herbert R.
La destruction de Guernica. - Paris : Ruedo Ibérico, 1975
xxiv, 535 p.. - Bibliography : p. 509-521
r6012099 X.700/16746
0618713

SOUTHWORTH, Herbert Rutledge
Guernica! Guernica!. - Berkeley ; London : University of California Press, 1977
xxvii,537p. - Bibliography: p.503-515. - Includes index
0-520-02830-9 X.800/26428
0618713

THOMAS, Gordon
The day Guernica died. - London : Hodder and Stoughton, 1975
319p : ill. - Bibliography: p.305-311. - Includes index
0-340-19043-4 X.800/11553
0618713

AIR RAIDS. Guernica. Spain
By Germany. Wehrmacht. Legion Condor, 1937
MAIER, Klaus A.
Guernica, 26.4.1937. - Freiburg : Rombach, 1975
166p.,plate. - (Einzelschriften zur militärischen Geschichte des Zweiten Weltkrieges ; 17)
3-7930-0176-8 S.F.441/48[no.17]
9127925

AIR RAIDS. Hampstead
By Germany. Luftwaffe, 1939-1945
HAMPSTEAD (Metropolitan Borough). Council
Hampstead, 1939-1945. - London : Camden History Society : Carlile House Press, 1977
40p : ill
0-904491-06-4 X.709/31958
1339125

AIR RAIDS. Koblenz region. West Germany
By Allied air forces, to 1945
SCHNATZ, Helmut
Der Luftkrieg im Raum Koblenz 1944/45. - Boppard am Rhein : Boldt, 1981
xiv,592p. : ill. - (Veröffentlichungen der Kommission des Landtages für die Geschichte des Landes Rheinland-Pfalz ; Bd.4). - Bibliography
3-7646-1774-8 X.0802/446(4)
8789711

AIR RAIDS. London
By Germany. Wehrmacht. V-1 flying bombs & V-2 rockets, 1944-1945
JOHNSON, David, 1950-
V for vengeance. - London : Kimber, 1981
203p : ill. - Includes index
0-7183-0188-9 X.800/31164
2542137

AIR RAIDS. London (City)
By Germany. Luftwaffe, 1940 (December)
JOHNSON, David
The city ablaze. - London : Kimber, 1980
217p : ill. - Includes index
0-7183-0267-2 X.800/29055
2208024

AIR RAIDS. Newton Abbot Railway Station
By Germany. Luftwaffe, 1940
The NEWTON Abbot blitz / [compiled] by A.R. Kingdom. - Oxford : Oxford Publishing Co., 1979
32p : chiefly ill. - (Transport topic series ; no. 1)
0-86093-054-8 X.629/13926 Woolwich
2067242

AIR RAIDS. Nuclear reactors. Osirak. Iraq
By Israel. Ḥel ăvir, 1981
PERLMUTTER, Amos
Two minutes over Baghdad. - [London] : Corgi, 1982
192p
0-552-11939-3 X.808/36795
2759292

AIR RAIDS. Nuremberg. West Germany
By Great Britain. Royal Air Force. Bomber Command, 1944
MIDDLEBROOK, Martin
The Nuremberg raid, 30-31 March 1944. - London : Allen Lane, 1973
xv,369p,[28]p of plates : ill. - Bibliography: p.360. - Includes index
0-7139-0612-x X.809/46249
0011134
The Nuremberg raid, 30-31 March 1944. - [London] : Fontana, 1975
iii-xv,369p,[16]p of plates : ill. - Bibliography: p.360. - Includes index
0-00-633625-6 X.708/20571
0011134

TAYLOR, Geoff, b.1920
The Nuremberg massacre. - London : Sidgwick and Jackson, 1980
xiii,176p,[16]p of plates : ill. - Bibliography: p.171
0-283-98646-8 X.809/45448
0011134

AIR RAIDS. Pearl Harbor. Hawaii
1939-1945
AIR raid / edited by Paul Stillwell. - Annapolis, Md : Naval Institute Press, [1981]
viii,299p : ill
0-87021-086-6 X.805/6274
904325x

LENGERER, Hans
Pearl Harbor 1941. - Friedberg : Podzun-Pallas, [1982]
159p. : ill
3-7909-0164-4 X.809/58344
904325x

TREFOUSSE, Hans Louis
Pearl Harbour. - Malabar, Fla : Robert E. Krieger, 1982
215p. - (An anvil original)
0-89874-261-7 X.808/40388
904325x

WILLMOTT, H. P. (Hedley Paul)
Pearl Harbor. - Repr. - New York : Galahad Books, 1982
64p : ill
0-88365-543-8 X.805/5718
904325x

AIR RAIDS. Pearl Harbor. Hawaii
By Japan. Army Air Force & Japan. Navy. Air Service, 1941. Influence on politics in United States, 1941-1946

BARTLETT, Bruce R.
Cover-up. - New Rochelle, N.Y. : Arlington House, c1978
189p
0-87000-423-9 X.809/51430
1463071

MELOSI, Martin V.
The shadow of Pearl Harbor. - College Station ; London : Texas A and M University Press, 1977
xiv,183p. - Bibliography: p.169-177. - Includes index
0-89096-031-3 X.800/26904
1463071

AIR RAIDS. Pearl Harbor. Hawaii
By Japan. Nihon Rikugun Koku-butai & Japan. Nihon Kaigun Kōkū-bu, 1941

PAŠEK, Petr
Tokio volá Honolulu. - 2. české vyd.. - Zürich : Konfrontation Verlag, 1975
143 p., 28 p. of plates : ill. - (Helvetia)
3-85770-017-3 X.808/10215
0890898

PRANGE, Gordon W. (Gordon William)
At dawn we slept. - London : Joseph, 1982, c1981
xvi,873p,[8]p of plates : ill. - Bibliography: p827-838. - Includes index
0-7181-2090-6 X.800/32651
0890898

At dawn we slept. - Harmondsworth : Penguin, 1983, c1982
xvi,873p,[16]p of plates : ill. - Bibliography: p827-838. - Includes index
0-14-006455-9 X.208/8642
0890898

TOLAND, John, 1912-
Infamy. - London : Methuen, 1982
xvi,366p,[16]p of plates : ill. - Includes index
0-413-49820-4 X.800/33873
0890898

AIR RAIDS. Peenemünde. East Germany
By Great Britain. Royal Air Force. Bomber Command, 1943

MIDDLEBROOK, Martin
The Peenemünde raid. - London : Allen Lane, 1982
265p,[16]p of plates : ill. - Bibliography: p258. - Includes index
0-7139-1335-5 X.800/34637
2834928

AIR RAIDS. Plymouth. Devon
By Germany. Luftwaffe, 1940-1944

WINTLE, Frank
The Plymouth blitz. - Bodmin : Bossiney, 1981
96p : ill. - Bibliography: p94
0-906456-51-7 X.809/47704
2405644

AIR RAIDS. Regensburg & Schweinfurt. West Germany
By United States. Army Air Force. Bomber Command, 1943

MIDDLEBROOK, Martin
The Schweinfurt-Regensburg mission. - London : Lane, 1983
363p,[16]p of plates : ill. - Bibliography: p353. - Includes index
0-7139-1522-6 X.809/57234
2987783

AIR RAIDS. Schweinfurt. West Germany
to 1945

GOLÜCKE, Friedhelm
Schweinfurt und der strategische Luftkrieg 1943. - Paderborn : Schöningh, [1980]
444p.,[1]folded leaf of plates : ill. - (Sammlung Schöningh zur Geschichte und Gegenwart)
3-506-77446-8 X.800/31824
8781974

AIR RAIDS. Schweinfurt. West Germany
By United States. Army Air Force. Air Force, 8th, 1943

BENDINER, Elmer
The fall of fortresses. - London : Pan, 1982, c1980
218p,[8]p of plates : ill
B83-19074 X.808/37879
2397102

COFFEY, Thomas M.
Decision over Schweinfurt. - London : Hale, 1978
ix,373p,[8]p of plates : ill. - Bibliography: p.359-364. - Includes index
0-7091-7152-8 X.800/27430
1758314

Decision over Schweinfurt. - London : Magnum Books, 1980
ix,373,[8]p of plates : ill. - Bibliography : p.359-364. - Includes index
0-417-04120-9 X.708/22390
1758314

AIR RAIDS. Schweinfurt. West Germany
By United States. Army Air Force. Air Force, 8th, 1943 - Biographies

BENDINER, Elmer
The fall of fortresses. - London : Souvenir Press, 1981, c1980
258p,[8]p of plates : ill
0-285-62478-4 X.629/15364 Woolwich
2397102

AIR RAIDS. Sheffield. South Yorkshire (Metropolitan County)
By Germany. Luftwaffe, 1940

WALTON, Mary
Raiders over Sheffield. - Sheffield : Sheffield City Libraries, 1980
xiv,169p,[24]p of plates : ill. - Includes index
0-900660-55-4 X.700/26360
2163934

AIR RAIDS. Southampton. Hampshire
By Germany. Luftwaffe, 1940-1944

BRODE, Anthony
The Southampton blitz. - Rev. ed. - Newbury : Countryside, 1982
96p
0-905392-15-9 X.808/36784
2760711

AIR RAIDS. York
By Germany. Luftwaffe, 1941 (June)

KESSLER, Leo
The Great York Air Raid. - Clapham via Lancaster : Dalesman, 1979
78p : ill. - Bibliography: p.78
0-85206-507-8 X.709/33410
1945807

AIR REED INSTRUMENTS
See Flute family instruments

AIR RESCUE SERVICES
See also
Helicopter rescue services

AIR RESCUE SERVICES. Royal Naval Air Station, Culdrose
Air sea rescue services, to 1976

PEARCE, Frank, 1909-
Mayday! Mayday! Mayday!. - Falmouth : Bantam, 1977
56p : ill
B78-18139 X.700/25680
1456830

AIR RIFLES
Spring operated air rifles. Internal ballistics

CARDEW, G.V.
The air gun from trigger to muzzle. - [Great Britain] : [The authors], 1976
2-96p : ill. - Includes index
B77-00310 SRL
0772496

AIR SEGMENTED CONTINUOUS FLOW CHEMICAL ANALYSIS. Water
Air segmented continuous flow automatic chemical analysis

PETTS, K. W.
Air segmented continuous flow automatic analysis in the laboratory, 1979. - London : H.M.S.O., 1980
23p : ill. - (Methods for the examination of waters and associated materials)
0-11-751470-5 SRL
2303299

AIR SERVICES
See also
Aerodromes
Airlines
Airports

AIR SERVICES
- Texts

EGYPT
[Treaties, etc. Great Britain, 1981 May 14]. Agreement between the Government of the United Kingdom of Great Britain and Northern Ireland and the Government of the Arab Republic of Egypt for air services between and beyond their respective territories. - London : H.M.S.O., [1981]
10p. - (Egypt ; no.1 (1981)) (Cmnd. ; 8348)
0-10-183480-2 BS.14/3440
2577887

GREAT BRITAIN
[Treaties, etc. Australia, 1958 Feb.7. Protocols, etc., 1975 Feb.13]. Exchange of notes between the Government of the United Kingdom of Great Britain and Northern Ireland and the Government of Australia replacing the schedule to the agreement for air services between and beyond their respective territories signed at London on 7 February 1958, London, 13-14 February 1975 ... - London : H.M.S.O., [1975]
6p. - (Treaty series ; no.67(1975)) (Cmnd.6079)
0-10-160790-3 BS.14/137
0374946

[Treaties, etc. Bulgaria, 1970 May 28. Protocols, etc., 1973 Aug.23]. Revised schedule to the air services agreement of 28 May 1970 between the Government of the United Kingdom of Great Britain and Northern Ireland and the Government of the People's Republic of Bulgaria (the revised schedule is deemed to have come into effect on 23 August 1973) ... - London : H.M.S.O., 1975
7p. - (Treaty series ; no.82(1975)) (Cmnd.6070)
0-10-160700-8 BS.14/137
0363871

[Treaties, etc. Italy, 1976 Nov.22]. Agreement between the Government of the United Kingdom of Great Britain and Northern Ireland and the Government of the Italian Republic for air services between and beyond their respective territories (with exchange of notes). Rome, 22 November 1976 ... - London : H.M.S.O., [1977]
12p. - (Italy ; no.1(1977)) (Cmnd.6873)
0-10-168730-3 BS.14/3011
1250418

[Treaties, etc. Ivory Coast, 1976 Dec.1]. Agreement between the Government of the United Kingdom of Great Britain and Northern Ireland and the Government of the Republic of the Ivory Coast for air services between and beyond their respective territories, London, 1 December 1976 ... - London : H.M.S.O., [1977]
12p. - (Ivory Coast ; no.1 (1977)) (Cmnd.6736)
0-10-167360-4 BS.14/2064
1084224

[Treaties, etc. Jamaica, 1970 Mar.25. Protocols, etc., 1974 Dec.31]. Exchange of notes between the Government of the United Kingdom of Great Britain and Northern Ireland and the Government of Jamaica revising the route schedule to the air services agreement signed at Kingston on 25 March 1970, Kingston, 31 December 1974 ... - London : H.M.S.O., [1975]
4p. - (Treaty series ; no.93(1975)) (Cmnd.6113)
0-10-161130-7 BS.14/137
0513725

[Treaties, etc. Liberia, 1975 Sept.26]. Agreement between the Government of the United Kingdom of Great Britain and Northern Ireland and the Government of the Republic of Liberia for air services between and beyond their respective territories, London, 26 September 1975 ... - London : H.M.S.O., [1975]
11p. - (Liberia ; no.1(1975)) (Cmnd.6312)
0-10-163120-0 BS.14/1929
0579580

[Treaties, etc. Tanzania, 1980 July 1]. Agreement between the Government of the United Kingdom of Great Britain and Northern Ireland and the Government of the United Republic of Tanzania for air services between and beyond their respective territories, Dar-es-Salaam, 1 July 1980. - London : H.M.S.O., [1981]
10p. - (Tanzania ; no.1 (1981)) (Cmnd. ; 8229)
0-10-182290-1 BS.14/3469
2498251

GREAT BRITAIN
[Treaties, etc. Union of Soviet Socialist Republics, 1974 Jan.29]. Exchange of notes between the Government of the United Kingdom of Great Britain and Northern Ireland and the Government of the Union of Soviet Socialist Republics designating the British Airways Board in agreements supplementary to the air services agreement of 1957, Moscow, 29 January/21 March 1974 ... - London : H.M.S.O., 1975
8p. - (Treaty series ; no.76(1975)) (Cmnd.6057)
0-10-160570-6 BS.14/137
0363863

[Treaties, etc. Zaire, 1974 May 17]. Agreement between the Government of the United Kingdom of Great Britain and Northern Ireland and the National Executive Council of the Republic of Zaire for air services between and beyond their respective territories ; Kinshasa, 17 May 1974. - London : H.M.S.O., 1974
11p. - (Zaire ; no.1(1974)) (Cmnd.5777)
0-10-157770-2 BS.14/1583
0238309

AIR SERVICES
Air freight transport services - Directories - Serials

AIRTRADE air freight directory. - Apr. to Sept. 1979-. - [London] : [Maclean Hunter Ltd.], 1979-
v.
Holdings: October 1981-March 1982, etc
B82-10123 P.823/117
2620936

AIR SERVICES
Air freight transport services - Serials

AIRTRADE gateway guide. - 1984- - London : Maclean Hunter, 1984-
v. : ill
B84-41666 P.625/492
3197204

AIR SERVICES
Air passenger transport services

CAUTER, Gaynor
International airport. - London : Octopus, 1980
127p : ill(some col.). - Includes index
0-7064-1295-8 X.622/9375 Woolwich
2331918

PETTY, Kate
On a plane. - London : Watts, 1984
30p : col.ill. - (Franklin Watts first library). - Includes index
0-86313-047-x X.990/23086
2765896

WILSON, Mike
Jet journey. - London : Marshall Cavendish, 1978
59p : col.ill. - (Woodpecker books). - Includes index
0-85685-336-4 X.622/13154
2765896

AIR SERVICES
Air passenger transport services - Atlases - Serials

ABC air travel atlas. - [No.1]- ; Dec. 1978/Apr. 1979-. - London : IPC Business Press, 1978-
B79-25456 P.703/625
1915908

AIR SERVICES
Air passenger transport services, to 1982

EMBLETON, Gerry
Passenger aircraft. - Hove : Wayland, 1982
64p : ill(some col.). - (Transport and society). - Bibliography: p62. - Includes index
0-85340-958-7 X.622/15152
2878100

AIR SERVICES
Air passenger transport services. Advance booking charter flights - Texts

GREAT BRITAIN
[Treaties, etc. United States, 1973 Mar.30. Protocols, etc., 1975 Apr.2]. Exchange of notes between the Government of the United Kingdom of Great Britain and Northern Ireland and the Government of the United States of America further amending the agreement concerning procedures for the implementation and enforcement of rules for advance booking charter flights, signed at Washington on 30 March 1973, London, 2-3 April 1975 ... - London : H.M.S.O., [1975]
4p. - (Treaty series ; no.88(1975)) (Cmnd.6107)
0-10-161070-x BS.14/137
0217123

[Treaties, etc. United States, 1973 Mar.30. Protocols, etc., 1975 June 4]. Exchange of notes between the Government of the United Kingdom of Great Britain and Northern Ireland and the Government of the United States of America further extending the agreement concerning procedures for the implementation and enforcement of rules for advance booking charter flights signed at Washington on 30 March 1973, London 4 June 1975 ... - London : H.M.S.O., [1975]
4p. - (Treaty series ; no.114(1975)) (Cmnd.6211)
0-10-162110-8 BS.14/137
0217123

AIR SERVICES
Air passenger transport services. Aeroplanes

TAYLOR, John, b.1922
Passenger aircraft and airlines. - London : Marshall Cavendish, 1975
[2],152p : ill(chiefly col)
0-85685-142-6 X.622/5370 Woolwich
0618950

AIR SERVICES
Air passenger transport services. Aeroplanes - Serials

AIRLINERS. - No.1-. - Hounslow : Airline Publications and Sales Ltd, [197-?]- : ill(some col)
Holdings: no.8, etc. [1976 etc.]
B76-28894 X.0619/15026 Woolwich
0720739

AIR SERVICES
Air passenger transport services. Catering services - Serials

TRAVEL food & duty free international. - Vol.3, no.2 (Mar./Apr. 1983)-. - Miami Springs ; London : A.C. Morton, 1983-
v. : ill(some col.)
B84-29331 P.2000/1144
2204258

AIR SERVICES
Air passenger transport services. Charter flights. Regulation

KAMP, Jaap
Air charter regulation. - New York : London : Praeger : [London] : [Distributed by Martin Robertson], 1976
xi,163p : 1 ill. - (Praeger special studies in US economic, social and political issues). - Bibliography: p.155-160. - Includes index
0-275-22890-8 X.520/10795
0674885

AIR SERVICES
Air passenger transport services. Great Britain-North America routes

COMBES, Peter
How to fly the Atlantic. - London : Kogan Page, 1980
160p : ill. - Includes index
0-85038-395-1 X.629/14897
2352672

AIR SERVICES
Air passenger transport services. Hygiene

WORLD HEALTH ORGANIZATION
Guide to hygiene and sanitation in aviation. - 2nd ed.. - Geneva : W.H.O. ; [London] : [H.M.S.O.], 1977
170p : ill. - Includes index
B78-14997 UNO.142/20
141545x
- [Another copy] SRL

AIR SERVICES
Air passenger transport services. Passengers. Injuries. Liability of airlines. Law

QUINTANA CARLO, Ignacio
La responsabilidad del transportista aéreo por daños a los pasajeros. - Salamanca : Universidad de Salamanca, 1977
xxxi,392p. - (Acta Salmanticensia. Derecho ; n.36). - Includes bibliographies
84-600-0858-4 Ac.153/3(3)
1978802

AIR SERVICES
Air passenger transport services. Safety aspects

JOHNSON, Daniel A.
Just in case. - New York ; London : Plenum, c1984
xvi,268p : ill. - Includes index
0-306-41576-3 X.629/24372
0762962

NORRIS, William, 1933-
The unsafe sky. - London : Arrow, 1981
223p
0-09-926600-8 X.629/16003 Woolwich
0762962

RAMSDEN, John Michael
The safe airline. - London : Macdonald and Jane's, 1976
viii,231p,[8]p of plates : ill. - Includes index
0-354-01029-8 SRL
0762962

AIR SERVICES
Air passenger transport services. Women stewards

STEWART, Anne
The air stewardess. - London : Hamilton, 1984
24p : col.ill. - (Cherrystones). - Includes index
0-241-11209-5 X.629/24056
3015858

AIR SERVICES
Benelux countries. Airports - For air services - Directories

AIR touring flight guide, Benelux. - 1975, Jan. / compiled and edited by Carolyn Pooley. - [Boreham Wood] : Airtour International Ltd, [1975]
136p : ill
Subsequent volumes, when received will be placed at this shelfmark
0-9502333-6-6 P.701/1521
0728993

AIR SERVICES
British international air passenger transport services. Policies of government

GREAT BRITAIN. Department of Trade
Future civil aviation policy. - London : H.M.S.O., [1976]
[2],17p. - (Cmnd.6400)
0-10-164000-5 BS.41/1063
0619337

AIR SERVICES
C-46 Curtiss aeroplanes

DAVIS, John Michael
The Curtiss C-46 Commando. - Hornchurch : Air-Britain, 1978
146p,[32]p of plates : ill. - (Monographs / Air-Britain). - Includes index
0-85130-065-0 X.611/8720 Woolwich
1833243

AIR SERVICES
De Havilland Comet aeroplanes, to 1975

COWELL, John Graham
DH Comet. - Hounslow : Airline Publications and Sales Ltd, 1976
viii,124p,[56]p of plates : ill
0-905117-04-2 X.619/16769 Woolwich
055183x

AIR SERVICES
Double taxation - Texts

GREAT BRITAIN
[Treaties, etc. Ethiopia, 1977 Feb.1]. Exchange of notes between the Government of the United Kingdom of Great Britain and Northern Ireland and the Provisional Military Government of Socialist Ethiopia for the avoidance of double taxation on profits derived from air transport, Addis Ababa, 1 February 1977 ... - London : H.M.S.O, [1977]
Folder(4p). - (Ethiopia ; no.1(1977)) (Cmnd.6857)
0-10-168570-x BS.14/2096
1145479

[Treaties, etc. Hungary, 1975 June 5]. Agreement between the Government of the United Kingdom of Great Britain and Northern Ireland and the Government of the Hungarian People's Republic for the avoidance of double taxation of air transport undertakings and their employees, Budapest, 5 June 1975 ... - London : H.M.S.O., [1975]
4p. - (Hungary ; no.1(1975)) (Cmnd.6209)
0-10-162090-x BS.14/1815
0531405

[Treaties, etc. Jordan, 1978 Mar.6]. Exchange of notes between the Government of the United Kingdom of Great Britain and Northern Ireland and the Government of the Hashemite Kingdom of Jordan for the avoidance of double taxation on profits derived from shipping and air transport, Amman, 6 March 1978 ... - London : H.M.S.O., [1978]
Folder (3p). - (Jordan ; no.1(1978)) (Cmnd.7223)
0-10-172230-3 BS.14/3173
1686585

[Treaties, etc. Venezuela, 1978 Mar.8]. Agreement between the Government of the United Kingdom of Great Britain and Northern Ireland and the Government of the Republic of Venezuela for the avoidance of double taxation in respect of shipping and air transport, Caracas, 8 March 1978 ... - London : H.M.S.O., [1978]
Folder(4p). - (Venezuela ; no.1(1978)) (Cmnd.7227)
B78-35152 BS.14/3174
1686569

[Treaties, etc. Zaire, 1976 Oct.11]. Agreement between the United Kingdom of Great Britain and Northern Ireland and the Republic of Zaire for the avoidance of double taxation on revenues arising from the business of shipping and air transport in international traffic, London, 11 October 1976 ... - London : H.M.S.O., [1977]
Folder(4 p). - (Zaire ; no.1 (1977)) (Cmnd.6717)
0-10-167170-9 BS.14/3028
1067605

AIR SERVICES
Economic forecasting. Applications of regression analysis

TANEJA, Nawal K.
Airline traffic forecasting. - Lexington, Mass. : Heath ; [Farnborough, Hants.] : [Distributed by] Teakfield Ltd, 1979
xvii,231p : ill. - Bibliography: p.217-224. - Includes index
0-669-02186-5 X.620/18484 Woolwich
1885898

AIR SERVICES
Effects of shortages of petroleum

SINGH, Karmjit
How airlines are coping on the fuel front. - London : Institute of Petroleum, [1983?]
14p. - (IP ; 83-002)
B84-31400 X.622/19921
3140555

AIR SERVICES
Europe. Airports - For air services - Directories

EUROPE flight guide. - 1980, Jan. / compiled and edited by Robert Pooley. - [Boreham Wood] : Airtour Associates International Ltd., [1980]
288p : ill
B80-25479 X.808/33363
0154083

AIR SERVICES
Finance - Serials

AIRFINANCE journal. - No.1 (Nov.1980)-. - Coggleshall : Hawkins, 1980-
v. : ill
B81-05590 P.513/2531
2352230

AIR SERVICES
Fokker F-27 aeroplanes
BARNES, F. G.
The Fokker F.27 Friendship. - Tonbridge : Air-Britain, 1979
76p : ill
0-85130-073-1 X.510/16401
1969226

AIR SERVICES
Great Britain. Airports - For air services - Directories
POOLEY'S flight guide United Kingdom and Ireland. - 1980. Mar. : [17th ed.] / compiled and edited by Robert Pooley ... with the assistance of the Civil Aviation Authority. - London : Airtour Associates International Ltd, [1980]
356p : ill
Subsequent volumes, when received, will be placed at this shelfmark
B80-28221 P.801/1537
0154105

AIR SERVICES
International air services
[CONVENTION for the Unification of Certain Rules Relating to International Carriage by Air. Protocols, amendments, etc., September 25, 1975]. Additional Protocol no.1 to amend the Convention for the Unification of Certain Rules Relating to International Carriage by Air signed at Warsaw on 12 October 1929, Montreal, 25 September 1975 ... - London : H.M.S.O., [1976]
7p. - (Miscellaneous ; no.12 (1976)) (Cmnd.6480)
0-10-164800-6 BS.14/1770
0678368

[CONVENTION for the Unification of Certain Rules Relating to International Carriage by Air. Protocols, amendments, etc., September 25, 1975]. Additional Protocol no.2 to amend the Convention for the Unification of Certain Rules Relating to International Carriage by Air signed at Warsaw on 12 October 1929 as amended by the Protocol done at The Hague on 28 September 1955, Montreal, 25 September 1975 ... - London : H.M.S.O., [1976]
8p. - (Miscellaneous ; no.15 (1976)) (Cmnd.6481)
0-10-164810-3 BS.14/1991
0678368

[CONVENTION for the Unification of Certain Rules Relating to International Carriage by Air. Protocols, amendments, etc., September 25, 1975]. Additional Protocol no.3 to amend the Convention for the Unification of Certain Rules Relating to International Carriage by Air signed at Warsaw on 12 October 1929 as amended by the Protocols done at The Hague on 28 September 1955 and at Guatemala City on 8 March 1971, Montreal, 25 September 1975 ... - London : H.M.S.O., [1976]
8p. - (Miscellaneous ; no.16 (1976)) (Cmnd. 6482)
0-10-164820-0 BS.14/1989
0678368

[CONVENTION for the Unification of Certain Rules Relating to International Carriage by Air. Protocols, amendments, etc., September 25, 1975]. Montreal Protocol no.4 to amend the Convention for the Unification of Certain Rules Relating to International Carriage by Air signed at Warsaw on 12 October 1929 as amended by the Protocol done at The Hague on 28 September 1955, Montreal, 25 September 1975 ... - London : H.M.S.O., [1976]
15p. - (Miscellaneous ; no.17 (1976)) (Cmnd.6483)
0-10-164830-8 BS.14/1993
0678368

AIR SERVICES
International air services. Political aspects
GIDWITZ, Betsy
The politics of international air transport. - Lexington, Mass. : Lexington Books, 1981, c1980
xii,259p. - Bibliography: p243-246. - Includes index
0-669-03234-4 X.622/10752 Woolwich
2531720

AIR SERVICES. Africa
1914-1953
JONES, David, b.1926
The time shrinkers, Africa. - London : Beaumont Aviation Literature, 1977
xii,273p,leaf of plate,[24]p of plates : ill. - Bibliography: p.266-268. - Includes index
0-902230-03-4 X.529/32545
158734x

AIR SERVICES. Argentina
to date
GÜIRALDES, Juan José
El poder aéreo de los argentinos. - Buenos Aires : Círculo de la Fuerza Aérea, Dirección de Publicaciones, 1979
208p : ill
r6701250 X.629/15751 Woolwich
8710376

AIR SERVICES. Australia
Air passenger transport services. Fares, to date
AUSTRALIA. Independent Public Inquiry into Domestic Air Fares
Domestic air fares. - Camberra : Australian Government Publishing Service, 1981
3v.. - (Parliament of the Commonwealth of Australia. Parliamentary Papers ; 1981, no.6, 7 and 8)
0-642-05993-4 CS.G.588/765
8840849

AIR SERVICES. British Airways
Air passenger transport services. Fares
BRITISH AIRWAYS
Air transport. - [London] : [British Airways], [197-]
3v. : col ill. - Includes index
r6032012 BS.420/14
1099248

AIR SERVICES. Canada
to 1941
MOLSON, K. M.
Pioneering in Canadian air transport. - 2nd ed. - Winnipeg : James Richardson, 1975
viii, 315 p., [2] leaves of plates : ill. - Bibliography : p. 301-302. - Includes index
0-919212-39-5 X.622/5815 Woolwich
1054961

AIR SERVICES. Canada
Lothian, George
LOTHIAN, George
Flight deck. - Toronto : McGraw-Hill Ryerson, [1979]
xiii,191p,plates
0-07-082975-6 X.629/20931
8976929

AIR SERVICES. China
1928-1949
WIETHOFF, Bodo
Luftverkehr in China 1928-1949. - Wiesbaden, 1975
380 p. : ill. - (Schriften des Instituts für Asienkunde in Hamburg ; Bd 39)
3-447-01577-2 Ac.9234.pu [Bd. 39]
0868019

AIR SERVICES. English Channel
Air passenger transport services - For yachting
The YACHTSMAN'S cross-Channel timetable guide. - 1978 / [compiled by] Patrick Blackwood. - Havant : K. Mason, [1978]
72p
Subsequent volumes, when received, will be placed at this shelfmark
0-85937-152-2 P.861/301
1590022

AIR SERVICES. Europe
Air passenger transport services. Fares
CIVIL AVIATION AUTHORITY
European air fares, a discussion document. - London : C.A.A., 1977
[2],70p : ill. - (CAP ; 409)
0-86039-053-5 BS.73/208
1308122

AIR SERVICES. Europe
International air passenger transport services. Fares
BRITISH AIRWAYS
Civil air transport in Europe. - [London] : [British Airways], 1977
[3],84,[55]p : ill(some col)
0-901515-03-5 BS.420/16
1060317

CIVIL AVIATION AUTHORITY. Airline Users Committee
European air fares. - London : Civil Aviation Authority, 1976
52p : ill
0-86039-023-3 BS.73/219
1060317

AIR SERVICES. Germany
to 1970
HEIMANN, Erich H.
Vom Zeppelin zum Jumbo-Jet. - Oldenburg : Stalling, 1971
224 p. : ill
3-7979-1929-8 X.800/11495
0861553

AIR SERVICES. Great Britain
Air passenger transport services. Fares. Discounts
GREAT BRITAIN
Report of the working party on discounted air fares. - London : The Department, 1978
[3],27p
B78-35991 BS.41/1157
1698982

AIR SERVICES. Great Britain
Aircraft. Maintenance & repair. Techniques
CIVIL AVIATION AUTHORITY
Condition monitored maintenance. - London : The Authority, 1978
v,33p : ill. - (CAP ; 418)
0-86039-069-1 BS.73/224
1946161
Condition monitored maintenance. - [1st ed. reprinted]. - London : The Authority, 1980
v,33p : ill. - (CAP. 418)
B80-35156 BS.73/255
1946161

AIR SERVICES. Great Britain
Delays. Avoidance. Use of simulations
PAULSON, G. A.
A technique to achieve pre-planned balancing of schedules and some results of its application. - London : Civil Aviation Authority, 1977
vi,30p : ill. - (Research papers / Civil Aviation Authority. Directorate of operational Research and Analysis ; no.7701) (Papers / Civil Aviation Authority ; 77005)
0-86039-039-x BS.73/170
1108654

AIR SERVICES. Great Britain
Domestic air services. Deregulation
FREEDOM of the air. - London : National Consumer Council, 1983
18p
0-905653-76-9 X.525/8412
2991500

AIR SERVICES. Great Britain
Domestic air services. Regulation
CIVIL AVIATION AUTHORITY
Domestic air services. - London : C.A.A., 1979
[2],29p. - (CAP ; 420)
0-86039-082-9 BS.73/230
1911163

AIR SERVICES. Great Britain
Industrial training. Organisations: Air Transport and Travel Industry Training Board
AIR TRANSPORT AND TRAVEL INDUSTRY TRAINING BOARD
A guide to the Air Transport and Travel Industry Training Board. - Staines : The Board, [1976]
16p : ill
B76-20403 X.519/25859
0664642

AIR SERVICES. Great Britain
International air passenger transport services. Fares
CRAYSTON TRAVEL RESEARCH
UK overseas air fares. - Richmond, Surrey : Crayston Travel Research, 1975
57 leaves
0-9504403-0-2 X.512/5061
0522600

AIR SERVICES. Great Britain
International air passenger transport services. Laker, Sir Freddie
MONCKTON, Christopher
The Laker story. - London : Christensen in association with Sunday telegraph, c1982
47p : ill(some col.)
0-9508007-0-8 X.955/2288
2185091

AIR SERVICES. Great Britain
International air passenger transport services. Laker, Sir Freddie - Biographies
BANKS, Howard
The rise and fall of Freddie Laker. - London : Faber, 1982
155p
0-571-11986-7 X.809/53974
2185091

EGLIN, Roger
Fly me, I'm Freddie!. - London : Weidenfeld and Nicolson, 1980
xi,238p,[8]p of plates : ill. - Bibliography: p.227-230. - Includes index
0-297-77746-7 X.809/45730
2185091

AIR SERVICES. Great Britain
Job analysis
COSTELLO, Martin
A survey of job description practices in air transport and travel industry. - [Staines] : Air Transport and Travel Industry Training Board, 1976
[5],30p. - (Occasional Studies Programme reports / Air Transport and Travel Industry Training Board ; no.1)
B78-11123 BS.23/860(1)
1378627

AIR SERVICES. Great Britain
Manpower. Forecasts for 1971-1974

MORGAN, R. G. T.
Manpower forecasts for the air transport industry, 1971-74. - Staines : Air Transport and Travel Industry Training Board, [1974]
165p in various pagings,6 leaves(4 fold) : ill. - (Research reports / Air Transport and Travel Industry Training Board ; 71/2)
B78-07365 BS.23/874
1344528

AIR SERVICES. Great Britain
Organisations: Civil Aviation Authority

BRITAIN'S Civil Aviation Authority. - [London] : Information and Publicity Division of the CAA, 1983
31p,[4]p of plates : col.ill
0-86039-177-9 X.629/20624
0060755

CIVIL AVIATION AUTHORITY
Britain's Civil Aviation Authority. - [London] : CAA, c1980
31p,[4]p of plates : col.ill. - List of films: p31
0-86039-088-8 X.629/15168
0060755

Britain's Civil Aviation Authority. - London : CAA, 1981
31p : ill(some col.). - (CAA documents ; no.225). - Bibliography: p30. - List of films: p31
0-86039-137-x BS.73/266
0060755

CIVIL AVIATION AUTHORITY
Britain's Civil Aviation Authority. - Cheltenham : CAA, 1984
31p,[4]p of plates : ill(some col.). - (CAA document ; no.225)
0-86039-203-1 BS.73/302
0060755

AIR SERVICES. Great Britain
Personnel. Training. Curriculum subjects: Industrial health & industrial safety

AIR TRANSPORT AND TRAVEL INDUSTRY TRAINING BOARD. Working Party on Health and Safety at Work
Training for health and safety. - Staines : The Board, 1977
[1],20,[1]p. - Bibliography: p.16-18
B78-09072 BS.23/826
1351982

AIR SERVICES. Great Britain
Redundant personnel. Redeployment

AIR TRANSPORT AND TRAVEL INDUSTRY TRAINING BOARD
Redeployment and retraining. - Staines : A.T.T.I.T.B., 1976
[2],18p
B76-27722 BS.23/835
0713376

AIR SERVICES. Great Britain
Redundant personnel. Retraining

AIR TRANSPORT AND TRAVEL INDUSTRY TRAINING BOARD
Redeployment and retraining. - Staines : A.T.T.I.T.B., 1976
[2],18p
B76-27722 BS.23/835
0713368

AIR SERVICES. Hong Kong
to date

DATABASE airlines 1983. - 1983-. - Hong Kong : DataBase, 1983-
v.
r6814289 X.0629/652
9073736

AIR SERVICES. Northern Norway
to date

AKSELSEN, Egil
Flyruteopplegg i Nord-Norge. - Bodø : Studieselskapet for nord-norsk næringsliv, 1979
194p
r6610993 X.629/21161
8995443

AIR SERVICES. Poland
to date

MIKULSKI, Mieczysław
Polski transport lotniczy. - Warszawa : Wydawnictwa Komunikacji i Kączności, 1980
491p : ill. - Includes bibliography
Pp 122, 124-125, 128-129 and 131 are blank and replaced with xerox copies
r6483701 L.49/2143
9047417

AIR SERVICES. Sovet ėkonomicheskoi vzaimopomoshchi countries
International air services, 1945-

KNEIFEL, J. L. (John L.)
Fluggesellschaften und Luftverkehrssysteme der sozialistischen Staaten UdSSR, Polen, CSSR, Ungarn, Bulgarien, Rumänien, Kuba, Jugoslawien und der VR China. - Nördlingen : Steinmeier, [1980]
318p. - Bibliography: p.279-318
3-921334-06-3 X.622/10897 Woolwich
8726426

AIR SERVICES. Sovet ėkonomicheskoi vzaimopomoshchi countries
International air services. Law, 1945-

KNEIFEL, J. L. (John L.)
Staatliches und internationales Luftverkehrsrecht der sozialistischen Staaten UdSSR, DDR, Polen, CSSR, Ungarn, Bulgarien, Rumänien, Kuba, Jugoslawien und der VR China. - Nördlingen : Steinmeier, [1980]
220p. - Bibliography: p.181-220
3-921334-05-5 X.622/10898 Woolwich
8726434

AIR SERVICES. Sweden

FREDÉN, Erland
Ett rundare Sverige. - [Luleå] : E. Fredén, 1976
72p : ill
91-7260-072-1 X.709/22422
0392057

AIR SERVICES. United States
Air passenger transport services. Deregulation. Effects

MEYER, John R. (John Robert)
Deregulation and the new airline entrepreneurs. - Cambridge, Mass. ; London : MIT, c1984
xvi,240p : ill. - (MIT Press series on the regulation of economic activity ; 9). - Includes index
0-262-13198-6 X.622/24118
330311x

AIR SERVICES. United States
Economic aspects

O'CONNOR, William Edmund
An introduction to airline economics. - New York ; London : Praeger, 1978
xiii,254,[1]p. - Bibliography: p.171-176. - Includes index
0-03-022416-0 X.520/13270
1767453

AIR SERVICES. United States
Regulation by government, to 1980

AIRLINE deregulation / John R. Meyer and Clinton V. Oster, Jr, editors ; [contributions by] John R. Meyer ... [et al.]. - Boston, Mass. : Auburn House, c1981
xx,287p : ill. - Includes index
0-86569-078-2 X.622/12149 Woolwich
2634554

AIR SPACE. Neutral countries
International law, to date

STECK, Eduard
Die Neutralität im Luftraum. - Nendeln : Kraus, 1977
99p.
r6303591 X.200/37152
9004378

AIR SPORTS
See also
Ballooning
Parachuting. Recreations
Racing. Aeroplanes

AIR SPORTS

CAREY, Keith
Flying for fun. - Cambridge : Stephens, 1984
151p : ill
0-85059-705-6 X.622/20957
0346268

WELCH, Ann
The book of airsports. - London : Batsford, 1978
135p,[4]p of plates : ill(some col). - Includes index
0-7134-1148-1 X.620/18547 Woolwich
0346268

AIR-SUPPORTED STRUCTURES

FIŘT, Vladimír
Statics, formfinding and dynamics of air-supported membrane structures. - The Hague ; London : Nijhoff, 1983
xvii,449p : ill. - (Mechanics of surface structures ; v.5). - Bibliography: p441-443. - Includes index
90-247-2672-7 SRL
2770482

AIR-SUPPORTED STRUCTURES
- Conference proceedings

SYMPOSIUM on air-supported structures. - London : Institution of Structural Engineers, 1980
207p : ill
B82-34143 SRL
2770482

AIR-SUPPORTED STRUCTURES. Sports buildings. Great Britain

SPORTS COUNCIL. Technical Unit for Sport
Air supported structures as sports buildings. - [London] : Sports Council, [1975]
[20]p : ill. - (TUS design note ; 6)
0-900979-20-8 BS.10/550
0668044

AIR TOURS. Great Britain
Air package tours. Marketing

YACOUMIS, John
Air inclusive tour marketing. - London : Economist Intelligence Unit, 1975
[3],71p : ill. - (ITQ special ; no.2)
0-900351-34-9 X.512/5303
0701777

AIR TOURS. West Germany
Air package tours. Marketing

YACOUMIS, John
Air inclusive tour marketing. - London : Economist Intelligence Unit, 1975
[3],71p : ill. - (ITQ special ; no.2)
0-900351-34-9 X.512/5303
0701785

AIR TRAFFIC
Control

FIELD, Arnold
International air traffic control. - Oxford : Pergamon, 1985
xviii,231p : ill(some col.). - Includes index
0-08-031312-4 X.622/23679
3053326

AIR TRAFFIC
Control. Psychological aspects

MATTSSON, Jan-Olof
Kognitiva funktioner vid flygtrafikledning. - Lund : Lund University, Department of Psychology, 1979
110p. - (Universitetet i Lund. Psykologiska institutionen. Dissertation series ; 1030)
r6029364 X.512/3250
2042029

AIR TRAFFIC
Control. Safety aspects

BROOKER, P.
Target levels of safety for controlled airspace. - London : Civil Aviation Authority, 1977
v,27,[9]p : 1 ill. - (Research papers / Civil Aviation Authority. Directorate of Operational Research and Analysis ; no.7509) (Papers / Civil Aviation Authority ; 77002)
0-86039-032-2 BS.73/170
105922x

AIR TRAFFIC
Control. Use of artificial satellites

[ARRANGEMENT between certain member states of the European Space Research Organisation and the European Space Research Organisation concerning the execution of an aeronautical satellite programme. Protocols, amendments, etc., November 20, 1975]. Protocol introducing amendments to the arrangement between certain member states of the European Space Research Organisation and the European Space Research Organisation concerning the execution of an aeronautical satellite programme, Neuilly-sur-Seine, 20 November 1975 ... - London : H.M.S.O., [1978]
16p. - (Miscellaneous ; no.1 (1978)) (Cmnd.7069)
0-10-170690-1 BS.14/3118
1408496

AIR TRAFFIC. Airports. Great Britain
Forecasting

GREEN, J. H. T.
United Kingdom air traffic forecasting. - London : Department of Trade, 1978
[5],24p
B78-10845 BS.41/1112
1376896

AIR TRAFFIC. Great Britain

REPORT of the Air Traffic Forecasting Working Party 1981. - London : Department of Trade, 1981
26p : 1ill
B83-14590 BS.41/1350
2867184

AIR TRAFFIC. Great Britain
Control

AIR traffic management in the United Kingdom. - London : Civil Aviation Authority, 1982
36p : ill(some col.)
0-86039-168-x BS.73/271
2395061

CIVIL AVIATION AUTHORITY
Air traffic control training manual. - 2nd ed. - London : C.A.A.
Section 2: Meteorology. - 1979
vii,86p(2 fold) : ill. - (CAP ; 390)
0-903083-84-1 BS.41/435
0123722
Section 3: Air navigation. - 1977
vi,118p : ill. - (CAP ; 390). - Includes index
0-86039-042-x BS.41/435
0123722
Section 4: Radar theory. - 1977
vii,90p : ill. - (CAP ; 390). - Includes index
0-86039-006-3 BS.41/435
0123722
Section 5: Approach radar control. - 1977
vi,42p : ill. - (CAP 390). - Includes index
0-86039-034-9 BS.41/435
0123722
- [Another copy.] X.629/13080 Woolwich

DUKE, G. R.
Air traffic control. - London : Ian Allan, 1984
80p : ill
0-7110-1379-9 X.629/24425
2395061

FIELD, Arnold
The control of air traffic. - [Eton] : [Eton].
Vol.1. - c1980
95p : ill(some col.). - Includes index
0-907159-01-x X.622/10101 Woolwich
2395061

AIR TRAFFIC. Heathrow Airport
Control

CIVIL AVIATION AUTHORITY
Air traffic control at Heathrow. - Cheltenham : C.A.A., 1978
24,[4]p : col ill. - (CAP ; 417)
0-86039-074-8 BS.73/225
1695657

AIR TRAFFIC. United States
Congestion - For private pilots

FOSTER, Timothy R. V.
Flying in congested airspace. - Blue Ridge Summit, Pa. : Tab, c1983
viii,184p : ill. - Includes index
0-8306-2358-2 X.629/23448
313766x

AIR TRAFFIC. Western Europe
International air traffic - Forecasts

BRITISH AIRPORTS AUTHORITY
Air traffic forecasts for Western Europe. - [London] : [B.A.A.], 1976
[1],108p : ill. - Bibliography: p.99-104
B77-17428 BS.401/6
1120018

BRITISH AIRPORTS AUTHORITY
Air traffic forecasts for Western Europe. - [London] : [B.A.A.].
Summary report. - 1976
[1],20p
B77-17429 BS.401/6(2)
1120018

AIR TRAFFIC CONTROL SYSTEMS. Great Britain
Very high frequency radio frequency wavebands

DAVIES, Ken
UK VHF airband guide. - Hounslow : Airline Publications and Sales Ltd, 1979
[2],28p : 1 ill
0-905117-68-9 X.629/13245 Woolwich
2108658

DAVIES, Ken, 19—
UK VHF airband guide. - 2nd ed. - Hounslow : Airline Publication and Sales, c1981
31p
0-905117-82-4 SRL
2108658

AIR TRAFFIC CONTROLLERS. Great Britain
Licensing. Examinations

CIVIL AVIATION AUTHORITY
The air traffic controller's licence and student air traffic controller's licence. - 9th ed. - London : C.A.A., 1977
viii,71p. - (CAP 160)
0-86039-026-8 BS.73/188
0057614

AIR TRAFFIC CONTROLLERS. United States
to date

BIGGS, Don
Pressure cooker. - New York : W.W. Norton, 1979
242p.
0-393-08815-4 X.809/48591
8716528

AIR TRANSPORT

ALTHEA
Flying in an aeroplane. - Over : Dinosaur Publications, 1977
[26]p : col ill. - (Althea's dinosaur books)
0-85122-140-8 X.619/18434 Woolwich
0736627

Flying in an aeroplane. - London : Evans Bros, 1978
[25]p : chiefly col ill
0-237-44892-0 X.992/3493 Woolwich
0736627

Flying in an aeroplane. - Cambridge : Dinosaur, c1981
[24]p : col.ill
0-85122-313-3 X.990/18952
0736627

BURRETT, Tony
Airports in action. - London : Ward Lock, 1976
2-47p : ill(some col). - Includes index
0-7063-5161-4 X.622/5632 Woolwich
0736627

KERROD, Robin
Air travel. - London : Macdonald Children's, 1985
[32]p : ill(some col.). - (My first library)
0-356-11174-1 X.622/23765
0736627

AIR TRANSPORT
- Conference proceedings

AEROSPACE enters a new era / sponsors: Financial Times, the Royal Aeronautical Society. - London : Financial Times Conference Organisation, [1982]
210p : ill. - (Financial times conferences)
B83-13215 SRL
2861410

WORLD AIRPORTS CONFERENCE (7th : 1983 : London)
Economics of air transportation. - London : Telford, 1984
129p : ill
0-7277-0191-6 X.525/10314
2861410

AIR TRANSPORT
to 1978

JACKSON, Aubrey Joseph
Air travel. - London : Macdonald Educational, 1979
2-46p : ill(chiefly col). - (Macdonald new reference library). - Bibliography: p.44. - Includes index
0-356-05814-x X.520/13771
1876600

WALTERS, Brian
The illustrated history of air travel. - London : Marshall Cavendish, 1979
185p : ill(some col.). - Includes index
0-85685-708-4 X.805/1521
2487748

AIR TRANSPORT
1909-1939

AIR transport before the Second World War / edited by John W.R. Taylor and Kenneth Munson. - London : New English Library, 1975
71p : ill(some col). - (History of aviation). - Includes index
0-450-02181-5 Cup.1285/53
0555355

AIR TRANSPORT
Air & space transport. Law. Documents on law of air & space transport in stock of libraries in Cambridge, London & Oxford - Catalogues

UNIVERSITY OF LONDON. Institute of Advanced Legal Studies
Union list of air and space law literature in the libraries of Oxford, Cambridge and London. - 2nd ed.. - London : The Institute ; [London] : [Distributed by Dillons], 1975
xii,[301]p. - (Union catalogue , no.4). - Includes index
0-901190-16-0 X.100/16046
0362239

AIR TRANSPORT
Air freight transport

[INTERNATIONAL news (Feltham)]. International news. - Vol.1, no.1 ([198-])-. - Feltham : Jardine Air Cargo (UK), [198-]-
v : ill
Holdings: Vol.2, no.4 (Aug.1984)-
r6863952 P.843/308
0189529

PEAK, D. W. (Douglas W)
Developments in the air cargo industry. - London : International Cargo Handling Coordination Association, c1981
iii,97p : ill. - Bibliography: p74-75
0-906297-17-6 SRL
0189529

AIR TRANSPORT
Air freight transport - Bibliographies

LAW, Peter R.
Air Cargo. - 2nd ed. - London : International Cargo Handling Co-ordination Association, 1980, c1979
19leaves. - (Information service library bibliography : no.2) (An ICHCA information access publication)
0-906297-18-4 2719.k.31
2787288

AIR TRANSPORT
Air freight transport. Penetration of Russian blockade of West Berlin, 1948-1949: Berlin Airlift

COLLIER, Richard
Bridge across the sky. - London : Macmillan, 1978
[13],239p,[8]p of plates : ill. - Bibliography: p.211-225. - Includes index
0-333-17802-5 X.800/27870
1904825

AIR TRANSPORT
By Great Britain. Royal Air Force, to 1958

GRANT, Roderick
But not in anger. - London : I. Allan, 1979
176p : ill. - Includes index
0-7110-0909-0 X.800/28227
1997971

AIR TRANSPORT
Civil aircraft - Serials

WORLD airline fleets monthly. - Edition no.1- ; [May 1977]-. - Hounslow : Airline Publications and Sales Ltd, [1977]- : ill(some col)
B78-19100 P.611/773
1462040

AIR TRANSPORT
Regional air transport - Conference proceedings

SYMPOSIUM ON THIRD LEVEL AND REGIONAL AIR TRANSPORT (1977 : London)
Third level and regional air transport, Wednesday 23 February 1977. - London : The Society, [1977]
105p in various pagings : ill
B78-00208 X.515/552
1272330

AIR TRANSPORT
Third level air transport - Conference proceedings

SYMPOSIUM ON THIRD LEVEL AND REGIONAL AIR TRANSPORT (1977 : London)
Third level and regional air transport, Wednesday 23 February 1977. - London : The Society, [1977]
105p in various pagings : ill
B78-00208 X.515/552
1272330

AIR TRANSPORT. Africa
- Serials

AFRICAN air transport. - Vol.1, no.1 (Jan./Feb.1980)-. - Winchester : Kingswood Publications, 1980-
v. : ill
B81-33284 SRL
2544539

AIR TRANSPORT. Dangerous goods. Great Britain
Air freight transport - Serials

DANGEROUS goods bulletin. - [No.]1- ; 1977-. - London : F.O.I., Dangerous Goods Section, 1977- : ill
B78-36116 BS.73/221
169698x

AIR TRANSPORT. Europe
Air freight transport - Serials

HEREFORD'S air cargo. Europe. - [No.1]- ; [1977]-. - London : Hereford Press Ltd, 1977- : ill
B79-00251 SRL
1741403

AIR TRANSPORT. Great Britain
Air freight transport

CIVIL AVIATION AUTHORITY
Air freight policy. - London : The Authority, 1977
[2],16p. - (CAP 405)
0-86039-045-4 BS.73/200
1246194

AIR TRANSPORT. Great Britain
Air freight transport. Demand
CIVIL AVIATION AUTHORITY
Air freight demand. - London : The Authority, 1977
[2],16,[21]p : ill. - (CAP 401)
0-86039-041-1 BS.73/186
1048570

AIR TRANSPORT. Great Britain
Energy. Conservation
ADVISORY COUNCIL ON ENERGY CONSERVATION
Civil aviation. - London : H.M.S.O., 1979
v,26p. - (Papers / Advisory Council on Energy Conservation ; 9) (Energy paper ; no.36)
0-11-410762-9 SRL
1967843

AIR TRANSPORT. Great Britain
Insurance. Abbreviations
SPURWAY, M. J.
Aviation insurance abbreviations, organisations and institutions. - London : Witherby, 1983
63p. - (Monument series)
0-900886-77-3 X.629/23920
3169278

AIR TRANSPORT. Middle East
Air freight transport - Serials
HEREFORD'S air cargo Middle East. - Vol.1, no.1- ; Jan. 1976-. - London : Hereford Press Ltd, 1976- : ill(some col)
B77-01335 SRL
0778605

AIR TRANSPORT. New York (City)
Air freight transport, to date
STEVENSON, Arthur J.
The New York-Newark air freight system. - Chicago : University of Chicago, Dept. of Geography, 1982
xvi,440p. - (Research paper / the University of Chicago, Department of Geography ; no.99-200). - Bibliography
0-89065-106-x Ac.2691.dxb(199-200)
8951373

AIR TRANSPORT. Newark. New Jersey
Air freight transport, to date
STEVENSON, Arthur J.
The New York-Newark air freight system. - Chicago : University of Chicago, Dept. of Geography, 1982
xvi,440p. - (Research paper / the University of Chicago, Department of Geography ; no.99-200). - Bibliography
0-89065-106-x Ac.2691.dxb(199-200)
8951381

AIR TRANSPORT. North America
Air freight transport - Serials
HEREFORD'S. North America. - Vol.1, no.1- ; June 1979-. - London : Hereford Press Limited, 1979- : ill
B80-02016 SRL
2005638

AIR TRANSPORT. United States
to 1981
BIEDERMAN, Paul
The U.S. airline industry. - New York : Praeger, 1982
xxiii,198p. - (Praeger studies in select basic industries). - Bibliography: p181-193
0-03-060324-2 X.520/27695
2742446

AIR TRANSPORT. Western Europe
- Conference proceedings
SYMPOSIUM ON AIR TRANSPORT IN EUROPE (1974 : London)
Symposium on Air Transport in Europe, 16th January 1974, proceedings. - London : The Society, [1975]
84p in various pagings : ill
0-903409-03-8 X.515/504
0573051

AIR TRANSPORT AND TRAVEL INDUSTRY TRAINING BOARD
AIR TRANSPORT AND TRAVEL INDUSTRY TRAINING BOARD
A guide to the Air Transport and Travel Industry Training Board. - Staines : The Board, [1976]
16p : ill
B76-20403 X.519/25859
0664642

AIR TRANSPORT AUXILIARY
1939-1945
CURTIS, Lettice
The forgotten pilots. - 2nd ed. - Twyford : E.L. Curtis, 1982
xii,337p,[16]p of plates : ill. - Includes index
0-9508157-0-5 X.809/56837
2983338

AIR TRANSPORT AUXILIARY
Women pilots, 1940-1945 - Biographies
DU CROS, Rosemary
Ata girl. - London : Muller, 1983
105p : ill
0-946569-00-2 X.800/36963
0296716

AIR TRAVEL
BARRETT, Frank
A consumer's guide to air travel. - London : Daily Telegraph in association with Woodhead-Faulkner, 1983
128p. - (A Family money-go-round special). - Includes index
0-901684-98-8 X.529/60829
1761250

BEATY, David
The complete skytraveller. - London : Magnum Books, 1979
236p : ill
0-417-03640-x X.619/19903 Woolwich
1761250

DUCKWORTH, Tessa
Flying without fear. - London : Sheldon, 1983
103p. - (Overcoming common problems). - Includes index
0-85969-400-3 X.529/62731
1761250
- [Another copy : paperback.] ISBN 0-85969-401-1 X.529/62732

HICKMAN'S world air travel guide / edited by R.H. and M.E. Hickman. - 1979. - London : Elm Tree Books, 1979
560p : ill
Subsequent volumes, when received, will be placed at this shelfmark
0-241-10171-9 P.701/1644
1761250

KOWET, Don
The jet lag book. - London : Futura, 1984, c1983
xv,138p : ill. - Includes index
0-7088-2648-2 X.529/67233
1761250

PARKER, Cliff
How to avoid flying. - Sevenoaks : New English Library, 1982
144p : ill
0-450-05412-8 X.529/48466
2756366

AIR TRAVEL
- Directories - Serials
HICKMANS' international air traveller. - [1982]-. - London : Mitchell Beazley, 1982-
v. : ill
B82-15768 X.629/17374 Woolwich
2527723

AIR TRAVEL
- Serials
COMMUTER world. - Vol.1, no.1 (Jan.-Apr. 1984)-. - Beaconsfield : Shephard Press, 1984-
v. : ill(some col.)
B85-21434 P.843/349
3280519

AIR TRAVEL
to 1978. Social aspects
HUDSON, Kenneth
Diamonds in the sky. - London : Bodley Head ; London : British Broadcasting Corporation, 1979
240p,[16]p of plates : ill(some col). - Bibliography: p.236-237. - Includes index
0-370-30162-5 X.800/28255
1894137

AIR TRAVEL
to 1981
A BOOK of air journeys / compiled by Ludovic Kennedy. - London : Collins, 1982
xvii,350p : ill
0-00-216377-2 X.950/18238
2839024

A BOOK of air journeys / compiled by Ludovic Kennedy. - [London] : Fontana, 1983, c1982
xvii, 350p : ill
0-00-636616-3 X.958/21932
2839024

AIR TRAVEL
Advertising, to date
MAIER, Lotte
Aufbruch zum Himmel. - Dortmund : Harenberg, c1982
99p : ill. - (Die bibliophilen Taschenbücher ; 328)
3-88379-328-0 X.429/15404
9047409

AIR TRAVEL
By British businessmen - Serials
WORLD of business travel. - June 1982-. - London : Business Magazines International on behalf of the External Affairs Division of British Caledonian Airways, 1982-
v. : ill
B83-04798 P.613/304
281983x

AIR TRAVEL
By businessmen to Great Britain. Econometric models
HANLON, J. P. (James Patrick)
Econometric estimates of the demand for air travel. - Birmingham : Department of Transportation and Environmental Planning, University of Birmingham, 1976
[2],52[i.e. 55]p
0-7044-0215-7 X.512/5233
0676268

AIR TRAVEL
International air travel
MOYNAHAN, Brian
Airport international. - London : Pan Books, 1978
336p : ill. - Includes index
0-330-25417-0 X.619/18673 Woolwich
1492209
Airport international. - London : Macmillan, 1978
336p : ill. - Includes index
0-333-24550-4 X.529/32423
1492209

AIR TRAVEL
Medical aspects
O'QUIGLEY, Seán
Health and travel. - Dublin : The O'Brien Press [for] Aer Lingus, 1979
xi,110p : ill. - Includes index
0-905140-81-8 X.529/37944
217779x

AIR TRAVEL
Science - For air travel
WOOD, Elizabeth Armstrong
Science from your airplane window. - 2nd revised ed. - New York : Dover Publications [etc.] ; London : Constable, 1975
xxiii,227,[2]p : ill(some col). - Bibliography: p.209-211. - Includes index
0-486-23205-0 X.619/16741
0650587

AIR TRAVEL. Special subjects. English humorous cartoons
- Collections
PUNCH in the air / edited by David Langdon. - London : Robson, 1983
[158]p : chiefly ill
0-86051-251-7 X.950/31940
3121526

AIR TRAVEL. West Germany
Companies: Lufthansa - Serials
LUFTHANSA Jahrbuch. - '84-. - Köln : Deutsche Lufthansa Ag, c1984-
r6310136 X.0629/674
9137793

AIR TRAVEL RESERVE FUND AGENCY
See Great Britain. Air Travel Reserve Fund Agency

AIR WAR GAMES
Naval air war games
SEA command. - Brooklyn : Enola Games ; Barnet : Navwar [Productions Ltd], [1980]
Leaf,ii,62p : ill
B80-41127 X.622/8891 Woolwich
2314665

AIR WAR GAMES BASED ON WORLD WAR 2
SPICK, Mike
Air battles in miniature. - Cambridge : Stephens, 1978
160p : ill. - Bibliography: p.160
0-85059-296-8 X.620/17735 Woolwich
1250051

AIR WARFARE
- Forecasts
BROWN, N. G.
Silver wings in the twilight. - [Birmingham] : University of Birmingham, 1983
26p
0-7044-0667-5 X.950/27399
3033015

AIR WARFARE
to 1970
WAR machines, air. - London : Octopus Books, 1975
2-143p : ill(chiefly col). - Includes index
0-7064-0417-3 X.622/5080
0516120

AIR WARFARE
to 1975

BROWN, David, 1938– (Sept.)
The Guinness history of air warfare. – London : Guinness Superlatives, 1976
viii,247p : ill(some col). – Bibliography: p.236–238. – Includes index
0–900424–61–3 X.620/16367 Woolwich
0748188

AIR WARFARE
to 1978

AERIAL warfare / edited by Anthony Robinson. – London : Orbis, 1982
384p : ill(some col.). – Includes index
0–85613–261–6 X.622/13259
1876937

CHANT, Christopher
The pictorial history of air warfare. – [London] : Octopus Books, [1979]
192p : ill(chiefly col). – Includes index
0–7064–0977–9 L.45/259
1876937

PIMLOTT, John, 1948–
Strategy & tactics of air warfare. – London : Marshall Cavendish, 1979
[4],80p : ill(some col)
0–85685–506–5 X.802/10856
1876937

AIR WARFARE
to date

GROEHLER, Olaf
Geschichte des Luftkriegs 1910 bis 1980. – 3. Aufl. – Berlin : Militärverlag der DDR, c1981
737p : ill
r6308763 X.800/37542
8619778

AIR WARFARE
Air antisubmarine warfare, to 1980

PRICE, Alfred
Aircraft versus submarine. – London : Jane's Publishing, 1980
xvi,272p,[28] of plates : ill. – Includes index
0–7106–0008–9 X.800/29184
2284103

AIR WARFARE
Fear by British government, 1932–1939. Influence on military policies

BIALER, Uri
The shadow of the bomber. – London : Royal Historical Society, 1980
[6],165p. – Includes index
0–901050–78–4 X.809/45894
2256983

AIRBASES. Greenwood. Nova Scotia
Great Britain. Royal Air Force. Operational Training Unit, No.36, 1942–1944 – Biographies

DARGUE, R. H. (Ralph H)
Just two years –. – [New Milton] : R.H. Dargue, 1982
35p : ill
B83–38583 X.629/21165
2954362

AIRBORNE ARTILLERY
Airborne, infantry & mountain artillery, 1939–1945

CHAMBERLAIN, Peter
Infantry, mountain and airborne guns. – London : Macdonald and Jane's, 1975
[4],64p. – (World War 2 fact files)
0–356–08225–3 X.622/5069
0504432
– [Another copy.] X.611/7053

AIRBORNE DISEASES
– Conference proceedings

The AERIAL transmission of disease / organised and edited by J.B. Brooksby. – London : The Royal Society, 1983
viii,604p : ill
0–85403–214–2 X.322/13036
3062988

AIRBORNE DISEASES. Crops
Movement. Influence of meteorological conditions

PEDGLEY, David E.
Windborne pests and diseases. – Chichester : Horwood, 1982
250p : ill. – Bibliography: p211–240. – Includes index
0–85312–312–8 SRL
2798190

AIRBORNE DUSTS. Mines
Control measures: Ventilation

RADCHENKO, G. A.
Ventilation for dust elimination in underground mines. – New Delhi : Amerind, c1976
xii,298p : ill
0–86249–369–2 X.622/17651
3030512

AIRBORNE DUSTS. Pollutants of environment. Great Britain

HAMILTON, Ernest Marcel
The identification of atmospheric dust by use of the microscope. – London : Central Electricity Generating Board, 1976
32,[1]p : ill(some col). – (Monographs / Central Electricity Generating Board. Research and Development Department)
B77–28651 SRL
120761x

AIRBORNE DUSTS. Quarries. Great Britain
Control measures

BRITISH QUARRYING AND SLAG FEDERATION. Plant Associates Section
Dust. – London : The Federation, 1977
49p : ill
B78–13635 SRL
1406787

GREAT BRITAIN. Health and Safety Executive
Airborne dust in quarries. – London : H.M.S.O., 1976
[2],21p : ill
0–11–880500–2 BS.385/288
1029312

AIRBORNE FORCES
See also
Allied Airborne Army. 1st Army
Great Britain. Army. Airborne Division, 1st
Great Britain. Army. Airborne Division, 6th
Parachute units
War games based on military operations during World War 2 of airborne forces

AIRBORNE FORCES
Allied airborne forces. Combined operations during D-Day, 1944 of Normandy campaign of World War 2

CROOKENDEN, Sir Napier
Dropzone Normandy. – London : Allan, 1976
304p : ill. – Bibliography: p.287–288. – Includes index
0–7110–0660–1 X.809/40765
061310x

AIRBORNE FORCES
American airborne forces. Army operations during D-Day, 1944 of Normandy campaign of World War 2

MARSHALL, S. L. (Samuel Lyman Atwood)
Night drop. – Nashville : Battery Press, c1962
425p,[8]p of plates : ill
0–89839–062–1 X.809/57469
3031217

AIRBORNE FORCES
Military operations, 1918–1945

GREGORY, Barry
Airborne warfare 1918–1945. – London : Phoebus, 1979
128p : ill(some col.). – Includes index
0–7026–0053–9 X.622/16013
2931338

AIRBORNE FORCES
Military operations, 1939–1977

AIRBORNE operations / [editor Philip de Ste Croix]. – London : Salamander Books, 1978
224p : ill(some col). – Includes index
0–86101–014–0 X.805/1058
1575864

AIRBORNE FORCES
Military operations, 1940–1977

AIRBORNE operations. – London : Salamander, 1978
221p : ill(some col.). – Includes index
0–86101–122–8 X.805/2557
2726483

AIRBORNE FORCES
Military operations during World War 2

CROOKENDEN, Sir Napier
Airborne at war. – London : I. Allan, 1978
144p : ill
0–7110–0771–3 X.802/10647
1548700

GAVIN, James M.
Airborne warfare. – Reprinted. – Nashville, Tenn. : Battery Press, 1980
xii,186p : ill. – (Airborne series ; 10)
0–89839–029–x X.800/34327
1548700

AIRBORNE FORCES. Germany. Heer
Military operations during World War 2

KUHN, Volkmar
German paratroops in World War II. – London : I. Allan, 1978
240p : ill. – Includes index
0–7110–0759–4 X.809/44008
1783270

AIRBORNE FORCES. Great Britain. Army
to 1982

NORTON, G. G. (Geoffrey Gordon)
The Red Devils. – New ed. – London : Leo Cooper in association with Secker & Warburg, 1984
310p : ill. – ([Famous regiments] / [Secker & Warburg])
0–436–31525–4 X.809/63821
3273695

AIRBORNE FORCES. Great Britain. Army
Generals. Military operations during World War 2

DOVER, Victor, 1919–
The sky generals. – London : Cassell, 1981
xiv,215p : ill. – Bibliography: p205. – Includes index
0–304–30480–8 X.809/51785
2729873

AIRBORNE FORCES. Great Britain. Army. Great Britain. Royal Air Force
Use of military aircraft, 1942–1946

GREAT BRITAIN. Air Ministry
RAF airborne forces manual. – London : Arms and Armour Press [etc.], 1979
288p : ill. – (RAF Museum series / Royal Air Force Museum ; vol.8)
0–85368–163–5 X.0622/507(8) Woolwich
1943871

AIRBORNE INTEGRATED DATA SYSTEMS
See Flight recorders

AIRBORNE MILITARY EQUIPMENT
to 1975

WEEKS, John
Airborne equipment. – Newton Abbot : David and Charles, 1976
192p : ill. – Bibliography: p.186–187. – Includes index
0–7153–7117–7 X.629/10574 Woolwich
0551023

AIRBORNE PESTS. Crops
Movement. Influence of meteorological conditions

PEDGLEY, David E.
Windborne pests and diseases. – Chichester : Horwood, 1982
250p : ill. – Bibliography: p211–240. – Includes index
0–85312–312–8 SRL
2798190

AIRBORNE RECREATIONS
See also
Gliding. Recreations

AIRBORNE RECREATIONS
– Serials

AMATEUR airsport. – Vol.1, no.1– ; Nov. 1978–. – Huntingdon : Welland Publications ; London : Distributed by S.M. Distribution, 1978– : ill(some col)
B79–14945 P.443/362
1836366

AIRBORNE SULPHUR COMPOUNDS. Pollutants of forests. Great Britain

DISCUSSION GROUP OF THE DEPARTMENT OF THE ENVIRONMENT AND THE NATURAL ENVIRONMENT RESEARCH COUNCIL
Effects of airborne sulphur compounds on forests and freshwaters. – London : H.M.S.O., 1976
vii,27p. – (Pollution paper ; no.7)
0–11–751066–1 BS.414/107
0683701

AIRBORNE SULPHUR COMPOUNDS. Pollutants of freshwater regions. Great Britain

DISCUSSION GROUP OF THE DEPARTMENT OF THE ENVIRONMENT AND THE NATURAL ENVIRONMENT RESEARCH COUNCIL
Effects of airborne sulphur compounds on forests and freshwaters. – London : H.M.S.O., 1976
vii,27p. – (Pollution paper ; no.7)
0–11–751066–1 BS.414/107
0683698

AIRBORNE WARFARE
to 1977

WEEKS, John
Assault from the sky. - Newton Abbot : Westbridge Books, 1978
192p : ill(some col). - Bibliography: p.186. - Includes index
0-7153-7564-4 X.622/6595 Woolwich
1408836

AIRBORNE WARFARE
to 1978

HICKEY, Michael
Out of the sky. - London : Mills and Boon, 1979
286p,[16]p of plates : ill. - Bibliography: p.274-277. - Includes index
0-263-06388-7 X.800/28250
200271x

AIRBORNE WARNING SYSTEMS. Military aircraft
Airborne early warning systems

HIRST, Mike
Airborne early warning. - London : Osprey, 1983
192p : ill. - Includes index
0-85045-532-4 X.622/20695
316473x

AIRBRUSHING

CAIATI, Carl
Airbrushing. - Blue Ridge Summit, Pa. : TAB, c1983
x,165p : ill(some col.). - Bibliography: p159-161. - Includes index
0-8306-0155-4 X.421/25499
2058626

CURTIS, Seng-gye Tombs
The airbrush book. - London : Orbis Publishing, 1980
160p : ill(some col.). - Bibliography: p159. - Includes index
0-85613-275-6 X.425/1836 Woolwich
2058626

MARTEN, Clement
The artist's airbrush manual. - Newton Abbot : David and Charles, 1980
80p : ill
0-7153-7997-6 X.421/21257 Woolwich
2058626

MARTIN, Judy
The complete guide to airbrushing. - London : Thames and Hudson, 1983
160p : ill(some col.). - Includes index
0-500-23372-1 X.421/24345
2058626

MISSTEAR, Cecil
The advanced airbrush book. - London : Orbis, 1984
160p : ill(some col.). - Bibliography: p157. - Includes index
0-85613-509-7 X.425/4966
2058626

PASCHAL, Robert
Airbrushing for fine & commercial artists. - New York ; London : Van Nostrand Reinhold, 1982
127p : ill (some col.). - Includes index
0-442-27508-0 X.425/3646
2058626

VERO, Radu
Airbrush. - London : Columbus, 1983
190p : ill(some col.). - Includes index
0-86287-031-3 X.425/4259
2058626

AIRBRUSHING
to date

RICHARD H. CHILDERS PRODUCTIONS
Air powered. - New York : Random House, [1979]
222p : ill mainly col
0-394-50755-x L.45/1392
8795991

AIRBRUSHING. Plastics models

GOLDMAN, Richard M.
Airbrushing for modellers. - 2nd ed. - London : Almark Publishing, 1977
48p : ill(some col)
B77-18916 X.319/17370 Woolwich
032390x

AIRCRAFT
See also
Aeronautical engineering
Aeronautics
Aeroplanes
Anti-aircraft artillery
Anti-aircraft cruisers
Biplanes
Civil aircraft
Commercial aircraft
Crashed aircraft
Crop spraying aircraft
Experimental aircraft
Flying bombs
Flying machines
Gliders
Hangars
Heavier than air aircraft

AIRCRAFT
See also
Home-built aircraft
Lighter than air aircraft
Man-powered aircraft
Military aircraft
Model aircraft
Presentation aircraft
Preserved aircraft
Second-hand aircraft
Sopwith aircraft
Vertical-lift aircraft

AIRCRAFT

AEROPLANES. - Maidenhead : Purnell, 1979
[28]p : of col ill. - (All in colour picture books)
0-361-04546-8 X.622/6873 Woolwich
189546x

ANGELUCCI, Enzo
World aircraft. - Maidenhead : Sampson Low. - (Sampson Low guides)
[Vol.1]: Origins - World War 1. - 1977
320p : ill(chiefly col). - Bibliography: p.318-320. - Includes index
0-562-00058-5 X.629/11622 Woolwich
1286269

[Vol.2]: 1918-1935. - 1977
319p : ill(chiefly col). - Bibliography: p.318-319. - Includes index
0-562-00059-3 X.629/11706 Woolwich
1286269

[Vol.3]: World War II. Part 1 : Great Britain, Germany, Italy, France, Czechoslovakia, Holland, Poland, Yugoslavia, Romania, Belgium, Greece, Hungary, Scandinavia. - 1978
320p : ill(chiefly col). - Bibliography: p.319-320. - Includes index
0-562-00095-x X.629/12081 Woolwich
1286269
- Bibliography: p.315-317. - Includes index
[Vol.4]: World War II. Part 2 : U.S.A., Japan, U.S.S.R. Australia, New Zealand, South Africa, China, Brazil. - 1978
317p : ill(chiefly col)
0-562-00096-8 X.629/12080 Woolwich
1286269

[Vol.5]: Commercial aircraft, 1935-1960. - 1979
317p : ill(chiefly col). - Bibliography: p.316-317. - Includes index
0-562-00125-5 X.629/12729 Woolwich
1286269
- Bibliography: p.317-318. - Includes index
[Vol.6]: Combat aircraft, 1945-1960. - 1980
318p : ill
0-562-00136-0 X.629/14085 Woolwich
1286269

CAVE, Ronald G. (Ronald George)
Aircraft. - London : Watts, 1982
30p : col.ill. - (What about?) (A Franklin Watts question and answer book). - Includes index
0-85166-979-4 X.629/17741 Woolwich
0151858

CONNELL, Stephanie
Aeroplanes and balloons. - London : Macdonald Educational, 1976
[1],32p : col ill. - (Readabout). - Includes index
0-356-05491-8 X.619/17065 Woolwich
0151858

GIRARD, Pat
Flying machines. - Oxford : Blackwell Raintree, c1981
32p : ill(some col.). - Includes index
0-86256-024-1 X.622/10075 Woolwich
0151858

GREEN, William, 19—
The observer's book of aircraft. - 1981 ed. 30th ed. - London : Warne, 1981
256p : ill. - (The Observer's pocket series ; 11). - Includes index
0-7232-1618-5 WP.12066/205
0100277

GREEN, William, 1927-
The new observer's book of aircraft. - 32nd ed. - London : Warne, 1983
256p : ill. - (The New observer's series ; N1). - Includes index
0-7232-1640-1 X.629/20666
0100277

The new observer's book of aircraft. - 33rd ed. - London : Warne, 1984
256p : ill. - (The New observer's series ; N1). - Includes index
0-7232-1673-8 X.629/23781
0100277

The observer's book of aircraft. - 31st ed. - London : Warne, 1982
256p : ill. - (The Observer's series ; 11). - Includes index
0-7232-1626-6 WP.12066/214
0100277

HUWS, Emily
Awyrennau. - Llandysul : Gwasg Gomer, 1983
[20]p : col.ill. - (Nawr gellwch chi ddarllen a dysgu)
0-85088-928-6 X.622/16900
2962586

MACKNIGHT, Nigel
Aircraft. - London : Hamlyn, 1979
32p : ill(chiefly col). - (Insight books). - Includes index
0-600-30400-0 X.611/9085 Woolwich
0151858

MILSOME, John
Machines in the air. - London : Burke, 1976
32p : col ill. - (The world of machines)
0-222-00501-7 X.629/10928 Woolwich
0151858

MUNSON, Kenneth
Aircraft. - 2nd ed. - London : Macdonald Educational, 1974, c1971
48p : ill(some col.). - (Visual books). - Includes index
B83-29403 X.622/15874
0151858

The OBSERVER'S book of aircraft. - 1976 ed. : 25th ed. / compiled by William Green ; with silhouettes by Dennis Punnett. - London : F. Warne, 1976
254p : ill. - (The observer's pocket series ; [11])
0-7232-1553-7 WP.12066/145
0251003

The OBSERVER'S book of aircraft. - 1977 ed. : 26th ed. / compiled by William Green ; with silhouettes by Dennis Punnett. - London : F. Warne, 1977
254p : ill. - (The observer's pocket series ; [11])
0-7232-1562-6 WP.12066/156
0251003

The OBSERVER'S book of aircraft / compiled by William Green ; with silhouettes by Dennis Punnett. - 1978 ed. : 27th ed. - London : F. Warne, 1978
254p : ill. - (The observer's pocket series ; [11])
0-7232-1577-4 WP.12066/170
0251003

The OBSERVER'S book of aircraft / compiled by William Green ; with silhouettes by Denis Punnett. - 1979 ed. : [28th ed.]. - London : F. Warne, 1979
254p : ill. - (The observer's pocket series ; [11])
0-7232-1591-x WP.12066/189
0251003

The OBSERVER'S book of aircraft / compiled by William Green ; With silhouettes by Dennis Punnett. - 1980 ed. : 29th ed. - London : F. Warne, 1980
256p : ill. - (The observer's pocket series ; [II]). - Includes index
0-7232-1604-5 W.P.12066/202
0251003

The SUPER book of aircraft. - London : Macdonald Educational, 1977
32p : ill(some col)
0-356-05592-2 X.611/7662 Woolwich
0151858

SWEETMAN, Bill
Aircraft 2000. - London : Hamlyn, c1984
158p : ill(some col.). - Includes index
0-600-34726-5 X.622/23778
3284905

TRYATT, Nigel
Cars, motorbikes & planes. - London : Usborne, 1980
192p : ill(some col.). - (Spotter's handbook). - Includes index
0-86020-419-7 X.629/18290
2079178

WALTERS, Brian
The young air traveller's handbook. - London : Hamlyn, 1980
224p : ill(chiefly col). - Includes index
0-600-30422-1 X.629/13858 Woolwich
0151858

AIRCRAFT -continued

WOOD, Derek
Jane's world aircraft recognition handbook. - London : Macdonald and Jane's, 1979
512p : ill. - Includes index
0-354-01221-5 SRL
2079178
- [Another copy: paperback]
ISBN 0-354-01261-4 X.619/22432 Woolwich
Jane's world aircraft recognition handbook. - New ed. - London : Jane's, 1982
559p : ill. - Includes index
0-7106-0202-2 SRL
2079178
Jane's world aircraft recognition handbook. - New ed. - London : Jane's, 1982
559p : chiefly ill. - Includes index
0-7106-0245-6 X.629/20681
2079178

AIRCRAFT
to 1900

BAUER, Melchior
Die Flugzeughandschrift des Melchior Bauer von 1765. - Rudolstadt : Greifenverlag, 1982
43p. : ill. - (Aus dem Greifenarchiv ; Nr.1)
r6307961 X.622/17567
9024255

AIRCRAFT
to 1920

HEWISH, Mark
Early aircraft. - Maidenhead : Purnell, 1977
[1],16[i.e.32],[1]p : ill(chiefly col). - (Little big books). - Includes index
0-361-03813-5 X.0319/1017(4) Woolwich
1340956

AIRCRAFT
to 1939

ALLWARD, Maurice
Early aircraft. - Glasgow : Collins, 1977
3-62p : ill(chiefly col). - (Collins transport series). - Includes index
0-00-106241-7 X.622/5960 Woolwich
1171135

AIRCRAFT
to 1970

AIRCRAFT / general editor Boswell Taylor ; ... illustrator John Young. - Leicester : Knight Books, 1977
[1],32p : of ill. - (Picture teach yourself. project books)
0-340-21554-2 X.619/17351 Woolwich
1044052

AIRCRAFT
to 1973

ALLEN, Roy
The story of flight. - Glasgow : Collins, 1975
4-61p : ill(chiefly col). - ([Collins prospect books]). - Includes index
0-00-106136-4 X.629/10210
0311774

TAYLOR, John W. R. (John William Ransom)
Aircraft, aircraft. - 4th ed. - London : Hamlyn, 1975
141p : ill(some col). - Includes index
0-600-30263-6 X.622/5559 Woolwich
0235806

AIRCRAFT
to 1974

BEAL, George
The quizzer book about flight. - London : Owlet Books, 1975
48p : ill(some col). - (Quizzer books)
0-263-05911-1 X.619/16655 Woolwich
063011x

AIRCRAFT
to 1976

AIRCRAFT / [compiled] by Christopher Pick. - London : Galley Press : [Distributed by] Cathay Books, 1979
64p : col ill. - Includes index
0-86178-015-9 X.622/8313 Woolwich
2258749

HEWISH, Mark
Aircraft. - Maidenhead : Purnell, 1978
[1],49p : ill(some col). - (Learn about). - Includes index
0-361-03832-1 X.615/2099 Woolwich
1254049

ILLUSTRATED history of aircraft / edited by Brendan Gallagher. - London : Octopus Books for W.H. Smith and Son, 1977
160p : ill(some col). - Includes index
0-7064-0600-1 X.622/6196 Woolwich
1334646

KERROD, Robin
Aircraft. - London : F. Watts, 1977
48p : ill(chiefly col). - (A first look book). - Includes index
0-85166-636-1 X.629/11504 Woolwich
1254049

AIRCRAFT
to 1976. Research

ROBERTSON, Bruce
Aviation archaeology. - Cambridge : Stephens, 1977
152p : ill
0-85059-255-0 X.421/10488 Woolwich
1206834

AIRCRAFT
to 1977

ANDREWS, Allen
Back to the drawing board. - Newton Abbot : David and Charles, 1977
168p : ill. - Includes index
0-7153-7400-1 X.622/6050 Woolwich
1246925

AIRCRAFT
to 1977 - Encyclopaedias

ENCYCLOPEDIA of aircraft / edited by Michael J.H. Taylor & John W.R. Taylor. - London : Weidenfeld and Nicolson, 1978
3-254p : ill(some col). - Includes index
0-297-77526-x SRL
1692992

AIRCRAFT
to 1978 - Encyclopaedias

The ILLUSTRATED encyclopaedia of aircraft / edited by David Mondey. - London : Hamlyn, 1979, c1978
320p : ill(some col.). - Includes index
0-600-30378-0 X.622/9317 Woolwich
2329026

AIRCRAFT
to 1979

AVIATION / general editor David Mondey. - London : Octopus Books, 1980
400p : ill(some col). - Includes index
0-7064-0962-0 X.622/8628 Woolwich
2276739

GUNSTON, Bill
The family library of aircraft. - London : Octopus, 1982, c1981
80p : ill(some col.). - Includes index
0-7064-1463-2 L.45/2188
2276739

KERSHAW, Andrew
The wonder book of aircraft. - London : Ward Lock, 1980
3-46p Ill(chiefly col), 2 ports. - Includes index
0-7063-6061-3 X.622/8142 Woolwich
2077191

MACKNIGHT, Nigel
The colour encyclopedia of aircraft. - London : Octopus, 1980
224p : ill(some col.)
0-7064-1303-2 L.45/854
2341301

AIRCRAFT
to 1980

HOARE, Robert J. (Robert John)
The story of aircraft. - London : A. & C. Black, 1983
96p : ill. - (Black's junior reference books). - Bibliography: p95. - Includes index
0-7136-2249-0 X.622/16641
2567946

HOYLE, Geoffrey
Flight. - Loughborough : Ladybird, 1984
51p : ill(some col.). - (Achievements). - Includes index
0-7214-0833-8 X.629/25191
2567946

MAYNARD, Christopher
Aircraft. - London : Pan, 1981
93p : ill(some col.). - (A Piccolo factbook). - Includes index
0-330-26416-8 X.629/16417 Woolwich
2567946

AIRCRAFT
to 1982

The ILLUSTRATED encyclopedia of aircraft. - London : Orbis, c1981-[1983?]
156v. : ill(some col.)
B82-12962 P.623/1168
2631695

AIRCRAFT
to date

[A collection of books on aircraft or spacecraft, printed or published in Hong Kong, not catalogued separately]. - Paris : Librairie Gründ, [1981]- : ill
r6211567 W.228
879944x

La CONQUÊTE de l'air / [Clive Hart ... et al.] ; [adaptation française de Marie-Pierre Bihet et d'Irene Boulloche-Morize. - Paris : Gründ, [1980]
253p : ill. - (Connaitre notre monde)
2-7000-5902-6 X.622/11930 Woolwich
879944x

DICORATO, Giuseppe
Storia dell'aviazione. - Novara : Ist. geografico De Agostini, 1978
320p : ill(chiefly col.)
r6505730 X.622/14193
879944x

GUNSTON, Bill
The world's greatest airplanes. - New York : Elsevier-Dutton, [1980]
253p : ill(some col)
0-525-70724-7 L.45/2022
879944x

MONDEY, David
An illustrated history of aircraft. - Sydney : Golden Press, 1981
62p,plates
0-85558-865-9 X.622/22726
879944x
- [Another issue]. - New York: Galahad Books, 1981. - ISBN: 0883654822 X.622/22727

AIRCRAFT
to date - Translations

[The WONDER book of aircraft. French]. Le monde passionnant de l'aviation ... / [Andrew Kershaw et Marilyn Gayle]. - Paris : Éditions G.P., [1981]
45p : ill
2-261-00840-6 X.622/14047
879944x

AIRCRAFT
1799-1909

EARLY flying machines, 1799-1909 / [compiled by] Charles Gibbs-Smith]. - London : Eyre Methuen, 1975
[128]p : chiefly ill. - (Picturefile)
0-413-33020-6 X.620/15073
0356212

AIRCRAFT
1870-1912

FLYING and ballooning from old photographs / introduction and commentaries by John Fabb. - London : Batsford, 1980
ix,[111]p : chiefly ill
0-7134-2015-4 X.622/8302 Woolwich
2237342

AIRCRAFT
ca 1900-1979

AVIATION photo album, 2 / compiled and edited by Michael J.F. Bowyer. - Cambridge : Stephen, 1980
96p : of cill
0-85059-410-3 X.622/9120 Woolwich
1941909

AIRCRAFT
ca 1910-1975

AEROPLANES in colour. - Shepperton : Allan, 1976
[64] : of col ill
0-7110-0709-8 X.622/5381 Woolwich
0623466

AIRCRAFT
1914-1976

AVIATION photo album / compiled and edited by Michael J.F. Bowyer. - Cambridge : Stephens, 1978
96p : chiefly ill
0-85059-297-6 X.611/8105 Woolwich
1258826

AIRCRAFT
Abandonment. Parachuting, to 1945 - Collections

BOWMAN, Gerald
Jump for it. - London : White Lion Publishers, 1975
198p,[12]p of plates : ill
0-85617-739-3 X.629/10484 Woolwich
0636142

AIRCRAFT
Abandonment. Parachuting, to 1976 - Collections

MACKERSEY, Ian
Into the silk. - Revised ed. - London : Mayflower, 1978
232p,[8]p of plates : ill. - Includes index
0-583-12793-2 X.708/21102
1400045

AIRCRAFT
Accidents

BEATY, David
Strange encounters. - London : Methuen, 1982
160p,[12]p of plates : ill. - Bibliography: p152-153. - Includes index
0-413-39760-2 X.520/29288
2848082

GREAT BRITAIN. Accidents Investigation Branch
BAC One-Eleven Type 518 G-AXMG, report on the accident at Stansted Airport, Essex on 27 February 1978. - London : H.M.S.O., 1979
[3],12p. - (Aircraft accident report ; 7/78)
0-11-512447-0 SRL
0000469

GREAT BRITAIN. Accidents Investigation Branch
Boeing 707 321C G-BEBP, report on the accident near Lusaka International Airport, Zambia, on 14 May 1977. - London : H.M.S.O., 1979
38,[4]p,[29] leaves of plates(4 fold) : ill(some col). - (Aircraft accident report ; 9/78)
0-11-512449-7 SRL
0000469

Boeing 707 Series 436 G-APFK, report on the accident at Prestwick Airport, Scotland on 17 March 1977. - London : H.M.S.O., 1978
[3],39p(4 fold) : ill. - (Aircraft accident report ; 6/78)
0-11-512446-2 SRL
0000469

!Pipe!Piper!. - London : H.M.S.O., 1979
[3],9p. - (Aircraft accident report ; 1/79)
0-11-512450-0 SRL
0000469

Piper PA-34 (Seneca II) G-BFKO. - London : H.M.S.O., c1979
29p,3leaves of plates(1folded) : ill. - (Aircraft accident report ; 7/80). - Bibliography: p29
0-11-513175-2 X.625/693 Woolwich
0000469

Sikorsky S.6 1N helicopter, G-BBHN, report on the accident in the North Sea, north east of Aberdeen on 1 October 1977. - London : H.M.S.O., 1979
[3],28p : ill. - (Aircraft accident report ; 8/78)
0-11-512448-9 SRL
0000469

SPAIN. Civil Aviation Accident Commission
British Airways Trident 1E G-AVYD, report on the accident at Bilboa Airport, Spain on 15 September 1975. - London : H.M.S.O., 1978
[2],22p,2p of plates,[2]leaves of plates(1 fold) : ill. - (Aircraft accident report ; 8/77)
0-11-511812-8 SRL
0000469

AIRCRAFT
Accidents. Investigation

BARLAY, Stephen
Aircrash detective. - [Revised ed.]. - London : Coronet, 1975
395p. - Bibliography: p.381-382. - Includes index
0-340-19890-7 X.619/15297 Woolwich
0539066

AIRCRAFT
Accidents. Investigation - Conference proceedings

SYMPOSIUM ON INTERNATIONAL AIRCRAFT ACCIDENTS INVESTIGATION
(1973 : London)
Symposium on International Aircraft Accidents Investigation, 15th January 1973. - London : Royal Aeronautical Society, [1975?]
58 leaves in various pagings
0-903409-01-1 X.105/500
0539066

AIRCRAFT
Aerodynamics

KÜCHEMANN, Dietrich
The aerodynamic design of aircraft. - Oxford : Pergamon, 1978
x,564p : ill. - (Pergamon international library). - Bibliography: p.515-555. - Includes index
0-08-020515-1 SRL
1448382

MORAN, Jack
An introduction to theoretical and computational aerodynamics. - New York ; Chichester : Wiley, c1984
xiii,464p : ill. - Includes index
0-471-87491-4 SRL
1448382

The PRINCIPLES of flight. - London : British Petroleum Educational Service, 1977
1wall chart : ill. (col.)
av00193 Cup.900.NN.15/60
2011778

SHEVELL, Richard S. (Richard Shepherd)
Fundamentals of flight. - Englewood Cliffs ; London : Prentice-Hall, c1983
xxiv,405p : ill. - Includes index
0-13-339093-4 X.322/12269
2011778

AIRCRAFT
Aeroelasticity

BISPLINGHOFF, Raymond Lewis
Principles of aeroelasticity. - New York : Dover Publications ; London : Constable, 1975
xi,527p : ill. - Includes index
0-486-61349-6 X.619/15388 Woolwich
0565415

AIRCRAFT
Application of chemicals - Conference proceedings

AGRICULTURAL Aviation Group all-day symposium jointly with Rotorcraft Section. - London : [The Society], [1979]
132p in various pagings : ill. - Includes bibliographies
B81-37730 SRL
2562030

AIRCRAFT
Archaeology

EPICS of aviation archaeology / compiled and edited by Bruce Robertson. - Cambridge : Stephens, 1978
156p : ill
0-85059-305-0 X.620/17893 Woolwich
1341235

ROBERTSON, Bruce
Aviation archaeology. - 2nd ed. - Cambridge : Stephens, 1983
168p : ill. - Includes index
0-85059-638-6 X.421/24706
3036863

AIRCRAFT
Automatic flight control systems

MCLEAN, D. (Donald)
Analytically redundant output feedback scheme for reduction of structural loads of a flexible transport aircraft. - [Loughborough] : Department of Transport Technology, Loughborough University of Technology, 1983
iv,71,14p : ill
B84-24581 X.622/19293
1901524

PALLETT, E. H. J. (Edwin Henry John)
Automatic flight control. - 2nd ed. - London : Granada, 1983
ix,286p : ill. - Includes index
0-246-12048-7 X.622/18063
1901524

PALLETT, Edwin Hery John
Automatic flight control. - London : Granada, 1979
vi,234p : ill. - Includes index
0-246-11167-4 X.620/18994 Woolwich
1901524

AIRCRAFT
Circumnavigation of world, to 1982

GLINES, Carroll V. (Carroll Vane)
Round-the-world flights. - New York ; London : Van Nostrand Reinhold, c1982
288p,[32]p of plates : ill. - Bibliography: p278-279. - Includes index
0-442-26140-3 X.622/17269
2990652

AIRCRAFT
Collisions into Empire State Building, 1945

WEINGARTEN, Arthur
The sky is falling. - London : Hodder and Stoughton, 1978
xv,260p,leaf of plate,[16]p of plates : ill. - Includes index
0-340-22932-2 X.800/27638
1383957

AIRCRAFT
Design

BENT, Ralph Daniel
Aircraft basic science. - 5th ed.. - New York ; London : McGraw-Hill, 1980
vii,374p : ill. - (Aviation technology series)
0-07-004791-x X.622/8332 Woolwich
225557x

AIRCRAFT
Design. Costigan, F J - Biographies

COSTIGAN, F. J.
Adventures of an aircraft designer. - Luton : Cortney Publications, 1980
viii,123p(1 fold),[14]p of plates : ill
0-904378-09-8 X.622/8799 Woolwich
2301067

AIRCRAFT
Digital automatic control systems - Conference proceedings

APPLICATION OF DIGITAL AVIONIC SYSTEMS IN AIRCRAFT
(Conference : 1974)
Joint RAeS-IEE Committee symposium on the application of digital avionic systems in aircraft, 12th December 1974. - [London] : [Royal Aeronautical Society], [1975]
109p in various pagings : ill
0-903409-05-4 SRL
0543888

AIRCRAFT
Dynamics

CLANCY, Laurence Joseph
Aerodynamics. - New York : Wiley ; London : Pitman, [1978]
xviii,610p : ill. - (Pitman aeronautical engineering series). - Includes index
0-273-01120-0 X.619/18789 Woolwich
0178497

AIRCRAFT
Dynamics. Mathematics - Conference proceedings

COMPUTATIONAL methods and problems in aeronautical fluid dynamics / edited by B.L. Hewitt ... [et al.]. - London : Academic Press, 1976
ix,525p : ill. - Includes bibliographies
0-12-346350-5 SRL
0702269

AIRCRAFT
Electrical systems

BENT, Ralph D. (Ralph Daniel)
Aircraft electricity and electronics. - 3rd ed. - New York ; London : Gregg, c1981
vii,359p : ill. - (Aviation technology series). - Includes index
0-07-004793-6 X.622/11153 Woolwich
2554046

AIRCRAFT
Electrical systems - Conference proceedings

SYMPOSIUM ON AVIONICS VERSUS ELECTRICS - WHO SHOULD DETERMINE FUTURE POWER SUPPLIES? (1977 : London)
Avionics versus electrics - who should determine future power supplies? Tuesday 15 March 1977. - London : Royal Aeronautical Society, [1977]
136p in various pagings : ill. - (RAeS/IEE)
B78-00963 SRL
2554046

AIRCRAFT
Electronic equipment

HELFRICK, Albert D.
Modern aviation electronics. - Englewood Cliffs ; London : Prentice-Hall, c1984
vii,312p : ill. - Includes index
0-13-587766-0 X.622/21093
3191028

AIRCRAFT
Electronic equipment - Serials

JANE'S avionics. - 1st ed. (1982-83)-. - London : Jane's Pub. Co., 1982-
v. : ill
B83-24922 SRL
2905310

AIRCRAFT
Electronic equipment. Vortella

SYKES, Egerton
An original concept in electronics. - Brighton : Markham House Press, 1978
8p. - Bibliography: p.8
B79-05165 X.619/19343 Woolwich
1773720

AIRCRAFT
Electronic warfare aircraft

STREETLY, Martin
World electronic warfare aircraft. - London : Jane's, 1983
127p : ill. - Includes index
0-7106-0166-2 X.622/17035
2967391

AIRCRAFT
Engines, to date - Catalogues

MUSEO STORICO DELL'AERONAUTICA MILITARE
Motori. - Bologna : Pàtron, 1980
195p : ill
r6507824 X.622/15917
895321x

AIRCRAFT
Flight

STEVER, H. Guyford
Flight. – Revised [ed.]. – New York : Time-Life Books, 1977
200p. – (Life science library)
r6188947 X.622/7155 Woolwich
1488155

AIRCRAFT
Flight, to 1974

MCKIM, Frank Robertson
By sea and air. – Oxford : Pergamon, 1975
[8],80p : ill. – (Projects in physics). – Bibliography: p.80
0-08-017052-8 X.619/15158
0500550

AIRCRAFT
Flight recorders. Use – Conference proceedings

AIDS today and tommorrow = RAeS/IEE. – [London] : [Royal Aeronautical Society], [1978]
[110]p : ill
B83-15447 SRL
2870452

AIRCRAFT
Flights, to 1925

PELLING, Jack
My book of flying to 1925. – [London] : Springwood Books, [1978]
[26]p : ill(chiefly col). – (Marvellous minis)
0-905947-50-9 X.629/11859 Woolwich
1457462

AIRCRAFT
Flights, to 1951

TAPPER, Oliver
The world's great pioneer flights. – London : Bodley Head, 1975
128p : ill. – (Putnam world aeronautical library). – Bibliography: p.126. – Includes index
0-370-01585-1 Cup.1281/49
0377244

AIRCRAFT
Flights, to 1958

GARRETT, Richard
Great air adventures. – London : Pan Books, 1978
127p : ill. – (Piccolo true adventures)
0-330-25439-1 X.708/21377
1577085

GARRETT, Richard
Great air adventures. – London : Severn House : [Distributed by Hutchinson], 1978
127p : ill
0-7278-0413-8 X.808/33241
1577085

AIRCRAFT
Flights, to 1970

NIVEN, Cyril
Man in flight. – Edinburgh : Holmes McDougall, 1971
47p : ill(chiefly col). – (Travels in time)
0-7157-0895-3 Cup.1281/785
1375075

AIRCRAFT
Flights, to 1973

FRIEDLANDER, Mark P.
Higher, faster, and farther. – New York : Morrow, 1973
349p : ill
0-688-00204-8 X.622/19784
0015725

AIRCRAFT
Flights, to 1974

HISTORY of aviation / [edited by] John W.R. Taylor and Kenneth Munson. – London : New English Library, 1975
511p : ill(chiefly col). – Includes index
0-450-02146-7 LR.425/15
0371718

AIRCRAFT
Flights, to 1975

SHEPHARD, Peter
Flight. – London : Ward Lock, 1976
92p : ill(chiefly col). – Includes index
0-7063-1860-9 X.622/5743 Woolwich
0791148

AIRCRAFT
Flights, to 1976

STORER, James Donald
Flying feats. – London : Beaver Books, 1977
127p : ill
0-600-33613-1 X.619/18319 Woolwich
135308x

AIRCRAFT
Flights, to 1977

The GUINNESS book of air facts and feats. – 3rd ed.. – Enfield : Guinness Superlatives, 1977
240p,[16]p of plates : ill(some col). – Bibliography: p.223-225. – Includes index
0-900424-34-6 X.620/17219 Woolwich
1267515

HISTORY of aviation / edited by John W.R.Taylor and Kenneth Munson. – 2nd revised ed. – London : New English Library, 1978
512p : ill(chiefly col). – Includes index
0-450-03582-4 LR.425/342
158880x

AIRCRAFT
Flights, to 1978

RECORD-BREAKING aircraft / compiled by Kenneth Munson, edited by John W.R. Taylor. – London : Macdonald and Jane's, 1978
264p : ill. – (Jane's pocket book ; 15). – Includes index
0-356-08423-x X.629/12308 Woolwich
1810057

AIRCRAFT
Flights, to 1983

TAYLOR, Michael J. H. (Michael John Haddrick)
Aircraft facts & feats. – [4th ed.]. – Enfield : Guinness Superlatives, 1984
256p : ill(some col.). – Includes index
0-85112-406-2 X.622/20604
3188213

AIRCRAFT
Flights, 1903-1913

WRAGG, David W. (David William)
Flight with power. – London : Barrie Jenkins, 1978
160p : ill. – Includes index
0-214-20484-7 X.620/18762 Woolwich
1936441

AIRCRAFT
Flights to Australia, 1930

CHICHESTER, Francis
Solo to Sydney. – Greenwich : Conway Maritime, 1982
208p,[9]p of plates : ill
0-85177-254-4 X.809/51456
2714086

AIRCRAFT
Flights to South Africa, 1947

LEWIS, Cecil
Gemini to Joburg. – [Harmondsworth] : Viking, 1984
128p
0-670-80062-7 X.809/63379
3259218

AIRCRAFT
Flying, to 1961

CHAMBERS, Aidan
Aidan Chambers' book of flyers and flying. – Harmondsworth : Kestrel Books, 1976
143p : ill. – Bibliography: p.143. – Includes index
0-7226-5087-6 X.622/5355 Woolwich
0616508

AIRCRAFT
Flying, to 1980

BROWN, Douglas, 19—
Flyers. – London : Hamlyn, 1981
93p : ill(some col.). – (History eye-witness). – Includes index
0-600-30494-9 X.622/10455 Woolwich
2494531

AIRCRAFT
Flying by Australian pilots, to 1975

COPLEY, Greg
Australians in the air. – Adelaide : Rigby, 1976
xiv,289p,[40]p of plates : ill(some col). – Bibliography : p281-282.-Includes index
0-7270-0139-6 X.802/10404
119559X

AIRCRAFT
Flying. Piggott, Derek – Biographies

PIGGOTT, Derek
Delta papa. – London : Pelham, 1977
xii,179p,[16]p of plates : ill
0-7207-0979-2 X.629/11346 Woolwich
1142429

AIRCRAFT
Flying. Problems – For pilots

MANNINGHAM, Dan
Crossfeed. – New York : London : Aronson, c1983
xii,304p : ill
0-87668-602-1 X.622/19680
3137295

AIRCRAFT
Flying. Simulations – Conference proceedings

50 YEARS OF FLIGHT SIMULATION (Conference : 1979 : London)
50 Years of Flight Simulation conference proceedings. Piccadilly Hotel, Piccadilly, London, England, April 23rd, 24th, 25th [1979]. – [Buckingham] : [Network] : [London] : [Royal Aeronautical Society].
Session 1. – [1979]
[1],iii,72[i.e.74] leaves : ill
0-904999-45-9 SRL
0508381
Session 2. – [1979]
[1],iii leaves,102p : ill. – Bibliography: p.85
0-904999-46-7 SRL
0508381

50 YEARS OF FLIGHT SIMULATION (Conference : 1979 : London)
50 Years of Flight Simulation conference proceedings. Piccadilly Hotel, Piccadilly, London, England, April 23rd, 24th, 25th [1979]. – [Buckingham] : [Network] : [London] : [Royal Aeronautical Society].
Session 3. – [1979]
[1],iii leaves,86p : ill. – Includes bibliographies
0-904999-47-5 SRL
0508381

FLIGHT SIMULATION SYMPOSIUM (2nd : 1973 : London)
Proceedings [of the] Second Flight Simulation Symposium. 16-17 May 1973. – London : Royal Aeronautical Society, [1975]
175p in various pagings : ill
0-903409-02-x SRL
0508381

AIRCRAFT
Fuels. Combustion. Pollution of atmosphere

BRAY, Kenneth Noel Corbett
Burning problems in aviation. – [Southampton] : University of Southampton, 1975
39p. – (Inaugural lectures / University of Southampton)
0-85432-140-3 X.619/15340 Woolwich
0567728

AIRCRAFT
Fuels. Conservation – Conference proceedings

FUEL ECONOMY IN THE AIRLINES (Conference : 1980 : London)
Fuel economy in the airlines. – [Sutton] : [IPC Business and Industrial Training Limited], [1980]
308p in various pagings : ill
B80-39581 SRL
2276712

AIRCRAFT
German reconnaissance aircraft, to 1945

NOWARRA, Heinz J. (Heinz Joachim)
Fernaufklärer 1915-1945. – Stuttgart : Motorbuch, 1982
207p : ill
3-87943-858-7 X.622/18399
9043039

AIRCRAFT
Hijacking. International legal aspects, to 1973

JOYNER, Nancy Douglas
Aerial hijacking as an international crime. – Dobbs Ferry, N.Y. : Oceana ; Leiden : Sijthoff, 1974
viii, 344 p.. – Bibliography : p. 321-338
0-379-00004-0 X.200/10674
0009512
– [Another copy.] OPL

MCWHINNEY, Edward
The illegal diversion of aircraft and international law. – Leyden : Sijthoff, 1975
vi, 123 p.
90-286-0155-4 X.100/16774
0009512

AIRCRAFT
Instruments

PALLETT, E. H. J. (Edwin Henry John)
Aircraft instruments. – 2nd ed. – London : Pitman, 1981
xvii,414p : ill. – Includes index
0-273-01539-7 SRL
2564718

AIRCRAFT
Large aircraft, to 1980

TAYLOR, Michael J. H. (Michael John Haddrick)
Giants in the sky. – London : Jane's, 1982
224p : ill
0-7106-0190-5 X.622/13126
2761734

AIRCRAFT
Maintenance
BENT, Ralph Daniel
Aircraft basic science. - 5th ed.. - New York ; London : McGraw-Hill, 1980
vii,374p : ill. - (Aviation technology series)
0-07-004791-x X.622/8332 Woolwich
2255588

AIRCRAFT
Navigational aids: Very high frequency omni directional radio range
BRAMSON, A. E. (Alan Ellesmere)
Flying the V.O.R.. - Shrewsbury : Airlife, [1981?]
vi,42p : col.ill
0-906393-00-0 X.629/16475 Woolwich
2569191

AIRCRAFT
Navigational aids: Very high frequency omni directional radio range & automatic direction finder
CASS, Martin
The VOR and ADF. - Shrewsbury : Airlife Publications, 1977
47],92p : ill(chiefly col)
0-9504543-3-8 X.629/11931 Woolwich
1514873

AIRCRAFT
Noise
MORGAN, Noël
A select bibliography of aircraft noise. - London : Airfields Environment Trust, c1984
[67]p. - Includes index
B85-11147 X.622/22261
0585130

AIRCRAFT
Noise. Control measures. Engineering aspects - Conference proceedings
AMERICAN INSTITUTE OF AERONAUTICS AND ASTRONAUTICS. Aero-acoustics Specialists Conference (1st : 1973 : Seattle)
Aeroacoustics, FAN, STOL, and boundary layer noise, sonic boom, aeroacoustic instrumentation : technical papers selected from the AIAA Aero-acoustics Specialists Conference, October 1973 subsequently revised for this volume. - New York : American Institute of Aeronautics and Astronautics ; Cambridge, Mass. ; London : M.I.T. Press, 1975
xvi,509p : ill. - (Progress in astronautics and aeronautics ; vol.38). - Includes bibliographies and index
0-262-14024-1 SRL
0707295

AIRCRAFT
Noise. Engineering aspects
BARDELL, N.
Aircraft fuselage design to minimise propeller-induced transmitted noise. - [Southampton] : University of Southampton, Department of Aeronautics and Astronautics, 1984
6leaves : ill. - (AASU memo. ; no.84/10)
r6863694 X.622/21852
0793159

AIRCRAFT
Noise. Engineering aspects - Conference proceedings
AIAA AERO-ACOUSTICS CONFERENCE (2nd : 1975 : Hampton, Va)
Aeroacoustics, acoustic wave propagation, aircraft noise prediction, aeroacoustic instrumentation : technical papers from AIAA 2nd Aero-acoustics Conference, March 1975, subsequently revised for this volume. - New York : American Institute of Aeronautics and Astronautics ; Cambridge, Mass. ; London : M.I.T. Press, 1976
xxxiv,307p : ill. - (Progress in astronautics and aeronautics ; vol.46). - Includes index
0-915928-10-8 SRL
0793159

AIRCRAFT
Performance. Prediction
ESHELBY, M. E.
Flight-path reconstruction for take-off and landing from inertial measurement data with Kalman filter. - [Caranfield] : [College of Aeronautics, Cranfield Institute of Technology], 1983
vi,158leaves : ill. - (Cranfield College of Technology report ; 8326)
0-902937-94-4 X.622/23125
3065014

AIRCRAFT
Piston engines, to 1980
SMITH, Herschel
Aircraft piston engines. - New York ; London : McGraw-Hill, c1981
xi,255p : ill. - Includes index
0-07-058472-9 SRL
2883953

AIRCRAFT
Remotely piloted aircraft
JANE'S pocket book of RPVs / [edited by] John W.R. Taylor, Kenneth Munson. - London : Macdonald and Jane's, 1977
239p : chiefly ill. - (Jane's pocket book ; 13). - Includes index
0-356-08404-3 X.619/18532 Woolwich
1449451

AIRCRAFT
Remotely piloted aircraft - Conference proceedings
REMOTELY piloted vehicles / sponsored jointly by the Royal Aeronautical Society and the Department of Aeronautical Engineering of the University of Bristol. - [Bristol] : [Department of Aeronautical Engineering, University of Bristol], [1979]
2v. : ill
0-906515-46-7 SRL
2590093

REMOTELY piloted vehicles / sponsored jointly by the Royal Aeronautical Society and the University of Bristol. - [Bristol] : [Department of Extra-Mural Studies, University of Bristol], [1981]
2v. : ill
0-906515-90-4 SRL
2590093

AIRCRAFT
Safety aspects - Serials
GENERAL aviation safety information. - 1/77- ; 21 Jan. 1977-. - Redhill : Civil Authority, Safety Data Unit ; [Cheltenham] : [Distributed by C.A.A.], [1977]-
B77-16411 BS.73/184
1113003

AIRCRAFT
Transonic aerodynamics
MOULDEN, T. H. (Trevor Holmes)
Fundamentals of transonic flow. - New York ; Chichester : Wiley, c1984
xvii,332p : ill. - Includes index
0-471-04661-2 X.622/23918
3287165

AIRCRAFT
Trials aircraft of British manufacturers. Identification marks, 1929-1977
REVELL, D. S.
Under B conditions. - Liverpool : Merseyside Aviation Society Ltd, 1978
103p : ill
0-902420-24-0 SRL
1526413

AIRCRAFT
Unmanned aircraft, to 1978
REED, Arthur
Brassey's unmanned aircraft. - London : Brassey's, 1979
ix,110p : ill. - Includes index
0-904609-32-4 SRL
2751739

AIRCRAFT
Unsuccessful aircraft, to 1975
GILBERT, James
The world's worst aircraft. - Walton-on-Thames : M. and J. Hobbs ; London : Joseph, 1975
192p : ill. - Bibliography: p.185-187. - Includes index
0-7181-1269-5 X.622/5159 Woolwich
0569178

The world's worst aircraft. - Sevenoaks : Coronet, 1978
xii,207p,8p of plates : ill. - Bibliography: p.199-201. - Includes index
0-340-21824-x X.619/18287 Woolwich
0569178

AIRCRAFT
Unusual aircraft, to 1980
BOWERS, Peter M.
Unconventional aircraft. - Blue Ridge Summit : Tab, c1984
ix,278p : ill. - Includes index
0-8306-2384-1 X.622/22565
2591758

TAYLOR, Michael J. H. (Michael John Haddrick)
Fantastic flying machines. - London : Jane's, 1981
142p : ill(some col.)
0-7106-0125-5 X.622/11620 Woolwich
2591758

AIRCRAFT
Use in agriculture - Conference proceedings
TECHNO-ECONOMIC TRENDS IN AIRBORNE EQUIPMENT FOR AGRICULTURE AND OTHER SELECTED AREAS OF THE NATIONAL ECONOMY (Conference : 1978 : Warsaw)
Trends in airborne equipment for agriculture and other areas. - Oxford : Pergamon for the United Nations, 1980
xvi,358p : ill
0-08-022425-3 SRL
1943987

AIRCRAFT
Use in crop spraying - Conference proceedings
DISTRIBUTION equipment and propulsion, 2 February 1977. - London : The Society, [1977]
73p in various pagings : ill
B78-03451 SRL
1294334

AIRCRAFT
Use in exploration of Andes
LARRA, Raúl
Desafio a los Andes. - Buenos Aires : Ánfora, 1978
157p : ill
r6101344 X.619/11192 Woolwich
2264374

AIRCRAFT. Air services. Great Britain
Maintenance & repair. Techniques
CIVIL AVIATION AUTHORITY
Condition monitored maintenance. - London : The Authority, 1978
v,33p : ill. - (CAP ; 418)
0-86039-069-1 BS.73/224
1946161

Condition monitored maintenance. - [1st ed. reprinted]. - London : The Authority, 1980
v,33p : ill. - (CAP. 418)
B80-35156 BS.73/255
1946161

AIRCRAFT. Airports. London
Noise at night
GREAT BRITAIN. Department of Trade
Night disturbance from aircraft noise at Heathrow and Gatwick. - London : [The Department], 1977
[2],ii,18,[2]p,[38]p of plates(18 fold),fold leaf of plate : col ill
B77-12472 BS.41/1054
1087312

AIRCRAFT. Ancient world
DEICHES, I. W. (I William)
"B.C. aeronautics - the world's greatest puzzle". - Brentwood : Time-Energy Information.
Pt.9: Plans-photographs-sketches of ancient aircraft from ancient Egypt, Tibet-Mexico-Columbia. - 1984
[8]leaves : ill
0-907863-17-5 L.49/3059
3150496

A collection of ancient aircraft, gliders, 'flying-chariots', 'magic carpets' insignia from the following countries, Ancient Egypt, Persia, India, Japan, Columbia and Mexico. - Brentwood : Time-Energy Information.
Pt.2. - 1984
leaves 23-41
0-907863-23-x X.622/20328
3150496

AIRCRAFT. Ancient world
- Serials
PLANS handbook. - No.1-. - Brentwood : Time-Energy Information International, 1984-
v. : chiefly ill
r6863650 P.843/300
3150496

AIRCRAFT. Andes. Mendoza Province. Argentina
Accidents, 1972. Survivors' experiences
READ, Piers Paul
Alive. - London : Pan Books, 1975
xii,318p,[16]p of plates : ill
0-330-24392-6 X.708/20174
0097020

AIRCRAFT. Antarctic
Accidents, to date
VETTE, Gordon
Impact Erebus. - Auckland : Hodder and Stoughton, 1983
347p : ill
0-340-32010-9 X.622/22256
9130438

AIRCRAFT. Austrian military forces
to 1938

HAUBNER, Fred
Die Flugzeuge der österreichischen Luftstreitkräfte vor 1938. – Graz : Weishaupt, 1982
245p : ill. – (Österreichs Luftfahrt in Einzeldarstellungen ; Bd.2)
3–900310–06–8 X.629/22711
9080295

AIRCRAFT. Barbados
1976 accidents

LEÓN COTAYO, Nicanor
Crimen en Barbados. – 2a ed. – La Habana : Editorial de Ciencias Sociales, 1978
295p : ill
r6700525 X.629/14876 Woolwich
2481936

AIRCRAFT. Brooklands. Surrey
Flying, 1907–1981

JOHNSON, Howard, 1916–
Wings over Brooklands. – Weybridge : Whittet, 1981
157p : ill. – Bibliography: p153. – Includes index
0–905483–20–0 X.622/11490 Woolwich
2579855

AIRCRAFT. Curriculum subjects. Students, 5–14 years. Schools. Great Britain
Aerodynamics – For teaching

SLACK, Derek
Flying starts here. – Basingstoke : Globe Education [for] West Sussex County Council, 1981
36p : ill. – (Science horizons. Level 2b). – Bibliography: p10. – List of films: p10
0–333–28541–7 X.622/10912 Woolwich
2536811

AIRCRAFT. England
Noise. Effects on residents of areas adjacent to airports

OLLERHEAD, J. B.
A comparison of annoyance caused by aircraft noise near London, Manchester and Liverpool airports. – [Loughborough] : Department of Transport Technology, Loughborough University of Technology, 1977
[1],iv leaves, 72p : ill. – (TT7706)
r6032119 X.615/1871 Woolwich
118704X

AIRCRAFT. England
Noise. Effects on residents of London Airport (Heathrow) region

OLLERHEAD, J. B.
A further survey of some effects of aircraft noise in residential communities near London (Heathrow) Airport. – Loughborough : Department of Transport Technology, Loughborough University of Technology, 1977
[1],viii leaves,139p : ill. – (TT7705)
r6032118 X.615/1872 Woolwich
1187031

AIRCRAFT. Exhibits. Museums. Great Britain

ELLIS, Ken, 1953–
British museum aircraft. – Liverpool : Merseyside Aviation Society Limited, 1977
166p : ill. – Bibliography: p.158–160. – Includes index
0–902420–15–1 X.619/17925 Woolwich
1247670

OGDEN, Bob
British aviation museums. – Stamford : Key, c1983
88p : ill. – (Flypast reference library). – Includes index
0–946219–06–0 X.629/24244
3167704

AIRCRAFT. Great Britain

WIDDOWSON, Kevin P.
Airguide Britain. – [Manchester] : [K.P. Widdowson], [1982]
51p : ill
B82–32987 X.629/18262
2767783

AIRCRAFT. Great Britain
Airmisses – Serials

GENERAL aviation airmisses / Joint Airmiss Working Group. – May 1980–. – London : Civil Aviation Authority, 1980–
v.
B81–02010 SRL
2327295

AIRCRAFT. Great Britain
Application of chemicals. Procedure

CIVIL AVIATION AUTHORITY
The Aerial Application Certificate. – London : C.A.A., 1978
1v.. – (CAP ; 414). – Bibliography(1p.)
0–86039–067–5 BS.73/227
0186996

AIRCRAFT. Great Britain
Components. Specifications. Documentation

VENDOR documentation guide. – London : Technical Section, Society of British Aerospace Companies, 1983
1v.(loose-leaf) : ill
B83–42579 X.525/8219
2969807

AIRCRAFT. Great Britain
Noise. Control

ACTION against aircraft noise / [prepared by the Department of Trade and the Central Office of Information]. – [London] : The Department, 1979
16p : ill(some col.)
B81–05897 BS.41/1243
2349264

AIRCRAFT. Great Britain
Noise. Control measures

GREAT BRITAIN. Department of Trade
Action against aircraft noise. – [London] : The Department, 1976
[2],25,[1]p : ill(some col)
0–85605–290–6 BS.41/1004
1106309

AIRCRAFT. Great Britain. Royal Flying Corps
1912–1918

BRUCE, J. M. (John McIntosh)
The aeroplanes of the Royal Flying Corps (Military Wing). – London : Putnam, 1982
xxv,642p : ill. – Includes index
0–370–30084–x SRL
3281329

AIRCRAFT. Great Britain. Royal Navy
to 1977

THETFORD, Owen
British naval aircraft since 1912. – 4th revised ed. – London : Putnam, 1978
480p : ill. – Includes index
0–370–30021–1 X.629/12203 Woolwich
1704958

AIRCRAFT. Great Britain. Royal Navy
to 1981

ELLIS, Paul, 1946–
Aircraft of the Royal Navy. – London : Jane's, 1982
176p : ill. – Includes index
0–7106–0135–2 X.622/12641 Woolwich
2727625

VICARY, Adrian
Naval wings. – Cambridge : Patrick Stephens, 1984
110p : ill. – Bibliography: p108. – Includes index
0–85059–660–2 X.622/21876
2727625

AIRCRAFT. Great Missenden region. Buckinghamshire
Departure routes from Luton Airport. Re-routing

GREAT BRITAIN. Department of Trade
Aircraft departing from Luton via Woodley, routing arrangements. – London : [The Department], 1977
[2],3p,fold plate
B78–07473 BS.41/1082
1344080

AIRCRAFT. Kent
Flying, to 1914

FLYING / [compiled] by David G. Collyer. – Rochester : North Kent Books, c1982
63p : ill
0–9505733–8–8 X.629/20408
2912589

AIRCRAFT. Mole Valley region. Surrey
Departure routes from London Airport (Heathrow). Re-routing

GREAT BRITAIN. Department of Trade
Mole Valley aircraft departure routes. – London : [The Department], 1977
[1],7p,[7]fold leaves of plates
B78–16707 BS.424/116
1441876

AIRCRAFT. Mole Valley region. Surrey
Revised departure routes from London Airports (Heathrow). Monitoring

GREAT BRITAIN. Department of Trade
Mole Valley routes. – London : The Department, 1976
7,3,[6]p(1 fold),[7]leaves of plates(3 fold) : ill
B79–06420 BS.41/1086
1780859

AIRCRAFT. Nairobi. Kenya
Accidents, 1974. Survivors' experiences

MOORHOUSE, Earl
Wake Up, it's a crash!. – London : Corgi, 1982, c1980
172p,[16]p of plates : ill. – Includes index
0–552–11932–6 X.808/36395
2149141

AIRCRAFT. Poland
Design

GLASS, Andrzej
Polskie konstrukcje lotnicze 1893–1939. – Warszawa : Wydawnictwa Komunikacji i Łączności, 1976
477p : ill
r6008555 X.625/183 Woolwich
1312790

SZEWCZYK, Witold
Samoloty na których walczyli Polacy. – Warszawa : Wydawnictwa Komunikacji i Łączności, 1978
177p : ill. – (Biblioteczka Skrzydlatej Polski)
r6097216 X.619/10900 Woolwich
1312790

AIRCRAFT. Soviet Union
[ZNAKOM'TES', samolet : raketa. English]. Soviet aircraft and rockets / N.A. Zhemchuzhin ... [et al.] ; translated from the Russian [by W.V. Nabar]. – New Delhi : Amerind, c1977
271p : ill
0–86249–269–6 X.622/21332
3178994

AIRCRAFT. Soviet Union
1917–1982 – Encyclopaedias

GUNSTON, Bill
Aircraft of the Soviet Union. – London : Osprey, 1983
414p : ill
0–85045–445–x SRL
3042936

AIRCRAFT. Special subjects. Children's short stories in English, 1837–1981
Flying – Anthologies

FLYING stories / edited by Hayden McAllister ; illustrated by Trevor Newton. – London : Octopus, 1982
304p : ill
0–7064–1734–8 X.950/17367
2818841

AIRCRAFT. Special subjects. English photography, 1900–1950

BROWN, Charles E. (Charles Edward)
Camera above the clouds. – Shrewsbury : Airlife, 1983
159p : chiefly ill(some col.)
0–906393–31–0 X.622/21029
3193977

AIRCRAFT. Special subjects. French humorous cartoons
Flights, ca 1910 – Collections

XAUDARO, J.
The perils of flight. – London : E. Benn, 1978
48p : chiefly col ill
0–510–00040–1 X.423/10711 Woolwich
174562x

AIRCRAFT. Special subjects. Picture postcards
– Serials

AVIATION postcard collector. – Issue no.1 (Aug. 1980)–. – Hounslow : A.P.C. Publications, 1980–
v. : ill
Holdings: issue no 3 (February 1981) etc
B82–09690 P.631/307
2617900

AIRCRAFT. Special subjects. Short stories in English, 1650–1970
Flying – Anthologies

AIRBORNE / compiled by Alan C. Jenkins. – Glasgow : Blackie, 1975
252p
0–216–90047–6 Nov.30784
0528617

AIRCRAFT. Stock. Duxford Airfield

IMPERIAL WAR MUSEUM
Duxford handbook. – [London] : The Museum, 1981
72p : ill
B82–03980 X.629/16804 Woolwich
2595087

AIRCRAFT. Switzerland. Kommando der Flieger- und Fliegerabwehrtruppen
1914–1973

URECH, Jakob
The aircraft of the Swiss Air Force since 1914. – Stäfa : Verlag Th. Gut + Co., 1975
367p : ill. – Bibliography : p.367.–Includes index
r6032030 X.622/5913 Woolwich
1122908

AIRCRAFT. Tulare County. California
Accidents, 1976. Survivors' experiences

ELDER, Lauren
And I alone survived. - London : Collins, 1978
xiv,188p
0-00-216781-6 X.809/43372
1544322
And I alone survived. - London : Pan Books, 1978
159p
0-330-25911-3 X.708/22262
1544322

AIRCRAFT. United States
Aerodynamics. Research, to 1935

HANLE, Paul A.
Bringing aerodynamics to America. - Cambridge, Mass. ; London : MIT, c1982
xiv,184p : ill. - Includes index
0-262-08114-8 X.629/18482
2783738

AIRCRAFT. United States. Navy

AIRCRAFT of the US Navy. - [Turin] : Delta, 1977
71p : of ill(some col). - (Air forces of the world ; Vol. 1)
r6032649 X.615/1972 Woolwich
1347403

AIRCRAFT. Western Europe
Flying. Safety measures

[EUROCONTROL Convention relating to Co-operation for the Safety of Air Navigation. Protocols, amendments, etc., November 21, 1978]. Protocol for the amendment of the additional protocol to the 'EUROCONTROL' International Convention relating to Co-operation for the Safety of Air Navigation dated 6 July 1970, Brussels, 21 November 1978 ... - London : H.M.S.O., [1979]
Folder(4p). - (Miscellaneous ; no.4(1979)) (Cmnd.7477)
0-10-174770-5 BS.14/3244
1965166

[EUROCONTROL Convention relating to Co-operation for the Safety of Air Navigation (1960 Dec.13). Protocols, etc., 1981 Feb.12]. Final Act of the diplomatic conference on the protocol amending the EUROCONTROL International Convention relating to Co-operation for the Safety of Air Navigation of 13 December 1960 with text of protocol, and a consolidated text of the provisions of the convention and the amendments and a multilateral agreement relating to route charges. - London : H.M.S.O., [1982]
55p. - (Cmnd. ; 8662) (Miscellaneous ; no.21 (1982))
0-10-186620-8 BS.14/3590
1965166

AIRCRAFT CARRIERS

PRESTON, Antony
Carriers. - London : Granada, 1984
48p : col.ill. - (Modern military techniques) (Dragon books). - Includes index
0-246-12255-2 X.622/20227
3169413

AIRCRAFT CARRIERS
to 1977

WATTS, Anthony J. (Anthony John)
A source book of aircraft carriers and their aircraft. - London : Ward Lock, 1977
128p : ill. - Includes index
0-7063-5538-5 X.629/11440 Woolwich
1198653

AIRCRAFT CARRIERS
to 1978

PRESTON, Antony
Aircraft carriers. - London : Hamlyn, 1979
192p : ill(some col). - Includes index
0-600-30437-x X.622/7245 Woolwich
2148196

AIRCRAFT CARRIERS
to 1980

CHESNEAU, Roger
Aircraft carriers of the world, 1914 to the present. - London : Arms and Armour, 1984
288p : ill. - Bibliography: p285. - Includes index
0-85368-636-x L.42/3102
2500469

FRIEDMAN, Norman, 1946-
Carrier air power. - Greenwich : Conway Maritime, 1981
192p : ill
0-85177-216-1 X.622/10581 Woolwich
2500469

WATTS, Anthony J. (Anthony John)
[A source book of aircraft carriers and their aircraft]. A source book of naval aircraft and aircraft carriers. - London : Ward Lock, c1980
128p : ill. - (Source books). - Includes index
0-7063-6052-4 X.629/14781
2335158
- Reprinted. - 1982 X.629/19536

AIRCRAFT CARRIERS
to date

HUMBLE, Richard
Aircraft carriers. - Seacaucas : Chartwell, [1982]
192p : ill
0-89009-503-5 X.805/4794
8799458

PRESTON, Antony
Aircraft carriers. - Greenwich, Conn. : Bison, 1982
64p : ill
0-86124-067-7 X.622/17508
8799458
[Aircraft carriers. French]. Histoire des porte-avions. - Paris : Fernand Nathan, [1980]
191p : ill
2-09-292040-5 X.622/11929 Woolwich
8799458

AIRCRAFT CARRIERS
1939-1945

BROWN, David, 1938- (Sept.)
Aircraft carriers. - London : Macdonald and Jane's, 1977
[1],64p : ill. - (World War 2 fact files)
0-354-01009-3 X.611/7932 Woolwich
1352342

AIRCRAFT CARRIERS
Naval air operations, 1945-1982

CARRIER air operations since 1945 / [compiled by] Paul Beaver. - London : Arms and Armour, [1983]
72p : chiefly ill(some col.). - (Warbirds illustrated ; no.19)
0-85368-598-3 X.0800/1646(19)
3063216

AIRCRAFT CARRIERS
World War 2. Naval air operations during Pacific campaigns

BELOTE, James H.
Titans of the seas. - New York ; London : Harper and Row, 1975
x,336p,[16]p of plates : ill. - Bibliography: p.305-322. - Includes index
0-06-010278-0 X.620/17740 Woolwich
1494600

AIRCRAFT CARRIERS. Great Britain. Royal Navy
to 1981

BEAVER, Paul
The British aircraft carrier. - Cambridge : Stephens, 1982
224p : ill. - Includes index
0-85059-493-6 X.622/12613 Woolwich
2725312
The British aircraft carrier. - 2nd ed. - Wellingborough : Stephens, 1984
240p : ill. - Includes index
0-85059-722-6 X.622/21411
2725312

AIRCRAFT CARRIERS. Great Britain. Royal Navy
Aircraft carriers called 'Ark Royal'. Naval operations, to 1975

APPS, Michael
The four 'Ark Royals'. - London : Kimber, 1976
256p,[24]p of plates : ill. - Includes index
0-7183-0344-x X.620/16166 Woolwich
0685232

AIRCRAFT CARRIERS. Great Britain. Royal Navy
Ark Royal (Ship), to 1978

BEAVER, Paul
Ark Royal. - Cambridge : Stephens, 1979
96p : ill
0-85059-381-6 X.800/29332
1825577

DALRYMPLE, James
HMS Ark Royal. - [Bodmin] : Bossiney Books, [1978]
48p : ill
0-906456-18-5 X.622/7462 Woolwich
1825577

AIRCRAFT CARRIERS. Great Britain. Royal Navy
Ark Royal (Ship), 1939-1941. Naval air operations during World War 2

POOLMAN, Kenneth
'Ark Royal'. - London : New English Library, 1978
173p
0-450-03953-6 X.708/21228
0122327

AIRCRAFT CARRIERS. Great Britain. Royal Navy
Illustrious (Ship). Naval air operations during World War 2 - Biographies

HANSON, Norman
Carrier pilot. - Cambridge : Stephens, 1979
255p : ill
0-85059-349-2 X.800/29289
1734326

AIRCRAFT CARRIERS. Great Britain. Royal Navy
Vindex (Ship). Naval air operations during World War 2

POOLMAN, Kenneth
Escort carrier. - London : Leo Cooper in association with Secker & Warburg, c1983
xx,216p : ill. - Includes index
0-436-37705-5 X.805/4379
2958600

AIRCRAFT CARRIERS. Japan. Kaigun
Aeroplanes. Naval air operations during Pacific campaigns of World War 2

FRANCILLON, René Jacquet
Japanese carrier air groups, 1941-45. - London : Osprey Publishing, 1979
48p : ill(some col). - (Airwar series ; 21)
0-85045-295-3 X.700/26057
1920065

AIRCRAFT CARRIERS. Japan. Kaigun
Naval operations during Pacific campaigns of World War 2, 1941-1942

THOMAS, David Arthur, b.1925
Japan's war at sea. - London : Deutsch, 1978
222p,[8]p of plates : ill. - Bibliography: p.195-202. - Includes index
0-233-97011-8 X.800/27264
1355392

AIRCRAFT CARRIERS. United States. Navy
to 1977

TERZIBASCHITSCH, Stefan
Aircraft carriers of the US Navy. - London : Conway Maritime Press, 1980
320p : ill. - Bibliography: p.320
0-85177-159-9 X.622/8027 Woolwich
2222728

AIRCRAFT CARRIERS. United States. Navy
1939-1945

HUMBLE, Richard
United States fleet carriers of World War II 'in action'. - Poole : Blandford, 1984
160p : ill. - Includes index
0-7137-1309-7 X.622/22384
324346x

AIRCRAFT CARRIERS. United States. Navy
Aeroplanes. Naval air operations during Pacific campaigns of World War 2

FRANCILLON, René Jacquet
US Navy carrier air groups, Pacific, 1941-45. - London : Osprey Publishing, 1978
48p : ill(some col). - (Airwar series ; 16)
0-85045-291-0 X.700/25748
1551418

SOWINSKI, Larry
The Pacific War. - London : Conway Maritime, 1981
208p : ill(some col.)
0-85177-217-x L.42/1217
257764x

AIRCRAFT CARRIERS. United States. Navy
Aeroplanes. Naval air operations during World War 2

KILDUFF, Peter
US carriers at war. - London : Ian Allan, 1981
128p : ill. - Bibliography: p128
0-7110-1077-3 X.805/1719
2538512

AIRCRAFT CARRIERS. United States. Navy
Escort aircraft carriers, to 1970

TERZIBASCHITSCH, Stefan
Escort carriers and aviation support ships of the US Navy. - Greenwich, London : Conway Maritime, 1981
208p : ill. - Bibliography: p192. - Includes index
0-85177-242-0 X.622/11583 Woolwich
2588412

AIRCRAFT CARRIERS. United States. Navy
Role

LEHMAN, John
Aircraft carriers. - Beverly Hills ; London : Sage Publications [for] the Center for Strategic and International Studies, 1978
83p : ill. - (The Washington papers ; vol.6. 52) (A Sage policy paper)
0-8039-1031-2 X.0709/603(52)
1766287

AIRCRAFT ENGINES INDUSTRIES
See also
Helicopter engine industries

AIRCRAFT ENGINES INDUSTRIES. Great Britain
Fedden, Sir Roy - Biographies
GUNSTON, Bill
By Jupiter. - London : Royal Aeronautical Society, 1978
ix,157p,[16]p of plates : ill. - Includes index
0-903409-07-0 X.619/19115 Woolwich
172908x

AIRCRAFT ENGINES INDUSTRIES. Great Britain
Military aircraft engines industries: Rolls-Royce, 1925-1971 - Biographies
HARKER, Ronald W.
Rolls-Royce from the wings. - Oxford : Oxford Illustrated Press, 1976
viii,168p : ill
0-902280-38-4 X.620/16528 Woolwich
0795917

AIRCRAFT ENGINES INDUSTRIES. Great Britain
Military aircraft engines industries: Rolls-Royce, 1939-1945
LLOYD, Ian
Rolls-Royce, the Merlin at war. - London : Macmillan, 1978
xvii,188p : ill. - Includes index
0-333-24016-2 X.809/44104
1383965

AIRCRAFT ENGINES INDUSTRIES. Great Britain
Rolls-Royce - Serials
The ROLLS-ROYCE magazine. - No.1 (June/Aug.1979)-. - London : Rolls-Royce, 1979-
v. : ill(some col)
B81-06391 P.623/974
2358883

AIRCRAFT ENGINES INDUSTRIES. Great Britain
Rolls-Royce, to 1978
HARKER, Ronald W.
The engines were Rolls-Royce. - New York : Macmillan ; London : Collier Macmillan, 1979
xxi,202p[32]p of plates : ill. - Includes index
0-02-548190-8 X.622/8729 Woolwich
2277840

AIRCRAFT ENGINES INDUSTRIES. Great Britain
Rolls-Royce, to 1980
DONNE, Michael
Leader of the skies. - London : Muller, 1981
158p : ill(some col.). - Bibliography: p158. - Includes index
0-584-10476-6 L.49/713
2376326

AIRCRAFT INDUSTRIES
See also
Aeroplane industries
Aircraft engines industries
Civil aircraft industries
Helicopter industries

AIRCRAFT INDUSTRIES
Companies, to 1979
GUNSTON, Bill
The plane makers. - London : New English Library, 1980
253p : ill(some col.). - Includes index
0-450-04754-7 L.45/904
233044x

AIRCRAFT INDUSTRIES. Brooklands. Surrey
1907-1981
JOHNSON, Howard, 1916-
Wings over Brooklands. - Weybridge : Whittet, 1981
157p : ill. - Bibliography: p153. - Includes index
0-905483-20-0 X.622/11490 Woolwich
2579863

AIRCRAFT INDUSTRIES. Ealing (London Borough)
Alliance Aeroplane Company, to ca 1945
LONDON BOROUGH OF EALING LIBRARY SERVICE
The story of Acton Aerodrome and the Alliance Factory, with notes on other early aviation in the Borough. - 2nd ed. - London : The Service, 1978
[7],31p
0-9502264-4-0 X.702/5915
1799193

AIRCRAFT INDUSTRIES. Germany
Manufacture of military aircraft for Germany. Luftstreitkräfte, 1914-1920. Economic aspects
MORROW, John H. (John Howard)
German air power in World War 1. - Lincoln, Neb. ; London : University of Nebraska Press, c1982
xii,267p : ill. - Bibliography: p245-261. - Includes index
0-8032-3076-1 X.800/34646
2835940

AIRCRAFT INDUSTRIES. Germany
Messerschmitt, Willy, to 1945
VAN ISHOVEN, Armand
Messerschmitt. - London : Omega : Galleon : [Distributed by Futura Publications], 1975
216p : ill. - Includes index
0-904442-04-7 X.611/7191 Woolwich
0260967

AIRCRAFT INDUSTRIES. Great Britain
De Havilland Aircraft Company, to 1981
SHARP, C. Martin (Cecil Martin)
D.H.. - Rev. and extended ed. - Shrewbury : Airlife, 1982
487p,[68]p of plates : ill. - Includes index
0-906393-20-5 X.809/55432
2915839

AIRCRAFT INDUSTRIES. Great Britain
Nationalisation. Law
GREAT BRITAIN. Parliament. Examiners of Petitions for Private Bills
Aircraft and Shipbuilding Industries Bill. - London : H.M.S.O., 1977
12p. - (1976/7 H.L.71)
0-10-407177-x BS.96/78
1387499

AIRCRAFT INDUSTRIES. Poland
Państwowe Zakłady Lotnicze
GLASS, Andrzej
Samoloty PZL 1928-1978. - Warszawa : Wydawnictwa Komunikacji i Łączności, 1980
178p : ill. - (Biblioteczka Skrzydlatej Polski)
83-206-0064-2 X.629/15994 Woolwich
8714061

AIRCRAFT INDUSTRIES. Rochester. Kent
Short Brothers (Firm), to 1964
PRESTON, J. M.
A short history. - Rochester [Kent] : North Kent Books, 1978
[40]p : ill
0-9505733-1-0 X.615/2272 Woolwich
1820591

AIRCRAFT INDUSTRIES. Soviet Union
ÎAKOVLEV, Aleksandr Sergeevich
Sovetskie samolety. - Izd. 2-e, perer. i dop.. - Moskva : Nauka, 1975
309p,[1]fold leaf of plate : ill
r6177221 X.629/8350 Woolwich
0481998

KOZLOV, Pavel Îakovlevich
ILy letîat na front. - Moskva : Izdatel'stvo DOSAAF SSSR, 1976
174p : ill
r6042940 X.809/23208
0481998

AIRCRAFT INDUSTRIES. Soviet Union
Il'iushin, Sergei Vladimirovich
UCHENYĬ konstruktor S.V. Il'îushin / [redaktšionnaîa kollegiîa : G.V. Novozhilov (otvetstvennyĭ redaktor) ... i dr.]. - Moskva : Nauka, 1978
207p,[24]p of plates
r6094186 X.620/8801 Woolwich
1980904

AIRCRAFT INDUSTRIES. Soviet Union
Lavochkin, Semen Alekseevich
ASTASHENKOV, Petr Timofeevich
Derzkie starty. - Moskva : Izdatel'stvo politicheskoĭ literatury, 1976
107 p., 4 p. of plates : ill. - (Geroi Sovetskoĭ Rodiny)
r6050895 X.708/19139
1024914

AIRCRAFT INDUSTRIES. Sweden
to date
BÖHME, Klaus-Richard
Svenska vingar växer. - Stockholm : Militärhistoriska förlaget, 1982
224p. - (Militärhistoriska studier ; nr 7). - Bibliography
91-85266-24-8 X.0809/2217(7)
8945950

AIRCRAFT INDUSTRIES. United States
to 1980
BLUESTONE, Barry
Aircraft industry dynamics. - Boston, Mass. : Auburn House, c1981
xv,208p : ill. - Bibliography: p192-197. - Includes index
0-86569-053-7 X.622/15007
2634538

AIRCRAFT INDUSTRIES. United States
to date
BRIGHT, Charles D.
The jet makers. - Lawrence : Regents Press of Kansas, c1978
xvii,228p. - Includes bibliography
0-7006-0172-4 X.322/13248
9065156

AIRCRAFT INDUSTRIES. United States
Lockheed Aircraft Corporation. Bribery of customers
BOULTON, David
The Lockheed papers. - London : Cape, 1978
xxiii,289p. - Includes index
0-224-01412-9 X.809/43659
1310070

CAMPOPIANO, Guido
Memoria di accusa contro l'onorevole Giovanni Leone ed altri scritti sull'Affare Lockheed. - Milano : SugarCo, 1978
243p. - (Fatti e misfatti ; 63)
r6195573 X.709/51437
1310070

AIRCRAFT INDUSTRIES. United States
Lockheed Aircraft Corporation. Role in political events in Italy, ca1966-1976
DE LUCA, Maurizio
Tutti gli nomini dell'antilope. - Milano : Mondadori, 1977
xiii,212p. - (L'immagine del presente ; 34)
r6085533 X.708/45913
2093227

AIRCRAFT MUSEUMS. Europe
- Directories
OGDEN, Bob
Aircraft museums directory. European edition. - London : Battle of Britain Prints International Ltd, 1978
64p : ill
0-900913-11-8 X.629/13988 Woolwich
2272156

AIRCRAFT MUSEUMS. Vigna di Valle. Italy
Museo storico dell'aeronautica militare - Catalogues
MUSEO STORICO DELL'AERONAUTICA MILITARE
Il lago degli aeroplani. - Roma : Santo Pietro, 1979
143p,[24]p of plates : ill(some col.). - (Collana d'arte e cultura)
r6505915 X.622/14435
8913684

AIRCRAFT TRADES
See also
Helicopter trades

AIRCREW WITH BURNS. Great Britain. Royal Air Force 1939-1945: Ex-patients of Queen Victoria Hospital (East Grinstead): Guinea Pig Club
WILLIAMS, Peter, 19—
McIndoe's army. - Leicester : Charnwood, 1981, c1979
222p : ill
0-7089-8016-3 Nov.44904 Woolwich
200416x

WILLIAMS, Peter, 1933-
McIndoe's army. - London : Sphere, 1981, c1979
xi,179p,[8]p of plates : ill. - Bibliography: p168. - Includes index
0-7221-4488-1 X.629/16925 Woolwich
200416x

AIRCREW WITH BURNS. Great Britain. Royal Air Force 1939-1945: Ex-patients of Queen Victoria Hospital (East Grinstead): Guinea Pig Club - Biographies
WILLIAMS, Peter, 1933-
McIndoe's army. - London : Pelham, 1979
176p,[8]p of plates : ill. - Bibliography: p.166. - Includes index
0-7207-1191-6 X.529/34239
200416x

AIRDRIE. Strathclyde Region. Scotland
Hand loom weaving industries. Organisations: Airdrie Weavers' Friendly Society, to 1968
AIRDRIE weavers / [compiled for the Local History Archives Project by Caroline Johnston, Jessie Shanks]. - [Glasgow] : [The Project], [1979]
[4] leaves,42p : ill. - Bibliography: p.42
B80-00618 X.615/2432 Woolwich
1999990

AIRDRIE WEAVERS' FRIENDLY SOCIETY
to 1968
AIRDRIE weavers / [compiled for the Local History Archives Project by Caroline Johnston, Jessie Shanks]. - [Glasgow] : [The Project], [1979]
[4] leaves,42p : ill. - Bibliography: p.42
B80-00618 X.615/2432 Woolwich
1999990

AIRDROPPED PROPAGANDA
1914–1968

The FALLING leaf. - Oxford : Museum of Modern Art, 1978
40p : ill. - Bibliography: p40
0-905836-09-x X.808/40576
316764x

AIRDROPPED PROPAGANDA. World War 1
British airdropped propaganda, 1916–1918: Facsimiles of correspondence from German prisoners of war in Great Britain

CATALOGUE of airdropped facsimile postal stationery World War 1 1916–1918 / compiled by R.G. Auckland. - [Sandridge] : Published for the Psywar Society [by R.G. Auckland], 1981
32p : ill. - (Blatter catalogue ; no.14)
B82-05812 X.629/16853 Woolwich
2591812

AIRE RIVER SYSTEM. Natural resources. West Yorkshire (Metropolitan County)
Pollution - Conference proceedings

RIVERS Aire and Calder (source to Castleford). - Leeds : Yorkshire Water Authority, [1974]
[2]p,66 leaves(3 fold),[2]fold leaves of plates : ill
0-905057-00-7 X.313/1259 Woolwich
0608807

AIRE VALLEY. West Yorkshire (Metropolitan County)
Description & travel

OGDEN, John
Yorkshire's River Aire. - Lavenham : Dalton, 1976
125,[3]p : ill. - Bibliography: p.125. - Includes index
0-900963-55-7 X.800/25751
0764086

AIRFIELDS
See Aerodromes

AIRFRAMES

CUTLER, John
Understanding aircraft structures. - London : Granada, 1981
v,170p : ill. - Includes index
0-246-11310-3 X.622/12016 Woolwich
2618087

PEERY, David J.
Aircraft structure. - 2nd ed.. - New York ; London : McGraw-Hill, c1982
ix,454p : ill. - Includes index
0-07-049196-8 X.622/13673
2618087

AIRFRAMES
Analysis

MEGSON, Thomas Henry Gordon
Aircraft structures for engineering students. - London : Edward Arnold, 1977
ix,485p : ill. - Includes index
0-7131-3393-7 X.619/19427 Woolwich
180278x

AIRFRAMES. Concorde aeroplanes
Materials

The SKIN of Concorde / prepared by Course Team. - Reprinted with corrections. - Milton Keynes : The Open University Press, 1974
84p.. - (An introduction to Materials ; Unit 16)
0-335-02669-9 P.613/68
0002720

AIRLINES

AIRLINES of the world / compiled by Michael J.H. Taylor and David Mondey. - London : Jane's, 1981
256p : ill. - (Jane's pocket book)
0-7106-0014-3 X.629/15400 Woolwich
2397692

METHOLD, Ken
Airline. - London : Evans, 1982
48p : ill(some col.). - (Talking shop)
0-237-29249-1 X.529/46644
2698390

SAMPSON, Anthony
Empires of the sky. - London : Hodder and Stoughton, 1984
254p : ill. - Includes index
0-340-34931-x X.622/21743
0618942

TAYLOR, John, b.1922
Passenger aircraft and airlines. - London : Marshall Cavendish, 1975
[2],152p : ill(chiefly col)
0-85685-142-6 X.622/5370 Woolwich
0618942

AIRLINES
- Directories

GREEN, William, 19—
The observer's world airlines and airliners directory. - 2nd ed. - London : F. Warne, 1980
416p : ill. - Includes index
0-7232-2417-x X.629/13139 Woolwich
0519979

GREEN, William, 1927-
The observer's world airlines and airliners directory. - London : F. Warne, 1975
384p : ill. - Includes index
0-7232-1547-2 X.629/10196
0519979

AIRLINES
Aeroplanes

BARRETT, Norman
Airliners. - London : Franklin Watts, 1984
32p : ill(some col.). - (Picture library). - Includes index
0-86313-134-4 X.622/23081
3272095

EUROPA. - 78 / compiled by A.B. Eastwood. - Luton : LAAS International, 1978
56p : ill
0-85075-015-6 X.700/26260
0713473

GREEN, William, 1927-
An illustrated guide to the world's civil airliners. - London : Salamander, c1982
279p : ill(some col.)
0-86101-107-4 X.629/22780
2543613

MCALLISTER, Chris
Planes and airports. - London : Batsford, 1981
64p : ill(some col.). - Includes index
0-7134-3911-4 X.629/16144 Woolwich
2543613

WORLD airline fleets. - 1977 / [compiled and edited] by Günter G. Endres. - Hounslow : Airline Publications and Sales Ltd, 1977
[7],436p : ill
Subsequent volumes, when received, will be found at SRL
0-905117-32-8 SRL
0713473

WORLD airline fleets, 'easy reference'. - 1978 / [edited] by Günter G. Endres. - Hounslow : Airline Publications and Sales Ltd, 1978
[2],92p
Subsequent volumes, when received, will be placed at this shelfmark
0-905117-43-3 P.511/5097
0713473

YOUNG, John, 1930-
Airliners. - [London] : Fontana, 1979
1folded sheet([12]p) : col.ill. - (Domino ; 12)
0-00-685455-9 Cup.911/129
240771x

AIRLINES
Aeroplanes - Directories

EUROPA. - 79 / compiled by A.B. Eastwood. - [Luton] : LAAS International Publications, [1979]
58p,[6]p of plates : ill
0-85075-028-8 X.629/13640 Woolwich
0713473

JP airline fleets. - 76 / by F.E. Bucher and U. Klee. - Hounslow : Airline Publications and Sales Ltd, 1976
[6],272p,[48]p of plates : ill
Subsequent volumes, when received, will be placed at this shelfmark
0-905117-08-5 P.621/710
0713473

TOMKINS, Nigel Milton
Airliner production list. - Hounslow : Airline Publications and Sales Ltd, 1978
[4],361p : ill
Subsequent volumes, when received, will be placed at this shelfmark
0-905117-42-5 P.621/801
0713473

WORLD airline fleets. - 1981. - Hounslow : Airline Publications & Sales, c1981
592p
0-905117-74-3 X.629/15623 Woolwich
0713473

WORLD airline fleets handbook. - 1978. - Tonbridge : Air-Britain, [1978]
201p : ill
Subsequent volumes, when received, will be placed at this shelfmark
0-85130-063-4 P.611/795
0713473

AIRLINES
Aeroplanes - Serials

AIRLINE data news. - Issue no.1 (June 1982)-. - Feltham : Aviation Data Centre, 1982-
v. : ill(some col.)
B83-33964 P.621/878
0713473

AIRLINES
Air America, to 1978

ROBBINS, Christopher
The invisible air force. - London : Macmillan, 1979
319p. - Bibliography: p.313-314. - Includes index
0-333-21847-7 X.809/44687
1945874

The invisible air force. - London : Pan in association with Macmillan, 1981, c1979
318p. - Bibliography: p311-312. - Includes index
0-330-26256-4 H.81/401 Woolwich
1945874

AIRLINES
Liability for injuries to passengers of air passenger transport services. Law

QUINTANA CARLO, Ignacio
La responsabilidad del transportista aéreo por daños a los pasajeros. - Salamanca : Universidad de Salamanca, 1977
xxxi,392p. - (Acta Salmanticensia. Derecho ; n.36). - Includes bibliographies
84-600-0858-4 Ac.153/3(3)
1978802

AIRLINES
Liveries

WORLD airline colour schemes. - Hounslow : Airline Publications and Sales Ltd.
Vol.1: [Airlines operating scheduled services to and from London's Heathrow Airport]. - [1977]
[2],73p : ill(some col)
0-905117-13-1 X.0619/15076(1) Woolwich
1486624

Vol.2: U.S. airlines. - 1977
72,[4] leaves : ill(some col)
0-905117-34-4 X.0619/15076 Woolwich
1486624

AIRLINES
Liveries - Encyclopaedias

MACKINTOSH, Ian
Encyclopedia of airline colour schemes. - Hounslow : Airline Publications and Sales Ltd.
Vol.1: North America. Part 1. - 1979
[160]p : col ill. - Includes index
0-905117-51-4 X.620/18478 Woolwich
1486624

AIRLINES
Management

TANEJA, Nawal K.
Airline planning. - Lexington, Mass. : Lexington Books, c1982
xiii,207p : ill. - Bibliography: p201-203. - Includes index
0-669-04346-x X.622/14304
2829231

AIRLINES
Organisation structure

LYNCH, James J. (James John)
Airline organization in the 1980s. - [Basingstoke] : Macmillan, 1984
xi,195 : ill. - (Globe industry report ; 3). - Includes index
0-333-38249-8 X.622/22440
3054195

AIRLINES. Germany
Lufthansa

SCHACK VON WITTENAU, Siegfried, Graf
Pionierflüge eines Lufthansa-Kapitäns. - Stuttgart : Motorbuch, 1981
221p : ill
3-87943-764-5 X.800/32542
8799717

AIRLINES. Great Britain

AIRLINES passenger & freight / industrial commentary by Roy Allen. - London : Jordan & Sons (Surveys), c1982
xxx,55p
0-85938-158-7 X.525/5954
077779x

AIRLINES. Great Britain
- Statistics

FINANCIAL results, United Kingdom airlines / [Civil Aviation Authority]. - 1968-1974. - London : C.A.A, 1975
v,45p(12 fold)
Subsequent volumes, when received, will be placed at this shelfmark
0-903083-77-9 BS.73/182
077779x

AIRLINES. Great Britain
- Statistics - Serials

UK airlines. - Jan. 1983-. - London : Civil Aviation Authority, 1983-
v.
Holdings: July 1983 etc
B85-13201 P.621/958
324735x

AIRLINES. Great Britain
Britain Airways - Serials

BRITISH AIRWAYS
Annual report and accounts. - 1972-73-. - [London] : [British Airways], [1973]-
r6000787 BS.420/13
0023280

AIRLINES. Great Britain
British Airways

ROAD, Alan
The facts about an airline. - [London] : [G. Whizzard], [1978]
[3],44p : ill(some col). - (Fact book)
0-233-96941-1 X.622/6607 Woolwich
173153x

AIRLINES. Great Britain
British Airways, to 1978

BRITISH AIRWAYS
The story of British Airways. - [London] : [British Airways], [1979]
44 leaves,[8]leaves of plates : ill
B79-26328 BS.420/20
1926551

AIRLINES. Great Britain
British Airways. Board. Loans. Default. Guarantees by Great Britain. Treasury

GREAT BRITAIN. Treasury
Statement of guarantee given by the Treasury on 22nd July 1977 in pursuance of section 11(1) of the British Airways Board Act 1977 for the payment of a debt incurred in United States dollars by the British Airways Board ... - London : H.M.S.O., [1977]
Sheet ([2]p). - ([1977-78 H.C.] 74)
Later statements are also placed at this shelfmark
0-10-207478-x BS.420/17
1409921

AIRLINES. Great Britain
British Airways. Air passenger transport services. Fares

BRITISH AIRWAYS
Air transport. - [London] : [British Airways], [197-]
3v. : col ill. - Includes index
r6032012 BS.420/14
1099248

AIRLINES. Great Britain
British Airways. Passenger transport services. Boeing 747 aeroplanes. Crews. In-flight duties on long journeys

STEWART, Stanley
Heathrow-Chicago. - Shepperton : Ian Allan, 1984
80p : ill. - (From the flightdeck ; 1)
0-7110-1434-5 X.622/23090
3274748

AIRLINES. Great Britain
British Airways. Passenger transport services. Boeing 747 aeroplanes. Supervisory stewards. In-flight duties on long journeys

COSTELLO, Martin
A study of supervisory grade cabin crew jobs on wide bodied aircraft. - Staines : Air Transport and Travel Industry Training Board, 1978
ix,187p. - (Research reports / Air Transport and Travel Industry Training Board ; 78/1)
0-86141-001-7 BS.23/979
1876503

AIRLINES. Great Britain
British Airways. Relations with British Caledonian Airways

THOMSON, Adam
Air transport, government subsidised or commercially viable?. - Edinburgh : Scottish Academic Press, 1975
23p : ill. - (Jubilee lectures / Edinburgh University Commerce Graduates' Association ; 1974)
0-7073-0119-x X.519/25623
0620122

AIRLINES. Great Britain
British Airways. Superannuation schemes: Airways Pension Scheme - Serials

AIRWAYS PENSION SCHEME
Annual report and accounts for the year ended 31 March. - [19-]-. - Hounslow : The Scheme, [19-]-
v. : ill
Holdings: 1983-
r6813663 P.823/274
2856573

The AIRWAYS Pension Scheme bulletin. - [19-]-. - Hounslow : The Trustees of the Airways Pension Scheme, [19-]-
v. : ill
r6863091 P.2000/816
2856573

AIRLINES. Great Britain
British Caledonian Airways - For petroleum industries - Serials

OIL flyer / British Caledonian Airways. - Apr. 1983-. - Crawley : British Caledonian Airways, 1983-
v. : ill
r6863373 P.613/320
2975750

AIRLINES. Great Britain
British Midland Airways, to 1979

CRAMP, B. G.
British Midlands Airways. - Hounslow : Airline Publications and Sales Ltd, 1979
208p : ill
0-905117-57-3 X.629/13605 Woolwich
2108968

AIRLINES. Great Britain
Imperial Airways, 1924

IMPERIAL AIRWAYS
Pilot's handbook and general instructions, 1924. - London : Ducimus Books [for] British Airways, 1974
96p : ill
0-903234-22-x X.619/16572 Woolwich
0605158

AIRLINES. Great Britain
Independent airlines, 1946-1975

JONES, Anthony Charles Merton
British independent airlines since 1946. - Uxbridge : LAAS International ; Liverpool : Merseyside Aviation Society.
Vol.2: (Brooklands Aviation to Irelfly). - 1976
p121-240 : ill
0-902420-08-9 X.0512/514(2)
0631345
- Uxbridge : LAAS International ; Liverpool : Merseyside Aviation Society Ltd.
Vol.3: Irish Air Charter to Sagittair. - 1976
p241-372 : ill
0-902420-09-7 X.0512/514
0631345
Vol.4: Scillonia Airways to Yorkshire Aeroplane Club. - 1977
p373-504 : ill
0-902420-10-0 X.0512/514
0631345

MERTON JONES, Anthony Charles
British independent airlines since 1946. - Uxbridge : LAAS International ; Liverpool : Merseyside Aviation Society.
[Vol.1]. - 1976
120p : ill. - Bibliography: p.3
0-902420-07-0 X.0512/514(1)
0631345

AIRLINES. Hong Kong
Cathay Pacific Airways

EATHER, Charles (Charles Edward James)
Syd's pirates. - Sydney : Durnmount, 1983
341p : ill. - Includes bibliography
0-949756-05-9 X.622/23020
9162496

AIRLINES. Ireland (Republic)
Aer Lingus, to 1980

RE-INTRODUCING Aer Lingus. - [Dublin] : Aer Lingus Public Relations and Publicity Departments, 1983
39p : ill
B84-22364 X.629/22793
3085236

AIRLINES. Latin America
to 1982

DAVIES, R. E. G. (Ronald Edward George)
Airlines of Latin America since 1919. - London : Putnam, 1984
xiv,698p : ill. - Includes index
0-370-30049-1 X.629/26276
3283879

AIRLINES. Netherlands
Luchtvaart Maatschappij, to 1977

ALLEN, Roy
Pictorial history of KLM, Royal Dutch Airlines. - London : I. Allan, 1978
192p : ill
0-7110-0845-0 X.620/17914 Woolwich
1578014

AIRLINES. Scandinavia
Scandinavian Airlines System

HAGRUP, Knut
SAS og den internasjonale luftfart. - Bergen : Norges Handelshøyskole, 1974
34 p.. - (Kristofer Lehmkuhl forelesning ; 1974)
r6004127 X.519/20958
088068X

AIRLINES. Scandinavia
Scandinavian Airlines System - Serials

SCANORAMA. - Vol.1, no.1- ; Feb.-Mar. 1972-. - Stockholm : Scandinavian Airlines System, 1972- : ill(chiefly col)
Holdings : vol.3.no.6- ; Dec. 1974/Jan. 1975-
r6032265 P.801/3089
1205307

AIRLINES. Scandinavia
Scandinavian Airlines System, to 1971

BURAAS, Anders
Fly over fly. - [Oslo] : Gyldendal, [1972]
311 p. : ill
82-05-00892-2 X.800/12044
1019422

AIRLINES. South Korea
Taehan Hanggong, Chusik, Hoesa. Passenger aeroplanes. Attack by Union of Soviet Socialist Republics. Voenno-vozdushnye sily, 1983. Fighter aeroplanes

ROHMER, Richard
[Massacre 747]. Massacre 007. - [Sevenoaks] : Coronet, 1984
213p
0-340-36447-5 X.808/40930
3023834

AIRLINES. United States

TANEJA, Nawal K.
The commercial airline industry. - Lexington, Mass. : D. C. Heath, 1976
xix 341 p. : ill. - Bibliography: p. 321-333. - Includes index
0-669-00129-5 X.520/11222
0782912

AIRLINES. United States
to date

FRADENBURG, Leo G.
United States Airlines. - Dubuque, Iowa : Kendall / Hunt, c1980
xx,446p. - (Aviation management series)
0-8403-2128-7 X.622/17618
9024247

AIRLINES. United States
1914-1981

DAVIES, R. E. G. (Ronald Edward George)
Airlines of the United States since 1914. - London : Putnam, 1972
xiii,746p : ill. - Bibliography: p578-580. - Includes index
0-370-30942-1 X.629/20436
291431x

AIRLINES. United States
American airlines. Economic aspects

AIRLINE economics / edited by George W. James. - [Lexington, Mass.] : Lexington Books, 1982
xxx,316p : ill. - Includes index
0-669-04909-3 X.520/27353
2751550

O'CONNOR, William E. (William Edmund)
An introduction to airline economics. - 2nd. ed. - New York : Praeger, 1982
xv,286p. - Bibliography: p172-178
0-03-059791-9 X.520/28226
2751550

AIRLINES. United States
Delta Air Lines

LEWIS, W. David
Delta. - Athens [Ga.] : The University of Georgia Press, c1979
xiii,503p,81p of plates : ill. - Bibliography
0-8203-0465-4 X.800/32182
8799709

AIRLINES. United States
Eastern Airlines

SERLING, Robert J.
From the captain to the colonel. - New York : Dial Press, [1980]
535p,16p of plates : ill
0-8037-4610-5 X.622/15426
8799709

AIRLINES. United States
Eastern Airlines. Aeroplanes. Ghosts. Investigation

FULLER, John G. (John Grant)
The ghost of flight 401. - London : Souvenir Press, 1978
319p
0-285-62356-7 X.809/43403
1490419

The ghost of flight 401. - London : Corgi, 1979
300p
0-552-11020-5 H.79/1989 Woolwich
1490419

AIRLINES. United States
Eastern Airlines. Aeroplanes. Ghosts. Investigation - Biographies

FULLER, Elizabeth
My search for the ghost of Flight 401. - London : Corgi, 1979
[3],216p
0-552-11220-8 H.80/569 Woolwich
1490419

AIRLINES. United States
Fleets. Re-equipment. Finance

MANDELL, Robert W.
Financing the capital requirements of the U.S. airline industry in the 1980s. - Lexington, Mass. : Heath ; [Farnborough, Hants] : [Distributed by] Gower Press, 1980
xviii,163p. - Bibliography: p.155. - Includes index
0-669-03215-8 X.622/8270 Woolwich
2235897

AIRLINES. Wales
Cambrian Airways, to 1976

STADDON, Tommy
History of Cambrian Airways. - Hounslow : Airline Publications & Sales Ltd, 1979
111p : ill
0-905117-56-5 X.519/30386
1993925

AIRMAIL
1920-1940

JACKSON, Donald Dale
Flying the Mail. - Alexandria, Va. : Time-Life, c1982
176p : ill(some col.). - (The Epic of flight). - Bibliography: p172
0-8094-3329-x L.42/3176
318014x

AIRMAIL. Great Britain
to 1934

GAMMONS, Tony
A handbook for the National Postal Museum [exhibition] Half way round the world. - London : The Museum, [1984]
20p : 1ill
B85-00785 X.429/16530
3195120

AIRMAIL. Great Britain
1936-1938

FIELD, John Charles Walsh
All up air mail. - Sutton Coldfield : F.J. Field, 1978
23p : ill. - (A popular pamphlet). - Bibliography: p.23
B78-29460 X.519/28370
157051x

AIRMAIL. Scandinavia
to 1970

LÜNING, Örjan
Luftpostens historia i Norden. - Stockholm : Sveriges filatelist-förbund, 1978
351p : ill(some col.). - (Specialhandbok ; nr 10)
r6698008 X.622/9167 Woolwich
2469294

AIRMAIL. Yorkshire
to 1982

BEITH, Richard
Yorkshire air mails. - Sheffield : Yorkshire Postal History Society, c1983
45p,[25]p of plates. - (Special series ; no.5). - Bibliography: p44-45
0-907022-08-1 X.622/16891
2960885

AIRMAIL POSTMARKS
Ethiopian airmail postmarks, to 1934

ETHIOPIA booklet no.VII. - Newbury : Philip Cockrill, [1982?]
48p : ill. - (Cockrill series booklet : no.19)
B85-12424 X.629/25435
3246655

AIRMAIL POSTMARKS
Hungarian airmail postmarks, 1918-1970

HOMOLYA, E.
Hungarian air mail postmarks. - [Grandborough] : [Magyar Philatelic Society of Great Britain], [1974]
26p : ill
0-9504552-0-2 X.519/25262
0519294

AIRMEN
See also
Paratroopers

AIRMEN
Great Britain. Royal Air Force. Aerodromes: Brawdy RAF Station - For airmen

ROYAL AIR FORCE, BRAWDY
Arrival information handbook. - Gloucester : British Publishing Co., [1977]
64p : ill
0-7140-1586-5 X.708/20863
1103482

AIRMEN
Great Britain. Royal Air Force. Marine stations: Mount Batten RAF Station - For airmen

MOUNT BATTEN RAF STATION
Information handbook. - Gloucester : British Publishing Co., [1977]
40p : ill
0-7140-1567-9 X.708/20862
1103466

AIRMEN
Serbian airmen. Role in World War 2

SPOMENICA palib Srba vazduhoplovaca 1941-1945 / uredio Miloš K. Aćin... - Lahr : Srpska Krila, 1975
297p : ill
r6046614 X.800/12690
1243772

AIRMEN
Soviet airmen. Air operations during Spanish Civil War

SMIRNOV, Boris Aleksandrovich
Ot Madrida do Khalkhin-Gola. - Izd. 2., ispr. i dop.. - Kuĭbyshev : Kuĭbyshevskoe knizhnoe izdatel'stvo
341 p. : ill
r6022909 X.809/23338
1009745

AIRMEN. Air operations, 1939. Mongolia (People's Republic)
Soviet airmen

SMIRNOV, Boris Aleksandrovich
Ot Madrida do Khalkhin-Gola. - Izd. 2., ispr. i dop.. - Kuĭbyshev : Kuĭbyshevskoe knizhnoe izdatel'stvo
341 p. : ill
r6022909 X.809/23338
1009737

AIRMEN. Belgium
Allied airmen. Escapes, to 1945

BUSSELS, J.
De doodstraf als risico. - [Hechtel-Eksel] : J. Bussels, 1981
339p : ill
r6330649 X.809/57021
9016333

AIRMEN. Denmark
Allied airmen, 1939-1945

BJØRNVAD, Anders
Faldne allierede flyvere 1939-1945. - Odense, 1978
218p : ill
87-7492-254-8 X.700/21258
213392X

AIRMEN. France
Allied airmen. Escapes. Role of resistance movements, 1944

RÉMY
Mission Marathon. - Paris : Librairie Académique Perrin, 1974
449 p., 16 p. of plates
r6002814 X.809/22436
0917370

AIRMEN. Soviet military forces
Kuibyshev airmen. Air operations during World War 2

OGNENNYE tarany / [sostavitel' B.M. Kosenkov]. - Kuĭbyshev : Kuĭbyshevskoe knizhnoe izdatel'stvo, 1978
100p
r6096777 X.708/46289
2094878

AIRMISSES. Aircraft. Great Britain
- Serials

GENERAL aviation airmisses / Joint Airmiss Working Group. - May 1980-. - London : Civil Aviation Authority, 1980-
v.
B81-02010 SRL
2327295

AIRO, AKSEL FREDRIK. Finland. Armeija
- Biographies

KUOSA, Tauno
A.F. Airo. - Porvoo : W. Söderström, 1979
268p,plates
951-00-9591-5 X.809/56463
8989567

AIRPORT EQUIPMENT INDUSTRIES. Great Britain
Civil airport equipment industries - Catalogues - Serials

BRITISH airport equipment catalogue. - 1st ed. (1980)-. - Farnborough : Combined Service Publications, 1980-
v. : ill
B81-04592 SRL
234064x

AIRPORT-RELATED EMPLOYMENT. Gatwick Airport region. South-east England

CRAWLEY. Borough Council
Airport related employment in the Gatwick area. - [Crawley] : [Crawley District Council], [1979]
41,[17]p : ill
0-900800-34-8 X.622/8851 Woolwich
2319845

INSTITUTE OF MANPOWER STUDIES
Gatwick airport and the labour market. - Brighton : Institute of Manpower Studies, 1983
vi,202p : ill
B84-00373 X.525/8296
2319845

AIRPORTS
See also
Aerodromes
Airport equipment industries
Areas adjacent to airports
Gatwick Airport region
Model airports

AIRPORTS

ALLEN, Roy
Major airports of the world. - London : I. Allan, 1979
128p : ill. - Includes index
0-7110-0918-x X.809/44640
1731092

Major airports of the world. - 2nd ed. - London : Ian Allan, 1983
124p : ill. - Includes index
0-7110-1267-9 X.629/21552
1731092

ASHFORD, Norman J.
Airport operations. - New York ; Chichester : Wiley, c1984
viii,476p : ill. - Includes index
0-471-89613-6 X.622/20295
1731092

BARTON, Byron
Airport. - London : MacRae, 1982
[24]p : chiefly col.ill
0-86203-111-7 L.49/1629
0176346

Airport. - London : Fontana, 1984, c1982
[32]p : ill. - (Picture lions)
0-00-662264-x X.629/23745
0176346

JAY, Michael, 1946-
Airports. - London : Watts, 1982
32p : col.ill. - (A First look book). - Includes index
0-85166-953-0 X.629/17711
0176346
- Reprinted - 1983 X.629/21152

KURTH, Heinz
At the airport. - Harmondsworth : Puffin Books, 1975
32p : ill(chiefly col). - (A new puffin picture book)
0-14-049121-x X.519/25333
0176346

MCALLISTER, Chris
Planes and airports. - London : Batsford, 1981
64p : ill(some col.). - Includes index
0-7134-3911-4 X.629/16144 Woolwich
1731092

AIRPORTS –continued

MILBURN, Constance
Airports. – Glasgow : Blackie, 1983
32p : col.ill. – (Inside story). – Includes index
0–216–91294–6 X.629/20302
0176346

PEACOCK, Frank
Let's go to the airport. – London : Watts, c1974
[36]p : col.ill. – (Let's go series)
B82–25841 X.629/17821 Woolwich
0176346

PETTY, Mike, 1947–
The airport. – London : Jackdaw [for] Thames Television, 1977
Portfolio : ill(some col). – (Magpies). – Bibliography: (1p.)
0–305–62135–1 Cup.911/43
1189409

PREGO DE OLIVER, Victoria
Airports. – Hove : Wayland, 1979
3–87p : ill(chiefly col). – (A read and build book)
0–85340–633–2 X.622/6842 Woolwich
0176346

ROSS
Ross's guide to airports. – London : Abelard [etc.], 1977
[28]p : chiefly col ill
0–200–72404–5 Cup.1281/710
0176346

RUTLAND, Jonathan
See inside an airport. – London : Hutchinson, 1977
4–29p : ill(chiefly col). – (See inside). – Bibliography: p.28. – Includes index
0–09–131480–1 X.622/6065 Woolwich
0176346

SIMPSON, Julie
Come and see the airport. – London : Longman, 1978
[32]p : ill(chiefly col). – (Come and see)
0–582–39013–3 X.619/19251 Woolwich
1758292

STROUD, John, b.1919
Airports of the world. – London : Putnam, 1980
xxiv,605p : ill. – Includes index
0–370–30037–8 X.809/45910
1731092

AIRPORTS
– Conference proceedings

WORLD AIRPORTS CONFERENCE (5th : 1976 : London)
Airports, the challenging future. – London : Institution of Civil Engineers, 1976
[3],251p : ill
0–7277–0017–0 X.525/2130
1731092

AIRPORTS
– Directories – Serials

AIRPORTS international directory. – 1981–. – Sutton : IPC Transport Press Ltd. for Airports international, c1981–
v.
B81–33288 P.521/3559
2544849

AIRPORTS
– Texts

SIMPSON, Julie
Come inside the airport. – London : Studio Vista, 1974
[27]p : ill(chiefly col)
0–289–70477–4 X.520/10309
0176346

AIRPORTS
Derelict aeroplanes & preserved aeroplanes

TOMKINS, Nigel Milton
Air transport hulks. – Hounslow : Airline Publications and Sales Ltd, 1979
126p : ill
0–905117–63–8 X.629/12876 Woolwich
2104490

AIRPORTS
Design

ASHFORD, Norman
Airport engineering. – New York ; Chichester : Wiley, 1979
ix,449p : ill. – Includes bibliographies and index
0–471–07255–9 SRL
2016427

AIRPORTS
Design & construction – Conference proceedings

WORLD AIRPORTS CONFERENCE (6th : 1979 : London)
Airports for the community. – London : Institution of Civil Engineers, 1980
[3],128p : ill
0–7277–0087–1 X.622/8676 Woolwich
2258544

AIRPORTS
Equipment – Serials

JANE'S airport equipment. – 1st ed. (1982–83)–. – London : Jane's Pub. Co., 1982–
v. : ill
B83–04821 SRL
2825503

AIRPORTS
Planning. Environmental aspects

ORGANISATION FOR ECONOMIC CO–OPERATION AND DEVELOPMENT
Airports and the environment. – Paris : OECD, 1975
290 p.
92–641–1377–0 UN.P.105/174
0194573

AIRPORTS
Planning. Political aspects

FELDMAN, Elliot J.
Technocracy versus democracy. – Boston : Auburn House, c1982
xlv,299p : ill. – Includes index
0–86569–063–4 X.622/12661 Woolwich
2729814

AIRPORTS
Regional planning

DE NEUFVILLE, Richard
Airport systems planning. – London : Macmillan, 1976
xvi,201p : ill. – Bibliography: p.182–194. – Includes index
0–333–17622–7 X.620/16316 Woolwich
0736589

AIRPORTS. Aberdeen. Grampian Region. Scotland
Aberdeen Airport. Development

The DEVELOPMENT of Aberdeen airport 1981–1995. – Glasgow : Scottish Airports, 1981
28p,[7]p,[5]leaves of plates(3 folded) : ill
B81–37626 BS.401/29
2560542

AIRPORTS. Aberdeen. Grampian Region. Scotland
Dyce Airport. Industrial relations. Disputes between Bristow Helicopters Limited & British Air Line Pilots Association

MCDONALD, Robert Howat, Lord McDonald
Report of a court of inquiry into a trade dispute at Dyce Airport, Aberdeen, between Bristow Helicopters Limited and members of the British Air Line Pilots Association, conducted by the Hon. Lord McDonald MC ... – London : H.M.S.O., [1977]
[1],54p. – (Cmnd.6951)
0–10–169510–1 BS.23/866
1257854

AIRPORTS. Belfast
Belfast Airport

BELFAST Airport / issued by authority of Northern Ireland Airports Ltd. – Wallington : Home Publishing, [1980?]
52p : ill
B83–05575 X.809/53893
2824167

NORTHERN IRELAND AIRPORTS
[Belfast Airport]. Belfast International Airport handbook. – Wallington : Home Publishing, 1984
56p : ill(some col.)
B84–40916 X.629/24356
2824167

AIRPORTS. Benelux countries
– For air services – Directories

AIR touring flight guide, Benelux. – 1975, Jan. / compiled and edited by Carolyn Pooley. – [Boreham Wood] : Airtour International Ltd, [1975]
136p : ill
Subsequent volumes, when received will be placed at this shelfmark
0–9502333–6–6 P.701/1521
0728993

AIRPORTS. Birmingham. West Midlands (Metropolitan County)
Birmingham Airport, to 1979

BIRMINGHAM Airport / editorial co-ordinated by B.J. Cheater. – Gloucester : British Publishing Co. [for] Birmingham Airport, [1979]
96p : ill(some col)
0–7140–1861–9 L.49/552
2273179

AIRPORTS. Blackpool. Lancashire
Blackpool Airport – Directories

BLACKPOOL Airport. – London : Burrow, [1977]
44p : ill
B77–30156 X.519/27267
1246445

BLACKPOOL Airport. – [New ed.]. – Gloucester : British Publishing Co., [1980]
64p : ill
Subsequent guides, when received, will be placed at this shelfmark
0–7140–1862–7 X.0700 / 3091
1246445

AIRPORTS. Bournemouth. Dorset
Bournemouth-Hurn Airport

BOURNEMOUTH-HURN airport. – Gloucester : British Publishing, c1981
52p : ill
B84–14479 X.800/37296
3060381

AIRPORTS. Brazil
Law

MOURA, Geraldo Bezerra de
Aspectos jurídicos dos aeroportos comerciais. – São Paulo : Editora Affonso, [1975?]
345p. – Bibliography : p.339–345
r6037902 X.100/26305
1323083

AIRPORTS. Canada
Policies of government

FELDMAN, Elliot J.
The politics of Canadian airport development. – Durham, N.C. : Duke University Press, 1983
xix,261p. – (Duke press policy studies)
0–8223–0479–1 X.520/32573
3066010

AIRPORTS. Castle Donnington. Leicestershire
East Midlands Airport

EAST Midlands Airport. – 7th ed.. – Carshalton : Home Publishing Co., [1976]
72 p. : ill
r6009715 X.519/26000
0688169

EAST Midlands Airport. – [New ed.]. – [Gloucester] : [British Publishing Co.], [1978]
96p
Subsequent editions, when received, will be placed at this shelfmark
0–7140–1800–7 X.0709/1084
0688169

AIRPORTS. Coventry. West Midlands (Metropolitan County)
Coventry Airport

COVENTRY Airport official handbook. – [London] : Burrow, [1975]
20p : ill
0–85026–598–3 X.519/26285
052686x

AIRPORTS. Crawley. West Sussex
Gatwick Airport, to 1980

KING, John, 1945–
Golden Gatwick. – [Gatwick?] : Royal Aeronautical Society, Gatwick Branch, c1980
[72]p : ill
B83–09185 X.629/19312
283703x

AIRPORTS. Crawley. West Sussex
Gatwick Airport. Civil aeroplanes. Arrival & departure

GATWICK schedules. – 1976 / [compiled by S.A. and R.A. [i.e. R.R.] Hillier]. – Hounslow : Airline Publications and Sales Ltd, [1976]
[1],17p : ill
Subsequent volumes, when received, will be placed at this shelfmark
0–905117–10–7 P.511/5020
0734829

AIRPORTS. Crawley. West Sussex
Gatwick Airport. Second terminal. Development. Proposals

SECOND terminal, Gatwick and other works / inspector, John Newey ; assessors, E.B. Haran, J.B. Veal ; secretary, Charles Nicholls. – London : H.M.S.O., 1981
iii,178p. – Includes index
0–11–751540–x BS.414/2072
2701898

AIRPORTS. Darlington. Durham (County)
Tees-side Airport

TEES-SIDE Airport Authority official handbook. – London : Burrow, [1979]
52p : ill
Later editions, when received, will be placed at this shelfmark
B79–37248 X.0709/1082
1996606

AIRPORTS. Eastleigh. Hampshire
Southampton (Eastleigh) Airport. Aviation, to 1946

NEW, Peter T. (Peter Talbot)
The Solent sky. - [Chandlers Ford] : [The author], [1976]
[8],118p : ill. - Bibliography: p.114-115
0-9504984-0-8 X.615/1608 Woolwich
0643785

AIRPORTS. England
London Airport. Third site

GREAT BRITAIN. Advisory Committee on Airports Policy
Report of the Advisory Committee on Airports Policy, the need for a third London airport. - London : H.M.S.O., 1979
v,53p : ill
0-11-512920-0 BS.41/1211
2100088

AIRPORTS. England
London Airport. Third site. Choice. Government decision making

TURRALL-CLARKE, Robert
Stansted. - [Cambridge] : Cambridge Academic Publications, [1984?]
95p in various pagings
B85-11208 X.629/25310
0136506

AIRPORTS. England
London Airport. Third site. Development. Potential effects on coastal ecology of Essex

BOORMAN, L. A.
Ecology of Maplin Sands and the coastal zones of Suffolk, Essex and North Kent. - Cambridge : The Institute, 1977
56p : ill. - Bibliography: p.55
0-904282-14-7 SRL
1707191

AIRPORTS. England
London Airport. Third site. Development. Potential effects on coastal ecology of North Kent

BOORMAN, L. A.
Ecology of Maplin Sands and the coastal zones of Suffolk, Essex and North Kent. - Cambridge : The Institute, 1977
56p : ill. - Bibliography: p.55
0-904282-14-7 SRL
1707183

AIRPORTS. England
London Airport. Third site. Location

BRINGING the third London airport down to earth / compiled by the Willingale Anti-Airport Group. - Ongar : Willingale Anti-Airport Group, [1979]
15leaves
0-9506718-1-9 X.805/3063
2100096

GREAT BRITAIN. Study Group on South East Airports
Report of the Study Group on South East Airports, possible sites for a third London airport. - London : H.M.S.O., 1979
v,116p,fold leaves of plates,[38]p of plates : ill
0-11-512921-9 BS.41/1212
2100096

AIRPORTS. England
Manual personnel. Remuneration. Determination

GREAT BRITAIN. Standing Commission on Pay Comparability
Municipal airport manual workers. - London : H.M.S.O., [1980]
v,18p. - (Report ; no.6) (Cmnd. 7852)
0-10-178520-8 BS.68 / 247(6)
2256975

AIRPORTS. England
Passengers. Origins & destinations

CIVIL AVIATION AUTHORITY
Passengers at major airports in the North, the North West and the south of England. - London : C.A.A., 1979
[2],64p : ill. - (CAP ; 423)
B80-21727 X.525/3189
2143798

AIRPORTS. Entebbe. Uganda
Entebbe Airport. Passengers: Israeli hostages of Palestinian Arabs. Rescue by Israeli military forces

BEN-PORAT, Yeshayahu
Entebbe rescue. - New York : Delacorte Press, [1977]
xvi,346p,16p of plates
0-440-02162-6 X.800/31955
0738581

OFER, Yehuda
Operation Thunder. - Harmondsworth : Penguin, 1976
iii-x,141p,[8]p of plates : ill. - (A Penguin special)
0-14-052321-9 12208.a.2/318
0738581

STEVENSON, William, 1924-
90 minutes at Entebbe. - Toronto ; London : Bantam, 1976
viii,216p,[16]p of plates : ill
0-552-60482-8 X.708/21263
0738581

WILLIAMSON, Tony
Counterstrike Entebbe. - London : Collins, 1976
[11],184p. - Includes index
0-00-216138-9 X.809/41253
0738581

AIRPORTS. Europe
- For air services - Directories

EUROPE flight guide. - 1980, Jan. / compiled and edited by Robert Pooley. - [Boreham Wood] : Airtour Associates International Ltd., [1980]
288p : ill
B80-25479 X.808/33363
0154083

AIRPORTS. Exeter. Devon
Exeter Airport

EXETER airport / issued by authority of Exeter Airport Ltd. - Wallington : Home Publishing, [1982?]
36p : ill
B82-36029 X.809/53046
2778408

AIRPORTS. Foulness. Essex
London Airport. Third site. Development. Law. Great Britain. Parliament. House of Commons. Maplin Development Bill. Session 1972-73

GREAT BRITAIN. Parliament. House of Lords. Select Committee on the Maplin Development Bill
Special report by the Select Committee of the House of Lords on the Maplin Development Bill, together with the proceedings of the Committee and minutes of evidence. - London : H.M.S.O., 1973 [i.e. 1974]
viii,342p. - (1973-74 H.L.179-I)
0-10-475673-x BS.96/70
0093653

AIRPORTS. France
to date - Serials

AÉROPORT DE PARIS
Rapport du Conseil d'administration. - Exercise (1978)-. - Paris : [l'Aéroport], [1979]-
r6206336 X.0622/643 Woolwich
8756988

AIRPORTS. Frankfurt am Main. West Germany
to date. Environmental aspects

ALLE Gewalt geht vom Staate aus. - Frankfurt am Main : Bunte Hilfe - Startbahn West, 1982
181p : ill
r6308953 X.529/36266
891656x

BUNTE HILFE
Einigkeit + Specht + Freiheit. - Frankfurt am Main : Bunte Hilfe, [1982]
[60]p : all ill
r6310156 X.429/16565
891656x

DANNEMANN, Christiane
Die Startbahn West ist überall. - München : Kaiser, 1982
216p
3-459-01467-9 X.809/57794
891656x

Die FALSCHE Richtung : Startbahn West / herausgegeben von Heinz Mees und W. G. Reinheimer. - Rüsselsheim : Edition Venceremos, 1982
300p : ill. - (Venceremos Taschenbuch)
3-88541-018-4 X.529/66257
891656x

KARASEK, Horst
Das Dorf im Flörsheimer Wald. - 2. Aufl. - Darmstadt : Luchterhand, 1981
208p : ill. - (Sammlung Luchterhand ; 368)
3-472-61368-8 X.629/20405
891656x

"KEINE Startbahn West" / Bürgerinitiative gegen die Flughafenerweiterung Frankfurt am Main (Herausgeber). - 2. aktualisierte Aufl. - Offenbach : Verlag 2000, [1981]
202p,plate : ill. - (Links pocket ; Bd. 7)
3-88534-306-1 X.808/38002
891656x

KUNST gegen Startbahn West / herausgegeben von Bruno Struif ; Autorengruppe : Alex Dill ... [et al.]. - Giessen : Anabas, 1982
167p : chiefly ill
3-87038-094-2 X.329/17429
891656x

NUR wer sich bewegt, spürt seine Fesseln / Bürgerinitiative gegen die Flughafenerweiterung Frankfurt Rhein-Main (Herausgeber) ; [Redaktion : Ingo Funk et al.]. - Offenbach : Verlag 2000, c1982
316p : ill. - (Links Pocket ; Bd.10). - Includes bibliography
3-88534-309-6 X.529/63402
891656x

OESER, Kurt
Es ist nie zu spät. - Dreieich : Bioverlag Gesundleben, [1981]
183p : ill
3-922434-10-x X.629/22434
891656x

TESTFALL Startbahn West / herausgegeben von Gottfried Orth und Adalbert Podlech. - Wuppertal : Jugenddienst-Verlag, 1982
158p. : ill
3-7795-7366-0 X.529/59730
891656x

AIRPORTS. Frankfurt am Main. West Germany
to date. Military aspects

MÜLLER, Harald, writer on defence
Startbahn für die Militärs?. - Frankfurt am Main : HSFK, 1982
iii,72p. - (Forschungsberichte aus der Hessischen Stiftung Friedens- und Konfliktforschung ; 1/1982)
r6310465 X.629/25796
915177x

AIRPORTS. Gravesend. Kent
Gravesend Airport, 1932-1958

CONNOLLY, P. A.
Gravesend Airport, 1932-1958. - [Rochester?] : [Kent County Library, Gravesham Division], [1979?]
[6]p. - (Local history pamphlet ; no.5)
B84-07118 X.809/57360
3038114

AIRPORTS. Great Britain

BARRETT, Sean D.
Airports for sale. - London : Adam Smith Institute, 1984
iv,27p. - Bibliography: p26-27
0-906517-52-4 X.529/66883
3225704

BRITISH AIRPORTS AUTHORITY
Airport information, Heathrow, Gatwick, Stansted, Glasgow, Edinburgh, Prestwick, Aberdeen. - Folkestone : [The Authority], 1977
96p : ill
B80-08706 BS.401/23
2044935

CONNER, Edwina
Working at an airport. - Hove : Wayland, 1982
70p : ill. - (People at work). - Includes index
0-85340-928-5 X.622/14652
2845288

TURNER, Paul St John
UK & Eire commercial airports. - Hounslow : Airline Publications and Sales Ltd, 1974
97p : ill
0-905117-17-4 X.519/26199
0741043

WRIGHT, Alan J. (Alan James)
British airports. - London : Ian Allan, 1980
80p : ill
0-7110-0967-8 X.629/16705 Woolwich
2044935

British airports. - 2nd ed. - London : I. Allan, 1983
80p : ill
0-7110-1290-3 X.629/21240
2044935

AIRPORTS. Great Britain
- For air services - Directories

POOLEY'S flight guide United Kingdom and Ireland. - 1980, Mar. : [17th ed.] / compiled and edited by Robert Pooley ... with the assistance of the Civil Aviation Authority. - London : Airtour Associates International Ltd, [1980]
356p : ill
Subsequent volumes, when received, will be placed at this shelfmark
B80-28221 P.801/1537
0154105

AIRPORTS. Great Britain
- Serials

BRITISH airports world. - [No.1]- ; Nov. 1978-. - London : British Airports Authority, [1978]- : ill(chiefly col)
B79-14640 BS.401/28
1836323

AIRPORTS. Great Britain
- Statistics - Serials

UK airports. - Jan. 1983-. - London : Civil Aviation Authority, 1983-
v.
Holdings: July 1983 etc
B85-13200 P.823/594
3247341

AIRPORTS. Great Britain
Air traffic. Forecasting

GREEN, J. H. T.
United Kingdom air traffic forecasting. - London : Department of Trade, 1978
[5],24p
B78-10845 BS.41/1112
1376896

AIRPORTS. Great Britain
Civil aeroplanes. Arrival & departure

UK and Ireland airport schedules. - 1979 / [compiled] by R. King and S.A. & R.R. Hillier. - [Hounslow] : [Airline Publications and Sales Ltd], [1979]
[3],92p : 1 ill
Subsequent volumes, when received, will be placed at this shelfmark
0-905117-59-x P.861/303
2062674

AIRPORTS. Great Britain
Civil aeroplanes. Arrival & departure - Serials

HEATHROW & Gatwick handbook. - 1983-. - Feltham : Aviation Data Centre, 1983-
v. : ill
B85-05542 P.861/702
2062674

AIRPORTS. Great Britain
Conditions of use

BRITISH AIRPORTS AUTHORITY
British Airports Authority airports. - From 1st April 1972-. - [London] : British Airports Authority, [1972]-
30cm
Holdings: 1972, 1977-
B78-14190 BS.401/22
141044x

AIRPORTS. Great Britain
Local authority airports - Statistics

CHARTERED INSTITUTE OF PUBLIC FINANCE AND ACCOUNTANCY. Airports Working Group
Financial costs and statistics of local authority airports. - [London] : [Chartered Institute of Public Finance and Accountancy], [1979]
49p
B80-41673 L.49/613
2317524

AIRPORTS. Great Britain
Location. Economic aspects

WHY London?. - Gatwick : External Relations, British Airports Authority, [1982?]
16p : col.ill
B83-05829 BS.401/31
2824817

AIRPORTS. Great Britain
Location. Regional planning

SEALY, Kenneth Royston
Airport strategy and planning. - London : Oxford University Press, 1976
iv,60p : ill. - (Theory and practice in geography). - Includes bibliographies
0-19-874041-7 X.519/25688
0632481

AIRPORTS. Great Britain
Luggage. Handling

AIR TRANSPORT AND TRAVEL INDUSTRY TRAINING BOARD
A survey of baggage handling. - Staines : The Board, [1977]
[7],30[i.e.40],[5]p : ill
B77-26256 BS.23/828
1194674

AIRPORTS. Great Britain
Passengers

CIVIL AVIATION AUTHORITY
Origins and destinations of passengers at United Kingdom airports. - London : The Authority, 1975
[2],103p : ill. - (CAP ; 363)
0-903083-63-9 BS.73/169
0523909

PASSENGERS at the London area airports in 1978. - London : Civil Aviation Authority, 1980
viii,70p : ill. - (CAP ; 430)
B83-13701 BS.73/282
0523909

AIRPORTS. Great Britain
Policies of government

GREAT BRITAIN. Department of Trade
Airports policy. - London : H.M.S.O., [1978]
iv,48p. - (Cmnd.7084)
0-10-170840-8 BS.41/1106
1373781

AIRPORTS. Great Britain
Railways, bus services & coach services - Serials

THOMAS Cook airport links. - Winter 1983/4 issue-. - Peterborough : Thomas Cook, 1984-
v.
B85-11090 P.843/258
3229505

AIRPORTS. Great Britain
Regional planning

ASSOCIATION OF BRITISH CHAMBERS OF COMMERCE
Airports in Great Britain. - London : Association of British Chambers of Commerce, 1977
24,[1]p
B78-02211 X.512/5733
1280287

COLVIN, Michael
Airport UK. - London : Centre for Policy Studies, 1982
64p. - Bibliography: p46
0-905880-44-7 X.805/3198
1280287

GREAT BRITAIN. Department of Trade
Airport strategy for Great Britain. - London : H.M.S.O.
Part 1: The London area. - 1975
[2],iii,78p : col ill
0-11-511679-6 BS.41/1051
0602655
Part 2: The regional airports. - 1976
[2],v,139p,leaf,[21]p of plates
0-11-511717-2 BS.41/1051
0602655

AIRPORTS. Great Britain
Security measures. Aviation Security Fund. Assets. Disposal

GREAT BRITAIN. Treasury
Treasury minute dated 16th January 1984 relative to the disposal of assets of the Aviation Security Fund. - London : H.M.S.O., 1984
2p. - (Cmnd. ; 9121)
0-10-191210-2 BS.44/684
3220583

AIRPORTS. Greater Manchester (Metropolitan County)
Manchester International Airport, to 1978

FIRST and foremost / [compiled by] R.A. Scholefield & S.D. McDonald. - Manchester : Manchester International Airport Authority, 1978
[98]p : chiefly ill
B79-01814 X.619/19235 Woolwich
1753207

AIRPORTS. Greater Manchester (Metropolitan County)
Manchester International Airport. Aviation, 1938-1951

WEBB, Raymond John
Early Ringway. - Bramhall : Webstar Graphics, 1978
230p : ill
0-9506172-0-2 X.611/8206 Woolwich
157034x

AIRPORTS. Hillingdon (London Borough)
Heathrow Airport

PARROTT, E. V. (Edmund Vaughan)
A pocket history and guide to Heathrow Airport, London. - Rickmansworth : EP Publications, [1977]
[14]p : ill
B78-14910 X.519/27872
1415417

AIRPORTS. Hillingdon (London Borough)
Heathrow Airport. Concorde aeroplanes. Noise, 1976-1977

CIVIL AVIATION AUTHORITY. Directorate of Operational Research and Analysis
Noise data from the first year of scheduled Concorde operations at Heathrow Airport, London. - London : C.A.A., 1977
v,10,[51]p : ill. - (Reports / Civil Aviation Authority. Directorate of Operational Research and Analysis ; 7703) (Papers / Civil Aviation Authority ; 77007)
0-86039-040-3 BS.73/170
1108646

AIRPORTS. Hillingdon (London Borough)
Heathrow Airport. Development. Proposals

BRITISH AIRPORTS AUTHORITY
Heathrow Airport, London master development plan report. - [New ed.]. - [London] : [B.A.A.], 1976
[1],18p,[12]leaves(1 fold). - Bibliography: p.18
0-903460-08-4 BS.401/11
0775525

REPORT of the Heathrow area working party. - London : Greater London Council, 1973
61,A.40p,[1]folded leaf of plates
0-7168-0579-0 X.622/15959
0775525

AIRPORTS. Hillingdon (London Borough)
Heathrow Airport. Road traffic. Planning

BRITISH AIRPORTS AUTHORITY
Heathrow: keeping traffic moving, second consultative paper. - [London] : [B.A.A.], 1976
[1],19p,fold plate : ill(some col)
0-903460-10-6 BS.401/17
0772305

AIRPORTS. Hillingdon (London Borough)
Heathrow Airport. Transfer passengers, 1975

BRITISH AIRPORTS AUTHORITY
1975 survey on transfer passengers, Heathrow. - [London] : [B.A.A.], 1976
[45]leaves(2 fold) : ill
0-903460-09-2 BS.401/9
0775096

AIRPORTS. Hillingdon (London Borough)
London Airport (Heathrow). Fourth terminal. Development. Proposals

GLIDEWELL, I. D. L.
Fourth terminal, Heathrow and associated road access. - London : H.M.S.O.. - Includes index
report of the public inquiry, 31 May - 15 December 1978 - 1979
[2],iii,156p
0-11-751438-1 BS.414/2020
2276380

AIRPORTS. Japan
Proposed sites: Agricultural land. Protest by agricultural communities

APTER, David E. (David Ernest)
Against the state. - Cambridge, Mass. ; London : Harvard University Press, 1984
ix,271p, [12]p of plates : ill. - Includes index
0-674-00920-7 X.520/35943
0480304

AIRPORTS. Kastrup. Denmark
Kobenhavns Lufthavn, to 1975

KØBENHAVNS LUFTHAVNSVÆSEN
København lufthavn 1925-1975. - [Kastrup] : Københavns Lufthavnsvæsen, [1975]
159 p. : ill
r6001512 X.525/1107
0878669

AIRPORTS. Leeds (District). West Yorkshire (Metropolitan County)
Leeds Bradford Airport

LEEDS Bradford Airport handbook. - London : Burrow, [1976]
99 p. : ill
r6010245 X.519/26432
0796158

LEEDS Bradford Airport handbook. - [New ed.]. - London : Burrow, [1978]
56p : ill
Subsequent editions, when received, will be placed at this shelfmark
r6092868 X.0709/1085
0796158

AIRPORTS. Liverpool. Merseyside (Metropolitan County)
Liverpool Airport, to 1983

BUTLER, P. H. (Philip Henry)
An illustrated history of Liverpool Airport. - Liverpool : Merseyside Aviation Society, c1983
78p : ill
0-902420-40-2 L.49/2014
2990008

AIRPORTS. Liverpool. Merseyside (Metropolitan County)
Liverpool Airport. Vickers Viscount aeroplanes. Preservation

LAWSON, J. G.
One summer at Speke. - High Wycombe : Skycol Publications, [1978]
[5],71p
B79-11041 X.708/21776
1825178

AIRPORTS. London
Aircraft. Noise at night

GREAT BRITAIN. Department of Trade
Night disturbance from aircraft noise at Heathrow and Gatwick. - London : [The Department], 1977
[2],ii,18,[2]p,[38]p of plates(18 fold),fold leaf of plate : col ill
B77-12472 BS.41/1054
1087312

AIRPORTS. London
Personnel

BRITISH AIRPORTS AUTHORITY
1975 survey on airport workers - Heathrow, Gatwick and Stansted. - [London] : [B.A.A.], 1976
[33],22 leaves,[19]p
0-903460-07-6 BS.401/15
0775630

AIRPORTS. Luton. Bedfordshire
Luton Airport. Development

HERTFORDSHIRE. County Council
The future of Luton and Stansted airports. - [Hertford] : [The Council], 1978
Folder(6p) : 1 ill
B78-40136 X.519/28813
1736590

STANSTED? / [NALGO. Luton Branch] - [Luton] : [NALGO. Luton Branch], [1982?]
5p.[1]leaf of plates
B83-05133 X.808/37473
2822261

AIRPORTS. Manchester
Manchester International Airport

BRITAIN'S northern gateway / editor: Mike Kiddey ; assistant editor: David Tierney. - Bradford : M.C.B., [1982?]
44p : ill
B82-36030 X.805/3070
2778416

AIRPORTS. Midlands. England
Passengers. Origins & destinations

CIVIL AVIATION AUTHORITY
Passengers at major airports in Scotland and central England. - London : C.A.A., 1976
[2],72p : ill. - (CAP ; 394)
0-86039-024-1 BS.73/176
1079948

AIRPORTS. Newcastle upon Tyne. Tyne and Wear (Metropolitan County)
Newcastle Airport

NEWCASTLE Airport. - 7th ed. - London : Burrow, [1980]
76p : ill
B83-06372 X.629/19164
1771418

NEWCASTLE Airport official handbook. - 6th ed. - London : Burrow, [1978]
76p : ill
B79-04248 X.619/19323 Woolwich
1771418

NORTH-EAST REGIONAL AIRPORT COMMITTEE
Newcastle international airport. - 8th ed. - London : Burrow, [1983]
72p : ill
B84-08242 X.809/57531
1771418

AIRPORTS. Ontario
to date

STEWART, Walter
Paper juggernaut. - Toronto : McClelland and Stewart, [1979]
207p
0-7710-8306-8 X.622/17795
9031014
- [Another copy] X.950/34675

AIRPORTS. Ostlandet. Norway
Location

NORSK INSTITUTT FOR BY- OG REGIONFORSKNING
Flyplasslokalisering i Østlandsområdet. - Oslo : Norsk institutt for by- og regionforskning, 1973
xvii,[200]p : ill. - (By- og regionforskning ; rapport 23)
r6079197 X.0802/100(23)
1508970

AIRPORTS. Providenciales. Turks and Caicos Islands
Development

GREAT BRITAIN
Second report from the Foreign Affairs Committee of the House of Commons, session 1982-3. - London : H.M.S.O., [1983]
6p. - (Cmnd. ; 8979)
0-10-189790-1 BS.14/3645
2996839

AIRPORTS. Rhoose. South Glamorgan
Cardiff-Wales Airport

CARDIFF. Wales Airport. - London : Burrows, [1984?]
36p : ill
B85-16117 X.629/25802
3258173

AIRPORTS. Rhoose. South Glamorgan
Glamorgan (Rhoose) Airport

GLAMORGAN (RHOOSE) AIRPORT JOINT COMMITTEE
Wales' airway to the world. - Carshalton : Home Publishing Co., [1978]
48,[1]p : ill
r6032700 X.519/27699
1354027

AIRPORTS. Scotland
Passengers

BRITISH AIRPORTS AUTHORITY
1975 Scottish airports origin and destination survey. - [London] : [B.A.A.], 1976
[5],26p,6 leaves : ill
0-903460-05-x BS.401/21
0630918

AIRPORTS. Scotland
Passengers. Origins & destinations

CIVIL AVIATION AUTHORITY
Passengers at major airports in Scotland and central England. - London : C.A.A., 1976
[2],72p : ill. - (CAP ; 394)
0-86039-024-1 BS.73/176
107993x

AIRPORTS. Scotland
Personnel

BRITISH AIRPORTS AUTHORITY
1976 survey on airport workers - Aberdeen, Edinburgh, Glasgow and Prestwick. - [London] : [B.A.A.], 1976
[3],23 leaves,[8]p
0-903460-11-4 BS.401/13
0775649

AIRPORTS. South-east England
Passengers

BRITISH AIRPORTS AUTHORITY
1975 South-east airports origin and destination survey. - [London] : B.A.A., 1975
[5],34p,4 leaves : ill
0-903460-04-1 BS.401/19
0589640

AIRPORTS. South Wales
Regional planning

CIVIL AVIATION AUTHORITY
Future airport development in South Wales and the south-west of England. - London : C.A.A., 1976
[3],13p. - (CAP380)
0-86039-000-4 BS.73/280
0607347

CIVIL AVIATION AUTHORITY. Economics and Statistics Division
Airport development in South Wales and the South West region of England. - London : C.A.A., 1975
[2],46p : ill. - (CAP 377)
0-903083-76-0 BS.73/175
0607347

AIRPORTS. South-west England
Regional planning

CIVIL AVIATION AUTHORITY
Future airport development in South Wales and the south-west of England. - London : C.A.A., 1976
[3],13p. - (CAP380)
0-86039-000-4 BS.73/280
0607339

CIVIL AVIATION AUTHORITY. Economics and Statistics Division
Airport development in South Wales and the South West region of England. - London : C.A.A., 1975
[2],46p : ill. - (CAP 377)
0-903083-76-0 BS.73/175
0607339

AIRPORTS. Stansted Mountfitchet. Essex
Stansted Airport - Serials

STANSTED & South East newsmagazine. - Aug. 1983-Oct./Nov. 1983. - [Stansted, Essex] : [s.n.], 1983-1983
3v. : ill
B84-27113 P.903/891
3134679

AIRPORTS. Stansted Mountfitchet. Essex
Stansted Airport. Development

HERTFORDSHIRE. County Council
The future of Luton and Stansted airports. - [Hertford] : [The Council], 1978
Folder(6p) : 1 ill
B78-40136 X.519/28813
1736604

AIRPORTS. Stansted Mountfitchet. Essex
Stansted Airport. Development. Proposals

BUCHANAN, Sir Colin, 1907-
No way to the airport. - Harlow : Longman, 1981
v,121p : ill. - Includes index
0-582-36123-0 X.529/43745
2566095

The CASE for the North / North of England Regional Consortium. - Manchester : City of Manchester Public Relations Office, [1982?]
1folded sheet([6]p) : col.ill
B82-36236 X.805/3085
2778653

AIRPORTS. Stansted Mountfitchet. Essex
Stansted Airport. Railways

GREAT BRITAIN. Working Party on Rail Access to Stansted Airport
Report of the Working Party on Rail Access to Stansted Airport. - [London] : Department of Transport, 1981
46p : ill
B82-40105 BS.43/616
2798093

AIRPORTS. Stockholm. Sweden
Bromma flygplats. Development

BENCKERT, Karl-Axel
Brommafrågan. - Rönninge : Vincent, 1975
85 p.
r6028830 X.510/9841
1026607

Bei Fragen zur Produktsicherheit wenden Sie sich bitte an:
If you have any questions regarding product safety,
please contact:

Walter de Gruyter GmbH
Genthiner Straße 13
10785 Berlin
productsafety@degruyterbrill.com